COLLINS BIRDS OF THE WORLD

Principal illustrators:
Norman Arlott and Ber van Perlo

Additional illustrations by:
Gustavo Carrizo, Aldo A. Chiappe, Luis Huber and
Jorge R. Rodriguez Mata

WILLIAM
COLLINS

William Collins
An imprint of HarperCollins*Publishers*
1 London Bridge Street
London SE1 9GF

WilliamCollinsBooks.com

HarperCollins*Publishers*
1st Floor, Watermarque Building, Ringsend Road
Dublin 4, Ireland

First published in Great Britain by William Collins in 2021

2022 2024 2023 2021
2 4 6 8 10 9 7 5 3 1

First published in the United States by Princeton University Press in 2021

A catalogue record for this book is available from the British Library

ISBN 978-0-00-817399-9

Set in Goudy Old Style and Univers

Edited, typeset and designed by D & N Publishing, Baydon, Wiltshire
Printed and bound in Bosnia and Herzegovina by GPS Group

FSC www.fsc.org
MIX
Paper from responsible sources
FSC™ C007454

Contents

Preface

It had always been an ambition of mine to produce a coloured checklist of every bird in the world, so I was more than happy when I was approached by Myles Archibald of HarperCollins to put together a book of the world's birds, drawing on their vast collection of bird-guide artwork.

The first step on the road was to decide which world bird list to follow – for there are, strangely, quite a few different ones. In this case it made sense to plump for the latest IOC list (version 8.2, as of January 2019), since it appears that many of the world's ornithological organisations are adopting this very comprehensive list. Therefore this publication follows that list and its nomenclature exactly, resulting in a book that includes 10,711 species, classified in 40 orders, 246 families and 2,313 genera.

The 301 plates are intended to give an 'at a glance' view of all the world's birds. The book is not meant to be a field guide, but hopefully the combination of illustration and concise text will help the reader to identify each species.

My most grateful thanks go to the artists who allowed me access to their work, as used in the various Collins guides. Apart from my own artwork I have drawn heavily on Ber van Perlo's work from Africa, New Zealand and South America, as well as other South American artwork produced by Jorge R. Rodriguez Mata, Gustavo Carrizo, Aldo A. Chiappe and Luis Huber.

Although it was my job to choose the artwork and design the plates, books are always a team collaboration, and this book would not have come to fruition without the great skill of David Price-Goodfellow and his team at D & N Publishing (designer: Namrita Price-Goodfellow; copy-editors: Susi Bailey, Hugh Brazier, Marianne Taylor; proofreader: Paul Stringer with assistance from Jan McCann; editorial assistants: Kai Price-Goodfellow, Ione Robinson. Layi Technologies and Geethik Technologies PVT Ltd in India contributed to plate creation and corrections). Thanks also to Paul Sterry, Nigel Redman, Marianne Taylor, Hazel Eriksson and Hugh Brazier for providing additional text.

Without encouragement (and a little pushing) from Myles Archibald and Hazel Eriksson of HarperCollins I would probably still be at the starting gate. I thank them both for allowing me the opportunity to be involved in this wonderful project.

Lastly, but most definitely not least, I must thank my wife Marie for all her help and support.

Norman Arlott, May 2021

SPECIES ACCOUNTS: ILLUSTRATIONS AND TEXT

Each of the 301 plates shows breeding plumages of males (♂), and females (♀) where female plumage differs substantially from that of the male, as well as a few colour morphs, where this was deemed appropriate and space allowed. To keep the book as simple as possible it was decided not to include any subspecies, even those that are very different from the nominate race. Species are shown approximately to scale within each plate.

The text for each species starts with the English and scientific names, as given in IOC 8.2, followed by the overall length of the bird (including tail streamers, unless otherwise specified). Where the sexes differ greatly in size, male and female lengths are given separately.

The description of each species is necessarily very brief. In some cases the plumage features and appearance of the bird are described, while in others the emphasis is on behaviour, or on other notable characteristics. Either way, the aim is to focus on the features that are most likely to be helpful in identifying the bird. Notes on the key points of difference between similar species are often included, but these are not comprehensive.

Unless otherwise indicated, the voice section (V) generally describes the territorial song of the species. Where space permits, some other calls are often included. For a few species, however, remarkably little is known.

The habitat section (H) cannot be comprehensive, because birds turn up in all sorts of unexpected places. The text indicates the main breeding habitat or habitats of each species, as well as notable wintering and migratory stopover habitats where appropriate.

Under distribution (D), the main range of the species is given, with breeding and non-breeding ranges described separately if appropriate. The IOC list was the main source for this information, supplemented from a range of other textbooks and online sources. The text refers to native distribution, but in some cases notable introduced populations are also mentioned. For places that are less likely to be familiar, including remote islands, the state or the island group is mentioned, or an indication of the location (e.g. 'SW Pacific' or 'Indian Ocean') is included (see world map on page 11). 'SE Asia' is used to refer to the mainland states of this region; where a species also occurs on the islands, these are noted separately.

ARTWORK SOURCES

Norman Arlott

Collins Field Guide to the Birds of the Palearctic: Passerines (William Collins, 2007)

Collins Field Guide to the Birds of the Palearctic: Non-Passerines (William Collins, 2009)

Collins Field Guide to the Birds of the West Indies (William Collins, 2010)

Collins Field Guide to the Birds of North America (William Collins, 2011)

Collins Field Guide to the Birds of India, Pakistan, Nepal, Bhutan, Bangladesh and Sri Lanka (William Collins, 2015)

Collins Field Guide to the Birds of South-East Asia (William Collins, 2017)

Collins Field Guide to the Birds of the Philippines, Sumatra, Java, Bali, Borneo, Sulawesi, the Lesser Sundas and the Moluccas (William Collins, 2018)

Birds of Australia and New Guinea (unpublished)

Ber van Perlo

Collins Illustrated Checklist: Birds of Western and Central Africa (William Collins, 2002)

Collins Field Guide to the Birds of Mexico and Central America (William Collins, 2006)

Collins Field Guide to the Birds of Eastern Africa (William Collins, 2009)

Collins Field Guide to the Birds of Southern Africa (William Collins, 2009)

Collins Field Guide to the Birds of New Zealand, Hawaii and the Central and West Pacific (William Collins, 2011)

Collins Field Guide to the Birds of South America: Passerines (William Collins, 2015)

Birds of Madagascar and South Africa (unpublished)

Jorge R. Rodriguez Mata, Gustavo Carrizo, Aldo A. Chiappe and Luis Huber

de la Peña, M.R. and Rumboll, M. *Collins Illustrated Checklist: Birds of Southern South America and Antarctica* (William Collins, 1998)

Mata, J.R.R., Erize, F. and Rumboll, M. *Collins Field Guide to the Birds of South America: Non-Passerines* (William Collins, 2006)

Bird Topography

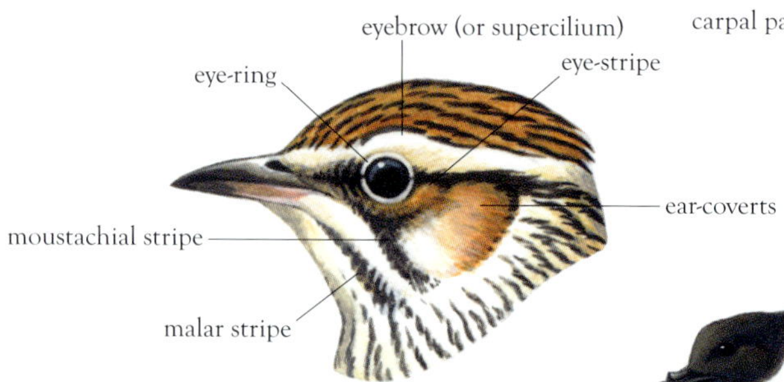

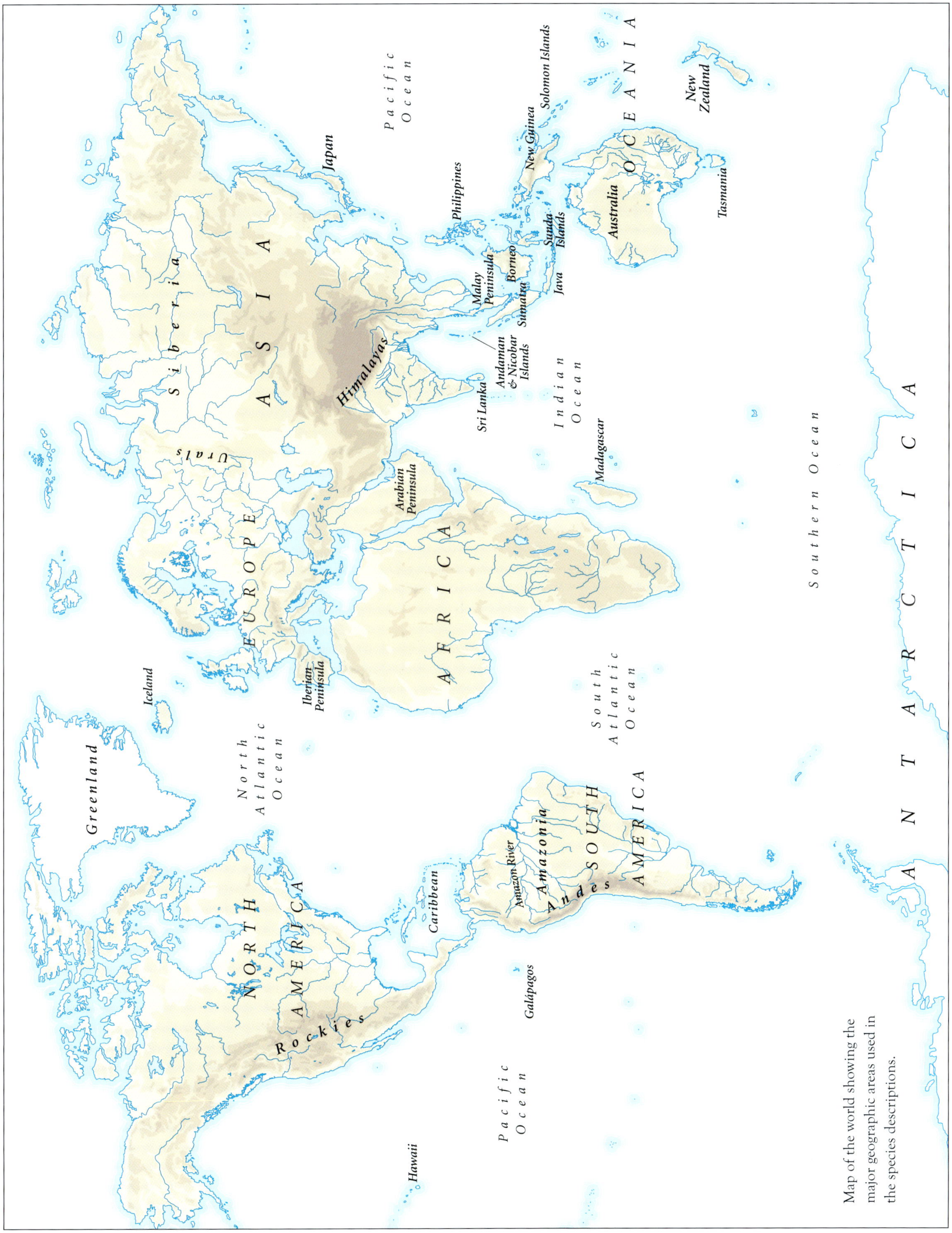

Map of the world showing the major geographic areas used in the species descriptions.

OSTRICHES STRUTHIONIDAE

1 COMMON OSTRICH *Struthio camelus* male 210–275cm, female 175–190cm Unmistakable, even at some distance. Flightless. **V** During breeding season males give a far-carrying, booming *ooom ooom booo-ooooo*. At other times usually silent, apart from various hissing and snorting utterances. **H** Dry open grassland, semi-desert with sparse vegetation. **D** W, C, E and SW Africa.

2 SOMALI OSTRICH *Struthio molybdophanes* male 210–275cm, female 175–190cm Unmistakable, even at some distance. Flightless. **V** During breeding season males give a far-carrying, booming *ooom ooom booo-ooooo*. At other times usually silent, apart from various hissing and snorting utterances. **H** Dry open grassland, semi-desert with sparse vegetation. **D** Somalia, Kenya.

RHEAS RHEIDAE

3 GREATER RHEA *Rhea americana* 170cm Huge, hefty flightless bird. Large wings with long, floppy, black-tipped feathers covering body. Body feathers short, thigh plumage dirty white. No tail. Powerful legs. Female usually less hefty, with base of neck paler. **V** Voice in breeding season is a deep, ventriloqual 'boom'. **H** Open plains, grassland, sparse woods, savannah, steppe, cultivated land. **D** SE South America.

4 LESSER RHEA *Rhea pennata* 140cm Smaller than Greater Rhea, with a short bill and dark iris. Body more rounded. Generally browner with no black on neck. Floppy wing feathers hang lower on thighs. White tips of feathers very visible in fresh plumage, wearing off with time. Body and thighs brownish white. Feathering extends onto tarsi. **V** During breeding season males make a low roaring sound. **H** Open terrain, steppe, arid and stony areas, brushland, plains to mountains, puna. Replaces Greater Rhea in arid regions. **D** S South America.

KIWIS APTERYGIDAE

5 SOUTHERN BROWN KIWI *Apteryx australis* 50cm Differentiated from North Island and Okarito brown kiwi by range. Sole brown kiwi to come out in daytime. **V** High, descending, shrieking yell *wee-oh* (male), that of female deeper and more hoarse. **H** Wet forest, subalpine tussockland. **D** South Island and Rakiura/Stewart Island, New Zealand.

6 NORTH ISLAND BROWN KIWI *Apteryx mantelli* 50cm Note brownish plumage. **V** High, rising, shrill whistle. **H** Prefers dense forest and shrubland; also in pine plantations, scrub, secondary growth, pastures. **D** North Island, New Zealand.

7 OKARITO BROWN KIWI *Apteryx rowi* 50cm Greyer than North Island Brown Kiwi. **V** At night, gives loud, shrill calls. **H** Dense lowland forest. **D** W South Island, New Zealand.

8 LITTLE SPOTTED KIWI *Apteryx owenii* 40cm Similar to Great Spotted Kiwi but smaller and otherwise differentiated by range. **V** Long, gradually ascending, high-pitched series of *wreeh* notes. **H** Mainly interior and at margins of forest with dense undergrowth; up to 1,000m. **D** South Island, New Zealand.

9 GREAT SPOTTED KIWI *Apteryx haastii* 55cm Differentiated from brown kiwi species by range and larger size. Strictly nocturnal and presence most easily detected by diagnostic calls. Like other kiwi, feeds by probing soft ground for invertebrates using long bill. **V** Slow, slightly ascending, mid-high series of *wraah* notes. **H** Mossy beech forest, tussock grassland, coastal scrub; up to 1,200m, mainly above 700m. **D** W South Island, New Zealand.

CASSOWARIES AND EMU CASUARIIDAE

10 SOUTHERN CASSOWARY *Casuarius casuarius* 130–170cm Length of adult wattles and casque very variable, and bare skin colours brighter in females. Shy, usually solitary. Presence often revealed by piles of orange or purplish droppings. **V** Various grunts, rumblings, roars and hollow-sounding booming. **H** Lowland and hill forest. **D** New Guinea; NE Australia; Aru Islands, Indonesia.

11 DWARF CASSOWARY *Casuarius bennetti* 100–110cm Similar plumage to Southern Cassowary, but smaller, with smaller, more triangular casque; wattle absent. Bare skin on neck blue, with small red patches at base. Female has larger casque and brighter bare parts. Varied diet. **V** Piping, booming and roaring sounds. **H** Lowland and hill forest; up to 3,500m depending on treeline. **D** New Guinea; New Britain, Papua New Guinea; Yapen Island, Indonesia.

12 NORTHERN CASSOWARY *Casuarius unappendiculatus* 120–150cm Impressive flightless bird with bluish-black adult plumage, similar to other cassowaries. Casque large and triangular. Largely featherless head purplish blue; bare neck blue with red or yellow on throat and red wattles. On average, female has larger casque and brighter colours. **V** Grunts and guttural roars. **H** Lowland rainforest, forested swamps. **D** New Guinea; Yapen and Salawati Islands, Indonesia.

13 EMU *Dromaius novaehollandiae* 150–190cm Umistakable flightless bird with shaggy, grizzled greyish-brown plumage. Capable of running at speeds of more than 40kph using long, powerful legs and three-toed feet. Head and neck sparsely feathered; throat and sides of neck blue. Bill stout and broad-based. Feeds on seeds, fruits and invertebrates depending on seasonal availability. **V** Female utters booming calls, male utters grunts; most vocal in breeding season. **H** Grassland, open scrub, woodland. **D** Australia.

1
♂
♂
♀
2
♀
♂
3
4
10
11
12
5
6
7
8
9
13
♂
♀

TINAMOUS TINAMIDAE

1 GREY TINAMOU *Tinamus tao* 52cm Overall bluish grey, upperparts with alternating thicker and finer black barring, less so on paler underside. Throat and line down side of neck white. **V** Call a short hoot, usually given singly, rarely a series of up to four, often repeated intermittently for several minutes. Mainly vocal late afternoon or at dusk. **H** Montane tropical, subtropical and cloud forest; up to 2,000m. **D** Amazonia.

2 SOLITARY TINAMOU *Tinamus solitarius* 50cm Dark brown above, barred blackish and with lighter speckling. Breast brownish grey, rest of underparts greyish buff and cinnamon buff with fine blackish barring. Line down side of neck ochreous buff. **V** At dusk emits three slow, mournful whistled notes, the last half a tone lower. **H** Dark floor of tropical and subtropical rainforest. **D** E Brazil to NE Argentina and E Paraguay.

3 BLACK TINAMOU *Tinamus osgoodi* 46cm Overall very dark. Upperparts brownish black, underparts blackish brown, undertail chestnut. Head and neck blackish slate; feathers edged black, giving a scaly effect. **V** Call a descending, deep whistle, similar to first note of song of White-throated Tinamou. Mainly vocal at dusk. **H** Upland tropical and subtropical rainforest; 1,500–2,500m. **D** SC Colombia, SE Peru.

4 GREAT TINAMOU *Tinamus major* 42cm Variable. Upperparts brownish olive, barred blackish. Underparts brownish buff, finely barred darker. Forehead, crown and nape (or whole head) chestnut. **V** Melodious song consisting of a pure whistle followed by a quavering trill, repeated several times. Mainly vocal in late afternoon and at dusk. **H** Montane tropical rainforest and cloud forest; up to 1,500m. **D** S Mexico through Amazonia.

5 WHITE-THROATED TINAMOU *Tinamus guttatus* 38cm Small. Noticeable white dots between dark barring on upperparts. Buffy brown below, darker on upper breast. Barred dark brownish on belly and undertail. Forehead and crown grey, rest of head and neck ochreous buff with black scaling. Horn-white legs. **V** Long, mournful hoot, *ooo* or *ooo-ooo*, up to 3 secs in duration; sometimes a long note followed by a shorter one. **H** Tropical rainforest at low elevation; 100–200m. **D** Amazonia.

6 HIGHLAND TINAMOU *Nothocercus bonapartei* 36cm Head slaty black, throat cinnamon. Upperparts and neck chocolate brown, underparts reddish cinnamon, all barred black. Buffy-white spots on sides, belly and lower upperparts. **V** Call a deep, throaty, slightly nasal *caw-oh* or *kooyoo*, repeated steadily. **H** Subtropical and temperate forests in foothills and mountains, cloud forest, humid grasslands; 500–2,500m. **D** Costa Rica to Peru.

7 TAWNY-BREASTED TINAMOU *Nothocercus julius* 38cm Head reddish chestnut, throat white. Upperparts dark brown, barred blackish; underparts rufous cinnamon. Buffy-white spots on sides. **V** Call a very distinctive fast, rhythmic series of almost chanting notes, *tree-er tree-er tree-er tree-er*, a little faster than one note per sec, which may continue for a minute or even longer, the phrases becoming shorter and lower pitched towards the end. **H** Subtropical and temperate rainforests in foothills, low open woodlands; up to 3,000m. **D** C Colombia and W Venezuela to SC Peru.

8 HOODED TINAMOU *Nothocercus nigrocapillus* 38cm Head grey, browner on forehead and cheeks; throat whitish. Upperparts dark brown, finely barred darker; underparts cinnamon brown. Buffy-white spots on sides and belly. Flight feathers dark brown, barred black. **V** Call a series of rising and falling *brreew* or *bwow* notes, averaging about one note per 10 secs. **H** Subtropical rainforest in foothills and mountains. **D** Peru, Bolivia.

9 BERLEPSCH'S TINAMOU *Crypturellus berlepschi* 30cm Very dark and uniform brownish black. Faint white streaking on sides of neck. Legs dirty pinkish brown. **V** Call a short, high-pitched (unusually so for a tinamou), piercing whistle, *teeeee*. **H** Wet forest, older secondary forest, hills, coastal lowlands. **D** W Colombia, NW Ecuador.

10 CINEREOUS TINAMOU *Crypturellus cinereus* 30cm Very dark, uniform brown. Cap cinnamon rufous, faintly streaked white on sides of neck. Legs dirty pinkish brown. **V** Calls resemble a tremulous police whistle. **H** Floodable lowland tropical forest; brush, thickets and secondary growth near rivers and swamps; várzea; savannah; regrowth; plantations. **D** Amazonia.

11 LITTLE TINAMOU *Crypturellus soui* 22cm Small. Almost uniform coloration, with no black markings. Many races. Typical female has greyish head, whitish throat, browny-chestnut upperparts, tawny-brown underparts; legs dirty cream or olive-washed cream. Typical male browner on neck and upperparts, underparts tawny brown; legs pale olive-grey. **V** Call a series of tremulous, penetrating whistles that rise in pitch and volume. **H** Humid or dry brushland and thickets, secondary growth, edge of transition woods and forests, cloud forest, clearings, regrowth, plantations. **D** S Mexico to E Brazil.

12 TEPUI TINAMOU *Crypturellus ptaritepui* 28cm Like a darker Little Tinamou but only the face is greyish; forehead, crown, nape and hindneck dark chestnut. Underparts dark brown, lacking black scaling on belly and undertail. **V** Call a long, high-pitched, pure-tone whistle, fading away at the end. **H** Dense subtropical cloud forest. **D** Local in SE Bolívar, Venezuela.

13 BROWN TINAMOU *Crypturellus obsoletus* 30cm Head greyish. Upperparts uniform chestnut brown; underparts buffy cinnamon, scaled black below. Legs yellowish. Eyes orange. **V** Call a succession of strident, accelerating whistles that rise in pitch until almost inaudible. **H** Lowland and mountain tropical and subtropical forest. **D** Widespread across South America.

14 UNDULATED TINAMOU *Crypturellus undulatus* 30cm Head, neck and upperparts dark chestnut brown, underparts pale brownish grey. All but breast and belly faintly vermiculated blackish. Throat whitish. Bill and legs pale grey. **V** Call a mournful four-note whistle, rising on slurred last note. **H** Humid tropical forest, gallery forest, várzea, woods, scrub, thickets, secondary growth, edges and clearings, transition woods, patches of woods, savannah, regrowth. **D** Amazonia.

15 PALE-BROWED TINAMOU *Crypturellus transfasciatus* 29cm Brow and throat whitish. Upperparts brown-barred, slightly scaled buff and black on lower back. Breast pale greyish brown; rest of underparts greyish cream, partly barred black. Legs reddish pink. Barring on lower back less evident in male. **V** Call a loud, resonant *whoooit*, generally given at intervals longer than 30 secs. **H** Open woods, transition woods, scattered trees, brushland and thickets in dry tropical areas. **D** W Ecuador, NW Peru.

16 BRAZILIAN TINAMOU *Crypturellus strigulosus* 30cm Virtually identical to Grey-legged Tinamou, differing only in brownish-grey breast and whitish belly. Legs pale greyish. **V** Single very long, mellow, whistled note lasting 6–8 secs, slightly tremulous towards end, repeated every few minutes. **H** Humid forest, woods, brushland, thickets. **D** E Peru and NW Bolivia through S Amazonian Brazil.

17 GREY-LEGGED TINAMOU *Crypturellus duidae* 30cm Large and hefty. Head, neck, breast and upper back rufous. Throat white. Rest of upperparts barred and scaled black and chestnut, to black and buff on lower back and uppertail coverts. Underparts tawny rufous, grading to buff, barred black below. Legs dirty cream, pale grey or creamy grey. Male dorsally uniform chestnut brown. **V** Slow, two-syllable whistle, *whoo-whoooooooooooo*. **H** Humid forest, brushland, thickets, open woods. **D** EC Colombia to S Venezuela.

18 RED-LEGGED TINAMOU *Crypturellus erythropus* 30cm Head, hindneck and upperparts brownish chestnut. Throat whitish. Foreneck and breast brownish grey. Rest of underparts tawny cinnamon, turning buff and barred blackish on flanks and undertail coverts. Legs reddish pink. Female strongly barred black and buff on lower upperparts and wings. **V** Call a clear trisyllabic whistle, mostly at dusk. **H** Dry and open woods, low deciduous woods, brushland, thickets, copses. **D** N South America.

19 YELLOW-LEGGED TINAMOU *Crypturellus noctivagus* 35cm Crown blackish, bordered by red. Back brownish black. Lower back and rump chestnut, finely barred black. Sides of head dusky, throat pale, neck grey, breast rusty. Belly buffish, lower belly and flanks barred darkish. Legs yellow. **V** Call distinctively low pitched, slow and mournful. Usually comprises four notes, *a-ooo, u-u* or *aa-oo oo-oo*, the first note highest and loudest. **H** Rainforest, woods, gallery forests. **D** E Brazil.

20 BLACK-CAPPED TINAMOU *Crypturellus atrocapillus* 30cm Head brownish black, throat cinnamon. Neck and upper breast slaty grey. Lower breast cinnamon chestnut, grading to buffish on undertail coverts. Barring on flanks. Rest of upperparts dark brown, vermiculated blackish in male, and barred buffy and blackish in female. Legs reddish pink. **V** Call a short whistle, *eeoee*, repeated every minute or so. Vocal throughout the day. **H** Humid tropical forests, secondary growth, bush and thickets, deciduous tropical forest. **D** Peru, Bolivia.

21 THICKET TINAMOU *Crypturellus cinnamomeus* 30cm Male rather grey with red legs. **V** High-level, fluting *pjeeuw*. **H** Dense scrub, secondary growth, forest edges; sea level to 1,000m. **D** WC Mexico to Costa Rica.

22 SLATY-BREASTED TINAMOU *Crypturellus boucardi* 30cm Note finely barred tawny underparts, dark grey neck and chest. Female has barred wings. **V** High, hollow, drawn-out, almost level *huuuh-huuuuuuh*, lasting 6 sec. **H** Dense forest undergrowth; <1,500m. **D** S Mexico to Costa Rica.

23 CHOCO TINAMOU *Crypturellus kerriae* 28cm Head slaty black, throat white. Neck and breast blackish brown. Upper belly and flanks brownish cinnamon. Female has blackish and buff barring on upperparts (almost absent in plain, dark brown male). Feet reddish pink. **V** Call a low-pitched, tremulous, resonant whistle about 1 sec long. **H** Montane tropical forest. **D** Rare and local. SE Panama, NW Colombia.

1
2
3
4
5
6
7
8
9
10
11
12
13
14
15
16
17
18
19
20
21
22
23

TINAMOUS ***CONTINUED***

1 VARIEGATED TINAMOU *Crypturellus variegatus* 28cm Long, straight bill. Head lead grey to blackish. Throat white, washed rufous. Neck and breast rufous. Upperparts barred rufous and black. Underparts tawny rufous, grading to buff, barred black below. Legs pale grey. **V** Call a long, plaintive, descending whistle, followed by 3–5 short, rising notes. **H** Tropical and subtropical forest in mountains, woods, open areas with scattered trees. **D** Amazonia, E Brazil.

2 RUSTY TINAMOU *Crypturellus brevirostris* 24cm Small. Short bill. Head chestnut, throat white. Neck and breast rufous. Upperparts barred chestnut and black. Underparts tawny rufous, grading to buff, somewhat barred black under belly. Legs greyish cream. **V** Call a series of 10–15 pure whistles, typically well spaced at the start and gradually accelerating, with a total length of 8 secs. **H** Tropical forest and várzea. **D** W and NE Amazonia.

3 BARTLETT'S TINAMOU *Crypturellus bartletti* 24cm Small. Short bill. Head blackish brown, throat white. Neck and breast dark brown. Upperparts barred black and brown. Underparts tawny rufous, grading to buff, somewhat barred lower. Legs greyish cream. Once considered a race of Rusty Tinamou. **V** Song a series of high-pitched, fluting notes, similar to that of Cinereous Tinamou (Plate 2) but not evenly paced. Vocal at dusk and at night. **H** Humid tropical forest. **D** W Amazonia.

4 SMALL-BILLED TINAMOU *Crypturellus parvirostris* 20–24cm Resembles Tataupa Tinamou, but bill smaller, red with greyish tip. Eyes pale orange. Legs coral red. Head, neck and breast browner. **V** Call a single *pip*, then a longish silence, then another *pip*, these accelerating until they fuse into churrs that descend to nothing. **H** Open areas with bush and scattered trees, tall grassland, open woods, forest edges. **D** SE Peru and S Amazonian Brazil to NE Argentina.

5 BARRED TINAMOU *Crypturellus casiquiare* 26cm Head rufous chestnut, throat very white. Neck, breast and flanks lead grey, paling to whitish belly. Lower belly and undertail coverts buffy, barred black. Back and rest of upperparts barred dark ochre and black. Legs pale greyish olive. **V** Call a whistle, followed by up to 30 shorter notes, rising at first, slowing at end. **H** Dense lowland, sandy-belt tropical forest near rivers. **D** E Colombia, S Venezuela.

6 TATAUPA TINAMOU *Crypturellus tataupa* 24cm Head, neck and breast lead grey, darker on head. Upperparts dark chocolate brown. Lower belly and undertail coverts black, feathers edged whitish, giving effect of scaling. Bill pinkish red, eyes dark brown, legs vinaceous red. **V** Call a series of loud, whistled *preeep* notes, rising and descending. **H** Tropical and subtropical forests, open woods, scrubland and thickets, savannah. **D** E, SE and SC South America.

7 RED-WINGED TINAMOU *Rhynchotus rufescens* 42cm Head and neck ochreous. Crown streaked black. Upperparts and flanks scaled ochreous brown, black and white. Breast plain ochreous brown. **V** Call a whistled *tooweeoo – who-who-who*, last notes falling in pitch and volume. **H** Open country with tall grass, often damp bottoms, sometimes with scattered trees. **D** E, SE and SC South America.

8 HUAYCO TINAMOU *Rhynchotus maculicollis* 42cm Resembles Red-winged Tinamou but head speckled and neck streaked black. Breast entirely barred. Some consider it the same species. **V** Call comprises two short notes, the first modulated and the second flatter and quieter. **H** Tall grass in high Andean valleys with brush and scattered small trees, cultivated land. **D** NW Bolivia, NW Argentina.

9 TACZANOWSKI'S TINAMOU *Nothoprocta taczanowskii* 36cm Resembles Ornate Tinamou. Long, heavy, downcurved bill. Face and neck dark greyish brown, crown and ear coverts blacker. Neck and breast greyish brown, barred and dotted blackish and whitish. Rest of underparts barred cinnamon and blackish. **V** When flushed, utters a loud *cuyy cuyy*. **H** Puna, semi-arid scrub, bunch grass, open woods, cultivated fields. **D** SC Peru.

10 ORNATE TINAMOU *Nothoprocta ornata* 35cm Robust. Upperparts brownish with contrasting black, whitish and ochre patterns. Head and neck with black spotting. **V** Call a short, sharp whistle, rising in pitch. **H** Puna, high mountain steppe, grass with patchy scrub, often near water. **D** Peru to N Chile and NW Argentina.

11 CHILEAN TINAMOU *Nothoprocta perdicaria* 30cm Brownish grey with contrasting black and whitish patterns on back. Crown dark. Foreneck and breast greyish, dotted and streaked whitish. Flight feathers cinnamon rufous, barred black. **V** Call a strident, far-carrying whistle that sounds double-syllabled, *sweee weeee*. **H** Grassland, scrub, cultivated land. **D** Chile.

12 BRUSHLAND TINAMOU *Nothoprocta cinerascens* 32cm Slender. Upperparts brownish grey with whitish, buff and black patterns. Black topknot raised when alarmed. Face and neck greyish white, speckled and streaked blackish. Breast grey, dotted white and faintly barred black. **V** Call a series of mournful, rising, whistled *wheeew* notes. **H** Open woods and brushland in dry areas. **D** S South America.

13 ANDEAN TINAMOU *Nothoprocta pentlandii* 26cm Greyish brown. Upperparts generally dark with few contrasting markings. Breast greyer, washed lavender and partly dotted whitish. Flanks barred and spotted whitish and brownish. Legs yellowish. **V** Call a sharp, ascending, piercing *wheeet*. **H** Ravines, slopes with brush, grass with scattered trees, vegetation along streams, cultivated land. **D** Ecuador to C Argentina.

14 CURVE-BILLED TINAMOU *Nothoprocta curvirostris* 28cm Upperparts brownish grey with contrasting black and whitish pattern. Neck whitish buff, streaked and dotted black. Breast and rest of underparts tawny with black and white markings on flanks and upper breast. Flight feathers barred cinnamon rufous and blackish. **V** Call a series of three whistled notes, the first pitched lower than the others. Repeated at 5 sec intervals. **H** Humid to semi-arid areas in puna and páramo, cultivated areas. **D** Ecuador, Peru.

15 WHITE-BELLIED NOTHURA *Nothura boraquira* 27cm Resembles Spotted Nothura but whiter below. Crown notably black when raised in alarm. Sides of breast and flanks more finely spotted and barred, respectively. Legs strong yellow. **V** Long, high-pitched whistle, dropping in pitch at end. **H** Open country, short-grass savannah, scrubby grassland with scattered trees, cultivated fields. **D** E Brazil to E Bolivia and NE Paraguay.

16 LESSER NOTHURA *Nothura minor* 18cm Small version of Spotted Nothura, with more rufous upperparts and flanks. Inner vane of flight feathers dark brown. **V** Utters a series of long, high, metallic whistles, *peee peeeeee-peeep peeep peeep peeeeee*, as well as shorter, faster notes. **H** Open short-grass savannah with scrub and trees, even tall grass, flooded areas, cultivated land. **D** SE Brazil.

17 DARWIN'S NOTHURA *Nothura darwinii* 21cm Pale, buffy version of Spotted Nothura, with less contrasting pattern on upperparts, breast and flanks. Flight feathers with inner vane all dark brown (wings seem darker). **V** Call a long series of *peep* whistles as in Spotted Nothura, but no coda at end. **H** Replaces Spotted Nothura in drier country. **D** Peru to SC Argentina.

18 SPOTTED NOTHURA *Nothura maculosa* 23cm Ochreous, black, buff and whitish pattern on upperparts. Neck spotted black, irregular barring on flanks. Brightness varies with race. Flight feathers all barred. **V** Calls a long, even trill, and a long series of *peep* whistles at rate of two per sec, ending in a rapid, descending *tiddle-toddle-too*. **H** Short grass in open country or with scattered trees, cultivated fields. **D** SE South America.

19 CHACO NOTHURA *Nothura chacoensis* 24cm Treated as subspecies of Spotted Nothura by some. Overall buffish brown. Back feathers have dark centres and pale margins, flight feathers barred. Usually secretive. **V** Territorial song a rapid trilling on one tone. Calls include peeps and long trills. **H** Open grassland with scrub, savannah. **D** NW Paraguay.

20 DWARF TINAMOU *Taoniscus nanus* 15cm Tiny. Like a miniature *Nothura*. Dumpier. Upperparts rufous ochreous, barred and scaled black and-whitish. Breast and flanks barred black, spotted white. **V** Insect-like call. **H** Open country with tall grass, sometimes with bushes and trees, Cerrado, savannah. **D** SE Brazil, NE Argentina.

21 ELEGANT CRESTED TINAMOU *Eudromia elegans* 39cm Long, pointed crest. White lines on face and upper neck. Upperparts brownish grey with varying creamy-white and black spotting. Blackish streaking on creamy foreneck and breast. Other races darker or paler. **V** Calls include a whistled, three-syllable *we? we do*, descending a tone on last note. **H** Scrub, open country, steppe, open woods, dry grassland, cultivated land. **D** S South America.

22 QUEBRACHO CRESTED TINAMOU *Eudromia formosa* 41cm Resembles Elegant Crested Tinamou but lines on neck reach base of neck. Upperparts brownish cinnamon. Blackish design on upperparts has bars, streaks and irregular patterns. Shyer and more easily flushed than Elegant Crested. **V** Most common call is a slow, descending double whistle, *foooo-ip foo-ip*, the first part more drawn out and with a third, quieter *foo-ip* occasionally added. Singing bouts may last 30 minutes or more and take place at dawn. **H** Dry Chaco woods, grasslands, scrub with scattered trees. **D** Paraguay, N Argentina.

23 PUNA TINAMOU *Tinamotis pentlandii* 40cm Striking white lines across face and down neck. Upperparts brown with olive wash, edged, dotted and streaked buffy white. Breast, belly and flanks barred whitish buff and brownish grey. Undertail deep rufous. Flight feathers barred. **V** Calls *kay-oo*, last syllable stressed. **H** Andean steppe with bunch grass and scattered low scrub, sandy or stony country; above 3,500m. **D** SC Peru, N Bolivia, N Chile, NW Argentina.

24 PATAGONIAN TINAMOU *Tinamotis ingoufi* 37cm Striking. White lines across face and down neck. Upperparts and flanks scaled ochreous, chestnut, grey and whitish. Undertail coverts ochreous. Flight feathers rufous. **V** Call a series of melodious *twee-oo* notes with variations. **H** Stony steppe with short, sparse grasses and low scrub. **D** S Chile, SW Argentina.

1
2
3
4
5
6
7
8
9
10
11
12
13
14
15
16
17
18
19
20
21
22
23
24

SCREAMERS ANHIMIDAE

1 HORNED SCREAMER *Anhima cornuta* 85cm Long, slender 'horn' angles forwards from forehead. Crown greenish black, lower neck and breast scaled white. Head and upper neck black. Belly white. Legs grey. Underwing-coverts white. **V** Series of strained, gasping honks. **H** Shores of rivers, lakes and swamps, oxbows, in forest or grassland. **D** Colombia and Venezuela through Amazonia to Bolivia.

2 NORTHERN SCREAMER *Chauna chavaria* 85cm Cap and nuchal crest grey. Throat and cheeks whitish. Upper neck blackish. Rest of plumage grey, darker on upperparts. Legs reddish pink. **V** Harsh, brief, rasping screams, repeated at short intervals. **H** Edge of swamps, lakes, marshes and rivers; open or wooded areas; bodies of fresh water with abundant emergent and floating vegetation. **D** N Colombia, NW Venezuela.

3 SOUTHERN SCREAMER *Chauna torquata* 85cm Head and nuchal crest grey. Face paler. Ring-collar white, choker velvety black. Rest of plumage grey, upperparts darker. Legs pinkish red. **V** Calls are solos or duets; one calls *aha*, the other interjects with soprano *oglik*. Another call is repeated very loud *chew-wheel, rrreap*. **H** Edge of lakes, marshes, temporary water, rivers, streams, roadside ditches, all preferably with emergent and floating vegetation; fields near water; open or wooded areas. **D** SE Peru and S Brazil to C Argentina.

MAGPIE GOOSE ANSERANATIDAE

4 MAGPIE GOOSE *Anseranas semipalmata* 70–90cm Unmistakable, mainly black and white plumage. Male larger than female. Often gathers in large flocks outside breeding season. Feeds on vegetation. Mainly sedentary but some movement related to seasonal food availability. **V** Loud, honking territorial calls. Pairs sometimes duet. Alarm calls include hisses and trills. **H** Swampy grassland, brackish marsh. **D** New Guinea, N Australia.

DUCKS, GEESE AND SWANS ANATIDAE

5 WHITE-FACED WHISTLING DUCK *Dendrocygna viduata* 45–53cm Often in very large flocks. **V** Whistled *tsri-tsri-trseeo*. **H** Freshwater marshes, lakes and rivers. **D** Widespread in South America and Africa.

6 BLACK-BELLIED WHISTLING DUCK *Dendrocygna autumnalis* 48–53cm In flight, shows a bold white bar on upperwing. Juvenile duller, with a grey bill. **V** High-pitched chattering whistle. **H** Freshwater marshes, shallow lakes, coastal lagoons. **D** E Texas to N Argentina.

7 SPOTTED WHISTLING DUCK *Dendrocygna guttata* 43cm Usually found in small flocks; regularly perches in trees. **V** Coarse, whistled *whu-wheouw-whi*. **H** Small lakes, swamps, tree-lined rivers, estuarine mangroves. **D** Philippines to New Guinea.

8 WEST INDIAN WHISTLING DUCK *Dendrocygna arborea* 48–58cm Feeds mainly at night and rests during daylight among swamp vegetation or mangroves. Often seen flying, at dusk, from roost areas. **V** Shrill, whistled *visisee*. **H** Wooded swamps and coastal mangroves, on Royal Palms (Roystonea regia), where it feeds on the fruits; also grazes on agricultural land. Breeds in any month; usually nests in a tree cavity but also in bushes and on the ground. **D** West Indies.

9 FULVOUS WHISTLING DUCK *Dendrocygna bicolor* 45–53cm Usually in small parties. **V** Whistling *k-weeoo* and a harsh *kee*. **H** Lowland freshwater lakes and marshes with fringing vegetation. **D** Tropical America, Africa and India.

10 PLUMED WHISTLING DUCK *Dendrocygna eytoni* 40–45cm Large, upright duck. Diagnostic pale eyes and striking plumes on flanks. **V** Piercing, whistling *whis-whee*. **H** Wet grassland, swamps, lakes. **D** Australia.

11 WANDERING WHISTLING DUCK *Dendrocygna arcuata* 40–45cm Gregarious, often in large flocks. Note white outer uppertail-coverts. **V** High-pitched twittering *pwit-wit-ti-t-t-t...* Also high-pitched whistles, usually in flight. **H** Various wetlands, including freshwater lakes and marshes. **D** Philippines to Australia.

12 LESSER WHISTLING DUCK *Dendrocygna javanica* 38–42cm Usually in flocks of 20–50. **V** In flight, utters a constant thin, whistled *whi-whee*. **H** Freshwater pools, lakes and swamps. **D** Tropical Asia.

13 WHITE-BACKED DUCK *Thalassornis leuconotus* 40cm Note white spot at bill base. White back and rump visible only in flight. **H** Secluded parts of lakes, dams and pools with some floating and fringing vegetation. **V** Extremely high *witleeet* or drawn-out *ooowouw*. **D** Widespread across Africa.

14 CAPE BARREN GOOSE *Cereopsis novaehollandiae* 75–100cm Mainly terrestrial. Visits water only with young or when moulting. **V** Grunts and hisses, uttered mainly in flight. **H** Grassland, scrub, beaches, water margins. **D** S Australia.

15 BRANT GOOSE *Branta bernicla* 55–66cm Small, short-billed, very dark goose. **V** Rolling, gargling *raunk raunk raunk*. **H** Breeds in high-Arctic coastal tundra with pools and inlets. Winters on coastal mudflats and grassland. **D** North America, Europe.

16 RED-BREASTED GOOSE *Branta ruficollis* 53–55cm Forms large winter flocks. **V** High-pitched *kik-yoik kik-yik*. **H** Breeds on tundra or open wooded tundra near rivers. Winters on salt-steppe, lowland pasture, crop and stubble fields. **D** Europe, SW Asia.

17 NENE *Branta sandvicensis* 64cm Note distinctive colour pattern of head and neck. **V** Call is a slightly plaintive *oweeh-oo* or high, goose-like cackling. **H** High, poorly vegetated volcanic slopes. **D** Hawaiian Islands.

18 CANADA GOOSE *Branta canadensis* 90–100cm Gregarious. **V** Varied musical honk. **H** Lakes, ponds, various types of grassland. **D** North America.

19 BARNACLE GOOSE *Branta leucopsis* 58–71cm Forms large winter flocks. **V** Flight notes likened to the yelping of a pack of dogs. **H** Breeds in high-Arctic coastal areas. Winters on coastal grasslands. **D** Greenland to NW Russia.

20 CACKLING GOOSE *Branta hutchinsii* 55cm Split from Canada Goose in 2004. Very small. **V** High-pitched cackling *yelk yelk a'lick a-lick*. **H** Breeds in Arctic coastal tundra. Winters on lakes, marshes, fields. **D** North America.

21 BAR-HEADED GOOSE *Anser indicus* 71–76cm Pale, with unique and distinctive head and neck pattern. **V** In flight, utters a nasal honking. **H** Breeds on high-altitude lakes and marshes. Winters on cultivated fields. **D** C Asia to India.

22 EMPEROR GOOSE *Anser canagicus* 66–89cm Plump, stocky small goose, very dark with contrasting white head and hind-neck. **V** In flight, gives a hoarse, high-pitched *kla-ha kla-ha* or a shrill *yang-yang*. **H** Breeds by coastal tundra lakes, and by pools on inland tundra. Winters on rocky seashores and estuaries. **D** W Alaska, NE Siberia.

23 ROSS'S GOOSE *Anser rossii* 53–66cm Similar to Snow Goose and often joins with flocks of this species. Dainty bill. **V** In flight, gives a grunt-like *kug* and a weak, cackling *kek ke-gak*. **H** Breeds on Arctic tundra. Winters mainly on farmland. **D** North America.

24 SNOW GOOSE *Anser caerulescens* 65–84cm Gregarious; large winter flocks numbering hundreds or thousands. Heavier bill than Ross's Goose. **V** In flight, utters a nasal, cackling *la-luk*, said to resemble the barking of a small dog. **H** Breeds in low tundra close to water. Winters on pastures and stubble fields, usually in lowland coastal areas. **D** North America, NE Siberia.

25 GREYLAG GOOSE *Anser anser* 75–90cm In flight, shows a distinctive pale grey forewing. **V** In flight, gives a deep, honking *aahng-ahng-ung*. All calls similar to those of the domesticated farmland goose. **H** Breeds on wetlands in open country with nearby damp grassland. Winters on farmland, estuaries, lakes, reservoirs. **D** Europe, N Africa, Middle East, Asia.

26 SWAN GOOSE *Anser cygnoides* 81–94cm Two-tone head pattern and black bill distinctive. Widely domesticated; domestic forms often show orange bill. **V** Various honking and cackling calls, uttered mainly in flight. All calls similar to those of the domestic goose. **H** Breeds on steppe and wooded steppe marshes, lakesides, river valleys, deltas. Winters on lowland marshes and wet cultivation. **D** C Asia to SE Siberia.

27 TAIGA BEAN GOOSE *Anser fabalis* 66–84cm Wary. Usually in smaller flocks than other grey geese. **V** *Wink-wink* and deep, nasal *hank-hank*; also typical goose cackling. **H** Breeds on damp tundra, coastal areas, islands, birch and coniferous forests. Winters in open country, including damp steppe and agricultural land. **D** N Europe to Siberia and E Asia.

28 PINK-FOOTED GOOSE *Anser brachyrhynchus* 60–75cm In flight, shows light grey forewing (not as pale as in Greylag Goose). **V** *Wink-wink* and high-pitched *ahng-ahng-ahng*; also a constant cackling from flocks in flight. **H** Breeds on tundra in mountain areas. Winters on salt marshes, lowland farmland. **D** Greenland to NW Europe.

29 TUNDRA BEAN GOOSE *Anser serrirostris* 65–80cm Similar to Taiga Bean Goose with larger bill and reduced orange on bill. **V** Nasal *hank-hank*, deeper than Taiga Bean Goose. **H** Breeds on Arctic tundra and taiga. Winters in open grassland, farmland, marshes. **D** NW Europe to E Asia.

30 GREATER WHITE-FRONTED GOOSE *Anser albifrons* 64–78cm Much larger than Lesser White-fronted Goose; lacks pale eye-ring and has variable black barring on belly. **V** In flight, call is a musical *lyo-lyok*; also typical goose cackling. **H** Breeds on lowland tundra, often by rivers and lakes. Winters on open steppe, grassland, stubble fields, salt marshes. **D** North America, Greenland, N Europe to NE Asia.

31 LESSER WHITE-FRONTED GOOSE *Anser erythropus* 53–66cm Smaller than Greater White-fronted Goose and shows pale eye-ring. Odd individuals sometimes found in flocks of other grey geese. **V** In flight, repeats a squeaky *kyu-yu-yu*. **H** Breeds on damp, bushy tundra, forest edges, mountain foothills. Winters in open areas, including salt steppe, meadows and arable farmland. **D** Eurasia.

32 COSCOROBA SWAN *Coscoroba coscoroba* 110cm Though swanlike, perhaps more closely related to whistling-ducks. Bill bright red. All white, black wing-tips. Pinkish-red feet. **V** Call an onomatopoeic trumpeted 3-syllable *cón currah*. **H** Often off shore. Open saline lagoons, freshwater lakes and marshes with reed-beds, estuaries, sheltered seashore. **D** S South America.

33 BLACK SWAN *Cygnus atratus* 115–140cm Unmistakable. In flight, shows white primaries and outer secondaries. **V** High-pitched musical bugling. **H** Lakes, flooded fields, coastal bays. **D** Australia, New Zealand.

34 BLACK-NECKED SWAN *Cygnus melancoryphus* 120cm Unmistakable. Bill greyish blue; reddish-pink knob. Head and neck black, white post-ocular line. Rest of body all white. Pale pink feet. **V** Call quiet and whistled *whews*. **H** Lakes and lagoons, open water, dense reed-beds with clearings, estuaries, edge of the sea and saline lakes. **D** S South America.

35 MUTE SWAN *Cygnus olor* 125–155cm In flight, wings make a distinctive musical throbbing noise. **V** Utters various hisses, grunts and snorts. **H** Rivers, lakes, reservoirs, large ponds, estuaries, sheltered coastal waters. **D** Europe to C Asia.

36 TRUMPETER SWAN *Cygnus buccinator* 150–180cm Long neck. Dark, long bill. Head and neck sometimes stained yellowish. Habits similar to those of Whooper Swan, which is often considered to be conspecific. **V** Deep, bugling *ko-hoh*. **H** Lakes, ponds, marshes. Winters on estuaries. **D** North America.

37 TUNDRA SWAN *Cygnus columbianus* 115–140cm Neck short. Gregarious, often in very large flocks. **V** In flight, gives a yelping *wow-wow-wow*. **H** Summers among marshy tundra ponds. Winters on coastal marshes, flooded land and pastures. **D** Breeds Alaska, N Canada. Winters E, W USA.

38 WHOOPER SWAN *Cygnus cygnus* 140–165cm Long neck. Elongated, yellow bill patch. **V** Various honking and bugling calls. In flight, gives a deep *hoop-hoop-hoop*. **H** Breeds by pools, lakes and rivers in northern taiga. Winters on farmland, flooded pastures, and occasionally coastal inlets and bays. **D** N Eurasia.

1
2
3
4
5
6
7
8
9
10
11
12
13
14
15
16
17
18
19
20
white phase
21
22
23
blue phase
white phase
24
blue phase
25
26
27
28
29
30
31
32
33
34
35
♂
♀
36
37
38

DUCKS, GEESE AND SWANS *CONTINUED*

1 FRECKLED DUCK *Stictonetta naevosa* 50–56cm Distinctive, freckled plumage. Forms flocks outside breeding season. Sifts water for plants and invertebrates. **V** Mainly silent. **H** Well-vegetated freshwater habitats. **D** SE and SW Australia.

2 BLUE DUCK *Hymenolaimus malacorhynchos* 53cm Unmistakable in its range and habitat. Well camouflaged when perching on and between boulders, but note striking yellow eyes and pale bill. **V** Female utters a low, grating *wrrrreh*, male a very high *feeeh*. **H** Fast mountain streams in forested areas. **D** New Zealand.

3 FLYING STEAMER DUCK *Tachyeres patachonicus* 68cm Smallest member of genus and the only one to fly. Wings reach base of longish tail. Male has a greyish-white head with a white post-ocular streak and eye-ring. Female darker, head dark brownish grey with a white post-ocular line; bill olive with a touch of yellow at base. **V** Male a *tishu tishu*; female a deep *ark ark*. **H** Inland freshwater and brackish ponds and lakes, rivers; sea coasts mostly in winter. **D** S Chile, S Argentina, Falkland Islands.

4 FUEGIAN STEAMER DUCK *Tachyeres pteneres* 80cm Flightless. Heaviest and heftiest member of genus. Wings barely reach base of rump. Generally paler than other species and sexes more uniform. Female has a slightly darker face. Bill of both male and female very thick at base, orangey yellow. **V** Pairs duet, female giving very deep croaks, male producing higher-pitched whooshing *whisha* notes, both accelerating into a rattle. **H** Sheltered waters of channels and fjords. **D** S Chile, S Argentina.

5 FALKLAND STEAMER DUCK *Tachyeres brachypterus* 70cm Flightless. Like Flying Steamer Duck but much larger and heftier. Heavy bill. Wings barely reach rump. Male has a grey head with a white eye-ring and post-ocular streak. Rusty wash on throat and breast, head paling with age to white. Female like female Flying Steamer. **V** Male a sneezy *tissue, tissue*: female rattles a deep *krok krok*. **H** Sea coasts, protected waters of inlets, bays and coves. **D** Falkland Islands.

6 CHUBUT STEAMER DUCK *Tachyeres leucocephalus* 70cm Flightless. Very similar to Falkland Steamer Duck. Dives and upends for marine invertebrates. **V** Whistles and clicks in breeding season. **H** Coasts, coastal pools. **D** Coasts of Chubut, SE Argentina.

7 TORRENT DUCK *Merganetta armata* 38cm Unmistakable. Slender. Long, rigid tail. Narrow red bill and red legs. Male with white head and neck, with black lines along crown and down hindneck, from eye down neck (S form) and across cheek to throat and foreneck. Breast and upper flanks black. Rest of underparts cinnamon, streaked black. Dorsal feathers black, edged white. Incredibly agile at water's surface and when diving. **V** high, whistled *prreeeps*. **H** Whitewater rivers and streams in the Andes. **D** South America.

8 SPUR-WINGED GOOSE *Plectropterus gambensis* 75–100cm In flight, shows a white forewing and extensive white on underwing. Female lacks forehead knob. **V** Mainly silent. **H** Marshes, lakes and rivers in open grassland. **D** Widespread across Africa.

9 COMB DUCK *Sarkidiornis sylvicola* 55–76cm Male has a disc-shaped knob on bill. Unmistakable; belly clean-white. Occurs in pairs or flocks; perches in trees. **V** Short, mid-pitched, rattly croak. **H** Rivers, marshes and lagoons with wooded banks. **D** N Colombia and Venezuela to N Argentina.

10 KNOB-BILLED DUCK *Sarkidiornis melanotos* 56–76cm Usually in small groups. **V** Mainly silent. **H** Lakes, swamps and rivers in open, lightly wooded areas. **D** Sub-Saharan Africa, S and SE Asia.

11 BLUE-WINGED GOOSE *Cyanochen cyanoptera* 70cm Often rests with neck tucked between fluffed-out mantle feathers. **V** Has varied calls including short barks and tuneful whistling notes. **H** Marshes, wet grassland, alpine moorland; >1,750m. **D** Ethiopia.

12 EGYPTIAN GOOSE *Alopochen aegyptiaca* 63–73cm Often perches on trees or buildings. Some individuals greyer than shown. In flight, shows extensive white forewing above and below. **V** Male utters a harsh, wheezy hiss; female has a noisy, braying *honk-haah-haah-haah*. **H** Freshwater lakes and rivers, nearby pastures or parkland. **D** Sub-Saharan Africa.

13 ORINOCO GOOSE *Neochen jubata* 55cm Upright stance. Neck thick. Head, neck and breast buffy cream. Rest of plumage tawny rufous, black and white. Legs reddish pink. Wings black with metallic-green coverts and a white triangle on secondaries. **V** Male makes a whistling *sheweep* and guttural honks; female utters a deep, growled *howg*. **H** Riverbanks, muddy beaches and islands in rivers in forest. **D** Amazonia to N Argentina.

14 ANDEAN GOOSE *Chloephaga melanoptera* 75cm Sexes alike. Neck thick. Bill and legs pink. Mostly white with a black-streaked upper back, and black lower back and tail. Conspicuous wing white with black primaries and a dark purply-green speculum. **V** Male whistles; female utters a gruff *prayer, prayer*. **H** High Andean lakes and streams, short grass; lower altitudes in winter. **D** Peru to S Chile and Argentina.

15 UPLAND GOOSE *Chloephaga picta* 65cm Wings of both sexes striking black and white with dark metallic-green speculum. **V** Male a series of *whee wheee* notes; female honks a loud *har har harrrr* with rolling 'r's at the end. **H** Short grass in valley bottoms near water; croplands in winter; steppe on migration. **D** S South America.

16 KELP GOOSE *Chloephaga hybrida* 60cm Male all white. Legs yellow in both sexes. Female has a pale pink bill and white eye-ring; head, neck and back blackish brown; breast, flanks and upper belly barred black and white; hindparts all white. In flight, female shows striking black-and-white wings with metallic-green speculum. Usually in pairs. **V** Male gives airy whistled *whsh* notes; female a querulous quack. **H** Rocky seashores, feeding on seaweeds at low tide. **D** S Chile, S Argentina, Falkland Islands.

17 ASHY-HEADED GOOSE *Chloephaga poliocephala* 55cm Sexes alike. Head and upper neck grey. Lower neck, upper back and breast reddish chestnut. Flanks barred black and white. Belly white. Tail black. Wings striking black and white with dark metallic-green speculum. **V** Male a series of strong whistled *wheels*; female a honky *crock crock*. **H** Edge of woods, short grass by water; often nests in trees. Agricultural land in winter; steppe on migration. **D** S South America.

18 RUDDY-HEADED GOOSE *Chloephaga rubidiceps* 52cm Like a small female Upland Goose. Head more rufous with finer barring, lower belly and undertail cinnamon, upperparts browner. Wings striking black and white with dark metallic-green speculum. **V** Male gives short, slightly squeaky whistled notes; female gives chattering, excitable quacks. **H** On short grass in valleys in open country, steppe, near water. Winters on agricultural land. **D** S Chile, S Argentina.

19 RAJA SHELDUCK *Radjah radjah* 48cm Generally found in pairs or small groups; often rests on mudbanks or tree branches. In flight from above, shows large white forewing; below, white with black primaries. **V** In flight, male utters a loud, upslurred whistle and female a harsh rattling or grunting. **H** Lakes, ponds, rivers, creeks, coastal flats. **D** New Guinea; N Australia; Maluku Islands, Indonesia.

20 COMMON SHELDUCK *Tadorna tadorna* 58–67cm Extensive white forewing above and below, making it appear black and white in flight. **V** Male utters thin, whistling *sliss-sliss-sliss-sliss*; female gives a rapid, nasal *gag-ag-ag-ag-ag-ak*. **H** Estuaries, sandy or muddy seashores, lakes and rivers. **D** W Europe to NE China.

21 RUDDY SHELDUCK *Tadorna ferruginea* 61–67cm In flight, shows extensive white forewing above and below. **V** Utters a honking *aakh* or *ah-onk*, also a repeated, trumpet-like *pok-pok-pok-pok*. **H** Lakes and rivers in open country, sometimes far from water on upland plateaux. Winters by lowland lakes and rivers. **D** S Europe to C Asia, NW Africa, Ethiopia.

22 SOUTH AFRICAN SHELDUCK *Tadorna cana* 65cm From Egyptian Goose by leg colour, less straight neck, black bill and thin black line through white on forewing. Note that male and female differ. **V** Male utters a mid-high, rather timid *raak raak raak*, female a *roh roh roh*. **H** Fresh and brackish waterbodies of any type and size; may breed away from water. **D** Namibia, Botswana, South Africa.

23 AUSTRALIAN SHELDUCK *Tadorna tadornoides* 55–72cm Flocks visit traditional grounds to moult outside breeding season. Very wary. Nests in holes in trees and banks. **V** Honking calls, given in flight. **H** Brackish lagoons, freshwater lakes. **D** S Australia.

24 PARADISE SHELDUCK *Tadorna variegata* 65cm No similar bird in range. White head of female especially distinctive. **V** Mewing, drawn-out quacks. **H** Farmland, pastures, dams, pools, mountain streams. **D** New Zealand.

25 CRESTED SHELDUCK *Tadorna cristata* 60–63cm In flight, shows extensive white forewing above and below. **V** Not recorded. **H** Thought to have bred by rivers in montane forests and wintered in coastal districts. **D** NE Africa. Possibly extinct.

26 PINK-EARED DUCK *Malacorhynchus membranaceus* 40cm Unmistakable by barred flanks and relatively large bill. **H** Normally on shallow water. **V** Very high, liquid, fluted chattering. **D** Australia.

27 SALVADORI'S TEAL *Salvadorina waigiuensis* 38–42cm Generally secretive. Favours fast-flowing torrents and pools. Upends and dives for freshwater invertebrates. **V** Mainly silent. **H** Mountain streams, lakes; usually 2,000–4,000m. **D** New Guinea.

28 MUSCOVY DUCK *Cairina moschata* 66–84cm In flight, both sexes show large white forewing and underwing. Domestic forms may have various white patches or are mostly white with a red face. **V** Usually silent. Male may give a low hiss, female a weak *quack*. **H** Wooded lakes, pools, marshes, rivers. **D** S Texas to N Argentina.

29 WHITE-WINGED DUCK *Asarcornis scutulata* 66–81cm Shy and chiefly nocturnal. In flight, shows extensive white forewing. Female as male but duller. **V** Generally silent. In flight, may utter a wailing honk, ending in a nasal whistle. **H** Streams and pools in tropical forests and nearby open swamp areas. **D** SE Asia.

30 HARTLAUB'S DUCK *Pteronetta hartlaubii* 50cm Differentiated from most other ducks in its region by habitat. Blue forewing diagnostic. **V** Very low, soft, raucous, muttered *rrarrarrar-ra*. **H** Well-wooded streams, especially in forests. **D** Sierra Leone to SW Uganda and N Angola.

31 WOOD DUCK *Aix sponsa* 43–51cm Eclipse male similar to female but with a pink bill. **V** Male utters a thin, rising *jeeeeee*; female has a sharp *cr-r-ek cr-r-ek* and squealed *oo-eek* when flushed. **H** Lakes, ponds and rivers in wooded country. **D** North America.

32 MANDARIN DUCK *Aix galericulata* 41–49cm Eclipse male similar to female but with a pink-red bill. **V** Generally silent. During display, male utters a whistle and a sharp *hwick* in flight; female sometimes gives short clucking notes. **H** Lakes and rivers with surrounding trees and bush. **D** SE Siberia, E Asia.

33 MANED DUCK *Chenonetta jubata* 50cm Note three parallel stripes over back and thin-necked jizz. **V** Low, grumbled shrieks in series: *wreeék - -*. **H** Short-grassed areas near smaller water bodies often surrounded by woodland or forest. **D** Australia.

34 AFRICAN PYGMY GOOSE *Nettapus auritus* 30cm In flight, note white secondaries. **V** Very high, fluted, short *feewee-feewee-tutukweet* or rapid, liquid *ripripripripri*. **H** Quiet, clear waters with some floating and emergent vegetation. **D** Sub-Saharan Africa.

35 COTTON PYGMY GOOSE *Nettapus coromandelianus* 30–37cm Male upperwing shows white primaries with black tips and white trailing edge to secondaries. Female upperwing lacks white primaries. **V** Male utters a sharp *car-car-carawak*; female gives a weak quack. **H** Vegetated lakes and pools. **D** Tropical Asia, New Guinea, Australia.

36 GREEN PYGMY GOOSE *Nettapus pulchellus* 32–36cm Usually seen in pairs or small groups, easily overlooked while swimming among floating vegetation. In flight, shows extensive white secondaries. **V** In flight, utters a shrill, whistled *tii-whit* or *whit*. **H** Freshwater lakes, swamps and rivers, especially with floating or submerged vegetation. **D** New Guinea; N Australia; west to Lesser Sunda Islands, Indonesia.

37 BRAZILIAN TEAL *Amazonetta brasiliensis* 41cm Bill and legs bright red. Body brown and buffy brown, tawnier on breast and with black spots on flanks. Cheeks and part of neck whitish. Crown and tail blackish. Wings black with startling green-and-blue sheen on upper side and contrasting white triangle on secondaries. **V** Male's call a whistled *wheeoh*; female a short, mid-pitched quack. **H** Water with floating vegetation, streams, flooded savannah, rice paddies. **D** SE Amazonia.

38 RINGED TEAL *Callonetta leucophrys* 36cm Male showy: bill light blue, legs pale pink. Crown, line down hindneck and half-collar black, framing buffy-cream face. Breast pinkish, dotted black. Flanks vermiculated grey. Back reddish chestnut. Rump, tail and undertail black with a large white patch on sides. Wings black with a metallic-green speculum and white oval on greater coverts. **V** Whistled, downslurred *wheer-oh* or more strained, lower-pitched *whrr-oh*. **H** Marshes with floating vegetation and nearby trees, flooded woods and savannah, roadside ditches, temporary ponds. **D** SE Bolivia to SE Brazil and Uruguay.

1
2
3
4
5
6
7
8
9
10
11
12
13
14
15
16
white form
17
18
19
20
21
22
23
24
25
26
27
28
29
white-headed form
30
31
32
33
34
35
36
37
38

DUCKS, GEESE AND SWANS *CONTINUED*

1 CRESTED DUCK *Lophonetta specularioides* 55cm Somewhat conspicuous nuchal crest. Dark mask. Mostly brownish buff with dark brown scaling. Lower upperparts dark. Conspicuous black-and-white wing with purply-copper wing speculum seen in flight. Primaries and tail blackish. **V** Male makes a *djeeeew* call; female quacks. **H** Inlets and sea, high mountain lakes and rivers, lagoons. **D** Peru, Chile, Argentina, Falkland Islands.

2 BRONZE-WINGED DUCK *Speculanas specularis* 52cm White 'spectacles' and 'choker'. Rest of head and upper neck dark chocolate. Flanks and upper back blackish brown. Breast and rest of underparts buffy brown. Wings black with greenish-blue gloss on coverts and purply-copper speculum with black-and-white trailing edge. **V** Female 'barks'. **H** Mountain tarns above treeline, lakes, fast-flowing rivers near woods. **D** S Chile, S Argentina.

3 BAIKAL TEAL *Sibirionetta formosa* 39–43cm Breeding male unmistakable; eclipse male shows 'shadow' of distinctive adult facial pattern. Forms large winter flocks, often mixed with other duck species. **V** Male utters a *wot-wot-wot*, often persistently; female utters a low quack. **H** Breeds by pools in taiga swamps and marshes at the edge of tundra. Winters in various fresh and brackish waterbodies. **D** E Siberia, E Asia.

4 GARGANEY *Spatula querquedula* 37–41cm In flight from above, male shows pale grey forewing, separated from green secondaries by a white wing-bar. At a distance, can look white-winged. Usually in pairs or small parties. After breeding forms large flocks. **V** Male utters a rattling *knerek*; female has a high, nasal quack. **H** Freshwater lakes and marshes with extensive vegetation. Winters on open freshwater lakes. **D** W Europe to Japan.

5 HOTTENTOT TEAL *Spatula hottentota* 35cm Note white cheek with dark smear and bluish (not red) bill. **V** Very high, dry, chuckled *kehehe-heheh*. **H** Shallow water with fringing reedy vegetation, sewage ponds, large open dams. **D** Sub-Saharan Africa, Madagascar.

6 PUNA TEAL *Spatula puna* 48cm Like Silver Teal but larger, and bill heavier and blue with a black culmen. Cap black, cheeks and neck creamy, flanks and tail vermiculated. **V** Unobtrusive, very dry rolling rattle. **H** High Andean lakes. **D** Peru to NW Argentina.

7 SILVER TEAL *Spatula versicolor* 41cm Bill blue and yellow with a black culmen. Cap brownish black. Rest of head and neck buffy cream, breast dotted black. Flanks barred black and whitish, more finely so on rump and tail. **V** Male calls a quiet non-vocal *rrrrrrrr*. **H** Marshes, ponds and lakes, streams with reeds and floating vegetation, flooded grasslands. **D** S Brazil to S Chile and S Argentina, Falkland Islands.

8 RED SHOVELER *Spatula platalea* 48cm Bill large, black and slightly spoon-shaped. Eyes white. Head and neck creamy, flecked blackish. Breast, flanks, belly and upper back tawny orange, dotted black. Long, sharp tail black and white. Female dark brown and buff, spotted and scaled. **V** Gives short, squeaky quacks and toneless dry rattles. **H** Lakes, marshes, temporary ponds, flooded grassland, streams, ditches. **D** S Peru and S Brazil to Tierra del Fuego.

9 CINNAMON TEAL *Spatula cyanoptera* 38–48cm Wing pattern of both sexes similar to that of Blue-Winged Teal. **V** Similar to Blue-winged Teal. **H** Shallow freshwater lakes and marshes. **D** W North America and Caribbean to S South America, Falkland Islands.

10 BLUE-WINGED TEAL *Spatula discors* 37–41cm Male unmistakable. Female has distinctive pale oval area at base of bill. In flight, both sexes show a pale, bright blue forewing. No white trailing edge to secondaries. **V** Males utter a thin *tsee-tsee*; female has a high-pitched quack. **H** Open country and grasslands with shallow lakes and pools. **D** S Canada , USA.

11 CAPE SHOVELER *Spatula smithii* 55cm Male and female differ from male Northern Shoveler by black bill and darker plumage. **V** Female makes a mid-high Mallard-like *week-wekwekwek*; male call is a low *rokrokrokrok*. **H** Sewage ponds, lagoons, estuaries; avoids large bodies of open water. **D** Namibia, Botswana, South Africa.

12 AUSTRALASIAN SHOVELER *Spatula rhynchotis* 50cm Breeding male easily separable from Northern Shoveler by facial pattern and scaled, not all-white breast and upper mantle; female from Northern by greenish dusky bill, missing orange tones at gape. **V** Irregular *quack* notes. **H** Large wetlands, floodplains and lakes. Occasionally on saline waters. **D** S Australia, New Zealand.

13 NORTHERN SHOVELER *Spatula clypeata* 44–52cm Male unmistakable. In flight, looks 'front heavy'. Upperwing shows pale blue forewing, separated from green secondaries by a white wing-bar; female wing duller, with a smaller wing-bar. **V** During display, male gives a hollow *sluk-uk* or *g-dunk*; female has a quacking *gak-gak-gak-ga-ga*. **H** Freshwater lakes and marshes; in winter also on estuaries and coastal lagoons. **D** North America, Europe.

14 GADWALL *Mareca strepera* 46–55cm In flight, both male and female show white secondary patch on upperwing. Usually found in small groups. Can be quite wary. **V** Usually fairly silent. Male gives a sharp *ahrk* and a low whistle; female has a mechanical Mallard-like quack. **H** Breeds and winters in lowland, open-country freshwater areas with fringing vegetation, less often on estuaries. **D** North America, Europe.

15 FALCATED DUCK *Mareca falcata* 48–54cm Sociable, forming large winter flocks that often include Eurasian Wigeon and Northern Pintail. **V** Male gives a low whistle, followed by a wavering *uit-trr*; female utters a throaty quack. **H** Breeds around lowland water meadows and lakes. Winters in similar habitats, as well as rivers and, to a lesser extent, estuaries. **D** E Siberia, N and E Asia.

16 EURASIAN WIGEON *Mareca penelope* 45–50cm In flight, male shows a large white patch on upperwing. Very gregarious outside the breeding season. Grazes more than most other ducks. **V** Male gives a clear, whistling *wheeooo*; female makes a growling *krrr*. **H** Breeds near lakes and marshes in open or light wooded country. Winters mainly on estuaries, flooded water meadows and, to a lesser extent, on internal freshwater lakes. **D** N Eurasia.

17 CHILOE WIGEON *Mareca sibilatrix* 46–56cm Sexes similar, with a white face, and black head and neck with greenish or purple sheen. Whitish patch on ear-coverts. Back black with sharply contrasting white edges to feathers. Breast barred blackish and white. **V** Male's call a strong, whistled *whooooeeeteeeooo*, louder and higher in the middle. **H** Shores of marshes, lakes and rivers. **D** S South America.

18 AMERICAN WIGEON *Mareca americana* 45–50cm In flight, upperwing pattern similar to that of Eurasian Wigeon. Centre of underwing white (usually greyer in Eurasian Wigeon). **V** Similar to that of Eurasian Wigeon but weaker and throatier. **H** Breeds in similar situations to Eurasian Wigeon; in winter prefers freshwater marshes. **D** N North America.

19 AFRICAN BLACK DUCK *Anas sparsa* 55cm Evasive; swims between and under fringing vegetation. In pairs. **V** Mid-high, loud Mallard-like *crick crick crick*. **H** Well-wooded montane streams, in open pools and streams at highest altitudes; 1,500–4,250m. **D** Widespread across Africa.

20 YELLOW-BILLED DUCK *Anas undulata* 55cm Yellow bill very striking and diagnostic. **V** As Mallard. **H** Lakes, dams, rivers, estuaries, pools, sewage ponds. **D** E, EC and S Africa.

21 MELLER'S DUCK *Anas melleri* 60cm Very similar to female Mallard but with a darker head. **V** Quacking notes, very like those of Mallard. **H** Freshwater bodies and swamps at any altitude. **D** Madagascar.

22 PACIFIC BLACK DUCK *Anas superciliosa* 47–60cm Usually encountered in pairs or small groups. In flight, shows extensive white underwing coverts. **V** Similar to that of Mallard, but hoarser. **H** Various wetlands, including lakes, ponds, marshes and estuaries. **D** Sunda Islands, Indonesia, to New Guinea, Australia and east to French Polynesia.

23 LAYSAN DUCK *Anas laysanensis* 41cm Very small and dark. No clear distinction between male and female, and amount of white to face sides varies in both sexes. Prefers to walk. **V** Fast, harsh-toned quacking *gak-gak-gak*. **H** Lagoon of Laysan, from where wanders throughout the island. **D** Laysan Island, Hawaiian Islands.

24 HAWAIIAN DUCK *Anas wyvilliana* 51cm Resembles female Mallard, but smaller, darker and with a rufous tinge, especially below. Speculum more green and less blue than in Mallard. **V** Mallard-like vocalisations, but slightly higher and thinner. **H** Any freshwater body. **D** Hawaiian Islands.

25 PHILIPPINE DUCK *Anas luzonica* 48–58cm Unmistakable. In flight from below, shows whitish coverts and dark greyish primaries and secondaries. **V** Much like that of Mallard but slightly more harsh. **H** Lakes, ponds, rivers, marshes, tidal creeks. **D** Philippines.

26 INDIAN SPOT-BILLED DUCK *Anas poecilorhyncha* 58–63cm In flight from above, shows large white patch on tertials; from below, dark primaries and secondaries contrast with white coverts. **V** Almost indistinguishable from Mallard. Generally silent. **H** Well-vegetated, shallow freshwater lakes and marshes, less often on rivers. **D** S and SE Asia, from Pakistan to SW Yunnan province, China.

27 EASTERN SPOT-BILLED DUCK *Anas zonorhyncha* 55–63cm In flight, shows less of a white patch on tertials than Indian Spot-billed Duck. **V** Calls similar to Mallard. **H** Well-vegetated lakes, pools and marshes. **D** China, SE Asia.

28 MALLARD *Anas platyrhynchos* 50–65cm Many of the 'odd' ducks on ornamental ponds are descendants of this species. **V** Male utters a rasped *kreep*; female's *quack-quack-quack* is probably one of the best-known bird calls. **H** Virtually any river, lake, pond or estuary. **D** Widespread across Europe, Asia, North America and N Africa.

29 MOTTLED DUCK *Anas fulvigula* 53–61cm Yellow bill. Pale sides of head and neck contrast with brown body. All habits and actions similar to those of Mallard, although usually occurs in smaller groups. **V** Similar to that of Mallard. **H** Coastal and inland marshes. **D** SC and SE North America.

30 AMERICAN BLACK DUCK *Anas rubripes* 53–61cm In flight, shows a striking white underwing. Beware, melanistic Mallards can look similar. **V** Similar to that of Mallard. **H** Similar to that of Mallard. **D** E Canada and NE USA.

31 MEXICAN DUCK *Anas diazi* 60cm Darker than female Mallard, with a more distinct eye stripe and streak over crown, and lacking whitish tail. Male has a conspicuous yellow bill. **V** Very like that of Mallard. **H** Ponds, marshes, fields. **D** SW USA to NC Mexican highlands.

32 CAPE TEAL *Anas capensis* 45cm Note pale head and black-based red bill. **V** Extremely high, smooth *feeeweeet* and Mallard-like *week-week-week*. **H** Mainly brackish waters, sewage ponds, alkaline lakes. **D** Chad, Sudan and Ethiopia to South Africa.

33 WHITE-CHEEKED PINTAIL *Anas bahamensis* 41–51cm Usually seen singly, in pairs or in small groups. **V** Male utters a low whistle; female has a descending series of quacks. **H** Freshwater or saltwater pools and lagoons, mangrove swamps, creeks, estuaries. **D** Widespread in South America, Caribbean, Galápagos Islands.

34 RED-BILLED TEAL *Anas erythrorhyncha* 45cm Unmistakable. Note dark cap and red bill. In flight, large, creamy speculum. Gregarious. **V** High, Mallard-like *pjehpjeh-pjeeh-pjeeh*. **H** Open, shallow freshwater lakes, dams and inundations with submerged, floating and fringing vegetation. **D** South Sudan and Ethiopia to South Africa.

35 YELLOW-BILLED PINTAIL *Anas georgica* 48cm Bill yellow with a black culmen. Brownish buff and dark brown, scaled especially on breast and flanks. Tail long and pointed. **V** Call a quiet, fast *piriweep* whistle. **H** Marshes, lakes, flooded grassland, stubble fields, roadside ditches, rice paddies, seashores. **D** W and S South America.

36 EATON'S PINTAIL *Anas eatoni* 35–45cm Recalls female Northern Pintail. Varied diet includes seeds, plant matter and invertebrates. Often feeds in flocks. **V** Wheezy whistle. **H** Freshwater pools, lakes. Coastal habitats in winter. **D** Kerguelen Islands, Indian Ocean.

37 NORTHERN PINTAIL *Anas acuta* 51–56cm (male with tail 61–66cm) In flight, long neck and tail make for elongated look. Very gregarious. **V** Male's call is a mellow *proop-proop*, similar to that of Eurasian Teal (Plate 7); female utters a series of weak quacks and a low croak when flushed. **H** Breeds in wet meadows, marshy lakes and by slow rivers. Winters on lakes, estuaries, coastal lagoons. **D** N Palearctic, North America, N Africa, tropical S Asia.

1
2
3 ♂
4 ♂
5
6
7
♀ 11 ♂
♀
8 ♂
9 ♂
10 ♂
♀
♀
♀
♀
12 ♂
13 ♂
14 ♂
15 ♂
♀
16 ♂
17 ♂
♀
18 ♂
♀
23 ♀
24 ♀ ♂
19
20
21
22
25
♀
27
♀
26 ♂
♀
30 ♂
28 ♂
29
31
32
33
34
35
36 ♂ ♀
37 ♂ ♀

DUCKS, GEESE AND SWANS *CONTINUED*

1 EURASIAN TEAL *Anas crecca* 34–38cm Forms very large winter flocks. Flight is rapid, with much twisting and turning. **V** Male utters a soft, high-pitched *preep-preep*; female rather silent, although often gives a nasal quack when alarmed. **H** Freshwater lakes and pools with fringing vegetation. In winter, also salt marshes, estuaries, sheltered coastal bays. **D** N and C Europe and Asia.

2 GREEN-WINGED TEAL *Anas carolinensis* 34–38cm Formerly considered a race of Eurasian Teal. Female inseparable from that species. **V** As Eurasian Teal. **H** As Eurasian Teal. **D** North America.

3 YELLOW-BILLED TEAL *Anas flavirostris* 38cm Bill yellow with black culmen, neck short. Brownish grey, darker on head, with slight occipital crest on male. Breast spotted black, flanks and undertail plain brownish grey. **V** Male's call is a high-pitched, fast *piriweep* whistle; female's a quiet *jziiiu*. **H** Lakes, ditches, streams, flooded grasslands. **D** South America.

4 ANDEAN TEAL *Anas andium* 35–45cm Resembles Yellow-billed Teal (of which it was previously considered a race) but note dark grey bill. **V** Clipped, mid-pitched single quacks. **H** High Andean lakes, swamps and streams. **D** Colombia, Ecuador.

5 SUNDA TEAL *Anas gibberifrons* 42cm Generally seen in pairs or small flocks. Male shows a 'bulge' on forehead. In flight, upperwing shows a white mid-wing panel. **V** Male utters a clear *pip*; female gives a wild, laughing cackle. **H** Small lakes, swamps, forest streams, rice fields, mudflats, mangroves. **D** Indonesia.

6 ANDAMAN TEAL *Anas albogularis* 37–47cm White facial markings variable: two extremes shown. In flight, upperwing shows a broad white wing-bar, underwing shows white axillaries. **V** Male utters a clear, low *preep*; female has a loud, laughing series of quacks. **H** Freshwater marshes and pools, paddyfields, mangrove swamps, coastal lagoons and estuaries. **D** Andaman Islands, Cocos Islands.

7 GREY TEAL *Anas gracilis* 42cm Usually occurs in pairs or small flocks. Very similar to Sunda Teal, but less brown; does not have a forehead 'bulge'. **V** Male gives a muted whistle; female utters a rapid series of harsh, laughing quacks. **H** Lakes, swamps, rivers, lagoons, shallow coastal waters. **D** New Guinea, New Caledonia, Australia, New Zealand.

8 CHESTNUT TEAL *Anas castanea* 35–45cm Filters mud while dabbling and upending. Usually nests in tree holes. **V** Whistles and laugh-like quacks. Noisiest in flocks. **H** Estuaries, coastal lagoons, marshes beside rivers. **D** Australia.

9 BERNIER'S TEAL *Anas bernieri* 40cm Pale reddish brown with a pink bill. In flight, shows a conspicuous white-and-black speculum. **V** Thin, dry, rather strained whistles. **H** Small, shallow saline lakes with adjacent vegetation and mangroves. **D** Madagascar.

10 BROWN TEAL *Anas chlorotis* 48cm Resembles Chestnut Teal but eyes dark with white eye-ring and wing pattern differs. **V** Male utters soft trills in alarm; female has rapid, high-pitched quacks and rasping growl. **H** Estuaries and wetlands with open water. **D** New Zealand.

11 AUCKLAND TEAL *Anas aucklandica* 48cm Separated from Brown Teal by range. Flightless. **V** Duet-like, fast chatter, partly sizzling and partly quacking. **H** Pools and creeks in moorland. **D** Auckland Islands, subantarctic SW Pacific Ocean.

12 CAMPBELL TEAL *Anas nesiotis* 40cm Separated from Brown Teal by range and small size. Flightless. **V** Long, high, shivering twitters. **H** Gullies and pools on tussock slopes. **D** Campbell Islands, subantarctic SW Pacific Ocean.

13 MARBLED DUCK *Marmaronetta angustirostris* 39–42cm Gregarious. Often difficult to see when in lakeside vegetation. Upperwing shows pale grey-buff secondaries and primaries, the latter with dark tips. **V** Generally silent. **H** Shallow freshwater and brackish lakes with developing and border vegetation. **D** S Spain and NW Africa to Pakistan.

14 PINK-HEADED DUCK *Rhodonessa caryophyllacea* 60cm Secretive. Feeds by both dabbling and diving. May perch in trees. **V** Male gives a low, weak, wheezy whistle; female utters a low quack. **H** Secluded marshes and pools in elephant-grass jungle. **D** NE Asia, E India, Myanmar. Probably extinct.

15 RED-CRESTED POCHARD *Netta rufina* 53–57cm Male in eclipse plumage has red bill, otherwise similar to female. Upperwings of both sexes show white flight feathers with a black trailing edge. **V** Generally silent. **H** Freshwater lakes with fringing vegetation, rivers, river deltas, coastal waters. **D** C and S Europe to N China.

16 ROSY-BILLED POCHARD *Netta peposaca* 55cm Male with a pinkish-red bill and caruncle. Head, neck, breast and upperparts black, the first two with a purply gloss. Wings with striking white bar. Female with a pale blue-grey bill. Brown, darker above. **V** Deep *aaark* call. **H** Lakes with reeds, marshes, streams, ditches, flooded grasslands, stubble fields, rice paddies. **D** Brazil and Paraguay to C Argentina and C Chile.

17 SOUTHERN POCHARD *Netta erythrophthalma* 48–51cm Usually in small groups, with larger concentrations after breeding. Upperwing of both sexes shows distinctive white flight feathers with a black trailing edge. **V** Generally silent. **H** Large freshwater and brackish lakes. **D** E, SE and S Africa, Amazonia.

18 CANVASBACK *Aythya valisineria* 48–61cm (Vagrant) Gregarious. In flight shows uniform grey upperwing. **V** Generally silent away from breeding sites. **H** Marshes with areas of open water. Winters on open lakes, estuaries, sheltered coastal waters. **D** North America.

19 REDHEAD *Aythya americana* 45–56cm Gregarious. In flight, upperwing of both sexes shows pale secondaries contrasting with rest of wing. **V** Generally silent away from breeding sites. **H** Freshwater lakes and marshes. In winter, also on tidal bays and brackish lagoons. **D** North America.

20 COMMON POCHARD *Aythya ferina* 42–49cm Gregarious. From above, both sexes show pale grey flight feathers contrasting with darker grey forewing. **V** Generally silent. **H** Lakes with surrounding vegetation. In non-breeding season also on open lakes, reservoirs and occasionally coastal waters. **D** W Europe to C Asia and N China.

21 HARDHEAD *Aythya australis* 42–59cm Usually occurs in flocks. Upperwing of both sexes shows a white bar across primaries and secondaries. **V** Generally silent. **H** Swamps and vegetated lakes. In winter, prefers large open lakes, rivers, coastal lagoons. **D** Australia, New Zealand, New Caledonia, Vanuatu.

22 MADAGASCAN POCHARD *Aythya innotata* 45cm No other pochard or duck with much white in wings occurs in Madagascar. **V** Not very vocal, a *crek crek crek*, when taking flight. **H** Freshwater bodies. **D** Madagascar.

23 BAER'S POCHARD *Aythya baeri* 41–46cm Occurs in pairs or small parties. In flight, upperwing of both sexes has white secondaries and inner primaries, contrasting with dark outer primaries, trailing edge and forewing. **V** Generally silent. **H** Freshwater lakes and pools with developing and fringing vegetation. **D** SE Siberia and N China.

24 FERRUGINOUS DUCK *Aythya nyroca* 38–42cm Occurs in pairs or small flocks. In flight, upperwing of both sexes has distinct wide white bar on secondaries and primaries. **V** Generally silent. **H** Freshwater lakes with surrounding and developing vegetation. Winters on open water and coastal lagoons. **D** W Europe and NW Africa to C Asia.

25 NEW ZEALAND SCAUP *Aythya novaeseelandiae* 40cm Normally unmistakable thanks to dark plumage (head and neck blackest) and black-tipped all-grey bill. Note white at bill base. In flight, shows restricted pale area on underparts. **V** High, shivering whistles and low quacks. **H** Open freshwater lakes. **D** New Zealand.

26 RING-NECKED DUCK *Aythya collaris* 37–46cm Occurs in small groups; larger flocks in winter. In flight, both sexes show pale grey secondaries contrasting with darker primaries and black forewing. **V** Generally silent. **H** Freshwater lakes and pools. Winters on large lakes, brackish lagoons and tidal bays. **D** North America.

27 TUFTED DUCK *Aythya fuligula* 40–47cm Gregarious. In flight, upperwing of both sexes shows distinct wide white bar on secondaries and primaries. **V** Usually silent. **H** Freshwater lakes, ponds and rivers with fringing vegetation. In winter, also sheltered coasts, brackish lagoons, estuaries. **D** Europe.

28 GREATER SCAUP *Aythya marila* 40–51cm Gregarious. In flight, upperwing of both sexes resembles that of Tufted Duck but with paler forewing. **V** Mostly silent. **H** Freshwater pools and lakes in tundra fringe areas. Winters on coastal waters and, locally, on freshwater lakes and reservoirs. **D** North America, Europe.

29 LESSER SCAUP *Aythya affinis* 38–46cm Gregarious. In flight, upperwing of both sexes shows wide white bar restricted to secondaries, grey inner primaries. **V** Generally silent. **H** Breeds on freshwater lakes and pools. Winters on lakes, estuaries and sheltered coastal waters. **D** North America.

4
1
3
2
9
5
7
6
8
12
11
10
13
15
16
14
18
19
17
21
20
22
23
24
25
26
27
28
29

DUCKS, GEESE AND SWANS *CONTINUED*

1 STELLER'S EIDER *Polysticta stelleri* 43–47cm Secondaries on female upperwing show a white bar and trailing edge. Male upperwing dark with a large white forewing patch. **V** Male relatively silent; female utters barks, growls and whistles. **H** Breeds on Arctic tundra. Winters on inshore coastal waters. **D** Siberia, Alaska.

2 SPECTACLED EIDER *Somateria fischeri* 52–57cm In flight, male wing pattern similar to that of Common Eider. **V** Male utters a weak crooning note; female has a harsh croak. **H** Breeds on coastal tundra, locally inland. Winters at sea. **D** Siberia, Alaska.

3 KING EIDER *Somateria spectabilis* 47–63cm In flight, male upperwing is black with a white forewing patch. **V** Similar to Common Eider; in flight, utters a low croak. **H** Breeds on high-Arctic tundra. Winters on open sea. **D** Arctic.

4 COMMON EIDER *Somateria mollissima* 50–71cm In flight, male has black flight feathers; rest of wing white. Female coloration varies from rufous to greyish brown. **V** Male utters a crooning *ahoooo*; female makes a grating *krrr*. **H** Breeds on coastal shores, including small islands. Winters mainly on coastal waters. **D** Arctic.

5 HARLEQUIN DUCK *Histrionicus histrionicus* 38–45cm Eclipse male as breeding female but with pale stripe on side of breast and white-edged scapulars. **V** Male utters a high-pitched whistle during display; female has various harsh calls. **H** Breeds by fast-flowing mountain rivers. Winters on coastal waters, especially rocky bays. **D** NE Siberia, NW North America, NE Canada, Greenland, Iceland.

6 SURF SCOTER *Melanitta perspicillata* 45–66cm (Vagrant) Gregarious, occurring in small parties up to large winter flocks. In Palearctic, stragglers found among Common Scoters. **V** Generally silent. **H** Breeds in freshwater bodies in boreal forests and tundra. Winters on shallow coastal waters. **D** Alaska, N and E Canada.

7 VELVET SCOTER *Melanitta fusca* 51–58cm Forms large flocks throughout the year. In flight, both sexes show white secondaries. **V** Male utters a loud piping; female makes a hoarse *braa-ah-braa-ah*. **H** Breeds by freshwater lakes. Winters on coastal waters. **D** N Europe, NW Siberia.

8 WHITE-WINGED SCOTER *Melanitta deglandi* 51–58cm Gregarious. In flight, both sexes show white secondaries. **V** Displaying male utters a thin whistle; both sexes give a harsh croak. **H** Lakes and large pools. Winters on inshore coastal waters and occasionally on inland lakes. **D** E and W coasts of North America.

9 COMMON SCOTER *Melanitta nigra* 44–54cm Gregarious all year; in winter usually forms very large flocks. Often seen flying low over sea in long, undulating lines. **V** Male gives a soft *pju*; female utters a grating *karr*. **H** Breeds by tundra lakes and pools. Winters mainly on coastal waters. **D** N Europe to NC Siberia.

10 BLACK SCOTER *Melanitta americana* 44–54cm Regularly in very large numbers mixed with other scoters. Often seen flying low over the sea in long, undulating lines. **V** Males gives a slurred *peeeew* and a whistled *cree*. **H** Tundra lakes and pools. Winters mainly on inshore coastal waters. **D** Alaska, NW and NE Canada, Siberia.

11 LONG-TAILED DUCK *Clangula hyemalis* 36–47cm (male with tail 48–60cm) Often forms large, often sexually segregated, winter flocks. **V** Male gives a yodelling *ow-ow-owlee... caloocaloo*; female has various weak quacks. **H** Breeds on Arctic tundra, lakes and pools, also by rivers and coastal inlets. Winters on coastal waters. **D** Siberia to NE Canada.

12 BUFFLEHEAD *Bucephala albeola* 32–39cm Unmistakable. Diagnostic white head patch shows well in flight. Wing pattern of male is similar to Common Goldeneye. **V** Generally silent apart from an occasional growl given by male; female utters a series of guttural notes. **H** Lakes, pools and rivers in wooded country. Winters on large lakes, rivers and coastal waters. **D** Alaska, Canada, NW USA. Winters south to Mexico.

13 COMMON GOLDENEYE *Bucephala clangula* 42–50cm Male has large white patch on upperwing secondaries and coverts. Female has same patch but split by two black bars on coverts. **V** During display male gives strange whistles and dry notes; female gives a purring *brra brra*. **H** Breeds in woodland close to rivers and lakes. Winters on lakes, reservoirs, coastal waters. **D** North America, Eurasia.

14 BARROW'S GOLDENEYE *Bucephala islandica* 42–53cm Male upperwing similar to that of male Common Goldeneye but with black bar dividing secondary and median coverts. Female with less white on upperwing coverts, which are split by one black bar. **V** During display male utters a soft *ka-kaa*; female has several low growling notes. **H** Breeds on lakes, pools, rivers. Often on inshore coastal waters in winter. **D** Canada, NW USA, Iceland.

15 SMEW *Mergellus albellus* 38–44cm Often in sexually segregated flocks in winter. Upperwing of both sexes shows a white wing patch. **V** During display male gives low croaks and whistles, otherwise fairly silent; female gives a low growling. **H** Breeds by lakes, pools and rivers in forest areas. Winters on lakes, estuaries, coastal bays. **D** N Eurasia.

16 HOODED MERGANSER *Lophodytes cucullatus* 42–50cm Unmistakable. **V** Generally silent, but displaying male utters a frog-like *crrrooooo*. **H** Forest lakes and rivers. In winter also on larger lakes and coastal lagoons. **D** Alaska to NW USA, SE Canada to SE USA.

17 BRAZILIAN MERGANSER *Mergus octosetaceus* 55cm Long and slender black bill with 'teeth'. Head, upper neck and drooping crest black with a green sheen. Upperparts black. Lower neck, breast, flanks and underparts vermiculated grey and black. Legs pinkish red. **V** Female gives harsh, crow-like rasping croak; male's call higher pitched. **H** Rivers and streams in dense forest and gallery forest. **D** SE Brazil, E Paraguay, NE Argentina.

18 COMMON MERGANSER *Mergus merganser* 58–72cm Male upperwing shows large white patch, this less extensive in female. **V** Displaying male gives a twanging *uig-a*; female utters various harsh notes. **H** Lakes and rivers in wooded areas. Winters mainly on freshwater lakes. **D** North America, Eurasia.

19 RED-BREASTED MERGANSER *Mergus serrator* 52–58cm Both sexes show much white on upperwing. **V** In display male gives a cat-like mewing and various soft notes, otherwise generally silent; female utters various harsh and grating calls. **H** Breeds on estuaries, lakes and rivers in wooded areas. Winters on inshore coastal waters. **D** North America, Europe.

20 SCALY-SIDED MERGANSER *Mergus squamatus* 52–62cm Both sexes show much white on upperwing. **V** As Red-breasted Merganser. **H** Breeds by rivers in mountain forest areas. Winters on rivers and lakes. **D** E Siberia, NE China, Korean Peninsula.

21 BLACK-HEADED DUCK *Heteronetta atricapilla* 40cm Floats low. Male with pale blue bill, pink at base. Head and upper neck black. Upperparts dark brown. Wings brownish black. Female with brownish-grey bill, dirty brown head, neck and upperparts. **V** Silent. **H** Dense vegetation in lakes, roadside ditches, marshes, open water, flooded areas. Parasitic nester. **D** C Chile to Paraguay and south to C Argentina.

22 MASKED DUCK *Nomonyx dominicus* 30–36cm Hides among aquatic vegetation. In flight, upperwings of both sexes show a large white patch. **V** Male utters a loud *kuri-kuroo*. Female gives low hisses and clucks. **H** Freshwater pools and marshes with extensive vegetation. **D** Central and South America, Caribbean.

23 RUDDY DUCK *Oxyura jamaicensis* 35–43cm Sociable. **V** Usually silent. During display taps bill against inflated chest, producing a bubbling of water around breast. **H** Lakes with bordering vegetation. **D** S Canada, USA, Caribbean.

24 ANDEAN DUCK *Oxyura ferruginea* 42cm Male with pale blue bill. Head and upper neck black, body deep chestnut; border between two colours slanted backwards. Undertail dirty white. Female with brownish-grey bill. Dark blackish brown, scaled and barred on breast and flanks. **V** Usually silent. Male produces tapping sound in courtship by striking bill against chest, accompanied by very low short groaning notes. **H** Open lakes, marshes, some reedbeds. **D** Colombia to Tierra del Fuego.

25 LAKE DUCK *Oxyura vittata* 37cm Male with blue bill. Black head and upper neck, rest of body reddish chestnut; border between two colours almost horizontal. Female with brownish-grey bill. Body brown, darker above, breast and flanks barred buff. **V** Generally silent, although displaying males make mechanical rustling noises and soft, popping *pup-pup-pup-pup* calls, and females give high-pitched squeaks during aggressive encounters. **H** Open water of marshes, lakes, roadside ditches, temporary ponds, dense reedbeds. **D** S South America.

26 BLUE-BILLED DUCK *Oxyura australis* 35–45cm Spine-like tail feathers sometimes held erect. Dives for invertebrates and seeds. Forms flocks outside breeding season. **V** Usually silent. **H** Well-vegetated pools; also open water in winter. **D** Australia.

27 MACCOA DUCK *Oxyura maccoa* 50cm Note uniform dark brown wings and shape and stance of tail. Female has pale horizontal lines below eye. **V** Very low, dry, drawn-out, rattling *purrrrrrrr*. **H** Shallow fresh (sometimes alkaline) lakes with emergent vegetation. **D** Ethiopia to South Africa.

28 WHITE-HEADED DUCK *Oxyura leucocephala* 43–48cm Usually in small parties; larger flocks in winter. **V** Male emits rattling and piping calls during display; female utters low, harsh notes. **H** Breeds and winters on small lakes with developing and fringing vegetation; also larger lakes and coastal lagoons after breeding. **D** S Europe and N Africa to C Asia and NW China.

29 MUSK DUCK *Biziura lobata* 60–70cm Dives well. Feeds mainly on invertebrates. Male has leathery lobe under bill. Has musky odour in breeding season. **V** Whistles and slaps feet when courting; otherwise silent. **H** Lakes, wetlands with open water. **D** SW and SE Australia.

1
2
3
4
5
6
7
8
9
10
11
14
12
13
15
16
17
18
19
20
21
22
23
24
25
26
27
29
28

MEGAPODES MEGAPODIDAE

1 AUSTRALIAN BRUSHTURKEY *Alectura lathami* 60–75cm Mainly dark plumage offset by red neck. Colour of male's neck and wattles intensifies in breeding season. Male builds mound nest. **V** Screams and gulping croaks; male 'booms' in breeding season. **H** Coastal rainforest. **D** E Australia.

2 WATTLED BRUSHTURKEY *Aepypodius arfakianus* 38–45cm Often seen in pairs. Flies up to tree when alarmed. Male builds nest mound. **V** Calls include harsh *kuee, kiu-kiu-kiu*. **H** Mainly mountain forest; sometimes lower than 750m. **D** New Guinea.

3 WAIGEO BRUSHTURKEY *Aepypodius bruijnii* 40–45cm Rare and poorly known. Bare parts more striking and colourful in male than female. Male builds mound nest. **V** Usually silent. **H** Dense, short cloudforest. **D** Waigeo Island, Indonesia.

4 RED-BILLED BRUSHTURKEY *Talegalla cuvieri* 45–55cm Head covered with bristle-like feathers. Usually runs from danger. Male builds nest mound from sticks and leaves. **V** Loud grunts and screeches, uttered day or night. **H** Lowland forests. **D** W New Guinea; Misool Island, Indonesia.

5 BLACK-BILLED BRUSHTURKEY *Talegalla fuscirostris* 50–58cm Powerful yellow legs aid identification. Usually runs from danger. Feeds on ground; diet includes fallen fruits and invertebrates. **V** Honks and screams. **H** Lowland rainforest, forested slopes. **D** New Guinea; Aru Island, Indonesia.

6 COLLARED BRUSHTURKEY *Talegalla jobiensis* 54–60cm Red legs and sparsely feathered red face aid identification. Male builds mound nest. **V** Loud braying calls, uttered day or night. **H** Moist forests; to 1,000m or more. **D** New Guinea.

7 MALLEEFOWL *Leipoa ocellata* 60cm Avoids detection by 'freezing'; cryptic plumage blends with background. Male builds nest mound filled with plant material and sand; decay generates heat for incubation. **V** Male utters territorial boom near nest mound. **H** Mallee scrub. **D** S Australia.

8 MALEO *Macrocephalon maleo* 55cm Usually in pairs feeding on ground; in tree branches when roosting or alarmed. **V** Nasal, vibrating braying, often answered by female with two-note *kuk-kuk*. **H** Lowland and hill forest. Breeds communally in sandy forest areas or on sandy beaches. **D** Sulawesi, Indonesia.

9 MOLUCCAN MEGAPODE *Eulipoa wallacei* 30cm Shies away from breeding areas; forages on ground, usually singly or in pairs. **V** Low, wavering *waaaaw*. **H** Hill and montane forest. Breeds communally on sandy beaches or clearings in coastal scrub. **D** Maluku Islands, Indonesia.

10 TONGAN MEGAPODE *Megapodius pritchardii* 33cm Very dark with pale head. **H** Sloping, forested areas, especially on innerside of volcano calderas. **V** Pairs duet with sharp, drawn-out whistling notes. **D** Tonga.

11 MICRONESIAN MEGAPODE *Megapodius laperouse* 29cm Brownish black with grey underparts. **H** Forest, forest remains, coconut groves, coastal scrub. **D** Palau Islands and Mariana Islands, W Pacific Ocean.

12 NICOBAR MEGAPODE *Megapodius nicobariensis* 43cm Usually in pairs or family groups. A ground forager, but will take to trees when alarmed. **V** Repeated, cackling *kuk-a-kuk-kuk* and a bullfrog-like *kiouk-kiouk-kok-kok-kok-kok*. **H** Dense forest undergrowth, usually near sandy beaches. **D** Nicobar Islands.

13 PHILIPPINE MEGAPODE *Megapodius cumingii* 32–38cm Secretive; usually encountered feeding on the ground, singly or in small groups. Generally flies only when alarmed. **V** Long, mournful, drawn-out scream, said to resemble an air-raid siren. During breeding issues chicken-like clucks and various *keeeee* or *kayooo* notes. **H** Coastal scrub, hill and montane forest. **D** Borneo; Philippines; Sulawesi, Indonesia.

14 SULA MEGAPODE *Megapodius bernsteinii* 35cm Forages on the ground, singly, in pairs or in small groups. Will fly short distances if alarmed. **V** Duets with a mournful, hoarse two-note braying, joined by second bird uttering a longer nasal chuckle. **H** Primary forest to coastal scrub. **D** Sula Islands, Indonesia.

15 TANIMBAR MEGAPODE *Megapodius tenimberensis* 35cm Shy; generally encountered singly or in pairs, usually heard rather than seen. **V** Loud, upslurred *kee-yu*, followed by a series of shorter, high-pitched *keyu* notes; second bird joins in with a series of rising *kwou* notes, repeated at varying intervals. **H** Primary, secondary and semi-evergreen forest. **D** Tanimbar Islands, Indonesia.

16 DUSKY MEGAPODE *Megapodius freycinet* 35cm Reasonably common. Pairs forage on ground. **V** Duets, the first bird uttering a chuckling, rising and then falling *kejowowowowowowowowow*, the second bird joining in with two rising *keyou* notes. **H** Various forests, including primary, tall secondary and swamp; also mangroves, swamps, plantations. **D** Maluku Islands, Indonesia.

17 BIAK SCRUBFOWL *Megapodius geelvinkianus* 35cm Often seen in pairs, foraging on ground for fruits and invertebrates. Red, stout legs aid recognition. **V** Loud screeching calls, sometimes in duet. **H** Moist, lowland forest. **D** New Guinea.

18 MELANESIAN MEGAPODE *Megapodius eremita* 35–40cm Forages on ground for fallen fruits and invertebrates. Breeds and often congregates near geothermal sites and sun-warmed beaches. **V** Harsh screeches, sometimes in duet. **H** Rainforest. **D** Bismarck Archipelago, Papua New Guinea; to Solomon Islands.

19 VANUATU MEGAPODE *Megapodius layardi* 42–44cm Nests communally. Digs burrows in sun-warmed or geothermally heated ground. Runs from danger. Forages on ground. **V** Screeches and croaks; most vocal at dawn and dusk. **H** Moist tropical forest. **D** Vanuatu.

20 NEW GUINEA SCRUBFOWL *Megapodius decollatus* 33–35cm Stout yellow legs and feet. Scratches ground for invertebrates and fallen fruit. Male builds nest mound. **V** Performs screeching duets, often at night. **H** Rainforest. **D** New Guinea.

21 ORANGE-FOOTED SCRUBFOWL *Megapodius reinwardt* 35cm Forages on ground, usually singly or in pairs; best located by calls. Runs when disturbed, but will fly to tree branches if alarmed. **V** Duets, starting with a series of melodious notes, followed by a long, descending, high-pitched, stammering *krrr-uk-uk-uk-uk-krrr*. **H** Various, from lower montane forest to coastal scrub and mangrove. **D** Lesser Sunda Islands, Indonesia; New Guinea; N Australia.

CHACHALACAS, GUANS AND CURASSOWS CRACIDAE

22 PLAIN CHACHALACA *Ortalis vetula* 55cm Separated from similar West Mexican and White-bellied Chachalacas by darker, tawny underparts and range. **V** High, excited chattering *chatatah-chatatah* in duet. **H** Open forest with dense brush, woodland. **D** S Texas to NW Costa Rica.

23 GREY-HEADED CHACHALACA *Ortalis cinereiceps* 50cm Red in wing is diagnostic. **V** Rather short, gull-like yelps. **H** Forest edges, secondary growth. **D** Honduras to N Colombia.

24 CHESTNUT-WINGED CHACHALACA *Ortalis garrula* 50cm Head, neck and breast cinnamon. Upperparts olive-brown. Primary flight feathers rufous. Tail blackish olive, outer feathers tipped white. Lower underparts white. **V** Creaky, upslurred wailing squeals. **H** Deciduous woods with shrubs, edges of humid forest, dry brushland, thickets, secondary growth, mangroves. **D** NW Colombia.

25 RUFOUS-VENTED CHACHALACA *Ortalis ruficauda* 53cm Head and upper neck greyish. Upperparts olive-brown. Tail blackish olive with greenish sheen. All outer-tail feathers tipped rufous. Underparts cinnamon-buff. Undertail coverts chestnutty rufous. **V** Excitable, chattering, creaky notes in rapid series. **H** Edges of woods and forest, brushland, deforested areas with scattered trees, riverine woods, gallery forest. **D** Colombia, Venezuela.

26 RUFOUS-HEADED CHACHALACA *Ortalis erythroptera* 52cm Head, neck and primary flight feathers rufous. Upperparts brownish olive. Tail blackish olive with green sheen. Outer-tail feathers tipped rufous. Underparts dirty whitish buff. Undertail coverts cinnamon. **V** Very throaty, rasping *crarr-cur, craa-craa-crarr-cur*. **H** Deciduous woods, savannah, brushland with trees. **D** W Ecuador, NW Peru.

27 RUFOUS-BELLIED CHACHALACA *Ortalis wagleri* 65cm Underparts and tail corners orange-rufous. **V** High, hoarse *truterut-tut* in chorus. **H** Dry woodland, thorn bush. **D** NW and W Mexico.

28 WEST MEXICAN CHACHALACA *Ortalis poliocephala* 65cm Separated from smaller White-bellied Chachalaca by tawny undertail coverts. Range differs from Plain Chachalaca. **V** Low, unpleasant pheasant-like nasal screeching in long, rhythmic series. **H** Dry woodland, thorn bush, montane forest. **D** W and SW Mexico.

29 CHACO CHACHALACA *Ortalis canicollis* 53cm Bare face reddish pink. Bill and feet creamy. Head and neck grey. Upperparts olive-brown. Tail blackish olive with green sheen. Underparts dirty buff, grading to cinnamon undertail coverts. **V** Very harsh rasping chatter *ca-ca-ca-CRAA-ca-ca*. **H** Dry woods and transition to forest, brushland with scattered trees, palm groves, savannah with patches of woodland. **D** C South America.

30 WHITE-BELLIED CHACHALACA *Ortalis leucogastra* 45cm White belly and undertail coverts diagnostic. **V** Low, harsh chicken-like *crat-crr-crarr* with last syllable ending in a trill, repeated and running into a cacophonous duet. **H** Woodland, forest edges. **D** S Mexico to Nicaragua.

31 SPECKLED CHACHALACA *Ortalis guttata* 47cm Cap blackish brown. Upperparts and tail dark olive-brown. Outer-tail feathers tipped rufous (all rufous when seen from below). Foreneck and breast spotted and marked whitish. Underparts buffy brown. **V** Harsh, rapid, slightly chicken-like chatter. **H** Brushland with scattered low trees, rainforest edges, gallery forest, clearings with secondary growth, regrowth and palm groves with bushes and low trees. **D** Amazonia.

32 EAST BRAZILIAN CHACHALACA *Ortalis araucuan* 47cm Like Speckled Chachalaca but has ochreous-cinnamon cap and hindneck. Upperparts lighter; lower underparts dirty white, more buff on undertail coverts. **V** Accelerating series of raspy, rather gull-like yelps. **H** Atlantic forests. **D** E Brazil.

33 SCALED CHACHALACA *Ortalis squamata* 47cm Like Speckled Chachalaca but has dark browny-cinnamon cap and hindneck. Underparts brown, slightly washed greyish. **V** Rapid, harsh chatter with fourth note higher pitched and emphatic; *cha-ca-cha-SHREE-ca*. **H** Forest. **D** SE Brazil.

34 COLOMBIAN CHACHALACA *Ortalis columbiana* 53cm Larger than Speckled Chachalaca with paler upperparts. Cap and hindneck pale slate grey, scaling on breast very fine. **V** A typical chachalaca chatter but with distinct pause after first note. **H** Forest and woodland. **D** NC Colombia.

35 LITTLE CHACHALACA *Ortalis motmot* 53cm Legs pinkish red. Head, upper neck and outer-tail feathers rufous. Upperparts olive-brown, central tail feathers brownish olive. Underparts brownish buff. **V** Three-note harsh chatter, sometimes with short, squeaking fourth note. **H** Edge of rainforest and cloud forest, brushland with low trees, cleared areas with scattered trees. **D** N Amazonia.

36 CHESTNUT-HEADED CHACHALACA *Ortalis ruficeps* 38cm Recently split from Little Chachalaca; similar in plumage but smaller. **V** Querulous hen-like chatter, similar to other chachalacas. **H** Forest. **D** NC Brazil.

37 BUFF-BROWED CHACHALACA *Ortalis superciliaris* 42cm Upperparts and tail dark olive-brown. Outer-tail feathers tipped rufous. Head and neck dark brown. Faint greyish brow and streaks on foreneck. Breast faintly scaled greyish, rest of underparts brownish buff. **V** Chattering song relatively high pitched, squeaky and excitable, with variable rhythm. **H** Edges of woodland and forest, brushland with small scattered trees, dry woods. **D** NE Brazil.

1
♂
♀
2
♀
♂
4
6
3
♂
♀
5
11
7
8
9
10
15
12
13
14
16
19
17
18
20
21
22
23
24
25
26
31
27
28
29
30
32
33
34
35
36
37

CHACHALACAS, GUANS AND CURASSOWS *CONTINUED*

1 BAND-TAILED GUAN *Penelope argyrotis* 60cm Smallish and relatively dumpy, with short legs. Brown with slight chestnut sheen. Scaled white on foreneck, upper back and underparts. **V** Rather shrill, yapping or clucking notes, also produces wing-rattling sounds. **H** Montane rainforest, cloud forest, tall secondary growth, plantations. **D** Colombia, Venezuela.

2 BEARDED GUAN *Penelope barbata* 58cm Dark brown with silvery speckles. Very like Andean Guan. Small red wattle. **V** Long series of clipped, squeaky piping notes, reminiscent of a small gull. **H** Moist montane rainforest. **D** S Ecuador, NW Peru.

3 BAUDO GUAN *Penelope ortoni* 65cm Uniform chestnut brown, scaled white on foreneck and breast. **V** Rather low, throaty and growling single notes, sometimes deeper and more abrupt. **H** Humid rainforest, especially at foot of mountains. **D** W Colombia, W Ecuador.

4 ANDEAN GUAN *Penelope montagnii* 58cm Smallish and dumpy, with short legs. Like Band-tailed Guan. Tail dark brownish chestnut. Tiny red wattle somewhat feathered. **V** Emphatic, yelping notes in fast series *ga-ga-ga...* increasing in pitch and speed when agitated. **H** Montane rainforest, edge, secondary growth, slopes with dense humid forest, dwarf forest, forest patches at greater elevations. **D** N and C Colombia, NW Venezuela.

5 MARAIL GUAN *Penelope marail* 63cm Legs pinky grey. Upperparts, breast and tail dark olive. Foreneck and breast scaled white. Rest of underparts dirty chestnut-brown. Long faint brow. **V** Drawn-out, upslurred rasping calls, becoming strained and squealing at the end of each phrase. **H** Rainforest, especially near water; secondary growth. **D** N South America.

6 RUSTY-MARGINED GUAN *Penelope superciliaris* 63cm Grey legs. Notable long brow whitish. Upperparts and tail dark olive. Foreneck and breast scaled whitish. **V** Gruff grunts and chicken-like clucks. **H** Forest in lowlands, secondary growth, plantation edges, groves in Cerrado, woods, Caatinga, gallery forest. **D** C and E South America.

7 RED-FACED GUAN *Penelope dabbenei* 70cm Bare face reddish orange. Legs short. Brown with slight olive sheen on neck and upperparts. Tail blackish brown. Faint scaling on foreneck, breast and upper back. **V** Call a squeaky, downslurred *squeer-yup*, finishing with low clipped grunt. **H** Alder woods, temperate montane forest. **D** SE Bolivia, NW Argentina.

8 CRESTED GUAN *Penelope purpurascens* 85cm Mostly dark brownish olive, contrasting with dark chestnut lower belly, tail coverts and tail. Foreneck, breast, wing coverts and upper back scaled white. **V** Series of yelping *keeow* notes, also wing-whirring. **H** Lowland rainforest, sometimes drier forest, hills. **D** Mexico to NW South America.

9 CAUCA GUAN *Penelope perspicax* 76cm Head and neck blackish grey, becoming browner lower on body and grading to rufous chestnut on wings, lower body and tail. Neck, breast and upper back feathers scaled white. **V** Series of querulous *quee* notes. **H** Humid forest. **D** W Colombia.

10 WHITE-WINGED GUAN *Penelope albipennis* 80cm Primary flight feathers white, very visible in flight. Overall dark brownish olive. Foreneck and breast scaled whitish. Bare face pinkish lilac. Bill pale blue tipped black. **V** Has repertoire of various high-pitched yelping, gull-like calls, and low creaking notes. **H** Dry woods and deciduous forest on mountain slopes, deep valleys with running streams, gallery forest, forested deltas. **D** NW Peru.

11 SPIX'S GUAN *Penelope jacquacu* 70cm Upperparts, breast and tail dark olive-brown. Rest of underparts and uppertail coverts brownish chestnut. Breast and upper back feathers scaled white. **V** Various growling and howling calls, varying greatly in pitch. **H** Rainforest and cloud forest, gallery forest, clearings with scattered trees. **D** Amazonia.

12 DUSKY-LEGGED GUAN *Penelope obscura* 73cm Legs dark violaceous grey. Generally dark blackish brown. Head and neck blackish. Foreneck, breast, wing coverts and upper back scaled white. **V** Two-note call, first part a harsh rasp, second more squeaky and drawn out, but can merge into single note. Also produces rasping barks. **H** Woods, forest, gallery forest in plains, deltas, lower montane forest, ravines. **D** SC and SE South America.

13 WHITE-CRESTED GUAN *Penelope pileata* 80cm White crest, feathers tipped chestnut. Neck and underparts chestnut. Contrasting blackish-olive back, wings and tail. Bare face edged black. Breast, back and wing coverts scaled white. **V** Very raucous chattering clucks and higher-pitched yelps. **H** Tropical lowland forest. **D** NC Brazil.

14 CHESTNUT-BELLIED GUAN *Penelope ochrogaster* 78cm Crest whitish and chestnut. Face edged black. Brow white. Neck and underparts chestnut, contrasting with dark olive-brown back, wings and tail. Breast, back and wing coverts scaled white. **V** Chicken-like clucks, pitch varying according to mood. Song combines clucks and raspier notes. **H** Lowland tropical forest, gallery forest, swampy forest, deciduous forest. **D** C Brazil.

15 WHITE-BROWED GUAN *Penelope jacucaca* 73cm Uniformly dark chestnut brown. Slight olive sheen on upperparts. Breast, back and wing coverts edged white. Bare face edged black. Long brow white. **V** Seldom recorded; produces yelping and harsh chattering notes. **H** Caatinga with stunted trees and shrubs, cactus and thorny thickets. **D** E Brazil.

16 TRINIDAD PIPING GUAN *Pipile pipile* 72cm Differs from White-browed Guan only in black-streaked cap and nape, and bare face washed pale blue. **V** Song a short series of whistled notes, call a soft *pee-oo*; also produces wing-clapping sounds. **H** Tree crowns in undisturbed primary forest, montane forest, steep terrain. **D** Trinidad.

17 BLUE-THROATED PIPING GUAN *Pipile cumanensis* 72cm Black. Greater wing coverts white. Bill base light blue, face white, triangular wattle blackish blue. Cap and nape faintly streaked. **V** Song an excitable series of downslurred whistles, also makes loud wing-whirring sounds. **H** Rainforest near rivers, várzea, terra firme forest, gallery forest, copses in palm groves. **D** W Guyana to SW Brazil.

18 RED-THROATED PIPING GUAN *Pipile cujubi* 72cm Differs from Trinidad Piping Guan in red wattle with small blue area under chin. Breast feathers bordered white. Face white. **V** Song a series of upslurred whistles, accelerating and rising in pitch. Also produces wavering whining calls and wing-rattles. **H** Rainforest, riverine forest in lowlands. **D** NC Brazil, Amazonia.

19 BLACK-FRONTED PIPING GUAN *Pipile jacutinga* 74cm Black. White wing patch, crown, nape and eye-ring, breast scaled white. Bill base blue, tip black. Wattle blue and red. **V** Song drawn-out slightly upslurred whistles, becoming louder and higher pitched. Also produces loud wing-rattles. **H** Rainforest near watercourses, mountains and hills. **D** SE Brazil, NE Argentina, SE Paraguay.

20 WATTLED GUAN *Aburria aburri* 71cm All black, body with bronzy-olive sheen. Head and neck with blue sheen. Blue bill tipped black. Long, fine wattle salmon pink. **V** Loud, slightly rattling wail that rises then falls in pitch. **H** Upper montane tropical forest, rainforest, cloud forest, secondary growth, copses. **D** Venezuela to Peru.

21 BLACK GUAN *Chamaepetes unicolor* 65cm Separated from Highland Guan by different range and blue naked face parts. **V** Dry, rattling *prr prrrrrr*. **H** Montane forest. **D** Costa Rica, Panama.

22 SICKLE-WINGED GUAN *Chamaepetes goudotii* 64cm No wattle. Head, neck, upperparts and tail brownish olive. Breast and rest of underparts contrasting chestnut rufous, with faint scaling on breast. Bare face and base of bill pale blue, eye red. **V** Slightly squeaky *prrree-up* notes, also whooshing and rattling wing sounds. **H** Montane rainforest, tall secondary growth, woods, coffee plantations. **D** Colombia to Bolivia.

23 HIGHLAND GUAN *Penelopina nigra* 60cm Black feathering and red chin wattle of male unmistakable. Female has red legs and delicate, dense barring. **V** Strange, rising to very high *fueuuuuuuuiiii* or descending *krrrrrrrr*. **H** Montane forest; 1,000–3,000m. **D** S Mexico to Nicaragua.

24 HORNED GUAN *Oreophasis derbianus* 80cm Red horn unmistakable. **V** Low, booming sounds; alarm call a very dry, low, rolling rattle. **H** Montane forest. **D** SE Mexico, Guatemala.

25 NOCTURNAL CURASSOW *Nothocrax urumutum* 60cm Heavy orange bill. Bare face colourful. Cap with long, curly black feathers forming an erectile crest. Chestnut, more cinnamon on belly. Upperparts finely barred black. Outer-tail feathers black, tipped buff. **V** Song a series of low, mellow, resonant hoots *buu buu buu...* in series of about seven notes with the last three slightly faster and higher pitched. **H** Dense humid forest, temporarily or permanently flooded, terra firme forest, usually near rivers. **D** W Amazonia.

26 CRESTLESS CURASSOW *Mitu tomentosum* 85cm Bill red. Cap of bristly, hair-like feathers. Glossy bluish black, feathers edged black. Lower underparts and tail tip chestnut. Legs wine red. **V** Single deep, resonant notes, like that made by blowing across the top of a bottle. **H** Riverine rainforest and terra firme forest, gallery forest, dense undergrowth. **D** N Amazonia.

27 SALVIN'S CURASSOW *Mitu salvini* 89cm Notable red bill with high, arched culmen. Glossy bluish black, feathers edged black. Lower underparts and tail tip white. Bushy erectile crest of straight feathers. Legs wine red. **V** Deep, booming, resonant notes that rise and fall in pitch, and vary in length. **H** Humid terra firme forest, forested ravines, hills and plains. **D** W Amazonia.

28 RAZOR-BILLED CURASSOW *Mitu tuberosum* 89cm Notable red bill, culmen highly arched. Glossy bluish black, feathers edged black. Lower underparts chestnut. Tail tipped white. Bushy erectile crest of straight feathers. Legs wine red. **V** Song is very low, short, booming or grunting notes; call resembles whip-crack. **H** Lowland forest. **D** Amazonia (south of the Amazon).

29 ALAGOAS CURASSOW *Mitu mitu* 83cm Like Razor-billed Curassow, but with a more moderate bill, bare area around ear, and tail narrowly tipped pale brown. **V** Chirping calls recorded in captivity. **H** Atlantic forest. **D** NE Brazil. Once extinct in the wild but efforts began in 2019 to reintroduce them using birds bred in captivity.

30 HELMETED CURASSOW *Pauxi pauxi* 90cm Bill red with fig-shaped, pale blue frontal knob. Glossy bluish black, feathers edged black. Lower underparts and tail tip white. Legs wine red. **V** Low breathy booms, and squeaky grunts. **H** Humid montane forest, dense cloud forest, steep slopes. **D** Venezuela, Colombia.

31 HORNED CURASSOW *Pauxi unicornis* 90cm Like Helmeted Curassow but with horn-shaped frontal knob. **V** Short grunting notes have been recorded. **H** Montane forest, steep slopes, ravines, usually with small streams. **D** C Bolivia.

32 SIRA CURASSOW *Pauxi koepckeae* 91cm Critically endangered. Has blue crest-like knob at base of pink bill. **V** Song low, breathy booms, also has abrupt toneless grunting call. **H** Cloud forest. **D** Cerros del Sira, C Peru.

33 GREAT CURASSOW *Crax rubra* 91cm Bill pale grey, cere yellow with one round knob above. Bushy crest of long, forward-curling feathers. Glossy black with faint blue sheen. Female has ivory bill. **V** Short hooting booms in song; has varied calls including shrill upslurred whistles. **H** Pristine humid lowland forest to foothills. **D** Mexico to Peru.

34 BLUE-BILLED CURASSOW *Crax alberti* 91cm Bill pale grey with two-lobed blue cere below. Long, bushy crest of forward-curling feathers. Glossy black with slight blue sheen. In female, blue cere lacks caruncles. **V** Song is typical currasow deep booming notes, also gives loud downslurred whistles and short slightly squeaky piping notes. **H** Lowland humid forest and foothills. **D** N Colombia. Not recorded in the wild since 1978.

35 YELLOW-KNOBBED CURASSOW *Crax daubentoni* 91cm Bill and eye-ring dark grey. Knobbly cere yellow. Long, forward-curling feathers form notable crest. Glossy black. Female with white dots in crest, no cere. **V** Long, downslurred whistle with vibrato at end. **H** Gallery forest, basal forest, rainforest, deciduous forest, savannah with scattered trees, semi-arid vegetation, often near rivers. **D** NE Colombia, N Venezuela.

36 BLACK CURASSOW *Crax alector* 90cm Bill pale grey, cere orange or orangy yellow. Bare eye-ring blue, lower edge yellow. Black crest of dense, forward-curling feathers. Glossy blackish blue. Female with white spots in crest. **V** Calls include whistles and whip-cracking notes. **H** Cloud forest, rainforest, gallery forest, terra firme forest, riverine brush and thickets. **D** N Amazonia

37 WATTLED CURASSOW *Crax globulosa* 90cm Bill and bare skin around eye blackish. Knobbly red cere with one round lobe above, two below. Bushy black crest of forward-curling feathers. Glossy black with slight blue sheen. Female lacks lobes on cere. **V** Downslurred whistle or lower-pitched wailing call. **H** Riverine forest, várzea near water, drier forest. **D** W Amazonia.

38 BARE-FACED CURASSOW *Crax fasciolata* 90cm Bill grey, cere yellow. Bare eye-ring grey. Glossy black. Forward-curling feathers form bushy crest. Female lacks cere and crest basally white. **V** Male produces a typical currasow booming song. **H** Gallery forest, woods. **D** C South America.

39 RED-BILLED CURASSOW *Crax blumenbachii* 90cm Bill and bare skin around eye blackish. Cere and one spherical knob below red. Dense crest of forward-curling feathers, shorter and more curly than in others of genus. Glossy black with slight blue sheen. Female lacks cere. **V** Call a rather explosive downslurring whistle. **H** Tall humid rainforest, river islands and spits, dense secondary growth, near water. **D** SE Brazil.

1
2
3
4
5
6
7
8
9
10
11
12
13
14
15
16
17
18
19
20
21
22
23
24
25
26
27
28
29
30
31
32
33
34
35
36
37
38
39

GUINEAFOWL NUMIDIDAE

1 WHITE-BREASTED GUINEAFOWL *Agelastes meleagrides* 40cm White collar diagnostic. **V** Foraging groups give chirping contact calls. **H** Forest undergrowth. **D** Sierra Leone to Ghana.

2 BLACK GUINEAFOWL *Agelastes niger* 40cm Separated from Helmeted, Plumed and Crested guineafowl by lack of white spots. **V** High-pitched, repeated *kweeh* notes. **H** Dense forest. **D** SE Nigeria to DR Congo.

3 HELMETED GUINEAFOWL *Numida meleagris* 55cm The only guineafowl found outside forest. **V** High, fluted *puweeet puweeet puweeet* (*pu-* barely audible). **H** Dry natural and cultivated areas with more or less tree cover, shrub and bush. **D** Widespread across Africa, but extirpated in Nigeria and Liberia.

4 PLUMED GUINEAFOWL *Guttera plumifera* 45cm Often difficult to separate from Crested Guineafowl, but erect crest longer and naked skin grey (not bluish purple). **V** Calls include short yelps and very harsh dry rasping notes. **H** Dense forest. **D** WC Africa.

5 CRESTED GUINEAFOWL *Guttera pucherani* 50cm Often difficult to separate from Plumed Guineafowl, but erect crest shorter and naked skin bluish purple (not grey). **V** Low, speeded-up cackles and rattles: *tetrut-tetrut-trrrueh*. **H** Dense undergrowth of forest and woodland. **D** W, C, E and SE Africa.

6 VULTURINE GUINEAFOWL *Acryllium vulturinum* 60cm Forages on ground. Forms flocks outside breeding season. Often adopts upright posture. **V** Rasping chatter, when alarmed or going to roost. **H** Dry bush and wooded areas, including forest edges. **D** Ethiopia and Somalia to Tanzania.

NEW WORLD QUAIL ODONTOPHORIDAE

7 STONE PARTRIDGE *Ptilopachus petrosus* 25cm Note bantam-like posture. **V** Upslurred tuneful short whistles, given in chorus by all members of a group. **H** Rocky hillsides with dense scrub, tall grass cover and more or less bush. **D** W Africa.

8 NAHAN'S PARTRIDGE *Ptilopachus nahani* 20cm Separated from Latham's Forest Francolin by red (not yellow) legs. **V** Short, upslurred whistles *wuit wuit wuit*. **H** Dense forests. **D** NE DR Congo, W Uganda.

9 BEARDED WOOD PARTRIDGE *Dendrortyx barbatus* 25cm Note head pattern. **V** Very high, shivering *wiwi-* (6×); very high, fluting *wuwufwuuh*. **H** Dense undergrowth of montane forest; 1,000–3,250m. **D** C Mexico.

10 LONG-TAILED WOOD PARTRIDGE *Dendrortyx macroura* 30cm Head pattern and broad red breast striping diagnostic. **V** High, sustained *uhju-uhju-uhju* (in duet) or *tidderah*. **H** Dense undergrowth of montane forest; 1,250–1,500m. **D** Mexico.

11 BUFFY-CROWNED WOOD PARTRIDGE *Dendrortyx leucophrys* 35cm Greyer than Long-tailed Wood Partridge and Bearded Wood Partridge; head pattern diagnostic. **V** Very high, rising and crescendoing *uh uhw uhwe uhwec*. **H** Dense undergrowth of montane forest; 500–3,000m. **D** S Mexico to Costa Rica.

12 MOUNTAIN QUAIL *Oreortyx pictus* 24–29cm Often in small parties. **V** Loud, repeated *plu-ark*; also a series of *cle* notes and squealing twitter when flushed. **H** Montane chaparral. **D** Mountain areas of W North America.

13 SCALED QUAIL *Callipepla squamata* 22–29cm Generally runs for cover when disturbed. After breeding, forms small to large coveys. **V** Hoarse, repeated *rrehh*; dry, nasal *chow-chowk chow-chowk*; and *pey-cos pey-cos*. **H** Dry, brushy grasslands. **D** SW USA, NC Mexico.

14 ELEGANT QUAIL *Callipepla douglasii* 25cm Crest and finely striped head pattern diagnostic. **V** High *shruck*; *tut-twueeh tut-twueeh*. **H** Scrubby woodland. **D** Mexico.

15 CALIFORNIA QUAIL *Callipepla californica* 24–28cm Scaled belly. Often in large non-breeding groups. **V** Loud, crowing *ka-kwa* and *ka ka-kwah*; also a loud *chi-ca-go*, cackles, chuckles and grunts. **H** Grassland, brushland, open woodlands. **D** Pacific coastal areas from SW Canada to NW Mexico.

16 GAMBEL'S QUAIL *Callipepla gambelii* 23–27cm No scaling on belly. Often in large parties after breeding. **V** Loud *ka-kya ka kah-ha* and querulous *chi-ca-go-go*; also grunts, cackles and chattering calls. **H** Arid desert scrub. **D** SW USA, NW Mexico.

17 BANDED QUAIL *Philortyx fasciatus* 20cm Unmistakable by belly barring. **V** Very high, wooden, sustained *puweeh puweeh*. **H** Dry thorn bush, scrub. **D** Mexico.

18 NORTHERN BOBWHITE *Colinus virginianus* 21–26cm Shy, typically in small groups that usually keep to cover. **V** Male gives a rising *bob-white* or *bob-bob-white*, female answers with a thin *a-loie-a-hee*. Also various other sounds, including a raucous squealing and whistled *hoy*. **H** Semi-open and open country, farmland, woodland edges. **D** North America, from EC Canada to SE Mexico.

19 YUCATAN BOBWHITE *Colinus nigrogularis* 20cm Black-scaled underparts and black throat of male diagnostic. **V** Ultra-high *wup whieet*. **H** Brushy woodland. **D** Mexico to Honduras.

20 SPOT-BELLIED BOBWHITE *Colinus leucopogon* 20cm Distinctive greyish-cinnamon underparts with white spotting of male. Head pattern less strong than in other bobwhites. Note distinctive eyebrow. **V** Very high *tjup tjip tjip*. **H** Brushy woodland. **D** Guatemala to Costa Rica.

21 CRESTED BOBWHITE *Colinus cristatus* 20cm Crest pale buff. Face and throat tawny or creamy (according to race). Hindneck, line down side of neck and collar black and white. Upperparts scalloped black, whitish and brown. Female duller, with smaller crest. **V** Three or four whistled notes, the last longer and rising in pitch, *whu whu whee-it*. **H** Open flat land, brush, grassland, thickets, plantations. **D** Costa Rica to N South America.

22 MARBLED WOOD QUAIL *Odontophorus gujanensis* 28cm Bare skin around eye reddish orange. Generally dark brown. Upperparts marked black, buff, rufous and whitish. Underparts greyish brown (or tawny brown according to race), lightly barred black. Head more chestnutty. Post-ocular stripe ochreous. **H** Humid forest, tall secondary growth, rainforest, cloud forest. **D** Costa Rica to Bolvia.

23 SPOT-WINGED WOOD QUAIL *Odontophorus capueira* 27cm Erectile crest brown. Bare face orangy red. All underparts brownish grey. Upperparts marked brown, black and buff, dotted white on wing coverts. Female duller with smaller crest. **V** Pairs sing in fast duet of alternating low and higher whistles. **H** Dense parts of shady forest, regrowth. **D** E Brazil to NE Argentina and E Paraguay.

24 BLACK-EARED WOOD QUAIL *Odontophorus melanotis* 25cm Note solid red breast. **V** High, wooden *tetjuuh tetjuuh* or very high *witwit wut*. **H** Forest. **D** Honduras to Panama.

25 RUFOUS-FRONTED WOOD QUAIL *Odontophorus erythrops* 29cm Forehead and face rufous chestnut, crown brownish black. Speckled white malar stripe, black bib, white collar. Upperparts dark brown, marked black and buff. Underparts chestnutty rufous. **V** Duetting pairs produce fast series of alternating higher- and lower-pitched querulous whistles *coo-wit coo-wit...*, gradually slowing and falling in pitch. **H** Humid forest and rainforest, foothills and lower slopes. **D** W Colombia, N Ecuador.

26 BLACK-FRONTED WOOD QUAIL *Odontophorus atrifrons* 28cm Forehead, face and throat black. Crown and nape chestnut. Upperparts dark brown, marked black, buff and rufous. Breast and belly buffy chestnut with white spots and slight blackish-brown barring. **V** Duetting song with rather harsh and throaty notes alternating in pitch. **H** Cloud forest. **D** NE Colombia, NW Venezuela.

27 CHESTNUT WOOD QUAIL *Odontophorus hyperythrus* 28cm Head, neck and upper breast chestnutty rufous. Ear patch white. Crown and nape dark brown. Lower belly and upperparts dark brown, marked black, ochre and whitish. Female with broad white line through face. Breast and belly brownish grey. **V** Pairs produce fast series of mid-pitched yelping calls. **H** Humid montane forest, dense edges, secondary growth. **D** Colombia.

28 DARK-BACKED WOOD QUAIL *Odontophorus melanonotus* 28cm Dark brown, faintly vermiculated blackish. Throat, neck and breast chestnutty rufous. **V** Very like that of Chestnut Wood Quail but higher pitched. **H** Subtropical montane forest. **D** SW Colombia, NW Ecuador.

29 RUFOUS-BREASTED WOOD QUAIL *Odontophorus speciosus* 29cm Black head. Faint speckled post-ocular streak down neck. Upperparts dark brown, marked white, ochre and black. Foreneck chestnut, breast and belly chestnutty rufous. Female with grey underparts. **V** Series of whistled rhythmic notes *whi-whi-wheeuh, whi-whi-wheeuh*. **H** Tropical and montane forest. **D** W Amazonia.

30 TACARCUNA WOOD QUAIL *Odontophorus dialeucos* 25cm Cap and bib blackish brown. White around eye, on chin and malar stripe. Pectoral collar white. Line from eye to hindneck ochre to chestnut. Body dark brown, vermiculated blackish. **V** Little recorded; typical wood quail duet of whistled notes. **H** Forest on slopes. **D** E Panama, NW Colombia.

31 GORGETED WOOD QUAIL *Odontophorus strophium* 26cm Head and neck black. Post-ocular streak, chin, malar stripe and collar white. Back and wings dark brown, marked black and buff. Underparts tawny rufous, breast neatly streaked white. Female duller with dark grey underparts. **V** Pleasant mid-pitched whistled series *whoo-ee-urr, whoo-ee-urr...* **H** Temperate and subtropical montane forest and oak woods. **D** Colombia.

32 VENEZUELAN WOOD QUAIL *Odontophorus columbianus* 28cm Cap blackish brown. Post-ocular line to neck cinnamon. Cheek black and speckled. Pectoral collar black. Throat and half-collar white. Upperparts dark brown, marked cinnamon and ochre. Underparts ochre-brown, breast with neat white markings. **V** Very like that of Gorgeted Wood Quail, but faster paced. **H** Cloud forest, edges of clearings at upper elevations. **D** NC Venezuela.

33 BLACK-BREASTED WOOD QUAIL *Odontophorus leucolaemus* 25cm Head and underparts distinctly patterned. **V** *Purrup'pupuk* (duet). **H** Montane forest; 750–1,500m. **D** Costa Rica, Panama.

34 STRIPE-FACED WOOD QUAIL *Odontophorus balliviani* 28cm Black face mask. Rufous cap. Ochre brow. Chin and malar stripe ochreous and rufous. Neck and upperparts dark brown, marked blackish, buff and rufous. Underparts dark chestnut with black-bordered white arrow marks. **V** Duet a very rapid, agitated series of mid-pitched whistles. **H** Dense montane forest, thickets, tree ferns and cane, humid cloud forest. **D** SE Peru, NW Bolivia.

35 STARRED WOOD QUAIL *Odontophorus stellatus* 29cm Bare skin around eye orangy red. Cap and long bushy crest chestnut. Cheeks, throat and neck grey. Upperparts dark brown, marked black, rufous, buff and white. Underparts chestnutty rufous with neat white spots on breast. **V** Duetting whistled song phrases are lower pitched than in other wood quails, *whoohoo-hoo-ee-ur...* **H** Tropical and riverine forest. **D** W Amazonia.

36 SPOTTED WOOD QUAIL *Odontophorus guttatus* 25cm Separated from Marbled Wood Quail and Black-eared Wood Quail by head pattern and colouring of breast, with white streaks below. **V** High, mellow, fluted *witterohwitterohweeweeweet*. **H** Dense forest undergrowth. **D** S Mexico to Panama.

37 SINGING QUAIL *Dactylortyx thoracicus* 20cm Head of male and female distinctly patterned. As Tawny-faced Quail, unobtrusive and shy forest dweller. **V** Very high, crescendoing and rising *fjuut*; melodious, fluted *weeterohweetwit-*. **H** Forest. **D** C Mexico to Honduras.

38 MONTEZUMA QUAIL *Cyrtonyx montezumae* 21–23cm Shy and wary, generally staying in cover. May dust-bathe in the open on dry tracks. **V** Descending, buzzing whistle; also a twittering *whi-whi whi-hu* given in alarm. **H** Open woodland with dense grass. **D** SW USA, Mexico.

39 OCELLATED QUAIL *Cyrtonyx ocellatus* 20cm As Montezuma Quail and with same clownish head pattern, but colouring below and range differ. **V** Song a series of sharp and piercing descending whistles. **H** Dry woodland; 1,000–3,000m. **D** SW Mexico to Nicaragua

40 TAWNY-FACED QUAIL *Rhynchortyx cinctus* 18cm Male separated from Singing Quail by white chin and grey breast. Compact jizz and belly barring of female diagnostic. **V** Very high, descending, fluted *pwuuuh* in chorus. **H** Dense forest undergrowth. **D** Honduras to Ecuador.

1
2
3
4
5
6
7
8
9
10
11
12
13
14
15
16
17
18
19
20
21
22
23
24
25
26
27
28
29
30
31
32
33
34
35
36
37
38
39
40

PHEASANTS AND ALLIES PHASIANIDAE

1 WILD TURKEY *Meleagris gallopavo* male 100–125cm, female 76–95cm Unmistakable. **V** Well-known 'gobbling'; also a liquid *cluk-cluck*, a yelping *keeow keeow keeow*, and a *putt* or *perk* when alarmed. **H** Deciduous and mixed woodland with extensive clearings. **D** S Canada, USA, N Mexico.

2 OCELLATED TURKEY *Meleagris ocellata* male 90cm, female 70cm Like Wild Turkey, but less black, more splendidly coloured green and blue, with white in wings, and in different habitat. **V** Hollow, accelerated bickering. **H** Forest, woodland. **D** SE Mexico, Belize, Guatemala.

3 RUFFED GROUSE *Bonasa umbellus* 43cm Male displays with tail fanned and ruff expanded. **V** Drumming sounds produced by beating open wings. Various hissing calls. **H** Open areas in dense woodland. **D** Alaska and N USA, Canada.

4 HAZEL GROUSE *Tetrastes bonasia* 35–40cm Feeds on the ground but often perches in trees. Some birds can be more tinged brown. In flight, shows black band at end of tail. **V** Repeated, high, penetrating *seeeeeeee-seee-see*. When alarmed gives a *plit plit* or *pitt pitt pitt*. **H** Mixed conifer and deciduous forest. **D** N Eurasia, from NE France to Japan.

5 CHINESE GROUSE *Tetrastes sewerzowi* 34cm Feeds on the ground and in trees. In flight, shows black bands on tail. **V** Little known. Hoarse *en er en er en er* given by male during a confrontation with another male and a *ze ze ze-dackdack* uttered by an alarmed female are only records. **H** Montane conifer forests with thickets of birch or willow. **D** NC, C and SC China.

6 SIBERIAN GROUSE *Falcipennis falcipennis* 38–43cm Can be very tame. Feeds on ground and in trees. In display, male cocks and fans tail while stretching neck upwards. **V** Displaying male gives a cooing whistle, followed by a clicking, before leaping into the air. **H** Coniferous forest, mixed forest with dense understorey. **D** E Siberia and Sakhalin Island, Russia.

7 SPRUCE GROUSE *Falcipennis canadensis* 38–43cm Male displays with tail fanned. Females vary from grey to rufous. **V** Low hoots during display. **H** Coniferous forest. **D** Alaska and N USA, Canada.

8 WESTERN CAPERCAILLIE *Tetrao urogallus* male 74–90cm, female 54–63cm Wary. During display male raises and fans tail while stretching neck upwards. **V** When agitated utters a harsh *koor krerk koroor*. During display male makes various knocking sounds, leading into a short 'drum roll', followed by a loud 'pop' and harsh wheezes. Female gives a chuckling *kok-kok*. **H** Mature coniferous and mixed forests. **D** NW and NC Eurasia, from Scotland to North Korea.

9 BLACK-BILLED CAPERCAILLIE *Tetrao urogalloides* male 90–97cm, female 69–75cm Actions and habits similar to Western Capercaillie. **V** *Tack-tack-tack*, rolling into a climactic *tr-r-rack*. **H** Mainly larch woods in mountains and plains. **D** E Siberia and Kamchatka Peninsula, Russia; Mongolia; N China.

10 BLACK GROUSE *Lyrurus tetrix* male 60cm, female 45cm In normal stance, tail long with outer feathers curved outwards, lyre-shaped. In flight, shows whitish underwing coverts. Groups of males perform display in a lek. **V** During display gives a low, bubbling crooning interspersed with sneezed *choo-EESH*. Various warning sounds recorded, including *guck guck* and *tuett-tuett-tuett*. **H** Forest edges, open woodland with clearings, moorland and heathland with tree cover. **D** N Eurasia, from Britain to NW Korean Peninsula.

11 CAUCASIAN GROUSE *Lyrurus mlokosiewiczi* male 50–55cm, female 37–42cm In flight, shows white underwing coverts. Groups of males perform display in a lek. **V** Generally silent. Lekking birds make wingbeat and bill-snapping sounds. **H** Low scrub on alpine meadows and slopes, above treeline. **D** Caucasus.

12 SAGE GROUSE *Centrocercus urophasianus* male 66–76cm, female 48–58cm Displaying male fans tail, fluffs out neck feathers, inflates neck sacs and raises thin head plumes. **V** Two wing-swishing sounds followed by two hooting or popping sounds. **H** Sagebrush plains. **D** SC Canada to WC USA.

13 GUNNISON GROUSE *Centrocercus minimus* male 55–64cm, female 45–50cm Displaying male resembles Greater Sage Grouse, but head plumes more prominent. **V** Nine low-pitched hooting sounds and three weak wing-swishes. **H** Sagebrush plains. **D** SW Colorado and SE Utah, USA.

14 DUSKY GROUSE *Dendragapus obscurus* 47–57cm Wary. Male displays with fanned tail and exposed purplish neck sacs. Female resembles Sooty Grouse female. **V** Series of low hoots. **H** Mainly conifer forests and forest edges. **D** Interior of W North America.

15 SOOTY GROUSE *Dendragapus fuliginosus* 47–57cm Male displays with fanned tail and exposed yellow neck sacs. **V** Series of low hoots. **H** Mainly conifer forests and forest edges. **D** Pacific coast of North America, from Alaska south to California.

16 SHARP-TAILED GROUSE *Tympanuchus phasianellus* 38–48cm Male displays with head down, tail cocked, wings spread and purplish neck sac inflated. **V** Various hoots, coos, barks and gobbling sounds. **H** Grassland, sagebrush and successional stages of forests. **D** NC and NW North America.

17 LESSER PRAIRIE CHICKEN *Tympanuchus pallidicinctus* 38–41cm During display inflates reddish neck sacs. Female pinnae short. **V** Bubbling hoot and descending clucking. **H** Dry grasslands, often mixed with dwarf shrubs. **D** C USA.

18 GREATER PRAIRIE CHICKEN *Tympanuchus cupido* 41–47cm During display inflates yellow-orange neck sacs. Female pinnae short. **V** Similar to Lesser Prairie Chicken, but generally lower pitched. **H** Prairie mixed with cropland. **D** SC Canada to Texas.

19 WHITE-TAILED PTARMIGAN *Lagopus leucura* 32cm In flight, shows white wings. **V** Rapid *pik pik pik pik pikkeea* and various clucking calls. **H** Barren, rocky tundra. **D** W and NW North America, from Alaska to New Mexico.

20 ROCK PTARMIGAN *Lagopus muta* 34–36cm In flight, shows striking white wings. During spring both sexes show white patches on mantle. **V** Dry *arr-arr-kakarr*, grating *kar-r-rk* and, during display-flight, *aa-ka-ka* followed by *ka-ka-ka* and with a *kwa-kwa* on landing. **H** Rocky areas on mountains and tundra. **D** N North America and N Eurasia.

21 WILLOW PTARMIGAN *Lagopus lagopus* 37–42cm In flight, shows striking white wings. During spring both sexes have patches of white on mantle. **V** Nasal, guttural *go-bak go-back go-back ak-ak-ak* and accelerating *ko-ko-ko-ko-ko-kokokokokokrrr*. During display flight gives an *aa*, followed by a *ka-ka-ka-ka-ka* and on landing *kohwa-kohwa-kohwa*. **H** Open tundra, moorland, heathland. **D** N North America and N Eurasia.

22 SNOW PARTRIDGE *Lerwa lerwa* 38–40cm Approachable. Occurs in pairs or small groups, with larger groups of 20–30 in winter. **V** Clear *jiju jiju jiju* that gradually quickens and rises in pitch. Also *huei huei* when flushed. **H** Grassy mountain slopes above treeline, with scattered shrub interspersed with scree and snow patches; 3,000–5,500m. **D** Himalayas to C China.

23 VERREAUX'S MONAL-PARTRIDGE *Tetraophasis obscurus* 48cm When disturbed walks off with tail cocked. Little else recorded. **V** Gives a 'loud cry' when flushed, otherwise calls reported to be similar to those of Szechenyi's Monal-partridge. **H** Mountains, on rocky slopes and in juniper and rhododendron scrub, meadows and ravines; 3,000–4,100m. **D** W China.

24 SZECHENYI'S MONAL-PARTRIDGE *Tetraophasis szechenyii* 50cm In non-breeding season usually encountered in family parties of 4–12, although larger groups are recorded. When disturbed tends to freeze or fly into forest cover, hiding in a tree until danger has passed. **V** Loud, far-carrying two- to three-note cackling interspersed with monosyllabic grating notes. **H** Coniferous, mixed coniferous, oak and rhododendron forest, rocky slopes with grass and scrub above the treeline; 3,200–4,875m. **D** Himalayas.

25 CAUCASIAN SNOWCOCK *Tetraogallus caucasicus* 52–56cm Shy and wary. Occurs in pairs or small parties of 3–9. Often walks with tail raised and undertail coverts fluffed out. **V** Mellow, Curlew-like *oolee-oooweeyuh*, and a low-pitched *pok-pok-pok-pok-pok* uttered by feeding parties or when flushed. **H** Grassy slopes with snow patches, rocky outcrops with ravines on mountains above treeline; 1,800–4,000m. **D** Caucasus.

26 CASPIAN SNOWCOCK *Tetraogallus caspius* 58–62cm Behaviour as Caucasian Snowcock. **V** Rising *oou-wee-eee-eee-et*, the last note very high pitched. When alarmed gives a *chok-chok-chok*. **H** Above treeline on mountain slopes with crags, ravines and open grassy areas; above 1,800m. **D** SW Turkey to Turkmenistan and SW Iran.

27 HIMALAYAN SNOWCOCK *Tetraogallus himalayensis* 58–62cm Behaviour similar to Caucasian Snowcock. **V** High-pitched *shi-er shi-er* and deeper *wai-wain-guar-guar*. When disturbed utters an accelerating *kuk kuk kuk*. **H** Sparsely vegetated scree or grassy patches among or below crags on mountain slopes between scrub and snowline; 3,500–6,000m. **D** Himalayas.

28 TIBETAN SNOWCOCK *Tetraogallus tibetanus* 50–56cm Habits much as Caucasian Snowcock. **V** Main breeding season call is a croaking *gu-gu-gu-gu*. Also makes a chuckling *chuck-aa-chuck-aa-chuck-chuck-chee-da da-da*. **H** High alpine pastures, bare or grassy mountain slopes up to the snowline; mainly 5,000–6,000m. **D** Himalayas.

29 ALTAI SNOWCOCK *Tetraogallus altaicus* 58cm Habits similar to Caucasian Snowcock. Very large groups have been recorded during harsh winters. **V** Not well documented. Said to be like others of group but perhaps more raucous than Himalayan Snowcock. **H** Open montane areas above treeline, including semi-desert, steppe and alpine meadows; 2,000–3,600m. **D** SW Siberia, NW Mongolia.

30 ROCK PARTRIDGE *Alectoris graeca* 32–35cm Parties usually number no more than 15 birds, sometimes larger during severe winter weather. **V** Rapidly repeated, grating *chitti-ti-tok chitti-ti-tok chitti-ti-tok*. When flushed gives a *wittoo-wittoo-witoo*. **H** Subalpine zone of mountains, rocky and grassy hillsides. **D** S and SE Europe.

31 CHUKAR PARTRIDGE *Alectoris chukar* 32–34cm Usually in small parties, but in cold winters has been encountered in vast flocks. When disturbed generally runs away. **V** Typically *chuck chuck chuck* or *chuck chuck chuck chuckarr chuckarr chuckarr*. When flushed often gives a repeated *wit-too-wittoo-wittoo*. **H** Mountain slopes with sparse cover, semi-arid hills, desert plains, sand dunes, forest clearings. **D** Eurasia, from Greece to C China.

32 PRZEVALSKI'S PARTRIDGE *Alectoris magna* 36–38cm Non-breeding season flocks average about 30, small summer parties are made up of unmated males. **V** Repeated *ga ga ga gela gela gela*. When flushed gives a *fei-ji fei-ji fei-ji* or *ja ja ja*. When alarmed at the nest utters a whistling *dirdir dirdir dirdir*. **H** Sparsely vegetated ravines and canyons of rocky mountain slopes and rugged plateaux. **D** NC China.

33 PHILBY'S PARTRIDGE *Alectoris philbyi* 34cm Occurs in pairs or small parties. **V** Repeated *chuk chuk-a-chuk-kar* or *chuk chuk chuk kar*. When disturbed gives a *chork chork chork*,a squealing *chuk-a-chuk-a-chuk* and a babbling *chuk-a-chuk-oo*. **H** Mountain plateaux and barren rocky slopes. **D** SW Arabian Peninsula.

34 BARBARY PARTRIDGE *Alectoris barbara* 32–34cm Usually occurs in pairs or small groups. Loath to fly, generally running away rapidly when disturbed. **V** Repeated *kutchuk kutchuk* with the odd *chukor* added. When flushed gives a squealing *kree-ah kree-ah* or a loud *chuckachew chew-chew*. **H** Bushy cover of desert wadis, hillsides and mountain sides, also orchards, coastal dunes, woodland clearings. **D** Canary Islands, N Africa.

35 RED-LEGGED PARTRIDGE *Alectoris rufa* 32–34cm Usually occurs in small parties of 6–10 birds, although larger groups have been recorded in winter. Tends to run away when disturbed. **V** Harsh, grating *go-chak-chak go-chak-chak go-chak-chak* and a *chuk-chuk-chukar-chukar*. When disturbed gives a squeaky *cheeragh cheeragh cheeragh*. **H** Dry open country, farmland, grassy hillsides and locally in mountains in south of range. **D** SW Europe.

36 ARABIAN PARTRIDGE *Alectoris melanocephala* 39cm Generally keeps to cover, in pairs or parties of up to 15. Mostly active in early morning and evening, when venturing to drinking sources. **V** Accelerating *kok kok kok kok kok chok-chok-chok-chok*, also a rapid *chuk-chuk-chuk-chuk-chuk*. When disturbed gives a *kerkow-kerkow-kerkow*. **H** Variable, including mountain slopes, rocky hillsides, sandy or stony plains with scattered bushes or trees. **D** S Arabian Peninsula.

♂
♀
1
2
rufous
3
grey
♀
rufous
♀ grey
4
♂
♀
♀
♂
5
6
♀
♂
7
♂
♂
8
♀
♂
9
♀
♂
10
♀
♀
13
♂
11
♂
♀
12
♂
15
♀
♀
♂
16
14
♀
17
18
♂
24
23
19
♀
♂
20
♀
♀
21
22
♂
♂
35
25
27
29
31
33
34
36
26
28
30
32

PHEASANTS AND ALLIES *CONTINUED*

1 SEE-SEE PARTRIDGE *Ammoperdix griseogularis* 24cm Occurs in small parties, although larger parties gather at drinking sites. When approached tends to run rather than fly. **V** Far-carrying, repeated *wheet-diu* or *hoe-it*. When alarmed utters a rapid, piping *bwuit-bwuit-bwuit*. **H** Barren stony hillsides, open stony land in semi-desert, wadis. **D** SE Turkey to Pakistan.

2 SAND PARTRIDGE *Ammoperdix heyi* 22–25cm Generally in parties of up to 10, although larger parties occur at drinking sites. When approached crouches before running, rather than flying, away. **V** Yelping *kew-kew-kew* or *watcha-watcha-watcha*, also a loud *quip* or *qu-ip*. When flushed utters an explosive *wuit-wuit-wuit*. **H** Rocky hillsides, ravines and wadis, generally with nearby water. **D** Arabian Peninsula, Egypt, N Sudan.

3 BLACK FRANCOLIN *Francolinus francolinus* 33–36cm Usually in pairs or family parties. Very shy, keeping to cover except when 'singing'. Reluctant to fly. **V** Generally silent apart from a strident, grating *clip gek-ge-gek gek-ge-gek* advertising call. **H** Lowland cultivated areas and grassland with bushy cover, lake edges with scrub and reeds. **D** Asia Minor to Bangladesh.

4 PAINTED FRANCOLIN *Francolinus pictus* 31–32cm Secretive, keeping to cover except when 'singing', which may be performed from low perch. Reluctant to fly, often squats in cover if disturbed. **V** High-pitched *click cheek-cheek-keray*, very similar to call of Black Francolin. **H** Dry grassland and scrub jungle interspersed with watercourses. **D** Most of peninsular India and SE Sri Lanka.

5 CHINESE FRANCOLIN *Francolinus pintadeanus* 31–34cm Very wary, but often 'sings' from prominent perch. **V** Harsh, metallic *kak-kak-kuich ka-ka* or *wi-ta-tak-takaa*, normally repeated after lengthy pauses. **H** Dry open forest and oak scrub-covered hills. **D** SE Asia.

6 GREY FRANCOLIN *Francolinus pondicerianus* 30–32cm Variable, southern birds with buff-orange throat, northern and northwestern races paler with whiter throat. Usually encountered in small groups. **V** Male has a rapid, repeated, strident *kat-ee-la kat-ee-la kat-ee-la*. Female gives a high-pitched, rising *tee-tee-tee*. When alarmed utters a sharp *kirr-kirr*. **H** Dry grassy plains with thorn scrub, stony semi-desert, cultivated areas. **D** SE Iran to India and Sri Lanka.

7 SWAMP FRANCOLIN *Francolinus gularis* 36–38cm Secretive, best located by call. Often feeds in the open early or late in the day. **V** Loud *kaw-care* or *ko-ko-care*, also a harsh *chuckeroo-chuckeroo-chuckeroo* preceded by a few harsh chuckles and croaks. **H** Tall, wet grassland, reedbeds, swamps, sugarcane fields. **D** N India, Nepal, Bangladesh.

8 LATHAM'S FRANCOLIN *Peliperdix lathami* 20cm Separated from Nahan's Partridge (Plate 11) by head pattern and buff underparts. **V** Accelerating series of about eight fluting whistles, falling in pitch. **H** Dense forests. **D** W and C Africa.

9 COQUI FRANCOLIN *Peliperdix coqui* 30cm Female separated from Grey-winged, Red-winged, Finsch's, Shelley's and Orange River Francolins by smaller size, barred underparts and, in flight, absence of red in wings. **V** Very high, sharp *keh-kwee-keh-kwee-keh-kwee* or very high *keh-krih kehkerriiiii*. **H** Grassy, more or less wooded and bushed areas. **D** S Africa.

10 WHITE-THROATED FRANCOLIN *Peliperdix albogularis* 25cm Red wings as Red-winged, Finsch's and Shelley's Francolins, but with other jizz and smaller bill. Note whitish chin. **H** Dry open areas with some grass cover, often near forest and water. **D** W and SC Africa.

11 SCHLEGEL'S FRANCOLIN *Peliperdix schlegelii* 25cm Separated from Coqui Francolin by different range, more extensive black barring below and darker, reddish upperparts. **V** Similar to that of Coqui Francolin, but faster and lower pitched. **H** Wooded grasslands. **D** Cameroon to Sudan.

12 RING-NECKED FRANCOLIN *Scleroptila streptophora* 25cm Note yellow legs, white chin and white-and-black barred collar and breast. **V** Long series of up to 10 high-pitched notes, with cooing tone. **H** Grassy, sparsely bushed and wooded rocky hillsides. **D** Cameroon, Uganda, Kenya, Tanzania.

13 GREY-WINGED FRANCOLIN *Scleroptila afra* 35cm Separated from Red-winged Francolin by 'reversed' neck and throat pattern. Belly finely spotted and barred black. Speckled throat diagnostic.**V** Very high, fluting *weeweeh-weeweeh*. **H** Grassy slopes at all altitudes. **D** Lesotho, South Africa.

14 RED-WINGED FRANCOLIN *Scleroptila levaillantii* 40cm Note tawny-rufous throat surrounded by speckled necklace, connected with speckled collar. **V** Very high, sharp *tiktiktiktiktik-let's-GO-then*. **H** Stony slopes, woodland with scrub, tall grass, reedy areas. **D** EC, SC and S Africa.

15 FINSCH'S FRANCOLIN *Scleroptila finschi* 35cm The only red-winged francolin without black head markings. **V** Series of sharp notes, similar to other francolins in its genus. **H** Grassland near forest and woodland, also on bare slopes at higher altitudes. **D** Gabon to SW DR Congo and W Angola.

16 SHELLEY'S FRANCOLIN *Scleroptila shelleyi* 35cm Separated from Orange River Francolin by black-and-white belly and darker upperparts. **V** Very high, sharp, fluting *wukwuk-weekweek-wukwuh weekweek*. **H** Stony, more or less wooded and bushy habitats, often with rocky outcrops. **D** E and SE Africa.

17 MOORLAND FRANCOLIN *Scleroptila psilolaema* 35cm Note red in wing and barring, scaling and spotting of underparts. **V** Typical of its genus, a very rapid series of shrill and squeaky notes. **H** Montane heath, grassland, moorland; 2,250–4,000m. **D** Ethiopia, Uganda, Kenya.

18 ORANGE RIVER FRANCOLIN *Scleroptila gutturalis* 35cm Note white chin, prominent double necklaces, and absence of black-and-white feathering on underparts. **V** Very high, piercing *turk-chéerup*. **H** Dry rocky, grassy, more or less wooded and bushy areas. **D** NE and S Africa.

19 CRESTED FRANCOLIN *Dendroperdix sephaena* 35cm Note prominent white eyebrow, red legs and, in flight, black tail. **V** Mid-high, loud, scraping *kurrk-kri-kurrk-kri*. **H** Dry, bushy and shrubby areas with some grass cover. **D** E and SE Africa.

20 SCALY FRANCOLIN *Pternistis squamatus* 30cm Note red legs and bill, and uniform overall colouring. **V** Very high, hoarse, running-up *skreetch-uh-skreetch-uh-skreetch*. **H** Tall grassy glades and forest edges. **D** C Africa.

21 AHANTA FRANCOLIN *Pternistis ahantensis* 35cm Rather unmarked appearance, striped white on dull brown. Orangish bill and legs. **V** Has barking and whistling contact calls. Song a fast chuckle *chuk-chukuruk*. **H** Forest edges and other dense growth. **D** W Africa.

22 GREY-STRIPED FRANCOLIN *Pternistis griseostriatus* 35cm Note all-rufous colouring, red bill and legs. **V** Said to be like that of Scaly Francolin. **H** Dense undergrowth of forest and woodland. **D** Angola.

23 HILDEBRANDT'S FRANCOLIN *Pternistis hildebrandti* 35cm Note scaly underparts of male, rufous underparts of female and red legs of both. **V** Series of high-pitched, grating cackles, typically starting as single notes before switching to double notes, *kek kek kek kerek kerek kerek*. Mainly vocal at dawn and dusk. **H** Rocky hillsides with rough grass. **D** E and SE Africa.

24 DOUBLE-SPURRED FRANCOLIN *Pternistis bicalcaratus* 30cm Note prominent white eyebrow. Separated from Hueglin's Francolin by absence of bare skin around eye and greenish (not orange) bill and legs. **V** Dry chuckling song *chuck-churr-uk*. **H** Grassland, farmland, woodland (not forest). **D** W and C Africa.

25 HEUGLIN'S FRANCOLIN *Pternistis icterorhynchus* 30cm Note yellowish bare skin behind eye, orange bill and unmarked centre of belly. **V** Song a short, agitated clucking phrase. **H** Open and bushy grassland and farmland. **D** S Chad and SW Sudan to N DR Congo and W Uganda.

26 CLAPPERTON'S FRANCOLIN *Pternistis clappertoni* 35cm Note bare red eye area, rufous crown and nape, scaly mantle and wings, and (in flight) pale buff wing patch. Separated from Red-necked Spurfowl (Plate 14) by absence of naked red throat. **V** Harsh, rolling, rasping chatter or trill. **H** Rocky hillsides and dry, lightly wooded and bushy habitats. **D** C and NE Africa.

27 HARWOOD'S FRANCOLIN *Pternistis harwoodi* 30cm Note bare red eye area and double-scaled breast and flanks. **V** Series of harsh, well-spaced, screeching notes. **H** Tall reedbeds with some trees along streams and adjoining (cultivated) areas. **D** Ethiopia.

28 SWIERSTRA'S FRANCOLIN *Pternistis swierstrai* 35cm Note diagnostic black band on breast that contrasts with white face and throat. **V** Crowing territorial call; clucking alarm call. **H** Montane forest, grassy and rocky mountain slopes. **D** W Angola.

29 MOUNT CAMEROON FRANCOLIN *Pternistis camerunensis* 35cm As Red-necked Spurfowl (Plate 14) except naked throat, and with different range and habitat. **V** Well-spaced, resonant hooting notes in a short series. **H** Dense forest undergrowth. **D** Cameroon.

30 HANDSOME FRANCOLIN *Pternistis nobilis* 35cm Separated from Scaly Francolin in same range by bare red eye-ring and rather rufous overall feathering. **V** Fast series of harsh clucking notes, accelerating. **H** Montane forest with dense undergrowth, bamboo, giant heath, moorland. **D** E DR Congo, Uganda, Rwanda.

31 JACKSON'S FRANCOLIN *Pternistis jacksoni* 35cm Note upright stance. **V** Song a raucous series of crowing and squeaking notes. **H** Dense forest undergrowth, bamboo, moorland, giant heath; 2,250–3,000m. **D** Kenya.

32 CHESTNUT-NAPED FRANCOLIN *Pternistis castaneicollis* 35cm Note black-edged white striping and white throat. **V** Short crowing phrases, notes speeding up into a rattle. **H** Tall grass in forest glades and dense undergrowth at forest edges; 1,000–4,000m. **D** NE Africa.

33 BLACK-FRONTED FRANCOLIN *Pternistis atrifrons* 40cm Usually adopts upright posture. Runs from danger. **V** Chuckling, cackling calls. **H** Wooded and scrub-covered rocky uplands. **D** S Ethiopia.

34 DJIBOUTI FRANCOLIN *Pternistis ochropectus* 35cm Occurs only in Djibouti. **V** Mid-toned harsh chatter. **H** Dense vegetation of dry watercourses. **D** Djibouti.

35 ERCKEL'S FRANCOLIN *Pternistis erckelii* 30cm Note dark face. **V** Song phrase begins with drawn-out, strained note, subsequent notes accelerate into a chatter. Alarm call duck-like quacking notes. **H** Steep forest glades with scrub, tall grass and herbage; 2,000–3,500m. **D** NE Sudan, Eritrea to C Ethiopia.

1
2
3
4
5
6
7
8
9
10
11
12
13
14
15
16
17
18
19
20
21
22
23
24
25
26
27
28
29
30
31
32
33
34
35

PHEASANTS AND ALLIES *CONTINUED*

1 HARTLAUB'S SPURFOWL *Pternistis hartlaubi* 25cm Note small size, short eyebrow, large bill, and black-streaked (male) or rufous (female) underparts. **V** Very high, excited, hurried *near-here near-here near-here.* **H** Slopes with large boulders. **D** Angola, Namibia.

2 RED-BILLED SPURFOWL *Pternistis adspersus* 35cm Note yellow eye-ring, black lore, red bill and pink legs. Less shy than other francolins. **V** Mid-high, loud, resounding, rising *gukgukgukgukkirrik*. **H** Low scrub, riverine thickets, woodland edges. **D** SC Africa.

3 CAPE SPURFOWL *Pternistis capensis* 40cm Note large size, pink legs, black-tipped pink upper mandible, white streaking on underparts and blackish tail in flight. **V** High, cackled *purwruut-purwruut-purpupperwrans.* **H** Coastal fynbos, riverine scrub. **D** South Africa.

4 NATAL SPURFOWL *Pternistis natalensis* 35cm Note pink bill and legs. Underparts paler than upperparts. **V** Very harsh, loud, excited *kekkerrek kekkerrek kekkerrek* or *kurruk kurruk*. **H** Rocky, wooded hillsides, riverine thickets. **D** SE Africa.

5 YELLOW-NECKED SPURFOWL *Pternistis leucoscepus* 35cm 33–35cm Note diagnostic bare yellow throat. Runs or flies from danger. **V** Grating *kek–kerruk kek–kerrek kek–kerruk*. **H** Forest edges and other wooded and bushy natural and cultivated areas. **D** Eritrea and E South Sudan to Somalia, Kenya and NE Tanzania.

6 GREY-BREASTED SPURFOWL *Pternistis rufopictus* 35cm Combines grey-brown legs and orange-pink bare throat. **V** Very harsh rolling *krrrraaaaaa*. **H** Grassy areas with scattered trees; dense undergrowth along streams. **D** NW Tanzania.

7 RED-NECKED SPURFOWL *Pternistis afer* 35cm Red legs and bare head parts diagnostic. **V** Mid-high, indignant, hurried, descending *kukukruk-kukukruk-kukukruk* or *corrupt-corrupt-corrupt*. **H** Forest patches, other bushy and wooded natural and cultivated areas with long grass. **D** C and SE Africa.

8 SWAINSON'S SPURFOWL *Pternistis swainsonii* 40cm Note black upper mandible and blackish legs. **V** Mid-high, loud, grating *korrah korrah korrah*. **H** Extensively used grasslands, thornveld, open woodland. **D** SC and SE Africa.

9 GREY PARTRIDGE *Perdix perdix* 29–31cm Apart from breeding season usually occurs in parties of 5–15. **V** Harsh, metallic *kierrr r-r-r-ik*, often likened to the squeak of a rusty gate. When flushed gives a rapid *skip skip kip kip kip-ip-ip-ip.* **H** Farmland, open grassland, semi-desert with scrub, locally in montane meadows. **D** Eurasia, from Scandinavia and Britain to SW Siberia and NW China.

10 DAURIAN PARTRIDGE *Perdix dauurica* 28–30cm During non-breeding season occurs in parties of 15–30, with much larger numbers recorded in winter. Habits little recorded, presumed as Grey Partridge. **V** As Grey Partridge. **H** Very varied, including lightly wooded areas adjoining grassland, wooded steppe, shrubby meadows, riverine scrub, farmsteads. **D** C Asia.

11 TIBETAN PARTRIDGE *Perdix hodgsoniae* 28–31cm In non-breeding season occurs in groups of 10–15. Usually runs for cover rather than fly. **V** Rattling *scherrrrrreck-scherrrrrreck*. When flushed utters a shrill *chee chee chee chee chee*. **H** Rocky mountain slopes and alpine meadows with scrub, descending to lower levels in winter. **D** Tibet, Bhutan, Nepal, N India, WC China.

12 LONG-BILLED PARTRIDGE *Rhizothera longirostris* 36–41cm When flushed flies to rest in trees. Little else recorded. **V** Bell-like *ti-ooah-whee* duet, repeated over long periods. **H** Lowland and montane forest, bamboo. **D** Malay Peninsula; Borneo; Sumatra, Indonesia.

13 HOSE'S PARTRIDGE *Rhizothera dulitensis* 30–35cm Very little recorded, actions presumed similar to those of Long-billed Partridge. **V** Not recorded. **H** Montane forest. **D** NE Borneo.

14 MADAGASCAN PARTRIDGE *Margaroperdix madagarensis* 26cm More a large quail than a partridge. Male's colour pattern is unmistakable. **V** Rather silent. **H** Mainly dry open areas. **D** Madagascar.

15 BLACK PARTRIDGE *Melanoperdix niger* 24–27cm Shy. Generally in pairs, but little else recorded. **V** Low, creaking contact call. **H** Evergreen forest. **D** Malay Peninsula; Borneo; Sumatra, Indonesia.

16 COMMON QUAIL *Coturnix coturnix* 16–18cm Shy and furtive, more often heard than seen. In flight, distinguished from other gamebirds and buttonquails by long, pointed wings. Some males have rufous-tinged face and throat. **V** Male utters a rapid *quip quip-ip*, often interpreted as *wet-my-lips*; female utters a low *bree-bree*. When flushed gives a shrill *tree-tree*. **H** Open grasslands, pastures and weedy waste areas. **D** Europe, Africa, Indian Ocean islands, C Asia.

17 JAPANESE QUAIL *Coturnix japonica* 17–19cm Action and habits similar to Common Quail. **V** Chattering *chrr-churrk-chrr*. When flushed call is similar to that of flushed Common Quail. **H** Rolling grassland, fields, forest clearings, montane foothills. **D** Mongolia, E Siberia, Japan, Korean Peninsula.

18 RAIN QUAIL *Coturnix coromandelica* 16–18cm Actions and habits very similar to those of Common Quail. **V** Male advertising call is a high-pitched *whit-whit whit-whit whit-whit whit-whit*. When flushed call is similar to that of flushed Common Quail. **H** Open grassland, cultivated fields, rice stubble, plantations, sometimes gardens. **D** Pakistan to Sri Lanka and Myanmar.

19 HARLEQUIN QUAIL *Coturnix delegorguei* 16–19cm Gregarious. Actions and habits similar to those of Common Quail. **V** Male gives a rapid *whit-whit*, *whit-whit-whit* or *tswic-tswic-tswic*; female may respond with a soft *quick-ik*. When flushed utters a squeaky *skreeee*. **H** Open grassland and fields. **D** Widespread across Africa.

20 STUBBLE QUAIL *Coturnix pectoralis* 16–20cm Nomadic, moving in response to drought, rainfall and food availability. Whirring sound created by wings on take-off. **V** Male song shrill *tee–tuleep*. **H** Grassland habitats, arable fields. **D** SE and SW Australia.

21 BROWN QUAIL *Coturnix ypsilophora* 19cm Generally seen in small coveys of 3–11 birds, typically elusive but occasionally recorded feeding along roadsides in the early morning or late afternoon. **V** Varies with race; includes an *er-errrrrrhh*, the drawn-out part with an upward inflection, and a loud *trrriup trrriup* when flushed. **H** Rank grassland, damp areas, light scrub, overgrown gardens. **D** Lesser Sunda Islands, Indonesia; New Guinea; Australia.

22 KING QUAIL *Excalfactoria chinensis* 12–15cm Shy, difficult to flush, preferring to squat or run for cover. Most often observed when running or dust-bathing on tracks. Generally occurs in pairs or family parties. **V** Piping *ti-yu ti-yu ti-yu ti-yu* or *ti-ti-yu*, the last note lower pitched. When flushed utters a weak *tir-tir-tir* or a sequence of sharp cheeps. **H** Shrubby and swampy grassland, marshes, paddyfields. **D** S and SE Asia, New Guinea, Australia.

23 BLUE QUAIL *Excalfactoria adansonii* 15cm The male from other quails by red wings, female from other female quails by barred (not streaked) underparts. Not gregarious. **V** Mid-high, descending, hoarse *wehwehweh-wehwehweh*. **H** Moist open grassland near swamps. **D** Sub-Saharan Africa.

24 SNOW MOUNTAIN QUAIL *Anurophasis monorthonyx* 26–28cm Feeds unobtrusively on ground. Barred plumage provides good camouflage. **V** Squealing call if disturbed; otherwise silent. **H** Alpine grassland; 3,000–3,800m. **D** New Guinea.

25 JUNGLE BUSH QUAIL *Perdicula asiatica* 17cm Generally occurs in parties of 6–20. Often encountered dust-bathing or walking on tracks, or feeding in shrubby grassland. Often flushed from underfoot with an explosion of whirring wings. **V** Harsh *chee-chee-chuck chee-chee-chuck*. Also a low, bubbling whistle, *tiri-tiri-tiri* or *whi-whi-whi-whi-whi*, and a low chuckle when flushed. **H** Grass areas in scrub jungle, grassy plains and hills with scrub. **D** India, Sri Lanka.

26 ROCK BUSH QUAIL *Perdicula argoondah* 17cm Very similar to Jungle Bush Quail. **V** Similar to that of Jungle Bush Quail. **H** Scrub-covered semi-desert plains and thinly vegetated rocky hills. **D** India.

27 PAINTED BUSH QUAIL *Perdicula erythrorhyncha* 18cm Actions and habits very similar to those of Jungle Bush Quail. **V** Pleasant *kirikee kirikee kirikee*. Reassembling coveys utter a soft, whistled *tu-tu-tu-tu-tu-tutu-tutu-tutu*, rising in pitch. When flushed utters a Common Quail-like whistle. **H** Open grassy hillsides with scattered scrub near forest edges, often interspersed with cultivation. **D** India.

28 MANIPUR BUSH QUAIL *Perdicula manipurensis* 20cm Best located by call. Occurs in coveys of 5–8, usually well ensconced in tall grass but will forage in open early and late, attracted to fresh shoots on burnt grassland. **V** Softly whistled *whit-it-it-it-it*, the notes getting progressively higher pitched, repeated three to four times, louder and higher each time. **H** Moist, tall elephant-grass areas, bogs, swamps. **D** NE India, Bangladesh.

29 HIMALAYAN QUAIL *Ophrysia superciliosa* 25cm Shy and elusive, keeping to thick cover. Prefers to run rather than fly when flushed. **V** Shrill whistle when alarmed. **H** Steep hillsides with scrubby thickets and tall grass. **D** N India. Probably extinct.

30 UDZUNGWA FOREST PARTRIDGE *Xenoperdix udzungwensis* 30cm Unobtrusive and shy; keeps to cover. Runs when disturbed. Roosts in trees. **V** Whistling cry when alarmed. **H** Forest interiors. **D** Udzungwa Mountains, S Tanzania.

31 RUBEHO FOREST PARTRIDGE *Xenoperdix obscuratus* 29–30cm Feeds on invertebrates and seeds, sometimes in pairs. **V** Soft whistle; loud scream if alarmed. **H** Dense montane forest. **D** Rubeho Mountains, S Tanzania.

1
2
3
4
5
6
7
8
9
10
11
12
13
14
15
16
17
18
19
20
21
22
23
24
25
26
27
28
29
30
31

PHEASANTS AND ALLIES *CONTINUED*

1 HILL PARTRIDGE *Arborophila torqueola* 28–30cm Feeds in forest leaf litter in groups of 5–10. **V** Mournful, repeated *pooo* or *pheaw*, followed by three to six rising double whistles *do-eat do-eat do-eat do-eat*. Also duets, female uttering *kwikwikwikwikwik* while male joins with series of *do-eat* calls. **H** Montane oak forest mixed with laurel and rhododendron. **D** W Himalayas to NW Vietnam.

2 RUFOUS-THROATED PARTRIDGE *Arborophila rufogularis* 26–29cm Actions and habits similar to those of Hill Partridge. **V** Far-carrying, mournful whistle, *wheeea-whu*, occasionally given in a series of two to three repeated three to four times, ascending and ending abruptly. **H** Oak forest with laurel and rhododendron, and thick undergrowth. **D** N India to C Vietnam.

3 WHITE-CHEEKED PARTRIDGE *Arborophila atrogularis* 25–27cm Actions and habits similar to those of Hill Partridge. **V** Far-carrying, quavering *prrrer prrrer prrrer prrrer*, ascending and accelerating, before ending abruptly, often followed by a number of *wi-chu* notes. **H** Bamboo and damp undergrowth in broadleaved evergreen forest, tea plantations, bushy grassland, scrub jungle, never far from forest edge. **D** NE India, Myanmar.

4 TAIWAN PARTRIDGE *Arborophila crudigularis* 27–28cm Forages on ground in small parties. Roosts in trees. **V** Soft whistle. **H** Thickets and damp undergrowth in broadleaved evergreen forest. **D** Taiwan.

5 CHESTNUT-BREASTED PARTRIDGE *Arborophila mandellii* 28–30cm Little known; actions and habits supposed similar to those of Hill Partridge. **V** Repeated *prrreet*, followed by series of ascending *prrr prrr-er-it* calls that end in crescendo. **H** Evergreen oak and rhododendron forest with thick undergrowth. **D** NE India to SE Tibet.

6 BAR-BACKED PARTRIDGE *Arborophila brunneopectus* 26–29cm Actions much as those of Hill Partridge. **V** Call is a whistled *ti-hu ti-hu ti-hu*, preceded by a series of throaty *brr* notes. Often duets, partner answering with a repeated *kew-kew-kew*. **H** Broadleaved evergreen forest. **D** SE Asia.

7 SICHUAN PARTRIDGE *Arborophila rufipectus* 29–31cm Actions and habits similar to those of Hill Partridge. **V** Slow *ho-wo ho-wo*; also a complex whistle. **H** Broadleaved forest with an understorey of bamboo and other shrubs. **D** SC China.

8 WHITE-NECKLACED PARTRIDGE *Arborophila gingica* 30cm Actions and habits presumed similar to others of genus. **V** Far-carrying, mournful, repeated *wooop* or quickly repeated *co-qwee*. **H** Dense, moist forest in foothills and mountains. **D** S China.

9 ORANGE-NECKED PARTRIDGE *Arborophila davidi* 27cm Forages among leaf litter in forest undergrowth. **V** Repeated *prruu*, accelerating and becoming higher pitched. Also a faster series of plaintive piping *tu* notes. **H** Broadleaved evergreen and semi-evergreen forests in low hills. **D** S Vietnam.

10 CHESTNUT-HEADED PARTRIDGE *Arborophila cambodiana* 28cm Actions similar to those of Hill Partridge, although said to be easier to approach than others in genus. **V** Unknown. **H** Broadleaved evergreen forest. **D** Cambodia.

11 SIAMESE PARTRIDGE *Arborophila diversa* 28cm No details recorded; actions as those of Hill Partridge male. **V** Whistled *tu-u...hu tu-u...hu*. **H** Tropical evergreen forest. **D** SE Thailand.

12 MALAYSIAN PARTRIDGE *Arborophila campbelli* 28cm Habits unrecorded, but presumed similar to others in genus. **V** Series of loud, whistled *pi-hor* notes, preceded by a soft, whistled *oii oii oii*. **H** Undergrowth in broadleaved evergreen forest. **D** Peninsular Malaysia.

13 ROLL'S PARTRIDGE *Arborophila rolli* 28cm Was, until recently, regarded as a subspecies of the previous species; actions are presumed to be similar. **V** Similar to Grey-breasted Partridge. **H** Dense forest undergrowth. **D** N Sumatra, Indonesia.

14 SUMATRAN PARTRIDGE *Arborophila sumatrana* 28cm Little known; actions presumed similar tothose of Roll's Partridge. **V** Very similar to that of Roll's Partridge. **H** Dense undergrowth in broadleaved foothill and mountain forest. **D** C Sumatra, Indonesia.

15 GREY-BREASTED PARTRIDGE *Arborophila orientalis* 28cm Little known; presumably forages among leaf litter in small parties. **V** Series of single whistles followed by a double chirping whistle, increasing in volume. **H** Undergrowth in primary montane forest. **D** E Java, Indonesia.

16 CHESTNUT-BELLIED PARTRIDGE *Arborophila javanica* 28cm Little recorded; presumably forages among leaf litter as other partridges. **V** Far-carrying series of double calls, increasing in volume and tempo. **H** Hill and montane forest. **D** Java, Indonesia.

17 RED-BILLED PARTRIDGE *Arborophila rubrirostris* 29cm Occurs in small groups, foraging in mossy gullies and dense undergrowth. **V** Loud, whistled *keow*, rising in pitch and volume. **H** Lower and upper montane forest. **D** Sumatra, Indonesia.

18 RED-BREASTED PARTRIDGE *Arborophila hyperythra* 27cm Nominate and grey eye-brow form (18b) shown. Forages in groups in thickets, on forest tracks and on river flats. **V** Repeated, ringing *chii*, which increases in volume and tempo, answered by a low double note. **H** Thickets, including bamboo in understorey of secondary and primary hill and montane forest. **D** Borneo.

19 HAINAN PARTRIDGE *Arborophila ardens* 26–28cm Forages among leaf litter, usually singly, in pairs or in small parties. **V** Repeated whistled *kwe-ho kwe-ho kwe-ho*, sometimes accelerating and rising in pitch. **H** Broadleaved evergreen tropical forest understorey, also monsoon evergreen forest. **D** Hainan, China.

20 CHESTNUT-NECKLACED PARTRIDGE *Arborophila charltonii* 26–32cm Shy; forages on ground at forest edges, usually in small parties. **V** Similar to that of Green-legged Partridge. **H** Dense lowland jungle and foothills with evergreen forest. **D** Myanmar; Thailand; Peninsular Malaysia; Borneo; Sumatra, Indonesia.

21 GREEN-LEGGED PARTRIDGE *Arborophila chloropus* 26–31cm Actions similar to those of others in genus. **V** Plaintive *tu-tu...tu-tu...tu-tu...tu tu tu tu tutututututututu chirra-chew-chirra-chew-chirra-chew* that ascends and then descends. **H** Dense evergreen and mixed deciduous primary forest with bamboo thickets. **D** SE Asia.

22 FERRUGINOUS PARTRIDGE *Caloperdix oculeus* 23–27cm Forages on forest floor, singly, in pairs or sometimes in larger parties. **V** Rising *pi-pi-pi-pipipipipipi*, repeated eight to nine times, followed by a clanging *dit-duit dit-duit*. **H** Broadleaved evergreen forest, bamboo and freshwater swamp forest. **D** Myanmar; Thailand; Peninsular Malaysia; Borneo; Sumatra, Indonesia.

23 CRIMSON-HEADED PARTRIDGE *Haematortyx sanguiniceps* 25cm Forages on forest floor; often encountered on forest trails. **V** Harsh, high-pitched *kro-krang*, repeated several times; also utters a harsh clucking. **H** Lower montane forest, heath forest and poor forest on sandy soils in valley bottoms. **D** Borneo.

24 CRESTED PARTRIDGE *Rollulus rouloul* 25cm Forages on ground among leaf litter, usually singly or in pairs. Has a brisk gait. **V** Long series of mournful *si-ul* whistles, the second note higher pitched. **H** Lowland rainforest. **D** Thailand; Peninsular Malaysia; Borneo; Sumatra, Indonesia.

25 MOUNTAIN BAMBOO PARTRIDGE *Bambusicola fytchii* 25–35cm Usually in family parties of 5–6. Wary; feeds early morning and late evening. **V** Rapidly repeated, cackling *che-chiree-che-chiree chirree chirree chirree*. When flushed, emits a scream. **H** Various open and scrubby areas, including tall grassland in damp areas and bamboo patches. **D** NE India to Myanmar and SC China.

26 CHINESE BAMBOO PARTRIDGE *Bambusicola thoracicus* 30–32cm Occurs in pairs or groups of up to 20, with larger groups recorded in winter. **V** Loud *gi-gi-gi-gi-gi-gigeroi-gigeroi*; also a *killl-killy e-put-kwai* duet. In non-breeding season utters *sih-mo-kuai sih-mo-kuai*, often written as *people pray people pray*. **H** Dense bushes (not necessarily bamboo), dry bush areas, parks. **D** S China.

27 TAIWAN BAMBOO PARTRIDGE *Bambusicola sonorivox* 30–32cm Field notes and **V** as for Chinese Bamboo Partridge, from which it was separated in 2014. **H** Dense bushes (not necessarily bamboo), dry bush areas, parks. **D** Taiwan.

28 RED SPURFOWL *Galloperdix spadicea* 35–38cm Variable. Secretive; more often heard than seen. When alarmed, runs rapidly into cover. **V** Rapidly repeated, crowing *k-r-r-r-kwek kr-kr-kwek kr-kr-kwek*. When flushed, utters a harsh, cackling *kuk-kuk-kuk-kukaak*. **H** Rocky foothills with scrubby bamboo thickets and dense secondary growth. **D** India, Nepal.

29 PAINTED SPURFOWL *Galloperdix lunulata* 30–34cm Secretive, keeping to cover. When disturbed, tends to run rather than fly; said to hide in holes or rock fissures if persistently pursued. **V** Loud, rapidly repeated *chur chur chur* and a fowl-like cackling. **H** Dry, rocky areas in thornbush or bamboo thickets. **D** India.

30 SRI LANKA SPURFOWL *Galloperdix bicalcarata* 30–34cm Secretive; best located by call. Occurs in pairs or family parties. Tends to run rather than fly when disturbed. **V** Rising series of trisyllabic notes, *yuhuhu yuhuhu yuhuhu yuhuhu yuhuhu yuhuheeyu*, the last note lower, similar in pitch to first. **H** Undisturbed lowland and hills. **D** Sri Lanka.

31 BLOOD PHEASANT *Ithaginis cruentus* 44–48cm Usually encountered in small groups, often tame, rarely flies. Very variable. **V** Repeated *chuck* or *chic*, and high-pitched, repetitive *see*. When maintaining contact, utters a loud *sree-cheeu-cheeu-cheeu* or high trill. **H** Forests or scrub at mid- to high altitudes; descends to lower forests during bad winters. **D** Himalayas.

32 WESTERN TRAGOPAN *Tragopan melanocephalus* male 68–73cm, female 60cm Wary and skulking. When disturbed may fly up to branches of nearby tree. During display inflates bare skin of throat and horns. **V** Bleating, repeated *khuwah*, said to sound like a lost goat, lamb or child. In alarm, gives a similar-sounding *waa waa waa*. **H** Mid-altitude oak-dominated forests with dense undergrowth; descends to lower altitudes in winter. **D** Himalayas.

33 SATYR TRAGOPAN *Tragopan satyra* male 67–72cm, female 58cm Generally very wary and skulking; disappears into nearby tree branches when alarmed. During display inflates bare skin of throat and horns. **V** Wailing *wah waah oo-ah oo-aaaa* repeated a dozen or so times. In alarm, utters a quiet *wak wak*. **H** High-altitude oak forests, or mixed conifer and broadleaved forests with dense undergrowth. **D** Himalayas.

34 BLYTH'S TRAGOPAN *Tragopan blythii* male 65–70cm, female 58cm Little known; actions and habits probably similar to those of other tragopans. **V** Male utters a loud, moaning *ohh-ohhah...ohaah-ohaaah...ohaaaha...ohaaaha-ohaaaha*. A resounding *gock gock gock* or *wak wak wak* also given during display. **H** Lush broadleaved forest with dense understorey. **D** Himalayas.

35 TEMMINCK'S TRAGOPAN *Tragopan temminckii* male 64cm, female 58cm Wary. Often feeds in trees. During spectacular display inflates bare skin of throat, showing off a dark blue oval, spotted paler blue and surrounded by a pale blue rim with red patches. **V** During breeding season gives an eerie *woh-woah-woah-woah-waah-waah-waah-waah-griiiik*. **H** Temperate and subalpine forests with dense undergrowth. **D** E Himalayas to SC China.

36 CABOT'S TRAGOPAN *Tragopan caboti* male 61cm, female 50cm Forages on ground and in trees in small groups, usually in early morning or late afternoon. **V** Territorial call resembles a baby's cry, *wa-r* followed by a few *gua* notes. When alarmed, utters a loud *gua-gua-gua*, often continuing for several minutes. **H** Evergreen and mixed forest; in summer also in open country above treeline. **D** S China.

1 ♂ ♀
2 ♂ ♀
3
4 ♂ ♀
5
6
7 ♂ ♀
8
9
10
11
12
13
14
15
16
17
18
18b
19
20
21
22
23 ♀ ♂
24 ♂ ♀
25
26
27
28 ♂ ♀
29 ♂ ♀
30 ♂ ♀
31 ♀ ♂
32 ♂ ♀ ♀
33 ♂
34 ♂ ♀
35 ♂ ♀
36 ♀ ♂

PHEASANTS AND ALLIES *CONTINUED*

1 KOKLASS PHEASANT *Pucrasia macrolopha* Male 56–64cm, female 52–56cm Wary. When alarmed, charges into undergrowth. Usually seen singly or in pairs; best observed early or late in day. **V** Early-morning call is loud *kok-kok-kok ko-kras* or similar. When disturbed, male gives a harsh *kwak kwak kwak* and female a musical *qui-quik qui-quik qui-quik*. **H** Coniferous and mixed forest with thick understorey. **D** Himalayas to SC China.

2 HIMALAYAN MONAL *Lophophorus impejanus* 64–72cm Plumage variable both in colour of gloss on breast, from green to purple, and in amount of white on back; some have no white. Usually occurs singly or in small, loose parties. Slightly less wary than others in genus **V** *Kur-lieu* or *kleeh-wick*; alarm note very similar. **H** Coniferous and broadleaved forest with thick understorey; alpine meadows in summer. **D** Himalayas.

3 SCLATER'S MONAL *Lophophorus sclateri* 64–68cm Little known; actions said to be similar to Himalayan Monal. Birds in W Arunachal Pradesh have all-white tail. **V** Loud, whistled *go-li*. When alarmed, gives a plaintive, shrill call. **H** Coniferous montane forests with thick undergrowth. **D** E Himalayas.

4 CHINESE MONAL *Lophophorus lhuysii* 75–80cm Action and habits as others in genus. **V** Repeated *guli* and whistled *guo-guo-guo* uttered every few minutes, starting high before dropping and fading to an end. When alarmed, gives a series of low *gee* notes. **H** Alpine and subalpine rocky meadows; 3,000–4,900m. **D** C China.

5 RED JUNGLEFOWL *Gallus gallus* male 65–78cm, female 41–46cm Generally encountered in small groups comprising a male and several females. Best observed early morning or late afternoon, foraging beside scrub-forest tracks. Roosts socially in trees. **V** *Cock-a-doodle-do*, similar to the typical farmyard cockerel call, although with shriller, more strangulated finish. **H** Forest undergrowth, scrub. **D** India, SE Asia.

6 GREY JUNGLEFOWL *Gallus sonneratii* male 70–80cm, female 38cm Wary, never far from cover. Usually encountered singly, although occasionally seen in groups of 5–6; sometimes in very large numbers where food is abundant. Roosts socially in trees or bamboo. **V** Loud, staccato *kuk ka kuruk ka* or *kuck kaya kaya kuck*, repeated up to five times a minute. **H** Wooded areas with thick understorey, scrub and bamboo thickets, overgrown or abandoned plantations. **D** India.

7 SRI LANKA JUNGLEFOWL *Gallus lafayettii* male 66–72cm, female 36cm Best observed in early morning and late afternoon, when it forages in open on forest tracks. Generally wary but can be confiding where not hunted. **V** Loud, staccato *chick chow-chik*, the final *ik* being higher than the rest. Female utters a high-pitched, metallic *kwikukk kwikkukkuk*. Both sexes also give a harsh *clock-clock*. **H** Various wooded areas, including primary and montane rainforests, scrub jungle, plantations. **D** Sri Lanka.

8 GREEN JUNGLEFOWL *Gallus varius* male 65–75cm, female 42–46cm Generally seen singly or in small groups, in early morning and late evening, feeding in open areas such as grain fields or tracks. **V** Male gives a three-note *cha-aw-awk* or *chow-a-aaaar* and a slow, cackling *wok-wok-wok*; female utters a rapid *kok-kok-kok*. **H** Woodland, woodland edges, plantations, grassland. **D** Java, Bali and W Lesser Sunda Islands, Indonesia.

9 KALIJ PHEASANT *Lophura leucomelanos* male 63–74cm, female 50–60cm Wary. Usually seen feeding early and late in small groups. Roosts in trees. **V** During breeding gives a loud chuckle and drums wings against body. Uses a *kurr-kurr-kurrchi-kurr* to keep contact, and in alarm gives a *koorchi koorchi* or *whoop-keet-keet*. **H** Temperate forests; 2,100–3,200m. **D** N Pakistan to W Thailand.

10 SILVER PHEASANT *Lophura nycthemera* male 80–125cm, female 56–70cm Encountered in small groups; forages on ground. Tends to run when disturbed. **V** Contact calls include a throaty *wutch-wutch-wutch* and a short *uwh* or *orh*. When alarmed, utters a grunting *uwerk* and *uwick*, combined with a sharp *sssiik* or *hssiik* and a rising *swiiieeik* and or *hwiiieeik*. **H** Broadleaved evergreen and mixed deciduous forests. **D** SE Asia.

11 EDWARDS'S PHEASANT *Lophura edwardsi* 58–65cm Little known. Said to be wary, keeping to dense cover. **V** Low, guttural *uk uk uk uk uk*. Males also indulge in wing-whirring. **H** Lowland evergreen forest with dense understorey. **D** C Vietnam.

12 SWINHOE'S PHEASANT *Lophura swinhoii* male 79cm, female 50cm Shy. Forages early morning or late afternoon in open areas of forest floor and among cover at road edges. Usually seen singly, sometimes in pairs and small groups. **V** Generally silent. Soft murmurings reported as birds forage. **H** Hardwood forest with dense undergrowth. **D** Taiwan.

13 HOOGERWERF'S PHEASANT *Lophura hoogerwerfi* 40–50cm Actions presumed much like those of Salvadori's Pheasant, with which it is often regarded as conspecific. **V** Not known, probably much like that of Salvadori's. **H** Montane forest. **D** N Sumatra, Indonesia.

14 SALVADORI'S PHEASANT *Lophura inornata* 46–55cm Rare and little recorded. Probably occurs in pairs. **V** Male utters a cluck in the breeding season. **H** Montane forest with dense undergrowth. **D** SW Sumatra, Indonesia.

15 CRESTLESS FIREBACK *Lophura erythrophthalma* male 47–50cm, female 42–44cm Wary. Forages on forest floor, in pairs or small groups. **V** Vibrating, throaty *purr* and repeated, croaking *tak-takuru* or *tooktaroo*. When alarmed, utters a loud *kak*. Wing-whirring takes place during display. **H** Secondary forest. **D** Peninsular Malaysia; Borneo; Sumatra, Indonesia.

16 CRESTED FIREBACK *Lophura ignita* male 65–70cm, female 56–57cm Wary. Usually encountered in small groups. Presence often given away by contact calls or male's wing-whirring sounds. **V** Male utters a subdued *woonk-k woonk-k*, often accompanied by loud wing-whirring. When alarmed, gives a sharp *chukun chukun*. **H** Lowland forest. **D** Peninsular Malaysia; Borneo; Sumatra, Indonesia.

17 SIAMESE FIREBACK *Lophura diardi* male 70–80cm, female 53–60cm Forages in cover early morning and late afternoon, singly or in small groups. Also recorded by forest tracks and roads. **V** Loud whistling, and a continual *pee-yu pee-yu*. Male produces wing-whirring sounds. **H** Lowland evergreen, semi-evergreen and bamboo forests. **D** SE Asia.

18 BULWER'S PHEASANT *Lophura bulweri* male 77–80cm, female 55cm Thought to occur in pairs or small family groups. In a spectacular display, male greatly extends wattles and raises and spreads tail. **V** In breeding season, utters a shrill, piercing cry. Also a metallic *kook kook* and distinctive *bek-kia*. **H** Primary hill and submontane forest. **D** Borneo.

19 WHITE EARED PHEASANT *Crossoptilon crossoptilon* 86–96cm Feeds on ground in small groups; larger groups of 30 or so form in winter. **V** Far-carrying, grating *gag gag gagerah gagerah gagerah gagerah*. **H** Subalpine birch, rhododendron, coniferous and mixed forests; 2,800–4,300m. **D** WC and SC China, Tibet.

20 TIBETAN EARED PHEASANT *Crossoptilon harmani* 75–85cm Usually in small parties feeding on ground. **V** Very similar to that of White Eared Peasant. **H** Subalpine meadows, alpine scrub, clearings in conifer and mixed forests; 3,000–5,000m. **D** Tibet.

21 BROWN EARED PHEASANT *Crossoptilon mantchuricum* 96–100cm Rare. Feeds on ground in small groups, with larger groups of up to 30 in winter. **V** Deep *gu-gu-gu-gu* and a *gu-ji-gu-ji* when foraging. **H** Deciduous and mixed conifer forests with shrubby understorey; 1,100–2,600m. **D** NE China.

22 BLUE EARED PHEASANT *Crossoptilon auritum* 96cm Feeds on ground in small groups, with larger groups of 50–60 in winter. **V** Loud, hoarse *ka ka...la*, *krip krraah krraah* or *wu wu wu*. When alarmed, utters *ziwo-ge ziwo-ge*. **H** Coniferous and mixed forests, juniper scrub, alpine meadows; 2,700–4,400m. **D** C China.

1
2
3
4
5
6
7
8
9
10
11
12
13
14
15
16
17
18
19
20
21
22

PHEASANTS AND ALLIES ***CONTINUED***

1 CHEER PHEASANT *Catreus wallichii* male 90–118cm, female 61–76cm Feeds on ground, usually close to cover. Roosts on rocky outcrops or trees. **V** Grating, accelerating *chir-a-pir chir-a-pir chir chir chirwa chirwa*; also a series of high whistles interspersed with harsh staccato and short *chut* notes. **H** Rocky and grassy hillsides with scrub. **D** Himalayas.

2 ELLIOT'S PHEASANT *Syrmaticus ellioti* male 80cm, female 50cm Forages mainly in morning and late afternoon in shy, small groups. **V** Low clucks and chuckles. Indulges in wing-whirring, usually followed by low-pitched *ge-ge-ge-ge-ge-ge*. **H** Evergreen broadleaved and conifer forests, bamboo and other dense scrub in mountains. **D** SE China.

3 MRS HUME'S PHEASANT *Syrmaticus humiae* male 90cm, female 60cm Generally forages in small groups, keeping to dense vegetation at forest edge. **V** Contact calls are a loud *chuck* and a low, muttering *buk-buk-buk-buk*. When alarmed, gives similar, even louder, calls and a noisy screech. **H** Evergreen broadleaved and mixed forest with patches of grass and bracken on steep rocky hillsides, oak–pine forest with scattered clearings, conifer plantations, scrub. **D** NE India to SW China, N Thailand.

4 MIKADO PHEASANT *Syrmaticus mikado* male 87cm, female 53cm Shy and elusive. Usually forages at dawn and dusk; seen more often during light rain or after heavy rain. Regularly takes to trees during heavy rain. **V** Mostly silent. During breeding season utters a short, rising squeal, preceded by a mellow *chup chup*. Gives a quiet, high-pitched *wok wok wok* when alarmed. **H** Primary forest with dense undergrowth of rhododendron and bamboo. **D** Taiwan.

5 COPPER PHEASANT *Syrmaticus soemmerringii* male 87–136cm, female 51–54cm Usually forages in deep cover. Roosts in trees. **V** Hoarse *ko ko ko*. **H** Coniferous and mixed forests with thick undergrowth. **D** Japan.

6 REEVES'S PHEASANT *Syrmaticus reevesii* male 150–210cm, female 70–80cm Forages at forest edge or on farmland early and late, otherwise keeps to forests. Roosts in trees. **V** Series of high chirps accompanied by wing-whirring. Also a soft *pu pu pu*. **H** Forests and areas of tall grass and bushes. **D** N and C China.

7 COMMON PHEASANT *Phasianus colchicus* male 66–89cm, female 53–63cm Familiar pheasant. Two main forms, green-necked and ring-necked, very variable, especially where introduced. **V** Harsh *korkk korrk ko ok korkk-kok*, often followed by wing-whirring. When alarmed, gives a rapid *kut-uk kut-uk kut-uk*. **H** Very varied, including farmland, open woodland, open country, riverine scrub. **D** EC Europe to E, C and S China and Korean Peninsula.

8 GREEN PHEASANT *Phasianus versicolor* male 66–89cm, female 53–63cm Regarded as subspecies of Common Pheasant by some. Male's greenish-blue sheen to neck and underparts diagnostic. **V** As Common Pheasant. **H** Varied habitats, including farmland, woodland edge. **D** Japan.

9 GOLDEN PHEASANT *Chrysolophus pictus* male 100–115cm, female 61–70cm Male unmistakable. Forages early and late on tracks and clearings, otherwise keeps to thick cover. **V** Loud *ka-cheek* or *cha-chak*. **H** Bamboo and scrub on rocky hills. Introduced population in Britain occurs mainly in rhododendron thickets. **D** C and S China.

10 LADY AMHERST'S PHEASANT *Chrysolophus amherstiae* male 105–120cm, female 60–70cm Male unmistakable. Skulking; usually keeps to thick cover. Forms large groups in winter. **V** Loud *cheek ker-chek* or *su-ik-ik*. **H** Forest, bamboo and thick scrub on hills and mountains. Introduced population in Britain frequents conifer plantations, mixed and deciduous woodland with undergrowth. **D** NE Myanmar, SE Tibet, SW China.

11 BRONZE-TAILED PEACOCK-PHEASANT *Polyplectron chalcurum* male 56cm, female 40cm Very shy and wary; presumably forages on the forest floor in pairs. **V** Male gives a harsh, loud, far-carrying *karau-karau-karau*. **H** Montane forest. **D** W Sumatra, Indonesia.

12 MOUNTAIN PEACOCK-PHEASANT *Polyplectron inopinatum* male 65cm, female 46cm Frequently encountered on tracks on or close to ridges, in small parties. Wary; if disturbed, quietly disappears into undergrowth. **V** Utters a burbling, descending whistle, also one to four loud, harsh clucks or squawks, repeated every few seconds. When alarmed, gives a chicken-like *cluck*. **H** Rugged mountain forest. **D** Peninsular Malaysia.

13 GERMAIN'S PEACOCK-PHEASANT *Polyplectron germaini* male 56–60cm, female 48cm Wary and difficult to observe; best located by calls. Usually runs if alarmed; flight is fast and low. **V** Chuckling or low purring, which becomes more rapid when responding to a rival. **H** Lowland and submontane evergreen and semi-evergreen forest, also secondary, swamp and bamboo forest. **D** Vietnam.

14 GREY PEACOCK-PHEASANT *Polyplectron bicalcaratum* male 56–76cm, female 48–55cm Extremely wary, creeping away through undergrowth at first sign of alarm. During display, male crouches and fans wings and tail, showing off ocelli. **V** Male utters a shrill, whistled *trew-tree*, *phee-hoi* or *taa-pwi* and a guttural, raucous *qua qua qua* or *wak wak wak*, the latter also given when alarmed. **H** Broadleaved evergreen and semi-evergreen forest with dense undergrowth. **D** SW China, Myanmar, N Thailand, Laos, CN Vietnam.

15 HAINAN PEACOCK-PHEASANT *Polyplectron katsumatae* male 53–65cm, female 40–45cm Habits presumed similar to those of Grey Peacock-pheasant, with which it is often thought to be conspecific. **V** Male utters a loud, melodious *guang-gui guang-gui*, female a more rapid *ga ga ga*. **H** Evergreen and semi-evergreen forests. **D** Hainan, China.

16 MALAYAN PEACOCK-PHEASANT *Polyplectron malacense* male 50–53cm, female 40–45cm Very wary, disappearing into undergrowth at slightest disturbance. Best located by calls or by cleared areas of leaf litter 'scrapes' used as display sites. **V** Male territorial call is a melancholy *puu pworr*; also a harsh cackle. **H** Broadleaved evergreen forest with a rich understorey. **D** SW Thailand, Peninsular Malaysia.

17 BORNEAN PEACOCK-PHEASANT *Polyplectron schleiermacheri* male 50cm, female 36cm Little recorded. Forages quietly on forest floor; spreads wings during display. **V** Melancholic *hor-hor* or *wu-wurh*. Contact call is a loud *cack cack*. **H** Primary lowland rainforest. **D** Borneo.

18 PALAWAN PEACOCK-PHEASANT *Polyplectron napoleonis* male 50cm, female 40cm Secretive; best located by calls or clean display 'scrapes'. Males recorded foraging singly and females in small groups. **V** Harsh, screeched *auukk*, *kratt* and *ka-reeeetch*; also a two-note, see-sawing *krotchh-kritchh*, likened to hitting a metal pipe with a piece of wood. **H** Forested slopes. **D** Palawan, Philippines.

19 CRESTED ARGUS *Rheinardia ocellata* male 190–235cm, female 64–75cm Wary; best located by calls or the presence of display 'scrapes'. **V** Series of up to eight loud *oowaaa* or *oowaaau* notes that vary in volume and length. Male gives a loud, resonant *woo-kia wau* during 'dancing' display. **H** Damp forest, primary and secondary evergreen forest, montane forest. **D** C Vietnam, E Laos, Peninsular Malaysia.

20 GREAT ARGUS *Argusianus argus* male 160–200cm, female 72–76cm Wary, moving into undergrowth at least disturbance. Best located by calls or display 'scrapes'. **V** Distinctive *kwoow-wow* and musical *wow*, repeated 70 times or more and getting higher pitched. **H** Broadleaved evergreen forest. **D** Peninsular Malaysia; Borneo; Sumatra, Indonesia.

21 INDIAN PEAFOWL *Pavo cristatus* male 180–230cm, female 90–100cm Unmistakable. During display, elevates and fans tail, showing off colourful ocelli. **V** Far-carrying, wailing, repeated *kee-ow kee-ow kee-ow* and a braying *ka-an ka-an ka-an*. When alarmed, utters a *kok-kok* or *cain-kok*. **H** Deciduous forest with understorey, scrub jungle, forest edges; where semi-feral, also in cultivated fields and around habitation. **D** India, Pakistan, Sri Lanka.

22 GREEN PEAFOWL *Pavo muticus* male 180–300cm, female 100–110cm Unmistakable. Timid and secretive. Best located when calling from roost sites. Tends to forage near watercourses. **V** Repeated, trumpeting *ki-wao or yee-ow*. Female utters a loud *aow-aa aow-aa*. When agitated, gives a repeated *tak tak ker-r-r-r oo oo ker-r-r-roo*. **H** Riverine forest and nearby open country. **D** SE Asia.

23 CONGO PEAFOWL *Afropavo congensis* 65cm Unmistakable. **V** Pairs sing in duet of *gowk* notes, male's higher pitched. **H** Undergrowth of tall forest. **D** DR Congo.

1
2
3
4
5
6
7
8
9
10
11
12
13
14
15
16
17
18
19
20
21
22
23

LOONS GAVIIDAE

1 RED-THROATED LOON *Gavia stellata* 53–69cm The smallest loon, with small, slightly uptilted bill giving distinctive profile. **V** Male utters a rolling, growling *oorroo-uh oorroo-uh*; female makes a slightly longer, higher-pitched *aarroo aarroo aarroo*. Also a barking and mewing. In flight, gives a goose-like *kah kah kah kah kah*. **H** By lakes and pools or marine inlets. Winters on shallow coastal waters and sometimes inland lakes. **D** N North America, N Eurasia.

2 BLACK-THROATED LOON *Gavia arctica* 58–73cm Often winters in small groups. **V** Loud, mournful *clowee-cok-clowee-cok-clowee*, a snoring *knarr-knorr-knarr-knorr* and a gull-like *aaah-owww*. **H** By large, deep lakes. Winters in coastal waters. **D** N Eurasia, W Alaska.

3 PACIFIC LOON *Gavia pacifica* 56–66cm Very similar to Black-throated Loon, but flanks uniformly dark. In non-breeding plumage often shows dark 'chin-strap'. **V** Loud, mournful *ooalee-koo ooalee-koo ooalee-koo*; also yodelling *o-lo-lee*, growls and croaks. **H** By freshwater tundra lakes. Winters in coastal waters. **D** NE Siberia, Alaska, N Canada.

4 COMMON LOON *Gavia immer* 69–91cm Large, heavy-billed loon, usually shows distinct forehead bulge. Bill black in breeding season, otherwise greyish. **V** Wailing *a-a-whoo-kwee-wheeooo-kwee-wheeooo*, manic *ho-yeyeyeyeyeye*, drawn-out howl and low moan. **H** By large lakes. Winters off coasts and sometimes on inland lakes. **D** N North America, W Europe.

5 YELLOW-BILLED LOON *Gavia adamsii* 76–91cm Similar to Common Loon but with distinctly larger, pale yellowish bill in all plumages. **V** Very similar to that of Great Northern Diver but louder and harsher. **H** Breeds by tundra lakes and rivers. Winters mainly at sea. **D** N North America, N Europe.

PENGUINS SPHENISCIDAE

6 KING PENGUIN *Aptenodytes patagonicus* 95–100cm Long bill black with orange-red patch along base of lower mandible. Feet black. Head black with golden ear-patches, the gold extending forward and paling on upper breast. Back blue-grey. Underparts white. Narrow black line along sides of breast. **V** Tinny trumpeting. **H** Colonies on beaches with no snow or ice and sea access. **D** Southern Ocean.

7 EMPEROR PENGUIN *Aptenodytes forsteri* 120cm Long bill black, feathered at base, with a red patch along base of lower mandible. Feet black. Head black with yellow ear-patches drooping onto neck. Chest and belly white with black line along sides of breast. **V** Gives single harsh downslurred calls, *hrrrraahh*. **H** Colonies on stable pack ice, shelf ice and land, usually sheltered. **D** Southern Ocean.

8 GENTOO PENGUIN *Pygoscelis papua* 80cm Bill red with black tip. Feet orange. Head black with band of white over crown from eyebrow to eyebrow. Back dark blue-grey, but brown when feathers old. Underparts white; flipper slate grey edged with white. **V** *Brrrrray* in a series. Also a *hhhaaaaaa* contact call at sea. **H** Colonies on shorelines or inland, preferably grassy areas **D** Southern Ocean.

9 ADELIE PENGUIN *Pygoscelis adeliae* 65cm Slightly raised transverse crest line between crown and nape. Bill black and short, feathered part way along. Feet pale pink. Head and throat black with white eye-ring. Back blue-black. Underparts white. Flipper black with white trailing edge. **V** On breeding grounds call starts with a series of staccato *kkk*, immediately followed by *kekekekek* and ending on a prolonged, harsh *air*. **H** Colonies on Antarctic coastlines. **D** Southern Ocean, Antarctica.

10 CHINSTRAP PENGUIN *Pygoscelis antarcticus* 60cm Bill black; feet pink. Forehead, crown and nape black; throat, cheeks, sides of neck and underparts white. Fine black 'chin-strap' from ear to ear. Back lead grey, flippers with white trailing edge. **V** At breeding colony, call is a hoarse, rapid *wayheyheyheyhey*, repeated three to four times. **H** Colonies on ice-free rocky coasts. **D** Southern Ocean, Antarctica.

11 FIORDLAND PENGUIN *Eudyptes pachyrhynchus* 55cm Note whitish cheek stripes (in most individuals), lack of bare skin at bill base and little black at tips of underflippers. **V** Very high, excited quacking. **H** Colonies in rainforest or in caves along rocky shores; forages in nearby seas. Probably pelagic outside breeding season. **D** S Australia, New Zealand.

12 SNARES PENGUIN *Eudyptes robustus* 60cm Note pink bare skin at bill base, robust bill and narrow yellow eyebrow. Black crown feathers not or hardly erectile. **V** Very high, hoarse quacking. **H** Colonies in muddy areas or on rocky flats with shading forest and other vegetation. Pelagic outside breeding season. **D** Snares Islands/Tini Heke, New Zealand.

13 ERECT-CRESTED PENGUIN *Eudyptes sclateri* 65cm Separated from Snares Penguin by broader yellow eyebrow, kept erect over eye. **V** Very high, nasal, shivering quacking. **H** Breeds bare rocky areas. Pelagic outside breeding season. **D** S New Zealand.

14 SOUTHERN ROCKHOPPER PENGUIN *Eudyptes chrysocome* 50cm Rather diminutive posture. Small bill. Yellow eyebrow spreading fanwise behind eye. **V** Low, Mallard-like quacking, sometimes in a dancing series. **H** Breeds in rocky places, gentle tussock slopes or under vegetation. Pelagic outside breeding season. **D** Cape Horn to Falkland Islands, Kerguelen Islands, subantarctic islands of New Zealand.

15 NORTHERN ROCKHOPPER PENGUIN *Eudyptes moseleyi* 50–60cm Similar to Southern Rockhopper Penguin but larger, with denser yellow crest. **V** Barking and braying calls when breeding. **H** Nests amongst boulders and tussock grass. Pelagic outside breeding season. **D** Tristan da Cunha, South Atlantic; Île Saint-Paul and Île Amsterdam, Indian Ocean.

16 ROYAL PENGUIN *Eudyptes schlegeli* 70cm Separated from other *Eudyptes* penguins by white cheeks. **V** Mid-high, drawn-out braying. **H** Colonies mainly on stony ground without vegetation. Pelagic after breeding. **D** Macquarie Island, subantarctic SW Pacific Ocean.

17 MACARONI PENGUIN *Eudyptes chrysolophus* 65cm Heavy bill sealing-wax red. Gape salmon pink. Head and upperparts black. Golden-yellow brow from mid-forehead ends in floppy tufts. Underparts white. **V** Like that of Royal Penguin. **H** Rocky shores to clifftops and sea. **D** Southern Ocean.

18 YELLOW-EYED PENGUIN *Megadyptes antipodes* 65cm Yellow eyes and streak from eye to hindcrown make it unistakable. **V** High-pitched *wrriturrriturrit*. **H** Breeds in dense vegetation on shores and slopes. Pelagic after breeding. **D** New Zealand.

19 LITTLE PENGUIN *Eudyptula minor* 42cm Note slaty-blue upperparts and lack of distinctive markings to head. Some variation in width of white flipper margins. **V** Varied, including soft, drawn-out, mumbled croaking or high, rather hoarse screams. **H** Small colonies on rocky flats and slopes; also in sand dunes. **D** S Australia, New Zealand.

20 AFRICAN PENGUIN *Spheniscus demersus* 65cm Bill black with a grey band. Feet black. Forehead, face, throat and upperparts black. Broad white brow turns down behind eye to lower throat. Underparts white with very few scattered black feathers. One black band across upper breast and down flanks. **V** Like braying of a donkey. **H** Breeds on offshore islands; sometimes seen near or on coast. **D** C Namibia to SE South Africa.

21 MAGELLANIC PENGUIN *Spheniscus magellanicus* 70cm Unmistakable thanks to head pattern and double breast-band. **V** During courtship and breeding, call is similar to a donkey's bray; almost explosive *ha-ha-ha-ha*, growing faster until notes join into a long, drawn-out *hhhaaaaaaa, hhaaaaaa*, ending with two to three short *ha-ha*. **H** Beaches, coasts, nearby areas for nesting, sea. **D** Chile and C Argentina to Cape Horn, Falkland Islands.

22 HUMBOLDT PENGUIN *Spheniscus humboldti* 70cm Bill and feet blackish. Upperparts slate grey. Throat and face black. Narrow white brow runs from base of bill and over eyes to lower throat, joining that from other side. Underparts white. Black band across breast and down sides of body to legs. **V** Slightly downslurred low mooing note; chicks produce shrill squeaks. **H** Coasts and offshore islands adjacent to the Humboldt Current. **D** N Peru to C Chile.

23 GALAPAGOS PENGUIN *Spheniscus mendiculus* 50cm Small. Head and upperparts black. Narrow white line runs from eye, down around cheeks and lower throat. Two pectoral collars, the upper less defined, the lower extending down flanks. Underparts white. **V** Downslurred bleating or honking note. **H** Beaches and rocky coasts, nearby sea. **D** Galápagos Islands.

AUSTRAL STORM PETRELS OCEANITIDAE

24 WILSON'S STORM PETREL *Oceanites oceanicus* 15–19cm Sooty black. Uppertail coverts and sides of rump white, extending onto vent. Often dangles feet and dances on sea surface when feeding. In flight, legs are longer than square tail. **V** Rapid 'chattering' occasionally uttered while feeding. **H** Maritime. Attracted to fishing vessels. **D** Temperate oceans, Southern Ocean.

25 ELLIOT'S STORM PETREL *Oceanites gracilis* 14cm Like Wilson's Storm Petrel but smaller. Bill and legs black, webs yellow. Mostly black, with white rump, lower flanks and lower belly. Large pale patch under wing. **V** Adult at nest gives harsh chatter, begging chick a high-pitched twitter. **H** Oceanic islands and sea. **D** Pacific Ocean.

26 PINCOYA STORM PETREL *Oceanites pincoyae* 15–16cm Similar to Wilson's and Elliot's Storm Petrels but with bolder white band on upperwing. Gathers in groups at good feeding spots. **V** Various chattering calls. **H** Nests in crevices; otherwise pelagic, mainly in Chilean fjords. **D** Reloncaví Sound, Chile.

27 GREY-BACKED STORM PETREL *Garrodia nereis* 39cm Unmistakable thanks to grey-and-dark-brown pattern of upperparts. Underparts and underwing white. Tail square. **V** Series of level, well-separated, hoarse, toneless shrieks. **H** Offshore. Nests in dense ground vegetation. **D** Southern Ocean.

28 WHITE-FACED STORM PETREL *Pelagodroma marina* 20–21cm Yellow webs between toes noticeable only at close range. Feeds in a series of swinging bounces, dangling feet at each bounce, often looking as though walking on water. **V** At breeding sites utters a slow *koo-koo-koo-koo*. **H** Breeds in burrows on islands, otherwise maritime. **D** Tropical and temperate oceans.

29 WHITE-BELLIED STORM PETREL *Fregetta grallaria* 20cm Feeds by hugging waves, legs often dangling and body swinging from side to side, bouncing from trough to trough. **V** Generally silent at sea. **H** Maritime. **D** Southern Ocean.

30 BLACK-BELLIED STORM PETREL *Fregetta tropica* 20cm Feeding actions as in White-bellied Storm Petrel. **V** Generally silent at sea. **H** Maritime. **D** Southern Ocean.

31 NEW ZEALAND STORM PETREL *Fregetta maoriana* 44cm Rediscovered in 2003. Separated from other storm petrels by streaked underparts. **V** Not recorded. **H** Offshore. **D** Pacific Ocean.

32 POLYNESIAN STORM PETREL *Nesofregetta fuliginosa* 54cm Polymorphic, pale morph most common throughout, but in Samoa only dark morph seen; all morphs occur on Phoenix Islands, Kiribati. Largest storm petrel. **V** High, well-separated *pew-pew*, second note high and staccato. **H** Offshore and open ocean. Nests on ground between boulders or in burrows under scrub or in crevices. **D** Pacific Ocean

1
2
3
4
5
6
7
8
9
10
11
12
13
14
15
16
17
18
19
20
21
22
23
24
25
26
27
28
29
30
31
32
pale morph
dark morph

ALBATROSSES DIOMEDEIDAE

1 LAYSAN ALBATROSS *Phoebastria immutabilis* 79–81cm No other dark-backed albatross with white head and underparts occurs in N Pacific. **V** Generally silent. **H** Breeds on islands with little or no vegetation; otherwise maritime. Occasionally follows ships. **D** Pacific Ocean.

2 BLACK-FOOTED ALBATROSS *Phoebastria nigripes* 68–74cm Some adults show pale greyish head and underparts. **V** Usually silent away from breeding sites. **H** Breeds on island beaches or slopes with little vegetation; otherwise maritime. Often around trawlers and ships. **D** Pacific Ocean.

3 WAVED ALBATROSS *Phoebastria irrorata* 90cm Unmistakable. Long bill yellowish ochre. Head whitish, washed yellow on nape and hindneck. Body, upperwings and tail greyish brown, finely vermiculated on breast. Underwing whitish bordered greyish brown, axillaries blackish brown. **V** Usually silent at sea. **H** Oceanic islands and tropical sea. Does not follow ships. **D** Pacific Ocean. Breeds on Galápagos Islands and coastal Ecuador.

4 SHORT-TAILED ALBATROSS *Phoebastria albatrus* 84–94cm Large bill is pale pink at all ages. Adult is only white-backed albatross in N Pacific. **V** Generally silent. **H** Breeds on islands with steep volcanic slopes, otherwise maritime. **D** Pacific Ocean.

5 WANDERING ALBATROSS *Diomedea exulans* 107–135cm Separation from royal albatrosses possible only at close range when lack of black line along cutting edge of upper mandible visible. Many adults have mainly black upperwing, with only very old birds attaining plumage shown. To take off, must run to gain some speed. So large it appears to move in slow motion. **V** Usually silent. **H** Maritime. Regularly follows ships. **D** Southern Ocean.

6 ANTIPODEAN ALBATROSS *Diomedea antipodensis* 110cm Marginally smaller than Wandering Albatross. Plumage lightens with age; overall darker in Antipodes Island subspecies *antipodensis* than Auckland Island subspecies *gibsoni*. **V** Usually silent. **H** Breeds on islands, otherwise pelagic. **D** Pacific Ocean. Breeds mainly on Auckland and Antipodes islands.

7 AMSTERDAM ALBATROSS *Diomedea amsterdamensis* 110–120cm Breeding plumage largely brown not white. **V** Usually silent. **H** Breeds on islands, otherwise pelagic. **D** Indian Ocean. Breeds on Île Amsterdam.

8 TRISTAN ALBATROSS *Diomedea dabbenena* 110cm Breeding-age plumage largely brown not white. **V** Usually silent. **H** Breeds on islands, otherwise pelagic. **D** Atlantic Ocean. Breeds on Tristan da Cunha islands, mainly Gough Island.

9 SOUTHERN ROYAL ALBATROSS *Diomedea epomophora* 120cm Separated from Wandering Albatross by black cutting edge of bill, white wing patch and all-white tail. **V** Grunts and mumblings. Also bill-rattling. **H** Nests sheltered by rocks or low vegetation, but needs gently sloping, exposed sites for take-off and landing. Otherwise open ocean. **D** South Pacific Ocean and Southern Ocean. Breeds on Campbell and Auckland islands.

10 NORTHERN ROYAL ALBATROSS *Diomedea sanfordi* 120cm To distinguish from Wandering Albatross, see Southern Royal Albatross. Separable from latter by solid black wing and range. **V** As Southern Royal Albatross. **H** As Southern Royal Albatross. **D** Pacific Ocean. Breeds on Chatham Islands.

11 SOOTY ALBATROSS *Phoebetria fusca* 85cm Overall sooty brown, slightly darker on head. Paler on back in worn plumage, especially on neck. Note line along lower mandible is pale yellow. **V** *Pee-oo* display call. **H** Vagrant over open sea. **D** Indian Ocean, South Atlantic Ocean.

12 LIGHT-MANTLED ALBATROSS *Phoebetria palpebrata* 78–79cm Very like Sooty Albatross, but with contrasting pale grey back. **V** Usually silent. **H** Maritime. Occasionally around fishing boats or following ships. **D** Southern Ocean.

13 BLACK-BROWED ALBATROSS *Thalassarche melanophris* 80–95cm Yellow bill and lack of grey to head diagnostic in adult. **V** Territorial braying and cackling. **H** Breeds on cliff ridges and ledges; otherwise offshore and pelagic. **D** Southern Ocean.

14 CAMPBELL ALBATROSS *Thalassarche impavida* 80–95cm Similar to Black-browed Albatross but note adult's pale yellow (not dark) eye and whiter underwing. **V** Territorial braying and cackling **H** Breeds on sea cliffs; otherwise pelagic. **D** South Pacific Ocean. Breeds on Campbell Island.

15 SHY ALBATROSS *Thalassarche cauta* 90–100cm White head and neck. Black mark in axillaries, where black leading edge of wings meets body, is diagnostic. **V** High, drawn-out, slightly shivering, lowing or toneless nasal twittering and braying. **H** Breeds on level places in rocky, broken areas; otherwise offshore. **D** Southern Ocean. Breeds in Tasmania and Auckland Islands.

16 CHATHAM ALBATROSS *Thalassarche eremita* 90cm Dark grey head and neck. **V** Bill-clacking, croaking and braying during breeding. **H** Nests on rocky ledges and steep slopes; otherwise pelagic. **D** Pacific Ocean. Breeds on The Pyramid, Chatham Islands.

17 SALVIN'S ALBATROSS *Thalassarche salvini* 90cm Head, throat and nape pale grey. **V** Braying and croaking during breeding, otherwise silent. **H** Breeds on rocky islands; otherwise pelagic. **D** Indian Ocean, Pacific Ocean. Breeds on Bounty Islands and Snares Islands/Tini Heke.

18 GREY-HEADED ALBATROSS *Thalassarche chrysostoma* 81cm Separated from yellow-nosed albatrosses by greyer head, more black to leading edge of underwing and yellow lower ridge to bill. Similar to Shy Albatross, but note that yellow starts narrowly at upper ridge of bill. **V** Mid-high nasal bleating. **H** Offshore and pelagic. **D** Southern Ocean.

19 ATLANTIC YELLOW-NOSED ALBATROSS *Thalassarche chlororhynchos* 81cm Separated from Indian Yellow-nosed Albatross by grey head, and from Shy Albatross by lack of black mark in axillaries. Lack of yellow line along lower mandible in adult is diagnostic. Note narrow black leading edge to underwings. **V** High, scratchy bleating or raspy *wuc-wuc-wuc*. **H** Inshore and offshore. **D** Atlantic Ocean. Breeds on Tristan da Cunha.

20 INDIAN YELLOW-NOSED ALBATROSS *Thalassarche carteri* 76cm As Atlantic Yellow-nosed Albatross, but head white. **V** As Atlantic Yellow-nosed. **H** As Atlantic Yellow-nosed **D** Indian Ocean.

21 BULLER'S ALBATROSS *Thalassarche bulleri* 209cm Note broad yellow base to yellow line along culmen. Also separable from Grey-headed Albatross by clear-cut black leading edge to underwings. **V** Braying notes and drawn-out lowing. **H** Breeds on slopes, ridges and tops of rocky slopes with sparse, low vegetation; otherwise inshore, offshore and open ocean. **D** Pacific Ocean. Breeds in New Zealand.

1
2
3
4
5
6
7
8
9
10
11
12
13
14
15
16
17
18
19
20
21

NORTHERN STORM PETRELS HYDROBATIDAE

1 EUROPEAN STORM PETREL *Hydrobates pelagicus* 14–17cm Feeds excitedly, repeatedly hovering or fluttering before dipping to seize food. **V** At breeding site utters a harsh, purring *arrrr-r-r-r-r-r-r-r-r,* ending with a grunt or hiccup-like sound. Generally silent at sea. **H** Breeds in crevices or burrows on rocky coasts and islands. Maritime after breeding. Follows ships and fishing boats. **D** NE Atlantic Ocean.

2 LEAST STORM PETREL *Oceanodroma microsoma* 13–15cm Very small. Sooty black. Pale brownish diagonal band on upperwing. Note short wedge-shaped tail. Flight swift and direct with deep wingbeats. **V** Silent. **H** Maritime. **D** NE Pacific Ocean.

3 WEDGE-RUMPED STORM PETREL *Oceanodroma tethys* 18–20cm Note large white rump. Flies with deep wingbeats, fast, direct and often high above waves. **V** Silent. **H** Maritime. **D** E Pacific Ocean.

4 BAND-RUMPED STORM PETREL *Oceanodroma castro* 19–21cm Methodically searches for food on even flight with glides, usually hovering without foot-pattering to take prey. **V** At breeding sites makes a low, prolonged purring *urr-rrr-rrr-rrr-rrr-rrr-rrr,* ending with a sharp *wika*; also a high-pitched *klair chuch-a chuk chuk chuk.* Usually silent at sea. **H** Breeds in rock crevices or burrows on islands. Forages at sea. **D** Pacific Ocean, Atlantic Ocean.

5 MONTEIRO'S STORM PETREL *Oceanodroma monteiroi* 19–21cm Very similar to Band-rumped Storm Petrel but tail longer and more deeply forked. **V** Chattering and purring, ending in yelps. **H** As for Band-rumped. **D** Atlantic Ocean. Breeds in the Azores.

6 CAPE VERDE STORM PETREL *Oceanodroma jabejabe* 19–21cm Very similar to Band-rumped Storm Petrel but has a proportionally long bill. **V** Elaborate chattering and purring. **H** As for Band-rumped. **D** Atlantic Ocean. Breeds in Cape Verde Islands.

7 SWINHOE'S STORM PETREL *Oceanodroma monorhis* 19–20cm Flight and feeding actions much as Leach's Storm Petrel, although some reports say it does not use foot-pattering. **V** Trilling chatter at breeding sites. Usually silent at sea. **H** Breeds on offshore islands, otherwise maritime. **D** Indian Ocean, Pacific Ocean.

8 LEACH'S STORM PETREL *Oceanodroma leucorhoa* 19–22cm Feeds erratically with bounding flight, then hovers with shallow wingbeats, often with foot-pattering, to seize food from sea surface. **V** At breeding sites utters a slow purring, ending in a high *whee-chaa*; also a screaming *par-kiki-kar-koo whuk-kuk-kuk-kuh-kuh.* Generally silent at sea. **H** Breeds in a wide variety of crevices or burrows on rocky coasts and offshore islands, otherwise maritime. **D** Pacific Ocean, Atlantic Ocean.

9 TOWNSEND'S STORM PETREL *Oceanodroma socorroensis* 19–22cm Distinguished from Leach's Storm Petrel by different vocalisations and morphology, and from sympatric Ainley's Storm Petrel also by seasonal isolation. **V** Call at nest is like that of Leach's Storm Petrel but faster and less wheezy. **H** Nests in crevices and burrows. Forages at sea. **D** Pacific Ocean. Breeds in summer on Guadalupe Island, Mexico.

10 AINLEY'S STORM PETREL *Oceanodroma cheimomnestes* 19–22cm Distinguished from Leach's Storm Petrel by different vocalisations and morphology, and from sympatric Townsend's Storm Petrel also by seasonal isolation. **V** Like that of Townsend's Storm Petrel. **H** As for Townsend's Storm Petrel. **D** Pacific Ocean. Breeds in winter on Guadalupe Island, Mexico.

11 MARKHAM'S STORM PETREL *Oceanodroma markhami* 23cm Sooty black. Diagonal band on upperwing whitish grey. Long, deeply forked tail. **V** Breathy, accelerating harsh chatter given at nest, otherwise silent. **H** Oceanic islands and maritime. **D** Pacific Ocean.

12 TRISTRAM'S STORM PETREL *Oceanodroma tristrami* 24–25cm Feeds in flight using foot-pattering as prey is snatched from sea surface. **V** Usually silent away from breeding colonies. **H** Breeds in recesses in scree or burrows on islands. Otherwise maritime. **D** Pacific Ocean.

13 BLACK STORM PETREL *Oceanodroma melania* 23cm Large, long-tailed. Flight like that of Black Tern (Plate 52). Often occurs in large numbers. **V** At breeding grounds utters a purring. Silent at sea. **H** Nests in crevices or burrows. Otherwise maritime. **D** Pacific Ocean. Breeds on islands off California and the Baja Peninsula.

14 MATSUDAIRA'S STORM PETREL *Oceanodroma matsudairae* 24–25cm Feeds on the wing, holding wings in a shallow 'V' while dipping down to pick food from sea surface. **V** Generally silent away from breeding site. **H** Nests in burrows on high ground. Otherwise pelagic. Follows ships. **D** Breeds on Volcano Islands, Japan. Winters in Indian Ocean.

15 ASHY STORM PETREL *Oceanodroma homochroa* 18–21cm Fluttering flight with shallow wingbeats. Feeds by dipping and snatching food from sea surface. Often occurs in large numbers. **V** At breeding areas utters a variable purring with an inhaled gasp. Generally silent at sea. **H** Breeds in rock crevices and burrows in colonies on offshore islands. Otherwise maritime. **D** Off California.

16 HORNBY'S STORM PETREL *Oceanodroma hornbyi* 21–23cm Dark cap. Feeds by pattering, dipping to snatch food from sea surface. Often gregarious. **V** Silent. **H** Maritime. **D** E Pacific Ocean, along coasts of Ecuador, Peru and Chile.

17 FORK-TAILED STORM PETREL *Oceanodroma furcata* 20–23cm Feeds by snatching food from sea surface when flying or sitting on water. **V** Generally silent away from breeding sites. **H** Breeds on offshore islands, among trees or in grassy areas on rocky hillsides. Otherwise maritime. Follows ships. **D** North Pacific Ocean.

PETRELS, SHEARWATERS AND DIVING PETRELS PROCELLARIIDAE

18 SOUTHERN GIANT PETREL *Macronectes giganteus* 86–99cm Separated from Northern Giant Petrel mainly by pea-green bill tip. Eyes normally brown but can be pale grey. **V** Low, nasal grumbles and croaks. **H** Oceanic islands, seas and coastal areas. Follows ships. **D** Southern Ocean, Antarctica.

19 NORTHERN GIANT PETREL *Macronectes halli* 90cm Very similar to Southern Giant Petrel but bill tip is red-brown. Eyes normally pale (sometimes brown), becoming even paler with age. **V** Low, nasal grumbles and croaks. **H** Oceanic islands and seas. Does not follow ships. **D** Southern Ocean.

20 NORTHERN FULMAR *Fulmarus glacialis* 45–50cm Intermediate forms occur. Much stiffer-winged in flight than gulls. **V** Guttural cackling, varying in speed. **H** Breeds mainly on sea cliffs. Winters at sea. **D** North Atlantic Ocean, North Pacific Ocean.

21 SOUTHERN FULMAR *Fulmarus glacialoides* 45–50cm Note diagnostic wing pattern, with white flash on outer wings more contrastingly patterned above than in Slender-billed Prion. Separated from gulls by gliding flight on stiff wings. **V** High, hoarse, subdued shrieks. **H** Offshore and open ocean. **D** Southern Ocean, Antarctica.

22 ANTARCTIC PETREL *Thalassoica antarctica* 43cm Note characteristic pattern of upperparts, with contrasting white secondaries and tail. **V** Low, hoarse croaks. **H** Subantarctic islands and seas. **D** Southern Ocean, Antarctica.

23 CAPE PETREL *Daption capense* 38–40cm Flight fulmar-like. Usually in large flocks. **V** Generally silent. **H** Maritime away from southern breeding sites. Follows ships. **D** Southern Ocean, Antarctica.

24 SNOW PETREL *Pagodroma nivea* 36–41cm No similar bird in the area. **V** Toneless croaks and bickering. **H** Offshore. **D** Southern Ocean, Antarctica.

25 BLUE PETREL *Halobaena caerulea* 28cm Separated from *Pterodroma* petrels by lack of dark bar on underwings, and from prions by white tail tip. **V** Mid-high cooing notes. **H** Breeds in burrows in coastal slopes with tussocks. Otherwise pelagic. **D** Southern Ocean.

26 BROAD-BILLED PRION *Pachyptila vittata* 25–30cm Seemingly large-headed. Bill black and larger than in other five prions. Note steep forehead. **V** Short, very low, rhythmic croaking. **H** Breeds in burrows or caves and crevices in cliffs. Otherwise offshore and open ocean. **D** Southern Ocean. Breeds in New Zealand and Tristan da Cunha.

27 SALVIN'S PRION *Pachyptila salvini* 29cm Separation from Broad-billed and Antarctic prions probably impossible at sea. In the hand, shows comb-like lamellae at base of closed bill. Bill grey. **V** Hoarse, rather excited croaking at different pitch. **H** Inshore and offshore. **D** Indian Ocean.

28 ANTARCTIC PRION *Pachyptila desolata* 28cm Bill smaller and bluer than in Broad-billed Prion. Collar more distinct than in Salvin's Prion, but not safely separable. **V** Low, hoarse and falsetto croaking (not unlike radio static). **H** Breeds in burrows and rock crevices. Feeds offshore and in open ocean. **D** Southern Ocean, Antarctica.

29 SLENDER-BILLED PRION *Pachyptila belcheri* 26cm 'M' on wings less distinct than in previous three species, especially in worn plumage. Note small, narrow bill and extensive, distinct white eyebrow, giving pale-faced expression. **V** Harsh clucking or croaking calls at nest. **H** Inshore, offshore and open ocean. **D** Southern Ocean.

30 FAIRY PRION *Pachyptila turtur* 23–28cm Note indistinct facial pattern and narrow 'M' on wings. At sea, not safely separable from Fulmar Prion. **V** High, excited croaking. **H** Breeds in burrows, rock crevices and caves. Feeds offshore and in open ocean. **D** Southern Ocean.

31 FULMAR PRION *Pachyptila crassirostris* 24–28cm Difficult to separate at sea from Fairy Prion; that species has similar pale face pattern but smaller bill. Best distinction is its 'looping' flight. **V** Low, hoarse croaking, at times bouncing. **H** Breeds in crevices and in tunnels in cliffs. At sea, inshore and offshore. **D** Southern Ocean. Breeds in New Zealand and on islands in S Indian Ocean.

32 KERGUELEN PETREL *Aphrodroma brevirostris* 30cm Note fairly thickset jizz. Distinctive underwing pattern with silvery underside to primaries and pale leading edge between body and wrist. Arcs higher up in flight than other petrels. **V** High, hoarse, trumpet-like shrieks or very high piercing screams. **H** Offshore and open ocean. Breeds in burrows, excavated in wet soil of marshes or lava ridges. **D** Southern Ocean.

33 GREAT-WINGED PETREL *Pterodroma macroptera* 42–45cm Separated from Providence Petrel by more uniform plumage and little or no pale flash on outer underwings. **V** Gives whistles and moaning calls in flight and at nest. **H** Offshore and open ocean. Nests in burrows or crevices under clumps of vegetation. **D** Southern Ocean.

34 WHITE-HEADED PETREL *Pterodroma lessonii* 40cm Note white head and pale grey rump and tail, contrasting with darker wings and mantle. **V** Low croaks and series sounding as if produced by a dry bicycle pump. **H** Breeds in burrows under tussocks, ferns, low shrub, often in open grassland. Otherwise offshore and open ocean. **D** Southern Ocean.

35 GREY-FACED PETREL *Pterodroma gouldi* 42–45cm Formerly classed as a subspecies of Great-winged Petrel. Very similar but with more white on face. **V** As for Great-winged. **H** As for Great-winged. **D** SW Pacific Ocean. Breeds on North Island, New Zealand.

36 ATLANTIC PETREL *Pterodroma incerta* 43cm In worn plumage chin and throat paler, face greyer and pale collar on nape. Flight typical of genus. **V** Silent at sea. **H** Maritime. Often follows ships. **D** Atlantic Ocean

37 PROVIDENCE PETREL *Pterodroma solandri* 40cm Head contrastingly darker. Note white markings at bill base and pale base of primaries on underwings. Very similar to Great-winged Petrel, and Murphy's, Kermadec and Herald Petrels (Plate 21). **V** High, upslurred *eeuuh-eeuuh.* **H** Nests in burrows or rock crevices, often within forest. Breeds also on sparsely vegetated cliffs. Otherwise offshore and open seas. **D** Disperses N into Pacific Ocean after breeding on Lord Howe Island off E Australia.

38 MAGENTA PETREL *Pterodroma magentae* 38cm Separated from Tahiti Petrel (Plate 21) by different underwing pattern, and from Phoenix Petrel (Plate 21) mainly by range. **H** Breeds in dense forest. Otherwise offshore. **D** Pacific Ocean.

1
2
3
4
5
6
7
8
9
10
11
12
13
14
15
16
17
18
normal morph
19
20
dark morph
light morph
21
22
23
24
white morph
25
26
27
28
29
30
31
32
33
34
35
36
37
38

PETRELS, SHEARWATERS AND DIVING PETRELS *CONTINUED*

1 MURPHY'S PETREL *Pterodroma ultima* 38–41cm Similar to dark form of Herald Petrel, but with whitish chin. Typical 'gadfly' flight. **V** Usually silent. **H** Maritime. Not known to follow ships. **D** Pacific Ocean.

2 SOFT-PLUMAGED PETREL *Pterodroma mollis* 33–36cm In the field very similar to Fea's and Zino's petrels, and probably not separable unless breast-band complete. **V** Generally silent. **H** Maritime. **D** Southern, Indian and Pacific oceans.

3 ZINO'S PETREL *Pterodroma madeira* 32–33cm Split from Soft-plumaged Petrel. Probably not possible to separate from Fea's Petrel in the field. **V** Much like Fea's Petrel. Silent at sea. **H** Breeds on cliffs. Otherwise maritime. **D** Atlantic Ocean.

4 FEA'S PETREL *Pterodroma feae* 36–37cm Split from Soft-plumaged Petrel. Flight typical of genus. Occasionally follows ships. **V** At breeding sites utters a cackling *gon-gon*; also a mournful wail ending with a hiccup. Silent at sea. **H** Breeds on cliffs. Otherwise maritime. **D** Atlantic Ocean. Breeds on Cape Verde Islands.

5 DESERTAS PETREL *Pterodroma deserta* 35–37cm Split from Fea's Petrel, which it closely resembles. **V** At breeding sites gives moaning and whimpering calls. **H** Breeds on cliffs, otherwise maritime. **D** Atlantic Ocean. Breeds on Desertas Islands.

6 BERMUDA PETREL *Pterodroma cahow* 38cm Typical 'gadfly' actions. **V** Generally silent at sea. **H** Maritime. **D** Atlantic Ocean. Breeds in Bermuda.

7 BLACK-CAPPED PETREL *Pterodroma hasitata* 35–46cm Some individuals show a much-reduced white collar. Typical 'gadfly' flight. Comes to breeding areas only at night. **V** At breeding grounds repeatedly utters a drawn-out *aaa-aw eek* or *ooow eek*; also a hurt, puppy-like yelp. **H** Mountain cliffs during breeding. Otherwise maritime. **D** Atlantic Ocean. Breeds from Cuba and Hispaniola to Martinique.

8 JUAN FERNANDEZ PETREL *Pterodroma externa* 43cm Note large size and narrow black bar to underwings. Often shows horseshoe-like white base to uppertail. **V** Short series of mewing and trumpet-like notes. Also drawn-out hooting. **H** Oceanic islands and sea. **D** Pacific Ocean. Breeds on Juan Fernández Islands, Chile.

9 VANUATU PETREL *Pterodroma occulta* 40cm Similar to White-necked Petrel but marginally smaller. First described in 2001. **V** Calls include a drawn-out squeaky croak, heard at burrow. Mainly silent at sea. **H** Nests in burrows, otherwise pelagic. **D** Pacific Ocean. Breeds on Vanuatu (island of Vanua Lava).

10 KERMADEC PETREL *Pterodroma neglecta* 38cm Polymorphic. All morphs generally difficult to separate from similar petrels, but white primary shafts on upperwings (hard to see) are diagnostic. **V** Long, drawn-out, mourning shrieks and bleating, changing in pitch. **H** Nests on cliffs or slopes with little vegetation. **D** Pacific Ocean.

11 HERALD PETREL *Pterodroma heraldica* 35–39cm Polymorphic. Intermediates between dark and pale morphs are most common. Note underwing pattern. **V** Drawn-out whinnying and bickering. **H** Offshore and open seas. Surface breeder at sheltered sites. **D** Pacific Ocean.

12 TRINDADE PETREL *Pterodroma arminjoniana* 35–39cm 'Gadfly' flight typical of the genus – beating wings to gain height, followed by long glides and wide, banking arcs. Much individual variation. **V** Generally silent. **H** Maritime. Occasionally follows ships. **D** Atlantic Ocean, Indian Ocean.

13 HENDERSON PETREL *Pterodroma atrata* 90cm Separated from dark Kermadec Petrel by lack of white primary shafts. **V** Undulating, shivering twitter. **H** Surface nester in dense forest. **D** Pacific Ocean.

14 PHOENIX PETREL *Pterodroma alba* 83cm Very similar to Magenta Petrel (Plate 20) and Herald and Tahiti petrels, but underwings solid brown except narrow white line along leading edge. **V** High, sharp bickering, downslurred at end. **H** Offshore and open ocean. Breeds on ground at sheltered sites under trees or bushes. **D** Pacific Ocean.

15 BARAU'S PETREL *Pterodroma baraui* 38cm Typical 'gadfly' flight, beating wings to gain height, followed by long glides and banking arcs. **V** Silent. **H** Maritime. **D** Indian Ocean.

16 HAWAIIAN PETREL *Pterodroma sandwichensis* 91cm Note characteristic dark cap and lack of 'M' pattern on upperwings. **V** High, hollow, three-syllabled hoots and mid-high, slightly squeaking or croaking calls. **H** Offshore and open seas. Breeds in burrows or crevices under vegetation high on mountain slopes. **D** Pacific Ocean. Breeds in Hawaii.

17 GALAPAGOS PETREL *Pterodroma phaeopygia* 43cm Note dark sides of neck, forming a semi-collar. Occasionally soars to a great height, hanging in wind. **V** Generally silent. **H** Oceanic islands and sea. **D** Pacific Ocean. Breeds in Galápagos Islands.

18 MOTTLED PETREL *Pterodroma inexpectata* 33–35cm Diagnostic broad bar across underwings and brownish belly. **V** High, thin shrieks in a short, slightly lowered series. **H** Nests in burrows in rocky ground and tussock grassland. **D** Pacific Ocean, Indian Ocean.

19 WHITE-NECKED PETREL *Pterodroma cervicalis* 43cm White neck diagnostic, but beware Juan Fernandez Petrel individuals with a pale grey neck. **V** Various wailing or shrieking notes. **H** Offshore and open seas. May nest in burrows (Kermadecs). **D** Pacific Ocean. Breeds on Kermadec, Norfolk and Phillip islands.

20 BLACK-WINGED PETREL *Pterodroma nigripennis* 28–30cm Separated from Gould's Petrel by paler cap and darker, broader underwing markings. **V** High, hollow, mourning *twoot* notes in series of 3–20; also high, nasal and falsetto shrieks. **H** Digs nest burrows under scrub and tussocks or breeds in rock crevices. Otherwise offshore and open seas. **D** E Pacific Ocean.

21 CHATHAM PETREL *Pterodroma axillaris* 67cm Diagnostic underwing pattern, in which carpal joint is connected by bar to rear edge where touching body. **V** Very high, rapid, staccato *tseetseetsee* in flight. **H** Breeds in burrows in soft clay under vegetation clumps and tussocks. Otherwise offshore and open ocean. **D** Pacific Ocean. Breeds in Chatham Islands.

22 BONIN PETREL *Pterodroma hypoleuca* 30cm Underwing pattern diagnostic. Typical bounding 'gadfly' flight. **V** Generally silent. **H** Breeds among vegetation in sandy areas or high slopes on oceanic islands. Otherwise maritime. Does not follow ships. **D** Pacific Ocean.

23 GOULD'S PETREL *Pterodroma leucoptera* 70cm Note small size, black cap, narrow underwing markings and dark 'M' across wings.**V** Very high, sharp, nasal shrieks *tfeet-tfeet*. **H** Offshore and open seas. **D** Pacific Ocean. Breeds on islands off New South Wales; New Caledonia; Raivavae, French Polynesia.

24 COLLARED PETREL *Pterodroma brevipes* 70cm Like Gould's Petrel but darker and has a complete collar, greyish belly and flanks. **V** Similar to that of Gould's Petrel. **H** Oceanic islands and sea. **D** Pacific Ocean.

25 COOK'S PETREL *Pterodroma cookii* 25–30cm Note uniform colouring, from pale grey crown to mantle. Flight rapid and erratic, bat-like with jerky wingbeats. **V** Mid-high nasal chatters. **H** Breeds in burrows near ridges or steep slopes in thick forest. Otheriwse maritime. **D** Pacific Ocean.

26 DE FILIPPI'S PETREL *Pterodroma defilippiana* 26cm Bill black, feet sky blue. Mask blackish. Crown, back, uppertail coverts and tail grey, outers white. Wings grey with black 'W'. Inner web of primaries white. Forehead, underparts and sides of rump white. Half-collar grey. **V** Generally silent at sea. **H** Oceanic islands and sea. **D** Pacific Ocean. Nests on islands off Chile.

27 STEJNEGER'S PETREL *Pterodroma longirostris* 26cm Weak underwing pattern. Crown darker than mantle. Flight rapid, weaving and banking, with jerky, bat-like wingbeats. **V** Generally silent. **H** Oceanic islands and sea. **D** Pacific Ocean. Nests on islands off Chile.

28 PYCROFT'S PETREL *Pterodroma pycrofti* 53cm Note small size and narrow black line from wrist to body. Generally darker than Cook's Petrel and with less extensive dark patch around eye than Stejneger's Petrel. **V** High *peepeep-puh* (last part much lower). **H** Offshore. Nests in burrows in forest. **D** Pacific Ocean. Breeds on islands off N New Zealand.

29 MASCARENE PETREL *Pseudobulweria aterrima* 90cm Note stout, dark bill and uniformly dark underparts. **V** Generally silent at sea. **H** Open sea. **D** Indian Ocean.

30 TAHITI PETREL *Pseudobulweria rostrata* 39cm Very similar to Phoenix Petrel. Stout bill, long body. Flight usually low over water with loose wingbeats. **V** Nasal, drawn-out shrieks at varying pitch. **H** Breeds in burrows on mountain slopes with thick forest cover. Forages at open sea. Solitary; rarely follows ships or fishing boats. **D** Pacific Ocean.

31 BECK'S PETREL *Pseudobulweria becki* 39cm Similar to Tahiti Petrel, from which it was split. **V** Generally silent at sea. **H** Breeds in burrows on mountain slopes with thick forest cover. Forages over open sea. **D** Pacific Ocean. Breeds on New Ireland, Papua New Guinea; Solomon Islands.

32 FIJI PETREL *Pseudobulweria macgillivrayi* 68cm Characterised by small size and uniform brown plumage. Separated from Jouanin's Petrel (Plate 22) by lack of paler wing-bar. **H** Breeding grounds unknown. Presumed to forage offshore. **D** Pacific Ocean. Recorded only on Gau Island, Fiji.

33 GREY PETREL *Procellaria cinerea* 48cm Bill dark ivory and black. Upperparts grey. Underparts white. Underwing blackish grey, paling towards tip; flight feathers grey. **V** Gives dry mechanical rattle and low honking calls at nest. **H** Subantarctic islands and seas. **D** Southern Ocean.

34 WHITE-CHINNED PETREL *Procellaria aequinoctialis* 56cm Bill ivory, with black lines along culmen and on mandible. Brownish black. Variable white spot on chin. **V** Screaming *tititititititi*, usually given when squabbling over food. **H** Oceanic islands, southern seas. **D** Southern Ocean.

35 SPECTACLED PETREL *Procellaria conspicillata* 56cm Formerly considered a race of White-chinned Petrel, but has white encircling face. **V** Similar to that of White-chinned Petrel but slightly lower pitched. **H** Oceanic islands, southern seas. **D** Southern Ocean.

36 BLACK PETREL *Procellaria parkinsoni* 46cm Like White-chinned Petrel but smaller. Bill ivory, tip blackish. All brownish black (no white chin). **V** Generally silent. **H** Oceanic islands and open ocean. Occasionally around fishing boats but does not follow ships. **D** Pacific Ocean. Breeds only on islands off N New Zealand.

37 WESTLAND PETREL *Procellaria westlandica* 52cm Like White-chinned Petrel but bill tipped black and no white chin: a large version of Black Petrel. **V** Harsh barking notes. **H** Seas. **D** Pacific Ocean. Breeds only on W coast of South Island, New Zealand.

1
2
3
4
5
6
7
pale
10
11
pale
8
9
inter-
mediate
10
pale
pale
dark
inter-
mediate
12
13
dark
dark
14
15
16
17
18
19
20
21
22
23
24
inter-
mediate
dark
pale
25
26
27
28
29
30
31
32
33
34
35
36
37

PETRELS, SHEARWATERS AND DIVING PETRELS *CONTINUED*

1 STREAKED SHEARWATER *Calonectris leucomelas* 48cm Often in large flocks. Sometimes shows pale crescent at base of tail. Flight action similar to Cory's Shearwater. **V** Generally silent. **H** Breeds on forested offshore islands. Winters at sea. Follows fishing boats. **D** Pacific Ocean

2 SCOPOLI'S SHEARWATER *Calonectris diomedea* 44–49cm Rather pale upperparts, tail darkest. Bill yellowish with dark tip. Like Streaked Shearwater, often has downwards-bending wings. **V** Series of low shrieks, each one inhaled and rising. **H** Inshore, offshore and open ocean. **D** Atlantic Ocean. Breeds near S Europe and W Africa.

3 CORY'S SHEARWATER *Calonectris borealis* 46–53cm Flight can appear lazy but is actually quite fast. **V** At breeding sites utters a harsh, wailing *keeoowwrrah*, otherwise silent. **H** Crevices in cliffs and rocky slopes on barren islands. Winters at sea. Often follows ships and trawlers. **D** Atlantic Ocean. Breeds in Macaronesia, Berlenga Islands.

4 CAPE VERDE SHEARWATER *Calonectris edwardsii* 42–47cm Smaller than Cory's Shearwater, from which it was split. **V** Gives cackling calls at breeding grounds, otherwise silent. **H** Crevices in rocky slopes and cliffs. Winters at sea. **D** Atlantic Ocean. Breeds on Cape Verde Islands.

5 WEDGE-TAILED SHEARWATER *Ardenna pacifica* 41–46cm Slow wing flaps followed by short glides. More bounding flight in strong winds. **V** Generally silent. **H** Breeds on flat ground on offshore islands. Otherwise maritime. Often around fishing boats. **D** Pacific Ocean, Indian Ocean.

6 BULLER'S SHEARWATER *Ardenna bulleri* 46cm Distinctive pattern of upperparts and black tail tip diagnostic.**V** Muttered, nasal cries. **H** Nests in burrows or crevices, preferring densely forested slopes. Otherwise offshore and open seas. Often scavenges around trawlers. **D** Pacific Ocean. Breeds on islands off New Zealand.

7 SOOTY SHEARWATER *Ardenna grisea* 40–46cm Note extensive white areas on underwings. Quick wing flaps followed by a long glide. **V** Low, pumping muttering. **H** Offshore. Nests in burrows under tussocks or dense scrub, sometimes in forest. **D** Worldwide.

8 SHORT-TAILED SHEARWATER *Ardenna tenuirostris* 41–43cm Like Fluttering Shearwater has white areas to underwings, but these mainly restricted to outer wings. Note protruding feet. Shorter bill than Sooty Shearwater. Similar in habits to Sooty Shearwater. Gregarious. **V** Hoarse shrieking and muttering. **H** Offshore. Breeds in burrows in open grassy or scrubby terrain. **D** Pacific Ocean.

9 PINK-FOOTED SHEARWATER *Ardenna creatopus* 48cm Colour dull overall. Feet and bill pink. Wings held straight out. Flight unhurried: high, rising wingbeats followed by long glides. Occurs alone or in small parties. **V** Strange, lamenting muttering. **H** Offhore and open seas. Nests in burrows in forested or open, hilly terrain. **D** Pacific Ocean. Breeds on islands off Chile, from where it disperses N.

10 FLESH-FOOTED SHEARWATER *Ardenna carneipes* 40–45cm Bill pinkish with dark tip, legs pink. Underside of flight feathers with silvery reflections. Lazy flaps followed by long glides on stiff wings. Gregarious. **V** High, slow, lamenting barks. **H** Offshore and open ocean. Nests in burrows in grassland or forest. **D** Pacific Ocean, Indian Ocean.

11 GREAT SHEARWATER *Ardenna gravis* 43–51cm Dark cap. Fast, strong wingbeats followed by long glides on stiff, straight wings. Gregarious. **V** Noisy, sounding like fighting cats when feeding around trawlers, otherwise generally silent. **H** Maritime. Often attracted to ships and fishing boats. **D** Atlantic Ocean.

12 CHRISTMAS SHEARWATER *Puffinus nativitatis* 35–38cm Buoyant flight with quick wingbeats followed by low, level glides. **V** Generally silent. **H** Maritime. **D** C Pacific Ocean.

13 MANX SHEARWATER *Puffinus puffinus* 30–38cm Rapid wingbeats followed by a low, swinging glide. Gregarious. **V** At breeding grounds emits raucous cackles and screams. Generally silent at sea. **H** Breeds in burrows on offshore islands and coastal cliffs. Otherwise maritime. Scavenges around fishing boats. **D** Atlantic Ocean.

14 YELKOUAN SHEARWATER *Puffinus yelkouan* 30–38cm Habits and flight actions much as those of Manx Shearwater, although said to be more 'fluttering'. **V** Said to be similar to that of Manx Shearwater. Generally silent away from breeding ground. **H** Breeds in burrows on offshore islands. Otherwise maritime. **D** Atlantic Ocean. Breeds in the Mediterranean.

15 BALEARIC SHEARWATER *Puffinus mauretanicus* 35–40cm Habits and flight actions similar to those of Manx Shearwater, although can appear more 'fluttering'. **V** Similar to that of Manx Shearwater. **H** Breeds in burrows and crevices. Otherwise maritime. **D** Atlantic Ocean.

16 BRYAN'S SHEARWATER *Puffinus bryani* 24–25cm Smallest known shearwater. New to science in 2011. Critically endangered; population fewer than 250 individuals. **V** Silent at sea. **H** Nests in crevices among rocky outcroppings, otherwise maritime. **D** Pacific Ocean. Breeds on Higashijima Island, possibly others nearby, Japan.

17 BLACK-VENTED SHEARWATER *Puffinus opisthomelas* 30–38cm Black undertail coverts. Resembles Pink-footed Shearwater, but smaller and less soaring, flying fast, close to water surface. Low flutter-and-glide flight. Often seen close to land. **V** Generally silent at sea. **H** Maritime. **D** Pacific Ocean. Nests on islands off California, and the Baja Peninsula.

18 TOWNSEND'S SHEARWATER *Puffinus auricularis* 31–35cm Low, fast flight with few glides. Note white flanks extending onto rump. Audubon's Shearwater has less black on upperparts. **V** Usually silent. **H** Offshore and open ocean. Breeds in burrows in areas with bush, grass and bracken, often at forest edges. **D** Pacific Ocean. Breeds on islands off W Mexico.

19 NEWELL'S SHEARWATER *Puffinus newelli* 33cm Split from Townsend's Shearwater. Feeds alongside tuna schools, which drive small squid and fish to surface. **V** Silent at sea. **H** Nests in burrows on remote islands; otherwise pelagic. **D** Pacific Ocean. Breeds in Hawaii.

20 RAPA SHEARWATER *Puffinus myrtae* 33cm Recent split from Newell's Shearwater. Critically endangered; population fewer than 250 individuals. **V** Silent at sea. **H** Nests in burrows; otherwise pelagic. **D** Pacific Ocean. Breeds in Austral Islands.

21 FLUTTERING SHEARWATER *Puffinus gavia* 33cm Separated from Hutton's Shearwater by less dark underwings, although note dark markings in axillaries. **V** Excited, rather rapid chattering. **H** Both inshore and offshore. Breeds in burrows in forest or grassland. **D** Pacific Ocean. Breeds in New Zealand. Migrates to Australia and Solomon Islands.

22 HUTTON'S SHEARWATER *Puffinus huttoni* 36–38cm Very similar to Fluttering, Manx and Little shearwaters, all four species with black at rump narrowed just after wings. Hutton's shows darkest underwings. **V** Excited, exhausted-sounding panting. **H** Offshore. Nests in burrows in mountain slopes with tussocks and scrub. **D** Pacific Ocean. Breeds in New Zealand.

23 AUDUBON'S SHEARWATER *Puffinus lherminieri* 27–33cm Note white markings around eyes and small size. Fluttering wingbeats followed by short, low glides. Often gregarious. **V** High, shrill, drawn-out rattling shrieks. **H** Mainly offshore. Breeds in hollows and crevices or on cliffs. **D** Atlantic Ocean, Caribbean.

24 PERSIAN SHEARWATER *Puffinus persicus* 30–35cm Brownish wash on underwing lining. In flight, wingbeats slower than in Tropical Shearwater, otherwise flight pattern similar. **V** Generally silent. **H** Maritime. **D** Indian Ocean, Arabian Sea, Persian Gulf.

25 TROPICAL SHEARWATER *Puffinus bailloni* 31cm Flight consists of fluttering wingbeats followed by short, low glides; in strong winds, glides rise and fall. **V** Mewing and twittering calls at breeding sites, otherwise silent. **H** Breeds in burrows or crevices in rocks, otherwise maritime. **D** Indian Ocean, Pacific Ocean.

26 GALAPAGOS SHEARWATER *Puffinus subalaris* 28cm Small. Bill blackish grey. Feet pinkish. White eye-ring. Upperparts dark brown. Underparts white. Undertail and tail blackish brown. Broad margin of underwing blackish. **V** Chattering, yelping and whooping notes when in groups. **H** Oceanic islands and sea. **D** Pacific Ocean. Breeds in Galápagos Islands.

27 BANNERMAN'S SHEARWATER *Puffinus bannermani* 27–32cm Formerly a race of Audubon's Shearwater. Seen mainly around Bonin and Volcano islands and presumed to breed there. **V** Silent at sea. **H** Probably nests in cliff burrows; otherwise pelagic. **D** Pacific Ocean.

28 HEINROTH'S SHEARWATER *Puffinus heinrothi* 35–38cm Sooty brown overall with pale underwing coverts. Long, slender bill and long tail. **V** Silent at sea. **H** Presumed to breed in highland burrows; otherwise pelagic. **D** Pacific Ocean.

29 LITTLE SHEARWATER *Puffinus assimilis* 27cm Upperparts greyish, wing coverts edged whitish. Face and sides of neck greyish, slightly scaled white. Underwings white without markings except darker tip and trailing edge. Flaps, rarely glides. **V** Low, hoarse panting and indignant muttering. **H** Inshore and offshore. Breeds in burrows or on ground, protected by scrub or rocks. **D** Pacific Ocean. Breeds in New Zealand, Norfolk Island and Lord Howe Island off E Australia and other islands off SW Australia.

30 SUBANTARCTIC SHEARWATER *Puffinus elegans* 27cm Like Little Shearwater but greyish cap extends neatly below eye. **V** Silent at sea. **H** Nests in burrows; otherwise pelagic. **D** Atlantic, Pacific and Indian oceans. Breeds on Tristan da Cunha, Gough, Chatham and Antipodes islands.

31 BAROLO SHEARWATER *Puffinus baroli* 25–30cm Fast, shallow wingbeats followed by a short, low glide; also fluttering flight close to sea surface. **V** At breeding grounds utters a high-pitched crowing, otherwise silent. **H** Breeds in crevices and burrows on islands, otherwise maritime. **D** Atlantic Ocean. Breeds in Macaronesia.

32 BOYD'S SHEARWATER *Puffinus boydi* 25–30cm Recent split from Barolo Shearwater; face is overall darker but otherwise very similar. **V** Silent at sea. **H** Nests in burrows; otherwise pelagic. **D** Atlantic Ocean. Breeds in Cape Verde Islands.

33 PERUVIAN DIVING PETREL *Pelecanoides garnotii* 22cm Bill black. Legs and feet blue with black webs. Upperparts black, some scapulars edged white. Wings black. Underparts white; flanks grey. The only species of diving petrel in its range. **V** Similar to that of Common Diving Petrel. **H** Nests in burrows in soft ground; otherwise pelagic. **D** Off W coast of South America.

34 MAGELLANIC DIVING PETREL *Pelecanoides magellani* 20cm Bill black. Legs and toes pale blue, webs black. Upperparts bluish black, underparts white. Scapulars edged with white. Wings brownish black. Diagnostic white 'gill' line curls up behind ear patch from throat. Dominant species around Fueguan waters. **V** Similar to that of Common Diving Petrel. **H** Nests in burrows in soft ground; otherwise pelagic. **D** S South America.

35 SOUTH GEORGIA DIVING PETREL *Pelecanoides georgicus* 20cm Bill black. Feet pale blue with black webs. Upperparts bluish black, underparts white. White along scapulars and tips to secondary flight feathers. No 'gill' line. Shape of nostrils differs from other species, but impossible to see in flight. **V** Various wailing or whooping calls near nest, otherwise silent. **H** Nests in burrows in soft ground; otherwise pelagic. **D** Southern Ocean.

36 COMMON DIVING PETREL *Pelecanoides urinatrix* 19cm Bill black. Feet pale blue with whitish webs. Flanks and sides of head grey, and a touch of grey on sides of breast, leaving hint of a white 'gill' line. Upperparts bluish black, underparts white. No white along scapulars. **V** Upslurring wail given near nest, otherwise silent. **H** Nests in burrows in soft ground; otherwise pelagic. **D** Southern Ocean.

37 BULWER'S PETREL *Bulweria bulwerii* 26–28cm Entirely dark brown with paler bar across upperwings. Note long-winged and long-tailed jizz. Usually flies low over water in erratic buoyant manner. Tail wedge-shaped when fanned. **V** At colonies utters a hoarse *hroo-hroo-hroo*. **H** Breeds in holes on rocky slopes or cliffs. Winters at sea. **D** Tropical and temperate oceans.

38 JOUANIN'S PETREL *Bulweria fallax* 30–32cm All dark with indistinct pale bar across upperwings. Note somewhat oversized bill. Faster, more swooping and towering flight than that of Bulwer's Petrel. **V** Silent. **H** Maritime. **D** Indian Ocean.

1
2
3
4
dark
5
pale
dark
6
7
8
9
10
11
12
13
14
15
16
17
18
19
20
21
22
23
24
25
26
27
28
29
30
31
32
33
34
35
36
37
38

GREBES PODICIPEDIDAE

1 LITTLE GREBE *Tachybaptus ruficollis* 25–29cm Usually encountered in pairs or small parties. In flight shows whitish secondaries. **V** High-pitched whinnying trill and various twitterings; also *beeh-ibto go on* contact call and, when alarmed, a metallic *whit whit*. **H** Vegetated lakes, ponds and rivers. In winter also on more open waters, including sheltered coastal bays. **D** Eurasia, Africa.

2 TRICOLORED GREBE *Tachybaptus tricolor* 25cm Usually in pairs or small groups. In flight, shows whitish secondaries. Darker than Little Grebe, from which it was split. **V** High-pitched whinnying trill and various twitterings. **H** Vegetated lakes, ponds and rivers, also coastal estuaries. **D** Indonesia, New Guinea, Solomon Islands.

3 AUSTRALASIAN GREBE *Tachybaptus novaehollandiae* 23–27cm Generally in pairs or small groups. In flight, shows much white on primaries and secondaries. **V** Series of rapid trills. **H** Freshwater lakes, ponds and swamps. **D** Indonesia, New Guinea, Solomon Islands, Vanuatu, New Caledonia, Australia, New Zealand.

4 MADAGASCAN GREBE *Tachybaptus pelzelnii* 18cm Similar to Little Grebe, but difference in breeding plumages is distinctive; in non-breeding plumage shows white wing panel when flying. **V** Rapid, laughing chatter given in courtship. **H** Any type of stagnant freshwater body, even slow-streaming rivers at any altitude. **D** Madagascar.

5 LEAST GREBE *Tachybaptus dominicus* 23–26cm Usually stays near thick cover. Feeds by diving for aquatic insects, amphibians or small fish. In flight, shows white patch on inner primaries and secondaries. **V** Nasal *teeen* or *weeek*, also a buzzy, descending *vvvvvvvvvvv*. **H** Freshwater cattail swamps, ponds with vegetation cover. **D** North, Central and South America.

6 PIED-BILLED GREBE *Podilymbus podiceps* 30–38cm Feeds mainly by diving for aquatic animals. When disturbed, often swims with body submerged. **V** Far-carrying, hollow cackle, starting fast, then slower and ending with mournful wail. **H** Shallow, well-vegetated freshwater lakes, ponds and rivers. In winter often on larger waterbodies and sheltered coastal areas. **D** North, Central and South America.

7 WHITE-TUFTED GREBE *Rollandia rolland* 25–32cm Head, neck and upperparts black. Large ear-tufts white. Flanks chestnut, faintly streaked blackish. **V** Short croaking note. **H** Lakes, marshes, roadside ditches, rivers, streams, with more or less vegetation. Also sea coasts in winter. **D** South America.

8 TITICACA GREBE *Rollandia microptera* 40cm Bill chestnut and yellow. Tufts on forehead and nape chestnut, streaked black. Sparse ear-tufts black. Throat, cheeks, foreneck and breast white. Hindneck chestnutty rufous. Upperparts and flanks blackish brown. **V** Sweet, brief downslurred whistle or piping note. **H** Lakes in high Andes with emergent vegetation. **D** Lakes Umayo, Titicaca and Poopo between Peru and Bolivia.

9 HOARY-HEADED GREBE *Poliocephalus poliocephalus* 29cm Separated from Horned and Black-necked grebes by range. Pale and slim, lacking chestnut tones in plumage (cf. New Zealand Grebe). **V** High, frog-like *purrreeh* in series of two to three. **H** Fresh and saline waters of open wetlands and estuaries. Occasionally on farm ponds. **D** Australia.

10 NEW ZEALAND GREBE *Poliocephalus rufopectus* 29cm Chestnut neck and silvery-streaked head. Male breeding plumage unmistakable. Non-breeding plumage separated from similar Australasian Grebe by more elongated jizz and lack of yellow at bill base. **V** Low, raspy *wro'wo'rough*. **H** Small, shallow waters such as farm dams and sewage ponds. **D** New Zealand.

11 GREAT GREBE *Podiceps major* 60cm Long bill black. Notable erectile crest and black line down hindneck. Rest of head dark grey, rest of neck chestnutty rufous. Upperparts and flanks blackish grey. **V** Short, squeaky piping note. **H** Lakes, marshes, rivers, edge of the sea. **D** South America.

12 RED-NECKED GREBE *Podiceps grisegena* 40–50cm In flight, similar to Great Crested Grebe but neck darker and shorter. **V** Wailing, braying and squeaking noises; also a grating *cherk-cherk-cherk*. **H** Lakes with surrounding vegetation. Winters on more open waters, including estuaries and sheltered coasts. **D** North America, Eurasia.

13 GREAT CRESTED GREBE *Podiceps cristatus* 46–51cm In flight, upperwing shows large white patches on secondaries and forewing. **V** Barking *rah-rah-rah*. Also various croaks, growls and slow nasal moaning. **H** Lakes and rivers with fringing vegetation. In winter also on large reservoirs, estuaries and sheltered coastal waters. **D** Eurasia, Africa, Australia, New Zealand.

14 HORNED GREBE *Podiceps auritus* 31–38cm Non-breeding birds have white cheeks and much like miniature Great Crested Grebe when in flight. **V** Far-carrying, rattling *joarrh*. Also an accelerating trill, similar to that of Little Grebe. **H** Vegetated lakes and ponds, more open waters in north. Winters on sheltered coastal waters, less so on inland waters. **D** North America, Eurasia.

15 BLACK-NECKED GREBE *Podiceps nigricollis* 28–34cm Often gregarious in winter quarters. In flight, upperwing shows large white secondary patch. **V** Flute-like *poo-eeet* and vibrant, trilled *tssrrrroooeep*. **H** Shallow lakes and ponds with fringing vegetation. Winters on more open waters, including sheltered coastal waters. **D** Europe, Asia, Africa, N South America, W and SW USA.

16 SILVERY GREBE *Podiceps occipitalis* 26cm Silvery white. Head lead grey. Slight crest and hindneck black. Ear-tuft golden yellow (blackish in Andean race). Eye golden yellow or orange. Upperparts blackish grey. Flanks somewhat streaked blackish. **V** Short piping notes, given in rippling series in song. **H** Patagonian and high Andean lakes with submerged or emergent vegetation. **D** South America.

17 JUNIN GREBE *Podiceps taczanowskii* 33cm Large version of Andean form of Silvery Grebe. Differs in proportionately longer bill, more elegantly slender head and neck. **V** Low, brief chirp. **H** Bays and channels at lake edge; nests in reedy margins. **D** Endemic to Lake Junin, central Peru.

18 HOODED GREBE *Podiceps gallardoi* 28cm White with black head and back. Forehead and brow white, crest rufous. Eye red eye with bare yellow eye-ring. **V** Chattering purrs and gull-like wails given in courtship. **H** Smallish open upland lakes in arid Patagonian steppe. Marine estuaries in winter. **D** SW Argentina.

19 WESTERN GREBE *Aechmophorus occidentalis* 51–74cm Dark eye surround in breeding birds. In winter, note dull yellowish bill. In flight, upperwing shows white on secondaries and inner primaries. **V** Two-note *kreed-kreet*. Also various creaking or high-pitched calls. **H** Summers on freshwater lakes with surrounding reeds. Winters mainly on coastal waters, less so on large inland lakes. **D** Summers in SW and SC Canada, W and NC USA. Winters along Pacific coast S to Mexico.

20 CLARK'S GREBE *Aechmophorus clarkii* 51–74cm Paler than Western Grebe, especially on flanks. White eye surround in breeding birds. In winter, note bright yellow bill. In flight, upperwing very similar to that of Western Grebe, but more white in primaries. **V** *Kreeed* or *kreee-eed*. Various other high or scratchy calls given on breeding grounds. **H** Similar to Western Grebe. **D** North America.

FLAMINGOS PHOENICOPTERIDAE

21 GREATER FLAMINGO *Phoenicopterus roseus* 120–145cm Unmistakable. **V** In flight, utters a honking *kla-ha*. Feeding flocks emit a constant, low, goose-like growling. **H** Salt lakes, sea bays, sometimes freshwater lakes. **D** Eurasia, Africa.

22 AMERICAN FLAMINGO *Phoenicopterus ruber* 107–122cm In flight, shows black primaries and secondaries, and extends neck and legs. Often feeds with head submerged while walking steadily forward. **V** Feeding flocks utter a constant low, babbling, goose-like chatter. In flight, gives a honking *ka-ha*. **H** Coastal estuaries and lagoons. Breeds colonially on lagoon borders. **D** Central America, Caribbean, South America, Galápagos Islands.

23 CHILEAN FLAMINGO *Phoenicopterus chilensis* 100cm Distal half of bill black, proximal ivory. Legs greyish blue, red at 'knee' and foot. Overall pink. Wing coverts and tertials scarlet, rest of flight feathers black. **V** Guttural and nasal. **H** Shallow and saline lakes, lagoons, estuaries, bays, sheltered waters. Freshwater marshes temporarily or on migration. **D** South America.

24 LESSER FLAMINGO *Phoeniconaias minor* 80–90cm Dark bill diagnostic. In flight, shows triangular pink patch on upperwing. Note shorter neck and small size. **V** In flight, utters a high-pitched *kwirrik*. Feeding flocks give a constant low murmuring. **H** Larger inland lakes, estuaries, lagoons. **D** Africa, NW India, SE Pakistan.

25 ANDEAN FLAMINGO *Phoenicoparrus andinus* 115cm Heftiest flamingo. Heavy bill; distal half black, proximal cream, red at base of mandible and spot before eye. Legs yellowish ivory. Head and neck bright pink, slightly tinged vinaceous. Rest pale pink, somewhat streaked darker pink. Flight feathers form a conspicuous black triangle over rump and tail. **V** Chattering calls, high pitched or low and goose-like. **H** High Andean saline lakes. Descends to plains in winter. **D** South America.

26 JAMES'S FLAMINGO *Phoenicoparrus jamesi* 90cm Small flamingo. Bill short and thick, tip black, rest bright yellow. Lores sealing-wax red. Legs red. Head and upper neck pink, lower neck whitish pink. Base of neck, breast and upper back streaked scarlet. Long dorsal plumes scarlet, covering most of black triangle of flight feathers. **V** Rolling, rattling chatter. **H** High Andean saline lakes; descends in winter. **D** South America.

TROPICBIRDS PHAETHONTIDAE

27 RED-BILLED TROPICBIRD *Phaethon aethereus* 90–105cm Flight pigeon-like on fluttering wingbeats followed by long glides. Catches fish by hovering and then plunge-diving on half-closed wings. **V** Shrill, rasping *kee-arrr*. **H** Breeds on offshore islands. Maritime after breeding. **D** Widespread across tropical oceans.

28 RED-TAILED TROPICBIRD *Phaethon rubricauda* 78–81cm Feeding and flight actions similar to those of Red-billed Tropicbird. **V** Harsh, rapid *keek-keek-keek-keek*. **H** Breeds on small oceanic islands. Maritime after breeding. **D** Indian Ocean, Pacific Ocean.

29 WHITE-TAILED TROPICBIRD *Phaethon lepturus* 70–82cm Flight consists of fluttering wingbeats followed by long glides. Feeds by hovering and then plunge-diving on half-closed wings. **V** Squeaky *chip-chip-chip*. **H** Pelagic and coastal waters. Breeds in rocky crevices or sheltered ground scrapes. **D** Widespread across tropical oceans.

1
2
3
4
5
6
7
8
9
10
11
12
13
14
15
16
17
18
19
20
21
22
23
24
25
26
27
28
29

STORKS CICONIIDAE

1 WOOD STORK *Mycteria americana* 85–100cm In flight, shows black primaries and secondaries and black tail. Flies with neck and legs outstretched. **V** Generally silent. At breeding grounds makes hissing and bill-clattering noises. **H** Swamps, coastal mudflats and lagoons, inland ponds and lakes. Breeds in mangroves. **D** SE USA to N Argentina.

2 MILKY STORK *Mycteria cinerea* 92–97cm Non-breeding birds have a pinkish-yellow bill. **V** Generally silent. During breeding makes screaming, hissing and bill-clapping sounds. **H** Tidal mudflats, mangroves, flooded forests. **D** Cambodia, Peninsular Malaysia, Indonesia.

3 YELLOW-BILLED STORK *Mycteria ibis* 95–105cm Yellow bill diagnostic. In flight, underwing has black primaries and secondaries with white coverts, tail black. **V** Generally silent away from breeding sites. **H** Any area with fish, especially larger waterbodies. **D** Sub-Saharan Africa.

4 PAINTED STORK *Mycteria leucocephala* 93–100cm In flight, shows all-dark underwing, apart from pale tips to coverts, tail black. Flies with neck outstretched. Roosts colonially in trees or on sandbanks. **V** During display, utters a weak 'fizzing' call along with bill-clapping; otherwise generally silent. **H** Marshes, lakes, ponds, rivers, coastal mudflats. **D** India, SE Asia.

5 ASIAN OPENBILL *Anastomus oscitans* 68cm In flight, underwing shows white coverts and black flight feathers, tail black. Gregarious. Breeds in colonies, often mixed with other storks and ibises. **V** During courtship display utters a series of hollow, nasal *hoo-hoo* calls. **H** Freshwater marshes, shallow lakes, lagoons, paddyfields. **D** India, SE Asia.

6 AFRICAN OPENBILL *Anastomus lamelligerus* 95cm Unmistakable, especially if 'open' bill is seen. In flight, all black below. Gregarious, especially in breeding season. **V** Generally silent except at breeding sites, where it utters a loud, raucous, sonorous *honks horrh-horrh*. Bill-clattering as in other storks physically impossible, but during copulation similar sounds are produced as pairs rattle bills sideways against each other. **H** Feeds on snails and freshwater mussels, which it opens with its specially adapted bill at water's edge, often in wooded areas. **D** Africa.

7 BLACK STORK *Ciconia nigra* 95–105cm In flight, underwing is black with white triangle on axillaries. **V** Generally silent. At nest site gives a rasping *shi-luu shi-luu shi-luu*. **H** Breeds in forested area with marshes, rivers and wet pastures. At other times may be found by lakes, rivers, estuaries or dry grassland. **D** S and E Africa, Eurasia.

8 ABDIM'S STORK *Ciconia abdimii* 69–81cm In flight from above, shows white rump and back. From below, resembles Black Stork but with larger white patch on wing. **V** High-pitched *peep peep peep*. **H** Dry plains and semi-deserts. **D** Africa.

9 WOOLLY-NECKED STORK *Ciconia episcopus* 75–92cm Usually encountered singly or small flocks. In flight, underwing black, with black tail often obscured by white undertail coverts. **V** Generally silent except for whistling greeting calls and bill-clapping at nest site. **H** Marshes, streams or ponds in open forest and freshwater swamp forest. **D** Sub-Saharan Africa, India to mainland SE Asia, Indonesia, Philippines.

10 STORM'S STORK *Ciconia stormi* 75–91cm Usually forages alone or in pairs. In flight, shows black underwing and tail, the latter often covered by white undertail coverts. **V** Short series of sibilant whistles and a *karau*. Bill-clapping occurs at nest site. **H** Freshwater peat-swamp forest, and rivers, streams and pools in forests. **D** Thailand; Peninsular Malaysia; Sumatra, Indonesia; Borneo.

11 MAGUARI STORK *Ciconia maguari* 120cm Bill horn-coloured, tip black. Face bare, legs red. White with black scapulars, flight feathers and tail. **V** Quiet. Young birds give soft clucking calls. **H** Marshes, flooded grass, fields, roadside ditches, streams, temporary water. **D** Colombia and Venezuela to C Argentina.

12 WHITE STORK *Ciconia ciconia* 100–115cm In flight, underwing has black primaries and secondaries with white coverts, tail white. **V** Generally silent. At nest site communicates mostly by bill-clapping. **H** Breeds in trees or on buildings in towns and villages, feeds in adjacent wet pastures, marshes. **D** Eurasia, Africa.

13 ORIENTAL STORK *Ciconia boyciana* 110–150cm In flight from below, similar to White Stork; from above, shows white edges to secondaries. **V** Generally silent. Communication during display is mainly through bill-clapping. **H** Islands of trees or scattered trees in river valleys, marshes and wet meadows, also lakes, marshes, coastal intertidal areas. **D** Siberia, E Asia.

14 BLACK-NECKED STORK *Ephippiorhynchus asiaticus* 129–150cm In flight, both upperwing and underwing white with wide black band on coverts, tail black. Flies with neck outstretched. Usually encountered singly or in small family groups. **V** Generally silent. **H** Freshwater swamps, shallow water on lakes and estuaries, all with nearby large trees. **D** India to Australia.

15 SADDLE-BILLED STORK *Ephippiorhynchus senegalensis* 145cm Eye yellow in female, dark in male. **V** Generally silent. At breeding sites adults perform bill-clattering during display. **H** Shallow waters, swampy areas. **D** Sub-Saharan Africa.

16 JABIRU *Jabiru mycteria* 140cm White. Huge bill slightly upturned, head bare, neck black. Bare red collar. **V** Generally silent. **H** Shallow marshes, ponds, wet pastures, also coastal estuaries, savannah, open areas with scattered trees. **D** S Mexico to N Argentina.

17 LESSER ADJUTANT *Leptoptilos javanicus* 110–120cm Usually seen singly or in small flocks. Generally breeds in small colonies. Flies with neck drawn back. In flight, underwing shows a small white triangle in axillary area. **V** During display, utters a series of high-pitched squeaks and cow-like moos. **H** Marshes, forest pools, swamps, floodplains, drying riverbeds. **D** Nepal and India through SE Asia to Greater Sunda Islands, Indonesia.

18 GREATER ADJUTANT *Leptoptilos dubius* 120–150cm Gregarious; often associates with vultures and kites during scavenging sorties. In flight, underwing shows a small white triangle in axillary area. Breeds singly or in small colonies. **V** At breeding sites utters various squealing and mooing sounds, along with much bill-clattering. **H** Marshes, lakes, cultivation, drying riverbeds, rubbish dumps. **D** N India.

19 MARABOU STORK *Leptoptilos crumenifer* 115–130cm Very large and unmistakable on ground. In flight, can look like Black Stork but flies with retracted head and neck (unlike other storks). **V** Generally silent. **H** Open savannas with lakes, pools, marshes, rivers. Also around rubbish dumps. **D** Sub-Saharan Africa.

IBISES AND SPOONBILLS THRESKIORNITHIDAE

20 AFRICAN SACRED IBIS *Threskiornis aethiopicus* 65–75cm Gregarious. In flight, shows black tips to primaries and secondaries. Bare pink-red 'arms' on underwing. **V** Squealing yelps and short barks at breeding sites, otherwise generally silent. **H** Freshwater marshes, rivers, coastal marshes, cultivated areas. **D** Sub-Saharan Africa, Iraq.

21 MALAGASY SACRED IBIS *Threskiornis bernieri* 65–75cm Similar to African Sacred Ibis but has pale eyes and lacks black trailing edges to wings in flight. **V** In flight, occasionally utters mid-high *kreh kreh*. **H** Wetlands, farmland, sewage ponds, coastal beaches, offshore islands. **D** Madagascar, Aldabra Islands.

22 BLACK-HEADED IBIS *Threskiornis melanocephalus* 65–75cm Usually in small flocks. Eastern counterpart of African Sacred Ibis. **V** Generally silent, apart from strange grunts at breeding sites. **H** Reed swamps, flooded grassland, reservoirs, coastal lagoons. **D** India, SE Asia.

23 AUSTRALIAN WHITE IBIS *Threskiornis molucca* 65–75cm Occurs singly, in pairs or in parties of up to 35 birds. Often perches high in isolated dead trees and on fallen riverside trees. **V** Generally silent, but occasionally gives harsh croaks and honks. **H** Swamps, moist and newly burned grassland, coastal mudflats, mangroves, margins of larger rivers. **D** Indonesia to New Guinea and Australia.

24 STRAW-NECKED IBIS *Threskiornis spinicollis* 60–75cm Nests colonially. Often seen in flocks. Eats terrestrial and freshwater invertebrates; also scavenges human refuse. **V** Various grunts and croaks, at colonies and in flight. **H** Grassland, farmland, wetland, parks. **D** New Guinea, Australia.

25 RED-NAPED IBIS *Pseudibis papillosa* 68cm Usually in small groups. White covert patch often concealed when perched or walking, but obvious in flight. Flies with neck outstretched. **V** Male utters a trumpet-like call when advertising or on the wing. **H** Grasslands, cultivated fields, lakesides, marsh edges, rubbish dumps. **D** India, Pakistan, Nepal.

26 WHITE-SHOULDERED IBIS *Pseudibis davisoni* 75–85cm In flight, shows a white patch on inner forewing. **V** Hoarse, screaming *errrrrrrroh* and subdued *ohhaaa ohhaaa* and *errr-ah*; also other screams and honking sounds. **H** Streams, ponds and marshy areas in open lowland forests. **D** Mainland SE Asia, Borneo.

27 GIANT IBIS *Pseudibis gigantea* 102–106cm Shy; probably best located by calls at dawn or dusk. Usually in pairs or small groups. In flight, upperwing greyish with black primaries, underwing all dark. **V** Repeated loud, ringing *a-leurk a-leurk*. **H** Marshy areas, ponds and streams in open lowland forest. Also around lakes and marshes, and on open plains. **D** Cambodia, S Laos, S Vietnam.

28 NORTHERN BALD IBIS *Geronticus eremita* 70–80cm Legs do not protrude beyond tail in flight. **V** At breeding sites utters a guttural *hrump* and hoarse *hyoh*, otherwise generally silent. **H** Breeds on cliffs in semi-arid hill country, feeding in nearby dry fields and along streams and rivers. **D** Morocco, Algeria.

29 SOUTHERN BALD IBIS *Geronticus calvus* 80cm Bald, bicoloured head and black plumage diagnostic. **V** In flight, occasionally utters a high, fluting *whuuhuh whuuh*. **H** Breeds on cliffs but feeds in short, overgrazed or burnt, upland grassland. **D** South Africa.

30 CRESTED IBIS *Nipponia nippon* 60cm In flight from below, tail and wings show strong pink tinge. **V** Mostly silent. Utters a *gak-gak-gak* when disturbed and, in flight, a single *gak*. **H** Breeds in forest patches in mountain valleys, feeding in nearby lakes, streams, ponds and paddyfields. Moves to lowland wetlands in non-breeding season. **D** NE China.

31 OLIVE IBIS *Bostrychia olivacea* 70cm Difficult to see in its forest habitat. **V** Mid-high *wa waa waa wa-waa*. **H** Open places in forests. **D** Africa.

32 SAO TOME IBIS *Bostrychia bocagei* 50cm Recalls Olive Ibis but smaller. Critically endangered; population fewer than 250 individuals. **V** Usually silent. **H** Primary forest, below 500m. **D** São Tomé.

33 SPOT-BREASTED IBIS *Bostrychia rara* 45cm Note crest and spotting. **V** Very noisy at dusk, *hakhak kakkah*. **H** In or close to forests. **D** WC Africa.

34 HADADA IBIS *Bostrychia hagedash* 75cm Note white cheek stripe and red ridge of bill. Very vocal in flight. **V** Mid-high, loud, raucous *haaah hahah haoh hahah*. **H** Cultivated areas, lawns, moist grassland, marshes, wooded streams. **D** Africa.

35 WATTLED IBIS *Bostrychia carunculata* 80cm Feeds in groups. **V** Harsh, croaky honks. **H** Grassland, farmland, marshes, open forests; roosts at cliffs. **D** Ethiopia, Eritrea.

36 PLUMBEOUS IBIS *Theristicus caerulescens* 73cm Bill grey. Legs pink. Plumage all grey. Bushy nuchal crest. **V** Calls include a low, muttering chuckle, and louder, high-pitched yelps. **H** Marshes, lakes, flooded savannah, open areas with scattered trees and patches of woods, rice paddies. **D** NE Bolivia and S Brazil to NE Argentina and Uruguay.

37 BUFF-NECKED IBIS *Theristicus caudatus* 73cm Like Black-faced Ibis, but silvery-white greater wing coverts and no pectoral collar. **V** Harsh, yelping calls. **H** Open woods and forest, savannah, open country, fields, small marshes, flooded areas, lakes, ponds, rivers, streams, also hill country. **D** South America.

38 BLACK-FACED IBIS *Theristicus melanopis* 73cm Bare face and wattles black. Legs pink. Head and neck buffy cinnamon. Upperparts grey. Pectoral collar grey. Underparts and tail black. **V** Abrupt, rather goose-like clucking honks. **H** Short grass in damp, low-lying areas, flooded areas, open woods. Ploughed fields and short grass on wintering grounds. **D** SW Peru, W Chile, S Argentina.

39 ANDEAN IBIS *Theristicus branickii* 75cm Similar to Black-faced Ibis but paler. **V** Harsh, clipped *ag-ag ag* notes. **H** Highlands. **D** Ecuadorian Andes to Peru, N Chile and N Bolivia.

1
2
3
4
5
6
7
8
9
10
11
12
13
14
♂
♀
15
♂
♀
16
17
18
19
20
21
22
23
24
25
26
27
28
29
30
31
32
33
34
35
36
37
38
39

IBISES AND SPOONBILLS *CONTINUED*

1 SHARP-TAILED IBIS *Cercibis oxycerca* 80cm Bill, bare face and legs reddish pink. Black with purple sheen on back. Thick nuchal crest. Long tail. **V** Goose-like honking calls, given in series by both members of a pair during courtship. **H** Lakes, rivers, marshes, fields, flooded areas in savannah. **D** E Colombia, Venezuela, Guyana, N Brazil.

2 GREEN IBIS *Mesembrinibis cayennensis* 55cm Stocky build. Note rather short tail and legs. **V** Loud, far-carrying, descending *cluk-clook-clook-clook-clook*. **H** Marsh in forested areas. **D** Honduras to SE Brazil and NE Argentina.

3 BARE-FACED IBIS *Phimosus infuscatus* 50cm Bill ivory. Bare face salmon pink. Shortish legs pale pink, and do not extend beyond tail in flight. Glossy black with bluish-green sheen. **V** Goose-like, rather rushing honking calls. **H** Fields, flooded grassland, marshes, lakes, swamps, streams, roadside ditches, rice paddies. **D** Widespread across South America.

4 AMERICAN WHITE IBIS *Eudocimus albus* 56–71cm Gregarious. In flight, neck and legs outstretched and shows black tips to primaries. **V** Harsh, nasal honking *hunk-hunk-hunk* given in flight and when disturbed. Also a murmuring *huu-huu-huu* when feeding. **H** Wetlands, including saltwater and freshwater areas. Breeds in mangroves. **D** S USA, Caribbean, N South America.

5 SCARLET IBIS *Eudocimus ruber* 56–61cm Breeding adult has a black bill. In flight, neck and legs outstretched and shows black tips to primaries. **V** Similar to that of White Ibis. **H** Coastal mangroves, swamps, lagoons. **D** Colombia and Ecuador to NE Brazil.

6 GLOSSY IBIS *Plegadis falcinellus* 55–65cm Often encountered feeding alongside herons and spoonbills. Flies with neck extended and legs protruding beyond tail. **V** Grunting *grru* or *graa*. Various grunting and croaking notes uttered at breeding sites. **H** Breeds in freshwater marshes, by lakes and rivers with surrounding vegetation. During migration also on wet meadows, paddyfields, coastal lagoons. **D** Worldwide.

7 WHITE-FACED IBIS *Plegadis chihi* 56–63cm Flies with neck outstretched, often in loose flocks or in undulating lines. Iris and legs red. **V** Generally silent although may utter a grunting sheep-like *grru* or *graa*. **H** Mainly freshwater marshes and ponds. **D** C USA to Argentina.

8 PUNA IBIS *Plegadis ridgwayi* 60cm Bill pinkish red at base, tip grey. Bare face deep red. Legs blackish. Head and neck chestnut, contrasting with blackish body, this with dark purple and green sheen. **V** Nasal quacking notes given in alarm or agitation. **H** High Andean lakes and marshes, nearby grassland. **D** N Peru to N Chile and NW Argentina.

9 MADAGASCAN IBIS *Lophotibis cristata* 50cm Unmistakable thanks to bicoloured crest (all black in *L. c. urschi*) and contrasting white wings. **V** Harsh, rather cawing notes. **H** Dry and humid forest, plantations, mangroves. **D** Madagascar.

10 EURASIAN SPOONBILL *Platalea leucorodia* 80–90cm Sociable, often mixing with herons and egrets. In non-breeding plumage loses yellow tinge to crest and breast. **V** Usually silent. **H** Brackish and freshwater lakes, reed swamps, coastal mudflats, lagoons. **D** Europe to N China, S Asia and N Africa.

11 BLACK-FACED SPOONBILL *Platalea minor* 60–84cm Habits similar to those of Eurasian Spoonbill. In non-breeding season loses yellow tinge on crest and breast. **V** Generally silent away from breeding sites. **H** Tidal mudflats, salt marshes, estuaries, inland lakes. **D** Korean Peninsula, NE China.

12 AFRICAN SPOONBILL *Platalea alba* 91cm Habits much as those of Eurasian Spoonbill. **V** Generally silent away from breeding sites. **H** Brackish and freshwater wetlands, coastal lagoons. **D** Sub-Saharan Africa.

13 ROYAL SPOONBILL *Platalea regia* 78cm Unmistakable. Note black bill and legs. **V** Low, hoarse grunts. **H** Large, shallow waterbodies, including estuaries, tidal mudflats, occasionally lagoons and inland wetlands. **D** Australia and New Zealand.

14 YELLOW-BILLED SPOONBILL *Platalea flavipes* 88cm Unmistakable. Note yellow bill and legs. **V** High, loud, strident *wheech* notes. **H** Normally inland wetlands, rarely coastal. **D** Australia.

15 ROSEATE SPOONBILL *Platalea ajaja* 66–81cm Forages in shallow water, sweeping bill from side to side while moving slowly forward. Flies with neck outstretched. **V** On breeding grounds utters a grunting *huh-huh-huh-huh* and coarse *rrek-ek-ek-ek-ek-ek*, otherwise usually silent. **H** Coastal lagoons, swamps, tidal pools. Breeds colonially in mangrove or other coastal trees. **D** S USA to N Argentina, Caribbean.

HERONS AND BITTERNS ARDEIDAE

16 FOREST BITTERN *Zonerodius heliosylus* 65–70cm Solitary, secretive and hard to observe well. Feeds at water's edge on invertebrates and small vertebrates. **V** Low, rumbling grunts. **H** Forested streams and pools; mainly lowlands. **D** New Guinea.

17 WHITE-CRESTED TIGER HERON *Tigriornis leucolopha* 65cm On first sight resembles a long-necked Eurasian Bittern. Differs from all other herons by finely barred plumage. Erectile crest normally hidden. Evasive. **V** Song a low, upslurring *uuuwuuuu*. **H** Marshy woodland, undergrowth near forest streams. **D** Senegal to Central African Republic and DR Congo.

18 RUFESCENT TIGER HERON *Tigrisoma lineatum* 72cm Bare face striking yellow. Head, neck and breast rufous chestnut. Two white lines down foreneck. Upperparts brown, finely vermiculated black. Belly vinaceous cinnamon. **V** Song a series of low, mournful moaning notes; also gives harsh calls. **H** Swamps, marshes, rivers, streams, temporary pools, all with trees on banks in forest or woods; mangroves; dense reedbeds in savannah and open areas. **D** Honduras to N Argentina.

19 FASCIATED TIGER HERON *Tigrisoma fasciatum* 68cm Bill and legs proportionately shorter than in other tiger herons. Cap black. Dark brownish grey from behind eye down side of upper neck. Neck, breast and upperparts blackish brown, finely barred buff. White lines down foreneck. Belly vinaceous cinnamon. **V** Song a short series of loud, deep groans, last note drawn-out. **H** Forest streams and rivers in mountains and hills. **D** Costa Rica to N Bolivia and NW Argentina.

20 BARE-THROATED TIGER HERON *Tigrisoma mexicanum* 78cm Black cap. Grey from behind eye down neck. Bare yellow throat. Neck and breast barred blackish and buff. Two lines down foreneck. Upperparts finely barred blackish brown and buff. Belly brownish rufous. **V** Gives a series of deep, hollow croaking grunts. **H** Watercourses in forest, swamps, mangroves, estuaries. **D** Mexico to Colombia.

21 AGAMI HERON *Agamia agami* 70cm Very long, slender yellow-and-black bill. Legs yellow. Head black, throat white. Neck and underparts chestnut. White and chestnut lines down foreneck. Base of neck and breast bluish grey, finely streaked white. Upper back, wings and tail bottle green. **V** Deep rattles and chuckles given near the nest. **H** Shady bogs, swamps, streams, ponds in dense pristine forest, mangroves. **D** E Mexico to Amazonia.

22 BOAT-BILLED HERON *Cochlearius cochlearius* 50cm Nocturnal. Very wide, heavy bill blackish. Cap black. Tuft of long feathers from occipital (longer in nuptial plumage). Forehead, face, neck and breast white. Upperparts grey. Underparts cinnamon, flanks black. **V** Hoarse, duck-like quacking calls. **H** Rivers and swamps in dense forest, gallery forest, mangroves. **D** WC Mexico to NE Argentina.

23 ZIGZAG HERON *Zebrilus undulatus* 32cm Short, thick neck. Bill black, mandible pinkish. Dense occipital crest. Upperparts blackish, irregularly and finely barred whitish buff. Buffish brown below, finely barred black. **V** Plaintive, downslurring single honking notes. **H** Forest streams, swamps with dense vegetation, waterholes. **D** Amazonia.

24 EURASIAN BITTERN *Botaurus stellaris* 70–80cm Secretive, more often seen flying over reedbeds. **V** Far-carrying booming, preceded by a short, muffled *up-rumbh* or *up-up-rumbh*, repeated at short intervals. In flight, gives a harsh, nasal *kau* or *krau*. **H** Freshwater and brackish reedbeds. **D** Europe to E Asia and S Africa.

25 AUSTRALASIAN BITTERN *Botaurus poiciloptilus* 71cm Secretive. Note cryptic colouring and distinctively tawny cheeks. **V** Series of extremely low, well-separated, resounding *boom* notes. **H** Wetlands with tall, dense vegetation. Occasionally in lagoons or estuaries. **D** SW and SE Australia, New Zealand.

26 AMERICAN BITTERN *Botaurus lentiginosus* 60–85cm Resembles Eurasian Bittern, but with speckled rather than barred plumage. **V** Booming *pump-er-lunk* call. **H** Freshwater and brackish marshes and bogs. **D** North America.

27 PINNATED BITTERN *Botaurus pinnatus* 70cm Like a young tiger heron. More streaked on upperparts. Bill and legs yellowish, feet big. Neck barred dark brown and ochreous. Throat white. Foreneck ochreous buff with tawny streaks forming lines. Underparts buffish white with irregular brown streaking. **V** Song comprises deep, gulping boom notes. **H** Lakes and marshes, riverside reedbeds, tall and dense emergent vegetation, flooded grassland, rice paddies. **D** E Mexico to NE Argentina.

28 STRIPE-BACKED BITTERN *Ixobrychus involucris* 32cm Tiny. Buff with black, cinnamon and white streaking on back. Foreneck white with lines of buff streaks. Underparts buffish white. **V** Various low-pitched honking and barking notes. **H** Lakes and marshes with dense reedbeds and other tall emergent vegetation. **D** Colombia to C Chile and C Argentina.

29 LEAST BITTERN *Ixobrychus exilis* 30cm Cap and back black. Neck tawny rufous to buff, foreneck white with tawny-buff lines. Upperwings tawny buff, primaries blackish. Underparts whiter. **V** Song a low, cawing chuckle, rising in pitch when agitated. **H** Swamps and lakes with very dense, tall emergent vegetation. **D** E and SW USA to SE Brazil.

30 LITTLE BITTERN *Ixobrychus minutus* 33–38cm Secretive. In flight, shows black back, tail and flight feathers contrasting with buff forewing. **V** Low, repeated *hoogh* or *grook*. When flushed utters a low *ker-ak* or *ker*, and when excited a loud, hoarse *eke-eke*. **H** Freshwater marshes, lakes, pools and rivers with surrounding reedbeds. **D** S and C Europe to C Asia, NW India and Africa.

31 BLACK-BACKED BITTERN *Ixobrychus dubius* 25–35cm Secretive and hard to observe. Recalls Little Bittern but flight feathers grey, not black. **V** Low, repeated croaking *crooah* uttered by territorial male. **H** Well-vegetated freshwater wetlands, mangrove swamps. **D** E and SW Australia.

32 YELLOW BITTERN *Ixobrychus sinensis* 30–40cm Skulking. Crepuscular or nocturnal feeder, usually occurring singly. In flight, shows black flight feathers and tail. **V** In flight, utters a sharp *kakak kakak*. **H** Vegetation surrounding marshes, lakes and flooded paddy fields. **D** India to Japan and Indonesia.

33 VON SCHRENCK'S BITTERN *Ixobrychus eurhythmus* 36cm Separated from Yellow Bittern by dark chestnut upperparts and different wing pattern. **V** Low, repeated *gup*. In flight, utters a low squawk. **H** Marshes, reedbeds, swamps, rice paddies. **D** Breeds in SE Siberia, Korean Peninsula, Japan, E China. Winters in SE Asia.

34 CINNAMON BITTERN *Ixobrychus cinnamomeus* 40–41cm Secretive. Generally forages alone, mainly at dawn and dusk, although also reported feeding during daytime, probably when young. **V** Low *kwok-kwok-kwok*, sometimes ending with two to three quieter notes. In flight, utters a croak. **H** Flooded paddy fields, swamps, reedbeds, overgrown ditches. **D** India to E China and Indonesia.

35 DWARF BITTERN *Ixobrychus sturmii* 25cm Note very small size, black eye, dark appearance. Largely nocturnal. **V** Mid-high, dull *roap-roap-roap*. **H** Thick reedy cover with some trees and shrub in or near marshes, lakes, streams, inundations, flooded woods. **D** Widespread across Africa.

36 BLACK BITTERN *Dupetor flavicollis* 54–66cm Secretive, skulking in dense cover. Peak feeding time is at dusk or dawn, but also feeds during the day in overcast weather. **V** In flight, utters a hoarse croak. **H** Dense reedy swamps. **D** S and SE Asia, New Guinea, Australia, Solomon Islands.

1
2
3
4
5
6
7
8
9
10
11
12
13
14
15
16
17
18
19
20
21
22
23
24
25
26
27
28
29
30
31
32
33
34
35
36

HERONS AND BITTERNS ***CONTINUED***

1 WHITE-EARED NIGHT HERON *Gorsachius magnificus* 54–56cm Crepuscular or nocturnal. Secretive. **V** Repeated, raspy *whoa*, said to sound like a large owl. **H** Dense undergrowth in marshy areas around forest streams. **D** S and E China, N Vietnam.

2 JAPANESE NIGHT HERON *Gorsachius goisagi* 49cm Skulks. Primaries with extensive rufous tips. **V** Deep *buo-buo*, usually at night. Emits a croak while feeding. **H** Dense undergrowth in mountain forests. **D** Japan.

3 MALAYAN NIGHT HERON *Gorsachius melanolophus* 49cm Shy. Generally crepuscular or nocturnal. In flight, shows white-tipped primaries. **V** Deep *oo oo oo oo*. Also croaks and a rasping *arh-arh-arh*. **H** Marshes and streams in dense forest. **D** S and SE Asia.

4 WHITE-BACKED NIGHT HERON *Gorsachius leuconotus* 55cm Note exceptionally large eyes. Separated from immature Black-crowned Night Heron by grey head and darker eyes. Nocturnal. **V** Low, crow-like *cra cra cra cra* or *craak-craak-craak-craak*. **H** Dense waterside undergrowth near slow streams and lakes. **D** Widespread across Africa.

5 BLACK-CROWNED NIGHT HERON *Nycticorax nycticorax* 58–65cm Adult unmistakable. Mainly crepuscular or nocturnal. Usually forages alone. **V** Various croaks given at breeding colonies. In flight, utters a frog-like croak. **H** Various watery areas, including marshes, lakes and rivers with extensive border vegetation, paddyfields, mangroves. **D** Worldwide except Australasia.

6 NANKEEN NIGHT HERON *Nycticorax caledonicus* 55–59cm Nocturnal. By day, rests in the cover of leafy trees. Best sighted in the early evenings as birds leave colonial roosts to visit feeding sites. **V** In flight, gives a loud *kyok* or *kwok*. **H** Swamps, ponds, lakes, mangroves, forest-lined creeks and rivers. **D** Java and Lesser Sunda Islands, Indonesia; Borneo; Philippines; New Guinea; Australia; New Zealand.

7 YELLOW-CROWNED NIGHT HERON *Nyctanassa violacea* 56–70cm Nocturnal, but will venture out at other times. **V** Squawking *kowk* or *kaow*. **H** Mangrove swamps, mudflats on coast or lakes, sometimes dry areas away from water. Breeds in trees, not necessarily near water. **D** C and E USA to Peru and E Brazil, Caribbean, Galápagos Islands.

8 GREEN HERON *Butorides virescens* 40–48cm Often stands, crouched and motionless, watching for prey. **V** When nervous utters a series of knocking *kuk-kuk-kuk* notes. When flushed delivers a squawking *skyeow*. **H** Water's edge in virtually any wetland area. **D** E Canada, E and C USA to Panama, Caribbean.

9 LAVA HERON *Butorides sundevalli* 40cm All lead grey. Chin spot white. In breeding plumage has long, pale grey feathers on back. Legs yellowish orange. **V** Usually silent, may give shrill *keowk* when alarmed. **H** Rocky seashores, inland waters. **D** Galápagos Islands.

10 STRIATED HERON *Butorides striata* 40–48cm Often forages alone, among vegetation on the banks of rivers or lakes; may use the same location for a number of days. **V** May utter a harsh *kyah* if disturbed. **H** Mangroves, rivers and streams in or near forests, lakes, coastal mudflats. **D** Widespread across South America, Africa, Indian Ocean, Asia to SE Siberia and Japan, New Guinea, Australia, Pacific Ocean.

11 SQUACCO HERON *Ardeola ralloides* 44–47cm In flight, appears largely white. Often skulks. **V** Harsh *kaahk*. **H** Well-vegetated marshes, lakes, ponds, rivers, coastal lagoons. Also paddyfields, flooded fields, etc. outside breeding season. **D** Widespread across Eurasia and Africa.

12 INDIAN POND HERON *Ardeola grayii* 42–45cm In flight, appears largely white. Regularly feeds in open. **V** Harsh squawk, similar to that of Squacco Heron. **H** Various wetlands, such as marshes, ponds, paddyfields, lagoons, etc. **D** Persian Gulf to Myanmar, Maldive Islands.

13 CHINESE POND HERON *Ardeola bacchus* 42–45cm In flight, appears largely white. Roosts and breeds in colonies, often with other heron species. **V** When alarmed utters a harsh croak. **H** Marshes, paddyfields, ponds, lagoons, etc. **D** E Asia.

14 JAVAN POND HERON *Ardeola speciosa* 45cm Crepuscular. Stands still waiting for prey. In flight, shows white wings and tail. **V** When flushed may give a harsh *kaa kaa*; in flight, utters a squawk. **H** Various inland and coastal wetlands. **D** SE Asia.

15 MALAGASY POND HERON *Ardeola idae* 45cm Less skulking than Squacco Heron, and in non-breeding plumage separated from that species by much heavier streaking. **V** Call like that of Squacco Heron but lower pitched and harsher. **H** Wooded marshes. **D** Madagascar, Aldabra Islands.

16 RUFOUS-BELLIED HERON *Ardeola rufiventris* 60cm Skulking. Note pale chin of female. **V** Rasping *graak* and other grunts. **H** Swamps, inundations, reedy edges of lakes and swamps. **D** Uganda and S Kenya to S Angola, N Botswana and E South Africa.

17 WESTERN CATTLE EGRET *Bubulcus ibis* 48–53cm Sociable. Often feeds on insects disturbed by grazing animals. **V** Usually fairly silent. **H** Colonial nester, often near water. Feeds in dry grassland, arable fields, marshes. **D** S Europe to Iran, Africa, Indian Ocean, and North, Central and South America.

18 EASTERN CATTLE EGRET *Bubulcus coromandus* 48–53cm Sociable, regularly seen feeding on insects disturbed by grazing animals. **V** At breeding sites utters a low croak and in flight may give a harsh, croaking *ruk* or *rik-rak*, otherwise usually silent. **H** Various wetlands and grasslands. **D** S and E Asia, Australasia.

19 GREY HERON *Ardea cinerea* 90–98cm Generally forages alone. Often stands motionless at water's edge or on a branch. Flies with arched wings and neck drawn back. Grey underwing coverts contrast with dark flight feathers. **V** At breeding sites utters harsh croaks; in flight, gives a harsh *frahnk*. **H** Variety of wetlands, including lakes, marshes, rivers, mangroves, tidal creeks. **D** Widespread across Europe, Africa and Asia.

20 GREAT BLUE HERON *Ardea herodias* 110–125cm Flies with large arched wings and head retracted. Often stands motionless, or walks stealthily in search of prey. **V** In flight, utters a deep, trumpeting *kraak*. **H** Lakes, ponds, marshes, estuaries. **D** North and Central America.

21 COCOI HERON *Ardea cocoi* 125cm Cap and occipital crest black. Neck and back greyish white. Line of black streaks on foreneck. Wings and tail darker grey. Belly and forepart of flanks black. Leggings and undertail coverts white. In breeding plumage crest, breast and back feathers very long. **V** Harsh, rasping, crow-like notes. **H** Freshwater bodies, estuaries, mangroves, seashores. **D** South America.

22 WHITE-NECKED HERON *Ardea pacifica* 91cm Unmistakable thanks to size and contrast between grey body and white neck and head. In flight, when seen head on, shows striking white 'headlights' at wrists. **V** High, hoarse, drawn-out shrieks. **H** Shallow water of wetlands and inundated grassland, swamps, watercourses, irrigation channels, ponds, occasionally estuaries. **D** Australia, S New Guinea.

23 BLACK-HEADED HERON *Ardea melanocephala* 90–95cm Black crown and legs diagnostic. In flight from below, shows prominent white wing lining. Often feeds in grassland far from water. **V** Harsh *kuark*. **H** Grassy areas, forest clearings, lakes, rivers, marshes, coastal areas, pastures. **D** Sub-Saharan Africa.

24 HUMBLOT'S HERON *Ardea humbloti* 105cm Upper- and underparts are uniform grey except black cap and chin. Bill and legs yellow in breeding birds. **V** Probably similar to that of Grey Heron. **H** Edges of freshwater or saltwater bodies. **D** Madagascar.

25 WHITE-BELLIED HERON *Ardea insignis* 127cm Feeds alone or in pairs. Flies with arched wings and neck drawn back, white underwing coverts contrasting with darker flight feathers. **V** Braying, croaking *ock ock ock ock urrrrr*. When disturbed utters a deep croak. **H** Marshes, lakes and rivers in forested areas. **D** E Himalayas.

26 GREAT-BILLED HERON *Ardea sumatrana* 115cm Generally shy and wary, foraging alone or in pairs. Flies with arched wings and neck drawn back, grey underwing coverts contrasting little with flight feathers. **V** Occasional loud, harsh croaks. **H** Mangroves, tidal mudflats, estuaries, coastal swamps, beaches. **D** SE Asia to N Australia.

27 GOLIATH HERON *Ardea goliath* 135–150cm Large, man-sized, much larger than superficially similar Purple Heron. Typically forages alone in the open. Flies with arched wings and neck drawn back. Underwing coverts extensively purple. **V** In flight, gives a deep, loud *kowoorrk-kowoorrk-woorrk-work-worrk*. **H** Lakes, rivers, marshes, coastal mudflats. **D** Sub-Saharan Africa, Iran and Iraq to India.

28 PURPLE HERON *Ardea purpurea* 78–90cm More secretive than others of genus. Flies with arched wings and neck drawn back, neck bulge more pronounced than in Grey Heron. **V** Various harsh utterances at breeding sites. Flight call like that of Grey Heron. **H** Marshes and lakes with dense aquatic vegetation; post-breeding often visits more open waters. **D** Widespread from S and C Europe to C, S and E Asia, Middle East, sub-Saharan Africa.

4
5
3
2
1
6
7
12
10
8
9
11
15
13
14
16
♀
♂
17
18
22
19
20
21
25
23
24
26
27
28

HERONS AND BITTERNS *CONTINUED*

1 GREAT EGRET *Ardea alba* 85–102cm At onset of breeding bill becomes black with yellowish base and facial skin turns greenish. Non-breeding birds lose long back plumes. **V** Dry, rattling *krr-rr-rr-rra*. **H** Breeds on well-vegetated freshwater lakes and marshes; visits other wetland areas, including coastal locations, in non-breeding season. **D** Worldwide.

2 INTERMEDIATE EGRET *Ardea intermedia* 65–72cm At onset of breeding facial skin becomes green and bill and tibia turn pinkish red. Non-breeding birds lose back and breast plumes. **V** Utters a *kwark* or *kuwark* when disturbed. **H** Freshwater lakes, rivers, marshes, less so in brackish or saltwater areas. **D** Sub-Saharan Africa, Japan to S India and Indonesia, New Guinea, Australia.

3 CAPPED HERON *Pilherodius pileatus* 60cm Bill pale blue with pink at mid-mandible. Forehead white, cap black. Long white occipital crest. Head and neck washed cream, rest white. **V** Low-pitched whooping notes. **H** Lakes, swamps, temporary pools, rivers and streams, in forest. **D** E Panama to SE Brazil.

4 WHISTLING HERON *Syrigma sibilatrix* 58cm Bill pink, tip black. Crown and occipital crest slaty grey. Neck buff. Upperparts grey. Wing coverts ochreous, edged black. Underparts whitish. **V** Gives thin, high-pitched, rather strained whistled notes. **H** Fields, short grass, islands of trees, scattered trees, always near fresh water. **D** N, SC and SE South America.

5 PIED HERON *Egretta picata* 50cm Agile forager, even seen to 'hover' before dropping onto prey. **V** In flight, utters a loud *awk* or *ohrk*. **H** Various wetlands, including lakes, mudflats and estuaries. **D** Sulawesi, Indonesia, to N Australia.

6 WHITE-FACED HERON *Egretta novaehollandiae* 61–74cm Occurs singly, in pairs or in small flocks; sometimes associates with other wading species. Active, stalking forager. **V** Generally silent; in flight, may utter a drawn-out, guttural croak. **H** Freshwater and brackish swamps, paddyfields, lake margins, rivers. **D** Australasia.

7 REDDISH EGRET *Egretta rufescens* 69–81cm Animated foraging actions, often with wings held open. Flies with head retracted. **V** Short grunt or soft groan. **H** Mainly sheltered, shallow coastal waters. **D** S USA, Mexico, Caribbean.

8 BLACK HERON *Egretta ardesiaca* 65cm Note black legs with yellow feet, and characteristic head profile with long plumes and black eyes. Unfolds wings repeatedly like a fast-opening umbrella to shut off glare when fishing. **V** Generally silent. **H** Lake margins, river edges, marshes, inundations, occasionally at seaside. **D** Sub-Saharan Africa.

9 SLATY EGRET *Egretta vinaceigula* 60cm Only black egret with all-yellow legs. Head plumes absent in non-breeding plumage. **V** Gives harsh *kraak* in flight. **H** Floodplains, shallow water with tall grass. **D** Zambia to NE Namibia, N Botswana and W Zimbabwe.

10 TRICOLORED HERON *Egretta tricolor* 63–68cm Forages in an active, dashing fashion. Flies with head retracted. **V** Nasal croak, usually given when disturbed. **H** Coastal marshes and mangroves, less common on freshwater lakes and marshes. Breeds in trees, often with other heron species. **D** SE USA to N South America.

11 LITTLE BLUE HERON *Egretta caerulea* 61–64cm Stealthy forager. Legs dark blue or greenish. **V** Various croaks at breeding sites. In flight, gives a harsh *gerr*. **H** Lakes, marshes, coastal pools, estuaries, inlets. **D** SE USA to S Brazil.

12 SNOWY EGRET *Egretta thula* 55–65cm Foraging actions much as Little Egret. Adults show more yellow on rear of tarsus compared to Little and Chinese egrets. **V** Rasping *graarr* or nasal *hraaa*. In flight, may utter a hoarse *charf*. **H** Freshwater swamps, rivers, saltwater lagoons. **D** Widespread across North, Central and South America.

13 LITTLE EGRET *Egretta garzetta* 55–65cm Facial skin and feet yellow or yellowish orange for a short period at onset of breeding. Loses head, back and breast plumes in non-breeding plumage. **V** Hoarse *aaah* or *kgarrk* when disturbed. **H** Lakes, rivers, marshland, coastal estuaries, saltpans, rice fields, occasionally dry grassland. **D** Europe, Africa, Asia, Australia, New Zealand.

14 WESTERN REEF HERON *Egretta gularis* 55–67cm Several forms with different plumages, from all white to dark grey. Habits much as Little Egret. Loses head, back and breast plumes in non-breeding plumage. **V** Guttural croak when feeding or alarmed, otherwise usually silent. **H** Rocky and sandy coasts, tidal mudflats, coastal lagoons and estuaries, less so inland. **D** W and NE Africa to W India and Sri Lanka.

15 DIMORPHIC EGRET *Egretta dimorpha* 60cm Dark and light forms known and rare intermediates with some darkish grey feathers. In flight, dark form may show some white patches at wrists like Western Reef Heron. Differs from Little Egret and Western Reef Heron by rather heavy, long all-black bill. **V** Harsh, low-pitched, Mallard-like quacking notes. **H** Mainly brackish and saltwater bodies. **D** E Africa, Madagascar.

16 PACIFIC REEF HERON *Egretta sacra* 58–66cm Breeding birds have a short, inconspicuous nape crest. Usual feeding action is lethargic, with a more rapid pursuit when potential prey is spotted. **V** Hoarse croak. When alarmed utters a harsh *arrk*. **H** Rocky shores, beaches, mudflats. **D** SE Asia to Australia, New Zealand and Pacific islands.

17 CHINESE EGRET *Egretta eulophotes* 65–68cm Non-breeding have yellowish-green facial skin and legs, and lose head and back plumes. Often feeds by dashing to and fro with wings held half-open and flapped. **V** Usually silent aside from low croaks. **H** Tidal mudflats, coastal bays, occasionally ponds and paddyfields. **D** E Asia.

HAMERKOP SCOPIDAE

18 HAMERKOP *Scopus umbretta* 48–56cm Unmistakable. Builds massive domed nest in trees or on cliffs. **V** Loud, yelping *yik-purrr yik-yik-yik-purr-purr-yik-yik*. In flight, utters a nasal *yip* or *kek*. **H** Wadis that have running water and are well vegetated. **D** Africa, Arabian Peninsula.

SHOEBILL BALAENICIPITIDAE

19 SHOEBILL *Balaeniceps rex* 150cm Unmistakable. **V** Usually quiet, may produce bill-rattling sounds. **H** Large swamps with high Papyrus (*Cyperus papyrus*) or reeds. **D** S Sudan, Uganda, and SW Ethiopia to SE DR Congo and N Zambia.

PELICANS PELECANIDAE

20 GREAT WHITE PELICAN *Pelecanus onocrotalus* 140–175cm Bill pouch pale yellow in non-breeding plumage. Often fishes cooperatively; forms a semi-circle to push fish into shallows, enabling each bird to scoop up a pouchful of fish. **V** In flight, may utter a deep croak. **H** Lakes, rivers, sheltered coastal waters. **D** E Europe to C Asia, sub-Saharan Africa.

21 PINK-BACKED PELICAN *Pelecanus rufescens* 125–132cm Non-breeding plumage duller, crest smaller and bill pouch pale yellow. Black loral spot diagnostic at all ages. **V** Gives various guttural calls at breeding colonies, otherwise generally silent. **H** Freshwater and saltwater lakes, mangroves, sheltered coastal waters. **D** Sub-Saharan Africa.

22 SPOT-BILLED PELICAN *Pelecanus philippensis* 140cm Non-breeding adult greyer, with paler bill and facial skin. Sociable, often feeding cooperatively like Great White Pelican. **V** At breeding sites utters squeaking, barking and bleating sounds. Fishing groups may utter a deep bleating noise. **H** Breeds colonially, preferring tall dead or bare trees within feeding range of large lakes, reservoirs and coastal waters. **D** India; Sri Lanka; Sumatra, Indonesia.

23 DALMATIAN PELICAN *Pelecanus crispus* 160–180cm In non-breeding plumage crest is smaller and bill pouch dull yellow. **V** Various barking, hissing and grunting noises at breeding colonies, otherwise fairly silent. **H** Breeds by shallow lakes and marshes with bordering vegetation. After breeding also on larger lakes and sheltered coastal waters. **D** SE Europe to China.

24 AUSTRALIAN PELICAN *Pelecanus conspicillatus* 150cm Gregarious, usually in small groups or large flocks. Often fishes in groups. Rests on sand bars, branches and man-made objects. Non-breeding birds have a paler pouch. **V** Generally silent. **H** Lakes, swamps, coastal lagoons, coral reefs. **D** Australia; New Guinea; Sumatra and Java, Indonesia; Fiji.

25 AMERICAN WHITE PELICAN *Pelecanus erythrorhynchos* 125–175cm Does not dive to catch food, instead dipping bill into water while swimming. **V** Generally silent. **H** Lakes, coastal bays and inlets. **D** SC Canada to C USA.

26 BROWN PELICAN *Pelecanus occidentalis* 105–150cm Feeds by plunge-diving, twisting with open wings before entering the water; can look quite dramatic. Gregarious, often seen flying in long lines, low over the water. **V** Generally silent. **H** Coastal bays, lagoons, occasionally inland lakes. Breeds on offshore cays. **D** North, Central and South America, Caribbean, Galápagos Islands.

27 PERUVIAN PELICAN *Pelecanus thagus* 125cm Huge yellowish-pink bill with reddish sides and tip. Pouch blackish with blue stripes. Yellow tuft on head and base of neck. **V** Gives harsh hisses and bill-snaps. **H** Coasts. **D** Peru, Chile.

1
3
5
6
dark morph
4
2
7
light morph
9
8
10
12
11
dark
morph
13
14
15
dark morph
light
morph
light
morph
dark morph
16
17
18
light
morph
19
21
23
20
22
26
24
25
27

FRIGATEBIRDS FREGATIDAE

1 ASCENSION FRIGATEBIRD *Fregata aquila* 89–96cm Male completely black, including upperwing and legs. Female has white-bellied and dark forms, both with black legs but red toes. **V** Generally silent at sea. **H** Open sea. **D** SE Atlantic Ocean.

2 CHRISTMAS FRIGATEBIRD *Fregata andrewsi* 90–100cm Actions and habits similar to Great Frigatebird. **V** Generally silent. **H** Maritime. **D** NE Indian Ocean.

3 MAGNIFICENT FRIGATEBIRD *Fregata magnificens* 95–110cm Scavenging actions similar to those of Great Frigatebird. Colonial breeder. Inflates gular pouch when displaying. **V** At breeding sites makes rattling and drumming sounds by vibrating and snapping bill. Generally silent at sea. **H** Breeds on rocky islets, usually in trees. Maritime. **D** Atlantic Ocean, Pacific Ocean.

4 GREAT FRIGATEBIRD *Fregata minor* 86–100cm During breeding season, male inflates red skin of throat. Pursues other seabirds, forcing them to regurgitate food. Scavenges around boats. **V** Braying, clapping and rattling calls at breeding sites, otherwise generally silent. **H** Mainly maritime. Nests on trees or flat ground. **D** Tropical oceans worldwide.

5 LESSER FRIGATEBIRD *Fregata ariel* 71–81cm Habits similar to those of Great Frigatebird. **V** Generally silent. **H** Maritime. **D** Tropical oceans worldwide.

GANNETS AND BOOBIES SULIDAE

6 NORTHERN GANNET *Morus bassanus* 87–100cm Feeds by plunge-diving vertically from as high as 30m. **V** At breeding sites and during communal feeding utters a harsh, grating *urrah*. **H** Breeds mainly on rocky islets, otherwise maritime. **D** N Atlantic Ocean.

7 CAPE GANNET *Morus capensis* 84–94cm Black tail and long gular stripe diagnostic. Actions as Northern Gannet. **V** Generally silent at sea. **H** Maritime. **D** S Africa coasts.

8 AUSTRALASIAN GANNET *Morus serrator* 84–91cm Reaches adult plumage in three years. Separated from Cape Gannet by white outer-tail feathers and shorter gular line; and from Masked Booby by facial pattern and white inner secondaries. **V** High, gull-like shrieking. **H** Normally inshore. Breeds on cliffs and small rocky islands. **D** Australia, New Zealand.

9 ABBOTT'S BOOBY *Papasula abbotti* 71cm Tends to have a more leisurely flight than other boobies, with slow flaps and languid glides. **V** Generally silent away from breeding grounds. **H** Pelagic post-breeding; breeds on trees. **D** NE Indian Ocean.

10 BLUE-FOOTED BOOBY *Sula nebouxii* 80cm Head and neck whitish, flecked dark brown. Back, wings and tail dark brown. Rump, uppertail coverts and underparts white, and white patch on upper back. Feet blue, bill and bare face bluish grey. **V** Gives harsh guttural croaks and rushing *shweeesh* calls close to nest. **H** Oceanic islands and sea, coasts. **D** EC Pacific Ocean.

11 PERUVIAN BOOBY *Sula variegata* 75cm Head, neck and underparts white. Back and tail dark brown, back feathers edged white. Feet blackish blue. **V** Generally silent at sea. **H** Coastal waters, offshore islands, coasts. **D** W coast of South America.

12 MASKED BOOBY *Sula dactylatra* 81–92cm Usually plunge-dives more vertically than other boobies. **V** Generally silent at sea. **H** Maritime. **D** Tropical oceans worldwide.

13 NAZCA BOOBY *Sula granti* 81–92cm Previously classed as a subspecies of Masked Booby. Separated from that species by range and orange-yellow bill (yellow in Masked). **V** Generally silent at sea. **H** Maritime. **D** EC Pacific Ocean.

14 RED-FOOTED BOOBY *Sula sula* 66–77cm Actions similar to those of Brown Booby. Variations on brown morph include a white-tailed form and a white-headed, white-tailed form. **V** Silent at sea; at nest sites utters a harsh squawk and guttural *ga-ga-ga-ga*. **H** Maritime away from breeding sites. **D** Tropical oceans worldwide.

15 BROWN BOOBY *Sula leucogaster* 64–74cm Feeding actions as Red-footed Booby. **V** Generally silent at sea. **H** Mainly maritime. May be found roosting in coastal trees or on buoys. **D** Tropical oceans worldwide.

CORMORANTS AND SHAGS PHALACROCORACIDAE

16 LITTLE PIED CORMORANT *Microcarbo melanoleucos* 55–65cm Usually occurs singly or in parties of up to 20 birds. Perches conspicuously, usually on waterside trees. In flight, outstretched neck shows a slight kink. **V** Generally silent; at breeding sites utter various cooing and clicking sounds. **H** Swamps, lakes, pools, rivers, coastal lagoons, mangroves, estuaries. **D** Australasia.

17 REED CORMORANT *Microcarbo africanus* 50–55cm Sociable. Non-breeding birds appear browner. **V** Hissing and cackling uttered at breeding colonies, otherwise silent. **H** Coastal islands, mangroves. **D** Widespread across Africa.

18 CROWNED CORMORANT *Microcarbo coronatus* 50cm Separated from Reed Cormorant by blacker upperparts year-round and slightly shorter tail. Solitary or in small groups. **V** Generally silent away from breeding colonies, where it cackles and hisses. **H** Only at sea coasts. **D** SW coast of Africa.

19 LITTLE CORMORANT *Microcarbo niger* 51cm In flight, outstretched neck shows a slight kink. Regularly sits with wings outstretched. **V** At breeding colonies utters various grunts, groans and roaring sounds, also a low-pitched *ah-ah-ah* and *kok-kok-kok*. **H** Various fresh and salt waters. **D** Widespread across Africa.

20 PYGMY CORMORANT *Microcarbo pygmeus* 45–55cm Sociable. Breeds colonially, often mixed with herons and egrets. **V** Grunts and croaks uttered at breeding colonies, otherwise mainly silent. **H** Breeds on lakes, marshes and rivers with dense vegetation, may visit brackish waters in winter. **D** SE Europe to Uzbekistan.

21 RED-LEGGED CORMORANT *Phalacrocorax gaimardi* 60cm Elegant. Grey, slightly paler below. Flight feathers and tail blackish. Bill yellow, foreface and feet bright red. **V** Pairs exchange various squeaking and groaning calls near nest. **H** Rias, tide rips, cliffs. **D** W coast of South America.

22 FLIGHTLESS CORMORANT *Phalacrocorax harrisi* 100cm Huge and hefty. Wings rudimentary. Blackish brown, lighter below. Iris green. **V** Dry, throaty rattles given near nest. **H** Seashores. **D** Galápagos Islands.

23 BANK CORMORANT *Phalacrocorax neglectus* 75cm Robust build. Pale eye in all-black face diagnostic. Feeds alone or in small groups. **V** Gives loud *wheee* call and makes foot-slapping sounds. **H** Maritime. **D** SW coast of Africa.

24 SPOTTED SHAG *Phalacrocorax punctatus* 69cm Separated from Pitt Shag by range and white stripe from eye down neck (faint in non-breeding plumage). **V** Low, groaning calls at breeding colonies. **H** Estuaries, inshore and offshore waters. **D** New Zealand.

25 PITT SHAG *Phalacrocorax featherstoni* 63cm Separated from Spotted Shag by solid black head and neck. **V** High, hoarse shrieks. **H** Inshore and offshore. **D** Chatham Islands.

26 BRANDT'S CORMORANT *Phalacrocorax penicillatus* 84–89cm Gregarious. In flight, outstretched neck shows slight kink. **V** Croaks and grunts at breeding sites, otherwise generally silent. **H** Coastal. Breeds on offshore rocks. **D** W coast of North America, from S British Columbia to California.

27 PELAGIC CORMORANT *Phalacrocorax pelagicus* 63–73cm Flies with neck held straight. **V** Silent away from breeding sites. **H** On rocky coasts; feeds and winters in coastal inlets, bays and open sea. **D** N Pacific Ocean coasts.

28 RED-FACED CORMORANT *Phalacrocorax urile* 79–89cm Flies with neck held straight. **V** Generally silent away from breeding colonies. **H** Exclusively marine. Breeds on rocky coasts. **D** N Pacific Ocean coasts.

29 NEOTROPIC CORMORANT *Phalacrocorax brasilianus* 63–69cm Tail longer than in Double-crested Cormorant, especially noticeable in flight. Often perches with wings open to dry feathers. **V** At breeding sites utters grunts and croaks, otherwise usually silent. **H** Inland and coastal waters. Generally found more often on fresh waters than Double-crested. Breeds colonially in trees and bushes near water. **D** S USA, Caribbean, Central and South America.

30 DOUBLE-CRESTED CORMORANT *Phalacrocorax auritus* 74–91cm In flight, outstretched neck shows a distinct kink. Gregarious. **V** Hoarse grunts at breeding sites, otherwise usually silent. **H** Lakes, rivers, coasts. **D** North and Central America.

1
♀
♂
2
♀
♂
3
♀
♂
4
♀
♂
5
♀
♂
6
7
8
9
♂
♀
10
11
12
13
14
light
morph
brown
morph
15
16
17
18
19
20
21
22
23
24
25
26
27
28
29
30

CORMORANTS AND SHAGS *CONTINUED*

1 EUROPEAN SHAG *Phalacrocorax aristotelis* 65–80cm Flies with neck held straight. **V** At breeding colonies utters various clicks and grunts, otherwise silent. **H** Breeds on rocky coasts. Occurs in other coastal areas, exceptionally inland freshwater sites, during non-breeding season. **D** W coasts of Europe and N Africa.

2 BLACK-FACED CORMORANT *Phalacrocorax fuscescens* 64–68cm Nests colonially. Often seen perched on rocks beside sea. **V** Various grunts and croaks heard at breeding colonies; otherwise silent. **H** Nests on coastal cliffs; otherwise in coastal waters. **D** S Australian coast.

3 INDIAN CORMORANT *Phalacrocorax fuscicollis* 63cm Flies with neck outstretched and slightly kinked. Regularly perches with wings outstretched. **V** At breeding colonies utters a sharp *kit-kit-kit-kit*. **H** Various fresh and salt waters. **D** India, SE Asia.

4 LITTLE BLACK CORMORANT *Phalacrocorax sulcirostris* 61cm Usually seen singly or in small or large flocks. Often perches conspicuously on dead trees near water. In flight, outstretched neck shows a slight kink. **V** Generally silent; at breeding sites utters various barks and croaks. **H** Lakes, ponds, coastal waters. **D** Australasia.

5 AUSTRALIAN PIED CORMORANT *Phalacrocorax varius* 65–85cm Large. Yellow bare skin in front of blue eyes diagnostic. **V** High, hoarse *ark ark*. **H** Subcoastal and inshore waters. Prefers sites with mangroves, some trees or high bush, but also with large rocks and cliffs. **D** Australia, New Zealand.

6 GREAT CORMORANT *Phalacrocorax carbo* 80–100cm In flight, outstretched neck shows slight kink. Often perches holding wings out to dry. **V** At breeding colonies utters various deep, guttural calls, otherwise generally silent. **H** Coastal cliffs, estuaries, inland lakes, reservoirs, rivers. **D** NE USA and E Canada to W Europe, C Europe to Japan, NW Africa, Australasia.

7 WHITE-BREASTED CORMORANT *Phalacrocorax lucidus* 90cm White throat and breast. Adult in breeding condition has white flank patches. **V** At breeding colonies utters various deep, guttural calls, otherwise generally silent. **H** Large lakes and rivers with open banks and shores; also marshes, inundations, coastal lagoons, estuaries. **D** Sub-Saharan Africa.

8 JAPANESE CORMORANT *Phalacrocorax capillatus* 92cm In flight, outstretched neck shows a slight kink. **V** Usually silent. **H** Mainly maritime, otherwise rocky cliffs. **D** NE Asia.

9 CAPE CORMORANT *Phalacrocorax capensis* 65cm Yellow gular pouch, short tail and absence of crest diagnostic. Large to very large flocks, which fly in long lines to fishing waters. **V** Deep, guttural rattle. **H** Coastal waters. Feeds at sea but may enter estuaries. **D** S coasts of Africa.

10 SOCOTRA CORMORANT *Phalacrocorax nigrogularis* 77–84cm Flies with neck held straight. Invariably found in large flocks. **V** Generally silent away from breeding colonies. **H** Coastal locations such as rocky islands, sandbanks and cliffs. **D** Persian Gulf.

11 ROCK SHAG *Phalacrocorax magellanicus* 65cm Head, neck and upperparts black, glossy in breeding plumage. Bare red face. Variably white ear patch, sometimes extending onto throat and foreneck. Lower breast and belly white. **V** Hoarse, doleful-sounding short rattle. **H** Rocky coasts and cliffs, coastal waters. **D** S coasts of South America.

12 GUANAY CORMORANT *Leucocarbo bougainvillii* 76cm Head, neck and upperparts black. Oval throat patch and underparts white. Face red, eye green. **V** At breeding colonies produces various croaking and barking sounds. **H** Coastal waters, coasts, offshore islands. **D** W coast of South America.

13 IMPERIAL SHAG *Leucocarbo atriceps* 75cm Nests in colonies. **V** Honks and hisses heard at breeding colonies; otherwise silent. **H** Breeds and roosts on coastal rock ledges and cliffs; feeds in inshore waters. **D** Coasts of S South America, Falkland Islands.

14 SOUTH GEORGIA SHAG *Leucocarbo georgianus* 75cm Best identified on basis of range. **V** Honks and hisses heard at breeding colonies; otherwise silent. **H** Breeds and roosts on coastal rock ledges and tussock grass; feeds in inshore waters. **D** South Georgia, South Sandwich and South Orkney islands, subantarctic Atlantic Ocean.

15 ANTARCTIC SHAG *Leucocarbo bransfieldensis* 75cm Unmistakable within geographical range. **V** Honks and hisses heard at breeding colonies; otherwise silent. **H** Breeds and roosts among boulders and on rock ledges; feeds in inshore waters. **D** South Shetland Islands, Antarctic Peninsula.

16 HEARD ISLAND SHAG *Leucocarbo nivalis* 75cm Gregarious. Has restricted and isolated geographical range. **V** Honks and hisses heard at breeding colonies; otherwise silent. **H** Breeds and roosts among boulders and tussock grass; feeds in inshore waters. **D** Heard Islands, subantarctic Indian Ocean.

17 CROZET SHAG *Leucocarbo melanogenis* 75cm Sedentary, with restricted geographical range. **V** Honks and hisses heard at breeding colonies; otherwise silent. **H** Breeds and roosts on vegetated, rocky cliffs and boulder beaches; feeds in inshore waters. **D** Prince Edward, Marion and Crozet islands, S Indian Ocean.

18 MACQUARIE SHAG *Leucocarbo purpurascens* 75cm Best diagnosed on basis of range. Note also facial pattern, especially of bare parts. **V** Very low croaks. **H** Inshore. **D** Occurs only on and near Macquarie Island, SW Pacific Ocean.

19 KERGUELEN SHAG *Leucocarbo verrucosus* 75cm Unmistakable within restricted range. Sedentary. **V** Honks and hisses heard at breeding colonies; otherwise silent. **H** Breeds and roosts on vegetated cliffs and among boulders; feeds in inshore waters. **D** Kerguelen Islands, subantarctic Indian Ocean.

20 NEW ZEALAND KING SHAG *Leucocarbo carunculatus* 76cm Best diagnosed on basis of range. Note also facial pattern, especially of bare parts. **V** Low grunts and high cries. **H** Islands off coast. **D** N South Island, New Zealand.

21 OTAGO SHAG *Leucocarbo chalconotus* 68cm Polymorphic, pied and bronze morphs shown. Best diagnosed on basis of range. Note also facial pattern, especially of bare parts. Shows a white dorsal patch. **V** High, shivering cries. **H** Islands off coast. **D** E South Island, New Zealand.

22 FOVEAUX SHAG *Leucocarbo stewarti* 68cm Split from Otago Shag in 2016. Also polymorphic, with pied and bronze morphs shown. Separated from Otago by more restricted range and lack of orange caruncles above bill base. **V** High, shivering cries. **H** Islands off coast. **D** S South Island, Stewart Island/Rakiura and Foveaux Strait, New Zealand.

23 CHATHAM SHAG *Leucocarbo onslowi* 63cm Best diagnosed on basis of range. Note also facial pattern, especially of bare parts. Shows a white dorsal patch. **V** Low, hollow grunts. **H** Inshore. **D** Chatham Islands, SW Pacific Ocean.

24 CAMPBELL SHAG *Leucocarbo campbelli* 63cm Best diagnosed on basis of range. Note also facial pattern, especially of bare parts, and black foreneck. **V** Barking notes given during courtship. **H** Inshore and offshore. **D** Campbell Island, SW Pacific Ocean.

25 AUCKLAND SHAG *Leucocarbo colensoi* 63cm Best diagnosed on basis of range. Note facial pattern, especially of bare parts. Black may be closed at foreneck or open. Shows a white dorsal patch. **V** Very low, hollow grunts. **H** Inshore and offshore. **D** Auckland Islands, SW Pacific Ocean.

26 BOUNTY SHAG *Leucocarbo ranfurlyi* 71cm Best diagnosed on basis of range. Note also facial pattern, especially of bare parts. **V** Similar to that of other species in its genus. **H** Inshore. **D** Bounty Islands, SW Pacific Ocean.

ANHINGAS AND DARTERS ANHINGIDAE

27 ORIENTAL DARTER *Anhinga melanogaster* 85–97cm In flight, holds neck in a distinct kink. Often sits on exposed perch with wings outstretched, and swims with only head and neck visible. **V** Breeding birds utter a loud *chigi chigi chigi* and various grunts and croaks, otherwise silent. **H** Inland and coastal waters, including lakes, marshes and mangroves. **D** Pakistan to SE Asia.

28 AFRICAN DARTER *Anhinga rufa* 85–97cm In flight, holds neck with a distinct kink. **V** Usually silent. Various rattles and grunts given during nesting. **H** Lakes, rivers and marshes, where trees and stumps provide perches. **D** Sub-Saharan Africa, Madagascar, Iraq, Iran.

29 AUSTRALASIAN DARTER *Anhinga novaehollandiae* 85–95cm Often perches and 'sunbathes' beside water with wings outstretched. When swimming, typically only head and snake-like neck are visible above water. **V** Usually silent. **H** Lakes, slow-flowing rivers, coastal lagoons. **D** New Guinea, Australia.

30 ANHINGA *Anhinga anhinga* 81–91cm Breeds colonially, often with cormorants, herons and ibises. Often swims with only neck protruding from water. Regularly perches, cormorant-like, with wings open to dry feathers. **V** Distinct, rapid series of clicking notes and guttural grunts. **H** Still, shallow waters, including lakes, estuaries and sheltered coastal areas. Nests in trees or bushes, usually above or near water. **D** SC and SE USA to N Argentina.

1
2
3
4
5
6
7
8
9
10
11
12
13
14
15
16
17
18
19
bronze
morph
22
pied
morph
23
20
bronze
morph
21
pied
morph
24
25
26
♀
28
♂
27
29
♂
♀
30
♀
♂

NEW WORLD VULTURES CATHARTIDAE

1 TURKEY VULTURE *Cathartes aura* 64–81cm Distinguished by bald red or pink head and neck. **V** At breeding sites utters various hissing and clucking sounds, otherwise generally silent. **H** Open areas and around human habitation and rubbish dumps. Nests in crevices, on cliff ledges and on the ground under thick vegetation. **D** North, Central and South America.

2 LESSER YELLOW-HEADED VULTURE *Cathartes burrovianus* 64cm As Turkey Vulture, but with yellow head and, typically, white shafts visible in primaries on upperwing. Slender. Dull brownish black. Flight wobbly, low and usually over grass, wings held above the horizontal. **V** May give hissing notes at nest, otherwise silent. **H** Marsh, grassland, river margins in woodland. **D** E Mexico to N Argentina.

3 GREATER YELLOW-HEADED VULTURE *Cathartes melambrotus* 78cm Like Lesser Yellow-Headed Vulture but bulkier. Deep black. Flight less wobbly, wings held horizontally. **V** May give hissing notes at nest, otherwise silent. **H** Only found in primary lowland tropical forests. **D** Amazonia.

4 BLACK VULTURE *Coragyps atratus* 56–68cm Separated from Turkey Vulture and Lesser Yellow-headed Vulture by six dark-tipped white primaries. Note short tail. Soars with flattened wings. **V** Guttural, almost dog-like calls, and various 'reptilian' hissing and wheezing calls. **H** Any place except true forest, including towns and rubbish dumps. **D** North, Central and South America.

5 KING VULTURE *Sarcoramphus papa* 80cm White, washed cream on upper back. Flight feathers and tail black. Head orange, black, grey and cream. Iris white. **V** May produce hisses and croaks, and bill-snaps during courtship. **H** Forest, woods, savannah. **D** Mexico to N Argentina.

6 CALIFORNIA CONDOR *Gymnogyps californianus* 117–134cm Soars with wings held level or slightly raised. **V** Generally silent, although may utter grunts and hisses. **H** Arid and sparsely wooded foothills and mountains. **D** C California and C Arizona.

7 ANDEAN CONDOR *Vultur gryphus* 120cm Huge and imposing. Head bare. Male with comb on forehead. White ruff on neck. Long, broad wings, primaries splayed. Upperwing coverts white. **V** Usually quiet but may give low, grunting rattle. **H** Mountains, down to sea coasts in some areas. **D** Venezuela to S Chile and Argentina.

SECRETARYBIRD SAGITTARIIDAE

8 SECRETARYBIRD *Sagittarius serpentarius* 90cm Extremely long legs, long tail and small head with 'secretary pens' diagnostic. **V** Deep, dry, rolling grunt. **H** Dry open, bushed and wooded, natural or cultivated areas. **D** Widespread across Africa.

OSPREYS PANDIONIDAE

9 WESTERN OSPREY *Pandion haliaetus* 56–61cm Head-on flight gull-like, with wings bowed. Feeds on fish, soaring over water before plunging feet first onto prey. **V** Series of mournful whistles; also a shrill, whistled *teeeeaa*. **H** Lakes, rivers, coastal lagoons, estuaries. **D** Worldwide except Australasia and Sulawesi, Indonesia.

10 EASTERN OSPREY *Pandion cristatus* 55–63cm Feeds on fish, hovering and then grabbing prey in feet-first plunge. **V** When alarmed gives a hoarse, sharp *kew-kew-kew-kew*. **H** Lakes, rivers, coastal lagoons, estuaries. **D** Australia, New Guinea, Indonesia, Philippines.

KITES, HAWKS AND EAGLES ACCIPITRIDAE

11 BLACK-WINGED KITE *Elanus caeruleus* 31–35cm Often hovers. In flight, shows large black patch on upperwing coverts. **V** Various, including a harsh *w-eeyah*, sharp *kree-ak* and piping *pii-uu*. **H** Grassland, semi-desert, open plains with scattered trees, forest fringes. **D** SW Iberian Peninsula to E China, SE Asia and New Guinea, Africa.

12 BLACK-SHOULDERED KITE *Elanus axillaris* 33–35cm Uses wind direction to remain stationary in air; sometimes hovers. Usually solitary when feeding but will roost communally. **V** Occasional high-pitched whistle; otherwise silent. **H** Open habitats, grassland, savanna. **D** Australia.

13 WHITE-TAILED KITE *Elanus leucurus* 38–43cm Hunts by hovering and then dropping to catch prey – usually small rodents. **V** High, rising whistle followed by a dry *sweeekrrkrr* note. When disturbed, utters a low, grating *karrrr*. **H** Parkland, fields and grassland with trees. **D** S and W USA to C Argentina.

14 LETTER-WINGED KITE *Elanus scriptus* 34–36cm Similar to Black-shouldered Kite but note black eye-surround and extensive black band on underwing. Irruptive; breeding associated with rodent 'plagues'. **V** Loud, shrill chatters and whistles. **H** Lightly wooded grassland; nests in woodland. **D** Australia.

15 PEARL KITE *Gampsonyx swainsonii* male 20cm, female 25cm Like a tiny falcon. Black and white with cinnamon-buff forehead and cheeks. Thighs cinnamon, flanks white or cinnamon according to race. **V** Series of high-pitched, twittering or warbling notes. **H** Open woods, savannah, palm groves, open land with trees, dry woods. **D** Nicaragua, N South America.

16 SCISSOR-TAILED KITE *Chelictinia riocourii* 35cm Tern-like in flight, often hovering. **V** High-pitched whinnying. **H** Semi-desert and open, bushy and wooded habitat with some grass cover. **D** Senegal and Gambia to Somalia and NE Kenya.

17 AFRICAN HARRIER-HAWK *Polyboroides typus* 60cm Raids weaver nests and extracts nestlings from tree holes with specially adapted, very flexible legs. **V** Very high, plaintive *phuweeee*. **H** Forest edges, moist wooded areas, suburbs. **D** Widespread across Africa.

18 MADAGASCAN HARRIER-HAWK *Polyboroides radiatus* 70cm No similar bird within its range. Separated from slightly darker African Harrier-hawk by broader black band along wings. **V** Sharp, short, descending *fiuuh* whistle. **H** Any habitat at any altitude with trees, from forest to open woodland and gardens. **D** Madagascar.

19 PALM-NUT VULTURE *Gypohierax angolensis* 60cm Distinctive round-winged, short-tailed flight silhouette. Note bare pink skin around eye. **V** Very high, drawn-out, whistled *feeeeeh*, often rising in pitch. **H** Open forest and plantations with oil palms (the fruits of which it eats). **D** Senegal and Gambia to South Sudan and E Kenya, south to Angola and NE South Africa.

20 BEARDED VULTURE *Gypaetus barbatus* 100–115cm Spends much time soaring, sometimes at great heights (recorded up to 8,000m). **V** Generally silent apart from a shrill *feeeee* uttered during aerial display. **H** Mountains with sheer crags. Hunts over plains, slopes and valleys. During hard winters resorts to lower areas. **D** S Europe to NE China, N, E and S Africa.

21 EGYPTIAN VULTURE *Neophron percnopterus* 60–70cm White with black flight feathers, although may be stained rusty brown by soil. **V** Usually silent. **H** Mountains, open arid areas, towns, villages, rubbish dumps. **D** W Africa to India, S Spain to C Asia.

22 MADAGASCAN SERPENT EAGLE *Eutriorchis astur* 65cm Very similar to Rufous-breasted Sparrowhawk (Plate 34), but has a 'less fierce' expression and plainer upperparts. **H** Forest interiors. **D** Madagascar.

23 GREY-HEADED KITE *Leptodon cayanensis* 50cm Short, rounded wings. Black underwing coverts and barred flight feathers. **V** Rising, rapid *wekwek* crescendo. **H** Forest and adjacent areas, mangroves. **D** EC Mexico to N Argentina, Caribbean

24 WHITE-COLLARED KITE *Leptodon forbesi* male 49cm, female 53cm Crown pale grey. Head, neck and underparts white. Upperparts blackish. Tail black with very broad whitish band. Underwing white. **V** Fast series of clucking *weka-weka-weka* notes, accelerating. **H** Atlantic rainforest. Very rare and endangered. **D** NE Brazil.

25 HOOK-BILLED KITE *Chondrohierax uncinatus* 38–43cm Loose, buoyant flight. Clambers in trees in parrot-like fashion searching for tree snails. **V** Rattling *kekekekekekeke* and quiet *huey*. **H** Dense, brushy woods. **D** S Texas to N Argentina.

26 CUBAN KITE *Chondrohierax wilsonii* 38–43cm Critically Endangered. Feeds on tree snails and slugs using its hooked bill. **V** Rattling *kekekekekekekeke*. **H** Riverside forests. **D** E Cuba.

27 EUROPEAN HONEY BUZZARD *Pernis apivorus* 52–60cm Very variable. Follows wasps to their nests to remove larvae and wax. Often occurs in large numbers on migration, especially around the Straits of Gibraltar and Bosporus. **V** Clear, melancholy *whee-oo* or *whi-whee-oo*. **H** Forests with open areas; more open areas on migration. **D** Europe and W Asia.

28 CRESTED HONEY BUZZARD *Pernis ptilorhynchus* 52–68cm Plumage very variable, especially underparts. Often seen soaring over forests, flying with deep, steady wingbeats interspersed with glides. **V** High-pitched *wee-wey-uho* or *weehey-weehey*, and a shrill, whistled *wheyeeee*. **H** Forested lowlands and hills. **D** E Europe and across Asia.

29 BARRED HONEY BUZZARD *Pernis celebensis* 50–58cm Usually seen singly, in pairs or occasionally in small groups; regularly soars high over forests, partly cleared ridges and valleys. **V** Sharp *whit-weee-oooo* or *weee-oooo*. **H** Forests, forest edges, partially cleared areas. **D** Sulawesi, Indonesia.

30 PHILIPPINE HONEY BUZZARD *Pernis steerei* 51–57cm Secretive; tends to 'hide' in dense foliage, although regularly soars high over forests. **V** Loud screaming or ringing calls. **H** Lowland to submontane forest and forest edges. **D** Philippines.

1
2
3
4
5
6
7
♂
♀
8
9
10
11
12
13
14
15
16
17
18
19
20
21
22
23
24
25
♂
♀
26
♀
♂
27
dark morph
28
29
30

KITES, HAWKS AND EAGLES *CONTINUED*

1 SWALLOW-TAILED KITE *Elanoides forficatus* 56–66cm Agile, graceful flier. **V** *Peat-peat-peat, klee-klee-klee* or soft whistling notes. **H** Damp woodlands, thickets, marshland. **D** S USA to NE Argentina.

2 SQUARE-TAILED KITE *Lophoictinia isura* 50–55cm Soars effortlessly, scanning ground for prey including invertebrates and small vertebrates. **V** Yelping calls near nest; otherwise mostly silent. **H** Open forest, lightly wooded grassland and farmland, scrub. **D** Australia.

3 BLACK-BREASTED BUZZARD *Hamirostra melanosternon* 50–60cm In flight shows long, parallel-sided wings and short tail. Hunting methods include quartering ground, speedy ambush flight and searching on foot. **V** Usually silent. **H** Dry woodland, savannah, grassy plains. **D** Australia.

4 AFRICAN CUCKOO-HAWK *Aviceda cuculoides* 40cm Note crest. Perches conspicuously and flies from tree to tree in a kite-like fashion. **V** Very high, gliding-down *pieuuh* and staccato *pi-pi-peh*, with *fieuuh* from a pair. **H** Forest edges, woodland, riverine belts, cultivation. **D** W, C, SC and SE Africa.

5 MADAGASCAN CUCKOO-HAWK *Aviceda madagascariensis* 40cm Note rounded head, short legs and greyish tail with three bars. In flight, shows strikingly and distinctively barred underwings. Perches conspicuously and flies from tree to tree in kite-like fashion. **V** Fast, yelping and staccato *pyeh-pyeh-pyeh*. **H** Forest, woodland, plantations, parks at any altitude. **D** Madagascar.

6 JERDON'S BAZA *Aviceda jerdoni* 41–48cm Hunts from a perch, making short sorties to grab prey – usually insects, lizards or frogs. **V** Plaintive *pee-ow*. During display utters an excited mewing *kip-kip-kip* or *kikiya kikiya*. **H** Tropical and subtropical broadleaved evergreen forests. **D** Widespread across Asia.

7 PACIFIC BAZA *Aviceda subcristata* 35–46cm Usually seen singly or in pairs, sometimes in groups. Often perches conspicuously in the tops of trees. Snatches insects and lizards from foliage. **V** Breezy, high-pitched *whee-chu*, often repeated and uttered more rapidly when excited. **H** Primary and tall secondary lowland and hill forest, forest edges, lightly wooded cultivation. **D** Widespread across Australasia.

8 BLACK BAZA *Aviceda leuphotes* 30–35cm Flies much like a crow, interspersed with level-winged glides. **V** Soft, quavering squeal or whistle, *tcheeoua*. **H** Deciduous or evergreen tropical forest clearings or wide forest streams. **D** Widespread across Asia.

9 LONG-TAILED HONEY BUZZARD *Henicopernis longicauda* 50–60cm Proportionately very long tail obvious in flight. Feeds on wasps and bees, also other invertebrates, lizards and small birds. **V** Loud, shrieking call. **H** Rainforest, lowlands to upland. **D** New Guinea.

10 BLACK HONEY BUZZARD *Henicopernis infuscatus* 48–50cm Recognised in flight by dark plumage and noticeably broad outer wings. Feeds on invertebrates and small vertebrates such as lizards. **V** Various piping calls. **H** Rainforest. **D** New Britain, Papua New Guinea.

11 HOODED VULTURE *Necrosyrtes monachus* 62–72cm Note thin bill and dark underparts. **V** Usually silent. **H** Savannah, woodland, towns and villages, also often around rubbish dumps. Rare outside main game reserves. **D** Widespread across Africa.

12 WHITE-BACKED VULTURE *Gyps africanus* 95cm Separated from Griffon Vulture by white underwing coverts and from Cape Vulture by darker, greyish-brown (not tawny-brown) colouring and blackish bare skin. First row of upperwing coverts only narrowly edged pale. In flight, white rump patch is diagnostic. **H** Open bushy and wooded habitats. **D** Widespread across Africa.

13 WHITE-RUMPED VULTURE *Gyps bengalensis* 75–85cm In flight from above, shows white rump and lower back. **V** When joining other birds at roost, or carcasses, gives a strident, creaky *kakakaka*. **H** Open country, towns and villages. **D** Iran through SE Asia.

14 INDIAN VULTURE *Gyps indicus* 89–103cm Separated from Slender-billed Vulture by whitish rump, paler undertail coverts and darker thighs. **V** Generally silent. **H** Cities, towns, villages, open wooded areas, rubbish dumps. **D** SE Pakistan, S India.

15 SLENDER-BILLED VULTURE *Gyps tenuirostris* 93–100cm Separated from Indian Vulture by slightly darker wings and mantle, and greyer rump. Pale thighs often show well in flight. **V** Generally silent. **H** Dry open country, open forested areas, rubbish dumps. **D** Himalayas through SE Asia.

16 RÜPPELL'S VULTURE *Gyps rueppelli* 85–95cm Distinctively scaled all over. **V** Various hisses, grunts, groans and guttural rattles, usually given during disputes over carcasses. **H** Mountains, rocky outcrops, hills, savannah, grassland. Prefers more arid areas than White-backed Vulture. **D** E, C and W Africa.

17 HIMALAYAN VULTURE *Gyps himalayensis* 120cm The largest *Gyps* species. Underwing shows pale lines on coverts. **V** Occasionally utters whistling and clucking noises. **H** High-altitude mountain areas. **D** Himalayas.

18 GRIFFON VULTURE *Gyps fulvus* 95–105cm The only member of the genus to occur in Europe. **V** At roosts or carcasses, gives various hisses and grunts. **H** Mountains and neighbouring grasslands. **D** S Europe and N Africa to C Asia and N India.

19 CAPE VULTURE *Gyps coprotheres* 115cm Note very large size and pale brown appearance. Separated from White-backed Vulture by pale eyes, pale blue bare skin and dark centres of first row of upperwing coverts. Unlikely to occur north of the equator. **H** Cliffs in otherwise flat country. **D** Namibia to Zimbabwe, Mozambique and S South Africa.

20 RED-HEADED VULTURE *Sarcogyps calvus* 85cm Unmistakable owing to prominent red head. **V** Hoarse croaks and screams usually given during disputes over carcasses. **H** Open country with nearby habitation and wooded hills. **D** Widespread across Asia.

21 WHITE-HEADED VULTURE *Trigonoceps occipitalis* 80cm Separated from Lappet-faced Vulture by much smaller size and pink bill and legs. **V** Usually silent but may give twittering calls while defending its space at a carcass. **H** Open bushy and wooded habitats. Rare outside main game reserves. **D** Widespread across Africa.

22 CINEREOUS VULTURE *Aegypius monachus* 100–110cm Generally found singly but may congregate in groups around carcasses. **V** Usually silent. **H** Forest areas on hills and mountains, semi-arid alpine meadows and grassland. **D** S Europe to C Asia, Pakistan and NW India. Non-breeding season in N Africa, India, China and SE Asia.

23 LAPPET-FACED VULTURE *Torgos tracheliotos* 95–105cm All dark above, with white thigh feathers. Bare head pink or pink and grey. **V** Usually silent, although may utter a growling *churr* and various metallic notes. **H** Semi-desert, savannah, grassland with scattered acacias. **D** Widespread across Africa, Arabian Peninsula.

24 CRESTED SERPENT EAGLE *Spilornis cheela* 50cm Soars above forest canopy. **V** In flight, utters a shrill *kwee-kwee kwee-kwee kwee-kwee-kwee*. **H** Open hill forests and plantations. **D** Widespread across Asia.

25 GREAT NICOBAR SERPENT EAGLE *Spilornis klossi* 38–42cm Encountered mostly in forest canopy. **V** Not described. **H** Mixed evergreen forest, grassland, areas of forest regeneration. **D** Great Nicobar, Little Nicobar and Menchal, Nicobar Islands.

26 MOUNTAIN SERPENT EAGLE *Spilornis kinabaluensis* 55–58cm Hunts along mountain ridges and dry riverbeds. Formerly considered a race of the previous species. **V** Long, thin scream and a drawn-out *kiillii*. **H** Stunted ridgetop montane forest. **D** Borneo.

27 SULAWESI SERPENT EAGLE *Spilornis rufipectus* 46–54cm Generally seen singly or in pairs. Noisy and conspicuous, often perching on exposed treetop branches. Regularly flies over forest canopy. **V** Variable, far-carrying, high-pitched whistle, occasionally ending in a weak chatter. **H** Primary lowland, hill and montane forest, tall secondary scrub woodland, forest edges, lightly wooded cultivation and adjacent grassland. **D** Sulawesi, Indonesia.

28 PHILIPPINE SERPENT EAGLE *Spilornis holospilus* 47–53cm Often perches at the forest edge. Soars high over forests and forest edges, usually calling. Hunts in the canopy. **V** Whistled *seee-ap weep weep*, which may also be interspersed with a *aahhe-e-a reep*. **H** Forest and forest edges. **D** Philippines.

29 ANDAMAN SERPENT EAGLE *Spilornis elgini* 51–59cm Dark chocolate brown with white ocelli. **V** Clear *kweep-kweep-kweep*. **H** Inland forests, preferring clearings or areas with scattered trees. **D** Andaman Islands.

30 PHILIPPINE EAGLE *Pithecophaga jefferyi* 90–100cm Occurs singly or in pairs. Frequently soars above forests on the lookout for small to medium animals. **V** Loud, plaintive, whistled *waaaauu waaaauu waaaauu* or *waa-leee-ahhh*. **H** Forest, logged forest and forest edges, from lowlands to mountains. **D** Philippines.

31 SHORT-TOED SNAKE EAGLE *Circaetus gallicus* 62–67cm Plumage variable. Head and upper breast pale grey-white to blackish, underparts virtually plain to barred black and white in the darker-headed forms. Often hovers when searching for prey. **V** Plaintive *weeo* or *weeooo*, also a gull-like *woh-woh-woh*. **H** Varied, including open country, scrub and semi-desert. **D** SW Europe to C Asia, NW China and India; Lesser Sunda Islands, Indonesia. Non-breeding season in E, C and W Africa and S and SE Asia

32 BEAUDOUIN'S SNAKE EAGLE *Circaetus beaudouini* 65cm Very similar to Short-toed Snake Eagle but note narrow, dense barring below and more distinct barring of tail and flight feathers. **H** Prefers more wooded habitats than other snake eagles. **D** Senegal to South Sudan, NW Kenya and Uganda.

33 BLACK-CHESTED SNAKE EAGLE *Circaetus pectoralis* 65cm Note grey cere and pale legs. Perches conspicuously. May hover. **V** High (or low), sometimes accelerated, whistled *fiuu-fiuu-fiuu*. **H** Dry, stony, lightly wooded plains and hillsides, often near water. **D** E Sudan and Ethiopia to South Africa.

34 BROWN SNAKE EAGLE *Circaetus cinereus* 70cm Note large size, broad face with large yellow eyes, and erect stance on top of tree or post. Like all snake eagles, shows broad wings in flight, sharply bent at wrists. **V** Very high, sharp, piercing *tjark tjark tjark* yelps. **H** Dense thornbush and woodland with some large trees. **D** Senegal and Gambia to Ethiopia and south to South Africa.

35 SOUTHERN BANDED SNAKE EAGLE *Circaetus fasciolatus* 55cm Note distinction between plain brown breast and prominent barring on belly. Underwings more densely barred and striped than in Western Banded Snake Eagle. **V** Mid-high, bouncing, rapid *ko-ko... koah* and high *kah-kah-kah-kah-kah-kah*. **H** Coastal forests. **D** Kenya to NE South Africa.

36 WESTERN BANDED SNAKE EAGLE *Circaetus cinerascens* 55cm Tail pattern diagnostic. Barring on underparts not very pronounced. Cere and gape yellow. Normally perches in (not on) dead trees or those with sparse foliage. **V** Mid-high, toy trumpet-like *uh kruruh* and high, descending *ko-ko-ko-ko-ko-koh*. **H** Riverine belts and forest patches in open country. **D** Senegal and Gambia to W Ethiopia and south to Namibia and Zimbabwe.

37 CONGO SERPENT EAGLE *Circaetus spectabilis* 50cm Note long tail, short wings, black throat and large eye. **V** Gull-like mewing calls. **H** Middle and ground strata of dense forest. **D** W and C Africa.

38 BATELEUR *Terathopius ecaudatus* 60cm Often dries feathers by holding wings straight out at sides and tipped vertically, following the sun. **V** Loud *yaaaow* and downslurred *wee weeye weeye weeye*. **H** Open savannah, thornbush grassland, open woodland. **D** Widespread across Africa.

39 BAT HAWK *Macheiramphus alcinus* 41–51cm Crepuscular or nocturnal hunter of bats. Flight is rapid, on shallow, stiff wingbeats. **V** High, yelping *kwik kwik kwik kwik*. **H** Open areas near broadleaved evergreen forest, near bat caves and around human habitation. **D** Widespread across Africa, and from Myanmar to Borneo and New Guinea.

1
2
3
4
5
6
♂
♀
7
8
9
10
11
12
13
14
15
16
17
18
19
20
21
♂
♀
22
23
24
25
26
27
28
29
30
31
32
33
34
35
36
37
38
♀
♂
39

KITES, HAWKS AND EAGLES *CONTINUED*

1 PAPUAN EAGLE *Harpyopsis novaeguineae* 75–90cm Hunts from perch, scanning ground below for mammals, birds and reptiles. Also disturbs prey in tree canopy by active pursuit. **V** Loud, abrupt, coughing *cowk-cowk...* **H** Rainforest, lowlands to uplands. **D** New Guinea.

2 CRESTED EAGLE *Morphnus guianensis* male 80cm, female 90cm Like Harpy Eagle but less massive. Dark and light morphs. Head and breast grey. Single crest. Upperparts blackish, marbled and irregularly barred greyish. Underparts white, faintly and finely barred tawny. In flight, very broad rounded wings. White axillaries. **V** Call a thin, drawn-out, slightly upslurred whistle. **H** Lowland tropical and subtropical forest, gallery forest. **D** Guatemala to NE Argentina.

3 HARPY EAGLE *Harpia harpyja* 86–107cm Large and very powerful looking, with huge claws. Black breast band of adult diagnostic. **V** Very high, stressed *tjiuw*. **H** Forest. **D** S Mexico to NE Argentina.

4 CHANGEABLE HAWK-EAGLE *Nisaetus cirrhatus* 58–77cm Uses a concealed forest perch from which it makes a short dash to capture prey. **V** Series of shrill whistles. **H** Broadleaved evergreen and deciduous forest. **D** Himalayas, C and S India and Sri Lanka to Philippines and Greater Sunda Islands, Indonesia.

5 FLORES HAWK-EAGLE *Nisaetus floris* 75–79cm Formerly considered a race of Changeable Hawk-eagle; actions presumed to be similar. **V** Unrecorded. **H** Submontane forest, lowland rainforest and nearby cultivation. **D** Lesser Sunda Islands, Indonesia.

6 MOUNTAIN HAWK-EAGLE *Nisaetus nipalensis* 67–86cm May soar above forest canopy but more often concealed among foliage. **V** Shrill, whistled *tlueet-weet-weet* and repeated *kee-kikik*. **H** Mountain and hill forests. **D** Himalayas to Taiwan and Japan.

7 LEGGE'S HAWK-EAGLE *Nisaetus kelaarti* 67–86cm Separated from Mountain Hawk-eagle in 2008 and best distinguished from that species by range. **V** Shrill, whistled *tlueet-weet weet* and repeated *kee-kikik*. **H** Mountain and hill forests. **D** SW India, Sri Lanka.

8 BLYTH'S HAWK-EAGLE *Nisaetus alboniger* 50–58cm Soars with wings held level. Hunts in the upper storey with an agile accipiter-like flight through trees. **V** Screaming *yhu yhu yip-yip-yip*, shrill *pik-wuee* and fast *wiii-hi*. **H** Broadleaved evergreen upland forest. **D** Malay Peninsula; Sumatra, Indonesia; Borneo.

9 JAVAN HAWK-EAGLE *Nisaetus bartelsi* 60cm Actions presumed similar to Changeable Hawk-eagle; preys on large birds, small mammals and lizards. **V** Harsh, high-pitched cries. **H** Tropical hill and mountain forest, and open wooded areas. **D** Java, Indonesia.

10 SULAWESI HAWK-EAGLE *Nisaetus lanceolatus* 55–64cm Usually seen singly. Perches in the tree canopy, sometimes conspicuously. **V** Descending *kluuu-kluuu-kluuu-kluuu-kluuu-kluuu*, preceded by an upward-inflected disyllabic note; also a rapid *kee-kee-kee-kee*. **H** Primary and tall secondary lowland, hill and montane forest, forest edges and nearby open country. **D** Sulawesi, Indonesia.

11 PHILIPPINE HAWK-EAGLE *Nisaetus philippensis* 56–67cm Often stays concealed in the canopy. Little recorded information; presumed to be similar to others of genus. **V** Screamed *wheeet whit* or *du-wheeet whit*. **H** Forest and advanced secondary growth, from lowlands to mountains. **D** Luzon, Philippines.

12 PINSKER'S HAWK-EAGLE *Nisaetus pinskeri* 54–61cm Formerly considered a race of Philippine Hawk-Eagle; actions presumed similar. **V** *Wheee-whit*, similar to previous species. **H** Mature forest, from lowland to low mountains. **D** Philippines except Luzon and Palawan.

13 WALLACE'S HAWK-EAGLE *Nisaetus nanus* 45–49cm Actions and habits probably much as Blyth's Hawk-eagle. **V** Shrill, high-pitched *yik-yee* or *kliit-kleeik*. **H** Broadleaved evergreen forest. **D** Malay Peninsula; Sumatra, Indonesia; Borneo.

14 BLACK HAWK-EAGLE *Spizaetus tyrannus* 58–70cm Separated from very rare dark form of Black-and-white Hawk-eagle by far less sharp demarcation between black breast and belly barring. Note also black underwing coverts. **V** Very high *pup-puphieuuw*. **H** Forest edges and clearings. **D** C Mexico to NE Argentina.

15 BLACK-AND-WHITE HAWK-EAGLE *Spizaetus melanoleucus* 50–60cm Unmistakable in its habitat, but see Semiplumbeous Hawk (Plate 36), which has a more restricted range. More *Buteo*-like than other eagles with, in flight, white leading edge to wings. **V** Whistled but sometimes slightly throaty *fweeee-uh*. **H** Forest edges and clearings. **D** S Mexico to NE Argentina.

16 ORNATE HAWK-EAGLE *Spizaetus ornatus* 56–68cm Bold colours and markings make it unmistakable. **V** Very high, fluted staccato *weet weet weetjr* or *pwu-pwu-pwipwipwipeh*. **H** Forest. **D** SE Mexico to N Argentina.

17 BLACK-AND-CHESTNUT EAGLE *Spizaetus isidori* male 70cm, female 85cm Very large. Head and upperparts brownish black. Underparts to feathered tarsi all chestnut, finely streaked black. Long tail whitish with wide black subterminal band. Huge wing. **V** High-pitched, yelping whistles. **H** Humid, undisturbed montane forest. **D** Venezuela to NW Argentina.

18 CROWNED EAGLE *Stephanoaetus coronatus* 80cm Note orange gape, large yellow toes with enormous claws, short wings, long tail. **V** Very high, excited *puweepuweepuwee*, undulating in pitch and volume, synchronised with up-and-down display flight. **H** Forest, wooded areas. **D** E South Sudan and W Ethiopia; Sierra Leone to S Kenya, E South Africa and N Angola.

19 RUFOUS-BELLIED EAGLE *Lophotriorchis kienerii* 53–61cm Soars high above forest canopy, from where it makes spectacular falcon-like stoops to capture prey. **V** Piercing scream. **H** Evergreen and moist deciduous forest. **D** S and SE Asia; Sulawesi, Indonesia.

20 MARTIAL EAGLE *Polemaetus bellicosus* 80cm One of the larger eagles. Fine black spotting on underparts and fully feathered legs. **V** In display flight, very high *fju-wirr* and *fwee-fwee-fwee*. **H** Semi-desert, more or less wooded and bushed areas. **D** Widespread across Africa.

21 LONG-CRESTED EAGLE *Lophaetus occipitalis* male 50cm, female 55cm Unmistakably long, loose crest and white 'socks'. **V** High, drawn-out *feeeeeh*. **H** Forest edges, swampy grassland, cultivation, suburban gardens. **D** Widespread across Africa.

22 BLACK EAGLE *Ictinaetus malaiensis* 69–81cm Often seen soaring low over forest canopy, with wings held in a shallow 'V', searching for eggs and nestlings. **V** Plaintive *kleeee-kee* or *hee-lee-leeuw*. **H** Lowland, hill and montane forest. **D** Widespread across Asia.

23 LESSER SPOTTED EAGLE *Clanga pomarina* 57–64cm Has a very rare pale morph, similar to pale Greater Spotted Eagle morph. In flight from above, shows white trailing edge to wings, small white patch at base of primaries and white lower rump. **V** High-pitched, yelping *k-yeep*. During display utters a whistled *wiiiik*. **H** Moist lowland woods near meadows, dry mountain forests. **D** E Europe.

24 INDIAN SPOTTED EAGLE *Clanga hastata* 59–67cm Soars on slightly arched wings. Yellow gape extends to back of, or slightly beyond, eye. **V** High-pitched laughing cackle. **H** Wooded areas interspersed with cultivation. **D** N India.

25 GREATER SPOTTED EAGLE *Clanga clanga* 59–69cm Pale morph very rare. **V** Yelping *kyak*, lower pitched than that of Lesser Spotted Eagle. **H** Woodland with nearby wetlands. In non-breeding season, wetlands with or without trees. **D** C Europe to E Siberia and N China. Non-breeding season in SW Asia and NE Africa.

26 WAHLBERG'S EAGLE *Hieraaetus wahlbergi* male 55cm, female 60cm Many colour forms. Four shown, of which darkest is most common. Separated from other eagles by more clearly visible barring of flight feathers and tail. Note also large black eyes and rather long tail. Flies with closed tail, giving cross-form to flight silhouette. **V** Call a drawn-out, high-pitched squeal or scream. **H** No preferred habitat but not in forest. **D** Widespread across Africa.

27 BOOTED EAGLE *Hieraaetus pennatus* 50–57cm Some variation occurs in both morphs, especially in colour of underwing coverts, which can be more rufous. **V** Shrill, chattering *ki-ki-ki*, also a buzzard-like *hiyaah*. **H** Forests mixed with open areas such as scrub, grassland, etc. **D** SW Europe and NW Africa to C Asia and N India, South Africa. Non-breeding season in Africa, S Asia.

28 LITTLE EAGLE *Hieraaetus morphnoides* 45–55cm Glides on flat wings, tail often fanned. Locates prey while soaring and from perch. Feeds on mammals, birds and reptiles. **V** Usually silent. **H** Open woodland, savannah, lightly wooded slopes. **D** Australia.

29 PYGMY EAGLE *Hieraaetus weiskei* 38–48cm Usually occurs singly. Glides on level wings, low over canopy or forest edges; often perches in the open on treetop branches. **V** Generally silent; gives a whistling *sip sip see* during display flight. **H** Primary rainforest, riparian and monsoon forest, forest edges. **D** New Guinea.

30 AYRES'S HAWK-EAGLE *Hieraaetus ayresii* male 45cm, female 55cm White forehead diagnostic but small crest often hidden. Note heavily barred underwings and tail. **V** Very high, pushed-out *fuweeh fuweeh*. **H** Forest and woodland canopy. **D** Widespread across Africa.

31 TAWNY EAGLE *Aquila rapax* 65–75cm Gape extends to middle of eye. General colour very variable, often appearing scruffy. Pale morphs usually show pale uppertail coverts. Dark form shown found in Himalayas. **V** In breeding season utters a repeated, barking *kowk*, otherwise normally silent. **H** Mountain forest and neighbouring plains and valleys. **D** Widespread across Africa, India.

32 STEPPE EAGLE *Aquila nipalensis* 67–87cm Gape extends to rear of eye. **V** In breeding season utters a barking *ow*, otherwise mainly silent. **H** Breeds in steppe and semi-desert in mountain or hill and lowland areas. **D** C Europe. Non-breeding season in S and SW Asia, Arabian Peninsula, E Africa.

33 SPANISH IMPERIAL EAGLE *Aquila adalberti* 72–85cm Often considered conspecific with Eastern Imperial Eagle, from which it is distinguished by more extensive white on shoulder. **V** As Eastern Imperial Eagle. **H** Forests with undergrowth brush, away from areas of human disturbance. **D** Spain, Portugal, Morocco.

34 EASTERN IMPERIAL EAGLE *Aquila heliaca* 72–84cm In flight, wing pattern similar to that of juvenile Tawny Eagle. **V** Deep, barking *owk*. **H** Upland and steppe forests, feeding on open plains or cultivated areas. **D** SE Europe to C Siberia. Non-breeding season in Asia, Arabian Peninsula.

35 GURNEY'S EAGLE *Aquila gurneyi* 74–85cm Generally seen singly, occasionally in pairs or threes. Hunts by quartering over forest or nearby open areas, gliding on flat or slightly raised wings. **V** Nasal, piping, downslurred note. **H** Primary lowland and hill forest, coastal forest, forest edges and plantations. **D** New Guinea; Maluku Islands, Indonesia.

36 GOLDEN EAGLE *Aquila chrysaetos* 76–93cm Note relatively long tail and pale nape patch. **V** Generally silent. Sometimes gives a fluty whistle in flight. **H** Mountains, steppe and, locally, marshes, preferring areas with sparse vegetation. **D** Widespread. Europe to NE Asia, N Africa to C Asia, North America.

37 WEDGE-TAILED EAGLE *Aquila audax* 85–105cm Recognised in flight by large size, dark plumage, diagnostic long, wedge-shaped tail. Prey includes birds and mammals to the size of wallabies. **V** Utters whistles near nest; otherwise silent. **H** Open and lightly wooded habitats. **D** Australia, S New Guinea.

38 VERREAUX'S EAGLE *Aquila verreauxii* 80–90cm In flight, all ages show diagnostic 'pinched-in' effect at wing base. **V** Usually silent, although various calls described, including *chorr-chorr-chorr*, upslurred *iiy-iii* and melodious *keee-uup*. **H** Desert mountains, especially those inhabited by hyraxes. Sometimes frequents more vegetated mountains. **D** Widespread across Africa, Arabian Peninsula.

39 CASSIN'S HAWK-EAGLE *Aquila africana* male 55cm, female 60cm Note short-winged and long-tailed jizz. Sparsely spotted and streaked (mostly around legs). In flight, shows clean white underparts, contrasting with dark underwing coverts. Separated from Black Sparrowhawk (Plate 34) by feathered lower legs. Rarely seen except soaring low over forest. **V** Shrill, protracted squealing call. **H** Forest canopy. **D** Widespread across Africa.

40 BONELLI'S EAGLE *Aquila fasciata* 65–72cm Often seen in pairs and unmistakable thanks to plumage, shape and flight. **V** Usually silent apart from a shrill *iuh* and whistled *eeeoo* given during display. **H** Dry hills and mountains with rocky gorges, either open or with sparse scrub or forest. After breeding visits open lowlands. **D** Spain to India and S China, Indochina, N Africa.

41 AFRICAN HAWK-EAGLE *Aquila spilogaster* male 65cm, female 80cm Note large size, feathered legs, rounded wings (in flight). Perches in cover, but can be seen soaring freely, especially in morning. **V** Slightly rising *kee-kee-kee*, repeated 6–7 times. **H** Open woodland, thornbush with some large trees, often near streams and rocky hillsides. **D** Widespread across Africa.

1
2
dark morph
light morph
3
4
5
6
7
8
9
10
11
12
13
14
15
16
17
18
19
20
21
22
light morph
23
24
light morph
dark morph
25
26
26
27
dark morph
28
dark morph
light morph
29
30
light morph
dark morph
31
32
33
34
35
36
37
38
39
♂
♀
41
40

KITES, HAWKS AND EAGLES *CONTINUED*

1 DOUBLE-TOOTHED KITE *Harpagus bidentatus* male 31cm, female 37cm Head grey. Throat white with black mid-line. Upperparts blackish brown. Breast rufous. Rest of underparts and thighs barred rufous and white. Perches among foliage. **V** Very high-pitched, drawn-out whistle, slurring down and then up in pitch. **H** Edges, cleared areas, rainforest, cloud forest, woods. **D** E Mexico to SE Brazil.

2 RUFOUS-THIGHED KITE *Harpagus diodon* male 30cm, female 36cm Head blackish. Throat white with blackish mid-line. Upperparts greyish black. Underparts pale grey. In topmost foliage. **V** High-pitched whistles. **H** Tropical and subtropical forests and woods. **D** E Amazonia, SE South America.

3 LIZARD BUZZARD *Kaupifalco monogrammicus* 30cm Note thickset, upright stance, dark eye and black median throat streak. Single white tail band diagnostic. **V** Drawn-out, slightly harsh, downslurring whistle. **H** Forest edges, woodland, cultivated areas, farmland. **D** Widespread across Africa.

4 GABAR GOSHAWK *Micronisus gabar* 30cm Rather thickset. Normal and dark forms shown. Note dark eye, pale grey breast, white rump, and white edge to secondaries and tail sides. Hunts from cover. **V** Series of sharp *plee-plee-plee* notes. **H** Woodland, open thornbush and scrub, suburban gardens, villages. **D** Widespread across Africa.

5 DARK CHANTING GOSHAWK *Melierax metabates* 38–48cm Upperparts, head and breast dark grey. Underparts white, finely barred black. Hunts from a perch. **V** In display gives a fluting *wheeu-wheeu-wheeu* or *kleeee-yeu*. **H** Open woodland, olive groves. **D** Widespread across Africa.

6 EASTERN CHANTING GOSHAWK *Melierax poliopterus* 50cm Note white upper tail coverts. Folded wing rather contrasting white and grey. **V** Mid-high, resounding, rapid *wuut-wuutwuutwuutwuutwutwutwut* (accelerated and in crescendo) and mid-high, descending, bouncing, rapid *wuuh-wicwicwickerrrrrrr*. **H** Often in more open, drier habitats than Dark Chanting Goshawk; <3,000m. **D** Ethiopia and Somalia to Tanzania.

7 PALE CHANTING GOSHAWK *Melierax canorus* 55cm Note erect stance on top of bush or post. Separated from Dark Chanting Goshawk by paler appearance, white rump and almost white secondaries in flight. **V** Mid-high, rapid *wuut-whut-wuutwuutwuutwuut*, slightly rising in pitch and volume, and a mid-high, descending *wooh-wikwikwikwikwirrrr*. **H** Open thornveld and woodland. **D** S Angola to Zimbabwe and South Africa.

8 LONG-TAILED HAWK *Urotriorchis macrourus* 60cm Unmistakable thanks to long tail and colour pattern. **V** Sharp, calmly descending, fluted note. **H** Forest canopies. **D** Liberia to Uganda and C DR Congo.

9 CHESTNUT-SHOULDERED GOSHAWK *Erythrotriorchis buergersi* 45–50cm Perches amongst foliage and scans for prey. Feeds mainly on birds, caught by active pursuit in forest interior. **V** Series of piercing calls. **H** Rainforest. **D** New Guinea.

10 RED GOSHAWK *Erythrotriorchis radiatus* 50–60cm Feeds mainly on birds, particularly parrots and pigeons. Prey located by concealed observation, then caught in active pursuit. **V** Shrieking *kip-kip-kip*... **H** *Coastal forest*. **D** N and E Australia.

11 DORIA'S GOSHAWK *Megatriorchis doriae* 55–70cm Seldom soars. Locates prey within tree canopy by watching from concealed perch. Diet includes birds, possibly mammals. **V** Occasional whistles but mainly silent. **H** Lowland rainforest, mangrove swamps. **D** New Guinea.

12 TINY HAWK *Accipiter superciliosus* male 25cm, female 30cm Note grey cheeks. **V** Low, hoarse *aaah-aaah*. **H** Forest, tall secondary growth. **D** Nicaragua to N Argentina.

13 SEMICOLLARED HAWK *Accipiter collaris* male 25cm, female 30cm Like Tiny Hawk but larger and with longer tail. Crown and nape blackish, cheeks and collar on hindneck white, barred black. Underparts white, broadly barred blackish. **V** Whistling calls, rather thin and hesitant. **H** Subtropical rainforest, cloud forest, edges and clearings in montane rainforest, generally at higher elevations than Tiny Hawk. **D** Venezuela to Peru.

14 CRESTED GOSHAWK *Accipiter trivirgatus* 30–46cm Usual flight consists of stiff wingbeats followed by short glides, often seen soaring high above forest canopy. **V** Shrill, prolonged, screaming *he he hehehehe*, and loud screams and deep croaks. **H** Deciduous and evergreen tropical and subtropical forests, well-wooded gardens. **D** Widespread across Asia.

15 SULAWESI GOSHAWK *Accipiter griseiceps* 28–37cm Usually seen singly. Favours forest interiors; hunts from a perch in dense foliage, swooping to capture prey on ground. **V** High-pitched, quite faint *tseee-tseee-tseee*. **H** Primary lowland, hill and montane forest, mangroves and occasionally tall secondary woodland. **D** Sulawesi, Indonesia.

16 GREY-BELLIED HAWK *Accipiter poliogaster* male 42cm, female 52cm Upperparts blackish. Underparts white or washed grey. Tail blackish with three grey bands and grey tip. **V** Rising, accelerating whistles *tuee-tuee-tweetweetwee*. **H** Tropical and subtropical rainforest, edges and clearings, riverine forest, woods. **D** Amazonia to N Argentina.

17 RED-CHESTED GOSHAWK *Accipiter toussenelii* 35cm Dark rufous below with little barring. **V** Gives brief chirping notes, also parrot-like series of yelps. **H** Primary and secondary forest. **D** S Cameroon to N and W DR Congo.

18 AFRICAN GOSHAWK *Accipiter tachiro* male 35cm, female 45cm Plumage, especially of female, often more brown than grey. Note white barring of tail. Skulking. **V** High, sharp *whip* in flight or from perch. Rather vocal. **H** In and near forests, dense thornbush and other woodland, riverine belts, suburban gardens. **D** E, SC and S Africa.

19 CHESTNUT-FLANKED SPARROWHAWK *Accipiter castanilius* male 30cm, female 35cm Note white spots to tail, not to rump. Rather like African Goshawk but flanks unbarred chestnut and underwing barred black on white. Also resembles Rufous-breasted Sparrowhawk, but that species lacks bars below, and has white eyebrows but no white on tail. **V** Not recorded. **H** Dense forests. **D** WC Africa.

20 SHIKRA *Accipiter badius* 30–36cm Hunts from hidden perch, taking prey (mainly lizards or birds) from trees or ground. Rarely indulges in aerial chases. **V** Piping *keeu-keeu-keeu* and shrill *kewik*. **H** Forest edges, open woodland, orchards, gardens. **D** Widespread across Africa and Asia.

21 NICOBAR SPARROWHAWK *Accipiter butleri* 30–34cm Upperparts pale grey, breast and flank barring rusty. Little information. **V** Shrill *kee-wick*. **H** Tree canopy. **D** Nicobar Islands.

22 LEVANT SPARROWHAWK *Accipiter brevipes* 33–38cm During migration often seen in large flocks. **V** Shrill *kee-wik kee-wik kee-wik*. **H** Deciduous forests, wooded plains, copses and orchards. **D** SE Europe to W Kazakhstan.

23 CHINESE SPARROWHAWK *Accipiter soloensis* 27–35cm Mainly catches prey on ground, including frogs. **V** Rapid, accelerating piping that descends in pitch. **H** Woodland, often near wetlands. **D** E Siberia, Korean Peninsula, China.

24 FRANCES'S SPARROWHAWK *Accipiter francesiae* male 30cm, female 35cm Very faintly barred orange below. Separated from Madagascan Sparrowhawk and Henst's Goshawk (Plate 34) by lighter belly. **H** Any wooded habitat. **D** Madagascar, Comoro Islands.

25 SPOT-TAILED SPARROWHAWK *Accipiter trinotatus* 26–30cm Perches in forest understorey, close to tree trunk, singly or in pairs. Active hunter of lizards, snakes, frogs, birds and insects. **V** Descending series of 4–6 *ke* notes. **H** Primary lowland, hill and lower montane forest, tall secondary forest, mangroves. **D** Sulawesi, Indonesia.

26 GREY GOSHAWK *Accipiter novaehollandiae* 45–55cm Watches for prey from concealed perch in tree foliage; diet comprises mainly birds, caught by surprise attack in flight. **V** Includes series of *kewick-kewick*... calls. **H** Forests. **D** Australia.

27 VARIABLE GOSHAWK *Accipiter hiogaster* 30–45cm Occurs singly or in pairs; frequently seen either soaring over forests or dashing along forest edges. **V** Series of high-pitched, upslurred notes. **H** Primary and tall secondary lowland and hill forest, forest edges, monsoon woodland, savannah scrub, plantations. **D** Maluku Islands, Indonesia, to New Guinea and Solomon Islands.

28 BROWN GOSHAWK *Accipiter fasciatus* 33–55cm Encountered singly, occasionally in pairs. Hunts from a concealed perch in lower canopy or mid-storey. **V** High-pitched *ki-ki-ki-ki*. **H** Monsoon forest and woodland, forest edges, lightly wooded cultivation and scrub. **D** Widespread across Australasia.

29 BLACK-MANTLED GOSHAWK *Accipiter melanochlamys* 35–40cm Unmistakable plumage. Diet mainly birds, also small mammals, amphibians and invertebrates. **V** Calls include shrill *kee-kee-kee*... notes. **H** Cloud forest, montane rainforest. **D** New Guinea.

30 PIED GOSHAWK *Accipiter albogularis* 35–40cm Watches for prey from concealed perch in tree foliage. Diet mainly birds, caught in rapid surprise attack; also lizards and invertebrates. **V** Mostly silent. **H** Forests. **D** Solomon Islands.

31 WHITE-BELLIED GOSHAWK *Accipiter haplochrous* 35–40cm Perches unobtrusively, scanning ground for prey including small mammals, invertebrates and birds. **V** High-pitched, rapid *ki-ki-ki-ki*... notes. **H** Rainforest. **D** New Caledonia.

32 FIJI GOSHAWK *Accipiter rufitorques* male 35cm, female 40cm Sole goshawk in Fiji and no other raptors except Swamp Harrier and Peregrine Falcon (both with very different jizz). Note pale underparts that lack markings. **H** Forest, woodland, city parks. **D** Fiji.

33 MOLUCCAN GOSHAWK *Accipiter henicogrammus* 37–43cm Occurs singly, in pairs, and occasionally in family groups. Inconspicuous, often perching in shaded areas of mid-storey, close to trunk. **V** Calls include a series of upslurred, slightly accelerating screams and a longer series of ascending whistles, the last few notes more shrieking. **H** Interior of primary lowland and hill forest, occasionally also forest edges. **D** N Maluku Islands, Indonesia.

34 SLATY-MANTLED GOSHAWK *Accipiter luteoschistaceus* 30–35cm Relatively short, round wings allow aerial manoeuvrability in pursuit of prey. Diet thought to include invertebrates, reptiles and small birds. **V** High-pitched *ki-ki-ki*... notes. **H** Forested and partly deforested habitats. **D** New Britain.

35 IMITATOR GOSHAWK *Accipiter imitator* 30–32cm Strikingly marked plumage. Perches unobtrusively in forest canopy. Diet presumed to include invertebrates, lizards and small birds. **V** Little known. **H** Tropical forests. **D** Solomon Islands.

36 GREY-HEADED GOSHAWK *Accipiter poliocephalus* 30–36cm Perches unobtrusively for long periods. Catches prey in rapid surprise attack. Diet includes reptiles and invertebrates. **V** Calls include series of thin whistles. **H** Rainforest. **D** New Guinea.

37 NEW BRITAIN GOSHAWK *Accipiter princeps* 40–45cm Diet includes reptiles and large invertebrates. Perches for extended periods, scanning surroundings; prey caught by surprise attack. **V** Unknown. **H** Rainforest. **D** New Britain.

38 RED-THIGHED SPARROWHAWK *Accipiter erythropus* male 25cm, female 30cm All black above except for conspicuous white rump. Note also small white spots on scapulars and on all but central tail feathers. Eye red (not yellow as in Little Sparrowhawk). **V** Fast, excitable chattered *ki-ki-ki*... call. **H** Dense forests. **D** W and WC Africa.

39 LITTLE SPARROWHAWK *Accipiter minullus* male 25cm, female 30cm Less black above than Red-thighed Sparrowhawk, and finely barred white and pink-rufous below. **V** Fairly rapid series of upslurring high-pitched whistles. **H** Woodland and thornbush. Less common in forests than Red-thighed. **D** S, SC and E Africa.

dark morph
1
2
3
4
5
6
7
8
9
10
11
12
13
14
15
16
17
18
19
20
21
22
23
24
25
26
27
28
29
30
31
32
33
34
35
36
37
38
39
white morph
♀
♂

KITES, HAWKS AND EAGLES *CONTINUED*

1 JAPANESE SPARROWHAWK *Accipiter gularis* 29–34cm Sometimes in small flocks during migration. Often perches hidden in tree foliage. **V** Shrill *kek-kek-kek*, repeated, squealing *tchew-tchew-tchew* and loud *ki-weer*. **H** Lightly wooded areas, forest edges, open country with patches of trees. **D** Breeds E Asia. Winters SE Asia, Indonesia, Borneo and Philippines.

2 BESRA *Accipiter virgatus* 29–36cm Typical sparrowhawk hunting technique, perching inconspicuously before giving chase to avian prey. **V** Rapid *tchew-tchew-tchew*. **H** Dense broadleaved forest, more open wooded areas post breeding. **D** Widespread in S and E Asia.

3 DWARF SPARROWHAWK *Accipiter nanus* 23–28cm Occurs singly or in pairs. Forages mainly in forest interior, also along forest tracks and roads. Stoops to capture large insects and small birds. **V** Thin, high-pitched *kiliu*, sometimes followed by sharp, rapid *ki-ki-ki-ki-ki*. **H** Primary hill and montane forest. **D** Sulawesi, Indonesia.

4 RUFOUS-NECKED SPARROWHAWK *Accipiter erythrauchen* 26–33cm Perches inconspicuously in dense canopy foliage, singly or occasionally in pairs. Dashes from cover to capture small birds. **V** Rapid series of high-pitched staccato notes. **H** Primary forest in lowlands and hills, tree plantations, logged lowland forest. **D** Maluku Islands, Indonesia.

5 COLLARED SPARROWHAWK *Accipiter cirrocephalus* 30–35cm Hunts small birds, launching surprise attack from concealed perch. Flight agile and rapid. **V** Shrill, rapid *ki-ki-ki...* in breeding season; otherwise silent. **H** Forested and wooded habitats, scrub. **D** Australia, New Guinea.

6 NEW BRITAIN SPARROWHAWK *Accipiter brachyurus* 30–35cm Recognised within range by unmarked white underparts and reddish nape. Perches unobtrusively and catches prey by surprise attack. **V** Little known. **H** Rainforest. **D** New Britain and New Ireland (Bismarck Archipelago).

7 VINOUS-BREASTED SPARROWHAWK *Accipiter rhodogaster* 26–33cm Usually solitary, sometimes in pairs. Often perches on exposed branches in the canopy, otherwise in forest interior. **V** Rapid *hihihihihi*. **H** Primary and tall secondary lowland and hill forest, lightly wooded cultivation, mangroves. **D** Sulawesi and the Sula Islands, Indonesia.

8 MADAGASCAN SPARROWHAWK *Accipiter madagascariensis* 35–40cm Very similar to smaller Frances's Sparrowhawk (Plate 33). **V** Typical call single, high-pitched squeaky *whik*; at nest calls *kwee-kwee-kwee*. **H** Mainly in native forests up to mid altitudes. **D** Madagascar.

9 OVAMBO SPARROWHAWK *Accipiter ovampensis* 30–35cm Note small head and dark-eyed appearance. Legs, cere and eyelids may be red, pink or yellow. Throat indistinctly barred grey. Small white streaks on tail shafts diagnostic even in rare black form. Hunts from cover or from exposed perch. **H** Open woodland, thornbush, eucalyptus and other exotic plantations. **D** Senegal and Gambia to Ethiopia, S to N South Africa.

10 EURASIAN SPARROWHAWK *Accipiter nisus* 28–38cm Surprises bird prey by dashing from a hidden perch or after a stealthy, twisting flight. **V** When alarmed utters rapid *kew-kew-kew-kew-kew*. **H** Open woodland, scrub forest, wooded cultivations. **D** Widespread in Eurasia.

11 RUFOUS-BREASTED SPARROWHAWK *Accipiter rufiventris* 30–35cm Rufous underparts and brown upperparts diagnostic. Secretive. **V** Sharp *kee kee kee* given when displaying. Chicks beg with mewing notes. **H** Mountainous forests, woodland, plantations and surrounding open country. **D** E, SE and S Africa.

12 SHARP-SHINNED HAWK *Accipiter striatus* 23–35cm Flight dashing and agile when tracking prey, soars (note square-ended tail) and glides with occasional flaps while exploring territory. **V** High-pitched *kew-kew-kew-kew-kew*. **H** Forests, more open woodland in winter. **D** Widespread in North America.

13 WHITE-BREASTED HAWK *Accipiter chionogaster* 30cm Similar to Sharp-shinned Hawk, but with mostly white underparts. **V** Sharp, well-spaced whistled notes. **H** Deciduous and pine woodland and forest. **D** S Mexico to Nicaragua.

14 PLAIN-BREASTED HAWK *Accipiter ventralis* 24–30cm Variable. Rufous and pale forms shown. Female larger than male. Prey mainly small birds, hunted in rapid-flight surprise attacks. **V** Shrill *kiu-kiu-kiu...* **H** Open forest, wooded slopes. **D** Venezuela to W Bolivia.

15 RUFOUS-THIGHED HAWK *Accipiter erythronemius* 24–33cm Variable. Crown and upperparts dark grey. Tail grey with four broad black bands. **V** Rapid shrill *kew-kew-kew* during breeding season. **H** Tropical, subtropical and temperate forest, edge and clearings, savannah, dry woods. **D** SE Bolivia and S Brazil to Paraguay, Uruguay, N Argentina.

16 COOPER'S HAWK *Accipiter cooperii* 35–51cm Hunting flight agile and dashing; when soaring, note rounded tail. **V** Nasal *cac cac cac*, also long *keeee*; female gives drawn-out *whaaaaaa*. **H** Mainly deciduous woodland with clearings; in winter uses other habitats, including urban areas. **D** S Canada to Honduras.

17 GUNDLACH'S HAWK *Accipiter gundlachi* 43–51cm In soaring flight note rounded end to tail. Behaviour similar to Sharp-shinned Hawk. **V** Cackling *kek-kek-kek-kek-kek...* and wailing squeal. **H** Mountain and lowland forests, swamps, coastal woodland. **D** Cuba.

18 BICOLORED HAWK *Accipiter bicolor* 35–43cm Crown and upperparts dark grey. Cheeks and underparts grey. Vent white, thighs rufous. Axillaries and underwing rufous or buffy-white according to race. Flight feathers barred. Tail dark grey with three bands and blackish tip. **V** During breeding season, repeated (10–20) nasal *keh* notes; female and chicks beg with squealing notes. **H** Woods and forest, gallery forest, savannah with palms, Cerrado, rainforest, *Araucaria* woods. **D** S Mexico to Argentina.

19 CHILEAN HAWK *Accipiter chilensis* 38–42cm Female larger than male. Perches for long periods. Feeds mainly on birds, caught by active pursuit after surprise attack. **V** Screaming *kek-kek-kek...* in breeding season; otherwise silent. **H** Montane forest. **D** C Chile and WC Argentina to Tierra del Fuego.

20 BLACK SPARROWHAWK *Accipiter melanoleucus* 45–55cm Separated from Cassin's Hawk-Eagle and Ayres's Hawk-Eagle (Plate 32) by dark eye, unfeathered lower legs and absence of crest. May soar, but in general rather secretive, though not shy. **V** Breeding male gives shrill *kyip*; female answers with lower *chep*, also *kow-kow-kow* if disturbed at nest. **H** Forests, woodland, exotic plantations, suburbs. **D** Widespread in Africa.

21 HENST'S GOSHAWK *Accipiter henstii* 50–60cm Mainly separable from Pale Chanting Goshawk (Plate 33) and Madagascan Serpent-Eagle (Plate 30) by size. **V** Repeated loud high-pitched *peer* notes given in territorial defence. **H** Primary and secondary forest up to high altitudes. **D** Madagascar.

22 NORTHERN GOSHAWK *Accipiter gentilis* 48–62cm White eyebrow. Agile hunting flight through trees in search of medium-sized birds and mammals. **V** Loud *kyee kyee kyee*, during display a gull-like *KREE-ah*. **H** Mixed or deciduous forest, preferring forest edge and clearings. **D** Widespread in North America, Central America and Eurasia.

23 MEYER'S GOSHAWK *Accipiter meyerianus* 48–56cm Generally seen alone, usually soaring low over forests; occasionally perched by forest clearing or hunting along forest streams. **V** Repeated, loud, nasal, upslurred *whi-i-yu*, also slurred *ka-ah*. **H** Primary lowland and hill forest. **D** New Britain, New Guinea, Solomon Islands.

24 WESTERN MARSH HARRIER *Circus aeruginosus* 48–56cm Generally flies very low, quartering reedbeds or open grassland, alternating flapping and gliding with wings held in shallow V. Drops feet-first onto prey. **V** Generally silent but when alarmed may give a cackling *chek-ek-ek-ek-ek*. **H** Reedbeds, grasslands, cultivated fields, salt marsh. **D** W and C Eurasia and Africa, some moving S in winter.

25 EASTERN MARSH HARRIER *Circus spilonotus* 47–55cm Behaviour similar to Western Marsh Harrier, which is often thought to be conspecific. Sometimes forms large flocks, especially near roosting sites. Intensity of head markings variable, often has blackish face or head. **V** Generally silent, may utter mewing *keeau* at roost sites. **H** Marshes, grasslands, paddyfields. **D** Breeds C Siberia to NE China and Japan. Winters SE Asia.

26 PAPUAN HARRIER *Circus spilothorax* 48–60cm Formerly considered conspecific with Eastern Marsh Harrier. Male's grey, black and white plumage is distinctive. Quarters ground looking for prey. **V** Mostly silent. **H** Marshes, grassland, rice fields. **D** C and E New Guinea.

27 SWAMP HARRIER *Circus approximans* 48–62cm Large raptor, with wings characteristically held in shallow V. No other harrier species or similar raptors in range. **V** High, sharp, mewing *vweeét* notes. **H** Wetlands, grassland. **D** Widespread in Australasia.

28 AFRICAN MARSH HARRIER *Circus ranivorus* 45cm Separated from Western Marsh Harrier by darker head and barred flight feathers and tail. **V** Gives various whistles, squeals and chattering notes around nest. **H** Larger swamps, grassland, farmland. **D** South Sudan and Kenya to South Africa.

29 REUNION HARRIER *Circus maillardi* 55cm Unmistakable by slender body and by colouring, especially of male. Flies with wings held in a V-shape. **V** Gives various wailing, chattering and chuckling calls near nest. **H** Wetlands and their surroundings, including woodland, up to high altitudes. **D** Réunion Island.

30 MALAGASY HARRIER *Circus macrosceles* 55–59cm Quarters ground in search of prey; diet includes small birds, mammals, reptiles and invertebrates. **V** Mostly silent; *kiu-kiu-kiu...* uttered when nesting. **H** Marshes, grassland, rice fields. **D** Madagascar, Comoro Islands.

31 LONG-WINGED HARRIER *Circus buffoni* 46–57cm Long broad wings and tail. Male has black hood, whitish face and white edge to facial disk. Underwing barred. Female has dark brown hood, cream face. Rump white. Tail buffy grey with three visible dark-brown bands. Underwing creamy, streaked brown. Glides low and drops onto small prey; seldom seen perched. **V** Shrill whistles and chatters. **H** Damp grasslands, reeds, marshes, rice paddies, cereal crops and stubble. **D** Colombia to the Guianas and NE Brazil, S to C Chile and C Argentina.

32 SPOTTED HARRIER *Circus assimilis* 50–61cm Occurs singly or in pairs. Hunts low over open country; regularly perches on isolated posts or dead trees. **V** Short, shrill whistle and a rapid chatter. **H** Grassland, paddyfields, scrub, cultivation. **D** Widespread in Australasia.

33 BLACK HARRIER *Circus maurus* 50cm Only all dark harrier, except for white unbarred underside of flight feathers. **V** Short, rather squeaky whistle. **H** Dry, open country, grassland, farmland. **D** Namibia, South Africa.

34 HEN HARRIER *Circus cyaneus* 45cm Bulkier and broader-winged than Pallid Harrier and Montagu's Harrier. Male separated from Pallid Harrier by better defined separation of grey chest from whitish belly and in flight by more extensively black primaries especially seen from above; also has more obvious white rump patch. Female very much like Pallid Harrier and Montagu's Harrier, but more heavily striped below and with less obvious white lower eyelids. In flight shows five (not four) projecting primaries, giving more rounded wing tip impression. **V** Rather squeaky, somewhat quavering whistles. **H** Dry open natural and cultivated plains. **D** BreedsEurope and C and N Asia. Winters S Asia, N Africa.

35 NORTHERN HARRIER *Circus hudsonius* 44–52cm Soars with wings level or in a shallow V. **V** During display utters *chuk-uk-uk-uk-uk*; when alarmed female gives twittering *chit-it-it-it-et-it* and male *chek-ek-ek-ek*. **H** Moorland, heathland, grassland, young conifer plantations, marshes, salt marsh, open country. **D** Breeds North America, Central America. Winters South America.

36 CINEREOUS HARRIER *Circus cinereus* 40–50cm Male head, neck, upper breast and upperparts ash-grey. Rump white. Flight feathers mostly white tipped black. Female head and upperparts brown faintly streaked buff. Rump white. Flight feathers grey barred black. Underwing white finely barred rusty, flight feathers whitish grey barred blackish. Tail whitish grey with three visible blackish bands. Flaps faster than Long-winged Harrier; flies low, wings in shallow V, perches in grass or on fences. **V** Whistles and sharp, shrill yapping notes and chatters. **H** Open grassland, low scrub. **D** C Colombia to Tierra del Fuego; Paraguay and SE Brazil to Tierra del Fuego.

37 PALLID HARRIER *Circus macrourus* 40–48cm Compared to Hen Harrier, has slimmer build and more buoyant, tern-like flight, with wings held in a shallow V. Hunting technique similar to Hen Harrier. **V** When alarmed gives a rapid *chit-er chit-er chit-it-it*. **H** Open grasslands, cultivations, marshes. **D** Breeds E Europe to C Asia. Winters Africa and SE Asia.

38 PIED HARRIER *Circus melanoleucos* 41–49cm Usually appears heavier in flapping flight than Pallid or Montagu's Harrier. **V** Female utters *chak-chak-chak-chak-chak* when alarmed. **H** Open grasslands, marshes, paddyfields, stubble fields. **D** Breeds SE Siberia, Korean Peninsula and NE China. Winters in SE Asia.

39 MONTAGU'S HARRIER *Circus pygargus* 50cm Wing tip reaches tail tip when perched. Separated from Pallid Harrier by shorter legs, greyer plumage, more black in wings (including wing bar). Some chestnut streaking on underparts and underwings. Female separated from female Pallid Harrier by paler facial disc, which is not separated by white line from streaks on sides of neck; also has paler secondaries below, while neatly arranged barring on underwing coverts is more chestnut. **V** Fast, emphatic high-pitched chatters, thin whistles. **H** Dry, open, natural and cultivated plains. **D** Breeds W Europe and NW Africa to C Asia. Winters Africa, India.

1
♂
♀
2
♀
♂
3
4
5
6
7
8
9
light morph
dark morph
10
♂
11
12
13
rufous
14
pale
15
16
17
18
light morph
dark morph
19
20
21
22
23
dark morph
light morph
24
♀
♂
25
♂
♀
26
27
♂
♀
28
29
♂
♀
30
♂
♀
31
♀
black morph
normal morph
♂
31
32
33
34
♂
♀
35
♀
♂
36
♀
♂
37
♀
♂
38
♀
♂
39
♀
♂

KITES, HAWKS AND EAGLES *CONTINUED*

1 RED KITE *Milvus milvus* 60–66cm Flight action much like Black Kite. **V** Mewing *peee-ooo*, followed by a drawn-out *peee-oooo-eee-ooo-eee-ooo-eee-ooo.* **H** Forests or scattered woodland with nearby grassland or wetland. **D** Europe and NW Africa to the Middle East, Cape Verde Islands.

2 BLACK KITE *Milvus migrans* 55–60cm Flight action often appears 'loose', with constant twisting of tail. Fork in tail often disappears when tail spread. **V** Whinnying *pee-errrr.* **H** Open country, often near wet areas, with or without woodland, also around human habitation. **D** Widespread across Europe, Africa and Asia.

3 YELLOW-BILLED KITE *Milvus aegyptius* 55–60cm Previously considered conspecific with Black Kite, from which it can be separated by all-yellow bill. **V** Whinnying *pee-errrr.* **H** Open country, often near wet areas, with or without woodland, also around human habitation and parks. **D** Widespread across Africa.

4 WHISTLING KITE *Haliastur sphenurus* 50–60cm Quarters ground for prey and sometimes hovers. Diet includes live prey but also scavenges roadkill and at bushfires. **V** Vocal and noisy; shrill whistling call typical. **H** Open woodland, marshes. **D** Widespread across Australasia.

5 BRAHMINY KITE *Haliastur indus* 45–51cm Often encountered sitting in a tall tree overlooking water, from where it swoops down to pick prey from the surface. **V** Wheezy *kyerrh* squeal. **H** Typically near water, such as lakes, rivers, marshes, flooded paddyfields, coastal lagoons, estuaries and fishing villages. **D** Pakistan to N Australia.

6 WHITE-BELLIED SEA EAGLE *Haliaeetus leucogaster* 66–71cm Grey upperparts contrast with white head, breast, underwing coverts and tail. Hunts near water. **V** Loud, honking *kank kank kank kank.* Also a faster *ken-ken-ken-ken* and *ka ka-kaaa.* **H** Typically near coastal waters. **D** India to Tasmania.

7 SANFORD'S SEA EAGLE *Haliaeetus sanfordi* 70–90cm Perches for long periods. Varied diet includes fish and other aquatic vertebrates; sometimes scavenges carrion. **V** Honking *kank-kank-kank* when nesting; otherwise mostly silent. **H** Coastal and riverine forest. **D** Solomon Islands.

8 AFRICAN FISH EAGLE *Haliaeetus vocifer* 74–84cm Resembles a Bald Eagle, being mostly brown with white head, breast and tail. **V** Ringing or yelping *kyow-kow-kow.* **H** Lakes, rivers, estuaries, lagoons. **D** Widespread across Africa.

9 MADAGASCAN FISH EAGLE *Haliaeetus vociferoides* 70cm No similar bird in its range. **V** Loud, gull-like mewed phrase *whee wheet-wheet-wheet.* **H** Forest and tall trees bordering sea, lakes and rivers. **D** Madagascar.

10 PALLAS'S FISH EAGLE *Haliaeetus leucoryphus* 76–84cm Distinctive white stripe on black tail is diagnostic. **V** Hoarse, barking *kvok kvok kvok.* **H** Lakes, rivers, extensive marshes. **D** C Eurasia, N India.

11 WHITE-TAILED EAGLE *Haliaeetus albicilla* 69–92cm Brown overall, with diagnostic wedge-shaped white tail. **V** Shrill *klee klee klee klee.* In alarm utters a lower *klek klek klek.* **H** Rocky islands, lakes, marshes and large rivers from desert to Arctic areas. **D** Europe to N Asia, Greenland.

12 BALD EAGLE *Haliaeetus leucocephalus* 70–102cm Soars with wings almost level. Mainly brown with distinctive white head and tail. **V** Cackling *kweek-kik-ik-ik-ik-ikik* and lower *kak-kak-kak-kak.* **H** Coasts, rivers, lakes. In winter may occur well away from water. **D** Widespread across North America.

13 STELLER'S SEA EAGLE *Haliaeetus pelagicus* 85–94cm Very dark brown to black with white shoulder patches and tail. **V** Barking *kyow-kyow-kyow.* Also a stronger *kra kra kra kra* when involved in disputes. **H** River valleys, river mouths, rocky coasts. **D** NE Asia.

14 LESSER FISH EAGLE *Haliaeetus humilis* 64cm Smaller than the similar Grey-headed Fish Eagle. **V** Plaintive *pheeow-pheeoow-pheeow.* During breeding utters a repeated *pheeo-pheeo.* **H** Mountain and foothills in the forested margins of rapid-flowing streams, rivers and lakes. **D** Widespread across Asia.

15 GREY-HEADED FISH EAGLE *Haliaeetus ichthyaetus* 69–74cm Separated from similar Lesser Fish Eagle by larger size, longer wings and darker underparts. **V** Squawking *kwok* or similar, harsh screams. During display utters a far-carrying *tiu-weeeu.* **H** Lowland forest with nearby water, lakes, rivers, coastal lagoons, sometimes near estuaries. **D** Widespread across Asia.

16 GRASSHOPPER BUZZARD *Butastur rufipennis* male 35cm, female 40cm Rather pot-bellied. Red wings, normally also visible when perched, diagnostic. **V** Series of high-pitched, rather thin short whistles. **H** More or less wooded areas. **D** Senegal and Gambia to Somalia and N Tanzania.

17 WHITE-EYED BUZZARD *Butastur teesa* 36–43cm White iris distinctive. Sits for long periods on a prominent perch, from which it drops onto ground-based prey. **V** Melancholic *pit-weer pit-weer.* **H** Dry open country with scattered trees and scrub. **D** Pakistan to Myanmar.

18 RUFOUS-WINGED BUZZARD *Butastur liventer* 35–41cm Solitary. Not shy. Regularly uses low-level perch. Glides and soars with wings held flat. **V** Shrill *pit-piu.* **H** Dry deciduous forest, secondary growth, savannah, paddyfields. **D** SE Asia; Java and Sulawesi, Indonesia.

19 GREY-FACED BUZZARD *Butastur indicus* 46cm Often forms large flocks during migration. Hunts frogs, lizards and rodents, typically using top of a dead tree as lookout site. **V** Tremulous *chit-kwee.* **H** Wooded country with nearby open areas. **D** Breeds in E Siberia, Japan, Korea, NE China. Winters in SE Asia.

20 MISSISSIPPI KITE *Ictinia mississippiensis* 35–38cm Captures insects in mid-air. Gregarious, feeding in groups and often nesting in loose colonies. In flight, shows white secondaries on upperwing. **V** High-pitched *pe-teew* whistle and an excited *pee-tititi.* **H** Woodland, usually close to water. **D** Breeds in SC and SE USA. Winters in SC South America.

21 PLUMBEOUS KITE *Ictinia plumbea* 34–37cm Long, sharp wings. Head and underparts grey. Upperparts blackish grey. Primaries mostly rufous. Tail black. **V** Shrill two-note whistle, second note drawn out, falling in pitch and fading, *pi-pwiiuuu.* **H** Forest, savannah, gallery forest, woods. **D** NE Mexico to N Argentina.

22 BLACK-COLLARED HAWK *Busarellus nigricollis* 45cm General colour and black throat patch make it unmistakable. **V** Low, dry, nasal *tetetjeeh.* **H** Clear, quiet waters in open or wooded country. **D** C Mexico to N Argentina.

23 SNAIL KITE *Rostrhamus sociabilis* 43–48cm Flight slow and buoyant. **V** Cackling *ka-ka-ka-ka-ka*, grating *krrkrrkrrkrr* and harsh *kerwuck.* **H** Marshes and swamps. Breeds in patches of vegetation or on dead tree stumps above water. **D** S Florida and Caribbean to N Argentina.

24 SLENDER-BILLED KITE *Helicolestes hamatus* 40cm Note yellow eyes and unmarked tail. **V** Very high, drawn-out *whi-eeeh*, like a toy trumpet. **H** Swampy woodland and forest. **D** E Panama through Amazonia.

25 CRANE HAWK *Geranospiza caerulescens* 50cm Slender, sluggish bird, specialised in extracting nestlings from hollow trees. Note long tail and white crescent across outer primaries. **V** Mid-high mewing *weeooo*, rising, then falling and tapering off in *Buteo*-like call. Also hollow, rattling *cuwokokokokoo.* **H** Open swampy woodland, mangrove, forest edges. **D** Mexico to Uruguay.

26 PLUMBEOUS HAWK *Cryptoleucopteryx plumbea* 35cm Note grey underparts, white underwings, black tail with single white bar. Does not soar. **V** Drawn-up, falling-off *weeeeeeh.* **H** Forest. **D** Colombia to Peru.

27 SLATE-COLORED HAWK *Buteogallus schistaceus* male 43cm, female 47cm Hefty. Dark slate grey, including underwing. Cere, bare face and longish legs bright orange. Iris yellow. Tail with mid-band and fine tip white. **V** Series of yelping whistles, slowing down towards end. Also shrill downslurred whistle *eeeurrrrrr.* **H** Lowland tropical and várzea forest, usually near water, mangroves, gallery forest. **D** Amazonia.

28 COMMON BLACK HAWK *Buteogallus anthracinus* 51–58cm All black except white tail band and tip. **V** Series of sharp whistles or screams and a harsh *haaaah.* **H** Montane forest. **D** SW USA to NW South America.

29 CUBAN BLACK HAWK *Buteogallus gundlachii* 51–58cm Previously considered conspecific with Common Black Hawk. Separated by range. **V** Series of sharp whistles or screams and a harsh *haaaah.* **H** Coastal forests, swamps and beaches. **D** Cuba.

30 RUFOUS CRAB HAWK *Buteogallus aequinoctialis* male 42cm, female 47cm Hood and upperparts blackish brown. Underparts and underwing cinnamon rufous, finely barred blackish. Short tail black with narrow white mid-bar and tip. **V** Series of sharp whistles, rising then falling in pitch, slowing towards end. **H** Mangroves, saline coastal marshes, wet savannah, river shores. **D** Coastal Venezuela to coastal NE Brazil.

31 SAVANNA HAWK *Buteogallus meridionalis* male 48cm, female 58cm Overall cinnamon, back and part of wing dark brown. Underparts finely barred blackish. Tail black with white mid-band and tip. **V** Plaintive, strongly downslurred whistle. **H** Open grassland and savannah with scattered trees or palms, marshes in woods, forest edges, riverine forest, mangroves. **D** Panama to N Argentina.

32 WHITE-NECKED HAWK *Buteogallus lacernulatus* male 40cm, female 43cm Head and underparts white. Back and wings black. Short tail black with very broad white subterminal band. **V** Series of shrill whistles, slowing and falling in pitch towards end. **H** Coastal mountainous Atlantic forest. **D** E Brazil.

33 GREAT BLACK HAWK *Buteogallus urubitinga* 60cm Note long tail and legs, double tail bars and grey lores. **V** Very high, shrill, lonely *fjeeeehir fjuuuuuh.* **H** Forest edges at water. **D** Mexico to N Argentina.

34 SOLITARY EAGLE *Buteogallus solitarius* 70cm Grey (not black) wing tip at or exceeding tail. Lore yellow. Rarely looks crested. **V** Very high, drawn-out *tjuuuwut*, like a train whistle. **H** Hilly and montane forest. **D** Mexico to W and N South America.

35 CHACO EAGLE *Buteogallus coronatus* male 75–79cm, female 80–84cm All grey with prominent crest. Shortish tail grey at base with broad whitish band halfway down, broad black subterminal band and narrow white tip. **V** Mournful rising and falling whistle, sometimes rather harsh-toned. **H** Savannah with trees, Chaco woodland. **D** E Bolivia, Paraguay and S Brazil to C Argentina.

36 BARRED HAWK *Morphnarchus princeps* male 53cm, female 59cm Head, neck, upper breast and upperparts blackish slate. Rest of underparts white, barred dark slate. Tail blackish slate, with three narrowish white bars on basal half. In flight, underwing all barred. **V** Wailing high-pitched downslurred whistle. **H** Montane forest to subtropical uplands, clearings, forest edges. **D** Costa Rica to N Ecuador.

37 ROADSIDE HAWK *Rupornis magnirostris* 33–41cm Relatively long tail with 4–5 grey bars. **V** Drawn-out *rreeeaew.* **H** Open and semi-open country. **D** Mexico to NE Argentina.

38 HARRIS'S HAWK *Parabuteo unicinctus* 46–56cm Brown with reddish shoulders. Tail reddish with white base and tip. **V** Harsh *raaaaaak* and a grating *keh keh keh keh keh.* **H** Semi-arid woodland, brushland, semi-desert. **D** SW USA to S Argentina and Chile.

39 WHITE-RUMPED HAWK *Parabuteo leucorrhous* male 33cm, female 37cm Mostly black, but rump, uppertail coverts, vent and two bands on tail white. Thighs cinnamon, faintly barred brown. Eye yellow. Rare. Feeds on small vertebrates and insects. **V** Sharp, piercing downslurred whistle. **H** Dense vegetation of cloud forest, rainforest, Chaco woods. **D** Venezuela to NW Bolivia, SE South America.

1
2
3
4
5
6
7
8
9
10
11
12
13
14
15
16
17
18
19
20
21
22
23 ♂
24
25
26
27
28
29 ♀
30
31
32
33
34
35
36
37
38
39

KITES, HAWKS AND EAGLES *CONTINUED*

1 WHITE-TAILED HAWK *Geranoaetus albicaudatus* male 52cm, female 60cm Upperparts and hood blackish grey, shoulders rufous, underparts white. Short tail white with black subterminal band. Wings appear much longer than tail in perched bird. **V** Harsh chattering *keha-keha-keha*. **H** Open grassland with trees, bushy steppe, savannah, open woods, plantations. **D** SW USA to C Argentina.

2 VARIABLE HAWK *Geranoaetus polyosoma* male 45cm, female 55cm Very variable with black, rufous, grey and intermediate forms. Wings shorter than in White-tailed Hawk, not extending beyond tail in perched bird. **V** Rapid series of yelping notes, also a chatter. **H** Sea level to high mountains in steppe and brushland, upland grass, temperate forest, gallery forest, páramo. **D** W South America, from Colombia to Tierra del Fuego, Falkland Islands.

3 BLACK-CHESTED BUZZARD-EAGLE *Geranoaetus melanoleucus* male 62cm, female 78cm Huge. Broad wings and very short tail giving delta-wing effect. Upperparts and breast grey, underparts whitish, tail grey. **V** Harsh yelps in fast series. **H** Mountains with trees and bushes, Chaco, savannah, steppe, lowlands, páramo. **D** Venezuela to S Chile, SE South America.

4 MANTLED HAWK *Pseudastur polionotus* male 47cm, female 56cm Head, neck and underparts white, upperparts black, flight feathers tipped white. Uppertail coverts edged white. Tail white, basal half black. **V** Shrill, slightly grating upslurred whistle. **H** Tropical forest in mountains. **D** E Brazil, Uruguay, Paraguay.

5 WHITE HAWK *Pseudastur albicollis* male 48cm, female 58cm Variable races. Head, neck and underparts white, or with streaked crown and neck. Back and wings black, feather tips variably edged white. Tail black, white at base and tip. **V** Short, clipped high-pitched whistles. **H** Rainforest, forest edges, open woods, lowlands, mountains. **D** S Mexico through Amazonia.

6 GREY-BACKED HAWK *Pseudastur occidentalis* male 49cm, female 57cm Upperparts blackish lead. Underparts white. Crown, cheeks and sides of the neck densely streaked. Tail white with broad black subterminal band. **V** Slow, doleful and harsh downslurring whistle. **H** Deciduous and evergreen forest, cloud forest. **D** W Ecuador, NW Peru.

7 SEMIPLUMBEOUS HAWK *Leucopternis semiplumbeus* 35cm White underparts and black tail with single white bar diagnostic. Does not soar. **V** From high, swept-up to ultra-high *wueuw*. **H** Lower storeys of forest and tall secondary growth. **D** Honduras to NW Ecuador.

8 BLACK-FACED HAWK *Leucopternis melanops* male 38cm, female 43cm Small. Lores and around eye black. Crown, hindneck and upper back white, streaked black. Upperparts black with white markings. Underparts white. Tail black with white medial band and narrow tip. **V** Piercing downslurred whistle. **H** Lowland rainforest, gallery forest, mangroves. **D** N Amazonia.

9 WHITE-BROWED HAWK *Leucopternis kuhli* male 37cm, female 45cm Small. Crown and face mask black, brow white. Back of cheek and sides of neck and breast white, streaked black. Upperparts dark slate grey. Underparts white. Tail black with white band and tip. **V** Similar to Black-faced Hawk. **H** Lowland tropical rainforest. **D** S Amazonia.

10 GREY HAWK *Buteo plagiatus* 41–43cm Swoops onto ground-dwelling prey from tree perch. **V** Series of long, plaintive whistles. **H** Streamside woods with nearby open land. **D** SW USA to NW Costa Rica.

11 GREY-LINED HAWK *Buteo nitidus* 38–45cm Feeds mainly on ground-dwelling reptiles. Prey located by prolonged observation from high perch. Banded tail striking in flight. **V** Series of piping notes and whistles. **H** Forested habitats. **D** Costa Rica to NC Argentina.

12 RED-SHOULDERED HAWK *Buteo lineatus* 40–61cm Red shoulders apparent in perched birds. Flies with wings slightly bowed. **V** Squealing, repeated *keeyuur*. Also a high *kilt* or *kilt kilt kilt*. **H** Wet woodland. **D** SE and C North America, NE Mexico.

13 RIDGWAY'S HAWK *Buteo ridgwayi* 36–41cm Critically Endangered due to persecution and habitat destruction. **V** Various squealing notes. **H** Undisturbed forests. **D** Now virtually confined to a small area in NE Dominican Republic.

14 BROAD-WINGED HAWK *Buteo platypterus* 35–41cm Shy. Flies with wings held level or slightly angled down. **V** Piercing, high-pitched whistle *teeteeeee*. **H** Deciduous and mixed woodland. **D** E and C North America.

15 WHITE-THROATED HAWK *Buteo albigula* male 38cm, female 47cm Tail longer and wings narrower than in Short-tailed Hawk. Upperparts brown, underparts white, tail with faint grey bars and subterminal band. Hunts birds and rodents. Flies agilely through vegetation and rises on thermals. **V** Rather thin high-pitched whistle. **H** Andean woods, from sea level to alpine areas, stunted vegetation. **D** Venezuela to C Chile.

16 SHORT-TAILED HAWK *Buteo brachyurus* male 35cm, female 40cm Long, broad wings and short tail notable in flight, wing tips bent upwards. Upperparts blackish, underparts white, tail with three black bars and broad subterminal band. Perches in open and hunts on the wing. **V** Sharp downslurring whistle *wheeeuhrrr*. **H** Forest and woods, mountains to plains, open land, scattered trees, thorny woods, often near water. **D** S Florida, E Mexico to N Argentina.

17 HAWAIIAN HAWK *Buteo solitarius* 45cm Pale and all-dark morphs occur in equal numbers. Compact jizz with short, rounded wings. **V** High, drawn-out *eeeoh* or series of excited *kiew* notes. **H** From farmland to forest; up to 2,700m. **D** Hawaiian Islands.

18 SWAINSON'S HAWK *Buteo swainsoni* 50cm Pale, dark and intermediate forms occur. Note diagnostic grey undertail and tawny undertail coverts, even in dark form. White underwing linings contrast with dark flight feathers. **V** Very high, fluted, drawn-out *tfuuuuh winuuuh*. **H** Over open country. **D** SW Canada, W USA and N Mexico. Winters in S South America.

19 GALAPAGOS HAWK *Buteo galapagoensis* male 50cm, female 55cm Sooty black. Very trusting, perching in open. Feeds on small vertebrates and insects. **V** Short, rather gull-like yelps. **H** Seashores to woods in mountains. **D** Galápagos.

20 ZONE-TAILED HAWK *Buteo albonotatus* 45–56cm Tail with 2–3 pale grey bars. **V** Screaming *meeeeeahhr*. **H** Wooded gorges, riparian woodland, scrubland. **D** SW USA to Paraguay.

21 RED-TAILED HAWK *Buteo jamaicensis* 48–64cm Regularly seen soaring. **V** Rasping scream, said to sound like a rusty hinge. **H** Mountains, woodland, open country, around human habitation. **D** Widespread across North and Central America.

22 RUFOUS-TAILED HAWK *Buteo ventralis* male 55cm, female 60cm Hefty. Short legs, powerful feet with long toes. Upperparts dark brown, underparts densely streaked and barred rusty and blackish. Tail from above rich rufous, barred black. Hunts birds up to duck size. **V** Downslurred yelping call, becoming grating like a squeaky hinge. **H** South Andean woods. **D** S Chile to SW Argentina.

23 FERRUGINOUS HAWK *Buteo regalis* 56–64cm Soars with wings slightly raised. Dark morph less common than pale morph. **V** Low, gull-like *kree-a* and a whistled *k-hiiiiiiiiw*. **H** Open arid grasslands. **D** SC Canada to WC USA. Winters in N Mexico.

24 ROUGH-LEGGED BUZZARD *Buteo lagopus* 50–60cm Variable, some with a pale belly. Also a rare dark morph with black body and underwing coverts. Actions and habits much as other buzzards. **V** Low-pitched, cat-like *peeeooo*. **H** Tundra, mainly treeless but also wooded areas in years of lemming and vole abundance. Winters in open country. **D** Widespread across North America and Eurasia.

25 UPLAND BUZZARD *Buteo hemilasius* 66–71cm Plumage variable. Soars with wings held in a deeper 'V' than Common Buzzard. **V** Prolonged mewing. **H** Open areas in hills and mountains. **D** Breeds in Ladakh, India. Winter visitor to Himalayas.

26 EASTERN BUZZARD *Buteo japonicus* 55cm Very variable, most with distinctive dark wrist patches, these normally less dark than in Rough-legged Buzzard. **V** In flight, very high, mewing, rapidly descending, whistled *niau*. **H** Forest edges and half-open and open areas with scattered trees or posts for perching. **D** C and S Siberia, Mongolia, NE China, Japan.

27 HIMALAYAN BUZZARD *Buteo burmanicus* 51–57cm Rufous-thighed buzzard from the Himalayas. Regularly perches in open on posts or trees. **V** Mewing *peeeeooo*. **H** Open country with scattered trees. **D** Himalayas.

28 LONG-LEGGED BUZZARD *Buteo rufinus* 50–65cm Variable. Three main colour morphs, medium morph generally rufous. Often sits on prominent perch for prolonged periods. **V** Mellow *aaah*. **H** Steppe, semi-desert, open woodland of plains, hills and mountains. **D** C Europe to C Asia, N Africa, Arabian Peninsula.

29 CAPE VERDE BUZZARD *Buteo bannermani* male 45cm, female 55cm Similar to Common Buzzard but geographical range aids identification. **V** In flight, very high, sharp, rapidly descending, whistled *niau*. **H** Open natural and cultivated areas with at least a few lookout posts. **D** Cape Verde Islands.

30 SOCOTRA BUZZARD *Buteo socotraensis* 51–57cm Separated from Common Buzzard in 2010. Breast and abdomen yellowish white. **V** Mewing *peeeooo*. **H** Woodland and woodland edges with adjacent open land, mountains. **D** Socotra Island, Yemen.

31 COMMON BUZZARD *Buteo buteo* 51–57cm Variable. Regularly perches in the open on telegraph poles, posts or trees, and also spends long periods soaring. **V** Mewing *peeeooo*. **H** Woodland and woodland edges with adjacent open land, mountains. In winter often occurs in more open country. **D** Widespread across Eurasia.

32 FOREST BUZZARD *Buteo trizonatus* male 40cm, female 45cm Normally brown without any rufous coloration. Underparts spotted (not streaked, striped or barred). Tail has indistinct terminal bar. **V** In flight, high, short, slightly descending *piuuh*. **H** In and around mountain forests. **D** S and E South Africa.

33 MOUNTAIN BUZZARD *Buteo oreophilus* male 40cm, female 45cm Rather small. Normally brown without any rufous coloration. Underparts spotted (not streaked, striped or barred). Indistinct terminal bar on underside of tail. **V** In flight, high, short, slightly descending *piuuh*. **H** In and around mountain forests. **D** Ethiopia to N Malawi.

34 ARCHER'S BUZZARD *Buteo archeri* 50–55cm In flight, note rufous body and broad, whitish wings with narrow dark trailing edge. Feeds on terrestrial small mammals, birds, reptiles. Detects prey either by soaring or watching from a perch for long periods. **V** Unknown. **H** Upland savannah and grassland. **D** N Somalia.

35 RED-NECKED BUZZARD *Buteo auguralis* male 40cm, female 45cm Note characteristic reddish glow. **V** Mewing call, similar to that of Common Buzzard. **H** Forest edges, heavily wooded areas. **D** Sierra Leone to Ethiopia, Uganda and Angola.

36 MADAGASCAN BUZZARD *Buteo brachypterus* 40cm Variable (bird shown is chestnut-brown form). Typical buzzard jizz (stocky, with wide head and broad, short wings); no other buzzard in its range. In flight from below, note pale median wing coverts, while lesser coverts are coloured like throat and upper breast. **V** Downslurred, quavering yelping whistle. **H** Any wooded area. **D** Madagascar.

37 AUGUR BUZZARD *Buteo augur* male 50cm, female 55cm Unmistakable. Dark morph form occasionally seen in NE DR Congo. **V** In flight, high *kjow-kjow-kjow-kjow*. **H** Open hilly areas, woodland, cultivation, suburbs. **D** Ethiopia and Somalia to Zimbabwe, C Angola to C Namibia.

38 JACKAL BUZZARD *Buteo rufofuscus* male 50cm, female 55cm Some individuals may lack the chestnut breast band but all have black underwing coverts. **V** In flight, mid-high jackal-like *iaueeh-iaueeh-iahueh*. **H** Open mountainous areas. **D** Namibia and South Africa.

rufous form
♂
2
♂
♀
3
4
5
6
7
8
1
9
16
17
dark morph
pale morph
10
11
12
13
14
15
18
dark morph
pale morph
19
20
21
22
dark morph
23
pale morph
24
dark morph
25
26
27
28
29
dark morph
30
31
32
33
34
35
36
♂
37
♀
38

BUSTARDS OTIDIDAE

1 GREAT BUSTARD *Otis tarda* male 105cm, female 75cm During display male appears to turn itself inside out, due to inflated neck, cocked tail and erected white wing feathers. **V** Low bark during disputes or when alarmed. During display gives a hollow drone. **H** Rolling or flat open grasslands, occasionally in Cork Oaks (*Quercus suber*) and Olive (*Olea europaea*) groves. **D** SW and C Europe and NW Africa to E Siberia, Mongolia and E China.

2 ARABIAN BUSTARD *Ardeotis arabs* male 100cm, female 75cm Upperwing flight feathers black with indistinct grey-white bars on inner primaries and secondaries. During display neck feathers are puffed out, tail cocked and spread, and wings drooped. **V** During display utters a *puk-puk*. **H** Semi-desert, dry grasslands, arid bush, cultivated fields. **D** W, C and NE Africa.

3 KORI BUSTARD *Ardeotis kori* 85cm Note predominantly grey wings in flight. Separated from other bustards by shaggy neck feathers. **V** Usually quiet but may give low, dry croak. **H** Dry, open plains with some trees and bush. **D** E and S Africa.

4 GREAT INDIAN BUSTARD *Ardeotis nigriceps* male 120cm, female 90cm Critically endangered. Unmistakable, large and weighty. Flies with neck outstretched. During display male inflates neck and struts around with tail cocked and wings drooped, calling with a deep booming moan. **V** Booming during display, also a gruff, barked *hook*. **H** Dry grassland with scattered bushes, also neighbouring open, dry deciduous forest. **D** India.

5 AUSTRALIAN BUSTARD *Ardeotis australis* male 120cm, female 90cm Unmistakable in geographical range. Mostly solitary; walks with stately gait. Feeds on invertebrates, small vertebrates, seeds and caper berries. **V** Croaking calls while feeding; displaying male utters roaring bark. **H** Grassland, open scrub, savannah. **D** Australia, S New Guinea.

6 HOUBARA BUSTARD *Chlamydotis undulata* male 65–75cm, female 55–65cm Generally shy, tending to hide when disturbed rather than fly. During display neck and head feathers are held erected and puffed out, as bird runs around display ground. **V** Generally silent. **H** Open semi-desert and dry steppe, usually with grass or low scrub. **D** N Africa, Canary Islands.

7 MACQUEEN'S BUSTARD *Chlamydotis macqueenii* male 65–75cm, female 55–65cm In flight, upperwing shows large white patch on base of outer primaries, inner primaries, secondaries and primary coverts black, the latter with whitish base patch. Flies with neck outstretched. **V** Generally silent. **H** Semi-desert with low scrub, sandy grasslands, crop fields. **D** Middle East to C China.

8 LUDWIG'S BUSTARD *Neotis ludwigii* 70cm Note darkish head, and small white spots on mantle and scapulars. Upper primary coverts mainly white. **V** Little documented. Dry knocking sounds recorded. **H** Dry areas, especially in the Karoo and Namib. **D** Angola, Namibia, South Africa.

9 DENHAM'S BUSTARD *Neotis denhami* male 100cm, female 80cm Upperwing flight feathers black with a white patch at base of inner primaries. Tips of secondary coverts white, forming a bar. Tail with two wide black-and-white bars. **V** Barking *kaa-kaa*. **H** Grassland, with or without bushes. **D** Widespread across Africa.

10 HEUGLIN'S BUSTARD *Neotis heuglinii* 75cm Female has grey (not rufous) neck in contrast to some other similar bustards. **V** Not known, probably mostly silent. **H** Desert and semi-desert with some occasional grass cover. **D** Eritrea to Somalia and N Kenya.

11 NUBIAN BUSTARD *Neotis nuba* male 70cm, female 50cm Upperwing flight feathers black with large white patch at base of primaries, contiguous with white bar at base of secondaries. Primary coverts white with broad black tips, black patch on bend of wing. **V** Low *wurk*, otherwise little documented. **H** Arid and semi-arid scrub or savannah. **D** Mauritania to E Sudan.

12 WHITE-BELLIED BUSTARD *Eupodotis senegalensis* 45cm Female separated from female Black-bellied Bustard by much shorter neck and legs, and pale underwing. **V** Series of short, dry rolling calls, alternating shorter deeper notes with higher, longer notes. **H** Grassy, lightly bushed and wooded parts of natural and cultivated areas, often near streams. **D** Widespread across Africa.

13 BLUE KORHAAN *Eupodotis caerulescens* 50cm Note blue colouring on underparts and upperwing. **V** Cacking, harsh song *kak-agagag*. **H** Open, short grassland. **D** E South Africa.

14 KAROO KORHAAN *Eupodotis vigorsii* 50cm Distinguished from Rüppell's Korhaan by having small black patches only on throat and nape (which can be puffed up and out in display). In flight, also shows much more dark brown on wing. **V** Deep, querulous quacking notes. **H** Dry scrubland with adjacent cultivated grassland and cropland. **D** Namibia, South Africa.

15 RÜPPELL'S KORHAAN *Eupodotis rueppelii* 50cm Note pale head and neck, and streaks down neck and throat. Separated from White-bellied Bustard by plain (not barred) upperparts and mainly cream upperwing in flight. **V** Frog-like or duck-like quacking croaks. **H** Vast, dry plains with sparse grass or scrub. **D** S Angola, Namibia.

16 LITTLE BROWN BUSTARD *Eupodotis humilis* 45cm Both male and female rather plain except for some black markings on nape and throat of male. Upperparts less distinctly striped and blotched than in other bustards. **V** Rattling, high-pitched call. **H** Dry, open bush. **D** E Ethiopia, N Somalia.

17 SAVILE'S BUSTARD *Lophotis savilei* 45cm Note black stripe down chin. Reddish crest normally concealed. **V** Sweet-toned short song of short whistles, with pause after first note, later notes falling in pitch. **H** Bushland and woodland with open areas. **D** Mauritania, Senegal and Gambia to C Sudan.

18 BUFF-CRESTED BUSTARD *Lophotis gindiana* 55cm Male has a distinct crest, no black behind eye and no white along black area of neck. Black-bellied male has no grey or black in face. **V** Powerful, pleasant-toned short whistles. **H** Dry, open, tall grassland, often with some shrubs and trees. **D** Ethiopia and Somalia to N Tanzania.

19 RED-CRESTED KORHAAN *Lophotis ruficrista* 50cm Reddish crest normally concealed. Both sexes separated from female Southern and Northern Black Korhaans by different pattern on upperparts (chevrons instead of fine barring). Note two conspicuous white patches within black area of breast. **V** Series of loud, pleasant-toned short whistles, slowing and falling in pitch. **H** Open woodland, locally in treeless grasslands. **D** Angola and Namibia to Mozambique, Swaziland and South Africa.

20 SOUTHERN BLACK KORHAAN *Afrotis afra* 45cm See Northern Black Korhaan. **V** Rushing, short series of whistles. **H** Extensive, lightly wooded grassland. **D** W and S South Africa.

21 NORTHERN BLACK KORHAAN *Afrotis afraoides* 45cm Male separated from male Southern Black Korhaan by white area in wing and darker crown. **V** Like that of Southern Black Korhaan. **H** Open, dry grassland, locally mixed with shrubland and woodland. **D** S Africa.

22 BLACK-BELLIED BUSTARD *Lissotis melanogaster* 55cm Note striking long, thin neck and legs, narrow black stripe running up front of neck and extensively white upperwing. Female separated from female White-bellied Bustard, Karoo Korhaan and Rüppell's Korhaan by blackish-brown (not partly creamy-buff) wings. **H** Tall grassland with some trees and lightly wooded farmland. **D** Widespread across Africa.

23 HARTLAUB'S BUSTARD *Lissotis hartlaubii* 60cm Note dark grey tail, grey-brown appearance and black (not mainly grey) face mask. Female separated from female Black-bellied Bustard by greyer neck. **V** Low clicking and popping sounds. **H** Arid open plains with scarce shrub and some acacias. **D** Sudan and South Sudan to Somalia and Kenya.

24 BENGAL FLORICAN *Houbaropsis bengalensis* 64–68cm Rare and declining. Flies with neck outstretched. During display male leaps into air, hovering momentarily at top of jump before gliding back down. **V** During display male emits a deep hum. When alarmed, utters a shrill, metallic *chik-chik-chik*. **H** Tall grassy plains with scattered scrub, feeding in more open areas in early morning. **D** S Nepal, NE India, S Vietnam.

25 LESSER FLORICAN *Sypheotides indicus* 46–51cm Endangered. Flies with neck outstretched. During display male leaps into air, showing off white wing patch before dropping down and repeating performance. **V** During display male utters a loud rattling. When alarmed, gives a short whistle. **H** Tall, dry grassland with scattered bushes, fields of cotton or millet. **D** India.

26 LITTLE BUSTARD *Tetrax tetrax* 43cm Flight feathers of male upperwing mainly white, with prominent black tips to outer primaries and black primary coverts appearing as a crescent. During display inflates neck feathers and leaps with wings held open. **V** Dry *prrit* and a low grunt when disturbed. In flight, wings make a whistling *sisisisisi* noise. **H** Open grasslands, rough plains, pastures, crop fields. **D** SW Europe and NW Africa to C Asia.

MESITES MESITORNITHIDAE

27 WHITE-BREASTED MESITE *Mesitornis variegatus* 31cm Separated from male Subdesert Mesite by crescent-shaped breast spots and straight bill. **V** High, rather sharp, fluted *peep-peep-(pip)-peep*, repeated 3–6 times. **H** Mostly undisturbed forest with some dense shrubby underlayer at lower altitudes. **D** Madagascar.

28 BROWN MESITE *Mesitornis unicolor* 30cm Unmistakable owing to rather uniform reddish-brown plumage, thick tail and white mark to neck sides. **V** Melodious *wuut-tjuw*, repeated 4–9 times. **H** Primary forest. **D** Madagascar.

29 SUBDESERT MESITE *Monias benschi* 32cm Unmistakable in range; note curved bill. **V** High, thin rattle or tickled *rit-rit-rit-tit*. **H** Spiny scrub to dry forest at low altitudes. **D** Madagascar.

SERIEMAS CARIAMIDAE

30 RED-LEGGED SERIEMA *Cariama cristata* 90cm Large. Notable untidy frontal crest. Bill and legs red. Bare blue around eye. Brownish-grey upperparts, slightly vermiculated. Underparts whitish, slightly streaked. Flight feathers barred black and white. **V** Pairs sing in duet, giving trumpeting notes that fall in pitch. **H** Open thorny woods and brush, Chaco, Caatinga, Cerrado, open areas with trees or bushes. **D** E Bolivia to E Brazil, south to N Argentina and Uruguay.

31 BLACK-LEGGED SERIEMA *Chunga burmeisteri* 70cm Lacks crest. Bill, eye-ring and legs black. Upperparts and tail grey, slightly vermiculated. Underparts paler. Flight feathers barred black and white. **V** Yelping, gull-like notes in series, falling in pitch. **H** Open dry woods. **D** S Bolivia and W Paraguay to C Argentina.

KAGU RHYNOCHETIDAE

32 KAGU *Rhynochetos jubatus* 55cm Unmistakable and heron-like. Diagnostic grey plumage and crest; red legs and bill. Forages in leaf litter for invertebrates and small vertebrates. Runs from danger. **V** Pairs duet, involving piping *wah wah-wah*, and *ko-ko* phrases. **H** Forests. **D** New Caledonia.

SUNBITTERN EURYPYGIDAE

33 SUNBITTERN *Eurypyga helias* 45cm No similar bird in its habitat. **V** High, pure, whistled descending note, sometimes with upwardly inflected disyllabic ending. Also a metallic rattle. **H** Vegetation along rivers and streams in forest, swamps. **D** Guatemala through Amazonia.

1
2
3
4
5
6
7
8
9
10
11
12
13
14
15
16
17
18
19
20
21
22
23
24
25
26
27
28
29
30
31
32
33

FLUFFTAILS SAROTHRURIDAE

1 GREY-THROATED RAIL *Canirallus oculeus* 30cm Unmistakable, but difficult to see clearly in its dense habitat. In flight, shows dark brown flight feathers, strikingly barred white. **V** Song a long series of hollow, drumming double-notes. **H** Undergrowth along forest streams. **D** Sierra Leone to Ghana, S Nigeria to Uganda and DR Congo.

2 MADAGASCAN WOOD RAIL *Canirallus kioloides* 28cm Note grey around eyes and rather long bill. **V** As Tsingy Wood Rail, but faster. **H** Undisturbed forests up to mid-high altitudes. **D** Madagascar.

3 TSINGY WOOD RAIL *Canirallus beankaensis* 30cm Separated from Madagascan Wood Rail by rufous colouring all around eyes, paler bill and different range. **V** Sustained, almost level series of resounding *tjuw*-notes. **H** Near rocks in or bordering dry forest at low altitudes. **D** W Madagascar.

4 WHITE-SPOTTED FLUFFTAIL *Sarothrura pulchra* 15cm Note pale eye. **V** High, hurried, hooted *poopoopoopoopoo*. Also a mid-high, fluted, rapid *poo-poo-poo-poo-poo* and very fast, high *tutitititi*. **H** Swampy forest and surrounding areas with tree cover. **D** W, WC and C Africa.

5 BUFF-SPOTTED FLUFFTAIL *Sarothrura elegans* 16cm Separated from White-spotted Flufftail by barred tail. **V** Low, hollow, level *booooo*, rising to a crescendo, and extremely low, hollow, rolling *rururuu*. **H** Forest, bamboo, dense bush, cultivation, gardens with dense undergrowth. **D** Widespread across Africa.

6 RED-CHESTED FLUFFTAIL *Sarothrura rufa* 16cm See Streaky-breasted Flufftail. Separated from female White-spotted and Buff-Spotted flufftails by brown (not rufous-orange) tail. **V** Mid-high hooted *hoo hoo hoo hoo*, often in duet with very high *piu piu piu* contra-song (*hoo* and *piu* alternating). Also a call resembling alarm cry of Black-tailed Godwit. **H** Marshy reedbeds and Papyrus (*Cyperus papyrus*) swamp, thick herbage and tall grass near rivers and pools. **D** Widespread across Africa.

7 CHESTNUT-HEADED FLUFFTAIL *Sarothrura lugens* 15cm Black-tailed male separated from Red-chested and Streaky-breasted flufftails by absence of rufous on breast. Female separated from other female flufftails by streaked (not barred or blotched) plumage. **V** Sequence of *oohoohooh* notes, rising in pitch and volume, and trailing off at end. Also mid-high *ooeh ooeh ooeh*. **H** Tall wet grassland near marshes, woodland or forest. **D** C and SC Africa.

8 STREAKY-BREASTED FLUFFTAIL *Sarothrura boehmi* 15cm Very short tail and pale lower mandible. Male distinguished from Red-chested Flufftail by whitish chin and belly. **V** Mid-high level, rhythmic *hooh hooh hooh*, also with shorter intervals and *uhooh-uhooh-uhooh* (*u-* as very low undertone). **H** Short wet grassland, inundations, river margins. **D** Cameroon to Kenya and south to Zimbabwe.

9 STRIPED FLUFFTAIL *Sarothrura affinis* 15cm Separated from Chestnut-headed Flufftail by habitat and red tail (male) or black-striped buff-orange tail (female). **V** Mid-high, thin, hollow *hooo hooo hooo*, each *hooo* crescendoing. Also sharp, mid-high rattle, ending in prolonged *wekwekwekwek*.**H** Dry grassland. **D** E and S Africa.

10 MADAGASCAN FLUFFTAIL *Sarothrura insularis* 14cm Very distinctively striped white in black. **V** Varied, including harsh, sharp, loud shrieks and cries, given singly or in long series. **H** Wet and dry habitats, including grassland, herbaceous marsh, brush, often at forest edges; at any altitude. **D** Madagascar.

11 WHITE-WINGED FLUFFTAIL *Sarothrura ayresi* 15cm Note brown mantle and white in wings. **V** Low *oow-oow-oow-oow* and other rhythmic pigeon-like calls. **H** Marshes, reedbeds, wet grassland. **D** Ethiopia, Zimbabwe, E South Africa.

12 SLENDER-BILLED FLUFFTAIL *Sarothrura watersi* 15cm Without, or with hardly any, marking except some barring across rufous tail. **V** Series of frog-like croaks. **H** Herbaceous marshes at higher altitudes. **D** Madagascar.

FINFOOTS HELIORNITHIDAE

13 AFRICAN FINFOOT *Podica senegalensis* 65cm Red bill (and feet) diagnostic. Swims with 'pumping' head movements, not submerged as African Darter (Plate 29). **V** Mid-high, dry, bouncing *crut crut-crur crut crut*. **H** Quiet rivers, streams, pools with fringing vegetation and overhanging trees. **D** Widespread across Africa.

14 MASKED FINFOOT *Heliopais personatus* 43–55cm Very elusive, although can be quite confiding. Most active at dawn or dusk. **V** High-pitched bubbling, grunting quack and, when alarmed, a *keek-keek-keek*. **H** Fresh or brackish water in dense forest and mangroves. **D** NE India to Sumatra, Indonesia.

15 SUNGREBE *Heliornis fulica* 30cm No similar bird in its habitat. **V** Irregular barking, descending, grebe-like *wek wek wek weh*. **H** Calm water with overhanging vegetation and branches in woodland, mangroves. **D** SE Mexico to NE Argentina.

RAILS, CRAKES AND COOTS RALLIDAE

16 NKULENGU RAIL *Himantornis haematopus* 45cm Note large size, black lore and red legs. **V** Low, rhythmic *boom tuckeh-heh boom tuckeh-heh*, actually an inseparable duet (common name of species is supposed to imitate its call). **H** Undergrowth along forest streams. **D** Sierra Leone to Uganda and DR Congo.

17 SWINHOE'S RAIL *Coturnicops exquisitus* 13cm Very secretive. In flight, shows white secondaries. **H** Wet meadows, grassy marshes or swamps. **V** *Tick-tick tick-tick-tick*. **D** E Siberia, NE China.

18 YELLOW RAIL *Coturnicops noveboracensis* 16–19cm Very secretive. In flight, has distinctive white secondaries. **V** *Tick-tick ticktictick tick-tick ticktictick*, said to sound like tapping two pebbles together. **H** Grassy marshes, damp grassland, salt marshes and rice fields. **D** E and C North America, from Canada to C Mexico.

19 SPECKLED RAIL *Coturnicops notatus* 13cm Tiny and dark. Mouse like. Very rare. Brownish black, spotted, barred and dotted white. In flight, shows white patch on secondaries. **V** When alarmed gives a quiet *coueee-cab* and *peep*. **H** Tall savannah grassland, dense marshy vegetation, rice paddies, alfalfa fields, wet edges of forest, ditches, flooded fields, tall grass by temporary puddles. **D** Paraguay and S Brazil to EC Argentina.

20 OCELLATED CRAKE *Micropygia schomburgkii* 14cm Very small. Overall rich ochre. Cinnamon wash on upperparts, which have drop-shaped spots with black borders. Legs light red. **V** Insect-like *prrrssss*. **H** Tall damp grass near water and bogs, dry grassland savannah. **D** N, SC and SE South America.

21 CHESTNUT FOREST RAIL *Rallicula rubra* 18–22cm Forages on forest floor and stream margins, mainly for invertebrates. Back marked with white spots (female). Creeps away from danger. **V** Harsh, repeated *keree-keree...* **H** Montane forest. **D** New Guinea.

22 WHITE-STRIPED FOREST RAIL *Rallicula leucospila* 20–22cm Forages on forest floor and stream margins, mainly for invertebrates. Back marked with white streaks (male). Creeps away from danger. **V** Repeated, croaking *who-a, who-a, who-a...* **H** Montane forest. **D** New Guinea.

23 FORBES'S FOREST RAIL *Rallicula forbesi* 20–25cm Seen in pairs or small parties. Forages on forest floor, mainly for invertebrates. Back is unmarked (male). Creeps away from danger. **V** Repeated, croaking *coo-ak, coo-ak, coo-ak...* **H** Montane forest. **D** New Guinea.

24 MAYR'S FOREST RAIL *Rallicula mayri* 20–22cm Rather uniform reddish-brown plumage. Male has only indistinct markings on back. Forages in streams and on forest floor, mainly for invertebrates. **V** Unknown. **H** Montane forest. **D** New Guinea.

25 RED-NECKED CRAKE *Rallina tricolor* 23–30cm Forages on forest floor, along streams or in shallow pools, singly or occasionally in pairs or small groups. **V** Repeated, descending *nark-nak-nak* or *kare-kare-kare*; also a monotonous *tock tock tock*. Contact calls include a low *um um um* and soft, repeated *plop*. **H** Primary and tall secondary lowland rainforest, monsoon forest, swamp forest. **D** Widespread across Australasia, from the Lesser Sunda Islands, Indonesia, through New Guinea to NW Australia.

26 ANDAMAN CRAKE *Rallina canningi* 34cm Skulking. Note bright green bill and barred underparts. **V** Deep, croaking *kroop kroop*. When alarmed, gives a sharp *chick chick*. **H** Forest marshes and streams, occasionally mangroves. **D** Andaman Islands.

27 RED-LEGGED CRAKE *Rallina fasciata* 23–25cm Skulking, difficult to observe or flush. **V** Loud *gogogogok*, a *girrr*, a nasal *pek pek pek* and a slow, descending trill. **H** Reedy swamps and marshes, paddyfields, watercourses. **D** S Myanmar to Lesser Sunda Islands, Indonesia, and the Philippines.

28 SLATY-LEGGED CRAKE *Rallina eurizonoides* 27cm Shy. Often flies into trees when flushed. **V** Persistent *kek-kek kek-kek* or *ow-ow ow-ow*. **H** Forest, forest edges. **D** Widespread across Asia, from NW Pakistan to Japan and south to the Sula Islands, Indonesia.

29 CHESTNUT-HEADED CRAKE *Anurolimnas castaneiceps* 20cm Tail very short. Bill short and greenish yellow. Head, neck, breast and upper belly reddish chestnut. Legs red. **V** Long, loud, antiphonal duet *kook tocock, kook tocock*. **H** Terra firme rainforest, secondary growth and thickets along stream banks, lowlands, plantations. **D** S Colombia to NW Bolivia.

30 RUSSET-CROWNED CRAKE *Laterallus viridis* 17cm No barring anywhere. Cap rufous chestnut. Face dark brownish grey. Breast and flanks dark rufous, undertail more tawny. Legs pinkish red. **V** Call like those of other *Laterallus* but more staccato. **H** Damp, partly flooded grass, dry grass with bushes, sometimes far from water. **D** Amazonia.

31 BLACK-BANDED CRAKE *Laterallus fasciatus* 19cm Head, neck and breast deep rufous chestnut. Flanks, rest of underparts and undertail barred black and pale rust. Legs pinkish red. **V** Very long (15 sec) rapid trilling call *trrrrrrrrrrrr...*, gradually falling in pitch. **H** Tall damp or slightly flooded grass, bogs, vegetation along stream banks and ponds. **D** W Amazonia.

32 RUFOUS-SIDED CRAKE *Laterallus melanophaius* 17cm Crown brownish grey. Cheeks, sides of neck and breast, and undertail tawny rufous. Flanks barred black and white. Throat to mid-belly white. Legs buffish olive. **V** Call a long, descending, trilled *prrrrrrrrrrrrr*. **H** Tall grass, reeds, canes, bushes in damp or partly flooded areas near small lakes, marshes, streams and rivers, with or without nearby trees. **D** Widespread across South America.

33 RUSTY-FLANKED CRAKE *Laterallus levraudi* 16cm No barring anywhere. Forehead, cheeks, sides of breast and rest of underparts rusty rufous. Crown and hindneck brownish grey. Throat, foreneck and centre of breast white. Legs pinkish red. **V** Three-sec rapid trill, initially harsh and shrill, falling in patch and becoming softer. **H** Tall grass and vegetation by swamps, marshes, lagoons, flooded or dry grassland. **D** Venezuela.

34 RUDDY CRAKE *Laterallus ruber* 15cm Colouring diagnostic: mainly red with black cap and grey face sides. **V** Very high, shivering, fast *srisrisri*. **H** Marshes, reedbeds, wet grassland. **D** E Mexico to Costa Rica.

35 WHITE-THROATED CRAKE *Laterallus albigularis* 15cm Head, neck and breast rufous and tawny. Flanks, belly and undertail barred black and white. Legs greyish green. **V** Calls like Rufous-sided Crake: a long, descending *churrrrrr* trill. **H** Tall grass, partially flooded areas, bogs, forest clearings, thickets by streams. **D** Nicaragua to Ecuador.

36 GREY-BREASTED CRAKE *Laterallus exilis* 15cm Diagnostic red patch on nape. **V** Very high *bic-bic* or shivering *srrrrr*. **H** Tall grass in wet areas. **D** Guatemala through Amazonia to SE South America.

37 BLACK RAIL *Laterallus jamaicensis* 14cm Very small and dark. Upperparts have discontinuous barring. Upper back plain, brownish chestnut. Breast grey. Flight feathers all blackish. **V** Calls insistently *tip trrrr tip trrrr*. **H** Tall grass near water, flooded grass in brackish areas, reeds with grass, cultivated fields. **D** USA and Greater Antilles to W South America.

38 GALAPAGOS CRAKE *Laterallus spilonota* 14cm Tiny and almost uniformly very dark. Few white dots and little barring. **V** Short, high-pitched slightly dry twitter. **H** Tall grass and scrub near water. **D** Galápagos.

39 RED-AND-WHITE CRAKE *Laterallus leucopyrrhus* 17cm Head, neck, upper back and sides of breast deep rufous. Throat to upper belly white. Flanks and part of belly barred black and white. Undertail white. Legs red. **V** Long, descending, trilled *trrrrrrr*. **H** Thickets, grassland, reedbeds, other dense vegetation between water and forest on edge of streams, rivers and lakes. **D** SE South America.

40 RUFOUS-FACED CRAKE *Laterallus xenopterus* 17cm Head to upper back deep rufous. Rest of upperparts black, barred white. Tail black. Sides of neck and breast tawny. Legs purply chocolate. **V** Long descending trill, about 6 sec with last 3 sec much fainter. **H** Tall damp or partially flooded grass, marshes bordered by gallery forest and Cerrado, dense thickets. **D** SC Brazil, C Paraguay.

1
2
3
4
5
6
7
8
9
10
11
12
13
14
15
16
17
18
19
20
21
22
23
24
25
26
27
28
29
30
31
32
33
34
35
36
37
38
39
40

RAILS, CRAKES AND COOTS *CONTINUED*

1 WOODFORD'S RAIL *Nesoclopeus woodfordi* 30cm Flightless rail with overall dark plumage. Feeds on invertebrates, amphibians, small reptiles. Forages on land and in water. **V** Loud squeals and repeated *kik-kik-kik...* **H** Moist lowland forest, marshes, freshwater margins. **D** Solomon Islands.

2 WEKA *Gallirallus australis* male 55cm, female 48cm Flightless, sturdy chicken-sized rail. **V** High, upslurred shrieks or high, liquid *wew*, or ascending series of *cohweetcohweet*. **H** Margins of any type of wetland, water or forest with some low cover, including sea beach and cultivated land. **D** New Zealand.

3 CALAYAN RAIL *Gallirallus calayanensis* 30cm Forages on the ground, singly or in small groups; tends to run away if disturbed. **V** Series of hoarse, staccato *ngeck* notes. **H** Primary and secondary forest on coralline limestone. **D** Calayan Island, N Philippines.

4 NEW CALEDONIAN RAIL *Gallirallus lafresnayanus* 46–48cm Flightless, long-billed rail. Critically endangered, possibly extinct but may survive in remote, inaccessible areas. Habits and behaviour largely unknown. **V** Unknown. **H** Humid forest habitats. **D** New Caledonia.

5 LORD HOWE WOODHEN *Gallirallus sylvestris* 32–42cm Flightless, long-billed, stout-legged rail. Hunted and predated to verge of extinction; small population now thought to be stable. **V** Loud, piercing whistle often delivered in duet. **H** Forest habitats, mainly in uplands. **D** Lord Howe Island, Tasman Sea.

6 OKINAWA RAIL *Gallirallus okinawae* 30cm Runs when flushed. **V** *Kwi kwi kwi ki-kwee ki-wee, kyip kyip kyip kyip* or *ki-kik-ki*. **H** Broadleaved evergreen forest with dense undergrowth and wet areas. **D** Okinawa.

7 BARRED RAIL *Gallirallus torquatus* 33–35cm Very shy. Usually found singly, in pairs or occasionally in small groups. Regularly forages along little-used tracks. **V** Loud, discordant, harsh croaking and screaming, often lasting several seconds. **H** Rank grassland at borders of primary and tall secondary lowland and hill forest, mangroves, mixed scrub, cultivation. **D** W New Guinea; Sulawesi, Indonesia; Philippines

8 PINK-LEGGED RAIL *Gallirallus insignis* 33cm Flightless, or nearly so. Often seen in vocal groups. Feeds mainly on invertebrates and some plant material. **V** Barking, grunting calls often delivered in duet. **H** Moist forest, cultivated fields. **D** New Britain.

9 BUFF-BANDED RAIL *Gallirallus philippensis* 30cm Distinguished by coloured head pattern, pink bill and yellowish legs, and markings on upperparts. Usually seen alone, occasionally in small, loose groups; forages along water's edge, tracks or roadsides. **V** Braying *coo-aw-ooo-aw-ooo-aw-ooo-aw*, creaking *swit-swit* and squeaking *krek-krek-krek*. **H** Drier parts of marshes, grassland, canefields, rank grassland, open hills. **D** From Cocos Keeling Islands in the Indian Ocean through New Guinea to Australia, New Zealand and Pacific islands; also Philippines.

10 GUAM RAIL *Gallirallus owstoni* 28cm Distinctive dark rail with unbanded upperparts except barred flight feathers. Flightless. **V** Gives various whistles, yelps and short *kip* calls. **H** Varied, including short grass, fern thickets, woodland, forest edges. **D** Guam.

11 ROVIANA RAIL *Gallirallus rovianae* 30cm Flightless or nearly so. Usually keeps to vegetated cover and runs from danger. Feeds on invertebrates, small vertebrates and plant material. **V** Repeated, squealing notes. **H** Forests, cleared habitats with secondary growth. **D** Solomon Islands.

12 SLATY-BREASTED RAIL *Gallirallus striatus* 29cm Very secretive. Some authorities now place it in genus *Lewinia*. **V** Sharp *terrik* or *trrrik*, which may be strung together as a 'song'; also a noisy *ka-ka-ka*. **H** Marshes, mangroves, paddyfields, marshy grassland. **D** Widespread across Asia, from India and Sri Lanka to S China and Taiwan, and Borneo and the Philippines.

13 MANGROVE RAIL *Rallus longirostris* 33cm Upperparts blackish brown, back feathers bordered buffish brown. Vinaceous cinnamon on breast. Flanks and rest of underparts barred black and white. Bill pinkish, culmen blackish. **V** Call is a series of loud, rapid *kak* or *kek* notes, descending and getting weaker at end. **H** Salt, brackish and freshwater marshes with tall emergent vegetation, flooded grass, mangroves. **D** Coasts of W Central America and N South America to NW Peru.

14 CLAPPER RAIL *Rallus crepitans* 31–40cm Mainly crepuscular. Back markings generally less contrasting and neat than those of King Rail. **V** Series of 'clapping' notes, accelerating and then slowing. **H** Salt marshes. **D** E coast of USA to Caribbean and coastal Belize.

15 RIDGWAY'S RAIL *Rallus obsoletus* 31–40cm Mainly crepuscular, more often heard than seen. Forages along higher reaches of mudflats. **V** Very similar to King Rail, with a series of accelerating and then slowing,'clapping' notes *kek kek kek kek*. **H** Mangroves and salt marshes. Nests among mangrove roots in Apr and May. **D** SW USA to Baja California, Mexico.

16 KING RAIL *Rallus elegans* 38–48cm Secretive. Generally bright and clearly marked. **V** Clapping *kek kek kek kek*, usually deeper than that of Clapper Rail. **H** Mainly freshwater marshes. **D** E North America and Cuba. Northern birds migrate south to winter in S USA and Mexico.

17 AZTEC RAIL *Rallus tenuirostris* 38–48cm Shy and secretive, feeding mostly at dawn and dusk. More often heard than seen. **V** Repeated *ket ket ket ket ket*. **H** Tall, dense vegetation in swamps, canals and flooded areas. Breeds Jun–Sep. **D** C Mexico.

18 PLAIN-FLANKED RAIL *Rallus wetmorei* 27cm Endangered. No barring. Upperparts dark brown, back feathers edged buffish brown. Underparts greyish brown. **V** Calls short, abrupt, explosive and toneless *krk krk krk...* **H** Mangroves, fresh and brackish coastal ponds, estuaries. **D** Venezuela.

19 VIRGINIA RAIL *Rallus limicola* 20–25cm Mainly crepuscular, shy and secretive. **V** Series of rapid *kid kid kidic kidic* notes; also a pig-like *wep wep wep wepwepwepwepwepppprrr* grunting, which usually descends and accelerates. **H** Freshwater or brackish and saltwater bodies with dense vegetation. **D** Breeds from S Canada to S Mexico. Winters from Mexico to Guatemala.

20 ECUADORIAN RAIL *Rallus aequatorialis* 24cm Upperparts blackish brown, back feathers edged buffish brown. Wing coverts chestnut. Face grey with slight whitish brow. Foreneck, breast and upper belly vinaceous cinnamon. Undertail barred. Bill and legs dirty coral pink, culmen blackish. **V** Call is a metallic clattering, loud, rapid *kid kid kirik kirik*. **H** Swamps and freshwater marshes with tall emergent vegetation, flooded grass. **D** SW Colombia to SW Peru.

21 BOGOTA RAIL *Rallus semiplumbeus* 27cm Endangered. Upperparts blackish brown, feathers edged buffish brown. Wing coverts chestnut. Face, neck, breast and upper belly grey. Undertail barred. Legs and bill dirty pink, culmen black. **V** Call is a clear, whistled *peeep*. **H** Marshes and reedbeds around lakes. **D** Colombia.

22 AUSTRAL RAIL *Rallus antarcticus* 22cm Differs from Ecuadorian Rail in that face, neck, breast and upper belly are very dark lead grey. Flanks, lower belly and undertail barred black and white. **V** Pairs produce duetting song of piping squeaks and low grunts. **H** Small lakes, streams and bodies of water with reedbeds and other vegetation. **D** S Chile and S Argentina.

23 WATER RAIL *Rallus aquaticus* 29cm Secretive, more often heard than seen, although will feed in the open if undisturbed. **V** Various pig-like squeals and grunts. **H** Dense reedbeds, marshes and overgrown ditches. **D** W Europe to N Africa and C Asia.

24 BROWN-CHEEKED RAIL *Rallus indicus* 23–28cm Secretive, but feeds in the open when undisturbed. **V** Long, clear, piping *kyu*; also a repeated, metallic, slurred *shrink shrink* call, about two per second. **H** Marshes. **D** Breeds in E Siberia and Japan. Winters from Nepal to SE Asia and Borneo.

25 AFRICAN RAIL *Rallus caerulescens* 30cm Separated from African Crake by long red bill. Secretive. **V** Very high, slowed-down, slightly lowered, piping trill, ending in mid-high *piu piu piu*. **H** Reedbeds of marshes, lake margins, stream banks. **D** E, SC, SE and S Africa.

26 MADAGASCAN RAIL *Rallus madagascariensis* 25cm Note long bill, dark plumage and white undertail coverts. **V** Song is a cackling rattle (in chorus?). **H** Thick herbaceous vegetation of wetlands. **D** Madagascar.

27 BROWN-BANDED RAIL *Lewinia mirifica* 21–22cm Little recorded information; habits presumed similar to those of Lewin's Rail. **V** Unrecorded. **H** Submontane and pine forest mixed with grasses and shrubs. **D** Philippines.

28 LEWIN'S RAIL *Lewinia pectoralis* 23–27cm Secretive and wary; mainly active at dusk and dawn, and on overcast days. Very difficult to flush, and uses runways in dense vegetation to escape when disturbed. **V** Loud, repeated *grr-eek grr-eek*; also a loud, metallic *jik-jik-jik-jik*. **H** Rank regrowth, dense grass and scrub, usually near water. **D** Lesser Sunda Islands, Indonesia; New Guinea; E Australia.

29 AUCKLAND RAIL *Lewinia muelleri* 23cm Note compact jizz and distinctive darkish colour pattern. **V** High *frueeh*. **H** Damp, densely vegetated areas with a (sub-)canopy at 1m above ground. **D** Auckland Islands, subantarctic SW Pacific Ocean.

30 WHITE-THROATED RAIL *Dryolimnas cuvieri* 32cm Note white throat, upper breast and lateral undertail coverts. **V** Hoarse shrieks and fluting calls. **H** Wetlands; in scrub on Aldabra. **D** Madagascar, Indian Ocean islands.

31 AFRICAN CRAKE *Crex egregia* 25cm Separated from Corn Crake by darker brown plumage with no rufous tones, and by seemingly longer head and bill. **V** Mid-high, loud, cackling *ceck-ceck-ceckceckceck*. **H** Moist grassland, inundations and swamps; often in dry habitats, but not in forests. **D** Widespread across Africa.

32 CORN CRAKE *Crex crex* 27–30cm Most active early or late in the day, and more often heard than seen. **V** Monotonous, dry *krek-krek-krek-krek*. **H** Grassland and cultivated areas. **D** Breeds from Europe to C Siberia and NW China. Winters in S Africa.

33 ROUGET'S RAIL *Rougetius rougetii* 30cm The only member of the genus. Declining owing to habitat modification. **V** Song a series of rolling, harsk shrieks, recalling a tern or small gull. **H** Tall grass, reeds and shrubbery near streams and ponds. **D** Ethiopia, Eritrea.

34 SNORING RAIL *Aramidopsis plateni* 29cm Forages in dense cover, singly, in pairs or in small family groups; reported to search for crabs in mountain forest streams. **V** Quiet snore, consisting of a brief wheeze followed by a drawn-out *ee-orrrr*; also utters a brief, deeply sighing *hmmmm*. **H** Primary and tall secondary lowland, hill and montane forest, dense wet secondary growth bordering forest. **D** Sulawesi, Indonesia.

35 INACCESSIBLE ISLAND RAIL *Atlantisia rogersi* 13–15cm World's smallest living flightless bird. Forages for invertebrates, like a small mammal. Usually keeps to cover of vegetation. **V** Trilling squeaks, often given in duet. **H** All habitats on Inaccessible Island in the S Atlantic Ocean. **D** Inaccessible Island, Tristan da Cunha.

36 LITTLE WOOD RAIL *Aramides mangle* 30cm Head grey, hindneck and upper back bluish grey. Foreneck, breast and flanks tawny rufous. **V** Repeated, rhythmic high yelps *eeyup eeyup eeyup...* **H** Coastal swamps, muddy shores, mangroves, thickets and nearby tall grass. **D** E Brazil.

37 RUFOUS-NECKED WOOD RAIL *Aramides axillaris* 30cm Separated from larger Grey-necked Wood Rail by rufous head. **V** High, rhythmic, repeated *tjuw*, each note downslurred. **H** Mangrove, wet places in forest. **D** Mexico to Ecuador, Belize and N South America.

38 GREY-NECKED WOOD RAIL *Aramides cajaneus* 39cm Head, neck and upper back lead grey, crown washed brownish. Breast and flanks tawny rufous. **V** Call is a loud *chiri-cot, chiri-cot...* with coos thrown in. **H** Swamps, tall riverine vegetation, mangroves, lakes, marshes, temporary ponds, rainforest, humid deciduous woods (sometimes far from water), tall grass, scrub, sugar cane and forestry plantations. **D** Costa Rica to N Argentina.

1
2
3
4
5
6
7
8
9
10
11
12
13
14
15
16
17
18
19
20
21
22
23
24
25
26
27
28
29
30
31
32
33
34
35
36
37
38

RAILS, CRAKES AND COOTS *CONTINUED*

1 RUFOUS-NAPED WOOD RAIL *Aramides albiventris* 40cm Note grey neck with rufous cap. **V** High, dry *eh-hoc ehhoc...chochochochoc*. **H** swamp, mangrove, at water edges in forest and woodland. **D** S Mexico to NE Costa Rica.

2 BROWN WOOD RAIL *Aramides wolfi* 36cm Dark grey head. Rest of body and wings dark chestnut. **V** Call a loud, repeated *kwee cor mwee*. **H** Mangroves, rivers, gallery forest, riverine and woodland swamp. **D** Colombia to Peru.

3 GIANT WOOD RAIL *Aramides ypecaha* 46cm Forehead, foreface, foreneck and breast grey. Back of head and hindneck vinaceous cinnamon. Sides of breast and flanks pale vinaceous. **V** Call a screamed *ha, ha, patterha, wok wok* duet, deranged and wild. **H** Edges of marshes, lakes and rivers, all with tall, dense vegetation; also tall grass and thickets, often with trees and shrubs, sometimes far from water on roads and fields. **D** E Paraguay to SE Brazil, Uruguay and NE Argentina.

4 SLATY-BREASTED WOOD RAIL *Aramides saracura* 36cm Forehead, crown and nape lead grey. Face brownish grey. Hindneck and upper back brownish cinnamon. Foreneck, breast and flanks bluish grey. **V** Call a loud, strident, usually duetted *wok...wahkwahk...wok...wahkwahk...* **H** Often far from water. Rainforest, dense vegetation near water, swamps and rivers in forest. **D** SE Brazil, E Paraguay, NE Argentina and Uruguay.

5 RED-WINGED WOOD RAIL *Aramides calopterus* 33cm Sides of neck and band in wing chestnut-red. Primaries black. Breast and belly bluish grey. **V** High-pitched, strongly downslurred whistle. **H** Swamps, near streams. **D** W Amazonia.

6 UNIFORM CRAKE *Amaurolimnas concolor* 22cm Short tail. Shortish bill greenish yellow. Upperparts brown. Cheeks and underparts tawny rufous. Legs red. **V** Call a long series of loud *tooee* notes, rising in volume, then descending. **H** Flooded thickets, secondary growth away from water, shady swamps near waterholes and streams, tall grass on edges of marshes and mangroves and on cultivated land, clearings in palm groves. **D** Widespread across Central and South America.

7 BLUE-FACED RAIL *Gymnocrex rosenbergii* 30cm Elusive. Occurs alone or occasionally in pairs. **V** Utters a quiet clucking. **H** Favours moist areas within primary and tall secondary lowland and hill forest, remnant patches of dense secondary forest and bushy abandoned rice fields. **D** Sulawesi, Indonesia.

8 TALAUD RAIL *Gymnocrex talaudensis* 33–35cm Little recorded information, but said to be extremely shy. **V** Series of 15 or more of rapid, high-pitched *peet-peet-peet* notes. **H** Patches of wet grass and rank vegetation at edges of lowland forest. **D** Karakelong Island, Talaud Islands, Indonesia.

9 BARE-EYED RAIL *Gymnocrex plumbeiventris* 30–33cm Usually solitary and secretive. Runs to and fro around forest floor in search of food, continually uttering grunting noises. **V** Continuous *uw uw uw uw...* grunting; also a loud, gulping *wow-wow-wow-wow*. **H** Primary and tall secondary lowland forest. **D** N Maluku Islands, Indonesia; New Guinea.

10 BROWN CRAKE *Amaurornis akool* 26–28cm Usually skulking. Feeds in the open more at dawn or dusk, but runs into cover at slightest alarm. **V** Long, vibrating trill. **H** Swamps, paddyfields and overgrown watercourses. **D** India to SE China.

11 ISABELLINE BUSH-HEN *Amaurornis isabellina* 35–40cm Generally shy, although regularly forages, quite openly, along edges of and within recently cleared areas of dense vegetation. **V** Jumble of sharp, slightly pulsating screeches and chattering notes, usually given by several birds together. **H** Dense, rank grassland, overgrown cultivation, low scrub mixed with grass, tall grass bordering forest. **D** Sulawesi, Indonesia.

12 PLAIN BUSH-HEN *Amaurornis olivacea* 30cm Generally seen alone, although may occur in groups in suitable habitats. More often heard than seen. **V** Raspy, growling *kaawww keerrr*. **H** Dry grassland, scrub, forest plantations, logged areas. **D** Philippines.

13 PALE-VENTED BUSH-HEN *Amaurornis moluccana* 30cm Shy and secretive. Usually keeps to dense cover; wades or swims in streams, generally keeping within shaded areas. **V** Series of shrieks or cat-like wails, decreasing in volume. **H** Dense rank grass and scrub bordering streams, swamps, marshy areas and forest. **D** Sangihe and Talaud islands (N of Sulawesi), Kai Islands (S Malukus) and Tanimbar Islands (E Lesser Sunda Islands), Indonesia.

14 TALAUD BUSH-HEN *Amaurornis magnirostris* 30–31cm Shy but inquisitive, although runs at the slightest disturbance. **V** Series of frog-like notes or low-pitched barks. **H** Forest with adjacent scrub, rank vegetation, swampy patches, overgrown cultivation. **D** Karakelong Island, Talaud Islands, Indonesia.

15 WHITE-BREASTED WATERHEN *Amaurornis phoenicurus* 28–33cm Often seen in the open; also climbs about in bushes and trees. **V** Loud grunts, croaks and chuckles, *kru-ak kru-ak kru-ak-a-wak-wak* or *krr-kwaak-kwaak krr-kwaak-kraak*. Contact call is a *pwik pwik pwik*. **H** Damp scrub, thick waterside vegetation. **D** Malaysia, Indonesia, Philippines.

16 BLACK CRAKE *Amaurornis flavirostra* 20cm Note yellow bill and red legs. Not secretive or shy. **V** High, excited chuckling, closely followed by a very low *ret-ret-ret-ret-ret* grunt. **H** Edges of reedbeds, Papyrus (*Cyperus papyrus*), shrubbery at lake margins, dams, ponds, streams. **D** Widespread across Africa.

17 SAKALAVA RAIL *Amaurornis olivieri* 19cm Combination of small size, dark plumage and green bill render it unmistakable. **V** Regularly repeated *tjuk*. **H** Marshes with dense reedy vegetation. **D** Madagascar.

18 BLACK-TAILED CRAKE *Porzana bicolor* 20–22cm Secretive. Will feed in the open at dawn or dusk, but at slightest disturbance retreats into cover. **V** Rasping *waak-waak*, followed by a descending trill. **H** Swamp areas and paddyfields in or near forests. **D** NE India to SC China.

19 LITTLE CRAKE *Porzana parva* 18–20cm Often feeds in quite deep water and tends to swim more than other crakes. Separated from Baillon's Crake by longer primary projection. **V** Far-carrying, accelerating *qwek qwek qwek qwek-qwek-qwek kwa kwa-kwa-kwa*. **H** Marshes and lakes with extensive, dense aquatic vegetation. **D** Breeds from S and C Europe to NW China. Winters in C Africa.

20 BAILLON'S CRAKE *Porzana pusilla* 17–19cm Forages at dawn or dusk, close to, or in, dense cover. Short primary projection compared to Little Crake. **V** Dry frog-like rattle. **H** Marshes, paddyfields, vegetation surrounding lakes and ponds. **D** Widespread across Europe, Africa, Asia and Australasia.

21 SPOTTED CRAKE *Porzana porzana* 22–24cm Secretive, although will feed in the open if undisturbed. **V** High-pitched, whiplash-like *whitt*; also a ticking *tik-tak* and a croaking *qwe-qwe-qwe*. **H** Marshes, bogs, wet meadows, paddyfields. **D** Breeds from W Europe to C Asia. Winters in Africa and W Asia.

22 AUSTRALIAN CRAKE *Porzana fluminea* 19–23cm Unobtrusive, usually feeding among dense vegetation but sometimes in the open. Diet includes invertebrates and plant material. **V** Various sharp ticking and rattling calls. **H** Well-vegetated margins of wetland habitats, freshwater and brackish. **D** Australia.

23 SORA *Porzana carolina* 20–23cm Secretive, but will feed in the open if not disturbed. In non-breeding adults black of throat is obscured by pale feather edgings. **V** Plaintive *ker-wee* and a high-pitched descending whinny. When alarmed gives a loud, sharp *keek*. **H** Thickly vegetated freshwater and saltwater swamps, including rice fields and mangroves. **D** Breeds across North America. Winters from S USA to N and NW South America.

24 DOT-WINGED CRAKE *Porzana spiloptera* 14cm Very small and dark. Separated from Black Crake by black and dark brown markings from crown to upper back. Flight feathers all blackish. **V** Rather low whistled two-note phrase *wee-prrouu*. **H** Flooded tall grassland and vegetation near fresh or brackish water, temporary marshes, dry grassland. **D** S Uruguay to NE Argentina.

25 ASH-THROATED CRAKE *Porzana albicollis* 25cm Bill greenish yellow. Upperparts blackish, feathers edged olive. Face to belly grey. Throat whitish. Legs greenish. **V** Male's call a series of loud *rrrrrow* or *krrrrow* notes (like a cow's moo); female joins in with *piri-piri*. **H** Tall grasses in marshes, flooded areas, lake edges by forest, reed beds, rice paddies. **D** N, SC and SE South America.

26 RUDDY-BREASTED CRAKE *Porzana fusca* 21–23cm Secretive. Occasionally feeds at edge of reeds. **V** Harsh *tewk*, often speeding up and ending with a grebe-like bubbling. **H** Marshes, paddyfields, dry bush by lakes. **D** Widespread across Asia.

27 BAND-BELLIED CRAKE *Porzana paykullii* 20–22cm Typical secretive crake behaviour; more often heard than seen. **V** Said to sound like intermittent drumbeats or the noise of a wooden rattle. **H** Marshes, wet meadows, damp woodlands, paddyfields. **D** Breeds in E Siberia, Korean Peninsula, NE China. Winters in E and SE Asia.

28 SPOTLESS CRAKE *Porzana tabuensis* 15–18cm Secretive. Forages in deep cover; sightings are usually of a bird darting from one patch of cover to another. Clambers in vegetation and along branches. **V** Loud, sewing machine-like *purr*, and a loud *pit* that is often interspersed with a harsh, nasal *harr*. **H** Wet grasslands, marshes. **D** Philippines, Australasia, Oceania.

29 HENDERSON CRAKE *Porzana atra* 18cm Uniform velvety black. **V** Drawn-out churring note, lowered at end. **H** Forest, thickets, coconut groves. **D** Henderson Island, Pitcairns.

30 YELLOW-BREASTED CRAKE *Porzana flaviventer* 14cm Feeds mainly at dawn and dusk; will feed in the open on floating vegetation or by wading in shallow water at swamp edges. **V** Harsh, rolling *k'kluk kurr-kurr* and a single or repeated *kreer* or *krreh*. **H** Freshwater swamps with fringing and floating vegetation. Breeds from March to June, nest placed among or on aquatic plants. **D** Breeds from Caribbean and S Mexico to SE South America. Winters in South America.

31 WHITE-BROWED CRAKE *Porzana cinerea* 19–22cm Forages early and late, regularly in the open, running for cover at the slightest disturbance; more often heard than seen. **V** Loud, nasal chattering *chika*, repeated 10–12 times; also a repeated nasal *hee* note. While feeding utters a sharp, loud *kek-kro*. **H** Well-vegetated lakes, marshes, overgrown ditches and paddyfields. **D** Malay Peninsula and Philippines to N Australia.

32 STRIPED CRAKE *Aenigmatolimnas marginalis* 20cm Note rufous undertail and flanks. **V** Mid-high, dry, fast, prolonged, mechanical *trrrrr* trill. **H** Dense vegetation in marshes, inundations and similar areas. **D** Ghana, Nigeria, and Cameroon to DR Congo and Gabon.

33 ZAPATA RAIL *Cyanolimnas cerverai* 29cm Critically Endangered. Appears to be almost flightless. Virtually unknown. **V** Low *cutucutu-cutucutu-cutucutu*, said to sound like a bouncing ball; also a loud *kwowk* or *kuck-kuck*. **H** Swamp with dense, tangled bush, low trees, sawgrass and cattails. **D** Zapata Swamp, S Cuba.

34 COLOMBIAN CRAKE *Neocrex colombiana* 15cm Like Paint-billed Crake but with plain buffish-white undertail and lower belly. **V** Little recorded; nasal croak given in alarm. **H** Damp or flooded grass. **D** Panama to Ecuador.

35 PAINT-BILLED CRAKE *Neocrex erythrops* 19cm Belly and undertail coverts broadly barred black. **V** Croaking *wrrot-wrrot*. **H** Marshes, wet grassland, rice fields. **D** Widespread across South America.

1
2
3
4
5
6
7
8
9
10
11
12
13
14
15
16
17
18
19
♂
♀
20
21
22
♀
♂
23
24
25
26
27
28
29
30
31
32
♀
♂
33
34
35

RAILS, CRAKES AND COOTS *CONTINUED*

1 SPOTTED RAIL *Pardirallus maculatus* 28cm Very secretive; more often heard than seen. **V** Grunt followed by rasping screech; also accelerating *tuk-tuk-tuk-tuk*, similar to call of Zapata Rail (Plate 40). **H** Freshwater swamps with emergent plants, rice fields. **D** Widespread across Central and South America, Caribbean.

2 BLACKISH RAIL *Pardirallus nigricans* 34cm Upperparts blackish brown, tinged olive. Tail black. Throat whitish. Underparts dark lead-grey. Bill yellowish green, more yellow at base. Legs bright red. **V** Call a high-pitched, whistled duet of *dweeeeet* and *chik chuk* notes. **H** Swamps, tall grass around marshes, flooded areas, rice paddies. **D** W Amazonia, E and SE South America.

3 PLUMBEOUS RAIL *Pardirallus sanguinolentus* 35cm Upperparts blackish brown, feathers edged dark brown, all washed olive. Underparts very dark lead grey. Bill greenish yellow, base of maxilla pale blue, base of mandible bright red. Legs red. **V** Call a duetted series of *wheeew rreeet* notes, punctuated by *teerreeet* notes and deep bass *doo doomph* notes. **H** Dense vegetation of marshes, streams, roadside ditches. **D** S and W South America.

4 CHESTNUT RAIL *Eulabeornis castaneoventris* 45–52cm Often seen in pairs. Diet comprises mainly invertebrates, mostly crabs and other crustaceans. **V** Grunts and pig-like squeals. **H** Mangrove swamps, estuary mudflats, tidal creeks. **D** Aru Islands, Indonesia; N Australia.

5 INVISIBLE RAIL *Habroptila wallacii* 35cm Usually solitary. Occasionally seen crossing open areas. **V** Reported to utter a low drumbeat sound, *wak wak wak*, given at the same time as a *tuk tuk tuk* sound made with wings. **H** Forest, dense swampy thickets (especially Sago *Metroxylon sagu* swamp), marsh edges, secondary growth, forest edges. **D** Halmahera, Indonesia.

6 NEW GUINEA FLIGHTLESS RAIL *Megacrex inepta* 36–38cm Has proportionately large head and long legs and toes. Flicks wings while walking; runs well. Typically stays close to cover when feeding. **V** Pig-like squeal. **H** Mangrove swamps, inundated forest. **D** New Guinea.

7 WATERCOCK *Gallicrex cinerea* 42–43cm Mainly a crepuscular, skulking forager, although readily swims across open water. **V** Long series of *kok* notes, followed by a series of hollow *utumb* notes and ending with a short series of *kluck* notes. **H** Reed or grassy swamps, paddyfields, vegetation alongside watercourses. **D** Widespread across Asia.

8 WESTERN SWAMPHEN *Porphyrio porphyrio* 45–50cm Unmistakable. Skulking. Clambers among dense reeds. **V** Low *chuk* or *chuk-chuk*; when alarmed gives a *toot*, said to sound like a toy trumpet. **H** Dense reedbeds, fringing vegetation by rivers and lagoons. **D** SW Europe, NW Africa.

9 AFRICAN SWAMPHEN *Porphyrio madagascariensis* 45cm Larger than most swamp birds. Further separated from Allen's Gallinule and Western Swamphen by heavy bill, concolorous with red frontal shield. **V** Repeated raucous, goose-like grunts, shrieks and cackles. **H** Large swamps and shallow lakes with reeds, Papyrus (*Cyperus papyrus*) and floating vegetation. **D** Africa, Madagascar.

10 GREY-HEADED SWAMPHEN *Porphyrio poliocephalus* 45–50cm Readily feeds in the open, although typically stays close to cover. Walks on floating water plants and clambers among reeds and bushes. **V** Very vocal, including a series of plaintive nasal rattles ending in a crescendo, a low *chuk-chuk* and, in alarm, a trumpeting *toot*. **H** Dense reedbeds, fringing vegetation of lakes, ponds and rivers. **D** Middle East to India, Malay Peninsula and S China.

11 BLACK-BACKED SWAMPHEN *Porphyrio indicus* 45–50cm Readily forages in the open, generally close to cover. Note large shield and black upperparts. **V** Series of plaintive, nasal rattles, ending in a crescendo; also a low *chuk-chuk* and a trumpeting *toot*. **H** Dense reedbeds and fringing vegetation by lakes, ponds and rivers. **D** Sumatra, Java and Sulawesi, Indonesia; Borneo; S Philippines.

12 PHILIPPINE SWAMPHEN *Porphyrio pulverulentus* 45–50cm Readily forages in the open, generally close to cover. Walks on floating vegetation and clambers among reeds and bushes. **V** Series of plaintive, nasal rattles, ending in a crescendo; also a low *chuk-chuk* and a trumpeting *toot*. **H** Dense reedbeds, fringing vegetation by lakes, ponds and rivers. **D** Philippines; Talaud Islands, Indonesia.

13 AUSTRALASIAN SWAMPHEN *Porphyrio melanotus* 46cm Unmistakable owing to size, long legs and wings, and jizz. Subspecies differ in plumage. **V** Peacock-like *uweeeh*. **H** Wetlands and adjoining open habitats such as grassland, cultivation, lawns, sports fields. **D** Widespread across Australasia and Pacific Ocean islands.

14 SOUTH ISLAND TAKAHE *Porphyrio hochstetteri* 63cm Flightless. Unmistakable. **V** Very high, yelping *weeh-weeh-weeh* in duet. **H** Tussocky alpine grassland and scrubland. **D** South Island, New Zealand.

15 ALLEN'S GALLINULE *Porphyrio alleni* 22–24cm Secretive. Swims, but not usually far from cover. Climbs among reeds and bushes, and often walks on floating vegetation. **V** Various noises, including a sharp *kik-kik-kik-kik-kik-kurr-kurr*, a dry *keek* and a *kli-kli-kli* uttered in flight. **H** Well-vegetated marshes and lakes. **D** Widespread across Africa.

16 PURPLE GALLINULE *Porphyrio martinica* 30–36cm Readily feeds in the open, although typically stays close to cover. Walks on floating plants, also clambers among reeds and bushes. **V** High-pitched *kyik*, also a wailing *ka-ka-ka* and a *kek-kek-kek* given in flight. **H** Well-vegetated lakes, pools and marshes. **D** SE USA to N Argentina.

17 AZURE GALLINULE *Porphyrio flavirostris* 23–26cm Skulking, sometimes seen perching on floating vegetation. **V** May utter a short trill. **H** Marshes. **D** N and SC South America, Amazonia.

18 MAKIRA WOODHEN *Gallinula silvestris* 26cm Creeps through dense vegetation on forest floor. Critically endangered, possibly extinct, no confirmed recent sightings. **V** Unknown. **H** Moist montane forest; mostly 500–1,000m. **D** Solomon Islands.

19 GOUGH MOORHEN *Gallinula comeri* 30cm Restricted to two small islands in the South Atlantic. Feeds on vegetable matter, invertebrates, carrion and eggs. **V** Short, quiet series of fast clucking notes. **H** Fern-bush, boggy areas, tussock grassland. **D** Gough Island, Tristan da Cunha.

20 COMMON MOORHEN *Gallinula chloropus* 32–35cm Often relatively tame. Regularly swims, with a jerky action. **V** Bubbling *krrrruk* or *kurr-ik*; also a *kik-kik-kik* given when alarmed. **H** Various wetland areas, including lake edges, ponds, rivers and ditches. **D** Widespread across Eurasia, Africa and Indian Ocean islands.

21 COMMON GALLINULE *Gallinula galeata* 22–23cm Secretive but often relatively tame. Forages in the water, often upending, or along edges of waterbodies. **V** Bubbling *krrrruk* or *kurr-ik*; also a cackling *kik-kik-kik-kik*. **H** Swamps, mangroves, lake edges, ponds, rivers and ditches, all with fringing vegetation. **D** Widespread across North, Central and South America, Caribbean and E Pacific Ocean islands.

22 DUSKY MOORHEN *Gallinula tenebrosa* 30cm Forages in the open and swims amid aquatic vegetation. **V** Various shrieks, screeches and yelps; also a nasal *kerk*. **H** Lakes, ponds and swamps with aquatic vegetation. **D** Borneo; Sulawesi, Indonesia; New Guinea; Australia.

23 LESSER MOORHEN *Paragallinula angulata* 22–23cm Shy. Actions similar to those of Common Moorhen. **V** A soft *pyup*, a rapid clucking and, when alarmed, a *tek*. **H** Reedbeds, marshes, lakes and rivers with fringing vegetation. **D** Widespread across Africa.

24 SPOT-FLANKED GALLINULE *Porphyriops melanops* 26cm Bill and frontal shield lime green. Flanks brownish, spotted white. Upperparts chestnut and olive-brown. **V** Fast clucking calls, also lower-pitched, single *teup* notes. **H** Lakes, marshes, temporary ponds and roadside ditches, with much floating and emergent vegetation. **D** E and S South America.

25 BLACK-TAILED NATIVEHEN *Tribonyx ventralis* 30–38cm Tail often raised and fanned like a bantam when walking. Usually feeds in groups. Swims well. **V** Usually silent. **H** Freshwater and brackish wetlands; moves in response to seasonal drought and rainfall. **D** Australia.

26 TASMANIAN NATIVEHEN *Tribonyx mortierii* 42–50cm Flightless. Lives in groups. Forages on the ground, diet comprising seeds and invertebrates. **V** Noisy. Calls include repeated raspy honks. **H** Short grassland, grazed agricultural pasture, wetlands. **D** Tasmania.

27 RED-KNOBBED COOT *Fulica cristata* 38–42cm In winter, red knob replaced by small red-brown spot. In all seasons note lack of pointed feather projection between bill and shield. **V** Wide variety, including a shrill *kik*, a double-clucking *klukuk*, a metallic *krrook* and a groaning *euh*. **H** Marshes and lakes with surrounding vegetation. **D** S Spain and Morocco to Ethiopia, Angola, Namibia, South Africa and Madagascar

28 EURASIAN COOT *Fulica atra* 36–38cm Gregarious. Often feeds on waterside grass. **V** Various metallic notes, including a short *kow* or *kowk* and a sharp *kick*. **H** Open water with fringing vegetation. **D** Widespread across Eurasia and Australasia.

29 HAWAIIAN COOT *Fulica alai* 39cm Note absence of pointed black wedge between frontal shield and upper mandible. Undertail coverts white. Two colour morphs differing in colour of beak and shield. **V** High, staccato *bic bic bic*. **H** Wide variety of wetlands, including estuaries, marshes, ponds, flooded land. **D** Hawaiian Islands.

30 AMERICAN COOT *Fulica americana* 38–40cm Highly gregarious, forming flocks numbering in the thousands. Omnivorous, eating mainly aquatic vegetation, but also aquatic animals, especially during the breeding season. **V** Varied; often metallic in tone but also deep grunts. **H** Reedbeds around lakes and ponds, marshes, slow-flowing watercourses. **D** North America, Central America and N South America, Caribbean, and Hawaiian Islands.

31 WHITE-WINGED COOT *Fulica leucoptera* 37cm Yellow bill. Rounded frontal shield varies with seasons from whitish, through yellow to yellowish orange. In flight, shows whitish-grey tips to flight feathers. Legs greyish olive. **V** Various cackling and goose-like honking notes. **H** Lakes, marshes with emergent vegetation, roadside ditches, temporary pools, dams, streams, rivers, flooded fields, brackish lagoons. **D** SC South America.

32 ANDEAN COOT *Fulica ardesiaca* 46cm Bill yellowish white. Prominent rounded shield purply chestnut (or white, washed yellow, in breeding condition in some populations). Legs greenish grey. **V** Goose-like honks and deeper-pitched rasping grunts. **H** High Andean lakes, sometimes with reeds and rushes, always with submerged vegetation. **D** Colombia to NW Argentina.

33 RED-GARTERED COOT *Fulica armillata* 43cm Bill and pointed frontal shield yellow, divided by an irregular reddish-chestnut band. When swimming, the 'knees' breaking the surface show the red 'garter'. Legs yellowish olive. Heftier than White-winged and Red-fronted coots, with which it is commonly found. **V** Various grunts, honks and squeals. **H** Lakes, marshes, roadside ditches, streams, rivers, reed-beds, temporary ponds and flooded fields, brackish lakes with little or no emergent vegetation. **D** S South America.

34 RED-FRONTED COOT *Fulica rufifrons* 38cm Bill rich yellow; base of bill and long, lanceolate frontal shield purply chestnut. Point of shield juts slightly above crown, forming a bump on head. Raises tail often to fan and show white undertail coverts. **V** Clucking and honking calls. **H** Lakes, marshes, roadside ditches, streams with much aquatic vegetation, especially reedbeds. **D** S South America.

35 GIANT COOT *Fulica gigantea* 52cm Huge and hefty. Bill reddish chestnut with yellow tip. Base of maxilla and central frontal shield white. Base of mandible and edge of shield bright yellow. Shield has groove up centre and ends in two lobes that push up into a 'double bump' on forehead. Legs red. **V** Wide range of grunting, squeaking and hissing notes. **H** Open high Andean lakes with submerged vegetation. **D** Peru to NW Argentina.

36 HORNED COOT *Fulica cornuta* 60cm Bill yellowish green, black caruncullation from shield (the 'horn') slopes down and forward along bill, giving a 'Roman nose' profile. Legs yellowish green. Head and neck black, rest of the bird slate grey, paler on undersides. Undertail coverts whitish. **V** Grumbling, barking and yelping calls. **H** Open lakes in high Andes; 3,500–4,800m altitude. **D** N Chile to NW Argentina.

4
6
1
2
3
5
♀
7
♂
8
9
10
12
11
13
14
15
18
16
17
19
22
♀
23
20
21
26
♂
27
24
25
30
28
31
33
29
36
32
34
35

TRUMPETERS PSOPHIIDAE

1 GREY-WINGED TRUMPETER *Psophia crepitans* 50cm As Pale-winged Trumpeter, but inner flight feathers grey, with ochre only on mid-back. **V** Loud call of 3–5 notes, *Oh-oh-oh-oh-oooooooh*, is used by adult birds to assert territoriality. **H** Humid tropical forest from lowlands to mountains, terra firme forests. **D** N Amazonia.

2 PALE-WINGED TRUMPETER *Psophia leucoptera* 50cm Black with greenish or blue sheen on breast, foreneck and some wing coverts. Inner flight feathers white (or ochreous, according to race). **V** Series of low-pitched, vibrating notes, each 1.5 secs long. Other individuals may join in with a series of equally low, short *woop* calls. **H** Dense, unmodified tropical forest away from human settlement, from lowlands to mountains. **D** W Amazonia.

3 DARK-WINGED TRUMPETER *Psophia viridis* 50cm Unlike other species in having green upper back and inner flight feathers. Sheen on breast and wing coverts purplish blue. One race has an ochreous-olive back and inner flight feathers, and a purple sheen on breast and coverts. **V** Series of low humming notes, given singly at first and then doubled. **H** Humid lowland tropical forest. **D** S Amazonia.

CRANES GRUIDAE

4 GREY CROWNED CRANE *Balearica regulorum* 95cm Separated from Black Crowned Crane by paler neck and upperparts, and by different cheek pattern. **V** Mellow honks, generally low pitched. **H** Roosts and nests in or near wet places, but feeds in open or wooded grassland and farmland. **D** E and SE Africa.

5 BLACK CROWNED CRANE *Balearica pavonina* 100cm Separated from Grey Crowned Crane by different colour pattern of cheeks. **V** As Grey Crowned Crane. **H** Breeds in marshes but feeds in natural and cultivated areas with tall grass. **D** NE, C and W Africa.

6 SIBERIAN CRANE *Leucogeranus leucogeranus* 140cm The most aquatic of cranes. Very rare. In flight from below, shows black primaries. **V** Soft, musical *koonk koonk*. **H** Breeds on tidal flats, bogs, marshes and open wetlands in the lowland tundra and taiga–tundra transition zone. On passage, occurs on large isolated wetlands. **D** NE and EC Siberia.

7 SANDHILL CRANE *Antigone canadensis* 100–120cm Gregarious. Older birds often stained rusty. In flight from below, shows black primaries and secondaries. **V** Vibrant, rolling *karr-rooo*. **H** Breeds in marshes, wet meadows and various open wetlands. Winters in similar habitats and on agricultural land and coastal wetlands. **D** Widespread across North America.

8 WHITE-NAPED CRANE *Antigone vipio* 125–150cm Gregarious; large flocks in winter. In flight from below, shows black primaries and secondaries, with a wide white bar on greater underwing coverts. **V** High-pitched bugling. **H** Breeds in marshes, wet meadows, vegetated lake shores; during passage and in winter on fields, wetlands, paddyfields, mudflats. **D** Breeds in SE Siberia, NE Mongolia and NE China. Winters in Korean Peninsula, S Japan and SE China.

9 SARUS CRANE *Antigone antigone* 156–176cm In flight from below, shows black primaries; flies with outstretched neck and legs protruding well beyond tail. **V** Loud trumpeting. **H** Watery areas such as marshes, floodplains, lakes, ponds, rivers, ditches, wet cultivations. **D** N India, SE Asia, N Australia.

10 BROLGA *Antigone rubicunda* 130cm Separated from Sandhill Crane by range, head pattern, darker primaries and narrow dark trailing edge to wings. **V** Excited, guttural trumpeting. **H** Normally found in wetlands and wet grassland. **D** S New Guinea, Australia.

11 WATTLED CRANE *Grus carunculata* 125cm Note black belly in flight. **V** High-pitched, piercing calls. **H** Shallow water of wet, wide-open areas. **D** DR Congo and Tanzania to Botswana; Zimbabwe; South Africa.

12 BLUE CRANE *Grus paradisea* 105cm In flight, separated from Wattled Crane by grey belly. **V** Low-pitched, raspy and broken calls. **H** Natural and cultivated open areas with or without low scrub. **D** South Africa.

13 DEMOISELLE CRANE *Grus virgo* 90–100cm Gregarious; often large flocks on migration. In flight from below, shows black primaries and secondaries. **V** Similar to Common Crane but drier and higher pitched. **H** Breeds on grassy steppe close to streams, shallow lakes or other wetlands, also semi-desert and desert with water nearby. **D** Breeds in NW Africa, E Turkey and SW Russia to N China. Winters in C Africa and India.

14 RED-CROWNED CRANE *Grus japonensis* 150cm Gregarious. More aquatic than other cranes. Rare. In flight from below, shows white primaries and black secondaries. **V** Penetrating, high-pitched trumpeting. **H** Breeds in marshes, bogs and wet meadows. In winter found in various freshwater wetlands, salt marshes, mudflats, paddyfields. **D** SE Siberia, NE China, Japan.

15 WHOOPING CRANE *Grus americana* 114–127cm Flies with neck and legs outstretched, showing black primaries. **V** Loud, rolling trumpeting. **H** Breeds in bogs, winters in marshes. **D** C Canada, C USA.

16 COMMON CRANE *Grus grus* 110–120cm Gregarious; forms large flocks post-breeding. In flight from below, shows black primaries and secondaries. **V** Far-carrying *krooh* and a repeated, harsh *kraah*. During breeding utters a musical duet: *krrroo* (male) *kraw* (female). **H** Breeds on bogs, marshes, damp heathland, swampy clearings in forests, steppe, semi-desert areas. Winters in open country with or without nearby lakes or marshes. **D** Breeds across Europe. Winters in N Africa and N Asia.

17 HOODED CRANE *Grus monacha* 100cm Occurs in small parties. In flight from below, wings dark with slightly darker primaries and secondaries. **V** Loud *krurrk*. **H** Breeds in high-altitude forested wetlands and isolated bogs. Non-breeders occur in open wetlands, grassland and agricultural fields, with wintering birds also using lake shores, river shores and paddyfields. **D** SE Siberia, N China.

18 BLACK-NECKED CRANE *Grus nigricollis* 115cm Gregarious. In flight from below, shows black primaries, secondaries and tail. **V** Series of loud trumpeting honks. **H** Breeds in high-altitude grassy wetlands, lakeside marshes and pastures; 2,950–4,900m. **D** Tibet, WC China.

LIMPKIN ARAMIDAE

19 LIMPKIN *Aramus guarauna* 66cm Secretive. Generally crepuscular, best located by call. Flight floppy, with neck and legs outstretched. **V** Loud, wailing *kwEEEeeer* or *gua-re-ao*; also a shorter *kwaouk*. **H** Marshes, swamps, lakes and wet wooded areas. Breeds in thick vegetation, generally near water. **D** SE USA to Argentina.

1
2
3
4
5
6
7
8
9
10
11
12
13
14
15
16
17
18
19

BUTTONQUAIL TURNICIDAE

1 COMMON BUTTONQUAIL *Turnix sylvaticus* 15–16cm Rare and secretive; when alarmed, runs rather than flies. In flight, shows short wings and buff upperwing coverts. **V** Female gives a low, droning *hoooo hoooo hooo*. Both sexes give a low-pitched *cree cree cree*. **H** Dry grassland, heaths, crop fields. **D** Widespread across Africa and Asia.

2 RED-BACKED BUTTONQUAIL *Turnix maculosus* 12–16cm Actions and habits similar to those of Common Buttonquail. **V** Deep, far-carrying, subdued *oom oom oom...* **H** Dense grassland, grassy scrub, cultivation. **D** Sulawesi, Indonesia, and New Guinea to N Australia.

3 HOTTENTOT BUTTONQUAIL *Turnix hottentottus* 14cm Separated from quails by more rufous plumage and creamy eye. Long buff streaks along scapulars diagnostic. In flight, looks long-necked and shows pale upperwing coverts. **V** Strange, very low, pumping, regular *wuh-wuh-wuh-wuh*. **H** Short, partly bare grassland, floodplains. **D** South Africa.

4 BLACK-RUMPED BUTTONQUAIL *Turnix nanus* 14cm Separated from quails by more rufous plumage and creamy eye. Long buff streaks along scapulars are diagnostic. In flight, looks long-necked and shows area of pale upperwing coverts. **V** Very low, pumping, regular *wuh-wuh-wuh-wuh*. **H** Short, partly bare grassland, and floodplains. **D** Widespread across Africa.

5 YELLOW-LEGGED BUTTONQUAIL *Turnix tanki* 15–18cm Shy and secretive. Actions and habits similar to those of Common Buttonquail. **V** Female has a human-like, moaning hoot and far-carrying *off-off-off*. Calls also include a *pook-pook*, probably uttered by male. **H** Scrub, grassland, slightly marshy areas, cultivated fields. **D** Widespread across Asia.

6 SPOTTED BUTTONQUAIL *Turnix ocellatus* 17–18cm Shy and secretive. Forages on the ground; if disturbed, tends to disappear into cover, but also recorded feeding unconcerned near observers. **V** Female said to utter a low booming, otherwise little known. **H** Very varied, including scrub, dry forest ravines, forest edges, gardens. **D** Philippines.

7 BARRED BUTTONQUAIL *Turnix suscitator* 15–17cm Secretive; actions and habits similar to those of Common Buttonquail. **V** *Groo groo groo drrr-r-r-r-r-r*, said to sound like a distant motorbike; also a far-carrying *hoon-hoon-hoon*. **H** Scrub, grassland, cultivated areas. **D** Widespread across Asia.

8 MADAGASCAN BUTTONQUAIL *Turnix nigricollis* 15cm Note pale eyes. Similar to Common, Blue and Harlequin quails (Plate 14), but these fly with faster wingbeat. **V** Low-pitched hum (3–8 sec). **H** Most habitats except dense forest and wetlands. **D** Madagascar.

9 BLACK-BREASTED BUTTONQUAIL *Turnix melanogaster* 17–19cm Female larger and more colourful than male; mates with several males, which incubate eggs. Forages in leaf litter, mainly for invertebrates. Freezes in face of danger. **V** Female utters a soft booming call. **H** Rainforest. **D** E Australia.

10 CHESTNUT-BACKED BUTTONQUAIL *Turnix castanotus* 15–20cm Female larger and more brightly marked than male. Often seen in groups. Usually runs from danger but sometimes flies. Diet includes seeds and invertebrates. **V** Female utters a soft booming call. **H** Woodland, savannah grassland. **D** N Australia.

11 BUFF-BREASTED BUTTONQUAIL *Turnix olivii* 19–21cm Unobtrusive and easily overlooked. Prefers to run from danger. Feeds on seeds and invertebrates. **V** Female utters a series of booming calls, with whistling response from male. **H** Grassy woodland. **D** N Australia.

12 PAINTED BUTTONQUAIL *Turnix varius* 18–22cm Usually in pairs. Scratches through leaf litter for seeds and invertebrates. **V** Female utters a series of booming calls, rising in pitch, usually at dawn or dusk. **H** Forest and woodland with dense leaf litter. **D** SW and E Australia.

13 WORCESTER'S BUTTONQUAIL *Turnix worcesteri* 12–14cm Little known; actions presumed similar to those of other buttonquails. **V** Unrecorded. **H** Probably grassland at high elevations. **D** Philippines.

14 SUMBA BUTTONQUAIL *Turnix everetti* 14cm Forages in pairs. Very little else recorded, but actions probably much as other buttonquails. **V** Unrecorded. **H** Scrubby grassland and fields. **D** Sumba, Lesser Sunda Islands, Indonesia.

15 RED-CHESTED BUTTONQUAIL *Turnix pyrrhothorax* 14–16cm Usually freezes when faced with danger. Female mates with several males. **V** Female utters a series of booming calls, rising in pitch. **H** Grassland, open grassy woodland, crop fields. **D** N, E and SE Australia.

16 LITTLE BUTTONQUAIL *Turnix velox* 12–16cm Looks proportionately large-headed. Female larger and more colourful than male. Usually seen in groups. Feeds on seeds and invertebrates. **V** Female utters a series of booming calls, often after dark. **H** Grassland, crop fields, open woodland. **D** Australia.

17 QUAIL-PLOVER *Ortyxelos meiffrenii* 12cm Said to resemble a small Cream-coloured or Temminck's courser (Plate 49) owing to its upright stance. In flight, not unlike a large butterfly. **V** Gives a very soft, low whistle. **H** Dry habitats with some grass cover and more or less bush. **D** Mauritania and Senegal to Sudan, Ethiopia and Kenya.

STONE-CURLEWS AND THICK-KNEES BURHINIDAE

18 EURASIAN STONE-CURLEW *Burhinus oedicnemus* 40–44cm Timid. Mainly crepuscular. During the day often stands motionless. **V** *Cur-lee* or *churrrreee*, often repeated. **H** Steppe, dry fields, heathland, semi-desert. **D** W Europe to N Africa and NW India.

19 INDIAN STONE-CURLEW *Burhinus indicus* 36–39cm Timid. Mainly crepuscular; during the day often stands motionless in shade. Can be hard to see against scrub or dry stony background. **V** *Cur-lee* or *churrreee*, usually given at night. **H** Open stony or scrubby barren areas, sand dunes, river sandbanks. **D** India to SE Asia.

20 SENEGAL THICK-KNEE *Burhinus senegalensis* 32–38cm Actions similar to those of Eurasian Stone-curlew, although tends to be less timid. **V** Similar to that of Eurasian Stone-curlew, but more nasal and metallic. **H** Riverbanks and islands, also open areas near water. In Egypt recorded nesting on flat roofs. **D** Senegal to Ethiopia and Kenya, and north to Egypt.

21 WATER THICK-KNEE *Burhinus vermiculatus* 40cm Note grey-and-white wing bar. **V** Very high, sharp *tjuutjuutjuu* (15–20×), first half speeded up and rising, second half decelerated and falling off; at night. **H** Riverbanks, lake shores, estuaries, lagoons. **D** SE and SC Africa.

22 SPOTTED THICK-KNEE *Burhinus capensis* 37–44cm Tends to stay near bush cover. Mainly nocturnal. **V** Plaintive *tche-uuuu* and rapid *pi-pi-pi-pi* when alarmed. **H** Bushy areas on broken ground and savannah grasslands, also rocky riverbeds. **D** Widespread across Africa.

23 DOUBLE-STRIPED THICK-KNEE *Burhinus bistriatus* 38–43cm Mainly crepuscular or nocturnal. In flight from above, shows white patch on inner primaries. **V** Chattering, strident *ca-ca-ca-ca-ca-ca-ca...*, descending in pitch, often given at dusk, dawn or night. **H** Semi-arid open grasslands, agricultural country. **D** S Mexico to N Brazil, also Hispaniola.

24 PERUVIAN THICK-KNEE *Burhinus superciliaris* 40cm White brow bordered above by black line. Upperparts brownish grey. Runs. **V** *Way ray kay kay*. **H** Arid and semi-arid regions, open land with short grass, cultivated land, scattered trees and bushes. **D** S Ecuador to S Peru.

25 BUSH STONE-CURLEW *Burhinus grallarius* 55–60cm Mainly nocturnal. In daytime, adopts a rigid, upright pose when nervous; relies on camouflage to avoid detection. **V** Mournful *whee-oo, whe-oo*, uttered at dusk and at night. **H** Open woodland, grassland, inland wetland margins. **D** Australia, except interior.

26 GREAT STONE-CURLEW *Esacus recurvirostris* 49–54cm Shy, mainly nocturnal or crepuscular. In flight, upperwing shows black secondaries, white inner primaries with a black subterminal black bar and black outer primaries with white subterminal spots on three outer feathers. **V** Wailing, whistled *see* or *see-ey*; also a harsh *see-eek* when alarmed. **H** Shingle and rocky riverbanks, rocky beaches, estuaries, reefs. **D** SE Iran through SE Asia.

27 BEACH STONE-CURLEW *Esacus magnirostris* 53–57cm Mainly crepuscular and nocturnal; spends day resting in shade. In flight, upperwing shows grey secondaries and black forewing and primaries, the latter with white subterminal spots on three outer feathers. **V** Harsh, wailing *wee-loo*. When alarmed, utters a weak, yapping *quip*, *peep* or rising *quip-ip-ip*. **H** Coastal shores, coral reefs. **D** Malay Peninsula and Philippines to N Australia.

SHEATHBILLS CHIONIDAE

28 SNOWY SHEATHBILL *Chionis albus* 38–40cm Note all-white plumage and pink skin around eyes. Scavenging diet includes food scraps, carcass remains and faeces. **V** Calls include coarse and harsh *errRGGHH, errRGGHH*. **H** Islands and coasts near seabird and seal colonies. **D** Breeds on Antarctic Peninsula and Antarctic islands including South Georgia. Winters in S South America.

29 BLACK-FACED SHEATHBILL *Chionis minor* 38–41cm Note all-white plumage, blackish facial skin and pink eye-ring. Diet comprises mainly penguin food scraps and faeces. **V** Calls include a rather squeakily harsh *uuh, uhh, uu-uuh-uh*. **H** Islands and coasts near seabird colonies, mainly of penguins. **D** Breeds on subantarctic islands in S Indian Ocean, including Crozet, Kerguelen and Prince Edward islands.

MAGELLANIC PLOVER PLUVIANELLIDAE

30 MAGELLANIC PLOVER *Pluvianellus socialis* 20cm Head, upperparts and breast ash grey. Underparts white. Bill black, eye red, legs reddish pink. **V** Longish sad, descending whistle. **H** Downwind on stony shores of inland lakes; seashores and tidal mudflats in winter. **D** S South America.

OYSTERCATCHERS HAEMATOPODIDAE

31 MAGELLANIC OYSTERCATCHER *Haematopus leucopodus* 44cm Upperparts, neck, tail and breast black. White underparts and secondary flight feathers, which do not curl in front of wing. Eye-ring and iris yellow. Legs pale pinkish. **V** Repeated piping *pee-pee*, and a distinctive, rather soft, plaintive, two-syllable whistle, *hoo-eep* or *pee-you*. **H** Seashores; in breeding season often inland on short-grass meadows. **D** Extreme S coasts of South America.

32 BLACKISH OYSTERCATCHER *Haematopus ater* 44cm All dirty black. Eye-ring red and iris yellow. Legs pinkish. **V** Repeated overslurred, piping whistle, *peep*. During display, one bird may utter these whistles while the second initiates a piping trill that increases in volume and becomes a series of longer whistles before trailing off, *prrrrrrrrr-pre-preee-pee-peee-peee-peee-purr-purr-prr*. **H** Rocky coasts. **D** S and W coasts of South America.

33 BLACK OYSTERCATCHER *Haematopus bachmani* 43–45cm Unmistakable. All black in flight. **V** Rapid *peep-peep-peep* or *pee-up*. **H** Rocky coasts, occasionally on adjacent shingle and sandy beaches. **D** W coast of North America.

34 AMERICAN OYSTERCATCHER *Haematopus palliatus* 40–44cm In flight from above, shows large white wing-bar and white lower rump. **V** Shrill, piping *kleep kleep kleep...* **H** Rocky shores and headlands. Breeds on remote beaches. **D** Widespread along coasts of North, Central and South America.

35 AFRICAN OYSTERCATCHER *Haematopus moquini* 45cm Unmistakable. **V** Loud, strident *kleep, kleep-a* or *klee-eep*, with a slower *kleep-a* given during butterfly display flight. **H** At or very near the seashore. **D** S and SW coasts of Africa.

36 EURASIAN OYSTERCATCHER *Haematopus ostralegus* 40–46cm In flight from above, shows broad white wing-bar and white rump. **V** Sharp *kleep* or *kle-eap*, also a quiet *weep*. In display, gives a trilling *ke-beep ke-beep ke-beep kwirrrrrr ee-beep ee-beep ee-beep*. In alarm, utters a *kip* or *pick*. **H** Salt marshes, beaches, inland farmland, by lakes and rivers. **D** Breeds in NW, C and E Europe. Winters in coastal Africa and Asia.

37 SOUTH ISLAND OYSTERCATCHER *Haematopus finschi* 46cm Separated from very similar pied morph of Variable Oystercatcher by different pattern of upperwing and back. **V** High, sharp *tuweet tuweet*. **H** Breeds inland on sandbanks and shingle beds in rivers and at lake sides; winters on sandflats and mudflats at coast and in estuaries. **D** New Zealand.

38 PIED OYSTERCATCHER *Haematopus longirostris* 42–50cm Occurs singly, in pairs or in small groups. Conspicuous and noisy, but wary. In flight, shows black primaries and mainly white secondaries, white rump and uppertail coverts, and mainly black tail. **V** Piping *kleep*. **H** Prefers remote sandy beaches and mudflats. **D** Maluku Islands, Indonesia; Australia.

39 VARIABLE OYSTERCATCHER *Haematopus unicolor* 48cm Polymorphic: all-black, intermediate (not shown) and pied morphs. Pied very similar to South Island Oystercatcher, but in flight shows a square-cut black or smudgy lower back, and wing-bars are narrow, not touching trailing edge. **V** High *tuweet*, like that of South Island Oystercatcher but more plaintive and gull-like. **H** Sandy coasts, especially near estuaries. **D** New Zealand.

40 CHATHAM OYSTERCATCHER *Haematopus chathamensis* 48cm Normally the sole oystercatcher in the Chatham archipelago, from where it does not wander. Note relatively short bill. **V** Like that of South Island Oystercatcher. Also long, meandering, very high rattling, varying in speed and pitch. **H** Rocky shores. **D** Chatham Islands, SW Pacific Ocean.

41 SOOTY OYSTERCATCHER *Haematopus fuliginosus* 46–50cm Feeds by probing substrate for soft-bodied invertebrates and hammering shelled molluscs off rocks. **V** Loud, piping whistle and series of shrill *peep* calls. **H** Sand and shingle beaches, rocky shores, estuaries. **D** Coastal Australia.

black morph
pied morph

CRAB-PLOVER DROMADIDAE

1 CRAB-PLOVER *Dromas ardeola* 38–41cm Adult unmistakable. **V** Barking *ka-how ka-how*. At breeding sites utters various sharp whistles, including *kew-ki-ki* and *ki-tewk*. **H** Sandy coasts, estuaries, lagoons, exposed coral reefs, mudflats. **D** Breeds from Madagascar to Andaman Islands. Winters from E Africa to India.

IBISBILL IBIDORHYNCHIDAE

2 IBISBILL *Ibidorhyncha struthersii* 39–41cm In flight from above, shows white patch on inner primaries and white spots near tips of other primaries. Feeds by probing among riverbed stones. **V** Ringing *klew-klew* and loud, rapid *tee-tee-tee-tee*. **H** Mountain river valleys with flat stony floodplains. **D** Himalayas to NW China.

STILTS AND AVOCETS RECURVIROSTRIDAE

3 BLACK-WINGED STILT *Himantopus himantopus* 35–40cm Usually seen in small groups. **V** Sharp *kek*, high-pitched *kikikikik* and yelping *kee-ack*. **H** Various wetlands, including saltmarshes, saltpans, lakes and marshes. **D** Widespread across Europe, Africa and Asia.

4 PIED STILT *Himantopus leucocephalus* 35–40cm Often considered a race of Black-winged Stilt; actions and habits similar. **V** Similar to that of Black-winged Stilt, but softer and more nasal. **H** Similar to those of Black-winged Stilt. **D** Widespread across Australasia.

5 BLACK-NECKED STILT *Himantopus mexicanus* 42cm Black bill long, slender and straight; legs bright pink. Nape, back of neck, scapulars and wings black; forehead, crown, foreneck, breast, belly, rump and tail white. **V** When alarmed, emits a repeated, penetrating *he*, like a little bark. **H** Edges of ponds, marshes and rivers, sometimes ploughed fields. **D** W and S USA to Peru and E Brazil, Hawaiian Islands.

6 WHITE-BACKED STILT *Himantopus melanurus* 34–39cm Unmistakable. In flight, shows black wings with a white tail and rump that extends as a 'V' on back. **V** Loud, repeated *kik kik kik kik*. **H** Various waterbodies, including mudflats, coastal salt marshes and freshwater pools. **D** Peru and NE Brazil to SC Argentina.

7 BLACK STILT *Himantopus novaezelandiae* 38cm Unmistakable. **V** When disturbed, utters a series of *whep* notes (higher pitched than in Pied Stilt). **H** Gravel and shingle banks in rivers, shallow ponds, other waterbodies. **D** New Zealand.

8 BANDED STILT *Cladorhynchus leucocephalus* 36–42cm Unmistakable, elegant wader. Feeds by probing and scything salt and saline water for crustaceans. Gathers in flocks when feeding is good. **V** Shrill, squeaky *keah*. **H** Saline wetlands. **D** S and SW Australia.

9 PIED AVOCET *Recurvirostra avosetta* 42–45cm Unmistakable. Gregarious. Readily swims, upending to feed like a dabbling duck. **V** Melodious *kluit-kluit-kluit*; harsher *kloo-eet* or *krrree-yu* given in alarm. **H** Shallow saline or brackish lakes, lagoons, saltpans, estuaries; in winter also on tidal mudflats, freshwater lakes, agricultural land. **D** Widespread across Europe and Africa.

10 AMERICAN AVOCET *Recurvirostra americana* 40–50cm Unmistakable. Readily swims, upending to feed, much like a dabbling duck. **V** High-pitched *kleet* or *kluit*. **H** Shallows of lakes, lagoons and tidal mudflats. **D** Breeds from SC Canada to W and SC USA and C Mexico. Winters in S USA and Central America.

11 RED-NECKED AVOCET *Recurvirostra novaehollandiae* 44cm Unmistakable, chestnut head diagnostic. **V** Long, irregular series of mid-high *wec* notes. **H** Prefers shallow inland wetlands. **D** Australia.

12 ANDEAN AVOCET *Recurvirostra andina* 47cm Long, slender, upturned bill black, legs and webbed feet bluish grey. White, with black back, wings and tail. **V** Bisyllabic *do-it*, nasal and penetrating. **H** Edges of lakes, streams and ponds; 3,000–4,000m. **D** Peru to NW Argentina.

PLOVERS CHARADRIIDAE

13 NORTHERN LAPWING *Vanellus vanellus* 28–31cm Unmistakable. In flight from above, looks black with white on uppertail and wing tips. **V** Plaintive *wee-ip* or *pee-wit*. In display flight gives a drawn-out medley of *coo-wee-ip* and *wee-willuch-coo-wee-up*. **H** Grasslands, sometimes on estuaries in cold winters. **D** Breeds across Europe. Winters in N Africa.

14 LONG-TOED LAPWING *Vanellus crassirostris* 30cm Unmistakable. **V** Repeated metallic clicking, *kick-k-k-k* or *kick-kick*. Particularly vocal during breeding season. **H** Floating vegetation at lake shores and calm riverbanks, large swamps, inundations. **D** C, E and SE Africa.

15 BLACKSMITH LAPWING *Vanellus armatus* 30cm Unmistakable. **V** Extremely high, metallic *tink-tink* (hence its common name). **H** Dry, muddy or marshy shores of lakes, dams, ponds, lagoons, swamps. **D** Angola to Kenya and south to South Africa.

16 SPUR-WINGED LAPWING *Vanellus spinosus* 25–28cm In flight, tail mainly black, rump white. Upperwing has black flight feathers and brown coverts split by a bold white wing-bar. Underwing black with white coverts. **V** When alarmed, utters a fast, repeated *kitt-kitt-kitt* or *tik-tik-tik*. Territorial call often transcribed as *did-ye-do-it*. **H** Lakes, marshes, edges of large rivers, coastal lagoons, irrigated fields. **D** SE Europe and Middle East, Mauritania to Guinea and east to Somalia, Kenya and Tanzania, north to Egypt.

17 RIVER LAPWING *Vanellus duvaucelii* 29–32cm Occurs singly, in pairs or in small groups. In flight, upperwing shows a broad white wing-bar contrasting with black flight feathers and grey-brown forewing; rump white and tail black. **V** Sharp *tip-tip* and longer *dip-dip-to-weet*. **H** Sand and shingle areas by rivers, also estuaries. **D** NC India through SE Asia.

18 BLACK-HEADED LAPWING *Vanellus tectus* 25cm In flight, upperwing similar to that of Spur-winged Lapwing except primary coverts white. **V** When disturbed, gives a piercing *kir*. Also utters a harsh *kwairr* and shrill *kiarr*. **H** Dry plains. **D** E, C and W Africa.

19 YELLOW-WATTLED LAPWING *Vanellus malabaricus* 26–28cm Generally occurs in pairs but may form small flocks after breeding. In flight, upperwing shows a prominent white bar on secondary coverts; lower rump white, tail black with white sides. **V** Plaintive *tchee-it*. When alarmed, gives a *chit-oo-eet* or sharp *whit-whit-whit*. **H** Dry grasslands, open dry country, wetland fringes. **D** India.

20 WHITE-CROWNED LAPWING *Vanellus albiceps* 30cm Separated from African Wattled Lapwing by all-white underparts and different upperwing pattern. **V** A single high-pitched, ringing and often repeated *pew* or *peeuw*, its quality reminiscent of the call of an oystercatcher. **H** Sandy riverbanks in forested areas. **D** Senegal to South Sudan and Angola; Tanzania, Zambia to Mozambique and South Africa.

21 SENEGAL LAPWING *Vanellus lugubris* 22cm Note sharp demarcation of white on forehead. Wing pattern diagnostic. **V** Clear, piping *tlu-wit* or *thi-wit* and a plaintive, trisyllabic whistle, *ti-ti-hooi*, the latter heard especially from migrants on moonlit nights. **H** Dry open short grassland with some shrubs. **D** SE, C and W Africa.

22 BLACK-WINGED LAPWING *Vanellus melanopterus* 25cm Note white forehead and eyebrow, gradually becoming grey on nape and cheeks. Separated from Senegal Lapwing by red (not blackish-red) legs. Note diagnostic upperwing and underwing pattern. Not in range of Brown-chested Lapwing. **V** *Ki-ki-kirreeek*, *kik* and *kikikikik* calls heard year-round, but significantly common during the breeding period, when birds are highly territorial. Mainly vocal during the day. **H** Grassland and burnt ground at higher altitudes. **D** E and SE Africa.

23 CROWNED LAPWING *Vanellus coronatus* 30cm Black-and-white head pattern and wing pattern distinctive. Very vocal and active, often performing flight displays in small groups. **V** High, loud, grating *kreeep* or *kree-kree-kree-kreeip*, especially in flight. **H** Drier areas with more or less short grass cover and some shrub and trees, e.g. airfields, golf courses. **D** E and S Africa.

24 AFRICAN WATTLED LAPWING *Vanellus senegallus* 35cm Note large size. Separated from White-crowned Lapwing by more uniform colouring. **H** Bare, muddy, sandy or short-grassed ground near marshes and lakes, ponds and rivers. **D** Widespread across Africa.

25 SPOT-BREASTED LAPWING *Vanellus melanocephalus* 34–35cm Distinctive breast markings and geographical range aid identification. Diet comprises mainly invertebrates, including those disturbed by cattle. **V** Calls include agitated, squeaky *eer-err-keer-ker-ker-r-rrr, eek, eek*. **H** Montane grassland and moorland, often near cattle; 1,750–4,250m. **D** Ethiopia.

26 BROWN-CHESTED LAPWING *Vanellus superciliosus* 25cm Small lapwing with bright yellow wattle and eye-ring; mostly grey and white with rufous on crown and lower breast. **V** In flight, gives three harsh, high-pitched calls in rapid succession, with a squeaky quality likened to a rusty hinge. **H** Dry short grassland, bare fields, often near riverbanks and lake shores. **D** Nigeria to N DR Congo.

27 GREY-HEADED LAPWING *Vanellus cinereus* 34–37cm In flight, shows tail with wide black tip and white rump. Upperwing shows large white area on secondaries and secondary coverts; rest of coverts brown, primaries black. Underwing white with black primaries. **V** Plaintive *chee-it* and rasping *cha-ha-eet* or *pink* when alarmed. **H** Marsh, swampy grassland and rice fields. **D** Breeds in E Siberia, Japan, NE and E China. Winters across Asia.

28 RED-WATTLED LAPWING *Vanellus indicus* 32–35cm In flight, similar to Spur-winged Lapwing but tail has a broad white tip. **V** Shrill *treent-trint teen-ty-too-int trinti-too-int*. When alarmed, gives a sharp, repeated *trint*. In display, gives a prolonged *trint-trint-tee-int tee-int trinti-too-int too-int too-int*. **H** Open areas such as farmland and grassland with nearby water. **D** Widespread across Asia, Middle East.

29 JAVAN LAPWING *Vanellus macropterus* 28cm Little recorded information. Not reported since 1940; possibly extinct. **V** Unrecorded. **H** Open areas near freshwater ponds, agricultural land, river deltas. **D** Java, Indonesia.

30 BANDED LAPWING *Vanellus tricolor* 26–29cm Feeds in a deliberate manner: walks slowly, bends to pick food from ground, then adopts an upright stance. Diet mainly invertebrates. Not dependent on water. Has benefited from overgrazing. **V** Nasal *ki-ke-kew*. **H** Short grassland, agricultural fields. **D** SW, S and SE Australia.

31 MASKED LAPWING *Vanellus miles* 33–37cm Noisy and conspicuous; usually encountered in small flocks. In flight, shows white uppertail coverts and base of tail. **V** Grating, staccato *keer-ki-ki-ki-ki* or *krik-krik-krik*, sharp *kek* and slurred, descending *kreerk-kreerk*. **H** Grassland near water, also lake and swamp margins. **D** Widespread across Australasia.

32 SOCIABLE LAPWING *Vanellus gregarius* 27–30cm Generally occurs in small flocks. In flight, shows white upperwing secondaries and secondary coverts, and black primaries and primary coverts; rest of wing grey-brown, lower rump white, tail black with white sides. **V** Harsh *kretch* or chattering *kretch-etch-etch*. **H** Dry grassland, stubble fields, scrub desert. **D** Breeds in C Asia. Winters in N India and NE Africa.

33 WHITE-TAILED LAPWING *Vanellus leucurus* 26–29cm Usually occurs in small flocks. In flight, upperwing shows brown forewing and black flight feathers split by a broad white wing-bar; tail white. **V** High-pitched *pet-ee-wit pet-ee-wit* and plaintive *pee-wick*. **H** Marshes, damp grasslands, shallows of lakes, pools and rivers. **D** Breeds in SW Asia. Winters in NE Africa.

34 SOUTHERN LAPWING *Vanellus chilensis* 35cm Unmistakable thanks to long, thin crest. **V** Very noisy, e.g. loud scolding metallic *keek-keek-keek*. **H** Short grass, swamps. **D** Widespread across Central and South America.

35 ANDEAN LAPWING *Vanellus resplendens* 32cm No occipital crest. Head, neck and breast pale greyish, blackish through face, throat whitish. Back and upperparts greyish brown; greenish and purple sheen on wing coverts. Underparts white. Base of bill, eye and legs reddish. **V** Repeated nasal *key-er*. **H** Short-grass meadows near water in high Andes, cultivated land, puna; above 3,000m. **D** Colombia to NW Argentina.

36 RED-KNEED DOTTEREL *Erythrogonys cinctus* 18cm No similar bird in range. **V** Mid-high *wrut wrutwrut*, with short, higher-pitched rattles. **H** Margins at wetlands with tussocks, reeds and rushes. **D** Australia, S New Guinea.

37 INLAND DOTTEREL *Peltohyas australis* 20–23cm Usually seen in small flocks. Runs from danger. Rests in shade during heat of day. Seen feeding on dirt roads at night. Diet includes seeds (typically daytime) and invertebrates (at night). **V** Usually silent. **H** Open inland arid habitats. **D** Australia.

38 WRYBILL *Anarhynchus frontalis* Unmistakable thanks to bill (curved right), narrow single breast-band and compact jizz. **V** Very high, thin, well-separated *weet* notes. **H** Breeds in large areas of shingle and sand in riverbeds. Winters at coast in shallow estuaries and mudflats. **D** New Zealand.

1
2
3
4
5
6
7
8
9
10
11
12
13
14
15
16
17
18
19
20
21
22
23
24
25
26
27
28
29
30
31
32
33
34
35
36
37
♀
38
♂

PLOVERS *CONTINUED*

1 EUROPEAN GOLDEN PLOVER *Pluvialis apricaria* 26–29cm Wing tips slightly projecting or level with tail. Underwing coverts white. **V** Mellow *too-ee* or *tloo*. **H** Breeds on dry tundra, moorland, heathland; winters on fields, short grasslands, salt marshes, coastal mudflats. **D** Breeds from Greenland to NC Russia. Winters in S Europe and N Africa.

2 PACIFIC GOLDEN PLOVER *Pluvialis fulva* 23–26cm Wing tips project beyond tail. Longer legs and longer tertials than both European Golden and American Golden plovers. Underwing coverts dusky grey. **V** Rapid *chu-wit* and drawn-out *klu-ee*. **H** Breeds on dry tundra; on migration occurs on coastal lagoons, mudflats, lake shores, river edges, sometimes fields. **D** Breeds in NC and NE Siberia, Alaska. Winters throughout Asia and Australasia.

3 AMERICAN GOLDEN PLOVER *Pluvialis dominica* 24–28cm Wings project beyond tail, tertials shorter than those of Pacific Golden Plover. Longer legs than European Golden Plover, slightly shorter legs than Pacific Golden Plover. Underwing dusky grey. **V** Sharp *klu-eet*, *kleep* or *klu-ee-uh*. Display call a repeated *wit wit weee wit weee* or *koweedl koweedl*. **H** Breeds in tundra; migrants occur on inland grassland and coastal mudflats. **D** Breeds in Alaska and N Canada. Winters in N South America.

4 GREY PLOVER *Pluvialis squatarola* 27–30cm In flight, shows white rump and white upperwing bar. Underwing white with black axillaries. **V** Mournful *tlee-oo-ee*. **H** Breeds on lowland tundra, usually in damp areas; on migration and in winter found mainly on coastal mudflats and beaches. **D** Breeds in N North America and N Europe. Winters worldwide.

5 NEW ZEALAND PLOVER *Charadrius obscurus* 27cm Large, thickset jizz with strong, slightly uptilted bill. S population in breeding plumage darker than N population. Nondescript pale non-breeding plumage. **V** Very high, sharp *wrut-wrut*. **H** Ocean beaches, estuarine mudflats, dunes, farmland. On exposed hilltops on Stewart Island/Rakiura during breeding. **D** New Zealand.

6 COMMON RINGED PLOVER *Charadrius hiaticula* 18–20cm In flight, upperwing shows a bold white wing-bar. **V** Mellow, rising *too-lee* and a soft *too-weep* when alarmed. **H** Shores of coasts, lakes and rivers. **D** N Europe, NE Canada.

7 SEMIPALMATED PLOVER *Charadrius semipalmatus* 17–19cm Yellow eye-ring slightly more obvious than in Common Ringed Plover but less so than in Little Ringed Plover; best distinguished from former by voice. Upperwing shows white bar. **V** Clear, thin *tu-wee* or *che-wee*, increasing in pitch. In alarm, utters a sharp *chip* or *kwiip*. During display gives a repeated *kerrwee-kerrwee*. **H** Breeds on tundra, coasts and margins of rivers, lakes and ponds; winters mainly on coastal mudflats. **D** Breeds in Alaska and N Canada. Winters in S North America, Central America and South America.

8 LONG-BILLED PLOVER *Charadrius placidus* 19–21cm Yellowish base to lower mandible, dark eye-ring. In flight, upperwing shows an inconspicuous white wing-bar. **V** Clear *pewee* and pleasant *tudulu*. **H** Stony margins of lakes and rivers; post-breeding also on paddyfields and coastal mudflats. **D** Breeds in SE Siberia, Japan, Korean Peninsula, NE and C China. Winters in SE Asia.

9 LITTLE RINGED PLOVER *Charadrius dubius* 14–17cm Usually seen singly or in small groups. Yellow eye-ring. In flight, upperwing shows inconspicuous white wing-bar. **V** Descending *pee-oo*; when alarmed, utters a *pip-pip-pip*. **H** Margins of lakes and rivers, marshes, mudflats and estuaries. **D** Breeds across Europe and Asia. Winters in Africa.

10 WILSON'S PLOVER *Charadrius wilsonia* 18–20cm In flight, upperwing shows a white wing-bar, edges of outer uppertail coverts white, forming a patch. Crown colour variable. **V** Musical, whistled *quit*, *whit* or *queet*. When alarmed, utters a sharp *dik* or *dik-ik*. **H** Coastal beaches, salt ponds, mudflats. **D** Coastal E and SW USA to E Brazil, Peru.

11 KILLDEER *Charadrius vociferus* 23–26cm Fairly typical plover foraging actions: short runs, then standing to 'look' before picking up food. Prominent white bar on upperwing, and chestnut-orange rump. **V** Shrill *kill-dee*, *kill-deeah* or variations such as *twill-wee-wee-wee*. **H** Wet grasslands, edges of lakes, pools and estuaries. **D** Alaska and C Canada to Peru.

12 PIPING PLOVER *Charadrius melodus* 17–19cm In flight from above, shows white uppertail coverts, black-tipped tail and wide white wing-bar. Typical plover feeding action: quick runs with frequent stops to pick up food. **V** Whistled *peep*, *peep-lo* or *peep peep peep-lo*. When alarmed, gives a series of soft whistles. **H** Coastal and lake shores. **D** Breeds in E and C North America. Winters in SE USA and Mexico.

13 MADAGASCAN PLOVER *Charadrius thoracicus* 13.5cm Elegant small bird with a characteristic pattern of black lines over its head. **V** Medium-high *pit* or *prip*, repeated every 2–3 secs, or a trilled *pipipipreeeeet*. **H** Coastal areas with more or less grass cover near water and mudflats. **D** Madagascar.

14 KITTLITZ'S PLOVER *Charadrius pecuarius* 12–14cm Diet comprises mainly terrestrial invertebrates. Usually solitary, occasionally in small groups. **V** Calls include a trilling phrase, slurred *tu-eep*, or *prit* given in alarm. **H** Lake margins, coastal lagoons, waterside grassland. **D** Widespread across sub-Saharan Africa, Nile Valley, NE African coast, coastal Madagascar.

15 ST. HELENA PLOVER *Charadrius sanctaehelenae* 14–15cm Recalls Kittlitz's Plover but note larger bill, even longer legs and isolated geographical range. Vulnerable due to predation by introduced mammalian predators, disturbance and habitat destruction. **V** Calls include a slurred, piping *tlu-eep*. **H** Open grassland, including grazed pasture. **D** St Helena, C Atlantic Ocean.

16 THREE-BANDED PLOVER *Charadrius tricollaris* 18cm Narrow white bar on upperwing and long, white-sided tail. **V** High-pitched *weee-weet* and *pi-peep*, and *wick-wick* when alarmed. **H** Edges of rivers, lakes and pools. **D** S and E Africa.

17 FORBES'S PLOVER *Charadrius forbesi* 19cm Eyebrows connected only at nape, not on forehead. Separated from adult Three-banded Plover by brown face sides and absence of wing-bar. Note long, tapered jizz. **V** Plaintive, piping *pee-oo*, which may be repeated. **H** Open, more or less grassy places away from water, occasionally on mudflats at lake and river sides. **D** Senegal and Gambia to South Sudan, Zambia and Angola.

18 WHITE-FRONTED PLOVER *Charadrius marginatus* 20cm No black on breast or chest. White hindcollar absent. Tail projects beyond wing tip. **V** Low-pitched, short *pwut* or *prrut*, and a dry, trilled *trrrr* or plaintive *pi-peep*. **H** Sandy and rocky seashores, lake shores and riverbanks. **D** Widespread across Africa.

19 KENTISH PLOVER *Charadrius alexandrinus* 15–17cm Often mixes with other small plovers, when its quicker movements may stand out. In flight, upperwing shows white wing-bar. **V** Soft, clear *pit* or *pit-pit-pit*; when alarmed, gives a hard *prrr*, *too-eet* or *pweep*. **H** Sandy shores of coasts, lagoons, saltpans and lakes. **D** Widespread across Europe, Africa and Asia.

20 SNOWY PLOVER *Charadrius nivosus* 14–15cm Forages by running with short stops to pick up food, generally in a quicker action than Common Ringed, Semipalmated and Piping plovers. In flight from above, shows white wing-bar and white sides to tail. **V** Low *ku-wheet*, hard *quip* and low *krut*. **H** Sandy beaches, saltflats, lake and lagoon edges. **D** C Canada to Panama, Peru to SC Chile.

21 JAVAN PLOVER *Charadrius javanicus* 15cm Formerly considered a race of Kentish Plover; actions and habits presumed to be similar. **V** Soft, rising *kweek kweek*. **H** Coastal lowlands. **D** Java, Indonesia.

22 RED-CAPPED PLOVER *Charadrius ruficapillus* 14–16cm Occurs singly, in pairs or in small groups. Active and tame. Formerly considered a race of Kentish Plover. **V** Rapid, hard trill; other calls similar to those of Kentish Plover. **H** Coastal and subcoastal sandy and shell beaches, brackish and saltwater mudflats and sandflats. **D** Australia.

23 MALAYSIAN PLOVER *Charadrius peronii* 14–16cm Usually encountered in pairs or small flocks. In flight, upperwing shows prominent white wing-bar, especially at base of primaries; outer-tail feathers white. **V** Soft *whit* or *twik*. **H** Undisturbed sandy, coral and shell beaches, also adjacent mudflats. **D** Malay Peninsula to the Philippines and Lesser Sunda Islands.

24 CHESTNUT-BANDED PLOVER *Charadrius pallidus* 15cm Narrow white bar on upperwing, white sides to tail and rump. **V** Soft *chup*, also a *drreet* or *dweeu*, and a *hweet* given in alarm. **H** Alkaline lakes, saltpans, coastal lagoons. **D** S Africa.

25 COLLARED PLOVER *Charadrius collaris* 14–15cm In flight, upperwing shows a very narrow white wing-bar and white sides to tail. **V** Sharp *peek*; also a short *kip* or *chit*. **H** Coastal mudflats, beaches, coastal lagoons and riverbanks. **D** Widespread across Central and South America.

26 PUNA PLOVER *Charadrius alticola* 18cm Black across crown, down to incomplete blackish collar; a second ill-defined pectoral collar brownish cinnamon. Crown and hindneck pale cinnamon. Upperparts pale brownish grey. Underparts white. Bill and legs black. **V** Short, emphatic *pit*. When excited, birds utter the note in rapid sequences. **H** On flats around lakes in puna; above 3,000m. **D** Peru to NW Argentina.

27 TWO-BANDED PLOVER *Charadrius falklandicus* 19cm Bill and legs black. Double pectoral collar and forecrown black. Crown, cheeks and hindneck cinnamon. Upperparts brownish grey. Underparts white. **V** Short, emphatic *pit* or a mellower *whit*. **H** Sandy beaches, dunes, short grass by water. **D** S South America.

28 DOUBLE-BANDED PLOVER *Charadrius bicinctus* 20cm Double breast-band diagnostic. **V** Very high, stressed *bic* in loose series. **H** Breeds on sand, shingle or gravel riverbanks; winters in variety of habitats, including coastal and inland wetlands, pastures. **D** Widespread across Australasia.

29 LESSER SAND PLOVER *Charadrius mongolus* 19–21cm Usually shows a narrow white bar on upperwing. **V** Short *drrit* and a sharp *chitik* or *chiktik*. **H** Mountain steppes and elevated tundra near water, coastal shingle, sand dunes. Winters on coasts. **D** Breeds in C and E Europe. Winters in Africa, Asia and Australasia.

30 GREATER SAND PLOVER *Charadrius leschenaultii* 22–25cm In flight, toes project beyond tail and shows an obvious white bar on upperwing. **V** Soft *trrri* and a melodious *pipruirr*. **H** Breeds in open, dry treeless areas with nearby water; winters mainly on coasts. **D** Breeds in C Europe. Winters in Africa, Asia and Australasia.

31 CASPIAN PLOVER *Charadrius asiaticus* 18–20cm In flight, toes project beyond tail; restricted white bar on upperwing. **V** *Tup*, *tik-tik-tik* and repeated *tyurlee* during song flight. **H** Saltpans on dry lowland plains, sometimes far from water. **D** Breeds from SW Russia through Kazakhstan to NW China. Winters in Africa.

32 ORIENTAL PLOVER *Charadrius veredus* 22–25cm Powerful flight. In flight, toes project beyond tail and upperwing uniform brown. **V** Trilling calls and a short, piping *klink*. In flight, gives a sharp, whistled *chip-chip-chip*. **H** Dry grasslands, dry areas near wetlands, coastal mudflats, estuaries. **D** Breeds from C Siberia to Mongolia and N China. Winters from Greater Sunda Islands to Australia.

33 EURASIAN DOTTEREL *Charadrius morinellus* 20–24cm In flight, outer primary has a white shaft, which can be conspicuous. **V** Soft *pweet-pweet-pweet* and *kwip-kwip*. When taking flight may give a trilling *skeer*. **H** Hills, mountains, farmland. **D** Breeds from N Europe to NE Siberia, NW Alaska. Winters in N Africa.

34 RUFOUS-CHESTED PLOVER *Charadrius modestus* 21cm Crown and upperparts dark brown. Long white brow. Face and upper neck grey. Breast rufous. Black pectoral band. Rest of underparts white. **V** In flight, gives a mellow *peeeu-dip*, sometimes shortened to *peeu*. Also has a wheezy *wheeerr* whistle. **H** Back of beaches, muddy shores, short-grass pastures, seashores, lakes. **D** S Chile and S Argentina, Falkland Islands.

35 MOUNTAIN PLOVER *Charadrius montanus* 23cm In flight, shows a small white bar on inner primaries and black patch at tip of tail. Often in large flocks post-breeding. **V** In flight, gives a harsh *grrrt*. During display gives a clear *wee-wee-wee*. **H** Breeds on dry upland grassland; winters on bare fields or in semi-desert. **D** Breeds in WC USA. Winters in SW USA and N Mexico.

1
2
3
4
5
6
7
8
9
10
11
12
13
14
15
16
17
18
19
20
21
22
23
24
25
26
27
28
29
30
31
32
33
34
35

PLOVERS *CONTINUED*

1 HOODED DOTTEREL *Thinornis cucullatus* 19–23cm Forms flocks outside breeding season. Forages along strandline and backwash from breaking waves, and feeds mainly on marine and brackish invertebrates. **V** Mostly silent. **H** Coastal beaches, estuaries, lake margins. **D** S Australia.

2 SHORE DOTTEREL *Thinornis novaeseelandiae* 20cm Adult unmistakable. Note compact jizz, long tail projection, bicoloured bill. **V** Single or vigorous series of *bic* notes. **H** Mainly rocky shores, boulder-strewn beaches. **D** Chatham Islands, SW Pacific Ocean.

3 BLACK-FRONTED DOTTEREL *Elseyornis melanops* 16–18cm In flight, shows pale grey central coverts contrasting with black primaries and darker grey secondaries and forewing. **V** Repeated, high-pitched, explosive *dip*, also a soft *tink-tink* and various clicking, buzzing and churring calls. **H** Margins of fresh or brackish water. **D** Widespread across Australasia.

4 TAWNY-THROATED DOTTEREL *Oreopholus ruficollis* 28cm Elegant and upright. Buff and grey. Black line through eye and large spot at mid-belly. Back streaked black and buff. Tawny bib. Legs dirty pink. Scattered pairs. Flocks in winter. **V** Melodious, whistled flight call. **H** Steppe and desert, short grass, stony terrain, usually far from water. **D** W and S South America.

5 DIADEMED SANDPIPER-PLOVER *Phegornis mitchellii* 16cm Small. Blackish head with white band around crown and half-collar. Hindneck and upper back rufous. Rest of upperparts dark brown. Underparts white, finely barred blackish brown. Legs pinkish orange. Shy; solitary or in pairs. **V** A clipped *pic* or *pic-pic* while foraging or in flight. **H** Puna, high mountain bogs, shores of streams and lakes. **D** Peru to SC Chile and Argentina.

6 PIED PLOVER *Hoploxypterus cayanus* 26cm Bill black, legs red. Pink metacarpal spur at bend of wing. Face and sides of head black, crown and nape greyish brown; white lines from forehead to lower nape separate the two. Throat, upper breast and underparts white, broad black band across lower breast Tail white with broad black terminal band. **V** *Kee-oo*, the second syllable lower pitched. **H** Margins of fresh or brackish water. **D** E Colombia to SE Brazil.

EGYPTIAN PLOVER PLUVIANIDAE

7 EGYPTIAN PLOVER *Pluvianus aegyptius* 19–21cm Unmistakable. In flight from above and below, shows broad black band on white flight feathers, upperwing coverts grey, underwing coverts white. **V** Harsh, high-pitched *cherk-cherk-cherk* and soft *wheeup*. **H** Riverbanks and islands. **D** Senegal and Gambia to Ethiopia, DR Congo and N Angola.

PAINTED-SNIPES ROSTRATULIDAE

8 GREATER PAINTED-SNIPE *Rostratula benghalensis* 24–28cm Diet comprises mainly invertebrates, probed for in soft ground. Most active at dusk. **V** Song (given by female) repetitive, piping *uOop, uOop, uOop, uOop* (one per sec). **H** Marshes, flooded fields, marshy open grassland, sewage pools. **D** Widespread across sub-Saharan Africa and Asia.

9 AUSTRALIAN PAINTED-SNIPE *Rostratula australis* 25–30cm Diet includes invertebrates and seeds. Usually solitary, sometimes seen in pairs. Mainly crepuscular and secretive. **V** Female utters booming *oOop, oOOp* calls (one per sec). **H** Vegetated freshwater wetlands, rice fields, sewage pools. **D** E and SE Australia.

10 SOUTH AMERICAN PAINTED-SNIPE *Nycticryphes semicollaris* 21cm Female upperparts, head and breast blackish brown. White medial line on forehead and crown. Ochreous 'V' on back starts white from sides of breast. Large white spots on wing coverts. Underparts white. Bill downcurved at the tip and legs yellowish olive. Male duller. Low, erratic flight. Shy. **V** Silent. **H** Emergent vegetation by lakes, marshes, streams, muddy shores. **D** Paraguay and S Brazil to C Argentina, C Chile.

JACANAS JACANIDAE

11 LESSER JACANA *Microparra capensis* 15cm Adult separated from immature African Jacana by much smaller size, paler upperparts and chestnut (not dark brown) nape. **V** Very high, fast, angry *didititititi*. **H** As African Jacana. **D** Mali to Ghana, east to Ethiopia and south to Namibia, Botswana and E South Africa.

12 AFRICAN JACANA *Actophilornis africanus* 30cm Walks with ease on floating vegetation, picking insects and other invertebrates. Much larger than Lesser Jacana. **V** High, often descending, yelping *wetwetwetwet-wetwet-wet* or mid-high *weh-weh-weh-weh*. **H** Swamps, lakes, ponds, slow streams with floating vegetation. **D** Widespread across sub-Saharan Africa.

13 MADAGASCAN JACANA *Actophilornis albinucha* 30cm Unmistakable in range by plumage pattern. **V** Medium-pitched, trilling *trrrt trrrt trrrt*, at rate of one note per sec, usually given in long sequences. **H** Waterbodies with floating vegetation. **D** Madagascar.

14 COMB-CRESTED JACANA *Irediparra gallinacea* 23cm Usually seen alone or in pairs. Walks with ease on floating vegetation. **V** In flight, utters a thin, twittering call. Also a nasal alarm call. **H** Freshwater lakes, ponds and swamps with abundant floating and emergent vegetation. **D** Philippines; Sulawesi, Indonesia, to N and E Australia.

15 PHEASANT-TAILED JACANA *Hydrophasianus chirurgus* 31–58cm Often gregarious. Forages by walking on floating vegetation, wading in shallow water or swimming. In flight, shows strikingly white wings. **V** In breeding season has a far-carrying, mewing *me-e-ou* or *me-onp*. Post-breeding flocks utter a nasal *tewn*. **H** Lakes and ponds with floating and emergent vegetation. **D** Widespread across Asia.

16 BRONZE-WINGED JACANA *Metopidius indicus* 28–31cm Forages by walking on floating vegetation or wading in shallow water. **V** Harsh grunt. When alarmed, utters a wheezy, piping *seek-seek-seek*. **H** Lakes and ponds with floating and emergent vegetation. **D** Widespread across Asia.

17 NORTHERN JACANA *Jacana spinosa* 19–23cm Forages principally by walking on floating vegetation. In flight, shows bright yellow primaries and secondaries. **V** Noisy cackling, usually given in flight. **H** Freshwater ponds, swamps and rivers with floating vegetation. **D** Caribbean, Mexico to W Panama.

18 WATTLED JACANA *Jacana jacana* 25cm Bill yellow with a red base and small red wattles, and red frontal shield. Legs green. Flight feathers a striking bright primrose-yellow, tipped with black. **V** Strident, sharp, raucous, penetrating series of *weck. weck weck. weck... weck*. **H** Floating vegetation of marsh edges and swamps. **D** Widespread across South America.

PLAINS-WANDERER PEDIONOMIDAE

19 PLAINS-WANDERER *Pedionomus torquatus* 15–19cm Enigmatic, well-camouflaged quail-like bird. Hard to observe in daytime; most observations are after dark, on roads. Diet mainly invertebrates, occasionally seeds. **V** Deep, repeated *hoo-ooo, hoo-ooo*. **H** Short grassland. **D** Inland E Australia.

SEEDSNIPES THINOCORIDAE

20 RUFOUS-BELLIED SEEDSNIPE *Attagis gayi* 30cm Head, neck, breast and upperparts tawny buff, scaled blackish. Underparts cinnamon. Underwing buffish white. Isolated race in Ecuador (*A. g. latreillii*) darker, with rufous ventral parts and underwing. **V** May call continuously in flight, but also sometimes when running: a melodic *gly-gly-gly* or *cul-cul-cul*. **H** Mountain bogs with hummocky vegetation, páramo, stony slopes, grassland, sheltered valleys. **D** Ecuador to S Chile and Argentina.

21 WHITE-BELLIED SEEDSNIPE *Attagis malouinus* 28cm Sexes alike. Like a large female Grey-Breasted Seedsnipe but no white on throat and no black or grey. Generally more ochreous, designs more contrasted. Underwing white. Occurs in pairs or small flocks. **V** In flight gives an excited *tu-whit tu-whit* or *too-ee too-ee* continually. **H** High mountains and plateaux, grassy or pebbly terrain. Descends altitude in winter. **D** S Chile, S Argentina.

22 GREY-BREASTED SEEDSNIPE *Thinocorus orbignyianus* 22cm Male has black-bordered white throat. Face, neck and breast grey, bordered by black pectoral band. White below. Crown and upperparts cryptically blackish, brown, cinnamon and buffish white. Underwing and axillaries blackish. Female head, neck and breast streaked, throat and underparts white. **V** Repetitive dove-like *coocoop coocoop coocoop*, at a rate of two notes per sec. **H** Mountains and hills with pebbly steppe, grass and small bushes, wet valley bottoms, peat bogs. **D** Peru to S Chile and Argentina.

23 LEAST SEEDSNIPE *Thinocorus rumicivorus* 18cm Like Grey-breasted Seedsnipe but smaller. In male, black border of throat and pectoral band joined by a line, forming an anchor. Female lacks grey and anchor is much narrower. **V** Utters a variety of repeated chuckling notes and whistles, including a low, doubled *cree-oo* and low *dzjuck* notes. In flight, the song's hollow-sounding notes are usually given in triplets, *pu-pu-HU'U'UUP*. **H** Flat pebbly steppe, mountains, dunes, cultivated lands; in winter, found short grasslands. **D** Ecuador to NW Argentina, S South America.

SANDPIPERS AND SNIPES SCOLOPACIDAE

24 UPLAND SANDPIPER *Bartramia longicauda* 28–32cm Usually occurs alone or in small groups. Regularly perches on posts or poles. Sometimes forages plover-like with short runs and sudden stops. **V** In flight, utters a piping *quip-ip-ip-ip*. In display, gives a strange bubbling *bububuLEEillayoooooooo*. **H** Grasslands. **D** Breeds in Alaska and NW Canada, and from SC Canada to C and NE USA. Winters in SE South America.

25 BRISTLE-THIGHED CURLEW *Numenius tahitiensis* 40–44cm In flight, shows buffish-cinnamon rump and uppertail. Underwing coverts bright cinnamon, barred black. **V** In flight, gives a whistled *chi-u-it* or *teeoip*. During display, gives a *wiiteew wiiteew...* followed by a *pidl WHIDyooooo*. **H** Barren tundra in mountains. **D** Breeds in W Alaska. Winters on C and S Pacific Ocean islands.

26 WHIMBREL *Numenius phaeopus* 40–46cm In flight, shows white rump and lower back. American race lacks white on rump and back. **V** On breeding grounds utters an accelerating bubbling that rises in pitch. A rippling *pupupupupupupu* is the most frequently heard call. **H** Breeds on moorland and tundra; on migration frequents coastal areas and nearby pastures. **D** Breeds in N North America and Europe. Winters worldwide.

27 LITTLE CURLEW *Numenius minutus* 29–32cm In display, flies to a considerable height while singing, then dives, producing a whistling from tail and wings. In flight, underwing coverts buff with dark bars. Wing tips level with tail. **V** Song is a rising *corr-corr-corr* followed by a level *quee-quee-quee*. Flight call is a whistled *te-te-te* or a rougher *tchew-tchew-tchew*. When alarmed, gives a harsh *kweek-ek*. **H** Breeds in clearings in larch woodland; at other times occurs on short grassland. **D** Breeds in NC and NE Siberia. Winters across Australasia.

28 ESKIMO CURLEW *Numenius borealis* 29–34cm Critically endangered, possibly extinct. Underwing coverts cinnamon, closely barred dark brown. At rest, wing tips project beyond tail. **V** Not well documented; noted calls include a rippling *tr-tr-tr* and a soft, whistled *bee-bee*. **H** Breeds on Arctic tundra; on migration occurs on short grassland. **D** Breeds in NW Canada. Winters in SE South America.

29 LONG-BILLED CURLEW *Numenius americanus* 51–66cm In flight, upperwing shows much cinnamon on inner primaries and secondaries; primary coverts and outer three primaries blackish; underwing coverts cinnamon. Feeds by deep probing and surface picking. **V** Loud, rising *cur-lee* or *coooLI*. **H** Coastal mudflats, marshes, lagoons, occasionally grassland. **D** Breeds from SW and SC Canada to SW and WC USA. Winters from S USA through Central America.

30 FAR EASTERN CURLEW *Numenius madagascariensis* 60–66cm Shy; may mix with flocks of Eurasian Curlews. In flight, lacks white rump and lower back. Female slightly larger and longer-billed than male. **V** Far-carrying *cour-lee*, flatter in tone than that of Eurasian Curlew. When disturbed, gives a strident *ker-ker-ee-ker-ee*. **H** Estuaries, beaches. **D** Breeds in E and SE Siberia, NE China. Winters across Australasia.

31 SLENDER-BILLED CURLEW *Numenius tenuirostris* 36–41cm Endangered. In flight, shows white rump and lower back. Slightly paler than Eurasian Curlew. **V** Short *cour-lee*, higher pitched than in Eurasian Curlew; also a sharp *cu-wee* given when disturbed. **H** Breeds on bogs and marshland in taiga forest; winters on fresh or brackish waters and nearby grassland. **D** Breeds in W Russia. Winters in N Africa.

32 EURASIAN CURLEW *Numenius arquata* 50–60cm In flight, rump and lower back white. Female slightly larger and longer-billed than male. **V** Far-carrying *cour-lee*, stammering *tutututu* when disturbed, and a sequence of bubbling or rippling notes given mainly during breeding season. **H** Breeds on moorland, heathland, grassy meadows; winters mainly at coastal sites but also visits pastures and arable land. **D** Breeds in C and W Eurasia. Winters across Africa and Asia.

1
2
3
4
5
6
7
8
9
10
11
12
13
14
15
16
17
18
19
20
21
22
23
24
25
26
27
28
29
30
31
32

SANDPIPERS AND SNIPES *CONTINUED*

1 BAR-TAILED GODWIT *Limosa lapponica* 37–41cm In flight, shows white from rump to upper back, no white wing-bar, barred tail, mainly white underwing. **V** High-pitched *kik* or *kiv-ik*, often repeated, and a nasal *ke-wuh*. Display calls include a rapid *a-wik a-wik a-wik* and *ku-wew*. **H** Lowland tundra, forest tundra, rolling uplands; winters mainly on estuaries and sandy or muddy coasts. **D** Breeds in N Eurasia and Alaska. Winters in Africa, Asia and Australasia.

2 BLACK-TAILED GODWIT *Limosa limosa* 40–44cm In flight, shows bold white bar on upperwing, white rump, black tail and mainly white underwing. Some races are darker chestnut. **V** *Kek, tuk* or *kip*, often repeated. During display gives an excited *wick-a wick-a-wick-a*, a *tititit* and a hoarse *wee-eeh*. **H** Damp moorland or pastures, lowland water meadows, grassy marshes; winters on estuaries, mudflats, lake shores, grassland. **D** Breeds across Eurasia. Winters in Africa, Asia and Australasia.

3 HUDSONIAN GODWIT *Limosa haemastica* 36–42cm In flight, shows narrow white bar on upperwing (shorter than in Black-tailed Godwit), white rump, black tail, black underwing coverts and axillaries. **V** Nasal *toe-wit, wit* or similar, and a soft *chow-chow*. **H** Marshland near coasts or rivers. **D** Breeds from W Alaska to C Canada. Winters in S South America.

4 MARBLED GODWIT *Limosa fedoa* 40–51cm In flight, superficially resembles Long-billed Curlew (Plate 46), but bill shorter and straight. **V** In flight, utters a harsh *cor-ack* or *kaaWEK*. **H** Breeds on grassy meadows near lakes and ponds; winters on coasts, salt marshes, coastal pools. **D** Breeds in C North America. Winters in S USA and Central America.

5 RUDDY TURNSTONE *Arenaria interpres* 21–26cm In flight from above, wing shows prominent white bar and white patch on innerwing coverts. Centre of back and lower rump white, split by dark band on upper rump. **V** Rapid, staccato *trik-tuk-tuk-tuk, tuk-e-tuk* or *chit-uk*. When alarmed, utters a sharp *chick-ik, kuu* or *teu*. **H** Stony coastal plains or lowlands; after breeding mostly stony and rocky shores. **D** Breeds in N North America and N Eurasia. Winters worldwide.

6 BLACK TURNSTONE *Arenaria melanocephala* 22–25cm In flight, very similar to Ruddy Turnstone but lacks any rufous. May roost alongside Ruddy Turnstones. **V** In flight, utters a trilling *keerrt*. When disturbed, gives a rattling *krkrkrkrkrkr...* **H** Breeds on coastal plains; winters mainly on rocky coasts. **D** Breeds in coastal W Alaska. Winters along W coast of North America.

7 TUAMOTU SANDPIPER *Prosobonia parvirostris* 16cm Pale morph shown, but many variations exist. Small, with warbler-like bill and long hind toe. Distinctive eyebrow. In flight, wings appear rounded. **V** Series of well-separated *beep* and *bic* notes, e.g. *beep beep bicbic bic beep*. **H** Ocean shores, lagoon beaches, bare gravel. **D** Tuamotu Islands, French Polynesia.

8 GREAT KNOT *Calidris tenuirostris* 26–28cm Usually in flocks, often in the company of other waders. In flight, upperwing shows narrow white wing-bar. Lower rump white, sparsely marked with dark specks (appears white), tail dark grey. **V** Low *nyut-nyut*, a harsh *chuker-chuker-chuker* and a soft *prrt*. **H** Mainly sandy or muddy coastal shores. **D** Breeds in NE Siberia. Winters across Asia and Australasia.

9 RED KNOT *Calidris canutus* 23–25cm Usually in small parties. In flight, upperwing shows a narrow white wing-bar; lower rump is white, barred blackish (appears grey). **V** Soft, nasal *knut, wutt* or *whet*. When alarmed, utters a *kikkiik*. **H** Sandy or muddy coastal shores. **D** Breeds in N North America and N Eurasia. Winters in South America, Europe, Africa and Australasia.

10 SURFBIRD *Calidris virgata* 23–26cm In flight, upperwing shows a white wing-bar and white tail with black tip. **V** In flight, gives a soft *iif iif iif*. Feeding flocks utter chattering nasal squeaks. **H** Breeds on rocky mountain ridges; winters mainly on rocky coasts. **D** Breeds in Alaska and NW Canada. Winters along W coasts of North and Central America.

11 RUFF *Calidris pugnax* 26–32cm Breeding males unmistakable but very variable (see examples on plate). Moulting birds have a non-breeding-type plumage splattered with dark blotches on breast. Females are known as Reeves. **V** Normally silent; migrating flocks may utter a shrill *hoo-ee*. **H** Lake, pool and river margins, marshes, wet grassland, coastal mudflats. **D** Breeds from Europe to E Siberia. Winters in Africa, SW Eurasia, India.

12 BROAD-BILLED SANDPIPER *Calidris falcinellus* 16–18cm Often forages among stint flocks. In non-breeding plumage, upperwing shows dark lesser wing coverts and leading edge, and white wing-bar. **V** Buzzing *chrrreet* or *trreet*. **H** Coastal creeks, lagoons, mudflats. **D** Breeds in N Eurasia. Winters in Middle East, Africa, Asia, Australasia.

13 SHARP-TAILED SANDPIPER *Calidris acuminata* 17–21cm In flight, shows a narrow white bar on upperwing and black-streaked white sides to lower rump and uppertail coverts. **V** Soft *wheep, pleep* or *trrt*, and a twittering *prrt-wheep-wheep*. During display, gives a low *hoop* and a muffled trill. **H** Tundra with wet peaty hollows and drier hummocks; after breeding, found on lakes, lagoons, wet grasslands, coastal mudflats. **D** Breeds in N Siberia. Winters in Australasia.

14 STILT SANDPIPER *Calidris himantopus* 18–23cm In flight, shows a white rump and plain upperwing; feet project well beyond tail. Often wades up to belly. **V** Soft *kirrr* or *drrr*, and a low *djew* or *toof*. During display, utters a series of nasal, buzzy trills. **H** Breeds in tundra; post-breeding favours inland and coastal wetlands. **D** Breeds in N Alaska and N Canada. Winters in SC South America.

15 CURLEW SANDPIPER *Calidris ferruginea* 18–23cm In flight, shows a prominent white bar on upperwing, and white lower rump and uppertail coverts. Regularly wades, often up to belly, in shallow water. **V** Rippling *chirrup*. In display, utters a series of chatters, trills and whinnies. **H** Arctic coastal tundra; post-breeding frequents coastal mudflats, lagoons, estuaries and salt marshes, also inland on marshes, lakes, rivers and flooded areas. **D** Breeds in N Asia. Winters in Middle East, Africa, Asia, Australasia.

16 TEMMINCK'S STINT *Calidris temminckii* 13–15cm Has a slow, deliberate foraging action, often among waterside vegetation. If alarmed, flees with a towering, jinking flight. At rest, white-sided tail projects beyond wing tips. Legs greenish yellow. **V** Rapid *tirrirrirrir* or trilled *tirrr*. **H** Marshes, lake and pond edges, paddyfields, saltpans, estuaries. **D** Breeds in N Europe and N Asia. Winters in Africa, Middle East, Asia.

17 LONG-TOED STINT *Calidris subminuta* 13–16cm Regularly forages among vegetation at water's edge. When alarmed, often stands upright with neck extended. If flushed, flees with a towering flight on weak, fluttery wingbeats. **V** Soft *prrt, chrrup* or *chulip*, and a sharp *tik-tik-tik*. **H** Marshes, edges of lakes and pools, coastal pools, estuaries. **D** Breeds in C and E Siberia. Winters across Asia and Australasia.

18 SPOON-BILLED SANDPIPER *Calidris pygmea* 14–16cm In flight, shows prominent white bar on upperwing, sides of rump white. Sweeps bill from side to side while feeding in shallow water. **V** Rolling *preep* or a shrill *wheet*. During display, gives a descending, cicada-like trill *preer-prr-prr*. **H** Coastal tundra, usually near freshwater lakes, pools and marshes. On passage occurs on muddy coasts and coastal lagoons. **D** Breeds in NE Siberia. Winters across Asia.

19 RED-NECKED STINT *Calidris ruficollis* 13–16cm In non-breeding plumage very similar to Little Stint, but rear end appears more attenuated and bill slightly blunter and shorter. Some breeding adults can also resemble Little Stint. Feeding action similar to that of Little Stint. **V** Coarse *chit, kreep, creek* or *chritt*. **H** Breeds on dry tundra; post-breeding frequents coastal or pool shores. **D** Breeds in N and NE Siberia. Winters in Asia and Australasia.

20 SANDERLING *Calidris alba* 20–21cm In flight from above, shows a broad white bar on upperwing and a dark leading edge when in non-breeding plumage. Feeds along water's edge, typically with rapid runs interspersed with quick dips to pick up prey. **V** In flight, gives a *twick* or *kip*, often repeated or forming a quick trill. **H** Barren stony tundra near water. Winters mainly on sandy or muddy shores. **D** Breeds in N North America and N Eurasia. Winters worldwide.

21 DUNLIN *Calidris alpina* 16–22cm Walks quickly, interspersed with short runs, probing and pecking vigorously. In flight, upperwing shows a prominent white wing-bar and white sides to rump and uppertail coverts. **V** Rasping *kreeeep* and a low *beep*. **H** Coastal mudflats, seashores, marshes, also lake and river sides. **D** Breeds in N North America and N Eurasia. Winters in S North America, Africa, S Eurasia.

22 ROCK SANDPIPER *Calidris ptilocnemis* 20–23cm In flight, shows prominent white wing-bar on upperwing and white sides to rump. Very variable. **V** Similar to those of Purple Sandpiper. **H** Breeds on upland tundra in coastal areas; post-breeding occurs on rocky and stony shores. **D** Breeds from E Siberia to Alaska. Winters in Japan and W coast of North America.

23 PURPLE SANDPIPER *Calidris maritima* 20–22cm Often considered conspecific with Rock Sandpiper. In flight, shows narrow white bar on upperwing and white sides to rump. **V** Short *whit* or *kut*. During display, utters various buzzing and wheezing trills, along with low moans. **H** Stony upland tundra; winters on rocky shores and man-made jetties etc. **D** Breeds in N Canada, Greenland, Iceland, Svalbard and Scandinavia to NC Siberia. Winters in E North America and W Europe.

24 BAIRD'S SANDPIPER *Calidris bairdii* 14–17cm At rest, wings project beyond tail. Usually occurs singly or in small groups. **V** Low *preeet*; also a grating *krrt* and a sharp *tsick*. **H** Mainly on fringes of lakes and pools that are bordered by grassland. **D** Breeds in NE Siberia, Alaska, N Canada and W Greenland. Winters in W and S South America.

25 LITTLE STINT *Calidris minuta* 12–14cm Difficult to separate from other stints, especially in non-breeding plumage, but note grey-sided tail, shape of bill and calls. Has a rapid feeding action. **V** Short *stit-tit*; other calls similar to those of Red-necked Stint. During display, utters a *swee-swee-swee*. **H** High-Arctic tundra near swampy areas; after breeding occurs on coastal mudflats and shores of lakes, pools and rivers. **D** Breeds in N Europe and N Asia. Winters in E Europe, Africa, India.

26 LEAST SANDPIPER *Calidris minutilla* 13–15cm Gregarious, often feeding alongside Semipalmated Sandpiper. Tends to adopt a crouched attitude when feeding. **V** Shrill, rising *trreee, prrrep* or *kreeep*; also a low, vibrant *prrrt*. **H** Coastal mudflats, marshes, lakes, pond fringes. **D** Breeds in Alaska and N Canada. Winters in S USA and Central America.

27 WHITE-RUMPED SANDPIPER *Calidris fuscicollis* 15–18cm At rest, wings project beyond tail. In flight, shows white uppertail coverts. **V** High-pitched *jeet*, and a short *tit* or *teet*. **H** Breeds on tundra; on migration occurs on a variety of inland and coastal wetlands. **D** Breeds in N Alaska and N Canada. Winters in SE South America.

28 BUFF-BREASTED SANDPIPER *Calidris subruficollis* 18–20cm Often very tame. In flight from below, shows dark tips to primary coverts, forming a crescent mark. **V** Generally silent, but occasionally utters a sharp *prrreet* when flushed. During display, makes rapid clicking sounds. **H** Arctic tundra; on passage frequents grasslands or dry mud surrounds of lakes and rivers. **D** Breeds in NE Siberia, N Alaska, N Canada. Winters in SE South America

29 PECTORAL SANDPIPER *Calidris melanotos* 19–23cm In flight, shows a narrow white bar on upperwing and clear white sides to lower rump and uppertail coverts. **V** Reedy *churk* or *trrit*. During display gives a hooting, repeated *oo-ah*. **H** Dry areas of vegetated wetlands in Arctic tundra; after breeding frequents coastal and inland wetlands. **D** Breeds in N Siberia, W and N Alaska, N Canada. Winters in Australasia and S South America.

30 SEMIPALMATED SANDPIPER *Calidris pusilla* 13–15cm Short, straight bill. Forages by running with head down, stopping frequently to probe for food. **V** In flight, gives a harsh *chrup, chirk* or *kreet*. Feeding flocks utter a rapid *tweed-do-do-do-do*. **H** Breeds on damp grassy tundra; on migration occurs on coasts, lake shores, pool sides. **D** Breeds in NE Siberia, W and N Alaska, N Canada. Winters in Caribbean and Central and South America.

31 WESTERN SANDPIPER *Calidris mauri* 14–17cm Compared with Semipalmated Sandpiper, bill is slightly more downturned and longer, with thicker base; also tends to forage more often in water. **V** Thin, high-pitched *jeet* or *cheet*. **H** Mainly coastal shores. **D** Breeds in NE Siberia and W and N Alaska. Winters in North and Central America and N South America.

32 ASIAN DOWITCHER *Limnodromus semipalmatus* 33–36cm Often associates with godwits. In flight, shows whitish lower back, barred and streaked dark grey; tail whitish, barred darker grey. **V** Yelping *chep-chep* or *chowp*, and a soft, moaning *kiaow*. **H** Coastal wetlands. **D** Breeds in C Asia and Mongolia to SE Siberia and NE China. Winters across Asia and Australasia.

33 LONG-BILLED DOWITCHER *Limnodromus scolopaceus* 27–30cm In non-breeding plumage resembles Short-billed Dowitcher, but breast slightly darker grey. In flight, shows white oval from upper rump to mid-back. White bars on tail usually much narrower than black bars, making tail appear dark. **V** In flight, utters a thin, high *keek* or *kik-kik-kik-kik*. **H** Shallow fresh or brackish waterbodies, coastal mudflats. **D** Breeds in NE Siberia, Alaska, N Canada. Winters in S USA and Central America.

34 SHORT-BILLED DOWITCHER *Limnodromus griseus* 25–29cm Difficult to separate from Long-billed Dowitcher, especially in non-breeding plumage; best distinguished by call. In flight, shows a white oval from upper rump to mid-back and pale secondaries. **V** In flight, utters a mellow, rapid *tutututu*. **H** Breeds in open marshes; on migration occurs on a wide variety of inland and coastal wetlands; winters mainly on coastal mudflats. **D** Breeds in N North America. Winters in S USA, Central America and N South America.

1
2
4
5
6
7
3
examples of
♂ plumage
variation
11
8
10
♀
11
9
13
14
♂
16
17
12
15
21
18
19
22
20
27
23
24
26
25
34
29
28
33
31
32
30

SANDPIPERS AND SNIPES *CONTINUED*

1 EURASIAN WOODCOCK *Scolopax rusticola* 33–35cm Usually encountered when flushed or during display flight (roding), when flies above territory with slow wingbeats. **V** During display flight, utters squeaks and grunts; also occasionally gives a snipe-like *schaap* or *schaap schaap*. **H** Breeds in mixed, coniferous and broadleaved woodland with glades and damp areas for feeding. **D** Breeds across Eurasia. Winters in N Africa and Asia.

2 AMAMI WOODCOCK *Scolopax mira* 34–36cm Often treated as conspecific with the very similar Eurasian Woodcock. **V** Little recorded; gives a continuous shrill *reep-reep-reep* during distraction display. **H** Evergreen forests and surrounding areas such as sugarcane fields. **D** Ryukyu Islands, East China Sea.

3 JAVAN WOODCOCK *Scolopax saturata* 30cm Nocturnal. Makes a roding display flight at dawn or dusk, usually over forest clearings. **V** Nasal *queet* and rapid, squealing *quo-quo-quo-quo*, both given during display flight and the latter also while perched. **H** Damp montane forest. **D** Sumatra and Java, Indonesia.

4 NEW GUINEA WOODCOCK *Scolopax rosenbergii* 30cm Roosts unobtrusively during the daytime. Feeds at night. Probes soft ground for soil invertebrates. **V** Series of soft grunts, followed by squeaky *purr-Whee*. **H** Montane cloud forest, alpine grassland; mostly 2,000–3,500m. **D** New Guinea.

5 BUKIDNON WOODCOCK *Scolopax bukidnonensis* 33cm Usually solitary and nocturnal. Makes a wide oval display flight over forest clearings, at dawn or dusk. **V** During display flight, utters a metallic, staccato, clicking *pip-pip-pip-pip-pip*, with a growling *gro-a gro-a* between the calls. **H** Mid-altitude and montane forest, especially mossy forest. **D** Philippines.

6 SULAWESI WOODCOCK *Scolopax celebensis* 30–35cm Skulks away from danger, reported making short flights over treetops at dusk; little other recorded information. **V** Unrecorded. **H** Dense montane woodland, especially where there is some open ground below the canopy; bamboo thickets. **D** Sulawesi, Indonesia.

7 MOLUCCAN WOODCOCK *Scolopax rochussenii* 32–40cm Elusive; best seen during dusk roding display flight, when flies on shallow wingbeats over treetops, watercourses or swampy areas. **V** Explosive trill during display flight. **H** Lowland, montane and slightly disturbed forest, near rivers or streams. **D** Maluku Islands, Indonesia.

8 AMERICAN WOODCOCK *Scolopax minor* 25–31cm Flies with fluttering wingbeats. Wings produce a whistling on take-off. **V** In ground display gives a nasal *beent* and a cooing *chako*. **H** Rich, moist woodland with open glades. **D** SC and SE Canada, E USA.

9 CHATHAM SNIPE *Coenocorypha pusilla* 19cm Like Subantarctic Snipe, but range differs. **H** Bush- and woodland, damp grassy habitats. **V** Level series of high, sharp whistles *tjew-tjew-tjew* (10–15 times). **D** Chatham Islands, SW Pacific Ocean.

10 SNARES SNIPE *Coenocorypha huegeli* 22–23cm Feeds on invertebrates, probed for in soil and leaf litter. Keeps to cover of dense vegetation. **V** Slurred, repetitive *tchuaa, tchuaa, tchuaa* (one per sec), delivered after dark. **H** Favours tussock grassland, low forest with dense undergrowth, tangled scrub. **D** Snares Islands/Tini Heke, SW Pacific Ocean.

11 SUBANTARCTIC SNIPE *Coenocorypha aucklandica* 22–24cm Probes for invertebrates in soil and leaf litter. **V** Repetitive *tchoaa, tchoaa, tchoaa* (one per sec) as if trying to build up a head of steam; delivered after dark. **H** Tussock grassland, low forest with dense understorey, scrub. **D** Antipodes, Campbell and Auckland islands, SW Pacific Ocean.

12 JACK SNIPE *Lymnocryptes minimus* 17–19cm Secretive, tends to wait until nearly trodden on before being flushed, flies away with less erratic movements than Common Snipe. **V** May utter a weak *gah* when disturbed. On breeding grounds gives a muffled *ogogok-ogogok-ogogok*. **H** Breeds in bogs of boreal forest and bushy tundra. Winters in marshes, flooded fields, wet grassy areas surrounding lakes, pools, rivers etc. **D** Breeds from N Europe to NE Siberia. Winters across Asia and Africa.

13 SOLITARY SNIPE *Gallinago solitaria* 29–31cm When flushed, zigzags away with a heavier flight than Common Snipe and drops into cover quite quickly. During display dives, stiff outer-tail feathers produce a bleating sound, more shrill than in Common Snipe. **V** When flushed, gives a harsh *kensh*. During display utters a deep *chok-a-chok-a*. **H** High-altitude marshes, bogs, watercourses. **D** Breeds in C and E Asia. Winters in S and E Asia.

14 LATHAM'S SNIPE *Gallinago hardwickii* 28–30cm Underwing with uniform dark barring. Tertials overlap primaries. Longer-tailed than Pin-tailed Snipe. **V** Gives a short *chak* when flushed. Utters a *ji-ji-ji-zubiykahk-zubiyahk-zubiyahk* during circling display flight, followed by a *ga-ga-ga* as bird dives. **H** Breeds on heathland and moorland, often amid light woodland; in winter, frequents freshwater margins. **D** Breeds in E Siberia and Japan. Winters in E Australia.

15 WOOD SNIPE *Gallinago nemoricola* 28–32cm Flight is direct and heavy. When flushed, rarely flies far. **V** When flushed, gives a low, croaking *chok-chok*. At breeding sites utters a nasal *check-check-check*. **H** Breeds in alpine meadows and dwarf scrub; winters in wet areas in forests. **D** S Himalayas to E India and N Vietnam.

16 PIN-TAILED SNIPE *Gallinago stenura* 25–27cm At rest, tertials overlap primaries and tail appears short. In flight, toes project well beyond tail and underwing is uniformly barred. Flushed flight is short and not as towering as in Common Snipe. Supercilium wider at base of bill than in Common Snipe. **V** When flushed, may utter a low-pitched, weak *scaap, squik* or *etch*. **H** Marshes, paddyfields, wet grassland, sometimes drier areas. **D** Breeds from NW Russia and Mongolia to E Siberia. Winters across Asia.

17 SWINHOE'S SNIPE *Gallinago megala* 27–29cm In flight, shows a small amount of toe projection. Underwing with uniform dark barring. Tertials shorter than primaries. **V** Sometimes utters a gruff *scaap* when flushed. During display, gives a repeated *tchiki*. **H** Breeds in swamps and damp forest clearings; after breeding occurs in similar habitats as Pin-tailed Snipe. **D** Breeds in C Asia and N Mongolia, SE Siberia. Winters across Asia and Australasia.

18 AFRICAN SNIPE *Gallinago nigripennis* 30cm Note very long bill, short wings, dark upperparts and contrasting white belly. White outer-tail feathers have some black barring. **V** When flushed, rises explosively and steeply, calling *tsjuk*, and zigzags away at low level before dropping into cover. During display flight, males produce quivering roars from spread tail. **H** Moorland, marshes, grassy lake shores, inundations at higher elevations. **D** E, SC and S Africa.

19 MADAGASCAN SNIPE *Gallinago macrodactyla* 30cm Sole snipe on Madagascar. Note dark mantle (with pale 'braces') and lack of white trailing edge to wings. **V** Rapid medium- to high-pitched *wikwikwikwikwikwikwikwik*, given at rate of 2–3 notes per sec and with each sequence lasting 2–4 secs. **H** Marshes, swamps and rice fields with short vegetation. **D** Madagascar.

20 GREAT SNIPE *Gallinago media* 27–29cm Usually more sluggish than Common Snipe. Extensive white on outer-tail, underwing uniformly dark barred. **V** When flushed, utters a weak *aitch-aitch-aitch*. When displaying, makes various bubbling, twittering and gurgling sounds. **H** Breeds on open damp grasslands, usually with nearby bushes, less often in dry woodland; on migration occurs in wet pastures, and marsh and swamp edges. **D** Breeds from N Europe to C Asia. Winters in Africa.

21 COMMON SNIPE *Gallinago gallinago* 25–27cm When flushed, flies off in an erratic manner. In flight, wings show a white trailing edge, underwing usually shows white mid-wing patch and white bar. **V** In display, utters a repeated *chipper-chipper-chipper-chipper*. When disturbed, gives a harsh *scaap*. **H** Breeds in various marshy or boggy places; after breeding spreads to flooded fields and ditches. **D** Breeds across Eurasia. Winters in S Europe, Africa, Asia.

22 WILSON'S SNIPE *Gallinago delicata* 25cm Very similar to Common Snipe. In flight, lacks white bars on underwing. Skulking. When flushed from among grass, zigzags away, calling before dropping into cover within 200–300m. **V** When flushed, utters a low, dry *critch*. During display, utters a very high, sharp, wooden *tretruttetrut*. **H** Grassy marshes, wet grass, floodplains. **D** North America and Central America to NW South America.

23 SOUTH AMERICAN SNIPE *Gallinago paraguaiae* 27cm Bold face pattern: three dark stripes on dirty white background. Back feathers edged white. Lower underparts white, barred on flanks. Flight feathers uniform greyish brown. Legs pale olive-grey. **V** Slow, excited *chwip chwip chwip* or *kek kek kek*, rising in volume in the middle and fading towards the end, with each note spaced at 1 sec intervals. **H** Damp grassland, peat bogs, marshy and boggy areas, flooded savannah. **D** Widespread across South America.

24 PUNA SNIPE *Gallinago andina* 22.5–25cm Resembles South American Snipe, but smaller and with a proportionally shorter bill and yellower legs. **V** Fast-paced but mellow-sounding *tip-tip-tip, cut-cut-cut* or *dyak dyak dyak*. **H** Damp grassland, peat bogs, marshy and boggy areas, flooded savannah. **D** Peru to NW Argentina.

25 NOBLE SNIPE *Gallinago nobilis* 30cm A hefty version of South American Snipe. Head, neck and breast darker buffish brown, pattern less marked, spotting more dense. Underwing and flight feathers darker. Legs proportionately shorter. Differs from Jameson's Snipe in having narrower bill base. **V** Male emits a reedy *keekeekee*. Female utters loud chipping notes, *kyuck-kyuck-kyuck-kyuck-kyuck*. **H** Bogs, streams and small lakes in páramo, damp grassland, savannah. **D** Venezuela, Colombia.

26 GIANT SNIPE *Gallinago undulata* 48cm Giant version of South American Snipe. Bill broad at base, seeming to flatten the forehead. Flight feathers dark brown, finely barred whitish. Unusually shy, hard to see. **V** In flight, makes sonorous, winnowing cries with a human timbre that fade away at the end, *HO-go, go* or *GA-ga, ga*. **H** Tall grass in bogs, flooded grassland (sometimes drier), marshes in savannah. **D** N and SE South America.

27 FUEGIAN SNIPE *Gallinago stricklandii* 30cm Generally cinnamon buff and fairly uniform. Back feathers edged cinnamon. Face and underparts ochre, barred dark brown on breast and flanks. Solitary or in small flocks **V** During nocturnal display, gives a series of repeated *cheep* notes, as well as a penetrating *char-woo*. **H** Bogs and damp transition zones with woods. **D** C and S Chile, S Argentina.

28 JAMESON'S SNIPE *Gallinago jamesoni* 30cm Resembles Fuegian Snipe but brow, throat, face and underparts whiter, belly more barred. **V** Male emits woodpecker-like *wee-kew* notes in a series lasting 30 secs. Female gives a strident, rising and falling *whi-whi-whi-whi-whi-whi* call in response. **H** Open areas with grass in páramo, boggy patches in woods and scrubby areas, coastal bogs. **D** Venezuela to Bolivia.

29 IMPERIAL SNIPE *Gallinago imperialis* 31cm Dark upperparts chestnut cinnamon with black streaking and bars. No pale edges to feathers. Cinnamon breast heavily streaked blackish. Lower underparts and vent with broad black-and-white barring. Flight feathers dark brown, underwing coverts black. Very rare and local. **V** Series of grating, far-carrying quacking notes, *keke-keh keke-keh kah-kah kuh kuhh kuhh*. **H** At treeline in mountains, elfin woods, bogs, damp places with tall grass, cane and tree-ferns. **D** Colombia, Peru.

30 TEREK SANDPIPER *Xenus cinereus* 22–25cm In flight, shows a wide white trailing edge. Often feeds in active, dashing manner. **V** In flight, gives a rippling *du-du-du-du-du* or mellow *chu-du-du*. During display, repeats a melodious *ka-klee-rree ka-klee-rree ka-klee-rree*. **H** Breeds by upland rivers and streams, otherwise usually encountered in a variety of wetland areas such as lakes, marshes, estuaries and saltpans. **D** Breeds from E Europe to NE Siberia. Winters in Africa, Asia, Australasia, Middle East.

31 WILSON'S PHALAROPE *Phalaropus tricolor* 22–24cm In flight, shows white patch on lower rump. Tends to swim less than other phalaropes, and when feeding on land walks quickly with a feverish pecking action. **V** Generally silent; sometimes utters a soft, grunting *aangh, wennf* or *vimp* in flight. **H** Shallow waterbodies. **D** Breeds in W and C Canada and N USA. Winters in S South America.

32 RED-NECKED PHALAROPE *Phalaropus lobatus* 18–19cm In flight, upperwing shows a white wing-bar. Feeding actions like those of Red Phalarope. **V** In flight, utters a harsh *twick*. **H** Breeds on tundra bogs; on migration occurs on lakes, pools and coastal waters. **D** Breeds in N Alaska, N Canada, N Europe, N Asia. Winters in W and E Pacific Ocean, N Indian Ocean.

33 RED PHALAROPE *Phalaropus fulicarius* 20–22cm In flight, upperwing shows wide white wing-bar. Regularly swims. Short, thick bill. **V** At breeding grounds, gives a buzzing *prrrt* and various chirrupings and twitterings. In flight, utters a sharp *pik*. **H** Breeds in coastal tundra marshes; on migration, sometimes found on coastal pools or lakes, otherwise strictly pelagic. **D** Breeds in N Alaska, N Canada, N Europe, N Asia. Winters in SE Pacific Ocean, SE Atlantic Ocean.

1
2
3
4
5
6
7
8
9
10
11
12
13
14
15
16
17
18
19
20
21
22
23
24
25
26
27
28
29
30
31
♀
♂
32
♂
♀
♂
33
♀

SANDPIPERS AND SNIPES *CONTINUED*

1 COMMON SANDPIPER *Actitis hypoleucos* 19–21cm In flight, shows a white bar on upperwing. Flies low with flicking wings. Bobs tail when walking. **V** In flight, gives a piping *tswee-wee-wee*. Alarm call is a *sweeet-eet*. During breeding display flight, often heard in winter and on migration, utters a repeated *kittie-needie*. **H** Breeds by upland rivers and streams, otherwise occurs on various freshwater and saltwater areas. **D** Breeds across Eurasia. Winters in Africa, Asia, Australasia, Middle East.

2 SPOTTED SANDPIPER *Actitis macularius* 19cm Resembles Common Sandpiper but with a white peak between carpal and breast; tail protrudes far beyond wing. Breeding plumage distinctive (note black-tipped orange bill). In flight, shows shorter white wing-bar than Common Sandpiper. **V** Descending *weetweetweet* and hurried *prrrrWeetprrrWeet*. **H** Sandy beaches, lagoons, mangroves. **D** Breeds across North America. Winters in S USA through South America.

3 GREEN SANDPIPER *Tringa ochropus* 21–24cm In flight, shows white rump and broad bars on tail. **V** Musical *tlueet-wit-wit* and a sharp *wit-wit-wit* when alarmed; variations of these used in display. **H** Damp areas in woodlands; after breeding frequents pools, lake edges, stream sides, ditches. **D** Breeds from N and C Europe to E Siberia and NE China. Winters in SW Europe, Africa and Asia.

4 SOLITARY SANDPIPER *Tringa solitaria* 18–21cm Walks slowly and deliberately, picking food delicately off the surface of water or ground. In flight, shows dark rump and white outer-tail feathers with broad dark bars. **V** Excited *peet, peet-weet-weet* or *tewit-weet*. Display call similar but more bell-like. **H** In breeding season, occurs on pools and marshes in woodland; post-breeding occurs on ponds, stream sides, ditches. **D** Breeds in N North America. Winters in the Caribbean and Central and South America.

5 WANDERING TATTLER *Tringa incana* 26–29cm In flight, appears plain grey with a white belly. When foraging, constantly bobs and teeters. Very similar to Grey-tailed Tattler, best distinguished by voice. **V** In flight, utters a ringing *pew-tu-tu-tu-tu-tu*, a rapid *lidididid* and a sharp *klee-ik*. Display call is a whistled *deedle-deedle-deedle- dee*. **H** Breeds by mountain streams; post-breeding primarily on rocky coasts. **D** Breeds in NE Siberia, Alaska and NW Canada. Winters in S USA and W Pacific Ocean islands.

6 GREY-TAILED TATTLER *Tringa brevipes* 24–27cm In non-breeding plumage shows slate-grey underwing that contrasts with white belly in flight. Bobs tail when walking. **V** In flight, utters a *tu-whip*, and when alarmed a *klee, klee-klee* or *weet-eet*. **H** Breeds on rocky upland rivers; on migration usually on sandy or muddy shorelines. **D** Breeds in NC and NE Siberia. Winters from Malay Peninsula through Australasia.

7 LESSER YELLOWLEGS *Tringa flavipes* 23–25cm Active feeder; often runs through water. Appears more delicate than larger Greater Yellowlegs. In flight, shows plain wings and square white rump, toes project well beyond tail. **V** Flat, harsh *tew-tew* or *tew*. **H** Breeds among forest bogs and grassy clearings; post-breeding occurs on a wide variety of coastal and inland wetlands. **D** Breeds from Alaska to C Canada. Winters from S USA to South America.

8 WILLET *Tringa semipalmata* 33–41cm Often the first from mixed groups to fly off in alarm. In flight, wings look distinctly black and white. **V** In flight, utters a ringing *kyaah yah* or harsh *wee-wee-wee*. In alarm, gives a loud *wik wik wik wik*. Territorial call is a rolling *pill-will-willet*. **H** E birds occur on salt marshes and coasts, W birds occur mainly on inland ponds and lakes; post-breeding birds usually found on coasts. **D** Breeds in E, C and SE North America. Winters from SW North America to South America.

9 COMMON REDSHANK *Tringa totanus* 27–29cm Wary, taking noisily to flight at the slightest disturbance. In flight, shows a white oval in centre of back and striking white secondaries and inner primaries. **V** Piping *teu-hu, teu-hu-hu* or similar. When alarmed, utters a loud *tli-tli-tli-tli*. During display flight gives a musical yodelling. **H** Wide variety of coastal and inland wetlands. **D** Breeds from W to C Eurasia. Winters in SW Europe, Africa and Asia.

10 MARSH SANDPIPER *Tringa stagnatilis* 22–25cm In flight, resembles a small Common Greenshank, but toes project well beyond tail. **V** Plaintive *keeuw* or *kyu-kyu-kyu*. When flushed, utters a loud *yip*. During display has a yodelling *tu-ee-u tu-ee-u tu-ee-e...* and a twittering *chip chip chip-ipepepepepepe*. **H** Open freshwater marshland; post-breeding frequents marshes, ponds, salt marshes, estuaries. **D** Breeds from E Europe to E Siberia and NE China. Winters in Africa, Asia and Australasia.

11 WOOD SANDPIPER *Tringa glareola* 19–21cm In flight, shows a white rump and narrow barring on tail. **V** High-pitched *chiff-iff-iff* and a *chip* or *chip-chip-chip* when alarmed. Gives a series of yodelling notes during display. **H** Forest bogs and lightly wooded marshes; post-breeding occurs on inland lakes, pools, flooded grasslands, marshes. **D** Breeds from N and C Europe to E Siberia and NE China. Winters in Africa, Asia and Australasia.

12 SPOTTED REDSHANK *Tringa erythropus* 29–32cm In flight, shows white oval in centre of back. Often wades up to belly. **V** In flight, gives a distinctive *chu-it*. When alarmed, utters a short *chip*. During display, gives a series of creaking, grinding notes. **H** Open Arctic tundra with nearby bogs and marshes; post-breeding occurs on upper reaches of estuaries, lagoons, lakes, marshes. **D** Breeds from N Europe to NE Siberia. Winters in SW Europe, Africa and Asia.

13 COMMON GREENSHANK *Tringa nebularia* 30–35cm In flight, shows white rump and back, toes project slightly beyond tail. **V** Ringing *chew-chew-chew* and a *kiu kiu kiu* when alarmed. During display, gives a repeated musical *too-hoo-too-hoo...* **H** Taiga and forest areas with clearings, wooded moorland, marshes, bogs; winters on a large variety of coastal and inland wetlands. **D** Breeds from N Europe to E Siberia. Winters in SW Europe, Africa, Asia and Australasia.

14 NORDMANN'S GREENSHANK *Tringa guttifer* 29–32cm In flight, shows white rump and back, pale grey tail; toes do not project beyond tail. **V** Piercing *keyew* and a harsh *gwark*. **H** Coastal lowland swamp and marshy areas in sparse larch forests; on passage occurs mainly on coastal mudflats, sandflats, lagoons. **D** Breeds in E Siberia. Winters in Asia.

15 GREATER YELLOWLEGS *Tringa melanoleuca* 29–33cm Very similar to smaller Lesser Yellowlegs. In flight, shows square white rump and plain wings, toes project well beyond tail. **V** Clear, slightly descending *teu-teu-teu*. **H** Breeds among forest bogs; during non-breeding season occurs on a wide variety of freshwater and brackish wetlands. **D** Breeds in S Alaska and from W Canada to E Canada. Winters from S USA to South America.

COURSERS AND PRATINCOLES GLAREOLIDAE

16 CREAM-COLORED COURSER *Cursorius cursor* 19–22cm Usually in pairs or small parties, foraging with rapid runs interspersed with pauses to pick up prey. In flight, underwing mainly black; above, buffish body and wings contrast with black primaries. **V** Harsh *praak-praak*. During display, gives a piping *quit quit quit* or *quit-quit-whow*. **H** Arid open desert or semi-desert, gravel plains, open fields, saltflats. **D** NE, C and W Africa and SW Eurasia.

17 SOMALI COURSER *Cursorius somalensis* 25cm Note gradual paling from throat to undertail coverts. **V** Poorly recorded, but a scratchy, descending, slurred *pyau* and a muffled *pip* have been reported. **H** Desert, semi-desert, open and bushy grassland and farmland. **D** NE Africa.

18 BURCHELL'S COURSER *Cursorius rufus* 21cm Note partly grey crown, grey band across belly, white secondaries. **V** Rather silent. **H** Arid areas with more or less grass cover. **D** Angola to South Africa.

19 TEMMINCK'S COURSER *Cursorius temminckii* 20cm Note orange (rufous) crown, small black breast patch, all-black flight feathers. **V** High, hooted, unhurried *hak hak hak zunk* (*zunk* slightly lower and inhaled). **H** Dry, more or less bushy habitats, bare fields, burnt ground. **D** Widespread across sub-Saharan Africa.

20 INDIAN COURSER *Cursorius coromandelicus* 23cm Foraging actions as Cream-coloured Courser. In flight from above, shows black primaries and inner secondaries that contrast with rest of wing and body; tail grey-brown, lower rump white. **V** Low *gwut* or *wut*. **H** Less arid areas than Cream-coloured Courser, including dry fields, stony plains, dry riverbeds. **D** Pakistan, Nepal, India, Sri Lanka.

21 DOUBLE-BANDED COURSER *Rhinoptilus africanus* 22cm Note two narrow black bands across breast and (in flight) rufous in wings. Mainly active at night. **V** Mid-high dry rattles; high, scolding, sharp *sheetsheet*; and very high, piercing *fififi* in several variations, all heard at night. **H** Arid and dry areas with little grass and occasionally some scrub. **D** SW and E Africa.

22 THREE-BANDED COURSER *Rhinoptilus cinctus* 25cm Note white underwing and complicated collar pattern. Nocturnal. **V** High, excited, staccato yelps, *wew wew wew*, heard after dark. **H** Dry, open woodland and thornveld. **D** E and SE Africa.

23 BRONZE-WINGED COURSER *Rhinoptilus chalcopterus* 25cm Note large red-rimmed eye. Beautiful purple tips of flight feathers not visible in folded wing. Nocturnal. **V** Strange high, peacock-like, repeated *miau-eh*, at night. **H** More or less wooded and bushy areas. **D** Widespread across sub-Saharan Africa.

24 JERDON'S COURSER *Rhinoptilus bitorquatus* 27cm Rare. Nocturnal; usually found in pairs or small parties. In flight, upperwing dark with a white subterminal patch on outer primaries and white mid-wing-bar; rump white, tail black. **V** Staccato *twick-too twick-too twick-too* or *yak-way yak-wak yak-wak*. **H** Scrub jungle with bare patches. **D** EC India.

25 AUSTRALIAN PRATINCOLE *Stiltia isabella* 24cm Usually occurs singly or in small groups. Forages on the ground or by hawking insects in the air. In flight, shows exceptionally long wings, white rump and short, squarish tail. Underwing coverts black. **V** Shrill, sweet *hoo-wee-too* or *hoo-wee*. **H** Open grassland. **D** Widespread across Australasia.

26 COLLARED PRATINCOLE *Glareola pratincola* 23–26cm In flight, shows narrow white trailing edge to secondaries and reddish-chestnut underwing coverts. In flight from above, dark primaries contrast with paler coverts and mantle. Tail deeply forked. **V** Harsh *kik* or *kirrik*, and a rolling *kikki-kirrik-irrik*. **H** Flat open areas, fields and steppe, usually near water. **D** Breeds across Africa and from S Europe to Pakistan. Winters in Africa.

27 ORIENTAL PRATINCOLE *Glareola maldivarum* 23–24cm Gregarious. In flight from above, shows all-dark wings, white rump and forked tail. Underwing coverts chestnut. **V** Sharp *kyik, chik-chik* or *chet*; also a rising *trooeet* and a *ter-ack*. **H** Bare flats or fields, often near wetlands. **D** Breeds in C and E Siberia, Japan, China and throughout Asia. Winters in Australasia.

28 BLACK-WINGED PRATINCOLE *Glareola nordmanni* 23–26cm In flight, shows all-dark underwing. Upperwing less contrasting than in Collared Pratincole. Tail deeply forked. **V** Low-pitched *chrr-chrr*, also a rapid *pwik-kik-kik* and a *kritt, krip* or *kikiip*. **H** Saline and alkaline steppe, grasslands, lake shores. **D** Breeds from SE Europe to C Asia. Winters in W and S Africa.

29 MADAGASCAN PRATINCOLE *Glareola ocularis* 24cm Note very short tail (when perched and in flight) and partly rufous belly. Further separated from Collared Pratincole by brighter, rufous underwing. **V** In flight, gives a descending *vik vik, vikavik, vikavik*, and while feeding on wing gives a *veet ee veet - veet ee veet - veet ee veet*. **H** Shores of open lakes and sea, muddy (not rocky) riverbanks. **D** Breeds in Madagascar. Winters in E Africa.

30 ROCK PRATINCOLE *Glareola nuchalis* 18cm Note white cheek and bar on underwing. **V** Alarm call is a faint *kip-kip-kip*. During the breeding season both sexes give a long series of short, sharp *ti-ti-ti... te-tic-te* notes. **H** Rocky river and lake shores; sandy and muddy beaches. **D** SC, WC and W Africa.

31 GREY PRATINCOLE *Glareola cinerea* 19cm Unmistakable thanks to pale colouring and facial pattern. Catches insects in swallow-like fashion; active at end of day. **V** Series of *zi* calls, with the intervals between notes becoming progressively shorter. **H** Sand bars in rivers and lakes. **D** Mali and Niger to Angola and DR Congo.

32 SMALL PRATINCOLE *Glareola lactea* 17cm In flight from below, shows white secondaries with black tips and white inner primaries; from above, has less white on inner primaries. Tail square-ended or shallowly forked. **V** *Tuck-tuck-tuck*. In flight, gives a high-pitched, rolling *prrip* or *tirirìt*. **H** Sand bars and shingle banks on rivers, coastal marshes and estuaries. **D** Afghanistan to Cambodia.

1
2
3
4
5
6
7
8
9
10
11
12
13
14
15
16
17
18
19
20
21
22
23
24
25
26
27
28
29
30
31
32

GULLS, TERNS AND SKIMMERS LARIDAE

1 BROWN NODDY *Anous stolidus* 38–45cm Feeds by hovering and dipping to pick prey from water. Underwing coverts paler than in Black Noddy. **V** Crow-like *kwok-kwok, karruuk* or *krao*. **H** Breeds on isolated islets; post-breeding, primarily pelagic. **D** Tropical oceans worldwide.

2 LESSER NODDY *Anous tenuirostris* 30–34cm Actions and habits similar to those of Black Noddy. Underwing coverts dark brown, concolorous with flight feathers, tail wedge-shaped and may appear forked when central tail feathers are moulted. **V** Purring *churr*; also rattling notes when alarmed. **H** Breeds on islets, otherwise maritime. **D** N and NW Indian Ocean.

3 BLACK NODDY *Anous minutus* 35–39cm Feeds by hovering and dipping to pick prey from water's surface, also foot-patters on surface. Underwing coverts brownish black, concolorous with flight feathers. Tail wedge-shaped; may looked forked during moult of central feathers. **V** Distinctive *tik-tikoree* and a staccato rattle. **H** Maritime. **D** Tropical Pacific and Atlantic oceans.

4 BLUE NODDY *Anous ceruleus* 25–28cm Often hovers and then dips to pick prey from water's surface; also paddles on water. Tail forked. **V** Generally silent, although sometimes gives a loud squeal. **H** Cliffs or rocky areas on islands. **D** W and C Pacific Ocean.

5 GREY NODDY *Anous albivitta* 30cm Separated from smaller Blue Noddy by white head and breast. **V** Upslurred rattling *purrr*. **H** Around rocky islands. **D** S Pacific Ocean.

6 WHITE TERN *Gygis alba* 25–30cm Dives to the water's surface to catch small fish as they leap out. Tail forked. **V** Guttural *heech heech*. **H** Breeds in bushes and trees on coral islands, otherwise maritime. **D** Tropical oceans worldwide.

7 BLACK SKIMMER *Rynchops niger* 40–51cm Unmistakable. Feeds by 'ploughing' water with lower mandible during skimming flight. Often nocturnal. **V** Soft, nasal *yep* or *yip*. **H** Sheltered coastal waters and lagoons. **D** Widespread across North, Central and South America.

8 AFRICAN SKIMMER *Rynchops flavirostris* 36–42cm Feeds by 'ploughing' water with lower mandible during skimming flight. Non-breeding plumage is browner with a whitish collar on lower hindneck. Rump and tail black, the latter forked and white-edged. **V** Sharp *kik kik kik*... or *kip kip kip*... and a harsh *kreeee*. **H** Lakes, coastal lagoons, rivers. **D** Widespread across Africa.

9 INDIAN SKIMMER *Rynchops albicollis* 38–43cm Feeds by 'ploughing' water with lower mandible during skimming flight. In flight, upperwing shows a broad white trailing edge, underwing white with dark greyish primaries. Tail forked, white with black central streak. **V** Nasal, yapping *kap kap kap*. **H** Sandbanks on large rivers. **D** E Pakistan to Myanmar.

10 SWALLOW-TAILED GULL *Creagrus furcatus* 51–57cm Tail forked, upperwing pattern similar to that of Sabine's Gull. **V** Harsh scream and a rattle. **H** Breeds on cliffs, sometimes on flatter areas; post-breeding, mainly pelagic. **D** Breeds in the Galápagos. Winters on W coasts of South America.

11 BLACK-LEGGED KITTIWAKE *Rissa tridactyla* 38–40cm Colonial breeder, sometimes in very large numbers. Outer primaries on upperwing and underwing black-tipped, underwing white. Tail slightly forked. At sea, often scavenges around fishing boats. **V** Nasal, wailing *kitt-i-waak kitt-i-waak*, harsh *vek-vek-vek* or *kek-kek-kek*, thin *zeep* and a short *kya*. **H** Breeds on coastal cliffs and islands, also locally on buildings and piers; winters at sea, unless blown inshore. **D** Breeds on N coasts of North America and Eurasia. Winters in the N Atlantic and N Pacific.

12 RED-LEGGED KITTIWAKE *Rissa brevirostris* 36–38cm Habits much as Black-legged Kittiwake. Wing pattern similar to that of Black-legged Kittiwake, although underwing grey. **V** Similar to that of Black-legged Kittiwake, although said to be higher pitched. **H** Breeds on cliffs on remote islands; winters at sea. **D** Breeds in the Aleutian and Pribilof islands, Alaska. Winters in the N Pacific Ocean.

13 IVORY GULL *Pagophila eburnea* 40–43cm Unmistakable. Aggressive scavenger, especially around animal carcasses. **V** Shrill, tern-like *kree-ar* or *keeer*. Other notes said to be like those of Black-headed Gull. **H** Breeds on sea cliffs, also rocky or flat shores; winters around pack ice. **D** Breeds on N Coasts of Eurasia and North America. Winters in the Arctic Ocean

14 SABINE'S GULL *Xema sabini* 27–33cm Distinctive tricoloured upperwing pattern: grey coverts, black outer primaries and primary coverts, white inner primaries and secondaries. Tail forked. **V** Grating *krrr*. **H** Breeds near marshes and pools in Arctic tundra; during migration, mainly pelagic. **D** Breeds on N Coasts of Eurasia and North America. Winters on coasts of NW South America and SW Africa.

15 SLENDER-BILLED GULL *Chroicocephalus genei* 42–44cm Often encountered in small parties. Upperwing shows black-tipped white outer primaries, similar to non-breeding Black-headed Gull, but Slender-billed always looks longer-necked. **V** Harsh rolling *krerrr*; other notes as for Black-headed Gull but lower pitched. **H** Breeds around saline and freshwater lakes, coasts and inland seas; winters mainly on coasts. **D** Iberian Peninsula through the Middle East to Pakistan and India; Mauritania, Senegal and Gambia.

16 BONAPARTE'S GULL *Chroicocephalus philadelphia* 28–30cm Note pale underwing. First-winter plumage very similar to that of Black-headed Gull, although with a slightly greyer hindneck, blackish bill and pink legs. **V** Grating, tern-like *gerrr* or *reeek*. **H** Coastal areas, including harbours. **D** Breeds from SW Alaska to SE Canada. Winters on coasts of USA and N Central America.

17 SILVER GULL *Chroicocephalus novaehollandiae* 41cm Red bill and legs, and yellow eyes. **V** Harsh, grating squeal, *kwarr* or *kwe-aarrr*. **H** Beaches, shores, inland fields, rubbish dumps. **D** Australia, New Zealand, New Caledonia.

18 BLACK-BILLED GULL *Chroicocephalus bulleri* 37cm Unmistakable. No black or brown markings to head. Reduced black to wing tips. **V** Hoarse screams (higher pitched than those of Silver Gull). **H** Breeds on shingle or gravel riverbanks; post-breeding, found on coasts, agricultural land, wet pastures. **D** New Zealand.

19 ANDEAN GULL *Chroicocephalus serranus* 48cm Black-headed version of Brown-hooded Gull. Bill and legs reddish black. Non-breeding birds have a white head and black ear patch. **V** Agitated, sometimes tremulous *yeeer*, and a hoarse *raggh-aggh-keeaagh*. **H** Breeds around high mountain waterbodies to 5,300m; winters lower, even to sea level on Pacific coast. **D** Ecuador to SC Chile and SC Argentina.

20 BROWN-HEADED GULL *Chroicocephalus brunnicephalus* 41–45cm Upperwing grey, outer primaries white with broad black tips, broken by white subterminal 'mirrors'. Underwing grey, primaries blackish, the outermost having white subterminal 'mirrors'. **V** Harsh *gek gek* or *grarhh*, a wailing *ko-yek ko yek* and a raucous *kreeak*. **H** Breeds on islands in high-altitude lakes or marshes; winters on coasts, large lakes and rivers. **D** Breeds in SC Asia, W China and Tibet. Winters in the Middle East and Asia.

21 BROWN-HOODED GULL *Chroicocephalus maculipennis* 37cm Dark blackish-brown hood. Incomplete white eye-ring, dark brown eye. White or pale grey upperparts. White wing tip, patterned with little black. Bill and legs dark red. Forms flocks, often huge. Follows the plough. **V** Shrill, tern-like *zhree* and a short *kip kip kip*. **H** Rubbish dumps, ports, inland water, sea coasts. **D** E and S coasts of South America.

22 BLACK-HEADED GULL *Chroicocephalus ridibundus* 37–43cm Upperwing grey with black-tipped white outer primaries. Underwing pale grey, leading primary white, tipped black, other outer primaries dark grey with black tips. **V** High-pitched *karr, kreeay* or *krreearr*, also a sharp *kek-kek*. **H** Brackish and freshwater lakes, lagoons and marshes; post-breeding, very cosmopolitan, including coasts, lakes, fields, grassland. **D** Breeds across Eurasia. Winters in Africa and Asia.

23 GREY-HEADED GULL *Chroicocephalus cirrocephalus* 38–43cm Occasionally shows a pinkish tinge to underparts. Upperwing grey; base of outer primaries white, otherwise black with white 'mirrors' on outermost pair. Underwing dusky, primaries blackish with white 'mirrors' on outermost pair. **V** Harsh *garr*, a querulous *kwarr* or a drawn-out *caw-caw*. **H** Mainly coasts, also inland lakes and rivers. **D** Widespread across South America and Africa.

24 HARTLAUB'S GULL *Chroicocephalus hartlaubii* 40cm Separated from Grey-headed Gull in breeding plumage by smaller size, black eyes, only traces of a grey hood; in non-breeding plumage by all-white head and less dark outer wings; black (not rosy) bill. **V** Harsh, nasal *kwaaarrr*. **H** Along sea coasts. **D** Coastal Namibia to South Africa.

25 SAUNDERS'S GULL *Chroicocephalus saundersi* 29–32cm Buoyant tern-like flight. Inner primaries form a black wedge, making underwing pattern diagnostic. Upperwing grey with black tips and white outer primaries. **V** Shrill *eek eek*. **H** Breeds on coastal wetlands; winters on coasts. **D** Breeds in coastal E China. Winters from Japan and Korean Peninsula to Vietnam.

26 LITTLE GULL *Hydrocoloeus minutus* 25–30cm Buoyant, tern-like flight. In flight, adult upperwing silvery grey, underwing dark grey, both showing a white trailing edge. **V** Short, nasal *keck* or *keck-keck-keck*. During display, gives a shrill *ke-kay ke-kay ke-kay*. **H** Breeds on well-vegetated lowland freshwater lakes, rivers and marshes; winters mainly on coasts or nearby lakes. **D** Breeds in E and N Europe to W Siberia, and in E Siberia. Winters in E North America and W Europe.

27 ROSS'S GULL *Rhodostethia rosea* 29–32cm In flight, which is light and buoyant, shows pointed tail and pale grey upperwing with black-edged outermost primary, mid-grey underwing with white secondaries and tips of inner primaries. **V** Melodic *a-wo a-wo a-wo* and *claw claw claw*. Also a squabbling *miaw miaw miaw* and soft *kew*. **H** Breeds on marshy areas in tundra and taiga; winters at edges of pack ice and at sea, occasionally blown into coastal areas. **D** Breeds on N coasts of North America and Eurasia. Winters in the Arctic Ocean.

28 DOLPHIN GULL *Leucophaeus scoresbii* 44cm Breeding birds grey, wings and back black. Tail white. Eye white. Bill and legs bright red. Non-breeding birds with blackish head. Bill and legs blackish red. Rowdy and aggressive at colonies of seabirds and mammals. **V** Guttural chatters and occasional high-pitched, piercing screams. **H** Sea coasts, dumps, beaches, ports. **D** Extreme S coasts of South America, Falklands.

29 LAVA GULL *Leucophaeus fuliginosus* 53cm Blackish hood, white eye-ring. Body and wings dark grey, belly and tail lighter. Outer flight feathers darker. Bill and legs black. **V** Single sharp *kow* notes, or a repeated *kow kow kow* in alarm. **H** Desert islands, rocky shores, ports, follows ships, sea-lion colonies. **D** Galápagos.

30 LAUGHING GULL *Leucophaeus atricilla* 36–41cm Upperwing has a white trailing edge and black outer primaries. **V** High-pitched, laughing *ka-ka-ka-ka-ka-kaa-kaa-kaaa-kaaa*; also a shorter *kahwi*. **H** Coastal areas. **D** Widespread on coasts of North, Central and South America.

31 FRANKLIN'S GULL *Leucophaeus pipixcan* 32–38cm In flight, shows grey centre to tail, upperwing has a white trailing edge, and broad white tips on outer primaries broken by a black subterminal band. **V** Short, soft *kruk, queel* or *kowii*. **H** Breeds on lakes and marshes in inland prairies and grasslands, otherwise also on coasts, fields and rubbish dumps. **D** Breeds in SC Canada and NC USA. Winters to W South America coasts.

32 GREY GULL *Leucophaeus modestus* 45cm Note grey rump and tail. Grey body and white head diagnostic. **V** High, trilling *rrrraargh* and soft mewing calls. Voice variable and not as harsh as in most gull species. **H** Sandy beaches. **D** N Chile.

33 RELICT GULL *Ichthyaetus relictus* 44cm Gregarious, breeds in large colonies. Has a similar upperwing pattern to Pallas's Gull. **V** Laughing *ka-ka ka-ka kee-aa*. **H** Shores or islands of saline mountain lakes. **D** Breeds from Kazakhstan to Mongolia and NC China. Winters in E China.

34 AUDOUIN'S GULL *Ichthyaetus audouinii* 48–52cm Colonial breeder. Usually feeds at sea on fish; less of a scavenger than other gulls. Upperwing shows black outer primaries with small white tips and tiny white subterminal 'mirror' on outermost feather. **V** Nasal *gleh-i-eh*, a low *ug-ug-uk* and a raucous *argh argh argh*... **H** Breeds on rocky islands; post-breeding occurs on coastal bays, estuaries, beaches. **D** Mediterranean and NW Africa coasts.

35 MEDITERRANEAN GULL *Ichthyaetus melanocephalus* 36–38cm Upperwing of adult very pale silvery white, primaries entirely white. Colonial. In some areas pairs frequently breed amid Black-headed Gull colonies. **V** *Yeah, jeeah* or *ga-u-a*, also a cooing *kyow*. **H** Breeds on brackish and freshwater lakes, lagoons and marshes; winters mainly in coastal areas. **D** Breeds across Europe. Winters along Mediterranean coasts.

36 PALLAS'S GULL *Ichthyaetus ichthyaetus* 57–61cm Flight action ponderous, with a prominent projecting head. Upperwing shows a narrow black subterminal band on white outer primaries. **V** Low *kyow-kyow* and a nasal *kraagh*. **H** Breeds in open country by freshwater or brackish lakes; in winter, mainly coastal. **D** Breeds from SW Russia to Mongolia and N China. Winters in the Middle East and India.

37 WHITE-EYED GULL *Ichthyaetus leucophthalmus* 39–43cm Sociable, often occurs with Sooty Gull. Upperwing and underwing dark grey with a white trailing edge. **V** Similar to that of Sooty Gull but less harsh and slightly lower pitched. **H** Coastal, including harbours. **D** Red Sea.

38 SOOTY GULL *Ichthyaetus hemprichii* 43–48cm Sociable, often occurs with White-eyed Gull. Upperwing and underwing very dark with a white trailing edge. **V** Mewing *kaarr, keee-aaar* or laughing *veeeaah ve vah veeeaah*; also a high-pitched *kee-kee-kee*. **H** Coastal, including harbours and ports. **D** Coasts of E Africa and the Middle East.

1
2
3
4
5
6
7
8
9
10
11
12
13
14
15
16
17
18
19
20
21
22
23
24
25
26
27
28
29
30
31
32
33
34
35
36
37
38

GULLS, TERNS AND SKIMMERS *CONTINUED*

1 PACIFIC GULL *Larus pacificus* 59–65cm Note massive bill, yellow legs and black back. Diet comprises mainly marine and intertidal fish and invertebrates; occasionally scavenges and takes young birds and eggs. **V** Calls include a moaning *Whoo-oaa, whoa, whoa* and shriller *kah-kah*. **H** Rocky coasts, beaches, harbours. **D** Coasts of W, S and SE Australia, Tasmania.

2 BELCHER'S GULL *Larus belcheri* 45–54cm In flight, shows black wings with white trailing edge and black subterminal band on tail. **V** High-pitched laughing call, reminiscent of a child's bugle. **H** Rocky shores. **D** WC coast of Peru and N Chile.

3 OLROG'S GULL *Larus atlanticus* 50–60cm Like a slightly larger Belcher's Gull, but black ring around bill more distinct. **V** Nasal and guttural calls. **H** Coastal waters. **D** NE coast of Argentina.

4 BLACK-TAILED GULL *Larus crassirostris* 44–47cm Often in large colonies. Upperwing primaries black with small white tips, outermost feather may have a small white subterminal 'mirror'. **V** Plaintive mewing. **H** Breeds on seashores, sea cliffs and rocky islands; winters in coastal areas. **D** E Siberia, Japan, Korean Peninsula, E China.

5 HEERMANN'S GULL *Larus heermanni* 43–49cm Flight buoyant on long wings. In flight, upperwing shows large white patch on primary coverts, tail black with white terminal band. **V** Low trumpeting nasal *yoww*; also a short *yek* and a *ye ye ye ye*... **H** Coastal areas. **D** NW coast of Mexico and SW coast of California.

6 MEW GULL *Larus canus* 40–46cm Often associates with other gulls, especially in winter. Upperwing shows black outer primaries with prominent white subterminal 'mirrors' on outermost feathers. **V** High-pitched, laughing *ke ke ke kleeeh-a... kleeeh-a... kay-a kay-a kay-a kay-a ke ke*, also a yelping *keea keea*. When alarmed, gives a persistent *glee-u glee-u glee-u*... **H** Breeds on coastal marshes and cliffs, beaches, inland moorland; winters in a wide variety of habitats, including coasts, lakes, reservoirs, farmland, town parks. **D** Breeds from W Europe to Siberia and NW North America. Winters worldwide.

7 RING-BILLED GULL *Larus delawarensis* 43–47cm Sociable, usually mixing with other gulls. Upperwing has black outer primaries with a small white subterminal 'mirror' on outermost feather. **V** Mellow *kowk*; other notes like those of American Herring Gull but higher pitched and more nasal. **H** Breeds on inland lakes and wet pastures; winters on estuaries, bays, coastal pools, lakes, fields, rubbish dumps. **D** Breeds in S Canada, and N USA. Winters S USA, Mexico and Caribbean.

8 CALIFORNIA GULL *Larus californicus* 47–54cm Gregarious scavenger. Actions, habits and wing pattern similar to those of American Herring Gull. **V** Scratchy *aow* and *uh-uh-uh*, also a squealing *kiarr*. Laughing call much as in American Herring, but higher pitched. **H** Breeds on inland lakes; winters on coasts, inland wetlands, farmland, towns and cities. **D** NC and W North America to S Mexico.

9 GREAT BLACK-BACKED GULL *Larus marinus* 64–78cm Can be aggressive and predatory. Heavy, powerful flight. Black upperwing shows white trailing edge and extensive white tips to outermost primaries. **V** Hoarse *oow-oow-oow*, also a deep *owk*. All notes gruffer than in other large gulls. **H** Coasts, locally on inland lakes. **D** Coastal E North America and W Europe.

10 KELP GULL *Larus dominicanus* 60cm In flight, wing tip not blacker than rest of wing. Note large, stocky build with short wings and colour of legs (paler than in Lesser Black-backed Gull); two out of every three birds have dark eye, rest have creamy eye as shown. **V** High, almost double-toned *tjik-tjiktjiktjik*. **H** Sea, beaches and nearby areas, but also increasingly seen inland. **D** S coasts of Australasia, South America and Africa.

11 GLAUCOUS-WINGED GULL *Larus glaucescens* 61–68cm Aggressive predator. Slow, powerful flight. Upperwing pattern resembles that of American Herring Gull but outer primaries grey. **V** Low *kak-kak-kak* or *klook-klook-klook*, a deep *kow-kow*, a high-pitched *keer-keer* and a screamed *ka-ka-ako*. **H** Mostly coastal, including ports and harbours, also visits rubbish dumps. **D** Coastal N Pacific.

12 WESTERN GULL *Larus occidentalis* 54–66cm In flight, shows white trailing edge and black primaries with a white subterminal spot on outermost feather. Third-winter birds have black spot near bill tip and pale pink legs. **V** Similar to, but lower than, American Herring Gull. **H** Coasts. **D** W coast of North America.

13 YELLOW-FOOTED GULL *Larus livens* 60–67cm Very similar to dark race of Western Gull but bill heavier and legs yellow. Second-winter birds (reaches adult plumage by third year) have black-tipped yellow bill. **V** Low *quock kuck kuck kuck*. **H** Mainly coasts. **D** Coastal NW Mexico and S California.

14 GLAUCOUS GULL *Larus hyperboreus* 64–77cm Can be an aggressive scavenger. In flight, broad-winged and more lumbering than the otherwise similar and smaller Iceland Gull. **V** Generally silent; most notes as in European Herring Gull but more hoarse. **H** Breeds on coastal and inland cliffs, also offshore islands and open land near coasts; winters mainly on coasts, including harbours, also inland on rubbish dumps. **D** Widespread along N coasts of Eurasia and North America.

15 ICELAND GULL *Larus glaucoides* 55–64cm Winter visitor. Often encountered within flocks of other gull species. Upperwing very pale grey with white outer primaries. **V** Similar to that of European Herring Gull but higher pitched. **H** Coastal areas, including harbours and ports, also on inland lakes and rubbish dumps. **D** Breeds from Canada to Greenland and Novaya Zemlya. Winters in N North America and N Europe.

16 EUROPEAN HERRING GULL *Larus argentatus* 55–67cm Gregarious. Upperwing shows black-tipped outer primaries with white subterminal 'mirrors' on two outermost feathers. **V** Loud, laughing *keeah-keeah-keeah-keah-kau-kau*..., also a short *keeah, keeow* or *kyow*. When alarmed, utters *gag-ag-ag*. **H** Breeds on coasts and nearby areas, also locally on inland moorland; post-breeding very cosmopolitan, including coasts, farmland, landfill sites, cities and towns. **D** Iceland and N Europe.

17 AMERICAN HERRING GULL *Larus smithsonianus* 55–67cm Most widespread gull of North America. Black primaries with distinct white subterminal spot on outer pair of feathers. **V** Loud, laughing *keeah-keeah-keeah-keah-kau-kau*... and a short *keeow*. **H** Coasts, inland lakes, rivers. **D** C Alaska to Newfoundland and the Great Lakes.

18 VEGA GULL *Larus vegae* 57–64cm In flight, upperwing shows black-tipped outer primaries with white subterminal spots on two outermost feathers. **V** Laughing *keeah-keeah-keeah-keah-kau-kau*... **H** Coasts and along large rivers. **D** NE and E Eurasia.

19 CASPIAN GULL *Larus cachinnans* 52–58cm Upperwing shows black outer primaries with large white tip to outer feather and white 'mirror' on next feather. All flight feathers tipped white. **V** Deep, nasal *keeah-keeah-keeah-keeah-kau-kau*... **H** Coasts and inland waters. **D** Breeds from C Eurasia to E China. Winters in SW and E Asia.

20 YELLOW-LEGGED GULL *Larus michahellis* 52–58cm Actions, habits and upperwing pattern very similar to European Herring Gull. **V** Similar to that of European Herring Gull, although slightly deeper and more nasal. **H** Breeds in coastal areas, also locally on inland lakes and rivers; post-breeding, frequents coasts, inland wetlands, fields, rubbish dumps. **D** W and S Europe, NW Africa, Mediterranean.

21 ARMENIAN GULL *Larus armenicus* 54–62cm Gregarious. Upperwing pattern similar to that of European Herring Gull but with one white subterminal 'mirror' on outermost primary feather. **V** Similar to that of Yellow-legged Gull. **H** Breeds on inland lakes, rivers, marshes; winters on coasts, inland wetlands. **D** Armenia to E Turkey and NW Iran.

22 SLATY-BACKED GULL *Larus schistisagus* 55–67cm Some individuals show a paler grey mantle. In flight, shows broad white trailing edge and often has an indistinct white band dividing slaty upperwing from black-tipped outer primaries, the latter with a white subterminal 'mirror' on outermost feather. **V** Similar to that of European Herring Gull. **H** Breeds on sea cliffs, rocky islets, sandy shores; winters in coastal areas. **D** Coastal NE and E Siberia, N Japan.

23 LESSER BLACK-BACKED GULL *Larus fuscus* 51–61cm Gregarious, often mixing with other gull species. Upperwing has white trailing edge, black outer primaries have white subterminal 'mirror' on outermost feather. **V** Similar to that of European Herring Gull, although usually deeper. **H** Breeds on coastal and inland wetlands; winters mainly in coastal areas, although often occurs inland on wetlands and farmland. **D** Breeds Greenland, Iceland, and from W Europe to NC Siberia. Winters N Africa, SW Europe to NW India, E North America.

24 GULL-BILLED TERN *Gelochelidon nilotica* 33–43cm Feeds by hawking insects or dipping to pick from surface of ground or water. Flight gull-like. In flight, mid- to outer primaries show a dark trailing edge above and below. Rump and forked tail pale grey. **V** Nasal *kay-did, kay-tih-did, gur-WICK, ger-erk* or *kay-vek*, also a metallic *kak-kak* and, when alarmed, a *kvay-kvay*. On breeding grounds utters a *br-r-r-r-r*... **H** Breeds in fresh and brackish lakes, salt marshes and rivers, often feeding over nearby grasslands; on migration frequents coastal areas. **D** Worldwide.

25 CASPIAN TERN *Hydroprogne caspia* 47–54cm Feeds mainly by hovering and then plunge-diving; also picks from surface of water and sometimes harasses food from other terns and gulls. Upperwing shows grey outer primaries with darker tips, underwing shows dark outer primaries. Tail has a shallow fork. **V** Loud, croaking *kraah, krah-krah* or *kree-ahk*. **H** Coastal mudflats and creeks, saltpans and lagoons, also inland lakes, rivers and marshes. **D** Worldwide.

26 ROYAL TERN *Thalasseus maximus* 45–50cm Feeds mainly by plunge-diving. Upperwing similar to that of Caspian Tern. Underwing pale with dark trailing edge to outer primaries. Tail deeply forked. **V** Low-pitched, grating *kerriup, krree-it* or *kirruck*. **H** Breeds on offshore islands; winters on coasts. **D** Coasts of North, Central and South America.

27 GREATER CRESTED TERN *Thalasseus bergii* 43–53cm Feeds mainly by plunge-diving, but also picks from surface of water. Underwing shows dark greyish tips to outer primaries. Rump and tail pale grey, concolorous with upperparts, tail deeply forked. **V** Grating *krrik* or *kee-rit*, also a high-pitched *kree-kree*. **H** Coastal waters. **D** Coasts of E Africa to the S Pacific Ocean.

28 LESSER CRESTED TERN *Thalasseus bengalensis* 36–41cm Feeds mainly by hovering and plunge-diving. Wing pattern similar to that of Sandwich Tern. Rump and tail pale grey, the latter deeply forked. **V** Harsh *krrrik-krrik* or *kerrick*. **H** Breeds on low-lying offshore islands; winters on coasts and at sea. **D** Widespread on coasts of Africa, Asia and Australasia.

29 CHINESE CRESTED TERN *Thalasseus bernsteini* 42cm Very rare. Actions little recorded, suspected to be much like those of Greater Crested Tern. Upperwing has outer webs of blackish outer primaries. Underwing white. Tail deeply forked. **V** Little recorded other than harsh, high-pitched cries. **H** Breeds on offshore islets; during migration, occurs on coasts. **D** Breeds on islands off China's E coast. Winters in the Philippines and Borneo.

30 SANDWICH TERN *Thalasseus sandvicensis* 36–41cm Usually occurs in noisy parties. Feeds mainly by plunge-diving, often from a considerable height (up to 10m). Inner webs of outer primaries on upperwing form a silver-grey wedge. Underwing primaries show dusky tips. Tail deeply forked. **V** Grating *kirrruk* or *kerRICK*, also a short *krik* or *krik krik*. **H** Breeds on undisturbed sandy shores and low coastal islands; during migration and in winter, mainly in coastal areas. **D** Breeds from N Europe to the Mediterranean, and Black and Caspian seas. Winters from S Europe to N Africa, Persian Gulf and India.

31 CABOT'S TERN *Thalasseus acuflavidus* 33–34cm Separated from Sandwich Tern by geographical range and proportionately slightly shorter bill. **V** Screeching flight call less obviously trisyllabic than Sandwich Tern's. **H** Inshore seas; nests on beaches. **D** Breeds on coasts from Virginia to French Guiana and Argentina. Winters along Pacific coast of North America from Mexico to Ecuador, and Atlantic coast from Florida to Uruguay.

32 ELEGANT TERN *Thalasseus elegans* 39–43cm Feeds by plunge-diving. Bill colour varies from deep yellow to orange-red. Outer primaries on upperwing often appear dark grey. Pale underwing has dark tips to outer primaries, with outermost feather mainly blackish. Tail deeply forked. **V** Nasal, rasping *karreek* or *ka-zeek*. **H** Breeds on sandy beaches and islands; post-breeding, occurs in estuaries, coastal lagoons, mudflats, harbours. **D** Breeds on coasts of SW USA and NW Mexico. Winters on W coast of South America.

1
2
3
4
5
6
7
8
9
10
11
12
13
14
15
16
17
18
19
20
21
22
23
24
25
26
27
28
29
30
31
32

GULLS, TERNS AND SKIMMERS *CONTINUED*

1 LITTLE TERN *Sternula albifrons* 22–28cm Very active when foraging; feeds by plunge-diving, with much hovering. Upperwing has dark outer pair of primaries. Tail deeply forked. **V** Rapid *kirrikikki kirrikiki* and sharp *kik-kik*, and a rasping *kyik* or *kriet* given in alarm. **H** Sand bars and shingle banks on coasts, lakes and rivers; on migration, mainly coastal. **D** Widespread across Europe, Africa, Asia and Australasia.

2 SAUNDERS'S TERN *Sternula saundersi* 20–28cm Actions and habits as Little Tern. Compared to Little Tern, upperwing has three blackish outer primaries and slightly darker grey upperparts, and lacks short white supercilium. **V** Like that of Little Tern, but less chattering. **H** Coastal waters. **D** Breeds in the Middle East. Winters to India and Malaysia.

3 LEAST TERN *Sternula antillarum* 22–25cm Feeds by plunge-diving, with much hovering. Tail and rump grey, tail deeply forked. **V** Rapid *kid-ick kid-ick...* and a rasping *zr-e-e-e-p*. **H** Beaches, lakes, rivers. **D** Breeds from NC North America to N South America. Winters south to Brazil.

4 YELLOW-BILLED TERN *Sternula superciliaris* 24cm Eye-line and cap black. Forehead white, as are underparts. Upperparts pale grey. Three (sometimes four or five) black outer primaries, forming a triangle. Legs and large, long bill yellow. Shortish forked tail. Solitary, or sometimes in small flocks. **V** Short nasal *kuk* and a slightly longer, raspy *rreh*, typically repeated in a series. **H** Preferably fresh water. Big rivers, marshes, estuaries, lagoons, on sandbanks and island beaches. **D** E Peru, C Colombia and N Venezuela to NE Argentina.

5 PERUVIAN TERN *Sternula lorata* 24cm Cap and eye-line black. Forehead, cheek and throat white. Upperparts grey. Underwing white. Three (sometimes two or four) black outer primaries with white shafts. Long slender bill black, dirty yellow at base. Legs orangey yellow. Tail forked. **V** Short, sharp *chirrio, chirrio*. **H** Coastal, on beaches and adjacent waters; nests on wide beaches, dunes and well inland in desert. **D** Ecuador to Chile.

6 FAIRY TERN *Sternula nereis* 25cm In breeding plumage separated from Little and Least terns by cut-off white above eyes and lack of black point to bill. **V** Goose-like snarls and very high, nasal *wiederweet*. **H** Coastal habitats such as reefs, estuaries, bays. **D** W and S Australia; New Caledonia; N North Island, New Zealand.

7 DAMARA TERN *Sternula balaenarum* 23cm Note long, slightly decurved bill, black cap (of breeding adult) and grey rump and tail. **V** Very high, almost sparrow-like *sree-sreee-ti-ti*. **H** Inshore coastal waters. **D** Coastal Angola, Namibia and South Africa.

8 ALEUTIAN TERN *Onychoprion aleuticus* 32–34cm Feeds by dipping to take food from water's surface. Underwing white with dark trailing edge to secondaries and dusky outer primaries; rump and deeply forked tail white. **V** Soft, wader-like *twee-ee-ee*. **H** Mainly maritime. **D** Breeds on N Pacific coasts. Winters in the South China Sea.

9 SPECTACLED TERN *Onychoprion lunatus* 38cm Feeds mainly by plunge-diving. Underwing shows grey-brown primaries and secondaries. Tail grey, white-edged and deeply forked, with long outer-tail feathers. **V** High-pitched screeching, less harsh than that of Sooty Tern. **H** Offshore and pelagic waters. **D** Islands of the SW Pacific Ocean.

10 BRIDLED TERN *Onychoprion anaethetus* 34–36cm Flight very buoyant. Feeds by plunge-diving and by dipping to pick food off water's surface. Underwing mainly white, with dusky tips to primaries and secondaries; rump and tail dark grey, the latter deeply forked and with long white outer-tail feathers. **V** Yapping *wep-wep* or *wup-wup*. **H** Mainly pelagic, coming to small islets only to roost or breed. **D** Tropical oceans worldwide.

11 SOOTY TERN *Onychoprion fuscatus* 36–45cm Feeds mainly by dipping and picking from water's surface; occasionally plunge-dives. Underwing white with dusky primaries and dusky tips to secondaries. Rump and tail blackish, the latter deeply forked with white outer-tail feathers. **V** Distinctive *ker-wacki-wah, ker-wacki-wack* or *wide-awake*; also a short *kraark*. **H** Mainly maritime. **D** Tropical oceans worldwide.

12 RIVER TERN *Sterna aurantia* 38–46cm Usually feeds by plunge-diving. In flight, whitish upperwing primaries form a conspicuous flash on outer wing. Tail deeply forked with long outer-tail feathers. **V** In flight, utters a shrill, staccato *kiuk-kiuk*. **H** Large rivers and lakes, rarely on coasts. **D** E Pakistan through India to S Vietnam.

13 ROSEATE TERN *Sterna dougallii* 33–41cm Feeding actions similar to those of Arctic Tern, although hovers less; flies 'into the water' often from a higher level than Common Tern. Tail deeply forked with very long tail streamers. **V** Short, soft *cher-vik* and rasping *kraak* or *zraaach*. **H** Coastal areas. Breeds in colonies. **D** Worldwide.

14 WHITE-FRONTED TERN *Sterna striata* 30cm Note pale upperparts, narrow black line along leading edge of wings, long, dark bill and white forehead. **V** Mid-high *tzit tzit*. **H** Nests mainly on shingle riverbanks, often at higher altitudes, but also seen on pastures, fields, lakes; post-breeding, occurs in sheltered coastal habitats. **D** Breeds in New Zealand. Winters in SE Australia.

15 BLACK-NAPED TERN *Sterna sumatrana* 34–35cm Feeds mainly by plunge-diving, but also by skimming low over water picking food from surface. Outer web of outermost primary blackish; tail deeply forked, with long outer-tail feathers. **V** Sharp *tsii-chee-chi-chip* and a *chit-chit-chit-er* when excited or alarmed. **H** Rocky islets, coastal bays, lagoons. **D** Breeds from coasts of E Indian Ocean to W Pacific and Australasia. Winters in Indian Ocean and South Pacific.

16 SOUTH AMERICAN TERN *Sterna hirundinacea* 40–41cm Bill and legs red in breeding plumage, black in non-breeding plumage. Top half of head black, back and wing coverts grey. Rump and deeply forked tail white. Tail extends beyond folded wings. Throat and breast pale greyish, rest of underparts white. **V** Nasal, high-pitched *care, careware* and aggressive *CAAARE*. **H** Sea islands and beaches. **D** Coastal Peru and SE Brazil to Tierra del Fuego, Falkland Islands.

17 COMMON TERN *Sterna hirundo* 37cm Feeds mainly by plunge-diving. In breeding plumage, cap black. Legs and bill bright red, this with black tip. Underwing shows dark outer primaries, the rest tipped blackish, forming a 'V'. Tail long and deeply forked. **V** All calls have a distinctive sharp, irritable timbre, including a sharp *kip*, a downslurred *keeur* and a harsh, shrill *kee-arr*. **H** Coasts and large inland waters. **D** Widespread across North America, Eurasia and Australasia.

18 WHITE-CHEEKED TERN *Sterna repressa* 32–34cm Feeds by plunge-diving and by dipping to pick food from water's surface. Rump and tail grey, concolorous with upperparts; tail deeply forked, with long outer-tail feathers. **V** Short *kep* or *keep*, also a rasping *kee-arrh*. **H** Breeds on offshore islands; post-breeding, occurs off coasts. **D** Breeds from coastal Kenya to coastal Pakistan. Winters in W India.

19 ARCTIC TERN *Sterna paradisaea* 35cm Note short legs; short bill; long tail (reaching wing tip or just beyond); narrow, dark trailing edge to primaries; 'translucent' flight feathers. In breeding plumage, bill all red and white cheek separated from greyer underparts. **V** Mid-high *eEEEhr*, lowered at end. **H** More offshore than inshore; may wander inland. **D** Breeds across N Eurasia and N North America. Winters in subantarctic oceans.

20 ANTARCTIC TERN *Sterna vittata* 37cm Short bill and legs. Breeding birds with bright red bill and legs, black cap, white stripe through cheeks, body grey, and white rump, undertail coverts and tail white, this slightly shorter than folded wing. Non-breeding birds whiter with blackish-red bill and legs, bill tipped black. **V** Noisy at nesting colonies. Gives a *trr-trr-kriah* contact call in flight and when fishing, and a *chrrrr* when defending nest against intruders. **H** Subantarctic islands and at sea; migrates far out at sea. **D** Southern Ocean.

21 KERGUELEN TERN *Sterna virgata* 33cm Similar to Antarctic Tern but sedentary, with geographically distinct range. **V** Harsh, screeching *kee-aaar*. **H** Breeds on volcanic islands; feeds on land, seashore and inshore seas. **D** Kerguelen, Prince Edward and Crozet islands, S Indian Ocean.

22 FORSTER'S TERN *Sterna forsteri* 33–36cm Feeding actions similar to Arctic Tern. Usually looks bulkier than Common and Arctic terns. Grey-centred tail deeply forked, with long streamers. **V** Rolling, nasal *kyarr, kwarr* or *kreerr*; also a rapid *kek-ke-kek...* **H** Coastal areas. **D** SC Canada through the USA to NE Mexico.

23 SNOWY-CROWNED TERN *Sterna trudeaui* 35cm In breeding plumage, bill yellow with blackish subterminal ring. Legs yellow. Head white with blackish eye-line. Rest of the body washed pale grey. Forked tail. Non-breeding birds generally whiter with yellow-tipped black bill. **V** Series of rapid notes, *je-je-je-je*, or a short, grating *jeeer*. **H** Freshwater lakes and marshes, sometimes seashores, especially estuaries, nearby lagoons. **D** SE Brazil to E Argentina, C Chile.

24 BLACK-BELLIED TERN *Sterna acuticauda* 33cm Feeds by plunge-diving and by surface-picking from water or land. On upperwing, whitish primaries contrast with grey of rest of wing; tail deeply forked, with long outer feathers (these shorter in non-breeding plumage). **V** Clear, piping *peuo*; also a shrill *krek-krek*. **H** Large rivers and lakes. **D** India through SE Asia.

25 BLACK-FRONTED TERN *Chlidonias albostriatus* 30cm Note striking orange-yellow bill and legs, white cheek streak and white rump. **V** High twittering *kiurrr* or *t-t-t-tirr*. **H** Breeds at higher altitudes on shingle riverbanks, occasionally also on lake shores; post-breeding, mainly in sheltered coastal habitats. **D** New Zealand.

26 WHISKERED TERN *Chlidonias hybrida* 23–29cm Buoyant foraging flight, regularly dipping to pick prey from water's surface; may also plunge-dive. Rump grey, paler in non-breeding plumage. Tail shallowly forked. **V** Rasping *cherk*; also a *kek* or *kek-kek*. **H** Vegetated lakes, marshes, paddyfields, coastal pools, mudflats. **D** Breeds across Europe, Africa, Asia and Australasia. Winters in Africa, Asia and Australasia.

27 WHITE-WINGED TERN *Chlidonias leucopterus* 23–27cm Actions similar to those of Black Tern. Underwing coverts black, rump white, tail pale grey and very shallowly forked. **V** Harsh, high-pitched *kreek*, a soft *kek*, and a rasping *kesch* or *chr-re re*. **H** Coastal lagoons, estuaries, marshes, paddyfields, lakes, pools, large rivers. **D** Breeds from E Europe to E Siberia and NE China. Winters across Africa, Asia and Australasia

28 BLACK TERN *Chlidonias niger* 25cm Non-breeding birds, bill and legs black; forehead, brow, neck and underparts white. Solitary or in flocks. Often feeds on insects in flight. **V** Relatively quiet compared with *Sterna* terns but vocal at breeding colonies, including a short *kip* or *kik* and a high-pitched *kik* or *keek* when alarmed. **H** Coastal waters and at sea, estuaries, inland lakes. **D** Widespread across North America and Eurasia. Winters south to N South America and Africa.

29 LARGE-BILLED TERN *Phaetusa simplex* 39cm Unmistakable. Huge yellow bill and legs. Part of forehead white. Cap and auricular black. Neck and underparts white. Back, rump and tail grey. Primaries black, rest of flight feathers and coverts white. Short tail slightly forked. **V** Raspy *kree* and *kew* notes, often given as a series. **H** Rivers, lakes, marshes, estuaries. **D** Widespread east of Andes to NC Argentina.

30 INCA TERN *Larosterna inca* 40cm Unmistakable. Overall slate grey. Cap and outer flight feathers blackish. Legs bright red. Heavy bill, base with bare yellow patch and downcurled, whiskery white feather. Gregarious. Feeds especially on anchovies, also scavenges from sea lions. **V** Vocal at nesting colonies. Calls include mewing and raucous cackling notes. **H** Cliffs, rocky coasts. **D** Coastal N Peru to C Chile.

SKUAS STERCORARIIDAE

31 CHILEAN SKUA *Stercorarius chilensis* 53–58cm Cinnamon brown on underparts and underwing. **V** Mainly vocal on breeding grounds. Long call is a series of 10–12 short nasal barks; other calls include short guttural grunts. **H** Post-breeding, mainly pelagic. **D** S coasts and oceans of South America.

32 SOUTH POLAR SKUA *Stercorarius maccormicki* 50–55cm Various morphs, grading from pale to dark, all lacing rufous in plumage. Much care is needed when trying to separate darker forms from other large skuas. **V** Generally silent during the non-breeding season. **H** Pelagic. **D** Breeds on Antarctic coasts; regular non-breeding visitor to N Atlantic and N Pacific.

33 BROWN SKUA *Stercorarius antarcticus* 61–66cm In flight, shows a prominent white flash at base of primaries. Very aggressive to seabirds. **V** Generally silent. **H** Coastal waters. **D** Southern Ocean waters and islands.

34 GREAT SKUA *Stercorarius skua* 53–58cm Very aggressive to seabirds, recorded killing victims when attempting to rob them of food. **V** On breeding grounds, gives a plaintive, wailing *piah-piah-piah-piah...* In alarm, utters a harsh *tuk* or *gek*, and when attacking intruders at breeding site gives a strangled *kayaya*. **H** Breeds on coastal moorland and offshore islands; post-breeding, mainly pelagic. **D** Breeds on coasts and islands in NW Europe. Winters in the N Atlantic.

35 POMARINE JAEGER *Stercorarius pomarinus* 46–51cm The bulkiest of all the 'smaller' skuas. Aggressively pursues other seabirds to steal food. **V** Generally silent away from breeding grounds, although occasionally utters barking sounds. **H** Primarily pelagic. **D** Breeds in N North America and N Eurasia. Winters in temperate and tropical oceans.

36 PARASITIC JAEGER *Stercorarius parasiticus* 41–46cm In flight, wings show a distinct white flash at base of primaries. Aerobatically chases and harries seabirds in an attempt to make them disgorge food. **V** Generally silent, although in flight may utter a nasal *gi-ooo*. **H** Coastal waters. **D** Breeds in N North America and N Eurasia. Winters in southern hemisphere oceans.

37 LONG-TAILED JAEGER *Stercorarius longicaudus* 48–53cm In flight, often gives the impression of being heavy chested. Less piratical than other skuas, but does chase terns. **V** On breeding grounds, gives a rattling *krr-krr-krr-kri-kri-kri* followed by a repeated, mournful *pheeeu*. When alarmed, utters a *krik* or *kreek*. **H** Breeds on coastal and inland tundra; post-breeding, mainly pelagic. **D** Breeds in N North America and N Eurasia. Winters in subantarctic oceans.

1
2
3
4
5
6
7
8
9
10
11
12
13
14
15
16
17
18
19
20
21
22
23
24
25
26
27
28
29
30
31
dark
morph
32
light
morph
33
dark
morph
34
35
36
36
light
morph
light morph with
dark morph above
37
light
morph

AUKS ALCIDAE

1 LITTLE AUK *Alle alle* 17–19cm Rapid, whirring flight. Underwing dusky, contrasting with white body. Breeds in large colonies. **V** At colonies gives a chattering *krrii-ek ak ak ak ak ak* or similar. When alarmed, utters a whinnying *whuwhuwhu*. Generally silent after breeding. **H** Breeds on rock scree of coastal slopes; winters at sea. **D** N North America, N Europe.

2 THICK-BILLED MURRE *Uria lomvia* 39–43cm Actions and habits similar to Common Murre. Underwing has white wing coverts and dark axillary area, and no dark bar on 'armpit'. Feet project slightly beyond tail. **V** Similar to that of Common Murre. **H** Coastal cliffs; winters at sea. **D** N North America, N Eurasia.

3 COMMON MURRE *Uria aalge* 38–43cm During breeding season often occurs in huge numbers at cliff-face colonies. Rapid, whirring flight. Underwing has white wing coverts with a dark axillary area and small dark 'armpit' bar. Feet project slightly beyond tail. **V** At colonies utters a *ha ha ha* that leads into a guttural *ha-rrrhr*, also a rumbling *mmm...* **H** Coastal cliffs; winters at sea. **D** N North America, N Eurasia.

4 RAZORBILL *Alca torda* 37–39cm Usually breeds in small groups on rocky cliffs or among boulders. Flight rapid and whirring. Underwing pattern similar to that of Thick-billed Murre. Feet do not project beyond tail. **V** At breeding sites utters a guttural *goarrr* or *urrr*. **H** Coastal cliff faces and rocky or boulder-pile areas below; winters at sea. **D** N Atlantic.

5 BLACK GUILLEMOT *Cepphus grylle* 30–36cm Rapid, whirring flight. In flight, shows white underwing coverts and large white patch on upperwing. Usually breeds in small scattered groups. **V** At breeding areas gives a high *peeeeeh* or *seeeeeuu*, also a *sipp-sipp-sipp* series. **H** Rocky coasts; winters at sea, usually near breeding sites, N populations often more dispersive. **D** N North America, N Europe.

6 PIGEON GUILLEMOT *Cepphus columba* 30–37cm Rapid, whirring flight. In flight, shows greyish-brown underwing coverts and a large white patch with broad black bar on upperwing. Usually breeds in small colonies. **V** Wheezy *peeeeee*. **H** Sea cliffs and rocky slopes; winters at sea near rocky coasts and in sheltered coves. **D** N Pacific.

7 SPECTACLED GUILLEMOT *Cepphus carbo* 38cm Rapid, whirring flight. Underwing blackish brown. Usually breeds in small colonies of <200 pairs. **V** Possibly similar to that of Pigeon Guillemot. **H** Sea cliffs, rocky slopes, boulder fields; winters at sea, usually near coasts. **D** Breeds in NE Eurasia. Winters in NW Pacific.

8 MARBLED MURRELET *Brachyramphus marmoratus* 24cm Recently split from Long-billed Murrelet. Flight rapid and whirring. In flight, winter birds show white scapulars. **V** High-pitched *peeeaah-peeeaah* or *meer-meer-meer*. **H** Breeds in coastal forests, feeds at sea; winters along sea coasts. **D** Breeds on NW coasts of North America. Winters in NE Pacific.

9 LONG-BILLED MURRELET *Brachyramphus perdix* 26cm Actions and habits similar to those of Marbled Murrelet, from which it has recently been split. Note pale white throat. **V** Possibly similar to that of Marbled Murrelet. **H** Mainly coastal forests; winters at sea. More dispersive than Marbled Murrelet. **D** Breeds on NE coasts of Eurasia. Winters in NW Pacific.

10 KITTLITZ'S MURRELET *Brachyramphus brevirostris* 22–23cm Underwing pattern similar to that of Marbled Murrelet. Flight wilder than in other *Brachyramphus*. **V** Low, groaning *urrrhhn*. **H** Coastal mountain slopes, feeds at sea; winters in seas near breeding areas. **D** N Pacific.

11 GUADALUPE MURRELET *Synthliboramphus hypoleucus* 23–25cm White chin. In flight, shows white underwing coverts. **V** Rattling trill. **H** Rocky islands; winters at sea near breeding grounds. **D** NW Baja California, Mexico.

12 SCRIPPS'S MURRELET *Synthliboramphus scrippsi* 23–25cm Lacks white crescent in front of eye of Guadalupe Murrelet. **V** High *seep seep seep...* **H** Rocky islands; winters at sea near breeding grounds. **D** Channel Islands off S California; islands off W Baja California, Mexico.

13 CRAVERI'S MURRELET *Synthliboramphus craveri* 21cm Black chin. In flight, shows plain mid-grey underwing. **V** High trill. **H** Offshore waters. **D** W coasts of North and Central America.

14 ANCIENT MURRELET *Synthliboramphus antiquus* 24–27cm Underwing coverts white, contrasting with grey flanks. Flight rapid and whirring. **V** Low whistles and chirping or clinking notes. **H** Islands, often with dense vegetation; winters off coasts. **D** N Pacific.

15 JAPANESE MURRELET *Synthliboramphus wumizusume* 26cm Actions and underwing pattern very similar to those of Ancient Murrelet. Colonial breeder. **V** Shrill whistles. **H** Rocky islets and coasts; winters at sea. **D** Japan.

16 CASSIN'S AUKLET *Ptychoramphus aleuticus* 20–23cm No marked difference in non-breeding plumage. Rapid flight. Underwing shows a pale wing bar. **V** Hoarse *RREP-nerreer* or similar. **H** Coastal islands; winters at sea. **D** NW coasts of North America.

17 PARAKEET AUKLET *Aethia psittacula* 23–25cm In flight, shows a dark underwing. Flight as in other auks, although often at greater height. Colonial breeder, often mixed with other auk species. **V** At breeding colonies gives a whistle or trill, rising in pitch. **H** Cliffs and rocky slopes on offshore islands; winters at sea. **D** N Pacific.

18 LEAST AUKLET *Aethia pusilla* 12–14cm Gregarious, often in very large flocks. **V** At breeding sites utters various chattering, twittering and squealing calls. **H** Rocky islands; winters at sea, usually near coasts. **D** N Pacific.

19 WHISKERED AUKLET *Aethia pygmaea* 17–18cm Flight rapid and whirring. Breeds in large colonies, often mixed with other auks. **V** High-pitched *eeaah ah ah ah...* or *eeaah ik eah*. **H** Rocky islands and sea cliffs; winters at sea, usually near coasts. **D** NC and NW Pacific.

20 CRESTED AUKLET *Aethia cristatella* 23–27cm Typical rapid, whirring flight. Gregarious, often in large flocks mixed with other auks. **V** Honking and grunting sounds. **H** Scree slopes and cliffs on islands; winters at sea. **D** N Pacific.

21 RHINOCEROS AUKLET *Cerorhinca monocerata* 35–38cm In winter often occurs in large groups close inshore. **V** At breeding sites utters growling and shrieking cries. **H** Grassy slopes on islands or coasts; winters at sea near coasts. **D** N Pacific.

22 ATLANTIC PUFFIN *Fratercula arctica* 26–36cm Rapid, whirring flight. Underwing dusky, contrasting with white body. Red-orange feet often conspicuous in flight. Colonial breeder. **V** At breeding sites utters a growling *arr* or *arr-uh*. Generally silent post-breeding. **H** Rocky and grassy slopes, on coasts and islands; winters at sea. **D** E Canada and NE USA, Greenland, Iceland and NW Europe to Novaya Zemlya, NW Russia.

23 HORNED PUFFIN *Fratercula corniculata* 36–41cm Typical auk flight, rapid and whirring. Underwing dusky. In flight, red-orange feet often conspicuous. Colonial breeder. **V** At breeding sites utters harsh grunting and growling notes. **H** Rocky coastal cliffs and offshore islands; winters at sea. **D** N Pacific.

24 TUFTED PUFFIN *Fratercula cirrhata* 36–41cm Unmistakable, especially in breeding plumage. Actions much as other puffins. Colonial breeder. **V** At breeding grounds utters soft grunts and growls. **H** Rocky coasts and islands; winters at sea. **D** N Pacific.

SANDGROUSE PTEROCLIDAE

25 TIBETAN SANDGROUSE *Syrrhaptes tibetanus* 40–48cm Gregarious. Unlike most sandgrouse does not make regular watering flights. In flight, upperwing coverts sandy, contrasting with black flight feathers; underwing mainly black; long central tail feathers. **V** Deep, loud *guk-guk* or *caga-caga*, also a *koonk-koonk*. **H** Stony or rocky areas of semi-desert or desert uplands. **D** Tajikistan through Tibet to WC China.

26 PALLAS'S SANDGROUSE *Syrrhaptes paradoxus* 40cm Flies to waterholes in small flocks, mainly during the morning. Post-breeding often forms flocks of several hundreds. In flight, underwing pale with dark trailing edge to secondaries. Note elongated central tail feathers. **V** Low-pitched *cu-ruu cu-ruu cu-ou-ruu*, also a rapid, bubbling *kukerik-kukerik* and a sharp *tchep* or *kep*. **H** Steppe and semi-desert with sparse, low vegetation, also fallow and abandoned fields. **D** W Kazakhstan to NE and EC China.

27 PIN-TAILED SANDGROUSE *Pterocles alchata* 31–39cm Often occurs in very large flocks. Flights to watering sites generally take place in the morning. In flight, underwing coverts white, contrasting with black flight feathers. Long central tail feathers. **V** Nasal *arrrh-arrrh*, a ringing *catar-catar* or *guettarr*, also a *ga-hg ga-hg arrr* and an *arrk-arrk-arrk*. **H** Arid and semi-arid plains with sparse vegetation. **D** Breeds in SW Europe, N Africa and Middle East. Winters east to Pakistan.

28 NAMAQUA SANDGROUSE *Pterocles namaqua* 25cm Note unmarked head and bi-coloured breast band of male. Female resembles female Double-banded Sandgrouse but has more rufous in barring. Drinks in mornings. **V** Flight call mid-high nasal *where is it now*. **H** Lower-altitude desert and semi-desert. **D** Angola, Namibia, Botswana, South Africa

29 CHESTNUT-BELLIED SANDGROUSE *Pterocles exustus* 31–33cm Small flocks join up to make large congregations at waterholes in mornings, with extra evening flights in hot, dry weather. In flight, appears dark due to blackish underwing and belly. **V** In flight, utters a chuckling *kt-arr kt-arr* or longer *whit kt-arr wit wit-ee-er kt-arrr-arr*, also a *chocka chocka* or *wot worp wot worp*. **H** Semi-desert with sparse vegetation, fallow fields, dry cultivations. **D** W Africa to India.

30 SPOTTED SANDGROUSE *Pterocles senegallus* 30–35cm Watering flights, generally in small to large flocks, usually take place in the morning, but also in evenings during hot weather. In flight, upperwing pale sandy with dark trailing edge on primaries; underwing coverts pale sandy buff with dark greyish flight feathers. **V** Musical *whitoo whitoo*, *wicko wicko* or *waqu waqu*. **H** Desert and semi-desert with sparse vegetation. **D** NW Africa to NW India.

31 BLACK-BELLIED SANDGROUSE *Pterocles orientalis* 33–39cm Small flocks travel to watering sites mainly in mornings. In flight, whitish underwing coverts contrast with black flight feathers and black belly. **V** Bubbling *tchowrrr rerr-rerr* or *churrll-urrll-urrll*. **H** Semi-desert with sparse vegetation and nearby fallow cultivation. **D** SW Europe and N Africa, east to NW China.

32 YELLOW-THROATED SANDGROUSE *Pterocles gutturalis* 30cm Note pale eyebrow and black-bordered creamy throat of male. Female separated from other female sandgrouse by thin, dark lore. Comes to drink within 1–3 hours after sunrise. **V** Flight call is a frog-like *chuwchuw* and *chieriet*. **H** Sparsely wooded grassland and farmland at higher altitudes. **D** E and SC Africa.

33 CROWNED SANDGROUSE *Pterocles coronatus* 27–29cm Watering flights usually made in mornings, with flocks of up to 50; much larger numbers recorded at waterholes. In flight, shows dark flight feathers and primary covert of upperwing, rest of coverts sandy. Underwing shows white coverts contrasting with dark flight feathers. **V** Rolling *ch-ga ch-gar-ra*, *que quet querrooo* or *chiruk-chirugaga*, also a *wok tu wok kock* and descending *wheeek*. **H** Stony desert and semi-desert, dry hills and mountains. **D** W Africa to Pakistan.

34 BLACK-FACED SANDGROUSE *Pterocles decoratus* 30cm Comes to water within an hour after sunrise. Note white underparts and black-mottled belly. **V** Flight call a rhythmic, three-note dove-like phrase, *what-wa-wha*. **H** Open spots in dry, open, flat or hilly areas with some grass and scrub; also dry bushland. **D** E Africa.

35 MADAGASCAN SANDGROUSE *Pterocles personatus* 35cm No similar bird in its range. Swift flyer. **V** In flight, utters a frog-like chatter. **H** Open areas at lower altitudes. **D** Madagascar.

36 LICHTENSTEIN'S SANDGROUSE *Pterocles lichtensteinii* 24–26cm Mainly crepuscular or nocturnal. Usually encountered in pairs or small groups; larger parties recorded during watering flights and at daytime roosts. In flight, upperwing shows pale buff wing-bar on greater coverts, underwing plain greyish buff. **V** Whistling *qewheeto*, *chee-weeup* or *witch-ouuu*. When alarmed, utters a *kerrek-kerrek-kerrek*, *krre-krre-kree* or *qua-qua-qua*. **H** Deserts, semi-deserts, scrubby hillsides, dry wadis. **D** NW Africa to Pakistan.

37 PAINTED SANDGROUSE *Pterocles indicus* 28cm Drinks after dusk, usually in pairs or small groups. In flight, shows golden-yellow upperwing coverts, underwing like Lichtenstein's Sandgrouse. **V** In flight, utters a *chirik-chirik*. When disturbed, gives a clucking *yek-yek-yek*. **H** Low hills with scattered thorn scrub, open grassy and rocky areas, firebreaks in forests. **D** India.

38 FOUR-BANDED SANDGROUSE *Pterocles quadricinctus* 30cm Drinks at sunset; otherwise mainly nocturnal. Note unmarked chin, throat and chest of female. **H** Open areas with some grass, scrubby coastal dunes. **D** Senegal and Gambia to Eritrea, Ethiopia and W Kenya.

39 DOUBLE-BANDED SANDGROUSE *Pterocles bicinctus* 25cm Comes to drink late after sunset. Note diagnostic black bar on forehead of male. Female separated from female Namaqua Sandgrouse by square-cut (not pointed) tail and pale-edged tertials and scapulars. **V** High, explosive, liquid *weeWeetitweeweeweeoh*. **H** Dry areas at lower altitudes. **D** S Africa.

40 BURCHELL'S SANDGROUSE *Pterocles burchelli* 25cm Separated from other sandgrouse by longer legs. Drinks 2–4 hours after sunrise. **V** Flight call is a high, mellow *tuweet tuweet tuweet*. **H** Dry, grassy areas with some scrub. **D** S Angola and Namibia through Botswana to Zimbabwe and South Africa.

1
2
3
4
5
6
7
8
9
10
11
12
13
14
15
16
17
18
19
20
21
22
23
24
25
26
27
28
29
30
31
32
33
34
35
36
37
38
39
40

PIGEONS AND DOVES COLUMBIDAE

1 ROCK DOVE *Columba livia* 31–34cm The ancestor of feral town pigeons. In flight, shows white rump, black subterminal band on tail and broad black bars on upperwing. Underwing white with dark trailing edge. **V** Moaning *oorh* or *oh-oo-oor*, also a hurried *oo-roo-coo t coo* given during display. **H** Sea and inland cliffs, caves, ruined buildings, deep wells. Feral populations use mainly city and town buildings as nest sites. **D** Widespread worldwide.

2 HILL PIGEON *Columba rupestris* 33–35cm Colonial, usually encountered in small flocks. Looks like a very pale version of Rock Dove with a whitish mid-band on tail. **V** High-pitched, rolling *gut-gut-gut-gut*. **H** Cliffs, gorges and caves in open rugged country, also in villages. **D** C and E Eurasia.

3 SNOW PIGEON *Columba leuconota* 31–34cm In flight, shows a white patch on upper rump; lower rump and tail black, the latter with a white mid-band. Often feeds in the company of Hill Pigeons; post-breeding may occur in large flocks. **V** *Hic hic cuck-cuck hic*, also a prolonged *coo-ooo-ooo*. **H** High-altitude rocky cliffs, steep gorges, snow fields. **D** SC and SE Eurasia.

4 SPECKLED PIGEON *Columba guinea* 40cm Shows striking white rump in flight. **H** Areas near cliffs and buildings, including towns and villages. **V** Mid-high (or high), slightly rising *coocoocoo* (10–20× 'coo') or *Roocopcoo Roocoocoo*. **D** Widespread across Africa.

5 WHITE-COLLARED PIGEON *Columba albitorques* 40cm Has white wing patch in opened wing. **H** Towns, villages and other settlements, woodland, grassland, farmland; 1,750–4,000m. **D** Ethiopia, Eritrea.

6 STOCK DOVE *Columba oenas* 32–34cm In flight, shows broad black terminal band on tail and two small black bars on inner part of upperwing. Underwing grey with dark trailing edge. Often occurs in flocks of Common Wood Pigeons. **V** Low *ooo-uh* or *ooo-er*. **H** Open country, cultivated areas, open woodland, forest edges, parks, all with old trees for use as nest sites. **D** W and SW Eurasia.

7 YELLOW-EYED PIGEON *Columba eversmanni* 29–31cm In flight, shows white upper rump, black bars on inner part of upperwing and broad black terminal band on tail. Underwing pale grey with dusky trailing edge. **V** Subdued *quooh quooh quooh-cuu-gooh-cuu-gooh-cuu-gooh*. **H** Light woodland in cultivated areas, soil cliffs, ruins; in winter, occurs on floodplains. **D** Breeds from Iran to NW China. Winters in NW India, Pakistan and Iran.

8 SOMALI PIGEON *Columba oliviae* 40cm Note pale colouring and purple crown. **V** Little information. Song said to be *wuk-wuk-wuk-oooh*, final note drawn out. **H** Cliffs and rocky hillsides near the seashore, with scarce grass and some shrubs. **D** Somalia.

9 COMMON WOOD PIGEON *Columba palumbus* 41–45cm Often in very large flocks. In flight from above, unmistakable due to white wing crescent that stretches back from alula area. **V** Mellow *coo COOO coo coo-coo cook* or similar, also a *coo ooo, coo coo coo* or *coo ke ke coo coo* given during display. **H** Forest and forest edges, cultivated areas with nearby woodland or hedgerows, town parks, gardens. **D** Europe to W Siberia and W China. Nominate race winters in NW Africa.

10 TROCAZ PIGEON *Columba trocaz* 38–40cm The only pigeon on Madeira, apart from feral Rock Doves. In flight, generally grey with darker flight feathers, tail with broad black tip and pale grey subterminal band. **V** Soft *coo-coo coooo cook*. **H** Cliffs and crags in laurel forest. **D** Madeira.

11 BOLLE'S PIGEON *Columba bollii* 35–37cm Often in fast flight over treetops, otherwise shy. In flight, generally dark grey with a pale subterminal band to dark-tipped tail. **V** Guttural *ruk ruk gruuuk guk* or *ruor ruor ruor rup*. **H** High-altitude laurel forest and nearby cultivated areas. **D** Canary Islands.

12 LAUREL PIGEON *Columba junoniae* 37–40cm White terminal tail band diagnostic. Often flies with a slow, floppy flight. **V** Hoarse *pu-pu-pooo* or *up-poooo*. **H** Crags and gorges in high-altitude laurel forest and nearby cultivated areas. **D** Canary Islands.

13 AFEP PIGEON *Columba unicincta* 35cm Large forest pigeon. In flight, shows a striking pale tail bar. **V** Hurried, low *oooh-oooh-oooh-oooh* (each 'oooh' falling off). **H** Forest canopies. **D** Guinea to Ghana; E Nigeria to Angola, DR Congo and Uganda.

14 AFRICAN OLIVE PIGEON *Columba arquatrix* 38–41cm In flight from above, looks uniform dark grey. Often skulks in trees. **V** Rolling *coo-coo croo-croo croo-croo croo-croo*. **H** Highland forest. **D** Angola; Ethiopia to South Africa.

15 CAMEROON OLIVE PIGEON *Columba sjostedti* 35cm Highland species of Mt Cameroon and Bioco. Note all-grey (not white-naped) head and black feet. Separated from African Oliver Pigeon by different range. **V** Deep, drawn-out growl, followed by a quavering series of low, muffled coos and, during display flight, a bleating sound. **H** Forest. **D** Cameroon.

16 SAO TOME OLIVE PIGEON *Columba thomensis* 40cm Very dark, including head and undertail. **V** Low-pitched, faint cooing repeated at long intervals, *cooohr-r-r-r... rhu-rhu-rhu-rhu-rhu... rhu-rhu-rhu-rhu... rhu-rhu-rhu-rhu...* **H** Forest. **D** São Tomé, Gulf of Guinea.

17 COMOROS OLIVE PIGEON *Columba pollenii* 40cm No similar pigeon in its range. **V** Calm, pumped *rooh rooh rooh*. **H** Forest and forest edges at higher altitudes, often near fruiting trees. **D** Comoros Islands off the NW coast of Madagascar.

18 SPECKLED WOOD PIGEON *Columba hodgsonii* 38–40cm Mainly arboreal. In flight, appears uniform dark grey-brown. Usually in pairs or small flocks. **V** Deep *whock-whrroo-whrrooo*. **H** Evergreen and semi-evergreen hill forests. **D** Himalayas to Myanmar.

19 WHITE-NAPED PIGEON *Columba albinucha* 40cm Note white-grey nape, tail pattern and unspotted wings. Feet red. **V** Rolling *coo-coo croo-croo croo-croo croo-croo*. **H** Canopies and mid-strata of montane forests, especially in palms. **D** Cameroon; E DR Congo, W Uganda, Rwanda and Burundi.

20 ASHY WOOD PIGEON *Columba pulchricollis* 31–36cm Mainly arboreal, usually occurring in pairs or small flocks. In flight, appears darkish grey with a paler head. **V** Deep *whuoo whuoo whuoo* or *coo coo coo*. **H** High-elevation, dense broadleaved evergreen forest. **D** Himalayas to S China, Taiwan.

21 NILGIRI WOOD PIGEON *Columba elphinstonii* 36–42cm Mainly arboreal, although regularly feeds on fallen fruit on the forest floor. Usually found in pairs or small parties. Female duller. **V** Deep *who who-who-who*. **H** Moist evergreen hill forests. **D** SW India.

22 SRI LANKA WOOD PIGEON *Columba torringtoniae* 33–36cm Mainly arboreal. Female shows less grey on head and neck. **V** Deep owl-like *hoo-hoo-hoo*. **H** Mainly hill forests. **D** Sri Lanka.

23 PALE-CAPPED PIGEON *Columba punicea* 36–41cm Generally in pairs or small groups. **V** Soft mew. **H** Broadleaved evergreen forest, secondary growth, mangroves and island forest; more open areas during local nomadic wanderings. **D** NE India through SE Asia.

24 SILVERY PIGEON *Columba argentina* 34–38cm Rare. Often associates with very similar Pied Imperial Pigeon (Plate 62); note black tip on undertail feathers is level across all feathers, while on Pied Imperial Pigeon black tip extends up central feathers. **V** Unrecorded. **H** Mangroves, woodland, coconut groves. **D** Islands off Sumatra, Indonesia; N Borneo.

25 ANDAMAN WOOD PIGEON *Columba palumboides* 36–41cm Generally seen in pairs or small parties; regularly flies from island to island in search of fruiting trees. **V** Mellow, purring *crroo-crroo*. **H** Dense broadleaved evergreen forests. **D** Andaman Islands.

26 JAPANESE WOOD PIGEON *Columba janthina* 37–43cm Unmistakable. Usually solitary, although sometimes occurs in small flocks. **V** Drawn-out *oo-woo oo-woo*. **H** Dense evergreen forest. **D** Islands south of Japan and South Korea.

27 METALLIC PIGEON *Columba vitiensis* 37–41cm Feeds in the canopy and on the ground on fallen fruit, usually in small flocks. **V** Deep *wuuuu woooo*. **H** Various forests, from lowlands to mountains. **D** Philippines to Samoa.

28 WHITE-HEADED PIGEON *Columba leucomela* 39–42cm Usually unobtrusive; seen singly or in pairs. Feeds on fruits and seeds, mainly in tree canopy. Nomadic, wandering according to food availability. **V** Hooting *whoo-ooo*. **H** Rainforest, secondary forest. **D** Coastal E Australia.

29 YELLOW-LEGGED PIGEON *Columba pallidiceps* 36–38cm Feeds on fruits and seeds, in tree canopy and on ground. Critically Endangered due to habitat loss, hunting and predation by introduced predators. **V** Unknown. **H** Forest. **D** Bismarck Archipelago, Solomon Islands.

30 EASTERN BRONZE-NAPED PIGEON *Columba delegorguei* 35cm Small, very dark pigeon. Male has a white neck patch. Female has a bronze head resembling that of Lemon Dove, but larger and habitat differs. **V** High, rising, accelerated and then descending *hoo-koo hoo koo koo-koo-kookookoo*. **H** Forest canopies. **D** E and SE Africa.

31 WESTERN BRONZE-NAPED PIGEON *Columba iriditorques* 30cm Note tail-pattern. **V** Mid-high *oohooroo*, high *oo-hooroo* and high *oohoo-oohoo-oohoo* (each 'oohoo' gliding down). **H** Forest canopies. **D** Guinea and Sierra Leone to Uganda, DR Congo and Angola.

32 ISLAND BRONZE-NAPED PIGEON *Columba malherbii* 30cm Separated from Lemon Dove by white eye and chestnut (not whitish) undertail coverts. **V** Two hoarse hoots, followed by a stuttering series of 15–20 notes that gradually become shorter and fade away, *rhuuuw... rhuuw... rhu-tu-tu-tu-tututu*. **H** Forests and plantations. **D** Islands in the Gulf of Guinea.

33 LEMON DOVE *Columba larvata* 25cm Female (not shown) duller than male. Another race has a more rufous than grey belly and grey (not coral-red) eye-ring. Note plain, unmarked wings. **V** High, hollow *woop woop* (2× 'woop' per 3 sec). **H** Undergrowth and floor of dense mountain forests, parks and suburban gardens. **D** Widespread across Africa.

34 WHITE-CROWNED PIGEON *Patagioenas leucocephala* 33–36cm Gregarious, often in very large roosts and breeding colonies. Mainly arboreal. **V** Loud, clear *Cruu cru cu-cruuu*, sounding like 'Who-took-two'. **H** Principally coastal woodlands and inland forests when food available. **D** S Florida to the Caribbean and E Central America.

35 SCALY-NAPED PIGEON *Patagioenas squamosa* 36–40cm Mainly arboreal, feeding alone or in small groups. Often seen flying over forests. **V** Strongly accentuated *who who hoo-oo-hoo*, the last three syllables sounding like 'who-are-you'. **H** Mainly mountain forests, also towns and villages. **D** Caribbean.

36 SCALED PIGEON *Patagioenas speciosa* 30–32cm Bill red. Head chestnut. Feathers on neck and breast 'scaled'. Violaceous sheen on shoulders, purplish on wing coverts. **V** Rhythmic series of very low-pitched, monotonous coos, *pu-whooo puh-whu-whooh puh-whu-whooh puh-whu-whooh*. **H** Subtropical rainforests. **D** S Mexico to S Brazil.

1
2
3
4
5
6
7
8
9
10
11
12
13
14
15
16
17
♀
♂
19
20
21
22
18
♂
23
24
25
26
27
♀
♂
28
29
♂
♀
30
♂
31
♀
32
♂
34
35
36
33
♀
♂

PIGEONS AND DOVES *CONTINUED*

1 PICAZURO PIGEON *Patagioenas picazuro* 36cm Vinaceous head, scaly on hindneck. Brownish back grading to lead grey on rump. Underparts vinaceous grey. Greater coverts with white tips forming a white line, visible in flight. **V** *Or whooo? who?who? or whoo?*, repeated 3–4×. **H** Forest, woodland, agricultural land. **D** S South America, expanding into areas where not previously known.

2 BARE-EYED PIGEON *Patagioenas corensis* 32cm Bare skin around eye sky blue. Head, neck and underparts pale vinaceous. Hindneck finely barred pale blue, black and white. **V** Rhythmic series of coos, typically starting with a hoarse single note followed by repeated, high-pitched triple notes, the latter clearly overslurred, *rwhoh woh-hu-whOAh woh-hu-whOAh woh-hu-whOAh*. **H** Deciduous woods, thorny thickets, cactus. **D** Colombia and Venezuela, Netherlands Antilles.

3 SPOT-WINGED PIGEON *Patagioenas maculosa* 33cm Grey bill. Head, neck, and underparts powder grey, darker on back to lead-grey rump. Wing coverts with white spots. **V** Like that of Picazuro Pigeon, but hoarse. **H** Woodland with clearings. **D** C Peru to SE Brazil and SC Argentina.

4 BAND-TAILED PIGEON *Patagioenas fasciata* 34–40cm In display, flies horizontally with tail and wings spread and neck extended, then glides in a circle, calling before beating wings with rapid, shallow beats. Feeds on the ground or in trees. **V** Low *coo-cooo...* and a grating *raaaaan*. **H** Coniferous and mixed woodland, parks, gardens. **D** SW British Columbia to NW Argentina.

5 CHILEAN PIGEON *Patagioenas araucana* 35cm Head, neck and breast vinaceous chestnut; white nuchal collar. Tail grey with black subterminal band. **V** Deep, doubled *hooo-HOOOO hooo-HOOOO hooo-HOOOO*, often preceded with a long, rolling *PHRRRRRRRRRR*. **H** Cool temperate forests. **D** Chile, W Argentina.

6 RING-TAILED PIGEON *Patagioenas caribaea* 38–48cm Gregarious, usually in small flocks. **V** Deep *crooo-crooo-crooooo*. **H** Mountain and hill forests. **D** Jamaica.

7 PALE-VENTED PIGEON *Patagioenas cayennensis* 32cm Crown and nape bronzy green. Cheeks lead grey. Forehead and forecrown, neck, breast and part of upperparts winy chestnut. Rest of underparts grey. Tail with paler terminal band. **V** Haunting coo, similar to the hooting of an owl, *kuk-kuk-croo-oo*; also a rising, hoarse, hissing growl, *rrhhhAHH*. **H** Rainforest, clearings, gallery forest, deciduous woods, open areas with scattered trees, dry woods, mangroves, cultivated land. **D** S Mexico to N Argentina.

8 RED-BILLED PIGEON *Patagioenas flavirostris* 32–37cm Mainly arboreal. In display flight, climbs with exaggerated wingbeats before gliding down, in circles, with wings held in a shallow 'V'. **V** Hoarse *Hhooo hwooo hwooooo hup hupA hwoooo hup hupA hwooooo...* **H** Forest and woodland with thick undergrowth. **D** S Texas to Costa Rica.

9 MARANON PIGEON *Patagioenas oenops* 33cm Red bill tipped yellow. Head, neck and breast vinaceous. Rest of underparts lead grey. Back chesnutty vinaceous. Primaries blackish, secondaries brownish, greater coverts narrowly edged white. Tail slate grey, darker at tip. **V** Repeated rhythmic series of coos, typically starting with a single note and followed by triple notes, *rwhoOoh pUh-hu-whoOOo pUh-hu-whoOOo pUh-hu-whoOOo*. **H** Arid subtropical woods and scattered trees. **D** Ecuador, Peru.

10 PLAIN PIGEON *Patagioenas inornata* 38–40cm Gregarious, mainly arboreal. Jamaican race darker, with white eye surrounded by red orbital skin. **V** Low *whoo wo-oo* or similar. **H** Lowland forests, mountain forests, mangroves, coastal scrub. **D** Cuba, Hispaniola, Jamaica, Puerto Rico.

11 PLUMBEOUS PIGEON *Patagioenas plumbea* 33cm Dark with no pattern. Head, neck and underparts dark greyish vinaceous. Whitish on throat and foreface. Violaceous wash on crown and hindneck. Rest of upperparts very dark olive-brown. **V** Repeated rhythmic series of 2–5 coos, and a purring, drawn-out, slightly overslurred *rrrrrow*. **H** Tropical and subtropical rainforest, montane cloud forest, tropical dryland woods, secondary growth. **D** Colombia to SE Brazil.

12 RUDDY PIGEON *Patagioenas subvinacea* 30cm Dark, no patterns. Upperparts dark olivaceous brown. Purply tint on crown, hindneck and upper back. Underparts dark greyish vinaceous. Cinnamon wash on forehead and lores. **V** Repeated rhythmic phrase of four high-pitched coos, *whu pU whu hu*, and a purring, drawn-out, slightly overslurred *rrrrrow*. **H** Tropical and subtropical rainforest, cloud forest, secondary growth. **D** Costa Rica to Amazonia.

13 SHORT-BILLED PIGEON *Patagioenas nigrirostris* 28cm Dark colours with no pattern. Back, wings and tail dark brown. Head, neck and breast vinaceous, lighter below. Hindneck with a violaceous wash. **V** Repeated rhythmic phrase of four coos, *whu... pwuh... whu... huw*, and a purring, drawn-out, slightly overslurred *rrrrrow*. **H** Tropical and subtropical rainforest. **D** SE Mexico to NW Colombia.

14 DUSKY PIGEON *Patagioenas goodsoni* 25cm Dark with no pattern. Head, foreneck and breast light bluish grey. Rest of underparts vinaceous grey. Hindneck with violaceous tint. Upperparts and tail very dark olive-brown. **V** Repeated phrase of three rather high-pitched coos, *whoah pup pup*, and a purring, drawn-out *rrrrow*. **H** Lowland and montane rainforest. **D** W Colombia, NW Ecuador.

15 MALAGASY TURTLE DOVE *Nesoenas picturatus* 30cm Darkish with cinnamon back and wings. Terrestrial. **V** Repeated coo, increasing in pitch and amplitude before fading away at the end. **H** Forest up to mid-high altitudes. **D** Madagascar region, Seychelles.

16 PINK PIGEON *Nesoenas mayeri* 35cm Unmistakable thanks to chestnut-orange tail, dark brown wings and pale body. **V** Repeated overslurred, low-pitched coo, preceded by a short introductory note, *cu cooOOoo cu cooOOoo*. **H** Remains of natural forest. **D** Mauritius.

17 EUROPEAN TURTLE DOVE *Streptopelia turtur* 26–28cm Feeds mainly on the ground. In flight, appears dark, uppertail dark grey with white outer edges and prominent white terminal band on all outer feathers. **V** Purring *turrrrrr turrrrrr turrrrrr*. **H** Open woodland, forest edges, copses, groves and hedgerows. **D** W and C Eurasia, N Africa, Middle East.

18 DUSKY TURTLE DOVE *Streptopelia lugens* 28–31cm In flight, appears very dark with dull greyish terminal band on outer-tail feathers. **V** Deep, slow *coo-oor coo-oor*. **H** Wooded areas, agriculture surrounded by trees, acacia wadis. **D** South Sudan and Ethiopia to Malawi; Saudi Arabia and Yemen.

19 ADAMAWA TURTLE DOVE *Streptopelia hypopyrrha* 30cm Separated from nominate race of European Turtle Dove by black neck patches, grey terminal tail band and rufous underparts. **V** As European Turtle Dove (which is normally silent in Africa). **H** In or near forest patches, woodland and stands of large wild or planted trees. **D** Senegal and Gambia; Nigeria; Cameroon.

20 ORIENTAL TURTLE DOVE *Streptopelia orientalis* 30–35cm Usually occurs singly or in pairs, sometimes in small flocks at rich feeding sites. In flight, tail shows grey sides and terminal band contrasting with blackish subterminal bar. **V** Mournful *coo-cooroo-coocoo* or *gur-grugroo*. **H** Open forest, secondary growth, scrub, cultivation. **D** C and E Asia through India and SE Asia.

21 ISLAND COLLARED DOVE *Streptopelia bitorquata* 31cm Usually seen alone or in pairs; feeds on the ground, rests in trees or on wires. **V** Repeated, throaty, purring *crrrrruw*. **H** Mangroves, open wooded areas, lightly wooded cultivation, forest edges. **D** Philippines; Java, Indonesia; Lesser Sunda Islands.

22 EURASIAN COLLARED DOVE *Streptopelia decaocto* 31–33cm Feeds mainly on the ground. Undertail coverts buffish grey. See African Collared Dove for differences from that species. **V** Loud *kook-koooo-kook* or similar; also a harsh *kreair* or *whaaa*, usually given on landing. **H** Urban areas. **D** W Europe to India and E China.

23 AFRICAN COLLARED DOVE *Streptopelia roseogrisea* 29–30cm Feeds mainly on the ground. In flight, appears very similar to Eurasian Collared Dove, but has white undertail coverts and paler overall. **V** Descending, rolling *crooo cro-cro-crococo, cruu currruuu* or *KOOK r-r-r-r-r-r-OOoooooooo*. Also a nasal *heh heh heh heh*. **H** Semi-desert and savannah with trees, mangroves, town parks. **D** NE, C and W Africa, SW Arabian Peninsula.

24 WHITE-WINGED COLLARED DOVE *Streptopelia reichenowi* 25cm Feeds on seeds and berries, in trees and on the ground. Seldom far from water. Usually rather wary. **V** Disyllabic, repeated *er-Wha, er-Wha...* **H** Dry woodland, especially with palms, often near streams. **D** SE Ethiopia, SW Somalia.

25 MOURNING COLLARED DOVE *Streptopelia decipiens* 30cm Note red-rimmed eye and white-edged black neck patch. In flight, separated from other similar doves by narrow white terminal tail band. **V** High *woodurrrr*, short *woo* or low *durrrr*, among others. **H** Dry wooded areas with some grass, often near streams, gardens. **D** Widespread across Africa.

26 RED-EYED DOVE *Streptopelia semitorquata* 30cm Feeds mainly on the ground. In flight, uppertail shows black base and broad grey-buff terminal band. **V** Slow, hoarse *coo coo coo-coo coo* or similar, often transcribed as 'I-am-a-red-eyed-dove'. **H** Well-wooded areas, usually near water. **D** Widespread in sub-Saharan Africa, Saudi Arabia, Yemen.

27 RING-NECKED DOVE *Streptopelia capicola* 25cm Note small pale black-eyed appearance and large white tail corners. Separated from African Collared Dove by range, grey rump and wider black half-collar. **V** Very high, sustained *wikwirrik-wuk*. **H** Forest edges, open woodland, suburbs. **D** E and S Africa.

28 VINACEOUS DOVE *Streptopelia vinacea* 25cm More pink than Ring-necked Dove. Tail feathers, except central pair, white with dark brown base. Belly and undertail coverts not as white as in Ring-necked. **V** Very high, almost rattling, hollow, fast *rurururu* and high, fast *weeweewee-weeweeru*. **H** Dry, more or less wooded and bushy natural and cultivated habitats. **D** Mauritania, Senegal and Gambia to Eritrea, Ethiopia and Uganda.

29 RED TURTLE DOVE *Streptopelia tranquebarica* 31–33cm Feeds mainly on the ground. In flight, tail shows white sides and greyish-white corners. **V** Deep *cru-u-u-u-u* or *groo-gurr-goo*. **H** Dry open country, scrub, cultivation. **D** Widespread across Asia.

30 SPOTTED DOVE *Spilopelia chinensis* 30cm In flight, shows white corners to dark tail. **V** Melodious *coo croo-oo croo-oo* or *coocoo croor-croor*. **H** Cultivated areas, open forests, around human habitation. **D** Widespread across Asia.

31 LAUGHING DOVE *Spilopelia senegalensis* 25–27cm In flight, uppertail appears mainly grey, outer feathers white with black bases. Feeds mainly on the ground. **V** Bubbling *do do dooh dooh do*. **H** Around human habitation, cultivated areas, savannah with scattered trees. **D** India and C Asia to S Arabian Peninsula and Africa.

32 BARRED CUCKOO-DOVE *Macropygia unchall* 37–41cm Usually seen in pairs or small flocks. Acrobatically clambers around tree branches foraging for small fruits. On the ground, carries tail slightly raised. **V** Booming *croo-oom*. **H** Broadleaved evergreen and semi-evergreen forest. **D** Widespread across Asia.

33 AMBOYNA CUCKOO-DOVE *Macropygia amboinensis* 35–37cm Usually seen in pairs or small groups. Feeds on small fruits, seeds and nuts, taken from shrubs up to the canopy; occasionally feeds or takes grit on the ground. **V** *Whoop-whoop*. **H** Rainforest, rainforest regrowth, woodland, scrubland. **D** C Maluku Islands, Indonesia, to New Guinea and its islands.

34 SULTAN'S CUCKOO-DOVE *Macropygia doreya* 33–37cm Occurs singly, in pairs or in flocks. Often perches on exposed branches and dead trees. **V** Nominate race not recorded; *M. d. albicapilla* utters a rising *kuoow* and a series of upslurred notes, *uwoop uwoop uwoop uwoop...* **H** Primary and tall secondary forest, forest edges, swamp forest, lightly wooded cultivation. **D** N Maluku Islands and Sulawesi, Indonesia, to N New Guinea and W Papuan islands.

35 RUDDY CUCKOO-DOVE *Macropygia emiliana* 30–37cm Forages from the mid-storey to the canopy, usually singly or in pairs. **V** Series of mournful *whu* notes. **H** Primary lowland and hill forest, tall secondary forest, open forest areas. **D** Java, Indonesia, to Lesser Sunda Islands.

36 ENGGANO CUCKOO-DOVE *Macropygia cinnamomea* 30–32cm Formerly considered a race of Ruddy Cuckoo-dove; habits presumed similar. **V** Loud *poh wa wao*. **H** Primary lowland and secondary forest, open forest areas. **D** Enggano Island off S Sumatra, Indonesia.

37 BARUSAN CUCKOO-DOVE *Macropygia modiglianii* 30–37cm Formerly considered a race of Ruddy Cuckoo-dove; habits presumed similar. **V** Unrecorded. **H** Primary and tall secondary forest, open forest areas. **D** Islands off W Sumatra (except Enggano), Indonesia.

38 TIMOR CUCKOO-DOVE *Macropygia magna* 44cm Generally inconspicuous unless calling. Usually seen singly, in pairs or in small flocks; forages in the mid-storey and lower canopy. **V** Series of three well-spaced notes: a long upslurred note followed by two shorter notes. **H** Primary and tall secondary monsoon forest, forest edges, secondary scrub. **D** E Lesser Sunda Islands.

39 TANIMBAR CUCKOO-DOVE *Macropygia timorlaoensis* 38–43cm Occurs singly, in pairs or in small groups, feeding in the mid-storey and lower canopy. **V** Slightly hoarse, mellow *koowuck-whuuuu*, repeated, sometimes monotonously. **H** Monsoon forest, secondary growth, forest edges. **D** Tanimbar Islands, Indonesia; E Lesser Sunda Islands.

40 FLORES SEA CUCKOO-DOVE *Macropygia macassariensis* 38–43cm Formerly considered a race of Timor Cuckoo-dove; habits presumed similar. **V** Unrecorded. **H** Monsoon forest, secondary growth, forest edges. **D** Sulawesi, Indonesia; and E Lesser Sunda Islands.

41 PHILIPPINE CUCKOO-DOVE *Macropygia tenuirostris* 38cm Usually seen in pairs or small flocks. Acrobatically clambers around tree branches foraging for fruits. On the ground carries tail slightly raised. **V** Deep *wua wu*. **H** Heavily wooded areas and ravines. **D** Philippines, Borneo.

42 BROWN CUCKOO-DOVE *Macropygia phasianella* 40–43cm Usually seen in pairs or small flocks. Feeds mainly on berries and fruits, usually in the tree canopy. Tail used for balance. Flies low and fast through trees. **V** Loud *whoop-er-whoop*. **H** Rainforest, wooded habitats. **D** Coastal E Australia.

1
2
3
4
5
6
7
8
9
10
11
12
13
14
15
16
17
18
19
20
21
22
23
24
25
26
27
28
29
30
31
32
33
34
35
36
37
38
39
40
41
42

PIGEONS AND DOVES *CONTINUED*

1 ANDAMAN CUCKOO-DOVE *Macropygia rufipennis* 39–40cm Feeds on small fruits, especially bird's-eye chillies. **V** Repeated, deep, subdued *o-o-o-o-ah*. **H** Dense evergreen and secondary forest. **D** Andaman and Nicobar islands.

2 BAR-TAILED CUCKOO-DOVE *Macropygia nigrirostris* 29–30cm Usually seen in pairs or small feeding groups. Diet comprises mainly fruits and berries, usually gathered in tree canopy. **V** Rapid series of *whup-whup-whup...* notes. **H** Forest habitats. **D** New Guinea, Bismarck Archipelago.

3 MACKINLAY'S CUCKOO-DOVE *Macropygia mackinlayi* 28–31cm Usually seen in pairs or small feeding flocks. Feeds on fruits, berries and seeds. **V** Soft, piping, repeated *whep-owor, whep-owor...* **H** Forested habitats. **D** Bismarck Archipelago, Papua New Guinea, to Vanuatu.

4 LITTLE CUCKOO-DOVE *Macropygia ruficeps* 30cm Usually occurs in pairs or small parties. Feeds on the ground and in trees, generally in the middle storey or lower canopy. **V** Soft, monotonous *wup-wup-wup-wup-wup-wup...* and a rapidly repeated *croo-wuck croo-wuck croo-wuck*. **H** Broadleaved evergreen forest, forest edges. **D** Widespread across Asia, from Myanmar to SW China and south to Lesser Sunda Islands.

5 GREAT CUCKOO-DOVE *Reinwardtoena reinwardti* 48–53cm Usually found singly, but may be seen in pairs or small flocks. Feeds in trees and sometimes on the ground. **V** Loud, clear *wh wuk-wu wuk-wu wuk-wu wuk-wu wuck-wu wuck-wu*. **H** Primary and secondary forest, forest edges, occasionally nearby gardens. **D** New Guinea and Maluku Islands, Indonesia.

6 PIED CUCKOO-DOVE *Reinwardtoena browni* 40–45cm Striking plumage and elongated shape aid identification. Usually solitary. Feeds on fruits and berries. **V** Soft, piping *er-Whoo*, delivered as if surprised. **H** Forested slopes. **D** Bismarck Archipelago, Papua New Guinea.

7 CRESTED CUCKOO-DOVE *Reinwardtoena crassirostris* 40–42cm Distinctive crest aids identification. Usually solitary or in pairs. Feeds mainly on fruits and berries, usually in trees. **V** Hooting, owl-like *whup-Awhoo*. **H** Forested habitats. **D** Solomon Islands.

8 WHITE-FACED CUCKOO-DOVE *Turacoena manadensis* 36–40cm Found singly or in pairs. Frequents the mid-storey and canopy, often feeding in the open on fruiting trees and shrubs. **V** Throaty *tik ko-koo* or *wok woo woo*. **H** Wooded gorges and patches of dense woodland in relatively open country. **D** Sulawesi, Indonesia.

9 SULA CUCKOO-DOVE *Turacoena sulaensis* 36cm Formerly considered a race of White-faced Cuckoo-dove; actions and habits presumed similar. **V** Untranscribed, although sonograph material suggests differences from White-faced Cuckoo-dove. **H** Wooded gorges and patches of woodland in relatively open country. **D** Sula and Banggai islands, Indonesia.

10 BLACK CUCKOO-DOVE *Turacoena modesta* 38–39cm Forages at the top of small trees and shrubs, usually alone or in pairs. **V** Series of upslurred *ho-wuk* notes that starts slowly and accelerates towards the end. **H** Primary and tall secondary monsoon forest, woodland, tall secondary growth. **D** Lesser Sunda Islands.

11 EMERALD-SPOTTED WOOD DOVE *Turtur chalcospilos* 20cm Separated from Blue-spotted Wood Dove by black bill. **V** As Blue-spotted but with more double notes, *ookoo ookoo ookoo*. **H** Woodland. **D** Ethiopia and Somalia to Angola, Namibia and South Africa.

12 BLACK-BILLED WOOD DOVE *Turtur abyssinicus* 20cm In flight, shows rufous inner webs to primaries. Usually solitary or in pairs, feeding mainly on the ground. **V** *Cuwoo cuwoo co-oo-cuwoo coo cuwoo cu-cu-cu-cu-cucucucucucucucucu...*, starting slowly then speeding up and descending in pitch. **H** Savannah woodland, woodland around marshes or water, woodland edges. **D** Mauritania, Senegal and Gambia to Eritrea, Ethiopia and Uganda.

13 BLUE-SPOTTED WOOD DOVE *Turtur afer* 20cm Separated from Emerald-spotted Wood Dove by yellow-tipped red bill. **V** *Oo oo oo oo oo oo oo-oo-oo-ookookookookoo*, starting high slow, lazy and hesitating, then slowly speeding up and descending with a bouncing finale. **H** Moist forest. **D** Senegal and Gambia to Eritrea, Ethiopia and Uganda, south to Angola, N South Africa and Mozambique.

14 TAMBOURINE DOVE *Turtur tympanistria* 25cm Shy and usually solitary, although sometimes in small family groups. Feeds on the ground on seeds and fruit. **V** As in Emerald-spotted Wood Dove, but whole sequence starts at a lower pitch. **H** Forest, scrubland, woodland, plantations, suburbs. **D** Guinea to Ethiopia, south to Angola, Zambia and South Africa.

15 BLUE-HEADED WOOD DOVE *Turtur brehmeri* 25cm Unmistakable thanks to orange colouring contrasting with grey head. In flight, note pale rump patch. **V** Mid-high series of *poe* notes ending in a slow trill. **H** Lower strata and floor of forest. **D** W and C Africa.

16 NAMAQUA DOVE *Oena capensis* 26–28cm In flight, underwing chestnut with dark trailing edge, upperwing flight feathers mainly chestnut with dark trailing edge. Usually in pairs or small groups, feeding mainly on the ground. **V** Soft, rising *hoo-ooooo*. **H** Thornbush, scrub, palm groves. **D** Sub-Saharan Africa, Arabian Peninsula, Madagascar.

17 COMMON EMERALD DOVE *Chalcophaps indica* 23–27cm Usually seen singly, in pairs or in small groups, feeding on the ground; often uses forest tracks. In flight, shows two white bars on black lower back. **V** Soft, drawn-out *tuk-hoop, hoo hoo* or *tk-hoon*. **H** Broadleaved, semi-evergreen and mixed deciduous forest. **D** Widespread from India to S China and south to New Guinea.

18 PACIFIC EMERALD DOVE *Chalcophaps longirostris* 23–28cm Occurs singly, in pairs or in small groups. Feeds on the ground on fallen fruit, roosts in trees. In flight, shows two pale bars on lower back. **V** Low, moaning *coo coo coo coo coo coo*, starting quietly and then rising; also utters a nasal *hoo-hoo-hoo*. **H** Forest, forest edges, secondary growth, scrub, thickets. **D** Australasia.

19 STEPHAN'S EMERALD DOVE *Chalcophaps stephani* 24–25cm Feeds on the ground on fallen fruit and seeds; occurs singly, in pairs or, at rich feeding sites, in small groups. In flight, shows two pale bars on lower back. **V** Soft *tuu* followed by a series of short notes, *du-du-duu-duu-duu-duu...* **H** Primary lowland and hill forest; occasionally secondary coastal woodland. **D** Sulawesi, Indonesia, and New Guinea.

20 NEW GUINEA BRONZEWING *Henicophaps albifrons* 34–40cm Feeds mainly on the ground, probing leaf litter for invertebrates and fallen fruit. Flight is low and fast. Solitary or in pairs. **V** Incessantly repeated, staccato, piping *wha-wha-wha-wha...* **H** Wet forested habitats. **D** New Guinea.

21 NEW BRITAIN BRONZEWING *Henicophaps foersteri* 38cm Feeds mainly on the ground, on invertebrates and fallen fruit. Flight is direct and low to the ground. Seen singly or in pairs. **V** Poorly known. **H** Undisturbed forest habitats. **D** Bismarck Archipelago, Papua New Guinea.

22 COMMON BRONZEWING *Phaps chalcoptera* 28–35cm Note metallic sheen on wings. Typically wary. Nomadic, moving in response to drought, rains and food availability. Diet comprises mainly seeds. **V** Deep, resonating *Whoo, Whoo, erWhooo...* **H** Wide range of wooded habitats, heaths, grassland. **D** Australia.

23 BRUSH BRONZEWING *Phaps elegans* 26–32cm Feeds on the ground. Seen singly or in pairs. Drinks at dawn and dusk. Diet includes berries, seeds and invertebrates. **V** Disyllabic, monotonous *who-oop, who-oop...* **H** Coastal heath, mallee scrub, open woodland. **D** Coastal S and SE Australia.

24 FLOCK BRONZEWING *Phaps histrionica* 28–30cm Well marked and distinctive. Feeds mainly on seeds and shoots. Nomadic, forming large flocks outside the breeding season. **V** Calls include soft cooing notes. **H** Arid plains, open grassland habitats. **D** Interior and NW Australia.

25 CRESTED PIGEON *Ocyphaps lophotes* 31–35cm Male's courtship display involves spreading wings and fanning tail. Forms flocks. Wings make whistling, soft-clapping sound on take-off. **V** Eider-like *a-Whooa, a-Whooa*, delivered as if surprised. **H** Open woodland, grassland, agricultural fields, parks. **D** Australia.

26 SPINIFEX PIGEON *Geophaps plumifera* 20–22cm Sedentary. Usually seen in small groups; larger flocks form when feeding is good. Regularly visits waterholes. **V** Soft, repeated three-note *oo-OO-oo*. **H** Arid, sparsely vegetated habitats. **D** W, N and C Australia.

27 SQUATTER PIGEON *Geophaps scripta* 28–32cm Strong facial markings aid identification. Feeds on the ground, mainly on seeds, occasionally invertebrates. Seen in pairs or small flocks. **V** Soft, repeated *oo-oo-oo*. **H** Open woodland, savannah, arid grassland. **D** NE Australia.

28 PARTRIDGE PIGEON *Geophaps smithii* 24–28cm Note colourful bare skin around eye. Feeds on the ground, mainly on seeds. Typically seen in flocks outside the breeding season. **V** Subdued, cooing *oooa-ou*. **H** Damp tropical forests and woodland. **D** N Australia.

29 THICK-BILLED GROUND PIGEON *Trugon terrestris* 32–35cm Feeds on the ground. Diet comprises mainly fallen fruit and seeds. Seen singly or in pairs. Takes off almost vertically if alarmed, with rapid wingbeats. Usually shy and hard to observe well. **V** Resonating *Whu-oop*. **H** Rainforest. **D** New Guinea.

30 WONGA PIGEON *Leucosarcia melanoleuca* 38–44cm Plump-bodied pigeon. Usually shy and wary. Feeds on the forest floor, bobbing head when walking. When disturbed, takes off with loud wing-claps. **V** Strident, piping and repetitive *whA-whA-whA...* **H** Rainforest, secondary woodland. **D** E Australia.

31 CHESTNUT-QUILLED ROCK PIGEON *Petrophassa rufipennis* 28–30cm Chestnut on flight feathers most striking in flight. Feeds on the ground, foraging for fallen seeds. If alarmed, flies off with loud wing-claps. **V** Soft, repeated cooing notes. **H** Rocky gorges, cliffs, sparsely vegetated grassland, desert. **D** N Australia.

32 WHITE-QUILLED ROCK PIGEON *Petrophassa albipennis* 28–30cm White on flight feathers most striking in flight, appearing as a panel. Feeds on the ground, foraging for fallen seeds. Often seen peering over rock ledges. **V** Soft, repeated cooing notes. **H** Rocky ravines, sparsely vegetated plateaux. **D** N Australia.

33 DIAMOND DOVE *Geopelia cuneata* 20–24cm Feeds on the ground, mainly on seeds. Nomadic, in search of food and water. Usually seen in pairs or small groups. Congregates near water. Walks with a waddle. **V** Purring, repeated *preow-pioo...* **H** Open woodland, arid grassland. **D** N Australia and arid interior.

34 ZEBRA DOVE *Geopelia striata* 21–22cm Feeds mainly on the ground. During display, bows with tail raised and fanned. **V** High-pitched, soft trilling that leads to a series of rapid *coo* notes. **H** Open country with scrub, cultivation, parks, gardens. **D** Malay Peninsula; Sumatra and Java, Indonesia.

35 PEACEFUL DOVE *Geopelia placida* 20–23cm Seen in pairs or small flocks. Feeds on the ground, mainly on seeds. Flight is rapid, wings making a whooshing sound. **V** Soft, repeated *whee-wuk* or *whee-we-wuk*. **H** Open woodland, grassland, urban parks. **D** New Guinea, N and E Australia.

36 BARRED DOVE *Geopelia maugeus* 21–24cm Found singly, in pairs or in small groups. Feeds on the ground. **V** Calls include a downslurred, purring *prrrrr*, a falsetto and a monotonous *ooo-loo*. **H** Lightly wooded open country, open scrubby monsoon woodland, edges of mangroves, grassland with scattered trees, cultivation. **D** SE Maluku Islands, Indonesia; Lesser Sunda Islands.

37 BAR-SHOULDERED DOVE *Geopelia humeralis* 26–30cm Well-marked dove. Seen in pairs and small groups; occasionally in large flocks. Feeds on the ground, mainly on seeds. **V** Loud, piping *wh-Wha-whe-whoa*. **H** Dense woodland and scrub, seldom far from water; commonest near coasts. **D** SC and SE New Guinea, N and E Australia.

38 INCA DOVE *Columbina inca* 19–22cm Ground feeder. In flight from above, shows rufous primaries and white-sided tail. Underwing rufous with dark markings on coverts. **V** Strong, high *COO-pup COO-pup COO-pup...* **H** Parks and gardens in suburban areas, farms, semi-desert. **D** SW USA to Costa Rica.

39 SCALED DOVE *Columbina squammata* 20cm Scaled black all over. Rufous primaries. White band on greater coverts. **V** Clear, repeated *ooo puru poo*. **H** Dry fields, scattered trees in scrubland, savannah, edge of copses, stubble fields, gardens, parks. **D** N and E South America.

40 COMMON GROUND DOVE *Columbina passerina* 15–18cm Common. Mainly a ground feeder, usually seen in pairs or small groups. **V** Repetitive *coo coo coo coo coo...*, *co-coo co-coo co-coo...* or *hoooip hoooip hoooip...* **H** Variety of open area, including cultivated land, woodland, savannah, urban gardens. **D** S USA to SE Brazil.

41 PLAIN-BREASTED GROUND DOVE *Columbina minuta* 14cm Tiny. Upperparts brownish grey, breast pinkish grey, wing coverts dotted blue, underwing reddish. **V** Series of evenly spaced, low-pitched, upslurred cooing notes, *whoop whoop whoop*. **H** Open and semi-open areas. **D** Widespread from S Mexico to NE Argentina.

1
2
grey
morph
3
4
5
6
7
8
9
rufous
morph
10
11
12
13
14
15
16
17
18
19
20
21
22
23
24
25
26
27
28
29
30
31
grey
morph
32
rufous
morph
33
34
35
36
37
38
39
40
41

PIGEONS AND DOVES *CONTINUED*

1 ECUADORIAN GROUND DOVE *Columbina buckleyi* 17cm Overall brownish grey, greyer on crown and hindneck. Black spots and lines on wing coverts. Flight feathers and outer-tail feathers blackish. Female more brownish buff. **V** Evenly spaced, low-pitched cooing, *huWOO huWOO huWOO*. **H** Dry tropical areas with trees and shrubs. **D** Ecuador, NW Peru.

2 RUDDY GROUND DOVE *Columbina talpacoti* 17cm Overall ruddy chestnut, underparts more vinaceous pink. Cap greyish blue. Black spots and lines on wing coverts. Female generally browner. **V** Monotonous *cooroo*. **H** Open areas in savannah with trees or shrubs; palm groves, often near water; open fields with scattered trees; edges of rainforest, gallery forest and cloud forest; cities. **D** Widespread from Mexico to S Argentina.

3 PICUI GROUND DOVE *Columbina picui* 17cm Upperparts pale brownish grey. Black line on lesser wing coverts. Underparts pale vinaceous cream, whitish below. White wing-bar. Outer-tail feathers notably white. Female browner. **V** High-pitched double *cooroo*. **H** Open areas, earth and sandy roads, scattered trees and bushes, dry land with thorny scrub and cactus, savannah, farmland, cities. **D** NE Brazil to C Chile and S Argentina.

4 CROAKING GROUND DOVE *Columbina cruziana* 18cm Bill yellow with a black tip. Notable purplish band across lesser coverts. Underparts vinaceous. Female's head and underparts greyish buff and whitish. **V** Deep, throaty *ghee gwa gwa*. **H** Trees and bushes in arid areas, valleys and oases, riverine woods, parks, gardens, farmland, plantations. **D** N Ecuador to N Chile.

5 BLUE-EYED GROUND DOVE *Columbina cyanopis* 16cm Head, upper-tail coverts and wing coverts chestnut rufous. Neck chestnut with a violaceous sheen. Black spotting on coverts. Tail black. Female paler. **V** Evenly spaced, soft cooing notes, *wah wah wah*. **H** Open areas and savannah. Very rare and local. **D** SC Brazil.

6 BLUE GROUND DOVE *Claravis pretiosa* 22cm Bill yellow. Feet pale pink. Upperparts light bluish grey, whiter on face and below. Outer-tail feathers black. Black markings on wing coverts. Female brownish with purple wing spots; rump and tail rufous. **V** Series of loud coos, usually in pairs, the second note slightly lower than the first: *POOP-POOP POOP-POOP*. **H** Rainforest, woods and woodland edges, savannah, plantations, cultivated land. **D** E Mexico to N Argentina and SE Brazil.

7 PURPLE-WINGED GROUND DOVE *Claravis geoffroyi* 24cm Upperparts darkish blue-grey, paler below. One dark blue and two dark purple bands on wing coverts. Outer-tail feathers white. Female brownish version of male. **V** Not recorded but said to be a two-note *oo-OOT*. **H** Rainforest, canebrakes, montane forest. **D** SE Brazil, E Paraguay and NE Argentina.

8 MAROON-CHESTED GROUND DOVE *Claravis mondetoura* 24cm Upperparts darkish blue-grey. Breast purply chestnut. Underparts grey, white below. One blue and two purple bands on wing coverts. Outer-tail feathers white. Female very like female Purple-winged Ground Dove but darker and with cinnamon foreface. **V** Series of low-pitched, slightly rising two-syllable coos, *cuWOOP cuWOOP cuWOOP*. **H** Cloud forest, open forest, upland woods in mountains. **D** SE Mexico to WC Bolivia.

9 BARE-FACED GROUND DOVE *Metriopelia ceciliae* 18cm Bare skin around eye orange. Pale scaling on upperparts. Breast tawny. Underwings rufous. Tail longish. Feeds on the ground, frequenting shady lower storeys. Flies low and weaves through vegetation. More often heard than seen. **V** Not well known, but two calls have been recorded: *eUup eUup eUup eUup eUup*, fading towards the end; and *Wo ow*. **H** High mountains (to sea level where arid), grasslands, stony steppe, cultivated land. **D** Peru to N Chile.

10 MORENO'S GROUND DOVE *Metriopelia morenoi* 17cm Bare skin around eye orange. All buffish grey. **V** Soft, nasal *cuEec cuEec cuEec*, and a faster and deeper *coo coo coo coo coo*. **H** High mountains, valleys, bushy and stony areas, steppe with sparse grass, puna. **D** NW Argentina.

11 BLACK-WINGED GROUND DOVE *Metriopelia melanoptera* 23cm Bare orange skin around whitish eye. Shoulders and leading edge of wing whitish. Legs grey. **V** Usually silent, but males coo incessantly from a tree mornings and evenings during the breeding season, *rrree-up rrree-up rrreee-up*. **H** Sandy and stony steppe, sparse grassland and scrub in high mountains, puna (usually near water). **D** Colombia to S Chile.

12 GOLDEN-SPOTTED GROUND DOVE *Metriopelia aymara* 17cm Dumpy, with short neck and tail. Buffish brown. Shiny golden-bronze spots on wing coverts. **V** Known to call, but no details of vocalisations have been recorded. **H** Sandy and stony steppe with sparse grasses in high mountains, near habitation, corrals and water. **D** SC Peru to N Chile.

13 LONG-TAILED GROUND DOVE *Uropelia campestris* 18cm Tail long. Forehead and lores greyish blue. Two purple, black and white bands on wing coverts. Outer feathers black, with outer vane and tip white. Female paler. **V** Long *teeeee* call. **H** Open areas in savannah, fields and grassland near water, sparse woods and woodland edges, scrubby parkland. **D** NC, C and WC Brazil; N and E Bolivia.

14 BLUE-HEADED QUAIL-DOVE *Starnoenas cyanocephala* 29–34cm Endangered. Forages on the ground, usually in pairs. **V** Hollow *huuup-up huuup-up*. **H** Forests with a thick overhead cover and an open floor with a quantity of leaf litter. **D** Cuba.

15 PURPLE QUAIL-DOVE *Geotrygon purpurata* 22–25cm Colourful, well-marked dove. Favours dense cover. Feeds on the ground, foraging for seeds and invertebrates in leaf litter. **V** Piping, repeated *whuk-Whoo*. **H** Moist forest. **D** NW Colombia to NW Ecuador.

16 SAPPHIRE QUAIL-DOVE *Geotrygon saphirina* 22cm Forehead, top half of cheek and belly white. Lower cheek black. Crown and nape very deep blue. Breast bluish grey. Hindneck shiny golden, violaceous and purple. **V** Monotonous two-note *hu huuuuu*. **H** Rainforest and tall secondary growth in lowlands, low mountains. **D** Colombia to SE Peru.

17 CRESTED QUAIL-DOVE *Geotrygon versicolor* 27–31cm Forages on the ground, generally in deep shade, although will venture onto tracks. **V** Low *woof-woo-wooo*. **H** Mountain forests. **D** Jamaica.

18 RUDDY QUAIL-DOVE *Geotrygon montana* 21–28cm Forages on the ground, usually under cover. **V** Low, fading *cooo*. **H** Mainly dense forests and coffee plantations on hills and mountains, also locally in coastal forests. **D** Greater and Lesser Antilles, Mexico to NE Argentina.

19 VIOLACEOUS QUAIL-DOVE *Geotrygon violacea* 24cm Bill red. Face greyish. Crown, hindneck and back purple, rump chestnut. Breast pale buffish vinaceous. Underparts creamy white. Female similar but hindneck and upper back brownish. **V** Single repeated high-pitched cooing notes. **H** Rainforest, valleys and slopes in mountains. **D** Nicaragua to Venezuela, E and C South America.

20 GREY-FRONTED QUAIL-DOVE *Geotrygon caniceps* 26–30cm Similar to White-fronted Quail-Dove but generally paler. Forages on the ground, often along tracks. Regularly calls from an elevated perch. **V** Low *hoot hoot hoot hoot*. **H** Tropical lowland forest, bordering swamps. **D** Cuba.

21 WHITE-FRONTED QUAIL-DOVE *Geotrygon leucometopia* 26–30cm Has a purer white forehead and is generally darker than Grey-fronted Quail-Dove. Forages on the ground, often along tracks. Regularly calls from an elevated perch. **V** Call similar to that of Grey-fronted Quail-Dove but with a sudden change to *coo-o-o*. **H** Mainly mountain forest and coffee plantations. **D** Dominican Republic.

22 KEY WEST QUAIL-DOVE *Geotrygon chrysia* 27–31cm Ground feeder, usually under the cover of bushes and trees. **V** Low, slightly descending *ooooo*; also an *ooowoo*, with the second syllable accentuated and slightly higher. **H** Dense woodland and thickets. **D** Caribbean.

23 BRIDLED QUAIL-DOVE *Geotrygon mystacea* 24–30cm Secretive. Forages on the ground, usually in deep cover. More often heard than seen. **V** Mournful *who-whooo*. **H** Dark undergrowth in dense forest and wooded ravines. **D** Lesser Antilles.

24 OLIVE-BACKED QUAIL-DOVE *Leptotrygon veraguensis* 23cm Very dark olive-brown. Lower underparts buffish. Forehead white and bluish. Broad, pure white line across face. **V** Single short, low-pitched note, *whOuw*, with a rather frog-like quality. **H** Rainforest and tall secondary growth in lowlands and foothills, valleys. **D** Costa Rica to NW Ecuador.

25 WHITE-TIPPED DOVE *Leptotila verreauxi* 28–30cm Secretive. Ground feeder. In flight, underwing shows grey flight feathers and rufous coverts. **V** Strong, low *poo pooorr* or *coo CRRRRRoo cup*, said to resemble the sound made by blowing across the top of a bottle. **H** Dense, shady woodland. **D** Widespread from S Texas to NC Argentina.

26 YUNGAS DOVE *Leptotila megalura* 30cm Upperparts earthy chestnut brown, washed violaceous on crown and hindneck. Face and underparts buffish vinaceous. **V** Three-syllable call, *coooo-co-coooo*. **H** Montane forest and woods, alder woods. **D** Bolivia, NW Argentina.

27 GREY-FRONTED DOVE *Leptotila rufaxilla* 28cm Forehead pale greyish blue. Eye-ring red. Upperparts dirty, dark chestnut brown, tinted purple on crown and neck. Face and underparts pinkish buff. **V** Call a low-pitched, resonant, descending *cooooo*. **H** Rainforest, gallery forest, várzea, montane forest. **D** Colombia to NE Argentina.

28 GREY-HEADED DOVE *Leptotila plumbeiceps* 26cm Virtually identical to Pallid Dove, but less white on forehead and deeper blue on crown and hindneck. **V** Double-note call, *who, who?* **H** Dry forest, secondary woods, woodland edges, copses. **D** E Mexico to W Colombia.

29 PALLID DOVE *Leptotila pallida* 26cm Forehead white. Crown, nape and hindneck lead blue. Face, fore neck and underparts creamy grey. Upperparts dull, dark chestnut brown. **V** Call a single *cooOOooo*, louder in the middle. **H** Rainforest, old secondary growth, forest edges. **D** Colombia, Ecuador.

30 AZUERO DOVE *Leptotila battyi* 25cm Separated from Grey-headed Dove mainly by rufous upperparts. **V** Mournful, constantly repeated two-syllable *whoo-oooo*. **H** Forest. **D** Panama.

31 GRENADA DOVE *Leptotila wellsi* 28–31cm Critically Endangered. Feeds on the ground. **V** Descending *oooo*. **H** Scrubby woods on hillsides and lowlands. **D** Grenada.

32 CARIBBEAN DOVE *Leptotila jamaicensis* 29–33cm Forages on the ground, often under the cover of bushes. **V** High-pitched *cu-cu-cu-oooo*. **H** Open areas, gardens and thick secondary forests in lowland and foothills. **D** Jamaica, Cayman Islands, Yucatán Peninsula.

33 GREY-CHESTED DOVE *Leptotila cassinii* 26cm Head, neck and chest light greyish blue, lower breast and belly creamy grey. Back, rest of upperparts and tail dirty, dark chestnut brown. **V** Call is a single deep long *cooooo*. **H** Edge of rainforest and woods, trails and clearings. **D** Guatemala to N Colombia.

34 OCHRE-BELLIED DOVE *Leptotila ochraceiventris* 26cm Forehead sky blue and white. Crown, nape and hindneck purple. Face tawny ochre. Breast vinaceous pink. Rest of underparts ochre. **V** Distinctive, slightly downslurred single note, *uOOOooo*, repeated every 8–10 secs. **H** Edges of tropical and subtropical forest on plains and foothills. **D** Ecuador, Peru.

35 TOLIMA DOVE *Leptotila conoveri* 26cm Forehead sky blue and whitish. Crown and hindneck dark violaceous blue. Cheeks, foreneck and underparts pinkish vinaceous. Flies low, weaving through vegetation, and seldom leaves the cover of trees. More often heard than seen. **V** Single slightly overslurred note, *wooOOOooo*, repeated every 6–8 secs. **H** Rainforest edges with bushes and thickets. **D** Colombia.

1
2
3
4
5
6
7
8
9
10
11
12
13
14
15
16
17
18
19
20
21
22
23
24
25
26
27
28
29
30
31
32
33
34
35

PIGEONS AND DOVES *CONTINUED*

1 TUXTLA QUAIL-DOVE *Zentrygon carrikeri* 30cm Note lack of white in tail and restricted range. **V** Very low *boom boom boom* in an irregular series, each strophe with a downward inflection and slightly choked. **H** Montane forest. **D** SE Mexico.

2 BUFF-FRONTED QUAIL-DOVE *Zentrygon costaricensis* 25cm Note plump jizz, plain tail and dark eyes. **H** Montane forest. **V** Low, hooting series of 6–9 strophes, each with an upward inflection *whoo whoo whoo*. **D** Costa Rica, Panama.

3 PURPLISH-BACKED QUAIL-DOVE *Zentrygon lawrencii* 25cm Separated from Tuxtla Quail-Dove by pale tail corners and range. **V** Very high, descending *tju-iow*. **H** Forest. **D** Costa Rica, Panama.

4 WHITE-FACED QUAIL-DOVE *Zentrygon albifacies* 30cm Whitish throat and cheeks contrasting with grey crown and nape is diagnostic. No malar stripe. Note striped neck sides. **V** High *wooh wooh*. **H** Montane forest. **D** E Mexico to Nicaragua.

5 WHITE-THROATED QUAIL-DOVE *Zentrygon frenata* 32cm Face vinaceous, crown and nape violaceous. Line around lower face black. Upperparts chestnut, scaled on hindneck with a green and violet sheen. Throat whitish, underparts greyish. **V** Single monotonous, low-pitched note, *whoOO*, repeated every 3–6 secs. **H** Rainforest and cloud forest in mountains, secondary growth, bushes and thickets. **D** W Colombia to NW Argentina.

6 LINED QUAIL-DOVE *Zentrygon linearis* 28cm Forehead and cheeks buffish cinnamon, crown rufous. Behind eye to nape greyish sky blue. Black line below cheek. Upperparts chestnut, violet sheen on upper back. Underparts cinnamon. **V** As White-throated Quail-Dove. **H** Montane rainforest and cloud forest, tall secondary growth, thickets on steep slopes. **D** N, C and E Colombia to NE Venezuela, Trinidad and Tobago.

7 CHIRIQUI QUAIL-DOVE *Zentrygon chiriquensis* 30cm Uniform reddish brown with grey crown. **V** High, hollow, drawn-out *whuooow*. **H** Undergrowth of montane forest. **D** Costa Rica, Panama.

8 RUSSET-CROWNED QUAIL-DOVE *Zentrygon goldmani* 32cm Differs from White-throated Quail-Dove in having a buffish-cinnamon face. Crown and nape rufous. **V** Single slightly overslurred *wOuu* note, sometimes sounding like two syllables, repeated every 4–6 sec. **H** Rainforest. **D** E Panama, NW Colombia.

9 MOURNING DOVE *Zenaida macroura* 30–31cm Feeds mainly on the ground. Female paler, with less iridescence on neck and less grey on head. **V** Mournful *oo-woo woo woo woo*, similar to that of Zenaida Dove. **H** Lowland open country, agricultural areas, dry coastal forests, frequently near fresh water. **D** Widespread from Canada to Panama and Caribbean.

10 SOCORRO DOVE *Zenaida graysoni* 30cm Rufous tawny below. **V** Low hoarse slow *eeeehwuoow ooch ochoch*. **H** Open forest. **D** Socorro Island W of Mexico, but extinct in the wild.

11 EARED DOVE *Zenaida auriculata* 22–25cm Feeds primarily on the ground, usually in pairs or small groups but occasionally in large flocks; often roosts communally. **V** Gentle *oooa-oo* or *u-ooa-oo*. **H** Lowland semi-arid brush. **D** Widespread across South America and Lesser Antilles.

12 ZENAIDA DOVE *Zenaida aurita* 25–28cm Primarily a ground feeder, although will feed in fruiting trees. In flight, secondaries have a white trailing edge and outer-tail feathers have greyish-white tips. **V** Mournful *coo-oo coo coo coo*, very similar to that of Mourning Dove. **H** Open coastal areas, urban gardens, open woodland, pine woods, scrub thickets. **D** Yucatán Peninsula, and Caribbean.

13 GALAPAGOS DOVE *Zenaida galapagoensis* 25cm Bill long, tail short. Eye-ring brilliant turquoise. Two black lines through face. Pale ear coverts. Head and neck brownish, the latter with a pinkish-bronze sheen on sides. Rest of underparts buffish vinaceous. Upperparts darker brown, patterned black and white. **V** Soft, deep *bob-bob-bob-rururur-bububurr*, repeated every 8–10 secs. **H** Scrub, arid bush, small trees, cacti. **D** Galápagos.

14 WHITE-WINGED DOVE *Zenaida asiatica* 28–30cm In flight, upperwings show a large white central patch. Feeds in trees and on the ground; usually gregarious. **V** Rhythmic *who hoo who hoo-oo* or *who hoo who hoo hoo-ah hoo-hoo-ah who oo*. **H** Open woodland, arid scrub, mangroves, gardens. Breeds colonially. **D** SW USA to Panama, Greater Antilles.

15 WEST PERUVIAN DOVE *Zenaida meloda* 30cm Bare patch around eye sky blue. Head greyish with small blackish spot below cheek. Neck and breast vinaceous. Belly and undertail coverts grey. White band on wing. Outer-tail feathers with broad white tip. **V** Low-pitched, rhythmic, monotonous cooing, *Who huLOO hu huLOO hu huLOO*. **H** Tropical and subtropical arid regions with trees or bushes, riverine woods, canyons, valleys, oases, parks, gardens, plantations. **D** Ecuador to N Chile.

16 NICOBAR PIGEON *Caloenas nicobarica* 40cm Usually feeds at dawn or dusk, mainly on the ground on fallen fruit. Female has smaller cere, shorter hackles and browner upperparts. **V** Usually silent, but a short, soft cooing and pig-like grunting during disputes both recorded. **H** Dense broadleaved evergreen forest. **D** Andaman Islands to Philippines and Solomon Islands.

17 SULAWESI GROUND DOVE *Gallicolumba tristigmata* 32–33cm Forages on the ground, singly or in pairs. Shy; tends to run when disturbed. **V** Similar to that of Eurasian Hoopoe (Plate 96), but softer. **H** Primary lowland, hill and mountain forest. **D** Sulawesi, Indonesia.

18 CINNAMON GROUND DOVE *Gallicolumba rufigula* 22–24cm Colourful little dove. Feeds on the ground on seeds, fallen fruit and invertebrates, foraged in leaf litter. Bobs head while walking. **V** Upslurred, purring trill. **H** Rainforest **D** New Guinea.

19 LUZON BLEEDING-HEART *Gallicolumba luzonica* 25–26cm Forages on the forest floor and along forest trails and roads. **V** Soft *aa-oooot* that rises at the end. **H** Forest and secondary growth, from lowlands to mid-altitudes. **D** N Philippines.

20 MINDANAO BLEEDING-HEART *Gallicolumba crinigera* 26–27cm Forages on the forest floor. **V** Soft *co-co-oooot* and an *a-ooooo*. **H** Primary and secondary forest. **D** S Philippines.

21 MINDORO BLEEDING-HEART *Gallicolumba platenae* 25–26cm Forages on the ground. Little reported. **V** Unrecorded, but suspected to be similar to that of Negros Bleeding-heart. **H** Lowland forest with dry floors and thick undergrowth. **D** Mindoro, WC Philippines.

22 NEGROS BLEEDING-HEART *Gallicolumba keayi* 25–26cm Little recorded, presumed similar to other members of the genus. **V** Series of 20–25 bubbling notes. **H** Primary and secondary forest. **D** WC Philippines.

23 SULU BLEEDING-HEART *Gallicolumba menagei* 28cm Feeds on the forest floor. **V** Unrecorded. **H** Primary and secondary forest. **D** Tawi-Tawi Island, Sulu Archipelago, Philippines.

24 WETAR GROUND DOVE *Alopecoenas hoedtii* 25cm Apparently solitary; little else reported. **V** Unrecorded. **H** Lowland and hill monsoon forest and woodland. **D** E Lesser Sunda Islands.

25 WHITE-BREASTED GROUND DOVE *Alopecoenas jobiensis* 24cm Feeds on the ground and in the tree canopy. Diet includes seeds, fruits and invertebrates. **V** Poorly known. **H** Rainforest, moist secondary forest. **D** New Guinea; Bismarck Archipelago, Papua New Guinea; Solomon Islands.

26 WHITE-FRONTED GROUND DOVE *Alopecoenas kubaryi* 28cm Separated from Polynesian Ground Dove mainly by range. Male and female similar. **V** Poorly documented, but said to utter a deep, moaning coo and a whistled call. **H** Ravines with thickets, plantations, natural forest. **D** Caroline Islands, WC Pacific.

27 POLYNESIAN GROUND DOVE *Alopecoenas erythropterus* 25cm Compare with White-fronted Ground Dove. Note that some males have all-white heads (as shown). Female much paler than male and with tawny underparts. **V** Repeated single, hoarse rolling coo, *rrruuh rrruuh rrruuh*. **H** Littoral scrub of some inhabited islands. **D** Tuamotu Archipelago, French Polynesia.

28 WHITE-THROATED GROUND DOVE *Alopecoenas xanthonurus* 26cm Unmistakable in range if seen well. Rather arboreal. **V** Single repeated mournful hoot, *whooooh whoooh*, and a grating snarl, *rrreeh*. **H** Forest canopy, but on Yap feeds largely on the ground. **D** Mariana and Caroline islands, Micronesia.

29 TONGAN GROUND DOVE *Alopecoenas stairi* 26cm Note range. Females occur in two colour morphs: some resemble male, while others lack white at breast shield. Both morphs occur in Fiji, while in Tonga and Samoa most females resemble males. **V** Series of short, evenly pitched hoots, *whuu whuu whuu whuu*. **H** Forest, bamboo thickets. **D** SC Polynesia.

30 SANTA CRUZ GROUND DOVE *Alopecoenas sanctaecrucis* 22–24cm Feeds mainly on the ground, occasionally in trees, on berries, seeds and invertebrates. Unobtrusive. Typically runs from danger. **V** Soft, repeated hoot. **H** Rainforest. **D** Santa Cruz Islands, Solomon Islands; Vanuatu.

31 MARQUESAN GROUND DOVE *Alopecoenas rubescens* 20cm Variable: width and length of white band in wings may vary, tail may be longer or narrower than shown, and head may be darker or almost white. **V** Poorly documented, but males are reported to utter a raspy, snarling note and females occasionally utter a *coo*. **H** Shrubland. **D** Marquesas Islands, French Polynesia.

32 BRONZE GROUND DOVE *Alopecoenas beccarii* 18–20cm Compact, plump dove. Sometimes cocks tail. Feeds on the ground and runs at speed. Diet includes, seeds, fallen fruit and invertebrates. **V** Repetitive, monotonous *hoo-hoo-hoo...* **H** Moist forest. **D** New Guinea; Bismarck Archipelago, Papua New Guinea; Solomon Islands.

33 PALAU GROUND DOVE *Alopecoenas canifrons* 22cm No similar bird occurs in Palau. **V** Series of 10–20 short, slightly upslurred hoots, *whuU whuU whuU whuU*. **H** Forest. **D** Palau, WC Pacific.

34 PHEASANT PIGEON *Otidiphaps nobilis* 45–50cm Tail laterally compressed and sometimes fanned. Feeds on the ground on seeds and fallen fruit. **V** Distinctive slurred, whistling *per-Weooooo*, rising in pitch and then slowly falling. **H** Hill and montane forest. **D** New Guinea.

1
2
3
4
5
6
7
8
9
10
11
12
13
14
15
16
17
18
19
20
21
22
23
♀
24
♂
♀
25
♂
♀
29
♀
♂
white-headed
morph
♀
♂
♂
♀
28
30
26
27
♂
♀
34
32
♂
33
31

PIGEONS AND DOVES *CONTINUED*

1 WESTERN CROWNED PIGEON *Goura cristata* 66–75cm Unmistakable. Forages on the ground, singly or in small groups; roosts in trees. **V** Very low, deep *oom*. **H** Lowland forest. **D** NW New Guinea; West Papua islands, Indonesia.

2 SCHEEPMAKER'S CROWNED PIGEON *Goura scheepmakeri* 72–78cm Feeds on the ground, mainly on seeds and fallen fruit. Sometimes seen in small groups. Runs from danger but generally tame. **V** Resonating, repeated *hoo-hoo, hoo-hoo* drumming notes. **H** Lowland rainforest. **D** SE New Guinea.

3 SCLATER'S CROWNED PIGEON *Goura sclaterii* 66–72cm Feeds on the ground. Seen in pairs or small groups. Diet includes seeds, fallen fruit and invertebrates. **V** Deep, booming, repeated *hoo-hoo, hoo-hoo*. **H** Lowland rainforest. **D** S New Guinea.

4 VICTORIA CROWNED PIGEON *Goura victoria* 68–74cm Usually seen in pairs or small groups. Feeds on the ground on seeds and fallen fruit. Flies up to the safety of a tree branch if disturbed. **V** Booming *oOoo, oOoo*, repeated several times. **H** Rainforest. **D** N New Guinea.

5 TOOTH-BILLED PIGEON *Didunculus strigirostris* 35cm Unmistakable owing to bill shape. **V** Male coos at dawn, a drawn-out *oooo* every 6–10 secs. Female has a similar voice, but higher pitched, and issues guttural growls and a plaintive *coo* in threat. **H** Mature forest. **D** W Samoa.

6 WHITE-EARED BROWN DOVE *Phapitreron leucotis* 23cm Occurs singly or in pairs; forages mainly in trees but will feed on the ground. Best located by its call. **V** *Hoot-ho hoot-ho hoot hoot hoot hoot-hoot-hoot*, accelerating and descending down the scale. **H** Dense woodland, open woodland, edges of cultivated fields. **D** Philippines.

7 AMETHYST BROWN DOVE *Phapitreron amethystinus* 26–27cm Found singly or in pairs foraging in the mid-storey or canopy. **V** Soft, deep, hollow *hoot hoot-hoot hoot hoot hoot*. **H** Primary and secondary forest, from lowlands to mountains. **D** Philippines.

8 TAWITAWI BROWN DOVE *Phapitreron cinereiceps* 26–27cm Frequents the mid-storey or canopy, alone or in pairs. **V** Deep, resonating *hoot hoot toot toot-toot-toot-toot-toot*, which accelerates and then tails off. **H** Lowland forest. **D** Tawi-Tawi Island, Sulu Archipelago, Philippines.

9 MINDANAO BROWN DOVE *Phapitreron brunneiceps* 26–27cm Frequents the mid-storey or canopy. Formerly considered conspecific with Tawitawi Brown Dove. **V** Short, fast *hoot toot-toot-toot-toot*. **H** Forest, from mid-altitudes to mountains. **D** Mindanao and Basilan islands, S Philippines.

10 CINNAMON-HEADED GREEN PIGEON *Treron fulvicollis* 25–27cm Said to feed mainly in small trees on fruit. **V** Similar to that of Little Green Pigeon, but less whining. **H** Freshwater swamp forest, coastal forest, secondary growth, mangroves. **D** Malay Peninsula; Sumatra, Indonesia; Borneo.

11 LITTLE GREEN PIGEON *Treron olax* 21–27cm Arboreal, feeding from the mid-storey to the canopy, usually in groups of up to eight birds. **V** High-pitched *wiiiiii-iiu-iiu-iiu-iiu iiui-iiuwu*, repeated after short intervals. **H** Broadleaved evergreen forest, freshwater swamp forest, secondary growth. **D** Malay Peninsula, Greater Sunda Islands.

12 PINK-NECKED GREEN PIGEON *Treron vernans* 26–32cm Feeds from the mid-storey to the canopy, usually in small flocks, with larger flocks at rich feeding sites. **V** Foraging groups utter a hoarse, rasping *krrak krrak...* **H** Forest and forest edges, plantations, bamboo, mangroves. **D** SE Asia to the Philippines; Sulawesi, Indonesia.

13 ORANGE-BREASTED GREEN PIGEON *Treron bicinctus* 29cm Actions and habits similar to those of Thick-billed Green Pigeon. May mix with other green pigeon species at trees bearing a large fruit crop. **V** Modulated, mellow whistle, a croaking note and a chuckling call. **H** Forests and well-wooded country. **D** Widespread from India to SE Asia.

14 SRI LANKA GREEN PIGEON *Treron pompadora* 27–28cm Actions and habits similar to those of Thick-billed Green Pigeon. **V** Low-pitched mellow whistles. **H** Forests and well-wooded areas. **D** Sri Lanka.

15 GREY-FRONTED GREEN PIGEON *Treron affinis* 27–28cm Actions and habits similar to those of Thick-billed Green Pigeon. **V** Low-pitched mellow whistles. **H** Forests and well-wooded areas. **D** SW India.

16 ASHY-HEADED GREEN PIGEON *Treron phayrei* 27–28cm Actions and habits similar to those of Orange-breasted Green Pigeon. **V** Low-pitched mellow whistles. **H** Broadleaved evergreen and semi-evergreen forest. **D** SE Asia, SW China.

17 ANDAMAN GREEN PIGEON *Treron chloropterus* 27–28cm Actions and habits similar to those of Thick-billed Green Pigeon. The only green pigeon on the Andaman and Nicobar islands. **V** Low-pitched mellow whistles. **H** Forests and well-wooded areas. **D** Andaman and Nicobar islands.

18 PHILIPPINE GREEN PIGEON *Treron axillaris* 28cm Occurs singly, in pairs or in small groups. Originally considered a race of the Sri Lanka Green Pigeon. **V** Mournful *coo* and a series of mellow whistles. **H** Primary and secondary evergreen forest and forest patches. **D** Philippines.

19 BURU GREEN PIGEON *Treron aromaticus* 28cm Occurs singly, in pairs or in small groups. Formerly considered a race of the Sri Lanka Green Pigeon. **V** Undescribed. **H** Forest, forest edges, tall secondary growth. **D** Buru, S Maluku Islands, Indonesia.

20 THICK-BILLED GREEN PIGEON *Treron curvirostra* 24–31cm Arboreal. Feeds on fruits and berries, usually in small flocks, with larger flocks at rich food sites. **V** Low-pitched, throaty whistles; also a hoarse *goo-goo* while feeding. **H** Various forests, including broadleaved evergreen, semi-evergreen and mixed deciduous; also mangroves and secondary growth. **D** Widespread from C Nepal to the Philippines and Borneo.

21 GREY-CHEEKED GREEN PIGEON *Treron griseicauda* 25cm Usually occurs in small or large flocks; generally feeds in the crown of tall trees, although regularly perches on exposed branches. **V** Jumble of melodious gurgles, coos and whistles; also utters a sad, howling *kuwu kuwu*. **H** Lowland forest, remnant forest patches, forest edges, lightly wooded cultivation, scrub with scattered trees, gardens. **D** Java, Bali and Sulawesi, Indonesia.

22 SUMBA GREEN PIGEON *Treron teysmannii* 29cm Found singly, in pairs or in small flocks. Generally quiet and inconspicuous. Forages mainly in the crown of fruiting trees. **V** Includes an *awop-awop-a-where-rup*, a mellow, coughing *korrr* and a *awhereu* advertising call. **H** Remaining areas of primary, selectively logged and tall secondary forest; also visits fruiting trees in parkland savannah. **D** Sumba, Indonesia; W Lesser Sunda Islands.

23 FLORES GREEN PIGEON *Treron floris* 29cm Wary and generally inconspicuous. Usually encountered in small flocks, with larger flocks in fruiting trees. **V** Generally silent, although low grunts have been reported. **H** Primary and tall secondary forest patches, coastal forest, woodland, lightly wooded cultivation, scrub. **D** W Lesser Sunda Islands.

24 TIMOR GREEN PIGEON *Treron psittaceus* 32cm Usually occurs in small flocks, also seen singly or in pairs; wary and inconspicuous. Forages in fruiting trees. **V** Series of descending, seesawing notes; also a complex jumble of moderately high-pitched bubbling and gargling notes. **H** Primary and tall secondary lowland monsoon forest. **D** E Lesser Sunda Islands.

25 LARGE GREEN PIGEON *Treron capellei* 35–36cm Forages in small groups of up to a dozen birds, often in fruiting figs. **V** Variable, deep nasal creaking *oo-oo-aah oo-oo-aah aa-aa-aah*; also *oooOOah oo-aah...* **H** Broadleaved evergreen and freshwater swamp forest, forest edges and clearings. **D** Malay Peninsula; Sumatra, Sulawesi, Java, Indonesia; Borneo.

26 YELLOW-FOOTED GREEN PIGEON *Treron phoenicopterus* 33cm Actions and habits similar to those of other green pigeons. Birds from Sri Lanka are duller and smaller. **V** Melodious whistling notes. **H** Forests, fruiting trees in cultivation, gardens, parks. **D** India to SE Asia.

27 BRUCE'S GREEN PIGEON *Treron waalia* 28–30cm Shy. Usually found in small groups feeding in the tree canopy; at fruiting fig trees may be seen in larger flocks. **V** Crooning whistle and a quarrelsome chatter. **H** Wooded wadis, open country with scattered trees, gardens with large trees. **D** Senegal and Gambia to Eritrea, Ethiopia, Somalia and Uganda; also SW Arabian Peninsula.

28 MADAGASCAN GREEN PIGEON *Treron australis* 30cm Range differs from that of African Green Pigeon. **V** Series of low, soft, mournful whistles, lasting 5–10 secs, *krrrrr tiuuu-kirriouu-tuwhip-kirriu-tuwhip-krrrr-whup-kirriwhip-kirriwhip-ker-whip-kurrr-whup-whup-kurrrrr-pop-whurrrr*. **H** Forest and woodland with fruiting trees. **D** Madagascar.

29 COMOROS GREEN PIGEON *Treron griveaudi* 32cm Unobtrusive and easily overlooked. Lives mostly in the tree canopy, foraging for fruit, notably figs (*Ficus* spp.). **V** Rasping *brrrr*, followed by sharp, upslurred *Whip-ooo*. **H** Evergreen forest. **D** Mohéli, Comoros Islands.

30 AFRICAN GREEN PIGEON *Treron calvus* 30cm The only green pigeon in the region. **V** Combination of mid-high frog-like grunts, fast rattles and other sounds. **H** Forest edges, cultivation, suburbs (wherever it finds wild figs to eat). **D** Widespread across Africa.

31 PEMBA GREEN PIGEON *Treron pembaensis* 30cm The only green pigeon on Pemba Island. **V** Drawn-out, melodious, modulated whistles, followed by short throaty growls. **H** Open forests, cultivation, parks, gardens. **D** Pemba Island, Zanzibar Archipelago, Tanzania.

32 SAO TOME GREEN PIGEON *Treron sanctithomae* 30cm Overall more dark grey than green. **V** Not well documented, but includes rattling, drawn-out whistling, 'hiccups' and short grunts. **H** Tree crowns of forests and plantations. **D** São Tomé, Gulf of Guinea.

33 PIN-TAILED GREEN PIGEON *Treron apicauda* 28–32cm Feeds with acrobatic, parakeet-like actions. Otherwise, habits generally much as for other green pigeons. **V** Deep, musical *oou ou-ruu oo-ru ou-rooou* and a mellow, whistled *ko-kla-oi-oi-oi-oilli-illio-kla*. **H** Evergreen broadleaved hill forests. **D** N India to SE Asia.

34 SUMATRAN GREEN PIGEON *Treron oxyurus* 34cm Usually occurs in nomadic flocks, but never far from forests. **V** Ringing *oo-oowao-oowau* or similar. **H** Thick hill and mountain forest. **D** Sumatra and W Java, Indonesia.

35 YELLOW-VENTED GREEN PIGEON *Treron seimundi* 26–28cm Feeds on fruit in the tree canopy. Also recorded flying across valleys at a great height. **V** High-pitched *pooaah po-yo-yo-pooaah*. **H** Broadleaved evergreen forest, forest edges. **D** SE Asia.

36 WEDGE-TAILED GREEN PIGEON *Treron sphenurus* 30–33cm Acrobatic forager in fruiting trees; occurs singly, in pairs or in small flocks. **V** Series of musical whistling or fluting notes; also a curious grunting. **H** Broadleaved evergreen forest and forest edges; post-breeding, sometimes occurs on plains in N Laos and C Vietnam. **D** N India to Java, Indonesia.

37 WHITE-BELLIED GREEN PIGEON *Treron sieboldii* 33cm Feeds mainly in fruiting trees, although also recorded feeding on the ground. Usually encountered in small parties. **V** Mournful *oaooh oaooh* or *o-vuuo-vuuo-vuuo-vouoou-oo*. **H** Broadleaved evergreen forest, forest edges and clearings. **D** Japan to SE Asia.

38 WHISTLING GREEN PIGEON *Treron formosae* 35cm Little recorded; actions and habits said to be much as for other members of the genus. **V** *Po-po-peh*, the last note higher pitched. **H** Subtropical broadleaved evergreen forests, trees surrounding cultivated areas, town gardens. **D** Islands south of Japan to the Philippines.

1
2
3
4
5
6
7
8
9
10
11
12
13
14
15
16
17
18
19
20
21
22
23
24
25
26
27
28
29
30
31
32
33
34
35
36
37
38

PIGEONS AND DOVES *CONTINUED*

1 BANDED FRUIT DOVE *Ptilinopus cinctus* 30–32cm Usually seen singly, in pairs and occasionally in small groups. Mainly frequents the mid-storey and lower canopy. **V** Series of evenly pitched, muted *whoo* notes. **H** Primary and tall secondary lowland, hill and montane forest, *Casuarina* groves, coastal woodland. **D** Bali, Indonesia; Lesser Sunda Islands.

2 BLACK-BANDED FRUIT DOVE *Ptilinopus alligator* 33–35cm Distinctive pigeon. Lives mostly in the tree canopy and feeds on fruits. Usually seen singly or in pairs. Take-off is accompanied by loud wing-clapping. **V** Repeated, pumping *whOOo*. **H** Rainforest. **D** N Australia.

3 RED-NAPED FRUIT DOVE *Ptilinopus dohertyi* 33cm Generally quiet and inconspicuous, although sometimes perches in the open at the top of tall trees, especially along limestone ridges. Frequents the mid-storey up to the canopy; often forages singly or in pairs. **V** Soft, deep *woo-oo*. **H** Remaining patches of primary and tall secondary forest. **D** Lesser Sunda Islands.

4 PINK-HEADED FRUIT DOVE *Ptilinopus porphyreus* 29cm Usually encountered singly or in pairs. Shy and inconspicuous. **V** Soft *hoo*. **H** Oak-laurel forest and montane heath forest. **D** Sumatra and Java, Indonesia.

5 FLAME-BREASTED FRUIT DOVE *Ptilinopus marchei* 34–38cm Generally frequents the forest canopy. Very shy; usually flies off, with much noisy wing-flapping, before being located. **V** Deep *hooot* or *hooot hooot*. **H** Montane mossy forest. **D** N Philippines.

6 CREAM-BREASTED FRUIT DOVE *Ptilinopus merrilli* 32cm Shy. Forages in the canopy and sometimes the understorey. **V** Soft, stuttering *crruuoop*, descending at the start and rising at the end, repeated in a pulsating sequence; also a soft, purring *rrrrrr*, repeated for several minutes. **H** Primary and selectively logged forest. **D** N Philippines.

7 YELLOW-BREASTED FRUIT DOVE *Ptilinopus occipitalis* 29–30cm Frequents the mid- and upper canopy, singly or in small flocks; larger flocks may occur in fruiting trees. **V** Deep, resonating *hhooott* or *hoorrrr hoorrrr*. **H** Lowland and mid-elevation primary forest. **D** Philippines.

8 RED-EARED FRUIT DOVE *Ptilinopus fischeri* 34cm Generally seen singly or in pairs, occasionally in small groups. Quiet and inconspicuous. Feeds on small fruits in the mid-storey and lower canopy. **V** Soft, rising, frog-like *oowup*. **H** Primary hill and montane forest. **D** Sulawesi, Indonesia.

9 JAMBU FRUIT DOVE *Ptilinopus jambu* 22–28cm Feeds in trees, although also recorded feeding on the ground on fallen fruit. **V** Soft, repeated *hooo*, but generally silent. **H** Broadleaved evergreen forest and occasionally mangroves. **D** Malay Peninsula, Greater Sunda Islands.

10 BANGGAI FRUIT DOVE *Ptilinopus subgularis* 33–36cm Usually seen singly, in pairs or in small flocks. Formerly considered conspecific with the next two species. **V** Series of 20 *whoop* notes. **H** Primary forest, also some secondary and disturbed forest. **D** Banggai Islands off E Sulawesi, Indonesia.

11 OBERHOLSER'S FRUIT DOVE *Ptilinopus epius* 33–36cm Forages in trees, feeding on fruit. Formerly considered a race of the previous species. **V** Series of 7–8 *whoop* notes. **H** Dense primary and secondary forest in lowlands and hills. **D** Sulawesi, Indonesia.

12 SULA FRUIT DOVE *Ptilinopus mangoliensis* 33–36cm Generally forages in the canopy of large fruiting trees. Formerly considered a race of the Banggai Fruit Dove. **V** Series of 11 *whoop* notes. **H** Dense primary and secondary forest. **D** Sula Islands, Indonesia.

13 BLACK-CHINNED FRUIT DOVE *Ptilinopus leclancheri* 26–28cm Feeds in fruiting trees. **V** Well-spaced *whoo-whoo-whoo*...; also a deep, hollow *rooooooo* and a long *whohooo*. **H** Patches of lowland and secondary growth. **D** Philippines, Taiwan.

14 SCARLET-BREASTED FRUIT DOVE *Ptilinopus bernsteinii* 28cm Generally seen singly or in pairs, foraging in the mid-storey and lower canopy. **V** Mournful, slow, hoarse *oo-oooh*; also a growling call. **H** Primary lowland, hill and montane forest, forest edges, lightly wooded cultivation, occasionally bamboo groves. **D** N Maluku Islands, Indonesia.

15 WOMPOO FRUIT DOVE *Ptilinopus magnificus* 35–45cm Unobtrusive despite colourful plumage. Gathers in large flocks when feeding is good. Diet comprises mainly fruit, occasionally invertebrates. **V** Resonating *uErr-Ooo*. **H** Moist forest. **D** Coastal E Australia, New Guinea, islands off Papua New Guinea.

16 PINK-SPOTTED FRUIT DOVE *Ptilinopus perlatus* 25–27cm Forms flocks outside the breeding season. Nomadic, wandering in search of food. Diet comprises mainly fruit, notably figs (*Ficus* spp.). **V** Haunting, upslurred *erWHOOoo*. **H** Forests. **D** New Guinea; islands off Papua New Guinea; Aru Islands, Indonesia.

17 ORNATE FRUIT DOVE *Ptilinopus ornatus* 23–26cm Forms sizeable flocks (several dozen strong) outside the breeding season. Nomadic, wandering in search of food. Feeds on fruits, preferring figs (*Ficus* spp.). **V** Series of upslurred *WHOoo* notes. **H** Rainforest. **D** New Guinea.

18 TANNA FRUIT DOVE *Ptilinopus tannensis* 28–30cm Usually seen singly or in pairs. Dispersive and slightly nomadic, wandering in search of food. Feeds on fruit. **V** Repeated piping *whooeh*, rising in pitch. **H** Rainforest, secondary woodland. **D** Vanuatu.

19 ORANGE-FRONTED FRUIT DOVE *Ptilinopus aurantiifrons* 22–24cm Arboreal dove with a colourful, well-marked head. Usually seen in pairs but sometimes forms flocks. Feeds on fruit. **V** Rapidly repeated *wu, wu, wu*... and *wer-Weo*... **H** Lowland forested habitats. **D** New Guinea.

20 WALLACE'S FRUIT DOVE *Ptilinopus wallacii* 26cm Usually occurs in groups of 3–50 or more. Shy and inconspicuous. **V** Generally silent, but may utter a low *hoo hoo hoo hoo*. **H** Primary and tall secondary lowland forest, forest edges, open secondary woodland, scrub, mangroves. **D** SW New Guinea; Maluku islands, Indonesia; Lesser Sunda Islands.

21 SUPERB FRUIT DOVE *Ptilinopus superbus* 23–24cm Inconspicuous; usually seen singly or in pairs in the mid-storey and canopy, often with other fruit-eating pigeons. **V** Repeated, slow series of upslurred *whup* notes that accelerates slightly. **H** Primary and tall secondary lowland, hill and lower montane forest, also remnant forest patches in secondary scrub and cultivation. **D** Sulawesi, Indonesia, to Solomon Islands and NE Australia.

22 MANY-COLORED FRUIT DOVE *Ptilinopus perousii* 23cm Male is unmistakable. Female may have a yellow or purple vent. Compare with Crimson-crowned Fruit Dove. **V** High *wooh wooh-oohpupooh*. **H** Forest, plantations, parks. **D** Fiji, Tonga, Samoa.

23 CRIMSON-CROWNED FRUIT DOVE *Ptilinopus porphyraceus* 23cm No purple in belly band and terminal tail-bar greyish. **V** High-pitched and rapidly delivered series of notes lasting 5–10 secs, roughly one note per sec, *coo, coo, cu-coo, cu-coo, cu-coo, cu-coo*. **H** Forest, mangrove, scrub. **D** Fiji, Tonga, Samoa.

24 PURPLE-CAPPED FRUIT DOVE *Ptilinopus ponapensis* 23cm Resembles the Crimson-crowned Fruit Dove, but with an indistinct belly band and yellow terminal tail-bar. **V** Not well documented, but includes a fast-paced series of cooing notes lasting up to 8–16 secs. **H** Forest, mangrove, scrub. **D** Chuuk and Pohnpei, Caroline Islands, Micronesia.

25 KOSRAE FRUIT DOVE *Ptilinopus hernsheimi* 22–24cm Seen singly or in pairs. Lives in the tree canopy and feeds on fruit. Unobtrusive and easily overlooked. **V** Churring note, followed by a repeated piping *tu-peh, tu-peh*... **H** Forest habitats. **D** Kosrae, Caroline Islands, Micronesia.

26 PALAU FRUIT DOVE *Ptilinopus pelewensis* 24cm No similar species occurs in Palau's islands. **V** Rhythmic cooing phrase, *wu-whOO wu-whOO wu-hu*... *whu-whu-whu-whu-whu-whu-whu-whu-whu-hu-hu*. **H** Forest. **D** Palau, WC Pacific Ocean.

27 LILAC-CROWNED FRUIT DOVE *Ptilinopus rarotongensis* 22cm Nominate race (shown) has a purplish breast patch. No similar fruit dove occurs in the Cook Islands. **V** Mid-high accelarating and slightly descending *cooh-cooh*. **H** Dense forest on Rarotonga, or wooded habitats (including plantations) elsewhere. **D** Cook Islands, Polynesia.

28 MARIANA FRUIT DOVE *Ptilinopus roseicapilla* 23cm No similar species occurs in the Mariana Islands. **V** Rhythmic cooing phrase, 10 secs long: 4–5 well-spaced emphatic coos, followed by a bubbling trill that changes into a series of 12–14 coos, these gradually slowing down. **H** Forest, secondary growth. **D** Mariana Islands, Micronesia.

29 ROSE-CROWNED FRUIT DOVE *Ptilinopus regina* 21–23cm Inconspicuous, best located by its call. Found singly, in pairs or in small groups of up to 15 birds. **V** Long series of upslurred *wuu* notes; also a loud, accelerating series of 10–20 or more *hoo* notes, starting slowly and ending with notes descending and running together. **H** Forest, monsoon woodland and scrub, coastal and riparian woodland, mangroves. **D** Lesser Sunda Islands, N and E Australia.

30 SILVER-CAPPED FRUIT DOVE *Ptilinopus richardsii* 20–22cm Lives in the tree canopy. Blends in well with dappled foliage. Sometimes seen in pairs, but forms flocks when feeding is good. Feeds on fruits and berries. **V** Mournful, repeated *wu-oo, wu-oo*... **H** Forested habitats. **D** Solomon Islands.

31 GREY-GREEN FRUIT DOVE *Ptilinopus purpuratus* 20cm No similar fruit dove occurs in the Society Islands. **V** Drawn-out, slightly rising, indignant, accelerating *woooow wooow-wooow*, repeated 9–10×. **H** Forest, plantations. **D** Society Islands, French Polynesia.

32 MAKATEA FRUIT DOVE *Ptilinopus chalcurus* 22cm No similar fruit dove occurs in its habitat. **V** Simple cooing phrase of 7–8 notes on an even pitch that gradually accelerate, lasting 4–5 secs. **H** Wooded areas. **D** Makatea Island, French Polynesia.

33 ATOLL FRUIT DOVE *Ptilinopus coralensis* 23cm No similar fruit dove occurs in the Tuamotu Archipelago. **V** Cooing phrase lasting 7–8 secs, *whoo whu-coo huwoo whu-coo hwoo-woo-hwoo-woo-hwoo-woo*. **H** Woodland, scrub, overgrown plantations. **D** Tuamotu Archipelago, French Polynesia.

34 RED-BELLIED FRUIT DOVE *Ptilinopus greyi* 22–24cm Blends in well with foliage despite its colourful plumage. Arboreal, feeding on fruits. Flight is fast and direct. **V** Rather strangled-sounding, repeated *per-WHeeer, per-WHeeer*... **H** Forest. **D** Solomon Islands to New Caledonia.

35 RAPA FRUIT DOVE *Ptilinopus huttoni* 31cm No similar fruit dove occurs on Rapa Iti Island. **V** Rather long series of descending, accelerating *wuh* notes, preceded by *OOH-rurr*. **H** Dense forest. **D** Rapa Iti, Bass Islands, French Polynesia.

36 WHITE-CAPPED FRUIT DOVE *Ptilinopus dupetithouarsii* 20cm White crown is diagnostic. **V** Mid-high, slightly downslurred *oowooh*. **H** Forest. **D** Marquesas Islands, French Polynesia.

1
2
3
4 ♀
♂
5
6
7
8
9 ♀
10
11
12
13 ♀
14 ♀
♂
♂
15
16
17
18
19
20 ♂
21 ♀
♂
22 ♀
23
24
25
26
27
28
29
♂
30
31
32
33
34
35
36

PIGEONS AND DOVES *CONTINUED*

1 HENDERSON FRUIT DOVE *Ptilinopus insularis* 23cm No similar fruit dove on Henderson Island. **V** Male's song a far-carrying, accelerating series of single *hoo* notes. **H** Forest. **D** Henderson Island, Pitcairns.

2 CORONETED FRUIT DOVE *Ptilinopus coronulatus* 18–20cm Colourful, well-marked pigeon. Arboreal. Feeds on fruits, mainly figs. Seen singly or in pairs. Congregates at good feeding spots. **V** Mournful, piping and repeated *oo-Whoo-oo*. **H** Forested habitats; mostly lowlands. **D** New Guinea, W Papuan Islands, Aru Islands.

3 BEAUTIFUL FRUIT DOVE *Ptilinopus pulchellus* 18–20cm Plumage lives up to its common name. Lives in tree canopy. Feeds on fruits. Seen singly or in pairs. **V** Repeated, rapid piping *Oo-oh, Oo-oh, Oo-oh...* **H** Rainforest, moist forested habitats; mainly 500–750m. **D** New Guinea, Papuan Islands.

4 BLUE-CAPPED FRUIT DOVE *Ptilinopus monacha* 18cm Occurs singly, in pairs and sometimes in small groups. Generally quiet. **V** Slow, monotonous *who-oo*, also single upslurred note and short series of accelerating upslurred notes. **H** Scrubby coastal woodland, tall secondary growth and forest edges, mangroves and coconut plantations with a mixed understorey of tall scrub. **D** N Maluku Islands, Indonesia.

5 WHITE-BIBBED FRUIT DOVE *Ptilinopus rivoli* 23–24cm Usually encountered singly, in pairs and occasionally in small groups; often with other fruit doves in mid-storey and canopy of fruiting trees. **V** Series of ascending, then descending, *hoo* notes. **H** Primary and tall secondary coastal, lowland, hill and montane forest. **D** Maluku Islands and New Guinea to Bismarck Archipelago.

6 YELLOW-BIBBED FRUIT DOVE *Ptilinopus solomonensis* 21–22cm Arboreal. Feeds in tree canopy on fruits. Usually seen singly or in pairs. Congregates at good feeding spots. **V** Repeated series of *who-Aa, who-Aa, who-Aa* phrases. **H** Forested habitats; mostly lowlands. **D** Solomon Islands, Bismarck Archipelago.

7 CLARET-BREASTED FRUIT DOVE *Ptilinopus viridis* 20–21cm Frequents canopy. Often perches conspicuously on high, exposed branches, especially early morning and late afternoon. Usually singly or in pairs. **V** Mournful *a-huh-hoo wo wo wo wo wo wo wo wo wo wo*, tailing off towards end. **H** Primary and tall secondary coastal, lowland and hill forest, forest edges, lightly wooded cultivation, small coastal islets. **D** S Maluku Islands, New Guinea, Solomon Islands.

8 WHITE-HEADED FRUIT DOVE *Ptilinopus eugeniae* 19–22cm Arboreal. Feeds on fruits and berries. Usually seen singly or in pairs. Distinctive markings allow easy identification. **V** Repeated cooing *whoo, u-Whoo, u-Whoo...* **H** Lowland forests, wooded habitats. **D** SE Solomon Islands.

9 ORANGE-BELLIED FRUIT DOVE *Ptilinopus iozonus* 20–22cm Feeds on fruits, notably figs. Arboreal. Usually seen singly or in pairs. **V** Repeated, piping and upslurred *weh-WEeoo, WEeoo, WEeoo...* **H** Forested habitats; mostly lowlands. **D** New Guinea, W Papuan Islands, Aru Islands.

10 KNOB-BILLED FRUIT DOVE *Ptilinopus insolitus* 22–24cm Red knob at front of head aids identification. Arboreal. Feeds on fruits. Seen singly or in pairs. **V** Rapid series of *Wheh, woo, woo, woo* phrases, accelerating during delivery. **H** Forested habitats; lowland to 1,000m. **D** Bismarck Archipelago.

11 GREY-HEADED FRUIT DOVE *Ptilinopus hyogastrus* 23–24cm Occurs singly, in pairs and sometimes in groups of 20 or more, especially in fruiting trees. Regularly perches on exposed treetop branches or overhead wires, particularly early morning and late afternoon. **V** Soft, slightly mournful, repeated *who-huu*. **H** Forest edges, secondary growth, grassland with scattered trees, lightly wooded cultivation and mangroves; occasionally near human habitation. **D** N Maluku Islands, Indonesia.

12 CARUNCULATED FRUIT DOVE *Ptilinopus granulifrons* 20–24cm Usually forages in twos or threes in canopy of fruiting trees, often along with other fruit-eating species. **V** Quiet, low-pitched creaking call has been noted. **H** Primary and secondary lowland and hill forest, forest edges, selectively logged forest, lightly wooded cultivation and coastal scrub. **D** Obi island, N Maluku Islands, Indonesia.

13 BLACK-NAPED FRUIT DOVE *Ptilinopus melanospilus* 21–27cm Singly, in pairs or in small flocks; rarely up to 100, usually at fruiting trees or roosts. Often mixes with other fruit-eating birds. **V** Call *hoo hoo*, repeated every few secs, often for several minutes. **H** Primary, selectively logged and tall secondary forest, forest edges, open woodland, lightly wooded cultivation, scrub and urban areas. **D** Java, Lesser Sunda Islands, Sulawesi and Philippines.

14 DWARF FRUIT DOVE *Ptilinopus nainus* 14–15cm World's smallest fruit dove. Arboreal. Feeds on fruits, berries and occasionally nectar. **V** Slightly strangled-sounding series of *oo-eerp, oo-eerp, oo-eerp...* phrases. **H** Forested habitats; lowlands to 1,000m. **D** New Guinea, Raja Ampat Islands.

15 NEGROS FRUIT DOVE *Ptilinopus arcanus* 16–17cm Known from only a single female specimen. Habits presumed to be as other fruit doves. May be extinct. **V** Unrecorded. **H** Specimen taken in forest; altitude 1,200m. **D** Negros, Philippines.

16 ORANGE FRUIT DOVE *Ptilinopus victor* 19cm Female distinguished from female Many-colored Fruit Dove (Plate 60) by all-green plumage. **V** Slow series of mechanical clicking or dripping notes. **H** Woodland, open forest; altitudes between 420 and 980m. **D** Fiji.

17 GOLDEN FRUIT DOVE *Ptilinopus luteovirens* 20cm Female distinguished from female Many-colored Fruit Dove (Plate 60) by all-green plumage. **V** Abrupt, well-spaced harsh barking notes. **H** Forest, secondary growth; altitudes between 60 and 2,000m. **D** Fiji.

18 WHISTLING FRUIT DOVE *Ptilinopus layardi* 20cm Female distinguished from female Many-colored Fruit Dove (Plate 60) by all-green plumage. **V** Bright, upslurred whistled note. **H** Forest, lowland bush, village gardens. **D** Fiji.

19 CLOVEN-FEATHERED DOVE *Drepanoptila holosericea* 30–32cm Colourful, distinctive, short-tailed pigeon with fluffy leg feathers. Arboreal. Feeds on fruits and berries. **V** Mournful, booming *whoo, whoo, whoo...* **H** Humid forest, savannah. **D** New Caledonia.

20 MADAGASCAN BLUE PIGEON *Alectroenas madagascariensis* 28cm Arboreal, usually in pairs or small groups. Unmistakable in its range. Note red tail. **V** Brief, low-pitched grunting calls recorded. **H** Forest and nearby wooded areas at any altitude. **D** Madagascar.

21 COMOROS BLUE PIGEON *Alectroenas sganzini* 30cm Behaviour similar to Madagascan Blue Pigeon. No similar bird in its range. Shows relatively little bare pink skin round eyes. **V** Call is a hollow *ou-wouh*. **H** Any wooded area. **D** Comoros.

22 SEYCHELLES BLUE PIGEON *Alectroenas pulcherrimus* 30cm Behaviour similar to Madagascan Blue Pigeon. No similar bird with this head pattern in its range. **V** Rapid series of gruff barking or clucking notes. **H** Any wooded area. **D** Seychelles.

23 PINK-BELLIED IMPERIAL PIGEON *Ducula poliocephala* 42cm Usually seen singly, in pairs or in small groups foraging in canopy. Shares roosts with other imperial pigeons. **V** Deep, booming *booouuum booouuum*. **H** Primary forest, undisturbed secondary growth and forest edges. **D** Philippines.

24 WHITE-BELLIED IMPERIAL PIGEON *Ducula forsteni* 43–52cm Occurs singly, in pairs and occasionally in large flocks, especially in fruiting trees. Generally vocal and conspicuous, regularly perching on exposed branches above canopy. Forages in lower canopy or mid-storey. **V** Very deep, far-carrying *uu-uum*. **H** Primary hill and montane forest. **D** Sulawesi and the Sula Islands, Indonesia.

25 MINDORO IMPERIAL PIGEON *Ducula mindorensis* 47cm Generally seen singly or in pairs, occasionally in small flocks; forages in understorey and canopy. **V** Deep, resonating, two-syllable note. **H** Montane forest and occasionally lowland forest. **D** Mindoro in the Philippines.

26 GREY-HEADED IMPERIAL PIGEON *Ducula radiata* 36–39cm Shy. Occurs singly or in flocks of up to 20 birds. Occasionally mixes with other pigeons in fruiting trees. Forages in mid-storey and lower canopy. **V** Subdued *huh*. **H** Primary hill and montane forest, forest edges and occasionally lowland forest. **D** Sulawesi.

27 SPOTTED IMPERIAL PIGEON *Ducula carola* 33–36cm Generally seen in small or large flocks, sometimes associating with flocks of Green Imperial Pigeons in lowlands. **V** Deep, descending *hu hu hu hu hu-hu-hu-huhu*, the last notes quieter and accelerating. **H** Lowland to mossy montane forest and forest edges. **D** Philippines.

28 GREEN IMPERIAL PIGEON *Ducula aenea* 43–47cm Usually in pairs or small parties; larger flocks at fruiting trees. **V** Deep, hollow *currr-whoo*. **H** Moist tropical broadleaved forest and mangroves. **D** W, E and S India and Nepal to S China, Thailand, also Sri Lanka, much of Indonesia and Philippines.

29 NICOBAR IMPERIAL PIGEON *Ducula nicobarica* 43–47cm Usually in pairs or small parties, sometimes larger flocks at fruiting trees. **V** Deep, hollow *currr-whoo*. **H** Mangroves and moist tropical broadleaved forest. **D** Nicobar and Andaman Islands.

30 SPECTACLED IMPERIAL PIGEON *Ducula perspicillata* 41cm Occurs singly, in pairs and in small flocks. Generally vocal and conspicuous, especially early morning and late afternoon, when tends to perch in the open. **V** Hurried series of 6–8 moderately low-pitched, short *hoo* notes. **H** Primary, coastal, lowland and hill forest, forest edges, tall secondary forest, lightly wooded cultivation and mangroves. **D** From Buru northwards in Maluku Islands, Indonesia.

31 SERAM IMPERIAL PIGEON *Ducula neglecta* 43cm Split from previous species; behaviour presumed to be similar. **V** Deep, rolling *wooo*. **H** Primary and selectively logged evergreen forest. **D** Seram and nearby Boano, Ambon and Saparua in S Maluku Islands, Indonesia.

32 ELEGANT IMPERIAL PIGEON *Ducula concinna* 45cm Usually in flocks of up to 40 birds, although sometimes alone or in pairs. Forages in upper mid-storey and canopy. Nomadic, recorded flying between islands. **V** Gruff, growled *urrauw*, interspersed with loud, drawn-out, upslurred growls. **H** Primary and secondary forest, forest edges and lightly wooded cultivation. **D** S Malukus and Lesser Sunda Islands.

33 PACIFIC IMPERIAL PIGEON *Ducula pacifica* 39cm Compare to Micronesian Imperial Pigeon. From Barking Imperial Pigeon (Plate 62) by enlarged cere, contrasting back, gloss to mantle and wings. **V** Slightly lowered *rrroooh* or low *oohóoroh*. **H** Forest canopy, littoral scrub. **D** Bismarck Archipelago to Cook Islands.

34 MICRONESIAN IMPERIAL PIGEON *Ducula oceanica* 44cm From Pacific Imperial Pigeon by range, and rufous belly. **V** Harsh, downslurred, growling *aaarooah*. **H** Canopy of forest and plantations. Also in mangrove. **D** Palau, Caroline Islands and Marshall Islands.

1
2
3
4
♀
♂
5
♀
♂
6
♀
♂
7
8
9
10
11
12
13
♀
♂
14
♀
♂
15
16
♀
17
♀
♂
18
♂
♀
19
♀
♂
20
21
22
23
24
25
26
27
♀
♂
28
29
30
31
32
33
34

PIGEONS AND DOVES *CONTINUED*

1 POLYNESIAN IMPERIAL PIGEON *Ducula aurorae* 51cm Unmistakable in range by large size. Note reddish eyes. **V** Interrogative *oo* or *whoo*. **H** Dense forest. **D** Tuamotu Archipelago and Society Islands, French Polynesia.

2 MARQUESAN IMPERIAL PIGEON *Ducula galeata* 55cm Not in range of Polynesian Imperial Pigeon. Note shape of enlarged cere. **V** Rasping, guttural crow-like note. **H** Upland forest. **D** Marquesas Islands, French Polynesia.

3 RED-KNOBBED IMPERIAL PIGEON *Ducula rubricera* 38–44cm Lives in tree canopy. Feeds on fruit and berries. Sometimes flies over forest canopy. Red bill-knob aids identification. **V** Calls include booming *arr-Oo-eo*. **H** Forested habitats, mainly lowlands. **D** Bismarck Archipelago, Solomon Islands.

4 SPICE IMPERIAL PIGEON *Ducula myristicivora* 41–43cm Perches on high, exposed branches early morning and late afternoon, alone, in pairs or occasionally in small flocks. **V** Harsh, far-carrying, short, upward-inflected, disyllabic growl; also short, harsh *koor*. **H** Primary and tall secondary lowland forest, mangroves. **D** N Maluku Islands, Indonesia.

5 PURPLE-TAILED IMPERIAL PIGEON *Ducula rufigaster* 34–38cm Usually solitary. Feeds on fruit in tree canopy. **V** Two clicks followed by long-drawn-out, wavering and slurred *oo-Ooo-uuerrr*, rising on second syllable then falling away. **H** Forest habitats; lowlands to 1,000m. **D** New Guinea, Papuan Islands.

6 CINNAMON-BELLIED IMPERIAL PIGEON *Ducula basilica* 41–42cm Presence revealed by loud calls. Singles, pairs or small flocks in mid-storey and lower canopy; also reported perching on tall emergent trees overlooking surrounding forest. **V** Unhurried, deep, throaty growl preceded by a brief, hoarse *oo* or *whoo*. **H** Primary lowland and hill forest, disturbed forest, lightly wooded cultivation. **D** NC Maluku Islands, Indonesia.

7 FINSCH'S IMPERIAL PIGEON *Ducula finschii* 36–38cm Lives in tree canopy. Feeds on fruit. Usually solitary. **V** Twin-note vibrating slur, pitch rising then falling; vaguely reminiscent of distant air-raid siren. **H** Forest habitats. **D** Bismarck Archipelago.

8 RUFESCENT IMPERIAL PIGEON *Ducula chalconota* 41–42cm Lives in mid-level tree canopy. Feeds on fruit. Usually solitary or in pairs. **V** Piping *whOoa*, increasing in volume, then fading; also clicking alarm call. **H** Montane forests. **D** New Guinea.

9 ISLAND IMPERIAL PIGEON *Ducula pistrinaria* 38–44cm Note rather uniform colouration. Arboreal. Feeds on fruit. Nomadic, wandering in search of food. **V** Gruff-sounding, descending *wuf-Woo-hoo-hoo*; also a trill. **H** Forest habitats; mainly lowlands. **D** Bismarck Archipelago, Solomon Islands.

10 PINK-HEADED IMPERIAL PIGEON *Ducula rosacea* 44cm Occurs singly, in pairs or in flocks of up to 20 birds, in canopy, often with other imperial pigeons. **V** Deep, muted, drawn-out *hoo-ooo*, also reported as *hoo hoo hoo-oo hoo-oo hoo-oo*. **H** Primary, selectively logged, tall secondary coastal, lowland and hill forest, forest edges, scrubby woodland, lightly wooded cultivation, small offshore islets. **D** Lesser Sunda Islands.

11 CHRISTMAS IMPERIAL PIGEON *Ducula whartoni* 42–45cm Forages mainly in canopy, sometimes in larger flocks at fruiting trees. **V** Series of slurred, cooing notes, *rroow... rroow... rroow*; also deep, booming call reported. **H** Inland primary and secondary rainforest, littoral forest when trees are in fruit. **D** Christmas Island, E Indian Ocean.

12 GREY IMPERIAL PIGEON *Ducula pickeringii* 40cm Usually singly or in pairs. May travel between islands. **V** Deep, gruff *o-oow* or *oot aaooooh*. **H** Primary and tall secondary coastal forest, lowland forest, degraded monsoon forest, lightly wooded cultivation. **D** Islands between Borneo and Philippines.

13 BARKING IMPERIAL PIGEON *Ducula latrans* 41cm Compare Pacific Imperial Pigeon (Plate 61). Cere not enlarged. Shows tawny vent and chestnut underwing (not slate-grey like Pacific Imperial Pigeon). **V** Characteristic, short barks *wuh wuh* in irregular series. **H** Mature forest. **D** Fiji.

14 CHESTNUT-BELLIED IMPERIAL PIGEON *Ducula brenchleyi* 40cm Lives in tree canopy. Feeds on fruits and particularly fond of figs. Usually solitary or in pairs. **V** Moaning, booming *ooEEoo*. **H** Remaining forested habitats; adversely impacted by deforestation. **D** S Solomon Islands.

15 VANUATU IMPERIAL PIGEON *Ducula bakeri* 40cm Lives mainly in tree canopy. Feeds on fruits. Usually seen singly or in pairs. Partly nomadic, wandering in search of fruiting trees. **V** Series of rapid-fire, accelerating hoots. **H** Montane rainforest. **D** Vanuatu.

16 GOLIATH IMPERIAL PIGEON *Ducula goliath* 50cm Note large size. Lives and feeds in tree canopy. Diet comprises mainly fruits. Usually seen singly or in pairs. **V** Owl-like, deep, booming *uh-Ooo*. **H** Primary montane forest. **D** New Caledonia.

17 PINON'S IMPERIAL PIGEON *Ducula pinon* 44–48cm Lives in tree canopy. Feeds on fruits and berries. Usually seen singly or in pairs. Small flocks gather at good feeding trees. **V** Low, booming *bu-Uo-u-oo*. **H** Forested habitats; mostly lowlands. **D** New Guinea, Papuan Islands.

18 BLACK IMPERIAL PIGEON *Ducula melanochroa* 38cm Uniformly dark plumage. Lives in tree canopy. Feeds on fruits, particularly figs. Usually seen singly or in pairs. **V** Low booming calls, undulating in pitch. **H** Forests; mostly 500–1,500m. **D** Bismarck Archipelago.

19 COLLARED IMPERIAL PIGEON *Ducula mullerii* 38–40cm Diagnostic black collar and white throat aid identification. Lives in tree canopy. Feeds on fruit. **V** Low, booming *ahOeeoo*, rising then falling in pitch during delivery. **H** Lowland rainforest, mangrove swamps. **D** New Guinea, Aru Islands.

20 ZOE'S IMPERIAL PIGEON *Ducula zoeae* 38–40cm Lives in tree canopy. Feeds on fruit. Usually seen in pairs, sometimes in small flocks. **V** Booming, tri-syllabic and rapidly slurred *u-oo-oo*. **H** Rainforest; moves between higher and lower ground depending on season. **D** New Guinea, Papuan Islands, Aru Islands.

21 MOUNTAIN IMPERIAL PIGEON *Ducula badia* 43–51cm Usually seen in pairs or small parties. **V** Clicking or clucking sound followed by deep, resonant, double-booming note. **H** Broadleaved evergreen forest. **D** India to SE Asia.

22 DARK-BACKED IMPERIAL PIGEON *Ducula lacernulata* 35–45cm Usually in canopy in small groups; occasionally found singly or in pairs. **V** Muted series of short *whu* notes. **H** Hill and montane forest, wooded cultivation, *Casuarina* forest in Flores. **D** Java and Lesser Sunda Islands.

23 TIMOR IMPERIAL PIGEON *Ducula cineracea* 39–45cm Forages in dense canopy foliage, alone, in pairs or in small groups. **V** Short series of rapid, deep, muted, resonant *hu* notes, producing quavering sound; deep *hoo hoo* call also reported. **H** Montane forest, preferring areas of monsoon woodland. **D** Timor and Wetar in Lesser Sunda Islands, Indonesia.

24 PIED IMPERIAL PIGEON *Ducula bicolor* 35–42cm Unmistakable. In flight shows black primaries and secondaries and a black terminal band on tail. Usually encountered in small parties. **V** Deep purring, also chuckling *hu-hu-hu* and *cru-croo*. **H** Island forest, coastal mainland forests, mangroves. **D** SE Asia to Philippines, W New Guinea.

25 SILVER-TIPPED IMPERIAL PIGEON *Ducula luctuosa* 37–38cm Gregarious; usually in small flocks, although very large flocks recorded, especially prior to dusk. **V** Growling *coo-cooo*, low *wh-hoo*, single loud, upslurred *whooo* and smooth *woooo*. **H** Forest edges, open wooded areas, lightly wooded cultivation, mangroves. **D** Sulawesi and Sula Islands, Indonesia.

26 TORRESIAN IMPERIAL PIGEON *Ducula spilorrhoa* 38–44cm Overall looks very pale. Lives in tree canopy. Feeds on fruit, usually singly or in pairs. Often nests colonially. **V** Upslurred, cooing *urrEeoo*. **H** Lowland rainforest, monsoon forest, swamp forest. **D** N and NE Australia, New Guinea, Aru Islands.

27 YELLOWISH IMPERIAL PIGEON *Ducula subflavescens* 38cm Plumage looks very pale, flushed buffish yellow. Feeds on fruits. Lives in tree canopy. Partly nomadic, wandering in search of fruiting trees. **V** Mostly silent but sometimes utters low, moaning *ooOOoo*. **H** Lowland rainforest. **D** Bismarck Archipelago.

28 TOPKNOT PIGEON *Lopholaimus antarcticus* 40–44cm Lives in tree canopy. Feeds on fruits. Nomadic, wandering in search of fruiting trees, sometimes forming flocks. **V** Various harsh, screeching and grating calls. **H** Rainforest, wet sclerophyll forest. **D** E Australia.

29 NEW ZEALAND PIGEON *Hemiphaga novaeseelandiae* 48cm Unmistakable by large size, sharp demarcation between white and dark green on breast and noisy swish in flight. **V** Sudden, descending *óoooh*. **H** Native forest, parks, gardens, plantations. **D** New Zealand.

30 CHATHAM PIGEON *Hemiphaga chathamensis* 48cm Very similar to New Zealand Pigeon but paler with more extensive green on breast and head. **V** Sudden, descending *óoooh*. **H** Native forest, parks, gardens, plantations. **D** Chatham Island, SW Pacific Ocean.

31 SOMBRE PIGEON *Cryptophaps poecilorrhoa* 46cm Usually singly, occasionally in pairs. Favours subcanopy in undisturbed forest. **V** Drawn-out, sonorous, descending *óoooh*. **H** Primary hill and montane forest. **D** Sulawesi, Indonesia.

32 PAPUAN MOUNTAIN PIGEON *Gymnophaps albertisii* 33–36cm Forages in canopy, singly, in pairs or in flocks. **V** Quiet, low-pitched, upslurred *wooooooo-m*; also soft, querulous whistles. **H** Montane forest. **D** N Maluku Islands, Indonesia; New Guinea.

33 BURU MOUNTAIN PIGEON *Gymnophaps mada* 33–39cm Occurs singly, in pairs or in small flocks. **V** Unrecorded. **H** Hill and montane forest, occasionally feeds in lowland and disturbed forest. **D** Buru in S Maluku Islands, Indonesia.

34 SERAM MOUNTAIN PIGEON *Gymnophaps stalkeri* 33–39cm Formerly considered a race of the previous species; behaviour presumed to be similar. **V** Unrecorded. Noisy swishing wingbeats. **H** Hill and montane forest. **D** Seram, S Maluku Islands, Indonesia.

35 PALE MOUNTAIN PIGEON *Gymnophaps solomonensis* 38cm Lives in tree canopy. Feeds on fruits and berries. Forms feeding flocks that move altitudinally in search of fruiting trees. **V** Low, moaning disyllabic *bu-whoo*, rising and falling in pitch. **H** Montane forests. **D** Solomon Islands.

1
2
3
4
5
6
♂ 7
♀
8
9
10
11
12
13
14
15
16
17
18
19
20
21
22
23
24
25
26
27
28
29
30
31
♀ 32
♂
♀
33
♂
♀ 34
♂
35

HOATZIN OPISTHOCOMIDAE

1 HOATZIN *Opisthocomus hoazin* 62cm Unmistakable appearance. Long untidy crest, bare blue face, red eye. Upperparts dark brown with buff streaks and feather-edges. Throat and breast buff. Rest of underparts and primaries chestnut. **V** Reptile-like huffing and puffing sounds. **H** Forest waterside thickets, shrub, low trees. **D** Amazonia.

TURACOS MUSOPHAGIDAE

2 GREAT BLUE TURACO *Corythaeola cristata* 75cm Unmistakable by size and colour pattern. **V** Long, level, rhythmic cackling. **H** Canopies of forests, riverine belts, forest remnants. **D** Guinea-Bissau and Guinea to W Kenya, NW Tanzania, N Angola.

3 GUINEA TURACO *Tauraco persa* 45cm Separated from Yellow-billed Turaco by white in front of eye, pink bill and blue tail. **V** High loud barking *wah-wah-wraah-wraah-wraah-wraah*. **H** Mid-strata and canopy of rather dense forests and woodland. **D** W and WC Africa.

4 LIVINGSTONE'S TURACO *Tauraco livingstonii* 40cm Separated from Knysna Turaco by black eyes, from Schalow's by shorter crest. **V** High, level barking. **H** Evergreen forest. **D** SE Africa.

5 SCHALOW'S TURACO *Tauraco schalowi* 40cm High crest makes it unmistakable. **V** Calm, high, hoarse, rather toneless series of *wroh* notes. **H** Forest, woodland. **D** Angola to SW Kenya, W Tanzania, Malawi.

6 KNYSNA TURACO *Tauraco corythaix* 40cm Crest shorter and more rounded than in Livingstone's and Schalow's turacos. **V** Series of *wroh* notes. **H** Evergreen forest. **D** SE Africa.

7 BLACK-BILLED TURACO *Tauraco schuettii* 40cm Separated from similar species by black bill. **V** Mid-high, sustained *wahweh-wahweh-wahweh* (recalling yapping of a small dog). **H** Forest canopy. **D** Central Africa.

8 FISCHER'S TURACO *Tauraco fischeri* 40cm Relatively drab olive-toned turaco with red-edged crest. **V** High, speeded-up, loud, barking *wQQ WQO Wah-Wah-W AHW AHWAH W AHW AH*. **H** Woodland, cultivation, gardens. **D** E Africa.

9 YELLOW-BILLED TURACO *Tauraco macrorhynchus* 40cm Rather dark overall with heavy bill. **V** Long series of raucous notes, starting fast then slowing and lowering in pitch. **H** Mature riverine forest. **D** WC and W Africa.

10 WHITE-CRESTED TURACO *Tauraco leucolophus* 40cm Unmistakable. **V** Mid-high *wahweh* followed by series of low-pitched *wroh* notes. **H** Leafy canopy at montane forest edges, woodland, bush, scrubland. **D** Nigeria, Cameroon and South Sudan to W Kenya.

11 BANNERMAN'S TURACO *Tauraco bannermani* 40cm No other red-crested turaco in its range. **V** High, hoarse and emphatic series of *wraoh* notes. **H** Mountain forests. **D** NW Cameroon.

12 RED-CRESTED TURACO *Tauraco erythrolophus* 40cm Separated from Bannerman's Turaco by white line along crest and different range. **V** Series of high, speeded-up, barking, goose-like *wah-wah* notes. **H** Forests. **D** Angola.

13 HARTLAUB'S TURACO *Tauraco hartlaubi* 40cm Distinctive black-and-white face pattern. **V** Series of low, brief, growling or clucking notes. **H** Forests, parks, suburban gardens. **D** C Kenya, E Uganda, N Tanzania.

14 WHITE-CHEEKED TURACO *Tauraco leucotis* 40cm Note dark crest. White neck patch not always present. **V** Series of harsh, upslurred whooping calls. **H** Forests, well-wooded streams. **D** NE Africa.

15 RUSPOLI'S TURACO *Tauraco ruspolii* 40cm No white in face except dirty white crest. **V** Series of throaty clucking notes, speeding up and raising in pitch in middle of series, then slowing and lowering. **H** Conifer forests with dense undergrowth, surrounding wooded areas. **D** S Ethiopia.

16 PURPLE-CRESTED TURACO *Tauraco porphyreolophus* 40cm Unmistakable by black-and-green crest. **V** Series that starts low and very rapid, then rises and slows, ending with very loud *thok* notes. **H** Well-wooded streams, forests, woodland, dense bushland, plantations, gardens. **D** SE Africa.

17 RWENZORI TURACO *Ruwenzorornis johnstoni* 40cm Colourful, with red breast-patch. **V** Fast, short, downslurred chuckle. **H** Montane forests and bamboo. **D** C Africa.

18 VIOLET TURACO *Musophaga violacea* 50cm Like Ross's Turaco, but with different head pattern and more southerly range. **V** Mid-high, harsh and drawn-out cooing notes. **H** Near streams in natural and cultivated areas with high trees. **D** Senegal and Gambia to Chad and Central African Republic.

19 ROSS'S TURACO *Musophaga rossae* 50cm Unmistakable by dark-blue, shining plumage and red crest. **V** High, short, slowing series of cackled *tut* notes. **H** Canopy of forest remnants, miombo, gardens. Often near streams. **D** Cameroon to South Sudan and W Kenya to Angola and Botswana.

20 GREY GO-AWAY-BIRD *Corythaixoides concolor* 50cm Unmistakable by uniform grey-brown plumage. **V** Mid-high, mewing, downslurred note. **H** Thornveld, dry open woodland. **D** S Africa.

21 BARE-FACED GO-AWAY-BIRD *Corythaixoides personatus* 50cm Unmistakable by crest and white neck. **V** Long series of excitable downslurred vibrating notes. **H** Wooded, natural and cultivated areas. **D** E Africa.

22 WHITE-BELLIED GO-AWAY-BIRD *Corythaixoides leucogaster* 50cm Unmistakable. **V** Mid-high, bleating bark *weecrh weh*. **H** Dry, wooded and bushed habitats. **D** South Sudan, Ethiopia and Somalia to N Tanzania.

23 WESTERN PLANTAIN-EATER *Crinifer piscator* 50cm Separated from Eastern Plantain-eater by plain tail and short stripes on upperparts. **V** Tuneful, rapid, downslurring series of hooting *wo-wo-wo* notes. Also a chicken-like cackle. **H** All types of wooded, natural and cultivated areas. **D** Senegal and Gambia to Central African Republic and W DR Congo.

24 EASTERN PLANTAIN-EATER *Crinifer zonurus* 50cm Note white windows in tail and upperwing. **V** Loud series of chuckling, whooping notes. **H** Open woodland, cultivated areas, suburban gardens. **D** Chad and W Sudan, Eritrea and Ethiopia to NE DR Congo and NW Tanzania.

CUCKOOS CUCULIDAE

25 GUIRA CUCKOO *Guira guira* 40cm Untidy appearance. Long 'uncombed' crest. White back and rump. **V** Series of slow descending *kee-ay kee-ay kee-eh kee-orr keeoh cure cure*; *keeerrrrrrr* rattle in alarm, slow quiet *yeus* in flight. **H** Trees in open spaces, savannah, plantations, bushes, thickets, edge of woods, urban gardens, parks. **D** E and S Brazil to Bolivia, Paraguay, N Argentina, Uruguay.

26 GREATER ANI *Crotophaga major* 45cm White eye. Shiny black with bright blue and green sheen. 'Keel' on top of arched bill. Noisy. **V** Call *roodle-toodle-toodle* with 'sizzling' background accompaniment. **H** Always low, by water. Edge of riverine vegetation, mangroves, flooded forest. **D** E Panama to N Argentina.

27 SMOOTH-BILLED ANI *Crotophaga ani* 30–33cm Usually occurs in small noisy flocks, walking on the ground, on branches and clambering through vegetation. Flight direct, with quick choppy wing-beats interspersed with short glides. **V** Ascending whistled *queee-ik* or *a-leep*; also a thin, descending *teeew*. **H** Various open lowland areas, with scattered trees or bushes. **D** S Florida to N Argentina; Bahamas, Greater and Lesser Antilles.

28 GROOVE-BILLED ANI *Crotophaga sulcirostris* 33–35cm Smaller-billed than other anis. Behaviour as Smooth-billed Ani. **V** Liquid *TEE-ho*, *TEET-way* or *TEEt*; also a sharp *pep pep*... **H** Farmland. **D** Mexico to Guyana and NW Argentina.

29 STRIPED CUCKOO *Tapera naevia* 28cm Cap tawny rufous striped black; bushy topknot. Whitish brow and post-ocular stripe. Upperparts brown and buff striped black. Underparts whitish, buffy-grey wash on neck and breast. **V** High-pitched two-note whistle, second note a semitone higher, insistently repeated. **H** Dry woods and scrub, clearings, scattered trees, forest edge, secondary growth. **D** S Mexico to N Argentina.

30 PHEASANT CUCKOO *Dromococcyx phasianellus* 36cm Throat and breast washed buff, streaked and spotted blackish brown. **V** Melancholy whistled *fee fee whirrrrrr* (trill or quaver on last note), also rising *sah, say, seesay*; clucking growl when agitated. **H** Forests, lowland flooded forest, thickets, undergrowth, lower cloud forest, forest edge, secondary woods, canebrakes. **D** S Mexico to N Argentina.

31 PAVONINE CUCKOO *Dromococcyx pavoninus* 27cm Throat, neck and breast uniform cinnamon-buff. Upperparts darker. **V** High-pitched rhythmic *we see thee titee*, rising a tone on last four notes. **H** Understorey thickets, dense tangled vegetation in forests, transition forests, secondary woods, seasonally flooded forest. **D** Guyana to N Argentina.

32 LESSER GROUND CUCKOO *Morococcyx erythropygus* 25cm Note greenish wash and black-bordered, naked blue skin round eye. Elusive. **V** Very high nasal yet mellow fluted *preuww preuww*..., drawn up to *djupdjupdjup*. **H** Dry woodland and forest edge. **D** W Mexico to NW Costa Rica.

33 GREATER ROADRUNNER *Geococcyx californianus* 56–59cm Unmistakable. Often perches on rocks or posts; chases prey on the ground. **V** Dove-like, descending *cooo cooo cooo cooo coo coo*; also a bill-rattling sound. **H** Dry open country with scattered brush. **D** SW USA, Mexico.

34 LESSER ROADRUNNER *Geococcyx velox* 50cm Throat and breast unmarked, not striped as Greater Roadrunner. **V** Low cooing descending *wooboo-wooboo*... **H** Arid open woodland. **D** W Mexico to Nicaragua.

35 RUFOUS-VENTED GROUND CUCKOO *Neomorphus geoffroyi* 50cm Similar to roadrunners, but in different habitat. Often at army-ant swarms. Note breast collar. **V** Explosive bill snaps. **H** Dense forest undergrowth. **D** Nicaragua to Colombia and S Brazil.

36 SCALED GROUND CUCKOO *Neomorphus squamiger* 50cm Differs from Rufous-vented Ground Cuckoo in lack of collar. Forehead, crown, cheeks, neck, breast and part of flanks cinnamon and buffy white, brown V-shaped scaling. Upperparts brownish with purply copper sheen. **V** Loud, sharp bill-snapping. **H** Forest around lower Tapajós river. **D** Brazil, SC Amazonia.

37 BANDED GROUND CUCKOO *Neomorphus radiolosus* 50cm Bill bluish black. Bare skin around eye pale blue. Tail dark blackish blue with green and deep purple sheen. Scaled buffy white on foreparts. **V** Drawn-out, sonorous hooting note, clicking bill-snaps. **H** Evergreen forest on foothills and lower slopes of Andes, tropical forest in lowlands. **D** SW Colombia, NW Ecuador.

38 RUFOUS-WINGED GROUND CUCKOO *Neomorphus rufipennis* 50cm Bill black with greenish-yellow tip. Bare skin around eye red. Crested hood blackish blue. Cheeks, throat and upper breast grey, scaled dark blue. Underparts pale greyish. **V** Dove-like rising and falling hooting note, bill-snaps. **H** Tropical forest in lowlands, foothills and lower slopes. **D** S Venezuela, Guyana, N Brazil.

39 RED-BILLED GROUND CUCKOO *Neomorphus pucheranii* 50cm Bill and bare skin around eye bright red. Cap dark blue. Cheeks and neck pale brownish grey. Breast scaled black. Black pectoral collar. Rest of underparts cinnamon or buffy. **V** Muffled bellowing calls, bill-snaps. **H** Lowland tropical forest. **D** W Amazonia.

1
2
3
4
5
6
7
8
9
10
11
12
13
14
15
16
17
18
19
20
21
22
23
24
25
26
27
28
29
30
31
32
33
34
35
36
37
38
39

CUCKOOS *CONTINUED*

1 BUFF-HEADED COUCAL *Centropus milo* 60–68cm Feeds mainly on ground. Diet comprises mostly invertebrates. **V** Harsh screeches and raspy, agitated *aaOOow*. **H** Moist forested habitats. **D** Solomon Islands.

2 WHITE-NECKED COUCAL *Centropus ateralbus* 42–48cm Very variable species. Three morphs are shown. Usually seen in pairs, sometimes family parties. Feeds mainly on ground. Diet includes invertebrates and small vertebrates. **V** Low, booming calls, often delivered in duet. **H** Forested habitats. **D** Bismarck Archipelago.

3 IVORY-BILLED COUCAL *Centropus menbeki* 60–66cm Feeds on ground and clambers through low vegetation, swinging tail. Diet includes invertebrates and small vertebrates such as lizards and frogs. **V** Penetrating, hooting *hoop*. **H** Forested habitats; mainly lowlands. **D** New Guinea, Aru Islands.

4 BIAK COUCAL *Centropus chalybeus* 44–46cm Feeds mainly on ground. Also clambers through vines. Diet mainly invertebrates. **V** Harsh, raucous and staccato *kra-kra-kra-kra-kra*. **H** Forested habitats. **D** Biak Island, New Guinea.

5 RUFOUS COUCAL *Centropus unirufus* 38–42cm Generally in small, noisy groups of up to 12. Forages in understorey in dense undergrowth, particularly favouring bamboo thickets. **V** Snapping *squip whip* or *squip whip whip whip*, often given by groups moving through trees. **H** Lowland and hill forest with thick undergrowth and bamboo. **D** Philippines.

6 GREEN-BILLED COUCAL *Centropus chlororhynchos* 43–46cm Secretive, behaviour similar to Greater Coucal. **V** Deep *hoop-poop-pooop*, last note lower-pitched, also short *hu-hu* and *chewkk*. **H** Humid, tall, evergreen forest with dense undergrowth, usually of bamboo or rattan cane. **D** Sri Lanka.

7 BLACK-FACED COUCAL *Centropus melanops* 42–48cm Skulks in dense foliage in middle and upper canopy, usually singly or in pairs. **V** Loud *wooooop woooop wooop*, also descending *boop boop boop boop*. **H** Lowland forest, forest edges. **D** Philippines.

8 BLACK-HOODED COUCAL *Centropus steerii* 46cm Secretive, more often heard than seen; skulks in dense foliage and vines in understorey and canopy. **V** Loud *hoooot hoot-hoot-hoot-hoot-hoot*..., descending and becoming softer. **H** Lowland primary forest. **D** Philippines.

9 SHORT-TOED COUCAL *Centropus rectunguis* 37cm Shy and skulking, looks very like Greater Coucal but much smaller and shorter tailed. Black underwing coverts. **V** Four or five resonant booming notes, *buup buup buup buup*, descending towards the end and repeated every few secs. **H** Broadleaved evergreen forest. **D** Extreme S Thailand; Peninsular Malaysia; Sumatra, Indonesia; Borneo.

10 BAY COUCAL *Centropus celebensis* 44–50cm Usually found in pairs or small groups, skulking in dense vegetation from undergrowth to lower canopy. **V** Accelerating, deep *hoo hoo hoo hoo hoo hoo hoo hoo hoo*, starting high then falling, slows and then rises with last two or three notes; also long series of upslurred *woop* notes. **H** Primary and tall secondary lowland and hill forest, forest edges, dense undergrowth. **D** Sulawesi, Indonesia.

11 GABON COUCAL *Centropus anselli* 50cm Note buff underparts. **V** Mid-high hooting slightly descending *foo-foo fooh fooh foo* or very high *fooh-fooh-fooh-fooh-wukwukwukwukwuk-wooh wooh wooh wooh* (*wukwuk* part very high). **H** Dense undergrowth of swampy forest, old cultivations. **D** Cameroon to Angola and C DR Congo.

12 BLACK-THROATED COUCAL *Centropus leucogaster* 50cm Only coucal with black chin and white belly. **V** Accelerating series of rather hollow *hoo hoo hoo* notes. **H** Forest undergrowth. **D** W and C Africa.

13 SENEGAL COUCAL *Centropus senegalensis* 40–42cm Often cumbersome. Feeds in trees and on ground, when can recall a long-tailed gamebird. **V** Hollow, hooting *hoo-hoo* then falling *hu-hu-hu-hu-hu-hu-hu*. **H** Thickets, cultivated land with bushes and trees, orchards, reedbeds. **D** Across W, C, E and S Africa.

14 BLUE-HEADED COUCAL *Centropus monachus* 45cm Separated from Senegal Coucal by larger size, heavier bill, less reddish-rufous upperparts; from Gabon Coucal, Coppery-tailed Coucal, and Burchell's Coucal by unbarred uppertail coverts. Note blue sheen on head and neck. **V** Mid-high, hollow, piping *poo-poo-Poo-PooPoo-Poo*. **H** Reedbeds, shrubbery, at riverbanks or in forest clearings. **D** NE, C and W Africa.

15 COPPERY-TAILED COUCAL *Centropus cupreicaudus* 50cm Undertail coverts palest area of underparts. Separated from Gabon Coucal by paler underparts, from Senegal Coucal and Burchell's Coucal by darker mantle gradually merging into neck and crown. **V** Low, piping, rapid, slightly descending series, like water running out of a bottle. **H** Reedbeds, papyrus, tall grass, shrubbery. **D** S DR Congo and S and SW Tanzania S to Angola, N Botswana and Malawi.

16 WHITE-BROWED COUCAL *Centropus superciliosus* 36–42cm Habits similar to Senegal Coucal. **V** Series of hollow notes, like water being poured from a bottle. Also various harsh *kak* and *hok* notes. **H** Dense scrub, palms; usually near water. **D** SW Arabian Peninsula, E and SC Africa.

17 BURCHELL'S COUCAL *Centropus burchellii* 45cm Only coucal in most of its range except Black Coucal. **V** As Coppery-tailed Coucal but at mid-high pitch and speeded up, or as White-browed Coucal. **H** Moist, grassy areas with some trees, gardens. **D** SE and S Africa.

18 SUNDA COUCAL *Centropus nigrorufus* 46cm Typical coucal skulking habits. **V** Deep, descending *hoop-hoop-hoop-hoop-hoop-hoop-hoop*. **H** Coastal marshes, thickets, swamps, mangroves, tall grasses near mangroves. **D** Java, Indonesia.

19 GREATER COUCAL *Centropus sinensis* 48cm Skulks in vegetation or walks with tail held horizontally when searching for food. Flight weak and clumsy. **V** Deep, descending then rising *hoop-hoop-hoop-hoop-hoop-hoop*. **H** Scrub, tall grassland, thickets, waterside vegetation, gardens. **D** Indian subcontinent and SE Asia.

20 MALAGASY COUCAL *Centropus toulou* 45cm No similar bird in its range. **V** Short or longer series of low, hollow *buc* notes, sometimes recalling Coppery-tailed Coucal. **H** Any habitat with thick, dense vegetation and undergrowth including hedges, gardens, wetlands. **D** Madagascar.

21 GOLIATH COUCAL *Centropus goliath* 62–70cm Usually in pairs or small groups, clambering in undergrowth to mid-storey and occasionally higher. **V** Persistent series of deep *ooom* notes; harsh, guttural *kcau* or *kcau-kuc* when alarmed. **H** Primary forest, forest edges, dense forest undergrowth. **D** N Maluku Islands, Indonesia.

22 BLACK COUCAL *Centropus grillii* 35cm Separated from dark Senegal Coucal by black belly and darker upperparts. **V** Mid-high, mellow *wukwuk wukwuk wukwuk*. **H** Marshes, inundations with some bush and shrub. **D** Widespread in Africa.

23 PHILIPPINE COUCAL *Centropus viridis* 41–43cm Skulks through dense vegetation or sits atop bushes and tall grass. All-black race occurs on Mindoro; rare white morph occurs on Luzon. **V** Various calls, including rapid *coo-coo-coo-coo*... or *boop-boop-boop-boop*..., and distinctive *chi-gook*, *chi-gook-gook* or *chi-go-go gook*. **H** Grassland, cultivation, secondary growth and thickets. **D** Philippines.

24 LESSER COUCAL *Centropus bengalensis* 31–33cm Actions and habits much like Greater Coucal. **V** Deep *whoot-whoot whoot-whoot kurook kurook kurook*... increasing in tempo and falling in pitch. **H** Tall grassland, reedbeds, thickets. **D** Indian subcontinent and SE Asia.

25 VIOLACEOUS COUCAL *Centropus violaceus* 64cm Clambers through vines and branches. Diet includes invertebrates and small vertebrates. **V** Calls include explosive, whooping *up-WHoo-up*. **H** Forested habitats; mostly lowlands. **D** New Britain and New Ireland (Bismarck Archipelago).

26 BLACK-BILLED COUCAL *Centropus bernsteini* 48–52cm Feeds on ground and clambers through vines. Diet includes invertebrates and small vertebrates such as frogs. **V** Calls include a loud, piping *boop*. **H** Lowland wooded and scrub habitats. **D** New Guinea.

27 PHEASANT COUCAL *Centropus phasianinus* 53–80cm Generally seen in pairs or singly. Forages by clambering around in vegetation, or walking or running. **V** Long, descending and then rising series of loud, hollow notes; also series of loud, harsh, hissing notes. **H** Swampy grassland, secondary monsoon forest, fringing grassland. **D** Timor in Lesser Sunda Islands, Indonesia.

28 ANDAMAN COUCAL *Centropus andamanensis* 45–48cm Actions and habits similar to Greater Coucal. **V** Rapid, deep *hoop-hoop-hoop-hoop-hoop*, starting weakly and slowly, increasing in intensity before ending abruptly. **H** Forests, cultivations, mangrove swamps, gardens. **D** Andaman Islands.

29 BORNEAN GROUND CUCKOO *Carpococcyx radiceus* 60cm Forages on ground, often following swarms of army ants, or pigs or bears, to prey on disturbed invertebrates. **V** Far-carrying *whoo-hooh whoo-hooh whoo-hooh*. **H** Alluvial and swamp forest, also undisturbed lowland riverine forest. **D** Borneo.

30 SUMATRAN GROUND CUCKOO *Carpococcyx viridis* 55cm Little recorded, very rare and shy ground forager. **V** Coughing *heh heh heh* and a loud *tock-tor*. **H** Forest in hilly areas. **D** Sumatra, Indonesia.

31 CORAL-BILLED GROUND CUCKOO *Carpococcyx renauldi* 65cm Terrestrial, but roosts in trees. Shy, runs when disturbed but has strong rapid flight. **V** Repeated, loud, mellow, moaning *woaaaah*, *wooaa* or *wohaaau*; also short *pohh-poaaah*, vibrant, rolling *wh ohh-whaaaaohu* and grumbling *grrro grrro*... **H** Broadleaved evergreen forest, secondary forest. **D** NW, NE and S Thailand, Cambodia, N, C and S Laos, C and N Vietnam.

32 CRESTED COUA *Coua cristata* 42cm Arboreal. Separated from Verreaux's Coua by less erect crest and orange-purple breast. **V** Varied, sometimes dog-like yapping or chicken-like clucking. **H** Primary and secondary forest, woodland, brushland, mangroves. **D** Madagascar.

33 VERREAUX'S COUA *Coua verreauxi* 36cm Compare larger Crested Coua. **V** Toneless, rather grating *khrarrr*. **H** Thorn scrub. **D** Madagascar.

34 BLUE COUA *Coua caerulea* 49cm No other large blue bird in the area. **V** Low, croaking *khorr*. **H** Forest, secondary growth, mangroves. **D** Madagascar.

35 RED-CAPPED COUA *Coua ruficeps* 42cm Slender, small-headed. **V** Rapid, descending series of high, fluted notes or low cooing notes. **H** Dry forest, woodland, gallery forest. **D** Madagascar.

36 RED-FRONTED COUA *Coua reynaudii* 39cm Separated from Red-breasted Coua by grey breast and distinctive orange front. **V** Single or cackling series of *wuc* notes, calm or rapid. **H** Dense undergrowth of forest and secondary growth. **D** Madagascar.

37 COQUEREL'S COUA *Coua coquereli* 42cm Normally without pinkish patch in blue area round eyes; compare much larger Giant Coua. **V** Rapid series of 5–7 loud, yet mellow *weew* notes. **H** Natural and secondary forest. **D** Madagascar.

38 RUNNING COUA *Coua cursor* 37cm Separated from Red-breasted Coua by pale, greyish-brown belly, from Red-capped Coua by olive, not lilac rump. Note orange area round throat. **V** High, single or doubled *weew*, answered by low, raucous but soft *roh*, singly or in series. **H** Spiny and thorny habitats, including dry woodland without undergrowth. **D** Madagascar.

39 GIANT COUA *Coua gigas* 60cm Note large size and purple spot below-left of eyes. **V** High, loud descending *tjee-tjee-tjee-tjeerow* with more or less peacock-like quality. **H** variety of forested and wooded habitats. **D** Madagascar.

40 RED-BREASTED COUA *Coua serriana* 42cm Very dark with a deep red breast. **V** Series of well-separated, downslurring *pee-ouw* notes. **H** Humid primary and secondary forest. **D** Madagascar.

1
2
2
3
4
5
6
7
12
8
9
10
11
13
normal form
dark morph
19
14
18
15
16
17
20
21
22
23
white
morph
black
morph
24
25
26
27
28
30
31
29
32
39
40
34
36
33
35
37
38

CUCKOOS *CONTINUED*

1 RAFFLES'S MALKOHA *Rhinortha chlorophaea* 32–35cm Moves steadily among middle-storey foliage and creepers. **V** Series of descending mewing notes, hoarse, strained *heeah* and harsh croaking sounds. **H** Broadleaved evergreen forest, forest edge, occasionally plantations. **D** Peninsular Malaysia; Sumatra, Indonesia; Borneo.

2 BLUE MALKOHA *Ceuthmochares aereus* 35cm Note heavy, yellow bill. Solitary and skulking. **V** Song starts with extremely high, staccato *tic* notes, then changes to mid-high, scolding *kah-kahkah* ending in trill. Many variations of this series. Also very high *fu Weeeh*. **H** Dense undergrowth at forest edges with connected riverine belts. **D** W and C Africa.

3 GREEN MALKOHA *Ceuthmochares australis* 33–34cm Usually skulking. Feeds mainly on insects, sometimes small vertebrates such as frogs. **V** Song an accelerating series of *tchiu-tchiu-tchiu* notes; calls include shrill, disyllabic *chip-chip*. **H** Forested habitats, dense scrub, often near water. **D** E Africa.

4 SIRKEER MALKOHA *Taccocua leschenaultii* 42cm Feeds mainly on ground; sometimes in shrubs and small trees; clambers among twigs or hops from branch to branch. **V** Generally silent, but may utter sharp *kek-kek-kek-kerek-kerek*. **H** Dry deciduous secondary forest, scrub, bush, thorn and grass jungle, dry stony hillsides. **D** Pakistan and India.

5 RED-BILLED MALKOHA *Zanclostomus javanicus* 42cm Behaviour much as other malkohas. **V** Frog-like *uc uc uc uc uc...* occasionally ends in quick *uc-uc-uc*; also a whistled *who-oo*, repeated every 10 secs. **H** Broadleaved evergreen forest, forest edge, secondary growth. **D** Peninsular Malaysia; Greater Sunda Islands.

6 YELLOW-BILLED MALKOHA *Rhamphococcyx calyorhynchus* 51–53cm Usually encountered singly, in pairs or small parties; forages in thick vegetation, creeping around squirrel-like. **V** Nasal rattle, accelerating then fading away; also drawn-out nasal whining like branch creaking in wind. **H** Primary and tall secondary lowland, hill and occasionally submontane forest, forest edges, open woodland, wooded cultivation, scrub. **D** Sulawesi, Indonesia.

7 CHESTNUT-BREASTED MALKOHA *Phaenicophaeus curvirostris* 42–49cm Often sits motionless in canopy watching for prey; moves around waving tail like a squirrel. **V** Low, clucking *kuk kuk kuk* and faster *kok-kok-kok...*; when agitated utters cat-like *miaou*. **H** Broadleaved evergreen forest, secondary growth, occasionally mangroves, plantations, gardens. **D** Peninsular Malaysia; Greater Sunda Islands.

8 RED-FACED MALKOHA *Phaenicophaeus pyrrhocephalus* 46cm Arboreal, works through foliage, hopping between branches and fluttering between trees, usually with other species. **V** Usually silent, although may utter some short yelping whistles and low *kra* or *kok*. **H** Dense forest. **D** Sri Lanka.

9 CHESTNUT-BELLIED MALKOHA *Phaenicophaeus sumatranus* 40–41cm Quiet, creeps through trees and thickets. **V** Low *tok tok...* also thin, high-pitched mewing. **H** Broadleaved evergreen forest, mangroves, secondary growth, plantations. **D** Peninsular Malaysia; Sumatra, Indonesia; Borneo.

10 BLUE-FACED MALKOHA *Phaenicophaeus viridirostris* 39cm Usually seen singly or in pairs, quietly but actively foraging in thickets. Weak flyer, generally only flies when travelling between thickets. **V** Low, croaking *kraa*. **H** Secondary woodland, thorn-scrub, bush. **D** Peninsular India, Sri Lanka.

11 BLACK-BELLIED MALKOHA *Phaenicophaeus diardi* 38cm Forages amidst dense undergrowth and creepers. **V** Gruff *gwuap*, a hurried *gwagaup* and a loud *pauk*. **H** Broadleaved evergreen forest, forest edge, secondary growth, plantations. **D** Peninsular Malaysia; Sumatra, Indonesia; Borneo.

12 GREEN-BILLED MALKOHA *Phaenicophaeus tristis* 38cm Skulking, forages in dense thickets. Best located when flying weakly from one thicket to another. **V** Frog-like croaking *ko ko ko ko*; also peculiar chuckle when flushed. **H** Dense broadleaved forest, thickets. **D** E and NE India, Bangladesh and Himalayas.

13 ROUGH-CRESTED MALKOHA *Dasylophus superciliosus* 41–42cm Travels slowly through dense forest understorey, with odd short flights, singly or in small groups. **V** Repeated, guttural, metallic *cheuk*. **H** Lowland forest, secondary growth, forest edges. **D** N Philippines.

14 SCALE-FEATHERED MALKOHA *Dasylophus cumingi* 40–41cm Usually seen singly or in small groups, foraging in dense understorey foliage or sometimes in canopy. **V** Explosive, high-pitched *quizzzzz-kid* or *whizzzzz-kid*. **H** Forest, forest edges, secondary growth. **D** N Philippines.

15 CHESTNUT-WINGED CUCKOO *Clamator coromandus* 41–47cm Generally stays hidden in canopy foliage, although regularly forages in low vegetation. **V** Metallic, whistled *thu-thu thu-thu thu-thu...*, also harsh *chee-ke-kek* or *crititititit*. **H** Forest clearings, scrub forest, mangroves, cultivation, gardens. **D** N India and Nepal to E China and SE Asia.

16 GREAT SPOTTED CUCKOO *Clamator glandarius* 35–39cm Presence of hidden bird often given away by Magpies (main parasitised host). **V** Loud rattling, cackling *cherr-cherr-che-che-che-che-che*. Female utters bubbling *gi-gi-gi-gi-gi-gi-gi-ku-ku-ku-ku*. Advertising call is clear *kleeok*. **H** Savannah-like heathland with cork oak and stone pine, also olive groves, cultivation with bushes and trees. **D** S Europe and Africa.

17 LEVAILLANT'S CUCKOO *Clamator levaillantii* 35cm Note striped throat and upper breast, relatively long tail. **V** Very high, piped *piu piu piu...* (each note trailing off) or high, chattered rattle, *chachacha*. **H** Forest edges, wooded swamps, dense bushland, gardens. **D** Sub-Saharan Africa.

18 JACOBIN CUCKOO *Clamator jacobinus* 34cm Mainly arboreal, often perches in open. In flight shows white patch at base of primaries. **V** Loud, fluting *piu piu pee-pee piu pee-pee-piu*. When alarmed, harsh *chu-chu-chu-chu*. **H** Open woodland, bushes, groves, gardens. **D** Africa and S Asia.

19 LITTLE CUCKOO *Coccycua minuta* 25cm Separated from much larger Squirrel Cuckoo by red orbital ring and less contrast between breast and belly. **V** Rising *puuuw*, very rapid *kikikik*. **H** Dense undergrowth at forest edge, often near water. **D** E Panama to NE Brazil and N Bolivia.

20 DWARF CUCKOO *Coccycua pumila* 20cm Face, throat and breast tawny rufous. Short and slender bill black. Rest of underparts creamy white. Shortish tail. **V** Calls *churr* and *trrrr*. **H** Gallery forest, rainforest, deciduous tropical forest, savannah, open woods, thorny woods, shrubby areas, scattered trees, parks, gardens. **D** Colombia and Venezuela.

21 ASH-COLORED CUCKOO *Coccycua cinerea* 23cm Black bill, red eye. Lead-grey head; back and underparts brownish grey, paler on throat and breast. Undertail yellowish. Tail with white tip. **V** Harsh *che-rro, che-rro, che-rro* repeated sporadically. **H** Savannahs, woodland. **D** S Brazil, Paraguay, N Argentina, Uruguay.

22 SQUIRREL CUCKOO *Piaya cayana* 47cm Huge tail. Greenish-yellow bill. Bare skin around eye pale green or red. Head and neck buffy vinaceous. Back, wings and upperside of tail rusty red. Underparts grey. Undertail black or rust-red and black, and white. **V** Calls *pit-wheew* and series of *wheep* notes. **H** Dense foliage in all forested and wooded habitats, including plantations, parks, gardens. **D** Mexico to N Argentina.

23 BLACK-BELLIED CUCKOO *Piaya melanogaster* 40cm Bill red. Eye red, bare skin around it yellow and pale green. Cap and nape grey. Throat, neck, breast and upperparts rufous. Belly and undertail coverts black. Underside of tail rufous with subterminal black, and white tip. **V** Explosive *djit, djitjitjit*, also a descending *yaaaaah* followed by dry rattle. **H** Humid forest, sandy soil forest, rainforest, savannah. **D** Colombia, Venezuela and the Guianas to N Bolivia and SC Brazil.

24 DARK-BILLED CUCKOO *Coccyzus melacoryphus* 27cm Skulking and secretive, keeping mostly to low branches of mangroves. In flight, upperwing is uniform brown. Undertail black with wide white tips. **V** Low *cu-cu-cu-cu-cu* or *cu-cu-cu-cu-cu-klop klop kulop*; also dry rattle. **H** Primarily mangrove swamps or nearby countryside. **D** Colombia and Venezuela to C Argentina and Uruguay.

25 YELLOW-BILLED CUCKOO *Coccyzus americanus* 28–32cm Skulking. In flight, upperwing shows distinct rufous bases to primaries. Undertail blackish with wide white tips. **V** Hollow *kuk-kuk-kuk-kuk-kuk-kuk...kow-kow-kow-kowlp-kowlp*; also dove-like *cloom...cloom...cloom*. **H** Woodland, forest edge, thickets, open country with bushes and trees. **D** Breeds S Canada to N Mexico and West Indies. Winters in South America.

26 PEARLY-BREASTED CUCKOO *Coccyzus euleri* 28cm Like Yellow-billed Cuckoo but with brownish-grey primaries. **V** Slow *kuope*, one per sec, also rattle followed by *tooks*, then *twops*. **H** Rainforest, gallery forest, forest edge, woodland on sandy soils, regrowth, scrub. **D** Colombia, Venezuela and the Guianas to E Paraguay and NE Argentina.

27 MANGROVE CUCKOO *Coccyzus minor* 28–30cm Shy, skulking, usually located by call. In flight, upperwing uniform brown. Undertail black with wide white tips. Light and dark morphs exist. **V** Low *gawk gawk gawk gawk gawk gawk*; also single *whit*. **H** Forests, dry scrub, mangroves, plantations, thickets. **D** S Florida, Caribbean and Mexico to NE Brazil.

28 COCOS CUCKOO *Coccyzus ferrugineus* 30cm Only cuckoo in its range. **V** Low-pitched, repeated *kcharr* notes. **H** Forest. **D** Isla del Cocos, Costa Rica.

29 BLACK-BILLED CUCKOO *Coccyzus erythropthalmus* 27–31cm Skulking. In flight, upperwing shows dull rufous-brown bases to primaries. Undertail grey, each feather with a narrow white tip. **V** Hollow *cu cu cu* or *cu cu cu cu*; also descending *k-k-k-k* or *kru-dru*. **H** Forests, woodlands, thickets, often along streams. **D** Breeds SC and SE Canada and C and E USA and S Canada. Winters in NW South America.

30 GREY-CAPPED CUCKOO *Coccyzus lansbergi* 25cm Contrast between grey crown and rufous upperparts diagnostic. **V** High rhythmic rapid wooden *jeWokokokokokokok*. **H** Wooded marsh, forest. **D** Colombia and Venezuela. Winters further south.

31 CHESTNUT-BELLIED CUCKOO *Coccyzus pluvialis* 48–56cm Often seen running along forest branches or gliding between trees; generally forages in middle storey or canopy. More often heard than seen. Underside of tail feathers black with broad white tips. **V** Guttural, accelerating *quawk-quawk-ak-ak-ak-ak-ak*. **H** Mid-elevation open wet forests, secondary forest, open woodland, thickets, gardens. **D** Jamaica.

32 BAY-BREASTED CUCKOO *Coccyzus rufigularis* 43–51cm Shy and secretive. Forages in middle storey and canopy, may be seen leaping between branches. Underside of tail feathers black with broad white tips. **V** Loud *cua...u-ak-u-ak-ak-ak-ak-ak-ak-ak*; also lamb-like bleating. **H** Mainly dry deciduous forests, also locally in mountain rainforest, arid lowlands. **D** Hispaniola.

33 JAMAICAN LIZARD CUCKOO *Coccyzus vetula* 38–40cm Forages from understorey to canopy, moving slowly along branches and among vegetation. More often heard than seen. Underside of tail feathers black with broad white tips. **V** Rapid, low *cak-cak-cak-ka-ka-k-k*. **H** Tropical forests, woodlands, wooded ravines, semi-arid country with trees and shrubs. **D** Jamaica.

34 GREAT LIZARD CUCKOO *Coccyzus merlini* 44–54cm Forages in middle storey and canopy, often descending to ground where runs with tail out straight, making it appear mongoose-like. Those on Cuba can be very tame. Underside of tail feathers grey with black subterminal bar and white tip. **V** Guttural *ka-ka-ka-ka-ka-ka-ka-kau-kau-ko-ko* and *tuc-wuh-h*. **H** Thickets with dense vegetation and vines, tropical lowland evergreen forest, abandoned coffee plantations, overgrown pastures. **D** Cuba and Bahamas.

35 PUERTO RICAN LIZARD CUCKOO *Coccyzus vieilloti* 40–48cm Retiring, habitually sits quietly among dense vegetation, more often heard than seen; forages mainly in the middle storey and canopy. Underside of tail feathers grey with black subterminal bar and broad white tip. **V** Emphatic *ka-ka-ka-ka-ka-ka-ka-ka-ka...*, accelerating and becoming louder. Also soft *caw*. **H** Tropical deciduous forest, tropical lowland evergreen forest, woodland, dry coastal forest, coffee plantations. **D** Puerto Rico.

36 HISPANIOLAN LIZARD CUCKOO *Coccyzus longirostris* 41–46cm Forages by striding or creeping along branches, often in thick vegetation, from canopy to near ground. Regularly sits quietly for several minutes. Underside of tail feathers black with broad white tips. **V** Similar to Great Lizard Cuckoo; also a harsh *tchk* and a *tick cwuh-h h*. **H** Tropical forests, woodlands, thickets, coffee plantations. **D** Hispaniola.

37 THICK-BILLED CUCKOO *Pachycoccyx audeberti* 35cm Note upright stance with drooping wings and short legs. **V** Extremely high, lashing *weeehwit weeehwit*. **H** Miombo, thornveld, riverine belts. **D** W, WC and SE Africa, Madagascar.

1
♀
♂
2
3
4
5
6
♂
7
♀
8
9
10
11
12
13
14
15
16
17
18
light
morph
black
morph
19
20
21
22
23
24
25
26
27
dark
morph
light
morph
28
29
30
31
32
33
34
35
36
37

CUCKOOS *CONTINUED*

1 DWARF KOEL *Microdynamis parva* 20–22cm Unobtrusive and easily overlooked. Keeps to cover of tree foliage. Diet includes fruits and berries. **V** Rapid, accelerating series of *ho, hoo, hooee-hooee-hooee...* notes, rising in pitch throughout delivery. **H** Forested habitats. **D** New Guinea.

2 ASIAN KOEL *Eudynamys scolopaceus* 40–43cm Unobtrusive, usually keeping to dense foliage; first sign is often of a bird flying from tree to tree. **V** Shrill *ko-el ko-el ko-el* that increases in pitch before ending abruptly; also descending, bubbling *wreep-wreep-wreep-wreep-wreepwreepwreep*. **H** Open woodland, forest edges, scrub, cultivation, parks, gardens. **D** Widespread in S Asia, SE Asia, Indonesia, Borneo and Philippines.

3 BLACK-BILLED KOEL *Eudynamys melanorhynchus* 36–44cm Occurs singly, in pairs and in small groups; usually shy, often heard rather than seen. Frequents mid-storey and canopy. **V** Various calls, including series of melancholy *kuOw* notes, starting low then becoming louder and higher in pitch; also loud bubbling, rising then quickly falling away, and constant-pitched *whu-wu-wu-ki-ki-ki-ki-ki-ki*. **H** Primary and tall secondary lowland, hill and submontane forest, riverine forest, forest edges, lightly wooded areas. **D** Sulawesi and Sula Islands, Indonesia.

4 PACIFIC KOEL *Eudynamys orientalis* 39–47cm Actions and habits similar to Asian Koel. **V** Loud, dreary *kooeei* or *ko-el*, rapid bubbling call, and high-pitched, nasal *keel*. **H** Coastal forest, secondary forest, forest edges, cultivation with trees, forest gardens. **D** S Maluku Islands, Indonesia, to N and E Australia.

5 PACIFIC LONG-TAILED CUCKOO *Urodynamis taitensis* 40cm Note barred and spotted upperparts and streaked underparts (unlike any other cuckoo). **V** Very high, hoarse, slightly upslurred *sreeeeuw'ft*. **H** Forest canopy, plantations, suburban areas. **D** Breeds New Zealand. Winters Polynesia.

6 CHANNEL-BILLED CUCKOO *Scythrops novaehollandiae* 58–65cm Usually seen singly or in flocks of up to 10; frequents canopy, occasionally seen in lower foliage. Noisy early and late. **V** Loud, drawn-out, rising scream, often repeated; also utters shorter, downslurred scream. **H** Forest edges, monsoon forest, lightly wooded areas, mangroves. **D** Australia.

7 ASIAN EMERALD CUCKOO *Chrysococcyx maculatus* 18cm Favours branches and foliage in canopy. Active forager, often hawking for flying insects. Flight fast and direct, very parrot-like. **V** Loud, descending *kee-kee-kee*, also *chweek* given in flight. **H** Mainly dense evergreen forest. **D** Himalayas to SC China and Peninsular Malaysia.

8 VIOLET CUCKOO *Chrysococcyx xanthorhynchus* 17cm Presumed to be similar to Asian Emerald Cuckoo. **V** Loud, sharp, repeated *tee-wit*, often uttered in flight; also descending trill *seer-se-seer seeseeseesee*. **H** Secondary evergreen forest, forest edge, orchards. **D** SE Asia to Philippines.

9 DIEDERIK CUCKOO *Chrysococcyx caprius* 18–20cm In flight from below shows black and white barred underwing and tail. Bill reddish. **V** Male calls with onomatopoeic *dee-dee-dee-dee-deederik*. Female utters *deea-deea-deea*. **H** Dry scrub and open woodland. **D** Widespread in Africa, also S Arabian Peninsula.

10 KLAAS'S CUCKOO *Chrysococcyx klaas* 18cm In flight from below has white underwing coverts and mainly white tail. **V** Mournful *whit-jeh whit-jeh whit-jeh*. **H** Open bushland, open woodland, forest edge. **D** Widespread in Africa.

11 YELLOW-THROATED CUCKOO *Chrysococcyx flavigularis* 20cm Note unbarred upperparts of female. Outer tail-feathers mainly white. **V** Very high, fluting *fee-wee-peepeepeepeepee*, slightly trailing off. **H** Forest canopies. **D** Sierra Leone to Ghana, Cameroon to E DR Congo and Gabon.

12 AFRICAN EMERALD CUCKOO *Chrysococcyx cupreus* 25cm Note regular barring of female. **V** Very high, mellow, liquid, affirmative *'helloh judy'*. **H** Forest glades and edges, woodland, bushland, cultivation, suburbs. **D** W, C, S and SE Africa.

13 LONG-BILLED CUCKOO *Chrysococcyx megarhynchus* 18cm Unobtrusive. Searches among tree foliage for insects and other invertebrates. **V** Whistling *tlueeeh*, rising in pitch then falling. **H** Moist, lowland forest. **D** New Guinea, Aru Islands, Waigeo Island.

14 HORSFIELD'S BRONZE CUCKOO *Chrysococcyx basalis* 16cm Forages at all levels. Sometimes occurs in small parties. **V** Descending, high-pitched, whistled *peeer* or *tseeeuw*, incessantly repeated. **H** Coastal scrub, coastal and subcoastal woodland. **D** Australia. E SE Asia in winter.

15 BLACK-EARED CUCKOO *Chrysococcyx osculans* 20cm Forages low down in shrubby vegetation; usually solitary. **V** Quiet, drawn-out, descending, plaintive *piiieeer*; also a cheerful *peeowee* or *cheeowee*. **H** Open woodland, open scrubland. **D** Australia. Winter visitor to Maluku Islands and E Lesser Sunda Islands.

16 RUFOUS-THROATED BRONZE CUCKOO *Chrysococcyx ruficollis* 16cm Sits unobtrusively in tree foliage, scanning for potential prey; diet mostly insects. **V** Repeated, piercing, downslurred and whistled *Peeoo-Peeoo-Peeoo...* **H** Montane forests. **D** New Guinea.

17 SHINING BRONZE CUCKOO *Chrysococcyx lucidus* 17cm Usually solitary. Inconspicuous forager from low scrub to treetops. **V** Generally silent; may utter upward-inflected *fleei*, rapidly repeated and often ending with descending *pee-eerr*. **H** Woodland, woodland edges, scrubby secondary growth, village gardens. **D** Widespread in Australasian region.

18 WHITE-EARED BRONZE CUCKOO *Chrysococcyx meyerii* 15–16cm Sometimes seen in mixed feeding flocks. Diet comprises mainly insects and other invertebrates. **V** Shrill, whistled series of *Teeo, Teeo, Teeo...* phrases. **H** Forest habitats. **D** New Guinea, W Papuan Islands.

19 LITTLE BRONZE CUCKOO *Chrysococcyx minutillus* 15–16cm Unobtrusive, tends to forage in dense foliage. **V** Descending *rhew rhew rhew rhew...* or *teu teu teu teu teu...* sometimes includes rising, screeching *wireg-reeg-reeg*; also utters high-pitched, drawn-out trill. **H** Coastal scrub, mangroves, secondary growth, forest edge, parks, gardens. **D** SE Asia to E Australia.

20 PALLID CUCKOO *Cacomantis pallidus* 31–32cm Often perches in the open; generally solitary, although recorded in small groups in Australia. **V** Eight-note ascending and accelerating whistle; female gives hoarse whistle. **H** Open woodland, clearings, open areas with trees. **D** Australia. Non-breeding visitor to N Maluku Islands and Lesser Sunda Islands.

21 WHITE-CROWNED CUCKOO *Cacomantis leucolophus* 32–33cm White crown stripe aids identification. Diet comprises mainly insects and other invertebrates. **V** Three-note piping call, *pee-pa-phoo*, downslurring. **H** Forested habitats. **D** New Guinea, W Papuan Islands.

22 CHESTNUT-BREASTED CUCKOO *Cacomantis castaneiventris* 24cm Feeds by gleaning foliage and watching patiently for movement while perched. Diet mainly insects. **V** Two-note strident whistle, *pee-puer*, second note upslurred. **H** Montane forest. **D** New Guinea, W Papuan Islands.

23 FAN-TAILED CUCKOO *Cacomantis flabelliformis* 26cm Unmistakable in range, although black morph known from Fiji. **V** Short, high, whistled, descending trill. **H** Lower levels of forest and woodland. **D** Widespread in Australasian region.

24 BANDED BAY CUCKOO *Cacomantis sonneratii* 24cm Frequents bare treetop branches. Calls with tail depressed. **V** Shrill *pi-pi-pew-pew*. **H** Forests and wooded country. **D** Widespread in S Asia, SE Asia, Greater Sundas, Borneo.

25 PLAINTIVE CUCKOO *Cacomantis merulinus* 23cm Mainly arboreal. Restless canopy forager, sometimes chasing flying insects or dropping to ground to collect prey. **V** Mournful *tay... ta...ta*, also repeated *tay... ta... tay*. **H** Forest, wooded country. **D** SE Asia to Philippines.

26 GREY-BELLIED CUCKOO *Cacomantis passerinus* 23cm Behaviour and habits very similar to Plaintive Cuckoo. **V** Clear *pee-pipee-pee...pipee-pee*. **H** Open woodland, secondary forest, bush, scrub, cultivated areas. **D** India. Winter visitor to Sri Lanka.

27 BRUSH CUCKOO *Cacomantis variolosus* 23cm Secretive, more often heard than seen. **V** Repeated series of 4–10 clear whistles, *weep weep weep weep weep...* **H** Forest, from coastal areas to mountains. **D** Widespread in E SE Asia, Indonesia, Borneo, Philippines and Australasia.

28 RUSTY-BREASTED CUCKOO *Cacomantis sepulcralis* 21–25cm Forages among foliage of trees and bushes. **V** Melancholy, descending *whi whi whi whi whi...* also accelerating series of *whi-wibu* notes. **H** Broadleaved evergreen forest, forest edge, secondary growth, mangroves, gardens. **D** Peninsular Malaysia to Greater and Lesser Sunda Islands and Sulawesi; SW Philippines.

29 MOLUCCAN CUCKOO *Cacomantis aeruginosus* 24cm Usually solitary, its presence often given away by calls. **V** Long series of rapidly repeated piping whistles that slows slightly towards the end; also short, high-pitched, whistled notes, repeated monotonously. **H** Montane forest. **D** N Maluku Islands, Indonesia.

30 DUSKY LONG-TAILED CUCKOO *Cercococcyx mechowi* 35cm Separated from Olive Long-tailed Cuckoo by slightly heavier belly baring and less buffy chest sides and from Barred Long-tailed Cuckoo by preference for lowlands; also by voice. **V** Very high rising, lashing *pu wee wir* (recalling Red-chested Cuckoo), fluted descending rapid *tijuwtijuwtijuw-tijuw-tijuw* and other whistles. **H** Mainly lowland forest. Undergrowth near streams. **D** Sierra Leone to Uganda, E DR Congo and N Angola.

31 OLIVE LONG-TAILED CUCKOO *Cercococcyx olivinus* 35cm Very difficult to separate from Dusky and Barred Long-tailed Cuckoos, but upperparts slightly more bronzy and underparts less heavily barred. **V** Mid-high fluted sustained *fiuw-fiuw-fiuw...* (each note with slight tremolo), or high very lazy *it-will-rain*. **H** Lowland wooded areas. **D** Guinea to Uganda and E DR Congo, Gabon, Angola and Zambia.

32 BARRED LONG-TAILED CUCKOO *Cercococcyx montanus* 35cm Boldy barred tail. **V** High loud *weet-weet-weet-weh* or long sequence of slowly rising *fjuu-fjuu-fjuu-fijuu-fijuu* notes (up to 20). **H** Montane forest edges. **D** E and SE Africa.

33 PHILIPPINE DRONGO-CUCKOO *Surniculus velutinus* 24–25cm Secretive. Arboreal. Sluggish while searching in tree foliage. **V** Ascending *wu wu wu waa waa wee* or *wi wi wi wi...* **H** Lowland forest, forest edges. **D** Philippines.

34 SQUARE-TAILED DRONGO-CUCKOO *Surniculus lugubris* 25cm Sluggish movements. Arboreal, occurs in canopy foliage; calls from bare treetop branch, when tends to perch more horizontally with wings drooped. **V** Rising *pip-pip-pip-pip-pip-pip-pip*. **H** Forests, well-wooded areas, plantations, orchards. **D** NW Himalayas to SE China, south to S Thailand; Malaysia; Greater Sunda Islands and Palawan, Indonesia.

35 FORK-TAILED DRONGO-CUCKOO *Surniculus dicruroides* 24–25cm Arboreal, sluggish while foraging in canopy foliage. Calls from bare treetop branch, in horizontal posture with drooping wings. **V** Rising *pip-pip-pip-pip-pip-pip*; also shrill, accelerating *phew phew phewphewphewphew* falling away at finish. **H** Broadleaved evergreen and deciduous forests, secondary growth, plantations, occasionally mangroves, parks. gardens. **D** Himalayan foothills, C and S India, Sri Lanka.

36 MOLUCCAN DRONGO-CUCKOO *Surniculus musschenbroeki* 24cm Presence usually given away by calls; generally seen singly or in pairs foraging in upper tree levels. **V** Rising, whistled *ki ki ki ki ki ki ki ki*, repeated and sometimes becoming hoarse and frantic. **H** Primary and tall secondary forest, forest edges. **D** Sulawesi to Halmahera, Indonesia.

1
♀
♂
2
♀
♂
♀
3
dark
morph
3
♀
♂
rufous
morph
4
♀
♂
5
6
7
♀
♂
8
♂
♀
9
♀
♂
10
♀
♂
11
♀
♂
12
♀
♂
13
♀
♂
14
15
16
17
18
♀
♂
19
♀
♂
20
21
22
23
24
25
♀
♂
26
♀
♂
27
28
29
30
31
32
33
34
35
36

CUCKOOS *CONTINUED*

1 MOUSTACHED HAWK-CUCKOO *Hierococcyx vagans* 28–30cm Best located by call. **V** Loud, monotonous *peu-peu*; also ascending sequence of mellow notes that accelerate to chatter then end abruptly. **H** Broadleaved evergreen forest, secondary growth. **D** SE Asia; Sumatra, Indonesia; Borneo.

2 LARGE HAWK-CUCKOO *Hierococcyx sparverioides* 38–40cm Secretive. Arboreal. **V** Shrill, screaming *pi-pee-ha*, increasing in speed and pitch leading to frantic climax. **H** High-elevation, open forests, especially oak. **D** Widespread in S and SE Asia, Indonesia, Borneo and Philippines.

3 DARK HAWK-CUCKOO *Hierococcyx bocki* 31cm Behaviour and habits presumed to be similar to those of Large Hawk-Cuckoo. **V** Two-note, whistled *pee-ha pee-ha* or *pi-phu pi-phu*. **H** Hill and montane forest. **D** Malaysian Peninsula; Sumatra, Indonesia; Borneo.

4 COMMON HAWK-CUCKOO *Hierococcyx varius* 33cm Arboreal. **V** Monotonous, screeching, high-pitched *wee-piwhit... wee-pwhit...* becoming even more frenetic than that of Large Hawk-Cuckoo. **H** Wooded country, groves, gardens. **D** India to Myanmar.

5 RUFOUS HAWK-CUCKOO *Hierococcyx hyperythrus* 28–30cm Behaviour similar to Malaysian Hawk-Cuckoo. **V** Shrieking *joo-ichi joo-ichi*. **H** Tropical and subtropical forests. **D** Breeds SE Siberia, Japan, Korean Peninsula, NE China. Winters SE Asia.

6 PHILIPPINE HAWK-CUCKOO *Hierococcyx pectoralis* 29cm Shy, generally seen singly or in pairs; forages at all levels. **V** Calls *wheet wheet wheet wheet tu*, repeated nine or 10 times, getting louder and faster, ending in frantic finish. **H** Lowland to montane forest and secondary growth. **D** Philippines.

7 MALAYSIAN HAWK-CUCKOO *Hierococcyx fugax* 28–30cm Arboreal, skulking, usually low down, moves higher when calling. **V** Shrill, insistent *gee-whizz... gee-whizz...* often followed by rapid *ti-tu-tu* phases. Accelerates to shrill crescendo and ends with slower *tu-tu-tu*. **H** Broadleaved evergreen and mixed deciduous forests. **D** Peninsular Malaysia; Sumatra and W Java, Indonesia; Borneo.

8 HODGSON'S HAWK-CUCKOO *Hierococcyx nisicolor* 28–30cm Behaviour as Malaysian Hawk-Cuckoo. **V** Much as Malaysian Hawk-Cuckoo but with rapid *trrrrr-titititititrrrtrrr* at the end. **H** Broadleaved evergreen and mixed deciduous forests. **D** Breeds Nepal and E Himalayas to Myanmar, Thailand, Hainan. Winters S to Greater Sunda Islands.

9 BLACK CUCKOO *Cuculus clamosus* 30cm Nominate from black morph of Jacobin Cuckoo (Plate 65) by absence of crest and wing bar. **V** Mid-high, fluted, unhurried *fiu-fiuu-fiuu* of male together with hurried, rising *wo-wikke-wikke...* of female. **H** Canopy at forest edges, woodland, bushland, scrubland, gardens. **D** Widespread in Africa.

10 RED-CHESTED CUCKOO *Cuculus solitarius* 30cm Chestnut breast of male may be unbarred and brighter rufous. From Black Cuckoo by pale upperparts, grey (not rufous) chin and throat. Often heard but seldom seen. **V** Very high, loud, fluted, staccato, unhurried *it-will-rain it-will-rain* (emphasis on first syllable). **H** Forest belts, cultivations, suburbs. **D** Widespread in Africa.

11 LESSER CUCKOO *Cuculus poliocephalus* 26cm Note small size and wider spaced barring on underparts. **V** Loud, cheery *pretty-peel-lay-ka-beet* or *that's-your-choky-pepper... choky-pepper*. **H** Broadleaved and pine forests, also secondary growth, scrub. **D** SE Siberia and Japan to S China, N SE Asia, Himalayas. Winters India and E Africa.

12 SULAWESI CUCKOO *Cuculus crassirostris* 38cm Shy, perching in leafy area of mid-storey, often close to the trunk; usually seen singly or in pairs, presence revealed by calls. **V** Far-carrying, mellow *ho-hoo* or *hoo-oo-oo*; also cooing *ko ko ku ku*, last note barely audible. **H** Primary and tall secondary hill forest, forest edges. **D** Sulawesi, Indonesia.

13 INDIAN CUCKOO *Cuculus micropterus* 33cm Note wide-spaced barring on underparts and broad blackish subterminal bar to tail. **V** Loud, hollow, persistently repeated four-note whistle, often transcribed as *crossword puzzle* or *one more bottle*. **H** Deciduous, evergreen and secondary forests. **D** Widespread in S, SE and E Asia, and Indonesia, Borneo and Philippines.

14 MADAGASCAN CUCKOO *Cuculus rochii* 30cm No rufous female form known. Often shows less undertail barring than Lesser Cuckoo. As Lesser Cuckoo with dark uppertail coverts. **V** High liquid yelping rapid *woopwoopwoop-wop* (lower in pitch than Lesser Cuckoo). **H** Forest, woodland, bushy areas. **D** Breeds Madagascar. Winters in Africa.

15 AFRICAN CUCKOO *Cuculus gularis* 35cm Similar to Common Cuckoo. Bill base often rather extensively yellow, but this is variable and not diagnostic. **V** Mid-high, rather level *ku-koo ku-koo ku-koo*. **H** Hilly, dry, more or less wooded and bushy areas. **D** Widespread in Africa.

16 HIMALAYAN CUCKOO *Cuculus saturatus* 30–32cm Only safely separated from Common Cuckoo by voice. **V** Distinctive, resonant *poo-poo-poo-poo*, also muted, hoopoe-like *poo-poo poo-poo poo-poo poo-poo*. **H** Forests, forest edge and clearings, open woodland. **D** Breeds Himalayas to N China. Winters Indonesia, Philippines.

17 ORIENTAL CUCKOO *Cuculus optatus* 30–33cm Forages in tree canopy. **V** Low *hoop hoop hoop hoop*; also harsh *gaak-gaak gak-ak-ak-ak*. **H** Evergreen, mixed forests, thickets. **D** Breeds NE Asia and Taiwan. Winters Indonesia to Australia.

18 SUNDA CUCKOO *Cuculus lepidus* 26–29cm Secretive, best located by voice. Forages mainly in canopy. **V** Loud *kuk hoo hoo hoo*, also mellow *pu-pu*; female has bubbling call. **H** Primary hill and mountain forest. **D** Peninsular Malaysia; Greater Sunda Islands and S Maluku Islands, Indonesia.

19 COMMON CUCKOO *Cuculus canorus* 32–34cm Often perches horizontally with tail cocked and wings drooped. Two colour morphs of the female occur. **V** Male gives far-carrying *cuck-oo*; female a bubbling call. **H** Open broadleaved forest, secondary growth, reedbeds, more open areas on migration. **D** Breeds widely in Eurasia. Winters Africa, S and SE Asia.

BARN OWLS TYTONIDAE

20 GREATER SOOTY OWL *Tyto tenebricosa* 38–43cm Strictly nocturnal. Preys mainly on arboreal mammals; occasionally also birds and insects. **V** Grating, raspy *shrrrrr-ip* lasting up to 2 secs. **H** Rainforest; lowlands, up to 3,500m in New Guinea. **D** E and SE Australia, New Guinea.

21 LESSER SOOTY OWL *Tyto multipunctata* 32–38cm Strictly nocturnal. Diet comprises mainly arboreal mammals; occasionally also birds and insects. **V** Grating, raspy and drawn-out *chrrrrourr* lasting 1–2 secs. **H** Lowland rainforest. **D** NE Queensland.

22 MINAHASSA MASKED OWL *Tyto inexspectata* 30cm Generally seen singly or in pairs; very rarely observed. **V** Recorded as being similar to Sulawesi Masked Owl, but weaker and less deep. **H** Primary and lightly disturbed hill and lower montane forest. **D** N Sulawesi, Indonesia.

23 TALIABU MASKED OWL *Tyto nigrobrunnea* 32cm Behaviour presumed to be similar to that of Minahassa Masked Owl. **V** Only records are of a hissing. **H** Rainforest, selectively logged lowland forest. **D** Taliabu Island, Sula Islands, Indonesia.

24 MOLUCCAN MASKED OWL *Tyto sororcula* 29–31cm Little recorded; behaviour presumed similar to that of Australian Masked Owl. **V** Harsh, grating screech and softer twitter. **H** Rainforest, and primary and mature monsoon forest. **D** S Maluku Islands, Indonesia.

25 MANUS MASKED OWL *Tyto manusi* 35–45cm Nocturnal, roosting during daytime in tree hollow. Diet comprises mainly terrestrial mammals, birds, amphibians and insects. **V** Rasping call. **H** Rainforest. **D** Manus Island (Admiralty Islands).

26 GOLDEN MASKED OWL *Tyto aurantia* 38–32cm Golden-buff plumage is diagnostic. Nocturnal. Uses tree hollow for daytime roost. Diet comprises mainly terrestrial mammals, birds and insects. **V** Rasping call. **H** Rainforest. **D** New Britain (Bismarck Archipelago).

27 AUSTRALIAN MASKED OWL *Tyto novaehollandiae* 34–43cm Nocturnal. Roosts during daytime in tree hollows or caves. Feeds on terrestrial prey including mammals, birds, reptiles and large insects. **V** Repeated, rasping *schrrrrrree*. **H** Open forest, lightly wooded habitats. **D** Coastal E and N Australia.

28 SULAWESI MASKED OWL *Tyto rosenbergii* 43–46cm Rarely seen in daylight. Generally hunts in clearings, cultivation and forest edges. **V** Hoarse, dry, almost hissing screech. **H** Lightly wooded cultivation, tall dead trees in open country, grassland, forest edges, plantations, suburban areas. **D** Sulawesi, Indonesia.

29 RED OWL *Tyto soumagnei* 30cm Very similar to larger Western Barn Owl but more orange, with more distinct black spotting and barring to wings. **V** Very similar to Western Barn Owl. **H** Forest edges, rice paddies. **D** Madagascar.

30 WESTERN BARN OWL *Tyto alba* 33–35cm Appears strikingly white in flight. Nocturnal, although often seen hunting in daylight. **V** Shrill, hoarse *shrrreeeeee*. **H** Open country and farmland with scattered trees, woodland edges, also around human habitation. **D** Widespread in Africa and Europe.

31 AMERICAN BARN OWL *Tyto furcata* 32–38cm Nocturnal, although often hunts by day; quarters with slow and buoyant flight. **V** Shrill *shrrreeeee* and loud clicks. **H** Open areas, open woodland, around urban dwellings. **D** Widespread throughout North America, Central America, South America, Caribbean.

32 EASTERN BARN OWL *Tyto javanica* 33–35cm Appears strikingly white in flight. Nocturnal, although readily hunts during daylight. **V** Shrill, hoarse *shrrreeeeee*; also various chuckling, snoring and hissing sounds. **H** Cultivations and around habitations. **D** Widespread from S and SE Asia to Australasia.

33 ANDAMAN MASKED OWL *Tyto deroepstorffi* 33–35cm More buff-brown on face and underside than Eastern Barn Owl. Nocturnal but will hunt in daylight. **V** Shrill, hoarse *shrrreeeeee*; also chuckling, snoring and hissing sounds. **H** Cultivations and around habitations. **D** Andaman Islands.

34 ASHY-FACED OWL *Tyto glaucops* 26–35cm Nocturnal. Foraging behaviour similar to Western Barn Owl. **V** Screeching *criiissssh* and clicks, followed by hissing cry. **H** Open woodland, forest, scrub, caves, agricultural land, urban areas. **D** Hispaniola.

35 AFRICAN GRASS OWL *Tyto capensis* 35cm Separated from Marsh Owl (Plate 73) by long-legged build and absence of dark wrist patches on underwing. **H** Moist grass- and moorland. **V** High shriek, *frieeet*, and low, Mallard-like *kweh kweh kweh kweh*. **D** Sub-Saharan Africa.

36 EASTERN GRASS OWL *Tyto longimembris* 32–38cm In flight upperwing shows ochre patch at base of primaries. Hunts from just before dusk to mid-morning. **V** Generally silent, although a screech, said to be similar to that of Western Barn Owl, has been recorded. **H** Open grassland. **D** S and SE Asia, SE New Guinea, N and NE Australia, New Caledonia.

37 CONGO BAY OWL *Phodilus prigoginei 30cm* Note reddish upperparts and short tail. **V** Possible recording exists, of series of mournful, drawn-out whistles. **H** Open montane forest. **D** E DR Congo, NW Tanzania.

38 ORIENTAL BAY OWL *Phodilus badius* 29cm Nocturnal. Hides in tree hollows during the day. Flight rapid. **V** Series of whistles, eerie and with upward inflection. **H** Dense evergreen and broadleaved forest, mangrove edges. **D** SE Asia to Java, Indonesia; Borneo.

39 SRI LANKA BAY OWL *Phodilus assimilis* 29cm Nocturnal. Similar in behaviour and appearance to Oriental Bay Owl, but darker-backed. **V** Series of whistles, eerie and with upward inflection. **H** Dense evergreen and broadleaved forest, mangrove edges. **D** Sri Lanka.

1
2
3
4
5
6
7
8
9
10
12
15
13 ♀
14
16 ♀
11
♀
♂
♂
♀
18
19
17
♂
20
21
rufous morph
♂
19
♀ normal form
22
23
24
25
26
27
28
29
30
31
33
35
32
34
36
37
38
39

OWLS STRIGIDAE

1 GIANT SCOPS OWL *Otus gurneyi* 30cm Usually forages high in understorey. Roosts by day in tree trunk forks. **V** Loud *wokkk* or *waookk*, or rasping *ouwwkkk*. **H** Forest and forest edges, from lowlands to mountains; also recorded in small clumps of trees in grassland. **D** Philippines.

2 WHITE-FRONTED SCOPS OWL *Otus sagittatus* 25–28cm Little known; behaviour probably like other scops owls. **V** Hollow, whistled *hooo*, like that of Reddish Scops Owl but more abrupt. **H** Broadleaved evergreen forest. **D** Peninsular Malaysia.

3 REDDISH SCOPS OWL *Otus rufescens* 20cm Behaviour presumably much as other scops owls. **V** Hollow, whistled *hoooo*, fading at end. **H** Broadleaved evergreen forest. **D** Peninsular Malaysia and Greater Sunda Islands.

4 SERENDIB SCOPS OWL *Otus thilohoffmanni* 17cm Nocturnal. Roosts on horizontal branch, camouflaged among dead leaves. **V** Hollow *whoor-u*, repeated at 20-sec intervals. **H** Dense rainforest. **D** Sri Lanka.

5 SANDY SCOPS OWL *Otus icterorhynchus* 20cm Separated from African Scops Owl and Eurasian Scops Owl by rather unmarked and more white-spotted than black-streaked plumage. **V** Series of descending whistles. **H** Lowland forest, scrub. **D** W and C Africa.

6 SOKOKE SCOPS OWL *Otus ireneae* 15cm Very variable, like most scops owls; rufous, brown and grey morphs exist. Ears prominent when in alert posture. **V** Single fluted hoots, given less than a second apart. **H** Dry forests, miombo. **D** SE Kenya and NE Tanzania.

7 ANDAMAN SCOPS OWL *Otus balli* 19cm Nocturnal, very little else known. More often heard than seen. **V** Loud, abrupt *hoot....hoot-curroo*. **H** Trees in semi-open areas, cultivations and around human habitations. **D** Andaman Islands.

8 FLORES SCOPS OWL *Otus alfredi* 19–21cm Little recorded information. Known from very few specimens. **V** Series of clucking *whup* notes, may run together into fast, chuckling twitter. **H** Presumed to occur in montane forest. **D** Flores Island, Lesser Sunda Islands, Indonesia.

9 MOUNTAIN SCOPS OWL *Otus spilocephalus* 17–21cm Nocturnal. Said to hunt below canopy, close to ground. **V** Far-carrying, metallic *plew plew* with 5–10-sec interval between notes. **H** Dense broadleaved montane forest. **D** Nepal, N India, SE and E Asia, N Greater Sundas and N Borneo.

10 RAJAH SCOPS OWL *Otus brookii* 23cm Little recorded information. Probably rare. Feeds mainly on insects. **V** Monotonous, explosive *whaooo*. **H** Montane rainforest. **D** Sumatra, Indonesia; Borneo.

11 JAVAN SCOPS OWL *Otus angelinae* 20cm Secretive; little recorded information. **V** Calls include soft *wook-wook*, hissing *tch-tschschsch* and *poo-poo* alarm call. **H** Montane forest. **D** Java, Indonesia.

12 MENTAWAI SCOPS OWL *Otus mentawi* 20cm Occasionally perches on exposed branches. **V** Series of *po-po* notes followed by descending, single *po* notes. **H** Lowland forest, secondary growth. **D** Mentawai Archipelago, Indonesia.

13 INDIAN SCOPS OWL *Otus bakkamoena* 20–24cm Nocturnal. Daytime roost is often on a branch close to the trunk or in a tree hollow. Only reliably separated from the very similar Collared Scops Owl by call. **V** Subdued, frog-like *wuk* or *whut*, repeated at 4–6-sec intervals. **H** Forests and well-wooded areas, groves and trees around human habitations, riverine forest. **D** S Pakistan, C and S India, Sri Lanka.

14 COLLARED SCOPS OWL *Otus lettia* 23–25cm Very similar to Indian Scops Owl; only reliably separated by call. **V** Soft, downward inflected *buuo*, repeated every 12–20 secs; also chattering when alarmed. **H** Forests, well-wooded areas. **D** W Himalayas to S China, N Indochina, Taiwan.

15 JAPANESE SCOPS OWL *Otus semitorques* 23–25cm Nocturnal. Behaviour similar to that of Indian Scops Owl. **V** Deep *whook* repeated at long intervals, also *koo* or *kwe* and repeated *kwee-kwee* or *pwe-u pew-u*. **H** Forests on hills and plains, in winter also in parks, gardens. **D** NE China; Korean Peninsula; Japan; Ryukyu Islands; Ussuriland, Russia.

16 SUNDA SCOPS OWL *Otus lempiji* 20cm Hunts from perch, dropping onto insects, occasionally small birds. **V** Soft upward-inflected *wooup*. **H** Secondary growth evergreen and deciduous forest, forest edge. Also open areas with scattered trees, parks, well-wooded gardens, tree-lined urban areas. **D** S Thailand; Malaya; Sumatra and Java, Indonesia; Borneo.

17 PALAWAN SCOPS OWL *Otus fuliginosus* 20cm Keeps to dense understorey, low to the ground. **V** Deep, harsh, growling *krarr-kruarr*, *wach grarhrhrh* or *wach waaarwwwhhh*. **H** Forest, secondary growth, trees in mixed cultivation in lowland. **D** Palawan, W Philippines.

18 PHILIPPINE SCOPS OWL *Otus megalotis* 23–28cm Usually seen singly or in pairs in the forest understorey. **V** Harsh *oiik oiik oiik oiik*. **H** Forest, forest edges. **D** N Philippines.

19 EVERETT'S SCOPS OWL *Otus everetti* 22–23cm Occurs singly or in pairs; presumed to feed on invertebrates. **V** Loud, repeated *wkuach*. **H** Humid lowland forest, forest edges, secondary growth. **D** S Philippines.

20 NEGROS SCOPS OWL *Otus nigrorum* 20cm Little recorded information; presumably similar to other Philippine Otus species. **V** Series of rapid *quick-quick* notes. **H** Dense lowland, montane tropical forest. **D** Negros and Panay, W Philippines.

21 WALLACE'S SCOPS OWL *Otus silvicola* 23–25cm Generally seen singly or in pairs; perches high, often in concealed position. **V** Gruff *rrow* and steady series of *whumph* notes. **H** Primary and tall secondary hill and montane forest, forest edges, degraded forest, cultivation, around human habitation. **D** Lesser Sunda Islands.

22 MINDANAO SCOPS OWL *Otus mirus* 19–20cm Little recorded information, apparently rare. **V** Far-carrying, whistled *paww piaww*. **H** Montane forest. **D** Mindanao, S Philippines.

23 LUZON SCOPS OWL *Otus longicornis* 18cm Can be approachable; usually perches in understorey. **V** Loud, far-carrying *whoo-hooo whoo-hooo*. **H** Submontane and montane mossy forest, mixed pine forest, rainforest. **D** Luzon, N Philippines.

24 MINDORO SCOPS OWL *Otus mindorensis* 17cm Quite common in suitable habitats. Roosts by day in tree hollows or dense foliage. **V** Soft, delicate *po-wo* or *wo-wo*. **H** Montane forest. **D** Mindoro, NC Philippines.

25 PALLID SCOPS OWL *Otus brucei* 21cm Nocturnal, although recorded hunting before dusk and by day. Roosts in rock crevices, tree hollows or in thick foliage. **V** Hollow, low-pitched *boo...boo...boo...* given at about 1-sec intervals. **H** Open areas with scattered trees and bushes, rocky foothills in semi-desert. **D** Breeds SW and C Asia. Winters NW India.

26 AFRICAN SCOPS OWL *Otus senegalensis* 20cm Like Eurasian Scops Owl, very variable in colour. Ear-tufts not always obvious. **V** Very high *prriur prriur...* at 4-sec intervals. **H** Wooded habitats including gardens. **D** Widespread in Africa.

27 ARABIAN SCOPS OWL *Otus pamelae* 16–19cm Nocturnal. Best distinguished from similar Eurasian Scops Owl by call. **V** Calls *da-pwoorp*, repeated at 12–20-sec intervals. **H** Hilly wooded areas. **D** Saudi Arabia, Oman and Yemen.

28 EURASIAN SCOPS OWL *Otus scops* 19–20cm Nocturnal. Often roosts on branch up against tree trunk. **V** Monotonous, plaintive whistle *tyuu* repeated about every 3 secs, often confused with frog or toad calls, although these usually briefer. **H** Broadleaved woodland, copses, orchards, groves, churchyards, large gardens. **D** Breeds W and C Europe. Winters in Africa.

29 CYPRUS SCOPS OWL *Otus cyprius* 19–20cm Very like Eurasian Scops Owl in appearance and behaviour. **V** Plaintive two-note whistle repeated about every 3 secs. **H** Broadleaved woodland, cultivated areas with trees, gardens. **D** Cyprus.

30 ORIENTAL SCOPS OWL *Otus sunia* 18cm Nocturnal. Hunts from perch or in flight, usually along forest edges. **V** Toad-like *wuk-tuk-tah*, *wut-chu-chraaii* or similar. **H** Broadleaved evergreen forest, mixed deciduous forest, forest edges, clearings, sometimes gardens. **D** Widespread in S, SE and E Asia and N Greater Sundas.

31 SOCOTRA SCOPS OWL *Otus socotranus* 21cm Nocturnal. Very like Pallid Scops Owl in appearance and behaviour. **V** Hollow, low-pitched whistling trill. **H** Open wooded areas, parks, large gardens. **D** Socotra, Gulf of Aden.

32 MOLUCCAN SCOPS OWL *Otus magicus* 21cm Usually encountered singly or in pairs perched high in trees; feeds on invertebrates and small vertebrates. **V** Harsh, raven-like croak. **H** Primary forest, coastal swamp forest, secondary forest. **D** Maluku Islands and Lesser Sunda Islands.

33 RINJANI SCOPS OWL *Otus jolandae* 22cm Presumed to be similar to other scops owls, resting in cover by day; calls from early evening and throughout the night. **V** Single, clear whistle. **H** Forest, from lowland to foothills, forest patches in more open country. **D** Lombok, Lesser Sunda Islands, Indonesia.

34 SULA SCOPS OWL *Otus sulaensis* 20cm Little recorded information; thought to hunt for insects and other invertebrates and small vertebrates. **V** Series of rapid, resonant notes or a long, churring song. **H** Primary and disturbed secondary forest. **D** Sula Islands, Indonesia.

35 SIAU SCOPS OWL *Otus siaoensis* 19cm Little recorded information. Known from a single specimen collected in 1866. **V** Unrecorded. **H** Forest. **D** Siau island, Sulawesi, Indonesia.

36 MANTANANI SCOPS OWL *Otus mantananensis* 18cm Hunts at woodland edges and in clearings. **V** Goose-like honk followed by three gruff, lower-pitched, gruff notes. **H** Lowland and foothill forest, woodland, plantations. **D** Mantanani island, Borneo; islets S of Palawan.

37 RYUKYU SCOPS OWL *Otus elegans* 20cm Nocturnal. Hunts along forest edges. **V** Repeated, hoarse, cough-like *uhu*, *kuru* or *u-kuruk*, or soft *poo-pup* or *pooo poo-pup*. **H** Dense evergreen forest. **D** Islands S of Japan to Philippines.

38 SULAWESI SCOPS OWL *Otus manadensis* 21cm Usually seen singly or in pairs, generally hiding in dense foliage. **V** Clear, plaintive whistle with rising inflection. **H** Primary and tall secondary lowland and hill forest, forest edges, lightly wooded cultivation, scrub, remnant forest patches. **D** Sulawesi, Indonesia

39 SANGIHE SCOPS OWL *Otus collari* 19–20cm Little recorded information. **V** High-pitched, downslurred, fluty whistle, *kleeeeer*. **H** Forest, mixed plantations, secondary growth, agricultural areas with trees and bushes. **D** Sangihe Islands, Indonesia.

40 BIAK SCOPS OWL *Otus beccarii* 24–25cm Nocturnal, roosts well-hidden in dense foliage in daytime. Feeds mainly on insects and other invertebrates. **V** Barking, croaking, crow-like *rrrruk*. **H** Dense, forested habitats. **D** Biak island, NW New Guinea.

1
2
3
4
5
6
rufous morph
7
8
7
grey morph
13
grey morph
normal form
brown morph
14
grey morph
15
9
10
11
12
rufous morph
18
19
20
21
22
23
rufous morph
16
17
grey morph
grey morph
32
brown morph
27
26
25
24
28
grey morph
29
30
rufous morph
tawny morph
31
38
40
rufous morph
36
rufous morph
37
grey morph
33
34
35
rufous morph
brown morph
39
grey morph

OWLS *CONTINUED*

1 SEYCHELLES SCOPS OWL *Otus insularis* 20cm No other scops owl on Seychelles except vagrant Eurasian Scops Owl, which has very different voice. **V** Very low, frog-like croaking. **H** Mixed forest with tall trees at higher altitudes. **D** Seychelles.

2 SIMEULUE SCOPS OWL *Otus umbra* 18cm Poorly studied, with little recorded information. **V** Two steady notes followed by higher, rising, inflected note, *pook-pook-pupook*; female gives a long whine. **H** Remnant forest patches, forest edges, clove plantations. **D** Simeulue island, Indonesia.

3 ENGGANO SCOPS OWL *Otus enganensis* 20cm Little recorded information. **V** Harsh croak, reported as similar to Moluccan Scops Owl (Plate 68). **H** Little recorded, presumed to be wooded areas, forest edges. **D** Enggano island, Indonesia.

4 NICOBAR SCOPS OWL *Otus alius* 19–20cm Behaviour probably similar to that of other SE Asian scops owls. **V** Repeated, melancholic, rising *ooo-m*. **H** Coastal forest. **D** Nicobar Islands.

5 PEMBA SCOPS OWL *Otus pembaensis* 20cm Not recently heard or seen. **V** Short, plaintive single hoots, given several secs apart. **H** Plantations with densely foliaged trees. **D** Pemba Island (Zanzibar), off NE coast of Tanzania.

6 KARTHALA SCOPS OWL *Otus pauliani* 20cm Only scops owl on Grande Comore. **V** Sustained, rather hurried series of *uhr* notes. **H** Areas with large trees. **D** Grand Comore island, Comoros.

7 ANJOUAN SCOPS OWL *Otus capnodes* 20cm Occurs in dark and brown morphs. No other scops owl on Anjouan. **V** High series of 3–4 fluted *tjuw* notes. **H** Mainly montane natural forest, but locally in plantations. **D** Anjouan island, Comoros.

8 MOHELI SCOPS OWL *Otus moheliensis* 20cm No other small owl on Mohéli; known as rufous and brown forms. **V** Short shriek *sreew*, given at irregular intervals. **H** Dense undisturbed and degraded forest. **D** Mohéli island, Comoros.

9 MAYOTTE SCOPS OWL *Otus mayottensis* 25cm Rufous form also known. No other small owl on Mayotte. **V** Calm series of 4–5 *hoot* notes. **H** Any wooded habitat. **D** Mayotte island, Comoros.

10 TOROTOROKA SCOPS OWL *Otus madagascariensis* 20cm No other small owl in W Madagascar. **V** Series of 3–6 barking notes. **H** Any wooded habitat. **D** W Madagascar.

11 RAINFOREST SCOPS OWL *Otus rutilus* 25cm Grey and rufous morphs known. No other small owl in E Madagascar. **V** Level series of 5–7 hooted notes, given rapidly yet well-separated. **H** Any wooded habitat at any altitude; prefers humid forest. **D** E Madagascar.

12 SAO TOME SCOPS OWL *Otus hartlaubi* 20cm Only small owl on São Tomé. Might occur on Principe. **V** Single plaintive downslurred hoots, given at long intervals. **H** Forests, plantations. **D** São Tomé, Gulf of Guinea.

13 FLAMMULATED OWL *Psiloscops flammeolus* 15–17cm Nocturnal. Dark eyes. Variable. **V** Hollow, low *hoop* singly or repeated at 2–3 sec intervals. **H** Mountain pine, or pine mixed with oak or aspen. **D** SW Canada to S Mexico.

14 EASTERN SCREECH OWL *Megascops asio* 16–25cm Nocturnal. Yellow eyes and yellowish bill. Variable; most are grey or rufous morphs. **V** Low-pitched, descending whinny; also long whistled trill. **H** Woodland, with clearings, orchards, parks, suburban gardens. **D** SE and SC Canada to NE Mexico.

15 WESTERN SCREECH OWL *Megascops kennicottii* 25cm Finer barred below than Eastern Screech Owl; note black bill. **V** Very high accelerated *ooh-ooh-ooh-oohohdear*. **H** Dry woodland. **D** W and S Alaska to S Mexico.

16 BALSAS SCREECH OWL *Megascops seductus* 25cm Dark but not black eyes. **V** High bouncing descending *ooh-ooh-ohohohrorr*. **H** Tall thorn, high cactus, dry open country. **D** SW Mexico.

17 PACIFIC SCREECH OWL *Megascops cooperi* 25cm In range of Whiskered Screech Owl and Vermiculated Screech Owl, but not in the same habitat. **V** High level loud *datdat*... or drawn-up *rrrorrorororororor*. **H** Dry open woodland. **D** S Mexico to NW Costa Rica.

18 WHISKERED SCREECH OWL *Megascops trichopsis* 16–18cm Nocturnal. Yellow eyes. Underparts appear spotted. **V** Steady *boo boo boo boo boo boo*...; also soft, descending *oooo*. **H** Dense oak and oak-pine forests. **D** SE Arizona to NC Nicaragua.

19 TROPICAL SCREECH OWL *Megascops choliba* 25cm Shares range only with Whiskered Screech Owl and Vermiculated Screech Owl (both forest species), separable only by voice and habitat. **V** High loud hooted *prrrruowHów*. **H** Open woodland, suburban areas. **D** Costa Rica to NE Argentina.

20 WEST PERUVIAN SCREECH OWL *Megascops roboratus* 21cm Like Tropical Screech Owl but streaks on breast broader at lower tip, drop-shaped. Cap blackish. Pale nuchal collar. **V** Harsh rising trill. **H** Dry woods, bushes with cactus, from coastal plains to foothills. **D** SW Ecuador, NW Peru.

21 KOEPCKE'S SCREECH OWL *Megascops koepckeae* 21cm Very similar to Tropical Screech Owl. **V** Fast, laughing trill, notes slowing towards the end. **H** Evergreen, high-altitude forest. **D** NW and C Peru.

22 BARE-SHANKED SCREECH OWL *Megascops clarkii* 25cm Rufous, lacks clear facial disk. Yellow iris. Crown and underparts have heavy bar-crossed streaks. Line of white spots on scapulars. Bare tarsi. **V** Call three deep *wouk* notes. **H** Montane forest. **D** Costa Rica to NW Colombia.

23 BEARDED SCREECH OWL *Megascops barbarus* 18cm No ear tufts. Separated from pygmy owls by different patterning of upperparts. **V** Very high rhythmic *how* (1 every 3 secs) or (accelerated) trill. **H** Mixed pine forest. **D** S Mexico to N Guatemala.

24 RUFESCENT SCREECH OWL *Megascops ingens* 28cm Rather uniform, vermiculated cinnamon and dark brown (or rufous morph). Eyes yellowish brown to dark brown. Faint streaking on underparts. **V** Long series of rapid *toototoo*... notes. **H** Dense humid cloud forest, scrub on steep slopes. **D** Venezuela, W Colombia to Bolivia.

25 CINNAMON SCREECH OWL *Megascops petersoni* 21–22cm Noctunal. Remains hidden in cover in daytime. Feeds mainly on insects. **V** Rapid-fire *boo-boo-boo-boo-boo-boo-boo*, pitch rising and falling during delivery. **H** Forested E slopes of Andes; mostly 1,600–2,500m. **D** S Ecuador, N Peru.

26 CLOUD-FOREST SCREECH OWL *Megascops marshalli* 23cm Dark rufous with contrasting streaks, bars and white markings. Facial disc rufous, framed black. Dark eyes. White spots around nape. **V** Series of *eee* notes. **H** Cloud forest with dense undergrowth. **D** C and S Peru.

27 TAWNY-BELLIED SCREECH OWL *Megascops watsonii* 22cm Plumage rather uniform, mostly vermiculated. Iris yellowish brown to blackish brown. Tawny-buff lower underside, finely streaked black; facial disc dark-edged. Several morphs, from brownish cinnamon to rufous and blackish brown. **V** Long call, accelerating, rising, then falling in pitch and volume. **H** Lowland rainforest, clearings and forest edges, gallery forest. **D** Amazonia.

28 MIDDLE AMERICAN SCREECH OWL *Megascops guatemalae* 25cm Unfeathered toes. Note inconspicuous ear tufts; little streaking below. **V** Very high trilling rising or level *drrr*... (6–7s). **H** Forest, scrubby woodland. **D** NW and NE Mexico to Nicaragua.

29 VERMICULATED SCREECH OWL *Megascops vermiculatus* 21cm Finely patterned, rather uniform. Yellow iris. Brown and rufous morphs. **V** Long trill growing in volume and ending abruptly. **H** Topical forest. **D** E Nicaragua, Costa Rica, W Panama.

30 FOOTHILL SCREECH OWL *Megascops roraimae* 21cm Facial disk indistinct. Pale brows. Underside finely and uniformly patterned. Yellow iris. **V** Long trill, falling in pitch and fading at end. **H** Humid tropical forest. **D** NE Colombia to N Venezuela, Guyana, Suriname and adjacent Brazil; also E slope of Andes from Colombia to Bolivia.

31 CHOCO SCREECH OWL *Megascops centralis* 25cm Naked toes, rather long face, inconspicuous ear tufts, little streaking below. **V** Very high, short trill, falling sharply in pitch at end. **H** Forest, scrubby woodland. **D** Panama to W Ecuador and W Colombia.

32 YUNGAS SCREECH OWL *Megascops hoyi* 23–24cm Upperparts brownish or rufous with white streaks and black and ochre spots. Grey face. Throat and upper breast grey with transverse dark bars. Rest of underparts with heavy streaks crossed by finer bars. Singly or in pairs. Roosts motionless in trees by day, ear tufts erect. **V** Low-pitched rolled trill *rrrrr-kooo* or *rrrrr-cuckOO*. **H** Forests, woods, gardens, savannahs. **D** S Bolivia to NW Argentina.

33 BLACK-CAPPED SCREECH OWL *Megascops atricapilla* 24cm Like Tropical Screech Owl but crown dark, dense streaks arrow-shaped. Brown or dull yellow iris. Dark and rufous morphs exist. **V** Very long, low-pitched trill, soft at beginning and end. **H** Rainforest, dense undergrowth, forest edges. **D** E Paraguay, SE Brazil.

34 LONG-TUFTED SCREECH OWL *Megascops sanctaecatarinae* 27cm Head and upperparts chestnut with black streaks on head, brown streaks on back. Breast chestnut with heavy black streaking, belly whitish with chestnut bars and streaks. Short ear-tufts, yellow irises. **V** Song *woo ii, ii, ii*..., with vibrato on first note. **H** Forests. **D** SE Brazil, NE Argentina, Uruguay.

35 PUERTO RICAN SCREECH OWL *Megascops nudipes* 23–25cm Nocturnal, roosts by day in dense vegetation or in tree hole. **V** Short trill or chatter, also whoop or maniacal laugh. **H** All types of forest and woodland. **D** Puerto Rico.

36 WHITE-THROATED SCREECH OWL *Megascops albogularis* 25cm No ear-tufts. Dark brown dotted and barred buff. Dark face. White throat. Yellow iris. **V** Very long series of *cheroo* and *choo* notes. **H** Humid montane and stunted forest, clearings and forest edges, scattered trees. **D** Venezuela to Bolivia.

37 SANTA MARTA SCREECH OWL *Megascops gilesi* 24–25cm The only screech owl species at that altitude, in that range. Nocturnal. Roosts in dense foliage or tree hole. **V** Piping, fluty trill lasting 1–2 secs. **H** Humid forested slopes; mainly 1,800–2,500m. **D** Sierra Nevada de Santa Marta, N Colombia.

38 PALAU OWL *Pyrroglaux podargina* 22cm Sole owl on Palau. **V** Excitable repeated short hoots or coos. **H** All types of forest. **D** Palau, WC Pacific Ocean.

39 BARE-LEGGED OWL *Margarobyas lawrencii* 20–23cm Nocturnal, roosts by day in tree holes or caves. **V** Low, repeated *cu-cu-cu-cucucu*. Females utter harsh scream. **H** Wooded areas. **D** Cuba.

40 NORTHERN WHITE-FACED OWL *Ptilopsis leucotis* 30cm Note orange eyes and pale grey feathering. **V** Mid-high cooing *bbbbb-woouow* (*bbbbb* as slow trill). **H** Large trees in woodland. **D** Senegal and Gambia to Somalia, S to N DR Congo, Uganda, C Kenya.

41 SOUTHERN WHITE-FACED OWL *Ptilopsis granti* 30cm Very similar to Northern White-faced Owl. **V** Understated short trill with longer final note. **H** Wooded areas with mature trees. **D** Gabon to S Uganda and SW Kenya, S to Namibia and South Africa.

brown morph
7
dark morph
rufous morph
8
brown morph
1
2
3
4
5
6
rufous
13
grey
rufous
14
grey
15
grey morph
14
rufous/ brown morph
15
9
10
11
12
brown
grey/ brown
16
17
18
21
23
24
20
22
19
rufous
cinnamon
28
grey
29
30
grey morph
25
27
brown morph
31
26
35
rufous morph
39
32
rufous morph
rufous morph
brown morph
33
grey morph
37
41
34
36
38
40
rufous morph
grey morph

OWLS *CONTINUED*

1 SNOWY OWL *Bubo scandiacus* 53–65cm Usually hunts at dusk or dawn, but by day too during summer. **V** Male gives booming *goo goo* or *gawh gawh*; female similar but higher pitched. When alarmed, male utters cackling *kre-kre-kre*, female gives loud whistling or mewing notes. **H** Tundra with rocks or hummocks and low vegetation. In winter, grassland, marshes, fields. **D** Breeds in Arctic North America and Eurasia. Winters further south.

2 GREAT HORNED OWL *Bubo virginianus* 53–58cm Mainly nocturnal. Very variable in plumage tone. **V** Deep, muffled *hooo hoo hoo*; during courtship, female answers with barking call. **H** Varied, including mountains, forests, parks, suburbs. **D** North America S of Arctic.

3 LESSER HORNED OWL *Bubo magellanicus* 55cm Large and hefty. Prominent ear-tufts. Plumage tones variable. Bright yellow iris. Upperparts irregularly patterned blackish brown, buffy and cinnamon. Underparts buffy barred blackish. **V** Powerful low-pitched hooted *who whowho... who who*, far-carrying. **H** Tropical forest, woods, scrub, grasslands, mangroves, parks, gardens. **D** C Peru to Tierra del Fuego.

4 EURASIAN EAGLE-OWL *Bubo bubo* 58–71cm Usually nocturnal or crepuscular, although in summer often hunts during daylight hours. **V** Male gives deep, muffled *HOO-o* or *BOO-ho*; female call higher pitched and more hoarse. Female also utters barking scream. When alarmed, gives shrill *ka-ka-kaKAYu*. **H** Rocky areas, cliffs, gorges and caves in grassland, open woodland, semi-desert, also locally around ruined buildings and farmland. **D** Widespread throughout Eurasia.

5 INDIAN EAGLE-OWL *Bubo bengalensis* 50–56cm Smaller than Eurasian Eagle-Owl, eyes orange-yellow. **V** Male gives deep two-note hoot, the second syllable longer. Female's call similar but higher pitched. **H** All habitat types from forest to open rocky areas and close to habitation. **D** Indian subcontinent.

6 PHARAOH EAGLE-OWL *Bubo ascalaphus* 60–75cm Pale and sandy with bold blackish markings, light orange eyes. **V** Single low hoot that falls in pitch and fades away; female's call higher pitched. **H** Rocky areas, cliffs, gorges and caves in semi-desert. **D** NW Africa to Arabian Peninsula.

7 CAPE EAGLE-OWL *Bubo capensis* 55cm Separated from Spotted Eagle-Owl by orange (not yellow) eyes, boldly blotched (not finely barred) upper- and underparts, coarse, lengthwise streaking on crown (not fine barring, concentric to facial disc). Often rests in full view on rocks or in trees by day. **V** Low, hooted *hooh hootooh*. **H** Rocky valleys and cliffs with some shrub and grass cover. **D** S, SE and E Africa.

8 SPOTTED EAGLE-OWL *Bubo africanus* 45cm Mainly nocturnal, usually feeds by dropping on prey from perch, also recorded catching bats in flight. **V** Soft *hoo-hoo-hoo* and fluty, nasal *wheeoo*. **H** Open woodland, rocky hills and ravines, locally around human habitation. **D** Africa S of Equator.

9 GREYISH EAGLE-OWL *Bubo cinerascens* 50cm Similar to Spotted Eagle-Owl but greyer-toned, and with dark not yellow eyes. **V** Low, short pressed-out *oohooh* calls, about 6 secs apart. **H** Rocky ravines, dry open woodland. **D** Senegal and Gambia to Ethiopia and N Kenya.

10 FRASER'S EAGLE-OWL *Bubo poensis* 55cm Note overall barring and boldly marked head. Separated from Shelley's Eagle-Owl by paler face, crown and upperparts. **V** High, crescendoing, plaintive *uuuuuuuh* and very low, pigeon-like rolling *rrrrru rrrrru*. **H** Forests. **D** Sierra Leone to Uganda, C DR Congo and N Angola.

11 USAMBARA EAGLE-OWL *Bubo vosseleri* 55cm Note overall striped markings and blue eyelids. **V** Low, fast, rolling chuckle. **H** Forests. **D** NE Tanzania.

12 SPOT-BELLIED EAGLE-OWL *Bubo nipalensis* 63cm Generally nocturnal. Hides in shady forest trees by day. **V** Deep *hoo...hoo*, and rising and falling mournful scream. **H** Dense broadleaved woodland. **D** S India and Sri Lanka, Himalayas to SE Asia.

13 BARRED EAGLE-OWL *Bubo sumatranus* 40–46cm Nocturnal and crepuscular, pounces on prey from perch, hops on the ground. **V** Loud *whooa-who whooa-who*, also quacking *gagagagogogo*. **H** Broadleaved evergreen forest, forest edge, clearings, mature plantations. **D** Peninsular Malaysia and Greater Sunda Islands.

14 SHELLEY'S EAGLE-OWL *Bubo shelleyi* 65cm Very large. Note dark, barred appearance. **V** Long, slightly downslurred, cat-like *whaaooa*. **H** Lowland forest. **D** Sierra Leone to Ghana, Cameroon to E DR Congo and Gabon.

15 VERREAUX'S EAGLE-OWL *Bubo lacteus* 65cm Unmistakable by size, grey colouring and overall fine barring. **V** Extremely low *prooh prooh prooh* (each note very short). **H** Well-wooded streams, other areas with some large trees. **D** Widespread in sub-Saharan Africa.

16 DUSKY EAGLE-OWL *Bubo coromandus* 58cm Usually in pairs. During the day hides in the shady areas of trees. **V** Deep, hollow *WO Wo wo-o-o-o*. **H** Trees with dense foliage, near water. **D** Pakistan and S Thailand, also E China.

17 AKUN EAGLE-OWL *Bubo leucostictus* 44cm Note yellow eyes. Underparts barred and blotched brown. Overall rather dusky. **V** Song a low, fast chuckle, call a high-pitched yowl. **H** Primary and secondary forest, gallery forest, cultivation with tall trees. **D** Sierra Leone to Ghana, Nigeria to E DR Congo and N Angola.

18 PHILIPPINE EAGLE-OWL *Bubo philippensis* 50–51cm Little recorded information. Reported roosting in trees by day; presumed to prey on small mammals or birds. **V** Deep, resonating *hoo-hoo-hoo*... **H** Forest and forest edges, often near water. **D** Philippines.

19 BLAKISTON'S FISH OWL *Bubo blakistoni* 60–72cm Feeds mainly on fish, crabs or frogs, dropping on prey from perch, hunts mostly at night or dusk. **V** Deep *boo-bo-voo*, *shoo-hoo* or *foo-fooroo*. **H** Forested rivers. **D** NE China, N Japan.

20 BROWN FISH OWL *Ketupa zeylonensis* 50–57cm Crepuscular and nocturnal. Swoops on fish from perch. **V** Deep *boom boom* or *boo-o-boom*, subdued *hu-who-hu* and harsh *we-aaah*. **H** Tree-lined rivers or lakes. **D** Middle East to SE Asia.

21 TAWNY FISH OWL *Ketupa flavipes* 61cm Crepuscular and nocturnal. Catches fish in talons, swooping from waterside perch. **V** Deep *whoo-hoo* and cat-like mewing. **H** Dense broadleaved forest by rivers and streams. **D** Himalayas to SE Asia.

22 BUFFY FISH OWL *Ketupa ketupu* 38–44cm Mainly nocturnal. Catches fish by swooping from perch, or wading in shallow water. **V** Loud, monotonous *kootookookootook*..., also ringing *pof pof pof*, musical *to-whee to-whee* and various hisses, mews and shrieks. **H** Forested streams and mangroves. **D** SE Asia to Java, Indonesia; Borneo.

23 PEL'S FISHING OWL *Scotopelia peli* 70cm Unmistakable. **V** Extremely low, sinister *hoop ho hoop ho*. **H** Well-forested streams and lakes. **D** Widespread in sub-Saharan Africa.

24 RUFOUS FISHING OWL *Scotopelia ussheri* 50cm Separated from Pel's Fishing-Owl by unbarred upperparts. **V** Very low grunting hoots. **H** Forest along rivers and lakes, mangroves. **D** Sierra Leone to Ghana.

25 VERMICULATED FISHING OWL *Scotopelia bouvieri* 48cm Note round head with indistinctly marked facial disk. Rather irregular blotchy streaks on upperside and underside. **V** Low, reverberating, foreboding *hoop* or *hoop-hoop*. **H** Rivers and pools in forest. **D** Nigeria to E DR Congo and N Angola.

26 SPOTTED WOOD OWL *Strix seloputo* 45–47cm Hunts from perch, preying mainly on small rodents, small birds and insects. **V** Deep *who*, usually preceded by rolling *huhuhuwhuwhu*; also resonant, rising *hoop-hoong* and a deep growling. **H** Broadleaved evergreen forest edge, logged forests, plantations, wooded parks, occasionally mangroves. **D** SE Asia; Sumatra and Java, Indonesia; Philippines.

27 MOTTLED WOOD OWL *Strix ocellata* 40–48cm Mainly nocturnal. By day rests in dense, shady trees. **V** Eerie, quavering *whaa-aa-aa-aa-ah*, also mellow hoot and occasional harsh screech. **H** Open woodland, thick foliaged trees on village outskirts, groves. **D** India and Myanmar.

28 BROWN WOOD OWL *Strix leptogrammica* 47–53cm Nocturnal. Shy, resting by day in dense foliage. **V** Low, hollow *tok tu-hoo tok-tu-hoo*; also eerie shrieks and chuckles. **H** All types of forest, from coasts to mountains. **D** Widespread in S, SE and E Asia, Greater Sundas and Borneo.

29 TAWNY OWL *Strix aluco* 37–39cm Rufous, brown and grey morphs occur. Nocturnal. Hunts from perch or by quartering over grassland, marshland or bushes, also chases prey in flight. **V** Shrill *kewick* and haunting *hooooo...hu huhuhu hoooooo*, also tremulous low trill. **H** Open forest or woodland, agricultural areas with trees, parks, large gardens. **D** W Eurasia.

30 HIMALAYAN OWL *Strix nivicolum* 37–39cm Nocturnal, during the day rests in foliage near tree trunk. Hunts mainly from perch, dropping onto small mammals, birds, amphibians, insects and worms. **V** Haunting *HU-HU* and shrill *kewick*. **H** Broadleaved evergreen and coniferous forests. **D** Himalayas to E Asia.

31 OMANI OWL *Strix butleri* 35–38cm Nocturnal. Hunts mainly from perch, often near tracks or roads. **V** Four-note hooting song. **H** Arid mountains and gorges in rocky deserts with nearby water and trees. **D** N Oman, E Iran.

32 DESERT OWL *Strix hadorami* 35–38cm Nocturnal. Behaviour similar to that of Omani Owl. **V** Hooting call comprises single long note followed by two short double notes. **H** Similar to that of Omani Owl. **D** Middle East and N Egypt.

1
♀
♂
2
3
4
5
6
7
8
9
10
11
12
13
14
15
16
17
18
19
20
21
22
23
24
25
26
27
28
grey
morph
29
rufous
morph
30
31
32

OWLS *CONTINUED*

1 SPOTTED OWL *Strix occidentalis* 41–48cm Nocturnal. Belly and flanks spotted and barred. **V** Barking *whoop-hu-hu-hooo.* **H** Wooded canyons, damp forests. **D** W North America, W Central America.

2 BARRED OWL *Strix varia* 48–51cm Mainly nocturnal. Barred breast and streaked belly and flanks. **V** Clear *hoo hoo ho-ho, hoo hoo ho-hoooooaw*, often transcribed as *who cooks for you, who cooks for you all.* **H** Coniferous and mixed woodland near open country. **D** SE Alaska and S Canada to SE, SC USA.

3 CINEREOUS OWL *Strix sartorii* 40–48cm Dark-eyed owl with greyish-buff plumage and yellow bill. Nocturnal. Roosts during daytime in dense foliage. **V** Mellow hooting *oh, oh, hu, Hu-Hu-hu-hu.* **H** Montane forest, cloud forest. **D** Mexico.

4 FULVOUS OWL *Strix fulvescens* 45cm Separated from Cinereous Owl by barred mantle. **V** Low barking *oh wow oh wow oh ooh-ooh.* **H** Forest. **D** S Mexico to El Salvador.

5 RUSTY-BARRED OWL *Strix hylophila* 36cm Very like Rufous-banded Owl but paler on head, back and breast, finer scaly pattern. **V** Raucous *hoo, hoo, HOO* followed by descending *HOO...hoo, hoo, hoo-hoo-hoo.* **H** Humid forest edge, woods, tropical mountain forest and lowlands, secondary growth, sometimes near human habitation. **D** SE Brazil, E Paraguay, NE Argentina.

6 CHACO OWL *Strix chacoensis* 40cm Upperparts barred dark greyish brown and buffy-white. Face paler, finely barred whitish and dark brownish grey. Underparts whitish barred dark greyish brown, becoming buffy, barred greyish brown. Iris brown. **V** Call *Crew crew crew (crew crew) CRAW CRAW.* **H** Dry chaco woods with cactus and scrub, Patagonian Andes' woods and forest. **D** S Bolivia, Paraguay, W Argentina.

7 RUFOUS-LEGGED OWL *Strix rufipes* 37cm Dark brown barred white with cinnamon tinge on belly. Thighs reddish cinnamon. **V** Loud and hoarse disyllabic *poorr, poorr.* **H** Woods, savannahs. **D** C Chile to Tierra del Fuego.

8 URAL OWL *Strix uralensis* 50–62cm Nocturnal, although often active during daytime. Can be very aggressive, especially when tending young. Hunts mostly from perch. **V** Deep *whooho... woohoo-uwoohoo*, also a rising *hoohoohoohoohoohoo*, falling away at end. When alarmed utters explosive, barking *waff.* **H** Coniferous, mixed or deciduous forests with clearings and nearby fields or bogs. **D** Widespread in Eurasia.

9 PERE DAVID'S OWL *Strix davidi* 58–60cm Similar to Ural Owl but smaller and overall darker; geographical ranges do not overlap. Prey include terrestrial mammals and birds. **V** Haunting, hooting *whoo-hoo.* **H** Montane forests; mostly 4,000–5,000m. **D** Qinghai and Sichuan provinces (China).

10 GREAT GREY OWL *Strix nebulosa* 65–70cm Mainly crepuscular. Feeds mostly on rodents, dropping on prey from low perch. **V** Soft, deep *hoo-hoo-hoo-hoo*, female answering with mellow *whoop* or *woo woo.* When alarmed gives growling *grrrrrrrrrrrok*, harsh *grrook-grrook-grrook* or high *kjah-kjah-kjah.* **H** Coniferous or mixed forests with clearings, nearby pastures or bogs. **D** Widespread in Eurasia and North America.

11 AFRICAN WOOD OWL *Strix woodfordii* 35cm Plumage variable, from pale brown via russet to dark brown, dark form shown. Note black eyes set in white face and barring below. **V** High and mid-high barking *woo-woo woorrrorrwoo.* **H** Forest, plantations, suburbs. **D** Widespread in Africa.

12 MOTTLED OWL *Strix virgata* 29–38cm Nocturnal. Streaked on breast. **V** Frog-like *gwho gwho gwho...* that increases in pitch and volume. **H** Forests and woodlands with open areas. **D** Widespread in Central America and South America.

13 RUFOUS-BANDED OWL *Strix albitarsis* 35cm Upperparts cinnamon-rufous barred and scaled blackish brown. Underparts cinnamon-buff fading to buffy-white, scaled blackish brown. Brow whitish. Iris orangy yellow. **V** Call *hoo... hoo-hoo-hoo HOOO.* **H** Cloud forest and humid forest, forest edges and clearings. **D** Venezuela to Bolivia.

14 BLACK-AND-WHITE OWL *Strix nigrolineata* 34cm Face, crown and upperparts black. Brow, hind-neck and underparts white, barred black. Tail black with fine white bars. Bill and legs yellow. Iris blackish brown. **V** Like that of Black-banded Owl. **H** Cloud forest, humid, gallery and deciduous forests, clearings and forest edges, other wooded habitats. **D** C Mexico to Venezuela and Peru.

15 BLACK-BANDED OWL *Strix huhula* 34cm All black, finely barred white. Bill and legs yellow. Eye blackish brown. **V** Call low-pitched *hoo hoo hooo HOOOO* and variations. **H** Terra firme, várzea and humid rainforest and clearings, *Araucaria* woods, plantations. **D** Widespread in South America.

16 MANED OWL *Jubula lettii* 40cm Note reddish glow and maned appearance. Larger than many other owls in the region but smaller than eagle-owls. **H** Forest, especially along rivers. **D** Liberia to Ghana, S Cameroon and Gabon to E DR Congo.

17 CRESTED OWL *Lophostrix cristata* 41cm Ear-tufts disproportionately long, starting at base of bill and rising through brow, partially white. Rufous face. Upperparts buff with white spots on wings. Underparts paler. A darker form also occurs. **V** Toad-like rolling *K-K-Krrrrrrrrr...* **H** Humid heavy forest, riverine woods, tall secondary growth. **D** S Mexico through Amazonia.

18 SPECTACLED OWL *Pulsatrix perspicillata* 50cm Large and hefty. Crown, head, upperparts and breast-band blackish brown or dark chocolate. 'Spectacles' and throat white. Underparts buffy. Iris yellow. **V** Motmot-like *boo boo boo boo boo...*, descending. **H** From humid tropical and subtropical to gallery and dry forest, savannah, transition woods, montane woods, swamps, secondary growth. **D** S Mexico to NE Argentina.

19 TAWNY-BROWED OWL *Pulsatrix koeniswaldiana* 42cm Smaller than Spectacled Owl. 'Spectacles' buffy-cinnamon. Underparts cinnamon-buff with variable rusty scaling. Tail blackish grey with two fine whitish bars. Iris brown. **V** Series of descending, diminishing, ventriloquial *brrrs* or *urrrs.* **H** Humid tropical and subtropical forest, *Araucaria* woods, forest edge and clearings. **D** SE Brazil, E Paraguay, NE Argentina.

20 BAND-BELLIED OWL *Pulsatrix melanota* 42cm Differs from Tawny-browed Owl in having underparts all barred buffy-white and rusty. Wing and scapulars barred buff. Pectoral collar has diffuse scaling. Some individuals look more like Tawny-browed Owl. **V** Short deep trill ending in 4–5 rapid *pop* notes; in Peru, deep *hoot* notes. **H** Humid tropical and montane forest, open woods. **D** SE Colombia to Bolivia.

21 NORTHERN HAWK-OWL *Surnia ulula* 36–39cm Mainly diurnal. Often perches on exposed treetops, also hovers to look for prey or chases prey like a Sparrowhawk. **V** Bubbling *ulululululululu...* When alarmed gives shrill *ki-ki-kikikikiki.* **H** Forest edges and clearings near bogs, moorland or cultivation. Some winters irrupts S. **D** N Eurasia, N North America.

22 EURASIAN PYGMY OWL *Glaucidium passerinum* 16–17cm Active during daytime, especially at dawn or dusk, when often located calling from top of tall tree. Often hunts shrike-like from prominent perch, feeding mainly on voles and birds. **V** Mellow, fluting *peeu* or *hyew* given at about 1–2 sec intervals, when excited often adds short, low vibrating hoots between normal notes. Also gives a squeaky *cheek-cheek-cheek...* in winter. **H** Coniferous and mixed forests. **D** Widespread in Eurasia.

23 COLLARED OWLET *Glaucidium brodiei* 16–17cm Crepuscular and diurnal. Bold, fierce hunter taking birds as large as itself. Has eye-like markings on rear of head. **V** Mellow, bell-like *hoo hoo-hoo hooo* or *toot-tootoot-toot.* **H** Broadleaved evergreen hill forest. **D** Widespread through Himalayas and SE and E Asia, also parts of Sumatra and Borneo.

24 PEARL-SPOTTED OWLET *Glaucidium perlatum* 20cm Note streaking (not barring) below, long tail, pseudo-face at back of head. Often seen by day. **V** High *puukpuukpuuk...*, sustained for 20–30 sec; also very high piercing descending *'piiiuuu'* (like fireworks). **H** Open woodland, bush with some trees. **D** Widespread in Africa.

25 NORTHERN PYGMY OWL *Glaucidium californicum* 16–18cm Most active at dawn and dusk. Back of head shows false eye markings. **V** Monotonous *hoo* or *hoo-hoo.* **H** Coniferous and deciduous forest, forest edge. **D** W North America from SE Alaska to S Baja California (Mexico).

26 MOUNTAIN PYGMY OWL *Glaucidium gnoma* 16cm In same range as Colima Pygmy Owl (which prefers lower altitudes), Tamaulipas Pygmy Owl (with more bi-coloured plumage and restricted range), Central American Pygmy Owl (grey-headed), Ferruginous Pygmy Owl (with shorter and barred tail; lower elevations). **V** Very high, slightly irregular, hurried and sustained *wegweg-wegweg...* **H** Mixed forest. **D** SW Arizona to C Mexico.

27 BAJA PYGMY OWL *Glaucidium hoskinsii* 16cm No other pygmy owl in its range. **V** High, long, drawn-out, irregularly spaced series of *phlew* notes, also trilling *pooo-pu-pu-pu-pu-pu-pu-pu-pu-pu.* **H** High-altitude pine forest. **D** Baja California (Mexico).

28 GUATEMALAN PYGMY OWL *Glaucidium cobanense* 15–16cm Red and brown morphs occur. Nocturnal. Remains concealed during daytime in tree hole or dense vegetation. **V** Repeated, electronic-sounding, disyllabic *pu-pu, pu-pu, pu-pu...* **H** Montane forest, cloud forest. **D** Central America.

29 COSTA RICAN PYGMY OWL *Glaucidium costaricanum* 15cm Note diagnostic barring below. **V** Varied high hurried *weetweet...* or *weeetrurrurrrrr.* **H** Montane, open forest and adjacent areas. **D** Costa Rica and Panama.

30 ANDEAN PYGMY OWL *Glaucidium jardinii* 15cm Very like Ferruginous Pygmy Owl (Plate 72) but smaller. Crown has white dots. Scaling on sides of breast, upper back and scapulars. Tail blackish with three visible white bars. Rufous morph also occurs. **V** Single or double low-pitched whistle. **H** Montane forest and woods, to tree line, grasslands, scrub. **D** W Venezuela to C Peru.

31 CLOUD-FOREST PYGMY OWL *Glaucidium nubicola* 16cm Like a rufous morph Andean Pygmy Owl, but all uniform chestnut-brown on sides of breast and upper back. Tail shorter, dark brown with three visible whitish bands. **V** Series of double whistles rapidly repeated. **H** Primary humid cloud forest. **D** Colombia, Ecuador.

32 YUNGAS PYGMY OWL *Glaucidium bolivianum* 16cm Generally reddish with round white spots on head, ochre dots on back. Dark false eye markings on back of head. Short white brow. Tail with four dark bands. **V** Two or three *yurr yurr* notes followed by two short whistles per sec; *too-too, too-too.* **H** Forests. **D** Peru to NW Argentina.

33 COLIMA PYGMY OWL *Glaucidium palmarum* 14cm Scapular spots weak, rather short tail. Note finely spotted crown. **V** High, slightly descending *wook-wook-wok-wok-wok.* **H** Thorn forest. **D** W Mexico.

34 TAMAULIPAS PYGMY OWL *Glaucidium sanchezi* 15cm Restricted range; best distinguished from Mountain Pygmy Owl by voice. **V** High plaintive hooting *fweet-fweet* or *oot-oot-oot.* **H** Forest. **D** NE Mexico.

2
3
5
1
4
12
7
6
8
9
11
15
13
10
19
20
14
18
dark
morph
16
17
30
31
22
24
light
morph
grey/
brown
morph
rufous
morph
grey
morph
32
21
23
25
29
27
26
28
33
34
grey
morph
rufous
morph
rufous
morph

OWLS *CONTINUED*

1 PERNAMBUCO PYGMY OWL *Glaucidium mooreorum* 13cm Critically endangered and possibly extinct, with no recent sightings. Nocturnal. Likely to feed on large insects. **V** Rapidly delivered series of five or so piping calls. **H** Humid, lowland forest. **D** NE Brazil.

2 CENTRAL AMERICAN PYGMY OWL *Glaucidium griseiceps* 16cm Head often greyer than unspotted mantle. **V** High hooting, slightly wooden or hurried *fjood-fjood-fjood*. **H** Forest. **D** S Mexico to W Colombia and NW Ecuador.

3 SUBTROPICAL PYGMY OWL *Glaucidium parkeri* 14–15cm Partly diurnal but also active after dark. Presumed to feed on large insects and small vertebrates such as lizards. **V** Repeated series of three piping *pu-pu-pu* notes. **H** High-altitude montane forests. **D** Andes, from S Colombia to N Boliva.

4 AMAZONIAN PYGMY OWL *Glaucidium hardyi* 14cm Partly diurnal. Inhabits dense tree canopy; difficult to observe. **V** Vibrating, rapid trilling *hu-hu-hu-hu-hu-hu*. **H** Lowland rainforest. **D** Amazonian regions of Brazil, Bolivia, Venezuela, Guyana, French Guiana, Suriname.

5 EAST BRAZILIAN PYGMY OWL *Glaucidium minutissimum* 13cm Tiny. Brownish-grey head with tiny white dots. Upperparts dark brown with whitish markings. Tail blackish brown with three visible white bands edged blackish. Underparts heavily streaked chestnut-rufous, especially towards flanks. Brown morph also occurs. **V** A series of double or triple low-pitched hoots. **H** Tropical humid forest, terra firme, várzea, subtropical evergreen forests on Andean slopes. **D** SE Brazil, E Paraguay.

6 FERRUGINOUS PYGMY OWL *Glaucidium brasilianum* 17cm Variable. Upperparts streaked and marked white. Upper back and scapulars uniform. Crown finely streaked. Underparts whitish and irregularly streaked, densest on breast and flanks. Tail blackish brown with 4–6 white bands. Some of rufous morph have unbarred tail. **V** Long series of whistles, three per sec. **H** Woods, forests, gardens. **D** Widespread in Central America and South America.

7 PACIFIC PYGMY OWL *Glaucidium peruanum* 16–17cm Partly diurnal. Lives and feeds in mid-canopy of trees. Diet includes insects and small vertebrates such as lizards. **V** Rapid, raptor-like piping *wik-wik-wik-wik-wik-wik* notes. **H** Forested W slopes of Andes. **D** Ecuador, Peru, Chile.

8 AUSTRAL PYGMY OWL *Glaucidium nana* 19.5–21cm Brown head with longish ochre streaks. Fine white brow. Back brown with touch of cinnamon; white spots on scapulars. Black and white false eye markings on nape. Face brown with buff spots. White throat. Brown collar. Breast cinnamon-brown with lines of ochre and white streaking. Brown wings with whitish dots. Tail brown with chestnut bands. Reddish morph also occurs. **V** Like that of Ferruginous Pygmy Owl. **H** Woods, forests. **D** S South America.

9 CUBAN PYGMY OWL *Glaucidium siju* 16–17cm Feeds by day or night. Travelling flight undulating, but swift and agile when in pursuit of prey. **V** Low, repeated *uh uh uh*; during breeding season calls *hui-hui-chiii-chiii-chi-chi-chi...*, which increases in strength. **H** Woods, plantations. **D** Cuba.

10 RED-CHESTED OWLET *Glaucidium tephronotum* 20cm Unmistakable by dark upperparts and reddish breast patches. May hunt by day. **V** High, mellow, gliding, hooting *too-too-too* (2–30 notes) or *tutju tutju...* or *huu-huu-huu-p*. **H** Forests. **D** W and C Africa.

11 SJÖSTEDT'S BARRED OWLET *Glaucidium sjostedti* 25cm From other owlets by reddish mantle, from Albertine Owlet by more and larger white spots on upperwing. **V** Series of short bubbling high-pitched hoots, falling in pitch. **H** Forest interior. **D** Cameroon to DR Congo and Gabon.

12 ASIAN BARRED OWLET *Glaucidium cuculoides* 22–25cm Mainly diurnal. Often seen in open, perched on bare branch or stump. Variable plumage. **V** Bubbling, whistled *wowowowowowowowo* and loud *hooloo hooloo hooloo kok kok* ending in shrill *chiurr*. **H** Broadleaved evergreen forest. **D** Himalayas to SE Asia.

13 JAVAN OWLET *Glaucidium castanopterum* 24cm Mainly nocturnal, but sometimes active by day. Pounces on insects and small vertebrates from perch. **V** Rapid trill, descending in pitch and increasing in volume. **H** Primary and secondary forest, trees in suburban areas. **D** Java, Indonesia.

14 JUNGLE OWLET *Glaucidium radiatum* 20cm Mainly crepuscular, although also recorded hunting by day. Roosts amid leafy branches or in tree hollows. **V** Loud *kao...kao...kao kao-kuk kao-kuk kao-kuk...* that quickens then fades at the end; also monotonous *cur-cur-cur-cur-cur-ur*. **H** Mixed deciduous forest, secondary growth. **D** India.

15 CHESTNUT-BACKED OWLET *Glaucidium castanotum* 19cm Diurnal, occurs in branches at top of tall trees, shy and wary. **V** Slow, far-carrying *kRaw kRaw kRaw kRaw kRaw*. **H** Dense forests. **D** Sri Lanka.

16 AFRICAN BARRED OWLET *Glaucidium capense* 25cm Plain mantle and spotted crown, darker in some races. **V** Very high, descending staccato *pjuipjuipjuipjui...* and very high *wruhwruhwruh*. **H** Forest, often near water. **D** SE, SC and W Africa.

17 ALBERTINE OWLET *Glaucidium albertinum* 20cm Separated from African Barred Owlet by buff, not white edge to mantle and fewer white spots on upperwing. **V** Said to be like that of African Barred Owlet. **H** Montane forest. **D** NE Congo, N Rwanda.

18 LONG-WHISKERED OWLET *Xenoglaux loweryi* 14cm Small and dumpy. Long whiskers from sides of facial disc. Upperparts and breast chestnut-brown, vermiculated darker. Brow, dotting and nape-band white. Paler below. **V** Single, well-spaced, abrupt and buzzing mid-pitch hoots. **H** Humid cloud forest with dense undergrowth. **D** N Peru.

19 ELF OWL *Micrathene whitneyi* 14–15cm Hunts at dusk and at night, mainly feeding on insects. **V** Sharp *pew* and a series of *pe pe pe* notes; also a descending, whistled *meeeew*. **H** Deserts, riparian woods, dry woods, wooded canyons. **D** SW USA to C Mexico.

20 LITTLE OWL *Athene noctua* 21–23cm Mainly crepuscular, although often seen in open, perched on posts, wires or buildings. **V** Sharp *KEE-ew*, also mellow *goooek* repeated every 5–10 secs. When alarmed utters loud *kyitt kyitt*. **H** Steppe, semi-desert, open woodland, farmland, around human habitation. **D** Europe, N and NE Africa.

21 SPOTTED OWLET *Athene brama* 19–21cm Nocturnal and crepuscular, although often seen basking in sun. **V** Medley of screeching and chattering notes; *chirurrr chirurrr chirurrr* interspersed with *cheevah cheevah cheevah*. Also rapid *kuerk-kuerk-kuerk* said to sound like fighting cats. **H** Agricultural areas, around human habitation, semi-desert. **D** S Iran to SE Asia.

22 BURROWING OWL *Athene cunicularia* 23–25cm Nocturnal, but hunts diurnally during breeding season, often by walking or hopping after prey on ground. Regularly perches, in daylight, at burrow entrance. **V** Soft, high-pitched *coo-cooo*; utters a clucking chatter when alarmed. **H** Open country, including grassland, prairie, golf courses. **D** Widespread throughout the Americas.

23 WHITE-BROWED OWL *Athene superciliaris* 26cm Note white eyebrows, barred underparts, spotted wings and mantle. **V** High *ruw*, sometimes preceded by lower *rrrrrr*. **H** All forested and wooded habitats, including villages. **D** Madagascar.

24 FOREST OWLET *Athene blewitti* 23cm Rare. Diurnal. Frequently perches on prominent bare branches. **V** Loud, mellow *uwwww* or *uh-wuwww*, also hissing *shreeee* or *kheek* and rising and falling series of *kwaak* notes. **H** Moist deciduous jungle and fairly open dry deciduous forest, especially teak. **D** C India.

25 BOREAL OWL *Aegolius funereus* 24–26cm Nocturnal. Feeds mainly on small rodents. **V** Soft, rapid *po-po-po-po-po-po-po...*, also nasal *kuwake* and short *chiak*. **H** Mature mixed forest with tall conifers, clearings and marshy areas. **D** N North America and Eurasia.

26 NORTHERN SAW-WHET OWL *Aegolius acadicus* 19–21cm Mainly nocturnal. Roosts in dense cover or cavities. **V** Repeated, low *poo poo poo*; also nasal *pew* and soft whistled *eeeooi*. **H** Dense conifer and mixed forests, wooded swamps. In winter a wider range of woodland, even in suburban areas. **D** Widespread in North America, also C Mexico.

27 UNSPOTTED SAW-WHET OWL *Aegolius ridgwayi* 20cm Range does not overlap with that of Northern Saw-whet Owl. **V** Very high sustained hooting *woot-woot...* (10–30 notes). **H** Mixed oak forest. **D** S Mexico to W Panama.

28 BUFF-FRONTED OWL *Aegolius harrisii* 21cm Dark chocolate-brown and cinnamon-ochre. White spots on wings and tail. **V** Call a warbled rolling *rrrrrrrr...* **H** Humid and montane forest, podocarp and alder woods to tree line, *Araucaria* and dry chaco woods, other wooded or bushy habitats. **D** E and C South America.

29 RUFOUS OWL *Ninox rufa* 40–52cm Diet includes terrestrial and arboreal mammals, birds and large insects. **V** Mournful, sonorous *Huoo-hoo*, the second element downslurred. **H** Rainforest, savannah woodland; mostly lowlands. **D** N Australia, New Guinea, Waigeo Island (W Papuan Islands), Aru Islands.

30 TOGIAN BOOBOOK *Ninox burhani* 25cm Usually seen in pairs, but singles and threes both recorded. Nocturnal; roosts in thick foliage. **V** Gruff, low-pitched *kok-ko-ro-ok*, usually repeated and occasionally preceded by single croaking note. **H** Disturbed and degraded lowland forest. **D** Togian Islands, Indonesia.

31 POWERFUL OWL *Ninox strenua* 50–65cm Large and imposing; Australia's largest owl. Diet includes arboreal mammals. **V** Resonant, booming *whoOO-uu*, first element upslurred. **H** Old-growth sclerophyll forest, planted parkland forest, secondary woodland. **D** E and SE Australia.

32 BARKING OWL *Ninox connivens* 40–45cm Usually seen singly or in pairs. Roosts by day, in foliage in mid-storey and canopy. Hunts and calls from open perch. **V** Two dog-like barking notes, *wuff-wuff*, given by male; female utters higher-pitched *wok-wok*. **H** Woodland, scrub and forest edges, often near watercourses and swamps. **D** Maluku Islands, Indonesia; New Guinea; Australia.

33 SUMBA BOOBOOK *Ninox rudolfi* 35–40cm Usually seen singly or in pairs, sometimes in small dispersed groups. Occasionally seen during daylight hours. **V** Series of monotonous, cough-like notes, uttered continuously. **H** Primary and tall secondary forest, forest edges, remnant forest patches. **D** Sumba, Lesser Sunda Islands, Indonesia.

34 SOUTHERN BOOBOOK *Ninox boobook* 27cm Occurs singly or in pairs. Roosts by day in thick foliage. **V** Simple *bru-bruk*. **H** Primary and tall secondary lowland forest, monsoon hill forest, monsoon woodland. **D** Lesser Sunda Islands; New Guinea; Australia.

35 MOREPORK *Ninox novaeseelandiae* 29cm Darker, larger-headed and longer-tailed than Little Owl. **V** High bassoon-like *more-pork* with slight tremolo. **H** Most wooded habitats, from rainforest to dry woodland. **D** Tasmania, New Zealand.

36 BROWN HAWK-OWL *Ninox scutulata* 27–33cm Crepuscular and nocturnal. Shortly before dusk, hunts insects in the manner of a nightjar. Tends to use same perch over prolonged period. **V** Soft *oo...ok oo...ok oo...ok*. **H** Forests, well-wooded areas. **D** E Asia to SE Asia.

1
2
3
4
5
grey (right), rufous (below) morphs
6
7
8
9
10
11
12
13
14
15
16
17
18
19
20
21
22
23
24
25
26
27
28
29
30
31
32
33
34
35
36

OWLS *CONTINUED*

1 NORTHERN BOOBOOK *Ninox japonica* 27–33cm Mainly nocturnal, but will hawk for insects shortly before dusk. Roosts deep in tree canopy. **V** Mellow, rising *coo-coo* repeated monotonously. **H** Woodlands, parks, gardens with large trees. **D** Far E Russia to Korean Peninsula, Japan, Taiwan.

2 CHOCOLATE BOOBOOK *Ninox randi* 27–33cm Formerly considered a race of Brown Hawk-Owl (Plate 72); actions and habits presumed to be similar. **V** Series of low-pitched *whoop* notes. **H** Primary lowland rainforest, secondary forest and mangroves. **D** Philippines.

3 HUME'S HAWK-OWL *Ninox obscura* 27–33cm Crepuscular and nocturnal. Recorded hawking for insects shortly before dusk. Tends to use same perch over prolonged period. **V** Soft *oo...ok oo...ok oo...ok*. **H** Forests, well-wooded areas. **D** Andaman Islands.

4 ANDAMAN HAWK-OWL *Ninox affinis* 25cm Little known; recorded hawking moths, presumably actions and habits are much like Hume's Hawk-Owl. **V** Loud *craw*. **H** Forest, secondary woodland, mangroves. **D** Andaman Islands, Nicobar Islands.

5 LUZON HAWK-OWL *Ninox philippensis* 20cm Roosts during daylight in darker parts of forest. Preys on insects and rodents. **V** Series of *cuk* notes, interspersed with softer *boo* or *boo-boo* notes, starting quietly, then accelerating and getting louder. **H** Primary and secondary rainforest, remnant forest patches. **D** N and C Philippines.

6 MINDANAO HAWK-OWL *Ninox spilocephala* 18cm Formerly considered a race of Luzon Hawk-Owl; behaviour presumed to be similar. Main prey probably insects. **V** Long series of low-pitched, mellow, dove-like double notes. **H** Primary lowland rainforest, secondary forest. **D** S Philippines.

7 MINDORO HAWK-OWL *Ninox mindorensis* 20cm Presumably nocturnal; roosts in dense cover. **V** Series of high-pitched whistles, often starting with high, tittering toots. **H** Primary and secondary forest, remnant forest patches, open woodland. **D** Mindoro, WC Philippines.

8 ROMBLON HAWK-OWL *Ninox spilonotus* 26cm Formerly considered a race of Luzon Hawk-Owl; behaviour presumed to be similar. **V** Series of short, hoarse whistles that fall in pitch, starting slowly then changing into three- or four-note phrases. **H** Remnants of primary and tall secondary forest. **D** Sibuyan and Tablas islands, C Philippines.

9 CEBU HAWK-OWL *Ninox rumseyi* 25cm Formerly considered a race of Luzon Hawk-Owl; behaviour presumed to be similar. Preys on rats and small birds. **V** Gruff, staccato *chuck* notes and plaintive, downslurred notes with occasional tree-frog-like *bwick* notes; also recorded are clear, bell-like *duit* notes and hoarse screeches. **H** Forests. **D** Cebu, W Philippines.

10 CAMIGUIN HAWK-OWL *Ninox leventisi* 25cm Formerly considered a race of Luzon Hawk-Owl; behaviour presumed to be similar. **V** Duets with short, low-pitched strophes, repeated with many rapid, irregular, barking notes per strophe. **H** Forest, forest edges. **D** Camiguin Island, C Philippines.

11 SULU HAWK-OWL *Ninox reyi* 19–20cm Formerly considered a race of Luzon Hawk-Owl; behaviour presumed to be similar. **V** Hollow, wooden knocking in short phrases, starting with low *cluck* notes, getting louder and higher in pitch, then slowing and dropping in pitch, before becoming louder and higher pitched again. **H** Subtropical or tropical moist lowland forest, also montane forest. **D** Sulu Archipelago, SW Philippines.

12 OCHRE-BELLIED BOOBOOK *Ninox ochracea* 29cm Usually seen singly or in pairs; recorded hunting from horizontal branches in mid-storey or lower canopy, usually overlooking open areas or forest roads. **V** Loud series of hoarse *kau* notes, slowing the end, and including *wuu kau* notes as it develops. **H** Primary and tall secondary lowland, hill and lower montane forest; also recorded in riverine forest. **D** Sulawesi, Indonesia.

13 CINNABAR BOOBOOK *Ninox ios* 22cm Nocturnal. Reported to make short sallies from exposed branch to capture flying insects. **V** Hard *wruck-wruck*. **H** Mid-elevation forest. **D** Sulawesi, Indonesia.

14 HANTU BOOBOOK *Ninox squamipila* 27–39cm Perches in mid-storey and lower canopy, often on exposed branches or stumps. Roosts during daylight in dense foliage. **V** Far-carrying, mellow *wooo wooo wu wu wu wu*. **H** Primary and tall secondary forest, selectively logged forest, forest edges in lowlands and hills. **D** S Maluku Islands, Indonesia.

15 HALMAHERA BOOBOOK *Ninox hypogramma* 27–39cm Formerly considered to be a race of Hantu Boobook; behaviour presumed to be similar. **V** Gruff two-note phrases, evenly pitched and far-carrying. **H** Tropical lowland rainforest, tall secondary forest. **D** N Maluku Islands, Indonesia.

16 TANIMBAR BOOBOOK *Ninox forbesi* 30cm Formerly considered a race of Hantu Boobook; behaviour presumed to be similar. **V** Call *ku-kuk ku-kuk ku-kuk*. **H** Primary and tall secondary lowland and hill forest. **D** Tanimbar Islands, Lesser Sunda Islands, Indonesia.

17 CHRISTMAS BOOBOOK *Ninox natalis* 26–29cm During daylight, hides in dense thickets or thick foliage in mid-canopy of trees. Preys on insects, small birds and rats. **V** Simple *bru-bruk*, repeated many times; also low, barking *ow-ow-ow*. **H** Tropical rainforest, monsoon forest, scrub. **D** Christmas Island, E Indian Ocean.

18 PAPUAN BOOBOOK *Ninox theomacha* 22–28cm Diet comprises mainly large insects, sometimes caught in flight. **V** Strident *Whao-Whao*, delivered with a barking quality. **H** Forested habitats, including gardens. **D** New Guinea, W Papuan Islands.

19 MANUS BOOBOOK *Ninox meeki* 25–30cm Roosts in tree cover in daytime. At night, perches on mid-level branches. Diet comprises mainly large insects. **V** Croaking, almost grunting *uh-uh-uh-uh* notes. **H** Forested and wooded habitats, including gardens. **D** Manus Island (Admiralty Islands).

20 SPECKLED BOOBOOK *Ninox punctulata* 27cm Usually encountered singly or in pairs. Has been reported foraging along small streams in forests. **V** Long series of *toi toi toi toi* or *wher wher wher wher* notes, rising and accelerating before ending with lower-pitched *toi*; also reported is a *toi-toi-toi-seeeet*. **H** Primary lowland and hill forest, forest edges and tall secondary forest; occasionally cultivation near human habitation. **D** Sulawesi, Indonesia.

21 NEW IRELAND BOOBOOK *Ninox variegata* 24–30cm Roosts in tree cover in daytime. At night, perches on mid-level branches. Eats mainly large insects. **V** Purring, croaking and vibrating series of notes *brr-crrr-crr-crr-crr-brr-brr*. **D** New Ireland (Bismarck Archipelago).

22 NEW BRITAIN BOOBOOK *Ninox odiosa* 22–24cm Presumed to feed on large insects and small vertebrates such as lizards. **V** Rapid series of hooting *hoo-hoo-hoo-hoo* notes. **H** Forested and wooded habitats including cultivation. **D** New Britain (Bismarck Archipelago).

23 SOLOMONS BOOBOOK *Ninox jacquinoti* 26–30cm Roosts by day in cover of foliage. At night perches on mid-level branches. Eats large insects and small vertebrates. **V** Distinctive, horn-like series of oUP, oUP, oUP, oUP, oUP, oUP notes. **H** Forests and wooded habitats, including gardens. **D** Solomon Islands.

24 LITTLE SUMBA HAWK-OWL *Ninox sumbaensis* 23cm Little recorded information; presumed to be similar to other *Ninox* owls. **V** Subdued, flute-like *duu* or *puu*. **H** Remnant patches of primary forest, disturbed primary and secondary forest. **D** Sumba, Lesser Sunda Islands, Indonesia.

25 PAPUAN HAWK-OWL *Uroglaux dimorpha* 30–32cm Roosts by day. Active after dark, eating mainly insects, small mammals and small birds. Perches on branches in mid-level canopy. **V** Repeated series of *hoo, hoo, hoo, hoo...* notes. **H** Rainforest; mostly lowlands. **D** New Guinea.

26 JAMAICAN OWL *Pseudoscops grammicus* 27–34cm Nocturnal. Arboreal, often has a regular roosting tree. **V** High, quivering hoot and throaty growl. **H** Open woodland, forest edges. **D** Jamaica.

27 STRIPED OWL *Pseudoscops clamator* 35cm Perches upright. Long ear-tufts. Whitish face framed blackish. Upperparts irregularly marked cinnamon, buff, whitish and blackish brown. Underparts paler with prominent blackish streaks. Iris pale or dark brown, or orange. **V** Loud, whistled, descending *pheeeeeew*, series of dog-like yaps. **H** Gallery forest, forest edge, marshes, grassland with scattered trees and bushes, savannah, patches of woods, agricultural areas, plantations, suburban areas. **D** Widespread in Central America and South America.

28 FEARFUL OWL *Nesasio solomonensis* 38–39cm Recalls an oversized Short-eared Owl with rich orange brown plumage. Diet includes medium-sized mammals and birds. **V** Drawn-out, mournful *WhaOooo*. **H** Forest, wooded habitats. **D** Solomon Islands.

29 STYGIAN OWL *Asio stygius* 41–46cm Nocturnal. Large size and dark plumage make it unmistakable. **V** Loud *uh*. During breeding season male gives low-pitched, repeated *fool*; female answers with high-pitched *niek*. **H** Dense deciduous and pine forests. **D** Widespread in Central America and South America.

30 LONG-EARED OWL *Asio otus* 35–37cm Nocturnal. When relaxed, ear-tufts lie flat. In flight outer primaries show four or five dark bars. **V** During breeding season gives series of 10 or more *hoo* notes. When alarmed utters barking *ooack ooack ooack*. **H** Woodland, copses, plantations in open country. In winter often on moorland and farmland with hedgerows. **D** Widespread in North America, Central America and Eurasia.

31 ABYSSINIAN OWL *Asio abyssinicus* 45cm Note long upright stance especially when disturbed but may perch also with fluffed-out feathers and no visible ear-tufts. **V** Not recorded, presumably similar to that of Long-eared Owl. **H** Montane forest edges, heaths. **D** E Africa.

32 MADAGASCAN OWL *Asio madagascariensis* 34cm Note ear-tufts and partially red facial disc. **V** Calls *wah*, notes given singly or in slowly rising and descending series. **H** Forest, plantations, parks. **D** Madagascar.

33 SHORT-EARED OWL *Asio flammeus* 35cm Active at dusk and dawn. Hunts low over vegetation, often hovers before pouncing onto prey. In flight, primaries look dark-tipped. Nomadic. **V** Short *hoo-hoo* or *bow-wow*; also emphatic *kee-ow*. **H** Open country. **D** Global, except Australasia.

34 MARSH OWL *Asio capensis* 29–36cm Nocturnal and crepuscular. Hunts mainly by quartering just above vegetation, often hovers or uses perch before dropping onto prey. **V** Croaking *kaaa-kaaa* or *quark-quark*. When alarmed gives high squeal. **H** Grassland, marshland, fields. **D** Widespread in Africa.

1
2
3
4
5
6
7
8
9
10
11
12
13
14
15
16
17
18
19
20
21
22
23
24
25
26
27
28
29
30
31
32
33
34

FROGMOUTHS PODARGIDAE

1 MARBLED FROGMOUTH *Podargus ocellatus* 38–48cm Nocturnal. Hard to locate when roosting thanks to cryptic plumage. Diet comprises mainly large insects. **V** Sharp, rather explosive series of *wHoa, wHoa, wHoa* notes ending with rapid-fire trill. **H** Rainforest, humid secondary forest. **D** N Australia, New Guinea, Aru Islands.

2 PAPUAN FROGMOUTH *Podargus papuensis* 45–60cm Nocturnal. Roosts in daytime among tree branches. Diet includes large invertebrates, small mammals, lizards. **V** Deep, booming *hooUP, hooUP, hooUP...* **H** Rainforest, secondary woodland, wooded gardens. **D** NE Australia, New Guinea, W Papuan Islands.

3 TAWNY FROGMOUTH *Podargus strigoides* 36–52cm Well-camouflaged when roosting in daytime, pressed against tree trunk. Often seen in pairs. Feeds at night. Diet includes large invertebrates, small mammals, reptiles. **V** Booming, upslurred trumpeting *who-uup, who-uup, who-uup...* **H** Forest, wooded parks and savannah, heath. **D** Australia.

4 SOLOMONS FROGMOUTH *Rigidipenna inexpectata* 36–38cm Remains motionless in daytime, often pressed close to tree trunk. Feeds at night. Diet includes large invertebrates and small vertebrates including mammals and lizards. **V** Shrill, penetrating, piping and whistled *Peee*. **H** Forest, secondary woodland, mature gardens. **D** N Solomon Islands.

5 LARGE FROGMOUTH *Batrachostomus auritus* 40–43cm Nocturnal. During daylight stays motionless in tree canopy. Makes sallies from perch to take insects. **V** Tremulous *prrrrooh prrrrooh prrrrooh...* **H** Broadleaved evergreen forest. **D** Peninsular Malaysia; Sumatra, Indonesia; Borneo.

6 DULIT FROGMOUTH *Batrachostomus harterti* 34–37cm Little recorded information; presumed to be nocturnal, with behaviour similar to other frogmouths. **V** Repeated, loud, trumpeting *whooooooaaah*. **H** Submontane forest. **D** Borneo.

7 PHILIPPINE FROGMOUTH *Batrachostomus septimus* 23cm Nocturnal; more often heard than seen. Typical frogmouth habits; perches upright, resembling a broken branch, hunts in flight or gleans invertebrates from leaves. **V** Harsh, growling *kaaoo, kaaww, paaww* or *pa-paaww*. **H** Forest, forest edges. **D** Philippines.

8 GOULD'S FROGMOUTH *Batrachostomus stellatus* 21–25cm Nocturnal. Sits motionless during daylight hours; little other recorded information. **V** Male utters eerie, weak *woah-weeo*, occasionally just *weeo*; female a growling, rapid, high-pitched yapping. **H** Primary lowland and hill forest. **D** Peninsular Malaysia; Sumatra, Indonesia; Borneo.

9 SRI LANKA FROGMOUTH *Batrachostomus moniliger* 23cm Nocturnal. Sits immobile, even when closely approached. Preys on insects. **V** Soft, rapid *kooroo kooroo kooroo*, also various chuckles and croaks. **H** Dense secondary growth in tropical and subtropical evergreen forests. **D** SW India, Sri Lanka.

10 HODGSON'S FROGMOUTH *Batrachostomus hodgsoni* 22–27cm Nocturnal. Behaviour little known. **V** Soft *gwaa gwaa gwaa...* also long, rising, then descending whistle. **H** Broadleaved evergreen, mixed coniferous and evergreen forest, secondary growth. **D** NE India to SE Asia.

11 SHORT-TAILED FROGMOUTH *Batrachostomus poliolophus* 20–22cm Little recorded; behaviour presumed to be similar to that of Philippine Frogmouth. **V** Downslurred, plaintive *weeooow*, repeated at irregular intervals. **H** Submontane primary forest, mixed pine forest. **D** Sumatra, Indonesia.

12 BORNEAN FROGMOUTH *Batrachostomus mixtus* 20–22cm Behaviour presumed to be similar to that of Philippine Frogmouth. **V** Pure, whistled *pwau* or *weeow*, repeated every 2–3 secs. **H** Submontane and montane forest. **D** Borneo.

13 JAVAN FROGMOUTH *Batrachostomus javensis* 19–25cm Typical frogmouth in appearance and behviour. Sits upright with bill pointing upwards while resting by day, usually near ground. **V** Hoarse *gwaa* notes, descending in pitch; also various barks, trills and whistles. **H** Moist lowland and hill forest. **D** Java, Indonesia.

14 BLYTH'S FROGMOUTH *Batrachostomus affinis* 23–24cm Rests by day, usually close to ground, in upright posture with head pointing skywards. **V** Male gives plaintive whistle; female maniacal laughing. **H** Broadleaved evergreen and mixed deciduous forest, forest edge, secondary growth. **D** Myanmar; N Thailand; C Vietnam to SE peninsular Thailand to Sumatra, Indonesia; Borneo.

15 PALAWAN FROGMOUTH *Batrachostomus chaseni* 21–22cm Nocturnal. Little recorded; behaviour presumed to be like that of other small frogmouths. **V** Plaintive, whistled *pheuuuuuuuuuu*, also harsh, mournful, growling *kawwrreerr*. **H** Forest, secondary growth, including dense tangles of vines and bamboo. **D** Banggi island (N of Borneo); Palawan, Philippines.

16 SUNDA FROGMOUTH *Batrachostomus cornutus* 23–28cm Plumage very variable. Reported resting by day on low branches, sometimes with bill wide open to lose heat. **V** Descending series of *gwaa* notes. **H** Secondary forest and forest edges. **D** Indonesia; Borneo.

OILBIRD STEATORNITHIDAE

17 OILBIRD *Steatornis caripensis* 46cm Rufous with white spots and narrow black bars. Nocturnal. Roosts and nests in caves. **V** Very dry, grating, rolling notes. **H** Tropical and subtropical montane forest down to sea level. **D** Panama to N and NW South America.

POTOOS NYCTIBIIDAE

18 GREAT POTOO *Nyctibius grandis* 52cm Paler than other potoos. Ash-grey, finely spotted and streaked in black, whitish and brown. Shoulders rusty. Eye dark brown. Seen high in large trees. **V** Long, harsh *wow*. **H** Open forest, savannah. **D** S Mexico to C Bolivia and SC and SE Brazil.

19 LONG-TAILED POTOO *Nyctibius aethereus* 55cm Large, with long, broad, graduated tail. Yellow eye. Sometimes more rufous, with whitish median wing coverts. **V** Plaintive *ra-OOH* repeated at intervals of 5–10 secs. **H** Deep, unspoilt forest, from very humid to relatively dry. **D** Amazonia and SE South America.

20 NORTHERN POTOO *Nyctibius jamaicensis* 43–46cm Nocturnal. Grey-toned, or sometimes more reddish brown. **V** Harsh, throaty *kwah waugh waugh waugh...* or hoarse *waark-cucu*. **H** Forests, palm groves, scrubland near open areas. **D** Mexico to Costa Rica and Greater Antilles.

21 COMMON POTOO *Nyctibius griseus* 35cm Variable; brownish grey to cinnamon-brown, with streaked, spotted and barred markings in black, chestnut, buff and whitish. Yellow eye. Blackish malar stripe. Dark shoulders. Median coverts sometimes very pale. **V** Mournful whistled notes (5–8), descending in pitch. **H** Forest, woods, mangroves, plantations. **D** Nicaragua to Uruguay.

22 ANDEAN POTOO *Nyctibius maculosus* 36cm Larger, mountain-dwelling relative of White-winged Potoo. **V** Similar to that of Long-tailed Potoo but higher pitched. **H** Humid cloud forest. **D** Venezuela to Bolivia.

23 WHITE-WINGED POTOO *Nyctibius leucopterus* 27cm Like Common Potoo but smaller. Wing coverts whitish with faint blackish streaks. Eyes yellow. **V** Single long, descending *feeeeooooo*. **H** Atlantic forest, montane forest. **D** Amazonia.

24 RUFOUS POTOO *Nyctibius bracteatus* 23cm Small. Cinnamon-rufous with white dots and black barring. Yellow eye. **V** Low-pitched, accelerating chuckle, falling in pitch. **H** Lowland rainforest, terra firme forest, swampy areas with palms. **D** Amazonia.

NIGHTJARS CAPRIMULGIDAE

25 SPOTTED NIGHTJAR *Eurostopodus argus* 30cm Found singly, in pairs or in small groups. Roosts by day on the ground. In flight, shows white patch on four outermost primaries. **V** Rapid series of ascending *whaw* notes, followed by bubbling gobble. **H** Savannah, grassland, rainforest edges. **D** Australia.

26 WHITE-THROATED NIGHTJAR *Eurostopodus mystacalis* 30–36cm Roosts by day in shade. Catches insects in flight after dark. **V** Song an accelerating series *who, wha, wup-up-up-up-up...* rising in pitch throughout delivery. **H** Rainforests, wooded savannah, dry sclerophyll forest. **D** E Australia. Winter range extends N to S New Guinea.

27 SOLOMONS NIGHTJAR *Eurostopodus nigripennis* 26–30cm Richly marked nightjar. Roosts during daytime in cover and on shaded branches. Feeds after dark, catching flying insects. **V** Series of rapid-fire clicking sounds. **H** Coastal forest and wooded habitats. **D** Solomon Islands.

28 NEW CALEDONIAN NIGHTJAR *Eurostopodus exul* 25–27cm Plumage overall pale grey with orange-flushed face and small white throat patch. Critically endangered, possibly extinct. Presumably feeds at night on flying insects. **V** Unknown. **H** Coastal wooded savannah. **D** New Caledonia.

29 SATANIC NIGHTJAR *Eurostopodus diabolicus* 26cm Feeds aerially, presumably mainly in twilight and at night, around small clearings. **V** Bubbling trill and *plip-plop* call. **H** Primary and selectively logged montane forest. **D** Sulawesi, Indonesia.

30 PAPUAN NIGHTJAR *Eurostopodus papuensis* 26–28cm Roosts in daytime in trees. Feeds after dark, catching mainly flying insects. **V** Accelerating series of whistled notes *whip-whip-whip-whipoo-whip...* **H** Lowland rainforest, secondary woodland. **D** New Guinea.

31 ARCHBOLD'S NIGHTJAR *Eurostopodus archboldi* 27–30cm Nocturnal, roosting in daytime in forest cover. Feeds after dark, catching flying insects over open ground. **V** Song an accelerating trill. **H** Montane forest, alpine heath. **D** New Guinea.

32 MALAYSIAN EARED NIGHTJAR *Lyncornis temminckii* 25–28cm Forages, in flight, over open areas. **V** Repeated *tut-wee-ow*. **H** Clearings in or near broadleaved evergreen forest. **D** Peninsular Malaysia; Sumatra, Indonesia; Borneo.

33 GREAT EARED NIGHTJAR *Lyncornis macrotis* 40cm Crepuscular and nocturnal. Hunts above forest clearings, flying with leisurely wingbeats, recalling small harrier. **V** Wailing *pee-wheeoo wheeoo wheeoo*. **H** Open areas in or near broadleaved evergreen and deciduous forest. **D** Widespread in SE Asia, also W India, Philippines, parts of Lesser Sundas.

grey morph
rufous
1
2
3
4
5
6
chestnut morph
7
8
dark morph
9
♀
♀
10
♂
♂
11
♀
brown morph
8
♂
14
♀
♀
15
♂
brown morph
16
♀
♀
12
♂
♂
13
♀
♂
♂
20
18
17
19
21
22
23
16
grey morph
♂
24
25
26
27
28
29
30
31
32
33

NIGHTJARS *CONTINUED*

1 COLLARED NIGHTJAR *Gactornis enarratus* 23cm Only nightjar in its range with a buff collar. **V** Not recorded. **H** Primary and secondary forest. **D** Madagascar.

2 NACUNDA NIGHTHAWK *Chordeiles nacunda* 30cm Large, unmistakable. Large rounded wings. White band at base of black primaries, also white triangle on throat. Lower underparts and underwing white. Female lacks white tail tips. Crepuscular, often also diurnal. Flies low in gentle wobbly flight. **V** Very low-pitched, brief, purring *churr*. **H** Savannah, open grassland near woods or water. **D** Widespread in South America.

3 LEAST NIGHTHAWK *Chordeiles pusillus* 17cm Tiny. Upperparts dark. White on throat. Secondaries blackish brown with buff trailing edge. From below, tips of central tail feathers white. White band on primaries. Underparts whitish buff barred black. **V** Stuttering *k k k k kuree*. **H** Edge of rainforest, savannah, open areas, Cerrado, Caatinga, rocky slopes. **D** N and E South America.

4 SAND-COLORED NIGHTHAWK *Chordeiles rupestris* 22cm Pale sandy with darker spots, bars and streaks on breast and back. Underparts white. Tail tipped black. Outer primaries black, forming contrasting triangle. **V** Drawn-out, low-pitched *churr*, falling sharply in pitch at end. **H** Dunes, rocky outcrops, sandbanks and river banks in open rainforest and marshes, near towns. **D** Amazonia.

5 LESSER NIGHTHAWK *Chordeiles acutipennis* 20cm Like Common Nighthawk but smaller, white band at mid-primaries. Barred underparts buffy. **V** Trill. **H** Open areas, desert, suburban areas, cultivated land, beaches, mangroves. **D** SW USA to N Argentina, partially migratory.

6 COMMON NIGHTHAWK *Chordeiles minor* 23cm Upperparts barred, spotted and streaked black, buff, greyish brown and cinnamon. White on throat. Tail slightly forked, tipped black with white subterminal band. White band near base of primaries. Perches on roofs or along cables and branches. High, erratic, fast flight. **V** Call an abrupt *tsip*, also has lower croaking call. **H** Most kinds of open habitats. **D** Breeds in North America and Central America. Winters in South America.

7 ANTILLEAN NIGHTHAWK *Chordeiles gundlachii* 20–21cm Often forages at dusk and dawn. At rest, wing tips do not protrude beyond tail. **V** Descending *que-re-be-bé*. **H** Open areas. **D** S Florida, Bahamas, Greater Antilles, Virgin Islands.

8 SHORT-TAILED NIGHTHAWK *Lurocalis semitorquatus* 24cm Swallow-like. Short square tail. Blackish. Tertials white, barred blackish. White triangle on throat. Lower underparts cinnamon-rufous, barred. Perches and nests in trees. **V** *Whit...whoo it...chew it* in flight. **H** Forests, woods, plantations, open areas with scrub. **D** Widespread in Central America and South America.

9 RUFOUS-BELLIED NIGHTHAWK *Lurocalis rufiventris* 23cm Swallow-like, short square tail. Similar to Short-tailed Nightjar but with unmarked rufous belly. Perches and nests in trees, hunts over cloud forest. **V** Series of evenly pitched or descending *kwa* notes. **H** High-altitude forests, other wooded habitats. **D** Venezuela to Bolivia.

10 BAND-TAILED NIGHTHAWK *Nyctiprogne leucopyga* 20cm Dark. No white on wing. White on throat split. Black tail with white bar midway. Perches in clusters on branches. **V** Triple-note *gole kwok quack*. **H** Savannah, near watercourses, rainforest, gallery forest, deforested areas, forest edge. **D** Amazonia.

11 PLAIN-TAILED NIGHTHAWK *Nyctiprogne vielliardi* 17cm Tiny. Chestnut-coloured with black and rufous streaks, scaling and barring. No white or black on wings. **V** Series of *wick-wicks*. **H** Caatinga along rivers, near dunes, rocky outcrops. **D** E Brazil.

12 BLACKISH NIGHTJAR *Nyctipolus nigrescens* 21cm Very dark. White on throat. Blackish flight feathers with short white bar on outer primaries. Outer three pairs of tail feathers tipped white. Female has no white on tail or wings. **V** Soft purring *crrrrrrc*. **H** Rocky outcrops along rivers, edge of rainforest, savannah. **D** Amazonia.

13 PYGMY NIGHTJAR *Nyctipolus hirundinaceus* 15cm Tiny. Pale with little contrast. White band on dark outer primaries. White tips to inner web of outer two pairs of tail feathers. Female has no white on tail, and smaller or absent wing-band. **V** Short, fluting and sweet-toned *weeoo* notes. **H** Open areas, woods, rocky areas. **D** NE and E Brazil.

14 PAURAQUE *Nyctidromus albicollis* 28cm In flight, male shows white bar on primaries and extensive white on tail. Female has buff primary bar and white on tail corners. **V** Buzzy *purwizheeeeer*; also slurred *po po po po po po pupppurrEEyeeeeeeerrrr*. **H** Woodland clearings with dense bushy cover. **D** Widespread in Central America and South America.

15 ANTHONY'S NIGHTJAR *Nyctidromus anthonyi* 22–24cm Feeds after dark on flying insects. Roosts in daytime in deep cover. Cryptic plumage provides good camouflage. **V** Song a squealing whistled and upslurred *cuEEo*. **H** Scrub, open woodland. **D** W Ecuador, N Peru.

16 TODD'S NIGHTJAR *Setopagis heterura* 20–22cm Small, well-marked nightjar. Feeds after dark on flying insects. Cryptic plumage provides good camouflage in daytime. **V** Song a rapid *pitchu, kbul-kbul-kbul*. **H** Woodland, open scrub; lowlands to 2,000m. **D** Colombia, Venezuela, Guyana.

17 LITTLE NIGHTJAR *Setopagis parvula* 20cm White on throat. Narrow malar stripe. Nuchal collar buffy-cinnamon. Outer flight feathers with white band. Tail feathers tipped white except central pair. Female has buffy wing-band, tail all barred. **V** Warbled *chorreee, froo-froo-froo-froo*. **H** Open dry woods, savannah, scrubby hills, plantations, parks. **D** E Peru to E Brazil, S to C Argentina and Uruguay.

18 RORAIMAN NIGHTJAR *Setopagis whitelyi* 22cm Very like Blackish Nightjar. Upperparts darker with less contrasting pattern. Female has narrow buffy wing-bar. **V** Buzzy, downslurring *choreohh*. **H** Wet forest. **D** E Venezuela, N Brazil, W Guyana.

19 CAYENNE NIGHTJAR *Setopagis maculosa* 23cm Known from only one specimen, a male taken in 1917. **V** Not known. **H** Specimen taken along Mana river. **D** French Guiana.

20 SICKLE-WINGED NIGHTJAR *Eleothreptus anomalus* 19cm Dumpy. Short tail and wings. In male, inner flight feathers much shorter than outers. Rather plain overall. Crown and scapulars streaked black with buffy-white edges. Breast streaked whitish. **V** Soft, dry, somewhat spluttering trill given in short phrases. **H** Edge of gallery forest, woods, transition zones, lightly wooded marshes, grassland and savannah. **D** SE Brazil, SE Paraguay to C Argentina and Uruguay.

21 WHITE-WINGED NIGHTJAR *Eleothreptus candicans* 21cm Pale brownish grey with dark face and strongly contrasting black and white streaks; underparts mostly white. Male has white flight feathers, primaries broadly tipped black; white outer tail. **V** Twittering call, also low grunting sound produced by wings in display flight. **H** Grassland, scattered trees, bushes, arid savannah, wood edge, Cerrado. **D** N Bolivia, SC Brazil, E Paraguay.

22 BAND-WINGED NIGHTJAR *Systellura longirostris* 20–27cm Variable in size and colour (from very pale to very dark) depending on race. **V** Thin *psee-wheet*. **H** Edge and clearings in forest and woods, scrub, steppe, páramo. **D** Widespread in South America.

23 TSCHUDI'S NIGHTJAR *Systellura decussata* 20–21cm Roosts on ground in daytime amongst low scrub. Feeds after dark on flying insects. Sometimes seen resting on roads at night. **V** Piping, wader-like whistled *pee-a, pee-a, pee-a...* **H** Arid open habitats. **D** Peru, N Chile.

24 SWALLOW-TAILED NIGHTJAR *Uropsalis segmentata* male 67cm, female 23cm Like Lyre-tailed Nightjar but smaller with shorter tail, browner, long feather tips more slender, inner webs finely barred whitish. Crown blackish brown flecked buff. Male has greatly elongated tail feathers. **V** Rather piercing, drawn-out notes, rising then falling in pitch. **H** Bushy slopes, edge of woods, to tree line and páramo, grassland with shrubs and canes, cloud forest (higher than Lyre-tailed Nightjar). **D** Colombia to C Bolivia.

25 LYRE-TAILED NIGHTJAR *Uropsalis lyra* male 90cm, female 26cm Unmistakable. Male has inordinately long lyre-shaped tail, three times body length, blackish tipped white. Overall dark, brownish black. Female has much shorter tail with slight fork. In courtship, several males follow one female. **V** Various rich, fluting warbled notes. **H** Rainforest and cloud forest in mountains, forest edge and clearings, especially near banks, rocks and caverns, open woods near water, alder woods, transition scrub. **D** Venezuela to S Bolivia.

26 WHITE-TAILED NIGHTJAR *Hydropsalis cayennensis* 22cm Upperparts very mottled. Male has white lores and throat, framing dark triangle. Hind-collar buffy-cinnamon. All but central tail feathers black and white; flight feathers with three white bands. Female more uniform. **V** High-pitched, thin *chip-wheeeeh*. **H** Grassland with scattered trees, edge of gallery forest, cloud forest, rainforest, arid scrub, towns. **D** Costa Rica to N South America.

27 SPOT-TAILED NIGHTJAR *Hydropsalis maculicaudus* 22cm Blackish crown and face. Long buffy-white brow. Nuchal collar tawny. Spotted breast. No white in wings. From above, male shows white tips to all but central pair of tail feathers. **V** High-pitched, thin *swit-sweeuh*. **H** Savannah, edge of forest and woods, secondary growth, slopes of dry hills. **D** Widespread in Central America and South America.

28 LADDER-TAILED NIGHTJAR *Hydropsalis climacocerca* 24–28cm Male's upperparts contrastingly patterned blackish, white and buff; underparts, underwing and tail mostly white. Outer primaries blackish with white band; rest of flight feathers barred blackish and white. Tail forked but with central pair as long as outermost pair. Tail feathers white, tipped greyish and barred blackish. Female plainer cinnamon-buff. **V** Repeated short *tchip* notes. **H** Gallery forest, rainforest, edge, rocky outcrops on rivers, sandbanks, other open lightly wooded habitats. **D** Amazonia.

29 SCISSOR-TAILED NIGHTJAR *Hydropsalis torquata* male 60cm, female 30cm Male's tail long, deeply forked, 1.5 times body length. Nuchal collar rufous. Face pale. Blackish triangle on lower cheek. Crown streaked black. Buffy-white line on coverts and scapulars. Flight feathers blackish, barred cinnamon-buff. **V** Short *tsip, tsip tsip*. **H** Edges and clearings in forest, woods, savannah, grassland with scattered trees and bushes, marsh edges, Cerrado, plantations, parks. **D** Amazonia and SC and SE South America.

30 LONG-TRAINED NIGHTJAR *Macropsalis forcipata* male 80cm, female 34cm Like Scissor-tailed Nightjar but larger and male's tail longer, three times body length. Crown and upper back speckled buff. Scapulars bordered buffy-white. Whitish on breast. **V** Calls *tsip tsip tsip*. **H** Forest and woods, edges and clearings, rolling and hilly country, secondary growth, suburban areas. **D** SE Brazil, NE Argentina.

31 LEAST POORWILL *Siphonorhis brewsteri* 17–20cm In flight, male shows narrow white tip to outer-tail feathers, replaced with buff on female. **V** Whistled, rising *toorrrrri*, warbled *tworrri* and throaty *torico torico*. **H** Coniferous forest, arid or semi-arid lowlands with cactus and thorn scrub. **D** Hispaniola.

32 CHOCO POORWILL *Nyctiphrynus rosenbergi* 20–21cm Dark chocolate-brown plumage is distinctive. Roosts on forest floor. Hunts flying insects after dark. **V** Song an ascending piping triplet *tor-ter-tay*. **H** Humid forested slopes; mostly lowlands. **D** W Colombia, N Ecuador.

33 EARED POORWILL *Nyctiphrynus mcleodii* 20cm Note white wing spots and grey scapulars. **V** Very high sweeping *píuw píuw...* **H** Mixed forest, dry rocky woodland. **D** W Mexico.

34 YUCATAN POORWILL *Nyctiphrynus yucatanicus* 20cm Note large white tail corners (tail looks all white below). **V** Very high, rather mellow *wheeow wheeow*. **H** Forest, scrub, woodland. **D** Yucatán Peninsula.

35 OCELLATED POORWILL *Nyctiphrynus ocellatus* 20cm Dumpy and large-headed. Blackish, finely barred chestnut (rufous form also exists). Narrow white collar on lower throat. White spots on wing coverts. Outer-tail feathers narrowly tipped white. **V** Repeated trilling *prreeeoh*. **H** Clearings in humid lowland forest, open understorey. **D** Widespread in Central America and South America.

36 COMMON POORWILL *Phalaenoptilus nuttallii* 18–21cm Variable in plumage tone. In flight, male shows white corners on tail; these markings smaller and duller on female. **V** Low, whistled *poor-will-ip*. **H** Arid or semi-arid country with scattered vegetation. **D** SW Canada to N Mexico, migratory in N of range.

dark
morph
rufous
morph

NIGHTJARS ***CONTINUED***

1 CHUCK-WILL'S-WIDOW *Antrostomus carolinensis* 27–34cm Nocturnal. In flight, male shows white inner webs on three outer-tail feathers; female has buffy tips to outer-tail feathers. Rufous phase similar to Rufous Nightjar. **V** Whistled *chip wido wido* or *chuck-wills-wid-ow*. **H** Woodlands. **D** Breeds in E North America. Winters in Central America and South America.

2 RUFOUS NIGHTJAR *Antrostomus rufus* 30cm Tinged rufous-cinnamon. Heavily dotted, barred, streaked black, buff and white. Cheek and throat barred. Buff half-collar. Flight feathers and tail barred. Male has white on inner web of three outer pairs of tail feathers, cinnamon from below; female's tail all barred. **V** Four-syllabled and insistent *chuck whipwhip wheeew*. **H** Forest and forest edge, gallery forest, secondary growth, open woods, grass and scrub, savannah, suburbs. **D** Widespread from Costa Rica S into South America.

3 CUBAN NIGHTJAR *Antrostomus cubanensis* 25–29cm Nocturnal. In flight, male shows broad white tips to outer-tail feathers, lacking in female. **V** Short trilled whistle and plaintive *gua bai ah ro*. **H** Dense forest, open woodland and edges of swamps. **D** Cuba.

4 HISPANIOLAN NIGHTJAR *Antrostomus ekmani* 26–30cm Nocturnal. Flight pattern similar to that of Cuban Nightjar, white tail tips narrower. **V** Call *tuc* followed by trilled, rising whistle. **H** Pine forests. **D** Hispaniola.

5 TAWNY-COLLARED NIGHTJAR *Antrostomus salvini* 25cm No other nightjar with similar tail pattern near its range except Dusky Nightjar (in Panama), which is much darker rufous. **V** Very high fluting hurried sustained *pivrow-pivrow*... **H** Dry dense woodland. **D** E Mexico.

6 YUCATAN NIGHTJAR *Antrostomus badius* 25cm As Tawny-collared Nightjar, but with darker face, paler body, larger white tail corners and different range. **V** Ultra-high, mellow, fluted *muh-miouiow*. **H** Dense woodland. **D** Yucatán Peninsula.

7 SILKY-TAILED NIGHTJAR *Antrostomus sericocaudatus* 29cm Like Rufous Nightjar but browner. Blackish cheeks. Flight feathers dark and almost uniform. Tail blackish, irregularly barred dark brown. In male, outer three pairs of tail feathers tipped white, finely edged buff; in female tail all barred and paler. **V** Trisyllabic *do, wheeu-weeu*. **H** Forest clearings, cane, secondary growth, forest edges. **D** Widespread in South America.

8 BUFF-COLLARED NIGHTJAR *Antrostomus ridgwayi* 21–25cm In flight, male shows white corners on tail; female buff. **V** Accelerating *cuk cuk cuk cuk cuk cuka-cheea* or accelerating and rising *tok tok tek tek tek teeka-teea*. **H** Rocky arid or semi-arid scrubland. **D** NW Mexico to Nicaragua.

9 EASTERN WHIP-POOR-WILL *Antrostomus vociferus* 23–26cm Nocturnal. In flight, male shows broad white tips to outer-tail feathers; buff in female. **V** Call *whip-poor-will*. **H** Dry, open woodlands. **D** SE USA and Central America.

10 MEXICAN WHIP-POOR-WILL *Antrostomus arizonae* 25cm Very similar to Eastern Whip-poor-will but with less white on tail. **V** As Eastern Whip-poor-will but lower and slower. **H** Mixed forest. **D** C Mexico to Honduras.

11 PUERTO RICAN NIGHTJAR *Antrostomus noctitherus* 22cm Nocturnal. In flight, male shows broad white tips to outer-tail feathers, buff in female. **V** Liquid *whilp whilp whilp*. **H** Dry, semi-deciduous forests. **D** SW Puerto Rico.

12 DUSKY NIGHTJAR *Antrostomus saturatus* 25cm Very dark upperparts, rufous and buff breast and belly. **V** Very high *prruh-purrieh*. **H** Open montane forest and adjacent areas. **D** Costa Rica and W Panama.

13 BROWN NIGHTJAR *Veles binotatus* 22cm Plain, dark red-brown with pale edge to scapulars. **H** Forest canopy. **V** Long series of *glu-glu-glu*... notes given from high branch, **D** Liberia to C DR Congo.

14 RED-NECKED NIGHTJAR *Caprimulgus ruficollis* 30–32cm Nocturnal. Flight pattern of male very similar to European Nightjar. Female has smaller white markings than male on wings and tail. **V** Prolonged *kuTOK kuTOK kuTOK*... **H** Dry open woodland or scrub, scrubby hillsides, semi-desert. **D** Breeds Iberia to NW Africa. Winters further S in Africa.

15 JUNGLE NIGHTJAR *Caprimulgus indicus* 28–32cm Nocturnal. In flight, male shows white spots on outer primaries and white tips to outer-tail feathers; female's wing and tail spots buff-tawny. **V** Slow *fwick-m fwick-m fwick-m*... **H** Forest clearings, scrubby slopes. **D** India and Sri Lanka.

16 GREY NIGHTJAR *Caprimulgus jotaka* 28–32cm Nocturnal. Male shows white spots on outer primaries and tips of outer-tail feathers; in female these are tawny-buff. **V** Rapid *tuk tuk tuk tuk*... **H** Open broadleaved evergreen and coniferous forest, secondary growth, open areas. **D** Breeds from Himalayas to E SE Asia. Winters in Indian subcontinent and SE Asia to Philippines.

17 PALAU NIGHTJAR *Caprimulgus phalaena* 21–23cm Previously considered conspecific with Jungle Nightjar. Roosts in deep cover in daytime. Feeds on flying insects after dark. **V** Agitated-sounding *wik-wik-wik* or *wik-wik*. **H** Lowland forest, mangrove forest. **D** Endemic to Palau, WC Pacific Ocean.

18 EUROPEAN NIGHTJAR *Caprimulgus europaeus* 26–28cm Nocturnal. Male has white throat patch; white spots on outer three or four primaries and white tips to outermost tail feathers. Female lacks white wing and tail markings and has smaller, more buff-tinged throat patch. **V** Prolonged *churr*, regularly changing in pitch. **H** Woodland edge, clearings, heathland, moorland with scattered trees or bushes, young forest plantations. **D** Breeds Europe, W and C Asia. Winters in Africa.

19 SOMBRE NIGHTJAR *Caprimulgus fraenatus* 25cm More brownish, less rufous than other nightjars. Note black-and-white scapulars. Very restricted range. **V** Low, rapid rattle *rrrrrrr*.... like a fast sewing machine. **H** Rocky outcrops in dry areas with scattered scrub and some grass cover. **D** Eritrea, Ethiopia and Somalia to NE Tanzania.

20 RUFOUS-CHEEKED NIGHTJAR *Caprimulgus rufigena* 24cm Separated from European Nightjar by tawny hind-collar. Little overlap with range of Fiery-Necked Nightjar. **V** Sings from ground (occasionally from perch); low sewing machine-like *rrrrrrr* sustained for several minutes, preceded by soft *tjuw tjuw tjuw*... **H** Dry woodland, thornbush, scrub desert. **D** Breeds S Africa. Winters WC Africa.

21 EGYPTIAN NIGHTJAR *Caprimulgus aegyptius* 24–26cm Nocturnal. Underwing pale, wing spots very small or lacking. Tail has small white or buff tips on outermost feathers. **V** Rapid, purring *kowrr-kowrr-kowrr*... slowing towards end. **H** Desert and semi-desert with scattered trees or scrub, often near water. **D** Breeds NW Africa to SC Asia. Winters in E, C and W Africa.

22 SYKES'S NIGHTJAR *Caprimulgus mahrattensis* 23cm Nocturnal. Male in flight shows white spots on outer primaries and white tips to outer-tail feathers; female wing and tail spots smaller and more buff. **V** Soft, even *churr*. **H** Mainly semi-desert with thorn-scrub. **D** Breeds SC Asia. Winters in India.

23 VAURIE'S NIGHTJAR *Caprimulgus centralasicus* 19cm Very little-known, only specimen recorded is thought to be an immature female. **V** Unrecorded. **H** Probably scrub covered sandy foothills, arid plains. **D** W China.

24 NUBIAN NIGHTJAR *Caprimulgus nubicus* 21–22cm Nocturnal. In flight male shows pale rufous underwing; broad white spots on four outer primaries and broad white tips to outermost tail feathers. Female wing and tail spots smaller. **V** Barking *ow-wow* or *ow-wow-wow*. **H** Desert areas with scattered scrub, often near water. **D** NE and N Africa; W and SW Arabian Peninsula.

25 GOLDEN NIGHTJAR *Caprimulgus eximius* 25cm Unmistakable. **V** Very long, low, monotonous *churr*. **H** Dry areas without trees. **D** NE, NC and W Africa.

26 JERDON'S NIGHTJAR *Caprimulgus atripennis* 28cm Crepuscular and nocturnal. Flight pattern similar to larger Large-tailed Nightjar. **V** Liquid, tremulous *ow-r-r-r-r*. **H** Forest edge. **D** SE India, Sri Lanka.

27 LARGE-TAILED NIGHTJAR *Caprimulgus macrurus* 33cm Crepuscular and nocturnal. Flies out from perches or ground to catch insects, also hawks flying insects. In flight, male shows white patch on outer primaries and on tail corners; female's patches buff. **V** Resonant *tok tok tok*... **H** Open forest edges, forest clearings, secondary growth, cultivation. **D** E India to N Australia.

28 MEES'S NIGHTJAR *Caprimulgus meesi* 25–29cm Crepuscular and nocturnal. Makes feeding sorties from exposed branch or possibly ground. Calls from exposed branches and ground. In flight, male shows white patch on outer primaries and tail corners; female's patches buff. **V** High-pitched *piok-piok* or *weelp-weelp*. **H** Dense scrubland with scattered small trees and forest edges. **D** Sumba and Flores, Lesser Sunda Islands, Indonesia.

29 ANDAMAN NIGHTJAR *Caprimulgus andamanicus* 28cm Crepuscular and nocturnal. Flight pattern as Large-tailed Nightjar. **V** Weak, rapid *tyuk tyuk tyuk*... **H** Open forest and open country with scattered trees. **D** Andaman Islands.

30 PHILIPPINE NIGHTJAR *Caprimulgus manillensis* 23–25cm Forages at twilight and by night, flying out from ground or an exposed perch. In flight, shows white patch on outer primaries and tail corners. **V** Harsh, pounding *chuck chur*, repeated, sometimes for several minutes. **H** Primary and secondary forest, pine forest, secondary growth, scrub, bamboo. **D** Philippines.

31 SULAWESI NIGHTJAR *Caprimulgus celebensis* 26cm Crepuscular and nocturnal. Hawks insects over small patches of grassland, alone or in pairs. Reported singing from palms and other trees at coastal scrub edges. In flight, shows white patch on outer primaries and tail corners. **V** Rapid, accelerating series of *chuck* notes, last note softer, fading away. **H** Dry coastal scrub and mangrove edges; in Sula Islands, also occurs along logging roads in lowland forest. **D** Sulawesi and Sula Islands, Indonesia.

32 DONALDSON SMITH'S NIGHTJAR *Caprimulgus donaldsoni* 17–19cm Small size and striking markings aid identification. Hunts after dark, often near water, sometimes near lights. **V** Series of piping, whistled *tuier-tur* notes, first phrase upslurred. **H** Arid scrub, semi-desert. **D** Ethiopia, Somalia, Kenya. Tanzania.

33 BLACK-SHOULDERED NIGHTJAR *Caprimulgus nigriscapularis* 25cm Rather rufous-brown colouring and dark forewing. **V** High, liquid *tjuuu tjuderrr*, second note short. **H** Rainforests, dense woodland, moist scrub. **D** E Ethiopia and NW Somalia to NE Tanzania.

34 FIERY-NECKED NIGHTJAR *Caprimulgus pectoralis* 24cm Note strikingly large head. Sings from tree perch. Male separated from Montane Nightjar by darker neck and collar, from Rufous-Cheeked Nightjar by different habitat and different voice. **V** High *tjuuu-tuwirrrrr*. **H** Wooded areas including gardens. **D** S, SE and SC Africa.

35 MONTANE NIGHTJAR *Caprimulgus poliocephalus* 25cm Note dark appearance. Male's two outer pairs of tail feathers mostly white (buff in female). Sings from dead tree branch or telephone line. **V** Very high fluting *pieuj pjurrr*. **H** Open forests, moist woodland, prefers higher altitudes. **D** SW Saudi Arabia, Ethiopia to Zambia and Angola.

36 INDIAN NIGHTJAR *Caprimulgus asiaticus* 24cm Crepuscular and nocturnal. In flight male shows white spots on outer-primaries and white tips to outer-tail feathers. **V** Far-carrying *chuck-chuck-chuck-chuck-k-k-roo*; in flight *quit-quit*. **H** Open dry forest, semi-desert, dry scrub. **D** SE Pakistan to SE Asia.

37 MADAGASCAN NIGHTJAR *Caprimulgus madagascariensis* 20cm Only nightjar in most of its range, sharing it only in forested areas with Collared Nightjar (Plate 75), which shows distinct buff collar; European Nightjar (vagrant in Madagascar) is much larger. **V** Short rattle preceded by *tjut*. **H** Open and partly open areas. **D** Madagascar.

38 SWAMP NIGHTJAR *Caprimulgus natalensis* 24cm Coarse blotching above and all-white (or buff in female) colouring of outer two pairs of tail feathers. Sings from ground. **V** High staccato *kukukuk*... like engine of small river boat, sustained for up to 30 secs, also high, level *wuwirrrrr*. **H** Damp grassy places near swamps, woodland. Often associated with palms. **D** Widespread in Africa.

39 NECHISAR NIGHTJAR *Caprimulgus solala* 24–25cm First discovered and described in 1990 on basis of a single wing from decomposing corpse. Anecdotal reports of subsequent sightings exist but reports not confirmed. **V** Unknown. **H** Inhospitable arid, rocky scrub. **D** Endemic to Ethiopia.

40 PLAIN NIGHTJAR *Caprimulgus inornatus* 25cm Rather unmarked: occurs in different colour forms in accordance with local soil colour (tawny, brown or grey). **V** Sustained *churr*. **H** Dry, wooded and bushy areas with some grass cover. **D** SW Arabian Peninsula, Mauritania and N Senegal to Ethiopia and Somalia.

1
2
3
4
5
6
7
8
9
10
11
12
13
14 ♂
15
16
17
18
19
20
21
22 ♂
23 ♂
24 ♂
25
26
27
28
29
30
31
32
33
34
35 ♂
36 ♂
37
38
39 ♂
40
♂ rufous morph
♂ grey morph

NIGHTJARS *CONTINUED*

1 STAR-SPOTTED NIGHTJAR *Caprimulgus stellatus* 25cm Note plain upperparts, buff belly and white throat spots. **V** Steady, repeated *pweu* notes. **H** Semi-desert, low-bushy areas with some grass cover. **D** Ethiopia and Somalia to C Kenya.

2 SAVANNA NIGHTJAR *Caprimulgus affinis* 23cm Crepuscular and nocturnal. In flight male shows white patch on outer primaries and white sides to tail. Drinks in flight, dipping over pools to scoop up water, much like swallows; also calls mainly in flight. **V** Repeated *chweep, chwip* or *dheet*. **H** Open woodland, scrubby hillsides, grassland with rocky outcrops. **D** Widespread in S Asia, SE Asia, Indonesia, Borneo and Philippines.

3 FRECKLED NIGHTJAR *Caprimulgus tristigma* 30cm Large, with dark uniform freckling, pale edge to scapulars. Rests by day on bare rock (never in trees); sings from ground. May nest on flat roofs in cities. **V** Very high staccato *whewwhew whewwhew whewhew*... sustained for up to 30 secs. **H** Bare (or sparsely overgrown) rocky outcrops and ravines. **D** Widespread in Africa.

4 BONAPARTE'S NIGHTJAR *Caprimulgus concretus* 20cm Crepuscular and nocturnal. Flies after insects from a perch; often forages along rivers. **V** Low, mournful, disyllabic *waouuuu*. **H** Primary lowland forest, often near rivers. **D** Sumatra, Indonesia; Borneo.

5 SALVADORI'S NIGHTJAR *Caprimulgus pulchellus* 24cm Crepuscular and nocturnal. Forages over forest clearings and near cliffs with slow, flapping wingbeats and frequent glides. Also makes short sallies from perch to capture insects. In flight, male shows white patch on outer primaries and tail corners; female's patches buff. **V** Five *tock* notes, in irregular pattern. **H** Montane and submontane forest, often on cliffs or cliff faces; occasionally in small marshes. **D** Sumatra and Java, Indonesia.

6 PRIGOGINE'S NIGHTJAR *Caprimulgus prigoginei* 20cm Very rare, only a female ever found. Small, rather rufous. **V** Not known. **H** Probably forest. **D** E DR Congo.

7 BATES'S NIGHTJAR *Caprimulgus batesi* 30cm Large, very dark rufous with grey-capped appearance. Female has rufous-brown (not grey) crown and lacks white in tail and wings. **V** Loud, rapid *whow-whow-whow*... (5–20 notes). **H** Probably forest. **D** Cameroon to Uganda and E DR Congo.

8 LONG-TAILED NIGHTJAR *Caprimulgus climacurus* 35cm White spots on outer primaries. Rather similar to Fiery-necked Nightjar (Plate 76) but more rufous, and with wider range. Sings from ground. **V** High (sometimes low) very fast rattled *rrrrr*..., faster than sewing machine. **H** Semi-desert, bushy and wooded grassland, extensive fields, open woodland. **D** C, WC and W Africa.

9 SLENDER-TAILED NIGHTJAR *Caprimulgus clarus* 30cm Separated from Black-shouldered Nightjar (Plate 76) by duskier, less rufous colouring, shorter tail and slower song tempo. Sings from ground. **V** High staccato sustained *ririririri*... like a rapid sewing machine. **H** Dry wooded and bushy areas. **D** Ethiopia and Somalia to C Tanzania.

10 SQUARE-TAILED NIGHTJAR *Caprimulgus fossii* 24cm All-white (buffy in female) outer-tail feathers (just terminal half white on next pair). Sings from ground (and maybe from tree perch). **V** Mid-high rapid staccato rattle, irregular, alternating between *ru* and *ri* notes; *rururuririririri*... sustained for up to three minutes. **H** Grassy and sandy areas, often at woodland edges near lakes and rivers. **D** SC and SE Africa.

11 STANDARD-WINGED NIGHTJAR *Caprimulgus longipennis* 25cm Note pale eyebrow. In non-breeding plumage looks small, no white in wing or tail. Buff scapulars and hind-collar. **V** Very high, sharp, sustained twitter *titititititititititi*... with slight decelerations. **H** Bushy and more or less wooded habitats. **D** Senegal and Gambia to Liberia, E to SW Sudan and Uganda.

12 PENNANT-WINGED NIGHTJAR *Caprimulgus vexillarius* 30cm Non-breeding plumage of male retains most of striking wing pattern but lacks pennants. Note absence of white or buff wing spots in female. Sings in flight or from low perch. **V** Extremely high, insect-like, shivering phrases of irregular length, *sisisisisisi*... **H** Woodland, often in stony, hilly areas. **D** Angola, S DR Congo and SW Tanzania to Namibia, Botswana and E South Africa.

OWLET-NIGHTJARS AEGOTHELIDAE

13 FELINE OWLET-NIGHTJAR *Aegotheles insignis* 29–30cm Recalls miniature frogmouth. Roosts by day. Hunts insects at night, in flight and on ground. Beautifully patterned plumage affords good camouflage. **V** Calls include harsh, screaming *woOoah* and other vaguely cat-like sounds. **H** Montane rainforest. **D** New Guinea.

14 STARRY OWLET-NIGHTJAR *Aegotheles tatei* 24–25cm Similar to but smaller than Feline Owlet-nightjar. Nocturnal. Feeds on insects, caught in flight and on ground. Roosts in cover in daytime. **V** Calls include various cat-like screams and yowling. **H** Lowland rainforest. **D** New Guinea.

15 MOLUCCAN OWLET-NIGHTJAR *Aegotheles crinifrons* 29cm Encountered singly, in pairs, occasionally small groups. Hunts small flying insects in short sallies from perches in mid-storey. **V** Weak, upslurred scream, followed by three more unhurried screams at same pitch; when alarmed gives various manic screams, cackles and cat-like yowling. **H** Primary and tall secondary lowland and hill forest, forest edges, selectively logged forest; occasionally in lightly wooded cultivation and coconut plantations. **D** Maluku Islands, Indonesia.

16 WALLACE'S OWLET-NIGHTJAR *Aegotheles wallacii* 20–22cm Nocturnal. Feeds on insects, caught with huge gape. **V** Calls include shrill, screeching almost budgie-like *pee-eh*. **H** Forests and wooded habitats including gardens; mainly lowlands. **D** New Guinea.

17 MOUNTAIN OWLET-NIGHTJAR *Aegotheles albertisi* 18–20cm Beautifully patterned plumage provides good camouflage when roosting. Feeds at night on insects. **V** Series of shrill and explosive notes, like squeaky toy. **H** Montane forest, secondary woodland. **D** New Guinea.

18 NEW CALEDONIAN OWLET-NIGHTJAR *Aegotheles savesi* 28cm Critically Endangered. No confirmed recent sightings and possibly extinct. Combination of large size, uniformly dark plumage and geographical range diagnostic. **V** Unknown. **H** Humid forest. **D** New Caledonia.

19 BARRED OWLET-NIGHTJAR *Aegotheles bennettii* 20–23cm Plumage marked with beautiful vermiculation. Formerly considered race of Vogelkop Owlet-nightjar. **V** Agitated-sounding loud squeaky notes including *kee, ke-ke*. **H** Lowland rainforest. **D** New Guinea.

20 VOGELKOP OWLET-NIGHTJAR *Aegotheles affinis* 23–24cm Like other owlet-nightjars, probably roosts mainly in tree holes. Hunts insects after dark. **V** Various shrill, almost parrot-like squeaky notes. **H** Hill forest; mostly 500–900m. **D** Vogelkop Mountains, New Guinea.

21 AUSTRALIAN OWLET-NIGHTJAR *Aegotheles cristatus* 22–25cm Nocturnal. Feeds at night by dropping from perch to catch insects on ground and in foliage. Also feeds on wing. **V** Loud, piercing shriek *keer* or *keer-ke-ke*. **H** Open woodland, deserts, mallee scrub, mangroves. **D** Widespread in Australia.

TREESWIFTS HEMIPROCNIDAE

22 CRESTED TREESWIFT *Hemiprocne coronata* 23–25cm Perches on branches, occasionally wires. In flight tail often held in 'spike', recalling larger version of Asian Palm Swift (Plate 80). At rest, tail extends past folded wings. **V** In flight utters harsh *whit-tuck whit-tuck* or *ti-chuck ti-chuck*. **H** Open deciduous forests, open areas. **D** India to SE Asia.

23 GREY-RUMPED TREESWIFT *Hemiprocne longipennis* 21–25cm Makes long sorties from favourite bare branch in tree canopy. At rest tail does not extend beyond folded wings. **V** Harsh, piercing *ki ki-ki-kew*, staccato *chi-chi-chi-chew* and *too-it*. **H** Well-exposed bare branches in forest and open wooded areas. **D** Peninsular Malaysia to Sulawesi, Indonesia.

24 WHISKERED TREESWIFT *Hemiprocne comata* 15–17cm In flight, from below shows dark underwing, white trailing edge, white undertail coverts. Spends much time perched, making only short flights to catch insects below tree canopy. **V** High-pitched, chattering *she-she-she-she-shoo-shee*, also plaintive *chew*, given when perched. **H** Clearings and edges of broadleaved evergreen forest. **D** Peninsular Malaysia; Sumatra, Indonesia; Borneo; Philippines.

25 MOUSTACHED TREESWIFT *Hemiprocne mystacea* 28–31cm Usually in pairs or loose groups, at dusk may form larger flocks. Makes aerial sallies from open branches after flying insects, generally more active at twilight. **V** Calls include short, hard *skip* and buzzing *krerr*; in flight, utters high-pitched, down-slurred *kiiee*. **H** Open areas with trees, forest edges, riverine areas, mangroves, coastal woodland. **D** Maluku Islands, Indonesia.

1
2
3
4
5
6
7
8
9
10
11
12
brown morph
13
rufous morph
14
15
rufous morph
16
brown morph
17
rufous morph
18
19
20
rufous morph
21
brown morph
dark morph
♀
22
♂
♀
23
♂
♂
24
♀
♂
25
♀

SWIFTS APODIDAE

1 SPOT-FRONTED SWIFT *Cypseloides cherriei* 14cm White dots at base of bill and behind eye, smaller in female. Belly faintly spotted paler; otherwise smoky blackish. Tail short and square, shafts emergent. **V** Abrupt, explosive shrill shrieks and chatters. **H** Forested mountains with deep valleys. **D** Costa Rica to N Venezuela and N Ecuador.

2 WHITE-CHINNED SWIFT *Cypseloides cryptus* 14cm Smoky black, white on chin hard to see or absent. Short square tail with emergent barbs. **V** Very high, fast, chuckling chatters. **H** Savannah, forest, mountains, lowlands. **D** Belize to Guyana, N Brazil and SE Peru.

3 WHITE-FRONTED SWIFT *Cypseloides storeri* 14cm Similar to White-chinned Swift but with paler forehead. **V** High, dry, accelerating chatter. **H** Over forest. **D** SW Mexico.

4 SOOTY SWIFT *Cypseloides fumigatus* 15cm Sooty black, square tail with emerging barbs. **V** Fast, clipped squeaks and chatters. **H** Over forest, cleared land, flooded land, hills, waterfalls. **D** E Bolivia, E Paraguay, NE Argentina, SE Brazil.

5 ROTHSCHILD'S SWIFT *Cypseloides rothschildi* 19cm All brown. Generally like Great Dusky Swift, but with different distribution. **V** Sharp chirping notes. **H** Savannahs, woods. **D** S Bolivia, S Peru, NW Argentina.

6 AMERICAN BLACK SWIFT *Cypseloides niger* 15–18cm Gregarious, often feeds high in the air in large flocks. Flight less erratic than smaller swifts, wingbeats shallow. **V** Soft *chip-chip*. **H** Mountain areas, often occurs near waterfalls, less common in lowlands and coastal areas. **D** Breeds in W North America and Central America. Winters South America.

7 WHITE-CHESTED SWIFT *Cypseloides lemosi* 15cm Sooty black with white patch on breast. Tail slightly forked. **V** Pleasant, sweet-toned, brief single notes and trilled chatter. **H** Upland and mountain grasslands, eroded foothills. **D** Upper Cauca valley, SW Colombia, NC Colombia, NW Peru.

8 GREAT DUSKY SWIFT *Cypseloides senex* 18cm Hefty. Sooty blackish brown, pale head. Black in front of eye. Square tail. **V** Very fast, rather low-pitched, sustained chatter. **H** Rainforest, farmland, over cities, associated with waterfalls. **D** C Brazil to NE Bolivia, E Paraguay and NE Argentina.

9 TEPUI SWIFT *Streptoprocne phelpsi* 16.5cm Like Chestnut-collared Swift but chin, throat, neck and breast orange-rufous (darker in female). Long tail slightly forked. **V** Fast, dry, irregularly spaced *chip-chip* notes. **H** Mountains near rock-faces and waterfalls. **D** South America.

10 CHESTNUT-COLLARED SWIFT *Streptoprocne rutila* 15cm Sooty black with throat, neck and breast chestnut (reduced in female), chin dusky. Long square tail. **V** Very dry and toneless insect-like chatter. **H** Mountains, forest, grassland, towns. **D** Widespread in Central America and South America.

11 WHITE-COLLARED SWIFT *Streptoprocne zonaris* 20–22cm Gregarious, often in large flocks; forages, with rapid and agile flight, above forests. **V** Shrill *sreee-screee*; also rapid *chip-chip-chip*. **H** Montane forest areas; visits lowland areas in bad weather. **D** Widespread throughout the Americas.

12 BISCUTATE SWIFT *Streptoprocne biscutata* 21cm Like White-Collared Swift but collar broken on sides of neck. Square tail. **V** Fast, tuneful *teu teu* notes, recalling a shorebird. **H** Over all habitats, including urban areas. **D** E South America.

13 WHITE-NAPED SWIFT *Streptoprocne semicollaris* 20cm Note white neck collar. Resembles White-collared Swift; heavy-looking. **V** Sharp, discordant notes given in uneven chatter. **H** Mainly in montane skies. **D** W Mexico.

14 GIANT SWIFTLET *Hydrochous gigas* 16cm Gregarious, often flies with other swifts. **V** Sharp *wicker* notes. **H** Mountain forest with nearby waterfalls. **D** Peninsular Malaysia; Greater Sunda Islands.

15 PLUME-TOED SWIFTLET *Collocalia affinis* 9–10cm Nests colonially in caves. Flight fast and gliding, interspersed with periods of fluttering when prey is abundant. **V** Shrill, excited twittering calls. **H** Feeds over moist lowland forest. **D** Malay Peninsula; Borneo; Sumatra, Indonesia; E Indian Ocean islands.

16 GREY-RUMPED SWIFTLET *Collocalia marginata* 9–10cm Formerly treated as race of Glossy Swiftlet. Nests in caves and rooves of buildings, making cup-shaped nest. **V** Shrill twitter. **H** Moist lowland forest. **D** Philippines.

17 RIDGETOP SWIFTLET *Collocalia isonota* 9–10cm Nests in buildings, sometimes in small colonies. Formerly treated as race of Glossy Swiftlet. **V** Shrill, disyllabic twittering whistle. **H** Feeds over open forested habitats. **D** Philippines.

18 TENGGARA SWIFTLET *Collocalia sumbawae* 9cm Small size and geographical location aid identification. Formerly treated as race of Glossy Swiftlet. **V** Sweet-sounding twitters. **H** Feeds over forested habitats. **D** W Lesser Sunda Islands.

19 DRAB SWIFTLET *Collocalia neglecta* 9–10cm Nesting associated with human habitation. Formerly treated as race of Glossy Swiftlet. **V** Excited, squeaky twitters. **H** Feeds over moist lowland forested habitats. **D** S, E and C Lesser Sunda Islands.

20 GLOSSY SWIFTLET *Collocalia esculenta* 10cm Gregarious. Banking and gliding flight interspersed with fluttering, bat-like wingbeats. **V** Sharp twitter. **H** Open and forested areas. **D** Sulawesi, Indonesia, to New Guinea and Solomon Islands.

21 SATIN SWIFTLET *Collocalia uropygialis* 9–10cm Flight combines glides and fluttering. Formerly treated as race of Glossy Swiftlet. **V** Harsh, shrill twitters. **H** Feeds over moist forested habitats. **D** New Caledonia; Loyalty Islands; Vanuatu; Santa Cruz islands, Solomon Islands.

22 BORNEAN SWIFTLET *Collocalia dodgei* 9–10cm Often considered a race of Cave Swiftlet; behaviour presumably similar. **V** Presumed to be similar to Cave Swiftlet. **H** Mountainous areas. **D** N Borneo.

23 CAVE SWIFTLET *Collocalia linchi* 10cm Flies low, often in loose flocks. **V** High-pitched *cheer-cheer*. **H** Over forest and open country. **D** Java, Indonesia; Borneo.

24 CHRISTMAS ISLAND SWIFTLET *Collocalia natalis* 9–10cm Formerly treated as race of Glossy Swiftlet. Often seen feeding in flocks. Nests in limestone caves. **V** Shrill twitter. **H** Feeds over lowland tropical rainforest. **D** Christmas Island, E Indian Ocean.

25 PYGMY SWIFTLET *Collocalia troglodytes* 9cm Forages in small groups. **V** Short, throaty *chirp*. **H** Lowland forests, inland waters. **D** Philippines.

26 SEYCHELLES SWIFTLET *Aerodramus elaphrus* 11cm No other similar small, uniform swift in range. **V** Normally silent outside breeding caves. **H** Over any sort of terrain. **D** Seychelles.

27 MASCARENE SWIFTLET *Aerodramus francicus* 11cm No similar bird in range, but may recall a larger Mascarene Martin (Plate 205), but with different wing shape. **V** Dry, quiet clicking notes. **H** Over all habitat types. **D** Mauritius, Réunion.

28 INDIAN SWIFTLET *Aerodramus unicolor* 12cm Gregarious, often in huge numbers at cave roosts. Typical banking and gliding flight, interspersed with bat-like fluttering wingbeats. **V** Twittering *chit-chit-chit* given at roosts, also harsh tern-like call note. **H** Rocky hills, scrub, dry forest with nearby cliffs. **D** SW India, Sri Lanka.

29 PHILIPPINE SWIFTLET *Aerodramus mearnsi* 10–11cm Forages above forests and in clearings. **V** Rhythmic series of *wi-ch-chew* notes. **H** Submontane forests with nearby caves. **D** Philippines.

30 HALMAHERA SWIFTLET *Aerodramus infuscatus* 10cm Usually forages in flocks of 10–30, often with other swiftlets. **V** Long, rhythmic series of *wee-chi-chi* notes. **H** Forests, disturbed areas, open areas. **D** N Maluku Islands, Indonesia.

31 SULAWESI SWIFTLET *Aerodramus sororum* 10cm Behaviour presumed to be similar to Halmahera Swiftlet, with which it was formerly considered conspecific. **V** Presumably similar to Halmahera Swiftlet. **H** Presumably similar to Halmahera Swiftlet. **D** Sulawesi, Indonesia.

32 SERAM SWIFTLET *Aerodramus ceramensis* 10cm Formerly considered to be a race of Halmahera Swiftlet; behaviour presumed to be similar. **V** Presumably similar to Halmahera Swiftlet. **H** Presumably similar to Halmahera Swiftlet. **D** Buru and Seram islands, S Maluku Islands, Indonesia.

33 MOUNTAIN SWIFTLET *Aerodramus hirundinaceus* 11–13cm Nests in caves. Cup-shaped nest made from plant material and saliva, stuck to cave wall. **V** Agitated-sounding series of squeaky notes that alternate in pitch. **H** Feeds over moist tropical forests. **D** New Guinea.

34 WHITE-RUMPED SWIFTLET *Aerodramus spodiopygius* 11cm Has paler rump than other swiftlets within its range. **V** High *weetweet* or *tjurreweet*. **H** Over wide range of habitats including grasslands, roads, forest. **D** Bismarck Archipelago to Samoa.

35 AUSTRALIAN SWIFTLET *Aerodramus terraereginae* 11–12cm Breeds in colonies in caves. Cup-shaped nest comprises plant material glued together with saliva and stuck to cave wall. **V** Squeaky, twittering calls. **H** Feeds over forests and open habitats; mostly sea level to 500m. **D** Queensland.

36 HIMALAYAN SWIFTLET *Aerodramus brevirostris* 13–14cm Regularly seen in flocks flying around mountain peaks and ridges. Flight often interspersed with bat-like fluttering. Forms large roosts, usually in caves. **V** Twittering *chit chit*, also low rattle. **H** Over wooded river valleys in mountainous areas. **D** NE India to SC China and Peninsular Malaysia.

37 VOLCANO SWIFTLET *Aerodramus vulcanorum* 14cm Occurs in fast-flying flocks around open peaks and ridges of highest mountains. **V** Piercing *teeree-teeree-teeree*. **H** Mountains, volcanic rims. **D** W Java, Indonesia.

38 WHITEHEAD'S SWIFTLET *Aerodramus whiteheadi* 14cm Little recorded; note broad head. **V** Unrecorded. **H** Forested mountain areas. **D** Philippines.

1
2
3
4
5
6
7
8
9
10
11
12
13
14
15
16
17
18
19
20
21
22
23
24
25
26
27
28
29
30
31
32
33
34
35
36
37
38

SWIFTS *CONTINUED*

1 BARE-LEGGED SWIFTLET *Aerodramus nuditarsus* 14cm Gathers in small flocks when feeding is good. Bare legs not useful for field identification. **V** Unknown but probably shrill twitters. **H** Feeds over moist tropical forests. **D** New Guinea.

2 MAYR'S SWIFTLET *Aerodramus orientalis* 14cm Overall dark plumage contrasts with paler grey rump. **V** Unknown but probably shrill twitters. **H** Feeds over highland forest. **D** New Ireland (Bismarck Archipelago), Guadalcanal, Solomon Islands.

3 MOSSY-NEST SWIFTLET *Aerodramus salangana* 12cm Forages over and sometimes within forests. **V** Shrill twitterings, chirrups, burbling notes. **H** Primary forest with nearby caves. **D** Sumatra and Java, Indonesia; Borneo.

4 UNIFORM SWIFTLET *Aerodramus vanikorensis* 13cm Gregarious, usually in small to large flocks, often alongside other swiftlet species. **V** Calls *zoo-zu-chee-chee*; also squeaky trill. **H** Forest, grassland, ponds, mangroves, with nearby caves. **D** Sulawesi, Indonesia, to Vanuatu.

5 AMELINE SWIFTLET *Aerodramus amelis* 13cm Considered by some authorities to be a race of the Uniform Swiftlet; actions presumed to be similar. **V** Presumably similar to Uniform Swiftlet. **H** Similar to Uniform Swiftlet. **D** Philippines.

6 PALAU SWIFTLET *Aerodramus pelewensis* 11cm Compare White-rumped Swiftlet (Plate 78). **V** Dry clicking call has been recorded. **H** Over canyons. **D** Palau, WC Pacific Ocean.

7 MARIANA SWIFTLET *Aerodramus bartschi* 11cm Compare White-rumped Swiftlet (Plate 78). Darker rump than similar swiftlets. **V** Has simple double-click call for echolocation in caves. **H** Over forest, forest remains, mangrove. **D** Mariana Islands, WC Pacific Ocean.

8 ISLAND SWIFTLET *Aerodramus inquietus* 11cm Compare White-rumped Swiftlet (Plate 78). Note dark rump, slightly paler than mantle. **V** Makes clicks for echolocation in caves. **H** Open and forested areas. **D** Caroline Islands, Micronesia.

9 TAHITI SWIFTLET *Aerodramus leucophaeus* 11cm Compare White-rumped Swiftlet (Plate 78). **V** Similar to other swiftlets. **H** Over rivers and forested valleys. **D** Society Islands, French Polynesia.

10 ATIU SWIFTLET *Aerodramus sawtelli* 10cm Compare White-rumped Swiftlet (Plate 78). Somewhat contrasting paler underparts. **V** Similar to other swiftlets. **H** Over forest and open areas. **D** Cook Islands, C Pacific Ocean.

11 MARQUESAN SWIFTLET *Aerodramus ocistus* 11cm Compare White-rumped Swiftlet (Plate 78). **V** High *tree-tuh*. **H** Forest canopy, forest edges. **D** Marquesas Islands, French Polynesia.

12 BLACK-NEST SWIFTLET *Aerodramus maximus* 12–13cm Most active in crepuscular period; gregarious, often with other swifts. **V** Similar to Himalayan Swiftlet (Plate 78). **H** Open areas, offshore islets, urban areas, occasionally over forests. **D** Peninsular Malaysia; Greater Sunda Islands.

13 EDIBLE-NEST SWIFTLET *Aerodramus fuciphagus* 12cm Gregarious, often with other swifts and swallows. Flight actions similar to Glossy Swiftlet. Grey rump sometimes indistinct. **V** Loud, metallic *zwing*. **H** Open areas, offshore islets, forests, mangroves. **D** Andaman Islands; Lesser Sunda Islands.

14 GERMAIN'S SWIFTLET *Aerodramus germani* 11–12cm Gregarious, often alongside other swifts and swallows. Colonial, breeds on cliff faces, sometimes on buildings. **V** Similar to Edible-nest Swiftlet; also utters various *chip* notes for echolocation. **H** Open areas, offshore islets, urban areas, occasionally over forests. **D** SE Asia, Borneo, Philippines.

15 THREE-TOED SWIFTLET *Aerodramus papuensis* 14cm Uniformly brown plumage lacks any contrast on rump. Only swift species to lack fourth toe. **V** Unknown, but probably shrill twitters. **H** Forests, secondary woodland, open country. **D** N New Guinea.

16 SCARCE SWIFT *Schoutedenapus myoptilus* 17cm Note tightly closed tail, narrow wings, contrasting black eye patch. Best identified in mixed flocks by voice. **V** Combination of high rapid *wutwutwut* with diagnostic *tiktik tikkerik*. **H** Rocky mountains and cliffs; feeds mainly over forests. **D** E and EC Africa.

17 SCHOUTEDEN'S SWIFT *Schoutedenapus schoutedeni* 17cm Generally similar to Scarce Swift, with which it may be conspecific, but with uniform dark plumage. Very rare. **V** Not known. **H** Probably a highland species. **D** E DR Congo.

18 PHILIPPINE SPINE-TAILED SWIFT *Mearnsia picina* 14cm Usually seen alone or in small groups, foraging above forests or nearby cleared areas. Sometimes seen with other swift species flying low, through passes or along logging tracks. **V** Unrecorded. **H** Forest areas. **D** S Philippines.

19 PAPUAN SPINE-TAILED SWIFT *Mearnsia novaeguineae* 11–12cm Flight very fast. In outline, note stocky body and short, rounded tail. **V** Thin, sharp *sweit sweit* calls. **H** Feeds over moist forest and fresh water in lowlands. **D** New Guinea.

20 MADAGASCAN SPINETAIL *Zoonavena grandidieri* 12cm No other small swift in range except occasional Little Swift (Plate 80) which shows more distinct white rump and throat. **V** Normally silent. **H** Forest. **D** Madagascar.

21 SAO TOME SPINETAIL *Zoonavena thomensis* 14cm Very small. Flutters around large trees, constantly twittering. **V** High-pitched piercing *tririririri*. **H** Forest clearings, plantations. **D** São Tomé, Príncipe, Gulf of Guinea.

22 WHITE-RUMPED SPINETAIL *Zoonavena sylvatica* 11cm Usually seen in small groups, hawking insects over forests. Underparts whitish, contrasting with dark breast and throat. **V** In flight gives twittering *chick-chick*, also transcribed as *swicky-sweezy*. **H** Broadleaved evergreen and moist deciduous forests. **D** Nepal, India.

23 MOTTLED SPINETAIL *Telacanthura ussheri* 14cm Note white belly-patch, connected with larger rump patch and wing shape. **V** Very high, dry *tchu-tchu tchrrrchrrr* or mid-high rattling *rururururururu*. **H** Forest edges, dry woodland, especially near baobab. **D** Widespread in Africa.

24 BLACK SPINETAIL *Telacanthura melanopygia* 14cm Note mottled throat and fluttering flight. **V** High, dry, sparrow-like *tsju tsjrrrsjrr*. **H** Forest edges, dry woodland, especially near baobab. **D** Sierra Leone to Cameroon and Gabon; SW Central African Republic and N DR Congo.

25 SILVER-RUMPED SPINETAIL *Rhaphidura leucopygialis* 11cm Flight fluttery, erratic. Forages around clearings and rocky outcrops. **V** High-pitched *tirrr-tirrr* and rapid chattering. **H** Broadleaved evergreen forests, clearings, secondary growth. **D** Peninsular Malaysia; Greater Sunda Islands.

26 SABINE'S SPINETAIL *Rhaphidura sabini* 12cm Note bluish sheen over upperparts. Black tail concealed between white coverts. **V** Extremely high, sharp *titititutu-tu-tu-tu*. **H** Rainforest edges, often over water. **D** Sierra Leone to W Uganda and W Kenya.

27 CASSIN'S SPINETAIL *Neafrapus cassini* 15cm Larger than Boehm's Spinetail, also with extremely short tail, but with narrower white band across tail base; different range. **V** Extra-high, soft, twittering notes. **H** Lakes, forest. **D** Ivory Coast to W Uganda.

28 BÖHM'S SPINETAIL *Neafrapus boehmi* 10cm Note diagnostic, seemingly tail-less appearance, white underparts and rump, characteristic wing shape. Slow, erratic, bat-like flight around trees. **V** Very high, twittered, fast trill. **H** Open woodland, especially near baobab trees. **D** E and SE Africa.

29 WHITE-THROATED NEEDLETAIL *Hirundapus caudacutus* 21–22cm Flight fast, powerful. White throat and white horseshoe-shaped area from undertail to flanks. **V** Weak, high-pitched twittering. **H** Forested and open areas. **D** Breeds E and S Asia. Winters Australasia.

30 SILVER-BACKED NEEDLETAIL *Hirundapus cochinchinensis* 20–22cm Usually over forests, in small groups. Fast, powerful flight. Note darker throat compared to White-throated Needletail. **V** Soft, rippling *trp-trp-trp-trp-trp*. **H** Forested and open areas, rivers near or in forests. **D** Widespread in SE Asia and Greater Sundas.

31 BROWN-BACKED NEEDLETAIL *Hirundapus giganteus* 23–25cm Typical powerful flight; wings make whooshing sound overhead. Hawks insects over forests and grasslands. Drinks at pools or rivers in evenings, scooping up water in flight. **V** In flight utters slow, rippling trill; also squeaky, repeated *cirrwiet* and thin *chiek*. **H** Forested and open areas. **D** SW India, Sri Lanka, SE Asia, Greater Sundas, Borneo.

32 PURPLE NEEDLETAIL *Hirundapus celebensis* 25cm Very fast flight, wings producing jet-like swoosh at close range. White horseshoe shape from undertail to flanks. **V** Unrecorded. **H** Forest in lowlands and hills, open country, urban areas. **D** Philippines; N Sulawesi, Indonesia.

33 LESSER ANTILLEAN SWIFT *Chaetura martinica* 11cm Gregarious, generally in flocks of 20–40, often associates with swallows and martins. **V** Soft twittering. **H** Mainly over mountain forests, occasionally lowland forests, open areas. **D** Lesser Antilles.

34 BAND-RUMPED SWIFT *Chaetura spinicaudus* 11cm Pale saddle sharply demarcated. **V** Thin, very high-pitched twitters and trills. **H** Over any type of country. **D** Panama to C Brazil.

35 COSTA RICAN SWIFT *Chaetura fumosa* 11cm Has more extensive pale rump than Band-rumped Swift. **V** High-pitched, slightly dry twittering. **H** Any type of country. **D** Costa Rica, Panama, N Colombia.

36 PALE-RUMPED SWIFT *Chaetura egregia* 10.5cm Like Grey-Rumped Swift (sometimes considered conspecific) but rump and tail coverts white, belly and flanks greyish brown. **V** Rapid, sweet and high-pitched twittering. **H** Over tropical lowland forest, hills, mountains, open areas. **D** W Amazonia.

37 GREY-RUMPED SWIFT *Chaetura cinereiventris* 11cm Gregarious, usually in small flocks of 20–30. May hover to take insects from treetops. **V** Light twittering. **H** Over forests. **D** Widespread in Central America, South America.

38 VAUX'S SWIFT *Chaetura vauxi* 12cm Western counterpart of Chimney Swift (Plate 80), generally paler on underparts and rump. May be seen together on migration, especially in California. **V** High-pitched, buzzy *tip tip tip tipto tipto tzeeeerip* or *chip chip chip cheweet-cheweet*. **H** Forests, urban areas. **D** W Canada to N South America.

1
2
3
4
5
6
7
8
9
10
11
12
13
14
15
16
17
18
19
20
21
22
23
24
25
26
27
28
29
30
31
32
33
34
35
36
37
38

SWIFTS *CONTINUED*

1 SICK'S SWIFT *Chaetura meridionalis* 14cm Note pale saddle and throat. **V** High-pitched *kip* notes, sometimes in accelerating chatter. **H** Over edges of lowland forest. **D** S Brazil to E Bolivia, Paraguay, N Argentina.

2 CHIMNEY SWIFT *Chaetura pelagica* 12–14cm Usually gregarious. Wingbeats rapid; bat-like fluttering when feeding low down. **V** Chattering *chip chip chip chip*... **H** Over woodland, fields, villages, towns. **D** Breeds SC, SE Canada and C, E USA. Winters Panama, South America.

3 CHAPMAN'S SWIFT *Chaetura chapmani* 13.5cm Short square tail with protruding barbs. Mostly uniform dark smoky brown; throat barely paler. Tail coverts and rump slightly paler and greyer. **V** Mid-pitched, chuckling twitters. **H** Over wooded hills, plains with trees, tropical forest, secondary forest, swamps, mangroves, coastal areas. **D** Panama to NE Brazil.

4 MATO GROSSO SWIFT *Chaetura viridipennis* 13–14cm Similar to Chapman's Swift with cigar-shaped body. Separated by geographical range. **V** High-pitched, thin tweets. **H** Feeds over moist lowland forest, secondary woodland, swamps. **D** Brazil, Bolivia, Colombia, Ecuador, Peru.

5 SHORT-TAILED SWIFT *Chaetura brachyura* 10cm Short-bodied but hefty. Very short square tail with protruding barbs. Black with lower body and tail pale brownish grey. Secondary flight feathers form aileron next to body. **V** High-pitched, sometimes sharp-sounding twitters and tweets. **H** Over forest and partially open areas in lowlands and mountains, grassland, semi-arid areas, coastal areas, beaches, mangroves. **D** Panama through Amazonia.

6 WHITE-THROATED SWIFT *Aeronautes saxatalis* 15–18cm Gregarious. Fast, dashing flight with swoops and quick changes in direction. White trailing edge to secondaries. **V** Descending *ki ki ki kir kir kiir kiir kirsh krrsh krrsh*. **H** Mountains, cliffs, rocky canyons. **D** W North America, Central America.

7 WHITE-TIPPED SWIFT *Aeronautes montivagus* 12.5cm Blackish. White bib extends in line down mid-breast to belly. White patch on lower belly and part of flanks. **V** Very fast, buzzing, nightjar-like trill. **H** Montane forest, semi-open areas, secondary growth and cleared areas, slopes, deep gorges, valleys. Occasionally in lowlands. **D** Venezuela to NW Argentina.

8 ANDEAN SWIFT *Aeronautes andecolus* 14.5cm Fast and rowdy. Long forked tail. Upperparts dark greyish brown. Collar, rump and underparts whitish, contrasting with greyish-brown underwing, tail and vent. **V** Rapid, rather grating twitter. **H** Rocky and arid mountains, scrub with bushes, open woods, cactus-studded slopes. **D** Peru to C Argentina.

9 ANTILLEAN PALM SWIFT *Tachornis phoenicobia* 10–11cm Flight rapid, bat-like, interspersed with glides. Usually in small groups, occasionally joined by swallows and martins. **V** Constant, high-pitched, weak twittering. **H** Open cultivated areas, urban areas, with nearby palms. **D** Greater Antilles.

10 PYGMY PALM SWIFT *Tachornis furcata* 10cm Like miniature Neotropical Palm Swift. Upper tail has white base. Dirty white on throat and mid-belly only. Very fast wingbeats. Uses nests as roost year-round. **V** Fast, high-pitched twitter. **H** Forest, open land, cleared areas, scattered trees, palm groves where it nests. **D** NE Colombia, W Venezuela.

11 NEOTROPICAL PALM SWIFT *Tachornis squamata* 14cm Small. Slender, deeply forked long tail. Smoky blackish upperside. Underparts whitish with variable dirty markings except on belly. Flanks and undertail coverts scaled white. Forms mixed flocks with swallows. Uses nest as roost year-round. **V** Dry *kir* notes, sometimes in accelerating series. **H** Open ground, forest, marshes, savannah, palm groves, urban areas. **D** Amazonia, E Brazil.

12 GREAT SWALLOW-TAILED SWIFT *Panyptila sanctihieronymi* 19cm Little overlap of range with Lesser Swallow-tailed Swift. **V** Shrill notes in series that accelerates then slows. **H** Mainly over montane areas. **D** SW Mexico to Honduras.

13 LESSER SWALLOW-TAILED SWIFT *Panyptila cayennensis* 13cm Long forked tail; feathers pointed. Blackish. Bib, nuchal collar, patch on lower flank and base of outer-tail feathers white. **V** Fast, high twitters interspersed with more drawn-out, downslurred notes. **H** Over forest, cleared areas, open fields in plains and mountains, slopes, foothills. **D** Mexico to N Bolivia, SE Brazil, Trinidad and Tobago.

14 AFRICAN PALM SWIFT *Cypsiurus parvus* 16cm Tail deeply forked but usually kept closed. Flight fast, fluttering, agile, glides with wings held below horizontal. **V** High-pitched *sisisi-soo-soo* or *skiiirrrrrrrr*. **H** Open country, towns and villages with nearby palms. **D** Widespread in Africa.

15 ASIAN PALM SWIFT *Cypsiurus balasiensis* 13cm Slim. Often in small active groups. Flight agile, rapid, with fluttering wingbeats and short glides. **V** Trilling *te-he-he-he-he* or *tititee*. **H** Open country, cultivation, urban areas, usually with nearby palms. **D** Widespread in S and SE Asia, Indonesia, Borneo and Philippines.

16 ALPINE SWIFT *Tachymarptis melba* 20–22cm Flight fast, powerful with long glides on slightly drooping wings, giving falcon-like impression. Usually associates with other swifts when foraging. **V** High-pitched twittering *trrr-titititititititititityiti-ti-ti-ti*..., rising and falling in pitch, accelerating then normally slowing at end. **H** Virtually any habitat but especially over mountain or hill areas. **D** Widespread in Eurasia and Africa.

17 MOTTLED SWIFT *Tachymarptis aequatorialis* 22cm Large, mottled below. Mantle darker than head and rump. **V** Extremely high sweeping *tcheeet tcheeet*. **H** As Alpine Swift. **D** Widespread in Africa.

18 CAPE VERDE SWIFT *Apus alexandri* 13cm Small, tail with shallow fork, rather pale below with dark saddle above. **V** Like that of Common Swift, but higher pitched. **H** Over any habitat. **D** Cape Verde islands.

19 COMMON SWIFT *Apus apus* 16–17cm Generally in small parties. Flight fast and powerful, rapid wingbeats interspersed with long high-speed glides. Forked tail not always apparent during fast direct flight. **V** Screaming *srrreeee*. **H** May forage over virtually any habitat. **D** Breeds widely in Eurasia. Winters in Africa.

20 PLAIN SWIFT *Apus unicolor* 15cm Small, streamlined, black with only obscure throat patch. More erratic flight than Common Swift. **V** Like that of Common Swift. **H** Sea cliffs. **D** Morocco, Canary Islands, Madeira.

21 NYANZA SWIFT *Apus niansae* 15cm Black with small, well-defined, sometimes finely striped, throat patch. Browner than other swifts (except Pallid Swift, which is bulkier and has different range). **V** Like that of Common Swift. **H** In and around rocky caves, buildings. **D** E Africa.

22 PALLID SWIFT *Apus pallidus* 15cm In mixed flocks looks paler than other swifts. Note large throat patch extending up to forehead, giving very pale head-on view. Mantle darkest part of upperside. Broader-winged and bulkier than other swifts. **V** Shorter shrieks than Common Swift; *sreeh sreeh*. **H** Prefers coastal areas in much of range. **D** Breeds S Europe, Arabian Peninsula, N Africa. Winters W Africa.

23 AFRICAN BLACK SWIFT *Apus barbatus* 18cm Separation of pale throat patch from breast shows graded scalloping. Belly, rump and underwing feathers slightly pale-fringed. Inner secondaries (seen from above) paler than rest of upperparts. **V** Extremely high sharp screams *tsjieeeah tsjieeeeah* (each longer than 4 secs and fading at end). **H** Rocky mountains and cliffs. **D** S and E Africa.

24 MALAGASY BLACK SWIFT *Apus balstoni* 16cm Smaller than African Black Swift; restricted to Madagascar. **V** Dry, downslurring screams and shorter, abrupt notes. **H** All habitats. **D** Madagascar.

25 FERNANDO PO SWIFT *Apus sladeniae* 18cm Separated from Common Swift by very black plumage, with very dark throat. **V** Very high, disyllabic short shrieks, *scréeh scréeh scréehscréeh*. **H** Mountainous areas. **D** SW Nigeria, W Cameroon, W Angola and Bioko.

26 FORBES-WATSON'S SWIFT *Apus berliozi* 20cm Throat patch rather large and ill-defined. General appearance rather pale. **V** Screams shorter and slightly lower than Common Swift. **H** Rocky mountains, coastal caves; feeds over all sorts of country. **D** E Africa.

27 BRADFIELD'S SWIFT *Apus bradfieldi* 17cm Slightly paler than Common Swift. Mottled below; primaries and tail darker than mantle and rump. **V** Very high screams at slightly lower pitch than Common Swift and African Black Swift, *sreeEEEeah*, with emphasis on middle part. **H** Like that of Alpine Swift. **D** SW Africa.

28 PACIFIC SWIFT *Apus pacificus* 17–18cm Forages over forests and open hilltops. Pale feather fringes below, especially on belly and undertail coverts. **V** High-pitched *skree-ee-ee*. **H** Mountain forests, forested areas, open areas. **D** Breeds E Asia. Winters SE Asia, Australasia.

29 SALIM ALI'S SWIFT *Apus salimalii* 18–19cm Note white rump and attenuated forked tail. Associated with human habitation when nesting. Feeds over fresh water and open ground. **V** Trilling, screaming *trrrrrrrr*. **H** Rugged montane regions. **D** Breeds Tibet; Sichuan, China; and possibly elsewhere. Wintering range unknown.

30 BLYTH'S SWIFT *Apus leuconyx* 17–18cm Formerly considered part of Pacific Swift complex. Separated by breeding range and altitude. **V** Chattering, insistent trill. **H** Cliffs, gorges, foothills. **D** Breeds Himalayas: Bhutan, N India, Nepal; mostly 2,000–3,000m. Winters India, Sri Lanka.

31 COOK'S SWIFT *Apus cooki* 17–8cm Formerly considered conspecific with Pacific Swift and allied species. Nests in limestone caves. Feeds over nearby forests and fresh water. **V** High-pitched *tsee-tsee-tsee*. **H** Gorges and cliffs. **D** Thailand, Myanmar, Cambodia, Vietnam.

32 DARK-RUMPED SWIFT *Apus acuticauda* 17cm Usually seen near breeding cliffs, although noted feeding above nearby forests. Belly can look very pale in strong light, contrasting with darker undertail. **V** High-pitched *tsee-tsee*. **H** Rocky cliffs, gorges. **D** NE India, Myanmar.

33 LITTLE SWIFT *Apus affinis* 12–13cm Square-ended tail, looks more rounded when spread. Fluttering, bat-like flight combined with short glides. From below appears black with prominent white throat. **V** High-pitched rippling trill and a rapid *siksiksiksiksik*... **H** Habitations, ruins, cliffs, ravines. **D** Widespread in Africa, SW Asia, India.

34 HOUSE SWIFT *Apus nipalensis* 15cm Has fluttering, bat-like flight, combined with short glides. Often nests in groups under building eaves, bridges or cliff faces. **V** Shrill, whickering scream. **H** Urban areas, open country, mountains, occasionally forested areas, often near water. **D** SE Asia and S Japan.

35 HORUS SWIFT *Apus horus* 15cm Looks stout, like fork-tailed version of Little Swift. White of rump extends to lower flanks. Often forages over water. **V** High, plover-like, fluted *piu-pgrrr piu-pgrrr*. **H** Prefers higher altitudes. Breeds in sand cliff tunnels excavated by other birds. **D** Widespread in Africa.

36 WHITE-RUMPED SWIFT *Apus caffer* 14cm Flight fast, agile with fluttering wingbeats. Tail deeply forked, although when held closed appears as single spike. From below appears quite dark, with pale tips to secondaries. **V** Rapid chattering. **H** Valleys and gorges in craggy hills and mountains, sometimes open country. **D** Widespread in Africa.

37 BATES'S SWIFT *Apus batesi* 14cm Small, with deeply forked tail and a strong gloss over upperparts. **V** Not known. **H** Rocky, hilly rainforest. **D** Sierra Leone to Ivory Coast; Cameroon to Gabon.

1
2
3
4
5
6
7
8
9
10
11
12
13
14
15
16
17
18
19
20
21
22
23
24
25
26
27
28
29
30
31
32
33
34
35
36
37

HUMMINGBIRDS TROCHILIDAE

1 SAW-BILLED HERMIT *Ramphodon naevius* 15cm Hefty. Black and yellow downcurved bill. Upperparts coppery brown. Rounded tail with cinnamon outer feathers. Bib tawny with black streaking on chin. Rest of underparts whitish streaked blackish. **V** High-pitched *tsk tsuk tsuk tsuk*. **H** Low in humid and shady forest with bromeliads. **D** SE Brazil.

2 WHITE-TIPPED SICKLEBILL *Eutoxeres aquila* 12cm Unmistakable bill. Tail feathers greyish brown with white tip. Underparts streaked. **V** Long series of variably squeaky notes in a regular rhythm. **H** Mountainous areas, dense thickets, wet forest, forest edges, near *Heliconia*. **D** Costa Rica to N Peru.

3 BUFF-TAILED SICKLEBILL *Eutoxeres condamini* 14cm Unmistakable bill. Throat blackish, underparts streaked. **V** Short phrase of usually three notes *swe-si-seé* repeated many times, occasionally with a short pause in between. **H** Humid montane forest, ravines, open areas, shady cane-brakes, cultivated areas, plantations. **D** SE Colombia to NW Bolivia.

4 HOOK-BILLED HERMIT *Glaucis dohrnii* 12.5cm Curved bill black, lower mandible yellow. Upperparts bronzy green. Tail rounded, tips white. Post-ocular brow and malar streak white. Underparts tawny. **V** High-pitched *tsk tsuk tsuk tsuk*, on descending scale. **H** Low in lowland primary forest with *Heliconia*. **D** SE Brazil.

5 RUFOUS-BREASTED HERMIT *Glaucis hirsutus* 10–12cm Generally solitary, aggressive, often seen chasing other hummingbirds through undergrowth. **V** High-pitched *sweep*, *sweep-sweep* or *sweep swee-swee*. **H** Montane forests, forest edges, plantations. **D** Panama to Brazil and NE Bolivia; also Grenada, Trinidad and Tobago.

6 BRONZY HERMIT *Glaucis aeneus* 9.5cm Small. Upperparts dark coppery with green sheen. Post-ocular and malar stripes whitish. Tail somewhat rounded, basally rufous, then black, tip white. Underparts tawny-rufous. **V** Song high-pitched descending phrase *tsee-tsee-tsi-tsi-tsitstitsi* irregularly altered with series of high-pitched *seee* notes. **H** Low in modified forest, forest edges, thickets, swamps, near streams, coastal areas. **D** Honduras to Ecuador.

7 BAND-TAILED BARBTHROAT *Threnetes ruckeri* 11cm Upperparts dark bronzy green. Outer-tail feathers bluish black with white base and tips. **V** Song fast high-pitched phrase of 5–10 notes repeated with intervals of several secs. Composition varies, typically a combination of some lower and higher notes: *tzi-tzi-tsee-ee-tsi-tzi-tzi*. **H** Low in disturbed forest, thickets, forest edges, plantations. **D** Guatemala to Venezuela and Ecuador.

8 SOOTY BARBTHROAT *Threnetes niger* 11cm Curved bill black, mandible yellow. All blackish with bronzy-olive sheen. **V** Unknown. **H** Tropical forest. **D** French Guiana, NE Brazil.

9 PALE-TAILED BARBTHROAT *Threnetes leucurus* 11cm Upperparts dark bronzy green. Central tail feathers as back, outers white, buffy or blackish. **V** Song fast, high-pitched phrase of 5–10 notes repeated at intervals of several secs. **H** Wet forest, scrub and thickets, plantations, regrowth, near rivers. **D** Amazonia, Guyana, Surinam.

10 BROAD-TIPPED HERMIT *Anopetia gounellei* 10cm Sides of nape rusty. Chin blackish, throat rusty. All tail coverts tawny-buff. Tail black with white tips, lacks long pointed central feathers. **V** Song a rising and falling bisyllablic series of *si-lew* notes repeated continuously (at rate of about one per sec). Sometimes altered with longer *suweesi*. **H** Thickets and damper areas in Caatinga and Cerrado. **D** E Brazil.

11 WHITE-WHISKERED HERMIT *Phaethornis yaruqui* 13cm Bill rather straight; white stripe down throat and breast. **V** Song continuous series of rather harsh *kree-u* notes, at rate of about one per sec. **H** Montane forest, dense secondary growth. **D** W Colombia, Ecuador.

12 GREEN HERMIT *Phaethornis guy* 14cm Dark green. Underparts grey, centre of belly ochreous. Throat tawny. Tail blackish blue. **V** Song loud, barking note repeated many times at regular intervals. **H** Humid montane forest, thickets, clearings, forest edges, woods, plantations. **D** Costa Rica to SE Peru.

13 WHITE-BEARDED HERMIT *Phaethornis hispidus* 15cm Mostly grey, darker on head and tail. Uppertail coverts and lower underparts paler. **V** Song continuous series of high-pitched single *seep* notes, repeated at about 1.5 notes per sec. **H** Wet, flooded and gallery forest, woods and plantations, forest edges, thickets, scrub. **D** W and C Amazonia.

14 LONG-BILLED HERMIT *Phaethornis longirostris* 15cm As Mexican Hermit, but not in same range. **V** Song continuous series of single, piercing, usually upslurred *sweeup* notes, at rate of about 1·5–2 notes per sec. **H** Forest interior and edges, lowlands. **D** S Mexico to N Colombia, W Ecuador and NW Peru.

15 MEXICAN HERMIT *Phaethornis mexicanus* 15cm Once treated as race of Long-billed Hermit. Note whitish breast. **V** Song continuous series of single, metallic notes, at rate of about 1–2 notes per sec; *chieh... chieh... chieh...* **H** Forest undergrowth and edges. **D** W and SW Mexico.

16 LONG-TAILED HERMIT *Phaethornis superciliosus* 14cm Like Great-billed Hermit but smaller, with shorter tail and whitish-tipped outer-tail feathers. Slightly buffier underparts. Throat buff and blackish. **V** Song continuous series of single *tsik* notes, about 1.6–2 notes per sec. **H** Understorey in lowland rainforest, forest edges, riversides and other forested habitats. **D** E Amazonia.

17 GREAT-BILLED HERMIT *Phaethornis malaris* 16cm Dark. Very long bill. Throat blackish. Upperparts blackish green with bronzy sheen, feathers edged dirty buff. **V** Song continuous series of single, bisyllablic rising, falling *slee-up* notes, c. 1–1.5 notes per sec. **H** Wet and upland forest, forest edges, cane-brakes, thickets, shrubs. **D** Amazonia, E Brazil.

18 TAWNY-BELLIED HERMIT *Phaethornis syrmatophorus* 14cm Lower back and rump feathers edged tawny. Underparts tawny, centrally paler. **V** Song continuous series of single, high-pitched, almost insect-like *tsi* calls, at rate of about 2–2.5 notes per sec. Sometimes notes are doubled. **H** Montane forest, cloud forest, thickets, forest edges. **D** Colombia to Peru.

19 KOEPCKE'S HERMIT *Phaethornis koepckeae* 14cm Recalls Tawny-bellied Hermit, but with straight bill and geographical ranges do not overlap. **V** Shrill, whistled *tzviit, tzviit, tzviit*. **H** Moist forested slopes; mainly 500–1,000m. **D** Cerros del Sira mountains, Peru.

20 NEEDLE-BILLED HERMIT *Phaethornis philippii* 13.5cm Bill only slightly curved. Small mask, no malar stripe. Uppertail coverts cinnamon. Outer-tail feathers tipped tawny-buff. **V** Song continuous series of single, upslurred high-pitched *tsee* notes, at rate of about 1.5–1.8 notes per sec. **H** Lowland rainforest, várzea, thickets, plantations. **D** W Amazonia.

21 STRAIGHT-BILLED HERMIT *Phaethornis bourcieri* 14cm Bill almost straight. Less contrast on face. Upperparts blackish grey. Underparts brownish grey, more whitish centrally. Undertail whitish. **V** Song is different from most large *Phaethornis* species; a high-pitched phrase repeated over and over, punctuated by short dull-sounding notes, *tsii'ti'ti'tsii...tip...tip...tsii'ti'ti'tsii....tip...tip... tip....* **H** Thickets, lowland forest, várzea forest, secondary growth, plantations. **D** N Amazonia.

22 PALE-BELLIED HERMIT *Phaethornis anthophilus* 13cm Crown and mask blackish grey. Throat scaled blackish grey. Upperparts bronzy green, rump and tail coverts edged ochreous. All tail feathers greenish black, tipped white. Underparts whitish grey. **V** Poorly known. Calls include rather piercing, high-pitched notes. **H** Semi-deciduous forest, dry woods, thickets, forest edges, gallery forest, plantations. **D** Panama to Venezuela.

23 SCALE-THROATED HERMIT *Phaethornis eurynome* 14cm Post-ocular brow tawny-buff. Rump and uppertail coverts blackish green edged coppery. Throat blackish, scaled white. Sides of body greyish, belly and undertail buffy. Tail feathers dark green and black with white tips. **V** Song a continuously repeated phrase that shows considerable variation (unlike most other *Phaethornis* species), but typically comprises 2–4 notes, *tsi-tseee... tsi-tseee*. **H** Low in forest, thickets, forest edges, regrowth. **D** SE South America.

24 PLANALTO HERMIT *Phaethornis pretrei* 14cm Underparts rich cinnamon. Uppertail coverts cinnamon-rufous. Throat buffy between brownish-grey sub-malar streaks. Outer-tail feathers tipped white. **V** Song variable: often a sequence of evenly spaced, alternating, single and double notes, *ti-tsi... tsi... tsi... tsi... ti-tsi... tsi... ti-tsi...tsi... tsi*, but also sometimes triple-noted phrases such as *chu-tsi-tsi... chu-tsi-tsi... chu-tsi-tsi...* **H** Gallery forest, dry forest, secondary growth, thickets, forest edges, parks, gardens. **D** E and SC Brazil to Bolivia, Paraguay and N Argentina.

25 SOOTY-CAPPED HERMIT *Phaethornis augusti* 14cm Throat finely flecked blackish grey. Rump and tail coverts rufous. Central tail feathers coppery green, outers black, tips white. **V** Song continuously repeated phrase of 2–3 high notes followed by 2–3 somewhat lower ones; *tsee-tsee-tsee-sew-sew... tsee-tsee-tsee-sew-sew...* **H** Montane and cloud forest, thickets in mountains, forest edges, plantations, secondary growth, parks and gardens. **D** N South America.

26 BUFF-BELLIED HERMIT *Phaethornis subochraceus* 11cm Dull. Upperparts dull bronzy green, rump and tail coverts edged cinnamon. Chin slightly flecked blackish. Underparts dirty buffy white. **V** Song continuously repeated high-pitched phrase, usually a short note followed by a rising one, or a short note followed by a falling one. **H** Dry forest in undergrowth, secondary growth, open woods. **D** C Bolivia to SW Brazil.

27 DUSKY-THROATED HERMIT *Phaethornis squalidus* 11cm Like Streak-throated Hermit but underparts browner. Chin and throat all dark brown. **V** Song a complex warbling high-pitched phrase repeated continuously. Phrase varies but typically ends with one or two noisy low-pitched notes; *tsi-teeé-tsa-tsa-tseé-CHAW-CHAW*. **H** Low in forest, thickets, open cleared areas. **D** SE Brazil.

28 STREAK-THROATED HERMIT *Phaethornis rupurumii* 11cm Very like Buff-bellied Hermit but throat blackish scaled grey; uppertail coverts cinnamon. **V** Male song incessant *tsi tsi jéb dé tsi tsi jéb dé* or *eesee-eesee-eesee, eesee-eesee-eesee-swur*, given from perches 1–1.5m high. **H** Low at rainforest edges, dryer or gallery forest, savannah, thickets, secondary growth, scrub. **D** N and E Amazonia.

29 LITTLE HERMIT *Phaethornis longuemareus* 9cm Black chin with faint streaking. Underparts buffy cinnamon, vent whiter. Tail shortish and broad, tipped whitish. **V** Song a high-pitched, chittering phrase, usually lasting 1–1.5 secs, *ee-wee tiddly weet*, uttered about every 2 secs. **H** Low in open forest, swamps, cleared areas, thickets, plantations. **D** NE Venezuela, the Guianas.

30 MINUTE HERMIT *Phaethornis idaliae* 8cm Small. Male dark. Throat blackish grading to chestnut on foreneck and upper breast. Female upperparts coppery olive, uppertail coverts and underparts rufous, throat brownish. **V** Song high-pitched phrases about 3 secs long, repeated incessantly. **H** Forest, thickets in craggy terrain in hills, forest edges, patches of forest. **D** SE Brazil.

31 CINNAMON-THROATED HERMIT *Phaethornis nattereri* 9.5cm Upperparts light bronzy olive. Uppertail coverts, hind-neck and tips of outer-tail feathers buffy rufous. Underparts all buffy. **V** Song a high-pitched phrase repeated incessantly without pauses between phrases, about one phrase per 2–2.5 secs. Phrase typically consists of some 3–5 similar single notes and ends in 1–3 lower-pitched, sometimes quite nasal-sounding, notes with a rhythmic pattern, *tsee... tsee... tsee... nya-ka-wee*. **H** Lowland thickets, secondary growth, forest edges, copses and gallery forest. **D** E Bolivia, SW and E Brazil.

32 REDDISH HERMIT *Phaethornis ruber* 7.5–9cm Variable. Some females lack black breast spot. Upperparts coppery olive. Smallest form has dark green upperparts, band on breast, and coppery-blackish tail with hardly any pale tips. **V** Song a high-pitched phrase repeated incessantly with clear pauses between phrases, about one phrase per 3–6 secs. **H** Forest, copses, forest edges, transition, woods, thickets, savannah, regrowth, parks, gardens. **D** Widespread in South America.

33 WHITE-BROWED HERMIT *Phaethornis stuarti* 9cm Differs from Reddish Hermit in that upperside of tail is bronzy olive and brownish, smaller black spot on breast, sometimes absent. **V** Song a descending, accelerating series of high-pitched notes, *tseee-tsee-tsee-tsi-ti-tututu*, interspersed with quiet *tewp* note. **H** Transition between várzea and terra firme forest, savannah, hills, thickets and secondary growth. **D** SE Peru and C Bolivia.

34 BLACK-THROATED HERMIT *Phaethornis atrimentalis* 8–9cm Similar to Reddish Hermit but note dark throat. **V** Thin, high-pitched *tsip, tsip, tsip*. **H** Lowland rainforest including várzea, secondary forest. **D** Amazonian Colombia, Ecuador, Peru.

35 TAPAJOS HERMIT *Phaethornis aethopygus* 9cm Unobtrusive, favouring shady forest understorey. Threatened by deforestation. **V** Thin, high-pitched *tsip, tsip, tsip* or *tzweip, tzweip, tzweip*. **H** Pristine rainforest, secondary forest. **D** SE Amazonian Brazil.

36 STRIPE-THROATED HERMIT *Phaethornis striigularis* 9cm Mainly brown and rufous with green mantle. **V** Song a high-pitched phrase repeated incessantly without pauses between, at rate of about one phrase per 2–5 secs. Phrase typically comprises several evenly spaced, slightly descending, single notes followed by a more complex warble; *tchee... tsee... tsup... tseecholeelee*. **H** Undergrowth of forest, secondary growth, mangrove, wooded swamp. **D** S Mexico to Venezuela and W Ecuador.

37 GREY-CHINNED HERMIT *Phaethornis griseogularis* 9cm Upperparts light coppery green. Black tail with white tip. Underparts all buffy. **V** Song a high-pitched series of notes repeated incessantly without pauses between, at rate of about one phrase per 2–3 secs. Each phrase typically comprises several, evenly spaced, slightly rising, single notes followed by a more complex warble; *tsi... tsee... tseeé... tseotsetsee*. **H** Rainforest and cloud forest, forest edges, clearings. **D** W and N Amazonia, NW Peru.

38 TOOTH-BILLED HUMMINGBIRD *Androdon aequatorialis* 13.5cm Very long upturned black bill with yellow lower mandible, some with hook at tip. Crown and rump coppery. Tail coverts white and blackish. Rounded tail grey, tip white. Underparts whitish streaked blackish. **V** Song a high-pitched, continuously repeated phrase, usually of three notes, *tsi-tsee-tsek...tsi-tsee-tsek...* with emphasis on second note. **H** Wet forest in lowlands and base of hills, low and at edges. **D** E Panama to NW Ecuador.

1
2
3
4
5 ♂ ♀
6
7
8
9
10
11
12 ♂ ♀
13
14
15
16
17
18
19
20
21
22
23
24
25
26
27
28
29
30
31
32
33
34
35
36
37
38

HUMMINGBIRDS *CONTINUED*

1 GREEN-FRONTED LANCEBILL *Doryfera ludovicae* 12cm Long, slender black bill, slightly upturned. Forehead shiny green. Crown to hind-neck coppery. Tail coverts blackish and bluish. Rounded tail bluish black, tip grey. **V** Usually silent. **H** Cloud forest with epiphytes in mountain terrain, low and on edges. **D** Costa Rica to Venezuela and NW Bolivia.

2 BLUE-FRONTED LANCEBILL *Doryfera johannae* 10.5cm Long, slender black bill slightly upturned. Forehead shiny violaceous blue. Nape coppery. Rest of upperparts greenish. Rounded tail bluish black. Underparts greenish black. Female is paler below, tail finely tipped greyish white. **V** Gives thin, dry, chittering notes while foraging, *chuert*. Song undescribed. **H** Edge of tropical and subtropical forest on lower slopes and ravines, adjacent lowlands. **D** N and NW Amazonia.

3 SCALY-BREASTED HUMMINGBIRD *Phaeochroa cuvierii* 12cm Scaled green below with white tail corners. **V** Song a loud, variable, continuous warble comprising phrases of 4–8 notes, each a mix of chips, sharp squeaks and short trills or sputters, without any apparent pattern. **H** Forest, gardens, plantations. **D** SE Mexico to N Colombia.

4 CURVE-WINGED SABREWING *Campylopterus curvipennis* 12–13cm Grey-green overall, with metallic bluish violet crown. **V** Chattering song; calls include staccato series of wren-like *chip-chip-chip*... notes. **H** Forest, secondary woodland, gardens. **D** Gulf slopes of Mexico.

5 WEDGE-TAILED SABREWING *Campylopterus pampa* 12cm Note form and patterning of outer primaries. As Long-tailed Sabrewing, but with shorter tail. **V** Song loud, prolonged, gurgling warble interspersed with squeaky chipping; begins with hesitant, nasal, reedy, insect-like chippering which may be continued for several minutes before full song begins. **H** Forest, woodland, gardens; mainly lowlands. **D** Yucután Peninsula to Guatemala, Belize and NE Honduras.

6 LONG-TAILED SABREWING *Campylopterus excellens* 13cm Separated from Wedge-tailed Sabrewing by longer tail. **V** Vocalisations are a wild jumble of tweets and gurgles. **H** Forest; mainly lowlands. **D** S Mexico.

7 GREY-BREASTED SABREWING *Campylopterus largipennis* 14cm Upperparts shiny metallic green. Central tail feathers dark green, outer-tail feathers blackish blue with greyish-white tips. Underparts grey, vent white. **V** Repeated short *chik* or *trzik* (the latter bisyllabic), occasionally a faster stuttering series. **H** Humid forest, secondary growth, woods, plantations, thickets. **D** Amazonia.

8 RUFOUS SABREWING *Campylopterus rufus* 13cm Note striking tail pattern with all-rufous outer feathers. **V** Song is varied, strong, squeaky chipping and chattering; also short, rich, warbled phrases. **H** Forest; mainly 1,000–2,000m. **D** SW Mexico to El Salvador.

9 RUFOUS-BREASTED SABREWING *Campylopterus hyperythrus* 11cm Dorsally bronzy green, underparts and outer-tail feathers cinnamon-rufous. **V** Weak, nasal squeak with strained quality. **H** Montane and cloud forest, scrub on tepuis. **D** NE South America.

10 VIOLET SABREWING *Campylopterus hemileucurus* 14cm Male unmistakable. Note white tail corners and purple chin of female. **V** Varied, loud, sharp chipping and warbles, often punctuated with fairly shrill, slightly explosive notes. **H** Forest, secondary growth, forest remains, gardens; 100–1,500(2,500)m. **D** S Mexico to Panama.

11 WHITE-TAILED SABREWING *Campylopterus ensipennis* 13cm Bright green, bib shiny dark blue. Outer-tail feathers white, black at the base. Female has blue throat, central underparts greyish white with green spots. **V** Song a persistently given, comparatively loud and slightly bisyllabic *tzchink*, uttered at a rate of about one note per sec, by males. **H** Montane cloud forest, clearings, forest edges, plantations. **D** NE Venezuela, Tobago.

12 LAZULINE SABREWING *Campylopterus falcatus* 12cm Dark green with large violaceous blue bib. Tail and undertail coverts rufous-chestnut. Female has blue throat, pale greyish underparts; tail paler than male's. **V** Song *Chik, it, chik, it splek, chat, seet, chik, seet, chik, it, chik, it*, or a simpler variant *sweep, tsit, tsuet, tsit* with first and third notes rising. **H** Montane forest, cloud forest to near páramo, plantations, gardens. **D** NE Venezuela to NE Ecuador.

13 SANTA MARTA SABREWING *Campylopterus phainopeplus* 13cm Shiny green, bib shiny dark blue. Tail blackish blue. Female whitish grey below with green flanks and vent, whitish-grey tips to outer-tail feathers. **V** A plaintive double *twit-twit*, both in flight and display. **H** Low in humid montane forest, forest edges, plantations. **D** Sierra Nevada de Santa Marta, N Colombia.

14 NAPO SABREWING *Campylopterus villaviscensio* 13cm Dark shiny green, bib dark blue. Tail dark violaceous blue. Female has grey underparts, narrow white tip to tail. **V** Song a long series of two notes repeated continuously at evenly spaced intervals; *tslip...tseek...tslip...tseek* or *trrip...tseek...trrip...tseek*. **H** Cloud forest. **D** S Colombia to NE Peru.

15 BUFF-BREASTED SABREWING *Campylopterus duidae* 11.5cm Upperparts bronzy green. Central tail feathers coppery, outers basally blackish coppery, distal half buff. Underparts dirty buffy grey. **V** Calls include somewhat buzzy chipping notes *chzzi*, at times a longer chattering series *chi-zizi* or *chizizizizizi*. **H** Montane forest, tepuis, thickets. **D** NE South America.

16 SOMBRE HUMMINGBIRD *Aphantochroa cirrochloris* 11cm All dull smoky brown. Greener on head and upperparts. Tail dull grey, blackish distal half. **V** Song a high-pitched *tchui-ui*, often repeated and sometimes doubled. **H** Humid forest, forest edges, secondary growth, small farms, plantations. **D** SE and E Brazil.

17 SWALLOW-TAILED HUMMINGBIRD *Eupetomena macroura* 20cm Shiny violaceous-blue hood. Long forked tail and undertail coverts dark blue. **V** Song a rapidly and repetitively delivered *tsuc tsuc tsuc*, with some individual variation. **H** Open savannah, gardens, secondary growth, plantations, forest, forest edges. **D** NE, E and C South America.

18 WHITE-NECKED JACOBIN *Florisuga mellivora* 11–12cm Favours high perches or forages above the canopy, usually solitary, although small parties have been recorded high above treetops, calling excitedly while indulging in rapid chases. **V** High-pitched *tit-tit-tit-tit*. **H** Various forests, mostly lowlands. **D** S Mexico through Amazonia.

19 BLACK JACOBIN *Florisuga fusca* 13cm Unmistakable. Black with lower flanks and tail white. **V** Song a series of high-pitched hissing notes above 10kHz, szee... szee... szee... szee... szee... szee, delivered somewhat irregularly at rate of 2–3 notes per sec. **H** Woods, gardens, open areas with scattered trees, secondary growth, plantations, forest edges. **D** SE Brazil to N Uruguay, E Paraguay and NE Argentina.

20 BROWN VIOLETEAR *Colibri delphinae* 11.5cm Greyish brown, paler below. Violet-blue ear-patch. Feathers on lower back and rump bordered buffy rufous. Throat scaled green and blackish. **V** Song typically comprises 4–7 loud, bisyllabic *chit* or *jit* notes, which may be uttered almost incessantly for long periods. **H** Hills and low mountain slopes, arid scrub, secondary growth, clearings, edges and canopy of forest, plantations. **D** Guatemala to N and W South America and E Brazil.

21 MEXICAN VIOLETEAR *Colibri thalassinus* 11–12cm Black band on tail. Can appear uniformly green. Female duller. **V** Repeated *chip-tsirr, tsip-tsup* or *chitik-chitik*. **H** Woods with clearings. **D** C Mexico to NC Nicaragua.

22 LESSER VIOLETEAR *Colibri cyanotus* 11cm Shiny dark green. Dark blue ear-patch. Blackish-blue subterminal band on tail. **V** Song is described as a vigorous but unmelodious two- to four-note phrase, *CHEEP chut-chut, chip CHEEP chut*, or *CHEET-chup*, repeated rapidly for minutes on end. **H** Edge of cloud forest, scrubby slopes, thickets, scattered trees in upland pastures, plantations, gardens. **D** Costa Rica to Bolivia.

23 SPARKLING VIOLETEAR *Colibri coruscans* 14cm Shiny dark green. Throat dark blue. Violaceous-blue ear-patch and upper belly. Large undertail coverts whitish. Blackish-blue subterminal band on tail. **V** Very vocal. Song a long series of monosyllabic metallic chips, *djit... djit... djit* or *tlik... tlik... tlik*, at rate of about two notes per sec. **H** Cloud forest edges, open woods, scrubby upland valleys, gardens, páramo. **D** N and W South America.

24 WHITE-VENTED VIOLETEAR *Colibri serrirostris* 13cm Shiny green. Purple-violet ear-patch. Lower belly and undertail coverts white. Blackish-blue subterminal band on tail. **V** Song a constantly repeated *tsilp, tsilp, zip, tsalp, tsalp* sometimes without break for about 25 mins. **H** Mountain scrub, open forest, woods, thickets, savannah, grassland. **D** Bolivia and N Argentina to E and SE Brazil and E Paraguay.

25 GREEN-THROATED MANGO *Anthracothorax viridigula* 12cm Differs from Green-Breasted and Black Throated Mangos in having green throat. Female virtually identical to female Black-throated Mango. **V** Unknown. **H** Thickets, shrubs, open woods, mangroves, coastal regions, marshy savannah, swampy places, open areas with scattered trees **D** NE Venezuela, the Guianas, NE Brazil.

26 GREEN-BREASTED MANGO *Anthracothorax prevostii* 11–12cm Feeds mainly at flowering trees, although often fly-catches, sallying from a perch, or hovers then darts after insect prey. **V** Song a buzzy *kazick-kazee*, usually rapidly repeated 3–4 times. Calls: repeated chipping, a high-pitched shrill note and a series of thin *see* notes. **H** Forest and forest edge, open woods, scrub and thickets, grassland with scattered trees, parks, gardens and mangroves. **D** E Mexico to N and NW South America.

27 BLACK-THROATED MANGO *Anthracothorax nigricollis* 12cm Shiny dark green. Huge black area from throat to belly. Undertail coverts blackish. Outer-tail feathers purple-violet. Female white below with wide black central streak, outer tail purple-violet and black, tip white. **V** Song comprises seven sibilant notes rendered *hsl-hsl-hsl-hsl*.... **H** Rainforest, secondary growth, clearings, shrubby areas, thickets, gallery forest, plantations, parks, gardens. **D** W Panama to NE Argentina.

28 VERAGUAN MANGO *Anthracothorax veraguensis* 12cm Male has little or no black on throat, female not safely separable from female Green-breasted Mango. **V** Song a repeated slightly buzzy phrase of about five notes *tzee-tzee-tzetzetzee*. **H** Open areas with trees and bush. **D** Panama.

29 ANTILLEAN MANGO *Anthracothorax dominicus* 11–12.5cm Feeds on nectar; also takes spiders and insects, the latter often caught in flight. **V** Sharp chipping notes and a thin trill. **H** Scrub and clearings in moist and dry areas, also plantations and gardens. **D** Hispaniola to Virgin Islands.

30 GREEN MANGO *Anthracothorax viridis* 11–14cm Feeds, from understorey to canopy, on nectar, spiders and insects, the latter often taken in flight. **V** Hard *tic*, a harsh chatter and a trill-like twitter. **H** Mountain forests and coffee plantations. **D** Puerto Rico.

31 JAMAICAN MANGO *Anthracothorax mango* 11–13cm Feeds on nectar from flowers including those of cacti; also takes insects. **V** Sharp, raspy *tic tic tic*.... **H** Forest edges, plantations and gardens. **D** Jamaica.

32 FIERY-TAILED AWLBILL *Avocettula recurvirostris* 8cm Bill upturned at tip. Shiny dark green. Black band on belly. Tail reddish chestnut. Female white from chin to upper belly, with wide central black streak, tail blue-black, outer feathers tipped white. **V** Song unknown. Call a series of *tsik* notes, delivered at an irregular pace, when hovering or in direct flight. **H** Forest, thickets, scrub, savannah, edges of secondary vegetation, near rivers. **D** Amazonia.

33 CRIMSON TOPAZ *Topaza pella* male 18cm, female 15cm Black hood. Glittering shiny pale green bib with golden sheen. Fiery crimson body. All tail coverts greenish gold. Rufous-chestnut tail with one pair of feathers black, curved and much longer than the rest. Female like female Fiery Topaz but two outermost pairs of tail feathers rufous-chestnut. **V** An irregular series of chattering chip notes given almost continually throughout the day. **H** Rainforest and trees, usually along watercourses. **D** NE Amazonia.

34 FIERY TOPAZ *Topaza pyra* male 18cm, female 15cm Like Crimson Topaz but tail vinaceous black. Female brilliant green, throat pale green, golden and fiery, tail coverts bright green, underwing coverts rufous, outer-tail feathers' outer web white. Central feathers shorter than in male. **V** Rich chatter decelerating into a series of *tchip* notes followed by high wiry *pseet-seet* notes. **H** Rainforest, trees in rocky areas, mostly near waterfalls and streams, edge of savannah. **D** W Amazonia.

1
2
3
4
5
6
7
8
9
10
11
12
13
14
15
16
17
18
19
20
21
22
23
24
25
26
27
28
29
30
31
32
33
34

HUMMINGBIRDS *CONTINUED*

1 PURPLE-THROATED CARIB *Eulampis jugularis* 11–12cm Feeds on flower nectar, spiders and insects, generally foraging from mid-level to canopy. **V** Sharp *chewp*, repeated rapidly when agitated. **H** Mountain forests, banana plantations. **D** Lesser Antilles.

2 GREEN-THROATED CARIB *Eulampis holosericeus* 11–12.5cm Feeding habits as Purple-throated Carib. **V** Sharp *chewp*. **H** Rainforests, parks, gardens. **D** NE Puerto Rico, Lesser Antilles.

3 RUBY-TOPAZ HUMMINGBIRD *Chrysolampis mosquitus* 8–9cm Feeds on nectar, spiders and insects. **V** High-pitched *tsip*. **H** Savannah vegetation, from sea level to hills. **D** E Panama to NE Argentina.

4 ANTILLEAN CRESTED HUMMINGBIRD *Orthorhyncus cristatus* 8–9.5cm Feeds on nectar, spiders and insects. **V** Various notes, which usually include a sharp *pit-chew*. **H** Open vegetation, parks, plantations, forest edges. **D** Puerto Rico, Lesser Antilles.

5 VIOLET-HEADED HUMMINGBIRD *Klais guimeti* 8cm Note distinctive blue crown, narrow white tips to tail feathers and white eye spot. **V** Male gives repeated high-pitched song comprising just two notes, *pip seet*. **H** Forest canopy, tall secondary growth, gardens; mainly lowlands. **D** E Honduras to W and N Amazonia.

6 GREEN-CROWNED PLOVERCREST *Stephanoxis lalandi* 8.5cm Unmistakable. Long straight crest. Large dark blue or black area on underparts. Female lacks crest; underparts all grey. **V** Male's song a long series of a repeated modulated note, *tsi-ling... tsi-ling... tsi-ling...* about 1–1.5 notes per sec. **H** Mountains and lowlands; forest, thickets, scrub, along watercourses. **D** E Brazil.

7 PURPLE-CROWNED PLOVERCREST *Stephanoxis loddigesii* 8.5–9cm Unmistakable. Note long straight crest and purple forecrown. Range does not overlap with Green-crowned Plovercrest. **V** Song high-pitched, insect-like trilling chirp. **H** Scrub, forest understorey. **D** Coastal SE Brazil, Paraguay, Argentina.

8 EMERALD-CHINNED HUMMINGBIRD *Abeillia abeillei* 7cm Note short black bill. **V** Song described as a high, thin, slightly squeaking chipping *tsin, tisn-tsin tsin-tsin tsin-tsin*. **H** Montane forest interiors. **D** SE Mexico to N Nicaragua.

9 TUFTED COQUETTE *Lophornis ornatus* 6.5cm Rufous crest of slender feathers. Tuft of very long cinnamon-rufous feathers with round, blackish-green tips forms a fan from cheeks. Female has bronzy green forehead and crown, cinnamon underparts, greenish on lower flanks and belly. **V** High *chik* is given while feeding. **H** Gallery forest, rainforest, savannah, scrub, thickets, plantations, cultivated land water. **D** N Venezuela and Trinidad, C Venezuela, the Guianas, NE Brazil.

10 DOT-EARED COQUETTE *Lophornis gouldii* 6.5cm Fan of long white feathers with greenish-black tips from cheek. Female indistinguishable from female Tufted Coquette. **V** Mostly silent. Short *tsip* given when feeding. **H** Edge of forest, scrub, thickets, savannah, Cerrado. **D** NE and C Brazil.

11 FRILLED COQUETTE *Lophornis magnificus* 6.5cm Fan of large white feathers edged black, like scales, from cheek. Female has rufous forehead, white bib with greenish spotting and rufous edge, white pectoral collar, rest of underparts dirty bronzy green. **V** Mostly silent. Short *tsip* given when feeding. **H** Edge of humid forest, scrub, secondary growth, Cerrado, regrowth, smallholdings, gardens, plantations. **D** SC, E and SE Brazil.

12 SHORT-CRESTED COQUETTE *Lophornis brachylophus* 7cm Note distinctive tail pattern. Lower belly tawny. **V** Mostly silent. High, sharp *tsip* given when feeding. Also quiet, dry chips; *chi-chi-chi*. **H** Forest; 1,000–2,000m. **D** SW Mexico.

13 RUFOUS-CRESTED COQUETTE *Lophornis delattrei* 6.5cm Rufous crest of long, slender feathers with black tips. Coppery ear-coverts and sides of neck. Bib coppery emerald. Female virtually the same as female Spangled Coquette but more cinnamon below with less marked pectoral collar. **V** Mostly silent. High, sharp *tsip* given when feeding. Also quiet, dry chips; *chi-chi-chi*. **H** Scrub, edges of woods, humid forest, clearings. **D** S Central America, W Amazonia.

14 SPANGLED COQUETTE *Lophornis stictolophus* 6.5cm Remarkable rufous crest, feathers with black tips. Bib emerald. Female has rufous forehead; whitish bib with rufous spotting; double pectoral collar green. **V** Mostly silent. A high, sharp *tsip* when feeding. **H** Forest, clearings, thickets, scrubby areas, edges of woods, hills. **D** W Venezuela to N Peru.

15 FESTIVE COQUETTE *Lophornis chalybeus* 7cm Big green fan covers cheek and neck, narrow feathers tipped white. Centre of dirty white breast streaked black. Female's underparts dirty greyish, faintly barred blackish; also has blackish chest-spot and white malar streak. **V** Mostly silent. A short *tsip* or *chip* when feeding. **H** Forest and forest edges, secondary growth, thickets, Cerrado, grassland with bushes. **D** SE Brazil.

16 PEACOCK COQUETTE *Lophornis pavoninus* 8.5cm Fan from cheek, broad green feathers with black spots at the tip. Underparts and uppertail coverts dark green. Female's throat, cheek and underparts streaked black and white. **V** Unknown. **H** Tepuis, montane forest, cloud forest and rainforest, forest edges, bushes, thickets. **D** NE South America.

17 BLACK-CRESTED COQUETTE *Lophornis helenae* 7cm Unmistakable. **V** Mostly silent. Song a clear, upslurred *tsuwee*, repeated. **H** Canopy in open forest, forest edges, plantations; mostly lowlands. **D** S Mexico to Costa Rica.

18 WHITE-CRESTED COQUETTE *Lophornis adorabilis* 7cm Unmistakable. **V** Soft liquid *tseping* when feeding. **H** Canopy of forest, secondary growth, plantations; mostly lowlands. **D** Costa Rica, Panama.

19 WIRE-CRESTED THORNTAIL *Discosura popelairii* male 12cm, female 8cm Very long slender crest. Black breast. Female lacks crest; has white malar area, white on sides of belly, blackish breast-patch, and short, white-tipped tail. **V** A quiet, somewhat liquid *tew*. **H** Humid forest, forest edges, thickets, bushy areas, groves at foot of hills. **D** EC Colombia to SE Brazil and NW Bolivia.

20 BLACK-BELLIED THORNTAIL *Discosura langsdorffi* male 12cm, female 8cm Emerald bib bordered orange below. Black breast, white on sides of belly. Female like female Wire-Crested Thorntail. **V** Mostly silent. A short *tsip* or *chip* given when feeding. **H** Rainforest, scrubby areas, gallery forest, montane and lowland forest edges. **D** W Amazonia, SE Brazil.

21 LETITIA'S THORNTAIL *Discosura letitiae* 9–10cm Little-known species. Possibly extinct. Sparse information based on a couple of museum specimens. **V** Unknown. **H** Presumed to include forest, secondary woodland, gardens. **D** Bolivia.

22 GREEN THORNTAIL *Discosura conversii* male 12cm, female 8cm All green with bluish wash. Long blue tail of pointed feathers. Female has short, white-tipped tail, malar area and sides of belly. **V** Usually silent, but may give soft, squeaky chipping calls. **H** Humid forest, forest edges, woods, clearings. **D** Costa Rica to W Ecuador.

23 RACKET-TAILED COQUETTE *Discosura longicaudus* male 10cm, female 7.5cm Long tail with 'racquets' on outer feathers. Female's tail lacks 'racquets' and is grey and purple-brown with white tips on outer feathers; also has buff band across rump, and black throat bordered white. **V** Unknown. **H** Fields with scattered trees, gallery forest, scrubby savannah, plantations. **D** C and S Venezuela, N and E Brazil, the Guianas.

24 RED-BILLED STREAMERTAIL *Trochilus polytmus* male 22–25cm, female 10.5cm Feeding habits similar to Black-billed Streamertail. **V** Similar to Black-billed Streamertail. **H** Forests, forest edges, plantations, parks, gardens. **D** Jamaica.

25 BLACK-BILLED STREAMERTAIL *Trochilus scitulus* male 22–24cm, female 10.5cm Feeds on nectar from various flowers, and insects. **V** Metallic *ting* or *teet*; also descending *twink-twink-twink-twink....* In flight, tail streamers create vibrating hum. **H** Humid forests, banana plantations, parks, gardens. **D** NE Jamaica.

26 BLUE-CHINNED SAPPHIRE *Chlorestes notata* 9cm Black bill with red at base of mandible. Dark shiny green, blue wash on throat. Square blue tail. Female dorsally paler and bronzy, with greyish-white underparts, heavily spotted green on breast, finer on throat, and all dark blue tail without white tip. **V** A monotonous, constantly repeated *tsip, tsip, tsip*. **H** Gallery forest, rainforest, forest edges, woods, bushy areas, savannah, plantations, fields, gardens. **D** Colombia to E Brazil and Amazonia.

27 GOLDEN-CROWNED EMERALD *Chlorostilbon auriceps* 8–10cm Long, deeply forked tail. Golden wash on crown and mantle. **V** Dry, rattling and chattering calls. **H** Dry areas. **D** W and C Mexico.

28 COZUMEL EMERALD *Chlorostilbon forficatus* 8–10cm As Golden-Crowned Emerald, but without golden wash. **V** Dry, rattling and chattering calls. **H** Woodland, woodland edges. **D** Islands off Yucatán Peninsula.

29 CANIVET'S EMERALD *Chlorostilbon canivetii* 8cm Bluish chin, rather short tail. **V** Dry, rattling and chattering calls. **H** Dry areas. **D** SE Mexico to Costa Rica.

30 GARDEN EMERALD *Chlorostilbon assimilis* 8cm Note all-black bill. **V** Dry, rattling and chattering calls. **H** Dry areas. **D** SW Costa Rica, Panama.

31 BLUE-TAILED EMERALD *Chlorostilbon mellisugus* 7.5cm Shiny dark green. Tail square and dark blue. Bill black. Female like female Red-billed Emerald but bill black, tail square. **V** Repeated *tsip* or *chwep* notes, with occasional rolls or twitters. **H** Gallery and deciduous forest, rainforest, savannah, thorny scrub, plantations, parks, gardens. **D** N and NE South America.

32 WESTERN EMERALD *Chlorostilbon melanorhynchus* 10–11cm Shiny emerald green. Straight red-tinged dark bill. **V** Thin, high-pitched *tsip, tsip* and chattering *tsrrr-t-t*. **H** Lowland forest, moist montane forest. **D** Colombia, Ecuador.

33 RED-BILLED EMERALD *Chlorostilbon gibsoni* 8cm Differs from Blue-tailed Emerald in having red on mandible. Longish, more forked tail. Female has pinkish lower mandible, white post-ocular streak and dark brown ear-patch, and is dorsally paler and bronzy with greyish-white underparts and dark blue, slightly forked tail with white tips to outer feathers. **V** Song a continuous series of wiry trilled notes, *wirrr... wirrr... wirrr* at rate of about three notes per sec. Calls include high-pitched penetrating *tseeee* and reedy *tzreee*. **H** Thorny woods and thickets, deciduous forest, rainforest, plantation, hill vegetation. **D** Colombia, NW Venezuela.

34 CHIRIBIQUETE EMERALD *Chlorostilbon olivaresi* 8.5–9cm Similar to Blue-tailed Emerald but larger, with longer bill. Male has shiny blue throat; female's underparts entirely whitish. **V** Yickering twitter. **H** Open habitats, scrub, savannah. **D** SE Colombia.

35 GLITTERING-BELLIED EMERALD *Chlorostilbon lucidus* 9cm Red bill, tip black. Shiny dark green, bib washed blue. Dark blue tail, slightly forked. Female has pale pinkish base of bill, is dorsally paler and bronzy with greyish-white underparts, and dark blue tail with white tips to outer feathers. **V** Song a high-pitched, cricket-like trill, repeated at intervals. Main call a short dry, scratchy rattle, *trrrr* or *krrr*, given repeatedly when feeding or hovering. **H** Woods, edges of forest, isolated trees, savannah, secondary growth, thickets and scrubby areas, plantations, parks, gardens. **D** E, SE and SC South America.

36 CUBAN EMERALD *Chlorostilbon ricordii* 9.5–11.5cm Feeds on the nectar of flowering trees and shrubs; also insects and spiders. **V** Squeaking twitter. **H** Woodland, swamp edges, coastal scrub forest, plantations, parks, gardens. **D** Cuba, Bahamas.

37 HISPANIOLAN EMERALD *Chlorostilbon swainsonii* male 9.5–10.5cm, female 8.5–9.5cm Feeds on nectar from flowering trees and shrubs; also hawks for insects. **V** A metallic *tic tic tic*. **H** Mountain forest and forest edges. **D** Hispaniola.

38 PUERTO RICAN EMERALD *Chlorostilbon maugaeus* 7.5–9.5cm Feeds on nectar; also spiders and insects. **V** Thin, rapid trill that ends with high-pitched buzz; also *tic tic tic* given at varying speeds. **H** Montane forests, forest edges, lowland woods, coffee plantations, mangroves. **D** Puerto Rico.

39 COPPERY EMERALD *Chlorostilbon russatus* 7.5cm Nape and upperparts coppery with golden and green sheen. Tail coppery rufous. Female has brown cap and ear-patch, underparts extensively greyish white, tail has dark subterminal band and white tip. **V** Poorly known. When feeding, a repeated short *tsik* or *trk*. **H** Montane forest and edges, scrub, rainforest, plantations. **D** NE Colombia, NW Venezuela.

40 NARROW-TAILED EMERALD *Chlorostilbon stenurus* 7.5cm Like Green-tailed Emerald but slightly forked green tail, feathers narrow and pointed. Female's upperparts paler and bronzy, ear-coverts brownish, underparts greyish white. Tail feathers unlike male's in shape, dark greenish blue with white tip. **V** Unknown. **H** Thickets and cloud forest in high mountains, secondary growth. **D** Venezuela, Colombia, NE Ecuador.

41 GREEN-TAILED EMERALD *Chlorostilbon alice* 7.5cm Very like Blue-tailed Emerald but square tail green. The bill is the most slender and shortest of the genus. Female very like those of Short-tailed Emerald and Narrow-tailed Emerald. **V** Poorly known. Thin, high, soft chittering while foraging. **H** Rainforest, cloud forest, forest edges, secondary growth, plantations, open areas. **D** N Venezuela.

42 SHORT-TAILED EMERALD *Chlorostilbon poortmani* 7.5cm Identical to Green-tailed Emerald but green tail slightly forked. Bill longer and heavier. Female like female Narrow-tailed Emerald. **V** Possible song a long series of high-pitched *tseep* notes at rate of about1.5 notes per sec. Calls include a repeated short *tsip* and longer downslurred plaintive *tsew... tsew*. **H** Humid forest, forest edges, open woods, secondary growth, plantations. **D** Colombia, NW Venezuela.

1
2
3
4
5
6
7
8
9
10
11
12
13
14
15
16
17
18
19
20
21
22
23
24
25
26
27
28
29
30
31
32
33
34
35
36
37
38
39
40
41
42

HUMMINGBIRDS *CONTINUED*

1 FIERY-THROATED HUMMINGBIRD *Panterpe insignis* 11cm Unmistakable by orange whiskers. **V** Calls include a repeated nasal, squeaky *kek... kek*, given in fast series when excited. Also a complex fast liquid twittering with sudden squeaky rises in pitch. **H** Montane forest and forest edges. **D** Costa Rica, W Panama.

2 WHITE-TAILED EMERALD *Elvira chionura* 8cm Separated from other green- and white-bellied hummers by diagnostic tail pattern and absence of red in wing. **V** Song a prolonged, thin, scratchy twittering mixed with buzzing or gurgling notes, rising and falling in pitch. Calls include soft scratchy chipping notes while foraging and high-pitched, buzzy notes during chases. **H** Montane forest, gardens. **D** Costa Rica, Panama.

3 COPPERY-HEADED EMERALD *Elvira cupreiceps* 8cm Unmistakable by head, rump and tail feathering. **V** Calls include a high, thin, liquid *quip* or rapid high sputtering in chases. **H** Montane forest. **D** Costa Rica.

4 OAXACA HUMMINGBIRD *Eupherusa cyanophrys* 11cm Blue cap diagnostic. **V** Song a high, rapid, slightly liquid to squeaky warbling, 2–8 secs in duration. Calls include a liquid to slightly buzzy, rolled *chip*, often run into rattled trills. **H** Montane forest. **D** SC Mexico.

5 WHITE-TAILED HUMMINGBIRD *Eupherusa poliocerca* 11cm Note tail pattern. **V** Song a high, rapid, slightly liquid to squeaky warbling, 2–8 secs in duration. Calls include a liquid to slightly buzzy, rolled *chip*, often run into rattled trills. **H** Montane forest, woodland, plantations. **D** S Mexico.

6 STRIPE-TAILED HUMMINGBIRD *Eupherusa eximia* 10cm Note tail pattern. Mainly seen in understorey. **V** A rapid, high, liquid, chipping warble. **H** Montane forest. **D** S Mexico to W Panama.

7 BLACK-BELLIED HUMMINGBIRD *Eupherusa nigriventris* 8cm Black face and belly diagnostic. Note all-white tail, with green central feathers in female. **V** High, thin, sputtering warble. **H** Montane forest. **D** Costa Rica, W Panama.

8 PIRRE HUMMINGBIRD *Goethalsia bella* 9cm Unmistakable, no other hummingbird has a sharply demarcated rufous tail. **V** Single sharp note repeated in a long series. **H** Forest interior and edges; mainly above 1,000m. **D** E Panama, NW Colombia.

9 VIOLET-CAPPED HUMMINGBIRD *Goldmania violiceps* 9.5cm Forehead and crown violaceous blue. Crimson-chestnut tail broadly tipped olive-brown. Female's forehead and crown green, underparts whitish spotted green. **V** Call includes a series of irregularly repeated, very nasal and fairly low-pitched notes, *kyek... kyek-kyek-kyek... kyek'*, given while hovering or feeding. Song a low, rapid chipping. **H** Humid forest, forest edges, foothills. **D** E Panama, NW Colombia.

10 DUSKY HUMMINGBIRD *Cynanthus sordidus* 10cm Note tail pattern and long postocular white stripe.**V** Dry, slightly buzzy chips and a quiet, dry, chippering warble. **H** Dry areas with some trees and scrub. **D** S Mexico.

11 BROAD-BILLED HUMMINGBIRD *Cynanthus latirostris* 9–10cm White mark behind eye on male often hidden. **V** Chattering *je-dit, tetek* or *tek*, similar to call of Ruby-crowned Kinglet (Plate 232). **H** Rocky canyons with springs or streams, and sycamore and mesquite. **D** SW USA to C Mexico.

12 DOUBLEDAY'S HUMMINGBIRD *Cynanthus doubledayi* 10cm Like Broad-billed Hummingbird but with more bluish underparts; ranges overlap only slightly. **V** Chatter call *je-dit*. **H** Dry open woodland, mainly lowlands. **D** S Mexico.

13 BLUE-HEADED HUMMINGBIRD *Cyanophaia bicolor* 9–11cm Feeds on nectar; also takes spiders and regularly hawks for insects over streams. **V** Metallic *click-click-click*; also shrill notes that rapidly descend in pitch. **H** Moist open areas in mountain forests, along mountain streams and wooded field edges. **D** Lesser Antilles.

14 MEXICAN WOODNYMPH *Thalurania ridgwayi* 10cm Note forked tail, blue crown, green gorget. **V** Song an irregularly repeated liquid note *tsip*, often in short fast series of 2–4 notes producing a liquid rattle. **H** Woodland. **D** W Mexico.

15 CROWNED WOODNYMPH *Thalurania colombica* 10cm Unmistakable by brilliant green and violet reflections. Note white throat, contrasting with greyish abdomen, and tail corners of female. **V** Monotonously repeated, single, plaintive squeaky *chip, ksit... ksit... ksit*, at rate of about 1.5 notes per sec. **H** Forest, often near streams. **D** Guatemala to Ecuador and NW Peru.

16 FORK-TAILED WOODNYMPH *Thalurania furcata* male 10.5cm, female 8.5cm Upperparts dark bronzy green. Blacker on cap. Bib brilliant green. Underparts purple-blue. Heavy forked tail blackish blue. Female coppery bronze on crown; underparts greyish white. **V** Song an incessant series of high-pitched bisyllabic, thin wiry notes; *see-tseet... see-tseet... see-tseet*. **H** Thickets, cloud forest, rainforest, woods, open areas with trees, lowlands and mountains, plantations, parks, gardens. **D** Widespread in South America.

17 LONG-TAILED WOODNYMPH *Thalurania watertonii* male 13cm, female 10cm Green. Back dark blue. Undertail coverts and long forked tail blackish blue. Female like female Violet-capped Woodnymph but tail longer, tail feathers narrower. **V** Very hard and comparatively loud chipping calls have been recorded, at varying speeds, often for relatively long periods. **H** Thickets, shrubs and copses on plains, coastal rainforest, Cerrado, plantations, parks, gardens. **D** E Brazil.

18 VIOLET-CAPPED WOODNYMPH *Thalurania glaucopis* male 11cm, female 8.5cm Green. Forehead and crown dark purple-blue. Blackish-blue tail, longish and forked. Female like female Fork-tailed Woodnymph but with green crown. **V** A monotonous rapid series of evenly spaced metallic chips occasionally accelerating to a rattle. Calls include short dry chips, given in rapid succession as dry trill or chatter in flight. **H** Forest, edges and clearings, bushes and thickets, secondary growth, parks, gardens. **D** SE Brazil to E Paraguay, N Argentina and N Uruguay.

19 VIOLET-BELLIED HUMMINGBIRD *Juliamyia julie* 8.5cm Head and back emerald. Rump bronzy. Breast and belly violaceous blue. Female's underparts greyish white spotted green, suggestion of breast-band. **V** Insect-like hiss, *vieiei veii veii veii*. **H** Humid forest, forest edges, thickets and scrub; lowlands and foothills. **D** Panama to NW Peru.

20 SAPPHIRE-THROATED HUMMINGBIRD *Lepidopyga coeruleogularis* 9.5cm Green. Violaceous-blue bib. Blackish-blue tail. Female's underparts whitish marked with green, tail tipped white. **V** Song high-pitched metallic rattle initiated by a separate note, *wi-didididididididididi*, usually descending, up to about 3 secs long and repeated at irregular intervals. Call is a short liquid rattle. **H** Edge of mangroves, arid scrub and woods; near sea level. **D** Panama, Colombia.

21 SAPPHIRE-BELLIED HUMMINGBIRD *Lepidopyga lilliae* 9.5cm Like Sapphire-throated Hummingbird but underparts shiny blue, throat more purple. Tail forked. Female unknown. **V** Series of short chatters. **H** Mangroves, thorny scrub. **D** Very restricted area in NC Colombia.

22 SHINING-GREEN HUMMINGBIRD *Lepidopyga goudoti* 9.5cm Green. Lower mandible red. Tail dark blue and forked. Undertail feathers edged white. Female's belly slightly marked white. **V** Song a short, thin lisping rattle initiated with an upslurred note, about 0.2 secs in length, and rapidly repeated 10–12 times, *pee-rrrr... pee-rrrr... peer-rrrr*. Calls include chips and a high-pitched descending trill. **H** Arid scrub, scattered trees, open woods, deciduous forest, edges of gallery forest, plantations, gardens. **D** Colombia and NW Venezuela.

23 BLUE-THROATED SAPPHIRE *Hylocharis eliciae* 9cm Like White-chinned Sapphire, but only bib blue, tail bronzy green, belly and undertail coverts buff. Female has whitish underparts. Throat dotted blue and breast spotted green. Tail bronzy green. **V** Song a phrase of 5–8 notes, the first a piercing *tseee*, followed by a series of single or double notes. **H** S Mexico to NW Colombia.

24 RUFOUS-THROATED SAPPHIRE *Hylocharis sapphirina* 9cm Rufous throat, blue bib. Tail rufous-chestnut. Female has pale rufous throat with rest of underparts whitish, spotted blue on foreneck. **V** Song a series of 4–7 bright, high-pitched notes, uttered at rate of about four notes per 2 secs and repeated every few secs, *sping... sping... sping* or more bisyllabic *sping... spewee... spewee*. Calls include a repeated short, dry trill and high-pitched seep. **H** Forest, forest edges, regrowth, savannah, brushland, woods, scattered trees, cleared land; sea level to hills. **D** Amazonia, SE Brazil, E Paraguay, NE Argentina.

25 WHITE-CHINNED SAPPHIRE *Hylocharis cyanus* 9cm Red bill. Violet-blue head and bib. White spot on chin. Uppertail coverts coppery chestnut. Square tail dark blue. Female has blue-spotted throat, coppery-chestnut uppertail coverts. **V** Song is insect-like, a series of squeaky phrases uttered at rate of 2–3 per sec, for up to several mins, *ca-seek* or *tweeh-chit*. **H** Woods, forest, regrowth, forest edges, scrub, thickets, plantations. **D** Amazonia, SE South America.

26 GILDED SAPPHIRE *Hylocharis chrysura* 9cm Red bill. All golden bronze with green sheen, female slightly greener. **V** Song repeated high-pitched, cricket-like trill of variable length. Calls include a short dry rattle, *trrrt*. **H** Gallery forest, woods, arid bush, secondary growth, thickets, forest edges, savannah, parks, gardens. **D** Bolivia to SE Brazil, Uruguay and N Argentina.

27 BLUE-HEADED SAPPHIRE *Hylocharis grayi* 9cm Red bill. Bronzy green. Blue head. Blue tail slightly forked. Female's underparts white spotted blue on throat, green on rest of underparts. **V** Song a repeated short warbled phrase that starts with a squeaky *tee* note, *tee... teetlitlitsee-chup... teetlitlitsee-chup*, with final note lower-pitched. Calls include short *chi* notes and a high-pitched descending rattle. **H** Humid woods and forest, mangroves, forest edges, arid bush, open woods, cultivated land. **D** W Colombia, N Ecuador.

28 HUMBOLDT'S SAPPHIRE *Hylocharis humboldtii* 9cm Separated from Sapphire-throated Hummingbird by blue front and black-tipped bill. **V** Poorly known. Possible song a repeated high-pitched phrase comprising hissing notes and trills, *tee-tsee-see-tsee-see-trtrt-tsee-see-trtrt* or similar. Calls include a descending squeaky twittering. **H** Forest edges, woodland, cultivation. **D** SE Panama to NW Ecuador.

29 GOLDEN-TAILED SAPPHIRE *Chrysuronia oenone* 9cm Cap blue. Tail coppery golden. Female has green cap, whitish underparts with green spotting. Tail like male's. **V** Quite vocal. Song rather variable, a repeated rhythmic phrase, often starting with a burry note, followed by several very squeaky or scratchy notes, and ending with repeated chips. Calls include a drawn-out metallic trill and shorter chipping notes. **H** Deciduous and gallery woodland, rainforest, edges and clearings, shrubs, thickets, scattered trees, plantations, gardens. **D** N South America, W Amazonia.

30 WHITE-THROATED HUMMINGBIRD *Leucochloris albicollis* 10.5cm Dark green with white throat, belly and undertail coverts. **V** Complex combinations of chips and squeaky warbles. **H** Regrowth, bush, thickets, edges of subtropical forest, trees, parks, plantations, gardens. **D** S Brazil, E Paraguay, NE Argentina, Uruguay.

31 WHITE-TAILED GOLDENTHROAT *Polytmus guainumbi* 10.5cm Reddish bill. White post-ocular and malar stripes. Upperparts coppery, underparts emerald. Tail rounded, outer feathers white, outer edge of each and subterminal band dark green. Female has lower underparts buffy, central tail feathers green, outers dark green with white outer edge and tip. **V** Fast-paced series of loud, excited spit notes. **H** Marsh edges, open country, deforested areas, tall savannah with scrub, edges of gallery and deciduous forests. **D** N, NE and SE South America.

32 TEPUI GOLDENTHROAT *Polytmus milleri* 12cm Upperparts coppery, stronger on cap. Rounded tail, outer feathers green, white band at base and white tips. Throat and belly speckled green and white. Female similar but has more white on throat and belly. **V** A loud series of *tsit* or *tizzie* notes given while foraging. **H** Cloud forest edges, open country with scattered bush and trees. **D** Tepuis of Venezuela, W Guyana and extreme N Brazil.

33 GREEN-TAILED GOLDENTHROAT *Polytmus theresiae* 8.5cm Post-ocular streak and lores white. Rounded tail bright green. Underparts green. Female has fine white speckling on throat and lower belly. **V** Song repeated, long whinnying series of 20–25 notes, first rising then slightly falling. **H** Forest edges, thickets and bushes, savannah, grassland. **D** NE and W Amazonia.

34 BUFFY HUMMINGBIRD *Leucippus fallax* 10cm Underparts buffy cinnamon. Central tail feathers light green, outers greyish brown with white tips. Female duller. **V** Song repeated short phrase of 2–4 notes, comprising in its simplest form just two, rather nasal squeaky notes, *tsik-tslee... tsik-tslee... tsik-tslee*. **H** Thorny trees and bushes, edges of mangroves. **D** NE Colombia, NW and N Venezuela.

35 TUMBES HUMMINGBIRD *Leucippus baeri* 9cm Dull. Upperparts pale golden green. Underparts greyish. Cap and cheeks brownish. Square tail green, outer feathers and subterminal band browner. **V** Complex series of chips and wheezing electric warbles. **H** Dry scrub, coastal and deciduous forest. **D** NW Peru, SW Ecuador.

36 SPOT-THROATED HUMMINGBIRD *Leucippus taczanowskii* 10.5cm Dull. Green spotting on white throat. Forehead and ear-coverts dirty brown. Underparts dirty whitish. Square tail green, outer feathers browner. **V** Complex series of chips and wheezing electric warbles. **H** Bush and scrub in arid uplands with cactus, plantations. **D** N and C Peru, west slopes of Andes.

37 OLIVE-SPOTTED HUMMINGBIRD *Leucippus chlorocercus* 10cm Underparts whitish with faint greyish-olive spotting, especially on throat and flanks. Square tail green, outer feathers browner. **V** Song a monotonous series of multisyllabic notes, *cliCHEW cliCHEW cliCHEW*. Calls include sharp *seek*, a wiry *seeuee*, a rich chatter and a hard *tcht*. **H** Secondary growth, shrub and thickets at humid forest edges, near rivers. **D** W Amazonia.

38 MANY-SPOTTED HUMMINGBIRD *Taphrospilus hypostictus* 13cm Hefty. Post-ocular and malar stripes white. Underparts dark green scaled white. **V** Song a quiet series of wheezy, electric warbles and gravelly sounding chatters. Calls include thin *chit* and wheezy *dew dew dew*. **H** Arid upland thickets, forest and forest edges, ravines in foothills. **D** Ecuador to Bolivia and SW Brazil.

1
2
3
4
5
6
7
8
9
10
11
12
13
14
15
16
17
18
19
20
21
22
23
24
25
26
27
28
29
30
31
32
33
34
35
36
37
38

HUMMINGBIRDS *CONTINUED*

1 WHITE-BELLIED HUMMINGBIRD *Amazilia chionogaster* 11cm Sexes alike. Upperparts golden-green, underparts white with green spots on throat and foreneck. Brown wings. Bronze-green tail with white tip. **V** Sings a persistent, high-pitched clear *tsing ... tsing ... tsing ...* from an exposed perch. **H** Thickets, woods. **D** N Peru to NW Argentina.

2 GREEN-AND-WHITE HUMMINGBIRD *Amazilia viridicauda* 9.3cm Upperparts and sides of breast brilliant green. Tail darker and olivaceous, slightly forked. Underparts white. **V** Song a repeated short phrase of typically three squeaky notes, *tseet-chew-chip... tseet-chew-chip.* **H** Arid upland bushy terrain to subtropical forest. **D** C Peru.

3 CINNAMON HUMMINGBIRD *Amazilia rutila* 10–11cm Female similar to male but has red only on lower mandible. **V** Squeaky, chipping notes, trills and thin chips. **H** Arid scrub, pastures, secondary growth, forest edges. **D** NW Mexico to Costa Rica.

4 BUFF-BELLIED HUMMINGBIRD *Amazilia yucatanensis* 10–11cm Green-bronze uppertail coverts and plain, dark wings. Can sometimes appear quite dark. **V** Hard chips that sometimes run into a rattle. **H** Lowland scrub. **D** S USA, E Mexico through Yucatán Peninsula.

5 RUFOUS-TAILED HUMMINGBIRD *Amazilia tzacatl* 9.2cm Bill red, tip black; or black with red only on mandible. Brilliant green, dirty white on lower belly. Tail and tail coverts rufous. **V** Song varied, high, thin, squeaky chirping; *tsi, tsi-tsi-tsit tsi-tsitsi tsi-si-si.* **H** Arid bush to cloud forest, rainforest, forest edges, open woods, cultivated land along streams, savannah, plantations. **D** C Mexico to Venezuela and W Ecuador.

6 CHESTNUT-BELLIED HUMMINGBIRD *Amazilia castaneiventris* 8.5cm Upperparts coppery. Tail reddish chestnut. Bib green. Underparts rufous-chestnut. **V** Song a short rhythmic phrase of three squeaky notes that sounds like a rusty door-hinge; *krey-ki-cheep... tsew... krey-ki-cheep.* **H** Lower montane humid forest, forest edges, bushy canyons. **D** NC Colombia.

7 AMAZILIA HUMMINGBIRD *Amazilia amazilia* 9cm Upperparts coppery. Forehead and crown more coppery green. Rufous-chestnut tail, barely forked. Large bib shining green with feathers edged white, or most of bib white. Rest of underparts cinnamon. **V** Song variable but typically descending series of 4–10 squeaky notes, repeated at intervals. Calls include short *tsip* and dry *zrrt* notes, sometimes in stuttering rattles. **H** Coastal arid scrub and open woods, plantations, parks, gardens. **D** Ecuador, Peru.

8 PLAIN-BELLIED EMERALD *Amazilia leucogaster* 9cm Upperparts and sides brilliant green. Uppertail coverts and central tail feathers brownish bronzy olive, outers bluish black. Throat and underparts white. **V** Song a long series of repeated *pseeee* notes. Calls include thin *tsink* and high-pitched stuttering series. **H** Open forest, forest edges, shrubs, mangroves, Cerrado, Caatinga, plantations, secondary growth, parks, gardens. **D** NE and E South America.

9 VERSICOLORED EMERALD *Amazilia versicolor* 8.3–9cm Upperparts bronzy green. Uppertail coverts and tail brownish bronzy olive. Dark brown subterminal band on tail. Bib white speckled emerald green. Rest of underparts bronzy green, white on central and lower belly and undertail coverts. Bib with less white, belly dirty white. **V** Song a long series of the same single note, which is quite variable, *pseee* or *tsew* or *tsidit*. Calls include a short *trr*. **H** Gallery forest, humid and cloud forest, savannah, open arid woods, forest edges, thickets and bushes, secondary growth, plantations, parks, gardens. **D** Amazonia and SC and SE South America.

10 WHITE-CHESTED EMERALD *Amazilia brevirostris* 9cm Upperparts and sides bronzy green. Uppertail coverts and tail coppery bronze, dark subterminal band seen from below. Throat and underparts white. **V** Song nasal and squeaky, and typically a repeated phrase, such as *tsri-lee... tsri-lee... tsri-lee.* **H** Edge of woods, rainforest, gallery and deciduous forest, secondary growth, semi-open scrub, savannah. **D** N South America.

11 ANDEAN EMERALD *Amazilia franciae* 9.2cm Dark blue cap (green in female). Rest of upperparts brilliant green. Tail and upper coverts bronzy olive. Throat and underparts white. **V** Song typically a repeated complex phrase of high-pitched squeaky whistles, trills and scratchy notes. Calls include a high-pitched *tsip* and scratchy squeaky chatters during social interactions. **H** Humid forest edges, wooded hills, thickets and scrub, secondary growth, arid brushy areas. **D** Colombia to N Peru.

12 WHITE-BELLIED EMERALD *Amazilia candida* 10cm Note white underparts, including throat and chin, green tail with faint subterminal bar; long, bent, strong bill. **V** Varied, high, thin, slightly shrill chipping, *tsi-si-sit... tsi-tsin.* **H** Forest. **D** E Mexico to Nicaragua.

13 AZURE-CROWNED HUMMINGBIRD *Amazilia cyanocephala* 11cm Only hummingbird in its range with blue crown and all-white underparts. Female lacks coppery sheen on upperparts like White-bellied Emerald. **V** Low, fairly hard, buzzy *dzzzrt*. Also has a fairly mellow, strong chipping that can run into a trill. **H** Forest edges, woodland. **D** E Mexico to Nicaragua.

14 VIOLET-CROWNED HUMMINGBIRD *Amazilia violiceps* 10–11cm Unmistakable. Crown often looks black. **V** Dry *tak* and descending *seew seew seew...* **H** Canyons with mesquite, sycamore or oak. **D** SW USA to SW Mexico.

15 GREEN-FRONTED HUMMINGBIRD *Amazilia viridifrons* 11cm Only hummingbird in its range with all-white underparts and dull-rufous tail. **V** Dry chattering. **H** Dry, open, mixed forest, riverine belts, suburban regions. **D** S Mexico.

16 CINNAMON-SIDED HUMMINGBIRD *Amazilia wagneri* 11cm Separated from Green-Fronted Hummingbird by pale rufous margin to green plumage areas. **V** Dry chattering. **H** Dry woodland. **D** C and S Oaxaca in Mexico.

17 GLITTERING-THROATED EMERALD *Amazilia fimbriata* 9cm Upperparts brilliant green, uppertail coverts bronzy green. Central tail feathers greenish bronze, outers blacker. Centre of belly and undertail coverts white (to throat in female). **V** Song a continuously repeated, single, high-pitched buzzy note, *tzee... tzee... tzee.* Calls include high-pitched *tsee* notes and soft chatters. **H** Gallery forest, woods, regrowth, mangroves, savannah, forest edges, thorny scrub in arid areas, plantations, parks, gardens. **D** Widespread in South America.

18 SAPPHIRE-SPANGLED EMERALD *Amazilia lactea* 9cm Upperparts brilliant dark green with bronzy sheen. Brilliant violet-blue bib. Central tail feathers dark green, bluer towards tip; outers dark blue, edged green. **V** Repeated buzzy phrase, *tzee... tzitzitzee... tzitzitzee... tzitzitzee* or similar. **H** Rainforest and gallery forest edges, bushes and thickets, secondary growth, clearings, parks, gardens. **D** Amazonia from SE Venezuela to NE Argentina and N Bolivia.

19 BLUE-CHESTED HUMMINGBIRD *Amazilia amabilis* 9cm Like Charming Hummingbird, but green area on crown smaller. **V** Song typically a long series of strident squeaky notes; *tsip-tsew-tsew-tseek-tsew.* Calls include metallic *tsink* and short *tsit*. **H** Forest edges, secondary growth, woodland, riverine belts, plantations; mostly lowlands. **D** Nicaragua to S Ecuador.

20 CHARMING HUMMINGBIRD *Amazilia decora* 10cm Note extension of green on head. Like Blue-Chested Hummingbird, with blue reflections to throat and rufous tail. **V** Song a repeated short phrase of high-pitched tinkling notes; *twee-ti-tlee... twee-ti-tlee.* Calls include dry *tsik* and rattling *trr*. **H** Woodland, secondary growth, gardens. **D** SW Costa Rica, W Panama.

21 PURPLE-CHESTED HUMMINGBIRD *Amazilia rosenbergi* 9cm Bronzy green with violaceous-blue chest. Central tail feathers brownish olive, outer-tail feathers blackish blue. Female whiter on throat and upper breast, with outer-tail feathers tipped white. **V** Song a repeated phrase of about 7–9 *tsee* notes, with emphasis on first two notes and the rest at a slightly faster pace; *tsee... tsee... tsee-tsee-tsee-tsee-tsee.* Calls include single *tsee* notes. **H** Humid forest edges, thickets, shrubs and regrowth, edges of woods, coastal scrubby areas. **D** W Colombia and NW Ecuador.

22 MANGROVE HUMMINGBIRD *Amazilia boucardi* 11cm No similar hummingbird in its range and habitat. **V** Soft *djt* sound given in rapid descending twitter. **H** Mangroves. **D** Costa Rica.

23 HONDURAN EMERALD *Amazilia luciae* 10cm Separated from White-bellied Emerald by bluish-green throat. Restricted range. **V** Metallic *tik tik tik* that can be given as a series of individual notes in succession, or as a rapid-fire, almost continuous *zzik-zzik-zzik* when more agitated. **H** Open, dry woodland in lowlands. **D** Honduras.

24 STEELY-VENTED HUMMINGBIRD *Amazilia saucerottei* 10cm Separated from Indigo-Capped Hummingbird by green, not blue, front and crown. **V** Calls include a high, sharp *tsit* or dry *chit*, sometimes uttered in sputtering series. **H** Savannah, open woodland, gardens. **D** NW South America.

25 INDIGO-CAPPED HUMMINGBIRD *Amazilia cyanifrons* 9cm Similar to Violet-capped Hummingbird. Note blue-black colouring of tail. **V** Song a repeated buzzy, squeaky phrase, *tzuk-keee... tsrp.* Calls include a high, sharp *tsit*, sometimes given in sputtering series. **H** Forest, woodland, plantations, gardens. **D** Colombia.

26 SNOWY-BELLIED HUMMINGBIRD *Amazilia edward* 10cm No rufous in wings; note difference in tail colouring between male (rufous) and female (blackish-blue). **V** Song a soft *bebeebee, d'beebee* or *tseer tir tir*, while calls include light *tip* or *tsip*. **H** Varied habitats, but not in forest. **D** Costa Rica, Panama.

27 BERYLLINE HUMMINGBIRD *Amazilia beryllina* 8–10cm Rufous in wings best distinguishing feature from similar Buff-bellied Hummingbird. Uppertail coverts dark violet. **V** Buzzy *drrzzzt* or *dzzzzir*; also loud *bob-o-lek* or similar. **H** Montane oak woods. **D** NW Mexico to Honduras.

28 BLUE-TAILED HUMMINGBIRD *Amazilia cyanura* 10cm Like Berylline Hummingbird, with rufous in wings, but with steel-blue tail. **V** Song a short twitter. Calls include a hard, raspy *bzzzrt*, hard chips and a high, sharp *siik* in flight. **H** Open forest, woodland, plantations. **D** S Mexico to Costa Rica.

29 BLUE-VENTED HUMMINGBIRD *Amazilia hoffmanni* 9–11cm Striking blue wings, tail and rump. Plumage otherwise green, tinged shiny copper. **V** Rapid-fire, tongue-smacking chatter and thin *tsip-tsip-tsip...* **H** Savannah scrub, secondary woodland, gardens. **D** W Nicaragua to C Costa Rica.

30 GREEN-BELLIED HUMMINGBIRD *Amazilia viridigaster* 9cm Overall brilliant dark green. Rump and all tail coverts bronzy olive. Tail vinaceous black. **V** Song a repeated rhythmic phrase of about four buzzy or squeaky notes, *tee-djew... djidjit*, the second note lower-pitched. Calls include buzzy *dzeee*, sometimes in long series, and *tsi* notes. **H** Forest, forest edges and clearings, scattered trees, foothills to uplands, secondary growth, plantations. **D** Colombia, W Venezuela, W Guyana, NC Brazil.

31 COPPER-RUMPED HUMMINGBIRD *Amazilia tobaci* 9cm Overall dark brilliant green. Back, rump, uppertail coverts and wing coverts coppery. Tail brilliant blackish blue. Undertail coverts coppery blackish. **V** Song a repeated phrase of three buzzy or squeaky, well-spaced notes, *tee-dee-dew* or *tee-dzee-djit*, the last note lower-pitched. Calls include high-pitched descending rattles while foraging. **H** Woods, rainforest, savannah, cloud forest, gallery forest, thorny brush, cleared areas, forest edges, plantations, parks, gardens. **D** Venezuela, Trinidad and Tobago.

32 SNOWCAP *Microchera albocoronata* 6cm Unmistakable; note tail colour of female. **V** Calls include soft, high-pitched, dry *tsip* and buzzy notes and chatters in aggressive interactions. **H** Forest canopy. **D** Honduras to Panama.

33 SANTA MARTA BLOSSOMCROWN *Anthocephala floriceps* 8cm Buff-white forehead, rufous-chestnut crown. Underparts greyish. Female has brownish crown and forehead. **V** Song a long series of repeated *tsip* notes, about 1.5 notes per sec. **H** Humid montane forest, open woods, thickets and bushes, secondary growth. **D** Sierra Nevada de Santa Marta, N Colombia.

34 TOLIMA BLOSSOMCROWN *Anthocephala berlepschi* 8.5cm Previously considered conspecific with Santa Marta Blossomcrown. Tips to outer-tail feathers white (buff in Santa Marta). **V** Harsh *tschak, tschak, tschak...* **H** Moist montane forest on E Andean slope; mostly 1,500–2,000m. **D** Magdalena Valley, C Colombia.

35 WHITE-VENTED PLUMELETEER *Chalybura buffonii* 11.5cm Hefty. Overall bright green, undertail coverts white. Slightly forked tail blackish blue. Female greyish from chin to belly, with white tips to outer-tail feathers. **V** Song unknown. Gives *chip* notes when foraging. **H** Rainforest, deciduous forest, dry forest, semi-open and secondary growth, woods, forest edges, bushes and thickets, swamps. **D** Panama to Venezuela and Colombia.

36 BRONZE-TAILED PLUMELETEER *Chalybura urochrysia* 11cm Red feet diagnostic. Note facial expression with unaccented lores. Female has no white spot behind ear. **V** Presumed song a soft, nasal, scratchy, trilled phrase, *ter-twee-ee-ee-ee-ee....ter-twee-ee-ee...* Chasing call is a drawn-out descending chattering trill. Other calls include loud *chup* and *chip* notes, as well as short chatters. **H** Edges of broken forest, tall secondary growth, adjacent areas. **D** Nicaragua to SW Ecuador.

37 BLUE-THROATED MOUNTAINGEM *Lampornis clemenciae* 13cm Large size, bronzy rump and dark bill separate from female of Broad-billed Hummingbird (Plate 84). **V** Repeated, quiet, hissing rattle *situtee trrrrrrttt*; also clear, monotonous *seek*. **H** Wooded canyons, especially near streams. **D** SW USA to C Mexico.

38 AMETHYST-THROATED MOUNTAINGEM *Lampornis amethystinus* 11cm Unmistakable. **V** Song a quiet chatter made up of a two-syllable introductory phrase followed by a mixture of notes arranged into very complex vocalisations. **H** Montane forest interior and edges. **D** C Mexico to Honduras.

39 GREEN-THROATED MOUNTAINGEM *Lampornis viridipallens* 11cm Note tail pattern. Female lacks throat spots. **V** Presumed song a complex warble of squeaky notes and buzzy, gurgling trills; *tsee-tsee-glr-tsee-glr-tsee-gluglugluglu-glr-glr-tsee.* Calls include a hard, buzzy *zzrrt*, a short *tsik* and a thin, high-pitched *see*. **H** Montane forest edges and clearings. **D** S Mexico to El Salvador.

1
2
3
4
5
6
7
8
9
10
11
12
13
14
15
16
17
18
19
20
21
22
23
24
25
26
27
28
29
30
31
32
33
34
35
36
37
38
39

HUMMINGBIRDS *CONTINUED*

1 GREEN-BREASTED MOUNTAINGEM *Lampornis sybillae* 11cm Separated from Green-throated Mountaingem by more extensive spotting to flanks and by blackish uppertail coverts. **V** Song a soft, scratchy warble, often with a trill at the end. Calls include short, buzzy *shrrrt* while feeding. **H** Montane forest interior and edges. **D** Honduras, Nicaragua.

2 WHITE-BELLIED MOUNTAINGEM *Lampornis hemileucus* 11cm Note blue chin, white belly, tail pattern. **V** Song a medley of squeaks, dry or liquid trills, and sputtering notes. Calls include a repeated, somewhat nasal deep, sometimes in rattling sequences, *de-de-drrrrr*, and squeaky sputtering trills during chases. **H** Forest canopy; 750–1,500m. **D** Costa Rica, Panama.

3 PURPLE-THROATED MOUNTAINGEM *Lampornis calolaemus* 11cm Purple throat diagnostic. Female as female Grey-tailed Mountaingem. **V** Song a high, thin and dry medley of sputtering and warbling notes. Calls include *trrrt*, and a sharp, penetrating, buzzy *zeet* or *zeep*. **H** Canopy of montane forest. **D** S Nicaragua to W Panama.

4 WHITE-THROATED MOUNTAINGEM *Lampornis castaneoventris* 11cm As Grey-tailed Mountaingem, but with blackish tail and grey belly. **V** Males have a sputtery, bubbly song. Calls include high-pitched *ziit* or *ziip*. **H** Undergrowth at forest edges and clearings. **D** W Panama.

5 GREY-TAILED MOUNTAINGEM *Lampornis cinereicauda* 11cm Formerly considered a race of White-throated Mountaingem. Chin white, belly green, tail pale grey. Note green tail and cinnamon-rufous underparts of female. **V** Song a rapid series of short, dry scratchy notes. Calls include single squeaky notes. **H** Montane forest edges and clearings. **D** S Costa Rica.

6 XANTUS'S HUMMINGBIRD *Basilinna xantusii* 8–9cm Unmistakable. Gleans insects from trees as well as feeding from blossoms. **V** Dry, chattering rattle and high *chi-ti* or *ti-tink*. **H** Arid scrub, brushy woodlands. **D** S Baja California (Mexico).

7 WHITE-EARED HUMMINGBIRD *Basilinna leucotis* 9–10cm Both sexes have a much bolder, white head stripe than shown by Broad-billed Hummingbird (Plate 84). **V** Clear, repeated *tink tink tink*... **H** Mountain canyons. **D** Mexico to Nicaragua.

8 GARNET-THROATED HUMMINGBIRD *Lamprolaima rhami* 11cm Unmistakable. Only hummingbird with all-rufous wings. **V** Song a soft, gruff, dry, crackling warble intermixed with nasal, gurgling notes. Calls include a nasal *nyik* and *choiw*, high-pitched chips and a sharp, slightly buzzy *tis-i-tyu-tyu*. **H** Montane forest interior and edges; 1,250–3,000m. **D** C Mexico to Honduras.

9 SPECKLED HUMMINGBIRD *Adelomyia melanogenys* 8.5cm Dull. Upperparts brownish green with bronzy sheen. Underparts spotted blue on throat. **V** Song a monotonous series of high-pitched *tsee* notes, occasionally interspersed with liquid twitters. Calls include repeated short dry rattling trills. **H** Montane forest, cloud forest, forest edges, thickets and bushes along mountain streams. **D** Venezuela to NW Argentina.

10 ECUADORIAN PIEDTAIL *Phlogophilus hemileucurus* 8cm Upperparts brilliant green. White collar, green half-collar below. **V** Song a series of typically three high-pitched buzzy notes followed by several twittering notes, *tzeeeee... tzeee-tzew-tititititi*, continuously repeated. Calls include a descending series of 3–4 high-pitched *see* notes, a repeated *tsik* note and high-pitched twittering. **H** Humid forest, foothills and slopes of mountains. **D** S Colombia to NE Peru

11 PERUVIAN PIEDTAIL *Phlogophilus harterti* 8cm Upperparts brilliant green with bronzy sheen. Rufous lores. Faint collar. White tail with diagonal blackish band across outer feathers. **V** Shrill, thin series of notes, which descend sequentially in pitch, *SIIII siii suuu*. **H** Rainforest at base of mountains, secondary growth, forest edges. **D** C and SE Peru.

12 BRAZILIAN RUBY *Clytolaema rubricauda* 12.3cm Brilliant green overall. Forehead and breast shining green. Throat ruby-red. Tail reddish rufous. Female's underparts cinnamon. **V** Song is *jig chrrrrrr... jig chrrrrrr... jig chrrrrrr*, while the most frequently heard call is a *jig jig jig*. **H** Forest, forest edges, thickets and bushes, parks, gardens. **D** SE Brazil.

13 VELVET-BROWED BRILLIANT *Heliodoxa xanthogonys* 11cm Very dark brilliant green. Cinnamon malar stripe. Female has white underparts, spotted green, tips of tail feathers white. **V** Repeated nasal *squank*. Also high-pitched drawn-out reeling trills, typically going up and down in pitch. **H** Tepuis, montane forest, forest edges, cleared areas. **D** NE South America.

14 PINK-THROATED BRILLIANT *Heliodoxa gularis* 11.5cm Central line of shining emerald on forehead and crown. Throat ruby-pink. Tail blackish or greenish bronze, slightly forked. **V** Presumed song a repeated nasal *keuw*, about one note per sec, also given as a single note in flight and while feeding. **H** Humid montane forest. **D** NW Amazonia.

15 RUFOUS-WEBBED BRILLIANT *Heliodoxa branickii* 11.8cm Overall dark brilliant green. Slightly forked tail blackish blue, central feathers bluish green. Inner flight feathers pale rufous. **V** Song a series of *chew* notes interspersed with a short, rapid, descending trill. Call is a sweet, descending *tew*. **H** Foothill rainforest, woods, forest edges, plantations. **D** SE Peru.

16 BLACK-THROATED BRILLIANT *Heliodoxa schreibersii* 12cm Overall dark brilliant green. Lower face and throat black. Foreneck shiny purple. Female has white malar stripe, throat blackish green, belly greenish grey, tail shorter than male's and less forked. **V** Presumed song a repeated, drawn-out, descending, reeling trill of about 4–5 secs. Also a single *chup*, repeated at a rate of about one note per sec. **H** Humid montane forest and scrub. **D** W Amazonia.

17 GOULD'S JEWELFRONT *Heliodoxa aurescens* 12cm Overall darkish brilliant green. Forehead purplish blue. Chin and upper throat black. Rufous-orange patch on breast. Tail reddish rufous. Female paler with bronzy green forehead; chin and throat speckled blackish and buffy green. **V** Series of high, thin *tseet* notes. **H** Humid rainforest, woods, damp places and along streams, sandy areas, várzea. **D** W and C Amazonia.

18 FAWN-BREASTED BRILLIANT *Heliodoxa rubinoides* 13cm Upperparts brilliant green. Pinkish-topaz patch on throat. Rest of underparts cinnamon-buff, faintly marked green. Longish forked tail olive-bronze from below. **V** Simple series of emphasised *tchik* notes. **H** Montane cloud forest, rainforest, forest edges, on slopes. **D** Colombia to Peru.

19 GREEN-CROWNED BRILLIANT *Heliodoxa jacula* 12cm Overall dark brilliant green. Forehead and crown shining deep emerald. Dark violet spot on throat. Longish forked tail violaceous black. Female like female Violet-fronted Brilliant. **V** Loud and squeaky *kyew* or *tyew*, often repeated when agitated. **H** Montane forest, cloud forest. **D** Costa Rica to Ecuador.

20 EMPRESS BRILLIANT *Heliodoxa imperatrix* 14.3cm Hefty. Long forked tail black. Violet throat-spot. Female has white throat and breast with lines of green dots. **V** A repeated single *tsit* note, about one note per sec, also given singly while hovering or feeding. **H** Lower cloud forest, forest edges, secondary growth, foot of mountains. **D** WC Colombia to NW Ecuador.

21 VIOLET-FRONTED BRILLIANT *Heliodoxa leadbeateri* 12cm Overall very dark green with bronzy sheen. Forehead and crown shining violet. Bib shining emerald green. Female has white underparts spotted green; tip of tail white. **V** Presumed song a continuous series of single strident *chup* or *tchep* notes, about one note per sec. **H** Montane forest, cloud forest and rainforest to open woods, forest edges, cleared areas, coffee plantations. **D** Venezuela to NW Bolivia.

22 RIVOLI'S HUMMINGBIRD *Eugenes fulgens* 11–13cm Male often appears all black. **V** Sharp *chip-chip*; also a buzzy warble. **H** Pine–oak forests. **D** SW USA to Honduras and Nicaragua.

23 TALAMANCA HUMMINGBIRD *Eugenes spectabilis* 12–13cm Paler gorget than Rivoli's Hummingbird, with which it was previously grouped as Magnificent Hummingbird. **V** Coarse, grating *tchh, tchh, tchh*... **H** Montane oak forests, secondary woodland. **D** Costa Rica, W Panama.

24 SCISSOR-TAILED HUMMINGBIRD *Hylonympha macrocerca* male 21cm, female 12cm Dark metallic green overall. Very long forked tail. Female has white throat and breast spotted green. **V** Presumed song a short, pulsating burst, repeated every 2–2.5 secs, *tsi-si-sip... tsi-si-sip... tsi-si-sip* or *tsi-sip... tsi-sip... tsi-sip*, the latter sounding almost monosyllabic to the human ear. **H** Cloud forest, woods, forest edges, clearings. **D** NE Venezuela.

25 VIOLET-CHESTED HUMMINGBIRD *Sternoclyta cyanopectus* 13cm Hefty. Overall brilliant dark green. Lower belly and undertail coverts brownish grey. Female's belly and vent buffy spotted green towards flanks. **V** Song series of sharp *chit! chit! chit!* notes, about one per sec, or slightly faster for periods of 10–20 secs. Also utters loud, staccato chipping notes while foraging. **H** Cloud forest, rainforest, coffee plantations, shady places with *Heliconia*. **D** N Venezuela.

26 WHITE-TAILED HILLSTAR *Urochroa bougueri* 15cm Dark green overall. Female has notable rufous malar stripe, shining blue throat, rest of underparts brownish grey. **V** Presumed song a continuous series of single *swit* or *tsit* notes, about 1–2 notes per sec. Calls include liquid *twit*, repeated in long sequences when alarmed. **H** Rainforest and cloud forest in mountains, forest edges, shrubs and thickets. **D** SW Colombia to NW Ecuador.

27 BUFF-TAILED CORONET *Boissonneaua flavescens* 12cm Overall bright green, scaled whitish buff on lower belly. **V** Presumed song a continuous series of single high-pitched *tsit* notes, about 0.6–1 note per sec. Also squeaky twittering, with rising piping notes and stuttering rattles, especially during social interactions. **H** Montane forest and cloud forest to upper edge, low stunted trees, bushy slopes, thickets and bush, forest edge. **D** Venezuela to Ecuador.

28 CHESTNUT-BREASTED CORONET *Boissonneaua matthewsii* 12cm Underparts rufous chestnut. Outer-tail feathers tawny-rufous, outer edge and tip greenish. **V** A high, thin, liquid *tip*, a rapid, sweet trill, and various squeaky notes. **H** Humid montane forest, treetops, forest edges. **D** Colombia to Peru.

29 VELVET-PURPLE CORONET *Boissonneaua jardini* 12cm Dark overall. Head and breast black. Outer-tail feathers white, outer edge and tip black. **V** Courtship song reported to be series of alternating harsh and soft whistles, *si, siii, si, siii, si, siii*. **H** Rainforest at foot of mountains, moss-hung trees, scrub on lower slopes, forest edges. **D** SW Colombia to NW Ecuador.

30 SHINING SUNBEAM *Aglaeactis cupripennis* 12cm Patch of buff-white spots on breast. Female duller, coloured patch on back smaller. **V** Song a repeated phrase of 3–5 high-pitched chirping notes, *tsip... chew... chew... tseep... tsip... chew*. Calls include similar-sounding squeaky twittering series, single upslurred *suweet* notes and single high-pitched *see* notes. **H** Edges of high montane woods, low and stunted trees, bushes, thickets, páramo, gardens. **D** Colombia to Peru.

31 PURPLE-BACKED SUNBEAM *Aglaeactis aliciae* 12cm Mostly dark chocolate-brown. Lower back and rump glittering purple. **V** Presumed song a repeated phrase of 3–5 high-pitched chirping notes, *tsip... tsee... chew... chew... tsip... tsee... tsee... chew-chew*, or similar. **H** Very high arid scrub, thickets, edges of montane woods. **D** N Peru.

32 WHITE-TUFTED SUNBEAM *Aglaeactis castelnaudii* 12cm Head, upperparts and breast-band dark chocolate-brown. Patch of white marks on breast. Rump shiny purple. **V** Calls include repeated *tzit* and during chases a twittering series, *titi-tsreet-tsreet-tsreet* or similar. Also a thin, high-pitched, downslurred, drawn-out *seeeuuu*. **H** Very high arid scrub, thickets and stunted trees, above montane woods and forest. **D** Peru.

33 BLACK-HOODED SUNBEAM *Aglaeactis pamela* 10cm Mostly black. Lower back to upper tail shining emerald. Tail and vent chestnut-rufous. Patch of white spots at mid-breast. **V** High-pitched *zeet-zeet-zeet*. Also gives very high pitched, slightly descending note, and lower pitched, sharper notes during antagonistic interactions. **H** Very high arid temperate scrub, thickets and stunted trees, edges of montane woods. **D** Bolivia.

34 ECUADORIAN HILLSTAR *Oreotrochilus chimborazo* 12cm Green throat present or not. Female has dull brownish-green upperparts. White bib dotted brownish. **V** Calls include repeated short *tsit* and strident *tseek*. Also fast melodious twittering with rising and falling sequences, probably a chasing call. **H** Arid rocky high mountain steppe, bush, thickets, grassland, scattered low trees, woods, montane forest. **D** S Colombia to C Ecuador.

35 ANDEAN HILLSTAR *Oreotrochilus estella* 12cm Like White-sided Hillstar but larger bib protrudes from sides of neck. Mid-belly stripe chestnut. Female has underparts pale greyish brown; base and tip of outer-tail feathers white. **V** Calls include a repeated short *tsip* or *swit*. Also rapid, melodious twittering during chasing. **H** Upland grassland and brush, near human habitations, edges of *Polylepis* woods and *Puya*. **D** SW Peru to NW Argentina.

36 WHITE-SIDED HILLSTAR *Oreotrochilus leucopleurus* 10cm Shining green bib does not protrude from side of neck. Underparts with blackish-blue median line on central breast and belly. Female dull; underparts whitish brown. **V** Poorly known. Calls include a repeated short *tsit*. Also fast twittering during chasing or display. **H** High arid mountain scrub, steppe with rocky outcrops near streams, sometimes with cactus. **D** S Bolivia to S Argentina and Chile.

37 BLACK-BREASTED HILLSTAR *Oreotrochilus melanogaster* 11cm Underparts black. Female pale above, greenish cap with coppery sheen, rest of underparts pale brownish grey; central tail dark green, outers blackish with white tip and base. **V** Poorly known. A fast squeaky twittering with rising and falling sequences during chases. **H** High mountain rocky steppe, grassland, thickets, scrub, temperate montane woods and forest, scattered trees. **D** C Peru.

38 WEDGE-TAILED HILLSTAR *Oreotrochilus adela* 12cm Breast and mid-belly black. Chestnut flanks. Female has white bib spotted brown, rest of underparts tawny-buff. **V** Song a medley of intense twittering notes interspersed with a distinct descending cadence. **H** High-mountain arid stony steppe, bushes, thickets, grassland. **D** C Bolivia to NW Argentina.

39 MOUNTAIN VELVETBREAST *Lafresnaya lafresnayi* 10cm Overall brilliant green with bronzy sheen. Female has buff underparts with scattered greenish markings. **V** Calls include a repeated high-pitched *tseee* or *pseeuw*, sometimes followed by a stuttering descending series. Also soft *tek* notes and a thin rattle. **H** Thickets and bushes in montane forest, edges of cloud forest, upland meadows with scattered bushes. **D** W Venezuela to C Peru.

1
2
3
4
5
6
7
8
9
10
11
12
13
14
15
16
17
18
19
20
21
22
23
24
25
26
27
28
29
30
31
32
33
34
35
36
37
38
39

HUMMINGBIRDS *CONTINUED*

1 BRONZY INCA *Coeligena coeligena* 14cm Blackish brown or cinnamon brown. Green sheen on back and rump. **V** Presumed song a continuous series of single rather sweet *tseet* notes, about 1–1.6 notes per sec. Calls include single *tsee* or *tzeet*, usually given in flight. **H** Cloud forest, montane forest, forest edges, coffee plantations, scattered trees and bushes. **D** N Venezuela to SE Bolivia.

2 BROWN INCA *Coeligena wilsoni* 13cm Dark brownish with greenish sheen. White patch on side of breast. **V** Possible song a repeated phrase comprising three notes; *tsip-tzreeew-tzrew*. Calls include strident, single *tsit* or doubled *tsi-tsit*, or longer series, *tsitsitsitsitsit... tsitsitsi... tsitsitsit*. Also gives short rattling *trrr*, short twitters and, in flight, a high-pitched *tzree... tzee... tzee... tzee*. **H** Cloud forest, montane forest, forest edges. **D** W Colombia to W Ecuador.

3 BLACK INCA *Coeligena prunellei* 14cm Blackish. White patch on sides of breast. **V** Mostly silent, but occasionally utters short *ick*. **H** Humid montane forest, open country and gallery forest. **D** Andes in NC Colombia (Quindío, SE Santander, W Boyaca, W Condinamarca).

4 COLLARED INCA *Coeligena torquata* 14cm Mostly blackish blue, or brilliant green. Crown violet, blue or green. Female's throat white or buffy. **V** Squeaky chatter, often a repeated phrase such as *tsi-tsi-tsiririt... tsi-tsi-tsiririt*. Also single notes, *tsit*, or a high-pitched *see*. **H** Cloud forest, humid montane forest, forest edges, near páramo. **D** W Venezuela and Colombia to Bolivia.

5 WHITE-TAILED STARFRONTLET *Coeligena phalerata* 14cm Brilliant dark green. Female's upperparts green, underparts tawny-rufous. Tail olive-green, tips buff. **V** Calls include a high-pitched chattering, *tsee-tsee-tsi-tsi-tsirrrrr* and lower-pitched short rattles. **H** Cloud forest and montane forest, forest edges, shrubs and thickets. **D** Sierra Nevada de Santa Marta, N Colombia.

6 GOLDEN-BELLIED STARFRONTLET *Coeligena bonapartei* 14cm Body green but belly and uppertail coverts golden olive. Female like Cinnamon Starfrontlet but tail, vent and uppertail coverts like male. **V** Short twitter and a more complex chatter that rises and falls in pitch. **H** Cloud forest, montane forest, scattered trees and bushes near páramo. **D** E Colombia to W Venezuela.

7 DUSKY STARFRONTLET *Coeligena orina* 14cm Previously treated as race of Golden-bellied Starfrontlet. Forehead glittering blue. **V** Tinkling notes and thin *tsip, tsip...* call. **H** Páramo, cloud forest; mostly 3,000–3,500m. **D** Cordeillera Occidental range, Colombia.

8 BLUE-THROATED STARFRONTLET *Coeligena helianthea* 14cm Overall greenish black. Female's head and upperparts bright green, throat cinnamon spotted green. **V** Rather silent and voice poorly known. Calls include single, strident *chit* notes. **H** Cloud forest, montane forest, scattered trees, forest edges, thickets, bushes, parks, gardens. **D** NE Colombia and W Venezuela.

9 BUFF-WINGED STARFRONTLET *Coeligena lutetiae* 14cm Upperparts black. Underparts dark green. Female green, with buffy cinnamon throat. **V** Thin, wiry chatter with rattles, often heard during high-speed aerial chases. Also a characteristic, nasal *unk* repeated at intervals. **H** Cloud forest, montane forest, forest edges, thickets and bushes. **D** Colombia to Peru.

10 VIOLET-THROATED STARFRONTLET *Coeligena violifer* 14cm Shiny violet patch on throat. Faint pale pectoral collar. Female lacks violet on throat. **V** A long series of single *tchit* notes given at a rate of about 1.6 notes per sec. Also a jerky mix of squeaky notes and rattles, often in flight. **H** Very high cloud forest and elfin forest, bushes and thickets. **D** S Ecuador to NW Bolivia.

11 RAINBOW STARFRONTLET *Coeligena iris* 14cm Several races differing in cap colour. Amount of green on underparts variable. Female duller. **V** Thin, wiry chatter, rising and falling, with rattles and squeaky notes, often emitted during high-speed aerial chases. Also single *tsit* or *tip*. **H** Upland cloud forest, forest edges, bushes, thickets, gardens. **D** Ecuador, Peru.

12 SWORD-BILLED HUMMINGBIRD *Ensifera ensifera* 20cm Inordinately long bill. Dark brilliant green overall, bronzy sheen on head. Female has whitish underparts spotted green. **V** A low, slightly trilled *trrr* is occasionally heard. **H** High mountain forest, forest edges, cloud forest, low trees and shrubs, scrubby slopes, scattered trees. **D** W Venezuela to NE Bolivia.

13 GREAT SAPPHIREWING *Pterophanes cyanopterus* 16.5cm Large and hefty. Brilliant dark green male. **V** High, thin, liquid chatter, as well as a drawn-out, piercing, high-pitched *zeee* and an agitated *tititititrrr*. **H** High mountains, scrubby slopes and thickets, scattered bushes and trees in páramo, edges of montane forest. **D** Colombia to N Bolivia.

14 GIANT HUMMINGBIRD *Patagona gigas* 22cm Huge and unmistakable. Upperparts browny olive with slight bronzy sheen. White rump. Female duller. **V** Mixture of squeaky piping whistles and trills. Calls include a single squeaky *squip* and a less diagnostic *teee* or *zrr*. **H** Arid mountains with cactus and scrub. **D** S Colombia to C Chile.

15 GREEN-BACKED FIRECROWN *Sephanoides sephaniodes* 9.5cm Cap fiery red. Mostly bronzy green. Female has bronzy green cap. **V** Song a series of high-pitched notes interspersed with gravelly trills and squeaky notes. Calls include single *psee* or *skee*. **H** S Andean woods, forest edges, clearings, thickets, parks, gardens. **D** S and C Chile, C and W Argentina.

16 JUAN FERNANDEZ FIRECROWN *Sephanoides fernandensis* 12cm All reddish chestnut. Female has green upperparts, underparts white spotted greenish black; cap blue. **V** Song a medley of high-pitched squeaky notes, dry gravelly trills and descending chatters. **H** Thickets, bushes, gardens. **D** Juan Fernandez archipelago, off WC Chile.

17 ORANGE-THROATED SUNANGEL *Heliangelus mavors* 9.5cm Large bib shining golden orange. Female has buffy cinnamon throat, flecked green. **V** Call a repeated, high-pitched, cricket-like, faint trill, given either from perch or in flight. **H** Cloud forest, montane forest and slopes with low and stunted trees and shrubs, forest edges, scrubby grassland. **D** Colombia, Venezuela.

18 AMETHYST-THROATED SUNANGEL *Heliangelus amethysticollis* 9.5cm Like Longuemare's Sunangel but with blackish chin and cinnamon-buff collar and underparts. Vent dirty white. Female throat tawny. **V** Call a repeated, high-pitched, cricket-like, short trill. **H** Cloud forest, upland woods with small or stunted trees, forest edges, open slopes with shrubs and scattered trees, scrubby ravines. **D** S Ecuador to NW Bolivia.

19 LONGUEMARE'S SUNANGEL *Heliangelus clarisse* 9.5cm Bib glittering rosy amethyst. Collar white. Vent blackish grey, feathers edged whitish. Slightly forked tail, central feathers bronzy green, outers blackish. Female has brown bib. **V** Call a repeated dry, upward-inflected *tsik* or *tsit*, given at rate of about 1–3 notes per sec. **H** Cloud forest, upland woods with low, stunted trees, bushy copses on páramo. **D** E Colombia to W Venezuela

20 MERIDA SUNANGEL *Heliangelus spencei* 11.2cm Formerly considered a subspecies of Amethyst-throated Sunangel. Note white collar and spot behind eye. **V** Cricket-like trilling call, lasting 1 sec. **H** Forested and scrub-covered slopes; mostly 2,000–3,000m. **D** Venezuela.

21 GORGETED SUNANGEL *Heliangelus strophianus* 9.5cm Like white-collared forms of Amethyst-throated Sunangel, but breast and belly green. Tail longer and somewhat forked. Female's throat blackish cinnamon, streaked white. **V** Song a repeated series of 2–3 high-pitched *pseee* notes followed by a slightly lower-pitched *tsip* note. Call is a dry chattering followed by a mellow note. Also a high-pitched twittering. **H** Humid montane forest, forest edges, scrubby ravines along watercourses. **D** Extreme SW Colombia (Nariño) and NW Ecuador.

22 TOURMALINE SUNANGEL *Heliangelus exortis* 10.3cm Dark green. Chin violaceous black, bib shining pink. Female's throat white, speckled blackish. **V** Song repeated series of single notes at a rate of about one note per sec. Calls include a short, dry, gravelly trilled *trrr*, repeated at intervals. **H** Montane forest, cloud forest, dwarf forest with cane, forest edges, upland slopes with trees and bushes, scrubby grassland. **D** Colombia and NW Ecuador.

23 FLAME-THROATED SUNANGEL *Heliangelus micraster* 9cm Dark green. Chin black, throat golden orange. Female like male but chin and upper throat white. **V** Call is a repeated dry *djit* or more gravelly *drrt*. **H** Montane temperate moss-laden woods, grassland with scrub, forest edges, slopes with scattered trees and shrubs. **D** Ecuador, Peru.

24 PURPLE-THROATED SUNANGEL *Heliangelus viola* 11.7cm Glittering green. Forehead shining blackish blue. Female lacks shining bib, has greenish-black throat. **V** Calls include a loose series of short, dry trills, *trr... trr... trr... trr*, and a repeated, drawn-out, dry, buzzy *bzzzrrr*. **H** Cloud forest, montane forest, forest edges, shrubs and thickets, alder woods. **D** Ecuador, Peru.

25 ROYAL SUNANGEL *Heliangelus regalis* male 10cm, female 7.5cm All brilliant dark blue. Female's underparts buffy cinnamon, small blackish eye-patch. **V** Calls include repeated short, dry and emphatic *tsik* or *tsawk* when feeding. **H** Low montane forest, cloud forest, dry grassland, scrub, wooded ravines, forest edges, stunted trees. **D** N Peru.

26 BLACK-BREASTED PUFFLEG *Eriocnemis nigrivestis* 9cm Critically Endangered. Blackish blue-green. Throat and vent violet. Female dark bronzy green, throat violaceous blue, breast scaled cinnamon, mid-belly scaled white. **V** Song a monotonous, metallic *tzeet... tzeet... tzeet*. When hovering, utters soft *tzik* notes. **H** Upland temperate woods, forest edges, bushes, thickets bordering páramo. **D** On ridges around Quito, NW Ecuador.

27 GORGETED PUFFLEG *Eriocnemis isabellae* 8–9cm Discovered in 2005. Critically Endangered. Male looks overall dark with bluish-violet throat. Female dull green. **V** Calls include a sharp *tvik*. **H** Páramo; mainly 2,500–3,500m. **D** Serranía del Pinche, Colombia.

28 GLOWING PUFFLEG *Eriocnemis vestita* 9cm Throat and vent violet. Female's throat scaled violet. Sides of throat and breast cinnamon spotted green. **V** A single metallic note, *tseek* or doubled *tsi-tseek*, repeated at irregular intervals. **H** High montane forest and cloud forest, forest edges, elfin woods, bushy areas, upland grasslands, copses in páramo. **D** NW Venezuela to N Peru.

29 BLACK-THIGHED PUFFLEG *Eriocnemis derbyi* 8.7cm Dark coppery green. All tail coverts brilliant light green. Outer-tail feathers pointed. Leg-puffs blackish. Female's underparts scaled white; leg-puffs greyish. **V** Short buzzy trilled *tzzrr*, repeated at irregular intervals. **H** Temperate montane forest and woods, forest edges, shrubby areas, thickets, neighbouring grassland, bushy slopes, ravines. **D** Colombia, NW Ecuador.

30 TURQUOISE-THROATED PUFFLEG *Eriocnemis godini* 9cm Coppery green. Throat and vent shining blue. Uppertail coverts pale green. Tail slightly forked. **V** Unknown. **H** Bushy upland slopes and ravines, edges of upland forest, thickets. **D** NW Ecuador.

31 COPPERY-BELLIED PUFFLEG *Eriocnemis cupreoventris* 9cm Dark green. Lower belly shining copper. Vent violet. Somewhat forked tail. **V** A single metallic *tseek* or doubled *tsi-tseek*, repeated at irregular intervals. **H** Páramo, scattered stunted trees and bushes, temperate montane forest, forest edges, thickets. **D** Colombia, Venezuela.

32 SAPPHIRE-VENTED PUFFLEG *Eriocnemis luciani* 12cm Dark green. Longish forked tail blackish blue. Forehead blue. Vent violet. Female has white scaling on bib and central belly. **V** Sharp *tirr tirr*. **H** Temperate forest to páramo, bushy slopes with grass, thickets and copses. **D** SW Colombia to W Ecuador.

33 GOLDEN-BREASTED PUFFLEG *Eriocnemis mosquera* 12cm Dark bronzy green. Coppery sheen on breast and face. Vent brownish grey. Longish forked tail blackish green. **V** *Trit* notes. **H** Low montane woods and bushes, dwarf copses, thickets at tree line in páramo. **D** W Colombia to NW Ecuador.

34 BLUE-CAPPED PUFFLEG *Eriocnemis glaucopoides* 9cm Dark green, bluer underparts. Forehead and vent dark blue. Slightly forked tail. Female has cinnamon bib, central underparts buffy, spotted green. **V** High-pitched *zee-zee*. **H** Montane forest, forest edges, bushes and thickets. **D** C Bolivia to NW Argentina.

35 COLORFUL PUFFLEG *Eriocnemis mirabilis* 8.5cm Emerald forehead. Belly dark shining blue. Female's throat and mid-breast white, spotted green. **V** Poorly known. Presumed song a repeated metallic short note *tsit*, about 1.5–2 notes per sec. **H** Montane forest, cloud forest, clearings and forest edges, bushes and thickets. **D** W slope of Andes, W Colombia.

36 EMERALD-BELLIED PUFFLEG *Eriocnemis aline* 7.5cm Huge pantaloons. Upperparts bronzy green. Shining emerald underparts and tail. White breast-patch spotted green. **V** Poorly known. Call a single, slightly buzzy, metallic *tzit*, repeated at irregular intervals. **H** Montane forest and cloud forest, less on forest edges. **D** Colombia to C Peru.

37 GREENISH PUFFLEG *Haplophaedia aureliae* 9cm Upperparts coppery green. Underparts dark green faintly scaled greyish. **V** Song an endlessly repeating double-noted *tur seet* or *tskut* at a rate of about one note per sec, given throughout much of day. **H** Cloud forest and montane forest, forest edges and clearings, bushes and thickets. **D** E Panama, Colombia, Ecuador.

38 BUFF-THIGHED PUFFLEG *Haplophaedia assimilis* 9cm Overall coppery green. Buff leg-puffs. **V** Repeated series of abrupt *tzip, tzip, tzip* notes and tongue-smacking chattering phrases. **H** Moist montane forest, secondary woodland. **D** Bolivia, Peru.

39 HOARY PUFFLEG *Haplophaedia lugens* 9cm Upperparts coppery green. Underparts blackish grey, scaled white on throat. **V** Single strident *tzik* or doubled *tsi-tsik*, repeated at irregular intervals. **H** Montane forest, cloud forest, forest at foot of mountains, forest edges, thickets. **D** SW Colombia to NW Ecuador.

40 PURPLE-BIBBED WHITETIP *Urosticte benjamini* 8–8.8cm Differs from Rufous-vented Whitetip in having lower half of bib purple. Female like female Rufous-vented Whitetip. **V** Mellow, nasal-sounding fast twittering *tweetweetweetwee*. **H** Montane forest, cloud forest and rainforest, forest edges. **D** W slopes of Andes, W Colombia to NW Ecuador.

41 RUFOUS-VENTED WHITETIP *Urosticte ruficrissa* 8.5–9.3cm Mostly brilliant green. Shiny dark green bib. Undertail coverts buffy. Female has short white malar stripe. **V** Mellow, nasal-sounding fast twittering *tweetweetweetwee*, also described as a laughing chatter. **H** Montane and humid forest, cloud forest, forest edges. **D** E slopes of Andes, SC Colombia, Ecuador, N Peru.

1
2
3
4
5
6
7
8
9
10
11
12
13
14
15
16
17
18
19
20
21
22
23
24
25
26
27
28
29
30
31
32
33
34
35
36
37
38
39
40
41

HUMMINGBIRDS *CONTINUED*

1 WHITE-BOOTED RACKET-TAIL *Ocreatus underwoodii* male 12cm, female 7.5cm Male unmistakable with long tail, outer feathers racquet-tipped. Female has buff undertail coverts, short and barely forked tail with outer feathers blackish blue, tipped white. **V** Diagnostic, descending, thin sweet trill, *ti-tlee-ee-ee*, usually repeated 2–3 times. Also single *tsit* and *trrt* notes. **H** Montane tropical to temperate, cloud and dwarf forests, forest edges. **D** N Venezuela to W Ecuador.

2 PERUVIAN RACKET-TAIL *Ocreatus peruanus* male 14cm, female 8cm Previously considered conspecific with White-booted Racket-tail. Only male has long tail feathers. **V** Calls include vibrating trill, pitch descending from start to finish. **H** Montane forest, cloud forest, secondary forest. **D** Ecuador, Peru.

3 RUFOUS-BOOTED RACKET-TAIL *Ocreatus addae* male 12cm, female 8cm Similar to Peruvian Racket-tail but note white freckling on throat. **V** Calls include thin vibrating trill, pitch descending from start to finish. **H** Montane forest, cloud forest, secondary forest. **D** Peru, Bolivia.

4 BLACK-TAILED TRAINBEARER *Lesbia victoriae* male 25cm, female 15cm Extremely long forked tail dark. Female has shorter forked tail, underparts white and buff spotted green. **V** Song a repeated gravelly, squeaky *skreee... skreee*. Calls include an upslurred *tseep* and a descending series of high-pitched notes accelerating into a trill. **H** Montane bushy terrain, slopes, banks and ravines, dwarf woods, scrubby grassland, plantations, bushy copses in páramo. **D** NE Colombia to SE Peru.

5 GREEN-TAILED TRAINBEARER *Lesbia nuna* male 16cm, female 11cm Very long forked tail green, except purple-black outer pair of feathers. Undertail coverts green. Female has shorter forked tail. **V** Song a repeated gravelly note; *drrrt... drrrt*. Calls include repeated *bzzzzt*, and a descending and accelerating series of high-pitched notes. **H** Bushes and thickets on edge of upper montane forest, grasslands with shrubs, scattered trees and parks. **D** NE Colombia to N Bolivia.

6 RED-TAILED COMET *Sappho sparganurus* male 18cm, female 13cm Tail long and forked, fiery red, feathers tipped black. Female has shorter tail. **V** Short, rapid, jumbled chatter. **H** Temperate and subtropical woods and bushy areas in mountainous and hilly country, near slopes and banks, scrubby grassland, copses, thickets, parks, gardens. **D** N Bolivia to C Chile.

7 BRONZE-TAILED COMET *Polyonymus caroli* 13cm Bronzy green. Long, deeply forked tail brilliant green and coppery purple. Female has white underparts spotted green and blackish, some carmine dots at mid-throat. **V** Dry, rapid chatter, *tcht* or *tchtcht*. **H** Edges of montane woods, copses of low trees and shrubs, arid temperate areas, thickets. **D** S and C Peru.

8 BLACK-BACKED THORNBILL *Ramphomicron dorsale* 9.5cm Tiny sharp bill. Female has bright green upperparts, purple-black outer-tail feathers, tipped white. **V** Calls include a short dry rattled *trrr*, repeated at intervals, and a long descending rattle starting and ending with one or a few single notes. **H** Edge of montane forest and cloud forest, dwarf woods at great elevations, shrubs and thickets on slopes, páramo, even above snowline. **D** NE Colombia.

9 PURPLE-BACKED THORNBILL *Ramphomicron microrhynchum* 8cm Tiny sharp bill. Upperparts shining violaceous purple. Female's upperparts brilliant green. **V** Song a continuous series of quiet scratchy notes interspersed with short buzzes or dry trills. Calls include a repeated short dry rattling *trrr*, a scratchy *krr-kit* and a long twittering rattle. **H** Edge of cloud forest and dwarf forest, bushes and thickets, scattered trees. **D** N Venezuela to Bolivia.

10 BEARDED MOUNTAINEER *Oreonympha nobilis* 16cm Large. Upperparts brown. Underparts brownish white. Outer-tail feathers mostly white. Female similar, duller, tail shorter, smaller beard. **V** A descending, squeaky series followed by a rich chatter, *swee swee chew-chew-chew*, and also a dry *dzzrt*. **H** Uplands, arid bush, thickets, copses, ravines. **D** C Peru.

11 GREEN-BEARDED HELMETCREST *Oxypogon guerinii* 11.5–12cm Usually feeds while perched on flower, not while hovering. Very short bill. **V** Calls include series of squeaky, tongue-smacking chatters. **H** High-altitude páramo; mostly 3,000–5,000m. **D** Colombia.

12 BLUE-BEARDED HELMETCREST *Oxypogon cyanolaemus* 11.5–12cm Note very short bill. Critically Endangered. **V** Calls include thin, high-pitched vibrating trill. **H** High-altitude páramo; mostly 3,000–4,000m. **D** Sierra Nevada de Santa Marta, N Colombia.

13 WHITE-BEARDED HELMETCREST *Oxypogon lindenii* 11.5cm Usually seen perched in flowering bushes. Typically feeds while clinging to flower, not hovering. Note short bill. **V** Thin, high-pitched *zrrp* or *tuip*. **H** High-altitude páramo; mostly 3,500–4,500m. **D** W Venezuela.

14 BUFFY HELMETCREST *Oxypogon stuebelii* 11.5–12.5cm Usually solitary. Feeds while perched on flower; sometimes catches invertebrates on ground. Note short bill. **V** High-pitched, abrupt *zip*. **H** High-altitude páramo. **D** Nevado del Ruiz, C Colombia.

15 TYRIAN METALTAIL *Metallura tyrianthina* 9cm Bronzy green with coppery sheen. Lower belly and undertail coverts cinnamon-buff. Female's outer-tail feathers tipped white. **V** Seldom-heard song a repeated series of weak, high-pitched, lisping notes. Frequently heard chase call includes alternating stuttering trills and repeated phrases of squeaky notes. **H** Cloud forest, dwarf forest, bushes in páramo. **D** N Venezuela to N Bolivia.

16 PERIJA METALTAIL *Metallura iracunda* 10cm Black with coppery and green sheen. Forehead and bib shining emerald. Tail shining rufous-red. Female like female Fiery-throated Metaltail but tail longer and square. **V** Unknown. **H** Bushes and grassland at great elevations, above montane forest. **D** N Colombia, Venezuela.

17 VIRIDIAN METALTAIL *Metallura williami* 10cm Very dark shining green. Brilliant purple-blue tail. Female's upperparts like those of male; lower belly and undertail coverts buffy cinnamon. Tail like male's. **V** A descending series of 3–5 squeaky notes, often followed by several jumbled dry buzzy trills; *trsee-seee-seee-sew... trrk-tsetswe-trrk*. Also repeated dry, buzzy or gravelly notes. **H** Bushy areas in ecotone between montane forest and páramo. **D** C Colombia to S Ecuador.

18 VIOLET-THROATED METALTAIL *Metallura baroni* 10cm Blackish coppery green. Shining violet bib. Slightly forked tail shining green, with purple-copper sheen dorsally. Female's bib buff, spotted shining violet, underpart feathers edged buffy cinnamon. **V** Descending series of 3–5 squeaky notes, followed by several jumbled ones; *trsee-seee-seee-sew... trr-tsee-tse-tsew... trr-tsee-tse-tsew*. **H** Edge of temperate woods and bushy areas at great elevations. **D** SC Ecuador.

19 NEBLINA METALTAIL *Metallura odomae* 10cm Like Fiery-throated Metaltail but fiery red bib much larger. Cheeks chestnut-copper. Upperside of tail blue, underside bronzy green. All outer-tail feathers tipped white. Female's bib whitish, spotted red. **V** Wiry chatter. **H** Above tree line, open páramo, dwarf copses. **D** S Ecuador, N Peru.

20 COPPERY METALTAIL *Metallura theresiae* 11cm Bright coppery purple. Violaceous purple tail, greener underside. Bib shining greenish gold. One race has blue undertail and reddish-bordered bib. **V** Series of frail *zeee* calls, often alternating with *ttrrrt* notes. **H** Thickets and edges of montane forest, moss-hung dwarf forest, edges of bogs. **D** Peru.

21 FIERY-THROATED METALTAIL *Metallura eupogon* 10cm Fiery red bib. Upperside of tail has bluish sheen. Female has feathers of lower underparts edged cinnamon-buff. **V** A descending series of 3–6 squeaky notes, followed by a repeated, buzzy, jumbled phrase; *trt-tsee-seee-seee-sew... trr-tsee-tsew... trr-tsee-tsew*. **H** Montane bushy areas and upper edges of forest. **D** Peru.

22 SCALED METALTAIL *Metallura aeneocauda* 10cm Bronzy green. Breast, upper belly and undertail coverts edged buff, giving slight scaly effect. **V** A descending series of 3–6 squeaky notes, followed by a repeated, buzzy, jumbled phrase; *trt-tsee-seee-seee-sew... trr-tsee-tsew... trr-tsee-tsew*. **H** Temperate montane forest, bushy areas, edges. **D** Peru, Bolivia.

23 BLACK METALTAIL *Metallura phoebe* 13cm Overall shining black. Tail shining purple-red. **V** Song a repeated high-pitched single note, *tsee... tsee... tsee*, given at a rate of one note every 2–5 secs. Also a short dry trill, *djrrt*, repeated at intervals, and longer dry chatter. **H** Semi-arid open scrub and woods in uplands, canyons, bushy slopes. **D** Peru.

24 RUFOUS-CAPPED THORNBILL *Chalcostigma ruficeps* 10cm Bronzy green. Pointed bib rainbow-coloured. Bronzy-olive tail somewhat forked. Vent and lower belly cinnamon-buff. **V** Calls include a short, dry double rattle, *trr-trr*, repeated at intervals, and a rising/falling twitter, *t-tr-tr-tsee-tseee-tititrrr*. **H** Bushes and edges of montane forest. **D** SE Ecuador to WC Bolivia.

25 OLIVACEOUS THORNBILL *Chalcostigma olivaceum* 14cm All brown, olive sheen on upperparts. Long beard rainbow-coloured. **V** Unknown. **H** Arid bush, thickets at high elevations, dwarf woods, copses, páramo. **D** WC Peru to WC Bolivia.

26 BLUE-MANTLED THORNBILL *Chalcostigma stanleyi* 14cm Blackish. Slight green sheen on shoulders. Female lacks violaceous tip to beard. **V** Call a descending squeaky twitter followed by a mellower upslurred note and a downslurred final note; *tseetsitsitsitsitr-whee-tsew*. **H** Thickets, dwarf woods and edges at great elevations. **D** Ecuador to WC Bolivia.

27 BRONZE-TAILED THORNBILL *Chalcostigma heteropogon* 13cm Bronzy green. Forehead and beard emerald, beard tipped bright violaceous-carmine. Somewhat forked tail bright bronzy olive. Belly and vent cinnamon-buff. **V** Calls include a repeated dull, short *tzk*. **H** Low and open montane woods, forest edges, bushy areas, copses, páramo, ravines. **D** W Venezuela, E Colombia.

28 RAINBOW-BEARDED THORNBILL *Chalcostigma herrani* 13cm Greenish black. Forehead and crown rufous, ending in an ochre crest. Long narrow beard rainbow-coloured. Rump coppery red. Tail blackish blue, broadly tipped white. Vent white. Female lacks beard; underparts dark brown. **V** Poorly known. Repeated low-pitched *cheet-dee-dee-cheet*, which may be a song. **H** Open or scattered woods, dwarf woods, copses, bushes, páramo. **D** Colombia to N Peru.

29 MOUNTAIN AVOCETBILL *Opisthoprora euryptera* 10cm Short bill, slightly upturned at tip. Green with coppery sheen on head, bronzy on upperparts. **V** Poorly known. Calls include a series of descending thin whistles, reminiscent of a piculet; *usee usee usee usee*. **H** Temperate montane forest at high elevations, bushes, thickets, forest edges, slopes in páramo. **D** Colombia to N Peru.

30 GREY-BELLIED COMET *Taphrolesbia griseiventris* 16cm Upperparts and tail bronzy green. Tail long and forked, bluish below. Female's upperparts and tail like male's though tail shorter. **V** Short, rapid series of *tsu-wit* notes. **H** Bushes and thickets in arid temperate zone of high elevations, hillsides, ravines. **D** NW Peru.

31 LONG-TAILED SYLPH *Aglaiocercus kingii* male 18cm, female 10.5cm Glittering dark green. Very long forked tail brilliant dark blue. Female's tail short, outer feathers tipped white. **V** Continuous series of buzzy notes, *bzzt... bzzt... bzzt*, at a rate of about one note per sec. Calls include a repeated, short, buzzy or raspy *dzzrt*, and a more drawn-out, higher-pitched *bzzeeew*. **H** From edge of humid forest in mountains to parks and gardens. **D** Venezuela to WC Bolivia.

32 VENEZUELAN SYLPH *Aglaiocercus berlepschi* male 21cm, female 10.5cm Only male has long violet and blue tail. Sometimes forays for insects from perch, **V** Rasping *chuup*. **H** Scrub and wooded slopes; mainly lowlands to 1,500m. **D** N Venezuela.

33 VIOLET-TAILED SYLPH *Aglaiocercus coelestis* male 18cm, female 9.7cm Differs from Long-tailed Sylph in having coppery-olive underparts. Female has short tail. **V** Continuous series of short notes, *psit... psit... psit*, about two notes per sec. Calls include a repeated, short, buzzy single or double *bzzt* or *bz-zzrt*, and a higher-pitched drawn-out *bzeee*. **H** Moss-hung montane forest and cloud forest, forest edges, scrub, copses. **D** Colombia, Ecuador.

34 HYACINTH VISORBEARER *Augastes scutatus* 8cm Front of face and bib shining emerald green. Female duller; head bronzy green, underparts grey flecked green, smaller bib, blue and white collars. **V** Males sing. Female observed giving a *tilp-tilp* when leaving or returning to nest and *ti-ti-chip-chip* or *chip-chip-chip* when perched a little distance away from it. **H** Stony areas with bushes, thickets and low vegetation in uplands. **D** SE Brazil.

35 HOODED VISORBEARER *Augastes lumachella* 9cm Front of face and bib shining golden green, bib ending in scarlet point. Body bronzy green with coppery sheen. Female similar, but no black. **V** Song an incessantly repeated short phrase, starting with an overslurred squeaky nasal note, followed by 1–2 burry notes, occasionally interspersed with a more rapid sequence. Calls are short *chup* and *chrr* notes. **H** Stony semi-arid upland terrain with cactus and shrubs. **D** E Brazil.

36 GEOFFROY'S WEDGEBILL *Schistes geoffroyi* 8.5cm Mask dark green. Violet and white marks on sides of neck. Coppery uppertail coverts. Female has white throat. **V** Song an insect-like series of regularly spaced, simple *tsit* or sibilant *sink* notes, given at a rate of 1.5–2.8 notes per sec. **H** Rainforest, dense cloud forest, forest edges, shrubs, thickets. **D** E Colombia, N Venezuela to E Peru and C Bolivia.

37 WHITE-THROATED WEDGEBILL *Schistes albogularis* 8.5–9cm Previously lumped with Geoffroy's Wedgebill and referred to as Wedge-billed Hummingbird. White collar more complete than Geoffroy's Wedgebill. **V** Sharp, tongue-smacking *tip, tip, tip-pp-pip*. **H** Cloud forest on W slopes of Andes; mainly 1,000–2,000m. **D** W Ecuador, W Colombia.

38 PURPLE-CROWNED FAIRY *Heliothryx barroti* 11cm Unmistakable by pure white underparts and peaked head shape. **V** Call includes a high, thin, slightly metallic *sssit*, which may be run into longer, rapid series. **H** Forest interior and edges in lowlands. **D** SE Mexico to SW Ecuador.

39 BLACK-EARED FAIRY *Heliothryx auritus* 12cm Black 'ear'. Upperparts and throat brilliant green. Rounded tail, central feathers blue, outers white. Female has white throat, spotted greyish. **V** Calls include a short high-pitched *tsit* and richer *tchip*, repeated at intervals. **H** Lowland forest, forest edges, secondary growth, thickets and bushes. **D** E South America, Amazonia.

40 HORNED SUNGEM *Heliactin bilophus* 11cm Shining rainbow-coloured 'horns'. Forehead glittering blue. Pointed black bib. Tail long and pointed. Female has bronzy-green cap, whitish throat. **V** Complex twittering comprising squeaky, burry and buzzy notes during chases. Calls include repeated *tsit* or *tseet*. **H** Savannah, gallery forest, bushes, thickets, gardens, hills, plains. **D** S Suriname and N, E and SE Brazil to E Bolivia.

1
2
3
4
5
6
7
8
9
10
11
12
13
14
15
16
17
18
19
20
21
22
23
24
25
26
27
28
29
30
31
32
33
34
35
36
37
38
39
40

HUMMINGBIRDS *CONTINUED*

1 MARVELOUS SPATULETAIL *Loddigesia mirabilis* male 17cm, female 9.5cm Very long racquet-tipped tail. Bib bluish green and golden. Ventral mid-line black. Female has dirty white throat and underparts, spotted on sides of breast and flanks, white undertail coverts extending beyond bluish-black tail; outer-tail feathers paddle-shaped. **V** Calls include an upslurred high-pitched sweet *tswee*, often given in fast series at a rate of about six notes per sec, as well as a repeated and more strident *tsik*. **H** Humid temperate and tropical montane forest, bushes and thickets, forest edges. **D** N Peru.

2 PLAIN-CAPPED STARTHROAT *Heliomaster constantii* 11.5–12.5cm Often makes low sallies to capture flying insects. **V** Loud *peek*; song transcribed as *chip chip chip chip pi-chip chip chip*..., or *chi chi chi chi whit-it chi*... **H** Shrubby, arid woodland, woodland edges, thickets. **D** W Central America.

3 LONG-BILLED STARTHROAT *Heliomaster longirostris* 10cm Brownish olive with bronzy sheen. Cap glittering emerald. Purple-red bib bordered white. Central underparts dirty white. Tail brownish olive, white tips. Female has brownish-olive cap, notable white malar stripe and throat spotted blackish. White mark on side of body. **V** Calls include a rich liquid *tseep* or *tsew*, while chasing call involves squeaky twitters. **H** Open woods, forest edges, copses, scattered trees, open areas, bushes, thickets, savannah, plantations, parks, gardens. **D** S Mexico to E Bolivia and NE Brazil.

4 STRIPE-BREASTED STARTHROAT *Heliomaster squamosus* 10cm Dark bronzy green. Cap shining blue. Gorget glittering purple-red. White stripe down mid-belly. Undertail coverts edged white. Female's underparts dirty white, throat white scaled blackish brown, white malar stripe and white spot on side of body; tail green with black subterminal band, tip white. **V** Song a subdued, buzzy, scratchy warbling *bzzzrrrr... bzzzrrr... chi-chi-chi-chi... bzrrr*. Calls include liquid *tsik*. **H** Tropical rainforest, forest edges, savannah, bushes and thickets, woods and scrub in hills, copses. **D** E and SE Brazil.

5 BLUE-TUFTED STARTHROAT *Heliomaster furcifer* 10cm Upperparts bronzy green, underparts dark blue. Gorget purple-red and shining dark blue. Vent dark green. Female's underparts plain dirty white, tail black, basally green, tipped white. **V** Call is a descending, pure-whistled *tseep*, repeated at intervals. **H** Dry woods, arid scrub, savannah, forest, forest edges, copses, parks, gardens. **D** Bolivia and SW Brazil to N Argentina, SE Brazil and Uruguay.

6 OASIS HUMMINGBIRD *Rhodopis vesper* male 13.5cm, female 10cm Upperparts bronzy olive. Uppertail coverts cinnamon. Bib shining purple-red. Tail longish and forked, blackish; feathers narrow and pointed. Central underparts whitish. Female's throat white, tail shorter that male's with broad outer feathers tipped white. **V** Melodious series of *tick* and *tzee* notes given at variable speeds and tone, as well as a rapid, thin, liquid chatter. **H** Oases in desert, lowlands and mountains, gardens, coastal hills. **D** NW Peru to N Chile.

7 PERUVIAN SHEARTAIL *Thaumastura cora* male 13cm, female 8cm Purple-red and shining blue bib. Broad white collar. Very long tail, central feathers slender. Mid-belly to vent white. Female's tail short and rounded, throat and rest of underparts buffy white, incomplete bronzy-green collar. **V** Squeaky, incessant, undirected song with at least four syllables, and lasting up to a minute or more. **H** Shrubs and thickets in ravines, oases and desert, lower slopes to coast. **D** NW Peru to N Chile.

8 SPARKLING-TAILED WOODSTAR *Tilmatura dupontii* 7–9cm Unmistakable by tail form and pattern. Note white tufts to rump sides, present also on Magenta-throated and Purple-throated Woodstars. **V** Usually silent, but rarely gives high, sharp, twittering chirps. **H** Forest, open woodland. **D** C Mexico to Nicaragua.

9 SLENDER SHEARTAIL *Doricha enicura* 9–12cm Distinctive long tail without rufous; note extent of rufous, and tail pattern of female. **V** Poorly known. Calls include fairly hard, rapid chips, often repeated steadily or slightly trilled. **H** Scrub, forest edges; 1,000–2,250m. **D** S Mexico to El Salvador

10 MEXICAN SHEARTAIL *Doricha eliza* 9–11cm Like Magenta-throated Woodstar, but underparts with more white, rounded tail tips, less rufous feathering; female has white in tail. **V** Call is fairly hard to slightly liquid, rapid chipping, often slightly rolled. **H** Low scrub, mangrove, gardens. **D** SE Mexico.

11 AMETHYST WOODSTAR *Calliphlox amethystina* male 9cm, female 6.5cm Forked tail, outer pair pointed. Shining purple-red bib. White collar. Female's throat buff, spotted green, collars green and buff, rest of underparts cinnamon-rufous. **V** Unknown. **H** Edge of humid and gallery forest, clearings, shrubs, thickets, savannah. **D** Widespread in South America.

12 BAHAMA WOODSTAR *Calliphlox evelynae* 8–9.5cm Female has buff tips on outermost tail feathers. Feeds on nectar and hawks insects. **V** Dry *prititidee prititidee prititidee*; also sharp *tit titit tit titit*, which often speeds into a rattle. **H** Mixed pine forests, forest edges, clearings, scrub and large gardens. **D** Bahamas, Caicos Islands.

13 INAGUA WOODSTAR *Calliphlox lyrura* 8cm Forecrown shiny (matt in related Bahama Woodstar). **V** Squeaky song and sharp *tchip* calls. **H** Varied, including parks, secondary woodland, mangroves, gardens. **D** Great Inagua island, Bahamas.

14 MAGENTA-THROATED WOODSTAR *Calliphlox bryantae* 7–9cm Separated from Purple-throated Woodstar by rufous, shorter tail; female similar. **V** Dry *chi* or territorial *chrrrrt*. **H** Forest edges and clearings, open woodland, cultivation. **D** Costa Rica, Panama.

15 PURPLE-THROATED WOODSTAR *Calliphlox mitchellii* 7–11cm See Magenta-throated Woodstar; separated from that species also by different range. **V** Poorly known. Calls include series of thin *chit* notes, often in bursts of 2–4 notes, and a squeaky *kyee-kyee-kyee-kyee*. **H** Forest canopy, mainly in lowlands. **D** E Panama to W Ecuador.

16 SLENDER-TAILED WOODSTAR *Microstilbon burmeisteri* male 8.5cm, female 7cm Strangely forked tail. Narrow shining purple-red gorget. Throat and mid-belly grey. Female's underparts buffy cinnamon. **V** Poorly known. Calls include a series of dull *chip* notes, sometimes doubled. **H** Shrubs and thickets in mountains, edges of montane forest. **D** C Bolivia to NW Argentina.

17 LUCIFER SHEARTAIL *Calothorax lucifer* 9–10cm Male has forked tail. Feeds on nectar and insects which are obtained by brief fly-catching sallies. **V** Twittering chips. **H** Desert areas with agave, mountain slopes, canyons. **D** SW USA to C Mexico

18 BEAUTIFUL SHEARTAIL *Calothorax pulcher* 9cm Very similar to Lucifer Sheartail, but with slightly shorter and straighter bill. Restricted range. **V** High, twittering chips. **H** Scrub, woodland, cultivation. **D** SC Mexico.

19 VERVAIN HUMMINGBIRD *Mellisuga minima* 6–7cm Feeds on nectar from various flowers. **V** Throaty buzz and high-pitched squeaks. **H** Open areas with small flowers. **D** Jamaica, Hispaniola.

20 BEE HUMMINGBIRD *Mellisuga helenae* 5–6cm The world's smallest bird. Feeds on nectar and small insects. Often uses a favourite perch for years. **V** Long, high-pitched twitter and low warbling. **H** Forest, woodland, swampland, gardens. **D** Cuba.

21 RUBY-THROATED HUMMINGBIRD *Archilochus colubris* 8–9.5cm Feeds mainly on nectar from flowering plants, also hawks for insects. **V** Squeaking *cric-cric*. **H** Woodland edges, copses, gardens. **D** Breeds SC and SE Canada, C and E USA. Winters Central America.

22 BLACK-CHINNED HUMMINGBIRD *Archilochus alexandri* 10cm Female is unmistakable when chin reflections can be seen. Somewhat longer-tailed than Ruby-throated Hummingbird. **V** Song a sweet, very high-pitched warble. **H** Dry country (summer); riverine belts, forest (winter). **D** Breeds SW Canada, W USA, N and NW Mexico. Winters W Mexico.

23 ANNA'S HUMMINGBIRD *Calypte anna* 10–11cm Feeds on nectar, and gleaned or hawked insects. **V** Song a jumble of squeaks and raspy notes. Call a sharp *chick*. **H** Chaparral, canyons, woodlands, coastal scrub. **D** SW Canada, W USA, NW Mexico.

24 COSTA'S HUMMINGBIRD *Calypte costae* 7.5–8.5cm As well as nectar, takes insects by hawking or hovering. **V** High, light, sharp *tik* or *tip*, often repeated to form a twitter. Song a thin, high, rising and falling buzz. **H** Dry areas with sages, ocotillo and yuccas. **D** Breeds SW USA, NW Mexico. Winters W Mexico.

25 BUMBLEBEE HUMMINGBIRD *Atthis heloisa* 7.5cm Bumblebee-like flight. Feeds mainly on nectar. **V** High chips. Song a high, thin *sssssssssssiu* or *seeuuuuu*, fading at the end. **H** Forest edges and clearings. **D** Mexico.

26 WINE-THROATED HUMMINGBIRD *Atthis ellioti* 7cm Separated from Bumblebee Hummingbird by range. **V** A series of high, slightly buzzy, squeaky chips that break into a warble. **H** Forest interior, edges and clearings. **D** S Mexico to Honduras.

27 PURPLE-COLLARED WOODSTAR *Myrtis fanny* male 9cm, female 7.5cm Longish forked tail, outer feathers curving inwards. Female's outer-tail feathers black, tips buffy white; underparts plain buffy cinnamon. **V** Calls include fast, dry *chi-chi-chi*, sometimes in longer rattling series, and a fast, high-pitched, descending twittering *ti-ti-ti-ti*. **H** Arid shrubs and thickets in tropical and temperate regions, coastal plains to high mountains. **D** Ecuador, Peru.

28 CHILEAN WOODSTAR *Eulidia yarrellii* male 8cm, female 7cm Bib shining purple-red and blue. Tail forked, outer feathers narrow, pointed. Female's throat and collar white, underparts mostly cinnamon. **V** Calls include slightly raspy *tsick* notes and high-pitched *zrrrrrrr* in display. **H** Shrubs, thickets and gardens in arid regions. **D** S Peru, N Chile.

29 SHORT-TAILED WOODSTAR *Myrmia micrura* 6.5cm Shining violaceous-red gorget. Very short rounded tail. Female's underparts buffy white. Tail tipped white. **V** Thin *tchi-tchi-tchi*. **H** Shrubs and thickets in arid tropical regions. **D** W Ecuador, NW Peru.

30 WHITE-BELLIED WOODSTAR *Chaetocercus mulsant* 7.5–8.5cm Shining purple-red bib. Collar and mid-belly white. Forked tail, feathers pointed. Vent white and green. Female's throat, collar and mid-belly buffy, flanks and vent rufous cinnamon, has incomplete green collar. **V** Calls include low-pitched *djup* and higher-pitched, dull *chip*. **H** Edge of humid forest, bushes, thickets, copses, grassland, cultivated areas in mountainous terrain. **D** Colombia to C Bolivia.

31 LITTLE WOODSTAR *Chaetocercus bombus* 6–6.8cm Bib shining red. Collar buffy. Tail forked, outer feathers thorn-like. Female's underparts cinnamon, incomplete green collar. **V** Song a mixture of chips, twitters and buzzy notes; *tsitsitsi... tzzeee-tzzeee... chichip*. Calls include a single dry *chip* or doubled *chichip* and a squeaky *kswee-kswee-ti-ti-ti*. **H** Bushes and thickets at edges of tropical and subtropical forest, mountains, valleys, plains. **D** Venezuela to W Ecuador.

32 GORGETED WOODSTAR *Chaetocercus heliodor* 5.7–6.4cm Gorget brilliant purple-red. Breast grey. Tail forked. Female's underparts cinnamon-rufous, incomplete green collar. **V** Calls include single dry *chit*, doubled *chichit* or tripled *chichichit*. **H** Montane and cloud forest, clearings, forest edges, bushes, thickets, copses, plantations. **D** Venezuela to W Ecuador.

33 SANTA MARTA WOODSTAR *Chaetocercus astreans* 7cm Male has green head, shiny blue body and attenuated reddish gorget. **V** High-pitched, disyllabic, repeated *tchu-ip*. **H** Montane forest, secondary woodland; mainly 1,000–2,000m. **D** Sierra Nevada de Santa Marta, N Colombia.

34 ESMERALDAS WOODSTAR *Chaetocercus berlepschi* 5.5–6.2cm Strange tail: outer feathers like long thorns. Female's underparts white, outer-tail tipped white. **V** Mostly silent. **H** Edge of subtropical and temperate montane forest, shrubs, thickets. **D** W Ecuador.

35 RUFOUS-SHAFTED WOODSTAR *Chaetocercus jourdanii* 6.5–7.2cm Broad white collar. Forked tail. Female has cinnamon and white underparts, white collar, incomplete green collar; somewhat rounded tail. **V** Song a rising 3–4-note *tssit tssit tssit tssit*. **H** Montane and cloud forest, forest edges, bushes, thickets, woods, coffee plantations. **D** N Venezuela, N Colombia.

36 BROAD-TAILED HUMMINGBIRD *Selasphorus platycercus* 10cm Feeds on nectar and insects, which are gleaned or hawked. Male's wings produce a loud trilling whistle. **V** Sharp *chip*. **H** Pine and juniper forests, riparian and dry pine-oak forest. **D** Breeds WC USA, Mexico, W Guatemala. Winters Central America.

37 RUFOUS HUMMINGBIRD *Selasphorus rufus* 10cm Small number of males have green upper back, making them virtually identical to Allen's Hummingbird. **V** Hard *tyuk*, excited *zee-chupity-chup* and various chipping and buzzing notes. **H** Coniferous forests, forest edges, clearings, mountain meadows. **D** Breeds SE Alaska, W Canada, NW USA. Winters S USA, N and C Mexico.

38 ALLEN'S HUMMINGBIRD *Selasphorus sasin* 10cm Female virtually identical to female Rufous Hummingbird. **V** Similar to that of Rufous Hummingbird. **H** Canyons, coastal chaparral, scrub, bushy slopes. **D** Breeds W North America. Winters C Mexico.

39 VOLCANO HUMMINGBIRD *Selasphorus flammula* 8cm Calliope Hummingbird has similar tail pattern. **V** Soft *chip* notes heard while birds are foraging. **H** Open disturbed habitats; below 1,750m. **D** Costa Rica, W Panama.

40 GLOW-THROATED HUMMINGBIRD *Selasphorus ardens* 7cm Both male and female have less rufous in tail than in Scintillant Hummingbird. **V** Unknown. **H** Forest edges and clearings; 750–1,750m. **D** WC Panama.

41 SCINTILLANT HUMMINGBIRD *Selasphorus scintilla* 7cm Note extent of rufous in tail. See also range of Glow-throated Hummingbird. **V** A liquid *tsip*. **H** Scrubby areas, forest edges, cultivation; 750–2,000m. **D** NC Costa Rica, W Panama.

42 CALLIOPE HUMMINGBIRD *Selasphorus calliope* 8cm Tends to feed on lower blossoms when taking nectar from plants. **V** Soft *tsip*. Song a thin *tseeeee-ew*. **H** Mountain glades, canyons. **D** Breeds SW Canada, W USA. Winters W Mexico.

1
2
3
4
5
6
7
8
9
10
11
12
13
14
15
16
17
18
19
20
21
22
23
24
25
26
27
28
29
30
31
32
33
34
35
36
37
38
39
40
41
42

MOUSEBIRDS COLIIDAE

1 SPECKLED MOUSEBIRD *Colius striatus* 35cm Rump concolorous with rest of upperparts. Some races have no or little white to cheeks. **V** Often silent but gives loud, explosive *siut* on finding food, similar quieter call by adult guiding young, contact call *siu-siu*, alarm call *pit* or *schiech... schiech* when confronting intruders. **H** Open forest, wooded grassland, orchards, suburbs. **D** Widespread in Africa.

2 WHITE-HEADED MOUSEBIRD *Colius leucocephalus* 30cm Greyish black, with white crest, crown and cheeks. **V** Not well known but reported to call *rit-rit*, also various plaintive notes. **H** Dry bushland with thickets, often near streams. **D** E Africa.

3 RED-BACKED MOUSEBIRD *Colius castanotus* 35cm Separated from Speckled Mousebird by chestnut rump and different range. **V** Twitters constantly when foraging in flocks; other calls include harsh *chee-chee chee*. **H** Dry wooded and bushy grasslands. **D** W Angola.

4 WHITE-BACKED MOUSEBIRD *Colius colius* 30cm Note pale, dark-tipped bill and white back. **V** Similar to the calls of Speckled Mousebird but more tuneful. **H** Dry, bushy and wooded areas. **D** Namibia to S Botswana and South Africa.

5 BLUE-NAPED MOUSEBIRD *Urocolius macrourus* 33–35cm Gregarious, first sighting is often of a flock flying from tree to tree. Works its way among tree branches like a rodent. **V** Plaintive *peeee peeeeeeeee* and shorter *pyee pyee pyee* or *pew t-lew pew t-lew*. **H** Semi-desert with bushes and trees. **D** E, C and W Africa.

6 RED-FACED MOUSEBIRD *Urocolius indicus* 35cm Note thin long tail and red bare parts. Eye may be black, pale blue or yellow. **V** Call melodious, bell-like *tee-ree-ree*, given in flight and at rest; also slurred *cheuwip* and various softer whistles and trills. **H** Wooded scrub and bushy areas, plantations, gardens. **D** S Africa.

TROGONS TROGONIDAE

7 EARED QUETZAL *Euptilotis neoxenus* 33–35cm More mobile than Elegant Trogon. Pairs move through trees in search of food. Underside of tail black with large white tips. **V** Squealing call with rising end note, *kweeeeeee-chk*. **H** Montane pine forests. **D** W Mexico.

8 PAVONINE QUETZAL *Pharomachrus pavoninus* 33cm Very like Golden-headed Quetzal but bill red and tail more graded, outer feathers very short. Female like female Crested Quetzal but bill dirty red. **V** Song described as series of five tuneful notes, *oh-ewoh-ewoh-ewoh-ewoh brr*; call downslurring whistle followed by short, loud single note, *heeeeear... CHOK*. **H** Terra firme and rainforest in lowlands. **D** Amazonia.

9 GOLDEN-HEADED QUETZAL *Pharomachrus auriceps* 33cm Differs from White-tipped Quetzal only in its entirely black tail. Yellow bill. Female differs from those of Crested Quetzal and White-tipped Quetzal in all-blackish tail and pinkish red reaching lower belly. **V** Gives drawn-out and far-carrying hawk-like whistles *whe-wheeoo, whe-wheeoo, whe-wheeoo*; also plaintive *ka-kaaaur*, and a fast, laughing *why-dy-dy-dy-drrr*. **H** Humid forest, cloud forest, dwarf copses, open country with scattered trees. **D** E Panama to N Bolivia.

10 WHITE-TIPPED QUETZAL *Pharomachrus fulgidus* 33cm Upperparts and breast bright emerald green. Head with strong golden sheen. Underparts bright red. Uppertail coverts brilliant green, longer than tail. Female's head greyish brown with golden-green gloss. Breast bright green. Vent pinkish red. **V** Song a series of three hooting whistles, *buu hu-hua*; call a loud *kirra*, becoming faster *kier, kip-kip-kip-a* when agitated. **H** Humid forest, cloud forest, forest edges, coffee plantations. **D** N Venezuela, N Colombia.

11 RESPLENDENT QUETZAL *Pharomachrus mocinno* male 95cm, female 35cm Unmistakable. **V** Very high *kiauw kiauw kiauw* or magpie-like *ketjauw ketjauw*. **H** Montane forest canopy and forest edges; 1,500–3,000m. **D** S Mexico to W Panama.

12 CRESTED QUETZAL *Pharomachrus antisianus* 30cm Crest from forehead. Brilliant emerald-green head, breast and upperparts. Underparts bright red. Central tail feathers black, outers white. Uppertail coverts green and longer than the tail. Female's head and underparts greyish brown, breast bright green, vent bright red; blackish tail with outer feathers partially barred and tipped white. **V** Call loud *way-way-wayo*; when alarmed calls *ka-ka-ka-ka*. **H** Cloud forest and forest edges. **D** Venezuela to N Bolivia.

13 CUBAN TROGON *Priotelus temnurus* 25–28cm Hovers, flycatcher-like, while feeding on flowers, buds or fruits. Usually in pairs. **V** Pleasant *toco-toco-tocoro-tocoro...* and short mournful note. **H** Shady areas in wet or dry forests. **D** Cuba.

14 HISPANIOLAN TROGON *Priotelus roseigaster* 27–30cm Usually in pairs. Feeds mainly on insects but also takes fruit and small lizards. **V** Repeated *toca-loro, coc ca-rao* or *cock-caraow*; also a puppy-like whimpering and cooing. **H** Mature pine and deciduous broadleaf montane forests, occasionally coastal mangroves. **D** Hispaniola.

15 LATTICE-TAILED TROGON *Trogon clathratus* 25cm White eyes and undertail pattern diagnostic. **V** High hurried *puhpuhuh...* (like hysterical laughter). **H** Forest; below 1,000m. **D** Costa Rica, Panama.

16 SLATY-TAILED TROGON *Trogon massena* 30cm Very like Black-tailed Trogon but lacks pectoral collar. Bill red in both sexes. Some specimens have slight white markings on outer vane and tip of outer-tail feathers. **V** Song a slow, steady series of low-pitched *cue* notes; calls include quiet clucking and a harsher, cackling chatter. **H** Humid rainforest, forest edges, scattered trees, clearings, secondary growth. **D** SE Mexico to NE Ecuador.

17 CHOCO TROGON *Trogon comptus* 28cm Like Black-tailed Trogon but smaller. White eye. Black eye-ring. No pectoral collar. Female like female Black-tailed Trogon but darker slate, especially on head and breast. Eye and eye-ring like male's. **V** Song very like that of Black-tailed Trogon but lower-pitched with slower delivery. **H** Humid rainforest, forest edges, foothills, broken terrain. **D** W Colombia, NW Ecuador.

18 ECUADORIAN TROGON *Trogon mesurus* 30cm Glossy green head, breast and back, with golden sheen. Face and throat blackish. Eye-ring orange-red. Bill yellow. Wing coverts finely vermiculated black and white. Central tail feathers green washed blue. Underparts bright red. Pectoral collar white. Undertail slaty. Female slate-grey, belly and vent pinkish scarlet, blackish bill, yellow mandible, orange eye-ring. **V** Song a series of slow *cow* notes, becoming faster and louder; calls include a rapid churring chatter. **H** Humid rainforest, forest edges, secondary growth. **D** W Ecuador and NW Peru.

19 BLACK-TAILED TROGON *Trogon melanurus* 30cm Black undertail diagnostic. Female similar to female Slaty-tailed Trogon but with yellow lower mandible. **V** Very high *wéjwéjwéj...* **H** Forest, tall secondary growth, mangroves. **D** Panama to Venezuela and Amazonia.

20 BLACK-HEADED TROGON *Trogon melanocephalus* 25cm Separated from Citreoline Trogon by black eyes and blue uppertail. Female undertail same as male's. **V** Very high shivering accelerated magpie-like chatter *wekwek-wekkering*. **H** Forest, tree stands, mangroves. **D** SE Mexico to N Costa Rica.

21 CITREOLINE TROGON *Trogon citreolus* 25cm Yellow eyes diagnostic. **V** High, dry, accelerated bickering. **H** Dry woodland, plantations, mangrove; below 1000m. **D** Mexico.

22 WHITE-TAILED TROGON *Trogon chionurus* 25cm Note white undertail of female. Separated from Amazonian Trogon by absence of white border to blue/brown breast. **V** Very high *wewwew...* (repeated 8–15 times). **H** Forest canopy. **D** Panama to W Ecuador.

23 BAIRD'S TROGON *Trogon bairdii* 30cm Unmistakable by colour combination. Note dark wing coverts of female. **V** Very high *wewwew...* rising to a bouncing trill. **H** Forest canopy. **D** Costa Rica, Panama.

24 GREEN-BACKED TROGON *Trogon viridis* 28cm Head and breast blue. Back bright green. Wing coverts black. Underparts rich yellow. Eye-ring pale blue. Female's head, breast and upperparts slate-grey. Underparts yellow. Eye-ring complete and narrow, pale blue. **V** Song fairly fast series of *kyow* notes, becoming louder; a slower variant of the song is also noted. **H** Montane forest, rainforest, deciduous and gallery forest, forest edges, scattered trees in deforested areas, plantations. **D** Trinidad, Amazonia, SE Brazil.

25 GARTERED TROGON *Trogon caligatus* 25cm Blue head and green tail diagnostic. Note also undertail pattern. Wing coverts pale grey. **V** Very high yelping *WefWefWef...* (repeated 6–7 times) or *titjèhtjèh...*, slowing down and falling away at end. **H** Forest edges, tall secondary growth, plantations. **D** E Mexico to NW South America.

26 AMAZONIAN TROGON *Trogon ramonianus* 24–25cm Previously lumped with Gartered and Guinan Trogons, and called Violaceous Trogon. **V** Song a throaty, repeated *wheE-ow, wheE-ow, wheE-ow* with first syllable upslurred. **H** Lowland rainforest (inundated and terra firme), secondary forest. **D** Amazonia.

27 GUIANAN TROGON *Trogon violaceus* 23cm Small version of White-tailed Trogon, but eye-ring orange, undertail barred and slight pectoral collar white. Central tail feathers bluish green or blue, according to race. Female differs from female White-tailed Trogon only in white eye-ring being incomplete and broader. **V** Song a long, fast series of hollow whistled *kyu* notes; calls include a rolling chatter. **H** Rainforest, forest edges, scattered trees in deforested areas, open woods, woods along streams, humid to arid regions. **D** S Venezuela, N Brazil, the Guianas, Trinidad.

28 BLUE-CROWNED TROGON *Trogon curucui* 25cm Face and throat blackish. Head and neck metallic blue with greenish sheen. Upperparts metallic green with bronzy sheen. Central tail feathers bluish green. Narrow pectoral collar white. Underparts bright red. Female like that of Surucua Trogon but with white pectoral collar and more extensive rosy-red underparts. **V** Similar to that of Green-backed Trogon, but song picks up speed throughout series; calls include a purring note. **H** Cloud forest, montane forest, várzea, terra firme and gallery forests, tall secondary growth. **D** S and E South America.

29 SURUCUA TROGON *Trogon surrucura* 26cm Head, neck and breast metallic blue. Upperparts metallic green with bronzy sheen. Central tail feathers green washed blue. Underparts bright red. Undertail white, base blackish. Female mostly dark slate-grey, darker on head, lighter on upper belly. Wings black, finely barred white. Lower belly and vent pinkish red. Undertail outer vane white, tip white. Bill yellowish. **V** Song long series of fairly high-pitched *diu* notes, becoming louder through series; alarm call *kiarr* or a fast rattle. **H** Montane and lowland forest, gallery forest, forest edges, Cerrado. **D** SE South America.

30 BLACK-THROATED TROGON *Trogon rufus* 25cm Face and throat black. Only trogon with head, breast and upperparts green. Underparts rich yellow. Wing coverts vermiculated black and white. Racial variation in green, blue or coppery central tail feathers. Female's head, breast and upperparts cinnamon-brown, underparts yellow. **V** Song two to five slow, mellow, descending *kyow* whistles; calls include a churring note. **H** Rainforest, gallery forest, cloud forest, upland open woods, edges. **D** Honduras to Ecuador, Amazonia and SE South America.

31 ELEGANT TROGON *Trogon elegans* 32cm Sits quietly with short sallies to catch insects or pick fruit. Underside of tail finely barred black and white, with white tips. **V** Soft, chattering *brr brr brr...*; also monotonous, croaking *co-ah*. **H** Dry woodland. **D** SW USA to Costa Rica.

32 MOUNTAIN TROGON *Trogon mexicanus* 30cm Separated from Elegant Trogon by different undertail pattern and by green, not brown central tail feathers. **V** Very high well-spaced *tjieuw tjieuw...* or rapid *tjuw-tjuw...* **H** Montane mixed forest. **D** Mexico to Honduras.

33 COLLARED TROGON *Trogon collaris* 25cm Head, breast and upperparts green with strong golden and bronzy sheen. Wings finely barred black and white. Pectoral collar white. Rest of underparts bright red. Eye-ring orange-red. Female: breast and upperparts tawny-brown. Face and throat dark brown. Pectoral collar white. Rest of underparts pale rosy red. **V** Song a short series of plaintive low whistled notes, in some regions higher-pitched with a stuttering introductory phrase; calls include snorting *chaaarrrr*. **H** Cloud forest, várzea, gallery forest, rainforest, forest edges, tall secondary growth. **D** E Mexico through Amazonia and SE Brazil.

34 MASKED TROGON *Trogon personatus* 25cm Very like Collared Trogon but upper tail green with golden or bronzy sheen. Wings darkly vermiculated. Undertail very finely barred black and white (appearing silvery in some), broader white tips. Female like that of Collared Trogon but bill all yellow, face, throat and forehead blackish, undertail barred, tail tipped white. **V** Song slow, quiet series of *kwa* notes; also gives a repeated descending trill. **H** Cloud forest, rainforest, forest edges, open woods at higher elevations. **D** Colombia to Peru and SC Venezuela.

1
2
3
4
5
6
7
8
9
10
11
12
13
14
15
16
17
18
19
20
21
22
23
24
25
26
27
28
29
30
31
32
33
34

TROGONS *CONTINUED*

1 NARINA TROGON *Apaloderma narina* 30cm Note white outer-tail feathers (especially striking in rapidly flapping flight). Sits motionless and hunched on perch from where it hawks insects in short sallies. **V** Loud, crescendoing, hooting *oooh-oo oooh-oo Oooh-Oo Oooh-Oo-* (6–12 times *oooh-oo*). **H** More or less open forests, large gardens, at all altitudes. **D** Sub-Saharan Africa.

2 BARE-CHEEKED TROGON *Apaloderma aequatoriale* 30cm Separated from Narina Trogon by yellow, more extensive bare parts on face sides and by some barring at base of tail feathers, seen from below. **V** Descending, crescendoing series of 6–8 notes, *ooh-ooh-ooh*. **H** Lowland rainforest. **D** SE Nigeria to DR Congo.

3 BAR-TAILED TROGON *Apaloderma vittatum* 30cm Distinguished from Narina and Bare-cheeked Trogons by darker head and uppertail and more distinct wing barring. Barred undertail diagnostic. Unobtrusive. **V** High loud crescendoing yelping, *hoo-hoo-hoo-hoo-hoo-hoo-hoo-hoo-hoo-hoo*. **H** Mid-strata of montane forests. **D** Sub-Saharan Africa.

4 JAVAN TROGON *Apalharpactes reinwardtii* 34cm Sits quietly on a shady perch before flying off to capture insect prey; also feeds on fruit. Often forms part of mixed-species feeding flocks. **V** Penetrating, hoarse *chierr chierr* or loud *turrr*. **H** Montane rainforest. **D** W Java, Indonesia.

5 SUMATRAN TROGON *Apalharpactes mackloti* 30cm Actions and habits similar to Javan Trogon, with which it was once thought to be conspecific. **V** High, whistled *wiwi wheeer-lu*, repeated every few seconds; other calls similar to Javan Trogon. **H** Montane rainforest, chiefly on the lower slopes. **D** Sumatra, Indonesia.

6 MALABAR TROGON *Harpactes fasciatus* 29–31cm Hard to see; generally the first indication is the call. Sits upright on a branch or tree stump, making short sallies to capture flying insects; feeds until well after sunset, often tagging on to mixed-species flocks. **V** Throaty, musical *cue-cue-cue-*. Also a low, rolling *krr-r-r-r* when alarmed. **H** Dense forests, often with an abundance of bamboo. **D** India, Sri Lanka.

7 RED-NAPED TROGON *Harpactes kasumba* 31–35cm Sits motionless for long periods, typical trogon behaviour. Tends to frequent the middle to upper storey. **V** Subdued, harsh *kau kau kau kau kau*; female utters a quiet whirring rattle. **H** Broadleaved evergreen and freshwater swamp forest, bamboo. **D** Malay Peninsula; Borneo; Sumatra, Indonesia.

8 DIARD'S TROGON *Harpactes diardii* 32–35cm Very unobtrusive, usually found in the middle storey, sitting still for long periods. **V** Descending *kau kau kau kau kau...* **H** Broadleaved evergreen forest. **D** Malay Peninsula; Borneo; Sumatra, Indonesia.

9 PHILIPPINE TROGON *Harpactes ardens* 30cm Perches in a shady location, making sallies to capture insects or pluck fruit from nearby trees. **V** Soft *nuu nu nu nu nu nu nu nuu*, accelerating at first and then gradually descending and slowing. **H** Understorey or primary and secondary forest. **D** Philippines.

10 WHITEHEAD'S TROGON *Harpactes whiteheadi* 29–31cm Perches in the higher branches of the understorey. **V** Harsh, even-pitched series of notes, recorded variously as *wark wark wark wark* or *poop poop poop poop poop*; also a soft rolling *rrrr*, sometimes followed by a loud *kekekeke* that drops in pitch. **H** Dark, wet patches of mountain forest. **D** N Borneo.

11 CINNAMON-RUMPED TROGON *Harpactes orrhophaeus* 25cm Very shy, frequents the lower to middle storey. Noted in mixed-species flocks. **V** Weak, descending *ta-aup ta-aup ta-aup*; also an explosive *purr*. **H** Broadleaved evergreen forest. **D** Malay Peninsula; Borneo.

12 SCARLET-RUMPED TROGON *Harpactes duvaucelii* 23–26cm Frequents the lower to middle storey and occasionally forest borders. Often a member of mixed-species flocks. **V** Male utters a rapid, accelerating, descending *yau-yau-yau-yau-yau-yau*. When alarmed gives a quiet whirring *kir-r-r-r*. **H** Broadleaved evergreen forest. **D** Malay Peninsula; Borneo; Sumatra, Indonesia.

13 ORANGE-BREASTED TROGON *Harpactes oreskios* 25–26cm Frequents the middle to upper storey. Recorded following mixed-species flocks. **V** Subdued, rapid *tu tu tau-tau-tau*. **H** Broadleaved evergreen, semi-evergreen and mixed deciduous forests, bamboo. **D** SW China, SE Asia, Greater Sunda Islands.

14 RED-HEADED TROGON *Harpactes erythrocephalus* 31–35cm Unobtrusive, sits motionless for long periods in the middle to upper storey. **V** Mellow, descending *tyaup tyaup tyaup tyaup tyaup*; when alarmed utters a chattering croak, *tewirr*. **H** Broadleaved evergreen forests. **D** Himalayas and S China to Sumatra.

15 WARD'S TROGON *Harpactes wardi* 35–38cm Sits motionless in cover, making occasional sallies to catch insects. Frequents lower storey, undergrowth and bamboo. **V** Mellow *klew klew klew klew klew...*, often slightly accelerating and dropping in pitch. **H** Montane broadleaved evergreen forest. **D** NE India, SW China, N Myanmar, N Vietnam.

CUCKOO ROLLER LEPTOSOMIDAE

16 CUCKOO ROLLER *Leptosomus discolor* 43cm Note characteristic head shape and small bill. **V** Slightly descending series of 3–5 whistling, raptor-like shrieks. **H** Primary and secondary forests and open areas with tall trees. **D** Madagascar, Comoros.

ROLLERS CORACIIDAE

17 PURPLE ROLLER *Coracias naevius* 35–40cm More sluggish than others of the genus. **V** Muffled *ga*. **H** Open savannah woodland, rocky and shrubby hillsides, farmland with trees. **D** Sub-Saharan Africa, South Africa.

18 INDIAN ROLLER *Coracias benghalensis* 30–32cm Pounces on prey from a prominent perch. Acrobatic rolling and diving display flight. **V** Short *rack*, chattering *rack rack rack rackrak-ak* and screeching *aaaarrr* given in warning. During display, loud rattling *ra-ra-ra-ra-raa-raa-aaaaaa-aaaaar*. **H** Open country with trees and bushes, coastal scrub, cultivations, parkland, gardens. **D** E Arabian Peninsula to India, SE Asia.

19 PURPLE-WINGED ROLLER *Coracias temminckii* 30–34cm Generally seen singly or in pairs; regularly perches in the open on a bare branch, telegraph wire or bush top. **V** Various harsh *krawk* notes, alternating with, or followed by, an upslurred, grating *tjorraa*. **H** Forest edges, swamp forest, secondary woodland, open scrub woodland, lightly wooded cultivation and savannah. **D** Sulawesi, Indonesia.

20 RACKET-TAILED ROLLER *Coracias spatulatus* 40cm Conspicuous elongated outer-tail feathers with paddle-shaped tips. In flight shows rufous of tertials extended onto coverts, merging into purple. **V** Loud, harsh, yelping notes exchanged between pair members; flight call loud screaming *kaiirsh*. **H** Open woodland. **D** Angola to S Tanzania, Mozambique.

21 LILAC-BREASTED ROLLER *Coracias caudatus* 40cm Unmistakable by breast colour and long sharp tail. **V** Loud, rasping *rak rak*; in display flight gives a raucous *kaaaa kaarsh*. **H** Dry scrub and woodland, cultivated areas, suburbs. **D** E, SE, SC Africa, from Eritrea to DR Congo and South Africa.

22 ABYSSINIAN ROLLER *Coracias abyssinicus* 28–30cm Habits similar to European Roller. **V** Loud *rack* in flight and screeched *aaaarh* when perched. During aerial display utters a strident *ra-ra-ra-ra-gaa-gaa-gaa-aaaaaar aaaaar*. **H** Semi-desert and savannah with scattered trees. **D** Mauritania and Ivory Coast to Somalia and Kenya.

23 EUROPEAN ROLLER *Coracias garrulus* 30–32cm Pounces on prey from a prominent perch such as a post, bare branch or overhead wire. Acrobatic rolling and diving display flight. **V** Short *rack*, chattering *rack rack rack rackrak-ak* and screeching *aaaarrr* given in warning. During display utters a loud rattling *ra-ra-ra-ra-raa-raa-aaaaaa aaaaar*. **H** Open country with scattered trees, open woods, cultivations. **D** Breeds N Africa, S and C Europe to W Himalayas and SW Siberia. Winters in Africa.

24 BLUE-BELLIED ROLLER *Coracias cyanogaster* 35cm Unmistakable, with brilliant blue wings contrasting with dark back and cream-coloured head. **V** Call a rapid *ga-ga-gaa* (3–5 notes per sec), or a rasping *keh-keh-keh rak*. **H** Mosaics of open spaces and large trees, forest edges, riverine belts. Often in palm trees. **D** From Senegal and Guinea to South Sudan and DR Congo.

25 BLUE-THROATED ROLLER *Eurystomus gularis* 25cm Separated from Broad-billed Roller by blue chin, brown undertail coverts and (in flight) paler-based flight feathers. **V** Very high slightly hoarse *hew hew hew...* **H** Less intensively cultivated areas with forest patches. **D** C and W Africa.

26 BROAD-BILLED ROLLER *Eurystomus glaucurus* 25cm Note short yellow bill, shallow-forked tail, purple underparts. Separated from Blue-throated Roller by azure-blue undertail coverts and absence of blue chin spot. **V** Low raucous rather frog-like *draaaf-draaaf*. **H** Forest glades and edges, wet woodland, grassland, farmland with scattered trees near streams and marshes. **D** Sub-Saharan Africa, Madagascar.

27 ORIENTAL DOLLARBIRD *Eurystomus orientalis* 25–28cm Often perches on topmost branches of a tall dead tree, making acrobatic sallies after flying insects. In flight shows a large pale blue patch on base of primaries. **V** Fast *krak-kak-kak-kak-kak* and hoarse *chak*. **H** Open broadleaved evergreen, semi-evergreen and deciduous forests, forest clearings and edge, plantations, mangroves. **D** E India, Sri Lanka, SE Asia, Greater and Lesser Sunda Islands to E Australia.

28 AZURE DOLLARBIRD *Eurystomus azureus* 27–35cm Uncommon, usually encountered in twos or threes sitting at the top of a tall tree; feeding actions similar to Oriental Dollarbird. **V** Harsh, grating calls given in a steady series. **H** Primary hill forest, forest edges, occasionally coconut groves and gardens. **D** N Maluku Islands, Indonesia.

GROUND ROLLERS BRACHYPTERACIIDAE

29 SHORT-LEGGED GROUND ROLLER *Brachypteracias leptosomus* 34cm Distinctive, with short legs, white breast-band and arboreal habits. **V** Long, slow series of hollow *ou* notes. **H** Humid parts of undisturbed forest. **D** Madagascar.

30 SCALY GROUND ROLLER *Geobiastes squamiger* 28cm Note colourful plumage, distinct facial pattern and long pink-orange legs. **V** Territorial call is a series of soft, hollow *whoo-oop* notes, each lasting 1 sec. **H** Dense parts of undisturbed forest. **D** Madagascar.

31 PITTA-LIKE GROUND ROLLER *Atelornis pittoides* 27cm Unmistakable by colouring and pattern of head. Note white throat. **V** Very slow series of *wou* notes (higher-pitched and much slower than Short-legged Ground-Roller). **H** Dense forest, nearby secondary growth and plantations. **D** Madagascar.

32 RUFOUS-HEADED GROUND ROLLER *Atelornis crossleyi* 26cm Note thin white lines across black necklace. Shows white wing band in flight. **V** Like Pitta-like ground roller but even higher-pitched and more melancholic. **H** Dense, damp parts of forest. **D** Madagascar.

33 LONG-TAILED GROUND ROLLER *Uratelornis chimaera* 40cm Unmistakable. **V** Short, descending series of *wut* notes. **H** Coastal areas with open scrub to dry forest. **D** Madagascar.

KINGFISHERS ALCEDINIDAE

34 GREEN-BACKED KINGFISHER *Actenoides monachus* 31cm A quiet, solitary bird, typically perching in the lower mid-storey, often close to a tree trunk; drops to the forest floor to pick up prey from the leaf litter. **V** Haunting series of long, mournful, ascending and descending whistled notes; also a shorter *huuuuwEEEEu*. Calls are usually given in the early morning or in overcast weather. **H** Primary and secondary forest. **D** Sulawesi, Indonesia.

35 SCALY-BREASTED KINGFISHER *Actenoides princeps* 24cm Usually solitary. Perches in shadowy areas of forest, from low levels to the understorey. **V** A series of soft, mournful whistles, the first notes rolling and the later ones rising in pitch and then falling. **H** Primary and tall secondary hill and montane forest. **D** Sulawesi, Indonesia.

36 MOUSTACHED KINGFISHER *Actenoides bougainvillei* 29–32cm Unmistakable, given facial markings and restricted geographical range. **V** Song a rising, squawking *tchr-tchr-tch-tch-tch*; call a chattering, downslurred trilling *tchRrrrrrrr* lasting 1 sec or so. **H** Forests, wooded habitats. **D** Bougainville and Solomon Islands.

37 SPOTTED WOOD KINGFISHER *Actenoides lindsayi* 25cm Perches quietly in shady areas of the understorey, generally singly or in pairs. **V** Loud, ringing, whistled *ptuuooo* precedes the main call, which is a stuttering, rising *tu-tu-tu-tu-tu* followed by a descending *tuuu tuu-a tuu-a tuu-a*; when alarmed, gives a rasping chatter. **H** Lowland and hilly forest. **D** Philippines.

38 HOMBRON'S KINGFISHER *Actenoides hombroni* 28cm Perches in dark areas in the forest understorey, singly or in pairs. **V** Calls include a loud, repeated *ki-aw* or *te-u* and a woodpecker-like *kaaa-a-a-a*; main call is similar to Spotted Wood Kingfisher. **H** Mid-montane and lower mossy forest. **D** S Philippines.

39 RUFOUS-COLLARED KINGFISHER *Actenoides concretus* 24–25cm Perches in the lower storey, keeping still apart from a slow wagging tail and alert head movements as the ground is scanned for prey. **V** Rising, whistled *kwee-i*, repeated every second for about 10 secs. **H** Broadleaved evergreen forest, generally close to water. **D** Malay Peninsula; Borneo; Sumatra, Indonesia.

1
2
3
4
5
6
7
8
9
10
11
12
13
14
15
16
17
18
19
20
21
22
23
24
25
26
27
28
29
30
31
32
33
34
35
36
37
38
39

KINGFISHERS *CONTINUED*

1 HOOK-BILLED KINGFISHER *Melidora macrorrhina* 27–28cm Distinctive plumage and bill shape. Recalls a miniature kookaburra. **V** Song a trisyllabic *hoo-he-tche-tche*, rising in pitch throughout delivery; call rasping and grating. **H** Rainforest, secondary woodland; mostly lowlands to 500m. **D** New Guinea, W Papuan Islands.

2 BANDED KINGFISHER *Lacedo pulchella* 21–25cm Often sits motionless for long periods. Hunts insects from a forest perch, sometimes hawks insects in the tree canopy. **V** Long whistled *wheeeoo* followed by up to 15 short *chi-wiu* whistles that gradually fade away. Calls include a sharp *wiak* or *wiak wiak*. **H** Broadleaved evergreen forest, mixed deciduous forest and bamboo. **D** SE Asia, Greater Sunda Islands.

3 COMMON PARADISE KINGFISHER *Tanysiptera galatea* 33–43cm Very variable. Perches, semi-concealed in dark forest understorey, darting to the ground or nearby foliage to take prey. **V** A series of separate, mournful whistles that accelerate into a short trill; also utters shrill squawks and a rasping chatter. **H** Primary and tall secondary forest, remnant forest patches, coastal scrub. **D** New Guinea; Maluku Islands, Indonesia.

4 KOFIAU PARADISE KINGFISHER *Tanysiptera ellioti* 32–34cm Unmistakable in geographical range. **V** Song a raucous *Tchrr, Tchrr, Tchrr...* often in duet; call a shrill *chip, chip, chip*. **H** Lowland primary forest, secondary woodland, gardens. **D** Kofiau Island, W Papua, E Indonesia.

5 BIAK PARADISE KINGFISHER *Tanysiptera riedelii* 36–37cm Unmistakable in geographical range. **V** Song a raucous *Tchrr, Tchrr, Tchrr...* parrot-like in tone; call a shrill *chip, chi-chip, chi-chi-chip*. **H** Lowland primary forest. **D** Biak Island and Supiori Island, Papua, E Indonesia.

6 NUMFOR PARADISE KINGFISHER *Tanysiptera carolinae* 35–38cm Unmistakable, given plumage and geographical range. **V** Song an accelerating *whuer, whe, whe, whhhhhhhh* rising in pitch throughout delivery; call a piping, upslurred *whuer-eEE*. **H** Lowland forest, secondary woodland. **D** Numfor Island, Papua, E Indonesia.

7 LITTLE PARADISE KINGFISHER *Tanysiptera hydrocharis* 30cm Like other paradise kingfishers, feeds on insects, gleaned from leaves or caught on ground. **V** Song a trilling, piping *kiu-kiu-kiu-ku-ku* descending in pitch throughout delivery; rasping call. **H** Lowland forest with dense understorey. **D** Aru Islands, lowland coastal S New Guinea.

8 BUFF-BREASTED PARADISE KINGFISHER *Tanysiptera sylvia* 30–36cm Very long tailed like other paradise kingfishers. Nests in termite nests. **V** Song a rasping, throaty *K-whe-er, K-whe-er, K-whe-er*; call a shrill, piping *kWhee-kWhee-kWhee*. **H** Primary forest with dense understorey; mostly lowlands, near water. **D** SE New Guinea, NE Australia.

9 BLACK-CAPPED PARADISE KINGFISHER *Tanysiptera nigriceps* 32–48cm Very long tailed like other paradise kingfishers. Perches on low branches, watching for insect prey. **V** Shrill, piping *kwhe-kwhi-kwi-kwi*, rising in pitch throughout delivery; call has similar tone to song. **H** Forested habitats; mostly lowlands. **D** New Britain, Duke of York Island and Umboi Island (Bismarck Archipelago).

10 RED-BREASTED PARADISE KINGFISHER *Tanysiptera nympha* 31–35cm Strikingly colourful plumage. Perches low and hunts mainly insects and other invertebrates. **V** Vibrating, trilling *Tchrrrrrrrrr* descending in pitch throughout delivery. **H** Lowland forest with dense understorey. **D** W and NE New Guinea.

11 BROWN-HEADED PARADISE KINGFISHER *Tanysiptera danae* 28–30cm Distinctive appearance aids identification. Usually unobtrusive. **V** Rapid-fire, accelerating *Pee-pu-pu-pe-pe-pe-ppppppp* descending in pitch throughout delivery. **H** Primary forest with dense understorey. **D** SE New Guinea.

12 LILAC KINGFISHER *Cittura cyanotis* 28cm Usually seen singly, perched at low level. Singing birds generally sit higher in the canopy. **V** Three or four descending, high, piping notes, repeated at 2–3-sec intervals; also falcon-like *ku-ku-ku-ku* and repeated *kebekek*. **H** Primary and tall secondary lowland and hill forest, also tree-crop plantations. **D** Sulawesi, Indonesia.

13 SHOVEL-BILLED KOOKABURRA *Clytoceyx rex* 32–34cm Massive bill used to dig and rummage in leaf litter for invertebrates. **V** Clear, distinct and monotonous piping *Peeup, Peeup, Peeup*. **H** Rainforest-covered slopes; mostly lowlands. **D** New Guinea.

14 LAUGHING KOOKABURRA *Dacelo novaeguineae* 40–42cm Diet comprises large invertebrates and small vertebrates such as lizards and frogs. **V** Choruses of iconic laughing, chattering calls involving two or more birds; most vocal dawn and dusk. **H** Primary forest, wooded habitats, parks, gardens. **D** E and SW Australia.

15 BLUE-WINGED KOOKABURRA *Dacelo leachii* 38–40cm Delicate and subtle dark markings on otherwise pale buff head and underparts. Distinctive blue on wings. **V** Agitated shrieks and squeals, often in duet. **H** Open woodland, wooded river margins, mangroves, parks. **D** N Australia, S New Guinea.

16 SPANGLED KOOKABURRA *Dacelo tyro* 32–34cm Unique black-and-white mosaic pattern on head. Perches and scans surroundings for insects. **V** Agitated, strident squawks and shrieks including *Who-er, Who-er*. **H** Savannah woodland, waterside forests. **D** S New Guinea, Aru Islands.

17 RUFOUS-BELLIED KOOKABURRA *Dacelo gaudichaud* 28–30cm Plumage strikingly marked and colourful. Forages for invertebrate prey in lower canopy. **V** Strident, squawking *Whi-up, Whi-up*, sometimes in duet. **H** Moist lowland rainforest, secondary woodland, mangroves. **D** New Guinea, W Papuan Islands, Aru Islands.

18 GLITTERING KINGFISHER *Caridonax fulgidus* 30cm Generally sits unobtrusively on a low forest branch, diving to the forest floor to capture prey. Usually seen singly, in pairs, or occasionally in small groups. **V** Series of rapid, harsh notes, *kuff-kuff-kuff*; also puppy-like yaps. **H** Primary and tall secondary moist forest, monsoon forest, lightly wooded cultivation, scrub with tall trees. **D** Lesser Sunda Islands.

19 STORK-BILLED KINGFISHER *Pelargopsis capensis* 35cm Sits quietly, fairly concealed, on waterside branch. Distinctive, with green back, blue wings and tail, olive-brown head, buff underparts. Large bright red bill. **V** Shrieking *ke-ke-ke-ke*; also a pleasant *peer peer peer*. **H** Shady waters in well-wooded country. **D** Widespread, from India through SE Asia to Indonesia and Philippines.

20 GREAT-BILLED KINGFISHER *Pelargopsis melanorhyncha* 35–37cm Perches on a partly concealed branch, singly or in pairs. **V** Loud *kak, kak-ka* or *ke-kak*. **H** Coastal woodland and scrub. **D** Sulawesi and Sula Islands, Indonesia.

21 BROWN-WINGED KINGFISHER *Pelargopsis amauroptera* 35cm Usually perches high in mangroves; feeds on fish and crabs. Azure rump conspicuous in flight. **V** Harsh *chak-chak-chak-chak-chak*, mournful and descending *tree treew-treew*. **H** Coastal, in mangroves. **D** India, Bangladesh, Malay Peninsula.

22 RUDDY KINGFISHER *Halcyon coromanda* 25cm Shy and secretive. Hunts from a perch, catching insects or diving for fish. **V** High-pitched, descending *titititititititi*, also a tremulous *pyorr pyorr pyorr...* **H** Watercourses in dense evergreen forests, mangroves and island forests. **D** E and SE Asia.

23 WHITE-THROATED KINGFISHER *Halcyon smyrnensis* 27–28cm Often encountered far from water. In flight from above shows large bluish-white patch on base of primaries. **V** Loud, rapid trilling *kilililililili...*, repeated, often quite incessantly. Also a cackling *chake ake ake-ake-ake-ake* as bird takes flight. **H** Very cosmopolitan, including roadside trees, plantations, bamboo, farmland, gardens, dams, ponds, canals, creeks, mudflats, industrial areas. **D** Widespread, from S Turkey and NE Egypt to SE Asia and Philippines.

24 JAVAN KINGFISHER *Halcyon cyanoventris* 27cm Perches on treetops, exposed side branches, wires or posts. In flight, shows white patches at base of primaries. **V** Loud scream, *tjie-rie-rie-rie-rie*, and a repeated *tjeu* or *tschrii*. **H** Various, including coastal mangroves, scrub, wooded and cultivated country. **D** Java and Bali, Indonesia.

25 CHOCOLATE-BACKED KINGFISHER *Halcyon badia* 20cm Unmistakable, with dark upperparts and pure white underparts. **V** Mid-high fluted unhurried *weeh fu-fee-fee-fee tenk fju fjuh* (central part highest-pitched). **H** Tropical rainforest, typically found in undergrowth beside streams. **D** From Sierra Leone and Ghana to South Sudan, Uganda, DR Congo.

26 BLACK-CAPPED KINGFISHER *Halcyon pileata* 28cm Shy, but will perch on exposed branches or overhead wires. In flight shows prominent white patches at base of primaries. **V** Ringing, cackling *kikikikikiki*, higher-pitched than White-throated Kingfisher. **H** Coastal and inland wetlands, cultivations, gardens. **D** From India and Sri Lanka through N Myanmar and China to Korean Peninsula. Winter visitor throughout SE Asia.

27 GREY-HEADED KINGFISHER *Halcyon leucocephala* 20cm Distinctive, with pale grey head, black mantle and back, bright blue rump, wings and tail, and chestnut underparts. Beak bright red. **V** Very high up-and-down warbled *tsi-tsi-tsi-tju-tji-tji-tji-tju* or *tsjurrr-tsjurrrr*. **H** Woodland, bush. **D** Sub-Saharan Africa from Cape Verde to South Africa; Arabian Peninsula.

28 BROWN-HOODED KINGFISHER *Halcyon albiventris* 20cm Note streaked head (not capped as in Striped Kingfisher) and buff underwing. Female browner above than male. **V** Very high short sharp fast chatter *wi-wuwuwu*. **H** Open forests, wooded areas, suburbs, especially near ponds and streams. **D** Sub-Saharan Africa.

29 STRIPED KINGFISHER *Halcyon chelicuti* 17cm Unobtrusive but strikingly blue in flight. Underwing of female plain grey without black bar or carpal patch. **V** Very high, short, fluted *tee, trrrrr* or *twee tweetwee*. Also a duet of male and female sounding as one short phrase, *teewuh*. **H** Open woodland, bushveld, often away from water. **D** Widespread from Mauritania and Senegal to Ethiopia and South Africa.

30 BLUE-BREASTED KINGFISHER *Halcyon malimbica* 25cm Separated from Woodland Kingfisher by blue breast-band and black carpal patch on underwing. **V** Extremely high loud fluted sharp *tiu-titititititiutiutiu*. **H** Forests, riverine belts. **D** C and W Africa.

31 WOODLAND KINGFISHER *Halcyon senegalensis* 21cm Separated from Blue-breasted Kingfisher by paler plumage without blue breast-band or black wrist patch on underwing. **V** Very high fast trill, *tiu trrrrrrrriueh*, and fast high chatter. **H** Dry open woodland, bush, often near streams. **D** Sub-Saharan Africa.

32 MANGROVE KINGFISHER *Halcyon senegaloides* 22cm Similar to Woodland Kingfisher, but note all-red bill. Preys on fish, crustaceans, lizards, insects. **V** Very high, descending, sharp *treet treet treet-treetreet treetreetrrrrt*. **H** Mangroves, forests, more or less wooded grassland, cultivation, lake margins, river banks. **D** E and SE Africa.

33 BLUE-BLACK KINGFISHER *Todiramphus nigrocyaneus* 22–24cm Strikingly colourful plumage. Probably feeds on waterside and aquatic invertebrates. **V** Penetrating, whistled *Wheee-hur*, first syllable upslurred, second downslurred. **H** Lowland forests with streams and pools. **D** W Papuan Islands, Indonesia.

34 WINCHELL'S KINGFISHER *Todiramphus winchelli* 25cm More often heard than seen; usually perches in the canopy, although will descend to the ground to pick up prey. **V** Loud, harsh, ascending *chup-chup-chep-chep-chep-chep-chep*, repeated every few seconds; also a faster *chup-chup-chup-chu-chu-chu-chu...*, rising at first, then descending and slowing. **H** Undisturbed or little-disturbed forest. **D** S Philippines.

35 BLUE-AND-WHITE KINGFISHER *Todiramphus diops* 19–21cm Perches in the open on branches or wires; generally found singly or in pairs, occasionally in small groups. **V** Warbling *tu-tu-tu-k-k* and a high, descending, three-note nasal whistle, repeated endlessly. **H** Secondary woodland, forest edges, mangrove edges, cultivated groves and wooded gardens. **D** N Maluku Islands, Indonesia.

36 LAZULI KINGFISHER *Todiramphus lazuli* 22cm Perches among shaded branches, dead branches and sometimes overhead wires. **V** Rapid, loud, high *ke-ke-ke-ke-ke-ke-ke*. **H** Partly cleared forest, forest edges, swampy woodland, occasionally mangroves. **D** S Maluku Islands, Indonesia.

37 FOREST KINGFISHER *Todiramphus macleayii* 20cm Perches conspicuously on bare branches or overhead wires, plunging to the ground to capture prey. **V** Harsh, strident *scissor-weeya scissor-weeya*, a chattering *kreek-kreek*, and various loud whistles and screeches. **H** Open woodland, forest edges, cultivated areas, savannah. **D** New Guinea, N and E Australia.

38 WHITE-MANTLED KINGFISHER *Todiramphus albonotatus* 16–18cm Often perches on low branch, scanning for insect prey in leaf litter. **V** Shrill, penetrating *Tchiu, tchu, tchu-tchu-tchu*. **H** Lowland rainforest, secondary woodland. **D** New Britain (Bismarck Archipelago).

1
2
3
4
5
6
7
8
9
10
11
12
13
14
15
16
17
18
19
20
21
22
23
24
25
26
27
28
29
30
31
32
33
34
35
36
37
38

KINGFISHERS *CONTINUED*

1 ULTRAMARINE KINGFISHER *Todiramphus leucopygius* 20–22cm Note deep blue plumage colour. Unobtrusive. Sits on shaded low branches scanning for invertebrates in leaf litter below. **V** Raucous screeching *Uerr, te-ch, tch-tch.* **H** Primary forest with dense understorey, secondary woodland. **D** Bougainville and Solomon Islands.

2 VANUATU KINGFISHER *Todiramphus farquhari* 20–22cm Unmistakable given striking plumage and restricted geographical range. **V** Piping, accelerating *whoa-wha, tchr, ch-ch-ch*, rising in pitch throughout delivery. **H** Lowland forest, secondary woodland, gardens. **D** Vanuatu.

3 SOMBRE KINGFISHER *Todiramphus funebris* 30cm Arboreal, perching inconspicuously in lower crown or mid-storey levels. Dives to the ground to capture prey. **V** Slow *ki... ki... ki...* **H** Open cultivated lowlands, coconut plantations, gardens. **D** N Maluku Islands, Indonesia.

4 COLLARED KINGFISHER *Todiramphus chloris* 23–25cm Conspicuous, bold and noisy, especially early in the day. **V** Harsh *krerk-krerk-krerk-krerk* that often ends with *jew-jaw* notes, also a shrieking *kick kyew kick kyew...* **H** Coastal wetlands, mangroves, cultivations, gardens, parks; occasionally large rivers and marshes. **D** Widespread from Red Sea coasts to SE Asia, Indonesia, Philippines.

5 TORRESIAN KINGFISHER *Todiramphus sordidus* 23–25cm Perches on low branches and scans for invertebrate prey in foliage and on ground. **V** Shrill *Ki, Ki* or *Ki ki-ki.* **H** Mangrove forests, coastal wooded habitats including gardens. **D** Coastal N Australia, S New Guinea, Aru Islands.

6 ISLET KINGFISHER *Todiramphus colonus* 24–25cm Diet of coastal individuals includes crabs, shrimps and small fish. **V** Piercing, shrill *KWhi-err* or *KWhi, err-err.* **H** Coastal habitats including mangroves. **D** Several islands in Louisiade Archipelago, Papua New Guinea.

7 MARIANA KINGFISHER *Todiramphus albicilla* 23–25cm Perches and scans for prey below. Diet comprises insects and small vertebrates in forest; crabs and shrimps taken on coast. **V** Rapid, shrill whistled *Kee, Kee, Kee, kiu-kiu-kiu-kiu*, latter half descending in pitch. **H** Lowland forest, mangroves, plantations. **D** Northern Mariana Islands.

8 MELANESIAN KINGFISHER *Todiramphus tristrami* 23–25cm Terrestrial diet comprises invertebrates and small vertebrates; crabs and shrimps taken on coast. **V** High-pitched *Pee, Pee, pee-pee-pu-pu-pu*, latter half descending in pitch. **H** Lowland forest, plantations. **D** Bismarck Archipelago and C Solomon Islands.

9 PACIFIC KINGFISHER *Todiramphus sacer* 23–25cm Often perches on bare branches, scanning below for movement of prey. **V** Strident, penetrating *Kiu, Kiu, Kiu, Kiu...* **H** Coastal forest, mangroves, forested slopes. **D** Numerous Pacific islands, including Solomon Islands, Vanuatu, Tonga, Fiji, American Samoa.

10 TALAUD KINGFISHER *Todiramphus enigma* 21cm Often seen perched in close proximity to a river. **V** Repeated *kekee-kekee-kekee-kekee-kekee.* **H** Forest and forest edges. **D** Talaud Islands, NE Indonesia.

11 GUAM KINGFISHER *Todiramphus cinnamominus* 20cm Iridescent blue back and cinnamon head. Extinct in the wild, surviving only in a captive breeding programme. **V** Harsh, rasping *kssh-skshh-kssh-kru-eee*, and mournful *kiu-kiu.* **H** Forest. **D** Guam.

12 RUSTY-CAPPED KINGFISHER *Todiramphus pelewensis* 20–21cm Diet includes invertebrates and small vertebrates including lizards. **V** Subdued chattering *tche, tche, tchee, tcheek, tcheek.* **H** Coastal forest, mangroves, forested slopes. **D** Palau, WC Pacific Ocean.

13 POHNPEI KINGFISHER *Todiramphus reichenbachii* 20cm Perches unobtrusively, scanning for prey including insects and other invertebrates, and small vertebrates. **V** Penetrating, screeching *Pee-ur, Pee-ur, Pee-ur.* **H** Lowland forest, secondary woodland, mangroves. **D** Pohnpei (E Caroline Islands).

14 BEACH KINGFISHER *Todiramphus saurophagus* 26–30cm Usually seen perched on a bare branch, rock, post or even driftwood, at the fringes of the seashore. Occasionally hovers before diving to catch fish or crabs. **V** Loud, deep *kill kill, kee-kee-kee* or *kiokiokiokio.* **H** Mangrove, coastal woodland. **D** Maluku Islands to Solomon Islands.

15 SACRED KINGFISHER *Todiramphus sanctus* 18–23cm Usually encountered in pairs, sitting on an exposed perch. Most prey is taken from the ground, although will dive for fish. **V** Rapid, high-pitched *kik-kik-kik-kik*; also a rasping *schssk schssk.* **H** Mangroves, open country, cultivation. **D** Australia, New Zealand, New Guinea, E Indonesia, Solomon Islands, other Islands in the region.

16 FLAT-BILLED KINGFISHER *Todiramphus recurvirostris* 22cm Sole kingfisher found in the islands of Samoa. **V** Typical call is a series of *kik* notes, rapid, abrupt and monotonous. **H** Wide variety of habitats, from montane forest to suburban areas. **D** Samoa.

17 CINNAMON-BANDED KINGFISHER *Todiramphus australasia* 21cm Usually perches in the mid- to upper storey in pairs or singly. **V** Rapid, descending trill, also a series of weak, wheezy *ch-w'hee* notes and single yapping notes. **H** Primary and tall secondary forest, forest edges, monsoon woodland, open forest, shady trees in cultivation or villages. **D** Lesser Sunda Islands.

18 CHATTERING KINGFISHER *Todiramphus tutus* 22cm Seen singly or in pairs. Feeds on insects and lizards taken on the wing or from the ground. Nests in tree cavities. **V** Typical call is a rapid series of *kee kee* notes, sometimes accelerating into a chatter, also *ke-kow ke-kow, shriii* and other cackling, chuckling and shrieking notes. **H** Highland streams in forest, secondary growth, plantations, gardens. **D** Cook Islands and Society Islands (French Polynesia).

19 MEWING KINGFISHER *Todiramphus ruficollaris* 22cm Sole kingfisher on Mangaia, where it is threatened by habitat loss and competition from introduced Common Mynas. **V** Loud, high *tjewtjewtjew...* **H** Woodland, scrub. **D** Mangaia (S Cook Islands).

20 SOCIETY KINGFISHER *Todiramphus veneratus* 21cm Dull-coloured male with weak frontal collar; female dark-brown. Sole kingfisher on Moorea. **V** Very high, rapid, sharp, trilling *teereereeh* in series of three. **H** Forest, plantations, secondary growth. **D** Moorea (Society Islands).

21 NIAU KINGFISHER *Todiramphus gertrudae* 20cm Separated from Marquesan Kingfisher by range. **V** Not known. **H** Woodland, plantations, villages. **D** Niau, French Polynesia.

22 MARQUESAN KINGFISHER *Todiramphus godeffroyi* 21cm White head and underparts, with broken blue eye-stripe and bright blue-green lower back, rump, tail and wings. Sole kingfisher on S Marquesas. **V** Single low *kiau*, sometimes given in speeded-up chatter; also a quiet *treee-teetee.* **H** Dense forest, plantations. **D** Marquesas Islands, French Polynesia.

23 RED-BACKED KINGFISHER *Todiramphus pyrrhopygius* 21–22cm Feeds mainly on invertebrates but also vertebrates including frogs and lizards. Partly nomadic or migratory within range. **V** Piping *Pi-up, Pi-up, Pi-up...* **H** Open woodland, mallee scrub, dry grassland. **D** Widespread across Australia.

24 YELLOW-BILLED KINGFISHER *Syma torotoro* 20cm Striking orange-red plumage and yellow bill aid identification. **V** Accelerating, rasping fluty *Tchee, tchrr, tchr-rrp-rrp-rrrp*, descending in pitch throughout delivery. **H** Lowland rainforest, mangrove, secondary woodland. **D** W Papuan Islands, Aru Islands, New Guinea, N Australia.

25 MOUNTAIN KINGFISHER *Syma megarhyncha* 23–24cm Perches unobtrusively in tree canopy. Insect prey caught on ground or gleaned from leaves. **V** Rapid-fire, vibrating piping trill; pitch rises and falls throughout delivery. **H** Rainforest, secondary woodland. **D** New Guinea.

26 AFRICAN DWARF KINGFISHER *Ispidina lecontei* 10cm Note diagnostic rufous crown and nape. Tiny forest species, normally away from water. **V** Extremely high, irregular *see see see.* **H** Dense forest undergrowth. **D** W and C equatorial Africa.

27 AFRICAN PYGMY KINGFISHER *Ispidina picta* 12cm From other small kingfishers by long, rufous-violet streak behind eye. Feeds on insects, taken from ground; rarely catches fish in water. **V** Very high, mumbled, fast trill, *trrt-trut trrrrrrrtrrrrrit*, or extremely high *tseet tseet.* **H** Ground strata of dry woodland, bush and scrubland, tall-grass forest glades, lake and river edges. **D** Widespread across sub-Saharan Africa.

28 MADAGASCAN PYGMY KINGFISHER *Corythornis madagascariensis* 14cm Unmistakable, with orange colouring and typical kingfisher bill. **V** Shrill, squeaky *treet-treet*; in alarm calls *cheiwp* or ringing *treee.* **H** Forest up to mid–high altitudes. **D** Madagascar.

29 WHITE-BELLIED KINGFISHER *Corythornis leucogaster* 13cm Blue, rufous and white. White stripe through underparts from throat to undertail. **V** Little known; a single high-pitched note has been recorded. **H** Dense swampy forests near streams. **D** W and C equatorial Africa.

30 MALACHITE KINGFISHER *Corythornis cristatus* 13cm From most other small blue kingfishers by blue cap touching eye. Normally not found in forest habitats. May show crest. **V** Song, given in duet, *ii-tiii-cha-cha chui chui chui tuichui*, finishing with a chuckled note. Calls include a clear rattled note, a short *tsip* or a *tsi-vit.* **H** Reedbeds, papyrus, shrubbery along lakes, dams, ponds, rivers, streams, creeks. **D** Widespread across sub-Saharan Africa.

31 MALAGASY KINGFISHER *Corythornis vintsioides* 15cm Dark blue upperparts, rufous underparts, blue-and-green barred crown. Bill black in both sexes. No similar kingfisher in range. **V** Short, shrill *seek* or longer, very high-pitched *treeee.* **H** Near water bodies including mangrove. Found up to mid–high altitudes. **D** Madagascar, Comoros, Mayotte.

32 CERULEAN KINGFISHER *Alcedo coerulescens* 13cm Usually seen singly or in pairs; fishes from a low perch at water's edge. Female slightly greener. **V** High-pitched, penetrating *tieh tieh.* **H** Water in low-lying country, including streams, canals, fish ponds, swamps, mangroves, tidal estuaries. **D** Sumatra, Java, Lesser Sunda Islands.

33 BLUE-BANDED KINGFISHER *Alcedo euryzona* 17cm Hunts from a low perch near forest streams; active, always moving from perch to perch. **V** High-pitched *cheep*, usually given in flight. **H** Streams in broadleaved evergreen forests. **D** Malay Peninsula, Greater Sunda Islands.

34 SHINING-BLUE KINGFISHER *Alcedo quadribrachys* 19cm From Half-collared Kingfisher by darker colouring and plain or less spotted wings. **V** Extremely high, irregular *sweep sweep sweep-sweep* (just within audible range). **H** Ponds, streams in forests, wooded and reedy fringes of lakes, rivers. **D** W and C Africa.

35 BLUE-EARED KINGFISHER *Alcedo meninting* 17cm Distinguished from Common Kingfisher by blue ear-coverts, more intense cobalt-blue upperparts and richer rufous underparts. Tends to fish from a low, shady perch overhanging a forest stream. **V** High-pitched, shrill *seet*; also thin, shrill contact calls. **H** Streams, small rivers and pools in dense forest; also creeks and channels in mangroves. **D** Nepal and India to SE Asia, Indonesia, Philippines.

36 COMMON KINGFISHER *Alcedo atthis* 16–17cm First sight is often just a blue flash flying low along a river giving a high-pitched call. Regularly uses a prominent perch from where it plunge-dives after small fish. Female has orange-red base to lower mandible. **V** Penetrating, high-pitched *tseee* or *tseee ti-ee ti-ee ti-ee*; when disturbed utters a harsh *shrit-it-it.* **H** Rivers, streams and ponds in open wooded areas, occasionally mangroves or estuaries; tends to avoid denser forests. **D** Widespread from W Palearctic to SE Asia, Sunda Islands, Philippines, New Guinea.

37 HALF-COLLARED KINGFISHER *Alcedo semitorquata* 19cm Note black bill, pale but brilliant colouring. From Shining-blue Kingfisher by overall paler colouring; from other similar kingfishers in region by blue cheeks. **V** Extremely high staccato *tseep tseep- tseep tseep.* **H** Rivers, streams, lakes, lagoons, normally with overhanging trees. **D** E and S Africa.

38 BLYTH'S KINGFISHER *Alcedo hercules* 22cm Dives after fish from a low concealed perch. Female has orange-red base to lower mandible. **V** In flight gives a loud *pseet.* **H** Streams and rivers in broadleaved evergreen forest and secondary growth. **D** E Nepal to C Vietnam.

♀
1
♂
♀
2
♂
4
♀
♂ 3
5
7
6
8
9
10
11
♀
13
♂
12
14
15
16
17
18
♀
♂
21
24 ♂
♀
25
♂
19
31
26
20
♀
23
22
27
28
29
30
32
33 ♀
♂
34
35
36
37
38

KINGFISHERS *CONTINUED*

1 ORIENTAL DWARF KINGFISHER *Ceyx erithaca* 14cm Perches low, in vegetation or on rocks, from where fish or insects are taken. **V** High-pitched shrill *tsriet-tsriet* or a soft *tjie-tjie-tjlie*. **H** Shady streams or ponds in damp broadleaved evergreen forests; occasionally in mangroves. **D** India, Sri Lanka, China, SE Asia, Sunda Islands.

2 PHILIPPINE DWARF KINGFISHER *Ceyx melanurus* 12–13cm Solitary; perches low down. **V** A high-pitched *zeeeep*. **H** Dense primary lowland and secondary forest. **D** Philippines.

3 SULAWESI DWARF KINGFISHER *Ceyx fallax* 12cm Uses a low perch from which to catch insect or amphibian prey. **V** High-pitched *seeee*. **H** Well-shaded lowland forest. **D** Sulawesi, Indonesia.

4 MOLUCCAN DWARF KINGFISHER *Ceyx lepidus* 14cm Usually solitary; uses a low perch as a vantage point to look for insects etc. **V** Shrill, wheezy *tzeeip*. **H** Primary and secondary forest. **D** Maluku Islands except Buru.

5 DIMORPHIC DWARF KINGFISHER *Ceyx margarethae* 14cm Variable; dark and light morphs occur, light morph depicted. Habits presumed similar to the Moluccan Dwarf Kingfisher. **V** Shrill *pe-teeeet pe-teeeet*. **H** Primary and secondary forest. **D** C and S Philippines.

6 SULA DWARF KINGFISHER *Ceyx wallacii* 14cm Habits presumed similar to Moluccan Dwarf Kingfisher. **V** Unknown. **H** Forested areas. **D** Sula Islands, Indonesia.

7 BURU DWARF KINGFISHER *Ceyx cajeli* 14cm Habits presumed similar to Moluccan Dwarf Kingfisher. **V** Unknown. **H** Forested areas. **D** Buru, S Maluku Islands, Indonesia.

8 PAPUAN DWARF KINGFISHER *Ceyx solitarius* 12cm Small size and striking plumage aid identification. Feeds on small invertebrates. **V** Thin, high-pitched whistling notes. **H** Lowland primary forest, secondary woodland. **D** W Papuan Islands, Aru Islands, New Guinea.

9 MANUS DWARF KINGFISHER *Ceyx dispar* 13–14cm Unobtrusive. Perches on low branches. Feeds mainly on terrestrial invertebrates. **V** Thin, high-pitched whistling notes. **H** Primary forest, secondary woodland, plantations. **D** Manus Island (Bismarck Archipelago).

10 NEW IRELAND DWARF KINGFISHER *Ceyx mulcatus* 13–14cm Rather shy. Often perches in dense foliage, making observation difficult. **V** Thin, high-pitched whistling notes. **H** Primary forest, secondary woodland, often near forest streams. **D** New Ireland, New Hanover, Lihir Island and Tabar Island (E Bismarck Archipelago).

11 NEW BRITAIN DWARF KINGFISHER *Ceyx sacerdotis* 13–14cm Unobtrusive. Often perches in dense cover. **V** Thin, high-pitched whistling notes. **H** Primary forest, secondary woodland; mainly lowlands to 1,000m. **D** New Britain and nearby islands (Bismarck Archipelago).

12 NORTH SOLOMONS DWARF KINGFISHER *Ceyx meeki* 13–14cm Yellow underparts and yellow feet are distinctive. **V** Thin, high-pitched whistling notes. **H** Lowland primary forest, secondary woodland with dense understorey. **D** Bougainville, Papua New Guinea; W and C Solomon Islands.

13 NEW GEORGIA DWARF KINGFISHER *Ceyx collectoris* 13–14cm Deep blue upperparts are distinctive. **V** Piercing, high-pitched whistling notes. **H** Lowland primary forest, secondary woodland; mainly lowlands. **D** W and C Solomon Islands.

14 MALAITA DWARF KINGFISHER *Ceyx malaitae* 13–14cm Feeds mainly on forest-floor invertebrates. **V** Thin, high-pitched whistling notes. **H** Lowland primary forest, secondary woodland; mainly lowlands. **D** Malaita Island, Solomon Islands.

15 GUADALCANAL DWARF KINGFISHER *Ceyx nigromaxilla* 13–14cm Perches unobtrusively in dense cover. Feeds mainly on terrestrial invertebrates. **V** Thin, high-pitched whistling notes. **H** Primary forest, secondary woodland; mainly lowlands. **D** Guadalcanal, Solomon Islands.

16 MAKIRA DWARF KINGFISHER *Ceyx gentianus* 13–14cm Has striking plumage with deep blue upperparts and clean white underparts. **V** Thin, high-pitched whistling notes. **H** Primary forest, dense secondary woodland, plantations. **D** Makira Island, Solomon Islands.

17 INDIGO-BANDED KINGFISHER *Ceyx cyanopectus* 14cm Perches on branches or rocks, diving to catch fish. **V** Thin, high-pitched single note. **H** Thickly forested streams and rivers, palm swamps, mangroves. **D** Philippines.

18 SOUTHERN SILVERY KINGFISHER *Ceyx argentatus* 15cm Perches on low streamside branches or rocks, diving to catch fish. **V** High-pitched *wheeet*. **H** Forested streams, rivers and pools. **D** S Philippines.

19 NORTHERN SILVERY KINGFISHER *Ceyx flumenicola* 14cm Behaviour presumed to be similar to Southern Silvery Kingfisher. **V** No known difference from Southern Silvery Kingfisher. **H** Forested streams, rivers, pools. **D** SC Philippines.

20 AZURE KINGFISHER *Ceyx azureus* 16–17cm Generally encountered singly; perches on low branches over water, diving to catch fish. **V** Repeated, shrill *tzeep* or *peeee*. **H** Wooded banks of streams and estuaries. **D** N Maluku Islands, Indonesia, and E Lesser Sunda Islands; New Guinea; N, E, SE Australia.

21 BISMARCK KINGFISHER *Ceyx websteri* 13–14cm More associated with forest streams and pools than its cousins. **V** Thin, high-pitched whistling notes. **H** Primary forest, secondary woodland; usually near water. **D** Bismarck Archipelago.

22 LITTLE KINGFISHER *Ceyx pusillus* 11cm Tiny blue and white kingfisher. Catches fish from a low perch. **V** High-pitched, repeated *tsee* or *tzweeip*. **H** Mainly mangroves, also forest streams, swamp forest. **D** Maluku Islands, Indonesia; New Guinea; Solomon Islands; N and E Australia.

23 AMERICAN PYGMY KINGFISHER *Chloroceryle aenea* 13cm Miniature version of Green-and-rufous Kingfisher. Very small (smaller than House Sparrow). Centre of belly and vent white. **V** Possible song a series of tuneful chirping notes; call *tik* or *tsweek*. **H** Small shady streams and ponds in forest, flooded forest, várzea, mangroves. **D** S Mexico to N Argentina.

24 GREEN-AND-RUFOUS KINGFISHER *Chloroceryle inda* 22cm Upperparts dark shining green. Spotted white on wings and tail. Throat and collar rufous ochre. Rest of underparts chestnut-rufous. Female has dark green breast-band, scaled white. **V** Song is a thin *week-week-week*. Calls include low *tootootoo*, harsh-*drrrt*, and twittering and crackling notes. **H** Shady forest streams, swamps and ponds, flooded forest, várzea, mangroves. **D** Nicaragua to Ecuador and SE Brazil.

25 GREEN KINGFISHER *Chloroceryle americana* 19cm Small, dumpy. Small version of Amazon Kingfisher but bill proportionately smaller. Upperwings dotted white. More chestnut-rufous on breast. Belt of green spots. Female has double breast-band of green dots. **V** Most common call a dry clacking rattle; twittering *kli-kli-kli-kli-kli* phrase may represent a song. **H** Lakes, marshes, roadside ditches, mangroves, watercourses with trees or reeds. **D** SW USA to N Argentina.

26 AMAZON KINGFISHER *Chloroceryle amazona* 28cm Medium-sized, with proportionately long and heavy beak. Erectile nuchal crest. Upperparts shining dark green. Throat, collar and underparts white. Breast chestnut-rufous. Flanks partly streaked green. Female has single green pectoral collar. **V** Wide range of calls including *tek* or *klek*, given singly or in chatter; series of clear *seee-seee... su su...* notes could be a song or greeting call. **H** Trees by lakes, marshes, streams, rivers, bodies of open water. Less in forest. **D** C Mexico to C Argentina.

27 CRESTED KINGFISHER *Megaceryle lugubris* 41–43cm Unmistakable. Shy; usually in pairs perched on rocks or branches in or by a river pool. **V** Loud *ping*. Also deep croaks and raucous grating notes. When disturbed utters a loud *kek*. **H** Fast-flowing rivers and streams in high-level forests. **D** NE Afghanistan and N India to Vietnam, China, Korean Peninsula, Japan.

28 GIANT KINGFISHER *Megaceryle maxima* 40cm Largest kingfisher in Africa, with shaggy crest, large black bill, white-spotted upperparts. Male less rufous below than female. **V** Mid-high loud resounding nasal shrieks, *kaai-kaai-kaai*. **H** Streams, ponds, lagoons with overhanging trees and shrubs, garden pools. **D** Widespread across sub-Saharan Africa.

29 RINGED KINGFISHER *Megaceryle torquata* 38–41cm Usually shy and solitary. Feeds by diving for fish, either from a perch or by hovering. Breeds from April to August. **V** Loud *kek, klek* or *klek-klek*; when alarmed, a rattling *klek-klek-klek-klek-klek....* **H** Edges of large streams and lakes. **D** SW USA, and throughout Central America and South America.

30 BELTED KINGFISHER *Megaceryle alcyon* 30cm Medium-sized, with erectile crest. Upperparts and pectoral band greyish blue. Throat, collar and underparts white. Female has chestnut-rufous band on upper belly. **V** The 'rattle call' is most frequently heard, but also utters a variety of other harsh and more warbling notes. **H** Swamps, trees bordering lakes, rivers and streams, freshwater bodies, sea coasts, estuaries, mangroves. **D** Breeds Canada, USA. Winters USA, Central America, N South America.

31 PIED KINGFISHER *Ceryle rudis* 25cm Usually encountered in pairs or small parties. Regularly hovers and dives to catch fish. **V** Noisy *kwik-kwik* or *chirruk-chirruk*, also a high-pitched *TREEtiti TREEtiti*. **H** Lakes, rivers, ponds, estuaries, tidal creeks, coastal lagoons. **D** Sub-Saharan Africa, Middle East, S and E Asia.

TODIES TODIDAE

32 CUBAN TODY *Todus multicolor* 10–11cm Feeds mainly by fly-catching, sallying out from exposed perch to pick insects off leaves or in mid-air; also feeds on caterpillars, spiders and small fruits. **V** Soft *pprreeee-pprreeee* and a short *tot-tot-tot-tot*. During courtship flights, wings make a rattling or cracking sound. **H** Forests, woodlands, thickets, gullies and stream sides. **D** Cuba.

33 BROAD-BILLED TODY *Todus subulatus* 11–12cm Feeding actions similar to Cuban Tody, although tends to feed higher up and on larger prey. **V** Monotonous, plaintive *terp terp terp*. Wings produce a cracking or rattling sound during courtship flights. **H** Mainly arid and semi-arid lowland scrub forest, second-growth forest, coffee plantations, mangroves. **D** Hispaniola.

34 NARROW-BILLED TODY *Todus angustirostris* 11cm Feeding techniques similar to Cuban Tody, although feeding flights can be more of a hop because it inhabits dense, tangled vegetation. **V** Chattering *chippy-chippy-chippy-chip* and a frequently repeated *chip-chee*. Produces wing noises similar to Cuban Tody. **H** Primarily ravines in dense wet forests, also coffee plantations. **D** Hispaniola.

35 JAMAICAN TODY *Todus todus* 9–11cm Feeding actions similar to Cuban Tody. Usually feeds in the understorey. **V** Rapid, throaty rattle, uttered during display; also a loud *beep* or *cherek*. Produces wing noises similar to Cuban Tody. **H** All types of forest. **D** Jamaica.

36 PUERTO RICAN TODY *Todus mexicanus* 11cm Feeding methods very similar to Cuban Tody. **V** Loud, harsh *beep, beep-beep* or *cherek*. Produces wing noises similar to Cuban Tody. **H** From sea-level to mountains in rainforest, arid scrub, dense thickets, shade coffee plantations. **D** Puerto Rico.

1
2
3
4
5
6
7
8
9
10
11
12
13
14
15
16
17
18
19
20
21
22
23
24
25
26
27
28
29
30
31
32
33
34
35
36

MOTMOTS MOMOTIDAE

1 TODY MOTMOT *Hylomanes momotula* 18cm Small; tail proportionately shorter than other motmots. Whitish line across face. Head rufous, brow blue. **V** Loud, resonant *kwakwakwakwa...* or long tremulous *coooooo-o-o-oo*. **H** Humid forest at foot of mountains. **D** S Mexico to NW Colombia.

2 BLUE-THROATED MOTMOT *Aspatha gularis* 25cm Face pattern diagnostic, with a black cheek spot that is usually more noticeable than the turquoise-blue throat. **V** Single, far-carrying *hoot* given every few seconds; also a longer series of yodelling hooting notes, possibly in duet. **H** Montane forest understorey. **D** S Mexico to Honduras.

3 RUSSET-CROWNED MOTMOT *Momotus mexicanus* 35cm Russet-rufous crown, greenish throat and green tail diagnostic. **V** Single rolling *krrrrp*; duets with a rhythmic series of *krr-uu krrr-uu* notes. **H** Open woodland, thorn bush, lowlands and foothills. **D** NW Mexico to SW Guatemala.

4 BLUE-CAPPED MOTMOT *Momotus coeruliceps* 40cm Separated from Lesson's Motmot by range and by all-blue crown. Red eyes distinctive. **V** Double *hooot-hoot*. **H** Forest edge, clearings, second-growth woodland. **D** NE Mexico.

5 LESSON'S MOTMOT *Momotus lessonii* Blue crown has dark top, unlike solid blue of Blue-capped Motmot. Red eyes distinctive. **V** Double *hoot-hoot*. **H** Forest edge, clearings, second-growth woodland. **D** S Mexico to W Panama.

6 WHOOPING MOTMOT *Momotus subrufescens* 38–42cm Sits motionless on shaded branch, watching for movement of prey. Diet includes invertebrates. **V** Song a hooting *WhoO-oop, WhoO-oop, WhoO-oop*; chattering, cackling calls. **H** Forested habitats with dense understorey. **D** S Panama, N Venezuela, NW Colombia, W Ecuador.

7 TRINIDAD MOTMOT *Momotus bahamensis* 38–42cm Long tail often swung like a pendulum. **V** Song a hooting *Whooup, Whooup, Whooup*; chattering calls. **H** Forested habitats with dense understorey. **D** Trinidad, Tobago.

8 AMAZONIAN MOTMOT *Momotus momota* 38–42cm Diet includes insects, caught in flight and pounced upon in leaf litter. **V** Song a piping *Whuup, Whuup, Whuup*; chattering calls. **H** Forested habitats with dense understorey. **D** Widespread across Amazonian regions of South America, from Venezuela to N Argentina.

9 ANDEAN MOTMOT *Momotus aequatorialis* 38–42cm Perches on overhanging branches, often near water. **V** Song a piping *Whudup, Whudup, Whudup*; chattering calls. **H** Montane forest, secondary woodland; mostly 1,000–2,000m. **D** Spine of Andean mountain range, from Colombia to Bolivia.

10 RUFOUS MOTMOT *Baryphthengus martii* 46cm Like Broad-billed Motmot but larger and lacking green chin. Rufous reaches belly. Tail seldom has 'racquet'. Broader mask. **V** Series of owl-like, low-pitched *hoot hoot....* **H** Rainforest, secondary growth, lowland forest and at the foot of mountains. **D** Honduras to SW Ecuador, W Amazonia.

11 RUFOUS-CAPPED MOTMOT *Baryphthengus ruficapillus* 44cm Cap rufous and ventral band cinnamon-rufous. Rectangular black mask. Tail with ill-defined 'racquets'. **V** Deep, guttural *hoorooroo hoorooroo* and long rolled *oorrrrrrrrrrrr*. **H** Rainforests and gallery forests. **D** SE Brazil, E Paraguay, NE Argentina.

12 KEEL-BILLED MOTMOT *Electron carinatum* 30cm Orange forehead, blue eye-stripe and chin. **V** Very similar to Broad-billed Motmot. **H** Forest; generally below 1,500m. **D** S Mexico to Costa Rica.

13 BROAD-BILLED MOTMOT *Electron platyrhynchum* 33cm Head and breast rufous, chin green, mask black, narrowing towards ear-coverts. Uppertail blue with 'racquets'. **V** Low-pitched, nasal, resonant *ornk* or *quoonk*. **H** Rainforest, secondary growth, lowland forest, foot of mountains. **D** E Honduras to SW Ecuador, W Amazonia.

14 TURQUOISE-BROWED MOTMOT *Eumomota superciliosa* 35cm Striking pale blue wing and tail pattern. Note long bare tail shafts. **V** High hoarse pushed-out *uhoooh*. **H** Dry forest, woodland, gardens. **D** S Mexico to Costa Rica.

BEE-EATERS MEROPIDAE

15 RED-BEARDED BEE-EATER *Nyctyornis amictus* 27–31cm Often frequents the lower canopy. Sits quietly, partly hidden, before making short sorties to catch winged insects. **V** Loud, hoarse *chachachacha...quo-qua-qua-qua* or a descending *kak kak-ka-ka-ka-ka*; also a deep *kwow* or *kwok*. **H** Broadleaved evergreen forest, occasionally well-wooded gardens. **D** Malay Peninsula; Borneo; Sumatra, Indonesia.

16 BLUE-BEARDED BEE-EATER *Nyctyornis athertoni* 31–34cm Catches insects in flight, but also clambers about in trees in search of insects. **V** Gruff *gga gga ggr*, a *kir-r-r kir-r-r-r* and a purring *grrew-grrew-grrew*. **H** Edges and clearings in dense broadleaved evergreen forest. **D** India, Nepal, Bangladesh, SE Asia.

17 PURPLE-BEARDED BEE-EATER *Meropogon forsteni* 25–26cm Perches on exposed branches in the open, at the mid-storey and upper canopy, making aerial sallies after insects. **V** Quiet, high-pitched *szit*, *peet* or *sip-sip*. **H** Edges of clearings and breaks in primary or tall secondary forest. **D** Sulawesi, Indonesia.

18 BLACK-HEADED BEE-EATER *Merops breweri* 33cm Unmistakable by all-black head. **V** Contact call a soft, bubbling or trilling *churruk-churruk*; alarm call short, fairly quiet repeated *wik*. **H** Along edges and open patches in forest, woodland, plantations. **D** W Africa, from Ivory Coast to NW Angola.

19 BLUE-HEADED BEE-EATER *Merops muelleri* 20cm Unmistakable, with ultramarine underparts, russet mantle and wings, red throat . **V** Generally quiet. Song combines various harsh and quieter well-spaced notes; calls include high-pitched, squeaky-toned *tsee-sup* and more musical *trrri-wit* or other trilled phrases. **H** Forest edge and interior. **D** S Cameroon to W Kenya.

20 BLUE-MOUSTACHED BEE-EATER *Merops mentalis* 19–21cm Previously considered conspecific with Blue-headed Bee-eater. Perches on dead branches. **V** Whistled, high-pitched *tsseip, tsseip, tsseip...* **H** Forest edge and interior. **D** W Africa, from Sierra Leone to Cameroon.

21 BLACK BEE-EATER *Merops gularis* 20cm From all other bee-eaters, except Blue-headed Bee-eater, by habitat. **V** Flight call *wik*; also has disyllabic, strident and repeated note. **H** Forest interior. **D** C and W Africa.

22 SWALLOW-TAILED BEE-EATER *Merops hirundineus* 20cm Unmistakable, with distinctive long blue forked tail. **V** Rather quiet, especially if alone; in flocks gives soft bubbling and rolling notes such as *prip... prip...* **H** Wooded and bushed grassland. **D** Sub-Saharan Africa.

23 LITTLE BEE-EATER *Merops pusillus* 16cm Mostly green, with yellow throat separated from buff underparts by black line. **V** Contact call a repeated *tsip*; other calls include thin *drrreeee*, and tinkling notes. **H** Bushland and forest edges, often near water. **D** Sub-Saharan Africa.

24 BLUE-BREASTED BEE-EATER *Merops variegatus* 20cm Similar to Little Bee-eater but yellow throat looks paler due to more white below black ear-coverts. **V** Similar to that of Little Bee-eater, but lower-pitched. **H** Generally wetter habitats than Little Bee-eater. Humid grassy areas at forest edges, lake sides. **D** SE Nigeria to Ethiopia and Tanzania.

25 CINNAMON-CHESTED BEE-EATER *Merops oreobates* 20cm Green wings, whitish cheeks, rufous underparts. Little or no rufous in tail. **V** Similar to that of Little Bee-eater, some notes more drawn-out. **H** More or less open montane forests, cultivation, roadsides, suburban gardens and parks. **D** S South Sudan to E DR Congo and N Tanzania.

26 RED-THROATED BEE-EATER *Merops bulocki* 20cm From White-fronted Bee-eater by green, not white forehead, all-pink throat and different range. **V** Has varied repertoire of melodious calls, including *wip* or *weep*, which may be extended into *wip-weep-piu piu terrip-terrip*. Alarm call a chattered *tic-ic-ic-ic-ic*. **H** Bushed grassland along streams near sand cliffs. **D** Senegal to Ethiopia and Uganda.

27 WHITE-FRONTED BEE-EATER *Merops bullockoides* 24cm Note striking, white forehead, white chin, pink throat and blue rear underparts. **V** Muffled, nasal, rolling *nyahh* or *gauuu*. **H** Wooded and bushed areas, swamp edges, other open areas, not too far from sand cliffs, where it breeds. **D** Gabon to Kenya, south to South Africa.

28 SOMALI BEE-EATER *Merops revoilii* 15cm Fulvous wing patches in flight. **V** Usually quiet, but may give loud *twee-twee-twee-te*, notes descending. **H** Arid areas with scarce bush and extensive cultivation. **D** Ethiopia, Kenya, Somalia.

29 WHITE-THROATED BEE-EATER *Merops albicollis* 19–21cm Bold white throat and distinctive head pattern. Very long central tail feathers. **V** Soft, high-pitched *prrrp prrrp prrrp....* **H** Breeds on hills, plains and wadis with trees and bushes. **D** Mauritania to Kenya and Somalia, SW Arabian Peninsula.

30 BÖHM'S BEE-EATER *Merops boehmi* 21cm Rich russet head. Not gregarious. Hawks insects in short sallies from perch. **V** Similar to that of Little Bee-eater. Contact call *sip* may be repeated to form a short (up to 5 secs) song. **H** Forest edges near rivers and streams. **D** Zambia, Tanzania, Malawi, Mozambique.

31 GREEN BEE-EATER *Merops orientalis* 22–25cm Darts from wire or branch to capture flying insects; also recorded using the backs of cattle as a moving perch. Roosts communally in trees, huddled together on a branch. **V** Quiet, trilling *trrr trrr trrr trrr*; also a sharp *ti-ic* or *ti-ti-ti* given when alarmed. **H** Open country with scattered trees, semi-deserts, grazing land. **D** Mauritania to Sudan, Middle East, S Asia, SE Asia.

32 BLUE-CHEEKED BEE-EATER *Merops persicus* 27–31cm Usually in pairs or small groups. Typical bee-eater action, sallying from exposed branch or wire to chase winged insects, gracefully circling back to same or nearby perch after capture. **V** Rolling *diririp*, mellow *tewtew*, and when alarmed a sharp *dik-dik-dik*. **H** Sandy areas with scrub and scattered trees, usually near water. **D** S Europe, N Africa, Middle East to Kazakhstan and W India. Winters in Africa.

33 OLIVE BEE-EATER *Merops superciliosus* 31cm From Little Bee-eater by rufous-brown throat and cheek, less brilliant green plumage, deep rufous underwings. Note dark brown cap. **V** Most common call a rolling *pr-reee... pr-reee...*, sometimes interspersed with a quiet *pup... pup...* **H** Sandy areas with scrub and scattered trees, usually near water. **D** Ethiopia to Namibia, Madagascar.

34 BLUE-TAILED BEE-EATER *Merops philippinus* 23–26cm Green with blue tail, thin black mask, rufous throat; bright rufous underwings visible in flight. Actions and habits similar to Blue-cheeked Bee-eater. **V** Similar to Blue-cheeked Bee-eater. **H** Wooded country, near water. **D** India, Sri Lanka, S China, SE Asia to New Guinea.

35 RAINBOW BEE-EATER *Merops ornatus* 19–21cm Yellow face with black mask, green and blue underparts, black tail streamers. Generally in pairs or small groups. **V** Rolling *prrrp prrrp*, *preee* or *drrrt*; when alarmed, *dip-dip* or *clip-lip-lip-lip*. **H** Grassy areas, forest clearings, scrub, forest edges, trees alongside rivers. **D** Australia. Non-breeding from Solomon Islands to Sulawesi and Lesser Sunda Islands.

36 BLUE-THROATED BEE-EATER *Merops viridis* 22–24cm Often in flocks, especially at roosts. Makes flying sorties from tall trees or power-lines. **V** A fast, short, trilled *brk brk*. In flight utters a loud *prrrp prrrp prrrp...* When alarmed gives a sharp *chip*. **H** Open country, river margins, forest edge, forest clearings, cultivations, parks and gardens. **D** S China, SE Asia, Philippines, Greater Sunda Islands.

37 CHESTNUT-HEADED BEE-EATER *Merops leschenaulti* 18–20cm Bright reddish head, yellow throat, green wings. Actions as other bee-eaters. **V** *Pruik* or *churit*. **H** Forests, often along watercourses, forest edge, coastal scrub, cultivations, mangroves, island forest. **D** India, Nepal, Sri Lanka to S China, Malay Peninsula; Sumatra to Bali, Indonesia.

38 EUROPEAN BEE-EATER *Merops apiaster* 27–29cm Multi-coloured. Sociable. Typically seen hawking from wires. Nests colonially in sandbanks. **V** Liquid, often repeated, *prruip*, *pruik* or *kruup*. **H** Open country, cultivated land, areas with scattered trees or bushes, rivers with steep banks, locally on woodland edge. **D** NW Africa, SW Europe to Middle East, C Asia; South Africa. Winters in sub-Saharan Africa.

39 ROSY BEE-EATER *Merops malimbicus* 28cm Unmistakable, with dark mantle and wings, white cheeks, pinkish-red underparts. **V** Alarm call *wic*; in flight gives rolling hoarse *crrrp*. **H** Forest and high woodland, normally not far away from large rivers and lakes. **D** Ghana to Gabon, DR Congo.

40 NORTHERN CARMINE BEE-EATER *Merops nubicus* 40cm Previously treated as subspecies of Carmine Bee-eater. Throat blue-green. Perches on dead branches. **V** Chattering *whrik, whrik...* calls, seldom heard in isolation, usually from flock. Flight call *chuk, chuk*. **H** Open savannah woodland, thorn bush, cultivated ground. **D** Broad equatorial band from Senegal to Kenya.

41 SOUTHERN CARMINE BEE-EATER *Merops nubicoides* 40cm Previously treated as subspecies of Carmine Bee-eater. Throat carmine-red. Perches on dead branches. **V** Whistled, high-pitched *tsseip, tsseip, tsseip...* **H** Open savannah woodland, thorn bush, cultivated ground. **D** Widespread in C sub-Saharan Africa, from Angola and Tanzania to N Namibia and N South Africa.

1
2
3
4
5
6
7
8
9
10
11
12
13
14
15 ♀
♂
16
17
18
19
20
21
22
23
24
25
26
27
28
29
30
31
32
33
34
35
36
37
38
39
40
41

HOOPOES UPUPIDAE

1 EURASIAN HOOPOE *Upupa epops* 26–32cm Unmistakable, even in flight, where it gives the impression of a giant butterfly. When agitated or alighting, crest is fanned. Forages mainly on the ground, usually singly or in pairs. **V** A low *hoop-hoop-hoop* or *poop-poop-poop*. When alarmed gives a harsh *schaahr*. **H** Open country with scattered trees, cultivations, around villages. **D** Widespread across Europe, Asia, N Africa, N sub-Saharan Africa.

2 AFRICAN HOOPOE *Upupa africana* 26–28cm Previously considered conspecific with Eurasian Hoopoe. **V** Soft, piping *hoo-poo-poo* or *hoo-whe-whe*. **H** Lightly wooded open country, savannah, thorn scrub. **D** Widespread in sub-Saharan Africa in suitable habitats.

3 MADAGASCAN HOOPOE *Upupa marginata* 30cm From Eurasian Hoopoe by range and paler colouring. **V** Song is a dove-like, crooning *rrrrooow*, lasting 1.5–2.5 secs and given at irregular intervals of a few secs; calls include various harsh growls and hisses. **H** Open country with some shrub and trees. **D** Madagascar.

WOOD HOOPOES PHOENICULIDAE

4 FOREST WOOD HOOPOE *Phoeniculus castaneiceps* 30cm Variable: brown-hooded in western part of range; green, brown, or whitish head further east. Note almost straight black bill. **V** Extra high, fast, trilling cackling *tritrutritritritritrutritri* and high, slightly hoarse fluting *wuut-wuut-wuut-wuut-wuut-wuut-wuut*. **H** Forest canopies, not on the ground. **D** W and C Africa, Liberia to Uganda.

5 WHITE-HEADED WOOD HOOPOE *Phoeniculus bollei* 35cm Variable, but always some white on face. Red bill and legs. **V** Very high, liquid chatters, *chachachachacha*. **H** Montane forest interiors and edges (canopies), often along streams. **D** W and C Africa, Liberia to Kenya.

6 GREEN WOOD HOOPOE *Phoeniculus purpureus* 40cm Note green iridescence on nape and mantle and red bill. Small parties. **V** Mid-high to high, magpie-like, fast, excited chatters, *cha-cha-cha-cha*. **H** Open woodland and bushland, well-wooded streams, suburbs. **D** Widespread across sub-Saharan Africa.

7 BLACK-BILLED WOOD HOOPOE *Phoeniculus somaliensis* 40cm Bill sometimes red-based. **V** Very like that of Green Wood Hoopoe. **H** Dense bush with some trees. **D** Eritrea, Ethiopia, SW Somalia, NE Kenya.

8 VIOLET WOOD HOOPOE *Phoeniculus damarensis* 40cm From Green Wood Hoopoe by mainly purple (not green, purple and blue) iridescence. Bill base may be black. Small parties. **V** Mid-high cackling, starting low and rising to excited climax, *rah-chah-rah-rah-rah...* with characteristic *ah* sound. **H** Dry bushland, thornveld, often near water. **D** Angola, Namibia.

9 GRANT'S WOOD HOOPOE *Phoeniculus granti* 40cm Bill sometimes black at base. Note dark purple-violet mantle and dark coppery head. **V** Mid-high, excited, cackling *racharararararahrahchara* (with typical *-ah-* sound). **H** Dry bush and scrubland, often near water. **D** Kenya.

10 BLACK SCIMITARBILL *Rhinopomastus aterrimus* 23cm From larger Green Wood Hoopoe and Violet Wood Hoopoe by black bill. **V** High, fluted, descending series of *weet* notes. **H** Dry woodland and bush. **D** W,C and E Africa: Senegal to Ethiopia, south to Angola.

11 COMMON SCIMITARBILL *Rhinopomastus cyanomelas* 30cm From Black Scimitarbill by strongly downcurved bill and less noisy behaviour. **V** Very high, fluted *pwuu-pwuu*. **H** Dry bushland, thornveld. **D** S and SE Africa: Kenya to Namibia and South Africa.

12 ABYSSINIAN SCIMITARBILL *Rhinopomastus minor* 25cm Race from Somalia, Ethiopia and N Kenya has white bar over primaries. **V** Harsh, chattering *kee-kee-kee*, rising and then falling in pitch; also a plaintive high-pitched *peuw-peuw-peuw...*, thought to be the song. **H** Bush and wooded areas. **D** E Africa: Djibouti and Ethiopia to Tanzania.

GROUND HORNBILLS BUCORVIDAE

13 ABYSSINIAN GROUND HORNBILL *Bucorvus abyssinicus* 105cm Adult male has bare blue skin around the eye and an inflatable patch of red skin on the neck and throat. Eyes are black. **V** Series of low, booming hooting notes, slightly higher-pitched and faster than those of Southern Ground Hornbill. **H** Arid or dry, open, wooded and bushed areas with more or less grass cover. **D** N sub-Saharan Africa, from Senegal to Kenya.

14 SOUTHERN GROUND HORNBILL *Bucorvus leadbeateri* 100cm Bare skin around eye is red. Note white primaries and primary coverts, which are only visible in flight. **V** Extreme low, booming *ohooh ohooh ohooh ohooh*, often in duet of male and female. **H** Wooded and bushed areas. **D** S sub-Saharan Africa, from Kenya to Angola and E South Africa.

HORNBILLS BUCEROTIDAE

15 TANZANIAN RED-BILLED HORNBILL *Tockus ruahae* 42–48cm Formerly treated as subspecies of Red-billed Hornbill. Note black eye patch and range. Female has less black on bill. **V** Raucous series of *We, We, We* notes and uncoordinated-sounding mix of chattering notes and screeches, often in duet. **H** Wooded savannah with baobab trees (*Adansonia* spp.). **D** Tanzania.

16 WESTERN RED-BILLED HORNBILL *Tockus kempi* 36–38cm Formerly treated as subspecies of Red-billed Hornbill. Note grubby eye patch and range. Female has less black on bill. **V** Raucous, accelerating series of agitated-sounding *Wa, Wa, Wa, wa-we, wa-we, wa-we, werte, werte...* notes. **H** Wooded grassland, savannah, thorn scrub. **D** Senegal, Gambia, Guinea-Bissau, W Mali.

17 DAMARA RED-BILLED HORNBILL *Tockus damarensis* 35–37cm Formerly treated as subspecies of Red-billed Hornbill. Note range and lack of eye patch. Female has less black on bill. **V** Raucous, accelerating series of agitated-sounding *Wa, Wa, Wa, wer-we, wer-we, wer-we* notes. **H** Thorn scrub, wooded savannah. **D** S Angola, NW Namibia.

18 SOUTHERN RED-BILLED HORNBILL *Tockus rufirostris* 35–38cm Formerly treated as subspecies of Red-billed Hornbill. Note range and pink flush around eye. Female has less black on bill. **V** Raucous, accelerating series of agitated *Wa, Wa, Wa, wer-we, wer-e-we, wer-e-we* notes. **H** Thorn scrub, wooded savannah. **D** Namibia, Botswana, Zimbabwe, Zambia, Mozambique, N South Africa.

19 NORTHERN RED-BILLED HORNBILL *Tockus erythrorhynchus* 40cm From Monteiro's Hornbill by smaller size, pale head and breast; from Southern Yellow-billed Hornbill by dark eye. Female has less black on bill. Feeds on the ground. **V** High to mid-high, excited, rapid, hoarse, yelping *wek-wek-wek-wek* (up to 25 secs), falling off at the end. **H** Dry, more or less wooded and bushed areas. **D** Mauritania through Somalia to NE Tanzania.

20 MONTEIRO'S HORNBILL *Tockus monteiri* 55cm From Crowned Hornbill and Bradfield's Hornbill by speckled, partly white wings and white outer-tail feathers, from Southern Yellow-billed Hornbill and Northern Red-billed Hornbill by dark throat and breast. Feeds mainly on the ground. **V** Low, mumbled *toktoktoktaktakkarrekarrekarrekarre*, rising in pitch, lowered at the end and starting afresh. **H** Desert and semi-desert with scarce scrub and trees. **D** Angola, Namibia.

21 VON DER DECKEN'S HORNBILL *Tockus deckeni* 50cm (male) 45cm (female) Feeds mainly on the ground. **V** Mid-high, chicken-like cackling, *tack tack tock-tock-tocktocktocktocktock*. **H** Dry, more or less bushed and wooded habitats. **D** Ethiopia and Somalia to Tanzania.

22 JACKSON'S HORNBILL *Tockus jacksoni* 50cm (male) 45cm (female) Feeds mainly on the ground. Only a few secondaries partly white. Bill mainly red with yellow tip. **V** Like that of Von der Decken's Hornbill but notes louder, slower and with more hollow sound. **H** Dry, wooded and bushed habitats. **D** S South Sudan, S Ethiopia, Uganda, Kenya.

23 SOUTHERN YELLOW-BILLED HORNBILL *Tockus leucomelas* 50cm From red-billed hornbills by yellow bill and fewer white spots on primaries. Feeds on the ground. **V** Mid-high, fast *wokwokwok-* (up to 30 secs), ending in excited yelps. **H** Dry, more or less wooded and bushed areas. **D** S Africa, from Angola and Zambia southwards.

24 EASTERN YELLOW-BILLED HORNBILL *Tockus flavirostris* 50cm Normally feeds on the ground. Naked skin at face sides is black. **V** Like that of Southern Yellow-billed Hornbill, but lower-pitched with hoarser quality. **H** Dry, more or less wooded and bushed habitats. **D** E Africa, from Eritrea to N Tanzania.

25 BRADFIELD'S HORNBILL *Lophoceros bradfieldi* 50cm From Crowned Hornbill by less dark upperparts and paler-edged feathers. Flight feathers and outer-tail feathers darker than rest of plumage. Arboreal but feeds also on the ground. **V** Very high, sharp, fluting *wik-wik-wik-wik-wik-wik-wik* (up to 10 secs), rising in pitch in the beginning. **H** Open woodland, thornveld. **D** S Angola and N Namibia to W Zimbabwe.

26 CROWNED HORNBILL *Lophoceros alboterminatus* 50cm Note white line between bill and head, yellow eye and all-brown upperparts. **V** Very high, shrieking, rapid *wutweetweetweetweet*, descending at the end. **H** More or less dense forests and woodland. **D** From Ethiopia to Angola and S to South Africa.

27 AFRICAN PIED HORNBILL *Lophoceros fasciatus* 50cm Mainly black, with white belly and tail tip. Long, curved black and red bill. Arboreal. **V** Very high rather short phrases, *fjuk-fjuk-fjuk-fjukfjuk-fjukfjukfjuk*. **H** Forest and surrounding areas. **D** W Africa from Senegal to Angola, and E to Uganda.

28 HEMPRICH'S HORNBILL *Lophoceros hemprichii* 60cm Feeds on the ground as well as in trees. Note tail pattern and dark red bill. **V** Long series of whistled *pi* notes, accelerating and shifting to lower pitch at end of sequence; contact calls single or double notes with similar piping quality. **H** Dry, rocky slopes, particularly with large candelabrum trees. **D** Eritrea, Djibouti, Ethiopia, Somalia, Uganda, Kenya.

29 AFRICAN GREY HORNBILL *Lophoceros nasutus* 50cm From Pale-Bellied Hornbill by white stripe through middle of rump and different bill pattern. **V** High, unhurried, piping *pipipipi-piu-piu-plee-plee-plee* (middle part extreme high) and high, fluted *tuuu-tjuh tuuu-tjeh*. **H** Dry, more or less wooded areas with some bush. **D** Widespread in sub-Saharan Africa.

30 RED-BILLED DWARF HORNBILL *Lophoceros camurus* 35cm Smallest hornbill. Chestnut, with white-spotted wing-coverts, white edges to flight feathers, white belly. Male has bright red bill. Female smaller, with black-tipped bill. **V** Very high tooting *toot-toot-toot-toot-toot-toot-toot-toot* (descending 4 notes from beginning to end). **H** Dense swampy lowland forests. **D** Sierra Leone to DR Congo and Uganda.

31 PALE-BILLED HORNBILL *Lophoceros pallidirostris* 45cm From African Grey Hornbill by yellow bill and plain brown rump. **V** Accelerating series of piping whistles, with the last notes lower-pitched and longer. **H** Dry miombo. **D** Angola and DR Congo to Tanzania and Mozambique.

32 PIPING HORNBILL *Bycanistes fistulator* 60cm Note dark, central patch on small-casqued bill and all-white outer-tail feathers. **V** Mid-high, rapid, fast, magpie-like chattering, *pipi-tji-tjak-tjak-tjak* (up to 5 secs). **H** Moist forests. **D** W Africa from Senegal to Angola, and E to Uganda.

33 TRUMPETER HORNBILL *Bycanistes bucinator* 65cm Note red orbital skin and all blackish bill. **V** Mid-high, loud, resounding, bleating, slightly lowered *weeeeee* and excited *wehwehwehweh*. **H** Montane rivers and connected riverine belts. **D** Angola to Kenya, S to E South Africa.

34 BROWN-CHEEKED HORNBILL *Bycanistes cylindricus* 60–70cm Usually seen in pairs. Groups gather where feeding good. Diet mainly fruits, but also invertebrates. **V** Metallic-sounding trumpeted *Wha-eek, Wha-eek, Wha-eek* with tones of honking goose and braying donkey. **H** Primary forest, secondary woodland. **D** W Africa, from Sierra Leone to Ghana.

35 WHITE-THIGHED HORNBILL *Bycanistes albotibialis* 70cm Note pale bill, white belly and thighs, and white-black-white pattern of tail. **V** Like that of Brown-cheeked Hornbill. **H** Dense forests. **D** Benin and S Nigeria to W Uganda, DR Congo, NW Angola.

36 BLACK-AND-WHITE-CASQUED HORNBILL *Bycanistes subcylindricus* 75cm From Trumpeter Hornbill by bi-coloured casque, purplish-black orbital skin, more white on wing, rump and tail. **V** Low, mallard-like chattering, *chaa-chaa-chaa* (up to 25 times). **H** Moist forests and surrounding areas. **D** W and C equatorial Africa, from Sierra Leone to Kenya.

37 SILVERY-CHEEKED HORNBILL *Bycanistes brevis* 75cm Bill darker than pale casque; note mainly black wings. **V** Loud *uHUH uHUH* in noisy wing-flapping flight. **H** Forests and riverine belts. **D** E Africa, from Ethiopia to Mozambique and Zimbabwe.

1
2
3
4
5
6
7
8
9
10
11
12
13
14
15
16
17
18
19
20
21
22
23
24
25
26
27
28
29
30
31
32
33
34
35
36
37

HORNBILLS *CONTINUED*

1 BLACK-CASQUED HORNBILL *Ceratogymna atrata* 80cm Very large, with cobalt-blue bare parts of head, dark brown upperparts, white at outer tail tips only. **V** Low, hoarse, chattered, fast *rhoa-rhoa-rhoa-rhoa-* (often partly breaking up two octaves) or *tet-tet rho-aauh-* or gull-like screams in flight. **H** Moist, lowland forest, riverine belts, plantations. **D** Sierra Leone to N Angola, S South Sudan to DR Congo.

2 YELLOW-CASQUED HORNBILL *Ceratogymna elata* 65cm From Black-casked Hornbill by smaller casque, different cheek colouring and more white in tail. **V** Drawn-out, nasal barking notes. **H** Forest, riverine tree belts, plantations. **D** Senegal to Cameroon.

3 BLACK DWARF HORNBILL *Horizocerus hartlaubi* 35cm Upperparts with bottle-green wash. Catches insects in rather flycatcher-like fashion. **V** High, mellow, staccato *week-week-week-week-week-week*. **H** Forest canopies. **D** W and C Africa.

4 WHITE-CRESTED HORNBILL *Horizocerus albocristatus* 65cm Different races show different amounts of white on head. **V** Very high yelping, gliding up to extra-high *eeeoouuww*. **H** Dense forests and surrounding areas. **D** W and C Africa.

5 WHITE-CROWNED HORNBILL *Berenicornis comatus* 75–85cm Forages in dense tangled growth in lower storeys or on the ground, spends much time digging in bark and debris searching for food. **V** Deep *hoo hu-hu-hu-hu-hu-hu* or *kuk kuk kuk kuk kuk* that often fades away. **H** Broadleaved evergreen forest. **D** Malay Peninsula; Borneo; Sumatra, Indonesia.

6 RHINOCEROS HORNBILL *Buceros rhinoceros* 91–122cm Arboreal. Usually encountered in pairs or small flocks. In flight, shows all-black wings and a white tail with black subterminal band. **V** Male utters deep *hok* notes and female gives a *hak*, often duetted as *hok-hak hok-hak...* **H** Primary lowland and hill forest. **D** Malay Peninsula, Greater Sunda Islands.

7 GREAT HORNBILL *Buceros bicornis* 95–105cm Usually encountered in pairs or small groups; may gather in larger flocks at fruiting trees or at communal roosts. Mainly arboreal, although may descend to the ground to pick up fallen fruit. In flight shows white tips to flight feathers, a buff-white wing-bar and white tail with a black subterminal band. **V** Loud, reverberating *tok.... tok.... tok*; a loud *ger-onk* in flight; various hoarse grunts, barks, roars. **H** Mature broadleaved forest. **D** W India, Himalayas, SE Asia, Sumatra.

8 RUFOUS HORNBILL *Buceros hydrocorax* 60–65cm Forages mainly in the canopy, occasionally descending to low bushes or the ground. Often seen flying across valleys or along mountain slopes in small groups. **V** Loud, deep *kaaww* or *aaww*. **H** Primary and tall secondary forest, from sea level to mountains. **D** Philippines.

9 HELMETED HORNBILL *Rhinoplax vigil* 110–120cm Usually occurs singly or in pairs, foraging in the canopy of tall trees in search of fruit, small animals, snakes and birds. **V** Series of loud, hollow *took* notes, quickening to *tee-poop* notes before ending in a manic laugh. In flight, utters a loud, clanking *ka-hank*. **H** Tall lowland and hill forest. **D** Malay Peninsula; Borneo; Sumatra, Indonesia.

10 PALAWAN HORNBILL *Anthracoceros marchei* 55cm Usually seen in small noisy groups foraging from the ground to the canopy. **V** Trumpeting *kuk, kuk-kak* or *kuk-kuk-ka*. **H** Primary and secondary evergreen forest, mangrove swamps, sometimes in cultivation. **D** Palawan, WC Philippines.

11 ORIENTAL PIED HORNBILL *Anthracoceros albirostris* 55–60cm Usually encountered in small groups. Mainly arboreal, although will descend to the ground to feed. In flight shows prominent white tips to flight feathers and large white tips to outer-tail feathers. **V** Various loud squeals and raucous cackles. **H** Evergreen forest edge, open deciduous forest, plantations. **D** Widespread from Nepal to Borneo.

12 MALABAR PIED HORNBILL *Anthracoceros coronatus* 65cm Generally occurs in small parties, but much larger flocks encountered at rich feeding sites which may include other hornbill species and fruit-eating birds. Mainly arboreal, but often descends to the ground to feed on fallen fruit or search for insects. In flight shows prominent white tips to all flight feathers and white outer-tail feathers. **V** Various loud, shrill squeals and raucous cackles. **H** Evergreen forest edge, open forests, plantations and fruiting trees around villages. **D** India, Sri Lanka.

13 SULU HORNBILL *Anthracoceros montani* 50cm Rare. Travels in pairs or small noisy groups, foraging mainly in the forest canopy. **V** Raucous *ghaakh*, uttered singly or in a long series of up to 20 notes, ending with a stuttering trill and two distinct, separate notes. **H** Lowland and hill forest. **D** Sulu Archipelago, SW Philippines.

14 BLACK HORNBILL *Anthracoceros malayanus* 76cm Usually found in pairs or small flocks, occasionally in much larger flocks. Forages in tangles in the lower to middle levels of forest; recorded catching bats emerging from caves at dusk. **V** Harsh grating growls and retching noises. **H** Broadleaved evergreen forest. **D** Malay Peninsula; Borneo; Sumatra, Indonesia.

15 MALABAR GREY HORNBILL *Ocyceros griseus* 45cm Generally occurs in pairs or small groups of 6–20, often in the company of other fruit-eating birds. In flight shows white tips to primaries and outer-tail feathers. **V** Raucous cackling *kyah kyah kyah*, also a maniacal, laughing *waa...waa...wa-wa-wa-wa*. **H** Evergreen and deciduous forest, usually near watercourses; plantations, gardens. **D** SW India.

16 SRI LANKA GREY HORNBILL *Ocyceros gingalensis* 45cm Arboreal, forages mainly in the foliage, below the canopy; may form small groups at fruiting trees. In flight shows white tips on primaries and white outer-tail feathers. **V** Loud *kaa...kaa...ka-ka-ka-ka* or *kuk...kuk-kuk-kuk ko ko kokoko*. **H** Evergreen forest and deciduous woodland, plantations, gardens. **D** Sri Lanka.

17 INDIAN GREY HORNBILL *Ocyceros birostris* 50cm Usually encountered in pairs or small parties, often in the company of mynas, green pigeons and bulbuls. Feeds mainly in trees, but descends to the ground to pick up fallen fruit or search for insects. In flight shows white tips to all flight feathers and outer-tail feathers. **V** Shrill *wheeee*, also a cackling *k-k-k-ka-e* and a rapid, piping *pi-pi-pi-pi-pipipieu-pipipieu-pipipieu*. **H** Deciduous woodland, open thorn forest with scattered fig trees, cultivations, gardens. **D** India, Nepal.

18 TICKELL'S BROWN HORNBILL *Anorrhinus tickelli* 60–65cm Actions and habits similar to Brown Hornbill. Often considered conspecific with Brown Hornbill. **V** Similar to Brown Hornbill. **H** Broadleaved evergreen forest, occasionally in nearby mixed deciduous. **D** S Myanmar, SE Thailand.

19 AUSTEN'S BROWN HORNBILL *Anorrhinus austeni* 60–65cm Mainly arboreal, often encountered in small noisy, restless flocks, often accompanied by other fruit-eating birds. In flight, males show white tips to primary feathers and outer-tail feathers. **V** Loud, yelping *klee-ah*; also various croaks, chuckles and screams. **H** Evergreen and deciduous forests. **D** Assam to SW China and Vietnam.

20 BUSHY-CRESTED HORNBILL *Anorrhinus galeritus* 65–70cm Usually found in flocks of 5–15 birds; forages in or just below the canopy. **V** Loud, excited *klia-klia-klia kliu-kliu* that rises and falls, often uttered by all members of a group and building to a crescendo. When alarmed, a loud *aak aak aak*. **H** Broadleaved evergreen forest. **D** Malay Peninsula; Borneo; Sumatra, Indonesia.

21 RUFOUS-NECKED HORNBILL *Aceros nipalensis* 90–100cm Usually seen in pairs or small groups feeding in the treetops, or flying across forested valleys. May descend to the ground to pick up fallen fruit. In flight shows white tips to outer primaries and white-tipped tail. **V** Short, repeated bark, said to resemble the noise made by an axe striking a sapling; also various loud roars, croaks or cackles. **H** Broadleaved evergreen forest. **D** NE India to W Thailand and NW Vietnam.

22 BLYTH'S HORNBILL *Rhyticeros plicatus* 65–85cm Found singly, in pairs or in small groups, foraging in the canopy of fruiting trees. **V** Various deep grunting and honking notes, uttered singly or in a sequence. **H** Primary lowland, hill and swamp forest. **D** Maluku Islands to Solomon Islands.

23 NARCONDAM HORNBILL *Rhyticeros narcondami* 45–50cm Generally encountered in small parties, with larger groups occurring in fruiting trees. The only hornbill on Narcondam Island. **V** A cackling *ka-ka-ka-ka-ka*. **H** Mature undisturbed forest with large trees. **D** Narcondam, Andaman Islands.

24 WREATHED HORNBILL *Rhyticeros undulatus* 75–85cm Generally in pairs or small parties, and in larger groups where food is plentiful or at communal roosts. Forages mainly in the canopy, although will descend to the ground to collect fallen fruit or take small animals. **V** A very loud, breathless *kuk-KWEHK*. **H** Tropical evergreen forest. **D** NE India to Greater Sunda Islands.

25 SUMBA HORNBILL *Rhyticeros everetti* 55cm Encountered singly, in pairs or in small groups, with much larger numbers at roosting sites. Forages mainly in the canopy in fruiting trees. **V** Short *erm-err* and a *kokokokokokokoko*. **H** Patches of primary forest, secondary forest and open parkland with fruiting trees. **D** W Lesser Sunda Islands.

26 PLAIN-POUCHED HORNBILL *Rhyticeros subruficollis* 65–70cm Generally found in pairs, occasionally in small groups. Forages mainly in the treetops, although will descend to the ground to collect fallen fruit or capture small animals. **V** Loud *keh-kek-kehk*. **H** Broadleaved evergreen and mixed deciduous forest. **D** Myanmar to Malay Peninsula.

27 KNOBBED HORNBILL *Rhyticeros cassidix* 70–80cm Seen singly, in pairs or in flocks; some non-breeding flocks can be large, with up to 50 birds. Forages in the forest canopy, often mixing with other fruit-eating species. **V** Deep, upslurred, honking *wha wha wha wha*. **H** Primary lowland, hill, montane, swamp and tall secondary forest, also forest stands within cultivated areas. **D** Sulawesi, Indonesia.

28 WALDEN'S HORNBILL *Rhabdotorrhinus waldeni* 60–65cm Generally occurs in small groups, foraging in the canopy; often seen flying over forest or crossing valleys. **V** Loud, stuttering, nasal *au-au-au-auk*. **H** Forest, from coast to hills, also large trees in clearings. **D** C Philippines.

29 WRITHED HORNBILL *Rhabdotorrhinus leucocephalus* 60–65cm Usually seen in noisy, conspicuous small groups foraging in the canopy of tall trees. **V** Loud, nasal *auk*, uttered singly or as a series of notes. **H** Primary lowland forest. **D** Philippines.

30 SULAWESI HORNBILL *Rhabdotorrhinus exarhatus* 45cm Usually encountered in noisy, active pairs or small groups, foraging in the mid-canopy. **V** Rapid series of harsh braying and honking notes. **H** Primary lowland and hill forest, tall secondary forest, swamp forest, forest edges. **D** Sulawesi, Indonesia.

31 WRINKLED HORNBILL *Rhabdotorrhinus corrugatus* 75cm Usually encountered in pairs or small flocks. Forages in the canopy of large emergent trees. **V** Sharp barking *kak kak-kak* or an echoing *wakowwakowkow* or *rowwrow*. **H** Broadleaved evergreen forest and freshwater swamp forest. **D** Malay Peninsula; Borneo; Sumatra, Indonesia.

32 LUZON HORNBILL *Penelopides manillae* 45cm Occurs in small groups of up to 15 birds. This and the following four species were all originally described as races of the Philippine 'Tarictic Hornbill'. **V** Repeated nasal *tuc* or *ta-ruc*, often extended to *ta-ruc-tuc-tuc-tuc-tuc*. **H** Primary evergreen lowland forest, riverine forest and even single fruiting trees in grassland. **D** N Philippines.

33 MINDORO HORNBILL *Penelopides mindorensis* 45cm Usually occurs in pairs or flocks of up to 20 birds. **V** Similar to Visayan Hornbill. **H** Primary evergreen forest, secondary forest, forest edges, isolated fruiting trees. **D** Mindoro, C Philippines.

34 MINDANAO HORNBILL *Penelopides affinis* 45cm Forages in the middle and lower forest levels, usually in small groups. *P. a. basilanicus* occurs on Basilan Island (SW Philippines). **V** Similar to previous species. **H** Primary evergreen forest, including clearings and edges, also secondary forest. **D** S Philippines.

35 SAMAR HORNBILL *Penelopides samarensis* 45cm Usually encountered in pairs or small flocks. Female similar to female Mindanao Hornbill. This species is often considered a race of Mindanao Hornbill. **V** Similar to Luzon Hornbill. **H** Moist lowland forest. **D** EC Philippines.

36 VISAYAN HORNBILL *Penelopides panini* 45cm Forages below the canopy and along forest edges. **V** Repeated, subdued, nasal *ta-ric* or *ta-ric-tic*. **H** Primary evergreen forest, secondary forest, isolated fruiting trees. **D** NC and C Philippines.

1
2
3
4
5
6
7
8
9
10
11
12
13
14
15
16
17
18
19
20
21
22
23
24
25
26
27
28
29
30
31
32
33
34
35
36

JACAMARS GALBULIDAE

1 WHITE-EARED JACAMAR *Galbalcyrhynchus leucotis* 20cm Somewhat like a kingfisher. Large pale pink bill. Overall strong chestnut and black. Cheeks white. Bare patch around eye bright pink. Iris red. **V** Single or repeated *keeew* or rising trill. **H** Edge of the várzea and gallery forest, secondary growth. **D** Amazonian Colombia, Ecuador, Peru, W Brazil.

2 PURUS JACAMAR *Galbalcyrhynchus purusianus* 19–21cm Reddish brown plumage and lack of white ear-patch aid identification. **V** Trilling, vibrating *tu, tu, tu ter, trrrrrrr*, accelerating and rising in pitch throughout delivery. **H** Inundated forest, swampy secondary woodland. **D** Amazonian Peru, Bolivia, W Brazil.

3 DUSKY-BACKED JACAMAR *Brachygalba salmoni* 17cm Overall slaty black with slight blue sheen. Throat white. Belly, flanks and undertail cinnamon with sparse black speckling. **V** High-pitched, insistent *pe pe pe peet* or *pe peet*, often in a long series. **H** Edge and clearings in rainforest, gallery forest, secondary growth with tall trees. **D** E Panama, NW Colombia.

4 PALE-HEADED JACAMAR *Brachygalba goeringi* 17cm The only jacamar with a buffy-white streak behind the eye and buffy-white collar. White throat. Dirty brown crown and cheeks, darker upper breast and back. Lower breast, lower belly and undertail white. Broad chestnut band across belly. **V** Single *weet*, or a series, rising and accelerating to a trill. **H** Edge and clearings in deciduous riverine forests, open woods, edge of savannah, stubble fields, scrub vegetation, bushes. **D** NE Colombia, NW Venezuela.

5 BROWN JACAMAR *Brachygalba lugubris* 17cm Mostly blackish brown. Crown feathers edged whitish, giving a scaly effect. Some streaking on cheeks. Whitish throat finely streaked dark brown. Buff-lined white triangle from centre of breast to lower belly. Undertail coverts black. **V** High, descending, accelerating insect-like stuttering trill. **H** Edge of forest plantations, scrubby and bush vegetation in rainforest, cloud forest and gallery forest, clearings, trees in open areas. **D** Amazonia, N and C South America.

6 WHITE-THROATED JACAMAR *Brachygalba albogularis* 17cm Ivory-yellow bill slightly downcurved. Face and throat white, creamier on cheeks. Overall blackish brown. Large chestnut spot at mid-belly. **V** Song is a series of high-pitched whistles, accelerating and then slowing, *pee-pipi-peeee tewee tewee tewee*. **H** Edges and clearings in forest, riverside forests, stubble fields, brushland and bushes at forest edge. **D** W Amazonia.

7 THREE-TOED JACAMAR *Jacamaralcyon tridactyla* 18cm Mostly dull slate-grey. White streaking on head. Breast and belly white, flanks and undertail feathers narrowly tipped whitish, giving effect of slight barring or large scales. Only three toes, unlike all other jacamars. **V** Series of high, short, whistled notes. **H** Forest along mountain rivers, patches of primary and secondary forest, forestry plantations; usually near earth banks of riversides. **D** SE Brazil.

8 YELLOW-BILLED JACAMAR *Galbula albirostris* 19cm Bill mostly yellow and appears slightly downcurved. Forehead, crown and part of cheeks brownish purple. Green above with bronze and coppery sheen, rufous below, paler on lower belly. Yellow eye-ring and lores. Male has white throat. Female has buffy-cinnamon throat. **V** Sharp *peek* or series accelerating to trill. **H** Bushes and scrub at edge of rainforest and várzea. **D** N Amazonia, from Colombia to the Guianas and N Brazil.

9 BLUE-NECKED JACAMAR *Galbula cyanicollis* 20cm Similar to Yellow-billed Jacamar. Bill straight. Forehead, crown and part of the cheeks dark blue. Male has rufous throat. **V** Indistinguishable from Yellow-billed Jacamar. **H** Edge of rainforest. **D** S Amazonia.

10 RUFOUS-TAILED JACAMAR *Galbula ruficauda* 23cm Upperparts, breast and central tail feathers metallic green with bronzy and purple sheen. Blue wash on forehead. Chin and throat white (male) or buffy cinnamon (female). Rest of underparts rufous; outer-tail feathers rufous with darker tip. **V** Sharp *pee-up*, accelerating *pee pee pe pe pe e e e...* ending in trill. **H** Edge and clearings in wet or dry forest, gallery forest, nearby bushes and thickets, copses, plantations, savannah, isolated trees in deforested areas and secondary growth. **D** Widespread, from S Mexico to Bolivia, N Paraguay, S Brazil.

11 GREEN-TAILED JACAMAR *Galbula galbula* 21cm Upperparts and broad band across breast metallic green with bronzy and coppery sheen. Forehead, crown and part of the cheeks bluish green. Throat white (male) or buffy cinnamon (female). Rest of underparts rufous. Tail green above, blackish blue below. **V** Repeated *peer*, or rhythmic *peep peep...* **H** Rainforest, gallery forest, riverine forest, tall secondary growth, scrub and bushes. **D** N Amazonia: Colombia to the Guianas and N Brazil.

12 COPPERY-CHESTED JACAMAR *Galbula pastazae* 23cm Similar to Bluish-fronted Jacamar but slightly larger; heavier bill slightly downcurved; narrow yellowish-orange eye-ring. Male has white chin spot, female chin and throat rufous. **V** Song is like that of other *Galbula* species, a series of gradually rising *pee* or *weee* notes. **H** Edge of montane rainforest. **D** S Colombia to E Ecuador.

13 WHITE-CHINNED JACAMAR *Galbula tombacea* 21cm Both sexes have white chin; throat and breast metallic green. Forehead, crown and part of the cheeks brown. Rest of underparts rufous, female paler. **V** Repeated *keelip*, and *pee pee pe pe...*, accelerating. **H** Edge and clearings of várzea, gallery forest, thickets. **D** NW Amazonia.

14 BLUISH-FRONTED JACAMAR *Galbula cyanescens* 22cm Plumage differs from White-chinned Jacamar only in bluish-green forehead, crown and part of the cheeks. Slightly larger than White-chinned Jacamar. Female paler below. **V** Series of shrill *kree* or *kree-ip* calls that may build into a long, chattered sequence with a final, falling, trilled flourish. **H** Humid primary forest edges, gallery forest, mature secondary growth, often close to water. **D** W Amazonia.

15 PURPLISH JACAMAR *Galbula chalcothorax* 20–22cm Previously treated as conspecific with Bronzy Jacamar. Favours clearings and stream margins. **V** Shrill, piping *wher-eep, wher-eep, wheep, wheep-wheep*; pitch rises incrementally throughout delivery. **H** Lowland riverine forest, terra firme rainforest. **D** S Colombia, Ecuador, Peru, SW Brazil.

16 BRONZY JACAMAR *Galbula leucogastra* 20cm Mostly coppery purple. Throat white in male, buffy cinnamon in female. Lower belly and undertail coverts white. Undertail blackish. **V** Song is a high-pitched and ascending series of whistles; calls include rising, very high-pitched *hew-hihi hew-hehe hew*; also gives sharp, abrupt single notes. **H** Edge of dryland rainforest, gallery forest, secondary growth, forest in sandy soils. **D** S Venezuela, the Guianas, W and C Brazil, N Bolivia.

17 PARADISE JACAMAR *Galbula dea* 30cm Metallic bluish black with white throat. Very long slender tail. **V** Descending *peep peep pee pee pe pe pe...* accelerating but quieter at end. **H** Rainforest, by bodies of water, in sandy soils, dryland forest, scattered trees in deforested areas, woods, savannah. **D** Widespread, Amazonia.

18 GREAT JACAMAR *Jacamerops aureus* 30cm Huge and hefty for the family. Large slightly downcurved black bill. Head, sides of the breast, back, tail and wings all glittery bright green with bronze, coppery and purple sheen. Underparts bright rufous. Male has white upper breast. Shy, keeps still for long periods in the foliage at all levels beneath the crowns. **V** Infrequent, long melancholy whistled *weeeeee...eeeeeer*, somewhat raptor-like. **H** Edge and clearings of rainforest and gallery forest, woodland and stubble fields. **D** Costa Rica to NW Ecuador, Amazonia.

PUFFBIRDS BUCCONIDAE

19 WHITE-NECKED PUFFBIRD *Notharchus hyperrhynchus* 25cm Huge, with heavy black bill. White head with black crown and mask. Broad black breast-band. Rest of underparts white, flanks barred. **V** Song is a monotonous bubbling trill, lasting up to 20 secs. Calls include soft growling and whistling notes. **H** Forest to open country with scattered trees, in treetops. **D** S Mexico to Bolivia and Brazil.

20 GUIANAN PUFFBIRD *Notharchus macrorhynchos* 25cm Similar to White-necked Puffbird but more restricted range. White head with black crown and mask. Broad black breast-band. Rest of underparts white, flanks barred. **V** Song is a series of whistles, ending with several repetitions of a three-note phrase *wi-di-dik*. Calls include tuneful, falling *duuur* and other melodious and harsh notes. **H** Forest to open country with scattered trees, in treetops. **D** Guianas, N Brazil.

21 BUFF-BELLIED PUFFBIRD *Notharchus swainsoni* 23–25cm Previously considered conspecific with White-necked and Guianan puffbirds. **V** Shrill, squeaky series of *We, We, we-er, weee-wee-wee* notes. **H** Humid Atlantic forest, secondary woodland. **D** SE Brazil, E Paraguay, NE Argentina.

22 BLACK-BREASTED PUFFBIRD *Notharchus pectoralis* 20cm Head black, white only on ear-patch and throat. Neck and breast black. Lower underparts white, flanks barred. **V** Whistled *whets, whews* and combinations. **H** From mid-storey upwards in rainforest, secondary growth, edge of forest. **D** Panama to NW Ecuador.

23 BROWN-BANDED PUFFBIRD *Notharchus ordii* 18cm Breast-band black over brown. Head white but for black cap. Black barring on white belly and flanks. Base and tips of tail feathers white. **V** Song is a long, varied series of loud, clear, whistled notes; begins with single notes, proceeding to couplets and triplets. **H** Rainforest. **D** Amazonia.

24 PIED PUFFBIRD *Notharchus tectus* 15cm Small. Cap black with white speckling, black mask, long white eye-stripe. Broad black breast-band. Scapular patch white. **V** High-pitched *pee-peed, pee-peedit, peedit, peedit, peeah, pee, pee, pee*. **H** Rainforest to mangroves, open areas with trees, stubble fields. **D** Costa Rica to Amazonia.

25 CHESTNUT-CAPPED PUFFBIRD *Bucco macrodactylus* 15cm Chestnut cap, whitish eye-stripe, black mask and neck-band. Cinnamon hindcollar. Underparts buffy, finely barred brown. Upperparts brown. Perches at lower levels especially near water and flooded areas. **V** Series of *pips* and *peeps* ending in *piz* twitter; series of *weeahs*, rising slightly. **H** Rainforest, its edge and woodland, gallery, dryland and várzea, secondary growth and thickets in clearings. **D** W and C Amazonia.

26 SPOTTED PUFFBIRD *Bucco tamatia* 18cm Brow and bib tawny. Cap and mask brown. Long white malar streak. Long black streak on lower cheek. Upperparts brown. Breast and rest of underparts white with black scaling simulating barring. Trusting, in lower branches. **V** Long series of whistled *chewys*, two per sec, ending emphatically with strong *petchowee*. **H** Edges of rainforest, várzea, sandy savannah along rivers, edge of woods and thickets. **D** Amazonia.

27 SOOTY-CAPPED PUFFBIRD *Bucco noanamae* 18cm Cap and mask sooty. Lores, brow and throat white. Broad black breast-band. Rest of underparts buffy, spotted dark, emulating barring. **V** Song is a series of 20–40 whistled notes, beginning rapidly and slowing. **H** Rainforest and secondary growth. **D** W Colombia.

28 COLLARED PUFFBIRD *Bucco capensis* 19cm Mostly rufous and white, with black breast-band. Bill orange with black. Perches quietly from mid-height to just below the crown of trees. **V** Song is a fast series of repeated, mewing *cua-will cua-will* or similar notes, sometimes given in duet. **H** Dry land and rainforest. **D** Colombia to the Guianas, Amazonia.

29 BARRED PUFFBIRD *Nystalus radiatus* 20cm Bill blackish and yellowish horn. Crown and rest of upperparts rufous. Face, hindneck and underparts buffy ochre, all finely barred blackish. **V** A long, slow *wheeeeeeet wheeeeeew* like a wolf-whistle, repeated twice a minute. **H** Rainforest, open damp areas with scattered trees, secondary growth. **D** Panama to W Ecuador.

30 WHITE-EARED PUFFBIRD *Nystalus chacuru* 19cm Bill orange. Face black, ear-coverts white. Underparts and hindcollar whitish. Typically seen on roadside telegraph wires and posts at the edge of forest, or on trees in fields. **V** Duets: *turutee turutay turutee turutay...*, descending. **H** Forest edge, secondary growth, transition forest. **D** E Peru, Bolivia, Brazil, Paraguay, N Argentina.

31 EASTERN STRIOLATED PUFFBIRD *Nystalus striolatus* 20cm Bill blackish and yellowish horn. Face and underparts ochre and buff streaked blackish. Upperparts barred. **V** Song is a melancholy whistled *whip, whiweeoo, weeooo*. **H** Edge of the forest, open woods, arid scrub with trees. **D** Amazonian Brazil.

32 WESTERN STRIOLATED PUFFBIRD *Nystalus obamai* 20–21cm Western counterpart of Eastern Striolated Puffbird. Favours clearings, usually near water. **V** Series of slurred *We, we-erwee, we-uu* notes. **H** Lowland forest, secondary woodland. **D** W Amazonian region, from S Colombia to Bolivia, SW Brazil.

33 CAATINGA PUFFBIRD *Nystalus maculatus* 19cm Bill orange and blackish. Lower throat and upper breast ochre. Belly and flanks white, streaked black. Crown and face streaked. Back and tail barred. **V** Whistled *to you, to you, to you, to you....* **H** Low, thorny arid scrub. Open woods. **D** NE and C Brazil.

34 CHACO PUFFBIRD *Nystalus striatipectus* 18–19cm Previously considered conspecific with Caatinga Puffbird, and referred to as Spot-backed Puffbird. **V** Shrill, whistled series of *K-wow, K-wow, kwow, kwo* notes. **H** Open woodland, savannah. **D** E Bolivia, CS Brazil, W Paraguay, N Argentina.

35 RUSSET-THROATED PUFFBIRD *Hypnelus ruficollis* 23cm Previously considered conspecific with Two-banded Puffbird. Usually trusting, perches low on outer branches of isolated trees or bush. **V** Noisy, usually a duetted rhythmic *oo-dooks*, ending with a half-speed solo. **H** Deciduous forest, dryland woods, thickets. **D** N Colombia, N Venezuela.

36 TWO-BANDED PUFFBIRD *Hypnelus bicinctus* 20–21cm Previously considered conspecific with Russet-throated Puffbird. Nests in arboreal termite nests. **V** Shrill, whistled series of *K-wow, K-wow, kwow, kwo* notes. **H** Llanos grassland, open secondary woodland, streamside scrub. **D** N Venezuela.

1
2
3
4
5
6
7
8
9
10
11
12
13
14
15
16
17
18
19
20
21
22
23
24
25
26
27
28
29
30
31
32
33
34
35
36

PUFFBIRDS *CONTINUED*

1 CRESCENT-CHESTED PUFFBIRD *Malacoptila striata* 22cm Lores, small brow and chin tawny. Moustache and band across foreneck white, pectoral collar black. Breast and flanks from tawny to buffy. Bill black. **V** Thin whistled *bee-ee, bee-ee...* and series of 10+ uniform *pee-u*. **H** Rainforest and edges, in shadier parts. **D** NE, SE Brazil.

2 WHITE-CHESTED PUFFBIRD *Malacoptila fusca* 20cm Lores, moustache and band on foreneck white. Breast and flanks streaked dark brown and white. Bill orange with black culmen and tip. Perches on branches from mid-levels to near the ground. **V** A thin whistled *seeee...* sometimes lasting up to 2 secs, descending in pitch. **H** Rainforest, plantations, dryland forest, secondary growth, edges. **D** N Amazonia.

3 SEMICOLLARED PUFFBIRD *Malacoptila semicincta* 20cm Practically identical to White-chested Puffbird but has a rufous half-collar on hindneck. Some consider this a race of White-chested Puffbird. **V** Series of mournful, high-pitched whistles, speeding up and dropping in pitch. **H** Rainforest. **D** W Amazonia: Peru, N Bolivia, W Brazil.

4 BLACK-STREAKED PUFFBIRD *Malacoptila fulvogularis* 22cm Similar to White-chested Puffbird and Semicollard Puffbird, but lores, chin and patch across throat buffy. Scapulars and wing coverts edged lighter, giving a scaly effect. Tail relatively longer. Bill black. **V** Very high-pitched whistle, drawn out (up to nearly 4 secs), rising in pitch. **H** Forested slopes, in damp understorey. **D** C Colombia to NW Bolivia.

5 RUFOUS-NECKED PUFFBIRD *Malacoptila rufa* 20cm Cap and ear-coverts grey, finely streaked white. Forehead and lores rufous cinnamon. Cheeks, sides of the neck and hindneck rufous. Moustache and band across upper breast white, the latter with slight dark barring below. **V** A rising phrase *ee, ee, ee, ee, ee*. **H** Rainforest and várzea. **D** S and C Amazonia.

6 WHITE-WHISKERED PUFFBIRD *Malacoptila panamensis* 20cm Sexes differ. Male mostly rufous; white moustache, fine streaking on cheeks; upper breast rufous; lower breast, flanks and belly streaked. Female greyish brown where male is rufous; upper breast ochre. **V** In alarm a thin *seee*; in territorial flight a high-pitched squeaky, repeated *seeet, hee, hee*. **H** Rainforest and secondary growth. **D** S Mexico to W Ecuador.

7 MOUSTACHED PUFFBIRD *Malacoptila mystacalis* 24cm Head, upperparts and tail brown, speckled buffy on back and wing coverts. Lores and moustache white. Breast tawny rufous, somewhat streaked brown, this extending onto the flanks. **V** High, thin peeping notes. **H** Rainforest, cloud forest and open or deciduous woods, in lower storeys. **D** NW Colombia, Venezuela, N Ecuador.

8 LANCEOLATED MONKLET *Micromonacha lanceolata* 13cm Very small. Black bill, downcurved at the tip. Upperparts brown, underparts white with blackish drip-shaped streaks. White around base of bill. Often overlooked. **V** Apparently silent. **H** Edge of rainforest, secondary growth near clearings. **D** Costa Rica to NW Bolivia.

9 RUSTY-BREASTED NUNLET *Nonnula rubecula* 15cm Bill black and yellowish. Upperparts and cheeks olive-brown; crown greyish. Lores buffy white. Throat, breast and flanks buffy cinnamon, lower belly lighter. **V** Song up to 20 mewing notes, becoming progressively louder and higher-pitched; calls include sharp *tick*. **H** Rainforest, várzea, gallery forests. **D** Amazonia, SE Brazil, NE Argentina.

10 FULVOUS-CHINNED NUNLET *Nonnula sclateri* 15cm Like Rusty-breasted Nunlet but face greyish, lores and chin tawny buff, crown olive-brown. **V** Song is a series of thin, weak *weep* notes with mewing quality. **H** Tropical rainforest. **D** W Amazonia: E Peru, W Brazil.

11 BROWN NUNLET *Nonnula brunnea* 16cm Bill black, bluish horn on mandible. Upperparts, cheeks and sides of neck dark brown. Lores and underparts tawny, lighter on belly. Eye-ring pink. **V** Song consists of 20–25 *treeuu* notes, becoming louder then fading away. **H** Tropical rainforest, dryland forest, lowland forest at the base of the Andes, secondary growth. **D** NW Amazonia: Colombia, Ecuador, N Peru.

12 GREY-CHEEKED NUNLET *Nonnula frontalis* 15cm Note contrasting grey face sides and sharp, slightly downcurved bill. **V** Series of very high fluty notes averaging 25, rising slightly, then monotone *plip plip plip plip - -*. **H** Lower forest storeys. **D** Panama, Colombia.

13 RUFOUS-CAPPED NUNLET *Nonnula ruficapilla* 15cm Bill black and blue. Face and sides of the neck all bluish grey. Cap rufous. Rest of upperparts and tail brown with olive wash. Throat, breast, flanks and rest of underparts tawny, ranging to white at lower belly. Eye-ring pink. **V** Song is a series of up to 30 upslurring short whistles, 2–3 per sec, becoming louder and then softer again. **H** Edge of rainforest and gallery forest, secondary growth in damp areas. **D** W and S Amazonia: E Peru, N Bolivia, Brazil.

14 CHESTNUT-HEADED NUNLET *Nonnula amaurocephala* 15cm Head and breast tawny rufous, belly and undertail coverts white. **V** Not known. **H** Tropical forests. **D** NW Brazil.

15 WHITE-FACED NUNBIRD *Hapaloptila castanea* 25cm Large. No streaking. Lead-grey cap. White around base of black bill. Black band across forehead. Rest of upperparts and tail dark brownish grey. Underparts orange-rufous. **V** Owl-like hoot, rising; sometimes a trill. **H** Treetops of rainforest or lower on the edge of forest or small clearings. **D** W Colombia to NW Peru.

16 BLACK NUNBIRD *Monasa atra* 28cm Nearly all black, belly slaty. 'Shoulders' and borders of upperwing coverts white. Bill bright coral-red. **V** Loud three-syllable *tree-de-lee*. **H** Rainforest, gallery forest, scrubby thickets, forest edge. **D** N Amazonia.

17 BLACK-FRONTED NUNBIRD *Monasa nigrifrons* 28cm Entirely slaty black, darker on face, cap and throat. Bright coral-red bill. **V** Quiet *keeor* or a long phrase of melodious whistles lasting several minutes. **H** Edge of rainforest, gallery forest, várzea, forested islands on rivers, secondary growth, scattered trees, palm groves. **D** S and W Amazonia to SE Brazil.

18 WHITE-FRONTED NUNBIRD *Monasa morphoeus* 28cm Like Black-fronted Nunbird but with white forehead and chin. The race *M. m. pallescens* has only forehead white. **V** Like Black-fronted Nunbird, in noisy choruses. **H** Rainforest, gallery forest, dryland forest. **D** Honduras to Amazonia and SE Brazil.

19 YELLOW-BILLED NUNBIRD *Monasa flavirostris* 24cm Yellow bill. Black with white 'shoulders', slatier on belly. Perches quietly on exposed limbs at upper levels of trees, sallying out to snap up prey from air or foliage. **V** Song is a series of fast, chuckling, melodious phrases, often including repeated *weekik-week weekik-week*. **H** Dryland forests, edges of forests, scattered trees in forest clearings, secondary growth, forest at the foot of hills. **D** W Amazonia.

20 SWALLOW-WINGED PUFFBIRD *Chelidoptera tenebrosa* 16cm Unlike any other puffbird. Long wings, short tail. Mostly black, lower belly cinnamon. Rump, tail coverts and underwing coverts white. Hunts insects on the wing in acrobatic flight. **V** Song is a twittering phrase; calls include *tsi-tsi-tsi*. **H** Perches in exposed positions in a variety of open habitats and edges of forest, rivers, savannah, woods. **D** NW Venezuela, Amazonia, E Brazil.

NEW WORLD BARBETS CAPITONIDAE

21 SCARLET-CROWNED BARBET *Capito aurovirens* 19cm Big yellowish-orange bib. Mask and upperparts dark brown. Belly and undertail brownish olive. Male has red cap; female has whitish cap. **V** Rapid, frog-like *crew crew....* **H** Várzea, edge, tall secondary growth. **D** W Amazonia.

22 SCARLET-BANDED BARBET *Capito wallacei* 19cm Red cap. White brow; black mask, upperparts and tail. Lower cheek, throat and foreneck white. Pectoral band red. Rest of underparts yellowish, browner on flanks. **V** Song is a rapid trill, recalling drumming of a woodpecker; calls include a throaty *graaak*. **H** Humid montane and stunted forest all dripping with mosses. **D** NC Peru.

23 SIRA BARBET *Capito fitzpatricki* 19–20cm Similar to Scarlet-banded Barbet. Note isolated range. Discovered in 2008. **V** Song purring, vibrating *crrrrrrrrr*, and range of grunting and croaking calls. **H** Montane cloud forest. **D** Cerros del Sira (C Peru).

24 SPOT-CROWNED BARBET *Capito maculicoronatus* 17cm Crown white, streaked black. Mask, upperparts and tail black. Male has malar region and throat white, breast yellow. Female has malar region, throat and breast black. **V** Song is a series of cough-like notes; some other recorded sounds possibly mechanical, for example bill-snapping. **H** Edge of humid rainforest, forest in hilly country, valleys, ravines, secondary growth. **D** Panama, NW Colombia.

25 ORANGE-FRONTED BARBET *Capito squamatus* 17cm Forehead orange. Crown and nape white, flecked black. Mask, upperparts and tail black. Underparts white, flanks streaked blackish. Male has white throat and yellow-washed breast. Female has black throat and breast. **V** Deep, fast, quiet purring trill, and a single or double *yit*. **H** Humid rainforest, lowland and hill forest, secondary forest, orchards, isolated fruiting trees, plantations, farms. **D** SW Colombia, W Ecuador.

26 WHITE-MANTLED BARBET *Capito hypoleucus* 18cm Forehead and forecrown red. Hindcrown, nape and upper back white, speckled black. Mask, rest of upperparts and tail black. Underparts white, pectoral collar pale brown. Flanks washed greyish yellow. **V** Deep croak. **H** Humid forest at base of hills. **D** NW Colombia.

27 BLACK-GIRDLED BARBET *Capito dayi* 17cm Male has red cap, female black cap. Mask, upperparts and tail black. White on side of upper back and on flight feathers. Black flanks and striking incomplete girdle. Breast and belly whitish buff. **V** Song is a long series of hollow, hooting *boo* notes; calls include grating and rattling notes. **H** Gallery forest, rainforest, forest in lowlands and slopes in foothills, secondary forest with fruiting trees, plantations. **D** S Amazonia.

28 BROWN-CHESTED BARBET *Capito brunneipectus* 18cm Forehead and pectoral collar brown. Crown, nape and throat yellowish buff. Rest of underparts dirty yellowish, streaked blackish grey. Mask, upperparts and tail black. Yellow line from nape to back. Female is duller, with throat dotted black. **V** Trilled *whooo*, rising and then ending *brrrrrr*. **H** Canopy of lowland rainforest. **D** NC Brazil.

29 BLACK-SPOTTED BARBET *Capito niger* 18cm Forehead, forecrown and throat red. Hindcrown and nape yellow, streaked black. Yellow line from brow to middle of back. Mask, upperparts and tail black. White bar on wing coverts. Underparts yellowish with drop-shaped spots on flanks. Female more heavily streaked. **V** Deep, frog-like, hollow, ventriloquial *hoot-oot*. **H** Rainforest, humid terra firme forest, swampy forest, secondary growth, plantations, occasionally várzea. **D** NE Amazonia.

30 GILDED BARBET *Capito auratus* 18cm Orange forehead and throat, orangey-yellow underparts and orange line from brow to back. Underparts yellowish with drop-shaped spots on flanks. Female more heavily streaked. **V** Deep, frog-like, hollow, ventriloquial *hoot-oot*. **H** Rainforest, humid terra firme forest, swampy forest, secondary growth, edge, plantations, occasionally in várzea. **D** NW Amazonia.

31 FIVE-COLORED BARBET *Capito quinticolor* 18cm Male black, white and yellow with red crown. Female more streaked, with cap streaked black and yellow and underparts spotted blackish. **V** Series of whooping notes for several seconds and guttural *churr*. **H** Rainforest, edge, tall secondary growth. **D** W Colombia, NW Ecuador.

32 LEMON-THROATED BARBET *Eubucco richardsoni* 15cm Variable. Males typically green-olive above and bright yellow below, with crimson crown and yellow bill. Variable amount of orange on breast. Females similar but duller, often with more grey in the head. **V** Series of fast *todoos* or *toodledoos*. **H** Terra firme forest and várzea, lowland forest, clearings, tropical hill forest, gallery forest. **D** W Amazonia, from Colombia to Bolivia.

33 RED-HEADED BARBET *Eubucco bourcierii* 16cm Hood and bib of male bright red; chin and around eye black; narrow, light blue hindcollar. Female has black forehead, pale greenish-blue face and throat, olive-green crown, orangey-yellow breast. **V** Song is a long, dry trill; calls include rattles, grunts and chatters. **H** Rainforest and transition with cloud forest, secondary growth, woodland edges. **D** Costa Rica to W Venezuela and Ecuador.

34 SCARLET-HOODED BARBET *Eubucco tucinkae* 17cm Male has bright red hood, pale orange hindcollar, bronzy-olive upperparts and tail. Female has olive crown and nape, yellow malar region and throat. **V** Longish series of rapid *oopoopoops*. **H** Thickets with cane in gallery forest, islands, tropical forest near rivers in the Andes. **D** SW Amazonia.

35 VERSICOLORED BARBET *Eubucco versicolor* 16cm Head of male mostly bright red with blue malar stripe and narrow collar; breast rich yellow; pale red band across upper belly. Crown and hindneck of female olive-green; forehead, face and throat blue; narrow red collar on foreneck; breast light green. **V** Low-pitched even trill. **H** Tropical forest, rainforest, secondary growth with epiphytes and mosses, dry forest on lower slopes. **D** Peru, Bolivia.

TOUCAN BARBETS SEMNORNITHIDAE

36 PRONG-BILLED BARBET *Semnornis frantzii* 17cm Unmistakable chunky brownish-yellow bird with thick silver-grey bill. **V** Low wooden cackling *wukwukwuk~* in chorus. **H** Montane forest and adjacent areas. **D** Costa Rica, Panama.

37 TOUCAN BARBET *Semnornis ramphastinus* 20cm Large and multicoloured. Cap and foreface black, white patch on hindcrown. Cheeks and throat bluish grey. Breast red. **V** Long series of fast, foghorn-like honks, sometimes duetted. **H** Montane forest, edge, secondary growth, isolated fruiting trees. **D** W Colombia, W Ecuador.

1
2
3
4
5
6
7
8
9
10
11
12
13
14
15
16
17
18
19
20
21
22
23
24
25
26
27
28
29
30
31
32
33
34
35
36
37

TOUCANS RAMPHASTIDAE

1 WAGLER'S TOUCANET *Aulacorhynchus wagleri* 30–34cm This and the next two species were previously treated as subspecies of 'Northern Emerald-toucanet'. Throat white, grading to blue. **V** Squawking, screeching and upslurred *o-Ak, o-Ak, o-Ak* notes. **H** Montane forest, forested slopes, secondary woodland. **D** SW Mexico.

2 EMERALD TOUCANET *Aulacorhynchus prasinus* 30–32cm Throat white. **V** High-pitched barking *wah, wah, wah*. **H** Montane forest, forested slopes, secondary woodland. **D** SE Mexico, Honduras, Nicaragua, El Salvador.

3 BLUE-THROATED TOUCANET *Aulacorhynchus caeruleogularis* 30–34cm Throat deep blue. **V** Harsh, screeching and monotonously repeated *wo-AK, wo-AK, wo-AK*. **H** Humid montane forest. **D** Costa Rica, Panama.

4 WHITE-THROATED TOUCANET *Aulacorhynchus albivitta* 30–36cm Previously considered conspecific with Black-throated Toucanet and treated as subspecies of 'Southern Emerald-toucanet'. Throat white or grey depending on race. **V** Monotonously repeated series of abrupt *waK, waK, waK...* notes. **H** Montane forest, cloud forest. **D** E Andean slopes of Venezuela, Colombia, N Ecuador.

5 BLACK-THROATED TOUCANET *Aulacorhynchus atrogularis* 30–36cm Previously considered conspecific with White-throated Toucanet and treated as subspecies of 'Southern Emerald-toucanet'. Throat black. **V** Croaking, monotonously repeated *wo-AK, wo-AK, wo-AK*. **H** Montane forest, cloud forest. **D** E Andean slopes of S Ecuador, Bolivia and N Peru.

6 GROOVE-BILLED TOUCANET *Aulacorhynchus sulcatus* 35cm Bill black and reddish chestnut, or black and yellow. Yellow-billed form previously considered a separate species. **V** Song is a series of variable harsh barking or growling notes, with bill-rattles. **H** Rainforest and cloud forest, plantations, copses, gardens. **D** NE Colombia, N Venezuela.

7 CHESTNUT-TIPPED TOUCANET *Aulacorhynchus derbianus* 35cm Bill mostly black with thin white line at base. Tip of tail chestnut. **V** Deep *guah...gawk...hawk...gahk* at the rate of 1 per sec. **H** Montane forest. **D** SE Colombia to C Bolivia.

8 TEPUI TOUCANET *Aulacorhynchus whitelianus* 34–40cm Previously considered conspecific with Chestnut-tipped Toucanet. Note separate and disjunct geographical range. **V** Monotonously repeated *wo-Ek, wo-Ek, wo-Ek...* **H** Humid forests, mountains and tepui. **D** Guyana, Suriname, N Brazil, Venezuela.

9 CRIMSON-RUMPED TOUCANET *Aulacorhynchus haematopygus* 42cm Bill black with broad reddish-chestnut longitudinal stripe on the maxilla. Bright red rump. Tip of tail chestnut. **V** Song is a series of low growling barks given in duet, with female's notes higher-pitched and clearer than male's. Calls include rattles and quieter vocalisations; also loud wing sounds. **H** Montane evergreen forests, basal forests, edges, isolated trees, gardens. **D** W Venezuela to Ecuador.

10 YELLOW-BROWED TOUCANET *Aulacorhynchus huallagae* 38cm Black and grey bill with ivory tip. Very like Blue-banded Toucanet but with white line at base of bill. Yellow post-ocular brow. Golden-yellow undertail coverts. **V** Long series of up to 30 croaking notes, about one per sec. **H** Cloud forest. **D** Very local, NE Peru.

11 BLUE-BANDED TOUCANET *Aulacorhynchus coeruleicinctis* 40cm Bill blackish and grey, ivory towards the tip, lacking white at base. Part of the cheeks, small eyebrow and throat white. Ill-defined blue band across upper belly. Greenish undertail coverts. Red rump. **V** Song consists of a variety of barking and growling notes. **H** Rainforest and cloud forests in mountains, edges and secondary growth. **D** Peru, Bolivia.

12 GREEN ARACARI *Pteroglossus viridis* 37cm Bill reddish and yellowish ochre with black mandible; serrated white line along cutting edge. Bare skin around eye red and blue. Hood black (male), dark chestnut (female). Both sexes have lemon-yellow underparts. **V** Rattled *ticker, ticker, ticker....* **H** Rainforests and gallery forests. **D** NE Amazonia.

13 LETTERED ARACARI *Pteroglossus inscriptus* 34–44cm Yellowish-ivory bill with irregular blackish 'teeth' along cutting edge. Bill of larger W race has all-black mandible. Hood black, with chestnut cheeks and foreneck in female. Underparts lemon-yellow. **V** Like a large kingfisher's explosive rattle and *tarack, rack-ack*. **H** Lowland evergreen forest, especially that which is seasonally flooded, secondary growth, gallery forest, other wooded habitats. **D** W and S Amazonia.

14 RED-NECKED ARACARI *Pteroglossus bitorquatus* 37cm Bill chrome-yellow, black and white, or entirely black (one race), or almost all white with black at the tip (another race); line of black along cutting edge. Head black. Back of neck, upper back and breast dark red. Pectoral collar and rest of underparts yellow. **V** *Wheat, wheat*. **H** Rainforest, gallery forest and edges, scattered trees. **D** S Amazonia.

15 IVORY-BILLED ARACARI *Pteroglossus azara* 38cm Bill ivory with variable amount of reddish chestnut; notable 'teeth'. Hood dark chestnut, male with even darker cap. Red chest. Black breast and upper belly, lower belly yellow. **V** Sharp *kissick* and rattled, croaked *co..ark... cro.. ark...* **H** Rainforests, edges, gallery forest, copses, plantations. **D** NW Amazonia.

16 BROWN-MANDIBLED ARACARI *Pteroglossus mariae* 36–44cm Previously treated as subspecies of Ivory-billed Aracari. Often seen in pairs. **V** Grating, chattery and harsh *chet, chet, w-chet, chet, chet...* **H** Inundated and terra firme forests, secondary woodland. **D** Bolivia, Brazil, Peru.

17 BLACK-NECKED ARACARI *Pteroglossus aracari* 46cm Maxilla and base of bill ivory; mandible, culmen and sub-basal line black. Hood bluish black. One red cummerbund across yellow underparts. **V** Tinkling high-pitched two-syllable *biditz*. **H** Rainforests, gallery forests, open woods, edges, scattered trees. **D** E South America, from Venezuela to SE Brazil.

18 CHESTNUT-EARED ARACARI *Pteroglossus castanotis* 46cm Bill mostly black with ochre-yellow band along maxilla, yellowish base and 'teeth'. Iris white. Head chestnut, cap black in some forms. Chest black. One red cummerbund across yellow underparts. Chestnut leggings. **V** Like scraping glass – high-pitched *tcheeeecheeek*. **H** Rainforest, gallery and várzea forest, clearings, savannah, secondary growth, plantations. **D** W Amazonia to SE Brazil.

19 MANY-BANDED ARACARI *Pteroglossus pluricinctus* 44cm Black and ivory bill with reddish tip, ivory line at base. Bare skin around eye bright blue. Hood black. Black bands across yellow underparts, flecked with red. **V** Jarring screech; also a high *pitchew*. **H** Rainforest, várzea. **D** NW Amazonia.

20 COLLARED ARACARI *Pteroglossus torquatus* 43cm Upper bill ivory with chestnut wash near base, blackish culmen and 'teeth'; black mandible; white line at base of bill. Hood black. Hindcollar chestnut. Bare skin around eye red. Underparts yellow with red flecking, black chest spot and cummerbund, both edged red. **V** High-pitched, sneezy *kiyik*. **H** Rainforest and edge, secondary growth, open forest and scattered trees. **D** Mexico to Venezuela.

21 STRIPE-BILLED ARACARI *Pteroglossus sanguineus* 43cm Very like Collared Aracari but has black stripe on maxilla, bare skin around eye is mostly blue, and it lacks collar. **V** High-pitched, sneezy *kiyik*. **H** Similar to those of Collared Aracari. **D** W Colombia, NW Ecuador.

22 PALE-MANDIBLED ARACARI *Pteroglossus erythropygius* 45cm Bill mostly ivory with black stripe along cutting edge of maxilla, black and blue tip of mandible. Underparts yellow with red flecking, black chest spot and cummerbund. Extensive red lower back, rump and uppertail coverts. **V** Like Collared Aracari. **H** Rainforest, edges, scattered trees. **D** W Ecuador.

23 FIERY-BILLED ARACARI *Pteroglossus frantzii* 45cm Partly red bill sides diagnostic. **V** High sustained *wreeet-wreeet-*. **H** Forest and clearings. **D** Costa Rica, Panama.

24 CURL-CRESTED ARACARI *Pteroglossus beauharnaesii* 44cm Multicoloured bill with chestnut, turquoise, ivory, white and ochre tip. Bare skin around eye pale blue. Curly cap-feathers black. Underparts yellow with wide red and black cummerbund. **V** Loud, gruff, low-pitched *rreeep* and *rree-rreee-rreee*. **H** Tropical rainforest, gallery forest, edge. **D** W Amazonia.

25 SAFFRON TOUCANET *Pteroglossus bailloni* 37cm Unpatterned. Bill greenish yellow with chestnut-red at base. Upperparts pale olive-green; rump red. Underparts greenish yellow with olive wash. **V** Whistled *gwee...gwee*; also quiet sneezed *spitz...spitz*. **H** Rainforests in lowland and rolling hills. **D** SE Brazil, E Paraguay, NE Argentina.

26 YELLOW-EARED TOUCANET *Selenidera spectabilis* 38cm Bill slate-grey, culmen broad yellow with a greenish wash. Only species with neither hindcollar nor cinnamon tip to tail. Crown and hindneck chestnut in female. **V** Slow, weak, rhythmic, 2-syllable *krek-ek*. **H** Rainforest; usually not in lowlands. **D** Honduras to SW Colombia.

27 GUIANAN TOUCANET *Selenidera piperivora* 35cm Bill black with dark red at the base, more so on mandible. Flanks greenish yellow. Male has black head and breast. Female has throat, lower cheeks, breast and upper belly lead-grey, chestnut on hindneck. **V** Rattly voice. **H** Tropical rainforest, lowlands. **D** NE Amazonia.

28 GOLDEN-COLLARED TOUCANET *Selenidera reinwardtii* 33cm Bill chestnut-red with black culmen, tip and cutting edge. Female is chestnut where male is black, with ear-tufts dull yellowish olive. **V** Longish series of froggy croaks. **H** Rainforests, sometimes in várzea. **D** W Amazonia.

29 TAWNY-TUFTED TOUCANET *Selenidera nattereri* 33cm Bill reddish chestnut, culmen and tip yellow, black spot at the base and 'teeth' along the white cutting edge. Female is chestnut where male is black, and lacks contrasting flank patch. **V** Series of deep-toned, quiet, frog-like notes. **H** Rainforest in sandy soils, tropical forest. **D** N Amazonia: Colombia, Venezuela, N Brazil.

30 GOULD'S TOUCANET *Selenidera gouldii* 33–34cm Previously considered conspecific with Spot-billed Toucanet. **V** Frog-like, croaking *wo-oh, wo-Ah, wo-Ah, wo-Ah...* **H** Inundated forest, Cerrado savannah, secondary woodland. **D** E Amazonia: NE Bolivia, Brazil.

31 SPOT-BILLED TOUCANET *Selenidera maculirostris* 33cm Shortish pale greyish-blue bill with yellower culmen and tip, and irregular black marks across it. Dull yellowish-olive ear-tufts. Female is chestnut where male is black. **V** Series of four or five deep, hollow *wawks*. **H** Rainforest. **D** SE Brazil, E Paraguay, NE Argentina.

32 GREY-BREASTED MOUNTAIN TOUCAN *Andigena hypoglauca* 48cm Bill red, with yellow and black below. Greyish blue hindcollar. Yellow rump. **V** Rising catlike *yaaaahs*, or repeated *eeeeeeks*. **H** High Andean cloud forest, woods up to near treeline. **D** Colombia, Ecuador, Peru.

33 PLATE-BILLED MOUNTAIN TOUCAN *Andigena laminirostris* 48cm Bill black with red basal half and buffy-white rectangle near base. Bare face yellow and green. Rump yellow, patch on flanks ochre-yellow. Active and vocal. **V** Series of nasal *waaahs*, each note rising. **H** Montane rainforest with many bromeliads and mosses. **D** SW Colombia to S Ecuador.

34 HOODED MOUNTAIN TOUCAN *Andigena cucullata* 48cm Bill yellow (or greenish), black tip and spot at base of mandible. Bare face blue. Hood black. Hindcollar greyish-blue. Rump and uppertail coverts olive-green. Shy. **V** Nasal *wa*, irregular *ke.. keke... ke...*, whining *ueeeee*. **H** Montane cloud forest with many epiphytes. **D** SE Peru, Bolivia.

35 BLACK-BILLED MOUNTAIN TOUCAN *Andigena nigrirostris* 51cm Black bill, white throat and cheeks, blue belly. Upperparts brownish with yellow rump. **V** Scraping, nasal *too-aaat*, rising. **H** Epiphyte-loaded cloud forest, rainforest, scattered trees. **D** W Venezuela to N Peru.

36 GREEN-BILLED TOUCAN *Ramphastos dicolorus* 45cm Yellow bill with greenish wash, slightly 'toothed'; black base. Face salmon-pink and yellow. Pale eye. Cheeks and throat yellow, foreneck orange. Chest-band yellow, rest of underparts and rump red. **V** Harsh tenor croaks, repeated *yehhhk*. **H** Rainforests and edges, plains, rolling hills. **D** SE Brazil, E Paraguay, NE Argentina.

37 CHANNEL-BILLED TOUCAN *Ramphastos vitellinus* 46cm Black bill with pale base (blue or yellow, depending on race). Throat and upper breast yellow or orange. Broad breast-band, undertail coverts and uppertail coverts red. **V** Grunting bisyllabic *yah, eah*. **H** Rainforest, edges, scattered trees, gallery forest, forestry plantations. **D** Amazonia, E and SE Brazil.

38 CITRON-THROATED TOUCAN *Ramphastos citreolaemus* 50cm Bill blackish purple, tip and culmen yellow, basally yellow and pale blue bands. Skin around the eye pale blue. Cheeks and foreneck white, with yellow wash on upper breast. Pectoral collar red. Uppertail coverts pale yellow. **V** Series of frog-like *creeop* croaks. **H** Rainforests and edges. **D** N Colombia, NW Venezuela.

39 CHOCO TOUCAN *Ramphastos brevis* 54cm A smaller version of Yellow-throated Toucan, with which it is largely sympatric. Bill relatively shorter (though this varies between individuals), but all have a higher arched culmen, which, seen from in front, is sharp, like an inverted V. **V** Croaking grunts in series. **H** Rainforest. **D** NW Colombia to SW Ecuador.

40 KEEL-BILLED TOUCAN *Ramphastos sulfuratus* 50cm Patterning of huge bill diagnostic. **V** High frog-like sustained *prut-prut-prrut-pruit*. **H** Forest canopy, tall secondary growth; descends for berries. **D** E Mexico to NW Venezuela.

41 TOCO TOUCAN *Ramphastos toco* 59cm Bill orangey yellow with black base and black spot at the tip; faint orange 'teeth'. Bare skin around eye orangey pink, eye-ring blue. Cheeks and bib white, as are uppertail coverts. **V** Deep, raucous croaking *aaaark*. **H** Rainforests, open areas with scattered trees, plantations, gardens. **D** From the Guianas and SE Peru to N Argentina.

42 WHITE-THROATED TOUCAN *Ramphastos tucanus* 58cm Dark part of bill varies from black to orangey chestnut, always with a black sub-basal band. Throat white. Uppertail coverts yellow. **V** Loud series of rhythmic, whistled yelps. **H** Rainforests, edges and scattered trees. **D** Amazonia.

43 YELLOW-THROATED TOUCAN *Ramphastos ambiguus* 58cm Huge bill black with broadly yellow culmen, which, seen from in front, is rounded, like an inverted U. Bare face pale blue. **V** Yelping *kyoss today today*, repeated at short intervals. **H** Rainforest and cloud forest, lower montane forest, mostly in the highest branches. **D** SE Honduras to W Ecuador, W Venezuela to Peru.

1
3
5
11
9
2
4
6
7
8
10
12 ♀
♂
13
14
16
18
♀
♂
15
19
20
17
♀
29
30 ♀
21
22
23
27
♂
24
♀
♀
28
♂
♂
♂
31
♀
25
♂
♀
32
34
35
26
♂
36
41
33
39
43
38
37
42
40

ASIAN BARBETS MEGALAIMIDAE

1 FIRE-TUFTED BARBET *Psilopogon pyrolophus* 29cm Black stripe on bill. Forages in pairs or small groups in the canopy of tall trees, acrobatic when attempting to reach fruit. **V** Cicada-like *dddzza-ddzza*, increasing to a *zz-zz-zz-zz*; also a whistled note and a squeak. **H** Primary and secondary forest, forest edges, forest patches. **D** Malay Peninsula; Sumatra, Indonesia.

2 GREAT BARBET *Psilopogon virens* 32–35cm Black head, massive pale bill. Usually seen singly or in small parties, with bigger groups where trees are in fruit. When not feeding, sits motionless in the topmost branches. **V** Mournful *piho piho piho*, a rapid *tuk tuk tuk* and a harsh *karr-r*. **H** Broadleaved evergreen, occasionally deciduous forests and gardens with fruiting trees. **D** NW India to Myanmar and SE China.

3 RED-VENTED BARBET *Psilopogon lagrandieri* 29cm Head mostly brown and grey. Arboreal; little recorded, probably forms groups at fruiting trees. **V** Male call is a throaty, strident *choa* or *chorwa*; female utters a descending *uk uk uk-ukukukukukuk*... When alarmed gives a harsh, high-pitched *grrric...grrric* or *brrret...brrret*. **H** Broadleaved evergreen and semi-evergreen forest. **D** Laos, Cambodia, Vietnam.

4 BROWN-HEADED BARBET *Psilopogon zeylanicus* 25–28cm Predominantly green and brown, with pale streaks on head and breast. Actions typical of the genus, usually seen singly or in small groups, with larger groups, mixed with other fruit-eating species, in trees laden with fruit. **V** Monotonous *kutroo kutroo kutroo*... or *kutruk kutruk kutruk*... **H** Broadleaved forests, wooded areas, plantations and trees around habitations. **D** India, Sri Lanka.

5 LINEATED BARBET *Psilopogon lineatus* 25–30cm Variable, dark and light forms shown. Usually seen singly or in small groups; larger groups occur at trees laden with fruit. **V** Monotonous *kotur kotur kotur*... or a trill followed by a long series of *poo-tok* notes. **H** Deciduous forests, open areas with scattered trees, coastal scrub and plantations. **D** NW India to Malay Peninsula.

6 WHITE-CHEEKED BARBET *Psilopogon viridis* 23cm Brown head with distinctive white facial pattern. Usually encountered in pairs or small parties. Often climbs trunks and branches like a woodpecker. **V** Trilling *prrrr-rrr* followed by a series of *tu-kowt* notes leading to shorter *t-kot* notes; also *tot-tot tot-tot*. **H** Forest and woodland, plantations, parks, gardens. **D** W and SW India.

7 GREEN-EARED BARBET *Psilopogon faiostrictus* 24cm Streaked head, yellow-green ear-coverts, grey bill. Forages in the canopy, feeding on fruits such as figs and berries. **V** Loud, rapidly repeated, throaty *took-a-prruk*; also a mellow, rising *pooouk*. **H** Broadleaved evergreen, semi-evergreen and mixed deciduous forest, open areas with scattered trees. **D** SE Asia.

8 BROWN-THROATED BARBET *Psilopogon corvinus* 26cm Brown head. Generally favours the forest canopy, where it forages singly or as part of mixed-species feeding parties. **V** Ringing *hoo-too-too-too too*, *boo toot-boo too-bootootoot* or *too-ta-ta-toot*, sometimes preceded by a fast trill. **H** Moist montane forest. **D** W Java, Indonesia.

9 GOLDEN-WHISKERED BARBET *Psilopogon chrysopogon* 30cm Bright yellow cheek. Forages mainly in the canopy, feeding on fruits. **V** Loud, rapid *too-tuk too-tuk too-tuk*... or *tehoop-tehoop-tehoop*...; also a repeated, low-pitched trill. **H** Broadleaved evergreen forest. **D** Malay Peninsula; Borneo; Sumatra, Indonesia.

10 RED-CROWNED BARBET *Psilopogon rafflesii* 25–27cm Intricate black-yellow-red-blue facial pattern. Forages in the canopy, feeding on fruits and insects. **V** Loud *took* followed by a rapid, repeated series of *tuk* notes. **H** Broadleaved evergreen forests. **D** Malay Peninsula; Borneo; Sumatra, Indonesia.

11 RED-THROATED BARBET *Psilopogon mystacophanos* 23cm Male has yellow, red, black and blue facial patches; female red and blue only. Forages mainly in the forest canopy, although will descend to rummage in dense cover in the understorey. **V** Slow, uneven series of deep notes, *chok... chok-chok... chok... chok-chok... chok*..., also *pooh pooh lentogok lentogok*. **H** Lowland forest. **D** Malay Peninsula; Borneo; Sumatra, Indonesia.

12 BLACK-BANDED BARBET *Psilopogon javensis* 26cm Black facial pattern. Often met with in more open or lightly wooded areas of forest. **V** Variable, including a repeated, ringing *tooloong-tumpook*, an accelerating *too-took too-took too-took*..., and singular *tyaap* or *tap* notes. **H** Lowland and hill forest. **D** Java and Bali, Indonesia.

13 YELLOW-FRONTED BARBET *Psilopogon flavifrons* 21cm Blue face and throat, yellow crown and moustachial stripe. Usually seen in pairs or small parties; in fruiting trees occurs in larger flocks, often accompanied by other fruit-eating species. **V** Ascending, rolling *kowowowowowo* leading to a repeated *kuiar kuiar kuiar*. **H** Hill forests, tree plantations and well-wooded gardens. **D** Sri Lanka.

14 GOLDEN-THROATED BARBET *Psilopogon franklinii* 23cm Black mask, red and yellow crown. Unobtrusive, forages in middle storey and canopy, sometimes descends to understorey. Large groups may gather in fruiting trees. **V** Monotonous, ringing *ki-ti-yook* or *pukwowk*; also a wailing *peeyu peeyu*. **H** Broadleaved evergreen forest. **D** Nepal to N Vietnam and Malay Peninsula.

15 NECKLACED BARBET *Psilopogon auricularis* 22–23cm Previously treated as conspecific with Golden-throated Barbet. Note black 'necklace' framing entirely yellow throat. **V** Repeated, hollow-sounding *k-Wheow, k-Wheow, k-Wheow*... **H** Lowland forest, montane forest on lower slopes. **D** SE Laos, S Vietnam.

16 BLACK-BROWED BARBET *Psilopogon oorti* 21–24cm Multicoloured facial pattern, with black above eye. Active forager in canopy and sub-canopy. **V** Loud, throaty *too-tuk-trrrrrk*, repeated about once a sec. **H** Broadleaved evergreen forest. **D** Malay Peninsula; Sumatra, Indonesia.

17 INDOCHINESE BARBET *Psilopogon annamensis* 21–24cm Sometimes considered conspecific with Black-browed Barbet. **V** Probably similar to Black-browed Barbet. **H** Broadleaved evergreen forest. **D** S Laos, SC Vietnam, E Cambodia.

18 CHINESE BARBET *Psilopogon faber* 21–24cm Crown mostly black. Frequents the middle and upper storey. **V** Hollow *tok-tr-trrrrrt*, repeated about 20 times a minute. **H** Subtropical and tropical forests. **D** S China (including Hainan).

19 TAIWAN BARBET *Psilopogon nuchalis* 21–24cm Pale crown. Active forager in upper and middle storey. **V** Probably similar to Black-browed Barbet. **H** Subtropical forests. **D** Taiwan.

20 BLUE-THROATED BARBET *Psilopogon asiaticus* 23cm Extensive blue throat. Usually encountered in pairs or small parties. Forages in the canopy, descends to understorey for fruits. **V** Rapid, harsh *took-a-rook took-a-rook*. **H** Broadleaved evergreen forest and secondary growth. **D** NE Pakistan to Myanmar, Vietnam, S China.

21 TURQUOISE-THROATED BARBET *Psilopogon chersonesus* 22–23cm Recalls Blue-throated Barbet but blue on face is much paler. Diet includes fruits and invertebrates. **V** Repeated *o-pe-whe-wer, o-pe-whe-wer, o-pe-whe-wer*... **H** Montane forest. **D** S Thailand.

22 MOUNTAIN BARBET *Psilopogon monticola* 20cm Relatively plain face, red streak on back of head. Forages at all levels, often forming part of mixed-species feeding parties. **V** Fast *tuk tuk tuk tuk* with the odd *tu-ruk*. **H** Lower montane and hill forest, forest edges, orchards and gardens. **D** Borneo.

23 MOUSTACHED BARBET *Psilopogon incognitus* 22–23cm Black eye-stripe and moustache. Tends to stay in forest cover. **V** Variable, including a repeated *u ik-a-ruk u ik-a-ruk*..., *tuk-uk* or *tuk-a-tuk* and a *tuk-tuk-trrrr*. **H** Broadleaved woodland. **D** SE Asia.

24 YELLOW-CROWNED BARBET *Psilopogon henricii* 21–23cm Blue throat, dark mask, yellow forehead. Spends much time foraging in the tree canopy, feeding on figs and other fruits. **V** Short, trilled *trrok* followed by a loud *tok-tok-tok-tok*, repeated for 10–20 secs. **H** Lowland and submontane forest. **D** Malay Peninsula; Borneo; Sumatra, Indonesia.

25 FLAME-FRONTED BARBET *Psilopogon armillaris* 20cm Orange forehead and breast-band. Common, often encountered as part of mixed-species feeding parties in fruiting trees. **V** Monotonous, repeated *trrk trrk trrk trrrk* or a series of short trilling notes, *t-t-t-rrrt*. **H** Primary lowland and hill forest, fruiting trees in plantations and gardens. **D** Java and Bali, Indonesia.

26 GOLDEN-NAPED BARBET *Psilopogon pulcherrimus* 20cm Striking blue throat and crown. Forages in the mid-storey and canopy. **V** Variable, including repeated, hollow *took-took-tarrrook*, rolling *trrr-trrr-trrrrrr* and *twaak twaak twaak*. **H** Montane forest. **D** Borneo.

27 YELLOW-EARED BARBET *Psilopogon australis* 16–17cm Yellow ear-coverts. Once considered conspecific with Blue-eared Barbet. Generally seen in pairs or small parties, with larger groups in fruiting trees. **V** Fast, endlessly repeated rattle, also a repeated shrill trill. **H** Primary forest, secondary forest, plantations. **D** Java and Bali, Indonesia.

28 BLUE-EARED BARBET *Psilopogon duvaucelii* 17cm Complex multicoloured facial pattern. Occurs in pairs or small groups, with larger groups in fruiting trees. **V** Endlessly repeated *tk-trrt tk-trrt*, *koo-turr koo-trr* or *too-rook too-rook*; also a whistled *teeow-teeow*... **H** Primary and secondary forest and cultivated areas. **D** NE India to Borneo; Sumatra, Indonesia.

29 BORNEAN BARBET *Psilopogon eximius* 15cm Bright yellow face spot, black throat, red and black crown. Generally forages in the canopy in fruiting trees. **V** Rapid series of *tiuk* notes, also a fast trill. **H** Montane forest. **D** Borneo.

30 CRIMSON-FRONTED BARBET *Psilopogon rubricapillus* 17cm Red forehead, yellow-orange around eye. Usually found in pairs or small parties; bigger groups, mixed with other fruit-eating species, occur in fruit-laden trees. **V** Slow *pop pop pop*..., also a rapid *popo-popo-popo-pop-po*. **H** Open wooded areas. **D** Sri Lanka.

31 MALABAR BARBET *Psilopogon malabaricus* 17cm Red over much of face and throat. Once considered conspecific with Crimson-fronted Barbet. **V** Like Coppersmith Barbet, but softer and quicker, *tik tik tik tik tik*..., *tunk tunk tunk tunk*... or *poop poop poop poop*... **H** Moist evergreen forest; also fig trees in coffee plantations. **D** SW India.

32 COPPERSMITH BARBET *Psilopogon haemacephalus* 17cm Streaked underparts, patterned face. Generally seen in pairs or small parties, with larger parties occurring in fruiting trees. **V** Monotonous, metallic *tuk tuk tuk tuk tuk tuk tuk*... **H** Open lowland forest and forest edges. **D** Widespread, NE Pakistan to Indonesia, Philippines.

33 BROWN BARBET *Caloramphus fuliginosus* 17cm Distinguished from other Asian Barbets by colour; from Sooty Barbet by range. Often encountered in groups, foraging on tree trunks and branches; calls constantly. **V** High-pitched, whistled squeak. **H** Primary and secondary forest. **D** Borneo.

34 SOOTY BARBET *Caloramphus hayii* 17–20cm Regularly found in small groups feeding from the understorey to the canopy, especially in fruiting trees; often forages acrobatically, like a tit. **V** Sibilant series of *pseeee* calls, also utters a thin *pseeoo*. **H** Lowland and swamp forest. **D** Malay Peninsula; Sumatra, Indonesia.

AFRICAN BARBETS LYBIIDAE

35 GREY-THROATED BARBET *Gymnobucco bonapartei* 18cm Pale bristle in nostrils diagnostic. Note black bill, grey head. **V** Very high, rapid, sharp chatters and trills. **H** Forest glades. **D** W and C Africa: Cameroon to W Kenya and N Tanzania.

36 SLADEN'S BARBET *Gymnobucco sladeni* 17cm Separated from Grey-throated Barbet where in the same range by dark eyes and naked head. **V** Calls include soft *piu* notes and various chatters and rattles. **H** Lowland forest. **D** DR Congo, S Central African Republic.

37 BRISTLE-NOSED BARBET *Gymnobucco peli* 17cm Very similar to Naked-faced Barbet, sometimes nesting in the same tree, but with different bill colour. **V** Five-second series of hooting *pyew* notes, similar to that of Naked-faced Barbet but higher-pitched. **H** Forest. **D** W Africa: Sierra Leone to Angola.

38 NAKED-FACED BARBET *Gymnobucco calvus* 16cm Blackish naked face. From Bristle-nosed Barbet by more yellow, less horn-coloured bill, white chin, yellow hair tuft under bill, not in nostril, darker rump. From Sladen's Barbet by pale bill, slightly paler underparts, which contrast more with dark wing. Breeds in colonies, often together with Bristle-nosed Barbet and Grey-throated Barbet. **V** Mid-high, hesitating, bouncing, accelerated *hoot-hoot-hootootoot*. **H** Forests. **D** W Africa: Sierra Leone to Angola.

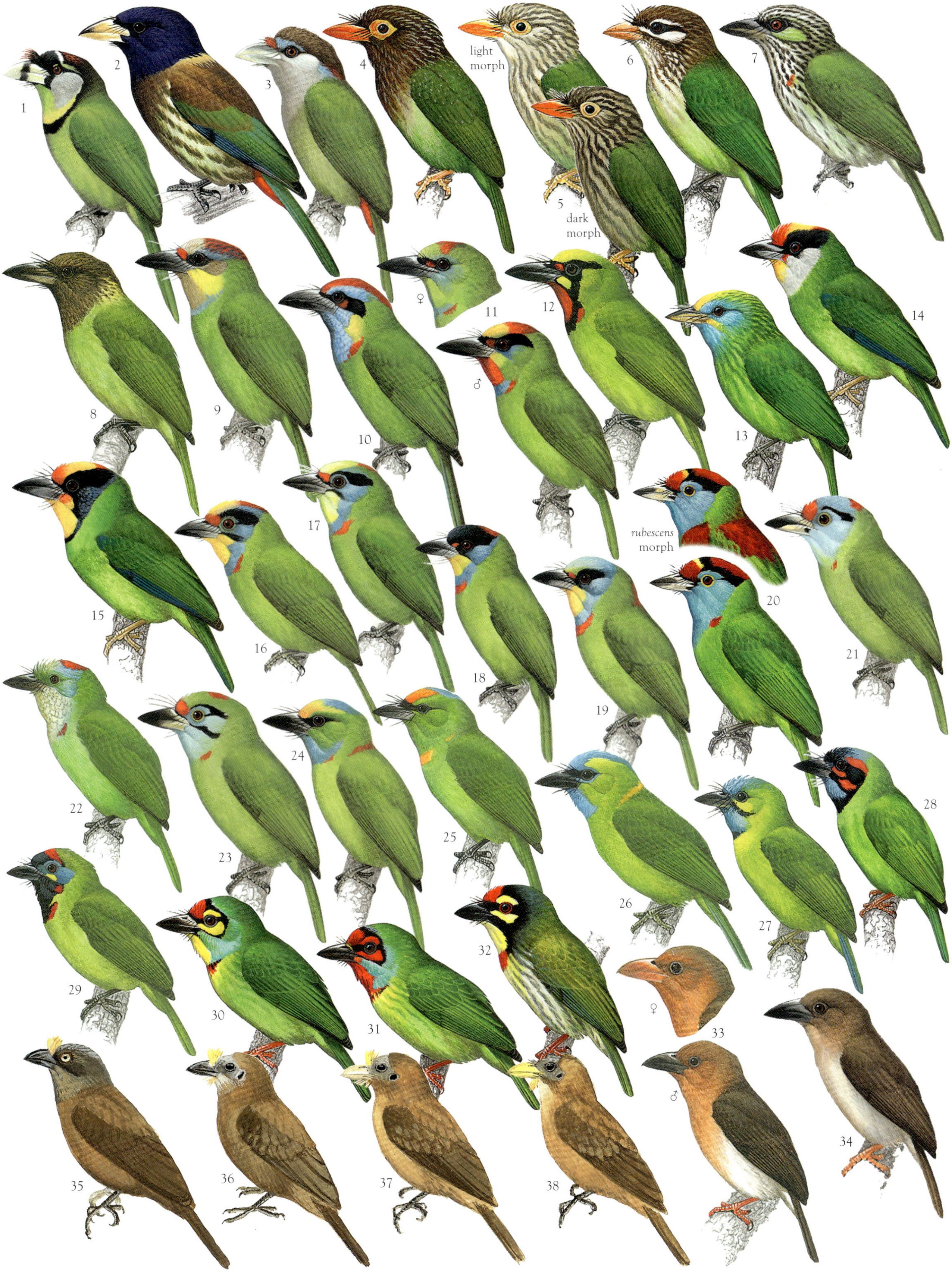

1
2
3
4
light morph
5
dark morph
6
7
8
9
10
♀
11
12
♂
13
14
15
16
17
18
19
rubescens morph
20
21
22
23
24
25
26
27
28
29
30
31
32
♀
33
♂
34
35
36
37
38

AFRICAN BARBETS *CONTINUED*

1 WHITE-EARED BARBET *Stactolaema leucotis* 17cm Note prominent white ear streak and belly. **V** Very high, shrieking *pee-pee-pee-pik-pik-pik-pik-per*, rather irregular in pitch. **H** Mainly in large forest trees. **D** E and SE Africa: Kenya to South Africa.

2 WHYTE'S BARBET *Stactolaema whytii* 18cm From Anchieta's Barbet by smaller white-yellow area to forehead, darker wing and white-streaked underparts. Double white wing patch diagnostic. **V** High hooted sighing, *hoo-hoo-hoo*. **H** Miombo woodland. **D** Tanzania, Zambia, Malawi, Mozambique, Zimbabwe.

3 ANCHIETA'S BARBET *Stactolaema anchietae* 17cm Note yellow face and single white wing patch. **V** Song is a series of *hoo* notes given at variable pace; calls include soft *kuut*, shrieking *keeew*, and various chatters, grating notes, and mechanical sounds. **H** Open woodland, riverine belts. **D** Angola, DR Congo, Zambia.

4 GREEN BARBET *Stactolaema olivacea* 17cm Note green wings and tail. **V** High, wooden, unhurried, slightly accelerated *tjip tjip tjip tjip tjip tjip-tjip-tjip tjip tjip*. **H** Canopy of forests, dense woodland. **D** E Africa: Kenya to Mozambique.

5 SPECKLED TINKERBIRD *Pogoniulus scolopaceus* 11cm Note pale eyes and chequered colouring. **V** Song is a series of *tikikik* notes; other vocalisations include *tok-tok*, piping trills, and nasal buzzes. **H** Forest edges and glades. **D** W and C Africa: Sierra Leone to Angola and W Kenya.

6 GREEN TINKERBIRD *Pogoniulus simplex* 8cm Differentiated from Moustached Tinkerbird by white faintly streaked forehead, green (not dusky) cheek, absence of moustache. **V** High, slow-trilled *prrruh prrruh prrruh*. **H** Dense forest, sometimes woodland. **D** E Africa: Uganda to Mozambique.

7 MOUSTACHED TINKERBIRD *Pogoniulus leucomystax* 9cm Differentiated from Green Tinkerbird by white moustache, plain forehead, dusky cheek. **V** High, loud, sharp, speedy *widu-duwi widi-diwi widuwi-*, sometimes uttered as rattles. **H** Forested slopes. **D** E Africa: E Uganda and W Kenya to Malawi.

8 WESTERN TINKERBIRD *Pogoniulus coryphaea* 9–10cm Yellow on crown, back and wings; white moustache. Diet comprises mainly fruit but some invertebrates. **V** Hollow-sounding, piping *wch-ch-ch-ch-ch-ch-ch-ch*. **H** Montane forested habitats. **D** Disjunct populations: Cameroon and Nigeria; DR Congo, Rwanda and Uganda; Angola.

9 RED-RUMPED TINKERBIRD *Pogoniulus atroflavus* 12cm Note unique pattern of white and yellow lines on face sides. No other tinkerbird with red rump in range. **V** Low, monotonous, hollow *hoot hoot hoot* or high *trri trri trri*. **H** Forest. **D** S Senegal to Uganda and DR Congo.

10 YELLOW-THROATED TINKERBIRD *Pogoniulus subsulphureus* 9.5cm Differs from Yellow-rumped Tinkerbird mainly by voice. **V** Very high, sustained, hurried *puck-puck-puck-* (faster than Red-rumped Tinkerbird; a *puck* in every 2–7 is left out). **H** Montane forest clearings. **D** Sierra Leone to Uganda and DR Congo.

11 YELLOW-RUMPED TINKERBIRD *Pogoniulus bilineatus* 10cm Bold white stripes on face, yellow-edged wing feathers, yellow rump. **V** Regular, slow or hurried, wooden *pook-pook-pook* (a *pook* in every 3–7 is often left out) or mid-high, hooting, trilled *prrru-prrru-prrr*. **H** Forest, woodland, suburbs. **D** W, C and SE Africa.

12 WHITE-CHESTED TINKERBIRD *Pogoniulus makawai* 11cm Black chin and absence of white eyebrow diagnostic. **V** Not known. **H** Dry forest. **D** NW Zambia.

13 RED-FRONTED TINKERBIRD *Pogoniulus pusillus* 9cm Distinguished from Yellow-fronted Tinkerbird by deeper yellow wing coverts. **V** Extremely high, irregular, rapid *bik-bik-bik-bik* (higher-pitched than Yellow-fronted Tinkerbird) and mid-high, monotonous *buk-buk-buk*. **H** Forest edges, woodland, gardens. In moister areas than Yellow-fronted Tinkerbird. **D** E and SE Africa, from Eritrea to South Africa.

14 YELLOW-FRONTED TINKERBIRD *Pogoniulus chrysoconus* 9cm Distinguished from Red-fronted Tinkerbird by faint (not bright) yellow wings. Yellow-fronted form distinctive; orange-fronted form difficult to distinguish from Red-fronted Tinkerbird. **V** High, regular, uninterrupted *puk-puk-puk*, sustained for very long periods with mid-high, mewing *neh-neh-neh-neh-neh* in duet. **H** Woodland and more or less wooded, cultivated areas. **D** Widespread across sub-Saharan Africa.

15 YELLOW-SPOTTED BARBET *Buccanodon duchaillui* 16cm Note red front and long, yellow eyebrow. **V** High, owl-like, hollow *rruu* and very high, loud, shivering, fast *bibibiwi bibibiwi*. **H** Canopy of montane forest. **D** W and C Africa, from Sierra Leone to W Kenya.

16 HAIRY-BREASTED BARBET *Tricholaema hirsuta* 15cm Striped crown browner and yellow-spotted in one race. **V** Mid-high, regular, fast, rapid or slow *hoothoothoothoot*. **H** Forests and nearby areas. **D** W and C Africa, from Sierra Leone to N Angola and N Tanzania.

17 RED-FRONTED BARBET *Tricholaema diademata* 13cm Note plain underparts and yellow and white eyebrow. **V** Song is a series of *hoop* notes delivered at variable pace, also incorporating *nyah* notes; calls include raspy *charr* and higher *tik-tik*. **H** Dry open wood- and bushland with dense thickets, sometimes along streams. **D** E Africa, from Eritrea to Tanzania.

18 MIOMBO PIED BARBET *Tricholaema frontata* 15cm Distinguished from Red-fronted and Yellow-fronted Tinkerbirds by spotted underparts. **V** Song consists of 6–30 *hoop* notes; secondary song a series of *nyeh* notes. Calls include grating and yelping notes. **H** Miombo woodland, grassland with some trees, other more or less open woodland. **D** Angola to Tanzania and Malawi.

19 ACACIA PIED BARBET *Tricholaema leucomelas* 15cm Note red forehead and striking facial pattern. From Tinkerbirds by larger size and black throat. **V** High, hooting, rapid *uhu-hu-hu-hu-huh*. **H** Dry thornveld, bushveld, often along (dry) streams. **D** From Angola, Zimbabwe and Mozambique to South Africa.

20 SPOT-FLANKED BARBET *Tricholaema lacrymosa* 13cm Distinguished from Miombo Pied Barbet, Acacia Pied Barbet and several tinkerbirds by pale eyes (male), lack of red on forehead, unspotted upperparts; only edges of flight feathers and tips of greater coverts are yellow, and flanks boldly spotted. **V** Song series of *hoop* notes, one or two per sec, also incorporating *nya* notes. Calls include grating notes; also makes mechanical sounds (wing-rustling, bill-wiping). **H** Woodland, bushveld, forest patches. **D** E Africa, from Uganda to Tanzania.

21 BLACK-THROATED BARBET *Tricholaema melanocephala* 13cm Striking black-and-white head pattern, yellow-fringed wing feathers. **V** Highly vocal. Song is a series of *kaaa* or *ptaw* notes; variety of calls include chatters, rattles, and single *kaaa* and *ptaw* notes. **H** Arid and dry areas with trees, bush, scrub and grass cover, often near water. **D** E Africa, from Eritrea to Tanzania.

22 BANDED BARBET *Lybius undatus* 15cm Heavily barred. Head mostly black, with red forehead. **V** Little known. Repeated *tock* notes are probable song; calls include *gr-gr-grgrgr*. **H** Woodland and scrubland, often near water. **D** Eritrea, Ethiopia.

23 VIEILLOT'S BARBET *Lybius vieilloti* 15cm Differentiated from Black-collared Barbet by different range and far less black around red face mask. **V** Strange, rather low piping, scops-owl-like *hoot-hoot hoot hoot-koot-hoot-hoot*, magpie-like chatters, high goose-like shouts. **H** Dry, more or less wooded and scrubbed, natural and cultivated areas. **D** From Mauritania to Cameroon, Sudan, Eritrea.

24 WHITE-HEADED BARBET *Lybius leucocephalus* 15cm Unmistakable, though variable in extent of black/brown and white in plumage. **V** High loud sharp chattering shrieks *peek-peek-peek-peek* or tinkling chattered *pi-pi-pi-pi-peh*. **H** Dry open woodland, bush, gardens, cultivation, often near water. **D** From Nigeria to South Sudan, and from Kenya to Angola.

25 CHAPLIN'S BARBET *Lybius chaplini* 15cm Note white thighs and black bill. Distinguished from White-headed Barbet by red forehead and white edge of mantle. **V** Members of group greet one another with buzzing, grating notes that build into a noisy, cackling song between two or three birds; calls include squawks and softer *wup* notes. **H** Cultivated and natural areas with a few trees, open woodland. **D** Zambia.

26 RED-FACED BARBET *Lybius rubrifacies* 15cm Mostly dark, with red face. Note brown unmarked wing coverts and chin. **V** Group greeting involves chatters and grating notes, which may develop into a duetting song of *kkaa* and *ko* notes, also with many mechanical sounds. **H** More or less wooded, natural and cultivated habitats. **D** Uganda, Rwanda, Burundi, NW Tanzania.

27 BLACK-BILLED BARBET *Lybius guifsobalito* 15cm Note striped wing coverts and red head, including chin. **V** Group greetings involve complex, sex-specific songs of chattering and grating notes. Calls include raspy *kek* and chatters as well as mechanical sounds. **H** Open, more or less wooded and bushed, natural and cultivated areas, mostly at higher elevations. **D** Eritrea and Ethiopia to NE DR Congo, Uganda, W Kenya, NW Tanzania.

28 BLACK-COLLARED BARBET *Lybius torquatus* 20cm Unmistakable, with pale belly. Not in same range as Viellot's Barbet and Black-billed Barbet. **V** Mid-high, miaowing, hurried *tjauw-tjauw-tjauw* or high loud bouncing *beedidder* often given in duet. **H** Woodland, grass- and farmland with scattered trees. **D** From South Africa N to Angola and Kenya.

29 BROWN-BREASTED BARBET *Lybius melanopterus* 15cm Note black thighs and pale bill. **V** Groups produce loud *nyekk* and dry *skzzzz* calls, also loud wing-rustling and bill-wiping. **H** Forest edges, cultivation with trees. **D** E Africa, from Somalia to Mozambique.

30 BLACK-BACKED BARBET *Lybius minor* 15cm Western populations have brown upperparts, while those further east have contrasting darker upperparts and white underparts. Note pale bill, red crown, pinkish belly with black belt. **V** Song consists of buzzing, trilled *kriiiiiii* notes. Calls include short *krek* and various screeching and hollow notes. **H** Woodland, bush and scrub, farmland with trees and hedges. **D** From Gabon to Angola, Tanzania, Zambia, Malawi.

31 DOUBLE-TOOTHED BARBET *Lybius bidentatus* 25cm Separated from Bearded Barbet by pink wing-bar, absence of black breast band, more extensive red underparts. **V** Mid-high, tinkerbird-like, sustained *pook-pook-pook*. **H** Understorey at forest edges, riverine belts, cultivation, more or less wooded natural and cultivated grasslands. **D** W and C Africa: Guinea-Bissau to Ethiopia, Angola to N Tanzania.

32 BEARDED BARBET *Lybius dubius* 26cm Separated from Double-toothed Barbet by black breast band and absence of wing-bar. Range generally just north of Double-toothed Barbet. **V** Calls include a rather grating *krawk* and a whirring *hurr-hurr*. **H** As Double-toothed Barbet, but more in upper storeys. **D** Senegal to Central African Republic.

33 BLACK-BREASTED BARBET *Lybius rolleti* 30cm White bill. Red restricted to centre of belly. **V** Little known; grating calls and loud wing-rustling sounds reported. **H** Dry open woodland, plantations and farmland with scattered trees, normally at higher elevations. **D** S Chad and SW Sudan to N Central African Republic, South Sudan, N Uganda.

34 YELLOW-BILLED BARBET *Trachyphonus purpuratus* 25cm Often perches quietly in un-barbet-like, upright stance. **V** Mid-high or very high, monotonous *hoot hoot* (2 secs between each *hoot*). **H** Dense forest undergrowth, often at edges and glades, near streams. **D** W and C Africa: Sierra Leone to Angola and Kenya.

35 CRESTED BARBET *Trachyphonus vaillantii* 23cm Unmistakable, with black-and-white wings and tail combined with yellow-orange head and underparts. **V** Very high, coppery, long trill, *prrrrrrrruh*, and similar, slightly hoarse, fluted *wuut-wutwut-wut-wut-wut*. **H** Dry, open woodland and bushland, riverine belts, even suburbs. **D** S and SE Africa, from Angola and Tanzania to South Africa.

36 RED-AND-YELLOW BARBET *Trachyphonus erythrocephalus* 25cm Note red head, orange-red bill and yellow underparts. **V** Mid-high, excited, descending, cackling *kehkerehkQkehkerehkQ-* in duet. **H** Dry, open, more or less wooded and bushed areas with some scrub and grass cover. **D** E Africa, from Ethiopia to Tanzania.

37 YELLOW-BREASTED BARBET *Trachyphonus margaritatus* 21–22cm Yellow head, black crown, reddish bill. Feeds on berries, fruit and invertebrates. **V** Trilling, cricket-like *tRrrrp, tRrrrp, tRrrrp...* **H** Acacia woodland, riverine scrub, savannah. **D** Broad south-Saharan range, from Mauritania and Mali to Sudan and Ethiopia.

38 D'ARNAUD'S BARBET *Trachyphonus darnaudii* 15cm Note red edge of uppertail coverts. **V** High, excited, up-and-down, rapid, cackling *keckkeckerec-keckeckerec* in duet. **H** Dry, open wood-, bush- and scrubland with some grass cover. **D** E Africa, from South Sudan to Tanzania.

1
2
3
4
5
6
7
8
9
10
11
12
13
14
orange-fronted form
yellow-fronted form
15
16
17
18
19
20
21
22
23
24
25
26
27
28
29
30
31
32
33
34
35
♂
♀
36
♀
♂
37
♀
♂
38

HONEYGUIDES INDICATORIDAE

1 CASSIN'S HONEYBIRD *Prodotiscus insignis* 9cm Differentiated from Brown-backed Honeybird by grey chin and by tail pattern. Like all honeybirds and honeyguides, rather pot-bellied and with white in tail normally concealed (but often visible in flight). Note short bill. **V** Little known; calls include chattering notes and a weak *ski-a* or *whi-hihi*. **H** Treetops in forests and surrounding areas. **D** W and C Africa: Sierra Leone to Angola and Kenya.

2 GREEN-BACKED HONEYBIRD *Prodotiscus zambesiae* 9cm Separated from Cassin's Honeybird by paler belly, thin white eye-ring and different range. May have small dusky tips to white tail feathers. **V** In display flight (singly or in pairs) gives series of falling *sch-sch* notes; other calls include chattering notes. **H** Forest edges, woodland. **D** E and S Africa, from Ethiopia to Angola and Mozambique.

3 BROWN-BACKED HONEYBIRD *Prodotiscus regulus* 13cm Note pale chin, dusky colouring and blackish T-pattern on tail. **V** Song is a short trill, also *zzzzz* notes in display flight; other vocalisations include *tsip* and *zeet-zeet*. **H** Woodland, plantations, cultivation. **D** Widespread in sub-Saharan Africa.

4 ZENKER'S HONEYGUIDE *Melignomon zenkeri* 13cm Distinguished from Yellow-footed Honeyguide by less yellow feet and bill, darker underparts, less white in tail and less bright upperparts. **V** Little known. Possible song is a long series of *psee* and louder *pseep* notes, becoming much softer at end. **H** Forests. **D** S Cameroon and N Gabon to W Uganda and E DR Congo.

5 YELLOW-FOOTED HONEYGUIDE *Melignomon eisentrauti* 13cm Olive-green head and upperparts, pale grey underparts, yellow bill and legs. **V** Song is a series of about 13 emphatic, strident *tuu-i tuu-i* notes; also a series of *wreew* notes. **H** Forests. **D** Sierra Leone, Liberia, S Ghana, SW Cameroon.

6 DWARF HONEYGUIDE *Indicator pumilio* 10cm Striking white loral spot, greenish crown and nape, white moustachial stripe. Flank striping faint. **V** Only known sound is a *tuutwi* call. **H** Moist montane forests. **D** E DR Congo, W Uganda, Rwanda, Burundi.

7 WILLCOCKS'S HONEYGUIDE *Indicator willcocksi* 11cm Faint malar stripe and no white between eye and bill. Note pale olive-green (not grey or brown) underparts and narrow dark stripes to lower flanks. **V** Extra-high, fluting *uweew-uweew-uweew* (each *uweew* slightly swept up). **H** Montane forests and connected well-wooded streams. **D** W and C Africa: Sierra Leone to Uganda.

8 PALLID HONEYGUIDE *Indicator meliphilus* 10cm Note plain golden-olive upperparts, pale loral spot, absence of moustachial stripe; shows faint or no streaks on lower flanks. Paler below and with less white in tail than most other honeyguides. **V** Calls from favourite tree, changing perches. **H** Miombo woodland, thorn scrub. **D** Uganda to Tanzania, Angola to Mozambique.

9 LEAST HONEYGUIDE *Indicator exilis* 11cm Note dark-centred, golden feathers of upperparts, moustachial stripe and streaks on flanks. **V** Extremely high, sharp *weew-weew-weew*. **H** Forest. **D** Senegal to Tanzania.

10 THICK-BILLED HONEYGUIDE *Indicator conirostris* 15cm Overall rather dark. No or faint loral spot. Faint black striping to lower flanks. **V** Song is a loud, far-carrying series of 10–30 *wit* notes. **H** Dense forests and connected well-wooded streams, normally at higher elevations. **D** Sierra Leone to Kenya.

11 LESSER HONEYGUIDE *Indicator minor* 15cm Note narrow dark moustachial stripe, sharply demarcated, grey cheek, tail pattern. **V** Very high, level, staccato *weew-weew-weew*. **H** Forest edges, wood- and scrubland, cultivation, grassland with scattered trees. **D** Widespread across sub-Saharan Africa.

12 SPOTTED HONEYGUIDE *Indicator maculatus* 18cm Striking white tail feathers especially visible in flight. Note plain forehead and crown, pale spotting and striping on underparts from chin to undertail coverts. **V** Little known; song is a fast 3-sec trill, also chatters and makes wing-rustling sounds. **H** Forest and surrounding areas. **D** Gambia to SW Sudan and N Angola.

13 SCALY-THROATED HONEYGUIDE *Indicator variegatus* 19cm Differentiated from Spotted Honeyguide by fine white streaks on head, whitish unstreaked belly and finely streaked undertail coverts. **V** Very high rather short drawn-up trill *rrrrrrruh* at 1-minute intervals. **H** Forest, woodland, wooded grassland. **D** E Africa, from Ethiopia and Somalia to E South Africa; also W through DR Congo and Zambia to Angola.

14 YELLOW-RUMPED HONEYGUIDE *Indicator xanthonotus* 15cm Dark brown, with yellow on head and rump. Feeds mainly on beeswax. May gather in numbers at exposed bees' nests. **V** Various calls, including a quiet *weet*, a chipping *tzt* and a *chaenp-chaenp*. **H** Mixed forests and wooded gorges. **D** NW India to N Myanmar.

15 MALAYSIAN HONEYGUIDE *Indicator archipelagicus* 16–18cm May sit motionless for long periods. Frequently seen in the vicinity of bees' nests. Olive-brown above and pale below. Females lack the small yellow shoulder patch. **V** Mewing, followed by an ascending rattle. **H** Broadleaved evergreen forests; occasionally in plantations and gardens. **D** Malay Peninsula; Borneo; Sumatra, Indonesia.

16 GREATER HONEYGUIDE *Indicator indicator* 20cm Note pale cheek and pink bill of male, yellow upperwing coverts and white-streaked rumps of male and female in flight. **V** Very high *weet-thrrree*, endlessly repeated at 1 minute intervals. **H** Well-wooded areas. **D** Widespread across sub-Saharan Africa.

17 LYRE-TAILED HONEYGUIDE *Melichneutes robustus* 20cm Unmistakable, when properly seen. **V** Soft extreme high shrill *feefeefeefeefeefeefee...*; also a mechanical 'song', produced by vibrations of its tail, sounding as a loud sequence of bleating toy trumpet-like *meheh-meheh-*, slightly speeded-up as the bird flies around over the forest canopy. **H** Forest, plantations. **D** Sierra Leone to Cameroon, DR Congo, W Uganda.

WOODPECKERS PICIDAE

18 EURASIAN WRYNECK *Jynx torquilla* 16–17cm Mottled shades of pale brown and grey with rufous and blackish bars and streaks. Feeds mainly on the ground. **V** Plaintive *quee-quee-quee-quee*. When alarmed utters a hard *teck*. **H** Breeds in open forests; post breeding found in open scrub and cultivations. **D** Breeds from SW Europe through Asia to NE China. Winters in sub-Saharan Africa, S Asia, SE Asia.

19 RED-THROATED WRYNECK *Jynx ruficollis* 19cm Distinctive rufous throat. Perches crosswise like Eurasian Wryneck. Forages more on ground than on tree trunks and branches. **V** High bleating *pjuuuut-pjuuut-pjuuut*. **H** Open more or less wooded natural and cultivated areas. **D** Sub-Saharan Africa, from Cameroon to Ethiopia and as far south as E South Africa.

20 SPECKLED PICULET *Picumnus innominatus* 10cm Olive-yellow wings, brown and white striped head, black-spotted breast and belly. Agile forager, hanging upside-down to poke in crevices, and flying or hovering in pursuit of flushed prey. **V** High-pitched *ti-ti-ti-ti-ti* and a squeaky *sik-sik-sik*. Drums on bamboo or dead branch with a persistent *brr-r-r brr-r-r...* **H** Bamboo and low bushes in moist deciduous and semi-evergreen forest. **D** From NE Afghanistan and Himalayas to India, China, SE Asia, Greater Sunda Islands.

21 BAR-BREASTED PICULET *Picumnus aurifrons* 8cm Underparts yellowish white, barred blackish on chest, streaked blackish below. Dark chocolate white-speckled cap, with yellow spots on forehead of male. **V** Calls include a very high, thin, hummingbird-like *tsirrit-sit-sit*. **H** Rainforest, várzea, terra firme forest, copses, bushes and thickets. **D** Amazonia.

22 LAFRESNAYE'S PICULET *Picumnus lafresnayi* 9cm Very like Orinoco Piculet but forehead of male dotted red and upperparts more noticeably barred. **V** Call, rarely heard, is a very high, thin *tseeyt* or *tsit*. **H** Very humid rainforest, secondary growth, clearings, thickets. **D** W and C Amazonia: Colombia, Ecuador, Peru, NW Brazil.

23 ORINOCO PICULET *Picumnus pumilus* 9cm Dark chocolate-brown cap with tiny white spots. Yellow spots on male's forehead. **V** Not known. **H** Humid forest, edge of gallery forest, dense thickets, savannah. **D** NW Amazonia: Colombia, Venezuela.

24 GOLDEN-SPANGLED PICULET *Picumnus exilis* 9cm Feathers of upperparts spangle-tipped black and yellow. Cap black with white spots. Red spots on male's forehead. **V** Call is a very high, thin *tsilit* or *tsirrr*. **H** Rainforest, cloud forest, savannah, open woods, mangroves, dense riverine vegetation with cane, regrowth. **D** Venezuela to E Brazil.

25 BLACK-DOTTED PICULET *Picumnus nigropunctatus* 9cm Very like Golden-spangled Piculet but lower breast and belly dotted black. May be a race of Golden-spangled Piculet. **V** Like Scaled Piculet. **H** Humid lowland tropical forest, várzea, terra firme forest, secondary growth, bushes and thickets, woodland. **D** NE Venezuela.

26 ECUADORIAN PICULET *Picumnus sclateri* 9cm Throat and breast barred, belly streaked, undertail barred, all black on white and contrasting. Black white-spotted cap with short yellow streaks on forehead of male. **V** Usually quiet. Possible song is a series of up to seven slow *swee* notes, falling in pitch. Calls include a high, thin *tseet* or *tseet-tseet* and stuttering notes. **H** Woods and edges, scrub with cactus, bushes and thickets in arid regions, more humid montane forest. **D** Ecuador, NW Peru.

27 SCALED PICULET *Picumnus squamulatus* 9cm Underparts white, feathers edged blackish, giving a scaly effect. Black cap with white spots, and very fine red streaks on male's forehead. **V** Call is a high, squeaky *chi-chi-chi-ch-e-chi*, last notes trilling. **H** Gallery forest, edge, secondary growth, deciduous woods, scattered trees, bushes, thickets. **D** Colombia, Venezuela.

28 WHITE-BELLIED PICULET *Picumnus spilogaster* 9cm Dirty white underparts, whitish face, brownish ear-coverts. Olivaceous-brown upperparts. Black cap with white spots. Male has dense red dotting on forecrown. **V** Song is a thin, high-pitched trill lasting 3 secs. **H** Rainforest, gallery forest, deciduous forest, bushes, thickets, scattered trees, mangroves. **D** Venezuela, the Guianas, NE Brazil.

29 ARROWHEAD PICULET *Picumnus minutissimus* 10cm Brown feathers of upperparts bordered buffy whitish. Underparts white, feathers finely edged blackish, giving a scaly effect. Black cap with white spots. Red forecrown in male. **V** Series of about 14 high, thin *ki ki ki* notes. **H** Secondary forest, vegetation near rivers and lagoons, plantations, montane forest, mangroves. **D** The Guianas.

30 SPOTTED PICULET *Picumnus pygmaeus* 10cm Chestnut-brown with white spots. Male has red forehead. **V** Call very high-pitched, squeaky *tsirrrr, tsi tsi tsi*. **H** Dry, open woods, Caatinga, thickets. **D** E Brazil.

31 SPECKLE-CHESTED PICULET *Picumnus steindachneri* 8.5cm Separated from other piculets in the region by dark breast marked with white spots and heavily barred belly. Male has red streaks on forecrown. **V** Song high-pitched, rapid trill that falls in pitch. **H** Montane rainforest, bushes and thickets, canebrakes. **D** Andes of N Peru.

32 VARZEA PICULET *Picumnus varzeae* 10cm Overall chestnut-brown, scaled blackish and whitish on cheeks, throat, neck, breast and flanks, fainter on mid-belly and undertail coverts. **V** Very high, thin trilling phrase that falls in pitch. **H** Várzea forest, forest on river islands, bushes and thickets. **D** C Amazonia.

33 WHITE-BARRED PICULET *Picumnus cirratus* 9cm Face and underparts barred blackish and white. **V** Call long, wavering high *tsirrrrr*, also *tsick, tsick*. **H** Woods, transition to arid zones in low mountains and lowlands, copses, scattered trees, gallery forest, savannah, bushes, thickets. **D** The Guianas, E Brazil, Bolivia to N Argentina.

34 OCELLATED PICULET *Picumnus dorbignyanus* 9cm Underparts whitish with black markings, arrow-pointed on breast. **V** High-pitched, fast trill that falls away in volume and pitch. **H** Humid montane forest, transition forest, tropical and subtropical regions. **D** E Peru, W Bolivia.

35 OCHRE-COLLARED PICULET *Picumnus temminckii* 9cm Face and hindneck ochre. Underparts strongly barred black and white. **V** High-pitched whistle. Drumming loud, in short bursts. **H** Humid forest, edge, tall canes and thickets, regrowth. **D** SE Brazil, E Paraguay, NE Argentina.

36 WHITE-WEDGED PICULET *Picumnus albosquamatus* 10cm Like Arrowhead Piculet but belly buffier, and breast whiter with feathers edged black, leaving white wedges, and below edged brown, giving a scaly effect. **V** A 1–2-sec rapid trill, falling in pitch. **H** Forest, gallery forest, edge, thickets, bush, savannah. **D** Bolivia, Paraguay, S Brazil.

37 RUSTY-NECKED PICULET *Picumnus fuscus* 9.5cm Underparts tawny buff, slightly scaled, paler centrally and on undertail coverts. Eye-ring and legs salmon-pink. Cap black. Male and female very similar, except for the orange-reddish feathers in the male's crown. **V** Not known. **H** Lowland riverine forest, várzea. **D** Very restricted range in NE Bolivia, WC Brazil.

38 RUFOUS-BREASTED PICULET *Picumnus rufiventris* 10cm Reddish chestnut with olive-brown back. Eye-ring grey. **V** Song is a slow, descending series of high-pitched *seep* notes. Calls include single *seep* and in flight fast *tsit-tsit*. **H** Terra firme forest and várzea, woods, secondary growth, scattered trees, clearings, bushes and thickets, climbers and creepers, canebrakes. **D** W Amazonia, from Colombia to Bolivia.

1
2
3
4
5
6
7
8
9
10
11
12
13
14
15
16 ♀
17
18
19
20 ♂ ♀
21 ♀ ♂
22 ♀ ♂
23
24 ♀ ♂
25 ♀ ♂
26 ♂
27 ♀ ♂
28 ♀ ♂
29 ♀ ♂
30 ♀ ♂
31 ♀ ♂
32 ♀ ♂
33 ♀ ♂
34 ♀ ♂
35 ♀ ♂
36 ♀ ♂
37 ♀ ♂
38 ♀ ♂

WOODPECKERS *CONTINUED*

1 OCHRACEOUS PICULET *Picumnus limae* 9cm Underparts greyish buff, slightly streaked paler on throat and breast. Back greyish brown. Black cap spotted white. Males have red forehead. **V** Thin, high *seeer, seeer* like a hummingbird's. **H** Semi-deciduous forest, Caatinga, in bushes, thickets and stunted trees, canes. **D** NE Brazil.

2 TAWNY PICULET *Picumnus fulvescens* 10cm Resembles Ochraceous Piculet, but underparts tawny with faint whitish streaks, and cheeks, hindneck and upperparts brown. **V** Song is a descending, very rapid series of high-pitched *driee* notes. **H** Atlantic forest, secondary thickets, palm groves, Caatinga. **D** E Brazil.

3 MOTTLED PICULET *Picumnus nebulosus* 10cm Ear-coverts and pectoral collar tawny brown, back brown. Malar region and throat whitish barred blackish. Underparts tawny buff, heavily streaked black. **V** Not well known. A humming *tsewrewt, sisisi* phrase noted, also single squeaking calls from begging juveniles; drums loudly with short pause after 2–4 rolls. **H** Woods, copses, savannah, edge, bushes, thickets, canebrakes. **D** SE Brazil, NE Argentina, N Uruguay.

4 PLAIN-BREASTED PICULET *Picumnus castelnau* 9.5cm Underparts dirty whitish, upperparts olivaceous brown. Hindneck barred whitish and blackish. Plain black cap (no white), forecrown dotted red in male. **V** Rapid, descending series of high-pitched notes. **H** Várzea, low islands, swampy forest, secondary growth, bushes, thickets. **D** W Amazonia.

5 FINE-BARRED PICULET *Picumnus subtilis* 9.5cm Like Plain-breasted Piculet, but with white spotting on crown. Underparts dirty whitish, faintly barred greyish on sides of breast. **V** Rapid series of sharp, high-pitched notes, falling in pitch. **H** Humid tropical forest; lowlands to foothills. **D** E Peru.

6 OLIVACEOUS PICULET *Picumnus olivaceus* 9cm Back and pectoral collar olive-brown. Cheeks, hindneck and throat cinnamon-buff scaled brown. Underparts yellowish cream streaked blackish brown. **V** Fast, soft trilling twitter, also sharp *sst, siip ssip*. Chicks in nest make a continuous buzzing sound. **H** Rainforest, cloud forest, open woods, deciduous forest, edge, scrub, cultivated land. **D** Guatemala to W Venezuela and NW Peru.

7 GREYISH PICULET *Picumnus granadensis* 9cm Dirty white underparts with very faint darker streaking. Upperparts and ear-coverts olivaceous brown. Cap has white dots, with yellow on forecrown of male. **V** Weak trilling phrase, without shifting in pitch. **H** Deciduous forest, rainforest edge, secondary growth, scrub, open woods. **D** NW Colombia.

8 CHESTNUT PICULET *Picumnus cinnamomeus* 10cm Overall reddish chestnut. Forehead and lores white. Pink eye-ring and legs. Cap speckled white (yellow and white in male). **V** Fast, descending trill of *ti* notes. **H** Rainforest, deciduous forest, scattered trees, arid scrubby areas, woods, mangroves. **D** N Colombia, NW Venezuela.

9 AFRICAN PICULET *Sasia africana* 8cm Only as long as a finger. The male from female by rufous forehead. Very active, difficult to follow. Only uses its feet, not its tail, for support. Climbs trunks, twigs, grass stems. **V** Extremely high, barely audible *srrrrrrreeh*. **H** Undergrowth at forest edges. **D** Sierra Leone to N Angola.

10 RUFOUS PICULET *Sasia abnormis* 8–10cm Olive above, orange below. Active forager, singly or in small parties. **V** High-pitched *kik-ik-ik-ik-ik*; also a sharp *tic* or *tsit*. **H** Primary and secondary lowland and hill forest. **D** Malay Peninsula, Greater Sunda Islands.

11 WHITE-BROWED PICULET *Sasia ochracea* 9–10cm Like Rufous Piculet with white eyebrow. Agile and restless forager, with regular, loud tapping of branches. **V** Short, sharp *chi*, also a fast, high-pitched trill *chi-rrrrrrrra* or *ti-iiiii*. A loud, tinny drumming, generally on bamboo. **H** Mixed semi-evergreen and deciduous secondary growth with bushes and bamboo. **D** Nepal to S China, SE Asia.

12 ANTILLEAN PICULET *Nesoctites micromegas* 14–16cm Olive-brown above, streaked below. Male has red on crown. Feeds primarily in the understorey. **V** Loud *kuk-ki-ki-ki-ke-ku-kuk* or *yeh-yeh-yeh-yeh*. **H** Dry and humid forests, mixed pine and broadleaved forests, thorn forests and mangroves. Breeds from March to July. **D** Hispaniola.

13 GREY-AND-BUFF WOODPECKER *Hemicircus concretus* 13–14cm Black-and-white scaling on the back. Male has bright red crown. Forages in canopy; best located by calls. **V** High-pitched, drawn-out *kiyow* or *kee-yew*, and a vibrating *chitter*. **H** Primary and secondary lowland and hill forest, plantations, occasionally gardens. **D** Malay Peninsula, Greater Sunda Islands.

14 HEART-SPOTTED WOODPECKER *Hemicircus canente* 16cm Black and white with large crested head. Female has white forehead. Generally seen among thin branches in treetops. **V** Squeaky, nasal *ki-yew*, *ch-yew* or *chirrick*, also a high-pitched *kee-kee-kee-kee*. **H** Broadleaved forests, bamboo and coffee plantations. **D** India, SE Asia.

15 WHITE WOODPECKER *Melanerpes candidus* 24cm Mostly white. Post-ocular streak, upperparts and tail black. Nape and mid-belly yellow (male) or white (female). Often in pairs and family groups. **V** Very loud and noisy *wheerrrr...keeerrr*. **H** Open woods, savannah, palm groves, scattered trees, copses, plantations. **D** Widespread, Guyana to N Argentina.

16 LEWIS'S WOODPECKER *Melanerpes lewis* 26–28cm Glossy green back, red face, grey collar, pink wash on belly. Regularly makes sallies from a prominent perch to catch insects; also takes insects from the ground. **V** Weak *teef*. Drumming weak, a roll followed by three or four taps. **H** Open woodland, orchards and farmland with trees. **D** W North America from British Columbia southward. Northern birds move south in winter.

17 GUADELOUPE WOODPECKER *Melanerpes herminieri* 24–29cm Black, sometimes with red chest and belly. Can often be shy and retiring. Forages mainly in the canopy. **V** Single *kwa*; when excited, a variable *wa-wa-wa* or *kakakakaka*. Drums in short, slow rolls. **H** All forest types from sea-level to tree line. **D** Guadeloupe.

18 PUERTO RICAN WOODPECKER *Melanerpes portoricensis* 23–27cm Dark blue back, red front. Often occurs in small parties. **V** *Wek wek wek-wek-wek-wek*, increasing in volume and speed; also a harsh *gurrr-gurrr*, a chicken-like *kuk* and a *mew*. Drumming weak and infrequent. **H** Woodland, from sea level to mountains, including mangroves and coffee plantations. **D** Puerto Rico.

19 RED-HEADED WOODPECKER *Melanerpes erythrocephalus* 23–25cm Brilliant crimson head, black back, large white wing patches, white belly. Forages on trees and on the ground; also takes insects on the wing. **V** Loud *queark*, *queeeah* or *krrrrr*. Drumming weak and short. **H** Mature lowland forest with open understorey, orchards, parks, gardens. **D** E and C North America, from S Canada to Texas and Florida.

20 ACORN WOODPECKER *Melanerpes formicivorus* 21cm Mask and area at base of bill black. Forehead, line before eye and bib yellow. Nape red. Back, wings, tail and breast black. Rest of underparts white, partly streaked black. Hoards acorns in treetrunks. **V** Has a large repertoire of squealing, rasping and chattering sounds. **H** Edge of forest, open woods, deforested areas, scattered trees, especially oaks. **D** NW USA to Colombia.

21 YELLOW-TUFTED WOODPECKER *Melanerpes cruentatus* 19cm Mostly black, with red belly and (in male) red forecrown. Brow yellow, ending in golden nape. Eye-ring yellow. Flanks and undertail coverts whitish barred black. Back white. Often seen in groups. **V** Double *wrack-up*. **H** Edge of and clearings in rainforest, secondary growth, scattered trees, gardens. **D** Amazonia.

22 YELLOW-FRONTED WOODPECKER *Melanerpes flavifrons* 19cm Mask and most of upperparts black. Forehead, lower face and foreneck bright yellow. Back and rump white. Upper breast and flanks pale olive-grey, barred black, lower half of breast red. Male has bright red crown and nape. Female has black crown and nape. **V** Noisy, giving a range of strident calls such as *kikiki, chlit* and in aggressive encounters *tweewetwee tweewetwee*. **H** Humid forest, gallery forest, edge, palm groves, secondary vegetation. **D** SE Brazil, E Paraguay, NE Argentina.

23 GOLDEN-NAPED WOODPECKER *Melanerpes chrysauchen* 19cm Male has crown, nape and mid-belly red, narrow hind-collar golden. Female has black crown, red only on nape. In both sexes, mask and rest of upperparts black, but white line down back. Forehead, lower face and rest of underparts olive-buff with black barring. **V** Resonant *churr*. **H** Humid forest, secondary growth, edge. **D** Costa Rica, Panama.

24 BEAUTIFUL WOODPECKER *Melanerpes pulcher* 18cm Previously treated as conspecific with Golden-naped Woodpecker. Both sexes have red, yellow and black on the head, but red more extensive in male. Feeds mainly on fruits. **V** Screeching *kuWhay, kuWhay, kuWhay...* **H** Forested slopes, secondary woodland; lowlands to 1,000m. **D** NC Colombia.

25 BLACK-CHEEKED WOODPECKER *Melanerpes pucherani* 19cm Crown, nape and belly patch red in male. Female has black hindcrown, red only on nape. Mask (with fine white brow) and rest of upperparts black, finely barred white on back and flight feathers. Black barring on breast, belly, flanks, undertail coverts. **V** Most frequent call is a short series of rattling trilled churrs; also gives a shrill *chirree* and a loud rattling *krrrr*. **H** Humid forest, edge, rainforest, secondary growth, scattered trees. **D** S Mexico to W Ecuador.

26 WHITE-FRONTED WOODPECKER *Melanerpes cactorum* 18cm White forehead and malar stripe. Mid-crown red in male. Neck, breast and rest of underparts buffy grey, barred blackish lower down. Whitish-grey line from nape to mid-back. **V** *Weep-weep* and *wee-beep*. **H** Thorny woods, scrub, cactus, palm groves. **D** Bolivia and W Paraguay to N Argentina.

27 HISPANIOLAN WOODPECKER *Melanerpes striatus* 20–25cm Grey face, yellow and black barred back. Often breeds in loose colonies. Feeds at all levels, mainly on insects and insect larvae, also scorpions, lizards, fruits. **V** Long rolling series interspersed with throaty notes. Also various short notes, *wup, ta* and *ta-a*. Drums infrequently. **H** Mountain forests, cultivations, wooded swamps, mangroves, coastal scrub. **D** Hispaniola.

28 JAMAICAN WOODPECKER *Melanerpes radiolatus* 24–26cm Whitish face, finely barred black and white back. Forages mainly at mid-crown level, feeding mostly on insects and fruit. **V** Loud *kaaa, kaaa-kaaa* or *kaaa-kaaa-kaaa*; also a parakeet-like *chee-ee-urp* or *wee-cha weecha*. Drums loudly. **H** Wide-ranging from lowland copses, including plantations, to mountain rainforest. **D** Jamaica.

29 GOLDEN-CHEEKED WOODPECKER *Melanerpes chrysogenys* 20cm Diagnostic black around eye. Male has flame-red crown patch and female has pale greyish crown. **V** Very high short chatter, *witwitwit*. **H** Forest, broken forest, open woodland. **D** W Mexico.

30 GREY-BREASTED WOODPECKER *Melanerpes hypopolius* 20cm Rather drab and greyish overall with a densely barred black and white back. Male has red crown patch and female has plain grey head. Note faint white eye-ring. **V** Calls include a nasal *yek-a yek-a* and *chuk*; drums loudly in territorial defence. **H** Dry, open areas with scarce tree stands and bush. **D** SW Mexico.

31 YUCATAN WOODPECKER *Melanerpes pygmaeus* 17cm Mainly light grey underparts, barred upperparts. Note yellow around bill base, small bill, small body. Investigates dead branches for tunnelling insect larvae. Typically solitary. **V** Harsh, trilling *trrrp, trrrp, trrrp...* **H** Lowland dry forest, secondary woodland. **D** Belize, Honduras, Yucatán Peninsula.

32 RED-CROWNED WOODPECKER *Melanerpes rubricapillus* 18cm Crown, nape and hindneck red in male. Female has red on hindneck only. Forehead yellowish. Central belly red. Upperparts and tail barred whitish and black. **V** Typical call is a wavering *churr churr krr-r-r-r*; other calls include *wicka wicka* when displaying. **H** Gallery forest, arid woods, thorny scrub with cactus, bushes and thickets, secondary growth, mangroves, scattered trees, gardens. **D** Costa Rica to NW Venezuela.

33 GILA WOODPECKER *Melanerpes uropygialis* 21–25cm In flight shows a barred rump and a white patch at base of primaries. Forages at all levels, from treetops to ground. **V** Loud, rolling *churr-ur-rr*, laughing *geet geet geet geet geet* and nasal, squeaky *kee-u kee-u kee-u*. Drums occasionally, with long regular rolls. **H** Arid areas with scattered trees and large cacti, riparian woodland, around habitations. **D** SW USA to Baja California and W Mexico.

34 HOFFMANN'S WOODPECKER *Melanerpes hoffmannii* 20cm Note white rump and uppertail coverts. Male has red crown and extensive yellow on nape. Female lacks red crown. **V** High bickering *titrrrrrih*. **H** Dry, open areas with tree stands and bush. **D** Honduras to Costa Rica.

35 GOLDEN-FRONTED WOODPECKER *Melanerpes aurifrons* 25cm Orange-yellow at bill base and on nape. Male has red crown. Separated from smaller Yucatán Woodpecker by barred uppertail coverts and lack of yellow on chin; from Hoffmann's Woodpecker by more orange on nape. **V** High sharp short whinnying *trrruh*. **H** Dry open woodland, forest edge. **D** SC USA to S Mexico.

36 VELASQUEZ'S WOODPECKER *Melanerpes santacruzi* 23–25cm Previously considered conspecific with Golden-fronted Woodpecker. Red crown continuous with orange nape. **V** Trilling *TtrrrRee* and excited, repeated screeches. **H** Arid, spiny scrub with cacti. **D** E Mexico to N Nicaragua.

37 RED-BELLIED WOODPECKER *Melanerpes carolinus* 24cm In flight shows a white rump, with a few black spots or bars and a white area at base of primaries. Feeds in trees, mainly by gleaning; occasionally uses fly-catching to capture insects. **V** Loud, rolling *churr* or *churr churr churr*; also a chuckling, descending *chig chigh chchchchchch*. Drumming weak, rolls with a steady rhythm. **H** Various open woodland, deciduous, mixed and swampy woodlands, suburban groves and trees. **D** Widespread in E USA.

38 WEST INDIAN WOODPECKER *Melanerpes superciliaris* 27–32cm Strongly barred back. Extensive red on nape of both sexes. Forages at all levels, including the ground. Feeds on insects, spiders, fruit, frogs, lizards. **V** Repeated, high-pitched *krruuu-krruu-kruu...*; also various low-pitched notes and *ke-ke-ke-ke-ke*. **H** Dry forests, forest and swamp edge, coastal forests, scrub, palm groves, gardens. **D** Cuba, Cayman Islands, Bahamas.

1
2
3
4
5
6
7
8
9
10
11
12
13
14
15
16
17
18
19
20
21
22
23
24
25
26
27
28
29
30
31
32
33
34
35
36
37
38

WOODPECKERS *CONTINUED*

1 WILLIAMSON'S SAPSUCKER *Sphyrapicus thyroideus* 21–23cm Male is mostly black with bold white wing patch, white rump, two white stripes on face, red throat, yellow belly. Female has brown head and intricate barring on back and wings, dark breast, white rump, yellow belly. Mainly arboreal. **V** Strong *queeeah*; also a low-pitched *k-k-r-r-r*. Drumming begins with a roll followed by 2–4 slow taps. **H** Pine forests. **D** W North America, from S British Columbia to Mexico. Northern and eastern birds migrate south.

2 YELLOW-BELLIED SAPSUCKER *Sphyrapicus varius* 19–21cm Male has red cap and throat; females have red cap and white throat. In flight, shows white rump and large white patch on upperwing. Feeds on insects, fruits and buds; also drills a series of holes, horizontally or vertically, to get at sap and insects attracted to it. **V** Nasal, squealing *neeah*; also a hoarse *wik-a-wika*.... Drumming consists of about five rapid taps then slower, with the odd double tap. **H** Forests, forest edge, woodlands, gardens. **D** Breeds from E Alaska and W Canada to E USA. Winters SE USA, Caribbean, Central America.

3 RED-NAPED SAPSUCKER *Sphyrapicus nuchalis* 19–21cm Formerly considered a race of Yellow-bellied Sapsucker. Very similar, but little overlap in range. **V** Calls and drumming similar to Yellow-bellied Sapsucker. **H** Forests, Aspen groves. **D** Breeds SW Canada to midwestern and W USA. Winters California, Arizona, New Mexico.

4 RED-BREASTED SAPSUCKER *Sphyrapicus ruber* 20–22cm Mostly red head and upper breast. Actions and habits similar to Yellow-bellied Sapsucker. **V** Calls and drumming similar to Yellow-bellied Sapsucker. **H** Mixed deciduous and conifer forest, especially if Aspen or Ponderosa Pine is present. **D** Breeds from S Alaska and W Canada to SW USA. Northern birds migrate south for the winter.

5 CUBAN GREEN WOODPECKER *Xiphidiopicus percussus* 21–25cm Only green-backed woodpecker on Cuba. Forages, usually in pairs, on trunks and branches and among vines or creepers, from low level to canopy, feeding primarily on insects. **V** Short, harsh *jorr-jorr-jorr*... and high-pitched *eh-eh-eh*. **H** Various types of forest, mangroves. **D** Cuba.

6 FINE-SPOTTED WOODPECKER *Campethera punctuligera* 22cm Differentiated from other similar woodpeckers by finely spotted sides of upper breast and spotted, not barred or striped, upperparts. **V** Excited, sharp, high, slightly barbet-like *weh-weh-weh-weh* in duet of male and female. **H** More or less open woodland and areas with scattered trees. **D** From SW Mauritania to N DR Congo.

7 BENNETT'S WOODPECKER *Campethera bennettii* 19cm Spotted underparts. Male has unmarked white cheeks and throat, and red moustachial streaks and crown, while female has brown throat, cheek, and eye-stripes. **V** Mid-high, excited, cackled *rrrrrrrutiwweetiweetiweeti* (*rrrrr* running up). **H** Open woodland and scrub. **D** S Africa, from Angola and Tanzania southwards.

8 SPECKLE-THROATED WOODPECKER *Campethera scriptoricauda* 19cm Differs from Bennett's Woodpecker by streaked ear-patches. Note white chin of female. **V** Churring and chattering notes. **H** Open woodland. **D** From Tanzania to Malawi and Mozambique.

9 NUBIAN WOODPECKER *Campethera nubica* 20cm Note heavily striped cheeks, barred upperparts and densely spotted sides of upper breast. Female has long white-spotted malar stripe. **V** Mid-high, loud, excited, accelerated *weetweetweetweet* (lasting 4 secs), often in duet. **H** Woodland and other more or less wooded habitats. **D** From Sudan and Ethiopia to Tanzania.

10 GOLDEN-TAILED WOODPECKER *Campethera abingoni* 20cm Separated from Bennett's Woodpecker by boldly streaked face and almost black throat and upper breast; from Nubian Woodpecker by paler, brighter colouring above and streaking (not scalloping) below. **V** Mid-high, bleating *puWEE* (*pu-* as undertone) or nasal, rapidly lowered *weeeeh*. **H** Forest, woodland, often near streams. **D** Widespread across sub-Saharan Africa.

11 MOMBASA WOODPECKER *Campethera mombassica* 22cm Feeds mainly on arboreal ants; occasionally other invertebrates. **V** Screeching *errr, errr, errr* recalling distant screaming monkey. **H** Lowland coastal tropical forest. **D** Somalia, Kenya, Tanzania.

12 KNYSNA WOODPECKER *Campethera notata* 20cm Note extensive dark scalloping below (extending to neck sides), dark, finely spotted upperparts, densely streaked and spotted face: female shows brown moustache. **V** Very high *siiiii*. **H** Forest, wooded scrubland, gardens. **D** South Africa.

13 GREEN-BACKED WOODPECKER *Campethera cailliautii* 16cm Note unmarked tail, short bill, diagnostic lack of moustachial stripe. **V** Very high, questioning *weeh*. **H** Forest edge, woodland, thorn scrub. **D** C Africa, from Ghana to S Ethiopia and Mozambique.

14 LITTLE GREEN WOODPECKER *Campethera maculosa* 16cm Upperparts yellowish-green or bronze-green. Flight feathers brown, barred buffish. Female lacks red in crest. **V** High sharp *puweeh puweeh puweeh* with emphasis on higher *weeh*. **H** Forest glades and edges. **D** W Africa, from Senegal to Ghana.

15 TULLBERG'S WOODPECKER *Campethera tullbergi* 20cm Note plain upperparts. Fine spotting below separates it from other green-backed woodpeckers in its montane habitat. **V** Quiet, but occasionally gives loud *kweek* note, singly or in a short series. **H** Canopy of moist montane forests. **D** Nigeria to Cameroon; N DR Congo to Tanzania.

16 BUFF-SPOTTED WOODPECKER *Campethera nivosa* 15cm Overall rather dark. Note densely spotted and barred underparts. **V** Short, soft rattle, *prruh prruh prruh*. **H** Forests. **D** W and C Africa, from Senegal to Uganda and N Angola.

17 BROWN-EARED WOODPECKER *Campethera caroli* 19cm Note dark colouring and diagnostic brown ear-coverts. **V** High drawn- and puffed-out rising then lowered *piuuuuiih*. **H** Forest, riverine belts, plantations. **D** W and C Africa, from Sierra Leone to NW Tanzania and NW Angola.

18 GROUND WOODPECKER *Geocolaptes olivaceus* 25cm Greyish-olive upperparts, pale eye, pink breast. Lives exclusively on the ground. **V** Song is a series of up to five *ree-chick* notes; calls include a rolling *kreee* and a loud, harsh *peer-peer-peer* in alarm. **H** Open, rock-strewn, hilly and mountainous areas. **D** South Africa, Lesotho.

19 SULAWESI PYGMY WOODPECKER *Yungipicus temminckii* 13cm Male has red flash on side of hindneck. Occurs singly or in pairs; forages from the mid-storey to the canopy. **V** Slightly descending nasal trill. **H** Primary and tall secondary forest, forest edges, wooded cultivation. **D** Sulawesi, Indonesia.

20 BROWN-CAPPED PYGMY WOODPECKER *Yungipicus nanus* 13cm Note size, pale eyes, brown-striped head pattern. Forages on smaller branches in the tops of tall trees; also has a liking for the lower woody stems of shrubs. Often part of mixed-species flocks. **V** Rapid *kikikikiki*. Soft, far-carrying drumming. **H** Light and secondary forest, bamboos and trees near cultivations and around villages. **D** NE Pakistan, Bangladesh, India, Sri Lanka.

21 GREY-CAPPED PYGMY WOODPECKER *Yungipicus canicapillus* 14–16cm Blackish above, pale with black streaking and buffy-grey wash below. Forages in treetops, bushes and saplings. Agile, often hanging upside-down in search of food. **V** Rattling *tit-tit-erh-r-r-r-r-h*; also a squeaky *kweek-kweek-kweek*. Call is a *kik* or *pit*, which often precedes the rattling song. **H** Broadleaved evergreen forest, deciduous forest, secondary growth and coastal scrub. **D** From N Pakistan through Nepal and N India to SE Siberia, China, SE Asia, Greater Sunda Islands.

22 PHILIPPINE PYGMY WOODPECKER *Yungipicus maculatus* 14cm Occurs singly or in pairs; regularly joins mixed-species feeding flocks. **V** Loud, stuttering, slightly descending trill. **H** Primary and secondary forest, cloud forest, forest edges, mahogany plantations. **D** Philippines.

23 SULU PYGMY WOODPECKER *Yungipicus ramsayi* 14cm Actions presumably similar to Philippine Pygmy Woodpecker, with which it was often considered conspecific. **V** Loud, staccato *kikikikikikikiki*. **H** Forests and mangroves. **D** Sulu Islands, SW Philippines.

24 SUNDA PYGMY WOODPECKER *Yungipicus moluccensis* 14–15cm Brownish with white-striped head, spotted back, and grey-streaked underparts. Forages in upper tree levels; sometimes a member of mixed-species feeding flocks. **V** High-pitched trill or a whirring *trrrrr-i-i*. **H** Various woodland types, including mangrove, primary and secondary forest, coastal forest and coastal gardens. **D** Peninsular Malaysia; Greater and Lesser Sunda Islands.

25 JAPANESE PYGMY WOODPECKER *Yungipicus kizuki* 13–15cm Round-headed, short-billed profile. Lightly streaked belly and white bars on the back. Forages among small branches and twigs, mainly in tree canopy, often hanging upside-down. **V** Buzzing *kzz kzz* and a sharp *khit* or *khit-khit-khit*. Drumming weak, short bursts in quick succession. **H** Many types of forests, also parks and gardens. **D** E Russia, NE China, Korean Peninsula, Japan.

26 EURASIAN THREE-TOED WOODPECKER *Picoides tridactylus* 21–24cm Two white bars on face. Male has yellowish crown. No red in plumage. In flight shows prominent white back. Forages low down, tends to favour dead trees and stumps. **V** Short, soft *kip* or *kyuk*; when alarmed gives a *kip-kip-kip-kip*... Drum bursts are loud and slightly accelerating. **H** Conifer and mixed forests, often in boggy areas. **D** Widespread from C and N Europe to Kazakhstan, NE China, E Russia, Japan.

27 AMERICAN THREE-TOED WOODPECKER *Picoides dorsalis* 21–23cm Face black with white stripes above and below the eye and white speckles on forehead. Males have yellow crown patch. Primarily arboreal. **V** High-pitched *kip*; also a rattling *klikliklikli*. Drumming rolls slow, faster at end. **H** Mature montane and boreal conifer forests, partial to burnt areas. **D** Boreal band from Alaska to E Canada, NE USA; also Rockies, south to New Mexico.

28 BLACK-BACKED WOODPECKER *Picoides arcticus* 24cm Distinguished from American Three-toed Woodpecker by black back with faint white markings on flight feathers only. Head mostly black with white moustachial stripe and throat. Mainly arboreal. Tends to forage on dead conifers and in less dense woodland than American Three-toed Woodpecker. **V** Sharp *kik*; also a grating variably pitched snarl. Drumming rolls long, accelerating at the end. **H** Conifer forests, partial to burnt and windfall areas. **D** Alaska, C and S Canada, N and W USA.

29 ARABIAN WOODPECKER *Dendrocoptes dorae* 18cm Grey-brown. Male has red hindcrown. Agile forager in canopy of acacia and other trees or on the ground. **V** Accelerating and descending *kek-kek-kek-ke-ke-kekekeke-ke-ke*, also a descending *keck-keck-keck-keck-keck* and a falcon-like *kik-kik-kik-kik*. Drumming feeble and only occasional. **H** Woodland, mainly acacia, also palm and fig groves, often close to habitation. **D** SW Arabian Peninsula.

30 BROWN-FRONTED WOODPECKER *Dendrocoptes auriceps* 19–20cm Brown forehead and yellow crown. Hindcrown red in male, black in female. Forages mainly in trees and bushes, often joins mixed-species feeding flocks. **V** Squeaky *chick* or *peek* also a *chitter-chitter-chitter-r-r-rh* or *cheek-cheek-cheek-rrrr*. Drums for long periods in spring. **H** Open temperate and pine forest, favours oaks, Deodar Cedar and mixed stands. **D** NE Afghanistan to Nepal.

31 MIDDLE SPOTTED WOODPECKER *Dendrocoptes medius* 20–22cm Both sexes have red crown. Moustachial stripe does not reach bill. Round-headed appearance. Mainly arboreal. Sometimes joins mixed-species feeding flocks. **V** Nasal *gueeah .. gueeah .. gueeah .. gueeah* also a short *teuk* and a rattling *kik keuk-keuk-keuk*. Rarely drums, although gives loud taps at nest hole. **H** Mature broadleaved forests and woodlands, orchards and parks. **D** W Europe (N Spain) to W Russia, Caucasus, W Iran.

32 YELLOW-CROWNED WOODPECKER *Leiopicus mahrattensis* 17–18cm Upperparts black, densely spotted white. Female has all-golden crown; male has red hindcrown. Forages singly or in pairs mainly in the crown of trees or on trunks. **V** Feeble *peek*, sharp *click-click* and rapid, repeated *kik-kik-kik-r-r-r-h*. **H** Deciduous woodland, open areas with scattered trees and scrub. **D** Pakistan, India, Sri Lanka, SE Asia.

33 BEARDED WOODPECKER *Chloropicus namaquus* 25cm Note large size and brownish-green, completely barred appearance (except head). Male more scalloped above than female. **V** High, loud, rapid, slightly descending *kluklukluklukluklukluklu*. **H** Dry woodland, bushveld, with some large trees. **D** E and S Africa, from Ethiopia to Namibia and South Africa.

34 YELLOW-CRESTED WOODPECKER *Chloropicus xantholophus* 23cm Male has yellow crown and rump, the latter also in female. Wing barring mainly visible in flight. **V** Very high shrill excited chattered trills like *tierrrr tier tu-tu-tu-tu-twi*. **H** Forest and surrounding areas. **D** C Africa, from SE Nigeria and Cameroon to Kenya.

35 FIRE-BELLIED WOODPECKER *Chloropicus pyrrhogaster* 24cm No similar woodpecker with black ears, red crown (male only) and red rump in its range. **V** Sharp *wip* or *wip-wi-di-di-di-di-di*; drums loudly but rolls tail off towards end. **H** Forests. **D** Guinea to W Cameroon.

1
2
3
4
5
6
7
8
9
10
11
12
13
14
15
16
17
18
19
20
21
22
23
24
25
26
27
28
29
30
31
32
33
34
35

WOODPECKERS *CONTINUED*

1 LITTLE GREY WOODPECKER *Dendropicos elachus* 13cm Very pale-coloured and short-tailed, with a red rump. **V** Typical call is a harsh grating rattled *skree-eek-eee-eee-eee-eee-eee-eee-eek*; also softer *tee-tee-tee* and loud series of *wik* notes. **H** Sparsely wooded, dry areas. **D** S Mauritania and Senegal to Mali and W Sudan.

2 SPECKLE-BREASTED WOODPECKER *Dendropicos poecilolaemus* 13cm Note plain golden-green mantle, unmarked belly and golden-yellow rump. **V** High, sharp, rasping, speedy *tethree-tethree-tethree*. **H** More or less wooded and bushed, natural and cultivated areas, normally at higher elevations. **D** Cameroon to Uganda, W Kenya and Rwanda.

3 ABYSSINIAN WOODPECKER *Dendropicos abyssinicus* 15cm Golden yellow mantle, red rump, barred wings and tail. Male has red nape and crown. Separated from Cardinal Woodpecker by brown stripe through eye and golden mantle. **V** Not known. **H** More or less wooded areas often with candelabrum trees. **D** Eritrea, Ethiopia.

4 CARDINAL WOODPECKER *Dendropicos fuscescens* 14cm Barred mantle and wings. Female has completely dark crown; male has brown forecrown and red hindcrown and nape. **V** Extreme high, rapid, sharp, twittered *tuteetee tletuteetutee* or dry, high, thrush-like *shreeshreeshree-*. **H** Forest, other wooded and bushed habitats. **D** Sub-Saharan Africa, widespread.

5 GABON WOODPECKER *Dendropicos gabonensis* 14cm Upperparts, including primaries and rump, all green. Forehead brown. Distinguished from Brown-eared Woodpecker (Plate 105) by different pattern of underparts. **V** Very high, level trill *rrrrrrri*. **H** Open forest, natural and cultivated areas with tree-stands and scattered trees. **D** S Nigeria to DR Congo, N Angola, W Uganda.

6 MELANCHOLY WOODPECKER *Dendropicos lugubris* 14cm Distinguished from Gabon Woodpecker by different patterning of face sides and underparts. **V** A drawn-out buzzing note, also a harsh rattle. Drums loudly. **H** Open forest, forest edges, clearings, secondary forest, woodland. **D** Sierra Leone to SW Nigeria.

7 STIERLING'S WOODPECKER *Dendropicos stierlingi* 16cm Note small, dark appearance and plain upperparts. Male has red mid-crown and nape. Female has brown crown with pale streaks and black nape. Underparts white or cream with dark arrow-shaped markings **V** Territorial call is a long series of wavering *pi-di-di* notes and similar phrases; drums loudly with about five rolls a minute. **H** Miombo woodland. **D** S Tanzania, S Malawi, NW Mozambique.

8 ELLIOT'S WOODPECKER *Dendropicos elliotii* 16cm Green back, striped buff or greenish underparts. The crown is black and red in males and all black in females. No similar species in its range. **V** Series of shrill *bwe* notes or softer *kiwik*; drumming quiet. **H** Dense forest. **D** W and WC Africa, from E Nigeria to Uganda and N Angola.

9 AFRICAN GREY WOODPECKER *Dendropicos goertae* 19cm Pale grey head and underparts, barred wings and tail. Male has red crown. **V** Very high angry sharp chattered *ti-ti-ti-tri-tri-tri-tri-tri*. **H** Open forests, more or less natural and cultivated areas, gardens. **D** Mauritania to Ethiopia and south to NW Tanzania, NW Angola.

10 EASTERN GREY WOODPECKER *Dendropicos spodocephalus* 20cm Note belly colour of male. **V** Very high, slightly descending sharp chatter, *tititritritritrrritrrri*. **H** Open forests and other, more or less wooded, natural and cultivated areas including gardens. **D** E Africa, from Ethiopia to Tanzania.

11 OLIVE WOODPECKER *Dendropicos griseocephalus* 17cm Separated from previous two species by lack of barring and streaking, and by olive (not grey) underparts. Males have a red cap, and both sexes show a red rump that is conspicuous in flight. **V** High, plaintive, pushed-out *wWeewit wWeewit wWeewit*. **H** Dense forest. **D** S and SC Africa, from Angola and SW Uganda to South Africa.

12 BROWN-BACKED WOODPECKER *Dendropicos obsoletus* 14cm Note neat face pattern. Rump barred white (not red or orange-yellow as in Little Grey, Speckle-breasted or Cardinal Woodpeckers). **V** Mid-high, hurried, sharp, nasal *tui-tui-truh-tui*. **H** Open, more or less wooded and bushed habitats. **D** W, C and EC Africa, from Senegal to Ethiopia and N Tanzania.

13 NUTTALL'S WOODPECKER *Dryobates nuttallii* 18–19cm Black-and-white barring on back, patterned flanks. Forages mainly in trees, occasionally feeds on the ground or by fly-catching sallies. **V** *Pitit* or *pit*, rattling *itititititit*. Drumming rolls last 1–3 secs. **H** Mainly open oak woods. **D** SW USA, NW Mexico.

14 LADDER-BACKED WOODPECKER *Dryobates scalaris* 18cm From Nuttall's Woodpecker by absence of black in crown (male), yellow wash over underparts. Black markings in face narrower, less distinct. **V** Very short rapid bickering *weetwitwit*. **H** Dry open country with scattered trees and cacti. **D** SW USA to NE Nicaragua.

15 DOWNY WOODPECKER *Dryobates pubescens* 15–17cm Smaller than Hairy Woodpecker, with noticeably shorter bill. Outer-tail feathers usually show black spots. Forages mainly in trees, often visits garden feeding stations. **V** Short, soft *pik*; also a prolonged *peet-peet-peet-peet-peet-pit-pit-pitpit*. Drums in short, slow rolls. **H** Deciduous and mixed forest, woods, parks, gardens. **D** North America, widespread.

16 CRIMSON-BREASTED WOODPECKER *Dryobates cathpharius* 17–19cm Usually forages low down in trees and bushes, also favours dead trees. **V** Loud, repetitive *chip* or *tchick*, also a short, rapid, descending rattle. **H** Broadleaved forests. **D** Nepal to NW Laos, NC China.

17 LESSER SPOTTED WOODPECKER *Dryobates minor* 14–16cm Prominent white barring on back and wings, whitish underparts, no red under tail. Male has red crown patch. Forages in trees, favouring smaller branches, bushes and plant stalks especially reeds. **V** Weak *kee kee kee...* or *piit piit piit piit piit...*; also a sharp *pik* or *chik*. Drumming high-pitched with short interval between bursts. **H** Open forests, often near water, forest edge, orchards, parks, gardens. **D** Widespread across Europe and Asia.

18 LITTLE WOODPECKER *Veniliornis passerinus* 15cm Face buffy brown. Throat and rest of underparts barred whitish and blackish grey. Upperparts olive with faint barring; pale spotting on wing coverts. **V** Series of high-pitched *ki ki ki ki* or *wi wi wi* notes, also rolling *wicka* and *wik-wik-wik*. **H** Gallery and terra firme forest, cloud forest, várzea, savannah, deciduous woods, secondary growth. **D** Widespread across South America, south to N Argentina.

19 DOT-FRONTED WOODPECKER *Veniliornis frontalis* 17cm Forehead whitish, streaked grey. Whitish cheeks faintly streaked brown. All underparts densely barred blackish and whitish. Upperparts dark olive, sparse yellow streaking on back, white dots on upperwing coverts. **V** Gives *jick* or *wick* notes, singly or in a series. **H** Montane and cloud forest, transition woods, upland subtropical and temperate zone. **D** Bolivia to NW Argentina.

20 WHITE-SPOTTED WOODPECKER *Veniliornis spilogaster* 17cm Dark blackish olive, mottled whitish below and barred yellowish white above. Forehead, face and throat whitish, ear-coverts streaked black. **V** Various sharp, mostly high-pitched calls, given singly or in a chatter, such as *pic*, *cheekit* and *reh-reh-reh-reh-reh*. **H** Humid forest, riverine forest, copses, open woods, parkland, thorny trees. **D** SE Brazil, E Paraguay, NE Argentina, Uruguay.

21 CHECKERED WOODPECKER *Veniliornis mixtus* 15cm Forehead, crown and nape black, the first two flecked white. White face, black ear-patch and malar stripe. Upperparts spotted and barred black and white. Underparts white, streaked and scaled black. Red mark on either side of nape in male. **V** Single *peek* notes and a trilling chatter. **H** Savannah, dry woods, riverine vegetation, trees and thorny bushes, copses, palm groves. **D** E and C South America, from E Brazil to C Argentina.

22 STRIPED WOODPECKER *Veniliornis lignarius* 18cm Large version of Checkered Woodpecker with red right across nape of male. **V** Similar to Checkered Woodpecker but lower-pitched and softer. **H** Southern beech woods in Patagonian Andes, edge, scattered trees in fields and slopes, gardens, plantations. **D** Bolivia, S Chile, SW Argentina.

23 SCARLET-BACKED WOODPECKER *Veniliornis callonotus* 14cm Unmistakable. Scarlet upperparts. White underparts, finely barred. **V** Rattling 1–2-sec call; also a sharp *ki-dik*, sometimes repeated in series. **H** Dry deciduous forest, arid scrub, scattered trees, cactus, dense waterside vegetation. **D** SW Colombia, Ecuador, NW Peru.

24 YELLOW-VENTED WOODPECKER *Veniliornis dignus* 18cm Crown, nape and hindneck red in male; red restricted to nape in female. Bold face pattern. Underparts yellow with dark barring, grading to all-yellow undertail coverts. **V** Rarely heard, but may give a fast, high-pitched nasal rattling call. **H** Montane forest and transition woods, subtropical and temperate zones. **D** SW Venezuela to Peru.

25 BAR-BELLIED WOODPECKER *Veniliornis nigriceps* 18cm Smoky olive, barred yellowish on underparts. Dense barring on underparts extends to vent. **V** Rarely heard single soft *chick* note; also a high, descending *kzzrr* and possible song of repeated *kee* notes. **H** Humid montane forest, cloud forest, clearings with dense thickets and canebrakes, temperate upland forest, stunted woods in high mountains. **D** N Colombia to Bolivia.

26 BLOOD-COLORED WOODPECKER *Veniliornis sanguineus* 14cm Unmistakable. Upperparts crimson in both sexes, and male has red cap. Face dark greyish brown. Underparts dark greyish brown barred white. **V** Call is a single *keek*; also rapid series of *wih* notes and drumming by both sexes. **H** Mangroves, swampy forest, coffee plantations, lowlands. **D** The Guianas.

27 RED-RUMPED WOODPECKER *Veniliornis kirkii* 16cm Red rump and golden-yellow nape in both sexes. Male has red crown. Underparts whitish barred blackish grey. Upperparts plain golden olive. **V** Nasal *keer* notes in series, also series of mewing notes. **H** Gallery forest, rainforest, secondary growth, scattered trees, mangroves, deciduous woods, savannah, thorny vegetation, plantations. **D** Costa Rica to Ecuador and SE Venezuela.

28 RED-STAINED WOODPECKER *Veniliornis affinis* 19cm Males have red crown and nape; female has yellowish nape. Underparts barred browny black and whitish. Upperparts olive, paler spotting on upperwing coverts. Wing coverts tinted reddish. **V** Rarely heard but may produce a series of 10–14 *ghi* or *kih* notes. **H** Humid forests, terra firme forest, várzea, secondary growth, edge, thickets. **D** Amazonia, E Brazil.

29 CHOCO WOODPECKER *Veniliornis chocoensis* 15–16cm More boldly marked than similar Red-stained Woodpecker; separated geographically from that species. Sometimes follows army ants. **V** Calls include squeaky *pueek, pueek*. **H** Humid primary forest, wooded slopes. **D** W Colombia, NW Ecuador.

30 GOLDEN-COLLARED WOODPECKER *Veniliornis cassini* 18cm Crown red in male, yellow-streaked in female. Hindneck yellow in both sexes. Face greyish, ear-coverts markedly streaked black. Underparts all densely and evenly barred blackish and whitish. **V** Very like Red-stained Woodpecker. **H** Rainforest, clearings, open country with trees, shrubberies, lowlands. **D** NE Amazonia.

31 YELLOW-EARED WOODPECKER *Veniliornis maculifrons* 17cm Olive-green back, yellow nape, grey face. Heavily barred below and on tail. Males have a red crown, green in females. **V** Long series of vibrating *ew* notes, rising then falling in pitch. **H** Forest, secondary growth, parks and gardens, montane forest. **D** E Brazil.

32 RED-COCKADED WOODPECKER *Leuconotopicus borealis* 22cm Boldly patterned black and white. Male's red spot at back of crown is tiny and difficult to see. Forages mostly on tree trunks, often in small groups, occasionally makes fly-catching sallies. **V** Raspy, nasal *sripp*, *shrrit* or high-pitched *tsick*; also a rattling *shirrp-chrchrchrchr...* **H** Mature pine forests. **D** SE USA.

33 SMOKY-BROWN WOODPECKER *Leuconotopicus fumigatus* 16cm Unmistakable. Crown and nape red in male, black in female. Otherwise uniform dark smoky brown, with paler face and undertail coverts. **V** Series of high-pitched, piping *keer* notes or more gravelly or squeaky sounds. Drums in long, fast rolls. **H** Humid montane forest, cloud forest, woods, secondary growth, bushes, thickets, coffee plantations. **D** Mexico to NW Argentina.

34 ARIZONA WOODPECKER *Leuconotopicus arizonae* 18–20cm Brown and white. Feeds mainly on tree trunks, occasionally makes fly-catching sallies. **V** Sharp *peep* or *keech*; also a descending rattle. Drumming long and loud. **H** Pine-oak woodland, riparian woodland. **D** SW USA, Mexico.

35 STRICKLAND'S WOODPECKER *Leuconotopicus stricklandi* 18cm Distinguished from Hairy Woodpecker by speckled and barred underparts; Arizona Woodpecker has plain back and is paler. **V** Very high variable unstructured jumble of strident whipped-up and trilled calls, e.g. *tsip-scrrr-crrr sip sip sip skrrr-skrrr*. **H** Pine woodland. **D** C Mexico.

36 HAIRY WOODPECKER *Leuconotopicus villosus* 20–23cm Black and white plumage similar to smaller Downy Woodpecker, but has longer bill and clean white outer-tail feathers. Forages by pecking, hammering, probing and gleaning, feeding mainly on insects, insect larvae, spiders, fruits and seeds. Common visitor to garden feeding stations. **V** Loud *keek* and a rapid whinny. **H** Woodland, gardens. **D** Widespread across North and Central America, from S Alaska and Newfoundland to Bahamas and Panama.

37 WHITE-HEADED WOODPECKER *Leuconotopicus albolarvatus* 24cm In flight shows a white patch at base of primaries. Forages mainly on lower trunk of conifers with visits to needle clusters and cones, occasionally making fly-catching sallies. **V** Sharp *chick-chick* or *pitik*; also a rattling *peekikikikikik*. Drumming fairly long with varying tempo. **H** Mixed conifer forests, in mountains. **D** SW Canada, W USA.

1
2
3
4
5
6
7
8
9
10
11
12
13
14
15
16
17
18
19
20
21
22
23
24
25
26
27
28
29
30
31
32
33
34
35
36
37

WOODPECKERS *CONTINUED*

1 RUFOUS-BELLIED WOODPECKER *Dendrocopos hyperythrus* 20–25cm Bright orange neck and belly. Red crown in male; black with white speckles in female. Usually seen singly or in pairs, occasionally joins mixed-species flocks. **V** Reeling *chit-chit-chit-r-r-r-h* and a fast *ptikitititititit* when alarmed. Drums in short fading rolls. **H** Open oak forests, pine forest, mixed broadleaved evergreen and coniferous forest; locally in deciduous forests. **D** N India to SE Asia, E Asia.

2 FULVOUS-BREASTED WOODPECKER *Dendrocopos macei* 18–19cm Underparts off-white with thin stripes on neck and breast. Favours tall trees, where usually encountered singly, in pairs or in small parties. **V** Sharp *tchick*, also a *pik-pik* or *chik-it-chik-it*. Drumming is weak, with short rolls. **H** Open forest and forest edge. **D** N Pakistan to N Myanmar.

3 FRECKLE-BREASTED WOODPECKER *Dendrocopos analis* 18–19cm A recent split from Fulvous-breasted Woodpecker; actions and habits therefore probably similar. Less red around vent. **V** Similar to Fulvous-breasted Woodpecker. **H** Open deciduous forest, secondary forest and open areas with scattered trees. **D** S Myanmar, Thailand, Cambodia, Laos, Vietnam; Andaman Islands; Java and Bali, Indonesia

4 STRIPE-BREASTED WOODPECKER *Dendrocopos atratus* 21–22cm Red on nape and more boldly streaked underparts separate this from the similar Fulvous-breasted Woodpecker. **V** Explosive *tchick* and a whinnying rattle. **H** Open oak and pine woodland in evergreen forest, edges of open broadleaved forest, cultivations. **D** S and E Asia, from NE India to Vietnam.

5 DARJEELING WOODPECKER *Dendrocopos darjellensis* 25cm Streaked yellowish belly, pale yellow-orange on neck and throat. Male has red patch on hindcrown which female lacks. Forages from ground to tree canopy, sometimes in mixed-species flocks. **V** Rattling *di-di-di-d-dddddt*, low *puk puk*, and when alarmed a *tsik tsik tsik*… **H** High-altitude forest and open woodland. **D** Nepal to SC China and N Vietnam.

6 HIMALAYAN WOODPECKER *Dendrocopos himalayensis* 23–25cm Male has red crown. Black mark below and behind eye often lacking, isolated or indistinct. Mainly arboreal, usually encountered singly or in pairs. **V** Sharp *kit*, rapid high-pitched *chissik-chissik* and fast *tri tri tri tri*… Drums in short bursts. **H** Dense mountain forests. **D** NE Afghanistan to W Nepal.

7 SIND WOODPECKER *Dendrocopos assimilis* 20–22cm Male has red crown. Forages among tree branches, often low down, on fallen trees and fence posts etc. **V** Explosive *ptik* a weak *chir-rir-rirrh-rirrh* and a rapid, repeated *wicka toi-whit toi-whit toi-whit*. **H** Riverine forests, thorn scrub, roadside trees and tree plantations. **D** SE Iran, Pakistan.

8 SYRIAN WOODPECKER *Dendrocopos syriacus* 23cm Note lack of black bar at rear of ear-coverts and pinkish undertail. Behaviour similar to Great Spotted Woodpecker. **V** Soft *cheuk*, *gipp* or when disturbed *gip-gip-gip*… Drumming like Great Spotted Woodpecker but longer, fading at end. **H** Broadleaved woodland, open country with scattered trees, orchards, vineyards, parks and gardens. **D** SE Europe to S Iran.

9 WHITE-WINGED WOODPECKER *Dendrocopos leucopterus* 22–23cm Little known. Actions probably much as Great Spotted Woodpecker. **V** *kewk* or *kig*, also rattling notes. **H** Poplar forests by streams, desert scrub, orchards and gardens. **D** Kazakhstan to N Afghanistan, W Mongolia, W China.

10 GREAT SPOTTED WOODPECKER *Dendrocopos major* 24cm Size and bold black and white plumage distinctive in most of range. Note big white shoulder patches, extensive red on vent, black crown. Male has red nape patch. **V** High-pitched *kik* or soft *chik*. Drumming loud, rapid and far-carrying. **H** Forest, woodland, parkland, gardens, farmland with hedges and trees. **D** Widespread across Europe and Asia.

11 OKINAWA WOODPECKER *Dendrocopos noguchii* 31–35cm Dark reddish brown with pale bill. In flight shows dull reddish rump. Forages mostly low down on tree trunks, stumps, bamboo and fallen logs. **V** Clear whistling *kwe kwe kwe kwe*, also a *kyu-kyu* or *kyu-kyu-kup* and a single *whit* when alarmed. Drumming can be in short or long bursts. **H** Mature subtropical forests. **D** Okinawa, East China Sea.

12 WHITE-BACKED WOODPECKER *Dendrocopos leucotos* 23–28cm In flight shows white lower back and rump. Actions similar to Great Spotted Woodpecker. **V** Soft, sharp *kiuk*, *kok* or *gig*; when alarmed gives a series of *kyig gyig* notes. Drumming loud, long and accelerating. **H** Deciduous and mixed mountain forests. **D** From Europe to China and Japan.

13 RUFOUS-WINGED WOODPECKER *Piculus simplex* 18cm Male has red crown and moustachial stripe; female has red on nape only. Pale eye diagnostic. **V** Series of downslurred *heew* notes; also a sharp, nasal *deeeah* and long bursts of drumming. **H** Lowland forest. **D** Honduras to W Panama.

14 STRIPE-CHEEKED WOODPECKER *Piculus callopterus* 17cm Distinguished from Rufous-winged Woodpecker (not in same range) by black eye and different head pattern. **V** Most frequently heard call is a nasal *nyah-wheet*; also a fast chatter. **H** Forest interior and edge. **D** Panama.

15 WHITE-THROATED WOODPECKER *Piculus leucolaemus* 20cm Golden line through face. Throat whitish faintly speckled olive. Olive-green upperparts, scaled breast, barred belly. Male has red crown and moustachial stripe; female has red only on hindcrown. **V** Usually quiet. Sometimes gives harsh slurring *pish* or *shreer*. **H** Gallery forest, humid forest in foothills, plains. **D** W Amazonia.

16 LITA WOODPECKER *Piculus litae* 18cm Formerly considered a form of Yellow-throated Woodpecker. Resembles that species, but has black throat and brown flight feathers. **V** Like White-throated Woodpecker. **H** Humid forest, edge, secondary growth, in lowlands and foothills. **D** W Colombia to NW Ecuador.

17 YELLOW-THROATED WOODPECKER *Piculus flavigula* 18–23cm Face and throat golden yellow. Forehead, crown, nape and malar region of male red; female has forehead and crown olive, malar region yellow. Upperparts olive, underparts whitish scaled olive. **V** Infrequent hissed *queer* or *shreer*, sometimes doubled. **H** Humid tropical forest, terra firme forest, gallery forest, edge, Caatinga. **D** Amazonia, SE Brazil.

18 GOLDEN-GREEN WOODPECKER *Piculus chrysochloros* 22cm Cap and malar region of male red; female has no red. Mask olive. Throat and line though face yellow. Upperparts olive. Underparts golden yellow, barred olive. **V** Shrill *shreeer*, given singly or in series. **H** Woods, savannah, humid terra firme forest, várzea, gallery forest, edge, deciduous woods, xerophytic vegetation, scattered trees. **D** Panama to Venezuela, Amazonia, SE Brazil, Paraguay, N Argentina.

19 YELLOW-BROWED WOODPECKER *Piculus aurulentus* 21cm Note complex head pattern: red crown on male (olive-green on female), yellow stripes above and below eye, red malar stripe. **V** Single shrill *piiiu* and series of descending *eeew* notes. Drumming rapid, in regularly spaced rolls. **H** Transition woods and forest in hills and plains. **D** SE Brazil, E Paraguay, NE Argentina.

20 GOLDEN-OLIVE WOODPECKER *Colaptes rubiginosus* 23cm Forehead and crown grey, nape red. Male also has red malar stripe. Face ivory-white, slightly barred on hind-cheeks. Upperparts golden olive. Throat finely speckled white on blackish. Rest of underparts greenish yellow, barred blackish. **V** Long, rising, rattling trill; also a single sharp *deeeeh* and various whirring and liquid notes. **H** Montane and cloud forest, clearings, secondary growth, open woods, lowland forest, deltas, edge, scattered trees. **D** Mexico to Guyana, NW Peru, NW Argentina.

21 GREY-CROWNED WOODPECKER *Colaptes auricularis* 17cm Grey crown, green back, barred underparts. Red only on malar stripe (male). **V** Descending *tjeeuw*. **H** Forest interior and edge. **D** W Mexico.

22 BRONZE-WINGED WOODPECKER *Colaptes aeruginosus* 20cm Formerly considered a race of Golden-olive Woodpecker. Distinguished from that species by absence of red eyebrow. **V** Very high series of 8–12 notes, *phwip phwip phwip*… **H** Forest interior and edge, plantations. **D** NE Mexico.

23 CRIMSON-MANTLED WOODPECKER *Colaptes rivolii* 28cm Large. Notably crimson. Yellowish-ivory face, black throat. Breast scaled crimson and black, rest of underparts yellow with markings on flanks. **V** Possible song a series of fast, monotone metallic notes. Calls include a rising, questioning *reee*. **H** Humid montane forest, cloud forest, dwarf forest, gnarled trees and bushes at páramo line. **D** NW Venezuela to Bolivia.

24 BLACK-NECKED WOODPECKER *Colaptes atricollis* 26cm Forehead and forecrown slate-grey. Nape, hindneck and malar region red in male; female has black malar region. Face creamy white. Throat, foreneck and breast black. **V** Series of *wic* notes; calls include short *peah* in alarm. **H** Semi-arid woods and cloud forest, riverside vegetation, irrigated areas, shrubby slopes, scattered bushes, plantations, gardens. **D** Peru.

25 SPOT-BREASTED WOODPECKER *Colaptes punctigula* 23cm Bold black spots on body and wings. Forehead and crown black, nape red. Males also show red malar stripe. Face creamy white. Throat white finely dotted black. Upperparts olive scaled and barred black. **V** Series of 8–12 high-pitched, weak and nasal *kah* or *keeh* notes; other calls include *ka-wick* and variants. **H** Gallery forest, rainforest, várzea, deciduous forest, open woods, edge, deforested areas, bushes and thickets, mangroves, palm groves, savannah, cultivated areas. **D** From Panama to Amazonia, Bolivia.

26 GREEN-BARRED WOODPECKER *Colaptes melanochloros* 27cm Mostly yellowish green, barred black on upperparts, spotted black on underparts. Forehead and crown black, nape red. Face whitish cream, throat streaked blackish. Malar stripe is red in male, blackish-streaked in female. **V** Ringing series of *kwik kwik kwik* notes; also a single *ker-wick*. **H** Edge of woods and forest, open country with trees, clearings, arid woods, scrub, canebrakes, savannah, Caatinga, palm groves, plantations, plains, mountains. **D** E, C, SE South America, from NE Brazil to Bolivia and C Argentina.

27 NORTHERN FLICKER *Colaptes auratus* 31–33cm In flight shows a white rump and yellow underwing. Mainly arboreal, although will descend to forage or dust bathe on the ground. **V** Calls and drumming similar to Gilded Flicker, although the former a little lower-pitched. **H** Open woodland, forest edge, open country with scattered trees, parks and gardens. **D** Widespread across North and Central America, from Alaska to Cuba and Nicaragua. Northern birds move south in winter.

28 GILDED FLICKER *Colaptes chrysoides* 26–30cm In flight shows a white rump and yellow underwing. Frequently feeds on the ground. Often considered a race of Northern Flicker, with which it is known to hybridise. **V** Descending *peah* or *klee-yer*; also a soft *wicka wicka wicka*… and a long *whit whit whit whit whit*… series. Drumming moderate to fast. **H** Arid scrub and desert with large cacti and yucca, riverside woodlands. **D** SW USA, NW Mexico.

29 FERNANDINA'S FLICKER *Colaptes fernandinae* 33–35cm Regularly forages on the ground, feeding on insects, larvae, worms and seeds. In flight, shows yellow underwing. Endangered due to habitat loss. **V** Loud *pic-pic-pic-pic-pic-pic*, slightly lower pitched than Northern Flicker; also gives a nasal *ch-ch-ch*. **H** Open woodland, low open country with palms. **D** Cuba.

30 CHILEAN FLICKER *Colaptes pitius* 30cm Generally barred blackish and whitish except grey cap, malar stipe flecked blackish, rest of face and throat greyish ivory. Female has no malar stripe. **V** Sharp, high-pitched *pitee-you*. **H** Cool temperate southern beech woods and ecotone, bushes, open fields. **D** Chile, SW Argentina.

31 ANDEAN FLICKER *Colaptes rupicola* 32cm Long bill. Cap grey. Malar stripe red and black in male, blackish-grey in female. Rest creamy, upperparts heavily barred black, underparts spotted black, especially on breast. **V** Varied calls, including loud, whistling *tew-tew-tew*, complex trilling call, and loud, harsh *kek* or *kek-kek-kek* in alarm. **H** Treeless upland steppe in puna, páramo scrub. **D** S Ecuador to NW Argentina.

32 CAMPO FLICKER *Colaptes campestris* 31cm Dark cap, yellow cheeks, yellow collar. Males have a red malar stripe, blackish in females. Rest of upperparts barred black and whitish, underparts more finely barred blackish and whitish. **V** Varied. Produces loud whistled notes in alarm. Possible song includes variations of *kya-wi*. **H** Open country, isolated trees, copses, palm groves, stony ground in hilly regions, cleared areas in forest, woodland edge, roadsides. **D** Suriname to Bolivia and E Argentina.

1
2
3
4
5
6
7
8
9
10
11
12
13
14
15
16
17
18
19
20
21
22
23
24
25
26
27
28
29
30
31
32

WOODPECKERS *CONTINUED*

1 CINNAMON WOODPECKER *Celeus loricatus* 22cm Head and upperparts cinnamon-chestnut, finely barred blackish. Malar region red in male. Underparts and tail buffy cinnamon, scalloped black. Flight feathers chestnut-cinnamon, barred black. **V** Loud call, accelerating, descending, *peee peeew pew poo puh*. **H** Rainforest, humid forest, secondary growth, clearings. **D** E Nicaragua to SW Ecuador.

2 WAVED WOODPECKER *Celeus undatus* 20cm Head and neck cinnamon, streaked black. Part of cheek and malar region red in male. Upperparts rufous cinnamon, markedly barred black. Underparts buffy cinnamon, densely scalloped black. Primary flight feathers and tail tip black. **V** Loud *whick coer*. **H** Dense rainforest, edge, riverine forest, dryland forest. **D** E Venezuela to NE Brazil.

3 SCALY-BREASTED WOODPECKER *Celeus grammicus* 21cm Head chestnut. Cheek and malar region red in male. Rest chestnut, finely barred and scaled blackish, but lower belly, rump and uppertail coverts plain tawny chestnut. Primary flight feathers and tail black. **V** Clear, sharp *curry-coo*. **H** Rainforest, humid forest and várzea, edge, secondary growth, savannah with scattered trees. **D** W, NW, C Amazonia.

4 CHESTNUT-COLORED WOODPECKER *Celeus castaneus* 25cm From Cinnamon Woodpecker by different crest shape, pale head, black tail and uniform basic colouring of body. **V** High nasal downslurred *piuw*. **H** Dense forest interior and edge. **D** SE Mexico to W Panama.

5 CHESTNUT WOODPECKER *Celeus elegans* 28cm Mostly chestnut, with shaggy crest. Red malar region in male. Some races have ochreous-yellow forehead, crown, crest and hind-half of body. Upperparts with small yellowish markings. Dark form is dark chocolate-chestnut, lacking yellow on head. There is a whole gradation between the extremes. **V** Descending *weewah ewewewewewewew* and mocking *ha hahahahaha*. **H** Tall, dense forest, gallery forest, várzea, edges, plantations. **D** N South America, Amazonia.

6 PALE-CRESTED WOODPECKER *Celeus lugubris* 27cm Big untidy bushy crest, warm ivory. Malar area red in male, brown in female. Brown around eye. Foreneck and rest of body brown, slightly scaled and barred ivory on top half of upperparts, wings barred chestnut. Rump and underwing warm ivory; flight feathers barred brown and dark ivory. **V** *wee-wee-week*. **H** Dense woods, patches, palm groves, savannah, open woods. **D** E Bolivia, Paraguay, S Brazil, NE Argentina.

7 BLOND-CRESTED WOODPECKER *Celeus flavescens* 27cm Head and prominent crest yellow. Male has red malar stripe. Otherwise mostly black or dark brown, scaled and barred yellow on upperparts. Yellow rump. **V** Series of long-spaced *weeps*, also *wicks*. **H** Humid forest and gallery forest, savannah, palm groves, secondary growth, edge, orchards. **D** E Brazil, E Paraguay, NE Argentina.

8 OCHRE-BACKED WOODPECKER *Celeus ochraceus* 26–27cm Previously treated as subspecies of Blond-crested Woodpecker. Note pale buff back and upperparts. **V** Shrill, piercing *keik, keik, keik, keik*; also rapid drumming. **H** Lowland forest, Caatinga semi-desert, savannah. **D** E Brazil.

9 CREAM-COLORED WOODPECKER *Celeus flavus* 26cm Unmistakable. Nearly all tawny cream. Male has red malar region. Wing coverts brown, edged cream. Flight feathers chestnut and blackish chestnut. Underwing cream, tail blackish. **V** Loud *pweer pweer purr paw, kew, kew...* and *wheejah*. **H** Humid forest, rainforest and várzea, all near water, mangroves, deciduous and open woods, secondary growth, plantations. **D** Amazonia to E Brazil.

10 RUFOUS-HEADED WOODPECKER *Celeus spectabilis* 27cm Head brownish chestnut, with red nuchal crest and malar region in male. Breast and tail black, contrasting with rest of yellowish-ochre body, scaled and barred black. **V** *squeear* followed by *klooklookloo*.... **H** Humid tropical forest, especially riverine forest, canebrakes. **D** W Amazonia, from E Ecuador to N Bolivia.

11 KAEMPFER'S WOODPECKER *Celeus obrieni* 30–38cm Previously treated as subspecies of Rufous-headed Woodpecker; preferred habitats differ. Feeds mainly on ants associated with bamboo. **V** Soft piping *Whe, weh-weh-weh-wa-wa*, first syllable upslurred; also rapid drumming. **H** Cerrado woodland; particularly associated with bamboo stands. **D** EC Brazil.

12 RINGED WOODPECKER *Celeus torquatus* 27cm Extent of black in plumage varies, but all races have solid black breast. **V** Series of five even-pitched *deeees*. **H** Rainforest, gallery forest, dryland forest, várzea, woods, tall secondary growth, clearings. **D** Amazonia, SE Brazil.

13 HELMETED WOODPECKER *Celeus galeatus* 31cm Bill greyish-horn. Cinnamon face with red crest. Male has red malar region. Line on neck, lower back and rump white. Neck, upperparts and tail black. Underparts barred whitish and blackish. **V** Series of up to 12 strident *keer* notes. **H** Rainforest, gallery forest. **D** SE Brazil, E Paraguay, NE Argentina.

14 BLACK-BODIED WOODPECKER *Dryocopus schulzii* 31cm Bill greyish-horn. Conspicuous red crest. Malar region red in male, black in female. Ear-patch grey, chin and throat whitish grey. White line across face and down sides of neck. Body black. Underwing and base of flight feathers white. **V** *Wick wick wick*.... **H** Unmodified dry Chaco woods. **D** SC Bolivia, W Paraguay, N Argentina.

15 LINEATED WOODPECKER *Dryocopus lineatus* 33cm Bill greyish-horn. Rounded crest red. Ear-coverts grey. White line through face, down side of black neck, onto back. Underparts whitish, scaled blackish. Underwing and base of flight feathers white. **V** *Wickwickwickwick* and *pickrrrr*. **H** Humid, transition and fairly dry forest, rainforest, patchy forest, clearings, gallery forest, thorn scrub, plantations, mangroves, gardens. **D** Widespread, from Mexico to S Brazil.

16 PILEATED WOODPECKER *Dryocopus pileatus* 40–43cm Dark underparts. In flight, from above, primaries show white bases, underwing white with black tips to primaries and secondaries. **V** Deep *wek* or *wuk* often given in flight, also in series. Drumming slow and powerful, accelerating and fading at end. **H** Mature forests. **D** S Canada, E and Pacific states of USA.

17 WHITE-BELLIED WOODPECKER *Dryocopus javensis* 40–48cm Black with white belly and pale eyes. Usually encountered singly, in pairs or in small groups in tall trees, on or near the ground, with a liking for dead trees, stumps, fallen logs and leaf litter. Flight crow-like, with leisurely, deliberate wingbeats, when white rump and underwing coverts are prominent. **V** Laughing *kek-kek-kek-kek-kek...* or *kiau-kiau-kiau-kiau...* Also a single loud, sharp *kiyow*, *kyah* or *keer*. Drumming loud and accelerating, lasting about 2 secs. **H** Forests and light secondary forest with tall trees. **D** Widespread, from S India to SE Asia, Indonesia, Philippines.

18 ANDAMAN WOODPECKER *Dryocopus hodgei* 38cm Black with red crest and pale eyes. Red malar area in male. Usually seen in pairs or loose parties. Actions and habits similar to White-bellied Woodpecker. **V** Loud, chattering *kuk-kuk-kuk* that ends with a whistled *kui*, also a loud, sharp *kik-kik-kik*. Drumming is loud and far-carrying. **H** Evergreen, open forest. **D** Andaman Islands.

19 BLACK WOODPECKER *Dryocopus martius* 45–47cm Entirely black apart from red crown (male) or red patch on hindcrown (female). Forages on lower parts of trunks, at base of trees and on the ground. Flight slow and loose, slightly undulating. **V** Loud *kwee kwee-kwee-kwee-kwee...*, also a clear *bleeeep* or *ke-yaa*. In flight gives a far-carrying *krry-krry-krry...* or softer *kruek-kruek-kruek...* Drumming loud, recalls a machine gun. **H** Mature forests of most types; after breeding often in more open areas with scattered trees. **D** Widespread from W Europe to Japan.

20 POWERFUL WOODPECKER *Campephilus pollens* 34cm Bill black. Mask, throat and neck black, split by white line through face, down sides of neck and along scapulars to form a V. Crest red in male, black in female. Upperparts and tail black. Central rump whitish. Underparts ochre, barred blackish. **V** Nasal *kyaaah* and *peeyour*, often repeated, also *kikikikikaw*. **H** Humid rainforest, cloud forest, montane forest and edge. **D** Colombia to Peru.

21 CRIMSON-BELLIED WOODPECKER *Campephilus haematogaster* 34cm Bill black. Both sexes have prominent red crest, black and yellow face, black throat. Female has yellow side of neck. Hindneck, underparts and rump crimson. Upperparts, wings and tail black. **V** Call is a loud *stk*; also gives song of loud, squeaky *eeer* notes. Drumming a loud, fast double knock. **H** Humid rainforest, edges, montane forests. **D** Panama to Peru.

22 RED-NECKED WOODPECKER *Campephilus rubricollis* 34cm Bill ivory. Head, neck and breast red, apart from black-bordered white triangle on face of female, small black-and-white ear-spot in male. Underparts, flight feathers and underwing chestnut. Upperparts black. **V** Explosive and nasal *nkaaah*, *kyah* and *kerra kerra*. **H** Rainforest, terra firme forest and várzea, cloud forest, edges, savannah, semi-open forest, riverine woods, light secondary growth. **D** Amazonia.

23 ROBUST WOODPECKER *Campephilus robustus* 37cm Bill ivory. Head and neck red. Black-and-white ear-spot in male; black-bordered white triangle on face of female. Buffy white along back, rump and uppertail coverts. Rest of upperparts black. Upper back, breast and belly barred whitish and blackish. **V** *Pseew, keeew* perched or in flight. **H** Humid forest, *Araucaria* woods, forest in hills, edges, canebrakes. **D** SE Brazil, E Paraguay, NE Argentina.

24 CRIMSON-CRESTED WOODPECKER *Campephilus melanoleucos* 36cm Bill bone-white. Head red, with white at base of bill, black and white ear-coverts in male, black and white below eye and black forecrown in female. Chin, throat and neck black. White line down sides of neck onto scapulars, forming a V. Rest of upperparts, rump and tail black. **V** *Queer queerer*. **H** Cloud forest, rainforest, gallery forest, deciduous forest open woodlands, savannah, secondary growth edges, semi-open country, palm groves, swampy areas, scattered trees. **D** Widespread from Panama through Amazonia to SE Brazil.

25 PALE-BILLED WOODPECKER *Campephilus guatemalensis* 35cm White bill diagnostic. Range probably does not overlap with that of Crimson-crested Woodpecker. **V** Double tap, followed by very high irregular rapid *tudurrut tudder*. **H** Forest, plantations, mangrove. **D** Mexico to W Panama.

26 GUAYAQUIL WOODPECKER *Campephilus gayaquilensis* 34cm Considered by some to be a race of Crimson-crested Woodpecker. Male lacks pale area around bill base. **V** *Quick quick quickerrrrr*. **H** Humid to dry deciduous forest, edge, secondary growth. **D** SW Colombia to NW Peru.

27 CREAM-BACKED WOODPECKER *Campephilus leucopogon* 34cm Mostly black with red head, ivory bill and buffy-white back. Female has cream moustachial stripe and black forehead. **V** *Pee-aw* or *kwee-or*. **H** Thorny woods, transition woods, savannah, palm groves, open woods, transition forest in mountains. **D** Bolivia, W Paraguay, NW Uruguay, N Argentina.

28 MAGELLANIC WOODPECKER *Campephilus magellanicus* 43cm The largest of the genus in South America. Bill black. Long, floppy, forward-curling crest. Head and neck red in male, largely black in female. Rest black but for white tertial flight feathers, underwing and axillaries. **V** Nasal *pisssaaah*. **H** Southern beech woods and forest. **D** Chile, SW Argentina.

29 IVORY-BILLED WOODPECKER *Campephilus principalis* 48–53cm In flight from above, shows distinctive white secondaries and inner primaries. Underwing also shows white coverts. **V** High-pitched, nasal *yank*. **H** Mature inaccessible forests. **D** SE USA. Almost certainly extinct.

30 IMPERIAL WOODPECKER *Campephilus imperialis* 60cm Very large, all-black underparts, male with red crest. **V** Said to have been surprisingly weak for a bird of its size, with the quality of a toy trumpet. **H** Open montane pine forest. **D** Mexico. Probably extinct.

31 BANDED WOODPECKER *Chrysophlegma miniaceum* 25–27cm Dark orangish face, dark chestnut wings, black-and-white-striped underparts, dark tail. Unobtrusive, usually encountered singly or in pairs foraging among vines and dense branches in the lower storey. **V** Screaming *kwee* or *chewerk-chewerk-chewerk-chewerk*; also utters a *keek* call note. **H** Broadleaved evergreen forest, secondary growth, plantations and occasionally mangroves. **D** Malay Peninsula; Borneo; Sumatra, Indonesia.

32 CHECKER-THROATED WOODPECKER *Chrysophlegma mentale* 26–30cm Moss-green with blood-red wings and neck sides, yellow crest on hindneck. Lively forager in lower and middle storey and higher parts of understorey. **V** Long series of *wi* notes; call includes a *kyick* and a *kiyee...kiyee...kiyee...* **H** Broadleaved evergreen forest and occasionally mangroves. **D** Malay Peninsula; Borneo; Sumatra, Indonesia.

33 GREATER YELLOWNAPE *Chrysophlegma flavinucha* 33–34cm Bright yellow nape, pale bill, grey underparts. Often encountered in small loose groups foraging on trunks and branches; also rummages on the ground searching for ants, termites and grubs. **V** Long, accelerating *kwee-kwee-kwee-kwee-kwee-kwee-kwee-kwee-kwi-kwi-kwi-kwi-wi-wi-wi-wik*, loud, plaintive *pee-u...pee-u...*, metallic *chenk*. **H** Broadleaved evergreen, deciduous and native pine forest. **D** Widespread from Himalayas and NE India through SE Asia and S China to Sumatra, Indonesia.

1
2
3
4
5
6
7
8
9
10
11
12
13
14
15
16
17
18
19
20
21
22
23
24
25
26
27
28
29
30
31
32
33

WOODPECKERS *CONTINUED*

1 LESSER YELLOWNAPE *Picus chlorolophus* 25–28cm Greenish, with white-barred chest and yellow hindcrest. Regularly appears in mixed-species feeding flocks that include woodpeckers, drongos and other insectivorous birds. Often feeds on the ground on ants and termites. **V** Loud, mournful *peee-ui*, *pee-a* or *pee-oow*. **H** Forest, secondary growth, plantations and well-wooded gardens. **D** India; Sri Lanka; Nepal; SE Asia; S China; Sumatra, Indonesia.

2 CRIMSON-WINGED WOODPECKER *Picus puniceus* 24–28cm Dark red crown and wings, yellow hindcrest. Favours tall trees; forages in the canopy on trunks and large branches. **V** Distinctive *pee-bee* or extended *pee-dee-dee-dee*; also a short *peep* and a *wee-eek*. **H** Broadleaved evergreen, secondary growth, plantations. **D** Malay Peninsula, Greater Sunda Islands.

3 STREAK-BREASTED WOODPECKER *Picus viridanus* 30cm Red cap in male, black in female. Regularly forages on the ground in search of ants. In flight shows a dull yellow-green rump. **V** *tcheu-tcheu-tcheu-tcheui*, explosive *kirrr*, squirrel-like *kyup*. **H** Mangroves. **D** Myanmar to Malay Peninsula.

4 LACED WOODPECKER *Picus vittatus* 30–33cm Distinguished from Streak-breasted Woodpecker by paler throat, yellow breast, more distinct black malar stripe. **V** Fast, low-pitched *pew-pew-pew*... also *keep* or *kee-ip*. **H** Broadleaved evergreen and deciduous forest, secondary growth, bamboo, plantations, coastal scrub, mangroves and gardens. **D** S China and E Myanmar through Thailand and Malay Peninsula to Sumatra and Java, Indonesia.

5 STREAK-THROATED WOODPECKER *Picus xanthopygaeus* 30cm Grey face, finely scaled underparts. In flight shows yellow rump. Generally solitary. Spends a lot of time on the ground. **V** Sharp *queemp*, otherwise rather silent. **H** Open deciduous, semi-evergreen and mixed bamboo forests, secondary growth, plantations. **D** India and Nepal to S Thailand and Cambodia.

6 SCALY-BELLIED WOODPECKER *Picus squamatus* 35cm Unmarked throat and breast, black-and-white scaly pattern from belly to tail. Feeds on trees and the ground. In undulating flight shows yellowish rump. **V** Ringing, melodious *klee-guh kleeguh* or *kuik-kuik-kuik*, also a high-pitched *kik* or a drawn-out nasal *cheenk*. Regularly drums during breeding season. **H** Mixed woodland, open country with copses, tamarisk scrub, orchards and groves. **D** SC Asia.

7 JAPANESE GREEN WOODPECKER *Picus awokera* 29–30cm Grey underparts with spotted belly. Feeds mainly in middle level of trees and occasionally on the ground. In flight shows a yellowish rump. **V** Loud *peoo peoo* and a *ket ket*. Drums in fairly long fast rolls. **H** Montane mixed and evergreen forest. **D** Japan.

8 EUROPEAN GREEN WOODPECKER *Picus viridis* 31–33cm Red crown, black face, pale eye. Feeds mostly on the ground. Flight undulating, showing prominent yellow rump. **V** Loud, laughing *klew-klew-klew-klew-klew*, also *kyu- kyu- kyuk* in flight. Rarely drums. **H** Forest edge, open woodland, copses, orchards, parks, large gardens. **D** NW Europe to Iran.

9 IBERIAN GREEN WOODPECKER *Picus sharpei* 31–33cm Iberian counterpart of European Green Woodpecker. Differs in having grey face. **V** Harsh *WHe-wa-wa-w-w-w-w* accelerating and descending in pitch towards end; screeching alarm call. **H** Open wooded grassland, arid ground, olive groves. **D** Iberian Peninsula.

10 LEVAILLANT'S WOODPECKER *Picus vaillantii* 30–32cm Female has mostly black cap. Actions much as Green Woodpecker, which is often considered conspecific, although drums more often. **V** Like Green Woodpecker but slightly quicker. **H** Mainly oak, poplar, cedar and pine forests. **D** N Africa, from Morocco to Tunisia.

11 RED-COLLARED WOODPECKER *Picus rabieri* 30cm Distinctive red pattern on head and neck. Generally forages low down on trunks or on the ground. **V** Not recorded. Drums in fast irregular rolls. **H** Primary and secondary semi-evergreen forest. **D** Laos, Vietnam.

12 BLACK-HEADED WOODPECKER *Picus erythropygius* 31–35cm Black face, yellow throat, green wings, red rump, white belly. Generally encountered in small, noisy groups. Forages in canopy and understorey, and on the ground. **V** Yelping *ka-tek-a-tek-a-tek-a-tek*... or *cha-cha-cha*...*cha-cha-cha*...; also gives a low double note. **H** Deciduous dry forest, open scrub. **D** Myanmar to Vietnam.

13 GREY-HEADED WOODPECKER *Picus canus* 25–26cm Grey face, thin black malar stripe. Regularly forages on the ground in search of ants, termites and grubs. **V** Musical, repeated *peeek peeek peeek peeek peeek* that fades at the end. When alarmed utters a chattering *kyakyakyak*. **H** Temperate and moist subtropical forests. **D** Widespread from N and C Europe to Siberia, Japan, China, SE Asia.

14 OLIVE-BACKED WOODPECKER *Dinopium rafflesii* 28cm Olive-brown body, black and white striped head and neck. Crest bright red in male, black in female. Forages from low to middle levels. **V** Slow *chakchakchakchakchak*...; also a single *chak*. **H** Dense, wet evergreen forests, mangroves. **D** Malay Peninsula; Borneo; Sumatra, Indonesia.

15 HIMALAYAN FLAMEBACK *Dinopium shorii* 30–32cm Brilliant orange back, black stripe through eye and cheek, black on back of neck. Systematically searches tree trunks for insects and grubs; also forages on the ground. **V** Rapid, tinny *klak-klak-klak-klak-klak*. **H** Mature deciduous and semi-evergreen forests. **D** N India to Myanmar.

16 COMMON FLAMEBACK *Dinopium javanense* 28–30cm Best told from very similar Himalayan Flameback by voice. Found at all forest levels, although prefers lower parts. Climbing is rapid and erratic. **V** Long, trilled *ka-di-di-di-di-di-di*..., a *kow* or *kow kow* and a *kowp-owp-owp-owp* in flight. **H** Open deciduous forest, scrub, gardens and occasionally mangroves. **D** SW India; NE India through SE Asia to Greater Sunda Islands.

17 SPOT-THROATED FLAMEBACK *Dinopium everetti* 28–30cm Actions presumably much as Common Flameback, with which it was formerly considered conspecific. **V** Probably similar to Common Flameback. **H** Open woodland. **D** W Philippines (Palawan and nearby islands).

18 BLACK-RUMPED FLAMEBACK *Dinopium benghalense* 26–29cm Dark throat. Unlike other members of the genus (which have three toes), this *Dinopium* has four toes. **V** Laughing *kyi-kyi-kyi* and strident *kierk*. **H** Open woodland, light forest, open country with trees, groves, plantations, gardens. **D** Indian subcontinent, NW Sri Lanka.

19 RED-BACKED FLAMEBACK *Dinopium psarodes* Both sexes have crimson upperparts. **V** Laughing *kyi-kyi-kyi* and strident *kierk*. **H** Open woodland, light forest, open country with trees, groves, plantations, gardens. **D** SE Sri Lanka.

20 BUFF-SPOTTED FLAMEBACK *Chrysocolaptes lucidus* 33cm This and the next four species were formerly considered races of Greater Flameback; actions and habits are presumed to be similar. Male has dark red crown and crest; female's crest is dark olive-brown to reddish-black with dark golden spots. **V** Single *kik* and monotonous metallic *di-di-di-di-di-di* or similar. **H** Deciduous and semi-evergreen forests, well-wooded areas. **D** EC and S Philippines.

21 LUZON FLAMEBACK *Chrysocolaptes haematribon* 28–34cm Male has dull red crown and crest; female's crown is dark, heavily spotted. Both sexes have spotted throat. **V** Presumed similar to Greater Flameback. **H** Primary and secondary forest, plantations with tall trees. **D** N Philippines.

22 YELLOW-FACED FLAMEBACK *Chrysocolaptes xanthocephalus* 28–30cm Distinctive golden-yellow face and underparts. Male has red crest. **V** Presumed similar to Greater Flameback. **H** Lowland forest, plantations. **D** C Philippines.

23 RED-HEADED FLAMEBACK *Chrysocolaptes erythrocephalus* 28–34cm Male has red head with black spot on ear-coverts. Female similar, but head is duller with pale-spotted crown. **V** Presumed similar to Greater Flameback. **H** Primary and secondary forest, plantations, clearings. **D** W Philippines (Palawan and nearby islands).

24 JAVAN FLAMEBACK *Chrysocolaptes strictus* 28–32cm Red crest in male, yellow in female. **V** Presumed similar to Greater Flameback. **H** Primary and secondary forest, coastal woodland, plantations. **D** E Java and Bali, Indonesia.

25 GREATER FLAMEBACK *Chrysocolaptes guttacristatus* 33cm Black and white bands on face. Male has red crest; female's is black with pale spots. Tends to prefer large trees, working up from the lower trunk in jerky spurts and spirals. Often seen as part of mixed-species feeding flocks. **V** Single *kik* and a monotone *di-di-di-di-di* or similar. **H** Coastal forest, open forest and forest edges. **D** Resident in W and C Sumatra, Java and Borneo.

26 CRIMSON-BACKED FLAMEBACK *Chrysocolaptes stricklandi* 29–31cm Colourful, well-marked species. Unmistakable within geographical range. **V** Screeching trill; drumming loud and accelerating. **H** Forested habitats, plantations; mainly lowlands. **D** Sri Lanka.

27 WHITE-NAPED WOODPECKER *Chrysocolaptes festivus* 29cm Large bill. Conspicuous black-bordered white nape. Usually encountered singly, in pairs or in small parties. Feeds on tree trunks and also on the ground. **V** Rattling, repeated *kwirri-rr-rr-rr-rr*, higher-pitched than calls of other flamebacks. **H** Deciduous forest, foothills with scrub and scattered trees. **D** India, Sri Lanka.

28 PALE-HEADED WOODPECKER *Gecinulus grantia* 25–27cm Dark body, pale head, reddish mantle and wings. Forages on bamboo, trees and fallen logs, sometimes on the ground. **V** Accelerating, nasal *chaik-chaik-chaik-chaik* or *kweek kwek-kwek* repeated four or five times. When alarmed, a rattling *kereki kereki*. Drums with fast steady rolls. **H** Bamboo jungle, mixed bamboo and secondary forest. **D** NE India, SE Asia, E China.

29 BAMBOO WOODPECKER *Gecinulus viridis* 25–26cm Plain coloration. Male has red crown. Best located by pecking when foraging, usually alone or in pairs. **V** Undulating, dry rattle and loud, clear *keep-kee-kee-kee-kee-kee*. Also *kweek-week-week-week-week* or similar during encounters with rivals. **H** Bamboo, broadleaved and deciduous forests. **D** Myanmar, Malay Peninsula.

30 MAROON WOODPECKER *Blythipicus rubiginosus* 23cm Dark reddish, with pale yellow bill. Generally forages low down, at the base of trunks, on fallen logs or on the ground. **V** Wavering, high-pitched *kik-kik-kik-kik-kik-kik-kik*... slowing towards the end; also a descending *chai-chai-chai-chai*. **H** Primary and secondary forest, plantations. **D** Malay Peninsula; Borneo; Sumatra, Indonesia.

31 BAY WOODPECKER *Blythipicus pyrrhotis* 27–30cm Rufous with black banding on wings and tail, paler brown head, bright yellow bill. Tends to forage low down on trunks, stumps and fallen logs. **V** Undulating, laughing *dit-d-d-di-di-di-di-dit-d-d-di-di-di*... or a *chake chake chake chake* that increases in tempo while dropping in pitch. **H** Broadleaved evergreen, occasionally semi-evergreen forest and mixed deciduous forest, nearby secondary growth and bamboo. **D** Nepal to S Vietnam and SE China.

32 ORANGE-BACKED WOODPECKER *Reinwardtipicus validus* 30cm Brightly coloured, crested, unmistakable. Forages on rotten logs, tree stumps, trunks and branches. Presence often given away by loud pecking and hammering. **V** Rapid, trilled *ki-i-i-i-i-i-ik*; also a squeaky *kit kit kit kit kit-it*, a *kee-wheet*, a *wheet wheet wheet wheet wheow*, and a chattering *cha-cha-cha*. **H** Forest, from the coast to mountains. **D** Malay Peninsula, Greater Sunda Islands.

33 RUFOUS WOODPECKER *Micropternus brachyurus* 25cm Brownish, with short dark bill. Usually seen in pairs, digging at tree-ant nests or foraging on fallen branches and termite nests; also feeds on fruit. **V** High-pitched *kenk kenk kenk*. Drumming said to sound like a stalling motorcycle engine. **H** Broadleaved evergreen forest, deciduous forest, forest edge and secondary growth. **D** From India and Sri Lanka to S China and Greater Sunda Islands.

34 BUFF-RUMPED WOODPECKER *Meiglyptes tristis* 17–18cm Tiny, with striped plumage and dark eye. Active forager from canopy down to smaller trees and saplings; generally encountered in pairs or mixed-species flocks. **V** Soft, rattled *drrrrrr*... also utters a *pit*, *chit* or *pit-pit* and a *peee* when alarmed. **H** Broadleaved evergreen forest, forest edge and secondary growth. **D** Malay Peninsula, Greater Sunda Islands.

35 BLACK-AND-BUFF WOODPECKER *Meiglyptes jugularis* 22cm Distinctive black and white patterning. Actions much as Buff-rumped Woodpecker, although not often seen in mixed-species flocks. **V** Nasal *ki-yew* and a rattling *titititit-week week week* or *titititititit weerk weerk weerk*... interspersed with a *ki-yew*. **H** Bamboo, open broadleaved evergreen and semi-evergreen forest. **D** SE Asia.

36 BUFF-NECKED WOODPECKER *Meiglyptes tukki* 21cm Lacks crest. Cream stripe on neck. Generally seen in the lower and middle storey, sometimes in mixed-species flocks. Often forages low down on stumps and fallen logs. **V** High, trilled *kirr-r-r* and a high-pitched *ti-ti-ti-ti-ti*. **H** Primary and secondary lowland forest. **D** Malay Peninsula; Borneo; Sumatra, Indonesia.

37 ASHY WOODPECKER *Mulleripicus fulvus* 39–40cm Long-tailed with grey upperparts. Male has red face. Forages on trunks and stout branches; also noted feeding on the ground or on rotting stumps. **V** Rapid, weak, even-pitched *kikikikikiki*..., laughing *hew-hew-hew-hew-hew-hew* and soft *twee twee twee*. **H** Primary and secondary forest, occasionally cultivated areas with trees, coconut groves, mangroves. **D** Sulawesi, Indonesia.

38 SOOTY WOODPECKER *Mulleripicus funebris* 34–35cm Mostly black or blackish. Male has dark red on face. Usually encountered singly or in pairs foraging in the upper storeys of tall trees. **V** Trilled *chil-lel-lel-lel-lel-lel*..., often repeated. **H** Various forest types, including secondary, mountain oak and pine, forest edges near plantations. **D** Philippines.

39 GREAT SLATY WOODPECKER *Mulleripicus pulverulentus* 45–51cm Largest Old World woodpecker, with a prehistoric appearance. Grey with bare yellowish throat patch. Forages mainly in tall trees. **V** Whinnying *woi-kwoi-kwoi-kwoi*, often given in flight; also utters a soft *whu-ick* and a single loud *dwot*, said to sound somewhere between the bleating of a goat and the barking of a dog. **H** Lowland primary forest, secondary growth, clearing edges, occasionally plantations. **D** From Nepal and N India to SW China, SE Asia, Greater Sunda Islands, Philippines.

1
2
3
4
5
6
7
8
9
10
11
12
13
14
15
16
17
18
19
20
21
22
23
24
25
26
27
28
29
30
31
32
33
34
35
36
37
38
39
♀
♂

CARACARAS AND FALCONS FALCONIDAE

1 BLACK CARACARA *Daptrius ater* male 43cm, female 48cm Black apart from bare orange face, yellow legs and broad white band at base of tail. Bill black. **V** Generally quiet, but may produce hoarse screaming and wailing calls. **H** Woods, savannah, edge, clearings, riverine forest, mangroves. **D** Amazonia.

2 RED-THROATED CARACARA *Ibycter americanus* male 47cm, female 55cm Bare face red and blue. Bill yellow. Legs orange-red. Lower underparts white. **V** Produces loud, raucous screeching calls when foraging in groups and as territorial defence. **H** Tropical lowland forest to mountains, deciduous forest, Cerrado. **D** S Mexico to S Brazil.

3 CARUNCULATED CARACARA *Phalcoboenus carunculatus* male 49cm, female 55cm Bare face and legs yellow-orange. Caruncles on sides of throat and behind eye. Slight topknot. Overall black. Foreneck, breast and upper belly irregularly streaked white. Lower belly, vent, uppertail coverts, thighs and underwing white. **V** Usually silent; when interacting with conspecifics gives grating squeals. **H** Páramos, grassy pastures, scattered bushes. **D** Ecuador, SW Colombia.

4 MOUNTAIN CARACARA *Phalcoboenus megalopterus* male 49cm, female 55cm Bare pinky-orange face. Legs yellow. Upperparts, neck, breast and flanks black. Rest of underparts, underwing coverts and tip of tail white. **V** Usually silent; may produce short, rasping barks. **H** Puna, high mountain desert, upland grassland, often near water. **D** Peru to C Chile.

5 WHITE-THROATED CARACARA *Phalcoboenus albogularis* male 49cm, female 55cm Yellow cere and legs. Upperparts black, underparts white, streaked on flanks. **V** Not known. **H** Upper mountain slopes to sea level, South Andean woods, rubbish dumps. **D** S Chile, S Argentina.

6 STRIATED CARACARA *Phalcoboenus australis* male 55cm, female 65cm Generally black with white streaking on neck and breast. Vent, thighs and underwing cinnamon-rufous. Tip of tail white. **V** Cat-like wails, screams and loud crow-like *kaaaa* in alarm. Courting pairs give low, throaty trills. **H** Rocky shores, open lowlands, to low coastal mountains, around colonies of seabirds and mammals. **D** Falkland Islands, several islands of Tierra del Fuego.

7 NORTHERN CRESTED CARACARA *Caracara cheriway* male 50cm, female 63cm Pale head with dark cap, dark body. Soars with wings slightly bowed. **V** Harsh, cackling *ca-ca-ca-ca*. **H** Palm savannahs, open country and pastures. Breeds, from February to December, mainly in palm trees. **D** S USA to N South America.

8 SOUTHERN CRESTED CARACARA *Caracara plancus* male 54–56cm, female 58–60cm Orange around bare face. Crown black and slightly crested at nape. White throat. Upperparts dark brown. Breast and underwings barred whitish. Long whitish tail finely barred black with broad blackish terminal band. **V** Courtship call *krakkrakkrak*, followed by *amrrrarrrrar r*. **H** Sparsely wooded open country, steppes, agricultural land, hills, marshes. **D** Widespread across South America, from Peru and N Brazil to Tierra del Fuego, Falkland Islands.

9 YELLOW-HEADED CARACARA *Milvago chimachima* 45cm Unmistakable by combination of buff head and body with dark mantle and wings. Walks around on the ground or perches low on fences and poles. **V** High piercing sibilant scream *sheeeargh*, sometimes in series. **H** Savannah, fields, ranchland. **D** Costa Rica to N Argentina.

10 CHIMANGO CARACARA *Milvago chimango* male 38cm, female 42cm Generally brown, paler below to whitish vent. Base of primaries whitish. Tail whitish, finely barred blackish, darkening towards tip. **V** Quiet outside the breeding season, when gives various squealing, chattering, growling and hissing notes. **H** Farmland, cultivated land, rubbish dumps, low in mountains, steppe, open woods, marshes, coastal regions, towns. **D** S South America, from S Brazil and Paraguay to Tierra del Fuego.

11 LAUGHING FALCON *Herpetotheres cachinnans* 50cm Unmistakable by black mask. Note upright stance. **V** High *haa-haah-hahhahhah* or high sustained *waak-waak-*, slowly rising in pitch. **H** Forest edge, forest remains, savannah. **D** Mexico to N Argentina.

12 BARRED FOREST FALCON *Micrastur ruficollis* 35cm Note yellow around large eyes. Brown and grey morphs. Finely barred underparts. **V** High sweeping *wew*. **H** Forest, forest edge. **D** S Mexico to N Argentina.

13 PLUMBEOUS FOREST FALCON *Micrastur plumbeus* 32–36cm Previously treated as subspecies of Lined Forest Falcon; differences include calls and single white tail band. **V** Calls include a shrill *peoow* and strident *Ke-ow*. **H** Humid primary forests, wooded slopes; mainly sea level to 1,500m. **D** SW Colombia, NW Ecuador.

14 LINED FOREST FALCON *Micrastur gilvicollis* male 32cm, female 38cm Like grey form of Barred Forest Falcon but cere and bare skin around eye orange, iris white, tail proportionately shorter with one or two very narrow white bands. Underparts less densely barred on breast and upper belly; white below, or all barred. **V** Nasal, slightly squeaky *kow-kah* or *kow-kaw-kaw-kaw*. **H** Primary lowland humid tropical forest, terra firme forest. **D** Amazonia.

15 CRYPTIC FOREST FALCON *Micrastur mintoni* 30–35cm Enigmatic. Compared to Lined Forest Falcon, note different call, shorter tail with broader white band, and orange skin surrounding whole eye. **V** Calls include a hooting, honking *ouP*, *ou-pouP*. **H** Lowland terra firme forest, várzea forest. **D** S and E Amazonian Brazil, Amazonian Bolivia.

16 SLATY-BACKED FOREST FALCON *Micrastur mirandollei* 45cm Large, with dark upperparts, more or less unmarked underparts, hazel eyes. **V** Plaintive rising loud *waah-waah-* (about 10 times). **H** Forest and adjacent areas. **D** Costa Rica to E Brazil.

17 COLLARED FOREST FALCON *Micrastur semitorquatus* 46–56cm Three colour morphs, two of which show pale collar across back of neck. Unmistakable by large eyes, long, white-barred tail, very short wings and habitat. Clambers around in trees, also runs on forest floor. **V** Far-carrying *cowh*. **H** Evergreen and deciduous forest. **D** C Mexico to N Argentina.

18 BUCKLEY'S FOREST FALCON *Micrastur buckleyi* male 40cm, female 50cm Resembles small version of pale-morph Collared Forest Falcon. Best identified by voice. **V** *Kawah, koh*, sometimes followed by a lower *koh*. **H** Forest in lowlands; also up to 1,800m. **D** East Ecuador, North East Peru, South West Colombia and adjacent Brazil.

19 SPOT-WINGED FALCONET *Spiziapteryx circumcincta* male 28cm, female 31cm Upperparts brownish grey, streaked blackish. White dots on wing coverts. White rump. Underparts pale greyish, streaked blackish. Brow, moustachial streak, throat and vent white. Rounded broad wings, longish rounded tail. **V** Series of nasal clucking notes, and loud, croaking *kronk-kronk-kronk*. **H** Dry and Chaco woods, dry savannah, plains and hills. **D** SE Bolivia and W Paraguay to C Argentina.

20 PYGMY FALCON *Polihierax semitorquatus* 18cm Back brown in female, grey in male. White rump conspicuous in flight. Perches shrike-like on top of tree or post. Roosts and breeds inside nests of Sociable Weavers. **V** Shrill, staccato yelps in series of three or four; also thin, squeaky *tsee-tsee* and purring *krr-krr-krr*. **H** More or less bushed areas and semi-desert where there are trees containing the large communal nests of weavers. **D** Sudan and Ethiopia to Tanzania; S Angola to NW South Africa.

21 WHITE-RUMPED FALCON *Polihierax insignis* 24–27cm Dark wings, clean white underparts. Male has a pale grey head; female is rusty orange on crown and back. Shrike-like actions: sits on high, bare branches and dives to the ground to take prey, which includes insects and frogs. Breeds in tree holes or nests of other birds. **V** Long, descending whistle. **H** Open deciduous woodland, clearings and forest edge. **D** SE Asia, from Myanmar to Vietnam.

22 COLLARED FALCONET *Microhierax caerulescens* 15–18cm Wide black eye-stripe, white collar, black upperparts, rufous beneath. Seeks prominent exposed perch from where to make short shrike-like sorties to capture prey. Slowly pumps tail and nods head while perched. **V** High-pitched *kli-kli-kli* or *killi-killi-killi*. **H** Edges and clearings in broadleaved tropical forest. **D** NE India and Nepal to SE Asia.

23 BLACK-THIGHED FALCONET *Microhierax fringillarius* 15–17cm Black above, white below, with rusty orange belly. Actions and habits similar to Collared Falconet. **V** Shrill, squealing *kweer*. **H** Broadleaved evergreen forest clearings, forest edge, wooded cultivations, parkland. **D** Malay Peninsula, Greater Sunda Islands.

24 WHITE-FRONTED FALCONET *Microhierax latifrons* 14–16cm Forecrown white in male, rusty orange in female. Perches on dead branches on a tall tree, from where dashing sorties are made to pursue insects or birds. **V** Undescribed, presumably similar to previous species. **H** Lowland primary forest and forest edges. **D** Borneo.

25 PHILIPPINE FALCONET *Microhierax erythrogenys* 15–18cm Black above, white below, no orange. Occurs singly, in pairs or in family groups. Favours the canopy or the upper branches of dead trees. Makes aerial forays after insects or small birds, picks lizards off trees or the ground. **V** High-pitched, rapid *chik-chik-erk-erk*, also a continuous squeaky *pew pew pew pew...* **H** Open forest, forest edges and clearings, from lowlands to mid-mountain. **D** Philippines.

26 PIED FALCONET *Microhierax melanoleucos* 18–20cm Black above, white below, black mask. Habits and actions similar to Collared Falconet. **V** Shrill whistle, a low chattering, also hissing sounds when agitated. **H** Forest edge and clearings, also recorded in tea plantations. **D** NE India to S China and C Vietnam.

27 LESSER KESTREL *Falco naumanni* 29–32cm A social bird, breeding colonially and migrating in flocks. Typically hovers less than Common Kestrel and in level flight wingbeats appear faster and shallower. Male has grey in wings and unspotted upperparts. White claws. **V** Rasping *chay-chay-chay*, usually given at roosts. When disturbed gives a trilling *keerrrl* or *kikikik*. **H** Open areas including cultivations. **D** S Europe and NW Africa to N China. Winters in sub-Saharan Africa.

28 COMMON KESTREL *Falco tinnunculus* 31–37cm Male has grey head, chestnut back, black-tipped grey tail. Female brown above, barred on back, wings, and tail. Frequently hovers when searching for prey or sits on post or other exposed perch from which it drops onto its quarry. Usually seen singly or in pairs. Black claws. **V** Shrill *kee-kee-kee-kee* and a trilling *vriii*. **H** Mountain slopes, hills, plains and cultivated areas. **D** Widespread across Europe, Africa, Asia. Migratory in N and E of breeding range, wintering in Africa and SE Asia.

29 ROCK KESTREL *Falco rupicolus* 35cm Rufous back and underparts, greyish head. Female has more strongly barred tail and more subdued colouring. **V** Like Common Kestrel. **H** Open or lightly wooded country, often near cliffs or tall buildings. **D** From Angola and S Tanzania to South Africa.

30 MALAGASY KESTREL *Falco newtoni* 25cm Variable: two colour forms, with rufous or white underparts. **V** Calls include fast, chattering *kikikiki* or *kree-kree-kree*, sometimes extended into 30-sec series. **H** Open areas, including settlement; up to high altitudes. **D** Madagascar and Aldabra Islands.

31 MAURITIUS KESTREL *Falco punctatus* 23cm No other resident raptor in range. Note irregular spotted, pale underparts and brown head. **V** Wailing *kreeeh* or *kyeeh*, abrupt *chip* and fast chattering calls. **H** Forest and scrub. **D** Mauritius.

32 SEYCHELLES KESTREL *Falco araeus* 20cm No other resident falcon in range; differs from other kestrels by small size. **V** Vocal. Calls include fast, thin, chattering *ki-ki-ki*. **H** Forest, cultivation, settlement. **D** Seychelles.

1
2
3
4
5
6
7
8
9
10
brown morph
11
12
13
14
15
16
17
3 colour morphs
17
grey morph
18
19
20
21
22
23
24
25
26
27
28
29
30
rufous morph
white morph
31
32

CARACARAS AND FALCONS *CONTINUED*

1 SPOTTED KESTREL *Falco moluccensis* 26–32cm Less sexually dimorphic than Common Kestrel. Perches on dead trees or hovers while searching for prey. **V** Shrill, repeated *kee-kee-kee-kee*, *kiek-kiek-kiek-kiek* or similar. **H** Grassland with scattered trees, lightly wooded cultivation, forest edges and clearings, and around human habitation. **D** Maluku Islands, Sulawesi, Java and Bali, Indonesia.

2 NANKEEN KESTREL *Falco cenchroides* 28–35cm Typical small kestrel, with actions similar to Common Kestrel. Usually occurs singly. **V** Rapid, shrill *ki-ki-ki*... and a slow *tek-tek-tek*. **H** Open country and farmland with scattered trees. **D** Australia. Non-breeding visitor to S New Guinea.

3 AMERICAN KESTREL *Falco sparverius* 23–30cm More colourful than most other kestrels. Watches for prey from high exposed perch or by hovering. **V** Shrill *killi-killi-killi-killi*. **H** Cosmopolitan, including forest edge, open areas, villages, towns and cities. **D** Widespread in North, Central and South America.

4 GREATER KESTREL *Falco rupicoloides* 34cm Pale eye of adult (visible at close range) diagnostic. Separated from other kestrels by more uniform rufous colouring, without cap or moustachial stripe, and by relatively longer wings, which reach tail tip. Normally hunts from perch, taking small prey from ground. **V** Quiet away from nest; alarm call *kwee-kwee*, also trilling notes and soft *kew*. **H** Dry, open country with scattered trees. **D** E and S Africa, from Ethiopia to Namibia and South Africa.

5 FOX KESTREL *Falco alopex* 40cm Underwing coverts very pale rufous with white (dark-tipped) flight feathers, the whole underwing contrasting strongly with dark-red body. **V** Around nest will give high-pitched shrieking *kree-kree-kree*. **H** Mountains and semi-desert with rocky hills. **D** Mauritania, Senegal and Gambia to Sudan, Ethiopia and Kenya.

6 GREY KESTREL *Falco ardosiaceus* 30cm Note absence of moustachial stripe, faint barring of flight feathers below. Wing tip shorter than tail when perched. Hunts at dusk from perch or in low flight over ground. **V** Shrill *kee-kee-kee*; also a rattling whistle and harsh twittering phrases. **H** More or less wooded grassland, often near palms and water. **D** Senegal to Ethiopia, south to Tanzania and west to Angola and Namibia.

7 DICKINSON'S KESTREL *Falco dickinsoni* 30cm Barred tail and white rump diagnostic. Head paler than breast and mantle. Hunts from perch. **V** In alarm gives a high-pitched *kee-kee-kee*; at nest a softer, mewing *ki-ki-ki-ki*. **H** Open woodland, plantations, edges of flood plains. Often near palms. **D** Angola and Namibia to Kenya and N Mozambique.

8 BANDED KESTREL *Falco zoniventris* 27cm Note pale eyes and blotch-barred plumage. Plunges into vegetation to catch prey. **V** Usually quiet, but has a high-pitched chattering call near the nest. **H** Forest edges and woodland. **D** Madagascar.

9 RED-NECKED FALCON *Falco chicquera* 30–36cm Rufous crown, nape, and moustachial stripe. Prominent yellow eye-ring. Finely barred white and grey underparts. Often hunts in pairs with a dashing low flight. **V** A strident *ki-ki-ki-ki-ki*, a harsh *yak yak yak* and a screaming *tiririri tiriririeee*. **H** Open country with patches of trees, cultivations, desert edge. **D** W, C and S Africa; Asia from Iran to India and Bangladesh.

10 RED-FOOTED FALCON *Falco vespertinus* 28–31cm Male dark grey with rufous 'trousers' and undertail coverts. Female has pale rufous head and underparts, whitish throat. Agile flyer when chasing flying insects. **V** Chattering *kekekeke*...; female gives a lower, slower *kwee-kwee-kwee*... **H** Open country with patches of trees. **D** Breeds C Europe to C Asia. Winters S Africa.

11 AMUR FALCON *Falco amurensis* 28–30cm Male sooty grey with rufous-orange thighs and undertail coverts. Female has well-marked whitish underparts. Highly manoeuvrable flight in pursuit of flying insects, also hovers. Gregarious. **V** Shrill *kew-kew-kew* when at roost. **H** Open country. **D** Breeds E Siberia, Korean Peninsula, NE China. Winters SE Africa.

12 ELEONORA'S FALCON *Falco eleonorae* 36–40cm Long-winged, long-tailed, elegant falcon. Pale and dark morphs illustrated. Fast acrobatic flier. Colonial nester. **V** In flight, nasal *kyeh-kyeh-kyeh-kyah*. **H** Islands and coastal cliffs. **D** Breeds S Europe, N Africa. Winters E Africa, Madagascar.

13 SOOTY FALCON *Falco concolor* 33–36cm Uniform grey plumage. Hunting flight energetic with fast glides and sudden stoops. **V** Loud *keee-keee-keee*...; also a plaintive chatter. **H** Desert, arid coastal areas, islands. **D** Breeds E Libya to SW Pakistan. Winters SE Africa, Madagascar.

14 APLOMADO FALCON *Falco femoralis* male 36cm, female 44cm Distinctive grey and white face pattern, white breast, black band across mid-belly, lower underparts tawny. Tail barred blackish and whitish. **V** Rather diverse range of chattering, chipping and wailing notes; most frequently heard is a repeated *kek*. **H** Anywhere except dense forest and mountain tops. **D** Widespread across Central and South America.

15 MERLIN *Falco columbarius* 25–30cm Dark above, paler and streaked below. Considerable geographical variation. In pursuit of prey, flight is often dashing with twists and turns, usually at low level. **V** Generally silent except in vicinity of nest. **H** Open country. **D** Breeds across North America from Alaska to Newfoundland, and right across N Europe and N Asia. Winters as far south as N South America, N Africa, Pakistan, S China.

16 BAT FALCON *Falco rufigularis* male 24cm, female 30cm Small and dark. Upperparts and cheeks black. Throat, lower cheeks and upper breast whitish or washed rufous. Breast, upper belly and flanks black, finely barred white. Lower underparts rufous. Tail black with four narrow white bands. Active at dusk. **V** Pairs call loudly, giving screaming *kee-kee-kee*; also sharp *kip*. **H** Lowland tropical and subtropical forest to mountains, edge and clearings, deforested areas with scattered trees. **D** N Mexico to NE Argentina.

17 ORANGE-BREASTED FALCON *Falco deiroleucus* male 33cm, female 39cm A larger and more powerful version of Bat Falcon. Upperparts and cheeks black. Upper breast rufous, streaked black. Lower breast and flanks black, barred rufous. Rest of underparts chestnut-rufous. Short black tail with fine white bars. **V** Harsh, rapid and very loud *key-key-key-key* when defending the nest. Pairs at nest exchange softer, chirping notes. **H** Forest in mountains and foothills, edge and clearings of tropical and subtropical forest, Chaco woodland, gallery forest. **D** S Mexico to NE Argentina.

18 EURASIAN HOBBY *Falco subbuteo* 30–36cm Grey upperparts, plain tail. Boldly streaked underparts with rufous 'trousers'. Fast acrobatic flier, catching prey, mainly insects and small birds, in flight. Regularly perches on isolated trees. **V** Rapid *kew-kew-kew-kew-kew*... **H** Breeds in well-wooded areas, resorts to more open areas after breeding. **D** Breeds from W Europe to Japan and S China. Winters sub-Saharan Africa, SE Asia.

19 AFRICAN HOBBY *Falco cuvierii* 30cm Darker above than European Hobby, with rich chestnut underparts. More graceful, less compact than Taita Flacon. **V** Gives rapid, high-pitched *kee-kee-kee* when defending nest or greeting partner; a screamed *keeeeee-ee* may be part of courtship vocalisation. **H** Forest edges, open landscapes with scattered trees, densely populated, rural areas. **D** Widespread in sub-Saharan Africa.

20 ORIENTAL HOBBY *Falco severus* 27–30cm Distinguished from other hobbies by solid rufous underparts. Actions and habits much like Eurasian Hobby. **V** Rapid *ki-ki-ki-ki*. **H** Open or lightly wooded hills. **D** From NW India through SE Asia to Solomon Islands.

21 AUSTRALIAN HOBBY *Falco longipennis* 33cm Perches conspicuously, usually singly. Swift, agile aerial hunter of birds, insects and bats. **V** Weak, high-pitched *ki-ki-ki-ki-ki*... **H** Lightly wooded grassland, open areas, wooded cultivation, edges of coastal and lowland monsoon forest. **D** Australia, New Guinea, Maluku Islands, Lesser Sunda Islands.

22 NEW ZEALAND FALCON *Falco novaeseelandiae* 45cm Unmistakable in its range. Streaky cream and brown breast, reddish-brown undertail coverts **V** Sharp, very high, hurried series *weetweetweet*. **H** Mosaic of forest and grassland, pastures, tussock land. **D** New Zealand.

23 BROWN FALCON *Falco berigora* 40–50cm Variable brownish plumage, ranging from dark chocolate brown to pale reddish. Dark moustache and finely barred tail. Hunts from perch, while hovering and in rapid pursuit flight. Diet includes both invertebrates and vertebrates. **V** Loud, shrill and vibrating *Kierr-ker-k-k-k* call. **H** Open habitats including savannah, grassland, agricultural land. **D** Australia, New Guinea.

24 GREY FALCON *Falco hypoleucos* 34–42cm Note grey plumage and striking yellow feet. Feeds mainly on birds; also small mammals, lizards and large invertebrates. Hunts from perch and in low-level pursuit. **V** Plaintive screaming *kee-er, kee-er*. **H** Open woodland, savannah, river-side scrub. **D** Australia.

25 BLACK FALCON *Falco subniger* 51cm Large falcon with uniformly dark plumage. Distinguished from darker individuals of Brown Falcon by proportionately smaller head, feathered legs, more pointed wings, less strongly marked face. **V** Quiet, but will give screaming and chattering sounds around nest. **H** Grassland and open woodland. **D** Australia.

26 LANNER FALCON *Falco biarmicus* 40–50cm Greyer and more slender than Saker Falcon. Rufous hindcrown. Hunts mainly by high-speed aerial chase, attacks often carried out by a pair of birds. **V** Rapid *kre-kre-kre*... and a plaintive *ueeh*. **H** Variable: forested mountains to dry open desert. **D** S Europe, Caucasus, Arabian Peninsula, widespread across Africa.

27 LAGGAR FALCON *Falco jugger* 43–46cm Darker than Lanner Falcon. Rusty crown, thin white supercilium. Waits on exposed perch, and when prey is sighted makes a swift low-level attack, victim taken in the air or on the ground. **V** Shrill *whi-ee-ee*. **H** Arid or semi-arid open country, cultivations, sand dunes. **D** Pakistan through India to Myanmar.

28 SAKER FALCON *Falco cherrug* 45–55cm Powerful falcon with pale head, narrow moustache, dark 'trousers'. Often encountered sitting on rocks. Usually hunts at low level; most prey, predominantly rodents, taken on the ground. Occasionally hovers or makes aerial stoops to catch birds. **V** Harsh *kek-kek-kek*... **H** Desert, semi-desert. **D** Breeds C Europe to C Asia, S Siberia, N China, Tibet. Winters S and SW Asia, Africa.

29 GYRFALCON *Falco rusticolus* 48–60cm Large falcon with a wide range of coloration, from white to dark sooty brown. All morphs tend to intergrade with each other. Mainly hunts at low level, prey usually taken on the ground. **V** Harsh *kak-kak-kak*... **H** Breeds in tundra and taiga, also occurs in cultivated and coastal areas during non-breeding season. **D** Arctic Europe, Asia, North America.

30 PRAIRIE FALCON *Falco mexicanus* 43–51cm Pale brown above, white below, with dark streaks and prominent moustache. Separated from Peregrine Falcon by dark axillaries. Hunts ground squirrels and other small mammals. **V** Shrill *kree-kree-kree*. **H** Mountain areas with nearby arid open country. **D** Widespread in W USA, north into SW Canada.

31 PEREGRINE FALCON *Falco peregrinus* 41–51cm Broad dark moustache. Bold barring on underparts. Prey usually killed in mid-air following high-speed pursuit and stoop. **V** Loud *ka-yak ka-yak ka-yak* or when alarmed a shrill *kek-kek-kek*. **H** Sea-cliffs, mountainous regions, open areas. Increasingly common in cities. **D** Worldwide, from tundra to tropics. Northern populations migrate south in winter.

32 BARBARY FALCON *Falco pelegrinoides* 35–42cm Often considered a race of Peregrine Falcon. Rufous nape may be hard to see. Prey usually taken in mid-air following a high-speed pursuit and stoop. **V** Scolding *kek-kek-kek-kek*... **H** Open desert or semi-desert with rocky hills. **D** N Africa, Arabian Peninsula to Pakistan and Mongolia.

33 TAITA FALCON *Falco fasciinucha* 30cm Small and compact appearance. Black above with grey rump and tail. Note large claws. Smaller and darker than Lanner Falcon. Chestnut nape patch diagnostic. **V** Silent except around nest, when gives loud, chattering *kek-kek-kek* and squealing *kree kree kree*. **H** Mountainous, dry, open country, normally near cliffs. **D** E Africa, from S Ethiopia to South Africa.

1
♂
♀
2
♂
♀
3
♂
♀
4
5
pale
morph
12
dark
morph
6
8
9
10
♀
11
♀
♂
7
♂
13
14
15
♀
♂
16
17
18
19
20
21
22
dark
morph
23
light
morph
24
25
26
27
28
29
29
3 colour
morphs
30
31
32
33

NEW ZEALAND PARROTS STRIGOPIDAE

1 KAKAPO *Strigops habroptila* 64cm Strictly nocturnal and very rare. Flightless. Unmistakable in the unlikely event of being seen. **V** Very low, resounding, mechanical *booh booh*, repeated. **H** Originally in a wide range of habitats: mossy forest, scrub, tussock grassland, peatlands, pastures. **D** New Zealand.

2 KEA *Nestor notabilis* 48cm Rare and almost exclusively restricted to mountains. From darker and smaller New Zealand Kaka by darker cap, and lack of crimson feather edges to feathers of vent and lower belly. **V** High, slightly descending, slightly mewing *keeaaah*, repeated. **H** Wide variety of habitats, especially at higher altitudes, from meadows to temperate rainforest; up to 2,100m. **D** South Island, New Zealand.

3 NEW ZEALAND KAKA *Nestor meridionalis* 45cm Note yellow-orange cheek feathers and orange wash to neck feathers. North Island race has darker cap. Rare, except on a few islands. **V** Varied, e.g. scratchy *kraakaha*, melodious, whistled *peerowuh'peeruwuh*, liquid *lokloklok*. **H** Prefers undisturbed native forest; up to 1,200m. **D** New Zealand.

COCKATOOS CACATUIDAE

4 COCKATIEL *Nymphicus hollandicus* 30–32cm Unmistakable. Forms large, nomadic non-breeding flocks. Visits water to drink. Feeds mainly on seeds. **V** Shrill whistling screech. **H** Arid, open woodland, savannah, cultivated ground. **D** Australia, generally interior.

5 RED-TAILED BLACK COCKATOO *Calyptorhynchus banksii* 50–65cm Note dark plumage except for red subterminal band on outer-tail feathers. Diet mainly fruits and nuts. **V** Screeching, trumpeting call with wildfowl- or crane-like tones. **H** Arid woodland, savannah, agricultural land. **D** Australia: disjunct populations in N and NE, interior, S, SE.

6 GLOSSY BLACK COCKATOO *Calyptorhynchus lathami* 48–50cm Red tail panels in male; orange-red in female. Specialist diet comprises mainly seeds of *Allocasuarina* (she-oaks). Small flocks gather where feeding is good. **V** Drawn-out, screeching, trumpeting call with crane-like tones. **H** Coastal forests with she-oaks, open woodland. **D** E Australia, Kangaroo Island (S Australia).

7 YELLOW-TAILED BLACK COCKATOO *Calyptorhynchus funereus* 55–60cm Yellow cheek patches and tail panels. Diet includes seeds (native species and introduced pines) and insects (notably colonial moth larvae). Flocks gather where feeding is good. **V** Harsh, drawn-out screeches and strident *wheEEer*. **H** Wooded habitats, heaths, pine plantations. **D** E and SE Australia, Tasmania.

8 BAUDIN'S BLACK COCKATOO *Calyptorhynchus baudinii* 55–60cm White cheek patches and tail panels. Diet includes seeds of native species of *Eucalpytus* and *Banksia*. Forms wandering post-breeding flocks. Sometime mixes with Carnaby's Black Cockatoo. **V** Screeching *wheEEouer*, often in duet or groups. **H** Eucalypt woodland, wooded farmland, parks. **D** Restricted range in SW Western Australia.

9 CARNABY'S BLACK COCKATOO *Calyptorhynchus latirostris* 55–60cm Distinguished from Baudin's Black Cockatoo by smaller upper mandible. Forms large, wandering post-breeding flocks. Diet includes seeds of native species including *Banksia*, and introduced Monterey Pine. **V** Screeching, drawn-out *wherEEouer*, often in duet or groups. **H** Eucalypt woodland, scrub, pine forest. **D** SW Australia.

10 PALM COCKATOO *Probosciger aterrimus* 55–60cm Massive bill. No red in tail. Usually seen singly or in pairs; forages in trees and occasionally on the ground to pick up fallen fruit. **V** Calls include peculiar whistles and a hoarse, nasal screech. **H** Lowland and hill forest and forest edges. **D** New Guinea, N Australia.

11 GANG-GANG COCKATOO *Callocephalon fimbriatum* 33–36cm Distinctive plumage. Altitudinal migrant. Breeds in mountains, winters in lowlands. Diets includes fruits, berries, some invertebrates. **V** Squeaky, creaky call, likened to cork being pulled from a bottle. **H** Wooded habitats, forested slopes, parks; seasonally, lowlands to 1,500m. **D** SE Australia.

12 GALAH *Eolophus roseicapilla* 36cm Unmistakable large grey parrot with pink underparts and pale cap. **V** Typical flight call a muffled two-note screech, *chee-chuh*. Alarm call is a harsh *khEH*. **H** Open woodland, savannah, shrub, agricultural land, parks, gardens. **D** Australia, widespread.

13 MAJOR MITCHELL'S COCKATOO *Lophochroa leadbeateri* 35–40cm Distinctive colour, markings and crest. Nomadic outside breeding season. Diet includes seeds; benefits from spilt agricultural grain. **V** Harsh, rather warbling screech; also subdued almost conversational squawk. **H** Arid open woodland, mallee scrub, tree-lined river banks. **D** Australia.

14 LONG-BILLED CORELLA *Cacatua tenuirostris* 38–30cm White plumage, red face and throat aid identification. Long bill used to excavate plant roots, bulbs and corms. **V** Excited, chattering screeching squawks; often in duets or groups. **H** Eucalypt woodland, grassland, cultivated land. **D** E, S and SW Australia.

15 WESTERN CORELLA *Cacatua pastinator* 38–44cm Uses bill to excavate soil for plant roots, including agricultural crops, but also invertebrates. **V** Hooting trisyllabic squawks, resonating rather like peacock's call. **H** Eucalypt forest, wooded grassland, agricultural land. **D** SW Australia.

16 LITTLE CORELLA *Cacatua sanguinea* 38–39cm Forms large non-breeding roosting flocks. Diet includes seeds, grain and underground plant roots. Regularly visits water to drink. **V** Shrill, insistent tri-syllabic screeches. **H** Wooded grassland, agricultural land. **D** Australia (widespread except W interior and NE), S New Guinea.

17 TANIMBAR CORELLA *Cacatua goffiniana* 31–32cm Often seen in large flocks. Regularly perches on bare branches at the top of tall trees. When crest is raised, pink feather bases are exposed. **V** Various loud, harsh screeches. **H** Primary and tall secondary lowland forest, forest edges, clearings, crop fields adjacent to forest. **D** Tanimbar Islands, Indonesia; introduced elsewhere.

18 SOLOMONS COCKATOO *Cacatua ducorpsii* 30cm Usually seen in pairs or small flocks. Diet includes seeds, fruits and invertebrates. Sometimes excavates underground tubers and bulbs. **V** Raucous *Whah, Whah-wheh, Whah, wheh* calls, often duetted or in groups. **H** Forested habitats; lowlands to 1,500m. **D** Solomon Islands.

19 RED-VENTED COCKATOO *Cacatua haematuropygia* 30cm Generally seen singly, in pairs or in flocks of up to 30 or so. Non-breeding groups nomadic; often raids crops. The only all-white land bird in the Philippines. **V** Various raucous calls, which can be very loud when groups are calling together. **H** Lowland, riverine and mangrove forest, forest edges and adjacent open ground. **D** Philippines.

20 SULPHUR-CRESTED COCKATOO *Cacatua galerita* 45–55cm Conspicuous large white cockatoo with spectacular yellow crest and dark bill. Occurs singly, in pairs or in small, noisy groups. Forages in trees, also on the ground when raiding crops. **V** Loud, raucous screech. **H** Forest, open woodland, forest edges and lightly wooded cultivation. **D** N, E, SE Australia; New Guinea.

21 BLUE-EYED COCKATOO *Cacatua ophthalmica* 45–50cm Note colourful blue skin around eye. Mainly arboreal. Feeds on fruits and nuts. **V** Trumpeted *Eeh, Eeh, Eeh, Eeh...* screeches and shrill, whistling calls. **H** Primary forest, secondary woodland; mainly lowlands. **D** New Britain (Bismarck Archipelago).

22 YELLOW-CRESTED COCKATOO *Cacatua sulphurea* 33cm White with yellow crest. Occurs singly, in pairs or in small groups, with larger groups at roosts. Forages mainly in the lower canopy, or on the ground when raiding crops. **V** Raucous screeching and a variety of whistles and squeaks. **H** Primary and tall secondary lowland and hill forest, forest edges, monsoon forest, tall scrubby woodland, wooded scrub, cultivation. **D** Sulawesi and Lesser Sunda Islands.

23 SALMON-CRESTED COCKATOO *Cacatua moluccensis* 46–52cm Salmon colour may be concealed if crest is not raised. Usually encountered singly, in pairs or in small groups. Inconspicuous, except when flying to and from roosts. Forages quietly in the upper mid-storey and canopy. **V** Harsh screeches and a nasal, staccato chatter. **H** Primary and tall secondary lowland and hill forest. **D** S Maluku Islands, Indonesia.

24 WHITE COCKATOO *Cacatua alba* 46cm Entirely white crest. Noisy and conspicuous; occurs singly, in pairs, in small groups or in large groups in late afternoon. Frequents the canopy. **V** Short, nasal, high-pitched screech. **H** Primary and tall secondary lowland and hill forest, forest edges, forest remnants. **D** N Maluku Islands, Indonesia.

AFRICAN AND NEW WORLD PARROTS PSITTACIDAE

25 GREY PARROT *Psittacus erithacus* 30cm Unmistakable grey parrot with white face and red tail. **V** Very varied whistles and screeches; capable of sophisticated mimicry, though this is rarely documented in the wild. **H** Forest. **D** SE Ivory Coast to Kenya, Tanzania and Angola; islands in Gulf of Guinea.

26 TIMNEH PARROT *Psittacus timneh* 28–32cm Previously treated as subspecies of Grey Parrot. Smaller and darker than that species, with maroon (not red) tail. **V** Abrupt screeches and whistles. **H** Primary forest, secondary woodland, plantations, agricultural land. **D** S Guinea, Sierra Leone, Liberia, W Ivory Coast; Príncipe Island, Gulf of Guinea.

27 RED-FRONTED PARROT *Poicephalus gulielmi* 30cm Note brilliant green plumage with darker face and wings, red forehead and shoulder, yellowish-green rump. **V** Typical call is an upslurred squealing whistle *kweet*; other screeching notes also recorded. **H** Forest canopy. **D** W and C Africa: Liberia to Ghana, Cameroon to NW Angola, DR Congo, Uganda.

28 YELLOW-FRONTED PARROT *Poicephalus flavifrons* 25cm Green, with yellow forehead, crown and upper cheeks. Flight feathers and tail dusky brown. **V** High-pitched, upslurred short *pEEh* or *kwEE-ee*, notes sometimes given in song-like series. **H** Forest. **D** Ethiopia.

29 BROWN-NECKED PARROT *Poicephalus fuscicollis* 35cm Similar to Red-fronted Parrot but paler head. Note pale yellowish bill. **V** Short, screeching, flat-pitched *cheeh*, and other short but piercing notes. **H** Woodland with baobab. **D** Senegal to Angola, E to Tanzania and NE South Africa.

30 CAPE PARROT *Poicephalus robustus* 35cm Note pale olive head, red 'socks', red on shoulder. Not all females have reddish forehead as shown. **V** Various rather sweet-toned but piercing whistled notes, upslurred or downslurred; also gives grating or buzzy notes. **H** Forest at mid–high altitudes and woodland. **D** E South Africa.

31 MEYER'S PARROT *Poicephalus meyeri* 22cm Note blue-green lower breast. From Niam-niam, Senegal and Rüppell's Parrots by smaller yellow area on underwing. Upperparts mainly brown. Yellow on forehead sometimes missing. **V** High-pitched, short whistled notes, screeches and squawks. **H** More or less wooded grassland. **D** C and SC Africa, from N Cameroon and Ethiopia to N South Africa.

32 RÜPPELL'S PARROT *Poicephalus rueppellii* 23cm Brown with yellow shoulder, yellow underwing coverts and blue belly without any green. Blue of female deeper. **V** Powerful, rather low-pitched *cheet*, and other shriller notes. **H** Dry grassy thornbush, woodland. **D** Angola, Namibia.

33 BROWN-HEADED PARROT *Poicephalus cryptoxanthus* 25cm Predominantly green with brown head and yellow underwing coverts. Note bicoloured bill. **V** Upslurred sharp and piercing *kwEEEt* or *krra-eet* and other similar-toned notes. **H** More or less wooded grassland, thornveld, plantations. **D** SE Africa, from SE Kenya to NE South Africa.

34 NIAM-NIAM PARROT *Poicephalus crassus* 23cm Mainly green, with grey head, extending to breast. Yellow underwings. **V** Little known, but a harsh, tern-like double screech has been recorded. **H** Wood patches and tree stands in open country. **D** Cameroon and Chad to South Sudan, NE DR Congo.

35 SENEGAL PARROT *Poicephalus senegalus* 25cm Grey head, green upperparts and breast, yellow belly. **V** Various drawn-out screeching, rasping and grating calls, also deeper and quieter nasal notes. **H** All types of woodland, especially with baobab. **D** W Africa, from Senegal to N Cameroon.

36 RED-BELLIED PARROT *Poicephalus rufiventris* 25cm Head, breast, upperparts and tail greyish brown, thighs and vent green. Male has orange-red upper belly and underwing coverts. Female has green belly and brown underwing coverts. **V** Has broad repertoire of grating screams and harsh notes, as well as some more complex higher-pitched phrases. **H** Dry bushland with some trees, especially baobab. **D** E Africa, from Ethiopia and Somalia to NE Tanzania.

1
2
3
4
5
6
7
8
9
10
11
12
13
14
15
16
17
18
19
20
21
22
23
24
25
26
27
28
29
30
31
32
33
34
35
36

AFRICAN AND NEW WORLD PARROTS *CONTINUED*

1 LILAC-TAILED PARROTLET *Touit batavicus* 15cm Parts of back and wings black and yellow. Tail lilac. Underwing very dark blue. **V** Penetrating *eee, eeeth.* **H** Montane deciduous woods to temperate, tropical, cloud, gallery and coastal forests. **D** Venezuela to French Guiana.

2 SCARLET-SHOULDERED PARROTLET *Touit huetii* 15cm Forehead black, upper cheeks dark blue. Crown and ear-patch yellowish. Leading edge of wing blue and red. Undertail bright yellow. Outer-tail feathers carmine (male) or light green (female). Underwing red. **V** *Touit.* **H** Tropical rainforest near rivers, várzea. **D** Amazonia.

3 RED-FRONTED PARROTLET *Touit costaricensis* 17cm Note red leading edge of wing. From Blue-fronted Parrotlet by head pattern. **V** Nasal *kyee kyee* given in flight; other calls include *kerr, kee-ree* and *chi-chi-kerrr.* **H** Montane forest. **D** Costa Rica to W Panama.

4 BLUE-FRONTED PARROTLET *Touit dilectissimus* 17cm Forehead and upper cheek blue. Lores and line under the eye red. Crown, nape and hindneck olivaceous brown. Red on leading edge of wing. Underwing and outer tail bright yellow in male, the latter greenish yellow in female. **V** High-pitched and nasal *too eet.* **H** Humid forest and cloud forest in mountains, tall secondary growth, foothills. **D** E Panama to Venezuela and NW Ecuador.

5 SAPPHIRE-RUMPED PARROTLET *Touit purpuratus* 16cm Dark brownish scapulars. Crown, ear-coverts and hindneck olive-brown. Lower back to rump very dark blue. Outer-tail feathers carmine, with green subterminal band in female. **V** *Kerreee, ke, ke.* **H** Tropical and subtropical forest in mountains, savannah, coastal forest, woods, várzea. **D** N and C Amazonia.

6 BROWN-BACKED PARROTLET *Touit melanonotus* 15cm Back and scapulars dark brown. Crown and hindneck brownish green. Outer-tail feathers carmine. **V** Harsh and sharp *tiree.* **H** Humid forest on lower mountain slopes. **D** SE Brazil.

7 GOLDEN-TAILED PARROTLET *Touit surdus* 16cm Scapulars brown. Forehead, lores and part of cheeks ochreous yellow. Outer-tail feathers golden yellow in male, yellowish green in female. **V** High-pitched *kree-kree-kree-kruh* or *wi-chi-chi-chi-chi.* **H** Lowland evergreen forest, coastal forest. **D** E Brazil.

8 SPOT-WINGED PARROTLET *Touit stictopterus* 17cm Wing coverts dark brownish grey with contrasting whitish spots in male, plain green in female. **V** a rough *ch, ch, ch.* **H** Tropical and subtropical montane forest. **D** S Colombia, Ecuador, N Peru.

9 GREY-HOODED PARAKEET *Psilopsiagon aymara* 18cm Cap brownish grey. Cheeks, foreneck, breast and part of belly very pale grey. **V** Clicking calls in flight. **H** Arid mountains, bushy slopes with grass and cactus, sheltered valleys. **D** N Bolivia to NW Argentina.

10 MOUNTAIN PARAKEET *Psilopsiagon aurifrons* 18cm Mostly green, paler below, flight feathers bluish. **V** Powerful *tee-teet* or *tee-tee-teet* and a low, guttural twittering phrase. **H** Arid mountains, puna, scrubby grass slopes with small trees, upland steppe, cultivated areas. **D** W South America, from Peru to C Chile and NW Argentina.

11 BARRED PARAKEET *Bolborhynchus lineola* 16cm Green, barred and scaled black all over. **V** Musical chatter. **H** Mountains with subtropical and temperate forest, cloud forest, canebrakes, open woods, plantations. **D** S Mexico to Colombia and Peru.

12 RUFOUS-FRONTED PARAKEET *Bolborhynchus ferrugineifrons* 18cm Broad tail. Overall dull green. Olivaceous wash on cheeks and breast. Narrow rusty forehead and lores. **V** Series of short, slightly nasal *gwuh* notes. **H** Mountains, temperate scrubby slopes at or above tree line. **D** C Colombia.

13 ANDEAN PARAKEET *Bolborhynchus orbygnesius* 17cm Broad tail. All dull green, with brighter tail. Head, breast and upper belly washed olive. Forehead pale green. **V** Short or long series of flat, toneless *juh-juh-juh* notes. **H** High mountains, stunted forest, *Polylepis* woods, bushy ravines in open country. **D** Peru to NC Bolivia.

14 TEPUI PARROTLET *Nannopsittaca panychlora* 14cm Tiny. Short square tail. Mostly dark green, paler below. Forehead, lores and line below the eye greenish yellow. Underwing dark green, flight feathers greyish green. **V** In flight utters a dry, high-pitched *scree* or less often a softer *dju-dju*; flocks give constant dry twittering. **H** Montane forest, cloud forest, lowlands around tepuis. **D** Venezuela, S Guyana, N Brazil.

15 MANU PARROTLET *Nannopsittaca dachilleae* 14cm Bright green, paler below. Crown and forehead blue. Black alula. Bill pinky cream, legs orange. Short tail. **V** Range of chirruping, sparrow-like calls. **H** Humid forest near big rivers. **D** SW Amazonia.

16 MONK PARAKEET *Myiopsitta monachus* 28–29cm Forehead, forecrown, cheeks, breast and upper belly grey, scaly on breast. Rest bright green, flight feathers bluish. Gregarious and noisy. Builds huge communal stick nests high in trees or pylons. **V** Loud, staccato shriek and high-pitched chattering. **H** Palm groves, woods, plantations, farmland, savannah, tree-lined avenues, urban parks, gardens. **D** SE South America, from S Bolivia to Argentina. Feral in North America, Europe.

17 CLIFF PARAKEET *Myiopsitta luchsi* 28–29cm Previously treated as subspecies of Monk Parakeet but with cleaner-looking unmarked underparts. Forms flocks. **V** High-pitched squeak recalling a child's squeaky toy. **H** Forested slopes. **D** Andean Bolivia.

18 TUI PARAKEET *Brotogeris sanctithomae* 17cm Tiny, with short, pointed tail. Pale green. Bill horn-grey. Forehead and ear-patch yellow. Underwing coverts pale yellow, flight feathers bluish green. **V** Call is a loud, dry, repeated *screek.* **H** Near water, secondary growth, gallery forest, várzea, edges, palm groves. **D** Amazonia.

19 PLAIN PARAKEET *Brotogeris tirica* 24cm Long pointed tail. Light green with brownish-olive shoulders and blue flight feathers. Underwing light green. **V** Usual calls, given in flight and when perched, are a shrill *kree* or *cra-creep* and a deeper *cra-cra-cra.* **H** Edge of the forest, palm groves, gardens and parks. **D** SE Brazil.

20 WHITE-WINGED PARAKEET *Brotogeris versicolurus* 23cm White inner primaries and secondaries, with yellow secondary coverts. Gregarious, occasionally in very large flocks. **V** High, scratchy *krere-krere*; high-pitched chattering from feeding flocks. **H** Coastal woodlands, hills and foothills of mountains, urban areas. **D** Amazonia. Introduced to California and Florida.

21 YELLOW-CHEVRONED PARAKEET *Brotogeris chiriri* 23–25cm Yellow wing coverts, with no white. Previously treated as conspecific with White-winged Parakeet, under the name 'Canary-winged Parakeet'. Feeds on fruits and nuts. Forms flocks. **V** Rapid and agitated chattering calls. **H** Open forest, wooded savannah, swamp forest, cultivated ground. **D** Bolivia, Brazil, Paraguay, NW Argentina. Introduced to California and Florida.

22 GREY-CHEEKED PARAKEET *Brotogeris pyrrhoptera* 19cm Short-tailed and dumpy. Mostly light green. Forehead and cheeks pearl-grey. Crown and nape pale blue. Chin spot and conspicuous underwing flame-orange. Flight feathers darker green. **V** Calls *chree, chree-chree* in flight and at rest; also gives rapid, chattering *cra-cra-cra.* **H** Evergreen forest and woods, tropical scrubby areas, open cultivated areas, plantations. **D** W Ecuador to NW Peru.

23 ORANGE-CHINNED PARAKEET *Brotogeris jugularis* 19cm Small orange chin spot hard to see. Shoulders conspicuously brown. Underwing contrasting yellow. Flight feathers blue-green. **V** Call in flight a harsh, continuous, single-note chatter, *ack, ack, ack...* **H** Dry woods, scrubby areas with trees, open forest, gallery forest, parks, plantations, gardens. **D** SW Mexico to W Venezuela.

24 COBALT-WINGED PARAKEET *Brotogeris cyanoptera* 19cm Orange chin, yellowish forehead. Underwing coverts green, flight feathers cobalt-blue. **V** Usual calls, given in flight and when perched, include a high-pitched *klee* or *chree* and a chattering *chichichichi.* **H** Lowland rainforest, edges, humid secondary growth, riverine forest, várzea, savannah, woods. **D** W Amazonia.

25 GOLDEN-WINGED PARAKEET *Brotogeris chrysoptera* 19cm Narrow forehead orange or brownish black. Chin orange. Orange on primary coverts. Underwing green. **V** Call *chill, chill, chill....* **H** Cloud forest, primary rainforest, coastal forest, savannah, gallery forest, woods, town parks. **D** N and C Amazonia.

26 PILEATED PARROT *Pionopsitta pileata* 22cm Bill black. Bare skin around eye dark. Male has red forehead, crown and part of face. Female has rusty ear-patch. Flight feathers, tail and underwing bluish. **V** Calls include *chee chee, cheely cheely* and *kloo-louie.* **H** Hill forest, copses, woods, rural areas. **D** SE Brazil, E Paraguay, NE Argentina.

27 BLUE-BELLIED PARROT *Triclaria malachitacea* 29cm Ivory bill. Very large and broad bluish squarish tail. Male has large blue area on mid-belly. **V** Male has fluted thrush-like *ee-ay ee-ay* and *jillid*; female joins in with *diot, diot.* **H** Mountain forests, adjacent drier woods, towns, plantations. **D** SE Brazil.

28 BROWN-HOODED PARROT *Pyrilia haematotis* 21cm Ivory bill. Dark brown hood. Nape and hindneck olive-washed. Red dot on ear-coverts, red half-collar on foreneck. Breast olive. Leading edge of wing, underwing coverts and flight feathers blue. Axillaries and part of flank bright red. Base of tail feathers when seen from below show rusty orange. **V** Calls in flight, *check* or *cheek.* **H** Cloud forest, wet lowland forest, piedmont forest. **D** SE Mexico to NW Colombia.

29 SAFFRON-HEADED PARROT *Pyrilia pyrilia* 21cm Hood yellow, orange wash on ear-coverts. Cere and bare skin around eye dark. Breast olive. Shoulders and leading edge of wing yellow and red. Underwing scarlet. **V** *Cheweek.* **H** Lowland humid forest to cloud forest, tall secondary growth. **D** E Panama to NW Venezuela.

30 ROSE-FACED PARROT *Pyrilia pulchra* 22cm Pink face, edged black and white. Formerly considered conspecific with Brown-hooded Parrot. **V** *Skreek skreek.* **H** Humid tropical and subtropical forest, tall secondary growth, clearings with scattered trees, plantations. **D** W Colombia, NW Ecuador.

31 ORANGE-CHEEKED PARROT *Pyrilia barrabandi* 23cm Bill black. Hood black, cheeks yellowish orange. Breast yellowish olive. Shoulders with orangey yellow and red. Leading edge of wing and underwing scarlet. **V** Distinctive bisyllabic *chew-it* or *choyet.* **H** Humid forest, várzea, isolated stands of trees, sandy-soil woods. **D** W Amazonia.

32 CAICA PARROT *Pyrilia caica* 23cm Greyish-ivory bill. Short black hood. Wide tawny-chestnut collar with darker scaling. **V** Strident *illit.* **H** Rainforest, tropical, gallery and piedmont forests. **D** E Venezuela, the Guianas, NE Brazil.

33 BALD PARROT *Pyrilia aurantiocephala* 23cm Differs from Vulturine Parrot in having featherless orange head with black bristles. Underwing coverts scarlet. **V** Makes wide range of sounds including nasal *chow-chow-chow*, higher *skee-skee-skee*, and various whistling, yelping and shrieking notes. **H** Rainforest at mid-Tapajos river, possibly also lower. **D** C Brazilian Amazonia.

34 VULTURINE PARROT *Pyrilia vulturina* 23cm Bill blackish and horn coloured. Bare black head with forehead and high collar yellow. Bend of the wing and leading edge red. Underwing coverts scarlet. Tail from below blue and yellow. **V** *Terek terek* and *tray, trayer.* **H** Lowland tropical forest, várzea, terra firme forest. **D** SE Amazonia.

35 RUSTY-FACED PARROT *Hapalopsittaca amazonina* 23cm Red face, darker on crown. Cheeks striped yellow. Golden wash on breast. Red shoulders and leading edge of wing. Tail dark carmine with blue terminal band. **V** Repeated *check.* **H** Subtropical humid and temperate forest, cloud forest, open, low woods near treeline. **D** Colombia, NW Venezuela.

36 FUERTES'S PARROT *Hapalopsittaca fuertesi* 23cm Differs from Rusty-faced Parrot in having yellow forehead, lores and ear-patch. Crown and nape bluish. Red patch at mid-belly. Underwing coverts red. **V** Grating, nasal *kraa*, given both in flight and when perched. **H** Upper temperate Andean cloud forest. **D** C Colombia.

37 RED-FACED PARROT *Hapalopsittaca pyrrhops* 23cm Like Rusty-faced Parrot but darker and duller, crown dark green. Tail green with blue terminal band, blue underside. Underwing coverts red. **V** In flight gives a soft, mellow *currrrt*; when perched gives a nasal *kyek* and more a powerful *creett*, along with other conversational notes. **H** Dripping cloud forest to transition shrubbery near páramo. **D** SW Ecuador, N Peru.

38 BLACK-WINGED PARROT *Hapalopsittaca melanotis* 24cm Mostly green, bluish on neck. Black spot on ear-coverts and black wing coverts. **V** Flight call 1–3 mellow *crrit* notes, sometimes finishing with a single more shrill note. **H** Humid temperate forest in Andes, cloud forest and stunted forests. **D** C Peru to Bolivia.

1
2
3
4
5
6
7
8
9
10
11
12
13
14
15
16
17
18
19
20
21
22
23
24
25
26
27
28
29
30
31
32
33
34
35
36
37
38

AFRICAN AND NEW WORLD PARROTS *CONTINUED*

1 DUSKY PARROT *Pionus fuscus* 27cm Bill blackish and pinky horn. Head dark dull blue. Upperparts blackish-brown scaled buff, wings and uppertail very dark blue. **V** In flight, 3–4 hoarse *craaaks*. **H** Tropical and subtropical montane forest, gallery forest, edges, clearings, coastal forests on dunes, savannah, plantations. **D** Venezuela, the Guianas, N and NE Brazil.

2 RED-BILLED PARROT *Pionus sordidus* 28cm Red bill. Head dark green with blackish scaling and dark blue forecrown. Blue breast, red vent. Underwing green. Upperparts olive brown, underparts greyer, somewhat scaled. **V** Very noisy *key-ank* or *pee-unt*. **H** Tropical and subtropical forest in mountains, rainforest, cloud forest, woods, partially cleared areas, scattered trees, plantations. **D** N and W South America, from Venezuela to Bolivia.

3 SCALY-HEADED PARROT *Pionus maximiliani* 28cm Bill blackish and ivory. All green, scaled blackish. Bluer on breast. Underwing green. **V** In flight, continuous resonant *crack, mytak, mytak* or *chack chok, choklock*. **H** Lowland forest and open woods, savannah, woods in marshy areas, humid montane forest. **D** C and E South America, from Bolivia to N Argentina and E Brazil.

4 PLUM-CROWNED PARROT *Pionus tumultuosus* 28cm Forehead, crown, nape and neck purply red. Scaled black on hindneck. Slightly scaled violaceous blue on breast. **V** In flight gives a low, harsh *kraah*; softer sounds made when perched. **H** Humid tropical and subtropical montane forest. **D** C Peru to Bolivia.

5 WHITE-CAPPED PARROT *Pionus seniloides* 27–29cm Previously considered conspecific with Plum-crowned Parrot (as 'Speckle-faced Parrot'). Cap pale, not speckled reddish purple. **V** Screeching *Whaa, wheh, whWha, whhr...* **H** Montane forest, páramo. **D** Andean spine of Venezuela, Colombia, Ecuador, N Peru.

6 BLUE-HEADED PARROT *Pionus menstruus* 28cm Bill black and red. Hood and breast cobalt-blue, ear-coverts blackish. Faint red scaling on foreneck. **V** In flight, double *key-wenk*. **H** Lowland tropical forest, gallery forest, rainforest, deciduous forest, scattered trees and copses in cleared areas, occasionally in tropical forest in mountains. **D** Costa Rica to SE Brazil.

7 WHITE-CROWNED PARROT *Pionus senilis* 28cm Whitish head with transverse blackish barring on crown. Forehead whiter. Speckled on cheeks and throat. Breast and upper belly greyish pink with black barring. **V** Low, laughing *chank*. **H** Subtropical and temperate forests in mountains, cloud forest and open areas with scattered trees near páramo. **D** NE Mexico to W Panama.

8 BRONZE-WINGED PARROT *Pionus chalcopterus* 28cm Mostly dark blue. Throat and breast pinkish white, scaled blue. Upperwing coverts brown. Underwing coverts dark blue, flight feathers blue. **V** Like Blue-headed Parrot. **H** Tropical and subtropical humid montane forests, edges and clearings with scattered trees, tropical dry forest. **D** Venezuela to NW Peru.

9 SHORT-TAILED PARROT *Graydidascalus brachyurus* 24cm Dumpy and very large-headed. Tail extremely short. Heavy black bill. Plumage uniformly bright green. **V** Gentle *kyerick* and strident *kia kia kia*. **H** Tropical forest, mostly near water, tall gallery forest, edge of várzea. **D** Amazonia.

10 YELLOW-FACED PARROT *Alipiopsitta xanthops* 26cm Head yellow, pectoral collar scaled green. Breast and upper belly yellowish orange. Flanks orange. Tail green, dull orange at base. **V** In flight, *cree-ay, gray-oh* and *totototo*. **H** Scattered low trees in Cerrado, or with cactus in Caatinga, gallery forest. **D** E Brazil to E Bolivia.

11 FESTIVE AMAZON *Amazona festiva* 34cm Lores and part of forehead deep red. Hind-brow blue. Lower back and rump bright red (lacking in some forms). **V** Nasal and laughing *whah whah*. **H** Gallery forest, várzea, rainforest, savannah, grassland with scattered trees, secondary growth, plantations. **D** Venezuela, W Amazonia.

12 VINACEOUS-BREASTED AMAZON *Amazona vinacea* 34cm Bill reddish with ivory tip. Red lores. Hind-collar pale blue, scaled black. Underwing green. Tail green, base of outer feathers carmine. **V** *Tay-oh* and *cryee-creeoh*. **H** Forest with *Araucaria*, isolated patches of woods in open country, *Araucaria* plantations, dry woods, humid forest with cane. **D** E Paraguay to SE Brazil.

13 TUCUMAN AMAZON *Amazona tucumana* 32cm All green, scaled blackish. Forehead and primary coverts red. Underwing green. Some consider this a race of Red-spectacled Amazon. **V** Various screeches, shrieks, barks and yelps; flight call a yelping *quiowk*. **H** Upper montane forest. **D** S Bolivia, NW Argentina.

14 RED-SPECTACLED AMAZON *Amazona pretrei* 32cm Forehead, crown and around eye red; leading edge of wing and 'garters' also red. Underwing green. **V** Strident whistles interspersed with *spee-ah, krek, clow* and *kayro*. **H** Forest, especially with *Araucaria*. **D** S Brazil.

15 BLACK-BILLED AMAZON *Amazona agilis* 25–26cm The smaller of the two parrots endemic to Jamaica. Plumage mostly emerald green, with some blue in flight feathers and red primary coverts. Small black bill. **V** In flight, bugling *tuh-tuh* and sharp screech; when perched or feeding, a low growl. **H** Wet mid-level forests on mountains and hills. **D** Jamaica.

16 WHITE-FRONTED AMAZON *Amazona albifrons* 30cm Separated from Yucatán Amazon by more extensive red on greater wing coverts. **V** Various screeches, grating notes and chatters; flight call a harsh *crek-crek*. **H** Forest, open woodland, mangrove. **D** W Mexico to Costa Rica.

17 YELLOW-BILLED AMAZON *Amazona collaria* 28–31cm Distinguished from Black-billed Amazon by rose-coloured throat, yellow bill and pale eye-ring. Generally occurs in small flocks feeding among the topmost branches. **V** In flight, a bugling *tuk-tuk-tuk-taaah*; high-pitched *tah-tah-eeeeep* when perched. **H** Forests on hills and mountains, mainly at mid-elevations; after breeding, may spread to more open areas, gardens, cultivations. **D** Jamaica.

18 CUBAN AMAZON *Amazona leucocephala* 28–33cm Distinctive rosy throat and white forehead. Usually occurs in small, noisy groups. **V** In flight, a harsh *squawk-squawk*; also a variety of harsh screeches and shrill shrieks when perched. **H** Primarily forests, at all elevations. **D** Cuba, Bahamas, Cayman Islands.

19 HISPANIOLAN AMAZON *Amazona ventralis* 28–31cm More restricted white forehead than Cuban Amazon, and green throat. Usually encountered in pairs, family groups or even quite large flocks. **V** In flight, an almost continual loud screech; at rest, or while feeding, groups often emit a low growl or chatter. **H** Forests, woodlands, palm savannah, scrub. **D** Hispaniola. Introduced Puerto Rico.

20 PUERTO RICAN AMAZON *Amazona vittata* 29–30cm Critically endangered, numbers less than 300 wild birds. Some birds may show a slight red tinge to the belly. **V** In flight, a loud, far-carrying *kar kar*; when feeding, often utters a low chuckling. **H** Mid-elevation wet forests. **D** E Puerto Rico.

21 LILAC-CROWNED AMAZON *Amazona finschi* 35cm Pale yellowish bill contrasting with dark red forehead. Lilac crown hard to see. **V** Calls include shrill and harsh screeches, also some lower-pitched notes. **H** Woodland, mangrove. **D** W Mexico.

22 RED-LORED AMAZON *Amazona autumnalis* 36cm Forehead red. Crown, nape and hindneck lilac-blue, slightly scaled blackish. **V** Flight calls harsh *cheekack cheekack* and *oorrack oorrack*. **H** Rainforest, gallery, open lowland and dry scrubby forests, scattered trees in cleared areas, plantations. **D** E Mexico to W Ecuador.

23 DIADEMED AMAZON *Amazona diadema* 36cm Formerly considered conspecific with Red-lored Amazon. Similar in appearance. **V** Flight calls harsh *cheekack cheekack* and *oorrack oorrack*. **H** Rainforest, gallery, open lowland and dry scrubby forests, scattered trees in cleared areas, plantations. **D** NW Brazil.

24 RED-CROWNED AMAZON *Amazona viridigenalis* 30–33cm In flight, shows distinctive red patch on secondaries. Usually seen in flocks, often noisy as they fly from tree to tree. **V** Harsh *kee-craw craw craw* or *keet kau-kau-kau-kau*. **H** Lowland wet forests and scrub. **D** NE Mexico. Introduced Puerto Rico.

25 YUCATAN AMAZON *Amazona xantholora* 30cm Note blue-black ear-patch and yellow lores. **V** Various calls, mostly loud and raucous, including short abrupt notes and longer trumpeting phrases; also low barks and rolling screeches. **H** Forest and forest edge. **D** Yucatán Peninsula.

26 BLUE-CHEEKED AMAZON *Amazona dufresniana* 37cm Bill grey and reddish. Forehead yellowish orange grading to scaled yellowish green on crown. Bluish cheeks. **V** Highly vocal. Has a gurgling song and various raucous, throaty and nasal calls. **H** Montane forests, irregularly coastal forests. **D** E Venezuela, the Guianas.

27 RED-BROWED AMAZON *Amazona rhodocorytha* 37cm Differs from Blue-cheeked Amazon in paler bill, reddish-orange forehead and crown, yellow to reddish lores and forepart of cheek. Rest of cheek bluish. **V** Clear *koiok, cow-ow* and *ullo ullo ullo*; *cray-oh* in flight. **H** Humid lowland forest, ranging into highlands. **D** E and SE Brazil.

28 RED-NECKED AMAZON *Amazona arausiaca* 40cm Endangered. In flight, shows red patch on secondaries. Forages mainly in the forest canopy, sometimes associating with the Imperial Amazon. **V** Drawn-out *rrr-eee*. **H** Rainforest, usually at mid-elevations. **D** Dominica.

29 ST. LUCIA AMAZON *Amazona versicolor* 42–46cm Endangered. Variable amount of red on breast and belly. In flight, shows red patch on secondaries. Forages in the forest canopy, generally in pairs or small parties, difficult to locate. **V** In flight, raucous screeching. **H** Mainly tropical moist forest. **D** St Lucia.

30 YELLOW-HEADED AMAZON *Amazona oratrix* 36cm Yellow on head. In flight, shows red patch on secondaries. Generally encountered in pairs. **V** Screeching *kurr-owk* and other raucous squawks. **H** Lowland forest. **D** S Mexico, Belize, Honduras.

31 TRES MARIAS AMAZON *Amazona tresmariae* 35–36cm Previously treated as subspecies of Yellow-headed Amazon. Feeds on fruits, berries and buds. **V** Wide range of screams and harsh *kraa-ah-ow* calls. Talented mimic. **H** Forested habitat, thorn scrub, cultivated ground. **D** Tres Marías Islands (W Mexico).

32 YELLOW-NAPED AMAZON *Amazona auropalliata* 34–36cm Grey bill, mostly green head with bright yellow nape. **V** Shrill, piping *pEer, pEer* and raucous *Wher, weer-wher*. **H** Arid forested habitats, scrub, savannah. **D** S Mexico to Costa Rica.

33 YELLOW-CROWNED AMAZON *Amazona ochrocephala* 38cm Forehead and crown yellow. Red patch on secondaries. Red shoulder. **V** *kerah* and *bow-wow*. **H** Tropical rainforest, gallery forest, várzea, swampy forest, arid open woods, scrubland, savannah, scattered copses, coastal forest, deforested areas, plantations. **D** W Panama to E Peru and N Bolivia.

34 YELLOW-SHOULDERED AMAZON *Amazona barbadensis* 35cm Extensive yellow on face and shoulders. **V** Various high-pitched screeching calls including *screet* or extended *scree-ee-eet*. **H** Savannah with scattered trees, thorny vegetation, brushland with cactus, plantations. **D** N Venezuela and nearby islands.

35 TURQUOISE-FRONTED AMAZON *Amazona aestiva* 37cm Extent of turquoise on forehead variable. Crown and face (sometimes almost entire head) yellow. Red and yellow shoulder very variable. **V** *Arrrow, how rrow, rrackow*. **H** Dry or humid forest, palm groves, gallery forest, savannah, plantations. **D** C South America, from Bolivia to NE Brazil, Paraguay, N Argentina.

36 SCALY-NAPED AMAZON *Amazona mercenarius* 34cm All scaly green (more scaled on nape and hindneck), with touch of yellow on leading edge of wing. **V** *Calee calee*. **H** Subtropical and temperate forest in Andes, cloud forest, partially open terrain at higher elevations. **D** Venezuela to Bolivia.

37 NORTHERN MEALY AMAZON *Amazona guatemalae* 38–40cm Northern counterpart of Southern Mealy Amazon, lacks yellow on forehead. Feeds on fruits and berries. Often seen in flocks. **V** Gruff, squawking screech and harsh *Whoaaaar*. **H** Lowland rainforest, secondary woodland. **D** SE Mexico to W Panama.

38 SOUTHERN MEALY AMAZON *Amazona farinosa* 41cm Yellow on forehead. End of tail pale green. **V** Low-pitched and musical *creek, cooreek* and *krop*. **H** Tall, dense tropical forest, gallery forest, coastal sand-ridge forest, palm groves, deciduous woods, semi-open secondary growth. **D** E Panama and Colombia to NE Bolivia, E Brazil, the Guianas.

39 KAWALL'S AMAZON *Amazona kawalli* 34–36cm Uniformly green body plumage, but note striking white skin at base of bill and lack of white eye-surround. **V** Squeaky *Pee-uk, Pee-uk* and notes reminiscent of child's electronic game. **H** Lowland forests, often near water. **D** Amazonian Brazil.

40 IMPERIAL AMAZON *Amazona imperialis* 46–51cm Endangered. In flight, shows a red patch on the secondaries. Usually occurs in pairs or small groups foraging in the tops of trees or flying over the forest canopy. **V** Metallic, trumpeting *eeeee-er*; various shrieks, squawks, whistles and bubbly trills. **H** Moist, mid- to high-elevation forest. **D** Dominica.

41 RED-TAILED AMAZON *Amazona brasiliensis* 37cm Forehead, crown and lores pinkish red (forehead only in female). Face pale mauve (less in female). Touch of red on leading edge of wing. **V** *Krah, clee, kalik, cree-oh*. **H** Coastal forest. **D** SE Brazil.

42 ORANGE-WINGED AMAZON *Amazona amazonica* 31–32cm In flight, shows orange-red patch on secondaries. Usually seen in pairs, although flocks also occur, especially at roosts. **V** Shrill *kee-ik kee-ik kee-ik* or *kweet kweet kweet kweet*; also a variety of screeches and whistles. **H** Lowland second-growth forests. **D** Colombia, Venezuela and the Guianas to Bolivia and S Brazil.

43 ST. VINCENT AMAZON *Amazona guildingii* 41–46cm Endangered. Very variable, two main colour morphs, one basically golden brown the other greenish. Usually occurs in small groups foraging in the forest canopy. **V** In flight, *gua gua gua* or *quaw quaw quaw*; while feeding, squeaks and squabbling noises. **H** Primarily mature moist mountain forest. **D** St Vincent.

1
2
3
4
5
6
7
8
9
10
♀
11
12
13
14
15
16
♂
17
18
19
♂
20
21
22
23
24
25
♀
26
27
28
29
30
31
32
33
34
35
36
37
38
39
40
41
♂
♀
42
43
green morph
brown morph

AFRICAN AND NEW WORLD PARROTS ***CONTINUED***

1 DUSKY-BILLED PARROTLET *Forpus modestus* 12cm Bill blackish. Dark green, washed olive. Dark blue lower back, rump and underwing. Female has paler yellowish-green underparts and face. **V** Usual call a buzzy *zeet*, given in flight and when perched. Flocks in flight give constant loud twittering. **H** Tropical rainforest, clearings with trees, thickets, gallery forest. **D** Amazonia.

2 MEXICAN PARROTLET *Forpus cyanopygius* 14cm Small, short-tailed, with bluish underwings. **V** Gives high-pitched *kree* or *kree-it* in flight. Flocks in flight produce a loud, tinkling chattering. **H** Forest, gallery forest, plantations. **D** W Mexico.

3 GREEN-RUMPED PARROTLET *Forpus passerinus* 12–13cm Uniform lime green, with blue wing patch in male. Pale bill. Gregarious. Flight swift with many twists and turns. **V** Shrill *chee chee chee*; also a *tsup-tsup*. Feeding flocks keep up a constant twittering. **H** Open country in dry lowlands and hills, forest edge, secondary growth, thornbush. **D** N Colombia, N Venezuela, Trinidad and Tobago, the Guianas, N Brazil, lower Amazonian Brazil.

4 TURQUOISE-WINGED PARROTLET *Forpus spengeli* 12–14cm Male has turquoise lower back and rump, and purple-blue underwing coverts and axillaries. Females lack blue markings. Previously considered conspecific with Blue-winged Parrotlet. **V** High-pitched, twittering squeaks. **H** Arid open woodland, scrub. **D** N Colombia.

5 BLUE-WINGED PARROTLET *Forpus xanthopterygius* 12cm Green, yellowish wash on underparts. Male has lower back and rump dark blue or turquoise, and blue underwing. Female's forehead and face yellowish. **V** *Tsip, tsip, tsip*. **H** Savannah, brushland and cactus, open semi-arid woods, edge of rainforest, palm groves, parks. **D** SE Peru, Bolivia, E Brazil, Paraguay, N Argentina.

6 LARGE-BILLED PARROTLET *Forpus crassirostris* 12–13cm Previously treated as subspecies of Blue-winged Parrotlet. Often seen in flocks. Feeds on fruits, seeds and buds. **V** Twittering squeaks and piping whistles. **H** Forested habitats, riverside woodland, secondary woodland. **D** S Colombia, Ecuador, N Peru, W Brazil.

7 SPECTACLED PARROTLET *Forpus conspicillatus* 12cm Dull green with bluish wash around eye. Forehead and cheeks brighter green. Lower back and rump dark blue. Underwing blue. **V** Calls are chortled and musical, recalling a finch. **H** Open tropical forest, gallery forest, semi-open woods, grassland with trees, plantations, agricultural land. **D** Panama, Colombia, Venezuela.

8 PACIFIC PARROTLET *Forpus coelestis* 12cm Forehead, face and crown bright green. Pale blue behind eye. Nape and neck greyish. Upper back and wings olivaceous grey. Lower back and rump blackish blue. **V** High-pitched chattering. **H** Arid scrub with scattered trees, open woods, gallery forest. **D** W Ecuador, NW Peru.

9 YELLOW-FACED PARROTLET *Forpus xanthops* 15cm Bright yellow on most of the head. Post-ocular streak, nape, neck and breast grey. Lower back and rump very dark blue. **V** Usual call a sharp *zit*, more passerine-like than parrot-like, given in flight and when perched. Groups in flight keep up a constant chattering. **H** Arid tropical montane scrub with cactus. **D** Marañon valley (NC Peru).

10 BLACK-HEADED PARROT *Pionites melanocephalus* 23cm Bill black. Cap black. Cheeks yellow, sides and back of the neck orange-ochre. Breast and mid-belly whitish. Leggings orange-ochre. Underwing in flight like White-bellied Parrot, axillaries orange. **V** High-pitched *cleeeoo, cleeeoo*. **H** Rainforest, open tropical forest, edges, woods, savannah, coastal forest. **D** N Amazonia.

11 WHITE-BELLIED PARROT *Pionites leucogaster* 23cm Bill ivory. Crown and hindneck tawny orange. Cheeks bright yellow. Breast and upper belly whitish. Leggings, lower belly green or yellow. Underwing coverts green, flight feathers blackish from below. **V** *Zrrree, zrrree*. **H** Gallery forest, várzea, dry forest. **D** S Amazonia.

12 RED-FAN PARROT *Deroptyus accipitrinus* 35cm Large crest of erectile blue-bordered red feathers. Forehead and crown speckled. Breast and upper belly also scaled red and blue. **V** Strident nasal calls, *kyah, ghee* and *he-ah*. **H** Terra firme forest, clearings, coastal dune forest, savannah, transition woods. **D** Amazonia.

13 OCHRE-MARKED PARAKEET *Pyrrhura cruentata* 29cm Ochre on sides and back of neck. Lores, ear-patch and rump reddish chestnut. Cap dark brown streaked even darker. Breast blue. Belly patch and underside of tail chestnut-red. **V** Short, nasal *kara kek kek kek*. Flocks are noisy on the wing but much quieter when perched. **H** Forest, in canopy. **D** SE Brazil.

14 BLAZE-WINGED PARAKEET *Pyrrhura devillei* 26cm Differs from Maroon-bellied Parakeet mainly in bright orange, red and yellow underwing coverts. Possibly a race of Maroon-bellied Parakeet, as they seem to interbreed in N Paraguay. **V** Shrill *kree kree kree* given in flight; calls softer when perched. **H** Savannah, woods, forest, scattered trees. **D** N Paraguay, SW Brazil.

15 MAROON-BELLIED PARAKEET *Pyrrhura frontalis* 26cm Green cap. Ear-patch pale brownish grey. Side of neck, throat and upper breast brownish ivory, scaled blackish. Belly and undertail reddish chestnut. Top of the tail green. Noisy flocks. Erratic fast flight through trees. **V** Harsh *kara kek kek kek* calls, given in flight and when perched; also a softer *krek*. Flocks chatter constantly. **H** Rainforest edges, upland forest, palm groves, *Araucaria* plantations, farmland, riverside forests. **D** E Brazil to SE Paraguay and N Argentina.

16 PEARLY PARAKEET *Pyrrhura lepida* 25cm Cap dark greyish brown. Cheek blue, ear-patch whitish brown. Scaling on neck and breast brown and whitish brown. Lower underparts all green. Underwing carmine. Tail chestnut-red. **V** Fast series of shrill, grating *kree kree kree* notes given in flight and when perched. Flocks produce constant harsh, piercing chattering. **H** Rainforest, gallery forest. **D** NE Brazil S of Amazon.

17 CRIMSON-BELLIED PARAKEET *Pyrrhura perlata* 25cm Startling crimson belly and underwing. Head and neck dark, scaled lighter; cheeks bluish to yellowish green. **V** Shrill, harsh *tieww...kri-tiew*. **H** Riverine forests. **D** S Amazonia, from Brazil to N Bolivia.

18 GREEN-CHEEKED PARAKEET *Pyrrhura molinae* 26cm Differs from Maroon-bellied Parakeet in having a dark brown cap, slightly scaled darker. Tail all chestnut-red. **V** In flight gives shrill, fast and grating *kreee kreee kreee* or single high-pitched *kuree*. Calls when perched are softer and more mellow-toned. **H** Edge and clearings in montane forests, wooded valleys, semi-open woods, copses. **D** Bolivia to SW Brazil and NW Argentina.

19 PFRIMER'S PARAKEET *Pyrrhura pfrimeri* 22–23cm Previously treated as subspecies of White-eared Parakeet. Seen in small flocks where feeding is good. **V** Shrill piping whistle, resembling raptor alarm call. **H** Arid scrub and Caatinga on limestone. **D** EC Brazil.

20 GREY-BREASTED PARAKEET *Pyrrhura griseipectus* 22–23cm Previously treated as subspecies of White-eared Parakeet. Endangered due to habitat destruction. Now known from just two small forested areas. **V** Calls include a shrill series of piping calls, rather like a squeaky toy. **H** Humid montane forests. **D** NE Brazil.

21 WHITE-EARED PARAKEET *Pyrrhura leucotis* 22cm Purply-chestnut cheeks. White ear-patch. Breast scaled green, black and white or just black and white. Dark red on belly, lower back and rump. Underside of tail feathers dull red. **V** Calls include shrill yelping notes and a softer *teet* when perched. **H** Tropical and subtropical forest, cloud forest, edges and clearings. **D** E and SE Brazil.

22 PAINTED PARAKEET *Pyrrhura picta* 22cm Blue forehead. Crown and nape dark brown. Cheeks purply chestnut and whitish. Markedly scaled on neck and breast. Red patch at mid-belly. Underside of tail dull red. **V** Usual call, given in flight, is a high-pitched *kyeek kyeek kyeek*; a softer *kek* when perched. **H** Tropical and subtropical forest and cloud forest, coastal forest, várzea. **D** SE Panama to the Guianas and N Brazil.

23 VENEZUELAN PARAKEET *Pyrrhura emma* 22–23cm Previously considered conspecific with White-eared Parakeet. **V** Chattering calls and repeated piping whistle, tone like raptor alarm call. **H** Humid forests, secondary woodlands; lowlands and foothill slopes. **D** N Venezuela.

24 SANTAREM PARAKEET *Pyrrhura amazonum* 22–23cm Distinguished from Painted Parakeet by narrower blue forehead and absence of red patch on shoulder. Seen in small flocks when feeding is good. **V** Chattering, rapid-fire squeaky whistle, with tone like raptor alarm call. **H** Lowland rainforest, both seasonally inundated and terra firme. **D** SC Amazonian Brazil, E Bolivia.

25 BONAPARTE'S PARAKEET *Pyrrhura lucianii* 21–23cm Previously treated as a subspecies of Painted Parakeet. Similar to Santarem Parakeet but ranges do not overlap. Diet includes fruits and seeds. **V** Chattering, squeaky screeches. **H** Rainforest, both terra firme and várzea (seasonally inundated). **D** W Amazonian Brazil.

26 ROSE-FRONTED PARAKEET *Pyrrhura roseifrons* 21–23cm Note rose-red face and forehead. **V** Shrill, high-pitched and insistent *Wee-ik, Wee-ik*. **H** Lowland rainforest, secondary woodland. **D** SE Ecuador, E Peru, NW Bolivia, W Brazil.

27 SANTA MARTA PARAKEET *Pyrrhura viridicata* 25cm Mostly green with two irregular bands across belly and flanks. Underwing, bend of the wing and narrow forehead all scarlet. Ear-patch chestnut, undertail chestnut-red. Bill ochreous ivory. **V** Gives loud screeching calls just before and during flight. **H** Humid subtropical montane forests. **D** Very restricted range: Sierra Nevada de Santa Marta (N Colombia).

28 FIERY-SHOULDERED PARAKEET *Pyrrhura egregia* 25cm Small ear-patch chestnut. Cream and black scales on breast. Patch of reddish chestnut on mid-belly. Underwing orangey yellow. Dark tail. **V** Usual call, given in flight and when perched, is a harsh *krreek krreek krreek*; flocks chatter constantly. **H** Forest on the slopes of the tepuis, tropical forest. **D** E Venezuela, Guyana.

29 MAROON-TAILED PARAKEET *Pyrrhura melanura* 25cm Cap scaled blackish on dark green. Cheeks all green. Extensive scaling on breast. Short red line on leading edge of wing. Tail reddish chestnut above, blackish below. **V** Calls *screet screet screet* in flight and when perched; flocks produce noisy chattering. **H** Tropical and subtropical forests from lowland to cloud forest, edges, semi-open areas. **D** NW Amazonia.

30 EL ORO PARAKEET *Pyrrhura orcesi* 22cm Ivory bill, red forehead and lores. Breast slightly scaled. Underwing green. **V** Most common call, given in flight and when perched, is a grating *creeet creeet creeet*; also *krrrr* when perched. Flocks produce noisy chattering. **H** Highland, very humid tropical forest. **D** Very local in El Oro province, SW Ecuador.

31 WHITE-BREASTED PARAKEET *Pyrrhura albipectus* 25cm Yellow-orange ear-patch. Whiter on neck forming a nuchal collar. Yellow breast. Basal half of uppertail green, distal half reddish chestnut, underside blackish. **V** Commonest call, given in flight and when perched, harsh *kree kree kree* or single *kurree*. Flocks chatter loudly. **H** Subtropical forest in mountains. **D** Local in Loja, SE Ecuador.

32 BLACK-CAPPED PARAKEET *Pyrrhura rupicola* 26cm Cap dark brown. Cheeks all yellowish green. Neck and breast yellowish white, broadly scaled blackish brown. Short red line on leading edge of wing. Tail green above, blackish below. **V** Usual call, given in flight and when perched, harsh *kree kree kree* or single *kurree* or *keee*. Flocks chatter loudly and constantly. **H** Tropical rainforest. **D** S Peru, extreme W Brazil, N Bolivia.

33 FLAME-WINGED PARAKEET *Pyrrhura calliptera* 25cm Bill ivory. Head scaly, with rusty ear-patch. Extensive blackish-brown scaling on foreneck and breast. Patch of chestnut-red markings at mid-belly. Yellow line on leading edge of the wing. Tail reddish. **V** Commonest call, given in flight and when perched, harsh *kree kree kree* or *kerr kerr kerr*. Also gives single notes when perched, and flocks chatter loudly. **H** Edge of subtropical and temperate forest in mountains, upland woods to páramo. **D** C Colombia.

34 BLOOD-EARED PARAKEET *Pyrrhura hoematotis* 25cm Ear-patch scarlet. Patch of reddish-chestnut markings at mid-belly. Tail all reddish chestnut. **V** In flight and when perched, gives harsh *krree krree krree*; when perched also a single *kurree, kurruk* or *kreekuk*. Flocks chatter loudly. **H** Subtropical montane forests, cloud forests, open areas, scattered trees in temperate zone. **D** N and NW Venezuela.

35 ROSE-CROWNED PARAKEET *Pyrrhura rhodocephala* 25cm Bill ivory. Scarlet cap. White line on leading edge of wing. Tail reddish chestnut. **V** Commonest call, given in flight and when perched, is a series of short, harsh, nasal *krrr kek kek kek* notes; also single *krree, kurree* and *krrr* notes. Flocks chatter loudly. **H** Subtropical forests in mountains, as high as páramo. **D** W Venezuela.

36 SULPHUR-WINGED PARAKEET *Pyrrhura hoffmanni* 25cm Head pattern and yellow in wings diagnostic. Note reddish tail. **V** Gives harsh *kree kree kree* in flight and when perched. Flocks chatter loudly. **H** Montane forest. **D** Costa Rica, Panama.

37 AUSTRAL PARAKEET *Enicognathus ferrugineus* 35cm Overall dark green, feathers edged dark, giving scaly effect. Forehead, belly patch and tail rusty red. Small bill blackish. **V** Rowdy in flight, silent when perched or feeding. **H** Subantarctic woodland. **D** S Chile, S Argentina.

38 SLENDER-BILLED PARAKEET *Enicognathus leptorhynchus* 41cm Long, slender bill. Overall dark green, feathers edged dark, looking like scales. Reddish forehead, tail and belly patch. **V** Series of nasal, grating *grrreh grrreh grrreh* notes; also a sweeter-toned *kreeh*. **H** Woods and farmland, forestry plantations, *Araucaria* woods. **D** C Chile.

39 BURROWING PARROT *Cyanoliseus patagonus* 46cm Dark olive-brown. Lower underparts, rump and lower back dirty yellow. Pale red at centre of belly. Dark race has dull olive-brown where former is yellow. **V** Raucous ugly screams. **H** Colonies at cliffs. Scrubland, Andean woods and brushland, agricultural land, orchards, scattered trees, riverbanks. **D** Argentina, C Chile.

1
2
3
4
5
6
7
8
9
10
11
12
13
14
15
16
17
18
19
20
21
22
23
24
25
26
27
28
29
30
31
32
33
34
35
36
37
38
39

AFRICAN AND NEW WORLD PARROTS *CONTINUED*

1 HYACINTH MACAW *Anodorhynchus hyacinthinus* 100cm The world's largest parrot. Overall cobalt-blue with hefty black bill. Bare skin around eye and at base of mandible bright yellow. **V** Very loud *kraah, kraah* (male); *wracker, wracker* (female). **H** Palm groves and water, forests and gallery forests nearby. **D** C Brazil to SE Bolivia, NE Paraguay.

2 LEAR'S MACAW *Anodorhynchus leari* 71cm General colour metallic blue, paler than Hyacinth Macaw. Bill black. Bare skin around eye and at base of mandible pale yellow. Critically endangered. **V** Harsh *arah, arah* and *trrarah*, relatively high pitched. **H** Deep canyons, where it roosts and nests in holes and on ledges; nearby palm groves, where it feeds on the fruit. **D** E Brazil, very restricted.

3 THICK-BILLED PARROT *Rhynchopsitta pachyrhyncha* 40cm Black bill, like Maroon-fronted Parrot, but with red front and different underwing pattern. **V** Gives far-carrying, laughter-like calls in flight; also gobbling calls and a sharp single alarm call that quickly spreads through the flock. **H** Pine forest. **D** W Mexico.

4 MAROON-FRONTED PARROT *Rhynchopsitta terrisi* 40cm Note black underwing with red leading edge. **V** Loud, far-carrying calls including rolling *kra-a-ak, rrahk* and laughing, chattering notes. **H** Pine forest. **D** E Mexico.

5 OLIVE-THROATED PARAKEET *Eupsittula nana* 22–26cm Brownish breast, white eye-ring. Gregarious, generally in small flocks. **V** High-pitched screeches. **H** Humid forest, deciduous woodland, crop fields, urban areas. **D** Mexico to W Panama, Jamaica.

6 ORANGE-FRONTED PARAKEET *Eupsittula canicularis* 23–25cm Bright orange forehead, yellowish eye-ring. Gregarious, usually in small flocks. **V** Raucous *can-can-can*, shrill screech; when feeding, quiet chattering. **H** Wooded pastures, ornamental trees in urban areas. **D** W Mexico to Costa Rica.

7 PEACH-FRONTED PARAKEET *Eupsittula aurea* 25cm Bill blackish. Forehead and feathering around eye yellowish orange. Cheek, foreneck and breast pale brownish grey. Underwing pale green. **V** In flight gives nasal, grating *rreh* or *rreh rreh*, also high-pitched, upslurred *krreek*. **H** Copses in open country, open woodland, savannah, scrub, cultivated areas. **D** Amazonia to S Bolivia, Paraguay, N Argentina.

8 BROWN-THROATED PARAKEET *Eupsittula pertinax* 25cm Green above with paler yellowish belly. Bill blackish-grey, bare skin around eye white. Cheek, foreneck and upper breast pale greyish brown. **V** Loud and penetrating *creeek*... **H** Open woods, scrubby dry areas, cactus, thorny woodland, transition to forest, grassland with scattered trees, mangroves, savannah, gallery forest, plantations, gardens. **D** Panama to Colombia, Venezuela, the Guianas, N Brazil.

9 CAATINGA PARAKEET *Eupsittula cactorum* 25cm Bill greyish ivory. Skin around eye white. Forehead, crown, lower cheek, foreneck and breast greyish brown. Ear-patch and upperparts green. Belly and flanks yellow. Underwing light green. **V** *Cree*... **H** Bushy areas with cactus, savannah, open woodland. **D** E Brazil.

10 DUSKY-HEADED PARAKEET *Aratinga weddellii* 28cm Bill black. Bare skin around eye white. Forecrown, forehead and face lead-grey. Underparts and underwing coverts light green. **V** Penetrating nasal *gee-eek*. **H** Edge of the forest, savannah, copses, transition woods, scrubby areas, várzea and gallery forests. **D** Colombia to W Brazil, Bolivia.

11 NANDAY PARAKEET *Aratinga nenday* 36cm Distinctive black hood and bill. Red thighs, blue on breast and in wings and tail. Underwing mostly black. **V** Screeching *kree-ah kree-ah*, generally given in flight; perched birds often utter a shrill chatter. **H** Gallery forest, deciduous forest, palm stands, moist Chaco woodland, dry scrub. **D** SW Brazil and Bolivia to Paraguay and N Argentina. Introduced S USA.

12 SUN PARAKEET *Aratinga solstitialis* 30cm Differs from Jandaya Parakeet in having bright yellow upperparts, wing coverts and leggings. Underwing coverts yellow. Some consider this a race of Jandaya Parakeet. **V** Shrill, grating *creeek* or *creeeek creeek*, given in flight and when perched. **H** Savannah, open woods and forests, edges, palm groves. **D** N Brazil, W Guyana.

13 SULPHUR-BREASTED PARAKEET *Aratinga maculata* 29–31cm Colourful and distinctive. Generally green and yellow, with orangish around eye. Diet includes fruits and flowers. Gathers in small flocks when feeding is good. **V** Raucous, screeching tern-like *kerrrr, kerrrr, kerrrr*. **H** Lowland forests, lightly wooded habitats on dry sandy soils. **D** NE Amazonian Brazil, S Suriname.

14 JANDAYA PARAKEET *Aratinga jandaya* 30cm Bill blackish. Bare skin around eyes greyish. Conspicuous orange and yellow head and body. Green on lower belly and back. Orange underwing coverts. **V** Loud, penetrating *creek, creeek*. **H** Edge of forest, palm groves, open woods. **D** NE Brazil.

15 GOLDEN-CAPPED PARAKEET *Aratinga auricapillus* 30cm Bill blackish grey. Bare skin around eye greyish. Forehead and upper face orange, grading to yellow at back of crown. Belly rusty orange, flanks and underwing coverts orange. **V** Loud, penetrating *kreee, kreeee*. **H** Edge of forest, secondary growth, savannah, open woods and cultivated areas. **D** SE Brazil.

16 SPIX'S MACAW *Cyanopsitta spixii* 57cm General coloration blue, pale greyish blue on head. Bill and bare skin around eye blackish. **V** *Kraah, kraah* and *kree, kree*. **H** Arid-zone palm groves, scattered trees in savannah, riverine woods with Caraiba trees. **D** NE Brazil. Extinct in the wild.

17 RED-BELLIED MACAW *Orthopsittaca manilatus* 46cm Bill black. Extensive bare face pale yellowish. Crown, part of the cheeks and forehead blue. Notable dull red patch on belly. **V** Calls include loud *chewiak* and *chee-ak*. **H** Riverine forests, savannah with palm groves and scattered palms. **D** Amazonia.

18 BLUE-HEADED MACAW *Primolius couloni* 41cm Bill black, greyish-ivory tip. Small bare patch around eye dark greyish. Head bluish. Tail reddish chestnut from above. **V** In flight gives mellow, rolling *graaa*, singly or in series. When perched, gives various mellow calls. **H** Wet tropical forest near water. **D** W Amazonia: E Peru, extreme W Brazil, N Bolivia.

19 GOLDEN-COLLARED MACAW *Primolius auricollis* 38cm Bill blackish, paler at tip. Bare white face. Black forehead and part of the cheeks. Crown and rest of the cheeks washed blue. Yellow hind-collar variable. Tail dorsally chestnut-red and blue. Pink legs. **V** Like *Aratinga* parakeets. **H** Forest and woodland, gallery forest, wooded swamps and clearings, montane forest. **D** C Brazil, SW Brazil and Bolivia to N Argentina and N Paraguay.

20 BLUE-WINGED MACAW *Primolius maracana* 43cm Bill black. Bare face creamy white. Forehead orange-red and blackish blue. Crown and part of cheek bluish. Irregular orange-red patch on belly. Topside of tail reddish chestnut, underside yellowish olive. Lower back red. **V** Similar to that of Blue-headed Macaw. **H** Edge of forest, swamps, gallery forest, palm groves. **D** C and E Brazil, E Paraguay, NE Argentina.

21 BLUE-AND-YELLOW MACAW *Ara ararauna* 85cm Unmistakable, with blue upperparts and bright yellow underparts. Bill black. Bare face white with lines of small black feathers. Black bib. **V** Loud, harsh, deep *ruaaark*. **H** Palm groves and scattered trees in savannah, often near rivers and marshes, gallery forest. **D** From E Panama to Brazil.

22 BLUE-THROATED MACAW *Ara glaucogularis* 85cm Like Blue-and-yellow Macaw but blue forehead, lines of blue feathers on bare face and a larger bib, which is blue, not black. Endangered. **V** Loud, raucous calls, similar to those of Blue-and-yellow Macaw but higher-pitched and softer. **H** Seasonally flooded grasslands with palms and small patches of forest. Once considered a race of Blue-and-yellow Macaw. Very localised in Bolivia, department of Santa Cruz. **D** N Bolivia.

23 MILITARY MACAW *Ara militaris* 75cm Mostly green, with blue flight feathers, tail red and blue above. Red pompom on forehead. Bare face pink with concentric lines of tiny black feathers. Bill blackish. **V** Harsh *crrraaahk*. **H** Montane forest and woods in arid and semi-arid zones, tall trees and open woods near cliffs. Only occasionally and seasonally in rainforest. **D** W Mexico; South America from NW Venezuela to Peru, Bolivia, NW Argentina.

24 GREAT GREEN MACAW *Ara ambiguus* 80cm Sometimes considered a race of Military Macaw. Very like Military Macaw, but bill heavier and paler at tip, lighter green faintly washed yellow, tail feathers paler, orangey. **V** Loud powerful *raaah*, deeper than Scarlet Macaw. **H** Rainforest in lowlands. **D** E Honduras to W Ecuador.

25 SCARLET MACAW *Ara macao* 85cm Mostly scarlet. Blue wings with yellow greater and median coverts. Ivory maxilla, black at the base, black mandible. Bare face white. **V** Deep, harsh and loud *rowaaark*. **H** Lowland and gallery forest, rainforest, copses and palm groves in savannah, large isolated trees near water. **D** SE Mexico to Colombia and Amazonia.

26 RED-AND-GREEN MACAW *Ara chloropterus* 90cm Differs from Scarlet Macaw in being slightly darker red and having concentric lines of tiny red feathers on face; greater and median wing coverts green. **V** Like Scarlet Macaw. **H** Crowns of the rainforest's tallest trees, edges, savannah with scattered trees, gallery forest; usually in the lowlands. **D** Amazonia, from Colombia to Brazil.

27 RED-FRONTED MACAW *Ara rubrogenys* 55cm Blackish bill. Small patch of bare skin around eye pink. Forehead, crown, auricular, shoulders and underwing strikingly orange. **V** Medium-pitched *crew-ack*. **H** Montane forests, wooded uplands, thorny woods. **D** C Bolivia.

28 CHESTNUT-FRONTED MACAW *Ara severus* 48cm Bill black, bare face white. Narrow forehead and chin blackish chestnut. Edge of wing scarlet. Underwing coverts scarlet and green. Flight and tail feathers reddish purple from below. **V** Medium-pitched *arrrrow* with very rolled r's. **H** Edge of forest near wetlands, swampy forest, gallery forest, copses in grassland, plantations, palm groves. **D** E Panama to the Guianas, SW Ecuador, N Bolivia, Amazonian Brazil.

29 GOLDEN-PLUMED PARAKEET *Leptosittaca branickii* 35cm Bare skin around eye white. Narrow forehead orange, ranging to yellow under eye to form a tufted ear-patch. Belly scaled dull orange. Underside of tail dull orange. Flanks and underwing green. **V** Shrill *kreeaah kreeaah*, in flight and when perched. **H** Temperate cloud forest, podocarp woods and above to treeline, dwarf woods, areas of brush. **D** Colombia to Peru.

30 YELLOW-EARED PARROT *Ognorhynchus icterotis* 43cm Bill black. Forehead, lores and ear-coverts bright yellow. Feathers of these are long and give the impression of ears. Lower cheeks and sides of the neck somewhat scaled. Underparts and underwing yellowish green. Tail dull orange beneath. **V** Distinctive, nasal, bisyllabic like a goose. **H** Subtropical and temperate forests in mountains, especially semi-open areas with palm groves. **D** Colombia, N Ecuador.

31 GOLDEN PARAKEET *Guaruba guarouba* 35cm Almost entirely bright yellow but for green flight feathers. Bill greyish ivory. Bare skin around the eye white. **V** Series of quietish *gra, gra, gra*. **H** Tropical forests, edges, palm groves, transition woods. **D** NE Brazil.

32 RED-SHOULDERED MACAW *Diopsittaca nobilis* 33cm Small. Bill ivory maxilla and black mandible, or all black according to race. Bare face white. Forecrown blue. Bend of the wing and primary underwing coverts bright red. **V** High-pitched for the genus, like *Aratinga* parakeets, *creek creek, airk airk*. **H** Sandy savannah with scattered trees and palm groves, scrubby areas, plantations, edge of the forest, cultivated clearings, coastal areas. **D** Venezuela to Bolivia and E Brazil.

1
2
3
4
5
6
7
8
9
10
11
12
13
14
15
16
17
18
19
20
21
22
23
24
25
26
27
28
29
30
31
32

AFRICAN AND NEW WORLD PARROTS *CONTINUED*

1 BLUE-CROWNED PARAKEET *Thectocercus acuticaudatus* 34–36cm Generally green but metallic pale blue on forehead, crown, lores, cheek, ear-patch and part of throat. Bare patch on face yellowish-white. Creamy beak with black tip and mandible. Yellowish-green on breast and belly. Sharp-pointed green tail with red on inner web of outer feathers. **V** Strident and rough. **H** Lowland dry forest. **D** Disjunct populations: N Colombia, N Venezuela; NE Brazil; Bolivia and S Brazil to N Argentina.

2 GREEN PARAKEET *Psittacara holochlorus* 32cm From Pacific Parakeet by smaller, less deep bill and more orange on throat. Little overlap of ranges. May or may not show the odd red feather on head or breast. **V** Raucous shriek, *krreh-krreh* or *kee-ik krii-krii kriir*; also a deeper *kreh kreh…* **H** Forests, plantations, arid pine–oak forests. **D** NW, E, S Mexico.

3 SOCORRO PARAKEET *Psittacara brevipes* 30cm Formerly treated as a race of Green Parakeet. Uniform green, with dull pink bill. No other parakeets in its range. **V** Not well known, but high-pitched, squeaky shrieks in a fast series have been recorded. **H** Forest. **D** Socorro Island (off W Mexico).

4 RED-THROATED PARAKEET *Psittacara rubritorquis* 30cm Formerly treated as a race of Green Parakeet. Orange throat diagnostic. Note purplish bare eye-ring. **V** Like that of Green Parakeet. **H** Mixed pine and semi-deciduous forest. **D** S Guatemala, Honduras, N Nicaragua.

5 PACIFIC PARAKEET *Psittacara strenuus* 30cm More olivaceous than Green Parakeet. May show a few flecks of red on head and neck. **V** Various harsh, sharp and squeaky notes; also noisy chattering. **H** Forest, woodland, plantations of the Pacific slope. **D** S Mexico to N Nicaragua.

6 SCARLET-FRONTED PARAKEET *Psittacara wagleri* 35cm Very like Mitred Parakeet but scarlet on forehead extends onto forecrown. Scarlet garters. Dorsally lighter green than Mitred Parakeet. **V** Squeaky notes, screeches and braying sounds. **H** Cloud forests, rainforest, semi-open areas of the Andes, valleys, plantations, foothills. **D** Colombia, N Venezuela.

7 CORDILLERAN PARAKEET *Psittacara frontatus* 38–40cm Previously treated as conspecific with Scarlet-fronted Parakeet. More red on head and on carpal area. Nests colonially in cliff-face holes. **V** Squeaky chattering calls, disorganised-sounding when coming from a flock. **H** Cloud forest, montane scrub. **D** W Ecuador, Peru.

8 MITRED PARAKEET *Psittacara mitratus* 38cm Scarlet on forehead and onto face, less so in the female, but always irregular and variable. A few scattered red spots on the head and body. Variable orangey red on shoulder. Underwing coverts green. Heavy bill buffy ivory. Bare skin around eye white. **V** Call like others of the genus, but strident and high-pitched. **H** Subtropical rainforest in mountains and adjacent valleys, upland woods, plantations. **D** Peru to NW Argentina.

9 RED-MASKED PARAKEET *Psittacara erythrogenys* 34cm Most of head bright scarlet. Underwing coverts scarlet, also visible on leading edge when wing is folded. Scarlet garters. **V** Wide variety of notes, including sharp *ke-REE* contact call, *eh eh-eh* in alarm, and various softer chuckling notes in courtship. **H** Dry woods. **D** W Ecuador, NW Peru.

10 FINSCH'S PARAKEET *Psittacara finschi* 36–40cm Forehead red, with a few red flecks elsewhere on head. White eye-ring. Edge of wing and outer underwing-coverts red, often tinged orange; greater underwing-coverts yellow. **V** A loud, shrill *sweet sweet swit*. **H** Hill forests. **D** Nicaragua to Panama.

11 WHITE-EYED PARAKEET *Psittacara leucophthalmus* 35cm Differs from Scarlet-fronted, Mitred and Red-masked Parakeets in having only a few scarlet spots on head. In flight, primary underwing coverts shine red and yellow. The red is also seen on the leading edge when wing is folded. **V** Series of penetrating, rough *cheerrry cheerrry*. **H** Open woods and forest, palm groves, savannah, mangroves, gallery forest, scattered trees, plantations, copses, rainforest, gardens, parks. **D** Venezuela, the Guianas, through Amazonia to N Argentina.

12 CUBAN PARAKEET *Psittacara euops* 24–27cm Scattered red feathers. In flight, which is swift and direct, shows red underwing coverts. Gregarious, generally occurs in small groups. **V** Loud, repeated *crick-crick-crick*, usually uttered in flight. When feeding or at rest, a low chattering. **H** Savannah with cabbage palms, forest edge, cultivated areas. **D** Cuba.

13 HISPANIOLAN PARAKEET *Psittacara chloropterus* 30–33cm In flight, which is swift and direct, shows red underwing coverts. Threatened by habitat loss, persecution as a crop pest, and pet trade. **V** Shrill screech, given when perched or in flight. **H** Mainly mountain forests but also arid lowlands, often attracted to crops. **D** Hispaniola. Introduced to Puerto Rico and Guadeloupe.

OLD WORLD PARROTS PSITTACULIDAE

14 PESQUET'S PARROT *Psittrichas fulgidus* 45–47cm Bare facial skin, bill shape and relatively long neck create vulture-like appearance. Feeds on fruits and flowers. **V** Harsh, croaking call, like a goose with a sore throat. **H** Montane primary forest, secondary woodland; mainly 500–1,500m. **D** New Guinea.

15 GREATER VASA PARROT *Coracopsis vasa* 50cm The largest African parrot; unmistakable by its size and brownish-black plumage. From smaller Lesser Vasa Parrot by its larger bluish area of skin around eyes. **V** Range of screeches, melodious whistles and harsh, grating notes. **H** All natural habitats with trees, cultivation, subdesert; at lower altitudes. **D** Madagascar, Comoros.

16 LESSER VASA PARROT *Coracopsis nigra* 38cm Compare Greater Vasa Parrot which is 25% larger. Bare skin round eyes hardly visible in the field. **V** Various loud notes such as *koo… keh… kiweek* and similar variants, chattering *kakakakakakakerkerker*, and soft croaking notes. **H** Forest, woodland and other habitats with some trees; up to high altitude. **D** Madagascar, Comoros, Seychelles.

17 SEYCHELLES BLACK PARROT *Coracopsis barklyi* 35–40cm Feeds on fruits and seeds. Roosts and breeds in hills. Ventures to lower elevations to feed. **V** Shrill, upslurred whistling calls. **H** Forested habitats with predominant palms, agricultural ground, gardens. **D** Seychelles (mainly Praslin, occasionally Curieuse).

18 YELLOW-CAPPED PYGMY PARROT *Micropsitta keiensis* 9cm Dusky yellow crown, shading to green at rear; upper face ochre. Occurs in pairs or small parties. Forages on trunks and large branches. **V** Short, weak, high-pitched notes. **H** All types of woodland areas. **D** New Guinea, W Papuan Islands, Kai Islands, Aru Islands.

19 GEELVINK PYGMY PARROT *Micropsitta geelvinkiana* 8–10cm Dark brownish crown and face, bluer on throat. Orange-yellow centre to underparts. Diet mainly lichens and fungi, also fruits, seeds and insects. Often seen in small flocks. **V** Thin, trilling whistled *tseee*. **H** Lowland primary forest, secondary woodland. **D** Biak Island, Numfoor Island (NW New Guinea).

20 BUFF-FACED PYGMY PARROT *Micropsitta pusio* 7–9cm Buff-ochre face, sometimes with yellowish eye-stripe. Crown and nape deep blue. Seen in pairs or small feeding flocks. Diet mainly lichens, fungi and insects. **V** Thin, high-pitched whistled *tsiii*. **H** Lowland primary forest, secondary woodland. **D** N New Guinea, Bismarck Archipelago, Louisiade Archipelago.

21 MEEK'S PYGMY PARROT *Micropsitta meeki* 9–11cm Head grey-brown with light barring. Nape and underparts yellow. Diet presumed to include lichens, fungi and insects, like other pygmy parrot species. Seen in pairs or small flocks. **V** Thin, high-pitched whistled *tsiip, tsiip*. **H** Primary forest, secondary woodland. **D** Admiralty Islands, St Matthias Islands (Bismarck Archipelago).

22 FINSCH'S PYGMY PARROT *Micropsitta finschii* 8–9cm Uniformly green plumage. Nests in arboreal termite nests. **V** Thin, high-pitched trilling whistles. **H** Primary rainforest, secondary woodland. **D** New Guinea, Bismarck Archipelago, Solomon Islands.

23 RED-BREASTED PYGMY PARROT *Micropsitta bruijnii* 8–9cm Male distinctively multi-coloured. Female has yellow face and blue crown. Frequents upper tree levels; acrobatic forager, seen singly, in pairs or in small parties. **V** A thin, high-pitched *tsee-tsee*. **H** Primary and disturbed hill and montane forest; occasionally lowland forest. **D** S Maluku Islands through New Guinea to Solomon Islands.

24 SUPERB PARROT *Polytelis swainsonii* 38–40cm Male has distinctive colours on head. Female almost entirely green. Feeds on eucalypt flowers and fruits; also grain and seeds. **V** Harsh, squawking *who-aak, who-aak-ak, who-aak* calls. **H** Dry eucalypt woodland, often beside rivers. **D** Interior SE Australia.

25 REGENT PARROT *Polytelis anthopeplus* 40cm Male has greenish-yellow body, black wings with red and yellow patches, red bill. Female duller and greener. Forms flocks, especially outside breeding season. Partly nomadic, following rains and resulting plant growth. Diet includes fruits, grain and flower buds. **V** Resonant, squawking *o-uWuek, o-uWuek, o-uWuek*. **H** Open eucalypt woodland, mallee scrub. **D** Disjunct populations: SW and interior SE Australia.

26 PRINCESS PARROT *Polytelis alexandrae* 45cm Pastel-toned parrot with pale blue crown, pink throat, bright green wing coverts. Female somewhat duller. Feeds mainly on seeds. Seasonally nomadic, following recent rains and resulting plant growth. **V** Screeching, squawking calls including *kyrrr, kyrrr, krrr, krrr*. **H** Arid grassland, sandy deserts, scrub. **D** Interior W and C Australia.

27 MOLUCCAN KING PARROT *Alisterus amboinensis* 35cm Head, nape and underparts red. Mantle, wings and tail dark green and blue. More often seen in flight; forages from mid-storey to canopy. **V** High-pitched, ringing, upslurred whistles; in flight, utters a dry *chack chack*. **H** Primary and secondary lowland and hill forest; occasionally gardens. **D** Sulawesi (Banggai Islands), Sula Islands, Maluku Islands, W New Guinea.

28 PAPUAN KING PARROT *Alisterus chloropterus* 35–37cm Similar to Moluccan King Parrot, but blue on hindcrown and nape, black mantle, greenish-yellow lesser wing coverts. Female much greener. Diet mainly seeds and berries. Usually seen in pairs or small flocks. **V** Shrill and penetrating whistled *kEe, kErr, kEe, kEe*. **H** Primary rainforest. **D** New Guinea.

29 AUSTRALIAN KING PARROT *Alisterus scapularis* 42–44cm Forms flocks. Sometimes mixes with other parrot species. Diet includes seeds, fruits and flowers. Altitudinal migrant in S of range. **V** Squeak, chattering *wEek, wiErk, wEek* calls. **H** Eucalypt woodland, primary rainforest, scrub, urban parks. **D** E and SE Australia.

30 JONQUIL PARROT *Aprosmictus jonquillaceus* 35cm Similar to female Red-winged Parrot but brighter head and underparts, darker back. Occurs singly, in pairs and occasionally in small groups, more often seen flying across forest openings; forages mainly from the mid-storey to the canopy. **V** Short, harsh, grating squawks. **H** Primary and tall secondary monsoon forest, savannah woodland, cultivation and scrubby secondary growth. **D** Maluku Islands, Lesser Sunda Islands.

31 RED-WINGED PARROT *Aprosmictus erythropterus* 30–32cm Slim-bodied, long-tailed parrot that shows blue rump in flight. Male has black back and large red wing patches. Female has green back and less red. Diet includes seeds, nuts and fruits. Gathers in flocks where feeding is good. **V** Sharp *wHit, wHit, wHit* calls, with tone of thrush alarm-call. **H** Eucalypt forest, scrub, mangroves (Australia), coastal scrub and forest (New Guinea). **D** N and E Australia, SC New Guinea.

1
2
3
4
5
6
7
8
9
10
11
12
13
14
15
16
17
18
19 ♀
♂ 20
21
♀ 22
♂
♂ 23 ♀
♂ 24
♀
♂ 25
♀
26
27
♂
28
♀
♂
29
♀
30
♂
♀
31

OLD WORLD PARROTS *CONTINUED*

1 BURU RACKET-TAIL *Prioniturus mada* 32cm Male has hindcrown, nape, mantle and shoulders purplish, shading to olive-green on back and wings. Female has less blue. Undertail coverts yellow. Tail spatules dark blackish blue. Forages in the crowns of forest trees, usually in small flocks. **V** Repeated pleasant whistling, a musical *si-quie*, a rapidly repeated *kwii kwii kwii* and a low-pitched *squr-squr*. **H** Primary lowland, hill and montane forest. **D** S Maluku Islands, Indonesia.

2 GOLDEN-MANTLED RACKET-TAIL *Prioniturus platurus* 28cm Orange stripe on male's back. Regularly seen flying over the forest canopy in small or large, noisy flocks; otherwise quite secretive and hard to see. **V** Harsh, nasal *kaaa*, a *krrrik* or *krrrri*, repeated nasal *quelie*. **H** Primary and tall secondary lowland and montane forest, mangroves and lightly wooded cultivation. **D** Sulawesi and Sula Islands, Indonesia.

3 MINDANAO RACKET-TAIL *Prioniturus waterstradti* 27cm Once considered a race of Montane Racket-tail, but less blue and no red on head. Actions presumed to be similar. **V** Presumably similar to the previous species. **H** Dense montane forest, ridgetop forest and stunted mossy forest. **D** Mindanao, S Philippines.

4 MONTANE RACKET-TAIL *Prioniturus montanus* 30cm Male has blue forehead and face and red mid-crown; tail spatules blackish. Female has green head. Occurs singly, in pairs or in small flocks; noisy when flying through or over forest. **V** Calls include a hacking *kak-kak-kak-kak ak ak ak...*, loud, harsh notes and a note like a swinging rusty gate. **H** Mid-montane forest. **D** N Philippines.

5 BLUE-HEADED RACKET-TAIL *Prioniturus platenae* 27cm Entire head of male light blue, greener in female. Occurs singly, in pairs or in small groups, foraging in the canopy. Noisy, especially in flight. **V** Similar to Montane Racket-tail. **H** Lowland forest, scrub, mangroves. **D** Palawan, W Philippines.

6 MINDORO RACKET-TAIL *Prioniturus mindorensis* 27cm Split from the previous species; habits presumed to be similar. Green forehead. **V** Presumably similar to the previous species. **H** Humid forest, mainly in the lowlands. **D** Mindoro, WC Philippines.

7 BLUE-WINGED RACKET-TAIL *Prioniturus verticalis* 30cm Male has red spot on crown. Noisy in flight; foraging actions similar to other racket-tails. **V** Calls include a harsh, rasping *aaaaak* and a squeaky *lee-aaack*. **H** Primary lowland forest, forest edges, mangroves. **D** Sulu Archipelago, SW Philippines.

8 YELLOW-BREASTED RACKET-TAIL *Prioniturus flavicans* 37cm Male has red spot on crown and olive-yellow from nape to mantle and breast. Usually seen singly, in pairs or in small groups; generally quiet and inconspicuous as it forages in the thick foliage of the crowns of tall trees. **V** Calls include a drawn-out screech with an alternating pitch, a series of three repeated nasal, barking screeches, and a higher-pitched nasal bugling or yodelling. **H** Primary lowland and hill forest, lightly wooded cultivation. **D** Sulawesi, Indonesia.

9 GREEN RACKET-TAIL *Prioniturus luconensis* 29cm Green all over, slightly yellower on head and underparts. Forages in the canopy and understorey, usually in pairs or small parties; very active in the late afternoon. **V** Calls include a harsh *aaaak*, a horse-like whinnying, a ringing *liiinng* and a *yuur-witt*, the last syllable rising sharply. **H** Lowland forest, forest edges and nearby cultivated areas. **D** N Philippines.

10 BLUE-CROWNED RACKET-TAIL *Prioniturus discurus* 27cm Crown pale blue; rest of head green. Occurs in small, noisy groups or pairs. **V** Various squeals, squeaks and harsh notes. **H** Primary and secondary forest, mangroves, orchards and banana plantations. **D** Philippines.

11 ECLECTUS PARROT *Eclectus roratus* 35–42cm Male bright green with red underwings, female bright red and blue-violet belly. Noisy and conspicuous; usually seen singly, in pairs and occasionally in small parties foraging in the canopy. Often perches on partially exposed branches. **V** Series of loud, hoarse, rapidly repeated screeches, generally given in flight; also a ringing, metallic squeal, uttered when perched. **H** Primary and tall secondary lowland and hill forest, mangroves and plantations. **D** Maluku Islands through New Guinea to Solomon Islands; NE Australia.

12 RED-CHEEKED PARROT *Geoffroyus geoffroyi* 21–30cm Male has red face, blue-purple crown, pale eye, reddish upper mandible. Female has brown head, pale eye, dark bill. Frequents the canopy, in small groups, in pairs or singly. Often sits prominently on an exposed branch. **V** Short, repeated *kee* notes. **H** Primary and tall secondary forest, monsoon forest, forest edges, coastal woodland, mangroves, scrub, parkland. **D** Maluku Islands, Lesser Sunda Islands, New Guinea, Solomon Islands, NE Australia.

13 BLUE-COLLARED PARROT *Geoffroyus simplex* 21–23cm Uniform plumage. Blue collar of male not useful in field identification. Diet includes seeds and nectar. Nomadic, forming flocks that search for good feeding trees. **V** Harsh, screeching calls and grating whistles. **H** Montane forests. **D** New Guinea.

14 SONG PARROT *Geoffroyus heteroclitus* 24–25cm Head and bill mostly yellow in male, grey in female. Feeds on seeds and fruits. Gathers in flocks where feeding is good. **V** Shrill, piping whistled *wheer, wheer, wheer, whe-er, whe-er*. **H** Primary forest, secondary woodland, plantations; mainly lowlands to 1,000m. **D** Bismarck Archipelago, Solomon Islands.

15 BLUE-RUMPED PARROT *Psittinus cyanurus* 18–19cm Male mostly green with orange bill and powder-blue head and rump. Female has brown head, dark bill. Pairs or small parties are usually seen foraging in the upper branches, or flying above the forest canopy. **V** Sharp *chi-chi-chi* and *chew-ee*, also a high-pitched *peep*. **H** Lowland forest, swamp forest, mangroves and cultivated areas. **D** Malay Peninsula; Borneo; Sumatra, Indonesia.

16 GREAT-BILLED PARROT *Tanygnathus megalorynchos* 33–43cm Note massive red bill. Seen singly, in pairs or in small parties; larger flocks at roosts. Nomadic, occasionally flying between small islands. **V** Harsh, reedy *ke-rarr ke-rarr*; repeated quavering, nasal notes. **H** Primary and secondary lowland and hill forest, forest edges, mangroves, coastal woodland, wooded cultivation, plantations and gardens. **D** Maluku Islands, Lesser Sunda Islands.

17 BLUE-NAPED PARROT *Tanygnathus lucionensis* 31cm Blue nape, orange-buff fringes on lesser and median wing coverts. Occurs singly, in pairs or in small parties; feeds on fruit and seeds in upper tree levels. **V** Piercing shriek and harsh *awwwk-awwwk*. **H** Forest, forest edges, remnant patches of forest within cultivation, wooded cultivation and coconut groves. **D** Philippines; N Maluku Islands and islands off N Borneo.

18 BLUE-BACKED PARROT *Tanygnathus sumatranus* 32cm Blue lower back and rump. Forages in the canopy, singly, in pairs or in small groups. **V** Harsh, squawking *nyak nyak...* followed by a screech, and a loud, piercing *kei*. **H** Forest and forest edges, swamp forest, tall secondary woodland, remnant forest patches, coconut plantations. **D** S Philippines; Sulawesi, Indonesia.

19 BLACK-LORED PARROT *Tanygnathus gramineus* 40cm Note black line through lores. Usually seen singly; more often heard than seen; little else recorded. Female has pink-grey bill. **V** Similar to Great-billed Parrot, but more drawn-out and higher-pitched. **H** Montane forest. **D** S Maluku Islands, Indonesia.

20 GREY-HEADED PARAKEET *Psittacula finschii* 36–40cm Paler grey head, upperparts with a slightly yellower wash and bluish tail separate this species from the very similar Slaty-headed Parakeet. Food includes fruit, grain and plant parts. Flight swift and agile, especially when flying among trees. Female lacks the maroon shoulder patch. **V** Loud, shrill *sweet sweet swit*. **H** Hill forests. **D** SE Asia.

21 SLATY-HEADED PARAKEET *Psittacula himalayana* 39–41cm Dark head, greener upperparts and yellow-tipped tail distinguish this species from the very similar Grey-headed Parakeet. Flight is agile and swift, especially when flying through trees. Feeds on fruit and seeds. Female lacks the maroon shoulder patch. **V** High-pitched *scree-scree* and a drawn-out *wee...eenee*. **H** Woodland and cultivated areas with large trees. **D** Himalayas.

22 BLOSSOM-HEADED PARAKEET *Psittacula roseata* 30–36cm Like paler-headed version of Plum-headed Parakeet. Actions and habits similar. **V** Similar to Plum-headed Parakeet. **H** Open forests and well-wooded country. **D** N India to Bangladesh and SE Asia.

23 PLUM-HEADED PARAKEET *Psittacula cyanocephala* 33–37cm Brilliantly coloured. Generally occurs in small parties, but found in much larger flocks where food is plentiful. Food includes fruit, seeds, buds, fleshy parts of plants. **V** High-pitched *tooi-tooi*. **H** Forests, well-wooded areas, cultivations. **D** India, Sri Lanka, widespread.

24 RED-BREASTED PARAKEET *Psittacula alexandri* 33–38cm Note black chin, black line through lores, pink breast. Usually occurs in small parties, with much larger flocks where food is plentiful. Food includes fruit, seeds, buds, flowers, nectar. **V** Short, nasal *kaink*, repeated by several birds when disturbed. **H** Moist deciduous forest, secondary growth, plantations. **D** N India to SE Asia, Greater and Lesser Sunda Islands.

25 LORD DERBY'S PARAKEET *Psittacula derbiana* 46–50cm Blue-purple face and underparts, black chin. Gregarious. Food includes fruit, seeds and ripening crops. **V** High-pitched shrill whistle; also a long, raucous, metallic cry. **H** Coniferous and mixed mountain forests. **D** SW China, NE India.

26 LONG-TAILED PARAKEET *Psittacula longicauda* 40–48cm Reddish cheeks, with black stripes above and below. Usually in small flocks, with larger flocks occurring at rich food sources. **V** High-pitched *pee-yo pee-yo pee-yo* and a nasal quavering *graak graak graak*. **H** Open broadleaved evergreen forest, freshwater swamp forest, plantations and mangroves. **D** Andaman Islands, Peninsular Malaysia, Singapore, Greater Sunda Islands.

27 BLUE-WINGED PARAKEET *Psittacula columboides* 36–38cm Male largely blue-grey with bright red upper bill and black and blue collar. Female greener with dark bill. Usually encountered in small parties; flight fast and agile. Feeds on grain, seeds and fruits. **V** Coarse *che-chwe*. **H** Tropical evergreen and moist deciduous forest, secondary growth, discarded plantations, nearby cultivations. **D** SW India.

28 LAYARD'S PARAKEET *Psittacula calthrapae* 29–31cm Green with grey hood and conspicuous black throat. Usually encountered in good-sized flocks on the edge or in open areas of woodlands. Food includes fruit, fleshy parts of plants and nectar. **V** Harsh, chattering scream. **H** Forest edges and clearings, plantations and gardens. **D** Sri Lanka.

29 ALEXANDRINE PARAKEET *Psittacula eupatria* 53–58cm Both sexes have red bill. Female lacks male's black and pink collar. Occurs in small flocks, with larger concentrations at rich food sources and roosts. Feeds on fruit, seeds, cereals and fleshy parts of plants. **V** Hoarse screaming *kii-e-rick*, *keeak* or *kee-ah*. **H** Well-wooded areas and plantations. **D** From Pakistan through India and Sri Lanka to SE Asia.

30 ROSE-RINGED PARAKEET *Psittacula krameri* 37–43cm Resembles larger Alexandrine Parakeet, but lacks dark red shoulder patch. Feeds on fruit, seeds and cereals. Flight fast and direct. **V** Screeching *kee-a* or *kee-ak*. **H** Open woodland, groves, parks, gardens. **D** Widespread across Africa from Mauritania to Djibouti, and across Asia from Afghanistan to Myanmar. Introduced Europe.

31 ECHO PARAKEET *Psittacula eques* 40cm Differs from Rose-ringed Parakeet by pale-rimmed eyes, less pointed tail and habitat. Note female's all-black bill. **V** Commonest call is described as a short, nasal *kaah*. **H** Natural forest. **D** Mauritius.

32 NICOBAR PARAKEET *Psittacula caniceps* 56–61cm Greyish head with prominent black frontal band and black throat; otherwise green. Occurs singly, in pairs or in small parties. Rests in foliage at the tops of trees when not feeding, where best located by noisy calls. Feeds mainly on the ripe fruit of *Pandanus*. **V** Wild screeching notes uttered while at rest or in flight: a crow-like, raucous *kraan...kraan*. **H** Tall forest. **D** Nicobar Islands, E Indian Ocean.

1
2
3
4
5
6
7
8
9
10
11
12
13
14
15
16
17
18
19
20
21
22
23
24
25
26
27
28
29
30
31
32

OLD WORLD PARROTS *CONTINUED*

1 BREHM'S TIGER PARROT *Psittacella brehmii* 23–25cm Beautifully patterned plumage. Male has chocolate-brown head with vertical yellow line on side of neck; nape, mantle, rump and uppertail coverts barred black and green. Female has barred breast and lacks yellow line. Diet includes seeds, fruits, berries and flower buds. **V** Piping, whistled *peup, peup, peup* calls. **H** Montane forest, secondary woodland; mainly 1,500–2,500m. **D** New Guinea.

2 PAINTED TIGER PARROT *Psittacella picta* 17–19cm Colourful and intricately patterned plumage. Diet includes seeds, berries and flower buds. **V** Distinctive, downslurred whistling *peeur, peeur, peeur* calls. **H** Montane forest, subalpine grassland and scrub; mainly 2,500–4,000m. **D** New Guinea.

3 MODEST TIGER PARROT *Psittacella modesta* 13–15cm Male's dark brown head aids identification. Feeds on fruits, berries and seeds. Seen in pairs or small flocks. **V** Upslurred, piping whistled *whuuiK, whuuiK*. **H** Montane primary forest, secondary woodland; mainly 2,000–2,500m. **D** New Guinea.

4 MADARASZ'S TIGER PARROT *Psittacella madaraszi* 13–15cm Male's head is paler and more speckled than Modest Tiger Parrot's. Feeds on fruits, berries and seeds. **V** Calls include upslurred, whistling *hu-wee*. **H** Montane primary forest, secondary woodland, cultivated ground; mainly 1,000–2,000m. **D** E New Guinea.

5 RED-RUMPED PARROT *Psephotus haematonotus* 26–28cm Red rump most noticeable in flight. Feeds on plant material including seeds. Seen in pairs or small flocks. **V** Chattering screeches, recalling high-pitched accelerated calls of House Sparrow. **H** Mallee scrub, open woodland, wooded grassland, agricultural land. **D** SE Australia.

6 EASTERN BLUEBONNET *Northiella haematogaster* 27–29cm Note yellow and red belly, deep blue face. Seen in pairs or small flocks. Feeds mainly on seeds, mostly on ground. **V** Abrupt *tchuup tchuup, tchip, tchip* and piping *pep-pee*. **H** Arid grassy scrub, open woodland, agricultural land. **D** Interior SE Australia.

7 NARETHA BLUEBONNET *Northiella narethae* 27–29cm Yellow belly, red undertail coverts. Feeds mainly on ground; diet includes seeds, fruits, berries and flowers. Visits water troughs. **V** Various screeching notes. **H** Open, dry woodland with *Acacia* and *Eucalyptus*, arid scrub. **D** Very local, S Australia.

8 MULGA PARROT *Psephotellus varius* 27–29cm Colourful plumage. Feeds mainly on ground; diet includes seeds, fruits, berries. **V** Shrill, rapid *chip-chip-chip*-chip and upslurred whistle. **H** Open dry woodland, arid mulga scrub, grassland. **D** Australia, from W to interior SE.

9 HOODED PARROT *Psephotellus dissimilis* 25–27cm Striking plumage. Male is turquoise with black cap, yellow wings, rosy undertail coverts, olive back. Female primarily pale green. Nests in terrestrial termite mounds. Feeds mainly on seeds, berries and fresh plant material. **V** Calls include a whistled *tsip, tsip, tsip*. **H** Open woodland, savannah grassland. **D** N Australia.

10 GOLDEN-SHOULDERED PARROT *Psephotellus chrysopterygius* 25–27cm Male's golden-yellow wing coverts are striking. Nests in terrestrial termite mounds. Feeds mainly on seeds. Rare and endangered. **V** Calls include a shrill *whip, whip, whip*. **H** Open eucalyptus forest with termitaria, usually near water. Visits coastal and estuarine mangroves. **D** Restricted to small area of N Queensland.

11 RED-CAPPED PARROT *Purpureicephalus spurius* 35–37cm Male's colourful plumage is distinctive. Female similar but slightly duller. Diet comprises mainly eucalypt seeds; occasionally flowers and invertebrates. **V** Calls include rasping *tchrr, tchrr, tchrr* notes. **H** Woodland, notably where Marri *Corymbia calophylla* dominates. **D** Coastal SW Australia.

12 GREEN ROSELLA *Platycercus caledonicus* 36cm Green and yellow with blue chin and wing coverts. Diet includes, fruits, berries, seeds, flowers; occasionally invertebrates. Seen in pairs or small groups. **V** Calls include distinctive and explosive paired or triplet squeaking *pTchee-pTchu* or *pTchee-pTchu-Tchu*. **H** Wooded habitats, including rainforest, parks, orchards. **D** Tasmania and Bass Strait islands.

13 CRIMSON ROSELLA *Platycercus elegans* 32–36cm Colourful plumage aids identification. In some races the red is replaced by red and yellow, or all yellow. Seen in pairs or small groups. Feeds on seeds, fruits, flowers, in trees and on ground. **V** Chirpy, whistling notes including *p-Pik-pu-pik*, each phrase jumping up and down an octave. **H** Coastal and adjacent mountain forests, woodland, farmland, suburban parks, gardens. **D** E and SE Australia. Introduced New Zealand.

14 NORTHERN ROSELLA *Platycercus venustus* 27–29cm Unmistakable plumage. Note black crown contrasting with white cheeks. Feeds on seeds, notably eucalpyts and grasses. Seen in pairs or small flocks. **V** Shrill, whistled *whip, whipt, whip* notes. **H** Eucalypt woodland, wooded savannah. **D** N Australia.

15 PALE-HEADED ROSELLA *Platycercus adscitus* 30–32cm Note white and yellow head. Feeds on seeds and fruits, including grasses, she-oaks and eucalypts. Nests in tree holes. **V** Abrupt series of chirping *tche-tche-tche-tche* notes. **H** Open woodland, savannah scrub. **D** NE Australia (Queensland to N New South Wales).

16 EASTERN ROSELLA *Platycercus eximius* 30–32cm Strikingly colourful plumage, with red head and white cheeks. Diet mainly seeds; occasionally also berries, buds, invertebrates. **V** Calls include wader-like series of rapid piping *te-tchu-tchu-tche-tche-tche* notes, pitch descending during delivery. **H** Various lightly wooded habitats including farmland, riversides, gardens, parks. **D** SE Australia, Tasmania. Introduced New Zealand.

17 WESTERN ROSELLA *Platycercus icterotis* 26–27cm Colourful plumage and restricted range aid identification. Seen in pairs or small groups. Diet comprises mainly grass seeds, grain, berries. **V** Calls include wader-like strident, piping whistles. **H** Open woodland, agricultural land. **D** SW Australia.

18 AUSTRALIAN RINGNECK *Barnardius zonarius* 34–44cm Yellow collar is distinctive. Nests in tree holes. Nomadic in times of drought. Diet includes fruits, seeds and underground tubers and bulbs. **V** Calls include strident piping *Peep-pe-pip-pip* notes. **H** Open woodland, mallee scrub, agricultural land. **D** W and interior Australia (absent from much of N and coastal E).

19 SWIFT PARROT *Lathamus discolor* 24–26cm Note red and yellow face. Feeds on pollen and nectar, especially from eucalypts, as well as insects, fruit, berries, seeds. Critically endangered. **V** Calls include shrill screeching phrases that 'dance' in tone and pitch. **H** Eucalypt forest, gardens, parks. In winter remnant open forest areas, parklands, suburbs **D** Breeds Tasmania. Non-breeding visitor to SE Australia.

20 CRIMSON SHINING PARROT *Prosopeia splendens* 45cm Distinguished from Maroon Shining Parrot by red (not maroon) head and underparts. **V** Short, harsh *erh*, grating *krreh*, and raucous *graaah*. **H** Forest, mangrove, secondary growth, cultivation. **D** Fiji.

21 MASKED SHINING PARROT *Prosopeia personata* 47cm Largely green, with black face and bill. No similar bird in range. **V** Mostly short, harsh notes with nasal or grating quality. **H** Forest, secondary growth, cultivation, mangroves. **D** Fiji.

22 MAROON SHINING PARROT *Prosopeia tabuensis* L 45cm Maroon head and underparts diagnostic. **V** Typical parrot-like *wrah wrah*. **H** Mangrove, forest. **D** Fiji, Tonga.

23 HORNED PARAKEET *Eunymphicus cornutus* 30–33cm Black face and bill, and red-tipped protruding crown feathers, aid identification. Usually seen in pairs. Feeds on seeds and berries. **V** Calls include series of hooting *wuut-wuut-wuut* notes. **H** Forested slopes with *Agathis* and *Araucaria* trees, savannah, scrub. **D** New Caledonia.

24 OUVEA PARAKEET *Eunymphicus uvaeensis* 30–33cm Diet comprises primarily flowers and fruits of forest trees. Often seen in pairs. **V** Calls include a rapid-fire series of harsh shrieks. **H** Primary forests, second woodland, agricultural land. **D** Ouvéa, Loyalty Islands (New Caledonia).

25 NEW CALEDONIAN PARAKEET *Cyanoramphus saisseti* 26–28cm Red crown, red line through eye. Previously considered conspecific with Red-crowned Parakeet. Diet includes buds, flowers, fruits, seeds. **V** Calls include shrill *wheit, wheit* notes. **H** Primary forest, secondary woodland, savannah scrub. **D** New Caledonia.

26 CHATHAM PARAKEET *Cyanoramphus forbesi* 26cm Restricted range. From Red-crowned Parakeet by yellow forehead. **V** Most common call a nasal, toy trumpet-like *neh-neh-neh*, also single screeching *krreeah*. **H** Undisturbed forest, scrub. **D** Chatham Islands, SW Pacific Ocean.

27 NORFOLK PARAKEET *Cyanoramphus cookii* 26–28cm Diet comprises seasonally available buds, flowers, fruits, seeds. Endangered. Population decline due to habitat destruction. **V** Rapid-fire shrill *ker-ke-ke-ke* and raucous nasal chatter. **H** Dwindling forested habitats, scrub. **D** Norfolk Island, SW Pacific Ocean.

28 ANTIPODES PARAKEET *Cyanoramphus unicolor* 30cm Unmistakable in range by wholly green head. **V** Varied, rapid chattering, hoarse screeches, cooing squeaks. **H** Tussock grassland, fernland, coastal swamp. **D** Antipodes Islands, subantarctic SW Pacific Ocean.

29 YELLOW-CROWNED PARAKEET *Cyanoramphus auriceps* 23cm Note yellow forecrown. Distinguished from Chatham Parakeet by range; from Malherbe's Parakeet by slightly broader, deeper orange forehead. **V** Very high, shivering chatter and other calls (higher-pitched than those of Red-crowned Parakeet). **H** Mainly in undisturbed native forest. **D** North and South Islands, New Zealand.

30 MALHERBE'S PARAKEET *Cyanoramphus malherbi* 20cm Similar to Yellow-crowned Parakeet, but note restricted range. **V** Sharp chatter and large diversity of musical and harsh notes, uttered singly or in series. **H** Forest edge. **D** New Zealand: confined to extreme N of South Island. Translocated to several islands off South Island (Chalky, Maud, Blumine) and North Island (Tuhua).

31 RED-CROWNED PARAKEET *Cyanoramphus novaezelandiae* 27cm Note distinctive red crown and small red patch behind eye. Blue leading edge to wing. **V** Varied, e.g. very high, hoarse chattering *ki-ki-ki-* and sharp screeches. **H** Rainforest, scrubland, open areas. **D** New Zealand (including Chatham Islands, Kermadec Islands).

32 REISCHEK'S PARAKEET *Cyanoramphus hochstetteri* 26–28cm Previously treated as subspecies of Red-crowned Parakeet. Terrestrial feeding habits. Diet includes seeds of grasses and sedges. **V** Shrill whistling calls and raucous nasal chatter. **H** Scrub, grassland. **D** Antipodes Islands, subantarctic SW Pacific Ocean.

33 EASTERN GROUND PARROT *Pezoporus wallicus* 29–30cm Upperparts green, mottled with black and yellow. Terrestrial habits. Feeds mainly on seeds, favouring button grass. **V** Extremely high-pitched series of *tutututu-tsee-tsee-se-se-se-se* notes, rising in pitch during delivery. **H** Coastal grassland, grassy heaths. **D** Coastal E and SE Australia, Tasmania.

34 WESTERN GROUND PARROT *Pezoporus flaviventris* 29–30cm Endangered. Terrestrial. Feeds mainly on seeds of sedges and grasses, favouring button grass. **V** High-pitched series of *tutututu-tsee-tsee-se-se-se-se* notes, rising in pitch during delivery. **H** Coastal heath, grassland **D** Coastal SW Australia.

35 NIGHT PARROT *Pezoporus occidentalis* 22–24cm Terrestrial, secretive, enigmatic and seldom observed. Diet includes seeds of spinifex (*Triodia*). **V** Short series of treefrog-like croaking *tink-tink* notes; uttered just after sunset. **H** Arid spinifex grassland. **D** Interior N and C Australia.

1
2
3
4
5
6
7
8
9
10
11
12
13
14
15
16
17
18
19
20
21
22
23
24
25
26
27
28
29
30
31
32
33
34
35

OLD WORLD PARROTS *CONTINUED*

1 BOURKE'S PARROT *Neopsephotus bourkii* 18–20cm Note pink underparts, blue undertail coverts and rump. Feeds mainly on seeds, particularly grasses. Nomadic, following rains and subsequent plant growth. Sometimes forms flocks outside breeding season. **V** Series of abrupt, shrill *tchuiP, tche-tchuiP* notes. **H** Arid scrub dominated by *Acacia*, especially mulga. **D** Interior W and C Australia.

2 BLUE-WINGED PARROT *Neophema chrysostoma* 21–22cm Wings mostly dark blue. Face yellow between eye and bill, with small yellow triangle behind eye and thin blue line across forehead. Breeds Tasmania and coastal SE Australia; most migrate N and spread inland outside breeding season. **V** Rapid chattering trills, tone reminiscent of hirundine contact calls. **H** Eucalypt woodland when nesting; scrub, open woodland, grassland, agricultural land at other times. **D** SE Australia, Tasmania.

3 ELEGANT PARROT *Neophema elegans* 22–24cm Very similar to Blue-winged Parrot, but brighter, with more yellow and less blue. Feeds on seeds of grassland plants. Partially nomadic in response to drought. **V** Shrill *Tsee-uk*, and rapid chattering twitters. **H** Eucalypt woodland, mallee scrub, grassland, dunes. **D** Disjunct populations: contiguous W and SW Australia, SC Australia.

4 ROCK PARROT *Neophema petrophila* 21–23cm More blue on face, less blue on wings, than precious two species. Feeds mainly on seeds and fruits. Nests in rock crevice or hollow. **V** Mostly silent; occasionally utters thin, high-pitched *tseet-tseet* calls. **H** Coastal woodland, scrub, dunes, mangroves. **D** Disjunct populations: SW Australia, SC Australia.

5 ORANGE-BELLIED PARROT *Neophema chrysogaster* 19–21cm Note orange patch on belly. Critically endangered. Feeds mainly on seeds and berries. **V** Trilling chattering notes with a buzzing quality. **H** Waterside woodland when nesting; salt marsh, dunes, coastal grassland at other times. **D** Breeds SE Tasmania. Winters S Australia.

6 TURQUOISE PARROT *Neophema pulchella* 20–21cm Brightly coloured, with blue face, green back, yellow belly. Male has chestnut-red on lesser wing coverts. Diet comprises flowers, fruits, seeds, especially grasses. Nests in tree hollows. **V** Shrill, high-pitched whistling *tseep* calls. **H** Open woodland, heath; typically, in foothills, rather than coastal habitats. **D** SE Australia.

7 SCARLET-CHESTED PARROT *Neophema splendida* 19–20cm Extremely colourful plumage. Nomadic and wandering, following rains and subsequent plant growth. Diet comprises mainly seeds. **V** Mostly silent. Occasionally utters subdued twittering calls. **H** Arid habitats including mallee scrub, open eucalypt woodland. **D** Interior S Australia.

8 PLUM-FACED LORIKEET *Oreopsittacus arfaki* 16–17cm Distinctive facial markings and colours aid identification. Diet includes nectar, flowers, fruits, berries. Usually seen in pairs. **V** Rapid-fire chattering twitters, tone reminiscent of hirundine contact calls. **H** Montane forests; mainly 1,500–3,500m. **D** New Guinea.

9 PALM LORIKEET *Charmosyna palmarum* 15–17cm Green, with small red patch on chin. Diet includes nectar, pollen, flowers, fruits. Gathers in flocks where feeding is good. Related to but geographically separated from Red-chinned Parakeet. **V** Thin, high-pitched twittering whistles. **H** Forested habitats; mainly 500–1,500m. **D** Solomon Islands, Vanuatu.

10 RED-CHINNED LORIKEET *Charmosyna rubrigularis* 17cm Diet includes palm flowers. Usually seen in pairs. **V** Thin, high-pitched whistling *tseit*. **H** Montane forest; mainly above 1,000m. **D** Karkar Island and New Britain, New Ireland, New Hanover (Bismarck Archipelago).

11 MEEK'S LORIKEET *Charmosyna meeki* 15–17cm Lacks red chin. Diet presumed to comprise flowers, fruits, berries. **V** Calls include thin, high-pitched twittering whistles. **H** Wooded foothills, montane forest; mainly 500–1,500m. **D** Solomon Islands (including Bougainville, Guadalcanal, New Georgia).

12 BLUE-FRONTED LORIKEET *Charmosyna toxopei* 16cm Pale blue forecrown. Usually found in small flocks. **V** A shrill *ti... ti... ti... ti... ti-ti-ti*. **H** Primary and secondary forest. **D** Buru, S Maluku Islands, Indonesia.

13 STRIATED LORIKEET *Charmosyna multistriata* 17–19cm Fine yellow streaks on undersides. Gathers in flocks when feeding is good. Diet comprises buds, flowers, fruits, berries. Threatened by deforestation. **V** Calls include shrill, *tsree-ik* notes and whistles. **H** Primary forest; mainly lowlands to 1,500m. **D** New Guinea.

14 PYGMY LORIKEET *Charmosyna wilhelminae* 12–13cm Male has purple-red rump. Usually seen in pairs. Diet comprises buds, flowers, fruits, berries. **V** Mostly silent. Occasional calls include thin, high-pitched whistles. **H** Montane forest, savannah; mainly 1,000–2,000m. **D** New Guinea.

15 RED-FRONTED LORIKEET *Charmosyna rubronotata* 16–18cm Forehead and breast sides red in male. Seen in pairs or small flocks. Mixes with other species when feeding is good. Diet presumed to comprise buds, flowers, fruits. **V** Thin, high-pitched chattering whistles. **H** Forested habitats, secondary woodland, plantations. **D** N and W New Guinea.

16 RED-FLANKED LORIKEET *Charmosyna placentis* 16–19cm Male has red facial patch, blue ear-coverts, red flanks. Female lacks red and has streaked blackish ear-coverts. Frequents upper canopy levels, feeding in flowering trees, usually in pairs or small flocks. **V** Dry *tst* and sharp *skeesk* notes. **H** Lowland forest, forest edges, coastal trees and plantations. **D** Maluku Islands, New Guinea, Bismarck Archipelago.

17 NEW CALEDONIAN LORIKEET *Charmosyna diadema* 18–19cm Critically endangered and possibly extinct; known only from a few verified and non-verified old specimens. Diet presumed to include forest flowers and fruits. **V** Unknown. **H** Montane forest, scrub. **D** Recorded only from Mt Ignambi, New Caledonia.

18 RED-THROATED LORIKEET *Charmosyna amabilis* 19cm Extensive red on lores, chin and upper breast, with thin yellow border below. Red thighs often concealed between feathers of lower belly. Does not occur on same islands in Fiji as Blue-crowned Lorikeet. **V** Gives short, shrill calls in flight. **H** Forest canopy. **D** Fiji.

19 DUCHESS LORIKEET *Charmosyna margarethae* 20–21cm Stunningly colourful plumage. Feeds on flowers, fruits; flocks gather where feeding is good. **V** Series of chattering *tche-che-che-che* whistling notes. **H** Forests, secondary woodland, plantations. **D** Solomon Islands.

20 FAIRY LORIKEET *Charmosyna pulchella* 18–20cm Extensive red on face, nape, underparts. Diet comprises nectar, pollen, flowers and fruits. Seen in flocks when feeding is good. **V** Series of thin, high-pitched *tsiip* whistles. **H** Forests, secondary woodland; mainly 500–1,500m. **D** New Guinea.

21 JOSEPHINE'S LORIKEET *Charmosyna josefinae* 24–25cm Black patch, mixed with some blue, from behind eye to nape. Diet includes buds, nectar, pollen, fruits. Usually seen in pairs. Small flocks form when feeding is good. **V** Series of thin, high-pitched chattering whistles. **H** Forests, secondary woodland; mainly 750–1,500m. **D** New Guinea.

22 PAPUAN LORIKEET *Charmosyna papou* 38–42cm Note colourful long tail streamers, which aid identification. Diet includes buds, nectar, pollen, fruits, seeds. **V** Series of high-pitched trilling whistles. **H** Montane forest; mainly 1,500–3,000m. **D** New Guinea.

23 BLUE-CROWNED LORIKEET *Vini australis* 19cm From Red-throated Lorikeet by blue crown and different range. **V** Sweet, whistled *seee* in flight and when perched; also more elaborate high-pitched whistled notes. **H** Plantations, gardens, scrub, forest. **D** SC Polynesia: Wallis and Futuna, Fiji, Samoa, American Samoa, Tonga, Niue.

24 KUHL'S LORIKEET *Vini kuhlii* 19cm Distinguished from Stephen's Lorikeet by blue crest, and shorter blue, green and red tail. **V** Very high, sizzling *sree* and *tjuh* notes. **H** Coconut plantations, forest, residential areas. **D** Rimatara (French Polynesia). Introduced on several other Pacific islands.

25 STEPHEN'S LORIKEET *Vini stepheni* 19cm Similar to Kuhl's Lorikeet. Only found on Henderson Island. **V** Very high *sreeeeh*. **H** Forest. **D** Henderson Island (Pitcairn Islands).

26 BLUE LORIKEET *Vini peruviana* 18cm Unmistakable by dark blue, almost black plumage with white bib. **V** Varied, e.g. high, rapid *siesiesie*. **H** Any type of wooded habitat, but especially in coconut trees; also in low, flowering plants. **D** Society Islands, W Tuamotu Archipelago (French Polynesia); Aitutaki (Cook Islands).

27 ULTRAMARINE LORIKEET *Vini ultramarina* 18cm Upperparts mostly dull turquoise-blue. No similar species in range. **V** Very high, thin, sizzling *srisisi*. **H** Any wooded habitat with flowering trees. **D** Marquesas Islands (French Polynesia).

28 COLLARED LORY *Phigys solitarius* 20cm Distinctive colourful plumage. No similar species in range. **V** Chattering buzzy *tzreet* given in flight and when perched; also agitated-sounding *tziririt*. **H** Mainly in forest, but also in cultivation and suburban areas. **D** Fiji.

29 YELLOW-BILLED LORIKEET *Neopsittacus musschenbroekii* 22–23cm Yellow bill. Olive-green head with russet on hindcrown. Yellowish streaking over much of head. Red breast and underwings. Usually seen in pairs. Diet includes buds, flowers, fruits. **V** Shrill, high-pitched *tsee, tsee-irp* whistling notes. **H** Montane forest, secondary woodland, cultivated ground; mainly 1,500–2,500m. **D** New Guinea.

30 ORANGE-BILLED LORIKEET *Neopsittacus pullicauda* 18–19cm Very similar to Yellow-billed Lorikeet, but orange bill aids identification. More red on underparts. Diet includes nectar and pollen. Usually seen in pairs. **V** Shrill, high-pitched *tsee, tsee-it* whistling notes. **H** Montane scrub, cloud forest; mainly 2,500–3,500m. **D** New Guinea.

31 LITTLE LORIKEET *Parvipsitta pusilla* 15–16cm Note red face mask. Forms flocks outside breeding season, sometimes mixing with other lorikeet species. Diet comprises mainly nectar and pollen, sometimes flowers themselves. **V** Series of extremely high-pitched chattering whistles. **H** Forest habitats. **D** E and SE Australia.

32 PURPLE-CROWNED LORIKEET *Parvipsitta porphyrocephala* 15–16cm Distinctive facial colours and pattern aid identification. Usually seen in small flocks. Sometimes mixes with other lorikeet species. **V** Hollow-sounding, harsh *tcheerp, tcheerp* notes. **H** Open woodland, mallee scrub, eucalypt forest. **D** S and SW Australia.

1
2
3
4
5
6
7
8
9
10
11
12
13
14
15
16
17
18
19
20
14
21
22
23
24
25
26
27
28
29
30
31
32

OLD WORLD PARROTS *CONTINUED*

1 CHATTERING LORY *Lorius garrulus* 30cm Usually seen in pairs, occasionally in larger groups when feeding in flowering trees. **V** Loud, disyllabic nasal bugle; in flight, gives a loud, harsh, quavering bray. **H** Primary and tall secondary lowland and hill forest, forest edges, occasionally coconut groves. **D** N Maluku Islands, Indonesia.

2 PURPLE-NAPED LORY *Lorius domicella* 28cm Generally seen singly or in pairs, foraging acrobatically to extract various seeds. **V** Recorded only as having a melodious call. **H** Submontane forest. **D** S Maluku Islands, Indonesia.

3 BLACK-CAPPED LORY *Lorius lory* 30–32cm Colourful plumage aids identification. Yellow on wing feathers most obvious in flight. Feeds on nectar, pollen, flowers and fruit. **V** Calls include trisyllabic *whe-uu-up*, descending in pitch during delivery. **H** Primary forest, secondary woodland; mainly lowlands to 750m. **D** New Guinea, West Papua islands, Biak Island.

4 PURPLE-BELLIED LORY *Lorius hypoinochrous* 26–28cm Note white cere. Diet comprises mainly flowers and fruits. **V** Calls include piercing, trumpeted *Whee, Whee-er* screeches. **H** Primary forest, secondary woodland, mangroves, plantations; mainly lowlands to 750m. **D** SE New Guinea and adjacent islands, Bismarck Archipelago, Louisiade Archipelago.

5 WHITE-NAPED LORY *Lorius albidinucha* 26–28cm White nape is distinctive. Diet comprises flowers and fruits (especially wild Oil Palm, *Elaeis guineensis*). **V** Calls include whistling *tsueet* notes. **H** Primary forest, secondary woodland; mainly 500–1,500m. **D** S and C New Ireland (Bismarck Archipelago).

6 YELLOW-BIBBED LORY *Lorius chlorocercus* 28–29cm Diet includes pollen, nectar, flowers and fruits. Small flocks gather where feeding is good. **V** Squeaky, chattering screeches and piercing, downslurred *TSeeeoor*. **H** Primary forest, secondary woodland, plantations. **D** E Solomon Islands.

7 BLACK LORY *Chalcopsitta atra* 32–34cm Purplish-black plumage is distinctive, contrasting with yellow-flushed underside to tail. Diet includes flowers and nectar. **V** Calls include squeaking, trilling screeches. **H** Lowland forested habitats, mangroves, plantations. **D** New Guinea, W Papua islands.

8 BROWN LORY *Chalcopsitta duivenbodei* 30–32cm Yellow feathering at base of bill contrasts with otherwise overall brown plumage. Orange-yellow underwing coverts obvious in flight. Diet includes flowers and fruits. **V** Calls include squeaking, trilling screeches. **H** Lowland primary forest, secondary woodland. **D** New Guinea.

9 YELLOWISH-STREAKED LORY *Chalcopsitta scintillata* 30–32cm Feeds on flowers and fruits. Forms flocks when feeding is good. **V** Calls include shrill, piercing *TSee, TSee* and *Tseeip-tsu*. **H** Primary forest, secondary woodland, mangroves; mainly lowlands and foothills. **D** S New Guinea; Aru Islands, Indonesia.

10 DUSKY LORY *Pseudeos fuscata* 25–26cm Feeds on flowers and fruits. Forms nomadic flocks outside breeding season. **V** Calls include raucous, screeching *Whah* or *Eaerrr*. **H** Primary forest, secondary woodland, agricultural land; mainly lowlands to 1,500m. **D** New Guinea; Salawati and Yapen islands, West Papua.

11 CARDINAL LORY *Pseudeos cardinalis* 30–32cm Uniform bright red colour aids identification. Usually seen in pairs. **V** Calls include buzzing *tsrrrp, chrrrr, cheee, cheee* series of trills. **H** Lowland primary forest, secondary woodland, mangroves, plantations. **D** Solomon Islands, Bougainville Island, E Bismarck Archipelago.

12 VARIED LORIKEET *Psitteuteles versicolor* 18–20cm Feeds on flowers and fruits. Nomadic, wandering in search of food and water. **V** Calls include trilling, screeching *tschrrrrp, chrrrrp, chrrrrp* notes. **H** Eucalypt forest, wooded habitats, waterside grassland. **D** N Australia, from N Western Australia to N Queensland.

13 IRIS LORIKEET *Psitteuteles iris* 20cm Usually occurs in small parties, occasionally in large flocks. Frequents flowering trees. **V** Various shrill screeches and whistles. **H** Primary and tall secondary monsoon, lowland, hill and submontane forest. **D** Lesser Sunda Islands.

14 GOLDIE'S LORIKEET *Psitteuteles goldiei* 18–20cm Diet includes flowers of flowering trees and lerps. Usually seen singly or in pairs. **V** Calls include silvery trilling phrases of 1 sec duration, delivered roughly 1 per sec. **H** Primary hill and montane forest; mainly 750–2,000m. **D** New Guinea.

15 RED-AND-BLUE LORY *Eos histrio* 31cm Usually found in pairs or small flocks, with much larger flocks at communal roosts. **V** Short, harsh, chattering screeches. **H** Primary lowland and hill forest, coconut plantations. **D** Islands north of Sulawesi, Indonesia.

16 VIOLET-NECKED LORY *Eos squamata* 23–27cm Found in pairs or small flocks, and larger flocks when feeding in flowering or fruiting trees. **V** Loud, shrill, harsh, rapidly repeated screeches. **H** Primary and tall secondary lowland and hill forest, forest edges, scrubby secondary growth, mangroves, coconut plantations. **D** N Maluku Islands, Indonesia.

17 RED LORY *Eos bornea* 30cm Generally seen in pairs or small parties, occasionally found in much larger groups. Feeds in flowering or fruiting trees. Noisy. **V** Screeches and more musical bell-like notes. **H** Primary and tall secondary lowland and hill forest, mangroves, coconut plantations, around human habitation. **D** S Maluku Islands, Indonesia.

18 BLUE-STREAKED LORY *Eos reticulata* 31cm Frequents flowering trees, usually in pairs or small flocks. **V** Drawn-out, nasal screech or screeches; also starling-like chatters and whistles. **H** Primary and secondary lowland forest, open forest, mangroves, wooded cultivation, coconut plantations. **D** Tanimbar Islands, Indonesia.

19 BLACK-WINGED LORY *Eos cyanogenia* 29–31cm Feeds on flowers and fruits. Nomadic. Striking markings and colours aid identification. **V** Calls include shrill, screeching *tseirp, tseeip* notes. **H** Lowland forested habitats, coconut plantations. **D** Biak, Num, Numfor and Manim islands, off NW New Guinea.

20 BLUE-EARED LORY *Eos semilarvata* 24cm Found singly, in pairs and in small groups; visits flowering trees and shrubs to feed. **V** Loud screech. **H** Primary montane forest, upper montane heath. **D** Maluku Islands, Indonesia.

21 ORNATE LORIKEET *Trichoglossus ornatus* 23–25cm Usually in pairs and small flocks, but often in larger flocks at flowering trees. **V** Short, squeaky, screeching notes, interspersed with rolling screeches. **H** Forest edges, tall secondary forest, coastal woodland, swamp forest, mangroves, wooded cultivation, coconut plantations. **D** Sulawesi, Indonesia.

22 SUNSET LORIKEET *Trichoglossus forsteni* 25–30cm See Coconut Lorikeet. **V** Similar to that of Coconut Lorikeet. **H** Most types of lower and montane woodland. **D** W Lesser Sunda Islands.

23 LEAF LORIKEET *Trichoglossus weberi* 25cm See Coconut Lorikeet. **V** Similar to that of Coconut Lorikeet. **H** Rainforest, *Casuarina* trees. **D** Flores, Indonesia.

24 MARIGOLD LORIKEET *Trichoglossus capistratus* 26cm See Coconut Lorikeet. **V** Similar to that of Coconut Lorikeet. **H** Primary and secondary forest, dry woodland and plantations. **D** E Lesser Sunda Islands.

25 COCONUT LORIKEET *Trichoglossus haematodus* 25–30cm Usually in pairs, groups or large flocks, especially at flowering trees. This and the Sunset, Leaf, Marigold and Red-collared lorikeets are often considered races of the Rainbow Lorikeet. **V** In flight, utters a repeated harsh screech; when perched, gives a strident *peaow peaow peaow*. **H** Lowland and lower montane wooded country, coconut and other plantations, suburbs. **D** S Maluku Islands to W New Guinea and east to the Loyalty Islands of New Caledonia.

26 BIAK LORIKEET *Trichoglossus rosenbergii* 26–30cm Previously treated as a subspecies of Coconut Lorikeet. Feeds on flowers and fruits. Forms flocks outside breeding season. **V** Calls include screeching *pe-errr, pe-err* phrases. **H** Lowland forest, wooded slopes, coconut plantations. **D** Endemic to Biak Island, off NW New Guinea.

27 RAINBOW LORIKEET *Trichoglossus moluccanus* 26–30cm Stunningly colourful. Feeds mainly on flowering tree nectar and pollen. Partly nomadic in search of food. **V** Calls include series of trilling, chirping phrases. **H** Rainforest, open woodland, mallee scrub, plantations, gardens, parks. **D** NE, E and SE Australia.

28 RED-COLLARED LORIKEET *Trichoglossus rubritorquis* 26cm See Coconut Lorikeet. **V** Similar to that of Coconut Lorikeet. **H** Presumed to be similar to that of Coconut Lorikeet. **D** E Lesser Sunda Islands to N Australia.

29 OLIVE-HEADED LORIKEET *Trichoglossus euteles* 25cm Occurs in small flocks, with larger flocks at flowering trees. **V** Rapid, buzzing trill, a drawn-out wheezy call, and a series of harsh squeaks, twitters and whistles. **H** Primary montane forest, forest edges, secondary growth, savannah woodland. **D** Lesser Sunda Islands.

30 CITRINE LORIKEET *Trichoglossus flavoviridis* 21cm Usually in pairs or small flocks; larger flocks at flowering trees. **V** Calls include a high-pitched screech, squeaky chatters, and dry *ksk* notes interspersed with thin whistles. **H** Primary hill and montane forest, forest edges. **D** Sulawesi and the Sula Islands, Indonesia.

31 MINDANAO LORIKEET *Trichoglossus johnstoniae* 18cm Usually occurs in pairs or noisy flocks; regularly seen in flight as flocks fly between trees. **V** Calls include a sharp *chick, chick-it* or *twick-it*. **H** Montane forest, forest edges. **D** S Philippines.

32 POHNPEI LORIKEET *Trichoglossus rubiginosus* 24cm No similar species occurs in its range. **V** Diagnostic *krr-EE-ah*, plus various notes and chattering. **H** Forest, plantations. **D** E Caroline Islands, Micronesia.

33 SCALY-BREASTED LORIKEET *Trichoglossus chlorolepidotus* 22–24cm Yellow 'scaly' markings on underparts aid identification. Feeds mainly on flowers, especially eucalypts. **V** Calls include piping whistles and chirping squawks. **H** Open woodland, heaths, suburban parks, gardens. **D** E, NE and SE Australia.

1
2
3
4
5
6
7
8
9
10
11
12
13
14
15
16
17
18
19
20
21
22
23
24
25
26
27
28
29
30
31
32
♀
♂
33

OLD WORLD PARROTS *CONTINUED*

1 MUSK LORIKEET *Glossopsitta concinna* 21–23cm Diet includes buds, flowers, pollen and nectar, mainly of eucalypts. **V** Calls include shrill piping notes and excited chattering squawks. **H** Eucalypt woodland, orchards, suburban parks, gardens. **D** E and SE Australia, Tasmania.

2 BUDGERIGAR *Melopsittacus undulatus* 18cm Popular cagebird that occurs in a variety of colours. In flight, shows a yellow wing-bar. **V** Subdued screech. In flight, commonly utters a pleasant warble. Feeding groups regularly utter quiet chattering. **H** Grassland. **D** Australia.

3 LARGE FIG PARROT *Psittaculirostris desmarestii* 18–20cm Usually seen in small flocks. Diet comprises mainly fruits and seeds on figs. **V** Calls include a shrill, squeaking *thcirrrp*. **H** Humid lowland forests, wooded slopes, mangroves. **D** New Guinea, West Papua islands.

4 EDWARDS'S FIG PARROT *Psittaculirostris edwardsii* 18–19cm Feeds primarily on fruits and seeds of figs. Gathers in large flocks where feeding is good. **V** Calls include series of high-pitched, piping *tsWeerp, tsWeerp* notes. **H** Primary lowland forest, secondary wooded slopes. **D** NE New Guinea.

5 SALVADORI'S FIG PARROT *Psittaculirostris salvadorii* 18–20cm Usually seen in pairs or small groups. Feeds primarily on fruits and seeds of figs. **V** Calls include series of high-pitched, piping *tsWeerk, tsWeerk* notes. **H** Lowland primary forest, lower forested slopes. **D** NW New Guinea.

6 ORANGE-BREASTED FIG PARROT *Cyclopsitta gulielmitertii* 12–13cm Note distinctive colours and markings on face and breast. Diet comprises fruits and seeds, including those of figs. **V** Calls include series of high-pitched, piping *tseeip* notes. **H** Rainforest, secondary woodland; mainly lowlands to 750m. **D** NW New Guinea.

7 DOUBLE-EYED FIG PARROT *Cyclopsitta diophthalma* 14–16cm Diet comprises fruits and seeds, including those of figs; also berries and invertebrates. Usually seen in pairs or small groups. **V** Calls include series of shrill, cheeping *sChip, sChip, sChip* notes. **H** Rainforest, wooded slopes, mangroves, suburban parks; mainly lowlands to 1,500m. **D** West Papua islands, New Guinea, NE Australia.

8 GUAIABERO *Bolbopsittacus lunulatus* 16–17cm Occurs singly or in pairs, with larger parties often seen in fruiting trees. **V** High-pitched *zeet* or *zeet zeet*. **H** Forest and forest edges, secondary growth, clearings with scattered trees, mangroves, orchards. **D** Philippines.

9 VERNAL HANGING PARROT *Loriculus vernalis* 14cm Sleeps hanging upside down, like a bat. Usually encountered in pairs or small parties feeding on fruit and berries, the birds' presence generally indicated by squeaking calls. **V** Squeaking *chi-chi-chee chi-chi-chee chi-chi-chee...* given in flight or when foraging. **H** Broadleaved evergreen and deciduous woodlands. **D** E India through SE Asia, Andaman Islands.

10 SRI LANKA HANGING PARROT *Loriculus beryllinus* 14cm Actions and habits similar to Vernal Hanging Parrot. Feeds on fruit, seeds, flowers and nectar. **V** Similar to that of the Vernal Hanging Parrot. **H** Wooded country, groves, plantations, gardens. **D** Sri Lanka.

11 PHILIPPINE HANGING PARROT *Loriculus philippensis* 14cm Occurs singly, in pairs, and in larger groups at feeding sites; favours flowering and fruiting trees. **V** Sharp twittering. **H** All types of forest, forest patches, gardens. **D** Philippines.

12 CAMIGUIN HANGING PARROT *Loriculus camiguinensis* 14cm Forages in the canopy, where it feeds on seeds of wild bananas, fruits, berries and blossoms. Considered by some authorities to be a race of the Philippine Hanging Parrot. **V** High-pitched *tziit-tziit-tziit*. **H** Upland forest. **D** Camiguin Island, S Philippines.

13 BLUE-CROWNED HANGING PARROT *Loriculus galgulus* 12–14cm Usually encountered in pairs or small parties. **V** Shrill *tsi, tsrri* or *tsi-tsi-tsi*. **H** Lowland forest, occasionally montane forest and plantations. **D** Malay Peninsula; Sumatra, Indonesia; Borneo.

14 GREAT HANGING PARROT *Loriculus stigmatus* 15cm Usually occurs singly, in pairs or in small parties; larger numbers are found feeding in flowering or fruiting trees. **V** High-pitched *tsu-tsee* or *tsu-tsee-tsee*. **H** Primary and tall secondary lowland and hill forest, forest edges, lightly wooded cultivation, mangroves, scrub, coconut plantations. **D** Sulawesi, Indonesia.

15 MOLUCCAN HANGING PARROT *Loriculus amabilis* 11cm Forages in the crowns of tall flowering and fruiting trees, singly, in pairs or in small parties. **V** Rapid, weak, high-pitched, buzzing call. **H** Primary and secondary lowland forest, coastal woodland, mangroves. **D** Maluku Islands, Indonesia.

16 SULA HANGING PARROT *Loriculus sclateri* 14cm Sometimes considered conspecific with the Moluccan Hanging Parrot; actions presumed to be similar. **V** Presumably similar to that of the Moluccan Hanging Parrot. **H** Primary and secondary forest, and partially cleared adjacent areas. **D** Sula and Banggai islands, Indonesia.

17 SANGIHE HANGING PARROT *Loriculus catamene* 12cm Usually found in pairs or small flocks; favours flowering coconut palms. **V** Dry, high, staccato, disyllabic whistle and an upslurred *sh-ui*. **H** Remnant forest patches, coconut and mixed tree-crop plantations. **D** Sangihe Islands, Indonesia.

18 ORANGE-FRONTED HANGING PARROT *Loriculus aurantiifrons* 10–11cm Diet comprises mainly flowers, nectar and pollen. Often feeds while hanging upside down. **V** Calls include shrill, high-pitched *tsseip, tsseip* notes. **H** Lowland primary forest, secondary woodland. **D** West Papua islands, New Guinea.

19 BISMARCK HANGING PARROT *Loriculus tener* 10–11cm Presumed to feed on flowers, nectar and pollen. Often feeds while hanging upside down. Usually seen in pairs. **V** Calls include high-pitched *tsseip, tsseip* notes. **H** Lowland primary forest, secondary woodland. **D** Bismarck Archipelago.

20 PYGMY HANGING PARROT *Loriculus exilis* 10–11cm Forages in the crowns of tall trees, feeding on flowers and fruits. **V** Short, dry, thin, weak *pssst*. **H** Primary lowland and hill forest; occasionally visits mangroves and trees in open country. **D** Sulawesi, Indonesia.

21 YELLOW-THROATED HANGING PARROT *Loriculus pusillus* 12cm Feeds on flowers, buds and small fruits. **V** In flight, gives a shrill, ringing *sree-ee*. **H** Rainforest, from sea level to mountains, and open areas with *Casuarina* trees. **D** Java, Indonesia.

22 WALLACE'S HANGING PARROT *Loriculus flosculus* 11–12cm Favours fruiting fig trees or other soft fruit and flowering trees. Usually seen in pairs or small groups. **V** Hoarse *chi-chi-chi-chi-chi*, given during display. In flight, utters a screeching *strrt strrt*. **H** Primary hill forest. **D** Lesser Sunda Islands.

23 GREY-HEADED LOVEBIRD *Agapornis canus* 15cm The only lovebird in Madagascar; the male's grey head is unmistakable. **V** Various whistles and squawks. **H** Forest, woodland and other habitats with some trees. **D** Madagascar.

24 RED-HEADED LOVEBIRD *Agapornis pullarius* 14cm Note green nape and neck, and pale blue rump. The only lovebird with red visible in folded tail (in all other lovebirds the red is normally visible only when tail is spread, such as in flight). **V** High-pitched double-note *ti-lee*. **H** Moist, wooded areas at lower altitudes. **D** W and C Africa.

25 BLACK-WINGED LOVEBIRD *Agapornis taranta* 15cm Note black primaries. **V** Rising, disyllabic *ksilee*, plus various calls, twitters and screeches. **H** Conifer forests and other more or less wooded habitats; 1,250–3,250m. **D** Ethiopia.

26 BLACK-COLLARED LOVEBIRD *Agapornis swindernianus* 13cm Note dark blue rump. There is also a yellow-necked form. **V** High, disyllabic *sri-leee*. **H** Forests, well-wooded streams. **D** W and C Africa.

27 ROSY-FACED LOVEBIRD *Agapornis roseicollis* 16cm Note faint, pale eye-ring, rosy cheeks and blue rump. **V** Disyllabic *chi-reep*, plus various screeches, chirrups and rattles. **H** Dry, wooded and open areas at higher altitudes. **D** SW Africa.

28 FISCHER'S LOVEBIRD *Agapornis fischeri* 14cm Note pinkish-purple rump. **V** Disyllabic *chi-reek* or screeching *krreek*, plus various screeches, chirrups and rattles. **H** Wooded areas with acacias and baobabs; 1,000–2000m. **D** Tanzania.

29 YELLOW-COLLARED LOVEBIRD *Agapornis personatus* 15cm Birds outside Tanzania are feral, with the possible exception of Kenya. **V** As for Fischer's Lovebird. **H** Bushland with scattered baobab and other trees; 1,000–2000m. **D** Tanzania.

30 LILIAN'S LOVEBIRD *Agapornis lilianae* 14cm Note yellow nape and neck, and pale green rump. **V** As for Fischer's Lovebird. **H** More or less wooded country. **D** Tanzania to Zimbabwe and Mozambique.

31 BLACK-CHEEKED LOVEBIRD *Agapornis nigrigenis* 14cm Note dark face and green rump. **V** As for Fischer's Lovebird. **H** Thornveld, open woodland. **D** S Zambia.

1
2
3
4
5
6
7
8
9
10
11
12
13
14
15
16
17
18
19
20
21
22
23
24
25
26
27
28
29
30
31

NEW ZEALAND WRENS ACANTHISITTIDAE

1 RIFLEMAN *Acanthisitta chloris* 8cm New Zealand's smallest bird. Note white eyebrow, pale wing-bar and white patch on tertials. **V** Very high-pitched, weak *srit*, single or in series. **H** Forest, pine plantations, locally in large hedgerows. **D** New Zealand.

2 NEW ZEALAND ROCKWREN *Xenicus gilviventris* 9.5cm Unmistakable in its range thanks to very short tail and general colour pattern. Hops and runs, avoiding flying. **V** Extremely high (for some people inaudible) *seet-sit-sit*. **H** Rocky habitats with low scrub; 920–2,900m. **D** New Zealand.

SAPAYOA SAPAYOIDAE

3 SAPAYOA *Sapayoa aenigma* 16cm Olive-green, paler below, with a broad bill and concealed yellow crown spot. **V** Soft nasal trill; little recorded. **H** Forest understorey. **D** E Panama to NW Ecuador.

BROADBILLS EURYLAIMIDAE

4 AFRICAN BROADBILL *Smithornis capensis* 14cm Note absence of rufous breast sides. Hawks insects from perch or gleans them from underside of leaves. **V** Low toy trumpet-like *prrrueh*. **H** Undergrowth of forest and dense woodland. **D** Widespread across Africa.

5 GREY-HEADED BROADBILL *Smithornis sharpei* 17cm Note rufous breast sides and absence of wing-bars. **V** Short, high-pitched *whee whee* or *huiiii* whistle. **H** Lower strata in forest. **D** C and WC Africa.

6 RUFOUS-SIDED BROADBILL *Smithornis rufolateralis* 12cm Female separated from female Grey-headed Broadbill by wing-bars. **V** High, rather level, toy trumpet-like *trrrui*. **H** Forest canopy, often near streams. **D** C, WC and W Africa.

7 GREEN BROADBILL *Calyptomena viridis* 15–17cm Quiet; forages mainly in lower levels and understorey of forest. Favours fruiting figs. **V** Soft, bubbling *toi toi-oi-oi-oi-oick* that starts quietly and then increases in tempo; also a *goik-goik*, a *goik-goik-doyik*, a loud *oik*, a frog-like *oo-turr*, and various wheezes, whines and cackles. **H** Broadleaved evergreen forest. **D** Malay Peninsula; Sumatra, Indonesia; Borneo.

8 HOSE'S BROADBILL *Calyptomena hosii* 20cm Forages at lower levels, singly or in pairs, or in small parties at fruiting trees. **V** Soft, dove-like cooing. **H** Hill and submontane forests. **D** Borneo.

9 WHITEHEAD'S BROADBILL *Calyptomena whiteheadi* 25cm Feeds in fruiting trees, singly, in pairs or sometimes in noisy groups. Occasionally joins mixed-species foraging flocks. **V** Loud, screeching *eek eek eek*, a *wark wark wark* and woodpecker-like rattles. **H** Montane forest, forest edges. **D** Borneo.

10 BLACK-AND-RED BROADBILL *Cymbirhynchus macrorhynchos* 20–24cm Unobtrusive; sits motionless for long periods. **V** Accelerating series of *parnk* notes and grating cicada-like notes; also various churrings, melodious and a monotonous, repeated *tyook*. When alarmed, utters a rapid *pip-pip-pip-pip...* **H** Broadleaved evergreen and semi-evergreen forests, forest edges near water, freshwater swamp forest, mangroves. **D** SE Asia; Sumatra, Indonesia; Borneo.

11 LONG-TAILED BROADBILL *Psarisomus dalhousiae* 28cm Regularly encountered in small, loose parties, moving from tree to tree. Forages by gleaning or making short sallies after flying insects. **V** Loud, sharp *tseeay-tseeay-tseeay-tseeay...*, or *pseew-pseew-pseew-pseew-pseew...* **H** Broadleaved evergreen and semi-evergreen forest. **D** Widespread from N India to S China, and through the Malay Peninsula to Borneo.

12 SILVER-BREASTED BROADBILL *Serilophus lunatus* 18cm Forages in pairs or small groups. Finds prey by gleaning from branches or leaves; also by making short aerial sallies after flying insects. **V** Soft, musical *chir-r-r-r*; also a squeaky *ki-uu*. **H** Mixed tropical evergreen, semi-evergreen and bamboo secondary forest. **D** Widespread from NE India to Hainan Island, and through the Malay Peninsula to Sumatra, Indonesia.

13 BANDED BROADBILL *Eurylaimus javanicus* 21–23cm Often encountered in small, slow-moving parties in the middle storey. **V** Sharp *wheeoo*, followed by a frantic, rising series of notes; also a nasal *whee-u*, a falling *kyeeow*, a rolling *keowrr* and a yelping *keek-eek-eek*. **H** Broadleaved evergreen, semi-evergreen and wet areas in mixed deciduous forest. **D** SE Asia to Java and Borneo.

14 BLACK-AND-YELLOW BROADBILL *Eurylaimus ochromalus* 13–15cm Usually encountered in small groups foraging in the middle to upper storey. Generally sits quietly, before making short sallies to capture insects from foliage. **V** Rapid, frantic series of notes, starting slowly and then gradually gaining speed; also a *kyeeow* and *keowrr*. **H** Various forest types, from lowlands to hills. **D** Malay Peninsula; Sumatra, Indonesia; Borneo.

15 WATTLED BROADBILL *Sarcophanops steerii* 16–17cm Forages in the middle and lower forest levels, singly, in pairs or in small flocks. Usually seen making short sallies to glean insects or capture them in flight. Sometimes a part of mixed-species feeding flocks. **V** Plaintive whistle. **H** Rainforest; dipterocarp, mixed dipterocarp and hillside secondary forest; occasionally mangroves and scrub forest. **D** Philippines.

16 VISAYAN BROADBILL *Sarcophanops samarensis* 16–17cm Formerly considered a race of the previous species; actions similar. **V** Insect-like *tik tik t-rrrrrrrr*. **H** Primary forest, often near limestone outcrops. **D** Philippines.

17 DUSKY BROADBILL *Corydon sumatranus* 25–29cm Small, noisy flocks occur in the upper storey, sitting quietly before making short sallies to pick insects off foliage. **V** Shrill, upward-inflected whistles and a shrill, falling *pseeoo*. **H** Broadleaved evergreen, semi-evergreen and wet areas in mixed deciduous forest. **D** SE Asia; Sumatra, Indonesia; Borneo.

18 GRAUER'S BROADBILL *Pseudocalyptomena graueri* 12cm Perches quietly, unlike *Agapornis* species, e.g. Red-headed Lovebird (Plate 122), and hawks insects. **V** *Cree-cree* or *tsi-tsi*, repeated 3–8 times. **H** Mainly mid-strata and canopies of montane forests, bamboo. **D** S Uganda, EC DR Congo.

19 VELVET ASITY *Philepitta castanea* 15cm Male is unmistakable owing to thickset jizz and colour. Note moustachial stripe of female. **V** Very high, sharp yet soft *seet*. **H** Near fruiting trees in interior, edges or neighbourhoods of forest. **D** Madagascar.

20 SCHLEGEL'S ASITY *Philepitta schlegeli* 13cm Male is unmistakable owing to glowing green-and-blue lappets. Note orange eye-ring of female. **V** Short, very high *tuh-de-weet-weet-wit*. **H** Forest. **D** Madagascar.

21 COMMON SUNBIRD-ASITY *Neodrepanis coruscans* 10cm Male separable from male Yellow-bellied Sunbird-asity by yellow edges to wing feathers and less clean underparts. Female distinguished from female Yellow-bellied by greener underparts. **V** Very high, short sizzles. **H** Rainforest to mid–high altitudes. **D** Madagascar.

22 YELLOW-BELLIED SUNBIRD-ASITY *Neodrepanis hypoxantha* 9.5cm See Common Sunbird-asity. **V** Very high sizzles. **H** Montane rainforest. **D** Madagascar.

PITTAS PITTIDAE

23 EARED PITTA *Hydrornis phayrei* 22cm Forages on the ground, in leaf litter and among rotting logs, progressing in short hops, although said to be more static than others of the family. **V** Whistled *wheeow-whit*. When alarmed, utters a dog-like whine or yelp. **H** Broadleaved evergreen and mixed deciduous forest, bamboo. **D** SE Asia.

24 BLUE-NAPED PITTA *Hydrornis nipalensis* 22–25cm Feeds on the ground, moving in short hops, stopping to turn over leaves or dig for food. **V** Sharp *chow-whit*, *uk-wuip* or *ip-wuiip*; feeding couples utter soft chuckles. **H** Usually near water, in tropical and subtropical secondary forest, bamboo, clearings with dense vegetation. **D** N India and Nepal to S China and C Vietnam.

25 BLUE-RUMPED PITTA *Hydrornis soror* 22–24cm Forages on the ground on snails, insects and earthworms; uses rocks and stones for cracking snail shells. Female slightly browner on mantle. **V** Full, repeated *weaoe* or *weeya*, and a frog-like *ppew* or *eau*. When agitated, may give a sharp *hwit* or *hwip*. **H** Broadleaved evergreen, semi-evergreen and deciduous forest, secondary forest. **D** S China, Laos, Vietnam, Thailand, Cambodia.

26 RUSTY-NAPED PITTA *Hydrornis oatesi* 21–25cm Forages on the ground among leaf litter. **V** Utters a sharp, repeated *chow-whit* and sometimes an explosive, descending *poouw*. When alarmed, gives a *tchick* or *chek* that lengthens to a *churr-r-r-t* or *werr-r-r-t* when an intruder approaches nest site. **H** Broadleaved evergreen forest, bamboo. **D** E Myanmar, Laos, Thailand, Vietnam, SE China.

27 SCHNEIDER'S PITTA *Hydrornis schneideri* 21–23cm Forages on the forest floor, turning over leaves in search of prey. **V** Low, soft, prolonged, tremulous, double-noted whistle, the first note rising and the second falling. **H** Primary mountain forest with dense undergrowth and tree-fall clearings, forest edges. **D** Sumatra, Indonesia.

1
2
3
4
5
6
7
8
9
10
11
12
13
14
15
16
17
18
19
20
21
22
23
24
25
26
27

PITTAS *CONTINUED*

1 GIANT PITTA *Hydrornis caeruleus* 25–29cm Forages among leaf litter, feeding on snails, earthworms, frogs, small snakes and large insects; uses rocks and stones to smash snail shells. **V** Slow, mournful *hwoo-er*; a soft, repeated *wheer*; and a falling *pheeeeeeoou*. **H** Bamboo and broadleaved evergreen forest. **D** Malay Peninsula; Sumatra, Indonesia (although possibly extinct here); Borneo.

2 BLUE-HEADED PITTA *Hydrornis baudii* 16–17cm Forages among leaf litter in search of worms, insects and arthropods. **V** Soft, descending *ppor-wi-iil* or *ppor-or*; female utters a drawn-out *hwee-ouu* when alarmed. **H** Primary lowland forest and old secondary forest; also logged forest, often with a river nearby. **D** Borneo.

3 BLUE PITTA *Hydrornis cyaneus* 22–24cm Forages on the ground; recorded scratching around, chicken-like, when searching for food. **V** Liquid *pleoow-whit*. **H** Damp ravines and scrubby undergrowth in evergreen and bamboo forests. **D** SE Asia, from NE India to S Vietnam.

4 BAR-BELLIED PITTA *Hydrornis elliotii* 19–21cm Forages by pecking at the ground or turns over leaves with bill in search of food. **V** Loud, repeated *chawee-wu*, *tu-wi-whil* or *per-ur-wu*; sometimes a mellow *hhwee-hwha*. When alarmed, utters a harsh, shrill *jeeow* or *jow*. **H** Bamboo, broadleaved evergreen, semi-evergreen and mixed deciduous forest; appears to prefer moist areas. **D** SE Asia.

5 JAVAN BANDED PITTA *Hydrornis guajanus* 20–23cm Forages on the forest floor using long hops with stops to pick over leaves in search of food. Formerly considered conspecific with Malayan and Bornean banded pittas. **V** Similar to that of Malayan Banded Pitta, but higher in pitch. **H** Primary forest and old secondary forest. **D** Java, Indonesia.

6 MALAYAN BANDED PITTA *Hydrornis irena* 21–24cm Forages on the ground, scratching about much like a chicken. Formerly considered a race of Javan Banded Pitta. **V** Short, repeated *pouw* or *poww*, and a whirring *kirrr* or *pprrr*. **H** Broadleaved evergreen and secondary forest. **D** Malay Peninsula; Sumatra, Indonesia.

7 BORNEAN BANDED PITTA *Hydrornis schwaneri* 20–23cm Forages on the forest floor, turning over leaf litter in search of food. This and the previous two species were formerly considered conspecific. **V** Repeated *pow pow pow* and *whrr whrr whrr*, also an explosive *pauk* and a gentle *kur-kur* or *purr*. **H** Forested slopes and hills, lowland forest on limestone hills. **D** Borneo.

8 GURNEY'S PITTA *Hydrornis gurneyi* 18–21cm Forages on the ground among leaf litter, tossing leaves aside with lateral flicks of bill. **V** Short, explosive *lilip*, repeated every 2–6 secs. When alarmed, utters a squeaky, falling *skyeew*. **H** Broadleaved evergreen forest, secondary forest, old rubber plantations near forests. **D** S Myanmar, S Thailand.

9 WHISKERED PITTA *Erythropitta kochi* 22–23cm Usually seen on or near the ground, often near wild pig diggings. Calls from an exposed perch or from the top of a small tree. **V** Monotonous, deep, mournful *haaawwww haaww haaww haaww haaw-r* or *goow-goow-goow-goow-goo*. **H** Montane and submontane forest, usually with dense undergrowth, and high-elevation mossy forest with *Rhododendron* or fern understorey. **D** Philippines.

10 PHILIPPINE PITTA *Erythropitta erythrogaster* 16–17cm Usually seen on or near the ground. Generally shy and inconspicuous, unless calling. Calls from an exposed rock, stump or high in a tree. **V** Two long, hollow, owl-like whistled notes, the first ascending and wavering, and the second descending. **H** Forests, secondary growth. **D** Philippines.

11 SULA PITTA *Erythropitta dohertyi* 16–18cm Forages on the forest floor, hopping with pauses to pick over leaves in search of insects, worms, snails, etc. Formerly considered a race of the Philippine Pitta. **V** Trisyllabic phrase, followed by a brief pause and then a descending series of five drawn-out notes that decrease in volume. **H** Lowland evergreen forest and degraded, selectively logged forest. **D** Sula and Banggai islands, Indonesia.

12 SULAWESI PITTA *Erythropitta celebensis* 16–17cm Formerly considered a race of the Philippine Pitta; actions presumed to be similar. **V** Presumed to be similar to that of the Philippine Pitta. **H** Subtropical and tropical moist lowland forests, plantations. **D** Sulawesi and the Togian Islands, Indonesia.

13 SIAU PITTA *Erythropitta palliceps* 16–17cm Formerly considered a race of the Philippine Pitta; actions presumed to be similar. **V** Presumed to be similar to that of the Philippine Pitta. **H** Subtropical and tropical moist lowland forest, secondary forest. **D** Siau, Tahulandang and Ruang islands, Indonesia.

14 SANGIHE PITTA *Erythropitta caeruleitorques* 16–17cm Formerly considered a race of the Philippine Pitta; actions presumed to be similar. **V** Presumed to be similar to that of the Philippine Pitta. **H** Subtropical and tropical moist lowland forest. **D** Sangihe Island, Indonesia.

15 SOUTH MOLUCCAN PITTA *Erythropitta rubrinucha* 16–17cm Formerly considered a race of the Philippine Pitta; actions presumed to be similar. **V** Presumed to be similar to that of the Philippine Pitta. **H** Tropical moist lowland forests. **D** Buru and Seram, S Maluku Islands, Indonesia.

16 NORTH MOLUCCAN PITTA *Erythropitta rufiventris* 16–17cm Formerly considered a race of the Philippine Pitta; actions presumed to be similar. **V** Presumed to be similar to that of the Philippine Pitta. **H** Subtropical and moist lowland forest. **D** N Maluku Islands, Indonesia.

17 LOUISIADE PITTA *Erythropitta meeki* 16–18cm Formerly treated as race of Philippine Pitta. Unobtrusive and generally secretive. Feeds on forest floor. Diet comprises mainly invertebrates. **V** Strangled-sounding, drawn-out *whooEE-whor*. **H** Lowland primary forest, secondary woodland. **D** Louisiade Archipelago.

18 BISMARCK PITTA *Erythropitta novaehibernicae* 16–17cm Formerly treated as race of Philippine Pitta. Rather shy and secretive. Forages for invertebrates in leaf litter on forest floor. **V** Drawn-out *whooEEorr* and various chattering calls. **H** Primary forest, secondary woodland; lowlands to 1,000m. **D** Bismarck Archipelago.

19 PAPUAN PITTA *Erythropitta macklotii* 16–17cm Formerly treated as race of Philippine Pitta. Shy, keeping to dense cover. **V** Drawn-out *whooEE-whor* with owl-like quality. **H** Lowland primary forest, secondary woodland. **D** New Guinea, Aru Islands, D'Entrecasteaux Archipelago, Raja Ampat Islands, NE Australia.

20 BLUE-BANDED PITTA *Erythropitta arquata* 15cm Little recorded information; presumed to forage on the ground, feeding on ants and other insects. **V** Monotonous, fluting whistle. **H** Forests with bamboo and many fallen trees. **D** Borneo.

21 GARNET PITTA *Erythropitta granatina* 15–16cm Forages on the ground, among leaf litter and around fallen branches and logs; occasionally searches on logs. **V** Drawn-out, monotone whistle that increases in volume. When agitated, utters a purring *prrr prrr prrr*. **H** Primary and logged lowland forests. **D** Malay Peninsula; Sumatra, Indonesia; Borneo.

22 GRACEFUL PITTA *Erythropitta venusta* 18cm Forages on the ground or on fallen logs. **V** Low, mournful whistle. **H** Moist, dark highland forest, with dense understorey and ravines. **D** Sumatra, Indonesia.

23 BLACK-CROWNED PITTA *Erythropitta ussheri* 15–16cm Forages by probing in leaf litter, or on damp ground or fallen logs. Formerly considered a race of Garnet Pitta. **V** Similar to that of Garnet Pitta but more prolonged. **H** Lowland primary and logged forest. **D** Borneo.

24 HOODED PITTA *Pitta sordida* 16–19cm Feeds on the ground; often perches on vines or branches while singing. **V** Loud *whew-whew*. **H** Primary riverine forest, secondary forest with thick understorey or scrub, wet or dry forests, various plantations. **D** Himalayas to New Guinea.

25 IVORY-BREASTED PITTA *Pitta maxima* 22–28cm Usually encountered singly, in pairs or occasionally in groups. Tends to spend more time off the ground than most pittas. **V** Distinctive, mournful *wu-whoouw*, repeated constantly. **H** Undisturbed and selectively logged hill forest on limestone, with an understorey of palms or dense thickets of spiny rattan; also occurs in damp primary lowland forest. **D** N Maluku Islands, Indonesia.

26 AZURE-BREASTED PITTA *Pitta steerii* 19cm Forages on the ground, on boulders and fallen logs. **V** Explosive *whirp whirp whirp whirp whirp* and a loud, repeated *kweioo*. **H** Forests with thick undergrowth on limestone, degraded forest. **D** Philippines.

27 SUPERB PITTA *Pitta superba* 21–22cm Endangered due to habitat destruction. Shy and unobtrusive. Forages for invertebrates in leaf litter. **V** Shrill, fluty, paired *WhaaOu, WhaaOu* phrases. **H** Lowland primary forest, secondary wooded slopes. **D** Manus Island (Bismarck Archipelago).

28 AFRICAN PITTA *Pitta angolensis* 20cm Distinguished from Green-breasted Pitta by buff (not green) breast, and in flight by slightly larger white wing spots. **V** Strange compressed *puwee* (*pu-* low and frog-like, *-wee* very fluting). **H** On the ground under dense shrub patches in wooded areas. **D** E, SE, WC and W Africa.

29 GREEN-BREASTED PITTA *Pitta reichenowi* 15cm See African Pitta. **V** High, fluting, pressed-out, short *fjull fjull fjull*. **H** Montane forest floors. **D** W Cameroon to S Uganda and C DR Congo.

30 INDIAN PITTA *Pitta brachyura* 19cm Feeds on the ground. When alarmed, often sits quietly on low branches or creepers with only a wagging tail giving away its presence. **V** Whistled *wheet-tieu*, *wieet-piyou* or *pree-treer*. **H** Light deciduous and evergreen forests with thick undergrowth. **D** India.

31 FAIRY PITTA *Pitta nympha* 19cm Forages on the ground among leaf litter. **V** Clear, whistled *kwah-he-kwa-wu*. When alarmed, gives a *kriaih* or *kahei-kahei*. **H** Broadleaved evergreen forest. **D** Breeds from Japan and Korean Peninsula to SE China. Winters in Borneo.

32 BLUE-WINGED PITTA *Pitta moluccensis* 18–21cm Forages on the ground, hops like a thrush. **V** Loud, clear *taew-laew taew-laew*, repeated every few seconds. When alarmed, utters a harsh *skyeew*. **H** Open broadleaved evergreen and mixed deciduous forest, secondary growth, bamboo, mangroves, parks and gardens. **D** Breeds from Bangladesh to New Guinea. Winters in Malay Peninsula; Sumatra, Indonesia; Borneo.

33 MANGROVE PITTA *Pitta megarhyncha* 20cm Forages on muddy areas around mangrove roots and nearby drier ground. **V** Loud *tae-laew* or *wieuw-wwieuw*. **H** Mangroves. **D** Bangladesh to Malay Peninsula and Sumatra, Indonesia.

34 ELEGANT PITTA *Pitta elegans* 19cm Occurs singly or in pairs. Shy and retiring, best located by calls. **V** *Kwweee-kwill*; variations have been noted on individual islands, such as a *ka-wha-kil* on Sumba, *kwuwik-kwk* in the Sula Islands and *perriew-priew* in the Banggai Islands. **H** Various forest types, including humid primary forest, dry monsoon forest, forest edges, and degraded and selectively logged forest. **D** Lesser Sunda Islands and S Maluku Islands, Indonesia.

35 RAINBOW PITTA *Pitta iris* 17–18cm Typically shy and secretive. Diet comprises mainly invertebrates, foraged in leaf litter. **V** Fluty, piping, paired *whor-whee, whor-whee* phrases. **H** Monsoon forests, densely wooded lowland thickets. **D** N Northern Territory and N Western Australia.

36 NOISY PITTA *Pitta versicolor* 20–21cm Shy and unobtrusive, heard more often than seen. Forages mainly for invertebrates in leaf litter. N Australia nesters are partial migrants, moving to New Guinea outside breeding season. **V** Fluty, piping, paired *Whoop, pe-Whoop* phrases. **H** Rainforest, dense secondary woodland, mangroves. **D** E and NE Australia, S New Guinea.

37 BLACK-FACED PITTA *Pitta anerythra* 16–17cm Unobtrusive and silent birds are easily overlooked. Forages in leaf litter, mainly for invertebrates. **V** Trilling, piping, paired *Wooerrr-whrreee* phrases. **H** Primary forest, dense secondary woodland; mostly lowlands. **D** Bougainville, Choiseul and Santa Isabel islands (Solomon Islands).

♀
1
♂
♀
2
♂
♀
3
♂
♀
4
♀
10
♀
5
♂
♀
6
♂
♀
7
♂
♀
8
♂
9
11
♀
13
12
14
15
16
17
18
19
20
21
22
23
24
25
26
27
28
29
30
31
32
33
34
35
36
37

OVENBIRDS FURNARIIDAE

1 CAMPO MINER *Geositta poeciloptera* 12cm No other miner occurs in its range. **V** Hurried, high, unstructured twittering. **H** Open dry grassland and Cerrado; 500–1,400m. **D** E Bolivia through SC Brazil.

2 COMMON MINER *Geositta cunicularia* 15cm Tail pattern varies depending on subspecies. **V** Rapid flow of *cheechiw* notes. **H** Sparse grassy plains, arid scrub, open mountain sides; <5,000m. **D** S and W South America.

3 PUNA MINER *Geositta punensis* 14cm Note tail pattern and unstreaked breast. **V** Series of *peep* notes, sung in flight. **H** Grass plains with some shrub; 3,200–4,900m. **D** S Peru to NW Argentina.

4 SHORT-BILLED MINER *Geositta antarctica* 15.5cm Separable from Common Miner by shorter, straighter bill and lack of rufous on wings. **V** Short, very high, sizzling twitters. **H** Arid plains with short grass and scattered bush; <1,000m. **D** S and W Argentina, S Chile.

5 SLENDER-BILLED MINER *Geositta tenuirostris* 18cm Long, slender, downcurved bill is diagnostic. **V** Continuous flow of *jee* notes, sung in flight. **H** Dry montane grassland and scrub; 2,500–4,600m. **D** C Ecuador to NW Argentina.

6 GREYISH MINER *Geositta maritima* 13cm Note tail pattern and lack of pale buff in wings during flight. **V** Short, level rattle. **H** Arid, barren or sparsely scrubbed areas; <2,900m, locally higher. **D** W Peru, N Chile.

7 COASTAL MINER *Geositta peruviana* 13cm Occurs at lower elevations than other miners except Greyish Miner, which is greyer above. Note black-tipped grey bill. **V** Call is a squeaky *cueet*. During display flight utters and unmusical *pjee-aww*. **H** Arid scrub, coastal desert; <700m. **D** W Peru.

8 DARK-WINGED MINER *Geositta saxicolina* 16cm Note buff forehead and sides of face, and uniform dusky-brown wings (lacking rufous). **V** Long rising and falling series of *peep* notes. **H** Rocky, grassy slopes; 3,700–4,900m. **D** C Peru.

9 RUFOUS-BANDED MINER *Geositta rufipennis* 17cm Wings and tail rufous with a black border. **V** Hurried, high, hoarse, level twitter, interspersed with short trills. **H** Rocky, sparsely vegetated slopes; 3,100–4,400m, almost down to sea level in the south of its range. **D** Bolivia, W Argentina, Chile.

10 CREAMY-RUMPED MINER *Geositta isabellina* 18cm Separable from Rufous-banded Miner by somewhat larger, more downcurved bill and lack of rufous in tail. **V** Long, descending, rapid series of *tree* notes. **H** Rocky, sparsely grassed mountain sides; 3,000–5,000m. **D** C Chile, WC Argentina.

11 THICK-BILLED MINER *Geositta crassirostris* 17cm Note rather heavy bill and pale legs. Mantle is slightly scaled. **V** Hurried series of harsh notes, rising at the start and lowered at the end (3–4 secs). **H** Arid slopes with scattered bushes and cacti; 600–3,000m. **D** W and SW Peru.

12 STRAIGHT-BILLED EARTHCREEPER *Ochetorhynchus ruficaudus* 18cm See Rock Earthcreeper. **V** Unstructured, high series of screeching notes. **H** Arid slopes with some scrub; 2,300–4,300m, locally (much) lower. **D** S Peru to C Chile and W Argentina.

13 ROCK EARTHCREEPER *Ochetorhynchus andaecola* 18cm Difficult to separate from Straight-billed Earthcreeper, but that species has a straighter bill, whiter throat and darker, broader stripes on underparts. **V** Loud, resounding series of *vee* notes, undulating rather randomly. **H** Arid slopes with scattered grass and scrub; 2,600–4,000m, locally higher. **D** Bolivia, N Chile, NW Argentina.

14 BAND-TAILED EARTHCREEPER *Ochetorhynchus phoenicurus* 17cm Note straight bill. Basal half of tail feathers except central pair rufous. **V** Fast, dry trill. Alarm call a repeated *wheet* or *suwee*. **H** Arid shrubby plains; <1,200m. **D** S Argentina.

15 CRAG CHILIA *Ochetorhynchus melanurus* 18.5cm Distinctly patterned and coloured; does not resemble any other bird in its range and habitat. Note straight bill. **V** Dry, short rattles. **H** Arid, sparsely vegetated slopes; 1,200–3,000m. **D** C Chile.

16 BUFF-BREASTED EARTHCREEPER *Upucerthia validirostris* 20cm Plain-looking earthcreeper, with dark brown upperparts, and a pale buffish supercilium and underparts. **V** High, very sharp, piercing, rising and lowered rattles. **H** High-altitude grassland and shrubland; 2,500–4,000m, locally higher. **D** C Peru to NW Argentina.

17 WHITE-THROATED EARTHCREEPER *Upucerthia albigula* 19cm Separable from Buff-breasted Earthcreeper by whiter throat and eyebrow. **V** Short, fast, slightly descending, unstructured twitter. **H** Dry, sparsely vegetated slopes and adjacent fields and grassland; 2,400–4,000m. **D** SW Peru, N Chile.

18 SCALE-THROATED EARTHCREEPER *Upucerthia dumetaria* 21cm Separable from Patagonian Forest Earthcreeper by slightly darker upperparts and fainter scalloping on upper breast, and by range. **V** Very high, sharp, hurried twitter, slightly rising at the start (2 sec). **H** Arid scrub and puna; <3,900m. **D** S South America.

19 PATAGONIAN FOREST EARTHCREEPER *Upucerthia saturatior* 21cm See Scale-throated Earthcreeper, which it resembles. **V** Hurried rising and falling twittering. **H** Shrubby steppe. **D** WC Argentina, C Chile.

20 STRIATED EARTHCREEPER *Geocerthia serrana* 20cm Finely streaked below, not scalloped or plain. **V** Unstructured twittering, including dry rattles. **H** Arid scrubby slopes and open woodland; 2,800–4,300m. **D** Peru.

21 BOLIVIAN EARTHCREEPER *Tarphonomus harterti* 17cm Separable from Chaco Earthcreeper by different head colouring and patterning, and by paler underparts, which contrast less with throat. **V** Rapid series of high, piercing, staccato notes (any length). **H** Arid scrub and low woodland; 1,400–3,000m. **D** Bolivia, NW Argentina.

22 CHACO EARTHCREEPER *Tarphonomus certhioides* 17cm See Bolivian Earthcreeper, whose range differs. **V** Sings as Bolivian Earthcreeper. **H** Dense undergrowth of woodland, scrub; <1,700m. **D** SC South America.

23 LONG-TAILED CINCLODES *Cinclodes pabsti* 21.5cm Separable from Buff-winged Cinclodes by more extensive wing edging. No other cinclodes normally occur in its range. **V** Very high, slightly rising, rather unstructured rattle, changing to a trill. **H** Open grassland and pastures; 750–1,700m. **D** SE Brazil.

24 BLACKISH CINCLODES *Cinclodes antarcticus* 19.5cm No similar bird occurs in its range and habitat. **V** Short, very high twitters. **H** Coastal gravel and rock beaches. **D** Tierra del Fuego, Falkland Islands.

25 BUFF-WINGED CINCLODES *Cinclodes fuscus* 17cm Note plain underparts, buff wing band and small, pointed bill. **V** Short, rather unstructured, very high twitter with embedded trill at start. **H** Grassy habitats, often near water; <5,000m. **D** C and S Chile, C and S Argentina.

26 CHESTNUT-WINGED CINCLODES *Cinclodes albidiventris* 17cm Note rufous tinge above and finely scalloped breast. Little or no white in wing band. **V** Long series of very high twittering, dominated by trills. **H** Grassy habitats, often near water; <5,000m. **D** N Andes from N Colombia and W Venezuela to Ecuador.

27 CORDOBA CINCLODES *Cinclodes comechingonus* 17cm Note deep rufous wing band and yellowish maxilla. **V** A trill, more complex than that of Buff-winged Cinclodes. **H** Open grassy, rocky areas; 1,600–2,400m, locally (much) higher. **D** NC Argentina.

28 CREAM-WINGED CINCLODES *Cinclodes albiventris* 17cm Wing band and undertail coverts whitish. **V** Long, undulating, very high series of notes, switching between twittering and trills. **H** Grassy habitats, often near water; <5,000m. **D** C Andes from N Peru to C Argentina.

29 OLROG'S CINCLODES *Cinclodes olrogi* 17cm White wing band; separable from White-winged Cinclodes by pale rufous tail corners. **V** High, sharp trill, slowing at the end to a few separate notes. **H** Open grassy and rocky areas near water; 1,600–2,800m. **D** NC Argentina.

30 STOUT-BILLED CINCLODES *Cinclodes excelsior* 20.5cm Note downcurved bill and scaling to breast. Darker below than other cinclodes. **V** Steeply rising, very fast series of notes, sometimes preceded or concluded by a few sparrow-like chirps. **H** Rocky grassland, montane scrub, *Polylepis* woodland. 3,200–5,200m. **D** Colombia to S Ecuador.

31 ROYAL CINCLODES *Cinclodes aricomae* 20.5cm Darker than any other cinclodes; separable from Buff-winged Cinclodes by heavier bill and buffer wingbar. **V** Nasal twittering, dominated by an embedded, steeply rising rattle. **H** Rocky slopes in *Polylepis* woodland; 3,600–4,600m. **D** S Peru.

32 WHITE-WINGED CINCLODES *Cinclodes atacamensis* 20cm Distinctive white wing band and tail corners. **V** Very short, hurried, harsh rattle, preceded by a single magpie-like note. **H** Puna grassland and rocky slopes; 2,200–4,500m, locally lower. **D** C Peru to WC Argentina.

33 WHITE-BELLIED CINCLODES *Cinclodes palliatus* 24cm No similar bird with white underparts occurs in its habitat. **V** Long, level series of harsh *zee* notes with small changes in intonation and tempo (4–7 secs). **H** Puna grassland; 4,400–5,000m. **D** C Peru.

34 GREY-FLANKED CINCLODES *Cinclodes oustaleti* 17cm Note fine bill and streaky spots to throat and breast. **V** Very high, sharp, dry trill. **H** Open rocky and grassy areas; <4,200m. **D** Chile, W Argentina.

35 DARK-BELLIED CINCLODES *Cinclodes patagonicus* 19cm Resembles Grey-flanked Cinclodes, but with a larger bill and less contrast between flanks and belly. **V** High, shrill trill, the first few notes slower. **H** Rocky beaches; <2,500m. **D** C and S Chile, W Argentina.

36 PERUVIAN SEASIDE CINCLODES *Cinclodes taczanowskii* 21.5cm Note faint eyebrows. **V** Short, sharp, descending series of notes. **H** Always at the edge of the ocean. **D** C and S Peru.

37 CHILEAN SEASIDE CINCLODES *Cinclodes nigrofumosus* 21.5cm Rather dark, with a striking white throat. Separable from Peruvian Seaside Cinclodes by range. **V** Hurried, short, almost level, sharp twitters. **H** Rocky sea beaches. **D** N and C Chile.

1
2
3
4
5
6
7
8
9
10
11
12
13
14
15
16
17
18
19
20
21
22
23
24
25
26
27
28
29
30
31
32
33
34
35
36
37

OVENBIRDS *CONTINUED*

1 LESSER HORNERO *Furnarius minor* 13cm Separable from Band-tailed, Pale-legged and Bay horneros by small size and slightly duller colouring. **V** Rapid series of very high, sharp, staccato notes with slight variations in pitch, intonation and tempo (4–5 secs). **H** Open riverine scrub; <300m. **D** Amazon and its tributaries.

2 BAND-TAILED HORNERO *Furnarius figulus* 16cm Throat contrasts less with breast than in Pale-legged Hornero. Often has black subterminal spots to tail feathers. **V** Rapid, level, staccato chattering. **H** Near water in open woodland, secondary growth, pastures, gardens. **D** SE Amazonia, E Brazil.

3 PALE-LEGGED HORNERO *Furnarius leucopus* 17cm See Band-tailed Hornero; also often has (much) paler eyes than that species. **V** Rapid, loud, level series of high, staccato notes, the first few notes higher and accelerating (8 secs). **H** Semi-open habitats, woodland, riverine belts, fields, parks. **D** W and S Amazonia, E Brazil.

4 PACIFIC HORNERO *Furnarius cinnamomeus* 19–20cm Recently split from Pale-legged Hornero. Note subtly larger size and grey crown. **V** Song a series of chirping *Tche-tche-tchee-tchee-tchu-tchu-chu-chu-chu* notes, wavering in pitch during delivery. Call a shrill, abrupt chirping *Tchiip, Tchiip, Tchiip*. **H** Open secondary woodland, agricultural land. **D** Pacific slope of Andes in Ecuador and N Peru.

5 CARIBBEAN HORNERO *Furnarius longirostris* 17–18cm Recently split from Pale-legged Hornero. Note grey crown. **V** Song a descending, chirping *Tche-tche-tchee-tchee-tchu-tchu-chu-chu chu* cascade of notes. **H** Open secondary woodland, agricultural land. **D** N Colombia, N Venezuela.

6 BAY HORNERO *Furnarius torridus* 17cm Underparts normally more saturated than in Band-tailed and Pale-legged horneros. **V** Loud, resounding, rapid, slightly descending series of ringing, staccato notes. **H** Riverine belts, river islands. **D** W Amazon and its tributaries.

7 RUFOUS HORNERO *Furnarius rufus* 19cm Less contrastingly coloured than Lesser, Band-tailed, Pale-legged and Bay horneros, and lacks distinct eyebrow. **V** Accelerating, resounding series of high, descending, dry, piercing notes. **H** Scrub, fields, pastures, gardens; often near houses. **D** E, SC and SE South America.

8 CRESTED HORNERO *Furnarius cristatus* 15cm Unmistakable due to crest. **V** Series of high, piercing, staccato notes, given in chorus. **H** Scrub and woodland; around houses. **D** S Bolivia, W Paraguay, N Argentina.

9 DES MURS'S WIRETAIL *Sylviorthorhynchus desmurii* 21cm Unmistakable. Note long, thin-feathered tail. **V** Song is a hurried mix of single high, sharp *peep* notes and ultra-short, dry rattles, or a very high, drawn-out, resounding trill. **H** Bamboo thickets in woodland and forest. **D** C and S Chile, SW Argentina.

10 THORN-TAILED RAYADITO *Aphrastura spinicauda* 14cm Unmistakable in its range owing to colour pattern and spiked tail. **V** Short, sharp, mid-high rattle. **H** Diverse habitats, from forest to tussock grassland on islands. **D** C and S Chile, SW Argentina.

11 MASAFUERA RAYADITO *Aphrastura masafuerae* 15cm Critically Endangered owing to habitat decline. Forages in the understorey and among leaf litter, gleaning for insects. **V** Low, churring *trrrt*. **H** Scrub and open woodland with ferns. **D** Alejandro Selkirk Island, Juan Fernández Islands, off the WC coast of Chile.

12 BROWN-CAPPED TIT-SPINETAIL *Leptasthenura fuliginiceps* 16cm Note rufous crown, normally shown as a bushy crest. **V** Short, fast, very high, thin twitter. **H** Montane scrub and *Polylepis* woodland. **D** Bolivia to NW and WC Argentina.

13 TAWNY TIT-SPINETAIL *Leptasthenura yanacensis* 16cm Note warm colouring. Wing pattern varies between that shown and (almost) completely rufous flight feathers. **V** More or less level, high, sharp twitter, varying in sound, tempo and strength (5–20 secs). **H** *Polylepis* woodland and scrub. **D** Peru to NW Argentina.

14 TUFTED TIT-SPINETAIL *Leptasthenura platensis* 16.5cm Separable from the smaller Plain-mantled Tit-spinetail by bushy crest, spot-striped throat and lack of white in tail. **V** Short, very high, thin trill. **H** Woodland and scrub. **D** N Argentina and C Paraguay to S Brazil, Uruguay and EC Argentina.

15 PLAIN-MANTLED TIT-SPINETAIL *Leptasthenura aegithaloides* 15cm Note unstreaked mantle. **V** Very high, thin, rattled trill of variable length and intonation. **H** Scrub and brush on rocky slopes. **D** S Peru to S South America.

16 STRIOLATED TIT-SPINETAIL *Leptasthenura striolata* 16cm No similar tit-spinetail occurs in its range. **V** Short, rapid, very high, sizzling trill, with an introductory note. **H** Forest, woodland and shrub. **D** SE Brazil.

17 RUSTY-CROWNED TIT-SPINETAIL *Leptasthenura pileata* 17cm Distinguished from White-browed Tit-spinetail by more extensive streaking on underparts. **V** Short, mid-high, rattled trill. **H** Scrub and woodland; 2,500–4,000m. **D** Peru.

18 WHITE-BROWED TIT-SPINETAIL *Leptasthenura xenothorax* 17cm Note plain underparts; streaking is restricted to throat. **V** Short, very high, rattled trill, varying in intonation. **H** *Polylepis* woodland; 3,600–4,500m. **D** S Peru.

19 STREAK-BACKED TIT-SPINETAIL *Leptasthenura striata* 16cm Similar to Rusty-crowned Tit-spinetail, but note rufous in folded wing. **V** Song a harsh chatter and melancholy, musical series of descending notes. Calls include a *twet* and *tcht*. **H** Montane scrub and *Polylepis* woodland; 2,000–4,000m. **D** W Peru to NW Chile.

20 ANDEAN TIT-SPINETAIL *Leptasthenura andicola* 17cm Note extensive streaking. Separable from Striolated Tit-spinetail by range. **V** Level, high, irregular chirping. **H** Scrub, páramo, *Polylepis* woodland; 3,000–4,500m. **D** W Venezuela to N Bolivia.

21 ARAUCARIA TIT-SPINETAIL *Leptasthenura setaria* 17cm Note rufous back. **V** Very high, sizzling trill with variations in speed and intonation (2–4 secs). **H** Normally found only in *Araucaria* trees in forest, woodland, plantations and parks; <1,900m, locally higher. **D** NE Argentina to SE Brazil.

22 PERIJA THISTLETAIL *Asthenes perijana* 19cm No similar thistletail occurs in its restricted range (Sierra de Perijá). **V** Song begins with 3–5 high-pitched *pee* notes, followed by descending trill. **H** Páramo, dense undergrowth at edge of cloud forest; 3,000–3,400m. **D** N Colombia, NW Venezuela.

23 WHITE-CHINNED THISTLETAIL *Asthenes fuliginosa* 19cm Note white chin, eye-ring and more or less distinct eyebrow. **V** Sharp, rising, rapid chatter, repeated 2–5 times. **H** Páramo, elfin forest, dense undergrowth at edge of cloud forest; 2,800–3,500m, locally higher or lower. **D** W Venezuela to C Peru.

24 VILCABAMBA THISTLETAIL *Asthenes vilcabambae* 18.5cm Rather dark and drab above. Note lack of eye-ring. **V** Song begins with three to five drawn-out high-pitched *pee* notes, rising in pitch, repeated notes that accelerate into a trill, which ends abruptly; calls include long, descending notes. **H** Treeline habitats, páramo, shrubby woodland; 2,800–3,600m. **D** Vilcabamba Mountains, S Peru.

25 AYACUCHO THISTLETAIL *Asthenes ayacuchensis* 18–20cm Previously treated as subspecies of White-chinned Thistletail (along with Vilcabamba Thistletail). Diet includes invertebrates. Solitary or seen in pairs. Unobtrusive. **V** Call similar to that of Vilcabamba Thistletail but higher pitched. **H** High-altitude woodland, scrub, puna; around 3,500m. **D** S Andean Peru.

26 OCHRE-BROWED THISTLETAIL *Asthenes coryi* 18cm Rufous face sides and chin are diagnostic. **V** Very high, rising series of 6–8 notes, starting slowly and accelerating to a short trill. **H** Páramo, dense undergrowth at edge of cloud forest; 3,400–4,100m. **D** W Venezuela.

27 MOUSE-COLORED THISTLETAIL *Asthenes griseomurina* 19cm Note distinct eye-ring and overall greyish plumage. **V** High, sharp, level series of notes, accelerating to a trill (2–3 secs). **H** Undergrowth of cloud forest and *Polylepis* woodland; 2,800–4,000m. **D** S Ecuador, N Peru.

28 EYE-RINGED THISTLETAIL *Asthenes palpebralis* 19cm Shows a distinctive eye-ring and chestnut upperparts. **V** Soft, rising series of notes, doubled after a tiny stop. **H** Páramo, dense undergrowth at edge of cloud forest; 3,000–3,600m. **D** C Peru.

29 PUNA THISTLETAIL *Asthenes helleri* 18cm Compare with the darker Vilcabamba Thistletail, which also has a distinct rufous throat patch. **V** Slightly rising, accelerating, thin rattle (2 secs). **H** Páramo, undergrowth and borders of woodland at the treeline; 2,800–3,600m. **D** S Peru, N Bolivia.

30 BLACK-THROATED THISTLETAIL *Asthenes harterti* 18cm Note distinct facial pattern and pale eyes. **V** Dry rattle, slightly rising and slightly lowered (1 sec). **H** Páramo, woodland undergrowth; 2,500–3,400m. **D** N and C Bolivia.

31 ITATIAIA SPINETAIL *Asthenes moreirae* 18cm No similar bird occurs in its range. Note long, pointed tail and pale orange throat patch. **V** Very high series of notes, starting with some slightly drawn-out notes that change to a small rattle, sometimes lowered at the end (1–2 secs). **H** Montane scrub, bamboo, tall grass; 1,850–2,800m. **D** SE Brazil.

32 SHARP-BILLED CANASTERO *Asthenes pyrrholeuca* 16cm Separable from Short-billed Canastero by longer tail, thinner bill and less distinct throat patch. **V** Short, rapid, very high, rising doubled twitter. **H** Scrub, rocky grassland, tall grass; <2,000m, locally higher. **D** S South America.

33 SHORT-BILLED CANASTERO *Asthenes baeri* 15cm Note distinct eyebrow. Outer two pairs of tail feathers are all rufous. **V** Short, level trill, changing in intonation. **H** Chaco scrub and woodland; <800m. **D** SE and SC South America.

34 CANYON CANASTERO *Asthenes pudibunda* 16cm Separable from Sharp-billed Canastero by thicker bill and more extensive rufous on wings. **V** High, sharp, rising, rattled series of notes, slightly lower at the end (2–3 secs). **H** Rocky slopes with scattered bush and stunted trees; 2,400–4,000m. **D** NW Peru to N Chile.

35 RUSTY-FRONTED CANASTERO *Asthenes ottonis* 18cm Note rufous front, chin and tail. **V** Short, rising series of notes, a few of them drawn out at the beginning. **H** Arid montane scrub and woodland; 2,500–4,000m. **D** SC Peru.

36 MAQUIS CANASTERO *Asthenes heterura* 16.5cm Separable from Rusty-fronted Canastero by range and relatively shorter tail. **V** Song includes a rapid series of 5–6 rising notes. **H** Arid montane scrub and low woodland; 3,000–4,200m, locally lower. **D** N Bolivia to NW Argentina.

37 CORDILLERAN CANASTERO *Asthenes modesta* 15cm Note rufous in tail and lack of streaking on back. **V** Shattering, sharp, slightly rising rattle (2–3 secs). **H** Grassland with scattered rocks and scrub; <4,500m. **D** Peru through S Argentina.

38 STREAK-THROATED CANASTERO *Asthenes humilis* 16cm See Streak-backed Canastero (Plate 127). Separable from Cordilleran Canastero by lack of rufous in tail. **V** Sings as Puna Canastero (Plate 127). **H** Boulder-strewn grassland; 3,000–4,800m. **D** Peru, N Bolivia.

39 RUSTY-VENTED CANASTERO *Asthenes dorbignyi* 16cm Separable from Berlepsch's Canastero (Plate 127) by wing pattern, and from Steinbach's Canastero (Plate 127) by grey or grey-brown (not rufous) wing coverts. Note rufous rump. **V** Sharp series of notes, starting with very high, piercing, staccato notes and rising or descending to a trill. **H** Arid scrubby habitats in hilly and mountainous areas; 1,800–4,000m. **D** Bolivia to NW Argentina.

40 DARK-WINGED CANASTERO *Asthenes arequipae* 15–17cm Previously treated as subspecies of Rusty-vented Canastero. Note dark wings. **V** Agitated, rapid-fire, wren-like trilling *tchrrrrrrrrrrr*. **H** High-altitude arid scrub, open woodland. **D** W Andean slopes of W Bolivia and SW Peru.

1
2
3
4
5
6
7
8
9
10
11
12
13
14
15
16
17
18
19
20
21
22
23
24
25
26
27
28
29
30
31
32
33
34
35
36
37
38
39
40

OVENBIRDS *CONTINUED*

1 PALE-TAILED CANASTERO *Asthenes huancavelicae* 15–16cm Previously treated as subspecies of Rusty-vented Canastero (Plate 126). Subtly smaller and paler than that species. **V** Song *tche-tche-tche-chu-chu chrrrrrrrrrrrrrrrr*, the first notes squeaky, the latter an agitated, mechanical-sounding trill. **H** High-altitude valleys with scrub-covered slopes. **D** W Andean slopes of WC and SC Peru.

2 BERLEPSCH'S CANASTERO *Asthenes berlepschi* 16cm Note wing pattern. **V** Song is an accelerating series of notes, starting with some toneless croaks and ending in a sharp, high, rattling trill. **H** Montane scrub and agricultural areas; 2,500–3,700m. **D** NW Bolivia.

3 CIPO CANASTERO *Asthenes luizae* 17cm Note fine black lines and small streaks on chin. No other canastero occurs in its range. **V** Level, sharp rattle (2–3 secs) or descending series of 10–12 *tseep* notes. **H** Rocky slopes with scattered bush and grass; 1,000–1,500m. **D** SE Brazil.

4 STREAK-BACKED CANASTERO *Asthenes wyatti* 18cm See Puna Canastero. Separable from Streak-throated Canastero (Plate 126) by rufous outer-tail feathers. **V** Mid-high, rising and lowered, sharp, rattled trill (1 sec). **H** Montane grassland and scrub; 3,000–5,000m, locally lower. **D** Venezuela to Peru.

5 PUNA CANASTERO *Asthenes sclateri* 18cm Separable from Streak-backed Canastero by black base to outer-tail feathers (not shown). **V** Song an accelerating, ascending trill. Call a *tzup*. **H** Grassland with tall bunchgrass and shrub; 1,800–4,000m. **D** Peru to NW Argentina.

6 AUSTRAL CANASTERO *Asthenes anthoides* 16.5cm Separable from Hudson's Canastero by range. **V** Series of 1–3 chopped notes, followed by a high, shrill trill. **H** Shrubby grass plains, sometimes with scattered scrub and low stands of trees; <1,500m. **D** S Chile, SW Argentina.

7 HUDSON'S CANASTERO *Asthenes hudsoni* 18cm No similar canastero occurs in its range. Note silvery edges to tail feathers. **V** Song comprises some very high, chopped notes, suddenly changing to an accelerating rattle. **H** Grassy areas in and around wetlands; <950m. **D** C and S Uruguay, NE Argentina.

8 LINE-FRONTED CANASTERO *Asthenes urubambensis* 16.5cm Has uniform brown upperparts (excluding crown), streaked underparts and an orange chin. **V** Series of 1–5 *weeh* notes, changing to a high, descending trill. **H** Treeline with scattered bush and *Polylepis* stands; 3,100–3,800m. **D** C Peru to W Bolivia.

9 MANY-STRIPED CANASTERO *Asthenes flammulata* 16cm Note distinctive 'looped' streaks to underparts. **V** Series of 1–5 *zreeh* notes, changing to a lowered, wheezing trill. **H** Grassland at the treeline with scattered bushes; 3,000–4,500m. **D** Colombia to C Peru.

10 JUNIN CANASTERO *Asthenes virgata* 17cm Separable from Many-striped Canastero by less rufous in wings and fainter streaking below. **V** Series of 1–5 jubilant yet hoarse notes, changing to a sharp, lowered trill. **H** Puna with scattered *Polylepis* patches; 3,300–4,500m. **D** C and S Peru.

11 SCRIBBLE-TAILED CANASTERO *Asthenes maculicauda* 17cm Note rufous forehead. Black scribbles are difficult to see. **V** Series of 1–5 *zreeh* notes, changing to a lowered, rattling trill. **H** Treeline grassland with low shrubs; 3,000–4,300m. **D** C Peru, Bolivia, NW Argentina.

12 DUSKY-TAILED CANASTERO *Pseudasthenes humicola* 15cm Note black tail (without rufous), fine throat streaking and rufous wing coverts. **V** High series of notes, accelerating to a rattle. **H** Arid scrub. **D** N and C Chile.

13 PATAGONIAN CANASTERO *Pseudasthenes patagonica* 15cm Separable from Short-billed and Sharp-billed canasteros (Plate 126) by lack of rufous throat patch. Note very fine black malar stripe, and in tail lack of rufous and black restricted to terminal half. **V** High, sharp, fast, level rattle (3 secs). **H** Low scrub. **D** SE Argentina.

14 CACTUS CANASTERO *Pseudasthenes cactorum* 14cm No similar bird occurs in its range and habitat. Note long bill. Orange throat patch absent, or at best very pale. **V** Dry, fast rattle of any length, sometimes interrupted by a tiny stop. **H** Arid scrub with tall cacti and sparse bush. **D** W Peru.

15 STEINBACH'S CANASTERO *Pseudasthenes steinbachi* 16cm Separable from Rusty-vented Canastero (Plate 126) by more rufous in tail, and browner cap and mantle. **V** Short, unstructured, sometimes shrill twittering. **H** Arid montane scrub; 1,500–2,500m, locally lower or higher. **D** W Argentina.

16 CHOTOY SPINETAIL *Schoeniophylax phryganophilus* 21cm Note black-bordered yellow chin. **V** Short, dry, croaking *cho* phrases. **H** Open areas with scattered trees, riverine belts, gardens. **D** C, SC and SE South America.

17 WHITE-WHISKERED SPINETAIL *Synallaxis candei* 16cm Similar to Hoary-throated Spinetail, but note black cheeks and white moustachial streak. Forages on the ground. **V** Song is a rising four-syllable chatter. **H** Arid scrub, scattered bush, overgrown fields, edges of mangrove. **D** N Colombia, W Venezuela.

18 HOARY-THROATED SPINETAIL *Synallaxis kollari* 15.5cm Separable from White-whiskered Spinetail by rufous cheeks and all-rufous tail. **V** Series of high, sharp *kee-ki* notes at slightly irregular intervals (<1 sec). **H** Riverine belts. **D** N Brazil.

19 OCHRE-CHEEKED SPINETAIL *Synallaxis scutata* 14.5cm Head is smartly patterned and coloured. **V** Series of four well-separated, piercing notes, the second higher. **H** Undergrowth and borders of forest and woodland. **D** E, C and SW Brazil, E Bolivia, NW Argentina.

20 RUFOUS SPINETAIL *Synallaxis unirufa* 17cm Separable from Black-throated Spinetail by range. **V** Series of endlessly repeated, high, nasal *kuh-weetweet* phrases. **H** Undergrowth and borders of montane forest; 1,700–3,700m. **D** NW Venezuela to C Peru.

21 BLACK-THROATED SPINETAIL *Synallaxis castanea* 17cm Note distinct black throat. **V** Series of endlessly repeated, sharp, nasal *ker-kik-kik* phrases. **H** Undergrowth and margins of montane forest; 1,300–2,200m. **D** N Venezuela.

22 RUSTY-HEADED SPINETAIL *Synallaxis fuscorufa* 17cm No similar spinetail occurs in its restricted range and habitat. **V** Series of endlessly repeated, sharp, nasal *dit-dit-dut* phrases, the last note lower. **H** Undergrowth and margins of montane forest; 2,000–3,000m, occasionally lower. **D** N Colombia.

23 RUFOUS-CAPPED SPINETAIL *Synallaxis ruficapilla* 16cm No similar spinetail occurs in its range, but see Bahia and Spix's spinetails. Note yellowish eyebrow. **V** Series of very short, nasal phrases composed from a small rattle and a squeak. **H** Undergrowth and borders of forest and secondary growth; <1,400m. **D** SE Brazil to E Paraguay and NE Argentina.

24 BAHIA SPINETAIL *Synallaxis cinerea* 16cm Separable from Rufous-capped Spinetail by range and darkish grey breast. **V** Hurried series of nasal *t'weet* notes. **H** Undergrowth and borders of montane forest and woodland; 500–1,000m. **D** E Brazil.

25 PINTO'S SPINETAIL *Synallaxis infuscata* 17cm Note dark grey underparts and white moustachial streak. **V** Series of high, nasal *weet* and *weetweet* notes. **H** Dense undergrowth and borders of forest and woodland. **D** E Brazil.

26 STRIPE-BREASTED SPINETAIL *Synallaxis cinnamomea* 14.5cm Note streaking below, the intensity of which differs by region. **V** Endlessly repeated, very high, nasal *chik-weeek*, the second note lower. **H** Undergrowth and tree-falls in forest, secondary growth, woodland, overgrown plantations. **D** N Venezuela, Colombia.

27 GREY-BELLIED SPINETAIL *Synallaxis cinerascens* 14cm Crown concolourous with mantle and underparts uniform grey. **V** Calm, very high series of thin, piercing *seeet seetseet* notes. **H** Dense undergrowth of forest and secondary growth. **D** SE Brazil to E Paraguay, NE Argentina and C Uruguay.

28 SILVERY-THROATED SPINETAIL *Synallaxis subpudica* 18cm Note olive-brown tail. **V** Very high, rapidly descending series of *weet* notes. **H** Undergrowth and edges of montane forest and secondary growth; 2,000–3,200m. **D** Colombia.

29 SOOTY-FRONTED SPINETAIL *Synallaxis frontalis* 15cm See Pale-breasted and Cinerous-breasted spinetails (both of which have a brown, not rufous, tail), and Azara's Spinetail. **V** Series of well-spaced, very high, sharp *ta-tawee* phrases. **H** Undergrowth and edges of forest, riverine belts, woodland, agricultural areas. **D** Uruguay.

30 AZARA'S SPINETAIL *Synallaxis azarae* 17cm Note indistinct buffish or whitish brow. Underparts paler in northern birds; eyebrow buffer in southern birds. **V** Series of well-spaced, mid-high, rising, disyllabic phrases. **H** Dense undergrowth and bamboo thickets in montane forest, woodland, clearings, hedgerows. **D** W Venezuela to NW Argentina.

31 APURIMAC SPINETAIL *Synallaxis courseni* 18cm Note brown tail and finely barred throat. **V** Series of well-spaced, mid-high, rising, nasal, disyllabic phrases. **H** Dense undergrowth with bamboo thickets and borders of forest and woodland; 2,450–3,500m. **D** C Peru.

32 PALE-BREASTED SPINETAIL *Synallaxis albescens* 15cm Separable from Dark-breasted and Cinerous-breasted spinetails by whiter underparts. Forages alone or in pairs in dense undergrowth. **V** Series of slightly hurried, rising, nasal, disyllabic phrases. **H** Cerrado, campo and extensive grassland with scattered bush and scrub. **D** Costa Rica to C Argentina.

33 RIO ORINOCO SPINETAIL *Synallaxis beverlyae* 15cm Separable from Pale-breasted Spinetail by its tiny range. **V** Series of 6–9 well-separated notes. **H** Scrubby vegetation on river borders and islands. **D** Venezuela, E Colombia.

34 DARK-BREASTED SPINETAIL *Synallaxis albigularis* 15.5cm Occurs only east of the Andes. Separable from Cinerous-breasted Spinetail (restricted overlap in range) by darker breast, and from Dusky and Cabanis's spinetails (Plate 128) by paler plumage. **V** Series of 3–4 fast, nasal notes that bounce down. **H** Scrub, tall grass, shrub, relict plantations, river islands. **D** W Amazonia.

35 CINEREOUS-BREASTED SPINETAIL *Synallaxis hypospodia* 15.5cm See Dark-breasted Spinetail, which it resembles; range differs from that of Spix's Spinetail. **V** Short, shivering series of high, sharp, slightly descending notes. **H** Shrub, woodland borders, often near water. **D** S Amazonia, E Brazil.

36 SPIX'S SPINETAIL *Synallaxis spixi* 16cm Separable from Cinerous-breasted Spinetail by more pointed tail, darker breast and more extensive black on throat. **V** Repeated series of high, sharp *weeet witwitwit* phrases. **H** Scrub, thickets, Cerrado, woodland borders. **D** SE Brazil to E Paraguay, NE Argentina and Uruguay.

37 RUDDY SPINETAIL *Synallaxis rutilans* 14cm Separable from Chestnut-throated Spinetail by black throat and voice. **V** Series of endlessly repeated, nasal *kee-kawow* phrases. **H** Dense undergrowth of terra firme forest. **D** Amazonia.

38 CHESTNUT-THROATED SPINETAIL *Synallaxis cherriei* 14cm See Ruddy Spinetail. **V** Series of endlessly repeated, high, nasal *trrr-tuuit* phrases. **H** Undergrowth and borders of lowland forest and secondary growth. **D** W and SC Amazonia.

39 RUFOUS-BREASTED SPINETAIL *Synallaxis erythrothorax* 16cm Note throat pattern and chequered throat. Forages alone or in pairs in dense undergrowth. **V** Very high, rising slightly shrieking *wrutuwit-wit-truw*. **H** Marsh, dense scrub. **D** SE Mexico to NW Honduras.

40 SLATY SPINETAIL *Synallaxis brachyura* 15.5cm The only darkish spinetail in its range with a rufous crown and wings. Forages alone or in pairs in dense undergrowth. **V** Short, descending chatter. **H** Scrub, dense undergrowth at borders of montane forest, gardens. **D** NC Honduras to N Peru.

1
2
3
4
5
6
7
8
9
10
11
12
13
14
15
16
17
18
19
20
21
22
23
24
25
26
27
28
29
30
31
32
33
34
35
36
37
38
39
40

OVENBIRDS *CONTINUED*

1 BLACKISH-HEADED SPINETAIL *Synallaxis tithys* 15.5cm Separable from Slaty Spinetail (Plate 127) by black or dark grey head pattern. **V** Short, sharp, rising rattle. **H** Undergrowth of forest, woodland, secondary growth. **D** SC Ecuador, NW Peru.

2 WHITE-BELLIED SPINETAIL *Synallaxis propinqua* 16cm See Plain-crowned Spinetail, which lacks black throat patch and is more buff on flanks. **V** Curious short, voiceless, hoarse rattle. **H** Early succession scrub, mainly on river islands. **D** W Amazon and tributaries.

3 MCCONNELL'S SPINETAIL *Synallaxis macconnelli* 16cm Rather dark grey, with extensive rufous on crown. **V** Short, dry rattle, often ending on a lower, soft note. **H** Undergrowth and edges of humid forest and woodland in hilly country and on tepuis. **D** NE and NC Amazonia.

4 DUSKY SPINETAIL *Synallaxis moesta* 16cm Note dusky forehead. **V** Series of low, croaking rattles (each 2–3 secs). **H** Dense undergrowth and edges of forest. **D** C Colombia to NE Peru.

5 CABANIS'S SPINETAIL *Synallaxis cabanisi* 16cm Forehead concolourous with rufous crown. **V** Utters dry, low *yap-yap* notes in an irregular series. **H** Dense undergrowth and edges of forest and secondary growth. **D** C Peru to N Bolivia, C Brazil.

6 PLAIN-CROWNED SPINETAIL *Synallaxis gujanensis* 16cm Lacks black on throat and rufous on crown. **V** Series of two well-separated notes, the second a higher double note. **H** Dense undergrowth of várzea and river-edge forest. **D** Amazonia.

7 MARANON SPINETAIL *Synallaxis maranonica* 15cm Note lack of black on throat and lack of rufous on head. No similar spinetail occurs in its restricted range. **V** Series of two well-separated, very high, nasal, squeezed-out notes, the second slightly higher. **H** Undergrowth and edges of forest, riverine belts, secondary growth, woodland. **D** S Ecuador, N Peru.

8 WHITE-LORED SPINETAIL *Synallaxis albilora* 16cm Grey head contrasts with brownish-ochraceous back and underparts. **V** Song is a *tjuuh-tuwiet* (the second part an upslurred double note) or a series of evenly spaced *keeip* notes. **H** Riverine belts. **D** C South America.

9 RUSSET-BELLIED SPINETAIL *Synallaxis zimmeri* 16.5cm Unmistakable due to grey throat and narrow white eye-ring. **V** Very short, dry trill followed by four higher, staccato *quik* notes. **H** Montane scrub; 2,100–2,800m, locally higher or lower. **D** W Peru.

10 NECKLACED SPINETAIL *Synallaxis stictothorax* 12cm Unmistakable due to white throat and finely streaked breast. **V** Fast, hurried, very high, thin twittering mixed with small rattles and trills (2–3 secs). **H** Arid scrub and woodland. **D** SW Ecuador, NW Peru.

11 GREAT SPINETAIL *Siptornopsis hypochondriaca* 18.5cm Separable from Necklaced Spinetail by larger size and longer tail. **V** Very fast, continuous series of dry, very high trills and rattles, switching in speed and pitch (2–8 secs). **H** Arid montane scrub; 1,700–2,800m. **D** N Peru.

12 RED-SHOULDERED SPINETAIL *Gyalophylax hellmayri* 18cm Note darkish milk-chocolate-brown plumage. **V** Mid-high accelerating, rising and falling trill that starts as a chatter and gradually fades away. **H** Low arid woodland. **D** E Brazil.

13 WHITE-BROWED SPINETAIL *Hellmayrea gularis* 13cm Note white face markings. **V** Ultra-high, thin, slightly descending rattle, sometimes with a small introductory roll or a few staccato notes (2 secs). **H** Dense undergrowth of montane forest at the treeline; 2,300–3,700m. **D** NW Venezuela to C Peru.

14 MARCAPATA SPINETAIL *Cranioleuca marcapatae* 16cm Note black line along rufous cap. **V** Series of rapid, very soft twitters, mixed with very high, loud, piercing mini-rattles. **H** Humid montane forest at the treeline; 2,400–3,500m. **D** SE Peru.

15 LIGHT-CROWNED SPINETAIL *Cranioleuca albiceps* 15cm Separable from Marcapata Spinetail by range and duskier sides of face. **V** Short, descending, jubilant series of notes. **H** Montane forest with bamboo; 2,300–3,300m. **D** S Peru to C Bolivia.

16 RUSTY-BACKED SPINETAIL *Cranioleuca vulpina* 14.5cm Uniform rufous above, including neck. Separable from Parker's Spinetail by less white throat. **V** High, nasal chattering, starting after some hesitating notes, often given by a pair. **H** Shrubby undergrowth and borders of forest at water. **D** Amazonia, C and E Brazil.

17 COIBA SPINETAIL *Cranioleuca dissita* 15cm No similar bird in its range. **V** Similar to that of Red-faced Spinetail. **H** Perches on lianas and small branches in forest and at forest edges. **D** Panama.

18 PARKER'S SPINETAIL *Cranioleuca vulpecula* 14cm See Rusty-backed Spinetail. **V** Very short, nasal series, starting with separate descending notes and changing to a small trill. **H** Undergrowth and borders of low woodland near water. **D** W Amazon and tributaries.

19 SULPHUR-BEARDED SPINETAIL *Cranioleuca sulphurifera* 15cm Distinguished by small yellow throat patch, buffy stripe in flight and habitat. **V** Series of short, nasal rattles, slightly rising at start and lowered at end to a trill (2–3 secs). **H** Marshes and surrounding scrub. **D** S Brazil, Uruguay, NE Argentina.

20 CRESTED SPINETAIL *Cranioleuca subcristata* 14cm No similar spinetail occurs in its range. Rarely displays crest. **V** Very high, rising and falling, thin trill with some staccato notes halfway (2–3 secs). **H** Forested habitats; <2,000m. **D** N Venezuela, Colombia.

21 STRIPE-CROWNED SPINETAIL *Cranioleuca pyrrhophia* 14.5cm Note wide white eyebrow and streaked crown. **V** Short, hurried series of very high, thin, staccato notes, ending as a trill (2–3 secs). **H** Shrubby parts of forest and woodland; <3,100m. **D** SC and SE South America.

22 BOLIVIAN SPINETAIL *Cranioleuca henricae* 14.5cm Note small rufous crown patch and broad white eyebrow, and also restricted range. **V** Very high, thin trill, slowing down at the end (5–10 secs). **H** Arid woodland. 2,500–2,900m. **D** C Bolivia.

23 OLIVE SPINETAIL *Cranioleuca obsoleta* 14cm Rufous restricted to wing coverts and tail. Range differs from that of Stripe-crowned Spinetail. **V** Series of very high trills and rattles, often lowered at end. **H** Humid forest, woodland, *Araucaria* stands; <1,400m. **D** Argentina.

24 PALLID SPINETAIL *Cranioleuca pallida* 14cm Striking rufous crown patch, tail and wing coverts. Note broad eyebrow. Range differs from that of Olive Spinetail. **V** Series of very high, thin notes and trills, mixed with short, high rattles and twittering. **H** Forest, woodland, tall secondary growth. **D** SE Brazil.

25 GREY-HEADED SPINETAIL *Cranioleuca semicinerea* 14cm Note pale grey head. No similar spinetail occurs in its range. **V** Song is a short or longer staccato series that changes to a downslurred trill (2 secs). **H** Forest and woodland. **D** E Brazil.

26 CREAMY-CRESTED SPINETAIL *Cranioleuca albicapilla* 17cm Note creamy-white crest and pale bill. **V** Short series of notes, starting as a rising chatter (2–3 notes) and then bouncing down. **H** Montane woodland and scrub, *Polylepis* woodland; 2,400–3,600m. **D** C and S Peru.

27 RED-FACED SPINETAIL *Cranioleuca erythrops* 14cm No similar spinetail with a rufous crown and face sides occurs in its range. **V** Fast series of high, thin notes, descending to a trill (1–3 secs). **H** Humid forest. **D** Costa Rica to W Ecuador.

28 TEPUI SPINETAIL *Cranioleuca demissa* 14.5cm Rather dark below; no other similar spinetail occurs in its range. **V** Fast series of high, thin notes, descending to a low trill (1–3 secs). **H** Montane and tepui forest, riverine belts. **D** SE Venezuela, WC Guyana, N Brazil.

29 STREAK-CAPPED SPINETAIL *Cranioleuca hellmayri* 14cm Note striking white eyes, and that black streaks in crown are often barely or not visible. No similar spinetail occurs in its range. **V** Fast, short series of high, thin notes, descending to a trill (1–2 secs). **H** Montane forest and woodland; 1,500–3,000m. **D** N Colombia.

30 ASH-BROWED SPINETAIL *Cranioleuca curtata* 14.5cm Note range and indistinct eyebrow. **V** Fast series of high, thin notes, descending to a low trill (1–3 secs). **H** Forest and woodland; 900–1,700m, locally higher or lower. **D** C Colombia to C Bolivia.

31 LINE-CHEEKED SPINETAIL *Cranioleuca antisiensis* 14.5cm Note narrow white eyebrow and restricted range. **V** Song comprises two high, nasal tones followed by a downslurred trill. **H** Montane forest, woodland, agricultural areas; 900–2,500m. **D** SW Ecuador, Peru.

32 SPECKLED SPINETAIL *Cranioleuca gutturata* 14cm Note pale eyes and streak-speckled underparts. See Scaled Spinetail, which is larger. **V** Series of 3–5 very high, thin, staccato notes followed by a few dry *truh* notes. **H** Lower storeys and borders of humid forest. **D** Amazonia.

33 SCALED SPINETAIL *Cranioleuca muelleri* 15.5cm Scaled appearance is diagnostic. **V** Song comprises two very high, thin notes followed by a short, loud chatter. **H** Várzea forest. **D** E Amazon.

34 YELLOW-CHINNED SPINETAIL *Certhiaxis cinnamomeus* 14cm Note yellow chin and overall red-and-white appearance. **V** Slightly descending, chattering trill, switching in intonation. **H** Marshes and adjacent areas, mangrove. **D** Colombia to Uruguay.

35 RED-AND-WHITE SPINETAIL *Certhiaxis mustelinus* 14cm Note long bill, white chin and pure white underparts. **V** Very high, barely audible, thin trill, constantly interrupted by 1–3 loud, hooted notes. **H** Marshy grass and shrub areas. **D** Amazon and tributaries.

36 ORINOCO SOFTTAIL *Thripophaga cherriei* 15cm Very restricted range. Note absence of rufous in crown. **V** Song a series of 30–40 staccato notes, accelerating to a crescendo, then falling slightly. Alarm call a *chrrr*. **H** Undergrowth of várzea. **D** SC Venezuela.

37 DELTA AMACURO SOFTTAIL *Thripophaga amacurensis* 17cm Separable from Striated Softtail by range, and from Grey-bellied Spinetail (Plate 127) by spotted crown. **V** Song is a mixture of rapid *tjuw* notes and short rattles, given in duet. **H** Seasonally flooded forest near rivers and streams. **D** N Venezuela.

38 STRIATED SOFTTAIL *Thripophaga macroura* 18cm Streaking and orange chin patch are diagnostic. See Pale-browed Treehunter (Plate 130), which shows a whitish chin. **V** Mixed series of strident rattles and shrieked notes. **H** Dense lower storeys and borders of humid forest. **D** SE Brazil.

39 PLAIN SOFTTAIL *Thripophaga fusciceps* 17cm Aptly named, with no similar species occurring in its range. **V** Harsh rattles in chorus. **H** In parts of humid várzea with dense vines. **D** W and C Amazonia.

40 RUSSET-MANTLED SOFTTAIL *Thripophaga berlepschi* 18cm Note pale ashy head, contrasting with rufous-orange body plumage. **V** Unstructured, very sharp, high chattering in chorus. **H** Lower growth and bamboo in forest at the treeline; 2,500–3,400m. **D** N Peru.

1
2
3
4
5
6
7
8
9
10
11
12
13
14
15
16
17
18
19
20
21
22
23
24
25
26
27
28
29
30
31
32
33
34
35
36
37
38
39
40

OVENBIRDS *CONTINUED*

1 RUFOUS-FRONTED THORNBIRD *Phacellodomus rufifrons* 16.5cm Separable from Little and Streak-fronted thornbirds by lack of rufous in tail. **V** Excited series of loud *chit* notes in chorus. **H** Fairly open scrub with some trees, often in agricultural areas. **D** Widespread in S Ecuador, N Peru, E Brazil.

2 PLAIN THORNBIRD *Phacellodomus inornatus* 16–17cm Previously treated as subspecies of Rufous-fronted Thornbird. Uniform upperpart colouration aids identification. **V** Squeaky, trilling *tchu-Tche-Tche-chtu-tu-tu-tu*, accelerating and falling, speeding up and then slowing during delivery. **H** Secondary woodland, open arid scrub, Llanos, wooded savannah. **D** N Venezuela.

3 LITTLE THORNBIRD *Phacellodomus sibilatrix* 14cm Note rufous in tail and on shoulders. **V** High, gradually descending series of sharp *cheep* notes (5–10 secs). **H** Forest, woodland, scrub. **D** S Bolivia and N Paraguay to CW Uruguay and C and E Argentina.

4 STREAK-FRONTED THORNBIRD *Phacellodomus striaticeps* 17cm Lacks a chin patch. Distinct rufous colouring on wings and tail sides. **V** Series of *cheet* and *tsidit* notes, often given as a duet. **H** Arid montane scrub with high cacti, *Polylepis* woodland, adjacent agricultural areas; 2,800–5,000m. **D** S Peru to N Argentina.

5 FRECKLE-BREASTED THORNBIRD *Phacellodomus striaticollis* 18cm Breast freckles are arrow-shaped. Less rufous on wings and tail than Greater Thornbird. **V** Unstructured, calm series of separate high, sharp notes. **H** Woodland, marsh thickets, almost always near water. **D** NE Argentina, Uruguay, S Brazil.

6 SPOT-BREASTED THORNBIRD *Phacellodomus maculipectus* 18cm Separable from Freckle-breasted Thornbird by range and shape of breast spots. **V** Unstructured series of high, dry or rasping and chopped notes of any length. **H** Scrub, woodland, forest, almost always near water; 1,000–3,100m. **D** C Bolivia to NW Argentina.

7 CHESTNUT-BACKED THORNBIRD *Phacellodomus dorsalis* 19.5cm Note bluish-grey iris. Upperparts rufous except contrasting head. **V** Calm series of sharp, descending, staccato notes (5 secs). **H** Dense scrub, low woodland; 2,000–2,700m. **D** N Peru.

8 GREATER THORNBIRD *Phacellodomus ruber* 20cm Note all-rufous wings and tail, and faint eyebrow. **V** Short, descending, slightly accelerating series of 5–6 sharp notes (sometimes much longer and accelerating). **H** Riverine belts and scrub. **D** W Bolivia to C Brazil, Paraguay and N Argentina.

9 ORANGE-EYED THORNBIRD *Phacellodomus erythrophthalmus* 17cm Rufous coloration is restricted to front and throat; note all-rufous tail. **V** Slow, gradually descending series of sharp, staccato notes (10–20 countable notes). **H** Undergrowth and borders of montane and lowland forest, secondary growth. **D** E Brazil.

10 ORANGE-BREASTED THORNBIRD *Phacellodomus ferrugineigula* 17cm Rufous extends to crown and upper breast. Central tail feathers brownish olive. Less rufous on wings than in Orange-eyed Thornbird. **V** Song as Orange-eyed Thornbird. **H** Undergrowth and borders of montane and lowland forest, secondary growth. **D** SE Brazil, N Uruguay, NE Argentina.

11 CANEBRAKE GROUNDCREEPER *Clibanornis dendrocolaptoides* 21cm Note striking eyebrow. Rarely seen on the ground despite name. **V** Series of loud, nasal *chet* notes. **H** Undergrowth, often with bamboo in forest. **D** E Paraguay, NE Argentina, S Brazil.

12 BAY-CAPPED WREN-SPINETAIL *Spartonoica maluroides* 13cm Secretive. Note rufous forecrown, dark markings on upperparts and long tail. **V** Very high, dry trill (2 secs). **H** Sedge, reeds and tall, wet grass; <4,300m. **D** S Brazil and Uruguay to C Argentina.

13 WREN-LIKE RUSHBIRD *Phleocryptes melanops* 14cm No similar bird occurs in its range. Note rather short tail with rounded feathers. **V** Rapid ticking (2–3 secs). **H** Reeds and rushes; <4,300m. **D** Coastal W Peru to C Chile and C Argentina.

14 CURVE-BILLED REEDHAUNTER *Limnornis curvirostris* 16cm Restricted to reedbeds. Note pale downcurved bill. **V** Fast series of notes, dominated by short, upslurred rattles. **H** Freshwater reedbeds. **D** NE Argentina, S Uruguay, S Brazil.

15 STRAIGHT-BILLED REEDHAUNTER *Limnoctites rectirostris* 16cm Separable from Curve-billed Reedhaunter by longer, straight bill and greyer crown. **V** Rolling rattle with some staccato notes halfway. **H** Reeds, shrubs and low trees in and around wet areas. **D** NE Argentina, S Uruguay, S Brazil.

16 FIREWOOD-GATHERER *Anumbius annumbi* 19cm Note long, white-margined tail and black-spotted malar stripes. Builds enormous stick nests. **V** Repeated short phrase of two staccato notes, followed by a short, descending trill. **H** Agricultural areas and Cerrado, often around farms and buildings. **D** Paraguay and C Brazil to C Argentina.

17 LARK-LIKE BRUSHRUNNER *Coryphistera alaudina* 16cm Unmistakable due to crest and face markings. Forages in groups on the ground. **V** Fast, thin, ultra-high trill with some small variations in pitch. **H** Arid scrub and woodland. **D** SC South America.

18 SPECTACLED PRICKLETAIL *Siptornis striaticollis* 12cm Note double-pointed tail, white eyebrow and incomplete eye-ring. Separable from xenopses (Plate 131) by its straight bill, lack of white malar stripe and lack of black wing band. **V** Short, very high, sharp, rapid series of notes that changes in intonation. **H** Canopy and borders of montane forest; 1,300–2,400m. **D** Colombia to N Peru.

19 ORANGE-FRONTED PLUSHCROWN *Metopothrix aurantiaca* 11.5cm Unmistakable due to its small size and general colour pattern. See Rust-and-yellow Tanager (Plate 292) and American Yellow Warbler (Plate 288). **V** Series of very thin *sie* notes just within earshot. **H** Forest and secondary growth near rivers. **D** W Amazonia.

20 DOUBLE-BANDED GREYTAIL *Xenerpestes minlosi* 11cm Note wing-bars. Distinctive in its range. **V** Evenly pitched, rapid, long, chattering trill (3–12 secs). **H** Canopy and borders of forest and tall secondary growth. **D** E Panama to N Colombia and NW Ecuador.

21 EQUATORIAL GREYTAIL *Xenerpestes singularis* 11.5cm Note streaking below and rufous forehead. **V** Barely audible ventriloquial trill (10–15 secs). **H** Canopy and borders of foothill and montane forest; 1,000–1,700m, locally higher. **D** Ecuador to N Peru.

22 RUSTY-WINGED BARBTAIL *Premnornis guttuliger* 14cm Tail rufous; wing feathers bordered by darkish rufous. Scalloped below. **V** Very high, soft but piercing, slow rattle, just audible (5–10 secs). **H** Montane forest; 1,600–2,300m, locally higher or lower. **D** NW Venezuela to C Peru.

23 SPOTTED BARBTAIL *Premnoplex brunnescens* 13.5cm No rufous on wings or tail. Paler-throated morph of nominate race also shown. **V** Very high trill (1 sec). **H** Montane forest; 900–2,500m, locally higher or lower. **D** Costa Rica to NW Venezuela and C Bolivia.

24 WHITE-THROATED BARBTAIL *Premnoplex tatei* 14cm Note rufous-brown upperparts. **V** Series of soft whistled notes. **H** Montane forest; 800–2,400m. **D** NE Venezuela.

25 RORAIMAN BARBTAIL *Roraimia adusta* 14.5cm Note white throat of this beautifully patterned species. **V** Rapidly ascending *tee-tee-tee-teuu-teuu-tuutuu*. **H** Undergrowth of montane and tepui forest; 1,000–2,500m. **D** SE Venezuela, W Guyana, extreme N Brazil.

26 PINK-LEGGED GRAVETEIRO *Acrobatornis fonsecai* 14cm Pink maxilla and legs are distinctive. Overall greyer than Double-banded and Equatorial greytails. **V** Long, sibilant rattles in duet. **H** Shading trees in cacao plantations, canopy and borders of forest and woodland. **D** E Brazil.

27 RUDDY TREERUNNER *Margarornis rubiginosus* 16cm Note whitish eyebrow and throat. **V** Very high, squeaking calls running into a descending trill, *kip kip krpkrpkrprprprprrr*. **H** Montane forest and adjacent areas. **D** Costa Rica, Panama.

28 STAR-CHESTED TREERUNNER *Margarornis stellatus* 15cm Overall rufous chestnut with a paler eyebrow and white throat. Note black-edged white feathers of necklace. **V** Rather silent, sometimes uttering a barely audible *tsit* or *tsisit*. **H** Higher levels in montane forest; 1,200–2,200m. **D** Colombia, Ecuador.

29 BEAUTIFUL TREERUNNER *Margarornis bellulus* 15cm Distinctively marked. **V** Short trill. **H** Mainly montane forest. **D** E Panama.

30 PEARLED TREERUNNER *Margarornis squamiger* 15cm Note straight, short bill, bright wings and tail, and rows of black-bordered white spots below. Resembles Beautiful Treerunner, but that shows a duller rufous (but browner) back and wings, and smaller spots below. **V** Short, soft, thin trill. **H** Montane forest, *Polylepis* woodland, secondary growth; 2,000–3,500m, locally higher or lower. **D** W Venezuela to Bolivia.

31 CAATINGA CACHOLOTE *Pseudoseisura cristata* 24cm Plumage is uniform orange-rufous. **V** Loud, resounding, slowly descending, chattering or chirping series of notes, sung in duet (1–6 secs). **H** Caatinga and semi-open areas, even where overgrazed, and around farm buildings. **D** E Brazil.

32 GREY-CRESTED CACHOLOTE *Pseudoseisura unirufa* 22cm Separable from Caatinga Cacholote by grey in crown and by range. **V** Rapid, slowly descending chatter, resumed a few times, sung in a duet. **H** Riverine belts, woodland stands, tree stands around buildings. **D** NC and E Bolivia, SW Brazil.

33 BROWN CACHOLOTE *Pseudoseisura lophotes* 26cm Head and tail are uniform rufous, more or less contrasting with irregular pale-scaled underparts and mantle. Crown is duskier. **V** Calm, slowly descending series of harsh, croaking notes, resumed a few times, sung in a duet. **H** Woodland, scrub. **D** SC and SE South America.

34 WHITE-THROATED CACHOLOTE *Pseudoseisura gutturalis* 23cm White throat bordered below with black is diagnostic. Crest not normally evident. **V** Rapid, raucous chatters in chorus. **H** Arid scrub; <2,900m. **D** Argentina.

35 BUFFY TUFTEDCHEEK *Pseudocolaptes lawrencii* 20.5cm Separable from Streaked Tuftedcheek by unstreaked back and occurrence at lower altitude. **V** Song includes a short, very high, thin trill, followed by a descending *sweee*. **H** Montane forest; 900–1,500m, locally higher or lower. **D** Costa Rica, Panama.

36 PACIFIC TUFTEDCHEEK *Pseudocolaptes johnsoni* 20–21cm Previously treated as subspecies of Buffy Tuftedcheek. Note more richly rufous back and wings. Forages for invertebrates in bromeliads. **V** Song a trilling, accelerating *tch-rrrrrrr*, followed by a wheezy *zeee*. Call a sharp, shrill *tchip, tchip*. **H** Montane forest, secondary woodland; mainly 1,000–1,500m. **D** W Andean slopes of Colombia and Ecuador.

37 STREAKED TUFTEDCHEEK *Pseudocolaptes boissonneautii* 21.5cm See Buffy Tuftedcheek. Note conspicuous tufts. **V** Song is varied and includes a very high, fast trill, slowing down to some thin, staccato notes. **H** Montane forest; 1,800–3,100m, locally higher or lower. **D** NW Venezuela to C Bolivia.

38 POINT-TAILED PALMCREEPER *Berlepschia rikeri* 20cm Unmistakable due to bold pattern and habitat. **V** Loud, sharp, rattled chatters (1–3 secs). **H** Palm stands. **D** Amazonia.

1
2
3
4
5
6
7
8
9
10
11
12
13
14
15
16
17
18
19
20
21
22
23
pale
normal
24
25
26
27
28
29
30
31
32
33
34
35
36
37
38

OVENBIRDS *CONTINUED*

1 SCALY-THROATED FOLIAGE-GLEANER *Anabacerthia variegaticeps* 16cm Facial pattern distinctive, with an ochraceous reclining '6' around eyes. Cheek pale, bordered blackish. Note range (W side of Andes). **V** Sluggish series of mid-high, sharp *tjik* notes (2 secs). **H** Forest and tall secondary growth; 700–2,000m. **D** SW Mexico to Ecuador.

2 MONTANE FOLIAGE-GLEANER *Anabacerthia striaticollis* 16.5cm Separable from Scaly-throated Foliage-gleaner by darker cheeks and range (E side of Andes). **V** Level, accelerating series of *zjik* notes (2 secs). **H** Montane and foothill forest; 900–2,100m. **D** NW Mexico to C Bolivia.

3 WHITE-BROWED FOLIAGE-GLEANER *Anabacerthia amaurotis* 15.5cm Note broad white eyebrows and unstreaked back. **V** Very thin, fast, slightly descending trill. **H** Understorey of humid forest; 600–1,500m. **D** SE Brazil, E Paraguay, NE Argentina.

4 GUTTULATE FOLIAGE-GLEANER *Syndactyla guttulata* 18.5cm Note upturned maxilla and extensive streaking on underparts. **V** High, level, croaking series of notes (2–3 secs). **H** Forest and secondary growth; 900–1,200m. **D** N Venezuela.

5 LINEATED FOLIAGE-GLEANER *Syndactyla subalaris* 19cm Note relatively shorter bill, brighter tail, and pronounced eyebrow and eye-ring. Forage alone, in pairs or in mixed bird flocks in the middle or lower forest storeys. **V** High nasal, almost trilling *wit-witwit-wtwt* (6 secs). **H** Mid–low forest storeys; >750m. **D** Costa Rica to NW Venezuela and C Peru.

6 BUFF-BROWED FOLIAGE-GLEANER *Syndactyla rufosuperciliata* 17.5cm Separable from Lineated Foliage-gleaner by lack of back streaking and from White-browed Foliage-gleaner by facial pattern (less contrast between eyebrow and cheek patch). **V** Rapid, nasal series of notes. **H** Undergrowth of forest and woodland; 1,000–2,500m. **D** S Ecuador to NW Argentina.

7 RUFOUS-NECKED FOLIAGE-GLEANER *Syndactyla ruficollis* 18cm Note rufescent neck and eyebrows. No similar bird occurs in its range. **V** Short series of notes, introduced by two stuttered, dry nasal notes and accelerating to a bouncing-down trill. **H** Montane forest with bamboo, secondary growth; 600–2,600m, locally higher. **D** SW Ecuador, NW Peru.

8 PLANALTO FOLIAGE-GLEANER *Syndactyla dimidiata* 17cm Note rich rufous plumage with browner upperparts. **V** Series of 9–10 loud, harsh notes, accelerating to a dry, downslurred trill. **H** Lowland forest, riverine belts; <1,200m. **D** NE Paraguay, SW and SE Brazil.

9 TEPUI FOLIAGE-GLEANER *Syndactyla roraimae* 18cm Facial pattern and colouring distinctive. White throat contrasts with underparts. **V** Short, very dry rattle. **H** Montane forest, stunted woodland, tepuis; 1,100–2,500m. **D** SE and SC Venezuela, W Guyana, N Brazil.

10 PERUVIAN RECURVEBILL *Simoxenops ucayalae* 19cm Brightly coloured, with a massive upturned bill. **V** Short, rising, dry rattle. **H** Undergrowth of humid forest; <1,300m. **D** S and SW Amazonia.

11 BOLIVIAN RECURVEBILL *Simoxenops striatus* 19cm Separable from Peruvian Recurvebill by smaller bill, and by streaked face sides, mantle and breast. **V** Short, ascending, very dry rattle. **H** Undergrowth of foothill and montane forest; 650–1,400m. **D** W and C Bolivia.

12 CHESTNUT-WINGED HOOKBILL *Ancistrops strigilatus* 18.5cm Contrasting plumage streaks are distinctive; bill is strong but not especially hooked. **V** Loud, level or gradually ascending trill that can last >30 secs. **H** Lowland forest; <900m. **D** W and C Amazonia.

13 EASTERN WOODHAUNTER *Hyloctistes subulatus* 17cm Only E of Andes. Note thin, long, pale bill. **V** Song comprises 2–3 loud, nasal *teeu* notes, followed by a short, dry trill. **H** Humid forest and woodland; <1,300m, locally higher. **D** Amazonia.

14 WESTERN WOODHAUNTER *Hyloctistes virgatus* 17–18cm Previously treated as subspecies of Eastern Woodhaunter. Forages for invertebrates in leaf litter, bromeliads and tree moss. **V** Song a piping *tchu-tchu-tchee-tchee-tchee*. Call a screeching *tchrrrp*. **H** Terra firme and várzea rainforest, dense secondary woodland. **D** Nicaragua, Costa Rica, Panama, W Colombia, W Ecuador.

15 RUFOUS-TAILED FOLIAGE-GLEANER *Philydor ruficaudatum* 17cm Separable from Rufous-rumped Foliage-gleaner by less distinct eye-ring, and more extensively flammulated underparts. Rump olive, not rufous. **H** Terra firme forest and locally in várzea; <850m, locally higher. **D** Amazonia.

16 SLATY-WINGED FOLIAGE-GLEANER *Philydor fuscipenne* 17cm Distinctively patterned and coloured. **V** Sharp, accelerating chatter. **H** Lower storeys of forest and woodland; <1,200m. **D** Panama to W Ecuador.

17 RUFOUS-RUMPED FOLIAGE-GLEANER *Philydor erythrocercum* 16cm Compare with Rufous-tailed Foliage-gleaner. Note striking eye-ring and eyebrow (forming a reclining '6'). **V** Rapid series of four level, high *seet* notes amid *shreet* calls. **H** Lower levels of humid forest; <1,300m, locally higher. **D** Amazonia.

18 CHESTNUT-WINGED FOLIAGE-GLEANER *Philydor erythropterum* 18.5cm Note contrasting rufous wings and tail, ochre-orange lores and chin, and lack of streaking. **H** Higher levels of terra firme forest, also in várzea in the north of its range; <900m. **D** Amazonia.

19 OCHRE-BREASTED FOLIAGE-GLEANER *Philydor lichtensteini* 18cm Separable from Buff-fronted Foliage-gleaner by greyish-brown forehead (concolourous with crown). **V** Short series of rising and falling nasal *wut* notes. **H** Higher levels of forest and woodland; <900m. **D** SE Brazil to E Paraguay and NE Argentina.

20 ALAGOAS FOLIAGE-GLEANER *Philydor novaesi* 18cm Compare with Black-capped Foliage-gleaner, which has blacker face markings. **V** Low-pitched rattle with changes in intonation. **H** Undergrowth of humid forest and woodland. **D** E Brazil.

21 BLACK-CAPPED FOLIAGE-GLEANER *Philydor atricapillus* 17cm Note distinctively patterned head. **V** Rapid rattle, slightly descending in pitch and changing in intonation (3–4 secs). **H** Undergrowth of lowland forest and secondary growth; <1,050m. **D** SE Brazil to E Paraguay and NE Argentina.

22 BUFF-FRONTED FOLIAGE-GLEANER *Philydor rufum* 18.5cm Compare with Ochre-breasted Foliage-gleaner and Brown Tanager (Plate 291). **V** As Black-capped Foliage-gleaner but less structured. **H** Higher levels of humid forest and woodland; <1,800m. **D** Costa Rica to NE Argentina.

23 CINNAMON-RUMPED FOLIAGE-GLEANER *Philydor pyrrhodes* 16.5cm No similar bird occurs in its range. Note strong, pale bill. **V** Song is a fast rattle, sounding at the start like a moped and then rising high, before ending abruptly or descending and fading out. **H** Terra firme and várzea forest; <700m. **D** Amazonia.

24 BAMBOO FOLIAGE-GLEANER *Anabazenops dorsalis* 18.5cm Note heavy bill and dusky cheeks. **V** Long, rising, dry rattle. **H** Bamboo undergrowth of forest and woodland; 200–1,300m. **D** W and S Amazonia.

25 WHITE-COLLARED FOLIAGE-GLEANER *Anabazenops fuscus* 19.5cm Note head shape and patterning; heavy bill is upturned. **V** Rapid series of 10–15 mid-high musical *jek* notes. **H** Montane forest and secondary growth; 350–1,200m. **D** SE Brazil.

26 PALE-BROWED TREEHUNTER *Cichlocolaptes leucophrus* 22.5cm No similar treehunter occurs in its range. More distinctly streaked below than on back. **V** Series of 5–7 loud, shrieking *wreeep* notes, starting suddenly. **H** Humid lowland and montane forest; <1,400m. **D** SE Brazil.

27 CRYPTIC TREEHUNTER *Cichlocolaptes mazarbarnetti* 21–22cm Enigmatic. Critically endangered, possibly extinct. **V** Song a rapid, mechanical-sounding trill, following by a squawky *whrrr, whrrip*. Call a shrill *te-pereep*. **H** Densely wooded slopes with abundant bromeliads and epiphytes. **D** NE Brazil (known from only two sites).

28 UNIFORM TREEHUNTER *Thripadectes ignobilis* 19cm Note fairly uniform plumage, eye-ring and eye-stripe. **V** High, sharp series of 6–10 rising and descending notes. **H** Undergrowth of foothill and lower montane forest; 700–1,700m, locally higher or lower. **D** Colombia, NW Ecuador.

29 STREAK-BREASTED TREEHUNTER *Thripadectes rufobrunneus* 20cm Note large size, heavy bill and rufous undertail coverts. **V** Hurried *tudrut* or *tiederruut*. **H** Montane forest, tall secondary growth. **D** Costa Rica, W Panama.

30 BLACK-BILLED TREEHUNTER *Thripadectes melanorhynchus* 21cm Separable from other treehunters by little or lack of streaking from breast down. **V** Slightly undulating series of 3–10 high, loud, piercing notes. **H** Undergrowth of foothill and lower montane forest; 900–1,700m. **D** Colombia to C Peru.

31 STRIPED TREEHUNTER *Thripadectes holostictus* 21cm Separable from Flammulated Treehunter by less contrasting streaking, especially on underparts. **V** Short, descending, shivering chatter. **H** Undergrowth with bamboo in montane forest; 1,500–2,500m. **D** C Colombia to C Bolivia.

32 STREAK-CAPPED TREEHUNTER *Thripadectes virgaticeps* 22cm Separable from Uniform Treehunter by cinnamon underparts and less distinct eye-ring. **V** Series of 6–7 rather high, piercing *chup* notes. **H** Undergrowth of montane forest; 1,300–2,100m, locally higher or lower. **D** N Venezuela to N Ecuador.

33 FLAMMULATED TREEHUNTER *Thripadectes flammulatus* 24cm Very distinctly striped. **V** Rising, rapid, sharp rattle, the last note lower (2 secs). **H** Undergrowth with bamboo in montane forest; 2,200–3,500m, locally lower. **D** W Venezuela to N Peru.

34 PERUVIAN TREEHUNTER *Thripadectes scrutator* 24cm Plumage of back and underparts rather tawny brown, not as greyish as in Black-billed, Striped or Flammulated treehunters. **V** Level, dry, shivering rattle, the last note lower (2–3 secs). **H** Undergrowth of montane forest; 2,450–3,300m. **D** N Peru to NC Bolivia.

35 BUFF-THROATED FOLIAGE-GLEANER *Automolus ochrolaemus* 18.5cm Note dark eyes and distinct eye-ring; shows a buffer throat and eye-ring than Olive-backed Foliage-gleaner, whose range overlaps. **V** Rapid series of 5–6 descending, nasal, almost mewing notes. **H** Terra firme and montane forest, várzea, secondary growth; <1,000m, locally higher. **D** S Mexico through Amazonia.

36 OLIVE-BACKED FOLIAGE-GLEANER *Automolus infuscatus* 19cm Note dark eyes and indistinct eyebrows. **V** Short, descending, shivering series of notes. **H** Undergrowth of humid forest; <700m. **D** W and N Amazonia.

37 PARA FOLIAGE-GLEANER *Automolus paraensis* 19cm Separable from Olive-backed Foliage-gleaner, with which it was previously considered conspecific, mainly by voice. **V** Rapid, very short, hoarse series of notes. **H** Undergrowth of humid forest; <700m. **D** SC Amazonian Brazil.

38 WHITE-EYED FOLIAGE-GLEANER *Automolus leucophthalmus* 19.5cm Note very distinctive white throat (compare with Pernambuco Foliage-gleaner). **V** Song resembles that of Pernambuco Foliage-gleaner. **H** Undergrowth of humid forest and woodland; <1,000m, locally higher. **D** E Brazil to NE Argentina.

39 PERNAMBUCO FOLIAGE-GLEANER *Automolus lammi* 19.5cm Separable from White-eyed Foliage-gleaner mainly by range. **V** Short series of 4–5 hoarse notes, the first higher or lower. **H** Undergrowth of humid forest and woodland; <1,000m. **D** NE Brazil.

1
2
3
4
5
6
7
8
9
10
11
12
13
14
15
16
17
18
19
20
21
22
23
24
25
26
27
28
29
30
31
32
33
34
35
36
37
38
39

OVENBIRDS *CONTINUED*

1 BROWN-RUMPED FOLIAGE-GLEANER *Automolus melanopezus* 18cm Note orange throat sides, red eyes and absence of breast flammulation. **V** Liquid series of two high notes, followed by a short, rolling twitter. **H** Undergrowth of humid forest, normally in bamboo thickets. **D** W and SW Amazonia.

2 RUDDY FOLIAGE-GLEANER *Automolus rubiginosus* 19cm Dark plumage, with some races even darker than the nominate shown. Note rufous throat sides. E Colombian ssp. *cinnamomeigula* has white eyes. **V** Song comprises one (or two) upslurred, plaintive notes. **H** Undergrowth and borders of humid forest. **D** S Mexico to N Bolivia, NE Amazonia.

3 CHESTNUT-CROWNED FOLIAGE-GLEANER *Automolus rufipileatus* 19cm Separable from Buff-throated Foliage-gleaner (Plate 130) by orange eyes, and from Brown-rumped Foliage-gleaner by rufous crown. **V** Short, rapid, bouncing-down series of notes. **H** In or near bamboo and cane in humid forest and woodland. **D** Amazonia.

4 SANTA MARTA FOLIAGE-GLEANER *Automolus rufipectus* 19cm Separable from Ruddy Foliage-gleaner by voice. **V** Three-syllable nasal rattle, given at irregular intervals. **H** Montane and lowland forest. **D** Sierra Nevada de Santa Marta, N Colombia.

5 HENNA-HOODED FOLIAGE-GLEANER *Hylocryptus erythrocephalus* 21cm No similar bird occurs in its range. Note long, pale bill. **V** Series of well-separated, very short, dry, nasal rattles. **H** On and near ground in forest and woodland. **D** SW Ecuador, NW Peru.

6 HENNA-CAPPED FOLIAGE-GLEANER *Hylocryptus rectirostris* 20.5cm Unmistakable due to its orange-rufous plumage and yellow iris. **V** Low, nasal chattering, sung in a duet or chorus. **H** Riverine belts and woodland. **D** SC Brazil, NE Paraguay.

7 TAWNY-THROATED LEAFTOSSER *Sclerurus mexicanus* 16cm Note lack of white on chin and throat. Separable from Grey-throated and Scaly-throated leaftossers by thinner bill, and from wrens by unbarred plumage. Distinctive rufous rump. **V** Ultra high *swéeeh-swéeeh-swéeh*. **H** Humid forest. **D** S Mexico to W Panama.

8 DUSKY LEAFTOSSER *Sclerurus obscurior* 15–17cm Previously treated as subspecies of Tawny-throated Leaftosser. Tosses leaves in search of invertebrates. **V** Song a high-pitched, descending series *tsii, tsiu, tsiu, tsiu, tsu*, ending in a vibrating trill. Call a shrill squeak. **H** Humid lowland rainforest. **D** E Panama to E Brazil.

9 SHORT-BILLED LEAFTOSSER *Sclerurus rufigularis* 16cm Separable from Tawny-throated Leaftosser by short, straight bill and white chin. **V** Song resembles that of Tawny-throated Leaftosser but slightly faster. **H** Humid forest. **D** Amazonia.

10 GREY-THROATED LEAFTOSSER *Sclerurus albigularis* 17cm Note grey throat contrasting with rufous chest. **V** Short, rising series of 5–6 upslurred, high notes. **H** Lowland, foothill and lower montane forest; 1,000–2,000m, locally higher or (much) lower. **D** Costa Rica to W and N South America.

11 BLACK-TAILED LEAFTOSSER *Sclerurus caudacutus* 18cm Note white chin and straight bill. **V** Short, fluted series of notes, starting very high and fast. **H** Terra firme forest. **D** Amazonia, E Brazil.

12 RUFOUS-BREASTED LEAFTOSSER *Sclerurus scansor* 18.5cm Separable from Tawny-throated Leaftosser and Black-tailed Leaftosser by rufous chest. **V** Very high, fast series of rising and lowered, chopped notes. **H** Lower montane forest; <1,500m. **D** NE Brazil to E Paraguay and NE Argentina.

13 SCALY-THROATED LEAFTOSSER *Sclerurus guatemalensis* 17cm Throat is distinctly scaled. **V** High-pitched series of notes that starts with some staccato notes, then lowers and rises again, finishing with a short trill. **H** Humid forest. **D** S Mexico to Colombia, W Ecuador.

14 SHARP-TAILED STREAMCREEPER *Lochmias nematura* 15cm Note long bill, white eye-stripe, short tail and bold pattern below. **V** Rapid, shivering series of thin notes that rises and descends. **H** Along streams in undergrowth of montane forest; <1,700m, locally higher. **D** N, W and SE South America, from E Panama to Uruguay.

15 SHARP-BILLED TREEHUNTER *Heliobletus contaminatus* 13cm Note small size and rather yellowish head. **V** Very high, sharp rattle (2 secs). **H** Humid lowland and montane forest and woodland; <1,800m. **D** SE Brazil, NE Argentina, E Paraguay.

16 RUFOUS-TAILED XENOPS *Microxenops milleri* 11cm Separable from Slender-billed, Plain and Streaked xenopses by straight bill and lack of white malar stripe. Tail is all rufous. **V** Very high, rising, piercing rattle, the last note lower. **H** Canopy and borders of humid forest. **D** Amazonia.

17 SLENDER-BILLED XENOPS *Xenops tenuirostris* 11cm Black in tail is normally visible. Separable from Streaked Xenops by bill shape and less streaking. **V** Very high series of 4–6 thin *tsip* notes. **H** Canopy and borders of humid forest. **D** Amazonia

18 PLAIN XENOPS *Xenops minutus* 12cm Separable from other xenopses by lack of streaking (check this carefully). **V** Short, level series of 4–5 very high, piercing notes. **H** Understorey of humid forest and woodland. **D** S Mexico through Amazonia to SE South America.

19 STREAKED XENOPS *Xenops rutilans* 13cm Streaking and habits as Olivaceous Piculet (Plate 104), but supporting itself on tail. Streaking quite different from Plain Xenops – see that species. Restricted range. **V** *Tjutju*; ultra-high, slightly descending *fjuhweetweetswat*. **H** Forest and forest edges. **D** Costa Rica to Venezuela, Bolivia and NE Argentina.

20 GREAT XENOPS *Megaxenops parnaguae* 16cm Unmistakable due to upturned bill and brightly coloured plumage. **V** Fast, short series of steeply rising, indignant, slightly liquid notes, the last one lower. **H** Caatinga and other woodland and forest. **D** E Brazil

21 WHITE-THROATED TREERUNNER *Pygarrhichas albogularis* 15cm Distinctively patterned. Note bicoloured bill. **V** Very high, piercing series of irregular single and paired *tsik* notes. **H** Tall forest. **D** C and S Chile, W Argentina.

22 TYRANNINE WOODCREEPER *Dendrocincla tyrannina* 25cm Larger and occurs at higher elevations than other plain woodcreepers. **V** Long, rather liquid rattle, starting very low and gradually rising, the last two notes separated (7–10 secs). **H** Montane forest; 1,800–2,700m, locally lower. **D** Colombia and W Venezuela to C Peru.

23 PLAIN-BROWN WOODCREEPER *Dendrocincla fuliginosa* 21cm Note pale lores and ear-coverts. Shows more or less dark malar stripe. **V** Nasal rattle, level at first and then lowered (4–5 secs). **H** Lower storeys of humid forest and woodland. **D** SE Honduras through Amazonia, E Brazil.

24 PLAIN-WINGED WOODCREEPER *Dendrocincla turdina* 21cm All brown with ochre streaks on crown, throat paler. Wings with chestnut hue. Tail chestnut. **V** Series of monotonous *tik* notes, lasting >1 min. **H** Rainforests. **D** E Brazil, E Paraguay, NE Argentina.

25 TAWNY-WINGED WOODCREEPER *Dendrocincla anabatina* 19cm No barring or streaking. Pale eyebrow, flight feathers tawny and contrasting. Short, straight bill, **V** Very high *tjuuw*; very high, sharp *tjutjutju* (5–60 secs), falling off at the end. **H** Forest, tall secondary growth, mangrove. **D** SE Mexico to W Panama.

26 WHITE-CHINNED WOODCREEPER *Dendrocincla merula* 20cm Small white chin patch is distinctive; face sides are mainly unmarked. **V** Very short, twittering rattles given at various intervals. **H** Undergrowth of humid forest. **D** Amazonia.

27 RUDDY WOODCREEPER *Dendrocincla homochroa* 20cm Plumage appears uniform orange-rufous except for grey lores. **V** Very long, harsh, chattered rattle, slightly undulating but generally descending (30–40 secs). **H** Undergrowth of humid forest and woodland. **D** S Mexico to NW Venezuela.

28 LONG-TAILED WOODCREEPER *Deconychura longicauda* 20cm Difficult to separate from several other small woodcreepers, but note length of bill, eye-stripe and extent of rufous coloration, which is restricted to uppertail coverts (not on rump as in Spot-throated Woodcreeper). **V** Slowly descending series of musical notes, each one slightly longer than the previous. **H** Lower storeys of mainly terra firme forest. **D** Honduras to Colombia, Amazonia.

29 SPOT-THROATED WOODCREEPER *Deconychura stictolaema* 17cm Separable from smaller Wedge-billed Woodcreeper by short, straight bill and relatively longer tail. **V** High, sharp, level trill (2 secs). **H** Undergrowth of terra firme forest. **D** Amazonia.

30 OLIVACEOUS WOODCREEPER *Sittasomus griseicapillus* 15.5cm Note fine, straight bill and lack of streaking. **V** Distinctive vocal variations, including rattles, trills and rolls. **H** Variety of forested and wooded habitats. **D** Mexico to NE Argentina.

31 WEDGE-BILLED WOODCREEPER *Glyphorynchus spirurus* 15cm Note small size and bill shape. **V** Short, fast, very high, warbling series of notes, the last one higher. **H** Undergrowth of humid forest and woodland. **D** S Mexico through Amazonia, E Brazil.

32 SCIMITAR-BILLED WOODCREEPER *Drymornis bridgesii* 31cm Note long, downcurved bill, and distinct eyebrow and malar stripe. **V** Utters 3–6 rapid, very high shrieks, sometimes as a rattle. **H** Chaco, scrub, savannah. **D** SE Bolivia and W Paraguay to C Argentina and W Uruguay.

33 LONG-BILLED WOODCREEPER *Nasica longirostris* 35cm Unmistakable due to long bill, slender jizz and white throat. **V** Slow series of 4–5 nasal, upslurred or downslurred notes. **H** Higher levels of humid forest. **D** Amazonia.

34 CINNAMON-THROATED WOODCREEPER *Dendrexetastes rufigula* 25cm Arrow-shaped white spots around collar and on breast are distinctive in nominate race (missing in other races). Note pale bill. **V** Very high, shivering, loud, nasal rattle, trailing off at end (3–4 secs). **H** Canopy and borders of terra firme forest and várzea. **D** Amazonia.

35 RED-BILLED WOODCREEPER *Hylexetastes perrotii* 29cm Note red bill and lack of barring. Nominate race shows a white malar stripe, which is missing in other races. **V** Sings as Bar-bellied Woodcreeper but slower. **H** Mainly terra firme forest. **D** NE Amazonia.

36 UNIFORM WOODCREEPER *Hylexetastes uniformis* 26–30cm Previously treated as subspecies of Red-billed Woodcreeper. Often associates with army ants. **V** Song a piercing, squeaky, upslurred whistle *hoo-EEE*, the second phrase louder and higher in pitch than first. Call a shrill disyllabic squeak. **H** Mainly terra firme rainforest. **D** SE Amazonia.

37 BRIGIDA'S WOODCREEPER *Hylexetastes brigidai* 25–30cm Previously treated as subspecies of Red-billed Woodcreeper. Associates with army ants. Diet includes invertebrates and small vertebrates. **V** Song a high-pitched upslurred *who-Eek, who-Err, who-Err*. Call a raspy squeak. **H** Mainly terra firme rainforest. **D** NC Brazil.

38 BAR-BELLIED WOODCREEPER *Hylexetastes stresemanni* 29cm Note red bill and barring on underparts. **V** Loud, resounding series of 4–5 *tou-weeh* notes. **H** Mainly terra firme forest. **D** W Amazonia.

1
2
3
4
5
6
7
8
9
10
11
12
13
14
15
16
17
18
19
20
21
22
23
24
25
26
27
28
29
30
31
32
33
34
35
36
37
38

OVENBIRDS *CONTINUED*

1 STRONG-BILLED WOODCREEPER *Xiphocolaptes promeropirhynchus* 30cm Has a heavier build and stronger bill than smaller Buff-throated Woodcreeper. **V** Descending series of 4–5 loud, staccato notes, superimposed on a continuous lower, softer note. **H** Humid, dry and cloud forests. **D** S Mexico through Amazonia.

2 WHITE-THROATED WOODCREEPER *Xiphocolaptes albicollis* 29cm Black streaking on head, olivaceous streaking on breast and mantle. Note blackish bill. **V** Song resembles that of Strong-billed Woodcreeper but lacks undertone and main note is *me-tjuu*, repeated 4×. **H** Montane forest, secondary growth. **D** NE Brazil to E Paraguay and NE Argentina.

3 MOUSTACHED WOODCREEPER *Xiphocolaptes falcirostris* 29cm No similar woodcreeper occurs in its range. **V** Song resembles that of Strong-billed Woodcreeper but slower, and undertone is missing or occurs only in part of series. **H** Woodland, Caatinga, riverine belts. **D** E Brazil.

4 GREAT RUFOUS WOODCREEPER *Xiphocolaptes major* 30cm Strikingly large in size. Note dark lores and face sides; some races show a whitish chin. **V** Descending series of 10–15 main notes (staccato, loud and harsh), superimposed on a lowered, mewing undertone. **H** Dry wooded habitats, riverine belts, Chaco, Cerrado. **D** SC South America.

5 NORTHERN BARRED WOODCREEPER *Dendrocolaptes sanctithomae* 27cm No other barred woodcreeper occurs W of Andes. **V** Short series of 3–4 upslurred notes, followed by a rapidly accelerating rattle. **H** Undergrowth in humid forest. **D** S Mexico to NW Ecuador.

6 AMAZONIAN BARRED WOODCREEPER *Dendrocolaptes certhia* 27cm Note reddish bill. **V** Short, ringing, descending rattle, preceded by two mumbled notes. **H** Undergrowth in humid forest and woodland. **D** Amazonia.

7 HOFFMANNS'S WOODCREEPER *Dendrocolaptes hoffmannsi* 28cm Rather uniform rufous coloration with faint barring on belly. Separable from similar *Dendrocolaptes* woodcreepers by blackish (not reddish or pale) bill. **V** Rapid, rather level series of liquid *wic* notes. **H** Undergrowth of terra firme forest. **D** SC Amazonia.

8 BLACK-BANDED WOODCREEPER *Dendrocolaptes picumnus* 27cm Note streaking on foreparts and barring on belly. **V** Rapid, descending series of jubilant notes (2–3 secs). **H** Forested habitats. **D** S Mexico to NW Argentina.

9 PLANALTO WOODCREEPER *Dendrocolaptes platyrostris* 26cm Separable from White-throated Woodcreeper by relatively shorter bill and less distinct malar stripe. **V** Rapid series of liquid yet nasal, slightly descending notes. **H** Undergrowth of humid forest and secondary growth. **D** NE Brazil to E Paraguay and NE Argentina.

10 STRAIGHT-BILLED WOODCREEPER *Dendroplex picus* 20cm Note straight white bill and squamate breast spots. **V** Fast, liquid, descending series of notes (5 secs). **H** Várzea and woodland. **D** Panama to E Brazil.

11 ZIMMER'S WOODCREEPER *Dendroplex kienerii* 22cm Separable from Straight-billed Woodcreeper by breast streaking, not spotting, and relatively longer tail; bill is also longer and slightly curved. **V** Sharp, dry rattle that may be downslurred (1–4 secs). **H** Várzea. **D** Amazon and its tributaries.

12 STRIPED WOODCREEPER *Xiphorhynchus obsoletus* 20cm Difficult to separate from Zimmer's Woodcreeper; check for its duller colouring, striping below and on back, and less straight bill. **V** Steeply rising, dry rattle (1–3 secs). **H** Várzea and woodland, especially near water. **D** Amazonia.

13 LESSER WOODCREEPER *Xiphorhynchus fuscus* 17cm Note slender bill and spotting on crown. **V** Song is a high, fast rattle, a series of high, staccato *djip* notes, or a combination of rattle and notes. **H** Forested habitats, especially in lowland and lower montane forest. **D** NE Brazil to E Paraguay and NE Argentina.

14 OCELLATED WOODCREEPER *Xiphorhynchus ocellatus* 22cm Note rather long bill with dark maxilla and unmarked, darkish mantle. **H** Undergrowth of humid forest. **D** NW Amazonia.

15 TSCHUDI'S WOODCREEPER *Xiphorhynchus chunchotambo* 22–24cm Previously treated as subspecies of Ocellated Woodcreeper. Note pale marks on mantle. **V** Song a confused-sounding series of squeaky notes, recalling noises of early electronic games. Call a hollow-sounding, trilling *Wtt-ttttttttt*, descending in pitch during delivery. **H** Humid forests and wooded slopes. **D** E Andean spine from SE Colombia to S Bolivia.

16 ELEGANT WOODCREEPER *Xiphorhynchus elegans* 21cm Separable from Ocellated Woodcreeper in most of its range by 'dotted' streaks on back; in SW birds these streaks are much weaker and breast spotting of Ocellated Woodcreeper is more distinctive. **V** Song is a rapid, descending series of jubilant staccato notes, often rising and lowering again, and often concluding with 1–2 'seducing' whistles. **H** Humid forest. **D** W and SC Amazonia.

17 SPIX'S WOODCREEPER *Xiphorhynchus spixii* 21cm Separable from Ocellated, Elegant, Chestnut-rumped and Cocoa woodcreepers by range, and from Buff-throated Woodcreeper by size. **V** Sharp, descending rattle, sometimes rising at the end (2–3 secs). **H** Undergrowth of humid forest. **D** SE Amazonia.

18 CHESTNUT-RUMPED WOODCREEPER *Xiphorhynchus pardalotus* 22cm Separable from Ocellated Woodcreeper by streaked mantle and from Buff-throated Woodcreeper by all-black bill. **V** Strong, level rattle, preceded by some staccato notes. **H** Lowland and lower montane forest and woodland. **D** NE Amazonia.

19 BUFF-THROATED WOODCREEPER *Xiphorhynchus guttatus* 26cm Note size, pale throat and lack of barring below. Pale-billed N nominate race is shown. **V** Rapid series of loud, slightly liquid notes, the first few somewhat lower. **H** Humid lowland forest, woodland, locally in mangroves. **D** Amazonia, E Brazil.

20 COCOA WOODCREEPER *Xiphorhynchus susurrans* 22cm Note range. **V** Resounding rapid, level series of strong *kwee* notes, the last one slightly lower. **H** Lower storeys and borders of humid forest, woodland, riverine belts, secondary growth. **D** E Guatemala to N Venezuela.

21 IVORY-BILLED WOODCREEPER *Xiphorhynchus flavigaster* 25cm Distinguished from smaller Olive-backed Woodcreeper by distinct striping on mantle. **V** High *tjouw* and very high, sharp, rapid, rising and falling-off *tititi-titjutju*. **H** Forest edges and clearings. **D** W and E Mexico to NW Costa Rica.

22 BLACK-STRIPED WOODCREEPER *Xiphorhynchus lachrymosus* 25cm Note bold patterning. **V** Long, descending series of *tjew tjiw...* (4 secs). **H** Forest interiors and edges. **D** E Nicaragua to NW Ecuador.

23 SPOTTED WOODCREEPER *Xiphorhynchus erythropygius* 23cm Occurs on W slope of Andes (compare Olive-backed Woodcreeper). **V** Short, rapid, shivering, descending series of notes. **H** Humid lowland, montane and cloud forest, montane pine woodlands. **D** S Mexico to SW Ecuador.

24 OLIVE-BACKED WOODCREEPER *Xiphorhynchus triangularis* 23cm Range differs from that of Spotted Woodcreeper, and unlike that species has a scaled (not lined) throat. **V** Song resembles that of Spotted Woodcreeper but longer. **H** Humid montane forest, cloud forest, older secondary growth. **D** Venezuela to Bolivia.

25 WHITE-STRIPED WOODCREEPER *Lepidocolaptes leucogaster* 20cm More marked with white than other woodcreepers. Note thin, downcurved bill. **V** Shrill, descending, trilling *trrr-iuuup*. **H** Pine and mixed forest. **D** Mexico.

26 STREAK-HEADED WOODCREEPER *Lepidocolaptes souleyetii* 20cm Note slender, downcurved bill, unstreaked back and streaked crown. Found at lower altitudes than Montane Woodcreeper. **V** Very high, descending, sharp yet liquid rattle, the last note higher. **H** Open woodland, riverine belts, arid scrub. **D** W Mexico to Guyana and NW Peru.

27 NARROW-BILLED WOODCREEPER *Lepidocolaptes angustirostris* 21cm Facial pattern, long downcurved bill and pale underparts are distinctive. **V** Descending, accelerating series of loud, nasal notes (2–3 secs). **H** Open woodland, scrub, Cerrado, agricultural areas. **D** E, C and SC South America.

28 SPOT-CROWNED WOODCREEPER *Lepidocolaptes affinis* 20cm Note white crown spots and red nape. Overall rather dark. **V** Ultra-high, loud, shrill, descending, very rapid *suseesusee-susee*. **H** Montane forest and adjacent areas. **D** NE Mexico to W Panama.

29 MONTANE WOODCREEPER *Lepidocolaptes lacrymiger* 19cm In its range, slender, downcurved bill is distinctive. **V** Song comprises unstructured, barely audible, very high, short, drawn-out, sizzling notes. **H** Montane forest and woodland; 1,500–3,000m, locally higher or lower. **D** W Venezuela to C Bolivia.

30 SCALED WOODCREEPER *Lepidocolaptes squamatus* 19cm Note neat streaking below. Separable from Scalloped Woodcreeper by buff-rufous neck. **V** Very high, stuttered or rolled shriek. **H** Humid forest, secondary growth, woodland. **D** E and SE Brazil.

31 SCALLOPED WOODCREEPER *Lepidocolaptes falcinellus* 19cm Compare Scaled Woodcreeper, whose range differs or barely overlaps. **V** Very high, descending, one-syllable shriek. **H** Humid forest and woodland. **D** E Paraguay, S Brazil, NE Argentina.

32 GUIANAN WOODCREEPER *Lepidocolaptes albolineatus* 19cm No similar woodcreeper (with a thin, downcurved bill) occurs in its habitat and range (although see Inambari Woodcreeper). **V** High, descending, accelerating, resounding series of fluted notes, ending as a small rattle. **H** Higher levels of forest and woodland. **D** E Venezuela, the Guianas, N Brazil.

33 DUIDA WOODCREEPER *Lepidocolaptes duidae* 17–19cm Shows slight supercilium. Usually feeds in forest canopy, foraging for invertebrates among epiphytes and under bark. **V** Song a woodpecker-like piping, accelerating *te-tu-tu-tu-tu-tttt*, descending in pitch towards end. **H** Favours terra firme rainforest. **D** NW Amazonia in Brazil, Venezuela, SE Colombia, E Ecuador, NE Peru.

34 INAMBARI WOODCREEPER *Lepidocolaptes fatimalimae* 19cm Separable from similar Guianan Woodcreeper by unspotted crown. Note lack of eye-stripe. **V** Very high, rapidly accelerating and simultaneously descending series of sharp, piercing notes. **H** Canopy of terra firme and foothill forest. **D** W Amazonia.

35 RONDONIA WOODCREEPER *Lepidocolaptes fuscicapillus* 17–19cm Feeds mainly in forest canopy. Diet includes mainly invertebrates. **V** Song a piping *pee-pee-piu-piu-piu-piu*, falling in pitch during delivery. **H** Mainly terra firme rainforest. **D** SW Amazonia in the Madeira–Tapajós interfluvial region.

36 LAYARD'S WOODCREEPER *Lepidocolaptes layardi* 16–18cm Searches for invertebrates under bark and in crevices, usually in tree canopy. **V** Call a shrill, downslurred *piu, piu, piu, piu, piu*, pitch falling during delivery. **H** Prefers terra firme rainforest. **D** SE Amazonian Brazil E of the Tapajós.

37 GREATER SCYTHEBILL *Drymotoxeres pucheranii* 28cm The only scythebill with a pale malar streak. **V** Unstructured series of 5–7 nasal notes. **H** Montane forest; 2,000–2,800m. **D** C Colombia to SE Peru.

38 RED-BILLED SCYTHEBILL *Campylorhamphus trochilirostris* 26cm Similar to Brown-billed Scythebill but habitat differs. **V** Short, rapid series of fluted yet sharp notes, the last one slightly lower. **H** Rather open wooded habitats, Chaco, Cerrado, Caatinga, riverine belts. **D** Panama to N Argentina.

39 BLACK-BILLED SCYTHEBILL *Campylorhamphus falcularius* 26cm Black bill is diagnostic. **V** Series of 3–7 rasping, high notes at any speed. **H** Lower storeys of humid forest and woodland. **D** SE Brazil to E Paraguay and NE Argentina.

40 BROWN-BILLED SCYTHEBILL *Campylorhamphus pusillus* 23cm Similar to Red-billed Scythebill but habitat differs and with a smaller range. **V** Descending, slightly plaintive series of loud, upslurred notes. **H** Lower montane and foothill forest. **D** Costa Rica to N Peru.

41 CURVEBILLED SCYTHEBILL *Campylorhamphus procurvoides* 23cm Compare Red-billed Scythebill. Nominate race (shown) lacks mantle streaking. **V** High, strong, slightly descending, shivering trill, sometimes introduced by a high *wee* note. **H** Mainly terra firme forest. **D** Amazonia.

1
2
3
4
5
6
7
8
9
10
11
12
13
14
15
16
17
18
19
20
21
22
23
24
25
26
27
28
29
30
31
32
33
34
35
36
37
38
39
40
41

ANTBIRDS THAMNOPHILIDAE

1 RUFOUS-RUMPED ANTWREN *Euchrepomis callinota* 10.5cm Resembles Chestnut-shouldered Antwren but found at higher altitudes. Note yellow shoulders (difficult to see in the field). **V** Very high series of notes, accelerating into a trill (1 sec). **H** Forest and woodland. **D** Costa Rica to Peru, the Guianas.

2 CHESTNUT-SHOULDERED ANTWREN *Euchrepomis humeralis* 10.5cm Compare Rufous-rumped Antwren. **V** Very high series of notes, accelerating into a rattle (1 sec). **H** Higher strata of lowland forest. **D** E Ecuador, E Peru, N Bolivia, W and WC Brazil.

3 YELLOW-RUMPED ANTWREN *Euchrepomis sharpei* 10.5cm Yellow rump is diagnostic. **V** Very high series of notes, accelerating into a slightly rising and falling trill (2 secs). **H** Canopy and borders of montane forest. **D** S Peru, W Bolivia.

4 ASH-WINGED ANTWREN *Euchrepomis spodioptila* 10cm Has a distinctive rufous rump; lacks yellow colour in plumage. **V** Very high series of notes, accelerating into a trill and often slowing down at the very end (1 sec). **H** Canopy and borders of montane forest. **D** N Amazonia.

5 WING-BANDED ANTBIRD *Myrmornis torquata* 15.5cm Unmistakable due to its odd shape. **V** Almost level series of 10–20 very high, ringing notes that starts slowly and accelerates towards the end. **H** Floor of humid forest. **D** Nicaragua, Panama, Amazonia.

6 SPOT-WINGED ANTSHRIKE *Pygiptila stellaris* 12.5cm Distinct jizz, with a very short tail and heavy, long bill. **V** Short, high, dry rattle, ending with a whistled *teeuw*. **H** Higher levels of forest. **D** Amazonia.

7 RUSSET ANTSHRIKE *Thamnistes anabatinus* 14cm Note strong bill and short tail (although longer than that of Spot-winged Antshrike). **V** Short, slightly accelerating series of very high, slightly upslurred notes. **H** Canopy and subcanopy of montane and lowland forest and secondary growth. **D** SE Mexico to C Bolivia.

8 RUFESCENT ANTSHRIKE *Thamnistes rufescens* 15–16cm Previously treated as subspecies of Russet Antshrike. Gleans invertebrate prey from leaves. Sometimes seen in mixed-species flocks. **V** Song a shrill, insistent *tsui-tsip-tsip-tsip-tsip*. **H** Forested slopes, montane secondary woodland. **D** E Peru to W Bolivia.

9 DOT-WINGED ANTWREN *Microrhopias quixensis* 11.5cm Tail and wing pattern are distinctive. Forages actively for insects. **V** Short, fast, warbling series of notes. **H** Dense growth and borders of humid forest. **D** S Mexico to S Amazonia.

10 BLACK BUSHBIRD *Neoctantes niger* 16cm Separated from Recurve-billed and Randonia bushbirds by range; rufous breast of female is unmistakable. **V** Long series of high, well-separated, rather staccato notes (10–15 secs). **H** At tree-falls, edges and streams in humid forest. **D** W Amazonia.

11 RECURVE-BILLED BUSHBIRD *Clytoctantes alixii* 16.5cm Separated from Black and Rondonia bushbirds by range. Note black throat and upper breast of male. **V** Short series of 1–3 high, loud notes, the last one resembling a very short rattle. **H** Undergrowth of forest, secondary growth and regenerating clearings. **D** NW Venezuela, N Colombia.

12 RONDONIA BUSHBIRD *Clytoctantes atrogularis* 17cm Separated from Black and Recurve-billed bushbirds by range. Male uniform black; female with black throat. **V** Male's song a far-carrying *tree-tree-tree*. **H** Undergrowth with dense tangles of terra firme forest. **D** SW Amazonia.

13 CHECKER-THROATED ANTWREN *Epinecrophylla fulviventris* 10.5cm The only *Epinecrophylla* antwren with a chequered throat W of Andes. **V** Slow or rapid series of 6–10 very high, piercing notes. **H** Understorey of humid forest and woodland. **D** SE Honduras to W Ecuador.

14 BROWN-BELLIED ANTWREN *Epinecrophylla gutturalis* 10.5cm No similar antwren occurs in its range. **V** Slightly descending, very high, piercing trill (2–3 secs). **H** Understorey of lowland forest. **D** NE Amazonia.

15 WHITE-EYED ANTWREN *Epinecrophylla leucophthalma* 11cm Mantle is brown (not rufous); paler overall than Foothill Antwren. **V** Very high, descending, plaintive series of sharp notes. **H** Dense understorey of lowland forest. **D** S Amazonia.

16 NAPO STIPPLE-THROATED ANTWREN *Epinecrophylla haematonota* 11cm Compare Ornate and Rufous-tailed antwrens. **V** Fast, very high, descending series of thin notes. **H** Understorey of lowland forest. **D** E Peru and W Brazil.

17 NEGRO STIPPLE-THROATED ANTWREN *Epinecrophylla pyrrhonota* 10–11cm Usually seen in pairs, often associated with mixed-species flocks. **V** Song a high-pitched, tinkling, trilling *tsee-tse-sssssssss*, falling in pitch during delivery. **H** Rainforest; mostly terra firme, occasionally várzea. **D** SE Colombia and S Venezuela to NE Ecuador, NE Peru and NW Brazil.

18 MADEIRA STIPPLE-THROATED ANTWREN *Epinecrophylla amazonica* 10–11cm Forages for invertebrates in clusters of dead leaves trapped in foliage. **V** Song a high-pitched, tinkling, trilling *tsu-tse-tsee-ssssssssss*, rising and falling in pitch during delivery. **H** Favours terra firme rainforest. **D** C and SC Brazil, N Bolivia.

19 YASUNI ANTWREN *Epinecrophylla fjeldsaai* 11cm Separable from Napo, Negro and Madeira stipple-throated antwrens by brown (not rufous) back. **V** Rapid series of very high, slightly descending, thin notes. **H** Undergrowth of lowland forest. **D** E Ecuador, N Peru.

20 FOOTHILL ANTWREN *Epinecrophylla spodionota* 11cm Normally at higher elevations than similar antwrens. **V** Short, fast series of very high, slightly descending, thin notes. **H** Forest undergrowth. **D** S Colombia to C Peru.

21 ORNATE ANTWREN *Epinecrophylla ornata* 10cm Distinctive NW nominate race is shown. This and other races differ from similar grey antwrens (e.g. Grey Antwren, Plate 134) in having wing spotting (not wing fringing). **V** Short, rapid series of high, thin, slightly descending notes. **H** Undergrowth of humid forest. **D** W and S Amazonia.

22 RUFOUS-TAILED ANTWREN *Epinecrophylla erythrura* 11cm Note lack of throat chequering and rufous tail and back. **V** Song is a high, rather unstructured but evenly spaced chirping. **H** Understorey of humid forest. **D** W Amazonia.

23 STRIPE-BACKED ANTBIRD *Myrmorchilus strigilatus* 16cm Distinctly patterned. No similar terrestrial antbird occurs in its range. **V** Very high, sharp pair of notes, the second drawn out. **H** Caatinga and Chaco woodland. **D** SC South America, E Brazil.

24 YAPACANA ANTBIRD *Aprositornis disjuncta* 13.5cm Resembles Silvered Antbird (Plate 138), but wings of male are more white-fringed, not spotted, and wings of female are unmarked. **V** Strange, toneless series of two grating, stretched notes. **H** Understorey and floor of dense woodland. **D** NC Amazonia.

25 GREY-BELLIED ANTBIRD *Ammonastes pelzelni* 13.5cm Note grizzled sides of face and large wing spots (compare Black-throated Antbird). **V** Hurried series of very high *zree* notes, increasing in sharpness. **H** On or near ground of lowland forest and secondary growth on poor soils. **D** NC Amazonia.

26 BLACK-THROATED ANTBIRD *Myrmophylax atrothorax* 14cm Note white throat of female (compare female Grey-bellied Antbird). **V** Level, decelerating series of high *chi* notes (2 secs). **H** Understorey of forest and secondary growth, often near river edges. **D** Amazonia.

27 MOUSTACHED ANTWREN *Myrmotherula ignota* 8cm See Pygmy Antwren. **V** Song differs from that of Pygmy Antwren in its slower pace, and can be described as a rapid, rather level series of sharp, high notes, accelerating to a short, falling-off trill (2 secs). **H** Higher levels of forest. **D** E Panama to NW Ecuador, W Amazonia.

28 PYGMY ANTWREN *Myrmotherula brachyura* 8cm Not reliably separable from Moustached Antwren but voice differs. **V** Rapid, rather level series of sharp, high notes, accelerating to a short trill (2 secs). **H** Higher levels of forest and secondary growth. **D** Amazonia.

29 GUIANAN STREAKED ANTWREN *Myrmotherula surinamensis* 9.5cm Separable from Amazonian Streaked Antwren by its range. Compare underparts of female with female Cherrie's and Klages's antwrens. **V** Song a sharp, short, 1-sec trill, slightly rising and falling at the end. **H** At water's edge in lower storeys of lowland forest. **D** NE Amazonia.

30 AMAZONIAN STREAKED ANTWREN *Myrmotherula multostriata* 9.5cm Note range; has a restricted overlap with Cherrie's and Klages's antwrens. **V** Sings as Guianan Streaked Antwren but longer. **H** At water's edge in lower storeys of lowland forest. **D** S and W Amazonia.

31 PACIFIC ANTWREN *Myrmotherula pacifica* 9.5cm No similar antwren occurs in its range. Female unstreaked below. **V** Short, sharp, staccato song, almost a rattle. **H** At edges of lowland forest, secondary growth, woodland, gardens. **D** Panama to W Ecuador.

32 CHERRIE'S ANTWREN *Myrmotherula cherriei* 10cm Note coarse streaking of female underparts. Best identified by voice. **V** Short, sharp, rattling trill. **H** Shrubby borders of forest, woodland and riverine belts. **D** NC Amazonia.

33 KLAGES'S ANTWREN *Myrmotherula klagesi* 10cm More finely streaked than Cherrie's Antwren. **V** High, repeated *cheedi*. **H** Higher levels of forest along rivers. **D** C Amazonia.

34 STRIPE-CHESTED ANTWREN *Myrmotherula longicauda* 10cm Note relatively long tail and fine chest streaking (coarser on Amazonian Streaked Antwren). **V** Series of *chidi* notes repeated 3×. **H** Forest, secondary growth. **D** W and SW Amazonia.

35 YELLOW-THROATED ANTWREN *Myrmotherula ambigua* 8.5cm Note yellow throat; female as female Pygmy Antwren but upperparts more distinct pale buff. **V** High, sharp, slightly descending series of 7–14 well-separated notes. **H** Higher levels of forests on poor soils. **D** NC Amazonia.

36 SCLATER'S ANTWREN *Myrmotherula sclateri* 8.5cm Note yellow underparts, including cheeks. **V** Short series of 4–8 well-spaced, high, piped notes. **H** Higher levels of lowland forest. **D** S Amazonia.

37 WHITE-FLANKED ANTWREN *Myrmotherula axillaris* 10cm Note distinct white flanks. **V** Short, hurried, descending series of chopped, whistled notes. **H** Understorey of forest and secondary growth. **D** SE Honduras through Amazonia.

38 SILVERY-FLANKED ANTWREN *Myrmotherula luctuosa* 10cm Resembles White-flanked Antwren, but black on underparts broadly fringed grey from breast to lower belly. **V** Short, hurried, descending series of chopped, whistled notes. **H** Understorey of forest and secondary growth. **D** E and SE Brazil.

39 SLATY ANTWREN *Myrmotherula schisticolor* 10.5cm At higher elevations than most other grey-black antwrens. **V** Series of 2–5 high, sharp notes. **H** Undergrowth of foothill and montane forest; 900–1,800m, locally higher or lower. **D** SE Mexico to N Venezuela and Peru.

1
2
3
4
5
6
7
8
9
10
11
12
13
14
15
16
17
18
19
20
21
22
23
24
25
26
27
28
29
30
31
32
33
34
35
36
37
38
39

ANTBIRDS *CONTINUED*

1 RIO SUNO ANTWREN *Myrmotherula sunensis* 9cm Difficult to separate from larger and longer-tailed Band-tailed Antwren. **V** Song as Slaty Antwren (Plate 133). **H** Lower storeys of lowland forest. **D** W Amazonia.

2 SALVADORI'S ANTWREN *Myrmotherula minor* 9cm Note black subterminal tail band of male. **V** Song as Slaty Antwren (Plate 133) but preceded by a short twitter. **H** Shaded lower storeys of forest and secondary growth. **D** SE Brazil.

3 LONG-WINGED ANTWREN *Myrmotherula longipennis* 10.5cm Note white mantle fringes of male. **V** Fast, rising series of whistled notes. **H** Lower storeys of lowland and foothill forest; mostly <700m, but higher in the Andes and on tepuis. **D** Amazonia.

4 BAND-TAILED ANTWREN *Myrmotherula urosticta* 9.5cm White terminal tail-band is diagnostic. **V** Ascending series of 3–5 nasal notes. **H** Undergrowth of humid forest. **D** E and SE Brazil.

5 IHERING'S ANTWREN *Myrmotherula iheringi* 9cm Note black throat and lores of male; wing pattern of female is distinctive. **V** Series of 10–15 slightly rising *peeu* notes. **H** Undergrowth of humid forest; <500m. **D** W and C Amazonia.

6 RIO DE JANEIRO ANTWREN *Myrmotherula fluminensis* 10cm Female undescribed. Male separable from male White-flanked Antwren (Plate 133) by grey flanks, and from Salvadori's and Long-winged antwrens by uniform grey tail. **V** Not recorded. **H** Not well known, probably edges of secondary forest. **D** SE Brazil.

7 YUNGAS ANTWREN *Myrmotherula grisea* 9.5cm Male uniform grey overall. Note unmarked wings and greyish crown of female. **V** Slightly hurried series of 3–8 *seeyr* notes. **H** Foothill and montane forest. **D** W Bolivia.

8 UNICOLORED ANTWREN *Myrmotherula unicolor* 9.5cm No other antwren with unmarked wings occurs in its range. **V** Hurried series of four sharp notes. **H** Lower storeys of lowland forest and secondary growth. **D** SE Brazil.

9 ALAGOAS ANTWREN *Myrmotherula snowi* 10cm Range differs from that of Unicolored Antwren. Underparts of female more rufescent than in female Unicolored. **V** Unstructured series of high, sharp, well-spaced *seeyr* notes. **H** Undergrowth of semi-humid forest. **D** E Brazil.

10 PLAIN-WINGED ANTWREN *Myrmotherula behni* 9cm Large black breast patch and plain, unmarked wings of male are distinct; female less grey above than female Grey Antwren. **V** Very high series of 3–4 sharp *seeyr* notes. **H** Lower storeys of foothill and montane forest. **D** N and NW Amazonia.

11 GREY ANTWREN *Myrmotherula menetriesii* 9cm Wags tail sideways, differing from wing-flicking of Long-winged Antwren. **V** Long, sharp series of 10–15 *sreeyr* notes, rising in pitch and volume. **H** Humid forest. **D** Amazonia.

12 LEADEN ANTWREN *Myrmotherula assimilis* 9.5cm Wings are basically grey, not black, and are marked with white. Note also habitat. **V** Sharp, rising and falling trill (2–3 secs). **H** Lowland forest, mainly along rivers. **D** C and S Amazonia.

13 STREAK-CAPPED ANTWREN *Terenura maculata* 10cm Note streaked head, contrasting with bright rufous back. **V** Very high, rhythmic chirping (2–3 secs). **H** Canopy and borders of humid forest and woodland. **D** E Paraguay, NE Argentina, SE Brazil.

14 ORANGE-BELLIED ANTWREN *Terenura sicki* 10cm Male distinctly patterned. Female separable from female Streak-capped Antwren by range. **V** Very high, hurried, descending warble. **H** Canopy and borders of humid forest and woodland. **D** E Brazil.

15 BLACK-AND-WHITE ANTBIRD *Myrmochanes hemileucus* 11cm No similar bird occurs in its restricted range and habitat. **V** Short, downslurred rattle. **H** Near water in várzea and swampy forest. **D** W Amazon and its tributaries.

16 NARROW-BILLED ANTWREN *Formicivora iheringi* 11.5cm Separable from White-Flanked Antwren (Plate 133) by finer bill. **V** Calm series of 3–19 staccato notes. **H** Lower forest storeys and woodland. **D** E Brazil.

17 BLACK-HOODED ANTWREN *Formicivora erythronotos* 11cm Rufous on back is distinctive. **V** Rapid, slightly lowered series of chopped notes (3 secs). **H** Scrubby woodland, young secondary growth, abandoned plantations. **D** SE Brazil.

18 SOUTHERN WHITE-FRINGED ANTWREN *Formicivora grisea* 12.5cm White fringe of male is distinctive. Note orange-buff underparts of female. **V** Mid-high series of 10–15 staccato *chup* notes. **H** Scrub, shrubby woodland, secondary growth, mangroves. **D** Guyana to S Amazonia and SE Brazil.

19 NORTHERN WHITE-FRINGED ANTWREN *Formicivora intermedia* 12–13cm Previously considered conspecific with Southern White-fringed Antwren. Invertebrate prey gleaned mainly from leaves and leaf litter. **V** Song a high-pitched burst of trilling notes. **H** Forested habitats, dense scrub, secondary woodland. **D** N Colombia, Venezuela, Caribbean islands of Tobago and Margarita, Pearl Islands in the Gulf of Panama.

20 SERRA ANTWREN *Formicivora serrana* 12.5cm Separable from Dusky-tailed Antbird (Plate 137) by rufous edging to tertials. **V** Rapid, mid-high series of *chup* notes (1–2 secs). **H** Scrub, secondary growth, stunted woodland, dry forest edges. **D** SE Brazil.

21 BLACK-BELLIED ANTWREN *Formicivora melanogaster* 12.5cm Compare male Southern and Northern White-fringed antwrens, which show wider white fringe to flanks. Female separable from female Serra Antwren by whiter underparts and wider white cheek. **V** Song resembles that of Southern White-fringed Antwren, but slower. **H** Lower storeys of forest, Caatinga scrub, woodland undergrowth. **D** E Brazil, C South America.

22 RUSTY-BACKED ANTWREN *Formicivora rufa* 12.5cm Note rusty upperparts. Female shows striped cheeks and breast. **V** Song is, e.g., a dry rattle (5 secs). **H** Shrubby habitats. **D** Amazonia, E Brazil, C South America.

23 SINCORA ANTWREN *Formicivora grantsaui* 12.5cm Restricted range. Male not reliably separable from male Southern or Northern white-fringed antwrens. Striping of female coarser than in female Rusty-backed Antwren. **V** Rather long, sometimes faltering series of dry *chup* notes (10 secs). **H** Rocky montane scrub. **D** NE Brazil.

24 MARSH ANTWREN *Stymphalornis acutirostris* 13.5cm No similar antwren occurs in the same habitat. **V** High series of repeated, short, chirping phrases. **H** Marshes with surrounding scrub, mangrove swamp. **D** SE Brazil.

25 BANDED ANTBIRD *Dichrozona cincta* 10cm Unmistakable plumage pattern and terrestrial behaviour. **V** Series of 10–15 high, sharp, drawn-out notes that increase in strength and pitch. **H** Open floor of terra firme forest. **D** S and W Amazonia.

26 STAR-THROATED ANTWREN *Rhopias gularis* 9.5cm Throat and front are mottled grey. Note broad wing-bars. **V** Short, hurried, descending series of shrill notes. **H** On or near ground of forest with dense understorey. **D** SE Brazil.

27 PLAIN-THROATED ANTWREN *Isleria hauxwelli* 9.5cm Note grey throat and large white tips to tertials. **V** Long, slowly descending and accelerating series of sharp notes (7–10 secs). **H** On and close to the ground in the understorey of lowland forest. **D** S and W Amazonia.

28 RUFOUS-BELLIED ANTWREN *Isleria guttata* 9cm Note striking wing markings and rufous belly. **V** Long, descending series of shrill notes, the first few well separated (10–15 secs). **H** On or near the ground in lowland forest. **D** NE Amazonia.

29 DUSKY-THROATED ANTSHRIKE *Thamnomanes ardesiacus* 14cm Note small black bib of male. Separable from Saturnine Antshrike by range (there are very small areas of overlap) and from Cinereous Antshrike by black bib. **V** Very high, accelerating, rising series of raspy notes. **H** Lower growth of terra firme forest. **D** N and W Amazonia.

30 SATURNINE ANTSHRIKE *Thamnomanes saturninus* 14cm Large bib of male diagnostic. Female difficult to separate from female Cinereous Antshrike. **V** Short, rising series of very high, piercing notes. **H** Lower growth of terra firme forest. **D** SC and SW Amazonia.

31 CINEREOUS ANTSHRIKE *Thamnomanes caesius* 14.5cm Tail is longer than in Dusky-throated Antshrike and Saturnine Antshrike. **V** Rising series of high, very piercing notes, accelerating to a trill (5–6 secs). **H** Lower storeys of lowland forest. **D** Amazonia, E and SE Brazil.

32 BLUISH-SLATE ANTSHRIKE *Thamnomanes schistogynus* 14cm Male is a distinctive dark bluish grey. Female shows rufous underparts contrasting with grey upper breast and throat. **V** High, descending series of sharp notes, starting slowly and accelerating to a trill (3–4 secs). **H** Lower storeys of lowland forest. **D** C Peru, SW Amazonia.

33 PEARLY ANTSHRIKE *Megastictus margaritatus* 13.5cm Note distinctive round wing spots and large white terminal tips to tertials. **V** Song starts with some low, querulous notes and is followed by rapid, raspy notes. **H** Understorey of forest and secondary growth. **D** W and C Amazonia.

34 BAHIA ANTWREN *Herpsilochmus pileatus* 11cm Range differs from that of larger, longer-tailed Black-capped Antwren. **V** Mid-high, clear, rising and falling rattle (2 secs). **H** Restinga woodland. **D** E Brazil.

35 CAATINGA ANTWREN *Herpsilochmus sellowi* 11cm Male separable from larger male Black-capped Antwren only by voice. Female shows no white/grey streaks in crown (unlike female Black-capped Antwren). **V** High, clear, rising and falling trill (2 secs). **H** Caatinga and restinga scrub. **D** E Brazil.

36 BLACK-CAPPED ANTWREN *Herpsilochmus atricapillus* 12cm Shares only part of its range with the smaller Roraiman Antwren (Plate 135), but can be further separated by habitat. **V** Mid-high, clear, rising and falling rattle (2 secs), decelerating at end. **H** Higher strata of forest. **D** E and SE Brazil to C Bolivia, NW Argentina and Paraguay.

37 ARIPUANA ANTWREN *Herpsilochmus stotzi* 11.5cm Separable from other *Herpsilochmus* antwrens by voice and range. Note that female shows less buff-orange to frontal face than Predicted Antwren. **V** High rattle, slowing down and lowered towards end, and without introductory notes (1 sec). **H** Campinarana forest and woodland. **D** Aripuanã–Machado interfluvial region of C Brazil.

38 PREDICTED ANTWREN *Herpsilochmus praedictus* 11cm Separable from other *Herpsilochmus* antwrens only by voice: **V** Mid-high or high purring trill that increases in force and falls off towards the end (1–2 secs). **H** Campinarana and forest near campina. **D** W Brazil.

39 CREAMY-BELLIED ANTWREN *Herpsilochmus motacilloides* 12cm Does not overlap with Ash-throated Antwren, or Dugand's or Ancient antwrens (Plate 135). **V** High, sharp, descending rattle (2 secs). **H** Higher strata and borders of montane forest. **D** E Peru.

40 ASH-THROATED ANTWREN *Herpsilochmus parkeri* 12cm Separable from Dugand's Antwren (Plate 135) by tail pattern and from Ancient Antwren (Plate 135) by less yellowish underparts. Note extensive buff-rufous areas of plumage on female. **V** High, sharp, descending rattle (2 secs), starting with a two-syllable stutter. **H** Higher strata of montane forest and woodland. **D** NC Peru.

1
2
3
4
5
6
7
8
9
10
11
12
13
14
15
16
17
18
19
20
21
22
23
24
25
26
27
28
29
30
31
32
33
34
35
36
37
38
39
40

ANTBIRDS *CONTINUED*

1 SPOT-TAILED ANTWREN *Herpsilochmus sticturus* 10.5cm Male difficult to separate from male Todd's Antwren, but forecrown solid black, not striped white. Female has a distinctive bright crown. **V** High, sharp, bouncing rattle (2 secs). **H** Higher strata of humid forest; <500m. **D** Venezuela, the Guianas, NE Brazil.

2 DUGAND'S ANTWREN *Herpsilochmus dugandi* 11cm Range differs from those of Spot-tailed and Todd's antwrens, with which it shares a similar tail pattern. **V** High, dry series of notes, accelerating to a rattle and slightly dropping in pitch. **H** Higher levels and borders of humid forest. **D** W Amazonia.

3 TODD'S ANTWREN *Herpsilochmus stictocephalus* 11cm Compare Spot-tailed Antwren. Note lack of rufous buff in forecrown of female. **V** High, short series of notes, falling in pitch and speed. **H** Higher strata of humid lowland forest. **D** Venezuela, the Guianas, NE Brazil.

4 SPOT-BACKED ANTWREN *Herpsilochmus dorsimaculatus* 11.5cm Pied back of both sexes is distinctive. **V** High, hurried, rising and falling rattle. **H** Higher levels and borders of humid forest. **D** NW Amazonia.

5 RORAIMAN ANTWREN *Herpsilochmus roraimae* 12.5cm Note relatively large size, black back streaking and 4–5 white tail-bars. **V** High, chopping rattle, falling off at end. **H** Canopy and borders of humid forest on tepuis. **D** Tepuis of SW and SE Venezuela, N and NW Brazil, and WC Guyana.

6 PECTORAL ANTWREN *Herpsilochmus pectoralis* 11.5cm Black crescent across breast of male diagnostic. Female separable from female Caatinga and Black-capped antwrens (Plate 134) by rufous cap. **V** Steeply rising, chattering rattle (2 secs). **H** Higher levels of Caatinga, secondary growth, restinga, gallery forest. **D** E Brazil.

7 LARGE-BILLED ANTWREN *Herpsilochmus longirostris* 12.5cm Note tiny black spots on breast sides of male. No similar female occurs in its range. **V** Mid-high, chopping rattle, decelerating and falling off at end. **H** Mainly in riverine belts, dry to deciduous forest, palm groves. **D** N Bolivia to C Brazil and south to SE Brazil.

8 ANCIENT ANTWREN *Herpsilochmus gentryi* 11cm Note pale yellow underparts. **V** Short series of high notes, starting fast and then lowering in speed and pitch. **H** Higher strata and borders of lowland forest. **D** E Ecuador, N Peru.

9 YELLOW-BREASTED ANTWREN *Herpsilochmus axillaris* 11.5cm Male separable from male Ancient Antwren by altitude. Female very different from female Ancient Antwren; note darkish face sides. **V** Short, high, descending rattle. **H** Canopy and borders of montane forest; 800–1,800m. **D** W Colombia to SE Peru.

10 RUFOUS-WINGED ANTWREN *Herpsilochmus rufimarginatus* 11.5cm Rufous wings diagnostic. Compare Rufous-winged Tyrannulet (Plate 144), which shows white edges to primaries. **V** Short series of high, descending notes, changing to a rattle. **H** Canopy of humid and montane forest, secondary growth. **D** Panama, N and W Amazonia, E and SE Brazil, E Paraguay, NE Argentina.

11 SPOT-BREASTED ANTVIREO *Dysithamnus stictothorax* 12cm Note spotting to sides of face and on throat. Note also rather pale eyes. Spotting or streaking on crown is diagnostic. **V** Short series of fluted notes, increasing and decreasing in pitch and volume. **H** Lower forest storeys. **D** SE Brazil.

12 PLAIN ANTVIREO *Dysithamnus mentalis* 11.5cm Dark ear-coverts are diagnostic. **V** Short series of high, plaintive notes, starting slowly and accelerating to a downslurred trill. **H** Wide variety of habitats, from evergreen forest in N of range to deciduous forest further S; also in dry savannah forest in NE Brazil. **D** SE Mexico to NE Argentina, Trinidad and Tobago.

13 STREAK-CROWNED ANTVIREO *Dysithamnus striaticeps* 11cm Separated from Spot-crowned Antvireo by more distinct breast streaking; spots on crown arranged in rows. **V** Very high *wutwutwutwutwutwuterrrrrr*. **H** Understorey and middle levels of forest. **D** SE Honduras to Costa Rica.

14 SPOT-CROWNED ANTVIREO *Dysithamnus puncticeps* 11.5cm Note spotted crown and pale eyes. **V** High, musical series that starts with some staccato notes, then accelerates to a trill. **H** Lower storeys of humid forest. **D** Costa Rica to NW Ecuador.

15 RUFOUS-BACKED ANTVIREO *Dysithamnus xanthopterus* 12cm Note distinctive rufous upperparts and streaked face sides. **V** Short, rapid series of sharp, descending notes. **H** Higher levels in forest. **D** SE Brazil.

16 BICOLORED ANTVIREO *Dysithamnus occidentalis* 13.5cm Male difficult to separate. Pale streaks to face sides and throat in female are distinctive. **V** Short, loud series of rapidly descending notes, accelerating to a very short rattle. **H** Undergrowth of montane forest. **D** SW Colombia, N and E Ecuador.

17 PLUMBEOUS ANTVIREO *Dysithamnus plumbeus* 12.5cm Note white shoulders of male; no similar antvireo occurs in its range. **V** Slow, rising and descending series of 7–8 low, rather plaintive, fluted notes. **H** Undergrowth of humid lowland forest. **D** SE Brazil.

18 WHITE-STREAKED ANTVIREO *Dysithamnus leucostictus* 12.5cm Note black throat of male. Female has distinctive bold streaking below. **V** Short series of whistles, falling in pitch. **H** Undergrowth of montane forest. **D** C Colombia and N Venezuela to Peru.

19 COLLARED ANTSHRIKE *Thamnophilus bernardi* 16.5cm Unmistakable in its small range due to collar of male and mask of female. **V** Very fast, rattling song (2 secs). **H** Dense undergrowth of woodland, arid scrub, mangroves. **D** W Ecuador, W Peru.

20 BLACK-BACKED ANTSHRIKE *Thamnophilus melanonotus* 15.5cm Compare Black-crested Antshrike (Plate 136). **V** Song is a mid-high, short roll. **H** Dense, viny thickets in woodland and arid scrub. **D** N Colombia, Venezuela.

21 BAND-TAILED ANTSHRIKE *Thamnophilus melanothorax* 16.5cm White terminal tail-band of male and black bib of female are diagnostic. **V** Accelerating series of hollow notes, bouncing down. **H** Undergrowth of humid forest, especially near tree-falls. **D** NE Amazonia.

22 BARRED ANTSHRIKE *Thamnophilus doliatus* 16cm Note yellow eyes and black crest. Compare underparts of Chapman's Antshrike. **V** Accelerating, descending, fast series of notes (magpie-like), with emphasised note at end (1–2 secs). **H** At edges and clearings in humid and dry forest, parks and gardens. **D** Widespread across Central and South America, from Mexico to SW Brazil.

23 CHAPMAN'S ANTSHRIKE *Thamnophilus zarumae* 15cm Note faintly streaked lower underparts of male. Female separable from female Bar-crested Antshrike and female Rufous-capped and Rufous-winged antshrikes (Plate 136) by different range. **V** High rattle, accelerating to a bouncing trill. **H** Undergrowth at forest edges and in clearings. **D** SW Ecuador, NW Peru.

24 BAR-CRESTED ANTSHRIKE *Thamnophilus multistriatus* 16cm Crown of male is spotted or barred. **V** Short, nasal series of notes, accelerating to a rattle. **H** Forest, secondary growth, cultivated areas, gardens; 2,200–2,900m, locally lower. **D** Colombia and W Venezuela.

25 LINED ANTSHRIKE *Thamnophilus tenuepunctatus* 16cm Male similar to male Barred Antshrike but width of black barring may make it appear blacker. Female not safely separable from female Chestnut-backed Antshrike, but ranges do not, or only marginally, overlap. **V** Short, accelerating, rattling series of notes with final mewing note. **H** Lower storeys of humid forest, scrubby clearings, secondary growth, dense tangles in parks. **D** NC Colombia to NE Peru.

26 CHESTNUT-BACKED ANTSHRIKE *Thamnophilus palliatus* 16.5cm Male is the sole antshrike with a black cap and barred underparts. Compare Lined Antshrike. **V** Song as Lined Antshrike. **H** Undergrowth of forest clearings, secondary woodland; <2,300m in Andes, elsewhere much lower. **D** S Amazonia, SE Brazil.

27 BLACK-HOODED ANTSHRIKE *Thamnophilus bridgesi* 17cm Unmistakable when sparse wing spotting can be seen. Female as Black Antshrike, but with olive upperparts. **V** High wooden *wekwekwekkerwikkerwek*, varying in speed and length. **H** Thickets at forest edges, in woodland, also mangroves. **D** Costa Rica, Panama.

28 BLACK ANTSHRIKE *Thamnophilus nigriceps* 16cm No other black antshrike without white wing spotting occurs in its range. Female unmistakable. **V** Rather level, accelerating series of mellow notes. **H** Shrubby understorey, edges and clearings in marshy forest and woodland. **D** E Panama, N Colombia.

29 COCHA ANTSHRIKE *Thamnophilus praecox* 16cm Separable from larger White-shouldered Antbird (Plate 138) by lack of bare blue skin around eyes. **V** Fast, almost level series of hollow notes. **H** Understorey of seasonally flooded forest. **D** NE Ecuador.

30 BLACKISH-GREY ANTSHRIKE *Thamnophilus nigrocinereus* 17cm Grey-headed female is distinctive. Note narrow fringing and small spotting to wings of male. **V** High, descending series of downslurred, nasal notes. **H** Lower storeys of humid forest along streams, locally in scrubby savannah woodland. **D** N Amazonia.

31 CASTELNAU'S ANTSHRIKE *Thamnophilus cryptoleucus* 16.5cm Male unmistakable. Female difficult to separate from other all-black antshrikes and antbirds, but note rather slender yet heavy bill. **V** Accelerating, downslurred series of hollow notes, the first few well separated. **H** Lower storeys of forest along rivers and streams. **D** W Amazonia.

32 WHITE-SHOULDERED ANTSHRIKE *Thamnophilus aethiops* 16cm Male shows more or less wing spotting. Female is uniform rufous brown, including sides of face. **V** Song is a slow, rising series of 5–6 nasal notes. **H** Understorey at gaps and streams in forest. **D** Amazonia.

33 UNIFORM ANTSHRIKE *Thamnophilus unicolor* 15cm No similar bird occurs in its montane range. **V** Short, accelerating series of nasal notes. **H** Dense undergrowth of montane forest; 1,200–2,300m, locally lower or higher. **D** Colombia to C Peru.

34 PLAIN-WINGED ANTSHRIKE *Thamnophilus schistaceus* 14.5cm Mainly grey, not black. Note rufous crown of female. **V** Rather short, accelerating series of nasal notes, concluding with a lower-pitched shriek. **H** Understorey of humid forest. **D** W and S Amazonia.

35 MOUSE-COLORED ANTSHRIKE *Thamnophilus murinus* 14.5cm Note hazel eyes and unobtrusive wing spotting. **V** Very short, mid-high series of notes, ending with a doubled final note. **H** Understorey of humid forest. **D** N and W Amazonia.

36 UPLAND ANTSHRIKE *Thamnophilus aroyae* 14.5cm Note sepia tinge on male's wings. Female tawnier below than female Plain-winged Antshrike. **V** Song a level series of 6–7 evenly spaced nasal notes, the last one lower. **H** Dense edges with bamboo in montane forest. **D** SE Peru, NW Bolivia.

37 BLACK-CROWNED ANTSHRIKE *Thamnophilus atrinucha* 14cm Note relatively long and heavy bill. No similar antshrike occurs W of Andes. **V** Song an accelerating, nasal series of notes, concluding with a high final shriek. **H** Undergrowth of forest, secondary growth and woodland. **D** S Belize to NW Venezuela and NW Peru.

38 NORTHERN SLATY ANTSHRIKE *Thamnophilus punctatus* 14cm Northern, Natterer's and Bolivian Slaty antshrikes, and Planalto and Sooretama slaty antshrikes (species on Plate 136) separated mostly on basis of voice and range (check range particularly). Difficult to separate from Amazonian Antshrike (Plate 136), but that species does not wag its tail. **V** Series of 12–16 nasal notes, starting slowly, then accelerating. **H** Understorey of lowland forest, woodland, gallery forest, secondary growth. **D** N and W Amazonia.

39 NATTERER'S SLATY ANTSHRIKE *Thamnophilus stictocephalus* 14.5cm See Northern Slaty Antshrike. **V** Series of 12–16 nasal notes, starting slowly, then accelerating to a downslurred rattle. **H** Understorey of forest and woodland. **D** S Amazonia.

40 BOLIVIAN SLATY ANTSHRIKE *Thamnophilus sticturus* 14.5cm See Northern Slaty Antshrike. **V** Rising, level or slightly descending series of sharp, high notes, accelerating to a shrieking rattle. **H** Lower storeys of dry and humid forest and woodland. **D** SW Amazonia.

1
2
3
4
5
6
7
8
9
10
11
12
13
14
15
16
17
18
19
20
21
22
23
24
25
26
27
28
29
30
31
32
33
34
35
36
37
38
39
40

ANTBIRDS *CONTINUED*

1 PLANALTO SLATY ANTSHRIKE *Thamnophilus pelzelni* 14.5cm See Northern Slaty Antshrike (Plate 135). Compare Variable Antshrike, with which it overlaps partially in Minas Gerais and Bahia, Brazil. **V** Song is a rising, cackling, accelerating series of notes. **H** Understorey of dry forest and riverine belts. **D** E and SC Brazil.

2 SOORETAMA SLATY ANTSHRIKE *Thamnophilus ambiguus* 14.5cm See Northern Slaty Antshrike (Plate 135). **V** Song is an accelerating, rising series of notes. **H** Understorey of lowland forest and woodland. **D** SE Brazil.

3 AMAZONIAN ANTSHRIKE *Thamnophilus amazonicus* 14.5cm Mantle usually blacker than those of Northern, Natterer's and Bolivian slaty antshrikes (Plate 135), Planalto and Sooretama slaty antshrikes, and Variable Antshrike. Note extensive orange on head of female compared with others in similar range. **V** Rather rapid rising and descending series of mid-high nasal notes. **H** Lower storeys of forest, savannah woodland, riverine belts, secondary growth. **D** Amazonia.

4 ACRE ANTSHRIKE *Thamnophilus divisorius* 16.5cm No similar antshrike occurs in its range. **V** Rapid, slightly descending and rising series of nasal notes, the last one accentuated as a full stop. **H** Stunted forest and woodland. **D** W Brazil, E Peru.

5 STREAK-BACKED ANTSHRIKE *Thamnophilus insignis* 16.5cm Note characteristic irregular white streaks on head and mantle. **V** Accelerating series of nasal notes, ending with an upslurred shriek. **H** Interior and edges of forest on tepuis. **D** NC Amazonia.

6 VARIABLE ANTSHRIKE *Thamnophilus caerulescens* 15cm Very variable indeed, but note that female shows rufous lower belly and vent. **V** Hurried, level series of 4–8 nasal notes. **H** Humid, dry or montane forest, especially at edges. **D** Peru; SE and WC Bolivia; NE, SE and SC Brazil; N Paraguay; NE and NW Argentina.

7 RUFOUS-WINGED ANTSHRIKE *Thamnophilus torquatus* 14cm Compare Rufous-capped Antshrike; male separable from male of that species by black cap. **V** Song as Rufous-capped but longer. **H** Open woodland and secondary growth. **D** SE Amazonia, E Brazil, NE Paraguay.

8 RUFOUS-CAPPED ANTSHRIKE *Thamnophilus ruficapillus* 16cm Note rufous cap and chequered tail pattern of male. Female not safely separable from female Rufous-winged Antshrike. **V** Short, accelerating series of loud, ringing, rather shrill notes. **H** Humid forest, forest patches and scrub, in most of range at higher elevations. **D** N Peru to C Bolivia, SE Brazil, NE Argentina and Uruguay.

9 BLACK-CRESTED ANTSHRIKE *Sakesphorus canadensis* 15cm Note conspicuous crest and greenish-brown mantle (male). Separable from Collared Antshrike (Plate 135) by range. **V** Rising and accelerating series of rather staccato notes (2 secs). **H** Savannah woodland, mangroves, shrubby forest borders. **D** W Amazonia, N South America.

10 SILVERY-CHEEKED ANTSHRIKE *Sakesphorus cristatus* 14cm Note spotted wing coverts in both male and female. **V** Falling and accelerating series of nasal, chopped notes (2 secs). **H** Low forest, deciduous woodland, arid scrub. **D** E Brazil.

11 GLOSSY ANTSHRIKE *Sakesphorus luctuosus* 17cm Note shaggy crest, white edges of mantle and white tips to tail. **V** Song is rather level, starting slowly but ending in an accelerating series of loud, chopped notes (3 secs). **H** Flooded forest, forested river islands. **D** C and E Amazonian Brazil.

12 WHITE-BEARDED ANTSHRIKE *Biatas nigropectus* 17.5cm Distinguishable by its white 'scarf'. **V** Mid-high series of 7–9 level notes, starting lower and trailing off at end. **H** Lower forest storeys. **D** E Paraguay, NE Argentina, SE Brazil.

13 FASCIATED ANTSHRIKE *Cymbilaimus lineatus* 17.5cm Note heavy, hooked bill and red eyes. Crown occasionally solid black. **V** Song comprises 4–6 solemn, melancholy notes. **H** Dense growth in lowland forest. **D** SE Honduras through Amazonia.

14 BAMBOO ANTSHRIKE *Cymbilaimus sanctaemariae* 17cm Note dense barring, crests (bicoloured in female) and dark eyes. **V** Song comprises 8–10 high, staccato notes. **H** Bamboo thickets in humid forest. **D** SW Amazonia.

15 GREAT ANTSHRIKE *Taraba major* 19.5cm Note red eyes and white underparts. Female shows rufous upperparts without white. **V** Gradually accelerating series of mid-high notes, bouncing to end and often concluding in a snarl. **H** Dense undergrowth at forest edges and woodland. **D** E Mexico to N Argentina.

16 LARGE-TAILED ANTSHRIKE *Mackenziaena leachii* 26cm Distinctly marked; note long tail. **V** Very high series of rising, piercing notes. **H** Dense undergrowth of forest and woodland. **D** S Brazil, E Paraguay, NE Argentina.

17 TUFTED ANTSHRIKE *Mackenziaena severa* 22cm Note upright crest and large size. **V** Slightly accelerating and rising series of 5–7 high, piercing notes. **H** Dense undergrowth, especially bamboo. **D** S Brazil, E Paraguay, NE Argentina.

18 BLACK-THROATED ANTSHRIKE *Frederickena viridis* 20cm Note large size, bushy crest and, in female, barred tail. **V** Series of melancholy notes, slowly rising and then levelling out. **H** Undergrowth of humid forest. **D** NE Amazonia.

19 UNDULATED ANTSHRIKE *Frederickena unduliger* 23cm Large, with a prominent crest. Male dark grey with wavy whitish barring; female rufous with wavy blackish barring (this varies on belly depending on subspecies). **V** Rather fast series of steadily rising notes (3 secs). **H** Dense undergrowth of lowland forest. **D** W Amazonia.

20 FULVOUS ANTSHRIKE *Frederickena fulva* 23cm Separable from Undulated Antshrike by darker, densely marked plumage. **V** Loud series of 10–12 rising notes. **H** Dense thickets and vine tangles in humid forest. **D** SC Colombia, E Ecuador, NE Peru.

21 SPOT-BACKED ANTSHRIKE *Hypoedaleus guttatus* 20.5cm Clean underparts make it unmistakable in its range. **V** Song is a fast series of slightly rising and falling sharp notes (3–4 secs). **H** Subcanopy, canopy and edges of forest. **D** E and SE Brazil, Paraguay, NE Argentina.

22 GIANT ANTSHRIKE *Batara cinerea* 32cm Very large size makes it unmistakable. **V** Fast series of mid-high rising notes preceded by a very short trill. **H** Understorey and mid-storey of various forest types and dense tree stands; locally to 2,600m. **D** SW Amazonia to SE Brazil, W Paraguay and NE Argentina.

23 SPECKLED ANTSHRIKE *Xenornis setifrons* 15.5cm Distinctively coloured and patterned. **V** Short, very high series of slightly upslurred, piercing notes. **H** Understorey of humid forest. **D** E Panama, NW Colombia.

24 WHITE-PLUMED ANTBIRD *Pithys albifrons* 12cm Unmistakable. **V** Very slow series of high, downslurred *tseeeu* notes. **H** Understorey of humid forest and secondary growth. **D** N and W Amazonia.

25 WHITE-MASKED ANTBIRD *Pithys castaneus* 14cm Note white face mask (without plume) and orange legs. **V** Single long, rising note. **H** Undergrowth of humid forest. **D** NE Peru.

26 OCELLATED ANTBIRD *Phaenostictus mcleannani* 19.5cm Unmistakable, especially due to large blue area at sides of face. **V** Steeply rising series of notes, accelerating to a twitter that suddenly changes to a short series of four notes at a calm tempo. **H** Understorey of humid forest and secondary growth. **D** Honduras to NW Ecuador.

27 BICOLOURED ANTBIRD *Gymnopithys bicolor* 15cm Unmistakable; note black cheeks. **V** Raucous *wrouw*; ultra-high *sweepsweep-pijeh*. **H** Forest undergrowth. **D** Honduras to W Ecuador.

28 WHITE-CHEEKED ANTBIRD *Gymnopithys leucaspis* 14cm No similar antbird with brown upperparts and white underparts exists. **V** Song a series of 5–6 loud, sharp, fluted notes, followed by a musical, descending chatter that becomes harsh at end (3 secs). **H** Undergrowth of humid lowland and foothill forest. **D** C Colombia to NE Peru, NW Amazonian Brazil.

29 RUFOUS-THROATED ANTBIRD *Gymnopithys rufigula* 13cm Very distinct bluish eye-rings. **V** Loud, fast chatter, rising and then suddenly descending. **H** Undergrowth of humid forest. **D** NC and NE Amazonia.

30 WHITE-THROATED ANTBIRD *Oneillornis salvini* 14cm Barred tail is diagnostic. **V** Song a descending series of 3–4 drawn-out, harsh notes. **H** Understorey of humid forest. **D** SW Amazonia.

31 LUNULATED ANTBIRD *Oneillornis lunulatus* 14cm Male separable from male White-throated Antbird by unbarred tail. Female separable from female White-throated Antbird by white face sides. **V** Song is a level series of high, fluted notes, distinctly slowing down and becoming very harsh at end. **H** Understorey of várzea. **D** W Amazonia.

32 BARE-EYED ANTBIRD *Rhegmatorhina gymnops* 14cm No similar bird occurs in its limited range. **V** High, descending, slightly accelerating series of fluted to wheezy notes. **H** Undergrowth of terra firme forest. **D** SC Amazonia.

33 HARLEQUIN ANTBIRD *Rhegmatorhina berlepschi* 15cm No similar bird occurs in its range. **V** High, descending, slightly accelerating series of wheezy notes, the first of which is drawn out. **H** Undergrowth of terra firme forest. **D** SC Amazonia.

34 WHITE-BREASTED ANTBIRD *Rhegmatorhina hoffmannsi* 15cm No similar bird occurs in its range. **V** Song resembles that of Bare-eyed Antbird. **H** Undergrowth of terra firme forest. **D** SC Amazonia.

35 CHESTNUT-CRESTED ANTBIRD *Rhegmatorhina cristata* 15cm No similar bird occurs in its limited range. Female resembles male but with some dotting on mantle. **V** Song resembles that of Bare-eyed Antbird but slightly slower. **H** Undergrowth of terra firme forest. **D** NW Amazonia.

36 HAIRY-CRESTED ANTBIRD *Rhegmatorhina melanosticta* 15cm Note fluffy crest. **V** Song resembles that of Bare-eyed Antbird, but first part is at a level pitch. **H** Undergrowth of terra firme forest. **D** W Amazonia.

37 BLACK-SPOTTED BARE-EYE *Phlegopsis nigromaculata* 17.5cm Note bare red skin around eyes. Separable from Reddish-winged Bare-eye by different pattern and colour of mantle and wings. **V** Slow, unhurried series of three descending, harsh, fluted notes. **H** Undergrowth of humid forest. **D** S and W Amazonia.

38 REDDISH-WINGED BARE-EYE *Phlegopsis erythroptera* 18cm Compare Black-spotted Bare-eye. **V** Descending, calm series of 3–6 hoarse notes that diminish in force. **H** Undergrowth of terra firme forest. **D** W and NW Amazonia.

39 PALE-FACED BARE-EYE *Phlegopsis borbae* 17cm Unmistakable, especially due to tufts in front of eyes. **V** Calm, slightly descending series of three drawn-out notes. **H** Understorey of humid forest. **D** SC Amazonia.

1
2
3
4
5
6
7
8
9
10
11
12
13
14
15
16
17
18
19
20
21
22
23
24
25
26
27
28
29
30
31
32
33
34
35
36
37
38
39

ANTBIRDS *CONTINUED*

1 COMMON SCALE-BACKED ANTBIRD *Willisornis poecilinotus* 13cm Birds of nominate race (illustrated) and other some others are distinctively scaled. See also Xingu Scale-backed Antbird. **V** Slow series of 9–11 sharp, drawn-out, high to ultra-high notes. **H** Undergrowth of humid forest. **D** Amazonia.

2 XINGU SCALE-BACKED ANTBIRD *Willisornis vidua* 13cm Male has a darker head than male Common Scale-backed Antbird. **V** Sings as Common Scale-backed Antbird. **H** Undergrowth of humid forest. **D** SC and SE Amazonian Brazil.

3 FERRUGINOUS ANTBIRD *Drymophila ferruginea* 13cm Not reliably separable from Bertoni's Antbird, but somewhat darker and with a blacker tail. **V** Very high series of irregular, repeated *jee-juwee* notes. **H** Undergrowth of forest, secondary growth. **D** SE Brazil.

4 BERTONI'S ANTBIRD *Drymophila rubricollis* 13.5cm Compare Ferruginous Antbird. **V** High, sharp, short series of 5–8 descending notes, the last one slightly higher. **H** Lower levels and edges of humid forest. **D** E Paraguay, NE Argentina, S and SE Brazil.

5 RUFOUS-TAILED ANTBIRD *Drymophila genei* 14cm Rufous tail and wings are diagnostic. **V** Level series of 5–6 snarly notes. **H** Bamboo thickets in montane forest. **D** SE Brazil.

6 OCHRE-RUMPED ANTBIRD *Drymophila ochropyga* 13cm Male streaked below, not scaled or solid rufous. Female has a black-streaked crown. **V** Simple two-syllable phrase, first syllable high and sharp, the second a hoarse snarl. **H** Undergrowth, borders and clearings with bamboo of montane forest. **D** E and SE Brazil.

7 DUSKY-TAILED ANTBIRD *Drymophila malura* 14cm Note distinct colour pattern and fine streaking. **V** High, sharp series of accelerating notes that bounce down. **H** Undergrowth and edges in lowland and foothill forest and secondary growth. **D** E Paraguay, NE Argentina, S and SE Brazil.

8 SCALED ANTBIRD *Drymophila squamata* 12cm Tail pattern is diagnostic. Note scaling of male and round buff spots of female. **V** Calm, very high series of 4–7 descending, sharp notes. **H** Undergrowth and borders of lowland and foothill forest and woodland. **D** E and SE Brazil.

9 STRIATED ANTBIRD *Drymophila devillei* 13.5cm Note long black tail adorned with large white dots on central feathers. **V** Series of three level, sharp, staccato notes followed by a downslurred warble. **H** Bamboo thickets in lowland and foothill forest. **D** S and SW Amazonia.

10 SANTA MARTA ANTBIRD *Drymophila hellmayri* 15–16cm Previously considered conspecific with East Andean Antbird. Note reddish (not grey) tail and isolated range. Gleans invertebrates from foliage. **V** Song a piping *tchup, tchup, t-chrrwerr, t-chrrwerr*. Call an abrupt *tChiup*. **H** Dense understorey of wooded slopes, bamboo thickets. **D** Sierra Nevada de Santa Marta, N Colombia.

11 KLAGES'S ANTBIRD *Drymophila klagesi* 15–16cm Previously considered conspecific with East Andean Antbird. Note limited streaking on throat and breast. **V** Song a shrill, grating *tchip, tchip, err-tchi-e-tchi*. Call an abrupt *tchiup-tchiup*. **H** Dense understorey of wooded slopes, bamboo thickets. **D** Venezuela, NE Colombia.

12 EAST ANDEAN ANTBIRD *Drymophila caudata* 15cm Separable from smaller Striated Antbird by solid dusky colour of tail; note also difference in habitats. **V** Song comprises 2–3 high, strident notes, followed by three hurried, hoarse notes. **H** Bamboo undergrowth of montane forest. **D** E Colombia.

13 STREAK-HEADED ANTBIRD *Drymophila striaticeps* 15–16cm Previously considered conspecific with East Andean Antbird. Vocalisation and subtle plumage differences aid identification. **V** Song a shrill *tchip, tchcip, t-cher-ip, t-cher-ip*. Call a shrill *tchip*. **H** Dense understorey of wooded slopes, bamboo thickets. **D** W and C Colombia, Ecuador, Peru and Bolivia along the main Andean cordillera.

14 GUIANAN WARBLING ANTBIRD *Hypocnemis cantator* 12cm This species, along with Imeri, Peruvian, Yellow-breasted, Rondonia and Spix's warbling antbirds, plus Yellow-browed Antbird, were formerly considered conspecific; they are now split mainly on the basis of voice. Note also range. **V** High series of hurried-up notes, becoming very harsh at end. **H** Undergrowth and borders of humid forest, secondary growth. **D** E Venezuela, the Guianas, NE Amazonian Brazil.

15 IMERI WARBLING ANTBIRD *Hypocnemis flavescens* 12cm Compare Guianan Warbling Antbird. **V** High series of notes, starting staccato and becoming very harsh. **H** Undergrowth and borders of humid forest, secondary growth. **D** S Venezuela, E Colombia, NW Brazil.

16 PERUVIAN WARBLING ANTBIRD *Hypocnemis peruviana* 12cm Compare Guianan Warbling Antbird. **V** Rather calm, loud series of notes, descending to rather harsh notes. **H** Undergrowth and borders of humid forest, secondary growth. **D** SE Colombia, E Ecuador, NE and E Peru, W Brazil.

17 YELLOW-BREASTED WARBLING ANTBIRD *Hypocnemis subflava* 12cm Compare Guianan Warbling Antbird. Separable from other warbling antbirds by yellow underparts combined with white eyebrow. **V** Loud series of 6–7 slightly descending, well-spaced notes. **H** Undergrowth and borders of humid forest, secondary growth. **D** C and SE Peru, N Bolivia.

18 RONDONIA WARBLING ANTBIRD *Hypocnemis ochrogyna* 12cm Compare Guianan Warbling Antbird. See also Manicore Warbling Antbird. **V** Sings as Yellow-breasted Warbling Antbird, but last few notes are harsher. **H** Undergrowth and borders of humid forest, secondary growth. **D** NE Bolivia and SC Amazonian Brazil.

19 SPIX'S WARBLING ANTBIRD *Hypocnemis striata* 12cm Compare Guianan Warbling Antbird. **V** Song resembles that of Yellow-breasted Warbling Antbird, but lower and longer. **H** Undergrowth and borders of humid forest, secondary growth. **D** C and EC Brazil.

20 MANICORE WARBLING ANTBIRD *Hypocnemis rondoni* 11.5cm Differs from similar *Hypocnemis* warbling antbirds by more distinct rufous edging to tail feathers. **V** Bouncing, accelerating, slightly descending series of loud, fluted notes (1–2 secs). **H** Understorey of terra firme forest. **D** Aripuanã–Machado interfluvial region, C Brazil.

21 YELLOW-BROWED ANTBIRD *Hypocnemis hypoxantha* 12cm Compare Guianan Warbling Antbird. **V** Song resembles that of Yellow-breasted Warbling Antbird, but more staccato. **H** Lower strata of lowland and foothill forest. **D** W and S Amazonia.

22 SOUTHERN CHESTNUT-TAILED ANTBIRD *Sciaphylax hemimelaena* 12cm Separable from Northern Chestnut-tailed Antbird by redder tail, range and voice. **V** Short, accelerating, descending series of *klee* notes. **H** Understorey and floor of lowland and foothill forest. **D** S Amazonia.

23 NORTHERN CHESTNUT-TAILED ANTBIRD *Sciaphylax castanea* 11.5cm Resembles Southern Chestnut-tailed Antbird, but never shows pale tips to tertials. **V** Short, accelerating, rising series of fluted notes. **H** Undergrowth and floor of humid forest. **D** W Amazonia.

24 WILLIS'S ANTBIRD *Cercomacroides laeta* 14cm See Dusky Antbird. Separable from that species by its very different song. **V** Short, very high, sharp series of 3–5 *peeur* notes. **H** Edges of humid forest and woodland, usually near water. **D** NC and SE Amazonia, E Brazil.

25 PARKER'S ANTBIRD *Cercomacroides parkeri* 14cm Separable from Dusky Antbird by its occurrence at higher altitudes and different song. **V** Short, high, sharp series of 6–7 *pee* notes, lowered at end. **H** Borders of montane forest and woodland; 1,100–1,900m. **D** WC Colombia.

26 BLACKISH ANTBIRD *Cercomacroides nigrescens* 14.5cm Not reliably separable from Black Antbird (male of which is blacker below, and female of which lacks rufous on forehead; see also ranges). **V** Short, rapid series of 4–8 chopped, harsh notes, preceded by a single *wor*. **H** Understorey with bamboo in humid forest, secondary growth, abandoned plantations. **D** S, W and NE Amazonia.

27 RIPARIAN ANTBIRD *Cercomacroides fuscicauda* 14–15cm Previously treated as a subspecies of Blackish Antbird. Vocalisation aids identification. **V** Song a shrill, vibrating *pu-CHEeerrrr*, second phrase downslurred. Call a buzzing *tcherrr*. **H** Favours várzea rainforest. **D** S Colombia, E Ecuador, E Peru, N Bolivia, SW Amazonian Brazil.

28 DUSKY ANTBIRD *Cercomacroides tyrannina* 14cm Not reliably separable from Willis's Antbird (which might show broader wing-bars), Parker's Antbird (which normally occurs at higher altitudes) or Blackish Antbird (which lacks white tail tips). **V** Fast, high, descending chatter. **H** Thick undergrowth and borders of forest and secondary growth; <800m, locally to 1,200m. **D** SE Mexico through N Amazonia.

29 BLACK ANTBIRD *Cercomacroides serva* 14.5cm See Blackish Antbird. **V** Loud, rising, slightly accelerating series of 5–10 sharp notes. **H** Undergrowth of humid forest and secondary growth. **D** E Ecuador, E Peru, N Bolivia, W Brazil.

30 MANU ANTBIRD *Cercomacra manu* 15cm See Blackish Antbird, which lacks this species' white tail tips. **V** Low, tuneless croaking notes in phrases of Black Antbird, and often in unstructured series. **H** Bamboo stands in humid forest. **D** SW and S Amazonia.

31 RIO DE JANEIRO ANTBIRD *Cercomacra brasiliana* 14cm No similar antbird occurs in its range. **V** Slow series of 4–7 repeated, low *kruh* notes. **H** Undergrowth of forest and secondary growth. **D** SE Brazil.

32 GREY ANTBIRD *Cercomacra cinerascens* 14cm Male SW ssp. *sclateri* and female NW nominate race are shown. Races differ in size of wing spots, these often missing in nominate. Large white tips to tail feathers are diagnostic. **V** Calm, rhythmic series of 7–9 *ch-krr* notes. **H** Mid-storey of humid forest with tangles of vines. **D** Amazonia.

33 MATO GROSSO ANTBIRD *Cercomacra melanaria* 16cm No similar antbird occurs in its range. **V** Slow, tuneless series of low, croaking *ker-cheeer-ker* phrases. **H** Viny understorey of riverine forest belts, savannah woodland near streams. **D** C and E Bolivia, N Paraguay, W Brazil.

34 BANANAL ANTBIRD *Cercomacra ferdinandi* 16cm No similar antbird occurs in its range. **V** Short, rhythmic series of low, croaked *oh-tchek* notes. **H** Thick undergrowth of riverine belts and islands. **D** C Brazil.

35 JET ANTBIRD *Cercomacra nigricans* 15cm Male is jet black. Female similar to female Bananal Antbird but ranges differ. **V** Calm, rhythmic series of croaking *oh-tchek* phrases. **H** Lower storeys of forest and secondary growth. **D** Panama to W Ecuador, NW Colombia and E Venezuela.

36 RIO BRANCO ANTBIRD *Cercomacra carbonaria* 15cm No similar antbird occurs in its range. **V** Calm, rhythmic series of croaking *oh-tchek* phrases (even slower than in song of Jet Antbird). **H** Lower storeys of riverine forest belts and edges of humid forest along streams. **D** N Brazil, SW Guyana.

37 FERRUGINOUS-BACKED ANTBIRD *Myrmoderus ferrugineus* 15cm Distinctively patterned and coloured. **V** Very high, hurried, slightly descending series of *wheehee* notes. **H** Floor of forest and tall secondary growth. **D** NE Amazonia.

38 CORDILLERA AZUL ANTBIRD *Myrmoderus eowilsoni* 14–15cm Described as a new species in 2016 and closely allied to the Ferruginous-backed Antbird, but with distinct plumage differences. Male with much more black below and female with russet breast and belly. **V** Song a shrill, whistled *te, ter, tu-tu*, pitch descending during delivery. Call a thrush-like chattering *tr-tch-ch*. **H** Humid montane forest; around 1,500m. **D** Cordillera Azul mountains, N Peru.

39 SCALLOPED ANTBIRD *Myrmoderus ruficauda* 14.5cm Distinctively patterned and coloured. Note lack of eyebrow. **V** Short, very high, piercing rattle. **H** Floor of humid forest and secondary growth. **D** E and SE Brazil.

40 WHITE-BIBBED ANTBIRD *Myrmoderus loricatus* 15cm Separable from Squamate Antbird by white throat. **V** Song resembles that of Scalloped Antbird but slower. **H** Floor of humid forest and secondary growth. **D** E and SE Brazil.

41 SQUAMATE ANTBIRD *Myrmoderus squamosus* 15cm See White-bibbed Antbird. **V** Short, very high, piercing series of notes, slightly falling off. **H** Floor of humid forest, secondary growth, restinga. **D** SE Brazil.

1
2
3
4
5
6
7
8
9
10
11
12
13
14
15
16
17
18
19
20
21
22
23
24
25
26
27
28
29
30
31
32
33
34
35
36
37
38
39
40
41

ANTBIRDS *CONTINUED*

1 BLACK-CHINNED ANTBIRD *Hypocnemoides melanopogon* 11.5cm Separable from Band-tailed Antbird by narrower white tail tips. **V** Rather level or descending series of high, sharp notes (2–3 secs). **H** Near or at water in undergrowth of várzea, riverine belts and savannah woodland. **D** C and N Amazonia.

2 BAND-TAILED ANTBIRD *Hypocnemoides maculicauda* 12cm Note pale eyes. Broad white tail-band is diagnostic. **V** Sharp, lowered rattle. **H** Near or at water in undergrowth of várzea, riverine belts and savannah woodland. **D** E Peru, N Bolivia, C and SE Brazil.

3 SPOTTED ANTBIRD *Hylophylax naevioides* 11cm Occurs only W of Andes. **V** Hurried series of very high, sharp *wheezee* notes (3–4 secs). **H** Undergrowth of humid and foothill forest. **D** Honduras to W Ecuador.

4 SPOT-BACKED ANTBIRD *Hylophylax naevius* 11cm Occurs only E of Andes. **V** Hurried series of very high, piercing *wheezee* notes, lowered at end (3–4 secs). **H** Undergrowth of humid lowland and foothill forest. **D** Amazonia.

5 DOT-BACKED ANTBIRD *Hylophylax punctulatus* 10.5cm Separable from Spot-backed Antbird by smaller dots on back and whiter sides of face. **V** Long series of well-interrupted, piercing *whee-beeyr* notes. **H** Undergrowth of forest and woodland, usually near streams. **D** Amazonia.

6 SILVERED ANTBIRD *Sclateria naevia* 15cm Compare with Yapacana Antbird (Plate 133), the male of which shows a bicoloured bill. **V** High, sharp, slightly undulating, rattling series of notes (4–5 secs). **H** Near water in understorey or on floor of forest and woodland. **D** Amazonia.

7 PLUMBEOUS ANTBIRD *Myrmelastes hyperythrus* 17cm Note combination of blue skin around eyes and wing spotting. **V** Sharp, rising chatter (2–3 secs). **H** Undergrowth and floor at forest edges. **D** W and SW Amazonia.

8 SLATE-COLORED ANTBIRD *Myrmelastes schistaceus* 14.5cm Male separable from male Spot-winged, Humaita, Brownish-headed, Rufous-faced and Roraiman antbirds by all-black bill. Female rather uniform rufous below with darker shaft streaks through crown. **V** Loud series of 8–10 sharp *pyeer* notes. **H** Understorey and floor of lowland forest. **D** W Amazonia.

9 SPOT-WINGED ANTBIRD *Myrmelastes leucostigma* 15cm This species and Humaita, Brownish-headed, Rufous-faced and Roraiman antbirds were formerly considered conspecific, but are now split based mainly on voice; they share wing spotting but not white fringing. Males are not safely separable; females are recognisable by colour of cheeks. **V** Very high, sharp, rattle-like song, downslurred at end (3 secs). **H** Understorey of lowland, foothill and montane forest. **D** N Amazonia.

10 HUMAITA ANTBIRD *Myrmelastes humaythae* 15cm See Spot-winged Antbird and range. **V** Slowly rising, shivering rattle (3 secs). **H** Understorey of lowland, foothill and montane forest. **D** W Amazonian Brazil, N Bolivia.

11 BROWNISH-HEADED ANTBIRD *Myrmelastes brunneiceps* 15cm See Spot-winged Antbird and range. **V** Very high, piercing, shivering rattle (2–3 secs), slightly downslurred at end. **H** Understorey of lowland, foothill and montane forest. **D** SW Peru, NW Bolivia.

12 RUFOUS-FACED ANTBIRD *Myrmelastes rufifacies* 15cm See Spot-winged Antbird and range. **V** Very high, accelerating, sharp rattle (2 secs). **H** Understorey of lowland, foothill and montane forest. **D** E Amazonian Brazil.

13 RORAIMAN ANTBIRD *Myrmelastes saturatus* 15cm See Spot-winged Antbird and range. **V** High, thin, downslurred rattle (2–3 secs). **H** Understorey of lowland, foothill and montane forest. **D** SE Venezuela, N Brazil, W Guyana.

14 CAURA ANTBIRD *Myrmelastes caurensis* 18cm Male larger than male Spot-winged, Humaita, Brownish-headed, Rufous-faced and Roraiman antbirds; note also red eyes. Female not reliably separable except by size. **V** Descending series of 10 piercing notes. **H** Understorey of lowland, foothill and montane forest. **D** NC Amazonia.

15 CHESTNUT-BACKED ANTBIRD *Poliocrania exsul* 14cm Note black eyes and bare blue skin. **V** High, loud, slow, descending series of 2–3 *chee* notes. **H** Undergrowth of forest and secondary growth. **D** Nicaragua to W Ecuador.

16 GREY-HEADED ANTBIRD *Ampelornis griseiceps* 13cm No similar antbird occurs in its range. **V** Short, slightly descending trill. **H** Montane forest and woodland. **D** SW Ecuador, NW Peru.

17 STUB-TAILED ANTBIRD *Sipia berlepschi* 14cm Best recognised by its accompanying female. **V** Slightly descending series of 6–7 high, piercing *chew* notes. **H** Understorey of wet lowland and foothill forest. **D** W Colombia to NW Ecuador.

18 ESMERALDAS ANTBIRD *Sipia nigricauda* 13.5cm Note red eyes. **V** Song is a slow, very high series of 6–7 piercing *psee* notes. **H** Understorey and floor of wet forest and secondary growth. **D** W Colombia, W Ecuador.

19 MAGDALENA ANTBIRD *Sipia palliata* 13.5cm Previously considered conspecific with Dull-mantled Antbird. Female paler than female Esmeraldas Antbird, and species ranges differ. **V** Calm, level series of very high *seet* notes. **H** Understorey and floor of wet forest. **D** C and N Colombia, NW Venezuela.

20 DULL-MANTLED ANTBIRD *Sipia laemosticta* 14cm Note narrow wing-bars. Female shows fine white spotting on chin. **V** *Dzing dzing* and ultra-high, loud *fjeet-fjeetuhfjeetfjeetfjeetweet*. **H** Forest undergrowth, often near streams. **D** E Costa Rica, Panama.

21 WHITE-BELLIED ANTBIRD *Myrmeciza longipes* 15cm Separable from other antbirds by its range and uniform rufous upperparts. **V** Fast, steeply descending series of shivering notes. **H** Undergrowth and floor of forest, riverine belts, secondary growth. **D** E Panama, N South America, Trinidad.

22 BLACK-TAILED ANTBIRD *Myrmoborus melanurus* 12cm Note reddish eyes. Male dark grey with a black head. Female has grey sides to face and lacks mask. **V** Series of high, harsh, ringing notes, bouncing down (2–3 secs). **H** Undergrowth of várzea. **D** W Amazonia.

23 WHITE-LINED ANTBIRD *Myrmoborus lophotes* 14.5cm Conspicuous crest is diagnostic. Note white belly of female. **V** Song comprises about 10 descending, fluted notes, the first note distinctively separated. **H** Undergrowth of river-edge forest, often in bamboo. **D** SW Amazonia.

24 BLACK-FACED ANTBIRD *Myrmoborus myotherinus* 13cm Distinct wing-bars, combined with black (or brown) mask, are diagnostic. **V** Descending, calm series of high, harsh, ringing notes (2–3 secs). **H** Undergrowth of terra firme forest, secondary growth and woodland. **D** Amazonia.

25 WHITE-BROWED ANTBIRD *Myrmoborus leucophrys* 13cm Bold eyebrow is distinctive. **V** Descending, rapid series of high, harsh, ringing notes (4 secs). **H** Undergrowth of woodland and forest borders. **D** Amazonia.

26 ASH-BREASTED ANTBIRD *Myrmoborus lugubris* 13cm Separable from White-browed Antbird by range; lacks distinct wing-bars shown by Black-tailed and Black-faced antbirds. **V** Rapid, rather steeply descending series of high, harsh, ringing notes (3 secs). **H** Undergrowth of forest along rivers and on islands. **D** Amazon and its tributaries.

27 BARE-CROWNED ANTBIRD *Gymnocichla nudiceps* 16cm Female separable from female Chestnut-backed Antbird by browner head, and from female Blue-lored Antbird by wing-bars. **V** High, slow, rather level series of 5–15 harsh, piercing notes. **H** Lowland and foothill forest borders and secondary growth. **D** SE Mexico to NW Colombia.

28 WHITE-BACKED FIRE-EYE *Pyriglena leuconota* 17cm Red eyes are distinctive. Separable from Fringe-backed and White-shouldered fire-eyes by range. **V** Loud, resounding, slightly falling chatter (2–3 secs). **H** Undergrowth of montane forest and woodland. **D** S Colombia to E Brazil.

29 FRINGE-BACKED FIRE-EYE *Pyriglena atra* 17cm Note pattern of white dorsal patch. Range differs from those of White-backed and White-shouldered fire-eyes. **V** Loud, resounding, slightly falling series of *peer* notes (2–3 secs). **H** Understorey of forest and secondary growth. **D** E Brazil.

30 WHITE-SHOULDERED FIRE-EYE *Pyriglena leucoptera* 17.5cm Usually only one wing-bar visible. Note white eyebrow of female. **V** Song like that of Fringe-backed Fire-Eye but slower. **H** Undergrowth of forest and woodland. **D** E Paraguay, NE Argentina, SE Brazil.

31 SLENDER ANTBIRD *Rhopornis ardesiacus* 19cm No similar bird occurs in its restricted range. **V** Series of 5–10 piercing *peer* notes. **H** Undergrowth of woodland, often with terrestrial bromeliads. **D** E Brazil.

32 BLACK-HEADED ANTBIRD *Percnostola rufifrons* 14.5cm Note black throat and bushy crest of male, and wing fringing (not spotting) of male and female. Range differs from those of Allpahuayo and White-lined antbirds. **V** Level series of 5–10 loud, sharp, fluted notes, faster in W birds. **H** Thick undergrowth of terra firme forest and woodland. **D** NE and NC Amazonia.

33 ALLPAHUAYO ANTBIRD *Percnostola arenarum* 14.5cm Male lacks crest. Crown of female concolourous with back. **V** Rather fast, slightly descending series of loud, harsh notes. **H** Understorey of lowland forest on wet, sandy soils. **D** NE Peru.

34 WHITE-SHOULDERED ANTBIRD *Akletos melanoceps* 18cm Darker blue around eye than Sooty Antbird. Female strikingly patterned. **V** Calm series of 5–7 mellow, fluted notes. **H** Undergrowth of várzea, riverine belts, secondary growth. **D** W Amazonia.

35 GOELDI'S ANTBIRD *Akletos goeldii* 17.5cm Separable from White-shouldered Antbird by lack of white on shoulder, dark bare skin around eyes and range. **V** Song resembles that of White-shouldered Antbird but slightly faster (first note stuttered). **H** Understorey and floor of forest, often near rivers and in bamboo. **D** SW Amazonia.

36 SOOTY ANTBIRD *Hafferia fortis* 18cm Male very similar to male White-shouldered Antbird, but blue around eye brighter; often shows some crest. **V** Loud, level, hurried, slightly accelerating series of *tjuh* notes (2 secs). **H** Understorey of lowland and foothill forest and secondary growth. **D** W Amazonia.

37 BLUE-LORED ANTBIRD *Hafferia immaculata* 18cm No similar all-black antbird with blue around eyes occurs in its range, but note Bare-crowned Antbird, which is similar but with white wing-bars. **V** High, calm series of 4–6 fluted *peer* notes. **H** Understorey of humid forest, secondary growth and woodland. **D** N Venezuela to C Colombia.

38 ZELEDON'S ANTBIRD *Hafferia zeledoni* 20cm Unmistakable in region due to large size and blue orbital ring. **V** Very high, loud *sreeet* (8×) or high *sruuh*. **H** Forest understorey. **D** S Nicaragua to W Ecuador.

1
2
3
4
5
6
7
8
9
10
11
12
13
14
15
16
17
18
19
20
21
22
23
24
25
26
27
28
29
30
31
32
33
34
35
36
37
38

ANTTHRUSHES FORMICARIIDAE

1 RUFOUS-CAPPED ANTTHRUSH *Formicarius colma* 18cm Separable from Black-faced Antthrush by rufous crown and black breast. **V** High, musical, descending and rising trill (1–5 secs). **H** Terra firme forest and woodland. **D** Amazonia, SE Brazil.

2 BLACK-FACED ANTTHRUSH *Formicarius analis* 17.5cm See Rufous-capped Antthrush. Forages for arthropods in leaf litter on the forest floor. **V** High phrase of four notes, first note separate and followed by three or more as a descending musical, trill-like series. **H** Humid forest and woodland. **D** Honduras through N Bolivia and Amazonian Brazil.

3 MAYAN ANTTHRUSH *Formicarius moniliger* 19cm Note grey underparts. Considered a race of Black-faced Antthrush by some authorities. **V** Rapid series of slightly descending, rich, fluted notes. **H** Forest floor. **D** S Mexico through NW Honduras.

4 RUFOUS-FRONTED ANTTHRUSH *Formicarius rufifrons* 18cm Rufous forehead is diagnostic; black on throat merges gradually into grey underparts. **V** Rapid, slightly descending series of fluted notes, with an embedded short rise and fall somewhere at beginning. **H** Swampy forest. **D** SW Amazonia.

5 BLACK-HEADED ANTTHRUSH *Formicarius nigricapillus* 18cm Unmistakable when seen properly. Like other antthrushes, does not hop, but walks; in general, much more easily heard than seen. **V** High, accelerated and decelerating, running-up *wukwukwuk...* (5 secs). **H** Forest floor. **D** Costa Rica to W Ecuador.

6 RUFOUS-BREASTED ANTTHRUSH *Formicarius rufipectus* 18.5cm Rufous breast is diagnostic. Note also range. **V** Short phrase of two hooted notes, the second note higher. **H** Wet forest and secondary growth. **D** Costa Rica to NW Venezuela and SE Peru.

7 SHORT-TAILED ANTTHRUSH *Chamaeza campanisona* 19.5cm Very similar to Cryptic, Rufous-tailed and Shwartz's antthrushes (although Rufous-tailed shows less rufous in and near tail); best recognised by voice. **V** Rapid series of low, hooted notes, slowly rising and with 4–6 very low, liquid *woop* notes. **H** Montane forest and open secondary growth. **D** Venezuela to Bolivia, E and SE Brazil, Paraguay, NE Argentina.

8 STRIATED ANTTHRUSH *Chamaeza nobilis* 22.5cm Shows a contrasting white throat. **V** Low series of *woop* notes that accelerates at first and then distinctly descends and slows down. **H** Terra firme forest. **D** W and C Amazonia.

9 CRYPTIC ANTTHRUSH *Chamaeza meruloides* 19cm See Short-tailed Antthrush. Note range. **V** Long, fast series of hollow, fluted, rising *cu* notes with an abrupt ending. **H** Montane forest. **D** SE Brazil.

10 RUFOUS-TAILED ANTTHRUSH *Chamaeza ruficauda* 19cm See Short-tailed Antthrush. **V** Fast, melodious series of rising notes (3 secs). **H** Dense undergrowth of montane forest and secondary growth. **D** SE Brazil and NE Argentina.

11 SCHWARTZ'S ANTTHRUSH *Chamaeza turdina* 19cm See Short-tailed Antthrush. **V** Very long, level series of strong *cu* notes. **H** Humid montane forest. **D** N Venezuela, WC Colombia.

12 BARRED ANTTHRUSH *Chamaeza mollissima* 20cm Barred, not streaked as other *Chamaeza* antthrushes. **V** Song resembles that of Rufous-tailed Antthrush but each note is doubled and whole series is much longer. **H** Dense undergrowth of humid forest. **D** Colombia to N Peru, SE Peru to C Bolivia.

ANTPITTAS GRALLARIIDAE

13 UNDULATED ANTPITTA *Grallaria squamigera* 22cm Note size and blackish malar stripe. **V** Slowly rising, hollow trill (5 secs). **H** Humid and wet forest, secondary growth, woodland; 1,850–3,800m. **D** W Venezuela to C Bolivia.

14 GIANT ANTPITTA *Grallaria gigantea* 24cm Very large – only head is shown to scale with other birds on this plate. Separable from Great Antpitta by lack of white throat and range. **V** Rising, hollow trill (3–4 secs). **H** Montane forest; 1,400–3,000m. **D** SW Colombia, N Ecuador.

15 GREAT ANTPITTA *Grallaria excelsa* 24cm See Giant Antpitta. **V** Sings as Giant Antpitta but rising to a higher pitch. **H** Montane forest; 1,700–2,300m. **D** Venezuela.

16 VARIEGATED ANTPITTA *Grallaria varia* 19cm Separable from Scaled Antpitta by larger size, shaft streaking on back and less warm colouring. **V** Level, somewhat irregular series of hollow *whoo* notes. **H** Humid forest and woodland. **D** Amazonia, SE Brazil, E Paraguay, NE Argentina.

17 MOUSTACHED ANTPITTA *Grallaria alleni* 16.5cm Not reliably separable from Scaled Antpitta, but that species shows hint of blackish 'tears' and normally occurs at lower elevations. **V** Short, low, hollow, slightly rising, trill-like song. **H** Dense undergrowth of wet forest; 1,850–2,500m. **D** SW Colombia, N Ecuador.

18 SCALED ANTPITTA *Grallaria guatimalensis* 19cm Unmistakable due to large size, scaled upperparts and throat pattern. **V** Low, hollow, turned-up *rrrurrrurur* (3 secs). **H** Forest floor, especially near water; 500–3,000m. **D** S Mexico to C Bolivia, Venezuela, N Brazil, Guyana, Trinidad.

19 TACHIRA ANTPITTA *Grallaria chthonia* 17.5cm Separable from Scaled Antpitta by whitish belly and grey-barred flanks. **V** Song is unrecorded. **H** Montane forest; 1,800–2,100m. **D** W Venezuela.

20 PLAIN-BACKED ANTPITTA *Grallaria haplonota* 16.5cm Note lack of rufous in plumage. **V** Calm, slightly rising series of hollow notes, the last pair lowered. **H** Humid foothill and montane forest. **D** N Venezuela, W Colombia to N Peru.

21 OCHRE-STRIPED ANTPITTA *Grallaria dignissima* 19cm No similar antpitta occurs in its range. Throat and chest rufous. **V** Mid-high series of two plaintive notes. **H** Terra firme forest. **D** W Amazonia.

22 ELUSIVE ANTPITTA *Grallaria eludens* 19cm Sides of face and breast are tawny, not rufous. **V** Calm series of two high, plaintive notes, the second higher. **H** Terra firme forest. **D** SW Amazonia.

23 CHESTNUT-CROWNED ANTPITTA *Grallaria ruficapilla* 18.5cm Separable from similar antpittas by rufous face sides. **V** Calm series of three high notes, the middle one lower. **H** Montane forest and woodland; 1,200–3,000m. **D** N and W Venezuela to C Peru.

24 WATKINS'S ANTPITTA *Grallaria watkinsi* 18cm Separable from Chestnut-crowned Antpitta by having paler rufous colouring restricted to crown and nape. **V** Calm series of 5–7 notes, the last three together sounding as *kew-ooheeh* (last note higher and accentuated). **H** Dry forest and secondary growth; <1,800m. **D** W Ecuador, NW Peru.

25 SANTA MARTA ANTPITTA *Grallaria bangsi* 18cm The only antpitta in its range with streaked underparts. **V** Series of two high, fluted notes, the second higher. **H** Montane forest and secondary growth. 1,300–2,400m. **D** N Colombia.

26 CUNDINAMARCA ANTPITTA *Grallaria kaestneri* 15.5cm Rather dull coloration, with a more tawny head and whitish underparts. **V** Series of three sharp, rising notes. **H** Montane forest and woodland; 1,800–2,500m. **D** C Colombia.

27 STRIPE-HEADED ANTPITTA *Grallaria andicolus* 16cm No other antshrike shows a streaked crown. **V** Upslurred frog-like rattle that starts with a stutter (2 secs). **H** Woodland; 3,000–4,300m. **D** Peru to W Bolivia.

28 GREY-NAPED ANTPITTA *Grallaria griseonucha* 16 cm The sole antpitta in its range. **V** Short, rapid, rising, hollow series of notes. **H** Montane forest; 2,300–2,900m. **D** Venezuela.

29 BICOLORED ANTPITTA *Grallaria rufocinerea* 16cm All-rufous head is diagnostic. **V** Song is a drawn-out, slightly rising, fluted note. **H** Montane forest and woodland; 2,400–3,100m. **D** C Colombia, NW Ecuador.

30 JOCOTOCO ANTPITTA *Grallaria ridgelyi* 21cm Unmistakable due to head pattern. **V** Song is a level, rising or irregular, short to very long series of owl-like *hoo* notes. **H** Montane forest with bamboo. **D** S Ecuador, N Peru.

31 CHESTNUT-NAPED ANTPITTA *Grallaria nuchalis* 20cm Separable from Bicoloured Antpitta by grey chin. **V** Short, accelerating and rising series of hooted notes. **H** Montane forest; 2,000–3,000m. **D** Colombia to N Peru.

32 PALE-BILLED ANTPITTA *Grallaria carrikeri* 19cm Pale bill is diagnostic. **V** Level series of high, hooted notes (2 secs). **H** Montane forest with bamboo; 2,350–2,900m. **D** Peru.

33 WHITE-THROATED ANTPITTA *Grallaria albigula* 18.5cm White throat, contrasting with rufous-orange head and grey breast, is distinctive. **V** Very short, level series of two hooted notes. **H** Humid montane forest; 800–1,700m. **D** SE Peru to NW Argentina.

34 YELLOW-BREASTED ANTPITTA *Grallaria flavotincta* 17cm No similar antpitta occurs in its range (see also White-bellied Antpitta). Yellow underparts, though not always distinctive, are diagnostic. **V** Song resembles that of White-throated Antpitta, but higher and more fluted. **H** Humid and wet montane forest; 1,300–2,350m. **D** Colombia to NW Ecuador.

35 WHITE-BELLIED ANTPITTA *Grallaria hypoleuca* 17cm Bicoloured plumage is distinctive. Occurs on E slope of Andes, not on W slope, where Yellow-breasted Antpitta is found. **V** Series of three high, hooted notes, the first slightly lower. **H** Humid forest and secondary growth; 1,400–2,300m. **D** Colombia to N Peru.

36 RUSTY-TINGED ANTPITTA *Grallaria przewalskii* 17cm Note greyish crown and orange throat and breast. **V** Series of three mid-high, hooted notes, the second note slightly lower. **H** Humid montane forest; 1,900–2,750m. **D** C Peru.

37 BAY ANTPITTA *Grallaria capitalis* 17cm Note pale mid-belly. Separable from Rusty-tinged Antpitta by range. **V** Series of three fluted notes in tempo 1, 2–3, the first note higher. **H** Humid montane forest and secondary growth; 2,000–3,000m. **D** C Peru.

38 RED-AND-WHITE ANTPITTA *Grallaria erythroleuca* 17.5cm Beautifully patterned orange-rufous and pure white bird. **V** Level, mid-high series of three fluted notes. **H** Humid forest and secondary growth with bamboo stands; 2,150–3,000m. **D** SE Peru.

1
3
4
7
6
5
11
8
2
9
10
12
13
14
15
16
17
18
20
22
19
21
23
24
25
26
27
28
29
30
31
32
33
34
35
36
37
38

ANTPITTAS *CONTINUED*

1 RUFOUS ANTPITTA *Grallaria rufula* 15cm Uniform rufous red. Note rather small eyes and bill, and pale eye-ring. **V** Song is very varied, e.g. Peru birds sing a high, fluted series of six notes, the first two well separated, the last one slightly lower. **H** Humid forest with bamboo; 2,200–3,650m. **D** Venezuela to C Bolivia.

2 CHESTNUT ANTPITTA *Grallaria blakei* 15cm Separable from Rufous Antpitta by darker plumage and lack of eye-ring. **V** Rapid, slightly descending series of notes, almost rattle-like (3–4 secs). **H** Montane forest and woodland; 2,100–3,100m. **D** C Peru.

3 TAWNY ANTPITTA *Grallaria quitensis* 17cm Tawny buff overall, paler below. **V** Calm series of three hollow notes, the first one slightly higher. **H** Shrub and scrub at high elevations; 2,200–4,500m. **D** Colombia to N Peru.

4 BROWN-BANDED ANTPITTA *Grallaria milleri* 16.5cm Separable from Tawny Antpitta by paler underparts (brown breast-band not distinctive or brown, but buffish). **V** Song comprises three rising, high, sharp notes. **H** Humid forest and woodland; 1,800–3,150m. **D** C Colombia.

5 URRAO ANTPITTA *Grallaria urraoensis* 16.5cm Note slaty-grey breast and belly. **V** Song comprises three piped, rising notes, the last two connected. **H** Montane forest; 2,500–3,200m. **D** NW Colombia.

6 RUFOUS-FACED ANTPITTA *Grallaria erythrotis* 18.5cm Note rufous face and neck sides. **V** Song comprises 2–3 slightly rising, piped notes. **H** Montane forest and woodland; 1,700–3,100m. **D** Bolivia.

7 STREAK-CHESTED ANTPITTA *Hylopezus perspicillatus* 13cm Range differs from that of Spotted Antpitta; note also single malar stripe and wing spotting. **V** Song comprises 7–9 piped notes, accelerating and slowing down, and rising and falling. **H** Humid forest. **D** E Honduras to NW Ecuador.

8 SPOTTED ANTPITTA *Hylopezus macularius* 14cm See Streak-chested Antpitta. Note double malar stripe and wing-bars. Breast is streaked rather than spotted. **V** Song comprises 3–6 slightly wavering, slightly rising and descending, hollow *ko* notes. **H** Humid forest. **D** N Amazonia.

9 SNETHLAGE'S ANTPITTA *Hylopezus paraensis* 14cm Separable from Streak-chested and Spotted antpittas by more distinct striping to mantle and less tawny-rufous flanks; also check ranges. **V** Song is a spooky series of *hoooh huhu-huhu* notes. **H** Floor and dense undergrowth of forest. **D** W and S Amazonian Brazil.

10 ALTA FLORESTA ANTPITTA *Hylopezus whittakeri* 14cm Separable from Snethlage's Antpitta by voice. **V** Level series of five hollow, evenly spaced notes. **H** Floor and dense undergrowth of forest. **D** SC Amazonian Brazil.

11 MASKED ANTPITTA *Hylopezus auricularis* 14cm Facial pattern is distinctive. **V** Plaintive rattle, shivering down (2–3 secs). **H** Humid forest and secondary growth near open areas. **D** N Bolivia.

12 THICKET ANTPITTA *Hylopezus dives* 14cm Separable from White-lored Antpitta by lack of white in face. **V** Hurried, rising series of 5–6 plaintive, fluted notes. **H** Dense edges of humid forest and woodland. **D** E Honduras to W Colombia.

13 WHITE-LORED ANTPITTA *Hylopezus fulviventris* 13cm Note grey crown and buff lower belly. **V** E.g. a very high, running-up, hurried *tututu...* (3 secs). **H** Dense undergrowth at forest edges, secondary growth. **D** W Amazonia.

14 AMAZONIAN ANTPITTA *Hylopezus berlepschi* 14.5cm Crown and mantle uniform olive-brown. **V** Slow series of three slightly nasal, hooted notes. **H** Dense undergrowth at edges of humid forest and woodland. **D** S Amazonia.

15 WHITE-BROWED ANTPITTA *Hylopezus ochroleucus* 13.5cm No similar bird occurs in its range and habitat. **V** Series of 5–6 slightly rising, pure, fluted notes, hesitant at the start. **H** Caatinga woodland. **D** E Brazil.

16 SPECKLE-BREASTED ANTPITTA *Hylopezus nattereri* 13.5cm Range differs from that of White-browed Antpitta. **V** Rapid series of 6–8 fluted notes that rise in strength and pitch. **H** Montane and humid forest, secondary growth. **D** E Paraguay, NE Argentina, S and SE Brazil.

17 THRUSH-LIKE ANTPITTA *Myrmothera campanisona* 15cm Note faintly streaked breast and generally plain plumage. **V** High, rapid series of 5–6 rising, plaintive notes. **H** Terra firme forest. **D** Amazonia.

18 TEPUI ANTPITTA *Myrmothera simplex* 16cm Note distinct white spot behind eye and white throat. **V** Song resembles that of Thrush-like Antpitta but longer (7–8 notes). **H** Dense forest and woodland. **D** Tepuis of SE Venezuela, N Brazil and W Guyana.

19 OCHRE-BREASTED ANTPITTA *Grallaricula flavirostris* 10cm Rather similar to Rusty-breasted Antpitta, but with ochre eye-ring (striking white in Rusty-breasted). **V** Single high, sharp, fluted shriek. **H** Humid and wet montane forest. **D** Costa Rica to C Bolivia.

20 SCALLOP-BREASTED ANTPITTA *Grallaricula loricata* 10cm Distinctive breast pattern. **V** Single downslurred, plaintive whistles. **H** Montane forest. **D** N Venezuela.

21 HOODED ANTPITTA *Grallaricula cucullata* 10cm Note all-rufous head and all-yellow bill. **V** Voice not recorded. **H** Dense, humid montane forest; normally 1,800–2,250m. **D** NW Venezuela, W Colombia.

22 PERUVIAN ANTPITTA *Grallaricula peruviana* 10cm Male separable from male Ochre-fronted Antpitta by heavy black scalloping (not streaking) below and rufous (not olive) crown. Female has pale lores. **V** Very high, piercing *seeeu* given at fairly long intervals. **H** Montane forest; 1,650–2,100m. **D** SE Ecuador, N Peru.

23 OCHRE-FRONTED ANTPITTA *Grallaricula ochraceifrons* 10.5cm See Peruvian Antpitta. **V** Very high, piercing *seeeu* given at fairly long intervals. **H** Montane forest; 1,900–2,400m. **D** C Peru.

24 RUSTY-BREASTED ANTPITTA *Grallaricula ferrugineipectus* 10cm Separable from Ochre-breasted Antpitta by lack of dark breast streaking/scalloping. Note distinct white eye-spot of male. **V** Very fast series of high, sharp *chip* notes. **H** Montane forest, often in dense bamboo; 600–2,200m, locally lower or (much) higher. **D** N Venezuela to Bolivia.

25 SLATY-CROWNED ANTPITTA *Grallaricula nana* 11cm Separable from Rusty-breasted Antpitta by grey crown. **V** Very high, short, sharp, shivering trill, slightly descending at end (2 secs). **H** Humid montane forest; 1,900–3,200, locally (much) lower. **D** Venezuela to N Peru, W Guyana.

26 SUCRE ANTPITTA *Grallaricula cumanensis* 11cm Separable from Slaty-crowned Antpitta by paler bill. **V** Series of evenly paced notes, rising to middle and then falling away slowly at end (1.6–2 secs). **H** Humid montane forest. **D** NE Venezuela.

27 CRESCENT-FACED ANTPITTA *Grallaricula lineifrons* 11.5cm Unmistakable due to facial pattern. **V** Very high series of piercing notes, rising at start (3 secs). **H** Humid montane forest and secondary growth; 2,900–3,400m. **D** C and S Colombia, Ecuador.

GNATEATERS CONOPOPHAGIDAE

28 RUFOUS GNATEATER *Conopophaga lineata* 13cm No other similar gnateater occurs in its range, only the very different Black-cheeked Gnateater. **V** Rattling, sharp series of notes, rising at the end (2 secs). **H** Woodland, riverine belts, secondary growth, humid forest. **D** E and C Brazil, Paraguay, Uruguay, NE Argentina.

29 CHESTNUT-BELTED GNATEATER *Conopophaga aurita* 12cm No similar gnateater occurs in its range, but check female Ash-throated Gnateater (with buff wing spots). **V** Song a high, sharp rattle (3 secs). **H** Undergrowth of humid lowland forest. **D** Amazonia.

30 HOODED GNATEATER *Conopophaga roberti* 12.5cm Note black crown of male and rufous crown of female. **V** Short, musical rattle, slightly upslurred at end. **H** Thick undergrowth of humid forest. **D** NE Brazil.

31 ASH-THROATED GNATEATER *Conopophaga peruviana* 12cm Note black marking of mantle and buff wing spots. **V** Series of well-separated, very short, trilling shrieks. **H** Densest parts of undergrowth in humid forest. **D** S and SW Amazonia.

32 CEARA GNATEATER *Conopophaga cearae* 12–14cm Previously treated as subspecies of Rufous Gnateater. Usually solitary but sometimes in pairs. Forages for invertebrates on forest floor. **V** Song a shrill, rapid-fire *chu-chu-cha-chachachichichi*, pitch rising during delivery. Call a rasping *Tchiup*. **H** Evergreen forest with dense understorey. **D** NE Brazil.

33 SLATY GNATEATER *Conopophaga ardesiaca* 13cm Male separable from male Ash-throated Gnateater by uniform wings. Female separable from female Chestnut-crowned Gnateater by lack of white in ear-tufts and grey breast. **V** Series of high, repeated *jeree* notes. **H** Densest parts of humid forest. **D** S Peru to S Bolivia.

34 CHESTNUT-CROWNED GNATEATER *Conopophaga castaneiceps* 13cm Male unmistakable due to rufous crown. Note female's bright orange-rufous lore and front. **V** High, piercing rattle, accentuated at the end by two separate notes. **H** Open spots in montane forest. **D** Colombia to C Peru.

35 BLACK-CHEEKED GNATEATER *Conopophaga melanops* 11.5cm See Rufous Gnateater, which shows plain wings and mantle. **V** High, sharp, slightly rising and falling trill (8–10 secs). **H** Undergrowth of forest and secondary growth, often along streams. **D** E and SE Brazil.

36 BLACK-BELLIED GNATEATER *Conopophaga melanogaster* 15cm Male is unmistakable. Note grey-white pattern of female. **V** Series of well-separated, very short rattles. **H** Dense undergrowth of humid forest. **D** SC and SE Amazonia.

37 BLACK-CROWNED ANTPITTA *Pittasoma michleri* 18cm Unmistakable due to black cap and general colour pattern. **V** Very long, slightly rising and slowing-down series of high *tu* notes (60 secs and longer). **H** Humid forest. **D** Costa Rica to NW Colombia.

38 RUFOUS-CROWNED ANTPITTA *Pittasoma rufopileatum* 17cm Unmistakable due to black eye-stripe. **V** Very long series of high, well-separated *keeyur* notes. **H** Humid and foothill forest. **D** W Colombia to NW Ecuador.

1
2
3
4
5
6
7
8
9
10
11
12
13
14
15
16
17
18
19
20
21
22
23
24
25
26
27
28
29
30
31
32
33
34
35
36
37
38

TAPACULOS RHINOCRYPTIDAE

1 OCELLATED TAPACULO *Acropternis orthonyx* 21.5cm Unmistakable due to its spectacular pattern. **V** High, descending *seeeuw* notes separated by rather long intervals. **H** Upper montane forest; 2,300–3,500m, locally lower or higher. **D** NW Venezuela to NW Peru.

2 CHESTNUT-THROATED HUET-HUET *Pteroptochos castaneus* 24cm Separable from Black-throated Huet-huet by chestnut throat. **V** Rising, accelerating, hurried series of dry *pruep* notes (5–6 secs). **H** Forest and woodland. **D** C Chile to WC Argentina.

3 BLACK-THROATED HUET-HUET *Pteroptochos tarnii* 23cm Striking pale eye-ring; scalloping restricted to lower flanks and belly. **V** Sings as Chestnut-throated Huet-huet but slower and longer. **H** Forest and woodland. **D** C and S Chile, SW Argentina.

4 MOUSTACHED TURCA *Pteroptochos megapodius* 23cm White face markings are diagnostic. **V** Descending, slow, slightly accelerating series of hollow *wook* notes. **H** Scrubby slopes. **D** N and C Chile.

5 WHITE-THROATED TAPACULO *Scelorchilus albicollis* 19cm Amount of scalloping varies. **V** Descending, rather rapid series of cooing notes. **H** Floor of dense scrub. **D** N and C Chile.

6 CHUCAO TAPACULO *Scelorchilus rubecula* 19cm Patterning and colouring are unmistakable. **V** Short, rapid series of six notes, the first and last two soft and low-pitched, and the middle three loud and cackling. **H** Forest and woodland floor. **D** Chile and Argentina.

7 CRESTED GALLITO *Rhinocrypta lanceolata* 21cm Bushy crest and cocked tail are distinctive. **V** Long series of dry *chorrok* notes with 1-sec intervals. **H** Scrub, open brush, woodland. **D** SE Bolivia, W Paraguay, N and C Argentina.

8 SANDY GALLITO *Teledromas fuscus* 18cm No similar bird that is so pale and uniform occurs in its range. **V** Rapid series of 10 high, dry *cho* notes. **H** Arid scrub; <3,500m, locally even higher. **D** C Argentina.

9 RUSTY-BELTED TAPACULO *Liosceles thoracicus* 19.5cm No similar bird occurs in its range. **V** Calm, accelerating series of slightly descending *toot* notes (3–4 secs). **H** Terra firme forest. **D** W and SW Amazonia.

10 SPOTTED BAMBOOWREN *Psilorhamphus guttatus* 13.5cm Distinctively spotted and coloured. Note tail pattern. **V** Long series of hollow *to* notes, slightly rising at start (15–25 secs). **H** Bamboo stands at borders of humid forest, woodland and secondary growth. **D** NE Argentina, SE Brazil.

11 SLATY BRISTLEFRONT *Merulaxis ater* 18.5cm Bristle in both sexes and chestnut rear body of male are distinctive. **V** Long, descending, fast series of sharp but musical notes, upslurred, then again continuing to fall at end (7–9 secs). **H** Humid forest woodland floor. **D** SE Brazil.

12 STRESEMANN'S BRISTLEFRONT *Merulaxis stresemanni* 19.5cm Separable from Slaty Bristlefront by range; also male lacks chestnut in plumage. **V** Song resembles that of Slaty Bristlefront but with a hollower quality. **H** Humid lowland forest floor. **D** E Brazil.

13 OCHRE-FLANKED TAPACULO *Eugralla paradoxa* 14.5cm Note size, yellow legs and rather heavy bill. Lacks flank barring. Female resembles male. **V** Short to very short, sharp rattle. **H** Dense undergrowth with bamboo in humid forest and woodland. **D** SC Chile, WC Argentina.

14 ASH-COLORED TAPACULO *Myornis senilis* 14cm Separable from *Scytalopus* tapaculos by longer tail. **V** Long, descending, accelerating, irregular series of chirped notes, slowing at end. **H** Thickets in humid montane forest; 2,300–3,500m, locally lower or (much) higher. **D** Colombia to Peru.

15 WHITE-BREASTED TAPACULO *Eleoscytalopus indigoticus* 11cm See Bahia Tapaculo. Separable from Brasilia Tapaculo by purer white underparts. Upperparts of female (not shown) somewhat browner than in male. **V** Slightly rising, purring trill (2 secs). **H** Undergrowth of humid forest and woodland; <1,500m. **D** SE Brazil.

16 BAHIA TAPACULO *Eleoscytalopus psychopompus* 11cm Separated from White-breasted Tapaculo by less or no flank barring and altitudinal range. Female resembles male. **V** Hollow trill, upslurred in second half (2 secs). **H** Undergrowth of humid lowland forest. **D** E Brazil.

17 MARSH TAPACULO *Scytalopus iraiensis* 11cm Habitat is diagnostic. **V** Accelerating and rising series of chopped *cher* notes (>1 min). **H** Tall grassy, seasonally inundated areas. **D** S Brazil.

18 MOUSE-COLORED TAPACULO *Scytalopus speluncae* 11cm Male uniform grey or with brownish grey on lower flanks (other tapaculos in range show some white below). **V** Fast, level, chopping series of *chek* notes (>1 min). **H** Undergrowth and edges of humid forest and woodland. **D** SE Brazil.

19 BOA NOVA TAPACULO *Scytalopus gonzagai* 12cm Similar to Mouse-colored Tapaculo. Subtle differences in plumage and vocalisation aid identification. **V** Song a repetitive series of *tchup-tchup-tchup-tchup-tchup...* phrases. Call a falcon-like *tche-te-te*. **H** Atlantic forests with dense understorey. **D** SE Brazil.

20 ROCK TAPACULO *Scytalopus petrophilus* 13cm Female resembles male. Range differs from those of Planalto Tapaculo and Diamantina Tapaculo (Plate 142). **V** Very long series of very high, chirping *chek* notes (>1 min). **H** Rocky areas with grass and shrub. **D** SE Brazil.

21 PLANALTO TAPACULO *Scytalopus pachecoi* 12cm Female resembles male. Range differs from those of Rock Tapaculo and Diamantina Tapaculo (Plate 142). Separable from Mouse-colored Tapaculo by different song. **V** Very long, rhythmic series of very high *chik* notes (>1 min), at a lower speed and less sharp than song of Mouse-colored Tapaculo. **H** Lower strata of secondary growth and forest. **D** S Brazil and NE Argentina.

22 BRASILIA TAPACULO *Scytalopus novacapitalis* 11cm Differs from White-breasted Tapaculo in having less barring on flanks and a different song. **V** Song starts with up to 30 well-separated, climbing, single chirps, which are suddenly transformed into a trill. **H** Undergrowth of riverine belts and woodland. **D** S Brazil.

23 BOLIVIAN WHITE-CROWNED TAPACULO *Scytalopus bolivianus* 12cm Sole tapaculo in its range (S of Northern White-crowned Tapaculo) with a white crown spot. **V** Slightly rising, sharp, piercing rattle (6–7 secs). **H** Dense undergrowth and borders of humid montane forest. **D** SE Peru to S Bolivia.

24 NORTHERN WHITE-CROWNED TAPACULO *Scytalopus atratus* 12cm Sole tapaculo in its range (south of Santa Marta Tapaculo) with a white crown spot. **V** Series of fast, short, twittering phrases. **H** Undergrowth and borders of humid montane forest. **D** W Venezuela to C Peru.

25 SANTA MARTA TAPACULO *Scytalopus sanctaemartae* 11cm Sole tapaculo in its range (N of Northern White-crowned Tapaculo) with a white crown spot. **V** High, descending trill (10 secs). **H** Dense undergrowth of humid forest. **D** Sierra Nevada de Santa Marta, N Colombia.

26 RUFOUS-VENTED TAPACULO *Scytalopus femoralis* 12.5cm Check range. Separable from Bolivian White-crowned and Northern White-crowned tapaculos by all-black crown. **V** Calm, level series of staccato *purp* notes (>1–2 min). **H** Undergrowth of humid forest. **D** C Peru.

27 LONG-TAILED TAPACULO *Scytalopus micropterus* 13.5cm Separable from Rufous-vented Tapaculo by range and from Choco Tapaculo by habitat at higher elevations; similar Magdalena Tapaculo has a very restricted range. Best distinguished by voice. **V** Rising series of jackdaw-like *chudok* notes (10–20 secs). **H** Undergrowth and borders of montane forest; 1,250–2,200m. **D** Colombia to N Peru.

28 NARINO TAPACULO *Scytalopus vicinior* 12cm Not reliably separable from Spillman's Tapaculo (Plate 142) except by voice. **V** Very long, sharp series of notes that starts as a rattle. **H** Undergrowth of montane forest. **D** W Colombia to C Ecuador.

29 EL ORO TAPACULO *Scytalopus robbinsi* 11cm Not reliably separable from Choco Tapaculo except by voice. **V** Very long, fast series of mid-high *tjow* notes. **H** Undergrowth of humid foothill and montane forest. **D** SW Ecuador.

30 TATAMA TAPACULO *Scytalopus alvarezlopezi* 12cm Vocalisation and subtle plumage differences allow separation from other similar tapaculo species. **V** Song a vibrating, trilling *Trrrrrrrrr*. Call a shrill, disyllabic, upslurred *tu-Wik*. **H** Mountain slopes covered with cloud forest; mainly 1,500–2,000m. **D** Pacific (W) slopes of Andean Colombia.

31 CHOCO TAPACULO *Scytalopus chocoensis* 11cm Occurs at lower elevations than other *Scytalopus* tapaculos on W flank of Andes. **V** Very long series of high, very sharp, staccato notes. **H** Undergrowth of foothill and montane forest; 250–1,200m. **D** E Panama to NW Ecuador.

32 MAGDALENA TAPACULO *Scytalopus rodriguezi* 11.5cm Reliably separable from Spillman's Tapaculo (Plate 142) only by voice. **V** Very long series of high, staccato notes with irregular, short interruptions. **H** Undergrowth and borders of montane forest; 2,000–2,300m. **D** C Colombia.

33 STILES'S TAPACULO *Scytalopus stilesi* 11.5cm Reliably separable from Narino Tapaculo and Magdalena tapaculos, and Spillman's Tapaculo (Plate 142), only by voice. Female paler than male. **V** Very long series of short, low rattles (each 1–2 secs). **H** Undergrowth and borders of montane forest; 1,400–2,100m. **D** C Colombia.

34 TACARCUNA TAPACULO *Scytalopus panamensis* 11.5cm Note white eyebrow and grey throat. **V** Very long series of high, sharp notes that is almost a trill at the start and gradually rises. **H** Undergrowth and borders of humid montane forest; 1,100–1,500m. **D** E Panama and NW Colombia.

35 SILVERY-FRONTED TAPACULO *Scytalopus argentifrons* 11cm Not in range of Tacarcuna Tapaculo. Note white forehead (however, white may be missing, even in most or all of eyebrow). **V** High, scolding, fast, staccato *tutrittrittrittrit*. **H** Forest, bamboo; 1,000–3,000m. **D** Costa Rica, Panama.

1
2
3
4
5
6
7
8
9
10
11 ♂ ♀
12 ♀ ♂
13
14
15
16
17
18
19
20
21
22
23
24
25
26
27
28
29
30
31
32
33 ♂ ♀
34
35

TAPACULOS *CONTINUED*

1 CARACAS TAPACULO *Scytalopus caracae* 11.5cm See Brown-rumped Tapaculo. **V** Series of repeated phrases that contain sharp notes like those of Brown-rumped, the second note higher. **H** Undergrowth and borders of montane forest and secondary growth; 1,200–2,000m. **D** N Venezuela.

2 MERIDA TAPACULO *Scytalopus meridanus* 11.5cm See Brown-rumped Tapaculo. **V** Very high, thin rattle (1 sec). **H** Undergrowth of montane humid forest; 1,600–4,000m. **D** NW Venezuela.

3 BROWN-RUMPED TAPACULO *Scytalopus latebricola* 11.5cm Caracas, Merida and Brown-rumped tapaculos can be separated only by range and song. **V** Sharp rattle that starts with some individual notes (4–7 secs). **H** Undergrowth and borders of montane forest and secondary growth; 2,000–3,600m. **D** N Colombia.

4 PERIJA TAPACULO *Scytalopus perijanus* 11–12cm Differences in vocalisations and restricted range allow separation from other similar tapaculo species. **V** Song a vibrating, trilling *tchueErrrrrrrr*, pitch and volume rising and then falling during delivery. Call a vibrating trill, similar to song. **H** Humid cloud forest; mainly 2,000–3,000m **D** Serranía de Perijá mountains of E Colombia and W Venezuela.

5 SPILLMANN'S TAPACULO *Scytalopus spillmanni* 12cm Separable from Narino Tapaculo (Plate 141) only by song. **V** Short, descending, shivering trill. **H** Undergrowth with bamboo in humid forest; 1,900–3,500m. **D** Colombia to Ecuador.

6 CHUSQUEA TAPACULO *Scytalopus parkeri* 11.5cm Separable from Spillman's Tapaculo only by song. **V** Short, level trill. **H** Undergrowth of montane forest; 2,250–3,300m. **D** S Ecuador, N Peru.

7 TRILLING TAPACULO *Scytalopus parvirostris* 11cm Rather dark grey above. Female resembles male. Note range. At lower elevations than Tschudi's Tapaculo. **V** Very high, sharp, irregularly rising and descending trill (4–10 secs). **H** Undergrowth of humid montane forest; 1,800–2,500m. **D** NC Peru to EC Bolivia.

8 TSCHUDI'S TAPACULO *Scytalopus acutirostris* 10.5cm Wings are browner and less rufous than those of Ancash and Neblina tapaculos. **V** Very long, calm series of high, piercing *d-drr* phrases. **H** Undergrowth and borders of montane forest; 2,500–3,700m. **D** C Peru.

9 UNICOLORED TAPACULO *Scytalopus unicolor* 11cm Mid-grey, paler below and with some rufous tingeing to rear underparts. Female resembles male but with slight brown wash. **V** Song of 4–6 notes, each becoming shorter and faster. **H** Dense undergrowth and edges of montane forest and woodland; 2,000–3,200m. **D** WC Peru.

10 PALE-BELLIED TAPACULO *Scytalopus griseicollis* 11cm Note bright rufous flanks with little barring. **V** High, sharp rattle (5–10 secs). **H** Undergrowth and borders of forest and woodland; 2,000–3,500m. **D** NC and C Colombia, W Venezuela.

11 PARAMILLO TAPACULO *Scytalopus canus* 11cm Note barred, buff-rufous flanks and range. **V** Fast, very high, shrill series of *zrut* notes. **H** Woodland and scrub at treeline; 3,050–4,000m. **D** W Colombia.

12 PARAMO TAPACULO *Scytalopus opacus* 11cm See Magallenic Tapaculo. Check range. **V** Piercing rattle (3–6 secs). **H** Woodland and scrub at treeline; 3,050–4,000m. **D** C Colombia, E Ecuador, N Peru.

13 ANCASH TAPACULO *Scytalopus affinis* 11cm The only tapaculo within its small range. Feeds on small invertebrates near the treeline. **V** Series of very high, piercing chirps (10–15 secs). **H** Woodland undergrowth and adjacent rocky grassland; 3,000–4,600m. **D** Ancash region of Peru on W flank of Andes.

14 NEBLINA TAPACULO *Scytalopus altirostris* 11cm Not safely separable from Tschudi's Tapaculo except by its range. **V** Series of sharp, high, staccato chirps (10 secs). **H** Undergrowth of montane forest; 3,000–4,000m. **D** C Peru on E flank of Andes.

15 VILCABAMBA TAPACULO *Scytalopus urubambae* 11cm Note lack of barring on flanks. **V** Very long series of high, sharp, rasping *churr* notes. **H** Undergrowth and borders of montane forest; 3,500–4,200m. **D** EC Peru.

16 DIADEMED TAPACULO *Scytalopus schulenbergi* 11cm Note white front and short eyebrow. **V** Fast, high, descending series of sharp *djirr* notes. **H** Undergrowth with bamboo in humid forest at treeline; 2,800–3,400m. **D** SE Peru to C Bolivia.

17 PUNA TAPACULO *Scytalopus simonsi* 11cm Note short white eyebrow of male. **V** Very long, level series of high sparrow-like *churr* notes. **H** Stunted forest; 3,000–4,300m. **D** SE Peru to C Bolivia.

18 ZIMMER'S TAPACULO *Scytalopus zimmeri* 11cm Separable from White-browed Tapaculo by range. **V** Very long series of twittered *chiti-chr* phrases. **H** Alder-dominated woodland and boulder-strewn slopes; 1,700–3,200m. **D** S Bolivia, NW Argentina.

19 WHITE-BROWED TAPACULO *Scytalopus superciliaris* 11cm Separable from Zimmer's Tapaculo by range. **V** Very long series of very high *tsit-tzeeu* phrases. **H** Alder-dominated woodland and boulder-strewn slopes; 1,500–3,000m. **D** NW Argentina.

20 MAGELLANIC TAPACULO *Scytalopus magellanicus* 10cm The original Magellanic Tapaculo is now split into 12 species: Magellanic, Tschudi's, Pale-bellied, Paramillo, Paramo, Ancash, Neblina, Vilcabamba, Puna, Zimmer's, White-browed and Dusky tapaculos. All are separable primarily on the basis of voice and range. **V** Song a very long, slightly lowered and climbing series of rhythmic *ka-chew* notes. **H** Humid forest, shrubby bogs, rocky grasslands. <1,000m up to 3,500m in Argentina. **D** C and S Chile, W Argentina.

21 DUSKY TAPACULO *Scytalopus fuscus* 11cm Separable from Magellanic Tapaculo by song. Never shows a white front. **V** Very long series of hoarse, upslurred *zriiip* notes. **H** Woodland; <800m. **D** C Chile.

22 BLACKISH TAPACULO *Scytalopus latrans* 11.5cm Note absence of rufous in plumage of male. Plumage of female races varies from uniform darkish grey to paler grey with faintly barred brown flanks (as in nominate, illustrated). **V** Short, chopping phrase, repeated and prolonged to a very long, often continuous series. **H** Humid undergrowth of montane forest and woodland; 2,000–3,500m. **D** Colombia to Peru.

23 JUNIN TAPACULO *Scytalopus gettyae* 11cm Vocalisation aids identification and separation from other similar tapaculo species. **V** Song a hooting, vibrating *oop, opp, Oopuwhererr, Oopuwhererr*. Call a shrill, nasal *o-rrr-errik*. **H** Humid montane forest with dense understorey; mainly 2,500–3,000m. **D** Junin region of C Peru on E Andean slopes.

24 LARGE-FOOTED TAPACULO *Scytalopus macropus* 14cm The sole tapaculo in its range that is all dark grey. **V** Slowly rising series of staccato *pur* notes (>1 min). **H** Undergrowth and borders of montane forest; 2,400–3,500m. **D** Peru.

25 DIAMANTINA TAPACULO *Scytalopus diamantinensis* 13cm Female resembles male. Range differs from those of Rock and Planalto tapaculos (Plate 141). **V** Very long series of high, staccato *chek* notes (>1 min). **H** Forest and secondary growth. **D** NE Brazil.

CRESCENTCHESTS MELANOPAREIIDAE

26 COLLARED CRESCENTCHEST *Melanopareia torquata* 14.5cm Separable from Olive-crowned Crescentchest by rufous neck collar and by range. **V** Long series of well-separated *tu* notes. **H** Stony Cerrado with scattered bush and stunted trees. **D** E and C Brazil, NE Paraguay.

27 DOUBLE-COLLARED CRESCENTCHEST *Melanopareia bitorquata* 14–15cm Subtle plumage differences and range aid identification. Diet comprises mainly invertebrates. Unobtrusive and easily overlooked in dense grassland. **V** Song a shrill, penetrating and piping *Tuip, Tuip, Tuip*. **H** Dry savannah, Cerrado grassland. **D** E Bolivia.

28 OLIVE-CROWNED CRESCENTCHEST *Melanopareia maximiliani* 14.5cm Compare Collared Crescentchest. **V** Fast, loud, rhythmic series of *tu* notes (4 secs). **H** Grassy areas with scattered bush; 1,200–3,000m. **D** Bolivia, W Paraguay, N Argentina.

29 MARANON CRESCENTCHEST *Melanopareia maranonica* 16cm Wing edging is white, not rufous. **V** Rapid, rhythmic series of *tu* notes (5–8 secs). **H** Arid scrub. **D** S Ecuador, N Peru.

30 ELEGANT CRESCENTCHEST *Melanopareia elegans* 14.5cm Note wing pattern. **V** Sings as Maranon Crescentchest. **H** Dry scrub and woodland; <2,300m. **D** W Ecuador, NW Peru.

TYRANT FLYCATCHERS AND CALYPTURA TYRANNIDAE

31 GREY-HEADED PIPRITES *Piprites griseiceps* 12cm Note distinctive eye-ring in grey head. **V** Very high *wut-wut wut*. **H** Middle storeys of forest, tall secondary growth. **D** E Guatemala to W Panama.

32 WING-BARRED PIPRITES *Piprites chloris* 13cm Note plump jizz with roundish head and large eyes. **V** Rapid, slightly rising series of short, nasal *wip* notes. **H** Forest and woodland with dense undergrowth. **D** N Colombia, Amazonia, E Brazil, E Paraguay, NE Argentina.

33 BLACK-CAPPED PIPRITES *Piprites pileata* 12cm Unmistakable due to yellow and rufous plumage and black crown stripe. Note yellow legs. **V** Rapid series of some *tjew* notes and fast, higher *weet* notes. **H** Montane forest. **D** SE Brazil, NE Argentina.

1
3
5
7
6
2
4
13
8
11
♀
9
12
18
♂
10
14
15
17
16
24
♀
22
20
19
♂
21
25
23
27
26
♀
29
♂
28
♀
33
32
♂
30
31

TYRANT FLYCATCHERS AND CALYPTURA *CONTINUED*

1 PLANALTO TYRANNULET *Phyllomyias fasciatus* 11.5cm Difficult to separate from other tyrannulets; note brownish crown. **V** Short, level series of three plaintive notes. **H** Humid forest and woodland. **D** Brazil, W and NE Bolivia, E Paraguay, NE Argentina.

2 YUNGAS TYRANNULET *Phyllomyias weedeni* 11cm Note lack of wing-bars. Separable from Sooty-headed Tyrannulet by grey (not sooty) crown. **V** Short, descending series of high *weeh* notes. **H** Humid and semi-humid forest. **D** SE Peru, NW Bolivia.

3 ROUGH-LEGGED TYRANNULET *Phyllomyias burmeisteri* 11.5cm Note distinctively bicoloured bill and dark red eyes. **V** Very high series of single or double *psee* notes (one per sec). **H** Humid forest. **D** E Paraguay, N Argentina, SE Brazil.

4 WHITE-FRONTED TYRANNULET *Phyllomyias zeledoni* 12cm Note grey head; pale lower mandible diagnostic. **V** Series of ultra-high, sharp *sheeet* notes. **H** Forest canopy, edges and clearings. **D** Costa Rica, Panama, NW South America.

5 GREENISH TYRANNULET *Phyllomyias virescens* 12cm Compare Rough-legged Tyrannulet (has a bicoloured bill), Reiser's Tyrannulet (check ranges) and Mottle-cheeked Tyrannulet (has a longer bill and speckled face sides). **V** Short, sharp rattle. **H** Humid forest and secondary growth. **D** E Paraguay, NE Argentina, S and SE Brazil.

6 REISER'S TYRANNULET *Phyllomyias reiseri* 11.5cm Separable from Greenish Tyrannulet by its different range and darker grey crown. **V** Short, hurried series of 4–5 very high, sharp notes, preceded and followed by a softer, lower note. **H** Dry forest, riverine belts. **D** EC Brazil, Paraguay.

7 URICH'S TYRANNULET *Phyllomyias urichi* 12cm Note whiter wing-bars than Greenish Tyrannulet. **V** Note recorded. **H** Montane forest and woodland. **D** NE Venezuela.

8 SCLATER'S TYRANNULET *Phyllomyias sclateri* 12cm Difficult to separate from Buff-banded Tyrannulet (with yellower wing-bars; Plate 144) and Plumbeous-crowned Tyrannulet (with yellower underparts). Underparts greyish white. **V** Calm series of mid-high, hoarse notes of any length. **H** Montane forest and woodland. **D** SE Peru to NW Argentina.

9 GREY-CAPPED TYRANNULET *Phyllomyias griseocapilla* 11cm Note bright yellow flanks and grey crown. **V** Short, calm, very high phrase of 3–4 fluted, descending *wheeuw* notes. **H** Forest and adjacent clearings. **D** SE Brazil.

10 SOOTY-HEADED TYRANNULET *Phyllomyias griseiceps* 10cm Compare Yungas Tyrannulet. **V** Very short, high, descending phrase of a single and a triple note. **H** Borders and clearings in humid forest. **D** E Panama to the Guianas and N Brazil, E Ecuador and E Peru, SW Colombia and NW Ecuador.

11 PLUMBEOUS-CROWNED TYRANNULET *Phyllomyias plumbeiceps* 11.5cm Note dark grey crown and yellowish-pink wing stripes. **V** Short, high, loud phrase of 3–4 staccato notes and a very short roll. **H** Higher levels of montane forest, especially in cloud forest. **D** C Colombia to S Peru.

12 BLACK-CAPPED TYRANNULET *Phyllomyias nigrocapillus* 11cm Note black cap and colour pattern in wings. **V** Very short, thin, sizzling series of notes. **H** Dense cloud forest and mossy woodlands, elfin forest; 1,500–3,400m, locally lower. **D** Venezuela to Peru.

13 ASHY-HEADED TYRANNULET *Phyllomyias cinereiceps* 11cm Note dark cheek patch, bluish-grey crown and breast streaking. Compare Marble-faced Bristle-tyrant (Plate 145), which has less distinct breast streaking. **V** Very high, sizzling phrase of one ultra-high note and a mini-trill. **H** Higher levels and borders of montane forest. **D** Venezuela to Peru.

14 TAWNY-RUMPED TYRANNULET *Phyllomyias uropygialis* 11cm Tawny rump diagnostic. Note dark brown crown. **V** Ultra-high phrase of one note followed by 1–3 triple notes. **H** Upper montane forest, often at edges with bamboo and in stunted elfin forest; 1,500–3,000m. **D** Venezuela to NW Argentina.

15 YELLOW-CROWNED TYRANNULET *Tyrannulus elatus* 10.5cm Small bird that perches upright; note dusky borders of yellow crown patch. **V** Irregular series of loud, clear, whistled two-syllable phrases. **H** Forest edges, secondary growth, plantations, gardens, parks. **D** S Costa Rica through Amazonia.

16 FOREST ELAENIA *Myiopagis gaimardii* 12.5cm Note yellow stripe dividing dusky crown, and vague breast streaking. **V** Very high, sharp, drawn-out *sreeeh* note. **H** Canopy and borders of forest and woodland, riverine belts, secondary growth. **D** Panama through Amazonia.

17 GREY ELAENIA *Myiopagis caniceps* 12.5cm Races and sexes vary. Females generally have greenish upperparts, and pale yellow belly and grey head. Male usually greyish with whitish wing-bars. **V** Very high, sizzling, descending series of notes, sometimes ending in a trill. **H** Canopy and borders of humid forest. **D** Panama, widespread through South America.

18 FOOTHILL ELAENIA *Myiopagis olallai* 12.5cm Note grey crown, white crown patch and distinct wing-bars. **V** Fast, very high, level rattle, sometimes introduced by a soft note or small trill (2–3 secs). **H** Edges of humid submontane forest. **D** NC Colombia, W Venezuela, Ecuador, Peru.

19 PACIFIC ELAENIA *Myiopagis subplacens* 14cm Note indistinct wing edging and blackish ear-coverts. **V** Sharp, piercing phrase of a double note and a harsh single note. **H** Undergrowth at borders of forest, woodland, riverine belts. **D** W Ecuador, NW Peru.

20 YELLOW-CROWNED ELAENIA *Myiopagis flavivertex* 13cm Overall rather dusky and greyish. **V** Varied short, high, harsh, explosive phrases, e.g. a single note and a short rattle. **H** Lower storeys of várzea and swampy forest. **D** Amazonia.

21 GREENISH ELAENIA *Myiopagis viridicata* 13.5cm Note indistinct wing edging and plain face. **V** Very high, thin *seeeh* note, followed by a short rattle. **H** Forest, woodland, tall secondary growth, plantations, riverine belts. **D** Widespread across Central and South America.

22 JAMAICAN ELAENIA *Myiopagis cotta* 12.5cm Forages from the understorey to the canopy, sallying out from a perch to pick insects off foliage while in flight. **V** Rapid, high-pitched *ti-si-si-sip*. **H** Middle elevations of wet forests, open woods, scrub, coffee plantations and dry forest. **D** Jamaica.

23 YELLOW-BELLIED ELAENIA *Elaenia flavogaster* 16.5cm Lacks any distinctive features (belly is not particularly yellow), so difficult to identify; compare Large and Lesser elaenias. Often has a raised bushy crest, in N of range revealing a white patch. **V** Very wide variations on *peeep* note. **H** Diverse types of more or less wooded habitats. **D** Widespread from SE Mexico across Central and South America.

24 CARIBBEAN ELAENIA *Elaenia martinica* 16–18cm Often sits quietly for long periods. Sallies out to capture prey in the air or from leaves. **V** Repetitious *jui-up wit-churr*; also a drawn-out *pee-wee-reecreeree*. **H** Mainly dry lowland forests, woods, scrub and gardens; in mountains in S Lesser Antilles. **D** West Indies.

25 LARGE ELAENIA *Elaenia spectabilis* 18cm Separable from Yellow-bellied Elaenia by third wing-bar and lack of white in crest. **V** Descending, pure, fluted *tuh* note, single or in a rapid series. **H** Edges and clearings in forest and woodland. **D** W Amazonia to E Brazil and C Bolivia.

26 NORONHA ELAENIA *Elaenia ridleyana* 17cm Sole elaenia on Fernando de Noronha Island. **V** Pure and scratchy variations of *puh* note. **H** Woodland, scrub and thickets around houses. **D** Fernando de Noronha Island, Atlantic Ocean.

27 WHITE-CRESTED ELAENIA *Elaenia albiceps* 15cm Often sits motionless for long periods in dense foliage. Gleans, including hover-gleaning, for berries, nectar and insects. **V** Generally silent post-breeding. **H** Woodland borders and scrub. **D** Widespread across South America.

28 CHILEAN ELAENIA *Elaenia chilensis* 14–15cm Previously treated as subspecies of White-crested Elaenia. Migratory, with subtly longer wings. **V** Song a repeated, shrill *Tchueeo, Tchueeo, Tchueeo*. Call a whistled *Whee-oo*. **H** Montane forests during breeding; wide range of wooded and scrub habitats at other times. **D** Breeds in the Andes, from S Bolivia to S Argentina and Chile. Outside breeding season, extends to Brazil and Peru.

29 SMALL-BILLED ELAENIA *Elaenia parvirostris* 14.5cm Crest absent or small; white crown spot normally visible. **V** Series of well-separated, short, 1–3-syllable phrases with varying intonation. **H** Range of wooded and forested habitats. **D** Colombia and Venezuela to C Bolivia and SE Brazil.

30 OLIVACEOUS ELAENIA *Elaenia mesoleuca* 15cm Separable from Small-billed Elaenia by lack of white in crest. **V** Series of 1–2-syllable phrases with a liquid quality. **H** Humid forest. **D** E Paraguay, NE Argentina, S and SE Brazil.

31 SLATY ELAENIA *Elaenia strepera* 15.5cm Slate-grey colouring distinctive. **V** Short, level or rising, dry rattles. **H** Woodland and forest borders. **D** Breeds in Bolivia, NW Argentina. Winters in Venezuela.

32 MOTTLE-BACKED ELAENIA *Elaenia gigas* 18cm Unmistakable due to its erect stance and crest patterning. **V** Short, high trill. **H** Shrubby clearings, river islands. **D** S Colombia to W Bolivia.

33 BROWNISH ELAENIA *Elaenia pelzelni* 18cm Lack of olive tinge to brown plumage distinctive. **V** Short, sharp phrases, e.g. a rising *kew-ik*. **H** Forest along rivers and river islands. **D** Amazon River and its tributaries.

34 PLAIN-CRESTED ELAENIA *Elaenia cristata* 14.5cm Crest (normally raised) lacks white. **V** Very short, rapid phrases of *djer* notes. **H** Scarcely wooded savannah, open woodland, Cerrado. **D** Venezuela, the Guianas, Brazil, SE Peru, Bolivia.

35 LESSER ELAENIA *Elaenia chiriquensis* 13.5cm Separable from Plain-crested Elaenia by white in less distinct crest and by call. **V** Series of hoarse, lowered *tsee-bur* notes. **H** Savannah, Cerrado, gardens, cultivated areas. **D** Costa Rica to NE Argentina.

36 COOPMANS'S ELAENIA *Elaenia brachyptera* 13–14cm Previously treated as subspecies of Lesser Elaenia. Vocalisation and subtle plumage differences aid separation from that species. **V** Calls include vibrating trills and a shrill, whistled *Pee-oo*. **H** Scrub, cleared open woodland. **D** SW Colombia, NW Ecuador.

37 RUFOUS-CROWNED ELAENIA *Elaenia ruficeps* 14.5cm Separable from Plain-crested and Lesser elaenias by some rufous on hindcrown and faint dusky streaking below. **V** Short, hurried, dry rattle, changing in intonation. **H** Scarcely bushed savannah, woodland edges, riverine belts. **D** N Amazonia.

1
2
3
4
5
6
7
8
9
10
11
12
13
14
15
16
17
♀
♂
18
19
20
21
22
23
24
25
26
27
28
29
30
31
32
33
34
35
36
37

TYRANT FLYCATCHERS AND CALYPTURA *CONTINUED*

1 MOUNTAIN ELAENIA *Elaenia frantzii* 14cm Without (or with only slight) crest and not showing any white on crown. **V** High, downslurred *sreee-ur* note. **H** Shrubby edges and clearings in forest and woodland; 750–3,000m. **D** Guatemala to Colombia and Venezuela.

2 HIGHLAND ELAENIA *Elaenia obscura* 17cm Rather large, with narrow wing-bars and a long tail. No crest or coronal patch. **V** Downslurred, fast *sree-ur*. **H** Undergrowth and borders of montane forest; 1,700–3,000m in Andes, <2,000m in E parts of range. **D** Ecuador to NW Argentina, SE South America.

3 GREAT ELAENIA *Elaenia dayi* 20cm Large elaenia whose darkish upperparts contrast with white eye-rings and wing-bars. **V** Short, nasal twittering in a duet. **H** Forest scrub, stunted woodland on tepuis; 1,800–2,600m, locally lower. **D** Tepuis of S Venezuela.

4 SIERRAN ELAENIA *Elaenia pallatangae* 14.5cm Note rather flat crest with a narrow white stripe. Difficult to separate from several other *Elaenia* flycatchers, e.g. White-crested Elaenia (which is whiter on belly; Plate 143) and Lesser Elaenia (which shows slightly more white in crest; Plate 143). **V** High, rapid, sharp two-syllable phrase. **H** Clearings and edges of montane forest, brushy pastures, secondary growth; 1,500–3,000m, locally higher or lower. **D** Colombia to Bolivia.

5 TEPUI ELAENIA *Elaenia olivina* 15–16cm Previously treated as a subspecies of Sierran Elaenia. Vocalisation and subtle plumage differences aid separation from that species. Varied diet includes fruit and insects. **V** Calls include a piping, whistled *Psee-up*. **H** Humid, forested tepui mountains and slopes. **D** Tepuis of Guyana and S and SE Venezuela.

6 GREATER ANTILLEAN ELAENIA *Elaenia fallax* 15cm White crown patch usually hidden. Often in pairs or with mixed-species flocks. Sallies from perch to take prey from leaves or the air; also feeds on fruit. **V** Harsh *pwee-chi-chi-chiup see-ere chewit-chewit*; also a repeated trill, given at dawn. **H** Humid mountain and lowland forest, pine forest, forest edges, thickets, and open country with scattered trees. **D** Jamaica, Hispaniola.

7 YELLOW-BELLIED TYRANNULET *Ornithion semiflavum* 9cm Note grey cap, short tail and indistinct greenish edging to wing feathers. **V** Very high, slightly running-down *peeuw peepeuwuohwut*. **H** Forest canopy, tall secondary growth. **D** SE Mexico to W Panama.

8 BROWN-CAPPED TYRANNULET *Ornithion brunneicapillus* 8cm Note tiny size, striking white eyebrow, short tail and lack of wing-bars. **V** Very high, thin, short twitter. **H** Humid forest, secondary growth and adjacent areas. **D** Costa Rica to NW Venezuela and NW Ecuador.

9 WHITE-LORED TYRANNULET *Ornithion inerme* 8cm Note white lores, eye-rings and wing-bars (broken up into 'drops'). **V** Very high, level, hurried series of three *pee* notes and a lower mini-trill. **H** Canopy and borders of humid forest and woodland. **D** Amazonia, E Brazil.

10 NORTHERN BEARDLESS TYRANNULET *Camptostoma imberbe* 11cm Secretive; best located by voice. Often flicks tail and wings. Feeds mainly by gleaning. **V** Thin, whistled *fleeeeer* or *fleeeee-rit*; also a descending *piti pi pi PEEE dee dee*. **H** Dense bushes or trees, usually near water. **D** SC and SW USA and Mexico to NW Costa Rica, Cozumel Island.

11 SOUTHERN BEARDLESS TYRANNULET *Camptostoma obsoletum* 9.5cm Includes about nine subspecies. Note bushy crest. **V** Varied, very short songs, e.g. a very high *pee-prrr*. **H** Wide range of dry and humid, open wooded and forested habitats. **D** Costa Rica to Uruguay.

12 SUIRIRI FLYCATCHER *Suiriri suiriri* 15.5cm Crown and back mostly grey, tail and wings darker, the latter with greyish-white wing-bars. Feeds on arthropods and small fruits. **V** Nasal twittering and trills. **H** Chaco, open woodland, Cerrado; <3,000m. **D** SC South America, Amazonia.

13 CHAPADA FLYCATCHER *Suiriri affinis* 16cm Separable from Suiriri Flycatcher by broad white tail tips. **V** Nasal twittering. **H** Campo, Cerrado. **D** SC Brazil, E Bolivia.

14 WHITE-THROATED TYRANNULET *Mecocerculus leucophrys* 14cm Includes 10 subspecies, but none has such broad or white wing-bars as White-banded Tyrannulet. **V** Song comprises one or two drawn-out, very high, thin *seee* notes. **H** Montane forest, woodland and adjacent clearings; 2,800–3,500m, locally lower. **D** Venezuela to N Argentina.

15 WHITE-TAILED TYRANNULET *Mecocerculus poecilocercus* 11cm Note pale rump and white outer-tail feathers. **V** Ultra-high series of 3–4 thin *see* notes. **H** Montane forest, especially cloud forest; 1,800–2,600m, locally lower. **D** Colombia to Peru.

16 BUFF-BANDED TYRANNULET *Mecocerculus hellmayri* 11cm Separable from White-tailed Tyrannulet by lack of white in tail, less pale rump and slightly buffer wing-bars. **V** Ultra-high series of 2–3 thin *see* notes. **H** Montane forest and woodland; 1,400–2,600m, locally lower. **D** SE Peru to NW Argentina.

17 RUFOUS-WINGED TYRANNULET *Mecocerculus calopterus* 11cm Unmistakable due to rufous on wings. See Rufous-winged Antwren (Plate 135). **V** Very short, nasal twittering. **H** Foothill and lower montane forest and woodland. **D** Ecuador, Peru.

18 SULPHUR-BELLIED TYRANNULET *Mecocerculus minor* 11.5cm Note yellow underparts and distinct wing-bars. **V** Series of 3–4 descending, nasal notes. **H** Montane forest and clearings; 1,600–2,800m. **D** Venezuela, Colombia to N Peru.

19 WHITE-BANDED TYRANNULET *Mecocerculus stictopterus* 12.5cm Note bold, contrasting wing pattern. Separable from White-throated Tyrannulet by lack of, or less, yellow in underparts. **V** Sharp *squeeyh*. **H** Montane forest and woodland, especially at the treeline; 2,400–3,500m. **D** Venezuela to Bolivia.

20 BLACK-CRESTED TIT-TYRANT *Anairetes nigrocristatus* 13cm Separable from smaller Pied-crested Tit-tyrant by larger white tail tips. **V** Rather steeply descending series of nasal *tri* notes, or given as 3–5 separate notes. **H** Arid scrub, *Polylepis* woodland; 2,300–3,900m, locally lower. **D** S Ecuador, N Peru.

21 PIED-CRESTED TIT-TYRANT *Anairetes reguloides* 11.5cm Compare with Black-crested Tit-tyrant. **V** Calm, descending series of 6–10 emphasised nasal notes. **H** Shrubby hillsides, hedgerows, open woodland; <3,000m. **D** WC Peru to NW Chile.

22 ASH-BREASTED TIT-TYRANT *Anairetes alpinus* 13cm No similar bird occurs in its habitat within its range. **V** Unstructured series of short rolls and nasal notes. **H** *Polylepis* woodland; 3,700–4,600m. **D** Peru, Bolivia.

23 YELLOW-BILLED TIT-TYRANT *Anairetes flavirostris* 11cm Separable from Tufted Tit-tyrant by black eyes and pale base to bill. **V** Calm, unstructured twittering with many sharp and a few hoarse notes. **H** Montane scrub, brush and *Polylepis* woodland; <4,000m. **D** Peru, NW Chile, Bolivia, NW Argentina.

24 TUFTED TIT-TYRANT *Anairetes parulus* 10cm Note pale eyes. Separable from Juan Fernandez Tit-tyrant by range. **V** Sharp trills (2–3 secs). **H** Montane brushy forest, elfin forest, cloud forest, *Polylepis* woodland, dry thorn scrub; 2,500–3,500m, locally down to sea level or up to 4,200m. **D** C Colombia to N Argentina, C Argentina, S Chile to SW Argentina.

25 JUAN FERNANDEZ TIT-TYRANT *Anairetes fernandezianus* 12.5cm Unmistakable in its range. **V** Not described. **H** Wooded habitats. **D** Robinson Crusoe Island and Juan Fernández Islands off WC coast of Chile.

26 AGILE TIT-TYRANT *Uromyias agilis* 13cm Note long, recurved crest (not curved up as in Tufted Tit-tyrant). **V** Chattering trill. **H** Undergrowth and borders with bamboo in montane forest; 2,600–3,500m, especially just below the treeline. **D** Venezuela to Ecuador.

27 UNSTREAKED TIT-TYRANT *Uromyias agraphia* 12.5cm Separable from Agile Tit-tyrant by range. **H** Forest undergrowth at margins of fast-flowing mountain streams; 700–2,800m, locally higher or lower. **D** Peru.

28 TORRENT TYRANNULET *Serpophaga cinerea* 11.5cm Distinctly patterned; note black head and black wings with contrasting wing-bars. **H** On rocks in and along fast-flowing mountain streams. **D** Costa Rica to Bolivia.

29 RIVER TYRANNULET *Serpophaga hypoleuca* 11cm Rather unmarked, brown-grey above and whitish below. Note habitat. **V** Short, very high, soft roll, preceded by a single *sip*. **H** Scrub, mainly on islands in large rivers. **D** Orinoco and Amazon rivers and their tributaries.

30 SOOTY TYRANNULET *Serpophaga nigricans* 12cm Brownish grey above, purer grey below. Usually seen at water's edge or on rocks in water. **V** Trill, preceded by single high *tsip* (2–3 secs in total). **H** Scrubby areas near small to large waterbodies. **D** S Bolivia, E Paraguay and SC Brazil to C Argentina and Uruguay.

31 WHITE-CRESTED TYRANNULET *Serpophaga subcristata* 11cm Note broad whitish wing-bars, semi-concealed white and black patch in crown, and yellowish belly and vent. **V** Very short, rattling trill. **H** Forest borders, riverine belts, Cerrado, scrub, gardens, cultivated areas. **D** SE Brazil and Uruguay, E Bolivia to SC Brazil, Paraguay, C Argentina.

32 WHITE-BELLIED TYRANNULET *Serpophaga munda* 11.5cm Separable from White-crested Tyrannulet by pure grey (not olive) upperparts and pure white (not yellowish) underparts. **V** Very short, high rattle. **H** Scrub, woodland. **D** NC Bolivia and SW Brazil to C Argentina and Uruguay.

33 STRANECK'S TYRANNULET *Serpophaga griseicapilla* 12cm Separable from White-bellied Tyrannulet by voice, olive tinge to upperparts, slightly buff wing stripes and lack of white crown patch. **V** High, dry *feet-thrrrree* rattle. **H** Arid lowland scrub. **D** Breeds in CW and NW Argentina. Winters in Bolivia, N Argentina, W Uruguay, Paraguay, possibly Brazil.

34 MOUSE-COLORED TYRANNULET *Phaeomyias murina* 12cm Note flesh-based lower mandible and pronounced eyebrow; brownish tone to upperparts. **V** Very high, accelerated, rising *wit-witwitwitwttr*. **H** Cultivation with scattered trees, savannah. **D** Panama to N Argentina.

35 TUMBESIAN TYRANNULET *Phaeomyias tumbezana* 12cm Difficult to identify, but note lack of crest, pale base to maxilla and indistinct eyebrow. Best recognised by its song. **V** Song a short, high trill, preceded by a higher inhaled note. **H** Scrub, woodland, riverine belts, Cerrado, parks, mangrove. **D** SW Ecuador, N Peru.

36 YELLOW TYRANNULET *Capsiempis flaveola* 11cm Yellow colouring distinctive. **V** Very high, loud, scolding *wurriwurriWecWicWic*. **H** Open areas with scrub, trees and hedges, often near water. **D** S Nicaragua to E Brazil.

37 BEARDED TACHURI *Polystictus pectoralis* 10cm Male distinctive due to finely streaked crown and 'beard'. Female separable from Rufous-sided Pygmy Tyrant (Plate 145) by less contrasting white throat. **V** Song comprises three rapid, very high, bouncing notes, followed by a small trill. **H** Variety of grassy habitats, including campo and Cerrado. **D** Widespread across South America.

38 GREY-BACKED TACHURI *Polystictus superciliaris* 9.5cm No similar bird occurs in its habitat within its range. **V** High note followed by a trill (4–5 secs). **H** Grassland with rocks and scattered bush, brushy edges of cloud forest. **D** SE Brazil.

39 COCOS FLYCATCHER *Nesotriccus ridgwayi* 13cm Only flycatcher on Cocos Island. Note long bill. **V** Explosive series of notes, often sung in duet. **H** Swampy forest, secondary growth. **D** Cocos Island, WC Pacific Ocean.

1
2
3
4
5
6
7
8
9
10
11
12
13
14
15
16
17
18
19
20
21
22
23
24
25
26
27
28
29
30
31
32
33
34
35
36
♂ 37
♀
38
39

TYRANT FLYCATCHERS AND CALYPTURA *CONTINUED*

1 DINELLI'S DORADITO *Pseudocolopteryx dinelliana* 11cm Separable from Warbling Doradito by paler cheeks and more olive (not rufescent) crown. **V** Short series of dry, voiceless *krek* notes, mixed with very high, thin, small warbles. **H** Reeds, wet grass, adjacent shrub. **D** C Paraguay and SW Brazil to C Argentina.

2 CRESTED DORADITO *Pseudocolopteryx sclateri* 11cm Black-and-white crest is distinctive. **V** A few ultra-high notes, just out of range of some people's hearing. **H** Marshes. **D** Venezuela, Guyana and Trinidad, Bolivia, E and S Brazil to NE Argentina.

3 SUBTROPICAL DORADITO *Pseudocolopteryx acutipennis* 11cm Bright green and yellow plumage with slightly duskier cheeks. **V** Short, ultra-high warbling. **H** Reedbeds, sedge and brush near water. **D** Colombia to NW Argentina.

4 WARBLING DORADITO *Pseudocolopteryx flaviventris* 11.5cm Compare with Ticking Doradito. Note rufescent crown and dusky cheeks. **V** Unstructured, soft series of sharp, squeaky notes. **H** Marsh and surrounding shrub. **D** Paraguay and S Brazil to EC Argentina.

5 TICKING DORADITO *Pseudocolopteryx citreola* 11.5cm Separable from Warbling Doradito by song and range. **V** Unstructured, short series of *krek* notes. **H** Fragments of brushy marshes. **D** C Chile, WC Argentina.

6 BRONZE-OLIVE PYGMY TYRANT *Pseudotriccus pelzelni* 11cm Overall dull olive; separable from female manakins by less green plumage and different behaviour (*Pseudotriccus* species forage alone, low in the understorey). **V** Very high *psee-see-see*, followed by a lower note and alternating with a small rattle (produced by wing-snap). **H** Undergrowth of foothill and montane forest; 300–2,100m. **D** Panama and Colombia to Peru.

7 HAZEL-FRONTED PYGMY TYRANT *Pseudotriccus simplex* 11cm Rufous on wings and head is less bright than in Rufous-headed Pygmy Tyrant; also check altitudes. **V** Short, ultra-high, descending, sibilant phrase. **H** Montane forest; 1,100–2,500m. **D** SE Peru, NW Bolivia.

8 RUFOUS-HEADED PYGMY TYRANT *Pseudotriccus ruficeps* 11cm Brightness of rufous coloration varies according to race. **V** Unstructured series of rising and falling rattles, alternated with wing-snapping. **H** Upper montane forest; 2,000–3,300m, locally lower. **D** Colombia to Bolivia.

9 RINGED ANTPIPIT *Corythopis torquatus* 14cm Iris dark or pale. Separable from Southern Antpipit (ranges overlap marginally) by less olive cheeks, whiter throat and browner mantle. **V** Snaps bill; song is a high, loud, shrill two-syllable *preer-preer*. **H** Mainly terra firme forest. **D** Amazonia.

10 SOUTHERN ANTPIPIT *Corythopis delalandi* 14cm Compare with Ringed Antpipit. **V** Snaps bill; song is a short series of very high, undulating Ringed Antpipit *peer* notes, the last note doubled. **H** Humid forest and woodland. **D** E Bolivia and C Brazil to E Paraguay, SE Brazil and NE and NW Argentina.

11 TAWNY-CROWNED PYGMY TYRANT *Euscarthmus meloryphus* 10cm No similar bird with rufous on crown or sides of face occurs in its habitat. **V** Long, undulating, rolling rattle. **H** Arid scrub, overgrazed grassland, waste areas, dry forest edges. **D** Widespread across South America.

12 RUFOUS-SIDED PYGMY TYRANT *Euscarthmus rufomarginatus* 11cm Separable from female Bearded Tachuri by white throat, central breast and belly. **V** Hurried, continuous series of short, dry rattles. **H** Campo, Cerrado, shrubby grassland. **D** Suriname, E Brazil to NE Bolivia and E Paraguay.

13 GREY-AND-WHITE TYRANNULET *Pseudelaenia leucospodia* 12.5cm Note white in spread crest. **V** Call is a dry, soft *pip*. **H** Arid scrub and thickets. **D** SW Ecuador, NW Peru.

14 LESSER WAGTAIL-TYRANT *Stigmatura napensis* 13.5cm Check range. Note tail pattern. **V** Excited chattering in short, descending phrases, sung in a duet. **H** Thick, tall grass along rivers, arid scrub. **D** Amazon and Orinoco rivers and their tributaries.

15 GREATER WAGTAIL-TYRANT *Stigmatura budytoides* 15cm Check range. Tail pattern similar to that of Lesser Wagtail-tyrant. **V** Liquid chattering in excited, short phrases. **H** Arid scrub, Chaco, dry riverine belts. **D** C and SE Bolivia, W Paraguay, NW and C Argentina, E Brazil.

16 PALTRY TYRANNULET *Zimmerius vilissimus* 11cm Greyish yellow below, prominent yellow edging to wing feathers. **V** Very high or ultra-high, sustained series of *swee* notes (20–30 secs). **H** Forest edges and clearings, plantations, areas with scattered trees. **D** S Mexico to El Salvador.

17 MISTLETOE TYRANNULET *Zimmerius parvus* 9–10cm Diet comprises mainly mistletoe berries, and associated insects. **V** Calls include upslurred, disyllabic and whistled *tse-iup, tse-iup, tse-iup*. **H** Wooded and forested habitats where mistletoes occur. **D** Honduras to NW Colombia.

18 SPECTACLED TYRANNULET *Zimmerius improbus* 11.5–12cm Diet comprises fruits, berries and associated insects. Usually solitary. **V** Calls include slurred, disyllabic and whistled *Whe-er, whe-er* phrases. **H** Montane forests and wooded slopes. **D** Colombia, Venezuela.

19 VENEZUELAN TYRANNULET *Zimmerius petersi* 11.5–12cm Diet comprises mainly berries (notably mistletoe) and associated insects. Seen singly, or occasionally in pairs. **V** Calls include shrill, trilling and grating *tchrr-tchrr-rrr-rrr-rr-r* phrases. **H** Coastal, hilly forested habitats. **D** N Venezuela.

20 BOLIVIAN TYRANNULET *Zimmerius bolivianus* 12cm Note pale eyes and lack of eyebrows. **V** Song is a very high, sharp *wifi*. **H** Montane forest, woodland, secondary growth. **D** SE Peru, C and W Bolivia.

21 RED-BILLED TYRANNULET *Zimmerius cinereicapilla* 11.5cm Separable from smaller Mishana Tyrannulet mainly by more extensive range. **V** Small, high rattle. **H** Humid foothill and montane forest. **D** Ecuador, Peru, Bolivia.

22 MISHANA TYRANNULET *Zimmerius villarejoi* 10.5cm Restricted range, overlapping with Red-billed Tyrannulet in Río Mayo area of Peru. See also Chico's Tyrannulet. **V** Lazy series of ultra-high, drawn-out *seee* notes. **H** Humid forest with stumpy trees and shrub. **D** N Peru.

23 CHICO'S TYRANNULET *Zimmerius chicomendesi* 11cm Separable from Mishana Tyrannulet only by voice and by range. **V** Very high, slightly rising *tuwee-teeh-three*. **H** Lowland woodland and scrub. **D** Rio Madeirinha in S Amazonas, Brazil.

24 SLENDER-FOOTED TYRANNULET *Zimmerius gracilipes* 10.5cm Often confused with other tyrant-flycatchers, but note grey head with short eyebrows. **V** Short, high, slightly rising rattle. **H** Swampy forest. **D** Amazonia.

25 GUIANAN TYRANNULET *Zimmerius acer* 11cm Separable from Slender-footed Tyrannulet by brown-grey (not pure grey) head. Ranges overlap marginally. **V** Short phrase of 2–3 rising notes. **H** Humid lowland forest. **D** The Guianas and NE Brazil.

26 GOLDEN-FACED TYRANNULET *Zimmerius chrysops* 11cm Difficult to separate from Choco and Peruvian tyrannulets, but note ranges. **V** Varied repertoire of calls, e.g. single *peuur* note or in a series. **H** Forest, woodland and adjacent clearings. **D** S Colombia and W Venezuela to Ecuador and N Peru.

27 COOPMANS'S TYRANNULET *Zimmerius minimus* 10.5–11.5cm Previously treated as conspecific with Golden-faced Tyrannulet. Diet includes mistletoe berries and associated insects. **V** Song a *tchu, tche, tchrrrrrr*, the latter phrase a vibrating trill. Calls include a shrill, downslurred *tSee-ur*. **H** Humid forest, secondary woodland. **D** NE Colombia, N Venezuela.

28 CHOCO TYRANNULET *Zimmerius albigularis* 11.5cm Difficult to separate from Golden-faced and Peruvian tyrannulets, but check ranges. **V** Call is a high *wit*. **H** Humid forest and adjacent clearings at mid-altitudes in mountains. **D** W Ecuador, SW Colombia.

29 LOJA TYRANNULET *Zimmerius flavidifrons* 10.5–11.5cm Previously treated as conspecific with Golden-faced Tyrannulet. Vocalisation aids identification. Diet includes mistletoe berries and associated insects. **V** Calls include a distinctive, shrill *Tsuu-Eeep*, the second phrase upslurred, overall sounding like an attention-seeking human whistle. **H** Humid forest, secondary woodland. **D** SW Ecuador.

30 PERUVIAN TYRANNULET *Zimmerius viridiflavus* 11cm Difficult to separate from Golden-faced and Paltry tyrannulets, but note ranges. **V** Simple two-syllable phrase: mid-high, slightly drawn-out *puuuh* note, abruptly rising to a high *wit*. **H** Forest and woodland. **D** Peru.

31 VARIEGATED BRISTLE TYRANT *Pogonotriccus poecilotis* 11.5cm Note colour of wing-bars. Separable from Chapman's Bristle Tyrant by grey (not green) crown, and from larger Slaty-capped Flycatcher (Plate 146) by bicoloured bill. **V** High, sharp note followed by a trill. **H** Lower storeys of montane forest. **D** N Colombia and NW Venezuela to Peru.

32 CHAPMAN'S BRISTLE TYRANT *Pogonotriccus chapmani* 12cm Compare Variegated Bristle Tyrant; crown is concolourous with back. **V** Not recorded. **H** Undergrowth of forest on tepuis. **D** Tepuis in S and SE Venezuela.

33 MARBLE-FACED BRISTLE TYRANT *Pogonotriccus ophthalmicus* 11.5cm Note distinct ear crescents and grey crown. Ashy-headed Tyrannulet (Plate 143) has a bluer crown and in Plumbeous-crowned Tyrannulet (Plate 143) ear crescent is less distinct. **V** High-pitched chatter ending in a trill. **H** Lower storeys and borders of montane forest. **D** Venezuela to Bolivia.

34 SPECTACLED BRISTLE TYRANT *Pogonotriccus orbitalis* 11cm Has a distinct eye-ring, grey crown and indistinct ear crescents. **V** Soft series of *tic* notes. **H** Interior of humid montane forest; 500–1,400m. **D** S Colombia to Bolivia.

35 VENEZUELAN BRISTLE TYRANT *Pogonotriccus venezuelanus* 11cm Separable from Marble-faced Bristle Tyrant by pale maxilla. **V** Call a *che-dip*; also a fast trill. **H** Lower storeys and borders of montane forest. **D** N Venezuela.

36 ANTIOQUIA BRISTLE TYRANT *Pogonotriccus lanyoni* 11cm Restricted range. Note wide wing-bars. **V** Song a short stutter. **H** Foothill forest and woodland. **D** NW Colombia.

37 SOUTHERN BRISTLE TYRANT *Pogonotriccus eximius* 11cm Pied facial pattern is diagnostic. **V** Short, thin trill, ending abruptly. **H** Lower storeys and borders of humid forest. **D** E Paraguay, NE Argentina, SE Brazil.

38 MOTTLE-CHEEKED TYRANNULET *Phylloscartes ventralis* 12cm No distinct features, but note horizontal posture and cocked tail. **V** Loose series of single staccato *whik* notes. **H** Humid forest and woodland; 1,000–2,200m. **D** Peru to NW Argentina.

39 ALAGOAS TYRANNULET *Phylloscartes ceciliae* 12cm No other *Phylloscartes* species occurs in its range. Note distinct white eyebrow. **V** Very high, piercing *sweek* and variations on this. **H** Canopy and borders of humid forest. **D** E Brazil.

40 RESTINGA TYRANNULET *Phylloscartes kronei* 12cm Resembles Mottle-cheeked Tyrannulet but note rather clean face sides and more distinct black line around cheek patch; also normally found at lower altitudes. **V** Very high, sharp *sit* notes, uttered singly or in a series. **H** Restinga and secondary growth at sea level. **D** SE Brazil.

1
2
3
4
5
6
7
8
9
10
11
12
13
14
15
16
17
18
19
20
21
22
23
24
25
26
27
28
29
30
31
32
33
34
35
36
37
38
39
40

TYRANT FLYCATCHERS AND CALYPTURA *CONTINUED*

1 BAHIA TYRANNULET *Phylloscartes beckeri* 12cm Range differs from those of Mottle-cheeked and Restinga tyrannulets (Plate 145). **V** Call is a high *tsik*. **H** Forest and old secondary growth. **D** E Brazil.

2 PANAMANIAN TYRANNULET *Phylloscartes flavovirens* 11cm Note distinctive eye-ring and wing-bars, and long tail. **V** Little recorded, but may include a short, high, rapid *pweet* call. **H** Forest and woodland canopy. **D** Panama.

3 OLIVE-GREEN TYRANNULET *Phylloscartes virescens* 12cm No other *Phylloscartes* species occurs in its range. Note slender jizz and distinct wing-bars (see Slender-footed Tyrannulet; Plate 145). **V** Song starts with a loud introductory note, followed by a series of high, reedy notes. **H** Canopy and borders of humid forest and tall secondary growth. **D** NE Amazonia.

4 ECUADORIAN TYRANNULET *Phylloscartes gualaquizae* 11.5cm Compare with Plumbeous-crowned Tyrannulet, which has a pale area behind ear crescent (Plate 143), and Rough-legged Tyrannulet (Plate 143) and Southern Bristle Tyrant (Plate 145), which both have a partially pale maxilla. **V** Short, dry, slightly descending rattle. **H** Humid foothill and montane forest. **D** S Colombia, Ecuador, N Peru.

5 BLACK-FRONTED TYRANNULET *Phylloscartes nigrifrons* 13cm No other *Phylloscartes* species occurs in its range. Note black front. **V** Dry rattle preceded by a *zip* note (2 secs). **H** Humid foothill and montane forest on tepuis. **D** Tepuis of Venezuela and W Guyana.

6 RUFOUS-BROWED TYRANNULET *Phylloscartes superciliaris* 12cm Separable from Rufous-lored, Cinnamon-faced and Minas Gerais tyrannulets by range. Note indistinct wing-bars. **V** High, sharp *speet*, followed by a rapid 2–5-syllable series of similar notes. **H** Humid montane forest, mainly in the interior. **D** Costa Rica to N Venezuela and Ecuador.

7 RUFOUS-LORED TYRANNULET *Phylloscartes flaviventris* 11.5cm Dark cheek and rufous eyebrow are distinctive. Separable from Cinnamon-faced and Minas Gerais tyrannulets by restricted range. **V** Short, fast series of thin, high notes. **H** Foothill and montane forest. **D** N Venezuela.

8 CINNAMON-FACED TYRANNULET *Phylloscartes parkeri* 12cm Separable from Rufous-lored Tyrannulet by range and less dark cheek. **V** Very high, downslurred trill. **H** Interior of montane forest. **D** C Peru to N Bolivia.

9 MINAS GERAIS TYRANNULET *Phylloscartes roquettei* 11.5cm No other *Phylloscartes* species occurs in its range. **V** Varied, very short rattles. **H** Riverine belts. **D** E Brazil.

10 SAO PAULO TYRANNULET *Phylloscartes paulista* 10.5cm Note dark ear patch and rather upright posture. **V** Short, sibilant rattle. **H** Undergrowth and borders of humid forest. **D** E Paraguay, NE Argentina, SE Brazil.

11 OUSTALET'S TYRANNULET *Phylloscartes oustaleti* 12.5cm Constantly quivers tail. Note lack of wing-bars. **V** Slow series of ultra-short nasal trills. **H** Understorey in interior of humid forest. **D** E and SE Brazil.

12 SERRA DO MAR TYRANNULET *Phylloscartes difficilis* 11.5cm Note striking white eye-ring and lore, bright upperparts and white-grey underparts. **V** Very short, nasal twitter. **H** Shrubby undergrowth of montane forest. **D** E and SE Brazil.

13 BAY-RINGED TYRANNULET *Phylloscartes sylviolus* 11.5cm Note pale eyes and pinkish rufous in face. **V** Very high series of notes, accelerating to a trill (1–2 secs). **H** Canopy and borders of humid forest. **D** E Paraguay, NE Argentina, SE Brazil.

14 STREAK-NECKED FLYCATCHER *Mionectes striaticollis* 13.5cm Separable from Olive-striped Flycatcher by greyish head and unstreaked mid-belly. Note altitudinal ranges of this and next species. **V** Calm series of very high, squeaky notes. **H** Montane forest and woodland; 1,300–2,500m, locally higher or lower. **D** Colombia to N Bolivia.

15 OLIVE-STRIPED FLYCATCHER *Mionectes olivaceus* 13.5cm Compare Streak-necked Flycatcher. **V** Continuous series of high, thin, squeaky notes in sets of three. **H** Undergrowth and borders of humid forest, woodland and plantations; 500–1,600m, locally lower or higher. **D** Costa Rica to N Venezuela and N Bolivia.

16 OCHRE-BELLIED FLYCATCHER *Mionectes oleagineus* 13cm Note olive head and rufous wing linings. **V** Unstructured series of *weet*, *prut* and other notes. **H** Undergrowth of forest and woodland. **D** S Mexico through Amazonia, E Brazil.

17 MCCONNELL'S FLYCATCHER *Mionectes macconnelli* 13cm Separable from Ochre-bellied Flycatcher by more uniform wings that lack clear bars and edgings. **V** Hurried series of *wreeh* notes. **H** Terra firme forest and montane forest. **D** E and SW Amazonia.

18 TEPUI FLYCATCHER *Mionectes roraimae* 12–13cm Previously treated as conspecific with McConnell's Flycatcher. Subtle differences in plumage and vocalisation aid identification. **V** Song is a vibrating, trilling *tch, tch, tcheerrrr*, the second phrase accelerating and rising in pitch during delivery. **H** Tepui and forested mountain slopes. **D** S and SE Venezuela, Guyana, N Brazil.

19 GREY-HOODED FLYCATCHER *Mionectes rufiventris* 13.5cm Note grey head and uniform brown wings. **V** Irregular, speeded-up series of nasal *kew* notes. **H** Undergrowth and borders of forest and woodland. **D** E Paraguay, NE Argentina, SE Brazil.

20 SEPIA-CAPPED FLYCATCHER *Leptopogon amaurocephalus* 13cm Sepia cap distinctive. **V** Very high, angry, bickering, slurred-down *tutrrrrrriit*. **H** Forest understorey, tall secondary growth, plantations. **D** Widespread from S Mexico to N Argentina.

21 SLATY-CAPPED FLYCATCHER *Leptopogon superciliaris* 13.5cm Separable from Inca Flycatcher by dark grey cheek mark and pale olive (not buff) chest. **V** Varied nasal rattles. **H** Humid forest. **D** Costa Rica to N Venezuela and N Bolivia.

22 RUFOUS-BREASTED FLYCATCHER *Leptopogon rufipectus* 13cm Pale rufous in face, throat and chest. **V** Slow series of very high, nasal *pee* notes. **H** Humid forest. **D** SW Venezuela and SE Colombia to N Peru.

23 INCA FLYCATCHER *Leptopogon taczanowskii* 13cm Separable from Rufous-breasted Flycatcher by grizzled grey-and-white face sides and by range, and from Slaty-capped Flycatcher by colour of breast. **V** High, thin, short, sibilant melody. **H** Humid montane forest, especially cloud forest. **D** Peru, S of Marañón River.

24 NORTHERN SCRUB FLYCATCHER *Sublegatus arenarum* 14cm Along with Amazonian and Southern Scrub flycatchers, differs from similar flycatchers in having a short all-black bill; note also their pale supraloral. This species is not safely separable from Amazonian or Southern scrub flycatchers, but check ranges and note slightly yellower belly. **V** Very high, thin *sree*. **H** Dry scrub, thorn woodland, savannah. **D** Costa Rica to the Guianas.

25 AMAZONIAN SCRUB FLYCATCHER *Sublegatus obscurior* 14cm See Northern Scrub Flycatcher. **V** Not very vocal. **H** Scrubby borders and edges of humid forest. **D** Amazonia.

26 SOUTHERN SCRUB FLYCATCHER *Sublegatus modestus* 13.5cm See Northern Scrub Flycatcher. **V** Short, rapid series of notes, concluding with a high *seeuw*. **H** Arid scrub, open woodland, Cerrado. **D** SC, E and SE South America.

27 SLENDER-BILLED INEZIA *Inezia tenuirostris* 9cm Separable from Plain and Amazonian inezias by range; note also very slender all-black bill and small size. **V** Short, dry trill. **H** Dry scrub, arid woodland, cactus desert, shrubby gardens. **D** NE Colombia, NW Venezuela.

28 PLAIN INEZIA *Inezia inornata* 10cm Note smooth grey crown. **V** High, fast, rising trill (2 secs). **H** Chaco, riverine belts, várzea. **D** W and SW Amazonia.

29 AMAZONIAN INEZIA *Inezia subflava* 12cm Separable from Pale-tipped Inezia by range and darker eyes. **V** Loud, level series of *chirrit* notes. **H** Shrubby woodland, usually near water. **D** C and SW Amazonia.

30 PALE-TIPPED INEZIA *Inezia caudata* 12cm See Amazonian Inezia. **V** Rapid or slow, level or descending series of high, sharp notes. **H** Shrubby woodland, scrub, mangroves. **D** Colombia to NE Brazil.

31 FLAVESCENT FLYCATCHER *Myiophobus flavicans* 12cm Separable from Orange-crested, Unadorned and Roraiman flycatchers by all-black bill. Has two wing-bars, one far more distinct than the other. **V** Unstructured, slow series of single, very high, thin *twik* notes. **H** Lower storeys and borders of humid forest. **D** Venezuela to Peru.

32 ORANGE-CRESTED FLYCATCHER *Myiophobus phoenicomitra* 11.5cm Very similar to Flavescent Flycatcher but breast striping is more distinct and maxilla is pale. **V** Ultra-high, slow, unstructured chirping. **H** Undergrowth and borders of forest. **D** Colombia to Peru.

33 UNADORNED FLYCATCHER *Myiophobus inornatus* 11.5cm Duller than Flavescent Flycatcher and wing-bars are more distinct. Note dark lores. **V** Very high, unstructured, hurried chirping, mixed with small rolls. **H** Lower storeys and borders of montane forest. **D** SE Peru to N Bolivia.

34 RORAIMAN FLYCATCHER *Myiophobus roraimae* 13cm Note brown mantle and large eyes. **V** Very high, hurried mixture of hoarse and sharp notes. **H** Understorey of montane forest. **D** Tepuis of S Venezuela, NW Brazil and W Guyana, and Colombia to Bolivia.

35 OLIVE-CHESTED FLYCATCHER *Myiophobus cryptoxanthus* 12cm Dull brownish grey; only lower belly has a yellow tinge. **V** Unstructured, high, loud, sharp *chwee* notes (one every 1–2 secs). **H** Shrubby clearings, pastures, open woodland. **D** E Ecuador, NE Peru.

36 BRAN-COLORED FLYCATCHER *Myiophobus fasciatus* 12.5cm Note rufous upperparts and distinctly striped underparts. **V** Short, sharp rattle, preceded by a higher, fluted note. **H** Shrubby clearings, pastures, open woodland. **D** Costa Rica to C Argentina.

37 HANDSOME FLYCATCHER *Nephelomyias pulcher* 11cm Crown greyer than mantle, and note wing pattern. Separable from Many-colored Rush Tyrant (Plate 147) by greyish (not pale rufous) face sides. **V** Series of three *tsi* notes (just within or out of human hearing). **H** Upper levels and borders of humid forest and secondary growth. **D** Colombia to Ecuador, SE Peru.

38 ORANGE-BANDED FLYCATCHER *Nephelomyias lintoni* 12.5cm No similar bird occurs in its restricted range. Note pale eyes and arrow-headed wing markings. **V** Unstructured, unhurried series of high nasal notes, short rattles and chatters. **H** Montane forest, mostly at the tree-line; 2,250–3,200m. **D** S Ecuador, NE Peru.

39 OCHRACEOUS-BREASTED FLYCATCHER *Nephelomyias ochraceiventris* 13.5cm Distinguishable from smaller, shorter-tailed Orange-banded Flycatcher by distinct separation between yellow cheeks and olive crown. **V** Series of *tjap* notes followed by three rapid, higher *see* notes. **H** Canopy and borders of montane forest and woodland; 2,800–3,500m. **D** N Peru to NW Bolivia.

1
2
3
4
5
6
7
8
9
10
11
12
13
14
15
16
17
18
19
20
21
22
23
24
25
26
27
28
29
30
31
32
33
34
35
36
37
38
39

TYRANT FLYCATCHERS AND CALYPTURA *CONTINUED*

1 ORNATE FLYCATCHER *Myiotriccus ornatus* 12cm Unmistakable due to its jizz and colouring. **V** Very high, sharp *pseeet* note. **H** Humid foothill and montane forest. **D** Colombia to Peru.

2 MANY-COLORED RUSH TYRANT *Tachuris rubrigastra* 11.5cm Beautiful bird that is unmistakable. **V** Song comprises a mixture of rattles, single high *tjuw* notes and low, dry croaks. **H** Marshes with reedbeds and adjacent grassy areas. **D** W, S and SE South America.

3 SHARP-TAILED GRASS TYRANT *Culicivora caudacuta* 10.5cm Separable from ovenbirds (e.g. *Leptasthenura* species, Plate 126; Chotoy Spinetail, Plate 127; Wren-like Rushbird, Plate 129; and Bay-capped Wren-spinetail, Plate 129) by black streaking, white edges to tertials, and buff flanks. **V** Calm series of single harsh notes (two per sec). **H** Campo, Cerrado. **D** W Bolivia to C Brazil, south to NE Argentina.

4 DRAB-BREASTED BAMBOO TYRANT *Hemitriccus diops* 11cm Separable from Brown-breasted Bamboo Tyrant by whiter lore and eye-ring. **V** Short, sharp, nasal rattle, sometimes doubled. **H** Undergrowth and edges with bamboo in humid forest and secondary growth; <1,300m. **D** E Paraguay, NE Argentina, SE Brazil.

5 BROWN-BREASTED BAMBOO TYRANT *Hemitriccus obsoletus* 11cm See Flammulated and Drab-breasted Bamboo tyrants; note altitudinal range. **V** Rapid, slightly undulating series of staccato notes (2–3 secs). **H** Undergrowth with bamboo in montane forest and woodland; 500–2,300m. **D** SE and S Brazil, NE Argentina.

6 FLAMMULATED BAMBOO TYRANT *Hemitriccus flammulatus* 11cm Separable from Drab-breasted and Brown-breasted Bamboo tyrants by range. **V** Series of 4–5 rapidly descending, hollow, nasal *tic* notes, sometimes given as a rattle. **H** Humid terra firme forest and secondary growth; <500m, locally much higher. **D** SW Amazonia.

7 SNETHLAGE'S TODY-TYRANT *Hemitriccus minor* 10cm Difficult to separate from other tyrants like White-bellied Tody-tyrant, and best identified by voice. See also Acre Tody-tyrant. **V** Hoarse, slightly lowered, trilling shriek. **H** Dense edges of várzea and secondary growth. **D** S Amazonia.

8 YUNGAS TODY-TYRANT *Hemitriccus spodiops* 11cm Note restricted range. **V** Song resembles that of Snethlage's Tody-tyrant but comprises a slightly longer series of notes that change in intonation. **H** Borders of foothill and lower montane forest. **D** C Bolivia.

9 ACRE TODY-TYRANT *Hemitriccus cohnhafti* 11cm Note buffish lores, streaked breast and pale panel in folded wing. **V** Song a series of well-separated, very short trills. **H** Secondary growth and forest edges. **D** Brazil–Bolivia border in SE Acre state and adjacent SE Peru.

10 BOAT-BILLED TODY-TYRANT *Hemitriccus josephinae* 11cm Lacks wing-bars. **V** Low, dry series of *puk* notes. **H** Lower storeys and edges of humid forest. **D** The Guianas, NC Brazil.

11 WHITE-EYED TODY-TYRANT *Hemitriccus zosterops* 11cm Occurs only N of the Amazon. **V** Hurried series of 5–6 *puk* notes. **H** Lower storeys of humid forest. **D** N Amazonia.

12 WHITE-BELLIED TODY-TYRANT *Hemitriccus griseipectus* 11cm Occurs only S of the Amazon. **V** Unstructured series of level, high, sharp, staccato *pik* notes. **H** Lower storeys of humid forest. **D** S Amazonia, E Brazil.

13 ZIMMER'S TODY-TYRANT *Hemitriccus minimus* 10cm Note brownish crown and yellow wing-bars. See also Snethlage's and White-eyed tody-tyrants. **V** Short, sharp rattle of 5–6 *tu* notes. **H** Campinarana. **D** Amazonia.

14 EYE-RINGED TODY-TYRANT *Hemitriccus orbitatus* 11.5cm Note large black eyes set in a distinct white eye-ring and lore, bordered below by a darkish line. **V** Series of ultra-short rattles of slightly different pitch. **H** Lower storeys of humid forest. **D** SE Brazil.

15 JOHANNES'S TODY-TYRANT *Hemitriccus iohannis* 11cm Note brownish ocular area, weak wing-bars and faint neck streaking. **V** Dry, slightly rising, high rattle (2 secs). **H** Dense riverside vegetation. **D** W Amazonia.

16 STRIPE-NECKED TODY-TYRANT *Hemitriccus striaticollis* 11cm Note white around eyes, white lores and distinct throat striping. **V** Series of three *pit* notes. **H** Humid secondary growth, riverine belts. **D** W and S Amazonia.

17 HANGNEST TODY-TYRANT *Hemitriccus nidipendulus* 9.5cm Unstreaked and bright green above, with white underparts. **V** High, very short, sharp trills, sung as triplets. **H** Restinga, humid forest, dense woodland. **D** E and SE Brazil.

18 PEARLY-VENTED TODY-TYRANT *Hemitriccus margaritaceiventer* 10.5cm Head greyish and wing-bars yellowish. Note red-rimmed eyes. **V** Small, sharp rattle, preceded by two higher, dry notes. **H** Arid scrub, woodland undergrowth, shrubby pastures. **D** N, SC and E South America.

19 PELZELN'S TODY-TYRANT *Hemitriccus inornatus* 9cm Note small size and whitish wing-bars. **V** Rapid series of 25 high *dic* notes. **H** Campinarana. **D** NW Brazil.

20 BLACK-THROATED TODY-TYRANT *Hemitriccus granadensis* 10.5cm Black throat is diagnostic. **V** Short, rapid, slightly undulating series of *whik* notes. **H** Humid forest, cloud forest, secondary growth; 2,000–3,000m, locally lower. **D** Venezuela to Peru.

21 BUFF-BREASTED TODY-TYRANT *Hemitriccus mirandae* 10cm No similar bird occurs in its restricted range. **V** Rising, rapid series of 3–5 sharp notes. **H** Highland forest and secondary growth. **D** NE and E Brazil.

22 CINNAMON-BREASTED TODY-TYRANT *Hemitriccus cinnamomeipectus* 10cm Note cinnamon face sides and breast, and characteristic pattern on tertials. **V** Short, high, sharp rattle. **H** Cloud forest. **D** SE Ecuador, N Peru.

23 KAEMPFER'S TODY-TYRANT *Hemitriccus kaempferi* 10cm Note brownish head and neck, white tail tips and partially white tertials. **V** Rapid series of 10–15 high, sharp *wit* notes. **H** Undergrowth at edges of forest and woodland. **D** SE Brazil.

24 BUFF-THROATED TODY-TYRANT *Hemitriccus rufigularis* 12cm Note pale eyes and plain wings; tertials are partly whitish. **V** Series of high, slightly upslurred notes (any length). **H** Higher levels of montane forest. **D** Ecuador to Bolivia.

25 FORK-TAILED TODY-TYRANT *Hemitriccus furcatus* 11cm Note all-brown head (with paler ocular region), colourful wings and notched, distinctly marked tail. **V** Very short, sharp, cranky rattle. **H** Humid forest and secondary growth. **D** SE Brazil.

26 EARED PYGMY TYRANT *Myiornis auricularis* 7cm Separable from White-bellied Pygmy Tyrant by range, greenish-yellow belly and striking face pattern. **V** Very short, mournful trill or series of chipped notes. **H** Lower storeys and borders of forest and woodland. **D** E and SE Brazil, SE Paraguay, NE Argentina.

27 WHITE-BELLIED PYGMY TYRANT *Myiornis albiventris* 7cm See Eared Pygmy Tyrant. **V** Short, mournful trill. **H** Lower storeys and edges of forest. **D** SE Ecuador, EC Peru, N Bolivia.

28 BLACK-CAPPED PYGMY TYRANT *Myiornis atricapillus* 6.5cm Separable from Short-tailed Pygmy Tyrant by range and the blackness of its crown. **V** Calm series of very high, chirping, single or double notes. **H** Lower storeys, edges and clearings in forest, secondary growth and plantations. **D** Costa Rica to NW Ecuador.

29 SHORT-TAILED PYGMY TYRANT *Myiornis ecaudatus* 6.5cm See Black-capped Pygmy Tyrant. **V** High, short, dry trills in sets of three. **H** Lower storeys and borders of humid forest. **D** Amazonia.

30 NORTHERN BENTBILL *Oncostoma cinereigulare* 9.5cm Separable from the smaller Southern Bentbill by greyish (not greenish) head. **V** Song is a downslurred gull-like note. **H** Scrubby open woodland, undergrowth at forest edges. **D** S Mexico to Panama.

31 SOUTHERN BENTBILL *Oncostoma olivaceum* 9cm See Northern Bentbill. **V** Song resembles that of Northern Bentbill, but more guttural and slightly longer. **H** Dense growth at forest borders. **D** Panama, Colombia.

32 SCALE-CRESTED PYGMY TYRANT *Lophotriccus pileatus* 9.5cm Note orange crest and yellow wing-bars. Separable from Double-banded Pygmy Tyrant by brighter crown. **V** Level, irregular series of 3–15 high, staccato notes. **H** Humid forest and older secondary growth. **D** Costa Rica to N Venezuela and SE Peru.

33 LONG-CRESTED PYGMY TYRANT *Lophotriccus eulophotes* 10cm Note long crest feathers. **V** Rapid, chipped series of 4–7 *pit* notes. **H** Bamboo undergrowth and borders of swampy forest and woodland. **D** SW Amazonia.

34 DOUBLE-BANDED PYGMY TYRANT *Lophotriccus vitiosus* 10cm Separable from Long-crested and Helmeted pygmy tyrants by distinct greenish wing-bars. **V** Song is a small, hoarse, fast rattle. **H** Terra firme and várzea forest, woodland. **D** W and NE Amazonia.

35 HELMETED PYGMY TYRANT *Lophotriccus galeatus* 10cm Separable from Long-crested Pygmy Tyrant by missing or obscure wing-bars. **V** Long series of *tic* notes. **H** Borders and undergrowth of forests and secondary growth. **D** N and SE Amazonia.

36 PALE-EYED PYGMY TYRANT *Atalotriccus pilaris* 9.5cm Separable from Pearly-vented Tody-tyrant by unstreaked throat. **V** Rattling, upslurred shriek. **H** Scrub, dry woodland, riverine belts. **D** Panama, N South America.

1
2
3
4
5
6
7
8
9
10
11
12
13
14
15
16
17
18
19
20
21
22
23
24
25
26
27
28
29
30
31
32
33
34
35
36

TYRANT FLYCATCHERS AND CALYPTURA *CONTINUED*

1 RUFOUS-CROWNED TODY-FLYCATCHER *Poecilotriccus ruficeps* 9.5cm Separable from Lulu's Tody-flycatcher by black-bordered cheeks and white throat. The ranges of this and Lulu's Tody-flycatcher are separated by the Marañón River. **V** Short, dry, very fast rattle. **H** Humid forest, especially várzea. **D** Venezuela to N Peru.

2 LULU'S TODY-FLYCATCHER *Poecilotriccus luluae* 9.5cm See Rufous-crowned Tody-flycatcher. **V** Short, rather liquid rattle. **H** Shrubby clearings and borders of montane forest. **D** N Peru.

3 WHITE-CHEEKED TODY-FLYCATCHER *Poecilotriccus albifacies* 9.5cm Male unmistakable; female not reliably distinguishable from female Black-and-white Tody-flycatcher, but separated by range. Note pattern formed by mainly white tertials. **V** Short, bouncing rattle. **H** Bamboo in humid forest. **D** SE Peru, NW Bolivia, extreme W Brazil.

4 BLACK-AND-WHITE TODY-FLYCATCHER *Poecilotriccus capitalis* 9.5cm See White-cheeked Tody-flycatcher. **V** Song comprises 3–5 staccato notes, followed by an excited rattle. **H** Dense undergrowth and borders of humid forest. **D** W and SW Amazonia.

5 BUFF-CHEEKED TODY-FLYCATCHER *Poecilotriccus senex* 9cm Separable from Smoky-fronted Tody-flycatcher by buff ocular area. **V** Series of high, short, dry trills. **H** Campina forest on poor sands. **D** NC Brazil.

6 RUDDY TODY-FLYCATCHER *Poecilotriccus russatus* 10cm Distinctive, richly coloured plumage. **V** Song is a short, bouncing-down rattle. **H** Undergrowth and borders of forest on tepuis. **D** Tepuis of SE Venezuela, N Brazil and W Guyana.

7 OCHRE-FACED TODY-FLYCATCHER *Poecilotriccus plumbeiceps* 9.5cm Bright face with grey cheeks and crown is diagnostic. **V** Hurried series of 1–6 dry rattles. **H** Thick undergrowth with bamboo at forest edges and in secondary growth. **D** Peru to N Argentina, Paraguay, and SE and E Brazil.

8 SMOKY-FRONTED TODY-FLYCATCHER *Poecilotriccus fumifrons* 9cm Separable from Rusty-fronted Tody-flycatcher by yellowish underparts and buff being restricted to lores. **V** Song is a series of shivering, downslurred rattles. **H** Thick woody vegetation near or in edges of forest. **D** E Amazonia, E Brazil.

9 RUSTY-FRONTED TODY-FLYCATCHER *Poecilotriccus latirostris* 9.5cm See Smoky-fronted Tody-flycatcher. Note buff shoulders. **V** Song resembles that of Smoky-fronted Tody-flycatcher. **H** Dense undergrowth in clearings and edges of forest with adjacent overgrown pastures, along riverbanks. **D** W and C Amazonia, C Brazil.

10 SLATY-HEADED TODY-FLYCATCHER *Poecilotriccus sylvia* 9.5cm Eye colour may vary, from dark to yellow. Note black loral patch. **V** Mid-high, dry, rattling *prrruh*. **H** Dense woody vegetation in forest edges, roadsides, secondary growth, riverine belts. **D** S Mexico to NE Brazil.

11 GOLDEN-WINGED TODY-FLYCATCHER *Poecilotriccus calopterus* 9cm Note striking wing pattern. Separable from Black-backed Tody-flycatcher by range. **V** Series of 4–6 dry, descending rattles. **H** Undergrowth and borders of forest, secondary growth, regenerating clearings. **D** W Amazonia.

12 BLACK-BACKED TODY-FLYCATCHER *Poecilotriccus pulchellus* 9.5cm Note head and wing pattern. **V** Song resembles that of Golden-winged Tody-flycatcher but may be even drier. **H** Undergrowth and borders of forest, secondary growth, regenerating clearings. **D** SE Peru.

13 BLACK-CHESTED TYRANT *Taeniotriccus andrei* 12cm Uniquely patterned and coloured. **V** Series of dry, slightly downturned *chew* notes (one per sec). **H** Dense undergrowth with bamboo in humid forest, especially várzea and woodland. **D** NC and E Amazonia.

14 SPOTTED TODY-FLYCATCHER *Todirostrum maculatum* 9.5cm Separable from Johannes's and Stripe-necked tody-tyrants (Plate 147) by more sharply defined striping and colouring. **V** Series of very high chirps (two per sec). **H** Dense growth along or at edges of waterbodies, including mangroves. **D** Amazonia.

15 YELLOW-LORED TODY-FLYCATCHER *Todirostrum poliocephalum* 9.5cm Separable from Common Tody-flycatcher by yellow lores. Often shows pale orange eyes. **V** Slow, unstructured series of very high chirps. **H** Humid forest edges and adjacent clearings and gardens. **D** SE Brazil.

16 COMMON TODY-FLYCATCHER *Todirostrum cinereum* 9cm Separable from similar tody-flycatchers by grey nape and neck, merging in green mantle. **V** Hurried series of dry *tic* notes, often sung in a duet. **H** In trees in a variety of open and semi-open areas; avoids dense forest. **D** S Mexico to NE Argentina.

17 MARACAIBO TODY-FLYCATCHER *Todirostrum viridanum* 9cm Paler and with a shorter tail than Common Tody-flycatcher. Note short eyebrow. **V** Series of very high *tic* notes. **H** Arid scrub and woodland. **D** NW Venezuela.

18 PAINTED TODY-FLYCATCHER *Todirostrum pictum* 9cm Note extensive streaking below. **V** Very high chirps (one per sec). **H** Humid forest and woodland with adjacent clearings. **D** NE Amazonia.

19 YELLOW-BROWED TODY-FLYCATCHER *Todirostrum chrysocrotaphum* 9cm Range does not overlap with that of Painted Tody-flycatcher. Yellow post-ocular stripe is diagnostic. **V** Very high series of *tic* notes (three every 2 secs). **H** Terra firme and várzea forest, secondary growth and adjacent clearings. **D** S and W Amazonia.

20 BLACK-HEADED TODY-FLYCATCHER *Todirostrum nigriceps* 8.5cm Note dark eyes, black head and white throat. Lacks breast streaking. **V** Very high *tic* notes (three every 2 secs). **H** Canopy and borders of humid forest, woodland and other wooded habitats, e.g. pastures with scattered trees. **D** Costa Rica to N Venezuela and Ecuador.

21 BROWNISH TWISTWING *Cnipodectes subbrunneus* 17cm Has a small head, yellow-buff wing-feather margins and shaggy plumage. **V** Song is a high, resounding double note. **H** Dense undergrowth of forest and woodland. **D** Panama through Amazonia.

22 RUFOUS TWISTWING *Cnipodectes superrufus* 17cm Separable from Brownish Twistwing by all-rufous plumage, darker on mantle and wings. **V** Mid-high, rapidly descending series of five nasal notes, the last one doubled. **H** Bamboo. **D** SE Peru, NW Bolivia.

23 EYE-RINGED FLATBILL *Rhynchocyclus brevirostris* 15cm Striking eye-ring is diagnostic. **V** Slow series of ultra-high *seee* notes. **H** Middle levels and borders of humid forest and secondary growth. **D** S Mexico to NW Colombia.

24 OLIVACEOUS FLATBILL *Rhynchocyclus olivaceus* 15cm Separable from Eye-ringed Flatbill by less distinct eye-ring. **V** Very high, accelerating series of piercing notes, sometimes given as a shiver. **H** Undergrowth in humid forest, especially in várzea and in woodland. **D** Panama through Amazonia, E Brazil.

25 PACIFIC FLATBILL *Rhynchocyclus pacificus* 15cm Separable from Fulvous-breasted Flatbill by greenish breast. **V** Very high series of 6–7 piercing *swee* notes. **H** Undergrowth of humid forest. **D** W Colombia, NW Ecuador.

26 FULVOUS-BREASTED FLATBILL *Rhynchocyclus fulvipectus* 15cm Pale fulvous breast is distinctive. **V** Irregular series of *tic* and *snip* notes. **H** Undergrowth of humid forest, especially near streams. **D** W Venezuela to NW Bolivia.

27 YELLOW-OLIVE FLATBILL *Tolmomyias sulphurescens* 14cm The 16 rather similar races in South America can be divided into those with a greenish crown (as shown) and those with a grey crown. Colour of iris varies. Grey-crowned birds can be separated from Yellow-margined Flatbill only in the hand and by voice. **V** Slow series of slurred, high, thin shrieks. **H** Lower storeys of forest, woodland and riverine belts. **D** S Mexico to SE Brazil.

28 ORANGE-EYED FLATBILL *Tolmomyias traylori* 13.5cm Orange-rimmed eyes and buff face and throat are diagnostic. **V** Slow, mounting series of 7–10 thin, shrill notes. **H** Higher levels of várzea forest, riverine belts and woodland. **D** W Amazonia.

29 ZIMMER'S FLATBILL *Tolmomyias assimilis* 13.5cm Separable from Yellow-olive Flatbill by darker eyes. Otherwise very similar; check habitats. See also Yellow-margined Flatbill. **V** Song resembles that of Orange-eyed Flatbill but series is shorter. **H** Higher levels of várzea, terra firme forest, woodland and neglected plantations. **D** Amazonia.

30 YELLOW-MARGINED FLATBILL *Tolmomyias flavotectus* 13.5cm Previously treated as conspecific with Zimmer's Flatbill. **V** Song a shrill, squeaky *tsuip, tsuip, tsuip*. Calls include a high-pitched *tsiip*, its tone reminiscent of song. **H** Rainforest, secondary woodland, plantations, mainly in lowlands. **D** Costa Rica to NW Ecuador.

31 GREY-CROWNED FLATBILL *Tolmomyias poliocephalus* 12cm Separable from Yellow-margined Flatbill by partially dark maxilla. **V** Slow, very high, sharp series of five mounting notes. **H** Higher levels and edges of humid forest and woodland. **D** Amazonia, E Brazil.

32 OCHRE-LORED FLATBILL *Tolmomyias flaviventris* 12cm Note yellow underparts and greenish upperparts with strikingly yellow-edged wing feathers. **V** Unstructured, level series of very high, sharp, single and double *see* notes. **H** Forest, woodland, riverine belts, restinga, Caatinga. **D** N and S Amazonia, E Brazil.

33 OLIVE-FACED FLATBILL *Tolmomyias viridiceps* 12–13cm Previously treated as conspecific with Yellow-margined Flatbill. Forages in tree foliage for invertebrates. **V** Song a whistled *tsueep, tsueep, tsueep*. Calls include a shrill *tsiip*. **H** Várzea rainforest. **D** W Amazonia.

34 KINGLET CALYPTURA *Calyptura cristata* 8cm Very small, with an unmistakable colour pattern. Might be the rarest bird of Brazil. **V** Not recorded. **H** Canopy of montane forest. **D** SE Brazil.

35 CINNAMON-CRESTED SPADEBILL *Platyrinchus saturatus* 9.5cm Overall rather plain, with a less distinct facial pattern than in other spadebills. **V** Calm series of *tjew* notes. **H** Undergrowth of terra firme forest. **D** Amazonia.

36 STUB-TAILED SPADEBILL *Platyrinchus cancrominus* 9cm As White-throated Spadebill, but with a paler breast and smaller yellow crown patch. **V** Hurried sharp, nasal *gifitme*. **H** Forest understorey. **D** Amazonia.

37 WHITE-THROATED SPADEBILL *Platyrinchus mystaceus* 10cm Note distinct facial pattern. **V** Song is a descending and then rising trill (2–3 secs). **H** Undergrowth of forest, woodland and riverine belts. **D** Costa Rica to Bolivia; N, E and SE South America.

38 GOLDEN-CROWNED SPADEBILL *Platyrinchus coronatus* 9cm Orange-yellow crown is diagnostic; other spadebills usually conceal their yellow or white crown stripe. **V** Short, very high, thin trill. **H** Undergrowth of humid forest. **D** Honduras to NW Ecuador, Amazonia.

39 YELLOW-THROATED SPADEBILL *Platyrinchus flavigularis* 10cm Note sparsely marked rufous head (with concealed white crown patch). **V** Short, rapidly rising trill. **H** Undergrowth of montane forest. **D** W Venezuela to E Peru.

40 WHITE-CRESTED SPADEBILL *Platyrinchus platyrhynchos* 11cm Note ochraceous underparts with a yellowish throat. **V** Nasal trill, ascending and then descending. **H** Undergrowth of humid terra firme forest. **D** Amazonia.

41 RUSSET-WINGED SPADEBILL *Platyrinchus leucoryphus* 12.5cm Unmistakable due to rufous wings. **V** Song resembles that of White-crested Spadebill, but descent is much shorter, more like a single note. **H** Undergrowth of humid forest and mature secondary growth. **D** E Paraguay, NE Argentina, SE Brazil.

1
2
3
♀
♂
4
♀
5
♂
6
7
8
9
10
11
12
♂
♀
♀
13
♂
14
15
17
19
16
18
20
21
22
23
25
24
26
27
29
31
34
28
30
32
33
36
35
37
38
39
40
41

TYRANT FLYCATCHERS AND CALYPTURA *CONTINUED*

1 CINNAMON NEOPIPO *Neopipo cinnamomea* 9.5cm Compare Ruddy-tailed Flycatcher (Plate 157). **V** Ultra-high, descending series of 6–8 sharp notes. **H** Undergrowth of terra firme forest. **D** W and NE Amazonia.

2 CINNAMON FLYCATCHER *Pyrrhomyias cinnamomeus* 13cm No other bird with such extensive rufous coloration occurs in its mountain habitat. **V** Very high, thin rattle (1–2 secs). **H** Borders and clearings in montane forest and woodland; 1,200–3,000m, locally higher or lower. **D** Venezuela to NW Argentina.

3 CLIFF FLYCATCHER *Hirundinea ferruginea* 18.5cm Unmistakable in its habitat. **V** Short rattle or small chatter, preceded by a *weew* note. **H** Stony and rocky vertical habitats, e.g. cliffs, canyon walls, road cuttings, buildings. **D** Widespread across South America.

4 EULER'S FLYCATCHER *Lathrotriccus euleri* 13cm Note bicoloured bill, brown upperparts and buffy wing-bars. **V** Series of repeated three-syllable *tjeewtjeewtjeew* phrases. **H** Lower levels and open borders of humid forest and woodland. **D** Widespread across South America.

5 GREY-BREASTED FLYCATCHER *Lathrotriccus griseipectus* 13cm Note distinct eye-ring and white wing-bars. Innermost tertial shows more distinct white margin than all other flight feathers. **V** Series of 2–4 notes, including a very high, downslurred *tjeew* note. **H** Undergrowth of humid forest and woodland. **D** W Ecuador, N Peru.

6 TAWNY-CHESTED FLYCATCHER *Aphanotriccus capitalis* 12cm Separable from Ochraceous Pewee by white lore and chin. **V** Very high, rapid *tututut Uureh*. **H** Forest streams and forest edges. **D** E Nicaragua, N Costa Rica.

7 BLACK-BILLED FLYCATCHER *Aphanotriccus audax* 13.5cm Note black bill, grey face sides and white margin to innermost tertial. **V** Very high, downslurred, rather harsh *tjeeuw* note. **H** Undergrowth near swamps and streams in forest. **D** Panama, N Colombia.

8 FUSCOUS FLYCATCHER *Cnemotriccus fuscatus* 14.5cm Note rufescent-grey plumage. Separable from Euler's Flycatcher by distinct eyebrow. **V** Very high, rising *whee-eu* note (one every 2–3 secs). **H** Thick undergrowth of forest, often near water. **D** Widespread across South America.

9 BELTED FLYCATCHER *Xenotriccus callizonus* 12cm Ochraceous breast-band sharply demarcated from yellow belly. Lore more distinct than in Ochraceous Pewee. **V** Ultra-high, rather hoarse *priuw* or bickering *tut-tutwrueet*. **H** Dry, scrubby woodland. **D** S Mexico, SW Guatemala, NW El Salvador.

10 PILEATED FLYCATCHER *Xenotriccus mexicanus* 14cm Note olive breast. Eye-ring more distinct than in other small flycatchers with crests. **V** High *tsjip* call; Whimbrel-like sounds included in song. **H** Scrubby forest. **D** SW Mexico.

11 EASTERN PHOEBE *Sayornis phoebe* 16.5–18cm Makes frequent fly-catching sallies from a prominent perch. Persistently pumps and spreads tail. **V** Distinctive *fee-be* and a sharp *chip* or *tsyp*. **H** Woodland, woodland edges, farmland, stream sides, farm and suburban buildings. **D** NC Canada to SC USA, SE USA and SE Canada.

12 BLACK PHOEBE *Sayornis nigricans* 15–18cm Actions and habits similar to those of Eastern Phoebe. **V** Two high, whistled phrases *sisee sitsew sisee sitsew...*, usually alternated; also a thin *tseew*. **H** Semi-open and shaded areas near water. **D** Breeds in North, Central and South America. Winters from W USA to NW Argentina.

13 SAY'S PHOEBE *Sayornis saya* 17–19.5cm Pumps and spreads tail. Perches on or near the ground, makes sallies to capture insects from the air, foliage or the ground. **V** Plaintive *pe-ee*; also a low *pidiweew pidireep pidiweew pidiireep...* **H** Open dry areas, including prairie, sagebrush and barren foothills. **D** Alaska to S Mexico.

14 NORTHERN TUFTED FLYCATCHER *Mitrephanes phaeocercus* 12–13.5cm Makes fly-catching sallies, regularly returning to the same perch, often shivering tail on landing. **V** Bright *tchwee-tchwee* or *turree-turree*. **H** Forest edges and clearings. **D** W Mexico to NW Ecuador.

15 OLIVE TUFTED FLYCATCHER *Mitrephanes olivaceus* 13cm Separable from Northern Tufted Flycatcher by greenish underparts. **V** Ultra-high, thin, lowered *seee* note. **H** Lower storeys and borders of montane forest. **D** Peru to NW Bolivia.

16 OLIVE-SIDED FLYCATCHER *Contopus cooperi* 18–19cm Makes fly-catching sallies, often using the same perch for weeks. **V** Sharp *quick three bears*; also a low *pep pep pep* or similar. **H** Coniferous forests. **D** Breeds from W Alaska to SW USA, NW Mexico, EC USA and SE Canada. Winters in N and C South America.

17 GREATER PEWEE *Contopus pertinax* 20cm Makes aerial sallies, generally returning to the same exposed perch. **V** Clear *ho-say ma-re-ah*; also a low *pip-pip-pip*. **H** Mountain pine-oak woodlands. **D** SW USA to Nicaragua.

18 DARK PEWEE *Contopus lugubris* 17cm Note uniform dark coloration and peaked crest. **V** High, loud, sharp *bic-bic-bic-bic bic-tic*. **H** Montane forest edges. **D** Costa Rica, W Panama.

19 SMOKE-COLORED PEWEE *Contopus fumigatus* 17cm Separable from other dark *Contopus* flycatchers by its size and grey throat. **V** Level series of 5–15 staccato *dit* notes. **H** Edges and large clearings in montane forest. **D** Venezuela to NW Argentina.

20 OCHRACEOUS PEWEE *Contopus ochraceus* 17cm Note yellow underparts, tinged ochraceous. Separable from Northern Tufted Flycatcher by pale, not cinnamon throat. **V** Sharp, repeated *pwit* or *pip pip pip*. **H** Canopy at forest edges and clearings; 2,000–3,000m. **D** Costa Rica, W Panama.

21 WESTERN WOOD PEWEE *Contopus sordidulus* 15–17cm Reliably distinguished from Eastern Wood Pewee only by voice. Makes sallies from perch to catch insects in the air or from foliage, returning to the same or a nearby perch. **V** Nasal, descending *peeyee* or *peeer*. **H** Woodland and river groves. **D** Breeds in W North and Central America. Winters in NW South America.

22 EASTERN WOOD PEWEE *Contopus virens* 16cm Feeding actions very similar to those of Western Wood Pewee. Separable with certainty only by voice. **V** Plaintive *pee-a-wee*, which slurs down and then up; also an upslurred *pawee* or downslurred *peeaaa*. **H** Forest edges, mixed woodland, coastal woodland, scrub, gardens. **D** Breeds from SC and SE Canada to SC and SE USA. Winters in NW South America.

23 TROPICAL PEWEE *Contopus cinereus* 14cm Reliably separable from other *Contopus* flycatchers only in the hand (and by voice, but normally silent on winter grounds). **V** Very high, sharp, slightly rolling *seew* note (one every 3–4 secs). **H** Forest, woodland, plantations, mangrove. **D** Widespread across Central and South America.

24 TUMBES PEWEE *Contopus punensis* 14cm Previously treated as conspecific with Tropical Pewee. Forays for insects from an exposed perch. **V** Song a series of shrill, whistled *pue-Pee, pue-Pee* phrases. Calls include a shrill *pee-Peep*. **H** Open woodland, forest clearings. **D** W Ecuador, W Peru.

25 WHITE-THROATED PEWEE *Contopus albogularis* 13cm Separable from Blackish Pewee by range. Note also contrasting white throat. **V** High *pik* note (two per sec). **H** Forest edges and small clearings. **D** NE Amazonia.

26 BLACKISH PEWEE *Contopus nigrescens* 13cm Separable from Smoke-colored Pewee by size and rounded head. **V** High *vi-duw* note (one every 2–3 secs). **H** Upper levels and edges of forest, often along streams. **D** Ecuador, Peru, E Amazonia.

27 CUBAN PEWEE *Contopus caribaeus* 15–16.5cm Generally uses a low perch from which it makes sallies to capture insects in the air, often returning to the same perch or one nearby. **V** Prolonged thin whistle; also a feeble *vi-vi* and a repeated *weet* or *dee*. At dawn, gives a squeaky *eeah ooweeah*; the Bahamian race adds a *dee-dee* to this. **H** Pine and broadleaved forest, forest edges, tree plantations, swamps, bushy scrub, mangroves. **D** Bahamas, Cuba.

28 HISPANIOLAN PEWEE *Contopus hispaniolensis* 15–16cm Feeding actions similar to those of Cuban Pewee. **V** Mournful *purr pip-pip-pip-pip*. At dawn gives a rapid *shurr pet-pit pit-pit peet-peet* that rises in pitch. **H** Various wooded areas, including pine and broadleaved forests, forest edges, shade coffee plantations, orchards. **D** Hispaniola.

29 JAMAICAN PEWEE *Contopus pallidus* 15cm Generally makes lengthy horizontal sallies, from an exposed perch, to take insects in the air; usually moves to a new perch after capturing prey. **V** Plaintive *pee*; also a rising and then falling *oeeoh* and, at dawn, two alternating *paleet weeleeah* phrases. **H** Mid-elevation to montane forest and forest edges. **D** Jamaica.

30 LESSER ANTILLEAN PEWEE *Contopus latirostris* 15cm Makes sallies from a low perch to take insects. **V** Rising *preee-e* and a high-pitched *peet-peet-peet*. **H** Mainly moist montane forests. **D** Lesser Antilles.

31 YELLOW-BELLIED FLYCATCHER *Empidonax flaviventris* 15cm Distinct pale eye-ring. Feeding actions similar to those of Willow Flycatcher. **V** Generally silent, but may utter an ascending *pee-wee*. **H** Dense vegetation in forests, secondary forest and forest edges, also plantations, copses, gardens. **D** Breeds from WC Canada to SE Canada and NE USA. Winters in Central America.

32 ACADIAN FLYCATCHER *Empidonax virescens* 12–14cm Slightly more distinct eye-ring and longer primary projection than in Willow Flycatcher. Feeding actions similar to those of Willow Flycatcher. **V** Generally silent, but may utter a soft *weet*. **H** Open woodlands, forest edges, copses, gardens. **D** Breeds in C and E USA. Winters from Nicaragua to Venezuela and W Ecuador.

33 WILLOW FLYCATCHER *Empidonax traillii* 14cm Reliably separable from *Contopus* flycatchers only in the hand and by voice (although normally silent on winter grounds). **V** Song comprises *fitz-bew*, *fizz-bew* and *creet* calls. **H** Forest borders, open woodland, pastures, scrub. **D** Breeds in E, C and W North America. Winters from W Mexico to NE Ecuador.

34 ALDER FLYCATCHER *Empidonax alnorum* 15cm Similar to Willow Flycatcher; best identified by voice. Feeding actions much as those of Yellow-bellied Flycatcher. **V** Harsh *rreeBEEa*, *fee-beeo* or *rrreep*; a clear *pew* or *peewi*, and a flat *pip*. **H** Alder and birch thickets in boggy areas. **D** Breeds from C Alaska to E Canada and NE USA. Winters in NW and WC South America.

35 WHITE-THROATED FLYCATCHER *Empidonax albigularis* 13cm Brownish with buff wing-bars. **V** Low, hoarse *tetruit*. **H** Swampy areas with trees and scrub. **D** Widespread across Central America.

36 LEAST FLYCATCHER *Empidonax minimus* 12.5–14cm Note distinct eye-ring and short primary projection. Feeding actions similar to those of Yellow-bellied Flycatcher. **V** Sharp, dry *pwit* or *pit*; also an emphatic, repeated *CHEbek* or *cheBIK*. **H** Mature forests, forest clearings and edges, Sugar Maple (*Acer saccharum*) woods. **D** Breeds from W Canada to NC and NE USA and SE Canada. Winters in Central America.

37 HAMMOND'S FLYCATCHER *Empidonax hammondii* 12.5–14.5cm Similar to American Dusky Flycatcher; best distinguished by voice and habitat. Feeding actions as Yellow-bellied Flycatcher. **V** Three-phrase *tsi-pik swi-vrk grr-vik*; also a sharp *peek* and a low *weew*. **H** Mountain conifers. **D** Breeds from C Alaska through W Canada to W USA. Winters from SW USA to Honduras.

38 AMERICAN DUSKY FLYCATCHER *Empidonax oberholseri* 13–15cm Distinguished from Hammond's Flycatcher by voice and habitat. **V** Three-phrase *sibip quwerrrp psuweet*; also a soft, dry *whit* or fuller *twip*. **H** Low chaparral, brush, small trees. **D** Breeds in W Canada and W USA. Winters in SW USA and Mexico.

39 AMERICAN GREY FLYCATCHER *Empidonax wrightii* 14–15.5cm Gentle downward tail-wag is unique in the genus (others flick tails, quickly upward). Feeding actions as Yellow-bellied Flycatcher. **V** Two-phrase *jr-vrip tidoo*; also a sharp *chivip* a low *weew* and a dry *whit*. **H** Arid woodland, brushland, open pine forests. **D** Breeds in SW Canada and W USA. Winters in Mexico.

40 PINE FLYCATCHER *Empidonax affinis* 14cm As Pacific-slope Flycatcher and Cordilleran Flycatcher (Plate 150) but with longer primary projection. **V** Call a *whip* or *pwip*. Song comprises 2–4 phrases, varying with region. **H** Open woodland, montane forest; 1,500–3,500m. **D** Mexico.

41 PACIFIC-SLOPE FLYCATCHER *Empidonax difficilis* 14–17cm Virtually identical to Cordilleran Flycatcher (Plate 150); best distinguished by voice. **V** Three-phrase *tsip kiseewii pTIK*. **H** Shady forest, often near watercourses. **D** Breeds in W North America. Winters in W Mexico.

1
2
3
4
5
6
7
8
9
10
11
12
13
14
15
16
17
18
19
20
21
22
23
24
25
26
27
28
29
30
31
32
33
34
35
36
37
38
39
40
41

TYRANT FLYCATCHERS AND CALYPTURA *CONTINUED*

1 CORDILLERAN FLYCATCHER *Empidonax occidentalis* 13–17cm Virtually identical to Pacific-slope Flycatcher (Plate 149), from which it has been split. Voice is the only sure way to differentiate the two. **V** Three-phrase *tsip kiseewii Ptik*. **H** Forests and woods. **D** Breeds in WC North America. Winters in Mexico.

2 YELLOWISH FLYCATCHER *Empidonax flavescens* 13cm Rich yellow below. **V** Very high *seeeh*. **H** Damp montane forest. **D** SE Mexico to Panama.

3 BUFF-BREASTED FLYCATCHER *Empidonax fulvifrons* 11.5–13cm Distinctive little flycatcher, with feeding methods similar to those of Yellow-bellied Flycatcher (Plate 149). **V** Sharp, musical *PIdew piDEW PIdew piDEW*; also a rolling *prrrew* or *pijrr* and a dry, sharp *pit*. **H** Open pine forests. **D** SW USA to Honduras.

4 BLACK-CAPPED FLYCATCHER *Empidonax atriceps* 12cm Unmistakable due to black cap. **V** Very high *wit*. **H** Montane forest and forest edges. **D** Costa Rica, Panama.

5 SCARLET FLYCATCHER *Pyrocephalus rubinus* 14cm Male upperparts, wings and tail blackish. Head and underparts bright red. Female upperparts greyish; underparts whitish, streaked brown. **V** Musical *pi pi pi pi prrreeee*, rising in pitch and volume, often uttered at night. **H** Open areas with scattered trees, woodland edges. **D** Breeds in SE Bolivia, Paraguay, SE Brazil to Argentina and Uruguay. Winters to Amazonia.

6 VERMILION FLYCATCHER *Pyrocephalus obscurus* 13–14cm Previously treated as conspecific with Scarlet and Darwin's flycatchers. **V** Song a rapid-fire squeaky trill, rising in pitch and accelerating during delivery. Calls include a shrill, whistled *tseip*. **H** Arid scrub, woodland, agricultural land. **D** SW USA to N Chile.

7 DARWIN'S FLYCATCHER *Pyrocephalus nanus* 13–14cm Previously treated as conspecific with Scarlet and Vermilion flycatchers. Unmistakable within its geographical range. **V** Song elements similar to those of both Scarlet and Vermilion flycatchers. Calls include a shrill *tchip-ip*. **H** Arid scrub, woodland. **D** Galápagos Islands.

8 AUSTRAL NEGRITO *Lessonia rufa* 12cm Male difficult to separate from male Andean Negrito, but that species has a paler mantle. Female separable from female Andean Negrito by distinctly paler and buffer plumage. Note also different altitudinal ranges. **V** Very thin *peep* notes, just within or beyond human hearing. **H** Open grassy or barren areas, usually near water; <100m, locally higher. **D** S South America.

9 ANDEAN NEGRITO *Lessonia oreas* 12.5cm See Austral Negrito. **V** Call a weak *tyt*. **H** Open areas near water; 3,000–4,000m. **D** B

10 CINEREOUS TYRANT *Knipolegus striaticeps* 13.5cm Note red eyes and grey (not black) plumage. Lacks white wing-bar in flight. Female separable from female Blue-billed Black Tyrant by white (not buff) wing-bars. **V** Very short, very high, double trill during courtship flight. **H** Chaco woodland. **D** Peru to NW Argentina.

11 HUDSON'S BLACK TYRANT *Knipolegus hudsoni* 15cm Shows a small white patch on flanks and a white patch on underwing. Female separable from female White-winged Black Tyrant by less buff plumage. **V** Very short, very high, dry tinkling during courtship flight. **H** Woodland and scrub. **D** Breeds in C Argentina. Winters to N Bolivia.

12 AMAZONIAN BLACK TYRANT *Knipolegus poecilocercus* 13.5cm Range and habitat differ from those of other black tyrants. Male lacks white in wing. **V** Unusually quiet. Male display call a faint *tic-dik*. **H** Undergrowth of várzea forest. **D** Amazonia.

13 JELSKI'S BLACK TYRANT *Knipolegus signatus* 16cm Note grey plumage of male and dusky plumage (especially below) of female. **V** Voice not recorded. **H** Undergrowth of montane forest and woodland. **D** N Peru.

14 PLUMBEOUS TYRANT *Knipolegus cabanisi* 15–17cm Southern counterpart to Jelski's Black Tyrant. Sometimes seen in pairs. Diet comprises mainly insects. **V** Song includes melancholy-sounding, vibrating, trilled *ti-up, wrrrrr-ttp, worrrr-tiup* phrases, with distinct pauses between each phrase. **H** Wooded slopes, often near water. **D** SE Peru, W Bolivia, NW Argentina.

15 BLUE-BILLED BLACK TYRANT *Knipolegus cyanirostris* 14.5cm Red eyes and lack of white in wing are diagnostic. Note buff wing-bars of female. **V** Irregular, ultra-high *see* note in courtship flight. **H** Borders of humid forest, woodland, riverine belts. **D** SE and S Brazil, E Paraguay, NE Argentina, Uruguay.

16 RUFOUS-TAILED TYRANT *Knipolegus poecilurus* 14.5cm Characterised by plain buff underparts and red eyes. **V** Very high, piercing *sweep* note. **H** Shrubby borders and clearings in humid forest, secondary growth, grassy areas with scattered trees. **D** W Venezuela to E Peru; tepuis of SE Venezuela, N Brazil and W Guyana.

17 RIVERSIDE TYRANT *Knipolegus orenocensis* 15cm Male (lacking white in wing) separable from male Amazonian Black Tyrant by habitat. Female separable from female Amazonian Black Tyrant by duller plumage, including duller rufous on rump and tail. **V** May utter a soft *bic* note. **H** Shrubby growth along rivers and near lakes. **D** Amazonia.

18 WHITE-WINGED BLACK TYRANT *Knipolegus aterrimus* 17cm Male separable from male Jelski's Black and Plumbeous tyrants by white patch in opened wing. **V** Very high, thin *sit-sueet* phrase. **H** Montane scrub, woodland and forest borders. **D** Peru to C Argentina.

19 SAO FRANCISCO BLACK TYRANT *Knipolegus franciscanus* 16cm Previously treated as a subspecies of White-winged Black Tyrant. Geographically isolated from that species. Often seen in pairs. **V** Poorly known. Probably similar to White-winged: high-pitched whistling notes. **H** Caatinga forest, associated with rocky outcrops. **D** E Brazil.

20 CRESTED BLACK TYRANT *Knipolegus lophotes* 20.5cm Male resembles female. Note long, thin crest. **V** Very high, shrill *swee* note. **H** Savannah and other grassy and shrubby areas with scattered groves of trees. **D** C and SE Brazil to Uruguay.

21 VELVETY BLACK TYRANT *Knipolegus nigerrimus* 18cm Separable from Blue-billed Black Tyrant by white band seen in flight. Rufous throat striping of female is diagnostic. **V** Normally quiet. **H** Grassy and scrubby areas with some trees. **D** E and SE Brazil.

22 SPECTACLED TYRANT *Hymenops perspicillatus* 15.5cm No similar bird occurs in South America. **V** Hurried *tee-rreb* note. **H** Grassy and shrubby areas near water. **D** S South America.

23 DRAB WATER TYRANT *Ochthornis littoralis* 13.5cm Nondescript bird, best identified by its habitat. **V** Excited, continuous, up-and-down chirping, sung in a duet or chorus. **H** Near water, especially on exposed roots and other perches along large rivers. **D** Amazonia.

24 YELLOW-BROWED TYRANT *Satrapa icterophrys* 16cm Striking yellow eyebrows and white wing-bars are distinctive. **V** Fast, rising, chattering phrase (three notes). **H** Edges of forest, lakes, fields, marshes. **D** N Venezuela, NE Brazil to Bolivia, C Argentina, Uruguay.

25 LITTLE GROUND TYRANT *Muscisaxicola fluviatilis* 13.5cm Note pale base to maxilla. Little overlap in range with Spot-billed Ground Tyrant. **V** Call a high-pitched, rising *peeeep*. **H** Sparsely vegetated sandbars, rocky beaches, adjacent grassy areas. **D** SW Amazonia.

26 SPOT-BILLED GROUND TYRANT *Muscisaxicola maculirostris* 15cm See Little Ground Tyrant, from which it differs by usually having buffer wing linings. Also shows a pale base to maxilla. **V** Short series of two *tip* notes and one *weetjuw* note. **H** Open rocky areas, short grass, ploughed fields. **D** Colombia to Chile and Argentina.

27 TACZANOWSKI'S GROUND TYRANT *Muscisaxicola griseus* 19cm Note distinct eyebrow, running over eye. Larger than Little and Spot-billed ground tyrants. **V** Call a soft, repeated *pip*. **H** Páramo, puna grassland, sparsely vegetated ground; 3,300–4,700m. **D** Peru, Bolivia.

28 PUNA GROUND TYRANT *Muscisaxicola juninensis* 16.5cm Crown has a slight tawny tinge. Compare to White-browed Ground Tyrant, which has a more distinctly coloured crown. **V** Call a high-pitched *peep*. **H** Puna grassland, often near rocky outcrops, boulders or cliffs, or around bogs, lakes or marshes; 4,000–5,000m. **D** Peru to N Chile and NW Argentina.

29 CINEREOUS GROUND TYRANT *Muscisaxicola cinereus* 16.5cm Smaller than the similar Taczanowski's Ground Tyrant and with a short eyebrow. **V** Call a soft, repeated *pip*. **H** Puna grassland, rocky pastures, often near rocks, bogs or around lakes; 2,500–4,500m. **D** Peru to N Chile and NW Argentina.

30 WHITE-FRONTED GROUND TYRANT *Muscisaxicola albifrons* 21cm Larger than most other ground tyrants, and with a distinct white front. **V** Unstructured series of very short, dry trills. **H** Bogs with cushion plants, barren rocky or grassy hillsides; 4,000–5,500m. **D** Peru to N Chile.

31 OCHRE-NAPED GROUND TYRANT *Muscisaxicola flavinucha* 20cm Shows a pale ochre nape, although this is not always prominent. Note long wings. **V** Ultra-high notes, just within or beyond human hearing. **H** Stony slopes with short grass and bogs; 3,000–4,500m, occasionally down to sea level. **D** Breeds in Argentina and Chile. Winters to Peru and W Bolivia.

32 RUFOUS-NAPED GROUND TYRANT *Muscisaxicola rufivertex* 17cm Rufous coloration on nape is more distinct than in Puna and White-browed ground tyrants. **V** Melodious, upslurred *wheet*. **H** Grassland, ploughed fields, rocky slopes, desert; 2,200–4,500m, locally much lower. **D** Peru to C Chile and W Argentina.

33 DARK-FACED GROUND TYRANT *Muscisaxicola maclovianus* 16cm Separable from Black-fronted Ground Tyrant by lack of white lores. **V** Soft, harsh chattering. **H** Open grassy habitats near forest and woodland; <2,500m. **D** SW South America.

34 WHITE-BROWED GROUND TYRANT *Muscisaxicola albilora* 17cm See Puna and Rufous-naped ground tyrants. **V** Irregular series of ultra-high *seet* notes. **H** Puna grassland and nearly barren rocky slopes; in winter, often seen near water; 1,200–4,000m. **D** Ecuador to S Chile and W Argentina.

35 PARAMO GROUND TYRANT *Muscisaxicola alpinus* 19cm Note distinct eyebrow (see Cinerous Ground Tyrant, whose eyebrow is smaller and less distinct). **V** Calm, unstructured series of very high *wuc* and *pic* notes. **H** Puna and páramo grassland; 3,300–4,700m, occasionally much lower. **D** Colombia, Ecuador.

36 CINNAMON-BELLIED GROUND TYRANT *Muscisaxicola capistratus* 17cm Head pattern is diagnostic. **V** Unstructured series of very high, sharp *tzip* notes. **H** Rocky montane areas, short grass; 2,000–4,500m. Breeds at much lower altitudes on humid steppes. **D** S Argentina, Chile, Peru.

37 BLACK-FRONTED GROUND TYRANT *Muscisaxicola frontalis* 18cm Note striking white lores and black front. **V** Not recorded. **H** Puna grassland, stony hillsides; 2,500–4,500m. **D** C Chile, WC Argentina, Peru.

38 BLACK-BILLED SHRIKE-TYRANT *Agriornis montanus* 24cm Separable from White-tailed and Great shrike-tyrants by all-black bill and pale eyes (although some variation is possible). As White-tailed Shrike-tyrant, has a mainly white tail. **V** Slow series of upslurred and downslurred *wee-tjuh* notes (one every 2–5 secs). **H** Open montane areas with scattered bushes, trees, fences and rocks, often near buildings; 2,000–4,500m. **D** Colombia to Chile and Argentina.

39 WHITE-TAILED SHRIKE-TYRANT *Agriornis albicauda* 27cm Note bicoloured bill. Flanks and belly brownish, not dull grey. **V** Calm series of *teeu* notes (1–5 secs). **H** Open areas with scarce shrubs and scattered rocks, often near buildings; 2,500–4,500m. **D** Ecuador to NW Argentina.

40 GREAT SHRIKE-TYRANT *Agriornis lividus* 28cm Only outer web of outer-tail feathers is white. Note size and bicoloured bill. Wing feathers have more distinct pale margins than in Black-billed Shrike-tyrant. **V** Very high *sit* note, given at long intervals. **H** Arid shrubby and agricultural areas; <1,500m, sometimes higher. **D** Chile, S Argentina.

41 GREY-BELLIED SHRIKE-TYRANT *Agriornis micropterus* 25cm Separable from larger Great Shrike-tyrant by longer eyebrow. **V** Usually silent. **H** Puna grassland with some bush and boulders; 2,000–4,000m in Andes, to sea level in the south of its range. **D** S South America.

42 LESSER SHRIKE-TYRANT *Agriornis murinus* 18cm Not much bigger than a House Sparrow. Perches in the open; note buffish tinge on cheeks and flanks. **V** Usually silent. **H** Shrubby steppes and scattered trees; <2,500m. **D** S South America.

1
2
3
4
5
6
7
8
9
10
11
12
13
14
15
16
17
18
19
20
21
22
23
24
25
26
27
28
29
30
31
32
33
34
35
36
37
38
39
40
41
42

TYRANT FLYCATCHERS AND CALYPTURA *CONTINUED*

1 FIRE-EYED DIUCON *Xolmis pyrope* 21cm Eye colour is distinctive. Note overall grey colour with brownish tinge to flanks. **V** Short, high series of three notes; first and last connected by a gliding middle note. **H** Edges of humid forest, woodland, clearings, gardens, orchards. **D** Chile, SW Argentina.

2 GREY MONJITA *Xolmis cinereus* 23cm Separable from mockingbirds (Plate 237) by red eye colour. **V** Very high, sharp, short *see* note, followed by an undulating second note. **H** Cerrado and other semi-open habitats; on migration also seen in cities (roofs and aerials). **D** E and C Brazil, SE South America.

3 BLACK-CROWNED MONJITA *Xolmis coronatus* 21cm Note black crown patch encircled by white. **V** Song a quiet and tuneful *whut-whut wheeer whut*. **H** Open and semi-open habitats with some shrub and trees. **D** N and C Argentina.

4 WHITE-RUMPED MONJITA *Xolmis velatus* 19.5cm Range barely overlaps with that of Black-crowned Monjita. Separable from White Monjita and Salinas Monjita by brownish (not white) mantle. **V** Nasal, downslurred *nzee*, given at long intervals. **H** Open areas with scattered trees and bush, often near buildings, along roads and near water. **D** C Brazil to Bolivia and N Paraguay.

5 WHITE MONJITA *Xolmis irupero* 17.5cm Unmistakable. Separable from Black-and-white Monjita by different wing and tail pattern. **V** Short, slightly rising, fluted *preep* note. **H** Open areas with scattered trees and bush, often near buildings, along roads and near water. **D** E and SE South America.

6 RUSTY-BACKED MONJITA *Xolmis rubetra* 19cm Compare to Salinas Monjita. **V** Little-known; adults with young recorded to make single short *pik* note. **H** Steppe with low bushes. **D** C and N Argentina.

7 SALINAS MONJITA *Xolmis salinarum* 16.5cm Separable from Rusty-backed Monjita by different wing pattern and lack of striping on breast. **V** Short, simple chipping or nasal notes. **H** In scrub around salty lakes. **D** NC Argentina.

8 BLACK-AND-WHITE MONJITA *Heteroxolmis dominicana* 21cm Note white scapulars of female. **V** Short series of 2–4 loud *weet* notes, separated by 1–2 soft, low *wuc* notes. **H** Grassy areas around swamps and marshes. **D** S Brazil, NE Argentina, Uruguay.

9 STREAK-THROATED BUSH TYRANT *Myiotheretes striaticollis* 23cm Separable from Santa Marta Bush Tyrant by noticeably larger size and preference for more semi-open habitat. Compare bill shapes of White-tailed Shrike-tyrant, Great Shrike-tyrant, and Grey-bellied Shrike-tyrant (Plate 150). **V** Series of high, loud, clear *weeuw* notes. **H** Shrubby and grassy habitats, including borders of forest and woodland. **D** W Venezuela to NW Argentina.

10 SANTA MARTA BUSH TYRANT *Myiotheretes pernix* 19cm Compare to Streak-throated Bush Tyrant, which also shows bright rufous in tail and wings in flight. Very restricted range. **V** Calm series of mid-high, clear *weeuw* notes. **H** Borders and adjacent clearings in montane forest and woodland. **D** N Colombia.

11 SMOKY BUSH TYRANT *Myiotheretes fumigatus* 20cm Note cinnamon in wings in flight. **V** Irregular series of high, clear *wuw* notes. **H** Higher levels of montane and elfin forest. **D** W Venezuela to Peru.

12 RUFOUS-BELLIED BUSH TYRANT *Myiotheretes fuscorufus* 18.5cm Separable from Streak-throated Bush Tyrant and Santa Marta Bush Tyrant by almost unmarked throat. **V** High, excited, staccato *t'tjuw* notes. **H** Borders of montane forest and woodland. **D** Peru, Bolivia.

13 RED-RUMPED BUSH TYRANT *Cnemarchus erythropygius* 23cm Distinctively coloured and patterned. **V** Very high, drawn-out *kyeee*. **H** Scrub, páramo, *Polylepis* woodland. **D** Colombia to Bolivia.

14 RUFOUS-WEBBED BUSH TYRANT *Polioxolmis rufipennis* 21cm Appears mainly grey when perched, and pale rufous in flight. **V** Very high, sharp *syeee* notes (one per sec). **H** Semi-open montane areas, especially grassland with scattered low woodland and shrub. **D** Peru to NW Argentina.

15 CHOCOLATE-VENTED TYRANT *Neoxolmis rufiventris* 23cm No similar species in its habitat. **V** High, sharp phrase of *tju* and higher *weee* notes. **H** Open grassy areas with scattered tussocks and low bushes; large fields in non-breeding season. **D** SE Brazil to Tierra del Fuego.

16 STREAMER-TAILED TYRANT *Gubernetes yetapa* 42cm Unmistakable. **V** Jubilant, loud phrases of *tuh* and *weeh* notes, sung in a duet. **H** Shrubby grassland and marshes, palm groves. **D** N and E Bolivia and C Brazil to NE Argentina.

17 SHEAR-TAILED GREY TYRANT *Muscipipra vetula* 22.5cm Unmistakable in its limited range by long forked tail. **V** Unstructured, nasal *pi-pup* phrases. **H** Borders of montane forest, woodland and secondary growth. **D** SE Brazil to E Paraguay and NE Argentina.

18 PIED WATER TYRANT *Fluvicola pica* 13cm Separable from Black-backed Water Tyrant by range, and by white scapulars. **V** Calm, unstructured series of *zreeo* phrases. **H** In and around marshes, ponds, lakes, adjacent open areas. **D** Panama to the Guianas and N Brazil.

19 BLACK-BACKED WATER TYRANT *Fluvicola albiventer* 14cm Compare to Pied Water Tyrant. **V** Very high series of *preeu* notes (one per sec). **H** Around marshes and at rivers and small lakes. **D** C, E and SE South America.

20 MASKED WATER TYRANT *Fluvicola nengeta* 14.5cm Note distinctive black eye-stripe and very long white uppertail coverts. **V** Repeated, short series of four rather musical notes, second note higher and stressed. **H** Around marshes and ponds, nearby shrubby areas, parks. **D** W Ecuador, NW Peru, E Brazil.

21 WHITE-HEADED MARSH TYRANT *Arundinicola leucocephala* 13cm Female separable from Pied Water Tyrant and Black-backed Water Tyrant by less marked contrast between upperparts and underparts. **V** Very short, high, thin, rapid *tirt-tirt* notes. **H** Marshy areas. **D** Widespread in South America.

22 COCK-TAILED TYRANT *Alectrurus tricolor* 14cm Unmistakable. **V** Usually silent. **H** Areas of tall grassland. **D** N Bolivia, Paraguay, SE Brazil.

23 STRANGE-TAILED TYRANT *Alectrurus risora* male 30cm, female 20cm Unmistakable. **V** Usually silent. **H** Damp grassland, marshy areas. **D** S Paraguay, NE Argentina, Uruguay.

24 TUMBES TYRANT *Tumbezia salvini* 13.5cm Unmistakable in its limited range. **V** Unstructured series of short, trilling *turrit* notes. **H** Arid scrub and woodland. **D** NW Peru.

25 CROWNED CHAT-TYRANT *Silvicultrix frontalis* 12.5cm Birds in the part of the range that overlaps with Jelski's Chat-tyrant and Golden-browed Chat-tyrant show no wing-bars. **V** Very high trill, preceded by single *seet* note (2–5 secs). **H** Undergrowth and borders of montane and elfin forest just below treeline. **D** Colombia, Ecuador. Mainly on E slope of Andes.

26 KALINOWSKI'S CHAT-TYRANT *Silvicultrix spodionota* 12–13cm Previously treated as race of Crowned Chat-tyrant. Diet comprises mainly insects. **V** Song a cricket-like trilling *t-trrrrrrrrrrrr-rrrr-urrr*. Call a sharp *tik*. **H** High-altitude montane forest and scrub, typically at or just below treeline. **D** Bolivia, Peru.

27 GOLDEN-BROWED CHAT-TYRANT *Silvicultrix pulchella* 12.5cm Eyebrow all yellow. **V** Call a very high, short *seet*. **H** Undergrowth of montane forest. **D** Peru, Bolivia.

28 YELLOW-BELLIED CHAT-TYRANT *Silvicultrix diadema* 12.5cm Extensive yellow on belly is diagnostic. **V** Hurried, short, high rattle. **H** Dense undergrowth of montane forest. **D** Peru, Bolivia.

29 JELSKI'S CHAT-TYRANT *Silvicultrix jelskii* 12.5cm Compare to Crowned Chat-tyrant. Separable from Golden-browed Chat-tyrant by partially (not fully) yellow eyebrow. **V** Excited mixture of very high, thin notes, ultra-short trills, and low chatters and rattles. **H** Borders of montane forest and secondary growth. **D** Colombia to N Peru; W slope of Andes.

30 SLATY-BACKED CHAT-TYRANT *Ochthoeca cinnamomeiventris* 12.5cm Distinctive bird. **V** Very high, thin, hurried series of exhaled *seetseet* notes, combined with softer, slightly lower, inhaled note. **H** At streams in edges of montane forest and secondary growth. **D** SW Ecuador, NW Peru.

31 BLACKISH CHAT-TYRANT *Ochthoeca nigrita* 12–13cm Previously treated as race of Slaty-backed Chat-tyrant. Often seen in pairs. Diet comprises mainly insects. **V** Song a slurred, whistled *tsurrrr*. Calls similar in tone. **H** Forested habitats, secondary woodland. **D** Andean Venezuela, Colombia, Ecuador, Peru.

32 MAROON-BELTED CHAT-TYRANT *Ochthoeca thoracica* 13–14cm Previously treated as race of Slaty-backed Chat-tyrant. Unobtrusive habits. Diet comprises mainly insects. **V** Song a piercing, high-pitched, downslurred *pseee-eup*. Calls include abrupt version of song. **H** Forested habitats, secondary woodland. **D** C and SE Peru, Bolivia

33 RUFOUS-BREASTED CHAT-TYRANT *Ochthoeca rufipectoralis* 13.5cm Rufous breast is distinctive. **V** Continuous, calm series of double note *seesee*, followed by lower *tjew* note. **H** Borders of montane forest and woodland. **D** N Colombia to W Bolivia.

34 BROWN-BACKED CHAT-TYRANT *Ochthoeca fumicolor* 15cm Note tawny underparts; separable from D'Orbigny's Chat-tyrant by partially tawny (not all-white) eyebrows. **V** Simple nasal twittering. **H** Páramo, shrub around pastures, montane woodland. **D** W Venezuela to C Bolivia.

35 D'ORBIGNY'S CHAT-TYRANT *Ochthoeca oenanthoides* 15cm Compare to Brown-backed Chat-tyrant; note all-white eyebrows. **V** Calls include a rhythmic *reeek a teek a* and sharp *kvee*. **H** Shrubby areas and woodland patches near the treeline. **D** Peru to NW Argentina.

36 WHITE-BROWED CHAT-TYRANT *Ochthoeca leucophrys* 15cm Compare to Piura Chat-tyrant, which is much smaller. Note slender impression. **V** Sharp, high chatter. **H** Dry montane forest, *Polylepis* woodland, thickets, hedgerows. **D** S Ecuador to NW Argentina.

37 PIURA CHAT-TYRANT *Ochthoeca piurae* 12.5cm Separable from White-browed Chat-tyrant by smaller size. **V** Hurried, sharp, nasal, up-and-down chattering. **H** Borders of montane woodland, scrub and thickets. **D** NW Peru.

38 PATAGONIAN TYRANT *Colorhamphus parvirostris* 12cm Separable from chat-tyrants by lack of eyebrows. **V** Very high, very sharp, drawn-out *seeet* note, given at 2-sec intervals. **H** Borders and shrubby openings in humid forest. **D** S Chile, SW Argentina.

39 LONG-TAILED TYRANT *Colonia colonus* male 25cm, female 20cm Unmistakable. **H** Canopy and borders in humid forest, secondary growth, adjacent clearings. **D** SE Honduras to NE Argentina.

40 SHORT-TAILED FIELD TYRANT *Muscigralla brevicauda* 11cm Unmistakable due to its short tail and long legs. **V** Mixture of very high *tic* notes and low grunts. **H** Open barren or sandy areas with scattered bush and sparse trees. **D** Ecuador to N Chile.

1
2
3
4
5
6
7
8
9
10
11
12
13
14
15
16
17
18
19
20
21
22
23
24
25
26
27
28
29
30
31
32
33
34
35
36
37
38
39
40

TYRANT FLYCATCHERS AND CALYPTURA *CONTINUED*

1 CATTLE TYRANT *Machetornis rixosa* 19.5cm Mainly seen on the ground but also sometimes on rooftops. **V** Slow series of ultra-high, thin, single to tripled *sit* notes. **H** Open habitats such as savannah, pastures, fields; often near houses, town parks, beaches. **D** Panama to SE South America.

2 PIRATIC FLYCATCHER *Legatus leucophaius* 14.5cm Note small size and yellowish tinge to flanks and lower belly. **V** Varied songs incorporating very high, sharp *feeeh* notes and *wit-wit-wit* phrases. **H** Forest borders, tall secondary growth, riverine belts, agricultural areas. **D** Widespread in Central America and South America.

3 WHITE-BEARDED FLYCATCHER *Phelpsia inornata* 17cm Separable from Lesser Kiskadee by shorter bill, from White-ringed Flycatcher by different range, from Rusty-margined Flycatcher by less pronounced rufous in wings, and from Social Flycatcher by lack of distinct wing-bars. **V** Call a loud, sharp chirrup, pairs give *chee-ter* in duet. **H** Habitats with scattered trees, woodland groves, riverine belts, trees near buildings. **D** N Venezuela.

4 RUSTY-MARGINED FLYCATCHER *Myiozetetes cayanensis* 17cm Separable from Social Flycatcher by lack of pale edging to wing feathers and by rufous in wings. **V** Series of long, plaintive, very high, drawn-out *seeeet* notes, given at intervals of any length. **H** Borders and clearings in humid forest, woodland and other habitats with some trees. **D** Panama to SE Brazil.

5 SOCIAL FLYCATCHER *Myiozetetes similis* 17cm Compare to Rusty-margined Flycatcher. Wing feathers distinctly pale-edged. **V** Varied songs, including very high series with some sharp *cheew* notes. **H** Wide variety of forested, wooded and other habitats with some trees, bushes, margins of waterbodies, gardens, pastures and fields. **D** Mexico to NE Argentina.

6 GREY-CAPPED FLYCATCHER *Myiozetetes granadensis* 17cm Pale greenish eyes are distinctive. **V** Varied songs such as a short, high, nasal series including *kew* notes. **H** Borders and clearings in humid forest and woodland. **D** Honduras to W Ecuador and W Amazonia.

7 DUSKY-CHESTED FLYCATCHER *Myiozetetes luteiventris* 14.5cm Head unmarked, with indistinct, paler front. Note coarse dusky breast striping. **V** Song comprises nasal, mewing *kew* notes. **H** Canopy and borders of terra firme forest. **D** Amazonia.

8 GREAT KISKADEE *Pitangus sulphuratus* 23cm Large, noisy and conspicuous. **V** Song varies, the best known a loud, harsh, rhythmic *kis-ka-dee-ka-dee*. **H** Wide variety of natural, cultivated and even urban habitats, but not in forest interiors. **D** SW USA to C Argentina.

9 LESSER KISKADEE *Philohydor lictor* 16cm Note long, slender bill. Distinguished from Great Kiskadee and Social Flycatcher by call. **V** Very high slurred-up hoarse *tjweeh*. **H** Wooded marsh, tree stands along slow streams. **D** Panama to SE Brazil.

10 WHITE-RINGED FLYCATCHER *Conopias albovittatus* 16.5cm Note eyebrows connect on nape. **V** High *weet* note, almost synchronous with a short, low rattle. **H** Canopy and borders of humid forest. **D** E Honduras to NW Ecuador.

11 YELLOW-THROATED FLYCATCHER *Conopias parvus* 16.5cm Separable from smaller, similarly yellow-throated Three-Striped Flycatcher by yellow crown patch (normally concealed) and darker mantle. **V** Irregular series of loud, short, chirped phrases. **H** Canopy and borders of humid forest. **D** N and W Amazonia.

12 THREE-STRIPED FLYCATCHER *Conopias trivirgatus* 14cm Compare to Yellow-throated Flycatcher. **V** Rapid, unstructured chirping. **H** Canopy and borders of humid forest. **D** Amazonia, SE South America.

13 LEMON-BROWED FLYCATCHER *Conopias cinchoneti* 16cm Note yellow eyebrows. **V** Repeated high, shrill, rattled trills. **H** Borders, clearings and tree-falls in montane forest. **D** W Venezuela to Peru.

14 GOLDEN-BELLIED FLYCATCHER *Myiodynastes hemichrysus* 20cm Note malar stripe, olive-green mantle, buff-yellow edges to flight feathers. **V** Very high sustained rather nasal *piuh piuh...* **H** Forest edge and clearings. **D** Costa Rica, Panama.

15 GOLDEN-CROWNED FLYCATCHER *Myiodynastes chrysocephalus* 21cm Note malar patch and coarse breast streaking. **V** Unstructured series of high nasal shrieks. **H** Borders of montane forest and woodland. **D** Panama and N Venezuela to NW Argentina.

16 BAIRD'S FLYCATCHER *Myiodynastes bairdii* 23cm Large flycatcher with a black mask that connects over bill. **V** Rapid phrase of very high *weet* notes, preceded by lower, short rolling *rrru*. **H** Arid scrub, woodland, settled areas. **D** SW Ecuador, NW Peru.

17 SULPHUR-BELLIED FLYCATCHER *Myiodynastes luteiventris* 19–22cm Frequently perches high in canopy. Hawks flying insects or makes sallies to glean insects from foliage. In flight shows a distinctive rufous tail. **V** Call *wee iz-uh* or *weez-ih*, sounding like a squeezy toy; also squeaky series *whee whee-whee-whee-whee-i-eezk*. **H** Broadleaf trees along mountain streams. **D** SE SW USA to Costa Rica.

18 STREAKED FLYCATCHER *Myiodynastes maculatus* 20cm Separated from very similar rufous-tailed Sulphur-bellied Flycatcher by whiter belly, white chin; streaking on flanks and belly further down; black malar does not reach chin. **V** High, sustained *wit-wit-wit...* **H** Forest interior, edges and clearings, plantations. **D** E Mexico to Uruguay.

19 BOAT-BILLED FLYCATCHER *Megarynchus pitangua* 25cm Note heavy bill; separated from Great Kiskadee by absence of rufous in wings and rump. **V** High canary-like twitter or *weeh weeh weeh*, or chattering slightly nasal *pirruhweck* or hurried *purpurpurreweh*. **H** Forest canopy and edge, plantations, open woodland. **D** Mexico to NE Argentina.

20 SULPHURY FLYCATCHER *Tyrannopsis sulphurea* 20cm Note grey head with darker cheeks and short stubby bill. **V** Unstructured series of *jeepee* phrases and other harsh chirping. **H** Borders of humid forest, savannah, parks, cultivated areas; often around palms. **D** Amazonia.

21 VARIEGATED FLYCATCHER *Empidonomus varius* 18.5cm Note broad eyebrow and rufous in wings and tail. **V** Ultra-high, thin *seee* note sung in an irregular series. **H** Borders and clearings in forest, secondary growth, riverine belts; also in savannah with some trees and bush, and occasionally in parks. **D** Widespread in South America.

22 CROWNED SLATY FLYCATCHER *Griseotyrannus aurantioatrocristatus* 18cm Note rather uniform greyish appearance contrasting with normally flat black crown. Yellow crown patch is normally hidden. **V** Ultra-high series of 1–8 *see* notes. **H** Open woodland, scrubby pastures, savannah with scattered trees. **D** Amazonia and SC and SE South America.

23 SNOWY-THROATED KINGBIRD *Tyrannus niveigularis* 19cm Separable from larger Tropical Kingbird by pure white throat, and from White-throated Kingbird by pale greenish upper breast. **V** Song a rapid and accelerating chatter. **H** Arid and semi-open habitats with some trees and bush, scattered patches of woodland; farmland. **D** SW Colombia to W Peru.

24 WHITE-THROATED KINGBIRD *Tyrannus albogularis* 21cm Compare to Snowy-throated Kingbird. **V** Song is very high *pip*, sometimes followed by falling trill. **H** Cerrado, savannah and other semi-open areas, often near water and palms. **D** Amazonia, C Brazil.

25 TROPICAL KINGBIRD *Tyrannus melancholicus* 18–23cm Longer-billed than Couch's Kingbird, but best distinguished by voice. Red crown patch usually concealed. Uses a prominent perch from which it launches aerial sallies after insects, returning to same perch or one nearby. **V** Twittering *tzitzitzitzitzi* or *pip-pip-pip-pip*. **H** Often near water in open and semi-arid scrubland with scattered trees, dry forest borders and areas of human disturbance, such as golf courses. **D** SW USA to C Argentina.

26 COUCH'S KINGBIRD *Tyrannus couchii* 20–24cm Shorter billed than Tropical Kingbird, but best distinguished by voice. Red crown patch usually concealed. Feeding technique similar to Tropical Kingbird. **V** Nasal *pik* and buzzy *kweeeerz*, *brrrear* or *brreeah*; also insect-like *dik dik dikweeeerz*. **H** Scrubby woodland, forest edge and plantations, often near water. **D** S Texas and E Mexico to Belize.

27 CASSIN'S KINGBIRD *Tyrannus vociferans* 20.5–23cm Pale tip to tail. Dark grey breast. Red crown patch usually concealed. Hawks insects from exposed perch, often at topmost branches of a tree. Also feeds at fruiting trees and shrubs. **V** Rough *sh-beehr*, *CHI-Vrrr* or *chi-bew*; also a nasal *breeahr breeahr...* **H** Varied, including open country with scattered trees, dry country thickets, pastures. **D** SW USA to Belize.

28 THICK-BILLED KINGBIRD *Tyrannus crassirostris* 20.5–24cm Red crown patch usually concealed. Hawks insects, often from a high, exposed perch. **V** Nasal *di-didweek*, *kidi-wik*, a buzzy *chweeer*, repeated and sometimes interspersed with clipped, nasal and bickering calls; also stuttering *ki-di di-di di-di-dee-yew* or similar. **H** Cottonwoods and mesquite, usually near water. **D** SW USA to Guatemala.

29 WESTERN KINGBIRD *Tyrannus verticalis* 21–22cm White outer-tail feathers. Red crown patch is usually concealed. Feeding methods similar to Cassin's Kingbird. **V** Sharp *whit* or *kit*; also rapid, rising *widik-pik-widi-pik-pik-pik* and lower *kdew-kdew-kdew-kdew*. **H** Variety of open areas, including grassland, sagebrush and urban areas with trees. **D** Breeds NW and NC Canada to N Mexico. Winters SW Mexico to Costa Rica.

30 SCISSOR-TAILED FLYCATCHER *Tyrannus forficatus* 31–38cm Underwing coverts bright pinkish-orange. Feeds by hawking flying insects, although tends to take prey from ground more often. **V** Low, flat *pik*, *pik-prrr* or *kopik*. At dawn utters *pup-pup-pup-pup-pup-perleep*. **H** Open areas with scattered bushes. **D** Breeds C USA to NE Mexico. Winters W Mexico to Panama.

31 FORK-TAILED FLYCATCHER *Tyrannus savana* 40cm (male), 30cm (female) Unmistakable due to its very long tail (shorter in female). Note that perched bird shown is not to scale. **V** Complex, very high, hurried series of hoarse trills, changing to bouncing rattles and vice versa. **H** Open grassy areas and pastures with scattered trees and bush, habitation, along rivers in forest, mangrove. See also Scissor-tailed Flycatcher. **D** S Mexico to Uruguay.

32 EASTERN KINGBIRD *Tyrannus tyrannus* 19–23cm White-tipped tail. Red crown patch usually concealed. Frequently perches in tall trees, hawks after flying insects. **V** Buzzy *kzeer*, and a series of spluttering notes ending in a descending buzz *kdik kdik kdik PIKa PIKa PIKa kzeeeer*. Early-morning song is a rapid rattling, building to a crescendo *kiu kittttttttiu ditide*. **H** Semi-open woodland, pastures with scattered shrubs and trees, orchards, shelterbelts, parks and gardens; often near water. **D** Breeds widely in North America. Winters in W Amazonia.

33 GREY KINGBIRD *Tyrannus dominicensis* 22–25cm Uses prominent perch, such as bare treetops, telephone posts and wires, to make sallies after flying insects; also feeds by hovering to glean from leaves; sometimes drops to ground to capture prey. Occasionally forms very large communal roosts. **V** Loud *pi-tirr-ri* and a musical *pi-ti-rée pi-ti-rro*. **H** Open areas with scattered trees in lowlands and mountains. **D** Breeds SE USA, Caribbean. Winters Panama to Guyana.

34 GIANT KINGBIRD *Tyrannus cubensis* 23cm Often found in pairs. Uses high exposed perches, feeds on flying insects, fruit and small lizards. **V** Loud *tooe-tooe-tooee-tooee-toee*; also a call of four distinct syllables. **H** Mixed pine and hardwood forests, semi-open woodland with tall trees, woodlands near rivers, swamps. **D** Cuba.

35 LOGGERHEAD KINGBIRD *Tyrannus caudifasciatus* 24–26cm Uses exposed perches from which to make sallies to capture insects, usually from foliage; also eats fruit and small lizards. Most races differ mainly in the amount of yellow-buff wash on the underparts and pattern of tail tip. **V** Variable, loud chattering such as *jo-bee-beep*. Also a bubbling, repeated *p-p-q*. **H** Forests, mangroves and swamp edges. **D** Greater Antilles.

36 GREYISH MOURNER *Rhytipterna simplex* 20cm Separable from larger Screaming Piha (Plate 155) by eye colour, which varies between almost white and dark red (not black), and by voice. Female has rufous in wings and tail. **V** Rising, rapid series of nasal, chipping notes, concluding with a loud *tchew* note. **H** Higher levels in humid forest. **D** Amazonia and E and SE Brazil.

37 PALE-BELLIED MOURNER *Rhytipterna immunda* 18.5cm Note rounded head. **V** Musical series of 3–5 rising, plaintive, ringing notes, the last one low. **H** Open woodland and scrubby várzea. **D** C and NE Amazonia.

38 RUFOUS MOURNER *Rhytipterna holerythra* 20.5cm Overall rich rufous coloration; separable from larger Rufous Piha (Plate 155) by longer tail. **V** Various plaintive, falling or rising *wheee* notes. **H** Higher levels in humid forest and mature secondary growth. **D** S Mexico to NW Ecuador.

1
2
3
4
5
6
7
8
9
10
11
12
13
14
15
16
17
18
19
20
21
22
23
24
25
26
27
28
29
30
31
32
33
34
35
♂
36
♀
37
38

TYRANT FLYCATCHERS AND CALYPTURA *CONTINUED*

1 SIBILANT SIRYSTES *Sirystes sibilator* 18.5cm White, grey and black plumage pattern is distinctive. **V** Resounding, rapid, slightly rising series of 4–8 *wheer* notes. Note slender jizz. **H** Canopy of forest and mature secondary growth. **D** E and SW Brazil to E Paraguay and NE Argentina.

2 WHITE-RUMPED SIRYSTES *Sirystes albocinereus* 18–19cm Perches on exposed branches. Often gleans insect prey while hovering. **V** Song a piping *whoa, chi-chi-chi-chi-uu*. Call a shrill *whoa, chi-uu*. **H** Forested habitats, secondary woodland; mainly lowlands. **D** Venezuela, Colombia, E Ecuador, E Peru, W Brazil, N Bolivia.

3 TODD'S SIRYSTES *Sirystes subcanescens* 18–19cm White rump most obvious in flight and when hovering while gleaning insect prey. **V** Song a shrill, whistling *wee-wee-wee-wu*. Calls include downslurred, whistled *wurroooh* and *we-wooh*. **H** Lowland primary forest, secondary woodland. **D** N Brazil, Guyana, Suriname, French Guiana

4 CHOCO SIRYSTES *Sirystes albogriseus* 17–19cm Diet comprises mainly insects, gleaned from foliage and pursued in flight. **V** Song comprises variations on shrill, piping *we-tchip, tchip, chu-per-whip* phrases; tone recalls thrush alarm call. Calls include harsh *tchuip, tchuip*. **H** Lowland primary forest, secondary woodland. **D** Panama, W Colombia, NW Ecuador.

5 RUFOUS CASIORNIS *Casiornis rufus* 18cm All rufous with a paler belly and slender jizz. Note partially pink bill. **V** Loud, high, mournful *wheer* note, single or doubled. **H** Cerrado, riverine belts, chaco woodland. **D** Bolivia to C Brazil, and S to N Argentina and SE Brazil.

6 ASH-THROATED CASIORNIS *Casiornis fuscus* 18cm Resembles Rufous Casiornis but with less uniform plumage coloration. **V** Series of 6–12 high, sharp *chew* notes. **H** Caatinga, Cerrado, campina. **D** E Brazil.

7 RUFOUS FLYCATCHER *Myiarchus semirufus* 20.5cm Note typical *Myiarchus* profile: relatively large head and slender jizz. **V** Excited, descending series of loud *tcheeer* notes, interspersed with softer notes. **H** Arid woodland, thorny desert, agricultural areas. **D** NE Peru.

8 YUCATAN FLYCATCHER *Myiarchus yucatanensis* 19cm Wing-bars indistinct, slightly peaked crown, white edges to tertials. **V** Very high *whueet whuéeet*, partly pressed-out. **H** Forest interior and edge, woodland. **D** Yucatán Peninsula.

9 SAD FLYCATCHER *Myiarchus barbirostris* 16.5cm Makes sallies from a perch, generally 3–9m from the ground, to pick insects from leaves; often returns to the same perch. **V** Forceful *pip pip-pip* or *pip-pip-pireee* and a single *huit*. **H** Mainly forests and woodland from lowlands to middle elevations. Scarce in semi-arid lowland forest and high-elevation forest. **D** Jamaica.

10 DUSKY-CAPPED FLYCATCHER *Myiarchus tuberculifer* 18cm Little or no rufous on tail. Inconspicuous wing-bars. Fly-catches or hover-gleans from foliage. **V** A plaintive whistle and a rolling *pree-pree-prrreeit*. **H** Shady woods, especially oak and sycamore in canyons. **D** SW USA to Amazonia and SE Brazil.

11 SWAINSON'S FLYCATCHER *Myiarchus swainsoni* 19cm Note lack of rufous in tail and wings. May look 'masked'. **V** Series of short, dry rattles, interspersed with rapid, undulating chattering. **H** Wide variety of forested and wooded habitats, occasionally in mangroves. **D** Widespread in S America.

12 VENEZUELAN FLYCATCHER *Myiarchus venezuelensis* 18.5cm Separable from other *Myiarchus* flycatchers by range and voice. **V** High, drawn-out, descending *weeew* note. **H** Clearings and borders of forest, woodland, plantations. **D** N Colombia, N Venezuela.

13 PANAMANIAN FLYCATCHER *Myiarchus panamensis* 18.5cm Not reliably separable from nominate race of Short-crested Flycatcher except by voice. **V** High, descending, fast series of *dee* notes. **H** Forest, open woodland, arid scrub, mangroves. **D** SW Costa Rica through Colombia.

14 SHORT-CRESTED FLYCATCHER *Myiarchus ferox* 18cm Separable from Venezuelan Flycatcher by range and voice. **V** High, descending, rolling trill, introduced by a *weeh* note. **H** Borders and clearings in humid forest and woodland. **D** Widespread in South America.

15 APICAL FLYCATCHER *Myiarchus apicalis* 18cm Tail tipped and edged whitish. Note habitat. **V** Undulating, rolling, nasal trill (2–3 secs). **H** Arid scrub, riverine belts, agricultural and suburban areas. **D** WC Colombia.

16 PALE-EDGED FLYCATCHER *Myiarchus cephalotes* 18cm Separable from Short-crested Flycatcher by whitish tail edges. **V** Slow, slightly descending series of loud, downslurred *weeer* notes. **H** Borders and clearings in forest, open woodland. **D** N Venezuela to Bolivia.

17 SOOTY-CROWNED FLYCATCHER *Myiarchus phaeocephalus* 18cm Note sooty crown and whitish tip and edges of tail. **V** Call an emphatic *wic* note. **H** Woodland, scrub. **D** W Ecuador, N Peru.

18 ASH-THROATED FLYCATCHER *Myiarchus cinerascens* 19–20.5cm Very pale yellow below. Inner webs of tail rufous, shows when tail spread. **V** Sharp *bik*, a *ki-brrrrnk-brr* and short *ka-wheer*. Song often transcribed as *tea-for-two*. **H** Thorn scrub, open woodland, brushy pastureland. **D** Breeds SW USA, N Mexico. Winters Central America.

19 NUTTING'S FLYCATCHER *Myiarchus nuttingi* 18–19cm Very like Ash-throated Flycatcher although secondaries generally brighter; best distinguished by voice. **V** Sharp *wheek, wheep, wih-ik whi-ik* or *kwee-week*. **H** Arid or semi-arid scrubby woodland, thorn forests, semi-open areas with small trees and bushes. **D** W Mexico to NW Costa Rica.

20 GREAT CRESTED FLYCATCHER *Myiarchus crinitus* 18–20.5cm Feeds by sallying from a perch to take insects in the air, from foliage or on the ground; also eats small berries and other fruit. **V** Harsh, rising *wheee-eep*, which is often part of a series *whee-up wheee whe-whe-whe-wheee-up*; also a rolling *whirr-r-r-r-r-up*. At dawn, gives a repeated *wheee-up whirr-r-r-r-r-up*. **H** Various wooded areas from semi-arid to humid. **D** Breeds SC and SE Canada to SC and SE USA. Winters S Mexico to Colombia and Venezuela.

21 BROWN-CRESTED FLYCATCHER *Myiarchus tyrannulus* 20cm Heavy bill, rufous on inner webs of tail shows when tail spread. **V** Sharp *h-whik, ha-wik* or*hwuik*, a longer *kir-ir-ik* and rolling *prEErrr-prdrdrrr wrrp-didider....* **H** Riparian woodland, thorn woodland, saguaro desert. **D** SW USA to Argentina.

22 GALAPAGOS FLYCATCHER *Myiarchus magnirostris* 16cm Compare to female Vermilion Flycatcher (Plate 150). **V** Calls are *wic* and *urr* notes. **H** Arid scrubland. **D** Galápagos islands.

23 GRENADA FLYCATCHER *Myiarchus nugator* 20cm Catches insects by sallying from a perch; often returns to the same perch, flicking its tail on landing. **V** Loud *quip* or harsh *queuk*. **H** Lowland scrub and open areas near settlements, usually near palms, secondary forest or tropical lowland forest. **D** Lesser Antilles.

24 RUFOUS-TAILED FLYCATCHER *Myiarchus validus* 24cm Sallies to catch insects from perches situated in dense foliage below forest canopy. Also feeds on fruit. **V** Descending, rolling *pree-ee-ee-ee-ee*; also *chi-chi-chiup* and *wick-up*. **H** Mainly moist forest, less common in secondary forests and dry scrub. **D** Jamaica.

25 LA SAGRA'S FLYCATCHER *Myiarchus sagrae* 19–22cm Captures caterpillars and insects during hovering flights in the understorey. **V** Plaintive *huit*. At dawn, and more often during the breeding season, utters whistled *tra-hee*. **H** Forests at all elevations, pine woodland, mixed woodland, thickets, mangroves. **D** Bahamas, Cuba

26 STOLID FLYCATCHER *Myiarchus stolidus* 20cm Hovers to snatch insects from a twig or foliage and to pluck small fruits. **V** Long, rolling *whee-ee-ee swee-ip bzzrt*. Birds on Hispaniola also give a plaintive *jui*. **H** Lowland forest, forest edge, arid woodlands, scrub, mangrove forest; also in pine forests on Hispaniola. **D** Jamaica, Hispaniola.

27 PUERTO RICAN FLYCATCHER *Myiarchus antillarum* 18.5–20cm Inconspicuous and inactive. Foraging methods not well documented although known to feed on insects, caterpillars, seeds and berries. **V** Plaintive, whistled *whee*. At dawn, utters *whee-a-wit-whee*; middle section may be given separately at other times of the day. **H** Tropical deciduous forest, tropical lowland evergreen forest, arid scrub, mangroves, coffee plantations, groves. **D** Puerto Rico.

28 LESSER ANTILLEAN FLYCATCHER *Myiarchus oberi* 19–22cm Feeds mainly by hovering to glean insects and pick small fruits. **V** Loud, whistled *peeu-wheeet* and short *oo-ee oo-ee* or *e-oo-ee* whistles. **H** Edges of forests, dense woodlands, tree plantations; also lower-altitude scrub or secondary growth. **D** Lesser Antilles.

29 FLAMMULATED FLYCATCHER *Deltarhynchus flammulatus* 16cm Thickset with broad bill. Note cinnamon edges to wing coverts. **V** Very high, slightly descending, drawn-out *tjúuuuuuh* (1 sec). **H** Dry scrubby woodland. **D** SW and S Mexico.

30 LARGE-HEADED FLATBILL *Ramphotrigon megacephalum* 13cm Compare to Rufous-tailed Flatbill and Dusky-tailed Flatbill, which it resembles in having pale base to maxilla and white eye-ring. Separable from Dusky-tailed Flatbill by distinct eyebrow and lack of rufous in tail. **V** Mournful *wooh* or *weh-wooh* notes given at long intervals. **H** Prefers dense bamboo at edge or interior of humid forest and woodland. **D** NW and S Amazonia and SE South America.

31 RUFOUS-TAILED FLATBILL *Ramphotrigon ruficauda* 16cm Separable from Large-headed Flatbill by darker head and extensive rufous in wings and tail. **V** Mournful, drawn-out, rising and lowered *whoooo* note (1 sec). **H** Lower storeys of lowland humid forest. **D** Amazonia.

32 DUSKY-TAILED FLATBILL *Ramphotrigon fuscicauda* 15.5cm Has a darker head and throat than similar Large-headed Flatbill and Rufous-tailed Flatbill. **V** Very high, sharp, plaintive series of 5–6 quick *piw* notes, preceded by a single downgliding *peeew* note. **H** Prefers dense bamboo in humid forest and woodland. **D** W and SW Amazonia.

33 RUFOUS-TAILED ATTILA *Attila phoenicurus* 18cm Grey coloration restricted to cheeks and crown; throat is yellow-orange. **V** Calm series of high, sharp, loud *weeh* notes, the last two close together, the first and last lower-pitched. **H** Higher levels in humid forest and woodland. **D** Breeds SE Brazil and Amazonia to NE Argentina. Winters to S Venezuela.

34 CINNAMON ATTILA *Attila cinnamomeus* 19.5cm Note lack of wing-bars. **V** Long, clear, single note that rises rapidly and descends slowly. **H** Swampy and wet places in várzea, woodland, and at lakes and streams. **D** Amazonia.

35 OCHRACEOUS ATTILA *Attila torridus* 21cm Brighter than rufous morph of Bright-rumped Attila and with dark eyes. **V** Song comprises 1–6 calm, rising, loud, fluted *wuu* notes, concluding with a 'full stop'. **H** Borders and clearings in humid forest and secondary growth. **D** SW Colombia to NW Peru.

36 CITRON-BELLIED ATTILA *Attila citriniventris* 18.5cm Front, cheeks and throat are faintly striped. **V** Rather rapid, level or slightly rising series of 4–8 loud, clear, fluted *whee* notes, the last one low. **H** Humid forest. **D** W Amazonia.

37 WHITE-EYED ATTILA *Attila bolivianus* 19cm Striking pale eyes are diagnostic. **V** Song resembles that of Citron-bellied Attila but slower. **H** Várzea forest, riverine belts. **D** S to SW Amazonia.

38 GREY-HOODED ATTILA *Attila rufus* 21cm Throat can be white and grey or rufous as shown. Compare to Rufous-tailed Attila, which has a shorter bill. **V** Song as Citron-bellied Attila but slower. **H** Higher levels and edges of forest. **D** E and SE Brazil.

39 BRIGHT-RUMPED ATTILA *Attila spadiceus* 19cm Variable; three colour morphs exist. Note wing-bars and orange eyes. **V** Song resembles that of Citron-bellied Attila but each note is doubled. **H** Higher levels and borders of humid forest, secondary growth, plantations, gardens, bushy patches in savannah. **D** W Mexico to Amazonia and E Brazil.

1
2
3
4
5
6
7
8
9
10
11
12
13
14
15
16
17
18
19
20
21
22
23
24
25
26
27
28
29
30
31
32
33
34
35
36
37
38
39
39
39
39
3 colour morphs

COTINGAS COTINGIDAE

1 RED-CRESTED COTINGA *Ampelion rubrocristatus* 21cm Note long erectile crest (normally semi-concealed). Vent and rump white, streaked black; shows white band across tail in flight. **V** Series of bouncing *k* notes. **H** Borders of montane forest and woodland; also in agricultural areas. **D** W Venezuela to Bolivia.

2 CHESTNUT-CRESTED COTINGA *Ampelion rufaxilla* 21cm Unmistakable due to chestnut collar and nuchal crest. **V** Very low, rasping or hissing, stuttered *rrèh*. **H** Borders and canopy of montane forest. **D** Colombia to Bolivia.

3 SWALLOW-TAILED COTINGA *Phibalura flavirostris* 22cm Races vary in amount of white around black cheek. Distinctive due to length and shape of tail. **V** Normally silent. **H** Forest, woodland, clearings, gardens. **D** SE Brazil to E Paraguay and NE Argentina.

4 PALKACHUPA COTINGA *Phibalura boliviana* 21–24cm Previously treated as race of Swallow-tailed Cotinga. More sedentary; note also subtle plumage differences. Catches flying insects, also eats fruits and berries. **V** Calls include subdued screeches. **H** Forested Andean foothills. **D** W Bolivia.

5 WHITE-CHEEKED COTINGA *Zaratornis stresemanni* 21cm Shows white cheeks and streaking above and below. **V** Irregular, dry, tuneless frog-like series of notes with a short, built-in rattle. **H** *Polylepis* woodland and forest near the treeline. **D** W Peru.

6 CHESTNUT-BELLIED COTINGA *Doliornis remseni* 21cm Separable from male Bay-Vented Continga by chestnut-rufous belly (concolourous with vent). **V** Usually silent. **H** Stunted woodland near the treeline. **D** Colombia, Ecuador.

7 BAY-VENTED COTINGA *Doliornis sclateri* 21.5cm Compare to Chestnut-Bellied Cotinga and check distribution. **V** Usually silent. **H** Cloud forest near treeline. **D** C Peru.

8 PERUVIAN PLANTCUTTER *Phytotoma raimondii* 18.5cm Male separable from male White-tipped Plantcutter by grey (not rufous) breast and range; female similar to female White-tipped Plantcutter. **V** Low, dry, nasal, tuneless *urrrrr*. **H** Desert scrub, barren dunes with sparse bushes. **D** NW Peru.

9 WHITE-TIPPED PLANTCUTTER *Phytotoma rutila* 19.5cm Compare to Peruvian Plantcutter. **V** Like that of Peruvian Plantcutter but in series of 3–6 short phrases. **H** Arid scrub and low woodland. **D** SC and SE South America.

10 RUFOUS-TAILED PLANTCUTTER *Phytotoma rara* 19cm No other plantcutter occurs in its range. Female more ochre-coloured than female White-tipped Plantcutter. **V** Song similar to that of Peruvian Plantcutter but with variations. **H** Thorny scrub, secondary growth, low woodland, agricultural areas. **D** Chile, W Argentina.

11 HOODED BERRYEATER *Carpornis cucullata* 23cm Distinctively patterned and coloured. **V** Soft, mellow phrase, described as *wurrit? what now?* **H** Higher levels in humid forest, palm groves. **D** SE Brazil.

12 BLACK-HEADED BERRYEATER *Carpornis melanocephala* 21cm Note red eyes, olive back and faintly barred underparts. **V** Mellow, steeply descending *njaau*. **H** Higher levels in humid forest. **D** E Brazil.

13 GREEN-AND-BLACK FRUITEATER *Pipreola riefferii* 18.5cm Note dark eyes and red bill and legs; separable from Band-tailed Fruiteater by range and all-green tail, and from Barred Fruiteater by pattern on wings and underparts. **V** Short, ultra-high, sibilant, bouncing note, accelerating to a trill. **H** Borders and lower storeys of montane forest and woodland. **D** N Venezuela to N Peru.

14 BAND-TAILED FRUITEATER *Pipreola intermedia* 19cm Has paler eyes than Green-and-black Fruiteater; note chevrons on flanks. **V** Series of very high, sibilant *ss* notes at various speeds. **H** Borders and understorey of montane forest; 2,300–3,000m (Peru), 1,100–2,800m (Bolivia). **D** Peru to Bolivia.

15 BARRED FRUITEATER *Pipreola arcuata* 23cm Separable from other fruiteaters by barring below and by larger size. **V** Ultra-high, thin *seeee* note, mounting until beyond human hearing. **H** Borders and understorey of montane forest. **D** W Venezuela to Bolivia.

16 GOLDEN-BREASTED FRUITEATER *Pipreola aureopectus* 17cm Male has bright yellow throat and breast; female separated from female Black-chested Fruiteater by narrow whitish tips to tertials, from Masked Fruiteater by range. **V** Like that of Barred Fruiteater. **H** Lower storeys of montane forest. **D** Colombia, Venezuela.

17 ORANGE-BREASTED FRUITEATER *Pipreola jucunda* 18cm Male separable from male Masked Fruiteater by all-black crown and orange breast merging into yellow underparts. Female similar to female Masked Fruiteater. **V** Ultra-high, slightly upslurred *see* note, repeated at long intervals. **H** Lower forest storeys. **D** Colombia, Ecuador.

18 BLACK-CHESTED FRUITEATER *Pipreola lubomirskii* 18cm Note grey legs and lack of yellow margin to bib. **V** Ultra-high, mounting *seee* note (lasting 2 secs). **H** Montane forest. **D** S Colombia to Peru.

19 MASKED FRUITEATER *Pipreola pulchra* 18cm Note lack of pale tips to tertials; female shows uniformly striped underparts. **V** Song resembles that of Black-chested Fruiteater but a slightly longer series. **H** Lower storeys of montane forest. **D** Peru.

20 SCARLET-BREASTED FRUITEATER *Pipreola frontalis* 16cm Male distinctive due to red throat; female shows a yellow front. Front barred in race *squamipectus*. **V** Very high *seee* note, changing to a fast twitter. **H** Higher levels in montane forest. **D** N Ecuador to Bolivia.

21 FIERY-THROATED FRUITEATER *Pipreola chlorolepidota* 13cm Male greener below than male of larger Scarlet-breasted Fruiteater; female shows green forehead. **V** Very high, thin *psi* note (one note every 2 secs). **H** Lower storeys of foothill forest. **D** S Colombia to Peru.

22 HANDSOME FRUITEATER *Pipreola formosa* 17cm No similar fruiteater (with an orange breast) occurs in its range; note small yellow patch on upper breast of female. **H** Lower storeys and borders of humid forest. **D** N Venezuela.

23 RED-BANDED FRUITEATER *Pipreola whitelyi* 16.5cm Unmistakable due to heavy build and colour pattern; note yellowish neck patches of female. **V** Rapid, ultra-high, decelerating series of thin *psi* notes. **H** Lower storeys of stunted secondary growth on tepuis. **D** NE South America.

24 SCALED FRUITEATER *Ampelioides tschudii* 19cm Very distinctive with heavy scaling and arrow markings below, yellow lores and pale eyes. Shows yellow band across wings. **V** High, mournful, drawn-out, rising and falling *wheeeur* note. **H** Higher levels of humid forest. **D** W Venezuela to Bolivia.

25 GUIANAN COCK-OF-THE-ROCK *Rupicola rupicola* 27cm Separable from Andean Cock-of-the-rock by range. Note also brown border line to crest and white patch in wings of male. **V** Utters wheezy and croaking notes at lek. **H** Humid forest near rocky outcrops. **D** N Amazonia.

26 ANDEAN COCK-OF-THE-ROCK *Rupicola peruvianus* 30cm Unmistakable in its range. Race *sanguinolentus* is shown; other races are orange like Guianan Cock-of-the-rock. **V** Utters harsh shrieks at lek. **H** Montane forest, especially near gorges and ravines. **D** W Venezuela to Bolivia.

27 GUIANAN RED COTINGA *Phoenicircus carnifex* male 22cm, female 24cm Separable from Black-necked Red Cotinga by range; male also separable from male Black-necked Red Cotinga by brown (not black) mantle. **V** At lek, utters *pee-che-weet* call. **H** Lower storeys of humid forest. **D** NE Amazonia.

28 BLACK-NECKED RED COTINGA *Phoenicircus nigricollis* male 22cm, female 24cm Compare to Guianan Red Cotinga. **V** At lek, utters explosive, sharp barks. **H** Lower storeys of humid forest. **D** W and SC Amazonia.

29 LOVELY COTINGA *Cotinga amabilis* 19cm Female lacks eye ring, not tawny below as Turquoise Cotinga; spotted above, not uniform as Blue Cotinga. **V** Usually silent. **H** Forest canopy and edges. **D** SE Mexico to Costa Rica.

30 TURQUOISE COTINGA *Cotinga ridgwayi* 18cm Female separated from Blue Cotinga by black neck spots. **H** Forest canopy, tall secondary growth. **D** Costa Rica, Panama.

31 BLUE COTINGA *Cotinga nattererii* 19cm Separable from other blue cotingas by range. Purple belly clearly demarcated from blue throat and upper breast. **V** Usually silent. **H** Canopy and borders of humid forest. **D** Panama to NW Ecuador.

32 PLUM-THROATED COTINGA *Cotinga maynana* 19.5cm Note plum-purple throat patch. Blue overall except flight feathers. Female separable from other female *Cotinga* species by paler, less marked plumage. **V** Usually silent. **H** Borders, canopy and adjacent clearings in humid forest. **D** W Amazonia.

33 PURPLE-BREASTED COTINGA *Cotinga cotinga* 18cm Separable from larger male Banded Cotinga by lack of blue breast band; no overlap in ranges. **V** Usually silent. **H** Canopy and borders of humid forest. **D** NE Amazonia.

34 BANDED COTINGA *Cotinga maculata* 20cm Note blue breast band of male. No similar cotinga occurs in its range. **V** Usually silent. **H** Forest canopy. **D** SE Brazil.

35 SPANGLED COTINGA *Cotinga cayana* 20cm Black bases of many feathers shine through, giving plumage a spangled effect; female overall rather dark. **V** Usually silent. **H** Canopy of humid forest. **D** Amazonia.

1
2
3
4
5
6
7
8
9
10
11
12
13
14
15
16
17
18
19
20
21
22
23
24
25
26
27
28
29
30
31
32
33
34
35

COTINGAS *CONTINUED*

1 THREE-WATTLED BELLBIRD *Procnias tricarunculatus* 30cm Unmistakable. **V** Resounding *prOingg*, like closing of an iron gate. **H** Forest; tall trees elsewhere. **D** Honduras to Panama.

2 WHITE BELLBIRD *Procnias albus* male 28.5cm, female 27cm Male White Bellbird, Bearded Bellbird and Bare-throated Bellbird are readily identifiable by facial wattles and naked skin. Females of the three are best distinguished by range. **V** Loud, resounding *kong-king* phrase. **H** Upper levels of foothill forest. **D** NE Amazonia.

3 BEARDED BELLBIRD *Procnias averano* male 28cm, female 26.5cm See White Bellbird. **V** *Bok* or *tic* notes, both like a hammer on iron, sung singly or in a series. **H** Upper levels of montane forest. **D** N and E South America.

4 BARE-THROATED BELLBIRD *Procnias nudicollis* male 28cm, female 26.5cm See White Bellbird. **V** Series of *bok* notes, slightly higher than those of Bearded Bellbird. **H** Canopy and borders of humid forest. **D** E Brazil to E Paraguay, S Brazil, N Argentina.

5 BLACK-AND-GOLD COTINGA *Tijuca atra* 27cm Male unmistakable due to yellow in wings; no similar greenish bird resembles female in its range. **V** Long, high, eerie, whistled note, first part level and then rising (3 secs). **H** Montane forest. **D** SE Brazil.

6 GREY-WINGED COTINGA *Tijuca condita* 24cm Separable from larger female Black-and-gold Cotinga by greyer wings and tail. **V** Ultra-high, piercing, double *teeh-teeh* note. **H** Humid elfin forest. **D** SE Brazil.

7 CHESTNUT-CAPPED PIHA *Lipaugus weberi* 23cm Has some rufous in crown and a pale cinnamon vent. **V** Short, very high, steeply rising *sreek*. **H** Mid-levels of montane forest. **D** NW Colombia.

8 DUSKY PIHA *Lipaugus fuscocinereus* 33cm Grey head and mantle, with olivaceous-brown tinge to wings, tail and underparts. **V** Upslurred, very high *weeh-oh-weeeh* phrase, like wolf-whistling. **H** Canopy and borders of montane forest, secondary growth, woodland. **D** Colombia to Peru.

9 SCIMITAR-WINGED PIHA *Lipaugus uropygialis* 30cm Distinctive due to rufous rump and vent, and long tail. **V** Loud, sharp parakeet-like shrieks, sung in chorus. **H** Mid-levels and borders of montane forest. **D** SE Peru, Bolivia.

10 RUFOUS PIHA *Lipaugus unirufus* 25cm Difficult to separate from Rufous Mourner (Plate 152), except by voice and size. **V** Croaking *krrrrrich*; high explosive resounding *piuWéeah* or *piuWaeh* (onomatopoeic). **H** Upper forest levels. **D** S Mexico to NW Ecuador.

11 SCREAMING PIHA *Lipaugus vociferans* 25cm Grey overall, somewhat darker on wings and tail, and paler on breast and belly. Compare to larger Dusky Piha and smaller Greyish Mourner (Plate 152). **V** Explosive, strident *week weeyou* phrase. **H** Lower storeys of humid forest. **D** Amazonia, E Brazil.

12 CINNAMON-VENTED PIHA *Lipaugus lanioides* 28cm Note brown tinge to vent. **V** Loud, hurried *weécou-weécou* phrase in a series. **H** Humid forest. **D** SE Brazil.

13 ROSE-COLLARED PIHA *Lipaugus streptophorus* 22.5cm Male is unmistakable; note rufous vent of female. **V** Very high, sharp, slightly upslurred *skeer* note. **H** Higher levels of forest and woodland on tepuis. **D** NE South America.

14 BLACK-FACED COTINGA *Conioptilon mcilhennyi* 23cm Note contrast between pale grey underparts and dark upperparts and head. **V** Upslurred, slightly shivering *hueee* note. **H** Higher levels of humid and swampy forest and woodland. **D** SW Amazonia.

15 GREY-TAILED PIHA *Snowornis subalaris* 23.5cm Note whitish eye-ring and pale grey tail. Has some semi-concealed black in crown. **V** High, drawn-out, ringing *cheeer* note. **H** Lower storeys of humid forest. **D** S Colombia to Peru.

16 OLIVACEOUS PIHA *Snowornis cryptolophus* 23.5cm Separable from Grey-tailed Piha by lack of grey in plumage. Note yellowish belly. **V** Short, mid-high, dry rattle. **H** Lower storeys of montane forest. **D** Colombia to Peru.

17 PURPLE-THROATED COTINGA *Porphyrolaema porphyrolaema* 18.5cm Unmistakable. Male distinguished by black and white pattern and purple throat patch, female by rufous throat and barring below. **V** High, loud, drawn-out *wheeer* note (1 sec). **H** Canopy and borders of humid forest. **D** W and SC Amazonia.

18 POMPADOUR COTINGA *Xipholena punicea* 19.5cm Male distinguished by crimson-purple plumage contrasting with white wings. No bird similar to female occurs across most of its range (note pale eyes). **V** Series of high *pip* notes (one every 1–2 secs). **H** Canopy of humid forest. **D** Amazonia.

19 WHITE-TAILED COTINGA *Xipholena lamellipennis* 20cm White tail of male is diagnostic; female resembles female Pompadour Cotinga but check ranges. **V** Single loud clucking *quip* notes. **H** Canopy of humid forest. **D** SE Amazonia.

20 WHITE-WINGED COTINGA *Xipholena atropurpurea* 19cm No similar bird (male or female) occurs in its range. **V** Similar to that of Pompadour Cotinga. **H** Canopy and borders of humid forest and secondary growth. **D** E Brazil.

21 BLACK-TIPPED COTINGA *Carpodectes hopkei* Male 24cm, female 23cm Male is unmistakable due to all-white plumage (except small black dots to tips of flight and tail feathers). Note red eyes of male and female. **V** Not known. **H** Canopy and borders of humid forest. **D** SW Panama to NW Ecuador.

22 SNOWY COTINGA *Carpodectes nitidus* 20cm From Yellow-billed Cotinga by black bill. **V** Little known; male heard to give series of 2–6 *chee or chiu* notes. **H** Forest canopy, tall trees elsewhere. **D** SW Honduras to W Panama.

23 YELLOW-BILLED COTINGA *Carpodectes antoniae* 20cm Yellow bill diagnostic. **V** Usually silent. **H** Forest, mangrove. **D** Costa Rica, Panama.

24 BARE-NECKED FRUITCROW *Gymnoderus foetidus* male 38cm, female 33cm Unmistakable; note small head and slender neck. **V** Normally silent. **H** Borders of humid forest; 700–2,500m (Brazil), <500m (N of Brazil). **D** Amazonia.

25 PURPLE-THROATED FRUITCROW *Querula purpurata* male 28cm, female 25.5cm Black overall, male with a distinctive purple throat; note pale bluish bill in both sexes. **V** Excited, low, level or descending *wah* notes, sung in chorus. **H** Humid forest, mature secondary growth. **D** S Nicaragua through Amazonia.

26 CRIMSON FRUITCROW *Haematoderus militaris* male 33cm, female 35cm Unmistakable due to heavy bill, red plumage and black wings. **V** Soft *bock* notes. **H** Upper levels of humid forest. **D** NE Amazonia.

27 RED-RUFFED FRUITCROW *Pyroderus scutatus* male 40cm, female 37cm Unmistakable due to size and colour pattern. **V** Weird, very deep, booming call. **H** Higher levels of montane forest; <850m (Brazil), <2,200m (Andes). **D** N, NW, E and SE South America.

28 CAPUCHINBIRD *Perissocephalus tricolor* 35cm Unmistakable due to bare face and crown. **V** Very weird song of long, drawn-out, very low notes, like chanting of Tibetan monks. **H** Higher levels of humid forest. **D** NE Amazonia.

29 BARE-NECKED UMBRELLABIRD *Cephalopterus glabricollis* 40cm Unmistakable. **V** Ultra low *wuhh* (2–5 secs). **H** Middle storeys of forest. **D** Costa Rica, Panama.

30 AMAZONIAN UMBRELLABIRD *Cephalopterus ornatus* male 50cm, female 42cm Note white eyes; male shows white frontal underside to 'umbrella'. **V** Song resembles that of Long-wattled Umbrellabird but lower. **H** Lowland forest along rivers and on islands (Brazil); humid forest at 900–1,300m (Andes). **D** W and C Amazonia. Occurs E of the Andes.

31 LONG-WATTLED UMBRELLABIRD *Cephalopterus penduliger* male 41cm, female 36cm Wattle of male might be even longer than shown. **V** Booms like a ship in the mist. **H** Canopy and borders of foothill and montane forest and secondary growth; also agricultural areas. **D** Colombia, Ecuador. Occurs W of the Andes.

1
2
3
4
5
6
7
8
9
10
11
12
13
14
15
16
17
18
19
20
21
22
23
24
25
26
27
28
29
30
31

MANAKINS PIPRIDAE

1 DWARF TYRANT-MANAKIN *Tyranneutes stolzmanni* 8cm Separable from Tiny Tyrant-Manakin by pale eyes. **V** Sharp, ascending *turwéet* note, repeated at 5-sec intervals. **H** Mainly terra firme forest. **D** W and S Amazonia.

2 TINY TYRANT-MANAKIN *Tyranneutes virescens* 7.5cm Compare to Dwarf Tyrant-Manakin. **V** High, thin, three-syllable *lili-marleen* phrase. **H** Lower storeys of humid forest. **D** NE Amazonia.

3 SAFFRON-CRESTED TYRANT-MANAKIN *Neopelma chrysocephalum* 13cm Separable from Pale-bellied Tyrant-Manakin by paler eye, from Wied's, Serra do Mar and Sulphur-bellied tyrant-manakins by range, and from Dwarf Tyrant-Manakin and Tiny Tyrant-Manakin by longer tail. **V** Accelerating series of five very low, nasal *wraang* notes. **H** Lower storeys of forest and woodland on sandy soils. **D** N Amazonia.

4 SULPHUR-BELLIED TYRANT-MANAKIN *Neopelma sulphureiventer* 13.5cm Note greyish head, and check range. **V** Low *wraah-wraah* phrase, repeated at 5–6-sec intervals. **H** Dense undergrowth of riverine belts and forest. **D** SW Amazonia.

5 PALE-BELLIED TYRANT-MANAKIN *Neopelma pallescens* 14cm Separable from Saffron-crested, Wied's, Serra do Mar, Sulphur-bellied, Dwarf and Tiny tyrant-manakins by creamy-white belly; also check range. **V** Unstructured series of croaking, low, double or triple *wraah* notes. **H** Lower storeys of woodland with dense vines. **D** E and C Brazil.

6 WIED'S TYRANT-MANAKIN *Neopelma aurifrons* 13cm Separable from Serra do Mar Tyrant-Manakin by range. **V** Rather hurried, often repeated series of 3–7 mid-high, nasal *choy* notes. **H** Lower storeys of humid forest. **D** SE Brazil.

7 SERRA DO MAR TYRANT-MANAKIN *Neopelma chrysolophum* 13.5cm Separable from Wied's Tyrant-Manakin by range. **V** Calm series of 3–6 mid-high *chip* and *zee* notes. **H** Lower storeys and borders of forest and secondary growth. **D** SE Brazil.

8 YELLOW-HEADED MANAKIN *Chloropipo flavicapilla* 12cm Distinctively coloured, male brighter than female. **V** Song probably comprises high, dry rattles. **H** Humid forest and mature secondary growth. **D** Colombia, Ecuador.

9 JET MANAKIN *Chloropipo unicolor* 12cm Male shows white underwing coverts; female has grey-toned, dark olive plumage. Separable from Black Manakin by range. **V** Irregular, calm series of dry, upslurred *dree* notes. **H** Undergrowth of montane forest. **D** Ecuador, Peru.

10 ARARIPE MANAKIN *Antilophia bokermanni* 15cm Male is unmistakable; female separable from female Helmeted Manakin by range. **V** Repeated *weewee weedeweet* phrase. **H** Forest undergrowth. **D** NE Brazil.

11 HELMETED MANAKIN *Antilophia galeata* 14cm Male is unmistakable; female separable from female Araripe Manakin by range. **V** Song resembles that of Araripe Manakin but concluded by an extra *weet* note. **H** Lower storeys of riverine belts and woodland. **D** E Bolivia to C Brazil and S to E Paraguay and SE Brazil.

12 LONG-TAILED MANAKIN *Chiroxiphia linearis* 21cm Male deeper black than White-Crowned Manakin (Plate 157), female with long reddish middle tail feathers. **V** Very high shrill *sweet* or mellow fluted *towheedo*. **H** Forest undergrowth, riverine belts, mangroves. **D** S Mexico to NW Costa Rica.

13 LANCE-TAILED MANAKIN *Chiroxiphia lanceolata* 13.5cm Separable from Blue-backed Manakin by elongated and pointed central tail feathers. **V** Short, mellow phrase of 1–4 *piu* notes at lek, and other varied series of notes, shrieks and low, rolling rattles. **H** Undergrowth of forest and woodland. **D** SW Costa Rica to N Venezuela.

14 BLUE-BACKED MANAKIN *Chiroxiphia pareola* 12cm Compare to Lance-tailed Manakin; separable from Yungas Manakin by range, and by colour of legs. N and E nom. **V** Unstructured *wir* and *tsir* notes. **H** Undergrowth of humid forest and woodland. **D** Amazonia, E Brazil.

15 YUNGAS MANAKIN *Chiroxiphia boliviana* 13cm Compare to Lance-tailed Manakin and Blue-backed Manakin. **V** Males at lek give loud *chew* notes in chorus. **H** Montane forest. **D** S Peru to S Bolivia.

16 BLUE MANAKIN *Chiroxiphia caudata* male 15cm, female 14cm Male unmistakable due to blue underparts. Both sexes have elongated central tail feathers. **V** Single or double *tjeow* note. **H** Undergrowth and borders of humid forest and woodland. **D** S and SE Brazil to E Paraguay and NE Argentina.

17 PIN-TAILED MANAKIN *Ilicura militaris* male 12.5cm, female 11cm Male unmistakable due to colour pattern and elongated tail feathers. Note tail shape and grey throat and cheeks of female. **V** Rapid, descending series of six ultra-high, thin *si* notes. **H** Undergrowth of humid forest. **D** SE Brazil.

18 GOLDEN-WINGED MANAKIN *Masius chrysopterus* 11cm Male unmistakable due to colour pattern of head and yellow in wings. Note isolated yellow throat patch of female. **V** Call is a low, frog-like croak. **H** Undergrowth and borders of wet montane forest. **D** NW Venezuela to N Peru.

19 WHITE-THROATED MANAKIN *Corapipo gutturalis* 9cm No similar manakin occurs in its range. Note white underparts of female. **V** Song resembles that of White-ruffed Manakin. **H** Mainly interior of humid forest. **D** NE South America, NE Amazonia.

20 WHITE-RUFFED MANAKIN *Corapipo altera* 9.5cm Not reliably separable from White-bibbed Manakin except by range. Note greyish chin of female. **V** Song is a very high, thin *seeuw* note. **H** Understorey of humid forest. **D** Honduras to Colombia.

21 WHITE-BIBBED MANAKIN *Corapipo leucorrhoa* 9.5cm See White-ruffed Manakin. **V** Various long, high-pitched trilling calls. **H** Understorey of humid forest. **D** N Colombia, NW Venezuela.

22 OLIVE MANAKIN *Xenopipo uniformis* 14cm Note dusky legs and bill; shows a very narrow, pale eye-ring. **V** Sharp, undulating, upslurred rattle, concluding with a few ultra-high *see* notes (2–3 secs). **H** Stunted montane woodland and forest. **D** N Amazonia.

23 BLACK MANAKIN *Xenopipo atronitens* 12cm Compare to Jet Manakin; note pale bill of male. Separable from Amazonian Black Tyrant (Plate 150) by smaller bill. **V** Varied, including short rattles and unstructured series of *kup* and *kew* notes. **H** Savannah woodland, scrub. **D** Amazonia.

24 GREEN MANAKIN *Cryptopipo holochlora* 12cm Female separable from female *Xenopipo* species by longer tail. **V** Calm series of very high, thin, upslurred *seee* notes. **H** Undergrowth of humid and wet forest. **D** E Panama to SE Peru.

25 BLUE-CROWNED MANAKIN *Lepidothrix coronata* 8.5cm Male black or green depending on race, but blue crown combined with dark eyes is diagnostic; female not safely separable from other female manakins. **V** Series of hoarse, ascending *deprée* phrases. **H** Undergrowth of humid forest and mature secondary growth. **D** Costa Rica to NW Ecuador and W Amazonia.

26 SNOW-CAPPED MANAKIN *Lepidothrix nattereri* male 9.5cm, female 9cm Male has distinctive white rump and crown. Female not reliably separable from other female manakins. **V** Rather hurriedly repeated, hoarse, ascending *deprée* phrase. **H** Mainly terra firme forest. **D** SC Amazonia.

27 GOLDEN-CROWNED MANAKIN *Lepidothrix vilasboasi* 8.5cm Green plumage with yellow crown of male is distinctive; female not reliably separable from other female manakins. **V** Repeated hoarse, ascending *deprée* phrase. **H** Humid forest. **D** C Brazil.

28 OPAL-CROWNED MANAKIN *Lepidothrix iris* 9cm Note pale blue cap, yellow belly and heavy, pale bill of male; female not reliably separable from other female manakins except by range. **V** Calm series of hoarse, ascending *deprée* phrases. **H** Humid forest and tall secondary growth. **D** SE Amazonia.

29 ORANGE-BELLIED MANAKIN *Lepidothrix suavissima* 9cm Male separable from male White-fronted Manakin by lack of orange patch on upper breast and by range; female not reliably separable from other female manakins. **V** Series of hoarse, ascending *deprée* phrases. **H** Undergrowth and borders of forest on tepuis. **D** N Amazonia.

30 WHITE-FRONTED MANAKIN *Lepidothrix serena* 9cm Male unmistakable due to colour pattern; female not reliably separable from other female manakins. **V** Calm series of high, fluting *wreee* notes (three every 2 secs). **H** Undergrowth and borders of humid forest. **D** NE Amazonia.

31 BLUE-RUMPED MANAKIN *Lepidothrix isidorei* 7.5cm Male unmistakable due to white crown. Female not reliably separable from other female manakins. **V** Calm series of high, ascending *deprée* phrases. **H** Humid forest. **D** Colombia to Peru.

32 CERULEAN-CAPPED MANAKIN *Lepidothrix coeruleocapilla* 8.5cm Male unmistakable due to blue cap and rump; female not reliably separable from other female manakins. **V** Calm series of high *pree* notes. **H** Forest undergrowth. **D** Peru.

33 ORANGE-CRESTED MANAKIN *Heterocercus aurantiivertex* 14cm Note pale grey cheeks. **V** Thin trill, also a hissing chatter. **H** Undergrowth of várzea and woodland. **D** E Ecuador to N Peru.

34 YELLOW-CRESTED MANAKIN *Heterocercus flavivertex* 14cm Male has olive-green face sides, darker at border with white throat. **V** Loud, whistled three-note call. **H** Undergrowth of várzea forest and riverine belts. **D** NC Amazonia.

35 FLAME-CRESTED MANAKIN *Heterocercus linteatus* 14cm Black head with narrow red crown stripe is distinctive. Compare range with Yellow-crested and Orange-crested manakins. **V** Descending *psieeuw* note (1 sec). **H** Undergrowth of várzea forest and riverine belts. **D** S Amazonia.

1
2
3
4
5
6
7
8
9
10
11
12
13
14
15
16
17
18
19
20
21
22
23
24
25
26
27
28
29
30
31
32
33
34
35

MANAKINS *CONTINUED*

1 WHITE-BEARDED MANAKIN *Manacus manacus* 11cm Male an unmistakable black-and-white manakin across most of range, but in some races black and yellow with grey-green underparts. Female not safely separable from other dark-eyed female manakins. **V** Slow series of nasal *tjuw* notes. **H** Thick undergrowth at edges of forest and woodland. **D** Widespread in South America.

2 WHITE-COLLARED MANAKIN *Manacus candei* 11cm Male unmistakable, female very like female Golden-collared Manakin but different range. **V** Pistol-like *tic tictic* or rasping explosive dry rattles in chorus. **H** Undergrowth at forest edge, plantations, tall secondary growth, often at streams. **D** SE Mexico to W Panama.

3 GOLDEN-COLLARED MANAKIN *Manacus vitellinus* 11cm Male unmistakable. Female not safely separable from similar red-legged White-collared Manakin (but different range), Orange-collared Manakin (slightly more yellowish below), Golden-headed Manakin (bill paler). **V** Repeated high, short, descending *krreow*, various *prrt-prrt* calls and short descending trills. **H** Undergrowth at forest edge, secondary growth. **D** Panama, Colombia.

4 ORANGE-COLLARED MANAKIN *Manacus aurantiacus* 10cm Male separated from Golden-collared Manakin by orange, not yellow, collar; for female see Golden-collared Manakin. **V** Pistol-like tic or rattling *trric* and *PafPafPaf...* in chorus. **H** Undergrowth at forest edge, plantations, gardens. **D** Costa Rica, Panama.

5 CRIMSON-HOODED MANAKIN *Pipra aureola* 11cm Separable from Band-tailed Manakin by all-black tail; also check ranges. **V** Thin, upslurred *seeuw* note. **H** Undergrowth of swampy forest. **D** NE and E Amazonia.

6 WIRE-TAILED MANAKIN *Pipra filicauda* male 16.5cm, female 13.5cm Unmistakable due to elongated tail feathers. **V** Slow series of nasal, descending *tjeeuw* notes. **H** Undergrowth of várzea and terra firme forest. **D** NW Amazonia.

7 BAND-TAILED MANAKIN *Pipra fasciicauda* 11cm White tail band is diagnostic. **V** Nasal, descending *sjeeuw* note. **H** Várzea forest and riverine belts. **D** C South America.

8 CLUB-WINGED MANAKIN *Machaeropterus deliciosus* 9.5cm Twisted secondaries of male are distinctive; female shows rufous malar stripe and white inner flags to tertials. **V** Slow series of short, nasal buzzes. **H** Montane forest and secondary growth. **D** SW Colombia to Ecuador.

9 KINGLET MANAKIN *Machaeropterus regulus* 9cm Male unmistakable due to orange-red crown and striped underparts. Female shows faint striping below and may show some orange-brown on breast sides. **V** Slow series of nasal, toy trumpet-like *tjuw* notes. **H** Undergrowth of forest and secondary growth. **D** SE Brazil.

10 STRIOLATED MANAKIN *Machaeropterus striolatus* 9–10.5cm Streaked underparts diagnostic among manakins in its geographical range. **V** Song and call a squeaky, whistled *ke-Chee* hiccup, second element upslurred. **H** Primary forest, mature secondary woodland; mainly lowlands. **D** Amazonian South America (W Brazil, N Peru, E Ecuador, Colombia, W Venezuela).

11 PAINTED MANAKIN *Machaeropterus eckelberryi* 9–10cm Distinguished from similar Striolated Manakin by differences in range, habitat preference and voice. **V** Song and call a shrill, disyllabic *tu-iip*, second element with upslurred inflection. **H** Andean foothill woodland with short trees. **D** Very restricted range in N Peru.

12 FIERY-CAPPED MANAKIN *Machaeropterus pyrocephalus* 9cm Male unmistakable due to broad yellow eyebrows. Female lacks any rufous-orange below and is very difficult to separate from other female manakins. **V** Calm series of *dzut* notes. **H** Humid forest and secondary growth. **D** Amazonia.

13 WHITE-CROWNED MANAKIN *Pseudopipra pipra* 10cm Male separable from similar white-capped or pale blue-capped manakins by red eyes; female shows grey head with red eyes. **V** Hoarse, lowered *zreeuw* (Brazil), concluded with a strong *weuw* in Andes birds. **H** Humid forest and tall secondary growth. **D** Costa Rica to Amazonia and SE Brazil.

14 SCARLET-HORNED MANAKIN *Ceratopipra cornuta* 12.5cm Separable from Round-tailed Manakin by range and tail shape. **V** Slow series of toneless *krets-krets* phrases. **H** Undergrowth of humid forest and tall secondary growth. **D** N Amazonia.

15 RED-CAPPED MANAKIN *Ceratopipra mentalis* 10cm Shows yellow thighs and yellow underwing coverts. Separable from Red-headed Manakin and Round-tailed Manakin by range. **V** Slow series of *tzik* notes. **H** Undergrowth of humid forest and tall secondary growth. **D** SE Mexico to W Ecuador.

16 ROUND-TAILED MANAKIN *Ceratopipra chloromeros* 11cm Has yellow thighs and black underwing coverts. Female not reliably separable from other female *Ceratopipra* manakins. **V** Slow series of notes including *tee* and *zzek*, sung separately or together at irregular intervals. **H** Lower storeys of humid forest and woodland. **D** SW Amazonia.

17 GOLDEN-HEADED MANAKIN *Ceratopipra erythrocephala* 9cm Colour of head is distinctive. Nominate race shows an almost orange head, while the more southerly *berlepschi* shows a paler yellow head. **V** Calm series of *prr* notes and an occasional small rattle. **H** Undergrowth of humid forest and secondary growth. **D** Panama, N South America, N Amazonia.

18 RED-HEADED MANAKIN *Ceratopipra rubrocapilla* 10cm Has red thighs and black underwing coverts. **V** Irregular, slow series of very high *see* notes and short dry rattles. **H** Humid forest and secondary growth. **D** S Amazonia, SE Brazil.

SHARPBILL, TITYRAS AND BECARDS TITYRIDAE

19 SHARPBILL *Oxyruncus cristatus* 17cm Note sharp-pointed bill, arrow-headed spotting below and barred head. **V** Long, descending *seeeeee*, like a flare (2 secs). **H** Borders and canopy of humid forest. **D** Widespread in South America.

20 AMAZONIAN ROYAL FLYCATCHER *Onychorhynchus coronatus* 16.5cm Unmistakable due to its hammerhead profile. Rarely fanned crest has a crosswise orientation. **V** Languid series of high, squeaking *peep* notes (one every 2 secs). **H** Lower storeys of forest, woodland, riverine belts and secondary growth, often near streams. **D** Amazonia.

21 NORTHERN ROYAL FLYCATCHER *Onychorhynchus mexicanus* 17–18cm Crest usually held flat and seldom displayed. **V** Song and call a strident, piping and repeated *tche-Uup, tche-Uup*, first element with upslurred inflection. **H** Forested habitats, often near water. **D** S and E Mexico, south to N Colombia, NW Venezuela.

22 PACIFIC ROYAL FLYCATCHER *Onychorhynchus occidentalis* 16–17cm Gleans insect prey from foliage and catches flying insects on the wing. **V** Song and call a strident, piping and repeated *tche-uup, tche-uup*, first element upslurred. **H** Deciduous forested habitats, often near water. **D** W Ecuador, NW Peru.

23 ATLANTIC ROYAL FLYCATCHER *Onychorhynchus swainsoni* 16–16.5cm Crest seldom displayed, usually held flat. **V** Song and call a strident, piping and repeated *tcheUk, tcheUk*, first element upslurred; less disyllabic than other royal flycatcher species. **H** Primary forests, often near water. **D** Coastal Atlantic forest regions of SE Brazil.

24 TAWNY-BREASTED MYIOBIUS *Myiobius villosus* 14cm Wings very dark, breast and flanks rather dark, tawny brown. Occurs at higher altitudes than Sulphur-rumped Myiobius and Black-tailed Myiobius. **V** Very high series of downslurred *teeuw* notes. **H** Lower storeys and borders of forest and woodland. **D** Panama and W Venezuela to NW Bolivia.

25 SULPHUR-RUMPED MYIOBIUS *Myiobius sulphureipygius* 12cm Note bright tawny breast. **V** Ultra high sweeping *twéet*. **H** Forest undergrowth. **D** SE Mexico to W Ecuador.

26 WHISKERED MYIOBIUS *Myiobius barbatus* 12.5cm Breast is olive in nominate race and less dark rufous in race *mastacalis*. Prefers more humid habitats than Black-tailed Myiobius. **V** Slow, ultra-high series of sharp single and double notes. **H** Lower storeys of humid forest, secondary growth, riverine belts. **D** Amazonia.

27 BLACK-TAILED MYIOBIUS *Myiobius atricaudus* 12.5cm Not reliably separable from Tawny-breasted Myiobius or Sulphur-rumped Myiobius. **V** Rapid, very high, slightly undulating series of very high, sharp notes (2 secs). **H** Undergrowth and borders of forest and woodland. **D** Costa Rica to Ecuador and E Brazil.

28 RUDDY-TAILED FLYCATCHER *Terenotriccus erythrurus* 10cm Note eye-ring, upright posture and broad bill. Separable from Cinnamon Neopipo (Plate 149) by more distinctive markings on wing and greyer mantle. **V** Very high, sharp, single or double notes at long intervals. **H** Lower storeys of terra firme and várzea forest, secondary growth. **D** SE Mexico through Amazonia.

29 BLACK-CROWNED TITYRA *Tityra inquisitor* 18.5cm Separable from both other tityras by absence of black to face sides. **V** Utters magpie-like calls and a three-syllable chattering. **H** Canopy, borders and clearings in tall humid forest and mature secondary growth. **D** E Mexico to NE Argentina.

30 BLACK-TAILED TITYRA *Tityra cayana* 21cm Separable from Masked Tityra by more black to face sides; note also striped plumage of female. **V** Low, croaking, tuneless *urt* given as single, double or triple notes. **H** Canopy and borders of humid forest, woodland, riverine belts, plantations. **D** Widespread in South America.

31 MASKED TITYRA *Tityra semifasciata* 20–24cm Often perches on a bare, exposed branch, feeds primarily on fruit; also pursues large insects in the air or from foliage. **V** Nasal, buzzing *zzzu rrk, zzr zzzrt* or *rr-rr-rrk*. **H** Forest edge, forest clearings, semi-open areas with scattered trees. **D** Mexico to NE Argentina.

1
2
3
4
5
6
7
8
9
10
11
12
13
14
15
16
17
18
19
20
21
22
23
24
25
26
27
28
29
30
31

SHARPBILL, TITYRAS AND BECARDS *CONTINUED*

1 VARZEA SCHIFFORNIS *Schiffornis major* 15.5cm Uniform orange-rufous, paler on rump and underparts, and with varying amounts of grey on face. Compare to Cinnamon Attila (Plate 153). **V** Short, fluted *tjuh-tuweet* phrase. **H** Várzea forest and woodland. **D** C and W Amazonia.

2 GUIANAN SCHIFFORNIS *Schiffornis olivacea* 16–16.5cm Diet includes berries, fruit, and insects caught by fly-catching sorties from low branch. **V** Song a shrill *puerr-Ee*, like a human attention-seeking whistle. Calls include abrupt chattering phrases. **H** Lowland terra firme rainforest. **D** Venezuela, Guyana, Suriname, French Guiana, N Amazonian Brazil.

3 NORTHERN SCHIFFORNIS *Schiffornis veraepacis* 16–16.5cm Diet includes berries, fruit and invertebrates. Feeds in understorey close to forest floor. **V** Song a whistled, trisyllabic but slurred *tu-eree-ter*. Calls include a shrill whistle. **H** Lowland terra firme rainforest. **D** S Mexico, through Central America to W Colombia, W Ecuador, NW Peru.

4 FOOTHILL SCHIFFORNIS *Schiffornis aenea* 16–17.5cm Feeds on berries and invertebrates, makes aerial sorties close to forest floor. Usually solitary. **V** Song a whistling *whoer-weer-Whi-pe-pe*, pitch rising and falling in slurred manner during delivery. **H** Forested slopes. **D** E Andean foothills of E Ecuador, N Peru.

5 RUSSET-WINGED SCHIFFORNIS *Schiffornis stenorhyncha* 16–17cm Feeds close to forest floor, using aerial sorties to capture insects and pluck berries. **V** Song a shrill, whistled *Tuer-whe-whe*. **H** Terra firme rainforest, mature secondary woodland. **D** E Panama, N Colombia, N Venezuela.

6 BROWN-WINGED SCHIFFORNIS *Schiffornis turdina* 16cm Rather nondescript bird, therefore resembling several other species. Note dark plumage, brownish wings and tail, and large eyes. **V** Song resembles that of Varzea Schiffornis but first part is drawn out and second part doubled: *tjuuuh-tuweet tuweet*. **H** Interior of humid forest. **D** E Guianas and S Venezuela through Amazonia and E Brazil.

7 GREENISH SCHIFFORNIS *Schiffornis virescens* 16cm Note pale eye-ring and dark legs. **V** Song resembles that of Varzea Schiffornis, but both parts are staccato. **H** Humid forest, tall secondary growth, riverine belts. **D** SE and S Brazil, E Paraguay, NE Argentina.

8 SPECKLED MOURNER *Laniocera rufescens* 20cm Note distinctive wing pattern. Yellow or orange pectoral tufts are usually hidden. **V** Long series of repeated *tuweet* notes. **H** Lower storeys of humid forest and woodland. **D** SE Mexico to Ecuador.

9 CINEREOUS MOURNER *Laniocera hypopyrra* 20cm Shows rows of wing spots and contrasting orange-yellow tips to tail and tertials. **V** Song resembles that of Speckled Mourner but lower-pitched. **H** Terra firme forest, savannah woodland, wooded sand ridges. **D** Amazonia, SE Brazil.

10 BUFF-THROATED PURPLETUFT *Iodopleura pipra* 9.5cm No similar bird occurs in its limited range. Tufts are normally hidden. **V** Call a very high, thin *sheee*. **H** Canopy and borders of forest, woodland, plantations. **D** E and SE Brazil.

11 DUSKY PURPLETUFT *Iodopleura fusca* 11cm Unmistakable in its range due to dark plumage with white medial stripe to underparts and fluffy vent. Male has dark tufts (white in female), but these contrast little with dark brown flanks or are hidden below wings. **H** Canopy and borders of humid forest. **D** E Venezuela, the Guianas, NE Brazil.

12 WHITE-BROWED PURPLETUFT *Iodopleura isabellae* 11.5cm Bold facial pattern is distinctive. Purple tufts (white in female) are usually hidden. **V** Ultra-high twittering with a drawn-out *pseee* note. **H** Canopy and borders of humid forest and secondary growth. **D** Amazonia.

13 BRAZILIAN LANIISOMA *Laniisoma elegans* 17.5cm Distinctively coloured and patterned. **V** Long, very high series of pushed-out *sueet* notes. **H** Humid forest and mature secondary growth. **D** SE Brazil.

14 ANDEAN LANIISOMA *Laniisoma buckleyi* 17–18cm Previously treated as race of Brazilian Laniisoma. **V** Song a whistled *Tsuuerr-Ee*, pitch falling then rising during delivery in slurred manner. Call a downslurred, whistled *tsiu*. **H** Humid forested habitats with dense understorey and epiphytes. **D** E Andean foothills of Colombia, Venezuela, Ecuador, Peru, Bolivia.

15 WHITE-NAPED XENOPSARIS *Xenopsaris albinucha* 13cm Separable from heavier Cinereous Becard by pure white plumage below and brownish wings. May show crest. **V** Very high, silvery warbling, interspersed with *seee* notes. **H** Woodland, riverine belts. **D** N and E Brazil to N Argentina.

16 GREEN-BACKED BECARD *Pachyramphus viridis* 14.5cm Note olive upperparts; female separable from female Barred Becard by wing pattern. **V** Series of high *tuwee* and *trutu-dee* phrases. **H** Borders and clearings in forest, open woodland, riverine belts. **D** N, E and C South America.

17 YELLOW-CHEEKED BECARD *Pachyramphus xanthogenys* 14–15cm Previously treated as race of Green-backed Becard. Note subtle differences in plumage. Usually seen in pairs. **V** Song an excited piping, raptor-like *tchu-tchu-tchu-tchu-tchee-che-che*; pitch rising slightly during delivery. Calls contain elements of song. **H** Margins of, and clearings in, montane forest, secondary woodland. **D** Ecuador, Peru.

18 BARRED BECARD *Pachyramphus versicolor* 13cm Note black upperparts of male. Female separable from female Green-backed Becard by wing pattern; faint barring below not always distinctive. **V** Very high, bouncing-down series of *see* notes. **H** Canopy and borders of humid forest, woodland, secondary growth. **D** Costa Rica to NW Bolivia.

19 SLATY BECARD *Pachyramphus spodiurus* 14cm Male separable from male One-coloured Becard by pale lores and wing edging; female not reliably separable from larger female One-coloured Becard (the latter sometimes has a dark smudge around eyes). **V** Very high, sharp, rolling series of notes, accelerating to a rattle (2 secs). **H** Woodland, scrubby areas, plantations with some shading trees. **D** W Ecuador, NW Peru.

20 CINEREOUS BECARD *Pachyramphus rufus* 14cm Male separable from male Black-and-white Becard by less white in wings and tail; female not reliably separable from female Slaty Becard. **V** Song is an almost level version of that of Slaty Becard. **H** Open woodland, forest borders and clearings. **D** Panama, N South America, Amazonia.

21 CHESTNUT-CROWNED BECARD *Pachyramphus castaneus* 14cm Note grey band between eye, running over nape. **V** Sings as Cinnamon Becard. **H** Borders and clearings in forest, woodland, riverine belts, plantations. **D** Widespread in South America.

22 CINNAMON BECARD *Pachyramphus cinnamomeus* 14cm Note pale supralorals. **V** Plaintive, slightly descending series of four *teeu* notes. **H** Borders and clearings in forest, woodland, plantations, riverine belts, mangroves. **D** SE Mexico to W Ecuador.

23 WHITE-WINGED BECARD *Pachyramphus polychopterus* 15cm Rather distinctive, almost or completely without white in supralorals. **V** Series of six strong *tew* notes, slightly accelerating except last note. **H** Borders and clearings in woodland, forest, riverine belts, old secondary growth. **D** Guatemala to Uruguay.

24 BLACK-CAPPED BECARD *Pachyramphus marginatus* 14cm Male separable from male White-winged Becard by white lores. Female separable from female White-winged Becard by rufous crown. **V** Sings as Black-and-white Becard. **H** Humid forest, tall secondary growth. **D** Amazonia, E Brazil.

25 BLACK-AND-WHITE BECARD *Pachyramphus albogriseus* 14.5cm Male shows distinctive white lores and grey mantle; note black band bordering rufous crown of female. **V** Variations on *teu-teu-weeh* phrase, the last note higher. **H** Canopy and borders of montane forest and tall secondary growth. **D** Costa Rica to N Venezuela and C Peru.

26 GREY-COLLARED BECARD *Pachyramphus major* 15cm Separated from White-winged, Black-and-white, Barred and Cinereous Becards by tail pattern. Note rufous female upperparts. **V** High wooden *pip-pipppijüeh*. **H** forest interior and edge, plantations. **D** Widespread in Central America.

27 GLOSSY-BACKED BECARD *Pachyramphus surinamus* 13cm Male unmistakable due to contrasting underparts and upperparts, and female due to grey nape and mantle. **V** Very high, sharp series of 4–8 *weeeh* notes, sometimes with emphasis on second note. **H** Canopy and borders of tall forest. **D** NE Amazonia.

28 ONE-COLORED BECARD *Pachyramphus homochrous* 16.5cm Very similar to smaller Slaty Becard but with a heavier bill, and in female often with a dark smudge around eye. **V** Call a short, undulating, nasal rattle. **H** Canopy and borders of humid forest and woodland. **D** Panama to N Venezuela and NW Peru.

29 PINK-THROATED BECARD *Pachyramphus minor* 17cm Note pink throat (often inconspicuous) in male, and contrasting grey crown and mantle of female. **V** Short, fast mixture of nasal rattles and shrieks. **H** Canopy and borders of terra firme forest and woodland. **D** Amazonia.

30 CRESTED BECARD *Pachyramphus validus* 18cm Note black crown of male (darker than mantle) and sooty crown of female. **V** Ultra-high, drawn-out *seee* note. **H** Canopy and borders of woodland and forest. **D** E and C South America.

31 ROSE-THROATED BECARD *Pachyramphus aglaiae* 17cm Note that nape is paler than crown. **V** Very high whistled *piweeeh* or unstructured excited twittering. **H** Canopy of forest edge, riverine belts. **D** Widespread in Central America.

32 JAMAICAN BECARD *Pachyramphus niger* 18cm Forages slowly below the forest canopy, occasionally hovering to pick insects off twigs; also hawks flying insects. **V** Hoarse *queeck queeck*, followed by a musical *co-ome and tell me what you hee-ear*, which rises in pitch then lowers pitch on last two syllables. **H** Tall open forests and forest edge in hills and mountains; has been seen in more closed forest, woodland, fields with scattered trees, gardens. **D** Jamaica.

1
2
3
4
5
6
7
8
9
10
11
12
13
14
15
16
17
18
19
20
21
22
23
24
25
26
27
28
29
30
31
32

LYREBIRDS MENURIDAE

1 ALBERT'S LYREBIRD *Menura alberti* 88–92cm Feeds on forest floor, using powerful feet to rake leaf litter. Diet assumed to comprise mainly invertebrates. **V** Varied mix of shrieking whistles, mechanical-sounding chattering phrases and clicks. A capable mimic. **H** Montane rainforest. **D** E Australia.

2 SUPERB LYREBIRD *Menura novaehollandiae* male 100cm, female 75–80cm Wiry, pheasant-like tail spread laterally in display. Powerful legs and feet used to rake leaf litter and run at speed. **V** Varied mix of whistles, shrieks, mechanical-sounding chattering phrases and clicks. A capable mimic. **H** Humid forested habitats; from sea level to montane slopes. **D** SE Australia.

SCRUBBIRDS ATRICHORNITHIDAE

3 RUFOUS SCRUBBIRD *Atrichornis rufescens* 17–19cm Endangered, in part due to habitat destruction. Feeds by raking leaf litter with feet; diet assumed to include invertebrates and fallen fruits. **V** Series of squeaking, tongue-smacking *ti, tip, tup, Tchiup, Tchiup, Tchiup* phrases. **H** Humid montane forests. **D** E Australia.

4 NOISY SCRUBBIRD *Atrichornis clamosus* male 22–23cm, female 19–20cm Endangered. Forages in leaf litter and in moss. Diet comprises invertebrates and small vertebrates. **V** Series of piercing, high-pitched *ti, tiu, tiu, tsi, tsiu, Tchiu-Tchiu-Tchiu-Tchiu* notes. **H** Coastal forested habitats with dense understorey. **D** SW Australia.

BOWERBIRDS PTILONORHYNCHIDAE

5 OCHRE-BREASTED CATBIRD *Ailuroedus stonii* 24–25cm Seen singly or in pairs, foraging in foliage. Diet includes fruits, berries, invertebrates and small vertebrates. **V** Song includes harsh cat-like hissing, growling notes. Call a high-pitching *pink*. **H** Humid forests; mainly lowlands. **D** SE New Guinea.

6 WHITE-EARED CATBIRD *Ailuroedus buccoides* 24–25cm Diet includes fruits, berries, invertebrates and small vertebrates. **V** Song includes harsh cat-like hissing, growling notes. **H** Humid lowland forests, wooded slopes with dense undergrowth. **D** W Papuan islands, NW New Guinea.

7 TAN-CAPPED CATBIRD *Ailuroedus geislerorum* 24–25cm Diet includes fruits, berries, invertebrates and small vertebrates. Sometimes joins mixed flocks of other fruit-eating species. **V** Song includes harsh cat-like hissing, growling notes. Call a disyllabic, frog-like croak. **H** Lowland humid forests, wooded slopes, monsoon forest. **D** N New Guinea.

8 GREEN CATBIRD *Ailuroedus crassirostris* 30–32cm Diet comprises mainly fruit, but also invertebrates and small vertebrates. Usually solitary or in pairs. **V** Agonised-sounding *EEeeOooArrr* screech, like poor rendition of screaming baby. **H** Humid eucalypt forest, rainforests, often near water. **D** Coastal E and SE Australia.

9 SPOTTED CATBIRD *Ailuroedus maculosus* 28–30cm Previously treated as race of Black-eared Catbird. Diet comprises mainly fruit, but also invertebrates and small vertebrates. **V** Agonised-sounding *eeRrOooRooAr* screech, cross between baby crying and cat having its tail trodden on. **H** Lowland humid forests, wooded slopes, monsoon forest. **D** NE Australia.

10 HUON CATBIRD *Ailuroedus astigmaticus* 28–30cm Previously treated as related to Spotted and Black-eared catbirds. Usually seen singly or in pairs. Diet comprises mainly fruits; occasionally also invertebrates. **V** Strange, strangled screech like a fighting, yowling cat. **H** Montane rainforest. **D** NE New Guinea (Huon Peninsula).

11 BLACK-CAPPED CATBIRD *Ailuroedus melanocephalus* 28–30cm Previously treated as race of Spotted Catbird. Note geographical separation and darker crown and chest. **V** Strangled-sounding *whereeorOoouaOoAr* screech, pitch modulated during delivery. **H** Montane rainforest. **D** SE New Guinea.

12 NORTHERN CATBIRD *Ailuroedus jobiensis* 28–30cm Previously treated as race of Spotted Catbird. Diet comprises mainly fruits and berries; occasionally invertebrates too. **V** Harsh-sounding *OoWher-Oowhe-ooer* screech. **H** Montane rainforest. **D** NC New Guinea.

13 ARFAK CATBIRD *Ailuroedus arfakianus* 28–30cm Previously treated as race of Spotted Catbird. Diet includes fruits and invertebrates. **V** Harsh-sounding *OoWher-Oowhe-ooer* screech, with a desperate-sounding tone. **H** Montane rainforest. **D** NW New Guinea.

14 BLACK-EARED CATBIRD *Ailuroedus melanotis* 28–30cm Previously treated as race of Spotted Catbird. Diet comprises mainly fruits and berries. **V** Harsh-sounding *Who-er-oOWher-Oowher* screech, pitch modulated during delivery. **H** Montane rainforest. **D** W New Guinea, Aru Islands, N Australia.

15 TOOTH-BILLED BOWERBIRD *Scenopoeetes dentirostris* 26–28cm Diet comprises fruits, berries, shoots. Bower decorated with fresh green leaves. **V** Varied song includes screeches, squawks, sharp *Tchuk* notes and mimicry of nearby sounds such as other birds, insects, tree-frogs. **H** Montane rainforest. **D** NE Australia.

16 ARCHBOLD'S BOWERBIRD *Archboldia papuensis* 36–38cm Varied diet includes fruits, invertebrates and small vertebrates. Bower floor decorated with ferns. **V** Two-phrase song, first element like two stones being knocked together, second a shrill *chwiuee*. **H** Montane rainforest. **D** W New Guinea.

17 VOGELKOP BOWERBIRD *Amblyornis inornata* 24–26cm Bower types include wigwam or hut of sticks decorated with colourful forest items including fruits, flowers, insect wings. **V** Calls include a harsh *chrruRRiek* and mournful screech. **H** Montane rainforest. **D** Vogelkop and Wandammen mountains, NW New Guinea.

18 MACGREGOR'S BOWERBIRD *Amblyornis macgregoriae* 25–27cm Bower comprises a conical wigwam arrangement of sticks. Diet includes fruits and invertebrates. **V** Song an upslurred, squawky *whrr-Rup*. Calls include a piping *whuu-iip*. **H** Montane rainforest. **D** C and W New Guinea.

19 STREAKED BOWERBIRD *Amblyornis subalaris* 23–25cm Male's orange crest striking only when displaying. Constructs hut-like bower decorated with colourful forest objects. **V** Calls include a hollow-sounding, barking *whuu-Uup*. **H** Montane rainforest. **D** SE New Guinea.

20 GOLDEN-FRONTED BOWERBIRD *Amblyornis flavifrons* 23–25cm Male constructs maypole bower decorated with colourful forest objects. **V** Songs and calls include varied range of whistles, croaks and whip-cracking sounds, plus mimicry of nearby sounds. **H** Montane rainforest. **D** Restricted to Foja Mountains, N New Guinea.

21 GOLDEN BOWERBIRD *Prionodura newtoniana* 24–25cm Male constructs maypole bower. Surroundings decorated with forest floor objects such as snail shells. **V** Song a hollow-sounding rattle, like mechanical reverberation. Call a harsh, squelchy *uRchh*. **H** Montane rainforest. **D** NE Australia.

22 MASKED BOWERBIRD *Sericulus aureus* 24–25cm Males constructs an avenue bower, surroundings decorated with colourful objects (particularly blue). Display involves shaking head and twisting wings and tail. **V** Harsh *tChrrrrruurr* lasting 1–2 secs. **H** Montane forests. **D** N New Guinea.

23 FLAME BOWERBIRD *Sericulus ardens* 25–26cm Male constructs an avenue bower. Displays nearby, twisting tail and spreading wings, then shaking head. **V** Rasping, rattling and mechanical-sounding screech lasting 1–2 secs. **H** Lowland rainforest, forested slopes, secondary woodland. **D** S New Guinea.

24 FIRE-MANED BOWERBIRD *Sericulus bakeri* 26–27cm Male constructs an avenue bower, bordered by rows of sticks. Surroundings decorated with blue and purple objects, including berries and plastic. **V** Rasping, rattling and mechanical-sounding screech lasting 1 sec. **H** Montane forest, dense secondary woodland. **D** Adelbert Mountains, NE New Guinea.

25 REGENT BOWERBIRD *Sericulus chrysocephalus* 24–25cm Male constructs an avenue bower. Surroundings decorated with forest-floor objects including blue plastic. **V** Squelchy, squawks and rattles, including sounds like radio static. **H** Coastal rainforest, sclerophyll woodland; mainly sea level and lowland slopes. **D** E Australia.

26 SATIN BOWERBIRD *Ptilonorhynchus violaceus* 32–33cm Male constructs bower comprising parallel rows of sticks framing display platform. Adorns surroundings with blue objects. **V** Strange, modulated squeaks and whistles, including sounds like wireless being tuned. **H** Rainforests, sclerophyll forest. **D** Disjunct populations: NE Australia, E and SE Australia.

27 WESTERN BOWERBIRD *Chlamydera guttata* 27–29cm Male constructs avenue bower. Surroundings decorated with green and white forest objects including snail shells, fruits, pebbles. Displaying male fans and flicks wings and tail, and shakes head. **V** Harsh, chattering *kerrrrrk* squawk. **H** Riverine woodland, scrub thickets in rocky ranges and gorges, parks, gardens. **D** C and W Australia.

28 GREAT BOWERBIRD *Chlamydera nuchalis* 34–36cm Male constructs avenue bower. Displaying male fans and shows off iridescent purple crest, often holding colourful object in bill. **V** Harsh, chattering *wherrrrup* squawk, lasting 1 sec. **H** Eucalypt forest, mature secondary forest, mangroves; mainly lowlands. **D** N Australia.

29 SPOTTED BOWERBIRD *Chlamydera maculata* 28–30cm Male constructs avenue bower. Surroundings decorated with stones, snail shells, eggshells and manmade objects. **V** Plaintive-sounding *eeerrr* and harsh squawks. **H** Dry sclerophyll woodland (typically comprising eucalypts and Brigalow *Acacia harpophylla*); often near rivers. **D** Interior E Australia.

30 YELLOW-BREASTED BOWERBIRD *Chlamydera lauterbachi* 26–28cm Male constructs avenue bower. Surroundings decorated with blue-grey pebbles, shells, fruits and manmade objects. **V** Abrupt, chattering *Tchiuup*. **H** Primary forest, secondary woodland; mainly lowlands. **D** C and N New Guinea.

31 FAWN-BREASTED BOWERBIRD *Chlamydera cerviniventris* 28–30cm Male constructs avenue bower. Surroundings decorated with sprays of green berries and fruits. **V** Harsh, rasping *wAAauk*, lasting 1 sec. **H** Forests, secondary woodland, mangroves, suburban parks. **D** E and NW New Guinea, NE Australia.

1
2
2
4
3
5
6
7
10
8
9
12
13
14
11
18
19
16
15
20
21
17
23
24
22
26
28
25
30
27
29
31

AUSTRALASIAN TREECREEPERS CLIMACTERIDAE

1 WHITE-THROATED TREECREEPER *Cormobates leucophaea* 15–17cm Forages for invertebrates, mostly ants, by probing bark crevices and flaky sections. **V** Delightful, descending series of piping whistles *See-pe-pe-pu-pu-pu-pu.* **H** Wide range of forest types and wooded habitats. **D** NE, E and SE Australia.

2 PAPUAN TREECREEPER *Cormobates placens* 14–15cm Probes for insects in tree bark, circling trunks and working its way upwards. **V** Series of penetrating, whistled *tse-tse, tsee, tsee, tsee, tsee, tsee* notes. **H** Humid montane forest habitats. **D** C, SE and NW New Guinea.

3 RED-BROWED TREECREEPER *Climacteris erythrops* 15–16cm Forages on tree bark for invertebrates, notably ants by gleaning, and probing crevices and flaky bark. **V** Rapid-fire, silvery and tinkling trill. **H** Forested habitats, open eucalypt woodland. **D** E and SE Australia.

4 WHITE-BROWED TREECREEPER *Climacteris affinis* 14–16cm Forages for invertebrates, notably ants, while climbing tree trunks and exploring branches. **V** Series of high-pitched *tsee-tse-tse-tsiu-tsiu-tsiu* whistles. **H** Arid open woodland with *Casuarina* and *Acacia*. **D** Disjunct populations: interior W Australia, C Australia, interior E Australia.

5 RUFOUS TREECREEPER *Climacteris rufus* 17–18cm Feeds mainly on ants, forages in tree bark, leaf litter and fallen branches. **V** Rapid-fire, silvery and tinkling trill, pitch descending during delivery. **H** Open woodland, mallee scrub, sclerophyll forest. **D** Disjunct populations: SW Australia, S Australia.

6 BROWN TREECREEPER *Climacteris picumnus* 15–18cm Diet comprises invertebrates, mainly ants. Forages by gleaning and probing bark crevices. **V** Repeated series of shrill *sWiip, sWiip, sWiip* whistles. **H** Wide range of eucalypt woodland types, mallee scrub. **D** NE and E Australia.

7 BLACK-TAILED TREECREEPER *Climacteris melanurus* 17–19cm Forages and gleans invertebrates on bark of tree trunks and branches. **V** Series of shrill, piercing *tuip, tuip, tuip* whistles. **H** Open eucalypt woodland, acacia scrub. **D** N and NW Australia.

AUSTRALASIAN WRENS MALURIDAE

8 WALLACE'S FAIRYWREN *Sipodotus wallacii* 11–13cm Forages in dense undergrowth for insects. Usually seen in small groups. **V** Series of rapid, high-pitched and tinkling *tsi-tsi-tsi-tsi-tsi* whistles. **H** Lowland rainforest, forested slopes. **D** New Guinea, Aru Islands.

9 BROAD-BILLED FAIRYWREN *Chenorhamphus grayi* 14–15cm Forages in small groups amongst tangled undergrowth for invertebrates. **V** Series of high-pitched and tinkling *tsi-tsi-tsi-tsi-tsi* whistles. **H** Primary forests with dense understorey; mostly lowlands. **D** N and NW New Guinea.

10 CAMPBELL'S FAIRYWREN *Chenorhamphus campbelli* 14cm Previously treated as race of Broad-billed Fairywren. Usually seen in small groups. **V** Rapid-fire, high-pitched bursts of tinkling *tsi-tsi-tsi-tsi-tsi* whistles. **H** Primary rainforest with dense understorey; mainly lowlands. **D** C and SE New Guinea.

11 EMPEROR FAIRYWREN *Malurus cyanocephalus* 14–16cm Diet comprises mainly invertebrates. Feeds in small groups. **V** Song includes varied chattering phrases including *purr-Wheek, purr-Wheek, purr-Wheek.* **H** Forest clearings, secondary woodland with dense undergrowth, mostly lowlands. **D** N and S New Guinea, Aru Islands, Biak Island.

12 LOVELY FAIRYWREN *Malurus amabilis* 12–13cm Usually seen in pairs or small groups. Feeds on invertebrates. **V** Song a high-pitched series of tinkling trills. Calls include vibrating, squeaky trills. **H** Lowland forest margins, secondary woodland with dense undergrowth. **D** N Australia.

13 PURPLE-BACKED FAIRYWREN *Malurus assimilis* 12–15cm Diet comprises mainly invertebrates; occasionally seeds. Forages in small groups. **V** Song a rapid, vibrating, rasping trill. Calls include a thin *tsii.* **H** Wide range of dense scrub habitats. **D** Widespread across Australia.

14 VARIEGATED FAIRYWREN *Malurus lamberti* 12–15cm Diet includes mainly invertebrates but also seeds. Usually seen in small groups. **V** Calls include a rasping, grating *tchrrrrr.* **H** Wide range of dense scrub habitats. **D** Coastal SE Australia.

15 BLUE-BREASTED FAIRYWREN *Malurus pulcherrimus* 14–15cm Diet comprises mainly insects. Often feeds on ground. Usually seen in small groups. **V** Song comprises a series of high-pitched rasping, trills. Calls include a short-burst grating trill. **H** Eucalypt woodland with dense understorey, heaths. **D** Disjunct populations: SW Australia, coastal SC Australia.

16 RED-WINGED FAIRYWREN *Malurus elegans* 14–15cm Often feeds on ground, foraging mainly for insects. Usually seen in small groups. **V** Song an accelerating high-pitched grating trill. Calls include a thin *tsii.* **H** Eucalypt woodland with dense understorey, heathland. **D** SW Australia.

17 SUPERB FAIRYWREN *Malurus cyaneus* 16–20cm Diet includes mainly insects, but also seeds and berries. **V** Song an accelerating high-pitched grating trill, pitch descending during delivery. Calls include a soft, grating *tsiup.* **H** Woodland with dense understorey, suburban parks and gardens. **D** SE and E Australia, Tasmania.

18 SPLENDID FAIRYWREN *Malurus splendens* 12–13.5cm Diet comprises mainly ants and other insects. Usually seen in small groups. **V** Song an animated, high-pitched grating trill, accelerating with pitch modulation during delivery. **H** Woodland with dense understorey, mallee scrub, shrubby heaths. **D** W, interior central and E Australia.

19 PURPLE-CROWNED FAIRYWREN *Malurus coronatus* 14–15.5cm Seldom strays far from water. Usually seen in small groups. Diet comprises invertebrates. **V** Song a varied *peer, tche-percu, tch, piu-tche,* tone rather chirping and sparrow-like. Calls include a squelchy *tcherr.* **H** Dense, scrubby waterside margins. **D** Disjunct populations in N Australia: W Northern Territory, W Queensland.

20 WHITE-SHOULDERED FAIRYWREN *Malurus alboscapulatus* 11–14cm Diet comprises mainly invertebrates. Usually forages in small groups. **V** Song a rapid series of squeaky and chattering notes, delivered in 1–2 sec bursts. Calls include a rapid-fire rattling *tchrrr-rrrrr.* **H** Riverside vegetation, secondary grassland. **D** New Guinea.

21 RED-BACKED FAIRYWREN *Malurus melanocephalus* 10–13cm Usually seen in small groups. Diet comprises insects and other invertebrates. **V** Song a series of chattering and twittering notes, ending in a silvery trill. Calls include a high-pitched, descending *tsi-i-i-u-u.* **H** Grassy habitats including scrub and open woodland. **D** N and E Australia.

22 WHITE-WINGED FAIRYWREN *Malurus leucopterus* 12–14cm Usually seen in pairs or small groups. Feeds mainly on insects. **V** Song a vibrating series of grating, squelchy notes, like sound produced by a bird-squeaker. Calls include a sharp *tsiip, tsiip.* **H** Low vegetation arid scrub. **D** Widespread across interior and W Australia.

23 ORANGE-CROWNED FAIRYWREN *Clytomyias insignis* 15–16cm Forages in small groups, gleaning insects from foliage. **V** Song a series of repeated *tsiu-tsiu-tsiu* whistles. Calls include a grating chatter and a lisping *tsee, tsee, tsee.* **H** Montane forests. **D** New Guinea.

24 SOUTHERN EMU-WREN *Stipiturus malachurus* 16–19cm Feeds amongst vegetation and often hard to observe well. **V** Song a trilling series of thin, high-pitched whistles. Call a high-pitched *tsiu, tsiu.* **H** Low vegetation of heaths, coastal scrub, marshland. **D** SW Australia, SE and E Australia, Tasmania.

25 MALLEE EMU-WREN *Stipiturus mallee* 14–15cm Diet includes insects and some seeds. Usually seen in small family groups. Endangered. **V** Song a series of thin, high-pitched whistles. Calls include a high-pitched *tsee-tsee* or *tsee-tsee-tsee* whistles. **H** Open eucalpyt woodland with spinifex (*Triodia*) grassland understorey. **D** Restricted inland range in SE South Australia.

26 RUFOUS-CROWNED EMU-WREN *Stipiturus ruficeps* 11–15cm Forages mostly within cover of vegetation. Diet includes insects and seeds. **V** Song an extremely high-pitched tinkling trill. Calls include a thin *tsii.* **H** Arid grassland with spinifex (*Triodia*). **D** Interior NC and NW Australia.

27 GREY GRASSWREN *Amytornis barbatus* 19–21cm Diet includes insects and seeds. **V** Calls include extremely high-pitched tinkling trills. **H** Inland floodplains with stands of lignum *Muehlenbeckia cunninghamii* and canegrass *Eragrostis australasica*. **D** Interior E Australia.

28 BLACK GRASSWREN *Amytornis housei* 19–21cm Note plump body and broad tail. Unobtrusive and hard to observe. Diet comprises insects and seeds. **V** Song and calls include short bursts of high-pitched chattering, warbling trills. **H** Scrub-covered sandstone outcrops. **D** Restricted range; Kimberley, NW Australia.

29 WHITE-THROATED GRASSWREN *Amytornis woodwardi* 20–22cm Diet includes insects and plant material. Usually seen in small groups. **V** Song comprises short bursts of squelchy phrases, whistles, clicks and trills; some elements recall tuning a shortwave radio. **H** Rocky outcrops with grassy tussocks. **D** Restricted range in N Australia.

30 CARPENTARIAN GRASSWREN *Amytornis dorotheae* 16–17.5cm Forages on ground and rock crevices for invertebrates. **V** Song comprises high-pitched, squeaky warbling phrases. Calls include a sharp, whistled *Tsi-up.* **H** Sandstone outcrops with low woodland, *Triodia* tussock grassland. **D** Restricted range; interior N Australia.

31 SHORT-TAILED GRASSWREN *Amytornis merrotsyi* 15–16cm Usually seen singly or in pairs. **V** Song comprises high-pitched, squeaky warbling phrases, ending with a trill. Calls include thin, high-pitched *tsee-ee-ee* whistles. **H** Rocky outcrops with *Triodia* tussock grassland and lightly wooded scrub. **D** Restricted range; interior S Australia.

32 STRIATED GRASSWREN *Amytornis striatus* 15–19cm Diet includes invertebrates and seeds; usually forages on ground. **V** Song comprises rapid-fire, high-pitched trills interspersed with squeaky notes. Calls include a disyllabic *tchurr-rr.* **H** Rocky ground with *Triodia* tussock grassland and open, scrubby woodland. **D** Interior SE Australia.

33 EYREAN GRASSWREN *Amytornis goyderi* 15–17cm Diet, comprising mainly seeds, is reflected in shape and size of bill. Feeds mainly on ground. **V** Song is a tinkling trill. Calls include a shrill, high-pitched *tseeup.* **H** Sand dune grassland, *Triodia* tussock grassland **D** C interior Australia.

34 WESTERN GRASSWREN *Amytornis textilis* 16–20cm Diet includes invertebrates, fruits and seeds; usually forages on ground. **V** Song comprises high-pitched, squeaky warbling phrases. Calls include a disyllabic, upslurred *pu-pwherr.* **H** Arid scrub. **D** Disjunct populations: W Western Australia; South Australia.

35 THICK-BILLED GRASSWREN *Amytornis modestus* 15–20cm Forages for invertebrates, fruits and seeds in leaf litter. Previously treated as conspecific with Western Grasswren. **V** Song comprises short bursts of chattering squeaks. Calls include slurred whistles. **H** Low shrubby habitats, on plains and dry river beds. **D** Interior SC Australia.

36 DUSKY GRASSWREN *Amytornis purnelli* 16–18cm Diet includes invertebrates, fruits and seeds; forages on ground. **V** Song comprises short bursts of high-pitched, warbling squeaks. Calls include slurred whistles. **H** Rocky ground and gorges with *Triodia* tussock grassland. **D** C Australia.

37 KALKADOON GRASSWREN *Amytornis ballarae* 15–16cm Diet comprises mostly insects and seeds. **V** Song comprises short bursts of high-pitched trills. Calls include a thin, high-pitched *tsi-tsi-tsi-tsi.* **H** Rocky outcrops and gorges with *Triodia* tussock grassland. **D** NC Australia.

1
2
3
4
5
6
7
8
9
10
11
12
13
14
15
16
17
18
19
20
21
22
23
24
25
26
27
28
29
30
31
32
33
34
35
36
37

HONEYEATERS MELIPHAGIDAE

1 BLACK HONEYEATER *Sugomel niger* 11–13cm Feeds primarily on nectar, occasionally insects. Seen singly or in small groups. **V** Song a shrill whistled *SEeep*, repeated at 1-sec intervals. Calls include notes similar to song. **H** Arid scrub, open woodland, eucalypt forest with flowering understorey. **D** Widespread in Australia although absent from humid coastlands and extreme deserts.

2 DRAB MYZOMELA *Myzomela blasii* 11–12cm Forages in the crowns of tall flowering trees; inconspicuous. **V** Very high-pitched single *pit* notes, sometimes given in a rapid chatter. **H** Lowland to montane forest. **D** S Maluku Islands.

3 WHITE-CHINNED MYZOMELA *Myzomela albigula* 13–14cm Diet comprises mainly nectar, probed for in native flowers. Usually seen singly or in small groups. **V** Song a dancing phrase of vibrant whistles. Calls include shrill *tsssp*. **H** Flower-rich scrub, open woodland. **D** Islands SE of New Guinea, including Louisiade Archipelago.

4 ASHY MYZOMELA *Myzomela cineracea* 15–16cm Diet includes nectar and insects. Usually seen singly. Sometimes mixes with other nectar-eating species where feeding is good. **V** Calls include shrill *tsip-tsip-tsip-tsip*. **H** Flower-rich woodland, scrub, gardens. **D** New Britain and Umboi islands (Bismarck Archipelago).

5 RUBY-THROATED MYZOMELA *Myzomela eques* 14–15cm Mixes with other species where feeding is good. Diet comprises mainly nectar, occasionally insects. **V** Calls include high-pitched *tchip*. **H** Rainforest, overgrown secondary woodland. **D** New Guinea, W Papuan islands.

6 DUSKY MYZOMELA *Myzomela obscura* 13cm Forages alone or in mixed-species groups, from sub-stage to mid-storey. **V** Thin, high *sut-sit sut-sit sut-sit*. **H** Primary and tall secondary forest, forest edges. **D** N Maluku Islands.

7 RED MYZOMELA *Myzomela cruentata* 12–13cm Diet comprises mainly nectar, occasionally insects. Unobtrusive. Usually seen singly. **V** Call a thin, whistled *tssiip*. **H** Rainforest, secondary woodland. **D** New Guinea; Duke of York Islands and New Britain (Bismarck Archipelago).

8 PAPUAN BLACK MYZOMELA *Myzomela nigrita* 10–12cm Feeds mainly in flowering trees, on nectar, occasionally insects too. Seen in small groups where feeding is good. **V** Call a thin, whistled *tssiip*. **H** Rainforest, secondary woodland. **D** New Guinea, W Papuan islands, Aru Islands, Louisiade Archipelago.

9 NEW IRELAND MYZOMELA *Myzomela pulchella* 11–12cm Diet presumed to include nectar and occasionally insects. **V** Calls include rapid-fire, shrill *tssrrr, tssrrr*. **H** Montane forest, secondary woodland. **D** New Ireland (Bismarck Archipelago).

10 CRIMSON-HOODED MYZOMELA *Myzomela kuehni* 11cm Forages from dense undergrowth up to canopy, singly, in pairs or within mixed-species feeding parties. **V** High-pitched, downslurred *tsiew tsiew...*; also thin, insect-like *tsii-tsii*. **H** Lowland monsoon forest, tall secondary woodland, overgrown cultivation, subcoastal scrub, gardens. **D** Lesser Sunda Islands.

11 RED-HEADED MYZOMELA *Myzomela erythrocephala* 12–13cm Feeds mainly on nectar, occasionally catches insects in flight. **V** Calls include raspy buzzing *tchrrrp*. **H** Mangrove forests, coastal woodland and scrub. **D** Coastal N Australia, coastal S New Guinea, Aru Islands.

12 SUMBA MYZOMELA *Myzomela dammermani* 11cm Generally encountered singly, in pairs and in small groups. Forages mainly in the canopy and mid-storey. **V** Unrecorded. **H** Primary forest; favours deciduous forest and forest edges. **D** Sumba (W Lesser Sunda Islands).

13 ROTE MYZOMELA *Myzomela irianawidodoae* 11cm Previously treated as related to Sumba Myzomela. Diet presumed to include nectar and insects. **V** Song a rapid-fire, shrill and squeaky *tch-ch-ch-ch-ch-ch-ch*. Calls include single-note squeaky *tch*. **H** Forest margins, secondary woodland. **D** Rote (Lesser Sunda Islands).

14 MOUNTAIN MYZOMELA *Myzomela adolphinae* 9–10cm Feeds in canopy of flowering trees. Diet comprises mainly nectar, occasionally insects. **V** Song a rapid *tch-t-t, tch-t-t*. Calls include single-note, strident *tcht*. **H** Primary forest, secondary woodland, overgrown farmland. **D** New Guinea.

15 BANDA MYZOMELA *Myzomela boiei* 9–12cm Seen singly, in pairs and sometimes in small parties; otherwise little recorded information. **V** Unrecorded. **H** Primary closed forest and tall secondary forest, secondary woodland, lightly wooded farmland, selectively logged forest and mangroves. **D** Lesser Sunda Islands.

16 SULAWESI MYZOMELA *Myzomela chloroptera* 9–12cm Forages in flowering trees in canopy or mid-storey. Usually found alone, in pairs or in small flocks; often part of mixed-species feeding parties. **V** Loud, repeated *peeeew*, a *treeu tree*, a fast *tuweedu*, and a brief, high-pitched warble or twittering. **H** Primary montane and tall secondary and moss forest. **D** Sulawesi and N Maluku Islands, Indonesia.

17 WAKOLO MYZOMELA *Myzomela wakoloensis* 9–12cm Seen singly, in pairs or in small groups; also forages with mixed-species parties in flowering trees. **V** Undescribed. **H** Primary and tall secondary forest, mangroves. **D** S Maluku Islands, Indonesia.

18 SCARLET MYZOMELA *Myzomela sanguinolenta* 10–11cm Diet comprises mainly nectar, occasionally insects. Usually solitary but gathers in groups when feeding is good. **V** Song a whistled variation of *tsi-su-tsi-su-se-se*. Call a shrill *tsiiup*. **H** Waterside scrub, open eucalypt woodland. **D** E Australia (east from Great Divide), N Queensland to Victoria.

19 NEW CALEDONIAN MYZOMELA *Myzomela caledonica* 10–12cm Feeds mainly on nectar. Usually seen singly or in pairs. **V** Song a shrill, piping, repeated *Tu-e-leh, Tu-e-leh*, roughly two phrases per sec. **H** Forests, open woodland, plantations. **D** New Caledonia.

20 CARDINAL MYZOMELA *Myzomela cardinalis* 12cm Unmistakable, no similar bird in range (Samoa). **V** Very high, thin *sreeeh* and other very high, piercing shrieks and twitters. **H** Wide range of habitats including gardens. **D** SW Pacific islands.

21 ROTUMA MYZOMELA *Myzomela chermesina* 12cm No similar bird in range (Rotuma and nearby islets). Note black face sides of males. **V** Undescribed. **H** Wide range of habitats. **D** Rotuma, north of Fiji.

22 MICRONESIAN MYZOMELA *Myzomela rubratra* 13cm Note all-red head and flanks of male. Females vary by race in extent and intensity of red. **V** Sharp, upslurred whistle or single chirp, song a series of chirping notes. **H** Forest, gardens, parks. **D** Micronesia.

23 SCLATER'S MYZOMELA *Myzomela sclateri* 11–12cm Feeds actively, diet comprising mainly nectar and insects. Usually seen singly. **V** Calls include thin *tsiuup*. **H** Primary forest, secondary woodland, plantations. **D** Small islands in Bismarck Sea, east of New Guinea.

24 BISMARCK BLACK MYZOMELA *Myzomela pammelaena* 11–13cm Usually solitary. Diet comprises mainly nectar and insects. **V** Calls include harsh, coarse chatters and squeaks. **H** Wide range of wooded and shrubby habitats. **D** Small islands in northern Bismarck Sea including Admiralty Islands, St Matthias Islands, islands off New Ireland (Bismarck Archipelago).

25 RED-CAPPED MYZOMELA *Myzomela lafargei* 12–13cm Diet comprises mainly nectar and insects. Usually seen singly or in pairs. **V** Calls include tongue-smacking *tssip*. **H** Primary forest, secondary woodland, mangroves, plantations. **D** N Solomon Islands.

26 CRIMSON-RUMPED MYZOMELA *Myzomela eichhorni* 12–13cm Feeds on nectar from forest canopy flowers. **V** Calls include whistled *tsiuup*. **H** Primary forest, secondary woodland, plantations, mature gardens. **D** Solomon Islands.

27 RED-VESTED MYZOMELA *Myzomela malaitae* 12–13cm Diet presumed to include nectar and insects. **V** Calls include sharp *tsiup*. **H** Primary forest, secondary woodland, plantations. **D** Malaita (Solomon Islands).

28 BLACK-HEADED MYZOMELA *Myzomela melanocephala* 12–13cm Feeds actively in tree canopy. Diet assumed to include nectar and insects. **V** Calls include shrill *tssiup*. **H** Primary forest, secondary woodland, plantations. **D** C Guadalcanal and other islands in Solomon Islands.

29 SOOTY MYZOMELA *Myzomela tristrami* 11–12cm Usually seen in small groups. Diet comprises mainly nectar, probably also insects. **V** Calls include squeaky, tongue-smacking *tchup, tchup, tchup*. **H** Primary forest, secondary woodland, plantations, gardens. **D** SE Solomon Islands.

30 SULPHUR-BREASTED MYZOMELA *Myzomela jugularis* 12cm Very distinctively colour-patterned. **V** Call a loud *chit-chit* or *sweet-sweet*, song a four-syllable phrase. **H** Wide range of habitats. **D** Fiji.

31 BLACK-BELLIED MYZOMELA *Myzomela erythromelas* 9–10cm Diet includes nectar, and presumably insects too. **V** Calls include shrill *tssiup*. **H** Primary forest, secondary woodland, plantations, gardens. **D** New Britain (Bismarck Archipelago).

32 BLACK-BREASTED MYZOMELA *Myzomela vulnerata* 10–11cm Active forager among flowering trees in the canopy or mid-storey, singly or in pairs. **V** Weak insect-like *sit-sit-sit...*, also rapidly repeated *tipa-tipa-tipa...* **H** Secondary monsoon forest and scrub, and primary forest. **D** Bismarck Archipelago.

33 RED-COLLARED MYZOMELA *Myzomela rosenbergii* 10.5–12.5cm Diet comprises mainly nectar, occasionally insects. Gathers in groups where feeding is good. **V** Song comprises short bursts of warbling whistles; *tu-leh-tu-tsee*. Call is a shrill chirping whistle. **H** Montane primary forest, alpine scrub. **D** New Guinea.

34 TAWNY-CROWNED HONEYEATER *Gliciphila melanops* 15–18cm Diet comprises mainly nectar, occasionally insects. Usually seen singly. **V** Song a series of separate piping whistle, each a different pitch; recalls somebody aimlessly whistling. Call a shrill squeak. **H** Heathland, open woodland, both with native flowering shrubs. **D** Disjunct populations: SW Australia; SE Australia; Tasmania.

35 GREEN-BACKED HONEYEATER *Glycichaera fallax* 11.5–12.5cm Feeds mainly on insects gleaned from foliage and flowers. Usually seen singly or in pairs. **V** Calls include dry, shrill *tchiip*. **H** Primary forest, secondary woodland, mainly lowlands. **D** New Guinea, W Papuan islands, Aru Islands, N Australia (N Queensland).

1
2
3
4
5
6
7
8
9
10
11
12
13
14
15
16
17
18
19
20
21
22
23
24
25
26
27
28
29
30
31
32
33
34
35

HONEYEATERS *CONTINUED*

1 LEADEN HONEYEATER *Ptiloprora plumbea* 14–15cm Diet includes invertebrates, gleaned from foliage and flowers, and nectar. Usually solitary. **V** Calls include shrill *tschwee, tschwee* notes. **H** Montane forests, secondary woodland. **D** New Guinea.

2 YELLOWISH-STREAKED HONEYEATER *Ptiloprora meekiana* 16–17cm Feeds on invertebrates, mainly insects gleaned from foliage and flowers, and nectar. Usually solitary and unobtrusive. **V** Calls include a shrill *tschuk, tschuk*. **H** Montane forests, secondary woodland. **D** New Guinea.

3 RUFOUS-SIDED HONEYEATER *Ptiloprora erythropleura* 16–17cm Feeds on invertebrates, nectar and fruit; gleans foliage and probes flowers. Often seen in pairs. **V** Song a slurred, rather mournful modulated *tchueeooo*, pitch rising then falling during delivery. Calls include a downslurred whistle. **H** Montane forest, secondary woodland. **D** New Guinea.

4 RUFOUS-BACKED HONEYEATER *Ptiloprora guisei* 16–18cm Diet includes mainly invertebrates, also nectar and fruit; gleans foliage and probes flowers. Seen singly or in pairs. **V** Song a shrill, whistled *tsweeooEE*, like attention-seeking human whistle. Calls include a shrill whistle. **H** Montane forest, secondary woodland. **D** E New Guinea.

5 MAYR'S HONEYEATER *Ptiloprora mayri* 19–20cm Feeds on invertebrates, also nectar and fruit. **V** Song a shrill, slurred and trisyllabic *tsuUEeeooo* whistle, pitch and volume rising then falling; delivered every 2–3 secs. **H** Montane forest, secondary woodland. **D** New Guinea.

6 GREY-STREAKED HONEYEATER *Ptiloprora perstriata* 19–20cm Diet includes invertebrates, nectar and fruit. **V** Song comprises series of shrill upslurred whistling *tsuEe, tsuEe, tsuEe* notes. Calls include shrill, trisyllabic, modulated *tseeuuee*. **H** Montane forest, alpine scrub. **D** New Guinea.

7 EASTERN SPINEBILL *Acanthorhynchus tenuirostris* 14–16cm Diet comprises nectar and insects, latter gleaned from foliage and by probing flowers. **V** Song a rapid-fire series of trilling whistles. Calls include mechanical-sounding piping *tik-tik-tik-tik*... notes. **H** Open eucalypt forest, heaths, both with native flowering shrubs. **D** E and SE Australia, Tasmania.

8 WESTERN SPINEBILL *Acanthorhynchus superciliosus* 13–16cm Diet comprises mainly nectar, probed from flowers; insects caught by gleaning and aerial sallies. **V** Song a series of repeated, chirpy, slurred *tchuerrrp* phrases. Calls include *tchueep* chirp, similar in tone to song. **H** Scrub, heath, open woodland, where flowering *Banksia* and eucalypts are plentiful. **D** SW Australia.

9 PIED HONEYEATER *Certhionyx variegatus* 16–20cm Diet comprises mainly nectar; also invertebrates, fruits, seeds. **V** Song a series of shrill, piping *wuee-we-we-we-we* whistles. Calls include whistles reminiscent of song. **H** Arid scrub, grassland, open woodland with flowering understorey. **D** Australia; widespread C and W Australia, absent N, E and parts of SW.

10 TUI *Prosthemadera novaeseelandiae* 30cm Unmistakable by reflections and details in plumage. **V** Very varied. Call a high, rising *huéeh*. Song a rapid, beautiful series of descending, pure, gong-like notes. **H** Forest, forest remains, suburbs; mainly at lower altitudes. **D** New Zealand.

11 NEW ZEALAND BELLBIRD *Anthornis melanura* 19cm Dark olive-green. On female, note white thin streak from bill down cheek. **V** Scolding *wec* or strong *tjuwtjuw*. Song a clear, rapid, liquid, ringing *pupuwéeh*. **H** Dense forest, secondary growth, subalpine scrub, gardens, parks. **D** New Zealand.

12 PLAIN HONEYEATER *Pycnopygius ixoides* 17–19cm Feeds mainly on nectar and fruits, occasionally insects too. Unobtrusive. Seen singly or in pairs. **V** Song a series of downslurred *tcheeurp* notes. Calls include harsh, squelchy whistles. **H** Forested slopes, secondary woodland. **D** New Guinea.

13 MARBLED HONEYEATER *Pycnopygius cinereus* 20–22cm Diet comprises nectar, fruit and insects. Usually solitary. **V** Song a series of disyllabic *tche-oop* phrases, first element upslurred, second downslurred. Calls include chirpy *tchirrup*. **H** Montane forest, secondary woodland. **D** New Guinea.

14 STREAK-HEADED HONEYEATER *Pycnopygius stictocephalus* 20–22cm Diet comprises mainly fruits, also nectar and insects. **V** Song a series of shrill, whistled notes, *tchu, tchuper-wheer*. Calls include *tchup-tchup-tchup*... notes. **H** Lowland forested and wooded habitats. **D** New Guinea, Aru Islands.

15 BANDED HONEYEATER *Cissomela pectoralis* 12–14cm Feeds mainly on nectar by probing flowers, also gleans insects from foliage. **V** Calls include harsh *tchak* notes followed by a shrill, upslurred whistle. **H** Eucalypt forest, savannah woodland, waterside woodland. **D** N Australia (N Western Australia to N Queensland).

16 SCALY-CROWNED HONEYEATER *Lichmera lombokia* 13–15cm Forages in flowering trees from sub-stage to canopy; alone, in pairs or in small parties; also forms part of mixed-species feeding flocks. **V** Series of rapidly repeated, chattering notes. **H** Primary and degraded forest, forest edges, lightly wooded cultivation, scrub. **D** Lesser Sunda Islands.

17 OLIVE HONEYEATER *Lichmera argentauris* 13–16cm Feeds in flowering trees, singly, in pairs or in small groups. **V** Harsh *zhip*. **H** Coastal *Casuarina* trees, coconut palms, scrub. **D** Maluku Islands, Indonesia; West Papuan islands.

18 INDONESIAN HONEYEATER *Lichmera limbata* 15cm Noisy and aggressive. Active, acrobatic forager in flowering trees and shrubs. Usually seen alone, in pairs or in small groups. **V** Variety of loud, staccato notes, rapid whistles and harsh chatters. **H** Monsoon woodland, secondary growth, mangroves, scrub, lightly wooded cultivation, gardens. **D** Bali and Lesser Sunda Islands.

19 BROWN HONEYEATER *Lichmera indistincta* 13–16cm Diet comprises mainly nectar, also invertebrates. **V** Song a series of shrill, chattering *Tchiip* and *tchuup* squeaky phrases. Calls include harsh, agitated-sounding chattering notes. **H** Wooded habitats including eucalypt forest, mangroves. **D** Widespread across Australia (except interior S, S and SE), S New Guinea, Aru Islands.

20 GREY-EARED HONEYEATER *Lichmera incana* 13–17cm Feeds mainly on nectar and fruit, also invertebrates. **V** Song a rapid-fire series of shrill, harsh *tch-tch-tch-tch-tchchchchchch* notes. Calls include shrill, rasping *tche-tchup-tchup* phrases. **H** Wide range of wooded and shrubby habitats. **D** New Caledonia, Loyalty Islands, Vanuatu.

21 SILVER-EARED HONEYEATER *Lichmera alboauricularis* 15cm Diet comprises mainly nectar probed from flowers, plus insects gleaned from foliage. **V** Song a series of squeaky *tswi-tswi-twsi*... notes. Call a rapid-fire accelerating series of *twch, twchchchchchchch* notes. **H** Waterside vegetation, including scrub, grassland, plantations. **D** New Guinea.

22 SCALY-BREASTED HONEYEATER *Lichmera squamata* 14–15cm Noisy and conspicuous. Forages at flowering trees, also gleans prey from undersides of leaves. **V** Downslurred *chirrup*, upslurred *chirrup* or *chisip*, whistled *tsi-tsi-tsi*, seesawing *whitcheo whitcheo whitcheo*... and a loud, descending trill. **H** Coastal and lowland secondary monsoon woodland, mangroves, lightly wooded cultivation, coconut plantations, villages with scattered trees. **D** Lesser Sunda Islands.

23 BURU HONEYEATER *Lichmera deningeri* 15–16cm Usually seen singly or in pairs; may associate with Wakolo Myzomela (Plate 161) when feeding at *Rhododendron* flowers or other flowering plants. **V** Soft *kew-kew-kew*. **H** Primary and selectively logged and disturbed forest, tall secondary forest, scrub. **D** Buru island, S Maluku Islands, Indonesia.

24 SERAM HONEYEATER *Lichmera monticola* 14–17cm Usually encountered in small flocks foraging in flowering trees and shrubs. **V** Repeated *tshoek tschoek tschoek*, very like the call of a Common Blackbird (Plate 244). **H** Montane forest, montane heath forest. **D** Seram island, S Maluku Islands, Indonesia.

25 FLAME-EARED HONEYEATER *Lichmera flavicans* 12–14cm Forages at flowers in upper mid-storey, alone or in pairs. **V** Series of short, very rapid, bubbling whistled notes, and soft, nasal *bzz*. Other phrases include an *ik-a-blik*, a *ze-plerk* and a *klook-klook*; also utters harsh, chortling song that includes various bell-like phrases. **H** Lowland to montane forests, riparian woodland, secondary growth, open *Eucalyptus* forest, forest edges. **D** Timor island, Lesser Sunda Islands.

26 BLACK-NECKLACED HONEYEATER *Lichmera notabilis* 13–15cm Occurs singly, in pairs or in mixed-species feeding flocks; forages at flowers in tree canopies and in vine tangles. **V** Unrecorded. **H** Lowland monsoon forest, coastal scrub, overgrown cultivation, tall secondary woodland and gardens. **D** Wetar island, Lesser Sunda Islands.

27 CRESCENT HONEYEATER *Phylidonyris pyrrhopterus* 15–17cm Feeds mainly on nectar, but also insects and fruit. **V** Song a rapid, squeaky *tu-tswe-tu-ser-tee*. Calls include shrill, disyllabic *tchSwe-chup*. **H** Eucalypt forest, open woodland, both with flowering understorey. **D** SE Australia, Tasmania.

28 NEW HOLLAND HONEYEATER *Phylidonyris novaehollandiae* 17–19cm Diet comprises mainly nectar, also insects and other invertebrates. **V** Calls include shrill, whistled *tu-tchip* or *tu-tchipipip*, and harsh *tchrrrp*. **H** Mallee scrub, heathland, both with flowering shrubs. **D** SW Australia, S and SE Australia, Tasmania.

29 WHITE-CHEEKED HONEYEATER *Phylidonyris niger* 17–20cm Probes flowers for nectar; gleans foliage and catches insects in flight. **V** Calls include harsh *tchaak* or *tchaa-ak*, and piping, shrill *tu-cherr, tu-cherr-err, tu-cheer*. **H** Heaths, scrub, open forest, all with flowering shrubs. **D** Disjunct populations: SW Australia; E and NE Australia.

30 WHITE-STREAKED HONEYEATER *Trichodere cockerelli* 15–17cm Diet comprises mainly nectar, also insects and other invertebrates. **V** Calls include piping whistles and harsh chattering notes. **H** Open woodland with flowering understorey, heaths. **D** N Queensland.

31 PAINTED HONEYEATER *Grantiella picta* 14–15cm Diet comprises mainly nectar and fruit of mistletoe species. **V** Song a sing-song, whistling *suee-suer, suee-suer, suee-suer*. Calls include shrill, seemingly breathless *tswee-te, tswee-te, tswee-te*, the first phrase upslurred. **H** Open woodland with mistletoe species. **D** E Australia.

32 STRIPED HONEYEATER *Plectorhyncha lanceolata* 22–24cm Unobtrusive. Usually solitary. Probes flowers for nectar; diet also includes fruit and invertebrates. **V** Song a fluty *tu-lu-leep*, repeated every few secs. Calls include a fluty, vibrating trill. **H** Open woodland, scrub, both with flowering trees and shrubs. **D** E Australia (coast and interior).

33 SPOTTED HONEYEATER *Xanthotis polygrammus* 15–17cm Diet comprises mainly invertebrates; also takes nectar and fruit. **V** Song a piping, fluty *whip-ur, wheep-er*, repeated every sec or so. **H** Primary rainforest, secondary woodland. **D** New Guinea, W Papuan islands.

34 MACLEAY'S HONEYEATER *Xanthotis macleayanus* 18–21cm Feeds mainly on insects; also consumes nectar and fruit. **V** Song a shrill *tu-tui-ti-weep*, repeated every sec or so. **H** Rainforest, open woodland, parks. **D** Coastal NE Australia (NE Queensland).

35 TAWNY-BREASTED HONEYEATER *Xanthotis flaviventer* 19–21cm Feeds mainly on insects; also nectar and fruit. **V** Song a shrill *tu-tui, tu-wee*, repeated every sec or so. **H** Lowland rainforest, secondary woodland, mangroves. **D** New Guinea, W Papuan islands, Aru Islands, N Australia (N Queensland).

36 KADAVU HONEYEATER *Xanthotis provocator* 18cm Unmistakable by pattern of face sides in tiny range (Kandavu). **V** Tuneful chirps and bubbling notes. **H** From coastal scrub to montane forest. **D** Kadavu island, Fiji.

1
2
3
4
5
6
7
♀
♂
8
♀
♂
9
♂
♀
10
♂
11
♀
12
13
14
15
16
17
18
♀
19
20
21
22
23
♂
24
25
26
27
♀
♂
28
29
30
31
32
33
34
35
36

HONEYEATERS *CONTINUED*

1 MEYER'S FRIARBIRD *Philemon meyeri* 21–22cm Feeds in tree canopy on nectar, fruit and insects. **V** Song a shrill and insistent *twi-tuer, twi-tuer,* repeated rapidly and increasing in volume during delivery. Calls include shrill, piping *tchuer.* **H** Rainforest, secondary woodland, mainly lowlands. **D** New Guinea.

2 BRASS'S FRIARBIRD *Philemon brassi* 21–22cm Feeds on nectar, fruit and insects in flowering waterside shrubs. **V** Calls include a shrill *tCHiu,* repeated every few secs. **H** Swamp grassland, riverside vegetation. **D** Restricted range, W New Guinea.

3 LITTLE FRIARBIRD *Philemon citreogularis* 26–27cm Feeds mainly on nectar and invertebrates; occasionally also seeds. **V** Song a strangled-sounding, repeated *erk-uerk, EEergh.* Calls include an abrupt, disyllabic wader-like *tchu-ip,* the second phrase upslurred. **H** Open eucalypt forest, riverside woodland, both with grassland understorey and flowering shrubs, parks, gardens. **D** Widespread in N and E Australia, S New Guinea.

4 GREY FRIARBIRD *Philemon kisserensis* 25cm Little recorded information; actions presumed to be similar to the previous species. Formerly considered to be conspecific with Little Friarbird. **V** Not known. **H** Primary and secondary monsoon forest, open woodland, open scrub and, occasionally, lightly wooded cultivation. **D** Islands off E Timor island, Lesser Sunda Islands.

5 TIMOR FRIARBIRD *Philemon inornatus* 24cm Noisy, conspicuous and aggressive. Forages in canopy, alone, in pairs or in small groups. **V** Song a medley of moderately loud notes, including rapid series of *t-chika-wook* notes; also utters series of seesawing notes that often precedes song. **H** Primary and secondary monsoon forest, open woodland, open scrub, occasionally lightly wooded cultivation. **D** Timor island, Lesser Sunda Islands.

6 DUSKY FRIARBIRD *Philemon fuscicapillus* 30cm Usually encountered singly or in pairs foraging in canopy of tall trees. **V** Unrecorded. **H** Primary and logged forest, secondary growth and coconut plantations. **D** N Maluku Islands, Indonesia.

7 SERAM FRIARBIRD *Philemon subcorniculatus* 35cm Forages in upper levels of forests, singly, in pairs or in small groups. Noisy and conspicuous. **V** Explosive *pprow* or *prrt;* also loud *gock* or *geck,* and single loud, nasal notes. **H** Forest, forest edges, mangroves, coastal coconut plantations. **D** Seram island, S Maluku Islands, Indonesia.

8 BLACK-FACED FRIARBIRD *Philemon moluccensis* 31–37cm Noisy and aggressive; forages in middle to upper levels of trees, singly or in pairs. **V** Calls include loud, fluid *yio-wheea,* short *ka wha* and hard *kawah.* **H** Lowland and montane forest, monsoon forest, secondary forest, plantations. **D** Buru island, S Maluku Islands, Indonesia.

9 TANIMBAR FRIARBIRD *Philemon plumigenis* 31–37cm Occurs singly and in pairs, foraging in middle to upper levels of trees. Noisy, aggressive and conspicuous. **V** Loud duet: first bird gives a *seeow,* followed by bubbling bugle from second bird. **H** All types of wooded areas, including plantations, selectively logged forests, mangroves. **D** Tanimbar Island, E Lesser Sunda Islands, and Kai Islands, S Maluku Islands, Indonesia.

10 HELMETED FRIARBIRD *Philemon buceroides* 32–36cm Forages in flowering and fruiting trees, singly, in pairs or in small parties. Noisy, conspicuous and aggressive. **V** Series of low-pitched, nasal notes. **H** All types of wooded areas, including around human habitation. **D** Lesser Sunda Islands to N Australia.

11 NEW GUINEA FRIARBIRD *Philemon novaeguineae* 33–36cm Previously treated as race of Helmeted Friarbird. Diet includes nectar, fruit, invertebrates. **V** Song a modulated, piping *eeer-uip-eh-ou.* Calls include a disyllabic *tchik-ow.* **H** Wide range of wooded habitats including primary rainforest, secondary woodland, mangroves, plantations. **D** New Guinea.

12 HORNBILL FRIARBIRD *Philemon yorki* 33–36cm Previously treated as race of Helmeted Friarbird. Diet includes nectar, fruit, invertebrates. **V** Song an extraordinary squawking and repeated *weh-Whii,* tone rather chicken-like. Calls include a harsh, shrill *kuer-weeik, kuer-weeik.* **H** Open woodland, wet eucalypt forest, mangroves, parks, gardens. **D** New Guinea.

13 NEW BRITAIN FRIARBIRD *Philemon cockerelli* 34–35cm Feeds mainly on nectar, also invertebrates. **V** Calls include a trilling, vibrating *ku-wer-rrp,* repeated every sec or so. **H** Wide range of wooded habitats including primary rainforest, secondary woodland, plantations. **D** Umboi, New Britain and Duke of York islands in Bismarck Archipelago.

14 NEW IRELAND FRIARBIRD *Philemon eichhorni* 32–33cm Diet includes nectar, fruit, invertebrates. **V** Calls include a three note *tu-pu-weep,* each note rising up scale. **H** Montane primary forest, secondary woodland. **D** New Ireland in Bismarck Archipelago.

15 MANUS FRIARBIRD *Philemon albitorques* 35–37cm Diet includes nectar and invertebrates. **V** Song an explosive, piping *CHUK-o* repeated every sec or so. Calls include *tchuk-rr, tche-chrr-ow.* **H** Favours secondary vegetation, gardens, cultivation. **D** Admiralty Islands (including Manus) in Bismarck Archipelago.

16 SILVER-CROWNED FRIARBIRD *Philemon argenticeps* 26–32cm Feeds on nectar and invertebrates. **V** Song a mix of squeaky, piping notes. Calls include a piping *peeUuup,* repeated every 2–3 secs. **H** Open eucalypt woodland with flowering shrubs and grass understorey, parks, gardens. **D** Widespread across N Australia.

17 NOISY FRIARBIRD *Philemon corniculatus* 30–35cm Diet includes nectar, fruit, invertebrates. **V** Song a curious, hiccupping *kee-koUUk,* repeated every 1–2 secs. Calls include a shrill *ke-Wuik,* repeated every 2–3 secs. **H** Open eucalypt forest, acacia woodland, coastal scrub with *Banksia.* **D** Widespread in E Australia (N Queensland to S Victoria).

18 NEW CALEDONIAN FRIARBIRD *Philemon diemenensis* 28–32cm Probes flowers for nectar. Insects caught by aerial sallies and by gleaning foliage. **V** Calls include intense, shrill piping *prrh, per-peah,* repeated every 1–2 secs. **H** Wide range of wooded habitats including gardens. **D** New Caledonia (including Loyalty Islands, Isle of Pines).

19 WHITE-STREAKED FRIARBIRD *Melitograis gilolensis* 23cm Feeds at flowering trees and shrubs, also gleans for invertebrate prey; usually seen singly, in pairs or in small groups. **V** Unmusical note, moderately high to high in pitch; also harsh, rasping note. **H** Lowland and montane forest, forest edges, remnant forest in grasslands and cultivation, selectively logged forest, regrowth forest, mangroves, coconut plantations. **D** N Maluku Islands.

20 BLUE-FACED HONEYEATER *Entomyzon cyanotis* 26–32cm Feeds on nectar and fruit; also insects, gleaned from foliage and on ground. **V** Calls include a shrill, piping *er-whEErp* or *wherEip* repeated every 1–2 secs. **H** Open eucalypt woodland, wooded parks and gardens. **D** SC New Guinea and widespread across E and N Australia.

21 BLACK-CHINNED HONEYEATER *Melithreptus gularis* 15–17cm Probes flowers for nectar, gleans insects from flowers, branches and foliage. **V** Song a shrill, whistling *ts-tse-tse, tchu-tchu-tchu-tchu.* Calls include rasping, squelchy phrases. **H** Open eucalypt forest, waterside *Acacia* woodland. **D** Widespread across N, E and SE Australia.

22 STRONG-BILLED HONEYEATER *Melithreptus validirostris* 15–16cm Diet comprises mainly invertebrates, foraged for in flowers, foliage and leaf litter. **V** Song a rapid-fire squeaky, warbling trill. Call a shrill *tchiup.* **H** Forested habitats, open woodland with shrub understorey. **D** Tasmania, Bass Strait islands.

23 BROWN-HEADED HONEYEATER *Melithreptus brevirostris* 12–14cm Varied diet includes invertebrates, nectar and fruit. Often seen in small groups. **V** Calls include an agitated squeaky warble and a rattling *tch-tch-tch-tch-tch-tch-tch.* **H** Open eucalypt woodland with shrubby understorey. **D** Disjunct populations: SW Australia; S, SE and E Australia.

24 WHITE-THROATED HONEYEATER *Melithreptus albogularis* 12–15cm Diet includes invertebrates, gleaned from foliage and caught while hovering; also drinks nectar. **V** Song a rapid-fire series of whistles. Calls include a *tsee-tsee,* like a squeaky toy. **H** Eucalypt forest, open woodland **D** N and NE Australia, S New Guinea.

25 WHITE-NAPED HONEYEATER *Melithreptus lunatus* 11–15cm Diet includes invertebrates and nectar. Associates with other honeyeater species when feeding is good. **V** Calls include a piping, repeated *tiu-tiu-tiu-tiu...* **H** Open woodland with shrub understorey including flowering *Banksia* and *Acacia.* **D** E and SE Australia.

26 GILBERT'S HONEYEATER *Melithreptus chloropsis* 11–15cm Diet includes invertebrates and nectar. Usually seen singly or in pairs. **V** Calls include a harsh, screeching *tcher-tcher-tcher-tcher...* **H** Open woodland with shrub understorey. **D** SW Australia.

27 BLACK-HEADED HONEYEATER *Melithreptus affinis* 14–15cm Feeds mainly on invertebrates; occasionally also nectar and fruit. **V** Calls include a shrill, whistled *tsiup, tsiup, tsiup.* **H** Open forest (including eucalypt woodland) with understorey of grass and flowering shrubs, parks, gardens. **D** Tasmania, Bass Strait islands.

28 POLYNESIAN WATTLED HONEYEATER *Foulehaio carunculatus* 20–21cm Feeds mainly on nectar; occasionally also fruit and invertebrates. **V** Calls include a fluty, abrupt *tchiUp, tchiUp, tchiUp* and harsh screeches. **H** Wide range of wooded habitats, plantations, parks, gardens. **D** Occurs widely in Polynesian islands including Tonga, Samoa, American Samoa, islands in E Fiji.

29 FIJI WATTLED HONEYEATER *Foulehaio taviunensis* 20–21cm Diet comprises mainly nectar; invertebrates also taken occasionally. **V** Calls include a fluty, whistled *tu-chew, tu-chew* and *whee-chu, whee-chu.* **H** Forested slopes. **D** Islands in NC Fiji.

30 KIKAU *Foulehaio procerior* 20–21cm Diet comprises mainly invertebrates and nectar. Gathers in small groups when feeding is good. **V** Song a repeated, fluty *tCHu-tCHow, tCHu-tCHow* phrases. Calls include harsh screeches and squeaks. **H** Wide range of forested habitats, mainly lowlands. **D** Islands in W and C Fiji.

31 WHITE-EARED HONEYEATER *Nesoptilotis leucotis* 17–21cm Feeds mainly on invertebrates; also nectar. Partial altitudinal migrant. **V** Song and calls include rapid-fire, mechanical-sounding *We-chchchchch* or chuckling, whiplash *we-o-whik.* **H** Open woodland including eucalypt forest, mallee scrub, coastal heaths, mainly lowlands. **D** SW, S and E Australia.

32 YELLOW-THROATED HONEYEATER *Nesoptilotis flavicollis* 18–22cm Feeds mainly on invertebrates; also nectar and fruit. **V** Song and calls include a squeaky introduction followed by rapid-fire, hollow-sounding trill. **H** Open forest with eucalypts and understory of flowering shrubs, coastal heath, parks, gardens. **D** Tasmania, Bass Strait islands.

33 GIBBERBIRD *Ashbyia lovensis* 12–14cm Diet comprises mainly insects, occasionally seeds. **V** Song comprises piping whistles, rising or falling in tone during delivery. Calls include a sharp *tssit.* **H** Stony desert plains (gibber) with sparse vegetation. **D** Interior C Australia (N South Australia, S Northern Territory, SW Queensland, NW New South Wales).

34 CRIMSON CHAT *Epthianura tricolor* 12–13cm Nomadic outside breeding season. Diet comprises mainly insects. **V** Song comprises silvery trills and piping whistles. Calls include a rapid-fire, strident trill. **H** Arid low scrub, grassland. **D** Breeds interior C and W Australia. Outside breeding season, widespread across interior Australia.

35 ORANGE CHAT *Epthianura aurifrons* 11–12cm Feeds mainly on insects, usually caught on ground. **V** Calls include a shrill, nasal squeak. **H** Arid, stony scrub, vegetated salt-lake margins. **D** Widespread across W and interior C Australia; mostly absent from SE, E and N.

36 YELLOW CHAT *Epthianura crocea* 11–12cm Feeds mainly on insects. Usually seen singly. **V** Song comprises various high-pitched whistles, including *tu-te-tu* phrase. Calls include a piping whistle. **H** Marginal vegetation of swamps, seasonally drying wetlands. **D** NC Australia (N Western Australia and Northern Territory to NE South Australia and Queensland).

37 WHITE-FRONTED CHAT *Epthianura albifrons* 12–13cm Feeds mainly on insects. Occasionally takes seeds. **V** Calls include nasal squeaks (like a squeaky toy) and a chattering warble. **H** Open habitats with low vegetation, coastal dunes, beaches, wetland margins. **D** S Australia, Tasmania, Bass Strait islands.

38 LONG-BILLED HONEYEATER *Melilestes megarhynchus* 21–23cm Varied diet includes invertebrates, small vertebrates, nectar, fruit. **V** Calls include a harsh screech, uttered every 2–3 secs. **H** Primary rainforest, secondary woodland, both with dense understorey, mainly lowlands. **D** New Guinea, W Papuan islands, Aru Islands.

1
2
3
4
5
6
7
8
9
10
11
12
13
14
15
16
17
18
19
20
21
22
23
24
25
26
27
28
29
30
31
32
33
♀
34
♀
35
♀
36
♀
37
38
♂
♂
♂
♂

HONEYEATERS *CONTINUED*

1 MACGREGOR'S HONEYEATER *Macgregoria pulchra* 39–41cm Feeds mainly on fruit; occasionally also invertebrates. **V** Calls include a thin, whistled *tseip* and rapid *tchip-tchip-tchip...* **H** Subalpine scrub and forest with small, epiphyte-laden trees and shrubs. **D** New Guinea.

2 ARFAK HONEYEATER *Melipotes gymnops* 21–22cm Diet comprises mainly fruit; occasionally invertebrates. **V** Calls include a strange, modulated screeching *whooeerrrr*, pitch rising and falling during delivery. **H** Montane forest. **D** NW New Guinea (restricted range, Wandammen and Vogelkop mountains).

3 COMMON SMOKY HONEYEATER *Melipotes fumigatus* 21–22cm Feeds mainly on fruits; occasionally also invertebrates. Usually solitary. **V** Calls include repeated, strident and piercing *wuiK, wuiK, wuiK, wuiK...* **H** Hill and montane forest, secondary woodland. **D** New Guinea.

4 WATTLED SMOKY HONEYEATER *Melipotes carolae* 21–22cm Similar to Common Smoky Honeyeater. Restricted range and size of pendant wattle aid identification. **V** Unknown. **H** Montane forest. **D** N New Guinea (restricted to Foja Mountains).

5 SPANGLED HONEYEATER *Melipotes ater* 29–31cm Diet presumed to comprise mainly fruit. Usually seen singly or in pairs. **V** Calls include a screeching, piping *weh-wuik-week-weik-weik*. **H** Montane forest, secondary woodland. **D** NE New Guinea (restricted to mountains in Huon Peninsula).

6 OLIVE STRAIGHTBILL *Timeliopsis fulvigula* 13–15cm Forages in dense cover for insects and other invertebrates. Seen singly or in pairs. **V** Calls include upslurred, piping *shree-shree-shreee...* phrases. **H** Montane forest. **D** New Guinea, W Papuan islands.

7 TAWNY STRAIGHTBILL *Timeliopsis griseigula* 18cm Varied diet includes invertebrates, fruit, nectar. **V** Calls include piping *weh-wuip, weh-wuip, weh-wuip...* **H** Primary rainforest, secondary woodland. **D** Disjunct range: E New Guinea, NW New Guinea.

8 RUFOUS-BANDED HONEYEATER *Conopophila albogularis* 13–15cm Feeds mainly on invertebrates; occasionally also nectar and seeds. **V** Song a mix of slurred notes, then sing-song wheezy whistles. **H** Open woodland, often near water, mangrove forests, parks, gardens. **D** New Guinea (mainly S), Aru Islands, N Australia.

9 RUFOUS-THROATED HONEYEATER *Conopophila rufogularis* 12–14cm Feeds mainly on invertebrates, gleaned from foliage and on ground; occasionally also fruit. **V** Song a squeaky, high-pitched *tswi-chu-per-chu-per-che*. Calls include a shrill, upslurred whistling *tchueee*. **H** Open woodland with understorey of shrubs and grasses. **D** N Australia.

10 GREY HONEYEATER *Conopophila whitei* 11–13cm Feeds mainly on insects but also takes nectar. **V** Song comprises slurred *tsueeoo* squeaks and warbling trill. Calls include a squeaky *tsueeoo*. **H** Open *Acacia* woodland with shrub understorey. **D** Interior W and C Australia.

11 BAR-BREASTED HONEYEATER *Ramsayornis fasciatus* 13–15cm Probes flowers for nectar; insects caught by gleaning foliage and in flight. **V** Calls include harsh, drawn-out, soft screeching *whoarrrr-tchkk*. **H** Forested wetland margins. **D** N Australia.

12 BROWN-BACKED HONEYEATER *Ramsayornis modestus* 12–13cm Probes flowers for nectar; insects gleaned from foliage and caught in flight. **V** Song a rapid *tchip-tchip-tchip-tchup-tchup-tchup*. **H** Forested wetland margins, mangrove forests. **D** N Australia (N Queensland), coastal N and S New Guinea, W Papuan islands, Aru Islands.

13 SPINY-CHEEKED HONEYEATER *Acanthagenys rufogularis* 23–26cm Feeds mainly on fruit but also nectar and invertebrates. **V** Calls include *tchwue* and assorted clicks, piercing whistles and piping notes. **H** Arid woodland and *Acacia* scrub with understorey of grasses and flowering shrubs. **D** Widespread across W and C Australia; absent from SE and N Australia.

14 LITTLE WATTLEBIRD *Anthochaera chrysoptera* 28–34cm Feeds on nectar (*Acacia* and *Banksia* important); occasionally invertebrates. **V** Song a squeaky, screeching *we-EErk-er-eErK*. Calls include harsh screeches and cackles. **H** Open woodland, heaths, mallee scrub, common factor being *Banksia*. **D** SE Australia, Tasmania.

15 WESTERN WATTLEBIRD *Anthochaera lunulata* 30–32cm Diet comprises mainly nectar and insects. **V** Song comprises mix of seemingly unstructured piping and gargling warbles, clicks and whistles. Calls include a harsh, crow-like screech. **H** Open woodland with *Eucalyptus* and understorey including *Banksia*, urban parks. **D** SW Australia.

16 RED WATTLEBIRD *Anthochaera carunculata* 34–36cm Diet includes nectar and some invertebrates. **V** Song is mix of unstructured clucking and gargling sounds, clicks and whistles. Calls include harsh, crow-like screeches. **H** Open woodland, shrubby heaths, parks, gardens. **D** Widespread across S Australia; absent from arid interior.

17 YELLOW WATTLEBIRD *Anthochaera paradoxa* 45–50cm Diet comprises mainly nectar; fruit and invertebrates also taken. **V** Song a harsh, explosive, disyllabic *krr-Whrrkk*. Calls include a screeching *chakk*. **H** Open eucalypt woodland with grass and *Banksia* understorey. **D** E and C Tasmania, King Island (Bass Strait).

18 REGENT HONEYEATER *Anthochaera phrygia* 21–24cm Critically Endangered. Diet comprises mainly nectar and invertebrates. **V** Song is strange mix of seemingly unstructured clucks, gargles, clicks and whistles. **H** Open eucalypt woodland **D** Restricted range in SE Australia; mostly inland slopes of Great Divide.

19 BRIDLED HONEYEATER *Bolemoreus frenatus* 20–22cm Varied diet includes nectar, fruit, invertebrates. **V** Song a mix of harsh screeches, squawks, clicks and whistles. Calls include rather plaintive screeches. **H** Primary rainforest, secondary woodland. **D** NE Australia (NE Queensland).

20 EUNGELLA HONEYEATER *Bolemoreus hindwoodi* 18–20cm Varied diet includes nectar, fruit and invertebrates. Usually seen singly or in pairs. **V** Song a mix of warbling and whistling screeches. **H** Upland rainforest. **D** Restricted range, NE Australia (Clarke Range).

21 YELLOW-FACED HONEYEATER *Caligavis chrysops* 15–17cm Varied diet includes nectar, seeds, fruits, insects. Usually seen singly, occasionally in pairs. **V** Song a mix of shrill whistling *tchiup, tchiup, tchiup* screeches. **H** Open eucalypt woodland with grass and shrub understorey, parks, gardens. **D** SE and E Australia.

22 BLACK-THROATED HONEYEATER *Caligavis subfrenata* 21–22cm Feeds mainly on nectar, also fruits and insects. **V** Song a bubbling trill, pitch rising and falling, and speed accelerating then slowing, during delivery. Calls include a thin *tssip*. **H** Primary montane forest, secondary woodland. **D** New Guinea.

23 OBSCURE HONEYEATER *Caligavis obscura* 18–19cm Diet comprises nectar, fruit and invertebrates. Usually seen singly or in pairs. **V** Song a series of shrill *tchip, tchip, tchup-tchu-tchu*. Calls include a thin *shree*. **H** Primary rainforest, secondary woodland. **D** New Guinea.

24 YELLOW-TUFTED HONEYEATER *Lichenostomus melanops* 17–21cm Feeds mainly on invertebrates, sometimes fruit and nectar. **V** Song comprises fluty, squeaky notes *u-tcheee, u-tchee, utchee-uchee-uchee*. Calls include an upslurred *sreee*. **H** Open forests, eucalypt woodland, both with dense shrub layer, often near water. **D** E and SE Australia.

25 PURPLE-GAPED HONEYEATER *Lichenostomus cratitius* 17–19cm Feeds mainly on invertebrates; also fruit and nectar. **V** Song a rattling trill of squeaky notes. Calls include an upslurred *wheip*. **H** Mallee heath and scrub, open eucalpyt woodland. **D** Disjunct range: S Western Australia, S South Australia and SE New South Wales.

26 BELL MINER *Manorina melanophrys* 18–19cm Diet comprises mainly nectar; occasionally also fruit and invertebrates. **V** Song a series of piercing, bell-like piping whistles. Calls include chattering squeaks. **H** Mosaic of open woodland and shrubby heath. **D** E and SE Australia.

27 NOISY MINER *Manorina melanocephala* 25–28cm Feeds on invertebrates, from foliage, branches and ground. Also nectar, fruits and seeds. **V** Calls include an agitated series of harsh, squeaky notes. **H** Open savannah woodland, parks, gardens. **D** E and SE Australia, Tasmania.

28 YELLOW-THROATED MINER *Manorina flavigula* 23–28cm Feeds mainly on invertebrates, nectars; occasionally fruit. **V** Calls include a series of shrill, raptor-like piping *kei-kei-kei-kei...* notes. **H** Wide range of wooded habitats including natural areas, farmland, parks, gardens. **D** Widespread across Australia, except coastal E, NE and SE.

29 BLACK-EARED MINER *Manorina melanotis* 24–26cm Endangered. Diet comprises mainly invertebrates; occasionally also nectar. **V** Calls include a series of shrill piping *weik-weiwik-wik-wwik...* notes, tone recalling wader alarm call. **H** Undisturbed mallee woodland with flowering shrubs and grasses. **D** Very restricted range: SE South Australia, NW Victoria.

30 MAKIRA HONEYEATER *Meliarchus sclateri* 26–28cm Diet comprises mainly insects, occasionally nectar. **V** Very vocal and noisy. Varied calls include croaks, whistles and an abrupt, harsh *wheeik*. **H** Mainly primary forest; occasionally mature secondary woodland and plantations. **D** Makira (Solomon Islands).

31 SOOTY MELIDECTES *Melidectes fuscus* 22–25cm Varied diet includes nectar, fruit, invertebrates. Usually solitary. **V** Calls include a series of harsh *waak-waak-waak-waak...* notes. **H** Primary montane forest, secondary woodland. **D** New Guinea.

32 GILLIARD'S MELIDECTES *Melidectes whitemanensis* 22–23cm Feeds mainly on invertebrates and nectar. Usually seen singly, feeding in tree canopy. **V** Song a series of slurred, disyllabic, thin whistles. Calls include chattering *waak* notes. **H** Montane forest. **D** New Britain (Bismarck Archipelago).

33 SHORT-BEARDED MELIDECTES *Melidectes nouhuysi* 27–28cm Diet includes nectar, fruit, invertebrates. **V** Song a series of slurred, disyllabic, thin whistles. Calls include soft chatters, usually the same note repeated several times. **H** Montane woodland and scrub. **D** C New Guinea.

34 LONG-BEARDED MELIDECTES *Melidectes princeps* 27–29cm Diet poorly known but presumed to include nectar and invertebrates. **V** Unknown. **H** Montane forest and scrub. **D** EC New Guinea.

35 CINNAMON-BROWED MELIDECTES *Melidectes ochromelas* 24–25cm Diet presumed to be varied and include invertebrates and nectar. **V** Song a series of fluty, piping *tui, tchu-tchu-tchu-tchu* notes. Calls include a hollow-sounding, dry rattle. **H** Montane forest. **D** New Guinea.

36 VOGELKOP MELIDECTES *Melidectes leucostephes* 25–26cm Diet presumed to be varied and to include nectar and invertebrates. **V** Calls include a screechy, hoarse *wher, weh-weh-weh-weh*. **H** Montane forest and open scrub. **D** NW New Guinea.

37 YELLOW-BROWED MELIDECTES *Melidectes rufocrissalis* 27–29cm Varied diet includes invertebrates, nectar, fruit. **V** Calls include a harsh, screechy and crow-like *whar, wa-wa-wa-weh*. **H** Hill and montane forest, secondary woodland. **D** EC New Guinea.

38 HUON MELIDECTES *Melidectes foersteri* 28–32cm Diet presumed to be varied and include nectar and invertebrates. **V** Calls include a shrill, squawking *whoh-wek-wek-wek*. **H** Montane forest, alpine scrub. **D** NE New Guinea (Huon Peninsula).

39 BELFORD'S MELIDECTES *Melidectes belfordi* 28–29cm Varied diet comprises invertebrates, nectar and fruit. **V** Calls include nasal, squeaky *kweek, kwik, kwo, kwik-wo*. **H** Montane forest, scrub. **D** New Guinea.

40 ORNATE MELIDECTES *Melidectes torquatus* 22–24cm Diet comprises mainly insects; nectar and fruit also consumed. **V** Song a repeated *kui-koo*, the first phrase upslurred, the second downslurred. Calls include a shrill, raptor-like *wah-kee-kee-kee-kee*. **H** Hill and montane forest. **D** New Guinea.

1
2
3
4
5
6
7
8
9
10
11
12
13
14
15
16
17
18
19
20
21
22
23
24
25
26
27
28
29
30
31
32
33
34
35
36
37
38
39
40

HONEYEATERS *CONTINUED*

1 WHITE-FRONTED HONEYEATER *Purnella albifrons* 14–18cm Diet comprises mainly nectar; occasionally also insects. **V** Song comprises varied mix of piping or fluty notes, each phrase comprising 5–7 elements, delivered at a different pitch and at different speeds. **H** Arid open woodland, mallee scrub, heathland. **D** W and C S Australia.

2 WHITE-GAPED HONEYEATER *Stomiopera unicolor* 20–22cm Varied diet includes nectar, fruits, invertebrates. **V** Song a rapid, squeaky *wer-wik-er-kkkk*. Calls include a squeaky chatter. **H** Waterside and swamp forest with understorey of grasses and flowering shrubs. **D** N Australia (N Western Australia, N Northern Territory, N and NE Queensland).

3 YELLOW HONEYEATER *Stomiopera flava* 17–18cm Feeds on nectar, fruit, invertebrates. **V** Song a shrill, whistled *we-Weeo* or *we-o-wee* recalling human wolf-whistle or attention-seeking whistle. Calls include a rattling chatter. **H** Swamp and waterside forest with understorey of flowering shrubs. **D** NE Australia (NE Queensland).

4 VARIED HONEYEATER *Gavicalis versicolor* 20–24cm Diet includes nectar and invertebrates. **V** Song is variation on a fluty, piping whistled *wu-chur-pa-we-chur* or elements thereof. **H** Mangroves, plantations, parks, gardens. **D** Coastal NE Australia, coastal S and N New Guinea and W Papuan islands.

5 MANGROVE HONEYEATER *Gavicalis fasciogularis* 19–20cm Diet includes nectar and invertebrates. Usually solitary. **V** Very vocal. Song comprises warbling, fluty whistling notes. Calls include various fluty whistles. **H** Mangroves, coastal forest, parks, gardens. **D** Coastal E Australia.

6 SINGING HONEYEATER *Gavicalis virescens* 17–23cm Diet comprises invertebrates, nectar and fruits. **V** Song *tchwee-koo-oo-oo-oo* the last section having an echoing quality. Calls include strident screeches and chirping notes. **H** Open woodland, scrub, parks, gardens. **D** Widespread across most of Australia, except N Queensland.

7 YELLOW-TINTED HONEYEATER *Ptilotula flavescens* 14–16cm Varied diet includes invertebrates, nectar, fruit, seeds. **V** Song a vibrant, piping *Tchi-tchu*. Calls include agitated screeches. **H** Open eucalypt forest, acacia woodland, mangrove, parks, gardens. **D** Patchy distribution in N Australia, S New Guinea.

8 FUSCOUS HONEYEATER *Ptilotula fusca* 14–17cm Diet comprises mainly invertebrates, some nectar and honeydew. **V** Song a vibrating trill of piping notes, preceded by a squelchy warble. Calls include a shrill *tchiup*. **H** Open woodland with eucalypts and shrub understorey, heathland outside breeding season. **D** E and SE Australia.

9 GREY-HEADED HONEYEATER *Ptilotula keartlandi* 14–17cm Feeds mainly on invertebrates and nectar. Seen in pairs or small groups. **V** Calls include a squeaky *tchip, tchip, tchip*... **H** Arid open woodland with shrub understorey. **D** C Australia (W Western Australia to C Queensland).

10 GREY-FRONTED HONEYEATER *Ptilotula plumula* 15–17cm Probes flowers for nectar, gleans and fly-catches invertebrates among foliage; occasionally also eats fruits and seeds. **V** Calls include a squelchy *tchwip, tchwip, tchwip*... **H** Mallee woodland with eucalypts, riverside forests. **D** Widespread across interior Australia.

11 YELLOW-PLUMED HONEYEATER *Ptilotula ornata* 15–19cm Feeds mainly on invertebrates and nectar; sometimes also fruits and seeds. **V** Song a rapid, trilling series of squeaky whistles. Calls include a squeaky *wer, whe-er*. **H** Eucalypt and acacia woodland. **D** Disjunct range: SW Australia, S and C SE Australia.

12 WHITE-PLUMED HONEYEATER *Ptilotula penicillata* 14–18cm Diet includes nectar, manna, fruit and invertebrates. **V** Song a rapid, shrill whistled *tchu-per-wee-wee*. Calls include coarse screeches. **H** Open sclerophyll woodland with eucalypts, parks, gardens. **D** Widespread across C Australia although absent from SW and N Australia.

13 MOTTLE-BREASTED HONEYEATER *Meliphaga mimikae* 16–17cm Varied diet includes invertebrates, fruits, nectar. Unobtrusive and usually solitary. **V** Song a rapid, trilling series of squeaky whistles; calls include piping notes. **H** Forested slopes, mainly lowlands. **D** C New Guinea.

14 FOREST HONEYEATER *Meliphaga montana* 17–18cm Diet is presumed to be varied and include nectar, fruits, seeds, invertebrates. Usually seen singly, sometimes in pairs. **V** Calls include a shrill, upslurred piping *whiiP*. **H** Primary montane forest, secondary woodland. **D** New Guinea.

15 MOUNTAIN HONEYEATER *Meliphaga orientalis* 14–16cm Feeds on nectars, fruit and invertebrates. Forages in tree canopy. **V** Calls include a shrill *pee-uu* or *tchu-ee* and sharp *tchup*. **H** Primary montane forest, secondary woodland, mature gardens. **D** New Guinea, W Papuan islands.

16 SCRUB HONEYEATER *Meliphaga albonotata* 18–19cm Diet comprises mainly invertebrates and nectar, occasionally also fruit. **V** Song a shrill *tu-ti-ti-ti-ti-ti*... recalling a raptor alarm call. Calls include a sharp *tchup*. **H** Secondary woodland, plantations, gardens. **D** C New Guinea.

17 MIMIC HONEYEATER *Meliphaga analoga* 16–19cm Diet includes invertebrates, nectar and fruit. **V** Song a shrill *we-we-we-we-we*... recalling a raptor alarm call. Calls include a sharp *tchiup*. **H** Primary forest, secondary woodland, mangroves, gardens. **D** New Guinea, W Papuan islands, Aru Islands.

18 TAGULA HONEYEATER *Meliphaga vicina* 16–17cm Poorly known, with affinities to Mimic Honeyeater. Diet presumed to include nectar and invertebrates. **V** Song a strident *wiip-uu*, first phrase upslurred. Calls include piping whistle, recalling a human attention-seeking whistle. **H** Forested habitats, mostly lowlands. **D** Tagula Island (Louisiade Archipelago).

19 GRACEFUL HONEYEATER *Meliphaga gracilis* 16–17cm Diet includes nectar, fruit and invertebrates. **V** Song a shrill *kee-kee-kee-kee*... Calls include a shrill, whistled *tchiup*. **H** Lowland primary forest, secondary woodland, mangroves. **D** N Australia (N and NE Queensland), S New Guinea, Aru Islands.

20 ELEGANT HONEYEATER *Meliphaga cinereifrons* 15–18cm Feeds mainly on nectar, fruit, invertebrates. **V** Calls include a shrill *tchiup* and *tu-tee-te-te-te*, recalling a woodpecker alarm call. **H** Lowland primary rainforest, swamp forest, secondary woodland, mangroves, gardens. **D** SE New Guinea.

21 YELLOW-GAPED HONEYEATER *Meliphaga flavirictus* 15–16cm Diet presumed to include nectar, as well as fruit and invertebrates. **V** Calls include a shrill *tchiup* and *tu-tee-te-te-te*, recalling a woodpecker alarm call. **H** Primary forest, secondary woodland, mainly lowlands. **D** New Guinea.

22 WHITE-LINED HONEYEATER *Meliphaga albilineata* 17–20cm Diet includes invertebrates, nectar and fruit. Unobtrusive and usually solitary. **V** Song comprises a series of thrush-like whistles *tcheeo, tcheeu, tcheeo*... Calls include sweet, whistling *tiu-teeoo*. **H** Scrub and woodland on sandstone cliffs, gorges and plateaux. **D** Restricted range in N Australia (W Arnhem Land in Northern Territory).

23 KIMBERLEY HONEYEATER *Meliphaga fordiana* 18–19cm Diet presumed to include nectar and invertebrates. **V** Song comprises shrill, thrush-like piping whistling phrases such as *wher-te-cha, wher-te-cha*. Calls include a tongue-smacking *tchiup*. **H** Sparse, arid scrub and woodland on sandstone cliffs and gorges. **D** Restricted range in NW Australia (Kimberley, NW Northern Territory).

24 STREAK-BREASTED HONEYEATER *Meliphaga reticulata* 15–16cm Noisy. Favours feeding at flowers of *Eucalyptus* trees, singly, in pairs or in small groups. **V** Series of plaintive, upslurred *wheep* notes, starting quickly and then slowing; also rapidly repeated, high-pitched *wik-wik-wik*... or *week-week-week*, and a lower-pitched *work-work*. **H** Primary and secondary deciduous and evergreen forest, degraded forest, scrub and semi-cultivated areas, urban gardens, mangroves. **D** Seram and Timor islands, Lesser Sunda Islands.

25 PUFF-BACKED HONEYEATER *Meliphaga aruensis* 17–19cm Feeds on fruits, seeds and invertebrates. Unobtrusive and shy. Usually solitary. **V** Calls include a sharp, clucking *Tchiup* and a vibrating trill. **H** Primary forest, secondary woodland, mainly lowlands. **D** New Guinea, W Papuan islands, Aru Islands.

26 YELLOW-SPOTTED HONEYEATER *Meliphaga notata* 17–20cm Diet includes fruits, nectar and invertebrates. Usually seen singly or in pairs. **V** Calls include a shrill, piping *tchiuk, tchiuk, tchiuk*... and a squeaky *tcheup, tcheup, tcheup*... with volume fading during delivery. **H** Primary rainforest, secondary woodland, mangroves, parks, gardens, mainly lowlands. **D** NE Australia.

27 LEWIN'S HONEYEATER *Meliphaga lewinii* 19–21cm Feeds on fruits, nectar and invertebrates. **V** Calls include a harsh *kah, kah, kah*... and a shrill *tuik, tuik, tuik*... volume fading during delivery. **H** Primary rainforest, secondary woodland, park, gardens, mainly lowlands. **D** E Australia, N Queensland to S Victoria.

28 GUADALCANAL HONEYEATER *Guadalcanaria inexpectata* 18–20cm Diet includes invertebrates; presumed to also take nectar and fruits. **V** Calls include squeaky contact calls and seemingly unstructured series of piping whistles. **H** Montane forest. **D** Guadalcanal (Solomon Islands).

29 ORANGE-CHEEKED HONEYEATER *Oreornis chrysogenys* 25–26cm Varied diet includes fruits, seeds and invertebrates. **V** Calls include shrill, rapid-fire *trrrrich* lasting around 1 sec. **H** Montane and subalpine forest, scrub and grassland. **D** W New Guinea.

30 YELLOW-BILLED HONEYEATER *Gymnomyza viridis* 28cm Timid, gregarious. Race *brunneirostris* has a dusky bill; bill of immature is yellowish with brownish tip. **V** Song **H** Prefers canopy of native forest. **D** Fijian islands of Taveuni and Vanua Levu.

31 GIANT HONEYEATER *Gymnomyza brunneirostris* 26–30cm Feeds mainly on nectar but also invertebrates. **V** Song a tirade of fluty notes tone reminiscent of emergency-service siren; sometimes delivered in duet. Calls include elements of song's notes. **H** Primary forest. **D** Viti Levu, W Fiji.

32 MAO *Gymnomyza samoensis* 28cm Unmistakable in restricted range by large size and dark plumage. Heard more than seen. **V** Duets: *mew wjuwjutwjut* (*mew* note nasal and downslurred) is answered by *wéetwoh*. **H** Forest, coconut plantations. **D** W Samoa.

33 CROW HONEYEATER *Gymnomyza aubryana* 36–42cm Critically endangered. Diet includes invertebrates and nectar. **V** Calls include harsh, crow-like croaks and piercing, high-pitched notes including *tchiiP*. **H** Forest habitats including remaining tracts of primary rainforest. **D** New Caledonia.

34 DARK-EARED MYZA *Myza celebensis* 17cm Feeds at flowering trees and gleans from vines, branches and foliage, mainly in canopy. Usually seen singly or in pairs. **V** Series of twittering squeaks, or thin, shrill *pst* or *pst-pst*; also sharp, harsh *tsreet*. **H** Montane forest. **D** Sulawesi, Indonesia.

35 WHITE-EARED MYZA *Myza sarasinorum* 20cm Active and conspicuous. Feeds at flowers and gleans arthropods from plants. **V** Calls include high-pitched *kik* or *kuik*, nasal *tuck* and 3–5 wheezy, high-pitched notes, given while feeding; song consists of series of high-pitched squeaks. **H** Upper montane forests. **D** Sulawesi, Indonesia.

36 BOUGAINVILLE HONEYEATER *Stresemannia bougainvillei* 17–18cm Diet includes fruits and invertebrates. Unobtrusive. Often seen in pairs. **V** Calls include harsh, rasping chirps and piping whistles. **H** Montane primary forest. **D** Bougainville Island (Solomon Islands).

37 BARRED HONEYEATER *Glycifohia undulata* 17–20cm Feeds on nectar, fruits and invertebrates. **V** Song includes fluty, oriole-like *kee-eup, kee-eup* notes and liquid-sounding trills, sometimes delivered in duet. **H** Primary forest, secondary woodland, gardens. **D** New Caledonia.

38 WHITE-BELLIED HONEYEATER *Glycifohia notabilis* 18–21cm Varied diet includes nectar, pollen, fruits and invertebrates. **V** Calls include a trisyllabic, fluty whistled *tche-uu-ee*. **H** Hill forests, wooded slopes, mangroves. **D** Vanuatu.

1
2
3
4
5
6
7
8
9
10
11
12
13
14
15
16
17
18
19
20
21
22
23
24
25
26
27
28
29
30
31
32
33
34
35
36
37
38

BRISTLEBIRDS DASYORNITHIDAE

1 EASTERN BRISTLEBIRD *Dasyornis brachypterus* 18–20cm Endangered due to habitat destruction. Feeds mainly on invertebrates and seeds. **V** Song comprises squeaky, whistling phrases such as *ti-tu-tswEE-tsu* delivered every 2–3 secs. Calls include a *tswuee, tswuee*. **H** Heathland, tussock-sedge stands, wetland scrub, woodland with dense understorey. **D** Coastal SE Australia (New South Wales and E Victoria).

2 WESTERN BRISTLEBIRD *Dasyornis longirostris* 16–18cm Endangered due to habitat loss. Diet mainly invertebrates and seeds. **V** Song comprises rapid, whistled phrases such as *tsit-t-tsi, tu-do-lee* or *tsichup tse-te*. **H** Coastal heathland comprising a dense, diverse array of shrubs. **D** Restricted range, SW Australia.

3 RUFOUS BRISTLEBIRD *Dasyornis broadbenti* 24–25cm Varied diet includes invertebrates, fruits and seeds. **V** Song a mix of shrill piping notes such as *tik-u, tik-u, tchu-de-wheuh*. Calls include a shrill, disyllabic *tShu-we*. **H** Coastal heathland and scrub, with mosaic of open ground and shrubs. **D** Coastal SE Australia (W Victoria).

PARDALOTES PARDALOTIDAE

4 SPOTTED PARDALOTE *Pardalotus punctatus* 9–10cm Diet includes invertebrates, manna and lerps. **V** Song comprises a piercing, whistled *chu-ee*, delivered every sec or so, or two or three phrases combined. **H** Wide range of eucalypt woodland. **D** Widespread across S and E Australia; mainly absent from interior.

5 FORTY-SPOTTED PARDALOTE *Pardalotus quadragintus* 9–11cm Endangered. Diet includes invertebrates, lerps and manna (latter encouraged by birds nipping stems). **V** Calls include nervous-sounding squeaky *tu-Urr* notes. **H** Eucalypt forest (mainly Manna Gum *Eucalpytus viminalis*). **D** Restricted range: coastal SE Tasmania and islands, Flinders Island (Bass Strait).

6 RED-BROWED PARDALOTE *Pardalotus rubricatus* 9–11cm Diet presumed to comprise mainly invertebrates. **V** Song comprises slightly accelerating series of piped, whistled *whu-pe-pe-pe-pe* notes. Calls include phrase-elements of song. **H** Wide range of wooded habitats. **D** Widespread across C and N Australia.

7 STRIATED PARDALOTE *Pardalotus striatus* 9–11cm Dispersive outside breeding season; Tasmanian birds migrate to mainland Australia. **V** Song a vibrant, fluty *whe-whe-wheo* or *whe-wo-wo*, each phrase delivered every 1–2 secs. **H** Eucalypt woodland. **D** Widespread across Australia except most arid regions of interior, Tasmania, Bass Strait islands.

AUSTRALASIAN WARBLERS ACANTHIZIDAE

8 GOLDENFACE *Pachycare flavogriseum* 11–12cm Gleans invertebrates in forest foliage and canopy. **V** Song *tchup-sisisisisi*, the first element shrill and fluty, the second a vibrant trill. **H** Montane forest, dense secondary woodland. **D** New Guinea.

9 FERNWREN *Oreoscopus gutturalis* 12–14cm Diet includes invertebrates and small vertebrates, found mainly in leaf litter. **V** Song a piercingly high-pitched *tsi-tsi-tsi-tsi-seeeeee*. Calls include a slurred, high-pitched *tsu-TSeee*. **H** Primary rainforest, humid dense secondary woodland. **D** Restricted range in NE Australia (NE Queensland).

10 PILOTBIRD *Pycnoptilus floccosus* 18–19cm Forages for invertebrates on ground; sometimes also eats seeds. **V** Song a whistled *su, see, tsuweSEUEee*, the first two elements subdued, the third piercing and shrill, the pitch modulated. **H** Rainforest, damp wooded habitats. **D** SE Australia.

11 SCRUBTIT *Acanthornis magna* 11–12cm Feeds on invertebrates, mainly insects by probing moss, bark, undergrowth and foliage. **V** Song a piping, whistled *tsee, tsu, tu-loo*. Call a shrill, vibrant trill. **H** Rainforest, humid wooded habitats, both with dense understorey of shrubs, ferns and mosses. **D** Tasmania and King Island (Bass Strait).

12 ROCKWARBLER *Origma solitaria* 13–15cm Diet comprises mainly invertebrates, particularly insects. Forages among rocks and on ground. **V** Song a series of fizzing, metallic, high-pitched *PSzzz* notes, delivered every sec or so. Calls include chattering notes and thin whistles. **H** Rocky outcrops with scrub. **D** Restricted range in coastal SE Australia (E New South Wales).

13 CHESTNUT-RUMPED HEATHWREN *Calamanthus pyrrhopygius* 13–16cm Feeds mainly on invertebrates. Forages on ground or in low vegetation. **V** Song a warbling series of high-pitched and chattering notes. Calls include thin whistling notes. **H** Heathland, open woodland with heath and shrub understorey, mainly lowlands. **D** SE Australia.

14 SHY HEATHWREN *Calamanthus cautus* 12–14cm Grey-brown with conspicuous rufous rump; despite name, not especially shy in manner. **V** Song a series of chirpy, warbled whistling notes such as *tili-tili, tuli-tuli, trrp-trrp*. Calls include harsh, tongue-smacking *wchtt*. **H** Mallee woodland and scrub with understorey of heathland shrubs. **D** S Australia except for far SE and far SW.

15 STRIATED FIELDWREN *Calamanthus fuliginosus* 13–14cm Forages for invertebrates on ground or in low vegetation. **V** Song a series of strident, warbled notes such as *tchr-chee-chee-ttttt-whchrr*. Calls include a tongue-smacking *tchutt*. **H** Heathland, low scrub, salt marsh, waterside grassland. **D** Coastal SE Australia, Tasmania.

16 WESTERN FIELDWREN *Calamanthus montanellus* 12–13cm Feeds mainly on invertebrates but also seeds, foraging on ground or in low vegetation. **V** Song a rapid, dancing series of piping whistles such as *piurr, tswee-tswee, tswe-tswe-terr-ter-we-er-we-er*. **H** Coastal heathland, scrub. **D** SW Australia.

17 RUFOUS FIELDWREN *Calamanthus campestris* 12–13cm Feeds on invertebrates and seeds, mainly on ground. **V** Song a series of piping whistles such as *whuik-whuik, te-te-tu-lee-oo*, pitch modulating during delivery. **H** Heathland, mallee scrub, gibber plains, saltbush habitats. **D** Disjunct populations from Western Australia to W New South Wales and W Victoria.

18 REDTHROAT *Pyrrholaemus brunneus* 11–12cm Diet comprises mainly invertebrates and seeds. Feeds on ground and in low vegetation. **V** Song a rapid warbling series of high-pitched notes. Calls include a sharp, tongue-smacking *tChtt*. **H** Arid scrub, chenopod habitats. **D** Widespread in Western Australia, C interior and S Australia.

19 SPECKLED WARBLER *Pyrrholaemus sagittatus* 11–12cm Feeds on invertebrates and seeds. **V** Song a series of shrill whistles, sometimes delivered in a warbling manner. Calls include a grating *tchrrr*. **H** Dry eucalypt woodland with understorey of grasses and shrubs. **D** E and SE Australia.

20 RUSTY MOUSE-WARBLER *Crateroscelis murina* 12cm Diet presumed to comprise mainly invertebrates. **V** Song a series of isolated, piercing whistles, pitch descending during delivery. Calls include a grating *tchrrr*. **H** Hill forest. **D** New Guinea, W Papuan islands, Aru Islands.

21 BICOLORED MOUSE-WARBLER *Crateroscelis nigrorufa* 12–13cm Diet presumed to include invertebrates, foraged for on ground and in low vegetation. **V** Calls include rapid, chattering *tch-kt* or agitated *tchrrrrr* and whistles. **H** Montane forest. **D** New Guinea.

22 MOUNTAIN MOUSE-WARBLER *Crateroscelis robusta* 12–13cm Feeds mainly on insects, foraged for on ground and in tangled, low vegetation. **V** Song a series of whistles *tuip, tee-tiu-tu-tu*, first phrase upslurred, thereafter pitch descending stepwise during delivery. Calls include rattling *tchttt*. **H** Montane forest. **D** New Guinea,

23 PALE-BILLED SCRUBWREN *Sericornis spilodera* 11–12cm Diet presumed to comprise mainly insects. **V** Song a series of wheezy, whistled *tchee-tuu, tchee-tuu...* phrases. Calls include a grating chatter. **H** Hill forest. **D** New Guinea, W Papuan islands, Aru Islands.

24 PAPUAN SCRUBWREN *Sericornis papuensis* 10–11cm Unobtrusive. Feeds mainly on insects and other invertebrates. **V** Song a shrill, warbling whistled *t-tchee-tchee, tur-tchee-tchiu*. Calls include a harsh *tchrrrr*. **H** Montane forest, dense secondary woodland. **D** New Guinea.

25 ATHERTON SCRUBWREN *Sericornis keri* 12–14cm Diet comprises mainly insects. Usually solitary. **V** Song a tinkling, warbling *t-tchiu-si, t-tchiu-si...* Calls include a fizzing, vibrating *tzzzzz*. **H** Dense rainforest. **D** NE Australia (NE Queensland).

26 WHITE-BROWED SCRUBWREN *Sericornis frontalis* 11–15cm Diet includes invertebrates, seeds, fruits. **V** Song a series of vibrating, high-pitched trills. Calls include a buzzing *t-t-tchttt*. **H** Wide range of habitats with dense undergrowth including woodland, scrub, heathland, gardens. **D** E, S and SW Australia.

27 TASMANIAN SCRUBWREN *Sericornis humilis* 13–14cm Diet comprises mainly invertebrates but also seeds. **V** Calls include high-pitched, thin whistled *tse-si-si-si-si* and grating, chattering notes. **H** Wooded habitats with damp, dense undergrowth. **D** Tasmania, King Island (Bass Strait).

28 YELLOW-THROATED SCRUBWREN *Sericornis citreogularis* 13–15cm Varied diet includes invertebrates, seeds and fruits. **V** Song a series of squeaky whistles delivered in a warbling manner. Calls include a tongue-smacking *tchttt*. **H** Shady, damp woodland with rich understorey. **D** Disjunct populations: coastal NE Australia (NE Queensland), coastal SE Australia.

29 LARGE-BILLED SCRUBWREN *Sericornis magnirostra* 11–12cm Diet comprises mainly invertebrates. **V** Song a series of squeaky *tchee-ur, tchee-ur* notes, each phrase alternating high and lower pitch. Calls include harsh chattering phrases. **H** Rainforest, mature damp secondary woodland. **D** Coastal E Australia (NE Queensland to S Victoria).

30 TROPICAL SCRUBWREN *Sericornis beccarii* 11–12cm Feeds mainly on invertebrates. **V** Song a series of shrill, strident whistling phrases. Calls include harsh chattering notes. **H** Rainforest, montane forest, mature secondary woodland. **D** N Australia (N Queensland), New Guinea, Aru Islands.

31 PERPLEXING SCRUBWREN *Sericornis virgatus* 11–12cm Diet presumed to include invertebrates, particularly insects. Often seen in pairs or small groups. **V** Song a rapid series of metallic, tinkling notes, alternating in pitch. Calls include a squelchy *tchhht*. **H** Primary forest, secondary woodland. **D** New Guinea.

32 LARGE SCRUBWREN *Sericornis nouhuysi* 12–13cm Diet comprises mainly invertebrates. Forages in small groups. **V** Song a series of sweet, whistling *tu-tuee, tu-tuee...* elements of each phrase alternating in pitch. Calls include a harsh, tongue-smacking *tchtt*. **H** Montane forest, secondary woodland. **D** New Guinea.

33 BUFF-FACED SCRUBWREN *Sericornis perspicillatus* 10cm Feeds mainly on insects. Forages in small groups. Mixes with other species in flocks. **V** Song a whistled *tu-tee, tu-tee, tu-tee...* gradually descending in pitch during delivery and ending with a trilled *tcht-tcht-tcht*. Calls include agitated chattering notes. **H** Montane forest. **D** New Guinea.

34 VOGELKOP SCRUBWREN *Sericornis rufescens* 10cm Forages for insects in small groups. Mixes with flocks of other insectivorous species. **V** Song comprises short bursts of whistling notes. Calls include an agitated *tcht-cht*. **H** Montane forest, secondary woodland. **D** NW New Guinea.

35 GREY-GREEN SCRUBWREN *Sericornis arfakianus* 9–10cm Diet presumed to comprise mainly insects. Sometimes seen in pairs or small groups. **V** Song an exuberant series of thin squeaky notes. Calls include a raspy *tchrr-rr*. **H** Hill forest, secondary woodland. **D** New Guinea.

36 WEEBILL *Smicrornis brevirostris* 8–10cm Diet comprises mainly invertebrates, sometimes seeds. **V** Song a series of shrill, whistled *t-Wheeo* or *t-Wheoee* phrases, upslurred and rising in pitch in middle of delivery. Calls include a soft *tchip-tchip*. **H** Wide range of wooded habitats. **D** Widespread across Australia.

1
2
3
9
♀
4
5
7
♂
8
♂
♀
13
6
10
12
11
18
♀
♂
15
17
16
14
19
♀
♂
21
22
23
20
26
♀
27
♀
24
25
♂
♂
♀
28
30
♀
♂
31
♂
29
32
33
34
35
36

AUSTRALASIAN WARBLERS *CONTINUED*

1 BROWN GERYGONE *Gerygone mouki* 9–11cm Diet comprises mainly insects. Usually feeds in pairs. **V** Song a series of whistled *tchr-s-s-s, tchr-s-s-s* phrases. Calls include a nasal *tchi-up*. **H** Primary rainforest, other wooded habitats outside breeding season. **D** Coastal E Australia (N Queensland to SE Victoria).

2 GREY GERYGONE *Gerygone igata* 10cm Small, greyish green, lacking eyebrow; distinctive tail pattern not easy to see in the field. **V** Cheerful, very high, warbling song, slightly descending, based on *turreweet* (in which *tur* low and *weet* very high pitched). **H** All types of native wooded habitats, from forest to shrublands and mangroves. **D** New Zealand.

3 NORFOLK GERYGONE *Gerygone modesta* 10–12cm Diet comprises mainly insects, gleaned from foliage. **V** Song a series of trilling, warbling notes, pitch descending during delivery. **H** Primary rainforest, dense secondary woodland, gardens. **D** Norfolk Island, SW Pacific Ocean.

4 CHATHAM GERYGONE *Gerygone albofrontata* 12cm Separated from Grey Gerygone by white or yellow sides of face, short eye-stripe and different range. **V** Short, very high, sharp series. *wutuwutwrwrwr*, last part a sharp rattle. **H** Dense forest and scrub. **D** Chatham Islands, SW Pacific Ocean.

5 FAN-TAILED GERYGONE *Gerygone flavolateralis* Forages for insects in pairs or small groups. Associates with mixed-species feeding flocks. **V** Song a series of tinkling *tink, te-ter-u* notes, tone recalling that of tree frog. **H** Primary forest, secondary woodland. **D** New Caledonia (including Loyalty Islands, Isles of Pines), Vanuatu (including Banks Islands).

6 BROWN-BREASTED GERYGONE *Gerygone ruficollis* Gleans insects from foliage, often in pairs or small groups. **V** Song a prolonged series of thin whistling notes, slurring into one another and slowly descending in pitch during delivery. Call is a thin, whistled *tsip*. **H** Upland primary forest, secondary woodland. **D** New Guinea.

7 GOLDEN-BELLIED GERYGONE *Gerygone sulphurea* 10–11cm Makes short aerial sorties to capture disturbed insects, also gleans insects from leaves in the tree canopy or mid-storey. **V** Various high-pitched whistles, a rising *chu-whee* and long series of wheezy *whiz* notes. **H** Mangroves, coastal scrub, freshwater swamp forest, plantations, parks, gardens. **D** Peninsular Malaysia; Greater and Lesser Sunda Islands; Philippines.

8 RUFOUS-SIDED GERYGONE *Gerygone dorsalis* 10cm Usually seen in pairs gleaning from foliage in the mid-storey and vine tangles. **V** Series of semi-trilled level-pitched notes, also jangle of unhurried, slightly warbled notes. **H** Coastal and lowland forest, secondary forest, forest edges, partially cleared areas, lightly wooded cultivation, mangroves. **D** Lesser Sunda Islands.

9 MANGROVE GERYGONE *Gerygone levigaster* 10–11cm Feeds mainly on insects, gleaned from foliage. **V** Song a series of shrill, whistled *toodlee, toodlee, toodlee*... notes or *tutee-dee, we-dee, wooh*. Calls include a soft, raspy *tchttt*. **H** Mangroves, coastal forest. **D** Coastal N and E Australia, S New Guinea.

10 PLAIN GERYGONE *Gerygone inornata* 10cm Active gleaner in mid-storey and lower canopy; often forms part of mixed-species feeding parties. **V** Melodious, descending series of 16–18 paired notes, *poo-pii, pee-pee, po-po, po-pu*, likened to rapid peal of bells. **H** Lowland and montane primary and secondary monsoon forest, woodland scrub, mangroves. **D** Lesser Sunda Islands.

11 WESTERN GERYGONE *Gerygone fusca* 9–11cm Diet comprises mainly invertebrates. **V** Song a series of slurred whistling notes, slurring into one another and modulating in pitch seemingly randomly during delivery. Calls include squeaky notes. **H** Dry, open woodland. **D** Widespread W Australia, inland C Australia, inland E Australia.

12 DUSKY GERYGONE *Gerygone tenebrosa* 10–12cm Diet comprises invertebrates, mainly insects. **V** Song comprises rather unenthusiastic squeaky whistles. Calls include various squeaky, chattering notes. **H** Mangrove habitats. **D** Coastal NW Australia.

13 LARGE-BILLED GERYGONE *Gerygone magnirostris* 10–12cm Diet comprises invertebrates. **V** Song includes chirping, chattering whistles, sometimes delivered in bursts; pitch of song's elements 'dance' up and down. **H** Mangroves, coastal forest. **D** N Australia (N Western Australia to E Queensland), New Guinea, W Papuan islands, Aru Islands.

14 BIAK GERYGONE *Gerygone hypoxantha* 10cm Diet presumed to comprised mainly insects and other invertebrates. Seen singly or in pairs. **V** Song comprises high-pitched whistles, the pitch of each modulating compared to previous note. **H** Forested habitats. **D** Biak and Supiori islands off NW New Guinea.

15 YELLOW-BELLIED GERYGONE *Gerygone chrysogaster* 10cm Feeds mainly on insects. **V** Song a series of shrill, piping whistles; *wher-wher, Tsi-Tsi-su-su-wer-wu*, pitch rising then falling stepwise during delivery. Calls include a shrill, trilled *wheurrrr* piping note. **H** Primary rainforest, lush secondary woodland, mainly lowlands. **D** New Guinea, W Papuan islands, Aru Islands.

16 GREEN-BACKED GERYGONE *Gerygone chloronota* 10–11cm Diet comprises invertebrates. **V** Song a series of trilling *teechu, teechu, teechu, teechu*... phrases. **H** Mangroves, eucalypt woodland, lush secondary woodland, gardens. **D** N Australia (N Western Australia, N Northern Territory), New Guinea, W Papuan islands, Aru Islands.

17 WHITE-THROATED GERYGONE *Gerygone olivacea* 10–12cm Feeds on invertebrates, mainly insects, gleaned from foliage. **V** Song includes rapid whistling *tsii, tsii* notes descending in pitch, followed by squelchy, buzzing chirps. **H** Wide range of open wooded habitats. **D** N, E and SE Australia.

18 FAIRY GERYGONE *Gerygone palpebrosa* 10–12cm Diet comprises invertebrates, mainly insects. Often joins mixed-species flocks. **V** Song comprises squeaky *tchiu, thipi-chu-ee* phrases. Calls include an agitated-sounding *tchtt*. **H** Forested clearings and margins. **D** NE Australia (NE Queensland), New Guinea, Aru Islands.

19 MOUNTAIN THORNBILL *Acanthiza katherina* 10–11cm Diet comprises invertebrates. Sometimes joins mixed-species flocks. **V** Song includes shrill, squeaky and squelchy whistles, and trilling phrases that descend in pitch during delivery. **H** Primary rainforest. **D** NE Australia (NE Queensland).

20 BROWN THORNBILL *Acanthiza pusilla* 10–11cm Diet comprises mainly invertebrates; seeds and fruit eaten occasionally. **V** Song comprises a few squeaky whistles followed by a vibrating, squeaky trill. Calls include a squeaky *tchttt*. **H** Heathland, open woodland with dense shrubby understorey, mature parks, gardens. **D** E and SE Australia, Tasmania, Bass Strait islands.

21 INLAND THORNBILL *Acanthiza apicalis* 10–11cm Feeds on invertebrates, mainly insects, gleaned from foliage. **V** Song comprises a *tsu-chuer, tsi-tsi*. Calls include grating, chattering *tchrrr* notes. **H** Wooded habitats with dense shrubby understorey. **D** S half of W Australia, interior E Australia.

22 TASMANIAN THORNBILL *Acanthiza ewingii* 10–11.5cm Diet comprises invertebrates, particularly insects. **V** Song comprises throaty, fluty notes and shrill whistles. Calls include grating, chattering *tchrr* notes. **H** Dense shrubby habitats including heathland and forest understorey. **D** Tasmania, Bass Strait islands.

23 NEW GUINEA THORNBILL *Acanthiza murina* 9–10cm Diet includes insects, nectar, fruits and seeds. Usually seen in small groups. **V** Song a shrill, disyllabic *Tchi-Tchup*. Calls include shrill, thin whistles. **H** Montane and subalpine forest; mainly above 2,000m. **D** New Guinea.

24 CHESTNUT-RUMPED THORNBILL *Acanthiza uropygialis* 10–11cm Diet comprises invertebrates, mainly insects, also occasionally seeds. **V** Song comprises thin, squeaky whistles, and grating warbling notes. Calls include thin whistled *tsee, tsee* notes. **H** Arid woodland with rich understorey, waterside vegetation. **D** Widespread across S half of Australia although absent from coastal SW, SE and E Australia.

25 BUFF-RUMPED THORNBILL *Acanthiza reguloides* 10–11cm Diet includes invertebrates, nectar and seeds. **V** Song a vibrating, whistling trill on one note. Calls include a thin, sharp *tssst*. **H** Open woodland with sparse understorey. **D** Widespread across E and SE Australia (Queensland to S Victoria and SE South Australia).

26 WESTERN THORNBILL *Acanthiza inornata* 8–11cm Feeds mainly on insects, gleaned and foraged from foliage and on ground. Often seen in pairs or small groups. **V** Song comprises squelchy shrill whistles and grating notes. Calls include a tinkling trill. **H** Heathland, open woodland with shrubby understorey, parks, gardens. **D** SW Australia.

27 SLENDER-BILLED THORNBILL *Acanthiza iredalei* 9–11cm Diet comprises mainly invertebrates, occasionally seeds as well. **V** Song and calls include trilling whistles and grating squeaks. **H** Salt marsh, chenopod shrubby habitats, heathland. **D** W and S Australia.

28 YELLOW-RUMPED THORNBILL *Acanthiza chrysorrhoa* 10–12cm Diet includes invertebrates and seeds. **V** Song comprises a whistled trill and squeaks. Calls include whistles and grating squeaks. **H** Shrubby habitats, open woodland with shrub understorey, parks, gardens. **D** Widespread across S half of Australia except arid E interior, Tasmania.

29 YELLOW THORNBILL *Acanthiza nana* 9–10cm Diet includes invertebrates, mainly insects, occasionally seeds. **V** Calls include grating notes and whistles, sometimes delivered in rapid trill. **H** Open woodland with shrubby understorey, mature parks, gardens, mangroves. **D** Widespread in E Australia.

30 GREY THORNBILL *Acanthiza cinerea* 9cm Feeds mainly on insects, caught by gleaning foliage and fly-catching. **V** Song a series of slurred, piercing whistling notes. Calls include buzzing *tzzzt* notes. **H** Montane forest. **D** New Guinea.

31 STRIATED THORNBILL *Acanthiza lineata* 9–11cm Diet comprises mainly invertebrates, occasionally seeds and manna. **V** Song comprises tuneful whistles. Calls include thin, metallic *tzzt-tzzt* or *tzzt, tzzt-tzzt*. **H** Open eucalypt woodland with shrubby understorey, mature parks, gardens. **D** E and SE Australia.

32 SLATY-BACKED THORNBILL *Acanthiza robustirostris* 10–11cm Diet comprises mainly insects. **V** Song a series of high-pitched squeaks delivered in a rapid, warbling manner. Calls include a thin, grating *tzzt* and piping *kvik*. **H** Arid scrub, dry open woodland, usually on bare rocky or stony ground. **D** Interior C Australia.

33 SOUTHERN WHITEFACE *Aphelocephala leucopsis* 11–12cm Feeds mainly on invertebrates. **V** Song comprises high-pitched, vibrating trilling whistles. Calls include a nasal *tvik-vik*. **H** Arid woodland with understorey of shrubs and grasses. **D** Widespread across S half of Australia although absent from SW and coastal SE and E.

34 CHESTNUT-BREASTED WHITEFACE *Aphelocephala pectoralis* 9–10cm Diet includes invertebrates and seeds. Often seen in pairs or small groups. **V** Song a whistled *tswee-swee-swee*. Calls include a piping, whistled *p-tu-tu-tu*. **H** Arid habitats with sparse shrubby vegetation, including gibber plains. **D** Restricted range in C South Australia.

35 BANDED WHITEFACE *Aphelocephala nigricincta* 9–10cm Diet includes insects and seeds. Usually seen in pairs or small flocks. **V** Song comprises slurred, whistling *su-see-see-see*. Calls include buzzing whistles. **H** Open, arid habitats with sparse grass and shrub vegetation. **D** C Australia.

AUSTRALASIAN BABBLERS POMATOSTOMIDAE

36 PAPUAN BABBLER *Garritornis isidorei* 24–25cm Diet presumed to include invertebrates and small vertebrates. **V** Song comprises a series of harsh, screeching *whoah, ch-ch-ch* phrases. Calls include a strident, screechy *tcha-tcha*. **H** Lowland primary forest, lush secondary woodland. **D** New Guinea, W Papuan islands.

37 GREY-CROWNED BABBLER *Pomatostomus temporalis* 24–26cm Forages on ground and in leaf litter for insects, other invertebrates and small vertebrates. **V** Song a series of slurred, screeching *whoo-e-ah, whoo-e-ah*... Calls include various strident screeching notes. **H** Open woodland. **D** NW, N NE and E Australia, S New Guinea.

38 HALL'S BABBLER *Pomatostomus halli* 20–21cm Presumed to feed mainly on insects, foraged and probed for on ground, under stones and behind bark. **V** Song comprises elements of calls including harsh, screeching notes and squeaky whistles. **H** Open woodland, often on barren ground with limited understorey. **D** Interior E Australia.

39 WHITE-BROWED BABBLER *Pomatostomus superciliosus* 18–22cm Varied diet includes insects, small vertebrates, seeds and fruits. Usually seen in small flocks. **V** Song comprises a series of piping notes, pitch descending during delivery. Calls include range of harsh screeches and croaks. **H** Open woodland with dense understorey of shrubs or heathland plants. **D** Widespread across S half of Australia, except coastal SE.

40 CHESTNUT-CROWNED BABBLER *Pomatostomus ruficeps* 20–23cm Diet comprises mainly insects but also fruits and seeds. Usually forages in small flocks. **V** Calls include a series of strident, piping whistles and trills comprising grating and chattering notes. **H** Open woodland with open shrub layer. **D** EC Australia.

1
2
3
4
5
6
7
8
9
10
11
12
13
14
15
16
17
♂
18
♀
19
20
21
22
23
24
25
26
27
28
29
30
31
32
33
34
35
36
37
38
39
40

LOGRUNNERS ORTHONYCHIDAE

1 PAPUAN LOGRUNNER *Orthonyx novaeguineae* 18–19cm Diet includes insects and other invertebrates, foraged for in leaf litter. **V** Song a series of jaunty, human-like whistles, pitch descending stepwise during delivery. Calls include elements of song. **H** Montane forest. **D** New Guinea.

2 AUSTRALIAN LOGRUNNER *Orthonyx temminckii* 19–21cm Feeds on insects and other invertebrates, foraged for by scratching and clearing leaf litter. **V** Song comprises triplets of screechy *whur-che, whur-che, whur-che* phrases. Calls include shrill squawks. **H** Lowland rainforest. **D** Coastal E Australia.

3 CHOWCHILLA *Orthonyx spaldingii* 25–28cm Diet includes invertebrates and small vertebrates, discovered by clearing leaf litter with feet. **V** Song a series of explosive, chattering chuckles and whistles, some notes with tone of electronic game; some renditions onomatopoeic. Calls include harsh squawks. **H** Primary rainforest, lush secondary woodland. **D** Coastal NE Australia (NE Queensland).

SATINBIRDS CNEMOPHILIDAE

4 LORIA'S SATINBIRD *Cnemophilus loriae* 21–22cm Diet comprises mainly fruits and berries; occasionally invertebrates. **V** Song comprises isolated piercing whistles. Calls include an upslurred whistled *tueerp*. **H** Montane forest, secondary woodland. **D** New Guinea.

5 CRESTED SATINBIRD *Cnemophilus macgregorii* 24cm Diet comprises fruits and berries. Usually solitary except where feeding is good. **V** Calls include harsh screeches and squawks. **H** Montane and subalpine forest, secondary woodland. **D** New Guinea.

6 YELLOW-BREASTED SATINBIRD *Loboparadisea sericea* 17–18cm Diet presumed to include fruits and berries. Sometimes seen in groups where feeding is good. **V** Calls include coarse, whistled screeches. **H** Montane forest. **D** New Guinea.

BERRYPECKERS AND LONGBILLS MELANOCHARITIDAE

7 OBSCURE BERRYPECKER *Melanocharis arfakiana* 11–12cm Diet presumed to comprise mainly berries and fruits. Usually seen singly or in pairs. **V** Call is a silvery, fizzing trill. Song comprises more prolonged version of call. **H** Forested slopes. **D** New Guinea.

8 BLACK BERRYPECKER *Melanocharis nigra* 11–12cm Diet includes berries and invertebrates. **V** Song a series of nasal, grating chattering notes. Calls include screeches and wheezy trills. **H** Primary forest, secondary woodland, mainly lowlands. **D** New Guinea, W Papuan islands, Aru Islands.

9 MID-MOUNTAIN BERRYPECKER *Melanocharis longicauda* 12–13cm Feeds on berries and invertebrates. Usually seen singly. **V** Song a series of high-pitched, silvery notes. Calls include agitated-sounding, harsh *tchik-tchik-tchik...* notes. **H** Primary hill forest, secondary woodland. **D** New Guinea.

10 FAN-TAILED BERRYPECKER *Melanocharis versteri* 14–15cm Forages actively in foliage for berries and invertebrates. Usually seen singly or in pairs. **V** Calls include a grating *tchet* and nasal *tchee-tchee-tchee*. **H** Montane forest, secondary woodland. **D** New Guinea.

11 STREAKED BERRYPECKER *Melanocharis striativentris* 13–14cm Diet presumed to include berries and invertebrates. Unobtrusive and hard to observe well. **V** Poorly known. **H** Montane forest, secondary woodland. **D** New Guinea.

12 SPOTTED BERRYPECKER *Rhamphocharis crassirostris* 12–14cm Diet presumed to include berries, small fruits and invertebrates. **V** Poorly known. **H** Montane forest, secondary woodland. **D** New Guinea.

13 DWARF LONGBILL *Oedistoma iliolophus* 11cm Diet includes insects, nectar, fruits. Usually solitary and constantly active. **V** Calls include a rapid-fire, mechanical-sounding rattle. **H** Primary forest, secondary woodland, mainly lowlands. **D** New Guinea, W Papuan Islands.

14 PYGMY LONGBILL *Oedistoma pygmaeum* 7–8cm Diet includes small invertebrates and nectar. Sometimes seen in pairs or small groups where feeding is good. **V** Song a series of chattering chirps. Calls include a soft *tchik*. **H** Primary forest, secondary woodland. **D** New Guinea, D'Entrecasteaux Archipelago.

15 YELLOW-BELLIED LONGBILL *Toxorhamphus novaeguineae* 12–13cm Diet includes small invertebrates and nectar. **V** Song a series of 6–7 shrill whistles, pitch descending stepwise during delivery. Calls include a resonating, metallic *kvink*. **H** Lowland primary forest, secondary woodland. **D** New Guinea, W Papuan islands, Aru Islands.

16 SLATY-HEADED LONGBILL *Toxorhamphus poliopterus* 12–13cm Feeds actively on small invertebrates (mainly insects) and nectar. **V** Calls include a shrill whistled *tiuk* or softer *tseeip*. **H** Primary hill forest, secondary woodland. **D** New Guinea.

PAINTED BERRYPECKERS PARAMYTHIIDAE

17 TIT BERRYPECKER *Oreocharis arfaki* 12–14cm Feeds mainly on berries. Usually seen in pairs or small flocks. **V** Calls include a thin, high-pitched *tssssseeur* and other thin calls. **H** Montane forest. **D** New Guinea.

18 CRESTED BERRYPECKER *Paramythia montium* 20–22cm Diet comprises small fruits. Usually seen in pairs. **V** Calls include a nasal, squeaky *tsee* and dry-sounding, crackly rustle. **H** Montane forest, alpine scrub. **D** New Guinea.

NEW ZEALAND WATTLEBIRDS CALLAEIDAE

19 NORTH ISLAND KOKAKO *Callaeas wilsoni* 37–39cm Near threatened due to habitat destruction. Recovering thanks to conservation efforts (predator control) and relocation projects. Diet comprises mainly plant material. **V** Calls and song include peculiar, discordant sounds, like badly played clarinet or saxophone notes. **H** Forest and wooded scrub. **D** North Island, New Zealand.

20 SOUTH ISLAND KOKAKO *Callaeas cinereus* 38cm No similar bird in New Zealand. Wattle colour varies by race. **V** Beautiful, calm stream of well-separated, loud, fluted notes of varying pitch. **H** Mainly in forests up to 900m. **D** South Island and Stewart Island in New Zealand.

21 NORTH ISLAND SADDLEBACK *Philesturnus rufusater* 24–26cm Near threatened due to habitat destruction and mammalian predators. Now mainly on predator-free offshore islands where relocated. Diet includes invertebrates and berries. **V** Song comprises a rapid-fire squeaky, penetrating *we-tu, tu-tu-tu-tu*. Calls include elements of song. **H** Forest habitats. **D** New Zealand (mainly islands off North Island).

22 SOUTH ISLAND SADDLEBACK *Philesturnus carunculatus* 25cm No similar bird in its range. **V** Nasal, shivering and chattering series, or high-pitched, sneezing *tseeh* notes. **H** Forest, coastal and montane shrubland. **D** South Island, New Zealand.

STITCHBIRD NOTIOMYSTIDAE

23 STITCHBIRD *Notiomystis cincta* 18cm Distinctively patterned and coloured. Note white in wings. **V** Call a rapid *tsee-tih*. **H** Prefers dense native forest. **D** New Zealand and several offshore islands.

WHIPBIRDS, JEWEL-BABBLERS AND QUAIL-THRUSHES PSOPHODIDAE

24 PAPUAN WHIPBIRD *Androphobus viridis* 16–17cm Diet comprises invertebrates, mainly insects, foraged for at ground level and in leaf litter. **V** Calls include shrill, screechy series of *wiik-wiik-wiik* notes. **H** Montane forest. **D** New Guinea.

25 EASTERN WHIPBIRD *Psophodes olivaceus* 26–30cm Diet comprises mainly invertebrates, occasionally small vertebrates. **V** Song a piping whistled *te-te-te*, rising in volume, ending in a whiplash *whuiiP* or *whuiiP, tchu-tchu*. **H** Forest habitats with dense understorey. **D** E and SE Australia.

26 BLACK-THROATED WHIPBIRD *Psophodes nigrogularis* 20–25cm Very similar to White-bellied Whipbird but has darker belly. Feeds mainly on invertebrates, foraged for on ground and in low vegetation. **V** Song a variation on a grating, screechy *tuer, tche-hu-te-te*. **H** Wooded and scrub habitats with dense understorey. **D** SW Western Australia.

27 WHITE-BELLIED WHIPBIRD *Psophodes leucogaster* 20–25cm Previously treated as conspecific with Black-throated Whipbird but has paler belly. Diet comprises invertebrates. **V** Song a mix of thrush-like notes, cheeps and vibrating whistles. **H** Wooded and scrub habitats with dense understorey. **D** SE South Australia.

28 CHIRRUPING WEDGEBILL *Psophodes cristatus* 19–20cm Diet comprises mainly insects and seeds. Usually seen singly or in pairs. **V** Song a chirruping, rasping *ti-tchue, ti-tchue, ti-tchue...* Calls include trilling chirrups. **H** Open shrubby habitats. **D** Interior E Australia.

29 CHIMING WEDGEBILL *Psophodes occidentalis* 20–22cm Diet includes invertebrates, mainly insects, and seeds. **V** Song a squeaky, wheezy *teh, ti-tu-er*. Calls include a soft piping *tchuu*. **H** Open, arid scrub. **D** Interior C and W Australia.

30 SPOTTED JEWEL-BABBLER *Ptilorrhoa leucosticta* 19–20cm Varied diet includes invertebrates, small vertebrates, fallen fruit. **V** Song a series of rapid, shrill and piping whistles, slurred into one another, increasing in volume throughout and ending abruptly. Calls include an abrupt whistled *kwik*. **H** Montane forest. **D** New Guinea.

31 BLUE JEWEL-BABBLER *Ptilorrhoa caerulescens* 21–22cm Feeds mainly on insects in leaf litter and forest debris. **V** Song a series of shrill, piping whistles, slurred into one another, increasing in volume throughout. Calls include a thin, high-pitched *tssiu*. **H** Primary lowland rainforest, lush secondary woodland. **D** New Guinea, W Papuan islands.

32 BROWN-HEADED JEWEL-BABBLER *Ptilorrhoa geislerorum* 22–23cm Previously treated as conspecific with Blue Jewel-babbler. Diet includes invertebrates and small vertebrates. **V** Song a series of shrill, piping *tssiu, tssiu, tsiuu...* whistles, each note distinct not slurred. **H** Forest habitats. **D** E New Guinea.

33 CHESTNUT-BACKED JEWEL-BABBLER *Ptilorrhoa castanonota* 22–24cm Turns leaf litter in search of invertebrates, mainly insects, and small vertebrates. **V** Song a tongue-smacking *Tsi, tChuk-tchuk*. Calls include a shrill *tch-tch*. **H** Hill forest. **D** New Guinea.

34 SPOTTED QUAIL-THRUSH *Cinclosoma punctatum* 24–30cm Diet comprises mainly insects; occasionally also small vertebrates and seeds. **V** Song a series of 14–15 shrill, whistled *Weik-Weik-Weik-Weik...* notes, gradually increasing in volume throughout. Calls include extremely high-pitched whistling *tsuee-tsee*. **H** Open woodland, often on stony ground. **D** E and SE Australia, Tasmania.

35 CHESTNUT QUAIL-THRUSH *Cinclosoma castanotum* 22–26cm Diet comprises mainly insects. **V** Song a piercing, vibrating and rapid-fire series of piping notes, gradually increasing in volume throughout. Calls include thin, high-pitched whistles. **H** Arid scrub habitats including mallee. **D** SE Australia.

36 COPPERBACK QUAIL-THRUSH *Cinclosoma clarum* 22–26cm Forages for insects, sometimes seeds, on ground. **V** Song a series a rapid-fire series of shrill, piping notes, gradually increasing in volume throughout and ending abruptly. Calls include thin, high-pitched whistles. **H** Arid scrub habitats including open woodland and mallee. **D** Scattered range across southern W and C S Australia.

37 CINNAMON QUAIL-THRUSH *Cinclosoma cinnamomeum* 20–22cm Diet includes invertebrates and seeds, foraged for and excavated on ground. **V** Song a series of 5–6 shrill, piping notes, gradually increasing in volume throughout. **H** Stony plains with sparse vegetation. **D** Interior C Australia.

38 NULLARBOR QUAIL-THRUSH *Cinclosoma alisteri* 19–20cm Forages for insects and seeds on ground, using bill to dig and excavate. **V** Song comprises subdued trilling phrases. Calls include a soft *tsiip*. **H** Stony plains with sparse vegetation. **D** Restricted to Nullarbor Plain (SE Western Australia, SW South Australia).

39 CHESTNUT-BREASTED QUAIL-THRUSH *Cinclosoma castaneothorax* 22–25cm Diet includes insects and seeds, forages on ground. **V** Song a piercing, whistled *tur, ter, lee* phrase, pitch rising stepwise during delivery. Calls include a vibrating trill, pitch descending during delivery. **H** Arid open woodland and scrub, both on stony ground. **D** Interior E Australia.

40 WESTERN QUAIL-THRUSH *Cinclosoma marginatum* 22–25cm Forages for insects and seeds on ground. **V** Song a piercing, vibrating 11- or 12-note whistle. Calls include squeaky, high-pitched whistles. **H** Arid open woodland and scrub, both on stony ground. **D** Interior W Australia.

41 PAINTED QUAIL-THRUSH *Cinclosoma ajax* 22–23cm Diet comprises mainly invertebrates. **V** Song a five-note, piercing whistle. Calls include repeated whip-cracking *uh-Whik, uh-Whik, uh-Whik...* phrases. **H** Primary rainforest, lush secondary woodland, mainly lowlands. **D** New Guinea.

1
2
3
4
5
6
7
8
9
10
11
12
13
14
15
16
17
18
19
20
21
22
23
24
25
26
27
28
29
30
31
32
33
34
35
36
37
38
39
40
41

BATISES AND WATTLE-EYES PLATYSTEIRIDAE

1 RWENZORI BATIS *Batis diops* 12cm No (or little) white in neck. Note white spot at lore, giving impression of second eye. Most similar to sympatric Ituri Batis, but larger and normally found at higher altitudes. **V** Series of simple whistles, repeated every few seconds. **H** Montane forests, bamboo. **D** E DR Congo, SW Uganda, W Rwanda, W Burundi.

2 MARGARET'S BATIS *Batis margaritae* 12cm Note very dark upperparts, without white in neck in male. Female has black breast-band and rufous wing-bar. **V** Call is a soft whistle, typically repeated many times. **H** Mid-strata of dry, dense forest and forest patches. **D** SC Africa.

3 FOREST BATIS *Batis mixta* 12cm Only males of Forest, Reichenow's and Dark Batis have red eyes (except some individuals of Woodward's Batis, which has different range). Female has predominantly brown upperparts. **V** Monotonous, level, slow series of *wuut*-notes. **H** Forest. **D** SE Kenya to NE Tanzania.

4 REICHENOW'S BATIS *Batis reichenowi* 9–10cm Previously treated as race of Forest Batis. Diet comprises mainly insects. **V** Poorly known but presumed to be similar to Forest Batis. **H** Lowland coastal forest. **D** SE Tanzania.

5 DARK BATIS *Batis crypta* 10cm Male not always safely separable from male Forest Batis but mantle is darker; female separated from female Forest Batis by less cinnamon wings. **V** Call a fluted *wuuh*. **H** Montane evergreen forest. **D** Eastern Arc mountains of Tanzania and NW Malawi.

6 CAPE BATIS *Batis capensis* 12cm Female separated from male Woodward's Batis by well-defined, white throat band. In male note buff-orange wing-bars (only male batis to show these within its range) and extensive white on wing coverts. **V** Mid-high, level, fluted, three-syllable *wuut-wuut-wuut*. **H** Forest, woodland, gardens. **D** S Africa.

7 WOODWARD'S BATIS *Batis fratrum* 12cm Note absence of breast-band in male and pale rufous underparts. Female separated from female Cape Batis by lack of white throat band. **V** Hesitant, slightly hoarse *ffuuuu*, sometimes in duet with female answering *woh* (four tones lower). **H** Forest, riverine belts, coastal bush. **D** S Malawi, C Mozambique, E Zimbabwe to E South Africa.

8 CHINSPOT BATIS *Batis molitor* 12cm Male (with white eyebrow extending well beyond the eye) separated from Pale Batis by broad breast-band. Note female's rich rufous breast and throat patch. **V** Very high, penetrating, sharp, slow, descending *it's so sad*. **H** Woodland. **D** C and S Africa.

9 SENEGAL BATIS *Batis senegalensis* 10cm The only batis in much of its range; female separated from other batises by different colouration. **V** Song is variable, sometimes brief and rather feeble, sometimes including repeated whistles and strong scolding notes. **H** All types of more or less wooded and bushy habitats, natural and cultivated. **D** S Mauritania to Sierra Leone and E to Niger, Chad and Cameroon.

10 GREY-HEADED BATIS *Batis orientalis* 10cm Very similar to the two black-headed batis species, but with slightly greyer crown and mantle. **V** Bell-like song resembles Chinspot Batis, but usually four notes rather than three. **H** Dry, more or less wooded areas. **D** C and NE Africa.

11 PALE BATIS *Batis soror* 11cm Note narrow breast-band of male. Female (with partly rufous eyebrow) separated from female Chinspot Batis by paler, rufous breast and throat patch. **V** Sustained, staccato, tooting *tioh-tioh-tioh* (*ti* note much higher but almost simultaneous with *oh*). **H** Woodland. **D** SE Kenya to E Zimbabwe and S Mozambique.

12 PRIRIT BATIS *Batis pririt* 12cm No other batis species in most of its range. Male separated from male Chinspot Batis by spots high on flanks, white eyebrow extending just beyond eye, different call; female by different colour pattern below. **V** High, slowly descending *djuu-djuu...* (28 times for a duration of 18 secs). **H** Thornveld. **D** SW Africa.

13 EASTERN BLACK-HEADED BATIS *Batis minor* 10cm Separated from most other batises in its range by black crown and mantle. **V** Long piping high-pitched song is slower than that of Western Black-headed Batis. Calls include a distinctive *ploop ploop*. **H** Forest patches, woodland, riverine belts. **D** S Somalia, SE Kenya and E Tanzania.

14 WESTERN BLACK-HEADED BATIS *Batis erlangeri* 10cm Separated from most other batises in its range by black crown and mantle. **V** Song is a monotonous penetrating whistle. **H** Forest patches, woodland, riverine belts. **D** Ethiopian plateau and Somalia W to Cameroon and Angola.

15 PYGMY BATIS *Batis perkeo* 8cm Note small size, dark brown flight feathers and short white eyebrow. **V** Song consists of up to 20 penetrating piping notes. Calls include harsh *churr* and weak *peep*. **H** Dry, wooded and bushed areas. **D** SE South Sudan, SE Ethiopia and Somalia to N Tanzania.

16 ANGOLAN BATIS *Batis minulla* 8cm Separated from other batises in its range by tiny size and limited white above lores. **V** A descending scale of weak, high-pitched pure notes. **H** Forest edges. **D** E Gabon, S Congo, W DR Congo, W Angola.

17 GABON BATIS *Batis minima* 8cm Similar to Fernando Po Batis, but female has grey, not chestnut breast-band. **V** Song is a series of thin short notes. **H** Forest edges and remnants. **D** S Cameroon, Gabon.

18 ITURI BATIS *Batis ituriensis* 9cm Note absence of grey colouring. Female has black chest collar. The only batis in its habitat and range. **V** Unknown. **H** Forest canopies. **D** DR Congo, W Uganda.

19 FERNANDO PO BATIS *Batis poensis* 9cm Only batis within its range, except Gabon Batis, which see. **V** Unknown, but may give quiet rasping notes. **H** All habitats with tall trees. **D** W and WC Africa.

20 WHITE-TAILED SHRIKE *Lanioturdus torquatus* 15cm No similar bird in range. Seen in parties of up to 20 on the ground. **V** Loud, hoarse shrieks. **H** Woodland, thornveld. **D** Angola and C Namibia.

21 WEST AFRICAN WATTLE-EYE *Platysteira hormophora* 10cm Diet presumed to comprise mainly insects. Sometimes joins mixed-species feeding flocks. **V** Song a series of piercing whistles, on average two per sec. Calls include a soft *pink* note. **H** Primary rainforest, lush secondary woodland, mainly lowlands. **D** Sierra Leone to Ghana.

22 CHESTNUT WATTLE-EYE *Platysteira castanea* 10cm See White-spotted Wattle-eye. **V** Song variable, but typically includes rather deep *gok, gok* or repeated monotonous *tuck* notes. **H** Forest undergrowth. **D** Nigeria to S Sudan, W Kenya and N Zambia.

23 WHITE-SPOTTED WATTLE-EYE *Platysteira tonsa* 9cm Note white rump and short tail. Separated from Chestnut Wattle-eye by full white collar (male only), higher eye wattle with white spot behind eye. **V** Song is a series of whistles, harsh churrs, grunts and explosive notes. **H** Forest canopy. **D** S Nigeria to Gabon, Congo and EC DR Congo.

24 BANDED WATTLE-EYE *Platysteira laticincta* 12–13cm Endangered. Feeds actively on insects. **V** Song a series of harsh chattering notes, speed increasing during delivery. Calls include a grating *tchak*. **H** Primary montane forest, lush secondary woodland. **D** Restricted to W Cameroon (Bamenda Highlands).

25 BLACK-THROATED WATTLE-EYE *Platysteira peltata* 13cm Separated from Brown-throated Wattle-eye by all-black wings. **V** High sharp lashing tit-like *tsja-tsja-weet-o-weet-o-weet-o...* **H** More or less wooded and bushed areas, including gardens. **D** S, E and SE Africa.

26 WHITE-FRONTED WATTLE-EYE *Platysteira albifrons* 11cm Paler than other wattle-eyes; female with completely white front. **V** Song unknown; calls include a rasping *chrr chrr*. **H** Thickets, riverine belts. **D** W Angola and SW DR Congo.

27 BROWN-THROATED WATTLE-EYE *Platysteira cyanea* 13cm Note red eye wattle, white wing-bar and (in female) deep-brown throat. **V** Very high, simple melody of sharp well-separated notes *fee-fee-fee-wee-wee-fee*. **H** Mangroves, forest edges, woodland, bush. **D** W and C Africa.

28 YELLOW-BELLIED WATTLE-EYE *Platysteira concreta* 9cm Races vary in colour pattern to underparts. Male may have black breast spot. **V** High fluted *whit-whit-tjuu-tjuu whip* (descending to lashed-up *whip*). **H** Forest. **D** W and C Africa.

29 RED-CHEEKED WATTLE-EYE *Platysteira blissetti* 9cm Male glossy and bottle-green, not brown as female. Separated from Jameson's Wattle-eye by wider brighter chestnut throat sides. **V** Song is a long series of ringing notes on an even pitch, three per sec, *peep-peep-peep-peep...* **H** Forest understorey. **D** Guinea and Sierra Leone to C Nigeria and NE Cameroon.

30 BLACK-NECKED WATTLE-EYE *Platysteira chalybea* 9cm Has striking blue eye wattle like Red-cheeked Wattle-eye and Jameson's Wattle-eye, but lacks chestnut to face sides (male) or throat (female). **V** Song includes a repeated *swee swee, swu swu*. **H** Forest understorey. **D** C Cameroon and Central African Republic to Gabon, Congo and WC Angola.

31 JAMESON'S WATTLE-EYE *Platysteira jamesoni* 9cm See Red-cheeked Wattle-eye and Black-necked Wattle-eye. **V** Song is a series of high-pitched whistles, *hee-hee-hee-hee...* Calls include a quiet *chawuk-chawuk-chawuk* and a soft *wuk, wuk, wuk*. **H** Forest understorey. **D** NE DR Congo and S South Sudan to Burundi and W Kenya.

1
2
3
4
5
6
7
8
9
10
11
12
13
14
15
16
17
18
19
20
21
22
23
24
25
26
27
28
29
30
31

BUSHSHRIKES MALACONOTIDAE

1 FIERY-BREASTED BUSHSHRIKE *Malaconotus cruentus* 25cm Note white around eye. No wing-bars visible in folded wing. Birds in SW Cameroon may have yellow underparts. **V** Song is a series of evenly spaced, resonant *hoh* notes. Calls include a sharp *krrriieek*. **H** Canopies and mid-strata of forests. **D** Guinea and Sierra Leone to NE DR Congo, W Uganda; south to Gabon, N Congo and W DR Congo.

2 MONTEIRO'S BUSHSHRIKE *Malaconotus monteiri* 22cm Separated from Grey-headed Bushshrike by white circle around eye, larger bill and pale blue (not yellow) eyes; separated from Fiery-breasted Bushshrike by wing-bars. **V** Little information, but a mournful whistle has been recorded. **H** Dense foliage of woodland. **D** SW Cameroon, NW Angola.

3 GREY-HEADED BUSHSHRIKE *Malaconotus blanchoti* 23cm Separated from Monteiro's Bushshrike by white restricted to lores, and yellow eye. **V** Mid-high fluted *fweeee fweeee fweeeep* (each note adding to slight crescendo). **H** Riverine belts in woodland. **D** Widespread in Africa.

4 LAGDEN'S BUSHSHRIKE *Malaconotus lagdeni* 20cm Note black, yellow and green wing pattern. **V** Song is a mournful, quavering, drawn-out *whoooooo* or *uuuuuuuh*, or a disyllabic *whooo-up*. **H** Montane forests. **D** W and C Africa.

5 GREEN-BREASTED BUSHSHRIKE *Malaconotus gladiator* 23cm Overall dark grey and olive green. **V** Song consists of 3–10 drawn-out, mournful, low-pitched whistles, audible from a great distance. **H** Montane forest. **D** SE Nigeria, W Cameroon.

6 ULUGURU BUSHSHRIKE *Malaconotus alius* 20cm Large, with all-dark crown and cheeks. **V** Distinctive series of strident, far-carrying, low-pitched disyllabic whistles, *huw-teew* or *wu-chiew*, second note higher and downslurred. **H** Forest canopies. **D** C Tanzania.

7 MOUNT KUPE BUSHSHRIKE *Chlorophoneus kupeensis* 18cm No similar bird in its restricted range and habitat. **V** Song is a loud, brief chatter, *tchec, tchec, kh-kh-kh*. **H** Montane forest. **D** SE Nigeria, SW Cameroon.

8 MANY-COLORED BUSHSHRIKE *Chlorophoneus multicolor* 18cm Occurs in different colour forms. Shows white above eye in most forms. **V** Song is a short, hollow whistle, *hoh* or *whoup*, or a disyllabic *huwo*, repeated at same pitch every few seconds. **H** Forest canopy and mid-strata. **D** W and C Africa.

9 BLACK-FRONTED BUSHSHRIKE *Chlorophoneus nigrifrons* 20cm Occurs in four colour forms. **V** Low, melodious *pwop* followed by miaowing, swept-up *twueet*; together as *pwop twueet*. **H** Canopies and mid-strata of forests. **D** E Africa.

10 OLIVE BUSHSHRIKE *Chlorophoneus olivaceus* 18cm Occurs in two colour forms. Grey or olive head, with blackish ear-coverts in male. **V** Very variable, includes melodious, rapid, fluted *pipip púupúupúurp*. **H** From coastal forests to montane shrub. **D** SE Africa.

11 BOCAGE'S BUSHSHRIKE *Chlorophoneus bocagei* 15cm Note pied plumage, not unlike that of puffbacks, but with different head shape. **V** Song variable, consisting of sharp ringing notes, e.g. *tp-tooi-tweeee, tewp-tewp-tewp, wewewewit* or *tyo-tyo-tyo-tyo-tyo...* **H** Areas with tall trees, including forests and gardens. **D** C Africa.

12 ORANGE-BREASTED BUSHSHRIKE *Chlorophoneus sulfureopectus* 17cm Note yellow eye stripe. **V** Fluted *fututùtèèh*, varied in pitch and speed. **H** Dense canopy and mid-strata at forest edges, dense parts of wooded and bushed areas. **D** Widespread in Africa.

13 GORGEOUS BUSHSHRIKE *Telophorus viridis* 18cm Separated from Doherty's Bushshrike by different colour of underparts. Female has black breast-band, narrower than male's, and green tail. **V** Duet a low, fluid, fast *pu-pu-pu-Wee*. **H** Dense forest undergrowth near streams. **D** W Africa to SE Kenya, E Tanzania S to South Africa and Swaziland.

14 DOHERTY'S BUSHSHRIKE *Telophorus dohertyi* 18cm From Gorgeous Bushshrike by uniform yellow lower breast and belly. Rare morph exists in which red at head is replaced by yellow. **V** Song variable, consisting of short phrases of loud piercing fluty whistles or liquid notes. **H** Thick undergrowth at forest edge. **D** E DR Congo, SW Uganda, Rwanda, Burundi, NW Tanzania and W Kenya.

15 BOKMAKIERIE *Telophorus zeylonus* 23cm Black-bordered bib diagnostic. Less shy than other bush shrikes. **V** Duets, for example high, rapid, liquid *wik-wik-wik-wik-wik* from male answered by very high, hurried *tree-tree-tree* from female. **H** Grassland and rocky areas with some scrub and bush. Often in gardens. **D** S Africa.

16 ROSY-PATCHED BUSHSHRIKE *Telophorus cruentus* 23cm Mainly terrestrial, although often perches on bush tops. Usually located by song. **V** Monotonous, duetted *twee-u-twee-u-twee-u...* or *tswee-ur tswee-ur tswee-ur...* **H** Acacia and thorn scrub in arid regions. **D** NE and SC Africa.

17 MARSH TCHAGRA *Bocagia minuta* 17cm Note black cap of male, split by white eyebrow in female; female differs from Black-crowned Tchagra by buff, not grey, underparts. **V** Includes high, very loud, rapid *weet-weet-weet*. **H** Tall grass and herbage near streams and in swamps. **D** W and C Africa.

18 BROWN-CROWNED TCHAGRA *Tchagra australis* 18cm Separated from Southern Tchagra by mainly buff eyebrow, bordered below and above by black line, and by row of dusky, triangular spots along edge of mantle. Normally forages on ground. **V** Includes loud, resounding, descending *weweetree-tree-tree tree*. **H** Open bush with dense thickets. **D** Widespread in Africa.

19 THREE-STREAKED TCHAGRA *Tchagra jamesi* 15cm Note narrow black crown-stripe between wide eyebrows. Mantle has no black edging. **V** Song, in display flight, is a series of downslurred whistles descending the scale, *wi-weo-weo-weo-weo* or *tui-tui-tui-tui-tui*. Typical call is a harsh *churr*. **H** Dry bushland. **D** NE Africa.

20 SOUTHERN TCHAGRA *Tchagra tchagra* 20cm Mainly white eyebrow, unbordered or very narrowly bordered with black above. **V** Very loud, sustained, staccato flutes; *piu-piu-piu-piu*. **H** Dense, coastal thornbush, riverine scrub, other thick cover. **D** S Africa.

21 BLACK-CROWNED TCHAGRA *Tchagra senegalus* 20cm Note black crown and grey underparts. **V** Mid-high, loud, unhurried, human-like whistles of male, interrupted by cackling laughter of female. **H** Dry, bushed and wooded areas with dense, weedy and thorny cover. **D** Widespread in Africa and Arabian Peninsula.

22 SABINE'S PUFFBACK *Dryoscopus sabini* 18cm Male separated from male Red-eyed Puffback by dark eyes and heavy bill; female separated from female Pink-footed Puffback by almost white legs and heavier bill. **V** Calm series of *fuut* notes, given rapidly, level or descending. **H** Forest edge and other areas with trees. **D** W and C Africa.

23 PINK-FOOTED PUFFBACK *Dryoscopus angolensis* 15cm Unmistakable when pink legs visible; female also separated from female Sabine's Puffback by less heavy bill. **V** Rather quiet. Utters a variety of harsh, clipped calls. **H** Forest. **D** C and E Africa.

24 RED-EYED PUFFBACK *Dryoscopus senegalensis* 16cm Note red eye of male and almost white underparts of female. **V** Explosive and frequently repeated *chyow* or *ptiou*, or a monotonous *chew-chew-chew...* **H** Open forest. **D** Nigeria to Central African Republic, SW South Sudan and W Uganda south to N Angola and E DR Congo.

25 BLACK-BACKED PUFFBACK *Dryoscopus cubla* 16cm Separated from Brubru (Plate 171) by larger size, red eye and all-black mantle. **V** Includes loud lashing whistles; rapid *weet-tee-weet-tee-weet-tee* (*tee* low and almost toneless). **H** Forest, riverine belts, woodland. **D** C, E and S Africa.

26 NORTHERN PUFFBACK *Dryoscopus gambensis* 18cm Male separated from Black-backed Puffback by grey rump. Note buff edges to wing feathers of female. **V** Unstructured, rapid chattering and series of staccato *djub* notes. **H** Forest edges, woodland, suburban gardens. **D** W and C Africa.

27 PRINGLE'S PUFFBACK *Dryoscopus pringlii* 13cm Note white belly and brown wings; also black crown. **V** Song is short, low-pitched, harsh and monotonous. Calls include a rasping, nasal *cheee-tzrrrr*. **H** Dry woodland and bushland. **D** S Ethiopia and Somalia to NE Tanzania.

1
2
3
4
5
6
7
8
8
8
8
9
9
9
9
4 colour forms
10
10
ruddy morph
olive morph
4 colour forms
11
12
13
14
15
16
17
18
19
20
21
22
23
24
25
26
27

BUSHSHRIKES *CONTINUED*

1 LOWLAND SOOTY BOUBOU *Laniarius leucorhynchus* 21cm Deep pure black. Often fluffs up its rump feathers. **V** Song consists of a series of low, almost pure hollow whistles, *ooooo, ooooo*. **H** Forest undergrowth. **D** Guinea and Sierra Leone to Ghana, and SE Cameroon to Uganda, C DR Congo and N Angola.

2 MOUNTAIN SOOTY BOUBOU *Laniarius poensis* 19cm Deep slaty black. Separated from Lowland Sooty Boubou by more bluish plumage and different range. **V** Duet of high fluid *sht-whit trrrril* (male), directly answered by high lashing *puwheet* (female). **H** Forest undergrowth, bamboo. **D** SE Nigeria, SW Cameroon, Bioko (Gulf of Guinea).

3 WILLARD'S SOOTY BOUBOU *Laniarius willardi* 19cm Deep pure black. Separated from Mountain Sooty Boubou by pale blue eyes and preference for lower altitudes. **V** Song pure *hoo-hoo-hoo-hoo*. Calls include harsh grating sounds. **H** Forest undergrowth in highlands. **D** Albertine Rift in Uganda, Burundi.

4 FÜLLEBORN'S BOUBOU *Laniarius fuelleborni* 19cm Deep slate-black. Separated from Willard's Sooty Boubou by different range. **V** Duet of high, fluid *sht-whit trrrril* (male), directly answered by high, lashing *ppWHéet* (female). **H** Forest undergrowth, bamboo. **D** EC Africa.

5 SLATE-COLORED BOUBOU *Laniarius funebris* 20cm Note silvery sooty appearance, deepest black in face. **V** Duet; low, running down, fast (almost trilling) *cierookookookoo* immediately answered with high *sweep*. **H** Thickets and undergrowth of dry, more or less wooded and bushed natural and cultivated habitats. **D** C Ethiopia and Somalia to Uganda and C Tanzania.

6 LÜHDER'S BUSHSHRIKE *Laniarius luehderi* 19cm Unmistak able. **V** Male's soft *prrrouu, prrrouu* is answered with a dry *k-k-k-k* by the female. **H** Dense undergrowth at forest edges, often near streams. **D** SE Nigeria and Cameroon to W DR Congo, E DR Congo to S South Sudan, W Kenya and C Tanzania.

7 BRAUN'S BUSHSHRIKE *Laniarius brauni* 18cm Brown crown colour a different shade to rufous breast. **V** Little known; calls include a nasal twang. **H** Forest undergrowth. **D** NW Angola.

8 GABELA BUSHSHRIKE *Laniarius amboimensis* 18cm Note all-white underparts; compare with Lühder's Bushshrike and Braun's Bushshrike. **V** Resembles Lühder's Bushshrike. Calls include a deep, guttural *worrrk* and a dry rattle. **H** Undergrowth of primary and secondary forest. **D** WC Angola.

9 RED-NAPED BUSHSHRIKE *Laniarius ruficeps* 15cm Hind crown bright fiery orange. **V** Solo song is a repeated low-pitched *kwoi* or *whooi*. In duet, a low-pitched *cheeo-oo* or guttural *gwaaar* is answered by a loud ticking, *gwaaar-tktktktktktktk*. **H** Ground and undergrowth of bush and shrubland. **D** NE Africa.

10 BLACK BOUBOU *Laniarius nigerrimus* 20cm All-black plumage, glossed greenish. **V** Various loud ringing calls, including repeated *wee-oo* and loud double *werk-werk*. **H** Thickets and woodland. **D** S Somalia and N coastal Kenya.

11 ETHIOPIAN BOUBOU *Laniarius aethiopicus* 25cm Secretive, black-and-white boubou with pink tinge to underparts and long white stripe on wing. **V** Duets; sounds like pure, hollow, piping whistle *puuu-wée*. **H** Forest undergrowth in highlands and lowlands, woodland and bushland, gardens, parks. **D** Eritrea, Ethiopia, NW Somalia, N Kenya.

12 TROPICAL BOUBOU *Laniarius major* 24cm Separated from Swamp Boubou by pale pinkish wash to underparts, and different habitat. **V** Varied duets, including raucous introduction of male answered by piping flutes from female, together *graagraagraa-wutwut*. **H** Forest edges, riverine belts, thickets, gardens. **D** W and C Africa south to S Zambia and N and C Mozambique.

13 EAST COAST BOUBOU *Laniarius sublacteus* 23cm Similar to Ethiopian Boubou but with no white in wing. **V** High, drawn-out, fluty *woo-wee-woo*, repeated frequently. **H** Coastal forest undergrowth and scrub, mainly in lowlands. **D** S Somalia, Kenya, NE Tanzania.

14 SOUTHERN BOUBOU *Laniarius ferrugineus* 22cm Separated from Tropical Boubou and Swamp Boubou by orange-rufous underparts (and little overlap in range). Skulking. **V** Varied duets with low, hollow, pure whistles from male answered by short, melodious flutes from female; *weep-weep-weep-fiufiufiu*. **H** Low down in dry bush, forest undergrowth. **D** S Africa.

15 SWAMP BOUBOU *Laniarius bicolor* 24cm Pure white below. **V** Magpie-like rattle *tjetjetjetjit*, immediately answered by a hollow pressed-out *fiuu*. **H** Papyrus, reedbeds, riverine thickets. **D** WC Africa.

16 TURATI'S BOUBOU *Laniarius turatii* 18cm Note white mantle spots and pink wash to underparts. **V** In synchronous duet, male utters a pure, soft, hollow whistle or a resonant metallic *hoo, hoou-hoou*, and female a nasal *dizhizhizhaaaah* or strident clicks. **H** High dense bush, woodland, forest edges. **D** Guinea-Bissau to Sierra Leone.

17 YELLOW-CROWNED GONOLEK *Laniarius barbarus* 18cm Unmistakable. **V** In duet, a pure, resonant, liquid *whee-ou* or *kweeho* is answered by a short sharp *ki-kik*. **H** Dense undergrowth in woodland and riverine bush, gardens. **D** W Africa.

18 PAPYRUS GONOLEK *Laniarius mufumbiri* 20cm Separated from Yellow-crowned Gonolek by white eye, wing-bar and undertail coverts. **V** In duet, a hollow, gong-like *pyo-pyop* is answered with a grating *zeetu* or buzzy *tzrrrr*. **H** Papyrus swamp. **D** Uganda to W Kenya, E DR Congo, Rwanda and Burundi.

19 BLACK-HEADED GONOLEK *Laniarius erythrogaster* 20cm Note pale eye and black crown. **V** Very high, pure, fluted *fee-fju-fee-ju*. **H** Dense parts of bush. **D** E Nigeria and N Cameroon to Eritrea, W Ethiopia, W Kenya, N Tanzania and E DR Congo.

20 CRIMSON-BREASTED SHRIKE *Laniarius atrococcineus* 23cm Underside vivid crimson or more rarely yellow. **V** Duets, for example *sree... djap-djap*. **H** Dry thornveld. **D** Angola, Zambia and Zimbabwe to C South Africa.

21 YELLOW-BREASTED BOUBOU *Laniarius atroflavus* 18cm No other similar bird in its range. **V** Noisy, with an extensive repertoire of whistling, swishing, rattling, harsh, explosive and stuttering notes; also sings in duet. **H** Dense undergrowth at forest edge in mountains. **D** SE Nigeria, W Cameroon.

22 BRUBRU *Nilaus afer* 13cm Some races show white eyebrow and rufous on flanks. **V** Very high, inquiring, short trill, *prrrrnu*. **H** Miombo, thornveld. **D** Widespread in Africa.

BOATBILLS MACHAERIRHYNCHIDAE

23 YELLOW-BREASTED BOATBILL *Machaerirhynchus flaviventer* 12cm Broad, flattened bill distinctive. Longish tail often held cocked. Crown and upperparts black in male, with two white wing-bars. Mainly yellow below. Female olive above with yellow supercilium. **V** Buzzing *chee-choo* calls and trilled variants. **H** Rainforest and secondary growth in lowlands and hills. **D** New Guinea, NE Australia.

24 BLACK-BREASTED BOATBILL *Machaerirhynchus nigripectus* 14cm Similar to Yellow-breasted Boatbill but larger and with black patch on breast in both sexes. **V** Series of descending *swee* notes. **H** Montane forest, secondary growth. **D** New Guinea.

VANGAS AND ALLIES VANGIDAE

25 RED-TAILED VANGA *Calicalicus madagascariensis* 14cm Separated from Red-shouldered Vanga by all dark eyes and brown, not rufous greater coverts. **V** Loud, level *wit-weet* or *wit-weet-weet*. **H** Primary forest. **D** Madagascar.

26 RED-SHOULDERED VANGA *Calicalicus rufocarpalis* 15cm See Red-tailed Vanga. **V** Short series of stutters, directly followed by drawn-out *weeeh*. **H** Dense *Euphorbia* scrub, low forest. **D** Madagascar.

27 HOOK-BILLED VANGA *Vanga curvirostris* 27cm Note long bill, grey-based tail and black and white pattern on wings. **V** Calm series of fluted notes. **H** Forest. **D** Madagascar.

28 BERNIER'S VANGA *Oriolia bernieri* 23cm Male has distinctive black plumage, white eyes and pale bill; note barred, reddish-brown plumage of female. **V** Sharp, scratchy *sreet* in irregular series. **H** Primary forest. **D** Madagascar.

29 LAFRESNAYE'S VANGA *Xenopirostris xenopirostris* 24cm Differs from Van Dam's Vanga and Pollen's Vanga by range and paler upperparts. **V** Call a short, fluted shriek. **H** Spiny forest and thickets at lower altitudes. **D** Madagascar.

30 VAN DAM'S VANGA *Xenopirostris damii* 23cm Note almost black upperparts and dark bill. **V** Call is downslurred *tjuw*. **H** Dry forest with little undergrowth at lower altitudes. **D** Madagascar.

31 POLLEN'S VANGA *Xenopirostris polleni* 24cm Note black throat of male and orange tinge to underparts of female. **V** Call is sweeping *tjuw* or nasal twitter. **H** Mainly primary forest. **D** Madagascar.

1
2
3
4
5
6
7
8
9
14
12
13
11
10
19
16
18
17
15
22
23
21
yellow form
20
normal form
28
24
25
29
30
26
27
31

VANGAS AND ALLIES *CONTINUED*

1 SICKLE-BILLED VANGA *Falculea palliata* 32cm Unmistakable, with long, curved bill. **V** Various hoarse shrieks. **H** Any habitat with trees. **D** Madagascar.

2 WHITE-HEADED VANGA *Artamella viridis* 20cm Unmistakable because of all-white head. Bill short and straight. **V** Short series of fluted notes; *tjeeuw* or *wuuh*. **H** Mainly in primary forest and nearby secondary growth. **D** Madagascar.

3 CHABERT VANGA *Leptopterus chabert* 14cm Note black, not blue, upperparts and blue eye-ring. **V** Call a sharp *chit-chit-chit-sit*. **H** Any habitat with trees. **D** Madagascar.

4 BLUE VANGA *Cyanolanius madagascarinus* 16cm Unmistakable with blue upperparts and pale blue iris. **V** Call a sharp, grating *grit-grit-grit*... in irregular series. **H** Higher storeys of primary and secondary forest. **D** Madagascar.

5 RUFOUS VANGA *Schetba rufa* 20cm Distinctive with rufous upperparts and black hood (male) or cap (female). **V** Short, voiceless rattle. **H** Primary forest. **D** Madagascar.

6 HELMET VANGA *Euryceros prevostii* 29cm Unmistakable by very large, blue bill. **V** Song a fluted, descending series, accelerating to a trill. **H** Primary forest. **D** Madagascar.

7 TYLAS VANGA *Tylas eduardi* 21cm Compare with female Pollen's Vanga (Plate 171), which has deep, but compressed bill. **V** Calm series of upswept *wu-weet* notes. **H** Humid forest. **D** Madagascar.

8 NUTHATCH VANGA *Hypositta corallirostris* 13cm Unmistakable by small size, bluish plumage, pink bill and tree-climbing behaviour. **V** Sequence of quiet hissing or squeaking notes, and a quiet trill increasing in volume. **H** In or near primary forest up to high altitudes. **D** Madagascar.

9 DARK NEWTONIA *Newtonia amphichroa* 12cm Browner upperparts than Common Newtonia. **V** Warbling sharp series of *fee-der-oh*..., 2–4 times, concluded by *weet*. **H** Understoreys of humid forest up to mid-high altitudes. **D** Madagascar.

10 COMMON NEWTONIA *Newtonia brunneicauda* 12cm Best separated from Dark Newtonia by voice. **V** Strong, rapid, rhythmic *tic-tac-tic-tac*... (1 sec). **H** Most forest types. **D** Madagascar.

11 ARCHBOLD'S NEWTONIA *Newtonia archboldi* 12cm Separated from Common Newtonia by rufous on forehead and around eyes; legs almost black. **V** Strong, rapid *tjic-tjic-tjic-tjah-tuwee*. **H** Thorn scrub; occasionally in dry forest. **D** Madagascar.

12 RED-TAILED NEWTONIA *Newtonia fanovanae* 12cm Distinctive with rufous tail, orange in wings, grey head and pale lower mandible. Eyes black. **V** Strong, rapid *weet-weet-weet*... *wit* and variations thereof. **H** Canopy of humid forest. **D** Madagascar.

13 WARD'S FLYCATCHER *Pseudobias wardi* 15cm No similar bird in range. Compare to white form of long-tailed Malagasy Paradise Flycatcher (Plate 185). **V** Series of high, dry trills, interspersed with some short warbling. **H** Undisturbed forest, secondary forest, plantations. **D** Madagascar.

14 CROSSLEY'S VANGA *Mystacornis crossleyi* 16cm Distinctive because of head pattern, long bill and short tail. **V** Ethereal, fluted *wuh-weeeh*, last syllable crescendoing. **H** Forest floor. **D** Madagascar.

15 WHITE-CRESTED HELMETSHRIKE *Prionops plumatus* 21cm Several races occur, mainly differing in size of crest and extent of white in wing. Occurs in parties of up to 20. **V** Very varied calls, including buzzing, chanting, chattering, whistling, churring, clicking and growling notes. **H** Miombo, bush, thornbush. **D** Widespread in Africa.

16 GREY-CRESTED HELMETSHRIKE *Prionops poliolophus* 26cm Larger than White-crested Helmetshrike with long grey crest and long white wing stripe. Differs from all races of White-crested Helmetshrike by lacking yellow eye-wattle and having black patches at sides of breast. **V** Calls similar to those of White-crested Helmetshrike but deeper. **H** Acacia bush. **D** SW Kenya, W Tanzania.

17 YELLOW-CRESTED HELMETSHRIKE *Prionops alberti* 20cm Unmistakable. **V** Not well known. A variety of bubbling, chattering, slurred and rolling notes. **H** Montane woodland, bamboo, forest. **D** EC DR Congo.

18 RED-BILLED HELMETSHRIKE *Prionops caniceps* 20cm Varies by race in colour of cheeks, and extent of rufous on underparts. **V** Song is a loud whistling, descending in pitch. Calls include a corvid-like *wrrraak, wrraak*. **H** Forest canopy. **D** W Africa.

19 RUFOUS-BELLIED HELMETSHRIKE *Prionops rufiventris* 21cm Unmistakable with white-grey head, black collar and rufous underparts. **V** Song is a liquid *tyooyoo*. Calls include a strident *chaja*, and various hoarse and whistling notes. **H** Forest canopy; up to mid-high altitude. **D** C Africa.

20 RETZ'S HELMETSHRIKE *Prionops retzii* 20cm In flight shows striking white tail corners. Occurs in parties of up to 20. **V** Song is a repeated whistle, *tweeoo* or *cheeeow, cheeee-ow*. **H** Woodland, riverine belts. **D** C Africa.

21 GABELA HELMETSHRIKE *Prionops gabela* 19cm Very similar to slightly larger Retz's Helmetshrike but compare distribution. **V** Almost unknown, apart from some churring and clicking notes. **H** Open forest, woodland. **D** W Angola.

22 CHESTNUT-FRONTED HELMETSHRIKE *Prionops scopifrons* 17cm Chestnut forehead diagnostic. Usually in groups of 3–30, occasionally mixed with Retz's Helmetshrikes. **V** Whistled song resembles Retz's Helmetshrike, *cheee-ouw*. **H** Canopy of forest and woodland. **D** E and SE Africa.

23 BAR-WINGED FLYCATCHER-SHRIKE *Hemipus picatus* 15cm Makes fly-catching sallies from habitually used perches. **V** Sharp *chisik* or *chir-up*; also a high-pitched trilling. **H** High open forests, forest edge, secondary forest, bamboo stands. **D** Widespread in S and SE Asia, Sumatra and Borneo.

24 BLACK-WINGED FLYCATCHER-SHRIKE *Hemipus hirundinaceus* 13–15cm Hovers to snatch insects from outer foliage and makes short sallies to catch flying insects. Often part of mixed-species feeding flocks. **V** Coarse *tu-tu-tu-tu hee-tee-tee-teet*, and *hee-too-weet* interspersed with a high *cheet-weet-weet-weet*. **H** Broadleaved evergreen forest, freshwater swamp forest and forest edge, occasionally plantations and mangroves. **D** Malay Peninsula; Sumatra and Java, Indonesia; Borneo.

25 LARGE WOODSHRIKE *Tephrodornis virgatus* 23cm Usually encountered in small parties, actively foraging in treetops, hopping from branch to branch in search of insects; occasionally makes aerial sallies after flying insects. After breeding may be found in larger flocks, often in association with other species. **V** Musical, repeated *kew-kew-kew-kew*; also various harsh notes and mellower *tra-a-a* and *thul thull*. **H** Evergreen forest, open deciduous secondary growth, various other well-wooded areas. **D** Widespread in S and SE Asia, Greater Sundas and Borneo.

26 MALABAR WOODSHRIKE *Tephrodornis sylvicola* 23cm Similar to Large Woodshrike but greyer overall. Grey throat with whitish submoustachial stripe. **V** Rapid *witoo-witoo-witoo-witoo*. **H** Deciduous forest and woodland. **D** SW India.

27 COMMON WOODSHRIKE *Tephrodornis pondicerianus* 18cm Occurs in pairs or small parties, foraging among foliage; sometimes takes invertebrate prey from ground or during aerial sallies. **V** Plaintive *weet-weet* followed by quick *whi-whi-whi-whee*; also soft trills during breeding season. **H** Light deciduous woodland, secondary growth, scrub, wooded gardens. **D** Widespread in S and SE Asia.

28 SRI LANKA WOODSHRIKE *Tephrodornis affinis* 18cm Only woodshrike in Sri Lanka. Male similar to Common Woodshrike but greyer with whitish submoustachial stripe. Female browner. **V** Descending, mournful *hooee-hooee-hooee-hooee*. **H** Woodland. **D** Sri Lanka.

29 RUFOUS-WINGED PHILENTOMA *Philentoma pyrhoptera* 16–17cm Sluggish. Frequents the lower to mid-storey; also occurs in low undergrowth and on the ground. Blue morph makes up 5% of population. **V** Mellow, piping *tu-tuuuuu* or *wee tooooo*, and harsh, scolding notes. **H** Primary and secondary forest, peat-swamp, heath, logged forest. **D** Myanmar to Borneo.

30 MAROON-BREASTED PHILENTOMA *Philentoma velata* 19–21cm Often sits motionless for long periods. Movements generally sluggish, but will make sallies after flying insects; frequents middle to upper storey. **V** Clear, bell-like *phu phu phu phu phu*..., strong, clear *chut-ut chut-up chut-ut*... and grating, metallic *churr*. **H** Broadleaved evergreen lowland and lower montane primary forest, logged forest, secondary forest. **D** Myanmar to Java (Indonesia) and Borneo.

31 AFRICAN SHRIKE-FLYCATCHER *Megabyas flammulatus* 15cm Quiet. Swings tail from side to side. Note upright posture. **V** Very high, very sharp *srit-sru-wit-sweet-sweet*. **H** More or less open forests with dense undergrowth. **D** W and C Africa.

32 BLACK-AND-WHITE SHRIKE-FLYCATCHER *Bias musicus* 15cm Active and conspicuous. **V** Sequence of mid-high to very high, short to very short phrases *tju-tjip-tjut-jip weet weet tjip-tju-weet*. **H** Open forest and woodland. **D** W, C and SE Africa.

1
2
3
4
5
6
7
8
9
10
11
12
13
14
15
16
17
18
19
20
21
22
23
24
25
26
27
28
29
30
31
32
blue
morph

BRISTLEHEAD PITYRIASIDAE

1 BORNEAN BRISTLEHEAD *Pityriasis gymnocephala* 26cm Usually seen in parties of 5–12 birds foraging in the middle level of trees; often forms part of mixed-species feeding flocks. Female has patchy red flanks, also sometimes shown by male. In flight, shows white wing-bar on primaries. **V** Various peculiar snorts and whistles, and strange nasal wheezes; also utters a *pit-pit-peeoo*, interspersed with a crow-like chatter. **H** Primary and secondary lowland forest, especially peat-swamp forest. **D** Borneo.

WOODSWALLOWS, BUTCHERBIRDS AND ALLIES ARTAMIDAE

2 ASHY WOODSWALLOW *Artamus fuscus* 19cm Gregarious; groups often perch on bare treetop branches, wires or palm leaf-stalks, from where they make aerial sallies to capture flying insects. **V** Harsh *chek-chek-chek* or *chake-chake-chake*; the latter sets off and finishes a pleasant twittering song, given during the breeding season. **H** Open wooded country with nearby palms. **D** Widespread in S and SE Asia.

3 WHITE-BREASTED WOODSWALLOW *Artamus leucorynchus* 19cm Gregarious, groups often perching together on bare treetop branches or wires, from where aerial sallies are made after flying insects. **V** Chattering, including mimicry; calls include rasping *wek-wek-wek*, sharp *pirt pirt* and *git*, *geet* or *geet geet*. **H** Open wooded country, forest clearings, open areas with scattered trees, plantations and near human habitation. **D** Widespread in Australasia, also Andaman Islands to Philippines.

4 FIJI WOODSWALLOW *Artamus mentalis* 17cm No similar bird in range. Like all woodswallows perches openly; normally in groups. **V** Song is a mixture of mellow whistles and chortling notes. Typical call is a rapid chatter. **H** Open habitats, but occasionally seen hawking over forest. **D** Fiji.

5 IVORY-BACKED WOODSWALLOW *Artamus monachus* 20cm Encountered in pairs or flocks of up to 20 birds. Perches on exposed branches in forest clearings or open areas close to forests, from where short sallies are made after flying insects; also hawks insects and soars above unbroken forests. **V** Series of monosyllabic notes, more metallic and penetrating than those of White-breasted Woodswallow. **H** Forest, forest edges and clearings. **D** Sulawesi, Indonesia.

6 GREAT WOODSWALLOW *Artamus maximus* 20cm Larger and darker than White-breasted Woodswallow with more extensive dark throat. **V** Rapid twittering with some mimicry. Nasal flight call similar to that of White-breasted Woodswallow. **H** Clearings and forest edges in mountains. **D** New Guinea.

7 WHITE-BACKED WOODSWALLOW *Artamus insignis* 20cm Striking pied woodswallow with white back. **V** High-pitched *zeep-zeep* in flight. **H** Forest edge and clearings in lowlands and hills. **D** Bismarck Archipelago, Papua New Guinea.

8 MASKED WOODSWALLOW *Artamus personatus* 19cm Note white-bordered face mask. **V** Series of high *tjew* notes, short chirps and rattles. **H** Dry open woodland; occasionally in low shrubland. **D** Australia.

9 WHITE-BROWED WOODSWALLOW *Artamus superciliosus* 19cm Unmistakable by colour pattern and white eyebrow. A yellow-eyed morph exists. **V** Series of high *tjew* notes, incorporating soft, nasal twitters. **H** Open woodland, grassland with sparse shrub and trees, orchards. **D** C Australia.

10 BLACK-FACED WOODSWALLOW *Artamus cinereus* 18cm Occurs singly, in pairs or in small flocks. Perches on low branches, fences, posts and overhead wires, from where it makes aerial sallies after flying insects; also hawks insects while in soaring flight. Tends to fly lower than other woodswallows. **V** Drawn-out *tchiff-tchiff-tchiff* contact call and soft, twittering song that often includes mimicry. **H** Open woodland, shrubland, cultivation. **D** Australia; Lesser Sunda Islands, SC New Guinea.

11 DUSKY WOODSWALLOW *Artamus cyanopterus* 18cm Dark brown woodswallow with white outer primaries, visible at rest and in flight. **V** Song soft and canary-like. Call harsh *chk-zweee*. Typical twittering and chattering in flight. **H** Open woodland, forests. **D** S and E Australia.

12 LITTLE WOODSWALLOW *Artamus minor* 13cm Like a small Dusky Woodswallow but lacks white outer primaries. **V** Similar calls to other woodswallows but less vocal. **H** Open woodlands, rocky outcrops, wetlands. **D** N and C Australia.

13 LOWLAND PELTOPS *Peltops blainvillii* 18cm Striking arboreal fly-catching bird, perching on high branches. Bluish-black plumage with white cheek-patch and red belly. **V** Harsh, mechanical single or disyllabic note, repeated in long series, sometimes given as *tik-a-tik, tik-a-tik, tik-a-tik*. **H** Lowland rainforest, forest edges, clearings. **D** New Guinea.

14 MOUNTAIN PELTOPS *Peltops montanus* 20cm Similar to Lowland Peltops, but with more extensive white on sides of face and more white on back. **V** Rapid trilling song given in short bursts. Call upslurred *hweeet*. **D** New Guinea.

15 BLACK BUTCHERBIRD *Melloria quoyi* 43cm Large all-black bird with bluish sheen. Bill blue-grey with black tip. **V** Loud yodelling song, deeper than other butcherbirds. **H** Coastal rainforests, mangroves. **D** N Australia.

16 AUSTRALIAN MAGPIE *Gymnorhina tibicen* 41cm Unmistakable by size, colour pattern, bill shape and noisy behaviour; white-backed and black-backed colour forms occur. **V** Short series of melodious, gurgling, fluted and guttural notes. **H** Open, grassy or partly bare habitats with some scattered trees and hedgerows. Also in bordering forest or woodland. **D** SC New Guinea and Australia.

17 GREY BUTCHERBIRD *Cracticus torquatus* 28cm Pied plumage, with black hood and white loral spot. Back mid-grey. **V** Loud varied song includes whistles and warbles, with much mimicry, often in duet. **H** Woodlands, forests, gardens. **D** C, S, E Australia.

18 SILVER-BACKED BUTCHERBIRD *Cracticus argenteus* 26cm Similar to Grey Butcherbird but smaller with paler grey back. Lacks loral spot. **V** Similar to Grey Butcherbird but less vocal. **H** Woodlands, forests. **D** N and NW Australia.

19 BLACK-BACKED BUTCHERBIRD *Cracticus mentalis* 26cm Similar to Grey Butcherbird but back black and wing-coverts more extensively white. Has broad white collar and lacks loral spot. **V** Similar to that of Grey Butcherbird. **H** Open forests, woodlands. **D** SE New Guinea, NE Australia.

20 PIED BUTCHERBIRD *Cracticus nigrogularis* 34cm Large pied butcherbird with complete hood extending onto upper breast. **V** Celebrated songster, with varied and complex whistles and warbles, often in duet and including some mimicry. **H** Open woodlands and wooded farmland. **D** Australia.

21 HOODED BUTCHERBIRD *Cracticus cassicus* 33cm Similar to Pied Butcherbird, but less white in wings (no overlap in distribution). **V** Pleasant series of fluty whistles, often in duet and including some mimicry. **H** Forest edges, secondary growth, gardens, savannah. **D** New Guinea.

22 TAGULA BUTCHERBIRD *Cracticus louisiadensis* 28cm Black, except for white vent and undertail-coverts, and white markings on wings and tail. Uncommon and poorly known **V** Typical butcherbird song with varied phrases and trills. **H** Forests. **D** Tagula Island, Louisiade Archipelago.

23 PIED CURRAWONG *Strepera graculina* 42–50cm Large, crow-like bird with yellow eyes and large bill. Mainly black with white vent, small white wing-patch and white tips to tail. **V** Loud rollicking call, often rendered as *curra-wong, curra-wong*. Also gives long whistles and scolding calls. **H** Forests, woodlands, farmland, urban areas. **D** NE and E Australia.

24 BLACK CURRAWONG *Strepera fuliginosa* 48cm Similar to Pied Currawong, but white restricted to wing-tips and tail-tips. **V** Distinctive series of cackling phrases. **H** Forests, woodlands, moorland, mainly in uplands. **D** Tasmania.

25 GREY CURRAWONG *Strepera versicolor* 45–50cm Mainly grey or blackish grey, often with darker face. Some races browner. All have white vent and white tail tips. **V** Distinctive loud ringing call, usually given as series of 2–4 notes. **H** Forests, woodlands, farmland, urban areas. **D** S Australia.

MOTTLED BERRYHUNTER RHAGOLOGIDAE

26 MOTTLED BERRYHUNTER *Rhagologus leucostigma* 16cm Elusive forest fruit-eater, formerly in whistler family but now elevated to its own family. Rather drab olive-brown plumage with rusty face and variably mottled underparts. Sexes similar in nominate race, but dimorphic in *obscurus* (female is brighter). Singly or in pairs. **V** Distinctive discordant series of slurred whistled notes. **H** Montane forests in mid-storey. **D** New Guinea.

IORAS AEGITHINIDAE

27 COMMON IORA *Aegithina tiphia* 14cm Acrobatic forager in the canopy, often making darting sallies to capture passing insects. Usually alone or in pairs; non-breeders often join mixed-species feeding parties. **V** Various strident, unmusical, whistling and whining songs. **H** Open forest, mangroves, peat-swamp forest, secondary growth, plantations, parks, gardens. **D** Widespread in S and SE Asia, Indonesia and Borneo.

28 MARSHALL'S IORA *Aegithina nigrolutea* 14cm Behaviour similar to that of Common Iora. **V** Slurred *tchoo-tchoo*, followed by loud, metallic *chee-tchoo-tchoo-tchee* or *chee-cho-chi-choo*; other calls include *wheeti wheeti*, *twsee-ku-kee* and *tswee-twsee-tee-dik*. **H** Thorny acacia, scrubby groves, sheesham woodland. **D** NW India, Sri Lanka.

29 GREEN IORA *Aegithina viridissima* 12–15cm Feeds on invertebrates in canopy foliage, in pairs or small parties, and as part of mixed-species foraging groups. **V** High-pitched *itsu tsi-tu tsi-tu*, chattering *tit-teeer* and subdued *chitititit*. **H** Primary and tall secondary forest, forest edges. **D** SE Asia; Sumatra, Indonesia; Borneo.

30 GREAT IORA *Aegithina lafresnayei* 15–18cm Dark morph has blackish mantle extending up neck to back of head. Long, powerful bill. Regular member of mixed-species feeding flocks. **V** Clear *chew-chew-chew*... **H** Broadleaved evergreen, semi-evergreen and mixed deciduous forest; also forest edge. **D** Widespread in SE Asia.

1
2
3
4
5
6
7
8
9
10
11
12
13
14
15
black-backed form
16
white-backed form
17
18
19
20
21
22
23
24
25
26 ♀ ♂
27 ♀ ♂
28 ♀ ♂
29 ♀ ♂
dark morph
30 ♂
30 ♀ ♂

CUCKOOSHRIKES CAMPEPHAGIDAE

1 WHITE-BELLIED MINIVET *Pericrocotus erythropygius* 15–16.5cm Usually found in pairs or small parties; generally forages low down. Regularly perches on bushes or long grass, dropping to ground to pick up prey; may also hover over grassland before dropping onto insects. **V** Soft *tchip* or *tsip-i-sip*; also various sweet, high-pitched notes such as *thi, tuee, chi, tschi* and *tchu-it*. **H** Open, dry scrub, grassland, savannah with scattered trees. **D** India.

2 JERDON'S MINIVET *Pericrocotus albifrons* 15–17cm Usually in pairs or small parties. Regularly perches on bushes or long grass from where it drops to the ground to pick up prey; also hovers over grassland before dropping onto prey. **V** Various sweet, high-pitched notes, such as *thi, tuee* and *tchu-it*; also a soft *tchip* or *tsip-i-sip*. **H** Open, dry scrub with scattered trees and dry cultivation. **D** C Myanmar.

3 FIERY MINIVET *Pericrocotus igneus* 15–16cm Often occurs in small groups or as a member of mixed-species feeding parties; forages in canopy. **V** Thin, rising, musical *swee-eet*. **H** Forests, forest edges, clearings, woodlands, scrub, occasionally mangroves. **D** Myanmar to Borneo and Palawan, Philippines.

4 SMALL MINIVET *Pericrocotus cinnamomeus* 16cm Active, flits about among foliage in search of insects, also makes short fly-catching sallies; often part of mixed-species feeding flocks. Plumage variable by race. **V** Constantly repeated, thin, drawn-out *tswee-swee*. **H** Foothill forests, scrub jungle, acacia, subtropical dry woodland. **D** Widespread in S and SE Asia, Java.

5 GREY-CHINNED MINIVET *Pericrocotus solaris* 17–19cm Behaviour similar to that of Scarlet Minivet. **V** Repeated *tzee-zip*, a slurred *swirrrririt*; also soft *trip* or stronger *trii-ti*. **H** Broadleaved evergreen forest, forest edge, occasionally pine forests. **D** Widespread in S and SE Asia, Greater Sundas and Borneo.

6 SUNDA MINIVET *Pericrocotus miniatus* 19cm Forages in canopy in small or large flocks. **V** Hard, shrill *chee-chee-chee* and drawn-out, loud *tsree-ee*. **H** Montane forest and forest edges, occasionally in cultivated areas. **D** Sumatra and Java, Indonesia.

7 SHORT-BILLED MINIVET *Pericrocotus brevirostris* 19cm Forages among foliage, usually in pairs or as part of mixed-species feeding flocks. **V** Distinctive thin, whistled *tsuuuit*; also dry *tup*. **H** Deciduous and evergreen forests, forest edges, secondary growth. **D** Himalayas to Laos.

8 LITTLE MINIVET *Pericrocotus lansbergei* 16cm Occurs in pairs or small groups, regularly among mixed-species feeding parties; feeds by gleaning and hover-gleaning from horizontal perches in mid-storey. **V** Quiet, trilling calls. **H** Open parkland-type forest with sparse understorey; also primary, tall secondary and riverine forests. **D** W Lesser Sunda Islands.

9 LONG-TAILED MINIVET *Pericrocotus ethologus* 20cm Forages mainly in canopy, gleans or hovers in front of foliage to take insects; also makes short fly-catching sallies. **V** Sweet, rolling, repeated *prrr wi* and *prrr i*; also thin, sibilant *swii-swii swii-swii-swii*. **H** Forests, well-wooded areas. **D** Himalayas to Laos.

10 ORANGE MINIVET *Pericrocotus flammeus* 20–22cm More orange than the closely related Scarlet Minivet. Gleans insects from foliage, also hovers or makes short sallies to capture flying insects. Forages mainly in canopy; regularly joins mixed-species feeding flocks. **V** Piercing *sweep-sweep-sweep-sweep* or *weep-weep-weep-wit-wip*. **H** Forests, wooded areas. **D** SW India, Sri Lanka.

11 SCARLET MINIVET *Pericrocotus speciosus* 20–22cm Slightly redder than Orange Minivet. Gregarious; often joins mixed-species foraging flocks. Feeds by gleaning or hovering to take invertebrates from foliage, and makes short aerial sallies to catch flying insects. **V** Loud, piercing *sweep-sweep-sweep-sweep, swee-et sweet-e* or *twii-it twii-it* and similar. **H** Forests, wooded areas. **D** Himalayas to Indonesia and the Philippines.

12 ASHY MINIVET *Pericrocotus divaricatus* 18cm Often in large flocks, forages in outer branches of canopy. Regularly joins mixed-species flocks. **V** Jangling metallic trill, usually uttered in flight. **H** Various woodlands, open areas with scattered trees, plantations and mangroves. **D** Breeds SE Siberia, NE China, Korean Peninsula, Japan and islands to Taiwan. Winters SE Asia.

13 RYUKYU MINIVET *Pericrocotus tegimae* 20cm Often regarded as race of Ashy Minivet. **V** As Ashy Minivet. **H** As Ashy Minivet. **D** S Japan.

14 SWINHOE'S MINIVET *Pericrocotus cantonensis* 20cm Forages in canopy, often part of mixed-species feeding flocks. **V** Metallic trill. **H** Broadleaved evergreen forest, semi-evergreen forest, deciduous forest, forest edge. **D** Breeds C and SE China. Winters SE Asia.

15 ROSY MINIVET *Pericrocotus roseus* 20cm More sluggish than other minivets; often sits quietly at the top of trees. May occur in large flocks post breeding. **V** Whirring trill. **H** Deciduous and semi-deciduous forest; occasionally open broadleaved evergreen forest. **D** Breeds Himalayas to S China, Myanmar, NW Vietnam. Winters C India, Thailand, Laos, S Vietnam.

16 MADAGASCAN CUCKOOSHRIKE *Ceblepyris cinereus* 24cm Varies by race in breast and belly colour. **V** Very high, sharp, hurried but slowing-down twitter. **H** Forest, scrub, mangroves. **D** Madagascar.

17 COMOROS CUCKOOSHRIKE *Ceblepyris cucullatus* 22cm Grey morph similar to Madagascan Cuckooshrike (formerly considered conspecific). Olive morph distinctive, with dark olive head and upperparts and pale yellowish underparts. **V** Similar to Madagascan Cuckooshrike but quieter. **H** Forests, in mid-storey and canopy. **D** Comoros.

18 GRAUER'S CUCKOOSHRIKE *Ceblepyris graueri* 20cm Unmistakable. Not found in same range as White-breasted Cuckooshrike. **V** Unknown. **H** Forest at higher elevations. **D** E DR Congo.

19 WHITE-BREASTED CUCKOOSHRIKE *Ceblepyris pectoralis* 25cm Female separated from female Grey Cuckooshrike by almost white underparts. **V** Very high *shree-shree witwit...* (*witwit* much lower pitched). **H** Open miombo and other woodland at lower altitudes. **D** Widespread in Africa.

20 GREY CUCKOOSHRIKE *Ceblepyris caesius* 21cm Note uniform grey plumage. **V** Extremely high *shreeu* with barely audible slow twittering. **H** Montane forest, well-wooded streams. **D** Widespread in Africa.

21 STOUT-BILLED CUCKOOSHRIKE *Coracina caeruleogrisea* 33–37cm Forages slowly and deliberately along large branches in upper levels of trees, usually in pairs or small parties. **V** Slightly musical, harsh *shhhy yun-eshhhyuway*; also soft, mewing slurred note, short *chirp*, a soft rasp and harsh buzzy notes. **H** Forest, forest edges, tall secondary growth, disturbed areas. **D** New Guinea.

22 HOODED CUCKOOSHRIKE *Coracina longicauda* 33cm Large grey cuckooshrike with black flight feathers and tail. Male has black hood, female has black face and throat. Generally encountered in pairs or small groups, mainly in canopy. **V** Series of dry rasping notes. **H** Montane forest. **D** New Guinea.

23 CERULEAN CUCKOOSHRIKE *Coracina temminckii* 30–31cm Generally encountered in pairs or small groups, foraging in mid-storey or crown of tall trees; often calls from high bare branches. **V** Drawn-out, nasal, rasping *ssssschu*, which is sometimes preceded by 2–3 dry, staccato notes; also utters whistled *pi pi pi pi pi* and short, high-pitched flight note. **H** Primary and tall secondary hill and montane forest, occasionally pine plantations. **D** Sulawesi, Indonesia.

24 PIED CUCKOOSHRIKE *Coracina bicolor* 31cm Usually seen in pairs or small groups foraging in dense leafy canopy. Best located by call. **V** Loud, ringing series of 3–5 trisyllabic notes followed by 4–8 louder single whistles, often preceded by several rasping notes. **H** Primary lowland and hill forest, forest edges and clearings, scrub, mangroves. **D** N Sulawesi, Indonesia.

25 GROUND CUCKOOSHRIKE *Coracina maxima* 35cm Large, long-legged grey cuckooshrike, with black wings and fine barring on belly. **V** Thin, high-pitched *chew* or *wi-chew*. **H** Open woodland, shrubs in arid areas. **D** C Australia.

26 BARRED CUCKOOSHRIKE *Coracina lineata* 23cm Medium-sized grey cuckooshrike with prominent yellow eyes, underparts strongly barred black and white. Gregarious, in flocks of up to 20. **V** Frequently calls with nasal whistles and trills, often in flight and by several birds. **H** Coastal rainforests. **D** New Guinea, E Australia, Solomon Islands.

27 BLACK-FACED CUCKOOSHRIKE *Coracina novaehollandiae* 32–35cm Generally seen singly or in small groups, usually perching on exposed branches; on alighting, repeatedly folds wings. **V** Nasal, plaintive *plee-urk*; also a creaky, nasal purring note, often given in flight. **H** Open forest, forest edges, clearings and secondary growth, savannah woodland, urban gardens. **D** Widespread in Australasia.

28 BOYER'S CUCKOOSHRIKE *Coracina boyeri* 22cm Uniform blue-grey cuckooshrike. Male has black face mask, female lacks mask and has plain face. Rufous underwing-coverts in flight. **V** Song long upslurred or downslurred whistle. Also gives variety of chirping or clucking calls. **H** Forests, secondary growth, mangroves; commoner at lower altitudes. **D** New Guinea.

29 BURU CUCKOOSHRIKE *Coracina fortis* 35cm Very little information; said to be unobtrusive. **V** Unrecorded. **H** Lowland, montane and monsoon forest, disturbed areas. **D** S Maluku Islands, Indonesia.

30 WALLACEAN CUCKOOSHRIKE *Coracina personata* 30–35cm Quietly forages in canopy and at tops of isolated trees, alone, in pairs or in threes. **V** Long, drawn-out *weeeer*, series of piping whistles, 2–3 upslurred buzzy notes and nasal squabbling notes. **H** Primary lowland and hill monsoon forest and woodland; also secondary woodland, scrub, savannah. **D** Lesser Sunda Islands.

31 NORTH MELANESIAN CUCKOOSHRIKE *Coracina welchmani* 32cm Large blackish grey cuckooshrike with dark eyes. Male has glossy black face and throat. **V** Long downslurred whistle and various other calls and chatters. **H** Montane forests, coastal forests and mangroves. **D** Solomon Islands.

1
2
3
4
5
6
7
8
9
10
11
12
13
14
15
16
17
18
19
20
21
22
23
24
25
26
27
28
29
30
31

CUCKOOSHRIKES *CONTINUED*

1 SOUTH MELANESIAN CUCKOOSHRIKE *Coracina caledonica* 35cm Very large slate-grey cuckooshrike with yellow eyes. Female slightly paler. **V** Loud upslurred whistle and series of chattering notes. **H** Forests, urban areas, mainly in lowlands. **D** Vanuatu, Loyalty Islands, New Caledonia.

2 BAR-BELLIED CUCKOOSHRIKE *Coracina striata* 28cm Noisy and conspicuous, often seen flying over forest clearings or between ridges, generally in small groups or as part of mixed-species feeding flocks. **V** Clear, whinnying *kliu-kliu-kliu* or shrill *kriiu-kriiu*, also harsh *klee kleep* and rising *see-up*. **H** Lowland forest, including mangroves, plantations, secondary growth. **D** SE Asia to the Philippines.

3 JAVAN CUCKOOSHRIKE *Coracina javensis* 28cm Forages singly or in pairs at tops of tall trees or at edges of forest clearings. **V** Loud, piercing, whistled *pee-eeo-pee-eeo* and *tweer* or *twee-eet*. **H** Open forest, forest edges and clearings, savannah. **D** Java, Indonesia.

4 LARGE CUCKOOSHRIKE *Coracina macei* 30cm On landing, flicks wings alternately. Usually forages in the tops of trees, in pairs or loose parties. **V** Loud, whistled *tee-treee, ti-eee or pee-eeo-pee-eeo*, often uttered while flying from tree to tree. **H** Open broadleaved evergreen forest and mixed deciduous forest, pine forest and open wooded country. **D** Widespread in S and SE Asia.

5 ANDAMAN CUCKOOSHRIKE *Coracina dobsoni* 26cm Keeps to tops of trees, often part of mixed-species feeding parties. **V** Clear whinnying *kliu-kliu-kliu-kliu* and grating *gree-ew gree-ew*. **H** Forests. **D** Andaman Islands.

6 SLATY CUCKOOSHRIKE *Coracina schistacea* 31cm Usually forages singly or in pairs, 10m or higher in trees. **V** Series of slightly downslurred, nasal, whistled notes; also short, unmusical, slightly metallic notes. **H** Primary and selectively logged lowland forest, degraded forest, secondary forest, lightly wooded cultivation, lightly wooded swampland, tall mangrove forest. **D** Sula Islands and Banggai Islands, Indonesia.

7 WHITE-RUMPED CUCKOOSHRIKE *Coracina leucopygia* 25–29cm Generally seen in pairs or small groups; forages in canopy, but will venture to ground to collect food. Regularly perches on exposed branches at tops of trees and shrubs. **V** Series of jangling notes; also weak, nasal, chattering notes. **H** Open wooded areas, lightly wooded cultivation, secondary woodland, scrub with trees, swamp forest. **D** Sulawesi, Indonesia.

8 SUNDA CUCKOOSHRIKE *Coracina larvata* 26–27cm Frequents canopy, singly or in pairs; sometimes found in mixed-species feeding flocks. **V** Harsh, ringing *eeooo-eeooo-eeooo, shreeok* and a curious loud, wheezy song. **H** Montane forests. **D** Sumatra and Java, Indonesia; Borneo.

9 WHITE-BELLIED CUCKOOSHRIKE *Coracina papuensis* 22–29cm Occurs singly, in pairs or in small groups; forages from mid-storey to the canopy, feeding by gleaning or making short aerial sallies. Often perches on branches of dead trees; has exaggerated wing-flicking habit while perched. **V** High-pitched, nasal, squealing *wheo-eeyu* or *witchew*. **H** Forest edges, secondary forest, lightly wooded cultivation, scrub with scattered trees, mangroves, coconut plantations. **D** Widespread in Australasia.

10 MANUS CUCKOOSHRIKE *Coracina ingens* 28cm Similar to White-bellied Cuckooshrike but larger with heavier bill and paler underparts. Sexes similar but female has duller mask. **V** Shrill, disyllabic *zh-zeek*. **H** Forests, secondary growth and disturbed habitats, in lowlands and hills. **D** Admiralty Islands (Bismarck Archipelago).

11 MOLUCCAN CUCKOOSHRIKE *Coracina atriceps* 32–35cm Often noisy and conspicuous. Forages in upper tree levels, singly, in pairs or in threes. **V** Harsh, staccato, dry chattering, given alternately by two individuals. **H** Primary lowland and hill forest, tall secondary and selectively logged forest, coastal *Casuarina* groves and woodland, lightly wooded cultivation, occasionally mangroves. **D** Maluku Islands, Indonesia.

12 BLACK CUCKOOSHRIKE *Campephaga flava* 18cm All-black form (common) and yellow-shouldered form (less common) occur. Note predominantly white (black-barred) underparts of female. Gleans insects from foliage. Floppy undulating flight. Often in mixed-species parties. **V** Extremely high silvery short insect-like trill *srrrrri*. **H** Open forests, plantations, other wooded and bushed habitats. **D** South Sudan and Kenya to Angola, Botswana and E South Africa.

13 RED-SHOULDERED CUCKOOSHRIKE *Campephaga phoenicea* 18cm Unmistakable, but some males rather similar to individuals of Black Cuckooshrike's yellow-shouldered form. Female separated from female Black-Cuckooshrike by greyer upperparts and less yellow on underside of tail feathers. **V** Song is a jumble of high-pitched whistles, squeaks and churrs. Typical call is a quiet *tchit*. **H** More or less wooded and bushed areas. **D** Senegal and Gambia to Sierra Leone, east to Eritrea, Ethiopia and Kenya.

14 PETIT'S CUCKOOSHRIKE *Campephaga petiti* 19cm Separated from Black Cuckooshrike by blue sheen. Female separated from female Black Cuckooshrike by yellower and less barred underparts. **V** Combinations of extreme high *shreeeeh* with much lower *weewee, dee-deeder*. **H** Tall trees at forest edges and surrounding areas. **D** SE Nigeria and SW Cameroon, SE Gabon to NW Angola, E DR Congo and W Uganda, W Kenya.

15 PURPLE-THROATED CUCKOOSHRIKE *Campephaga quiscalina* 18cm Separated from Black Cuckooshrike and Petit's Cuckooshrike by deep purplish glow over underparts, female by absence of (or much reduced) barring. **V** Song is a rapid series of melodious whistles, *slueet-slueet-swit-wit-slueet-slueet-swit-wit-tluw-tluweew*. Calls include *whit-whit-whit* and *tsee-up*. **H** Montane forests. **D** W and C Africa.

16 WESTERN WATTLED CUCKOOSHRIKE *Lobotos lobatus* 18cm Not unlike some orioles, but with black bill, orange wattles and different shape and manner. **V** Mainly silent. Flight call is *zit* or *tsik*. **H** Rainforest. **D** Sierra Leone to Ghana.

17 EASTERN WATTLED CUCKOOSHRIKE *Lobotos oriolinus* 18cm Separated from Western Wattled Cuckooshrike by green rump, yellow underparts and no yellow in wings; also different range. **V** Mainly silent and inconspicuous, apart from *zit* or *tsik* flight calls. **H** rainforest. **D** SE Nigeria, S Cameroon and SW Central African Republic to Gabon, NW Congo and E DR Congo.

18 GOLDEN CUCKOOSHRIKE *Campochaera sloetii* 20cm Striking colourful cuckooshrike with white in wings. Plumage mainly black and golden-yellow. Male has black face and throat. Female duller with grey face. Often in small flocks. **V** Song excited bubbling whistle or series of *chip-chip-chip* notes. **H** Forest and clearings in foothills and lowlands. **D** New Guinea.

19 MCGREGOR'S CUCKOOSHRIKE *Malindangia mcgregori* 21cm Often seen in groups or mixed-species feeding flocks; forages in canopy and understorey, and at forest edges. **V** Rasping or snapping *chu chu chu zwheeeeeeet-zhuuuuuuuu*, likened to sound made while tuning a radio; also *zwheeet zwheeet zwheeet*. **H** Mossy montane forest, forest edges. **D** Mindanao, S Philippines.

20 NEW CALEDONIAN CUCKOOSHRIKE *Edolisoma anale* 28cm Dark grey cuckooshrike with rufous vent and rufous underwing-coverts. Sexes alike. Usually in mid-storey or canopy. **V** Various loud whistles and squeals. **H** Forest in hills and mountains. **D** New Caledonia.

21 WHITE-WINGED CUCKOOSHRIKE *Edolisoma ostentum* 25cm Usually found in small flocks or as part of mixed-species feeding parties, foraging in canopy or at forest edges. **V** Loud, downward-inflected *schiirp* or *pi-ieu*; also *chu chu chu chu pi-ieuu ki chu*. **H** Lowland and mid-montane forest, forest edges, secondary forest. **D** WC Philippines.

22 BLACKISH CUCKOOSHRIKE *Edolisoma coerulescens* 25–26cm Generally occurs in small groups or mixed-species feeding parties; forages in canopy or upper understorey of forests or secondary growth. **V** Loud, harsh *peeeeeuuuu* or *peeeeeuuu peeuu-tip-tip-peeuu*..., the latter snapping and almost twittering; also utters harsh, twittering *tip-tip-tip-tip*. **H** Subtropical or tropical moist lowland forest, secondary growth. **D** Philippines.

23 BLACK-BELLIED CUCKOOSHRIKE *Edolisoma montanum* 24cm Male mainly blue-grey above and glossy black below. Female mainly grey with black face. Usually in pairs or small groups, mainly in canopy. **V** Duets. Male gives downslurred *chewoooo*, answered immediately by female's harsh rattle. **H** Forest, clearings, gardens in mountains. **D** New Guinea.

24 PALE-SHOULDERED CICADABIRD *Edolisoma dohertyi* 20–24cm Usually occurs in pairs, or sometimes alone or in threes. Forages in mid-storey and lower canopy of large trees. **V** Song consists of three grating cicada-like notes; also utters harsh, rasping *queep*. **H** Primary and tall secondary evergreen forest, monsoon forest, *Casuarina* woodland. **D** Lesser Sunda Islands.

25 KAI CICADABIRD *Edolisoma dispar* 21–25cm Generally seen alone or in pairs. Active forager, mainly from mid-storey to lower canopy, gleaning from undersides of leaves. **V** Repeated, unmusical, downslurred note; also utters single *chuk*. **H** Forest, forest edges, secondary woodland. **D** Kai Islands in SE Maluku Islands, Indonesia.

26 GREY-HEADED CUCKOOSHRIKE *Edolisoma schisticeps* 22cm Strongly sexually dimorphic. Male mainly slate-grey with darker face and wings. Female rufous-brown with grey head. Usually in pairs or small parties. **V** Mournful upslurred and downslurred whistle *hoowit-huuu*. **H** Forest and forest edges in lowlands and hills. **D** New Guinea.

27 PALE CICADABIRD *Edolisoma ceramense* 24–25cm Often seen in mixed-species feeding parties foraging in dense foliage of crowns of tall trees; gleans from branches and undersides of leaves. **V** Rapid, dry chatter and short, staccato *tick*; in flight, utters distinctive nasal chatter. **H** Forests, including selectively logged forest, secondary growth, occasionally mangroves. **D** S Maluku Islands, Indonesia.

28 BLACK-BIBBED CICADABIRD *Edolisoma mindanense* 23cm Inconspicuous forager, mainly in canopy; occurs singly, in small groups or as part of mixed-species feeding parties. **V** No information. **H** Forest and secondary growth. **D** Philippines.

29 MAKIRA CICADABIRD *Edolisoma salomonis* 23cm Male slate-grey with black lores, flight feathers and tail. Female has rufous-orange underparts. Usually in pairs, sometimes in mixed flocks. **V** Song decelerating series of downslurred whistles *tseeuu- tseeuu- tseeuu*. **H** Forests, secondary growth. **D** Makira, Solomon Islands.

30 SOLOMONS CUCKOOSHRIKE *Edolisoma holopolium* 21cm Small grey cuckooshrike. Male has black underparts. Female all grey except for black wings and tail. Usually in pairs or groups of three. **V** Mournful downslurred series of whistles with rasping quality. **H** Forests and secondary growth, from lowlands to hills. **D** Solomon Islands.

31 SULAWESI CICADABIRD *Edolisoma morio* 23–25cm Occurs in pairs or small groups, and as part of mixed-species feeding parties; forages in canopy and subcanopy. **V** Slow sequence of buzzy, nasal, ringing cicada-like notes, alternating with clear descending whistle; also piercing, excited *ki* or *ke* calls, rapid nasal chatter and low-pitched *chup*. **H** Primary and tall secondary lowland, hill and montane forest, forest edges, small forest patches, secondary growth. **D** Sulawesi, Indonesia.

yellow-shouldered form

CUCKOOSHRIKES *CONTINUED*

1 BLACK-SHOULDERED CICADABIRD *Edolisoma incertum* 22cm Male mainly dark blue-grey, with blackish face and throat. Female paler with dark lores. Usually in pairs or small groups. **V** Series of fast disyllabic notes *chewee- chewee- chewee- chewee*, repeated many times. **H** Forest, forest edges and secondary growth, in hill and lower montane areas. **D** New Guinea.

2 GREY-CAPPED CICADABIRD *Edolisoma remotum* 25cm Closely related to Common Cicadabird and may be conspecific. Male dark blue-grey with darker mask. Female has rufous upperparts, grey crown and unbarred pale rufous underparts. **V** Series of rapid chattering notes and longer downslurred whistles. **H** Forest, forest edges and secondary growth from lowlands to hills. **D** Bismarck Archipelago.

3 SULA CICADABIRD *Edolisoma sula* 20–24cm Usually seen singly, in pairs or in small groups, often forming part of mixed-species feeding parties. Forages mainly in lower storeys of forests, up to 15m or so from ground. **V** Series of short, explosive metallic notes, 2–3 per sec. **H** Primary forest, woodland, plantations, scrub. **D** Islands east of Sulawesi, Indonesia.

4 COMMON CICADABIRD *Edolisoma tenuirostre* 24–27cm Inconspicuous forager in dense leafy crowns of tall trees, best located by voice. Generally seen alone or in pairs. **V** Jumbled cicada-like *ch-ch-ch*, followed by downslurred nasal note. **H** Primary, secondary and selectively logged forest, forest edges. **D** Widespread in Australasia and S Pacific islands.

5 ADMIRALTY CICADABIRD *Edolisoma admiralitatis* 26cm Closely related to Common Cicadabird; may be conspecific. Male dark slate-grey. Female has rufous-brown upperparts, grey crown and pale rufous underparts with barring from throat to belly. **V** Series of nasal disyllabic notes *chewee-chewee-chewee*. **H** Forest, forest edges and secondary growth from lowlands to foothills. **D** Manus, Admiralty Islands (Bismarck Archipelago).

6 PALAU CICADABIRD *Edolisoma monacha* 23cm Closely related to Common Cicadabird; formerly conspecific. Male dark slate-grey. Female has rufous-brown upperparts, grey crown and buffy underparts with barring from throat to belly. **V** Series of mellow, disyllabic, upslurred notes. **H** Forest, forest edges, secondary growth. **D** Palau, WC Pacific.

7 YAP CICADABIRD *Edolisoma nesiotis* 25cm Closely related to Common Cicadabird; formerly conspecific. Male slate-grey. Female has rufous-brown upperparts, grey crown and pale rufous underparts with faint barring. **V** Irregular series of short squeaky notes. **H** Forest, forest edges, secondary growth. **D** Caroline Islands, WC Pacific.

8 POHNPEI CICADABIRD *Edolisoma insperatum* 25cm Closely related to Common Cicadabird; may be conspecific. Male dark slate-grey. Female has dark rufous-brown upperparts, dark grey crown and unbarred cinnamon-rufous underparts. **V** Series of irregular, nasal squeaks and whistles. **H** Forest, forest edges, secondary growth. **D** Caroline Islands, WC Pacific.

9 BLACK CICADABIRD *Edolisoma melas* 23cm Male all black. Female rufous-brown above with paler supercilium, and plain rufous-buff below. Often in mixed flocks in mid-storey and canopy. **V** High-pitched downslurred whistle *teeew-teeew-teeew*, repeated regularly. **H** Forest, forest edges and secondary growth in lowlands and foothills. **D** New Guinea.

10 HALMAHERA CUCKOOSHRIKE *Edolisoma parvulum* 25cm Generally found in pairs, foraging in canopy or perched at top of dead trees, from where sallies are made after flying insects. **V** Series of rapidly repeated staccato, chattering notes, slowing and descending before tailing off. **H** Primary and selectively logged hill forest. **D** Halmahera, N Maluku Islands, Indonesia.

11 PYGMY CUCKOOSHRIKE *Celebesica abbotti* 20cm Forages sluggishly in crowns of tall trees, climbs nuthatch-like on thick branches; usually encountered singly or in pairs, occasionally in small groups or as part of mixed-species feeding parties. **V** Song consists of two quiet notes followed by four-note warble; calls include high-pitched descending *thip* and sharp *tseet*, given in flight. **H** Montane forest. **D** Sulawesi, Indonesia.

12 BLUE CUCKOOSHRIKE *Cyanograucalus azureus* 20cm Unmistakable, but can look black in shade. **V** Song is clear and powerful, *chup-peeeoo* or *pooeet-pooit-pooeet-peeoo*, interspersed with fast chattering notes. **H** Canopy of forest and woodland. **D** Sierra Leone and Liberia to E and SW DR Congo.

13 POLYNESIAN TRILLER *Lalage maculosa* 16cm Varies by race; birds from Samoa, Tonga and Niue are clean black and white (male) or brown and white with some barring to breast sides and flanks (female); birds from Fiji have barred underparts and those from Rotuma are buffy below. **V** High, loud, rapid *weetweetweet*...(5 secs) or *tutjew* and other sharp or melodious notes and strophes. **H** Forest, gardens, parks. **D** Fiji, W Samoa.

14 SAMOAN TRILLER *Lalage sharpei* 13cm Unmistakable by whitish eyes, orange bill and lack of white in wings. No similar species in range. **V** Short, nasal, upslurred squeaks. **H** Forest, wooded farmland. **D** W Samoa.

15 WHITE-SHOULDERED TRILLER *Lalage sueurii* 18cm Usually conspicuous. Forages singly, in pairs or in small groups, on ground in short grass or on bare areas. Perches in isolated trees, and on fence posts and stumps. **V** Harsh, rattling series of rapid, staccato notes that almost run together; song consists of 12 clear, rapidly repeated musical notes. **H** Woodland, savannah, open forest, mangroves, overgrown or lightly wooded cultivation, scrub, gardens. **D** E Java, Bali, the Lesser Sunda Islands and Sulawesi.

16 LONG-TAILED TRILLER *Lalage leucopyga* 18cm Six island races (one extinct). Males pied with white wing patch, females browner. **V** Series of strident double notes *too-wee too-wee too-wee*; also other calls and trills. **H** Forest, forest edges, scrub. **D** Solomon Islands, Vanuatu, Loyalty Islands, New Caledonia.

17 WHITE-WINGED TRILLER *Lalage tricolor* 17cm Separated from Polynesian Triller by range and inconspicuous eyebrow. **V** High, vigorous rattle, slightly descending at the end (10–15 secs). **H** Open forest, woodland, wooded farmland, shrubland. **D** S New Guinea, Australia.

18 RUFOUS-BELLIED TRILLER *Lalage aurea* 18–20cm Usually found singly, in pairs or in small groups; forages from mid-storey to canopy. **V** Series of seven staccato, piping notes. **H** Primary and secondary logged forest, forest edges, open secondary forest, scrub, coastal woodland, mangroves. **D** N Maluku Islands, Indonesia.

19 BLACK-BROWED TRILLER *Lalage atrovirens* 18cm Male pied with white shoulder and wing-bar. Female similar but with narrow barring below. Usually in pairs, foraging in canopy. **V** Series of strident whistled notes *wheet-wheet-wheet-wheet* with many variations. **H** Forest edges, secondary growth and gardens in lowlands and foothills. **D** New Guinea.

20 WHITE-BROWED TRILLER *Lalage moesta* 18–19cm Forages, singly or in pairs, from mid-storey to canopy; feeds by gleaning from leaves and taking small fruits. **V** Muted, unmusical, staccato *nit nit* notes. **H** Secondary forest, mangroves, low open forest, occasionally tall forest. **D** S Maluku Islands, Indonesia.

21 VARIED TRILLER *Lalage leucomela* 18cm Widespread pied triller with many races. Males have prominent supercilium and variable fine barring below with cinnamon vent. Usually in pairs or small flocks, foraging in canopy. **V** Varies across range, often including nasal trilling in short descending series. **H** Forest edges, mangroves, secondary growth in lowlands and foothills. **D** New Guinea, Bismarck Archipelago, N, E Australia.

22 MUSSAU TRILLER *Lalage conjuncta* 17cm Typical pied triller, with white rump and lacking supercilium and barring. Lower abdomen strongly cinnamon. Sexes alike. **V** More vocal than Varied Triller, with longer, high-pitched phrases. **H** Forest, forest edges and secondary growth in lowlands and hills. **D** St Matthias islands (Bismarck Archipelago).

23 BLACK-AND-WHITE TRILLER *Lalage melanoleuca* 21–22cm Generally encountered in canopy or at forest edges, singly or as part of mixed-species feeding parties; often noisy. **V** Loud *cha-chi cha-chi cha-chi*... or *ka-choo ka-choo ka-choo*; also staccato *chi chi chi chi*. **H** Forest, forest edges. **D** Philippines.

24 PIED TRILLER *Lalage nigra* 18cm Usually occurs singly, in pairs or in small parties. Forages at all levels and occasionally hunts on ground in short grass. **V** Various nasal chuckles, *chaka-chevu* or similar; also rattling *wheek chechechecheche-chuk*, and *kew kew* followed by 2–8 *kyhek* or *chack* notes. **H** Open woodland, forest edges, cultivation, plantations, coastal woodland, parks, gardens. **D** Peninsular Malaysia to the Philippines.

25 WHITE-RUMPED TRILLER *Lalage leucopygialis* 19cm Forages from near ground to crowns of small or large trees, singly, in pairs or in small groups. **V** Loud, swelling series of rapid, hard, clear chattering notes. **H** Forest edges, heavily disturbed forest, secondary forest, lightly wooded cultivation and mangroves. **D** Sulawesi, Indonesia.

26 BLACK-WINGED CUCKOOSHRIKE *Lalage melaschistos* 24cm Arboreal. Active and conspicuous, forages singly or in pairs; regularly joins mixed-species feeding flocks. **V** Slow, descending *twii-twii-weeo-weeow* or similar. **H** Broadleaved evergreen forest; wintering birds occasionally in deciduous forests, open woods and gardens. **D** Himalayas through to SE Asia.

27 BLACK-HEADED CUCKOOSHRIKE *Lalage melanoptera* 18cm Usually seen singly, in pairs or in small parties; regularly part of mixed-species feeding flocks. **V** Mellow whistling followed by rapid, repeated *pit-pit-pit*. **H** Broadleaved forest and secondary growth. **D** India, Sri Lanka.

28 INDOCHINESE CUCKOOSHRIKE *Lalage polioptera* 21–22cm Slow, deliberate forager in middle to upper storey; sometimes joins mixed-species feeding parties. **V** High-pitched, descending *wi-wi-wi-wi-wu* or similar; also nasal *uh uh uh uh-ik*. **H** Deciduous, semi-deciduous and pine forests, locally also in peat-swamp forest. **D** SE Asia.

29 LESSER CUCKOOSHRIKE *Lalage fimbriata* 19–21cm Forages in canopy and is frequently part of mixed-species feeding flocks. **V** Loud, clear *whit-it-it-chui-choi*; also rapid *whit-whit-whit-whit-whit-whit*. When alarmed utters squeaky, nasal *wherrrh-wherrrh-wherrrh*... and high *whit-weei*. **H** Broadleaved evergreen forest, secondary growth, plantations. **D** SE Asia.

30 MAURITIUS CUCKOOSHRIKE *Lalage typica* 22cm No other cuckooshrike in its range. Darker upperparts than Reunion Cuckooshrike. **V** Sharp, rattling chatter (1 sec). **H** Humid forest with surrounding wooded and scrubby areas. **D** Mauritius, Indian Ocean.

31 REUNION CUCKOOSHRIKE *Lalage newtoni* 22cm No other cuckooshrike in its range. Note uniform grey upperparts with dark wing and tail feathers of male and barring below in female. **V** Very high, strong phrase of 4–6× *weet* notes. **H** Restricted to a small forest area at about 1,250m altitude. **D** Réunion, Indian Ocean.

1
2
3
4
5
6
7
8
9
10
11
12
13
14
15
16
17
18
19
20
21
22
23
24
25
26
27
28
29
30
31

WHITEHEADS MOHOUIDAE

1 YELLOWHEAD *Mohoua ochrocephala* 15cm Unmistakable. Note thickset appearance and strong, black legs and feet. Forms noisy groups in canopy. **V** Very high, rapid *tsitsitsi...* (up to six notes). **H** Prefers tall beech forest. **D** New Zealand.

2 WHITEHEAD *Mohoua albicilla* 15cm Separated from Yellowhead by white head and different range. Forages at all forest levels. **V** Varied, including short series beginning with descending chirrup, ending with *reetseesee*. **H** Tall, open, native forest. **D** New Zealand.

3 PIPIPI *Mohoua novaeseelandiae* 13cm Note grey face sides and short, white stripe behind eye. **V** Short, descending series of fluted notes, including *tuut-tuweet-tuweet-tooh*. **H** Forest, secondary growth, pine plantations. **D** New Zealand.

SITTELLAS NEOSITTIDAE

4 VARIED SITTELLA *Daphoenositta chrysoptera* 10–14cm Highly variable, with five races across range. Easily identified by compact size, short tail and slightly upturned bill. Agile tree-climbing behaviour recalls northern-hemisphere nuthatches (Plate 236). Social and active, often in restless groups. **V** Simple vocalisations with variety of squeaky *chip* notes, uttered randomly or in excited series. **H** Eucalypt forests, woodlands. **D** Australia.

5 PAPUAN SITTELLA *Daphoenositta papuensis* 11cm Formerly conspecifc with Varied Sitella. Variable with six races across range. Best identified by distinctive behaviour (see Varied Sitella). Males have variable head coloration. Females have white heads. **V** High-pitched twittering, repeated constantly. **H** Montane forest, forest edges. **D** New Guinea.

6 BLACK SITTELLA *Daphoenositta miranda* 12cm Black with pink face and buffish tail tip. Usually in canopy. Habits similar to Papuan Sitella, but lives at higher elevations. **V** Flocks give high-pitched twittering. **H** Montane forest. **D** New Guinea.

PLOUGHBILL EULACESTOMATIDAE

7 WATTLED PLOUGHBILL *Eulacestoma nigropectus* 13cm Unique small compact bird with distinctive stout bill. Male has large fleshy pink wattle at base of bill. Female lacks wattle. Usually singly or in pairs, sometimes in mixed flocks. Keeps low. **V** Long, mournful whistle *eeeeeeee*. **H** Montane forest edges, bamboo. **D** New Guinea.

AUSTRALO-PAPUAN BELLBIRDS OREOICIDAE

8 RUFOUS-NAPED BELLBIRD *Aleadryas rufinucha* 17cm Small inhabitant of understorey with short erectile crown feathers (often hidden). Mainly olive-green with rufous nape patch and yellow throat. Sexes similar. **V** Series of rapid whistles *choo-choo-choo-choo-choo*. Also raspy grating sounds and strident shorter calls. **H** Monatne forest, secondary growth. **D** New Guinea.

9 PIPING BELLBIRD *Ornorectes cristatus* 25cm Shy inhabitant of understorey. Larger than Rufous-naped Bellbird, also with short erectile crown feathers. Dark rufous-brown above, paler below. Sexes alike. **V** Distinctive, continuous pulsating *poo-poo-poo-poo-poo-poo*, sometimes lasting for many minutes. **H** Forest in lowlands and foothills. **D** New Guinea.

10 CRESTED BELLBIRD *Oreoica gutturalis* 21cm Male has striking face pattern and erectile crest (usually hidden). Female much plainer. Unobtrusive, feeds on ground. **V** Distinctive song consists of 3–5 pure whistles followed by whip-like sound: *hoo hoo hoo-hoo whip*. **H** Woodland and scrub in arid areas. **D** Australia.

WHISTLERS AND ALLIES PACHYCEPHALIDAE

11 CRESTED SHRIKETIT *Falcunculus frontatus* 15–19cm Colourful thickset bird with very stout bill. Striking black-and-white head pattern and short erectile crest. Sexes similar. Usually in pairs or small parties. **V** Slow, mournful, descending series of whistles *hee-u hee-u hee-u hee-u*. **H** Eucalypt forest, woodland. **D** Australia.

12 MAROON-BACKED WHISTLER *Coracornis raveni* 15cm Shy and skulking. Usually seen in pairs or singly, foraging from sub-stage to lower mid-storey. **V** Repeated, loud, explosive whip-crack call. **H** Montane forests. **D** Sulawesi, Indonesia.

13 SANGIHE SHRIKETHRUSH *Coracornis sanghirensis* 17–19cm Forages in middle and upper storey, also in dense rattan undergrowth and occasionally on ground. **V** Song loud, consists of repeated 10-sec phrases; also lisping *chweep chweep*. **H** Lower montane primary forest, well-established secondary forest. **D** Sangihe Island, north of Sulawesi, Indonesia.

14 BLACK PITOHUI *Melanorectes nigrescens* 23cm Stout, heavy-billed inhabitant of lower levels. Male all-black. Female olive-brown above and tawny below. Singly or in pairs. **V** Variety of calls including fast ringing series of whistles on same note, and slower series of mournful downslurred notes. **H** Lower montane forest. **D** New Guinea

15 OLIVE WHISTLER *Pachycephala olivacea* 21cm Drab, thickset whistler with whitish throat. Sexes similar. Favours undergrowth. **V** Considerable variation in calls, comprising powerful drawn-out whistles. **H** Thickets, scrub, woodland, forests. **D** SE Australia.

16 RED-LORED WHISTLER *Pachycephala rufogularis* 21cm Plain brownish-grey whistler with rufous lores and throat. Sexes similar. Shy and elusive, keeping to cover. **V** Series of loud rich whistles *cheeoo-cheeoo-cheeoo*. **H** Mallee scrub with dense undergrowth. **D** SE Australia.

17 GILBERT'S WHISTLER *Pachycephala inornata* 20cm Similar to Red-lored Whistler but male has black lores, and rufous restricted to throat. Female lacks rufous and is plain-faced. Note white undertail-coverts. Feeds on ground or in low cover. **V** Series of rasping double whistles. **H** Dry open woodland and forest with shrubs. **D** S Australia.

18 MANGROVE WHISTLER *Pachycephala cinerea* 17cm Unobtrusive and sluggish; gleans insects from branches and trunks. **V** Loud, whistled *oo-oo-oo-oo chew-it* or similar; in W Philippines, utters a whistled *peeee pur-purr chiaoonkk* or *peee pur chiaoonkk*, last note ending abruptly. **H** Mangroves and nearby vegetation, forest, riverine vegetation, plantations, wooded gardens. **D** Widespread in coastal S and SE Asia, Indonesia and Borneo.

19 GREEN-BACKED WHISTLER *Pachycephala albiventris* 16cm Shy, foraging quietly in canopy or understorey; seen singly, in pairs or in mixed-species feeding groups. **V** Loud, metallic *peee* or *peee-wit*, repeated frequently. **H** Forests, from lowlands to high mountains. **D** Philippines.

20 WHITE-VENTED WHISTLER *Pachycephala homeyeri* 16–17cm Inconspicuous. Generally occurs singly or as part of mixed-species feeding parties. **V** Various whistles, including *oo-wichee, yump, yump-wit* and *u-wichee-u...*; also high-pitched *wheeu tu tu tu* or similar. **H** Forests, from sea-level to mountains. **D** Philippines and islands off NE Borneo.

21 ISLAND WHISTLER *Pachycephala phaionota* 16cm Skulking. Forages in shady understorey, singly or in pairs. **V** Loud *weet-chuw-weeEEee*, busy *chup* calls and nasal notes. **H** Mangroves, coastal woodland, beach scrub, secondary scrub, plantations. **D** Maluku Islands and West Papuan Islands, Indonesia.

22 RUSTY WHISTLER *Pachycephala hyperythra* 15cm Shy and unobtrusive; usually seen singly. Grey head, white throat and pale rufous underparts. Sexes alike. **V** Series of musical whistles that speeds up at the end *tooee-tooee-tooee-tooee chew-chew*. **H** Forest and forest edges in lowlands and foothills. **D** New Guinea.

23 BROWN-BACKED WHISTLER *Pachycephala modesta* 14cm Small, rather drab unobtrusive whistler of montane forest. Usually singly or in pairs; joins mixed flocks. **V** Typical whistler song with series of whistled *wheet-wheet* notes, ending with explosive *chew-chew*. **H** Montane forest and forest edges, clearings, gardens. **D** E New Guinea.

24 YELLOW-BELLIED WHISTLER *Pachycephala philippinensis* 15–16cm Inconspicuous. Forages from understorey to canopy, singly, in pairs or as part of mixed-species feeding flocks. **V** Repeated, rising *peeeeeup*; also forceful *hu-i-yu wit weeu*. **H** Forests, from sea-level to mountains. **D** E Philippines.

25 SULPHUR-VENTED WHISTLER *Pachycephala sulfuriventer* 14–15cm Encountered singly, in pairs or as part of mixed-species feeding parties; forages mainly in mid-storey, on trunks or large branches. **V** Whistled *wiwiwiwiwiwiuWHIT*, decreasing in pitch and increasing in volume; also short, staccato *wheeu tu tu tu*. **H** Primary and secondary forest, forest edges. **D** Sulawesi, Indonesia.

26 BORNEAN WHISTLER *Pachycephala hypoxantha* 16cm Forages in middle to upper canopy, often as part of mixed-species feeding parties. Captures prey by gleaning or during fly-catching sallies. **V** Whistled *dee-dee-dee-dee-dit*, last note like a whip-crack. **H** Montane forests. **D** Borneo.

27 VOGELKOP WHISTLER *Pachycephala meyeri* 15cm Small-billed whistler restricted to the Vogelkop Peninsula. Grey head and yellow belly. Sexes alike. Usually seen singly. **V** Pleasant series of powerful descending whistles. **H** Montane forest and thickets, at lower altitudes. **D** NW New Guinea

28 GREY WHISTLER *Pachycephala simplex* 15–16cm Forages in mid-storey and canopy, mainly gleaning from foliage or branches. Often forms part of mixed-species feeding flocks. **V** Series of five whistled notes, the first two brief and the last three bisyllabic. **H** Lowland monsoon forest. **D** N Australia, New Guinea.

29 FAWN-BREASTED WHISTLER *Pachycephala orpheus* 14cm Occurs singly or in pairs, foraging from understorey to mid-storey. **V** Series of 10–16 clear, whistled notes, beginning softly and then increasing in volume and speed, the notes at the end running together before ending in sharp, explosive double note. **H** Monsoon forest, secondary growth, beach and hill scrub, mangroves. **D** Lesser Sunda Islands.

30 SCLATER'S WHISTLER *Pachycephala soror* 15cm Colourful widespread whistler with some racial variation. Males have black head and white throat bordered below with black necklace (which females lack). Usually singly or in pairs but joins mixed flocks. **V** Typically series of accelerating whistles, ending in a flourish. **H** Lower montane forest, forest edges, secondary growth. **D** New Guinea.

31 RUSTY-BREASTED WHISTLER *Pachycephala fulvotincta* 16–19cm Usually seen singly or in pairs. Forages mainly in middle levels, although may sing from canopy. **V** Loud, clear whistles. **H** Primary and secondary forest, scrub, wooded cultivation, occasionally mangroves. **D** Java and the Lesser Sunda Islands.

32 YELLOW-THROATED WHISTLER *Pachycephala macrorhyncha* 16–19cm Behaviour presumed to be similar to that of Rusty-breasted Whistler. **V** Loud, clear whistles. **H** Presumed to be similar to Rusty-breasted Whistler. **D** Lesser Sunda Islands and S Maluku Islands, Indonesia.

33 BALIEM WHISTLER *Pachycephala balim* 16cm Similar to Sclater's Whistler, but male has yellow hind-collar and female has finely scalloped throat. **V** Loud explosive whistles typical of genus. **H** Forest and forest edges at 1,500–2,500m. **D** Baliem Valley, WC New Guinea.

34 BLACK-CHINNED WHISTLER *Pachycephala mentalis* 16–19cm Behaviour presumed to be similar to that of Rusty-breasted and Yellow-throated whistlers. **V** Loud, clear whistles. **H** Presumed to be similar to that of Rusty-breasted and Yellow-throated Whistlers. **D** N Maluku Islands, Indonesia.

1
2
3
4
5
6
7
8
9
10
11
12
13
14
15
16
17
18
19
20
21
22
23
24
25
26
27
28
29
30
31
32
33
34

WHISTLERS AND ALLIES *CONTINUED*

1 AUSTRALIAN GOLDEN WHISTLER *Pachycephala pectoralis* 17cm Striking colourful whistler with much racial variation, although many races now elevated to full species (see below). Strong sexual dimorphism. Male has black hood, white throat, black breast-band and yellow underparts. Female rather plain but much variation. Usually singly or in pairs, but will join mixed flocks. **V** Series of pure slow whistles *hoo hoo hoo hoo* followed by whiplash sound, sometimes delivered faster. **H** Wide range of forest and woodland. **D** E Australia.

2 WESTERN WHISTLER *Pachycephala occidentalis* 17cm Western counterpart of Australian Golden Whistler and formerly conspecific. Usually singly or in pairs, but will join mixed flocks. **V** Similar to that of Golden Whistler. **H** Forest, woodland. **D** SW Australia.

3 BISMARCK WHISTLER *Pachycephala citreogaster* 16cm Formerly conspecific with Australian Golden Whistler. Male has broad black breast-band. Female yellowish below. Usually singly or in pairs, but will join mixed flocks. **V** Short explosive notes, often ending with a whiplash. **H** Forest, secondary growth. **D** Bismarck Archipelago.

4 ORIOLE WHISTLER *Pachycephala orioloides* 17cm Formerly conspecific with Australian Golden Whistler. Nine races, with much variation. All males have yellow throat with variable black breast-band. Usually singly or in pairs. **V** Variable, comprising loud explosive melodious whistles ending in a whiplash. **H** Forests. **D** Solomon Islands.

5 LOUISIADE WHISTLER *Pachycephala collaris* 16cm Formerly conspecific with Australian Golden Whistler. Long-billed whistler with greenish wings. Male has pale rusty breast band. **V** Rich melodious whistles given in short bursts. **H** Forests. **D** Louisiade Archipelago (SE New Guinea).

6 RENNELL WHISTLER *Pachycephala feminina* 15cm Formerly conspecific with Australian Golden Whistler, but no colourful plumage in male. Olive-rufous above and dull yellow below. Sexes alike. Unobtrusive, often seen close to ground. **V** Loud melodious song. **H** Forests. **D** Rennell (Solomon Islands).

7 MELANESIAN WHISTLER *Pachycephala chlorura* 17cm Formerly conspecific with Australian Golden Whistler. Male has narrow collar. Usually singly or in pairs, but will join mixed flocks. **V** Typical series of notes ending in a whiplash but also squeakier bursts of song. **H** Forest, forest edges, gardens. **D** Vanuatu, Loyalty Islands.

8 NEW CALEDONIAN WHISTLER *Pachycephala caledonica* 17cm Formerly conspecific with Australian Golden Whistler. Male has grey head, white throat and narrow black breast-band. Usually singly or in pairs, but will join mixed flocks. **V** Simple short whistles including a whiplash. **H** Forest, forest edges, gardens. **D** New Caledonia.

9 FIJI WHISTLER *Pachycephala vitiensis* 15cm Formerly conspecific with Australian Golden Whistler. Ten races with much variation. Most males have yellow throats. Black breast-band varies from narrow to incomplete. **V** Simple series of whistles ending in a whiplash. **H** Forest, forest edges, gardens. **D** Fiji islands.

10 TEMOTU WHISTLER *Pachycephala vanikorensis* 17cm Formerly conspecific with Australian Golden Whistler. Male olive or black above, sometimes with golden wing-edgings. Female has brownish breast-band and yellowish underparts. **V** Song squeaky melodious warble. **H** Forest, forest edge and gardens. **D** Santa Cruz Islands (Solomon Islands).

11 TONGAN WHISTLER *Pachycephala jacquinoti* 18cm Male unmistakable with all-black hood and broad yellow collar; female much duller, with grey crown washed cinnamon, and greyish-white chin and throat. **V** Typical song comprises series of whistles ending in explosive whiplash. **H** Undergrowth of forest, secondary growth. **D** Tonga.

12 MANGROVE GOLDEN WHISTLER *Pachycephala melanura* 16cm Male similar to Australian Golden Whistler but has narrow yellow collar and grey wings. Female has pale yellow belly. **V** Similar to Australian Golden Whistler. **H** Mangroves, rainforest, scrub. **D** W New Guinea, N Australia.

13 SAMOAN WHISTLER *Pachycephala flavifrons* 17cm Male blackish above and yellow below with variable head pattern. Forehead and short supercilium white or yellow, throat white or yellow with variable mottling. **V** Short, thrush-like strophes. **H** Mature forest, secondary growth, gardens. **D** W Samoa.

14 HOODED WHISTLER *Pachycephala implicata* 16cm Male similar to Bougainville Whistler but much duller. Head and breast blackish grey, upperparts dark olive, underparts yellowish olive. Female similar to Bougainville Whistler. **V** Loud explosive series of notes, given in short bursts. **H** Montane forest. **D** Guadalcanal (Solomon Islands).

15 BOUGAINVILLE WHISTLER *Pachycephala richardsi* 16cm Male has black hood and breast, but lacks yellow hind-collar. Female similar but head grey and throat greyish white. **V** Short varied bursts of song, repeated frequently. **H** Montane forest. **D** Bougainville Island (Bismarck Archipelago).

16 BARE-THROATED WHISTLER *Pachycephala nudigula* 19–20cm Usually encountered singly. Very vocal; its calls are one of the characteristic sounds of montane forests in Flores. **V** Complex, beautiful, random series of loud, clear notes, including trilled whistles, bell-like bugles, squawks and squeaks. **H** Primary and tall secondary moist hill and montane forest. **D** Sumbawa and Flores, Lesser Sunda Islands.

17 LORENTZ'S WHISTLER *Pachycephala lorentzi* 16cm Small-billed whistler with greyish-white throat and breast, and yellow belly. Sexes alike, resembling female Regent Whistler. Usually seen singly. **V** Sweet accelerating series of whistles. **H** Montane forest. **D** WC New Guinea.

18 REGENT WHISTLER *Pachycephala schlegelii* 16cm Male striking with very broad breast-band, broad yellow collar, small white throat-patch and orange wash on belly. Female like female Lorentz's Whistler. **V** Typical whistler vocalisations but more varied than most species. **H** Montane forest and forest edge. **D** New Guinea.

19 GOLDEN-BACKED WHISTLER *Pachycephala aurea* 16cm Similar to Australian Golden Whistler and its close relatives, but back bright yellow. Sexes alike. Usually singly or in pairs. **V** Rich, melodious song typical of group, but delivered at rather leisurely speed. **H** Secondary growth, thickets and scrub in lowlands and foothills. **D** S New Guinea.

20 RUFOUS WHISTLER *Pachycephala rufiventris* 17cm Distinctive white-throated whistler with grey upperparts, black breast-band and rufous underparts. Female lacks strong head pattern. **V** Rather varied, comprising pleasant whistles and typical whistler vocalisations. **H** Open forest, woodlands. **D** Australia, New Caledonia.

21 BLACK-HEADED WHISTLER *Pachycephala monacha* 16cm Unusual whistler, with black head and breast, blackish upperparts and white belly. Female greyer. **V** Varied series of whistles at different speeds and pitch, sometimes ending in a whiplash. **H** Forest edge, clearings, gardens, in foothills. **D** C New Guinea.

22 WHITE-BELLIED WHISTLER *Pachycephala leucogastra* 15cm Male similar to Black-headed Whistler, but has white throat and black breast-band. Female greyer and lacks breast-band. Usually singly or in pairs. **V** Typical series of whistles with an explosive ending. **H** Forest edge, plantations and mangroves, in lowlands and foothills. **D** SE New Guinea.

23 WALLACEAN WHISTLER *Pachycephala arctitorquis* 14cm Forages in upper or middle levels of trees, singly or in pairs. **V** Slowly accelerating series of five upslurred whistles, ending with high *whee*; also rising and falling, loud *chitchuwit* and *chuuuu-wit*, followed by two short notes and then nine bisyllabic notes. **H** Mangroves, forest, forest edges, plantations, gardens. **D** E Lesser Sunda Islands.

24 DRAB WHISTLER *Pachycephala griseonota* 14–16cm Forages from understorey to canopy, singly, in pairs or as part of mixed-species feeding groups. Best located by its calls. **V** Loud, prolonged, cheery warble; also six staccato notes with explosive finish. **H** Primary and tall secondary forest, selectively logged forest, lightly wooded cultivation. **D** Maluku Islands and Sula Islands, Indonesia.

25 CINNAMON-BREASTED WHISTLER *Pachycephala johni* 14–16cm Forages from understorey up to canopy, sometimes in mixed-species feeding parties. **V** Presumed to be similar to that of Drab Whistler. **H** Primary and tall secondary forest and lightly wooded cultivation. **D** Obi Island, N Maluku Islands, Indonesia.

26 WHITE-BREASTED WHISTLER *Pachycephala lanioides* 19cm Striking whistler, grey and white with chestnut collar extending around below black breast-band. Female rather plain without head pattern. **V** Varied repertoire includes typical series of whistles ending with a whiplash. **H** Mangroves. **D** W, N Australia.

27 MORNINGBIRD *Pachycephala tenebrosa* 19cm Dull-coloured with rather large head and dark eyes. Skulks in undergrowth. **V** Song is a slow and deliberate jumbled warble of liquid chirrups and whistles. **H** Forest undergrowth. **D** Palau, WC Pacific.

28 WHITE-BELLIED PITOHUI *Pseudorectes incertus* 22cm Nondescript rufous-brown with plain face and mottled breast. Sexes alike. Usually in small groups. **V** Pleasant melodic warbling, preceded by quiet ticking notes. **H** Seasonally flooded lowland forest. **D** S New Guinea.

29 RUSTY PITOHUI *Pseudorectes ferrugineus* 26–28cm Rusty thrush-like bird of lower levels of forests. Note pale eye. Bill may be black or yellowish. Sexes alike. Usually in pairs or small groups; joins mixed flocks. **V** Rich melodic warbling. More often heard than seen. **H** Rainforest, secondary growth and plantations, in lowlands and foothills. **D** New Guinea.

30 BOWER'S SHRIKETHRUSH *Colluricincla boweri* 20cm Medium-sized, with robust black bill. Note lightly streaked throat and rufous underparts. Singly or in pairs, in mid-storey. **V** Rich series of powerful calls, often preceded by quiet ticking notes. **H** Rainforest, usually above 400m. **D** NE Australia.

31 SOOTY SHRIKETHRUSH *Colluricincla tenebrosa* 19cm Formerly known as Sooty Whistler. All-dark brown or blackish brown, unobtrusive and little-known. Sexes alike. Singly or in pairs. **V** Not known. **H** Montane forest and forest edge. **D** C New Guinea.

32 LITTLE SHRIKETHRUSH *Colluricincla megarhyncha* 17–19cm Small shrikethrush with many similar races, some with pale bill and others with dark bill. Nondescript plumage. Sexes alike. **V** Rich melodious whistles, which vary across range. **H** Rainforest, secondary growth. **D** New Guinea, N, E Australia.

33 GREY SHRIKETHRUSH *Colluricincla harmonica* 23cm Large widespread shrikethrush with five races. Mostly grey but some races have brownish backs. Males usually have white lores. Sexes similar. Usually in pairs. **V** Fine songster; rich melodious and varied. **H** Wide range of habitats. **D** E New Guinea, Australia.

34 SANDSTONE SHRIKETHRUSH *Colluricincla woodwardi* 25cm Large slender shrikethrush with long tail. Sexes similar. Forages on rocks. **V** Loud melodious whistles given from prominent perch. **H** Sandstone gorges, escarpments. **D** N Australia.

1
2
3
4
5
6
7
8
9
10
11
12
13
14
15
16
17
18
19
20
21
22
23
24
25
26
27
28
29
30
31
32
33
34

SHRIKES LANIIDAE

1 YELLOW-BILLED SHRIKE *Corvinella corvina* 30cm Very large with long tail. **V** Song sometimes transcribed as *may we wait, may we wait, may we*. Calls include a variety of whistles, buzzes and rasping sounds. **H** Bushy and wooded areas. **D** E, C and W Africa.

2 MAGPIE SHRIKE *Urolestes melanoleucus* 45cm Very long tail. Usually seen in parties of 3–10. **V** Territorial calls include a variety of loud, repeated whistles, *needle boom needle boom* or *come here come here*. Pairs call in duet, the male's *teeloo* answered by a higher-pitched *teeu* from the female. **H** Open, wooded grassland, thornveld. **D** C and E Africa.

3 NORTHERN WHITE-CROWNED SHRIKE *Eurocephalus ruppelli* 25cm Loose groups of 2–12, hunting small prey from top of bush or perch. Note conspicuous, white crown. **V** Song is a complex mix of noisy, harsh, nasal notes. Typical call is a short rather high-pitched squawk. **H** More or less wooded and bushy areas. **D** SE South Sudan, C Ethiopia and C Somalia to C Tanzania.

4 SOUTHERN WHITE-CROWNED SHRIKE *Eurocephalus anguitimens* 20cm Very similar to Northern White-crowned Shrike. **V** High, mewing *peew-peew pee-peew-pu-pieuw*. **H** Wooded country. **D** C Africa.

5 TIGER SHRIKE *Lanius tigrinus* 17cm Perches in cover more than most shrikes. Captures insect prey from leaves and branches. **V** Calls include repeated *tcha* and sharp *tchik*. **H** Open forest, forest edges, clearings. **D** Breeds SE Siberia, E and NE China, Korean Peninsula and Japan. Winters SE Asia to Greater Sunda Islands.

6 SOUZA'S SHRIKE *Lanius souzae* 18cm White edge to dull-brown mantle diagnostic in adult. **V** Various chattering and whistling calls, but generally rather quiet. **H** Miombo and other woodland. **D** C Africa.

7 BULL-HEADED SHRIKE *Lanius bucephalus* 19cm Behaviour much as others of genus. In flight, shows white flashes at base of primaries. **V** Loud, consisting of clear whistles and mimicry of other birds and insects such as cicadas. Calls include chattering *ju-ju-ju* or *gi-gi-gi*, trilling *kürrrririri* and short *kew* repeated 3–4 times when a male first notices a female. **H** Light hill forest, tree-lined meadows, light riverine forests. In the Kuril Islands and Japan, frequents clearings in taiga forest and bamboo, cultivated areas with scattered trees, roadside trees and shrubs, parks, gardens. **D** E Asia.

8 BROWN SHRIKE *Lanius cristatus* 18cm Uses prominent perch to launch aerial sallies to capture prey on or near the ground. **V** Harsh *chr-r-r-ri*. **H** Forest edges, scrub, open cultivation. **D** Breeds E Asia. Winters S and SE Asia.

9 RED-BACKED SHRIKE *Lanius collurio* 17cm Usually perches prominently on bush tops or other vantage points from where it pounces on insects or small vertebrates. In flight shows grey rump and white bases on outer-tail feathers. **V** Harsh *shack* or *shak-shak*; also harsh *churruck-churruck*. **H** Dry country with bushes, also cultivation edges. **D** Breeds Europe to W Siberia, W Kazakhstan, Turkey and NW Iran. Winters E and S Africa.

10 ISABELLINE SHRIKE *Lanius isabellinus* 17cm Perches on bush tops or other vantage points, drops down to catch large insects or small vertebrates. Also takes some flying insects in the air. Impales prey items. In flight, shows white flashes at base of primaries. **V** Babbling and warbling, including melodious whistles and harsh notes. Very similar to Red-backed Shrike. **H** Arid areas with scattered bushes, cultivation edges, bushy areas in lowlands and hills. **D** Breeds C Asia. Winters S Asia and E Africa.

11 RED-TAILED SHRIKE *Lanius phoenicuroides* 17cm Actions and habits similar to Red-backed Shrike. Formerly considered conspecific with Isabelline Shrike. **V** Babbling and warbling with harsh and melodious whistles; also harsh grating call. **H** Open dry scrub. **D** Breeds C Asia to Pakistan. Winters SW Asia and E Africa.

12 BURMESE SHRIKE *Lanius collurioides* 20cm Confiding; feeding actions similar to Bull-headed Shrike. **V** Sweet, with musical and grating notes; when alarmed utters rapid, harsh *chikachikachitchit* or similar. **H** Open woodland, forest edge, clearings, secondary growth, cultivations. **D** SE Asia.

13 EMIN'S SHRIKE *Lanius gubernator* 15cm Note chestnut (not grey) rump and brown tail feathers narrowly edged white (not black and white). **V** Song is a short series of whistles, *tweet-u-wee-u-weet* or *trip-tu-trip-srtp*, interspersed with hoarse *chweeeh* notes. Calls include hisses, twitters, whistles and clear notes. **H** Lightly wooded and bushy areas. **D** S Mali and N Ivory Coast to S South Sudan, N Uganda and NE DR Congo.

14 BAY-BACKED SHRIKE *Lanius vittatus* 18cm Territorial, often occupying same prominent perches day after day. Actions similar to Red-backed Shrike. In flight shows white rump, white patches at base of primaries and white outer-tail feathers. **V** Pleasant loud warble interspersed with mimicry of other birds' calls; also scolding *chur-r* or *chee-urr*. **H** Open dry country with scattered trees and scrub. **D** SC Asia.

15 LONG-TAILED SHRIKE *Lanius schach* 24cm Noisy, restless and aggressive, recorded robbing other birds of food; otherwise behaviour much as Red-backed Shrike. In flight shows white bases to primary feathers and a rufous rump. **V** Metallic warble, often including mimicry of other birds' calls and songs; also utters harsh *tchick*, buzzing *grennh* and coarse *rrrre*. **H** Open wooded country, cultivations, gardens. **D** Widespread from S and SE Asia to Australasia.

16 GREY-BACKED SHRIKE *Lanius tephronotus* 22cm Usually seen singly, or in widely separated pairs; feeding behaviour much as Red-backed Shrike. In flight shows rufous rump. **V** Prolonged, subdued, melodious warble interspersed with mimicry of others birds' calls; also harsh grating calls. **H** Forest clearings, high-altitude open scrub; winters in lower-altitude scrub and cultivations. **D** Breeds Himalayas, C China. Winters SE Asia.

17 MOUNTAIN SHRIKE *Lanius validirostris* 20–23cm Generally found singly or in pairs. Dives down, from exposed perch, to capture insect prey from or near ground. **V** Series of harsh, loud, whistled *piaaoo* or *chaaoo* notes. **H** Montane forest clearings, open secondary growth, forest edges, scrub in grassland. **D** Philippines.

18 MACKINNON'S SHRIKE *Lanius mackinnoni* 20cm Separated from Lesser Grey Shrike by white eyebrow and scapulars. No or only very little white shows on closed wings. **V** Song recalls a *Sylvia* warbler (Plate 227), a mixture of hoarse notes and intense warbles, with much mimicry. **H** More or less bushy natural and cultivated areas. **D** SE Nigeria and Cameroon to NW Angola and NE Congo, and N DR Congo to W Kenya and N Tanzania.

19 LESSER GREY SHRIKE *Lanius minor* 20cm Pounces on prey from prominent perches; tends to hover and pounce more than other shrikes. In flight shows white patch on primaries and white outer-tail feathers. **V** Babbling chatter that includes mimicry; also harsh *kerrib-kerrib*. **H** Open semi-desert and cultivation with scattered bushes. **D** Breeds S Europe to S Russia, S to Syria, Iraq and Afghanistan. Winters S Africa.

20 LOGGERHEAD SHRIKE *Lanius ludovicianus* 23cm In flight shows large, white patch at base of primaries. Drops from exposed perches onto quarry; sometimes hovers while searching for prey. **V** Harsh *jaaa* and grating *teen raad raad raad raad raad*; song a sharp *krrDI krrDI krrDI...* **H** Quite variable, including open areas with low vegetation, small shrubs and trees, pastures, woodland edges. **D** Widespread in North America and Central America.

21 NORTHERN SHRIKE *Lanius borealis* 24cm In flight shows large white patch at the base of the primaries. Perches prominently, often using same vantage point for long periods, often hovers when searching for small vertebrate prey. Regularly pumps tail. **V** Harsh *shraaaa*; also repeated *kdldi kdldi kdldi...* **H** Open scrub, western birds favouring feltleaf willow. **D** Breeds N North America, N Asia. Winters N China, Japan.

22 GREAT GREY SHRIKE *Lanius excubitor* 24cm Perches prominently, often using same vantage point for long periods. Feeds mainly on small vertebrates, especially voles. In flight, shows large white patches at base of primaries. **V** Song *trrr-turit-trrr-turit...*; call a ringing *shreee*; also often repeated nasal *shack*. **H** Various types of open country with scattered trees, bushes or scrub. **D** Widespread in Europe, Asia and N Africa.

23 STEPPE GREY SHRIKE *Lanius pallidirostris* 25cm Perches prominently on bushes, rocks or wires, from where sallies are made to capture prey on or near ground. **V** Calls include repeated *kwi-wide*, *sheenk-sheenk* or *shihk-shihk*. **H** Open country with scattered bushes. **D** Breeds S Russia and S Kazakhstan to Mongolia and N China, S to NW Iran, Afghanistan and Pakistan. Winters SW Asia and NE Africa.

24 IBERIAN GREY SHRIKE *Lanius meridionalis* 25cm Feeding actions similar to Great Grey Shrike. In flight shows extensive white patches on base of primaries and on outer-tail feathers. **V** Repetition of two melodious notes, the second higher-pitched, often interspersed with a long subdued warble; also harsh *tscheee*. **H** Semi-desert scrub, open thornbush, cultivation edge. **D** Breeds S France, Iberian Peninsula. Winters to NW Africa.

25 CHINESE GREY SHRIKE *Lanius sphenocercus* 29cm Behaviour similar to Great Grey Shrike. In flight, shows large white wing-bar. **V** Repetition of two melodious notes, *tschriii*, the first lower in pitch; call a harsh, nasal *tscheee*. **H** Open country with scattered trees and bushes, open steppe, semi-deserts. In Japan, frequents cultivated plains, open woodland, reclaimed land. **D** E Asia.

26 GREY-BACKED FISCAL *Lanius excubitoroides* 25cm Gregarious, usually in parties of up to 20. Confiding. In flight, shows white patches at base of primaries and large black tip to white-sided tail. Feeding behaviour much as others of genus. **V** Very varied: *kyoir, kyoi, kyo-ooh*, also screaming and chattering notes when moving from tree to tree. **H** Wooded savannah, woodland, open bush with scattered thorn trees. **D** C and NE Africa.

27 LONG-TAILED FISCAL *Lanius cabanisi* 30cm Note plain black (not white-edged) mantle and long all-black tail. Female has chestnut on flanks but this is often hidden. **V** Simple series of 3–5 notes. **H** Wooded and bushy natural and cultivated areas; <1,750m. **D** SE Somalia to C Tanzania.

28 TAITA FISCAL *Lanius dorsalis* 20cm Resembles Somali Fiscal but has no white on secondaries. **V** Song has liquid and rhythmic quality. **H** Dry, more or less wooded and bushy areas. **D** SE South Sudan, S Ethiopia and C Somalia to NE Tanzania.

29 SOMALI FISCAL *Lanius somalicus* 20cm Black crown, grey back and white scapulars excludes all except Taita Fiscal. Note white tips to secondaries (best seen in flight). **V** Song more scratchy than that of Taita Fiscal. **H** Bush and savannah in arid and semi-arid areas. **D** Ethiopia, Somalia, N Kenya.

30 NORTHERN FISCAL *Lanius humeralis* 20cm Similar to Southern Fiscal. Female has chestnut flank patch, which is often hidden. **V** Song, given by both sexes, consists of a disjointed medley of short clear whistles, trills and warbles, grating notes and mimicry. **H** More or less wooded and bushy country, including gardens; often on telephone lines. **D** Ethiopia and sub-Saharan Africa S to Angola, SE DR Congo, Zambia, Malawi, N Mozambique.

31 SOUTHERN FISCAL *Lanius collaris* 22cm Similar to Northern Fiscal. Some races show a white eyebrow. Female has chestnut flank patch, which is often hidden. Watches from prominent perches for small vertebrate and insect prey. **V** Song like that of Northern Fiscal. **H** Any type of country. **D** Namibia, SW Angola E to S Mozambique and S South Africa, also NE, C and SW Tanzania and N Malawi.

32 SAO TOME FISCAL *Lanius newtoni* 19cm The only shrike on São Tomé. **V** Territorial song is a series of loud, far-carrying, well-separated *tiu* or *juurt* notes. **H** Open forest. **D** São Tomé, Gulf of Guinea.

33 WOODCHAT SHRIKE *Lanius senator* 18cm Typical shrike, using mainly a perch-and-pounce feeding method. In flight shows white on rump, primary bases and outer-tail feathers. **V** Rattling *trr-trr-trr, gek-gek-gek*, short *crex* and *kwikwik*. **H** Open woodland, woodland edges, scrubby country. **D** Breeds SW Eurasia. Winters C Africa.

34 MASKED SHRIKE *Lanius nubicus* 17cm Often perches partly hidden. Usually feeds in typical shrike fashion, although will often take flying insects by making aerial sallies. In flight, shows large white patches at base of primaries. **V** Song a long, chattering warble; calls include various harsh notes such as *keer-keer-keer*. **H** Open woodland, woodland edge, bushy hillsides, orchards, olive groves. **D** Breeds E Greece and Bulgaria through S Turkey to Israel and Iran. Winters C and NE Africa.

1
2
3
4
5
6
7
8
9
10
11
12
13
14
15
16
17
18
19
20
21
22
23
24
25
26
27
28
29
30
31
32
33
34

VIREOS, GREENLETS AND SHRIKE-BABBLERS VIREONIDAE

1 RUFOUS-BROWED PEPPERSHRIKE *Cyclarhis gujanensis* 15cm Diagnostic rufous eyebrow. **V** Song comprises varied phrases, including *purre-purre-weet-weet* (4–5-sec intervals), repeated many times before switching to another phrase. **H** Borders and clearings in woodland and forest. **D** Widespread in Central America and South America.

2 BLACK-BILLED PEPPERSHRIKE *Cyclarhis nigrirostris* 15cm Differs from Rufous-browed Peppershrike in having darker eyes and eyebrows. Most have pale base to maxilla. **V** Sings as Rufous-browed Peppershrike. **H** Canopy and borders of montane forest. **D** Colombia, Ecuador.

3 CHESTNUT-SIDED SHRIKE-VIREO *Vireolanius melitophrys* 17cm Unmistakable by face pattern. **V** Mewing *pwîêeeh*; drawn-up *piwueúww*. **H** Pine and mixed forest. **D** C Mexico to Guatemala.

4 GREEN SHRIKE-VIREO *Vireolanius pulchellus* 14cm Heavy jizz, strong bill, grass-green above, yellow below. **V** Very high clear loud *twiwíwiwí*. **H** Forest canopy, woodland. **D** SE Mexico to Panama.

5 YELLOW-BROWED SHRIKE-VIREO *Vireolanius eximius* 13.5cm Separable from Slaty-capped Shrike-Vireo by darker eyes. **V** Endlessly repeated, high, rapid *weet-weet-weet* phrase. **H** Canopy of humid forest and tall secondary growth. **D** E Panama, N Colombia, NW Venezuela.

6 SLATY-CAPPED SHRIKE-VIREO *Vireolanius leucotis* 14cm Distinctive pale blue eyes. **V** Long series of single nasal *tjeeu* notes (one per sec). **H** Canopy of humid forest. **D** Amazonia.

7 SLATY VIREO *Vireo brevipennis* 12cm Unmistakable, but rather skulking. Like other vireos, less active and lively than warblers. **V** Very high sharp decisive *titjíistjúuiet titjúitsjúuiet*. **H** Forest edge, scrub, thickets. **D** S Mexico.

8 WHITE-EYED VIREO *Vireo griseus* 12.5cm Sluggish and rather secretive when moving around in dense vegetation. **V** Song a repeated *chip-a-tee-weeo-chip* or similar, often incorporating mimicry; calls include *rik* or *rikrikrikrik-rik-rik-rik-rik*. **H** Bushy woodlands, undergrowth, scrub, coastal thickets. **D** E North America, E Central America.

9 THICK-BILLED VIREO *Vireo crassirostris* 13cm Slow and secretive, more often heard than seen. Underpart colour varies, being yellow on southern Bahamas, greyish on the northern Bahamas and Cayman Islands, and buff on Île Tortue. **V** Song and calls very similar to that of White-eyed Vireo, but generally slower; calls usually longer. **H** Woodland edge, bushes, undergrowth. **D** West Indies.

10 MANGROVE VIREO *Vireo pallens* 12cm Eye colour and plumage may vary regionally (two morphs shown), but eye yellowish, never as white as White-eyed Vireo, yellow more uniformly spread over underparts than White-eyed, Bell's, Yellow-throated and white-chinned Yellow-winged vireos. **V** High rattling *tjatjatjatjah*. **H** Mangroves and adjacent scrub. **D** Widespread in Central America.

11 PROVIDENCIA VIREO *Vireo approximans* 13cm Formerly conspecific with Mangrove Vireo. Note white eye-ring and yellow supraloral stripe. **V** Typically series of 3–6 harsh notes, or more rapid trill. **H** Woodland, mangroves. **D** Colombian islands of Isla de Providencia and Isla Santa Catalina (Caribbean).

12 COZUMEL VIREO *Vireo bairdi* 12cm Identified by restricted range. **V** High sharp rapid rhythmic *wietwietwiet* (repeated 6–7 times). **H** Scrubby woodland. **D** Cozumel, off NE Yucatán.

13 SAN ANDRES VIREO *Vireo caribaeus* 12cm Active forager in bushes and trees, from ground level up to 10m. **V** Three different songs: chatter of repeated single-syllable notes, 2–20 or more times; a two-syllable phrase *se-wi se-wi se-wi*... repeated up to 15 times; and a variable three-syllable call. **H** Mangroves, bushes, scrubby pasture. **D** San Andrés, Caribbean.

14 JAMAICAN VIREO *Vireo modestus* 13cm Active but secretive, keeps to dense vegetation. **V** Various repeated phrases such as *sewi-sewi*, *twee-weet-weet-wuu* or *pee-eu*; also gives rapid, scolding *chi-chi-chi-chi-chi*. **H** Forests, forest edge and thickets, primarily in the arid lowlands. **D** Jamaica.

15 CUBAN VIREO *Vireo gundlachii* 13cm Sluggish, usually encountered in pairs or among mixed-species feeding flocks, often alongside Yellow-headed or Oriente warblers. **V** High-pitched, repeated *wi-chivi wi-chivi wi-chivi*, and guttural *shruo* given when agitated. **H** Brushland, forest edge, thickets and dense scrub, mainly in lowlands but also in hills and mountains. **D** Cuba.

16 PUERTO RICAN VIREO *Vireo latimeri* 13cm Inactive, best located by frequent calls. **V** Melodious three- or four-syllable whistle; also rattling *chur-chur-churr-rrr*. Other calls include *tup tup* and a grating, cat-like *mew*. **H** All types of forest including mangroves, shade coffee plantations and coastal scrub. **D** W Puerto Rico.

17 FLAT-BILLED VIREO *Vireo nanus* 12–13cm Forages slowly through bushes, occasionally feeds on ground or pursues insects in flight. **V** High-pitched, chattering *weet-weet-weet-weet-weet-weet-weet*. Also slower, repeated version. **H** Mainly semi-arid scrub and undergrowth. **D** Hispaniola.

18 BELL'S VIREO *Vireo bellii* 12cm Flicks and bobs tail in a nervous manner. **V** Nasal, rapid *sheh-sheh* or *chih-chih*. Song is chatty *chewed jechewide cheedle jeeew*. **H** Dense, low bushes near water. **D** SW and SC USA to Nicaragua.

19 BLACK-CAPPED VIREO *Vireo atricapilla* 11cm Distinctive head pattern. Active and acrobatic, although often hard to see when foraging in tangled understorey. **V** Harsh, rising *zhreee*, *sherr-sherr* or *dr-dr-dri*; also a dry *tidik*. Song consists of hurried, scratchy, warbled phrases. **H** Semi-arid scrub in oak-juniper stands. **D** SC USA and Mexico.

20 DWARF VIREO *Vireo nelsoni* 11cm Note white ring, broken above eye, whiter wing bars than Black-capped Vireo. **V** Very high rapid loud twitters, each starting with nasal *èèh*. **H** Dry scrub. **D** SW Mexico.

21 GREY VIREO *Vireo vicinior* 14cm Distinct pale eye-ring. Active forager, flicks tail as it moves through foliage. **V** Low, rasping *charr*, *jerr* or *jerr-jerr-jerr-jerr*; also rough *cherrcherr*... Song is a hesitant *ch-ree ch-ruh chee-r ch-ree*... **H** Primarily dry hillsides with scattered juniper-pinyon areas. **D** SW USA, NW Mexico.

22 BLUE MOUNTAIN VIREO *Vireo osburni* 13cm Secretive, heard more often than seen due to its habit of foraging in dense vegetation. **V** Slightly descending, trilling whistle. When agitated gives harsh, descending *burr*. **H** Primarily moist and humid mountain forest; also upland woods and shaded coffee plantations. **D** Jamaica.

23 YELLOW-THROATED VIREO *Vireo flavifrons* 13cm Usually forages alone in canopy. **V** Song a slow, slurred *rrreeyoo rreeooee three-eight* or *de-ar-ie come-here three-eight*, on average repeated every 3 secs. Call a harsh, descending *chi-chi-chur-chur-chur-chur-chur* or *ship-shep-shep-shep-shep-shep-shep*. **H** Various types of forests, woodlands, coastal scrub and secondary growth. **D** Breeds SC and SE Canada to SC and SE USA. Winters N South America.

24 PLUMBEOUS VIREO *Vireo plumbeus* 15cm Feeds steadily in high parts of trees and shrubs. **V** Call and song similar to that of Yellow-throated Vireo, song often ending with a rising *zink*. **H** Deciduous and mixed woods. **D** W North America, W Central America.

25 CASSIN'S VIREO *Vireo cassinii* 14cm Behaviour similar to that of Plumbeous Vireo. **V** Song like Plumbeous Vireo, but slightly higher pitched. Calls like Yellow-throated Vireo. **H** Deciduous and mixed woods. **D** Breeds W North America. Winters Central America.

26 BLUE-HEADED VIREO *Vireo solitarius* 14cm Behaviour similar to that of Plumbeous Vireo. **V** Song high-pitched, clear *see you cheerio be-seein-u so-long seeya* given, on average, every 2.5 secs. **H** Deciduous and mixed woods. **D** Breeds E North America. Winters Costa Rica, Cuba.

27 YELLOW-WINGED VIREO *Vireo carmioli* 11cm Only vireo with yellow wing bars in its range. **V** Very high thin *tjuwéeh tútjeeh*... **H** Montane forest canopy. **D** Costa Rica, Panama.

28 CHOCO VIREO *Vireo masteri* 11cm No similar vireo occurs in its range. **V** Very high, descending, warbling phrases. **H** Canopy and borders of tall forest. **D** Colombia.

29 TEPUI VIREO *Vireo sclateri* 12cm No similar vireo occurs on the tepuis. **V** Very high, sharp *seet-suweet* or *suuweet*, often repeated. **H** Canopy and borders of humid montane forest on tepuis. **D** Venezuela, Guyana, N Brazil.

30 HUTTON'S VIREO *Vireo huttoni* 13cm Incomplete eye-ring. Looks like a chunky Ruby-crowned Kinglet (Plate 232). Often part of mixed-species feeding parties, foraging in mid- to upper levels. **V** Rising, nasal *reeee dee de* or laughing *rrrreeeee-dee-dee-dee-dee*; also high, harsh mewing *shhhhhrii shhhri shhr shhr* and dry *pik*. Song a simple mix of two phrases: *trrweer trrweer trrweer, tsuwiif tsuwiif*... **H** Moist woodlands, especially oak. **D** SW Canada to W Guatemala.

31 GOLDEN VIREO *Vireo hypochryseus* 13cm Separated from warblers by heavier bill. **V** Very high *wicwicwictst*; very high fluted *tuweetwit-weetwit*. **H** Forest edge, woodland, plantations, dry scrub thornbush. **D** W Mexico.

1
2
♀
3
♂
4
5
6
7
8
grey morph
9
10
grey morph
buff morph
yellow morph
9
11
12
yellow morph
13
14
15
16
♀
19
♂
20
17
18
24
21
22
23
25
26
27
28
29
30
31

VIREOS, GREENLETS AND SHRIKE-BABBLERS *CONTINUED*

1 WARBLING VIREO *Vireo gilvus* 13–15cm Forages in mixed-species groups, from low down to upper levels. Western birds can be dingier on underparts. **V** Harsh, nasal mewing *meeerish*, *nyeeah* or *rreih*; also a dry *ch* or *ch-ch-ch*... Song is a long, rapid husky warble. **H** Deciduous and mixed woodland. **D** Widespread in W Canada, south of Arctic, most of USA apart from much of Texas and the SE. Winters Central America.

2 BROWN-CAPPED VIREO *Vireo leucophrys* 12cm Brown crown and narrow eye-stripe are diagnostic. **V** Rather short bursts of fast, very high, meandering and yet level warbling. **H** Humid forest, tall secondary growth, plantations. **D** Mexico to Venezuela and N Bolivia.

3 PHILADELPHIA VIREO *Vireo philadelphicus* 13cm Some birds can be less bright than that shown, when they can look very similar to Warbling Vireo. Forages in the low to middle level in trees; recorded hovering, or fluttering to pick insects from vegetation. **V** Descending *weeej weeezh weeezh weeezh*. Song very similar to Red-eyed Vireo but weaker and higher-pitched. **H** Broadleaf forests, especially edges and clearings, also thickets and parkland. **D** Summers across S Canada and NE USA. Winters S Mexico to Panama.

4 RED-EYED VIREO *Vireo olivaceus* 14–15cm Note dark stripes through and above eye. Active but with heavy movements; forages mainly in tree canopy. The South American race (Chivi Vireo) is now widely recognised as a separate species. **V** Rambling, warbled phrases that often end abruptly, *teeduee-tueedee-teeudeeu*... or *here-I-am in-the-tree look-up at-the-top*..., each phrase given every 2 secs. Calls include a soft mewing *meerf* and a nasal *tshay*. **H** Wet and dry forests, open woodlands, scrub and gardens. **D** Widespread across Canada and USA, apart from SW. Winters South America. Chivi Vireo widespread across South America.

5 NORONHA VIREO *Vireo gracilirostris* 14cm Note dull plumage and slender bill. No similar bird occurs on Fernando de Noronha Island. **V** Song *tet-tjee*, repeated at short intervals. **H** Woodland and scrub. **D** Fernando de Noronha Island (off NE Brazil).

6 YELLOW-GREEN VIREO *Vireo flavoviridis* 14–15cm Central American counterpart of Red-eyed Vireo, with less contrasting head pattern. Sluggish, forages in trees at the mid- to upper levels. **V** Song like Red-eyed Vireo but shorter and more rapid. Calls include a dry chatter and a rough mewing. **H** Woodland, scrubby forest edge and plantations. **D** Breeds Mexico to Panama. Winters Colombia to Bolivia.

7 BLACK-WHISKERED VIREO *Vireo altiloquus* 15–16cm Note black throat stripe. Often sits motionless; best located by song. **V** A monotonous *chip-john-phillip chiip-phillip*...*chillip phillip*. Calls include a nasal mew, a thin *tsit* and a nasal chatter. **H** All types of forest. Breeds in May and June. **D** Breeds throughout Caribbean and in S Florida. Winters S Caribbean, N South America.

8 YUCATAN VIREO *Vireo magister* 15cm Duller plumage than Red-eyed Vireo. Forages slowly in thick vegetation; more often heard than seen. **V** Song consists of varied rich phrases, given in a tentative manner, *chu-ree chu-i-chu ch-weet ch ee chu ch oo choo-choo*... Calls include a soft, dry chatter *shit chi-chi-chi-ch*..., a sharp *peek*, *beenk* or *peenk peenk*. **H** Scrubby woodland, woodland edge and mangroves. **D** SE Mexico, Belize, Grand Cayman Island.

9 RUFOUS-CROWNED GREENLET *Hylophilus poicilotis* 12.5cm Separable from Grey-eyed Greenlet by yellower underparts. Note pointed bill. **V** Level, high, sharp *tsee-tsee-tsee* phrase. **H** Humid forest, woodland, secondary growth, scrub. **D** E Paraguay, NE Argentina, SE Brazil.

10 GREY-EYED GREENLET *Hylophilus amaurocephalus* 13cm Separable from Rufous-crowned Greenlet by less yellow underparts and less marked cheeks. **V** Varied, hurried, musical 1-2-3 phrases, repeated at short intervals. **H** Forest, woodland, Caatinga, scrub. **D** E and SW Brazil, N Bolivia.

11 LEMON-CHESTED GREENLET *Hylophilus thoracicus* 12cm Separable from Grey-chested Greenlet by grey nape. Note white eyes and olive-yellow forecrown. **V** High series of 5–10 *seerr* phrases. **H** Borders and canopy of humid forest, tall secondary growth, woodland (SE Brazil); <1,000m. **D** Venezuela and the Guianas to Peru and N Bolivia; SE Brazil.

12 GREY-CHESTED GREENLET *Hylophilus semicinereus* 12cm Separable from Lemon-chested Greenlet by grey breast. **V** Rapid series of 10–30 clear, whistled *weet* notes. **H** Canopy and borders of humid forest, secondary growth, scrubby várzea. **D** Amazonia.

13 ASHY-HEADED GREENLET *Hylophilus pectoralis* 12cm Separable from Lemon-chested Greenlet by dark eyes and all-grey head. **V** High, clear *tweet-tweet-trrill*, often repeated. **H** Forest, woodland, mangrove, plantations, gardens. **D** N and C Amazonia.

14 BROWN-HEADED GREENLET *Hylophilus brunneiceps* 11.5cm Separable from Dusky-capped Greenlet by less yellow underparts and different habitat. **V** High, sharp series of downslurred, interconnected *teeeu* phrases. **H** Scrubby forest on sandy soils, savannah, woodland. **D** S Venezuela, E Colombia, NW Brazil.

15 RUFOUS-NAPED GREENLET *Hylophilus semibrunneus* 12.5cm Note distinct facial pattern and orange-rufous breast sides. **V** Fast *sisideeweet* phrase, often doubled. **H** Canopy and borders of montane forest. **D** NW Venezuela to E Ecuador.

16 GOLDEN-FRONTED GREENLET *Hylophilus aurantiifrons* 11.5cm Note buffy-brown head and underparts. No similar greenlet occurs in its habitat. **V** Continuous stream of very high, very fast, short, chirped phrases. **H** Woodland, scrub, riverine belts, gardens. **D** E Panama to N Colombia, N Venezuela, Trinidad and Tobago.

17 DUSKY-CAPPED GREENLET *Hylophilus hypoxanthus* 11.5cm Note brown (not orange-rufous) head. **V** Song resembles that of Golden-fronted Greenlet but even sharper. **H** High levels of humid forest. **D** Amazonia.

18 BUFF-CHEEKED GREENLET *Hylophilus muscicapinus* 11.5cm Buff is restricted to face sides and throat. Rufous lores and forehead. Rear crown and nape grey. **V** Song resembles that of Golden-fronted Greenlet but individual phrases are slightly longer. **H** Higher levels of humid forest. **D** N and E Amazonia.

19 SCRUB GREENLET *Hylophilus flavipes* 11.5cm Dull greenish and yellowish plumage with pale eyes. **V** Long, fast, rhythmic series of *peer* notes. **H** Scrub, open woodland, riverine belts. **D** Costa Rica to N Colombia, N Venezuela.

20 OLIVACEOUS GREENLET *Hylophilus olivaceus* 12cm Separable from Scrub Greenlet by range. **V** Long, fast, rhythmic series of *weet* notes. **H** Edges and clearings in humid forest and secondary growth. **D** Ecuador, Peru.

21 TAWNY-CROWNED GREENLET *Hylophilus ochraceiceps* 12cm Note brownish upperparts. **V** Song is based on a long, very high, drawn-out *weeh* note (one every 3 secs). **H** Undergrowth of humid forest. **D** S Mexico through Amazonia to Bolivia and N Brazil.

22 LESSER GREENLET *Hylophilus decurtatus* 10cm Note habitat, plump jizz and long pinkish bill. **V** Series of very high, hurried *wichee-wee* phrases (one every 2 secs). **H** Canopy and borders of humid forest and woodland. **D** E Mexico to NW Peru.

23 WHITE-BELLIED ERPORNIS *Erpornis zantholeuca* 11cm Yellowish green above, white below, with prominent crest. Agile, acrobatic forager in the lower canopy and higher bushes. **V** High-pitched, trilling *si-i-i-i-i-i*, or a rising and falling *ss-ss-ss-se-se-se-se*. Calls include a loud, nasal *jeeer-jeeer-jeeer* interspersed with dry trills and chitterings. **H** Broadleaved evergreen forest, rhododendrons and secondary growth. **D** Himalayas to S China, SE Asia, Borneo.

24 BLACK-HEADED SHRIKE-BABBLER *Pteruthius rufiventer* 18–20cm Male has black head and wings. Female has green wings and grey head, with darker crown. Forages from low undergrowth to the canopy; regularly part of mixed-species feeding flocks. **V** A mellow *wip-wu-yu*, repeated every few seconds. Calls include a scolding *rrrrt-rrrrt-rrrrt*... and a quick *ukuk-wrrrrrii-yiwu*. **H** Dense moss-covered oak and evergreen forest. **D** Himalayas to N Myanmar, S China, NW Vietnam.

25 PIED SHRIKE-BABBLER *Pteruthius flaviscapis* 16cm Note broad white stripe behind eye of male. Forages slowly, shuffles sideways along branches to search mosses and lichens for insects; feeds mainly in the canopy, often in the company of other species. **V** Loud, far-carrying *chu-wip-chip-chip* or *cha-cha chip*. Calls include a quickly repeated harsh grating and a short *pink*. **H** Mid-storey and canopy of montane and submontane forest. **D** Java, Indonesia.

26 HIMALAYAN SHRIKE-BABBLER *Pteruthius ripleyi* 16cm Formerly treated as conspecific with Blyth's Shrike-babbler. Similar to Blyth's but tertials are darker chestnut in both sexes. **V** Vocalisations similar to Blyth's Shrike-babbler. **H** Broadleaved evergreen forest and mixed evergreen–coniferous forest, from foothills to mountains. **D** W Himalayas to C Nepal.

27 BLYTH'S SHRIKE-BABBLER *Pteruthius aeralatus* 16cm Male is white below, black above with a grey back, a white stripe behind eye, and bright orange tertials. Female resembles a faded version of the male. **V** Variable, repeated, strident series, e.g. *ip ch-chu ch-chu*, *ip chip chip chip ch-chip* or *ip chu ch-chu*; calls include a harsh grating and a short *pink*. **H** Broadleaved evergreen forest and mixed evergreen–coniferous forest. **D** E Himalayas through SE Asia to Sumatra and Borneo.

28 DALAT SHRIKE-BABBLER *Pteruthius annamensis* 16cm Formerly treated as conspecific with Blyth's Shrike-babbler. Similar to Blyth's but has smaller white tips to primaries. Male has darker mantle and is whiter below. Female has more chestnut on tertials. **V** Strident series of repeated notes, similar to Blyth's Shrike-babbler. **H** Broadleaved evergreen forests. **D** S Vietnam.

29 GREEN SHRIKE-BABBLER *Pteruthius xanthochlorus* 13cm Greyish head, olive upperparts, whitish-grey throat and breast. Actions much like a sluggish leaf warbler; regular member of mixed-species feeding parties. **V** Rapid, monotonous, repetition of a single note. Calls include a repeated *whit*, a nasal *nyeep nyeep* and a high *jerri*. **H** Subalpine, mixed forests. **D** Himalayas to NE Myanmar, SE China.

30 BLACK-EARED SHRIKE-BABBLER *Pteruthius melanotis* 11cm Note white eye-ring and black on rear of ear-coverts. Arboreal, mainly in the canopy; regularly found among mixed-species foraging parties; actions sluggish and methodical. **V** Bright *tew wee tew we tew-wee*. Calls include a short *chid-it* and a high *t-cheer-cheer chee chee chee chee*. **H** Broadleaved evergreen forest. **D** Himalayas to SE Asia, Peninsular Malaysia.

31 TRILLING SHRIKE-BABBLER *Pteruthius aenobarbus* 11–12cm Resembles Black-eared Shrike-babbler, but lacks black on ear-coverts. Arboreal; sluggish forager in the canopy and mid-storey. **V** Monotonous *chip-chip-chip* or similar, and tinny trill; calls include a buzzy *jer-jer-jer* and a sharp *pwit*. **H** Broadleaved evergreen forest and forest edges. **D** Java, Indonesia.

32 CLICKING SHRIKE-BABBLER *Pteruthius intermedius* 11–12cm Formerly treated as conspecific with Trilling Shrike-babbler (and called Chestnut-fronted). Very similar to Trilling Shrike-babbler, but male has less extensive chestnut on breast. Arboreal; sluggish forager in the canopy and mid-storey. **V** Harsh couplets, *chu-wip chu-wip chu-wip*, repeated monotonously. **H** Broadleaved evergreen forest and forest edges. **D** NE India, SE Asia.

1
2
3
4
5
6
7
8
9
10
11
12
13
14
15
16
17
18
19
20
21
22
23
24
25
26
27
28
29
30
31
32

FIGBIRDS AND ORIOLES ORIOLIDAE

1 GREEN FIGBIRD *Sphecotheres viridis* 26cm Male has olive-green throat and breast. Forages in fruiting trees and shrubs, singly, in pairs or in groups, and with other fruit-eating species. **V** Burred, metallic trill, consisting of two very brief phrases. **H** Primary and tall secondary monsoon forest, remnant forest patches, woodland, wooded cultivation, mangroves and scrub. **D** Timor and Rote Island, E Lesser Sunda Islands.

2 WETAR FIGBIRD *Sphecotheres hypoleucus* 26cm Male has white underparts. Little recorded information. **V** Muted series of harsh, chirpy, nasal and unmusical notes. **H** Primary and secondary deciduous monsoon forest, woodland and lightly wooded scrub. **D** Wetar, E Lesser Sunda Islands.

3 AUSTRALASIAN FIGBIRD *Sphecotheres vieilloti* 27–30cm Male has grey chin, throat and breast. Feeds mainly in fruiting trees, often in flocks of 20 or so birds. **V** High-pitched, descending whistles and a short, weak, rustling trill. Song consists of 3–5 simple whistles, the final note downslurred. **H** Coastal secondary forest and adjacent cultivation. **D** SE New Guinea, N and E Australia.

4 NORTHERN VARIABLE PITOHUI *Pitohui kirhocephalus* 23–25cm Rusty-plumaged frugivore, with nine races. Three colour morph groups: black-headed, grey-headed and brown-headed. Sexes usually alike but some forms are sexually dimorphic. Usually in groups but will join mixed flocks. **V** Rich fluty warbling, with some variation between races. **H** Forest edge, secondary growth and gardens from lowlands to foothills. **D** New Guinea.

5 RAJA AMPAT PITOHUI *Pitohui cerviniventris* 23–25cm Paler than other pitohuis, with grey-brown head and upperparts and pale rusty underparts. Sexes alike. **V** Rich fluty warbling, similar to Northern Variable Pitohui but sweeter and more mellow. **H** Lowland and hill forest. **D** West Papuan islands, New Guinea.

6 SOUTHERN VARIABLE PITOHUI *Pitohui uropygialis* 23–25cm Similar to Northern Variable Pitohui, with five races. Head and wings black or brown, rest of plumage rusty to deep chestnut. **V** Rich fluty warbling, with some variation between races. **H** Forest edge, secondary growth and gardens from lowlands to foothills. **D** New Guinea.

7 HOODED PITOHUI *Pitohui dichrous* 23cm Striking rufous and black plumage. Smaller than Variable Pitohuis and generally allopatric. Sexes alike. **V** Rich melodious warbling, often slowly delivered, but also a rapid bubbling call. **H** Forest, forest edge, secondary growth. **D** New Guinea.

8 BROWN ORIOLE *Oriolus szalayi* 26–28cm Dull brown oriole with strong reddish bill, and prominently streaked head and breast. Somewhat similar to New Guinea Friarbird (Plate 163) but lacks bill casque. **V** Rather typical oriole-like warbling, rich and melodious. **H** Wide range of forested habitats, mostly in the canopy, from lowlands to hills. **D** New Guinea.

9 DUSKY-BROWN ORIOLE *Oriolus phaeochromus* 26cm Forages in the crowns of tall trees, singly, or sometimes in small flocks or as part of mixed-species feeding parties. A visual mimic of Dusky Friarbird. **V** Liquid *k k-wheeou*, repeated at 8-sec intervals. **H** Primary and mature secondary lowland and hill forest, cultivated land. **D** Halmahera, N Maluku Islands, Indonesia.

10 GREY-COLLARED ORIOLE *Oriolus forsteni* 31–32cm Occurs singly, in pairs, or sometimes as a member of mixed-species feeding parties; forages in the lower canopy and mid-storey of tall trees. A visual mimic of Seram Friarbird. **V** Calls include 3–4 ascending, liquid, musical notes and a fluty *whee-who*. **H** Primary lowland and montane forest. **D** Seram, S Maluku Islands, Indonesia.

11 BLACK-EARED ORIOLE *Oriolus bouroensis* 23–32cm Forages alone or in pairs, often in the company of Black-faced Friarbirds; as the oriole is a visual mimic of the friarbird, it is difficult to separate the two. **V** Not certainly recorded; possibly fluid *yio-wheea*. **H** Lowland and montane forest, deciduous monsoon forest, secondary woodland, mangroves, lightly wooded cultivation. **D** Buru, S Maluku Islands, Indonesia.

12 TANIMBAR ORIOLE *Oriolus decipiens* 23–32cm Forages alone or in pairs. Formerly considered a race of Black-eared Oriole. A visual mimic of the Tanimbar Friarbird. **V** Rich, loud, clear, down-slurred musical note; also a clear, pure whistle, low at first and rising at the end. **H** Moist montane and lowland forests, and mangroves. **D** Tanimbar, E Lesser Sunda Islands.

13 OLIVE-BROWN ORIOLE *Oriolus melanotis* 25cm Usually best located by its calls. Occurs singly or in pairs, foraging from the mid-storey to the canopy. Female is a visual mimic of Helmeted Friarbird. **V** Variable, liquid, musical, three-syllable yodel, *ti-ti-lu-i*; also a harsh, nasal scolding call. **H** Remnant patches of monsoon forest, secondary monsoon forest, open woodland, partially wooded cultivation and mangroves. **D** E Lesser Sunda Islands.

14 OLIVE-BACKED ORIOLE *Oriolus sagittatus* 27cm Occurs singly, in pairs or in small family parties. Noisy – usually heard before it is seen. Forages in the leafy canopy on arthropods and fruit. **V** Rolling, mellow, repeated *olly, olly-ole, urry, orry-ole* or *olio*; also utters a harsh, scolding note. **H** Open woodland, wooded cultivation and forest edges. **D** S New Guinea, N and E Australia.

15 GREEN ORIOLE *Oriolus flavocinctus* 25–30cm Favours the canopy. Forages alone, in pairs or in small groups. **V** Loud, bubbling *yok-yok-yoddle*, clear *peek-kwwek*, harsh *scarab*, soft warbling subsong. **H** Monsoon forest and, occasionally, lightly wooded cultivation. **D** E Lesser Sunda Islands, S New Guinea, N Australia.

16 DARK-THROATED ORIOLE *Oriolus xanthonotus* 20–21cm Male has striking black, yellow and white plumage; female is olive-green. Usually forages from mid-storey to canopy. **V** Melodious, fluty *tu-u-liu* or *peu-peu-peu-poh*, the last note a descending rasp. Call is a high-pitched, piping *kyew, pheeu* or *ti-u*. **H** Broadleaved evergreen forest, forest edge, secondary growth. **D** Malay Peninsula, Sumatra, Borneo, Philippines.

17 PHILIPPINE ORIOLE *Oriolus steerii* 20–21cm Sexes alike. Forages singly, in pairs and, occasionally, in mixed-species feeding flocks, usually in the canopy. **V** Trumpeted, high-pitched *per-jek* that speeds up and is repeated; also trumpeting *eeerk eeerk eeerk* and *hooooo-op*. **H** Forest, forest edge, secondary growth. **D** S and C Philippines.

18 WHITE-LORED ORIOLE *Oriolus albiloris* 20–21cm Formerly considered a race of the previous species; actions presumed to be similar. **V** Slightly mournful *chow-wooooo*, rapid, staccato *chup-chup-chup*. **H** Forest, forest edge, secondary growth. **D** N Philippines.

19 ISABELA ORIOLE *Oriolus isabellae* 20–21cm Very rare. Forages from mid-storey to canopy, singly, in pairs or in small groups, and as part of mixed-species feeding parties. **V** Unrecorded. **H** Lowland rainforests, especially thick bamboo; also secondary growth and forest edges. **D** N Philippines.

20 EURASIAN GOLDEN ORIOLE *Oriolus oriolus* 24cm Male unmistakable. Arboreal, often stays hidden in foliage and can be hard to see, in spite of bright plumage. **V** Mellow, fluty *weela-weeoo*, also a harsh *kweeaahk* and a fast *gigigigigi*. **H** Open, well-wooded country. **D** Breeds from W Europe to S Siberia, Mongolia, NW China. Winters sub-Saharan Africa.

21 INDIAN GOLDEN ORIOLE *Oriolus kundoo* 25cm Mainly arboreal, attracted to fruiting trees. Often in mixed-species parties with jungle babblers, drongos, flycatchers. **V** Very similar to Eurasian Golden Oriole. **H** Open woodland, groves, cultivations and urban gardens with large trees. **D** Breeds from Kazakhstan to N India. Winters S India, Sri Lanka.

22 AFRICAN GOLDEN ORIOLE *Oriolus auratus* 25cm Note yellow-edged wing feathers of male and female. **V** High, melodious, fluted *weet-weet-oh-wee-o-weer*. **H** Forest edges, woodland and open areas with large trees. **D** Widespread across sub-Saharan Africa.

23 SLENDER-BILLED ORIOLE *Oriolus tenuirostris* 23–26cm Similar to Black-naped Oriole, but thinner bill and narrower black band through eyes. Arboreal, usually stays hidden in foliage; attracted to fruit-bearing trees. **V** Drawn-out *wheeow* or a liquid *chuck-tarry-you*; also a high-pitched *kich* and a cat-like *miaow*. **H** Pine forests, open woods, plantations and open country with scattered trees. **D** E Himalayas to SE Asia.

24 BLACK-NAPED ORIOLE *Oriolus chinensis* 27cm Actions and habits similar to Slender-billed Oriole. **V** Liquid, fluty *luwee – wee – wee-leeow* or similar; also a harsh scolding *kyerrr*. **H** Open woodland, plantations, mangroves, parks and gardens. **D** Breeds E Asia, Myanmar, Malay Peninsula, Indonesia, Philippines. Northern populations winter in India and SE Asia.

25 GREEN-HEADED ORIOLE *Oriolus chlorocephalus* 24cm Unmistakable by green head and mantle. Wings grey, not black. **V** Mid-high, liquid, short *wit-wit-oh-Weeoh* (first part *wit-wit* very fast) or simple, liquid, high *wheet* (preceded by barely audible *piupiu-*). **H** Montane forest. **D** E Africa, from Kenya to Mozambique.

26 SAO TOME ORIOLE *Oriolus crassirostris* 24cm Only oriole on São Tomé. No similar bird within its range. **V** Song is variable, a long, mellow downslurred *tyeeow, way-whee-ya, chip-aw-hah-aw* or *wit-wuay-kuwow*, or an ascending *ko-ku-waayoo*; slower and deeper than other African orioles. **H** Mainly forest. **D** São Tomé, Gulf of Guinea.

27 WESTERN ORIOLE *Oriolus brachyrynchus* 22cm From Black-headed Oriole by more grey wing coverts, green- (not yellow-) edged tertials, paler middle tail feathers. **V** Extensive repertoire of rich, mellow and fluty notes. **H** More restricted to true forest than Black-headed Oriole. **D** W and equatorial Africa, from Guinea-Bissau to W Kenya.

28 ETHIOPIAN ORIOLE *Oriolus monacha* 25cm Tail pattern exactly like Western Oriole but separated geographically. Some individuals have no black in the tail. **V** Song is higher-pitched and less rich and fluty than other African orioles. Scolding call is a short *grraaa-grraaa*. **H** Forest, wooded grassland. **D** Ethiopia, Eritrea.

29 MOUNTAIN ORIOLE *Oriolus percivali* 24cm Extensive black in wings. No green in tail. Wing feathers edged yellow, not white or green. **V** Song consists of short liquid phrases, higher-pitched than Western Oriole. Commonly sings in duet, with the male's *weeka-ku-weeu* answered in a higher pitch by the female. **H** Forest, plantations, cultivation with high trees. **D** E DR Congo to C Kenya, W Tanzania.

30 BLACK-HEADED ORIOLE *Oriolus larvatus* 24cm As Western Oriole but central tail feathers greener and edges of wing feathers paler grey. Tail from below all yellow, without black at base. **V** Extensive vocabulary of rich and liquid whistles, often including mimicry of many other birds. **H** Woodland, plantations, suburbs. **D** E and S Africa, from Sudan to Angola and E South Africa.

31 BLACK-WINGED ORIOLE *Oriolus nigripennis* 24cm From Western and Black-headed orioles by green mantle, scapulars and wing coverts, and different tail pattern. **V** Varied repertoire of fluty notes, delivered with frequent changes in pitch and intensity. **H** Forest. **D** W and C Africa, from Sierra Leone to South Sudan, W Uganda, DR Congo, NW Angola.

32 BLACK-HOODED ORIOLE *Oriolus xanthornus* 25cm No similar oriole in its range. Mainly arboreal, usually singly or in pairs, sometimes in small parties post breeding. **V** Melodious, fluty *why-you* or *why-you-you*, often interspersed with harsh *cheeahs* and *kwaaks*. **H** Open broadleaved forest, well-wooded areas, cultivations, parks and wooded gardens. **D** India, Sri Lanka, SE Asia, Borneo.

33 BLACK ORIOLE *Oriolus hosii* 21cm Male all black apart from chestnut undertail coverts. Female has grey belly. Forages on fruits, berries and invertebrates in the canopy. **V** Clear whistles with a downward inflection. **H** Montane forest, mossy transitional forest. **D** N Borneo.

34 BLACK-AND-CRIMSON ORIOLE *Oriolus cruentus* 23–25cm Male has red breast patch and primary coverts. Forages from understorey to canopy; regularly joins mixed-species feeding parties. **V** Short, melodious call and a hard *kek kreo* or *ee-oo*; also a cat-like *keeeeu* and a strained *hhsssu*. **H** Broadleaved evergreen forest. **D** Peninsular Malaysia; Borneo; Sumatra and Java, Indonesia.

35 MAROON ORIOLE *Oriolus traillii* 24–28cm Note pale eye in both sexes. Arboreal, tends to keep to the tops of trees, often part of mixed-species flocks. **V** Harsh *kee-ah* followed by a rich, fluty *pi-lo-lo*. **H** Broadleaved evergreen forest, forest edge and, occasionally, deciduous forest. **D** Himalayas to SW China, SE Asia.

36 SILVER ORIOLE *Oriolus mellianus* 28cm Coloration unlike any other oriole. Usually forages in the canopy, often in mixed-species flocks. **V** Fluty whistles and a cat-like call. **H** Broadleaved evergreen and semi-evergreen forests. **D** Breeds SC China. Winter visitor to S Thailand, W Cambodia.

1
2
3
4
5
6
7
8
9
10
11
12
13
14
15
16
17
18
19
20
21
22
23
24
25
26
27
28
29
30
31
32
33
34
35
36

DRONGOS DICRURIDAE

1 SQUARE-TAILED DRONGO *Dicrurus ludwigii* 18cm Small with dull purplish gloss and forked tail. Perches less exposed than Fork-tailed Drongo. **V** Varied song, with repeated disyllabic calls, *tyip-tyip* or *toylu, toylu*; also explosive whistles and hard buzzing notes. **H** Forest, riverine belts. **D** Sub-Saharan Africa, widespread.

2 SHINING DRONGO *Dicrurus atripennis* 22cm From Fork-tailed Drongo by black, not whitish, underwing, stronger gloss and different habitat. **V** Noisy, in flocks, often together with other bird species. **H** Lowland forest. **D** Sierra Leone to E and C DR Congo and S Gabon.

3 FORK-TAILED DRONGO *Dicrurus adsimilis* 24cm From Square-tailed Drongo by larger size and deeply forked tail, from Shining Drongo by different habitat. Hawks insects from an exposed perch. **V** Extremely varied repertoire of sharp calls, short whistles and squeaky, liquid, grating and scratchy notes, often rather discordant. **H** Forest edge, dry woodland, farmland with some trees. **D** Sub-Saharan Africa, widespread.

4 VELVET-MANTLED DRONGO *Dicrurus modestus* 26cm Very similar to Fork-tailed Drongo but less glossy especially on mantle. **V** Very high *weet-weet* and variations. **H** Wide range of wooded habitats with some large trees. **D** W and C Africa, from Nigeria to Kenya and NW Angola.

5 GRAND COMORO DRONGO *Dicrurus fuscipennis* 28cm No other drongo in its small range. Note brown tail and flight feathers without gloss. **V** Varied repertoire like other drongos, with squeaks, clicks, trills, whistles and grinding sounds, as well as softer *wit wit* notes. **H** Forest interior and edge. **D** Grande Comore, Comoros.

6 ALDABRA DRONGO *Dicrurus aldabranus* 23cm No other drongo in its range. Note tufts at base of bill. **V** Very noisy, uttering a range of squawks, whistles, scolding chatters, harsh chuckles and scratchy calls. **H** Mangrove, dense scrub, woodland. **D** Aldabra, Indian Ocean.

7 CRESTED DRONGO *Dicrurus forficatus* 28cm No other drongo in its range. Long crest diagnostic. **V** Harsh, short chattering. **H** Forest and woodland. **D** Madagascar and nearby islands.

8 MAYOTTE DRONGO *Dicrurus waldenii* 36cm No other drongo in its range. Note long, deeply forked tail. **V** Short, rather harsh, yet not unmelodious chatters. **H** Mainly in forest with large trees. **D** Mayotte, one of the Comoros Islands.

9 BLACK DRONGO *Dicrurus macrocercus* 30cm Uses a prominent perch from where it makes fly-catching sallies. Captures prey in the air or pounces on insects on the ground; attracted to areas where grazing animals disturb insects. **V** Harsh *ti-tui*, rasping *jeez*, *cheece* or *cheece-cheece-chichuk*. **H** Open country, cultivations, scrub, roadsides. **D** From Pakistan through India, Sri Lanka, SE Asia to NE China and Taiwan; also Lesser Sunda Islands. NE populations winter to the south.

10 ASHY DRONGO *Dicrurus leucophaeus* 30cm Ash-grey rather than black. Makes fly-catching sorties from the tops of trees. **V** Harsh *cheece-cheece-chichuck*, querulous *kil-kil-kil-kil* or *tililili*, loud *tchik wu-wit tchik wu-wit*, wheezy *phuuuu* or *hieeeeer*. **H** Forest clearings, forest edge, secondary growth, mangroves, coastal scrub. **D** Breeds Himalayas to NE China, SE Asia, Greater Sunda Islands, SW Philippines. Winters India and in south of breeding range.

11 WHITE-BELLIED DRONGO *Dicrurus caerulescens* 24cm White belly diagnostic (but Sri Lankan birds may have very little white). Makes fly-catching sallies from the tops of trees; often a member of mixed-species parties. **V** Three or four pleasant whistling notes, similar to Black Drongo but less harsh. **H** Clearings and edges of open forest and well-wooded areas. **D** India, Sri Lanka.

12 CROW-BILLED DRONGO *Dicrurus annectens* 28cm Glossier than Black Drongo, with less deeply forked tail. Keeps to tall undergrowth and lower tree branches. **V** Loud, musical whistles, chatterings and churrs. **H** Broadleaved evergreen forest, mixed deciduous forest, mangroves, coastal scrub, secondary growth, plantations. **D** Breeds from N India and Nepal to S China, Laos, N Vietnam. Winter visitor Malay Peninsula and Greater Sunda Islands.

13 BRONZED DRONGO *Dicrurus aeneus* 24cm Glossy black with relatively short tail. Arboreal, mainly in treetops. **V** Loud, clear musical whistles. **H** Broadleaved evergreen, semi-evergreen and deciduous forest, forest edge and secondary growth. **D** India through S China, Taiwan, SE Asia, Malay Peninsula, Borneo, Sumatra.

14 LESSER RACKET-TAILED DRONGO *Dicrurus remifer* 40cm Distinctive elongated outer-tail feathers. Arboreal; makes bold, dashing pursuits after flying insects, mainly in the canopy. **V** Range of loud, metallic, musical whistles; also mimics other birds. **H** Broadleaved evergreen and semi-evergreen forests. **D** S Himalayas to S China, SE Asia; Sumatra and Java, Indonesia.

15 BALICASSIAO *Dicrurus balicassius* 26–27cm Noisy and conspicuous. Usually seen in small groups, sometimes in mixed-species feeding parties. **V** Mixture of clear and screechy whistles, interspersed with *chuck*, *chunk* or similar; often mimics other species. **H** Subtropical or tropical moist lowland forest, sometimes forest edge. **D** Philippines.

16 HAIR-CRESTED DRONGO *Dicrurus hottentottus* 32cm Note twisted tail feathers. Takes insects, mainly by fly-catching sorties; also feeds on nectar. **V** Calls include a loud *chit-wiii*, a single *wiii* and a *tsit-wit wuu*. **H** Forest, forest edges, lightly wooded areas, scrub woodland, mangroves. **D** From India to China, SE Asia, E Indonesia, Philippines.

17 TABLAS DRONGO *Dicrurus menagei* 35–36cm Long and deeply forked tail. Captures prey by hawking, and by foraging in foliage and on trunks. Favours the mid-canopy, often near streams. **V** Rasping, cicada-like calls, *dzak-tess-k* and *tsee-ik*. **H** Mature closed-canopy forest, clearings. **D** Tablas Island, C Philippines.

18 SUMATRAN DRONGO *Dicrurus sumatranus* 29cm Deep ultramarine iridescence. Tail has shallow fork. Regularly joins mixed-species feeding parties. **V** Melodious *tee-tiyeeah* and harsh notes. **H** Submontane and lower montane dry primary and tall secondary forest, forest edge. **D** Sumatra, Indonesia.

19 WALLACEAN DRONGO *Dicrurus densus* 28–38cm Occurs singly, in pairs and occasionally in small groups. Perches in the under-canopy, from where sallies are made to catch passing insects. **V** Scratchy grinding notes, short phrases of chopped percussive notes, twittery and liquid, starting with a harsh, rapid churr. **H** Primary and tall secondary forest, open woodland, lightly wooded cultivation, mangroves. **D** SE Maluku and Lesser Sunda Islands.

20 SULAWESI DRONGO *Dicrurus montanus* 28cm Usually encountered singly, in pairs or as part of mixed-species foraging parties. **V** Duet: first bird utters a harsh, grating *trrsh-trrsrh-trrrsh*, second bird answers with a soft, low-pitched *twuu* or *tee-twuu*. **H** Hill and montane forest. **D** Sulawesi, Indonesia.

21 SPANGLED DRONGO *Dicrurus bracteatus* 30–32cm Favours the mid-storey and canopy; feeds by hawking flying insects from an exposed perch, and sometimes eats fruit. **V** Penetrating, long, nasal, high-pitched upslur. Duets with a collection of disjointed phrases. **H** Lowland and hill forest, forest edges, and secondary and littoral woodland. **D** Maluku Islands, Indonesia; New Guinea; Solomon Islands; N and E Australia.

22 PARADISE DRONGO *Dicrurus megarhynchus* 50–62cm Large glossy drongo with very long twisted outer-tail feathers (hence alternative name of Ribbon-tailed Drongo). Tail streamers may be broken or missing. Sexes alike. **V** Loud melodic *chu chu-chu chu*. Also variety of other mechanical and shrieking sounds. **H** Lowland and montane forest, including secondary growth. **D** New Ireland (Bismarck Archipelago).

23 ANDAMAN DRONGO *Dicrurus andamanensis* 32cm Arboreal. Often clings to trunks, woodpecker-like, when searching for insects: also makes fly-catching sallies. **V** Variety of sharp metallic notes. **H** Forests. **D** Andaman Islands.

24 GREATER RACKET-TAILED DRONGO *Dicrurus paradiseus* 30–65cm (depending on tail length) A large drongo with two long tail streamers. Gleans insects from foliage or flowers; also makes fly-catching sallies. **V** Monotonous *kit-kit-kit-kit*, mainly pre-dawn. Regularly mimics other birds. **H** Broadleaved, bamboo and secondary forest. **D** Widespread from India to SE Asia and Greater Sunda Islands.

25 SRI LANKA DRONGO *Dicrurus lophorinus* 34cm Actions as Greater Racket-tailed Drongo. **V** Explosive mix of fluty, bell-like and grinding notes; also a nasal *urdle-eee* or a quicker *urd-lee*. A frequent mimic. **H** Wet-zone forests. **D** Sri Lanka.

FANTAILS RHIPIDURIDAE

26 MINDANAO BLUE FANTAIL *Rhipidura superciliaris* 16cm Forages in the understorey, often as part of mixed-species feeding parties. Best located by its distinctive call. **V** Repeated, rapid, ascending *woo-woo-woo-woo-woo...*; also a sharp, raspy *whickkk whickkk*. **H** Shady areas of forests, also forest edges. **D** S Philippines.

27 VISAYAN BLUE FANTAIL *Rhipidura samarensis* 16cm Formerly considered a race of Mindanao Blue Fantail; actions presumed to be similar. **V** Presumed similar to previous species. **H** Forests, forest edges. **D** C Philippines.

28 BLUE-HEADED FANTAIL *Rhipidura cyaniceps* 18cm Conspicuous and noisy. Forages in the understorey, usually in small parties or as part of mixed-species feeding flocks. **V** Sharp, metallic, staccato *chip-chip-chip-chip chip chip*, slowing, interspersed with sharp *chick* notes. **H** Primary oak and pine forests, and heavily disturbed secondary growth. **D** N Philippines.

29 TABLAS FANTAIL *Rhipidura sauli* 18cm Formerly considered conspecific with the previous species; actions presumed to be similar. **V** Probably similar to previous species. **H** Mature and semi-mature lowland forest. **D** Tablas Island, C Philippines.

30 VISAYAN FANTAIL *Rhipidura albiventris* 18cm Actions presumed to be similar to Blue-headed Fantail, with which it was formerly considered conspecific. **V** Presumed similar to Blue-headed Fantail. **H** All types of woodland. **D** C Philippines.

31 WHITE-THROATED FANTAIL *Rhipidura albicollis* 17cm Restless forager, often working up and down main tree trunk or nearby branches; also makes fly-catching sallies. Tail often fanned and held erect. **V** Descending *tut-tut-tut-sit-sit-sit-sit* or *tsu sit tsu sit sit sit sit-tsu*; call is a sharp *cheep, jick* or *chuck*. **H** Broadleaved evergreen forest, cultivations, parks, wooded gardens. **D** S Himalayas, NE India, SE Asia, S China, Greater Sunda Islands.

32 WHITE-SPOTTED FANTAIL *Rhipidura albogularis* 17cm Grey breast band with white spots. Actions and habits as White-throated Fantail. **V** As White-throated Fantail but slower. **H** Wooded country, secondary forest. **D** S and C India.

33 WHITE-BELLIED FANTAIL *Rhipidura euryura* 18cm Broad white supercilium, blue-grey throat and breast. Active feeder on small flying insects, often joins mixed-species foraging parties. **V** Excited, squeaky *cheet-cheet*. **H** Montane forest. **D** Java, Indonesia.

34 WHITE-BROWED FANTAIL *Rhipidura aureola* 17cm Forages restlessly, low down in bushes and undergrowth and is often part of mixed-species feeding flocks. Makes fly-catching sallies. **V** Rising then descending series of tinkling notes, *chee-chee-cheweechee-vi*; harsh *chuck-chuck*. **H** Forests, wooded areas, areas with scattered trees and groves. **D** Pakistan, India, Sri Lanka, SE Asia.

35 MALAYSIAN PIED FANTAIL *Rhipidura javanica* 17–20cm Constantly on the move in vegetation, interspersed with fly-catching sallies; reported to follow domestic animals or monkeys, feeding on disturbed insects. Occurs singly, in pairs or in small groups, and joins mixed-species feeding parties. **V** High-pitched, squeaky *chee-chee-wee-weet*; various squeaky chattering and squawking calls. **H** Open wooded areas, secondary forest, mangroves and gardens. **D** SE Asia, Greater and Lesser Sunda Islands.

36 PHILIPPINE PIED FANTAIL *Rhipidura nigritorquis* 19cm Conspicuous and noisy; occurs alone or in pairs. Formerly considered conspecific with Malaysian Pied Fantail. **V** Metallic chime-like *pip pip chop siitt chop why-su-weet*, repeated several times a minute. **H** Early secondary growth, bamboo thickets, mangroves, parks and gardens. **D** Philippines.

37 SPOTTED FANTAIL *Rhipidura perlata* 17–18cm Active forager in the middle and upper storey; makes regular sallies after flying insects. Regular member of mixed-species feeding flocks. **V** Melodious *chilip pechilip-chi*. **H** Primary and old secondary forest with tall trees. **D** Malay Peninsula; Borneo; Sumatra, Indonesia.

38 WILLIE WAGTAIL *Rhipidura leucophrys* 21cm Almost entirely black above, white below. Searches for insects on or near the ground; regularly runs around flashing wings and tail, presumably to flush insects. **V** Various squeaky, whistled notes that rise and fall; harsh, rattling chatter when alarmed. **H** Grassy areas, beaches, mangroves, cultivation and urban areas. **D** From Maluku Islands to New Guinea, Solomon Islands, widespread in Australia.

39 BROWN-CAPPED FANTAIL *Rhipidura diluta* 16–17cm Forages in the dense understorey, sometimes in mixed-species flocks. **V** Sweet and unmusical, discordant, staccato notes; also a high-pitched *chingk*. **H** Primary and degraded forest, mangroves, bamboo thickets, coastal savannah, shrubland with grass cover. **D** W Lesser Sunda Islands.

1
2
3
4
5
6
7
8
9
10
11
12
13
14
15
16
17
18
19
20
21
22
23
24
25
26
27
28
29
30
31
32
33
34
35
36
37
38
39

FANTAILS *CONTINUED*

1 CINNAMON-TAILED FANTAIL *Rhipidura fuscorufa* 18cm Forages by gleaning and by making aerial sallies; active, constantly fanning tail. Regular member of mixed-species feeding parties. **V** Song consists of short, discordant, whistled notes, interspersed with a high-pitched *cheep*. **H** Forest, forest edges and mangroves. **D** E Lesser Sunda Islands.

2 NORTHERN FANTAIL *Rhipidura rufiventris* 16–19cm Conspicuous, often perching upright when at rest. Favours open areas in the mid-storey. Plumage very variable. **V** Series of halting, high, piping notes. **H** Primary and secondary forest, forest edges, lightly wooded cultivation, open scrub and mangroves. **D** Maluku and E Lesser Sunda Islands; New Guinea and surrounding islands; N Australia.

3 COCKERELL'S FANTAIL *Rhipidura cockerelli* 17cm Perches upright and fans tail; rather flycatcher-like, inhabiting mid-storey and canopy. Seven races with variable white on tertials and variable white spotting on breast. **V** Song is a short series of mournful whistled notes. **H** Primary and secondary forest, forest edge. **D** Solomon Islands.

4 SOOTY THICKET FANTAIL *Rhipidura threnothorax* 17–18cm Large, reclusive fantail of the understorey. Blackish-brown above with bold white supercilium, white throat and white-spotted breast. No white tips to tail. **V** Irregular *chick* notes followed by a loud, penetrating *chow-chow-chow*. More often heard than seen. **H** Primary and gallery forest in lowlands and hills. **D** New Guinea.

5 BLACK THICKET FANTAIL *Rhipidura maculipectus* 18–19cm Large, reclusive fantail of the understorey. Mainly blackish with short white supercilium and white malar patch. Note broad white tips to tail feathers. **V** Series of stuttering squeaky notes, accelerating and ending in a flourish. **H** Dense forest, swamp forest, mangroves, secondary growth in lowlands. **D** New Guinea.

6 WHITE-BELLIED THICKET FANTAIL *Rhipidura leucothorax* 18cm Large, reclusive fantail of the understorey. Similar to Black Thicket Fantail but belly is white. Tail often fanned and raised. Very active. **V** Series of harsh and scratchy notes which accelerate and end in a loud pure *chu-whee*. **H** Dense scrub and thickets, mainly in the lowlands; also in swampy areas and gardens. **D** New Guinea.

7 BLACK FANTAIL *Rhipidura atra* 17cm Male is all black except for white spot above eye. Female is rufous with darker wings and central tail feathers. Singly or in pairs. **V** Typical song is a repetitive *whit-too whit-to whit-too*, followed by a fast, jumbled series of warbles. **H** Forest and secondary growth in mountains, inhabiting the lower and mid-storey. **D** New Guinea.

8 CHESTNUT-BELLIED FANTAIL *Rhipidura hyperythra* 14–15cm Active, dark fantail with white head markings and wing-bars. Underparts mostly rufous. Sexes alike. Characteristically cocks tail and droops wings. **V** Rapid series of identical slurred notes, which gradually slow down and descend in pitch. **H** Forest in hills and lowlands. **D** New Guinea.

9 FRIENDLY FANTAIL *Rhipidura albolimbata* 14–15cm Similar to Chestnut-bellied Fantail but lacks rufous on underparts. Throat, belly and vent white. **V** Thin, discordant, high-pitched series of staccato notes, with regional variations. **H** Forest, forest edge, clearings; generally above the altitude range of Chestnut-bellied Fantail. **D** New Guinea.

10 GREY FANTAIL *Rhipidura albiscapa* 16cm Mostly grey and white with narrow black breast-band and two narrow wing-bars. Sexes alike. Eight races across wide range, with subtle variations. **V** A sweet high-pitched jumble of notes, with harsh scolding. **H** Most habitats, including forests, woodlands, parks, gardens. **D** Australia, Solomon Islands, Loyalty Islands, Vanuatu, New Caledonia.

11 NEW ZEALAND FANTAIL *Rhipidura fuliginosa* 16cm No other arboreal fantail in New Zealand. Two morphs: pied and dark. Dark morph uncommon on North Island, rather commoner on South Island. **V** High, unstructured series, partly clear warbling, partly nasal twittering, ending in some *tfeet* notes. **H** All habitats with trees or scrub from forest and plantations to gardens and wooded farmland. **D** New Zealand.

12 MANGROVE FANTAIL *Rhipidura phasiana* 15cm Similar to Grey Fantail but smaller and paler with narrower, paler breast-band. **V** Thin, high-pitched trilling. Call notes similar to Grey Fantail. **H** Mangroves. **D** S New Guinea, N Australia.

13 BROWN FANTAIL *Rhipidura drownei* 16cm Nondescript, grey-brown fantail. Note whitish throat, lightly streaked breast and buff tips to tail. **V** Song is a rich melodious warble. **H** Montane forest. **D** Bougainville and Guadalcanal (S olomon Islands).

14 MAKIRA FANTAIL *Rhipidura tenebrosa* 17cm Dark brown fantail, with small white throat, white tips to tail and double golden-brown wing-bars. In pairs or mixed-species flocks. **V** Song is a sweet twittering. Call is a hard tik. **H** Forest, in lowlands and hills. **D** Makira (Solomon Islands).

15 RENNELL FANTAIL *Rhipidura rennelliana* 16cm Grey-brown above, paler below. Note buff wing-bars and white outer-tail feathers. In pairs or mixed-species flocks. **V** Song is an unmelodic staccato series of notes. **H** Forest and forest edge. **D** Rennell (Solomon Islands).

16 STREAKED FANTAIL *Rhipidura verreauxi* 17cm Grey-brown fantail with short supercilium and white throat. Note grey breast with bold white spots. Five races with some variation. Often in mixed-species flocks. **V** Call is a harsh *zhik* or a longer, slurred *zhweek-zhweek*. **H** Closed forest and secondary growth, in lowlands and hills. **D** Fiji to New Caledonia.

17 KADAVU FANTAIL *Rhipidura personata* 15cm The only fantail in its range. Grey-brown above and white below with prominent black breast-band. Often in mixed-species flocks. **V** Song is a long series of high-pitched notes all at the same pitch, which accelerates and then slows down. Call is a harsh *zhik*. **H** Forest undergrowth. **D** Kadavu and Ono, Fiji.

18 SAMOAN FANTAIL *Rhipidura nebulosa* 14.5cm Mostly dark grey, with some white on ear-coverts, throat and undertail coverts. **V** Short, very high, thin, twittered strophes *tjee-tjee-tjuh-rrrruh* or *pipirueh*. **H** Dense undergrowth of forest, hedgerows, gardens. **D** Samoa.

19 RUFOUS-TAILED FANTAIL *Rhipidura phoenicura* 17cm Chestnut tail and rump contrasting with grey back. Forages close to the ground in the dense understorey; often joins mixed-species feeding parties. **V** Song rendered as *he-tee-tee-tee-oh-weet*. **H** Mature forests with dense thickets and bushes. **D** Java, Indonesia.

20 BLACK-AND-CINNAMON FANTAIL *Rhipidura nigrocinnamomea* 16cm Conspicuous, actively flitting through the forest understorey, singly, in pairs or as part of mixed-species feeding parties. **V** A sharp *squeek* or *chick*, likened to the sound of a squeaky toy. **H** Mid-montane and montane forest. **D** Mindanao, S Philippines.

21 DIMORPHIC FANTAIL *Rhipidura brachyrhyncha* 16cm Brownish fantail with rufous back and rump. Two morphs: dark morph has dark brown tail with broad rufous tips; pale morph is paler with all-grey tail. Singly or in pairs. **V** Song is a short discordant medley of pure high-pitched notes preceded by several harsh notes. **H** Montane forest, usually in understorey. **D** New Guinea.

22 PALAU FANTAIL *Rhipidura lepida* 18cm Striking, large rufous fantail with white throat and broad black breast-band. Tail black with broad orange-buff tips. No similar bird in its range. **V** Song is a fast twittering medley. Call is a soft high-pitched *zip*. **H** Forest, forest patches, secondary growth, occasionally mangroves. **D** Palau, W Pacific.

23 STREAK-BREASTED FANTAIL *Rhipidura dedemi* 13–14cm Very conspicuous and a regular member of mixed-species foraging parties. **V** Attractive, rapidly repeated, medium-pitched musical notes. **H** Forest undergrowth in lowland and montane forest. **D** Seram, S Maluku Islands, Indonesia.

24 TAWNY-BACKED FANTAIL *Rhipidura superflua* 14cm Usually occurs singly or in pairs; forages in the understorey. **V** Short, quiet series of tinkling notes. **H** Hill and montane forest, forest edges. **D** Buru, S Maluku Islands, Indonesia.

25 SULAWESI FANTAIL *Rhipidura teysmanni* 14cm Active forager from the mid-storey to the lower canopy. Reported following malkohas to capture disturbed insects. **V** Rising and falling series of 3–5 subdued, high-pitched notes, interspersed with *churr* and *tzic* notes. **H** Hill and montane forest; occasionally in selectively logged lowland forest. **D** Sulawesi, Indonesia.

26 TALIABU FANTAIL *Rhipidura sulaensis* 14cm Formerly considered a race of Sulawesi Fantail, and differs only subtly morphologically. Vocalisations distinct. **V** Song is a high-pitched tinkling preceded by short *tseep* notes. **H** Submontane and montane forest. **D** Sula Islands, Indonesia.

27 LONG-TAILED FANTAIL *Rhipidura opistherythra* 17cm Shy and skulking; often part of mixed-species foraging flocks. **V** Song consists of a series of high-pitched squeaky notes; calls are short, unmusical repeated notes. **H** Primary and secondary forest, forest edges and, occasionally, mangroves. **D** Tanimbar Islands, Indonesia.

28 RUFOUS-BACKED FANTAIL *Rhipidura rufidorsa* 14cm Small active fantail of the mid-storey and canopy. Sexes alike. Often joins mixed-species flocks. **V** Song is a series of four descending slurred notes, *see-see-see-seew*, repeated at regular intervals. **H** Forest and secondary growth in the lowlands and foothills. **D** New Guinea.

29 BISMARCK FANTAIL *Rhipidura dahli* 14cm Small, dull rufous fantail with a dark subterminal tail-band. Sexes similar. Often joins mixed-species flocks. **V** Song is a series of rising and falling nasal notes. **H** Inhabits undergrowth in montane forest. **D** Bismarck Archipelago.

30 MUSSAU FANTAIL *Rhipidura matthiae* 15cm Striking fantail with black head and breast, white forecrown and malar stripe, and rufous wings and tail. Sexes alike. Often joins mixed-species flocks. **V** Song is a fast, quiet medley of scratchy notes. **H** Undergrowth in forest, forest edge, secondary growth. **D** Mussau Island (Papua New Guinea).

31 MALAITA FANTAIL *Rhipidura malaitae* 16cm Dull rufous fantail with darker wings and prominent dark eye. Sexes alike. Often joins mixed-species flocks. **V** Song is not known, but call is a quiet *seep*. **H** Montane forest. **D** Malaita (Solomon Islands).

32 MANUS FANTAIL *Rhipidura semirubra* 14cm Bright rufous above, with darker wings and tail, latter with white tips. Striking head pattern with narrow white eye-ring, dark face and white throat. Note black spots on breast. Sexes alike. Very active. **V** Song is a fast series of descending scratchy notes preceded by a few slow notes. Call is a high-pitched *seep*. **H** Forest and scrub. **D** Admiralty Islands (but extinct on Manus Island).

33 RUFOUS FANTAIL *Rhipidura rufifrons* 15–18cm Note rufous forehead and rump. Forages from the ground to the upper canopy; feeds by searching foliage, fly-catching and, occasionally, hover-gleaning. **V** Calls include chips, buzzes and scolding notes. **H** Lowland, hill and montane forests. **D** Maluku Islands, Indonesia; New Guinea; Solomon Islands; E Australia.

34 POHNPEI FANTAIL *Rhipidura kubaryi* 16.5cm Like Rufous Fantail but with no or restricted rufous in plumage. No other fantail in range. **V** Poorly known. Song is a series of high-pitched staccato notes moving randomly up and down the scale. **H** Forest undergrowth. **D** Pohnpei, Caroline Islands, W Pacific.

35 ARAFURA FANTAIL *Rhipidura dryas* 16–17cm Often joins mixed-species feeding parties foraging from the ground to the canopy, favouring mainly the understorey to lower middle levels. **V** Song consists of piping and bubbling notes and a jangle of short notes. **H** Mangroves, coastal woodland, monsoon woodland, primary and secondary lowland, hill and montane forest, and forest edges. **D** Lesser Sunda Islands, SC New Guinea, N Australia.

36 TAVEUNI SILKTAIL *Lamprolia victoriae* 12cm Unmistakable, with striking white rump, short tail and sparkling black plumage. Forages on the forest floor. **V** High-pitched whistles, a whistling trill, and low chattering squeaks. **H** Mature rainforest, secondary growth. **D** Taveuni (Fiji).

37 NATEWA SILKTAIL *Lamprolia klinesmithi* 10cm Similar to Taveuni Silktail but smaller and more iridescent. **V** Poorly known, but presumably similar to Taveuni Silktail. **H** Mature rainforest, secondary growth. **D** Vanua Levu (Fiji).

38 DRONGO FANTAIL *Chaetorhynchus papuensis* 20cm Aberrant sooty-black 'fantail' with a square-ended tail. Formerly considered to be a drongo (Pygmy Drongo). Perches upright like a large monarch (Plates 185–187), in mid-storey. **V** Song is a rapid jumbled series of squeaky notes. Also makes loud, rasping, chattering sounds. **H** Hill forest, from foothills to lower mountains. **D** New Guinea.

1
2
3
4
5
6
7
♀
♂
8
9
10
11
dark morph
pied morph
12
13
14
15
16
17
18
19
20
21
pale morph
dark morph
22
23
24
25
26
27
28
29
30
31
32
33
34
35
36
37
38

MONARCHS MONARCHIDAE

1 BLACK-NAPED MONARCH *Hypothymis azurea* 16cm Note male's black nuchal tuft. Active; gleans insects from foliage. Also hovers or makes aerial sallies to capture flying insects. **V** Ringing *wii-wii-wii-wii-wii-wii*; also a high-pitched, rasping *sweech-which* or *che-chwe*. **H** Broadleaved evergreen, semi-evergreen, deciduous and peatswamp forest, secondary growth, overgrown plantations. **D** From India and Sri Lanka to SE Asia, Indonesia, S China.

2 PALE-BLUE MONARCH *Hypothymis puella* 15–17cm Occurs singly, in pairs or in small groups. Forages mainly by gleaning from leaves. Often sits in an upright position. **V** Song consists of a series of whistled notes, including a *s-si-si SI-SI-SII*, a bubbling *whiwhiwhiwhiwhi...* and an upslurred *sweei-sweei-sweei...* **H** Lowland, hill and secondary forest, scrub, lightly wooded cultivation. **D** Sulawesi and nearby islands, Indonesia

3 SHORT-CRESTED MONARCH *Hypothymis helenae* 14cm Male has black around base of bill. Generally seen alone, in pairs or as part of mixed-species foraging parties. **V** Rapid, high-pitched, metallic *pi-pi-pi-pi-pi-pi-pi-pi* that increases in volume; also a rasping *tzeet-zip-zip* or *swee-sip*. **H** Forest understorey. **D** Philippines.

4 CELESTIAL MONARCH *Hypothymis coelestis* 15–16cm Forages in the middle and upper canopy, singly or in mixed-species feeding flocks. **V** Loud, ringing *pwee pwee pwee* and a repeated, raspy *pee-chittt*. **H** Lowland forest, forest edges, secondary growth. **D** Philippines.

5 CERULEAN PARADISE FLYCATCHER *Eutrichomyias rowleyi* 18cm Forages mainly in the canopy and subcanopy in small groups, occasionally joining mixed-species feeding flocks. Feeds by fly-catching or gleaning from vegetation. **V** Loud, rasping *chew chew chew chew chew chew*; a single *tuk*; a loud, descending, trilled *chreechreechreechree*; and a high, fizzing *streeeeee*. **H** Broadleaved tropical hill rainforest; favours sheltered forest in valleys. **D** Sangihe Island, N of Sulawesi, Indonesia.

6 BLUE-MANTLED CRESTED FLYCATCHER *Trochocercus cyanomelas* 14cm Like all small African flycatchers, very active, constantly cocking and fanning its tail. Separated from other flycatchers by white wing patches. **V** Varied medley of well-separated, instrumental phrases: high, nasal *weetwit-weetwit*, low, liquid *weetweetweetweet*, fluted trill, *wrrrrrrh*. **H** Dense, lower storeys of forests. **D** E and SE Africa, from Somalia to South Africa.

7 BLUE-HEADED CRESTED FLYCATCHER *Trochocercus nitens* 12cm Not unlike other small tail-fanning flycatchers, but note sharp demarcation of white and black on breast. **V** Song is a rapidly delivered, far-carrying, hollow *hohohohohohohoho*. **H** Dense forest undergrowth. **D** W and C Africa, from Guinea to Uganda and NW Angola.

8 BEDFORD'S PARADISE FLYCATCHER *Terpsiphone bedfordi* 18cm Grey with black head. Note blue eyelid wattle. All paradise flycatchers are known to hybridise easily within their genus, which makes identification not always easy. **V** Song unknown. Calls include a *zre-zre*. **H** Forest undergrowth. **D** E DR Congo.

9 RUFOUS-VENTED PARADISE FLYCATCHER *Terpsiphone rufocinerea* Male 23–28cm, female 18cm Separated from African Paradise Flycatcher by rufous undertail coverts, smaller crest and less conspicuous eye wattle. Female lacks tail streamers. **V** Song is a mellow *huit-huit-huit* or *lulululu...* **H** Forests, riverine belts. **D** S Cameroon to NW Angola.

10 RED-BELLIED PARADISE FLYCATCHER *Terpsiphone rufiventer* Male 30–32cm, female 18–21cm Many races occur, all with rufous underparts. Female lacks tail streamers. **V** Song is a short, ringing *huee-huee-huee-huee*. **H** Forest interiors. **D** W and C Africa.

11 ANNOBON PARADISE FLYCATCHER *Terpsiphone smithii* 18cm Short-tailed paradise flycatcher with bluish-black head, brown wings and tail, and bright rufous body. Central tail feather of male only extends c. 1cm. **V** Similar to other paradise flycatchers. **H** Forest, secondary growth, cultivation. **D** Annobón, Gulf of Guinea.

12 BATES'S PARADISE FLYCATCHER *Terpsiphone batesi* 20cm From Rufous-vented Paradise Flycatcher by rounded head without crest. **V** Song is short and cheerful, very like that of Red-bellied Paradise Flycatcher. Typical calls are harsh and rasping. **H** Forests. **D** C Africa, from S Cameroon to E DR Congo and NW Angola.

13 AFRICAN PARADISE FLYCATCHER *Terpsiphone viridis* Male 36cm, female 20cm Rufous and white forms. Rufous males sometimes have white tail streamers; white wing patch variable in size or occasionally absent (see plate) . **V** Pleasant thrush-like *twee-twoo-twoo-twoo-twoo*, harsh *tsveit*, *scheep* and *tscaeae-tseaeaet*. **H** Savannah woodland, plantations, wadis with trees. **D** Widespread across sub-Saharan Africa; S Arabian Peninsula.

14 INDIAN PARADISE FLYCATCHER *Terpsiphone paradisi* Male 45cm, female 20cm Male has rufous and white morphs; female similar to rufous-morph male. Generally hunts from a perch in the lower part of tree canopy, usually in pairs or mixed-species flocks. **V** Clear rolling *chu-wu-wu-wu-wu-wu...*, also a loud *chee-tew*, a harsh *tst*. When mobbing utters a *weep-poor-willie-weep-poor-willie*. **H** Open forests, bushes, groves and gardens. **D** Breeds N Afghanistan and W China to Bangladesh, SW Myanmar, India, Sri Lanka. Northern populations winter in Indian subcontinent.

15 BLYTH'S PARADISE FLYCATCHER *Terpsiphone affinis* Male 45cm, female 20cm Formerly considered conspecific with Indian Paradise Flycatcher. Male has rufous and white morphs. Generally hunts from a perch in lower canopy, usually in pairs. **V** Clear, rolling *chu-wu-wu-wu-wu...*, a loud *chee-tew*, a *zhee* and a whistled *whit-it* or *pop-it*. **H** Primary and secondary lowland and hill monsoon forest, open woodland and lightly wooded cultivation. **D** NE India, SE Asia, Borneo, Indonesia.

16 AMUR PARADISE FLYCATCHER *Terpsiphone incei* Male 45cm, female 20cm Formerly considered conspecific with Blyth's Paradise Flycatcher. Very similar to Blyth's Paradise Flycatcher but darker chestnut on upperparts and tail, and black hood is more sharply demarcated from white underparts. Note white vent. **V** Similar to Blyth's Paradise Flycatcher. **H** Various wooded habitats including parks and gardens. **D** Breeds China, far E Russia, N Korea. Non-breeding mainly in SE Asia.

17 JAPANESE PARADISE FLYCATCHER *Terpsiphone atrocaudata* Male 30–35cm, female 18cm A dark paradise flycatcher with a short or medium-length crest. Actions similar to Indian Paradise Flycatcher. **V** Whistled *tsuki-hi-hoshi-hoi-hoi-hoi*; call is a querulous *jouey*. **H** Forest and forest edge; on migration found in mangroves, parks and wooded gardens. **D** Japan, Korean Peninsula, Taiwan, islands off N Philippines. Winters Philippines, Malay Peninsula, Sumatra.

18 BLUE PARADISE FLYCATCHER *Terpsiphone cyanescens* Male 22cm, female 19cm Male mostly grey-blue, with black lores, chin, flight feathers and underside of tail. Female has grey-brown wings and rufous mantle, rump and tail. **V** Loud, staccato trill, lasting about 4 secs and repeated several times; also a mild *chh-chh-chh*. **H** Primary lowland forest and secondary growth. **D** W Philippines.

19 RUFOUS PARADISE FLYCATCHER *Terpsiphone cinnamomea* Male 30–31cm, female 21–22cm Note distinctive coloration. Forages singly, in pairs and in mixed-species feeding parties. **V** Continuous series of 30 or more strident *schweet* whistles, and a harsh, raspy *tre-chee*. **H** Forest and secondary growth in lowlands and hills. **D** Philippines; Talaud Islands, Indonesia.

20 SAO TOME PARADISE FLYCATCHER *Terpsiphone atrochalybeia* Male 25–28cm, female 18cm Only paradise flycatcher on São Tomé. Male is glossy blue-black; female has grey head and is otherwise brown and white. **V** Song is a whistled *tzup-tewee, tewee*. **H** All types of habitat, wherever there are trees or shrubs. **D** São Tomé, Gulf of Guinea.

21 MALAGASY PARADISE FLYCATCHER *Terpsiphone mutata* Male 38cm, female 18cm No similar bird in the area. Variable, with rufous and white forms, plus intermediates. **V** Short, rapid, melodious, descending strophes, mixed with twitters. **H** Forest, plantations, gardens. **D** Madagascar, Comoros islands.

22 SEYCHELLES PARADISE FLYCATCHER *Terpsiphone corvina* Male 36cm, female 20cm Male is all black with blue iridescence and very long tail. Female strikingly different, and lacks elongated central tail feathers. **V** Irregular series of chirping, loud *weet-wit* notes. **H** Woodland at low altitudes near marshes. **D** Seychelles.

23 MASCARENE PARADISE FLYCATCHER *Terpsiphone bourbonnensis* 17.5cm Note rounded head, and sharp demarcation in male between black hood and grey throat. **V** high, sharp twittering. **H** Forest, secondary growth, plantations. **D** Mauritius, Réunion.

24 KAUAI ELEPAIO *Chasiempis sclateri* 14cm Formerly considered conspecific with Hawaii Elepaio, but vocally and genetically distinct. Differs from Hawaii Elepaio in grey-brown crown and mantle, and white chin and throat. **V** Various squeaky call notes and a 2–3-note whistle, *huu yu-yu*. **H** Dense wet forest in mountains, but also in drier forest, sometimes with introduced vegetation. **D** Kauai (Hawaii).

25 OAHU ELEPAIO *Chasiempis ibidis* 13cm Formerly considered conspecific with Hawaii Elepaio, but vocally and genetically distinct. Differs from Hawaii Elepaio in not having much black on throat. **V** Pleasant five-note whistle with emphasis on second and fourth notes. **H** Mixed wet forest with tall canopy and dense understory; also in drier forest. **D** Oahu (Hawaii).

26 HAWAII ELEPAIO *Chasiempis sandwichensis* 14cm Variable, but unmistakable by its cocked tail, white wing bands and white rump. **V** Twittering, rapid *witwitwit* or *wit-wit-wit* or *weetjur weetjur* or *weetjur wír*. **H** Closed-canopy forest, woodland, savannah. **D** Hawaii (Hawaii).

27 RAROTONGA MONARCH *Pomarea dimidiata* 15cm Unmistakable in very restricted range. Mostly slate-grey above, white below. **V** Rapid *tjuhwéetwéet* or *rapraprap*. **H** Undergrowth of upland native forest. **D** Rarotonga, Cook Islands.

28 TAHITI MONARCH *Pomarea nigra* 15cm Male and female similar. Distinguished from dark morph of Tahiti Reed Warbler (Plate 212) by shorter grey-blue bill. **V** Powerful, toneless *kratchkratchkratch* followed by e.g. melodious *feeohweeh*. **H** Dense native forest. **D** Tahiti, French Polynesia.

29 MARQUESAN MONARCH *Pomarea mendozae* 17cm Male entirely black. Female unmistakable. **V** Varied e.g. two-noted *Tòh-tjep* and other, partly inhaled phrases. **H** Forest and degraded forest at all elevations. **D** Mohotani, Marquesas Islands, French Polynesia.

30 UA POU MONARCH *Pomarea mira* 17cm Formerly considered conspecific with Marquesan Monarch. Male all black, like male Marquesan. Female differs in being all black except for white tail and much white in wing. Critically endangered, possibly extinct. **V** Not known. **H** Native forest, most recently in montane areas. **D** Ua Pou, Marquesas Islands, French Polynesia.

31 IPHIS MONARCH *Pomarea iphis* 17cm Male partly white below, female with distinctive eye-ring. **V** Very high, rapid, sharp *sweesweeswee*. **H** Forest, dry scrub. **D** Uua Huka, Marquesas Islands, French Polynesia.

32 FATU HIVA MONARCH *Pomarea whitneyi* 19cm Sexes similar. From other Marquesan monarchs by stiff, very black frontal feathers (see Marquesan Monarch) or black underparts (see Iphis Monarch); does not share island ranges of the other species. **V** Song not recorded. Calls include a shrill cat-like *cri-ri-a-rik*. **H** Dense native forest, wooded thickets. **D** Fatu Hiva, Marquesas Islands, French Polynesia.

1
2
3
4
5
6
7
8
9
10
11
12
13
14
15
16
17
18
19
20
21
22
23
24
25
26
27
28
29
30
31
32
white and rufous (right) morphs
rufous morph
white morph
white morph
white morph
rufous morph
white morph
rufous
♂
♀

MONARCHS *CONTINUED*

1 VANIKORO MONARCH *Mayrornis schistaceus* 14cm Mainly slate-grey. Tail black with white tips to outer feathers. Cocks and fans tail. **V** Song is reported to be a trisyllabic wavering whistle; also gives harsh chattering and scolding notes. **H** Forest, forest edge and secondary growth. **D** Vanikoro (SE Solomon Islands).

2 VERSICOLORED MONARCH *Mayrornis versicolor* 12cm Distinctively colour-patterned. **V** Typical call is a far-carrying *tsic*; sometimes uttered in series, followed by an upslurred whistle. **H** Forest interior and edge. **D** Fiji.

3 SLATY MONARCH *Mayrornis lessoni* 13cm Note whitish lores and eye-ring. **V** Scolding *tchic tchic tchic tchic*; also a harsh churr and a high-pitched descending whistle. **H** Dense forest, tall trees in parks and gardens. **D** Fiji.

4 BUFF-BELLIED MONARCH *Neolalage banksiana* 15cm Unmistakable, striking monarch with a white face and yellowish underparts. Often half-cocks tail. **V** Song is reported to be a thin wavering whistle; also gives harsh chattering and scolding notes. **H** Forest and scrub at all altitudes. **D** Vanuatu.

5 SOUTHERN SHRIKEBILL *Clytorhynchus pachycephaloides* 19cm Large brown monarch with stout grey bill. Note white tips to outer-tail feathers. Sexes alike. Often in mixed-species flocks. **V** Song is a loud quavering whistle, sometimes slurred upwards or downwards; also gives harsh scolding notes. **H** Wet forests, in understorey and subcanopy. **D** Vanuatu and New Caledonia.

6 FIJI SHRIKEBILL *Clytorhynchus vitiensis* 19cm Rather nondescript. Unmistakable in range by bill shape. **V** High, slightly shivering, downslurred *pipriiuuh* or mid-high *titjuwuuuh* or parrot-like twittering *chatchatchatter*. **H** Forest with thick undergrowth, woodland, thick scrub. **D** Fiji, Tonga, Samoa.

7 BLACK-THROATED SHRIKEBILL *Clytorhynchus nigrogularis* 21cm Male unmistakable; female sepaerated from Fiji Shrikebill by heavier bill. **V** Long, drawn-out wavering whistle, *hu-hu-hu-hooo*. **H** Dense forest. **D** Fiji.

8 SANTA CRUZ SHRIKEBILL *Clytorhynchus sanctaecrucis* 19cm Large monarch with stout blue-grey bill. Male is black and white with prominent white ear-coverts. Female is all brown. Sometimes joins mixed-species flocks. **V** Song is a long whistle, often upslurred; also harsh scolding notes. **H** Primary forest, in the understorey. **D** Nendö (Santa Cruz Islands, Solomon Islands).

9 RENNELL SHRIKEBILL *Clytorhynchus hamlini* 19cm Large brown monarch with a very long blue-grey bill. Rufous-brown plumage with a black face. Sexes similar, but female duller. Often in mixed-species flocks. **V** Song is a wavering whistle or varied 2–3-note whistles, repeated at regular intervals. **H** Forest and secondary growth, in the understorey. **D** Rennell (Solomon Islands).

10 CHUUK MONARCH *Metabolus rugensis* 20cm Male unmistakable, almost entirely white with glossy blue-black face and throat. Most females have some reddish feathers at random through plumage. **V** Various upslurred or downslurred whistles. **H** Undisturbed forest; also in mangroves. **D** Chuuk Islands, Micronesia.

11 BLACK MONARCH *Symposiachrus axillaris* 16cm Glossy blue-black monarch with white pectoral tufts. Sexes similar, but female duller. Very similar to unrelated Black Fantail (Plate 184), hence alternative name of Fantail Monarch. **V** Song is a harsh series of grating rattles, repeated at irregular intervals. **H** Lower montane forest, secondary scrub. **D** New Guinea.

12 SPOT-WINGED MONARCH *Symposiachrus guttula* 15cm Distinctive grey-and-white monarch with black face mask. Sexes alike. Often in mixed-species flocks. **V** Harsh, scolding grating sounds, often ending in a fast high-pitched 4–5-note whistle. **H** Rainforest and tall secondary growth, in lowlands and hills. **D** New Guinea and nearby islands.

13 BLACK-BIBBED MONARCH *Symposiachrus mundus* 16cm Occurs singly, in pairs or in small groups; also joins mixed-species feeding parties. Actively gleans from the underside of leaves. When excited, cocks and fans tail. **V** Song consists of a series of well-spaced whistled *swih* notes, increasing in volume and accelerating, followed by a series of high, rising notes that run together. **H** Primary semi-evergreen forest, mangroves. **D** E Lesser Sunda Islands.

14 FLORES MONARCH *Symposiachrus sacerdotum* 15–16cm Forages singly, in pairs or in small mixed-species feeding parties. **V** Upward-inflected whistle; also buzzing notes, a nasal *schr schr schr* followed by a flute-like whistle, and a scolding *sjay-sjay*. **H** Primary moist semi-evergreen hill forest, and old secondary and partially degraded forest. **D** Flores, Indonesia.

15 BLACK-CHINNED MONARCH *Symposiachrus boanensis* 16cm Very rare. Occurs singly, in twos or in threes, and a regular member of mixed-species feeding parties. Active forager in undergrowth near the ground. **V** Clear *tjuuu tjuuu*, immediately followed by a soft, monotonous, fading, buzzing trill. **H** Remnant patches of dense semi-evergreen secondary forest in gorges and valleys in foothills. **D** S Maluku Islands, Indonesia.

16 SPECTACLED MONARCH *Symposiachrus trivirgatus* 14–16cm Grey upperparts, with black mask and chin, rufous-orange throat and upper breast. Forages from the sub-stage to the lower canopy, singly or in pairs; also joins mixed-species feeding parties. **V** Series of harsh, muted, rasping notes, a churring *prrrrt* and a melodious *whit*. **H** Primary and secondary forest, forest edge, mangroves. **D** Lesser Sunda Islands, S Maluku Islands, Louisiade Archipelago, Torres Strait islands, NE and E Australia.

17 MOLUCCAN MONARCH *Symposiachrus bimaculatus* 14–16cm Singles or pairs forage from the sub-stage to the lower canopy; also forms part of mixed-species feeding groups. **V** Rapid, rasping, slightly nasal, staccato notes, often with an ascending or descending whistle; also various harsh, nasal, squabbling notes. **H** Primary and secondary forest. **D** N Maluku Islands, Indonesia.

18 WHITE-TAILED MONARCH *Symposiachrus leucurus* 15–16cm Gleans from leaves in the mid-storey, alone, in pairs and in mixed-species feeding parties. **V** Series of four clear, descending, high-pitched whistles; a series of slow, dry, rasping, double-note scolding calls and a series of staccato, slightly musical, harsh chattering notes. **H** Primary and secondary coastal lowland monsoon forest, hill forest, lightly wooded cultivation. **D** Kai Islands, Indonesia.

19 WHITE-TIPPED MONARCH *Symposiachrus everetti* 14cm Noisy. Forages in pairs or occasionally in small groups; also in mixed-species feeding parties. Frequently cocks and half-fans tail. **V** Slightly tremulous, plaintive whistle; also harsh, scolding notes. **H** Forest, scrub, mangroves with scattered large trees. **D** Tanah Jampea, Indonesia (Flores Sea).

20 BLACK-TIPPED MONARCH *Symposiachrus loricatus* 18cm Forages in the understorey and in secondary growth, alone or in pairs. **V** Rich, descending *teoow teoow teoow*. **H** Lowland and submontane primary forest, shrubby secondary growth, edges of logged areas and cultivation. **D** Buru, S Maluku Islands, Indonesia.

21 KOFIAU MONARCH *Symposiachrus julianae* 16cm Resembles Spot-winged Monarch but is darker above and lacks wing spots. Sexes alike. Singly or in pairs. **V** High-pitched trisyllabic note, similar to Spot-winged Monarch. **H** Coastal and lowland forests. **D** Kofiau, West Papuan islands, Indonesia.

22 BIAK MONARCH *Symposiachrus brehmii* 17cm Distinctive monarch with creamy crescent around ear-coverts and cream patch on wings. Female has white throat. Singly or in pairs. **V** Utters a variety of loud musical whistles. **H** Rainforest and forest fragments, in lower and mid-storey. **D** Biak, West Papuan islands, Indonesia.

23 HOODED MONARCH *Symposiachrus manadensis* 16cm Similar to Kofiau Monarch but has more extensive black hood. Sexes alike. Usually singly but will join mixed-species flocks. **V** Long, drawn-out, plaintive whistle, sometimes tremulous. **H** Lowland and hill forest. **D** New Guinea.

24 MANUS MONARCH *Symposiachrus infelix* 15cm Striking pied monarch with large white patch on side of neck and on wing coverts. Sexes alike. **V** Long, drawn-out, quavering whistle, sometimes upslurred at end; also dry rasping scolding notes. **H** Mature forest, occasionally in secondary growth. **D** Manus Island and nearby islands (Bismarck Archipelago).

25 MUSSAU MONARCH *Symposiachrus menckei* 15cm Distinctive monarch with mostly white plumage. Sexes alike. Singly or in pairs. **V** Long, drawn-out, quavering whistle, similar to Manus Monarch. Also gives harsh scolding notes. **H** Forest and secondary growth in lowlands and hills. **D** Mussau Island (Bismarck Archipelago).

26 BLACK-TAILED MONARCH *Symposiachrus verticalis* 16cm Similar to Mussau Monarch but more black on head, throat and mantle. Sexes alike. Singly or in pairs. Often in mixed-species flocks. **V** Thin, high-pitched whistle, *teee-weeee teee-weeee*. **H** Primary forest, less commonly in secondary growth and montane forest. **D** Bismarck Archipelago.

27 SOLOMONS MONARCH *Symposiachrus barbatus* 15cm Similar to Black-tailed Monarch but more black on head. Sexes alike. Singly or in pairs. Often in mixed-species flocks. **V** Long drawn-out whistle like others in the genus, but lower-pitched and more tremulous, almost trilling at times. Also gives raspy scolding notes. **H** Primary forest, mainly in hills, but sometimes lowland forest. **D** Solomon Islands.

28 KOLOMBANGARA MONARCH *Symposiachrus browni* 15cm Typical pied monarch. Sexes alike. Four races, with differences in the extent of the black bib and the white in the wings and tail. **V** Thin trilling whistle, lower-pitched than others in the genus. Also gives raspy scolding notes. **H** Lowland primary forest. **D** New Georgia island group, Solomon Islands.

29 WHITE-COLLARED MONARCH *Symposiachrus vidua* 15cm Typical pied monarch with black hood and complete white collar. Sexes alike. Often in mixed-species flocks. **V** Thin drawn-out whistle, similar to others in the genus. Also gives raspy scolding notes. **H** Primary forest, sometimes in secondary growth and forest edge, in lowlands and hills. **D** Makira (Solomon Islands).

30 RUFOUS MONARCH *Monarcha rubiensis* 18cm Mainly rufous plumage. Male has black throat, female is entirely rufous. Singly or in pairs; often joins mixed-species flocks. **V** Poorly known, but gives buzzy scolding notes and whistles similar to others in the genus. **H** Lowland and hill forest, in lower to mid-storey. **D** New Guinea.

31 ISLAND MONARCH *Monarcha cinerascens* 16–19cm Forages singly, in pairs or in small groups, from near the ground to the canopy, feeding mainly by gleaning. Noisy. **V** Rapid, loud, slurred *weeweeweeweeweeweewee...*; calls include harsh, scolding chatters, followed by a jangle of squeaky, twittering notes. **H** Lowland forest and scrub. **D** Sulawesi to Solomon Islands.

32 BLACK-FACED MONARCH *Monarcha melanopsis* 18cm Distinctive pale grey upperparts, rufous underparts, black face and throat. Sexes alike. **V** Song is a mellow whistled *wheeuuu whit-chew, wheeuuu whit-chew* repeated frequently. **H** Rainforest and other wet forests; occasionally mangroves. **D** Breeds E Australia. Winters extreme NE Australia, S New Guinea.

33 BLACK-WINGED MONARCH *Monarcha frater* 18cm Differs from Black-faced Monarch in having black wings and tail. Underparts deeper rufous. **V** Call is an upslurred whistle. Also gives a repeated *weeuu-witchew*, similar to Black-faced Monarch but less strident. **H** Lowland rainforest, in lower to mid-storey. **D** New Guinea, NE Australia. Australian breeding birds probably winter in S New Guinea.

34 BOUGAINVILLE MONARCH *Monarcha erythrostictus* 17cm Glossy black head, throat and upperparts, contrasting with rufous belly and vent. Male has white spot in front of eye; female has rufous spot. Singly or in pairs; often joins mixed-species flocks. **V** Long mournful whistle at the same pitch, which may ascend or descend the scale slightly. **H** Primary forest, occasionally secondary growth, lowlands and hills. **D** Bougainville, Solomon Islands.

35 CHESTNUT-BELLIED MONARCH *Monarcha castaneiventris* 17cm Similar to Bougainville Monarch but lacks spot in front of eye. Sexes are similar. Often in mixed-species flocks. **V** Song is a loud, strong whistle, given as a single note or downslurred or upslurred, and varying in pitch. **H** Primary and secondary forest in lowlands and hills. **D** Solomon Islands.

2
3
4
5
1
♀
9
♀
6
8
10
♂
♀
7
♂
♂
15
11
12
13
14
16
♀
17
19
22
♂
18
20
21
23
24
25
26
27
28
31
32
♂
34
♀
29
♀
30
♂
33
35

MONARCHS *CONTINUED*

1 WHITE-CAPPED MONARCH *Monarcha richardsii* 17cm Striking large monarch, mainly blue-black above with white crown and nape, and chestnut belly. Often in small flocks or mixed-species flocks. **V** Song is a short tremulous whistle. Calls are typical harsh scolding notes, often in threes. **H** Forest and scrub. **D** New Georgia island group, Solomon Islands.

2 YAP MONARCH *Monarcha godeffroyi* 15cm Unmistakable in restricted range. **V** Song is a series of loud clear whistles, *we're here, we're here*. Calls include a rapid rasping *chick-chick-cher-dee* and a squeaky *we'er*. **H** Forest, secondary growth, scrub, mangrove. **D** Yap, Micronesia.

3 TINIAN MONARCH *Monarcha takatsukasae* 15cm Unmistakable by contrasting white plumage parts; restricted range. **V** Song is a short whistled *tee-tee-wheeeooo*. Calls include low rasping notes. **H** Forested and wooded habitats. **D** Tinian, Mariana Islands, W Pacific.

4 WHITE-EARED MONARCH *Carterornis leucotis* 14cm Distinctive pied monarch with complex head pattern. Sexes alike. Forages in the canopy, singly or in pairs. **V** Typically a four-note whistle with the final two notes downslurred. Also other whistles and harsh scolding notes. **H** Forest edge, mangroves and thickets. **D** NE Australia, south to extreme NE New South Wales.

5 WHITE-NAPED MONARCH *Carterornis pileatus* 14–15cm Active. Feeds by gleaning or hover-gleaning in the canopy or mid-storey, generally in small groups or mixed-species flocks. **V** Series of semi-musical, nasal, squabbling and chattering notes, followed by a single harsh downslurred note. **H** Lowland and hill forest, selectively logged forest and forest edges. **D** Maluku Islands, Indonesia.

6 GOLDEN MONARCH *Carterornis chrysomela* 13cm Small yellow and black monarch with nine races. Sexually dimorphic. Males vary in extent of yellow on back and wings, and intensity of golden-yellow colour on head. Females usually duller. Both sexes have white spot in front of eye. **V** Varies regionally. Usually includes a rapid jumble of whistles and scolding notes. **H** Rainforest and other forest types in lowlands and hills. **D** New Guinea, Bismarck Archipelago.

7 OCHRE-COLLARED MONARCH *Arses insularis* 16cm Striking monarch with blue eye-wattles and erectile feathers on nape forming a frill. Sexually dimorphic. Male has ochraceous breast and broad collar on hindneck. Female is browner. **V** Rapid series of mechanical notes or whistles, often increasing in volume. **H** Forest in lowlands and hills. **D** N New Guinea.

8 FRILLED MONARCH *Arses telescopthalmus* 16cm Striking monarch with blue eye-wattles and erectile feathers on nape forming a frill. Sexually dimorphic. Male resembles Ochre-collared Monarch but is entirely black and white. Female is similar to Ochre-collared but has blacker head and more rufous upperparts. **V** Similar trills to Ochre-collared Monarch, but usually slower and lower-pitched. **H** Rainforest and secondary growth in lowlands and hills. **D** New Guinea and nearby islands.

9 FRILL-NECKED MONARCH *Arses lorealis* 15cm Striking monarch with blue eye-wattles and erectile feathers on nape forming a frill. Male closely resembles Frilled Monarch. Female is like male (unlike female Frilled) but has white chin and lores. **V** Similar to Frilled Monarch. **H** Rainforest and gallery forest. **D** Cape York Peninsula, NE Australia.

10 PIED MONARCH *Arses kaupi* 15cm Striking monarch with blue eye-wattles and erectile feathers on nape forming a frill. Male is similar to male Frill-necked Monarch but has broad black breast-band. Female is like male but breast-band joins hood. **V** Similar to Frill-necked Monarch but softer and slower. **H** Rainforest and other wet forests. **D** NE Australia.

11 MAGPIE-LARK *Grallina cyanoleuca* 26cm Forages mainly on the ground, but also perches on trees, from where occasional sallies are made after flying insects. **V** Duet: *tee-hee* from first bird, followed by *pee-o-wee* or *pee-o-wit* from second bird. Also a liquid, mellow *cloop cloop cloop* or *clue-weet clue-weet*, and an emphatic *pee pee pee* when alarmed. **H** Open vegetation, savannah, farmland, parks, gardens, urban areas; near surface water. **D** Widespread across Australia; E Lesser Sunda Islands, S New Guinea.

12 TORRENT-LARK *Grallina bruijnii* 20cm Striking black and white inhabitant of mountain torrents. Forages on boulders and close to water. Sexually dimorphic. Male has white patch on side of neck and black belly. Female has broad white supercilium and white belly. **V** Call is a single raspy, buzzing sound, often repeated and audible above the sound of running water. **H** Fast-flowing mountain streams throughout highlands. **D** New Guinea.

13 OCEANIC FLYCATCHER *Myiagra oceanica* 15cm No similar bird in the small forest remnants on Chuuk Is. **V** Song consists of upslurred or downslurred whistles, *twee-twee-twee-twee*. Calls include a distinctive *quick-eRICK*. **H** Forest, forest edge. **D** Chuuk Islands, Micronesia.

14 PALAU FLYCATCHER *Myiagra erythrops* 15cm No similar small, crested bird on Palau islands. **V** Song is an evenly pitched *pee-pee-pee*; also a rapid, flat-pitched trill, *titititititititi*. **H** Forest, forest edge, open areas with scattered trees. **D** Palau, W Pacific.

15 POHNPEI FLYCATCHER *Myiagra pluto* 15cm No similar small, all-black flycatcher in range. Note sepia-brown breast of female. **V** Repeated whistles, low-pitched and melodious, at a variable pace, *whee...whee...whee...whee*. **H** Forest, forest edge, open areas with scattered trees. **D** Pohnpei, Caroline Islands, W Pacific.

16 MOLUCCAN FLYCATCHER *Myiagra galeata* 14cm Forages mainly in the canopy, or lower down when in mixed flocks. Often raises crown feathers and shivers tail. **V** Loud, whistled *teu-teu-teu-teu-teu-teu...*, rapid, unmusical *wik* notes. **H** Primary and secondary lowland and hill forest, logged forest, forest edges, forest patches in cultivation, mangroves, coconut groves. **D** Maluku Islands, Indonesia.

17 BIAK BLACK FLYCATCHER *Myiagra atra* 13cm Small, sexually dimorphic, restricted-range flycatcher. Male is all glossy blue-back. Female is grey and paler below. Male Shining Flycatcher is larger and avoids forest; female is very different. **V** Song is a loud whistle, *wee-choo-wee-choo-wee-choo wee* or *wee-wee-wee-wee-wee-wee*. **H** Forest and secondary growth. **D** Biak, N West Papua, Indonesia.

18 LEADEN FLYCATCHER *Myiagra rubecula* 13–16cm Typical *Myiagra* flycatcher. Sexually dimorphic. Male has blue-grey upperparts with darker throat and white belly. Female has rufous throat and upper breast. Six races with minor differences. Perches upright, frequently erects crown feathers and shivers tail. **V** Song is a loud whistle, *hoowee hoowee hoowee* or *hwee hwee hwee hwee*. **H** Dry forests, coastal forests, forest edge, mangroves, scrub. **D** N and E Australia, S New Guinea, Torres Strait islands.

19 STEEL-BLUE FLYCATCHER *Myiagra ferrocyanea* 14cm Male is similar to male Leaden Flycatcher but is glossy black above. Note angle of lower border of breast-band. Female is grey above with rufous-brown wings and tail; no rufous on throat. **V** Song is a loud, drawn-out, upwards-inflected whistle, *hoooowee hoooowee hoooowee*. **H** Forest and degraded wooded habitats. **D** Bougainville, Solomon Islands.

20 MAKIRA FLYCATCHER *Myiagra cervinicauda* 14cm Male is similar to male Steel-blue Flycatcher. Female has grey head, pale eye-ring, brown upperparts and deep orange throat and breast. In pairs or joins mixed-species flocks. **V** Song is a thin high-pitched whistle, *hwee hwee hwee hwee*. **H** Forest. **D** Makira (Solomon Islands).

21 MELANESIAN FLYCATCHER *Myiagra caledonica* 14cm Typical *Myiagra* flycatcher, reminiscent of Makira Flycatcher. Sexually dimorphic. Five races, differing mainly in colour of upperparts and extent of white tail markings. **V** Song is a mournful whistle, *hoo-wee hoo-wee hoo-wee*. Also gives buzzy call notes. **H** Forest, secondary growth, mangroves. **D** Solomon Islands, Vanuatu, New Caledonia.

22 VANIKORO FLYCATCHER *Myiagra vanikorensis* 13cm From Blue-crested Flycatcher by black bill and pale orange underparts. **V** Hoarse, almost toneless squeaks *twittweettweet*, in rapid series of three. **H** Forest, secondary growth, wooded farmland, gardens. **D** Fiji, Vanikoro in the SE Solomon Islands.

23 SAMOAN FLYCATCHER *Myiagra albiventris* 14cm No similar small bird in restricted range. **V** High, tit-like *titjitjitjjitji*. **H** Multi-structured forest, wooded farmland, mangroves. **D** Samoa.

24 AZURE-CRESTED FLYCATCHER *Myiagra azureocapilla* 16cm Male is dark slate-blue above, with azure crest and cheek patch. Female mostly brown-grey above. Orange bill. **V** Song is a rapid, evenly pitched *weet-weet-weet-weet*; may also include a longer, drawn-out *weeeer*. **H** Lower levels of dense forest. **D** Taveuni (Fiji).

25 CHESTNUT-THROATED FLYCATCHER *Myiagra castaneigularis* 15cm Formerly considered conspecific with Azure-crested Flycatcher. Male differs from Azure-crested in having a rufous throat and white tail tips. **V** Song is a whistled *weet-weet-weet-weet*, similar to Azure-crested Flycatcher. **H** Mature forest in hills and mountains. **D** Viti Levu and Vanua Levu (Fiji).

26 BROAD-BILLED FLYCATCHER *Myiagra ruficollis* 15cm Forages from middle to lower storeys, usually alone or in pairs. Captures prey by gleaning or aerial sallies. **V** Musical *weeoo weeoo weeoo*, interspersed with *chwik chwik chwik*; calls include a harsh, raspy *zzzt* or *zsh-wit* and a nasal *bzzzsh*. **H** Mangroves, deciduous forest, secondary monsoon forest and lightly wooded cultivation. **D** Lesser Sunda Islands, S New Guinea, Flores Sea islands, N Australia.

27 SATIN FLYCATCHER *Myiagra cyanoleuca* 17.5cm Distinctively patterned. Note slightly peaked head and plain wings (superficially similar Tomtit has white wing-bars). **V** Call: raspy *shrueeeh*. Song: *tchooweéh tchooweéh*. **H** Migrants in open country, woodland, parks, gardens, forest clearings. **D** E Australia, Tasmania. Non-breeding as far north as NE New Guinea.

28 SHINING FLYCATCHER *Myiagra alecto* 16–17cm Usually encountered in pairs, foraging in dense thickets. **V** Song consists of a series of clear, trilled, whistled notes; calls include a loud, long, down-slurred, rasping note. **H** Mangroves, forest edges, secondary growth and scrub. **D** From Maluku Islands to New Guinea, Bismarck Archipelago, N Australia.

29 VELVET FLYCATCHER *Myiagra hebetior* 16cm Male is very similar to Shining Flycatcher, but is slightly smaller with a more rounded head, and with a shorter tail and shorter bill. Female is duller than female Shining with a grey head. **V** Song is a rapid descending series of whistles or a more buzzy series at the same pitch. **H** Rainforest, forest edge and thickets, in lowlands and hills. **D** Bismarck Archipelago.

30 PAPERBARK FLYCATCHER *Myiagra nana* 18cm Distinctive *Myiagra* flycatcher with glossy black upperparts and entirely white underparts. Sexes similar, but female duller with grey lores. Sweeps tail from side to side. **V** Song is a musical whistle, *choo-wee choo-wee*. Call is a loud buzzy drawn-out note, often given as a double note. **H** Riverine scrub, swampy areas, mangroves. **D** SC New Guinea, N Australia.

31 RESTLESS FLYCATCHER *Myiagra inquieta* 20cm Similar to Paperbark Flycatcher but larger and slightly paler-backed. Female has buff wash on upper breast. Noisy and conspicuous. Sweeps tail from side to side. **V** 'Scissors-grinder' call is unique to this species. Other calls are similar to Paperbark Flycatcher. **H** Open woodland, farmland and mallee. **D** SW, EC, E Australia.

1
2
3
4
5
6
7
8
9
10
11
12
13
14
15
16
17
18
19
20
21
22
23
24
25
26
27
28
29
30
31

CROWS AND JAYS CORVIDAE

1 CRESTED JAY *Platylophus galericulatus* 31–33cm Unmistakable, with long crest and white neck patch. Generally encountered in pairs or small parties, foraging in the low to middle canopy. **V** Excited, stacatto, chattering rattle; also a single *chik* usually delivered while foraging in foliage. **H** Broadleaved evergreen forest. **D** Malay Peninsula, Greater Sunda Islands.

2 BLACK MAGPIE *Platysmurus leucopterus* 39–41cm Usually found in pairs or small parties. Forages in trees from the lower to upper canopy. While perched, repeatedly bobs and bows head. **V** Noisy, varied vocabulary, including a loud, discordant *keh-eh-eh-eh-eh*, a bell-like *tel-ope* and a *kontingka-longk*; also utters a xylophone-like *tok-tok terlingk-klingk-klingk...*, a repeated *kip* and a high-pitched mewing. **H** Lowland forest, secondary forest, swamp woodland, forest edges. **D** Malay Peninsula; Borneo; Sumatra, Indonesia.

3 SIBERIAN JAY *Perisoreus infaustus* 30cm Usually wary, though will visit forest picnic sites. Recorded hanging upside-down, feeding in tit-like manner. **V** Various quiet whistles, trills and creaky notes, with some mimicry. Harsh *hearrr-hearrr*, hoarse *skaaaak* and buzzard-like mewing calls. **H** Mainly dense coniferous forest; in the far north may frequent birch forest. **D** From Scandinavia through Siberia to NE China.

4 SICHUAN JAY *Perisoreus internigrans* 30cm Note pale bill. Can be quite inquisitive, but usually unobtrusive; forages in the dense foliage of conifers at mid-storey level. Best located by mewing calls. **V** High *kyip-kyip*, sometimes repeated to make a longer series; also a buzzard-like mewing note. **H** Steep mountainside spruce and pine forest. **D** WC China.

5 GREY JAY *Perisoreus canadensis* 27–31cm Generally in pairs or small groups, can become quite bold. Also known as Canada Jay. **V** Soft, whistled *wheeeoo* and *weef weef weef*; also a musical, husky *chuf-chuf-weeff* and a harsh, grating *cha-cha-cha-cha* given in alarm. **H** Primarily conifer forest, sometimes in mixed forest. **D** Boreal Alaska and Canada south through the Rockies and the Cascades, also into NE USA.

6 BLACK-COLLARED JAY *Cyanolyca armillata* 32cm Cobalt blue with black mask and narrow black breast band. Can appear blackish in poor light. Throat darker blue than in Turquoise Jay and forecrown less white. **V** Very high, sharp, piercing *sree*, single or doubled. **H** Mossy cloud forest. **D** Venezuela to Ecuador.

7 WHITE-COLLARED JAY *Cyanolyca viridicyanus* 34cm Note diagnostic narrow white collar. **V** Very high, sharp *jeetjeet*. **H** Cloud forest, elfin forest. **D** Peru, Bolivia.

8 TURQUOISE JAY *Cyanolyca turcosa* 32cm Similar to Black-collared Jay, and some overlap in range. **V** Varied calls, e.g. a downslurred *cheeur*. **H** Cloud forest, elfin woodland. **D** S Colombia to NW Peru.

9 BEAUTIFUL JAY *Cyanolyca pulchra* 27cm Similar to Black-collared Jay and Turquoise Jay, but note white crown and absence of black breast band. **V** Varied calls, e.g. a low *cratch*. **H** Andean cloud forest. **D** Colombia, Ecuador.

10 AZURE-HOODED JAY *Cyanolyca cucullata* 30cm Pale hood contrasting with otherwise dark plumage diagnostic, but can be hard to see. **V** Very varied, e.g. low *verdriet-verdriet* or high *shreek shreek* or high dry rapid ticking. **H** Montane forest. **D** Mexico to W Panama.

11 BLACK-THROATED JAY *Cyanolyca pumilo* 25cm Distinctive head pattern, with thin white upper border to black face and throat. **V** High dry *reek-reek* or *rupreprep*. **H** Montane forest. **D** S Mexico to NW Honduras.

12 DWARF JAY *Cyanolyca nanus* 25cm Distinguished from White-throated and Silvery-throated Jays by blue crown. **V** Very high swept-up *ireet-ireet*. **H** Montane pine–evergreen forest. **D** S Mexico.

13 WHITE-THROATED JAY *Cyanolyca mirabilis* 25cm Paler than Silvery-throated Jay with cheek bordered white all around. **V** Very high shrill warbling *srieoh-srieoh*. **H** Montane forest. **D** SW Mexico.

14 SILVERY-THROATED JAY *Cyanolyca argentigula* 25cm Very dark azure-blue with prominent silvery-white throat and eye-stripe. **V** High *wehwehwehweh*. **H** Montane forest. **D** Costa Rica, W Panama.

15 BUSHY-CRESTED JAY *Cyanocorax melanocyaneus* 30cm Separated from San Blas Jay, Yucatan Jay and Purplish-backed Jay by black legs, more distinct crest and different range. **V** Very high *sraksraksraksrak*. **H** Open woodland. **D** Guatemala to Nicaragua.

16 SAN BLAS JAY *Cyanocorax sanblasianus* Note yellowish legs. Highly sociable. Best distinguished from Yucatan and Purplish-backed jays by different range. **V** Very high scratchy *shriiik-shriiik-shriiiik*. **H** Dry woodland, plantations. **D** Pacific coast of C Mexico.

17 YUCATAN JAY *Cyanocorax yucatanicus* 30cm Dark eye. Best separated from San Blas and Purplish-backed jays by range. **V** Varied, e.g. high rapid dry chatters of varying length. **H** Forest, secondary growth, plantations. **D** Yucatán Peninsula, Mexico.

18 PURPLISH-BACKED JAY *Cyanocorax beecheii* 35cm Dark purplish-blue above. Legs brighter yellow than San Blas Jay, glossy colour of upperparts more violet. **V** High hurried *shrek-shrek-shrek*. **H** Dry woodland in coastal lowlands. **D** NW Mexico.

19 VIOLACEOUS JAY *Cyanocorax violaceus* 36cm Predominantly dark violet-blue, with black mask and white nape. **V** High downslurred *peeyur*. **H** Forest edges, open woodland. **D** Venezuela, SW Guyana, W Amazonia.

20 AZURE JAY *Cyanocorax caeruleus* 38cm Blue body, black head, no white. More brightly coloured than Purplish Jay. Might show a small crest. **V** Varied calls, e.g. mewing, hoarse *ew* notes. **H** *Araucaria* woodland, restinga, isolated tree stands in grassy areas. **D** SE Brazil, E Paraguay, NE Argentina.

21 PURPLISH JAY *Cyanocorax cyanomelas* 37cm Compare Violaceous Jay, which has a white neck. Separable from Azure Jay by range. **V** Single or hurried crow-like series of *raa* notes. **H** Rather dry forest, open woodland, scrub, riverine belts. **D** S Peru, Bolivia, SW Brazil through Paraguay to N Argentina.

22 CURL-CRESTED JAY *Cyanocorax cristatellus* 35cm Dark eyes and stiff crest are diagnostic. Note also white underparts and white distal half of tail. **V** Series of raucous *ew* notes. **H** Cerrado, riverine belts, grassland with sparse trees. **D** C Brazil to E Bolivia, E Paraguay.

23 TUFTED JAY *Cyanocorax dickeyi* 35cm Unmistakable, with spiky crest, yellow eyes and striking head pattern. No similar jay in (restricted) range. **V** High loud *tjatatat* and other chatters. **H** Mixed montane forest. **D** W Mexico.

24 BLACK-CHESTED JAY *Cyanocorax affinis* 30cm Black head and whitish belly diagnostic. **V** High staccato rapid *tjeowtjeow*. **H** Tall secondary growth in upper levels of forest, woodland, riverine belts, plantations. **D** Costa Rica to NW Venezuela.

25 WHITE-TAILED JAY *Cyanocorax mystacalis* 33cm Note white nape, undersides and outer-tail feathers. No similar jay occurs in its range. **V** Series of 2–3 high, nasal *tjat* notes. **H** Thick growth at borders of forest and woodland. **D** SW Ecuador, NW Peru.

26 CAYENNE JAY *Cyanocorax cayanus* 33cm No similar jay occurs in its range. A cooperative breeder, like many other jays. **V** Varied calls, e.g. a downslurred *tjew*. **H** Woodland, forest borders, scrub. **D** N Amazonia: E Venezuela, N Brazil, the Guianas.

27 AZURE-NAPED JAY *Cyanocorax heilprini* 34cm Note white vent and tail; shows a pale blue nape. **V** As Cayenne Jay. **H** Woodland on sandy soils. **D** NW Amazonia: E Colombia, S Venezuela, NW Brazil.

28 PLUSH-CRESTED JAY *Cyanocorax chrysops* 34cm Note pale whitish-yellow underparts, black bib, black crest. **V** Various single-note nasal calls or a magpie-like *cho-cho-cho*. **H** Woodland and forest. **D** Brazil south of the Amazon, Bolivia, Paraguay, NE Argentina, Uruguay.

29 WHITE-NAPED JAY *Cyanocorax cyanopogon* 35cm Note brown upperparts and pale eyes. Distinctive in its range. **V** High, nasal *tuw* note. **H** Caatinga, Cerrado. **D** E Brazil.

30 GREEN JAY *Cyanocorax luxuosus* 25–28cm Unmistakable. Generally shy, but can be quite bold and inquisitive, especially when in groups. **V** Very varied, including mewing, chattering, rattling, buzzing and squeaking notes. **H** Bushy thickets, shrubby woodland and forest edge. **D** S Texas to Honduras.

31 INCA JAY *Cyanocorax yncas* 28cm No similar bird occurs in South America. **V** Song varied, e.g. triplets of low, dry rattles. **H** Montane forest. **D** Venezuela to Bolivia.

32 BROWN JAY *Psilorhinus morio* 38–44cm Forages in small groups. **V** High-pitched, nasal *peeeeah*; also makes a clicking or hiccupping sound. **H** Open woodland and riverine forest. **D** Extreme S Texas through E Central America to Panama.

33 BLACK-THROATED MAGPIE-JAY *Calocitta colliei* 70cm Two colour forms shown. From White-throated Magpie-Jay by longer tail and black, not white, front and lores. **V** Very high mellow yelping *wow wew-wew* and other sounds such as a mewing *weeeeeh*. **H** Dry woodland, areas with scattered tree stands. **D** W Mexico.

34 WHITE-THROATED MAGPIE-JAY *Calocitta formosa* 50cm Note white around eyes. **V** High shrill rasping *shreer-shreer* or sharp *trreeer*. **H** Dry woodland, areas with scattered tree stands, cultivation. **D** S Mexico to Costa Rica.

1
2
3
4
5
6
7
8
9
10
11
12
13
14
15
16
17
18
19
20
21
22
23
24
25
26
27
28
29
30
31
32
33
34
2 colour
morphs

CROWS AND JAYS *CONTINUED*

1 BLUE JAY *Cyanocitta cristata* 26–28cm Unmistakable. Usually forages alone or in pairs; may form small flocks post breeding. **V** Very varied, including short, shrill *peeeah peeeah* and high-pitched *too-doodle-up-to*. **H** Deciduous and mixed woodland, parks, gardens. **D** C and E USA, CS and NE Canada.

2 STELLER'S JAY *Cyanocitta stelleri* 28–32cm Forages in pairs or small groups. Usually shy and wary, but can be quite tame around campsites. **V** Very varied, including a harsh *shaaak shaaak*, a mellow *klook klook* and a mewing *hidoo*. **H** Open conifer and mixed pine–oak forests in mountain areas. **D** From S Alaska to Nicaragua.

3 MEXICAN JAY *Aphelocoma wollweberi* 29cm Often encountered in noisy groups, feeding among leaf litter. **V** Ringing *wink-wink-wink*... **H** Mountainsides and canyons with oak or pine–oak–juniper woodland. **D** From SW USA to C Mexico.

4 TRANSVOLCANIC JAY *Aphelocoma ultramarina* 28–32cm Previously considered conspecific with Mexican Jay. **V** Similar to Mexican Jay. **H** Montane pine and pine–oak woodland. **D** SC and SW Mexico.

5 UNICOLORED JAY *Aphelocoma unicolor* 35cm Rich blue overall, much darker than Pinyon Jay, which does not occur in the same range. **V** Very high sweeping rapid *wheetwheet*. **H** Forest interior and edge. **D** C Mexico to W Honduras.

6 CALIFORNIA SCRUB JAY *Aphelocoma californica* 28–30cm Blue upperparts apart from grey-brown back. Light grey underparts. Flies with quick wingbeats followed by a glide. **V** Harsh, rising *shreeeenk*, a rapid *wenk wenk wenk*... or *kkew kkew kkew*...; also a pounding *sheeyuk sheeyuk*. **H** Mixed oak, pine and juniper woodlands, scrubland and chaparral. **D** W North America, from Washington to Baja California.

7 WOODHOUSE'S SCRUB JAY *Aphelocoma woodhouseii* 28–30cm Formerly considered conspecific with California Scrub Jay (and called Western Scrub Jay). Differs from California Scrub Jay in being duller and paler, with only a faint breast-band and greyer underparts. **V** Similar to California Scrub Jay but, lower and more hoarse. **H** Dry, open oak and juniper woodlands. **D** Interior W North America, from Oregon to S Mexico.

8 ISLAND SCRUB JAY *Aphelocoma insularis* 33cm Unmistakable. Very like California Scrub Jay, but the only jay on Santa Cruz Island. **V** Very similar to California Scrub Jay. **H** Low-growing oak chaparral, oak woodland and relict pine forest. **D** Santa Cruz Island, off SW California.

9 FLORIDA SCRUB JAY *Aphelocoma coerulescens* 26cm Usually encountered in pairs or small family parties. Feeds mainly on the ground near cover. **V** Calls include a low *kereep*, a sweet *ch-leep* and a rising *kreesh*. **H** Scrubby oak thickets. **D** C Florida.

10 PINYON JAY *Gymnorhinus cyanocephalus* 25–27cm Gregarious, often forages in large flocks. **V** Varied, including a high-pitched *kwa-kwa-kwa-kwa*, a slow *kura kura kura* and a mewing *kraa-aha*. **H** Pinyon–juniper forests. **D** Mountain and plateau areas of interior W USA.

11 EURASIAN JAY *Garrulus glandarius* 34cm Unmistakable when seen well, but wary. Harsh alarm calls and white rump, as bird flies from tree to tree, are often the first signs of bird's presence. **V** Song a quiet, often unnoticed warble. Call is a harsh *skaaaak-skaaaak*; also a weak *piyeh*. **H** Deciduous, mixed and coniferous forests, parks and large gardens. **D** Widespread from W Europe to Japan and SE Asia.

12 BLACK-HEADED JAY *Garrulus lanceolatus* 33cm Actions similar to Eurasian Jay, usually less wary, often recorded feeding on scraps around houses in isolated villages and camps. In flight shows a pinkish-buff rump. **V** Dry *skaaaak*, thinner than similar call of Eurasian Jay. **H** Montane forests. **D** Himalayas, from E Afghanistan and N Pakistan to Tibet, Nepal, N India.

13 LIDTH'S JAY *Garrulus lidthi* 38cm A large jay, predominantly dark purple-blue and chestnut. Sociable, often forming large parties in winter. **V** Harsh, variously pitched *kraah*. **H** Coniferous, deciduous and mixed forest. **D** Ryukyu Islands, off S Japan.

14 AZURE-WINGED MAGPIE *Cyanopica cyanus* 34cm Sociable, forming small family parties in breeding season, larger parties in winter. **V** Shivering *screep*, often uttered by roving parties, also a harsh chatter, a sharp *wee-wee-wee-u*; a harsh *karrah* when alarmed, often followed by a series of *kwink* notes. **H** Mixed and deciduous woodland, forest edge, parks and large gardens. **D** E Siberia, Mongolia, N and C China, Korean Peninsula, Japan.

15 IBERIAN MAGPIE *Cyanopica cooki* 31–35cm Formerly considered conspecific with Azure-winged Magpie. Very similar to Azure-winged Magpie but mantle and underparts browner and less grey. Often in small or large groups outside the breeding season. **V** Similar to Azure-winged Magpie but less harsh. **H** Woodlands and orchards. **D** Iberian Peninsula.

16 SRI LANKA BLUE MAGPIE *Urocissa ornata* 42–47cm Rare and endangered. Mainly a bird of the tree canopy, moving through forests in noisy small parties and regularly foraging in a tit-like acrobatic style. **V** Very varied: includes a far-carrying *chink-chink* or *cheek-cheek*, a rasping *crakrakrakrak* and a loud *whee-whee*. **H** Dense evergreen broadleaved forests. **D** Sri Lanka.

17 TAIWAN BLUE MAGPIE *Urocissa caerulea* 69cm Note black hood, pale eye, long graduated tail. Shy and wary. Usually encountered in small parties foraging in the canopy. **V** Soft *kwee-eep* or *swee-eee*, usually uttered when relaxed; when alarmed gives a cackling *kyak-kyak-kyak-kyak*. **H** Deciduous hill forests, lower forests in winter. **D** Taiwan.

18 YELLOW-BILLED BLUE MAGPIE *Urocissa flavirostris* 63cm Forages in trees and on the ground and is regularly a member of mixed-species flocks. Generally wary, but recorded feeding on human food scraps in and around isolated hill stations and villages. **V** Wheezy *bu-zeep-peck-peck-peck pop-unclear pu-pu-weer* and a high-pitched *clear-clear*. **H** Deciduous and mixed mountain forests. **D** Himalayas from N Pakistan to N Myanmar; N Vietnam.

19 RED-BILLED BLUE MAGPIE *Urocissa erythroryncha* 66cm Generally bluer and whiter than Yellow-billed Blue Magpie, with more white on crown. Social, mainly arboreal, regularly forages in the canopy of fruiting trees; also recorded feeding on food scraps around hill stations. **V** Very varied, including a piercing *quiv-pig-pig*, a soft, repeated *beeee-trik* and when alarmed a rapid *penk-penk-penk-penk*. **H** Broadleaved forests in foothills. **D** Himalayas, China, SE Asia.

20 WHITE-WINGED MAGPIE *Urocissa whiteheadi* 45–46cm Note bold white and blackish wing pattern. Very social, forages in parties of 20–25 birds. **V** Harsh, rising *errreep errreep*, a hoarse, rising *shureek*, a low hoarse *churreee* and a liquid, rippling *brrriii brrriii*... **H** Broadleaved evergreen forest, forest edge, secondary forest. **D** S China to Hainan, Laos, Vietnam.

21 COMMON GREEN MAGPIE *Cissa chinensis* 37–39cm Inconspicuous, presence usually given away by its whistled calls. Generally forages low down in shrubbery or in forest understorey; post breeding forms small flocks which often join mixed-species feeding flocks. **V** Loud *peep-peep* or *kik-wee*; also a rich, squealing whistle. **H** Tropical and subtropical evergreen forest, with a liking for bamboo thickets and shrubbery along watercourses. **D** Himalayas to S China, SE Asia, Malay Peninsula, Sumatra, Borneo.

22 INDOCHINESE GREEN MAGPIE *Cissa hypoleuca* 31–35cm Yellower than Common Green Magpie, with shorter tail. Actions similar to Common Green Magpie. **V** Variable, including a loud, shrill *peeeoo-peeeoo-peeeoo-peeeoo*..., a clipped, shrill *peu-peu-peu*, an abrupt *weep* and a long piercing *peeeeooo*. **H** Broadleaved evergreen forest, semi-evergreen forest, bamboo. **D** S China, Hainan, S and E Thailand, Cambodia, Laos, Vietnam.

23 JAVAN GREEN MAGPIE *Cissa thalassina* 31–33cm Hunts in the lower levels of forests, in small, noisy groups. **V** Unrecorded, presumed to be as other green magpies. **H** Foothill and montane forest, occasionally lowland forest and adjacent cultivated areas. **D** Java, Indonesia.

24 BORNEAN GREEN MAGPIE *Cissa jefferyi* 32cm Often part of mixed-species foraging flocks. Sometimes feeds on the ground around mossy tree trunks. **V** Harsh whistles and chatterings; also an incisive, rapid *swe-swi-swee-swi-swe sweet* or similar. **H** Montane forests. **D** Borneo.

25 RUFOUS TREEPIE *Dendrocitta vagabunda* 46–50cm Usually encountered in pairs or small parties, with much larger groups where food is plentiful; also joins other species to feed in fruiting trees. **V** Flute-like *ko-ki-la*, often mixed with a harsh rattle; also a variety of harsh metallic and mewing notes. **H** Open wooded areas, cultivations, parks, gardens. **D** Pakistan and India to SW China, Vietnam.

26 SUMATRAN TREEPIE *Dendrocitta occipitalis* 40cm Forages in small, noisy parties in the forest canopy. **V** Bell-like calls, presumed to be much like those of the previous species. **H** Mountain and foothill forests, favouring open forest with plantations and bamboo thickets. **D** Sumatra, Indonesia.

27 BORNEAN TREEPIE *Dendrocitta cinerascens* 40cm Noisy and conspicuous; sociable, usually encountered in small parties foraging in the canopy. **V** Variable bell-like *choing*, also a harsh *shraank*. **H** Montane forests, preferring foothill and valley forests with clearings and cultivation; also secondary scrub jungle and bamboo thickets. **D** Borneo.

28 GREY TREEPIE *Dendrocitta formosae* 36–40cm Regularly found in small parties. Forages mainly in trees but will descend to feed on the ground; often forms part of mixed-species flocks. **V** Loud, rapid *klok-kli-klok-kli-kli*; also a variety of short harsh and musical calls. When alarmed gives a Magpie-like chatter. **H** Deciduous mountain forests, usually open and near scrub or cultivation. **D** Widespread from Himalayas to E and NE India, N Myanmar, N Thailand, Laos, N Vietnam, China, Taiwan.

29 WHITE-BELLIED TREEPIE *Dendrocitta leucogastra* 48cm Separated from Rufous Treepie by black face and white underparts. Generally an arboreal forager; often encountered in mixed-species flocks, especially with Greater Racket-tailed Drongos (Plate 183). **V** Often mimics Greater Racket-tailed Drongo; also utters a throaty *chuff-chuff-chuff* and various other creaking, quacking, clicking and dove-like calls. **H** Humid evergreen hill forest and secondary growth. **D** SW India.

30 COLLARED TREEPIE *Dendrocitta frontalis* 38cm Arboreal; forages in small parties in dense forest. Recorded making fly-catching sorties much like a drongo. **V** Little recorded; said to have typical range of treepie calls. **H** Dense mixed humid evergreen forests with bamboo thickets. **D** N India to NW Vietnam.

31 ANDAMAN TREEPIE *Dendrocitta bayleii* 32cm Arboreal forager, usually in pairs or small parties although big groups of 20 or more are not unusual. Tags on to mixed-species feeding parties, especially those containing Andaman Drongos (Plate 183). **V** Little recorded; said to have various harsh and melodious notes. **H** Dense evergreen forests. **D** Andaman Islands.

32 RACKET-TAILED TREEPIE *Crypsirina temia* 30–33cm Note long spatulate tail. Generally found singly, in pairs or in small parties. Agile; forages in shrubbery. **V** Rasping *churg-churg*, *grasp-grasp*, *chrrrk-chrrrk* or *chrrrk-churrrk*. **H** Various open lowland forests, including mixed deciduous woodland, bamboo, mangroves. **D** SE Asia; Java, Indonesia.

33 HOODED TREEPIE *Crypsirina cucullata* 29–31cm Usually encountered in pairs or small parties. **V** Quiet purring *drrrriii-k*; also various discordant, harsh call notes. **H** Open deciduous woodland, thorn-scrub jungle, bamboo, cultivation borders. **D** Myanmar.

34 RATCHET-TAILED TREEPIE *Temnurus temnurus* 32cm Unobtrusive, usually seen singly or in pairs; sluggish. Best located by persistent calls. **V** Calls include a ringing *clee-clee-clee*, a harsh *graak-graak* and a squeaky, rising *eeup-eeup-eeup*. Also utters a ringing *clipeeee*, a hollow *pupueeee* and a rasping, rippling *rrrrrrr*. **H** Broadleaved evergreen forest, forest edge, bamboo and secondary growth. **D** Hainan, Vietnam, C Laos, W Thailand, S Myanmar.

1
2
3
4
5
6
7
8
9
10
11
12
13
14
15
16
17
18
19
20
21
22
23
24
25
26
27
28
29
30
31
32
33
34

CROWS AND JAYS *CONTINUED*

1 EURASIAN MAGPIE *Pica pica* 45–50cm Usually in pairs or small parties. Forages mainly on the ground; rests in trees, on rocks or on buildings. **V** Chattering *chack-chack-chack-chack-chack…*, also an enquiring *ch-chack* and a squealing *keee-uck*. **H** Very varied; avoids treeless landscapes and dense forest. **D** Widespread, from W Europe to Iran, Russia, N China, E Siberia.

2 MAGHREB MAGPIE *Pica mauritanica* 45cm Formerly considered conspecific with Eurasian Magpie. Differs from Eurasian in having a small patch of bare blue skin behind the eye and less white on the scapulars. **V** Similar to Eurasian Magpie. **H** Open woodlands and scrub, from lowlands to mountains. **D** NW Africa (Morocco and Algeria).

3 ASIR MAGPIE *Pica asirensis* 50cm Formerly considered conspecific with Eurasian Magpie. Larger than Eurasian, with a stouter bill, shorter tail, reduced white on the scapulars and less gloss. **V** Poorly known, but reported to be a mournful long screech. **H** Juniper woodland and open hillsides with scattered trees in Asir Mountains. **D** SW Arabian Peninsula.

4 BLACK-RUMPED MAGPIE *Pica bottanensis* 48cm Formerly considered conspecific with Eurasian Magpie. Differs from Eurasian in its black rump, shorter tail, stout bill and less gloss. **V** Similar to Eurasian Magpie. **H** Open woodland and cultivated valleys. **D** C Bhutan and WC China.

5 ORIENTAL MAGPIE *Pica serica* 45cm Formerly considered conspecific with Eurasian Magpie. Similar to Eurasian but is smaller and darker with a shorter tail and greener gloss. **V** Similar to Eurasian Magpie. **H** Open country with trees. **D** E and S China and Taiwan to SE Asia.

6 BLACK-BILLED MAGPIE *Pica hudsonia* 45–60cm Very similar to Eurasian Magpie. Southern birds show bare grey-black skin around eye. **V** Various chattering and rattling calls. **H** Open country with scattered trees. **D** From S Alaska through W and S Canada to inland W and C USA.

7 YELLOW-BILLED MAGPIE *Pica nuttalli* 43–54cm Apart from yellow bill and facial skin, very similar in all respects to Black-billed Magpie. **V** Very like Black-billed Magpie. **H** Wooded rangelands and foothills. **D** C California.

8 STRESEMANN'S BUSHCROW *Zavattariornis stresemanni* 30cm Distinctive pale grey corvid with black wings and tail. **V** Very vocal, with a range of calls including a single *keh*, a metallic *kaw-kaw*, a soft *kwak* and a rapid *wah-wah-wah-wah*. **H** Bushland. **D** S Ethiopia.

9 HENDERSON'S GROUND JAY *Podoces hendersoni* 28cm Shy. Ground feeder, digging for seeds and insects. Usually runs; when put to flight, shows conspicuous white primary wing patch. **V** Poorly recorded. Harsh whistles and *clack-clack-clack*, said to sound like a wooden rattle. **H** Flat stony or gravel desert with scattered trees. **D** Mongolia, NC China.

10 BIDDULPH'S GROUND JAY *Podoces biddulphi* 28cm Shy, actions typical of genus. In flight, shows white wing patch, white trailing edge to secondaries and a mainly white tail. **V** Includes *chui-chui-chui*, last syllable higher than rest; also a succession of low whistles in a rapidly descending scale. **H** Sandy wastes and semi-deserts with scattered trees. **D** NW China.

11 PANDER'S GROUND JAY *Podoces panderi* 25cm Feeding actions typical of genus. In flight, wings appear mainly white. **V** Ringing *chweek-chweek-chweek*. **H** Sandy deserts with scattered saxaul shrubs. **D** Turkmenistan, Uzbekistan, Kazakhstan.

12 PLESKE'S GROUND JAY *Podoces pleskei* 24cm Little studied, actions suspected to be much as Pander's Ground Jay. In flight, shows large white wing patch and white trailing edge to secondaries. **V** Clear, rapid, high-pitched *pee-pee-pee-pee-pee…* **H** Sandy desert with scattered shrubs. **D** Iran.

13 CLARK'S NUTCRACKER *Nucifraga columbiana* 29–31cm In flight shows black wings with white secondaries and black tail with extensive white on outer feathers. **V** Nasal *kraaaa kraaa kraaaa*. **H** Montane conifer forests. **D** W USA and SW Canada.

14 SPOTTED NUTCRACKER *Nucifraga caryocatactes* 32–35cm Usually shy and wary. Mainly arboreal, although will descend to feed on pine nuts on the forest floor. First sightings are often of birds flying lazily from one treetop to another. **V** Dry *kraaaak*, often repeated to form a discordant rattle; also utters a weak *zhree*. **H** Coniferous forests. **D** Widespread from C and N Europe to Himalayas, N Myanmar, E Siberia, Japan, Taiwan.

15 LARGE-SPOTTED NUTCRACKER *Nucifraga multipunctata* 35cm Formerly considered conspecific with Spotted Nutcracker. Differs in being more profusely spotted with larger spots, giving whitish appearance overall. Note diagnostic spotting on rump and uppertail coverts. **V** Similar to Spotted Nutcracker. **H** Coniferous forest in mountains. **D** W Himalayas (E Afghanistan, N and W Pakistan, NW India).

16 RED-BILLED CHOUGH *Pyrrhocorax pyrrhocorax* 38cm Note fine downcurved red bill. Sociable, forming large winter flocks that forage on pastures or crop fields, often in association with Alpine Choughs. **V** Far-carrying *chee-aw*, *chaow* or *chi-ah*. **H** Mountains, alpine pasture, coastal grassland, cultivations. **D** Widespread, from W Europe, NW Africa to Iran, N India, China, Mongolia; also Ethiopia.

17 ALPINE CHOUGH *Pyrrhocorax graculus* 38cm Regularly encountered in flocks, usually at higher elevations than Red-billed Chough; often scavenges around ski centres and climbing camps; generally fairly tame. **V** Descending, thin *sweeeoo*, a rippling *preep* and a rolling *churr*. **H** High mountains, mountain pastures and cultivations. **D** S and C Europe, N Africa, to Turkey, Iran, C Asia.

18 PIAPIAC *Ptilostomus afer* 35cm Note red eye and long tail. Associated with cattle. **V** Contact call, uttered constantly by a feeding flock, is a loud and rather shrill *pee-ac, pee-ac, pee-ac*. **H** Lightly wooded grassland with palms. **D** W and C Africa, from Senegal to S Ethiopia, Uganda, Kenya.

19 WESTERN JACKDAW *Coloeus monedula* 34cm Note pale eye and grey nape. Sociable, often forming large flocks; regularly associates with other crows and starlings. **V** Abrupt *chjak* often repeated seven or so times; also a low, drawn-out *chaairurr* accompanied by *chak*. **H** Open country with scattered trees, farmland, parkland, towns, villages, coastal and inland cliffs. **D** Widespread, from W Europe and NW Africa to C Asia.

20 DAURIAN JACKDAW *Coloeus dauuricus* 32cm Sociable, usually in small flocks. Often feeds among grazing cattle and regularly associates with other crows. **V** In flight, a short *chak*. **H** Open meadowland with stands of trees, riverine plains and foothills and near human habitations. **D** Breeds from S and E Siberia through Mongolia to W China. Non-breeding as far south as Korean Peninsula, S Japan, E China.

21 HOUSE CROW *Corvus splendens* 40cm Bold, very sociable. Note large bill, and pale collar contrasting with glossy black back and wings. **V** Flat, dry *kaaa-kaaa*. **H** Villages, towns and cities, often common around ports; also cultivated areas. **D** From Pakistan through India, Maldives and Sri Lanka to Myanmar and SW Thailand.

22 NEW CALEDONIAN CROW *Corvus moneduloides* 41cm Bulky, glossy black crow with distinctive slightly upturned bill. Usually in pairs or small groups. Flicks wings when perched. **V** Gives series of nasal barks, *wak-wak, wak-wak*, usually in couplets, but also in a continuous series. **H** Forest and savannah. **D** New Caledonia (and Loyalty Islands), SW Pacific.

23 BANGGAI CROW *Corvus unicolor* 39cm Little information, known only from two old specimens. Possibly a secretive bird of forested hills. **V** Unknown. **H** Presumed to be hill forest. **D** Banggai Islands, off E Sulawesi, Indonesia.

24 SLENDER-BILLED CROW *Corvus enca* 43–47cm Generally encountered in small flocks feeding in, or flying above, the canopy. **V** Variable, including a high-pitched *ahk-ahk-ahk*, a nasal *werk-werk-werk* and a *whaaa-whaaa-whaaa*. **H** Forests, forest edge, secondary growth. **D** Peninsular Malaysia, Indonesia, Borneo, Philippines.

25 VIOLET CROW *Corvus violaceus* 40–42cm Usually found in pairs or small parties; often raids ripening maize fields. Formerly considered a race of Slender-billed Crow. **V** Short nasal barks, given singly or in a 2–3-note series. **H** Forests, semi-open country, farmland. **D** Seram, S Maluku Islands, Indonesia.

26 PIPING CROW *Corvus typicus* 19–20cm Shy and very nervous, flying away with noisy, swooshing wingbeats. Regularly occurs in small parties. **V** Raucous, shrill screech; also three rising whistles followed by a crowing sound, and a variety of creaking and trilling notes. **H** Forest edge, open woodland with clearings. **D** Sulawesi, Indonesia.

27 FLORES CROW *Corvus florensis* 40cm Usually seen singly or in pairs, occasionally found in small groups; frequents the canopy or subcanopy. **V** High-pitched, downward-inflected, rasping *cwaaa, cawaraa* or *waak*; also a popping or gurgling *pol-ok* or *burr-ok*. **H** Primary, tall secondary, and disturbed moist and semi-deciduous lowland and hill forest; occasionally in relict patches of forest and lightly wooded areas. **D** Flores, Indonesia.

28 MARIANA CROW *Corvus kubaryi* 40cm Note slender bill. The only crow on Rota; probably extinct on Guam. Critically endangered. **V** Poorly known. Nasal *hi* or *caw*, or more drawn-out *caaaw*. **H** Forest. **D** Rota, Mariana Islands, W Pacific.

29 LONG-BILLED CROW *Corvus validus* 46–48cm Note pale eye. Usually occurs in pairs or small parties, mainly in the tree canopy; often calls from the topmost branches. **V** Loud, short, dry, croaked *cruk… cruk… cruk… cruk*. **H** Forest. **D** N Maluku Islands, Indonesia.

30 WHITE-BILLED CROW *Corvus woodfordi* 41cm Bulky black crow with strongly downcurved culmen. Massive white bill with black tip. Note blue-grey eyes and short tail. **V** Loud, high-pitched rolling *kraaow*, frequently repeated. **H** Forest and forest edge, in lowlands and hills. **D** Choiseul, Santa Isabel and Guadalcanal (Solomon Islands).

31 BOUGAINVILLE CROW *Corvus meeki* 41cm Bulky black crow with strongly downcurved culmen. Massive black bill. Note dark eyes and short tail. **V** Similar to White-billed Crow, but notes shorter and lower-pitched. **H** Forest and forest edge, in lowlands and hills. **D** Bougainville, Solomon Islands.

32 BROWN-HEADED CROW *Corvus fuscicapillus* 45cm Large crow with brownish head and breast, long stout bill and strongly downcurved culmen. Bill black in males, yellow with a black tip in females. Singly or in pairs. **V** Call is a harsh, hoarse *harh-harh*, the second note often slightly higher-pitched. **H** Rainforest and mangroves, in lowlands and hills. **D** W New Guinea, including West Papuan and Aru Islands.

33 GREY CROW *Corvus tristis* 51–55cm Blackish-brown or greyish plumage with distinctive bare pink face and large bill. Note rather long tail and pinkish legs and feet. Commonly in small flocks. **V** Distinctive high-pitched cries, *wah wah*, loud and ringing. **H** Forest, forest edge, secondary growth, in lowlands and hills. **D** New Guinea.

34 CAPE CROW *Corvus capensis* 45cm Note long, slender bill. Usually in flocks. **V** Harsh *kraa-kraa-kraa*, sometimes ending with a downslurred *krooaaauw*. **H** Grassland with some trees. **D** NE, E and S Africa.

35 ROOK *Corvus frugilegus* 47cm Baggy 'trousers' compared to other crows. Usually in flocks of sometimes many hundreds of birds, often accompanied by other crows. **V** Dry *kraah* and higher-pitched *kraa-a*. **H** Mainly agricultural land with stands of trees. **D** Widespread, Europe to E Asia.

1
2
3
4
5
6
7
8
9
10
11
12
13
14
15
16
17
18
19
20
21
22
23
24
25
26
27
28
29
30
31
32
♂
♀
33
34
35

CROWS AND JAYS *CONTINUED*

1 AMERICAN CROW *Corvus brachyrhynchos* 43–53cm Most widespread North American crow, usually in pairs or small parties. May form larger groups at good food sources. **V** Short *ahhh*, *caaw* or *carr*, a rapid, rattling *tatatato* and a soft *prrrk*. **H** Varied, including cities, farmland, woodland, shorelines. **D** Most of North America south of the sub-tundra zone, but absent from much of SW USA. Northern birds move south in winter.

2 NORTHWESTERN CROW *Corvus caurinus* 42–45cm Very sociable. Only reliably identified by range. **V** As American Crow, maybe slightly lower and hoarser. **H** Shorelines, islands, coastal villages. **D** Pacific coast of North America, from Alaska to S Washington.

3 TAMAULIPAS CROW *Corvus imparatus* 40cm Not in range of Northern Raven. Noticeably glossy plumage. Voice distinctive. **V** Very low croaking *wehih-wehih-wehweh*. **H** Areas with scattered tree stands and hedges, cultivations, settlements, towns. **D** NE Mexico, S Texas.

4 SINALOA CROW *Corvus sinaloae* 35cm From American Crow by call, size, gloss. **V** Very high mewing *tjouw tjouw-tjouw*. **H** Areas with scattered tree stands and hedges, cultivations, settlements, towns. **D** NW Mexico.

5 FISH CROW *Corvus ossifragus* 36–41cm Usually gregarious. Very similar to American Crow, best distinguished by voice. **V** Short, nasal *cah* or *cah-ah*; also a throaty rattle, higher-pitched than American Crow. **H** Coastal marshes, riversides and seashores. **D** E USA, from New England to Texas.

6 HISPANIOLAN PALM CROW *Corvus palmarum* 43cm Forages in small to medium-sized groups in trees and on the ground, feeding mainly on fruit, seeds, insects and small lizards. Flight similar to Cuban Palm Crow. Formerly considered conspecific with Cuban Palm Crow, under the name 'Palm Crow'. **V** Harsh, nasal *aaar aaar aaar...* **H** Pine forests, swamp and dry plains with wooded ravines. Breeds from March to May. **D** Hispaniola.

7 CUBAN PALM CROW *Corvus minutus* 38cm Usually occurs in pairs or small groups. Forages primarily in trees, but will feed on the ground. Eats mainly fruit, seeds, insects and small lizards. Flies with rapid, flapping wingbeats. **V** A harsh *craaao*. **H** Forest, scrub and palm savannah. **D** Cuba.

8 JAMAICAN CROW *Corvus jamaicensis* 38cm Mainly arboreal; usually occurs in pairs or small groups that forage in the canopy in search of fruit and invertebrates, the latter found by probing bark and bromeliads. Flight slow and heavy with deliberate wingbeats. **V** Loud *craa-craa* and various bubbling, chuckling and gobbling noises strung together in garbled outbursts. **H** Hill and mountain forests, descending to lower levels during dry season. **D** Jamaica.

9 CUBAN CROW *Corvus nasicus* 40–46cm Usually occurs in small noisy parties; forages in trees and on the ground, feeding on fruit, seeds, lizards and frogs. Flight is unhurried, with deep wingbeats. **V** High, nasal *caah-caaah*; also a bubbling or turkey-like gobbling. **H** Primarily wooded areas, but also in urban areas where trees are plentiful. **D** Cuba, Caicos Islands.

10 WHITE-NECKED CROW *Corvus leucognaphalus* 48–51cm White bases of body feathers only show during display. Occurs in pairs or small flocks, used to gather in very large flocks; leaves mountain forest roosts to forage in lowland forest on fruit, berries, nestlings and small toads. Flight is graceful, and often soars high up. **V** Variable, including babbling and squawking, sounding much like a parrot. **H** Mountain and lowland forests and woodlands; also open areas with scattered trees. **D** Hispaniola.

11 HAWAIIAN CROW *Corvus hawaiiensis* 49cm No other crow in Hawaii, but probably not surviving in the wild. **H** Forest, woodland, grassy areas with scattered trees. **V** Very varied, e.g. high, raucous, slightly upslurred *Aoow* or falsetto, trumpet-like *Rrruah*. Generally higher-pitched than other crow species. **D** Hawaii (Hawaii).

12 CARRION CROW *Corvus corone* 49cm Distinguished from Rook (Plate 190) by head shape, fully feathered face, less feathering around thighs. Wary, usually in pairs or small groups. **V** A vibrant *kraaa*, often repeated; sometimes utters a hollow *konk-konk*. **H** Huge variety of open country, from moorland to farmland and urban areas. **D** W Europe; C and E Asia.

13 HOODED CROW *Corvus cornix* Black head, wings and tail; otherwise grey. Formerly treated a subspecies of Carrion Crow, and sometimes hybridises with that species. **V** Vibrant *kraaa*, often repeated; sometimes utters a hollow *konk-konk*. **H** Open country, cultivated areas. **D** N and E Europe to Middle East and C Asia.

14 COLLARED CROW *Corvus torquatus* 54cm Little recorded. Said to be less of a scavenger than Carrion Crow. Sometimes forms mixed flocks with Large-billed Crow. **V** More hoarse, less rolling, than Carrion Crow; a loud *kaaarr*, also a *kaar-kaar*. **H** Open lowland cultivation with scattered trees, farmland, parks, large gardens, edges of towns and villages. **D** E, C China, N Vietnam.

15 LARGE-BILLED CROW *Corvus macrorhynchos* 41–49cm Usually seen singly, in pairs or in small parties. This and the next two species were formerly treated as a single species, the 'Jungle Crow'. **V** Hoarse *kyarrh kyarrh* or *kyearh kyearh*. **H** Wooded areas, open country, also near habitations. **D** Afghanistan, Pakistan, N India through Tibet to E Russia, China, Japan, Taiwan, Philippines, Indonesia, S Malay Peninsula.

16 EASTERN JUNGLE CROW *Corvus levaillantii* 41–49cm Similar to Large-billed Crow; best separated by range. **V** Typical crow *kaaah-kaaah*; vocabulary also said to include a nasal yapping with a duck-like quality. **H** Open forest and woodland, mangroves, open country and near habitations. **D** N India and Nepal to Bangladesh, Myanmar, N Thailand, N Malay Peninsula.

17 INDIAN JUNGLE CROW *Corvus culminatus* 41–49cm Distinguished from House Crow (Plate 190) by absence of grey neck. **V** Extensive vocabulary, with at least 10 different calls recorded. **H** Various, including wooded areas, open country and near habitations, at lower altitude than Large-billed Crow. **D** Peninsular India, Sri Lanka.

18 TORRESIAN CROW *Corvus orru* 48–54cm Occurs singly, in pairs or in small flocks. Regularly scavenges at road-kills. **V** 8–14 nasal *uk* notes, ending with an *uk-ohw*; also a variety of harsh cawing or *uk* notes. **H** Open coastal areas, lightly wooded cultivation, and woodland near villages or towns. **D** Maluku Islands, E Lesser Sunda Islands, New Guinea, Torres Strait islands, W, C and N Australia.

19 BISMARCK CROW *Corvus insularis* 41cm Formerly considered conspecific with Torresian Crow, but larger-billed and has distinct voice and habits. Glossy black with pale blue eyes. Large flocks gather at communal roosts. **V** High-pitched series of clipped notes, *ka-ka-ka-ka-ka*. **H** Open habitats including forest edge, plantations, gardens. **D** Bismarck Archipelago.

20 LITTLE CROW *Corvus bennetti* 45–48cm Similar to Torresian Crow but slightly smaller. Gregarious and nomadic, occurring in large flocks, mainly in the interior of Australia. **V** Call is a nasal, medium-pitched *kaaah kaaah kaaah*. **H** Occurs in varied inland habitats including outback towns; also coastal areas in W Australia. **D** W and C Australia.

21 FOREST RAVEN *Corvus tasmanicus* 52–54cm Differs from Australian Raven mainly by call. Note slightly shorter tail. **V** Call is a slightly tremulous *krraah krraah krraah*, deeper and harsher than other Australian corvids. **H** Occurs in most habitats in Tasmania. On mainland, occurs in forests, plantations and farmland. **D** SE Australia (including Tasmania).

22 LITTLE RAVEN *Corvus mellori* 48–50cm Slightly smaller than other corvids in region except Little Crow. Head more rounded and bill less massive. Nomadic, often in large flocks. **V** Call is a strangled *wah wah wah*, deeper and shorter than Australian Raven. **H** Open woodlands and farmland. **D** SE Australia.

23 AUSTRALIAN RAVEN *Corvus coronoides* 48–54cm Large, glossy black corvid with long throat hackles. Common and territorial. **V** Typical call is a high-pitched drawn-out wail, *waaaaaahh*. **H** Open woodlands, farmland and urban areas. **D** Australia.

24 PIED CROW *Corvus albus* 48cm Often scavenges around human habitations and rubbish dumps. **V** Harsh *karrh-karrh-karrh*; also a knocking *kla-kla-kla*. **H** Open country with scattered trees, cultivation, also around towns and villages. **D** Sub-Saharan Africa, Madagascar.

25 BROWN-NECKED RAVEN *Corvus ruficollis* 53cm Desert counterpart of Northern Raven. Small gatherings form at feeding locations, which often include rubbish dumps; larger ones often occur at communal roosts. **V** Dry, rising *aarg-aarg-aarg* and an abrupt croak. All calls appear to be less harsh than those of Northern Raven. **H** Arid desert and semi-desert plains and foothills, also around oases and human settlements. **D** Desert regions of N Africa, Middle East, C Asia.

26 SOMALI CROW *Corvus edithae* 45cm Note brown wings and mantle (but this colouring not always so distinct). **V** Similar to Pied Crow but higher-pitched and more nasal, a dry *karrh* that becomes harsher and shriller in excitement. **H** Desert, semi-desert. **D** Eritrea, Djibouti, E and S Ethiopia, Somalia, SE South Sudan, N Kenya.

27 NORTHERN RAVEN *Corvus corax* 51–69cm Huge black crow with massive bill and shaggy throat. In flight shows a distinct wedge-shaped tail. Usually seen singly, in pairs, or in larger groups where food is plentiful. Often wary. **V** Deep, hollow *pruk-pruk-pruk*, also various other croaks and a guttural rattle. **H** Extreme generalist, occurring everywhere from Arctic tundra and highest mountains to forests and subtropical desert; every habitat except tropical rainforest. **D** Occurs throughout most of the Holarctic, across North America, Europe, Asia, N Africa.

28 CHIHUAHUAN RAVEN *Corvus cryptoleucus* 48–50cm In flight shows slightly wedged tail. Sometimes white feather bases show when preening or in a wind. **V** Rising *graak*, usually higher-pitched than Northern Raven. **H** Primarily flat, arid grassland. **D** SW USA to N Mexico.

29 FAN-TAILED RAVEN *Corvus rhipidurus* 47cm In flight, distinctly short-tailed. Often in quite large gatherings. Regularly associates with Brown-necked Ravens. **V** High-pitched *craah-craah*, also guttural croaks, quiet high-pitched croaks, squeals, trills and clucks. **H** Desert cliffs and ravines with nearby oases. **D** Mali and Niger to Ethiopia, Kenya, Somalia, Arabian Peninsula, Israel, Jordan.

30 WHITE-NECKED RAVEN *Corvus albicollis* 55cm Unmistakable, with huge bill and white neck patch. **V** High-pitched rolling *kraaw, kraaw*. **H** Rocky, bushed and wooded mountainous areas. **D** E and SE Africa, from Uganda to South Africa.

31 THICK-BILLED RAVEN *Corvus crassirostris* 65cm Note white patch on upper nape and enormous bill, creating a front-heavy appearance. **V** Hoarse, metallic, bubbling croak; also a deep grunt, *urrrrk*. **H** Rocky areas. **D** Eritrea, Ethiopia, Somalia.

AUSTRALIAN MUDNESTERS CORCORACIDAE

32 WHITE-WINGED CHOUGH *Corcorax melanorhamphos* 43–47cm Crow-sized, all-black bird with strong legs, curved bill, long rounded tail and red eyes. In flight note broad wings and extensive white in primaries (not visible at rest). Forages on the ground. Gregarious; usually in family groups. **V** Mournful descending whistles, usually given by several birds together. **H** Open forest, woodland and parks, preferring areas with sparse ground cover. **D** S and E Australia.

33 APOSTLEBIRD *Struthidea cinerea* 29–33cm Mostly grey with a long rounded blackish tail and a short stubby black bill. Longer feathers on head and breast give a somewhat shaggy appearance. Gregarious; usually in family groups. Forages on the ground. **V** Gives a variety of harsh, raucous and scratchy notes, in a constant chatter. **H** Arid and semi-arid open woodlands, usually near water. **D** E Australia

MELAMPITTAS MELAMPITTIDAE

34 LESSER MELAMPITTA *Melampitta lugubris* 18cm All black, pot-bellied and short-tailed. Male has red eyes. Mainly terrestrial and skulking bird of dense cover. Singly or in pairs. **V** Distinctive call is a penetrating, sharp *tlik!* repeated at regular intervals. **H** Montane forest. **D** New Guinea.

35 GREATER MELAMPITTA *Megalampitta gigantea* 29cm Large all-black bird with a heavy hook-tipped bill. Shy, rare and little known. Largely terrestrial, more often heard than seen. **V** Loud ringing song, *choo-wee choo-wee*, repeated for long periods. **H** Lives in areas of limestone formations in dense rainforest. **D** New Guinea.

IFRIT IFRITIDAE

36 BLUE-CAPPED IFRIT *Ifrita kowaldi* 16cm Plump, short-tailed bird now placed in its own family. Iridescent blue cap in both sexes. Forages on mossy branches and logs. Usually in pairs or small groups. **V** Call is a loud squeaky *chu-wee*, repeated frequently. Also gives longer sequences of similar squeaky notes. **H** Montane forest. **D** New Guinea.

1
2
3
4
5
6
7
8
9
10
11
12
13
14
15
16
17
18
19
20
21
22
23
24
25
26
27
28
29
30
31
32
33
34
♀
♂
35
36
♀
♂

BIRDS-OF-PARADISE PARADISAEIDAE

1 PARADISE-CROW *Lycocorax pyrrhopterus* 40–44cm Crow-like bird-of-paradise, with smallish head and broad rounded tail. Blackish-brown plumage lacks gloss (except in race obiensis). Singly, in pairs or small groups. Does not lek. **V** Makes a variety of clipped mechanical sounds, often in duet. **H** Forest, forest edge and plantations, in mid-storey or canopy. **D** Maluku Islands, Indonesia.

2 GLOSSY-MANTLED MANUCODE *Manucodia ater* 32–43cm All black, glossed purplish and greenish, with relatively long tail and rounded head. Red eyes. Sexes alike. Singly or in pairs. **V** Long, high-pitched whistle, reminiscent of a tuning fork. **H** Forest edge, secondary growth, wet forests and mangroves, in lowlands and hills. **D** New Guinea.

3 JOBI MANUCODE *Manucodia jobiensis* 30–36cm Smaller and shorter-tailed than Glossy-mantled Manucode. 'Head bumps' inconspicuous. Sexes alike. **V** Poorly known. **H** Rainforest and swamp forest in lowlands and foothills. **D** N New Guinea.

4 CRINKLE-COLLARED MANUCODE *Manucodia chalybatus* 33–37cm Differs from Glossy-mantled Manucode in having conspicuous 'head bumps' (tufts of feathers above the eyes). The ruffled feathers of the throat and breast are only obvious at close range. **V** Typical call is a short, sharp *chuk*. **H** Forest, forest edge and secondary growth, in lowlands and hills. **D** New Guinea.

5 CURL-CRESTED MANUCODE *Manucodia comrii* 43–44cm Largest manucode, with prominent curly mop of feathers on the crown, crinkled neck and breast feathers, and twisted central tail feathers. **V** Typical call is a rasping *chek*. Also a fast bubbling *hoo-hoo-hoo-hoo...* **H** Forest, coastal woodland and gardens. **D** D'Entrecasteaux Archipelago and Trobriand Islands (W Papua New Guinea).

6 TRUMPET MANUCODE *Phonygammus keraudrenii* 28–33cm Small manucode with prominent head plumes and neck hackles. Six races, differing in size, altitude range, colour of gloss and vocalisations (may represent more than one species). **V** Very diverse, with some geographical variation. **H** Forest, forest edge and woodland, from lowlands to mountains. **D** New Guinea, NE Australia.

7 LONG-TAILED PARADIGALLA *Paradigalla carunculata* 35–37cm All black with a long, graduated tail and slender bill. Prominent yellow and blue facial wattles. Rare and little known. Sexes similar. Singly or in pairs. **V** Calls include a long, drawn-out *wheeeee*, and a rising whistle. **H** Mid-montane forest in Arfak mountains. **D** NW New Guinea.

8 SHORT-TAILED PARADIGALLA *Paradigalla brevicauda* 22–23cm Similar to Long-tailed Paradigalla but stocky and short-tailed. Sexes similar. Usually seen singly. **V** Melodious, bell-like *zhee*, and a series of clear whistled notes. **H** Montane forest and forest edge. **D** W and C New Guinea.

9 ARFAK ASTRAPIA *Astrapia nigra* male 75cm, female 50cm Bulky body with very long tail. Female duller with shorter tail. Poorly known. Usually seen singly. **V** Poorly known. Short, sharp *chup*, reminiscent of two pebbles being struck together. **H** Montane forest in Vogelkop. **D** NW New Guinea.

10 SPLENDID ASTRAPIA *Astrapia splendidissima* male 39cm, female 37cm Shorter-tailed than other astrapias, with white base to the tail. Male has spatulate tips to tail feathers. Sexually dimorphic. **V** Series of evenly paced *tchuk* notes, or a rapid trisyllabic *tchuk tck-tck*. **H** Montane forest, forest edge, secondary growth. **D** W and C New Guinea.

11 RIBBON-TAILED ASTRAPIA *Astrapia mayeri* male 125cm, female 53cm Male has tuft of feathers at base of bill and extraordinarily long white central tail feathers. Female has shorter dark tail, resembling other female astrapias. **V** Makes varied raucous mechanical sounds. **H** Upper montane forest and forest edge. **D** EC New Guinea.

12 PRINCESS STEPHANIE'S ASTRAPIA *Astrapia stephaniae* male 84cm, female 53cm Male has two very long central tail feathers that broaden towards the end. Female resembles female Ribbon-tailed Astrapia. **V** Various high-pitched whistles and squawks. **H** Montane forest, forest edge, secondary growth. **D** New Guinea.

13 HUON ASTRAPIA *Astrapia rothschildi* male 66cm, female 47cm Male has very long tail feathers that broaden towards the end. Most closely resembles Arfak Astrapia but ranges do not overlap. Female is rather dark. **V** Poorly known, but probably similar to other astrapias. **H** Montane forest and forest edge. **D** SE New Guinea.

14 WESTERN PAROTIA *Parotia sefilata* 30–33cm Mostly black with white forehead tuft and iridescent emerald-green breast-shield feathers. Like all male parotias, has six fine plumes on the head. Sexually dimorphic, with female mostly brown above with a darker head and fine barring below. **V** Makes various harsh, nasal scolding sounds. **H** Mid-montane forests and secondary growth. **D** NW New Guinea.

15 QUEEN CAROLA'S PAROTIA *Parotia carolae* 25–26cm Male has different forehead pattern, pinkish-violet breast-shield, white flank plumes and pale eyes. Female has a striped face pattern. **V** Shrill, scratchy whistle of 3–4 notes; also a tremulous double *trrrrr-trrrrr*. **H** Mid-montane forests and secondary growth. **D** W and C New Guinea.

16 BRONZE PAROTIA *Parotia berlepschi* 25–26cm Formerly considered conspecific with Queen Carola's Parotia. Differs in black forehead, no eye-ring, golden-bronze breast-shield and overall more bronzy coloration. Female very similar to Queen Carola's. **V** Poorly known, but a shrill *wee-deet* is unlike call of Queen Carola's. **H** Mid-montane forest in Foya Mountains. **D** NC New Guinea.

17 LAWES'S PAROTIA *Parotia lawesii* 25–27cm Male is similar to Western Parotia but has less white on forehead (restricted to a white tuft above the nostrils). Breast-shield feathers golden-green. Female rufous-brown with a dark head. **V** Raucous, nasal *rah rah rah rah*. Also higher-pitched, plaintive *wah*. **H** Mid-montane forest including disturbed areas. **D** E and SE New Guinea.

18 EASTERN PAROTIA *Parotia helenae* 25–27cm Formerly considered conspecific with Lawes's Parotia. Male differs in having reduced frontal crest (which is golden-brown, not white). Female is very similar to Lawes's. **V** Not known. **H** Mid-montane forest including disturbed areas. **D** SE New Guinea.

19 WAHNES'S PAROTIA *Parotia wahnesi* 36–43cm Longer-tailed than all other parotias. Male is noticeably larger than female. Female has prominent post-ocular and malar stripes. **V** Poorly known, but includes typical harsh parotia nasal sounds. **H** Mid-montane forest. **D** SE New Guinea.

20 KING OF SAXONY BIRD-OF-PARADISE *Pteridophora alberti* 20–22cm Male is unmistakable, with two long pale blue head plumes. Female is grey-brown above and heavily barred or spotted below, with ochraceous undertail coverts. Males are often in the canopy, females usually lower down. **V** Male gives distinctive jangly sequence of hurried electrostatic notes. **H** Mid-montane and upper montane forest. **D** C New Guinea.

21 GREATER LOPHORINA *Lophorina superba* 25–26cm Formerly called Superb Bird-of-paradise, together with following two newly split species. Male is mainly black with a blue-green breast-shield and an extendable iridescent green cape of elongated neck feathers. Female is barred below with a prominent supercilium and face pattern. Males usually solitary, but females may gather in small groups at fruiting trees. **V** Male gives harsh series of *scher scher scher* notes. **H** Mid-montane forest and secondary growth. **D** New Guinea.

22 CRESCENT-CAPED LOPHORINA *Lophorina niedda* 25–26cm Formerly considered conspecific with Greater Lophorina. Male is almost identical. Black-headed female is whiter below and more heavily barred. **V** Plaintive series of harsh, evenly spaced downslurred notes, *hiu hiu hiu*. **H** Mid-montane forest and secondary growth. **D** NW New Guinea.

23 LESSER LOPHORINA *Lophorina minor* 25–26cm Formerly considered conspecific with Greater Lophorina. Male is almost identical. Female is much darker above than other female lophorinas, with supercilium reduced to a post-ocular streak or absent. **V** Not known. **H** Mid-montane forest and secondary growth. **D** SE New Guinea.

24 PARADISE RIFLEBIRD *Ptiloris paradiseus* 28–30cm Long, slender, downcurved bill. Sexually dimorphic. Male is velvety black with iridescent blue-green breast-shield, crown and tail. Female is brown above and white below with scaling; note white supercilium. Male displays on horizontal branch. **V** Unmusical, long rasping call, repeated frequently. **H** Subtropical and temperate rainforests. **D** NE Australia.

25 VICTORIA'S RIFLEBIRD *Ptiloris victoriae* 23–25cm Smaller than Paradise Riflebird. Male is very similar. Female is tinged cinnamon below. Male displays from top of broken vertical stump. **V** Various rasping calls, similar to Paradise Riflebird. **H** Tropical rainforest. **D** NE Australia.

26 MAGNIFICENT RIFLEBIRD *Ptiloris magnificus* male 32cm, female 28cm Largest riflebird, with larger iridescent breast-shield in male. Note wispy flank plumes. Female is rufous-brown above with prominent malar stripe. **V** Long downslurred and upslurred whistles, sometimes like a 'wolf-whistle'. Also harsh rasping sounds. **H** Lowland and hill forest. **D** W and C New Guinea, NE Australia.

27 GROWLING RIFLEBIRD *Ptiloris intercedens* M32cm, F28cm Formerly considered conspecific with Magnificent Riflebird. Very similar to Magnificent, but voice is diagnostic and no overlap in range. Male has shorter flank plumes. **V** Harsh, growling *hrrraaak hrrraaak*, quite unlike the whistles of Magnificent Riflebird. **H** Hill forest. **D** E New Guinea.

1
2
3
4
5
6
7
8
9
10
11
12
13
14
15
16
17
18
19
20
21
22
23
24
25
26
27

BIRDS-OF-PARADISE *CONTINUED*

1 BLACK SICKLEBILL *Epimachus fastosus* male 110cm, female 55cm Long, downcurved bill and very long tail. Sexually dimorphic. Male is all dark with iridescent blue patches. Female is brown with rufous crown, and finely barred below. Note red eyes. **V** Rapid, gunfire-like *ta-tk ta-tk*, often with long intervals between calls. **H** Hill and lower montane forest. **D** WC to SE New Guinea.

2 BROWN SICKLEBILL *Epimachus meyeri* male 96cm, female 52cm Similar to Black Sicklebill but slightly smaller and bill more downcurved. Male is browner below. Female is similar to female Black Sicklebill. Note pale eyes in both sexes. **V** Typical call is a machine-gun chatter; also short high-pitched trumpet-like notes. **H** Montane forest and forest edge. **D** WC to SE New Guinea.

3 BLACK-BILLED SICKLEBILL *Drepanornis albertisi* 36cm Small sicklebill with long downcurved bill and rounded tail. Note pale blue bare skin behind eye. Sexually dimorphic. Usually singly in forest interior. **V** Fast series of piping notes, vaguely reminiscent of a Whimbrel (Plate 46). **H** Montane forest. **D** New Guinea.

4 PALE-BILLED SICKLEBILL *Drepanornis bruijnii* 35cm Similar to Black-billed Sicklebill but bill is pale. Note lavender bare skin behind eye. Sexually dimorphic. Usually singly but will join mixed-species flocks. **V** Melodic series of *whik-hoo whik-hoo whik-hoo* notes, with variations. **H** Lowland rainforest. **D** New Guinea.

5 MAGNIFICENT BIRD-OF-PARADISE *Diphyllodes magnificus* 19cm (26cm incl. tail) Sexually dimorphic. Male is spectacular and unmistakable. Female is largely brown with blue post-ocular streak. Male maintains open display ground on forest floor. **V** Series of downslurred *chow chow chow chow chow* notes. **H** Hill forest, preferring lower and middle storey. **D** New Guinea.

6 WILSON'S BIRD-OF-PARADISE *Diphyllodes respublica* 16cm (21cm incl. tail) Sexually dimorphic. Male is spectacular and unmistakable. Female is largely brown with blue skullcap. Favours forest interior. Male maintains open display ground on forest floor. **V** Calls resemble Magnificent Bird-of-paradise, but notes are purer. **H** Hill and mid-montane forest. **D** Batanta and Waigeo, West Papuan islands, Indonesia.

7 KING BIRD-OF-PARADISE *Cicinnurus regius* 16cm (31cm incl. tail) Sexually dimorphic. Male is spectacular and unmistakable. Female is brown with rufous wings. Usually singly. Males display in the canopy. **V** Noisy at display sites, with varied vocalisations including raucous *kow kow kow kow kow*. **H** Lowland and hill forest, including secondary growth. Favours vine tangles in mid-storey and canopy. **D** New Guinea.

8 STANDARDWING *Semioptera wallacii* 25–30cm Flat-headed and short-tailed with tuft of feathers at base of upper mandible. Adult male has two long white plumes on each wing ('standards'). Note orange legs. Males lek in groups in subcanopy. **V** Noisy at leks. Males give a wide variety of loud throaty whistles and squawks, including typical *wow-wow-wow-wow-wow* calls. **H** Closed-canopy forest. **D** Halmahera and Bacan, N Maluku Islands, Indonesia.

9 TWELVE-WIRED BIRD-OF-PARADISE *Seleucidis melanoleucus* 32–34cm Long-billed and short-tailed. Sexually dimorphic. Male unmistakable, with lemon-yellow underparts and 12 wire-like flank plumes. Female is russet above with a black head and neck. Male displays from dead tree stump or other exposed perch. **V** Loud, resonant *kow*, repeated at regular intervals or in a slow series. **H** Swamp forest and wet forests in the lowlands. **D** New Guinea.

10 GREATER BIRD-OF-PARADISE *Paradisaea apoda* male 43–46cm (excl. tail wires), female 35cm Spectacular and sexually dimorphic. Male has elongated yellowish and white flank plumes and two long, wire-like central tail feathers. Female entirely dark brown. Has communal leks in large emergent trees. **V** At leks, males give raucous bugling *wok wok wok wok wok* calls, audible at long range. **H** Lowland and hill forest. **D** S New Guinea.

11 RAGGIANA BIRD-OF-PARADISE *Paradisaea raggiana* 33–36cm (excl. tail wires) Male is similar to Lesser Bird-of-paradise but has red or orange flank plumes. Female is brown with a dark face. Has communal leks in large emergent trees. **V** Calls of males at leks are more nasal and high-pitched than Greater Bird-of-paradise, a loud far-carrying cry increasing in intensity and/or pitch, *wau wau wau wau wau waauu waauu waaauuu*. **H** Wide range of habitats including forest, forest edge, secondary growth from lowlands to mountains. **D** NE and SE New Guinea.

12 LESSER BIRD-OF-PARADISE *Paradisaea minor* 32–33cm (excl. tail wires) Male is similar to Greater Bird-of-paradise, but smaller and has a golden back. Female has a dark maroon head and white underparts. Has communal leks in large emergent trees. **V** Males at leks give similar calls to Raggiana Bird-of-paradise, but shriller and high-pitched. **H** Lowland and hill forest, secondary growth. **D** W and N New Guinea.

13 GOLDIE'S BIRD-OF-PARADISE *Paradisaea decora* 29–33cm (excl. tail wires) Restricted range. Male resembles Raggiana Bird-of-paradise but is smaller and has a paler breast. Female has yellowish crown and hindneck. Has communal leks in large emergent trees. **V** Distinctive, upslurred *huipp huipp huipp huipp huipp* and harsher variations. **H** Hill forest, forest edge, secondary growth. **D** D'Entrecasteaux Archipelago, W Papua New Guinea.

14 RED BIRD-OF-PARADISE *Paradisaea rubra* 30–33cm (excl. tail wires) Restricted range. Male has unique green cushion of feathers above each eye, golden-yellow breast and scarlet flank plumes. Female has blackish-brown face and throat. Has large communal leks in large emergent trees. **V** Males at lek give loud raucous *wark wark wark wark* calls, remininscent of Raggiana Bird-of-paradise. **H** Lowland and hill forest. **D** Batanta and Waigeo, West Papuan islands, Indonesia.

15 EMPEROR BIRD-OF-PARADISE *Paradisaea guilielmi* 31–33cm (excl. tail wires) Restricted range. Male is distinctive, with bottle-green face, throat and breast, and white flank plumes. Female has nape and mantle pale yellowish. Has communal leks in large emergent trees. **V** Loud raucous *wuk wuk wuk* calls, similar to others in the genus. **H** Hill and montane forest in Huon Peninsula. **D** SE New Guinea.

16 BLUE BIRD-OF-PARADISE *Paradisaea rudolphi* 30cm (excl. tail wires) Spectacular and unmistakable. Note broken white eye-ring. Sexually dimorphic. Male has narrow, elongated central tail feathers and cinnamon-blue flank plumes. Usually seen singly. Performs extraordinary display in canopy. **V** Male gives a loud nasal *kweh kweh kweh*. **H** Montane forest. **D** C and SE New Guinea.

1
2
3
4
5
6
7
8
9
10
11
12
13
14
15
16

AUSTRALASIAN ROBINS PETROICIDAE

1 ASHY ROBIN *Heteromyias albispecularis* 15–18cm Large-headed, shy robin. Eastern races have broad white supercilium behind the eye. Sexes alike. Seen singly or in pairs, in the understorey. **V** Soft piping *hoo-hoo-hoo-hoo*. **H** Montane forest. **D** New Guinea.

2 GREY-HEADED ROBIN *Heteromyias cinereifrons* 16–18cm Similar to Ashy Robin but browner with more white in the wing. Sexes alike. Inhabits the understorey. **V** Monotonous piping whistle, slower than Ashy Robin. **H** Lower-montane rainforest. **D** NE Australia.

3 BLACK-CHINNED ROBIN *Poecilodryas brachyura* 14–15cm Striking pied robin with long white supercilium. Note black chin. Sexes alike. Singly or in pairs, in lower or middle storey. **V** Soft plaintive medley of whistles. **H** Lowland and foothill forest. **D** NW and N New Guinea.

4 BLACK-SIDED ROBIN *Poecilodryas hypoleuca* 13–15cm Similar to Black-chinned Robin but has white chin, shorter and narrower supercilium, and black patches on sides of breast. Sexes alike. **V** Loud emphatic *pit-chew* or *whit-chew*. **H** Lowland forest, swamp forest, secondary growth. **D** New Guinea.

5 WHITE-BROWED ROBIN *Poecilodryas superciliosa* 14–16cm Note long white supercilium, grey and white ear-coverts and brownish-grey upperparts. Sexes alike. Inhabits lower storey. **V** Clear piping whistles; also a loud *chew chew*. **H** Rainforest, coastal woodland. **D** NE Australia.

6 BUFF-SIDED ROBIN *Poecilodryas cerviniventris* 16–18cm Similar to White-browed Robin but larger and browner. Ear-coverts all dark; flanks and undertail coverts tawny-rufous. Sexes alike. Cocks tail and flicks wings. **V** Loud whistles, similar to White-browed Robin. **H** Riverine forest, monsoon rainforest. **D** N Australia.

7 BANDED YELLOW ROBIN *Poecilodryas placens* 14–15cm Striking, colourful robin with dark grey head and yellow underparts with broad olive-green breast-band. Sexes alike. Inconspicuous and shy, inhabiting the understorey. **V** Slow series of unevenly spaced bell-like notes, *tew tew-tew tew*. **H** Primary forest in lowlands and foothills, but fragmented distribution. **D** New Guinea.

8 BLACK-THROATED ROBIN *Poecilodryas albonotata* 18–19cm Large, dumpy robin with black throat and thin white line on sides of neck. One race has grey underparts. Sexes alike. Favours middle and upper storey. **V** Series of thin high-pitched whistles or a much longer single whistle at the same pitch. **H** Montane forests. **D** New Guinea.

9 WHITE-WINGED ROBIN *Peneothello sigillata* 14–15cm All black with white inner secondaries. One race has a white patch on side of breast. Sexes alike. Singly, pairs or small family groups. Keeps low down. **V** Call is a distinctive fast trill. Song is a series of rapid jumbled notes. **H** Montane forest, up to treeline. **D** C and E New Guinea.

10 SMOKY ROBIN *Peneothello cryptoleuca* 14–15cm Plumage all smoky-grey, with paler belly (whitish in some races). Sexes alike. Inhabits understorey. **V** Song is a series of very short *too-chip* notes, with considerable variation. **H** Montane forest, forest edge, thickets (only in west). **D** W New Guinea.

11 SLATY ROBIN *Peneothello cyanus* 14–15cm Plumage all blue-grey, with darker face. Sexes alike. Inhabits understorey. Singly, pairs or small family groups. **V** Rather variable whistles and trills. **H** Lower montane forest and secondary growth. **D** New Guinea.

12 WHITE-RUMPED ROBIN *Peneothello bimaculata* 13–14cm Sooty-black with white rump and variable white patch on sides of breast. Nominate race has white belly. Eastern race *vicaria* has black belly. Sexes alike. Singly or in pairs. Inhabits dense undergrowth. **V** Mournful series of *wee prr-prr-prr-prr-prr*, the first note higher-pitched or absent. **H** Foothill and lower montane forest. **D** New Guinea.

13 MANGROVE ROBIN *Peneoenanthe pulverulenta* 14–15cm Smoky grey above with blackish wings and tail; note white bases to outer-tail feathers. Mainly white below with greyish breast-band. Sexes alike. Seen singly or in pairs. Keeps low. **V** Plaintive piping notes, uttered rather randomly. **H** Mangroves and riparian vegetation. **D** NC and S New Guinea, N Australia.

14 WHITE-FACED ROBIN *Tregellasia leucops* 12–13cm Distinctive small robin with grey head, olive-green upperparts and yellow underparts. Ten races, mainly varying in amount of white on face and throat. Sexes alike. Seen singly or in pairs. **V** Variety of rather harsh grating sounds, but racial variation poorly known. **H** Moist lowland forests and lower montane forest. **D** New Guinea, NE Australia.

15 PALE-YELLOW ROBIN *Tregellasia capito* 12–13cm Similar to White-faced Robin but plainer with paler grey head and white lores. Sexes alike. **V** Typically gives series of harsh grating squeaks. **H** Coastal and inland rainforests with dense understorey. **D** E Australia.

16 EASTERN YELLOW ROBIN *Eopsaltria australis* 14–16cm Attractive grey and yellow robin, not dissimilar to Pale-yellow Robin, but note dark lores. In flight, shows bright yellow rump and whitish wing-bar. Sexes alike. **V** Gives short series of clear piping whistles in morning and evening. **H** Wide variety of forests and woodlands with undergrowth. **D** E and SE Australia.

17 WESTERN YELLOW ROBIN *Eopsaltria griseogularis* 14–16cm Western counterpart of Eastern Yellow Robin, differing mainly in having a grey breast, with yellow restricted to belly and undertail. Sexes alike. **V** Gives short series of clear piping whistles, longer and lower-pitched than Eastern Yellow Robin. **H** Wide variety of forests and woodlands with undergrowth. **D** SW Australia.

18 WHITE-BREASTED ROBIN *Eopsaltria georgiana* 15–16cm Grey and white robin, reminiscent of Mangrove Robin, but lacks white sides to base of tail. Sexes alike. Singly or in pairs. **V** Song is a loud, whistle-like *see-ow*, sometimes repeated. Call is a harsh *chit-chit*. **H** Dense coastal scrub, riverine vegetation, wet woodlands. **D** SW Australia.

19 HOODED ROBIN *Melanodryas cucullata* 15–17cm Striking pied robin. Male has black hood and upperparts with white stripes on scapulars and wings. Female much greyer, lacking hood. Usually in pairs or small groups. **V** Song is a jumbled squeaky medley of hurried notes. **H** Wide variety of wooded habitats with suitable understorey. **D** Australia.

20 DUSKY ROBIN *Melanodryas vittata* 16–17cm Rather sombre, with olive-brown upperparts; paler below. Sexes alike. Singly or in pairs. Endemic to Tasmania. **V** Song is a plaintive descending whistle, *piu piu-piu*. **H** Forest edge, open woodlands, parks and gardens. **D** Tasmania.

21 GREEN-BACKED ROBIN *Pachycephalopsis hattamensis* 15cm Mainly olive-green with grey head, and rusty-brown wings, uppertail coverts and tail. Note white lores and chin, and pale eyes. Sexes alike. Seen singly or in pairs. **V** Poorly known. Call is an upslurred *poo-wee*, or a downslurred *seeoo*. **H** Montane forest, in damp understorey. **D** W and C New Guinea.

22 WHITE-EYED ROBIN *Pachycephalopsis poliosoma* 15–16cm Similar to Green-backed Robin, but entirely grey except for whitish throat. Note white eyes. Sexes alike. Seen singly or in pairs; shy and elusive. **V** Poorly known. Typical song is a series of 4–10 loud notes. **H** Hill and lower montane forest, in understorey. **D** C and E New Guinea.

23 TORRENT FLYROBIN *Monachella muelleriana* 14cm Striking riverine inhabitant with white forehead and black cap. Perches on boulders and branches alongside rivers. Sexes alike. In pairs or small groups. **V** Song is a high-pitched trilling, given in a long series, or in short bursts. **H** Fast-flowing rocky rivers in hills and lower mountains; at lower altitudes in New Britain. **D** New Guinea, Bismarck Archipelago.

24 CANARY FLYROBIN *Microeca papuana* 12cm Olive-green above and bright yellowish below. Rather small, and formerly called Canary Flycatcher. Sexes alike. Seen singly or in pairs. **V** Thin, high-pitched notes given in short bursts of song. **H** Montane forest, forest edge, clearings. **D** New Guinea.

25 YELLOW-LEGGED FLYROBIN *Microeca griseoceps* 12cm Inconspicuous inhabitant of the upper strata with a distinctly grey head and whitish lores and throat. Sexes alike. Singly or in pairs. Occurs at lower elevations than Canary Flyrobin and higher than Olive Flyrobin. **V** Short, squeaky, high-pitched bursts of song. **H** Hill forest, sometimes in lowlands. **D** New Guinea, NE Australia.

26 OLIVE FLYROBIN *Microeca flavovirescens* 13–14cm Similar to Canary Flyrobin, but paler yellow below. Sexes alike. Seen singly or in pairs. Occurs at lower elevations than Canary Flyrobin. **V** Song is a mournful series of 4–6 whistler-like notes, sometimes repeated for long periods. **H** Forest, from lowlands to hills. **D** Aru Islands, West Papuan islands, mainland New Guinea.

27 LEMON-BELLIED FLYROBIN *Microeca flavigaster* 12cm Similar to other flyrobins, but occurs in drier habitats, not inside forest. Sexes alike. Singly or in pairs. **V** Variable. In Australia, gives a rather flat, short warble, repeated at regular intervals; more strident in New Guinea. **H** Forest edge, secondary growth, savannah and mangroves, from lowlands to hills. **D** New Guinea, N Australia.

28 YELLOW-BELLIED FLYROBIN *Microeca flaviventris* 14–15cm Typical dumpy flyrobin, with white throat, grey breast and yellow belly. Note indistinct dark breast-band. Sexes alike. Unobtrusive, usually in pairs. **V** Gives a loud warble of repeated short phrases. **H** Wet forests, occasionally drier forests. **D** New Caledonia.

29 GOLDEN-BELLIED FLYROBIN *Microeca hemixantha* 12cm Usually uses exposed branches or overhead wires to launch short sallies to capture flying insects; also forages with mixed-species feeding parties. **V** Medley of 12–14 sweet, warbled notes. **H** Mangroves, forest, forest edges and open woodland. **D** Tanimbar Islands, Indonesia.

30 JACKY WINTER *Microeca fascinans* 14cm Nondescript grey-brown robin with prominent white outer-tail feathers. Sexes alike. Singly or in pairs. Forages low, but relatively tame. **V** Loud monotonous *wit-a-wit-a-wit-a-wit* or *wit-wit-wit-wit-wit-wit-wit*. **H** Forest edge, open woodlands, scrub, parks, gardens. **D** SE New Guinea, Australia.

31 GARNET ROBIN *Eugerygone rubra* 11cm Small active robin, mainly in mid-storey. Sexually dimorphic. Male has maroon-red upperparts; female is bright olive-green. Flicks one wing at a time, and waves tail from side to side. **V** Song is a thin, high-pitched tinkling jumble of notes. **H** Montane forest, forest edge, secondary growth. **D** New Guinea.

1
2
3
4
5
6
7
8
9
10
11
12
13
14
15
16
17
18
19
♀
♂
20
21
22
23
24
25
26
27
28
29
30
31
♀
♂

AUSTRALASIAN ROBINS *CONTINUED*

1 ROSE ROBIN *Petroica rosea* 11–12cm Small, sexually dimorphic robin. Male has dark grey head and upperparts and rose-pink breast. Female is much paler, with wing-bars and a hint of pink on the breast. Seen singly or in pairs. Partially migratory. **V** Short bursts of song with a buzzy quality, especially the last two or three longer notes. **H** Wet forest and rainforest with dense gullies. **D** SE Australia.

2 PINK ROBIN *Petroica rodinogaster* 12–13cm Sexually dimorphic. Male has sooty-black head and upperparts contrasting with lilac-pink underparts. Female is brown and buff. Seen singly or in pairs. Partially migratory. **V** Song is a loose rattling warble. **H** Gullies and undergrowth of tall eucalypt forest and rainforest. **D** SE Australia.

3 SNOW MOUNTAIN ROBIN *Petroica archboldi* 14cm Little-known, restricted-range robin. Sexes similar. Male is slaty or blackish with bright red spot on breast. Note small white spot on forehead. Singly or in pairs. Reported to be inactive. **V** Poorly known. A medley of chirps and scolding notes; also a series of fast churrs. **H** Bare rocky slopes and cliffs above the treeline in the Snow Mountains; usually above 3,800m. **D** WC New Guinea.

4 MOUNTAIN ROBIN *Petroica bivittata* 11cm Small black and white robin. Note white spot on forehead. Sexes similar. Generally seen singly or in pairs, in the canopy. **V** Song is a fast twittering series of notes, continuing for long periods without much variation. **H** High montane forest and subalpine scrub. **D** New Guinea.

5 FLAME ROBIN *Petroica phoenicea* 12–14cm Sexually dimorphic robin with much white in the wing. Male is grey above with extensive red on the breast and belly; female mainly brown. Partially migratory, often occurring in small flocks in winter. **V** Song is a short, rich musical medley, gradually descending in scale. **H** Moist eucalypt forests with open understorey, wintering in more open habitats. **D** SE Australia.

6 PACIFIC ROBIN *Petroica pusilla* 12–14cm Formerly considered conspecific with Norfolk and Scarlet Robins. Small, dumpy black and white robin with red on the breast in both sexes. Thirteen races across wide range, differing in colour and extent of wing markings and forehead spot, and colour of chin. Males of some races are browner. Usually in pairs, in mid-storey or canopy. **V** Variable according to race. Song is typically a short warble. **H** Forest and forest edge; in mountains on Solomon Islands. **D** Solomon Islands to Samoa.

7 NORFOLK ROBIN *Petroica multicolor* 12cm Formerly considered conspecific with Pacific and Scarlet Robins. Smaller and longer-billed than Scarlet Robin, with less white in wings. Male has white extending onto forecrown. **V** Song is a pleasant, short, trilled warble. **H** Native rainforest, but also other habitats. **D** Norfolk Island, SW Pacific Ocean.

8 SCARLET ROBIN *Petroica boodang* 12–14cm Formerly considered conspecific with Pacific and Norfolk Robins. Small robin with extensive white in the wings and large white forehead spot. Singly or in pairs; does not form flocks in winter. **V** Song is a plaintive short, high-pitched warble, repeated frequently. **H** Open eucalypt forests and woodlands with open understorey. **D** S and E Australia.

9 RED-CAPPED ROBIN *Petroica goodenovii* 11–12cm Small black and white robin with diagnostic red forecrown. Note white sides to scarlet breast. Female is mostly brown but has rusty forecrown. Seen singly or in pairs. Mainly sedentary. **V** Song is a scratchy warble, often incorporating short trills. **H** Varied inland habitats with trees or shrubs, often in more arid habitats than other robins. **D** C Australia.

10 TOMTIT *Petroica macrocephala* 13cm From larger North Island and South Island robins by white wing-bar and wholly white or yellowish belly. **V** High-pitched, simple song, slightly descending *tufec-tufec-tufec* or *teerohteerohweeh*. **H** Forest, secondary growth, subalpine scrub, tussock grassland, pine plantations. **D** New Zealand, including several outlying islands.

11 NORTH ISLAND ROBIN *Petroica longipes* 18cm A dark grey-black bird with a pale area on breast and belly and pale streaking on upperparts. Can be very confiding. **V** Very like South Island Robin; a series of repeated notes, *ti-ti-ti-ti*, *tonk-tonk-tonk*, *too-too-too*, varying in speed and pitch. **H** Natural forest. **D** North Island and nearby islands, New Zealand.

12 SOUTH ISLAND ROBIN *Petroica australis* 18cm Note all-blackish wings, upright stance, long legs. Male has clearer division than female between dark breast and white belly. **V** Melodious series of high to very high trills and rattles at varying speed. **H** Forest, tall scrub, pine plantations, up to treeline. **D** South Island and Stewart Island, New Zealand.

13 BLACK ROBIN *Petroica traversi* 15cm Both sexes entirely black. No similar bird on Chatham Islands. **V** Short, high, warbling *tutuwituwiweet-* (*wi* lower-pitched) or meandering-down *fut tweetfeet futd'r'r'r*. **H** Forest, scrub. **D** Chatham Islands, SW Pacific Ocean.

14 PAPUAN SCRUB ROBIN *Drymodes beccarii* 20cm Formerly considered conspecific with Northern Scrub Robin, differing in voice and habits. Darker plumage than Northern Scrub Robin. Usually seen singly. Elusive and largely terrestrial. **V** Song is a long, mournful, drawn-out whistle, usually downslurred. **H** Dense thickets and understorey in forest, from lowlands to lower montane areas. **D** New Guinea.

15 NORTHERN SCRUB ROBIN *Drymodes superciliaris* 21–22cm Long legs and long tail, latter often cocked. Bold face pattern with black vertical stripe through eye, black-and-white wing pattern, and rusty upperparts. Less terrestrial than Papuan Scrub Robin. **V** Similar whistles to Papuan Scrub Robin, but shorter and usually upslurred. Also gives harsher scolding calls. **H** Rainforest and monsoon forest, in Cape York Peninsula. **D** NE Australia.

16 SOUTHERN SCRUB ROBIN *Drymodes brunneopygia* 21–23cm Plainer and less well marked than Northern Scrub Robin. Mainly terrestrial, but sings from elevated perches. Singly or in pairs. **V** Gives a thin, high-pitched whistle similar to other scrub robins, and also a cheerful, whistled *did-you-eat?* **H** Woodlands and shrublands with sparse ground cover, including dense mallee. **D** S Australia.

17 GREATER GROUND ROBIN *Amalocichla sclateriana* 20cm Large, heavy-billed ground robin. Shy and elusive; mainly terrestrial. More often heard than seen. **V** Song is a rich mournful whistle, usually 3–4 notes, with last note longer and upslurred. Longer series are interspersed with more scratchy notes. **H** High montane forest and subalpine shrubs. **D** C and SE New Guinea.

18 LESSER GROUND ROBIN *Amalocichla incerta* 14–15cm Smaller and more widespread than Greater Ground Robin, with small bill and long legs. Shy and elusive; mainly terrestrial. More often heard than seen. **V** Song is a sweet series of discordant whistles, with some variation. **H** Montane forest, at lower altitudes than Greater Ground Robin. **D** New Guinea.

ROCKFOWL PICATHARTIDAE

19 WHITE-NECKED ROCKFOWL *Picathartes gymnocephalus* 32cm Unmistakable, with bare head, and yellow face contrasting strongly with black patch behind eye. **V** Mostly silent, but has a harsh *aaoow* alarm call and a whistling contact call. Roosting birds churr and grunt. **H** Primary or secondary forest in rocky terrain. **D** W Africa, from Guinea to Ghana.

20 GREY-NECKED ROCKFOWL *Picathartes oreas* 32cm Unmistakable. Bare head is pale blue, black and red. **V** Mostly silent, but a quiet *wheet* call is said to resemble the sound of furniture being pushed across a wooden floor. **H** Near large rocks, caves and cliffs in rainforest. Keeps to the undergrowth. **D** S Nigeria and Cameroon to Gabon and N Congo.

ROCKJUMPERS CHAETOPIDAE

21 CAPE ROCKJUMPER *Chaetops frenatus* 24cm Male has a dark rufous belly, a black throat, and a white malar stripe; female's plumage is similar but more muted. Pairs and small groups forage for insects on the ground. **V** Very high *fiufiufiufiutipuptuetiptiptuetiptiptue*, repeated 3–6 times. **H** Rocky mountain slopes, fynbos. **D** W South Africa.

22 DRAKENSBERG ROCKJUMPER *Chaetops aurantius* 21cm From Cape Rockjumper by paler underparts and largely different range. **V** Sequence of very high, frequently repeated variations on *sreesreesreesree-*. **H** Rocky slopes at high altitudes. **D** E South Africa, Lesotho.

RAIL-BABBLER EUPETIDAE

23 RAIL-BABBLER *Eupetes macrocerus* 29cm Forages on the forest floor. Walks with a nodding head, much like a chicken; also dashes over ground and fallen branches in pursuit of invertebrates. **V** Long, drawn-out, monotonous whistle. **H** Broadleaved evergreen forest. **D** Malay Peninsula; Borneo; Sumatra, Indonesia.

WAXWINGS BOMBYCILLIDAE

24 BOHEMIAN WAXWING *Bombycilla garrulus* 20cm Note white tips to secondaries and primary coverts. Forms flocks in winter. **V** Ringing *sirrrr*. Song consists of quiet trilled phrases. **H** Breeds in boreal forest. In winter spreads to a wide variety of woodland, including parks and gardens, in search of berries. **D** Breeds from Alaska to C Canada, and from N Scandinavia through Russia to NE Siberia. Winters south to C USA, C and E Europe, C Asia, Japan.

25 JAPANESE WAXWING *Bombycilla japonica* 16cm Crest held erect more often than Bohemian Waxwing. Gregarious. **V** High-pitched, lisping *hee-hee-hee*. **H** Cedar and larch forest. Wanderers in winter can be found feeding on berry bushes and trees, from mountain woodland to suburban parks. **D** Breeds SE Russia, NE China. Non-breeding E China, Korean Peninsula, Japan.

26 CEDAR WAXWING *Bombycilla cedrorum* 18cm Much plainer wings than Bohemian Waxwing. White undertail coverts. Forms large flocks in winter. **V** High-pitched *sreee*. Song a series of high *screee* notes in an irregular rhythm. **H** Breeds in conifer and mixed woods. In winter spreads to a wider range of woodland and suburban areas in search of berries. **D** Breeds across North America, from S Alaska to Newfoundland as far south as Oregon and Virginia. Winters throughout the USA and south to Mexico.

SILKY-FLYCATCHERS PTILIOGONATIDAE

27 BLACK-AND-YELLOW PHAINOPTILA *Phainoptila melanoxantha* 20cm Male is unmistakable, black with yellow rump, breast and flanks, and grey belly. Female is olive and grey with a black cap. Rather thrush-like in behaviour. **V** No song is known. Calls include a variety of high-pitched *tsit* and *tseep* notes. **H** Montane forest. **D** Costa Rica, W Panama.

28 GREY SILKY-FLYCATCHER *Ptiliogonys cinereus* 18–21cm Often perches conspicuously on tall trees, making sallies to catch flying insects. Usually occurs in pairs or small flocks. **V** Dry *chi-che-rup che-chep* and a sharp *chureet*, *chu-leep* and *ch-tuk*. **H** Pine, oak and juniper forest, forest edge, open areas with scattered trees. **D** NW Mexico to Guatemala.

29 LONG-TAILED SILKY-FLYCATCHER *Ptiliogonys caudatus* 19cm Distinctive blue-grey and yellow plumage, long tail. Unmistakable in its range. **V** Very high *tjirp tjirp tjirp*. **H** Canopy of high-altitude montane forest and adjacent areas. **D** Costa Rica, W Panama.

30 PHAINOPEPLA *Phainopepla nitens* 18–21cm In flight, male shows extensive white on primaries. Makes sallies, often with erratic changes of direction, after flying insects. **V** Soft, rising *wurp*. Song is a short warble. **H** Desert with trees and shrubs. **D** SW USA, N and C Mexico. Northern birds move south in winter.

HYPOCOLIUS HYPOCOLIIDAE

31 GREY HYPOCOLIUS *Hypocolius ampelinus* 23cm Gregarious, usually occurring in small parties or larger groups of up to 20. Forages in bushes and sometimes drops to the ground to pick up insect prey. **V** Various mewing and whistling notes, a low, harsh *chirr* and a continuous *kirrrkirrrkirrr*. **H** Semi-desert with scattered scrub and palm groves. **D** Breeds Iraq, Iran, S Turkmenistan, S Afghanistan. Winters from Arabian Peninsula to S Pakistan, W India.

PALMCHAT DULIDAE

32 PALMCHAT *Dulus dominicus* 20cm Gregarious, active forager in the upper levels of trees; occasionally hawks after flying insects. **V** Variety of short, harsh notes; also a musical whistle that drops in pitch, given when alarmed. **H** Royal palm savannahs and open areas with scattered trees. **D** Hispaniola.

HYLOCITREA HYLOCITREIDAE

33 HYLOCITREA *Hylocitrea bonensis* 14–15cm Inconspicuous. Forages in the mid-storey and understorey, singly, in pairs or as part of mixed-species feeding flocks. **V** Thin, high-pitched buzzy notes; also a loud, piping call. **H** Mountain forest, especially moss forest. **D** Sulawesi, Indonesia.

1
2
3
4
5
6
7
8
9
10
11
12
13
14
15
16
17
18
19
20
21
22
23
24
25
26
27
28
29
30
31
32
33

FAIRY FLYCATCHERS STENOSTIRIDAE

1 YELLOW-BELLIED FANTAIL *Chelidorhynx hypoxanthus* 13–14cm Restless, tail often raised and fanned. Regular member of mixed-species feeding flocks. **V** Thin, high-pitched *sip-sip* followed by a feeble trill. **H** Broadleaved evergreen forests. **D** Himalayas to SW China, Myanmar, NW Thailand, N Laos, N Vietnam.

2 FAIRY FLYCATCHER *Stenostira scita* 11cm Diagnostic white wing-bar, distinct facial pattern. Constantly fans its tail. **V** Very high, thin twitter with some inhaled notes. **H** Woodland, thornveld, plantations, suburbs. **D** South Africa, Lesotho.

3 GREY-HEADED CANARY-FLYCATCHER *Culicicapa ceylonensis* 12–13cm Returns to the same perch after making aerial sallies after flying insects. **V** Soft *pit - pit - pit*, clear *kitwik, kui-whi-whi* and quiet *chichictrr*. Song is a sweet, clear whistle, *tyissi-a-tyi*. **H** Forests and wooded ravines. **D** Himalayas, India, Bangladesh, Sri Lanka, SE Asia, Greater and Lesser Sunda Islands.

4 CITRINE CANARY-FLYCATCHER *Culicicapa helianthea* 11–12cm Favours the mid-storey and lower canopy, and the tops of secondary-growth shrubbery; feeds by fly-catching or gleaning from foliage. **V** Series of 4–5 loud, clear notes of varying pitch or with rising and falling buzzing phrases, such as *tsu-si-tchu-si-si* or *sweet su sweet*, finished with a short trill. **H** Lowland to montane forest, forest edges, secondary growth and edges of cultivation. **D** Philippines; Sulawesi, Indonesia.

5 AFRICAN BLUE FLYCATCHER *Elminia longicauda* 14cm Distinguished from White-tailed Blue Flycatcher by bluer colouring and lack of white in tail. **V** Very high, calm, unstructured twittering. **H** Lower storeys of forest, moist woodland, bushveld, gardens. **D** W and C Africa, from Senegal east to South Sudan and south to NW Angola.

6 WHITE-TAILED BLUE FLYCATCHER *Elminia albicauda* 14cm Note tail pattern. **V** Unhurried, unstructured series of very high *tsip* and *tsee* notes. **H** Montane, open forest, riverine belts. **D** Angola and S DR Congo to S Uganda, Tanzania, Malawi.

7 DUSKY CRESTED FLYCATCHER *Elminia nigromitrata* 9cm Dark grey, with underparts not obviously paler than upperparts. **V** Song is a long series of highly variable phrases, a mixture of whistles and trills, interspersed with chattering and higher-pitched *tweeet* notes. **H** Forest undergrowth and moist bush. **D** W and C Africa, from Liberia to Cameroon, Gabon to W Kenya, NW Tanzania.

8 WHITE-BELLIED CRESTED FLYCATCHER *Elminia albiventris* 9cm Note white belly centre and all-black tail. **V** Song is a soft, slow-paced melodious warble with frequent pauses. **H** Montane forest undergrowth. **D** Nigeria, Cameroon, N DR Congo, Rwanda, W Tanzania.

9 WHITE-TAILED CRESTED FLYCATCHER *Elminia albonotata* 13cm Note tail pattern. **V** Very high, hesitating, wagtail-like *tweet treet treetoweet* with high, fluted *weet tsio-weet*. **H** Montane forests and forest remains. **D** E Africa, from E DR Congo and Kenya to Mozambique and E Zimbabwe.

TITS AND CHICKADEES PARIDAE

10 FIRE-CAPPED TIT *Cephalopyrus flammiceps* 10cm Active and agile forager in tall trees and occasionally in bushes. **V** High-pitched, constantly repeated *tsit*. Song variable, including high-pitched *tink-tink-tink-tink*, slow *pitsu-pitsu*, quicker *pissu-pissu-pissu* and vibrant *psing-psing-psing*. **H** Moist temperate mixed or deciduous forests, orchards, poplar stands. **D** Himalayas to SW China. Winters C and NE India, N Thailand, N Laos.

11 YELLOW-BROWED TIT *Sylviparus modestus* 10cm Active and acrobatic forager in tree foliage, often part of mixed-species feeding flocks. **V** Probable song is a shrill *zee-zi-zee-zi-zee-zi*... Calls include *tsip, tchup, tszizizizizizizizi*..., *pli-pli-pli-pli*. **H** Broadleaved evergreen forest. **D** Himalayas to SC China, SE Asia.

12 SULTAN TIT *Melanochlora sultanea* 20.5cm A large tit, strikingly black and yellow. Acrobatic forager in foliage of trees and bushes. Crest is raised when excited. **V** Song consists of a mellow, whistled *piu-piu-piu-piu-piu*. Calls include a rattling *chi-dip tri-trip* and a fast, squeaky *tria-tria-tria* or *squear-squear-squear*. **H** Evergreen and deciduous forests, especially along paths and roads; also large trees near cultivations. **D** E Himalayas to S China, SE Asia, Malay Peninsula.

13 RUFOUS-NAPED TIT *Periparus rufonuchalis* 13cm Black crest, white cheeks. Agile forager in trees and bushes, and on the ground. **V** Calls include a *cheep*, a deep *chut-chut* and a squeaky *trip-ip-ip*. Two song types, a trilling and a whistled *tsi-tsi-peedduw*. **H** Conifer and mixed forests; also poplars, willows and cultivations in NW Pakistan. **D** W, NW Himalayas, from Kyrgyzstan and E Afghanistan to Nepal.

14 RUFOUS-VENTED TIT *Periparus rubidiventris* 12cm Lacks Coal Tit's wing-bars; more rufous than Rufous-naped Tit. Habits much as others of genus. **V** Rattled *chi-chi-chi-chi-chi-chi*..., also some slurred whistles and trills. **H** High-altitude forests, including oak, birch, pine, rhododendron. **D** Himalayas to Myanmar, SC China.

15 COAL TIT *Periparus ater* 11cm Considerable geographical variation, but always has double white or yellowish wing-bar. Active, restless feeder. Gregarious in winter. **V** Song *teechu-teechu-teechu*... or *chickwee-chickwee-chichwee*... Call *tsuu* or *hseeoo*, high *psit* or *psitisit*, buzzing *tsee-tsee-tsee-tsee*... **H** Coniferous and mixed woodland, parks, gardens. **D** Widespread, from W Europe and NW Africa through Asia to Himalayas, China, E Siberia, Japan.

16 YELLOW-BELLIED TIT *Pardaliparus venustulus* 10cm Forages in middle to low levels in trees and in undergrowth; usually in pairs or small parties. **V** Song is a series of metallic phrases, *swi-swi-swi, suwi-suwi-suwi, sipu-sipu-sipu*, occasionally interspersed with more complex phrases. **H** Open broadleaved deciduous and evergreen forests, bamboo, plantations. **D** E, C China.

17 ELEGANT TIT *Pardaliparus elegans* 11–12cm Usually encountered in pairs or small flocks; an active forager from the middle levels to the canopy. **V** Variable; includes a *tweet chuck-z-chuck-z-chuck-z*, a monotonous *sweet sweet sweet sweet*..., a *chi-bow sweet-zee-sweet-zee-sweet-zee* and a faster *swee-zee-swee-swee-zee-zoo*. **H** Dense primary evergreen forest, edges of secondary forest, scattered trees at edge of cultivation. **D** Philippines.

18 PALAWAN TIT *Pardaliparus amabilis* 12cm Entire head black. Active forager in the mid-storey and canopy. **V** Very vocal; calls include harsh *chuwi-chuwi-chuwi*..., *wichi-wichi-wichi-wichi*, simple accelerating rattles, musical *tui-tui-tui-tui*... and silvery, descending trill. **H** Forest, forest edges, wooded fringes of swamps and secondary growth. **D** Palawan Archipelago, W Philippines.

19 EUROPEAN CRESTED TIT *Lophophanes cristatus* 11.5cm Broad white fringes on crest feathers. Restless, flicks wings and tail. Usually found singly or in pairs, forms small parties in the winter. **V** Song is rapid *seeh-i-i-burrurrlt-seeh-i-i-burrurlt*... Call: low *brrrrrr-t'brrrrrr*, often preceded by thin *zizizi*... **H** Conifer forests in north; in south frequents a wider variety of woodland, from mixed forest to Cork Oak. **D** Europe, from Iberian Peninsula to S Scandinavia, W Russia.

20 GREY CRESTED TIT *Lophophanes dichrous* 12cm Plainer grey than European Crested Tit. Shy. In pairs or small parties. **V** Song *whee-whee-tz-tz-tz*. Calls include high *zai* and rapid *ti-ti-ti-ti-ti*. **H** Various forests, including oak, fir, birch. **D** Himalayas to Myanmar and WC China.

21 BRIDLED TITMOUSE *Baeolophus wollweberi* 13cm Unmistakable, with crest and striking face pattern. Forms small parties after breeding; also occurs in mixed-species flocks. **V** Rapid, harsh chatter *ji ji ji ji ji* or *jedededededed*. Song is a rapid, whistled *pidi pidi pidi pidi pidi* or *pipipipi*.... **H** Primarily mountain evergreen oak, pine and juniper woods. **D** SW USA, Mexico.

22 OAK TITMOUSE *Baeolophus inornatus* 15cm More often heard than seen. Generally forages in the upper storey. Formerly lumped with Juniper Titmouse under the name 'Plain Titmouse'. **V** Call *si si si chrr* or *pi pi pi peeew*. Song variable, including *chuwi-chuwi-chuwi-chuw* and *chit it-chit it-chit it*... **H** Dry open woods, especially of oak. **D** W USA and Mexico, from SW Oregon to Baja California.

23 JUNIPER TITMOUSE *Baeolophus ridgwayi* 15cm Action and habits similar to Oak Titmouse. **V** Rapid *sisisi-ch-ch-ch-ch, si-ch-ch-ch* or similar. Song lower and faster than Oak Titmouse, *jijiji jijiji jijiji*... **H** Juniper–pinyon woodland, scrub oak and Ponderosa Pine, alder and willow in winter. **D** WC USA to N Mexico.

24 TUFTED TITMOUSE *Baeolophus bicolor* 17cm Note rusty flanks and black forehead patch. Generally occurs singly or in pairs, foraging at all levels, including on the ground. **V** Calls include *chick-a-dee*, harsh *tsee-day-day-day* and *sit-sit-sit*. Song is a fast or slow *peter-peter-peter-peter* or variants. **H** Deciduous and mixed woodland, wooded farmland, parks, suburban gardens. **D** E and SE USA.

25 BLACK-CRESTED TITMOUSE *Baeolophus atricristatus* 17cm Often considered a race of Tufted Titmouse. Actions and habits similar; intergrades with dark grey crown occur where ranges overlap. **V** Similar to Tufted Titmouse, but generally louder and more nasal. **H** Dry forest and scrub. **D** SC USA (S Oklahoma, Texas), NE Mexico.

26 VARIED TIT *Sittiparus varius* 12cm Secretive. Feeds in canopy. The next three species were formerly considered races of Varied Tit. **V** Song variable, *peee-spit'tit - peee-spit'tit*..., *tsre-tsre-peee-triri-peee-triri-peee-triri*; always with the *peee* whistle. **H** Coniferous, deciduous and mixed forest, often with thick undergrowth. **D** NE China, Korean Peninsula, Japan.

27 OWSTON'S TIT *Sittiparus owstoni* 15cm Formerly considered conspecific with Varied Tit. Similar to Varied Tit, but forehead, cheeks and breast all rusty-orange. Note stout bill. **V** Song and calls similar to Varied Tit, but slightly lower-pitched. **H** Evergreen broadleaf forest, woodland, gardens. **D** S Izu Islands, off EC Japan.

28 IRIOMOTE TIT *Sittiparus olivaceus* 13cm Formerly considered conspecific with Varied Tit. Slightly smaller and markedly paler than Varied Tit, with rusty plumage replaced with tawny-buff. **V** Song and calls similar to Varied Tit. **H** Evergreen broadleaf forest. **D** Yaeyama Islands, S Ryukyu Islands, SW of Japan.

29 CHESTNUT-BELLIED TIT *Sittiparus castaneoventris* 12–14cm Forages mostly in the upper storeys, collecting food from foliage and trunks; occasionally catches insects in flight. **V** Calls include weak *tsuu tsuu tsuu*, high, drawn-out *spit-see-see-see* and scolding *chi-chi-chi*; song variable, monotone whistles combined with a ringing *peee*. **H** Open deciduous and mixed forest. **D** Taiwan.

30 WHITE-FRONTED TIT *Sittiparus semilarvatus* 13–14cm Mostly blackish, with white patch on forehead and lores. Forages in the middle storey and the canopy of tall trees. **V** Calls include sharp, high *psit*, plaintive, rolling *tsuit*, high-pitched *tsi-tsi-tsi-tsi-tsi* and *tseeeh tsi-tsi-tsi-tsi-tsi-tsi-ts ts ts ts*. **H** Forest, forest edges, remnant forest patches, secondary growth, scrub. **D** Philippines.

31 WHITE-BROWED TIT *Poecile superciliosus* 13–15cm Feeds low down, typical active tit habits. **V** Variable, complex song, *tsee-tsee-tsee-pwi-pee, tchip-tchip-pwi-pee, tsir'r'r'r'r'r-pee-pee-pee, peta-peta-peta-peta-peta*. **H** Dwarf alpine shrubs, scrub and bushes, usually near watercourses; post breeding occurs in taller vegetation and open areas of spruce forests. **D** WC China.

32 SOMBRE TIT *Poecile lugubris* 14cm Usually in pairs. Less acrobatic than many of genus. **V** Piping, buzzy *be-zoo-be-zoo-be-zoo*... and *doodle-lu-doodle-lu-doodle-lu*... Call: harsh, churring *zi-zi-zi-z-chrrr*. **H** Open mixed or deciduous woodland or tall scrub, mostly in hilly, rocky areas. **D** SE Europe and Middle East, from Balkans to N Israel and Iran.

33 PERE DAVID'S TIT *Poecile davidi* 12–13cm Note cinnamon underparts. Acrobatic feeder, mainly in the canopy of tall trees. Often occurs in small flocks post breeding. **V** Calls include *tsip, psit* and a drawn-out *chi-it-it*; song *tsip zee zee*. **H** Mature mixed forest. **D** NC China.

34 MARSH TIT *Poecile palustris* 11.5cm Agile feeder. Considerable overlap with Willow Tit, both in range and in appearance. Generally appears smaller-headed than that species, with a smaller, neater black bib. **V** Liquid, monotonous *tchip-tchip-tchip*..., *tew-tew-tew*... or a faster *chipchipchipchip*... Calls include explosive *pichay* and scolding *chicka-dee-dee-dee-dee*. **H** Mature deciduous woodland, mixed woodland, riverine woodland. **D** From W Europe to S Scandinavia, W Russia, Caucasus; from C Siberia to E Asia.

35 CASPIAN TIT *Poecile hyrcanus* 13cm Formerly considered conspecific with Sombre Tit but differs in morphology, genetics and voice. Slightly smaller than Sombre Tit, with dark brown cap and whitish underparts. Excavates own nest hole, like Willow Tit. **V** Song is reported to be similar to Willow Tit, a simple series of *tiu tiu tiu* notes. Calls are unlike those of Sombre Tit. **H** Montane broadleaf woodland. **D** S Azerbaijan, N Iran.

36 BLACK-BIBBED TIT *Poecile hypermelaenus* 11–12cm Forages from low down in scrub to the mid-storey and canopy; occurs singly or in small flocks. **V** Calls include a high thin *stip*, a thin *si-si* and an explosive *psiup*, also a chattering *chrrrrrr* and a scolding *chay*. **H** Open broadleaved evergreen and pine forest, forest edge and scrub. **D** NC and SW China, NW Myanmar.

37 WILLOW TIT *Poecile montanus* 11cm Often hard to distinguish from Marsh Tit. Best separated by voice. **V** Song variable, *tsiu-tsiu-tsiu-tsiu*..., *du-duu-duu-duu*... etc. Call is a harsh, nasal *tchar-tchar-tchar* or *tsi-tsi-chay-chay*, also a buzzing *zi-zi-zeerr-zeerr*. **H** Thickets, deciduous and mixed woodland, preferably willow, alder or birch, with undergrowth. Frequents coniferous forests in northern and montane regions. **D** Europe, north to N Scandinavia; across Siberia to NE China, Japan, Kamchatka.

1
2
3
4
5
6
7
8
9
10
11
12
13
14
15
16
17
18
19
20
21
22
23
24
25
26
27
28
29
30
31
32
33
34
35
36
37

TITS AND CHICKADEES *CONTINUED*

1 SICHUAN TIT *Poecile weigoldicus* 13cm Often occurs in small parties, regularly feeds on catkins in spring. **V** Calls include nasal, scolding *dzee*, thin *psit* or *psit-dzee* and *chick-a-dee*. **H** Coniferous and mixed forests. **D** WC and SW China.

2 CAROLINA CHICKADEE *Poecile carolinensis* 11–12cm Very similar to Black-capped Chickadee, but is smaller-headed, shorter-tailed, has less distinct pale wing edges and is usually duller. **V** Call is a high-pitched, rapid *chikadeedeedeedee*. Song is a *fee-bee fee-bay, sufee-subee* or similar. **H** Broadleaf woodlands, especially along watercourses, swamp forest, parks and wooded gardens. **D** SE USA.

3 BLACK-CAPPED CHICKADEE *Poecile atricapillus* 13–15cm Large-headed with distinct whitish wing edges. Tame and confiding; regular member of winter mixed-species flocks. **V** Sharp *chik* leading to a slow *chick-a-dee-dee-dee*. Song *fee-bee* or *fee-bee-ee*. **H** Open woodlands, clearings, parks and gardens. **D** Widespread across North America, from boreal Alaska and Canada south over north half of USA.

4 MOUNTAIN CHICKADEE *Poecile gambeli* 13cm When worn, white eyebrow can become indistinct. **V** Harsh *chick-adee-adee-adee*. Typical song is a *fee-bee-bay* or *fee-bee fee-bee*. **H** Montane conifer forest; moves to lower level forest or scrub in winter. **D** Mountainous areas of W North America.

5 MEXICAN CHICKADEE *Poecile sclateri* 13cm Large black bib and dark grey flanks. **V** Buzzy *sschleeeer*, also buzzy trills followed by a hissing *tzee tzee tzee shhhh shhhh*. In song gives a *peeta peeta peeta*. **H** Open montane conifer forests and deciduous woodland. **D** SW USA, Mexico.

6 GREY-HEADED CHICKADEE *Poecile cinctus* 14cm Known in Europe as Siberian Tit. Very similar to Boreal Chickadee, but longer-tailed and rear of cheeks white. **V** Calls include a thin *chit-sit, si-si, tsit* and a sharp *chik*, like tapping stones together. **H** Mature willows along watercourses, spruce forest and in winter in aspen and alder. **D** The only tit to occur in both Old and New Worlds, from Norway east across Eurasia into Alaska and Arctic Canada.

7 BOREAL CHICKADEE *Poecile hudsonicus* 14cm Rufous flanks, brown cap, grey rear cheeks. Occurs in small flocks after breeding, often part of mixed-species flocks. **V** Nasal *tseek-a-day-day, tsi-jaaaay* or *tsi ti jaaaay jaaay*. Song is a simple trill, *p-twee-titititititititititi*. **H** Primarily dense boreal conifer forest. **D** Boreal Alaska, Canada, extreme NE USA.

8 CHESTNUT-BACKED CHICKADEE *Poecile rufescens* 12cm Distinctive, with dark brown to blackish cap and rufous-chestnut back and rump. **V** Rapid, hoarse *tseek-a-dee-dee*, a nasal *tsidi-tsidi-tsidi-cheer-cheer* and a weak *tsity ti jee jee*. **H** Coniferous and mixed woods. **D** Extreme W North America, from Alaska to California.

9 AFRICAN BLUE TIT *Cyanistes teneriffae* 12cm Formerly considered conspecific with Eurasian Blue Tit, but differs in head pattern, with darker cap (often blackish) and broader eye-stripe. Some races lack a wing-bar. Strong racial variation on Canary Islands (five races). **V** Song of simple repeated phrases recalls Great Tit or Coal Tit (Plate 196). Calls often similar to Eurasian Blue Tit. **H** Woodlands, orchards and gardens. **D** North Africa and Canary Islands.

10 EURASIAN BLUE TIT *Cyanistes caeruleus* 12cm Note blue cap, white face with dark line through eye. Very acrobatic. Often tame. **V** Song variable, e.g. *tee-tee - see-see - tee-tee - see-se - psi-dada - psi-dada* or *pee-pee-ti - sihihihihiihi - pee-pee-ti - sihihihihi*. Many calls, e.g. scolding *churrrrit* and thin *sissississississ*. **H** Deciduous woodlands, evergreen forests, orchards, hedgerows, parks, gardens. **D** Widespread across Europe as far east as W Russia and Caucasus.

11 AZURE TIT *Cyanistes cyanus* 13.5cm Active, noisy; feeds in cover of low trees, bushes and reeds. Hybridises with Eurasian Blue Tit. **V** Variable, includes a descending trill, *tii-tsi-dji-daa-daa-daa, tee-tee-tee-chup-chup* and a repeated *tea-cher*. Various calls: slurred *tstirrup*, nasal *tsee-tsee-dze-dze*, scolding *chr-r-r-r-rit*. **H** Willow, birch and poplar thickets, preferably near water, also light deciduous and mixed woodland and reedbeds in winter. **D** From Belarus through S Siberia, Mongolia to NE China; Kazakhstan to NE Afghanistan.

12 GROUND TIT *Pseudopodoces humilis* 19cm Looks and acts much like a wheatear: bounces along and when at rest flicks wings and tail. Usually shy and wary but can become very confiding, especially around human settlements. **V** Short *chip* followed by a rapid, whistled *cheep-cheep-cheep-cheep*. **H** Open grassy steppe and plains with scattered boulders and bushes, above the treeline. **D** Himalayas.

13 GREAT TIT *Parus major* 14cm Active and agile, but less acrobatic than the smaller tits. Usually encountered in pairs, small parties or mixed-species flocks. **V** Calls include a *tsee tsee tsee* and a harsh churring when alarmed. Song very variable, including *weeter-weeter-weeter*, repeated *wheet-chee-chee* and clear *zwink zwink*. **H** Forests and well-wooded country, orchards, groves, urban parks, gardens. **D** Europe, N Africa, Middle East, east to Mongolia.

14 JAPANESE TIT *Parus minor* 13–14cm Bold and less agile than many of the smaller tits. Resembles a faded Great Tit. **V** Song is a shrill *shi-ju shi-ju shi-ju*; calls similar to Cinereous Tit. **H** Open woodland, stands of pines in hill evergreen forest (Thailand), light forest, wooded cultivations and gardens. **D** From SE Russia to China, Korean Peninsula, Japan, Myanmar, N Thailand, N Laos, N Vietnam.

15 CINEREOUS TIT *Parus cinereus* 13–14cm Bold, less agile than many of the smaller tits. This and Japanese Tit were formerly treated as races of Great Tit. **V** Song is a rapidly repeated *chew-a-ti chew-a-ti chew-a-ti* or similar. Calls include a harsh *tcha-tcha-tcha* and a ringing *pink-pink-pink*. **H** Deciduous forest mixed with pines, dry teak forests, bamboo and mangroves. **D** Himalayas to India and Sri Lanka, through SE Asia to Indonesia, Hainan.

16 GREEN-BACKED TIT *Parus monticolus* 12–13cm Forages from the ground up to the canopy, post breeding may occur in small flocks. **V** Variable song includes a *seta-seta-seta*, a *tu-weeh-tu-weeh* and a *whit-ee whit-ee*. Calls include a rapid *si-si-si-si*, a harsh *shick-shick-shick* a clear *te-te-whee* and a musical *pling-pling-pling-tee-eurr*. **H** Evergreen, deciduous and mixed forests. **D** Himalayas to C China, Myanmar, Laos, Vietnam, Taiwan.

17 WHITE-NAPED TIT *Machlolophus nuchalis* 12cm Strongly patterned black-and-white plumage. Usually found in pairs or small family parties, foraging in the canopy and shrub layer. **V** Song is a thin, high-pitched *tiu-sut-sut-sut*. Calls include repeated *ti pee-pee-pee* or *teep whee whee whee whee* and bold *whew whew whew whew whew*. **H** Thorn-scrub forest. **D** NW and S India.

18 YELLOW TIT *Machlolophus holsti* 13cm Usually encountered in the canopy or upper storey, singly or in pairs. May also join mixed-species foraging flocks. **V** Song consists of a variety of trisyllabic ringing notes, *tu-wich-ch* or similar; calls include a thin *si-si-si*, a sibilant *tzee-tzee-tzee* and a scolding *dz-za-za-za-za*. **H** Broadleaved and occasionally primary mixed forest and open secondary growth. **D** Taiwan.

19 HIMALAYAN BLACK-LORED TIT *Machlolophus xanthogenys* 13cm Arboreal; forages in the canopy, occasionally makes fly-catching sallies after flying insects. When agitated raises crest and flicks wings and tail. **V** Song consists of repeated phases, such as *pui-pui-tee pui-pui-tee..., tsi-teuw tsi-teuw...* or *tsi-eheeah-wheeah...* Calls include a *si-si, tzee-tzee-wheep-wheep-wheep, tsi-tsi-pit-tui* and a rapid *ch-chi-chi-chi-chi*; when alarmed, *tst-reet*. **H** Mainly light open subtropical forest, forest edge. **D** Himalayas, from N Pakistan to E Nepal.

20 INDIAN BLACK-LORED TIT *Machlolophus aplonotus* 13cm Formerly considered conspecific with Himalayan Black-lored Tit, differing in having white (not yellowish) wing-bars. Often joins mixed-species flocks. **V** Vocalisations distinct, comprising short phrases of varied notes unlike the repetitive phrases of Himalayan Black-lored Tit. **H** Well-wooded areas including evergreen forest and plantations. **D** Peninsular India.

21 YELLOW-CHEEKED TIT *Machlolophus spilonotus* 14cm Actions similar to Great Tit. **V** Song is a ringing, rapidly repeated *chee-chee-piu chee-chee-piu chee-chee-piu* or *dzi-dzi-pu dzi-dzi-pu...* **H** Mainly light oak or pine forests, also rhododendrons and secondary growth with scattered trees. **D** Himalayas to SE China, SC Vietnam.

22 WHITE-SHOULDERED BLACK TIT *Melaniparus guineensis* 15cm Note white eye. Unmistakable. **V** Song is a series of whistles, *tip-tu-wip, tu-wip, tu-wip, yu-wip*. Calls include a variety of harsh *churrr* and *tchip* notes. **H** Open woodland, thornbush. **D** S Mauritania to S Sudan and Eritrea, south to Cameroon and N DR Congo.

23 WHITE-WINGED BLACK TIT *Melaniparus leucomelas* 15cm Dark eye. From Carp's Tit by black undertail coverts and only 1–3 visible black centres in greater wing coverts. **V** Song is a whistled *wheee-to-trrrr-tooee*. Calls include a sharp *chut* and a buzzing *zzzzz-rr*. **H** Open woodland. **D** SC Africa, from Gabon, Congo and Angola to Uganda, Tanzania, Malawi, Zambia.

24 SOUTHERN BLACK TIT *Melaniparus niger* 15cm From Carp's Tit and White-Winged Black Tit by more than six visible black centres in greater coverts, white-tipped undertail coverts and broadly white-edged tail feathers. **V** Calls include *jee-jee-jee, tit-tit-chacha* and a grating *jrrr-jrrr-jrrr*, as well as a softer *peeoo-pu-peeoo*, etc. Song consists of combinations of call notes. **H** More or less wooded areas including plantations and suburbs. **D** SE Africa, from Zambia to E South Africa.

25 CARP'S TIT *Melaniparus carpi* 15cm From White-winged Black Tit by grey undertail coverts, white tip to all tail feathers, 5–6 visible black centres in greater coverts. **V** Similar to Southern Black Tit: *tsi-tsi-jer-jer, churia-churia, piu-piu-piu*, etc. **H** Thornveld, scrubland. **D** SW Angola to C Namibia.

26 WHITE-BELLIED TIT *Melaniparus albiventris* 14cm White belly makes this black tit unmistakable. **V** Calls include a nasal *tsi-chah-chah-chah* and a sweeter *pee-pee-purr-purr*. Song is made up of combinations of call notes. **H** Montane forest edges and more or less wooded and bushed grassland, including gardens. **D** SE Nigeria, Cameroon; S South Sudan, E Uganda, Kenya, Tanzania.

27 WHITE-BACKED BLACK TIT *Melaniparus leuconotus* 13cm Mainly black, with a distinctive off-white patch on the mantle, often extending to the nape. **V** Wide range of calls, including a typical *bzz-bzz-bzzz*. **H** Wooded valleys in mountainous regions. **D** Eritrea, Ethiopia.

28 DUSKY TIT *Melaniparus funereus* 14cm Note conspicuous red eye. **V** Loud, lowered, sometimes hoarse *vi-uuuh* or questioning *vreeeeh?* **H** Forest, riverine belts, plantations. **D** C Africa, from Guinea to Uganda, Kenya, DR Congo, W Angola.

29 RUFOUS-BELLIED TIT *Melaniparus rufiventris* 14cm Note pale eye, black head, rufous underparts. **V** Calls include a thin *tsit, tsit* and a longer *tsitsit-chaa-chaa-chaa*; also a sweeter *peh-teepio, peh-teepio*. Typical song is a combination of call notes. **H** Miombo, other woodland. **D** SC Africa: Congo, DR Congo, Angola, Namibia, Zambia, Malawi, Botswana.

30 CINNAMON-BREASTED TIT *Melaniparus pallidiventris* 14cm Note dark eye. Previously considered conspecific with Rufous-bellied Tit. **V** Said to be indistinguishable from Rufous-bellied Tit. **H** Miombo woodland. **D** SE Africa: Tanzania, Malawi, Zimbabwe, Mozambique.

31 RED-THROATED TIT *Melaniparus fringillinus* 12cm Grey with rufous face, dark cap, white in wings. **V** Variety of *plikk, chitt, chik-chik-chik, prtt-tchay-tchay-tchay* and *brzz-zzz-zzz* notes. **H** More or less wooded and bushed grassland. **D** S Kenya, N Tanzania.

32 STRIPE-BREASTED TIT *Melaniparus fasciiventer* 12cm Note black head and breast, black stripe down belly. No other similar tit within its range and habitat. **V** Song resembles Great Tit. Calls include a variety of harsh notes, as well as a softer *whit* or *tweep* contact call. **H** Montane forests, giant heath. **D** Extreme E DR Congo, W Uganda, Rwanda, Burundi.

33 ACACIA TIT *Melaniparus thruppi* 12cm Note whitish cheeks and narrow mid-breast stripe. **V** Song is a softly whistled *tu-tu-wee*, or a harsher *tsi-tsi-tchuerr-tchuerr*. **H** Dry, more or less wooded and bushed habitats, especially along wooded streams. **D** Ethiopia, Somalia, Kenya, E Uganda, NE Tanzania.

34 MIOMBO TIT *Melaniparus griseiventris* 15cm From Grey Tit by blue-grey mantle, small white neck spot, more white in wing. From Ashy Tit and Grey Tit by greyish cheeks, concolorous with sides of breast. **V** Calls include a soft *whup* and an insistent *chrrrr-chrrrr-chrrrr*. Song contains phrases such as *tsi-trreeeoo-tweetoo* and *twee-too-you, twee-too-you*. **H** Miombo woodland. **D** Angola to W Tanzania, Malawi, Zimbabwe.

35 ASHY TIT *Melaniparus cinerascens* 15cm Flanks concolorous with mantle. Isolated narrow white cheek. **V** Song consists of a series of mellow trills, *tlu-tlu-tlu-tlu-tlu, chi-chi-chi-chi-chi, tri-tri-tri*. **H** Thornveld, riverine belts, Kalahari scrub. **D** SW Angola, Namibia, Botswana, Zimbabwe, N and C South Africa.

36 GREY TIT *Melaniparus afer* 14cm From Miombo Tit by white tips to tail feathers, from Miombo Tit and Ashy Tit by pale, tawny underparts. **V** Calls include a sibilant *tsi-cha-cha-cha*; alarm call harsher, *tzi-chre-chre-chre*. Song is a series of repeated loud phrases, *we-toolee-woo, klee-klee-klee, piet-tiou-tiou*, etc. **H** Rocky areas with scrub and small thorn trees. **D** SW Namibia, South Africa, Lesotho.

1
2
3
4
5
6
7
8
9
10
11
12
13
14
15
16
17
18
19
20
21
22
23
24
25
26
27
28
29
30
31
32
33
34
35
36

PENDULINE TITS REMIZIDAE

1 EURASIAN PENDULINE TIT *Remiz pendulinus* 11.5cm Agile feeder. Unobtrusive, often first detected by call, but can be quite confiding. **V** Thin *tsssseeoo* with variations, e.g. *ssss-lu-lu-lu* and a buzzing *tizzz*. **H** Reedbeds, rank vegetation with patches of bushes and small trees. **D** S and E Europe to Caucasus, Kazakhstan. Northern and eastern populations move south in winter.

2 BLACK-HEADED PENDULINE TIT *Remiz macronyx* 11cm Formerly considered conspecific with Eurasian Penduline Tit, and some hybridisation in zones of contact; differs in having all-blackish head and throat. **V** Similar to Eurasian Penduline Tit. **H** Reedbeds and bulrushes, along lakes and rivers. **D** C Asia (presumed resident).

3 WHITE-CROWNED PENDULINE TIT *Remiz coronatus* 11cm Male distinguished from Eurasian and Chinese Penduline tits by larger black mask, paler crown, more contrasting plumage pattern. Agile forager in trees and reeds. **V** Thin high-pitched *tsee, tseeuh* or *tee-tsee-tsee*. **H** Reedbeds, riverside acacias and irrigated plantations. **D** Breeds C Asia to Mongolia. Winters Iran to NW India.

4 CHINESE PENDULINE TIT *Remiz consobrinus* 10–11cm Similar to Eurasian Penduline Tit. Agile, active forager on reed heads and scrub foliage; sometimes in flocks. **V** Repeated, see-sawing *si si tiu si si tiu*; calls include a thin *tseee* or *pseee*, a *piu* and a series of *siu* notes. **H** Reedbeds, scrub and rank vegetation. **D** Breeds N China. Winters S China, Korean Peninsula, S Japan.

5 SENNAR PENDULINE TIT *Anthoscopus punctifrons* 7cm Distinguished from Yellow Penduline Tit by whitish belly. **V** Extra-high, thin, piercing *fie fie fie fie*, alternating with a low, dry trill. **H** Open bush. **D** S Mauritania to Eritrea.

6 YELLOW PENDULINE TIT *Anthoscopus parvulus* 7cm Smaller than any other yellow bird and with different bill form. Note black frontal spots. **V** Contact call is a thin *tsi-tsi-tsi-tsi*. Song is a tuneless, rapid, urgently repeated *ska-ska-ska-ska*. **H** Dry acacia steppe. **D** S Mauritania to South Sudan, NE DR Congo.

7 MOUSE-COLORED PENDULINE TIT *Anthoscopus musculus* 7cm Note pale eyebrow and slightly greyish upperparts. **V** Typical call is a high-pitched *tit, tit, tit*; also a nasal *tzeee*. Song includes a rattling *di-di-di-di-di-di...* and a ringing *we-cha-we-cha-we-cha*. **H** Dry, rocky, more or less wooded areas. **D** E Africa, from Ethiopia and South Sudan to N Tanzania.

8 FOREST PENDULINE TIT *Anthoscopus flavifrons* 7cm Mostly plain olive-green with paler undersides. Yellow-orange spot on forehead often hard to see. **V** Very high-pitched calls may be mostly beyond the range of human hearing. **H** Canopy and mid-levels of forest. **D** WC Africa, from Liberia to N DR Congo.

9 GREY PENDULINE TIT *Anthoscopus caroli* 8cm Separated from Cape Penduline Tit by more extensive white from chin to lower breast. **V** Extremely high *wit wit wit witterdewitterdewitterdewit*. **H** Open forest, woodland. **D** From Kenya to NE South Africa, and from Angola to Mozambique.

10 CAPE PENDULINE TIT *Anthoscopus minutus* 9cm Separated from Grey Penduline Tit by spotted forehead and grey crown. **V** Very high *puweepuwee* and extremely high *see-swee-swee*. **H** Thornveld. **D** W Angola, Namibia, Botswana, Zimbabwe, South Africa, Lesotho.

11 VERDIN *Auriparus flaviceps* 10–11cm Extent of orange-yellow on face varies. Active, usually forages alone but also in pairs or family parties. **V** Sharp *tschep, tschik* or *chip*, the latter often repeated. Song is a whistled *chee-chee-chee...* or similar. **H** Dry scrubland. **D** SW to SC USA, N Mexico.

NICATORS NICATORIDAE

12 WESTERN NICATOR *Nicator chloris* 21cm Note large bill. Solitary and secretive. **V** High, hurried *weet-weet-weet-weet-weet kisch kisch* (last part loud and explosive). **H** Thickets, understorey of woodland, riverine belts. **D** W and C Africa, from Senegal to DR Congo and Uganda.

13 EASTERN NICATOR *Nicator gularis* 21cm Distinguished from Western Nicator by dark brown cap and by range. **V** Mid-high, partly harsh, partly melodious *tjurk tjurk kwik-kor kwikker-kweet*. **H** Dense forest undergrowth, scrub. **D** E and SE Africa, from Somalia to NE South Africa.

14 YELLOW-THROATED NICATOR *Nicator vireo* 14cm Distinguished from Western Nicator by smaller size, thinner bill, yellow throat patch. **V** Song, typically given from dense cover, is loud and resonant: *po-pweee-toh-toh-toh-twee-tut-tut* or similar. **H** Canopy and middle strata of forest. **D** From Cameroon to NE DR Congo, N Angola.

BEARDED REEDLING PANURIDAE

15 BEARDED REEDLING *Panurus biarmicus* 14–17cm Unmistakable. Best located by pinging call; often feeds at the base of reeds. **V** Ringing *ping* or *pching*, buzzing *tjipp*, soft *pitt*. **H** Reedbeds. **D** W, C, S Europe through C Asia to N China.

LARKS ALAUDIDAE

16 GREATER HOOPOE-LARK *Alaemon alaudipes* 18–20cm Forages on the ground, runs about like a courser. In flight black and white wing-pattern is striking. **V** Characteristic desert sound; given during an acrobatic display flight; melodious piping notes that start slowly, accelerate during climb, a short trill at peak of ascent, reverting to piping notes, lowering in tone and speed during descent. Call is a buzzing *zee*. **H** Desert and semi-desert with low dunes or clay flats. **D** From Morocco and Mauritania through N Africa, Arabian Peninsula to Iran, Pakistan.

17 LESSER HOOPOE-LARK *Alaemon hamertoni* 20cm Note dark primaries. Breast striping more or less forms a breast-band. **V** Undescribed. **H** Open plains with rough grass. **D** Somalia.

18 BEESLEY'S LARK *Chersomanes beesleyi* 12cm White tail-corners, long, decurved bill, white throat contrasting with light rufous breast. **V** High, un-lark-like, more wader-like, rapid, bouncing *tseeptseeptseeptseep*. **H** Dry areas with more or less short grass cover. **D** NE Tanzania.

19 SPIKE-HEELED LARK *Chersomanes albofasciata* 15cm Note short tail with white corners, thin bill, erect stance. **V** Sings in short phrases just after take-off, plover-like, sharp, very fast *wut-wut-wut*. **H** Grassland, shrubland, gravel plains with sparse grass cover. **D** S and SW Africa, from Angola and Botswana southwards.

20 GRAY'S LARK *Ammomanopsis grayi* 14cm Note very pale colouring, sometimes appearing white in the field. **V** Sings only before dawn, sharp, extremely high *witweet* or *ee-ee-weee-ee-*. **H** Gravel-covered deserts with rocky outcrops. **D** W Namibia, SW Angola.

21 SHORT-CLAWED LARK *Certhilauda chuana* 19cm Note rust-red rump and 'capped' appearance (due to long creamy supercilium). **V** Extremely high to very high, drawn-out, whistled *feeeeuh-weetweetweeuuuuh*. **H** Dry thornveld and scrubland. **D** SE Botswana, N South Africa.

22 KAROO LONG-BILLED LARK *Certhilauda subcoronata* 20cm Note colouring of upperparts, with mantle more rufous than neck and crown. **V** Similar to Eastern Long-billed Lark. **H** Rocky scrubland and grassland. **D** SW Africa, from Namibia to W South Africa.

23 BENGUELA LONG-BILLED LARK *Certhilauda benguelensis* 19cm Best separated from other long-billed larks by range. **V** High, resounding, drawn-out, descending *wip-weeeer*. **H** Arid stony slopes and plains. **D** SW Angola, NW Namibia.

24 EASTERN LONG-BILLED LARK *Certhilauda semitorquata* 18cm Smaller than the other long-billed larks, and the least streaked. **V** High, drawn-out *pfeeer*. **H** Grassland, rocky slopes. **D** E South Africa, Lesotho.

25 CAPE LONG-BILLED LARK *Certhilauda curvirostris* 20–24cm Distinguished from Short-clawed Lark by brown (not contrasting rust-red) rump. **V** When displaying rises steeply to about 20m, calls high, sharp, resounding *pjiier pjiier*, and descends with folded wings. **H** Grass- and scrubland with stony outcrops and ridges. **D** SW Namibia, W South Africa.

26 AGULHAS LONG-BILLED LARK *Certhilauda brevirostris* 18–20cm Distinguished from Cape Long-billed Lark mainly by smaller size and by habitat. **V** Triple-syllabled *pfit-tih-weeer*. **H** Waste and croplands, dry scrub. **D** SW South Africa (Western Cape).

27 DUSKY LARK *Pinarocorys nigricans* 18cm Note dark appearance with striking face pattern. Flicks wings frequently when foraging on the ground. **V** In flight, high rasping *sreeeeh-sreeeh-sreeeh*. **H** Open grassy thornbush, woodland. **D** Breeds Angola, S DR Congo, Zambia, W Tanzania. Non-breeding Namibia to Mozambique.

28 RUFOUS-RUMPED LARK *Pinarocorys erythropygia* 20cm Note distinctive facial pattern and rufous tail, rump and flanks. **V** High, sharp, fluted, descending *tuuuu-tuuuu* and *tee-tiuu-tee-tui-tee*. **H** Wooded grassland, burnt ground, bare fields. **D** Mali, Guinea, Sierra Leone to E Sudan, South Sudan, NW Uganda.

29 THICK-BILLED LARK *Ramphocoris clotbey* 17cm When feeling threatened will often run, rather than fly. Forms large winter flocks. **V** Song is a quiet medley of tinkling and warbling notes, given from the ground or in flight. **H** Breeds in flat stony desert with scattered low vegetation, sometimes in more vegetated areas. In winter often spreads to more cultivated areas and arid coastal plains. **D** Mauritania and Morocco to Libya; Arabian Peninsula.

30 DESERT LARK *Ammomanes deserti* 17cm Forages on the ground, generally in pairs or small parties; after breeding often in larger flocks. Song given from the ground or in a hesitating, undulating song flight. **V** Trilled *trreooee*. Calls include a *chu* and a *chee-lu* flight note. **H** Desolate, barren country, stony hill slopes, and fallow fields in desert-canal cultivations. **D** From Morocco through N Africa to Somalia, Arabian Peninsula, Iran, Afghanistan, Pakistan, NW India.

31 BAR-TAILED LARK *Ammomanes cinctura* 14cm Forages on the ground; post breeding frequently encountered in small flocks. Song often given during undulating display flight. **V** Repeated, thin fluty notes interspersed with a louder *see-oo-lee*. **H** Barren gravelly plains and low stony hills. **D** Cape Verde; Morocco through N Africa to Arabian Peninsula, Iran, Afghanistan, S Pakistan

32 RUFOUS-TAILED LARK *Ammomanes phoenicura* 16cm Forages on the ground; in song flight, at about 30m, bird circles on deeply flapping wings before making steep, stepped dives to the ground. **V** Random *juu juuh tcherre tcherre tcherre...juuh...tchwerrwe...* sometimes mixed with short warbling notes. **H** Dry open areas with sparse vegetation. **D** NE Pakistan, widespread over India.

33 BLACK-EARED SPARROW-LARK *Eremopterix australis* 13cm Male is largely black with chestnut fringes to feathers of wings and mantle. Female can be distinguished from other female sparrow-larks by absence of black line and patch on belly. **V** In flight, high *trutru tree tree trutru* (*trutru* mellow, *tree* higher and sharper). **H** Sand and stone plains with some shrub. **D** Namibia, S Botswana, South Africa.

34 MADAGASCAN LARK *Eremopterix hova* 13cm No similar lark or pipit in its range. **V** Very high sharp *zreeeh*. **H** Open country. **D** Madagascar.

35 BLACK-CROWNED SPARROW-LARK *Eremopterix nigriceps* 13cm Forages on the ground. Sings from a low perch or during a low, drifting, wing-fluttering display flight. **V** Repeated *wit-ti-weee* or similar. Calls include a *tchip* and a *zree* when alarmed. **H** Sandy deserts. **D** Cape Verde; Mauritania to NW India.

36 CHESTNUT-BACKED SPARROW-LARK *Eremopterix leucotis* 12cm Note chestnut upperparts of male and buff-and-chestnut wing coverts of female. **V** In flight, unstructured flow of very fast *wuwuchi tuwuee*. **H** Short open grassland, semi-arid savannah. **D** Widespread in Africa, from Senegal to Somalia and N Tanzania, from Angola to Mozambique and South Africa.

37 ASHY-CROWNED SPARROW-LARK *Eremopterix griseus* 12cm In spectacular display flight, soars to about 30m, sings while hovering in wide circles, then dives toward the ground, pulls up with a few rapid wingbeats. **V** Short fluty notes followed by a drawn-out whistle. **H** Dry open areas with scattered low vegetation. **D** Pakistan, India, Bangladesh, Sri Lanka.

38 CHESTNUT-HEADED SPARROW-LARK *Eremopterix signatus* 12cm Outside breeding season forms flocks of 40 or so. **V** Simple twittering given from low perch or in flight. Call: sharp *chip-up*. **H** Semi-arid or arid grassy plains. **D** Ethiopia, Somalia, Kenya.

39 GREY-BACKED SPARROW-LARK *Eremopterix verticalis* 13cm Male has a black head with bold white ear patches that reach the nape. Female has a dark belly patch and is generally paler than other female sparrow-larks. **V** In flight, very high, sharp, pleasant variations of *wee-weee-wur-wee*. **H** Grassy plains and shrubland in semi-arid regions. **D** W Angola to Botswana, S Zambia, Zimbabwe, South Africa.

40 FISCHER'S SPARROW-LARK *Eremopterix leucopareia* 12cm Note rusty-tawny, black-bordered crown of male, spotted mantle of female. **V** Song is a rather feeble and tuneless warble. **H** Dry areas with some short grass. **D** Kenya to N Mozambique.

1
2
3
4
5
6
7
8
9
10
11
12
13
14
15
16
17
18
19
20
21
22
23
24
25
26
27
28
29
30
31
32
33
34
35
36
37
38
39
40

LARKS *CONTINUED*

1 SABOTA LARK *Calendulauda sabota* 14cm Bill shape variable; may be thinner or heavier than shown. Note prominent, long, white eyebrow, lack of red in wings, dark malar stripe and bold streaking on upper breast. **V** Normally sings from perch, occasionally in hovering, fluttering flight, unhurried, melodious twitters interspersed with extremely high *seee* notes. **H** Bushveld, open woodland, especially on rocky slopes. **D** S Africa, from SW Angola southwards.

2 PINK-BREASTED LARK *Calendulauda poecilosterna* 15cm Note pale slim appearance, pinkish-buff face sides and breast streaking. **V** Accelerated, descending trill given from perch. **H** Bushed and wooded habitats. **D** SE South Sudan to NE Tanzania.

3 FOXY LARK *Calendulauda alopex* 14cm General colouring rufous. White eye-ring and eyebrow. Breast markings in the form of small black triangles. **V** Very high, happy, descending, accelerated *fi fi-fi-fifjuweehweehjuh*. **H** Open and more or less wooded areas with some grass cover and scrub. **D** E Ethiopia and Somalia to Tanzania.

4 FAWN-COLORED LARK *Calendulauda africanoides* 14cm Note white underparts, light breast streaking, white line above and below eye (both starting at bill). Upperparts often darker and more heavily streaked than shown. **V** Short, sharp, ascending twitter. **H** More or less shrubby areas. **D** S Africa, from S Angola and SW Zambia southwards.

5 KAROO LARK *Calendulauda albescens* 17cm Extensive streaking of underparts diagnostic. Note dark tail and white eyebrow. From Monotonous Lark by more slender bill and (where in same area) by more streaking on upperparts. **V** Sings from bush top or in flight, rapid, short rattle concluding with *titi-tjuh*. **H** Stony, open shrub land and dunes. **D** W South Africa.

6 RED LARK *Calendulauda burra* 19cm Note heavy bill, dark wings and tail, rather plain upperparts, restricted range. Hovers shortly before settling. **V** Sings from bush top or in flight, high, very fast *tuh-Wee-dudu*, second syllable higher and stressed. **H** Red sand dunes sparsely covered with grass, shrublands. **D** WC South Africa.

7 DUNE LARK *Calendulauda erythrochlamys* 17cm Note pale appearance and dark feathers in tail. **V** Sings in flight or from perch: short dry rattle, preceded by double *tjiptjip*. **H** Scrubby places in Namib desert. **D** Namibia.

8 BARLOW'S LARK *Calendulauda barlowi* 18cm Distinguished from Red Lark by smaller bill and different range, from Spike-Heeled Lark (Plate 198) by darker, browner colouring and different range, and from Karoo Lark (with which it shares only a small part of its range) by less extensive streaking below and different voice. **V** Series with sustained phrases, preceded by 2–5 (not 1–2) introductory notes. **H** Scrubland and other sparsely covered areas. **D** SW Namibia, NW South Africa.

9 RUDD'S LARK *Heteromirafra ruddi* 14cm Note buffy crown stripe, short tail, erect stance. From Spike-heeled Lark (Plate 198) by shorter, conical bill and different tail pattern. **V** Alternating flapping and gliding display flight, giving short whistles, *witurweeh*. **H** Montane grassland. **D** E South Africa.

10 ARCHER'S LARK *Heteromirafra archeri* 14cm Note short-tailed appearance. No wing patch. **V** Song is a series of hurried warbling notes with a buzzy trilling quality. **H** Open rocky areas with some short grass cover and sparse bush. **D** E Ethiopia, SW Somalia.

11 EASTERN CLAPPER LARK *Mirafra fasciolata* 15cm Differs from Cape Clapper Lark by slightly larger size, slightly heavier bill, different range. **V** Mainly long ascending *seeeeh* notes. **H** Dry, rocky areas often with tall grass, Kalahari scrub. **D** SW Zambia, N and E Namibia, Botswana, C and E South Africa.

12 CAPE CLAPPER LARK *Mirafra apiata* 14cm Separated from Flappet Lark by redder tinge and pale, almost white, outer-tail feathers. **V** Flaps at the summit of an upward display flight before descending, giving a loud whistle, *fuuiiii titjitjirrrr*. **H** Boulder-strewn sloping grassland, dry shrublands, Kalahari scrub. **D** SW Namibia, W and SW South Africa.

13 RED-WINGED LARK *Mirafra hypermetra* 25cm Often perches on bush. Note large size, black breast striping and rufous wing. **V** Long, loud, whistled phrases. **H** Sparsely bushed, rough grassland. **D** South Sudan, Ethiopia, Somalia to N Tanzania.

14 RUFOUS-NAPED LARK *Mirafra africana* 17cm Distinguished from Angolan Lark by buff, not white, outer-tail feathers. Crest and rufous wings diagnostic. **V** Extremely high, short, sharp, yet sweet whistle *tee-tjuih* or *tiu-uweek*, given from boulder or bush top. **H** Open and bushed, partly bare or overgrazed grass and pasture land. **D** Sub-Saharan Africa, widespread.

15 FLAPPET LARK *Mirafra rufocinnamomea* 14cm Note buff (not white) outer-tail feathers and red in wing. Much smaller than Rufous-naped Lark. **V** Dry, short burst of wing flaps (like distant motorbike exhaust) during high aerial cruise, sometimes with very high, short *witweetrieweeh*. **H** Woodland, wooded grass plains. **D** Sub-Saharan Africa, widespread, south to Angola and N South Africa.

16 ANGOLAN LARK *Mirafra angolensis* 17cm Distinguished from Rufous-naped Lark by darker plumage, white outer-tail feathers, brown (not rufous) secondaries, barred tertials. **V** Song is a distinctive series of trills and clear notes, *zi-zi-zi, zu-zu-zu-zu, trrp trrp trrp tiu-tuu, tiu*, with some longer notes, *peeee-oo*. **H** Montane grassland, moist valleys. **D** Angola to SW Tanzania.

17 WILLIAMS'S LARK *Mirafra williamsi* 13cm Note dark, slightly streaked upperparts, black breast streaking, rusty wing patch and white outer-tail feathers. **V** Sequence of thin scratchy notes, ending with a distinctive *tsir-eeet tsir-eeet*. **H** Dry, short grass with some occasional bush patches. **D** Kenya.

18 MONOTONOUS LARK *Mirafra passerina* 14cm From Melodious Lark and Sabota Lark by absence of white eyebrow, from Sabota Lark by red wings and white outer-tail feathers, from Sclater's Lark (Plate 200) by red wings and mainly different range. **V** Sings from top of bush or in short display flight, very high, fluted, short *uppertjeetjuuk*. **H** Stony patches in dry bushveld and woodland. **D** W Angola to C Zambia and south to N South Africa.

19 MELODIOUS LARK *Mirafra cheniana* 12cm From Monotonous Lark by sharper eyebrow, more buffy underparts, different habitat. **V** Sings for long periods from high perch or in fluttering display flight in a continuous flow of short, repeated phrases, trills, fluted interludes and imitations of other birds. **H** Grassland, cultivated fields. **D** C Zimbabwe, E South Africa.

20 HORSFIELD'S BUSH LARK *Mirafra javanica* 13–15cm Forages on the ground, singly, in pairs or small loose groups. **V** Sings from a perch or in towering song flight; repeated short varied phrases, often including mimicry. **H** Short grassland with bushes, dry marshland edge and paddyfield stubble. **D** S China, SE Asia, Philippines, Indonesia, New Guinea, Australia.

21 SINGING BUSH LARK *Mirafra cantillans* 13–14cm Forages on the ground, usually singly, in pairs or in small scattered parties. During song flight, rises to about 30m before hovering, then drops to ground or bush. **V** Various short chirps, whistles and buzzes that end in a buzzing trill; often includes mimicry. **H** Dry bush-covered plains, grassland, fallow cultivation, scrubby semi-desert. **D** Senegal to Kenya, Arabian Peninsula, Pakistan, India, Bangladesh.

22 BURMESE BUSH LARK *Mirafra microptera* 13–15cm Mainly terrestrial, also perches on trees or overhead wires. **V** Short hurried jingle of varied high-pitched whistles, given from an elevated perch or in high song flight. **H** Semi-desert scrub with scattered trees and cultivation. **D** C Myanmar.

23 BENGAL BUSH LARK *Mirafra assamica* 16cm Note stout bill and rufous wing panel. Forages on the ground. **V** Monotonous repeated, squeaky bisyllabic note, usually given in flight. Also a jingle of short varied notes, with some mimicry, generally given from a low perch or the ground. **H** Open grassy areas on plains and plateaux, edge of open forest. **D** N India to W Myanmar.

24 INDOCHINESE BUSH LARK *Mirafra erythrocephala* 15cm Note whitish supercilium. Mainly terrestrial, also perches on trees and overhead wires. **V** Quick series of thin, clear, mainly drawn-out notes, given in 2–8-sec strophes, from an elevated perch or the ground. **H** Open dry areas with scattered trees and bushes, cultivation and forest edge with scrub, bamboo and trees. **D** SE Asia, from S Myanmar to Vietnam.

25 INDIAN BUSH LARK *Mirafra erythroptera* 14cm Forages on the ground; regularly perches on bushes or posts. Sings from perch or song flight, in the latter fluttering upward for about 10m before parachuting down to a nearby perch. **V** Clear whistles, alternating in length and pitch. **H** Arid, scrubby and rocky areas. **D** Pakistan, N and C India.

26 JERDON'S BUSH LARK *Mirafra affinis* 14cm Similar to Indian Bush Lark. Sings mainly from a perch, on the ground or in a song flight similar to that of Indian Bush Lark. **V** Dry, drawn-out, rattling *zizizizezezezezezezezezezezezeze-zezezeze*. **H** Fallow fields edged with bushes or trees, scrub-covered rocky ground, scrubby areas of open forest. **D** E and S India, Sri Lanka.

27 GILLETT'S LARK *Mirafra gilletti* 14cm Note beautiful buff-tawny appearance. Wings paler than mantle, underparts white. **V** Not well known. A simple *dzee-dzit* call has been recorded, and a brief four- or five-note song. **H** Hilly, sandy and stony areas with rocky outcrops and some grass and scrub. **D** Horn of Africa: SE Ethiopia, Somalia, NE Kenya.

28 RUSTY BUSH LARK *Mirafra rufa* 13cm Rather plain and darkish without the typical well-defined rufous *Mirafra* wing patch. Note the reddish, striped breast. **V** Unknown. **H** Open grassy plains, rocky areas with some bush, open woodland. **D** C Sahel zone, from Mali to Sudan.

29 COLLARED LARK *Mirafra collaris* 13cm Distinguished by white throat, black collar, rufous upperparts. **V** Plaintive whistle. **H** Dry areas with some grass cover, bush and trees. **D** Ethiopia and Somalia to C Kenya.

30 ASH'S LARK *Mirafra ashi* 14cm Note thin bill, darkish upperparts, clear rufous wing patch and white tail edges. **V** Unknown. **H** Rocky, rough grass plains near seashore. **D** E Somalia; very restricted range.

31 SOMALI LARK *Mirafra somalica* 20cm A large rufous lark with a long slender bill and white tail edges. **V** Undescribed. **H** Arid areas with some grass. **D** Somalia.

32 FRIEDMANN'S LARK *Mirafra pulpa* 13cm A small colourful lark with white outer-tail feathers. **V** Song is a distinctive whistle, *hooeeeooo*, repeated every 1–2 secs, very different from other small larks in its range. **H** Sparsely bushed grassland. **D** S Ethiopia, Kenya, NE Tanzania.

33 KORDOFAN LARK *Mirafra cordofanica* 15cm Lightly streaked golden-rufous upperparts, pale face, whitish-buff underparts. **V** High long-lasting song flight. Much like Singing Bush Lark with a variation of whistles and trills, but less repetitive. **H** Arid desert or semi-desert with sparse grass and scattered bushes. **D** Sahel zone: Mauritania and Senegal to Niger; E Chad, S Sudan, N South Sudan.

34 WHITE-TAILED LARK *Mirafra albicauda* 13cm Note black-marked upperparts, bicoloured bill, white outer-tail feathers and rufous in wing. **V** Long sequences of harsh and mellow whistles, without trills. **H** Open and sparsely bushed habitats. **D** Scattered distribution, in Chad, Sudan, South Sudan, Ethiopia, Uganda, Kenya, Tanzania.

35 WOODLARK *Lullula arborea* 15cm Note broad wings, short tail, long white supercilium. Readily perches on trees, bushes or wires. **V** Beautiful song starts slowly then quickens with a repeated *lee - lee-lee-leeleeleeleelululululul...ee-lu-ee-lu-ee-luee-lu-eelulululu...* **H** Open forest, heathland, young plantations, open country with trees; in winter, often found in more open cultivation, stubble-fields. **D** Widespread across Europe, N Africa, east to W Russia, Iran, Turkmenistan.

rufous morph
brown morph
1
2
3
4
5
6
7
8
9
10
11
12
13
14
15
16
17
18
19
20
21
22
23
24
25
26
27
28
29
30
31
32
33
34
35

LARKS *CONTINUED*

1 OBBIA LARK *Spizocorys obbiensis* 13cm Note face pattern and narrow white tail margins. **V** Song unknown. Flight call *tip-tip-tip*. **H** Sandy, partly grassed and scrubbed plains and hills near coast. **D** C Somalia.

2 SCLATER'S LARK *Spizocorys sclateri* 14cm Overall colour variable. Note large head with distinctive 'teardrops', broad white tail edges, absence of crest. **V** Sequence of more or less connected, fluted, melodious or harsh notes. **H** Dry, stony, sparsely grassed and shrubbed areas. **D** S Namibia, W South Africa.

3 STARK'S LARK *Spizocorys starki* 14cm Note very pale colouring, peaked or crested crown, streaked upperparts. **V** Sings on ground or in high display flight, hardly varied series of *wee*, *tjuh* and *sree* notes. **H** Dry grassland. **D** W Angola to South Africa.

4 SHORT-TAILED LARK *Spizocorys fremantlii* 14cm Note bold face pattern and long bill. **V** Sings from the ground, a rather slow series of mournful whistles, both upslurred and downslurred. **H** Open areas often with some grass cover, scrub and trees, near rocky outcrops and settlement. **D** Ethiopia and Somalia to N Tanzania.

5 MASKED LARK *Spizocorys personata* 15cm Note face mask and white throat contrasting with buff breast. **V** Calls include a rolling *chew-chi chew* or *tiu tiu tu-tiu tiu*. **H** Bare sparsely grassed areas with some small scrub and bush. **D** Ethiopia, Kenya.

6 BOTHA'S LARK *Spizocorys fringillaris* 12cm From Pink-billed Lark by less conical bill, white outer-tail feathers, white chin, lower breast and belly. **V** In low flight, very high *sireee sireeh*. **H** Upland, short grasslands. **D** NE South Africa.

7 PINK-BILLED LARK *Spizocorys conirostris* 13cm General tinge varies from very pale to rather dark. Distinguished from Botha's Lark by larger bill, from Stark's Lark by absence of crest, and from both of these by buff outer-tail feathers. **V** Very high, short, more or less connected twitters, *sreee-sreee*, given from ground or in flight. **H** Montane grasslands. **D** SC Africa, from SW Zambia south.

8 WHITE-WINGED LARK *Alauda leucoptera* 18cm Wing with broad white trailing edge. Forms large flocks in winter. **V** Song resembles Eurasian Skylark but sweeter, higher-pitched and jerkier, given in flight or from ground. **H** Natural steppe, dry heath; also cultivated areas in winter. **D** S Ukraine to Kazakhstan and SC Russia. Winters south to Caucasus and N Iran.

9 RASO LARK *Alauda razae* 14cm The only lark on Raso Island. **V** Song like Eurasian Skylark but less complicated, with longer gaps between phrases, delivered in flight or on ground. **H** Mostly on flat, decaying, lava plain with sand patches that support low vegetation; disperses to other areas in non-breeding season. **D** Ilhéu Raso (Cape Verde).

10 ORIENTAL SKYLARK *Alauda gulgula* 16cm Shorter primary projection and shorter-tailed than Eurasian Skylark, and lacks white trailing edge to wing. Sings from a low perch or more often in a very high display flight. **V** Prolonged mix of warbling, twittering and short whistles. Calls include a buzzy *baz-baz* or *baz-terr* and a *twip*. **H** Grassland, crop fields, grass bordering saltpans and coastal mudflats, playing fields in urban areas. **D** Widespread from W Himalayas to India, Sri Lanka, China, SE Asia, Philippines.

11 EURASIAN SKYLARK *Alauda arvensis* 16–18cm In flight, shows white or whitish trailing edge to wing. Forms winter flocks, often with other ground feeders. **V** Various, lengthy, warbling and trilling phrases mixed with mimicry and the odd call note, from high display flight or low perch. Call is a liquid *chirrip*. **H** Various grassy areas, e.g. cultivated fields, moorland and alpine pastures; disperses in winter to include salt marshes and coastal beaches. **D** Widespread from W Europe to extreme E Asia, mostly further N than Oriental Skylark.

12 SYKES'S LARK *Galerida deva* 14cm Spiky crest is not always obvious. Forages on the ground, usually alone, in pairs or small groups. Song delivered in a soaring, hovering and wandering flight. **V** Various harsh and clear notes mixed with mimicry. **H** Stony, sparse scrub areas and dry cultivations. **D** C India.

13 SUN LARK *Galerida modesta* 14cm Note striking long eyebrow, short crest, rather stout bill, black moustachial stripe and heavy breast streaking. Some rufous in opened wing. **V** Very high, rather sharp, twittering *sritisrititwee*. **H** Open ground, often with rocky outcrops and occasional stands of large trees. Also pasture, bare fields, airstrips. **D** Senegal to South Sudan and NE Uganda.

14 LARGE-BILLED LARK *Galerida magnirostris* 18cm Note heavy bill with pale base of lower mandible. May show a crest. Heavily marked above and below, without rufous in wings or tail. **V** Sings in flight or from perch, short, pleasant, high little phrases, *wip-wip-wu-reeh*. **H** Dry montane grassland, shrubland, cultivated fields. **D** S Namibia, South Africa, Lesotho.

15 THEKLA'S LARK *Galerida theklae* 17cm Underwing greyish. Perches more readily on bushes than Crested Lark. **V** Prolonged loud fluting, interspersed with short whistles, trills and mimicry, often from a low perch. **H** Mainly rocky slopes with bushes, mountain slopes, dry steppe; less often lowland cultivation. **D** Iberian Peninsula, N Africa, Eritrea, Ethiopia, Somalia, N Kenya.

16 CRESTED LARK *Galerida cristata* 18cm Note prominent crest. Forages on the ground, although regularly perches on bushes, posts and wires. **V** Song is given mainly during a simple display flight, a clear pleasant warble mixed with call notes and mimicry. **H** Desert, semi-desert, dry cultivated areas, dry coastal mudflats. **D** Widespread across Europe, Africa, Asia.

17 MALABAR LARK *Galerida malabarica* 16cm Actions similar to Crested Lark. Sings mainly from the ground or low perch, but also in flight. **V** Rich mixture of piping and melancholy notes along with some mimicry. **H** Open sparse scrub, forest clearings, grass-covered stony hillsides, cultivations, and grassy edges of coastal mudflats. **D** W India.

18 MAGHREB LARK *Galerida macrorhyncha* 18–20cm Formerly considered conspecific with Crested Lark, but bill noticeably longer and stouter. **V** Song is similar to Crested Lark, but simpler and slower, with longer and lower-pitched notes. **H** Interior deserts and mountains, typically in more arid habitats than Crested Lark. **D** NW Africa, from Morocco and Algeria to Mauritania.

19 HORNED LARK *Eremophila alpestris* 16–19cm Note distinctive facial pattern. Small winter flocks often include buntings and finches. **V** Sings from perch or in flight, simple rippling trills followed by a short chatter. **H** Breeds on dry tundra; winters on salt marshes, beaches, stubble fields. **D** Holarctic: widespread across North America, N Europe, Asia. Northern populations move south in winter.

20 TEMMINCK'S LARK *Eremophila bilopha* 15cm Often confiding. Loose flocks include other lark species. **V** Sings in flight or from ground, monotonous warbling and twittering interspersed with short whistles. **H** Barren, flat stony desert or semi-desert with sparse vegetation. **D** NW Mauritania and Morocco to Syria and Iraq.

21 HUME'S SHORT-TOED LARK *Calandrella acutirostris* 14cm Forages restlessly on the ground; forms large winter flocks. Song given during wandering display flight, when bird soars, hovers, then dives to the ground. **V** Series of monotonous, faint, disjointed notes. **H** High-altitude semi-desert and rocky hills; winters in fallow cultivation. **D** Breeds Pakistan to Himalayas. Winters S Nepal, N India.

22 MONGOLIAN SHORT-TOED LARK *Calandrella dukhunensis* 15cm Formerly considered conspecific with Greater Short-toed Lark, but differs in genetics and vocalisations; only slight morphological differences. **V** Song is a continuous fast ramble of variable notes, with whistles, clicks and mimicry, given in song flight. **H** Open dry sandy steppe with short grasses and scattered bushes. **D** E Mongolia to W China.

23 ERLANGER'S LARK *Calandrella erlangeri* 15cm Formerly considered conspecific with Red-capped Lark (together with Blanford's Lark). Boldly marked, with deep rufous cap and large black-and-chestnut patches on sides of breast. **V** Song is a short series of rapid rising and falling notes. **H** Short dry grasslands and fields in highlands. **D** Ethiopia and Eritrea.

24 BLANFORD'S LARK *Calandrella blanfordi* 14cm Rufous crown. Actions similar to Greater Short-toed Lark. **V** Song given in flight, *chew-chew-chew-chew* with call notes and more fluid phrases added. **H** High, open, bare or grassed stony areas. Lower ground in winter. **D** SW Arabian Peninsula, Eritrea, N Somalia.

25 RED-CAPPED LARK *Calandrella cinerea* 15cm Note brick-red crest (not always erect) and diagnostic shoulder patch. **V** Very high unstructured unhurried connected short twitters including *sweeeh* and sharp descending *tweeeeeh*. **H** Dry bare more or less grassed and scrubbed areas. **D** Sub-Saharan Africa.

26 GREATER SHORT-TOED LARK *Calandrella brachydactyla* 15cm Dark patch on breast side. Crouched feeding posture. Large flocks out of breeding season. **V** Song given mainly in flight, contains accelerating simple notes, several hesitant notes, finishing with long series of repeated bubbling phrases. **H** Dry plains, dry cultivation, semi-desert. **D** Breeds S Europe, N Africa, C Asia. Winters south to Sahel zone, Arabian Peninsula, India.

27 BIMACULATED LARK *Melanocorypha bimaculata* 16–18cm In flight, underwing greyish, white-tipped tail. **V** Like Calandra Lark but with fewer harsh notes, given in flight or perched. **H** Stony sparsely grassed plateaux in mountainous areas, dry grass areas on plains. **D** Breeds Turkey to Kazakhstan. Winters NE Africa, Arabian Peninsula, N India.

28 CALANDRA LARK *Melanocorypha calandra* 18–20cm Underwing looks blackish with distinct white trailing edge; white-sided tail. Sociable. **V** Like Eurasian Skylark but louder and more complex with harsher, stronger notes, usually given in flight. **H** Steppe, open cultivation, wasteland. **D** N Africa, S Europe to C Asia.

29 BLACK LARK *Melanocorypha yeltoniensis* 20cm Fresh-plumage male has pale fringes, giving bird a scaly look. Worn females becomes darker above, including head, often blotched darker below. **V** Like Eurasian Skylark but with shorter frantic phrases, higher-pitched and including mimicry, given in display flight or from low perch. **H** Grassy and shrubby steppe, often by lakes; cultivation edge in winter. **D** S Russia, Kazakhstan. In winter further west and south.

30 MONGOLIAN LARK *Melanocorypha mongolica* 18cm In flight shows a very wide white trailing edge to wing. **V** Song is sustained twittering, mixed with dry notes, much like Bimaculated Lark. **H** Prefers dry grassland, but also found in damper areas and rocky uplands. **D** Mongolia, C China.

31 TIBETAN LARK *Melanocorypha maxima* 21–22cm In flight, tail shows much white on tips and outer feathers, and wing shows a white trailing edge. **V** Mix of slow-flowing phrases interspersed with grating notes and mimicry. **H** High-altitude grassy or marshy areas by lakes or rivers. **D** Tibetan Plateau from NW India to C China.

32 DUPONT'S LARK *Chersophilus duponti* 17cm Note long legs, long neck, thin slightly curved bill. A skulker, often runs off when encountered. **V** Twittering and buzzing notes, given from ground or during towering song flight. **H** Open arid or semi-arid flat land with low scrub or tufted grass. Outside breeding season often frequents cereal fields. **D** N Africa, Iberian Peninsula.

33 DUNN'S LARK *Eremalauda dunni* 14–15cm Note sandy coloration and heavy bill. Forms small, nomadic flocks in non-breeding season. **V** Rapid series of chirping and dry notes, given from a low perch or in flight. **H** Flat, gravelly or sandy desert or semi-desert. **D** Sahara, from Mauritania to Sudan; Middle East, from Syria to Arabian Peninsula.

34 ATHI SHORT-TOED LARK *Alaudala athensis* 14cm Formerly considered conspecific with Somali Short-toed Lark. Differs in greyer, less sandy plumage. **V** Similar to Somali Short-toed Lark. **H** Open plains with short grass cover. **D** S Kenya, N Tanzania.

35 ASIAN SHORT-TOED LARK *Alaudala cheleensis* 13cm Very similar to Lesser Short-toed Lark. Shows a long primary projection. Forages on the ground. **V** Dry buzzing *prritt* or *chirrick*. **H** Open sandy semi-desert and stony foothills. **D** C and E Asia, from Kazakhstan and Uzbekistan to Mongolia and China.

36 SOMALI SHORT-TOED LARK *Alaudala somalica* 14cm Smallish lark with pale supercilium and eye-ring, and pale pinkish bill. Note fine streaking on breast. **V** Complex song consists of a long series of rising and falling notes, with mimicry. **H** Open plains with short grass cover. **D** N Somalia, E and S Ethiopia, N Kenya.

37 LESSER SHORT-TOED LARK *Alaudala rufescens* 14cm Shows more primary projection than Greater Short-toed Lark. Low, circling display flight with slow wingbeats. **V** Song richer, more varied and faster-paced than Greater Short-toed Lark. **H** Dry plains, steppe, cultivation, saline flats, semi-desert. **D** S Europe, N Africa, Middle East, Arabian Peninsula to Kazakhstan.

38 SAND LARK *Alaudala raytal* 12cm Forages on the ground; sings in flight or on the ground. During song flight soars high, flying aimlessly around with an intermittent series of rapid wing-flaps before parachuting to the ground in a series of steps. **V** Song short, with disjointed, dry rattling phrases. **H** Sandy banks and islets in rivers or lakes. **D** From Iran through N India to Myanmar.

1
2
3
4
5
6
7
8
9
10
11
12
13
14
15
16
17
18
19
♀
♂
20
21
22
23
24
25
26
27
28
29
♀
♂
30
31
32
33
34
35
36
37
38

BULBULS PYCNONOTIDAE

1 BARE-FACED BULBUL *Nok hualon* 19cm Newly described in 2009. Generally encountered singly, in pairs or small groups; predominantly arboreal, although frequently seen on limestone cliff edges or jagged crags on steep hillsides. **V** Rising, bubbling trill, short dry bubbling notes and a harsh churring. **H** Sparse open deciduous forest and shrubs on limestone. **D** Laos.

2 CRESTED FINCHBILL *Spizixos canifrons* 22cm Noisy bulbul with prominent crest. Often found in small groups. **V** Bubbling *purr-purr-prruit-prruit-prruit*; calls include a buzzy *grz-grz-grz*. **H** Open woodland, secondary growth, overgrown forest clearings and scrub. **D** NE India, SW China, Myanmar, NW Thailand, N Laos, N Vietnam.

3 COLLARED FINCHBILL *Spizixos semitorques* 22cm Note dark head, white patch at bill base, white streaks on ear-coverts, white half-collar. Usually in small groups, often fly-catches for insects. **V** Hurried whistles, *ji-de-shi-shei - ji-de-shi-shei - shi-shei*. **H** Forest clearings and edges, undergrowth and bamboo; up to 3,000m. **D** S and C China, Taiwan, N Vietnam.

4 STRAW-HEADED BULBUL *Pycnonotus zeylanicus* 29cm Usually encountered in small family groups, foraging from the ground up to the canopy. **V** Melodious, rich warbling; weak chattering and gurgling while foraging. **H** Broadleaved evergreen forest, secondary growth, scrub, plantations and occasionally mangroves. **D** Malay Peninsula to Borneo.

5 STRIATED BULBUL *Pycnonotus striatus* 23cm Note streaked plumage and long crest. Generally forages in treetops. **V** Series of pleasant warbling notes. Calls include a sharp, repeated *tyiwut* and a loud *pyik...pyik*. **H** Broadleaved evergreen forest, moist oak–rhododendron forest, deciduous forest. **D** C Himalayas through NE India to N Vietnam.

6 CREAM-STRIPED BULBUL *Pycnonotus leucogrammicus* 17–18cm Forages on fruit and insects in the mid-storey or subcanopy. Usually seen singly or in pairs, occasionally in small groups; regular member of mixed-species feeding flocks. **V** *tree-troo-troo*. **H** Hill and montane evergreen forest, forest edge, secondary growth. **D** Sumatra, Indonesia.

7 SPOT-NECKED BULBUL *Pycnonotus tympanistrigus* 16cm Conspicuous yellowish patch behind ear-coverts. Usually seen in pairs or small groups; sometimes joins mixed-species foraging parties. Tends to favour the middle storey and canopy. **V** Bisyllabic *tdip-diew*; also a repeated, emphatic *jret-jret-jtry*. **H** Evergreen forest, forest edges and secondary growth. **D** Sumatra, Indonesia.

8 BLACK-AND-WHITE BULBUL *Pycnonotus melanoleucos* 18cm Unmistakable, with blackish plumage and contrasting white wing coverts. Most often encountered in the forest canopy, but also forages in low shrubs. **V** Generally quiet; occasionally utters a tuneless *pet-it* or *tee-too* and a longer *cherlee-chlee-chlee-chee-chee*. **H** Broadleaved evergreen forest and forest edge. **D** Malay Peninsula; Borneo; Sumatra, Indonesia.

9 GREY-HEADED BULBUL *Pycnonotus priocephalus* 19cm Note olive-green mantle, grey head, peaked crown. Forages at all levels, singly, in pairs or in small parties. **V** Calls include incessant shrill notes, a thin metallic wheezy *jzhwink* and a jarring *chraink*. **H** Moist broadleaved evergreen forest with dense undergrowth, scrub in abandoned clearings. **D** SW India.

10 BLACK-HEADED BULBUL *Pycnonotus atriceps* 18cm Variable, but most morphs show a bright yellow inner-wing panel. Usually found in pairs or small parties. **V** The song is a disjointed series of rising and falling *tink* notes; calls include an emphatic *chew* or *chewp*. **H** Lowland primary and secondary forest, wooded gardens. **D** NE India, Bangladesh, SE Asia, Greater Sunda Islands to W Philippines.

11 ANDAMAN BULBUL *Pycnonotus fuscoflavescens* 14–17cm Usually forages singly, in pairs or in small groups. **V** A short series of high-pitched *tsit-tsit-tsit* notes. Calls include a *tsit* and an upward-inflected *shrinkit*. **H** Forest, forest edge, thick secondary growth. **D** Andaman Islands.

12 BLACK-CAPPED BULBUL *Pycnonotus melanicterus* 19cm Note solid black head. Arboreal; usually in pairs or small groups. **V** Various sweet, mellow piping whistles and sharper notes. Calls include a rapid whipcrack note. **H** Luxuriant forest, secondary growth, and forest edge. **D** Sri Lanka.

13 BLACK-CRESTED BULBUL *Pycnonotus flaviventris* 22cm Crest is permanently raised. Arboreal, generally seen singly or in pairs. **V** Song is a sweet *weet-tre-trippy-weet*. **H** Forest with undergrowth, abandoned clearings, scrub around cultivations and orchards. **D** Himalayas, NE India to SE Asia.

14 RUBY-THROATED BULBUL *Pycnonotus dispar* 17–20cm Note rich orange-yellow on upper breast. Forages for fruits and insects. **V** Includes *tee-tee-wheet-wheet*, *whit-wheet-wit* or *hii-tii-hii-tii-wiit*. **H** Open woodland, shrubby areas and plantations. **D** Sumatra, Java and Bali, Indonesia.

15 FLAME-THROATED BULBUL *Pycnonotus gularis* 18–19cm Pale eye, orange-red throat. Arboreal, generally in the canopy. **V** Low, constant churring. Song consists of six or so tinkling notes. **H** Evergreen forest edge, thickets, overgrown clearings. **D** SW India.

16 BORNEAN BULBUL *Pycnonotus montis* 17–18cm Often perches in the open at forest edges. Forages in fruiting trees, making regular aerial sallies after passing insects. **V** Constantly repeated *yek yek*, also a whistled *grrrt grrrt*. **H** Evergreen forest, secondary forest, shrubby regrowth of abandoned cultivation. **D** Borneo.

17 SCALY-BREASTED BULBUL *Pycnonotus squamatus* 14–16cm Usually found in the forest canopy, singly or in small groups; sometimes in larger gatherings in fruiting trees. **V** Calls include a high-pitched *wit* or *tit*, a thin *tree*, a persistent *trrip trip* and a pretty trill. **H** Submontane primary and secondary forest. **D** Malay Peninsula; Borneo; Sumatra and Java, Indonesia.

18 GREY-BELLIED BULBUL *Pycnonotus cyaniventris* 16–17cm Usually keeps to the canopy, in pairs or small parties, unobtrusive. Agile, often hanging tit-like to gather berries. **V** Sweet *pi-pi-pwi... pi-pi-pwi...*, trilled, bubbling *pi-pi-pi-pi-pi-pi-pi* and subdued *wit-wit-wit*. **H** Primary and secondary forest, forest edge and clearings by roads and rivers. **D** Borneo; Sumatra, Indonesia.

19 RED-WHISKERED BULBUL *Pycnonotus jocosus* 20cm Black crest, black-and-white face, red ear-patch, red undertail coverts. **V** Lively, varied musical phrases, e.g. *wit-ti-waet - queep-kwil-ya - queek-kay*; call is a rolling *prroop*. **H** Forest, gardens, orchards, cultivated areas. **D** India and Nepal to S China, SE Asia.

20 BROWN-BREASTED BULBUL *Pycnonotus xanthorrhous* 20cm White throat contrasts sharply with brown breast-band. Often occurs in groups, picking fruit from various trees and plants. **V** Simple, repeated *chirriwu-i-whi-chu whirri-ui*; calls include a harsh *chi* or *brzzp* and a thin *ti-whi*. **H** Secondary growth, scrub, tall grass, thickets, clearings, streamside vegetation, cultivations, gardens. **D** C and S China to Myanmar, N Thailand, N Laos, N Vietnam.

21 LIGHT-VENTED BULBUL *Pycnonotus sinensis* 19cm Striking head pattern. Regularly perches prominently on bush tops. Often occurs in large flocks post breeding. **V** Range of cheerful phrases and peculiar notes; calls with a loud *tocc-tocc-tocc*. **H** Open woodland, cultivations and scrub. **D** E China to Taiwan, N Vietnam.

22 STYAN'S BULBUL *Pycnonotus taivanus* 19cm Usually seen singly or in pairs, forms large flocks post breeding. **V** Loud, variable *qiao-keli qiao-keli*. **H** Secondary growth, farms and developed areas in coastal lowlands. **D** S Taiwan.

23 HIMALAYAN BULBUL *Pycnonotus leucogenys* 20cm Can be very confiding. Usually in small parties; restless, constantly flicks tail and wings when foraging, and often bows and postures while perched on bush top. **V** Song often transcribed as *tea for two* or *take me with you*. Call: when agitated, a *pit-pit*, *pit-lo* or a chattering *pit-pit-it-it- it*. **H** Hillsides with scattered wild fruit bushes, open scrub, hedgerows, often around human habitations. **D** Himalayas.

24 WHITE-EARED BULBUL *Pycnonotus leucotis* 18–19cm Resembles Himalayan Bulbul, but slightly smaller, lacks of crest, and has thicker and shorter bill. **V** Like Himalayan Bulbul. **H** Dry woodland, thorn-scrub, semi-desert, agricultural land, orchards, palm groves, mangroves. **D** Arabian Peninsula, S Iraq to NW India.

25 RED-VENTED BULBUL *Pycnonotus cafer* 20cm Black head, scaly upperparts, crimson vent. In flight shows pale rump. Noisy and flock-loving. **V** Cheery, often transcribed as *be-careful*. Call: when alarmed, a sharp, loud, repeated *peep*; also a chattering *peep-a-peep-a-lo*, a rapid *pitititit* and a slower *peet-wit-wit-wit wit*. **H** Gardens, orchards, cultivation, scrub, sometimes around buildings. **D** From Pakistan through India, Sri Lanka, Bangladesh to Myanmar.

26 SOOTY-HEADED BULBUL *Pycnonotus aurigaster* 19–21cm Black crown and face, white ear-coverts. Often occurs in large flocks post breeding. **V** Chatty *whi-wi-wiwi-wiwi*, *u whi hi hu* or *wh-i-i-wi*. **H** Forest clearings, secondary growth, scrub, cultivations. **D** S China; SE Asia; Java and Bali, Indonesia.

27 WHITE-SPECTACLED BULBUL *Pycnonotus xanthopygos* 20cm Note whitish eye-ring and yellow undertail coverts. Often confiding, sometimes gathers in noisy loose groups. **V** Song consists of variable fluty, bubbling, repeated phrases, *whee-too-too* or similar. Call is a sharp scolding *weck* or a harsh *tscheck*; also a strong *wit-wit-wit*, *teewit* or *tew*. **H** Gardens, palm groves, bush and various fruit-growing plantations. **D** S Turkey and Syria to S Arabian Peninsula.

28 AFRICAN RED-EYED BULBUL *Pycnonotus nigricans* 17cm Slightly peaked crown. Differs from Cape and Common bulbuls in its eye-ring of bare orange-red skin. **V** High, liquid *wuut-weeroh-weet* like Common Bulbul. **H** Dry woodland, savannah, semi-arid shrubland, orchards, gardens. **D** SW Africa, from Angola to South Africa.

29 CAPE BULBUL *Pycnonotus capensis* 17cm Note bare white skin around eye. **V** High, sharp *tut three-toweet tuweet*. **H** Any area with bushes and trees, including gardens. **D** W and S South Africa.

30 COMMON BULBUL *Pycnonotus barbatus* 18cm Ubiquitous, rather nondescript bird with dark head and white vent. **V** High happy up-and-down whistled *I'm coming home*. **H** Forest edges, gardens, more or less wooded and bushed areas. **D** Widespread across N and sub-Saharan Africa.

31 SOMALI BULBUL *Pycnonotus somaliensis* 18cm Formerly considered conspecific with Common Bulbul. Very similar to Common Bulbul but has a small white patch on side of neck and a scaly breast. **V** Similar to Common Bulbul. **H** Bushed and wooded areas and gardens. **D** Djibouti, NW Somalia, NE Ethiopia.

32 DODSON'S BULBUL *Pycnonotus dodsoni* 18cm Formerly considered conspecific with Common Bulbul. Note white patch on side of neck and scaly breast like Somali Bulbul, but has yellow vent. **V** Similar to Common Bulbul but faster and more shrill. **H** Bushed and wooded areas, usually in more arid country. **D** N Somalia and SE Ethiopia to EC Kenya.

33 DARK-CAPPED BULBUL *Pycnonotus tricolor* 18cm Formerly considered conspecific with Common Bulbul. Very similar to that species but has yellow vent. **V** Similar to Common Bulbul. **H** Forest edges, gardens, more or less wooded and bushed areas. **D** C, E and SE Africa, from E Cameroon to S Ethiopia and south to South Africa.

34 PUFF-BACKED BULBUL *Pycnonotus eutilotus* 20–22cm Forages in trees after fruits and insects, usually singly or in pairs; sometimes joins mixed-species feeding parties. Note white tips on undertail. **V** Loud, high-pitched, quavering warble. **H** Lowland and hill primary forest, peat swamp forest, secondary growth and forest edges. **D** Malay Peninsula; Borneo; Sumatra, Indonesia.

35 BLUE-WATTLED BULBUL *Pycnonotus nieuwenhuisii* 18cm Little information; known from only two specimens. Status unclear; may be a hybrid. **V** Unknown. **H** Lowland dipterocarp forest in Borneo and low shrubbery in Sumatra. **D** Sumatra and Borneo.

36 YELLOW-WATTLED BULBUL *Pycnonotus urostictus* 19cm Note bare fleshy yellow eye-ring. Occurs singly, in pairs or in small parties; favours fruiting trees. **V** Variable, loud, tri-syllabic, whistled *wee-we-weeee* or *wee-eeee-ee*; also high-pitched, squeaky whistles and longer, more complicated whistles ending with a *wee-eeee-ee*. **H** Open woods, secondary growth, forest edge. **D** Philippines.

37 ORANGE-SPOTTED BULBUL *Pycnonotus bimaculatus* 20cm Orange at base of bill, yellow ear-coverts. Generally seen alone or in small parties; favours open glades and forest edges. **V** Various loud, harsh songs, including a *toc-toc-toc-toroc*, a clear *chewlk-chewlk-chewlk* or *chuk-chuk-chooh-chewlewlewlew*; also a sharp *tik*. **H** Montane forest edges, tall secondary growth, scrub and elfin forest near summits. **D** Sumatra, Java and Bali, Indonesia.

38 ACEH BULBUL *Pycnonotus snouckaerti* 20–21cm Favours small scrubby clearings dominated by ferns and high grasses. Formerly considered a race of Orange-spotted Bulbul. **V** Song similar to the previous species, but with a more metallic or dry, reedy quality. **H** Submontane forest with clearings. **D** N Sumatra, Indonesia.

39 STRIPE-THROATED BULBUL *Pycnonotus finlaysoni* 19–20cm Face and forecrown olive, streaked yellow. Usually seen singly or in pairs. **V** Throaty *whit-chu whic-i*, *whit-tu-iwhit-whitu-tu* or similar. **H** Forest clearings, forest edge, secondary growth. **D** SE Asia.

1
2
3
4
5
6
7
♀
8
♂
9
normal
10
grey morph
10
intermediate morph
11
12
13
14
15
16
17
18
19
20
21
22
23
24
25
26
27
28
29
30
31
32
33
34
35
36
37
38
39

BULBULS *CONTINUED*

1 YELLOW-THROATED BULBUL *Pycnonotus xantholaemus* 20cm Yellow head contrasts with whitish-grey underparts. Skulking, but restless; in pairs or small groups. **V** Rich, low burbling and warbling. Calls include a brief *teetle-lerp* and a nasal *rhid-tu-tu*. **H** Rocky wooded hillsides. **D** S India.

2 YELLOW-EARED BULBUL *Pycnonotus penicillatus* 20cm Striking black, white and yellow head pattern. Shy; forages mainly in undergrowth, although higher if trees are in fruit. **V** Explosive, ringing *Swink-Swink-Swink* and a low *crr-crr*. **H** Forests and nearby gardens. **D** Sri Lanka.

3 FLAVESCENT BULBUL *Pycnonotus flavescens* 22cm Note small white patch in front of eye, and peaked crown. Shy; forages mainly in undergrowth, higher if trees are in fruit. **V** Rising and falling, chuckling *brzk bzeek zink-zenk-zink* and a tumbling *johi bwiki-bwiki-bwik*; both song types sometimes combined. Calls include a rasping *brzzk brzzk*. **H** Open forest, forest edge, secondary growth, scrub. **D** NE India, SE Asia, Borneo.

4 YELLOW-VENTED BULBUL *Pycnonotus goiavier* 20–21cm Note broad white supercilium. Forages singly, in pairs or in small parties. Occasionally makes fly-catching sallies from an exposed perch. **V** Song consists of a series of variable choppy notes; calls are cheerful and loud, including a bubbling *chic-chic-chic, tiddloo-tiddloo-tiddloo* or *tid-liu tid-liu tid-liu* and similar. **H** Open areas, secondary growth, cultivation, roadsides, gardens. **D** SE Asia, Malay Peninsula, Indonesia, Philippines.

5 WHITE-BROWED BULBUL *Pycnonotus luteolus* 20cm Separated from Yellow-throated and Yellow-eared bulbuls by head pattern. Shy; more often heard than seen. Forages in thick cover. **V** Song is a loud burst of lively, discordant, bubbling, spluttering notes. Calls include *churr* and *krr-kurr*. **H** Dry open scrub, thickets, forest edge, clearings, gardens. **D** Peninsular India, Sri Lanka.

6 OLIVE-WINGED BULBUL *Pycnonotus plumosus* 20–21cm A plain and unobtrusive bulbul, usually seen low down, singly or in pairs. Joins other bulbuls in fruiting trees. **V** Variable, soft, liquid chirping, transcribed as *quick-doc-tor-quick*; calls include a throaty *whip-whip* and a purring *wrrh-wrrh-wrrh*. **H** Secondary growth, coastal scrub, mangroves. **D** Malay Peninsula, Greater Sunda Islands.

7 ASHY-FRONTED BULBUL *Pycnonotus cinereifrons* 19–20cm Formerly considered to be a race of the Olive-winged Bulbul. Favours fruiting trees. Usually seen singly or in pairs, and sometimes in small groups; often a member of mixed-species feeding parties. **V** Bubbling, repeated *chop wit chu do-de-do-de-du*; calls are presumed to be much like those of the Olive-winged Bulbul, including a *whip-whip* and a purring *wrrh wrrh wrrh*. **H** Open country, forest edges and secondary growth. **D** Palawan, W Philippines.

8 AYEYARWADY BULBUL *Pycnonotus blanfordi* 17–20cm White-streaked ear-coverts may be hard to see. Generally found singly, in pairs or small groups foraging in trees; also makes fly-catching sallies. **V** Calls include a rasping *which-which-which*, a piping *brink-brink-brink* and a harsh *chu-chu-chu*. **H** Open mixed deciduous forest, scrub, cultivations, gardens. **D** Myanmar.

9 STREAK-EARED BULBUL *Pycnonotus conradi* 17–20cm Formerly treated as conspecific with Ayeyarwady Bulbul. Differs genetically from Ayeyarwady Bulbul, and also in more yellowish underparts and vent, and whitish or greyish eyes (reddish in Ayeyarwady). **V** Very similar to Ayeyarwady Bulbul. **H** Open forest, scrub, cultivations, gardens. **D** SE Asia.

10 CREAM-VENTED BULBUL *Pycnonotus simplex* 18cm A nondescript but noisy and conspicuous bulbul. Occurs singly, in pairs or as part of mixed-species foraging flocks. Favours fruiting trees. **V** Subdued, quavering *whi-whi-whi-whi*..., interspersed with a low *pru-pru prr* and *prr-pru*; also harsh call notes and short phrases such as *quik chop, quik-plik chop* and a longer *bee-quik* or *pee dee kew*. **H** Broadleaved evergreen forest, forest edge, secondary growth. **D** Borneo; Sumatra and Java, Indonesia.

11 ASIAN RED-EYED BULBUL *Pycnonotus brunneus* 19cm Arboreal, usually occurs singly or in pairs, occasionally joins mixed-species feeding flocks. **V** High-pitched, bubbling, rising *pri-pri-pri-pri-pri-pit-pit*; also brief trills and a strident *chirrup* or *whit-it*. **H** Broadleaved evergreen forest, forest edge, secondary growth. **D** Malay Peninsula; Borneo; Sumatra, Indonesia.

12 SPECTACLED BULBUL *Pycnonotus erythropthalmos* 16–18cm Very similar to Asian Red-eyed Bulbul, but note orange eye-ring. Usually encountered singly or in pairs, often with other bulbul species. **V** High-pitched, mechanical *pip-pippidi* or *wip-wip-wi i i i*; also short phrases, such as *willy-nilly, pippi-dippi* or *willy-nilly-no*. **H** Primary and mature secondary lowland forest, swamp forest, secondary growth. **D** Malay Peninsula; Borneo; Sumatra, Indonesia.

13 SHELLEY'S GREENBUL *Arizelocichla masukuensis* 17cm Uniform dull olive greenbul with whitish eye-ring. Feeds on arthropods and small fruits. **V** Rather quiet *wit-wit-wit*. **H** Montane forest and surrounding well-wooded areas. **D** Tanzania, N Malawi.

14 KAKAMEGA GREENBUL *Arizelocichla kakamegae 17cm* Formerly considered conspecific with Shelley's Greenbul. Grey crown contrasts with olive-green upperparts. Forages on mossy trunks, probing for insects. **V** No song is known. **H** Montane forest interior. **D** E DR Congo to W Kenya and W Tanzania.

15 CAMEROON GREENBUL *Arizelocichla montana* 15cm Dark with paler yellowish chin and lower underparts. From other bulbuls in same habitat by uniform plumage and dark brown (not rufous) tail. **V** Song is a quiet nasal babble ending with a faster phrase, *chur-chur-chur-chipurchipurcherr*. **H** Forest undergrowth, often near clearings. **D** SE Nigeria, W Cameroon.

16 WESTERN GREENBUL *Arizelocichla tephrolaema* 20cm Note white eye-ring and striped cheeks. Often occurs in mixed-species flocks. **V** High, three-note *it's-not-so-chilly*. **H** Montane forest and bamboo. **D** SE Nigeria, W Cameroon.

17 OLIVE-BREASTED GREENBUL *Arizelocichla kikuyuensis* 18cm Formerly considered conspecific with Mountain Greenbul. Distinguished by grey head and yellow-green underparts. Note narrow broken eye-ring and fine lines on ear-coverts. **V** Song is a throaty warble. Calls are harsh churrs. **H** Highland forest and secondary growth. **D** E DR Congo, SW and SE Uganda, Rwanda, Burundi, C Kenya.

18 MOUNTAIN GREENBUL *Arizelocichla nigriceps* 18cm Note blackish crown or supercilium and greyer underparts, compared to Olive-breasted Greenbul. **V** Song is similar to Olive-breasted Greenbul, but includes cat-like calls. **H** Highland forest and secondary growth. **D** S Kenya, N Tanzania.

19 ULUGURU GREENBUL *Arizelocichla neumanni* 19cm Formerly treated as conspecific with Mountain Greenbul. Lacks broken eye-ring and fine lines on ear-coverts. **V** Song is a deep chattering warble with a flourish at the end. **H** Highland forest in Uluguru Mountains. **D** E Tanzania.

20 BLACK-BROWED GREENBUL *Arizelocichla fusciceps* 18cm Formerly treated as conspecific with Mountain Greenbul. Note greyer crown, lack of streaks on ear-coverts and yellowish undertail coverts. **V** Song is a 3–4-note warble with the last two notes louder, higher and slurred together. **H** Highland forest. **D** SW Tanzania to NE Zambia, Malawi and WC Mozambique.

21 YELLOW-THROATED GREENBUL *Arizelocichla chlorigula* 18cm Formerly considered conspecific with Mountain Greenbul. Note dark grey head and yellowish throat and vent. **V** Song is similar in structure to Black-browed Greenbul. **H** Forest, secondary growth and cultivation. **D** EC and S Tanzania.

22 STRIPE-CHEEKED GREENBUL *Arizelocichla milanjensis* 19cm Plain olive-green greenbul with grey head and white streaking on ear-coverts. Note diagnostic white upper eyelid. Eyes pale grey or brown. **V** Mid-high *tjah tjah tjekkertjektjek*. **H** Forest edges and adjacent areas. **D** SE Malawi, extreme E Zimbabwe, WC Mozambique.

23 OLIVE-HEADED GREENBUL *Arizelocichla olivaceiceps* 19cm Formerly considered conspecific with Stripe-cheeked Greenbul. Duller and darker than that species, with greener underparts. Eyes brownish or yellow. **V** Song is a short staccato warble with a deep throaty quality. **H** Variety of forest types including forest edge and scrub, mainly in the highlands. **D** SW Tanzania, Malawi, NW Mozambique.

24 STRIPE-FACED GREENBUL *Arizelocichla striifacies* 19cm Formerly considered conspecific with Stripe-cheeked Greenbul. Olive-green above and olive-yellow below (lacking grey head of Stripe-cheeked). Eyes usually pale (bluish-grey or yellow) but sometimes reddish-brown. **V** Song consists of short phrases of rich melodic notes. Call is a mechanical, monotonous series of scolding notes. **H** Variety of forest types including forest edge and scrub, mainly in the highlands. **D** SE Kenya to SW Tanzania

25 SLENDER-BILLED GREENBUL *Stelgidillas gracilirostris* 17cm The only forest greenbul to show pale grey underparts and pale yellowish undertail coverts. **V** Very high, descending, repeated *tjiuuuu tjuuu* and mid-high, drawn-up *nieau nieau*. **H** Upper strata of montane forest. **D** W and C Africa, from Senegal to W Kenya and NW Angola.

26 LITTLE GREENBUL *Eurillas virens* 15cm Separated from Little Grey Greenbul by yellowish (not greyish) chin, absence of eye-ring. **V** Hurried strophes like mid-high, babbling, up-and-down *prrrrk-prrrrk-jterp-fuwit*. **H** Mid-strata and undergrowth of forest. **D** W, C and E Africa, from Senegal to N Zambia, Tanzania.

27 LITTLE GREY GREENBUL *Eurillas gracilis* 15cm Note narrow eye-ring, greyish throat, yellowish lower belly. Forages rather high in trees. **V** Very high up-and-down *wee-tup-to-wup tjut*. **H** Forest. **D** W and C Africa, from Sierra Leone to N Angola, W Kenya.

28 ANSORGE'S GREENBUL *Eurillas ansorgei* 15cm Similar to Little Grey Greenbul, but more greyish green, less warm brown. Note white eye-ring, dark grey head and greenish edges of flight feathers. Tail more reddish brown than rest of body. **V** Very high, rapid *weet-wju-twee*. **H** Mid-strata of montane forest. **D** W and C Africa, from Guinea to W Kenya.

29 PLAIN GREENBUL *Eurillas curvirostris* 16cm Note well-defined grey throat, small, short bill, pale grey eyelids. **V** Mid-high, fluted *tuweetudirrrr* or *weetoweehoh*. **H** Forest understorey. **D** W and C Africa, from Sierra Leone to N Angola, W Kenya.

30 YELLOW-WHISKERED GREENBUL *Eurillas latirostris* 18cm Mostly dark plumage, with conspicuous yellow whiskers. **V** Mid-high, unstructured, up-and-down, scratching, unhurried *tjash tjesh tjesh-tjush tresh-tjash*. **H** Forest, forest remains, plantations, large gardens. **D** W and C Africa, from Senegal to W Kenya, W Tanzania.

31 SOMBRE GREENBUL *Andropadus importunus* 22cm Note creamy-white eye and unmarked face sides. **V** Very high, fast *fweeh twitwitwitwiti* or *torritrreeweehweeh*. **H** Forest, coastal bush. **D** E and SE Africa, from Somalia to South Africa.

32 GOLDEN GREENBUL *Calyptocichla serinus* 15cm Note pale, flesh-coloured bill. From other similar bulbuls in its habitat by keeping to the forest canopy. **V** Sharp, high *tjut-tweee tjut-Tweee*, last part higher and accentuated. **H** Forest and other habitats with high trees. **D** Sierra Leone to Ghana; SE Nigeria to Central African Republic and extreme NW Angola.

33 HONEYGUIDE GREENBUL *Baeopogon indicator* 18cm Differs from honeyguides in darker plumage without any yellow tones. **V** Very high, melodious *pi-pi-piuuuh*, or *pipiwiuuuuh* or *wee wee wee piuwweeh*. **H** Open forest and forest edges. **D** W and C Africa, from Sierra Leone to N Angola and W Kenya.

34 SJÖSTEDT'S GREENBUL *Baeopogon clamans* 18cm Note absence of black tips to outer-tail feathers. Distinguished from Honeyguide Greenbul also by more buffish underparts, paler throat and harsher call. **V** Nasal very high *tjuw tjuw* or rapid *tititjuweeh*. **H** Along streams and rivers in forest. **D** SE Nigeria to Central African Republic, DR Congo.

35 SPOTTED GREENBUL *Ixonotus guttatus* 17cm Triangular white feather tips on wings and lower back are distinctive. **V** Very high *tji-terk*. **H** Forest canopy. **D** W and C Africa, from Sierra Leone to N Angola and Uganda.

36 JOYFUL GREENBUL *Chlorocichla laetissima* 20cm Separated from Yellow-bellied Greenbul by more uniform and bright green upperparts and yellow underparts. **V** Low hoarse rather excited *tjuut-tjuwt-tjuut-tuu-terre-tuut*. **H** Forest. **D** From S South Sudan to E DR Congo, SW Kenya.

37 PRIGOGINE'S GREENBUL *Chlorocichla prigoginei* 19cm Separated from Joyful Greenbul by white chin and lore. **V** Not recorded. **H** Forest patches at higher elevations. **D** E DR Congo.

38 YELLOW-BELLIED GREENBUL *Chlorocichla flaviventris* 21cm Note olive-brown upperparts, yellowish underparts, broken eye-ring which is broader above eye. **V** Mid-high miaowing *tjuk tjuk tjeeh-tjeeh tjee tutje tji-tji-tji-tji-tji*. **H** Forest, riverine belts, coastal shrub. **D** From Angola to S Somalia and N South Africa.

1
2
3
4
5
6
7
8
9
10
11
12
13
14
15
16
17
18
19
20
21
22
23
24
25
26
27
28
29
30
31
32
33
♂
♀
34
35
36
37
38

BULBULS *CONTINUED*

1 FALKENSTEIN'S GREENBUL *Chlorocichla falkensteini* 18cm No other mainly green, yellow-throated greenbul in its range and habitat. **V** Mid-high miaowing chatters like *tu-tju-tji-wiuuh*. **H** Thick bush outside forest. **D** Cameroon to Angola.

2 SIMPLE GREENBUL *Chlorocichla simplex* 19cm Note broken eye-ring and white throat. Shy skulker in foliage. **V** High short hurried miaowing chatter *mew-mau-mi-chet-chet*. **H** Forest edge, savannah scrub, thickets, abandoned cultivation. **D** W and C Africa, from Guinea Bissau to W South Sudan, W Uganda, DR Congo, N Angola.

3 YELLOW-THROATED LEAFLOVE *Atimastillas flavicollis* 21cm From Simple Greenbul by pale eye, absence of eye-ring, greener upperparts. Puffs out throat feathers when singing. Noisy flocks. Not shy. **V** Song is harsh, nasal and grating. **H** Open forest, riverine belts, abandoned cultivation, large gardens. **D** W and C Africa, from Senegal to W Ethiopia; Angola, Zambia.

4 SWAMP PALM BULBUL *Thescelocichla leucopleura* 20cm Note white-streaked ear-coverts and white tail tips. Noisy flocks in swamps. **V** Low, slightly rising, rapid miaowing chatters *whit-tau-wit-tau-wit-wit*. **H** More or less wooded swamps with palms. **D** W and C Africa, from Senegal to DR Congo.

5 RED-TAILED LEAFLOVE *Phyllastrephus scandens* 20cm Note large size, grey head, pale breast (but darker than rest of underparts), buff edging of flight feathers. **V** Mid-high, melodious, jackdaw-like chatters *piu-piu piew-pu-pu-pu-pu*. **H** Forest, often near streams and swampy places; also surrounding areas. **D** W and C Africa, from Senegal to Uganda and DR Congo.

6 TERRESTRIAL BROWNBUL *Phyllastrephus terrestris* 18cm Note conspicuous white chin. Often on the ground in small, noisy groups. **V** Low, scratchy, froglike notes, uttered singly or in series. **H** Dense forest undergrowth, riverine belts, dense woodland. **D** E and SE Africa, from S Somalia to E and S South Africa.

7 NORTHERN BROWNBUL *Phyllastrephus strepitans* 16cm Note rather rufous appearance with pale eye-ring. Smaller than Terrestrial Brownbul and not so olive-brown. **V** Sharp, meandering, chattering series. **H** Forest edge, open woodland, wooded grassland, thickets, riverine scrub. **D** E Africa, from Ethiopia to Tanzania.

8 GREY-OLIVE GREENBUL *Phyllastrephus cerviniventris* 16cm Similar to Northern Brownbul and Terrestrial Brownbul, but with pale bill. **V** Song is mellow but rather rasping, including elements such as *chup, chup, cherry-yeck* and a rapid *chewki-chiki-chew-chew*. **H** Forest undergrowth, riverine belts. **D** Kenya to S DR Congo, Zambia, Mozambique.

9 PALE-OLIVE GREENBUL *Phyllastrephus fulviventris* 16cm From Terrestrial Brownbul by more pronounced contrast between upperparts and (uniform) underparts, and rather striped eye area. **V** Buzzy, nasal chattering. **H** Forest undergrowth. **D** S Congo, extreme W DR Congo, W Angola.

10 BAUMANN'S OLIVE GREENBUL *Phyllastrephus baumanni* 15cm Overall drab brown with pale underparts. Tail only slightly rufous. From similar Little Greenbul (Plate 202) in same habitat by much longer bill. **V** Calls include *week* and *wik* notes, varying in volume. **H** Forest, woodland and other habitats with trees and bush. **D** W Africa, from Guinea to SE Nigeria.

11 TORO OLIVE GREENBUL *Phyllastrephus hypochloris* 15cm Very similar to Little Greenbul and Slender-billed Greenbul (Plate 202), but without eye-ring, paler below and with finer, longer bill. **V** Song begins with a raspy *whee-oo-whee* and ends with a descending *shree-shree-shree-shree*. **H** Forest. **D** S South Sudan, E DR Congo through Uganda to W Kenya, N Tanzania.

12 SASSI'S OLIVE GREENBUL *Phyllastrephus lorenzi* 15cm A small greenbul with a well-defined dark cap. **V** Unknown. **H** Forest. **D** NE and E DR Congo.

13 FISCHER'S GREENBUL *Phyllastrephus fischeri* 17cm Pale eye diagnostic. Distinguished from Cabanis's Greenbul by white (not creamy) throat and by voice. **V** Reed-warbler-like notes uttered singly or in rapid level series. Rhythmic, level, scratching series. **H** Forest undergrowth. **D** E Africa, from S Somalia to NE Mozambique.

14 CABANIS'S GREENBUL *Phyllastrephus cabanisi* 17cm Note well-defined creamy throat and creamy-white eye-ring. Eye colour may be pale grey. Small groups. **V** Harsh series of bickering notes. **H** Highland forest. **D** Angola to DR Congo, S South Sudan, Kenya, Tanzania, N Zambia.

15 PLACID GREENBUL *Phyllastrephus placidus* 15cm Note rather dark appearance with striped cheek, whitish eye-ring and buff wing-feather edges. Underparts are pale brownish grey, contrasting with white puffed-out throat. Eyes normally very pale, but Nairobi birds have dark eyes as shown. **V** Often sings in duet, with a *churr* answered by a whistled series of double notes, *chirriu, chorriu, chirriu, chorriu*. **H** Forest undergrowth. **D** E Kenya through Tanzania to NE Zambia, Malawi, NW Mozambique.

16 CAMEROON OLIVE GREENBUL *Phyllastrephus poensis* 20cm No similar greenbul in its restricted range and habitat. Note face pattern. **V** Song is a rather unmusical *chiup chop chop chip chiup chip cher*, and sometimes a loud *kweep-kweep-kwerr*. **H** Montane forest. **D** SE Nigeria, W Cameroon.

17 ICTERINE GREENBUL *Phyllastrephus icterinus* 15cm Underparts uniform pale yellow without greenish breast-band. Eye colour variable. **V** High, rapid, miaowing *weck-weck weck-weck* mixed with low, drawn-out miaowings. **H** Forest undergrowth, surrounding wooded areas. **D** W and C Africa, from Guinea to DR Congo.

18 XAVIER'S GREENBUL *Phyllastrephus xavieri* 15cm Note pale lore, rather yellow underparts and rather dark breast. **V** Very high, staccato, fluted *tju-tee-tu* and *tee-tju-tee*. **H** Forest undergrowth and nearby wooded habitats. **D** Cameroon and Gabon to W Uganda, NW Tanzania.

19 WHITE-THROATED GREENBUL *Phyllastrephus albigularis* 16cm Note striking white throat. Distinguished from Red-tailed Greenbul by overall less bright colouring, more slender build, smaller size. **V** High melodious bubbling *wit-wit-weet-wit-juu tjuu-tjuu* (first part rapidly chattered). **H** Forest edges, gardens. **D** W and C Africa, from Senegal to W Uganda.

20 YELLOW-STREAKED GREENBUL *Phyllastrephus flavostriatus* 19cm Note strikingly long bill. Yellow streaking on underparts not apparent in the field. Not shy. Constantly raises or flicks a wing on one side. **V** Mid-high, lazy *terrettjup terrettjup tjup-tjup-tjup-tjup* or high *tsjirrup-tsjirrup*. **H** Mid- and higher strata of forest. **D** E and SE Africa, from NE DR Congo to E South Africa.

21 SHARPE'S GREENBUL *Phyllastrephus alfredi* 19cm Formerly treated as conspecific with Yellow-streaked Greenbul. Olive-brown above (including head) with rufous-brown wings and tail. Underparts lack yellow streaking. **V** Similar to Yellow-streaked Greenbul, but notes are more musical. **H** Forest, especially in the highlands. **D** SW Tanzania, NE Zambia, N Malawi.

22 GREY-HEADED GREENBUL *Phyllastrephus poliocephalus* 19cm No other similar greenbuls in its restricted range and habitat except Western Greenbul (Plate 202), which is duller below. Note yellowish corners to tail. **V** Similar to Yellow-streaked Greenbul: a repeated *churp* or *tyow*, and a fast *tsjirrup*. **H** Forest. **D** SE Nigeria, W Cameroon.

23 LOWLAND TINY GREENBUL *Phyllastrephus debilis* 14cm In size and habits rather warbler-like. Note long, thin, pale bill, grey face sides, pale (but occasionally dark) eye. **V** Low, ascending, nasal, fast *mau-mau-mau-mau-kirrik*. **H** Coastal shrub, forest, forest remains. **D** SE Kenya to Mozambique.

24 MONTANE TINY GREENBUL *Phyllastrephus albigula* 14cm Formerly treated as conspecific with Lowland Tiny Greenbul. Darker above and greyer below, with olive-green crown. Eyes yellowish-orange. **V** Poorly known. Song is a rich churring, varying in pitch. **H** Montane forest. **D** Usambara and Nguru Mountains, N Tanzania.

25 RED-TAILED BRISTLEBILL *Bleda syndactylus* 21cm Note yellow underparts, rufous tail and bare, blue skin above eye. Very shy. **V** Mid-high, hurried, mewing *kiau-kiau-kiau* and many other repeated, thrushlike or magpie-like whistles, changing in speed, pitch and volume, often with crescendo. **H** Forest, riverine belts, other habitats with dense undergrowth. **D** Sierra Leone to DR Congo, N Angola, S South Sudan, W Kenya, NW Zambia.

26 GREEN-TAILED BRISTLEBILL *Bleda eximius* 20cm Note olive-yellow lores and yellow tail corners. **V** Very high, twittered, fluted *titjituu-tuu* continued with rolls, trills and clear whistles. **H** Forest edge and interior. **D** W Africa, from Sierra Leone to Ghana.

27 YELLOW-LORED BRISTLEBILL *Bleda notatus* 20cm Unmistakable by facial pattern with yellow lores. Note also yellow tail corners. **V** Rapid series starting with a chatter that turns to a nice burble. **H** Humid forest, forest patches, secondary forest. **D** C Africa, from SE Nigeria to Central African Republic, DR Congo, Uganda.

28 GREY-HEADED BRISTLEBILL *Bleda canicapillus* 20cm Distinguished from Green-tailed Bristlebill by grey head and grey (not yellow) lores. Keeps close to the ground, often at ant swarms. **V** Melodious, fluid, rapid *peeweeweeweeweeh wik wik puweeweehweeh*. **H** Forest interior. **D** W Africa, from Senegal to Cameroon.

29 WESTERN BEARDED GREENBUL *Criniger barbatus* 20cm Separated from Eastern Bearded Greenbul by yellow 'beard', often fluffed out, and more yellow undertail coverts. **V** High, fluid *piuuh-piuuh-piuuh-piuuh* or *woo-whiuuuh* (*whiuu*-part higher). **H** Forest, open forest, gallery forest. **D** W Africa, from Sierra Leone to S Nigeria.

30 EASTERN BEARDED GREENBUL *Criniger chloronotus* 20cm Note whitish 'beard' of often fluffed-out throat feathers contrasting with dark head, grey-brown breast and reddish tail. **V** Mid-high, trilled whistle *tu-tu-tu-tjiuu*. **H** Forest. **D** C Africa, from Cameroon to Central African Republic, DR Congo.

31 RED-TAILED GREENBUL *Criniger calurus* 17cm Differs from White-bearded Greenbul by dark lores and voice. Eastern Bearded, Red-tailed and White-bearded greenbuls show pure white 'beard' contrasting with cheek and breast, bare blue skin around eye, red tail. **V** High, fluted, short, series of stressed *tjuh*-notes. **H** Lower level of forest, riverine belts. **D** W and C Africa, from Senegal to DR Congo, W Tanzania.

32 WHITE-BEARDED GREENBUL *Criniger ndussumensis* 18cm Separated from very similar Red-tailed Greenbul (in same habitat) by paler lore, buffier undertail coverts and different call. **V** Call is a hard trill. **H** Forest. **D** C Africa, from SE Nigeria to E DR Congo and extreme NW Angola.

33 YELLOW-BEARDED GREENBUL *Criniger olivaceus* 16cm All-green underparts separate this species from other yellow-throated greenbuls (except Falkenstein's Greenbul, which has a different range). Quiet and unobtrusive. **V** High, level, rather staccato *tu-tu-tjuu*. **H** All levels in rainforest. **D** W Africa, from Sierra Leone to Ghana.

34 FINSCH'S BULBUL *Alophoixus finschii* 16–17cm Note creamy-yellow chin. Forages mainly in the middle storey, comes lower to feed on fruit at lower tree levels and shrubs. Occasionally makes short sallies after flying insects. **V** Calls include a subdued *wek* or *twut*, a loud nasal *hwuiikt* and a grating *scree*. **H** Broadleaved evergreen forest. **D** Malay Peninsula; Borneo; Sumatra, Indonesia.

35 WHITE-THROATED BULBUL *Alophoixus flaveolus* 22cm Note pointed crest, elongated white throat feathers and bright yellow underparts. Creeps and clambers about undergrowth in chattering parties. **V** Strident, nasal *nyak nyark nyark*, higher-pitched *yap* and shrill *shree-shree-shree*. **H** Undergrowth in evergreen forest and secondary jungle. **D** E Himalayas to Myanmar and W Thailand.

36 PUFF-THROATED BULBUL *Alophoixus pallidus* 23cm Large, noisy, conspicuous bulbul with distinct crest and elongated throat feathers, often puffed out. Occurs in small active parties, regularly forms part of mixed-species feeding flocks. **V** Calls include a harsh, abrupt *churt churt churt...chutt-chutt-chutt...* or *chutt-chutt chik-it chik-it*, also a weak *twee twee twee*. **H** Broadleaved evergreen forest, open woodland in lowlands and foothills. **D** SE Asia.

37 OCHRACEOUS BULBUL *Alophoixus ochraceus* 19–22cm Similar to Puff-throated Bulbul but slightly smaller and duller, with shorter crest. Usually encountered in loose, noisy parties feeding in fruiting trees and shrubs. **V** Song consists of warbled sequences and chattering notes; calls include a harsh *chrrt-chrrt-chrrt-chrrt... chik-chik-chik-chik* or *chi-it-chit-it-chit-it-chitit*, all often preceded by a fluty, nasal *eeyi* or *iiwu*. **H** Montane broadleaved evergreen forest, tall secondary forest with bamboo. **D** SE Asia, Sumatra, Borneo.

1
2
3
4
5
6
7
8
9
10
11
12
13
14
15
16
17
18
19
20
21
22
23
24
25
26
27
28
29
30
31
32
33
34
35
36
37

BULBULS *CONTINUED*

1 GREY-CHEEKED BULBUL *Alophoixus bres* 21–22cm Noisy and conspicuous; usually seen singly, in pairs or small groups foraging in fruiting trees and shrubs. **V** Variable; song consists of a mournful *whi-u wiu iwi* followed by a high-pitched *ii-wi-tchiu-tchiu*. **H** Broadleaved evergreen forest. **D** Malay Peninsula, Greater Sunda Islands.

2 PALAWAN BULBUL *Alophoixus frater* 21–22cm Formerly considered a race of Grey-cheeked Bulbul; actions presumed to be similar. **V** Loud *chip-pu chu-chu-chu-chu cha-wheeet*. **H** Forest, forest edge, secondary growth. **D** Palawan, W Philippines.

3 YELLOW-BELLIED BULBUL *Alophoixus phaeocephalus* 20–21cm A large, brightly coloured, relatively shy bulbul. Frequents lower storey, regularly descending to feed on the ground. **V** Subdued, buzzy *whi-ee whi-ee whi-ee* or a rasping *cherrit-berrit*; also a variety of grating notes. **H** Broadleaved evergreen forest. **D** Malay Peninsula; Borneo; Sumatra, Indonesia.

4 YELLOW-BROWED BULBUL *Acritillas indica* 20cm Note bright yellow underparts and contrasting dark eye. Found in pairs or small parties. **V** Low-toned, pleasant warble. Calls include a clear, mellow double whistle and a harsh jarring *chaink-chaink*. **H** Forest edge, secondary forest, shade coffee plantations, occasionally gardens. **D** Peninsular India, Sri Lanka.

5 HOOK-BILLED BULBUL *Setornis criniger* 20cm Noisy. Note distinctive face pattern and strong bill with hooked tip. **V** Harsh *carrrk* or *currrk*, and a harsh *chow-cho cho*. **H** Lowland primary forest, stunted ridgetop forest and abandoned plantations. **D** Borneo; Sumatra, Indonesia.

6 HAIRY-BACKED BULBUL *Tricholestes criniger* 16–17cm Note pale yellowish plumage on lores and around eye. Forages for fruit or invertebrates, and often makes short sallies to take flying insects. Unassuming but can be inquisitive. **V** Scratchy, chattering warble, interspersed with a quavering *whirrrh*; calls include a high-pitched, rising *whiiii*. **H** Primary, secondary and coastal forest, mangroves. **D** Malay Peninsula; Borneo; Sumatra, Indonesia.

7 OLIVE BULBUL *Iole viridescens* 19cm Shy and unobtrusive; forages in high bushes and in the middle and top storeys of trees. **V** Upslurred, strident, mewing *whee-ik*. **H** Broadleaved evergreen forest, semi-evergreen forest and tall secondary growth. **D** Myanmar, Malay Peninsula.

8 CACHAR BULBUL *Iole cacharensis* 19cm Formerly treated as conspecific with Olive Bulbul. Uniform brownish-olive above, dull olive below with yellowish throat and belly. Vent cinnamon-tinged. **V** Poorly known. Gives a harsh, upslurred *whhee-ik*, similar to Olive Bulbul. **H** Broadleaved forest, tall secondary growth, wooded ravines. **D** NE India, SE Bangladesh.

9 GREY-EYED BULBUL *Iole propinqua* 17–19cm Rather dull plumage, and generally unobtrusive. Seen singly or in pairs, with larger parties at fruiting trees or shrubs. **V** Calls include a nasal, loud *uuu-wit*, *beret* or *prrrit*, also a *whi-it* and a flatter *wowh* or *weeao*. **H** Broadleaved evergreen forest, forest edge, secondary growth. **D** SE Asia, from S China to Myanmar and S Vietnam.

10 BUFF-VENTED BULBUL *Iole crypta* 20–21cm Slight crest is often raised. Forages in the canopy and middle storey, sometimes descending to lower levels to feed in fruiting shrubs. **V** Calls include a musical *cherrit*, a *whe-ic* and a flatter *whirr*. **H** Broadleaved evergreen forest, forest edges and secondary growth. **D** Malay Peninsula; Sumatra, Indonesia.

11 CHARLOTTE'S BULBUL *Iole charlottae* 20cm Formerly treated as conspecific with Buff-vented Bulbul. Differs from Buff-vented in having darker ear-coverts and more yellowish below. **V** Calls include a series of harsh *wher-wher-wher* notes, but varies regionally. **H** Broadleaved evergreen forest, forest edges and secondary growth. **D** Borneo.

12 SULPHUR-BELLIED BULBUL *Iole palawanensis* 18cm Occurs singly, in pairs or in small groups. **V** Calls include a *chirrup*, *chirrrup-chi* or *chip*. **H** Evergreen forest, forest edge, secondary growth. **D** Palawan, W Philippines.

13 NICOBAR BULBUL *Ixos nicobariensis* 20cm Rather drab, lacking a crest. Often forages in noisy flocks. **V** Chattering babble. **H** Forests and secondary growth. **D** Nicobar Islands.

14 MOUNTAIN BULBUL *Ixos mcclellandii* 21–24cm A large and conspicuous bulbul with square tail, bright olive upperparts, chestnut underparts, elongated crown feathers frequently raised in a crest, elongated throat feathers often puffed out. Pairs or small parties forage in the tops of trees, although descend to feed on fruiting bushes. **V** Screechy *chirrut chewt chirrut chewt*; calls include a downward inflected *tsiuc*. **H** Broadleaved evergreen forest and secondary growth. **D** From Himalayas to China and SE Asia.

15 STREAKED BULBUL *Ixos malaccensis* 23cm Note uniform creamy-white belly. Forages mainly in the canopy, singly or in pairs; occasionally makes sorties after flying insects. **V** High-pitched, descending *chiri-chiri-chu*, or longer *ka-jee ka-jee ka-jee jueee* that sometimes runs into a delicate, descending trill; calls include harsh rasping or rattling notes. **H** Lowland and hill primary and secondary forest. **D** Malay Peninsula; Borneo; Sumatra, Indonesia.

16 SUNDA BULBUL *Ixos virescens* 20cm Note streaking on flanks. Sluggish forager in the middle storey and the canopy, singly, in pairs or in small noisy groups. **V** Calls include a *chiit-chiit* and a sharp *twink*. **H** Evergreen montane forest, forest edge, secondary growth. **D** Sumatra and Java, Indonesia.

17 SERAM GOLDEN BULBUL *Thapsinillas affinis* 22–24cm A conspicuous bulbul with olive and yellow plumage. Forages from the understorey to the subcanopy, and occasionally feeds among leaf litter on the forest floor. **V** Mournful series of sweet minor-key notes, essentially slow and leisurely, occasionally interspersed with harsh *chrek* or *kruk* notes; also a raucous, parakeet-like *ki-ki-ki-ki-kreee-kreee-kreee-kre kr kr kr*. **H** Forests, from sea-level to submontane areas. **D** Seram, S Maluku Islands, Indonesia.

18 NORTHERN GOLDEN BULBUL *Thapsinillas longirostris* 21–22cm Formerly considered conspecific with Seram Golden Bulbul. Usually encountered in small groups, foraging for fruit and invertebrates. **V** Emphatic husky, chattering notes, often accelerating into a repetitive, loud, melodious phrase; also a meandering sequence of clipped notes and a lorikeet-like *twe-twe* call. **H** Primary and secondary forest, from lowlands to montane areas; also secondary growth, plantations and scrub at the edge of disturbed forest. **D** Islands E of Sulawesi and N Maluku Islands, Indonesia.

19 BURU GOLDEN BULBUL *Thapsinillas mysticalis* 22–24cm Formerly considered conspecific with Seram and Northern Golden bulbuls; actions presumed to be similar. **V** Begins with a series of *kek-kek* notes, followed by a series of upslurred or warbled, whistled notes. **H** Broadleaved evergreen forest, light woodland. **D** Buru, S Maluku Islands, Indonesia.

20 ASHY BULBUL *Hemixos flavala* 20–21cm A distinctive combination of grey, black, white and golden-olive. Forages in the middle to upper storey. Sociable, especially post breeding. **V** Metallic, perky *skrink-er rink-er-rink*; calls include twangy buzzes and musical chatters. **H** Broadleaved evergreen forest. **D** From Himalayas to SE Asia.

21 CINEREOUS BULBUL *Hemixos cinereus* 20–21cm Browner and more uniform than Ashy Bulbul. Forages mainly in the middle storey and canopy; often gregarious. **V** Brief, sweet *beelee-bear beelee burlee*; calls include a loud, ringing *tree-tree-tree*, a whining *whear*, a sharp *chiap* and a whistled *wheesh-wheesh*. **H** Broadleaved evergreen forest and secondary growth. **D** S Thailand; Peninsular Malaysia; N Borneo; Sumatra, Indonesia.

22 CHESTNUT BULBUL *Hemixos castanonotus* 21–22cm Note chestnut upperparts, white throat and belly. Noisy and conspicuous, often in large post-breeding flocks. **V** Clear, simple *whi-wi-wu* or *to-to te-wee*, usually interspersed with churring notes. **H** Broadleaved evergreen forest, forest edge and secondary growth. **D** S China, Hainan, N Vietnam.

23 SEYCHELLES BULBUL *Hypsipetes crassirostris* 24cm Note bright orange bill. No other bulbul in the Seychelles. **V** Noisy. Song consists of a mixture of clucks, squawks, whistles and chatters. **H** Woodland, plantations, gardens. **D** Seychelles, Indian Ocean.

24 REUNION BULBUL *Hypsipetes borbonicus* 22cm Pale eye diagnostic. **V** Song is loud, short and monotonous, a series of chuckling phrases. Calls also include mewing notes and a sharp *kek*. **H** Natural forest at higher altitudes. **D** Réunion, Indian Ocean.

25 MAURITIUS BULBUL *Hypsipetes olivaceus* 25cm Pinkish bill, dark crown, dark eyes. No similar bulbul on Mauritius. **V** Similar to Reunion Bulbul. **H** Natural forest. **D** Mauritius, Indian Ocean.

26 MALAGASY BULBUL *Hypsipetes madagascariensis* 24cm Not easily distinguished from Grand Comoro Bulbul where ranges overlap, but has colder and greyer underparts; elsewhere no similar bulbul. **V** Highly vocal, with a range of mostly unmusical notes. A downslurred *tireet* or *tyeeu* is often heard. **H** Any habitat with trees. **D** Madagascar, Comoros, Aldabra, Indian Ocean.

27 GRAND COMORO BULBUL *Hypsipetes parvirostris* 25cm Larger than Malagasy Bulbul, and may show a greenish wash above and a yellowish wash below. **V** Repertoire resembles other bulbuls of the Indian Ocean. Song is a series of rather monotonous calls, e.g. *kwok-kwuk-chuk-chik-waa-waa-chuk-chak-chuk-chik*. **H** Mainly forest at higher altitudes. **D** Grande Comore, Comoros, Indian Ocean.

28 MOHELI BULBUL *Hypsipetes moheliensis* 25cm Differs from Grand Comoro Bulbul in thicker bill and darker underparts. **V** Song is lower-pitched, slower and more deliberate than that of Grand Comoro Bulbul. **H** Mainly forest. **D** Mohéli, Comoros, Indian Ocean.

29 BLACK BULBUL *Hypsipetes leucocephalus* 23–27cm Forages mainly in the topmost branches; often launches sorties after flying insects. Forms large post-breeding flocks. Some races have black head. **V** Song is a series of discordant and high notes; calls include a mewing *hwiiii* and an abrupt, nasal *ber-ber-bic-ber – ber*. **H** Broadleaved evergreen and mixed deciduous forest. **D** From Afghanistan through Himalayas to China, Taiwan, SE Asia.

30 SQUARE-TAILED BULBUL *Hypsipetes ganeesa* 22cm Habits and actions similar to Black Bulbul, which was formerly considered conspecific. **V** Similar to Black Bulbul, but harsher and less nasal. **H** Broadleaved evergreen forests, shade trees in plantations. **D** SW India, Sri Lanka.

31 PHILIPPINE BULBUL *Hypsipetes philippinus* 22cm Noisy and conspicuous. Usually seen singly or in pairs, and sometimes occurs in small parties. Forages from the understorey to the canopy. **V** Variable, including a rising and falling *deut doo doo doo da-leee-eut*. **H** Broadleaved evergreen forest, forest edges, shrubby clearings, secondary growth, cultivated areas with trees, coconut groves. **D** Philippines.

32 MINDORO BULBUL *Hypsipetes mindorensis* 22cm Formerly considered a race of Philippine Bulbul; actions and habits presumed to be similar. **V** Teetering *we-to we-to we-to-to*. **H** Presumed to be similar to Philippine Bulbul. **D** Mindoro, WC Philippines.

33 VISAYAN BULBUL *Hypsipetes guimarasensis* 22cm Actions presumed to be similar to the Philippine Bulbul, with which it was formerly considered conspecific. **V** Starts *deut deut deut*, accelerating into a complex series of musical notes. **H** Presumed to be similar to Philippine Bulbul. **D** WC Philippines.

34 ZAMBOANGA BULBUL *Hypsipetes rufigularis* 24cm Distinguished from Philippine Bulbul by unstreaked rufous chin. Forages in the middle and upper levels of trees, usually in small groups. **V** Song includes *chigur-chigur-chigur-chigur-chigur* and faster *chli-chli-chli...*; calls include *chuk*, *tweer* and *choik*. **H** Forest and forest edges. **D** Basilan (Sulu Archipelago, SW Philippines).

35 STREAK-BREASTED BULBUL *Hypsipetes siquijorensis* 26–27cm Note dark cap. Forages at all forest levels, usually in small groups. **V** Strident *whit pu whit tooer whit whee-er* or *pu-er whit pu-er*. **H** Forest, forest edges, secondary growth and clearings. **D** Cebu, Tablas Island, Romblon, Siquijor, C Philippines.

36 YELLOWISH BULBUL *Hypsipetes everetti* 24–26cm Note olive-green crown and yellow belly. Generally forages from the understorey to the mid-storey, in small groups or in mixed-species flocks. **V** Long *peeeee* or *peeeeee-yuk*; song is a melodious *doo dee-dee-dee-dee-dee dee-boy-aay dee-boy-aaaay*. **H** Forest, forest edges and secondary growth. **D** Philippines.

37 BROWN-EARED BULBUL *Hypsipetes amaurotis* 28cm Note chestnut ear-coverts and white-streaked underparts. Noisy and conspicuous; usually in groups or in mixed-species flocks. Forages at all forest levels, in forest corridors and in secondary-growth patches. **V** Harsh *pee-yuk* and a descending, accelerating *pee-yuk p-uk p-uk*... **H** Deciduous, mixed and evergreen forest. **D** Sakhalin, Korean Peninsula, Japan to N Philippines.

38 WHITE-HEADED BULBUL *Cerasophila thompsoni* 20cm Distinctive smoky-grey bulbul with pure white head. Arboreal, foraging in loose flocks. **V** Rhythmic *chit-chirui chit-chiriu*... **H** Secondary forest, scrub, forest edge and grassy areas with scattered trees. **D** Myanmar, NW Thailand.

39 BLACK-COLLARED BULBUL *Neolestes torquatus* 16cm Unmistakable, with white throat and cheeks bordered by a black line. Often seen in groups feeding on fruits. **V** High, hurried, nasal, mewing *weetu-wee-tree-treetu-weet treetreet*. **H** Wooded country. **D** C Africa, from Gabon and N Angola to DR Congo, Rwanda.

1
2
3
4
5
6
7
8
9
10
11
12
13
14
15
16
17
18
19
20
21
22
23
24
25
26
27
28
29
30
31
32
33
34
35
36
37
38
39

SWALLOWS AND MARTINS HIRUNDINIDAE

1 AFRICAN RIVER MARTIN *Pseudochelidon eurystomina* 15cm Note blue-black plumage with green sheen on back, red eyes, orange bill. **V** Song is a jingling series of twitters. **H** Over and near large waters and rivers; nests in sand bars and grassy plains near water. **D** Gabon, Congo.

2 WHITE-EYED RIVER MARTIN *Pseudochelidon sirintarae* 24cm Little known, may be extinct. Plumage mostly black with white rump; very long central tail feathers. Said to have a buoyant, graceful flight. **V** Not recorded. **H** Roosts in reeds of marshy reservoirs. **D** SE Asia.

3 SQUARE-TAILED SAW-WING *Psalidoprocne nitens* 13cm Note absence of gloss on dark brown to blackish-brown plumage, and square tail. **V** Very high, single *tjirps*. **H** Forest interior, forest edge. **D** W and C Africa, from Guinea to DR Congo, NW Angola.

4 MOUNTAIN SAW-WING *Psalidoprocne fuliginosa* 17cm Uniform dark blackish brown. Distinguished from Black Saw-wing (where ranges overlap) by black, not pale or dark grey, underwing coverts. **V** Song is melodious and rather quiet, *deeu-deeu-deeop-diuop-diuop*. Flight calls include *tsee-tsu* and *chik-chuk*. **H** Forest, forest clearings, natural and cultivated areas with large trees. **D** W Cameroon, Bioko (Gulf of Guinea).

5 WHITE-HEADED SAW-WING *Psalidoprocne albiceps* 18cm Note distinctive head pattern and long, shallowly forked tail. Appears dark dusky brown without noticeable gloss. Flight slow, often low over vegetation. **V** Has a twittering call, but generally rather quiet. **H** Forest glades, woodland, sparsely wooded areas. **D** E and C Africa, from S South Sudan to N Zambia, NE Angola.

6 BLACK SAW-WING *Psalidoprocne pristoptera* 18cm Flight slow and steady with much low gliding over vegetation. Also hawks from perch. **V** Very high variations on *sreeeuw*. **H** Forest glades and edges, more or less open woodland, marshes, often near settlement. **D** Widespread across C, E and S Africa, from Eritrea to South Africa.

7 FANTI SAW-WING *Psalidoprocne obscura* 20cm Note green gloss (difficult to see in flight). Individuals without long tail streamers often seen. **V** Subdued squeaks and twitters may be too quiet to hear except at very close range. **H** Areas near forests. **D** W Africa, from Senegal to Cameroon.

8 GREY-RUMPED SWALLOW *Pseudhirundo griseopyga* 14cm Note grey rump, weak blue gloss, dark brown head, absence of white in tail. **V** Mid-high, shrieking *shree free wih shree*. **H** Sheltered areas with some grass, often near water. **D** Widespread across sub-Saharan Africa.

9 WHITE-BACKED SWALLOW *Cheramoeca leucosterna* 15cm Unmistakable pied swallow with white mantle. Head mainly white with long black eye-stripe. **V** Song is a continuous twittering. Flight call is a soft *chek*. **H** Open country, often near watercourses. **D** Australia.

10 MASCARENE MARTIN *Phedina borbonica* 13cm Distinguished from Brazza's Martin by forked tail, larger size, different range. Streaking on underparts difficult to see from some distance. Nests in small groups at cliffs or in buildings. **V** Very high drawn-out *sreeeuw*. **H** Open woodland. **D** Madagascar, Mauritius, Réunion.

11 BRAZZA'S MARTIN *Phedina brazzae* 11cm Note dark underwing coverts and striped underparts (looking uniform dark from some distance). **V** Unknown. **H** Riverine belts with sand banks. **D** Congo, DR Congo, NE Angola.

12 BROWN-THROATED MARTIN *Riparia paludicola* 12cm Variable, with pale or brown underparts. Weak, fluttering flight that can recall a small bat. **V** Soft twittering, often uttered in flight. Calls include a low *chrrr* and a harsh *svee-svee*. **H** Rivers, streams and lakes; also forages over grasslands. **D** Widespread across Africa, Madagascar.

13 GREY-THROATED MARTIN *Riparia chinensis* 12cm Gregarious. Weak fluttering flight that can recall a small bat. Occasionally some birds show a paler throat. **V** Soft twittering, often uttered in flight. Calls include a low *chrrr* or *skrr*, and a harsh *svee-svee*. **H** Rivers, streams and lakes; also forages over grasslands. **D** From Tajikistan to S China, Taiwan, SE Asia.

14 CONGO MARTIN *Riparia congica* 13cm Restricted range. Separated from Sand Martin by less well-defined breast-band, from pale-bellied form of Brown-throated Martin by pale throat. **V** Unknown. **H** Over large rivers and nearby forests. **D** Congo River.

15 SAND MARTIN *Riparia riparia* 12cm Upperparts uniform grey-brown, underparts white with brown breast-band. Rapid, light flight usually low over water or ground. Nests colonially, excavating tunnels in sand cliffs. Known in North America as 'Bank Swallow'. **V** Short, harsh, often-repeated *tschr*, *chirr* or *shrrit*. Song is a repeated *wit wit dreee drr drr drr*. **H** Open lowland country, especially near water; requires banks for nest holes. **D** Breeds across North America, Europe, Asia. Winters South America, sub-Saharan Africa, SE Asia.

16 PALE MARTIN *Riparia diluta* 13cm Very similar to Sand Martin, but paler and greyer, with breast-band less well defined. Fast fluttery flight, often in the company of other swallows. **V** Short grating twittering. Calls include a short *ret* or *brrit*. **H** Mainly open country with nearby water; regularly roosts in reedbeds. **D** Breeds C Asia to Mongolia, S Siberia, China. Winters Pakistan, India, S China.

17 BANDED MARTIN *Riparia cincta* 16cm Distinguished from Sand Martin by darker, more contrasting upperparts, very dark eye area, whitish underwing coverts. Narrow, white eyebrow diagnostic. **V** Mid-high hurried twitters in phrases of any length. **H** Open, bushed grassland near rivers, marshes, dams, ponds. **D** Sub-Saharan Africa, widespread. Breeding visitor to South Africa; non-breeding visitor to W Africa; mainly resident in central part of range.

18 TREE SWALLOW *Tachycineta bicolor* 13–15cm Iridescent blue-black upperparts, white underparts. Usually alone or in small groups. Flight light, often straight and direct with sudden dips or turns to catch prey. **V** High chirping or twittering; when alarmed, a harsh chatter. Song starts with three descending notes followed by a liquid warble. **H** Open woodland, near water. Nests in tree cavities. **D** Widespread in breeding season over much of North America. Winters S USA to Central America and Caribbean.

19 MANGROVE SWALLOW *Tachycineta albilinea* 11–12cm Feeds alone or in pairs. Flight generally direct with fast wingbeats and some gliding, normally low over water. **V** Chirping *chiri-chrit*, *chrit* or *chriet*. **H** Coastal areas, including beaches and mangroves; inland over various water bodies such as lakes, marshes, rivers. **D** Mexico to Panama.

20 TUMBES SWALLOW *Tachycineta stolzmanni* 12.5cm Note grey shaft streaking in whitish areas of plumage. **V** Soft, hoarse shrieks. **H** Arid scrub, overgrazed areas. **D** SW Ecuador, NW Peru.

21 WHITE-WINGED SWALLOW *Tachycineta albiventer* 14cm White wing patches diagnostic. Flies low over water or ground, hawking insects. Often perches on branches overhanging water. **V** Shrill *urreeeet* or *chirrup*. **H** Open, wet, lowland areas, primarily mangroves, rivers, lakes. **D** Widespread across South America, from Colombia to N Argentina.

22 WHITE-RUMPED SWALLOW *Tachycineta leucorrhoa* 13.5cm Separable from smaller Chilean Swallow by white on forehead. **V** Mixture of small rolls and inhaled *see* notes. **H** Open and semi-open areas near water and woodland edges. **D** Bolivia to C Brazil, Paraguay, Uruguay, C and E Argentina.

23 CHILEAN SWALLOW *Tachycineta leucopyga* 12.5cm Very like White-rumped Swallow, but lacks white forehead. **V** Twittered *tzee* notes. **H** Open and semi-open areas, woodland, forest edges, human habitation. **D** Breeds S Chile, S Argentina. Winters to N Argentina.

24 GOLDEN SWALLOW *Tachycineta euchrysea* 12cm Iridescent bronzy-green to copper-bronze upperparts. Flies low, darting after insects, alone or in small groups. Also seen perched on tall dead pines. **V** Soft, repeated *tchee-weet*. **H** On Hispaniola usually in open country, mountain pine forests and rainforests. On Jamaica occurs over open areas such as sugarcane fields. **D** Jamaica, Hispaniola.

25 VIOLET-GREEN SWALLOW *Tachycineta thalassina* 12–13cm Upperparts green and violet, with white of ear-coverts extending to above eye. Females generally duller than males. Flight rapid and direct including some gliding. Forages in small groups or loose flocks. **V** Includes *chee-che*, *chilip* or *chip-lip*; also creaking *twee tsip-tsip-tsip...* **H** Open coniferous, deciduous or mixed woodland and around human habitations; feeds over open country, water or forest canopy. **D** Breeds W North America, from Alaska to Mexico. Winters Mexico to Costa Rica.

26 BAHAMA SWALLOW *Tachycineta cyaneoviridis* 15cm Active mainly in the evening or during overcast weather, chasing insects either high up, where gliding flight seems to be the norm, or low over the ground in a rapid darting flight. **V** Metallic *chep* or *chi-chep* and a plaintive *seew-seew-seew-sew*. **H** Pine forests, woodland clearings, open fields and urban areas. **D** Bahamas, Cuba.

27 PURPLE MARTIN *Progne subis* 19–22cm Male glossy blue-black; female paler and more streaked. Forages high in the air, frequently at 50m or more. Less manoeuvrable than smaller swallows, alternates flapping flight with gliding on outstretched wings. **V** Rich *cherr* and a melodious whistle; also a gurgling croak uttered by the male, while the female gives various chortles. **H** Open areas, often near water; villages and towns. **D** Breeds across North America, from S Canada to N and C Mexico. Winters South America.

28 CUBAN MARTIN *Progne cryptoleuca* 18–19cm Foraging techniques similar to Purple Martin, which means high-flying birds are virtually impossible to identify. Sometimes considered to be a race of Purple Martin. **V** Melodious warble and a gurgling that includes a high-pitched *twick-twick*, similar to, but said to be distinct from, Purple Martin. **H** Lowland open areas, swamp borders, towns and cities. **D** Breeds on Cuba. Presumed to winter in South America.

29 CARIBBEAN MARTIN *Progne dominicensis* 20cm A large metallic blue swallow with a white belly. Forages at high and low level; chases insects that are disturbed by cattle; flights consist of gentle flapping interspersed with gliding. **V** Very similar to Purple and Cuban martins. **H** Open and semi-open areas, usually near water, coasts, cliffs, and towns. **D** Breeds on Caribbean islands other than Cuba. Believed to winter in NE South America.

30 SINALOA MARTIN *Progne sinaloae* 18cm A poorly known species. Not reliably separable from Caribbean Martin, but white breast area is broader. **V** Soft, nasal shrieks. **H** Pine–oak woodland. **D** Breeds W Mexico. Presumed to winter in South America.

1
2
3
4
5
6
7
8
9
10
11
12
pale-bellied morph
dark morph
13
14
15
16
17
18
19
20
21
22
23
24
25
26
27
28
29
30

SWALLOWS AND MARTINS *CONTINUED*

1 GREY-BREASTED MARTIN *Progne chalybea* 16–18cm Note greyish breast and white belly. Forages at medium heights; spends much time gliding, interspersed with fast flapping when chasing insects; also recorded feeding on the ground when there is an abundant source of insects. **V** Variable, including a *cheur*, a rattle and a *zurr*. **H** Lowland woodland, forest clearings, savannah, farmland, mangroves, around human habitations. **D** Central and South America, from Mexico to C Argentina.

2 GALAPAGOS MARTIN *Progne modesta* 15cm Smaller than Purple Martin (Plate 205, uncommon in Galápagos). Male steel-blue; female rather uniformly dark. **V** Poorly known; includes a twittering *tchur-tchur*. **H** Arid lowland scrub. **D** Galápagos.

3 PERUVIAN MARTIN *Progne murphyi* 17cm Male dusky steel-blue; female ashy grey with darker back and wings. **V** Mixture of scratchy and rolling notes and twitters. **H** Natural and cultivated open and semi-open areas, mainly along coasts. **D** W Peru, N Chile.

4 SOUTHERN MARTIN *Progne elegans* 17cm Male iridescent bluish-purple, but often appears almost black; female barred grey below. Forages alone or in small groups, at high or low level. Flight is slow and weak with much gliding. **V** Variety of chirruping calls and trills. **H** Grasslands, dry forest, around human habitations. **D** S South America.

5 BROWN-CHESTED MARTIN *Progne tapera* 16cm Sandy upperparts, white underparts with indistinct brown breast-band. Forages alone or in small groups. Flight is fast and low over vegetation or water, slower and weaker around trees or open ground. **V** Contact call is *chu-chu-chip*. **H** Open or semi-open areas with trees, often near water; also around human habitations. **D** Widespread from Panama to C Argentina.

6 BLUE-AND-WHITE SWALLOW *Notiochelidon cyanoleuca* 12cm Unmistakable. Steel-blue upperparts and white underside, with contrasting black undertail coverts. **V** Very high slightly mournful drawn-out *psweet psweet*, tailing off. **H** Montane areas near towns and villages. **D** Costa Rica, Panama, throughout South America.

7 BROWN-BELLIED SWALLOW *Notiochelidon murina* 13.5cm Note entirely black and blue upperparts and dusky underparts. **V** Nasal twittering. **H** Natural and cultivated open areas. **D** W Venezuela to C Bolivia.

8 PALE-FOOTED SWALLOW *Notiochelidon flavipes* 12cm Rufous chin and black vent are diagnostic. **V** Sharp *zeet* note. **H** Edges and clearings in cloud forest and elfin forest of the northern Andes. **D** W Venezuela to C Bolivia.

9 BLACK-CAPPED SWALLOW *Notiochelidon pileata* 13cm Resembles Blue-and-white Swallow, but note blue restricted to forehead, brown spotting to throat sides and brown flanks. **V** Song is a buzzy chatter. **H** Woodland, towns, villages. **D** S Mexico to Honduras, El Salvador.

10 ANDEAN SWALLOW *Haplochelidon andecola* 13.5cm Note grey throat, grading towards pale belly and whitish vent. **V** Short, hoarse and musical twitters. **H** Open areas near cliffs and habitation in the high Andes. **D** C Peru to NW Argentina.

11 WHITE-BANDED SWALLOW *Atticora fasciata* 14.5cm Unmistakable, with glossy blue-black plumage, contrasting white breast-band, deeply forked tail. **V** Musical mini-rattles. **H** Along forest rivers and broad streams. **D** Amazonia.

12 BLACK-COLLARED SWALLOW *Atticora melanoleuca* 14.5cm Full black collar and black vent are diagnostic. **V** Very high, sweet and scratchy twittering. **H** Over rivers, especially near rapids and waterfalls. **D** E Colombia through C Venezuela, the Guianas, N Brazil, SE Amazonia, NE Argentina.

13 WHITE-THIGHED SWALLOW *Neochelidon tibialis* 12cm Dark uniform brown plumage distinctive. White thighs difficult to see. **V** Constant *tseet-tseet* uttered by feeding birds. **H** Forest edge and clearings, often at rivers. **D** Panama to W Ecuador, the Guianas; SE Brazil.

14 NORTHERN ROUGH-WINGED SWALLOW *Stelgidopteryx serripennis* 13cm Greyish-brown upperparts, creamy-white underparts, with pale greyish-brown on breast and flanks. Lacks distinct breast-band of Sand Martin (Plate 205). Flight direct with leisurely but purposeful wingbeats; usually forages low over water or land. **V** Low, harsh *prrit*; also a rising *frrip-frrip-frrip*... **H** Open areas, wetlands. **D** Breeds across North and Central America, from SE Alaska and S Canada to Costa Rica. Winters S USA to Panama.

15 SOUTHERN ROUGH-WINGED SWALLOW *Stelgidopteryx ruficollis* 13cm Dusky brown with pale rump, rusty throat, pale yellowish belly. **V** Rolling, hoarse *zreet* note and other calls. **H** Semi-open areas, especially near water. **D** Central and South America, from Costa Rica to N Argentina.

16 TAWNY-HEADED SWALLOW *Alopochelidon fucata* 12cm Slightly smaller than Southern Rough-winged Swallow, with distinctive all-rufous head. **V** Mixture of *seet* notes and very high and low twittering. **H** Savannah, other open areas. **D** Bolivia to SC Brazil and south to N Argentina.

17 BARN SWALLOW *Hirundo rustica* 18cm Note metallic blue-black upperparts, rufous forehead and throat, very long outer-tail feathers. Underparts vary from white to deep chestnut. An agile, fast flier that twists and turns to capture flying insects. Post breeding often occurs in large flocks. **V** Song is a melodious twittering interspersed with a grating rattle, given in flight or when perched. Calls include a *vit-vit*, a sharper *flitt-flitt* and a *chir-chir* when agitated. **H** Open country and cultivations, usually not far from water. **D** Widespread, worldwide. Breeds across North America, Europe, N Africa, Asia. Winters South America, sub-Saharan Africa, S and SE Asia.

18 RED-CHESTED SWALLOW *Hirundo lucida* 15cm Note large chestnut throat patch, dusky underwing coverts, white underparts and large white tail spots. **V** Song resembles Barn Swallow, but is slower and less varied. **H** Montane grassland, marshes, lakes, rivers. **D** W, C and NE Africa: SW Mauritania to Togo and Benin; Congo and DR Congo; Ethiopia.

19 ANGOLAN SWALLOW *Hirundo angolensis* 15cm Similar to Barn Swallow, but with brownish underparts. Note dusky underwing coverts. **V** Song is a rather weak twitter. **H** Open and bushed habitats, river edges, lakes, swamps, forest, human settlements. **D** Gabon and Angola to Uganda, W Kenya, Tanzania, N Malawi, N Zambia.

20 PACIFIC SWALLOW *Hirundo tahitica* 13cm Dark glossy steel-blue upperparts; rufous forehead, ear-coverts, throat and upper breast. Flight fast with glides and frequent swerving and banking. **V** Twittering song and *titswee* call. **H** Mainly coastal, near human habitations. **D** SE Asia to New Guinea, Pacific islands.

21 HILL SWALLOW *Hirundo domicola* 13cm Resembles Pacific Swallow, and was formerly considered conspecific, but has greener gloss on upperparts. **V** Twittering song and a *titswee* call. **H** Grassy hills near plantations. **D** S India, Sri Lanka.

22 WELCOME SWALLOW *Hirundo neoxena* 15cm Very similar to Barn Swallow but not in same range. Note absence of black breast-band. **V** Song is a rhythmic series of trills, nasal twitters and mewing notes. **H** Open habitats near water; common in settled areas. **D** Australia, New Zealand.

23 WHITE-THROATED SWALLOW *Hirundo albigularis* 16cm Note white throat, narrow black breast-band, chestnut forehead. **V** Extreme high, soft *sreeeeeeh sreeeeeeh* in flight. **H** Open grass plains, inundations, dams, rivers, settlements. **D** Breeds Botswana, Zimbabwe, South Africa. Winters north to Angola, S DR Congo, Zambia.

24 ETHIOPIAN SWALLOW *Hirundo aethiopica* 13cm Note incomplete dark breast-band. Low-flying fast flight, with few glides. **V** Song is a melodious weak twittering. Call is a repeated *cheep* or *chi*, and *preut* when alarmed. **H** Open areas, open woodland, coastal cliffs, human habitations. **D** Senegal to Ethiopia, Somalia, Kenya, N Tanzania.

25 WIRE-TAILED SWALLOW *Hirundo smithii* 14–21cm Distinctive rufous forehead and crown. Long thin tail streamers may be hard to see. Flight fast, often low over water. **V** Song is a twittering *chirrickweet-chrrickweet*. Calls include *chit-chit* and, when alarmed, *chichip-chichip*. **H** Grassland, cultivations, urban areas, usually near water. **D** Widespread across sub-Saharan Africa, SC and SE Asia.

26 BLUE SWALLOW *Hirundo atrocaerulea* 21cm Note very long thin tail streamers (shorter in female) and blue-black plumage. **V** High rapid *weetweetweetweetweet*... mixed with short chirps and twitters. **H** Open bushed and wooded grassland near swamp or forest edges, often at high altitudes. **D** Patchy distribution from SE DR Congo and C Tanzania to E South Africa.

27 WHITE-BIBBED SWALLOW *Hirundo nigrita* 15cm Note white throat and large white windows in almost square tail. **V** Song is a weak twitter. Calls include *hweet* and *hwit*. **H** Along forest streams, rivers. **D** Sierra Leone to W Uganda, S DR Congo, N Angola.

28 PIED-WINGED SWALLOW *Hirundo leucosoma* 13cm Unmistakable, with large white patches on inner wing coverts. **V** Rather silent. Song and call both consist of quiet *chwit* notes. **H** Dry, grassy plains with large trees, around settlements, open forest. **D** Senegal to Nigeria.

29 WHITE-TAILED SWALLOW *Hirundo megaensis* 15cm Note white inner tail feathers. **V** Not well known, but twittering flight calls like those of other swallows have been recorded. **H** Arid plains with scarce grass, bush or trees. **D** S Ethiopia.

1
2
3
4
5
6
7
8
9
10
11
12
13
14
15
16
17
18
19
20
21
22
23
24
25
26
27
28
29

SWALLOWS AND MARTINS *CONTINUED*

1 BLACK-AND-RUFOUS SWALLOW *Hirundo nigrorufa* 15cm Often appears black in the field. Similar to Red-breasted Swallow but lacks rufous rump. **V** Song is a shrill warble. Calls include a high-pitched *tsek*. **H** Grass plains, open woodland, marsh edges. **D** C Angola through S DR Congo to E Zambia.

2 PEARL-BREASTED SWALLOW *Hirundo dimidiata* 14cm Distinguished from Common House Martin by lack of white rump, less strong gloss on upperparts. **V** Mid-high, nasal *sreet*, *sreet-sreet* or *sriauw-sriauw*. **H** More or less wooded and bushed habitats, often near swamps and settlements. **D** Angola to Mozambique and south to South Africa.

3 EURASIAN CRAG MARTIN *Ptyonoprogne rupestris* 15cm Shows pale spots on spread tail. Powerful, slow but agile flight with much gliding. Usually encountered in small parties; larger flocks occur post breeding. **V** Quiet, throaty, rapid twitter. Calls include a *prrit*, a warning *zrrr*, a plaintive *whee*. **H** Cliffs, gorges and old buildings. **D** S Europe and N Africa through S Asia to Mongolia and E China. Northern and eastern populations migratory, wintering in Africa, India.

4 PALE CRAG MARTIN *Ptyonoprogne obsoleta* 13cm Flight slow with regular gliding; white spots show when tail spread. Usually in small groups, with larger flocks post breeding. Considered by some authorities to be a race of Rock Martin. **V** Low twitter. Calls include a high-pitched *twee*, a rapid *chir-chir-chir* and a low *chirp*. **H** Cliffs, gorges and ravines in dry country. **D** N Africa, Arabian Peninsula, east to Pakistan.

5 ROCK MARTIN *Ptyonoprogne fuligula* 13cm Separated from dark form of Brown-throated Martin (Plate 205) by paler underparts (contrasting more distinctly with upperparts), white spots in tail (difficult to see unless spread tail is seen against dark background). Occurs singly or in small colonies. **V** Very high, shrill *sreeh* and *tu-sjir*. **H** Near cliffs, houses, bridges. **D** Africa, widespread.

6 DUSKY CRAG MARTIN *Ptyonoprogne concolor* 13cm Mostly dark sooty brown, with pale spots on spread tail. Flight slow with frequent periods of gliding; usually in pairs or small groups, with larger parties post breeding. **V** Twittering song and a soft *chit-chit* contact call. **H** Mountainous and hilly areas with cliffs, caves and gorges; also old buildings and urban areas. **D** India, SE Asia.

7 COMMON HOUSE MARTIN *Delichon urbicum* 13cm White rump. Flies with much gliding and soaring, often at a great height. Gregarious, often in the company of other swallows and swifts. **V** Song consists of a rapid twittering interspersed with a dry rattling. Calls include an abrupt *prrt*, a longer *pri-pit* and a *za-za-za*. **H** Open areas, coastal cliffs, cultivation, human habitations, including urban areas. **D** Widespread, breeding from Europe and N Africa to E Asia. Winters Africa, SE Asia.

8 ASIAN HOUSE MARTIN *Delichon dasypus* 12cm Square-ended tail. Flight contains frequent gliding, swooping and banking, often at a great height. Regularly accompanied by other swallows and swifts. **V** Shrill flight call and a soft trilling song. **H** Gorges, valleys and around villages in hilly or mountain areas, generally at higher altitude than Common House Martin. **D** Breeds Himalayas to China, Korean Peninsula, Japan, Taiwan. Winters SE Asia, Greater Sunda Islands, Philippines.

9 NEPAL HOUSE MARTIN *Delichon nipalense* 13cm Rump and underparts greyer and tail shallower-forked than Common House Martin. Flight includes much gliding and swooping at a great height or along cliffs and above treetops. Regularly joins with other swallows and swifts. **V** In flight utters a high-pitched *chi-i*. **H** Wooded mountain ridges with cliffs and river valleys; also around mountain villages. **D** Himalayas to NW Vietnam. Resident.

10 GREATER STRIPED SWALLOW *Cecropis cucullata* 19cm Note glossy deep-blue upperparts, chestnut cap and rump, pale cheek and throat, finely streaked underparts. **V** Mid-high, unhurried *trrir-trrit-treet-treet*. **H** Open grassland. Avoids woodland and forest. **D** Breeds S Africa, from Namibia and Zimbabwe to South Africa. Winters north to DR Congo and Tanzania.

11 LESSER STRIPED SWALLOW *Cecropis abyssinica* 15–19cm Note boldly streaked underparts. Flight erratic with much fluttering and gliding, although can be swift and direct. **V** Song: *chip-chip-chwip-kreek-kree-kree-kreep-chwip-kreee*, more vigorous and squeaky than Barn Swallow (Plate 206). Call: *tee-tee-tee*. **H** Forest edge, open woodland, grassland, savannah and around villages and towns. **D** Widespread across sub-Saharan Africa.

12 RED-BREASTED SWALLOW *Cecropis semirufa* 24cm Distinguished from Black-and-rufous Swallow by red rump, longer tail streamers. **V** Mid-high, explosive, rather sharp *tit ti-tri-UOOH*. **H** Open and bushed habitats with some scattered trees, often near water. **D** Senegal to South Sudan and South Africa. Resident near equator; breeding visitor to north and south of range.

13 MOSQUE SWALLOW *Cecropis senegalensis* 21cm Similar to Red-rumped Swallow, but larger and lacks black undertail coverts. Note white throat and white underwing coverts. **V** Song is rambling and discordant. Call is a short, nasal *hank*. **H** Open and bushed habitats, often near water. Avoids the most arid areas. **D** Widespread in sub-Saharan Africa except in south.

14 RED-RUMPED SWALLOW *Cecropis daurica* 16–17cm Distinguished from most other swallows by chestnut collar and rump, black undertail coverts, no white in tail, no breast-band. Flight slow and graceful, with much gliding and soaring. **V** Song is shorter and quieter than Barn Swallow (Plate 206). Calls include *djuit*, aggressive *krr* and territorial mewing. **H** Open hilly country, valleys and gorges where cliffs, caves, old buildings or bridges can be used for nest sites. **D** Widespread across S Europe, Africa, Asia. Northernmost populations winter in Africa, S and SE Asia, N Australia.

15 SRI LANKA SWALLOW *Cecropis hyperythra* 17cm Formerly treated as a race of Red-rumped Swallow. Cheeks and underparts entirely dark rufous (darker than Red-rumped) with shorter tail. Lacks rufous hind-collar. **V** Song is a rambling medley, distinct from Red-rumped Swallow. Call is a loud *chweet*. **H** Open areas and lightly wooded hillsides, in lowlands and foothills. **D** Sri Lanka.

16 WEST AFRICAN SWALLOW *Cecropis domicella* 18cm Sometimes treated as a race of Red-rumped Swallow. Note rufous neck isolating black cap from mantle, and black undertail coverts. **V** Mid-high nasal *weet-weet*. **H** Rocky hillsides. **D** Senegal to E Sudan.

17 STRIATED SWALLOW *Cecropis striolata* 19cm Differs from Red-rumped Swallow in heavier streaking and lack of clear collar. Flight is slow and buoyant, usually low over the ground or around cliffs. **V** Soft twittering song; calls include a drawn-out *quitsch*, a *pin* and a repeated *chi-chi-chi*. **H** Open areas, often near water, cliffs and river gorges. **D** SE Asia, Taiwan, Philippines, Indonesia.

18 RUFOUS-BELLIED SWALLOW *Cecropis badia* 19–20cm Similar to Striated Swallow, and often considered conspecific, but underparts deep rufous with only faint streaking. **V** Twittering warble; calls include a *tweep*, a trembling *schwirrr* and a *chi-chi-chi*. **H** Open hilly country, rocky outcrops, over open country and forest and, occasionally, around habitations. **D** Malay Peninsula.

19 RED-THROATED CLIFF SWALLOW *Petrochelidon rufigula* 13cm Differs from South African Cliff Swallow in white spots in tail, darker, plain, rufous rump, blue-black crown and neck, clean throat, less obvious breast-band. Ranges barely overlap except in N Zambia and maybe in Cabinda. **V** Twittering song, and high-pitched *chrr, chrr* flight call. **H** Lowland grassland, cultivation, often near water and cliffs. **D** Gabon to DR Congo, south to C Angola and Zambia.

20 PREUSS'S CLIFF SWALLOW *Petrochelidon preussi* 14cm From Red-throated Cliff Swallow and South African Cliff Swallow by absence of rufous on throat, rump and undertail coverts. From Common House Martin by less 'clean' appearance. Red spot behind eye difficult to see. **V** Noisy twittering song, and *prrrp-prrrp* flight call. **H** Natural and cultivated plains with wood near streams and/or settlement. **D** W and C Africa, from Guinea Bissau to NE DR Congo.

21 RED SEA CLIFF SWALLOW *Petrochelidon perdita* 15cm Distinctive cliff swallow with grey rump, blackish throat and white underwing coverts. Note narrow rufous breast-band and rufous wash on undertail coverts. Known only from a single male specimen found on the Red Sea coast of Sudan. **V** Not known. **H** Not known. **D** E Sudan.

22 SOUTH AFRICAN CLIFF SWALLOW *Petrochelidon spilodera* 14cm Similar to Red-throated Cliff Swallow; absence of tail spots diagnostic. **V** Very high *sree-titirrrrrr*. **H** Grasslands not too far from buildings, bridges or buildings. **D** Namibia, Zimbabwe, SE Botswana, South Africa. Winters in S Congo, DR Congo.

23 FOREST SWALLOW *Petrochelidon fuliginosa* 14cm Note compact build, mostly blackish-brown plumage with darker cap and rusty brown throat. Distinguished from saw-wings (Plate 205) by less fluttering, more swift and purposeful flight. **V** Not very vocal. Flight calls include *hweet* and *pitchree*. **H** Rainforest. **D** SE Nigeria to Gabon and NW Congo.

24 STREAK-THROATED SWALLOW *Petrochelidon fluvicola* 11–12cm Dull chestnut crown. Weak, fluttering flight. Often occurs in flocks and in the company of other swallows. **V** Twittering song and a sharp *trr-trr* call. **H** Open country, foothills, cultivation and around human habitations; often near water. **D** NE Afghanistan, Pakistan, India.

25 FAIRY MARTIN *Petrochelidon ariel* 11cm Note rufous crown. Generally in small groups; often mixes with other swallows. Flight is slow and fluttery, usually high up. **V** Calls include a short *chrrr* or *prrrt-prrrt*. **H** Open woodland and open country, usually near water. **D** Australia; resident.

26 TREE MARTIN *Petrochelidon nigricans* 13cm Differs from Fairy Martin in having deep blue crown. Flight is swift, with frequent twists and turns; often hawks for insects over water. Usually in groups of 10–30 birds. **V** A short, thin, high-pitched twittering, a low-pitched twitter, and a high-pitched, repeated, hard, nasal *zzit* or *tsweet*. **H** Open areas, including open woodland, grassland, human habitations, including towns and cities. **D** Australia. Migratory in SE (wintering N to New Guinea); mostly resident elsewhere.

27 AMERICAN CLIFF SWALLOW *Petrochelidon pyrrhonota* 13–15cm Separated from Cave Swallow by dark throat, white forehead and (mostly) different range. Usually in flocks, hawking insects both high in the air and near to the ground; frequently soars and glides. Nests in large colonies. **V** Soft, husky *verr*, *purr* or *chur*. **H** Variety of habitats, including grasslands, towns, broken forest, water's edge. **D** Breeds North and Central America, from Alaska to Nova Scotia and Mexico. Winters South America.

28 CAVE SWALLOW *Petrochelidon fulva* 12.5–14cm Note square tail and pale orange rump. Usually in loose flocks; flight strong with frequent periods of gliding. Nests colonially. **V** Calls include a *chu-chu*, *weet*, *cheweet* and a short *choo*. Song is a series of squeaks and a warble ending with a series of two-toned notes. **H** Caves or artificial structures such as bridges or churches. **D** S USA, Mexico, Caribbean.

29 CHESTNUT-COLLARED SWALLOW *Petrochelidon rufocollaris* 12cm Rufous neck is connected to rufous breast. Note whitish throat. **V** High twitters. **H** Natural, cultivated and urban semi-open areas. **D** SW Ecuador, W Peru.

1
2
3
4
5
6
7
8
9
10
11
12
13
14
15
16
17
18
19
20
21
22
23
24
25
26
27
28
29

CUPWINGS PNOEPYGIDAE

1 SCALY-BREASTED CUPWING *Pnoepyga albiventer* 9–10cm Tiny and almost tail-less. Dark rufous-brown upperparts, underparts scaled either white or coffee-tan. Nervously flicks wings while foraging in dense tangled undergrowth and among fallen mossy logs; usually in pairs, but generally solitary in winter. **V** Rapid, high-pitched, jumbled warble. Call is an explosive *tschik* or *tchik*. **H** Undergrowth in damp, shady broadleaved, fir, hemlock and birch forests, forest edge; usually near water. **D** Himalayas to Myanmar.

2 CHINESE CUPWING *Pnoepyga mutica* 9–10cm Formerly considered conspecific with Scaly-breasted Cupwing, but differs genetically and vocally. Habits like Scaly-breasted Cupwing. **V** Rapid, high-pitched, jumbled warble, but lower-pitched than Scaly-breasted Cupwing. **H** Undergrowth in damp, shady forests, forest edge; usually near water. **D** SC and C China.

3 TAIWAN CUPWING *Pnoepyga formosana* 8–9cm Similar to Scaly-breasted Cupwing, but separated by range. **V** Very high-pitched whistles; calls with a wheezy, querulous *pwshhhhhht*. **H** Dense tangled undergrowth and bamboo in montane broadleaved evergreen and mixed broadleaf–conifer forests. **D** Taiwan.

4 NEPAL CUPWING *Pnoepyga immaculata* 10cm Separated from Scaly-breasted and Pygmy Cupwings by less streaked upperparts. Hops about in low herbage or about boulders. **V** Piercing *tsi-tsu-tsi-ts-si-tsu-tsi-tsi*. Call is a sharp *tchit tchit*. **H** Vegetation and undergrowth with rocks and boulders near streams or rivers, open areas and edge of broadleaved evergreen forest, overgrown clearings, secondary growth and scrub in gulleys. **D** Himalayas.

5 PYGMY CUPWING *Pnoepyga pusilla* 9cm Forages on or near the ground, in leaf litter, tangled vegetation and on mossy logs or branches. **V** Slow, drawn-out *se-e-e-s-w*, each syllable lasting 1 sec with a 2-sec interval. Call is a repeated *tchit*. **H** Undergrowth in broadleaved evergreen forest, dense vegetation in forest ravines, mossy rocks and fallen logs. **D** C and E Himalayas, China, SE Asia, Indonesia.

CROMBECS AND AFRICAN WARBLERS MACROSPHENIDAE

6 MOUSTACHED GRASS WARBLER *Melocichla mentalis* 19cm Large and reddish. Skulking. **V** Very high unhurried basically staccato phrases with sudden accelerations *tjup tjup priweetweetweet tjup tjuwèh*. **H** Tall herbage along streams in wooded landscapes. **D** Sub-Saharan Africa, south to Angola and Mozambique.

7 CAPE GRASSBIRD *Sphenoeacus afer* 19cm Note long, pointed tail, rufous cap and black malar stripe. Rather skulking. **V** Very high, hurried, running-up twitter, *si-ri-si-ti-si-ti ... suuh-wee*. **H** Areas with long grass and scrub from coast to mountain slopes. **D** E Zimbabwe, W Mozambique to South Africa.

8 ROCKRUNNER *Achaetops pycnopygius* 17cm Note streaked head and breast, rufous belly. No similar bird in its rocky habitat. **V** Mid-high, melodious, bulbul-like descending warble and other pleasant phrases. **H** Keeps to the ground on and between rocks and grass in hilly country. **D** SW Angola, N Namibia.

9 YELLOW LONGBILL *Macrosphenus flavicans* 13cm Note white chin, throat and upper breast, and grey legs. **V** Rapidly descending, slow, fluted *fjuu fjuu fjuu fjuu fjuu fjuu-fjuu* (starting very high). **H** Dense forest undergrowth. **D** C Africa, from SE Nigeria to N Angola, DR Congo, Tanzania.

10 KEMP'S LONGBILL *Macrosphenus kempi* 13cm Warm brown above, rusty red on flanks. **V** Partly as Yellow Longbill. Also *dee-dee-dee* (8–12 times), starting rapidly, but slowed and slurred down. **H** Undergrowth in secondary forest, depleted primary forest, forest edge. **D** W Africa, from Sierra Leone to W Cameroon.

11 GREY LONGBILL *Macrosphenus concolor* 13cm Fluffier than Kretschmer's Longbill. Underparts not or only slightly paler than upperparts. Legs brownish flesh. **V** Very high, hurried, crescendoing twitter, descending or with repeated phrases such as *weet-turu-weet-*. **H** Dense forest undergrowth. **D** W and C Africa, from Guinea and Sierra Leone to Uganda, Rwanda and E DR Congo.

12 PULITZER'S LONGBILL *Macrosphenus pulitzeri* 13cm Distinguished from Yellow Longbill by pink legs, less bright colouring and buffish (not white) chin and throat. **V** Not well known, but a cheerful *ch-pwrrr* call has been recorded. **H** Dry forest. **D** W Angola.

13 KRETSCHMER'S LONGBILL *Macrosphenus kretschmeri* 14cm Distinguished from most greenbuls (Plates 202, 203) by small size and long bill. Note grey (not pinkish) legs. **V** High *ti*, directly followed by compressed *ti-der-rie dip*. **H** Forest undergrowth. **D** E Tanzania, N Mozambique.

14 NORTHERN CROMBEC *Sylvietta brachyura* 7cm Note pure grey upperparts and white belly. **V** Very high, descending, short, hurried twitter *tu-tjeet-ter-tsjee-tu-wee*. **H** Dry woodland, bush. **D** Mauritania to Somalia and Tanzania.

15 RED-FACED CROMBEC *Sylvietta whytii* 9cm Crown and upperparts grey, face and underparts tawny-buff, paler on belly. Face and underparts concolorous. **V** Very high, soft, well-separated cicada-like trills (each 2–3 secs). **H** Forest edges, woodland. **D** E Africa, from Ethiopia to Zimbabwe.

16 PHILIPPA'S CROMBEC *Sylvietta philippae* 9cm Note dark eye-stripe and reddish legs. **V** Song consists of repeated *ti-chirr-cheeziz*, rising and then falling in pitch. Calls include *chirr* and a loud *chink*. **H** Dry bushland. **D** E Ethiopia, Somalia.

17 LONG-BILLED CROMBEC *Sylvietta rufescens* 12cm Distinguished from Red-faced Crombec by broken white eyebrow and throat. **V** Very high, short, soft yet shrill *tuwee three-wee three-wee three*. **H** Dry bushveld, semi-desert with some shrub. **D** SC and S Africa, from SE DR Congo to South Africa.

18 SOMALI CROMBEC *Sylvietta isabellina* 10cm A pale, mostly grey crombec with a long and slightly downcurved bill. **V** Repeated *tichit-tichit-tiri-chirichiri chirichiri*. **H** Dry woodland and bushland. **D** Ethiopia, Somalia through Kenya to NE Tanzania.

19 RED-CAPPED CROMBEC *Sylvietta ruficapilla* 12cm Note rufous ear-coverts. Separated from Burnt-necked Eremomela (Plate 219) by lack of tail and larger size. **V** Very high, hurried, warbling *tititiuwreetuwreeThree* (rising in pitch at the end). **H** Miombo woodland. **D** SC Africa, from Gabon to Angola and Mozambique.

20 GREEN CROMBEC *Sylvietta virens* 9cm Note dull colouring with pale tawny eyebrow, cheek and throat. **V** Extreme high descending sharp fast twitter *si-see-sree-si-si-wee*, often followed by softer and lower syllables. **H** Undergrowth of forest, riverine belts. **D** W and C Africa, from Senegal to South Sudan, Tanzania, NE Angola.

21 LEMON-BELLIED CROMBEC *Sylvietta denti* 9cm Note mottled cheeks, buffy throat and pale rufous stripe down breast. **V** Extreme high siffled decelerated *seeseeseeseesee-see-see see*. **H** Canopy at forest edge, deserted cultivation. **D** W and C Africa, from Guinea to W Uganda, DR Congo, NE Angola.

22 WHITE-BROWED CROMBEC *Sylvietta leucophrys* 8cm Note striking white eyebrow and rufous-brown upperparts. **V** Song is a repeated high-pitched phrase ending in a trill, e.g. *CHIR-wi-wee-chrrrrrrr*. **H** Undergrowth of montane forests and bamboo. **D** Uganda, W Kenya to E DR Congo, Tanzania.

23 VICTORIN'S WARBLER *Cryptillas victorini* 16cm Note orange eye. **V** Very high *tutuwlEtututwlETutuWlET*. **H** Rank vegetation at higher altitudes. **D** S South Africa.

CETTIA BUSH WARBLERS AND ALLIES CETTIIDAE

24 YELLOW-BELLIED WARBLER *Abroscopus superciliaris* 9–11cm Note pale supercilium and lemon-yellow underparts. Forages mainly in the understorey, gleans from foliage or makes fly-catching sallies. Frequently flicks wings. **V** Halting, tinkling high-pitched whistled *dee-dee-dir-rit-tit-deweet* that rises at the end; repeated 3–4 times, each time higher than the last; also subdued twittering. **H** Undergrowth in evergreen forest, especially bamboo. **D** C Himalayas to S China, SE Asia, Malaysia, Indonesia.

25 RUFOUS-FACED WARBLER *Abroscopus albogularis* 8cm Note orange face and dark stripes on sides of crown. Nervous and very active; flicks wings and occasionally fans tail. **V** Repetitive, high-pitched, drawn-out, plaintive whistle; also shrill twittering. **H** Undergrowth in evergreen forest, especially bamboo. **D** E Himalayas, NE India, SE Asia.

26 BLACK-FACED WARBLER *Abroscopus schisticeps* 9cm Striking grey, black and yellow head pattern. Nervous and highly active; frequently flicks wings and tail. **V** Song is a thin, high-pitched tinkling. Call is a subdued *tit*; when alarmed gives a high-pitched *tz-tz-tz-tz-tz-tz*. **H** Moist hill and mountain forests with moss-covered trees, bamboo thickets, thick undergrowth. **D** C Himalayas to NW Vietnam.

27 MOUNTAIN TAILORBIRD *Phyllergates cucullatus* 11–12cm Note bright yellow belly. Elusive, forages in low thickets; also makes aerial sorties after flying insects. **V** Thin high-pitched 4–6-note whistle, *pee-pee-peeeeeeeeee-pee-pee* or similar. Calls include a dry descending trill, a low buzzy *kiz-ki*, a thin *trit* and a long nasal chatter. **H** Bushy thickets and undergrowth in broadleaved evergreen forest. **D** E Himalayas to S China, N Philippines, SE Asia, Indonesia.

28 RUFOUS-HEADED TAILORBIRD *Phyllergates heterolaemus* 11cm Usually seen in pairs or small parties, foraging in thick, tangled vegetation; occasionally shows itself when flitting out after a passing insect. **V** High-pitched *tee-tee lee-oot tee-lee-leee*; when alarmed, utters a rapid, trilled *trip-p-p-p-p-p-p-p*, often repeated continuously. **H** High-elevation forest and secondary scrub. **D** Mindanao, S Philippines.

29 BROAD-BILLED WARBLER *Tickellia hodgsoni* 10cm Note chestnut crown, dark stripe through eye, yellow belly. Very active; post breeding joins mixed-species parties. **V** Long series of very high-pitched notes; also a rapid metallic *witiwiwititw-chu- witiwitiwiti*. **H** Undergrowth and bamboo thickets at the edge of broadleaved evergreen forest. **D** E Himalayas to Myanmar, N Vietnam.

30 PHILIPPINE BUSH WARBLER *Horornis seebohmi* 12cm Note long greyish supercilium. Forages on or near ground, in thick cover; usually alone. **V** Rhythmic *pee chap chop, pee chap chop chi or do doooooor dor-pro-dee-dor-pee*; calls include a repeated *chekk*. **H** Open deciduous and pine forest, with understorey, thickets and low vegetation. **D** Luzon, N Philippines.

31 JAPANESE BUSH WARBLER *Horornis diphone* 18cm More prominent supercilium than Philippine Bush Warbler. Inconspicuous forager in thick foliage. **V** Calls include a dry *tchet-tchet-tchet* and a rattling *trrt*. **H** Bamboo thickets and brush. **D** S Sakhalin, Japan. Northern populations move south in winter.

32 MANCHURIAN BUSH WARBLER *Horornis canturians* 18cm Very like Japanese Bush Warbler, but with slightly more rufous crown. Often remains immobile in thick foliage. **V** Loud, fluty *wrrrrr-whuciuchi*; calls include a dry *tchet-tchet-tchet* and rattling *trrt*. **H** Dense thickets, scrub, secondary forest. **D** Breeds C and NE China, SE Siberia, Korean Peninsula. Winters S China, SE Asia, N Philippines.

33 PALAU BUSH WARBLER *Horornis annae* 15cm Note white supercilium, pale orange legs and long, thin bill. **V** A long penetrating flute-like whistle, sometimes followed by a sequence of warbling notes; may be a duet. **H** Dense forest undergrowth. **D** Palau, WC Pacific.

34 TANIMBAR BUSH WARBLER *Horornis carolinae* 12–13cm Note bright rufous crown and long buffish supercilium. Forages in dense undergrowth, alone or as part of mixed-species groups. **V** Drawn-out whistle, followed by a short, jumbled warble; calls include a soft *cherr* or *chuck*. **H** Primary and monsoon forest, secondary woodland, bamboo, scrub and forest edges. **D** Tanimbar Islands, Indonesia.

35 SHADE BUSH WARBLER *Horornis parens* 12cm Small brown forest warbler with paler supercilium. Forages in the understorey or in the canopy. **V** Song is a pure loud whistle, starting with a long, drawn-out note and ending in a flourish of 3–4 rapid notes. **H** Forest with dense undergrowth in hills and mountains. **D** Makira, S Solomon Islands.

36 BOUGAINVILLE BUSH WARBLER *Horornis haddeni* 13cm Dumpy forest warbler with rufous head, rufous-brown upperparts and grey underparts. Forages mostly on the ground. **V** Song is a series of melancholy, high-pitched whistles, in phrases of 2–3 notes, rising up the scale. **H** Montane forest. **D** Bougainville, N Solomon Islands.

37 FIJI BUSH WARBLER *Horornis ruficapilla* 13cm Dark rufous crown, rather dark upperparts with rufous tinge, especially to wing and tail feathers. Pale pink legs. **V** Typical song is a flute-like whistle, followed by a warbled *fiddle-dee-dee, tchik-tchik*. **H** Dense forest undergrowth and secondary growth. **D** Fiji.

rufous morph
1
pale morph
rufous morph
2
pale morph
rufous morph
3
pale morph
rufous morph
4
pale morph
rufous morph
5
pale morph
6
7
8
9
10
11
12
13
14
15
16
17
18
19
20
21
22
23
24
25
26
27
28
29
30
31
32
33
34
35
36
37

CETTIA BUSH WARBLERS AND ALLIES *CONTINUED*

1 BROWN-FLANKED BUSH WARBLER *Horornis fortipes* 12cm Note relatively plain face. Skulks in thick cover; more often heard than seen. **V** Sustained, rising *weeeee* followed by an explosive *chiwiyou*; calls include a harsh *chuk* or *tchuk tchuk*. **H** Undergrowth and bamboo clumps in broadleaved evergreen forest, overgrown clearings, forest edge. **D** Himalayas to E China, Taiwan.

2 HUME'S BUSH WARBLER *Horornis brunnescens* 11cm Skulks in thick cover, usually heard rather than seen. **V** Rasping *brr* and a sharp *tik*. Song is a high-pitched *see-saw see-saw see-saw...* **H** Bamboo thickets. **D** Himalayas.

3 YELLOW-BELLIED BUSH WARBLER *Horornis acanthizoides* 9.5–11cm Note yellower belly than Hume's Bush Warbler. Skulks; best located by voice. **V** Three or four thin drawn-out whistles, followed by a short series of fast see-sawing notes; calls include a rasping *brrr* and a sharp *tik-tik-tik*. **H** Bamboo and undergrowth in broadleaved and mixed forest. **D** C and SE China, Taiwan.

4 SUNDA BUSH WARBLER *Horornis vulcanius* 12–13cm Forages alone or in pairs in thick undergrowth or on the ground. **V** Rising and falling *chee-heeoow cheeoow-wee-ee-eet*; calls include a sharp, repeated *trr-trr*. **H** Stunted montane forest, with dense undergrowth, ferns or tall grasses. **D** Indonesia; Borneo; W Philippines.

5 ABERRANT BUSH WARBLER *Horornis flavolivaceus* 12–14cm Upperparts yellowish olive-green, supercilium pale yellowish, underparts buff to olive-yellow. Skulking, sometimes inquisitive; constantly flicks wings and tail. **V** Series of thin rising notes followed by a long whistle; calls include a soft *brrt-brrrt* and a sharp *chick*. **H** Thickets and tall grass clumps at forest edge or in clearings; also bamboo and undergrowth in forests. **D** Resident in SW China, W, N, E Myanmar, N Laos and N Vietnam. Winter visitor in NW Thailand.

6 GREY-BELLIED TESIA *Tesia cyaniventer* 9–10cm Note pale green supercilium. Restless forager, generally on or near the ground. Dances to and fro on branches like a clockwork toy. **V** Begins with a few high-pitched notes, followed by a series of loud slurred phrases. Call is a loud, rattling *trrrrrrk*. **H** Dense tangled undergrowth in thick forest, usually near streams. **D** Himalayas to S China, SE Asia.

7 SLATY-BELLIED TESIA *Tesia olivea* 9–10cm Note dark stripe behind eye. Secretive, although can be inquisitive; forages on or close to the ground, often tossing leaf-litter into the air. **V** Begins with 4–11 single whistled notes, followed by an explosive jumble of tuneless notes. Calls include a sharp *tchirik* spluttering *trrrrt trrrrt trrrrt...* **H** Dense undergrowth in moist evergreen forest, often near water. **D** NE India to SW China, N Thailand, N Vietnam.

8 RUSSET-CAPPED TESIA *Tesia everetti* 9–10cm Plain brown upperparts and russet cap. Forages on the ground, in undergrowth, understorey and among creepers on tree trunks. **V** Loud, staccato series of ascending and other whistled notes, alternating up and down the scale. **H** Damp forest, degraded woodland, secondary forest, scrub. **D** Lesser Sunda Islands.

9 JAVAN TESIA *Tesia superciliaris* 7cm Forages on or close to the forest floor; favours dense undergrowth, thickets in clearings, forest edge. **V** Loud, explosive, fast song of about 15 notes. **H** Montane forest. **D** Java, Indonesia.

10 CETTI'S WARBLER *Cettia cetti* 14cm Fairly uniform rich brown upperparts with greyish supercilium. Skulks, more often heard than seen, although sometimes sings in the open. **V** Sharp *chip*, a *tsuk* and when alarmed a rattling *twik-ik-ik-il*. Song is a loud and explosive *chee-weechoo-weechoo-weechoo*. **H** Reedbeds, swampy undergrowth, tall grass. **D** W Europe and N Africa to C Asia.

11 CHESTNUT-CROWNED BUSH WARBLER *Cettia major* 13cm Note rich chestnut crown. Shy and skulking, creeps on the ground. **V** Song is a short whistle followed by a rapid shrill warble. **H** Forest undergrowth, forest bushes, rhododendron shrubberies; winters in tall grass and reedbeds. **D** Himalayas to NE India, C China.

12 GREY-SIDED BUSH WARBLER *Cettia brunnifrons* 10–11cm Smaller and more slender than Chestnut-crowned Bush Warbler, with whitish (not rufous-buff) supercilium in front of eye. Skulking, but sometimes ventures to the edge of cover. **V** Song is high-pitched *dzit-su-ze-sizu* followed by a nasal *bzeeuu-bzeeuu*. **H** Subalpine bamboo and bushes in forest clearings, dense thickets in open coniferous forests; winters at lower altitudes in scrub and forest undergrowth. **D** Himalayas, NE India, N Myanmar, S China.

13 CHESTNUT-HEADED TESIA *Cettia castaneocoronata* 8–10cm Chestnut head, olive-green upperparts, largely yellow underparts. Forages, wren-like, in and around dense undergrowth or moss-covered logs or rocks. **V** Loud, explosive *cheep-cheeu-che-wit* or rapid *ti-tisu eei* followed by a slower *tis-tit-ti-wu*. **H** Undergrowth in broadleaved evergreen forest and secondary growth, often near water. **D** Himalayas to Myanmar, N Vietnam, SW China.

14 ASIAN STUBTAIL *Urosphena squameiceps* 10–11.5cm Note prominent pale supercilium and dark eye-stripe. Skulks mainly on the ground, foraging among undergrowth. **V** Song is a high-pitched, insect-like *see-see-see-see-see-see-see-see*, final notes getting progressively louder. **H** Coniferous and deciduous montane forests with thick undergrowth. **D** Breeds E Siberia, NE China, Korean Peninsula, Japan. Winters Taiwan, S China, SE Asia.

15 BORNEAN STUBTAIL *Urosphena whiteheadi* 10cm Creeps about mouse-like in thick undergrowth and on the ground. Usually alone. **V** Drawn-out, high-pitched, quiet, bat-like, squeaky *tzi-tzi-tzeeee*. **H** Montane forest. **D** Borneo.

16 TIMOR STUBTAIL *Urosphena subulata* 10cm Forages in leaf litter and around roots, fallen trees or branches. **V** Thin, high-pitched, slightly rising whistle. **H** Primary and tall monsoon forest and hillside scrub. **D** Timor and nearby islands.

17 PALE-FOOTED BUSH WARBLER *Urosphena pallidipes* 11–13cm Skulking, flits through bushes or grasses; best located by voice. **V** Explosive jumble of chattering notes. **H** Grassy areas, scrub, bracken-covered slopes and, occasionally, in open broadleaved evergreen and pine forests. **D** Himalayas to SW China, SE Asia; Andaman Islands.

18 NEUMANN'S WARBLER *Urosphena neumanni* 13cm Note large head and short tail. Forages mainly near or on the ground. **V** Loud and explosive song, *tee-tiyoo-tee*. **H** Undergrowth of mountain forests. **D** E DR Congo, SW Uganda, Rwanda, Burundi.

STREAKED SCRUB WARBLER SCOTOCERCIDAE

19 STREAKED SCRUB WARBLER *Scotocerca inquieta* 10cm Pale sandy brown with dark eye-stripe and long tail. Mouse-like actions, usually difficult to see well, although can be quite inquisitive. **V** Song variable, usually starting with a series of staccato piping notes followed by a scratchy warble. **H** Dry stony hillsides with low scrub. **D** NW Africa to Pakistan.

YELLOW FLYCATCHERS ERYTHROCERCIDAE

20 LITTLE YELLOW FLYCATCHER *Erythrocercus holochlorus* 9cm Note olive-yellow upperparts, bright yellow underparts, dark eye in plain face. **V** Song is a hurried high-pitched warbling, often combined with sharp *chip* notes. **H** Forest, moist bushland. **D** SE Somalia, E Kenya, NE Tanzania.

21 CHESTNUT-CAPPED FLYCATCHER *Erythrocercus mccallii* 10cm Bright chestnut crown distinctive. Active with much tail flirting. **V** Very high, sharp, unstructured twitter. **H** Forest. **D** Sierra Leone to NW Angola, W Uganda.

22 LIVINGSTONE'S FLYCATCHER *Erythrocercus livingstonei* 10cm Note subterminal black spots on rufous tail. **V** Combination of hurried, short, low twitters and very high, fluted, sharp, ascending twitter. **H** Open woodland, riverine belts, coastal bush. **D** S Tanzania to Mozambique, S Zambia, N Zimbabwe.

GRAUERIA, HYLIA AND *PHOLIDORNIS* *INCERTAE SEDIS*

23 GRAUER'S WARBLER *Graueria vittata* 15cm Dull greenish-olive with pale barring on face and underparts. **V** Song is a purring *wrrrrrrrrrrrrr*, rising in pitch, lasting 1–2 secs and repeated every few seconds. **H** Dense montane forest undergrowth. **D** E DR Congo, SW Uganda, Rwanda, Burundi.

24 GREEN HYLIA *Hylia prasina* 12cm Darker than Willow Warbler (Plate 210) with more pronounced eyebrow. Longer tail than Neumann's Warbler. Often in mixed flocks. **V** Song is a penetrating double whistle, *twee-twee*. **H** Undergrowth and mid-strata of forest and bush. **D** W and C Africa, from Senegal to Angola and Kenya.

25 TIT HYLIA *Pholidornis rushiae* 8cm Note tiny size and yellow rear part of body. **V** Song is a brief double trill, typically from high in the canopy. **H** Forest canopies. **D** W and C Africa, from Sierra Leone to NW Angola and Uganda.

BUSHTITS AEGITHALIDAE

26 LONG-TAILED TIT *Aegithalos caudatus* 14cm Always on the move, often in large or mixed-species flocks. **V** Very vocal, with various *tup* and *ssrit* contact notes and a thin *tsee-tsee-tsee* as flocks move through trees. **H** Mixed woodland, undergrowth, scrub, hedges, heathland. **D** Widespread, from W Europe to E Asia, Japan.

27 SILVER-THROATED BUSHTIT *Aegithalos glaucogularis* 14cm Formerly considered conspecific with Long-tailed Tit. Distinct, with grey upperparts and buffish underparts. Note black lateral crown stripes and dark bib. **V** Similar range of vocalisations to Long-tailed Tit, but trills are harder. **H** Mixed and deciduous forest, woodland and scrub. **D** C, NE, EC China.

28 WHITE-CHEEKED BUSHTIT *Aegithalos leucogenys* 11cm Typical bushtit; agile and acrobatic forager in trees and bushes. **V** A *si-si-si-si*, a fuller *see-see-see*, a *seeup* and a buzzing *trrrp*. Song is a weak jumble of squeaky and piping notes. **H** Open scrubby forest and riverside tamarisk bushes. **D** Afghanistan, Pakistan.

29 BLACK-THROATED BUSHTIT *Aegithalos concinnus* 10.5cm Note distinctive face pattern with black throat spot and black mask. Gregarious, often forms large flocks. **V** Calls include *si-si-si-si* and rattling *churr-trrrt-trrrt*. **H** Deciduous forest edge, bush, secondary growth, bamboo, gardens. **D** Himalayas to S Vietnam.

30 WHITE-THROATED BUSHTIT *Aegithalos niveogularis* 11cm Note contrast between white throat and cinnamon breast-band. Active and acrobatic, usually in pairs, with larger flocks post breeding. **V** Calls include *tze-tze-tze* and *wi*. **H** Mixed conifer and birch forests, alpine shrubs. **D** W Himalayas.

31 RUFOUS-FRONTED BUSHTIT *Aegithalos iouschistos* 11cm Typical bushtit behaviour, active and agile. **V** Constant *see-see-see-see* and *tup* or *trrup*; when alarmed a shrill *zeet* and *trr-trr-trr*. **H** Forest undergrowth and low trees. **D** C and E Himalayas.

32 BLACK-BROWED BUSHTIT *Aegithalos bonvaloti* 11cm Often considered conspecific with Rufous-fronted Bushtit. Active and agile forager. **V** Very like Rufous-fronted Bushtit. **H** Open broadleaved evergreen and mixed forest, pine forest, forest edge, secondary growth. **D** SW China, N Myanmar.

33 BURMESE BUSHTIT *Aegithalos sharpei* 11cm Often considered conspecific with Black-browed Bushtit. Typically agile and active. **V** Calls include a high-pitched *tsi-si-si-si*, a soft *tsup* and a slurred, rattling *tsirrup*. **H** Open pine forest, occasionally high-altitude oak and rhododendron forest. **D** Resident in W Myanmar (Mt Victoria).

34 SOOTY BUSHTIT *Aegithalos fuliginosus* 11.5cm Note clear-cut dark breast-band in otherwise rather dull plumage. Forms family flocks post breeding. **V** Rolling *sir-rrup*, a thin *si-si-si - si-si-si - si-si-si* and a low *trrr*. Call: thin *chit* repeated at irregular intervals. **H** Mixed forest with willow, birch, rhododendron or bamboo. **D** NC China.

35 PYGMY BUSHTIT *Aegithalos exilis* 8–9cm Nondescript bushtit with stubby bill. Forages in small flocks, often low down. **V** High-pitched *si, sisi* or *sililili*; also a high-pitched, grating *srrr* and a soft *chip, tchip* or *sip*, these sometimes combining to make longer phrases. **H** Montane forest, forest edges and plantations. **D** Java, Indonesia.

36 WHITE-BROWED TIT-WARBLER *Leptopoecile sophiae* 10cm Male has patches of violet and blue; female more uniform brown. Acrobatic forager in thick undergrowth, often a member of mixed-species feeding flocks. **V** Calls include a sibilant, trilling *sirrrr*, a hissing *psee*, a drawn-out *pseee* and an abrupt, buzzy *psrit*. **H** Montane scrub and thickets; winters at lower elevations in valley-bottom thickets. **D** SE Kazakhstan to N India, C China.

37 CRESTED TIT-WARBLER *Leptopoecile elegans* 10cm Male has pale grey crown, rich rufous nape, face and breast sides, indigo mantle. Female similar but duller. Very active, foraging high in canopy. **V** Calls include a thin *pseee* and a shrill wren-like chatter. **H** Coniferous forest, dwarf alpine juniper and birch scrub above treeline. **D** SE Tibet, W China.

38 AMERICAN BUSHTIT *Psaltriparus minimus* 11cm Considerable geographical variation in coloration and head markings. Male has dark eye, female has pale eye. Active and confiding. Forms flocks after breeding. **V** Emphatic *pit* or *tsit*, a *skrrti ti ti*; travelling groups sometimes utter a trilling *pit pit pit sre-e-e-e-e-e-e* or *srrit srrit srrit sisisi*. **H** Dry open woodland, woodland edge scrub and chaparral. **D** SW Canada through W USA to Guatemala.

1
3
5
2
4
6
7
10
8
9
14
11
12
13
16
18
15
17
19
20
21
22
24
23
25
26
27
28
29
30
31
33
35
♀
37
♂
32
34
♀
36
♂
♂
38
♀

LEAF WARBLERS AND ALLIES PHYLLOSCOPIDAE

1 WOOD WARBLER *Phylloscopus sibilatrix* 12.5cm Bright green upperparts, well-defined supercilium, yellow breast, white belly. Feeds in upper layers of woodland canopy, often fly-catching or hovering to capture insects. **V** Distinctive song, starting slowly, accelerating into a shivering trill, *zip - zip - zip - zip - zip-zip-zip-zipzipzipzipzvurrrurrrr...* Calls include plaintive *pew* and soft *wit-wit-wit*. **H** Mature deciduous and mixed woodland. **D** Breeds from W Europe through N Kazakhstan and C Russia. Winters W and C Africa.

2 WESTERN BONELLI'S WARBLER *Phylloscopus bonelli* 11.5cm Note yellowish rump and whitish underparts. Actions much as others of genus, including fly-catching. **V** Dry trill, *twee-wee-wee-wee-wee-wee...* Call is *hoo-eet* or *doo-eeo*. **H** Open deciduous, mixed and coniferous forests. **D** Breeds W Europe, NW Africa. Winters WC Africa.

3 EASTERN BONELLI'S WARBLER *Phylloscopus orientalis* 11.5cm Formerly considered conspecific with Western Bonelli's Warbler. Very similar in appearance. **V** Much as Western Bonelli's Warbler, but shorter and less vigorous. Call is a short, hard *tsiup* or *chip*. **H** As Western Bonelli's Warbler. **D** Breeds E Europe, Middle East. Winters EC Africa.

4 BUFF-BARRED WARBLER *Phylloscopus pulcher* 10cm Long supercilium, two wing-bars, pale rump. Forages mainly in canopy. **V** Song consists of one or more call notes followed by a fast trill. Call is an often repeated *tsip* or *twick*. **H** Coniferous forests, oak and birch woods, rhododendron and other high-altitude scrub. **D** Breeds Himalayas, C China. Winters SE Asia.

5 ASHY-THROATED WARBLER *Phylloscopus maculipennis* 9–10cm Greyish crown stripe and yellow rump. Active forager, generally in the tree canopy; in winter also frequents bushes. **V** Thin, *sweet wee-ty-wee-ty-weet-ty* or *du-ze-zuu-ze-za*. Call is a repeated *zip* or *zit*. **H** Open mixed forests with thick undergrowth. **D** Resident in Himalayas, N India, SW China, N SE Asia.

6 HUME'S LEAF WARBLER *Phylloscopus humei* 10cm Very similar to Yellow-browed Warbler, but generally slightly duller. Arboreal, active, restless forager, with much wing and tail flicking. **V** Drawn-out, descending *eeeeeeeeezzzzzzzzzzzzz* or a repeated *wesoo*. **H** Subalpine forests; winters in forests, woodland, orchards, gardens. **D** Breeds SC Russia and E Kazakhstan to NW and C China, Nepal. Winters Indian subcontinent, SE Asia.

7 YELLOW-BROWED WARBLER *Phylloscopus inornatus* 10–11cm Note prominent supercilium. Often first detected by call. Active and agile, flicks wings and tail. Forages mainly in tree canopy, often in mixed-species flocks. **V** Calls include a rising *suu-eet*, *swee-ooo*, *seweest* or *weest*. **H** Open forests, woodland, plantations. **D** Widespread breeding population from C Russia across Siberia to NE China. Winters SE Asia.

8 BROOKS'S LEAF WARBLER *Phylloscopus subviridis* 9–10cm Indistinct yellowish rump and crown stripe. Restless forager, generally in tree canopy. **V** Series of single notes followed by a reeling *tr-r-r-r-r-r-r*. Call is a strident *chwee*, *chewy* or *psee*. **H** Coniferous or mixed forests; after breeding in bushes, scrub, olive groves, orchards, plantations and gardens. **D** Breeds E Afghanistan to NW India. Winters C Pakistan, NC India.

9 CHINESE LEAF WARBLER *Phylloscopus yunnanensis* 10cm Forages mainly in the canopy, hovers and makes sallies after insects. **V** Mechanical, monotonous *tsiridi-tsiridi-tsridi...* Calls include a scolding *tueet-tueet-tueet...* or similar, and a single *tueet* or *swit*. **H** Broadleaved evergreen forest and secondary growth. **D** Breeds NC China. Winters SE Asia.

10 LEMON-RUMPED WARBLER *Phylloscopus chloronotus* 9–10cm Distinct pale yellowish rump and crown stripe. Frequently flicks tail and wings; hovers to capture insects. **V** Stuttering, endless *tsi-ts-tsi-tsi-tsu-tsu-tsi-tsi-tsu-tsi-tsi-tsidididididididi-tsi-tsi-tsu-tsu...* or drawn-out rattle followed by rapid notes, *tsirrrrrrrrrrrrrrrrrrr-tsi-tsi-tsi-tsi-tsi-tsi-tsi...* **H** Mixed or coniferous forest with thick undergrowth; also secondary growth in winter. **D** Himalayas to C China. Winters at lower altitudes.

11 SICHUAN LEAF WARBLER *Phylloscopus forresti* 9–10cm Formerly treated as conspecific with Lemon-rumped Warbler. Darker, less brown above, brighter yellow rump and all-black bill (no pale base). **V** Song is similar to Lemon-rumped but slower and with a wider frequency range. **H** Montane forests. **D** SC China. Winters south to NW Thailand and N Vietnam.

12 GANSU LEAF WARBLER *Phylloscopus kansuensis* 9–10cm Formerly treated as conspecific with Lemon-rumped Warbler, but differs genetically, vocally and in choice of habitat. **V** Song starts with several very high-pitched *tsi* notes and ending in a loud trill. Call is a short disyllabic *ch-di*. **H** Montane deciduous broadleaved forest. **D** C China. Presumably winters in SC China.

13 PALLAS'S LEAF WARBLER *Phylloscopus proregulus* 10cm Note pale yellowish central crown stripe and yellow rump. Frequently flicks wings and tail. Often hovers around foliage in search of insects. **V** Song loud, with vibrant whistles and trills. Has been called 'the canary of the taiga'. **H** Mainly mixed and coniferous forest, with thick undergrowth. **D** Breeds E Siberia, Mongolia, NE China. Winters mainly SE China, SE Asia.

14 TYTLER'S LEAF WARBLER *Phylloscopus tytleri* 10–12cm Note prominent supercilium and absence of wing-bars. Arboreal, acrobatic forager, best located by song. **V** Even-paced, high-pitched *pi-tsi-pi-tsu*. Call is a plaintive *sooeet* or *tsee-it*. **H** Coniferous forest, dwarf willow and birch; winters in broadleaved forest. **D** Breeds NE Afghanistan, N Pakistan, NW India. Winters SW India.

15 YELLOW-STREAKED WARBLER *Phylloscopus armandii* 13–14cm Slow, deliberate forager in bushes and lower tree branches, and on the ground. **V** Series of rapid, husky, slurred undulating phrases, sometimes introduced by a few *zick* notes. Call is a sharp *zick*. **H** Low bushes, thickets, subalpine spruce forest; winters in low bushes in plains and hills and forest edge. **D** Breeds C China. Winters SE Asia.

16 RADDE'S WARBLER *Phylloscopus schwarzi* 12cm Rather stocky leaf warbler. Forages with quite slow, lumbering movements; frequently flicks wings and tail. **V** Clear trill, followed by one or two quiet notes. Calls include a soft *chek* and a nervous *pwit*. **H** Dense undergrowth, bushy forest edge and clearings, waterside thickets. **D** Breeds E Kazakhstan to SE Russia, North Korea. Winters SE Asia.

17 SULPHUR-BELLIED WARBLER *Phylloscopus griseolus* 11cm Usually a ground forager, running or hopping over rocks; becomes more arboreal in winter. **V** Short, rapid *tsi-tsi-tsi-tsi-tsi*. Call is a *quip* or *pick*. **H** Stunted mountain shrubs on dry stony slopes; winters in open forests, rocky open ground, old buildings. **D** Breeds Afghanistan and Pakistan to S Russia, W Mongolia, C China. Winters India.

18 TICKELL'S LEAF WARBLER *Phylloscopus affinis* 10cm Skulking, feeds low down in thick vegetation, or on ground. Appears not to flick wings or tail. **V** Rapid *tchip-chi-chi-chi-chi-chi-chi*. Call is a sharp *chep*, and rapidly repeated *tak-tak* when alarmed. **H** Open, bushy, alpine scrub, low vegetation in rocky valleys, dwarf bamboo and bushes in upland cultivation. **D** Breeds Himalayas. Winters India to Myanmar, S China.

19 ALPINE LEAF WARBLER *Phylloscopus occisinensis* 10–11cm Usually seen singly or in pairs, with small parties post breeding. Sometimes treated as conspecific with Tickell's Leaf Warbler. **V** Fast *chip-chi-chi-chi-chi-chi*. **H** Rocks and bushes on dry barren mountains. **D** Breeds WC China. Winters India.

20 SMOKY WARBLER *Phylloscopus fuligiventer* 11cm Skulking, keeps to low vegetation, often among rocks; nervously flicks wings and tail, the latter often cocked. **V** Repetition of a single note, *tsli-tsli-tsli-tsli-tsli...* Calls include a soft *stup*, a sharp *chek* and a *tzik* or *tsrr*. **H** Rocky alpine pasture, low alpine scrub; winters in dense scrub, usually near water. **D** Breeds E Himalayas, Tibetan Plateau. Winters Indian subcontinent.

21 DUSKY WARBLER *Phylloscopus fuscatus* 11.5cm Drab brown with prominent supercilium. Skulking but active feeder, usually low in thick vegetation or on ground. **V** Song *chewee-chewee-chewee-chewee*, with a concluding trill. Call is a sharp *chett*. **H** Birch and willow thickets, often near water, scrub in open forest, dwarf willows, birch scrub and low conifers in subalpine zone. **D** Breeds C and E Siberia to Mongolia, NE China. Winters NE India, SE Asia, S China.

22 PLAIN LEAF WARBLER *Phylloscopus neglectus* 9–10cm Active; forages in trees and bushes, often hovering to glean insects from leaves. **V** Variable short rising warble, including *pt-toodla-toodla* or *chit-chuwich-chissa*. Calls include a harsh *churr* or *trrr trrr* and a nasal *chit* or *chi-ip*. **H** Juniper thickets, open oak woods, orchards. **D** Breeds Iraq and Iran to Afghanistan and Pakistan. Winters S to Arabian Peninsula and NW India.

23 BUFF-THROATED WARBLER *Phylloscopus subaffinis* 11–12cm Active, nervous forager, low down in thick vegetation or on the ground. **V** Soft *tuee-tuee-tuee-tuee*; calls include a weak *trrup* or *trip*. **H** Montane forest, open hillsides with scrub; winters in grassland and bushes in foothills. **D** Breeds C and S China, N Vietnam. Winters S China, N SE Asia.

24 WILLOW WARBLER *Phylloscopus trochilus* 11.5cm Relatively long-winged and slender leaf warbler. Plain plumage with prominent supercilium. **V** Distinctive series of sweet notes, faint at first, then increasing in volume before drifting to a quieter, descending flourish. Call is bisyllabic *hoo-eet*. **H** Birch, willow and alder forests, secondary growth, scrub. **D** Widespread, breeding from W Europe to E Siberia. Winters sub-Saharan Africa.

25 MOUNTAIN CHIFFCHAFF *Phylloscopus sindianus* 11cm Actions and habits similar to Common Chiffchaff. **V** Repeated *chit-chiss-chyi-chiss-chit-chiss-chyi-chip-chit-chyi*. Calls include a bisyllabic *tiss-yip* and a *pseeu* or *pee-oo*. **H** Willow and poplar groves along streams, scrub and orchards; winters in waterside trees and bushes. **D** Breeds Turkey to Himalayas. Winters S Asia.

26 CANARY ISLANDS CHIFFCHAFF *Phylloscopus canariensis* 11cm Formerly considered conspecific with Common Chiffchaff, but differs vocally, morphologically and genetically. Note longer supercilium and pale yellowish vent. **V** Song is similar to Common Chiffchaff but is distinctly richer, deeper and faster. **H** Forest, woodland, scrub, gardens and cultivation. **D** Canary Islands.

27 COMMON CHIFFCHAFF *Phylloscopus collybita* 11cm Shorter-winged than Willow Warbler, with less obvious supercilium. Forages from ground level to tree canopy, with nervous flicking of wings and tail. **V** Song is *chiff-chaff-chiff-chaff-chiff-chaff...*, often for 20 secs or more. Call is a monosyllabic *hweet*. **H** Open deciduous, mixed and coniferous forests, scrubby thickets, hedgerows, gardens. **D** Widespread, breeding from Europe to E Siberia. Winters S Europe, N Africa, C Africa.

28 IBERIAN CHIFFCHAFF *Phylloscopus ibericus* 11cm Subtly different from Common Chiffchaff, with whiter underparts and yellowish vent, but best separated by voice. **V** Song is a series of *chi-chi-chi-chu-chu-chu* notes, ending in a short trill. Call is a downslurred *siu*. **H** Open forest with scrub in hilly areas; also riverine woodland and Mediterranean maquis (mainly in winter). **D** Breeds SW France, Portugal, Spain, NW Africa. Some birds winter in tropical W Africa.

29 EASTERN CROWNED WARBLER *Phylloscopus coronatus* 11–12cm Single wing-bar, long supercilium. Very active forager from undergrowth to canopy. **V** Call is a *phit-phit*, *swe-zueee* or *zweet*. **H** Mixed and evergreen forests, low trees and bushes. **D** Breeds SE Russia, NE China, Korean Peninsula, Japan; C China. Winters SE Asia, Indonesia.

30 IJIMA'S LEAF WARBLER *Phylloscopus ijimae* 12cm Often acts like a flycatcher to take insects, otherwise most actions typical of genus. **V** Mechanical *swee-seww-swee-swee-swee*, usually ending in *chew*. **H** Forest undergrowth, alder thickets, bushes around habitations. **D** Breeds Izu Islands off SC coast of Japan. Winters south to Philippines.

31 PHILIPPINE LEAF WARBLER *Phylloscopus olivaceus* 11–13cm Forages alone or as part of mixed-species feeding party; picks insects from the outer foliage. **V** Frequently repeated, short, sharp *prrrr-chi*. **H** Lowland and middle-elevation forest and forest edges. **D** S and C Philippines.

32 LEMON-THROATED LEAF WARBLER *Phylloscopus cebuensis* 11–12cm Forages alone or in mixed-species flocks. **V** Loud, repeated *chi-chi-oo* or *chip chi-u*. **H** Lowland forest and forest edge. **D** WC and N Philippines.

33 YELLOW-THROATED WOODLAND WARBLER *Phylloscopus ruficapilla* 9cm Distinctive reddish-brown forehead and crown. Extent of yellow on throat and breast variable. **V** Extremely high, bouncing, hurried *wee-turi wee-turi wee-turi weet* and other short strophes. **H** Montane forests, bamboo. **D** E and SE Africa, from Kenya to South Africa.

34 BROWN WOODLAND WARBLER *Phylloscopus umbrovirens* 11cm Active, energetic. In worn plumage upperparts become paler grey-brown. **V** Loud, fast and slow phrases often ending with a fast trill, transcribed as *titititititi-wit-vuit-vuit-vuit-wit-wit-wit-wit-tu-vi-tu-vi-tuvi-twit-diu-diu-diu*. Call: descending *dzziieep*; also a low *psew* and *swee-vik* when alarmed. **H** Hillside and highland trees and bushes, wadis and gorges with lush growth. **D** E Africa, Arabian Peninsula.

35 RED-FACED WOODLAND WARBLER *Phylloscopus laetus* 9cm Note green upperparts and buff face and breast. **V** Song consists of a series of bubbling phrases in quick succession, *putreeputriputri, pitjupitju...piyjupiyju*. **H** Mountain forests, bamboo. **D** E DR Congo, Uganda, Rwanda, Burundi.

36 LAURA'S WOODLAND WARBLER *Phylloscopus laurae* 10cm Yellow underparts interrupted by white between lower breast and undertail coverts. **V** High-pitched and variable descending phrases, *chirrijee-chirrijee-prichirri-prichirri-chichee, titiwee-titiwee-titiwee* or similar. **H** Forest canopy. **D** Angola, S DR Congo, N and E Zambia, SW Tanzania.

37 BLACK-CAPPED WOODLAND WARBLER *Phylloscopus herberti* 11cm Note distinctive head pattern. **V** Song is a short and variable sequence of high-pitched warbles, ending with a flourish. **H** Montane forest. **D** SE Nigeria, W Cameroon, Equatorial Guinea.

38 UGANDA WOODLAND WARBLER *Phylloscopus budongoensis* 10cm Note darkish green colouring, slightly yellow-streaked breast, short primary projection. **V** Song is a short snatch of high-pitched notes, up and down the scale. **H** Mainly mid-strata of dense mountain forests. **D** Equatorial Guinea, Cameroon, Uganda, W Kenya.

1
2
3
4
5
6
7
8
9
10
11
12
13
14
15
16
17
18
19
20
21
22
23
24
25
26
27
28
29
30
31
32
33
34
35
36
37
38

LEAF WARBLERS AND ALLIES *CONTINUED*

1 WHITE-SPECTACLED WARBLER *Phylloscopus intermedius* 11–12cm Note black stripe on side of crown. Active, forages in trees and undergrowth; often makes fly-catching sallies. **V** Song starts hesitantly then accelerates, *uee-tiu uee-tiu-chu-weet-chu-chu-weet-chu-weet...* **H** Humid broadleaved evergreen forest; winters at lower elevations in mixed forests. **D** NE India to SE China, Vietnam.

2 GREY-CHEEKED WARBLER *Phylloscopus poliogenys* 10–11cm Grey head, broad white eye-ring. Less obvious crown-stripe than White-spectacled Warbler. Forages in the understorey. **V** Song consists of varied whistled phrases, often quick and trilling. Call is a high-pitched *ueest*. **H** Bamboo jungle and dense undergrowth in evergreen hill forests. **D** E Himalayas to S Vietnam.

3 GREEN-CROWNED WARBLER *Phylloscopus burkii* 11cm Arboreal, forages from mid-storey down to undergrowth; flicks wings, cocks and fans tail. **V** Similar to Grey-crowned Warbler. Call is a high-pitched, soft *huit*. **H** Forest, secondary growth. **D** Breeds Himalayas. Winters N India.

4 GREY-CROWNED WARBLER *Phylloscopus tephrocephalus* 11–12cm Agile forager in foliage or undergrowth. Post breeding often joins mixed-species flocks. **V** A few single notes followed by a trill; call is a soft *trrup* or *turup*. **H** Broadleaved evergreen forest; in winter also forest edge, bamboo and scrub. **D** Breeds NE India to N Vietnam, C China. Winters N SE Asia.

5 WHISTLER'S WARBLER *Phylloscopus whistleri* 11–12cm Very similar to Green-crowned Warbler, but has complete eye-ring and less clearly defined lateral crown-stripe, and may show more obvious wing-bar. **V** Simple, whistled *chu chu-weet-tu-chu-wee...chu chu-weet-tu-chu-wee...* Call is a soft *chip* or *tiu-du*. **H** Lush undergrowth in temperate forest. **D** Himalayas to W Myanmar.

6 BIANCHI'S WARBLER *Phylloscopus valentini* 11–12cm Central crown-stripe greyer than in Whistler's Warbler. Forages in the understorey and the canopy of low trees, often making short fly-catching sallies. **V** Similar to Whistler's Warbler but lower-pitched. Calls include a soft *tiu* or *tiu tiu*. **H** Broadleaved evergreen forest with lush undergrowth, secondary growth. **D** Breeds C and SE China, N Vietnam. Winters S China, N SE Asia.

7 ALSTRÖM'S WARBLER *Phylloscopus soror* 11–12cm Relatively large bill and short tail. Forages mainly in the understorey. **V** Simple *chip chu-se-sis-chu-se-sis...* Calls include a high-pitched *tsrit*. **H** Broadleaved evergreen forest with lush undergrowth. **D** Breeds C and SE China. Winters S China, N SE Asia.

8 MARTENS'S WARBLER *Phylloscopus omeiensis* 11–12cm Head pattern generally less distinct than in Grey-crowned Warbler. Forages mainly in the understorey. **V** Various short, abrupt phrases, some ending in trills, *chu-si-tsu-chu-si-tsu...pi tsu-pi-tsu-hueetse...huee-huee-tse-tse-tse-tse...* **H** Broadleaved evergreen forest. **D** Breeds C China. Winters N SE Asia.

9 GREEN WARBLER *Phylloscopus nitidus* 10–11cm Formerly considered a race of Greenish Warbler. Restless, forages mainly in the upper canopy; flicks wings and tail. **V** Song very like Greenish Warbler. Calls include a *che-wee*, a *chirr-ir-ip* and a longer *chi-ru-weet*. **H** Dense forest, orchards, parks, gardens. **D** Breeds N Turkey, Caucasus, N Iran, NW Afghanistan. Winters India, Sri Lanka.

10 TWO-BARRED WARBLER *Phylloscopus plumbeitarsus* 12cm Note double white wing-bar. Arboreal, forages mainly in middle levels. **V** Call is a *tissheep* or *chi-ree-wee*. **H** Forests, parks, gardens. **D** Breeds C Russia to NE Mongolia, E Russia, NE China, North Korea. Winters S China, SE Asia.

11 GREENISH WARBLER *Phylloscopus trochiloides* 10–11cm Distinguished from Arctic Warbler by pale supercilium extending to forehead. Mainly arboreal, flitting among foliage or hovering to capture insects. **V** Song variable, consisting of high-pitched accelerating phrases that culminate in an abruptly ending trill. Calls include *ch-wee*, *chirree* and *chis-weet*. **H** Deciduous, mixed and coniferous forests with rich undergrowth; in winter also woodland, gardens. **D** Breeds E Europe to C Siberia, N Pakistan, C China. Winters India, Sri Lanka, SE Asia.

12 EMEI LEAF WARBLER *Phylloscopus emeiensis* 11–12cm Note prominent supercilium and wing-bars. Forages at all levels. Often gives quick wing flicks. **V** Clear, slightly quivering trill; call is a soft *tu-du-du* or *tu-du-du-di*. **H** Subtropical broadleaf forest. **D** Breeds S China. Probably winters SE Asia.

13 LARGE-BILLED LEAF WARBLER *Phylloscopus magnirostris* 12–13cm Conspicuous long supercilium. Shy; forages in the upper canopy along boughs rather than among foliage. **V** Descending *tee-ti-tii-tu-tu*. Calls include a *dir-tee* and an ascending *yaw-wee-wee*. **H** Open areas and glades in montane coniferous or mixed forests, usually near water. **D** Breeds Himalayas to C China. Winters India, Sri Lanka.

14 SAKHALIN LEAF WARBLER *Phylloscopus borealoides* 11.5cm Little studied. Note long whitish supercilium and broad dark-brown eye-stripe. **V** Repeated series of high-pitched whistles, *hee-tsoo-kee*. **H** Dark coniferous or mixed montane forest. **D** Breeds Sakhalin, Japan. Winters SE Asia.

15 PALE-LEGGED LEAF WARBLER *Phylloscopus tenellipes* 12–13cm Note pale pink legs. Forages in undergrowth and lower tree branches. **V** Calls with a metallic *tik-tik* or loud *peet*. **H** Various forests, secondary growth, mangroves and gardens. **D** Breeds E Russia, NE China, North Korea. Winters SE Asia.

16 JAPANESE LEAF WARBLER *Phylloscopus xanthodryas* 12–13cm Formerly considered a race of Arctic Warbler; actions and habits are similar. **V** Song *tree'diret-tree'diret-tree'diret...* or *tschiritschiritschiri...* **H** Woodlands, secondary forest. **D** Breeds Japan (except Hokkaido). Winters Taiwan, Philippines, Borneo, Java.

17 KAMCHATKA LEAF WARBLER *Phylloscopus examinandus* 12–13cm Formerly considered a race of Arctic Warbler; actions and habits are similar. **V** Song is a rolling *ch-chiri ch-chiri ch-chiri ch-chiri*. Call is similar to Arctic Warbler. **H** Deciduous, coniferous and mixed forests. **D** Breeds Kamchatka, Sakhalin, Kuril Islands, N Japan. Winters Philippines to Lesser Sunda Islands.

18 ARCTIC WARBLER *Phylloscopus borealis* 12cm Mainly arboreal, urgent feeder, dashing in and around cover, flicking wings and tail; also hovers to catch insects. **V** Shivering trill *dyryryryryryry...*, often interspersed with call notes. Call is a sharp, metallic *dzik* or *dzrt*. **H** Open birch or coniferous woodland, taiga willow scrub, tundra edge, often near water. **D** Breeds from N Scandinavia through Siberia to W Alaska. Winters SE Asia, Philippines, Indonesia.

19 CHESTNUT-CROWNED WARBLER *Phylloscopus castaniceps* 9.5cm Note chestnut crown, dark lateral crown-stripe, double wing-bar, yellow belly and undertail coverts. Very active, forages in the outer foliage of upper canopy. **V** Upward-inflected, thin, high-pitched *see see see-see-see-see-see*. Calls include quiet *chik* and *chee-chee*. **H** Oak forest. **D** Himalayas to S China and SE Asia; Sumatra, Indonesia.

20 SUNDA WARBLER *Phylloscopus grammiceps* 10cm Chestnut, grey and white. Active forager in the understorey and lower canopy of forest edges. **V** Ringing, high-pitched *chee-chee-cheechee*; calls with a buzzing *turrr*. **H** Subtropical humid montane forest with well-developed understorey. **D** Sumatra, Java and Bali, Indonesia.

21 YELLOW-BREASTED WARBLER *Phylloscopus montis* 9–10cm Forages mainly in foliage of the understorey or mid-canopy at forest edges; feeds by gleaning, hovering or fly-catching. **V** Very high-pitched, rising and then fading *zizizizizizi-azuuuu*, interspersed with *chit chit* notes. **H** Subtropical humid forest with a well-developed understorey and stands of bamboo. **D** Sumatra, Borneo, Palawan (W Philippines), Lesser Sunda Islands.

22 LIMESTONE LEAF WARBLER *Phylloscopus calciatilis* 11cm Yellow and black head-stripes, yellow underparts. Actions like Sulphur-breasted Warbler. **V** Short series of 7–9 soft whistled notes on a slightly falling scale. Calls include a short, soft *pi-tsiu*, *pi-tsu*. **H** Broadleaved evergreen and semi-evergreen forests on limestone karst. **D** N and C Vietnam, N and C Laos.

23 SULPHUR-BREASTED WARBLER *Phylloscopus ricketti* 11cm Entirely yellow underparts. Active canopy forager, often making aerial fly-catching sallies. **V** High-pitched series of short notes speeding up towards the end. Call is a subdued *wi-chu* or *pi-chu*. **H** Mainly broadleaved evergreen and mixed deciduous forests. **D** Breeds EC and S China. Winter visitor to SE Asia.

24 YELLOW-VENTED WARBLER *Phylloscopus cantator* 10cm Note white belly. Active forager in bushes or the low to middle level of trees. When calling, spreads tail and flicks it upwards. **V** High *sit weet weet seep seep seep si-chu-chu to-you*. Calls include a soft *see-chew* and *see-chew-chew*. **H** Broadleaved evergreen and semi-evergreen forests. **D** E Nepal to C Laos.

25 WESTERN CROWNED WARBLER *Phylloscopus occipitalis* 11–13cm Two wing-bars, but median bar can be indistinct. Restless, flicks wings and tail. **V** Rapid, piercing *stic-swee-swee-swee-swee-swee-swee*, *chwi-chwi-chwi-chwi-chwi-chwi* or *tityu-tityu-tityu-tityu-tityu-tityu*. Calls are a *chiwee*, *chip-wee* and a repeated *stic* or *stic-swick*. **H** Well-wooded country; in winter also uses deciduous and conifer hill forests. **D** Breeds W to C Himalayas. Winters India.

26 BLYTH'S LEAF WARBLER *Phylloscopus reguloides* 10–12cm Acrobatic forager, often clinging upside down like a nuthatch. When agitated, slowly flicks one wing at a time. **V** Call is a *pit-chew*; song is a trilling extension of the call. **H** Broadleaved, conifer and mixed forests; winters in open forest and forest edge. **D** Himalayas to SE Asia.

27 CLAUDIA'S LEAF WARBLER *Phylloscopus claudiae* 11–12cm Formerly considered conspecific with Blyth's Leaf Warbler. Best separated by voice. **V** Song begins with 1–2 notes followed by a single-note rapid trill. **H** Forest. **D** Breeds C and E China. Winters S China, N SE Asia.

28 HARTERT'S LEAF WARBLER *Phylloscopus goodsoni* 11–12cm Actions similar to Blyth's Leaf Warbler, which was formerly considered conspecific. **V** Presumed similar to Blyth's Leaf Warbler. **H** Broadleaved evergreen forest. **D** Breeds SE China. Winters S China, Hainan.

29 KLOSS'S LEAF WARBLER *Phylloscopus ogilviegranti* 10–11cm Actions similar to Davison's Leaf Warbler, which was formerly considered conspecific. **V** Song is a short warbled phrase, e.g. *ti-see-chee-sui*, lacking obvious repetition. **H** Deciduous woodland or mixed conifer woodland and bamboo. **D** C and S China to C Vietnam.

30 HAINAN LEAF WARBLER *Phylloscopus hainanus* 10–11cm Note yellow head-stripes, wing-bars and underparts. Active forager in the middle and upper canopy. **V** High-pitched, variable phrases, *tsitsitsui-tsitsui...titsu-titsui-titsui...titsu-titsui-titsui*. **H** Montane secondary forest, forest edge, scrub. **D** Hainan (S China).

31 DAVISON'S LEAF WARBLER *Phylloscopus intensior* 11–12cm Generally forages in the canopy. When agitated, flicks both wings simultaneously. **V** Single, high call note followed by a *tit-sui-titsui-titsui* or *see-chee-wee see-chee-wee see-chee-wee*. Calls include *pitsiu*, *wit-see* or *pitsitsui*. **H** Deciduous woodland or mixed conifer woodland and bamboo. **D** S China, SE Asia.

32 GREY-HOODED WARBLER *Phylloscopus xanthoschistos* 10cm Note contrast between grey head and yellow underparts. Agile forager in trees and shrubs; fans and cocks tail. **V** Repeated *tsi-tsi-tsi-weetee*. Call is a high *psit-psit* or a plaintive *tyee-tyee*. **H** Forest, secondary growth. **D** Himalayas to NW Myanmar.

33 MOUNTAIN LEAF WARBLER *Phylloscopus trivirgatus* 10–11cm Favours the middle level and canopy, feeding on insects picked from among foliage or by hovering to take them from outer leaves. Forages alone, in pairs or in small groups, and as part of mixed-species flocks. **V** Unmusical *tisiwi-tsuwiri-swit* or faster *tsee-chee-chee-weet*. Calls include *cheecheechee* and jangling *tersiwit*. **H** Submontane and montane forest, and forest edges. **D** Malay Peninsula, Borneo, Indonesia.

34 NEGROS LEAF WARBLER *Phylloscopus nigrorum* 10–11cm Formerly considered conspecific with Mountain Leaf Warbler. Regularly seen in flocks, gleaning from leaves and branches. **V** Variable, including an explosive *chi-chi-chi-chi-pa-aw*, a *chi-cherru* and a *sweet sweet chu*. **H** Submontane and montane forest. **D** Philippines.

35 TIMOR LEAF WARBLER *Phylloscopus presbytes* 11cm Occurs singly, in small groups or in mixed-species flocks; forages in the middle level and the canopy, gleaning insects from leaves. **V** Short, staccato jangle of sweet high-pitched notes; in Flores, a high-pitched rising and falling warble. **H** Lowland and hill semi-evergreen forest and woodland; birds on Flores occur in primary and slightly degraded montane forest, secondary scrub and *Casuarina* forest. **D** Timor and other islands of Lesser Sunda Islands.

36 MAKIRA LEAF WARBLER *Phylloscopus makirensis* 10cm Formerly considered conspecific with Island Leaf Warbler. Whitish throat and lemon-yellow underparts. Inhabits the canopy. **V** Hurried jangle of high-pitched notes. **H** Hill forests. **D** Makira (S Solomon Islands).

37 SULAWESI LEAF WARBLER *Phylloscopus sarasinorum* 11cm Gleans and hover-gleans from foliage in the mid-storey and canopy, alone, in pairs, in small groups, or in mixed-species flocks. **V** Weak, variable, rapid series of high-pitched, repetitive warbles and trills. **H** Montane moss forest and forest edges. **D** Sulawesi, Indonesia.

38 KOLOMBANGARA LEAF WARBLER *Phylloscopus amoenus* 11cm Dark olive, short-tailed warbler. Note dark crown and stout bill. Inhabits the understorey; often on the ground. Joins mixed-species flocks. **V** Song is a soft high-pitched warble. Call is a sharp *sweet*. **H** Montane moss forest. **D** Kolombangara (N Solomon Islands).

39 ISLAND LEAF WARBLER *Phylloscopus maforensis* 10–11cm Note grey crown and face. Active, gleaning from leaves in the mid-storey and the crowns of small trees. **V** Rapid series of sweet, sibilant, tinkling notes or a sweet, melodious warble. **H** Hill and montane primary and secondary forest. **D** Maluku Islands, Indonesia; New Guinea; Solomon Islands.

1
2
3
4
5
6
7
8
9
10
11
12
13
14
15
16
17
18
19
20
21
22
23
24
25
26
27
28
29
30
31
32
33
34
35
36
37
38
39

REED WARBLERS AND ALLIES ACROCEPHALIDAE

1 MALAGASY BRUSH WARBLER *Nesillas typica* 17.5cm Distinguished from Subdesert Bush Warbler by darker plumage. Note short supercilium. **V** Short, dry rattles, given singly or in series. **H** Ground cover in forest. **D** Madagascar, Comoros islands.

2 SUBDESERT BRUSH WARBLER *Nesillas lantzii* 17cm Note pale underparts, short supercilium (more a loral spot), pale upper eyelid. **V** Dry tics, given in rapid series. **H** Thorn scrub, secondary growth, cultivation. **D** Madagascar.

3 ANJOUAN BRUSH WARBLER *Nesillas longicaudata* 17.5cm Differs from Malagasy Brush Warbler in range and slightly paler underparts. **V** Dry trills, each one slightly longer than Malagasy Brush Warbler. **H** Dense undergrowth of forests. **D** Anjouan (Comoros).

4 GRAND COMORO BRUSH WARBLER *Nesillas brevicaudata* 15.5cm The only *Nesillas* brush warbler on Grand Comoro. **V** Unstructured staccato squeaks. **H** Dense forest understorey at mid- and higher altitudes. **D** Grande Comore (Comoros).

5 MOHELI BRUSH WARBLER *Nesillas mariae* 15.5cm Distinguished from larger Malagasy Brush Warbler by greenish wash to upperparts and plain face. **V** Soft squeaking *tjirps*. **H** Dense undergrowth of forest. **D** Mohéli (Comoros).

6 BASRA REED WARBLER *Acrocephalus griseldis* 17cm Elusive, less clumsy than other large *Acrocephalus* reed warbler. **V** Song like Great Reed Warbler but quieter, less guttural and grating. Call is a harsh *chaarr*. **H** Waterside vegetation, especially reeds. **D** Breeds S Iraq. Winters E Africa.

7 CAPE VERDE WARBLER *Acrocephalus brevipennis* 14cm Grey or greyish plumage, especially on head. Elusive, especially out of breeding season. Sings from within cover. **V** Rich, vibrant rattle of 3–5 notes. Calls include throaty *pitchow*, sharp *chuk* and croaking *churr*. **H** Trees, bushes, giant cane, reeds; plantations e.g. sugarcane, banana, coffee. **D** Cape Verde.

8 GREATER SWAMP WARBLER *Acrocephalus rufescens* 17cm Drab brown above, brown-grey below, long bill, short wing, ill-defined supercilium. **V** Low to very high, unhurried *tjiptjip-WHaaro-PjUPjUPjU*. **H** Papyrus. **D** W and C Africa.

9 LESSER SWAMP WARBLER *Acrocephalus gracilirostris* 15cm Note bright orange gape and black legs. Short-winged, white below, faint supercilium. Warmer coloured than Greater Swamp Warbler. **V** Mid-high sustained thrush-like medleys of *weet-weetweet-* with loud rapid melodious rattles, flutes and oriole-like warbles. **H** Reedbeds, sedges, papyrus. **D** E and S Africa.

10 MADAGASCAN SWAMP WARBLER *Acrocephalus newtoni* 18cm Note breast streaks, orange eyes and long tail. **V** Song combines melodious, resonant phrases with lower, throaty notes. Calls include sharp single or repeated *chak* notes. **H** Reedbed areas with herbaceous vegetation; up to high altitudes. **D** Madagascar.

11 SEYCHELLES WARBLER *Acrocephalus sechellensis* 14cm No similar birds in range except other, vagrant warblers. **V** Song consists of repeated, rich, thrush-like whistled phrases. Call is a rapid chatter. **H** Wide range of habitats, including dense woodland and marshy areas. **D** Seychelles.

12 RODRIGUES WARBLER *Acrocephalus rodericanus* 13.5cm No similar bird in range. **V** Soft, melodious song of repeated *ch-chwee ch-chwee* phrases; alarm call a harsh chatter. **H** Thickets, woodland, forest. **D** Rodrigues, Indian Ocean.

13 GREAT REED WARBLER *Acrocephalus arundinaceus* 19cm A large unstreaked *Acrocephalus* warbler with relatively heavy bill. Long primary projection. Clumsy actions while foraging often give away bird's presence. **V** Song is loud and frog-like, *karra-karra-karra gurk gurk gurk chirrr-chirrr*. Calls include harsh *tchack* and croaking *churr*. **H** Reedbeds. **D** Breeds Europe and N Africa to W Mongolia, NW China. Winters sub-Saharan Africa.

14 ORIENTAL REED WARBLER *Acrocephalus orientalis* 18cm Very similar to Great Reed Warbler but may show more olive-grey on upperparts, slightly whiter underparts, more streaked throat. Clambers among reed stems or in bushes. **V** Deep, guttural churring and croaking, interspersed with warbling phrases. Calls include a quiet *krak*, a *chichikarr* and a hoarse *si-si-si-si*. **H** Reedbeds and scrub, usually not far from water. **D** Breeds Mongolia, SE Russia, Korean Peninsula, Japan, N and E China. Winters SE Asia, Indonesia.

15 CLAMOROUS REED WARBLER *Acrocephalus stentoreus* 18cm Short primary projection and long slender bill. Clambers clumsily among reed stems and in bushes. **V** Loud combination of harsh grating, chattering, squeaky and sweet notes, *karra-karra-kareet-kareet-kareet, skareet-skareet-skareet* or *prit-prit-pritik*. Calls include a harsh *chack* and a low *churr*. **H** Reedbeds, papyrus, waterside scrub and vegetation. **D** Widespread, breeding from Egypt across Asia to Indonesia, Philippines. Some sedentary; C Asian populations winter in Indian subcontinent.

16 AUSTRALIAN REED WARBLER *Acrocephalus australis* 17cm Resembles Oriental Reed Warbler, but smaller, with shorter bill. **V** Song is a long series of varied musical, chattering notes. Calls include dry rattling or churring in territorial disputes. **H** Low weedy vegetation in and around any type of wetland. **D** Australia, Lesser Sunda Islands, New Guinea, Solomon Islands.

17 MILLERBIRD *Acrocephalus familiaris* 13cm The only Old World warbler in the Hawaiian Islands. No similar bird in its range. **V** High, nasal, sustained *cheep cheep cheep* with many variations. **H** Bushy hillsides. **D** Hawaii.

18 SAIPAN REED WARBLER *Acrocephalus hiwae* 18cm Unmistakable, with extremely long and thin bill. **V** Song very loud, rather slow-paced, combining tuneful fluting phrases with drier rasping sounds. Calls include a loud *chuck*. **H** Reedbeds, thickets, forest undergrowth. **D** Saipan, Alamagan, Northern Mariana Islands, NW Pacific.

19 NAURU REED WARBLER *Acrocephalus rehsei* 15cm 'Average', brownish reed warbler, unmistakable in its restricted range. **V** Very accomplished vocalist with powerful, fluting thrush-like song. **H** Forest, scrub, gardens. **D** Nauru, W Pacific.

20 CAROLINIAN REED WARBLER *Acrocephalus syrinx* 15cm Rather dark overall with long bill (but not as long as Saipan Reed Warbler). **V** Song complex and melodic, with trills and warbled phrases interspersed with short pauses. **H** Forest, secondary growth, gardens, stands of tall grass. **D** Caroline Islands, W Pacific.

21 BOKIKOKIKO *Acrocephalus aequinoctialis* 15cm Unmistakable in its range. Note pale grey plumage. Only unmated males sing. **V** Low, hoarse, rather toneless *srrruh*. **H** Dense brush, open, sparsely wooded areas. **D** Line Islands, Kiribati, C Pacific.

22 NORTHERN MARQUESAN REED WARBLER *Acrocephalus percernis* 18cm A large reed warbler with greenish upperparts and yellow underparts. **V** Melodious mixture of whistles, warbles, churring and scolding notes. **H** Dense undergrowth of forests, plantations, coconut groves. **D** N Marquesas Islands, French Polynesia.

23 TAHITI REED WARBLER *Acrocephalus caffer* 18cm Unmistakable in range. Distinctively long bill. **V** Series of high, single, strong *tjew* notes alternated with melodic *weetjewhat*. **H** Forest, secondary growth, bamboo thickets, coconut plantations. **D** Tahiti.

24 SOUTHERN MARQUESAN REED WARBLER *Acrocephalus mendanae* 18cm Note yellow underparts. Unmistakable in range. **V** Strange, rapid series of alternating short, melodious notes, hoarse croaks and high-pitched twitters. **H** Forest, plantations, brush. **D** S Marquesas Islands, French Polynesia.

25 TUAMOTU REED WARBLER *Acrocephalus atyphus* 18cm Plumage variable, generally brownish to greyish. **V** Song *sree*, preceded and followed by some mellow or staccato, high, finch-like notes. **H** Woodland, bush, plantations, gardens. **D** Tuamotu Archipelago, French Polynesia.

26 COOK REED WARBLER *Acrocephalus kerearako* 16cm Unmistakable in range. **V** Call: *wreet* or *wrut* or *feetjeeh*. Song: energetic stream of high, loud notes and trills, often introduced by *twit*. **H** Woodland, reeds, gardens. **D** Cook Islands, C Pacific.

27 RIMATARA REED WARBLER *Acrocephalus rimitarae* 17cm Unmistakable in range. Confined to one tiny island, and critically endangered. **V** Irregular series of *cheeck* notes. **H** Forest, reedbeds. **D** Rimatara, French Polynesia.

28 HENDERSON REED WARBLER *Acrocephalus taiti* 17cm Unmistakable in range. Head normally all-white with some blackish feathers in crown. **V** Unmelodic, scratchy *sruuuh* or *sreeeh*. **H** Forest with thick undergrowth. **D** Henderson Island, Pitcairn Islands, SC Pacific.

29 PITCAIRN REED WARBLER *Acrocephalus vaughani* 17cm Unmistakable in range. Formerly treated as conspecific with Rimatara and Henderson Reed warblers. **V** High, unmelodic *sreeeh sreeeh*. **H** Prefers tall forest. **D** Pitcairn Island, SC Pacific.

30 BLACK-BROWED REED WARBLER *Acrocephalus bistrigiceps* 12cm Note pale supercilium with prominent brownish-black stripe above. Shy and skulking, forages in thick cover, sings from exposed perches during breeding season. **V** Short phrases interspersed with rasping and churring notes. Calls include a soft *dzak* and a harsh *chur*. **H** Emergent vegetation near marshes or watery areas. **D** Breeds E Mongolia, SE Russia to E China, Japan. Winters SE Asia.

31 MOUSTACHED WARBLER *Acrocephalus melanopogon* 12–13cm Note strong head pattern and clean white throat. Usually stays out of sight, feeding among reed debris near water surface. Cocks tail when alarmed. **V** Song is *lu-lu-lu-lu-lu* followed by a scratchy warble. Calls include a churring *trrrp* and a *tac-tac* contact note. **H** Reedbeds, waterside scrub. **D** Breeds SW Europe to C Asia. Winters Mediterranean region, Middle East, Indian subcontinent.

32 AQUATIC WARBLER *Acrocephalus paludicola* 13cm Separated from Sedge Warbler by pale central crown-stripe and more heavily streaked mantle. Shy and retiring. Sings from exposed perch or during short song flight. **V** Less complex and more monotonous than Sedge Warbler, *trrr...dew-dew-dew...churr...di-di-di...* Calls include *tuk, chuck, cher-cherr*. **H** Open, waterlogged sedge meadows. On passage uses a wider range of waterside vegetation. **D** Breeds C and E Europe. Winters W Africa.

33 SEDGE WARBLER *Acrocephalus schoenobaenus* 13cm Note distinct pale supercilium. Furtive, but often shows well, especially when singing. **V** Medley of sweet and harsh phrases. Calls include a *tuc* or *tuc-tuc-tuc* when alarmed, and a soft *churr*. **H** Dense vegetation, usually near water. **D** Breeds N and C Europe to C Russia. Winters sub-Saharan Africa.

34 SPECKLED REED WARBLER *Acrocephalus sorghophilus* 13cm Forages low down in vegetation, climbs reed stems and sings from a prominent perch. **V** Series of rasping, churring notes. **H** Reedbeds and boggy grassland, often near water. **D** Breeds NE China. Winters SE China, Philippines.

35 BLUNT-WINGED WARBLER *Acrocephalus concinens* 13cm Skulking, although regularly calls while foraging in cover. **V** Repeated slurred whistles, clear and buzzing notes. Calls include a quiet *tcheck* and a soft *churr*. **H** Reedbeds, rank vegetation and scrub in swamps and marshes; also drier open areas in mountain valleys. **D** Breeds Afghanistan to NW India; NE India to Myanmar; C and E China. Winters Bangladesh, SE Asia.

36 MANCHURIAN REED WARBLER *Acrocephalus tangorum* 13–15cm Not well known; noted climbing to reed tops with tail cocked. **V** Series of warbled phrases interspersed with squeaky notes. Calls include a sharp *chi-chi*, a harsh *chr-chuck* and a slurred *zack-zack*. **H** Reedbeds and emergent vegetation in marsh and lake borders. **D** Breeds NE China. Winters SE Asia.

37 LARGE-BILLED REED WARBLER *Acrocephalus orinus* 13cm Little recorded; actions probably much like others of the genus. **V** Song variable, consisting of some short, scratchy *wjitchety-wjitchety-wjitchety*, or *we-chuck we-chuck* phrases as well as some beautiful bulbul-like notes. **H** Breeds in waterside scrubby bush; post-breeding habitat not documented. **D** Breeds NE Afghanistan. Winters N India, Thailand.

1
2
3
4
5
6
7
8
9
10
11
12
13
14
15
16
17
18
19
20
21
22
23
24
25
26
27
28
29
30
31
32
33
34
35
36
37

REED WARBLERS AND ALLIES *CONTINUED*

1 PADDYFIELD WARBLER *Acrocephalus agricola* 13cm An unstreaked *Acrocephalus* warbler with a long pale supercilium. Constantly cocks and flicks tail. Forages low in cover and sometimes on the ground. **V** Hurried series of melodious, chortled notes interspersed with mimicry. Calls include a sharp *chik-chik*, a harsh *chr-chuck* and a slurred *zack zack*. **H** Waterside vegetation, paddyfields. **D** Breeds E Europe to Mongolia, WC China. Winters S Asia to Myanmar.

2 BLYTH'S REED WARBLER *Acrocephalus dumetorum* 13cm Note relatively uniform plumage and short primary projection. Best identified by voice. Mainly arboreal, often fans and flicks tail. **V** Sings mainly at night, very varied mix of harsh and clear notes, often including mimicry. Calls include a clicking *thik*, a harsh *tchirr* and a hard *chack*. **H** Scrub, rank vegetation in dry or wet locations, cultivation, forest edge. **D** Breeds NE Europe to C Russia, NW Mongolia. Winter visitor to Indian subcontinent, Myanmar.

3 EURASIAN REED WARBLER *Acrocephalus scirpaceus* 13cm Long primary projection. Furtive while foraging in reeds or bushes. **V** Song is a series of grating and squeaky notes, *kerr-kerr chirruc chirruc chirruc kek-kek-kek chirr-chirr*. Calls include *churr* and harsh *tcharr*. **H** Reedbeds, bushes near water. **D** Breeds Europe to C Asia; Arabian Peninsula, NE Africa. Winters sub-Saharan Africa.

4 AFRICAN REED WARBLER *Acrocephalus baeticatus* 13cm Very similar to Eurasian Reed Warbler. Note general rusty-reddish colouring and brown wings and tail. **V** High, loud, chattered, sustained, rapid syllables, *krits-krit-tjitji weetkarreweetweet-tri–*, each syllable repeated twice. **H** Tall grass, reedbeds, other lush herbage, not necessarily near water. **D** Breeds Iberian Peninsula, NW Africa, sub-Saharan Africa. Winters sub-Saharan Africa.

5 MARSH WARBLER *Acrocephalus palustris* 13cm Eye-ring dominates supercilium. May show raised crest. **V** Very high, powerful, rapid warble, with perfect mimicry, e.g. *weet-weet-o-kree-kree-wee-tic-tic-tic...* **H** Thickets, reedbeds, tall herbage, not necessarily near water. **D** Breeds C and NC Europe to Turkey and C Russia. Winters SE Africa.

6 THICK-BILLED WARBLER *Arundinax aedon* 18–19cm Elusive; clumsy while foraging in vegetation and bushes. **V** Song *chok chok* followed by loud, hurried, chattering warble. **H** Tall grass, marshy areas with reeds and bushes. **D** Breeds Siberia, Mongolia, NE China. Winters India, SE Asia.

7 AFRICAN YELLOW WARBLER *Iduna natalensis* 14cm Distinguished from Mountain Yellow Warbler by darker cap, from Icterine Warbler and Willow Warbler by brighter underparts and longer-tailed, upright profile. **V** Very high, melodious, staccato trills and rolls connected by low, dry *tektek tek*. **H** Skulks in rather dense undergrowth. **D** Sub-Saharan Africa, widespread.

8 MOUNTAIN YELLOW WARBLER *Iduna similis* 13cm Crown and mantle uniform olive-green. **V** High, crescendoing *tjuutjuuTjuuTjuuTJuuTJuu-* and other high, fluted notes. **H** More mountainous than African Yellow Warbler but habitats overlap. **D** E Africa.

9 BOOTED WARBLER *Iduna caligata* 12cm Very similar to Sykes's Warbler, but also resembles a *Phylloscopus* warbler (Plates 210–211). **V** Dry *chek* and short trilled *tr-r-rk*. **H** Herbaceous plants and shrubs, often near water. **D** Breeds E Europe to C Asia. Winters India.

10 SYKES'S WARBLER *Iduna rama* 12–13cm Forages mainly in bushes, with brief visits to pick food from the ground; while moving about twitches tail upwards or sideways. **V** Rattling, scratchy warble. Calls are a *chek* and a fuller *tslek*. **H** Arid areas with scattered bushes. **D** Breeds C Asia, Iran, Pakistan. Winters Indian subcontinent.

11 EASTERN OLIVACEOUS WARBLER *Iduna pallida* 12cm Much paler lores than *Acrocephalus* warblers. Note absence of any yellow or green tinge to plumage, short wings and faint wing panel. **V** High rasping warble, sounding like a slightly speeded-up Reed Warbler and includes mimicry. **H** As Icterine Warbler but often in drier areas. **D** Breeds SE Europe through Middle East to Iran, Kazakhstan; also scattered populations in N Africa. Winters sub-Saharan Africa.

12 WESTERN OLIVACEOUS WARBLER *Iduna opaca* 13cm Clambers through foliage, stretching neck to gather insects. Pumps tail up and down. **V** Reed Warbler-like chattering, faster and more monotonous, with musical notes along with rattles and trills. Call is a sharp *tec* and a *tick-tick-tick*; also sparrow-like chatter. **H** Open bushy areas in dry country; olive groves, orchards, parks and gardens. **D** Breeds Iberian Peninsula, NW Africa. Winters W Africa.

13 PAPYRUS YELLOW WARBLER *Calamonastides gracilirostris* 13cm Supercilium ill-defined. From Mountain Yellow Warbler and African Yellow Warbler also by very narrow bill. **V** Song is a simple, short phrase of up to five rattling or metallic *chowweet* or *t-tslowee* notes. **H** Papyrus, reeds. **D** E Africa, from E DR Congo and W Kenya to N Zambia.

14 UPCHER'S WARBLER *Hippolais languida* 14cm Clambers, often clumsily, while foraging through foliage, regularly stretches neck to reach food; tail often appears 'unhinged' when moved up, down or sideways. **V** Scratchy warble. Call said to resemble two stones being struck together, *chuk-chuk*. **H** Open bushy hillsides. **D** Breeds Turkey to Pakistan. Winters E Africa.

15 OLIVE-TREE WARBLER *Hippolais olivetorum* 15cm Note powerful bill. Slight downward tail-flicks. **V** Song resembles a melodious Great Reed Warbler (Plate 212), *chak - chu - chi-chak - chira - chuk - chi-chi - chak-era - chak - chu...* **H** Open broadleaf forest, orchards, olive groves, almond plantations, vineyards, hillside scrub. **D** Breeds SE Europe to Israel. Winters E and SE Africa.

16 MELODIOUS WARBLER *Hippolais polyglotta* 13cm Skulking, lethargic, methodical feeder, fluttering flight between cover. Shallow tail-flick. **V** Faster, more rattling than Icterine Warbler, lacking repeated nasal notes. Call: Soft *tuk*, a harsh rattled *trrrrr* and a chattering *chret-chet*. **H** Much as Icterine Warbler, but also in more bushy areas. **D** Breeds SW Europe. Winters W Africa.

17 ICTERINE WARBLER *Hippolais icterina* 13cm Shallow tail-flick. Active feeder; strong, dashing flight between cover. **V** Resembles Marsh Warbler, but slower and higher-pitched with typical call note *chi-chi-vooi* and nasal *geea-geea* added. **H** Broadleaved woodland, forest edge, parkland, orchards, hedgerows, gardens. **D** Breeds C Europe to C Russia, Kazakhstan. Winters S Africa.

GRASSBIRDS AND ALLIES LOCUSTELLIDAE

18 CORDILLERA GROUND WARBLER *Robsonius rabori* 20–22cm Forages among leaf litter or wood debris by walking or hopping; also said to run through the forest like a small rail. Often cocks tail. **V** Variable, very high-pitched insect-like *tsui-ts sii uu ee, tsui-ts ssuuu ee, tit-tuuits ts ssu eeet* or similar. **H** Lowland forest, secondary growth. **D** Luzon, N Philippines.

19 SIERRA MADRE GROUND WARBLER *Robsonius thompsoni* 20–22cm Recently discovered. Forages on the ground, walking and running around in search of prey. **V** High-pitched and ventriloquial, therefore very difficult to locate. **H** Forest with a dense understorey. **D** Luzon, N Philippines.

20 BICOL GROUND WARBLER *Robsonius sorsogonensis* 20–22cm Walks, runs or hops while foraging among leaves on the forest floor; often fans and cocks tail. **V** High-pitched *tit–tsuuuu-tsiiiiii*, and a *tssuit swiieeii, tssuu sit suuiee, tssiiuu sweeiieet* or *tit-tsuee-swiieet*. **H** Broadleaved evergreen forest, forest edge, secondary growth, bamboo; especially limestone areas. **D** Luzon and Catanduanes, N Philippines.

21 SAKHALIN GRASSHOPPER WARBLER *Helopsaltes amnicola* 16–18cm Formerly treated as conspecific with Gray's Grasshopper Warbler. Differs genetically from Gray's but very similar morphologically. Very secretive. **V** Song is similar to Gray's but distinctly slower. **H** Undergrowth and bushes in lowland forest. **D** Breeds Sakhalin, S Kuril Islands, N Japan. Winters Wallacea (Indonesia).

22 GRAY'S GRASSHOPPER WARBLER *Helopsaltes fasciolatus* 16–18cm Secretive, foraging in deep cover or among grass stems. **V** Song is rich and fluty, starting softly, then getting louder and finishing with a rapid, falling-away *u-chic-toi-tu-tee-chee*. **H** Thickets, forest edges and tall grass. **D** Breeds SE Russia, Mongolia, NE China, Korean Peninsula, N Japan. Winter visitor to Philippines, Indonesia, New Guinea.

23 MARSH GRASSBIRD *Helopsaltes pryeri* 12–13cm Streaked upperparts. Skulking, clambers nimbly among grass or reeds. Sings from prominent reed perch or in flight. **V** Low-pitched *djuk-djuk-djuk*; call is a low *chuk*. **H** Dense grass or reedbeds by rivers or lakes. **D** Breeds E Mongolia, SE Russia, E China, Japan. Winters south to C China.

24 PALLAS'S GRASSHOPPER WARBLER *Helopsaltes certhiola* 13–14cm Note white-tipped tail. Skulks and creeps among reeds and tall grass. **V** Calls include a thin *tik-tik-tik*, a sharp *chirk* or *chuck* and a trilling *chi-chirr*, similar to Lanceolated Warbler but less urgent and quieter. **H** Reedbeds, paddyfields. **D** Breeds Siberia and Kazakhstan to N China, Sakhalin. Winters India, Sri Lanka, SE Asia, Indonesia.

25 STYAN'S GRASSHOPPER WARBLER *Helopsaltes pleskei* 15–16cm Skulking, but often ventures into the open to forage. **V** Song *chip-chit-chip-chir-chit-chi-schwee-schee*; call is a short *chit*. **H** Coastal mangroves, water vegetation, reedbeds, seashore rocks. **D** Breeds on islands off far E Russia, Korean Peninsula, S Japan. Winter visitor to coastal S China, N Vietnam.

26 MIDDENDORFF'S GRASSHOPPER WARBLER *Helopsaltes ochotensis* 15cm Skulker, but exposed when singing either from a perch or looping song flight. **V** Short, jarring warbles. **H** Open thickets in damp grassland, dwarf bamboo, waterside scrub and reedbeds. Winters mainly in reedbeds and mangroves. **D** Breeds far E Russia, N Japan. Winters Philippines, Borneo.

27 LANCEOLATED WARBLER *Locustella lanceolata* 12cm Note heavy streaking above and below. Very secretive; tends to run through vegetation, very reluctant to fly. **V** Song is a continuous thin reeling, slightly higher-pitched than Common Grasshopper Warbler. Calls include a loud, urgent *chirr-chirr* or *chi-chirr* and a low-pitched *chk*. **H** Dense bush, reedbeds, tall grassy vegetation. **D** Breeds from E Europe to far E Russia. Winters SE Asia, Greater Sunda Islands, Philippines.

28 BROWN BUSH WARBLER *Locustella luteoventris* 13–14cm Skulking, although occasionally sings from a prominent perch. **V** Dry reeling; calls include a hard *tack* and a higher *tink-tink-tink*. **H** Dense grass, weedy areas or scrub near forest edge or clearings. **D** E Himalayas to C China, N Vietnam; mostly resident.

29 LONG-BILLED BUSH WARBLER *Locustella major* 13cm Very skulking; runs through vegetation like a small rodent, hard to flush. **V** Monotonous, rapid reeling. **H** Grassy slopes with bushes, waste areas, forest clearings, upland cultivation. **D** W Himalayas, W China.

30 COMMON GRASSHOPPER WARBLER *Locustella naevia* 13cm Streaked above but relatively plain below. More often heard than seen. **V** Song is a dry insect-like reeling, which may continue for several minutes. **H** Scrubby thickets or rank vegetation on heathland, moorland, downland, marshy areas, forest edge. **D** Breeds from W Europe to C Asia. Winters W Africa, Indian subcontinent.

31 CHINESE BUSH WARBLER *Locustella tacsanowskia* 13–14cm Keeps well hidden, creeping about in tangled vegetation. **V** Rasping, insect-like *raaasp...raaasp...raaaaaasp... raaaaaaasp...raaasp...* **H** Grassy upland meadows with shrubby thickets and tall grasses. Winters in lowland grasslands, reeds and scrubby edges of cultivation. **D** Breeds SC Siberia to C China. Winters SE Asia.

32 BAMBOO WARBLER *Locustella alfredi* 14cm Note white chin and grey breast and flanks. Separated from Cinnamon Bracken Warbler (Plate 214) in same habitat by less warm colouring, faint throat spots and song. **V** Monotonous *tsjuu-kuh tsjuu-kuh tsjuu-kuh...* **H** Undergrowth of bamboo, montane forest. **D** C Africa, from Ethiopia to S DR Congo.

33 RIVER WARBLER *Locustella fluviatilis* 14cm Note pale tips to dark undertail coverts. Sings from exposed perch, otherwise shy and elusive. **V** Rapid, cricket-like *zre-zre-zre-zre-zre...* delivered for long periods. **H** Various types of dense bush, usually near water or other damp areas. Not necessarily near water in winter. **D** Breeds C Europe to Kazakhstan, SW Siberia. Winters SE Africa.

34 SAVI'S WARBLER *Locustella luscinioides* 14cm Very plain, unstreaked. Skulks, with mouse-like actions. Sings from exposed perch. **V** Quiet ticking followed by vibrant, insect-like, buzzy reeling. **H** Reedbeds and dense waterside vegetation. **D** Breeds W Europe to C Asia. Winters sub-Saharan Africa.

35 FRIENDLY BUSH WARBLER *Locustella accentor* 15cm Pale throat may be hard to see in forest gloom. Forages on or close to the ground, alone or in pairs. Often follows people in the forest, picking up disturbed invertebrates. **V** Repeated *dzhee-dzhee-zeeeeee-ah* or *trrp trrrp trzzzzz*. **H** Damp, dark rainforest undergrowth with moss-covered dead tree trunks or branches. **D** N Borneo.

36 CHESTNUT-BACKED BUSH WARBLER *Locustella castanea* 14–15cm Note rich dark brown upperparts. Occurs singly or in pairs; forages on or close to the ground, gleaning insects from foliage and twigs. **V** Buzzy, rising and falling three-note trill, repeated at 5-sec intervals. Calls include a *tsp* or (on Seram) a soft, repeated *zit-oh-zit*. **H** Dense undergrowth, vines and tangles in montane forest, forest edge, upland grassland. **D** Sulawesi and S Maluku Islands, Indonesia.

1
2
3
4
5
6
7
8
9
10
11
12
13
14
15
16
17
18
19
20
21
22
23
24
25
26
27
28
29
30
31
32
33
34
35
36

GRASSBIRDS AND ALLIES *CONTINUED*

1 LONG-TAILED BUSH WARBLER *Locustella caudata* 16–18cm Fairly uniform dark plumage, and can appear darker in forest gloom. Forages alone or in pairs, on or close to the ground; when disturbed, runs for cover like a rodent. **V** High-pitched, insect-like *to-to-zeeee, trp trp twzzz, zeeeuuu* or *zeee-zuuu*. **H** Forest undergrowth and dense secondary growth in forest clearings. **D** Philippines.

2 BAIKAL BUSH WARBLER *Locustella davidi* 12cm Formerly considered conspecific with Spotted Bush Warbler but differs vocally, morphologically and genetically. **V** Song is a series of short, buzzy insect-like trills, each trill about 1 sec long. Call is a harsh *tchik*. **H** Scrubby thickets, reedbeds, damp grassland. **D** SC Siberia to C China. Winters SE Asia.

3 SPOTTED BUSH WARBLER *Locustella thoracica* 11–13cm May show rufous forecrown. Secretive, especially post breeding; generally keeps to thick cover. **V** Song consists of a series of clicks and buzzing sounds. Calls include a harsh *shtak* and a low *tuk*. **H** Scrubby thickets near forests, high-level meadows; winters in rank vegetation by watercourses. **D** Breeds C Himalayas to C China. Winters at lower altitudes and south to N SE Asia.

4 WEST HIMALAYAN BUSH WARBLER *Locustella kashmirensis* 12cm Formerly considered conspecific with Spotted Bush Warbler but differs vocally, morphologically and genetically. **V** Song is a rhythmic series of buzzy, insect-like couplets, with the final note slightly longer. **H** Breeds in alpine scrub. Winters at lower altitudes. **D** Breeds W Himalayas. Non-breeding range largely unknown.

5 TAIWAN BUSH WARBLER *Locustella alishanensis* 13cm Fairly uniform dull brown. Skulking, forages low down. **V** Loud, clear monotonous notes ending with two or three clicks. Calls with a scratchy *ksh ksh ksh*. **H** Open grassy slopes, thickets, forest clearings, undergrowth in cultivations. **D** Taiwan.

6 RUSSET BUSH WARBLER *Locustella mandelli* 13–14cm Variable; may show whitish throat and pale supercilium. Typical skulking *Locustella* behaviour. **V** Rapid *cre-ut cre-ut cre-ut cre-ut*... Calls include a short *shtuk*. **H** Low vegetation in forest clearings, forest edge, grass, scrub. **D** E Himalayas to C and E China, N SE Asia.

7 DALAT BUSH WARBLER *Locustella idonea* 13–14cm Formerly considered conspecific with Russet Bush Warbler. Similar to that species, but darker above and lighter below. **V** Song and call are similar to Russet Bush Warbler. **H** Low vegetation in forest clearings, forest edge, grass, scrub. **D** SC Vietnam.

8 BENGUET BUSH WARBLER *Locustella seebohmi* 14cm Little information; presumed to have the typical skulking behaviour of the genus. **V** Metallic zipping sound. **H** Steep, grassy valleys. **D** N Philippines.

9 JAVAN BUSH WARBLER *Locustella montis* 14–15cm Fairly large bush warbler with rusty brown upperparts, greyish face and underparts. Hops around on low branches or on boulder-strewn slopes. **V** Monotonously repeated *klpklpzeeep*. **H** Bushes and grassy areas at forest edges, cleared slopes with scattered trees and scrub. **D** Java and Bali, Indonesia; Timor.

10 SICHUAN BUSH WARBLER *Locustella chengi* 13–14cm Recently described. Differs from Russet Bush Warbler in plumage and song. **V** Rhythmic series of notes repeated monotonously. **H** Bushes and scrub at forest edge and in clearings, at mid-elevation. **D** C China.

11 FLY RIVER GRASSBIRD *Poodytes albolimbatus* 14–15cm Smaller than Tawny Grassbird with shorter, more rounded tail. Endemic to Trans-Fly region. Skulking, usually keeping low. **V** Call is a thin mournful whistle. Song is poorly known. **H** Wet clumps of reeds and floating grass. **D** SC New Guinea.

12 SPINIFEXBIRD *Poodytes carteri* 15–16cm Slim, unstreaked warbler with long broad tail. Note rufous cap and buff supercilium. Seen singly or in pairs, usually skulking. **V** Song is a short melodious warble. Call is a hard *chup*. **H** Tall clumps of spinifex in arid regions. **D** C Australia.

13 NEW ZEALAND FERNBIRD *Poodytes punctatus* 18cm Note bold streaking above and below; no similar bird in range. Skulking in dense cover; weak flyer; supports itself on tail when clambering around. **V** Contact call is duet, started by male, answered by female, together as *TU'Weeh*. Also heard is rapid *Turrit* or *Trit*, short, dry rattle or fluted *Tuwt*. **H** Dense low vegetation in fresh and saline wetlands. **D** New Zealand.

14 LITTLE GRASSBIRD *Poodytes gramineus* 14cm Smaller and shorter-tailed than Tawny Grassbird, with darker and greyer plumage. Secretive, often first detected by song. **V** Song is a plaintive, far-carrying trisyllabic whistle, *pe-peeeeee-peeeee*. **H** Reedbeds, swamps, wetlands, salt marshes. **D** New Guinea (rare), Australia.

15 MALIA *Malia grata* 28–29cm Olive-green above, greenish-yellow below. Usually occurs in small, noisy, active parties, foraging from the ground to the lower canopy. **V** Group calls include a repeated, high-pitched whistle, accompanied by a harsh, rapid *tsut-sut KA-KA* and a cacophony of guttural, warbling notes. **H** Primary montane forest and, occasionally, disturbed forest. **D** Sulawesi, Indonesia.

16 BROWN SONGLARK *Cincloramphus cruralis* male 24–26cm, female 18–20cm Sexually dimorphic, with male much larger than female. Long-legged bird of treeless habitats. Mainly terrestrial. Partially migratory (within Australia). **V** Rich melodious song includes scratchy notes, often given in diagnostic song flight. **H** Treeless plains, grassland, farmland. **D** Australia.

17 RUSTY THICKETBIRD *Cincloramphus rubiginosus* 19cm Orange-rufous with long graduated tail. Often cocks and fans tail. Secretive and mainly terrestrial. **V** Series of clear whistles, *see too too-woo-wee*, with the last notes slurred together. **H** Lowland and hill forest with dense undergrowth. **D** Bismarck Archipelago.

18 NEW BRITAIN THICKETBIRD *Cincloramphus grosvenori* 18cm Dark rusty-brown above with dark mask; paler orange-buff below. Rare and little known. Skulking. **V** Not known. **H** Undergrowth with bamboo in montane forest. **D** New Britain (Bismarck Archipelago).

19 BUFF-BANDED THICKETBIRD *Cincloramphus bivittatus* 18–19cm Note chestnut crown and broad buff supercilium. Skulks low down in dense thickets. **V** Staccato *zi-ka-cheet* or *tswi-chit*, a harsh, descending trill and a high-pitched rattle. **H** Patches of monsoon forest, secondary woodland, scrub. **D** Timor, Lesser Sunda Islands.

20 RUFOUS SONGLARK *Cincloramphus mathewsi* male 18–19cm, female 14–16cm Sexes similar, but male is larger. Similar to Brown Songlark but smaller and has diagnostic rufous rump. Lacks dark belly patch. Mainly terrestrial. Partially migratory (within Australia). **V** Melodious medley of whistles and trills, given in song flight or from a perch, often for long periods. **H** Open grassy woodlands and scrub. **D** Australia.

21 PAPUAN GRASSBIRD *Cincloramphus macrurus* 23cm Larger and longer-tailed than Tawny Grassbird with less rufous crown. Seven races; five montane, two lowland. Skulking. Singly or in pairs. **V** Less varied than Tawny Grassbird; a rich loud chirrup, often preceded by single, short, sharp notes. **H** Grasslands and forest edge. **D** New Guinea, Bismarck Archipelago.

22 TAWNY GRASSBIRD *Cincloramphus timoriensis* 18–20cm Skulking; often uses the tops of grass stems as a look-out. **V** Song contains ascending and descending, rattling, ringing notes. Calls include *tsip, chir-up* or *threr-up*. **H** Reedbeds and grassland with scattered trees and shrubs. **D** Philippines to N and E Australia.

23 MELANESIAN THICKETBIRD *Cincloramphus whitneyi* 17cm Dark ground warbler with long graduated tail and rufous supercilium. Secretive. Forages on or close to the ground. **V** Alarm call is a piercing series of fast, high-pitched notes. Voice of Guadalcanal race not known. **H** Montane forest with dense undergrowth. **D** Guadalcanal (Solomon Islands), Espiritu Santo (Vanuatu).

24 NEW CALEDONIAN THICKETBIRD *Cincloramphus mariae* 18cm Long-tailed with prominent white supercilium, dark mask and white throat. Singly or in pairs; secretive. **V** Song is a fast series of rich notes sounding like a rattle or trill. Call is a harsh *tchak*. **H** Scrubby habitats, clearings and tall grass savannah in foothills. **D** New Caledonia.

25 LONG-LEGGED THICKETBIRD *Cincloramphus rufus* 19cm Dark rufous with prominent eye-stripe. Terrestrial beneath dense vegetation. **V** Song consists of a short, loud, variable series of clear notes. Calls include an explosive scolding chatter. **H** Rainforest. **D** Fiji.

26 BOUGAINVILLE THICKETBIRD *Cincloramphus llaneae* 17cm Dark brown above with orange-rufous supercilium and underparts. Long graduated tail. Skulking and secretive. **V** Clear melodic whistle of 1–2 notes followed by a higher note. **H** Montane forest. **D** Bougainville (N Solomon Islands).

27 STRIATED GRASSBIRD *Megalurus palustris* 22–28cm Large and conspicuous, with long tail and strongly streaked upperparts. Forages in reeds and bushes, sometimes on the ground; often perches prominently on reeds or bush tops, flicking the tail up and down. **V** Strong, rich warble; also a subdued whistle followed by a loud *wheeechoo*. Calls include an explosive *pwit* and a harsh *chat* or *tzic*. **H** Tall damp grasslands, reedbeds, tamarisks. **D** Pakistan to N India, S China, SE Asia, Indonesia, Philippines.

28 SRI LANKA BUSH WARBLER *Elaphrornis palliseri* 16cm Note buffish throat. Secretive; constantly on the move and continually flicking tail. Post breeding often forms loose flocks. **V** Song consists of squeaky scattered notes. Call is a low, explosive *quitz* or *queek*. **H** Dense undergrowth in damp forests. **D** Sri Lanka.

29 BROAD-TAILED GRASSBIRD *Schoenicola platyurus* 18cm Small head and broad tail with long undertail coverts, resulting in 'rear-heavy' appearance. Skulking, forages low in thick tangled vegetation; best located while singing, either from a reed stem or during display flight. **V** High-pitched, metallic, piping *twink twink twink* ending in a rattling *cheep-cheep-cheep*; also a soft ringing *tseenk tseenk*. Calls include a repeated, trisyllabic *jur-jur-jur* and a harsh *chick*. **H** Grass, reed-swamp, bracken-covered hillsides. **D** W India.

30 BRISTLED GRASSBIRD *Schoenicola striatus* 19–21cm Note short strong bill. Generally elusive, forages through reeds and tall grass, sometimes on the ground; sings from prominent perch or during song flight. **V** Repeated, rising then falling *trew treuw* or *ji-jee jee-ji*. Calls include a harsh *cha* and a sharp *zip*. **H** Lowland grassland with scattered bushes, paddyfields. **D** Breeds NE Pakistan to Bangladesh. Winter visitor to EC India.

31 FAN-TAILED GRASSBIRD *Catriscus brevirostris* 16cm Note disproportionately small head and heavy tail. **V** Extremely high, unstructured, weak yet piercing *sweeh sweet sweeh sweeh sweeh*. **H** Lush vegetation near or away from water. **D** Sub-Saharan Africa, widespread.

32 KNYSNA WARBLER *Bradypterus sylvaticus* 14cm Distinguished from Barratt's Warbler by pinkish legs. **V** High-pitched *teet teet-teet-tee-ti-tjsrrrrrr*. **H** Forest undergrowth. **D** S and E South Africa.

33 BANGWA FOREST WARBLER *Bradypterus bangwaensis* 14cm Brightly coloured, with long pale supercilium, dark rufous-brown upperparts, whitish throat and belly contrasting with rufous breast-band and flanks. **V** Rhythmic series of repeated notes, increasing in volume, *piya-piya-piya-PIYA-PIYA*. **H** Dense undergrowth, montane forest edge, clearings, secondary growth. **D** SE Nigeria, W Cameroon.

34 BARRATT'S WARBLER *Bradypterus barratti* 15cm Note white chin, dark legs. **V** Very high, sweeping whistles, *fifitutrrrrrrrt*. **H** Vegetation at edge of or under montane forests. **D** SE Africa.

35 EVERGREEN FOREST WARBLER *Bradypterus lopezi* 14cm Note overall dark appearance with little or no white. Normally no other *Bradypterus* warbler in its habitat. **V** High hurried rattled sharp *tjutjutjutjutjut* and crescendoing *titju-titju-titju-titju*. **H** Forest interior. **D** Scattered distribution, from Cameroon to Kenya, from Angola to Mozambique.

36 CINNAMON BRACKEN WARBLER *Bradypterus cinnamomeus* 15cm Crown darker than mantle, supercilium well defined. **V** Very high, powerful, melodious rattles, *fjuut-rrrrr* or *tjuut-wee-tweet-weet*. **H** Undergrowth of montane forests, bamboo, moist thickets. **D** E Africa, from Ethiopia to NE Zambia, N Malawi.

CISTICOLAS AND ALLIES *CONTINUED*

1 SOCOTRA WARBLER *Incana incana* 13cm Note long thin bill, warm brown upperparts and near-white margins of primaries. **V** Little known. Song consists of a short, soft, hesitant trill. Call is a series of scolding *chip* notes; alarm call a rolling low-pitched trilling *churr*. **H** Open and bushed grassland. **D** Socotra, Arabian Sea.

2 STRIATED PRINIA *Prinia crinigera* 16cm Rather drab, with long graduated tail. Usually skulking, but in breeding season sings from a prominent perch on trees or bushes. **V** Jaunty, wheezy *chitzereet-chitzereet-chitzereet-chitzereet-chitzereet-chitzereet*; the last few notes speed up, making a slight climax. **H** Open grassy mountainsides or hillsides with scattered shrubs, rank herbage in cultivations. **D** Afghanistan and Pakistan through Himalayas to Myanmar, China, Taiwan.

3 BROWN PRINIA *Prinia polychroa* 16cm Distinguished from Striated Prinia by absence of streaking on underparts. Keeps to thick cover; if disturbed tends to creep through vegetation rather than fly. **V** Wheezy *chirt-chirt-chirt* or *chook-chook-chook*. Call is a loud *twee-ee-ee-eet*. **H** Grassland with low scrub, cultivations, open deciduous forest with low shrubby growth. **D** SE Asia; Java, Indonesia.

4 BLACK-THROATED PRINIA *Prinia atrogularis* 17cm Active; forages in thick cover, constantly flicks tail. **V** Scraping, buzzing *szelik szelik szelik* or similar. Calls include a soft *tp-tp-tp* and a scolding *chrrr-chrrr-chrrr*. **H** Grassy mountainsides and hillsides with scattered shrubs, scrub in forest clearings, thick vegetation in cultivations. **D** E Himalayas to W Myanmar.

5 HILL PRINIA *Prinia superciliaris* 17cm Note unstreaked upperparts and whitish supercilium. Noisy and active in grass and low vegetation. **V** Loud, piercing *cho-ee cho-ee cho-ee*. **H** Bracken-covered slopes, grass, scrub, overgrown clearings. **D** NE India; S China; SE Asia; Sumatra, Indonesia.

6 GREY-CROWNED PRINIA *Prinia cinereocapilla* 11cm A small prinia with a relatively short tail. Rufous forehead merging into grey crown. Shy and elusive; forages in thick tangles, constantly twitches tail. **V** Twittering *tweetoo-wee-too-weetoo-weeto* followed by a tinny *ti-ti-ti-ti-ti*. Call note is a repeated *tzit*. **H** Bushes and scrub in forest clearings, secondary growth and scrub in cultivations. **D** Himalayan foothills.

7 RUFOUS-FRONTED PRINIA *Prinia buchanani* 12cm Distinguished from Plain Prinia by white tail tips and different song. Active, tail constantly twitched; forages in tangled vegetation. **V** Musical, reeling *chid-le-weeest*, the last note with a rising inflection, followed by a repeated *chid-le-ee...* **H** Semi-desert areas with scrub and coarse grasses, clearings in dry forests. **D** Pakistan, India.

8 RUFESCENT PRINIA *Prinia rufescens* 11cm Note supercilium extending to just behind eye. Very secretive, occasionally in small parties; tail regularly waved up and down or sideways. **V** Rhythmic *chewp-chewp-chewp-chewp*. Calls include a *chip chip chip* and a buzzy *peez-peez-peez*. **H** Grassy areas in open forest, grassland with scattered trees. **D** NE India, through SE Asia to Peninsular Malaysia, S China.

9 GREY-BREASTED PRINIA *Prinia hodgsonii* 11cm White throat and grey breast-band. Active, moves quickly through bushes or undergrowth; often in parties post breeding. **V** Squeaky, warbling *chiwee-chiwee-chewi-chip-chip-chip*. Calls include a tinkling *zeee-zeee-zeee* and a thin *chew-chew-chew*. **H** Scrub jungle, bushes in open forest or forest edge, also bushes in cultivations and gardens. **D** Pakistan, India, Sri Lanka through SE Asia to Vietnam.

10 GRACEFUL PRINIA *Prinia gracilis* 11cm Greyish, with indistinct supercilium. Twitches tail and wings, the latter often producing a 'snapping' sound, especially in aerial display. **V** Monotonous repetition of a thin *tzeee-bit* or *zerwit*. Calls include a trilling *bleep* and a flat *jit*. **H** Very varied, from low vegetation in arid areas to grassy swamp, reedbeds, cultivations, gardens. **D** From Egypt through Middle East and Arabian Peninsula to NE India, Bangladesh.

11 JUNGLE PRINIA *Prinia sylvatica* 13cm Forages in low vegetation; sings from a prominent perch or during an undulating display flight. **V** Loud, rapid *pit-pretty pit-pretty pit-pretty...* Calls include a rapid chatter and a quickly repeated *pit pit pit*. **H** Scrubby bush jungle mixed with boulders and grassland. **D** India, Sri Lanka.

12 BAR-WINGED PRINIA *Prinia familiaris* 13cm Note double white wing-bar. Forages from the ground to the treetops, usually alone or in pairs; post breeding, occurs in noisy parties. **V** Loud, high-pitched *chweet-chweet-chweet*. **H** Mangroves, secondary growth, plantations, parks, shrubby gardens. **D** Sumatra, Java and Bali, Indonesia.

13 YELLOW-BELLIED PRINIA *Prinia flaviventris* 13cm Yellow belly, no wing-bars. Restless forager among grasses. Sings from a prominent perch or during a short display flight. **V** Song is flute-like, starting with a *chirp* followed by a descending trill, repeated many times. Calls include a *chink chink*.... and a plaintive *twee-twee*. **H** Wet areas with tall grassland and small bushes, waterside reeds. **D** From Pakistan through N India to SE Asia, Indonesia, S China, Taiwan.

14 ASHY PRINIA *Prinia socialis* 13cm Distinctive blue-grey, rufous and buff plumage. Skulking but active forager low down in tall grass or scrub. Sings from prominent perch or during display flight. **V** Loud, cheerful, wheezy *jimmy-jimmy-jimmy*. Calls include a sharp, nasal *tee-tee-tee* and a cracking *kit kit kit*. **H** Tall grassland and scrub along wetlands, open scrub jungle, mangroves, cultivations, gardens. **D** Pakistan, India, Bangladesh, Sri Lanka.

15 TAWNY-FLANKED PRINIA *Prinia subflava* 12cm Distinguished from cisticolas and other warblers by slim build and sideways tail waving. Note supercilium and rusty edges of wing and tail feathers. **V** Extremely high staccato hurried *weet-weet-weet-weet* or a series of rapid *tjiep-tjiep-tjiep* of irregular length. **H** Bush, open areas in forest, woodland, bush, swamp edges. **D** Sub-Saharan Africa, widespread.

16 PLAIN PRINIA *Prinia inornata* 13cm A small unstreaked prinia with whitish supercilium and underparts and very long tail. Unobtrusive, keeps low in vegetation. **V** Insect-like, wheezy *jirt jirt jirt jirt...* Calls include a buzzy *zzpink* or *bzzp* and a clear *clact*. **H** Scrubby grassland, reedbeds, mangroves and cultivations. **D** From E Afghanistan through India, Sri Lanka, SE Asia to China, Taiwan; Java, Indonesia.

17 PALE PRINIA *Prinia somalica* 12cm Note short supercilium. **V** Song is a series of buzzing dry notes, recalling a cricket; also gives high-pitched chatter and snapping sounds. **H** Dry, open areas with some grass and bush. **D** South Sudan, Ethiopia, Somalia, Kenya.

18 RIVER PRINIA *Prinia fluviatilis* 12cm Distinguished from Tawny-flanked Prinia by greyer upperparts and less tawny flanks, and by wetter habitat. **V** Song is a fast series of mournful, downslurred, whistling *plieu* notes. **H** Swamps, rank vegetation along streams. **D** NW Senegal and SW Mauritania to Chad; NW Kenya.

19 BLACK-CHESTED PRINIA *Prinia flavicans* 14cm Breeding adult may be white below. Non-breeding adult distinguished from Tawny-flanked Prinia by uniform brownish wings; from Karoo Prinia, Drakensberg Prinia and Namaqua Warbler by unmarked underparts. **V** Extremely high, rapid *weet-weet-weet*, *djap-djap-djap* or rattling *sisisisisisi*. **H** Dry areas with bush and scrub. **D** Angola and W Zambia to South Africa.

20 KAROO PRINIA *Prinia maculosa* 14cm Distinguished from Namaqua Warbler by white tail feather tips and prominent supercilium. **V** Extremely high, sharp *sreetsreetsreet*, or high *tretretretret*. **H** Montane forest edges and bracken-covered slopes. **D** S Namibia, South Africa, Lesotho.

21 DRAKENSBERG PRINIA *Prinia hypoxantha* 14cm Separated from Karoo Prinia by less streaked underparts, from Namaqua Warbler by white tail feather tips and prominent supercilium. **V** Extremely high, sharp *sreer-sreet-sreer*, or high *tre-tre-tre-tret* and other strophes. **H** Montane forest edges and bracken-covered slopes. **D** E South Africa, Eswatini.

22 SAO TOME PRINIA *Prinia molleri* 12cm Dark upperparts with bright tawny cheeks. **V** Repeated *tsee* or *tsi-tsip tsi-tsip tsi tsip...* **H** All natural and cultivated habitats on São Tomé, except true forest. **D** São Tomé, Gulf of Guinea.

23 BANDED PRINIA *Prinia bairdii* 12cm Note strongly barred underparts, and white tips of tail feathers, tertials and wing coverts. **V** High, hurried, sharp, staccato *wit-wit-wit* (sustained for up to 30 secs). **H** Forest edges, often along streams. **D** C Africa, from SE Nigeria to NW Angola, and east to W Uganda.

24 BLACK-FACED PRINIA *Prinia melanops* 12cm Sometimes treated as a race of Banded Prinia. Note white tips to tail feathers, tertials and wing coverts. **V** High, hurried, sharp staccato *witwitwit* (sustained for up to 30 secs). **H** Forest edges, often along streams. **D** EC Africa, from E DR Congo through Uganda, Rwanda and Burundi to W Kenya.

25 WHITE-CHINNED PRINIA *Schistolais leucopogon* 13cm Striking white throat diagnostic. **V** Unsynchronised duet of extremely high, fast warbling, *fififi*, with very high *weep-weep-weep*. Also individual *frifrifrifrifrifri*. **H** Tall herbage and shrubbery in forest glades and cultivation. **D** C Africa, from SE Nigeria to South Sudan, W Kenya, N Angola, S DR Congo.

26 SIERRA LEONE PRINIA *Schistolais leontica* 12cm Note very striking pale eye. **V** Soft, unstructured duet. **H** Forest edge and other places with thick cover. **D** Guinea, Sierra Leone, Ivory Coast, Liberia.

27 NAMAQUA WARBLER *Phragmacia substriata* 13cm Separated from Karoo and Drakensberg prinias by grey supercilium, absence of white tail tips, white throat and breast. **V** Very high, sharp, ascending then descending, bouncing trill, *ti-tri-tri-trrrrrreh*. **H** Vegetation along streams in the Karoo. **D** S Namibia, South Africa.

28 ROBERTS'S WARBLER *Oreophilais robertsi* 14cm Note dark upperparts, pale eyes, tawny rear underparts. Skulking. **V** Very high, rapid chatter, *chachacha* (in chorus). **H** Thick herbage at edges of montane forest. **D** E Zimbabwe, W Mozambique.

29 RED-WINGED WARBLER *Heliolais erythropterus* 13cm Note undertail pattern, long bill and rufous wings. **V** High, rapid *fiu-fiu-fiu-fiu* (or *fit-fit-fit*) by male with toneless twittered *trtrtr* of female in duet. **H** Tall-grassed places in woodland. **D** Senegal to Cameroon, Central African Republic, South Sudan, NW Uganda, Ethiopia; Kenya to E Zimbabwe, Mozambique.

30 VISAYAN MINIATURE BABBLER *Micromacronus leytensis* 7–8cm Tiny, with bright yellow underparts and supercilium, and whitish spiny feathers projecting beyond tail from flanks and back. Forages from the undergrowth to the canopy, in small groups or mixed-species parties. **V** Unknown. **H** Montane broadleaved evergreen forest and forest edges. **D** EC Philippines.

31 MINDANAO MINIATURE BABBLER *Micromacronus sordidus* 7–8cm Little information; noted feeding in the canopy in mixed-species flocks. **V** Unknown. **H** Montane broadleaved evergreen forest. **D** Mindanao, S Philippines.

32 GREEN LONGTAIL *Urolais epichlorus* 15cm Long tail diagnostic. No similar warbler in its restricted range and habitat. **V** Said to resemble that of Common Chiffchaff (Plate 210). **H** Montane forest. **D** SE Nigeria, SW Cameroon, Bioko (Gulf of Guinea).

33 BLACK-COLLARED APALIS *Oreolais pulcher* 11cm Unmistakable. In small flocks, normally not together with other species. **V** Very high, rapid, complaining *weet-weet-weet-weet* or *witwitwitwitwitwit*. **H** Undergrowth at forest edges and around clearings. **D** SE Nigeria to South Sudan and Kenya; SE DR Congo.

34 RWENZORI APALIS *Oreolais ruwenzorii* 10cm Note all-grey upperparts including tail, and pale salmon throat and belly. **V** Very high, squeaking *weeweeweeweeweeweewee*. **H** Undergrowth of montane forests and bamboo. **D** E DR Congo, S Uganda, Rwanda, Burundi.

1
2
3
4
5
6
7
8
9
10
11
12
13
14
15
16
17
18
19
20
21
22
23
24
25
26
27
28
29
30
31
32
33
34

CISTICOLAS AND ALLIES *CONTINUED*

1 RED-WINGED GREY WARBLER *Drymocichla incana* 13cm Distinguished from Red-winged Warbler (Plate 217) by less extensive red in wing. Active and restless, gleaning insects in shrubs and lower parts of trees. **V** Song is short, loud, querulous, *chwee* or *chwit weer-weer-weer*, and other member of pair may join in with *tchwee*. **H** Open woodland, often in thickets and near water. **D** E Nigeria and Cameroon to South Sudan and Uganda.

2 CRICKET WARBLER *Spiloptila clamans* 11cm Note long graduated tail and yellowish rump. Active, often in small family groups. **V** Usually a lively, rapid, tinkling trill. Calls include a ringing whistle and a sharp *zzt*. **H** Scattered scrub and grass on desert or semi-desert plains. **D** S Mauritania and N Senegal to Eritrea.

3 BUFF-BELLIED WARBLER *Phyllolais pulchella* 9cm Note pale grey upperparts, white outer-tail feathers and absence of eye-stripe. **V** Extra-high, dry, short trills *srrrrr*. **H** Open acacia woodland. **D** S Niger and N Nigeria to Eritrea, Ethiopia, Kenya, Tanzania.

4 BAR-THROATED APALIS *Apalis thoracica* 11cm Black breast-band, yellow eye, white outer-tail feathers. **V** Mid-high, loud *wrut-wrut-wrut-* of male; pauses filled in by female with extremely high *srrr*. **H** Woodland. **D** E and S Africa, from Kenya to South Africa.

5 YELLOW-THROATED APALIS *Apalis flavigularis* 11cm Formerly considered conspecific with Bar-throated Apalis. Greenish above and bright yellow below with black breast-band. **V** Voice of both sexes is similar to Bar-throated Apalis. **H** Forest and forest edge at mid to high altitudes. **D** S Malawi.

6 TAITA APALIS *Apalis fuscigularis* 11cm Formerly considered conspecific with Bar-throated Apalis. Brownish-grey above with blackish throat and breast, and almost no yellow on underparts. Critically endangered. **V** Voice of both sexes is similar to Bar-throated Apalis. **H** Forest remnants in the Taita Hills. **D** SE Kenya.

7 NAMULI APALIS *Apalis lynesi* 11cm Formerly considered conspecific with Bar-throated Apalis. Similar to Yellow-throated Apalis but throat all black. **V** Voice of both sexes is similar to Bar-throated Apalis. **H** Montane forest and secondary growth. **D** N Mozambique.

8 RUDD'S APALIS *Apalis ruddi* 13cm Distinguished from Bar-throated Apalis by green (not grey) tail without white edges, narrow, white supercilium and black eye. **V** Song is a duet, combining male's dry *trrt* phrases with female's faster, higher-pitched, *pip* or *prp* notes. Contact call is a soft *chirg*; snaps bill in agitation. **H** Coastal woodland and scrub with adjoining lowland areas. **D** SE Africa, from Malawi and Mozambique to E South Africa.

9 YELLOW-BREASTED APALIS *Apalis flavida* 11cm Grey crown merges into olive-green back. Female lacks male's black breast spot. **V** High, dry *tju-tju-tju*, *terruterruterru* or sustained, rapid *tjuuptjuuptjuup*. **H** Woodland, bush, scrub. **D** Sub-Saharan Africa, widespread.

10 LOWLAND MASKED APALIS *Apalis binotata* 10cm Differs from Black-throated Apalis in all-green wings and shorter tail. Female has more white on neck than male, and smaller black bib. **V** Very high, rapid *fiufiufiu-* or high, fast *krrrk krrrk keelerkeeler*. **H** Riverine belts. **D** S Cameroon, Gabon, NW Angola, NE DR Congo, NW Tanzania, Uganda.

11 MOUNTAIN MASKED APALIS *Apalis personata* 10cm Distinguished from Lowland Masked Apalis by black area running further down from throat to belly and white spot (not stripe) on side of neck. **V** Similar to Lowland Masked Apalis (which normally occurs in lowlands), but more clearly trisyllabic and higher-pitched. **H** Montane forest. **D** E DR Congo, Rwanda, Burundi, W Uganda.

12 BLACK-THROATED APALIS *Apalis jacksoni* 12cm Separated from Lowland Masked Apalis and Mountain Masked Apalis by yellow-green underparts, including breast. **V** Duets *tuuttuwt-tuuttuut-tuuttet*, often just out of synchrony. **H** Montane forest edges. **D** SE Nigeria, Cameroon, South Sudan to N Tanzania, E DR Congo, N Angola.

13 WHITE-WINGED APALIS *Apalis chariessa* 12cm Brightly plumaged, with prominent white panel on folded wing. **V** Pairs sing in duet, giving 6–7 bright *tee-lu dee-lu* notes per sec for 2–3 secs. **H** Forest, riverine belts. **D** E Kenya; C Tanzania to Malawi and N Mozambique.

14 BLACK-CAPPED APALIS *Apalis nigriceps* 10cm Male's black face and crown contrast strongly with white throat and yellow-olive nape. Female has grey crown. **V** Extra-high, dry trill *srrrrrr srrrrr srrrrr* (each one 1.5 secs). **H** Canopies of montane forest. **D** Sierra Leone to SW Central African Republic and Gabon; E DR Congo and Uganda.

15 BLACK-HEADED APALIS *Apalis melanocephala* 15cm Dark above, light below. Note that the eye may be either yellow or red-brown. **V** Very high *swee-swee-swee* of male often together with mid-high *shra-shra-shra* of female in unsynchronised duet. **H** Forest riverine belts, coastal bush. **D** E Africa, from coastal S Somalia to Malawi highlands and Mozambique.

16 CHIRINDA APALIS *Apalis chirindensis* 13cm Resembles all-grey female Gosling's Apalis, but different range. **V** Very high, loud, sharp *tsjeeptsjeeptsjeeptsjeep*. **H** Forest. **D** Zimbabwe, Mozambique.

17 CHESTNUT-THROATED APALIS *Apalis porphyrolaema* 13cm Mainly grey, with chestnut throat. Sexes alike. Distinguished from Buff-throated Apalis and Sharpe's Apalis females by white tips to tail feathers. **V** Extra-high, insect-like double-buzz *turrrree-turrree*. **H** Montane forest interiors and edges. **D** E DR Congo to Kenya and N Tanzania.

18 KABOBO APALIS *Apalis kaboboensis* 12cm Formerly considered conspecific with Chestnut-throated Apalis. Similar to Chestnut-throated Apalis but throat and breast blackish. Sexes alike. **V** Song is an insect-like trill, similar to Chestnut-throated Apalis. **H** Canopy of montane forest; restricted to Mount Kabobo. **D** E DR Congo.

19 CHAPIN'S APALIS *Apalis chapini* 12cm Note distinctive brown-rufous face, throat and breast. **V** Song is a sibilant *psi-psi-psi*, given in duet; in south of range female's song is a faster *tik-ik-ik-ik*. **H** Canopy of montane forest. **D** Tanzania, NE Zambia, Malawi.

20 SHARPE'S APALIS *Apalis sharpii* 11cm Male is overall dark sooty grey. Female is paler, with chestnut throat and grey breast, differing from female Buff-throated Apalis in grey (not brown) upperparts. Note also absence of white in tail. **V** Very high, rapid, staccato *trittrittrit...* **H** Lowland forest and riverine belts. **D** Guinea, Sierra Leone to Togo and Benin.

21 BUFF-THROATED APALIS *Apalis rufogularis* 11cm Female has buff-orange throat. Males vary, with western races having dark throats, while those further east and south have white throats. **V** High rapid rattling *wrreet-wrreet-wrreet*. **H** Forest. **D** W and C Africa, from Nigeria to W Kenya and south to Angola.

22 KUNGWE APALIS *Apalis argentea* 13cm Male is grey above, paler below. Female has more olive mantle, rump and wings. **V** Song consists of repeated *prui-tju prui tju* phrases; female may join in with *tjup* notes. **H** Forests and bamboo. **D** E DR Congo, Rwanda, Burundi, W Tanzania.

23 KARAMOJA APALIS *Apalis karamojae* 12cm Note distinctive white wing panel. **V** Pairs perform duet of musical, liquid notes in phrases lasting up to a min or more. Contact call is a soft *we we we*, alarm call a rasping *jirrr jiirr*. **H** Thickets, low shrub. **D** NE Uganda, SW Kenya, N Tanzania.

24 BAMENDA APALIS *Apalis bamendae* 11cm The only apalis with an all-rufous head. **V** Very high, rapid *weet-witwitweet-witwitweet* (*weet* higher). **H** Montane forest. **D** Cameroon.

25 GOSLING'S APALIS *Apalis goslingi* 11cm Note short tail. Female has paler face than male. **V** High, sharp, rapid, tinkling *tingc-tingc-tingc*. **H** Along forest streams. **D** Cameroon, Equatorial Guinea and Gabon to NE and E DR Congo.

26 GREY APALIS *Apalis cinerea* 12.5cm Note white outer-tail feathers. **V** Mid-high, rapid *rap-rap-rap - - -* or very high, insect-like *trrrr*. **H** Dense forest at higher altitudes. **D** SE Nigeria, Cameroon, Bioko (Gulf of Guinea), W Angola; E DR Congo and South Sudan to C Kenya and NW Tanzania.

27 BROWN-HEADED APALIS *Apalis alticola* 12cm Sometimes treated as conspecific with Grey Apalis. Has browner head and less white in tail. **V** Song consists of dry, monotonous repetitions of *krrip* or *krrip-ip*, female sometimes contributing squeaky notes and trills. **H** Forest and forest remains. **D** From N Angola and S DR Congo to Malawi, Tanzania, SW Kenya.

28 RED-FRONTED WARBLER *Urorhipis rufifrons* 12cm Note black, often cocked tail, white wing-bars, rufous-buff crown. Often in parties of 10–14. **V** Song consists of *stiuck* notes, interspersed with cheeping and chipping sounds and wing-snapping. Typical call is a soft *tseep*, alarm call similar but louder. **H** Dry bush and acacia. **D** Chad to Somalia, SE Kenya, N Tanzania.

29 RUFOUS-EARED WARBLER *Malcorus pectoralis* 15cm Brick-red around eye to ear-coverts. Forages mainly on the ground, is even reluctant to fly. **V** High or extremely high, sharp, hurried *sreet-sreet-sreet-* with sudden stops and interruptions by lower *piu* (at a sustained speed). **H** Dry, sparsely covered areas. **D** Namibia, Botswana, South Africa.

30 ORIOLE WARBLER *Hypergerus atriceps* 20cm Feathers of head, neck, throat and upper breast black with silvery fringes. Unmistakable, but shy and secretive. **V** Very high, sharp, whistled *puweeweeh pwik pwik* or *puwee pwik*, going up and down. **H** Undergrowth at forest edge, thickets, mangrove. **D** From Senegal to Central African Republic and N DR Congo.

31 GREY-CAPPED WARBLER *Eminia lepida* 15cm Unmistakable, with grey cap, black band through eyes and chestnut throat, but rather secretive; more often heard than seen. **V** Extra-high, fast, rattling trill *tirrrrrrr*, high song phrases like *tutu-tu tu* and other trills. **H** Undergrowth, tall herbage near water, suburban gardens. **D** S South Sudan, E DR Congo, Rwanda, Burundi, Kenya, N Tanzania.

32 GREEN-BACKED CAMAROPTERA *Camaroptera brachyura* 10cm Grey head; wings, mantle and tail are green. **V** Mid-high, very dry, loud, staccato *treet-treet-treet* (3–5 times) and mid-high, mewing *meh meh-meh*. **H** Dense thickets, coastal bush, forest edges. **D** E and SE Africa, from Kenya to E South Africa.

33 GREY-BACKED CAMAROPTERA *Camaroptera brevicaudata* 10cm Similar to Green-backed Camaroptera (and sometimes treated as conspecific), but mantle is grey, not green. **V** Rapid series of staccato *triwt-triwt-triwt*. **H** Dense vegetation in woodland, forest, gardens. **D** Sub-Saharan Africa, widespread.

34 HARTERT'S CAMAROPTERA *Camaroptera harterti* 10cm Formerly treated as conspecific with Grey-backed Camaroptera, differing in its green tail and whiter underparts; also differs genetically. **V** Rapid *chi-chi-chi*; also a plaintive bleating, similar to Grey-backed Camaroptera. **H** Dense forest understorey, thickets, secondary growth. **D** WC Angola.

35 YELLOW-BROWED CAMAROPTERA *Camaroptera superciliaris* 9cm Distinguished from Laura's Woodland Warbler (Plate 210) by pale eye, white chin and throat, longer bill, shorter tail. **V** Mid-high loud mewing, *piauw-piauw piauw-piauw* or *ee-weh-weh ee-wek-weh*. **H** Undergrowth at forest edge, old cultivation, riverine belts. **D** Guinea and Sierra Leone to Uganda, DR Congo, NW Angola.

36 OLIVE-GREEN CAMAROPTERA *Camaroptera chloronota* 10cm Short tail is often cocked. Some races lack tawny face. **V** Very long, rapid sequences of very high, penetrating *weetweetweet* or *wuutwuutwuut*, often very slowly descending, lasting several seconds. **H** Dense undergrowth of forest and riverine belts. **D** Senegal to Cameroon, Gabon and Congo; DR Congo to SW Kenya, NW Tanzania.

1
2
3
4
5
6
7
8
9
10
11
12
13
14
15
16
17
18
19
20
21
22
23
24
25
26
27
28
29
30
31
32
33
34
35
36

CISTICOLAS AND ALLIES *CONTINUED*

1 GREY WREN-WARBLER *Calamonastes simplex* 13cm Note long legs, dark grey plumage with very faint barring (if any) on underparts. Frequently waves its tail. **V** Song is a monotonous repeated tapping *clack-clack*, sometimes *click-clack*, with the bill snapped audibly on each note. **H** Thick undergrowth of woodland and bushland. **D** Ethiopia and Somalia to SE South Sudan, NE Uganda, Kenya, NE Tanzania

2 MIOMBO WREN-WARBLER *Calamonastes undosus* 14cm Formerly considered a single species with Stierling's Wren-Warbler. Note faint barring of rear underparts. **V** Song consists of loud chirping notes repeated monotonously, about one per sec; also repeated *chi* notes. Call is a bleating *maa*. **H** Miombo and other woodland. **D** SW Congo, DR Congo, Angola, Zambia, Rwanda, Burundi, Tanzania, Kenya.

3 STIERLING'S WREN-WARBLER *Calamonastes stierlingi* 14cm Formerly considered conspecific with Miombo Wren-Warbler. Regular fine barring of underparts diagnostic. **V** Extremely high *tweet-tweet tweet*. **H** Bushveld. **D** SE Angola and NE Namibia to Zambia, E Tanzania, Mozambique, Botswana, NE South Africa.

4 BARRED WREN-WARBLER *Calamonastes fasciolatus* 14cm Breeding male distinguished from Stierling's Wren-warbler by darker breast-band, female and non-breeding male by fainter barring and buff (not white) underparts. **V** Very high, piercing *turwiet-turwiet-turwiet* (like a badly oiled wheel). **H** Thornveld, dry bushland. **D** S Angola, Namibia through Botswana to SW Zimbabwe, N South Africa.

5 CINNAMON-BREASTED WARBLER *Euryptila subcinnamomea* 13cm Dark brown warbler with grey throat and breast, often with cocked tail. **V** Very high, hurried *weet-weet-weet-weet* or drawn-up *uWeeet-uWeeet*. **H** Rocky hillsides. **D** S Namibia, W South Africa.

6 BLACK-HEADED RUFOUS WARBLER *Bathmocercus cerviniventris* 13cm Unmistakable. Male has black head, neck and centre of breast; female has off-white on chin and sides of throat. **V** Loud, whistled *weeh-oh deee* or similar, often repeated. **H** Forest undergrowth, especially along streams. **D** Sierra Leone to Ghana.

7 BLACK-FACED RUFOUS WARBLER *Bathmocercus rufus* 13cm Unmistakable. Male has rufous upperparts with black face and throat, with black extending to centre of upper belly; female similarly marked, but grey instead of rufous. **V** Extra-high, sharp *tjeeeee-tjeeeee-tjeeeee*. **H** Forest undergrowth. **D** S Cameroon and Gabon to NW DR Congo; South Sudan and E DR Congo to W Kenya, NW Tanzania.

8 WINIFRED'S WARBLER *Scepomycter winifredae* 15cm Also known as Mrs Moreau's Warbler. Olive-brown with rufous from head to breast. **V** Mid-high *fjuuuuuufjuuuui* and *tuui-tuui-tuui-tio-tui-fi'e*. **H** Dense forest undergrowth. **D** Tanzania.

9 RUBEHO WARBLER *Scepomycter rubehoensis* 15cm Recently described species (formerly considered to be a race of Winifred's Warbler). Differs from Winifred's in being brighter on head and breast. **V** Male gives a single mournful drawn-out whistle, repeated frequently; female often responds in duet with a repeated piping. **H** Dense forest undergrowth. Restricted to Rubeho–Ukaguru Mountains. **D** E Tanzania.

10 COMMON TAILORBIRD *Orthotomus sutorius* 10–14cm Greyish green, with rufous cap. Males in the breeding season develop long central tail feathers. Forages in cover or on the ground under cover, often in pairs. **V** Rapid, repetitive, loud *pitchik-pitchik-pitchik* or *cheeyup-cheeyup-cheeyup*. Calls include a sharp *cheep cheep cheep cheep cheep* and a quickly repeated *pit-pit-pit-pit.....* **H** Forest and cultivation edge, bushy cover in urban areas and mangroves. **D** Widespread from Pakistan through India, Sri Lanka, SE Asia to S China; also Java, Indonesia.

11 DARK-NECKED TAILORBIRD *Orthotomus atrogularis* 11–12cm Note dark breast-band in male. Shy; restless forager, low in the centre of undergrowth, more often heard than seen. **V** Repeated, high, shivery *pirra pirra pirra...* and a trilled *kri-kri-kri*. Also a *titrrrrrt-churrit* duet between partners. **H** Heavy scrub and edges of evergreen forest. **D** Bangladesh and NE India through SE Asia; Borneo; Sumatra, Indonesia.

12 CAMBODIAN TAILORBIRD *Orthotomus chaktomuk* 10–11cm Very similar to Dark-necked Tailorbird. Elusive, generally in pairs foraging in dense vegetation. **V** Multiple repeated trilled phrases, some rising, some falling and some slurred (rising and falling). Calls include a nasal squeak, occasionally repeated in quick succession. **H** Dense, humid evergreen scrub, sometimes mixed with tall grasses or trees. **D** Cambodia.

13 PHILIPPINE TAILORBIRD *Orthotomus castaneiceps* 13cm Noisy. Forages in low, tangled undergrowth, usually near the ground. **V** A *twee-pee twee-doo* and a more complex series of similar notes. **H** Lowland forest, forest edge, clearings with secondary growth and dense, tangled undergrowth. **D** C Philippines.

14 TRILLING TAILORBIRD *Orthotomus chloronotus* 13cm Formerly considered a race of Philippine Tailorbird; actions are similar. **V** Bubbly, pulsing *tuuut tuuutt p-p-p-p*. Calls include a scolding *speeee*. **H** Lowland forest, especially forest edges and clearings with secondary growth and associated tangled undergrowth. **D** N Luzon, Philippines.

15 RUFOUS-FRONTED TAILORBIRD *Orthotomus frontalis* 12cm Note restricted extent of rufous-chestnut on head. Forages in low undergrowth. **V** Drawn-out *de de drer-r-r-rw*, likened to a bouncing table-tennis ball. **H** Lowland forest, forest edge, clearings with rank secondary growth. **D** EC and S Philippines.

16 GREY-BACKED TAILORBIRD *Orthotomus derbianus* 11cm Keeps hidden while foraging in thick undergrowth, and can appear very dark in forest gloom. **V** Loud, pulsing *chew-ee-peee, twee-purrr* or simple *purr*, and a buzzing *eeeeh eeeeh eeeeh*. **H** Lowland secondary forest, forest edge, clearings with tangled undergrowth. **D** S Luzon, Philippines.

17 RUFOUS-TAILED TAILORBIRD *Orthotomus sericeus* 12–14cm Note bright rufous cap and tail, white throat. Shy and skulking; keeps to dense foliage, usually alone or in pairs. **V** Variety of loud, rapid, bisyllabic notes; calls include a harsh *terr-terr*, a wheezy *tzee-tzee-tzee*, a sharp *twip-twip-twip* and a scolding *speee-speee-speee*. **H** Forest edge, overgrown clearings, shrubby cultivations, overgrown gardens. **D** Malay Peninsula; Borneo; Sumatra, Indonesia; W and SW Philippines.

18 ASHY TAILORBIRD *Orthotomus ruficeps* 11–12cm Occurs in pairs or in small parties, foraging from the undergrowth up to the canopy. **V** Repeated *chip-wee chip-wee...* or similar; calls include a *trrree-yip*, a plaintive *cho-choee*, a spluttering *prrrt*, a rolling *prit-u* and a harsh *thieu*. **H** Secondary forest, bamboo thickets, mangrove swamps, coastal scrub, peat-swamp forest, roadside vegetation. **D** Malay Peninsula; Borneo; Sumatra and Java, Indonesia.

19 OLIVE-BACKED TAILORBIRD *Orthotomus sepium* 12cm Distinguished from Ashy Tailorbird by greenish tone of breast and upperparts, pale yellow undertail coverts. Active forager from the understorey to the canopy. **V** Various single notes, rhythmically repeated monotonously, such as *chew-chew-chew...* or *turr-turr-tsee-weet – tsee-weet*; also a quavering nasal call and a loud, trilled *tree-yip*. **H** Inland secondary forest, forest edge, bamboo thickets, bushy edges of cultivation. **D** Java, Bali and Lombok, Indonesia.

20 WHITE-EARED TAILORBIRD *Orthotomus cinereiceps* 12–13cm Note distinctive white ear-spot. Usually seen in pairs foraging in dense undergrowth; rarely seen in the open. **V** Similar to the two previous species, but less explosive, *dededede-dep-dep-deep-deeup deee-up deee-up deee-up deee-up*; also an agitated *chat-chat-chat-chat-chat*. **H** Dense, tangled undergrowth in lowland forest clearings or forest edges. **D** W Mindanao, Philippines.

21 BLACK-HEADED TAILORBIRD *Orthotomus nigriceps* 12cm Head pattern is distinctive. Shy and elusive forager in dense undergrowth, rarely appearing in the open. **V** Long, descending, stuttering trill, very similar to that of the previous species but more metallic and ringing; also utters an agitated, descending *ssiirrrrpppppp* and a ringing *key-e ei*. **H** Dense forest undergrowth in lowlands. **D** SE Philippines.

22 YELLOW-BREASTED TAILORBIRD *Orthotomus samarensis* 12cm Distinctive tailorbird with black head and yellow underparts. Shy forager in dense undergrowth, usually in pairs; rarely comes into the open. **V** Long, descending, stuttering trill that slows, levels out and then becomes a monotonous throbbing, *dededede-de-de-de-de-dep-dep-dep-deep-deep-deep-deeup-deeup-deeup...* **H** Dense forest undergrowth in lowlands and forest edge; sometimes along watercourses, gullies, dry streambeds. **D** EC Philippines.

23 LONG-BILLED FOREST WARBLER *Artisornis moreaui* 11cm Distinguished from Red-capped Forest Warbler by longer bill, less orange face sides, greyer upperparts. **V** Song is a steady, rhythmic repeated *tcheu tcheu* or disyllabic *t'wee t'wee*. Probable contact call is a rasping note. **H** Canopy of montane forests. **D** Tanzania, Mozambique.

24 RED-CAPPED FOREST WARBLER *Artisornis metopias* 10cm Similar to Long-billed Forest Warbler, but with more rufous head **V** Song given as duet, one bird giving high, thin *swee* notes and the other short buzzing *dzz-dzz-dzz* phrases. Calls consists of various low, soft notes. **H** Understorey of montane forests. **D** Tanzania, Mozambique.

25 WHITE-TAILED WARBLER *Poliolais lopezi* 12cm Note short, square, white-sided tail. **V** Bisyllabic, extra-high *tju tjeeh*, second part lower and slurred down, or long series of *tjeeh-tjeeh-tjeeh–* or *tjuh-tjuh-tjuh–*. **H** Undergrowth of montane forests. **D** SE Nigeria, SW Cameroon, Bioko (Gulf of Guinea).

26 YELLOW-BELLIED EREMOMELA *Eremomela icteropygialis* 10cm Extent of yellow on underparts variable. Distinguished from Karoo Eremomela by well-defined supercilium. **V** Very high *titiWeeh-titiWeeh* or sharp, descending, hurried, warbled *seesirititititu-Wee*. **H** Desert, woodland, bushveld. **D** Sub-Saharan Africa, widespread.

27 SALVADORI'S EREMOMELA *Eremomela salvadorii* 11cm Bright yellow underparts sharply demarcated from grey breast. Note greenish rump. Hardly any overlap in range with Yellow-bellied Eremomela. **V** Song bright high-pitched *chiri-chu-chit-chit-chuwee*, middle notes higher pitched; call high *chit*, alarm call upslurred whistle.. **H** Open areas with some bush. **D** SE Gabon to S DR Congo, Angola, W Zambia.

28 YELLOW-VENTED EREMOMELA *Eremomela flavicrissalis* 9cm Very similar to Yellow-bellied Eremomela but slightly smaller and with yellow restricted to vent. **V** Song is a fast medley of notes, scratchier and less varied than Yellow-bellied Eremomela. **H** Dry bush and semi-desert areas. **D** Somalia and S Ethiopia to Kenya and NE Uganda.

29 SENEGAL EREMOMELA *Eremomela pusilla* 8cm Rather similar to Yellow-bellied Eremomela, but with green, not grey back, yellow undertail coverts and pale eye. **V** Song is a bright *chi-chi-chi-chi* that rises and falls in pitch; also a higher-pitched *pirp-pirp-pirp-pirp*. Calls include a dry insect-like rattle and a soft *tsep*. **H** Forest edge, more or less wooded areas, including gardens and mangrove. **D** S Mauritania to Sierra Leone and east to S Chad, Central African Republic.

30 GREEN-BACKED EREMOMELA *Eremomela canescens* 10cm Note black mask. Hybridises with Senegal Eremomela, where ranges meet. **V** Several together give a 'foraging song' of cheery *reelu reelu* or *tureetree* notes. Typical call is a thin *see-see*. **H** Hilly, wooded and bushed habitats, often near streams. **D** S Chad, Central African Republic, east to Eritrea, Ethiopia, W Kenya.

31 GREEN-CAPPED EREMOMELA *Eremomela scotops* 10cm Note pale, red-rimmed eye and bright yellow breast-band. **V** High, sustained, rather staccato *trrit-trrit-trrit*. **H** Open woodland, riverine belts. **D** SC Africa, from SE Gabon to W Kenya, south to NE Namibia, NE South Africa.

32 KAROO EREMOMELA *Eremomela gregalis* 12cm Note sharp demarcation between grey-brown face sides and white underparts, pale eyes and yellow undertail coverts. **V** Very high, hurried, single-noted *ti-ti-ti-ti*. **H** Karoo and semi-desert. **D** S Namibia, W and SW South Africa.

33 BURNT-NECKED EREMOMELA *Eremomela usticollis* 10cm Note pale eye and rufous on cheeks and ear-coverts (rufous on throat often absent). **V** Extremely high, wader-like, fast *weeweeweewee* or very high, tit-like *tjut-tjut-tjut*. **H** Thornveld and dry woodland. **D** S Angola to Malawi, south to Namibia and NE South Africa.

34 RUFOUS-CROWNED EREMOMELA *Eremomela badiceps* 10cm Chestnut crown, black mask, white throat. Distinguished from Turner's Eremomela by rufous extending to crown. **V** Extremely high hurried siffled *sisisisi–* or very high soft fluted warbling very varied in pitch. **H** Canopy at forest edges, riverine belts. **D** Sierra Leone to N Angola and east to W Uganda.

35 TURNER'S EREMOMELA *Eremomela turneri* 10cm Resembles Rufous-crowned Eremomela, but rufous restricted to forehead. **V** Extra-high, complex, unmusical siffling. **H** Canopies of montane forest edges. **D** EC DR Congo to SW Uganda; W Kenya.

36 BLACK-NECKED EREMOMELA *Eremomela atricollis* 11cm Distinctive yellow and black head pattern. **V** Song is a harsh, grating, low-pitched *pree pree-prurrut pree-prurrut*, with short pauses between repetitions of phrase. Flocks give harsh *zwut zwut* calls. **H** Miombo, bushveld. **D** S DR Congo to C Angola and C Zambia.

1
2
3
4
5
6
7
8
9
10
11
12
13
14
15
16
17
18
19
20
21
22
23
24
25
26
27
28
29
30
31
32
33
34
35
36

BABBLERS AND SCIMITAR BABBLERS TIMALIIDAE

1 LARGE SCIMITAR BABBLER *Pomatorhinus hypoleucos* 26–28cm Skulking; forages on or near the ground, in an ungainly manner. **V** Three short, hollow notes, often from dueting pairs, e.g. *wiu-pu-pu – wup-up-piu*. Calls include a grating rattle and a hard *puh*. **H** Broadleaved evergreen forest, mixed deciduous forest, bamboo, scrub-jungle, reeds and elephant grass. **D** Bangladesh to Malay Peninsula, S China, Hainan.

2 RUSTY-CHEEKED SCIMITAR BABBLER *Pomatorhinus erythrogenys* 22–27cm Orange-rufous extends from side of head to flanks and vent. Forages on the ground and occasionally in trees. **V** Dueted *whi-u-ju-whi-u...iu-chu-ip-iu-chu*...or *yu-u-yi-yu-u*..., etc. Also a high *pu* or *ju*, a repeated, rolling *jrr-jrr-jrr* and, when alarmed, a rattling *whih-whihihihihi*. **H** Thick scrub and dense undergrowth at forest edge, scrub in open pine forest and secondary growth. **D** Himalayas to Myanmar.

3 BLACK-NECKLACED SCIMITAR BABBLER *Pomatorhinus erythrocnemis* 23–25cm Note white chin and black streaked 'necklace'. Forages mainly on the ground, rummaging amongst leaf litter. **V** Male gives a hurried *wiu'wi* or *whiu'whi* quickly answered by a female with a *wuu* or *woh*. **H** Upper undergrowth and lower storey in foothill and montane forests. **D** Taiwan.

4 BLACK-STREAKED SCIMITAR BABBLER *Pomatorhinus gravivox* 21–25cm Rufous ear-coverts, flanks and vent, bold black streaks on breast. Forages on or near the ground. **V** Quick, husky *whi-wip*, a rapid *whi-chu* or a *whi-tu*; calls include a *whup-which'ch'ch'ch'ch'ch*, *whoip-tut'ut'ut'ut'ut'ut* or *whoi-t't't't't*. **H** Open forest, forest edge, scrub-jungle, bamboo, secondary growth, thickets, scrub, abandoned cultivations. **D** C China to N SE Asia.

5 SPOT-BREASTED SCIMITAR BABBLER *Pomatorhinus mcclellandi* 25cm Breast is spotted ash-brown. Forages mostly on the ground, but also in trees. **V** Duets, two fluty notes quickly followed with a sharp note: *wi-wru-pi* or *wi-wwu-jrr*. When alarmed, *wi-wi-chitit*. **H** Forest undergrowth, scrub jungle and thickets in forest clearings. **D** NE India to W Myanmar.

6 GREY-SIDED SCIMITAR BABBLER *Pomatorhinus swinhoei* 22–24cm Rufous mantle and wings, but grey flanks. Forages on the ground often amongst dead leaves. **V** Whip-like *whi'ru-chi-whi'ru* or *whi'wu-pi-whi'wu*; call is a harsh *chut'ut'ut'ut'ut'ut*... **H** Dense cover in deciduous forest, damp summit forest, brush-covered hillsides, thickets near woods. **D** SE China.

7 INDIAN SCIMITAR BABBLER *Pomatorhinus horsfieldii* 22cm Note white supercilium, dark eye, yellow bill. Forages on the ground or on mossy trunks and branches. **V** Mellow, fluty *oo-pu-pu-pu* or *wot-ho-ho-ho*. Calls include various chirps and rattles, hoarse hoots and a guttural *woch-wohorro*. **H** Forests, secondary growth, bamboo patches, thorn-scrub, dense bush. **D** Peninsular India.

8 SRI LANKA SCIMITAR BABBLER *Pomatorhinus melanurus* 19–21cm Only scimitar babbler in Sri Lanka. Often works up tree trunks like a woodpecker. **V** An *oop-oop-oop-oop* followed by a *kraa-kree kraa-kree*, the latter presumed to be uttered by the female. **H** Moist deciduous and evergreen forest. **D** Sri Lanka.

9 WHITE-BROWED SCIMITAR BABBLER *Pomatorhinus schisticeps* 22cm Has been treated as conspecific with Indian and Sri Lanka Scimitar Babblers; separated by range. **V** Short, low whistled *tji-u-u-u-u* or *woot-a-ah-hoot*. Calls include a *tjoo-tjoo* or *gouk-gouk* and some loud and shrill notes. **H** Forest undergrowth, thick secondary growth, scrub jungle. **D** Himalayan foothills to SE Asia.

10 CHESTNUT-BACKED SCIMITAR BABBLER *Pomatorhinus montanus* 19–21cm Note contrast between dark grey crown and rich rufous back. Usually forages in undergrowth and the lower middle storey. **V** Quick *wu-pwi*, followed by a second bird uttering a *wu-pu pu pu pu pu pu*. **H** Broadleaved evergreen forest. **D** Malay Peninsula; Borneo; Sumatra and Java, Indonesia.

11 STREAK-BREASTED SCIMITAR BABBLER *Pomatorhinus ruficollis* 19cm Dull olive-brown streaks on breast. Forages on the ground, in bushes or low in trees. **V** Fast, mellow *win-wun-wun*. Contact call is a raspy *wreep* or *wreep-wreep*; when alarmed gives a scolding rattle. **H** Thick forest, open forest with thick undergrowth and hillside scrub. **D** Himalayas to S China.

12 TAIWAN SCIMITAR BABBLER *Pomatorhinus musicus* 19–21cm Usually in pairs or small parties foraging on or near the ground, often around tree trunks. **V** Husky *wuh-wuh-wuh-wree* answered with a jolly *wu-wii*, also a piping *uh-pu-pu-pu* answered with a harsh *wheer-wu'wu*. Calls include a rich, musical *tui-tui* and a burry *jrrr-jeee*. **H** Undergrowth in foothill, submontane and montane forest. **D** Taiwan.

13 RED-BILLED SCIMITAR BABBLER *Pomatorhinus ochraceiceps* 23cm Note long downcurved orange bill. Forages in trees and bushes, and on the ground. **V** Hurried, piping *wu-wu-woi*, also a human-like whistle. **H** Dense undergrowth in evergreen or mixed forest and bamboo jungle. **D** NE India to SE Asia.

14 CORAL-BILLED SCIMITAR BABBLER *Pomatorhinus ferruginosus* 23cm Underparts are rich rufous in west of range, paler to the east. Elusive; forages on the ground. **V** Fluty *oo-pu-pu* or *oo-pu-pu-pu*. Calls include grating *churr*, *weeitch-oo* and shrill *wheep-wheep*. **H** Dense shrubbery, bamboo jungle, dense undergrowth and forest edge. **D** Himalayas to C Vietnam.

15 SLENDER-BILLED SCIMITAR BABBLER *Pomatorhinus superciliaris* 20cm Note strikingly long thin bill. Shy, restless and noisy; forages on or near the ground. **V** Rapid, hollow piping, which may be accompanied with a high-pitched *u-wi* or *ti-wee*. **H** Broadleaved evergreen forest with thick undergrowth, bamboo and thick secondary growth. **D** Himalayas to SW China, N Vietnam.

16 RUFOUS-THROATED WREN-BABBLER *Spelaeornis caudatus* 9cm Small, brown, short-tailed, with rusty throat and breast, dark white-spotted belly. Very elusive, forages on the ground. **V** Powerful *witchu-witchu-witchu-witchu-witchu*; when alarmed, a low *birrh birrh birrh*... **H** Dense thickets in broadleaved evergreen forest, especially where ferns and mossy rocks occur. **D** E Himalayas.

17 RUSTY-THROATED WREN-BABBLER *Spelaeornis badeigularis* 9cm Very like Rufous-throated Wren-Babbler, but more contrasty throat and breast. **V** Variable; includes short three- or four-note whistles and a combination of short warbles and trills. **H** Dense undergrowth in secondary broadleaved evergreen forest; in winter in moist subtropical forest. **D** NE India.

18 BAR-WINGED WREN-BABBLER *Spelaeornis troglodytoides* 10cm Barred wings and tail. More arboreal than many wren-babblers; clambers on bamboo stems and mossy tree trunks. **V** Five- to eight-note husky, rapid, rolling warble. **H** Dense undergrowth and bamboo in moist temperate forest. **D** Himalayas to C China.

19 NAGA WREN-BABBLER *Spelaeornis chocolatinus* 10cm Note strong buff streaks on brown flanks. Active, often forages in vegetation well off the ground. **V** Undescribed. **H** Undergrowth in montane broadleaved evergreen forest, steep hillsides with rocks covered with moss and thick vegetation, forest edge. **D** NE India.

20 GREY-BELLIED WREN-BABBLER *Spelaeornis reptatus* 10cm Forages in low, thick vegetation. **V** Loud, strident, accelerating trill that changes into a descending warble. Contact call is a soft, repeated *pt*. **H** Thick undergrowth in broadleaved evergreen forest and wet, well-vegetated ravines and gulleys. **D** NE India to E Myanmar, SW China; W Thailand.

21 CHIN HILLS WREN-BABBLER *Spelaeornis oatesi* 10cm Differs from the once conspecific Naga Wren-Babbler by black spots on throat and breast. Forages in low vegetation. **V** Loud undulating warble that is quickly repeated: *chiwi-chiwi-chiwi-chew* or *witchi-witchi-witchi-wu*. Calls include soft *tuc tuc tuc*, a *chit-chit-chit* and quiet *ik ik ik*. **H** Understorey of broadleaved evergreen forest, forest edge, secondary growth, scrub. **D** NE India, W Myanmar.

22 PALE-THROATED WREN-BABBLER *Spelaeornis kinneari* 11–12cm Off-white throat contrasts with dark malar area. Forages close to the ground. **V** Trills that either slow towards the end or are stressed at the end with a stuttering middle. **H** Broadleaved evergreen forest. **D** N Vietnam.

23 TAWNY-BREASTED WREN-BABBLER *Spelaeornis longicaudatus* 10cm Brown above, ochre-buff below. Creeps and climbs in low vegetation and on rocks. **V** Loud, short, shrill warble. **H** Undergrowth in broadleaved evergreen forest, secondary growth, steep hillsides with rocks and boulders. **D** NE India.

24 SIKKIM WEDGE-BILLED BABBLER *Sphenocichla humei* 18cm Blackish brown, with white streaks. Relatively arboreal, often clambers on tree trunks; also forages in thick undergrowth near streams. **V** Loud, piping and slurred whistles, often given in duet. Calls include a strident set of whistles; when alarmed gives a low *hrrrh hrrrh hrrrh hrrr it hrrrh hrrh*... **H** Understorey of broadleaved evergreen forest and bamboo brakes. **D** E Himalayas.

25 CACHAR WEDGE-BILLED BABBLER *Sphenocichla roberti* 18cm Habits and actions similar to Sikkim Wedge-billed Babbler. **V** Melodious, fluty *uu-wii-wuu-yu*. **H** Evergreen forest and bamboo on the edge of dense jungle. **D** NE India to SW China.

26 WHITE-BREASTED BABBLER *Stachyris grammiceps* 12cm Rufous-brown upperparts, mostly white below. Forages from the undergrowth to the middle levels. **V** Loud, rattling trill that increases in volume; calls include a *chrr chrr*, a throaty *cheek-cheek*, a soft *tik* and *chup chup* flight notes. **H** Primary lowland and hill forest, shrubs along forest trails and wooded streams. **D** Java, Indonesia.

27 SOOTY BABBLER *Stachyris herberti* 16–18cm Entirely dark apart from slightly paler throat. Forages on rocks, in undergrowth and in the canopy. Runs rapidly along vines with tail fanned. **V** Calls include soft, repeated *tip* or *tu-ip* and subdued, metallic *cheet cheet cheet*. **H** Broadleaved evergreen forest on limestone, near cliffs and outcrops. **D** Laos, Vietnam.

28 NONGGANG BABBLER *Stachyris nonggangensis* 16–18cm Note white crescent on ear-coverts. Forages on the ground, on rocks or low branches. **V** Unrecorded. **H** Seasonal rainforest on limestone karst. **D** SW China, N Vietnam.

29 GREY-THROATED BABBLER *Stachyris nigriceps* 12cm Distinctive head pattern. Very active, constantly on the move; usually encountered in small parties foraging in low growth. **V** High-pitched, quavering, rising *ti tsuuuuuuuueee*. Calls include *chi chi chi* and *chrrrt* or *chrrrrr-rrr-rrt*. **H** Secondary growth and bamboo in light or dense forest. **D** Himalayas to SW China, SE Asia; Borneo; Sumatra, Indonesia.

30 GREY-HEADED BABBLER *Stachyris poliocephala* 13–15cm Reddish brown, with dark grey hood. Forages in undergrowth and in trees. **V** Clear high-pitched, repeated phrases, e.g. *chit-tiwi-wioo-iwee* or *yit-uip-ui-wiee* and a higher *chu-chi-chiee* or similar. Calls include a descending *dji-dji-dji-du*, a scolding *chrrrrtutut* and soft *tip-tip-tip* contact calls. **H** Broadleaved evergreen forest, secondary growth. **D** Malay Peninsula; Borneo; Sumatra, Indonesia.

31 SPOT-NECKED BABBLER *Stachyris strialata* 15–17cm Note white throat and black-and-white spotting on side of neck. Shy and skulking. **V** High-pitched, whistled *tuh tih tih, tuh tih tuh* or *tuh-tih* sometimes accompanied by a rattling *whiiii-titititititi*. Calls include scolding *tirrrirrirr* and high-pitched *tip* notes. **H** Broadleaved evergreen forest, secondary growth, scrub and grass. **D** SE Asia; Sumatra, Indonesia.

32 SNOWY-THROATED BABBLER *Stachyris oglei* 15cm White throat, black mask, white supercilium. Forages in fast-moving flocks, in thick cover. **V** Thin high-pitched rattle. **H** Broadleaved evergreen forest, secondary growth, bamboo and scrub in ravines. **D** NE India, N Myanmar.

33 CHESTNUT-RUMPED BABBLER *Stachyris maculata* 17–19cm Note chestnut rump and dense black streaks on underparts. Forages in the lower and middle storey, in noisy, busy groups. **V** Variable, including a persistent, loud *whup whup whup*, *wu wup-wwuhup-wup-wuhup* or similar, often mixed with harsh trills and quiet conversational notes. **H** Lowland forest. **D** Malay Peninsula; Borneo; Sumatra, Indonesia.

34 WHITE-NECKED BABBLER *Stachyris leucotis* 14–15cm White supercilium linked to white spots on side of neck. Forages on or near the ground, singly, in pairs or in small groups, occasionally in the company of other small babblers. **V** Simple, whistled *uu-wi-u-wi, uu-wi-uwi-u* or *uui-wi-oi-wi*. **H** Primary evergreen and logged forest. **D** Malay Peninsula; Borneo; Sumatra, Indonesia.

35 BLACK-THROATED BABBLER *Stachyris nigricollis* 15–16cm Note distinctive white-spotted head pattern. Generally forages in low vegetation, sometimes in the company of other small babblers. **V** Well-spaced, monotonous, piping *pu-pu-pu-pu-pu-pu* or faster *pupupupupupupupupu*, followed by a low churring from a female. Calls include a slow rattle, a harsh, descending *chi-chi-chew-chew* and a high *tchi-tchu*. **H** Primary and secondary evergreen, swamp forest, overgrown plantations. **D** Malay Peninsula; Borneo; Sumatra, Indonesia.

36 WHITE-BIBBED BABBLER *Stachyris thoracica* 18cm Distinctive combination of blackish throat and white breast-band. Forages in undergrowth and thickets, generally in small groups. **V** Calls include a low, pulsating *chr r r r r*..., a shorter *chrr chrr chrr chrr* or an abrupt *chrrrp*. **H** Broadleaved evergreen forest. **D** Java, Indonesia.

37 CHESTNUT-WINGED BABBLER *Stachyris erythroptera* 12–14cm Small babbler with rich chestnut and dark grey plumage. Forages in the middle storey in small parties, gleaning from foliage and twigs. **V** Mellow, quite quick, variable piping whistle of 7–10 notes, e.g. *hu-hu-hu-hu-hu-hu*. May be accompanied by low *churrs*, presumably given by the female. Calls include a scolding *trrrrrt-trrrrrt* and a soft *wip* or *wit*. **H** Broadleaved evergreen forest, secondary growth. **D** Malay Peninsula; Borneo; Sumatra, Indonesia.

38 CRESCENT-CHESTED BABBLER *Stachyris melanothorax* 13cm Black crescent across breast. Forages in parties and occasionally with mixed-species groups. **V** Fast, rolling, piping trill, accompanied by harsh, low, churring calls. **H** Dense thickets and scrub at forest edges, monsoon forest and wooded areas around villages. **D** Java and Bali, Indonesia.

1
2
3
4
5
6
7
8
9
10
11
12
13
14
15
16
17
18
19 ♀
20 ♀
21
22 ♀
♂
♂
♂
23
24
25
26
27
28
29
30
31
32
33
34
35
36
37
38

BABBLERS AND SCIMITAR BABBLERS *CONTINUED*

1 DEIGNAN'S BABBLER *Stachyridopsis rodolphei* 12cm Very similar to Rufous-fronted Babbler and now considered to be conspecific. Duller and less rufescent than Rufous-fronted with greyish (not white) underwing coverts. **V** Not known but probably similar to Rufous-fronted Babbler. **H** Montane bamboo forest on Doi Chiang Dao. **D** NW Thailand.

2 RUFOUS-FRONTED BABBLER *Stachyridopsis rufifrons* 12cm Restless, forages through undergrowth and in tops of bamboo clumps, usually in pairs or small groups, often with other babblers. **V** Monotonous, piping *per pe-pe-pe-pe-pe* or *tuh tuh-tuh-tuh-tuh-tuh*. Calls include a short rolling *wirrri*, a fast *wu-yu-yu-yu-yu-yu-yi* and a soft *wit* or *wi* contact note. **H** Thick undergrowth in open forests, dense forest in ravines, bamboo, scrub jungle. **D** Myanmar, Thailand, Malay Peninsula; Borneo; Sumatra, Indonesia.

3 BUFF-CHESTED BABBLER *Stachyridopsis ambigua* 12cm Sometimes treated as conspecific with Rufous-fronted Babbler; very similar, but underparts warmer. **V** Very similar to Rufous-fronted Babbler and often indistinguishable. **H** Thick undergrowth and bamboo scrub in open forests. **D** E Himalayas to S Laos.

4 RUFOUS-CAPPED BABBLER *Stachyridopsis ruficeps* 12cm Note bright rufous crown. Forages low in undergrowth, actions tit-like; post breeding often found with other small babblers. **V** Similar to Rufous-fronted Babbler, but lower in tone and without the space between the first and second note. When alarmed, a harsh, scolding *trrrrt-trrrrt-trrrrt*. **H** Dense bushes in forest clearings and bamboo jungle. **D** E Himalayas to Laos, Vietnam, C and S China, Taiwan.

5 BLACK-CHINNED BABBLER *Stachyridopsis pyrrhops* 10cm Black lores and chin. Forages on or near the ground, in undergrowth; sometimes ascends trees. **V** Bell-like, mellow *wit-wit-wit-wit* repeated 7–8 times. Calls include a soft *chir*, a slow *pee-ve-ve* and a scolding *tchhirrirrr*, *irr-wir-wee* or *irr-wir-wir-wee*. **H** Open and secondary forest with low undergrowth, forest edge, scrub jungle, thickets, bamboo, hedgerows. **D** Himalayas.

6 GOLDEN BABBLER *Stachyridopsis chrysaea* 10cm Note striped crown and bright yellow underparts. Active, agile forager in tangled undergrowth and in tree foliage. **V** Level-toned *pee pi-pi-pi-pi-pi-pi*, very like Rufous-fronted Babbler. Calls include a soft twittering, a *chirik-chirik* and a scolding *chrrrrr*. **H** Dense bushes, bamboo, new undergrowth, thickets in moist secondary jungle, deserted cultivations. **D** Himalayas to SE Asia, SW China; Sumatra, Indonesia.

7 TAWNY-BELLIED BABBLER *Dumetia hyperythra* 13cm Striking rufous underparts. Forages mainly among tall grasses or on the ground; also noted taking nectar from flowers of *Erythrina* trees. **V** Clear, pleasant whistling that starts with a lark-like *psssi-yu* or *ssiiu* followed by *tit-ut-swit-it*, *whit-it-it-it* or *whit-ut*. Calls include a feeble *tit* and a soft *tack-tack*. **H** Open wooded areas, grassland with scattered scrub, bamboo clumps, thorn-scrub, deciduous scrub jungle, wasteland near forests. **D** India, Sri Lanka.

8 DARK-FRONTED BABBLER *Rhopocichla atriceps* 13cm Groups forage in undergrowth, often in the company of other species of babbler. **V** Rattling *kt t t t kt kt t t*, subdued *chur-r chur-r*, used when on the move; also various squeaks and, when alarmed. a rattling *chur-r*. **H** Sholas and thickets near streams, dense marshy and bamboo jungle, plantations, dense undergrowth, scrub. **D** SW India to Sri Lanka.

9 PIN-STRIPED TIT-BABBLER *Macronus gularis* 11cm Narrow streaks on breast. Noisy, actions tit-like; post breeding often in parties of 12 or more foraging in bushes or trees. **V** Repetitive *chaunk-chaunk-chaunk*... Calls include a harsh *chrrrt-chrr* and a scolding *tseep*. **H** Forest with bushes and undergrowth, bamboo jungle, long grass. **D** Himalayan foothills and E India through SE Asia to SW China; Sumatra, Indonesia; Palawan, W Philippines.

10 BOLD-STRIPED TIT-BABBLER *Macronus bornensis* 12cm Chestnut upperparts, boldly striped underparts. Creeps around in dense undergrowth, in small parties. **V** Duets: male utters a *chuhu-chuhu-chuhu-chuhu*, while female utters a harsh, crackling *chuk uk uku*. Calls include a *chrr chrr chrr*, which may develop into a chatter, with an occasional explosive *chirrr*. **H** Secondary forest, wooded riverbanks, coastal scrub, scrub jungle. **D** Borneo and nearby islands; Java, Indonesia.

11 GREY-CHEEKED TIT-BABBLER *Macronus flavicollis* 13–14cm Forages in tangled vegetation in the lower and middle storey, usually in small groups. **V** Monotonous series of 4–25 bell-like notes, repeated after short pauses. **H** Open forest, secondary growth, dry forest patches, edge of beach forest. **D** Java, Indonesia.

12 GREY-FACED TIT-BABBLER *Macronus kelleyi* 14cm Forages mainly in trees singly or in pairs; post breeding in small parties, which may include other species. **V** Soft, even, well-spaced *tuh-tuh-tuh-tuh-tuh-tuh-tuh-tuh*... Calls include a harsh *chrrrrii-chrrruu-chrrrii-chru* and a harsh, squeaky *trrrrrrt-trrrrrt*... **H** Broadleaved evergreen and secondary forest. **D** E Cambodia, Laos, Vietnam.

13 BROWN TIT-BABBLER *Macronus striaticeps* 13cm Note bold white streaks on head. Forages in dense, tangled vegetation from the lower to middle storey. **V** Chattering *we-chu-we-chu we-chu*, *pe-we-chu pe-we chu*. Calls include a quiet *jit* or *gug* and a distinctive *fshhht*, *fssstut-fssstut*. **H** Primary and secondary evergreen forest, logged forest, forest edge, secondary growth, scrub. **D** S Philippines.

14 FLUFFY-BACKED TIT-BABBLER *Macronus ptilosus* 16–17cm Dark brown, with chestnut crown and black throat. Forages in foliage, usually pairs. **V** Repeated, low, long, liquid *puh-puh-puh-puh*, a slower *wuh wu-hu wu-hu*, *wuh-wuh hu-wu hu-wu* or a *poop-poop...poop-poop*, often accompanied by husky churring, presumably by a female. Calls include a *gertcha* or a harsh *ker*. **H** Broadleaved evergreen forest edge, freshwater swamp forest, secondary forest, bamboo. **D** Malay Peninsula; Borneo; Sumatra, Indonesia.

15 CHESTNUT-CAPPED BABBLER *Timalia pileata* 15–17cm Note rufous crown, white supercilium and throat. Elusive, threads through tangles of tall grass and bushes. **V** Husky phrases ending with thin, metallic notes, e.g. *wher-wher witch-it-it* or similar, also a *tseeen-weer-skrich-richrichit* or a *whinnying tweer'r'r'r'r'r'r'r'r'r'r*, fading at the end. Calls include a short *tzit*, a harsh *chrrt* and various chuntering notes. **H** Tall grass, reeds, scrub and secondary growth. **D** Nepal, Bangladesh, NE India through SE Asia to S China; Java, Indonesia.

FULVETTAS AND GROUND BABBLERS PELLORNEIDAE

16 RUFOUS-VENTED GRASS BABBLER *Laticilla burnesii* 17cm Keeps low in grass clumps, best located by song. **V** Loud, clear liquid warble. Calls include a wheezy *feez* and a quiet, rapid, nasal rattle. **H** Tall grassland or tall grass mixed with scrub, reedbeds. **D** Pakistan, NW India, SE Nepal.

17 SWAMP GRASS BABBLER *Laticilla cinerascens* 17cm Usually in pairs or small parties; tends to stay well hidden while foraging on or near the ground. Sometimes considered a race of Rufous-vented Grass Babbler. **V** Song consists of a rising and falling, rich, mellow warble interspersed with short trills. Typical call is a low *chisp chisp*. **H** Tall grass, especially in swamps. **D** NE India, N Bangladesh.

18 GOLDEN-FRONTED FULVETTA *Alcippe variegaticeps* 10–12cm Note bright yellow forehead and distinctive face pattern. Forages in undergrowth and bamboo. **V** Chattering alarm call. **H** Broadleaved evergreen, mixed deciduous and pine forest, with bamboo cover. **D** W and C China.

19 YELLOW-THROATED FULVETTA *Alcippe cinerea* 10–11cm Frequents dense undergrowth and understorey, often in small flocks. **V** Series of thin notes ending in a complex, rapid jumble. Also a thin *si-si-si-si-si-si-si'si'si* interspersed with *tit* notes. **H** Broadleaved evergreen forest. **D** Himalayas to NC Vietnam.

20 RUFOUS-WINGED FULVETTA *Alcippe castaneceps* 10cm Note black and chestnut panels in wing. Occurs in highly active, large winter flocks. **V** Rich, undulating and slightly descending *ti-du-di-du-di-du-di*. Calls include a three-note crescendo *tu-twee-twe*, a high wheezy *tsi-tsi-tsi-tsirr* and a quiet *chip*. **H** Thick evergreen undergrowth at forest edge. **D** Himalayas to SE Asia.

21 BLACK-CROWNED FULVETTA *Alcippe klossi* 12–13cm Crown is black with white streaks. Forages on mossy trunks in the under storey. **V** Long series of thin, shrill notes, trailing at the end; also a grating *hht't't't't'it'it*... mixed with hard *tid* and *tid-rr* notes. **H** Broadleaved evergreen forest, secondary growth. **D** SC Vietnam.

22 RUFOUS-THROATED FULVETTA *Alcippe rufogularis* 12cm Note black, white, grey and chestnut face pattern. Skulks in deep cover near the ground, usually in small parties mixed with other babblers. **V** Loud, shrill *wi-chuw-i-chewi-cheeu*, repeated every few seconds. **H** Evergreen forest undergrowth, bamboo, secondary growth. **D** E Himalayas to SE Asia.

23 RUSTY-CAPPED FULVETTA *Alcippe dubia* 13cm Forages in dense undergrowth, close to or on the ground; regularly occurs in small parties mixed with other babblers. **V** Short, sweet, fluty warble. Calls include a low rattle, buzzy trills and a quiet *peet-seet-seet*. **H** Forest undergrowth. **D** Himalayas to SC China, C Vietnam.

24 DUSKY FULVETTA *Alcippe brunnea* 13–14cm Note long black lateral crown-stripe. Skulking, forages on or near the ground, in small parties. **V** Hurried, shrill *whi wi-wi wich'uu, uu wi-witchi-chuu*. Foraging birds utter a low *jurt*. **H** Broadleaved evergreen forest, grass-jungle. **D** China, Hainan, Taiwan.

25 BROWN FULVETTA *Alcippe brunneicauda* 14–15cm Very plain. Forages in the lower middle storey and in low vegetation, in pairs or small parties. **V** Slow, high-pitched *hi-tu-tu ti-tu ti-tu*. Calls include a stressed *whit* and short, harsh rattles. **H** Broadleaved evergreen forest, often near streams. **D** Malay Peninsula; Borneo; Sumatra, Indonesia.

26 BROWN-CHEEKED FULVETTA *Alcippe poioicephala* 15cm Sides of head are greyish, underparts buff. Forages from undergrowth up to canopy. **V** Pleasant, whistled *tui-tui-tui-tuwee-twee-tuee*. Call is a harsh, buzzy, spluttering rattle. **H** Forest undergrowth. **D** India to SE Asia, S China.

27 JAVAN FULVETTA *Alcippe pyrrhoptera* 14–15cm May show very faint dark lateral crown-stripe. Forages in small parties or occasionally in mixed-species feeding flocks. **V** Loud, clear *ti-ti chi chi chew*. Calls include an explosive *bhip* or *jesip*, and sometimes a loose rattle. **H** Submontane and montane broadleaved evergreen forest. **D** W Java, Indonesia.

28 MOUNTAIN FULVETTA *Alcippe peracensis* 14–16cm Prominent black lateral crown-stripe contrasts with slate-grey crown. Forages in pairs or small parties, often in the company of other species. **V** Pleasant *iti-iwu uwi-u wheer wheer*. Calls include a harsh *chrr'rr'r*. **H** Montane broadleaf forest. **D** Laos, Vietnam, Peninsular Malaysia.

29 BLACK-BROWED FULVETTA *Alcippe grotei* 15–17cm Formerly considered conspecific with Mountain Fulvetta. Forages in the mid-storey and undergrowth, in parties and mixed-species flocks. **V** Pleasant *yu-chi-chiwi-chu-woo*. Calls include a rasping, spluttering *wit-it-itrrrt*, *witchititit* or *err-rittirrirrrt*. **H** Broadleaved evergreen forest, forest edge, secondary growth, bamboo, scrub. **D** SE Thailand, E Cambodia, Laos, Vietnam.

30 GREY-CHEEKED FULVETTA *Alcippe morrisonia* 12–14cm Grey head, smudgy dark lateral crown-stripe, white eye-ring. Forages in the mid-storey and low undergrowth, in parties and mixed-species flocks. **V** Repeated, high-pitched *it chi wi-wi, yu yu wi wi-you*. Calls include a *chr'rr'r* and a harsh *chitititit*. **H** Montane broadleaf evergreen forest, secondary forest, bamboo, scrub. **D** Taiwan.

31 DAVID'S FULVETTA *Alcippe davidi* 12–14cm Formerly treated as conspecific with Grey-cheeked Fulvetta. Differs in lack of dark lateral crown-stripes. **V** Similar to Grey-cheeked Fulvetta. **H** Montane broadleaf evergreen forest, secondary forest, bamboo, scrub. **D** SC and S China to NW Vietnam.

32 YUNNAN FULVETTA *Alcippe fratercula* 12–14cm Formerly treated as conspecific with Grey-cheeked Fulvetta. Differs in darker buff underparts. **V** Similar to Grey-cheeked Fulvetta. **H** Montane broadleaf evergreen forest, secondary forest, bamboo, scrub. **D** S China to SE Asia.

33 HUET'S FULVETTA *Alcippe hueti* 12–14cm Formerly treated as conspecific with Grey-cheeked Fulvetta. Similar to David's Fulvetta but head and throat greyer. **V** Similar to Grey-cheeked Fulvetta. **H** Montane broadleaf evergreen forest, secondary forest, bamboo, scrub. **D** SE China, Hainan.

34 NEPAL FULVETTA *Alcippe nipalensis* 12cm Agile forager in undergrowth and the tops of low trees. Post breeding often in flocks of up to 20 individuals; also joins mixed-species feeding flocks. **V** Simple *chu-chui-chiwi*. Calls include *chr'rr'r*. **H** Broadleaved evergreen forest, forest edge, secondary growth, bamboo, scrub. **D** Himalayas to SW Myanmar.

1
2
3
4
5
6
7
8
9
10
11
12
13
14
15
16
17
18
19
20
21
22
23
24
25
26
27
28
29
30
31
32
33
34

FULVETTAS AND GROUND BABBLERS *CONTINUED*

1 BORNEAN WREN-BABBLER *Ptilocichla leucogrammica* 15–16cm Usually seen in pairs or small groups foraging on the forest floor among leaf litter and rotting, fallen branches. **V** Mournful *doo-dee* or *doo-doo-dee*. Calls include a spluttering *churr* or *prr prr*. **H** Primary, secondary and logged forest, peat-swamp, upland heath forest. **D** Borneo.

2 STRIATED WREN-BABBLER *Ptilocichla mindanensis* 13–14cm Note bold white supercilium. Found singly, in pairs or in small parties foraging in low vegetation or among leaf litter. **V** Descending *hi hi hi uu uu uu uu-u* and a penetrating *hiuu-hiuu-hiuu-hiuu...* Calls include *trrr-t, trrr trrr* and *wh rrp*. **H** Primary and logged forest. **D** Philippines.

3 FALCATED WREN-BABBLER *Ptilocichla falcata* 19–20cm Long white stripes on upperparts and underparts; relatively long tail. Forages by walking or hopping on the ground, or scrambling around in dense thickets and fallen trees. **V** Loud, mournful, undulating *hiuuu-huu oo*. **H** Primary broadleaved evergreen forest, often near streams; also bamboo forest. **D** Palawan, W Philippines.

4 RUSTY-BREASTED WREN-BABBLER *Napothera rufipectus* 18–19cm Whitish throat. Forages on or near the ground, generally under dense vegetation cover. **V** Clear, piping *hi-hi-hi-hi-hi-huuuuh* or *ip ip ip ip ip ip-puuuuuh*; during duets, one bird utters a continuous *ip ip ip ip ip...* while the second bird utters an undulating *hu wip ii* or *hu-wip ee*. **H** Montane primary broadleaved evergreen forest, dense roadside vegetation in conifer plantations. **D** Sumatra, Indonesia.

5 BLACK-THROATED WREN-BABBLER *Napothera atrigularis* 18cm Forages mainly on the ground among leaf litter, beneath tree roots and by probing in tree-trunk crevices. Usually found in small parties. **V** Duets, one bird giving a long series of clear, bell-like notes, while the second adds a slightly descending *iuuh-iuh-iuh... iuuh-uh-uh*. Calls include a *we-ah we-ah we-ah* and a nasal, coarse *krar krar*. **H** Lowland primary broadleaved evergreen forest, old secondary forest, logged forest. **D** Borneo.

6 LARGE WREN-BABBLER *Napothera macrodactyla* 19cm Blackish breast-band is variable, and absent in some races. Forages singly, in pairs or in small parties, on or close to the ground; sings from higher perches. **V** Very variable, consisting of short, loud, clear, whistled phrases, repeated every few seconds. **H** Broadleaved evergreen, selectively logged and bamboo forest. **D** Malay Peninsula; Sumatra and Java, Indonesia.

7 MARBLED WREN-BABBLER *Napothera marmorata* 21–22cm Prominent scaling above and below. Shy and secretive, foraging on or near the ground, in undergrowth and often in damp areas. **V** Clear, whistled *puuu-chiiii, pyuuu-jhiiii, puuui-jhiiii, piuuu-whiiii, uuuu-jhiii* or a single *piuuu*. **H** Lower montane broadleaved evergreen forest. **D** Malay Peninsula; Sumatra, Indonesia.

8 LIMESTONE WREN-BABBLER *Napothera crispifrons* 15–16cm Forages among tangled vegetation and rocks. The white-throated morph occurs in S Myanmar. **V** Rapid series of uneven, harsh slurred notes. Calls with harsh scolding rattles. **H** Evergreen and mixed deciduous forest and scrub in limestone hill country. **D** SE Asia.

9 STREAKED WREN-BABBLER *Napothera brevicaudata* 12cm Forages in pairs or small groups, on the ground in low, tangled vegetation or among rocks and boulders. **V** Various loud, clear, melancholy ringing whistles, including a *chee-oo, peee-oo, pu-ee, chiu-ree, cheweе-chui, pee-wi* or a single *pweee*. Calls include a hard *churk-urt-churk-urt* and a buzzy *trrreeettt*. **H** Moist and shady hill forest, broken up by rocky ravines and steep slopes. **D** NE India to SE Asia.

10 MOUNTAIN WREN-BABBLER *Napothera crassa* 14cm Whitish throat, rufous belly, pale supercilium. Forages in low vegetation and sometimes at higher levels. Skulking but can be very tame. **V** Shrill, piping *hi-hi-hi hu hi, hi hi hi-hu-hi* or *hi-hi-hi hu hi-hu-hu*; during duets, one bird sings song while the second utters a repeated, descending *hii-hii-hiiu*. Calls include a quick *whit, whik* or *whiti chrrh*. **H** Broadleaved evergreen forest, favouring dense, dark areas broken by steep slopes and rocky ravines; also forest edges, low bamboo, and densely vegetated banks and beds of small streams. **D** Borneo.

11 EYEBROWED WREN-BABBLER *Napothera epilepidota* 10cm Note very short tail. Forages on the ground, turning over leaves in search of insects; also creeps about in low vegetation and among mossy logs or rocks. *N. e. roberti* occurs south of the Brahmaputra River. **V** Thin, clear, falling *cheeeoo, cheeeeeu* or *piiiiiu*, repeated every few seconds. Calls include a loud, repeated *chyurk*, a low *pit pit pit* and a rattling *prrrt-prrrt-prrrt*. **H** Dense, dark forests with natural openings, especially where strewn with mossy logs or boulders. **D** NE India through SE Asia to S China, Indonesia.

12 WHITE-HOODED BABBLER *Gampsorhynchus rufulus* 23cm Longish tail, white head and breast. Forages mainly in the mid-storey of trees or the canopy of bamboo; gregarious. **V** Loud, hollow cackling or accelerating, hollow, laughing *khurk khurk khurk-khurk-khurk-khurk*. Contact call is *wit, wet* or *wyee*. **H** Bamboo, broadleaved evergreen forest, secondary growth, scrub and vegetation at forest edge. **D** E Himalayas, NE India to C Myanmar, SW China.

13 COLLARED BABBLER *Gampsorhynchus torquatus* 23–24cm Formerly considered conspecific with White-hooded Babbler. Slightly more rufous, with less extensive white hood. **V** Calls include a harsh, stuttering rattle, e.g. *rrrrtchu-rrrrtchu-rrrrtchu*. **H** Bamboo, broadleaved evergreen forest and secondary growth. **D** SE Asia, from E Myanmar to Vietnam and Peninsular Malaysia.

14 SPOTTED THRUSH-BABBLER *Ptyrticus turdinus* 22cm Note triangular spots on breast and otherwise white underparts. **V** Mid-high, melodious, fluted *wuWutju-tjitjutuditjuw* and other mellow phrases. **H** Swampy woodland, riverine belts. **D** Cameroon, South Sudan, DR Congo, E Angola, NW Zambia.

15 BLACKCAP ILLADOPSIS *Illadopsis cleaveri* 14cm Resembles Brown-chested Alethe (Plate 249), but note uninterrupted grey around cheek. **V** Very high, sharp, fluting *tuh-teeeh* or *tutu-teeh*. **H** Forest and forest remnants. **D** Sierra Leone to Central African Republic, Congo.

16 SCALY-BREASTED ILLADOPSIS *Illadopsis albipectus* 14cm Note pale legs and scaly breast (though some individuals have reduced or no scaling). Solitary or in pairs. **V** Very high, slightly rising, sharp *piuu-piuu-peeh* or very high, sharp, whistled *fjuu-fjee-fjee*. **H** Forest. **D** C Africa, from SE Central African Republic and S South Sudan through DR Congo to W Kenya, NW Angola.

17 RUFOUS-WINGED ILLADOPSIS *Illadopsis rufescens* 14cm Note contrasting white underparts with ill-defined partial breast band. **V** Rhythmic *ti-ti-ti-hu-hu-hu*. **H** Forest undergrowth. **D** Gambia, Guinea, Sierra Leone to Togo.

18 PUVEL'S ILLADOPSIS *Illadopsis puveli* 15cm Distinguished from Rufous-winged Illadopsis by full (though ill-defined) breast band. Note grey around eye. **V** Very high, babbling, robin-like *peehpuhwripeepee*. **H** Forest, bush, thickets. **D** Senegal to Cameroon, S South Sudan, W Uganda.

19 PALE-BREASTED ILLADOPSIS *Illadopsis rufipennis* 14cm Similar to Brown Illadopsis, but with more extensive white belly. **V** Very high, pure, short, level whistle with very slight tremolo and crescendo **H** Dense forest undergrowth. **D** Sierra Leone to Tanzania.

20 BROWN ILLADOPSIS *Illadopsis fulvescens* 16cm Sharp division between grey cheeks and white chin, and uniform underparts. Less on the ground than other illadopsises. Typically in pairs or small family groups. **V** Three-toned, soft, unhurried, mellow, fluted phrase in duet. Also high *fweeeeh* or *due* with varied *pic* as counter-song. **H** Forest. **D** Senegal to W Kenya, Tanzania, N Angola.

21 MOUNTAIN ILLADOPSIS *Illadopsis pyrrhoptera* 14cm Note very dark colouring with whitish chin, rufous-brown wings and tail. Solitary or in small groups. **V** Very high fluted descending *one-two-three four* (*four* is very low answer in duet of male and female). **H** Montane forest. **D** E DR Congo to W Kenya, W Tanzania, N Malawi.

22 SHORT-TAILED SCIMITAR BABBLER *Jabouilleia danjoui* 18–19cm Note long downcurved bill. Forages on or close to the ground. **V** Series of short, monotone, clear, high-pitched whistles. Calls include a scolding *chrrr-chrrr-chrrr...* **H** Broadleaved evergreen forest, secondary forest, bamboo. **D** Laos, Vietnam.

23 NAUNG MUNG SCIMITAR BABBLER *Jabouilleia naungmungensis* 18–19cm Sometimes considered conspecific with Short-tailed Scimitar Babbler. **V** Probably much like Short-tailed Scimitar Babbler. **H** Primary and secondary broadleaved evergreen forest, rocky areas. **D** N Myanmar.

24 LONG-BILLED WREN-BABBLER *Rimator malacoptilus* 11–12cm Skulking; forages on the ground and in undergrowth, usually in pairs. **V** Short, clear, whistled *chiiuuh* or *fyeeer* that falls in pitch but rises in volume and is given every few seconds. **H** Forest undergrowth and dense scrub in steep broken country. **D** NE India to N Myanmar, S China.

25 SUMATRAN WREN-BABBLER *Rimator albostriatus* 13cm Forages on the forest floor or on fallen trees or branches. **V** Short, clear, bell-like *puu* or *puh*; also a series of *whipu ip* or *whipu-wip* notes. Calls include a low *wrrrt, trrrp* or *trr h*. **H** Lower montane oak and laurel forest. **D** W Sumatra, Indonesia.

26 WHITE-THROATED WREN-BABBLER *Rimator pasquieri* 11–12cm Skulks in undergrowth, rummages among fallen leaves. **V** Whistled *chiiii'uh* or *tiiiii'u*, often interspersed with short *pit'wip* or *pi-wip* notes; call is a low *prrp* or *prrt*. **H** Broadleaved evergreen forest, secondary growth and bamboo. **D** N Vietnam; very restricted range, endangered.

27 ABBOTT'S BABBLER *Malacocincla abbotti* 15–17cm Forages among leaf litter, alone or in pairs. **V** Series of 3–4 rich, fluty, liquid, whistled notes. Calls include mewing, purring and an explosive *cheu* interspersed with nervous *wer* notes. **H** Mangroves, coastal forest, riverside forest, secondary forest, forest edges and scrub. **D** From E Himalayas and E India through SE Asia to Indonesia.

28 HORSFIELD'S BABBLER *Malacocincla sepiaria* 15–17cm Very like Abbott's Babbler, but darker plumage. Forages on the ground, in low vegetation and, occasionally, at higher levels. **V** Strident 3–4-note *wi-cho-teuu* or *tip top tiu*; Javan birds, give a plodding *wip-chup-chu-puiii*. **H** Broadleaved evergreen forest, usually near water. **D** Malay Peninsula; Borneo; Sumatra, Java and Bali, Indonesia.

29 BLACK-BROWED BABBLER *Malacocincla perspicillata* 15–16cm Formerly known from only one specimen collected in the 1840s and long feared extinct, but spectacularly rediscovered in October 2020. **V** Unrecorded. **H** Lowland forest. **D** SE Borneo.

30 SHORT-TAILED BABBLER *Malacocincla malaccensis* 13–15cm Short tail is often cocked. Forages on or close to the ground, usually in pairs. **V** Series of 6–7 loud, rich, descending, whistled notes, introduced by a dry trill. Calls include a harsh, mechanical *chutututututututut...*, soft *yer, fit-zweet* and *pew*. **H** Broadleaved evergreen forest, peat-swamp forest and secondary growth, thickets and scrub. **D** Malay Peninsula; Sumatra; Borneo and nearby islands.

31 ASHY-HEADED BABBLER *Malacocincla cinereiceps* 13cm Often seen hopping on the ground or foraging in low vegetation, usually alone. **V** Series of 4–9 husky, nasal *jhew, chiew* or *jhew* notes, sometimes introduced by 1–4 stuttering notes. **H** Lowland and mid-elevation forest, secondary growth and scrub. **D** Palawan, W Philippines.

32 MOUSTACHED BABBLER *Malacopteron magnirostre* 18cm Dark malar stripe not always clear. Forages mainly in the middle storey. **V** Clear, sweet *tii-tu-ti-tu* or *ti-tiee-ti-ti-tu*. Calls include a buzzy *bzzii* and an explosive *whit*. **H** Broadleaved evergreen forest. **D** Malay Peninsula; Borneo; Sumatra, Indonesia.

33 SOOTY-CAPPED BABBLER *Malacopteron affine* 15–17cm Gleans insects from the foliage of small trees, bushes and vine tangles. **V** Series of 6–9 rising and falling whistles, sometimes introduced by scratchy or jumbled notes. Calls include a sharp *which-it*. **H** Broadleaved evergreen forest, swamp forest and forest edges, often with nearby water. **D** Malay Peninsula; Borneo; Sumatra, Indonesia.

34 SCALY-CROWNED BABBLER *Malacopteron cinereum* 14–16cm Active and agile forager in the mid-storey, often in parties. **V** Variable, *dit-dit-dit-dit-dit-dit-du-du-du-du-du-phu-phu-phu-phu-w-wiwiwiwiwi-wi-wi-wi-wu*. Calls include a subdued *chit-chit* or *chreu-chreu*, and a shrill *chit, tcheu* or *titu*. **H** Primary broadleaved evergreen forest, secondary growth, lowland forest and coastal forest. **D** SE Asia, Borneo, Indonesia.

35 RUFOUS-CROWNED BABBLER *Malacopteron magnum* 18–20cm Forages in the middle storey and undergrowth, in pairs, small parties or as part of mixed-species flocks. **V** Three-part song, transcribed as *phu-phu-phu-phi chuwee-chuwee-chuwee-chuwu-chuwu chut-chut-chut-chut-chut*. **H** Primary broadleaved evergreen, mixed dipterocarp, logged, upland heath and stunted ridgetop forest; also mangroves and abandoned plantations. **D** Malay Peninsula; Borneo; Sumatra, Indonesia.

36 MELODIOUS BABBLER *Malacopteron palawanense* 17–18cm Forages in the middle and lower storeys, alone or in small parties; noted picking insects from dead leaves caught in bamboo canopy. **V** Loud series of mournful, clear, alternating, high and low whistles that increase in volume, accompanied by *whit-whit-whit-whit* or *chi-chi-chi-chi-chi*. **H** Lowland primary and old secondary evergreen forest, bamboo and forest edges. **D** Palawan, W Philippines.

37 GREY-BREASTED BABBLER *Malacopteron albogulare* 14–16cm Usually forages in the understorey. **V** Long, subdued, discordant *whu-whi, whit-whu* and *uu-whi-u* phrases, with variants. Calls include *trr* and persistent churring. **H** Broadleaved evergreen and freshwater swamp forests. **D** Malay Peninsula; Borneo; Sumatra, Indonesia.

1
2
3
4
5
6
7
8
white-throated morph
9
10
11
12
13
14
15
16
17
18
19
20
21
22
23
24
25
26
27
28
29
30
31
32
33
34
35
36
37

FULVETTAS AND GROUND BABBLERS *CONTINUED*

1 WHITE-CHESTED BABBLER *Trichastoma rostratum* 15–17cm Forages on or near the ground, and on roots, rocks or branches at the water's edge. **V** Variable, including a repeated, high-pitched *wi-ti-ti*, *chui-chwi-chew* or *chr chr oo-iwee*. Calls with harsh rattles. **H** Riverine broadleaved evergreen forest, secondary and freshwater peat-swamp forest, sometimes mangroves. **D** Malay Peninsula; Borneo; Sumatra, Indonesia.

2 SULAWESI BABBLER *Trichastoma celebense* 15–16cm Relatively featureless. Coloration varies. Forages mainly on or near the ground with mouse-like actions. **V** Loud, cheerful phrase of 2–5 liquid, fluty, melancholy whistles, sometimes answered, presumably by female, with a down-slurred *kiew kiew kiew*. **H** Primary forest, secondary woodland, thickets at forest edge. **D** Sulawesi, Indonesia.

3 FERRUGINOUS BABBLER *Trichastoma bicolor* 16–18cm Rufous-brown upperparts with brighter rufous tail. Forages mainly in the lower to middle storey, in pairs or loose parties. **V** Repeated, loud, sharp *u-wit* also variable jolly phrases, e.g. *wit wi-ti-tu-tu*. Calls include explosive *wit* notes and dry rasping sounds. **H** Broadleaved evergreen forest and secondary forest. **D** Malay Peninsula; Borneo; Sumatra, Indonesia.

4 STRIPED WREN-BABBLER *Kenopia striata* 14–15cm Note white face and prominent streaking of mantle and flanks. Forages on or close to the ground among leaf litter and fallen branches. **V** Short, monotonous *chuuii*, *chiuuu* or *chi-uuu*. Calls include a soft *pee-pee-pee*, a soft, frog-like *churrh-churrh-churrh* and short, twangy nasal notes. **H** Broadleaved evergreen forest, lightly logged forest, upland heath forest. **D** Malay Peninsula; Borneo; Sumatra, Indonesia.

5 INDIAN GRASSBIRD *Graminicola bengalensis* 18cm Skulking; occurs low down in reeds or thick grass, best seen while in song from reed top or during song-flight. **V** Strong downslurred note followed by a series of clipped, rhythmic notes; also includes rattles and short wheezy notes. Calls include a mewing and a nasal screech. **H** Lowland wet grassland, reedbeds and other vegetation near water. **D** N and NE India, S Nepal, Bhutan, Bangladesh.

6 CHINESE GRASSBIRD *Graminicola striatus* 19–21cm Generally elusive, forages through reeds and tall grass, sometimes on the ground; sings from prominent perch or during song-flight. **V** Repeated, rising then falling *trew treuw* or *ji-jee jee-ji*. Calls include a harsh *cha* and a sharp *zip*. **H** Lowland grassland with scattered bushes; also paddyfields. **D** S China (including Hainan), Myanmar, Thailand, Vietnam.

7 SPOT-THROATED BABBLER *Pellorneum albiventre* 14–15cm Olive-brown above and below. Very skulking, forages near the ground. **V** Rich mix of short whistles and hard ringing notes. Calls include a buzzing *chrrr-chrrr-chrrr-chrrrit...* and a quickly repeated *tip* or *tchip*. **H** Undergrowth in pine forest, overgrown clearings, scrub, bamboo, secondary growth. **D** NE India to SE Asia, SW China.

8 MARSH BABBLER *Pellorneum palustre* 15cm Note whitish throat and heavy streaking on breast. Very skulking, best located by call. **V** Song variable, including *grgrgrgrgr chew-hwee chichi-chu-hee*. Calls include a loud *chi-chew*, an aggravated *chik-chik-tuwheeu* and clattery bursts combined with clear notes. **H** Reeds and coarse grass alongside swamps and rivers, tall grass and bushes on marshy ground. **D** NE India.

9 PUFF-THROATED BABBLER *Pellorneum ruficeps* 15cm Distinctive head pattern, broad streaks on flanks. Forages on the ground, usually in pairs or small parties. **V** Repeated loud, shrill *whi-chu*, *wi-ti-chu* or *pre-tee-deer*. Calls include a nasal *chi* and a rasping *rrrrrit*. **H** Forest undergrowth, secondary growth, bamboo and thickets in ravines and by watercourses. **D** Himalayas, India, Bangladesh, SE Asia.

10 BROWN-CAPPED BABBLER *Pellorneum fuscocapillus* 16cm Distinctive combination of rich rufous underparts and dark brown crown. Forages on or near the ground. **V** Monotonous, clear *tu-weee-deyuuu*. Calls include a sharp *wit*, a *quit-it-it* and a low *chr chrr chrr*. **H** Dense thickets, undergrowth in and near evergreen broadleaved forest, scrub and overgrown areas near villages. **D** Sri Lanka.

11 BUFF-BREASTED BABBLER *Pellorneum tickelli* 13–16cm Note absence of grey on face. Skulking; forages close to the ground. **V** Loud, quickly repeated *wi-twee* or *wi-choo*. Calls include a rattling *prree* and an explosive *whit*. **H** Broadleaved evergreen forest, secondary growth, bamboo, occasionally mixed deciduous forest. **D** NE India, E Bangladesh, SE Asia.

12 SUMATRAN BABBLER *Pellorneum buettikoferi* 15cm Short tail, grey face. Skulks in undergrowth or forages on the ground. **V** Loud, repeated *pwiy pii biyo* or *pii byopwiyu*. **H** Primary and logged forest with undergrowth, forest edge. **D** Sumatra, Indonesia.

13 TEMMINCK'S BABBLER *Pellorneum pyrrogenys* 15cm Forages on the ground in undergrowth, although reported working up vines to the canopy, searching for insects. **V** Shrill *witichew witichew witichew*, *wi-chu wi-chu* or single *chew*; also utters a skipping *tip ip ip ip ip-zrriiu*. **H** Undergrowth and dense vegetation at the edge of clearings in lower montane and montane primary forest. **D** Borneo; Java, Indonesia.

14 BLACK-CAPPED BABBLER *Pellorneum capistratum* 16–17cm A brown and rufous babbler with a striking head pattern. Walks or runs while foraging among leaf litter. **V** Loud *ti-tuu* or *pi-tuu*; Bornean birds utter a thinner *ti-ii-uu* or *ti iiuu*, while Sumatran birds sing with a high-pitched *teeu*. Calls include a *yeryeryer*, a nasal *nwit-nwit-nwit* and a liquid, rising *pooeet*. **H** Primary broadleaved evergreen forest, logged forest, peat-swamp forest, overgrown plantations, and bamboo and palm thickets. **D** Malay Peninsula; Borneo; Sumatra and Java, Indonesia.

LAUGHINGTHRUSHES AND ALLIES LEIOTHRICHIDAE

15 CAPUCHIN BABBLER *Phyllanthus atripennis* 20cm Blackish-maroon with grey and black head. Extent of black on head variable. **V** Chorus song, given by group, consists of cackling and trilling notes, becoming piercing squeals when agitated. Calls include grunts and soft notes, with a louder *kaa kaa kaw* when alarmed. **H** Dense lowland forest undergrowth. **D** Senegal to SW Cameroon, S Central African Republic, NE DR Congo, W Uganda.

16 WHITE-THROATED MOUNTAIN BABBLER *Kupeornis gilberti* 20cm Note striking white throat and cheeks. Typically seen in noisy flocks high up in trees. **V** Call is a harsh explosive *chak*, given singly or repeated; flocks chatter loudly. **H** Edges of montane forest. **D** SE Nigeria, SW Cameroon.

17 RED-COLLARED BABBLER *Kupeornis rufocinctus* 20cm Unmistakable, when properly seen in canopy. **V** Harsh chattering, rustling *chuck* notes. **H** Montane forest. **D** E DR Congo, Rwanda, Burundi, SW Uganda.

18 CHAPIN'S BABBLER *Kupeornis chapini* 20cm Note chestnut panel in wings. Normally occurs in flocks, often together with other bird species. **V** Harsh chattering calls, often given by group in chorus. **H** Forest canopy. **D** E DR Congo.

19 SPINY BABBLER *Turdoides nipalensis* 25cm Forages mainly on the ground, under bushes; usually in pairs, with small groups post breeding. **V** Variable, including a rising *ter-ter-ter-ter-ter* and a descending *tee-tee-ker-chee-ker-chee-ker-chee*. Calls include musical *churr* and screaming notes. **H** Dense secondary scrub, thick scrub and bracken on hillsides. **D** Nepal.

20 IRAQ BABBLER *Turdoides altirostris* 22cm Gregarious, usually in groups of up to 10, sometimes furtive, sometimes inquisitive; feeds on or near ground, often stands with tail raised. Groups often first catch the eye as they fly 'follow-my-leader' from cover to cover. **V** Commonest contact note is a drawn-out, whistled *pherrrreee*, also a chattering *pherr-pherr-pherr...* When alarmed, a squeaking *phsioe* and during aggression a series of *phist* or *phic* notes. **H** River, canal and marsh margins, frequenting various trees, scrub and reeds, spreading into nearby thickets, palm groves, edges of cultivation. **D** SC Turkey, Syria, Iraq, SW Iran.

21 COMMON BABBLER *Turdoides caudata* 23cm Upperparts streaked, underparts only lightly so. Usually encountered in parties of up to 20 birds, hopping rapidly on the ground or scuttling rat-like under vegetation; regularly cocks tail. **V** Various loud, descending whistles and a squeaky *qwe-e-e qwe-e-e* or *qwee* given when alarmed. **H** Thorn-scrub, scrub in plains and cultivations. **D** Indian subcontinent.

22 AFGHAN BABBLER *Turdoides huttoni* 23cm Formerly considered conspecific with Common Babbler. Differs in being larger and greyer, with a longer bill. **V** Series of downslurred whistles, similar to Common Babbler. **H** Scrubby river valleys with tamarisks and cultivation. **D** SE Iraq to SW Pakistan.

23 STRIATED BABBLER *Turdoides earlei* 21cm More strongly streaked than Common Babbler. Gregarious, clambering in reeds and grass. **V** Repeated *tiew-tiew-tiew-tiew*, interspersed with *quip-quip-quip* calls from other flock members. **H** Tall grass and reedbeds in marshy areas. **D** Pakistan through N India to Myanmar.

24 WHITE-THROATED BABBLER *Turdoides gularis* 25–27cm Note striking white throat. Forages on the ground, under scrub and in small trees. **V** Flocks utter a sibilant, low *trrrr trrrr trrrr* and a louder, constantly repeated *chr'r'r'r'r'r'r* or *whir'r'r'r'r'r'r*. **H** Scrub and bushes in semi-desert, cultivation edge, thickets, thorn hedges, patches of bamboo. **D** Myanmar.

25 SLENDER-BILLED BABBLER *Turdoides longirostris* 23cm Skulking, gregarious and noisy; forages on the ground and on reed or grass stems. **V** Variable; includes a high-pitched series of shrill notes; also a strident *chiu-chiu-chiu-chiu*, a discordant *tiu-tiu-tit-tit-tu-tu* and a buzzy call that leads into an irregular rattle. **H** Tall grass and reeds, usually near water. **D** Nepal to SW Myanmar.

26 LARGE GREY BABBLER *Turdoides malcolmi* 28cm Large, long-tailed greyish babbler with pale forehead. Forages on the ground or in low vegetation; gregarious, often in quite large groups. **V** Monotonous, drawling *kay-kay-kay-kay*. Gives a noisy chatter when alarmed. **H** Open dry scrub, cultivations, gardens. **D** Indian subcontinent.

27 ARABIAN BABBLER *Turdoides squamiceps* 27cm Long-tailed grey-brown babbler with dark streaks and pale scaling on head and mantle. **V** Song is a quiet, high-pitched warble. Calls include loud, piercing whistles, *peee - peee - peee...* or *piu - piu - piu...*, a hoarse trill and a loud *pew*; a soft whinnying when groups are at roost. **H** Open acacia woodland, scrub thickets, palm-grove undergrowth, edges of reedbeds and gardens. **D** NE Egypt, Israel, Jordan, Arabian Peninsula.

28 FULVOUS BABBLER *Turdoides fulva* 24cm Sandy brown, with whitish throat. Said to be more secretive than other babblers. **V** Subdued squeaking and chirruping. Call: series of piping notes, *peeoo-peeoo-peeoo-peeoo-peeoo-peeoo*, also a hollow rattle and a sharp *pwit*; a clear *peep* in flight. **H** Open acacia woodland, scrub, thickets and palm groves with undergrowth. **D** N Africa, from Morocco to Senegal and east to Sudan, Eritrea.

29 SCALY CHATTERER *Turdoides aylmeri* 20cm Some faint scaling of throat and breast, varying according to race. **V** Not well known. Foraging groups make constant chorus of mouse-like squeaks. **H** Dry woodland and bushland. **D** E Africa, from Ethiopia, Somalia to Tanzania.

30 RUFOUS CHATTERER *Turdoides rubiginosa* 20cm Note brown upperparts and contrasting tawny underparts. Scaly breast and face sides. **V** Chorus song is a series of up to nine twittering notes. Calls include a shrill *tschyee* and a soft *pew*. **H** Open places in dense bush and shrub. **D** E Africa, from South Sudan, Ethiopia, Somalia to Tanzania.

31 RUFOUS BABBLER *Turdoides subrufa* 25cm Note black-and-yellow bill, dark lores. Forages on or near the ground in thick cover; more often heard than seen; gregarious, usually in small parties. **V** Shrill, whistling *tree-tree-tree* interspersed with harsh squeaks. **H** Dense undergrowth, especially with coarse grass and bamboo, forest edge, abandoned clearings. **D** SW India.

32 JUNGLE BABBLER *Turdoides striata* 25cm Note pale eye, whitish lores, yellowish bill, dull grey-brown plumage. Very gregarious; forages mainly on the ground. **V** Harsh *ke-ke-ke* that may break into a squeaking and chattering. Calls include a nasal chortling and a low buzzy *churweeur*. **H** Deciduous forests, cultivations, gardens. **D** Pakistan, India, Bangladesh.

33 ORANGE-BILLED BABBLER *Turdoides rufescens* 25cm Note warm colouring, orange-yellow bill, yellow legs. Gregarious; forages in trees or undergrowth, gleaning insects from foliage and branches. **V** Constant chattering, squeaking and chirping. **H** Broadleaved evergreen forest, thickets and bamboo scrub in well-wooded areas. **D** Sri Lanka.

34 YELLOW-BILLED BABBLER *Turdoides affinis* 23cm Pale head and yellow bill are distinctive. Forages mainly on the ground, but often gleans insects from tree branches and foliage. **V** High-pitched tinkly chittering that varies in pitch and volume. **H** Open forest, secondary woodland, dry scrub, cultivations, gardens. **D** S India, Sri Lanka

35 BLACK-FACED BABBLER *Turdoides melanops* 30cm Note yellow eye circled in black. **V** Chorus song nasal; consists of chattering notes that build to a crescendo. Calls include a low *juk* and a high *jizi-jizi-jizi*. **H** Bushed and wooded areas. **D** S Angola, N Namibia, NW Botswana.

36 BLACK-LORED BABBLER *Turdoides sharpei* 25cm Note scaled feathering of head, breast and mantle. **V** Chorus song consists of wild, laughter-like babbling *chookaaah* notes. Calls are mostly loud, harsh notes, single or repeated. **H** Bushed and wooded areas. **D** Uganda, Kenya, Tanzania, Rwanda, Burundi, E DR Congo.

37 DUSKY BABBLER *Turdoides tenebrosa* 20cm Note black, sharply demarcated lores. **V** Little known. Recorded calls include a hoarse *chow* and a loud, nasal *what-chow*. **H** Dense forest undergrowth near streams. **D** N Central African Republic, S Sudan; NE DR Congo, S South Sudan, W Uganda; E South Sudan, W Ethiopia.

1
2
3
4
5
6
7
8
9
10
11
12
13
14
15
16
17
18
19
20
21
22
23
24
25
26
27
28
29
30
31
32
33
34
35
36
37
darker-faced morph (above)

LAUGHINGTHRUSHES AND ALLIES *CONTINUED*

1 BLACKCAP BABBLER *Turdoides reinwardtii* 24cm Unmistakable, with solid black hood and spotted breast. Moves in flocks through dense vegetation. **V** Chorus song consists of scraping *jarrrrr* notes interspersed with raspy, staccato chatters. Calls include a harsh, short *dzwit* and similar single notes. **H** Thickets, dense patches of bush and tree stands. **D** Senegal to Central African Republic, N DR Congo.

2 BROWN BABBLER *Turdoides plebejus* 25cm Note absence of black before and around eye. **V** Chorus song consists of chuckling bursts of grating notes, loud and wooden-sounding. Calls include a twangy *caa* and a buzzing *chay-o* when alarmed. **H** Wooded and bushed, mostly dry areas, including cultivations and gardens. **D** S Mauritania to Kenya.

3 WHITE-HEADED BABBLER *Turdoides leucocephala* 20cm Note naked blue-grey eye area and brown rump, tail and mantle. **V** Loud chorus song, *waaaak-waaaak*. Calls include a raspy *kwik*, various churrs and similar notes. **H** Dense scrub, dry acacia woodland. **D** E Sudan, Eritrea, NW Ethiopia.

4 ARROW-MARKED BABBLER *Turdoides jardineii* 24cm Note red-rimmed yellow eye, arrow-marked feathering, uniform brown upperparts. **V** Chorus song consists of harsh, chattering notes that build to a whirring crescendo. Calls include harsh *chack-chack-chack* and loud *chow-chow-chow*. **H** Any type of woodland, riverine belts, acacia bush, suburbs. **D** Gabon to SE Kenya and south to N Botswana, NE South Africa.

5 SCALY BABBLER *Turdoides squamulata* 20cm Note dark (red) eye and brown (not white or tawny) rump. **V** Sings in chorus, an extremely harsh and toneless *gagagagagagaga*. Groups make constant rasping sounds, and other calls include a harsh *chack* and other throaty notes. **H** Dense bush. **D** S Ethiopia, S Somalia, coastal Kenya, NE Tanzania.

6 WHITE-RUMPED BABBLER *Turdoides leucopygia* 20cm Note white rump. Generally very variable in extension of white and black, especially on head, which may be all white. **V** Loud and very vocal. Chorus song and typical calls both have a harsh, chattering tone. **H** Damp wooded and bushed grassland. **D** E Sudan, Eritrea, Ethiopia, Somalia.

7 HARTLAUB'S BABBLER *Turdoides hartlaubii* 25cm Note pale scaling on head, mantle and throat, white rump, red eye. **V** Very noisy. Chorus song is a long series of raucous, high-pitched babbling sounds. Calls include a variety of harsh and shrill notes. **H** Reedbeds, other wet habitats, nearby woodlands. **D** E DR Congo and Rwanda to Angola, N Botswana, W Zimbabwe.

8 HINDE'S BABBLER *Turdoides hindei* 20cm Note tawny rump and flanks. Otherwise variable. Tawny feathers and white areas can appear asymmetrically anywhere on head and body. **V** Chorus song is a noisy chattering that builds to long screeching calls. Other calls include braying and chattering notes. **H** Bush and shrubland especially in small valleys. **D** Kenya.

9 NORTHERN PIED BABBLER *Turdoides hypoleuca* 25cm Brown above, white below, with full or partial brown breast-band. **V** Chorus song includes loud *teeya-teeya-teeya* and *skare-skare-skare* notes. Calls include harsh chatters, churrs and chuckling sounds. **H** Wooded and bushed natural and cultivated areas, gardens, parks. **D** Kenya, Tanzania.

10 SOUTHERN PIED BABBLER *Turdoides bicolor* 25cm Unmistakable black and white plumage. **V** Loud, harsh chorus song of high-pitched chattered notes. Calls include crowing and purring sounds. **H** Dry bushveld, more or less open woodland. **D** S Angola, Namibia to Zimbabwe, N South Africa.

11 BARE-CHEEKED BABBLER *Turdoides gymnogenys* 24cm Unmistakable, with white head and contrasting bare black cheek. **V** Chorus song *kerrakerra-kek-kek-kek*, very similar to Arrow-marked Babbler. Calls include harsh *jeee-jeee-jee* and soft *lull lull lull* notes. **H** Dry, rocky areas with wooded outcrops. **D** Angola, Namibia.

12 CHINESE BABAX *Babax lanceolatus* 28cm Skulking; forages on the ground or in low scrub, although will ascend trees to feed in the topmost branches. **V** Wailing *ou-phee-ou-phee*, repeated several times; groups keep up a continual flow of soft musical notes, with some harsher outbursts. Calls include a sound like a creaky gate hinge. **H** Thin scattered forest, open hillsides with a covering of bracken, brambles and grass. **D** Myanmar to SE China.

13 MOUNT VICTORIA BABAX *Babax woodi* 28cm Formerly considered conspecific with Chinese Babax. Differs in having black moustachial stripe and bolder streaking above. **V** Series of loud upslurred whistles, *hu-Wheet! hu-Wheet!* repeated frequently, and sometimes given by multiple individuals simultaneously. **V** Open evergreen forest, forest edge, secondary growth, scrub. **D** NE India, W Myanmar.

14 GIANT BABAX *Babax waddelli* 31cm Heavily marked with brown and grey streaks. Skulks in dense bushes; forages on the ground, turning over dead leaves in a search for food. **V** A series of quavering, whistled, thrush-like notes. Call is a harsh grating. **H** High-altitude arid scrub and thickets. **D** Tibet.

15 TIBETAN BABAX *Babax koslowi* 28cm Relatively plain, rufous. Actions and habits said to be much as others of genus. **V** Undescribed. **H** Scrubland, rocky areas and abandoned agricultural fields. **D** Tibet.

16 CHINESE HWAMEI *Garrulax canorus* 23cm Unmistakable, with bold white eye-ring and post-ocular streak. Rather uniform warm brown plumage. Head and neck slightly streaked. **V** Song is a rapid, thrush-like fluting with repetitions. Call is a high, shrill, sustained *sreeh-sreeh*. **H** Dense forest, parks, gardens. **D** S and EC China to Vietnam.

17 TAIWAN HWAMEI *Garrulax taewanus* 21–24cm Forages on the ground, singly, in pairs or small parties. **V** Similar to Chinese Hwamei but less complex, containing fewer syllables but more repeated phrases. **H** Secondary growth and lower tree strata. **D** Taiwan.

18 WHITE-CRESTED LAUGHINGTHRUSH *Garrulax leucolophus* 28cm Chestnut and brown, with white crest and breast and black mask. Gregarious, often in the company of other species; noisy, constantly uttering contact chuckles. Forages on the ground and in the lower and middle storeys. **V** Rapid, laughing, chattering cackle; also a subdued staccato *ker-wick-erwick*. **H** Broadleaved evergreen forest, secondary growth, bamboo jungle. **D** Himalayas to SE Asia.

19 SUMATRAN LAUGHINGTHRUSH *Garrulax bicolor* 24–28cm Noisy. Forages in groups in the middle and lower storeys, and occasionally feeds on the ground. **V** Starts with chattering notes, followed by loud, cackling, melodious laughter. **H** Broadleaved evergreen forest. **D** Sumatra, Indonesia.

20 WHITE-NECKED LAUGHINGTHRUSH *Garrulax strepitans* 28–32cm Olive-brown crown, dark rufous ear-coverts, blackish-brown face and breast, whitish neck sides. Forages mainly among leaf litter. Often found in large parties. **V** Cackling laughter combined with rapid chattering, interspersed or preceded with clicking *tick* or *tekh* notes, the latter also used as contact calls. **H** Broadleaved evergreen forest. **D** E Myanmar, Thailand, NW Laos, SW China.

21 CAMBODIAN LAUGHINGTHRUSH *Garrulax ferrarius* 28–30cm Note white spot at rear of ear-coverts. Forages mainly on the ground in dense undergrowth, often in small flocks. Formerly considered conspecific with White-necked Laughingthrush. **V** A rapid cackling laughter. **H** Broadleaved evergreen forest. **D** SW Cambodia.

22 BLACK-HOODED LAUGHINGTHRUSH *Garrulax milleti* 28–30cm Forages from the ground up to the lower canopy, usually in flocks of 3–10 individuals. Occasionally found with other laughingthrushes. **V** Loud outbursts of extended, rapid, cackling laughter interspersed with rattling and tinkling notes. **H** Broadleaved evergreen forest. **D** Laos, Vietnam.

23 GREY LAUGHINGTHRUSH *Garrulax maesi* 28–31cm Forages mainly on the ground in leaf litter. Gregarious, sometimes with other laughingthrushes. **V** Sudden outbursts of loud cackling laughter combining rapid chattering and repeated double-note phrases preceded by a few subdued *ow* notes. **H** Broadleaved evergreen forest. **D** Tibet to S China and N Vietnam.

24 RUFOUS-CHEEKED LAUGHINGTHRUSH *Garrulax castanotis* 28–31cm Habits and actions similar to Grey Laughingthrush, which is sometimes regarded as conspecific. **V** Outbursts of cackling, rapid chattering and repetitive double-note phrases. **H** Submontane broadleaved evergreen forest. **D** Laos, Vietnam, Hainan (S China).

25 SNOWY-CHEEKED LAUGHINGTHRUSH *Garrulax sukatschewi* 28cm Usually in small parties that feed mainly on ground. Birds shake head, flick tail and quiver feathers when calling. **V** Repeated, shrill *hwii-u, hwii-u* or *h'wi-i, h'wi-i*. **H** Coniferous hill forest and scrub. **D** NC China.

26 MOUSTACHED LAUGHINGTHRUSH *Garrulax cineraceus* 21–24cm Note black crown and malar stripe. Generally in pairs or small parties; forages mostly on the ground. **V** Repeated, high-pitched, upslurred *pr r r r ip* that may be interspersed with hard, chuckling staccato notes. **H** Thick bushes in moist forest, thick scrub and secondary growth near cultivations. **D** NE India to E China.

27 RUFOUS-CHINNED LAUGHINGTHRUSH *Garrulax rufogularis* 23–26cm Note black scaling above and below. Skulking; forages on the ground or in low bushes. **V** Repeated, clear, husky *whi-whi-whu-whi* or *whi-whi-whi-whi*; also short grating rattles, a low buzzing *jzzzzz* and a twangy *gshwee*. **H** Dense undergrowth in subtropical forest. **D** Himalayas to Myanmar, NW Vietnam.

28 CHESTNUT-EARED LAUGHINGTHRUSH *Garrulax konkakinhensis* 24cm Skulking; forages in the understorey, singly, in pairs or in small parties. **V** Rambling series of well-spaced and stressed notes and mimicry. Calls include a low grumbling *rrreeek rrreeek rrreeek*. **H** Broadleaved forest, forest edge, secondary growth. **D** Vietnam, SE Laos.

29 BARRED LAUGHINGTHRUSH *Garrulax lunulatus* 23cm Upperparts barred black and buff, underparts scaled white. Forages in groups. **V** Song is *wu-chi - wi-wuoou*, repeated at short intervals. **H** Bamboo understorey in broadleaved and coniferous forest. **D** C China.

30 WHITE-SPECKLED LAUGHINGTHRUSH *Garrulax bieti* 26cm Little known. Similar to Barred Laughingthrush, which has been treated by some authors as conspecific. **V** Undescribed. **H** Bamboo thickets in conifer and secondary forests. **D** SW China.

31 GIANT LAUGHINGTHRUSH *Garrulax maximus* 30–35cm Large, with dark cap and chestnut face. Secretive; mainly a ground forager, often in the company of other laughingthrushes. **V** Jerky rattling, also some shrill notes. **H** Dense, dry subalpine forest with undergrowth and glades. **D** W and NC China.

32 SPOTTED LAUGHINGTHRUSH *Garrulax ocellatus* 30–33cm Forages on the ground or in bushes; usually in pairs or small parties. **V** Rich, mellow, fluty *wu-it wu-u wu-u wi-u wi-u w'you w'you uu-i w'you uu-i...* or *fuwit-fuwee fuwit-fuweet*, often mixed with a nasal *fu-u-uwheen*. Calls include a screechy *schuwee*. **H** Undergrowth in high-elevation forest, open mixed forest with undergrowth and thick rhododendron scrub, bushes at the edge of fields. **D** Himalayas to Myanmar and S China.

33 ASHY-HEADED LAUGHINGTHRUSH *Garrulax cinereifrons* 24–25cm Plain chestnut above, buffy brown below, with grey head and pale eye. Usually found low in vegetation or on the ground. Flocks keep up a constant flow of squeaks and chatters. **V** Continuous high-pitched squeaky notes and harsh churring sounds interspersed with sharp *chit* notes or higher-pitched *pieu-pieu*. **H** Dense humid forest undergrowth and bamboo thickets. **D** Sri Lanka.

34 SUNDA LAUGHINGTHRUSH *Garrulax palliatus* 24–25cm Note large pale eye-ring. Forages in low vegetation or on the ground, usually in small flocks or in mixed-species feeding parties. **V** Group songs start gently, then speed up to a raucous, bubbling, tumbling chaos of screeching, chattering laughter, before easing into a series of repeated *wiku* notes or a flowing *wipiwuwipiwpiwu*. **H** Broadleaved evergreen forest and sometimes in secondary growth. **D** Borneo; Sumatra, Indonesia.

35 RUFOUS-FRONTED LAUGHINGTHRUSH *Garrulax rufifrons* 27cm Usually found in noisy groups or as part of mixed-species foraging flocks, feeding mainly in the understorey. **V** Calls include a repeated, subdued, harsh *kheh, queck* or *hii-tii hii-tii-tii...* When agitated, a chuckling or tinkling *hi tu tu tu tu tu tu...* or *hihi hu hu hu hu hu hu...* **H** Primary montane forest. **D** W Java, Indonesia.

36 MASKED LAUGHINGTHRUSH *Garrulax perspicillatus* 30cm Note large black or blackish mask. Mainly a ground feeder, turning over leaf litter. Usually in small parties. **V** Contact and alarm call loud and piercing, *jhew* or *jhow*, also a harsh chattering. **H** Bamboo thickets, thick scrub, reeds, cultivation and parks. **D** E and C China to Vietnam.

37 WHITE-THROATED LAUGHINGTHRUSH *Garrulax albogularis* 28–30cm Gregarious; in winter often forms large flocks, also joins mixed-species feeding parties; forages in trees and undergrowth, and on the ground. **V** While feeding gives a constant, low *teh teh* combined with a subdued chattering. Other calls include shrill, wheezy whistles and a gentle *chrrr*. **H** Broadleaved evergreen forest, deciduous forests, coniferous forests, open secondary growth and light jungle. **D** Himalayas to NW Vietnam, C China.

1
2
3
4
5
6
7
8
9
10
11
12
13
14
15
16
17
18
19
20
21
22
23
24
25
26
27
28
29
30
31
32
33
34
35
36
37

LAUGHINGTHRUSHES AND ALLIES *CONTINUED*

1 RUFOUS-CROWNED LAUGHINGTHRUSH *Garrulax ruficeps* 27–29cm Similar to White-throated Laughingthrush (Plate 224) but has orange-rufous crown. Forages in middle strata, usually in pairs or flocks. **V** Song similar to White-throated Laughingthrush. Calls include a thin, shrill *tswiiiii* or *dziiiii*. **H** Primary oak, fir and cedar forest, secondary growth, scrub. **D** Taiwan.

2 LESSER NECKLACED LAUGHINGTHRUSH *Garrulax monileger* 27cm Narrow black 'necklace' extends from ear-coverts. Gregarious, foraging mainly on ground. **V** Mellow, repeated *u-wi-uu*, a more subdued *ui-ee-ee-wu*, *wiu-wiu-wiu* or *ui-ui-ui* and a downslurred *tieew ti-tiew...* **H** Broadleaved evergreen and deciduous forests, secondary growth. **D** Himalayas to SE Asia, SE China.

3 GREATER NECKLACED LAUGHINGTHRUSH *Garrulax pectoralis* 29cm Very like Lesser Necklaced Laughingthrush, but larger, with dark eye and bolder 'necklace'. **V** Repeated, clear, ringing *kleer-eer-eer-eer...* or an upslurred, mellow *tu-tweetu-tweetu-twee...* Also a clear, rapid *chit-it*. **H** Dense forest, secondary growth, bamboo. **D** E Himalayas to N SE Asia, SE China.

4 CHESTNUT-BACKED LAUGHINGTHRUSH *Garrulax nuchalis* 23–26cm Forages mainly on or near the ground, often mixes with other laughingthrush species. **V** Song includes mellow whistles mixed with higher and lower slurred notes, also a slightly different type with hardly a pause, *whit-oo-whit-oo-whit, wheeoo-wheeoo-wheeoo, tiu-whit-tiu tiu-whit-tiu tiu-whit-tiu, whit-oo-whit-oo*. **H** Secondary growth, thickets, scrub jungle, rocky ravines with scrub, overgrown cultivations, forest edge. **D** NE India, N Myanmar.

5 BLACK-THROATED LAUGHINGTHRUSH *Garrulax chinensis* 23–30cm Generally in pairs or small parties; skulking forager in trees and bushes. **V** Repetitive, rich fluty song mixed with course *wraah* notes and squeaky whistles. **H** Broadleaved evergreen and mixed deciduous forest, secondary growth, scrub, grass. **D** SE Asia, S China.

6 WHITE-CHEEKED LAUGHINGTHRUSH *Garrulax vassali* 26–29cm Gregarious, often in large parties; forages mainly on or near the ground. **V** Simple, repeated *whii-u*, foraging flocks utter quick *whi* notes and harsh rattles. **H** Broadleaved evergreen forest, secondary growth, scrub, grass. **D** Laos, Cambodia, Vietnam.

7 RUFOUS-NECKED LAUGHINGTHRUSH *Garrulax ruficollis* 22–27cm Note black face, rufous neck patch and undertail coverts. **V** Repeated, jolly, whistled *wiwi wi-whu whi-yi-ha*, and clear, shrill *krkrkrkeeerkrookeerkoo*, building to a loud crescendo before falling at the end. **H** Forest and cultivation edge, scrub, secondary growth, bamboo, tall grass, reeds. **D** Himalayas to N Myanmar, SW China.

8 YELLOW-THROATED LAUGHINGTHRUSH *Garrulax galbanus* 23–25cm Forages on or near the ground; usually encountered in pairs or small parties; often mixes with Rufous-necked Laughingthrush. **V** Feeble chirping. **H** Tall grass with trees and shrubs, edge of dense broadleaved evergreen forest. **D** NE India, Myanmar, Bangladesh.

9 BLUE-CROWNED LAUGHINGTHRUSH *Garrulax courtoisi* 24–25cm Blue-grey crown, black mask, yellow throat. Critically endangered. **V** Continuous, tittering, also pleasant *piiuu, djew* or *djoh* notes and, occasionally, a louder *dju-dju-dju-dju...* **H** Mixed evergreen and deciduous forest, forest patches and nearby bushy areas. **D** Two very restricted populations in E and S China.

10 WYNAAD LAUGHINGTHRUSH *Garrulax delesserti* 23–26cm Slate-grey head, white throat. Gregarious, forages mainly on the ground. **V** Strident *tseeurp*, clanging *jhur-jhur-jheer-jheer-jheer-jheer* and a slower, piping *jheer-jheer-jheer*; also a nasal *tree-tree-true*, various squeaks and rattles and a rasping *churr*. **H** Humid rainforest with dense undergrowth, thorny canebrakes, cardamom sholas. **D** SW India.

11 RUFOUS-VENTED LAUGHINGTHRUSH *Garrulax gularis* 23–26cm Note black mask and primrose-yellow throat. Gregarious and skulking. **V** Sweet chiming whistles, upslurred then downslurred. Calls include harsh rattling churrs, interspersed with high-pitched whistled phrases. Flocks utter mellow squabbling and chattering notes. **H** Broadleaved evergreen forest, secondary growth and scrub. **D** Bhutan to C Laos.

12 PLAIN LAUGHINGTHRUSH *Garrulax davidi* 23cm Grey-brown with yellow bill, grey wing panel. **V** Repeated series of short notes, starting with weak whining notes and followed by lower, weaker notes, *wia - wa - WIKWIKWIK-woitwoitwoitwoit*. **H** Mountain thickets and scrub. **D** China.

13 GREY-SIDED LAUGHINGTHRUSH *Garrulax caerulatus* 27–29cm Rufous above, whitish below, with grey flanks. Forages on the ground or in bushes. **V** Clear, loud, spaced whistled phrases. Calls include a gruff *zhyurt-zhyurt*, a mewing *zhaow-zhaow* and a nasal rattle. **H** Undergrowth in dense forest and bamboo thickets. **D** Himalayas through NE India to Myanmar.

14 RUSTY LAUGHINGTHRUSH *Garrulax poecilorhynchus* 27–29cm Forages in small groups in undergrowth and lower parts of trees. **V** Slow, melodic whistling. Calls include various chirring notes and low cat-like contact notes. **H** Broadleaved evergreen or deciduous forest, mixed forest, often along watercourses. **D** Taiwan.

15 BUFFY LAUGHINGTHRUSH *Garrulax berthemyi* 27–29cm Formerly considered conspecific with Rusty Laughingthrush. Forages mainly on the ground and in undergrowth, occasionally venturing to mid-storey. **V** Loud, melodious and variable whistles. Calls include mewing notes and odd loud whistles. **H** Broadleaved evergreen forest, bamboo and low ground cover in conifer plantations. **D** SW, SC and SE China.

16 WHITE-BROWED LAUGHINGTHRUSH *Garrulax sannio* 22–24cm Note distinctive face pattern. Forages on the ground or in low bushes; occurs in small noisy parties after breeding. **V** Harsh, shrill *tcheu...tcheu...tcheu*; also harsh *tcheurrrr* and buzzy *dzwee* notes. **H** Undergrowth in dense forest, secondary growth, bamboo thickets, hillsides covered with bracken and bramble scrub. **D** NE India through N SE Asia to S China.

17 CHESTNUT-CAPPED LAUGHINGTHRUSH *Garrulax mitratus* 22–24cm Note yellow bill and white eye-ring. Forages in pairs or small parties among creepers and thick foliage in the lower and mid-storey. **V** Subdued, shrill *wi wu-wi-wu-wi, wi-wu-wi, wi-wu-wiu-wu-wi* or *wiu-wu-wui-wi etc*. Calls include sibilant *ju-ju-ju-ju-ju* and cackling *wikakakaka*. **H** Broadleaved evergreen forest and forest edge. **D** Extreme S Thailand; Peninsular Malaysia; Sumatra, Indonesia.

18 CHESTNUT-HOODED LAUGHINGTHRUSH *Garrulax treacheri* 22–24cm Forages from the ground to the canopy, in small parties or mixed-species flocks. **V** Thin, high *chu-wu chwi-wi-wi-wi-wiee-wu* or similar. Calls include harsh, scolding notes and a descending *ah-ah-ah-ah*. **H** Broadleaved evergreen forest, disturbed and secondary forest, adjacent cultivation. **D** Borneo.

19 SPOT-BREASTED LAUGHINGTHRUSH *Garrulax merulinus* 22cm Skulking, noisy and gregarious; forages mainly on the ground. **V** Various rich, melodious phrases with some mimicry; also a repeated coughing chuckle. **H** Dense undergrowth in damp forests, overgrown clearings, bamboo. **D** NE India to N Vietnam.

20 ORANGE-BREASTED LAUGHINGTHRUSH *Garrulax annamensis* 24–25cm Note black throat and neat black streaks on rufous-orange breast. **V** Loud, rich, musical, rambling phrases and mimicry. **H** Broadleaved evergreen forest, forest edge, overgrown clearings, secondary growth. **D** SC Vietnam.

21 STRIATED LAUGHINGTHRUSH *Garrulax striatus* 28cm Note bushy crest, stout bill and white-streaked plumage. Arboreal, foraging from undergrowth to canopy. **V** Repeated, vibrant *prrit-you prrit-pri-prii-u* or *krrrrwhit kwit-kwit-wheeuuw*. Calls include a high, soft, nervous *wer-wer-wer-wer-wer*, a rising *wu-wiw* and a grumbling *greip-greip-greip*. **H** Broadleaved evergreen forest, secondary forest, scrub jungle, thickets. **D** Himalayas to N Myanmar.

22 BLACK LAUGHINGTHRUSH *Garrulax lugubris* 25–27cm Black plumage, orange or reddish bill, bare patch behind eye. Typically forages close to the ground. **V** Hollow, whooping *huup-huup-huup* and a rapid *okh-ohk-okh-okh-okh-okh...*, accompanied by harsh *awk* or *aak*. **H** Broadleaved evergreen forest, forest edges and secondary growth. **D** Extreme S Thailand; Peninsular Malaysia; Sumatra, Indonesia.

23 BARE-HEADED LAUGHINGTHRUSH *Garrulax calvus* 25–26cm Favours the lower and middle storeys. Often acrobatic, hanging upside down like a giant tit. **V** Long series of flat, resonant *ooh* notes; in a duet, *ooh* notes are combined with *yow-yow* notes, *ooh-yow-yow-yow ooh-yow-yow*. **H** Broadleaved evergreen forest and secondary growth. **D** Borneo.

24 STREAKED LAUGHINGTHRUSH *Trochalopteron lineatum* 18–20cm Forages on the ground, in the open or under cover. **V** Similar to Bhutan Laughingthrush but longer. Calls include a *tsip tsip tsip* and high-pitched buzzy notes. **H** Hillside scrub, bushes in open forest, forest edge and around human habitations. **D** E Uzbekistan, S Tajikistan through Afghanistan, Pakistan to C Himalayas.

25 BHUTAN LAUGHINGTHRUSH *Trochalopteron imbricatum* 19–20cm Forages mostly on the ground, usually in pairs or small parties. **V** Rapid trill followed by a ringing whistle. Calls include a buzzy *bzzrt-bzzrt*. **H** Bushes, long grass, thick scrub and thickets in open secondary forests, edge of cultivations. **D** Bhutan, NE India.

26 STRIPED LAUGHINGTHRUSH *Trochalopteron virgatum* 23cm Fine white streaks are prominent on upperparts. **V** Clear, hurried *chwi-pieu* and loud, staccato, rattling trill often preceded by *chrrru-prrrrrt*. Calls include a harsh *chit* and a *chrrrrr*. **H** Thick scrub and ground cover near broadleaved evergreen forest, forest edge, secondary growth. **D** NE India, NW Myanmar.

27 BROWN-CAPPED LAUGHINGTHRUSH *Trochalopteron austeni* 24cm Note scaly appearance of underparts. Skulking, foraging on the ground or in low vegetation. **V** Repeated, clear, jolly *whit-wee-wi-weeoo*. Calls include a harsh *grrrret-grrrret-grrrret* when alarmed. **H** Oak and rhododendron forest, secondary forest, forest edge, bushes in ravines, clearings, bamboo. **D** NE India, NW Myanmar.

28 BLUE-WINGED LAUGHINGTHRUSH *Trochalopteron squamatum* 22–25cm Skulking; forages close to the ground, usually in pairs or small groups. **V** Rich *cur-white-to-go* or *free-for-you*. Calls include a buzzy *jrrrrr-rrr-rrr*, a harsh *cher-cherrrru* and a thrush-like *chuk*. **H** Dense undergrowth in open broadleaved evergreen forest, secondary growth, scrub near forest, bamboo. **D** Himalayas to NW Vietnam.

29 SCALY LAUGHINGTHRUSH *Trochalopteron subunicolor* 23–25cm Dark scaling above and below, yellowish wing panel. Forages on the ground or in tangled vegetation. **V** Clear shrill 'wolf-whistle'. Calls include a squeaky chatter and a shrill alarm note. **H** Broadleaved evergreen forest with thick undergrowth, secondary growth, bramble thickets and rhododendron shrubberies. **D** Himalayas to NW Vietnam.

30 ELLIOT'S LAUGHINGTHRUSH *Trochalopteron elliotii* 23–26cm Note golden-olive fringes to wing and tail feathers. Forages mainly on the ground, in pairs or small parties. **V** Wavering *whi-pi-piu*. Calls include a subdued, high-pitched chattering. **H** Thickets, undergrowth and bamboo, at or above the treeline; also in open broadleaf forest, mixed and juniper forests, sometimes close to human habitation. **D** SW and SC China.

31 BROWN-CHEEKED LAUGHINGTHRUSH *Trochalopteron henrici* 24–26cm Note blue-grey wings and chocolate-brown mask. **V** Repeated, clear *wichi-pi-choo, wi-pi-choo* or *whi-choo-it*. **H** Scrub bordering forests. **D** S and SE Tibet.

32 BLACK-FACED LAUGHINGTHRUSH *Trochalopteron affine* 24–26cm Note white malar patch. Forages on or near the ground, but also in trees. **V** Repeated, shrill *wiee-chiweeoo, wiee-chweeiu* or *wiee-weeoo-wi*. Calls include a high rattle, a low chuckle and wheezy purrs and whines. **H** Forest undergrowth, scrub above the treeline. **D** Himalayas to SC China, NW Vietnam.

33 WHITE-WHISKERED LAUGHINGTHRUSH *Trochalopteron morrisonianum* 25–28cm Striking head pattern and olive-gold wing panel. Forages on the ground and up to mid-storey. **V** Rich, clear, whistled *wit-chi'wi* or *wip chi'rri*, a repeated, laughing *hee-hee-hee-hee hee-hee* and a bell-like *di di di...* also various quiet whistles and churrs. **H** Undergrowth and low trees in coniferous and mixed deciduous forests, open forest with clearings, coniferous scrub, bamboo, juniper, scrub above the treeline. **D** Taiwan.

34 VARIEGATED LAUGHINGTHRUSH *Trochalopteron variegatum* 24–26cm Skulking, forages on the ground, in bushes and trees; generally in pairs, with larger parties post breeding. **V** Loud, musical, whistled *weet-a-weer* or *weet-a-woo-weer*. When alarmed utters muttering and squealing notes. **H** Undergrowth, rhododendron and other bushes in open oak and mixed forests. **D** Himalayas, from NE Afghanistan to Nepal.

35 CHESTNUT-CROWNED LAUGHINGTHRUSH *Trochalopteron erythrocephalum* 24–28cm Skulking; forages on the ground or in low cover, usually in small parties. **V** Various repeated short phrases, transcribed as *pearl-lee to-reaper to-real-year you-reap*. Calls include a *pheeou*, a loud *wee-ou-wee-whip* and a soft, musical *twi-ee-you*. When alarmed gives a grating *m-u-r-r-r*. **H** Dense undergrowth, scrub in gulleys and by cultivation, forest edge, bamboo. **D** Himalayas, from Kashmir to NE India.

36 ASSAM LAUGHINGTHRUSH *Trochalopteron chrysopterum* 23–25cm Forages on or near the ground, occasionally ascends moss- or lichen-covered tree branches. **V** Similar to Chestnut-crowned Laughingthrush, but lower, mellower and more complex. Calls include a low purring *squar-squar-squar*. **H** Understorey and bamboo in broadleaved evergreen, pine and mixed forest. **D** NE India, W Myanmar, SW China.

37 SILVER-EARED LAUGHINGTHRUSH *Trochalopteron melanostigma* 26cm Differs from Assam Laughingthrush in unspotted back, dark primary coverts, more extensive silvery edges on ear-coverts, black chin, olive-tinged lower underparts. **V** Loud, liquid *wi-wiwioo, wu-weeeoo* or *tu-tweeoo*, sometimes a fast *wiu-wip*; also utters clear mewing notes. **H** Broadleaved evergreen, pine and mixed broadleaf–pine forests, secondary growth, bamboo, scrub. **D** N SE Asia, SW China.

1
2
3
4
5
6
7
8
9
10
11
12
13
14
15
16
17
18
19
20
21
22
23
24
25
26
27
28
29
30
31
32
33
34
35
36
37

LAUGHINGTHRUSHES AND ALLIES *CONTINUED*

1 GOLDEN-WINGED LAUGHINGTHRUSH *Trochalopteron ngoclinhense* 27cm Chestnut rear crown, broad golden fringes to wing feathers. Forages singly or in pairs. **V** Double-noted cat-like mewing. **H** Understorey and bamboo in montane broadleaf evergreen forest. **D** W Vietnam.

2 MALAYAN LAUGHINGTHRUSH *Trochalopteron peninsulae* 25–27cm Mostly dull chestnut-brown, with grey head and bright chestnut crown. Forages on or near the ground. **V** Clear *wip-weeoo, wiw-weeoo* or a quicker *wip-wi-eeoo.* **H** Understorey and bamboo in broadleaved evergreen forest, secondary growth, scrub, grass. **D** S Thailand, Peninsular Malaysia.

3 COLLARED LAUGHINGTHRUSH *Trochalopteron yersini* 26–28cm Strikingly patterned. Forages on the ground or low down in dense vegetation, in pairs or small parties. **V** Loud, high-pitched rising *wueeeeoo, u-weeeeoo, uuu-weeoo* etc. Whistles are sometimes answered (presumably by females) with harsh mewing calls, but these are sometimes given on their own. **H** Understorey of primary broadleaf evergreen forest, thick regrowth, scrub at forest edge. **D** S Vietnam.

4 RED-WINGED LAUGHINGTHRUSH *Trochalopteron formosum* 27–28cm Skulking; forages in pairs or small parties, on or close to the ground. **V** Thin, clear, whistled *chu-weevu* or a rising *chiu-wee*; also a *chiu-wee – u-weeoo*, which may be duetting. **H** Broadleaved evergreen forest, secondary growth, scrub near forests. **D** S China, NW Vietnam.

5 RED-TAILED LAUGHINGTHRUSH *Trochalopteron milnei* 26–28cm Note rufous-chestnut crown, whitish ear-coverts, crimson wings and tail. **V** Clear, whistled *uuu-weeoo, eeoo-wee, uuuwi* or *uuu-hiu hiu*, the latter with slight rising introduction and faster, soft laughter at the end. **H** Understorey of broadleaf evergreen forest, dense secondary growth, bamboo, scrub and grass near forest. **D** Myanmar to S China, C Vietnam.

6 HIMALAYAN CUTIA *Cutia nipalensis* 20cm Forages nuthatch-like; creeps along branches and tree trunks; post breeding forms small parties, often mixed with shrike-babblers and other babblers. **V** Long series of ringing notes. Calls include a light *chick-chick-chick...*, a loud *chip* and a low *jert*. **H** Broadleaved evergreen forest, particularly larger trees festooned with moss and epiphytes. **D** Himalayas to SW China, SE Asia, Peninsular Malaysia.

7 VIETNAMESE CUTIA *Cutia legalleni* 17–20cm Very similar to Himalayan Cutia, but has more strongly barred underparts. **V** Variable, including a distinctive *wuyeet wu wi wi wi woo*, a *wipwi-weei-weei-weei* and a fast, loud, high-pitched *wei-wuu-wei-wuu*. **H** Broadleaved evergreen and mixed broadleaf–pine forest. **D** SE Laos, Vietnam.

8 BLUE-WINGED MINLA *Minla cyanouroptera* 14–15cm Note blue highlights on wings and tail. Forages in the tops of bushes and trees, often in large groups or mixed-species flocks. **V** Repeated, high-pitched *psii sii suuu*. Calls include a short *whit* or *bwik* and dry staccato buzzes. **H** Broadleaved evergreen forest and secondary growth. **D** Himalayas to S China, SE Asia.

9 BAR-THROATED MINLA *Minla strigula* 16cm Note ragged yellowish-rufous crown and barred throat. Forages in high bushes and medium-height trees. **V** High-pitched, quavering *tui-twi ti-tu, twi ti-u* or *twi-twi-twi-twi*. Calls include a sharp *kip*, a soft *yeep* and a trilling buzz. **H** Broadleaved evergreen and mixed broadleaf–conifer forest. **D** Himalayas to SW China, SE Asia.

10 RED-TAILED MINLA *Minla ignotincta* 14cm Note prominent white supercilium and white throat. Acrobatic, methodical forager in treetop branches and on mossy trunks. **V** High-pitched, repeated *wi ti wi-wu* or a slurred, falling *si-swee-sweeeuuuu*. Calls include a harsh *wih-wih-wih-wih*, a short *wit* and a fast *witti-wi-wrrh*. **H** Humid dense forest, especially oak and rhododendron. **D** Himalayas to S China, SE Asia.

11 RED-FACED LIOCICHLA *Liocichla phoenicea* 21–23cm Unobtrusive; forages on the ground, in undergrowth and occasionally in trees; post breeding may form small parties. **V** Various clear, beautiful phrases, including *chewi-ter-twi-twitoo, chi-cho-choee-wi-chu-chooee* and *chiu-too-ee*. Calls include a rasping *chrrt-chrrt* and a buzzy, upslurred *grssh grssh*. **H** Dense undergrowth in broadleaved evergreen forest, dense thickets of secondary growth near cultivations or streams. **D** Nepal, NE India to SW China.

12 SCARLET-FACED LIOCICHLA *Liocichla ripponi* 21–23cm Differs from Red-faced Liocichla in having brighter and more extensive scarlet on face, less black above eye. **V** Clear, repeated *chu'u-wiu-wwuu* or similar. Calls include a mewing and a falling, mellow *tyuuuuu*. **H** Ravines, open forest, bamboo jungle near swamps. **D** E Myanmar to N Vietnam.

13 EMEI SHAN LIOCICHLA *Liocichla omeiensis* 19–21cm Wing panel orange-red in male, duller yellowish in female. Skulking; forages in thick vegetation. **V** Weak, shrill whistled phrases of long slurred notes, e.g. *chwi-weeiee-eeoo* or *chui-weeiee-ieeoo-ueeoo*. **H** Undergrowth in secondary broadleaved evergreen forest, dense secondary growth, bamboo, scrub. **D** SW China.

14 BUGUN LIOCICHLA *Liocichla bugunorum* 22cm Forages from ground to treetops, creeps through tangled vines and undergrowth and clambers on tree trunks. Critically endangered. **V** Variable fluty phrases, including a descending *wieu-u-wee i-tuu i-tuu uw-tu oow* and a shorter *weee-keew*; also an accompanying dry *trrrr-trii-trii*, presumed to be made by the female. **H** Forest edge, disturbed hillsides and ravines with dense shrubbery. **D** NE India.

15 STEERE'S LIOCICHLA *Liocichla steerii* 17–19cm Note bright yellow loral patch. Forages in low vegetation. **V** Variable song consists of high-pitched, quavering and buzzy notes. Calls include a rasping *djr* or *djr drrrrrrr*. **H** Forest, forest edge, tangled shrubbery, orchards. **D** Taiwan.

16 RUSTY-FRONTED BARWING *Actinodura egertoni* 23cm Forages from undergrowth to the canopy, often among epiphytic growth; regularly joins mixed-species feeding parties. **V** Sweet, piping *tsit-tsit-seet-seeetsuut*, repeated every few seconds. Calls include a high-pitched rattle and a harsh, buzzy *gursh-gursh...* **H** Dense undergrowth, forest edge, shrubbery and scrub in broadleaved evergreen forest. **D** Himalayas to NE Myanmar.

17 SPECTACLED BARWING *Actinodura ramsayi* 23–25cm Note white eye-ring. Forages in pairs or groups. **V** Repeated, high-pitched, descending *iee-iee-iee-iuu*, sometimes followed by a high-pitched *ewh ewh ewh*, presumably uttered by a female. Calls include a low, harsh *baoh*. **H** Broadleaved evergreen forest, bamboo, forest edge, scrub, grass. **D** N Myanmar to NE Laos, SW China.

18 BLACK-CROWNED BARWING *Actinodura sodangorum* 24cm Forages in the canopy and on mossy trunks and branches, singly or in pairs. **V** Starts quickly, more paced at end, *tutututu'tu'tudi-duuu* or *tutututututu'tu'tee-tuuu*, occasionally followed by a grumbling *hwerr'rr'rr*, presumably given by the female. **H** Broadleaved evergreen forest, forest edge, secondary growth, tall grass and scrub near forests. **D** SE Laos, C Vietnam.

19 HOARY-THROATED BARWING *Actinodura nipalensis* 21cm Arboreal; usually in small parties, foraging for insects on mossy branches and tree trunks. **V** Variable, including a whistled *tui-whee-er*, a clear quavering *wiu-iu* and *duit-duwee-duweer*, repeated every few seconds. When alarmed, a repeated, rapid *je-je-je-je-je...* **H** Oak, conifer and rhododendron forest with rich mixed undergrowth. **D** Nepal to Bhutan.

20 STREAK-THROATED BARWING *Actinodura waldeni* 20–22cm Clambers about mossy branches and tree trunks, pulling moss apart to find insects; often forms part of mixed-species foraging flocks. **V** Strident, rising *tchrrrr-jo-jwiee, dddrt-juee-iwee* or a shorter *jorr-dwidu*. Calls include a nasal, grumbling *grrr-ut grrr-ut* and a *grr-grr-grr-grr-grr*. **H** Mossy broadleaved evergreen and mixed forest. **D** NE India, Myanmar, SW China.

21 STREAKED BARWING *Actinodura souliei* 21–23cm Similar to Streak-throated Barwing, but has darker lores, more silvery ear-coverts and nape, heavier streaking above and below. **V** Calls include soft contact notes and harsh churrs. **H** Broadleaved evergreen and semi-evergreen forest, open fir forest with bamboo. **D** SC China, N Vietnam.

22 TAIWAN BARWING *Actinodura morrisoniana* 18–19cm Distinctive dark rusty-brown hood, streaked grey collar and breast. Forages in the canopy and middle storey. **V** Loud, clear *whit chiwewii*, a quavering *chiriririrt* and a short *hiu* or *huu*. **H** Broadleaf evergreen, broadleaf deciduous and mixed broadleaf–coniferous forest. **D** Taiwan.

23 SILVER-EARED MESIA *Leiothrix argentauris* 15–17cm Unmistakable when seen well. Forages in bushes and trees; post breeding forms parties of 5–30 individuals **V** Cheerful, descending *che tchu-tchu cher-it* or *che-chu chiwi chwu*. Calls include a flat, piping *pe-pe-pe-pe-pe* and a harsh chattering. **H** Bushes and undergrowth in forests, forest edge, secondary growth and scrub. **D** Himalayas to SE Asia, S China; Sumatra, Indonesia.

24 RED-BILLED LEIOTHRIX *Leiothrix lutea* 14–15cm Post breeding often encountered in small, noisy parties; regular member of mixed-species foraging groups; feeds mainly low down in undergrowth or on the ground. Known in the pet trade as Pekin Robin. **V** Fluty warble of up to 15 notes. Calls include a nasal *zhirk*, a *shreep* and a rattling *zhri-zhri-zhri...* When alarmed utters a buzzy *zhriti-zhriti-zhriti...* **H** Undergrowth in open broadleaved evergreen, pine and mixed forests, forest edge, secondary growth and scrub. **D** Himalayas to E China and NE Vietnam. Introduced Hawaii and elsewhere.

25 GREY-CROWNED CROCIAS *Crocias langbianis* 22cm Distinctive bold dark streaks on whitish underparts. Arboreal, forages from the canopy down to undergrowth. **V** Loud, repeated *wip'ip'ip-wiu-wiu-wiu-wiu-wiu-wiu-wiu*. Call is a buzzing *pzah-pzah-pzah-pzah*. **H** Broadleaved evergreen forest. **D** Vietnam.

26 SPOTTED CROCIAS *Crocias albonotatus* 20cm Generally seen in small groups or mixed-species foraging flocks, in the canopy or at forest edges. **V** A husky series of *jhew* or *jhrr* notes, and a series of repeated *whi tu* or *whi tui* notes. **H** Broadleaved evergreen forest. **D** W Java, Indonesia.

27 RUFOUS-BACKED SIBIA *Heterophasia annectens* 18cm Distinctive black, white and rufous plumage. Forages in the canopy, searching mosses and lichens. **V** Pretty, descending warble. Calls include a harsh chattering and a buzzy chortling. **H** Broadleaved evergreen forest. **D** Himalayas to SW China, SE Asia.

28 RUFOUS SIBIA *Heterophasia capistrata* 24cm Arboreal, lively feeder, foraging amongst mossy branches. Often forms part of mixed-species feeding flocks. **V** Fluty *tee-dee-dee-dee-dee-o-lu*. Call is a rapid *chi-chi*; a harsh *chrai-chrai-chrai...* when alarmed. **H** Mixed forests. **D** Himalayas.

29 GREY SIBIA *Heterophasia gracilis* 23–25cm Forages among epiphytes and on mossy trunks. **V** Repeated, descending *tu-tu-ti-ti-ti-tu*. Calls include a soft *ti-tew*, a squeaky *witwit-witarit* and a grating *trrit-trrit*. **H** Broadleaved evergreen forest. **D** NE India to SC China.

30 DARK-BACKED SIBIA *Heterophasia melanoleuca* 21–23cm Forages mainly in the treetops. **V** High-pitched wavering whistle with a drop in pitch at the end. Calls include a harsh *trr-trr-trr*. **H** Broadleaved evergreen forest. **D** Myanmar, Thailand, Laos.

31 BLACK-HEADED SIBIA *Heterophasia desgodinsi* 21–25cm Distinguished from Dark-backed Sibia by sharp demarcation between black crown and grey back. Forages among mossy epiphytes. **V** Slightly descending *hi wi-wi wi wi*; contact calls consist of thin *tsrrri* notes. **H** Broadleaved evergreen forest. **D** N Myanmar to SC China, Laos, Vietnam.

32 WHITE-EARED SIBIA *Heterophasia auricularis* 22–24cm Note prominent white eye-stripe. Forages mainly in the middle to upper storey. **V** Resonant *weep-weeo-weep-weeeooo*. Calls include a tailing-off *sirrrrrrrr*. **H** Broadleaved evergreen forest. **D** Taiwan.

33 BEAUTIFUL SIBIA *Heterophasia pulchella* 24cm Uniform blue-grey with diffuse black mask and black wing coverts. **V** Shrill *ti-ti-titi-tu-ti*, descending slightly towards end. Calls include a variety of notes, including a series that is said to sound like the jingling of a bunch of keys. **H** Mossy forests. **D** NE India to SE Tibet, N Myanmar, SC China.

34 LONG-TAILED SIBIA *Heterophasia picaoides* 28–35cm Note long tail and white wing patch. Forages mainly in the canopy, usually in pairs or small parties; also a regular member of mixed-species flocks. **V** Rich, six-note, whistled phrase. Calls include a high *tsip-tsip-tsip*, sometimes interspersed with a dry rattling or trilling. **H** Broadleaved evergreen forest, forest edge, open scrub with large trees and forest clearings. **D** E Himalayas to SE Asia; Sumatra, Indonesia.

35 NILGIRI LAUGHINGTHRUSH *Montecincla cachinnans* 20cm Forages on the ground or in low vegetation. Can be hard to see, but easily detected by loud call. Formerly abundant; now endangered. **V** Includes a nasal *wi-yu wi-yu wi* followed by a low *dhu-dhu-dhu*, a rising series of *aingk* notes, and a nasal *whur-whaink whur-whaink*; also a short *gruk* and chittering *chink-chink-chink*. **H** Shola forests, forest edge, thick scrub, plantations. **D** SW India (Western Ghats, in Nilgiri Hills).

36 BANASURA LAUGHINGTHRUSH *Montecincla jerdoni* 20cm Formerly considered conspecific with Nilgiri Laughingthrush. Resembles that species but ear-coverts, throat and breast grey. **V** Loud nasal *whir-wir wee, whir-wir wee*, with the second note lower and the final note higher. **H** Streamside thickets, plantations, scrub, secondary forest. **D** SW India (Western Ghats, in SW Karnataka)

37 PALANI LAUGHINGTHRUSH *Montecincla fairbanki* 20cm Forages in bushes, undergrowth and occasionally on the ground. **V** Short, mellow clear notes. Calls include a dry *ptr trit* or *ptr tr tr trit*. **H** Streamside thickets, plantations, scrub, secondary forest. **D** SW India (Western Ghats, in Palani Hills).

38 ASHAMBU LAUGHINGTHRUSH *Montecincla meridionalis* 20cm Formerly considered conspecific with Palani Laughingthrush. Differs in having shorter supercilium and white belly, with rufous restricted to flanks. Note grey-streaked breast. **V** Short, mellow clear notes, similar to Palani Laughingthrush. **H** Dense undergrowth in evergreen forest, secondary growth and plantations. **D** SW India (Western Ghats, in S Kerala and adjacent S Tamil Nadu).

1
2
3
4
5
6
7
8
9
10
11
12
13
14
15
16
17
18
19
20
21
22
23
24
25
26
27
28
29
30
31
32
33
34
35

SYLVIID BABBLERS *CONTINUED*

1 RUFOUS-TAILED BABBLER *Moupinia poecilotis* 15cm Brown upperparts, whitish-buff underparts, with rusty wings and long tail. Rather sluggish forager in very dense undergrowth. **V** Clear, quickly delivered *phu pwii*, occasionally introduced by a short *chit* or similar. When agitated, various combinations of short stuttering notes. **H** Grass, scrubby hillsides, thickets near streams. **D** C China.

2 WHITE-BROWED FULVETTA *Fulvetta vinipectus* 11cm Note white supercilium and throat. Forages in low trees, bushes and undergrowth. **V** Faint *chit-it-it-or-key*. Calls include an incessant *chip-chip* and a *churr* when alarmed. **H** Subalpine scrub and bamboo in forest. **D** Himalayas to SC China, N Vietnam.

3 CHINESE FULVETTA *Fulvetta striaticollis* 11–12cm Streaks on throat and breast; otherwise rather drab. Forages low down in vegetation. **V** Simple *ti tsew* or *ti chuu*. Calls include a low rattle. **H** High-level scrub. **D** Tibet, SW China.

4 SPECTACLED FULVETTA *Fulvetta ruficapilla* 10–11cm Forages low down or in the canopy of small trees. **V** Unrecorded. **H** Broadleaved evergreen oak forest and secondary growth. **D** C and S China.

5 INDOCHINESE FULVETTA *Fulvetta danisi* 11–12cm Similar to Spectacled Fulvetta but with darker crown, plainer wings, darker streaks on breast. Forages singly, in pairs or in small parties. **V** Calls with a quickly repeated, rapid *chrrrit* or *chrrt-chrrt-chrrt*... **H** Broadleaved evergreen forest, bamboo, scrub. **D** Laos, Vietnam.

6 BROWN-THROATED FULVETTA *Fulvetta ludlowi* 11cm Distinctive brown-and-white streaking on throat and breast. Usually in pairs or small parties, often with mixed-species flocks. **V** Thin *see-see-spir r r*. **H** Bamboo thickets and bushes in rhododendron forests. **D** SE Tibet to NW Myanmar.

7 GREY-HOODED FULVETTA *Fulvetta cinereiceps* 12cm Extent of grey on head varies. Actions tit-like. **V** High, thin *ti wuu wiiuu*. Call is a thin, churring rattle. **H** Undergrowth and bamboo in forests, secondary growth. **D** China.

8 MANIPUR FULVETTA *Fulvetta manipurensis* 11cm Formerly treated as a race of Grey-hooded Fulvetta. Forages low down, sometimes in trees or on the ground. **V** High-pitched *ti ti si-su, si-swu* or *see si-wu*. Calls include a low *tirrru* a high *swi-swi-swi-sw*... and a dry metallic rattle. **H** Broadleaved evergreen forest, forest edge, bamboo, scrub. **D** NE India, SW China, N Myanmar, N Vietnam..

9 TAIWAN FULVETTA *Fulvetta formosana* 12cm Formerly considered a race of Grey-hooded Fulvetta. Forages, mouse-like, on or near the ground. **V** Repeated *ti tuuu*. **H** Undergrowth and bamboo in coniferous and deciduous forest. **D** Taiwan.

10 YELLOW-EYED BABBLER *Chrysomma sinense* 18cm Elusive; sometimes climbs to exposed perches, where yellow legs may be visible, but soon dives back into cover. **V** Short, powerful whistled *tit-toowhee-twitchoo*, often mixed with trills or nasal notes. Calls include a dry trill and a subdued *stik*. **H** Scrub, bush-covered grassy hillsides, secondary growth, bamboo, sugarcane fields. **D** Pakistan to Sri Lanka, S China, N Vietnam.

11 JERDON'S BABBLER *Chrysomma altirostre* 17cm Highly skulking; best seen while singing from the top of a reed stem. **V** Weak *chi-chi-chi-chew-chew-chew*. Calls include a high-pitched *tic* or *ts-ts-tsik*. **H** Tall grassland and reed-beds. **D** Pakistan, S Nepal, NE India, Bhutan, Myanmar.

12 BEIJING BABBLER *Rhopophilus pekinensis* 16cm Active. Frequently runs with long tail slightly cocked. **V** Sweet *dear-dear-dear-dear*, each first syllable starting high, falling, then rising again before start of second syllable. Call: a mellow *chee-anh*. **H** Scattered bushes on dry, stony hill- or mountainsides. **D** NC to NE China, N Korea.

13 TARIM BABBLER *Rhopophilus albosuperciliaris* 17cm Formerly treated as a race of Beijing Babbler. Differs from that species in having dark brown (not whitish) eyes, and in being paler and less well streaked. **V** Song is a sweet musical chattering, slower than that of Beijing Babbler. **H** Scattered bushes on dry, stony hillsides. Prefers drier, more desert-like habitats than Beijing Babbler. **D** NW China.

14 WRENTIT *Chamaea fasciata* 16–17cm Typically heard before seen. Pale eye. Tail often cocked as it hops from twig to twig, much like a wren. Pumps tail as it makes short, weak flights from bush to bush. **V** Song is a series of accelerating, staccato notes developing into a descending trill. Typical call is a dry scolding *trrrk*. **H** Dense chaparral, bushy forest margins, parks, gardens. **D** Pacific coast of USA, from Oregon to NW Mexico.

15 GREAT PARROTBILL *Conostoma aemodium* 28cm Much larger than other parrotbills. Forages on or near the ground. **V** Loud, repeated *whip whi-uu* or similar. Calls with nasal wheezes, cackling, churring and squeals. **H** Open evergreen forest, bamboo and rhododendron thickets. **D** Himalayas, NE Myanmar to C China.

16 THREE-TOED PARROTBILL *Cholornis paradoxus* 23cm Resembles Great Parrotbill, but smaller, and has prominent white eye-ring. Outer toe is vestigial. **V** High-pitched, plaintive *tuwi-tui* or *tuii-tew*, also a weaker *tidu-tui-tui* and a low chuntering, interspersed with high *tuwii, tuwii* or *tuuu*. Calls include a harsh *chah*. **H** Bamboo thickets in broadleaf and conifer forests. **D** C China.

17 BROWN PARROTBILL *Cholornis unicolor* 21cm Skulking; acrobatic, often hanging upside-down when foraging. **V** Loud *ii-wuu-iiew* or *it ik ik - ii-wuu-iiew*. Calls include a shrill *whi-whi-whi*, a low *brrh* and a cackling *churrrh*. **H** Dense bamboo thickets. **D** Himalayas to S China.

18 SPECTACLED PARROTBILL *Sinosuthora conspicillata* 14cm Note white eye-ring. Forages low down in bamboo, in small, active flocks. **V** Twangy, high-pitched *triiih-triiih-triiih*..., also shorter *triit* notes. **H** Bamboo layer in montane forests. **D** C China.

19 VINOUS-THROATED PARROTBILL *Sinosuthora webbiana* 11–13cm Rufous-brown crown merges into mid-brown mantle. Forages in fast-moving flocks, in the understorey or the tops of small flowering trees. **V** Quick introductory notes followed by rapid notes, ending with high, thin notes, *ri riti ri ri chididi wii-ssi-tssu*. Flocks call with a rapid, subdued chattering interspersed with thin, high and chuntering notes. **H** Scrub, thickets, bamboo, secondary growth, reedbeds, plantations. **D** SE Russia, China, Korean Peninsula, Taiwan, N Vietnam.

20 ASHY-THROATED PARROTBILL *Sinosuthora alphonsiana* 12–13cm Chestnut crown more strongly demarcated than in Vinous-throated Parrotbill. Forages in fast-moving flocks. **V** Loud, high-pitched *tssu-tssu-tssu-tssu, tsser-tsser-tsser-tsser*. Calls include a harsh chuntering *twer-trr'ir'ir'irrit, twi'it'ti, trr'it'it'it tcher'der'der, chip-ip-ip-ip*, and *tu, twi* or *du* contact notes. **H** Scrub, grass thickets, plantations. **D** C China to N Vietnam.

21 BROWN-WINGED PARROTBILL *Sinosuthora brunnea* 12–13cm Extensive chestnut on head contrasts with warm brown upperparts. Usually occurs in fast-moving flocks. **V** Continuous twittering while feeding; no other information. **H** Montane scrub and grass thickets, sometimes open forest or forest edge. **D** SW China, Myanmar.

22 GREY-HOODED PARROTBILL *Sinosuthora zappeyi* 12–13cm Forages in low vegetation; usually found in small parties. **V** High, piercing *ss-ss-su-si*. Calls include rasping *trr'ik* and *trrrh* notes. **H** Bamboo and bushes in open mountain-top conifer forest and mixed fir–rhododendron forest. **D** W China (SC Sichuan, NW Guizhou).

23 PRZEVALSKI'S PARROTBILL *Sinosuthora przewalskii* 13cm Distinctive small parrotbill with chestnut breast, ash-grey head and long black supercilium. **V** Contact call consists of short rattles, interspersed with thin, high notes. **H** Montane scrub, bamboos, grasses in open larch forests. **D** SC China.

24 FULVOUS PARROTBILL *Suthora fulvifrons* 12–13cm Post breeding forages in fast-moving flocks of 20–30, occasionally more. **V** High-pitched *si-tsiiii chuu, si-si juu* or *si-ti ti tituuuu-jhiiu*. Calls include a husky *chew-chew-chew*, a *cher-cher-cher* and a spluttering *trrrip*. **H** Bamboo thickets, in or near forests. **D** Himalayas, SW China, N Myanmar.

25 BLACK-THROATED PARROTBILL *Suthora nipalensis* 11–12cm Distinctive black, white and brown patterning. Acrobatic forager in small to large flocks. **V** Wheezy, nasal *chu-irrr-diii-dirrr* or similar. Contact calls are a soft *tu, ti, tit* or *tip*, creating a constant twittering from flocks. **H** Bamboo and undergrowth in forests, secondary growth and forest edge. **D** Nepal to Vietnam.

26 GOLDEN PARROTBILL *Suthora verreauxi* 11–12cm Predominantly rich rufous-brown, with black throat. Forages in pairs or small parties, also found in mixed-species flocks. **V** Weak *chuur-dii* and a high-pitched, wispy *hsu-ssu-ssu-ssi*. Calls include a low, spluttering *trr'it* or *trr'eet*. Foraging flocks utter a jumble of *it, twit* and *tip* notes. **H** Broadleaved evergreen forest edge and bamboo. **D** China, Taiwan, E Myanmar, Laos, N Vietnam.

27 SHORT-TAILED PARROTBILL *Neosuthora davidiana* 10cm Short tail creates a bulbous-headed appearance. Forages in bamboo or occasionally trees. **V** Thin, high-pitched, rapid ascending *ih'ih'ih'ih'ih'ih'ih* or *zu'zu'zu'zu'zu'zu'*... also a *tit tiwit tit tit-tew hiuuuu-ti-di'di'di*. Flocks call with *tip* or *tut* notes that can run into a twittering. **H** Bamboo, grass and edge of broadleaved evergreen forest. **D** S China, E Myanmar, N Thailand, N Laos, N Vietnam.

28 PALE-BILLED PARROTBILL *Chleuasicus atrosuperciliaris* 15cm Note dark mark above eye. Acrobatic forager, regularly in small parties or in larger mixed-species groups. **V** Chirping *tik-tik-tik-tik-tik-tik*. Flocks give a chattering interspersed with harsh metallic notes. **H** Bamboo, forest edge and forest undergrowth. **D** Himalayas to N Thailand.

29 WHITE-BREASTED PARROTBILL *Psittiparus ruficeps* 18cm Clambers acrobatically among small twigs and branches; usually in pairs or small parties and in the company of other species. **V** High, slightly descending *he-he-he-hew-hew*. Calls include a rattling *trrrrt trrrrrrrrrt trrrrrrrrrt* and a twangy *jhew*. **H** Bamboo, forest edge and tall grass. **D** NE India, Bhutan, adjacent S China.

30 RUFOUS-HEADED PARROTBILL *Psittiparus bakeri* 18cm Formerly considered conspecific with White-breasted Parrotbill. Darker than that species, with more buff on underparts. Differs also in song. **V** Repeated, breathless *wi-wi-wi-wu* interspersed with *jhaowh* call notes. **H** Bamboo in or near broadleaved evergreen forest and forest edge. **D** NE India through Myanmar, SW China, to N Vietnam.

31 GREY-HEADED PARROTBILL *Psittiparus gularis* 21cm Distinctive black and grey head pattern. Forages from treetops down to undergrowth, occasionally on the ground. **V** Shrill *eu-chu-chu* or *eu-chu-chu-chu* and a high *wi-wuu* or *wi-wuu-wuu-wuu*. Calls include a harsh, slurred *jiew* and a scolding *chit-it-it-it-it-it-it*... **H** Broadleaved evergreen forest, secondary growth, scrub and bamboo. **D** Himalayas to S and SE China, N SE Asia.

32 BLACK-HEADED PARROTBILL *Psittiparus margaritae* 15–16cm Black cap and indistinct dark cheek patch. Forages from shrub level up to the canopy. **V** Husky *jhu'jhu'jhu* or *eu'ju'jhu'jhu*. Calls similar to Grey-headed Parrotbill. **H** Broadleaved evergreen forest, secondary growth, scrub. **D** S Vietnam, E Cambodia.

33 BLACK-BREASTED PARROTBILL *Paradoxornis flavirostris* 19cm Forages low down, but regularly sings from the top of reeds or grass stems; usually occurs in small parties. **V** Clear, high *woi-woi-woi-woi-woi-woi, whii-whii-whii-whii* or a huskier *jhor-jhor-jhor-jhor-jhor-jhor*. Calls include a *wu-wi-wi* and a nasal *uh-uh-uh-uh-uh-uh*. **H** Reeds and tall grass. **D** NE India.

34 SPOT-BREASTED PARROTBILL *Paradoxornis guttaticollis* 19cm Separated from Black-breasted Parrotbill by spotted, not black, chin and breast. Noisy, usually in small parties, which may contain other species. **V** Staccato *whit-whit-whit-whit*... Calls include a low *ruk-ruk, ruk-uk-uk* or *rut-rut-rut-rut*. **H** Secondary growth, scrub, tall grass. **D** NE India through N SE Asia to S China.

35 REED PARROTBILL *Paradoxornis heudei* 18–20cm Note grey head with black supercilium. Forages in reedbeds in small or large flocks. **V** Long series of *chut, chu, hui* or *tui* notes, delivered at varying rates. Calls include a gentle, low *u uui-uui-uui-uui*, a rapid thin chirping and a subdued *jhew-jhew*. **H** Reedbeds by rivers or estuaries. **D** E Mongolia to NE China, SE Russia.

2
4
5
1
3
11
6
8
9
7
10
16
14
13
12
15
18
17
19
20
21
22
27
24
25
23
33
26
32
29
35
28
30
31
34

WHITE-EYES ZOSTEROPIDAE

1 STRIATED YUHINA *Yuhina castaniceps* 13cm Agile forager in the foliage of high bushes and low trees; often in parties of up to 30 birds. Crest may be brown or grey, depending on race, but ear-coverts always warm brown. **V** Series of high-pitched, shrill *tchu, tchi* or *tchi-chi* notes. Flocks utter an incessant twittering or cheeping. **H** Undergrowth, scrub and understorey in broadleaved evergreen forest. **D** Himalayas to NW Thailand.

2 INDOCHINESE YUHINA *Yuhina torqueola* 14–15cm White-spotted brown of ear-coverts extends around nape. Forages in high bushes and low trees. **V** Loud *tu'whi*. Calls include a continuous chattering, high squeaks and dry trills. **H** Scrub and undergrowth in broadleaf forest. **D** S China, N Thailand, Laos, N Vietnam.

3 CHESTNUT-CRESTED YUHINA *Yuhina everetti* 14cm Usually found in fast-moving flocks, foraging in the tops of small or large trees. **V** Calls include a low, quick *whit whit whit* or similar, and a rattling *chr r r r r t*; flocks produce a hurried chatter. **H** Submontane and montane broadleaved evergreen forest, moss forest, forest edge, secondary growth. **D** Borneo.

4 WHITE-NAPED YUHINA *Yuhina bakeri* 13cm Note prominent crest and white nape. Forages in bushes and treetops in small parties; post breeding occurs in larger flocks, often with other species. **V** Series of high thin notes, repeated every few seconds. Calls include a short metallic *tsit*, a falling *seep* and a *seet-chuut*. **H** Broadleaved evergreen forest, secondary growth. **D** E Himalayas to N Myanmar.

5 WHISKERED YUHINA *Yuhina flavicollis* 13cm Prominent crest is grey and blackish brown. Actions tit-like, also fly-catches from bush tops; forages in small noisy parties, often mixed with tits, warblers, nuthatches and small babblers. **V** Repeated high-pitched *tzii-jhu ziddi* or *twe-tyurwi-tyawi-tyawa*. Calls include a thin squeaky *swii-swii-swii*, a buzzy *jhoh* or *fzee-tzzup jhoh*. **H** Broadleaved evergreen forest, oak and open deciduous forest, secondary growth. **D** Himalayas to C Laos.

6 BURMESE YUHINA *Yuhina humilis* 12–13cm Lacks warm brown colouring. Forages in pairs or small groups, often in flowering trees. **V** Flocks utter a low *chuck-chuck* or *chir-chir-chir-chir*. **H** Broadleaved evergreen forest, forest edge, secondary growth. **D** E Myanmar, N Thailand, C Laos.

7 STRIPE-THROATED YUHINA *Yuhina gularis* 14cm Arboreal, actions tit-like but slower; usually encountered in small parties or in mixed-species foraging flocks. **V** Descending, nasal, mewing *mher* or *wher*, which is sometimes followed by a hurried *whu-whu-whu-whi-whi-whi*; also utters a short *wiht*. **H** Temperate oak, birch and rhododendron forest or mixed rhododendron and conifer forest, also bamboo and low scrub. **D** Himalayas to S China, Vietnam.

8 WHITE-COLLARED YUHINA *Yuhina diademata* 14–18cm Forages in bushes and treetops, usually in parties, with larger parties post breeding. **V** Hurried, high *tsu'tsu'tsu, tsu'tst* or *tsu'tsu'tsut*. Calls include a *seep* and a *seet-chuut*. **H** Primary and secondary broadleaf evergreen forest. **D** S China, N Myanmar, N Vietnam.

9 RUFOUS-VENTED YUHINA *Yuhina occipitalis* 13cm Note rufous-orange on back of crest, and rufous undertail coverts. Forages in the canopy and on moss-covered trunks and branches. **V** High-pitched *swi-si-si-su-su swi-si-si-si-si-sisu-su-su...* Calls include buzzy and nasal notes, *bee, beebee* or *bzzzee*. **H** Broadleaved evergreen forest, especially oak and rhododendron. **D** C Himalayas through N Myanmar to S China.

10 TAIWAN YUHINA *Yuhina brunneiceps* 11–12cm Head pattern is very distinctive. Forages from low strata to canopy. **V** Jaunty *to-meet-you* or *so-pleased-to-meet-you*. **H** Forests, forest edge and clearings. **D** Taiwan.

11 BLACK-CHINNED YUHINA *Yuhina nigrimenta* 11cm Black in crest more noticeable than black chin. Active, agile, noisy forager in the canopy of tall trees and in low shrubs. **V** Clear, thin sequence of ringing whistles. Calls include a staccato chattering and a metallic *pik pik pik* that is often mixed with other notes. **H** Broadleaved evergreen forest, secondary jungle and overgrown clearings. **D** Himalayas to SE China, N SE Asia.

12 CHESTNUT-FACED BABBLER *Zosterornis whiteheadi* 15cm Generally forages from low down up to the canopy; regularly seen in small groups or as part of mixed-species feeding parties. **V** Busy, twittering *chip chip chip chip*, interspersed with *pe-chu* notes. **H** Forest, secondary growth, scrub, tangled high grass and small mountainside trees. **D** N Philippines.

13 LUZON STRIPED BABBLER *Zosterornis striatus* 13–14cm Heavily streaked underparts, white eye-ring. Forages slowly and methodically from the lower to upper storey. **V** High-pitched *tsi* notes, accelerating to a trill and ending with a *zeeep zep*. Calls include a staccato, trilled *tiptiptiptiptiptip*. **H** Primary, secondary, logged and bamboo forest, forest edge, secondary growth, overgrown clearings. **D** N Philippines.

14 PANAY STRIPED BABBLER *Zosterornis latistriatus* 15cm Note extensive white on face. Generally forages low down. **V** Ascending, staccato, trilling *chi chi-chi-chi-chi...*, occasionally interspersed with sharp *tsik* notes or chattering. **H** Montane evergreen and mossy forest. **D** WC Philippines.

15 NEGROS STRIPED BABBLER *Zosterornis nigrorum* 14–15cm Similar to Panay Striped Babbler, but note black extending from forehead over eye. Forages mainly in understorey bushes and trees. **V** Whistled *pli-hi pli-hy pli-hy* or *plea-he plea-he plea-he plea-he-plea-hu*. Calls include a soft *tsip tsip*, a short *tzi* and a high *weeet*. **H** Montane, secondary and degraded forest, forest edges, adjacent plantations. **D** WC Philippines.

16 PALAWAN STRIPED BABBLER *Zosterornis hypogrammicus* 14–15cm Like other striped babblers, but with orange-buff crown. Forages mainly in the canopy, gleaning insects from leaves. **V** Distinctive *zeep zeep zeep zeep*, also trills and bubbling notes. **H** Montane evergreen forest. **D** Palawan, W Philippines.

17 GIANT WHITE-EYE *Megazosterops palauensis* 14cm Note yellow bill (with dusky culmen) and greyish, finely white-streaked cheeks. **V** Song is a peculiar combination of thin downslurred and upslurred whistles and rhythmic rattling, giving impression of two birds singing at once. Calls include harsh chatter, grating notes and a short downslurred whistle. **H** Forest, thickets. **D** Palau, W Pacific.

18 BONIN WHITE-EYE *Apalopteron familiare* 13.5cm Note distinctive triangular black eye patch. Feeds on fruits of papaya, acacia, etc., and various flowers. **V** Melodious warble. Calls include a soft *pee-you, weet* and *pit*, an explosive *tit-tit*, a harsh *weet-weet* and a *zhree-zhree*. **H** Secondary forest, forest edge, bushes, plantations, gardens. **D** Japanese Bonin Islands, NW Pacific.

19 GOLDEN WHITE-EYE *Cleptornis marchei* 14cm Unmistakable in restricted range. **V** Song is a protracted warble of repeated phrases, *can-you-seeeee-meeeee*. Calls include strident twangs and whistling notes. **H** Any wooded habitat. **D** Northern Mariana Islands, NW Pacific.

20 TEARDROP WHITE-EYE *Rukia ruki* 14cm Note orange legs. Eye-ring reduced to white segment under eye. **V** Song is a loud warbled phrase. **H** Canopy at edge of dense forest. **D** Caroline Islands, W Pacific.

21 LONG-BILLED WHITE-EYE *Rukia longirostra* 15cm Note long, decurved bill and narrow eye-ring. **V** Song begins with downslurred whistle, followed by a short warbling phrase and a dry chattering phrase. Contact calls are downslurred whistles with a nasal tone. **H** Undergrowth of palm and broadleaf forest. **D** Caroline Islands, W Pacific.

22 FLAME-TEMPLED BABBLER *Dasycrotapha speciosa* 13cm Note yellow, orange, black and white head pattern. Forages in undergrowth, understorey bushes, vine tangles and the middle storey. **V** Pleasant, slurred song that starts faintly and then gets louder. Calls include a soft *yir, ju, chu* or *ju-jrrr*. **H** Primary, secondary and degraded secondary forest. **D** WC Philippines.

23 MINDANAO PYGMY BABBLER *Dasycrotapha plateni* 10cm Occurs from the middle storey up to the lower canopy; acrobatic forager, occasionally making fly-catching sallies. **V** Calls include a *tchik*, a *chik-chik-chik-chik-chik*, and a quiet *tsieu-tsieu...* **H** Primary and secondary forest, forest edge, secondary growth, fruiting trees in abandoned cultivation. **D** S Philippines.

24 VISAYAN PYGMY BABBLER *Dasycrotapha pygmaea* 10cm Less rufous than Mindanao Pygmy Babbler. Forages in the middle storey or occasionally higher. **V** Unknown; may be similar to Mindanao Pygmy Babbler. **H** Forest, forest edge, secondary growth. **D** EC Philippines.

25 GOLDEN-CROWNED BABBLER *Sterrhoptilus dennistouni* 13–14cm Black bill contrasts with yellow forehead and throat. Forages alone, in pairs or in small parties, usually fairly low down. **V** Noisy, fast *pilit-pilit-pilit poo poo poo poo poo wi-wi-wi-wi-prr-prr-prrt*, and a quiet *wit-wit, pi* or *pri-pri-pri*. **H** Primary forest, logged and degraded forest, forest edges, bamboo and areas of tall grass. **D** N Philippines.

26 BLACK-CROWNED BABBLER *Sterrhoptilus nigrocapitatus* 13–14cm A slow, methodical, acrobatic forager in the lower branches of trees. **V** Level, descending *wi-wi-wi-wu-wiu-wiu* or similar. Calls include a quiet *whit, whut* or *wue wue*. **H** Undergrowth in forest and forest edges. **D** C Philippines.

27 RUSTY-CROWNED BABBLER *Sterrhoptilus capitalis* 14–15cm Note bold whitish streaks on head and upperparts, rusty crown and throat. Forages fairly low down. Acrobatic gleaner, although often sits motionless. **V** Soft, bubbling *whit whit chew-chew-chew*. **H** Primary and logged forest, and forest edges. **D** EC and S Philippines.

28 RUFESCENT DARKEYE *Tephrozosterops stalkeri* 12–13cm Forages alone, in pairs or in mixed-species parties. **V** Unrecorded. **H** Dense secondary growth at forest edges or in clearings, overgrown cultivation and scrub. **D** Seram, S Maluku Islands, Indonesia.

29 GREY-HOODED WHITE-EYE *Lophozosterops pinaiae* 14cm A large white-eye with a prominent eye-ring. Forages in the thick foliage of tree crowns and among dense arboreal epiphytes, often in mixed-species flocks. **V** Unknown. **H** Montane forest. **D** Seram, S Maluku Islands, Indonesia.

30 MINDANAO WHITE-EYE *Lophozosterops goodfellowi* 13–14cm Note white throat. Forages at all levels of the forest, in small parties or mixed-species flocks. **V** Musical, whistled *tu-pik, chu-beer* or *su si deer*. **H** Submontane and montane forest, forest edge. **D** S Philippines.

31 STREAK-HEADED WHITE-EYE *Lophozosterops squamiceps* 12cm Forages from the understorey to the canopy, in pairs or small groups, and as part of mixed-species flocks. **V** Repeated, warbled series of loud, clear, sweet, sibilant, high-pitched notes; also a harsh, chirruping trill. **H** Primary montane forest, forest edges and secondary growth. **D** Sulawesi, Indonesia.

32 MEES'S WHITE-EYE *Lophozosterops javanicus* 13cm Mainly found in the canopy or high up in forest edges, in pairs, small flocks or as part of mixed-species groups. **V** Song comprises melodious whistles and call notes. Calls include a long, drawn-out, high cheeping, and a throaty *turr, teerrr-teerrr* or *chee-ee-wheet-chee-ee-weeeet*. **H** Forest, dense secondary growth and neglected plantations. **D** Java and Bali, Indonesia.

33 CREAM-BROWED WHITE-EYE *Lophozosterops superciliaris* 13cm Yellowish supercilium extends from eye-ring. Generally encountered foraging in small groups, usually in the mid-storey. **V** Rapid series of high-pitched, bubbling, trilled and warbled notes, occasionally interspersed with *tchee-tchee* notes; also utters a ringing *peu-peu*. **H** Upper montane primary forest, semi-evergreen rainforest, *Casuarina* forest, logged and degraded forest, forest edges, thin secondary growth and scrub. **D** Sumbawa and Flores, Indonesia.

34 CRESTED WHITE-EYE *Lophozosterops dohertyi* 12cm Favours the understorey and dense scrub, alone, in pairs or in small groups; also joins mixed-species feeding parties. **V** Moderately rapid series of 14 clear, sweet whistles; also gives a soft *tsip-tsip*. **H** Moist primary forest, degraded forest, tall secondary forest, lightly wooded cultivation and scrub. **D** Sumbawa, Satonada and Flores, Indonesia.

35 SPOT-BREASTED HELEIA *Heleia muelleri* 13–14cm Usually found in pairs or compact, small groups; also joins mixed-species foraging flocks. **V** Unmusical rattle of 10–20 rapid, mechanical notes, initially rising and then falling; also gives a harsh, weak, grating noise. **H** Primary and tall secondary monsoon and evergreen forest. **D** Timor, Lesser Sunda Islands.

36 THICK-BILLED HELEIA *Heleia crassirostris* 13–14cm Generally encountered in small groups, alone or in pairs, foraging from the understorey to the lower middle storey. Often joins mixed-species flocks. **V** Rapid series of loud whistles and trilled notes on an even pitch. Calls include a deep, quiet *chup... chup...* **H** Primary and degraded semi-evergreen rainforest and moist deciduous monsoon forest, forest edges and scrub. **D** Sumbawa and Flores, Indonesia.

37 PYGMY WHITE-EYE *Oculocincta squamifrons* 9–10cm Fast-moving small groups forage in the crowns of tall trees, and in the middle level at forest edges. **V** A *chit-chit-chit* or high-pitched *tsee-tsee...* **H** Hill and lower montane forest, forest edges, secondary growth, dense scrub and sandy forest with dwarf vegetation. **D** Borneo.

1
2
3
4
5
6
7
8
9
10
11
12
13
14
15
16
17
18
19
20
21
22
23
24
25
26
27
28
29
30
31
32
33
34
35
36
37

WHITE-EYES *CONTINUED*

1 BARE-EYED WHITE-EYE *Woodfordia superciliosa* 14cm Dumpy, grey-brown white-eye with long horn-coloured bill and short tail. Note white line surrounding face mask. Usually in small groups. **V** Song is a short fast warble. Flocks make soft whistled contact notes. **H** Forest, forest edge, gardens. **D** Rennell (Solomon Islands).

2 SANFORD'S WHITE-EYE *Woodfordia lacertosa* 15cm Dumpy, rufous-brown white-eye with paler face and underparts. Note long yellowish bill and silvery eye-ring. **V** Song not known. Calls with harsh nasal notes. **H** Primary and secondary forest. **D** Nendö (Solomon Islands).

3 MOUNTAIN BLACKEYE *Chlorocharis emiliae* 11–12cm Olive-green, with black lores and black eye-ring. Forages from near the ground to treetops, in pairs or in small flocks. **V** Melodious *wit-a-wit wit wit wheer*. Calls include a twittering *tut-u*, *stweet-u* or *te-wio*; in flight, utters jangling notes and a stuttering *gujuguju*. **H** Montane moss forest and stunted growth at higher elevations. **D** Borneo.

4 FERNANDO PO SPEIROPS *Zosterops brunneus* 12cm Unmistakable in very limited range. **V** Calls include a soft *peep*, a fast *trik-trik-trik* and a drawn-out, trilling *trrrrruuu*. **H** Forest clearings, heathland scrub, open woodland. **D** Bioko (Gulf of Guinea).

5 FOREST WHITE-EYE *Zosterops stenocricotus* 10cm Formerly treated as a race of African Yellow White-eye (Plate 232). **V** Similar to African Yellow White-eye (Plate 232) but lacks buzzy quality. **H** Forest edge and interior, giant heath, all types of woodland, suburban gardens. **D** SE Nigeria to SW Central African Republic, Equatorial Guinea, N Gabon.

6 PRINCIPE SPEIROPS *Zosterops leucophaeus* 13cm Pale grey head contrasts with dark grey upperparts. **V** Not well known. Possible song is a fast, tuneless rattled phrase; calls include trills and rattles. **H** Forest, bush, plantations. **D** Príncipe (Gulf of Guinea).

7 BLACK-CAPPED SPEIROPS *Zosterops lugubris* 14cm Dark grey with contrasting black cap and white eye-ring. In pairs or small flocks, often in mixed-species flocks. **V** Song is a loud warble. Calls include fast twitterings and trills. **H** All wooded habitats, at all altitudes. **D** São Tomé (Gulf of Guinea).

8 MOUNT CAMEROON SPEIROPS *Zosterops melanocephalus* 13cm Note dark cap and white at base of bill. **V** Song consists of loud, staccato, toneless notes in a 2-sec phrase; calls include rattled *trrrr* and a soft cheeping. **H** In all areas with trees. **D** Mount Cameroon (SW Cameroon).

9 CHESTNUT-FLANKED WHITE-EYE *Zosterops erythropleurus* 11–12cm Chestnut flanks are distinctive, but may be rather faint. Gregarious, often in mixed-species feeding flocks along with Japanese White-eye. **V** Twittering *dze-dze*. **H** Broadleaved evergreen forest and secondary forest. **D** Breeds NE China, extreme SE Russia. Winters SE Asia.

10 JAPANESE WHITE-EYE *Zosterops japonicus* 10–12cm Gregarious, restless forager in treetop foliage; often part of mixed-species feeding flocks. Note that the classification of many of the *Zosterops* white-eyes is in a state of flux. **V** Calls include a staccato *tsip-tsip-tsip*, a *chi-i chi-i*, a *jeet-jeet* and a cicada-like trill. **H** Forest, secondary growth, cultivations, parks and gardens. **D** Breeds Sakhalin, Japan, Korean Peninsula, N and E China, Taiwan, Hainan, N Vietnam. Winter visitor across SE Asia.

11 LOWLAND WHITE-EYE *Zosterops meyeni* 10–12cm Noisy. Forages in groups or mixed-species flocks. **V** A *swit* or *swit-tzee* and a complicated series of twittering and wheezy notes. **H** Forest, forest edge, bamboo thickets, scrub, cultivated areas, gardens. **D** N Philippines, Taiwan.

12 ORIENTAL WHITE-EYE *Zosterops palpebrosus* 9–11cm Restless, agile forager among tree foliage and blossoms. Often part of mixed-species feeding parties. **V** Wispy, made up of slurred call notes. Calls include a twittering *dzi-da-da* and a raspy, down-slurred *djeeeer*. **H** Broadleaved evergreen, deciduous and swamp forest, secondary growth, mangroves, cultivation, parks and gardens. **D** Widespread from Afghanistan through India, S China and SE Asia to Indonesia.

13 SRI LANKA WHITE-EYE *Zosterops ceylonensis* 11cm Restless, agile forager in bushes, undergrowth and trees; generally found in pairs with larger, scattered parties after breeding. **V** Jingling song, said to sound like the shaking of a bunch of keys. Calls include a constant *cheep* or a reedy *chisip*. **H** Forests and nearby plantations and gardens. **D** Sri Lanka.

14 ROTA WHITE-EYE *Zosterops rotensis* 10cm Entire underparts are yellow. Bill mainly yellowish-brown. **V** Song not known. Calls include a rather sparrow-like *tshrip*. **H** Native forest with tall trees in hills. **D** Rota (Northern Mariana Islands, NW Pacific).

15 BRIDLED WHITE-EYE *Zosterops conspicillatus* 10cm Note combination of pale throat and yellow underparts. **V** Song combines buzzing *zip* and *zeep* notes in a lilting series. Calls include sparrow-like chirps. **H** Forest, thickets. **D** Mariana Islands, NW Pacific.

16 CITRINE WHITE-EYE *Zosterops semperi* 10cm Note white in front of eye and all-yellow underparts. Eyes may be light chestnut. **V** Calls include a high-pitched, clear whistling *tee dee, tee dee, tee dee* and a squeaky note. **H** Forest, scrub. **D** Palau and Caroline Islands, W Pacific.

17 PLAIN WHITE-EYE *Zosterops hypolais* 10cm Pale eye with very narrow eye-ring. Rather plain with seemingly large head. **V** Song is a series of short chirping phrases at a slowish pace. Calls include a thin *chee* and a buzzing *zee-up*. **H** From forest canopy to grassy fields. **D** Caroline Islands, W Pacific.

18 ENGGANO WHITE-EYE *Zosterops salvadorii* 10cm Seen foraging in small parties; little other information. **V** Reported as similar to that of the Oriental White-eye. **H** Lowland wooded areas, coconut groves. **D** Enggano, off SW Sumatra, Indonesia.

19 BLACK-CAPPED WHITE-EYE *Zosterops atricapilla* 9–10cm Front of face black, breast and undertail coverts yellow. Acrobatic forager in trees and bushes; regularly encountered in small flocks and as a member of mixed-species feeding flocks. **V** Calls include a trembling note and a loud, twinkling series of notes. **H** Hill and lower montane forest, montane bush, alpine ericaceous meadows. **D** Borneo; Sumatra, Indonesia.

20 EVERETT'S WHITE-EYE *Zosterops everetti* 11–12cm Gregarious, usually in flocks of 5–20 individuals, although higher numbers reported; forages in treetops. **V** Series of 8–13 metallic, raspy, whistled notes and a series of sweet, weak, twittering notes. Calls include a *tsee-tsee*, a metallic *spreet* or *peeet* and a buzzing *dzee* or *dzee-ap*. **H** Evergreen forest. **D** Thailand to Philippines.

21 YELLOWISH WHITE-EYE *Zosterops nigrorum* 10–12cm Note yellow patch on lores. Forages in noisy flocks of up 20 individuals or in mixed-species parties. **V** Series of *pit-it tit*, *sit-it sit-it* or *sip-it sip-it* notes. **H** Forest, forest edges, secondary growth, clearings and lower storeys of dipterocarp forest. **D** Philippines.

22 MOUNTAIN WHITE-EYE *Zosterops montanus* 11–12cm Gregarious, often seen in small or large flocks foraging in the upper tree levels. **V** Variable: melodious song in Sumatra, a *peet-peet* followed by a trill in Sulawesi. Contact calls are soft and high-pitched. **H** Primary montane forest, forest edges, secondary growth, *Casuarina* stands and wooded cultivation. **D** Philippines; Sumatra and Sulawesi, Indonesia.

23 YELLOW-RINGED WHITE-EYE *Zosterops wallacei* 11–12cm Note orange-yellow forehead. Forages in the middle and upper levels, singly, in pairs or in small flocks. **V** Variable: on Sumba, a descending, tinkling warble; on Komodo, a rapid series of warbled notes, starting with two short notes, a single upslurred whistle and a jumble of high-pitched sweet notes; on Flores, two short insect-like notes followed by a series of warbled notes. **H** Primary and secondary forest, forest edges, dry scrub, cultivation. **D** Lesser Sunda Islands.

24 JAVAN WHITE-EYE *Zosterops flavus* 9–10cm Arboreal; forages in tight gleaning groups. **V** Calls include a *trrieew*, a short, soft *trrip* and a *wiwiwiwi* when alarmed. **H** Mangroves, coastal forest edge, low waterside trees, groves, scrub, gardens. **D** Borneo; Java, Indonesia.

25 LEMON-BELLIED WHITE-EYE *Zosterops chloris* 11–12cm Generally encountered in small flocks; restless forager at all levels. **V** Mix of rich, beautiful, high-pitched, seesawing notes, along with short *si-si* notes, repeated, often monotonously. Calls include a *shilp* or *chiew*. **H** Secondary forest, open woodland, scrub, mangroves, coastal woodland and thickets, cultivations, gardens. **D** Sulawesi, Lesser Sunda Islands and Maluku Islands.

26 ASHY-BELLIED WHITE-EYE *Zosterops citrinella* 10–11cm Belly almost pure white. Active, foraging in the foliage of the outer canopy in small groups or as a member of mixed-species flocks. **V** Series of weak, high-pitched, twittering notes, interspersed with rapidly repeated warbles, trills and slurs. **H** Primary and secondary forest, degraded forest, forest edge, open woodland, monsoon forest, mangroves, coastal woodland, cultivations, scrub. **D** Lesser Sunda Islands, Torres Strait islands.

27 PALE-BELLIED WHITE-EYE *Zosterops consobrinorum* 11–12cm Forages mainly in the understorey and forest-edge thickets, and in the canopy of tall trees, usually in pairs, small groups or mixed-species flocks. **V** Attractive, pleasing song that lacks trilling notes. **H** Remnant patches of lowland forest, forest edges, forest plantations, cultivations, scrub, gardens. **D** Sulawesi, Indonesia.

28 PEARL-BELLIED WHITE-EYE *Zosterops grayi* 13cm Forages singly, in pairs and in mixed-species feeding parties in the mid-storey. **V** Series of strident, squeaky, chattering notes interspersed with three sharp, high-pitched notes. Calls include a series of rapid, unmusical chattering or bubbling notes, and a *pipip* and *trrr*. **H** Primary and secondary forest, open woodland, gardens. **D** Kai Islands, S Indonesia.

29 GOLDEN-BELLIED WHITE-EYE *Zosterops uropygialis* 12–13cm Distinguished from Pearl-bellied White-eye by dusky forehead, almost invisible eye-ring, bright yellow underparts. Encountered in pairs or small parties; gleans and hover-gleans in the foliage of the canopy. **V** Calls include squeaky, nasal chatters; harsh, rasping squeaks; short, bubbling notes; and a mellow, nasal *chow*. **H** Forest and cleared areas with scattered trees. **D** Kai Islands, S Indonesia.

30 BLACK-RINGED WHITE-EYE *Zosterops anomalus* 12cm Eye-ring is black, not white. Found singly, in pairs, in small groups or in mixed-species parties, from understorey vegetation to the canopy. **V** Muted series of teetering, chattering notes and a *chewchicheruit-chewticheru-i-uu rrr*. Calls with a quivering whistle. **H** Scrubby deforested hills, secondary forest, forest edge, orchards, gardens. **D** Sulawesi, Indonesia.

31 CREAM-THROATED WHITE-EYE *Zosterops atriceps* 12cm Dark face; white throat and belly. Relatively skulking; forages in thickets, the mid-storey or the canopy. **V** Thin, sweet, whistled notes, alternating up and down, ending with 1–2 *tu-wit* notes. **H** Lowland and hill primary and secondary forest, forest edge, cultivations. **D** Maluku Islands, Indonesia.

32 SANGIHE WHITE-EYE *Zosterops nehrkorni* 10–12cm Forages in small parties in dense canopy and subcanopy foliage. **V** Thin, tinkling and trailing away. Calls include a *swiit... swiit... swiit*. **H** Primary broadleaved ridgetop forest, with dense *Pandanus*. **D** Sangihe Islands, off Sulawesi, Indonesia.

33 BLACK-CROWNED WHITE-EYE *Zosterops atrifrons* 11–12cm Forages from low down up to the canopy, in small groups or sometimes in much larger parties. **V** Variable: a shrill warble, a rapid series of high-pitched sweet notes and a clear, high-pitched series of rolling, sweet, whistled notes. Calls include a high-pitched *peee*, *tiu* or *teew*. **H** Primary and secondary lowland and hill forest, logged and degraded forest, forest edge. **D** Sulawesi, Banggai and Sula islands, Indonesia.

34 TOGIAN WHITE-EYE *Zosterops somadikartai* 11cm Note grey eye-ring. Forages in dense, low shrubs, generally in pairs or small flocks. **V** Thin, sweet warble; moving flocks utter twittering chirrups. **H** Low bushes near mangroves, coconut groves, secondary scrub in logged forest, gardens. **D** Togian Islands, off NW Sulawesi, Indonesia.

35 SERAM WHITE-EYE *Zosterops stalkeri* 12–13cm Forages alone, in pairs or in mixed-species parties. **V** Unrecorded. **H** Dense secondary growth at forest edges or in clearings, overgrown cultivation and scrub. **D** Seram (S Maluku Islands, Indonesia).

36 BLACK-FRONTED WHITE-EYE *Zosterops minor* 11cm Note yellow throat and undertail-coverts in all races. Some races have black forehead and width of eye-ring varies. **V** Song is a sweet descending warble, but may vary locally. **H** Hill forest and forest edge. **D** New Guinea.

37 TAGULA WHITE-EYE *Zosterops meeki* 11cm Olive-green above with blackish forecrown. Note white throat. Only white-eye on Tagula. **V** Poorly known. Call is a soft *sueep*. **H** Forest. **D** Tagula (Louisiade Archipelago, SE Papua New Guinea).

1
2
3
4
5
6
7
8
9
10
11
12
13
14
15
16
17
18
19
20
21
22
23
24
25
26
27
28
29
30
31
32
33
34
35
36
37

WHITE-EYES *CONTINUED*

1 BISMARCK WHITE-EYE *Zosterops hypoxanthus* 11cm Olive-green above and yellow below with blackish or brownish hood. In pairs or small flocks; often in mixed-species flocks. **V** Song is a simple sweet warble; also gives nasal contact calls. **H** Forest, forest edge, secondary growth and scrub, from lowlands to mountains. **D** Bismarck Archipelago.

2 BIAK WHITE-EYE *Zosterops mysorensis* 11cm Only white-eye in New Guinea that lacks an eye-ring. Note longish bill and white throat contrasting with greyish underparts. Joins mixed-species flocks. **V** Call is a thin downslurred note, typical of the genus. **H** Lowland forest. **D** Biak, NW New Guinea.

3 CAPPED WHITE-EYE *Zosterops fuscicapilla* 11cm Note yellowish-olive underparts and blackish forecrown and face. Usually in flocks. **V** Sparrow-like chirping notes and trills. Call is a thin downslurred note. **H** Lower montane forest. **D** W and C New Guinea.

4 BURU WHITE-EYE *Zosterops buruensis* 11–12cm Note small yellow patch on lores. Forages in mixed-species flocks in dense foliage in the canopy or subcanopy. **V** While foraging, utters a quiet, quickly repeated *tsu-tsu-tsu-tsu-tsu*; also a *chewit chewit chewit*. **H** Primary and selectively logged forest, secondary growth, scrub. **D** Buru (S Maluku Islands, Indonesia).

5 AMBON WHITE-EYE *Zosterops kuehni* 12cm Active forager in tree crowns, especially when in flower; occurs in pairs or small groups. **V** Song is a musical warble. Calls with a sibilant *teeu*. **H** Lowland forest, remnant patches of secondary forest and woodland, lightly wooded cultivation, scrub, gardens. **D** Ambon and Seram (S Maluku Islands, Indonesia).

6 PAPUAN WHITE-EYE *Zosterops novaeguineae* 11cm Seven races vary in minor plumage details and in width of eye-ring. Note yellow throat and dark lores. Usually in pairs or flocks; often joins mixed-species flocks. **V** Song is a descending series of 6–7 pure notes. Call is a rather squeaky single note. **H** Mid-montane forest, forest edge, secondary growth, gardens. **D** New Guinea.

7 YELLOW-THROATED WHITE-EYE *Zosterops metcalfii* 11cm Note yellow throat and narrow white eye-ring (race *floridanus* has no eye-ring). Usually in pairs or flocks; often joins mixed-species flocks.**V** Song is a melodious series of 6–9 notes. Call is a loud *tchirup*. **H** Forest, secondary growth and gardens from lowlands to hills. **D** Bougainville, Choiseul, Santa Isabel and Nggela Island (Solomon Islands).

8 CHRISTMAS WHITE-EYE *Zosterops natalis* 12–14cm Ear-coverts are grey. Forages from low levels to the canopy; gregarious. **V** Song is *yerr yerr weet yerr yerr tyerr weet...* Calls include twittering and chirping notes, a thin *ts-ee-sect... tsee-eet...*, a *tsirr-tsirr* when alarmed, and a scolding *cheeuw cheeuw cheeuw*. **H** Woodland, shrubs, forest edges, open country with trees and bushes. **D** Christmas Island, E Indian Ocean.

9 CANARY WHITE-EYE *Zosterops luteus* 11cm Entire underparts bright yellow. Note yellow supraloral band. In pairs or small flocks.**V** Song is a melodious medley of whistles and trills. **H** Mangroves and riverine vegetation in coastal areas. **D** N Australia.

10 LOUISIADE WHITE-EYE *Zosterops griseotinctus* 12cm Rather plain with prominent eye-ring. Four races, varying in plumage tones and colour of bill and legs. Usually in small flocks; joins mixed-species flocks. **V** Song is a loud melodious warble. Call is a musical trill. **H** All types of forest including plantations and gardens. **D** Louisiade Archipelago (SE Papua New Guinea) and Bismarck Archipelago

11 RENNELL WHITE-EYE *Zosterops rennellianus* 12cm Plain olive white-eye without a white eye-ring. Bill and legs dull orange. Note creamy axillaries. In pairs or small flocks, sometimes with mixed-species flocks. **V** Rarely heard song is a fast trill or warble. Chattering call notes are rather squeaky. **H** Forest, forest edge, secondary growth and scrub. **D** Rennell (Solomon Islands).

12 VELLA LAVELLA WHITE-EYE *Zosterops vellalavella* 12cm Note yellow throat, broad white eye-ring and olive breast-band. Bill and legs dull orange-yellow. In pairs or small flocks in the canopy; sometimes joins mixed-species flocks. **V** Song is a series of downslurred notes. Call is a high-pitched *tse-tse*. **H** Forest, forest edge, secondary growth and plantations. **D** Vella Lavella (Solomon Islands).

13 GIZO WHITE-EYE *Zosterops luteirostris* 12cm Underparts entirely yellow with green wash on breast and flanks. Bill and legs bright orange-yellow. In pairs or small flocks in the canopy; sometimes joins mixed-species flocks. **V** Song is a slow series of loud notes, more melodious than other white-eyes. Call is a loud *sweep!* **H** Forest edge, secondary growth and plantations. **D** Gizo (Solomon Islands).

14 RANONGGA WHITE-EYE *Zosterops splendidus* 12cm Resembles Gizo White-eye but bill is black and eye-ring is broader. In pairs or small groups. **V** Song is a simple series of 3–10 clear notes. Call is a simple tsieu. **H** Forest, forest edge and secondary growth. **D** Ranongga (Solomon Islands).

15 SOLOMONS WHITE-EYE *Zosterops kulambangrae* 12cm Plain olive white-eye with narrow white eye-ring and yellowish legs. In pairs or small flocks; often joins mixed-species flocks. **V** Song is a descending series of downslurred notes. Call is a sparrow-like chirrup. **H** Forest, forest edge and gardens, in lowlands and hills. **D** New Georgia (Solomon Islands).

16 DARK-EYED WHITE-EYE *Zosterops tetiparius* 12cm Formerly treated as a race of Solomons White-eye. Dark skin around eye (no white eye-ring). Belly either whitish or yellow. In pairs or small flocks; often joins mixed-species flocks. **V** Song is similar to Solomons White-eye, but is less musical. **H** Forest, forest edge and gardens. **D** Tetepare and Rendova (Solomon Islands).

17 KOLOMBANGARA WHITE-EYE *Zosterops murphyi* 12cm Plain olive white-eye with very broad white eye-ring. Bill and legs dark. In pairs or flocks, sometimes quite large. **V** Song is a short warbling of pure, sweet notes. Flocks give a constant chattering. **H** Montane forest. **D** Kolombangara (Solomon Islands).

18 GREY-THROATED WHITE-EYE *Zosterops rendovae* 12cm Olive above and silvery-grey below. Three races, varying in plumage details. Two races have yellow undertail-coverts (white on Makira). Race *hamlini* on Bougainville has greenish throat and narrow white eye-ring. In pairs or small flocks; often joins mixed-species flocks. **V** Song is a rolling series of mellow notes. Flocks call with loud *cheep* notes. **H** Forest, forest edge and secondary growth, mainly in hills and mountains. **D** Bougainville, Guadalcanal and Makira (Solomon Islands).

19 MALAITA WHITE-EYE *Zosterops stresemanni* 13cm Plain olive-green with yellow wash on throat, belly and vent. No white eye-ring. Note stout pale bill. Small flocks sometimes join mixed-species flocks. **V** Song is a melodious thrush-like warble. Call is a high-pitched *tsip*. **H** Forest edge, secondary growth, bushes and gardens, mainly in hills and mountains. **D** Malaita (Solomon Islands).

20 SANTA CRUZ WHITE-EYE *Zosterops sanctaecrucis* 12cm Plain olive white-eye with dark bill, lores and eye-ring, creating small mask. In pairs or small flocks, in the canopy. **V** Song is a rich melodious warble. Call is a sparrow-like *cheep*. **H** Forest, secondary growth and gardens. **D** Nendö (Solomon Islands).

21 VANIKORO WHITE-EYE *Zosterops gibbsi* 12cm Plain olive with long black bill and pale orange legs. Note pale grey eye-ring. Usually in pairs in the canopy. **V** Song is a series of melodious notes, slower and shorter than Santa Cruz White-eye. Single nasal call notes repeated frequently. **H** Forest edge and secondary growth in lowlands, but inhabits primary forest in mountains. **D** Vanikoro (Solomon Islands).

22 SAMOAN WHITE-EYE *Zosterops samoensis* 10cm The only white-eye on Savai'i. Note pale legs and eyes. **V** Very high, soft *suhsuh* or *wuut-tjeetjee*. **H** Canopy of native forest, open scrub. **D** Savai'i (Samoa).

23 FIJI WHITE-EYE *Zosterops explorator* 10cm Restricted to Fiji. White eye-ring is complete, and black of lores extends beneath eye. **V** Song is a rather nondescript series of unvarying notes. Calls include a thin, reedy *siu-siu* and a short *zick*. **H** Forest, woodland, gallery forest, wooded farmland. **D** Fiji.

24 VANUATU WHITE-EYE *Zosterops flavifrons* 12cm Yellowish-olive white-eye with broad white eye-ring. Seven races, varying in minor plumage details. Note yellow forehead and dark lores in most races. In pairs, small groups or large flocks. **V** Song is a melodious warble. Calls are typical of the genus. **H** All types of forest, including scrub, bushes and gardens, at all altitudes. **D** Vanuatu.

25 SMALL LIFOU WHITE-EYE *Zosterops minutus* 10cm Small size, with yellow forehead and underparts, contrasting with silvery flanks. Note small bill and broad white eye-ring. In pairs or small flocks, often with other species. **V** Song is a loud fast warble. Calls include scratchy notes and trills. **H** Forest, forest edge, secondary growth and gardens. **D** Lifou (Loyalty Islands, SW Pacific).

26 GREEN-BACKED WHITE-EYE *Zosterops xanthochroa* 11cm Note broad white eye-ring, yellowish throat and vent, and greyish flanks. In pairs and small flocks in the canopy. **V** Song is a high-pitched medley of whistles. Flocks give plaintive nasal notes. **H** Forest and forest edge at all altitudes; less common in scrub and gardens. **D** New Caledonia.

27 SILVEREYE *Zosterops lateralis* 11cm Grey back and chestnut flanks diagnostic, but flanks paler in female, and greyer in some races. No other white-eye in New Zealand. **V** Song: high to very high twitters such as *feefeefeetjuweeh*. Call: tinkling *sree*. **H** Shrub, heathland, woodland, forest, mangrove, orchards. **D** Australia, New Zealand, W Pacific islands.

28 SLENDER-BILLED WHITE-EYE *Zosterops tenuirostris* 13cm Large plump white-eye with long slender bill. Note yellow throat and vent. Unobtrusive. **V** Call is a high-pitched chatter. **H** Rainforest and palm forest. Less common in secondary habitats. **D** Norfolk Island, SW Pacific.

29 LARGE LIFOU WHITE-EYE *Zosterops inornatus* 14cm Distinctive dull olive white-eye with long heavy bill. Lacks white eye-ring. Note grey mantle and underparts, and pink legs. Singly, in pairs or small groups. **V** Song is a rich melodious warble. Flocks do not constantly call. **H** Forest, forest edge, secondary growth and gardens. **D** Lifou (Loyalty Islands, SW Pacific).

30 KOSRAE WHITE-EYE *Zosterops cinereus* 10cm Note entirely grey plumage and very narrow or absent eye-ring. **V** Has a variety of chirping notes, some loud and sharp, others thinner and nasal-toned. **H** Most habitats with trees and/or shrubs. **D** Kosrae (Caroline Islands, W Pacific).

31 GREY-BROWN WHITE-EYE *Zosterops ponapensis* 10cm Formerly conspecific with Kosrae White-eye. Resembles Kosrae White-eye, but is browner above and paler below. In pairs or small flocks. **V** Call is a single harsh squeaky note. **H** Most habitats with trees or bushes, including gardens, at low altitudes. **D** Pohnpei (Caroline Islands, W Pacific).

32 OLIVE-COLORED WHITE-EYE *Zosterops oleagineus* 13cm Largish dark olive white-eye with broad white eye-ring, dark lores and robust bill. Usually in pairs or small groups. **V** Poorly known. Call is a high-pitched squeaky note. **H** All types of forest including mangroves and non-native vegetation. **D** Yap (Caroline Islands, W Pacific).

33 DUSKY WHITE-EYE *Zosterops finschii* 10cm Lacks white eye-ring and is darker than any other small bird on Palau. **V** Flocks are noisy, giving frequent *cheee* notes, extended into *chee-che-che-che* in flight. **H** Any wooded habitat. **D** Palau, W Pacific.

34 ABYSSINIAN WHITE-EYE *Zosterops abyssinicus* 12cm Note pale yellow throat and undertail coverts, grey lower breast and belly. **V** Squeaky and descending slurs in a random series. Calls include a soft, fine *tilu* or *teuu*, and a low *waouw*. **H** Trees (mainly acacia), wooded mountain slopes, gardens. **D** Eritrea to Tanzania; S Arabian Peninsula.

35 CAPE WHITE-EYE *Zosterops virens* 11cm Variable white-eye with dark lores and split white eye-ring. Belly varies from grey (in SW) to green or yellow (in N and E). All birds have yellow throat and vent. Usually in small flocks. **V** Song is a loud warble. Call is a slightly trilled *preee*, given by members of a flock to maintain contact. **H** Forest, woodland, thickets and gardens. **D** SE Botswana and SW Mozambique to South Africa (except NW).

36 ORANGE RIVER WHITE-EYE *Zosterops pallidus* 11cm Formerly treated as conspecific with Cape White-eye. Similar to that species, but paler above and has peachy-buff flanks. **V** Similar to Cape White-eye but higher-pitched and more trilling. **H** All types of wooded and bushed habitats, including riparian woodland. **D** Namibia to W and C South Africa.

1
2
3
4
5
6
7
8
9
10
11
12
13
14
15
16
17
18
19
20
21
22
23
24
25
26
27
28
29
30
31
32
33
34
35
36

WHITE-EYES *CONTINUED*

1 AFRICAN YELLOW WHITE-EYE *Zosterops senegalensis* 10cm Distinguished from Cape White-eye (Plate 231) by brighter colouring above and especially below. **V** Song is more mellow than Cape White-eye, a high *pipiwee-piriwee*. Contact call is a sharp *sieeeh sieeh*. **H** Forest, riverine belts, woodland, suburbs. **D** Sub-Saharan Africa, south to N Angola and NE South Africa.

2 MONTANE WHITE-EYE *Zosterops poliogastrus* 11cm Variable highland white-eye with a broad white eye-ring. Underparts vary from grey to greenish or yellowish. Six races recognised, which may represent six different species. **V** Song is mournful warble. Call is a downslurred *seeuu*. **H** Forest, forest edge, secondary growth and gardens, in highlands. **D** SE Sudan and Eritrea south to N Tanzania.

3 KIKUYU WHITE-EYE *Zosterops kikuyuensis* 11cm Formerly treated as a race of Montane White-eye. Differs in more extensive yellow forehead, greenish flanks and broader eye-ring. **V** Song is a rising and falling medley of slurred notes. Calls are similar to Montane White-eye. **H** Forest, forest edge, secondary growth and gardens, in highlands. **D** C Kenya.

4 TAITA WHITE-EYE *Zosterops silvanus* 10cm Formerly considered conspecific with Montane White-eye. Differs in lacking yellow on forecrown and having mainly grey underparts. Note very broad white eye-ring. **V** Song is similar to Kikuyu White-eye, but notes are less pure. Calls are similar to Montane White-eye. **H** Highland forest in the Taita Hills. **D** SE Kenya.

5 REUNION GREY WHITE-EYE *Zosterops borbonicus* 11cm Distinctive white-eye with white rump, no eye-ring, and a tail that is frequently cocked. Variable in colour, with grey and brown morphs. **V** Song is a complex, varied combination of warbles and call notes, with some mimicry, rising and falling in pitch over up to 30 secs. Calls include *chip* and *chee chee*. **H** Any habitat with trees or shrubs. **D** Réunion, Indian Ocean.

6 MAURITIUS GREY WHITE-EYE *Zosterops mauritianus* 11cm Formerly considered conspecific with Reunion Grey White-eye. Similar to grey morph of that species, but paler grey above, and paler below with rufous-tinged flanks. **V** Similar to Reunion Grey White-eye but lower-pitched and more squeaky. **H** Any habitat with trees or shrubs. **D** Mauritius, Indian Ocean.

7 PRINCIPE WHITE-EYE *Zosterops ficedulinus* 10cm Rather plain white-eye with yellow supraloral stripe and narrow white eye-ring. Pale greyish below. **V** Flocks give constant chittering notes, *prrip prrip, pink pink pink*. **H** Forests of the hilly interior. **D** Príncipe (Gulf of Guinea).

8 SAO TOME WHITE-EYE *Zosterops feae* 10cm Formerly considered conspecific with Principe White-eye. Similar to that species, but greener above and darker below. **V** Song is a trilling warble. Flocks give constant chittering notes. **H** Forest, secondary growth, plantations and bushy habitats in more degraded areas, mainly in the hills. **D** São Tomé (Gulf of Guinea).

9 ANNOBON WHITE-EYE *Zosterops griseovirescens* 12cm The only white-eye on Annobón. Note tawny-buff flanks and greyish ear-coverts. **V** Song is a quiet and melodious series of jumbled notes. Calls include *tsip* and trilled *chrrr*. **H** All areas with trees. **D** Annobón (Gulf of Guinea).

10 MALAGASY WHITE-EYE *Zosterops maderaspatanus* 11cm The only white-eye in most of it range. **V** Song consists of thin, reedy notes going up and down scale, lasting 2–8 secs. Calls include *tew, ter-tew* and a repeated querulous *wrrrri*. **H** All habitats with trees. **D** Madagascar, Comoros, SW Seychelles.

11 KIRK'S WHITE-EYE *Zosterops kirki* 11cm Formerly considered conspecific with Malagasy White-eye. Underparts mainly yellow with greenish wash on flanks. **V** Song is a medium-pitched warble. Calls are a series of high-pitched *tew* notes. **H** All habitats with trees, but absent from higher altitudes, where replaced by Karthala White-eye. **D** Grande Comore (Comoros).

12 MAYOTTE WHITE-EYE *Zosterops mayottensis* 11cm Very distinctive, with bright yellow forehead and underparts, and chestnut-brown flanks. **V** Not well documented, but described as fairly typical of its genus. **H** All habitats with trees. **D** Mayotte, Mozambique Channel.

13 SEYCHELLES WHITE-EYE *Zosterops modestus* 10cm Note absence of yellow and bright green. No similar white-eye in range. **V** Song is a pleasant series of complex high-pitched phrases. Calls include soft and chattering trills and a sharp *tik tik*. **H** Woodland, orchard, gardens. **D** Seychelles.

14 KARTHALA WHITE-EYE *Zosterops mouroniensis* 11cm Rather uniformly coloured. **V** Song is a pleasant warble. Calls include whistles and buzzing notes. **H** Woodland and heathland at higher altitudes. **D** Comoros.

15 REUNION OLIVE WHITE-EYE *Zosterops olivaceus* 11cm Note black lores, striking white eye-ring and curved bill. **V** Song is loud and warbling, including *tu* and *tchip* notes. Calls include a clipped *chip-chip* and, in flight, *chuck chuck*. **H** Forest at higher altitudes. **D** Réunion, Indian Ocean.

16 MAURITIUS OLIVE WHITE-EYE *Zosterops chloronothos* 10cm Note curved bill and pale orange legs. **V** Song is a short, nondescript combination of trills and warbles notes. Calls include metallic *plik plik* and *pit*. **H** Forest and scrub. **D** Mauritius, Indian Ocean.

17 PEMBA WHITE-EYE *Zosterops vaughani* 10cm Bright yellowish white-eye with a broad white eye-ring and dark lores. The only white-eye on Pemba Island. **V** Song is a melodious warble of slurred notes. **H** All wooded habitats. **D** Pemba Island (off NE Tanzania).

DAPPLE-THROAT AND ALLIES MODULATRICIDAE

18 SPOT-THROAT *Modulatrix stictigula* 14cm A babbler-like passerine with long bill, reddish rump and generally warm brown colouring. **V** Song is a loud, shrill series of slurring whistled notes. Alarm call is a harsh *jerr jerr jerr*. **H** Forest ground strata, gardens. **D** Tanzania, Malawi.

19 DAPPLE-THROAT *Arcanator orostruthus* 14cm Note long bill, buff and greenish head. **V** Song consists of sweet, melodious phrases; also a repetitive chatter. Calls include a fluting *hooo-reee* and tinkling notes. **H** Undergrowth of moist forests. **D** Tanzania, Mozambique.

20 GREY-CHESTED BABBLER *Kakamega poliothorax* 15cm Unmistakable, with beautiful rufous upperparts. **V** High, descending thrush- or oriole-like *pju-pju-pjuwi* (*j* falsetto). **H** Mountain forest undergrowth. **D** Disjunct populations: SE Nigeria, W Cameroon and Bioko; E DR Congo, W Uganda, Rwanda, Burundi and W Kenya.

SUGARBIRDS PROMEROPIDAE

21 CAPE SUGARBIRD *Promerops cafer* male 37cm, female 24cm Long downcurved beak and dramatic tail. Distinguished from Gurney's Sugarbird by prominent malar stripe, streaked chest and longer tail. **V** Hoarse chatter. **H** Fynbos shrub, suburban gardens. **D** S South Africa.

22 GURNEY'S SUGARBIRD *Promerops gurneyi* male 25cm, female 23cm Note russet crown and breast. **V** Rather unstructured chattering, less hoarse than Cape Sugarbird. **H** Montane scrub, similar to Cape Sugarbird but different range. **D** E Zimbabwe, W Mozambique to E South Africa.

FAIRY-BLUEBIRDS IRENIDAE

23 ASIAN FAIRY-BLUEBIRD *Irena puella* 25cm Generally encountered in treetops. Keeps on the move, hopping from branch to branch and flying from tree to tree. **V** Percussive, liquid *weet-weet be-quick peepit whats-it*, usually repeated every few seconds. In flight utters a sharp *chichichichik*. **H** Evergreen and moist deciduous forests. **D** W India through SE Asia to Indonesia, Philippines.

24 PHILIPPINE FAIRY-BLUEBIRD *Irena cyanogastra* 23–28cm Forages in fruiting canopy trees, alone or in small loose groups. Sexes alike, but female slightly duller. **V** Snapping, whip-like *weep-weep pul paaawwww*; also a repeated, fluty *hu wee-u whip-tip hu wee-u*, the *whip* emphasised and the *tip* quiet. **H** Tall lowland to mid-elevation forest and forest edges. **D** Philippines.

GOLDCRESTS AND KINGLETS REGULIDAE

25 COMMON FIRECREST *Regulus ignicapilla* 9cm Tiny, with striking head pattern. Often feeds at low levels in bushes and shrubs. **V** High-pitched, accelerating *zi-zi-zi-zi-zi-zi-zi-zi-zi-zi-zizizit*. Calls include *ze-ze-zeep*; also a single, or repeated, *zeep*. **H** Deciduous, coniferous and mixed woodland, all with rich undergrowth; scrubby Evergreen Oak, Cork Oak, parks, large gardens. **D** W, C and S Europe east to Caucasus; NW Africa.

26 FLAMECREST *Regulus goodfellowi* 9cm Very colourful. Black eye patch contrasts with white on side of face. Agile, active forager in tree foliage. **V** High-pitched *seeh-seeh-seeh*. Call is a quiet *seeeh*. **H** Coniferous and montane forest. **D** Taiwan.

27 GOLDCREST *Regulus regulus* 9cm Agile forager, always on the move; flicks wings and tail and frequently hovers to catch insects; often gives away presence by continual calling. **V** Song is a rapid trill. Call is a high, persistent *tsi-tsi-tsi-tsi*. **H** Mainly coniferous forest. **D** Widespread, from W Europe to SE Siberia, C Asia, Korea, Japan; Canary Islands and Azores.

28 MADEIRA FIRECREST *Regulus madeirensis* 9cm Formerly treated as a race of Common Firecrest, but differs in minor plumage details, vocalisations and genetics. Lacks long white supercilium of Common Firecrest. **V** Song is more reminiscent of Goldcrest than Common Firecrest. Distinctive call is an upslurred *wheez*. **H** Forests. **D** Madeira.

29 GOLDEN-CROWNED KINGLET *Regulus satrapa* 10cm When foraging, constantly on the move, flitting from twig to twig. **V** Song rising then ending in a tumbling chatter, *see see see si si si titichichichichi*. Call is a very high-pitched *see-see-see*. **H** Primarily coniferous woodland. **D** Breeds from S Alaska to Newfoundland, mountain regions of W and E USA, S Mexico, Guatemala. Canadian birds move south in winter to become widespread in USA.

30 RUBY-CROWNED KINGLET *Regulus calendula* 10cm Restless acrobatic forager. Females lack the red crown patch, and in males it is usually concealed. **V** Song is a series of high, then descending notes ending in a warble, *sii si sisisi berr berr pudi pudi pudi see*. Dry *jidit, jit* or *jit jit jit...* when alarmed. **H** Woodland, thickets, scrub. **D** Breeds Alaska, Canada, Rocky Mountains, extreme NE USA. Northern birds migrate to winter in S USA, Mexico.

ELACHURA ELACHURIDAE

31 SPOTTED ELACHURA *Elachura formosa* 10cm Brown with whitish speckles. Russet on wings and tail. **V** High drawn-out tinkling. Call is a spluttering trill. **H** Broadleaved evergreen forest, scrub and weeds in steep-sided gulleys. **D** Himalayas to Myanmar, Vietnam, SE China.

HYLIOTAS HYLIOTIDAE

32 YELLOW-BELLIED HYLIOTA *Hyliota flavigaster* 12cm Distinguished from Southern Hyliota by buff (not white) rear underparts, short white edges to tertials, stronger overall gloss. **V** Unstructured sequence of well-separated very high scratchy notes, *treetreet twitreetreet wietwiet wit wit treetreet*. **H** Miombo woodland. **D** Sub-Saharan Africa, from Senegal to Ethiopia, Angola, Mozambique.

33 SOUTHERN HYLIOTA *Hyliota australis* 11cm Rather dull black above. **V** Very high short slightly scratchy phrases, *tutree-tutree-twee-twee- twee*. **H** Woodland. **D** NE DR Congo; Angola to Mozambique, NE South Africa.

34 USAMBARA HYLIOTA *Hyliota usambara* 11cm Formerly treated as a race of Southern Hyliota. Similar to that species, but glossy blue-black above and richer rufous-orange below. Sexes alike. **V** Song is a squeaky medley preceded by introductory chipping notes. **H** Forests in the Usambara Mountains. **D** NE Tanzania.

35 VIOLET-BACKED HYLIOTA *Hyliota violacea* 13cm Glossy violet-black upperparts. White in wing absent in W part of range. Gleans insects from foliage and bark. **V** Song is a series of short whistles. Calls include a sharp *tit-tit*. **H** Canopy of forest, especially at edges; not in miombo or woodland. **D** Sierra Leone to E DR Congo, Rwanda.

1
2
3
4
5
brown morph
grey morph
6
7
8
9
10
11
12
13
14
15
16
17
18
19
20
21
♂
♀
22
23
♂
♀
24
25
♀
♂
26
27
♀
28
♀
29
♂
♀
30
31
32
♀
♂
33
♀
♂
34
35
♀
♂

WRENS TROGLODYTIDAE

1 WHITE-HEADED WREN *Campylorhynchus albobrunneus* 18.5cm Unmistakable, with contrasting plumage and all-white head. **V** Low, harsh, scratchy chattering in chorus. **H** Dense borders of humid forest. **D** Panama to SW Colombia.

2 BAND-BACKED WREN *Campylorhynchus zonatus* 18.5cm Separable from Stripe-backed Wren, Fasciated Wren and Thrush-like Wren by ochrous flanks and belly. **V** Dry, sharp chatters. **H** Canopy, borders and clearings in humid lowland and montane forest. **D** Mexico to Ecuador.

3 GREY-BARRED WREN *Campylorhynchus megalopterus* 18cm Distinguished from Yucatan Wren by barred mantle, spotted breast, different range. **V** Very high magpie-like *sreetsreet*. **H** Montane forest interior and edge. **D** SC Mexico.

4 STRIPE-BACKED WREN *Campylorhynchus nuchalis* 17.5cm Note yellow eyes and striped (not barred) mantle. **V** Mid-high cackling chatters. **H** Open forest, woodland, riverine belts, habitations. **D** Colombia, Venezuela.

5 FASCIATED WREN *Campylorhynchus fasciatus* 19cm Separable from Band-backed Wren by boldly barred and spotted whitish flanks and belly. **V** Hoarse chattering in chorus. **H** Arid scrub, cultivated areas, woodland. **D** W Ecuador, W Peru.

6 GIANT WREN *Campylorhynchus chiapensis* 20cm Long white supercilium and distinctive white-spotted tail. Restricted range. **V** Very high mewing *peeohweeoh*. **H** Open areas with tree stands, hedges, scrub, gardens, cultivations. **D** S Mexico.

7 BICOLORED WREN *Campylorhynchus griseus* 22cm Unmistakable, with colour contrast of upperparts and underparts, and bold black and white head pattern. **V** Mellow, rolling cackling. **H** Open woodland and scrub, often near palm groves. **D** Colombia to the Guianas, N Brazil.

8 RUFOUS-NAPED WREN *Campylorhynchus rufinucha* 17cm Note lightly spotted underparts. Separable from Spotted Wren by black crown, and from Rufous-backed and Sclater's wrens by different range. **V** High croaking *wefwuf* or staccato *wic* or *weoh*. **H** Arid and semi-arid areas with dry scrub, open areas with bushes and cacti. **D** E Mexico.

9 SCLATER'S WREN *Campylorhynchus humilis* 15cm Formerly considered conspecific with Rufous-naped Wren, differing in smaller size, mainly rufous crown and unspotted underparts. **V** Series of rich chortling phrases in a rhythmic pattern. **H** Arid and semi-arid areas with dry scrub, open areas with bushes and cacti. **D** SW Mexico.

10 RUFOUS-BACKED WREN *Campylorhynchus capistratus* 17–19cm Formerly considered conspecific with Rufous-naped Wren, differing in more extensive rufous on mantle and unspotted underparts. **V** Series of rich chortling phrases, but varies geographically. Sexes duet. **H** Arid and semi-arid areas with dry scrub, open areas with bushes and cacti. **D** SW Mexico to NW Costa Rica.

11 SPOTTED WREN *Campylorhynchus gularis* 17cm Rufous cap and barred, buff flanks diagnostic. **V** Chicken-like *whec* or *erc* or *chatchatchatchat*. **H** Scrubby woodland, open areas with scattered bush, cacti. **D** W and EC Mexico.

12 BOUCARD'S WREN *Campylorhynchus jocosus* 18cm Not in range of Spotted Wren, Yucatan Wren or Cactus Wren. **V** High rather rasping *uwrEetit* or *uhweet*. **H** Dry areas with scrub and cacti. **D** S Mexico.

13 YUCATAN WREN *Campylorhynchus yucatanicus* 18cm No similar wren in range. **V** Mewing *ewktjic-*. **H** Dry scrub, gardens. **D** N Yucatán Peninsula, Mexico.

14 CACTUS WREN *Campylorhynchus brunneicapillus* 18–19cm Conspicuous and noisy. Forages low in vegetation or on the ground. **V** Staccato *tek tek tek tek*... Song is a harsh *cha cha cha*... **H** Arid areas with cacti. **D** SW USA, Mexico.

15 THRUSH-LIKE WREN *Campylorhynchus turdinus* 20.5cm Separable from Band-backed Wren, Stripe-backed Wren and Fasciated Wren by range. Note distinct eyebrow. **V** Rather musical, short, cackling phrases. **H** Humid forest. **D** Amazonia from S Colombia to Paraguay, extreme N Argentina.

16 GREY-MANTLED WREN *Odontorchilus branickii* 12cm Fairly distinctive, with grey upperparts, a tawny crown and a barred tail. **V** Very high, soft, rhythmic trill. **H** Humid forest on Andean slopes. **D** Colombia to N Bolivia.

17 TOOTH-BILLED WREN *Odontorchilus cinereus* 12cm Resembles Grey-mantled Wren but more brownish grey, and tawny coloration of crown extends to face sides. **V** Very high, level, rhythmic trill or rattle. **H** Canopy and borders of humid forest. **D** SW Amazonia, from C Brazil to NE Bolivia.

18 ROCK WREN *Salpinctes obsoletus* 14–15cm Greyish-brown upperparts are speckled white and black. Forages among rocks, habitually bobs up and down. **V** Song consists of variable phrases, *cheer-cheer-cheer-cheer, deedle-deedle-deedle-deedle, tur-tur-tur-tur*. Calls include a buzzy, trilled *dee-dee, dee-dr-dr-dr-dr-dr*. **H** Rocky hillsides and gullies. **D** SW Canada to Costa Rica. Northern birds move to winter in SW USA.

19 CANYON WREN *Catherpes mexicanus* 13–15cm Note white throat and breast. Unobtrusive forager among rock crevices, more often heard than seen. **V** Song is a series of clear descending and slowing whistles followed by a nasal hissing. Typical call is a loud, buzzy *jeet*. **H** Mainly canyons. **D** SW Canada, W USA, Mexico.

20 SUMICHRAST'S WREN *Hylorchilus sumichrasti* 16cm Separable from Nava's Wren by less white on throat and by different range. **V** Short decelerating loud warble or loud sharp *sweeep*. **H** Limestone outcrops in forest. **D** SC Mexico.

21 NAVA'S WREN *Hylorchilus navai* 16cm Terrestrial bird, with a very restricted range that does not overlap with that of Sumichrast's Wren. **V** Very high fluted unhurried *teeteeteetjuw* or *teetutteeturweeweeh*. **H** Limestone outcrops in forest. **D** SC Mexico (further E than Sumichrast's Wren).

22 RUFOUS WREN *Cinnycerthia unirufa* 16.5cm Brighter plumage than Sepia-brown Wren, with less distinct barring. **V** Rapid, hurried series of repeated notes, phrases and trills. **H** Undergrowth with bamboo in montane forest. **D** Venezuela and Colombia to Peru.

23 SEPIA-BROWN WREN *Cinnycerthia olivascens* 16cm Similar to Rufous Wren. Some birds show a white frontlet. **V** Unbelievably rich musical repertoire, containing long series of strong phrases, trills, notes and rattles. **H** Wet forest with mosses and bamboo. **D** Colombia to Peru.

24 PERUVIAN WREN *Cinnycerthia peruana* 15cm Note faint eyebrow. May show varying amounts of white on front of face. **V** Very rich musical repertoire, resembling that of Sepia-brown Wren but perhaps with more harsh notes. **H** Undergrowth of montane forest. **D** Peru.

25 FULVOUS WREN *Cinnycerthia fulva* 14.5cm Separable from Peruvian Wren by range and by dark eye-stripe. Some birds have much white on face. **V** Song repertoire resembles that of Peruvian Wren, but often with shorter rattled or trilled phrases. **H** Undergrowth of wet forest. **D** Peru, Bolivia.

26 SEDGE WREN *Cistothorus stellaris* 11cm Very secretive, best located by song. **V** Song is a sharp staccato trill or chatter. Calls include a rich *chip* or *chip-chip* and a sharp *chadt*. **H** Tall-grass meadows with scattered bushes. **D** Breeds Canadian Prairies and NE USA. Winters SE USA, NE Mexico.

27 MERIDA WREN *Cistothorus meridae* 10cm Separable from Sedge Wren by distinct eyebrow and barring to flanks. **V** Dry cricket-like trills. **H** Páramo, especially near water or large boulders. **D** Venezuela.

28 APOLINAR'S WREN *Cistothorus apolinari* 12cm Crown not streaked. Note habitat. **V** Duet of a sustained rolling rattle contra double notes, e.g. a fluted *fu-fu*. **H** Marshes, boggy páramo, vegetation at lakesides. **D** Colombia.

29 GRASS WREN *Cistothorus platensis* 10.5cm Note streaked and spotted mantle and crown. **V** Rattles and *zit* notes at very different pitches. **H** Savannah, grasslands. **D** Central and South America, widespread.

30 MARSH WREN *Cistothorus palustris* 11.5–12.5cm Secretive, forages low in marsh vegetation. **V** Song consists of gurgling, rattling, buzzing and trilling notes, introduced by a few *tek* notes. Typical call is a sharp *tek*. **H** Reed swamps. **D** Breeds over much of North America. Winters S USA and on E and W coasts of USA, and in Mexico.

31 BEWICK'S WREN *Thryomanes bewickii* 13cm Whitish supercilium, pale underparts, longish tail. Forages low, in vegetation or on the ground. **V** Song is variable, incorporating a high-pitched, thin buzzing and slow trills. Calls include a scolding buzzy *shreeee*, a soft *wijo* and a high-pitched, rising *zrink*. **H** Dense brushy places, including thickets, woodland clearings and suburban areas. **D** SW Canada, W, C and S USA, Mexico.

32 ZAPATA WREN *Ferminia cerverai* 16cm Very secretive, best located by song. Feeds low down in vegetation and on the ground. Tail is regularly cocked but is held down while singing. **V** Utters a clear series of gurgling whistles interspersed with harsh churring notes, which can continue for over a minute; also gives a low, harsh *chut-chut, churr-churr* or similar. **H** Sawgrass marshes. Breeds from January to July. **D** Cuba.

33 BLACK-THROATED WREN *Pheugopedius atrogularis* 18cm Forages mainly on the ground among leaf litter, beneath tree roots and by probing in tree-trunk crevices. Usually found in small parties. **V** Duets, one bird giving a long series of clear, bell-like notes, while the second adds a slightly descending *iuuh-iuh-iuh... iuuh-uh-uh*; calls include a *we-ah we-ah we-ah* and a nasal, coarse *krar krar*. **H** Lowland primary broadleaved evergreen forest, old secondary forest, logged forest. **D** Nicaragua to Panama.

34 SOOTY-HEADED WREN *Pheugopedius spadix* 15cm Black throat is sharply demarcated from rufous breast. **V** Repeated phrases, *Why can't I see you?* **H** Undergrowth of wet forest. **D** Panama, Colombia.

35 BLACK-BELLIED WREN *Pheugopedius fasciatoventris* 15cm Note large white bib contrasting with faintly white-barred black belly. **V** Beautiful short phrases of mellow, sometimes gurgling notes at different pitches. **H** Thick vegetation beside streams. **D** Costa Rica to Colombia.

1
2
3
4
5
6
7
8
9
10
11
12
13
14
15
16
17
18
19
20
21
22
23
24
25
26
27
28
29
30
31
32
33
34
35

WRENS *CONTINUED*

1 PLAIN-TAILED WREN *Pheugopedius euophrys* 16cm Note grey and white markings on head, and absence of barring on tail and wings. **V** Song resembles that of Sooty-headed Wren (Plate 233), given in a duet, *weehweetjeh-witwit*. **H** Dense forest undergrowth, forest edge. **D** Colombia to Peru.

2 INCA WREN *Pheugopedius eisenmanni* 15cm Boldly streaked below; tail and wings not barred. **V** Pleasant but unstructured, whistled phrases. **H** Bamboo thickets in undergrowth of montane forest and woodland. **D** Peru.

3 MOUSTACHED WREN *Pheugopedius genibarbis* 15.5cm Note narrow 'whiskers' and bold tail barring. Compare Plain-tailed Wren and Whiskered Wren. **V** Series of complicated repeated, rapid, mellow phrases mixed with ultra-high *see* notes. **H** Dense undergrowth at borders of forest and woodland. **D** Amazonia.

4 WHISKERED WREN *Pheugopedius mystacalis* 16cm Note broad 'whiskers'. Separable from Moustached Wren by range and song. **V** Short, mellow, gurgling phrases with a hurried end. **H** Dense forest borders, secondary growth. **D** Venezuela, Colombia, Ecuador.

5 CORAYA WREN *Pheugopedius coraya* 14.5cm Face mostly black, with white supercilium and variable extent of white markings. **V** Combinations of high, gliding notes and a strong rattle. **H** Humid forest, secondary growth. **D** N Amazonia.

6 HAPPY WREN *Pheugopedius felix* 13cm Distinctive facial pattern with white throat, rest of plumage unmarked or faintly marked. May have almost white underparts. **V** Very high rapid fluted *tsjuweeh wupwupwup*. **H** Dry forest interior and edge, secondary growth, plantations. **D** W Mexico.

7 SPOT-BREASTED WREN *Pheugopedius maculipectus* 13cm Warm brown back and spotted breast diagnostic. **V** Very high *weetweetweetuwweet* or buzzing two-part trill. **H** Dense vegetation at forest edge, plantations and rivers. **D** E Mexico to Costa Rica.

8 RUFOUS-BREASTED WREN *Pheugopedius rutilus* 14cm Speckled face sides and throat are sharply demarcated from orange-rufous breast. **V** Short, rapid phrase of strong whistles at different pitches. **H** Undergrowth and borders of woodland and forest. **D** Costa Rica to Colombia, Venezuela, Trinidad and Tobago.

9 SPECKLE-BREASTED WREN *Pheugopedius sclateri* 14cm Dull rufous-brown upperparts, white underparts with black spots or bars on cheeks, breast and flanks, a buffy wash to the rear flanks, and a black-barred tail. **V** Song resembles that of Rufous-breasted Wren but with 2–10 slightly varied phrases connected in a continuous series. **H** Undergrowth and borders of woodland. **D** Colombia to Peru.

10 BANDED WREN *Thryophilus pleurostictus* 14cm Pattern of flanks, neck and face distinctive. Resembles Riverside Wren, but lacks distinctly marked tail, and range is different. **V** Hurried tickling or series of hurried ultra high rattles. **H** Dry forest, forest edge. **D** S Mexico to Costa Rica.

11 RUFOUS-AND-WHITE WREN *Thryophilus rufalbus* 15cm Distinctly two-toned (more so than Buff-breasted Wren). Separable from Niceforo's Wren by orange (not ochrous) upperparts. **V** Musical, low, whistled phrases with a high exclamation note at the end. **H** Undergrowth of dry forest, riverine belts, palm thickets. **D** Mexico to Colombia and Venezuela.

12 ANTIOQUIA WREN *Thryophilus sernai* 14cm Distinguished from Rufous-and-white Wren by browner upperparts, and from Niceforo's Wren by unmarked mantle. First described in 2012. **V** Calm stream of very musical, fluted single and two-syllable notes. **H** Dry forest and scrub, vegetation along watercourses. **D** NW Colombia.

13 NICEFORO'S WREN *Thryophilus nicefori* 14.5cm Formerly treated as conspecific with Rufous-and-white Wren. **V** Beautiful, contrasting series of several level, single, low, mid-high and very high notes. **H** Dense acacia scrub. **D** NC Colombia.

14 SINALOA WREN *Thryophilus sinaloa* 13cm Separable from warmer-coloured Happy Wren in same range by weaker face pattern. **V** Very high short warbles with small rattles. **H** Dry forest interior, forest edge, secondary growth, plantations. **D** W Mexico.

15 CABANIS'S WREN *Cantorchilus modestus* 13cm Formerly lumped with Canebrake and Isthmian wrens, under the name 'Plain Wren'. Prominent eye-stripe, buff underparts, unbarred undertail coverts. **V** Very high *tjuptjupprrrrrrrreer* or ultra-high drawn-up *treehweewitwitwit*. **H** Forest edge, scrubby woodland, weedy vegetation, neglected fields, grassland. **D** S Mexico to Costa Rica.

16 CANEBRAKE WREN *Cantorchilus zeledoni* 14cm Formerly considered conspecific with Cabanis's Wren. Darker and duller than Cabanis's and Isthmian wrens. **V** Similar to Cabanis's Wren but slower and lower-pitched. **H** Broadleaf forest, secondary growth and gardens, in understorey. **D** E Nicaragua, E Costa Rica and NW Panama.

17 ISTHMIAN WREN *Cantorchilus elutus* 13cm Formerly considered conspecific with Cabanis's Wren. Paler than Cabanis's Wren with more prominent barring on wings and tail. **V** Song is a fast, high-pitched series of slurred notes, repeated several times. **H** Broadleaf forest, secondary growth, scrub and gardens, in understorey. **D** SW Costa Rica and W Panama.

18 BUFF-BREASTED WREN *Cantorchilus leucotis* 14.5cm Has little black in face; note barred wings. **V** Calm to hurried warbling with repetitive elements. **H** Thickets, forest edge, arid scrub, usually near water. **D** From Panama through Amazonia to Pagaguay.

19 SUPERCILIATED WREN *Cantorchilus superciliaris* 14.5cm Unmistakable, with white face sides and breast. **V** Sharp, hurried warbling with repetitive elements. **H** Dry woodland, hedges, arid scrub, agricultural areas. **D** W Ecuador, NW Peru.

20 FAWN-BREASTED WREN *Cantorchilus guarayanus* 13.5cm Note narrow white supercilium behind eye, mottled grey, white and blackish ear-coverts, unstreaked crown and upperparts. **V** Sharp, rapid warbling with repeating phrases. **H** Várzea forest, secondary growth. **D** Bolivia, SW Brazil, N Paraguay.

21 LONG-BILLED WREN *Cantorchilus longirostris* 15cm Long bill and rufescent upperparts are distinctive. **V** Forceful, unstructured, sometimes repetitive warbling. **H** Dense vegetation at forest and woodland borders, mangrove, Caatinga. **D** E Brazil.

22 GREY WREN *Cantorchilus griseus* 11.5cm Small and overall dusky grey. **V** Strong series of 2–3 rich, staccato notes. **H** Dense growth of várzea and woodland. **D** W Brazil.

23 RIVERSIDE WREN *Cantorchilus semibadius* 13cm Distinctive, with strong barring and unmarked crown and mantle. **V** Very high, hurried, sustained *sweet sweet - -*. **H** Forest edge, dense vegetation near streams. **D** C Costa Rica to W Panama.

24 BAY WREN *Cantorchilus nigricapillus* 14.5cm No other wren in its South American range has the combination of mainly black head, white throat, and rufous and/or heavily barred underparts. Extent of barring and rufous coloration on underparts varies geographically. **V** Rapid, repetitive mixture of dry rattles and hurried, forceful, short warbling. **H** Forest edges and secondary growth, usually near water. **D** Nicaragua to Ecuador.

25 STRIPE-BREASTED WREN *Cantorchilus thoracicus* 13cm Striped breast diagnostic. Note also barred wings. **V** Very high mellow fluted *weetweettjuwehweet* or *here we-are-again*. **H** Forest interior and edge, damp secondary growth. **D** E Honduras to Panama.

26 STRIPE-THROATED WREN *Cantorchilus leucopogon* 12cm Small, with faint barring on wings and tail. **V** Continuous series of 2–10 short, slightly varied phrases of 3–4 notes at very different pitches. **H** Dense undergrowth of humid woodland and forest. **D** Panama to Ecuador.

27 CAROLINA WREN *Thryothorus ludovicianus* 13–14cm Forages in undergrowth. **V** Song is a loud, ringing *tea-kettle tea-kettle tea-kettle* or similar. Calls include a harsh *zhwee zhwee...*, a low *dip* or *didip* and a descending trill. **H** Woodlands, forest edge, urban areas. **D** SE Canada and E USA to Guatemala and Nicaragua.

28 EURASIAN WREN *Troglodytes troglodytes* 10cm Restless; forages mouse-like amongst low vegetation, showing itself only in fleeting glimpses; more prominent when singing from an exposed perch. **V** Song is very loud, a mixture of trills and rattling warbles. Calls include a harsh *tek* or *tac* and loud churrs. **H** Very varied: forest, woodland, farmland, treeless offshore islands, coastal cliffs, suburban parks, urban gardens. **D** Widespread, from Iceland, W Europe through C Asia to Himalayas, China, SE Siberia, Japan.

29 WINTER WREN *Troglodytes hiemalis* 10cm Very similar to Eurasian Wren. Secretive. Forages in woodland undergrowth. **V** Song is a series of high trills and thin buzzes. Call is a hard *jip-jip*, also a rapid series of staccato notes when alarmed. **H** Coniferous forest, woodland with dense undergrowth. **D** Breeds SE Canada, NE USA, mountains of West Virginia to Georgia. Canadian birds move south to winter in SE USA.

30 PACIFIC WREN *Troglodytes pacificus* 10cm Formerly considered conspecific with Eurasian Wren and Winter Wren. Darker and more rufous than Winter Wren, and has distinctly different song and calls. **V** Tumbling series of notes. More staccato than Winter Wren, and harsher. Calls include a sharp *chat-chat*. **H** Wide range of habitats, from treeless islands to dense coniferous forest. **D** Pacific North America, from westernmost Aleutian Islands, Alaska, to California

31 CLARION WREN *Troglodytes tanneri* 13cm The only wren on Clarión. **V** High chattering *wekekekekrrwek*. **H** Dry open areas with some scrub. **D** Isla Clarión, W Pacific (Mexico).

32 HOUSE WREN *Troglodytes aedon* 11.5cm Brown to greyish-brown, with indistinct barring, spotting and scaling on upperparts, sometimes extending to flanks and belly. Western birds are greyer than eastern birds. **V** Short, very diverse bubbling and gurgling warbles, twitters and rattles. **H** Very varied habitat, from forest edges, scrub and hedgerows, to cultivation and suburban gardens. **D** Breeds from S Canada to Tierra del Fuego. Northern populations winter in S USA.

33 COBB'S WREN *Troglodytes cobbi* 13cm Similar to House Wren but larger with a longer bill, paler upperparts and shorter tail. **V** Similar to House Wren but slower and lower-pitched. **H** Tussock grass on small outlying islands. **D** Falkland Islands.

34 SOCORRO WREN *Troglodytes sissonii* 12cm The only wren on Socorro. **V** Ultra-high sharp chattering hurried *wuutwutweetweetohweet*. **H** Dry open areas with some scrub and trees. **D** Isla Socorro, W Pacific (Mexico).

35 RUFOUS-BROWED WREN *Troglodytes rufociliatus* 11cm Note warm brown plumage. No similar wren in its range and habitat. **V** Ultra-high shrill hurried *sweetsreeweetohweet*. **H** Humid montane pine and mixed forest. **D** S Mexico to Nicaragua.

1
2
3
4
5
6
7
8
9
10
11
12
13
14
15
16
17
18
19
20
21
22
23
24
25
26
27
28
29
30
31
32
grey
morph
rufous
morph
33
34
35

WRENS *CONTINUED*

1 OCHRACEOUS WREN *Troglodytes ochraceus* 10cm Distinguished from House Wren (Plate 234) by short tail and richer coloration, with buff-orange supercilium. **V** Ultra-high, hurried, sharp descending warble. **H** Montane forest, tall secondary growth, adjacent areas. **D** Costa Rica, Panama.

2 MOUNTAIN WREN *Troglodytes solstitialis* 11cm Distinct pale supercilium varies from whitish to buff. **V** Ultra-high, soft, rhythmic twitter. **H** Interior and borders of montane forest, including cloud forest. **D** Andes, from W Venezuela to NW Argentina.

3 SANTA MARTA WREN *Troglodytes monticola* 11.5cm Separable from Mountain Wren by boldly barred grey rear parts. **V** Typical call is a brief *di di*. **H** Stunted and elfin forest near the treeline. **D** Sierra Nevada de Santa Marta, N Colombia.

4 TEPUI WREN *Troglodytes rufulus* 12cm Separable from Flutist Wren by lack of scaling. **V** Calm series of very high *zeet* notes. **H** Humid forest on tepuis. **D** S Venezuela, Guyana, extreme N Brazil.

5 TIMBERLINE WREN *Thryorchilus browni* 10cm Note prominent facial markings, white supercilium, white wing patch. **V** Ultra-high sustained *tehsreehweeh-*. **H** Bamboo and scrub at montane forest edge, at higher altitude than Ochraceous Wren. **D** Costa Rica, Panama.

6 WHITE-BELLIED WREN *Uropsila leucogastra* 10cm Note faint barring to wings, tail and undertail coverts. **V** Very high short rapid warble *dibliyou*. **H** Understorey of dry forest, woodland. **D** C Mexico to Honduras.

7 WHITE-BREASTED WOOD WREN *Henicorhina leucosticta* 11cm Warm brown upperparts, whitish supercilium, heavily white-and-black-streaked ear coverts, mainly white underparts. Separable from Grey-breasted Wood Wren by more extensive white on underparts, without grey. **V** Varied, hurried series of a musical phrase, typically repeated three times. **H** Undergrowth of humid forest, normally at lower altitude than Grey-breasted Wood Wren. **D** S Mexico to Peru, the Guianas, N Brazil.

8 GREY-BREASTED WOOD WREN *Henicorhina leucophrys* 11cm Note greyish-white supercilium, grey breast and belly. **V** Hurried series of repeated, very high phrases. **H** Undergrowth and borders of montane forest. **D** Mexico to N Bolivia.

9 HERMIT WOOD WREN *Henicorhina anachoreta* 11cm Formerly considered a race of Grey-breasted Wood Wren. Occurs at higher altitude. More easily heard than seen. **V** Higher frequency than Grey-breasted Wood Wren, with a greater frequency range. **H** High-altitude montane forest. **D** Sierra Nevada de Santa Marta, N Colombia.

10 BAR-WINGED WOOD WREN *Henicorhina leucoptera* 11cm White wing-bars are diagnostic. **V** Varied, rapid series of musical notes or phrases repeated 2–4 times. **H** Humid forest on mountain ridges, fern-covered slopes. **D** S Ecuador, N Peru.

11 MUNCHIQUE WOOD WREN *Henicorhina negreti* 11cm Very similar to slightly paler Grey-breasted Wood Wren; best distinguished by voice. **V** Rapid, high series of merry musical phrases. **H** Very wet, stunted cloud forest. **D** W Colombia, very restricted range.

12 NORTHERN NIGHTINGALE-WREN *Microcerculus philomela* 11cm Almost black with long bill. Forages on the ground, bobbing its rear end and short tail; more often heard than seen. **V** High, meandering, fluted, well-separated *wuh wuh uuweeh tuwee-wuh*, like human whistling. **H** Forest undergrowth. **D** S Mexico to Costa Rica.

13 SOUTHERN NIGHTINGALE-WREN *Microcerculus marginatus* 11cm Much paler underparts than Northern Nightingale-Wren. Sometimes known as Scaly-breasted Wren. **V** Very high, fluted *uuh* and ultra-high *feeeh* notes, given at a very calm pace or in a rapid, slightly mounting series. **H** Floor of dense forest. **D** Costa Rica through NW Ecuador, W Peru, W Amazonia to N Bolivia.

14 FLUTIST WREN *Microcerculus ustulatus* 11.5cm Uniform rufous-brown with varying amounts of scaling on flanks. **V** Song is a very beautiful, calm, high, clear, fluted series of notes, like a song whistled by a person. **H** On or near the ground in dense montane forest on tepuis. **D** Venezuela, Guyana, extreme N Brazil.

15 WING-BANDED WREN *Microcerculus bambla* 11.5cm Unmistakable, with white tips to greater coverts forming a prominent wing-bar. **V** Single or rapidly repeated note, starting very high and sometimes descending, with a pure, clear quality. **H** On or near the ground in lowland rainforest. **D** E Ecuador, E Peru; Venezuela, the Guianas, N Brazil.

16 CHESTNUT-BREASTED WREN *Cyphorhinus thoracicus* 15cm Note deep orange-brown throat and breast. Occurs at higher elevations than Musician Wren and Song Wren, both of which show wing and tail barring. **V** Calm series of slow, very pure, fluted notes, interrupted by sudden rattles and trills. **H** Wet montane forest. **D** Colombia to Bolivia.

17 MUSICIAN WREN *Cyphorhinus arada* 12.5cm Note white streaking on rear ear-coverts and nape, rufous front, barring on wings. **V** Song is like that of Song Wren, but phrases may be longer. **H** Lower storeys of humid forest, including várzea. **D** Amazonia.

18 SONG WREN *Cyphorhinus phaeocephalus* 13cm Separable from Chestnut-breasted Wren by distinct bare bluish eye-ring and barring on wings. **V** Short phrases of happy whistling. **H** On or near ground in lowland humid forest and secondary growth. **D** Honduras to Ecuador.

GNATCATCHERS POLIOPTILIDAE

19 COLLARED GNATWREN *Microbates collaris* 10.5cm Note brown crown, black and white facial pattern, black breast patch. **V** Very high and ultra-high drawn-out, slightly descending, plaintive *peee* notes. **H** Undergrowth of wet forest. **D** N Amazonia, from E Ecuador through S Colombia, S Venezuela, N Brazil to the Guianas.

20 TAWNY-FACED GNATWREN *Microbates cinereiventris* 10.5cm Note distinctive facial pattern, with russet cheeks and white throat. **V** Song is like that of Collared Gnatwren, but sharper. **H** Undergrowth of lowland humid forest. **D** Nicaragua to E Peru.

21 LONG-BILLED GNATWREN *Ramphocaenus melanurus* 12cm Long bill and tail are diagnostic. Tends to stay hidden in undergrowth. **V** Varied, mid-high and high rattles and vibrating trills. **H** Lower storeys of humid forest and woodland. **D** SE Mexico through Amazonia to S Brazil.

22 BLUE-GREY GNATCATCHER *Polioptila caerulea* 11cm Male distinguished from female by dark supercilium. Active, forages in trees and tall bushes, often fly-catches by making short flits from branches. **V** A thin *zpee-zpee*, *pwee* or *zeef-zeef*; also a series of soft warbles. **H** Woodland, scrub, mangroves, gardens. **D** Breeds from SE Canada to S Mexico; Bahamas. Northern birds winter south to Honduras, Cuba, Cayman Islands.

23 BLACK-TAILED GNATCATCHER *Polioptila melanura* 11cm Note black and white undertail pattern. Active, often hovers to glean insects. **V** Various, including a *psssh*, *gee-gee* and a *ch-ch-ch-ch...* **H** Arid brush. **D** SW USA, N Mexico.

24 CALIFORNIA GNATCATCHER *Polioptila californica* 11cm Darker grey underparts than Black-tailed Gnatcatcher, with less white in tail, but best separated by voice. **V** A mewing *mee-eew*, a soft *dear dear...* and a harsh *tssshh*. **H** Coastal brush. **D** SW California, NW Mexico.

25 CUBAN GNATCATCHER *Polioptila lembeyei* 10.5cm Black crescent behind eye is generally less distinct in females. A tame, very active forager in scrub; often occurs in family groups. **V** A loud, four-note whistle followed by a trill and a variable whisper, *pss-psss-psss-psss-tttiizzzt-zzzz-ttizzz-tzi-tzi-tzi*; also *speee* and *pit*. **H** Mainly dense, coastal thorn scrub, occasionally inland in thick scrub. **D** Cuba.

26 WHITE-LORED GNATCATCHER *Polioptila albiloris* 11cm Shows much white on tertials and secondaries. Male has black cap and black tail. Female has white supercilium. **V** Very high inhaled sizzled *shreezupzup*. **H** Dry scrub, open woodland, forest edge. **D** W Mexico to Costa Rica.

27 BLACK-CAPPED GNATCATCHER *Polioptila nigriceps* 11cm Glossy black cap of male contrasts with pale eye-ring and blue-grey back. Actions as other gnatcatchers. **V** A buzzy *jehrr* and a sharp *chip chip chip*. **H** Thickets along desert streams. **D** N Mexico, extreme SE Arizona.

28 TROPICAL GNATCATCHER *Polioptila plumbea* 11cm Note head and wing pattern, pure white underparts and range. **V** Ultra-high, rapid or slow trill (1–2 secs). **H** Wide variety of habitats, from humid forest and overgrown pastures to arid scrub and mangrove. **D** S Mexico to Peru and E Brazil.

29 CREAMY-BELLIED GNATCATCHER *Polioptila lactea* 11cm Note creamy underparts of both male and female. **V** Ultra-high, sizzled, slow rattle. **H** Canopy and borders of humid forest. **D** SE Brazil, NE Argentina, E Paraguay.

30 GUIANAN GNATCATCHER *Polioptila guianensis* 11cm Similar to Inambari and Iquitos Gnatcatchers, but has whitish throat and different range. **V** Fairly simple song, consisting of repeated high, sharp notes. **H** Canopy and borders of terra firme forest. **D** E Venezuela, the Guianas, N Brazil.

31 RIO NEGRO GNATCATCHER *Polioptila facilis* 11cm Formerly considered conspecific with Guianan Gnatcatcher, but lacks pale eye-ring and white supraloral, and has less white in the tail. **V** Similar to Guianan Gnatcatcher but slightly softer. **H** Canopy and borders of humid primary forest. **D** E Colombia, S Venezuela to N Brazil.

32 PARA GNATCATCHER *Polioptila paraensis* 11cm Formerly considered conspecific with Guianan Gnatcatcher, but is paler and has a longer tail. **V** Similar to Guianan Gnatcatcher. **H** Canopy and borders of humid primary forest. **D** Amazonian Brazil, south of the Amazon.

33 INAMBARI GNATCATCHER *Polioptila attenboroughi* 11cm A newly discovered species, first described in 2013. Separable from Guianan Gnatcatcher by greyer throat, and from other gnatcatchers by range. **V** Song is a series of level notes like that of Guianan Gnatcatcher, but with a slightly lower tempo. **H** Canopy of upland forest. **D** SW Amazonian Brazil.

34 IQUITOS GNATCATCHER *Polioptila clementsi* 11cm Note grey throat. Known only from a few sites near Iquitos, NE Peru. **V** Very high, sizzling *see-see* trill. **H** Tall forest. **D** NE Peru, very restricted range.

35 SLATE-THROATED GNATCATCHER *Polioptila schistaceigula* 11cm White belly sharply demarcated from grey breast. **V** Very high, dry trill. **H** Canopy and borders of humid forest and secondary growth. **D** E Panama to Ecuador.

36 MASKED GNATCATCHER *Polioptila dumicola* 12.5cm Male has a black mask. Female has a distinct black crescent behind the eye. **V** Varied repertoire, e.g. a very high, sharp, descending series of six *see* notes. **H** Cerrado, Chaco, riverine belts, savannah. **D** Bolivia to C Brazil, Paraguay, Uruguay, N Argentina.

1
2
3
4
5
6
8
10
7
9
11
12
13
15
14
16
17
18
20
19
21
24
23
25
22
26
29
28
31
27
30
33
36
32
34
35

NUTHATCHES SITTIDAE

1 EURASIAN NUTHATCH *Sitta europaea* 14cm Typical nuthatch, with blue-grey upperparts and black eye-stripe. Agile feeder on tree trunks and larger branches; in winter, often visits bird tables or nut feeders. **V** Song variable, from a slowish *pee-pee-pee...* to a rapid *wiwiwiwiwiwiwi...* Calls include a liquid *dwip*, a shrill *sirrrr*, a harsh *trah* and a thin *tsee-tsee-tsee*. **H** Deciduous and mixed forest, parks, large gardens. **D** Widespread from W Europe through Asia to SE Siberia, Japan, Taiwan.

2 SIBERIAN NUTHATCH *Sitta arctica* 14cm Formerly considered a race of Eurasian Nuthatch. Differs from that species in multiple morphological characters, including shorter black eye-stripe and more white in outer tail. **V** Similar to Eurasian Nuthatch. **H** Deciduous and mixed forest. **D** C and NE Siberia.

3 CHESTNUT-VENTED NUTHATCH *Sitta nagaensis* 12.5–14cm Deep rusty red flanks, reddish scaling on undertail coverts. Regularly feeds on the ground. **V** Song is a rapid *chichichichichi...*, or a slower *chi-chi-chi-chi-chi...* Calls include *sit* or *sit-sit*, a *quir* and a *tsit* or *tsit-tsit-tsit...* **H** Evergreen hill forest, pine forests, mixed light deciduous forest. **D** Tibet to SC Vietnam.

4 KASHMIR NUTHATCH *Sitta cashmirensis* 14cm Note buffy-orange underparts. **V** A *tsi-tsi*, a rapid *pee-pee-pee-pee* and a harsh *kraaa*. **H** Various deciduous and pine forests. **D** NW Himalayas.

5 INDIAN NUTHATCH *Sitta castanea* 13cm Note contrast between white throat and chestnut breast and belly. Often considered conspecific with Chestnut-bellied Nuthatch. **V** Rapid, descending trill. Calls include a short *stit* that sometimes accelerates into a fast trill. **H** Deciduous forest, groves and gardens. **D** India, SW Nepal, SW Bangladesh.

6 CHESTNUT-BELLIED NUTHATCH *Sitta cinnamoventris* 14cm Very like Indian Nuthatch, but has slightly darker crown, more orange (less brown) underparts, white scallops on undertail coverts. **V** Calls include a mellow *tsup*, a thin *sit* or *sit-sit*, and a screechy *chreet chreet chreet*. **H** Mainly open deciduous forest. **D** N Pakistan through Himalayas to N SE Asia.

7 BURMESE NUTHATCH *Sitta neglecta* 13cm Usually forages on the upper trunk or upper branches. Formerly considered a race of Chestnut-bellied Nuthatch. **V** Song is a mellow, wavering trill; calls include a screechy *chreet-chreet* and a hard explosive rattle. **H** Dry dipterocarp and pine forest. **D** Myanmar to Vietnam.

8 WHITE-TAILED NUTHATCH *Sitta himalayensis* 12cm Note bright white spot at base of tail, and unspotted undertail coverts. Forages mostly on mossy branches in the upper parts of trees. **V** Songs include a fast, rising *dwi-dwi-dwi...* and a slower *tui-tui-tui-tui...* Calls include *nit*, *shree*, *tschak* and *chak-kak*. **H** Mixed forests of pine, oak, maple, rhododendron. **D** Himalayas to NW Vietnam.

9 WHITE-BROWED NUTHATCH *Sitta victoriae* 11–12cm Note largely whitish face and small bill. Forages on the smaller, outer branches of trees. **V** Song is a crescendo of 9–12 notes, *whi-whi-whi-whi-whi-whi...* **H** Oak and mixed oak and rhododendron forests. **D** W Myanmar, very restricted.

10 PYGMY NUTHATCH *Sitta pygmaea* 11cm Note grey-brown cap and ill-defined dark eye-stripe. Restless forager. Occurs in small to large flocks after breeding. **V** High pitched, including *bip-bip-bip*, *kip* and a *wee-bee*. **H** Pine forests, especially ponderosa pines. **D** Mountain regions from SW Canada to NW Mexico.

11 BROWN-HEADED NUTHATCH *Sitta pusilla* 9.5–11cm Resembles Pygmy Nuthatch, but crown is browner, and there is no range overlap. **V** Weak, high-pitched chattering. **H** Pine barrens. **D** SE USA, Bahamas.

12 CORSICAN NUTHATCH *Sitta whiteheadi* 12cm Note prominent white supercilium. Feeds tit-like among branches, or feeds by hovering or fly-catching. **V** Rapid, quavering *hidididididid...* and an ascending *dew-dew-dew...*, sometimes combined. Calls include *pu*, a thin *tsi-tsi-tsi* and a nasal *pink*; *chay-chay-chay* when agitated. **H** Pine forest. **D** Corsica.

13 ALGERIAN NUTHATCH *Sitta ledanti* 13.5cm Very like Corsican Nuthatch, but underparts more pinkish-buff, and male has black on forecrown only. **V** Fluty *quair-di - quair-di - quair-di* or fast *du-wid-du-wid...* Calls include a soft *kna*, *quuwee*, a rasping *schrr-shrr-schrr* and a loud *chwa-chwa-chwa*. **H** Mainly Atlas Oak and cedar. **D** N Algeria.

14 KRÜPER'S NUTHATCH *Sitta krueperi* 12.5cm Both sexes have rusty breast patch. Black forehead less pronounced on female. When agitated, sits upright, flicking wings. **V** Various trills, *pip-pip-pip-pip...*, *veet-veet-veet...* or *yu-di - yu-di - yu-di - yu-di...* **H** Coniferous forests. **D** E Greece, Turkey, Caucasus.

15 YUNNAN NUTHATCH *Sitta yunnanensis* 12cm Forages, tit-like, among needle clusters of pine trees. **V** Calls include a nasal *nit*, an abrupt *chit-chit*, a piping *pi-pi-pi-pi...* and a scolding *schri-schri-schri...* **H** Open mature pine forests. **D** SW China.

16 RED-BREASTED NUTHATCH *Sitta canadensis* 11.5cm Female duller on crown and flanks. Often forages head-down on trunks and branches, regularly gleans from outer twigs. **V** Song is a clear, rising *eeeen eeeen eeeen...* **H** Coniferous and mixed forest. **D** S Canada, NE USA, mountains of W and E USA. Winters across whole of USA.

17 CHINESE NUTHATCH *Sitta villosa* 11–12cm Forages among small twigs and branches at the top of trees; occasionally makes fly-catching sorties. **V** Song consists of a series of upward-inflected whistles, *tsi-pui-pui-pui-pui*. **H** Coniferous forests. **D** C to NE China, Korean Peninsula, SE Siberia.

18 WHITE-CHEEKED NUTHATCH *Sitta leucopsis* 13cm Black cap and white face. Shy, best located by call; forages mainly in the treetops. **V** Gives a constant, bleating *kner-kner* or similar. Song consists of prolonged, rapid, wailing squeaks. **H** Coniferous forests or mixed forests. **D** W Himalayas.

19 PRZEVALSKI'S NUTHATCH *Sitta przewalskii* 12–13cm Formerly treated as a race of White-cheeked Nuthatch. Forages mainly in treetops. **V** Calls include a muffled *chip*, a whistled *dweep* or *dweep-eep* and a thin *pee-pee-pee*. **H** Spruce and fir forests. **D** SE Tibet, WC China.

20 WHITE-BREASTED NUTHATCH *Sitta carolinensis* 13–15cm Female has a greyer crown. Western birds have darker grey flanks. **V** Call variable: eastern birds utter a *yenk*, western birds a *eeern* and interior birds a *yidi-yidi-yidi*. Song is a soft *whi-whi-whi...* **H** Open woodland, especially oak and pine. **D** S Canada, USA, Mexico, widespread.

21 WESTERN ROCK NUTHATCH *Sitta neumayer* 13cm Restless, feeds on ground as well as rocks. Often perches upright. **V** Song is a loud trill, *iti-tit - tuit-tuit-tuit-tuit...* Calls include *chik*, a harsh *charr* and a screeching *creea*. **H** Rocky slopes, gorges, etc., in dry or arid areas. **D** SE Europe to Iran.

22 EASTERN ROCK NUTHATCH *Sitta tephronota* 15cm Forages with jerky hops, on cliff faces and boulders and in rock crevices. **V** Loud trilling calls given all year round; also a *ch-ch-ch* when agitated. Song very like Western Rock Nuthatch. **H** Rocky valleys, with broken cliffs and boulders. **D** SE Turkey to W Pakistan and Kazakhstan.

23 VELVET-FRONTED NUTHATCH *Sitta frontalis* 12.5cm Distinctive, with red bill, yellow eye, black forehead, violet-blue upperparts. Very active; forages from undergrowth to canopy. **V** Song is a rattling *sit-sit-sit-sit...* Calls include a hard *chat* and a thin *sip* or *tsit*. **H** Open forests and well-wooded areas. **D** India, Sri Lanka, Himalayas to SE Asia, Indonesia, W Philippines.

24 YELLOW-BILLED NUTHATCH *Sitta solangiae* 12–14cm Found singly or in small parties, often part of mixed-species flocks, foraging from undergrowth to the upper storey. **V** A stony *chit* that often lengthens into a *chit-it-it-it-it-it...* Song is a fast *sit-ti-ti-ti-ti-ti...* **H** Broadleaved evergreen forest. **D** Hainan, Laos, Vietnam.

25 SULPHUR-BILLED NUTHATCH *Sitta oenochlamys* 12–13cm Forages on main tree trunks and nearby branches in the upper tree levels. **V** Calls include a *chit*, a thinner *sit* or *sit-sit-sit-sit*, a squeaky *snii* and a winding rattle. **H** Pine and evergreen forests, forest edges and clearings, remnant patches of forest and secondary growth. **D** Philippines.

26 BLUE NUTHATCH *Sitta azurea* 13–14cm Often appears black and white, with noticeable pale eye-ring. Forages mainly in the upper storey, on trunks and main branches. **V** A short *chit*, which when excited lengthens into a *chi-chit*, a *chit-chit-chit*, a *chir-ri-rit* or a trilling *tititititititik*. **H** Broadleaved evergreen forests. **D** Peninsular Malaysia; Sumatra and Java, Indonesia.

27 GIANT NUTHATCH *Sitta magna* 20cm Actions much as other nuthatches, but less restless. Occasionally cocks tail; in flight wings give out a whirring sound. **V** Calls include a chattering *gd-da-da*, *dig-er-up* or *get-it-up*, also a *ge-de-ku* that sometimes becomes a gamebird-like *gu-drr gu-drr gu-drr*. **H** Open mature pine and mixed oak and pine forest. **D** SC China, NW Thailand, Myanmar.

28 BEAUTIFUL NUTHATCH *Sitta formosa* 16–17cm Forages up, down and along branches; usually high up in tall trees covered with epiphytes; more sluggish than other nuthatches. **V** A soft, liquid *plit* and an explosive *chit* that sometimes lengthens to a *chit-it chit-it chit-it* or *chit-it-it chirririt*. Song is a *chi it it it it it it it...* **H** Broadleaved evergreen and semi-evergreen forests. **D** Himalayas to NW Vietnam.

WALLCREEPER TICHODROMIDAE

29 WALLCREEPER *Tichodroma muraria* 16.5cm Wary. Forages on cliff faces and in crevices, regularly flicking wings open to reveal red and white markings. **V** Song is a drawn-out, piping *tu-tuee-zreeeeeu* or *chewee-cheweeooo*. Calls include a whistled *tseooo*, a rapid twitter, a buzzing *zree*, a *tui* and a *chup*. **H** Cliffs and rock faces; in bad weather often forced lower to old buildings and rocky riverbeds. **D** S Europe to C China.

TREECREEPERS CERTHIIDAE

30 EURASIAN TREECREEPER *Certhia familiaris* 12.5cm Brown and buff streaked upperparts, whitish supercilium, white underparts. Typically feeds by climbing tree trunks and large branches. **V** Song is a very high-pitched *tsee-tsee-tsi-tsi-si-si-si-si-sisisisisi*. Calls include *tsrree* or *tsee*. **H** Deciduous and mixed woodland, occasionally pure conifer forests. **D** Widespread, from Ireland through Europe to China, Japan.

31 HODGSON'S TREECREEPER *Certhia hodgsoni* 11–12cm Note mottled black, brown and white ear-coverts. Creeps up and around trees in jerky movements. **V** Song is a high-pitched *tzee-tzee-tzizizi*. Call is *tsree* or *tsree-tsree-tsree*. **H** Mixed conifer and birch forests. **D** Himalayas from N Pakistan to NE Myanmar, SW China.

32 BROWN CREEPER *Certhia americana* 13–15cm Forages, with a jerky creeping action, from base of trees upwards. Occurs in brown and greyish morphs. **V** Song variable, often transcribed as *trees trees pretty little trees*. Typical call is a thin *seee* or *sreee*. **H** Coniferous, mixed and deciduous forests. **D** Resident across mountainous regions of North and Central America, from Alaska to Newfoundland to Nicaragua. Occurs in winter throughout USA.

33 SHORT-TOED TREECREEPER *Certhia brachydactyla* 12.5cm May be impossible to distinguish from Eurasian Treecreeper in the field. **V** Song is a series of loud, plaintive whistles, *seet-seet-seet-e-roi-deitt*. Calls include an explosive shrill *zeet* and a tremulous *tsrree*, like Eurasian Treecreeper, but lower. **H** Deciduous, mixed and coniferous woodland, parks, orchards, large gardens. **D** W, C and SE Europe, Turkey, Caucasus, NW Africa.

34 BAR-TAILED TREECREEPER *Certhia himalayana* 14cm Forages by climbing up tree trunks and along branches; sometimes fly-catches or occasionally feeds on the ground. **V** Song is a lilting *tsee tsu-tsu tsut tut tut li tee* or *ti ti tu-du-du-du-du*. **H** Montane coniferous forest. **D** From Kyrgyzstan through Afghanistan to W Nepal; W Tibet and N Myanmar to C China.

35 RUSTY-FLANKED TREECREEPER *Certhia nipalensis* 14cm Note contrast between off-white throat and breast and cinnamon lower underparts. **V** Song is a high-pitched, accelerating trill. **H** Breeds in mixed forests; winters at lower elevations in broadleaved forests. **D** Himalayas to NE Myanmar.

36 SIKKIM TREECREEPER *Certhia discolor* 14cm Separable from Rusty-flanked Treecreeper by less contrasty underparts. Creeps up mossy trunks and branches. **V** Song is a monotonous *chit-it-it-it-it-it-it-it...* Calls include a *tchip*, a *tsit* and a rattling *chi-r-r-it*. **H** Broadleaved evergreen and mixed broadleaved and pine forest. **D** Nepal, Tibet, NE India.

37 HUME'S TREECREEPER *Certhia manipurensis* 14cm Formerly considered conspecific with Sikkim Treecreeper; actions thought to be similar. **V** Song is a monotonous, slow rattle. Calls include a explosive *chit* or *tchip* which sometimes turns into a rattle. **H** Moist evergreen hill forest, mixed pine–broadleaf forest. **D** E Himalayas to SC Vietnam.

38 SICHUAN TREECREEPER *Certhia tianquanensis* 14cm Actions typical of genus. Forages on trunks and branches in the upper storey. **V** Song is a loud, rapid trill, starting explosively before tailing off and falling in pitch. **H** Open stands of mature conifers. **D** SC China.

39 INDIAN SPOTTED CREEPER *Salpornis spilonota* 15cm Spotted and barred plumage is distinctive. Forages on trunks and branches. **V** A strident, rising *tui-tui-tui*. Song is a short series of plaintive whistled notes. **H** Open deciduous forest, well-wooded country and groves. **D** W and C India.

40 AFRICAN SPOTTED CREEPER *Salpornis salvadori* 15cm Very distinctive if seen well. Climbs tree trunks without support of tail. **V** Extremely high, narrow, piercing *seeeeseeeeseeeeseeee*. **H** Miombo woodland. **D** Senegal and Ethiopia south to Zimbabwe, Mozambique.

1
2
3
4
5
6
7
8
9
10
11
12
13
14
15
16
17
18
19
20
21
22
23
24
25
26
27
28
29
30
31
32
33
34
35
36
37
38
39
40

MOCKINGBIRDS AND THRASHERS MIMIDAE

1 GREY CATBIRD *Dumetella carolinensis* 23cm Grey with black crown and reddish undertail coverts. Skulks in thick cover, near or on the ground. More often heard than seen. **V** An explosive *kak-kak-kak* and a soft, cat-like *mew*. Song consists of sweet, varied phrases interspersed with mewing and harsh notes. **H** Dense, low, streamside thickets, woodland, garden shrubbery. **D** Breeds across S Canada, C and E USA. Winters E coastal USA, Florida and Gulf coast, Caribbean, Mexico.

2 BLACK CATBIRD *Melanoptila glabrirostris* 20cm Other all-black birds in Central America are larger, with stout bills and less skulking. **V** Long sustained very high wailing flutes with *weet* and *sjuut*. **H** Forest edge, woodland, dense scrub. **D** S Mexico (Yucatán Peninsula) to Honduras.

3 NORTHERN MOCKINGBIRD *Mimus polyglottos* 24–28cm Conspicuous and often aggressive. In flight shows large white wing patch and white outer-tail feathers. **V** A series of melodious phrases, each repeated several times; often mimics the calls of other birds. When agitated, a harsh *tchack*. **H** Thickets, copses, shelterbelts and gardens. **D** Resident across North America, from S Canada to S Mexico, Caribbean.

4 TROPICAL MOCKINGBIRD *Mimus gilvus* 23–24cm Similar to Northern Mockingbird, but less white in wings. Forages mostly on the ground or low in vegetation, also hawks for insects. **V** Repeated musical whistles and phrases, similar to Northern Mockingbird but slightly more harsh with less mimicry. When agitated, gives a harsh *chuck*. **H** Open areas near habitations, lowland scrub and cultivations. **D** S Mexico to Lesser Antilles, N and E South America

5 BAHAMA MOCKINGBIRD *Mimus gundlachii* 28cm Browner and more streaked than Northern Mockingbird. Forages mostly on the ground or in tall dense vegetation. **V** Repeated phrases, less variable than Northern Mockingbird, and lacks mimicry. **H** Woodland, semi-arid scrub, urban areas. **D** Bahamas, Turks and Caicos Islands, N Cuba, S Jamaica.

6 CHILEAN MOCKINGBIRD *Mimus thenca* 27cm Separable from Patagonian Mockingbird by brownish-black malar stripe and less distinct wing markings. **V** Irregular series of very high, fluted shrieks and low, short, rhythmic chatters. **H** Low coastal scrub, semi-desert, savannah. **D** Chile, WC Argentina.

7 LONG-TAILED MOCKINGBIRD *Mimus longicaudatus* 29cm Note facial pattern and, in flight, white tail corners. **V** Unstructured series of mostly scratchy or croaking notes, chatters and rattles. **H** Desert scrub, arid woodland, agricultural areas. **D** SW Ecuador, W Peru.

8 CHALK-BROWED MOCKINGBIRD *Mimus saturninus* 26cm Note distinctive broad white supercilium and white tail tips. **V** Stream of upslurred trills, rattles and fluted notes, most of which are repeated 3–4 times. **H** Savannah, open woodland, farmland, Cerrado, Caatinga. **D** Suriname, S Amazonia, NE Brazil to C Argentina.

9 PATAGONIAN MOCKINGBIRD *Mimus patagonicus* 24cm Relatively plain plumage with whitish supercilium and two narrow wing-bars. **V** Series of short elements, e.g. a repeated, very high *peep*, or low croaks, rattles and trills. **H** Low desert shrub, low woodland. **D** S, C and NW Argentina, S Chile.

10 WHITE-BANDED MOCKINGBIRD *Mimus triurus* 23cm Note distinctive white wing panel. **V** Rather high series of sharp rattles, shrieks, staccato notes, trills. **H** Low woodland, scrubby areas, farmland, habitation. **D** Bolivia through Paraguay, S Brazil, Uruguay to C Argentina.

11 BROWN-BACKED MOCKINGBIRD *Mimus dorsalis* 25cm Uniform rufescent brown above; less white in wings than White-banded Mockingbird. **V** Bursts of rattles, chirps and staccato notes. **H** Arid scrub, brush, habitation. **D** Bolivia, NW Argentina.

12 GALAPAGOS MOCKINGBIRD *Mimus parvulus* 25cm The four mockingbirds of the Galápagos Islands are best separated by range. This is the only one of the four that occurs on multiple islands. Note dark mask and fairly unmarked underparts. **V** Unstructured, gurgled twittering. **H** Arid coastal scrub, low woodland, mangrove. **D** Galápagos.

13 FLOREANA MOCKINGBIRD *Mimus trifasciatus* 25cm Note black patch on side of breast. **V** Song is loud and melodious, without mimicry. **H** Desert scrub, low vegetation at beach. **D** Islets off Floreana, Galápagos.

14 ESPANOLA MOCKINGBIRD *Mimus macdonaldi* 27cm Note long downcurved bill. **V** Song is loud and strident, without mimicry. **H** Arid scrub, dry woodland. **D** Española, Galápagos.

15 SAN CRISTOBAL MOCKINGBIRD *Mimus melanotis* 25cm Intermediate in appearance between Galapagos and Espanola Mockingbirds, with dark mask and blotched breast. **V** Varied musical repertoire, dominated by an endlessly repeated mini-phrase. **H** Most habitats on San Cristóbal, except dense forest, tall woodland and grassland. **D** San Cristóbal, Galápagos.

16 SOCORRO MOCKINGBIRD *Mimus graysoni* 25cm Distinguished from Northern Mockingbird by pale stripes to flanks and absence of white wing patch. **V** Very high short sharp *wup wippip-wup* and other strophes. **H** Wooded slopes. **D** Isla Socorro, W Pacific (Mexico).

17 SAGE THRASHER *Oreoscoptes montanus* 20–23cm Forages on the ground. Sings from the tops of bushes. **V** Song is a continuous series of warbled phrases. Calls include a low *chup*, a high *churr* and a whistled *whee-er*. **H** Sagebrush plains. **D** Breeds W USA and extreme SW Canada. Winters SW USA, Mexico.

18 BROWN THRASHER *Toxostoma rufum* 28cm Rufous upperparts, coarsely streaked whitish underparts, two wing-bars. Forages primarily in leaf litter under bushes and trees. **V** Song consists of rich musical phrases; also a quieter low warble. Typical call is a loud *tschek* or *chip*. **H** Bushy woods, shelterbelts, copses, shrubby gardens. **D** Breeds SC and SE Canada, USA east of the Rockies. Winters SE USA.

19 LONG-BILLED THRASHER *Toxostoma longirostre* 26–29cm Grey-faced, with slightly downcurved bill. Actions similar to Brown Thrasher. **V** Song consists of rich musical phrases, harsher and less rambling than Brown Thrasher. Calls are a sharp *chak*, a mellow *cheeop* and a rattled *chttr*. **H** Dense thickets. **D** Resident in S Texas, NE Mexico.

20 COZUMEL THRASHER *Toxostoma guttatum* 25cm No other thrasher on Cozumel. Critically endangered. **V** Song is rich and varied, with warbles and scratchy notes and little repetition. **H** Scrubby woodland. **D** Cozumel, off Yucatán Peninsula.

21 GREY THRASHER *Toxostoma cinereum* 25cm Note arrow-shaped spots on white underparts. **V** Song consists of loud, warbling scratchy phrases that are often repeated. **H** Dry areas with scrub, bush, trees, cacti. **D** Baja California.

22 BENDIRE'S THRASHER *Toxostoma bendirei* 23–25cm Note relatively uniform olive-brown plumage and yellow eye. Forages mainly on the ground. Often cocks tail when running. **V** Song is a continuous, flowing, sweet and husky warble. Call is a low *chuk*. **H** Open farmland, grassland, brushy desert. **D** SW USA, NW Mexico.

23 OCELLATED THRASHER *Toxostoma ocellatum* 30cm Bold spotting on white underparts diagnostic. **V** Very long stream of very varied very high notes and strophes, sharp flutes, repetitions and mimics. **H** Dry scrubby open woodland. **D** SC Mexico.

24 CURVE-BILLED THRASHER *Toxostoma curvirostre* 26–28cm Orange eye. Plumage variable. Forages primarily on the ground. **V** Song consists of trills, warbles and rattles. Calls include a sharp, liquid *wit-WEET-wit*, a dry *pitpitpitpit* and a low *chuk*. **H** Open desert with emergent trees, cacti and thorn scrub. **D** SW USA, Mexico.

25 CALIFORNIA THRASHER *Toxostoma redivivum* 28–32cm Forages on the ground, usually under cover of shrubs and around fallen branches. Runs swiftly between feeding sites, with tail slightly raised. **V** Song is a series of variable, harsh and mellow phrases. Calls include a loud *churreep*, a soft *shtupp* and when alarmed a sharp *chack*. **H** Primarily chaparral-covered slopes. **D** W California, Baja California.

26 CRISSAL THRASHER *Toxostoma crissale* 30cm Greyer than California Thrasher and LeConte's Thrasher, with dark rusty undertail coverts. Note pale eye. **V** Series of hurried rather nasal twittering with *wehwehweet*. **H** Dry scrubby areas. **D** SW USA, N and C Mexico.

27 LECONTE'S THRASHER *Toxostoma lecontei* 30cm Note overall pale colouring, black tail and malar stripe in otherwise weakly patterned face. **V** Rather short streams of very high sharp fluted syllables and notes, each repeated 2–3 times. **H** Desert with some low bush and scrub. **D** SW USA, NW Mexico.

28 WHITE-BREASTED THRASHER *Ramphocinclus brachyurus* 20–23cm Forages mainly on the ground, usually among leaf litter; also feeds in bushes and trees, especially when they are in fruit. **V** Repeated short, mellow phrases of several syllables; also utters a *chek-chek-chek* on Martinique and a *tschhhhhh* on St Lucia. **H** Dry woodland, coastal thickets, wooded stream valleys and scrub forest. **D** Martinique and St Lucia.

29 BLUE MOCKINGBIRD *Melanotis caerulescens* 25cm Dark red eye in black mask diagnostic. Skulking, though sings from an exposed perch. **V** Series of well-separated rather short fluted strophes, *fufisheetwusheetwee*. **H** Pine forest undergrowth, scrubby woodland. **D** Mexico.

30 BLUE-AND-WHITE MOCKINGBIRD *Melanotis hypoleucus* 25cm Deep blue upperparts, black mask, white underparts, long tail. **V** Very high rattling *fjujiweetweetwettjutju*. **H** Dry scrub, woodland, forest edge. **D** S Mexico to Honduras and El Salvador.

31 SCALY-BREASTED THRASHER *Allenia fusca* 23cm Retiring. Mainly arboreal, feeds on fruits. **V** Repeated phrases much like those of the Tropical Mockingbird, but less vigorous. **H** Moist and semi-arid forests and woodlands. **D** Lesser Antilles.

32 PEARLY-EYED THRASHER *Margarops fuscatus* 28–30cm Arboreal, often forages in small groups from the middle level to the canopy. Takes a large variety of food, including fruits, berries, frogs, lizards, land crabs, eggs, nestlings and occasionally adult birds. **V** A slow series of one- to three-syllable phrases with lengthy pauses between segments; also a *craw-craw* or *chook-chook*. Often sings late in the day and occasionally at night. **H** Forests, woodlands, thickets, mangroves, coastal palm groves, urban areas. **D** Bahamas to Lesser Antilles.

33 BROWN TREMBLER *Cinclocerthia ruficauda* 23–26cm Arboreal, searches for food among epiphytic vegetation and tree hollows. Frequently cocks tail over back. The name 'trembler' refers to the way these birds quiver when they sing. **V** A rasping *yeeak* and a series of melodic and harsh phrases. **H** Wet forests, secondary forests, drier woodland, scrub. **D** Lesser Antilles, more widespread than Grey Trembler.

34 GREY TREMBLER *Cinclocerthia gutturalis* 23–26cm Greyer and less rufous than Brown Trembler, with paler underparts. Actions and foraging methods similar. **V** Repeated wavering, whistled notes and phrases; also harsh scolding notes. **H** Mainly high-level mature, moist forests. Less common in open woodlands, secondary growth and dry scrub. **D** Martinique and St Lucia.

1
2
3
4
5
6
7
8
9
10
11
12
13
14
15
16
17
18
19
20
21
22
23
24
25
26
27
28
29
30
31
32
33
34

STARLINGS AND RHABDORNIS STURNIDAE

1 METALLIC STARLING *Aplonis metallica* 21–26cm Gregarious. Favours the crowns of flowering or fruiting trees. **V** Loud, unmusical, downslurred note; also a harsh *squeeow*. **H** Lightly wooded cultivation, disturbed forest, forest edge, plantations, mangroves. **D** Maluku Islands (Indonesia) through New Guinea to Solomon Islands, NE Australia.

2 VIOLET-HOODED STARLING *Aplonis circumscripta* 21–26cm Formerly considered a race of Metallic Starling; habits and actions similar. **V** Unknown; presumed to be similar to Metallic Starling. **H** Forest edges, lightly wooded cultivation, disturbed forest, secondary woodland, mangroves. **D** Tanimbar Islands, Indonesia.

3 YELLOW-EYED STARLING *Aplonis mystacea* 20cm Similar to Metallic Starling but has prominent forehead tuft and shorter tail. Note yellow eye. Usually in small flocks, sometimes mixed with Metallic Starling. **V** Similar to Metallic Starling but deeper and harsher. **H** Lowland forest. **D** New Guinea.

4 SINGING STARLING *Aplonis cantoroides* 19cm All black with a blue-green gloss. Note red eye and short square tail. Breeds in holes in dead trees. **V** Calls include a clear downslurred *tseeeu* or a lower-pitched trilled *trrree*. **H** Open habitats with trees and bushes, at all altitudes but favours lowlands. **D** New Guinea to Solomon Islands.

5 TANIMBAR STARLING *Aplonis crassa* 20cm Dark charcoal-black glossed metallic green, slightly paler on underparts. Regularly seen in the crowns of flowering or fruiting trees. **V** Repeated piping, metallic notes; flocks utter a jumbled cacophony of piping, metallic calls. **H** Wooded areas, including secondary forest and woodland edges. **D** Tanimbar Islands, Indonesia.

6 ATOLL STARLING *Aplonis feadensis* 20cm All black with green and blue gloss. Note short square tail. Differs from Metallic and Singing starlings by yellow eye. **V** Calls are squeaky and high-pitched. **H** Forest, forest edge, plantations and gardens. Only on small islands. **D** Bismarck Archipelago, Solomon Islands.

7 RENNELL STARLING *Aplonis insularis* 17cm Plump short-tailed starling with yellow eye. Note rounded wings in flight. Singly or in pairs. **V** Call is a thin high-pitched *zeek*. **H** Open forest, forest edge and gardens. **D** Rennell (Solomon Islands).

8 LONG-TAILED STARLING *Aplonis magna* 40cm All black with dark eye and very long tail. In pairs and small flocks. **V** Song is an unmusical medley of rich notes. Call is a high-pitched downslurred *tseeeu*. **H** Lowland and hill forest, secondary growth and gardens. Endemic to islands in Geelvink Bay. **D** Islands off NW New Guinea.

9 WHITE-EYED STARLING *Aplonis brunneicapillus* 32cm All black with elongated central tail feathers (often broken or missing). Note white eye and heavy bill. Usually in small flocks. **V** Call is a loud whistled *kwee* or *kwee-kwee*. Also gives a series of whistled and harsh notes. **H** Forest, forest edge and secondary growth, in lowlands and hills. **D** Solomon Islands.

10 BROWN-WINGED STARLING *Aplonis grandis* 25cm Bulky all-black starling with brown wings. Usually in pairs. **V** Calls include high-pitched whistles, as well as clicks and pops. **H** Forest edge, secondary growth and gardens, in lowlands and hills. **D** Solomon Islands.

11 MAKIRA STARLING *Aplonis dichroa* 20cm Glossy black starling with brown wings and short tail. Note red eye. Singly, in pairs or small groups; sometimes in mixed-species flocks. **V** Song is a repeated series of slurred notes. Also gives various whistles and squeaks. **H** Forest edge and secondary growth, especially in hills. **D** Makira (Solomon Islands).

12 RUSTY-WINGED STARLING *Aplonis zelandica* 19cm Grey-brown starling with rufous-chestnut wing patch and rump. Underparts paler. In pairs or small groups. **V** Calls include thin high-pitched whistles and harsh clicks. **H** Forest, forest edge, secondary growth and gardens, from lowlands to hills. **D** Nendö and Vanikoro (Solomon Islands), N Vanuatu.

13 STRIATED STARLING *Aplonis striata* 18cm Male is glossy black with brownish wings and heavy bill. Note red or orange eye. Female is grey, usually pale but sometimes much darker. In pairs or small flocks. **V** Gives variety of whistles and squeaks. **H** Forest, savanna and gardens. **D** New Caledonia (including Loyalty Islands).

14 MOUNTAIN STARLING *Aplonis santovestris* 17cm Plump rufous-brown starling with white eye. Singly or in pairs in the understorey. **V** Song is a series of whistles and trills. Calls include a thin whistle and soft clicks. **H** Montane forest, at the highest altitudes. **D** Espiritu Santo (Vanuatu).

15 ASIAN GLOSSY STARLING *Aplonis panayensis* 20cm Note greenish gloss, red eye, black bill and legs. Mainly arboreal. **V** Series of metallic squeaks. **H** Forest edge, secondary growth, coastal forest, plantations, urban areas. **D** Bangladesh and NE India through W Myanmar to Malay Peninsula, Greater and Lesser Sunda Islands, Philippines.

16 MOLUCCAN STARLING *Aplonis mysolensis* 20cm Gregarious. Generally occurs in small or large groups, often seen high in fruiting trees.. **V** Unmusical, nasal, slurred whistle, and various chattering and squealing notes. **H** Mangroves, coastal woodland, primary and degraded forest, forest edge, cultivation, urban areas. **D** Islands E of Sulawesi, Maluku Islands and West Papuan islands, Indonesia.

17 SHORT-TAILED STARLING *Aplonis minor* 18cm Gregarious; forages in small flocks, with larger concentrations at fruiting trees. **V** Calls include a plaintive *seep* and a slurred, metallic *chilanc*. **H** Forest, forest edge. **D** Lesser Sunda Islands, Sulawesi, S Philippines.

18 MICRONESIAN STARLING *Aplonis opaca* 24cm Black plumage and yellow eyes diagnostic. **V** Song is a tuneful rising phrase of call notes. Calls include gurgles, whistles and chipping notes, also a harsh rolling *brleep*. **H** Varied habitats at all elevations. **D** Caroline Islands, Palau, Mariana Islands, W Pacific.

19 POHNPEI STARLING *Aplonis pelzelni* 16cm Very small; plumage is sooty black without reflections. Note dark eyes. **V** Call is a shrill, ringing *see-ay*. **H** Montane forest. **D** Pohnpei (Caroline Islands, W Pacific).

20 POLYNESIAN STARLING *Aplonis tabuensis* 19cm Note relatively short bill and brown or brownish plumage. **V** Rather toneless *tjuh-wrueeeh* and loud, sharp *tiew-tiew-tiew* or *weet-weet-weet*. **H** Mainly forest edge; any wooded habitat on smaller islands. **D** Many Polynesian islands.

21 SAMOAN STARLING *Aplonis atrifusca* 30cm Dark overall, almost black below. **V** High *tju-wrrreh*. **H** Forest clearings, plantations, coastal fringe, gardens. **D** Samoa.

22 RAROTONGA STARLING *Aplonis cinerascens* 21cm Mouse-grey, mottled paler. **V** Varied, with melodious, well-spaced notes such as *prrruh*, *wrihwrih* and *rrreeet*. **H** Native forest. **D** Rarotonga (Cook Islands, S Pacific).

23 YELLOW-FACED MYNA *Mino dumontii* 25cm Glossy black plumage with large orange-yellow patch around eye. Note heavy orange-yellow bill and white wing patch. Usually in pairs or small flocks. **V** Strange frog-like croaks and squawks. **H** Forest, secondary growth and gardens, in lowlands and hills. **D** New Guinea.

24 LONG-TAILED MYNA *Mino kreffti* 28cm Formerly considered conspecific with Yellow-faced Myna. Similar to that species, but differs in shape of face wattle and slightly longer tail. No overlap in range. Usually in pairs or small flocks. **V** Gives a variety of downslurred whistles and squeaks. **H** Forest, secondary growth and gardens, in lowlands and hills. **D** Bismarck Archipelago, Solomon Islands.

25 GOLDEN MYNA *Mino anais* 24cm Striking black and yellow myna, with black head, yellow eye and yellow breast and rump. Some races have yellow crown. Usually in pairs or small flocks. **V** Calls include a bell-like *skew-wee*, and various squawks and whistles. **H** Lowland forest, forest edge and clearings. **D** New Guinea.

26 SULAWESI MYNA *Basilornis celebensis* 23–27cm Distinctive, with helmet-like crest and white patches on ear-coverts and breast sides. **V** High-pitched whistles and drawn-out nasal notes; also brief squeaks and *chip* notes. **H** Primary and disturbed forest, forest edge, secondary woodland, lightly wooded cultivation. **D** Sulawesi, Indonesia.

27 HELMETED MYNA *Basilornis galeatus* 24–25cm Generally seen in pairs; small flocks occur in fruiting tees. **V** Booming *poo poo poop*, repeated every few seconds. **H** Primary and degraded forest, tall secondary forest, overgrown cultivation, mangroves. **D** Banggai and Sula islands, E of Sulawesi, Indonesia.

28 LONG-CRESTED MYNA *Basilornis corythaix* 24–26cm Noisy. Favours fruiting trees. **V** Calls include a series of five piercing, rising whistles and various ascending and descending nasal notes, interspersed with occasional short, piping notes. **H** Tall secondary forest, forest edge, overgrown cultivation, gardens. **D** Seram (S Maluku Islands, Indonesia).

29 APO MYNA *Basilornis mirandus* 28–30cm Note wispy crest and bare yellow skin around eye. Perches on high, exposed branches. **V** Various tinkling and metallic notes, interspersed with snaps, gurgles and squeaks. **H** Montane forest, forest edge. **D** Mindanao, S Philippines.

30 COLETO *Sarcops calvus* 26–28cm Grey and black, with bare skin on most of face. Frequently seen perched at the top of dead trees. **V** Metallic click, followed by a high-pitched *kliing-kliing*; also chiming notes mixed with mechanical gurgles and splutters, like an orchestra warming up. **H** Forest, forest edge, secondary growth, clearings with isolated trees, coconut groves. **D** Philippines.

31 WHITE-NECKED MYNA *Streptocitta albicollis* 42–50cm Distinctive pied plumage. Generally seen in pairs or small groups; joins mixed-species flocks. **V** Clear, whistled *towee*, 2–3 twangy, nasal notes, a drawn-out rasping and a penetrating *keee* alarm call. **H** Primary and secondary forest, swamp forest, wooded bushland, isolated thickets and groups of trees. **D** Sulawesi, Indonesia.

32 BARE-EYED MYNA *Streptocitta albertinae* 42–45cm Usually seen in pairs, occasionally alone or in trios. Tends to favour the middle- to upper levels of tall trees. **V** Call recalls the sound made by a squeaky gate, consisting of a descending series of five notes. **H** Degraded, selectively logged and open forest, lightly wooded cultivation and lightly wooded swamps. **D** Sula Islands, Indonesia.

33 FIERY-BROWED STARLING *Enodes erythrophris* 27cm Note orange lateral crown-stripe, yellow rump and undertail coverts. Forages acrobatically in the canopy. **V** Metallic *pik*, *tik*, *tsiit*, *sip* or *ssii*; also a rapid, sharp, hard *tik-tik-tt-tt-tt*, and a high-pitched *peeep* interspersed with various guttural notes. **H** Mature forest, tall secondary forest, selectively logged forest and forest edges. **D** Sulawesi, Indonesia.

34 GROSBEAK STARLING *Scissirostrum dubium* 17–21cm Note massive yellow bill and yellow legs. Breeds colonially. **V** Calls include a whistled *urriu*, a loud, clear, high-pitched *swee*, a nasal chatter and a liquid *chiruip* flight note. **H** Forest edges, lightly wooded country, plantations. **D** Sulawesi and Banggai, Indonesia.

35 SPOT-WINGED STARLING *Saroglossa spilopterus* 19cm Usually in tree canopy, feeding on insects, fruit and nectar; often forms large flocks alongside other starling species. **V** Noisy chattering, an aggressive *chek-chek-chek* and a soft *chik-chik*. Song consists of a musical warbling and dry discordant notes. **H** Open hill forest. **D** Breeds NC India. Winters NE India to Thailand.

36 GOLDEN-CRESTED MYNA *Ampeliceps coronatus* 19–21cm Arboreal, regularly perches in exposed tops of tall trees. In flight shows yellow primary patches. **V** Calls include a metallic bell-like note; otherwise similar to Common Hill Myna, but higher-pitched. **H** Broadleaved evergreen and deciduous forest, forest edge and clearings. **D** NE India, SE Asia.

37 SRI LANKA HILL MYNA *Gracula ptilogenys* 25cm Arboreal. In flight shows white primary patches. **V** A variety of piercing whistles, croaking and guttural notes. **H** Forests, well-wooded areas, plantations, gardens. **D** Sri Lanka.

38 COMMON HILL MYNA *Gracula religiosa* 25–29cm Arboreal, generally encountered in pairs or small groups. In flight shows white primary patches. **V** Various *chip* notes, soft *um* sounds, whisper-whistles and many types of whistles, croaks and wails. **H** Moist forest, forest edge and clearings; also plantations. **D** Widespread from N and NE India through SE Asia to Indonesia.

39 SOUTHERN HILL MYNA *Gracula indica* 24cm Arboreal, usually in pairs or small parties; larger flocks occur in flowering trees. In flight shows white primary patches. **V** Similar to Common Hill Myna, but higher-pitched and less variable. **H** Evergreen forest, well-wooded cultivations. **D** SW India, Sri Lanka.

40 NIAS HILL MYNA *Gracula robusta* 32cm Usually found in small flocks. In flight, shows a white patch at base of primaries. **V** Said to be an excellent mimic of human speech; no other information. **H** Hill forests. **D** Islands off NW Sumatra, Indonesia.

41 ENGGANO HILL MYNA *Gracula enganensis* 27cm In flight, shows a white patch at base of primaries. Actions presumably much like those of Common Hill Myna. **V** Unreported; probably like Common Hill Myna. **H** Forested areas. **D** Enggano, off SW Sumatra, Indonesia.

1
2
3
4
5
6
7
8
9
10
11
12
13 ♂ ♀
14
15
16
17
18
19
20
21
22
23
24
25
26
27
28
29
30
31
32
33
34
35 ♀ ♂
36 ♂ ♀
37
38
39
40
41

STARLINGS AND RHABDORNIS *CONTINUED*

1 GREAT MYNA *Acridotheres grandis* 25cm Note bright yellow bill and white undertail coverts. Forages mostly on the ground. In flight shows white primary patches. **V** Song and calls similar to Common Myna. **H** Open country, cultivation, urban areas. **D** NE India, SE Asia.

2 CRESTED MYNA *Acridotheres cristatellus* 25cm Usually occurs in flocks foraging on the ground. Sometimes forms large post-breeding flocks that roost colonially. In flight shows white flashes at base of primaries. **V** Similar to Common Myna, but song is more whistled, less fluty. **H** Open areas, scrub, cultivation, urban areas. **D** S China to SE Asia, Taiwan.

3 JAVAN MYNA *Acridotheres javanicus* 24–25cm Forages mainly on the ground in pairs or small groups, and larger flocks occur where food is plentiful. In flight, shows white patches at base of primaries. **V** Very similar to Common Myna. **H** Open country, cultivation, urban areas. **D** Java and Bali, Indonesia; widely introduced elsewhere in E and SE Asia.

4 PALE-BELLIED MYNA *Acridotheres cinereus* 25cm Mostly greyish, with charcoal-black forehead and glossy black crown. Usually encountered in small flocks foraging on the ground, often around cattle. **V** Random variety of dry, unmusical, nasal, squeaky rattles and whistles. **H** Dry, fallow paddyfields, cattle pastures, lightly wooded cultivation, urban areas. **D** Sulawesi, Indonesia.

5 JUNGLE MYNA *Acridotheres fuscus* 23cm Gregarious. Forms large groups at roosts; forages mainly on the ground. In flight shows white primary patches and much white on tail. **V** A repeated *tiuck-tiuck-tiuck*; other calls and song similar to Common Myna. **H** Open wooded areas, plantations, around villages. **D** N Pakistan through Himalayas and India to Malay Peninsula.

6 COLLARED MYNA *Acridotheres albocinctus* 23cm White collar is distinctive. Forages mainly on the ground. In flight shows white primary patches. **V** Undescribed. **H** Tall grassland and cultivated areas. **D** NE India, Myanmar, SW China.

7 BANK MYNA *Acridotheres ginginianus* 23cm Often tame. Gregarious. In flight shows ochre primary patches and underwing. **V** A *wheek* and a harsh note when agitated. Song consists of tuneless gurgles and whistles. **H** Agricultural and urban areas. **D** Pakistan, India, Bangladesh.

8 COMMON MYNA *Acridotheres tristis* 23cm Very tame. In flight shows white primary patches. **V** Song is a tuneless mixture of gurgling and whistled phrases. Calls include a querulous *kwerrh* and many gurgling and chattering notes; when alarmed, a harsh *chake-chake*. **H** Around human habitations and agricultural areas. **D** Widespread, from Kazakhstan to India, Sri Lanka, SE Asia, S China. Widely introduced across the world.

9 BLACK-WINGED STARLING *Acridotheres melanopterus* 22–24cm Usually encountered in pairs or small parties foraging on the ground or in trees and bushes. Breeds and roosts colonially. **V** Throaty *chok*, a harsh *kaar* and a drawn-out *keeer*; in flight gives a high, whistled *tsoowit* or *tsoowee*. **H** Open areas with scrub. **D** Java and Bali, Indonesia; introduced Singapore.

10 VINOUS-BREASTED STARLING *Acridotheres burmannicus* 21–24ccm Mainly a ground forager, favours short, well-watered grassland. **V** Harsh *tchew-it, tchew-tchieuw* or *tchew iri-tchew iri-tchieuw...* **H** Semi-desert, dry open country, scrub, cultivation, forest clearings. **D** SW China, SE Asia.

11 RED-BILLED STARLING *Spodiopsar sericeus* 22cm Male is grey with darker wings and tail, and white head; female is duller and browner. Gregarious, often in large flocks. Forages on the ground and in trees. **V** Song is sweet and melodious; flocks make chattering calls. **H** Cultivated areas with scattered trees, scrub, gardens. **D** S and SE China.

12 WHITE-CHEEKED STARLING *Spodiopsar cineraceus* 22cm Note white patches on head, and white rump. Gregarious. Mainly forages on the ground. **V** Monotonous, creaking *chir-chir-chay-cheet-cheet...* **H** Open country, cultivations, open woodland. **D** Breeds SE Russia, Korean Peninsula, Japan, C China. Winters South Korea, Japan, SE China.

13 BLACK-COLLARED STARLING *Gracupica nigricollis* 26–30cm A large starling with distinctive pied plumage. Forages on the ground, occasionally among grazing cattle. **V** Mixture of dry rattles and melodic notes, also a harsh *kraak-kraak*. **H** Open grasslands, scrub, cultivations, urban areas. **D** S China to SE Asia.

14 PIED MYNA *Gracupica contra* 22–25cm Forages on the ground, usually in pairs or small parties with larger flocks post breeding. **V** Prolonged series of shrill churrs with a few croaking and buzzing notes; sometimes includes mimicry of other birds. Calls include a loud *staar-staar*, a shrill *shree-shree* and various chuckles, warbles and whistles. **H** Open, usually moist areas with scattered trees, cultivations, urban areas. **D** From Pakistan through India to SE Asia, S China, Indonesia.

15 DAURIAN STARLING *Agropsar sturninus* 17cm Gregarious, roosting communally; often associates with other starling species. Resting flocks often sit on exposed dead branches, preening. **V** Calls include loud crackling sounds and a soft *prrp*, given in flight. **H** Coastal forest, forest edges, secondary growth, open areas with trees, cultivation, gardens. **D** Breeds E Mongolia, SE Russia, North Korea to C China. Winters SE Asia.

16 CHESTNUT-CHEEKED STARLING *Agropsar philippensis* 16–18cm Male has white head with brick-red patch. Female duller and browner, without cheek patch. Mainly arboreal. Gregarious, roosting communally; often mixes with other starling species. **V** Calls include an *airr* or *tshairr*, and a penetrating *tshick* when alarmed; flight call is a melodious *chrueruchu*. **H** Secondary growth, open areas with trees, cultivation, gardens. **D** Breeds S Sakhalin, Japan. Winters Philippines, Borneo.

17 WHITE-SHOULDERED STARLING *Sturnia sinensis* 17cm Usually gregarious, forages in trees and on the ground. **V** Harsh *kaar* and soft *preep* given in flight. **H** Open areas with scattered trees, scrub, cultivation, urban areas. **D** Breeds SC China to N Vietnam. Winters Taiwan, S China, Hainan, SE Asia.

18 CHESTNUT-TAILED STARLING *Sturnia malabarica* 20cm Gregarious; forages mainly in trees. Eastern race is paler, with less extensive rufous on underparts. **V** Song is a series of short hard notes and low squeaky churrs, the same notes given in rapid subdued outbursts. Calls include short buzzes and whistles. **H** Open woodland and open country with scattered trees. **D** Widespread from India to SE Asia.

19 WHITE-HEADED STARLING *Sturnia erythropygia* 21cm Note extent of greyish white in plumage. Gregarious; forages in trees and shrubs. **V** Song is a mess of discordant musical, snarling, snorting, squeaky and rattling notes intermingled with mimicry of other bird species. **H** Forest clearings, forest edge, secondary woodland, open cultivated and grassland areas. **D** Andaman and Nicobar Islands.

20 MALABAR STARLING *Sturnia blythii* 21cm Formerly considered conspecific with Chestnut-tailed Starling. Male differs in having a white head and breast. Female is duller with white restricted to face. **V** Song is a medley of squeaks, squawks and whistles. **H** Forest edge and open country with scattered trees. **D** SW India.

21 BRAHMINY STARLING *Sturnia pagodarum* 21cm Distinctively coloured starling with longish crest. Forages mainly on the ground, often with cattle; also feeds in fruiting and flowering trees. **V** Song is short, a nasal *slurr* followed by a bubbling yodel; also includes mimicry. Calls are croaking and chattering notes, and short grating churrs when alarmed. **H** Open deciduous forests, thorn-scrub, cultivations, around human habitations. **D** E Afghanistan, India, Bangladesh.

22 WHITE-FACED STARLING *Sturnornis albofrontatus* 19–22cm Arboreal; usually forages in the tree canopy. **V** Typical call is a high downslurred *cheewp*, *cheow* or *chirp*. Song said to be soft and sweet. **H** Rainforest and adjacent fruiting trees. **D** Sri Lanka.

23 BALI MYNA *Leucopsar rothschildi* 25cm Crown feathers elongated to form erectile crest. Mainly arboreal. Usually occurs in pairs; roosts communally. **V** Loud, harsh whistles and chattering notes. **H** Open woodland with a grassy understorey. **D** Bali, Indonesia.

24 ROSY STARLING *Pastor roseus* 21cm Combination of pale pink and black with green and purple iridescence is unmistakable. Gregarious. **V** Flocks emit a constant chatter; flight calls short and harsh or a clear *ki-ki-ki*. **H** Open country, wooded areas including orchards. **D** Breeds SE Europe to C Asia. Winters India, Sri Lanka.

25 COMMON STARLING *Sturnus vulgaris* 22cm In breeding plumage, note purple and green iridescence. After post-breeding moult, extensive buff or whitish tips to body feathers. Gregarious, foraging mainly on the ground. **V** Song is a medley of clicks, creaks, chirrups and warbles, interspersed with drawn-out whistles and mimicry. Calls include a harsh *tcherrr*, a hard *kyik* and a grating *schaahr*. **H** Grassland, agricultural areas, suburban and urban areas. **D** Widespread from W Europe through C Asia to Pakistan, Mongolia, S Siberia. Northern and eastern populations move south and west in winter, as far as N Africa and India. Introduced worldwide.

26 SPOTLESS STARLING *Sturnus unicolor* 22cm Very like Common Starling, but far less whitish-buff spotting. Usually in smaller flocks. **V** Song Common Starling-like, perhaps more melodious. Calls like Common Starling. **H** Similar to Common Starling. **D** SW Europe, N Africa.

27 WATTLED STARLING *Creatophora cinerea* 21cm Gregarious. Ground feeder. In flight, shows pale rump. **V** High, squeaky *tsirrit-tsirrit-tseep - seeet-seeereeet*, a harsh *graaaah* when alarmed, and a harsh three-syllable flight note. **H** Open bush and savannah. **D** From Eritrea to Angola and South Africa.

28 BLACK-BELLIED STARLING *Notopholia corusca* 21cm Note dark appearance and absence of contrasting black wing spots. **V** Very high, melodious, scratchy, rapid series. **H** Forest edge, woodland, riverine belts, suburbs. **D** E Africa, from SE Somalia to E South Africa.

29 PURPLE-HEADED STARLING *Hylopsar purpureiceps* 20cm Note compact appearance, dark plumage and dark eyes. **V** Song is a series of short tuneful phrases, sometimes given by several birds in chorus. In flight, a metallic *twink*. **H** Lowland forest. **D** Nigeria to DR Congo and Uganda.

30 COPPER-TAILED STARLING *Hylopsar cupreocauda* 20cm No green in plumage, and with coppery tail. Pale eye separates this species from Purple-headed Starling. Often together with Splendid Starling. **V** Not well documented. Song described as jumbled harsh notes, calls harsh and grating. **H** Canopy of lowland forest. **D** Sierra Leone to Ghana.

31 CAPE STARLING *Lamprotornis nitens* 22cm Black ear-coverts not well-defined. **V** High, mewing yet fluting *meh-juwee meh-meh-juwee tju-pu-ruhe*. **H** Woodland, bushveld, suburbs. **D** Angola to Zimbabwe, S Mozambique, South Africa.

32 GREATER BLUE-EARED STARLING *Lamprotornis chalybaeus* 22cm Distinguished from Cape Starling by blue (not bluish-green) belly, larger wing spots and more slender, brighter appearance. **V** Very high, mewing, reed warbler-like series. **H** Woodland, bushveld, suburbs. **D** Senegal to Eritrea; Ethiopia to Namibia, N South Africa.

33 LESSER BLUE-EARED STARLING *Lamprotornis chloropterus* 19cm Patch on ear-coverts smaller than in Greater Blue-eared Starling. Gregarious in loose flocks, often together with other glossy starlings. **V** High, miaowing, throaty, rather staccato *reh piu tu tju pri tju tui*. **H** Savannah woodland, farmland, bushveld. **D** Senegal and Guinea to Eritrea, Ethiopia and Kenya.

34 MIOMBO BLUE-EARED STARLING *Lamprotornis elisabeth* 19cm Formerly conspecific with Lesser Blue-eared Starling and adult is very similar to that species. No overlap in range. **V** Calls are long and harsh, repeated frequently. **H** Miombo woodland. **D** S Tanzania to N Botswana and Zimbabwe.

35 BRONZE-TAILED STARLING *Lamprotornis chalcurus* 20cm Note purple (not greenish) middle tail feathers. Shorter-tailed than Greater Blue-eared Starling. **V** Song combines whistles, nasal notes and chatters, often given in chorus. Contact call *ju-wee-yrr*. **H** Bushed areas. **D** Senegal to W Kenya.

36 SPLENDID STARLING *Lamprotornis splendidus* 28cm Arboreal species with diagnostic coppery patch on neck sides and black bar across secondaries. **V** Low mewing *kiauw-kjew*. **H** Forest, cultivation, gardens. **D** Senegal to Ethiopia, DR Congo, W Tanzania, N Zambia, Angola.

37 PRINCIPE STARLING *Lamprotornis ornatus* 30cm Looks blackish; less colourful than Splendid Starling, which also occurs on Príncipe. **V** Song consists of short, disjointed phrases. **H** Wooded areas. **D** Príncipe (Gulf of Guinea).

38 EMERALD STARLING *Lamprotornis iris* 18cm Mostly brilliant emerald-green, with purple ear-coverts and belly glossed purple. Note absence of wing spots. **V** Calls include squeaks and wheezes; alarm call *pee-pee-pee* noted in captivity. **H** Any area with bush or trees, except forest. **D** Guinea, Sierra Leone, Ivory Coast.

39 PURPLE STARLING *Lamprotornis purpureus* 25cm Green gloss restricted to wings and back. **V** Song consists of prolonged chattering. Contact call *twee-twee*, alarm call *shree*. **H** More or less wooded and bushed natural and cultivated habitats, including gardens. **D** Senegal to W Kenya.

1
2
3
4
5
6
7
8
9
10
11
12
13
14
15
16
17
18
19
20
21
22
23
24
25
26
27
28
29
30
31
32
33
34
35
36
37
38
39

STARLINGS AND RHABDORNIS *CONTINUED*

1 RÜPPELL'S STARLING *Lamprotornis purpuroptera* 35cm Predominantly glossy blue-purple, with some blue-green on head. No wing spots, long tail. **V** Song is a long, varied series of musical whistles, harsh notes and some mimicry. Calls include *swi-chew* and *kwerr.* **H** More or less wooded and bushed areas, normally near settlements. **D** E Africa, from Eritrea to Tanzania.

2 LONG-TAILED GLOSSY STARLING *Lamprotornis caudatus* 50cm The only long-tailed glossy starling in its range. **V** Calls include a lilting *ek-e-le-le* and a throaty *chu-chu-chu.* **H** Woodland, cultivation, suburban areas, mangroves. **D** S Mauritania to Guinea-Bissau and east to Sudan and South Sudan.

3 GOLDEN-BREASTED STARLING *Lamprotornis regius* 35cm Distinctive colourful starling with long graduated tail. Can look very dark in the field. **V** Song is a long warbled series. Alarm call is a harsh *chairrr;* in flight, *cheeo cheeo.* **H** More or less wooded and bushed areas. **D** Ethiopia and Somalia to N Tanzania.

4 MEVES'S STARLING *Lamprotornis mevesii* 30cm Separable from Burchell's Starling by longer tail, dark-looking appearance, less reflective plumage, especially on wings. **V** Mid-high, mewing *sreetireet reereeh.* **H** Mopane woodland, riverine belts. **D** S Angola to Malawi and N South Africa.

5 BURCHELL'S STARLING *Lamprotornis australis* 30cm Note dark eye, long graduated tail, large-winged appearance. **V** Low, throaty *ririruak rirituah ririria ria tjouwtjouw.* **H** Woodland, thornveld. **D** S Angola, Namibia, Botswana, N South Africa, Eswatini.

6 SHARP-TAILED STARLING *Lamprotornis acuticaudus* 23cm Note red (male) or orange (female) eye, pale underwing and pointed tail. **V** High, lashing *sreepsreep tjisreep mewi-sreep.* **H** Dry woodland. **D** Angola and N Namibia, through Zambia and S DR Congo to W Tanzania.

7 SUPERB STARLING *Lamprotornis superbus* 20cm Note white band between glossy blue-green upper breast and chestnut-brown belly. **V** Song consists of a long, quiet, varied and rambling series of musical notes. Alarm call *chirrr,* flight call *skrrrrri.* **H** Towns, villages, grassy bush and wooded areas. **D** Ethiopia and Somalia to Tanzania.

8 HILDEBRANDT'S STARLING *Lamprotornis hildebrandti* 20cm Rufous belly darker than breast. Note distinct large black wing spots and greenish hind-collar. **V** Song is slow and steady, a series of low-pitched notes. Alarm call *chu-ee,* flight call a whistling *chule.* **H** More or less bushed and wooded areas. **D** Kenya, Tanzania.

9 SHELLEY'S STARLING *Lamprotornis shelleyi* 15cm Similar to Hildebrandt's Starling, but note uniformly dark rufous underparts and small black wing spots. **V** Song is a series of short whistles, nasal and scratchy phrases. Calls include a two-note *ja-raah.* **H** Dry wooded and bushed areas. **D** Ethiopia and Somalia to SE South Sudan and N Kenya. Non-breeding as far south as N Tanzania.

10 CHESTNUT-BELLIED STARLING *Lamprotornis pulcher* 20cm Crown and ear-coverts are charcoal-grey with faint gloss, upperparts glossy bronze-green. Shows pale wing patch in flight. **V** Song consists of soft, tuneful notes. Calls include *whirri whirri* when flocks are on the move, *churr-churr* when mobbing an enemy. **H** Wooded and bushed areas, often near settlements. **D** Sahel region, from Mauritania to Eritrea, N Ethiopia.

11 ASHY STARLING *Lamprotornis unicolor* 30cm Slender, uniformly ashy-grey starling with slight green sheen and very long tail. **V** Song is a series of low chuckles and scratchy notes, with phrases often repeated. Calls include a harsh *charrr* in alarm, and a squeaky two-note contact call, *kuri kiwera.* **H** More or less wooded and bushed areas. **D** Tanzania.

12 FISCHER'S STARLING *Lamprotornis fischeri* 20cm Note white eye and white underparts. Cap or whole head rather paler than breast. **V** Song is a short phrase of squeaky, metallic notes. Flocks in flight give a shrill whistled call and a repeated wheezing *cree-wee-creewoo.* **H** Dry bush, bushed grassland. **D** S Ethiopia and Somalia to NE Tanzania.

13 PIED STARLING *Lamprotornis bicolor* 25cm Note yellow wattle at corner of gape, striking pale eye and white undertail coverts. **V** Song, lasting several minutes at times, combines warbles and harsher notes. Flight call loud is a *reek-reek;* gives harsh squawk when alarmed. **H** Grassland, farmyards, dry riverbeds, roadsides. **D** S Botswana, Lesotho, South Africa.

14 WHITE-CROWNED STARLING *Lamprotornis albicapillus* 25cm Note white eye, and whitish crown, lower belly, thighs and undertail coverts. Breast and flanks dull greyish brown with pale streaks. **V** Song is a series of upslurred whistles; calls include shrill and harsh notes. **H** Dry bush. **D** Ethiopia, Somalia, extreme N Kenya.

15 MADAGASCAN STARLING *Hartlaubius auratus* 20cm Slender, dark, distinctive starling; note white in wings and slender bill. **V** Song consists of high-pitched repeated *chee chreetee* phrases, and may last several minutes. **H** Forest, scrub. **D** Madagascar.

16 VIOLET-BACKED STARLING *Cinnyricinclus leucogaster* 18cm Male has strikingly iridescent upperparts and throat, violet or blue depending on the light. Female has dark brown-streaked upperparts. Often in small parties. Fairly shy, feeds mainly in trees. **V** Loud, gurgling warble. Calls include a long, ringing, grating, musical squeal with a rising inflection, ending in a quiet chuckle. Chuckles also used when flushed. **H** Plains, hills, wadis with trees. **D** Sub-Saharan Africa, widespread; SW Arabian Peninsula.

17 RED-WINGED STARLING *Onychognathus morio* 30cm Male differs from Slender-billed, Chestnut-winged and Waller's starlings in having blue (not green) gloss to head. Note dark eye of female. **V** Song is a series of sweet whistles and warbles, given by both sexes. Calls include a two-note *twee-twoo* as contact call, and harsher notes when alarmed or mobbing intruders. **H** Any more or less wooded area near rocky outcrops or buildings (including town centres). **D** E Africa, from Eritrea to South Africa.

18 SLENDER-BILLED STARLING *Onychognathus tenuirostris* 25–30cm Note slender bill, greenish gloss to head of male, and scalloped feathering of female. **V** Mid-high, somewhat rasping, rapid *pju pjee-pjeepji.* **H** Montane forest, moorland and cultivation. **D** Eritrea to E DR Congo, Tanzania, N Malawi.

19 CHESTNUT-WINGED STARLING *Onychognathus fulgidus* 28–35cm Note greenish gloss to head of male. Larger and longer-tailed than Waller's Starling. **V** Song is a varied series of low-pitched and screeching notes. Flight call a whistling *ti-ew,* alarm call a harsh note. **H** Lowland forest. **D** Guinea through Cameroon to Uganda, NW Angola.

20 WALLER'S STARLING *Onychognathus walleri* 20cm Red on primaries in folded wing does not extend beyond tertials. Female less extensive grey on head than Red-winged Starling. **V** Very high, whistled *fiu-fiu-fjee-fiee.* **H** Forest. **D** SE Nigeria and W Cameroon; S Sudan and E DR Congo to Kenya, Tanzania, N Malawi.

21 SOMALI STARLING *Onychognathus blythii* 28cm Note long tail of male and plain (not striped or scaled) head of female. **V** Calls include a high-pitched, tuneful *tleep* given in flight, a harsh *skrach* alarm call, and *tee tee* from flocks on the move. **H** Rocky ravines in wooded and bushed areas. **D** Ethiopia, Eritrea, Djibouti, Somalia, Socotra.

22 SOCOTRA STARLING *Onychognathus frater* 25cm Uniformly glossy black except for reddish-brown primaries, which show prominently in flight. Sexes alike. **V** Calls include far-carrying whistled *pee-oo,* rasping alarm call, and soft single notes exchanged between paired birds. **H** Rocky areas, woodland, urban gardens. **D** Socotra, Arabian Sea.

23 TRISTRAM'S STARLING *Onychognathus tristramii* 25cm Gregarious, noisy and fairly tame, often seen at picnic areas. In flight, shows bright orange-red primaries. **V** Parties keep up a constant flow of wolf-whistle-like notes, *dee-oo-ee-o* or *o-eeou;* also a mewing *vu-ee-oo.* **H** Rocky hills and ravines, semi-desert, various urban areas. **D** Arabian Peninsula.

24 PALE-WINGED STARLING *Onychognathus nabouroup* 25cm Uniformly glossy black except for pale rufous panel on primaries. Sexes alike. **V** High, warbling babbles in phrases of any length, *sree-wuwmweh freefrufruruririweh.* **H** Cliffs and rocky slopes in dry areas. **D** SW Angola to South Africa.

25 BRISTLE-CROWNED STARLING *Onychognathus salvadorii* 40cm Large dark starling with very long graduated tail, and bristly feathers on forecrown and around bill. Female slightly paler than male, with shorter tail. **V** Song is a series of musical whistles. Calls include chattering notes, loud *swi-chit* in flight, and harsh *schwaa* in alarm. **H** Cliffs and ravines in arid and dry areas. **D** Ethiopia to Somalia, Kenya, NE Uganda.

26 WHITE-BILLED STARLING *Onychognathus albirostris* 25cm White bill is diagnostic. Female has paler head. **V** Song is composed of chattering and whistling phrases. Call at nest is a repeated *kwit-kwit,* alarm call a harsh *charrr.* **H** Cliffs and ravines in montane moor- and grassland. **D** Eritrea, Ethiopia.

27 NEUMANN'S STARLING *Onychognathus neumanni* 28–33cm Large with long graduated tail. Eyes red or brown. Female has streaky grey head. In pairs or small flocks. **V** Gives a variety of soft whistles and squeaks. **H** Rocky outcrops, crags and cliffs. **D** W Africa, from Senegal and Mauritania to Chad and Central African Republic.

28 STUHLMANN'S STARLING *Poeoptera stuhlmanni* 17cm Distinguished from Narrow-tailed Starling by shorter tail, less gloss, darker wing and narrow yellow eye-ring. **V** Song consists of a series of whistles and chuckling phrases. Contact call is a pleasant *prtlee.* **H** Forest. **D** Ethiopia to W Kenya, E DR Congo, W Tanzania.

29 KENRICK'S STARLING *Poeoptera kenricki* 15cm Note small size (sparrow-sized). No gloss. Rufous inner flight-feather webs of female not visible in folded wing. Eyes grey or yellow. **V** Flocks produce constant tuneful babbling; possible flight call *peleep.* **H** Montane forest. **D** Kenya, Tanzania.

30 NARROW-TAILED STARLING *Poeoptera lugubris* 18cm Note pale wing of male, strong gloss and long, pointed tail. Sparrow-sized. **V** Flocks in flight give tuneful cheeping and melodious whistled notes. **H** Forest. **D** Sierra Leone to W Uganda, E and C DR Congo, N Angola.

31 SHARPE'S STARLING *Poeoptera sharpii* 15cm Note buff belly and undertail coverts, and pale eye. **V** Song is a rising and falling series of very thin, sharp, high-pitched whistles. Contact call is a soft whistle, flight call a high-pitched *chink.* **H** Forest canopy. **D** C Ethiopia to E DR Congo, Tanzania.

32 ABBOTT'S STARLING *Poeoptera femoralis* 15cm Combination of glossy blue-black upperparts and breast, white belly and pale eye is distinctive. No white in wing. **V** Song is high-pitched and squeaky, with twanging notes. Possible contact call a whistled six-note phrase, other calls metallic and very high-pitched. **H** Forest canopy. **D** S Kenya, NE Tanzania.

33 WHITE-COLLARED STARLING *Grafisia torquata* 20cm Male unmistakable; note pale eye of all-brown, slightly pale-scaled female. **V** Song recalls those of the *Lamprotornis* glossy starlings. Typical call consists of three short whistles; also chirruping notes. **H** From woodland to dry areas with sparse trees. Not on the ground. **D** Cameroon, Central African Republic, N DR Congo.

34 MAGPIE STARLING *Speculipastor bicolor* 20cm Note orange eye and white wing panel. **V** Song is a series of soft babbles with some harsher notes, and may last several minutes. Flight call a shrill whistle, alarm call a harsh repeated *ti-chuck chuck chuck.* **H** Wooded and bushed areas. **D** E Ethiopia and N Somalia to N Uganda, Kenya, NE Tanzania.

35 BABBLING STARLING *Neocichla gutturalis* 20cm Distinctive plumage, with black wedge-shaped mark in centre of breast. **V** Song is a long series of plaintive notes combined with upslurred squeals. Calls varied, shrill and parrot-like. **H** Miombo woodland. **D** S Angola, Zambia, Tanzania, Malawi.

36 STRIPE-HEADED RHABDORNIS *Rhabdornis mystacalis* 15–16cm The three *Rhabdornis* species resemble treecreepers (Plate 236) in structure and actions. Stripe-headed Rhabdornis occurs in active groups from the mid-storey to the canopy, hopping and jumping in search of food, or making short fly-catching sorties. Often a member of mixed-species feeding parties. **V** Unmusical, high-pitched *tsee tsee wick-tsee,* the first two notes hardly audible; also a series of *zeet* notes, uttered when perched or feeding. **H** Forest, forest edge, clearings. **D** Philippines.

37 STRIPE-BREASTED RHABDORNIS *Rhabdornis inornatus* 15–19cm Tends to forage in small groups, in the middle and upper canopy of forest trees. Active, searching among branches with hops, jumps and stretches; also makes sallies after passing insects. May join mixed-species feeding parties. **V** High-pitched *tzit* or *tzit-tzit-tzit-tzit-tzit.* **H** Submontane and montane forest. **D** C and S Philippines.

38 GRAND RHABDORNIS *Rhabdornis grandis* 16–18cm Forages after insects and small fruits by hopping and jumping along branches. Usually seen singly or in small groups; also joins mixed-species feeding flocks. Often perches at the top of dead trees. **V** High-pitched *zip zip zeet zip,* some notes more trilled. **H** Mid- and high-elevation primary and secondary dipterocarp and hardwood forest. **D** N Philippines.

OXPECKERS BUPHAGIDAE

39 YELLOW-BILLED OXPECKER *Buphagus africanus* 22cm Distinguished from Red-billed Oxpecker by different eye colour, different bill shape and colour, and pale rump. **V** Calls include buzzing, peeping and hissing notes; alarm call is an intense hissing *tchissss.* **H** Savannah, game reserves and other places with large ungulates. **D** Sub-Saharan Africa, widespread.

40 RED-BILLED OXPECKER *Buphagus erythrorynchus* 21cm Note that rump is brown, concolorous with mantle. **V** Song combines soft call notes, trills and whistles; female has a simpler song than male. Alarm call is a sharp *ksss.* **H** Savannah, game reserves. **D** E Africa, from Eritrea to NE South Africa.

1
2
3
4
5
6
7
8
9
10
11
12
13
14
15
16
17
18
19
20
21
22
23
24
25
26
27
28
29
30
31
32
33
34
35
36
37
38
39
40

THRUSHES TURDIDAE

1 RED-TAILED ANT THRUSH *Neocossyphus rufus* 20cm Thrush-like in stance and feeding habits. Note all-rufous tail. **V** Extra-high, trilling *fi-fi-fi-fi-fi-fi-sruu fjui weee* (*-sruu* rapidly descending, *weee* as a high, loud, level whistle). **H** Forests, moist bushland and scrubland. **D** S Cameroon and Gabon to N DR Congo and W Uganda.

2 WHITE-TAILED ANT THRUSH *Neocossyphus poensis* 20cm Note white tail corners and red bar in flight, formed by red bases of all flight feathers. **V** Includes a very high *fueeeet* or mid-high *weeet-weeet-weeet*. **H** Middle strata and undergrowth of forest. **D** W and C Africa.

3 FRASER'S RUFOUS THRUSH *Stizorhina fraseri* 20cm Note flycatcher-like stance and characteristic tail-flicking (showing red outer-tail feathers). Separated from White-tailed Ant Thrush, in flight, by smaller red patch formed by red bases of primaries only, red rump and by red outer-tail feathers. **V** Mid-high, slightly drawn-out whistle, *uweeeet-uweet-tereet* (drawn up at end) and mid-high *weeet-wreet-wreet*. **H** Middle and higher levels of forest. **D** C Africa.

4 FINSCH'S RUFOUS THRUSH *Stizorhina finschi* 19cm Separated from White-tailed Ant Thrush by different stance, paler plumage and less red in wings. **V** Song similar to that of Fraser's Rufous Finch, but slower and lower in pitch. Calls include a *wurd-wurd-wurd*. **H** Lower storeys of very dense forest. **D** Sierra Leone to Nigeria.

5 SLATY-BACKED THRUSH *Geokichla schistacea* 16–17cm Forages mainly on or near the forest floor; when disturbed, often flees to a perch in the subcanopy. Usually seen alone, in pairs or in small groups. **V** Exquisite melody of eight clear, sweet notes, ending with a long, upslurred, high-pitched whistle; also a thin *tsee*. **H** Primary and secondary lowland forest. **D** Tanimbar Islands, E Lesser Sunda Islands.

6 BURU THRUSH *Geokichla dumasi* 17cm Forages on the ground. Little recorded information. **V** Call is a high *pthhhhhh*. **H** Primary lower montane rainforest. **D** Buru, S Maluku Islands, Indonesia.

7 SERAM THRUSH *Geokichla joiceyi* 17cm Forages on the ground, in deep undergrowth, usually alone or in pairs. **V** Call is a thin *tseep*. **H** Montane rainforest. **D** Seram, S Maluku Islands, Indonesia.

8 CHESTNUT-CAPPED THRUSH *Geokichla interpres* 17–19cm Shy; usually seen singly or in pairs. Forages on the ground up to the mid-storey. **V** Rising, flute-like whistles, interspersed with chirrups, transcribed as *see it-tu-tu tyuu*. Calls include a harsh *tac* and a high-pitched, falling *tsi-i-i-i*. **H** Broadleaved evergreen forest. **D** S Thailand to S Philippines.

9 ENGGANO THRUSH *Geokichla leucolaema* 16–19cm Little information; presumed to be similar to the Chestnut-capped Thrush, with which it was formerly considered conspecific. **V** Song said to resemble a juvenile begging call. **H** Rainforest. **D** Enggano, off SW Sumatra, Indonesia.

10 RED-BACKED THRUSH *Geokichla erythronota* 20cm Shy and secretive, foraging mainly on the ground in shady forest areas, usually singly. **V** Thin, high-pitched, upslurred note. Song consists of a series of liquid notes. **H** Tropical lowland and mid-elevation broadleaved evergreen forest. **D** Sulawesi, Indonesia.

11 RED-AND-BLACK THRUSH *Geokichla mendeni* 20cm Formerly considered a race of the previous species; actions and habits presumed to be similar. **V** Unrecorded. **H** Selectively logged forest. **D** Sulawesi, Indonesia.

12 CHESTNUT-BACKED THRUSH *Geokichla dohertyi* 16–18cm Generally seen singly; occasionally occurs in small groups at fruiting trees, otherwise forages on the ground. **V** Song comprises 3–7 mellow, sweet whistles, interspersed with more complex phrases. Calls include a thin whistle and high-pitched, squeaky whistle. **H** Primary hill and montane forest. **D** Lesser Sunda Islands.

13 PIED THRUSH *Geokichla wardii* 22cm Shy when breeding, less so post breeding. Forms small flocks during migration. Feeds on the ground and in fruiting trees. **V** Short, repeated two-note warble. Call is a sharp, spitting *ptz-ptz-ptz-ptz*. **H** Broadleaved forest and secondary growth. **D** Himalayas to Sri Lanka.

14 ASHY THRUSH *Geokichla cinerea* 19cm Forages on or near the ground, usually in open patches. **V** Wheezy, up-and-down *wheeezizizizi zizizizi*. **H** Primary, secondary and selectively logged forests, forests with limestone outcrops, remnant ridgetop forest patches, mossy forest, forest with an open understorey dominated by rattans. **D** Philippines.

15 ORANGE-SIDED THRUSH *Geokichla peronii* 19–22cm Usually encountered singly or in pairs. Small groups gather in fruiting trees, otherwise forages mainly on the ground, with occasional forays into the mid-level or canopy. **V** Beautiful, rich, clear whistled notes, interspersed with occasional short, harsh notes and long, high-pitched notes. **H** Closed-canopy monsoon forest, degraded forest patches. **D** Lesser Sunda Islands.

16 ORANGE-HEADED THRUSH *Geokichla citrina* 20–23cm Shy; may form small flocks post breeding. **V** Loud, clear series of lilting phrases. Calls include a soft *chuk* and a screeching *kreeee*. **H** Moist, shady areas in forests. **D** Widespread across Asia, from the Himalayas to S and E China and N Borneo.

17 SIBERIAN THRUSH *Geokichla sibirica* 22cm Secretive, feeding on the ground or in fruiting trees. **V** Weak *tseee*, soft *zit* and a gruff squawk when alarmed. **H** Hill and montane forests. **D** Breeds in E Asia. Winters in SE Asia.

18 ABYSSINIAN GROUND THRUSH *Geokichla piaggiae* 20cm Feeds on invertebrates and fruit, foraging on the ground and in low vegetation. **V** Very high, rather sharp, fluted, slightly hurried *pi-pi-pji-pju*. **H** Ground strata of montane forests and bamboo. **D** C and E Africa.

19 CROSSLEY'S GROUND THRUSH *Geokichla crossleyi* 22cm Note: may have white tail tips. Separated from Oberländer's Ground Thrush by broad black band through eye, and from Grey and Black-eared ground thrushes by absence of black ear mark and generally more orange colour. **V** Song a series of mellow, rising whistles, *hor-her-hiiwo-chichiwo-tsitsi*. In alarm, utters a high *siiip* call. **H** Forest. **D** C and WC Africa.

20 ORANGE GROUND THRUSH *Geokichla gurneyi* 22cm Differs from Olive Thrush by facial pattern, black bill and wing-bars. **V** High, loud, piping, repeated whistles, including *tutu-tutir-tjirr* and other variations (lacking double-toned quality of other thrushes). **H** Montane forests. **D** E and S Africa.

21 OBERLÄNDER'S GROUND THRUSH *Geokichla oberlaenderi* 20cm Note rufous face and overall warm colouring. **V** High, unstructured, up-and-down *peeeuu-wip to-three* and *riu-riu-reez reep-reep*. **H** Ground strata of forests. **D** E DR Congo and W Uganda.

22 BLACK-EARED GROUND THRUSH *Geokichla camaronensis* 20cm Separated from Grey Ground Thrush in same range by tawny underparts. Note rather pale brown colouring of upperparts and pale chin and throat. **V** Song not described. Calls include a high *ssreee*. **H** Ground strata of lowland forests. **D** C Africa.

23 GREY GROUND THRUSH *Geokichla princei* 22cm Note white tail corners. Separated from Black-eared Ground Thrush by slightly larger size, heavier bill and longer tail. **V** Song not described. Call a trilling *tsssrrr*. **H** Dense parts of forest, often near streams. **D** W and C Africa.

24 SPOTTED GROUND THRUSH *Geokichla guttata* 22cm Note facial pattern and spotted wing-bars. **V** High, loud, plaintive, fluted whistles, including *twee-oh-twee-twit-tirrik*. **H** Ground strata of humid forest and woodland. **D** E and SE Africa.

25 SPOT-WINGED THRUSH *Geokichla spiloptera* 23cm Best located while singing. Retiring, keeping to the cover of thick vegetation. **V** Variable, rich series of short, whistled phrases. Contact call is a soft *tzseee*. **H** Undergrowth in dense rainforest. **D** Sri Lanka.

26 GEOMALIA *Zoothera heinrichi* 28–30cm Shy and secretive. Forages on the ground, among decaying vegetation in thick undergrowth. **V** Intermittent, thin, high-pitched, insistent whistle. **H** Primary montane forest with thick undergrowth. **D** Sulawesi, Indonesia.

27 EVERETT'S THRUSH *Zoothera everetti* 23cm Forages on the ground, often in moist areas and near streams; quickly disappears into thick undergrowth if disturbed. White face markings are variable. **V** Soft, musical song; little other information. **H** Deciduous submontane forest. **D** Borneo.

28 SUNDA THRUSH *Zoothera andromedae* 25cm Extremely shy and skulking. Forages on the ground or in low, dense cover; flees to perch in mid-storey when disturbed. **V** Not described. **H** Hill forest, montane mossy forest. **D** Sumatra and Java, Indonesia; Philippines; Lesser Sunda Islands.

29 ALPINE THRUSH *Zoothera mollissima* 26cm Feeds mainly on the ground. Generally shy and unapproachable. **V** Variable, rich musical phrases. Calls include a thin *chuck* and a rattling alarm. **H** Rocky alpine areas with juniper and dwarf rhododendron. Winters at lower altitudes. **D** NW Himalayas to S China.

30 SICHUAN THRUSH *Zoothera griseiceps* 26cm Generally shy and unapproachable; feeds mainly on the ground. Note that very similar Himalayan Thrush may occur in extreme SW China. **V** Variable, rich musical phrases, including *plee-too ti-ti-ti* or *plee-too ch-up-ple-ooop*. Calls include a thin *chuck* and a rattling alarm. **H** Rhododendron and coniferous forest, low vegetation and rocky areas above the treeline. **D** SC China to NW Vietnam.

31 HIMALAYAN THRUSH *Zoothera salimalii* 26cm Little known. Very like Alpine and Sichuan thrushes but bill heavier. **V** Rich, variable phrases, more musical than those of Alpine Thrush with few or no scratchy notes. **H** Coniferous montane forest with rhododendrons, other forest types with dense understorey, scrubby areas above the treeline. **D** E India (West Bengal and Arunachal Pradesh) to S China (Yunnan).

32 LONG-TAILED THRUSH *Zoothera dixoni* 27–28cm Secretive; feeds on the ground in thick vegetation. **V** Often begins with a *w-t-it*, followed by a slow, slurred *wu-ut - cheet-sher - wut-chet-shuur*, interspersed with twitters and *too-ee* phrases. **H** Dense forests above or along the treeline. Winters in lower-altitude thick forest or open areas with bushes. **D** Breeds in C and E Himalayas to SC and SW China. Winters in SE Asia.

33 WHITE'S THRUSH *Zoothera aurea* 27cm Large golden-yellow thrush with scaly markings on upperside and underside. Mainly feeds on ground, taking invertebrates and berries. **V** Song a simple, repeated 1.5-sec whistle. Calls include thin *sreet* and brief *chik*. **H** Shady coniferous or mixed forest; in boreal zone, riverine forest. Outside of breeding season may visit more varied habitats, including parks, gardens. **D** Breeds in E Asia. Winters in Taiwan, SE Asia.

34 AMAMI THRUSH *Zoothera major* 28cm Very shy. Little recorded about habits, but presumed to be similar to those of Scaly Thrush (Plate 242). **V** Series of slow, varied, flute-like notes, rising and falling, e.g. *piri-piri-kyo-kyo* or *chirrup-chewee - cheue - wiow-we-ep - chewee-wiop*. Call as Scaly Thrush. **H** Uncut primary forest and selectively cut mature forest. **D** Ryukyu Islands, off S Japan.

1
2
3
4
5
6
black-winged
morph
7
8
9
10
11
12
13
♀
14
15
16
♀
♂
♂
17
♀
♂
18
19
20
21
22
23
24
25
26
27
28
29
30
31
32
33
34

THRUSHES *CONTINUED*

1 SCALY THRUSH *Zoothera dauma* 27cm Usually remains hidden. Mainly a ground feeder, in moist areas of forests. **V** Repeated slow phrases, e.g. *pur-loo-tree-lay... dur-lee-dur-lee... drr-drr-chew-you-we-eeee*, sometimes interspersed with low squeaks or twitters. Call is a soft *tsi*, drawn-out *tzeep* or *seeh*, and *chuck - chuck* when alarmed. Generally silent, apparently reluctant to call, even when alarmed. **H** Various forest types, including mature oak, dense spruce in the taiga zone, and montane and submontane forest, with tangled undergrowth. **D** Himalayas to Sumatra and Java, Indonesia.

2 NILGIRI THRUSH *Zoothera neilgherriensis* 29cm Large thrush. Upperparts golden and underparts whitish, both with black scaling. **V** Song very like that of Amami Thrush (Plate 241). **H** Dark and wet forest ravines and similar areas. **D** SW India.

3 SRI LANKA THRUSH *Zoothera imbricata* 24cm Forages low down, mainly on the ground. **V** Series of eight or more single, rich whistles. Calls include a high-pitched, repeated *tchiss*. **H** Dense moist forest. **D** Sri Lanka.

4 FAWN-BREASTED THRUSH *Zoothera machiki* 21–22cm Skulks in dense scrub, although regularly forages in the open on the forest floor, along streams, on forest roads and in recently burnt areas; usually alone. **V** Unknown. **H** Lowland monsoon and gallery forest, degraded and secondary forest, forest edges and scrub. **D** Tanimbar Islands, E Lesser Sunda Islands.

5 RUSSET-TAILED THRUSH *Zoothera heinei* 25cm Long-billed, rather light-coloured *Zoothera* thrush, with typical black scaling on upperparts and underparts. **V** Song a series of double whistles. Calls include a thin, soft *tsep* and, in alarm, harsh chacking. **H** Hillside rainforest with deep leaf litter. **D** New Guinea to Solomon Islands, NE Australia.

6 BASSIAN THRUSH *Zoothera lunulata* 28cm Fairly large thrush with a strong bill and shortish tail. Forages in leaf litter, feeding mainly on earthworms. **V** Song comprises three loud, whistling, trilled phrases. Calls include a thin *seep* and abrupt *chi-lit*. **H** All kinds of cool, damp forests, pine plantations. **D** NE, SE and S Australia, Tasmania.

7 BLACK-BACKED THRUSH *Zoothera talaseae* 23cm Dark grey upperparts, whitish underparts with light scaling on flanks, prominent double whitish wing-bar. Forages on the ground. **V** Unknown. **H** Montane mist forest. **D** Bismarck Archipelago, Solomon Islands.

8 MAKIRA THRUSH *Zoothera margaretae* 23cm Dark olive-brown upperparts, whitish underparts with dark scaling, thin double wing-bar. **V** Song combines simple, well-spaced whistles with grating and clicking notes. Calls include a very thin *tseep* and sharp *chik*. **H** Montane ridge forest, upper zones of gulleys, sometimes overgrown gardens. **D** Makira, Solomon Islands.

9 GUADALCANAL THRUSH *Zoothera turipavae* 20cm Little known. Small, short-tailed, dark brown *Zoothera* thrush. Sings mainly at dawn from one of several favourite high perches. **V** Series of loud, tuneful, slurred whistles. Calls include thin, high-pitched *tssss*. **H** Moist forest. **D** Guadalcanal, Solomon Islands.

10 LONG-BILLED THRUSH *Zoothera monticola* 27cm Shy and retiring. Ground feeder in dense forest undergrowth. **V** Usually sings from the top of a tall tree: loud, melancholic series of up to three plaintive whistles, *te-e-uw*, *sew-a-tew-tew* or similar. Alarm call a loud *zaaaaaaa*. **H** Thick undergrowth, often with moist mossy areas, in dense, high-altitude fir forests. Winters at lower levels. **D** Himalayas, Myanmar, Vietnam.

11 DARK-SIDED THRUSH *Zoothera marginata* 25cm Very shy. Often forages in moist stream beds. **V** Thin whistle. Calls include a low *chuck* and, when alarmed, a high-pitched *pit-pit-pit*. **H** Dense forest with nearby streams. **D** Himalayas to Thailand.

12 VARIED THRUSH *Ixoreus naevius* 23cm Shy; feeds low down or on the ground in dark, shaded areas. **V** Weak *chuk*; also a thin *woooeee*. Song consists of vibrant, eerie, melancholic and sustained notes, varying rapidly from high to low pitch; also a buzzing trill. **H** Dense coniferous forests with undergrowth of dogwood and wild currant. **D** W North America.

13 AZTEC THRUSH *Ridgwayia pinicola* 24cm Forages on the ground or low down in bushes or thick vegetation. Best detected by call. **V** Thin, slightly quavering *wheeerr*; also an upslurred, buzzing *zrrip*, *prrip* or *prreep* and a nasal *sweee-uh*. **H** Pine or pine–oak forests in ravines. **D** W Mexico, SW USA.

14 SULAWESI THRUSH *Cataponera turdoides* 20–25cm Usually encountered alone or in small groups; occasionally joins mixed-species foraging parties. Hops along horizontal branches searching for prey in epiphytic growth. **V** A chatter and a pleasant melody of notes, generally given at dawn or dusk. **H** Montane evergreen forest and moss forest with dense undergrowth. **D** Sulawesi, Indonesia.

15 GRANDALA *Grandala coelicolor* 19–23cm Forages mainly on the ground; posture much like that of a rock thrush. Winter flocks act starling- or wader-like, wheeling and circling, before dropping onto the top branches of trees or onto cliffs. **V** Ringing *tji-u*. Song is a subdued *tju-u tiu-u ti-tu tji-u*. **H** Alpine meadows, bare rocky areas above the treeline, open forests. **D** Himalayas to C China.

16 EASTERN BLUEBIRD *Sialia sialis* 15–16.5cm Uses an exposed perch to pounce upon insects on the ground; occasionally makes short fly-catching sallies. **V** Musical *chur-lee*, *tu-a-wee* or *jeew wiwi*. Song consists of soft, mellow whistles. **H** Open country with hedgerows, woodland edges, roadsides. **D** E and C North America to Nicaragua.

17 WESTERN BLUEBIRD *Sialia mexicana* 17cm Note brown mantle. **V** Short, sparrow-like chatter and a high, sweet, clear *wheet wheetwitwheer*. **H** Open pine woodland, open bush; 1,500–3,000m. **D** W North America to C Mexico.

18 MOUNTAIN BLUEBIRD *Sialia currucoides* 17–18cm Makes darting flights from a branch or rock perch to catch flying insects; also hovers in pursuit of insect prey. **V** Soft *feeer* or muffled *perf*; also a harsh *chik* or *chak*. Song is a short, clear warble. **H** High-elevation open areas with scattered trees. **D** Breeds in W North America. Winters in W Mexico.

19 OMAO *Myadestes obscurus* 19cm Note grey forehead. Closely resembles Puaiohi and Olomao, but range does not overlap with that of Olomao. **V** Song is liquid, rapid, varied *weetjohwih*. Hoarse *wreeeh* or high police whistle-like *rreeh*. **H** Forest, scrub and savanna above 1,000m and open scrub above 2,000m on Mauna Loa. **D** Hawaii, Hawaii.

20 PUAIOHI *Myadestes palmeri* 17cm Critically Endangered; restricted to high-altitude forest on Kauai. **V** Song comprises high, short, descending phrases, *ohcomenowyou*. Call a very high, sizzling *ssssh*. **H** Streams in undergrowth of dense forest. **D** Kauai, Hawaii.

21 OLOMAO *Myadestes lanaiensis* 17cm Buff patch at base of primaries more distinct than in other thrushes. Critically Endangered. **V** Song slurred and fluty. Alarm call a cat-like rasp; other calls vary widely in tone, amplitude and repetition. **H** Montane forest. **D** Oahu, Lanai and Molokai, Hawaii.

22 TOWNSEND'S SOLITAIRE *Myadestes townsendi* 21–23cm In flight, shows white outer-tail feathers and buff wing-bar. Makes darting fly-catching sallies from a high perch. **V** High-pitched *eek* or *heeh*. Song is a complex, prolonged warble, very finch-like in quality. **H** Montane coniferous forests. **D** W Canada to C Mexico.

23 BROWN-BACKED SOLITAIRE *Myadestes occidentalis* 20cm Note brown-olive upperparts and white chin with black malar stripe. **V** Very high, rusty, descending warble. **H** Forest, often along streams. **D** NW Mexico to Honduras.

24 CUBAN SOLITAIRE *Myadestes elisabeth* 19cm Perches high in trees, from where it makes sallies to catch flying insects or to pick prey from vegetation; also hovers to pluck fruits. **V** Short whistles and a loud, flute-like, far-carrying song, said to sound like a wet finger rubbed on the rim of a fine porcelain cup. **H** Dense, humid mountain and hill forests. **D** Cuba.

25 RUFOUS-THROATED SOLITAIRE *Myadestes genibarbis* 19–20.5cm Often sits motionless and silent in dense vegetation. **V** Drawn-out *teut* or *toot*; also a series of semi-discordant whistles and trills, highly ventriloquial. **H** Moist, montane forests. **D** Hispaniola, Jamaica, Dominica, Martinique, St Lucia, St Vincent.

26 BLACK-FACED SOLITAIRE *Myadestes melanops* 17cm The only solitaire in its range. Note face, tail and wing pattern. **V** High, well-separated *tjuuh fjoh fjeehohfjeehoh*, second part very high and shrill, like a badly oiled gate. **H** Bamboo and undergrowth in montane forest. **D** Costa Rica, Panama.

27 VARIED SOLITAIRE *Myadestes coloratus* 18cm Unmistakable due to pattern on wings and face, and all-orange bill and legs. **V** Song comprises drawn-out notes or fluty phrases, interspersed with distinct pauses. **H** Montane cloud forest. **D** Panama.

28 ANDEAN SOLITAIRE *Myadestes ralloides* 18cm Plumage is two-toned. In flight, shows flashing silvery wing-bar and outer-tail feathers. **V** Short mixtures of 1–2 sharp notes and 1–2 musical notes, repeated three times every 2 secs. **H** Lower storeys and edges of humid forest and secondary growth. **D** N Venezuela to W Bolivia.

29 SLATE-COLORED SOLITAIRE *Myadestes unicolor* 20cm Separated from Black-faced Solitaire by range, and wing and face pattern. **V** Very high, mellow yet rusty *fjeehfjeeh-turruwwheeohwhee*. **H** Forest. **D** S Mexico to Nicaragua.

30 RUFOUS-BROWN SOLITAIRE *Cichlopsis leucogenys* 20.5cm Note bicoloured bill and unstreaked, light chestnut throat. **V** Calm phrases of three fluted, sharp or rolling notes. **H** Humid and wet forest, dense secondary growth. **D** W, N and SE South America.

31 BLACK-BILLED NIGHTINGALE-THRUSH *Catharus gracilirostris* 15cm Bill colour distinctive; note grey crown. **V** Very high *uueeeh ueeeh* or drawn-up, ultra-high *seeesususwier*. **H** Open places in montane forest; >2,150m. **D** Costa Rica, Panama

32 ORANGE-BILLED NIGHTINGALE-THRUSH *Catharus aurantiirostris* 16cm Note bright eye-ring; orange bill and legs distinctive. **V** Very high *fteewih tih* and very high, strong *tjuptjuptjeehohwrrr*. **H** Dry woodland undergrowth, forest edges, gardens. **D** NW Mexico to N Colombia and C Venezuela.

33 SLATY-BACKED NIGHTINGALE-THRUSH *Catharus fuscater* 18cm Head slightly darker than upperparts; blotched (not spotted) below. **V** Variations of a calm, fluted phrase of 2–3 pure notes, e.g. *too-tee*. **H** Undergrowth of humid and wet forest. **D** Costa Rica to N Bolivia.

34 RUSSET NIGHTINGALE-THRUSH *Catharus occidentalis* 17cm Lower mandible has a flesh base. Mantle more rufous russet and bill with darker tip than in Ruddy-capped Nightingale-Thrush (Plate 243). **V** Very high, short metallic whistle *sreeohweer*. **H** Open forest undergrowth. **D** Mexico.

1
2
3
4
5
6
7
8
9
10
11
12
13
14
15
16
17
18
19
20
21
22
23
24
25
26
27
28
29
30
31
32
33
34

THRUSHES *CONTINUED*

1 RUDDY-CAPPED NIGHTINGALE-THRUSH *Catharus frantzii* 17cm Note contrasting russet of crown and nape. **V** High, clear, sometimes rusty phrases, e.g. *wirohwirrwirr*. **H** Montane forest undergrowth, tall secondary growth. **D** S Mexico to Panama.

2 BLACK-HEADED NIGHTINGALE-THRUSH *Catharus mexicanus* 15–16.5cm Forages among low branches or on the ground, where it progresses using a series of springing hops. **V** Sharp, ascending *seeet*, plaintive mewing, a rising *chowr* and a buzzing *chrrr* when alarmed. **H** Forest undergrowth. **D** EC Mexico to Panama.

3 GOULD'S NIGHTINGALE-THRUSH *Catharus dryas* 18cm Head contrasts with black-spotted yellow underparts. **V** Variations of a calm, fluted phrase of 2–3 notes, e.g. *tee-too-tee* (sometimes two notes uttered together). **H** Undergrowth of humid forest. **D** S Mexico to Honduras.

4 SCLATER'S NIGHTINGALE-THRUSH *Catharus maculatus* 18cm Recently split from Gould's Nightingale-Thrush; closely resembles that species but is darker. **V** Presumed to be like that of Gould's Nightingale-Thrush. **H** Undergrowth of montane cloud forest. **D** E Colombia to N Argentina.

5 VEERY *Catharus fuscescens* 17cm Secretive. Forages primarily on the ground. Western birds are more earthy above, with darker spotting on breast. **V** Song a descending *da-vee-ur - vee-ur - veer - veer*. Calls include a fluted *phew*, *whee-uu* and a slow, slurred *wee-oo*. **H** Undergrowth in deciduous or mixed forests. **D** Breeds across North America. Winters across Central America and as far south as Brazil.

6 GREY-CHEEKED THRUSH *Catharus minimus* 16–20cm Wary. Forages on or near the ground. **V** Song a series of high-pitched, repeated notes interspersed with a sharp *chee-chee* and ending with a descending *wee-oh wee-oh*. Call a downslurred *wee-ah*; also a short *chuck* and a light *pheeeu*. **H** Dense conifer, mixed and open forest, shrubby thickets. **D** Breeds in NE Siberia and N North America. Winters in Central America and N South America.

7 BICKNELL'S THRUSH *Catharus bicknelli* 16–17cm Wary. Forages on or near the ground, occasionally making short sallies to catch flying insects. **V** Song like that of Grey-cheeked Thrush, but higher pitched. Call a *wee-ooo*, *pee-oo* or *psee-uuu*; also a sharp *shrip* or *chirp*. **H** High-elevation broadleaved forests, woods, gardens with large trees. **D** Breeds in NE USA and SE Canada. Winters on Hispaniola.

8 SWAINSON'S THRUSH *Catharus ustulatus* 16–20cm Shy. Forages on the ground and in fruiting trees. Western coastal birds are slightly more rufous above. **V** Song a musical *whip-poor-will-a-will-e-zee-zee-zee*. Calls include a liquid *whit* and soft *whup*. **H** Shaded or damp understorey of dense spruce forests. **D** Breeds across North America. Winters in Central and South America as far south as Peru.

9 HERMIT THRUSH *Catharus guttatus* 19cm Shy. Forages on the ground. Constantly flicks wings and tail. **V** Song a series of flute-like notes at different pitches. Calls include a low *chuck*, a ringing *cheeee* and a harsh *pay*. **H** Mixed or coniferous forest, forest clearings with shrubby undergrowth. **D** Breeds in North American boreal zone and along the Rockies. Winters in the USA, on Pacific and Atlantic coasts and in southern states, south to Mexico and Guatemala.

10 WOOD THRUSH *Hylocichla mustelina* 20cm Usually shy and retiring. Mainly a ground feeder. **V** Song consists of fluty phrases interspersed with call notes and ending with a soft trill. Calls include a *pit-pit-pit* and a low *tuck-tuck*. **H** In native N America: deciduous woodland undergrowth, often near water. **D** Breeds in C and E USA, SE Canada. Winters in Central America.

11 BLACK SOLITAIRE *Entomodestes coracinus* 23cm Unmistakable due to contrasting white in face, wings and tail. **V** Single, rather low, pure, fluted notes given at long intervals. **H** Wet cloud forest, foothill forest, secondary growth. **D** Colombia, Ecuador.

12 WHITE-EARED SOLITAIRE *Entomodestes leucotis* 24cm Note distinctive bold plumage pattern. **V** Curious single notes, as if from a hoarse toy trumpet, given at long intervals. **H** Borders and mid-levels of humid forest. **D** Peru, Bolivia.

13 GROUNDSCRAPER THRUSH *Turdus litsitsirupa* 21cm Separated from Dusky Lark (Plate 198) by larger size and uniformly coloured wings. May occasionally flick one wing to show yellow-buff panel on underwing. **V** Short, mid-high, melodious, slightly harsh whistles, e.g. *tiktiktweedsrohweeth*. **H** Suburbs, cultivation, open woodland, moorland. **D** E and S Africa.

14 YELLOW-LEGGED THRUSH *Turdus flavipes* 22cm Body is grey, and head, wings and tail are black, contrasting with yellow bare parts. **V** Strong phrases of a loud, musical, inhaled or metallic quality. **H** Borders and interior of humid forest, secondary growth, adjacent clearings, plantations. **D** N and SE South America.

15 PALE-EYED THRUSH *Turdus leucops* 20cm Note contrasting pale eyes without an eye-ring. **V** Series of loud, muttered, sharp, squeaky and sizzled notes. **H** Lower storeys and borders of humid and wet montane forest and secondary growth. **D** Venezuela and Colombia to Bolivia.

16 AFRICAN THRUSH *Turdus pelios* 21cm Separated from Olive Thrush by different range, and from Kurrichane Thrush by less prominent moustache and by voice. **V** Very high, calm, sustained, far-carrying, fluted *peeweepeeweetuweetuwee* (each part repeated 2–3 times). **H** Forests, woodlands, gardens. **D** W and C Africa.

17 BARE-EYED THRUSH *Turdus tephronotus* 20cm Note bare orange eye area, pale orange underparts, olive breast and distinct throat striping. **V** Song resembles that of Olive Thrush, but louder and slower. Calls include a rattling *chrrrr*. **H** Arid bush, cultivation. **D** Ethiopia and C Somalia through Kenya to C Tanzania.

18 KURRICHANE THRUSH *Turdus libonyana* 22cm Separated from Olive Thrush by prominent malar stripe and white underbelly. **V** Very high, short, fluted, far-carrying whistles, e.g. *weetoweet* or *weetoweethree*. **H** Suburbs, woodland, bushveld. **D** C and S Africa.

19 SAO TOME THRUSH *Turdus olivaceofuscus* 23cm Previously considered conspecific with Principe Thrush. Forages on the ground, feeding on invertebrates and fruit. **V** Song a series of mellow whistles and trills. Alarm calls include a low *chup* and flight calls a *sip*. **H** Forest, plantations, woodland with tall trees. **D** São Tomé, Gulf of Guinea.

20 PRINCIPE THRUSH *Turdus xanthorhynchus* 23cm Dusky brown thrush with heavily dark-scaled white underparts. Bill yellow, eyes pale blue. Confiding ground forager. **V** Similar to that of the closely related Sao Tome Thrush but less rich in tone. **H** Lowland and mid-elevation primary forest with dense understorey. **D** Príncipe, Gulf of Guinea.

21 OLIVE THRUSH *Turdus olivaceus* 24cm Underparts vary from orange to dusky (birds at higher altitudes). **V** High, full, up-and-down, short whistles, e.g. an almost level *weet-oh-weet-three*. **H** Suburbs, riverine belts, plantations, forests. **D** SE Africa

22 USAMBARA THRUSH *Turdus roehli* 22cm Upperparts dark blackish brown, underside mid-brown to white on belly, with bright reddish flanks. Forages in leaf litter, using bill and feet to move leaves. **V** Said to be very like that of the closely related Abyssinian Thrush. **H** Primary and secondary forest and forest edges with open understorey. **D** Pare and Usambara mountains, NC Tanzania.

23 ABYSSINIAN THRUSH *Turdus abyssinicus* 22cm Less extensive orange on flanks and less distinctive throat striping than in Olive Thrush. **V** Short, varied phrases, often with some trilling notes. **H** Suburbs, riverine belts, plantations, forests. **D** N Somalia to N Tanzania.

24 KAROO THRUSH *Turdus smithi* 22cm Pale orange (almost) absent or restricted to lower belly. **V** Song comprises short phrases, including *weet-oh-weet*. **H** Suburbs, riverine belts, plantations, forests. **D** S Namibia, SE Botswana, N South Africa.

25 SOMALI THRUSH *Turdus ludoviciae* 25cm Note throat streaking. **V** Song resembles that of Olive Thrush. **H** Montane forests. **D** N Somalia.

26 TAITA THRUSH *Turdus helleri* 21cm Blackish with a white belly and reddish flanks, bill bright orange-yellow. Forages in leaf litter. Sings from a well-concealed perch in mid-storey. **V** Song rich and slow, given mainly at dawn and dusk. Calls include a thin, high-pitched *tseee*, with a soft *chook* given in alarm. **H** Montane forests. **D** SE Kenya.

27 YEMEN THRUSH *Turdus menachensis* 23cm Shy. Ground feeder in dense scrub. **V** Song a series of soft, low- and high-pitched phrases, *tissik-tissik-tssechup*, a stuttering trill or a soft croaking note. Calls include a quiet *chuk-chuk*, a thin *seep* and a more explosive *chuck-chuck-chuck*. **H** Scrub on steep rocky slopes and hillsides; cultivated wadis, tree-lined terraces. **D** S and SW Arabian Peninsula.

28 COMOROS THRUSH *Turdus bewsheri* 23cm Ssp. *comorensis* is illustrated; nominate race is similar but with scaled underparts. No similar thrush or other bird occurs in its restricted range. **V** Short phrases of three slow notes. **H** Evergreen forest and other wooded habitats at mid–high altitudes. **D** Comoros islands, Mozambique Channel.

29 GREY-BACKED THRUSH *Turdus hortulorum* 23cm Very shy. Ground feeder among leaf litter, but also spends long periods in foliage of trees. Forms loose flocks in winter. **V** Not well documented. Said by some authorities to be superlative, consisting of a series of loud, fluty whistles. Calls include a harsh *chack-chack*, a low chuckle and, in alarm, a shrill *cheee*. **H** Dense oak woods, often near water, and mixed woods of various species. Winters in open woods, bamboo, scrub woodland. **D** Breeds in SE Russia and NE China. Winters in SE Asia.

30 TICKELL'S THRUSH *Turdus unicolor* 21cm Feeds mainly on the ground. In winter sometimes forms loose flocks. **V** Song is a weak, monotonous series of phrases. Calls include a loud *juk-juk*, a chattering, decelerating *juh juk-juk juk*, and a short, high buzz. **H** Breeds in mixed forests with clearings, open broadleaved woodland, groves and orchards. Winters in groves and well-wooded areas. **D** Breeds in W and C Himalayas. Winters in N India and Pakistan.

31 BLACK-BREASTED THRUSH *Turdus dissimilis* 23cm Forages on the ground and in fruiting trees and bushes. **V** Song a sweet, lilting series of short-spaced notes. Calls include a thin *seeee* and a resounding *tuc tuc tuc*. **H** Breeds in oak and conifer forest, broadleaved evergreen forest and secondary growth. Post breeding also occurs in scrub jungle and mangroves. **D** NE India to N Vietnam.

32 JAPANESE THRUSH *Turdus cardis* 22cm Usually shy. In winter often joins with flocks of Siberian thrushes (Plate 241). **V** Song a series of repeated fluty trills and whistles, *see-tew - see-tew - see-tew - titupi-tit - seea-tyew - see-a-tyu - se-a- tyu - tilit-tilit-tilit - tyu-tyatyew - tyatyew - tullut-tullut-tillit...* and variants. Call a thin *tsweee* or *tsuuu*. **H** Usually dense mixed or deciduous woods and forests on hillsides or mountainsides. **D** Breeds in Japan and C China. Winters in SE Asia.

33 WHITE-COLLARED BLACKBIRD *Turdus albocinctus* 28cm Generally wary. Feeds in fruiting trees and shrubs and on the ground. **V** Song a mellow series of descending whistles, *tew-i-tew-u-tew-o* or similar, occasionally with some hissing or squeaking notes. Call a throaty *tuck-tuck-tuck*. **H** Various forest types, including deciduous, open conifer with good ground cover, mixed rhododendron scrub above the treeline. **D** Breeds from Himalayas to SC China. Winters in SE Asia.

34 RING OUZEL *Turdus torquatus* 26cm Actions like those of Common Blackbird (Plate 244) but generally more shy. **V** Song comprises several loud, repeated, melancholy, fluty phrases, sometimes rounded off with a rattling chuckle. Call a hard *tak-tak-tak*. **H** Mountains, rocky uplands, hilly areas with stunted trees; in the Middle East, often down to sea level. **D** Breeds in W and SW Eurasia, from N Europe to Iran. Winters from S Europe to N Africa and SW Asia.

1
2
3
4
5
6
7
8
9
10
11
12
13
♂
14
♀
♂
15
♀
16
17
18
19
20
21
22
23
24
25
26
27
28
♀
29
♂
30
♀
♀
31
♂
♀
32
♂
♀
33
♂
♂
♀
34
♂

THRUSHES *CONTINUED*

1 GREY-WINGED BLACKBIRD *Turdus boulboul* 29cm Generally shy. May form small winter flocks, these often including White-collared Blackbirds (Plate 243). **V** Song rich and melodious, usually consisting of a soft opening note followed by four high notes descending in tone. Calls include a *chuck-chuck*, a *chook-chook* and a *churi* contact note. **H** Moist broadleaved forest of oak and rhododendron with thick cover, conifer forest clearings and dry scrub on hillsides. **D** Himalayas to N Vietnam.

2 COMMON BLACKBIRD *Turdus merula* 27cm Usually conspicuous. Feeds on the ground and in fruiting bushes and trees. **V** Song comprises beautiful clear, rich, fluty notes that merge into short continuous phrases, not like the repeated phrases of a Song Thrush. Call a low *chuck-chuck-chuck* or similar, rapidly repeated when alarmed. Also a high, drawn-out *tseee*. **H** Very varied, including all types of forest and woodland, scrub, farmland with hedges, parks, gardens. **D** Widespread across Europe, NW Africa and Asia to N Afghanistan.

3 CHINESE BLACKBIRD *Turdus mandarinus* 28cm Very similar to Common Blackbird in appearance and habits. **V** Similar to that of Common Blackbird, but song phrases lack high-pitched ending. **H** All kinds of open woodland, also farmland with hedges, plantations, parks, gardens. **D** E and C China.

4 TIBETAN BLACKBIRD *Turdus maximus* 26–29cm Forages on the ground; attracted to soft areas at edges of melting snow. **V** Song a series of phrases that includes rapid metallic notes, squeaks, wheezy sounds, guttural caws and a pure whistle. Calls include a low *chut-ut-ut* and harsh *chak-chak-chak-chak*. **H** Subalpine rocky, grassy slopes with dwarf juniper. Winters at lower altitudes. **D** Himalayas, from N Pakistan to SE Tibet.

5 INDIAN BLACKBIRD *Turdus simillimus* 25cm Forages on the ground and in fruiting trees. **V** Song loud and melodious, with much mimicry. Call a *kack-kack*. **H** Moist forest and wooded ravines. **D** India, Sri Lanka.

6 ISLAND THRUSH *Turdus poliocephalus* 21–25cm Generally shy, mostly foraging on the ground among leaf litter in dense cover. Plumage very variable, with many races described (around 50 worldwide). **V** Song a subdued melody of flute-like whistles. Calls include a *chook-chook*, a rapid, chattering *tchick-tchick-tchick* or *tchick-chink-chink*, and a *tchook-tchook-toaweet-oweet-toaweetoweet* when excited or alarmed. **H** Montane forest. **D** From Greater Sunda Islands and the Philippines to Fiji.

7 CHESTNUT THRUSH *Turdus rubrocanus* 24–28cm Shy. Perches in treetops, but mainly a ground feeder. In winter, often forms small flocks, which may include other thrushes. **V** Song comprises short, warbled phrases, repeated up to eight times, then changing to a Song Thrush-like *yeee-bre - yee-bre - diddiyit-diddiyit-diddiyit - yip-bru - yip-bru...* Calls include a *chuck-chuck-chuck* and, when alarmed, a rapid *kwik-kwik - kwik-kwik*. **H** Conifer and mixed forest with ground cover; 2,300–3,300m, but lower in winter. **D** E Afghanistan to Myanmar.

8 KESSLER'S THRUSH *Turdus kessleri* 28cm Gregarious, even in the breeding season. **V** Song rarely heard, but said to be a short series of Mistle Thrush-like melancholy phrases. Calls include a soft *squack*, harsher when alarmed, also a soft *dug-dug*, a high *swi-swi-swi-swi* and *chock-chock-chock*. **H** Dwarf juniper and conifers, various scrub types; 3,600–4,500m. **D** W China to SE Tibet.

9 GREY-SIDED THRUSH *Turdus feae* 23cm Shy. Mainly a ground feeder. Forms winter flocks, which often include Eyebrowed Thrushes. **V** Song little recorded; said to consist of series of repeated, short double- or triple-note phrases, *sit-tweoo - wet-too - chit-to-loo*. Call a thin *zeeee*. **H** Broadleaved montane forest. **D** Breeds in NE China. Winters in SE Asia.

10 EYEBROWED THRUSH *Turdus obscurus* 22cm Wary during the breeding season, less so on migration and in winter, when often in flocks with other thrushes. Feeds on the ground, and in fruiting trees and bushes. **V** Song consists of two or three mournful phrases, *teveteu - trryutetyute - trrryutetyutyu*, a pause, then a lower twittering, warbling and chattering, interspersed with pauses. Calls include a soft *chuk*, a hard *tack-tack* and a *shree*; also a *dzee* flight call. **H** Dense spruce and fir forests, high-altitude mixed forests. Winters in open forests, open country, parks and gardens. **D** Breeds from C to SE Russia. Winters in SE Asia.

11 PALE THRUSH *Turdus pallidus* 23cm Shy and wary. Mainly a ground feeder, in undergrowth. In winter, forms flocks. **V** Far carrying. Song comprises various repeated double or triple whistles, *tuvee-tulee - tulee-tevee, tve-tveeeu-weet-weet-tveeu-trrrsss* or similar. Calls include a *chook*, *tuck-tuck* and a thin *tsee* or *tsee-ip*; alarm call a Common Blackbird-like *think-think*. **H** Montane and submontane forests. In winter and on migration, occurs in more open areas. **D** Breeds in SE Russia, NE China, Korean Peninsula. Winters in Japan, Taiwan, SE China.

12 BROWN-HEADED THRUSH *Turdus chrysolaus* 23cm Feeds low down, in or near the cover of bushes, or at all levels in fruiting trees. Often in large flocks. **V** Song a repeated *krurr krurr krr-zeee* or *kiron kiron tsee*. Calls include a *chuck-chuck-chuck* and, when alarmed or in flight, a thin *zeeee*. **H** Broadleaved forests, secondary growth, clearings near cover, cultivations, well-wooded parks and gardens. **D** Breeds on Sakhalin and Kuril Islands (Russian N Pacific), Japan. Winters in Japan, including Ryukyu Islands.

13 IZU THRUSH *Turdus celaenops* 23cm Actions similar to those of others in the genus. **V** Song usually a repeated high-pitched rattle, *tche-e-e-e-e - tche-e-e-er* or similar. Call a series of dry, grating notes. **H** Deciduous or mixed woods, orchards, large gardens. **D** Islands south of Japan.

14 BLACK-THROATED THRUSH *Turdus atrogularis* 26cm Gregarious. Forages on the ground and in bushes. **V** Calls include a thin *seet*, a squeaky *qui-kwea* and a throaty chuckling when alarmed. **H** Grassy scrubby hillsides, forest edges, cultivation. **D** Breeds from WC Russia to NC Siberia and NW Mongolia. Winters in SW and S Asia.

15 RED-THROATED THRUSH *Turdus ruficollis* 24cm Ground feeder, with actions like those of a Fieldfare. Forms large winter flocks, often with other thrush species. **V** Song a rambling, simple *chooee - whee-oo-ee - oo*. Calls include a soft *chuk*, a weak *seep* and a harsh *chak*; also a Blackbird-like chatter. **H** Mainly sparse mountain forest and mossy scrub tundra. **D** Breeds in SC Russia, N Mongolia, NW China. Winters from N India to N Vietnam.

16 NAUMANN'S THRUSH *Turdus naumanni* 23cm Mainly a ground feeder; acts much like a Fieldfare. Forms flocks in winter, often in association with Black-throated Thrush. **V** Song a melodious, fluty *tvee-tryuuu-tee - tvee-tryuuuu-tvee*, ending in a faint trill. Calls include a shrill *cheeh-cheeh*, often repeated, a harsher *ket-ket-ket* and a chuckling *chak-chak*. **H** Open forests and woods. Winters in more open areas, including woodland edges, scrub, stubble-fields, parks and gardens. **D** Breeds in SC Russia. Winters from SE Russia to E China.

17 DUSKY THRUSH *Turdus eunomus* 23cm Forages on the ground and in fruiting bushes and trees. **V** Calls include a low staccato *chuck* and a rhythmic *chek-chek-chek*. **H** Open broadleaved evergreen forest, forest edges, open areas with scattered trees, scrubby hillsides, winter stubble fields, parks and gardens. **D** Breeds from NC to NE Russia. Winters from India to Japan.

18 FIELDFARE *Turdus pilaris* 25cm Note white underwing coverts. Feeds on the ground, but also attracted to fruiting trees and bushes. **V** Calls include a *chach-chack*, a nasal *tseee* and, when anxious, a *chetchetchetje* or *trt- trrrrt-trrt*. **H** Open conifer, deciduous or mixed forest, tundra scrub. Winters in more open areas. **D** Breeds from W Europe through E Russia. Winters to N Africa and Iceland.

19 REDWING *Turdus iliacus* 21cm Underwing coverts rusty. Forms large winter flocks, often in the company of other thrushes, especially Fieldfares. **V** Song comprises a few short notes, *chirre-cherre-churre* or similar, followed by a prolonged fast, variably pitched twittering. Calls include a thin *seeeh* or *seeip*, and, when alarmed, a rattling *trrrt-trrrt-trrt* or *jip-jip*. **H** Birch woodland, birch and willow scrub, mixed conifer woodland, locally in parks and gardens. **D** Breeds from Iceland and N Europe to C Siberia. Winters to N Africa and SW Asia.

20 SONG THRUSH *Turdus philomelos* 23cm Forages mostly on the ground; specialises in feeding on snails, which it usually smashes on stones or rocks. In flight, shows ochre underwing coverts. **V** Song comprises clear repeated phrases. Calls include a *sip* or *zip*, and a loud *chick* when alarmed. **H** Open woodland with extensive undergrowth. **D** Breeds across Europe to N Iran and C Siberia. Winters in N Africa and S Asia.

21 CHINESE THRUSH *Turdus mupinensis* 23cm Usually in undergrowth; shy. **V** A pleasant *drip-dii-du dudu-du-twi dju-wi-wi chu-wii-wr up chu-wi i-wu-wrrh dju-dju-wiii u...*, plus shorter twitterings and *se-wee* phrases. **H** Mixed or deciduous montane forest, plantations and woodlands with rich undergrowth. **D** C China.

22 MISTLE THRUSH *Turdus viscivorus* 27cm Feeds in the open. Flight undulating. Underwing coverts white. **V** Several clear, far-carrying, fluted phrases, e.g. *chuwee - trewuu - tureetruruu - truwutru - truwuwutru...*, interspersed with short pauses and some shorter phrases. Calls include a harsh rattle and a *tuck-tuck-tuck*. **H** Various types of woodland, usually with open areas; also hedgerows, parks and gardens. Often moves to more open areas in winter. **D** Breeds from Europe to W Siberia and N Iran, and SC Asia to W Nepal. Winters in N Africa and S Asia.

23 GREAT THRUSH *Turdus fuscater* 31cm Note large size. Separable from Chiguanco Thrush by yellow eye-ring. **V** Continuous series of closely interconnected, short, melodious phrases of 3–6 notes. **H** Borders of forest and woodland, fields and gardens. **D** W Venezuela to W Bolivia.

24 CHIGUANCO THRUSH *Turdus chiguanco* 27cm See Great Thrush. **V** Song consists of phrases of 4–5 notes, starting with a level *woot-woot-woot* and ending with an inhaled mini-twitter. **H** Mainly arid areas like fields, thickets and gardens, woodland. **D** S Ecuador to W Argentina.

25 SOOTY THRUSH *Turdus nigrescens* 25cm Eye, bill and leg colouring distinctive. **V** High, simple *tjilping wit wit wit titjerrup*. **H** Open montane areas. **D** Costa Rica, Panama.

26 BLACK THRUSH *Turdus infuscatus* 25cm Male distinctive; female with uniform underparts, including unstriped or very faintly striped throat. **V** Very high, varied *sreek tujotujo wrrr ssssi...* **H** Forest edges. **D** Mexico to Honduras.

27 GLOSSY-BLACK THRUSH *Turdus serranus* 25cm Male is the blackest thrush in South America; female uniform brown. **V** High, loud, sharp, short *zeet-it* phrases, often extended to *zeet-t-t-it*. **H** Humid and wet montane forest. **D** Venezuela to NW Argentina.

28 ANDEAN SLATY THRUSH *Turdus nigriceps* 20cm Male is distinctively coloured; note yellow eye-ring and brownish breast of female. **V** Unstructured twittering with many inhaled notes (5–8 secs). **H** Montane forest and woodland. **D** S Ecuador to N Argentina.

29 EASTERN SLATY THRUSH *Turdus subalaris* 22cm Male lead grey, underparts paler grey; throat white with black streaks; bill yellow, legs orange. Female upperparts, wings and tail brown; throat buff, streaked dark; breast brownish, belly whitish. **V** High-pitched 'scraping' song. **H** Canopy of woods and forests. **D** E Paraguay, NE Argentina, SE Brazil.

30 PLUMBEOUS-BACKED THRUSH *Turdus reevei* 23cm Note pale eyes and blue-grey upperparts. **V** Calm series of simple fluted phrases given in (almost) similar sets of two. **H** Forest, woodland, scrub, adjacent clearings. **D** W Ecuador, NW Peru.

31 BLACK-HOODED THRUSH *Turdus olivater* 23cm Note contrasting black head and breast of male. **V** Slow series of high, loud, sharp, single- and double-syllable phrases. **H** Interior and borders of humid forest and secondary growth. **D** Colombia to S Guyana.

32 MARANON THRUSH *Turdus maranonicus* 22cm No other thrush has such heavy spotting below. **V** Song is much as that of Black-billed Thrush (Plate 245). **H** Woodland, arid scrub, gardens. **D** N Peru.

33 CHESTNUT-BELLIED THRUSH *Turdus fulviventris* 24cm No other thrush with a rufous belly occurs in its range. **V** Series of short, high-pitched, fluted phrases. **H** Montane forest and woodland. **D** NE Venezuela to Peru.

34 RUFOUS-BELLIED THRUSH *Turdus rufiventris* 24cm Brown with contrasting rufous-orange belly. **V** Continuous stream of strong, fluted notes of varying pitch. **H** Open woodland, Chaco woodland, agricultural areas with scattered trees, gardens. **D** E and SC South America.

1
2
3
4
5
6
7
8
9
10
11
12
13
14
15
16
17
18
19
20
21
22
23
24
25
26
27
28
29
30
31
32
33
34

THRUSHES *CONTINUED*

1 AUSTRAL THRUSH *Turdus falcklandii* 25cm Note contrasting dark head. Eye-ring, bill and legs yellow. **V** Slow, unstructured series of single or double fluted notes of various intonations. **H** Undergrowth of open forest, plantations, secondary growth, agricultural areas with some trees and hedges, gardens. **D** S South America.

2 PALE-BREASTED THRUSH *Turdus leucomelas* 24cm Note greyish crown and nape. Separable from Creamy-bellied Thrush by pale (not dark) lores. **V** Groups of 3–5 short, almost similar, mid-high, fluted phrases. **H** Forest borders and adjacent clearings, woodland, habitats with scattered trees, gardens. **D** N and SC South America.

3 CREAMY-BELLIED THRUSH *Turdus amaurochalinus* 23cm No similar thrush has such dark lores and a yellow bill (dusky in female). **V** Slow, continuous series of single, principally harsh notes. **H** Cerrado, open forest, light woodland, farmland, gardens. **D** Breeds from Bolivia and S Brazil to C Argentina. Winters in N South America through Amazonia.

4 MOUNTAIN THRUSH *Turdus plebejus* 25cm Dark brown colouring with black bill is distinctive. **V** Song a very high, rather hurried *tjow-tjew-tjow-tjeu-tjew*... Call a *tjur*. **H** Montane forest. **D** SE Mexico to Panama.

5 BLACK-BILLED THRUSH *Turdus ignobilis* 23cm Blackish bill is diagnostic. **V** Series of short phrases of 2–4 rather hoarse, hurried notes. **H** Mainly in semi-open and open habitats, e.g. savannah with tree stands, Cerrado, gardens. **D** N and W Amazonia.

6 LAWRENCE'S THRUSH *Turdus lawrencii* 22cm Note orange eye-ring and bill. **V** Beautiful, very varied phrases with clear or sharp notes given at 6–8-sec intervals. **H** Borders and canopy of humid forest. **D** W Amazonia.

7 COCOA THRUSH *Turdus fumigatus* 23cm Feeds primarily on the ground, although often attracted to fruiting trees. **V** Variable series of musical notes, consisting mostly of rapid, repeated phrases; also utters a descending *weeo weeo weeo* or *wee-a-wee-a-wee-a*. **H** Forest, cacao plantations, farmland with scattered trees. **D** S Lesser Antilles, N Amazonia.

8 PALE-VENTED THRUSH *Turdus obsoletus* 22cm Has little or no throat streaking. Note pure white vent. **V** Calm series of rapidly delivered fluted phrases, many of them variations of *wee-oh-wee*. **H** Lower storeys of forest, secondary growth, riverine belts. **D** Costa Rica to Ecuador.

9 HAUXWELL'S THRUSH *Turdus hauxwelli* 23cm Compare with Cocoa Thrush. Note white vent. **V** Song is similar to that of Cocoa Thrush. **H** Lower storeys of humid forest. **D** W Amazonia.

10 UNICOLORED THRUSH *Turdus haplochrous* 23cm Separable from Hauxwell's Thrush by less rufescent upperparts, brighter yellow bill and dull orange eye-ring. **V** Song resembles that of Ecuadorian Thrush but delivered with a bit of hesitation. **H** Woodland, várzea forest. **D** N Bolivia.

11 CLAY-COLORED THRUSH *Turdus grayi* 25cm Note greenish bill. Undertail coverts concolorous with rest of underparts. **V** High, simple *pruweeh-puweeh-weejoh-juju*... **H** Areas with scattered tree stands, bush, hedges, gardens. **D** Mexico to Colombia.

12 SPECTACLED THRUSH *Turdus nudigenis* 23–25cm Forages in trees and on the ground, feeding mainly on berries and fruit. Often aggressive towards other thrushes. **V** Loud, liquid *cheerily cheer-up cheerio* or similar; also utters a high *miter-ee*. **H** Open lowland woods, forest edges, secondary growth, plantations. **D** Lesser Antilles, N Amazonia.

13 VARZEA THRUSH *Turdus sanchezorum* 23cm Note colour of bill and throat streaks. **V** Lazy series of 4–10 melodious yet slurred, drawn-out *tuweee* notes. **H** Várzea and other open forest. **D** W Amazonia.

14 ECUADORIAN THRUSH *Turdus maculirostris* 23cm No similar thrush occurs in its range. **V** Series of 4–6 melodious phrases. **H** Forest, woodland, clearings with scattered trees. **D** W Ecuador, NW Peru.

15 WHITE-EYED THRUSH *Turdus jamaicensis* 23cm Usually forages in dense vegetation from the ground to treetops; shy and secretive. **V** Song varied and musical, with repeated phrases such as a bell-like *hee-haw* and trilling whistles and chimes. Call a shrill *dzeee* or *dzaw*. **H** Wet mountain forests and shade coffee plantations. **D** Jamaica.

16 WHITE-THROATED THRUSH *Turdus assimilis* 22–26.5cm Shy. Forages mainly in mid-level or canopy of fruiting trees, much less so on the ground. **V** Song a rich warble with repeated phrases. Call a nasal *rreeuh* or *rreuh*; also a clucking *ch-uhk* and a thin *ssi* given in flight. **H** Forest, forest edges, plantations. **D** Widespread from NW Mexico to Panama.

17 DAGUA THRUSH *Turdus daguae* 22–26cm Recently split from White-throated Thrush; it closely resembles that species but is a little darker. **V** Song like that of White-throated Thrush but faster paced; has a similar varied array of calls. **H** Forest, forest edges, plantations. **D** Panama to NW Ecuador.

18 WHITE-NECKED THRUSH *Turdus albicollis* 22cm Separable from White-throated Thrush by warmer colouring, darker head and rufous flanks. **V** Slightly hurried series of 4–5 melodious phrases. **H** Undergrowth of humid forest and woodland. **D** Colombia to E Brazil and N Argentina.

19 RUFOUS-BACKED THRUSH *Turdus rufopalliatus* 23.5cm Forages on the ground and in fruiting trees. **V** Song a slow series of rich, warbled notes. Call a plaintive, mellow *peeeoooo*, *cheeoo* or *teeeuu*; also a throaty *chuck chuck chuck*. **H** Deciduous and mixed forest, woodland edges, scrub and large gardens. **D** Mexico.

20 RUFOUS-COLLARED THRUSH *Turdus rufitorques* 25cm Unmistakable. Note dark hood and scapulars of female. **V** Very high *tjurrup trur wih* (little variation). **H** Open woodland, human habitation. **D** S Mexico to Honduras.

21 AMERICAN ROBIN *Turdus migratorius* 23–25cm Feeds mainly on the ground but also in trees and bushes, where it is attracted to berries and fruit. **V** Song loud and liquid, often transcribed as *cheerily cheer-up cheerio* or *cheerily-cheery-cheerily-cheery*. Call a *tut-tut-tut*; also a descending *shheerr* and an excited *kli quiquiquiqul*... **H** Forests, woods, thickets, meadows with hedges, urban areas. **D** Alaska and Canada to Guatemala.

22 LA SELLE THRUSH *Turdus swalesi* 26cm Forages mainly on the ground. **V** A deliberate *tu-re-oo* or *cho-ho-cho*, and a strident *poo-ip poo-ip* or *wheurry-wheurry wheury*; also utters low chuckling or gurgling notes. **H** Dense understorey of wet forests, including mountain pine forests. **D** Hispaniola.

23 WHITE-CHINNED THRUSH *Turdus aurantius* 24cm Active, especially after rain; forages mainly on the ground. **V** Song slow and lilting, including shrill whistles that are often given on their own; also utters a clucking *kek*. **H** Mountain and hill woodlands, plantations, roadsides, gardens. **D** Jamaica.

24 RED-LEGGED THRUSH *Turdus plumbeus* 26–27cm Forages mainly on the ground. Generally shy but becomes more conspicuous during the breeding season. **V** Call a high-pitched *weecha weecha weecha*, *cha-cha-cha* or *chu-week chu-week chu-week*. Song a melodious, monotonous series of 1–3-syllable phrases. **H** Woodlands, thick undergrowth, scrub, shade coffee plantations, gardens. **D** West Indies.

25 FOREST THRUSH *Turdus lherminieri* 24–27cm Shy, often running into cover when disturbed. Forages from the ground to tree canopy. **V** Song a fluty, whistled cadence that may include some harsh notes. Call a loud *chuck-chuck*, often extended to a harsh chatter. **H** Moist mountain forests. **D** Lesser Antilles.

26 TRISTAN THRUSH *Turdus eremita* 22cm Unusual small, very dark thrush with dark bill and eyes; paler underparts heavily marked with diffuse dark streaking. Omnivorous and opportunistic, scavenging carrion and preying on storm-petrel chicks; will even attack and kill adult storm-petrels. **V** Song a combination of single and double whistled phrases. Calls include a wheezing *swee* and a soft, chirping chatter. **H** Wet heath, grasslands, shorelines, gardens. **D** Tristan da Cunha, S Atlantic Ocean.

27 PURPLE COCHOA *Cochoa purpurea* 25–28cm Secretive. Generally arboreal, feeding in fruiting trees, but will also forage on the ground. **V** Song a pure, whistled *peeeee* or *peeee-you-peeee*. Calls include a thin *sit* or *tssri*, a chuckling *nyerr* and a soft, high *pink-pink-trrrrrew*. **H** Dense, humid broadleaved evergreen forest. **D** Himalayas to N Vietnam and N Thailand.

28 GREEN COCHOA *Cochoa viridis* 25–28cm Lethargic forager in trees and undergrowth and on the ground; frequently in pairs or small groups. **V** A pure, thin monotonous whistle. Calls include a harsh note and a high, thin *pok*. **H** Dense, moist broadleaved evergreen forest, often near streams, also locally in dry evergreen forest and tall forests in limestone valleys. **D** Himalayas to Vietnam.

29 SUMATRAN COCHOA *Cochoa beccarii* 28cm Mainly seen in the middle and upper storeys – look for short flights between perches. Forages for fruit in tall trees and possibly also searches for invertebrates on the forest floor. **V** Long, thin, high, mournful whistle. In flight, also gives a *sip*. **H** Lower montane forest. **D** Sumatra, Indonesia.

30 JAVAN COCHOA *Cochoa azurea* 23cm Often sits silently for long periods or moves about quietly, making detection difficult – watch for short flights from perch to perch. Favours fruiting forest trees. **V** Thin, high, whistled *siiiit*. Calls include a scolding *cet-cet-cet*, usually given when alarmed. **H** Lower and upper tropical montane rainforest. **D** Java, Indonesia.

31 FRUITHUNTER *Chlamydochaera jefferyi* 22–23cm Usually seen in small wandering groups and mixed-species parties. Forages in the mid-storey and on the ground in leaf litter. **V** Quiet, high-pitched *seep*. **H** Tall montane forest at lower altitudes and nearby gardens. **D** Borneo.

1
2
3
4
5
6
7
8
9
10
11
12
13
14
15
16
17
18
19
20 ♀
21 ♀
♂
22
♂
23
24
25
26
27 ♀
♂
28 ♀
♂
29 ♀
♂
30 ♀
♂
31 ♀
♂

CHATS AND OLD WORLD FLYCATCHERS MUSCICAPIDAE

1 WHITE-TAILED ALETHE *Alethe diademata* 17cm Note orange crown with contrasting grey face sides. **V** Simple, repeated, strong mid-high whistle, *fuuu fuuu fuuu* , falling-off at the end, or *fju weh*. **H** Lowland forests, riverine belts, plantations. **D** W Africa, from Senegambia to Togo.

2 FIRE-CRESTED ALETHE *Alethe castanea* 19cm Note pale orange centre to crown, and grey on sides of face, breast and flanks. Up to 50 may follow a swarm of army ants. **V** Very like that of White-tailed Alethe but song comprises two rather than three notes. Gives a *chip chip* call at ant swarms. **H** Undergrowth of mature forest, mainly in lowlands. **D** W and C Africa, from Nigeria to Uganda.

3 KAROO SCRUB ROBIN *Cercotrichas coryphoeus* 17cm Separated from other scrub robins absence of rufous in plumage and tail, which is not normally cocked. Note also plain wings. **V** Very high, simple-structured flow of slightly harsh, whistled notes, *fiu-fri-fri-rutrit-reeh*. **H** Dry, shrubby areas. **D** S Namibia and South Africa.

4 FOREST SCRUB ROBIN *Cercotrichas leucosticta* 17cm Separated from Miombo Scrub Robin by much darker plumage and brown-black tail. **V** Song a melodious, warbling series of varied phrases. Alarm call a rapid *chit-chit-chit*. **H** Forest interiors, often near termite mounds. **D** W and C Africa.

5 BEARDED SCRUB ROBIN *Cercotrichas quadrivirgata* 17cm More saturated plumage colours than Miombo Scrub Robin, and with more extensive white on underparts. **V** Continuous flow of mid- to very high, clear, fluted notes and short phrases, with very fast, repeated parts, *tie tjutie-tie-tjutie-fuwie-fuwie-fuwie-ririri-tututjie*. **H** Undergrowth of dry forest, woodland, bushveld, gardens. **D** E, SE and S Africa.

6 MIOMBO SCRUB ROBIN *Cercotrichas barbata* 17cm Separated from Forest Scrub Robin by paler plumage, red cheek, rufous (not brown or grey) flanks, and different habitat and range. **V** Very high, sharp strophes. **H** Miombo woodland, thick bush and scrub. **D** Angola and S DR Congo to W Tanzania, Malawi and S Zambia.

7 BLACK SCRUB ROBIN *Cercotrichas podobe* 19cm Tail often cocked and spread over back. **V** Song a series of melodious whistles, occasionally interspersed with harsher *tew-ti-heat* notes. Call a hoarse squeak or liquid chatter. **H** Flat or rolling sandy plains with dry scrub, bushy oases and gardens. **D** Breeds from W Africa to Arabian Peninsula. Winters in Arabian Peninsula.

8 RUFOUS-TAILED SCRUB ROBIN *Cercotrichas galactotes* 16cm Forages mainly on the ground. **V** Song includes rich, varied ringing notes. Calls include a hard *tek-tek*. **H** Dry scrub jungle. **D** Breeds from Iberian Peninsula to W Pakistan and south to Senegal, Gambia and N Somalia. Winters in W, E and NE Africa.

9 KALAHARI SCRUB ROBIN *Cercotrichas paena* 16cm Note absence of white in wing, buff-orange mantle and tertials, and orange-rufous rump and tail. **V** Song a sequence of very high, very short warbler-like phrases, each repeated 2–10 times, *weet weet weettietjittietjittietjit*, all parts well separated. **H** Dry thornveld with open patches. **D** SC and SW Africa.

10 BROWN-BACKED SCRUB ROBIN *Cercotrichas hartlaubi* 18cm Note dull plumage and lack of reddish breast-band and flanks. Separated from Forest Scrub Robin and Miombo Scrub Robin by different range; overlaps only marginally with range of Bearded Scrub Robin. **V** Rather short, clear whistles, each one starting with very high *feee* and then descending. **H** Dense undergrowth. **D** SE Nigeria and S Cameroon to SW Central African Republic; NE DR Congo to C Kenya and Burundi; NW Angola.

11 WHITE-BROWED SCRUB ROBIN *Cercotrichas leucophrys* 15cm In most of its range, the only scrub robin with a streaked breast. **V** Very high, loud, fluted whistles, repeated 2–3 times, e.g. *weeduwietweeduwiet weeduwiet*. **H** Dense grassy thornbush patches in more or less wooded and bushy areas. **D** C, E and S Africa.

12 BROWN SCRUB ROBIN *Cercotrichas signata* 18cm Note dull plumage and lack of reddish breast-band and flanks. Separated from Forest Scrub Robin and Miombo Scrub Robin by different range; overlaps only marginally with range of Bearded Scrub Robin. **V** Rather short, clear whistles, each one starting with very high *feee* and then descending. **H** Dense undergrowth. **D** SE and S Africa.

13 INDIAN ROBIN *Copsychus fulicatus* 16cm Confiding; hops or runs with tail regularly carried erect or well over back. **V** Repeated *wheech* or a harsh *chu-r - chu-r*. Song is a short, high-pitched, cheery warble. **H** Dry stony foothills with scrub, edges of cultivation, in and around human habitation. **D** India, Pakistan, Sri Lanka.

14 ORIENTAL MAGPIE-ROBIN *Copsychus saularis* 20cm Confiding; usually conspicuous but can become more secretive in the non-breeding season. **V** Song a varied musical warbling, alternating with churrs and sliding whistles. Calls include a clear rising whistle; when alarmed, issues a harsh *che-e-e-e-h*. **H** Open forests, groves, parks and gardens. **D** Widespread across Asia, from NE Pakistan to S and E China, and south to SE Borneo.

15 RUFOUS-TAILED SHAMA *Copsychus pyrropygus* 25cm Sits still for long periods; regularly cocks tail. Forages on the ground and in trees, gleaning from leaves and the ground. **V** Song is a loud, whistled *whi-iii whi-iii whi-uuu*. Call is a scolding *tchurr*. **H** Primary lowland and hill broadleaved evergreen forest. **D** S Thailand; Malaysia; Sumatra, Indonesia; Borneo.

16 MADAGASCAN MAGPIE-ROBIN *Copsychus albospecularis* 18cm Male distinctive; note wing pattern of female. **V** Song a series of high, rather sharp phrases comprising six notes. **H** Wooded habitats, from evergreen forest to coastal scrub. **D** Madagascar.

17 SEYCHELLES MAGPIE-ROBIN *Copsychus sechellarum* 19cm No similar bird occurs in its tiny range. **V** Song a repetitive series of mellow and squeaky phrases, including mimicry, interrupted by short pauses. **H** Forested areas, human settlement. **D** Seychelles.

18 PHILIPPINE MAGPIE-ROBIN *Copsychus mindanensis* 19–20cm Noisy. Regularly flicks and fans tail when perched. **V** Melodious, repetitive, clear whistles. **H** Secondary growth, cultivated areas, scrub, bamboo thickets. **D** Philippines.

19 WHITE-RUMPED SHAMA *Copsychus malabaricus* 25cm Retiring; usually forages on the ground or in low cover. Uses wings to make a clicking sound when flying across open ground. **V** Song comprises rich, melodious phrases. Alarm call harsh and scolding; also utters a musical *chir-chur* or *chur-chi-churr*. **H** Forest undergrowth. **D** Widespread from India to S China and south to Borneo.

20 ANDAMAN SHAMA *Copsychus albiventris* 25cm Considered by some authorities to be a race of White-rumped Shama; resembles that species but belly is white rather than rufous. **V** Reported as being similar to that of White-rumped Shama. **H** Dense forest, scrub, gardens. **D** Andaman Islands.

21 WHITE-CROWNED SHAMA *Copsychus stricklandii* 25cm Considered by some authorities to be a race of White-rumped Shama. Generally forages on the ground or in low cover. **V** Reported as being similar to that of White-rumped Shama. **H** Primary lowland and hill forest. **D** Borneo.

22 WHITE-BROWED SHAMA *Copsychus luzoniensis* 17–18cm Forages in forest undergrowth close to the ground. Best located by its calls. **V** Pleasant series of melodious phrases, including gurgles and whistles, rising and falling, and often finishing on a high note. **H** Primary forest and secondary growth. **D** Philippines.

23 WHITE-VENTED SHAMA *Copsychus niger* 18–22cm Occurs singly or in pairs. Best located by its loud song. **V** Various repeated, loud, melodious, rising and falling phrases. **H** Lowland forest, forest edges, secondary growth, scrub. **D** Philippines.

24 BLACK SHAMA *Copsychus cebuensis* 20cm Forages on or close to the ground in the understorey. Secretive; best located by its calls. **V** Long series of melodious rising and falling whistles. **H** Primary forest, dense scrub, thickets, bamboo groves. **D** Philippines.

25 FRASER'S FOREST FLYCATCHER *Fraseria ocreata* 14cm Found in flocks of up to 20. Active and noisy. **V** Song melodious. Calls buzzing and harsh. **H** Forest interiors. **D** W and C Africa.

26 WHITE-BROWED FOREST FLYCATCHER *Fraseria cinerascens* 17cm Differs from Fraser's Forest Flycatcher by white supraloral spot and less neatly scaled breast. Normally seen in pairs, not in groups. **V** Song a series of thin trills. Calls include a high-pitched *tseeee*. **H** Forest streams. **D** W and C Africa.

27 GREY-THROATED TIT-FLYCATCHER *Myioparus griseigularis* 11cm Note uniform grey colouring below. In behaviour, very much like a warbler: active, moving through the foliage gleaning insects while constantly wagging its tail. **V** Very high, hoarse, rising three-note *reeh-reeh-reeh*. **H** Forest and nearby areas. **D** C Africa.

28 GREY TIT-FLYCATCHER *Myioparus plumbeus* 14cm Separated from Grey-throated Tit-Flycatcher by paler underparts and constant fanning (not only wagging like Grey-throated) of tail. Active insect hunter. **V** Very high *srrieeeeh* (last part descending). **H** Open woodland, riverine belts. **D** Widespread W, C, E and S Africa.

29 ANGOLAN SLATY FLYCATCHER *Melaenornis brunneus* 15cm No similar flycatcher with brown upperparts and a white belly occurs in its restricted range. **V** Not recorded. **H** Montane forest, woodland patches. **D** Angola.

30 WHITE-EYED SLATY FLYCATCHER *Melaenornis fischeri* 16cm Note broad whitish-grey eye-ring. **V** Song a high-pitched *tsp, swee-oo-sweetoo*. Call a harsh *screescreet*. **H** Montane forest edges, woodland, gardens. **D** C and E Africa.

31 ABYSSINIAN SLATY FLYCATCHER *Melaenornis chocolatinus* 15cm Note stout bill. Eyes pale yellow. **V** Little information; said to resemble that of White-eyed Slaty Flycatcher. **H** Forest edges. **D** Ethiopia, Eritrea.

32 NIMBA FLYCATCHER *Melaenornis annamarulae* 18cm Separated from very similar Northern Black Flycatcher by more bluish, less dull black plumage and very restricted, different range. **V** Usually silent. Song varied, including a whistling *peeew-peeew-peeew-peeew-toytoytoy*. Calls loud and strident. **H** High up in trees around forest clearings. **D** Sierra Leone to Ivory Coast.

33 YELLOW-EYED BLACK FLYCATCHER *Melaenornis ardesiacus* 20cm Note striking yellow eye. **V** Calls include a harsh *tch-tchec* and *raap, raap*. **H** Dense undergrowth at edges of montane forest. **D** E DR Congo, W Uganda, Rwanda, Burundi.

34 NORTHERN BLACK FLYCATCHER *Melaenornis edolioides* 20cm Note sooty-black colouring. **V** Song a series of melodious and buzzing phrases interspersed with pauses, sometimes sung in a duet. Calls include a *tseeeu* and *tsik*; alarm call a deep *kchchew*. **H** Woodland, bush, gardens. **D** W, C and E Africa.

35 SOUTHERN BLACK FLYCATCHER *Melaenornis pammelaina* 20cm Glossy black. Tail only slightly forked (compare with Shining Drongo, Square-tailed Drongo, and Fork-tailed Drongo; Plate 183). **V** High, loud, melodious, fluted *feeh-tjuutjuu* (*feeh-* extremely high). **H** Wooded and bushy natural and cultivated areas. **D** E and S Africa.

1
2
3
4
5
6
7
8
9
10
11
12
13 ♀
14 ♀ ♂
15 ♀ ♂
16 ♂ ♀
17 ♂
18 ♀ ♂
19 ♂
19 ♀
20
21
22 ♀ ♂
23
24 ♂ ♀
25
26
27
28
29
30
31
32
33
34
35

CHATS AND OLD WORLD FLYCATCHERS *CONTINUED*

1 PALE FLYCATCHER *Melaenornis pallidus* 16cm Separated from Chat Flycatcher by distinct lore and less bright edges to wing feathers. Perches on low branches. **V** Mostly silent. Calls include a soft *churr*. **H** Woodland, bushveld. **D** Widespread across Africa.

2 CHAT FLYCATCHER *Melaenornis infuscatus* 20cm Underparts not white as in Pale and Marico flycatchers, and lacks dark loral stripe. Perches upright on top bushes (not on low perches like Pale and Marico flycatchers). Often seen with partly spread wings. **V** Unstructured, twittering *treet-treet-twitwitwit*, punctuated with some fluting notes. **H** Dry scrubland and bushland. **D** Angola, Namibia, Botswana, South Africa.

3 AFRICAN GREY FLYCATCHER *Melaenornis microrhynchus* 13cm Note white forehead between bill and striped crown. **V** Song rarely heard. Call low, thin and tweeting. **H** Dry wooded and bushy areas. **D** E Africa.

4 MARICO FLYCATCHER *Melaenornis mariquensis* 18cm Conspicuous white below. Note narrow eye-ring and brown loral area, which is uniform with forehead and ear-coverts. **V** Short, hoarse, rapid *retjehtjeh* and *freeh*, sometimes multiplied to an unstructured song with well-spaced parts. **H** Thornveld. **D** S Africa.

5 FISCAL FLYCATCHER *Melaenornis silens* 19cm Separated from Southern Fiscal (Plate 179) by absence of white edge to mantle, smaller bill and different tail pattern. **V** Extremely high whistling, together with very high, thin flutes, *sreesreetuweeweh*. **H** Open country with some trees, suburbs. **D** Botswana, South Africa.

6 SILVERBIRD *Empidornis semipartitus* 18cm Upperparts very striking silver-grey, underparts orange. Catches insects by gleaning or pursuit on the ground; also makes aerial sallies. **V** Song a repeated three-note phrase of rich, tuneful thrush-like notes. **H** Dry bushy grassland. **D** W Sudan to W Kenya and N Tanzania.

7 SPOTTED FLYCATCHER *Muscicapa striata* 14cm Makes fly-catching sallies in a sweeping circle, returning to its favoured perch or one nearby. **V** Song consists of a series of squeaky notes, often with lengthy intervals between phrases. Calls include a squeaky *zeee* or *chick*, and a *zee-zucc* when agitated. **H** Juniper and open forests, especially pine. **D** Breeds from Europe to C Siberia and Mongolia. Winters in E and S Africa.

8 MEDITERRANEAN FLYCATCHER *Muscicapa tyrrhenica* 14cm Very like Spotted Flycatcher in appearance and habits, but warmer brown on upperparts and more lightly marked on underparts. **V** Like that of Spotted Flycatcher. **H** Like that of Spotted Flycatcher. **D** Breeds on the Balearic Islands, Corsica, Sardinia. Winters in W and SW Africa.

9 GAMBAGA FLYCATCHER *Muscicapa gambagae* 12cm Habits similar to those of Spotted Flycatcher. **V** Song a quiet, high-pitched squeaking, combined with trills and creaking notes. Calls include a *chick* and *zick-zick-zick*. **H** Wooded highlands and bushy, mainly acacia, scrub. **D** Mali and Ivory Coast to Somalia and Kenya, also SW Arabian Peninsula.

10 GREY-STREAKED FLYCATCHER *Muscicapa griseisticta* 14cm Usually solitary. Makes rapid dashes after flying insects, frequently returning to the same treetop perch. **V** Calls include a loud *chipee*, *tee-tee* or *zeet zeet zeet*. **H** Open woodland, forest edges and clearings, grassy areas with trees. **D** Breeds in SE Russia and NE China. Winters from the Philippines to New Guinea.

11 DARK-SIDED FLYCATCHER *Muscicapa sibirica* 13cm Makes darting aerial sallies after flying insects from a prominent perch. **V** Song a series of repetitive thin notes with trills and whistles. Call a tinkling *chi-up-chi-up-chi-up*. **H** Open broadleaved, coniferous, mixed deciduous and rhododendron forests and secondary growth. **D** C and E Asia to SE Asia.

12 ASIAN BROWN FLYCATCHER *Muscicapa dauurica* 13cm Actions similar to those of Dark-sided Flycatcher. **V** Song a faint, squeaky, melodious whistle. Calls include a soft rattling, a short *tzi* and a soft *churr*. **H** Open forests, secondary growth, mangroves, parks and gardens. **D** C and E Asia to India and SE Asia.

13 SULAWESI STREAKED FLYCATCHER *Muscicapa sodhii* 12–14cm Makes aerial sallies after flying insects from a perch in the understorey or higher. Occasionally seen in mixed-species foraging parties. **V** Thin, high whistles, chirps, twitters, trills and buzzy notes. **H** Primary lowland and submontane broadleaved evergreen forests. **D** Sulawesi, Indonesia.

14 BROWN-STREAKED FLYCATCHER *Muscicapa williamsoni* 14cm Actions as Dark-sided Flycatcher. **V** Song similar to that of Asian Brown Flycatcher. Calls include a thin *tzi* and a harsh, slurred *cheititit*. **H** Open broadleaved evergreen and semi-evergreen forests, parks and gardens. **D** S Myanmar, S Thailand, N Peninsular Malaysia, NE Borneo.

15 ASHY-BREASTED FLYCATCHER *Muscicapa randi* 12–13cm Forages low down in the understorey or lower levels at forest edges or clearings. Usually alone or in pairs. **V** High-pitched, repeated *wee-tit* and a longer *zeeeeee-tip-tip-zee zizizizi*. **H** Montane and selectively logged forest. **D** Philippines.

16 SUMBA BROWN FLYCATCHER *Muscicapa segregata* 13cm Forages in the understorey and lower levels of trees; often sits unobtrusively for long periods. **V** Jumbled series of rapid, high-pitched, whistled, trilled and harsh notes, often ending with a trill. **H** Lowland primary forest, secondary forest, forest edges. **D** Sumba, W Lesser Sunda Islands.

17 BROWN-BREASTED FLYCATCHER *Muscicapa muttui* 14cm Secretive. Frequents low vegetation, from where it makes short sallies to catch flying insects. **V** Song is feeble but pleasant. Calls include a soft, low note that is uttered while rapidly vibrating half-open wings. **H** Dense thickets in broadleaved evergreen forest and riverside forests. **D** Breeds from NE India to NW Thailand. Winters in SW India, Sri Lanka.

18 FERRUGINOUS FLYCATCHER *Muscicapa ferruginea* 12.5cm Retiring; actions much like those of Spotted Flycatcher. **V** Soft trilling *si-si-si*. Song is probably a high-pitched *tsit-tittu-tittu*. **H** Broadleaved or fir forest and dense mixed jungle. **D** Breeds from the Himalayas to C China, Taiwan. Winters in SE Asia, Philippines.

19 ASHY FLYCATCHER *Muscicapa caerulescens* 14cm Note strong bill with paler lower mandible, uniform upperparts (including tail) and white eye-ring. Hawks from mid-stratum perch. **V** High *peepeepeepu-peeweh*, sometimes changing to a twittering song. **H** Woodland, riverine belts, coastal bush. **D** Widespread W, C, E and S Africa.

20 SWAMP FLYCATCHER *Muscicapa aquatica* 13cm Note white chin and dark brown upperparts. **V** Very high, short, sweet, thin twitters. **H** Lake shores, swamps. **D** W and C Africa.

21 CASSIN'S FLYCATCHER *Muscicapa cassini* 13cm Rather long-tailed. Lacks eyebrow and has only a faint breast-band. The only grey flycatcher that feeds along forest streams. **V** Song includes chirps, whistles, trills and buzzes. Call a buzzing, high-pitched note. **H** Hawks from a perch over forest streams. **D** Guinea and Sierra Leone, east to Uganda and south to Angola and Zambia.

22 OLIVACEOUS FLYCATCHER *Muscicapa olivascens* 13cm Rather featureless, but normally the only flycatcher living in the forest canopy. **V** Courtship song a nasal twitter. Calls include *wit-wit* and *seeee*. **H** Forest interiors. **D** W and C Africa.

23 CHAPIN'S FLYCATCHER *Muscicapa lendu* 10cm Note yellowish gape, stumpy bill, brown eye and plain upperparts. **V** Calls include a thin *tsseet, tsseet*. **H** Montane forests. **D** NE DR Congo, SW Uganda, W Kenya.

24 ITOMBWE FLYCATCHER *Muscicapa itombwensis* 12cm Very like Chapin's Flycatcher in appearance and habits, but slightly darker and with stronger bill. **V** Like that of Chapin's Flycatcher. **H** Dense montane forests. **D** E DR Congo.

25 AFRICAN DUSKY FLYCATCHER *Muscicapa adusta* 12cm There is usually no similar flycatcher in its habitat. Note smudgy underparts and thickset jizz. Often an orange tinge to broken eye-ring and lore. **V** Song is an extremely high, thin, unstructured twitter. Call is an extremely high *sieeeh*. **H** Forest edges, riverine belts, dense woodland, suburbs, in highlands and coastal regions. **D** WC, E and S Africa.

26 LITTLE GREY FLYCATCHER *Muscicapa epulata* 10cm Very small, with a blotched breast. **V** Song a series of high-pitched notes and rattling calls. Calls include a high-pitched *tsee-tsee*. **H** Forest, deserted plantations, cultivation with trees. **D** Sierra Leone to Togo, C Nigeria, S Cameroon and SW Central African Republic to NW Angola, NE DR Congo.

27 YELLOW-FOOTED FLYCATCHER *Muscicapa sethsmithi* 9cm Note yellow feet, white chin and pale grey breast. **V** Thin, high-pitched, buzzing notes. **H** Forests. **D** SE Nigeria to SW Republic of the Congo, NE and E DR Congo, Uganda.

28 DUSKY-BLUE FLYCATCHER *Muscicapa comitata* 12cm Note dark colouring, thin eyebrow, conspicuous white throat and pale brown undertail coverts. **V** Song unobtrusive. Calls include a soft *tsseet, tsseet*. **H** Forest edges. **D** W and C Africa.

29 TESSMANN'S FLYCATCHER *Muscicapa tessmanni* 12cm Very similar to Dusky-blue Flycatcher, with which it shares its range, but whiter below, especially on undertail coverts, and lacks white eyebrow. **V** Song like that of *Turdus* thrush; medley of sweet and buzzing notes. **H** Forest, working and overgrown plantations. **D** Sierra Leone to Ghana, Cameroon, NE DR Congo.

30 SOOTY FLYCATCHER *Muscicapa infuscata* 12cm Very dark, with mottled underparts. Hunts from a high, exposed position. **V** Mostly silent; song rarely heard. Calls include a harsh *tsiew*. **H** Forest edges and nearby areas. **D** S Nigeria to C and E DR Congo and Angola, Uganda.

31 USSHER'S FLYCATCHER *Muscicapa ussheri* 12cm Very similar to Sooty Flycatcher, but blackish rather than dark brown. **V** Mostly silent. Calls include a *tssrip*. **H** Normally seen high in the canopy, perching on a bare branch, often several together. **D** Guinea Bissau and Guinea to Nigeria.

32 BÖHM'S FLYCATCHER *Muscicapa boehmi* 12cm Note very upright stance and triangular spots below. Unobtrusive percher. **V** Usually silent. Song includes repeated *tch-tee* notes. **H** Miombo. **D** Angola to Tanzania, Malawi and S Zambia.

33 WHITE-GORGETED FLYCATCHER *Anthipes monileger* 11.5–13cm Occurs low down in dense undergrowth. Captures insects by making short aerial sallies or from the ground. **V** Song consists of a weak high-pitched whistle. Call a scolding, rattling short whistle. **H** Thick forest undergrowth and bamboo. **D** Himalayas to Thailand.

34 RUFOUS-BROWED FLYCATCHER *Anthipes solitaris* 12–13cm Spends long periods in the lower storey, resting on a shady perch. Makes short aerial fly-catching forays; also collects insects from the ground. **V** Song thin and tremulous, transcribed as *three-blind-mice*. Calls include a thin *tseep*, a sharp *tchik* and a harsh churring. **H** Broadleaved evergreen forest and bamboo. **D** Myanmar to Sumatra.

1
2
3
4
♀
♂
5
6
7
8
9
10
11
12
13
14
15
16
17
18
19
20
21
22
23
24
25
26
27
28
29
30
31
32
33
34

CHATS AND OLD WORLD FLYCATCHERS *CONTINUED*

1 HAINAN BLUE FLYCATCHER *Cyornis hainanus* 13–14cm Forages in the middle level of forest trees. **V** Song often transcribed as *hello mummy*. Calls include a series of soft *tic* notes. **H** Broadleaved evergreen, semi-evergreen and mixed deciduous forests, bamboo forest, mangroves. **D** Myanmar to S China and Thailand.

2 PALE BLUE FLYCATCHER *Cyornis unicolor* 18cm Makes aerial sorties after flying insects; also forages in the middle and upper levels of trees. **V** Song rich and melodious, and very thrush-like. Alarm call a soft *tr-r-r-r*. **H** Moist primary and secondary broadleaved forest, bamboo. **D** Himalayas through Java (Indonesia) and Borneo.

3 RÜCK'S BLUE FLYCATCHER *Cyornis ruckii* 17cm Little information; known only from two specimens collected in 1917 and 1918. **V** Unrecorded. **H** Logged lowland forest. **D** Sumatra, Indonesia.

4 BLUE-BREASTED BLUE FLYCATCHER *Cyornis herioti* 15cm Encountered singly, in pairs or in mixed-species foraging parties, from the understorey to mid-levels of forest trees. **V** Whistled *seeeep wheeu* or *seeeep seeeep wheeu*, interspersed with chattering and buzzy notes. **H** Lowland and submontane primary and selectively logged forest. **D** Philippines.

5 WHITE-BELLIED BLUE FLYCATCHER *Cyornis pallidipes* 15cm Forages in thick undergrowth, with occasional sallies after flying insects. At rest, often lifts and spreads tail. **V** Song a rambling, faltering series of unmelodious squeaky and slurred notes. Call a low *tsk-tsk*. **H** Undergrowth in dense broadleaved evergreen forest, patches of dense or tangled hillside vegetation. **D** SW India.

6 PALE-CHINNED BLUE FLYCATCHER *Cyornis poliogenys* 14cm Forages from undergrowth to canopy. **V** Song is a loud, rising and falling series of high-pitched notes, often interspersed with chuckling and harsh notes. Calls include a grating rattle and repeated *tik*. **H** Open forests. In winter also occurs in open country with scrub. **D** Himalayas to S Myanmar, EC India.

7 HILL BLUE FLYCATCHER *Cyornis banyumas* 14cm Unobtrusive; hawks insects from a low perch. **V** Song similar to that of Tickell's Blue Flycatcher, but descends overall. Calls include a hard *tac* and scolding *trrt-trrt-trrt*. **H** Dense humid forest with abundant undergrowth. **D** SW China and N Myanmar to SE Asia and Borneo.

8 LARGE BLUE FLYCATCHER *Cyornis magnirostris* 15cm Actions probably similar to those of Hill Blue Flycatcher, which some authorities consider to be conspecific. **V** Unrecorded, possibly similar to that of Hill Blue Flycatcher. **H** Broadleaved evergreen forest. **D** Breeds in E Himalayas. Winters in Myanmar, Thailand, Peninsular Malaysia.

9 PALAWAN BLUE FLYCATCHER *Cyornis lemprieri* 16cm Usually encountered singly or in pairs. Forages in the forest understorey. **V** Soft, pleasant, descending and then rising *de do da da do de*, and a *da de do poy*. **H** Lowland and submontane dry primary and secondary forest. **D** Philippines.

10 TICKELL'S BLUE FLYCATCHER *Cyornis tickelliae* 14cm Forages from undergrowth up to forest middle levels, regularly making sallies to capture flying insects. **V** Song comprises a short metallic trill of 6–10 notes, first descending and then ascending. Calls include a harsh *tac* or *kak*, a *tik-tik* and a sharp, churring *trrt-trrt*. **H** Open dry forest and wooded areas. **D** Widespread across S Nepal, India, Sri Lanka, Bangladesh, Myanmar, Malay Peninsula.

11 SUNDA BLUE FLYCATCHER *Cyornis caerulatus* 14cm Forages by fly-catching from a low, exposed perch; generally seen in pairs. **V** Thin, metallic, rising and falling *si-si-tiuuuw*. **H** Lowland mixed dipterocarp forest. **D** Sumatra, Indonesia; Borneo.

12 BORNEAN BLUE FLYCATCHER *Cyornis superbus* 15cm Usually seen alone or in pairs. Fly-catches from a low perch, and hovers to pick insects from branches. **V** High-pitched *hiu-te-hie*. **H** Submontane primary and secondary forest; occasionally lower forests. **D** Borneo.

13 BLUE-THROATED BLUE FLYCATCHER *Cyornis rubeculoides* 14cm Forages low down in undergrowth; regularly makes aerial sallies to capture flying insects. **V** Song similar to that of Tickell's Blue Flycatcher but more trilling, higher pitched and more rapid. Calls include a soft *tac* or *check* and, when agitated, a harsh *trrt* or *trrt-trrt*. **H** Dense undergrowth in dry broadleaved evergreen and mixed deciduous forests. **D** Himalayas to Thailand.

14 CHINESE BLUE FLYCATCHER *Cyornis glaucicomans* 14–15cm Forages low down in undergrowth; also makes sallies after flying insects. Previously considered conspecific with Blue-throated Blue Flycatcher. **V** Song rich, with varied warbling notes. Calls include a soft *tac* and a harsh *trrt* or *trrt-trrt*. **H** As Blue-throated Blue Flycatcher. **D** Breeds in S and SE China. Winters in SE Asia.

15 MALAYSIAN BLUE FLYCATCHER *Cyornis turcosus* 13–14cm Makes aerial pursuits of insects from a low perch. **V** Song a soft, whistled *diddle diddle dee diddle dee*. Calls include a harsh *chrrk*, and a hard *tik-tk-tk* when alarmed. **H** Primary lowland and secondary forest, often near rivers or streams in Sumatra. **D** Malay Peninsula; Sumatra, Indonesia; Borneo.

16 MANGROVE BLUE FLYCATCHER *Cyornis rufigastra* 14–15cm Forages low down, and frequently sallies out from a secluded perch to capture passing insects. **V** Song a slow, low, short, metallic trill, first descending and then ascending. Calls include a repeated dry *psst* and a sharp, staccato *chi-chik-chik-chik*. **H** Mangroves, coastal forest, scrub, overgrown edges of plantations. **D** Malay Peninsula to the Philippines and Borneo.

17 TANAHJAMPEA BLUE FLYCATCHER *Cyornis djampeanus* 14–15cm Formerly considered a race of the Mangrove Blue Flycatcher; actions presumed to be similar. **V** Unknown. **H** Closed-canopy forest. **D** Islands in the Flores Sea.

18 SULAWESI BLUE FLYCATCHER *Cyornis omissus* 14–15cm Often seen in roadside vegetation at dawn, otherwise favours understorey and mid-storey of forest trees. Usually encountered singly, in pairs or occasionally as part of mixed-species feeding parties. Formerly considered a race of the Mangrove Blue Flycatcher. **V** Short, weak 4–5-note warble, or a clear, whistled 4–6-note warble followed by 3–4 loud, clear whistles. **H** Hill, lower montane and tall secondary forest, montane secondary scrub. **D** Sulawesi, Indonesia.

19 TIMOR BLUE FLYCATCHER *Cyornis hyacinthinus* 16cm Frequents lower to middle levels of trees. Sallies after flying insects, frequently returning to the same high perch. **V** Song a monotonous, rapid series of babbled rising and falling notes. Calls include a short, subdued bubbling. **H** Remnant patches of primary and secondary monsoon forest and woodland, hillside shrubbery, borders of degraded forests and plantations. **D** E Lesser Sunda Islands.

20 BLUE-FRONTED BLUE FLYCATCHER *Cyornis hoevelli* 15cm Forages from the undergrowth to the middle levels of forest trees, gleaning insects from foliage. Usually seen singly, in pairs or, occasionally, in mixed-species feeding parties. Inconspicuous; best located by its voice. **V** Song a pleasant, rich sequence of 10–20 notes; also a medley of 4–5 discordant notes and a loud *tsat-tsat-tsat*. **H** Montane rainforest, moss forest. **D** Sulawesi, Indonesia.

21 MATINAN BLUE FLYCATCHER *Cyornis sanfordi* 14–15cm Inconspicuous. Frequents the undergrowth and lower to middle levels of forests, making fly-catching sallies after insects; usually alone or in mixed-species foraging flocks. **V** Song a series of rapid, thin, clear or subdued notes, varying in pitch. **H** Primary hill and montane broadleaved evergreen forest. **D** Sulawesi, Indonesia.

22 WHITE-TAILED FLYCATCHER *Cyornis concretus* 18cm Forages low down in undergrowth or lower branches of trees; often spreads tail to reveal white panels. **V** Song a variable series of penetrating and sibilant whistles. Calls include a soft *pweee* and harsh *scree*. **H** Dense forest, often near streams. **D** NE India to Borneo.

23 RUSSET-BACKED JUNGLE FLYCATCHER *Cyornis oscillans* 13–15cm Forages in thick undergrowth and to middle levels of forest trees; pounces on insects or pursues them in flight. Usually seen singly or in mixed-species feeding parties. May raise and spread tail while perched. **V** Song a loud series of jumbled, high-pitched, slightly staccato notes. Calls include a harsh *tak* or *chek*. **H** Primary hill and montane semi-evergreen forest. **D** Lesser Sunda Islands.

24 BROWN-CHESTED JUNGLE FLYCATCHER *Cyornis brunneatus* 15cm Frequents lower canopy and bushes; catches insects in flight but also forages among leaf litter on the forest floor. **V** Song a series of loud, descending whistles. Calls include a harsh *churr*. **H** Broadleaved evergreen forest, bamboo thickets. Post breeding, frequents semi-evergreen and mixed deciduous forest, mangroves and scrub. **D** Breeds in SE China. Winters in Malay Peninsula.

25 NICOBAR JUNGLE FLYCATCHER *Cyornis nicobaricus* 15cm Fairly large flycatcher with plain, dull brown upperparts and grey-white underparts with extensive brownish wash on breast. Bill rather long with a hooked tip, eyes large. **V** Song a series of loud, descending whistles. Calls include a harsh *churr*. **H** Primary and secondary forest and forest edges, bushes, thickets, sometimes gardens. **D** S Nicobar Islands, Indian Ocean.

26 FULVOUS-CHESTED JUNGLE FLYCATCHER *Cyornis olivaceus* 15cm Usually found in the lower and middle storeys. Makes sallies after flying insects. Occasionally cocks and fans tail. **V** Song a rapid series of short phrases, alternating in pitch, consisting of musical, scratchy, tinkling and churring notes. Calls include a low *tchuck-tchuck*, a *trrt* and a harsh *tac*. **H** Lowland broadleaved primary and secondary forest, plantations, forest edges. **D** Malay Peninsula; Sumatra, Java and Bali, Indonesia; Borneo.

27 GREY-CHESTED JUNGLE FLYCATCHER *Cyornis umbratilis* 15cm Forages from the undergrowth up to the middle canopy; occasionally makes aerial sallies after insects. Cocks and spreads tail. **V** Song a thin, sweet, descending *si ti-tu-ti tlooeeu*. Calls include a scolding *churr-churr-churr*, a *trrrt it it it* and a clicking *tchk-tchk*. **H** Lowland primary and secondary forest, peat-swamp forest, plantations. **D** Malay Peninsula; Sumatra, Indonesia; Borneo.

28 RUFOUS-TAILED JUNGLE FLYCATCHER *Cyornis ruficauda* 15cm Forages in the understorey and along forest edges; occasionally hawks insects from a low perch. **V** Series of up to three repeated, high-pitched *chirr* notes or a more musical *cheep cheep chirr*, sometimes followed by a buzzing note or a trill. **H** Lowland and lower montane forests and forest clearings. **D** Borneo to the Philippines.

29 SULA JUNGLE FLYCATCHER *Cyornis colonus* 14cm Occurs singly, in pairs or in small groups. Unobtrusive. Little other information. **V** Jumble of shrill, sweet, clear, high-pitched notes, including single rising notes, hurried descending whistles and trills, and alternate high and low notes. **H** Lowland primary, secondary and degraded forest, selectively logged forest. **D** Sula Islands, Indonesia.

30 BANGGAI JUNGLE FLYCATCHER *Cyornis pelingensis* 14cm Plain olive-brown flycatcher with large eyes. Quiet and unobtrusive. **V** Song a soft jumble of shrill and sweet, clear, high-pitched whistles and trills, and alternate high and low notes. **H** Undisturbed lowland primary, secondary and degraded forest. **D** Peleng Island, Banggai Islands, Indonesia.

31 FUJIAN NILTAVA *Niltava davidi* 18cm Forages mainly in dense undergrowth, sitting quietly before pouncing on prey. **V** Song a repeated, high-pitched *sssew* or *ssiiiii*. Calls include a sharp, metallic *tit-tit-tit* and a harsh *trrt-trrt-trrt-tit-tit-trrt-trrt*. **H** Broadleaved evergreen woodland. On migration, also in scrub, parks, gardens. **D** C and E China to SE Thailand.

32 RUFOUS-BELLIED NILTAVA *Niltava sundara* 18cm Unobtrusive. Sits on a low perch, from where it makes darting sallies after flying insects; also drops to the ground to capture prey. Constantly flicks and spreads tail. **V** Rasping *zi-i-i-f-cha-chuk*, hard *tic*, thin *see* and soft *chacha*. **H** Undergrowth and bushes in broadleaved evergreen or mixed forests. **D** Himalayas to Thailand.

33 RUFOUS-VENTED NILTAVA *Niltava sumatrana* 15cm Forages on the ground, in dense undergrowth and in the lower to middle storeys of forest trees. **V** Song a monotonous series of clear, undulating whistles; also a series of rapid, scratchy, slurred notes. Calls include a hard *chik*. **H** Broadleaved evergreen forest. **D** Peninsular Malaysia; Sumatra, Indonesia.

34 VIVID NILTAVA *Niltava vivida* 17.5cm Forages on large branches in the middle or upper storey of trees; also makes aerial sallies after flying insects. **V** Song consists of slow, mellow whistles, interspersed with some scratchy notes, transcribed as *beu-wii-riu-chrt-trrt-heu-wii-tiu-wii-u...* Call a whistled *yiyou-yiyou*. **H** Broadleaved evergreen or mixed forests. **D** Himalayas to China, Taiwan.

35 LARGE NILTAVA *Niltava grandis* 21cm Skulks and flits about in low bushes; occasionally feeds on the ground. Less agile than most flycatchers. **V** Song a whistled, ascending *whee-whee-wip tee-ti-tree* or *uu-uu-du-di*. Calls include a nasal *dju-ee*, a loud *trr-k trr-k* and a harsh rattle. **H** Broadleaved evergreen forest. **D** Himalayas to Sumatra, Indonesia.

36 SMALL NILTAVA *Niltava macgrigoriae* 13cm Forages in shady undergrowth and bushes, regularly making sallies after flying insects. **V** Song a high-pitched, rising and falling *twee-twee-ee-twee*. Calls include a high-pitched *see-zee* and various metallic scolding and churring notes. **H** Forest edges and clearings, bushes along tracks. In winter, also occurs in dense reeds and tall grass with scattered trees. **D** Himalayas to S China and Thailand.

1
2
3
4
5
6
7
8
9
10
11
12
13
14
15
16
17
18
19
20
21
22
23
24
25
26
27
28
29
30
31
32
33
34
35
36

CHATS AND OLD WORLD FLYCATCHERS *CONTINUED*

1 BLUE-AND-WHITE FLYCATCHER *Cyanoptila cyanomelana* 17cm Forages in the canopy; also visits lower levels to glean insects from foliage or branches, and makes fly-catching sorties from a prominent perch. **V** Calls include a harsh *tchk-tchk* and a soft *tic* or *tac*. **H** Primary and secondary lowland and submontane forest, coastal woodland and scrub. **D** Breeds in E Asia. Winters in SE Asia through Java and the Philippines.

2 ZAPPEY'S FLYCATCHER *Cyanoptila cumatilis* 16cm Little known. Formerly considered conspecific with Blue-and-white Flycatcher, which it closely resembles. **V** Similar to that of Blue-and-white Flycatcher, but song lower and more evenly pitched. **H** Coniferous and deciduous submontane forest. **D** C China.

3 DULL-BLUE FLYCATCHER *Eumyias sordidus* 15cm Forages low down in trees or undergrowth, also on the ground. **V** Song a series of mournful, downslurred warbling notes. Call a series of four or five *chip* notes. **H** Edges of forest and plantations, well-wooded areas, large gardens. **D** Sri Lanka.

4 VERDITER FLYCATCHER *Eumyias thalassinus* 16cm Makes aerial sallies after flying insects from an exposed perch on trees, wires or buildings. **V** Song a pleasant trilled *pe-tititi-wu-pititi-weu*. Call a *tze-ju-jui*. **H** Open forests, forest clearings and edges, groves, gardens. **D** Himalayas to SE Asia and Sumatra, Indonesia..

5 TURQUOISE FLYCATCHER *Eumyias panayensis* 14cm Usually seen singly or in pairs; regularly joins mixed-species feeding parties. Captures prey by sallying or gleaning. **V** Song a monotonous warble, consisting of about 20 notes. **H** Submontane forest, forest clearings and edges. **D** Philippines; Sulawesi and Maluku Islands, Indonesia.

6 NILGIRI FLYCATCHER *Eumyias albicaudatus* 15cm Forages in bushes and trees, with occasional aerial sallies after flying insects. **V** Song a rambling, mournful warble of about eight notes. Call a series of *chip* notes. **H** Evergreen hill forest, plantations, forest edges and clearings, thick streamside vegetation. **D** SW India.

7 INDIGO FLYCATCHER *Eumyias indigo* 14cm Forages low down, often in dark undergrowth. Captures small invertebrates by gleaning or making short aerial sallies; also feeds on small berries. Occurs singly, in pairs or in mixed-species parties. **V** Song a squeaky, ringing *fee-foo-fu-fee-fee-fee* or *chit chwit choo wee tooo*. Calls include a harsh, rattling *turrr-tur*, and a *tzit-tzit-tzit* when alarmed. **H** Submontane and montane forest. **D** Sumatra, Java and Borneo.

8 STREAK-BREASTED JUNGLE FLYCATCHER *Eumyias additus* 15cm Active flycatcher; little other information known. **V** Unrecorded. **H** Lowland, montane and selectively logged forest, forest edges. **D** Maluku Islands, Indonesia.

9 EUROPEAN ROBIN *Erithacus rubecula* 14cm Wary, but can become quite confiding. **V** Melodic warbling and trilling phrases, usually beginning with a few high-pitched, drawn-out notes. Call a hard *tick* or, *tick-tick-tick-tick...*, also a plaintive *seeh*. **H** Woodland, hedgerows, parks and gardens. **D** Breeds in W and C Eurasia. Winters to N Africa.

10 RED-THROATED ALETHE *Chamaetylas poliophrys* 15cm Restricted range. Separated from similar thrush-like birds by small area of orange on throat. **V** Single downslurred note. **H** Undergrowth of montane forest and bamboo. **D** EC Africa.

11 BROWN-CHESTED ALETHE *Chamaetylas poliocephala* 15cm Note narrow white eyebrow and reddish-brown upperparts. **V** High, fluted, descending, slow, whistled *fju-fju-fju-fju-fju-fju-fjufju*. **H** Forest undergrowth and nearby areas. **D** W and C Africa.

12 WHITE-CHESTED ALETHE *Chamaetylas fuelleborni* 17cm Strikingly white below. **V** High, monotonous, sustained, fluted *uWeeet uWeeet*. **H** Montane forest. **D** Tanzania to Malawi, C Mozambique.

13 THYOLO ALETHE *Chamaetylas choloensis* 16cm Note diagnostic white chin and tail corners. **V** Songs consists of 3–4 notes, *tyerr... wor-tyer-chii*. Calls include a downslurred *piiiyuu*. **H** Dense undergrowth near forest edges. **D** S Malawi, NC and WC Mozambique.

14 WHITE-BELLIED ROBIN-CHAT *Cossyphicula roberti* 14cm Very small. Note white eyebrow and dark middle tail feathers. **V** Very high, plaintive song, lasting about 1 min. **H** Forest understorey. **D** C Africa.

15 MOUNTAIN ROBIN-CHAT *Cossypha isabellae* 16cm Separated from White-bellied Robin-Chat by larger size, orange belly and black cheek. **V** Extremely high, shrill *wheeteetshreeh* and similar variations. **H** On the ground and in understorey of montane forest. **D** E Nigeria, W and SW Cameroon.

16 ARCHER'S GROUND ROBIN *Cossypha archeri* 15cm Note absence of black on head, warm overall rufous-brown colouring and white eyebrow. **V** Very high *firra-tweet-tweet-tweet*. **H** Undergrowth of montane forests, bamboo and giant heath. **D** EC Africa.

17 OLIVE-FLANKED GROUND ROBIN *Cossypha anomala* 15cm Note white throat and dark rufous tail and rump. **V** Song a phrase of 4–7 high-pitched simple whistles. Calls include a *bairk* and *chop*. **H** Undergrowth of forest near glades and streams. **D** Malawi, Tanzania, NE Zambia.

18 CAPE ROBIN-CHAT *Cossypha caffra* 17cm Note interruption of orange on underparts between breast and undertail coverts. **V** Short, fluted whistles, *tuh-twitwi-rurutweereeree*, immediately followed by other strophes. **H** Suburbs, cultivation, open forests, scrubland. **D** E, SE and S Africa.

19 WHITE-THROATED ROBIN-CHAT *Cossypha humeralis* 17cm Unmistakable due to combination of pied plumage and orange-and-black tail. **V** Short, well-separated, fluted whistles. **H** Dry bush, thornveld, riverine belts. **D** Zimbabwe to N South Africa and S Mozambique.

20 ANGOLAN CAVE CHAT *Cossypha ansorgei* 19cm Unmistakable in its restricted range. **V** Song resembles that of a Woodlark (Plate 199). Calls include a soft *ui ti ti, ui ti ti*. **H** Rocky hillsides, often with cave mouths near forest patches. **D** Angola.

21 GREY-WINGED ROBIN-CHAT *Cossypha polioptera* 16cm Separated from White-browed Robin-Chat by shorter eyebrow, grey crown and all-red tail; and from Bocage's Akalat by white eyebrow. **V** Very high, sharp whistling of varied pitch, interspersed with clicks and short tempo changes, often resembling happy, random human whistling. **H** Forest and well-wooded streams at middle elevations. **D** W and C Africa.

22 BLUE-SHOULDERED ROBIN-CHAT *Cossypha cyanocampter* 20cm Note shining, pale orange underparts and blue shoulder. **V** Mid-high, liquid, rich, unhurried whistles, repeated 1–3 times (like a person happily whistling). **H** Wet forest undergrowth, especially along streams. **D** W and C Africa.

23 RÜPPELL'S ROBIN-CHAT *Cossypha semirufa* 20cm Separated from White-browed Robin-Chat by darker middle tail feathers, slightly greyer mantle and incomplete orange neck collar. **V** High, melodious *puweecpuweet puweetpuweetpuweet* and many other variations and imitations, mainly of other birds. **H** Forest patches and gardens with dense shrub. **D** E Africa.

24 WHITE-BROWED ROBIN-CHAT *Cossypha heuglini* 19cm Note diagnostic long white eyebrows. **V** High, melodious, crescendoing phrase, endlessly varied with small, skilful tempo changes and interwoven with imitations, mainly of other birds. **H** Well-planted gardens, bushland, open forest. **D** C, E and S Africa.

25 RED-CAPPED ROBIN-CHAT *Cossypha natalensis* 18cm Unmistakable due to all-orange-rufous head. **V** Song a stream of rather short phrases with frequently repeated, rich, fluted notes that incorporate adapted imitations. Call a *weeeh rurr*. **H** Suburbs, forest undergrowth, coastal bush, riverine belts. **D** C, E and S Africa.

26 CHORISTER ROBIN-CHAT *Cossypha dichroa* 19cm Note absence of white on head. **V** Mid-high, rich, mellow stream of fluted phrases, interwoven with perfect imitations of other bird songs. **H** Mid-strata of lowland forests. **D** South Africa.

27 WHITE-HEADED ROBIN-CHAT *Cossypha heinrichi* 23cm Unmistakable. Feeds on insects. **V** Unrecorded. **H** Riverine belts. **D** W DR Congo, N Angola.

28 SNOWY-CROWNED ROBIN-CHAT *Cossypha niveicapilla* 23cm Separated from White-crowned Robin-Chat by chestnut on chin reaching bill. **V** Very rich, liquid, sustained whistling with perfect imitations, mainly of other birds. **H** Forest edges, undergrowth of moist woodland and gardens. **D** S Mauritania, Senegal and Gambia to Ethiopia, Kenya, DR Congo, NE Angola and Tanzania.

29 WHITE-CROWNED ROBIN-CHAT *Cossypha albicapillus* 25cm Separated from Snowy-crowned Robin-Chat by larger size, and darker lower belly and undertail coverts. **V** Rich, rather sharp and slightly hurried fluted phrases, normally without imitations. **H** Swampy woodland thickets. **D** W, C and E Africa.

30 SWYNNERTON'S ROBIN *Swynnertonia swynnertoni* 14cm Note all-dark tail and conspicuous white throat patch. **V** Extremely high, plaintive, monotonous, descending *pfut-pfut-pfee*. **H** Undergrowth of montane forest. **D** E Zimbabwe, W Mozambique, C Tanzania.

31 WHITE-STARRED ROBIN *Pogonocichla stellata* 16cm Separated from Swynnerton's Robin by partly orange-yellow outer-tail feathers. White spot in edge of grey on throat normally concealed or barely visible. **V** Very high, clear, single or repeated short phrases, often sounding like carefree human whistling. **H** Coastal and montane forest edges and clearings. **D** E and SE Africa.

32 FOREST ROBIN *Stiphrornis erythrothorax* 12cm Olive wash on upperparts, intense orange breast and throat. Note black face and white spot before eye. **V** Extra-high, sharp, hurried whistle *feetjeetowee*, *feetjeeweeteewo* and *sri-sweep-sreepper-sreep-sreepsreep-sweep*. **H** Undergrowth of swampy forests. **D** W and C Africa.

33 BOCAGE'S AKALAT *Sheppardia bocagei* 14cm Note dark rufous rump and tail. **V** Very high, fluted, loud, sharp *fuu-tji-fuu-tu-weeh*. **H** Montane forests. **D** SC Africa.

34 LOWLAND AKALAT *Sheppardia cyornithopsis* 13cm Similar to Equatorial Akalat, but distinguished by altitude and song. Note short, stumpy bill, greyish cheeks, and orange flanks and rump. **V** Very high, varied, calm, warbling strophes (2–3 secs) with melodious rattles and rolls. **H** Undergrowth of more or less swampy forests. **D** W and C Africa.

35 EQUATORIAL AKALAT *Sheppardia aequatorialis* 13cm Similar to Lowland Akalat but separated by altitude and voice. **V** Mid-high, simple, rolling, fluty *prurr prurrr*. **H** Montane forest undergrowth. **D** E and C Africa.

36 SHARPE'S AKALAT *Sheppardia sharpei* 13cm Small white eyebrow and uniform upperparts diagnostic. **V** Song a repeated, tinkling *chii chiddly chiddly*. Calls include a metallic *pink*. **H** Montane forest undergrowth, bamboo. **D** Tanzania, Malawi.

♂
1
♀
♂
2
♀
3
♂
4
♀
5
6
♀
♂
7
8
9
10
11
12
13
14
15
16
17
18
19
20
21
22
23
24
25
26
27
28
29
30
31
32
33
34
35
36

CHATS AND OLD WORLD FLYCATCHERS *CONTINUED*

1 EAST COAST AKALAT *Sheppardia gunningi* 13cm Note small size, dark tail, and grey eyebrow and wings. Very robin-like in behaviour. **V** Very high, hesitating, slightly hoarse whistle, *rrur-Hee*. **H** Dense, moist forest undergrowth. **D** E and SE Africa.

2 GABELA AKALAT *Sheppardia gabela* 13cm Small colourless robin with pale brown breast-band and very restricted range. **V** Song presumed to comprise a soft, descending whistles. **H** Dense forest undergrowth. **D** Angola.

3 RUBEHO AKALAT *Sheppardia aurantiithorax* 14cm Recalls a dusky version of European Robin (Plate 249), but with less orange on face. Feeds mainly on the ground; may attend swarms of driver ants. **V** Calls include dry, nasal rattles. **H** Undergrowth in montane forest. **D** Tanzania.

4 USAMBARA AKALAT *Sheppardia montana* 13cm Note general olive-brown colouring and warm brown rump. May show rufous spot in front of eye. **V** Song a series of thin, evenly pitched, *ii siibii ii hii-hii ii siibii hii lichii siibii*. Alarm call a nasal *jahh jah jah jah*. **H** Forest undergrowth. **D** Tanzania.

5 IRINGA AKALAT *Sheppardia lowei* 13cm Note more or less uniformly coloured upperparts, more rufous throat and rather whitish belly. **V** Song a series of whistles or tinkling trills. Calls include a *ra-a-a-a-a-a-a-ah* alarm call. **H** Dry forests. **D** Tanzania.

6 COLLARED PALM THRUSH *Cichladusa arquata* 18cm Separated from Rufous-tailed Palm Thrush by dark line around throat patch and different range. **V** Unhurried sequence of fluted notes and phrases of any length, *weet weet weet-weet-wu-weet*. **H** Riverine palm thickets, palm savannahs. **D** C DR Congo, Uganda and Kenya to NE South Africa.

7 RUFOUS-TAILED PALM THRUSH *Cichladusa ruficauda* 17cm Separated from Collared Palm Thrush by absence of breast collar. Note upright stance. Forages mainly on the ground. **V** Unhurried sequence of fluted notes varying in pitch, *fiufiuflehweedjubdjubdjubweetweet-(-djubdjubdjub-)*, almost a short rattle. **H** Palm thickets, palm plantations, gardens. **D** S Central African Republic to N Namibia.

8 SPOTTED PALM THRUSH *Cichladusa guttata* 15cm Note black arrow markings on creamy-white underparts and whitish supercilium. **V** High, continuous, 4–5-toned, up-and-down song with short whistles. **H** Dense thickets along dry streams. **D** E Africa.

9 GREAT SHORTWING *Heinrichia calligyna* 17–18cm Very shy. Hops along the ground foraging among leaf litter, moss and lichen, and around small stones. **V** Song consists of three loud, clear, upslurred disyllabic whistles, ending with a shrill, high-pitched note. **H** Primary montane forest, with rocky gullies and overgrown stream sides. **D** Sulawesi, Indonesia.

10 BAGOBO BABBLER *Leonardina woodi* 19–20cm Very secretive. Forages on or close to forest floor. **V** Very high-pitched, tinkling *seeeeep seepseep seep*, repeated every 4–5 secs. **H** Primary broadleaved evergreen and montane evergreen forest, forest edges, ridgetop forest, transition (lowland/montane) forest. **D** Philippines.

11 GOULD'S SHORTWING *Heteroxenicus stellatus* 13cm Mouse-like forager among tangled roots and fallen branches. **V** High-pitched *tssiu – tssiu – tssiu – tsitsitsiutssiutssiutsitssitsiu...* Utters a *tik-tik* when alarmed. **H** Boulder-strewn areas in rhododendron, bamboo and conifer forests, also broadleaved evergreen forest with nearby streams. **D** Himalayas to NW Vietnam.

12 RUSTY-BELLIED SHORTWING *Brachypteryx hyperythra* 13cm Skulking. Forages on the ground in thick undergrowth. **V** A *tu-tiu* that leads into a fast warble of slurred notes. **H** Dense thickets and forest undergrowth. **D** E Himalayas, NE India.

13 LESSER SHORTWING *Brachypteryx leucophris* 13cm Often holds tail erect while foraging among dead leaves on forest floor. **V** Song a single note followed by a melodious, sibilant warble, with a jumbled finish of buzzy, rich, melodious notes. Calls include a thin whistle and a harsh *tack*. **H** Dense, damp undergrowth in broadleaved forest. **D** Himalayas to S China; SE Asia; Sumatra and Java, Indonesia; Lesser Sunda Islands.

14 WHITE-BROWED SHORTWING *Brachypteryx montana* 14cm Skulking; forages on the ground. Actions much those of a European Robin (Plate 249). **V** Variable; generally starts slowly with a few single notes, speeds to a plaintive babble and then ends abruptly, or similar. Calls include a hard *tack*, and a *tt-tt-tt* when alarmed. **H** Dense undergrowth and thickets in forests. **D** Himalayas to the Philippines; Sumatra and Java, Indonesia; Flores, Lesser Sunda Islands; Borneo.

15 EYEBROWED JUNGLE FLYCATCHER *Vauriella gularis* 15cm Occurs alone, in small groups or in mixed-species feeding parties. Forages low down or on the ground. Inquisitive. **V** A churring call, and a sharp *prrrt* when alarmed. **H** Montane forest. **D** Borneo.

16 WHITE-THROATED JUNGLE FLYCATCHER *Vauriella albigularis* 16–17cm Forages in the forest understorey. Unobtrusive. **V** Song a series of very high-pitched, almost inaudible phrases, interspersed with low churring notes. Calls include a high-pitched, rapid, rising and accelerating *ti-ti-ti-ti-tip*. **H** Lowland, lower montane and secondary forest. **D** Philippines.

17 WHITE-BROWED JUNGLE FLYCATCHER *Vauriella insignis* 16–20cm Very secretive, foraging in the understorey and in shady lower levels of trees. **V** Thin, high-pitched phrases. **H** Primary montane moss forest, secondary forest close to primary oak forest. **D** Philippines.

18 SLATY-BACKED JUNGLE FLYCATCHER *Vauriella goodfellowi* 17–18cm Usually alone or in pairs, sitting quietly in the forest understorey. **V** Song a series of up to 16 high-pitched, rapidly repeated *tsi* notes, rising and then levelling out, before accelerating and then descending at the end. **H** Montane forest. **D** Philippines.

19 INDIAN BLUE ROBIN *Larvivora brunnea* 14cm Forages on the ground or in low growth; runs rapidly and flicks wings and tail. **V** Song starts with a few introductory whistles, followed by a short, sweet jumble of hurried phrases. Calls include a high-pitched *tsee* and a hard *tuk-tuk* when alarmed. **H** Dense undergrowth in forests. **D** Breeds from Himalayas to C China. Winters in S India, Sri Lanka..

20 SIBERIAN BLUE ROBIN *Larvivora cyane* 14cm Shivers tail while foraging on the ground or in low cover; said to run and hop like a small crake. **V** Calls include a subdued *tak*, a louder *se-ic* and, when alarmed, a *chuck-chuck-chuck*. **H** Broadleaved evergreen and mixed deciduous forest, bamboo, secondary growth, mangroves, parks and gardens. **D** Breeds in E Asia. Winters in SE Asia and to Greater Sunda Islands.

21 RUFOUS-TAILED ROBIN *Larvivora sibilans* 13cm Skulks on or near the ground; shivers tail. **V** Song a repeated accelerating trill, falling in pitch towards the end. Calls include a *chirp* or *chirrup*. **H** Broadleaved and semi-evergreen forest. **D** Breeds from C Russia to SE Russia, NE China and N Japan. Winters in SE Asia.

22 RUFOUS-HEADED ROBIN *Larvivora ruficeps* 15cm Skulks, usually on the ground in thick cover; often cocks tail. **V** Rich, powerful, melodious phrases, preceded by a single note, transcribed as *ti-chulululu – ti-chewtchwetchew – tititichewtchewtchew...* Calls include a deep *tuc* or *toc* and a thin *si*. **H** Dense scrubby subalpine forest. **D** NC China.

23 RYUKYU ROBIN *Larvivora komadori* 14cm Forages on the ground or in low bushes. **V** Song similar to that of Japanese Robin but weaker and with more variation. Calls include a penetrating *tsiiii* and a *kirrick* when alarmed. **H** Dense undergrowth in broadleaved evergreen forest, often near small streams. **D** Ryukyu Islands, off S Japan.

24 JAPANESE ROBIN *Larvivora akahige* 14cm Skulks on or near the ground, tending to keep to dense undergrowth. **V** Song is a repeated *tsee – chararararararar* or similar. Calls include a thin *tsip*. **H** Broadleaved evergreen forest, parks and gardens. **D** Sakhalin and Kuril islands (Russian N Pacific), Japan.

25 BLUETHROAT *Luscinia svecica* 14cm Forages on the ground or in low cover. Chestnut tail bases often 'flash' as bird flits into cover. **V** Vigorous, with a bell-like *ting-ting-ting* and a throaty *torr-torr-torr-torr*; often mimics other birds or insects. Calls include a *tucc-tucc*, a croaky *turrc-turrc* and a plaintive *hweet*. **H** Breeds in waterside scrub. Winters in scrub and tall grass. **D** Breeds across Eurasia from N Europe to C China, also NW Alaska. Winters to N Africa, S and SE Asia, and E China.

26 WHITE-BELLIED REDSTART *Luscinia phaenicuroides* 19cm Retiring; forages mainly on the ground. Tail often held vertically and spread. **V** Whistled song transcribed as *he-did-so*. Calls include a *chack* and, when alarmed, a *tsiep-tsiep-tk-tk* or *tck-tck-sie*. **H** Rhododendron and coniferous forest, dense low scrub above the treeline. Winters at lower levels in dense low scrub or undergrowth. **D** Himalayas to C China and Myanmar.

27 THRUSH NIGHTINGALE *Luscinia luscinia* 17cm Note large black eye and pale eye-ring. Yellowish mouth corner gives face a characteristic expression. Tail is slightly russet brown. **V** Very rich stream of highly variable, well-separated single or repeated liquid notes and harsh, fluted or trilling rattles (lacking famous crescendo of extra-limital Common Nightingale). **H** Thickets, woodland with leafy undergrowth. **D** Breeds from N and C Europe to C Russia. Winters in SE Africa.

28 COMMON NIGHTINGALE *Luscinia megarhynchos* 16cm Extreme skulker. **V** Loud, melodic phrases mixed with trills and rattles, the most striking being a fluty *pew-pew-pew-pew-pew...* and rapid, low *jug-jug-jug-jug...* Calls include a deep *grrrrr*, a thin *seeeee*, a hard *tacc-tacc* and a harsh *tucc-tucc*. **H** Thick undergrowth in deciduous woodland, dense thickets. **D** Breeds from W and C Europe to NW China. Winters in Africa.

29 WHITE-THROATED ROBIN *Irania gutturalis* 17cm Skulks in ground cover. **V** Song comprises loud, clear, bell-like notes, interspersed with harsher notes; given from perch or in flight, as bird glides down hillside with wings and tail fully spread. Calls include a loud *tji-thyt* or a soft *teck*. **H** Stony, scrubby hillsides and valleys. On migration, also in dense lowland scrub. **D** Breeds from Turkey to Kyrgyzstan, Afghanistan and Iran. Winters in NE and E Africa.

30 HIMALAYAN RUBYTHROAT *Calliope pectoralis* 15cm Secretive, although often more exposed when singing. Regularly cocks tail while foraging on the ground. **V** Song a complex series of shrill, undulating, warbling trills and twitters. Calls include a *tchuk* and, when alarmed, a *siiii-siiii*. **H** Dwarf rhododendron, juniper and scrub above the treeline. Post breeding, also in marshy grassland and scrub. **D** Breeds from Tien Shan mountains to the Himalayas. Winters to S Asia.

31 CHINESE RUBYTHROAT *Calliope tschebaiewi* 15cm Male slaty grey on upperparts, white on belly, with a scarlet throat and broad black band below. Female dark grey-brown, rather nondescript. **V** Song a series of loud, warbling phrases. Calls include a *skut fweep* and, when alarmed, a *ke*. **H** Alpine and subalpine scrub with boulders and bare scree, often near water. Scrubby lowlands post-breeding. **D** Kashmir, Tibet, C China, N Myanmar.

32 SIBERIAN RUBYTHROAT *Calliope calliope* 15cm Shy, skulking in dense vegetation but more exposed when singing; cocks tail. **V** Song a loud, varied, sustained warbling, interspersed with harsh notes. Calls include a *chak-chak* and a falling, whistled *ee-uk*. **H** Scrub thickets, forest edges, grassy areas with bushes, occasionally gardens. **D** Breeds in Siberia, N Mongolia, NE China, N Korea. Winters in SE Asia, Philippines.

33 FIRETHROAT *Calliope pectardens* 14cm Flicks tail; skulks in thick cover. **V** Song is loud, long, sweet and varied, each note repeated several times and including some mimicry, transcribed as *wiu-wihui wi – wi chu-wi chu – whi-iiii – wi-chudu chudu – t sii-sii – wi chu-wi chu-wi chu – chu tsri sri...* Call is a hard *tok*. **H** Broadleaved evergreen forest, bamboo, dense scrub. **D** Breeds in C China. Winters in SE Asia.

34 BLACKTHROAT *Calliope obscura* 14cm Skulks in thickets; flicks tail. **V** Song comprises shrill, cheerful phrases alternating with purring trills. Contact call is a subdued *tup*. **H** Bamboo thickets, scrub, secondary growth. **D** Breeds in NC China. Winters in SE Asia.

35 WHITE-TAILED ROBIN *Myiomela leucura* 18cm Shy, foraging in dark thickets. Often spreads tail, revealing white bases. White neck spot is usually concealed. **V** Song consists of clear, liquid, separated phrases, transcribed as *tey-tlee-i-ta-wey-i*. Calls include a low *tuc* and a thin whistle. **H** Dense undergrowth in broadleaved evergreen montane forest. In winter, descends to lower-level forest. **D** Himalayas to Thailand and Taiwan.

36 SUNDA ROBIN *Myiomela diana* 15cm Forages in undergrowth; little other information. **V** Song a simple 2–5-note, melancholic, sweet warble. **H** Montane forests with undergrowth. **D** Sumatra and Java, Indonesia.

37 NILGIRI BLUE ROBIN *Sholicola major* 15cm Secretive; forages on the ground. **V** Song a short jumble of whistles, twangy buzzes and harsh notes. Calls include a harsh rattle and an indrawn whistle. **H** Undergrowth in sheltered woods. **D** SW India.

38 WHITE-BELLIED BLUE ROBIN *Sholicola albiventris* 15cm Forages on the ground. **V** Song a series of beautiful phrases comprising rich, slurred whistles and buzzy notes. Calls include a high-pitched whistle and a loud chatter. **H** Wet areas of undergrowth in forest patches and densely wooded ravines. **D** SW India.

1
2
3
4
5
6
7
8
9
10
11
12
13
14
15
16
17
18
19
20
21
22
23
24
25
26
27
28
29
30
31
32
33
34
35
36
37
38

CHATS AND OLD WORLD FLYCATCHERS *CONTINUED*

1 WHITE-BROWED BUSH ROBIN *Tarsiger indicus* 15cm Forages on or near the ground in thick cover. **V** Song a rapidly repeated, sharp, bubbling *shri-de-de-dew... shri-de-de-dew*. Calls include a sweet *heed* or *tuit-tuit*, a croaking *churr* and a clucking *tukukukukukukukuk*. **H** Mixed subalpine forest. **D** Himalayas to SC China, Taiwan.

2 RUFOUS-BREASTED BUSH ROBIN *Tarsiger hyperythrus* 13cm Forages on or near the ground, where it often adopts an upright stance. **V** Song a lisping warble, *zeew-zeew-zeew...* Calls include a low *duk-duk-duk-tseak*. **H** Forest-edge bushes, streamside vegetation. **D** Himalayas to SC China.

3 COLLARED BUSH ROBIN *Tarsiger johnstoniae* 12cm Forages in shade, on the ground and in the lower storey. **V** Song a short, jolly *wiwi s-wizuwu wiwi s-wu-wi wiwi s-wu-srr*. Calls include a piping *pi-pi-pi...* and a grating *sipsipsip grrgrrgrr sipsip grrgrr*. **H** Mountain forest, forest edges. **D** Taiwan.

4 RED-FLANKED BLUETAIL *Tarsiger cyanurus* 14cm Cover-loving, but not an intense skulker. **V** Song a *tetee-teeleee-tititi*, trailing away at end. Calls include a *tic-tic*, a soft *huit* and a guttural *kerrr*. **H** Moist mixed and coniferous forest with undergrowth. On migration, all types of woodland, orchards or gardens. **D** Breeds in N Eurasia, from Finland to Japan. Winters in SE Asia, Japan, Taiwan.

5 HIMALAYAN BLUETAIL *Tarsiger rufilatus* 14cm Forages on the ground and in low cover. Slightly brighter than Red-flanked Bluetail, from which it has been split. **V** Song a soft, weak *churrh-cheee* or similar. Calls include a deep, croaking *tack-tack*. **H** Forest understorey, bushes in forest clearings. **D** W Himalayas to C China.

6 GOLDEN BUSH ROBIN *Tarsiger chrysaeus* 15cm Forages on the ground or in low cover; often holds tail cocked. **V** Song a wispy *tse-tse-tse-tse-tse-chu-r-r* or similar. Calls include a croaky *trrr* and scolding *chirik-chirik*. **H** Scrub above the treeline, thickets on boulder-covered alpine slopes, forest undergrowth. **D** Himalayas to C China.

7 LITTLE FORKTAIL *Enicurus scouleri* 12–14cm Constantly wags tail up and down, and rapidly opens and closes it in a scissor-like movement. **V** Loud, thin *ts-youeee*. **H** Mountain streams, rivers and waterfalls. **D** Himalayas to E China and NW Vietnam, Taiwan.

8 SUNDA FORKTAIL *Enicurus velatus* 16cm Little information; presumed similar to other forktails. **V** Hard, shrill *chee* or *hie-tie-tie*. **H** Boulder-strewn rivers and fast-flowing streams in hill and mountain forests. **D** Sumatra and Java, Indonesia.

9 CHESTNUT-NAPED FORKTAIL *Enicurus ruficapillus* 18–20cm Wary; searches for food along stream edges and on rocks in water. Flicks tail on landing. **V** Utters a series of thin, shrill whistles and a high *dir-tee*. **H** Rivers, streams, waterfalls. **D** Malay Peninsula; Sumatra, Indonesia; Borneo.

10 BLACK-BACKED FORKTAIL *Enicurus immaculatus* 20–25cm Forages among rocks or along the water's edge, constantly wagging tail. **V** A hollow *huu*, often followed by a shrill *zeee*; also a squeaky *weeng*. **H** Fast-flowing waters with exposed rocks in damp dense forests. **D** Himalayas to NW Thailand.

11 SLATY-BACKED FORKTAIL *Enicurus schistaceus* 22–25cm Hops or flits from rock to rock in search of food; flight undulating, much like that of a wagtail. Sways tail slowly up and down. **V** A mellow *cheet* and a metallic *teenk*. **H** Fast-flowing rocky rivers. **D** Himalayas through SE Asia.

12 WHITE-CROWNED FORKTAIL *Enicurus leschenaulti* 25–28cm Forages on rocks or along the water's edge; when disturbed, often disappears into nearby forest cover. **V** Song an elaborate series of sweet, high-pitched whistles. Calls include a harsh *tssee* or *tssee-chit-chit-chit*. **H** Fast-flowing rocky streams and rivers in dense evergreen forest. **D** NE India through SE Asia to Java, Indonesia.

13 BORNEAN FORKTAIL *Enicurus borneensis* 28cm Actions and habits similar to those of White-crowned Forktail, with which it is often considered conspecific. **V** A shrill double whistle. **H** Rocky streams in primary montane forest. **D** Borneo.

14 SPOTTED FORKTAIL *Enicurus maculatus* 26–27cm Actions and habits much as those of Slaty-backed Forktail. **V** A sharp, creaky *cheek-chik-chik-chik-chik* and a shrill, rasping *kreee* or *tseek*. **H** Rocky streams in dense mountain forests. Winters at lower elevations, often on wider watercourses. **D** E Afghanistan to SC Vietnam.

15 SRI LANKA WHISTLING THRUSH *Myophonus blighi* 20cm Very shy and elusive; forages on the ground, usually at stream edges. **V** A rich mixture of tinkling, chortling and buzzing notes; also a down-slurred buzzy *sriii* or *sriii-sriii*, often preceded by a shrill warbling series. **H** Near fast-flowing streams in dense mountain forests. **D** Sri Lanka.

16 SHINY WHISTLING THRUSH *Myophonus melanurus* 24–29cm Shy. Forages on, or close to, the forest floor, often near running water; usually alone or in pairs. **V** A high-pitched screech; little other information. **H** Primary mossy hill and montane forest with rivers and streams. **D** Sumatra, Indonesia.

17 JAVAN WHISTLING THRUSH *Myophonus glaucinus* 24–27cm Forages on the ground and in lower storeys. Regularly fans tail. **V** Song a series of loud, clear whistles. Calls include a screeching, a pleasant whistled note and, when alarmed, an *ooweet-ooweet-teet* followed by a *truuu-truuu*, *cheet* or *tee-ee-eee... tee-ee-ee-eet*. **H** Montane forest with dark caves and crevices. **D** Java, Indonesia.

18 BORNEAN WHISTLING THRUSH *Myophonus borneensis* 25cm Forages on the ground among leaf litter. Frequently flicks tail open, like a fan. **V** Calls include a long, drawn-out whistle, a screech, a high-pitched ringing whistle and a long chittering. **H** Hill and montane forest, near rocky cliffs or streams. **D** Borneo.

19 BROWN-WINGED WHISTLING THRUSH *Myophonus castaneus* 25cm Generally forages in the middle and upper levels of trees, especially those in fruit; occasionally on rocks in streams. Regularly fans tail. **V** A grating *waach*; little other information. **H** Hill and montane forest, mainly alongside watercourses. **D** Sumatra, Indonesia.

20 MALAYAN WHISTLING THRUSH *Myophonus robinsoni* 25–26cm Very shy. Forages on the ground and in the lower storey of trees; occasionally seen on mountain roadsides at dawn or dusk. **V** A soft mix of fluty and scratchy notes, similar to that of Blue Whistling Thrush; also a high-pitched *tseee*. **H** Broadleaved evergreen forest, usually near streams. **D** Peninsular Malaysia.

21 MALABAR WHISTLING THRUSH *Myophonus horsfieldii* 25cm Forages on the ground or in shallow water at stream edges; readily perches in trees. **V** A series of rambling, rich, mournful whistles; also a high-pitched, descending *kree-ee*. **H** Rocky hill streams in forests, secondary growth, plantations; also recorded on forest paths and in gardens. **D** S India.

22 TAIWAN WHISTLING THRUSH *Myophonus insularis* 28cm Forages on the ground, usually close to rivers or streams. Wary; when disturbed, flees into cover with shrieking alarm calls. **V** A screeching *zi* or *sui yi*. **H** Dense broadleaved evergreen forest with bamboo, near rivers or streams. **D** Taiwan.

23 BLUE WHISTLING THRUSH *Myophonus caeruleus* 33cm Feeding actions much like those of a Common Blackbird (Plate 244). Generally seen feeding from rocks in strong-flowing streams, dipping to collect food from water surface. **V** Song a disjointed string of melodious, high-pitched, human-like whistles; sometimes includes mimicry. Calls include a far-carrying *tzeet-tze-tze-tzeet* or *tzeet-tzuit-tzuit-zuit* and a shrill *skreee*. **H** Broadleaved evergreen and mixed deciduous forest; often in gorges and ravines. **D** C Asia to E China, SE Asia and Sumatra and Java, Indonesia.

24 BLUE-FRONTED ROBIN *Cinclidium frontale* 19cm Forages by clambering among bamboo and probably also on the ground. **V** Song a series of short melodious phrases, transcribed as *tuee-be-tue* and *tuu-buudy-doo*. Calls include a shrill *shraak*, a faint *tch-tch-tch-tch-tch* and a harsh, buzzy *zschwick* when alarmed. **H** Broadleaved evergreen forest, bamboo. **D** Nepal to SE Asia.

25 RUSTY-TAILED FLYCATCHER *Ficedula ruficauda* 14cm Unobtrusive; forages in trees, snapping up insect prey while flitting from branch to branch. **V** Song comprises 3–4 loud, clear notes repeated at short intervals. Call a *tee-peup tee-peup*; alarm call a plaintive, ceaselessly repeated *peup*, followed by a soft *churr*. **H** Mixed coniferous broadleaved forest. Winters in broadleaved evergreen forest. **D** Breeds from Uzbekistan and Kyrgyzstan to Nepal and N India. Winters in SW India.

26 EUROPEAN PIED FLYCATCHER *Ficedula hypoleuca* 13cm Makes fly-catching sallies, usually darting out from a hidden perch in middle or top of tree and returning to a similar position nearby or to cover some distance away. **V** Sweet warble, individually variable and variously transcribed, e.g. *chee-chee-chee-tsri-tsri-chee* or *zi-vreezi-vreezi-vreezi-tsu-tsu-chu-vee-chu-vee-zi-zi-zi*. Calls vary, e.g. *hweet* and *tic*, often combined as *whit-tic* or *whee-tic*. **H** Deciduous and mixed woodland, parks and large gardens. **D** Breeds from N and W Europe to SC Siberia. Winters in W Africa.

27 ATLAS PIED FLYCATCHER *Ficedula speculigera* 13cm Very like European Pied Flycatcher. In male, black more intense and velvety; in both sexes, white wing patch larger. **V** Very similar to that of European Pied Flycatcher. **H** Most woodland and forest habitats. **D** Breeds from Morocco to Tunisia. Winters in W Africa.

28 COLLARED FLYCATCHER *Ficedula albicollis* 13cm Actions and habits similar to those of European Pied Flycatcher. **V** Song like that of European Pied, but slower and interspersed with thin, high notes. Call a far-carrying, drawn-out *eep* and a soft *tsrr*. **H** Deciduous woodland with glades and clearings, orchards, parks and gardens. **D** Breeds from C and SC Europe through W Russia. Winters in SC Africa.

29 SEMICOLLARED FLYCATCHER *Ficedula semitorquata* 13cm Actions and habits similar to those of European Pied Flycatcher. **V** Song similar to that of Collared Flycatcher with repeated European Pied-style phrases. Call a whistling *eep* or *tseep*, or a *tuup* when alarmed. **H** Deciduous and mixed woodlands, orchards, riverside copses. **D** Breeds from Greece to Azerbaijan and Iraq. Winters in EC Africa.

30 YELLOW-RUMPED FLYCATCHER *Ficedula zanthopygia* 13cm Forages in the foliage of trees and undergrowth, often making short sallies after flying insects. **V** A dry rattling *tr-r-r-rt*. **H** Forests and undergrowth along rivers and streams, coastal scrub and mangroves. **D** Breeds from E Mongolia and SE Russia to E China. Winters from SE Asia to Java, Indonesia.

31 NARCISSUS FLYCATCHER *Ficedula narcissina* 13cm Makes sallies from middle storey and canopy. **V** Repeated warbles and whistles, transcribed as *o-shin-tsuk-tsuk* and often including some mimicry of other bird calls. **H** Deciduous, mixed or coniferous forests with dense undergrowth, on hills and mountains. **D** Breeds on Sakhalin and Kuril Islands (Russian N Pacific) and Japan. Winters from SE Asia to Java, Indonesia.

32 GREEN-BACKED FLYCATCHER *Ficedula elisae* 13–14cm Previously considered conspecific with Narcissus Flycatcher; actions and habits presumed similar. **V** Calls include a low *tok tok tok*, a sharp *tek tek* and a high-pitched *pee*. **H** Forest, plantations, parks and gardens. **D** Breeds in NE China. Winters in Thailand, Peninsular Malaysia, possibly South Korea.

33 MUGIMAKI FLYCATCHER *Ficedula mugimaki* 13cm Unobtrusive forager in the middle and upper canopy. Often flicks and spreads tail. **V** Calls include a soft, rattled *trrrr*. **H** Forest, forest edges. **D** Breeds from N Mongolia and SE Russia through NE China to the Korean Peninsula. Winters from SE Asia to Java, Indonesia; Philippines.

34 SLATY-BACKED FLYCATCHER *Ficedula hodgsonii* 13cm Sallies after flying insects from a tree-canopy perch; occasionally takes insects from the ground. **V** Song is a ripple of descending, whistling notes. Calls include a hard *tchat* and a rattled *terrht*. **H** Damp broadleaved forest, shrubberies, bamboo. **D** Himalayas to E India and SE Asia.

35 RUFOUS-CHESTED FLYCATCHER *Ficedula dumetoria* 11–12cm Forages low down in dense vegetation, often near streams; regularly catches insects in flight. **V** Song a distinctive high-pitched, rising and falling *sii wi-sii si-wi-si-ii* and *si-wi-oo*. Calls include a soft *sst-sst*. **H** Lowland to lower montane primary forest, with bamboo and understorey. **D** Malay Peninsula; Java and Sumatra, Indonesia; Borneo; Lesser Sunda Islands.

36 TANIMBAR FLYCATCHER *Ficedula riedeli* 11–12cm Forages in the forest understorey and bamboo. Formerly considered a race of Rufous-chested Flycatcher. **V** Presumed to be similar to that of Rufous-chested Flycatcher. **H** Lowland and lower montane forest. **D** Tanimbar Islands, E Lesser Sunda Islands.

37 RUFOUS-GORGETED FLYCATCHER *Ficedula strophiata* 14cm Flits and spreads tail when agitated, showing white tail sides. Forages mainly in undergrowth or low down in trees. **V** Song a spirited *tin-ti-ti*. Call a low *tik-tik* or *pink*, and a croaking *churr* when alarmed. **H** Dense or open forests, thick secondary scrub, forest edges. **D** Himalayas to N and C China and SE Asia.

1
2
3
4
5
6
7
8
9
10
11
12
13
14
15
16
17
18
19
20
21
22
23
24
25
26
27
28
29
30
31
32
33
34
35
36
37

CHATS AND OLD WORLD FLYCATCHERS *CONTINUED*

1 RED-BREASTED FLYCATCHER *Ficedula parva* 11.5–12.5cm Forages among foliage. Makes occasional aerial sorties after flying insects, when white outer tail bases are prominent. **V** Calls include a *chick* or *chick-chick*, also a rattled *serrrt*. **H** Open forest, secondary growth, orchards, urban trees. **D** Breeds from C and N Europe to W Russia and Iran. Winters in Pakistan, India.

2 TAIGA FLYCATCHER *Ficedula albicilla* 11–12cm Forages among foliage. Makes occasional sallies after flying insects, when white outer tail bases are prominent. **V** A buzzing *drrrrrt* and a harsh *zree*. **H** Open forests, forest edges, plantations, parks and gardens. **D** Breeds from W Russia to E Russia and south to N Mongolia and NE China. Winters from SE Asia through Borneo.

3 KASHMIR FLYCATCHER *Ficedula subrubra* 13cm Actions similar to those of Rufous-gorgeted Flycatcher (Plate 251). **V** Song is a rising twitter. Calls include a sharp *chack*, a subdued, harsh *purr* and a dry rattle. **H** Deciduous forests. Winters in plantations, forest edges, gardens. **D** Breeds in N India. Winters in Sri Lanka.

4 SNOWY-BROWED FLYCATCHER *Ficedula hyperythra* 11–13cm Forages low down in scrub and thickets; runs along the ground much like a shortwing. **V** Song is a wheezy, shrill *tsit-sit-si-sii tsi-sii-swrri* or *tsi-sit-i*. Calls include a thin *sip* and an upslurred *seep*. **H** Hill and montane forest. **D** Widespread across Asia except the Philippines, from the C Himalayas to SC China and south to Borneo and Indonesia.

5 LITTLE SLATY FLYCATCHER *Ficedula basilanica* 12–13cm Usually secretive, foraging on the ground and in tangled understorey; best detected by its calls. **V** Song a series of chattering whistles, sometimes ending in a trill. Calls include a soft, whistled *pee-pawww pee-pee-hi-hi*. **H** Primary forest, secondary growth. **D** Philippines.

6 RUFOUS-THROATED FLYCATCHER *Ficedula rufigula* 11–12cm Generally forages on the ground in dense rattan thickets, usually in pairs. **V** Song an up-and-down *si-si-si* or *swee-wee-seee*. Calls include a quick *chik* and a staccato chattering. **H** Primary lowland and hill forest, swamp forest. **D** Sulawesi, Indonesia.

7 CINNAMON-CHESTED FLYCATCHER *Ficedula buruensis* 11–12cm Inconspicuous. Favours the shady parts of dense forest understorey. **V** A buzzing, nasal *tup tup tezeeew*. **H** Lowland and hill forest. **D** S Maluku Islands, Indonesia.

8 DAMAR FLYCATCHER *Ficedula henrici* 12–13cm Generally seen alone or in pairs. Uses an exposed branch to make drops onto insects; also gleans prey. **V** A far-carrying whistle. **H** Lush evergreen forest. **D** Damar Island, E Lesser Sunda Islands.

9 SUMBA FLYCATCHER *Ficedula harterti* 11cm Shy and secretive. Forages in the forest undergrowth, usually alone or in pairs. Continuously pumps tail. **V** Song a mournful series of three descending, whistled notes, followed by a sharp *zip*. Calls include an insect-like, buzzing *tszzz*, a *chik-ik* and a *chikikikikikiki*. **H** Primary and secondary forest, forest edges, bushy thickets, secondary growth. **D** Sumba, Lesser Sunda Islands.

10 PALAWAN FLYCATCHER *Ficedula platenae* 11–12cm Shy and secretive, favouring dense tangles of rattan and climbing bamboo; usually alone. **V** A varied, raspy, high-pitched *zee-zawww, zawwp-zeeepp, zee zeet zeeett* or *zeee-zawwwpppp*, followed by a short chatter. **H** Lowland primary forest, secondary growth. **D** Philippines.

11 CRYPTIC FLYCATCHER *Ficedula crypta* 11–12cm Secretive, usually favouring undergrowth and secondary growth; occurs singly or in pairs. **V** A series of high-pitched phrases. **H** Submontane primary moss forest and secondary growth. **D** Philippines.

12 BUNDOK FLYCATCHER *Ficedula luzoniensis* 11–13cm Secretive, foraging in the forest understorey. Formerly considered to be a race of the Snowy-browed Flycatcher. **V** Little information; may be similar to that of Snowy-browed. **H** Montane forest. **D** Philippines.

13 FURTIVE FLYCATCHER *Ficedula disposita* 11–12cm Favours dense climbing bamboo in the forest understorey. Secretive; usually seen alone or in pairs. **V** A faint, high-pitched, whistled *wan he, wau he hu, he haaww* or *he-u-heee*, repeated every 4–5 secs; also utters a sharp, repeated *zeet zeet*. **H** Dense lowland secondary forest. **D** Philippines.

14 LOMPOBATTANG FLYCATCHER *Ficedula bonthaina* 10–11cm Frequents the dense forest understorey, on or close to the ground. **V** Unknown. **H** Primary montane moss forest. **D** Sulawesi, Indonesia.

15 LITTLE PIED FLYCATCHER *Ficedula westermanni* 11–12cm Forages in treetops, constantly on the move; also makes short sallies after flying insects. **V** Song variable, including a flute-like *sueep* or *seeeup*, interspersed with a rapid, vibrating trill, and *tee-dee* notes followed by a long warbling. Calls include a mellow *tweet*, a low *chur*, and a high note followed by soft, trilled *trrrt* notes. **H** Mossy, *Casuarina* and other forest. Post breeding may use more open woodlands. **D** Widespread from the W Himalayas to SC China and south to the Philippines, Sulawesi and Lesser Sunda Islands.

16 ULTRAMARINE FLYCATCHER *Ficedula superciliaris* 12cm Forages in the foliage of low trees and bushes. **V** Song a repeated *chi-chi-purr*. Calls include a soft *tick* and a low rattling *trrrt*. **H** Open mixed forests. In winter, also occurs in orchards and gardens. **D** E Afghanistan to S China, C India and SE Asia.

17 SLATY-BLUE FLYCATCHER *Ficedula tricolor* 12–13cm Secretive. Forages in undergrowth and lower tree branches; also takes insects from the ground. **V** Song is a three-note whistle, *zieth-ti-zietz*. Calls include an *ee-tik* and a rapid *ee-tick-tick-tick-tick*. **H** Evergreen mountain forests, shrubberies, forest edges. In winter, also in reedbeds and tall grass. **D** Himalayas east to N Vietnam.

18 SAPPHIRE FLYCATCHER *Ficedula sapphira* 11cm Forages in undergrowth and trees; occasionally picks food from the ground or makes aerial sallies after flying insects. **V** Song a series of several high-pitched notes followed by short rattles. Call a low, rattled *tit-tit-ti*. **H** Open broadleaved evergreen forest. **D** Himalayas to C China and C Laos.

19 BLACK-AND-ORANGE FLYCATCHER *Ficedula nigrorufa* 11cm Forages in shady undergrowth. **V** Song is a high, insect-like *chiki-riki-chiki*. Calls include a low *pee* and a *zit-zit* when alarmed. **H** Shola forest with dense undergrowth, damp riverine thickets, plantations. **D** SW India.

20 BLACK-BANDED FLYCATCHER *Ficedula timorensis* 11cm Shy and elusive, favouring dense undergrowth. Feeds by snatching or sallying for insects. Generally seen singly or in pairs. **V** Song consists of a series of muted, high-pitched disyllabic whistles, repeated every 2–3 secs. Calls include a soft, low-pitched *buzz-buzz-buzz*, often repeated and sometimes interspersed with a short, piercing, descending whistle. **H** Monsoon hill forest with limestone boulders and rocky scree slopes. **D** Timor, E Lesser Sunda Islands.

21 PYGMY FLYCATCHER *Muscicapella hodgsoni* 10cm Active leaf warbler-like forager in tree foliage. Usually in pairs or alone; also joins mixed-species feeding parties. Often flicks wings and cocks tail. **V** Song a high-pitched *tzzit-che-che-che-heeee* or similar. Calls include a feeble *tsip, tip* or *tup* and a low *churr*. **H** Dense montane broadleaved forest. **D** Himalayas to Sumatra (Indonesia) and Borneo.

22 PRZEVALSKI'S REDSTART *Phoenicurus alaschanicus* 15cm Drops from a twig or rock to pick up insects; flicks tail up and down. **V** A whistled *few-eet* and a croaking *gre-er*. **H** Montane coniferous forest, dense bushes, rocky slopes. **D** NC China.

23 EVERSMANN'S REDSTART *Phoenicurus erythronotus* 15cm Drops from a rock or twig to capture insects. Flicks tail up and down. **V** A whistled *few-weet* and a croaking *gre-er*. **H** Dry scrub areas on hills and valleys. **D** C Asia.

24 BLUE-CAPPED REDSTART *Phoenicurus coeruleocephala* 15cm Drops to the ground, or makes aerial sallies to capture insects. Shakes tail. **V** Song consists of loud, fast, high-pitched, ringing jingles. Calls include a *tik-tik* and, when alarmed, a piping *tit-tit-tit*. **H** Open montane forests with rocky slopes. Winters in lower-altitude open forest and scrubby areas. **D** E Afghanistan to C Himalayas.

25 BLACK REDSTART *Phoenicurus ochruros* 15cm Flits from a low perch to catch insects from the ground or catches flying insects by making short aerial sallies. **V** Song is a rapid warble, interspersed with a rattle, and ending with a rushed burst of ringing notes. Calls include a *tsip*, a *tucc-tucc* and, when alarmed, a *tititicc*. **H** Steppe. Winters in plantations and cultivations. **D** Breeds from W Europe to C China. Winters in N Africa, S Asia.

26 COMMON REDSTART *Phoenicurus phoenicurus* 15cm Shivers tail. Usually sits in cover; shows itself when dropping to feed on the ground. **V** Song variable; *hooeet* followed by a weak, melancholic warbling. Calls include a *hooeet* and a *tek*, often combined. **H** Deciduous and mixed forest, orchards, parks and gardens. On migration can occur in any woods and scrub. **D** Breeds from Europe to N Africa and C Asia. Winters in NE and E Africa.

27 HODGSON'S REDSTART *Phoenicurus hodgsoni* 15cm Chases insects in short sallies from a rock or branch perch; in taller trees often acts like a flycatcher. **V** Song consists of short phases with tinny, tinkly notes. Calls include a *prit*, a *trr* and, when alarmed, a *tschrrr*. **H** Tree-lined stony riverbeds, open scrub jungle, cultivated areas with bushes. **D** Himalayas to C China.

28 WHITE-THROATED REDSTART *Phoenicurus schisticeps* 15cm Restless; often feeds in a flycatcher-like manner. **V** Song a series of dry, trilled phrases, usually accelerating towards the end. Call is a drawn-out *zieh* followed by a rattling note. **H** Thick scrub in subalpine conifer forests. In winter descends to lower elevations. **D** Himalayas to C China.

29 DAURIAN REDSTART *Phoenicurus auroreus* 15cm Forages in trees and bushes, often in the manner of a flycatcher. Shivers tail. **V** Song is a series of cheerful, whistled notes. Calls include a *wheep* and a soft *tac-tac* that is preceded by a series of soft whistles. **H** Bushy areas, cultivation. **D** Breeds in C and E Asia. Winters in S China, NE India.

30 MOUSSIER'S REDSTART *Phoenicurus moussieri* 13cm Perches in the open. Actions much like those of a stonechat. **V** Song a short scratchy warble. Calls include a thin *wheet, hiip-hiip* or *psew*. **H** Dry stony or grassy hills or mountains with scrub, coastal forest and thickets. On winter moves to lower, often desert, areas. **D** Morocco to NE Libya.

31 GÜLDENSTÄDT'S REDSTART *Phoenicurus erythrogastrus* 18cm Rather shy, but showy during quivering-wing display flight, when white wing patches are prominent. **V** Song a series of clear notes, followed by a burst of short wheezy notes. Call a hard *tik* or *tek*. **H** High-altitude boulder-strewn slopes, rocky meadows, often near water; forced to lower elevations during severe weather. **D** Breeds in Caucasus; E Afghanistan to SC Siberia, W Mongolia and NW and C China. Winters to S China.

32 BLUE-FRONTED REDSTART *Phoenicurus frontalis* 15cm Descends from a rock or branch to capture insects. Flicks tail up and down. **V** Song similar to that of Black Redstart, but less wheezy. Calls include a *tic* and, when alarmed, a repeated *ee-tit - ti-tit*. **H** Subalpine scrub. Winters in bushes and open forest at lower altitudes. **D** E Afghanistan to C China.

33 PLUMBEOUS WATER REDSTART *Phoenicurus fuliginosus* 14cm Continuously opens and shuts tail; also wags tail up and down. Makes sallies from rocks or low branches. **V** Song a rapid *streee-treee-tree-treeeh*. Calls include a sharp *ziet-ziet* or a threatening *kree*. **H** Fast-flowing streams and rivers. **D** E Afghanistan to E China, Taiwan.

34 LUZON WATER REDSTART *Phoenicurus bicolor* 14–15cm Often seen in small groups perching on streamside rocks and feeding on invertebrates at the water's edge. **V** A fairly loud, repeated *seep*. **H** Clear, fast-flowing, rocky-sided mountain streams and rivers running through montane forests. **D** Philippines.

35 WHITE-CAPPED REDSTART *Phoenicurus leucocephalus* 19cm Continuously flicks tail; sits on rocks in or close to water. **V** Song a weak, undulating whistle. Call a plaintive *tseeit-tseeit*. **H** Mountain streams and rivers. **D** Breeds from C Asia through the Himalayas to C and NE China and N SE Asia. Winters from N India to Vietnam.

1
2
3
4
5
6
7
8
9
10
11
12
13
14
15
16
17
18
19
20
21
22
23
24
25
26
27
28
29
30
31
32
33
34
35

CHATS AND OLD WORLD FLYCATCHERS *CONTINUED*

1 WHITE-WINGED CLIFF CHAT *Monticola semirufus* 20cm Little known. Male glossy blue-black with a rufous lower breast and belly. Female dusky blackish with a rufous undertail. Both sexes have a white patch at base of primaries. **V** Song described as fluty with some rasping notes. **H** Stony ground, cliffs, rocky outcrops, often near forests. **D** Ethiopia, Eritrea.

2 CAPE ROCK THRUSH *Monticola rupestris* 21cm Grey restricted to head of male; female is uniform orange below. **V** High, loud, sweet, fluting *wee-weet-wree-tuwie weet-weet.* **H** Rocky and grassy slopes with some bush. **D** SE Botswana and SW Mozambique to S South Africa.

3 SENTINEL ROCK THRUSH *Monticola explorator* 18cm Separated from Short-toed Rock Thrush by uniform grey crown and mostly different range. **V** High, slightly sharp, up-and-down *tjit-si-sirup-tsit-sit-sirri-weer-weeh.* **H** Open rocky mountain sides. **D** E and S South Africa, Lesotho to S Mozambique.

4 SHORT-TOED ROCK THRUSH *Monticola brevipes* 18cm Note pale crown of male (reduced in non-breeding plumage). Separated from Sentinel Rock Thrush by mostly different range. **V** High, fluting *ni-wi-wi-phrir-weer-wee,* rising and then descending. **H** Rocky outcrops in open natural and cultivated areas. **D** W Angola, Namibia, NW and C South Africa, SE Botswana.

5 MIOMBO ROCK THRUSH *Monticola angolensis* 17cm Male separated from male Cape, Sentinel, Short-toed and Common rock thrushes by spotted crown and mantle; female by well-defined white moustache. **V** Rather short mid-high *fiuugwirwir.* **H** Miombo; at higher altitudes than Cape Rock Thrush. **D** S Africa.

6 COMMON ROCK THRUSH *Monticola saxatilis* 20cm Usually wary. Generally forages on the ground, although sometimes feeds in trees or makes short aerial sorties to capture flying insects. Frequently wags tail. **V** A series of soft, clear melodic phrases; also a low *chak-chak* and a clear *diu.* **H** Open rocky hillsides. **D** S Europe to SC China, NW Africa.

7 LITTLE ROCK THRUSH *Monticola rufocinereus* 15cm Usually feeds by dropping onto insects from a bush or tree perch, or by making fly-catching sallies. Quivers tail when perched. **V** Song comprises scratchy notes interspersed with low- and high-pitched fluty notes, *tryyh-rrr-twi-rirp-tschak-tshak.* Call a soft *tyyt* or, when alarmed, a *trrrt.* **H** Rocky slopes, cliffs and gorges in highland regions, always with some trees and bushes. **D** E Africa, W Arabian Peninsula.

8 BLUE ROCK THRUSH *Monticola solitarius* 20cm Forages on the ground, or drops on prey from a low perch; sometimes makes aerial sorties after flying insects. **V** Song a loud, fluty, melodic *tju-sri - tjurr-titi - wuchi - trr-trrt-tri;* may also include some mimicry. Calls include a deep *chak-chak,* a plaintive *see* and a *wit-wit.* **H** Barren rocky hills and steep hillsides. In winter, occurs in a wider range of rocky locations, including cliffs, rocky seashores, quarries, old habitations. **D** Widespread from SW and SC Europe to NW Africa and east to Japan and Sumatra, Indonesia. Winters in N Africa, India, Greater Sunda Islands.

9 CHESTNUT-BELLIED ROCK THRUSH *Monticola rufiventris* 23cm Forages mainly on the ground; occasionally makes fly-catching sorties from high trees. **V** Song a pleasant warble and a sharp whistle followed by an upslurred *fweeeur-fweet,* usually delivered from a treetop. Calls include a *quock,* a coarse *quach* and a shrill *tick.* **H** Open broadleaved evergreen and coniferous forests on rocky hillsides, forest edges, scrub, rocky outcrops. **D** Himalayas through SE Asia.

10 BLUE-CAPPED ROCK THRUSH *Monticola cinclorhyncha* 17cm Mainly arboreal. Picks insects off trunks and branches; also feeds on the ground among leaf litter. **V** A fluty *tew-li-di - tew-li-di - tew-li-di* or *tra-trr-treee-treea...,* usually given from the topmost branch of a tree or during a song flight. **H** Open pine and oak forests and rocky grass-covered slopes with scattered trees. Winters in moist forest and well-wooded areas. **D** Afghanistan to Myanmar and S India.

11 WHITE-THROATED ROCK THRUSH *Monticola gularis* 18cm Forages on the ground or in small trees. **V** Song comprises melancholic, flute-like, rising whistles, combined with a *chat-at-at* call. Calls include a sharp *tack-tack,* a soft *queck-queck* and a thin *tsip,* the latter given in flight. **H** Open deciduous and broadleaved evergreen forest, plantations, secondary growth. **D** Breeds from E Mongolia and SE Russia to C China. Winters in S China, SE Asia.

12 LITTORAL ROCK THRUSH *Monticola imerina* 16cm Separated from Forest Rock Thrush by much paler plumage and grey (male) or brownish (female) tail. **V** Hurried series of 2–3 *wreet* notes, immediately followed by a rattle or rapid twittering. **H** Sandy shrubland. **D** Madagascar.

13 FOREST ROCK THRUSH *Monticola sharpei* 16cm Male has a blue-grey head, otherwise rufous. Female rufous-brown with a white throat and white streaking on underside. Often perches motionless for long spells on a rock or other raised perch; approachable. **V** Song a varied series of clear, tuneful whistles with long pauses in between. Calls include a soft *tseet-tak-tak* and, when alarmed, a *taktaktak* or low *krrr.* **H** Evergreen humid forest with open understorey, other wooded habitats, especially near streams and creeks. **D** Madagascar.

14 WHINCHAT *Saxicola rubetra* 13cm Usually perches less upright than a stonechat, with less wing- and tail-flicking. **V** Song a short, scratchy series of warbling phrases. Calls include a *tek-tek* or *whuk-tek-tek.* **H** Open country with scattered bushes, including heathland, upland pastures, hillsides, young conifer plantations. **D** Breeds from Europe to C Russia and NW Mongolia. Winters in W, C and NE Africa.

15 WHITE-BROWED BUSH CHAT *Saxicola macrorhynchus* 15cm Actions much like those of Siberian Stonechat, although tends to spend more time hopping on the ground. In flight, white tail-base pattern looks like that of a wheatear. **V** Song a low, musical *twitch-chhe chee chee.* Calls include a sharp *chip-chip* and a deep *prupp prupp.* **H** Sandy, semi-desert scrubland. **D** NW India.

16 WHITE-THROATED BUSH CHAT *Saxicola insignis* 15cm Actions much as Siberian Stonechat, although may forage more on the ground. **V** A metallic *tek-tek.* **H** Tall riverside vegetation and cane fields. **D** Mongolia.

17 CANARY ISLANDS STONECHAT *Saxicola dacotiae* 13cm Actions as for European Stonechat. **V** Song a scratchy *bic-bizee-bizeeu,* similar to that of European Stonechat. Call a *chut,* a little sharper than that of European Stonechat. **H** Arid rocky ground with scattered scrub; generally avoids open plains. **D** Canary Islands.

18 EUROPEAN STONECHAT *Saxicola rubicola* 13cm Sits atop a prominent perch, flicking wings and tail, dropping to the ground to pick up food and then returning to same perch or one nearby. At times feeds aerially, much like a flycatcher. **V** Song a thin, scratchy warble, more melodious when given during song flight. Calls include a *chak* or *wheet,* often combined as *wheet-tak-tak.* **H** From bushy coastal cliffs to heathland, moorland, steppe and scrubby mountainsides. **D** W and S Europe, N Africa.

19 SIBERIAN STONECHAT *Saxicola maurus* 13cm Sits atop a prominent perch, flicking wings and tail, with frequent sallies to capture insects on or near the ground; at times makes flycatcher-like aerial sallies to catch flying insects. **V** Song a thin, scratchy warble, often given in a short song flight. Call a *chak* or *wheet,* often combined as *wheet-tak-tak.* **H** Open country with bushes. Winters in scrub, reedbeds, cultivation. **D** Breeds in E Caucasus, SE Turkey and W Iran, C Siberia, C Asia, Himalayas to W and C China. Winters in NE Africa, S Asia.

20 STEJNEGER'S STONECHAT *Saxicola stejnegeri* 14cm Actions similar to those of Grey Bush Chat. **V** At the time of writing, no difference known from calls of Siberian Stonechat. **H** Moist meadows with rich grass, shrubby marshes, dry pinewoods. **D** Breeds from E Siberia and E Mongolia to Korean Peninsula and Japan. Winters in SE Asia.

21 AFRICAN STONECHAT *Saxicola torquatus* 14cm Note all-black tail and white neck patches. Female has only an inconspicuous eyebrow. **V** Short, nervous rattle, *sisititirrut-trirut.* **H** Areas with some scattered bush and shrub, moorland, swamp edges. **D** SW Arabian Peninsula, sub-Saharan to S Africa.

22 MADAGASCAN STONECHAT *Saxicola sibilla* 14cm Sometimes considered conspecific with African Stonechat; closely resembles that species but male has rufous restricted to a very narrow band below black throat. **V** Like that of African Stonechat. **H** Remote areas of slash-and-burn agriculture within rainforest, also montane scrub and stunted montane forest. **D** Madagascar.

23 REUNION STONECHAT *Saxicola tectes* 13cm No similar bird on Réunion; note white throat. Variable, some with white eyebrow as shown, other individuals lacking this. **V** Short, high whistles with small trills. **H** Any high-altitude habitat with trees or tall heath and without undergrowth. **D** Réunion, Indian Ocean.

24 WHITE-TAILED STONECHAT *Saxicola leucurus* 12–14cm Actions and habits very similar to those of Siberian Stonechat. **V** Song a series of lark-like phrases, consisting of rapid, scratchy, squeaky notes and ending on a high, slurred note. Calls include a short *peep-chaa,* a hard, dry *kek-kek-kek* and warning *pseep.* **H** Tall grassland and reeds. **D** Pakistan, N India.

25 PIED BUSH CHAT *Saxicola caprata* 13–14cm Actions similar to those of Siberian Stonechat. **V** Song a brisk, whistled *chip-chepee-chewee-chu.* Calls include a repeated, plaintive *chep-chep-hee* or *chek-chek-trweet* and a scolding *chuh* when alarmed. **H** Stony hillsides with low scrub, open country with scattered bushes, tall grass and reeds, cultivated fields, tamarisk growth. **D** NE Iran through SE Asia to New Guinea.

26 JERDON'S BUSH CHAT *Saxicola jerdoni* 15cm Actions and habits very similar to those of Siberian Stonechat. **V** Song comprises a series of sweet, clear, mellow, warbled phrases, often ending with a trilled flourish. Calls include a high, nasal, downslurred *heeew;* a short, plaintive, high-pitched *chirr* or *chit-churr;* and, when alarmed, a high, dry, rapid ticking. **H** Tall grassland. **D** N India to N Thailand.

27 GREY BUSH CHAT *Saxicola ferreus* 15cm Sits on a prominent perch, making frequent sallies to capture insects on or near the ground; occasionally makes aerial fly-catching sorties. **V** Song a short, feeble trill, ending with a rolling whistle. Calls include a soft *zizz,* a clear *hew* and a sharp *tak-tak-tak-tak.* **H** Open woodland, open scrubby and bush-covered hillsides, cultivation, parks and gardens. **D** Breeds from the Himalayas to C and S China and C Vietnam. Winters in N India, S Myanmar to S Vietnam.

28 WHITE-BELLIED BUSH CHAT *Saxicola gutturalis* 15–17cm Usually seen in pairs; gleans and makes short sallies after insects. Favours the canopy or tall understorey shrubs, and sings from dense canopy of tall trees. **V** Song a series of sweet, unhurried phrases. Calls include a *tchk-tchk.* **H** Remnant patches of monsoon forest, monsoon woodland, secondary growth. **D** E Lesser Sunda Islands.

29 BUFF-STREAKED CHAT *Campicoloides bifasciatus* 16cm All-black wings and tail of male diagnostic. Female separated from Blackstart, Familiar Chat and Brown-tailed, Sombre and Brown rock chats (Plate 254) by black tail, contrasting with orange-buff rump. **V** Short, hurried, very high phrases, preceded by a compact *click-click.* **H** Rocky and stony mountain slopes. **D** South Africa.

30 SICKLE-WINGED CHAT *Emarginata sinuata* 15cm Separated from Tractrac and Karoo chats by rufous-edged wing feathers and larger, black wedge on a pale rufous (not white) tail. **V** Combination of fluting and harsh notes, *tsreeetsreeetfjuwhich-which-which.* **H** Bare or short-grassed slopes; virtually sea level to high altitudes. **D** W, C and S South Africa, Lesotho, Namibia.

31 KAROO CHAT *Emarginata schlegelii* 16cm Note white outer-tail feathers. **V** Mid-high, harsh, well-separated *cratch-cratch-crut-crut-cree.* **H** Dry scrubby areas with sparse bush. **D** NW, W and C South Africa, SW Angola, Namibia.

32 TRACTRAC CHAT *Emarginata tractrac* 14cm Separated from Karoo Chat by only partly white outer-tail feathers. **V** Combination of a high, rolling *preeee* and a low, sparrow-like *tsja-wits-jaw.* **H** Dry stony plains, sand dunes with very little grass, desert scrub. **D** SW Angola, Namibia, W South Africa.

33 MOORLAND CHAT *Pinarochroa sordida* 15cm Note white tail with inverted black 'T'. **V** Song a loud piping. Call a metallic chirp. **H** Rocky alpine grassland and moorland; 2,250–4,500m. **D** E Africa.

34 MOCKING CLIFF CHAT *Thamnolaea cinnamomeiventris* 21cm Separated from *Cossypha* robin-chats (Plate 249) by black breast and rump. **V** Very varied, extremely high whistles, sunbird-like twitters and melodious nightingale-like strophes. **H** Well-wooded sloping areas near cliffs, rocky outcrops, buildings. **D** E, C and S Africa.

35 WHITE-CROWNED CLIFF CHAT *Thamnolaea coronata* 20cm Very like Mocking Cliff Chat, with which it is often considered conspecific. Male has a white crown, while female has more extensive rufous than female Mocking Cliff Chat. **V** Like that of Mocking Cliff Chat. **H** Rocky areas with some trees, also around buildings. **D** WC Africa.

1
2
3
4
5
6
7
8
9
10
11
12
13
14
15
16
17
18
19
20
21
22
23
24
25
26
27
28
29
30
31
32
33
34
35

CHATS AND OLD WORLD FLYCATCHERS *CONTINUED*

1 SOOTY CHAT *Myrmecocichla nigra* 16cm Separated from immature Arnot's Chat by smaller, white wing patch, shorter tail and more open habitat. **V** Very high, sharp, calm, 4–5-toned thrush-like phrase, *tu-tu-tjee-tu-tju*. **H** Open, short grassland with termite mounds. **D** Nigeria, Central African Republic and South Sudan to Angola, Zambia and Tanzania.

2 ANTEATER CHAT *Myrmecocichla aethiops* 20cm In flight, note striking white wing patches. **V** High, loud, rather slow whistle, *pee-pee-pju tju-fju*. **H** Termite mounds in dry, short grassland at higher elevations. **D** E, C and W Africa.

3 CONGO MOOR CHAT *Myrmecocichla tholloni* 17cm Note white rump and wing patch. **V** Song a series of melodious whistles, interspersed with rolled *chiurrr* calls. Alarm call a shrill *piip*. **H** Open grassland with some trees, bushes and other lookout posts. **D** Central African Republic; Gabon to SW DR Congo and C Angola.

4 ANT-EATING CHAT *Myrmecocichla formicivora* 17cm Separated from Sooty Chat by brown (not black) plumage and large white primary patch. Usually lacks white shoulder patch. **V** Unhurried sequence of separate, slightly rasping *wrah* notes at different pitches, between recurring, high *fu-fu-weet*. **H** Dry, open grassy areas with termite mounds. **D** S Angola, Namibia and Botswana to S South Africa.

5 RÜPPELL'S BLACK CHAT *Myrmecocichla melaena* 20cm Entirely black except for white patch on inner primaries, revealed in flight. **V** Song a soft warbling. Call a short, piercing whistle. **H** Cliffs, gorges and other rocky habitats, often near water. **D** Ethiopia, Eritrea.

6 MOUNTAIN WHEATEAR *Myrmecocichla monticola* 18cm Various colour forms, but all males have a white shoulder. Separated from Ant-eating and Arnot's chats by white rump and partly white tail. **V** High, melodious yet rasping *sishraw titjuruwirwir*. **H** Short-grassed rocky slopes. **D** SW and W Angola, W and S Namibia, South Africa.

7 ARNOT'S CHAT *Myrmecocichla arnotti* 17cm Note black tail. **V** Very high, hurried canary-like twittering, with some rasping notes added to recurring, very high *feee*. **H** Bare patches in miombo and other woodland. **D** SC Africa.

8 RUAHA CHAT *Myrmecocichla collaris* 17cm Male similar to male Arnot's Chat; female generally less white on chest. **V** High, fluted, repetitive series of notes, e.g. *teeh-tjeeh-tjir*. **H** Dry grassland and savannah. **D** W Tanzania, E Rwanda, Burundi, N Zambia, E DR Congo.

9 NORTHERN WHEATEAR *Oenanthe oenanthe* 14.5–15.5cm In flight, shows a white rump and white outer tail bases. Flicks wings and tail. **V** In display flight, sometimes gives a brief scratchy warble. Calls include a hard *chak* and a *wheet*, often combined as *wheet-chak-chak*. **H** Varied habitats, from tundra and moorland to coastal islands and sand dunes. **D** Widespread across Eurasia, N North America, NW Africa.

10 CAPPED WHEATEAR *Oenanthe pileata* 17cm Unmistakable due to white throat and broad black breast collar. **V** Unhurried sequence of high, unstructured, rasping and chattered notes. **H** Semi-desert, montane moorland, open coastal shrub. **D** SC and S Africa.

11 RED-BREASTED WHEATEAR *Oenanthe bottae* 17cm In flight, shows a white rump and white outer tail bases. Distinctive downward tail-wagging. **V** Song consists of fluty and scratchy notes, often uttered in low flight. Alarm call a *tjeet*. **H** Open hillsides with sparse vegetation, high-plateau cultivation. **D** S Arabian Peninsula, Ethiopia.

12 HEUGLIN'S WHEATEAR *Oenanthe heuglini* 14cm Similar to Red-breasted Wheatear, but smaller, with a narrower white eyebrow and more brown-toned underparts. Timid. Sometimes sings at night. **V** Song consists of long, jumbled phrases (30–60 secs), often given in display flight. Calls include a hard *chack*. **H** Flat stony ground with short turf, burnt-over and overgrazed areas. **D** Mauritania, Guinea and Cameroon to Sudan and NW Kenya.

13 ISABELLINE WHEATEAR *Oenanthe isabellina* 17cm Often has a very upright stance. Jerky, emphatic tail-wagging. In flight, shows a white rump and outer tail bases. **V** Song loud, with various croaks, whistles and mimicry. Calls include a piped *weep* or *dweet*, a high-pitched *wheet-whit* and a quiet *cheep*. **H** Plains and plateaus with sparse vegetation. Winters in sandy semi-desert. **D** Ukraine to Mongolia and N China, and south to Israel, Iran and Pakistan.

14 HOODED WHEATEAR *Oenanthe monacha* 17.5cm Usually wary. In flight, shows an extensive white rump and tail. Makes prolonged aerial pursuits of flying insects. **V** Song a sweet medley of whistles and thrush-like notes. Call is a harsh *zack* or low *wit-wit*. **H** Barren cliffs and ravines in hot, arid, desolate areas. **D** E Egypt to SW Pakistan.

15 DESERT WHEATEAR *Oenanthe deserti* 14–15cm Wary; often perches in low vegetation. Usually sings from the top of a bush. In flight, shows a white rump and mainly black tail. **V** Song a descending *swee-you*, occasionally interspersed with rattles or trills. Calls include a low *chuck*, a whistled *peeeeoo* and a *trrr*. **H** Arid areas with scattered scrub. **D** From N Africa through the Middle East to the Himalayas, W China and Mongolia.

16 BLACK-EARED WHEATEAR *Oenanthe hispanica* 14.5cm Pale-throated and dark-faced forms occur. In flight, shows a white rump and white outer tail bases, the latter often white to near tail tip. **V** Sings from a perch or during display flight, a rich, cheerful warbling with some buzzy and scratchy notes and some mimicry. Calls include a clicked *zack-zack* or *chep-chep*. **H** Steppe with rocky outcrops; rocky mountainsides; arid, lightly wooded country. **D** SW Europe and NW Africa to the Middle East.

17 CYPRUS WHEATEAR *Oenanthe cypriaca* 14cm Rump, tail pattern and actions similar to those of Pied Wheatear. **V** Song a cicada-like purring, *bizz-bizz-bizz-bizz...* Call a harsh *zack-zack*. **H** As Pied Wheatear. **D** Cyprus.

18 PIED WHEATEAR *Oenanthe pleschanka* 14.5cm Forages on the ground; also makes short vertical sallies after flying insects. In flight, shows white rump and outer tail bases, with dark-edged outer-tail feathers. **V** Song is a repeated buzzy, trilling phase with much mimicry. Calls include a harsh *zack-zack* or *chep-chep* and a *psyiep*. **H** Stony lowlands with scattered small trees. **D** Romania and Ukraine to SC Russia, Mongolia and C China.

19 WHITE-FRONTED BLACK CHAT *Oenanthe albifrons* 14cm Separated from Sooty Chat by lack of white cap, and Arnot's and Ruaha chats by different wing pattern. **V** Very high series of swept-up notes, interwoven with imitations, mainly of other bird species. **H** Bare stony places in bushy and wooded areas. **D** W, C and E Africa.

20 SOMALI WHEATEAR *Oenanthe phillipsi* 14cm Grey, black and white wheatear, female slightly duller than male. Scans for insect prey from a raised perch, then flies down. **V** Calls include metallic clicks, low whistles and a double buzzing note. **H** Open stony areas, semi-desert, grassland, light bush. **D** Somalia, Ethiopia.

21 RED-RUMPED WHEATEAR *Oenanthe moesta* 16cm In flight, shows ochre/rufous rump and small rufous area on outer tail bases. Rump can wear and become very pale buff or off white. **V** Song a sweet, throaty warbling, interspersed with whistles. Calls include a dull *trrrp* and a short *k-wik*. **H** Flat, stony semi-desert or desert with scattered shrubs and an abundance of rodent burrows in which to nest. **D** N Africa to NW Saudi Arabia, Syria and W Iraq.

22 BLACKSTART *Oenanthe melanura* 15cm Active; feeds by fly-catching or dropping onto insects from a low perch. On alighting, frequently half-spreads wings and fans tail. **V** Song a repeated mellow *chee-yu-chee-yu* or *cheee-yu-chee*. Calls include a piping *cher-u*, a weak *chirp* and, in alarm, a short *tzeetch-eetch*. **H** Scrub in semi-desert, desert, foothills, wadis, isolated buildings and settlements. **D** Arabian Peninsula through E, C and W Africa.

23 FAMILIAR CHAT *Oenanthe familiaris* 15cm Feeding methods much as Blackstart. Frequently flicks wings and trembles tail. In flight, shows rufous rump and extensive rufous on outer-tail feathers. **V** Song a random series of soft whistles and quiet chattering notes. Calls include a scolding *whee-chuck-chuck* and a shrill *swiiip-swiiip-swiiip*. **H** Areas of rocky ground in scrub or light woodland. **D** Widespread across W, C, E and S Africa.

24 BROWN-TAILED ROCK CHAT *Oenanthe scotocerca* 13cm Note dark brown tail and upright stance. **V** Song a rapid series of thin, liquid, chirruping notes. Calls include a loud *chuck-chuck* and a brief, tuneful trill. **H** Dry bushy areas with rocky outcrops. **D** NE and E Africa.

25 SOMBRE ROCK CHAT *Oenanthe dubia* 14cm Note rather dark appearance. Tail and upperparts uniformly brown. Sings from a prominent perch. **V** Song a monotonous series of *chip* notes. **H** Not known, but supposed to be rocky areas with some bush. **D** Somalia, Ethiopia.

26 BROWN ROCK CHAT *Oenanthe fusca* 17cm Drops from a low perch to capture prey from the ground. Flexes legs and spreads and raises tail. **V** Song consists of a sweet thrush-like warbling. Calls include a whistled *chee*, a mournful *pseeu* and a harsh *tchk-tchk-tchk* when alarmed. **H** Low rocky hills, sandstone cliffs, old buildings. **D** Pakistan, N and C India.

27 VARIABLE WHEATEAR *Oenanthe picata* 15cm Sings from a prominent perch or in flight. In flight, shows a white rump and outer tail bases. **V** Song is scratchy, consisting of low-pitched *chott* notes, whistles, chirrups and trills. Call a loud *chek-chek*. **H** Arid boulder-covered hills with scattered vegetation. Winters in stony desert foothills, cultivated areas. **D** Turkmenistan, Uzbekistan and Kyrgyzstan to S Iran and S Pakistan.

28 BLACK WHEATEAR *Oenanthe leucura* 18cm Wary. In flight, shows a white rump and extensive white outer tail bases. **V** Song a melodious *chokereu-keu-keke*, or similar. Calls include a plaintive *pee-pee-pee* and a scolding *chak*. **H** Steep rocky areas, ravines, sea cliffs, ruins, etc., with sparse vegetation. **D** Iberian Peninsula, S France, NW Africa.

29 ABYSSINIAN WHEATEAR *Oenanthe lugubris* 15cm Similar to Mourning Wheatear but male mostly black with a grey crown and a white rump, tail sides and undertail. Female dull grey-brown with faint diffuse streaking. **V** Like that of Mourning Wheatear. **H** Rocky desert country with caves and sparse vegetation. **D** E Africa.

30 WHITE-CROWNED WHEATEAR *Oenanthe leucopyga* 17cm Usually wary. In flight, shows a white rump and extensive white on outer tail. **V** Song very variable, combining warbles, discordant notes and much mimicry of birds and mammals. Calls include a grating *dzik* and a far-carrying *hwee-weet*. **H** Some of the most desolate desert mountain areas, including ravines, wadis, rocky hill areas and around settlements. **D** N Africa to N and C Arabian Peninsula.

31 HUME'S WHEATEAR *Oenanthe albonigra* 17cm Shows an extensive white rump and outer tail bases. **V** Song is a cheerful, rising *chew-de-dew-twit*. Calls include a whistled *triki-treet* or *trooti-trooti-tree* and a harsh *chack-chack*. **H** Barren rocky slopes with sparse vegetation. **D** Iraq to Afghanistan, Pakistan and Oman.

32 FINSCH'S WHEATEAR *Oenanthe finschii* 15cm In flight, shows an extensive white rump and large white outer tail bases. On landing, cocks tail and then spreads and lowers it slowly. **V** Song, often given during display flight, a variable warble. Call a harsh *chak-chak*. **H** Rocky steppe, barren ravines and semi-desert with scattered vegetation. **D** Breeds from SC and E Turkey to Israel, Iran, Kazakhstan and W Pakistan. Winters in Cyprus, Egypt, SW Asia.

33 MOURNING WHEATEAR *Oenanthe lugens* 14cm In flight, shows a white rump and extensive white outer tail bases. **V** Song a pleasing warble, often combined with call notes. Calls include a quiet *chack-chack* and a harsh *zeeb*. **H** Gorges, rocky wadis and rolling hills with sparse vegetation in semi-arid and desert areas. **D** N Africa through Arabian Peninsula to Iran, Jordan and Syria.

34 ARABIAN WHEATEAR *Oenanthe lugentoides* 14cm Rump, tail pattern and actions similar to those of Mourning Wheatear. **V** Song comprises short bubbling phrases, also a musical *to-too-too*. Calls include a *chuck-a-do* and a rasping, repeated *kaak*, frequently combined with a high-pitched *seek*. **H** Rocky hillsides and mountainsides with sparse vegetation, less frequently in areas with bushes or thick juniper scrub near cultivation. **D** S Arabian Peninsula.

35 KURDISH WHEATEAR *Oenanthe xanthoprymna* 14.5cm In flight, shows a chestnut rump and black tail with a small white area on outer tail bases. Often perches on bushes; flicks wings and spreads tail. **V** Song a short throaty warble, *see - wat-shew - eeper* or *wee-chu-chree*. Calls include a low *chek-chek*, *zvee*, *zvee-tuk* and a soft *thrrr-thrrr-thrrr*. **H** Arid rocky hillsides and mountainsides with scattered boulders and sparse vegetation. **D** Turkey, Iran.

36 RED-TAILED WHEATEAR *Oenanthe chrysopygia* 14.5cm Wary, often running for cover. In flight, shows a chestnut rump and outer tail bases. **V** Song a loud warble with much mimicry. Call a low *thrrr-thrr-thrr*. **H** Dry rocky slopes, often with nearby streams. Winters in semi-desert areas with scattered scrub. **D** Armenia to C Iran, Tajikistan and Afghanistan.

37 BOULDER CHAT *Pinarornis plumosus* 24cm In flight, note conspicuous spotted wing-bar and white tail corners. **V** Sequence of drawn-out, very high to extremely high notes. **H** Rocks under tree cover in hilly country. **D** E Zambia and Malawi to E Botswana and S Zimbabwe.

38 HERERO CHAT *Namibornis herero* 17cm Note narrow white eyebrow and tail pattern. Hunts from a low perch, catching insects from the ground. **V** Rather oriole-like *twi-te-deelee-doo* or warbler-like variations on this phrase. **H** Dry, shrubby slopes with rocks and boulders. **D** SW Angola through W Namibia.

39 HUMBLOT'S FLYCATCHER *Humblotia flavirostris* 14cm Note diagnostic yellow bill and legs. **V** High, hurried, descending three-note *reeh-reeh-reeh*. **H** Wooded habitats with some tall trees. **D** Comoros.

1
2
3
4
5
various morphs
6
6
6
7
8
9
10
11
12
13
14
15
16
16
17
pale-throated morph
18
dark-throated morph
18
pale-throated morph
19
20
21
22
23
24
25
26
27
27
3 colour morphs
28
29
30
31
32
32
pale-throated morph
dark-throated morph
33
34
35
36
37
38
39

DIPPERS CINCLIDAE

1 WHITE-THROATED DIPPER *Cinclus cinclus* 18cm Swims and walks underwater, feeding on stream or river bottoms. Bobs whole body while perched. **V** Song a sustained, rippling warble. Calls include a loud, rasping *zink* or *zrets*. **H** Fast-flowing rivers and streams, chiefly in upland areas. Sometimes visits lower areas in winter. **D** Widespread across Eurasia, NW Africa.

2 BROWN DIPPER *Cinclus pallasii* 20cm Actions similar to those of White-throated Dipper. **V** Song short and rich, more musical than that of White-throated. Calls include a buzzing *zzit-zzit* or *dzit-dzit*. **H** Mountain streams and small lakes. **D** C and E Eurasia.

3 AMERICAN DIPPER *Cinclus mexicanus* 19cm Bobs whole body while perched. Swims and walks underwater, feeding on stream or river bottoms. **V** Song is a bubbling *k-tee k-tee wij-ij-ij treeoo treeoo tsebrr tsebrr tsebrr tsebrr...* Calls include a metallic *zeet* and a rapid *dzik-dzik*. **H** Fast-flowing rivers or streams. **D** W Canada to Panama.

4 WHITE-CAPPED DIPPER *Cinclus leucocephalus* 15.5cm White crown is diagnostic. **V** Call is a *dzit* sound. **H** Rocky mountain streams and rivers. **D** Colombia to Bolivia.

5 RUFOUS-THROATED DIPPER *Cinclus schulzii* 15.5cm Unmistakable due to habitat and throat patch. **V** Rapid series of 3–4 *tuweet* notes. **H** Rocky mountain streams and rivers. **D** NW Argentina, SE Bolivia.

LEAFBIRDS CHLOROPSEIDAE

6 PHILIPPINE LEAFBIRD *Chloropsis flavipennis* 18–19cm Forages in the canopy, singly or in pairs. Usually difficult to see; best located by calls. **V** A whistled *chick weeeeep* and a whistled *chick-ur-treet*. **H** Lowland evergreen forest and secondary growth, forest edges. **D** Philippines.

7 YELLOW-THROATED LEAFBIRD *Chloropsis palawanensis* 15–17cm Hard to see; best located by calls. Forages in the canopy, singly or as part of a mixed-species feeding party. **V** A warbled *zo-o zo-o* and a rhythmic *twick err treet*. **H** Lowland evergreen forest edges, secondary growth. **D** Philippines.

8 GREATER GREEN LEAFBIRD *Chloropsis sonnerati* 20–23cm Frequents the middle to upper storey, singly or in pairs; sometimes forms part of mixed-species feeding parties. **V** Musical whistles, interspersed with brief chattering notes; also indulges in mimicry. **H** Lowland evergreen forest, peat-swamp forest, well-grown secondary forest, occasionally tall mangroves. **D** Thailand to Sumatra (Indonesia) and Borneo.

9 LESSER GREEN LEAFBIRD *Chloropsis cyanopogon* 16–19cm Forages in the canopy, singly or in pairs; often a member of mixed-species feeding flocks. **V** Loud, varied, sequence of rich, warbling phases, including deep, mellow notes. **H** Broadleaved evergreen forest, open forest, forest edges. **D** Myanmar to Sumatra (Indonesia) and Borneo.

10 BLUE-WINGED LEAFBIRD *Chloropsis cochinchinensis* 16–18cm Acrobatic searcher of insects, fruit and nectar; often a member of mixed-species feeding parties. **V** Various sweet, musical notes; also mimics other species. **H** Deciduous and broadleaved evergreen forests, forest edges, secondary growth. **D** Widespread from NW India to S China, Sumatra and Java (Indonesia), and Borneo.

11 BORNEAN LEAFBIRD *Chloropsis kinabaluensis* 17cm Forages alone, in pairs or in small groups; feeds in the canopy of fruiting trees. **V** High-pitched twittering and a rapidly repeated *chit chit chit*. **H** Mature and well-regenerated lower and upper montane forest. **D** Borneo.

12 JERDON'S LEAFBIRD *Chloropsis jerdoni* 20cm Forages in trees, acrobatically searching for insects, fruit and nectar; often part of mixed-species feeding flocks. **V** Random, varied combination of whistles, buzzing and rich, sharp notes. **H** Open forests, secondary growth, orchards, wooded gardens. **D** India, Sri Lanka.

13 GOLDEN-FRONTED LEAFBIRD *Chloropsis aurifrons* 19cm Acrobatic forager in the thick foliage of trees. **V** A musical *swich-chich-chich-weee*; also a repeated *tzik* and a *chup-chaw*. Mimics the calls of other species. **H** Broadleaved forests, secondary growth. **D** Widespread from E Himalayas to SW China, SE Asia and south to Sumatra, Indonesia.

14 SUMATRAN LEAFBIRD *Chloropsis media* 17–19cm Reported feeding on figs in the tree canopy, otherwise little recorded information. **V** Songs of captive birds recorded as loud and richly melodic. **H** Lowland evergreen and secondary forest, plantations, orchards. **D** Sumatra, Indonesia.

15 ORANGE-BELLIED LEAFBIRD *Chloropsis hardwickii* 20cm Forages in tree foliage, acrobatically probing flowers for nectar and leaves for insects. **V** Said to have the sweetest song of all the leafbirds. Various ringing and melodious calls recorded, including a soft *tilu-tilu-tilu-tilu-ti*, a low *tp-tp-tp-tp-tp*, a *tshiwatshishi-watshishi-watshishi* and a loud, rapid, repeated *ti-ti-tsyi*. **H** Broadleaved forest. **D** Himalayas to SE China and W Malaysia.

16 BLUE-MASKED LEAFBIRD *Chloropsis venusta* 14cm Forages in the canopy, usually in pairs, although also recorded in small groups. **V** Undescribed. **H** Mature lowland and lower montane forest, forest edges and clearings. **D** Sumatra, Indonesia.

FLOWERPECKERS DICAEIDAE

17 OLIVE-BACKED FLOWERPECKER *Prionochilus olivaceus* 9cm Forages in the understorey, especially in flowering and fruiting trees. Often joins mixed-species feeding parties. **V** A repeated, loud, precise *peeit* or *peeith*, a high-pitched *tsoo-eet* and a rattling trill. **H** Forest, forest edges, secondary growth. **D** Philippines.

18 YELLOW-BREASTED FLOWERPECKER *Prionochilus maculatus* 10cm Forages in the middle to upper storey, in pairs or alone; favours flowering and fruiting trees. **V** A high-pitched *tswik*, a hoarse *tsweet-tsweet* and harsh, metallic, chittering calls. **H** Lowland and hill dipterocarp, peat-swamp and secondary forest, forest edges, scrub. **D** Malay Peninsula, Borneo.

19 CRIMSON-BREASTED FLOWERPECKER *Prionochilus percussus* 10cm Forages in the middle and lower storeys, especially in fruiting and flowering trees. **V** A *see-sik* and a fast *weg*. **H** Primary, swamp and secondary forest, forest edges, old plantations, scrub. **D** Malay Peninsula; Sumatra and Java, Indonesia; Borneo.

20 PALAWAN FLOWERPECKER *Prionochilus plateni* 9cm Forages at all levels, alone or occasionally in mixed-species flocks, usually in flowering and fruiting trees. **V** A repeated, high-pitched, metallic *seep-seep*. **H** Forest, secondary growth, scrub, gardens. **D** Philippines.

21 YELLOW-RUMPED FLOWERPECKER *Prionochilus xanthopygius* 9cm Forages at all levels; favours fruiting and flowering trees. **V** Calls include a *tsee-oo, ship-ship, ship-ship-ship* and *tsik-tsik*; also utters 7–9 descending notes and, in flight, a high-pitched chittering. **H** Dipterocarp, peat-swamp, heath and secondary forest, forest edges, clearings, plantations. **D** Borneo.

22 SCARLET-BREASTED FLOWERPECKER *Prionochilus thoracicus* 9–10cm Forages at all levels; sometimes climbs tree trunks like a nuthatch. **V** A metallic, clicking twitter, a very high-pitched, insect-like *seek* and a harsh *chink*. **H** Forest, forest edges, secondary growth, coastal vegetation. **D** Malay Peninsula, Borneo.

23 GOLDEN-RUMPED FLOWERPECKER *Dicaeum annae* 9–10cm Favours the nectar, fruit and pollen of mistletoes; occurs singly, in pairs or in small groups. **V** A repeated, thin, level-pitched *see-see-see-see seeee seeee seeee*. **H** Deciduous, semi-evergreen and secondary forest, woodland, cultivation. **D** Lesser Sunda Islands.

24 THICK-BILLED FLOWERPECKER *Dicaeum agile* 10cm Restless; habitually twitches tail from side to side. Attracted to flowering or fruiting trees and shrubs, especially if infested with mistletoe. **V** Song comprises 6–8 notes of differing pitches mixed with dry trills. Calls include a sharp *chik-chik-chik-chik*, a rattling *titititili* and a high-pitched *chit-chit*. **H** Dry or moist deciduous or evergreen forest, orchards, groves, gardens. **D** Widespread from NE Pakistan to SW China, SE Asia and south to Borneo and the Lesser Sunda Islands.

25 STRIPED FLOWERPECKER *Dicaeum aeruginosum* 10–11cm Forages in the canopy, mainly in fruiting and flowering trees, singly or in single-species or mixed-species groups. Wags tail from side to side. **V** Unrecorded. **H** Forest, forest edges, secondary growth. **D** Philippines.

26 BROWN-BACKED FLOWERPECKER *Dicaeum everetti* 10cm Forages at all levels, feeding on spiders, insects and flowers, and possibly also nectar and fruits. **V** A sharp, metallic *chip-chip*. **H** Coastal keranga and secondary forest, forest edges. **D** Peninsular Malaysia, Borneo.

27 WHISKERED FLOWERPECKER *Dicaeum proprium* 9cm Forages in flowering and fruiting trees, especially those with mistletoe flowers and fruits. **V** Various raspy, snappy, insect-like notes; occasionally a series of notes such as *zaach-zee-peew*; and high, buzzy notes that sometimes run into a 'song'. **H** Forest, forest edges, secondary growth. **D** Philippines.

28 YELLOW-VENTED FLOWERPECKER *Dicaeum chrysorrheum* 10cm Active forager at all levels of vegetation; attracted to mistletoes and small figs. **V** Calls include a *zeet*, a repeated *chip-a-chip-tree* and a *zit-zit-zit*; also utters various soft squeaks. **H** Open jungle, forest edges, orchards. **D** Himalayas to Borneo.

29 YELLOW-BELLIED FLOWERPECKER *Dicaeum melanoxanthum* 11–13cm Elusive. Noted making fly-catching sallies from dead branches; often sits upright. **V** A harsh, agitated *zit-zit-zit-zit*. **H** Pine forest, tall trees in open forest clearings in dense forest, forest edges. **D** Himalayas to Thailand.

30 LEGGE'S FLOWERPECKER *Dicaeum vincens* 10cm Acrobatic forager, mainly in treetops. **V** Song said to consist of very high-pitched notes alternating in pitch. Calls include a *tchip tchip-twee-see*, a high *tee-too* and a *wheep-wheep-wheep*. **H** Tall trees and creepers in rainforest, dry forest, plantations, occasionally gardens. **D** Sri Lanka.

31 YELLOW-SIDED FLOWERPECKER *Dicaeum aureolimbatum* 8–9cm Forages in pairs, singly or in small groups in flowering or fruiting trees. **V** Calls include a *s-uit*, 5–6 dry, staccato *tuk* notes and a sharp, clear *zit-zit-zit...* **H** Primary and tall secondary forest, forest edges, woodland, plantations, scrub. **D** Sulawesi, Indonesia.

32 OLIVE-CAPPED FLOWERPECKER *Dicaeum nigrilore* 9–10cm Favours flowering and fruiting trees and mistletoes. **V** A high-pitched, scratchy *zuti-zuti-zuti-zuti...*, repeated up to 15 times, a high *tseep-tseep* and a rapid rising and falling, high-pitched trill. **H** Submontane and montane forest. **D** S Philippines.

1
2
3
4
5
6
7
8
9
10
11
12
13
14
15
16
17
18
19
20
21
22
23
24
25
26
27
28
29
30
31
32

FLOWERPECKERS *CONTINUED*

1 FLAME-CROWNED FLOWERPECKER *Dicaeum anthonyi* 9–10cm Forages at all levels, especially in flowering and fruiting trees. Occurs singly, in pairs and in mixed-species feeding parties. **V** High-pitched, sharp *srrreep*. **H** Mossy forest, forest edges. **D** Philippines.

2 BICOLORED FLOWERPECKER *Dicaeum bicolor* 9cm Forages in the canopy, alone, in pairs, in small parties or in mixed-species flocks. Favours flowering and fruiting trees. **V** A *swip-swip*... that develops into a rapid trill and gets lower and slower. **H** Forest, forest edges, secondary growth. **D** Philippines.

3 RED-KEELED FLOWERPECKER *Dicaeum australe* 10cm Forages mainly in the canopy, especially in fruiting and flowering trees and mistletoes. Occurs in small groups, in pairs and alone; also joins mixed-species feeding flocks. **V** High-pitched, insect-like, trilling *suit-sui*... and a high, rising and falling *tik-tik* that evolves into a trill. **H** Forest, forest edges, secondary growth, coconut groves, fruiting shrubs in open country. **D** Philippines.

4 BLACK-BELTED FLOWERPECKER *Dicaeum haematostictum* 10cm Forages alone or in pairs, small flocks or mixed-species feeding parties; favours fruiting and flowering trees and mistletoes. **V** A *seet-seet* that occasionally develops into a trill, a *chip*, rapid tinkling notes and a *chip-seet-seet* followed by a short trill. **H** Forest, forest edges, secondary growth, cultivation, coconut groves, fruiting bushes in open country. **D** Philippines.

5 SCARLET-COLLARED FLOWERPECKER *Dicaeum retrocinctum* 10cm Favours flowering mistletoes and flowering and fruiting trees, generally in the higher levels; seen alone, in pairs or in small flocks, and also joins mixed-species flocks. **V** Continuous *tipk-tipk-tipk-tipk* or *tip-chik zeet zeet zeet*. **H** Closed-canopy forest, forest edges, secondary growth, coconut groves, cultivation, fruiting bushes in open country. **D** Philippines.

6 CEBU FLOWERPECKER *Dicaeum quadricolor* 9cm Attracted to flowering mistletoes, trees and vines, and fruiting trees. **V** A *seep-seep-seep tik tik tik tik*, a series of *tsip-tsip* or *trik-trik* notes that occasionally develops into a trill, a high insect-like *see-ip* and a *sit-sit-sit*. **H** Open and closed-canopy forest. **D** Philippines.

7 ORANGE-BELLIED FLOWERPECKER *Dicaeum trigonostigma* 9cm Forages mainly in the tops of trees, especially those in flower or fruit. **V** Song a rising series of rapid, upslurred notes. Calls include a *swit* or *swit-szee*, a drawn-out *zeeee*, a series of wheezy and twittering notes, a *zit-zit-zit*, a *chik-chik-chik* and a high-pitched *zeeeep-zeeeep*. **H** Forest glades and edges, secondary growth, gardens. **D** Bangladesh to SE Asia, Borneo and the Philippines.

8 BUZZING FLOWERPECKER *Dicaeum hypoleucum* 8–9cm Forages at all levels, feeding on fruits, nectar and pollen. Occurs singly, in pairs, in small parties or as part of mixed-species flocks. **V** High-pitched, buzzing *bzeeeppp*, uttered singly or as a series, occasionally followed by a trilled *cheenjet*; also utters a metallic *chimp chimp*, which develops into a trill. **H** Forest, forest edges, cultivation, scrub. **D** Philippines.

9 PALE-BILLED FLOWERPECKER *Dicaeum erythrorhynchos* 8cm Restless; usually keeps to the tops of trees. Attracted to mistletoe fruits. **V** Song a series of twittering notes or a reel. Calls include a repeated high-pitched *pit* and a sharp *chik-chik-chik*. **H** Deciduous forest, plantations, groves, orchards, sometimes mangroves. **D** Himalayas to Sri Lanka and Myanmar.

10 NILGIRI FLOWERPECKER *Dicaeum concolor* 8–9cm Active, agile forager; especially attracted to mistletoes. **V** Song a high-pitched trill. Calls include twitterings and a *tik-tik-tik*. **H** Broadleaved forests, forest edges, well-wooded areas. **D** SW India.

11 PLAIN FLOWERPECKER *Dicaeum minullum* 7–9cm Active, agile forager at all levels; attracted to flowers and fruits of mistletoes. **V** Song a high-pitched trill. Calls include twitterings, a sharp *chek*, a short, ticking *chrik* and a *tik-tik-tik*. **H** Open broadleaved evergreen, semi-evergreen and deciduous forest, secondary growth. **D** E Himalayas to Taiwan, W Indonesia.

12 ANDAMAN FLOWERPECKER *Dicaeum virescens* 8–9cm Very similar to Plain Flowerpecker in appearance and behaviour, but brighter green. **V** Like that of Plain Flowerpecker. **H** Broadleaved evergreen, semi-evergreen and deciduous forest, secondary growth, plantations and other cultivation. **D** Andaman Islands.

13 PYGMY FLOWERPECKER *Dicaeum pygmaeum* 8cm Forages high in the understorey or in the canopy among flowering mistletoes; encountered singly, in small groups or in mixed-species flocks. **V** Loud, sharp, irregularly spaced *tip tip*..., repeated almost continuously; also a high *schenk-schenk*... and a *zip-zip*... *zip-zip*... *zip-zip*. **H** Forest, forest edges, secondary growth. **D** Philippines.

14 CRIMSON-CROWNED FLOWERPECKER *Dicaeum nehrkorni* 8–9cm Forages mainly in the canopy in fruiting trees, singly or in pairs, and as part of mixed-species feeding flocks. **V** A sharp *zit-zit*, a repeated, hard *tit* and a high-pitched insect-like trill. **H** Hill and montane forest, forest edges. **D** Sulawesi, Indonesia.

15 FLAME-BREASTED FLOWERPECKER *Dicaeum erythrothorax* 9cm Occurs in small parties or alone, foraging in the canopy in fruiting trees. **V** Repeated *tcheep tcheep tcheep*. **H** Lowland forest and scrub in riverine areas. **D** Buru, S Maluku Islands, Indonesia.

16 HALMAHERA FLOWERPECKER *Dicaeum schistaceiceps* 9cm Formerly considered a race of the previous species; habits and actions presumed to be similar. **V** Presumed to be similar to that of the previous species. **H** Primary, degraded, open and secondary forest, riverine forest edges, scrub, cultivation with trees. **D** Halmahera, N Maluku Islands, Indonesia.

17 ASHY FLOWERPECKER *Dicaeum vulneratum* 8–9cm Forages from low down in shrubs to the treetops, singly, in pairs or as a member of mixed-species flocks. **V** Song consists of high-pitched, metallic disyllabic or trisyllabic notes. Call a hard, staccato *tst*. **H** Hill and coastal forest, forest edges, plantation edges, old gardens. **D** Maluku Islands, Indonesia.

18 OLIVE-CROWNED FLOWERPECKER *Dicaeum pectorale* 9cm Light olive-grey flowerpecker. Male has a red breast-patch. **V** Short buzzing note, recalling an insect, also a drawn-out *chew*. **H** Canopy of primary forest and secondary growth. **D** NW New Guinea.

19 RED-CAPPED FLOWERPECKER *Dicaeum geelvinkianum* 8–9cm Both sexes have an orange-red crown and rump; male also has an orange-red breast patch. **V** Call a buzzing, downslurred *bszzt*. **H** Canopy and edges of forest, secondary growth, plantation edges, old gardens. **D** New Guinea.

20 LOUISIADE FLOWERPECKER *Dicaeum nitidum* 9cm Similar to Red-capped Flowerpecker but has a more silvery tone to back, wings and tail. Forages alone or in pairs. **V** Calls include buzzing insect-like sounds and an upslurred high-pitched note. **H** Forest and secondary growth. **D** Louisiade Archipelago, SE Papua New Guinea.

21 RED-BANDED FLOWERPECKER *Dicaeum eximium* 8–9.5cm Upperparts dull grey-brown apart from red rump, underparts whitish; male has a small red breast patch. **V** Calls include a *tzick* and various squeaky notes. **H** Forest and forest edges away from coasts and mountains, larger gardens. **D** Bismarck Archipelago.

22 MIDGET FLOWERPECKER *Dicaeum aeneum* 8cm Small slate-grey flowerpecker with whitish underparts flushed yellowish on flanks; male has a large red breast patch. Forages in canopy, often hovering. **V** Calls varied, including a fast *tiktiktik* and very high-pitched chips and whistles. **H** All kinds of forest and scrub habitats. **D** Solomon Islands.

23 MOTTLED FLOWERPECKER *Dicaeum tristrami* 8–9cm Brown flowerpecker with a speckled head and light greyish underparts. Male has a dark head, female has a white eyebrow and sides of cheeks. Searches for insects by hovering close to foliage. **V** High-pitched, abrupt *tschip-tschip*. **H** Primary forest, secondary growth, forest edges, mainly in lowlands. **D** Solomon Islands.

24 BLACK-FRONTED FLOWERPECKER *Dicaeum igniferum* 9cm Occurs singly, in pairs, in small groups or as part of mixed-species flocks. Favours flowering trees. **V** Song comprises a very rapid, descending series of short, thin, dry notes. Call a rapidly repeated, high-pitched *see-saw*. **H** Semi-deciduous, degraded or secondary forest, coastal monsoon scrub forest, lightly wooded cultivation, clearings. **D** Lesser Sunda Islands.

25 BLUE-CHEEKED FLOWERPECKER *Dicaeum maugei* 8–10cm Forages alone or in pairs; often seen around mistletoes and in fruiting trees. **V** A high-pitched *tsit* and a high-pitched 2–3-note whistle. **H** Primary, secondary and degraded forest, woodland, plantations, wooded cultivation, bamboo. **D** Lesser Sunda Islands.

26 MISTLETOEBIRD *Dicaeum hirundinaceum* 9cm Especially fond of mistletoe; forages from low down up to the canopy. Generally in pairs or alone; occasionally joins mixed-species flocks. **V** A dry *tick* note and sibilant calls. **H** Forest, forest edges, mangroves, village trees. **D** Maluku Islands, Indonesia; Lesser Sunda Islands; Australia.

27 GREY-SIDED FLOWERPECKER *Dicaeum celebicum* 8–10cm Favours mistletoe in the canopy; forages alone or in pairs. **V** Calls include a thin, upslurred *seeei*, a repeated *tjjit*, a dry *trri-tri* and a high-pitched *tsip*; in flight, gives a sharp *chip chip chip*... **H** Primary and tall secondary forest, forest edges, lightly wooded cultivation, gardens. **D** Sulawesi and nearby islands, Indonesia.

28 BLACK-SIDED FLOWERPECKER *Dicaeum monticolum* 8cm Forages at low levels and sometimes visits the canopy; especially fond of mistletoe fruits. **V** A sharp, piercing, metallic *zit*, a repeated *tit*, a rapid ticking and a *tsweet-tsweet*. **H** Hill dipterocarp, montane and heath forest, scrub. **D** Borneo.

29 FIRE-BREASTED FLOWERPECKER *Dicaeum ignipectus* 9cm Restless, agile forager, mainly in treetops; attracted to mistletoes. **V** Song a high-pitched *titty-titty-titty* or *see-bit see-bit see-bit see-bit*. Calls include a buzzing *zeeep*, a metallic *chip* and a rattling trill. **H** Montane and hill forests. **D** Himalayas to Sumatra (Indonesia), Taiwan and the Philippines.

30 BLOOD-BREASTED FLOWERPECKER *Dicaeum sanguinolentum* 8–10cm Favours mistletoe in the canopy; forages alone or in pairs. **V** Song consists of 4–5 jerky, thin, high-pitched, sweet notes. Calls include high-pitched clicks and a hard, sharp, buzzing double note. **H** Montane, hill and secondary forest, forest edges, *Casuarina* groves, open and degraded woodland, lightly wooded cultivation. **D** Java, Indonesia; Lesser Sunda Islands.

31 SCARLET-BACKED FLOWERPECKER *Dicaeum cruentatum* 9cm Usually forages in the tops of trees among clumps of mistletoes. **V** Song includes a rising and falling *see-sip-see-sip-see-sip* and a ringing *chipi-chipi-chipi dzee-dzee-dzee*. Calls include various twitterings, a *chip-chip*, a clicking *tchik-tchik-tchik* and a high-pitched *chizee*. **H** Open forests, forest edges, secondary forest, mangroves. **D** Himalayas to Sumatra (Indonesia) and Borneo.

32 SCARLET-HEADED FLOWERPECKER *Dicaeum trochileum* 8–9cm Forages alone or in pairs, especially around mistletoe at upper tree levels. **V** Song contains sweet, high-pitched, rising and falling double notes. Calls include a short, high-pitched *zit-zit-zit* and a buzzing *seeeeep*... *seeeeep*. **H** Woodland, mangroves, cultivation, open scrubby areas, gardens. **D** SE Sumatra, Java, Bali and Lombok, Indonesia; Borneo.

1
2
3
4
5
6
7
8
9
10
11
12
13
14
15
16
17
18
19
20
21
22
23
24
25
26
27
28
29
30
31
32

SUNBIRDS NECTARINIIDAE

1 RUBY-CHEEKED SUNBIRD *Chalcoparia singalensis* 10–11cm Forages mainly in the upper storey, probing flowers for nectar and gleaning from leaves. **V** Song a shrill, trilled *tirr-tititirr tir tir* and a rapid, high *switi-ti-chi-chu tusi-tit swit-swit*... Calls include a shrill *seet-seet*, a *tweest-wit* and a soft *chi-wip*. **H** Forest, secondary growth, coastal scrub, mangroves. **D** Nepal to Java (Indonesia) and Borneo.

2 FRASER'S SUNBIRD *Deleornis fraseri* 11cm Note narrow eye-ring. Male separated from all other greenish male and female sunbirds by (often concealed) scarlet tufts. **V** Song a series of high-pitched *tsserr-tseep* notes, repeated every sec. Call a squeak. **H** Forest canopy. **D** W and WC Africa.

3 GREY-HEADED SUNBIRD *Deleornis axillaris* 11cm Note grey head. **V** Long trill of rising and falling notes. **H** Mid-strata of forests. **D** E DR Congo, Uganda, Rwanda.

4 PLAIN-BACKED SUNBIRD *Anthreptes reichenowi* 10cm Female separated from Fraser's and Little Green sunbirds, and Bates's Sunbird (Plate 259), by different range. No tufts. **V** Very high, lowered, Willow Warbler-like song. **H** Forest canopy. **D** E and SE Africa.

5 ANCHIETA'S SUNBIRD *Anthreptes anchietae* 10cm No other sunbird in its range has a similar colour pattern. **V** Song a series of complex notes, these repeated up to 40 times. **H** Miombo. **D** Angola to Tanzania and Mozambique.

6 PLAIN SUNBIRD *Anthreptes simplex* 13cm Forages much like a leaf warbler, gleaning from leaves. **V** Metallic chips and trills, and a high-pitched *seep*. **H** Various forests, forest edges, secondary forest and scrub, including coastal scrub and mangroves. **D** Malay Peninsula; Sumatra, Indonesia; Borneo.

7 BROWN-THROATED SUNBIRD *Anthreptes malacensis* 14cm Forages singly or in pairs, mainly in the canopy but also at all other levels; gleans from leaves and twigs. **V** Song a *sweet-sweet, swit-swit-sweet* or *wee-chew-chew-wee*. Calls include a much-repeated *kelichap*, a hard *chip*, a drawn-out, high-pitched *siiewei* and a shrill *whiiu*. **H** Forest edges, mangroves, freshwater swamp forest, secondary growth, coastal scrub, plantations, gardens. **D** Myanmar to Java and Sulawesi, Indonesia; Lesser Sunda Islands and the Philippines.

8 GREY-THROATED SUNBIRD *Anthreptes griseigularis* 12–13cm Formerly regarded as conspecific with the previous species; actions presumed to be similar. **V** Unknown. **H** Forest, mangroves, plantations, scrub, groves, gardens. **D** N and SE Philippines.

9 RED-THROATED SUNBIRD *Anthreptes rhodolaemus* 12cm Gleans from foliage, usually in the canopy but also at lower levels. **V** Song a high-pitched *sit-sit-sit-see* or slurred *sit-sit-sit-swe-er*. Calls include various chirps and trills. **H** Forest, forest edges, secondary growth, plantations, coastal vegetation. **D** Malay Peninsula; Sumatra, Indonesia; Borneo.

10 MANGROVE SUNBIRD *Anthreptes gabonicus* 10cm Note white in face. **V** Song a sequence of twitters, e.g. *tser-tser-tsew-tsi-tsi-tsi-tsi-tsi-tsi-tseuuur*. Call a quiet *tserr*, *wit-wit-sqee-witter-witter* or *tsurp-tseep-tseep*. **H** Mangroves, tree belts along rivers, swamp thickets. **D** Senegal and Gambia to Cameroon and south to Angola.

11 WESTERN VIOLET-BACKED SUNBIRD *Anthreptes longuemarei* 13cm Male separated from male Uluguru Violet-backed Sunbird by more reddish-purple colouring above, some bluish reflections on rump and whiter flanks. Note white eyebrow of female. **V** High, unstructured, reed warbler-like twittering. **H** Miombo and other woodland. **D** W, C, E and SC Africa.

12 EASTERN VIOLET-BACKED SUNBIRD *Anthreptes orientalis* 12cm Mid-purplish above, with white belly. Note green reflections in rump. **V** Song a series of rising and falling twitters, *too-wit-woo-tweu, zeet-zeet*. Call one or two *chwee* notes. **H** Dry, more or less wooded and bushy areas. **D** Ethiopia and C Somalia to South Sudan, Rwanda and S Tanzania.

13 ULUGURU VIOLET-BACKED SUNBIRD *Anthreptes neglectus* 13cm Note grey flanks and belly of male. Female resembles male but with a white chin. **V** Call a *tsssp* or *sweep-sweep-sweep*, repeated up to four times. **H** Coastal forest. **D** SE Kenya to EC Mozambique.

14 VIOLET-TAILED SUNBIRD *Anthreptes aurantius* 11cm Note green wings. **V** Little known. Flight call high-pitched. **H** Wooded and overgrown streams. **D** Cameroon to E DR Congo and south to N Angola.

15 LITTLE GREEN SUNBIRD *Anthreptes seimundi* 9cm Separated from Bates's Sunbird (Plate 259) by paler underparts. Note narrow yellowish eye-ring. **V** Squeaky *twip twip twip*, rapidly repeated up to eight times; also a *pse-ee* or *pss-upp*, repeated every sec. **H** Open forest. **D** W and C Africa.

16 GREY-CHINNED SUNBIRD *Anthreptes rectirostris* 9cm Separated from Collared Sunbird by whitish chin and paler belly. **V** Calls include a high-pitched *tsi-tseeet* (last note rising in pitch), repeated every 1–2 secs. **H** Forest canopy. **D** W and C Africa.

17 BANDED GREEN SUNBIRD *Anthreptes rubritorques* 9cm Note whitish (not green) throat. **V** Song by male from high in tree, *chip* or *teuu*, repeated every sec for up to a min. Calls include a *thk-eeer* and far-carrying *shwerp*. **H** Canopy at forest edges. **D** Tanzania.

18 COLLARED SUNBIRD *Hedydipna collaris* 10cm Unmistakable, but see Pygmy Sunbird. Note green rump and short bill. **V** Very high, fast *tu-ju-ti-ju-ti-ju-ti-ju* and other unstructured whistles and twitters. **H** Forest, riverine belts, coastal bush, suburbs. **D** W, C, E and S Africa.

19 PYGMY SUNBIRD *Hedydipna platura* 13cm In non-breeding plumage (without tail streamers), separated from Variable Sunbird (Plate 259) and Collared Sunbird by all-green (not purple-banded) chest and purple (not blue or green) rump. **V** Song 3–4 secs long, comprising varied phrases. Calls include *cheek-cheek, twee-weet, tsuup-tsuup-tsuup* and *tsei*. **H** Dry, bushy and wooded areas. **D** Mauritania to Guinea Bissau and east to Sudan, South Sudan and Uganda.

20 NILE VALLEY SUNBIRD *Hedydipna metallica* 16cm Actions as for Pygmy Sunbird. Often in small parties in non-breeding season. **V** High-pitched warble, *pruiit-prruiit-pruiit-tiririririri-tiririri*. Call a grating *pee* or *pee-ee*. **H** Scrub (usually acacia), gardens. **D** Egypt to South Sudan, Ethiopia and Somalia; SW Arabian Peninsula.

21 AMANI SUNBIRD *Hedydipna pallidigaster* 8cm Note bottle-green upperparts, bluish-green head and purple breast-band. **V** Extra-high, fast, amusical twitter. **H** Forest canopies. **D** Tanzania, SE Kenya.

22 PURPLE-NAPED SUNBIRD *Hypogramma hypogrammicum* 14–15cm Forages mainly in the understorey, often flicking and fanning tail. **V** Song a high, strong *sweet sweet sweet sweet*. Calls include a strident *schewp, tsit-tsit, tchu* or *chip*. **H** Broadleaved evergreen forest, freshwater swamp forest, secondary growth. **D** Myanmar to Sumatra (Indonesia) and Borneo.

23 REICHENBACH'S SUNBIRD *Anabathmis reichenbachii* 11cm Note yellow undertail coverts. Tail graduated, each feather with pale tips. May show yellow tufts. **V** Song comprises high-pitched trills and whistles, delivered from a high perch. Calls include a *chirr-up*. **H** Forest edges, cultivation, gardens; prefers habitats along rivers. **D** Liberia to E DR Congo and south to N Angola.

24 PRINCIPE SUNBIRD *Anabathmis hartlaubii* 14cm The only sunbird on Príncipe. **V** Song is repetitions of *peek-oo-wee, peek-oo-wee* or *pstes-to-tchou, pstes-to-tchou*. Calls include *dreee* and *wee wee*. **H** Forest, woodland, plantations, gardens. **D** Príncipe, Gulf of Guinea.

25 NEWTON'S SUNBIRD *Anabathmis newtonii* 9cm The only small sunbird on São Tomé. **V** Song a 2-min burst of ascending and descending high-pitched notes of varied sounds. Calls similar to song notes, including a *tsee-ee, jit-jit-jit* and *bink*. **H** Forest, woodland, plantations, gardens. **D** São Tomé, Gulf of Guinea.

26 GIANT SUNBIRD *Dreptes thomensis* 18cm Unmistakable. **V** Song a series of three notes each 1 sec long, *chee-cheep-eeep, tsweet-chut-uu* or *huet-tsip-tsuit*. Call a loud *cheep*. **H** Forest and immediate surroundings. **D** São Tomé, Gulf of Guinea.

27 ORANGE-BREASTED SUNBIRD *Anthobaphes violacea* 14cm Female differs from female *Cinnyris* sunbirds by strong orange wash on underparts. **V** High, soft, fast, up-and-down, chattered song phrase. **H** Coastal hillsides. **D** South Africa.

28 GREEN-HEADED SUNBIRD *Cyanomitra verticalis* 14cm Male separated from male Bannerman's Sunbird by more brilliant reflections, especially on throat and chin. Female separated from female Bannerman's by green (not blue) cap. **V** Song a repetition of *tsk* every 1–5 secs, then a rapid burst of 5–11 *tse* notes. Calls include a single *tsk*, also a plaintive *chi-u-wee*. **H** Forest, bamboo, bushveld, gardens, cultivation. **D** W and C Africa.

29 BANNERMAN'S SUNBIRD *Cyanomitra bannermani* 14cm Resembles Green-headed Sunbird, but see that species; also has a shorter bill and ranges don't overlap. **V** Song a sequence of *chuk* notes followed by a wheezing *purz-urr-wee*. Call a *chuk*. **H** Forest and forest remains, especially along streams. **D** Angola, S DR Congo, N Zambia.

30 BLUE-THROATED BROWN SUNBIRD *Cyanomitra cyanolaema* 14cm Note brown upperparts and blue restricted to forehead and throat. Male may show yellow tufts. Female shows diagnostic white lines above and below eye. **V** Song an extended series of descending trills and twitters involving *ptssew* and *psit* notes. Calls include a *tschuk* and a *tsit*. **H** Forest canopy. **D** W and C Africa.

31 CAMEROON SUNBIRD *Cyanomitra oritis* 12cm Separated from other similar sunbirds in its range by yellowish (not grey) underparts. **V** Song consists of bursts of metallic notes and a quiet warble interspersed with *tsi* notes, each burst lasting 2–4 secs. Calls include repetitive ticking and a *pseep*. **H** Forest, especially along streams. **D** SE Nigeria; W Cameroon; Bioko, Gulf of Guinea.

32 BLUE-HEADED SUNBIRD *Cyanomitra alinae* 14cm Note saffron mantle of male and female, and dark underparts of male. **V** Song a quiet twittering. Calls include a rising and falling sequence ending with distinctive *tcii tcii tcii yehu*. **H** Montane forest canopies. **D** EC Africa.

33 OLIVE SUNBIRD *Cyanomitra olivacea* 14cm Separated from Fraser's and Little Green sunbirds, and Bates's Sunbird (Plate 259), by olive (not green) upperparts and much longer bill. Southern birds (not shown) have orange wash on throat and breast. **V** Song a repetition of rising and falling whistling notes. Calls include a scolding *tsk-tsk-tsk* and *tsick*. **H** Forest, woodland, riverine belts, gardens. **D** Widespread across W, E and S Africa.

34 GREY SUNBIRD *Cyanomitra veroxii* 12cm May show red pectoral tufts. **V** Rather staccato, warbler-like *twee-twee-twee-tritri*, starting very high and then gradually lowered. **H** Undergrowth of coastal wood- and bushland, mangroves, gardens. **D** E, SE and S Africa.

1
2
3
4
5
6
7
8
9
10
11
12
13
14
15
16
17
18
19
20
21
22
23
24
25
26
27
28
29
30
31
32
33
34

SUNBIRDS *CONTINUED*

1 BUFF-THROATED SUNBIRD *Chalcomitra adelberti* 11cm Unmistakable. **V** Song a complex twittering. Calls include a *chip, tseep* and *pitchew*. **H** Forest and adjacent areas, cultivated areas with trees. **D** W Africa.

2 CARMELITE SUNBIRD *Chalcomitra fuliginosa* 13cm Looks all black in the field. Rump non-reflecting. Differs from Amethyst Sunbird by range. Female separated from female Green-throated Sunbird by absence of visible eyebrow. **V** Song a 0.5-sec *pttrreeeee*, repeated every 2–3 secs. Call a *tsit*. **H** Open wooded country, cultivation, gardens. **D** W Africa.

3 GREEN-THROATED SUNBIRD *Chalcomitra rubescens* 13cm Note dense streaking on female underparts, standing out against pale background. **V** Male song complex, delivered from a vantage point. Call a typical *chip*. **H** Open forest. **D** W and SW Africa.

4 AMETHYST SUNBIRD *Chalcomitra amethystina* 14cm Looks all black in the field with, occasionally, a small reflecting spot. Male separated from male Green-throated Sunbird by green-reflecting (not blue) forehead and crown. Female separated from female Green-throated by more distinct eyebrow. **V** Mid-high, sustained, rather slow *tuut puwietpuwiet tututut fuwie*. **H** Forest edges, woodland, mangroves, suburbs. **D** E, SE and S Africa.

5 SCARLET-CHESTED SUNBIRD *Chalcomitra senegalensis* 14cm Female heavily marked below; separated from female Green-throated Sunbird by sharply demarcated cheek. **V** Mid-high, slow *tuut twit-tuut twuut*. **H** Woodland, bushveld, suburbs. **D** W, C, E and S Africa.

6 HUNTER'S SUNBIRD *Chalcomitra hunteri* 14cm Note black chin and purple rump; also, blue-green colouring of forehead reaches nape. Bill more curved than in Scarlet-chested Sunbird, and little or no overlap in range or habitat with that species. Female separated from female Scarlet-chested by pale supercilium compared with plain head of that species. **V** Song complex, comprising rising and falling notes, a whistle and a trill. Call a typical *chip*. **H** Hot, dry country with scrub. **D** Ethiopia and Somalia to NE Tanzania.

7 SOCOTRA SUNBIRD *Chalcomitra balfouri* 11.5–14cm Upperparts greyish, underparts white with a black throat patch and black speckling on breast and upper belly; bright yellow pectoral tufts. Very active in pursuit of small invertebrates; also takes nectar. **V** Song a series of fast, jangling notes, may include some mimicry. Calls shrill and rasping. **H** All vegetated areas. **D** Socotra, Gulf of Aden.

8 PURPLE-RUMPED SUNBIRD *Leptocoma zeylonica* 10cm Active, acrobatic forager among flowers and foliage. **V** Song a sharp, twittering *tityou tityou tityou trr-r-r-tit tityou...*, a weak *sisiswee-sisiswee...* or a *sit-sit tseet-tseet-tseet tsut-tsut-tsut-tsut*. Calls include a constant *sweety-swee sweety-sweety-swee* and a high, metallic *tzip*. **H** Open jungle, secondary jungle, cultivation, gardens. **D** Indian subcontinent.

9 CRIMSON-BACKED SUNBIRD *Leptocoma minima* 8cm Active, acrobatic forager among flowers and foliage; favours the blossoms of *Erythrina* and *Loranthus*. **V** Song a squeaky, repeated *see-see-whi-see-see-siwee...* or a tinkling *tseet-tsut-tseet*. Calls include a constant *chik*, a metallic ticking and a chittering. **H** Evergreen forest, plantation shade trees, secondary growth, gardens. **D** SW India.

10 PURPLE-THROATED SUNBIRD *Leptocoma sperata* 10cm Active and agile. Usually forages in treetops, hovering to take insects or water from foliage. **V** Song a *psweet psweet psweet psweet psweet psweet... psit-it pstit psweet psweet...* Calls include a weak *chip chip*, a sharp *si-si-si*, an upslurred *psweeet*, a *fut-chit* and short, high trills. **H** Forests, gardens. **D** Philippines.

11 VAN HASSELT'S SUNBIRD *Leptocoma brasiliana* 10cm Usually forages in the treetops, hovering to take insects or water from foliage and nectar from flowers. **V** Song a series of discordant *psweet* notes. Calls include a weak *chip chip*, a sharp *si-si-si*, an upslurred *psweet*, a *fut-chit* and short, high trills. **H** Various forests, forest edges, secondary growth, coastal scrub, cultivation, gardens. **D** NE India to SE Asia and Greater Sunda Islands.

12 BLACK SUNBIRD *Leptocoma aspasia* 11–12cm Active, foraging by gleaning or hover-gleaning. Sometimes occurs in small groups. **V** Song a rapid, sweet, tinkling cadence. Calls include a shrill *zi-zi-zi-zi*, a downslurred *swee*, and various high-pitched sibilant or harsh notes. **H** Primary and secondary forest, forest edges, mangroves, scrub, village trees, gardens. **D** Sulawesi and Muluku Islands, Indonesia, through Bismarck Archipelago.

13 COPPER-THROATED SUNBIRD *Leptocoma calcostetha* 12–13cm Forages mainly in mangroves, feeding on nectar and small invertebrates. **V** A high trill and a deep, melodious trill. **H** Mangroves, coastal scrub, secondary growth, cultivation. **D** Myanmar to the Philippines and Java, Indonesia..

14 BOCAGE'S SUNBIRD *Nectarinia bocagii* 20cm Male dark iridescent violet above, reflecting blue-green. Female yellowish below. No non-breeding plumage and no tufts. **V** Call a loud *wiep-wiep*. **H** Woodland. **D** S DR Congo to C Angola.

15 PURPLE-BREASTED SUNBIRD *Nectarinia purpureiventris* 20cm Unmistakable. **V** Song an explosive, rapid series of wheezy notes, squeaks and twitters. Calls include *tsi-tsi-tsi-tsi*. **H** Open montane forests. **D** E DR Congo, S Uganda, Rwanda, Burundi.

16 TACAZZE SUNBIRD *Nectarinia tacazze* male 22cm, female 15cm Male dark with a green gloss on head shading to violet-purple on wings and breast; tail long. Female drab olive-green. Bill strongly curved. Feeds mainly on nectar, also fly-catches. **V** Song a twittered *sweet-siuswitter tseu tsit-tsit-tist* or similar. Calls include harsh chatters and a low *chup-chup-chup see chup*. **H** High-altitude forest in mountains. Moves to lower slopes post-breeding. **D** EC Africa.

17 BRONZY SUNBIRD *Nectarinia kilimensis* 22cm Very dark, with bronze and green reflections. Female has diffuse streaks below. No non-breeding plumage and no tufts. **V** Very high, excited, rapid, sustained twittering. **H** Montane forest edges, wooded and bushy areas. **D** C and E Africa.

18 MALACHITE SUNBIRD *Nectarinia famosa* 24cm Female separated from most other *Nectarinia* females by rather yellow-greenish underparts and well-defined moustachial stripe. Non-breeding male similar to female, often with a few reflecting feathers. **V** Very high, unstructured, slow tinkling, built around a high *cink-cink*. **H** More or less open, wooded and scrubbed areas, gardens. **D** E and SE Africa.

19 SCARLET-TUFTED SUNBIRD *Nectarinia johnstoni* 28cm Reflecting (not shining, as in Malachite Sunbird) green with black undertail coverts. Female dusky with faintly, finely barred throat. Non-breeding male as female but retaining reflecting wing coverts and rump feathers, black flight feathers and elongated middle tail feathers. **V** Calls include metallic *tspk* and a fast *tsp-tk tsp-tk*. **H** Montane and alpine areas. **D** E Africa.

20 GOLDEN-WINGED SUNBIRD *Drepanorhynchus reichenowi* 20cm Unmistakable. **V** Chattering song includes bursts of twitters mixed with high *chi-chi-chi* notes. Calls include quiet *jwee* and *tweep* notes **H** Montane forests, bamboo, bushy areas, gardens, cultivation, grassland with flowers. **D** E Africa.

21 OLIVE-BELLIED SUNBIRD *Cinnyris chloropygius* 10cm Note blue rump and dusky underparts of male. Female has streaked underparts and a distinct eyebrow. **V** Song 3–4 secs, beginning with *pisr-pisr*, then fast rising and falling notes. Calls include repetitions of *chip*. **H** Forest edges, moist bushland, cultivation. **D** W and C Africa.

22 TINY SUNBIRD *Cinnyris minullus* 8cm Similar to Olive-bellied Sunbird, but much smaller. Note narrow red chest-band and dark wings. **V** Song a high-pitched succession of squeaky *tsi-tsi-tsi* notes. Calls include a *chip*. **H** Forests. **D** Sierra Leone and Liberia to Uganda, DR Congo and Gabon; Bioko Island, Gulf of Guinea.

23 EASTERN MIOMBO SUNBIRD *Cinnyris manoensis* 10cm Male separated from male Eastern Double-collared Sunbird by olive rump, narrower red breast-band and paler whitish belly. Female separated from female Eastern Double-collared by cleaner underparts. Range differs from that of Olive-bellied Sunbird. **V** Very high, whistled chattering (lower pitched than that of Southern Double-collared Sunbird). **H** Montane forest, miombo. **D** SC and SE Africa.

24 WESTERN MIOMBO SUNBIRD *Cinnyris gertrudis* 12cm Separated from Northern Double-collared Sunbird in same range by paler belly, longer bill and preference for higher altitudes. **V** Song similar to those of other *Cinnyris* sunbirds but ends with a drawn-out, descending trill. Calls include a *chip*. **H** Normally at high altitudes in mountains. **D** C Angola and S DR Congo to SW Tanzania and N Malawi.

25 SOUTHERN DOUBLE-COLLARED SUNBIRD *Cinnyris chalybeus* 10cm Differs from other double-collared sunbirds by range, except Greater Double-collared Sunbird (see that species). **V** Extremely high, fast whistling, interrupted by very high, staccato *zikzik zikzik*. **H** Forest edges, coastal scrub, gardens. **D** South Africa.

26 NEERGAARD'S SUNBIRD *Cinnyris neergaardi* 10cm Note red breast-band and uniform brown-black underparts of male, and plain, faintly greenish underparts of female. **V** Extremely high, whistling *si-si-si-si*, turning to a fast rattle. **H** Coastal woodland. **D** SE Mozambique to Swaziland and E South Africa.

27 RWENZORI DOUBLE-COLLARED SUNBIRD *Cinnyris stuhlmanni* 12cm Separated from Northern Double-collared Sunbird in same range by paler belly, longer bill and preference for higher altitudes. **V** Warbling song, *chee-oo, che che, chee-oo, che che, se, se, se, se, se, se, ce chit, che chit-che chit*. Calls include a *tsp*. **H** Normally at high altitudes in mountains. **D** SC Africa.

28 WHYTE'S DOUBLE-COLLARED SUNBIRD *Cinnyris whytei* 12cm Formerly considered a subspecies of Ludwig's Double-collared Sunbird. Male has a brighter red breast-band and wider blue breast-band than that species; female darker. **V** Song a high-pitched *ti ti titi pitsi pitsi chu chu chu tititi* and similar. Calls include ticking and chipping notes. **H** Montane forest, forest edges, grassland. **D** EC Africa.

29 PRIGOGINE'S DOUBLE-COLLARED SUNBIRD *Cinnyris prigoginei* 13cm Note narrow red breast-band. **V** Song a high-pitched warble. Calls various cheeps and ticking notes. **H** Montane forest and fields. **D** SE DR Congo.

30 LUDWIG'S DOUBLE-COLLARED SUNBIRD *Cinnyris ludovicensis* 12cm Very similar to Rwenzori Double-collared Sunbird but differs in range. **V** Song comprises bursts of high notes, *ti,ti, titi pitsy-pitsy, chu-chu-chu pitsy-pitsy, tititititipitsy, pichew, chichichichi...* Calls include various ticking notes. **H** Montane forest and grassland. **D** Angola.

31 NORTHERN DOUBLE-COLLARED SUNBIRD *Cinnyris reichenowi* 10cm Separated from similar Olive-bellied Sunbird in same range by purple breast-band and uppertail coverts. See also Rwenzori Double-collared Sunbird. **V** Song a burst of *chup-chup-ch-ch-ch-ch* or *tsup-tsup* notes preceding a run of warbles. Call a high-pitched *zeet* or *chip*. **H** Montane forest, bamboo, gardens. **D** WC and E Africa.

32 GREATER DOUBLE-COLLARED SUNBIRD *Cinnyris afer* 12cm Male separated from male Southern Double-collared Sunbird by longer bill and broader red-breast band; female separated by bill size. Separated from Eastern and other double-collared sunbirds by different range. **V** High to very high, pleasant warbler-like twittering. **H** Forest edges, bush, gardens. **D** E and S South Africa.

33 REGAL SUNBIRD *Cinnyris regius* 11cm Separated from Rockefeller's Sunbird by yellow belly. **V** Song a repeated rapid twitter of rising and falling notes, followed by a *tsi*. Calls include a *djer* and *dzit*. **H** Montane forest and bamboo. **D** EC Africa.

34 ROCKEFELLER'S SUNBIRD *Cinnyris rockefelleri* 11cm Very restricted range. Note red undertail coverts. **V** Song undescribed. Calls reported as harsh *schick schick*. **H** Montane forest, bamboo. **D** E DR Congo.

35 EASTERN DOUBLE-COLLARED SUNBIRD *Cinnyris mediocris* 10cm Note green rump, orange (not red) breast and yellowish underparts of male. **V** Song likened to sound made by shaking a metal chain. Calls include a *chek chek* and *tse*. **H** Forest edges, bamboo, gardens. **D** W and C Kenya, N Tanzania.

36 USAMBARA DOUBLE-COLLARED SUNBIRD *Cinnyris usambaricus* 12cm Very like Eastern and Forest double-collared sunbirds, but has a longer bill than the former, and male has a narrower red breast-band than the latter. **V** Song little known but appears to be shorter and simpler than those of Eastern and Forest double-collareds. **H** Montane forests, bamboo, gardens. **D** SE Kenya, NE Tanzania.

37 FOREST DOUBLE-COLLARED SUNBIRD *Cinnyris fuelleborni* 12cm Note red (not maroon) breast-band in combination with dark belly. **V** High, rapid, unstructured twittering around a high, slightly thrush-like rattle *tutututututut*. **H** Miombo, woodland, bush, gardens. **D** C and S Tanzania, Malawi, NE Zambia, N Mozambique.

38 MOREAU'S SUNBIRD *Cinnyris moreaui* 12cm Note moss-green edges of wing feathers. **V** Song a fast outburst of high-pitched notes. Calls consist of regularly spaced chirps. **H** Forest. **D** C Tanzania.

1
2
3
4
5
6
7
8
9
10
11
12
13
14
15
16
17
18
19
20
21
22
23
24
25
26
27
28
29
30
31
32
33
34
35
36
37
38

SUNBIRDS *CONTINUED*

1 LOVERIDGE'S SUNBIRD *Cinnyris loveridgei* 12cm Note yellow (not red) undertail coverts and orange (not red) breast. **V** Song a fast, short (3–4 secs) warble of high-pitched *tsi* notes given in such quick succession, like the sound of an insect. Calls a metallic *tsk* or *pzit*. **H** Undergrowth and mid-strata at forest edges. **D** EC Tanzania.

2 BEAUTIFUL SUNBIRD *Cinnyris pulchellus* 20cm Unmistakable. **V** Song a warble of rising and falling high-pitched notes. Call a repeated *chip-chip*. **H** More or less wooded natural and cultivated areas. **D** W, C and E Africa.

3 MARICO SUNBIRD *Cinnyris mariquensis* 14cm Separated from Purple-banded Sunbird by long, downcurved bill and less brilliant colouring, and from Orange-tufted Sunbird by green forehead and chin and different range. **V** High to very high, very fast warbling. **H** Miombo, dry thornveld. **D** E, SE and S Africa.

4 SHELLEY'S SUNBIRD *Cinnyris shelleyi* 12cm Note red (not maroon) breast-band. Separated from Neergaard's Sunbird (Plate 258) by longer bill and different range; female from female Neergaard's by streaked (not plain) underparts. **V** High, rapid, unstructured twittering around a high, slightly thrush-like rattle, *tu-tu-tu-tu-tu-tu-tut*. **H** Miombo, woodland, bushland, gardens. **D** Tanzania to Zambia, N Zimbabwe and N Mozambique.

5 HOFMANN'S SUNBIRD *Cinnyris hofmanni* 12cm Note all-green (not partly purplish) head and orange-red chest-band (narrower than in Shelley's Sunbird). **V** Like that of Shelley's Sunbird. **H** Miombo, woodland, bushland, gardens; often along streams. **D** E Tanzania.

6 CONGO SUNBIRD *Cinnyris congensis* 20cm Unmistakable. Bluish green with a long tail. **V** No information available. **H** Forest edges, especially along rivers or at clearings. **D** Congo region.

7 RED-CHESTED SUNBIRD *Cinnyris erythrocercus* 15cm The only black-bellied green sunbird with elongated central tail feathers except the larger, green-rumped Congo Sunbird. Note also maroon (not scarlet) breast-band. **V** Song a brief, twittering *si-si-sip-see-see-swee*. Calls include a repeated *jik* or *trink*. **H** Shrubbery and undergrowth near water, swamps and gardens. **D** Extreme S South Sudan to W and C Tanzania.

8 BLACK-BELLIED SUNBIRD *Cinnyris nectarinioides* 13cm Like black-bellied form of Beautiful Sunbird but with shorter tail streamers and orange (not red) breast. **V** Song comprises 6–20 *tsk*, *chip* or *che* notes, repeated 6–7 times per sec, followed by a brief warble of rising and falling *tsi* notes. Calls use song notes. **H** Dry woodland and bushland. **D** E Africa.

9 PURPLE-BANDED SUNBIRD *Cinnyris bifasciatus* 11cm Note maroon (not red) breast-band and rather short bill. **V** Very high *tit-tit-trrri-ri-trrr*. **H** Dense riverine belts, forest, coastal scrub, mangroves. **D** WC, E and SE Africa.

10 TSAVO SUNBIRD *Cinnyris tsavoensis* 11cm Note bluish-green colouring and purple breast-band. **V** Song a rapid *tsusitiseesee, chuchiti-tsi-tsi-tsi-tsi-sitisee-see-see-see-chitisee*. **H** Woodland, bushland, dense shrubland, mangroves, gardens. **D** E Kenya and NE Africa.

11 VIOLET-BREASTED SUNBIRD *Cinnyris chalcomelas* 11cm Note beautiful purple-blue-green colouring. **V** Song a trilling *chrrrrrrrrssssuwwizzzzlllle*, ending with *chee-per-chichi-woo-per-chichi-chee-dzurr*. Calls include a *chip* and *chut*. **H** Woodland and bushland, including gardens. **D** EC Somalia to SE Kenya.

12 PEMBA SUNBIRD *Cinnyris pembae* 9–10cm Male all dark with a green gloss on head, throat, upper breast and back, shading to blue on wings and tail, plus a glossy violet breast-band. Female mid grey-brown on upperparts, whitish on underside with a short white eyebrow, and with a blue gloss on dark tail. **V** Song a series of up to 15 *tslink* notes, ending with a short warble. Call a shortened version of first part of song. **H** Most well-vegetated habitats, including gardens. **D** Pemba Island, Zanzibar Archipelago.

13 ORANGE-TUFTED SUNBIRD *Cinnyris bouvieri* 10cm Note purple forehead of male. Female lightly streaked below. **V** Song starts with a quiet *tswee*, followed by before a louder *tsi* and then a warble terminating in *tsi-pu, tsi-pu, tsi-pu, tsi-pu, tsi-pu, tsi-pu tsi* or *chieta-chieta-chieta-chit*. Calls include a *cheep* and *ziet*. **H** Open woodland. **D** Cameroon and the Central African Republic to Uganda, N Zambia and N Angola.

14 PALESTINE SUNBIRD *Cinnyris osea* 11cm Typical sunbird actions and habits. Erects orange pectoral tufts during display. **V** Fast, rambling, high-pitched trill, *dy-vy-vy-vy-vy-vy*, or rising *tveeit-tveeit-tveeit* or similar, often ending in Serin-like trill. Call: Variable, includes a thin *ftift* and a hard *tsak*, also a loud *te-weeit-teweeit* and a Siskin-like *tiu*. **H** Woodland, places with some vegetation in deserts. **D** NC Africa, S and W Arabian Peninsula.

15 SHINING SUNBIRD *Cinnyris habessinicus* 13cm Typical sunbird, active and agile. **V** Fast, fluty, trilling and whirring *tuu-tuu-tuu-tuu-vita-vita-vita-vita-du-du-du-du*, often ending in a Eurasian Wren-like trill. Call a hard *tshak, dzit* or *chewit-chewit*; also a fast, dry *tje-tje-tje-tje*. **H** Flowering trees and plants on hills, slopes, coastal cultivation. **D** NE and E Africa, S and W Arabian Peninsula.

16 SPLENDID SUNBIRD *Cinnyris coccinigastrus* 13cm Note irregular red breast-patch, surrounded by blue. **V** Most common song a series of 6–9 clear, descending *chip-chee-cho-cho-choo-choo-choo-choo* whistles. Calls include a *chee-iip, chip, tschup* and *choo*. **H** Forest edges and other wooded habitats, including gardens. **D** Senegal and Gambia south to Liberia and east to NE DR Congo and NW Uganda.

17 JOHANNA'S SUNBIRD *Cinnyris johannae* 14cm Separated from Superb Sunbird by green chin, strikingly short tail and yellow pectoral tufts. Note striped underparts of female. **V** Song a high-pitched warble of rising and falling whistling notes. Calls include a *wit*. **H** Forest, plantations, suburbs. **D** W and C Africa.

18 SUPERB SUNBIRD *Cinnyris superbus* 14cm Maroon underparts of male diagnostic. Note long bill and orange and yellow tinges on underparts of female. **V** Song comprises sequences of *weet-choo weet-choo witchoo witch* and variants. Calls include a *chip* and *weeet*. **H** Forest canopy. **D** W and C Africa.

19 RUFOUS-WINGED SUNBIRD *Cinnyris rufipennis* 12cm Male has a glossy blue-black head, tail, forewings and upper breast, and bright chestnut breast-band and fringes to flight feathers. Female dull sandy brown but shares male's rufous wing feather fringes. **V** Song a shrill trilling (female often accompanies male with chirping notes). Calls include a *csee-it* and trilling *tiddit*. **H** Moist, lush montane forest. **D** C Tanzania.

20 OUSTALET'S SUNBIRD *Cinnyris oustaleti* 10cm Male separated from White-bellied Sunbird by less bluish plumage; female by indistinct, dusky throat streaks. **V** Song a series of ascending and descending high-pitched notes. Calls include a repeated *cheep*. **H** Miombo. **D** SC Africa.

21 WHITE-BELLIED SUNBIRD *Cinnyris talatala* 11cm Male separated from Oustalet's Sunbird (which has a restricted range) by blue (not purplish-maroon) breast-band and longer, more curved bill; female by whiter throat and chest. **V** Combination of a very high *uweet-uweet*, a fast, wren-like trill and other short twitters. **H** Woodland, bushveld, gardens. **D** S Tanzania to S Angola, N Botswana, N and E South Africa, and Mozambique.

22 VARIABLE SUNBIRD *Cinnyris venustus* 10cm Male separated from Collared Sunbird (Plate 257) by blue rump and longer bill. Note blue-black tail of male and female. **V** Very high, descending, very fast warbler-like chatter and very high, level, fast twittering. **H** Forest edges, riverine belts, bushveld, gardens. **D** W, C and E Africa.

23 DUSKY SUNBIRD *Cinnyris fuscus* 10cm Female (as male) predominantly white below. Non-breeding male dusky brown above with a broad, untidy black stripe from chin to belly, leaving throat and breast sides white. **V** Song a combination of mid-high trills, canary-like warbles and reed warbler-like rattles. **H** Very dry, open areas with scrub and rocks. **D** SW Africa.

24 URSULA'S SUNBIRD *Cinnyris ursulae* 9cm No similar bird occurs in its range and habitat. **V** Songs include a descending high-pitched trill, a whistling warble and trills. Calls include a *tsit-tsit* and *tche-tchu*. **H** Forest. **D** Cameroon.

25 BATES'S SUNBIRD *Cinnyris batesi* 11cm Very similar to Little Green Sunbird (Plate 257), but with shorter tail, narrower bill and darker underparts. Lacks tufts. **V** Song a quiet trill followed by nasal *ts-tsp*. Calls include a *tsk* and *weet*. **H** Upper forest strata, cultivation. **D** Ivory Coast and S Liberia east to E and S DR Congo.

26 COPPER SUNBIRD *Cinnyris cupreus* 11cm No similar sunbird. Note black tail of female. **V** Song is a combination of its sparrow-like call, *dut-dut-dut-dut*, and slow, high, nasal twittering. **H** Forest edges, open woodland, bushveld, cultivation, gardens. **D** W, C, E and SC Africa.

27 PURPLE SUNBIRD *Cinnyris asiaticus* 10cm Active, agile forager; very attracted to flowering trees and shrubs. **V** Song an excited *cheewit-cheewit...cheewit-cheewit...cheewit-cheewit... cheewit-cheewit...*, delivered from a high, exposed perch. Calls include a *chip*, a *chweet*, a *sweep*, a hard *zik* and a crackling alarm note. **H** Deciduous forests, thorn scrub, cultivation, gardens. **D** E Arabian Peninsula to SE Asia.

28 OLIVE-BACKED SUNBIRD *Cinnyris jugularis* 11–12cm Feeds on invertebrates, nectar and small fruits. **V** Song a feeble twittering. Calls include a *chip* and a nasal *sweei*, and female utters a persistent *sweep*. **H** Forest, forest edges, mangroves, coastal vegetation, scrub. **D** S China to NE Australia.

29 APRICOT-BREASTED SUNBIRD *Cinnyris buettikoferi* 11cm Forages singly or in pairs in the middle and upper storeys. **V** Long, sweet, high-pitched, tinkling warble and a *wee-chew wee-chew wee-chew wee-chew-wee*; in flight, utters a sharp, high *chee* or *sip*. **H** Secondary forest, forest edges, cultivation, scrub. **D** Sumba, Lesser Sunda Islands.

30 FLAME-BREASTED SUNBIRD *Cinnyris solaris* 11cm Favours foraging in flowering trees, singly, in pairs or in small groups. **V** Song consists of 3–4 high-pitched, halting notes that go up and down the scale. **H** Secondary forest, forest edges, woodland, plantations, *Eucalyptus* savannah, scrub, gardens. **D** Lesser Sunda Islands.

31 SOUIMANGA SUNBIRD *Cinnyris sovimanga* 10cm Note maroon breast-band. **V** Song comprises short warbling phrases. Calls include loud chirps. **H** Any type of wooded habitat up to very high altitude. **D** Madagascar.

32 ABBOTT'S SUNBIRD *Cinnyris abbotti* 10cm Recently split from Souimanga Sunbird; similar to that species but male has a dark brown rather than white belly. **V** Like that of Souimanga Sunbird. **H** Forest and similar habitats. **D** Aldabra Islands, Indian Ocean.

33 MALAGASY GREEN SUNBIRD *Cinnyris notatus* 14cm Souimanga and Humblot's sunbirds by size, longer bill and lack of maroon breast-band. **V** Song, delivered from cover, a quiet, slow warbling (based on a single description). Calls include a loud, penetrating, nasal *chew-chew-chew* or *chip-chip-chip-chip*. **H** Any wooded habitat, including gardens. **D** Madagascar, Comoros.

34 SEYCHELLES SUNBIRD *Cinnyris dussumieri* 12cm No other sunbird in its range. Note that gloss is restricted to throat and upper breast. **V** Male's song a high-pitched squeaky jumble of 6–20 or more notes, beginning as a slow *dze-dze-dze-dze* and accelerating to *der-tseet-tseet-tsit-tsit-tsit*; female's song quieter and less harsh. Calls include a *pseeeu* repeated at 2-sec intervals. **H** Any type of habitat, from forest to settlements. **D** Seychelles.

35 HUMBLOT'S SUNBIRD *Cinnyris humbloti* 11cm Unmistakable due to orange glow and colouring. Only other sunbird in its range is the very different Malagasy Green Sunbird. **V** Song a jumbled series of chipping notes. **H** All habitats with trees and scrub. **D** Comoros.

36 ANJOUAN SUNBIRD *Cinnyris comorensis* 10cm No other sunbird in its range. **V** Song a variable jumbled series of notes. Calls include a sharp *pit pit*. **H** Any habitat with trees or shrub. **D** Comoros.

37 MAYOTTE SUNBIRD *Cinnyris coquerellii* 10cm The most colourful sunbird in the Madagascar region. **V** Various harsh chipping notes. **H** Any type of habitat. **D** Comoros islands.

38 LOTEN'S SUNBIRD *Cinnyris lotenius* 13cm Agile, active forager after insects, spiders and nectar. **V** Song a quickly repeated *cheewit-cheewit-cheewit* or *ti-ti-ti twink-it-twink*; also gives a *twink-it-tee*. Calls include a metallic *chit-chit*, a clipped *twink-tuwink*, *cheerit* and a *chee-chee-teer*. **H** Deciduous woodland, open country with trees, cultivation, tea plantations, gardens. **D** Sri Lanka, India.

1
2
3
4
5
6
7
8
9
10
11
12
13
14
15
16
17
18
19
20
21
22
23
24
25
26
27
28
29
30
31
32
33
34
35
36
37
38

SUNBIRDS *CONTINUED*

1 GREY-HOODED SUNBIRD *Aethopyga primigenia* 10–11cm Forages alone, in pairs or in mixed-species flocks; favours the flowers of banana plants. **V** Repeated, high-pitched *pink-pink-pink*, level in pitch or ascending, and sometimes turning into *see-see-see…*; also *seck-seck*, repeated up to six times. **H** Submontane and montane forest and forest edges. **D** Mindanao, S Philippines.

2 APO SUNBIRD *Aethopyga boltoni* 12cm Active and noisy. Forages alone, in pairs or in mixed-species flocks, especially in flowering trees and shrubs. **V** Ascending, metallic *twip twip twip twit twit…*, repeated after a short pause; also gives rapid, continuous, snapping *twit twit twit…* or *whit whit whit…*, occasionally changing to clear, whistled *whirp*. **H** Montane forest. **D** Mindanao, S Philippines.

3 LINA'S SUNBIRD *Aethopyga linaraborae* 10–11cm Forages in middle and upper storeys, alone, in pairs or in mixed-species flocks. **V** Long series of high-pitched, squeaky, twittering notes with repeated sequences. Calls include high-pitched *suweet suweet suweet*, upward-inflected *su-weet* or *tsoo-eet*, and metallic *tip-tip-tip-tip…* **H** Montane mossy forest. **D** E Mindanao, S Philippines.

4 FLAMING SUNBIRD *Aethopyga flagrans* 9–10cm Forages alone, in pairs or in mixed-species flocks. **V** Repeated, short, high-pitched, rising *tsweet*. **H** Forest, forest edges, secondary growth. **D** N Philippines.

5 MAROON-NAPED SUNBIRD *Aethopyga guimarasensis* 9–10cm Formerly considered a race of Flaming Sunbird; behaviour presumed to be similar. **V** Presumed to be similar to that of Flaming Sunbird. **H** Forest, forest edges, secondary growth. **D** WC and SC Philippines.

6 METALLIC-WINGED SUNBIRD *Aethopyga pulcherrima* 9–10cm Forages alone, in small groups or with mixed-species flocks. **V** Song slow, well-spaced notes, followed by trill and then further series of slow, spaced notes; calls include repeated *squeak* and repeated high-pitched *zeeeep*; also utters sharp *see* or *see-see-tsik tsik*, which may turn into rising trill. **H** Forest, forest edges, secondary growth, plantain plantations. **D** C and S Philippines.

7 LUZON SUNBIRD *Aethopyga jefferyi* 9–10cm Formerly considered a race of Metallic-winged Sunbird; behaviour presumed to be similar. **V** Presumed to be similar to that of Metallic-winged Sunbird. **H** Forest, forest edges, secondary growth, plantations. **D** Luzon, Philippines.

8 BOHOL SUNBIRD *Aethopyga decorosa* 9–10cm Formerly considered a race of the Metallic-winged Sunbird; behaviour presumed to be similar. **V** Presumed to be similar to that of Metallic-winged Sunbird. **H** Forest, forest edges and plantations. **D** Bohol, SC Philippines.

9 ELEGANT SUNBIRD *Aethopyga duyvenbodei* 12cm Gleans insects from leaves and spider webs; occurs in mixed-species flocks, in small groups, in pairs or alone. **V** Song short insect-like trill, and high trill of chipping notes; calls include high-pitched, rasping *treek* or *tseeek*, and *tit* or *tit-tit-tit-tit*. **H** Mixed plantations near remnant forest patches, secondary forest, bamboo, tree ferns, scrub. **D** Sangihe, N Indonesia.

10 LOVELY SUNBIRD *Aethopyga shelleyi* 10–11cm Feeds on nectar, insects and larvae, in pairs, alone or as a member of mixed-species flocks. **V** Seesawing *zuep-ziip* that continues for several seconds. **H** Forest, forest edges, cultivation with flowering trees. **D** W Philippines.

11 HANDSOME SUNBIRD *Aethopyga bella* 9cm Feeds on insects, larvae and nectar, foraging singly, in pairs or in mixed-species parties. **V** Rapid *sit-sit-sit-sit-tee-tee-tee* and lower-pitched *tsit-tsit-tit-it*. **H** Forest, forest edges, thickets, cultivation. **D** Philippines.

12 MRS. GOULD'S SUNBIRD *Aethopyga gouldiae* male 14–15cm, female 10cm Restless, agile forager, attracted to flowers of rhododendrons and *Loranthus*. **V** Described as strong seesawing sound. Calls include *tzit-tzit*, *squeeeee* that rises in the middle and *tshi-stshi-ti-ti-ti* given when alarmed. **H** Forests, scrub jungle. **D** Himalayas to S Vietnam.

13 GREEN-TAILED SUNBIRD *Aethopyga nipalensis* male 14–15cm, female 10cm Typical active agile forager. During the winter, in Nepal, regularly seen with Fire-tailed, Black-throated and Crimson sunbirds feeding among flowering trees and shrubs. **V** Song lively, twittering *swit-it-it-it-it-it-it* mixed with high notes and a dry metallic trill; calls include sharp *dzii* or *reet*, regularly repeated *tee-tzree-tzweeeet* and series of staccato *stip* notes. **H** Forests, woodland, scrub jungle, secondary growth, orchards, gardens. **D** Himalayas to S Vietnam.

14 WHITE-FLANKED SUNBIRD *Aethopyga eximia* 13cm Favours flowering trees and vines in lower and middle levels. Seen alone, in pairs or in small parties. **V** Clear, precise *tee-tee-tee-leet* or similar. **H** Forest, forest edges, clearings and alpine scrub above the treeline. **D** Java, Indonesia.

15 FORK-TAILED SUNBIRD *Aethopyga christinae* 10cm Attracted to flowering trees and shrubs. **V** Song accelerating *pe-et pe-et pit pit*; calls include *twisk* and *chip-chip* that develops into trill. **H** Forest, forest edge. **D** C China to S Vietnam.

16 BLACK-THROATED SUNBIRD *Aethopyga saturata* male 14–15cm, female 10cm Active, agile forager from treetops to undergrowth; especially attracted to flowering trees and shrubs. **V** Song twittering, containing sharp, high-pitched notes mixed with rapid metallic trills; calls include repeated high-pitched *tit tit-tit* or *ttiss-it* and *tu-ti-tee-tee*. **H** Dense scrub and bushes in forests, secondary growth, open areas with scattered bushes. **D** Himalayas to S Vietnam.

17 CRIMSON SUNBIRD *Aethopyga siparaja* male 12–14cm, female 10cm Acrobatic forager among tree and shrub blossom. **V** Song loud chirping trill; calls include *zit-zit* and soft *siesiep-siepsiep*. **H** Dense evergreen forest, pine forest, open deciduous and scrub jungle, orchards, gardens. **D** Himalayas through SE Asia and Indonesia to Sulawesi.

18 MAGNIFICENT SUNBIRD *Aethopyga magnifica* 15cm Encountered singly, in pairs, in small groups or as part of mixed-species foraging parties; favours flowering trees. Formerly considered a race of Crimson Sunbird. **V** Call metallic *zit-zit*. **H** Forest, secondary growth, cultivation, scrub. **D** WC Philippines.

19 VIGORS'S SUNBIRD *Aethopyga vigorsii* male 14–15cm, female 11cm Active agile forager after nectar or small insects. **V** Sharp, harsh *chi-wee* and *shwing*. **H** Evergreen and moist deciduous forest. **D** W peninsular India.

20 JAVAN SUNBIRD *Aethopyga mystacalis* 12cm Forages in noisy pairs in upper levels of forest trees, favouring mistletoe flowers. **V** Soft, ringing *tseep-tzeep cheet-cheet*. **H** Lower montane forest, hill dipterocarp forest, secondary forest, forest edges. **D** Java, Indonesia.

21 TEMMINCK'S SUNBIRD *Aethopyga temminckii* male 13cm, female 10cm Forages mainly in canopy, especially among mistletoe clumps. Occurs alone or in pairs, and occasionally found in small parties. **V** Soft *cheet-cheet* and rhythmic *tit-ti tit-it tit-it tit-it…* **H** Lowland, hill and lower montane forest, mountain gardens. **D** Peninsular Malaysia, Sumatra, Borneo.

22 FIRE-TAILED SUNBIRD *Aethopyga ignicauda* male 15–20cm, female 10cm Very active around flowering bushes. **V** Song series of descending high-pitched notes; calls include repeated, high-pitched *dzdzi-dzidzidzidzi* and rapid staccato twittering. **H** Open coniferous forest with rhododendron understorey; also juniper scrub above treeline. **D** Himalayas.

23 LITTLE SPIDERHUNTER *Arachnothera longirostra* 16cm Restless, acrobatic forager, attracted to banana blossoms. **V** Song monotonous metallic *which-which*, given twice per sec for about two mins; calls include harsh *cheep*, *chee-chee-chee* or loud *sheep*, latter repeated up to 25 times. **H** Dense forest, forest edges and glades, wild and cultivated banana patches. **D** Widespread in SE Asia, Indonesia and Borneo.

24 ORANGE-TUFTED SPIDERHUNTER *Arachnothera flammifera* 13–16cm Formerly considered conspecific with Little Spiderhunter; behaviour presumed to be similar. **V** Presumed to be similar to that of Little Spiderhunter. **H** Forest, forest edges, secondary growth, banana plantations. **D** Philippines.

25 PALE SPIDERHUNTER *Arachnothera dilutior* 13–16cm Behaviour presumed to be like that of Little Spiderhunter, with which it was formerly considered conspecific. **V** Presumed to be similar to that of Little Spiderhunter. **H** Forest, forest edges, secondary growth, cultivation. **D** Palawan, W Philippines.

26 THICK-BILLED SPIDERHUNTER *Arachnothera crassirostris* 16–17cm Forages from canopy down to understorey, singly or in pairs; feeds on small invertebrates and nectar. **V** Hard, nasal *chit-chit* or *chissie-chissie*, also *tch-tch* and *chek-chek-chek* or similar. **H** Forest, forest edges, secondary growth, banana plantations, gardens. **D** Peninsular Malaysia, Sumatra, Borneo.

27 LONG-BILLED SPIDERHUNTER *Arachnothera robusta* 21–22cm Forages mainly in canopy; solitary and aggressive. **V** Song rising *choi choi choi choi…*; calls include harsh *chuu-luut chuut-luut*, given from high perch, and high-pitched *chit-chit chit-chit* flight call. **H** Broadleaved evergreen forest, forest edges. **D** Peninsular Malaysia, Sumatra, Java, Borneo.

28 SPECTACLED SPIDERHUNTER *Arachnothera flavigaster* 21–22cm Feeds on small invertebrates, nectar and small fruits. Forages in middle and upper levels, singly or occasionally in pairs; small groups may occur in fruiting trees. **V** High-pitched *chit-chit*, and explosive *tak*, *cha-tak*, *cha-ta-tak* or variants. **H** Open forest, secondary forest, secondary scrub, plantations, gardens. **D** Peninsular Malaysia, Sumatra, Borneo.

29 YELLOW-EARED SPIDERHUNTER *Arachnothera chrysogenys* 17–18cm Acrobatic. Forages mainly in canopy, searching for invertebrates in bark and broken branches; also feeds on nectar. Occurs in pairs, alone and occasionally in small groups. **V** Rough *chit* and high-pitched *twit-twit-twit-twee-ee* flight call. **H** Broadleaved evergreen forest, forest edges, secondary growth. **D** Peninsular Malaysia, Sumatra, Java, Borneo.

30 NAKED-FACED SPIDERHUNTER *Arachnothera clarae* 17cm Noisy and conspicuous, seen singly and in pairs. Favours foraging among plantains. **V** Loud, raspy, repeated *serp-rp-rp-rp-rp-rp-rp*, insect-like *seee*, low, croaking *crrr*, rapid *trrrik* and various rapid trills. **H** Forest, forest edges, clearings, scrub. **D** Philippines.

31 GREY-BREASTED SPIDERHUNTER *Arachnothera modesta* 17–18cm Active. Forages at all forest levels, mostly in low to middle storeys; occurs in pairs or alone. **V** Continuous *tee-chu*, first note rising and second falling. **H** Broadleaved evergreen forest, secondary growth, cultivation, gardens. **D** S Myanmar to S Vietnam, Peninsular Malaysia, Sumatra, Borneo.

32 STREAKY-BREASTED SPIDERHUNTER *Arachnothera affinis* 21cm Forages at higher levels in forest, alone or occasionally in pairs. **V** Calls include *chee-wee-dee-weet… tee-ree chee chee-chur*, and piercing, ringing and raucous calls. **H** Lower montane forest, forest edges, around plantations. **D** Java, Bali, Indonesia.

33 BORNEAN SPIDERHUNTER *Arachnothera everetti* 21cm Feeds on nectar of banana and Ginger (*Zingiber officinale*) flowers, and on small invertebrates. Occurs alone or in pairs. Formerly considered race of Streaky-breasted Spiderhunter. **V** Presumed to be similar to that of Streaky-breasted Spiderhunter. **H** Lowland, lower montane and submontane forest. **D** Borneo.

34 STREAKED SPIDERHUNTER *Arachnothera magna* 19cm Noisy agile forager from treetops to low levels. **V** Song begins with soft *vijvitte vij* that accelerates then becomes rapid and monotonous; calls include chirruping *chiririrík* or *chirik chirik*; in flight utters loud musical trill. **H** Dense evergreen forest, abandoned cultivations where wild banana and plantain trees grow, gardens with flowering shrubs. **D** Himalayas to S Vietnam.

35 WHITEHEAD'S SPIDERHUNTER *Arachnothera juliae* 16–18cm Usually found foraging in tops of trees, especially among orchid clusters and other flowers growing high up on tree trunks. Forages alone, in pairs or in small groups. **V** Song high pitched and squeaking; calls include loud shrieking, prolonged twittering, nasal *swee-urr*, buzzing *erz dee erz*, *tech-tech-wee* and *tee-tee-swee-eee*. **H** Hill dipterocarp forest, montane forest, forest edges. **D** Borneo.

1
2
3
4
5
6
7
8
9
10
11
12
13
14
15
16
17
18
19
20
21
22
23
24
25
26
27
28
29
30
31
32
33
34
35

OLD WORLD SPARROWS AND SNOWFINCHES PASSERIDAE

1 CINNAMON IBON *Hypocryptadius cinnamomeus* 15cm Active, gleaning insects from foliage and branches. Regular member of mixed-species foraging flocks. **V** Soft, whistled *chuuu, pee chuuu* or *pee chuuu chuuu*, often given by a group of birds together. **H** Submontane and montane mossy forest, forest edges. **D** Mindanao, S Philippines.

2 SAXAUL SPARROW *Passer ammodendri* 15cm Wary, often stays hidden in foliage of saxaul or other shrubs. In winter, regularly forms mixed flocks with House, Spanish and Eurasian Tree sparrows. **V** Melodic chirps and a short whistle. **H** Oases and riverbeds in deserts and foothills, usually with saxaul, tamarisks or poplar thickets nearby, also at edges of desert settlements. **D** C Asia.

3 HOUSE SPARROW *Passer domesticus* 15cm In winter occurs in large flocks. **V** Song excited series of mostly call notes, *chirrup-chirrup-cheep-chirp-chirrup…*; calls include *chirrup, chirp* and *chissick*, also soft *swee-swee* and rolling *chur-r-r-it-it* given when alarmed. **H** Urban areas, including towns and cities; after breeding also in cultivations and scrub jungle. **D** Widespread native in Eurasia and Africa, widely introduced elsewhere.

4 ITALIAN SPARROW *Passer italiae* 16cm Male intermediate in plumage between Spanish and House sparrows; female very like female House Sparrow. Interbreeds with both species where ranges meet. **V** Male's advertising call most like that of Spanish Sparrow; other vocalisations very like those of both House and Spanish. **H** Cultivated and urban areas, rocky areas, light woodland. **D** Italian peninsula, Corsica, Sicily, Crete.

5 SPANISH SPARROW *Passer hispaniolensis* 15cm Usually in very large flocks and often in the company of House sparrows. **V** Song similar to that of House Sparrow but more rhythmic; calls include *chweeng chweeng*, squeaky *cheela-cheela, chirrup* and *chee-chee-chee*; all higher-pitched and more metallic than those of House Sparrow. **H** Cultivated areas, semi-desert, marshes, reedbeds. **D** SW and SC Europe, SW and S Asia, N Africa.

6 SIND SPARROW *Passer pyrrhonotus* 13cm Sociable, usually in flocks of 20 or so, occasionally with House Sparrow flocks. **V** Song House Sparrow-like, but softer with warbling twitters and a high wagtail-like note; calls also House Sparrow-like but softer and less strident. **H** Bushes and trees alongside rivers and swamps. **D** S Iran, Pakistan, N India.

7 SOMALI SPARROW *Passer castanopterus* 13cm Male with brighter rufous tones than male House Sparrow; female very like female House Sparrow. **V** Soft, House Sparrow-like chirrup. **H** Dry open areas. **D** E Africa.

8 RUSSET SPARROW *Passer cinnamomeus* 14–15cm Behaviour similar to that of House Sparrow. **V** Song frequently repeated *cheep-chirrup-cheweep* or *chwe-cha-cha*; calls include *cheep* or *chilp*, also *swee-swee* and rapid *chit-chit-chit* given when alarmed. **H** Mountain or upland forests, scrub near cultivation and villages; in winter also in grassland and fields. **D** Afghanistan to Japan and SE Asia.

9 PLAIN-BACKED SPARROW *Passer flaveolus* 13–15cm Forages mainly on ground, in loose groups with larger flocks post breeding. **V** Song and calls similar to those of House Sparrow, although slightly louder and harsher. **H** Open woodland, coastal scrub, dry open areas, cultivations, near human habitations. **D** Myanmar to S Vietnam.

10 DEAD SEA SPARROW *Passer moabiticus* 12cm Forages on ground or among foliage of bushes; usually in small flocks. **V** Song rhythmic, high-pitched *tweeng-tweeng-tweeng* or *chilung-chilung-chilung*; calls include high *trrirp*, a *chet-chet-chet-chet* and usual sparrow chirps, churrs and rattles. **H** Tamarisks and other bushes near water. **D** S Turkey to SW Afghanistan.

11 IAGO SPARROW *Passer iagoensis* 13cm Behaviour much like that of House Sparrow. Social, occurs in small, loose breeding colonies and larger flocks at other times. **V** Song loose series of call notes such as *cheep-chirri-chip-cheep-chirri-chip-chip*; call *chirrp* like House Sparrow but lower, also nasal *cheesp* or *chew-weep*. **H** Barren, arid country, stony plains, gorges, rocky cliffs, also villages and towns. **D** Cape Verde.

12 GREAT SPARROW *Passer motitensis* 15cm Both sexes more brightly coloured than House Sparrow. Note striking rufous rump in flight. Male's bib smaller in non-breeding plumage. Note pale brown upper mandible and yellowish lower mandible outside breeding season. **V** Male's call at nest a low, twanging *cheeurr*. Pairs 'converse' together with various soft calls such as *cheewee* and *chee-ti-cheet*. **H** Dry areas away from settlements. **D** S Africa.

13 SOCOTRA SPARROW *Passer insularis* 15cm Note grey rufous-edged mantle. The female has no clear eyebrow. **V** Various dry chirps and chirrups. **H** Dry scrub, woodlands, around settlements. **D** Socotra, Gulf of Aden.

14 ABD AL-KURI SPARROW *Passer hemileucus* 12.5cm Mostly pale, resembling a Sind Sparrow, but with boldly contrasting black markings on back and wings, and, in male, also black lores, eye-stripe and bib. **V** Like that of Socotra Sparrow. **H** Bushy slopes and gullies, villages and towns. **D** Abd al-Kuri island, W of Socotra, Gulf of Aden.

15 KENYA SPARROW *Passer rufocinctus* 14cm Note grey crown and chestnut rump. **V** Song combines low chirps with metallic notes. Male's call at nest loud and deeper than House Sparrow's; *chuweep* or *chreep*; other calls include nasal *jerwey* and sharp *tsui*. **H** More or less wooded and bushy natural and cultivated areas including gardens. **D** C Kenya to N Tanzania.

16 SHELLEY'S SPARROW *Passer shelleyi* 13cm Predominantly grey sparrow with an orange-buff eyebrow, rump and sides of back; female slightly duller than male. **V** Musical chirps and drier scratchy notes. **H** Mainly grassy savannah with trees, also edges of cultivation but not usually around settlements. **D** South Sudan, S Ethiopia, N Uganda, NW Kenya, NW Somalia.

17 KORDOFAN SPARROW *Passer cordofanicus* 15cm Both sexes more brightly coloured than House Sparrow. Note striking rufous rump in flight. Male's black chin spot reduced in non-breeding plumage. **V** Not well studied but said to be indistinguishable from that of Great Sparrow. **H** Dry areas away from settlements. **D** E Chad, WC Sudan.

18 CAPE SPARROW *Passer melanurus* 15cm Male unmistakable by head pattern and by rufous scapulars. Female separated from Southern Grey-headed and Swahili Sparrows by white eyebrow. **V** Song jerky, melodious for a sparrow, comprising repeated elements of calls. Male's call at nest a loud *tweeng*; other calls include typical sparrow chirps and churrs. **H** Rural areas, villages, towns. **D** S Africa.

19 NORTHERN GREY-HEADED SPARROW *Passer griseus* 15cm Separated from Southern Grey-headed Sparrow and Swahili Sparrow by contrasting white chin, yellowish-grey underparts, rusty-brown back, rather dark head. Upper mandible pale brown and lower mandible yellowish outside breeding season. **V** Male at nest produces various chipping and chirping notes, sometimes combined with more liquid *twee* notes to form a song; alarm call dry rolling *chrr-it-it*. **H** Bushy and wooded areas including cultivation and suburbs. **D** W and C Africa.

20 SWAINSON'S SPARROW *Passer swainsonii* 16cm Rather drab grey sparrow with brown wings, rufous shoulders and rump, and short white wing-bar; very like Northern Grey-headed Sparrow. **V** Male at nest utters *chirrip* call, like that of Northern Grey-headed Sparrow but more musical. **H** All kinds of open countryside, most abundant in villages and towns. **D** Eritrea through Ethiopia to E South Sudan, N Kenya and N Tanzania.

21 PARROT-BILLED SPARROW *Passer gongonensis* 18cm Resembles other African grey-headed sparrow species but is the largest, with a distinctively thick, arched bill. Usually seen in pairs or small groups. **V** Call a low *choop* that may be given in series. **H** Open country, bush, light woodland, rarely in villages but will frequent rural game lodges. **D** E South Sudan, S Ethiopia and C Somalia through Kenya to E Uganda and N Tanzania.

22 SWAHILI SPARROW *Passer suahelicus* 16cm Head, mantle and underparts concolorous grey. Has pale brown upper mandible and yellowish lower mandible outside breeding season. **V** Song a varied sequence of notes, such as *chup chu-weew chop wor chu-weet chip* and so on. **H** Like that of Northern Grey-headed Sparrow. **D** SW Kenya to S Tanzania.

23 SOUTHERN GREY-HEADED SPARROW *Passer diffusus* 16cm Underparts off-white, head mid-grey, mantle paler and more buff than wings. Has pale brown upper mandible and yellowish lower mandible outside breeding season. **V** Song a series of simple chirrups. Calls similar notes given singly. **H** Like that of Northern Grey-headed Sparrow. **D** SE and S Africa.

24 DESERT SPARROW *Passer simplex* 14cm Shy and retiring, spends much time hidden, especially in high heat of the day, usually in tops of palm trees. Feeds on ground. **V** Song a musical trill; calls include soft *chup*, high-pitched *chip-chip* and House Sparrow-like chirps. **H** Sandy plains with scattered trees or scrub, oases, wadis and in and around human settlements. **D** N Africa.

25 ZARUDNY'S SPARROW *Passer zarudnyi* 13.5–15cm Rather toneless grey sparrow with typical black facial markings; female similar but buffer. A ground feeder, taking more insects than most other sparrows. **V** Like that of Desert Sparrow. **H** Bare rocky areas, sand plains with sparse trees and bushes. **D** Turkmenistan, Uzbekistan.

26 EURASIAN TREE SPARROW *Passer montanus* 13–14cm Usually in pairs, with larger flocks post breeding; forages on ground and in trees and shrubs. **V** Song series of call notes interspersed with *tsooit, tsveet* or *tswee-ip* notes; calls include high *chip*, sharp *tet* and dry *tet-tet-tet* flight call. **H** Around habitation, including towns and cities, and in lightly wooded areas and cultivation. **D** Widespread in Europe and Asia.

27 SUDAN GOLDEN SPARROW *Passer luteus* 13cm Behaviour typical of genus. Highly gregarious. Breeds colonially. **V** Song repetition of call notes, *chirp-chirp-chirp*; call *chirp, schilp* or *tchirrup* and, in flight, rhythmic *che-che-che*. **H** Arid scrub, often near cultivation. **D** Mauritania and Senegal to Sudan, Eritrea and Ethiopia.

28 ARABIAN GOLDEN SPARROW *Passer euchlorus* 13cm Highly gregarious, breeds and roosts colonially, forms small flocks in breeding season and much larger flocks in non-breeding season. **V** Flocks give constant, whispering twitter, also noted is a subdued *chirp*. **H** Arid thorn and acacia savannah; in non-breeding season visits cereal crops and edges of towns. **D** SW Arabian Peninsula, N Somalia.

29 CHESTNUT SPARROW *Passer eminibey* 12cm Both sexes very distinctive. **V** Song a thin, high-pitched combination of *tchi, weza* and *see* notes. Calls include a dry *chrrrit* and, in flight, a ringing *chew chew*. **H** Dry wooded areas near papyrus swamp, settlements. **D** Sudan and Ethiopia to N Tanzania.

30 PALE ROCKFINCH *Carpospiza brachydactyla* 14cm Feeds on ground. In non-breeding season, often forms very large flocks. **V** Song distinctive, rising buzzing, somewhat like cicada, *tss-tss-tss-tsseeeeeeei* or *tee-zeeeze-zeeezeeei*; calls include high, nasal *twee, zweee* or *twee-oo*, also soft trill or *churr* in flight. **H** Grassy plains and hillsides with rocky outcrops, arid, stony semi-deserts, wadis, sometimes in cultivation. **D** E Turkey, Armenia and Azerbaijan to Turkmenistan, Afghanistan and Iran.

31 ROCK SPARROW *Petronia petronia* 14cm Sociable, usually in small groups. Breeds colonially. Mainly a ground feeder. **V** Song collection of repeated call notes, usually delivered from high vantage point; calls varied, vaguely similar to House Sparrow with some added metallic notes such as *cheeooee* or *pee-uoo-ee, dliu, viep* or *vi-viep*; utters *sup* or *doui* in flight. **H** Varied, including barren hills and mountain regions with rocky outcrops, gorges, ravines, desert edges, vineyards, villages and, in winter, farmland. **D** SW and SC Europe, SW Asia to C and E Asia, NW Africa.

32 YELLOW-THROATED PETRONIA *Gymnoris superciliaris* 15cm Long, conspicuous, white eyebrow diagnostic. Yellow throat patch normally concealed. **V** High, fast *tsjee-tsjee-tsjee*. **H** Woodland, thornveld, riverine belts. **D** SC and S Africa.

33 BUSH PETRONIA *Gymnoris dentata* 13cm Often sits for long periods in top of tree or bush. Usually feeds in bushes or trees, less so on ground. **V** Song fast twittering *triup-triup-triup-triup* or bunting-like *chu-chu-chu-chu*; call (in flight) soft *chewee*. **H** Wadis and terraced slopes with scattered trees. **D** Mauritania to Guinea and east to Eritrea, Ethiopia and South Sudan, also SW Arabian Peninsula.

34 YELLOW-SPOTTED PETRONIA *Gymnoris pyrgita* 13cm Note white chin and overall grey tone. Males and some females have a small concealed yellow throat patch. **V** Song a series of low *chiew* notes. Calls include a shrill *tyeet* and lower-pitched chirps. **H** More or less wooded and bushy natural and cultivated areas. **D** W, C and E Africa.

35 YELLOW-THROATED SPARROW *Gymnoris xanthocollis* 14cm Forages mostly on ground; in winter forms large flocks, often with House Sparrows and Black-headed Buntings (Plate 279). **V** Song constantly repeated, rhythmic *chilp chalp cholp*; calls include *chilp, chirrup* and harsh churring notes. **H** Open country and low hills with trees, scrub jungle, groves and trees near cultivations and villages. **D** Turkey to India and Sri Lanka.

1
2
3
4
5
6
7
8
9
10
11
12
13
14
15
16
17
18
19
20
21
22
23
24
25
26
27
28
29
30
31
32
33
34
35

OLD WORLD SPARROWS AND SNOWFINCHES *CONTINUED*

1 WHITE-WINGED SNOWFINCH *Montifringilla nivalis* 17cm Often tame and confiding. Feeds on ground in small loose flocks; forms larger flocks during winter, often with other snowfinches and mountain finches. Has a circular display-flight. **V** Monotonous *sitticher-sitticher-sitticher...*; call sharp, nasal *pschieu* or *pchie* often given in flight, also *tsee*, soft *pruuk*, and *pchurrt* when alarmed. **H** Rocky slopes and gullies, between the treeline and snowfields of high mountains; often visits buildings in ski resorts. **D** S Europe east to S and C Asia.

2 TIBETAN SNOWFINCH *Montifringilla henrici* 17cm Handsome brown-and-white snowfinch with black-and-white wing and face markings. Similar to White-winged Snowfinch in behaviour. **V** Like that of White-winged Snowfinch. **H** High-altitude steppe between the treeline and snowline. **D** Tibet.

3 BLACK-WINGED SNOWFINCH *Montifringilla adamsi* 17cm Forages on ground, walks and runs in lark-like fashion; in flight shows much white on secondaries, secondary coverts and sides of tail. **V** Song consists of a single note repeated monotonously; calls include sharp *pink-pink* and soft mewing; flocks utter a constant twittering. **H** High stony plateaux, cliffs and rocky slopes, cultivated fields near villages. **D** Tibet, Nepal.

4 WHITE-RUMPED SNOWFINCH *Onychostruthus taczanowskii* 17cm Forages on ground; often occurs in large winter flocks. **V** Song short, loud *duid-ai-duid duid-duid-ai*; call sharp, resounding *duid duid*. **H** High stony plateaux. Regularly found around pika colonies; uses their burrows to roost, breed and shelter. **D** Tibet, W China.

5 PERE DAVID'S SNOWFINCH *Pyrgilauda davidiana* 15cm Regularly flicks wings while perched atop low stone or earth mound. Usually occurs in large flocks after breeding. **V** Flocks utter constant twittering. **H** Stony mountains and semi-desert with sparse grass; usually occurs among colonies of small burrowing animals. **D** C Asia.

6 RUFOUS-NECKED SNOWFINCH *Pyrgilauda ruficollis* 15cm Forages on ground; during winter encountered in small flocks, often with other snowfinches. **V** Calls include soft *duuid* or *doooid* and a magpie-like chattering. **H** High, barren, stony steppe, grassy plateaux. **D** Tibet, Nepal, W China.

7 BLANFORD'S SNOWFINCH *Pyrgilauda blanfordi* 15cm Runs around, mouse-like, while foraging on ground. In winter occurs in flocks, often in the company of finches, sparrows and other snowfinch species. **V** Constant twittering. **H** Dry, stony steppe, hillsides with stunted grass. **D** Tibet, Nepal, W China.

8 AFGHAN SNOWFINCH *Pyrgilauda theresae* 14cm Little known. Ground feeder. In winter, joins flocks of White-winged snowfinches, Horned larks (Plate 200) and Rock sparrows (Plate 261). **V** Undescribed. **H** High mountains, open stony hillsides and plateaux. **D** Afghanistan.

WEAVERS AND WIDOWBIRDS PLOCEIDAE

9 WHITE-BILLED BUFFALO WEAVER *Bubalornis albirostris* 25cm Unmistakable. White bill of male turns to black in non-breeding plumage. **V** Song *chiew-chiew-chiew-chiew-tcheep*. Calls include a shrill *tyip* and various chirping notes. **H** Wooded and bushy areas. **D** Mauritania and Senegal to Sudan, Eritrea, Ethiopia and W Kenya.

10 RED-BILLED BUFFALO WEAVER *Bubalornis niger* 23cm Breeds in large, untidy group nests, several in a large tree. **V** Mid-high, fast chatter, slightly falling away, *cha-cha-cha...* **H** Dry, wooded areas. **D** E, SE and S Africa.

11 WHITE-HEADED BUFFALO WEAVER *Dinemellia dinemelli* 24cm Unmistakable by colour pattern and red rump. **V** Song slow and varied, comprising trills and piercing nasal notes. Calls include a powerful *kiyerr* and ringing *tew*. **H** Wooded habitats, savannah. **D** E and EC Africa.

12 WHITE-BROWED SPARROW-WEAVER *Plocepasser mahali* 17cm Striking, shows white rump in flight. Builds untidy, communal nests in thorn trees. Feeds on ground. **V** Mid-high, pleasant nasal chattering,*'tjiau-tji-tjauw-tjauw-ti-ti-ti*. **H** Thornveld. **D** E and S Africa.

13 CHESTNUT-CROWNED SPARROW-WEAVER *Plocepasser superciliosus* 17cm Note rufous rump. **V** Song a short series of jumbled notes with a final trill. Calls include a soft *tik*. **H** Dry wooded and bushed areas. **D** Senegal and Gambia to Eritrea, Ethiopia and W Kenya.

14 DONALDSON SMITH'S SPARROW-WEAVER *Plocepasser donaldsoni* 15cm Note heavy bill. **V** Song varied and rambling, including whistles, trills and harsh notes. Calls include a loud *chink chink* and quieter *cluk cluk*. **H** Dry rocky areas with some grass and scattered trees. **D** S Ethiopia to C Kenya and S Somalia.

15 CHESTNUT-BACKED SPARROW-WEAVER *Plocepasser rufoscapulatus* 18cm Note whitish crown. Behaviour like that of White-browed Sparrow-Weaver. **V** Song a loud jumble of varied notes. Call a high-pitched *srp*. **H** Miombo. **D** Angola and S DR Congo to Zambia and Malawi.

16 RUFOUS-TAILED WEAVER *Histurgops ruficauda* 20cm Large, with pale eyes and scaly plumage; rufous rump and tail. **V** Song a prolonged series of nasal slurring and grating notes. Call a harsh *skweuk*. **H** More or less wooded areas in dry hilly country. **D** Tanzania.

17 GREY-CAPPED SOCIAL WEAVER *Pseudonigrita arnaudi* 13cm Note whitish-grey cap. **V** Song combines about 12 high-pitched notes; in territorial dispute gives a loud *cheep*. **H** Wooded and bushy areas with some grass cover. **D** E and C Africa.

18 BLACK-CAPPED SOCIAL WEAVER *Pseudonigrita cabanisi* 13cm Small, top-heavy weaver, with crisply demarcated black cap, white underside, heavy pale bill and bright red eyes. **V** Sparrow-like chirps and chatters. **H** Dry wooded and bushy areas. **D** Ethiopia and Somalia to NE Tanzania.

19 SOCIABLE WEAVER *Philetairus socius* 14cm Breeds in large, communal nests (50 pairs or more) which are used for many years. **V** High, excited, collective chattering near nest. **H** Woodland, thornveld. **D** Namibia to Botswana and South Africa.

20 SCALY-FEATHERED WEAVER *Sporopipes squamifrons* 13cm Delicately marked. Occurs in small groups. Compare with Sociable Weaver. **V** Song a repeated two-note phrase, *kreep krop, kreep krop...* Contact call a rolling *sirrip*. **H** Dry thornveld, near settlements. **D** S Africa.

21 SPECKLE-FRONTED WEAVER *Sporopipes frontalis* 13cm Small. Always seen in flocks, breeds in colonies. **V** Song a fast series of high-pitched *tsi* and *tee* notes. In flight, calls *tsip-tsip-tsip*. **H** Bushy and wooded areas, often near water. **D** W, C and E Africa.

22 THICK-BILLED WEAVER *Amblyospiza albifrons* 18cm Occurs in pairs or loose colonies in tall reeds and elephant grass. **V** Song combines dry clicks and high trills. Contact call a loud *chuck*. **H** Marshes, swamps, wet places in forest and woodland. **D** Widespread in Africa.

23 BAGLAFECHT WEAVER *Ploceus baglafecht* 14cm The only black-cheeked weaver with yellow chin and throat. Shows some racial variation in extent of black on head. **V** Song a two- or three-part series of mixed tuneful and hissing sounds. Calls include *pseet*, *swee-chit* and *zwenk*. **H** mMontane forest edges. **D** C, E and SE Africa.

24 BANNERMAN'S WEAVER *Ploceus bannermani* 14cm Separated from Spectacled Weaver by black cheeks and all-yellow breast. Female similar to male. **V** Song a series of accelerating tuneful, high-pitched notes, ending in a long wheezing sound. Contact call a sharp *prit*. **H** Montane forest. **D** Nigeria, Cameroon.

25 BATES'S WEAVER *Ploceus batesi* 14cm Male has unique cherry-red mask. Note dark eye of male and female. **V** Not known. **H** Forest. **D** Cameroon.

26 BLACK-CHINNED WEAVER *Ploceus nigrimentus* 15cm Separated from most other male weavers except Black-necked Weaver and Dark-backed Weaver (Plate 263) by black mantle. **V** Song a short *whit-pu pui trr pui* phrase. Calls include a harsh *chichichip*. **H** Woodland. **D** E Gabon, Congo, C Angola.

27 BERTRAM'S WEAVER *Ploceus bertrandi* 15cm Note orange-brown forecrown of male and plain mantle in both sexes. **V** Song begins with a squeaky phrase followed by swishing and clicking sounds, petering out at end. Contact call a sparrow-like chirp. **H** Forest edges, woodland, cultivation and often along streams, in montane areas. **D** N Zambia and C Tanzania through Malawi to N Mozambique.

28 SLENDER-BILLED WEAVER *Ploceus pelzelni* 12cm Note small size and thin bill. Female has uniform yellow face sides and black bill. Forms small colonies. **V** Song combines chattering and hissing or swishing sounds, plus *si-si-si* notes. Calls include a buzzy *bzzzzt*. **H** Marshes, swamps, damp woodland. **D** C and W Africa.

29 LOANGO WEAVER *Ploceus subpersonatus* 15cm Overall golden-yellow with black head and uniform green mantle and wings. **V** Song a quiet jumbled mixture of typical, rather toneless weaver notes. Calls include a *chreep* and *dseerp*. **H** Open coastal forest near water. **D** Coastal Gabon, Congo and DR Congo.

30 LITTLE WEAVER *Ploceus luteolus* 12cm Small with short bill; lacks orange tones. **V** Song begins with a few sweet notes and then becomes a fast, dry chatter. Calls include a quiet *tsip* and a shriller note given in alarm. **H** Woodland, thornbush, gardens. **D** W, C and E Africa.

31 SPECTACLED WEAVER *Ploceus ocularis* 14cm Female separated from male Orange Weaver by lack of black in wing. Skulking. Not colonial. **V** Very high, lowering, fast, wader-like *fwee-fwee-fwee-fwee-fwee*. **H** Dense vegetation at forest edges, riverine belts, suburbs. **D** C, E and SE Africa.

32 BLACK-NECKED WEAVER *Ploceus nigricollis* 14cm Note extensively black upperparts of male and female and black, narrow bib of male. Not colonial. **V** Song a short *prr-it tew tew twee* series. Calls include a metallic *treeng treeng*; alarm call a sharp, twanging note. **H** Forest, moist woodland, riverine belts. **D** W, C and E Africa.

33 STRANGE WEAVER *Ploceus alienus* 14cm Combines rufous bib with black (not dark brown) head and olive-green mantle. **V** Song a short, rhythmic *wee chow-chow-chow* phrase. Calls include a short phrase of descending whistles. **H** Montane forest edges, bamboo. **D** SW Uganda, E DR Congo, Rwanda, Burundi.

34 BLACK-BILLED WEAVER *Ploceus melanogaster* 13cm Unmistakable. **V** Song a short series of tuneful notes, ending with a buzz. Contact call a rapid *zhink-zhink-zhink*. **H** Forest, near the ground. **D** C Africa.

35 CAPE WEAVER *Ploceus capensis* 15cm Note large size, pale eye and strong, rufous wash on head. Non-breeding male like female, but with pale eye. Female separated from other female weavers by size and heavy bill. Forms small colonies of up to 10 pairs. **V** Song a long (up to 27 secs) series of monotone chatters and wheezes. Calls include a *chuck* and, when alarmed, a harsher *chak*. **H** Trees and reedbeds in woodland, suburbs. **D** South Africa.

36 BOCAGE'S WEAVER *Ploceus temporalis* 13cm Note pale eyes of male, olive-green cheeks and narrow, rufous bib. Non-breeding male like female. Forms small colonies. **V** Song a prolonged series of chattering notes. Calls include a *chyap*. **H** Riverine belts, reedbeds, tall grass; found at middle to high altitudes. **D** Angola, S DR Congo, N Zambia.

37 EASTERN GOLDEN WEAVER *Ploceus subaureus* 13cm Male uniform golden-yellow, female greenish above with faint streaks and dark brown eyes. Non-breeding male like female but with light, rufous wash on chin. Colonial. **V** Song begins with a few *tsip* notes, then becomes prolonged swishing, rolling sound. Alarm call a *tzit*. **H** Reedbeds, thornveld, open woodland, riverine belts. **D** E and S Africa.

38 HOLUB'S GOLDEN WEAVER *Ploceus xanthops* 16cm Note large size, creamy eye, black bill, uniform colouring. Not colonial. **V** Song a short chatter, ending with a trill and some squeaking notes. Contact call a brief *chirp*. **H** Reeds, bushy wetland areas. Sometimes in suburbs. **D** Gabon to Uganda and Kenya south to N Namibia, N Botswana and E South Africa.

39 ORANGE WEAVER *Ploceus aurantius* 14cm Note golden-yellow colouring, fine bill, black lores in male. Male and female have dark red eyes. Colonial. **V** Song a pleasant rambling series of musical notes that may be long or short. Calls include a loud *kew-kewp*. **H** Forest edges, swamps, reedbeds. **D** W and C Africa.

40 HEUGLIN'S MASKED WEAVER *Ploceus heuglini* 14cm Not highly social. Note pale yellow eye and yellow-orange forehead. **V** Song a long series of chattering and hissing sounds, but variable. Calls include a quiet *charr* and grating *chuk*. **H** Dry woodland. **D** Senegal and Gambia, Mali to Ivory Coast and east to Uganda and W Kenya.

1
2
3
4
5
6
7
8
9
10
11
12
13
14
15
16
17
18
19
20
21
22
23
24
25
26
27
28
29
30
31
32
33
34
35
36
37
38
39
40

WEAVERS AND WIDOWBIRDS *CONTINUED*

1 GOLDEN PALM WEAVER *Ploceus bojeri* 15cm Social. Note red (not pale) eye, saffron head and edge around yellow throat. **V** Song a series of creaking sounds; birds at colonies maintain constant low chattering. Calls include a harsh *tjip*, a sweet *ti-tye* and an emphatic *kik kik*. **H** Bushy and wooded areas. **D** E Africa.

2 TAVETA WEAVER *Ploceus castaneiceps* 14cm Social. Note saffron edge of 'missing' mask. **V** Song low-pitched, rather tuneless and harsh. Contact call a grating *jick*. **H** Reedbeds, papyrus and shrubbery along streams and swamps. **D** Kenya, Tanzania.

3 PRINCIPE WEAVER *Ploceus princeps* 15cm Note orange head. One of the most common birds on Príncipe. **V** Song begins with chirping notes that accelerate into a buzz and then end with slower squeaky notes. Calls include a *zeet* as a contact call and a *ksuip*. **H** Forest edges, bush, villages. **D** Príncipe, Gulf of Guinea.

4 NORTHERN BROWN-THROATED WEAVER *Ploceus castanops* 14cm Social. Note heavy bill, small rufous mask and plain mantle. **V** Song combines chattering and squeaking sounds; occurs in long and short forms. **H** Breeds in papyrus and reedbeds, but wanders to forest edges and other wooded areas when not breeding. **D** Uganda and E DR Congo to W Kenya and N Tanzania.

5 SOUTHERN BROWN-THROATED WEAVER *Ploceus xanthopterus* 14cm Male separated from male Cape Weaver (Plate 262) by brown eye and plain mantle; female from all other female weavers by lack of pale eyebrow. Non-breeding male as female. Colonial. **V** Song a series of chattering, buzzing, nasal and hissing sounds. Calls include hissing notes and a sharp *jick*. **H** Reedbeds, riverine belts, swamp edges. **D** S Africa.

6 KILOMBERO WEAVER *Ploceus burnieri* 15cm Social. Note sharply demarcated mask. Less saffron round mask edge than in Northern Masked Weaver. **V** Song a rambling, variable series of chipping, squeaking and other notes. Contact call a *chuk*. **H** Riverside swamp. **D** Tanzania.

7 RÜPPELL'S WEAVER *Ploceus galbula* 14cm Gregarious, frequently in large flocks, often accompanied by sparrows. Breeds colonially, nest suspended from tree branches. **V** Song wheezy chatter, ending in insect-like hiss; call dry *cheee-cheee*. **H** Savannahs and wadis with acacia or other bushes, also palm groves and crops. **D** Sudan to Somalia and extreme N Kenya, SW Arabian Peninsula.

8 NORTHERN MASKED WEAVER *Ploceus taeniopterus* 15cm Very social. Black mask merging gradually into rufous of crown and breast. **V** Song a series of unmusical notes with hesitant, jerky delivery. Alarm call harsh and grating. **H** Reedbeds and nearby bushes. **D** EC Africa.

9 LESSER MASKED WEAVER *Ploceus intermedius* 12cm Note absence of orange tones and diagnostic pale eye in both sexes. Female is more yellow than most other female weavers. Seen in pairs and large colonies, often together with Village Weaver. **V** Song comprises swishing and chattering sounds with some other elements and a final flourish. Calls include *pew pew pew* and *tsop-tsop*. **H** Nests in trees over water or in reedbeds. Open thornbush, woodlands. **D** E, SE and S Africa.

10 SOUTHERN MASKED WEAVER *Ploceus velatus* 12cm Separated from Lesser Masked Weaver and Tanzanian Masked Weaver by narrow black on forehead, red eye, faintly streaked mantle; female separated from other female weavers by yellowish throat, buffish breast, whitish underparts. Non-breeding male like female. Small to large colonies. **V** Song a long jangling chatter with some buzzy and chattering elements. Calls include a sharp *chik*. **H** More or less wooded areas, often away from water; suburbs. **D** S Angola to Mozambique and south to S and E South Africa; São Tomé, Gulf of Guinea.

11 KATANGA MASKED WEAVER *Ploceus katangae* 14cm Very similar to Vitelline Masked Weaver, but not in same range and with different habitat. **V** Song a prolonged chatter. Contact call a *chuk*. **H** Swamps. **D** SC Africa.

12 LUFIRA MASKED WEAVER *Ploceus ruweti* 14cm Separated from Vitelline Masked Weaver by habitat, restricted range and orange-rufous breast. **V** Song combines wheezing sounds with hard tacking notes. Calls mostly brief, but territorial call includes a short trill. **H** Swamp. **D** DR Congo.

13 TANZANIAN MASKED WEAVER *Ploceus reichardi* 14cm Male separated from Southern Masked Weaver by more extensive, rufous colouring, female by fainter streaking above. **V** Song an unmusical series of chirping, chattering and trilling notes. Calls include *click* and a fast *chut-chut*. **H** Swampy areas. **D** Tanzania, NE Zambia.

14 VITELLINE MASKED WEAVER *Ploceus vitellinus* 14cm Less social than most masked weavers. Note greenish (not yellow) neck. Demarcation between black and saffron-rufous on crown normally rather sharp and in front of eye. **V** Song combines swishing, whistling, rasping and chirping notes. Calls include a sharp *pink*. **H** Wooded and bushy natural cultivated areas. **D** W, C and E Africa.

15 SPEKE'S WEAVER *Ploceus spekei* 15cm Social. Combines black face mask and heavy mantle streaking. **V** Song a long series of nasal notes that accelerate into a chatter. Contact call a sharp *tseep*. **H** Wooded and bushy areas, including cultivation and gardens. **D** Ethiopia and Somalia to N Tanzania.

16 FOX'S WEAVER *Ploceus spekeoides* 15cm Social. Note almost completely dark mantle and wings. **V** Not known. **H** Wet wooded and bushy habitats. **D** Uganda.

17 VILLAGE WEAVER *Ploceus cucullatus* 17cm Gregarious when foraging and breeding, colonies can reach up to 100 nests in one tree. **V** High-pitched chatter interspersed with squeaks and churrs. **H** Bushes and scrub near water, rice fields, open woodland, trees around urban areas. **D** Widespread in Africa.

18 GIANT WEAVER *Ploceus grandis* 21cm Resembles an unusually large Village Weaver. **V** Song a long chatter, finishing with a wheezing note. In flight, call a repeated *prric*. **H** Forest, woodland. **D** São Tomé, Gulf of Guinea.

19 VIEILLOT'S BLACK WEAVER *Ploceus nigerrimus* 14cm Male unmistakable. Female greenish, not black as female Maxwell's Black Weaver and Yellow-Legged Weaver (Plate 264). Very social. **V** Song a rapid, stuttering series of tweets and chatters. Calls include a *chak*, an upslurred *way-yee* and a whistling *sweet*. **H** Forest edges, woodland. **D** W and C Africa.

20 WEYNS'S WEAVER *Ploceus weynsi* 15cm Separated from Village Weaver by dark brown back and pale eye. **V** Song a very high-pitched hissing, sizzling phrase. Calls include a high-pitched *chip*. **H** Forests. **D** Uganda and E DR Congo.

21 CLARKE'S WEAVER *Ploceus golandi* 13cm Social and often in mixed-species flocks; breeding habits unrecorded. **V** Flocks produce chattering, hissing and sharp chipping notes. **H** Sokoke Forest. **D** Kenya.

22 JUBA WEAVER *Ploceus dichrocephalus* 15cm Social. Note mottled transition from head to rest of body. **V** Song a rambling jumble, with squeaks and buzzing notes. **H** Wooded and bushy areas near swamps and water. **D** C Ethiopia, extreme N Kenya, S Somalia.

23 BLACK-HEADED WEAVER *Ploceus melanocephalus* 14cm Note yellow neck and plain mantle. Non-breeding male like female. Colonial. **V** Song includes wheezes, and grating and creaking sounds, and has long and short forms. Alarm call a harsh *chak*. **H** Riverine belts, lake edges, swamps. **D** W, C and EC Africa.

24 GOLDEN-BACKED WEAVER *Ploceus jacksoni* 15cm Very social. Note contrasting yellow mantle. **V** Song a prolonged, complex mixture of sounds, including hisses and buzzes. **H** Wooded and bushy areas near swamps and water. **D** S South Sudan to Tanzania.

25 CINNAMON WEAVER *Ploceus badius* 14cm Male distinctive, with a black head and chestnut-and-yellow body plumage. Female very like female Northern Masked Weaver. Gregarious and colonial. **V** Similar to that of Northern Masked Weaver; song clearer and (at colonies) given more frequently. **H** Grassland with trees and bushes, near rivers. **D** Sudan, South Sudan.

26 CHESTNUT WEAVER *Ploceus rubiginosus* 14cm Note brownish (not greenish as most other female weavers) plumage of female. Colonial breeder. **V** Song comprises loud, sizzling notes. Calls include a high-pitched *pyewp* and *chik-chik*. **H** Dry woodland, bushland. **D** E and SW Africa.

27 GOLDEN-NAPED WEAVER *Ploceus aureonucha* 16cm Separated from Preuss's Weaver by greenish underparts. No yellow stripe through mantle in female. **V** When flocking, gives a weak chattering. **H** Forests. **D** DR Congo.

28 YELLOW-MANTLED WEAVER *Ploceus tricolor* 14cm Unmistakable by tricoloured pattern; yellow in both sexes restricted to mantle. **V** Song varied, including chattering notes and tuneful two-note whistles and swishing sounds. Alarm call a rasping note. **H** Forest canopy. **D** W and C Africa.

29 MAXWELL'S BLACK WEAVER *Ploceus albinucha* 13cm Shows racial variation in presence or absence of a white or greyish neck patch. Social. **V** Possible song a soft, swishing sizzling phrase. Contact call a *chick chick*. **H** Forests. **D** W and C Africa.

30 NELICOURVI WEAVER *Ploceus nelicourvi* 15cm Plumage pattern distinctive by black mask (male, ghosted in female), yellow collar, grey underparts and rufous undertail coverts. **V** Calls include swishing sounds and a nasal *tiang tiang*. **H** Mainly found in humid forests. **D** Madagascar

31 SAKALAVA WEAVER *Ploceus sakalava* 15cm Yellow head of male and pinkish-orange head of female very distinctive. **V** Song a repeated *chee-chee*. In flight, calls *treeyoo*. **H** Dry, open country. **D** Madagascar.

32 ASIAN GOLDEN WEAVER *Ploceus hypoxanthus* 15cm Gregarious, breeding in small colonies. **V** Song comprises chattering notes, ending with long rattle; call harsh *chit*. **H** Marshes, paddyfields, flooded grassland. **D** Myanmar to Java, Indonesia.

33 COMPACT WEAVER *Ploceus superciliosus* 12cm Note solid brown upperparts. Pairs or small groups. **V** Song a long series of chirping *cheerr* and *chrray* notes. Calls include a ticking note in flight, and a two-note *cheewery*. **H** Forest, swamps, woodland, bushveld, cultivation. **D** Senegal to Liberia, east to Ethiopia and south to Angola and S DR Congo.

34 BLACK-BREASTED WEAVER *Ploceus benghalensis* 14cm Variable face pattern; intermediates occur with black ear-coverts and white throat. Gregarious, often with other weavers, waxbills and starlings. **V** Song soft series of sibilant notes ending with a low buzzing; in flight utters quiet *chit-chit*. **H** Tall grass or reedy areas near water. **D** Pakistan to SC China.

35 STREAKED WEAVER *Ploceus manyar* 14cm Gregarious. Forages and roosts in flocks, breeds in small scattered colonies. **V** A trill of high-pitched whistles that ends with a wheezy note. Calls include a *re tre cherrer cherrer*, usually given by displaying males, and a *chirt-chirt* given in flight. **H** Reedbeds, reed swamps, tall grass and seasonally flooded areas. **D** Widespread in S and SE Asia, also Java and Bali, Indonesia.

36 BAYA WEAVER *Ploceus philippinus* 15cm Highly gregarious. Roosts and forages in large flocks, often with other weavers, waxbills, sparrows and starlings. **V** Chittering followed by wheezy whistle, a buzz and some chirps. **H** Generally open areas with nearby water, including grassland scrub with scattered trees, paddyfields, cultivated areas and mangroves. **D** Pakistan and India to Indonesia.

37 FINN'S WEAVER *Ploceus megarhynchus* 17cm Gregarious, often with other weavers. **V** Loud, harsh chatter ending with a wheezy note; flight call *twit-twit*. **H** Grassland with scattered trees, paddyfields, tall grass, sugarcane. **D** N India, Nepal, S Bhutan.

38 DARK-BACKED WEAVER *Ploceus bicolor* 13cm Chin finely speckled. Not colonial. **V** Lashing *weep* followed by delightful, toy trumpet-like *pwee pwee pu-weep*. **H** Forest, moist woodland. **D** Widespread in Africa.

39 PREUSS'S WEAVER *Ploceus preussi* 15cm Note black chin and throat. Woodpecker-like feeding behaviour. **V** A harsh *chwep* has been documented. **H** Forests. **D** Guinea to Ghana, Cameroon to Gabon, E DR Congo.

white-eared morph

WEAVERS AND WIDOWBIRDS *CONTINUED*

1 YELLOW-CAPPED WEAVER *Ploceus dorsomaculatus* 14cm Separated from similar black-chinned Preuss's Weaver (Plate 263) by yellow throat. **V** Not known. **H** Forests. **D** Cameroon to Central African Republic, Gabon and Congo, E DR Congo.

2 OLIVE-HEADED WEAVER *Ploceus olivaceiceps* 14cm Male separated from Bocage's Weaver (Plate 262) by red, not yellow, eye and different range. Non-breeding male like female. Not colonial. **V** Song a short phrase with four descending notes at end. Calls include a quiet *tssp tsssp, twee-wheep-wheep* and chattering given by flocks. **H** Miombo. **D** Tanzania, E Zambia, Malawi, Mozambique.

3 USAMBARA WEAVER *Ploceus nicolli* 14cm Not social. Brown (not black) head merges in darker mantle. **V** Song combines *swi-iri swi-iri swi-iri* phrases in a long, rhythmic series. Calls a *chip* and *chak*. **H** Forest. **D** Tanzania.

4 BROWN-CAPPED WEAVER *Ploceus insignis* 13cm Separated from Yellow-capped Weaver and Preuss's Weaver (Plate 263), which have similar patterned upperparts, by brown or black crown. **V** Song includes descending whistles combined with swishing notes. Contact call a piercing *siip siip*. **H** Forests. **D** E Nigeria and W Cameroon, W Angola, S South Sudan to E DR Congo and W and N Tanzania.

5 BAR-WINGED WEAVER *Ploceus angolensis* 12cm Unmistakable by wing-bars, yellow rump, thin bill. Not colonial. Often in mixed-species parties. **V** Song a series of fast, melodious notes that increase in volume. Contact call a single squeaky *tyoo*, sometimes extended with additional lower-pitched notes to *tyoo-vo-vo-vo*. **H** Woodland. **D** Angola, S DR Congo, Zambia.

6 SAO TOME WEAVER *Ploceus sanctithomae* 14cm Nuthatch-like feeding behaviour. **V** Song begins with slow, pure notes and accelerates into a chatter. Flocks are noisy and give frequent contact calls, *tchim-tchim-tcholo, psink* and, when alarmed, a harsher *chuk*. **H** Moist wooded and bushy grassland. **D** São Tomé, Gulf of Guinea.

7 YELLOW-LEGGED WEAVER *Ploceus flavipes* 15cm Faintly green-scaled. Yellowish legs diagnostic. Usually in pairs or small groups. **V** Not known. **H** Forests. **D** DR Congo.

8 RED-CROWNED MALIMBE *Malimbus coronatus* 16cm Red of male restricted to crown. **V** Song a series of very high-pitched trills. At colonies, birds vocalise constantly with chattering notes and, in confrontations, a sharp *pi-tsit-tsit*. **H** Forests. **D** Cameroon to E DR Congo.

9 CASSIN'S MALIMBE *Malimbus cassini* 16cm Separated from Red-vented Malimbe and Ibadan Malimbe by narrower red-orange breast band and black undertail coverts. Female all black. **V** Song begins with harsh notes and develops into a nasal sizzling sound. Calls by groups include *teuc teuc* and *tsip tsip*. **H** Wet parts of forests. **D** Cameroon to W and E DR Congo.

10 RACHEL'S MALIMBE *Malimbus racheliae* 16cm From Golden-naped Weaver (Plate 263) by yellow undertail coverts. Female separated from female Gola Malimbe by more orange on breast. **V** Song brief with a buzzing final sound. Contact call a harsh *zhep-zhep*. **H** Forest. **D** SE Nigeria to C Gabon.

11 GOLA MALIMBE *Malimbus ballmanni* 16cm Male has unique yellow ruff at neck. Note yellow undertail coverts. **V** Both sexes produce a monotonous song of *cheg* and *chig* notes, male's version ending in a long wheeze. Contact call *chch chchch*. **H** Forests. **D** E Sierra Leone through Liberia to Ivory Coast.

12 RED-VENTED MALIMBE *Malimbus scutatus* 16cm Like Ibadan Malimbe but with more extensively red undertail coverts. Female separated from Blue-billed Malimbe by red undertail coverts and different bill colour. **V** Song a series of melodious notes and phrases. Calls include nasal chirping, a rasping *zee-zee-zee* and a grating *tuk-tuk-tuk*. **H** Forests. **D** W Africa.

13 IBADAN MALIMBE *Malimbus ibadanensis* 17cm Separated from Red-vented Malimbe (with more extensive range and different habitat) by less or no red to undertail coverts. Note narrow red girdle over breast of female. **V** Both sexes sing, uttering a short *chup ee wrr zzzzzzzz* phrase. **H** Open woodland, cultivation. **D** Nigeria.

14 BLUE-BILLED MALIMBE *Malimbus nitens* 16cm Note pale bill. See also female Red-vented Malimbe. **V** Male's song combines warbling phrases with shrill notes, its composition varying according to whether male is defending territory or actively courting. Calls include a *zheep* and *zhep-zhep*. **H** Moist forest often along streams. **D** Senegal and Gambia to Liberia and east to Uganda and E DR Congo, south to N Angola.

15 RED-HEADED MALIMBE *Malimbus rubricollis* 17cm Unmistakable but compare Cassin's Malimbe. Usually seen in pairs. **V** Song combines sizzling and whistling notes; female may join male in duet. Calls include a *chee* and harsh *zheet*. **H** Forest, woodland, dense bushveld. **D** W and C Africa.

16 RED-BELLIED MALIMBE *Malimbus erythrogaster* 16cm Unmistakable. **V** Song combines phrases such as *tya-tya-tya* and *tsee-tsee-tsee-chu-way*. Contact call a dry *ptsik*. **H** Forest. **D** SE Nigeria to Congo and east to Uganda and E DR Congo.

17 CRESTED MALIMBE *Malimbus malimbicus* 16cm Male has a small crest. Separated from other malimbes by black neck. **V** Song may be very short or longer than 30 secs; includes twitters and sizzling notes. Contact call a harsh *scree scree scree*. **H** Forest, dense thickets. **D** W and C Africa.

18 RED-HEADED WEAVER *Anaplectes rubriceps* 14cm Non-breeding male like female. In breeding season occurs in small groups of one male with several females. Nomadic outside breeding season. **V** Song includes high-pitched, fast squeaky and sizzling sounds. In flight, calls include *tsi* or *tsi-woo*, and alarm call a high-pitched *chi-chi*. **H** Woodland, thornbush. **D** Widespread in Africa.

19 CARDINAL QUELEA *Quelea cardinalis* 10cm Male separated from Red-headed Quelea by gradual transition of red from head to rest of body; female by yellowish chin. Gregarious (but less so than Red-bellied Quelea). **V** Song begins with some sharp notes, then combines buzzes and sizzling sounds and ends with a nasal whistle. Contact call a quiet *zeet*. **H** Grassland, farmland. **D** E and EC Africa.

20 RED-HEADED QUELEA *Quelea erythrops* 11cm Note sharply defined red of male mask and in female dark mark near base of bill, yellow eyebrow and white (not yellow) chin. Small flocks. **V** Song a series of churring notes. Flocks produce constant chattering and *tyap* contact calls. **H** Marsh shrubbery, moist grassland. **D** Senegal and Gambia to Ethiopia and south to E South Africa.

21 RED-BILLED QUELEA *Quelea quelea* 12cm Variable, but red bill diagnostic in both sexes and in all colour forms. Very gregarious (flocks of far more than 10,000 may occur). **V** Song a short phrase, beginning with a chatter, followed by warbling and ending with a long whistle. Calls include *chirt chirt chirt, tseep* and a harsh *chak*. **H** Dry bushy natural and cultivated areas. **D** Widespread in Africa.

22 RED FODY *Foudia madagascariensis* 13cm All-red male unmistakable, but variations and moulting birds often resemble other fody species. Spends more time on ground than other fodies. Female not safely separable from other female fodies; differs from female House Sparrow by smaller size. **V** Very high *tsip*, often in rapid series. **H** Many types of habitat but less frequent in closed forest at any altitude. **D** Madagascar.

23 COMOROS FODY *Foudia eminentissima* 13cm Bill slightly larger and wing-bars more distinct than in Red Fody. **V** Like that of Red Fody but slightly lower. **H** Wooded areas. **D** Comoros islands.

24 ALDABRA FODY *Foudia aldabrana* 13cm Male has a red face, breast and rump, and black lores and eye-stripe. Omnivorous and opportunistic, joining mixed foraging flocks. **V** Varied combination of metallic, trilling and fizzing notes. **H** Mixed scrub, mangroves, coconut groves, woodland. **D** Aldabra Islands, W Indian Ocean.

25 FOREST FODY *Foudia omissa* 14cm Difficult to separate from variations of Red Fody, but underparts are slightly yellower. **V** Very high thin siffling. **H** Humid forest and other types of woodland; up to high altitudes. **D** Madagascar

26 MAURITIUS FODY *Foudia rubra* 14cm Not safely separable from variations of Red Fody, but bill slimmer and tail slightly shorter. **V** Series of high *tjip* notes. **H** Most types of habitat, especially with softwoods. **D** Mauritius.

27 SEYCHELLES FODY *Foudia sechellarum* 12cm If present, yellow on face is diagnostic in male; note sharp bill. Darker individuals and females not safely separable from variations of Red Fody. **V** Series of high, sharp *tjirps*. **H** Prefers humid forest but now found in most habitats. **D** Seychelles.

28 RODRIGUES FODY *Foudia flavicans* 12cm Male unmistakable by yellow head and breast. Brown variants differ from other fodies by smaller, sharper bill and by call. **V** Series of very high, slightly hoarse *tjirps*. **H** Prefers habitats with tall trees and closed canopy. **D** Rodrigues, northeast of Mauritius.

29 BOB-TAILED WEAVER *Brachycope anomala* 12cm Unmistakable by short tail, even shorter than in Compact Weaver (Plate 263), but beware of other moulting weavers. **V** Song a short series of liquid chirps; call a harsh *jit*. **H** Open spaces along forest streams. **D** SE Cameroon and S Central African Republic through Congo and DR Congo.

30 YELLOW-CROWNED BISHOP *Euplectes afer* 12cm Generally occurs in flocks. Non-breeding males look very similar to breeding females. **V** Unmusical, monotonous chipping, given perched or during display flight. **H** Reeds and tall grass near fresh water. **D** W, C, E and S Africa.

31 FIRE-FRONTED BISHOP *Euplectes diademätus* 10cm Note orange forehead patch. **V** Song a short, thin sizzling phrase. Calls include a liquid *tilly* and *plee-pleew*; contact call *chweep*. **H** Dry bushy areas, also rice fields. **D** C Somalia through Kenya to NE Tanzania.

32 BLACK BISHOP *Euplectes gierowii* 12cm Female shows dark-mottled moustachial area and buff-coloured breast. Non-breeding male rather dark with blackish wings. **V** Song a long buzz followed by a short sequence of shorter buzzy notes. Calls said to be high-pitched, sizzling notes. **H** Bushy tall grassland, sugarcane. **D** EC, E and SE Africa.

33 ZANZIBAR RED BISHOP *Euplectes nigroventris* 10cm Note black breast of breeding male. Female separated from females of Southern and Black-winged Red Bishop by darker ear-coverts. **V** Song comprises quiet sizzling and chattering. Calls include *tsip* and, when alarmed, a harsh *drrt*. **H** Open and bushy grassland. **D** SE Kenya to S Tanzania, EC Mozambique.

34 BLACK-WINGED RED BISHOP *Euplectes hordeaceus* 14cm Female from female Southern Red Bishop by rich tawny face sides and breast-band. Non-breeding male has much darker wing than males of Northern and Southern Red Bishop. Often in flocks together with Golden-backed Bishop. **V** Song a three-part series: first and third parts ticking notes, with a sizzling sound in the middle. Calls include twitters, high-pitched notes and, in alarm, *chak*. **H** Reedbeds, elephant grass, sugarcane. **D** W, C E and SE Africa.

35 SOUTHERN RED BISHOP *Euplectes orix* 12cm Male separated from male Northern Red Bishop by black forehead (not extending to crown) and from male Black-winged Red Bishop by red, not whitish, undertail coverts. **V** Song a long, sizzling phrase, ending with series of *tsip* notes. Contact call *chip*, alarm call *chuk*. **H** Reedbeds, elephant grass, sugarcane. **D** S and E DR Congo and SW Kenya to Mozambique, Angola, W Zambia, N Botswana, Namibia and South Africa.

36 NORTHERN RED BISHOP *Euplectes franciscanus* 12.5cm Forages in small to large flocks, searching out seeds on grass stems or the ground; probably also feeds on grain crops and rice. Non-breeding males look very like breeding females. **V** Simple tuneless chipping and trilling, given perched or in display flight. **H** Grassy borders of sugarcane fields. **D** W, C and E Africa.

37 GOLDEN-BACKED BISHOP *Euplectes aureus* 14cm Separated from Yellow Bishop by more extensive yellow upperparts and white undertail coverts. **V** Song a series of loud metallic *zik* notes. **H** Reedbeds, tall grass. **D** Angola, São Tomé in the Gulf of Guinea.

38 YELLOW BISHOP *Euplectes capensis* male 14cm, female 12cm Note rather heavy streaking below and (in flight) yellowish rump of female. **V** Song a long series of chipping notes followed by hissing sounds. Calls include a thin *tseep* and, when alarmed, a harsh *chak*. **H** Tall montane grassland with some bushes, forest edges. **D** C, E and S Africa.

1
2
3
4
5
6
7
8
9
10
11
12
13
14
15
16
17
18
19
20
21
21
22
23
3 colour
morphs
24
25
26
27
28
29
30
31
32
33
34
35
36
37
38

WEAVERS AND WIDOWBIRDS *CONTINUED*

1 FAN-TAILED WIDOWBIRD *Euplectes axillaris* male 17cm, female 13cm Note red and cinnamon epaulets of breeding male (orange-yellow and cinnamon in some races), with no white in wing. Female and non-breeding male show some rufous at wing shoulder. **V** Song is a long series of high-pitched scratchy, sizzling and twittering notes. Calls include rolling *tseek-wirra-wirra* given in flight, and contact call *zip-zip*, with a harsher version of this when alarmed. **H** Swamp edges, long grass, sugarcane. **D** Widespread across sub-Saharan Africa, from Mali to Ethiopia and south to South Africa.

2 YELLOW-MANTLED WIDOWBIRD *Euplectes macroura* male 20cm, female 13cm Male in breeding plumage has variable amount of yellow on mantle. Note yellow shoulders of female and non-breeding male. **V** Song combines rustling and hissing sounds, sometimes beginning with *tsee-piti, tsee-piti*; flight call a whistle. **H** Moist, open grassland at mid–high altitudes. **D** From Senegal to Ethiopia and south to Zimbabwe.

3 MARSH WIDOWBIRD *Euplectes hartlaubi* male 30–36cm, female 15cm Male in breeding plumage resembles Fan-tailed Widowbird but has longer tail (much longer in southern race). Female, like other female widowbirds, is streaked and short-tailed. **V** Song begins with rolling notes that rise in pitch, continues into buzzing trill. Contact call is a brief *drrt*. **H** Marshes, swamp edges, moist grassland. **D** Nigeria and Cameroon to Uganda and W Kenya; Angola, S DR Congo, N Zambia, W Tanzania.

4 MONTANE WIDOWBIRD *Euplectes psammacromius* male 32cm, female 15cm Note yellow traces in female's shoulder; non-breeding male like female but with wing of breeding plumage. **V** Song is a complex series of very high-pitched notes; calls include whistles and rattles. **H** Montane grassland. **D** C Tanzania to NE Zambia and N Malawi.

5 WHITE-WINGED WIDOWBIRD *Euplectes albonotatus* male 16–20cm, female 12cm Length of tail varies, but breeding male always has yellow lesser coverts and prominent white wing patch. Female has some (often concealed) rufous at wing bend and shows more or less a white wing flash in flight. **V** Song consists of a short sizzling phrase; flight call a simple chipping note. **H** Bushed areas with tall grass. **D** From Gabon to Ethiopia and south to South Africa.

6 RED-COLLARED WIDOWBIRD *Euplectes ardens* male 27cm, female 12cm Some males have red on crown and nape, as well as on foreneck. Female is nondescript. Non-breeding male retains blackish wings and more or less elongated tail. **V** Song is a prolonged, sibilant trilling phrase. Calls include hisses and rattles, and a husky *sskip* when alarmed. **H** Marshy open grassland with some shrub. **D** Senegal to South Sudan; Eritrea to South Africa.

7 LONG-TAILED WIDOWBIRD *Euplectes progne* male 50–70cm, female 18cm Tail of breeding male extraordinarily long. Note large size and plain colouring of female; non-breeding male retains colourful wing patches. **V** Song *twi-twi-twi-twi-zi-zi-zi*, high-pitched and repetitive. Calls include *zik* and a powerful *tsek tsek* when alarmed. **H** Damp, open grassland. **D** Kenya, Angola, S DR Congo, Zambia, SE Botswana to E South Africa.

8 JACKSON'S WIDOWBIRD *Euplectes jacksoni* male 30cm, female 14cm Male in breeding plumage almost entirely black, with buff-fringed brown flight feathers. Distinct 'cape' on nape. Tail long and curved. **V** When perched, sings a series of sibilant notes; during jumping display, song is more wheezing. **H** Open rough grassland. **D** W and C Kenya to N Tanzania.

WAXBILLS, MUNIAS AND ALLIES ESTRILDIDAE

9 WOODHOUSE'S ANTPECKER *Parmoptila woodhousei* 9cm Note rufous face and scaly brown and white underparts. Warbler-like actions. Very secretive. Feeds on ants. **V** Repeated high-pitched whistling *tcseeep*. **H** Ground and understorey at forest edge. **D** SE Nigeria and Cameroon to S DR Congo, N Angola.

10 RED-FRONTED ANTPECKER *Parmoptila rubrifrons* 9cm Similar in behaviour to Woodhouse's Antpecker, but note uniform rufous-chestnut underparts of male and brown, white-speckled cheeks of female. **V** Call *pee-yoo*. **H** Mid-strata and undergrowth of forest edge. **D** Sierra Leone to Ghana.

11 JAMESON'S ANTPECKER *Parmoptila jamesoni* 11cm Similar to Red-fronted Antpecker, but male's red coloration more extensive; also note red cheeks and throat of female. **V** Little known; *wheet* call noted. **H** Damp riverine forest, forest edge, gallery and partly logged forests. **D** DR Congo, W Uganda, extreme NW Tanzania.

12 WHITE-BREASTED NIGRITA *Nigrita fusconotus* 10cm Resembles a *Sylvia* warbler (Plate 227), with glossy black cap and white underparts, but noticeably smaller. Sexes alike. **V** Song is a varied phrase of slow, slurred and tuneful whistles, such as *wee-hyoo-hyoo*, sometimes accelerating into a trill. **H** Forest edge and clearings, riverine belts. **D** Guinea to W Kenya and N Angola.

13 CHESTNUT-BREASTED NIGRITA *Nigrita bicolor* 10cm Very dark, with slate-grey upperparts and reddish underparts. Skulking. Sexes similar. **V** Song combines various musical whistles; calls include *chi-chi-hooeee*. **H** Forest edge and clearings. **D** Senegal to W Uganda and N Angola.

14 PALE-FRONTED NIGRITA *Nigrita luteifrons* 10cm Male has whitish to yellowish forehead, grey crown and back, black face and underparts. Female has black mask and grey underparts. **V** Song is a short, simple series of descending whistles. Calls include a descending, whistled *fu-fi-fi-fi*; contact call is a soft *choo*. **H** Forest, lower in canopy than Grey-headed Nigrita. **D** Sierra Leone and Liberia to W Uganda and NW Angola.

15 GREY-HEADED NIGRITA *Nigrita canicapillus* 13cm Separable from Pale-fronted Nigrita by white wing spots and black forehead. Sexes similar. **V** Song consists of three whistled notes, *hooee hooeeee hoo*, sometimes beginning with a soft bubbling note and harsh chatter. **H** Montane forest edges and clearings. **D** Guinea to Kenya and NW Angola.

16 SHELLEY'S OLIVEBACK *Nesocharis shelleyi* 9cm Note black head and grey nape of both sexes. Best separated from White-collared Oliveback by different range. **V** Song is a series of 4–8 buzzing trills; calls include a thin *tsi-tsi* and short trills. **H** Montane forest, clearings, plantations. **D** SE Nigeria, SW Cameroon, Bioko, Gulf of Guinea.

17 WHITE-COLLARED OLIVEBACK *Nesocharis ansorgei* 10cm Very similar to Shelley's Oliveback. Has longer tail, but best separated by range. **V** Song begins with two clear slurring whistles, followed by a trill; calls include a lisping *tssep*. **H** Wet, wooded and bushed areas with thick grass and shrubbery at higher elevations. **D** E DR Congo and W Uganda to NW Tanzania.

18 GREY-HEADED OLIVEBACK *Nesocharis capistrata* 12cm Note black bib and yellow flanks. **V** Song is a clear whistled phrase, notes becoming shorter and lower-pitched; call a lisping *tsssp*. **H** Undergrowth of forest edge and wet, wooded and bushed areas. **D** Senegal to S South Sudan, W Uganda, N DR Congo.

19 RED-BILLED PYTILIA *Pytilia lineata* 13cm Very like Red-winged Pytilia, but bill is red, not black. **V** Song is a short series of squeaking notes followed by rattling *titititititititi*. Alarm and flight call *chwick*. **H** Open bushed and wooded areas. **D** Ethiopia.

20 RED-WINGED PYTILIA *Pytilia phoenicoptera* 13cm Note red tail, red wings and barred underparts. **V** Song is a series of rising whistles, followed by a rising buzzy whistle and a falling whistle. Calls include a double whistling chatter; contact call *seee*. **H** Tall grass of bushed and wooded areas. **D** Senegal to S South Sudan, N Uganda.

21 YELLOW-WINGED PYTILIA *Pytilia hypogrammica* 13cm Male has red face. In both sexes, note combination of grey upperparts, yellow (or orange-yellow) wings, red and black in tail. **V** Song varied, including chattering and buzzing phrases. Contact call *seeee*, also croaks and chatters. **H** Open woodland, bush, cultivation. **D** Guinea and Liberia to S Chad, Central African Republic.

22 ORANGE-WINGED PYTILIA *Pytilia afra* 11cm Note deep orange in wings. **V** Mid-high *trip trip trip* or extremely high *tseep tseep tseep*. **H** Forest, miombo and other woodland, thorn-bush. **D** S South Sudan and Ethiopia south to NE South Africa and west to S DR Congo and W Angola.

23 GREEN-WINGED PYTILIA *Pytilia melba* 12cm Separable from Orange-winged Pytilia by all-green wings and more strongly barred underparts. Forages near or on the ground. **V** Song is a quiet, unstructured sequence of well-separated notes, *fjuw prri sreeh juut-tjuw-tjih*. Call is lowered *fieeuw*. **H** Dry woodland, thornveld, thickets, grassland, cultivation. **D** Mauritania to Eritrea and south to N South Africa.

24 RED-HEADED FINCH *Amadina erythrocephala* 14cm Female can be separated from female Cut-throat Finch by plain upperparts, regular, rather scalloped barring of underparts, brownish (not grey) bill. Forages on the ground, typically in small flocks. **V** Song is a quiet stream of churring or buzzing notes. Calls include chirping *shep* and *zree-zree* in flight, and *tek* when alarmed. **H** Wooded thornbush, semi-desert grassland, savannah. **D** From W Angola through Namibia, Botswana, S Zambia, Zimbabwe to South Africa.

25 CUT-THROAT FINCH *Amadina fasciata* 12cm Named for the distinctive red throat of the male. Note zig-zag barring of underparts of both sexes. Often forages on the ground, in pairs or flocks. **V** Song soft, low, unobtrusive buzzing and bubbling phrase; calls include chirping notes and *eee-eee-eee* in flight. **H** Dry woodland, bushveld, cultivation, near waterholes. **D** Sahel zone from Mauritania to Eritrea, and south through E Africa to N South Africa.

26 GREEN TWINSPOT *Mandingoa nitidula* 10cm A tiny bird. Male has red face; both sexes have black belly with white rounded spots. Often on the ground. **V** Extreme high cicada-like trill *trrrrr*. **H** Undergrowth of forest edges, moist thickets. **D** Sierra Leone to Ethiopia and south to E South Africa, Eswatini.

27 RED-FACED CRIMSONWING *Cryptospiza reichenovii* 12cm Generally browner, less grey than Abyssinian Crimsonwing, the male with red sides to face. Range also differs. **V** Extreme high siffled *srrreeeeh*. **H** Undergrowth of montane forest, often near streams. **D** SE Nigeria, Cameroon, NW Angola, Bioko, Gulf of Guinea; E DR Congo to Mozambique.

28 ABYSSINIAN CRIMSONWING *Cryptospiza salvadorii* 12cm Rather pale. Note narrow red eye-ring of male. **V** Song consists of soft, mournful phrases, *dee-goo-goo-dee*; calls include soft *tsip tsip*. **H** Montane forest edges and bamboo. **D** Ethiopia to E DR Congo, N Tanzania.

29 DUSKY CRIMSONWING *Cryptospiza jacksoni* 12cm Much darker than Red-faced Crimsonwing, so reasonably easy to separate even where ranges overlap. **V** Song includes elements of call notes, such as trilled *geegeegeegeegee* and *tzeeek* or *tsit* notes. **H** Dense montane forest undergrowth. **D** E DR Congo, Rwanda, Burundi, W Uganda.

30 SHELLEY'S CRIMSONWING *Cryptospiza shelleyi* 13cm Note entirely red head and pink bill of male, and mainly pink bill of female. **V** Typical call is a twittering *tu-tu-tu-ti-ti-ti*. **H** Dense undergrowth of montane forest edges. **D** E DR Congo, Rwanda, Burundi, W Uganda.

31 BLACK-BELLIED SEEDCRACKER *Pyrenestes ostrinus* 15cm Male resembles male Red-headed Bluebill, but has uniform dark grey bill and more red on tail. Female is browner, with less red. **V** Song is a highly varied phrase of sweet whistles, each male having several unique variants. Calls include low *peenk*, and *terr* when alarmed. **H** Tall grass and shrubbery at forest edge, gardens. **D** Ivory Coast to S South Sudan, W Kenya, and south to Zambia and N Angola.

32 CRIMSON SEEDCRACKER *Pyrenestes sanguineus* 14cm Best separated from Black-bellied Seedcracker by different range. **V** Song is a melodious *titi-tsee-chweta-toi-ti-tsee*, or a squeaky phrase that rises in pitch over first four notes and tails off on final two. Calls include a clicking *tsut-tsut*. **H** Wet bush, marshes, dense scrub. **D** Senegal to Ivory Coast.

33 LESSER SEEDCRACKER *Pyrenestes minor* 13cm Both sexes resemble female Black-bellied Seedcracker, but can be identified by size, relatively smaller bill, and different range. **V** Song is a soft, twittering trill; calls include *tzeet* and explosive *chat-chat-chat*, also a clicking note when alarmed. **H** Dense undergrowth at forest edge, often near water. **D** Tanzania, Malawi, E Zimbabwe, N and C Mozambique.

34 GRANT'S BLUEBILL *Spermophaga poliogenys* 14cm Male like Red-headed Bluebill but with black nape and smaller bill. Note red throat and white-spotted grey belly of female. **V** Song is a stream of downslurred and upslurred whistles; calls include coarse *chip* and soft *thac-thac* in flight. **H** forest undergrowth. **D** N Congo through DR Congo to W Uganda.

35 WESTERN BLUEBILL *Spermophaga haematina* 15cm Note glossy black upperparts extending to black crown and forehead of male. Female has maroon forehead and face. **V** Song begins with clicks and progresses to musical whistling warbles. Calls include thin, high-pitched *seeee*; also sharp *tsip*, and *tswink tswink* when alarmed. **H** Forest edge, thickets, scrub. **D** Senegal to DR Congo.

36 RED-HEADED BLUEBILL *Spermophaga ruficapilla* 15cm Head entirely red in both sexes. Male resembles male Black-bellied Seedcracker but has all-black tail, with red restricted to rump. **V** Two song types occur, one fluting with a final trill and the other a faster jumble of rattles, clucks and whistles. Calls include *pik*, nasal *tewk*, and soft, high *seep*. **H** Tall grass and shrubbery, normally away from settlements. **D** SE Central African Republic to Uganda, Kenya, Tanzania; S DR Congo, N Angola.

1
2
3
4
5
6
7
8
9
10
11
12
13
14
15
16
17
18
19
20
21
22
23
24
25
26
27
28
29
30
31
32
33
34
35
36

WAXBILLS, MUNIAS AND ALLIES *CONTINUED*

1 BROWN TWINSPOT *Clytospiza monteiri* 13cm Underparts rufous-cinnamon (not black), spotted white. White spots on belly arranged in bars. Note also red rump. **V** Song is a mixture of long rising and falling whistles and chatters, trills and chuckles. Calls include *vay vay vay* and *chk*. **H** Tall grass and shrubbery in montane forest glades, moist bush. **D** SE Nigeria and Cameroon to S South Sudan, W Kenya; Gabon to W Angola and C DR Congo.

2 PINK-THROATED TWINSPOT *Hypargos margaritatus* 13cm Male similar to Red-throated Twinspot, with crimson throat and breast, but differs in dull pink (not crimson) face sides and rump, and brown (not grey) crown, female by grey (not pink) breast. **V** Song a trill followed by a buzz. Contact call a soft, rising *seesee*, and *zirrrr* when alarmed. **H** Tall-grassed areas, forest edges, woodland, bushveld. **D** S Mozambique, NE South Africa, Eswatini.

3 RED-THROATED TWINSPOT *Hypargos niveoguttatus* 13cm Male resembles Pink-throated Twinspot, but is deeper red, with grey crown. Female is mostly olive-brown above, with grey face and yellowish chin. **V** Extremely high, descending to mid-high and then rising again, *fififififitjehtjeh-sreeeh*. **H** Tall-grassed areas, forest edges, woodland, bushveld. **D** S Somalia, SE Kenya to Zambia, Zimbabwe, Mozambique

4 DYBOWSKI'S TWINSPOT *Euschistospiza dybowskii* 13cm Note purple-red mantle and all-grey head and breast. **V** Song loud, combining churrs, whistles and bubbling trills. Contact call *kek* or, at greater distance, *tsit*. **H** Open woodland, forest edge, cultivated land. **D** Guinea to Ivory Coast; C Nigeria to W Uganda.

5 DUSKY TWINSPOT *Euschistospiza cinereovinacea* 12cm Dark upperparts and red flanks and rump distinctive. **V** Song rather nondescript. Contact calls include a chirping *tsyip-tsyip*, and *tswilip* when alarmed. **H** Tall-grass glades in montane forest. **D** W Angola; E DR Congo, SW Uganda, Rwanda, Burundi.

6 BLACK-BELLIED FIREFINCH *Lagonosticta rara* 13cm Mantle of male is concolorous with head and breast. Female largely brown above. **V** Song is a series of whistles, some notes flat or buzzing, others mournful and musical. Calls include a nasal *keeyh* or a louder *squeer* in flight; alarm call is a sharp *chek*. **H** Tall grass and herbage in wooded and bushed areas. **D** Senegal to Kenya.

7 BAR-BREASTED FIREFINCH *Lagonosticta rufopicta* 10cm Very similar to Red-bellied Firefinch, but upperparts uniform brown, not tinged red. White spots form irregular bars. Rump and upper tail coverts more crimson-red than breast and head. **V** Song is a 4-sec jingling phrase. Calls include *kewp* or *kewp-kewp*, and *tik* when alarmed. **H** Herbaceous places in wooded and bushed habitats, often near water. **D** Senegal to Ethiopia, Uganda, W Kenya.

8 BROWN FIREFINCH *Lagonosticta nitidula* 10cm Separable from all other firefinches by brown (not pink) rump. **V** Song consists of jingling notes with some metallic and low nasal sounds. Contact call *kewp*, repeated chattering *trrrritit* when alarmed. **H** Thick vegetation near water. **D** C Angola, C DR Congo and SW Tanzania through Zambia to NE Namibia, N Botswana, NW Zimbabwe.

9 RED-BILLED FIREFINCH *Lagonosticta senegala* 10cm Male distinguished from other firefinches by pink legs, pale brown flanks and lower belly. In the female, the pink loral spot is distinctive. **V** Unstructured song of slightly hoarse notes and trills around basic *tit-weet*. **H** Open thornveld, cultivation, suburbs. May enter houses. **D** Sub-Saharan Africa, widespread.

10 ROCK FIREFINCH *Lagonosticta sanguinodorsalis* 11cm Very restricted range, where it is the only firefinch with a red back. Note also the grey crown. **V** Calls include various different trilled and whistled notes. **H** Open woodland. **D** Nigeria, NW Cameroon.

11 CHAD FIREFINCH *Lagonosticta umbrinodorsalis* 11cm Dark firefinch with pale grey crown and nape (entire head in female). Forages in pairs on the ground. **V** Very like that of Rock Firefinch. **H** Rocky areas with scattered trees, thickets and grass, sometimes along streams and rivers. **D** SW Chad, N Cameroon.

12 MALI FIREFINCH *Lagonosticta virata* 10cm Separable from African Firefinch by fewer white spots on flanks and brownish-grey, not red forehead. Female similar to male, but paler. **V** Song is a variable series of trills, sometimes with upslurred or downslurred whistles. Calls include a whistling *fiu* and a rattling *churrchurr churrr* in alarm; various soft notes exchanged between pairs in courtship. **H** Rocky hillsides with sparse grass and scrub. **D** E Senegal, S Mali.

13 AFRICAN FIREFINCH *Lagonosticta rubricata* 11cm Note greyish-brown crown and nape, and brown (not red) mantle. **V** Extreme high (or very high), shrill warbling or silvery-rattling trills. **H** Tall grass and shrubbery in moist forest edges, woodland, bushveld, thornbush. **D** S Senegal to Eritrea, and south to SE South Africa.

14 LANDANA FIREFINCH *Lagonosticta landanae* 11cm Note male's red forehead and crown, and partly pink upper mandible in both sexes. Tail and undertail coverts are black. **V** Like that of African Firefinch. **H** Dry, grassy areas, thornveld. **D** Gabon to W DR Congo and NW Angola.

15 JAMESON'S FIREFINCH *Lagonosticta rhodopareia* 11cm Note male's mid-brown, pink-washed upperparts and indistinct eye-ring. **V** Mid-high, warbled, loud (or soft), rapid (or very fast) trills. **H** Tall grass and shrubbery in bushveld and woodland. **D** South Sudan and Ethiopia to S DR Congo, Angola, Zambia, Zimbabwe, Mozambique, NE South Africa.

16 BLACK-FACED FIREFINCH *Lagonosticta larvata* 11cm Note extensive dark mask of male. **V** Song consists of a short series of rising, plaintive whistles. Contact call a shrill *seeseee*, alarm call a sharp *pitpitpit*. **H** Areas with tall grass, thickets, open woodland. **D** Senegal to Ethiopia, Uganda.

17 BLUE WAXBILL *Uraeginthus angolensis* 12cm Male differs from female in having bright blue extending to lower flanks. Female not safely separable from female Red-cheeked Cordon-Bleu, but male and female of both species are normally always seen in pairs, making confusion unlikely. **V** Song varies by region but is a short phrase including slurred and sibilant whistles and harsh notes. Calls include quiet *swip swip* or louder *chee-chee-chee*; alarm call is a sharp chattering note. **H** Thickets in woodland, bushveld. **D** São Tomé, W DR Congo to NW Namibia, Tanzania and Zambia to E South Africa.

18 RED-CHEEKED CORDON-BLEU *Uraeginthus bengalus* 13cm Male has grey-brown upperparts, bright turquoise to cobalt-blue underparts, and a red spot on the ear-coverts. Female very like female Blue Waxbill. **V** Song unique to each individual male, but includes sharp notes, tuneful upslurred whistles and buzzes. Female also sings, giving a simpler phrase. Calls include rising, thin *tseek* or *tsee-tsee-tsee*; alarm call slow, low chatter. **H** Woodland, bush, often near settlements. **D** S Mauritania to Eritrea, and south to N Zambia.

19 BLUE-CAPPED CORDON-BLEU *Uraeginthus cyanocephalus* 13.5cm Note pink bill of male and female. Blue crown of male diagnostic; female not safely separable from Blue Waxbill and Red-cheeked Cordon Bleu, but note pink bill. **V** Song unique to each male (and female); includes whistles, trills and jumbled notes. Calls include a thin *tsee*; alarm call is a sharp *tchek*. **H** Dry bushed areas. **D** S South Sudan and S Ethiopia to Tanzania.

20 VIOLET-EARED WAXBILL *Uraeginthus granatinus* 14cm Male is unmistakable, with violet face, red bill and eye-ring, black throat, blue rump. Female is like a washed-out version of the male. **V** High, slightly lark-like phrases mixed with soft, hoarse rattles. **H** Woodland, thornveld. **D** S Angola and Namibia to SW Zambia, Botswana, Zimbabwe, S Mozambique, N South Africa.

21 PURPLE GRENADIER *Uraeginthus ianthinogaster* 14cm Note male's cobalt- or purple-blue mask and belly, and red bill of both sexes. **V** Both sexes sing and have individually unique song form, comprising whistles and soft whispering or buzzing notes with trills. Calls include a thin *wis* between pairs, a trilling contact call, and a sharp alarm call, *tsek*. **H** Ground of more or less wooded and bushed areas with thickets, gardens, parks. **D** Ethiopia and Somalia to C Tanzania.

22 YELLOW-BELLIED WAXBILL *Coccopygia quartinia* 9cm Note yellow (not creamy-white) flanks, short tail and black-and-red (not all-red) bill. Female is slightly less yellow than male. **V** Song is a quiet but penetrating drawn-out *tuuuueeeet* note or similar. Calls include a soft *see-see* and a sharper *tswuee*. **H** Forest glades with grass, other wooded areas. **D** Eritrea to Mozambique, E Zimbabwe.

23 SWEE WAXBILL *Coccopygia melanotis* 9cm Male has distinctive black ear-coverts and chin, with rather indistinct white band on upper breast. Female has entire face grey but is otherwise similar. Note two-tone red and black bill of both sexes. **V** Song similar to that of Yellow-bellied Waxbill; call a soft *swee-swee*. **H** Mainly tall grass at edges of montane forest, wooded streams. **D** S Zimbabwe and SW Mozambique to S South Africa.

24 ANGOLAN WAXBILL *Coccopygia bocagei* 9–10cm Very like Swee Waxbill, but note male's yellow belly. **V** Not known, but probably like that of Swee Waxbill. **H** Tall grass and undergrowth at forest edges and along rivers and mountain streams. **D** Angola.

25 LAVENDER WAXBILL *Estrilda caerulescens* 10cm Separable from Grey Waxbill and Cinderella Waxbill by red undertail coverts. Note red bill. **V** Various high-pitched, thin piping notes. **H** Dry, more or less wooded areas. **D** Senegal to Central African Republic.

26 GREY WAXBILL *Estrilda perreini* 11cm Distinguished from Cinderella Waxbill by grey (not pink) bill, black tail and undertail coverts, different range. **V** Contact call is a thin but explosive whistled *pseeu pseeu*; also a drawn-out *fweeee*. **H** Dense undergrowth at forest edge, woodland with thick, tangled grass. **D** Gabon to N Angola and east to S Tanzania, Malawi, Mozambique, E South Africa.

27 CINDERELLA WAXBILL *Estrilda thomensis* 11cm Note partly pink bill base and partly red (not all-black) tail feathers. **V** Song is a whistling *see-eh see-eh see-eh*. Calls include harsh *brrt brrt* and *swee-swee*. **H** Dry miombo woodland, bush, thorn scrub, riverine belts. **D** W Angola, extreme NW Namibia.

28 ANAMBRA WAXBILL *Estrilda poliopareia* 12cm Pale eye is diagnostic. **V** Not well known. Calls include *tzzt*. **H** Long grass in swamps or at the edge of forest. **D** S Benin, S Nigeria.

29 FAWN-BREASTED WAXBILL *Estrilda paludicola* 12cm Longer-tailed than Yellow-bellied and Swee Waxbills, with all-red bill. Note also black tail, olive-brown back and red rump, which is conspicuous in flight. **V** Song is a rhythmic series of hard, ringing *tek* and *tektri* notes. Calls include various chattering notes and a downslurred *sieu*. **H** Wet grassland, swamps, forest edge, cultivated land. **D** Patchily distributed, in Central African Republic, DR Congo, Congo, Gabon, Uganda, Kenya, Tanzania, Angola, Zambia.

30 ABYSSINIAN WAXBILL *Estrilda ochrogaster* 9–10cm Very like Fawn-breasted Waxbill, from which it has been split; more yellowish on underside. Forages in pairs, small parties or flocks. **V** Song is a rhythmic *tek tek tek teketri teketri*... Calls include *sieu* and *tjip*. **H** Tall grass in woodland clearings, riversides, edges of cultivation. **D** SE Sudan, Ethiopia.

31 ORANGE-CHEEKED WAXBILL *Estrilda melpoda* 10cm Combination of orange face and red bill distinctive. Actions and habits much as Common Waxbill. **V** Song a collection of short notes, *de-de-de-sweea - sweea - sweea*, or similar. Call is a high-pitched *sieu*. **H** Cultivation, grassy and weedy areas. **D** S Mauritania to Central African Republic, DR Congo, N Angola, N Zambia. Introduced Japan, Spain, Hawaii, Caribbean.

32 CRIMSON-RUMPED WAXBILL *Estrilda rhodopyga* 10cm Note bicoloured bill, red in wings and red rump. **V** Song *tchek-er cherr, tche-chaeer*, increasing in volume. Contact call is a quiet *sspt-sspt*; also grating and nasal notes. **H** Tall grass and shrubbery in woodland, bush. **D** Sudan to Somalia and south to Malawi.

33 ARABIAN WAXBILL *Estrilda rufibarba* 11cm Mainly grey, with fine barring all over the body, crimson mask, black tail and rump. Social. Feeds on ground or on reed and grass seeds. **V** Buzzing *dzit*, and *chee-chee-chee*... in flight. **H** Reed patches, weedy thickets, rocky hillsides, cultivation. **D** SW Arabian Peninsula.

34 BLACK-RUMPED WAXBILL *Estrilda troglodytes* 10cm Similar to Arabian Waxbill, but with red bill. Forages in flocks, never far from cover. **V** Song is an explosive *tche-tcheer che-eeer*. Calls include *cheu-cheu* and *chihooee*. In flight, *tiup-tiup-tiup*. **H** Grassy woodland, dry grassland, abandoned cultivation, rank grass near water, swamp edges. **D** S Mauritania to Eritrea, Ethiopia, Kenya. Introduced S Europe, Canary Islands, Caribbean, Hawaii.

35 COMMON WAXBILL *Estrilda astrild* 11cm Note fine barring on upperparts and underparts, and red bill. Male has more black on underparts than female. **V** Mid-high, soft, nasal song, *tju-tu-tu-wee-teesj*. **H** Tall grass and shrubbery in woodland, bushveld, often near water or settlement. **D** Widespread across sub-Saharan Africa. Introduced throughout the world.

36 BLACK-LORED WAXBILL *Estrilda nigriloris* 11cm Black lores diagnostic. **V** Not known. **H** Grassland with some trees or bush. **D** S DR Congo.

37 BLACK-CROWNED WAXBILL *Estrilda nonnula* 10cm Note white undertail coverts and creamy-white underparts. **V** Song is a series of *speet* and *seet* notes. Calls include a buzzing *tseee*; alarm call *srree srreee*. **H** Moist tall grass near forest edges, woodland, and bush, cultivation and gardens. **D** Cameroon and Equatorial Guinea to S South Sudan, W Kenya, NW Tanzania

38 BLACK-HEADED WAXBILL *Estrilda atricapilla* 10cm Similar to Black-crowned Waxbill, but note more extensive red on flanks and black undertail coverts. **V** Calls include high, thin *tsree* and *psee* notes, also short *psit*. **H** Montane forest, bamboo. **D** Cameroon to DR Congo, N Angola.

1
2
3
4
5
6
7
8
9
10
11
12
13
14
15
16
17
18
19
20
21
22
23
24
25
26
27
28
29
30
31
32
33
34
35
36
37
38

WAXBILLS, MUNIAS AND ALLIES *CONTINUED*

1 KANDT'S WAXBILL *Estrilda kandti* 10cm Black cap and tail, red flanks and rump, otherwise mid-grey on upperparts (suffused brown in female), whitish grey on underparts. May flock with Black-crowned Waxbill (Plate 266). **V** Song is a twittering warble combining *cheep* and *wee* notes. Calls include a very thin *tee-tee-tee*. **H** Montane forest edges and clearings. **D** E DR Congo, W Uganda, Rwanda, Burundi, C Kenya.

2 BLACK-FACED WAXBILL *Estrilda erythronotos* 13cm Combination of black face and overall reddish colour (especially on flanks) is distinctive. Feeds on the ground. **V** Calls include a high-pitched *tip-tip*, alarm call similar but more explosive, distant contact call a whistled *teeee*. **H** Thornveld. **D** Uganda, Kenya, Tanzania; Angola and Namibia to Zimbabwe and N South Africa.

3 BLACK-CHEEKED WAXBILL *Estrilda charmosyna* 13cm Note pale (not black) chin. Generally paler than Black-faced Waxbill. **V** Little known, but song is said to be pleasant warbling; flocks give dry chatters. **H** Dry, rocky bush and shrub land. **D** Somalia and S Ethiopia to N Uganda, Kenya.

4 RED AVADAVAT *Amandava amandava* 10cm Male unmistakable, largely red with white spots. Forages on the ground or on grass-heads, often in small flocks. **V** Song is a weak, high-pitched warble combined with sweet twittering notes. **H** Swampy grassland, sugarcane fields, reedbeds, grass and scrub near cultivation. **D** Pakistan to SE Asia, Indonesia.

5 GREEN AVADAVAT *Amandava formosa* 10cm Mostly olive-green above, with red bill and black and white striped flanks. Forages on the ground, usually near cover; gregarious. **V** Prolonged twitter that ends with a loud trill. Calls include a constant *seee* or *swee swee* and some *cheeps* or *chirps*. **H** Tall grassland, sugarcane fields, boulder-strewn scrub jungle, open dry woodland. **D** C India.

6 ORANGE-BREASTED WAXBILL *Amandava subflava* 10cm Fairly plain olive-green to olive-grey upperparts, red rump, yellow to orange underparts. Actions much as Common Waxbill (Plate 266). **V** Short *cheep* or *chirp*; also soft clinking *zink-zink*, usually given in flight. **H** Cultivation, trees and scrub. **D** Sub-Saharan Africa (widespread), and SW Arabian Peninsula.

7 QUAILFINCH *Ortygospiza atricollis* 10cm Note male's black face and reddish bill. Skulking; does not perch but lives exclusively on the ground. **V** High *ti-tjeeterik-tjeeterik* when flushed. **H** Wet grassland, occasionally drier areas. **D** Sub-Saharan Africa (widespread).

8 LOCUST FINCH *Paludipasser locustella* 9cm A ground dweller, like Quailfinch. Note red wings and pale eyes. **V** Calls include *tissep*, *see-see-see*, and rattling and downslurred notes. **H** Wet grassland, swamps, open woodland. **D** Patchily distributed, from Nigeria and Cameroon to Angola, W Kenya, Mozambique, Zimbabwe, Botswana.

9 PAINTED FINCH *Emblema pictum* 11cm Male has red face, rump and central belly, underparts otherwise black with bold white spots, upperparts grey-brown. Female has more restricted red and more extensive black. Mainly feeds on ground; preferred food is seed of spinifex grass. **V** Song is a wheezy chattering of *che* notes. Calls include a harsh *trut-chek-chek*. **H** Arid areas, stone deserts, gorges and gulleys, rocky hills. Often near water. **D** C Australia.

10 BEAUTIFUL FIRETAIL *Stagonopleura bella* 11cm Plumage appears mainly grey (actually whitish with very fine blackish vermiculations). Lores and (in male) hind belly black, rump and bill red. Usually in flocks of up to 12. **V** Song is a whistled *pee-oo pee-oo pee-oo*. Calls include plaintive upslurred *weeee* and sequence of two or three high notes followed by falling *tee tee te te te te*. **H** Shrubby areas, woodland, wilder gardens. **D** SE and S Australia.

11 RED-EARED FIRETAIL *Stagonopleura oculata* 12cm Buff-grey with red bill, ear-covert patch and rump, belly and flanks blackish with white spotting. Forages alone or in pairs or family groups, feeding mainly on ground. **V** Song is *oweee u u u u*. Calls include a single *oweee* and *twit-twit*. **H** Wet evergreen forest, scrub, coastal heath, thickets. **D** SW Australia.

12 DIAMOND FIRETAIL *Stagonopleura guttata* 12cm Light grey with solid black lores and breast-band, flanks black with white spots. Hops on ground and climbs on grass stems to feed. Forms flocks of up to 40. **V** Song consists of low-pitched raspy notes. Calls include a penetrating *twoo-hee*. **H** Open woodland with grassy understorey, parks, gardens. **D** SC, SE and E Australia.

13 MOUNTAIN FIRETAIL *Oreostruthus fuliginosus* 13cm Rather plain grey-brown firetail with red rump and red flush to underparts (orange in female). Feeds inconspicuously on grass stems. **V** Calls include *howee* and, when agitated, an explosive *pit*. **H** Montane forest and forest edge. **D** SC New Guinea, N Australia.

14 RED-BROWED FINCH *Neochmia temporalis* 11.5cm Grey with leaf-green back and wings, dark tail, red rump and eye-stripe. May form large flocks, and mixes with related species. **V** Song comprises repeated call notes given in rhythmic sequence; calls include *ssitt-ssitt* and *see-see*. **H** Open woodland, mangroves, swamps, gardens, parks. **D** E, SC Australia.

15 CRIMSON FINCH *Neochmia phaeton* 12cm Male crimson-red with blackish hood and black or white (depending on race) centre of belly; female grey-brown with red face and tail. Forages in vegetation and on ground, fly-catches. **V** Song low and raspy, ending with three descending tuneful notes. Calls include *tsee-tsee-tsee* and *che-che-che*. **H** Grasslands (wet and arid) and swampy wetland areas. **D** SC New Guinea, NE Australia.

16 STAR FINCH *Neochmia ruficauda* 11.5cm Grey-green with greyer, white-spotted breast and flanks, yellowish unmarked belly, face red with white speckles, tail dark red. Feeds mainly on grass seeds, also fly-catches. **V** Song short, quiet; calls include penetrating *ssit*, *seet* and a rattling note given when alarmed. **H** Damp areas with tall grass, sedges, rushes and scattered trees, also open savannah woodland. **D** NW, N, NE Australia.

17 PLUM-HEADED FINCH *Neochmia modesta* 11cm Forecrown dark red (restricted to small patch, leaving white eyebrow, in female), lores and small bib black, bill grey. Feeds on ground and climbs plant stems, flocks with other estrildids. **V** Song begins with chirrups, then throaty sounds followed by fluted notes; calls include *tlip* and sharp *tyait*. **H** Open woodland, tall riverside vegetation sometimes gardens. **D** E Australia.

18 MASKED FINCH *Poephila personata* 12cm Grey-brown with long black tail, also black hind-belly and front of face, white rump and undertail. Very stout yellow bill. Sexes alike. Usually seen in flocks, sometimes mixed with Long-tailed or Black-throated finches. **V** Song is a short, soft phrase of nasal notes. Calls include loud *tiat* and softer *tuat tuat*. **H** Open woodland with grass and shrub understorey, open plains with bushes. **D** N and NE Australia.

19 LONG-TAILED FINCH *Poephila acuticauda* 15.5cm Greatly elongated central tail feathers. Also note bright yellow bill and black lores and bib. Feeds on the ground, on grass-heads or in vegetation, often flocks with other species. **V** Song is soft, notes becoming longer and rising in pitch towards end. Calls include soft *tet* and louder *peew*; also has chattering alarm call. **H** Dry scrubby grassland with rivers. **D** N Australia.

20 BLACK-THROATED FINCH *Poephila cincta* 11cm Stocky and short-tailed with black bill, bib, lores, tail and hind belly-band. Feeds mainly on grass seeds, with some insects. **V** Similar to Long-tailed Finch but lower-pitched. **H** Open woodland with grassy or scrubby understorey, edges of cultivation. **D** NE and E Australia.

21 ZEBRA FINCH *Taeniopygia guttata* 10cm Note bright orange bill and black-and-white 'moustache'. Gregarious, often in large flocks of up to 100. Feeds on the ground, on grass-heads or in vegetation. **V** Song comprises nasal call notes interspersed with chattering trills; calls include a sharp *tya*, *teea* or *tcheea*. **H** Dry grassland with scattered trees or scrub, lightly wooded cultivation, monsoon forest edge. Often near water. **D** Australia, Lesser Sunda Islands.

22 DOUBLE-BARRED FINCH *Taeniopygia bichenovii* 10cm Neat black ring framing whitish face, narrow black breast-band. Otherwise light brown, paler on underparts, tail black and wings black with white spotting. Often flocks with Zebra Finches. **V** Song comprises buzzing notes and trills; calls include high-pitched *tat-tat* and squeaky *tiatt tiatt*. **H** Grassy woodland and scrubland, riverside vegetation, parks, gardens. **D** Australia.

23 TAWNY-BREASTED PARROTFINCH *Erythrura hyperythra* 10–11cm Note green upperparts, tawny-buff underparts, blue forecrown. Forages on the ground on seeds and small fruits, usually in small flocks. **V** Series of soft notes followed by four bell-like notes. Calls include a hissing *tzit-tzit* contact call, given in flight. **H** Primary forest, bamboo, forest edge and adjacent grassland. **D** Philippines, Peninsular Malaysia, Greater and Lesser Sunda Islands.

24 PIN-TAILED PARROTFINCH *Erythrura prasina* male 15cm, female 11–12cm Both sexes have a pointed tail, longer in male than in female. Forages singly or in pairs; also in small parties, especially when bamboo is in seed. **V** Series of abrupt clinking or chirping notes. Calls include a sharp *teger-teter-terge*, a high-pitched *tseet-tseet*, and a shrill *tzit-tzit tzit* in flight. **H** Bamboo thickets, rice fields, forest edge, secondary growth. **D** SE Asia, Greater Sunda Islands, W Philippines.

25 GREEN-FACED PARROTFINCH *Erythrura viridifacies* 12–13cm Entirely green, apart from red rump and tail. Occasionally seen in large flocks at rich feeding sources. **V** Song comprises a soft *deedeedeedee...* followed by a chattering *day day day* and ending with a harsh *grey-grey-grey-ray-day-lay-grey*. **H** Montane forest, forest edge, bamboo, grassland and scrub near forest. **D** Philippines.

26 TRICOLORED PARROTFINCH *Erythrura tricolor* 10cm Underparts are almost entirely blue. Feeds on the ground, in low scrub, bamboo and trees, singly, in pairs, in small flocks or occasionally in larger flocks. **V** Shrill, high trill. Calls include a soft *ti-ti-ti-ti... ti-ti-ti*. **H** Monsoon forest, forest edge and clearings, woodland, secondary growth, thickets, bamboo, cultivated areas. **D** C and E Lesser Sunda Islands.

27 RED-EARED PARROTFINCH *Erythrura coloria* 10–11cm Note red, blue and green on head. Forages on the ground, in low vegetation and occasionally up to the forest canopy, usually alone or in pairs. **V** Calls include a sharp, repeated *tik*, a sharp *prrrt* and a trilled *tik-tik-tik-tik*. **H** Montane forest, forest edge, secondary growth, shrubs, thickets, dense understorey, tall grass. **D** S Philippines.

28 BLUE-FACED PARROTFINCH *Erythrura trichroa* 12cm Blue face extends to crown in male; more restricted in female. Forages alone, in pairs or in small groups, from the ground to the canopy. **V** Calls include a thin, high-pitched *tsit-tsit*, *ti-tu ti-tu-tu* or *t-t-t-t*. **H** Montane forest, rainforest, forest edge, bamboo, scrub, dense secondary growth. **D** Sulawesi and Maluku Islands (Indonesia) through New Guinea to Caroline Islands, Solomon Islands, other W Pacific islands, NE Australia.

29 PAPUAN PARROTFINCH *Erythrura papuana* 15cm Bright green parrotfinch with blue face and red tail and rump. Usually alone or in pairs, foraging in trees and near ground level on seeds of herbaceous plants. **V** Song is a high-pitched trill; calls include a thin, high-pitched *tsit-tsit*. **H** Evergreen forest and forest edge, secondary growth, mainly at high altitude. **D** New Guinea.

30 RED-THROATED PARROTFINCH *Erythrura psittacea* 12cm Green with bright red face, upper breast, rump and tail. Mainly feeds on the ground, alone, in pairs or in small groups. **V** Song is a high-pitched trill; calls include a high-pitched *seet*. **H** Forest, forest edge and clearings, plantations, mangroves. **D** New Caledonia.

31 RED-HEADED PARROTFINCH *Erythrura cyaneovirens* 10cm Colours of underparts less saturated than Fiji Parrotfinch and black on chin and around face mask missing. **V** Song a fast metallic trill, falling in pitch over its 1.5-secs duration; call a thin *seep*. **H** Rainforest, secondary growth, plantations. **D** Samoa.

32 ROYAL PARROTFINCH *Erythrura regia* 10cm Small, short-tailed parrotfinch. Male mainly deep blue (darkest on throat) with red face and tail; female similar but greener on back and wings. Feeds on seeds and fruits, especially figs, which it grips with its feet and tears open to reach seeds, insects, probably also nectar. **V** Song fast, metallic trill; calls include a thin, high-pitched *seep*. **H** Upland and lowland forest with fig trees. **D** Vanuatu.

33 FIJI PARROTFINCH *Erythrura pealii* 10cm Male and female similar (mask of female slightly more orange). Note black chin and black line bordering lower red cheek. **V** Song a repeated, whistled double-note; calls include a metallic *peep* or *seep*, repeated persistently. **H** Grassland, rice fields, parks, gardens. **D** Fiji.

34 PINK-BILLED PARROTFINCH *Erythrura kleinschmidti* 11cm Unmistakable, with black face, strong pink bill and short tail. **V** Calls include high *cheee cheee cheeee* and a dry clicking. **H** Mainly mature, wet forest. **D** Viti Levu (Fiji).

35 GOULDIAN FINCH *Erythrura gouldiae* 15cm Male stunningly multicoloured; female has same colours but more muted. Head may be black (commonest), red or orange (rarest). Gregarious, climbing on grass-heads to feed. **V** Song soft, complex series of whispering and hissing notes; calls include *sit* or *ssit-ssit* at variable volume. **H** Semi-arid woodland and forest with grasses, requires tree cavities for nesting. **D** N Australia.

36 MADAGASCAN MANNIKIN *Lepidopygia nana* 9cm Distinctive tiny bird with black bib and lores. Feeds on the ground and perched on grass stems. **V** Song is a soft, purring rattle; calls with a soft *tsip* or *tsip-tsirip* in flight. **H** Open areas with or without trees and bush, cultivated land, gardens and cultivation. **D** Madagascar.

37 AFRICAN SILVERBILL *Euodice cantans* 11cm Note creamy colouring and black rump and tail. Sociable. When perched, flicks, waves and spreads tail. **V** Song is a rapid repetition of rising and falling single and then slurred double notes. Calls include a *cheep*, a soft *seeip*, a harsh *tchwit* and a trilling *zip-zip*. **H** Dry open areas, often near settlement. **D** Mauritania to Sudan and south to Tanzania; Arabian Peninsula.

38 INDIAN SILVERBILL *Euodice malabarica* 10–11cm White rump shows in flight. Forages on the ground or on grass seeds. Gregarious, breeding and roosting communally. **V** Song is a short series of rambling twittering notes. **H** Dry cultivated country, grassland, light secondary jungle. **D** NE Egypt, Israel, Arabian Peninsula to India, Bangladesh, Sri Lanka.

1
2
3
4
5
6
7
8
9
10
11
12
13
14
15
16
17
18
19
20
21
22
23
yellow
morph
24
24
25
26
27
28
29
30
31
32
33
normal
form
34
orange
morph
normal
form
35
35
36
37
38
red
morph

WAXBILLS, MUNIAS AND ALLIES *CONTINUED*

1 GREY-HEADED SILVERBILL *Odontospiza caniceps* 12cm Note blue-grey head with white spots from forehead to side of face and upper throat, brown mantle, white rump. **V** Song is a whispering phrase, becoming slightly louder near the end. Calls include a high, thin *tsi* and a tuneful trill. **H** Dry, more or less wooded and bushed areas. **D** S South Sudan to Tanzania.

2 BRONZE MANNIKIN *Lonchura cucullata* 9cm Distinctively patterned and coloured. In flight resembles groups of dancing dots. **V** Song is a soft, steady 2-sec phrase *chi, chu, chi, chu, chiri-it, chu*. Calls include a wheezy *tsek* or *whit-whit*. **H** Long grass in woodland, bushveld, cultivation, marshes, suburbs. **D** Sub-Saharan Africa, widespread.

3 BLACK-AND-WHITE MANNIKIN *Lonchura bicolor* 9cm Head, upperparts and upper breast glossy black. Underparts otherwise white. Southern birds have dark brown back. **V** Song is a medley of soft call notes. Calls include nasal *sip* and *seep*, and louder *pee* and *pew*. **H** Forest edge, woodland, bush, grassland, cultivation, suburbs. **D** Guinea-Bissau to Cameroon, east to South Sudan, south to NW Zambia.

4 RED-BACKED MANNIKIN *Lonchura nigriceps* 9cm Resembles Black-and-white Mannikin, but with chestnut back. Note also barred flight feathers. **V** Like that of Black-and-white Mannikin. **H** Forest edge, woodland, bushveld, grassland, cultivation, suburbs. **D** C Kenya and S Somalia to E Angola, Zambia, Zimbabwe, E South Africa.

5 MAGPIE MANNIKIN *Lonchura fringilloides* 11cm Back is more dark brown than black. Some pale rufous on flanks. Often together with other mannikins. **V** Song consists of soft, bubbling notes. Calls include loud *pyoo-wee*, high-pitched *chee* in flight. **H** Forest edges, moist shrubby woodland, gardens. **D** Senegal to Liberia and east to Ethiopia, Kenya, and south to South Africa.

6 WHITE-RUMPED MUNIA *Lonchura striata* 11–12cm Upperparts are dark brown with pale shaft streaks. White rump is distinctive. Forages on the ground or on seedheads. Gregarious. **V** A series of rising and falling twittering notes. Calls include a plaintive *peep* and a twittering *tr-tr-tr, prrrrit* or *brrt*. **H** Lightly wooded areas, open dry scrub, forest edge and clearings, cultivations and gardens. **D** Nepal, India, Sri Lanka, Andaman Islands, SE Asia, Sumatra, S and SE China, Taiwan.

7 JAVAN MUNIA *Lonchura leucogastroides* 10–11cm Forages on the ground or low down in vegetation, usually in small flocks. **V** Calls include a short *tit*, a *p-tit*, a *peteet* or a *chirrup*. **H** Cultivated areas, mangrove edges, grassland at forest edges, woodland, gardens. **D** S Sumatra, Java, Bali and Lombok, Indonesia.

8 DUSKY MUNIA *Lonchura fuscans* 11cm Almost wholly dusky brown. Quite shy and retiring, foraging on the ground and low down in vegetation. **V** Calls include a shrill *pee pee*, a thin *chirrup*, and a low *teck teck* in flight. **H** Paddyfields, grassland, grassy riverbanks, forest edge, secondary scrub, cultivation, gardens. **D** Borneo, Philippines.

9 BLACK-FACED MUNIA *Lonchura molucca* 10–11cm Note black face and upper breast, and fine barring on lower breast and belly. Feeds on grass seeds and seeds on the ground. **V** Song contains a run of wheezing *peep* and *whee* notes. Calls include a short *tri* and a buzzy *tissip*. **H** Bushy and grassy areas, paddyfields and cultivated areas. **D** Sulawesi and Muluku Islands, Indonesia, Lesser Sunda Islands.

10 SCALY-BREASTED MUNIA *Lonchura punctulata* 10–12cm Mainly dark brown, with strong scaling on underparts. Forages on the ground or on grasses, usually in small groups, with bigger flocks post breeding. **V** Song is a series of quiet notes, followed by whistles and churrs, ending with a long slurred whistle. Contact calls are a repeated *tit-ti* and a *kit-teee kit-teee*. **H** Open country with scrub and trees, bushy hillsides, secondary growth, cultivation, gardens. **D** Widespread, from N Pakistan through Nepal, India, Sri Lanka, SE Asia, Indonesia, Philippines to S China, Taiwan.

11 BLACK-THROATED MUNIA *Lonchura kelaarti* 12cm Black face and throat, dark underparts with whitish markings. Forages on the ground. **V** Calls include a high-pitched, nasal *tay* and a *chirp*. **H** Scrub, grassland, forest clearings, plantations, gardens. **D** SW and E India, Sri Lanka.

12 WHITE-BELLIED MUNIA *Lonchura leucogastra* 11cm White belly is distinctive. Forages in pairs or small groups, feeding on small seeds. **V** Calls include a *twyrt*, a *tee-tee-tee*, and a strong *tik* or *tchek* when alarmed; song is transcribed as *di-di-ptchcee-pti-pti-ti-pteep*. **H** Lowland and hill forest, forest edges, paddyfields, grassland, scrub and around rural houses. **D** Malay Peninsula, Greater Sunda Isands, Philippines.

13 STREAK-HEADED MANNIKIN *Lonchura tristissima* 10cm Dark brown, shading to blackish on belly and tail, head with fine white streaking, rump whitish. Feeds on grass seeds, often in flocks of 20–30. **V** Song complex, combining buzzing and toy-trumpet notes; calls include *tik, zeeee*, and louder buzzing *toot*. **H** Grassy stream edges in forest, marshes, reedbeds, bamboo. **D** New Guinea.

14 WHITE-SPOTTED MANNIKIN *Lonchura leucosticta* 10cm Mid-grey upperparts, buff belly shading to whitish on breast. Head finely streaked and back finely spotted with white. Behaviour like Streak-headed Mannikin. **V** Song a buzzy sequence of *zwee* and *tik* notes; calls include soft *tik* and louder *toot*. **H** Tall grassland and savannah, forest edge, reedbeds. **D** SC New Guinea.

15 FIVE-COLORED MUNIA *Lonchura quinticolor* 11–12cm Despite its name, this is a largely chestnut and white bird. Forages in pairs and small groups, feeding on the ground or on seedheads. **V** Song *te te te te weeee weeee pti-ti-pti-pti*. Calls include a loud *triprip* and a single *peet*. **H** Grassland, paddyfields, lightly wooded savannah, mixed scrub, forest clearings. **D** Lesser Sunda Islands.

16 TRICOLORED MUNIA *Lonchura malacca* 11–12cm Note glossy black head and upper breast, chestnut back, white belly with black patch. Forages on the ground, usually in pairs or small parties; bigger flocks after breeding, often with other species. **V** Calls include a weak *peekt* or *pee-eet*, a *veet-veet* and a *chirp-chirp-chirp* given in flight. **H** Edges of marshes or swamps, reedbeds, grassland, cultivation edge. **D** Peninsular India, Sri Lanka.

17 WHITE-CAPPED MUNIA *Lonchura ferruginosa* 11–12cm Black throat contrasts with white head. Feeds on the seeds of grass-heads, especially rice; often seen in large flocks. **V** Song is a very quiet series of clicks and wheezing notes, followed by a long, drawn-out *wheeee*. **H** Grassland, paddyfields, wetland areas with reeds, grasses and sedges. **D** Java and Bali, Indonesia.

18 CHESTNUT MUNIA *Lonchura atricapilla* 11–12cm A combination of glossy black and rich chestnut-brown. Forages on the ground, usually in pairs or small parties; bigger flocks post breeding. **V** Calls include a loud *pink pink* and a clear *pee* or *peet* contact note. **H** Edge of marshes or swamps, reedbeds, grassland, cultivation edge. **D** N India through SE Asia to Greater Sunda Islands, Philippines, S China, Taiwan.

19 WHITE-HEADED MUNIA *Lonchura maja* 11cm Note lack of black throat. Feeds on small seeds on the ground or from seedheads, usually in pairs or small groups. **V** Song consists of a series of clicks and then a drawn-out *weeeee heeheeheeheeheehee*; calls include a soft *preet* or *prit*. **H** Open grassland, marshes, reedbeds, paddyfields, cultivation, village gardens. **D** Malay Peninsula; S Vietnam; Sumatra, Java and Bali, Indonesia.

20 PALE-HEADED MUNIA *Lonchura pallida* 11cm Similar to White-headed Munia, but range differs. Forages in small or large flocks, feeding on seeding stems and on the ground. **V** Song is a high-pitched, chattering *weeeeeeeee*. **H** Open grassland, paddyfields, grassy scrub. **D** Sulawesi and Lesser Sunda Islands.

21 GREAT-BILLED MANNIKIN *Lonchura grandis* 12cm Black head, breast and centre of belly, otherwise warm rufous-brown. Note very large grey bill. Feeds in vegetation and on the ground, taking various types of seeds. **V** Song a quiet series of *tk-tk* notes, ending with descending drawn-out *whhheeeeeeee*. Calls include a metallic *tink* from female in reply to the male's loud *quip*; also a loud *kweer*. **H** All kinds of marshy grassland. **D** New Guinea.

22 GREY-BANDED MANNIKIN *Lonchura vana* 10cm Head white, shading to grey on breast, otherwise brown (brighter on belly). Note whitish breast-band separating grey breast from rufous belly. Forages in small flocks; behaviour little known. **V** Call is a high, thin *ts ts ts*. **H** Wet grasslands, marshes, cultivation edges. **D** New Guinea.

23 GREY-HEADED MANNIKIN *Lonchura caniceps* 11cm Head white, shading to mid then dark brown, tail blackish with contrasting bright rufous rump. Usually in small groups, rarely in flocks of up to 100. **V** Call nasal whistling *tee* or *too*. **H** Grassland, damp savannah and other grassy areas. **D** New Guinea.

24 GREY-CROWNED MANNIKIN *Lonchura nevermanni* 11.5cm Head white with black streaks forming solid black throat, also black undertail, underparts otherwise rufous, upperparts grey-brown. Climbs along grass seed-heads when feeding. **V** Song begins with whispered notes, proceeds to ticking sounds and ends with ringing *wheeeeeeeee*. Calls include *tseet* or *dooreet*. **H** Grassy marshland, riversides, floating grass mats in flooded areas. **D** New Guinea.

25 HOODED MANNIKIN *Lonchura spectabilis* 10cm Small, stocky mannikin with black head and brown upperparts (paler on rump and tail). Underparts white or warm rufous depending on race. Feeds on seeds, may form flocks of up to 40. Breeding females visit beaches to eat mussel shells for calcium. **V** Song combines clicking and high-pitched *wee* notes; call *peep* or *seee*. **H** All kinds of grassy habitats, cultivation, sometimes beaches. **D** New Guinea, Bismarck Archipelago.

26 FORBES'S MANNIKIN *Lonchura forbesi* 11.5cm Stout brown mannikin (lighter and more rufous on underparts) with black head and undertail. Feeds on seeds of grasses and other plants. **V** Not known. **H** Grassland, including small patches. **D** Bismarck Archipelago.

27 HUNSTEIN'S MANNIKIN *Lonchura hunsteini* 11cm Black with white speckling on crown and nape, tail and rump rufous. Forms large flocks. **V** Calls include *peep-peep* and fluting *pee-up* or *pee*. **H** Grassy areas, gardens, cultivation. **D** Bismarck Archipelago.

28 YELLOW-RUMPED MANNIKIN *Lonchura flaviprymna* 11cm Pale with whitish-grey head shading to light buff underparts, rump and tail, undertail and thighs black, wings and back grey-brown. Feeds on seeding stems and on ground, often in mixed flocks. **V** Song is a rhythmic series of clicks and tweets, ending with chucking notes; calls include *teet* and *weee*. **H** Grassland, reedbeds, rice fields, scrub, sometimes gardens. **D** NW Australia.

29 CHESTNUT-BREASTED MANNIKIN *Lonchura castaneothorax* 11cm Note black breast-band separating light rufous upper breast from white belly. Head dark brown; in some races, as shown, crown and nape grey with fine dark streaking. Will hybridise with other mannikin species. **V** Song consists of soft clicks ending with a *weeeeeee* note; calls include elements of song. **H** Grassland, savannah, mature gardens. **D** New Guinea, N and E Australia.

30 BLACK MANNIKIN *Lonchura stygia* 11cm Black apart from light buff rump and tail. Climbs among grasses feeding on seeds, forms small flocks and will flock with other estrildids. **V** Song a series of soft burbling notes; calls include quiet *tyu-tyu* and louder *tseeu*. **H** Savannah, marshland, riverside vegetation. **D** New Guinea.

31 BLACK-BREASTED MANNIKIN *Lonchura teerinki* 11cm Black face, breast and undertail, white belly, dark brown upperparts with paler tail. Behaviour little studied but probably similar to that of other mannikins. **V** Song little known, includes *huueet* phrases. Calls include a soft, clear *tseep*; groups maintain a constant quiet twittering. **H** Grasslands, clearings, cultivation, mainly at high elevations. **D** WC New Guinea.

32 EASTERN ALPINE MANNIKIN *Lonchura monticola* 11.5cm Note white breast and low black breast-band extending in blotches over flanks. Often flocks with Grey-headed Mannikins. **V** Thin *see see see* flight call; also gives rattling buzz. **H** Alpine scree and grasslands. **D** New Guinea.

33 WESTERN ALPINE MANNIKIN *Lonchura montana* 11cm Similar to Eastern Alpine Manikin but flanks white with narrow black barring, breast light buff. **V** Call series of *tyu* notes. **H** Alpine grassland and scrub edges. **D** New Guinea.

34 BUFF-BELLIED MANNIKIN *Lonchura melaena* 11cm Mostly black; lower belly and flanks buff with black barring, tail and rump rufous. Feeds on seeds including sugar cane. **V** Call short, high-pitched note. **H** Grassland, open forest, visits beaches for calcium sources. **D** Bismarck Archipelago, Solomon Islands.

35 TIMOR SPARROW *Lonchura fuscata* 12cm Note distinctive white cheeks. Forages in pairs or small groups, mainly on the ground; regularly perches on stumps and posts and in trees. **V** Rising *chip chip chip chip chipchipchipchip*. Calls include *tchik* and *wheeee*. **H** Degraded monsoon forest and grassy areas with scattered bushes and trees. **D** Timor, Lesser Sunda Islands.

36 JAVA SPARROW *Lonchura oryzivora* 15cm Note pink bill and pinkish-red eye-ring. Forages on the ground and clings to grass stems to feed on seedheads; usually in small flocks. **V** Bell-like trilling and clicking notes ending with a drawn-out whistle. Call is a *tup, t'luk* or *tack*. **H** Cultivations, reedbeds, gardens. **D** Java and Bali, Indonesia.

37 PICTORELLA MANNIKIN *Heteromunia pectoralis* 11cm Black face and black-scaled white breast, otherwise buff underparts, grey-brown upperparts. Feeds on ground and on plant stems. **V** Call short *chip* or *pik*; flocks very noisy. **H** Grassy areas in woodland or in open, reedbeds, rice fields. **D** N Australia.

1
2
3
4
5
6
7
8
9
10
11
12
13
14
15
16
17
18
19
20
21
22
23
24
♀
♂
25
26
27
28
29
30
31
32
33
34
35
36
37

INDIGOBIRDS AND WHYDAHS VIDUIDAE

1 VILLAGE INDIGOBIRD *Vidua chalybeata* 11cm It is almost impossible to separate the 10 short-tailed indigobirds in the field, with the exception of the Zambezi Indigobird, but note bill and leg colour and basic colour to gloss in male plumage. Also compare their ranges. All are brood parasites, and they incorporate parts of their host's song into their own. Village Indigobird has a green to steel-blue gloss, and parasitises Red-billed Firefinch (Plate 266). **V** Song combines dry chatters, churrs, scratchy notes, and mimicry of host species. **H** Woodland, thornveld, cultivation. **D** Mauritania to Sierra Leone and through Sahel zone to Ethiopia, and through E Africa to E South Africa; Angola, N Namibia.

2 PURPLE INDIGOBIRD *Vidua purpurascens* 12cm Note white bill and legs. Male has purplish gloss. Parasitises Jameson's Firefinch (Plate 266). **V** Song is a series of whistles, trills, chatters and churrs, including mimicry of host species. Males match song types with those in neighbouring territories. **H** Thornveld. **D** Kenya to Angola, Botswana, NE South Africa.

3 JAMBANDU INDIGOBIRD *Vidua raricola* 11cm Male has bright green gloss. Parasitises Black-bellied Firefinch (Plate 266) and Orange-breasted Waxbill (Plate 267). **V** Song consists of churring, grating, scratchy notes combined with mimicry of host species; calls also mimic host species. **H** Woodland and bush with tall grass. **D** Guinea to Sudan, South Sudan, NE DR Congo.

4 BARKA INDIGOBIRD *Vidua larvaticola* 11cm Male has blue to green-blue gloss. Parasitises Black-faced Firefinch (Plate 266). **V** Song consists of churring and scratchy notes combined with mimicry of one of host species; calls also mimic host species. **H** Open woodland with tall grass. **D** Guinea-Bissau; Ivory Coast to Ethiopia.

5 DUSKY INDIGOBIRD *Vidua funerea* 12cm Male plumage is not very glossy, black with a dull bluish-purple to purplish-blue cast. Parasitises African Firefinch (Plate 266). **V** Song includes metallic and dry notes, churrs and scratchy notes combined with mimicry of host species; calls also mimic host species. **H** Forest edge, thornbush, riverine belts, gardens. **D** W Kenya and Tanzania to N Angola, Zambia, Mozambique, N and E South Africa.

6 ZAMBEZI INDIGOBIRD *Vidua codringtoni* 12cm The only male indigobird with no white in the wing. The female can be separated from other females by white bill. Parasitises Red-throated Twinspot (Plate 266). **V** Song is churring and scratchy, incorporating mimicry of host species; calls also mimic host species. **H** Open woodland, thornveld. **D** Tanzania to Zimbabwe and W Mozambique.

7 WILSON'S INDIGOBIRD *Vidua wilsoni* 11cm Male has purple to bluish-purple gloss. Parasitises Bar-breasted Firefinch (Plate 266). **V** Song combines churring, chattering and scratchy notes with infrequent mimicry of host species; calls also mimic host species. **H** Woodland, bush. **D** Senegal and Gambia to Sierra Leone and east to Ethiopia, S South Sudan, NE DR Congo.

8 QUAILFINCH INDIGOBIRD *Vidua nigeriae* 11cm Male has dull green gloss. Parasitises Quailfinch (Plate 267). **V** Song is churring and scratchy, combined with accurate mimicry of the complex vocalisations of its host species; calls also mimic host species. **H** Wet grassland. **D** Senegal and Gambia, Guinea, Mali, Nigeria and Cameroon, S South Sudan.

9 JOS PLATEAU INDIGOBIRD *Vidua maryae* 11cm Male has green to blue-green gloss. Note very restricted range. Parasites mainly African Firefinch (Plate 266). **V** Song consists of churring, chattering and scratchy notes combined with mimicry of host species; calls also mimic host species. **H** Forest edge, woodland. **D** Jos Plateau, C Nigeria; also N Cameroon.

10 CAMEROON INDIGOBIRD *Vidua camerunensis* 11cm Male has blue or bluish-green gloss. Parasitises Black-bellied Firefinch, African Firefinch, Dybowski's Twinspot and Brown Twinspot (all Plate 266). **V** Song churring and scratchy, combined with mimicry (each male only ever mimics one of the many host species); calls also mimetic of that host species. **H** Forest edge, woodland, cultivation. **D** Guinea-Bissau to Sierra Leone and east to South Sudan, Congo, NE DR Congo.

11 PIN-TAILED WHYDAH *Vidua macroura* male 31cm, female 11.5cm Breeding-plumage male unmistakable. Non-breeding male similar to breeding female, but with stronger face markings and a red bill. Parasitises waxbills (Plates 266–267) and other species. **V** Song is a jumble of harsh squeaks and chirps that form a rhythmic twittering. Calls include a loud *sweet*. **H** Short grass areas, including fields and garden lawns. **D** Sub-Saharan Africa, widespread. Introduced Puerto Rico.

12 STEEL-BLUE WHYDAH *Vidua hypocherina* male 30cm, female 10cm Male unmistakable, with all-black plumage and long tail; note dark markings of female. Main host is Black-faced Waxbill (Plate 267). **V** Song is a series of short, varied phrases, including chatters but also twittering or plaintive whistling. **H** Dry, more or less bushed areas. **D** Ethiopia and Somalia to Tanzania.

13 STRAW-TAILED WHYDAH *Vidua fischeri* male 31cm, female 10cm Note pale crown of breeding male. Parasitises mainly Purple Grenadier (Plate 266). **V** Mimics songs and calls of its host; calls also include fast and slow chatters and a buzzy *chuz*. **H** Dry, more or less bushed areas. **D** South Sudan, Ethiopia, Somalia to Tanzania.

14 SHAFT-TAILED WHYDAH *Vidua regia* male 32cm, female 12cm Breeding-plumage male has black crown and central tail feathers very thin with a flag at the tip. Female separable from female indigobirds by redder bill and more buff colouring overall. Parasitises Violet-eared Waxbill (Plate 266). **V** Songs include mimetic and non-mimetic forms; calls include harsh buzzing chatters. **H** Dry areas with thornbush and more or less grass cover. **D** Angola, Zambia, Zimbabwe, Botswana to N South Africa, S Mozambique.

15 LONG-TAILED PARADISE WHYDAH *Vidua paradisaea* male 37cm, female 15cm Long-tailed, Sahel, Exclamatory, Togo and Broad-tailed Paradise whydahs can be separated by range, length and form of longest tail streamers and intensity of neck colouring. Breeding male of this species separable from Broad-tailed Paradise Whydah by longer, tapered tail, female by blacker markings on head and upperparts. Non-breeding male as female but often with some scattered black feathers on head or mantle. Parasitises Green-winged Pytilia (Plate 265). **V** Song is mimetic of its host, also of Blue-capped Cordon-bleu (Plate 266), which is not confirmed to be a host. Non-mimetic calls include a high *weeee* or a buzzing *zzzeee*. **H** Open woodland, bushland, thornveld. **D** SE Sudan, Ethiopia, Somalia to N South Africa.

16 SAHEL PARADISE WHYDAH *Vidua orientalis* male 31cm, female 14cm Note pale nape and tail that does not taper until near the end. Parasitises mainly Green-winged Pytilia (Plate 265). **V** Song and calls are mimetic of its host; calls also include chattering notes and single *chuff*. **H** Open woodland, bushland, thornveld. **D** Sahel region, from Senegal to Eritrea.

17 EXCLAMATORY PARADISE WHYDAH *Vidua interjecta* male 38–40cm, female 14cm Long tail of breeding male is about the same width as the bird's body, tapering only at the end. Parasitises mainly Crimson-winged Pytilia (Plate 265). **V** Song and calls are mimetic of one of the two host species, depending on region; calls also include varied chattering notes. **H** Open woodland, bushland, thornveld. **D** Senegal to Ethiopia.

18 TOGO PARADISE WHYDAH *Vidua togoensis* male 40–43cm, female 14cm Longest tail feathers of breeding male are of even width almost to the tip, and narrower than the bird's body. Parasitises mainly Orange-winged Pytilia (Plate 265). **V** Song and calls are mimetic of its host; calls also include non-mimetic chattering notes and phrases. **H** Open woodland, bushland, thornveld. **D** Guinea to Togo.

19 BROAD-TAILED PARADISE WHYDAH *Vidua obtusa* male 27cm, female 15cm Note tail shape of breeding male. Non-breeding male as female. Parasitises Orange-winged Pytilia (Plate 265). **V** Song and calls are mimetic of its host; calls also include chattering notes and flight call *whooee*. **H** Open woodland, bushland, thornveld. **D** Angola to E DR Congo, Tanzania, Botswana, Zimbabwe, Mozambique.

20 CUCKOO-FINCH *Anomalospiza imberbis* 12cm Note short, conical bill and combination of clean yellow head and streaked green upperparts. Non-breeding male less bright. Breeding female has streaked breast sides and flanks. Brood parasite, using cisticolas (Plates 215–216) as hosts. **V** High, soft chattering. **H** Moist, open and bushed, natural and cultivated areas. **D** Sierra Leone and S Mali to Ethiopia and south to E South Africa.

OLIVE WARBLER PEUCEDRAMIDAE

21 OLIVE WARBLER *Peucedramus taeniatus* 13cm Only distantly related to other New World warblers. Note dark mask and double white wing-bar. Often forages in pine-needle clumps. **V** Song is a loud two-note whistle, *pee-ter pee-ter pee-ter pee-ter*. **H** Montane conifer forests. **D** SW USA to Nicaragua.

ACCENTORS PRUNELLIDAE

22 ALPINE ACCENTOR *Prunella collaris* 15.5–17cm Note black scaling on white throat. Forages on the ground, among rocks, stones and vegetation. Song given from a rock or during a short display flight. **V** Song is a variable warble of low-pitched trills and ripples with fluty whistles. Calls include a rolling *chirrup* and a low *chit-chittur*. **H** Rocky mountain slopes with sparse vegetation; often descends in winter. **D** From SW Europe and NW Africa through C Asia to China, E Siberia, Japan.

23 ALTAI ACCENTOR *Prunella himalayana* 15cm Note rufous streaking on upper breast and flanks. Gregarious, often encountered with other accentors and mountain finches. **V** Calls include a finch-like *tee-tee* and a low twitter. **H** Grassy, rocky hillsides, valleys at lower altitudes. **D** E Kazakhstan, NE Afghanistan, S Russia to Mongolia and NW China.

24 ROBIN ACCENTOR *Prunella rubeculoides* 16–17cm Combination of plain grey head and rufous breast is distinctive. Occurs in small flocks in winter. **V** Song is a sweet and short chirping *si-tsi-si-tsi - tze-e-you*. **H** Dwarf scrub, sedge tussocks around lakes; in winter frequents stony and rocky ground and around upland villages. **D** Himalayas to C China.

25 RUFOUS-BREASTED ACCENTOR *Prunella strophiata* 15cm Supercilium is white in front of the eye, becoming rufous behind. Wing-twitching makes bird appear nervous as it forages, mouse-like, under bushes. **V** Long, melodious, wren-like warbling and trilling. **H** Forest and scrub near or above the treeline; in winter at lower altitudes. **D** E Afghanistan through Himalayas to C China and N Myanmar.

26 SIBERIAN ACCENTOR *Prunella montanella* 14–15cm Supercilium is long, broad and creamy-yellow. Forages mainly on the ground, but also in bushes and trees. **V** Typical call is a *tsee-ree-see* or *dididi*. **H** Thickets and scrub usually along watercourses. **D** Breeds across Siberia, from west of the Urals to the far east. Winters N China, Korean Peninsula.

27 BROWN ACCENTOR *Prunella fulvescens* 15cm Predominantly olive-brown, with whitish supercilium. Often forages in the open; in winter regularly encountered feeding on village refuse or muck heaps. **V** Song is a short low warble with trills. Call is a bunting-like *ziet-ziet-ziet*. **H** Dry, rocky boulder-strewn hillsides with low scrub; in winter often around villages. **D** E Afghanistan and E Kazakhstan through Himalayas, Mongolia, China.

28 RADDE'S ACCENTOR *Prunella ocularis* 15–16cm Note white supercilium, black-streaked tawny-brown upperparts, orange-tinged breast. In worn plumage may show a band of darkish spots across upper breast. **V** Song is a gentle, sweet twittering or trembling. Calls include a slurred *tseer*, a *ti-ti-ti* and a *tseep*. **H** Low scrub in dry rocky mountains; moves to slightly lower areas in winter. **D** SW Russia and Turkey to Turkmenistan, Afghanistan, Iran.

29 ARABIAN ACCENTOR *Prunella fagani* 15cm Easily identified, as the only accentor in its range. **V** Song is a short, fast *drsi-drsi-drsi-dy-dy-dy*, often with a scratchy ending; also a more trilling *dri-drrriii-tyi-driivivivi*. **H** Bushy areas in rocky mountains. **D** SW Arabian Peninsula.

30 BLACK-THROATED ACCENTOR *Prunella atrogularis* 15cm Blackish ear-coverts separated from black throat by thin white malar stripe. Forages on the ground, like other accentors. **V** Call is a *teeteetee*. **H** Sandy semi-desert near cultivation, damp valleys with dense vegetation, scrub jungle and cultivations. **D** Breeds Urals and C Asian mountains. Winters from Iran to Kazakhstan and south to N Indian subcontinent.

31 KOZLOV'S ACCENTOR *Prunella koslowi* 15cm The palest and plainest of the accentors. **V** Song is a quite warble. **H** Juniper scrub and grassland on dry mountain slopes; in winter, arid scrub in sandy desert. **D** Mongolia, N China.

32 DUNNOCK *Prunella modularis* 14.5cm Grey head and breast contrast to varying degree with streaky chestnut back and wings. Wing twitching makes it appear nervous as it creeps, mouse-like, in undergrowth. **V** Song is a fairly slow jangling. Call is *seep* or *seeh*, also a quieter *ti-ti-ti*. **H** Gardens, open woodland, heath, spruce and juniper forest; in south of range, more a treeline species. **D** Widespread across Europe east to Urals and Caucasus. Birds from N and E of range move S and W in winter.

33 JAPANESE ACCENTOR *Prunella rubida* 14cm Similar to Dunnock, but generally browner. Never far from cover. **V** High-pitched slow trill, like a squeaky bicycle wheel. **H** Dwarf pines and birch up to treeline on high mountains; in winter, at lower elevations. **D** Japan.

34 MAROON-BACKED ACCENTOR *Prunella immaculata* 14–15cm Distinctive unstreaked accentor, predominantly grey, brown and black, with pale eye. **V** Call is a weak, high-pitched *zieh-dzit* or *tzip*. **H** Undergrowth in humid conifer and rhododendron forests; winters in secondary forest and forest edge. **D** Himalayas to N Myanmar and C China.

1
2
3
4
5
6
7
8
9
10
11
12
13
14
15
16
17
18
19
20
21
22
23
24
25
26
27
28
29
30
31
32
33
34

WAGTAILS AND PIPITS MOTACILLIDAE

1 FOREST WAGTAIL *Dendronanthus indicus* 16–18cm Note distinctive wing and breast pattern. Runs or walks rapidly; when flushed often flies into tree canopy. Sways tail and rear of body from side to side. **V** Song is a repeated, high-pitched, squeaky *zlic-zhee zlic-zhee zlic-zhee...* Call is a hard, shrill *pick* or *pick-pick*. **H** Tracks, clearings and edges of forests, plantations and bamboo jungle. **D** Breeds far E Russia, E China, Korean Peninsula, S Japan. Winters India, Sri Lanka, SE Asia, S China, Indonesia.

2 WESTERN YELLOW WAGTAIL *Motacilla flava* 16cm Very variable. Head may be largely grey, black, white or yellow, depending on race. Nominate race (illustrated) often referred to as Blue-headed Wagtail. In winter and on migration very gregarious. **V** Simple, repeated, harsh *tsre-tsree - tsre-tweeo* or *tsre - tsre - tsre*. Calls include a loud *pseet* and a longer *swee-eep*. **H** Meadows, cornfields, grazed marshes, tundra, moist steppe and large clearings in forest. On migration and in winter, almost any open habitat. **D** Breeds across Europe and Asia, as far east as Mongolia and NC Siberia. Winters sub-Saharan Africa, Indian subcontinent.

3 EASTERN YELLOW WAGTAIL *Motacilla tschutschensis* 16–18cm Similar to Western Yellow Wagtail, and formerly considered conspecific. **V** Song is a short fast twittering of 3–6 notes, faster than song of Western Yellow Wagtail. **H** Favours damp habitats. **D** Breeds far E Siberia, N China, Alaska. Winters Taiwan, S China, SE Asia through Philippines and Indonesia to W New Guinea and N Australia.

4 CITRINE WAGTAIL *Motacilla citreola* 17–20cm Yellow head of breeding male contrasts with black hindneck and upper back. Actions much like White Wagtail, but tail-wagging less pronounced; regularly perches on vegetation, posts and wires. **V** Song consists of call notes interspersed with warbling phrases. Calls include a *dzreeip* or *tzreep* and a soft *tslee*, *tselee* or similar. **H** High-altitude wet grassland; winters in freshwater wetlands. **D** Breeds from E Europe across Asia to Mongolia, N China. Winters from Iran to SE Asia and S China.

5 CAPE WAGTAIL *Motacilla capensis* 19cm Separable from Grey Wagtail by pale wing-bars, from immature African Pied Wagtail by less white in wing. **V** Song is transcribed as *sweep tededjeh* (*sweep* very high, *tededjeh* rattling and low). **H** Wet open woodland, grassland, pasture, swamp, lake edges. **D** E DR Congo to Kenya; Angola and S DR Congo to South Africa.

6 MADAGASCAN WAGTAIL *Motacilla flaviventris* 19cm Note narrow black breast-band and yellow belly. **V** Song is a three-note whistled phrase, *tree treeoo*. Calls include a similar but buzzier phrase. **H** Open areas (including gardens) at any altitude, normally near water. **D** Madagascar.

7 SAO TOME SHORTTAIL *Motacilla bocagii* 13cm An enigmatic bird, once placed with warblers. No other similar bird in its range and habitat. **V** Typical call is a thin, whistled *tseep* or *tsuuit*. **H** On the ground along forest streams. **D** São Tomé, Gulf of Guinea.

8 GREY WAGTAIL *Motacilla cinerea* 17–20cm Female lacks black throat of male. Constantly pumps rear body and tail. **V** Song consists of short notes, often with added high *si-si-si-siu*. Call is a high-pitched *zit-zit*. **H** Mountain streams; also lowland waterways with weirs, lock gates or mill races; winters by lowland waters. **D** Widespread across Europe (including Azores, Madeira, Canary Islands) and Asia (east to Japan). Southern populations resident; others winter in Africa, Middle East to SE Asia, Philippines, New Guinea.

9 MOUNTAIN WAGTAIL *Motacilla clara* 19cm No yellow in plumage. Delicately marked with black and pure grey above. **V** Extremely high, descending *sreeeeeh* or *treederup*. **H** Mountain streams. **D** Sub-Saharan Africa.

10 WHITE WAGTAIL *Motacilla alba* 17–18cm Colour of mantle varies from pale grey to black, depending on race and sex. Constantly wags tail up and down as it walks. **V** Song is a twittering interspersed with call notes. Calls include *chissik* and *ts-lee-wee*. **H** Wide range of open habitats, farmyards, often near water. **D** Breeds across Europe, NW Africa, Asia, W Alaska. Southernmost breeding populations resident; otherwise winters from Africa through Middle East and S Asia to Philippines.

11 AFRICAN PIED WAGTAIL *Motacilla aguimp* 20cm Striking black and white plumage. Actions as White Wagtail. **V** Song simple and monotonous to rich and varied, often including complex nasal buzzing and rattling notes. Call is a loud metallic *tzink*, *tzchip* or similar. **H** Lake shores, sandbanks, riversides, around human habitations. **D** Sub-Saharan Africa, widespread.

12 MEKONG WAGTAIL *Motacilla samveasnae* 17–18cm Regularly forages in bushes near water. Sings from a bush top or boulder. **V** Rapid series of thin, high-pitched harsh notes, interspersed with long pauses. In flight utters a harsh *dzeer*. **H** Broad, lowland rivers with sandbars, emergent rocks and bushes. **D** Lower Mekong River and tributaries in S Laos, NE Cambodia; marginally in adjacent parts of Thailand, Vietnam.

13 JAPANESE WAGTAIL *Motacilla grandis* 21cm Upperparts and breast black, apart from white forehead, supercilium, chin and tiny crescent below eye. Actions as White Wagtail. **V** Series of simple notes and phrases, *chzeeo... birzzchle-birzzchle... chzleeo... birzzchelou... chzleeo-pzchy...* Call is a shrill *bzzr*, *tzzr* or *tzreh*. **H** Edges of streams, rivers, lakes, ponds; in winter spreads to wider variety of habitats. **D** Japan, South Korea.

14 WHITE-BROWED WAGTAIL *Motacilla maderaspatensis* 21–24cm Resembles Japanese Wagtail, but ranges do not overlap. Wags tail up and down. In winter sometimes forms flocks that roost in tamarisks or reeds. **V** Very simple *tchi-tchi-tchi-tchUU* or a mixture of harsh, shrill and melodious notes with some drawn-out harsh rolling notes. **H** Embankments of watercourses; villages and gardens with nearby water. **D** Indian subcontinent.

15 GOLDEN PIPIT *Tmetothylacus tenellus* 15cm In flight, male appears all yellow with black wing tips. **V** Series of flute-like whistles interspersed with a hurried, grating warbling, given during fluttering display-flight. **H** Dry acacia bush, open savannah, scrubby grassland. **D** Somalia to SE South Sudan and C Tanzania.

16 SHARPE'S LONGCLAW *Macronyx sharpei* 15cm Looks like a small slim yellow pipit. Lacks the distinct dark necklace of other longclaws. **V** Song is a series of four or five thin whistles, rising in pitch, sometimes a longer, more complex variant. Calls include *tsip* given in flight, or sharper *tswit*, also descending *seeooo*. **H** Rough grassland. **D** W Kenya.

17 ABYSSINIAN LONGCLAW *Macronyx flavicollis* 20cm Note orange-yellow throat bordered by black necklace, and yellow forehead. **V** Song is a clear trill; call a piping note. **H** Grassland, moorland. **D** Ethiopia.

18 FÜLLEBORN'S LONGCLAW *Macronyx fuelleborni* 19cm Separable from Yellow-throated Longclaw by less white in outer-tail feathers, absence of streaks below necklace and off-yellow underparts. **V** Mid-high, loud, fluted notes, *fiu feew feeuw*. **H** Short grassland, marshes. **D** Angola through S DR Congo and Zambia to SW Tanzania.

19 CAPE LONGCLAW *Macronyx capensis* 19cm Upperparts are less strongly marked than in Yellow-throated Longclaw. Note also deep orange throat. **V** High series of nasal *dweet* notes. **H** Short grassland. **D** Zimbabwe and S Botswana to South Africa.

20 YELLOW-THROATED LONGCLAW *Macronyx croceus* 19cm Distinguished from Fülleborn's Longclaw by all-white outer-tail feathers. **V** Very high *mieeeeh mioweeh* combined with low *chachachacha*. **H** Moist, sparsely wooded grassland. **D** Senegal and Gambia to South Sudan and Kenya, south to Angola and South Africa.

21 PANGANI LONGCLAW *Macronyx aurantiigula* 20cm Throat patch more orange-yellow than supercilium and breast. Black collar rather narrow and streaks on breast rather extensive. **V** Song is a long series of repeated, descending notes. Calls are varied, including a whistling *ooeee* and a dry *chrry*; alarm call a plaintive *k-lee*. **H** Dry bushed and slightly wooded areas with some grass. **D** S Somalia to NE Tanzania.

22 ROSY-THROATED LONGCLAW *Macronyx ameliae* 19cm Tail corners and outer-tail feathers all white. Male separable from Grimwood's Longclaw by red belly, white margins to wing feathers and solid black necklace; female by rosy (not buffy) underparts. Rather skulking. **V** High *weetjeh-weetjeh*. **H** Rough grass near lakes, dams, inundations, marshes. **D** Kenya and Tanzania to Angola, Botswana, Zimbabwe; S Mozambique, Eswatini, E South Africa.

23 GRIMWOOD'S LONGCLAW *Macronyx grimwoodi* 20cm Upperparts darker than Rosy-throated Longclaw. Not also narrow black streaks instead of a solid black necklace. **V** Song said to be a series of whistles, similar to that of Fülleborn's Longclaw. **H** Open, grassy hills. **D** SW DR Congo to C Angola, NW Zambia.

24 RICHARD'S PIPIT *Anthus richardi* 17–20cm Forages on the ground; when agitated often stands very upright with neck stretched. Flight powerful and undulating, with long dips; regularly hovers above grass before landing. **V** Simple, grinding *tschivu-tschivu-tschivu-tschivu-tschivu...* given in flight. Calls include a harsh *schreep* or a longer *sherrreeep* given as bird takes flight; also a short *chup*, a subdued *chirp* and a *r-rump*. **H** Open country, grassy areas and cultivations. **D** Breeds SE Russia to Kyrgyzstan and China. Winters Indian subcontinent, S China, SE Asia.

25 PADDYFIELD PIPIT *Anthus rufulus* 15cm Forages on the ground. Call note is the best way to distinguish from the similar but larger Richard's Pipit. **V** Similar to Richard's Pipit, but faster and higher-pitched. Calls include a hard *chep* or *chep-chep*, a thin *pipit* and harsh *chwist*. **H** Open country, short grasslands, paddyfields, stubble fields and cultivations. **D** From Afghanistan through Indian subcontinent to SE Asia, SW China, Indonesia, Philippines.

26 AUSTRALIAN PIPIT *Anthus australis* 17–18cm Recently split from New Zealand Pipit; plumage more yellowish in tone. **V** Like New Zealand Pipit. **H** All kinds of short grasslands, roadsides, dunes, playing fields, gardens. **D** Australia and New Guinea.

27 NEW ZEALAND PIPIT *Anthus novaeseelandiae* 18cm No other pipit in New Zealand. **V** Song is *uswee* or descending *sreeeuw*, or very high, sharp series, *weet-witwitwitwit*. **H** Open habitats, occasionally open woodland. **D** New Zealand.

28 AFRICAN PIPIT *Anthus cinnamomeus* 16cm Note erect, slender stance, yellow lower mandible, white outer-tail feathers. From Wood Pipit by more distinct striping, from Long-Billed Pipit by diffuse buff (not white) supercilium. **V** Very high, liquid, rapid, level *tjeetjeetjee*. Song is a level, fast twittering *tweetweehweehweehweehweeh*. **H** Any open area. **D** S, C and E Africa.

29 MOUNTAIN PIPIT *Anthus hoeschi* 18cm Similar to Richard's Pipit but less slender. Outer-tail feathers buff. **V** Song as Richard's Pipit but shorter. Call is a very high *tuwee tuwee*. **H** Montane grasslands. **D** Breeds Lesotho, E South Africa. After breeding, moves north to E Angola, S DR Congo, NW Zambia.

30 BLYTH'S PIPIT *Anthus godlewskii* 16cm Actions similar to the slightly larger Richard's Pipit, best distinguished from it by voice. Generally has back more heavily streaked than Paddyfield Pipit. **V** Calls include a loud *chup*, *chep* or *chep-chep* and a longer *pshee* or *pshee-chep-chep*. **H** Grasslands, dry paddyfields and edges of cultivations. **D** Breeds Mongolia, SE Russia, N China. Winters Indian subcontinent.

31 TAWNY PIPIT *Anthus campestris* 16–17cm Similar to Richard's and Paddyfield Pipit, but with dark lores. Forages on the ground, but often perches on bushes, posts or wires. Frequently pumps tail, especially when agitated. **V** Calls include an explosive *chilp* or *cherleep* and a loud *tseep* or *tseuc* given at take-off. **H** Open, stony country with scattered scrub, semi-desert, fallow fields and pastures. **D** Breeds from SW Europe through C Asia to Mongolia. Winters Africa (Sahel zone, Nile valley and south into Somalia), Arabian Peninsula, S Asia.

32 LONG-BILLED PIPIT *Anthus similis* 17–20cm Forages on the ground, tends to creep about; when flushed regularly lands on rocks, bushes or trees. **V** Series of monotonous, unmusical, well-spaced phrases, *tjup-threeee-tjup-tjup-threeee* or *chreep-shreep-chew-ee*. Calls include a *klup* and a loud *che-vlee*. **H** Rocky hillsides with sparse cover; winters on grassy plains, sparsely scrubbed country, fields and sand dunes. **D** Scattered distribution through sub-Saharan Africa, and from Middle East to Myanmar.

33 WOOD PIPIT *Anthus nyassae* 18cm Note well-defined white supercilium, general warm colouring, grey neck. **V** High, monotonous, well-spaced *fee trit trit fee free fui fee*. **H** Miombo woodland, lightly wooded savannah. **D** Gabon to Angola and east to S Tanzania, Mozambique.

34 BUFFY PIPIT *Anthus vaalensis* 18cm Note very erect stance, unstreaked upperparts and faint streaking on breast. **V** Very high, rapid *triuh* as part of an unstructured, well-spaced sequence. **H** Dry, bare or sparsely grassed areas. **D** From Angola, S DR Congo, S Tanzania to S South Africa.

1
2 ♀
♀
♀
5
♂
3
♂
4
♂
7
9
6
8
♀
10
11
12
♂
♀
♂
13
14
15
16
♂
♀
17
18
19
20
21
♀
♂
23
♀
24
22
♀
25
26
27
28
29
30
31
32
33
34

WAGTAILS AND PIPITS *CONTINUED*

1 PLAIN-BACKED PIPIT *Anthus leucophrys* 17cm Separated from Long-Legged Pipit by fainter breast streaking and buff (not pale brown) outer-tail feathers, from Buffy Pipit (Plate 270) by greyer, more two-toned colouring, from African Rock Pipit by paler underparts and buff (not yellowish) edges to wing feathers. **V** High, unstructured *tjee tjup tree swee tjup*. **H** Dry, hilly areas with short grass. **D** Widespread in Africa.

2 LONG-LEGGED PIPIT *Anthus pallidiventris* 17cm Note white underparts, pale brown outer-tail feathers, yellow base to lower mandible. Separated from Plain-Backed Pipit and African Rock Pipit by different range, from Buffy Pipit (Plate 270) by darker upperparts and paler underparts. **V** Extremely high *pweet pweet pirriweet*. **H** Clearings in forest and woodland. **D** WC Africa.

3 MEADOW PIPIT *Anthus pratensis* 15cm Tone of olive-buff upperparts and paler underparts varies with wear. Often perches on low wires after flushing. **V** Given mainly during a display flight that ends in a parachuting descent, consists of a series of accelerating *seep-seep-seep-seep* notes, rising in pitch, mixed with longer *tseut* phrases and ending in a long trill. Call a squeaky *seep* or *seep-seep-seep*. **H** Open grassland from coast to alpine meadows; heathland and moorland. In winter often by lakesides and beaches. **D** Breeds Iceland through NW Europe to W Russia. Winters N Africa, S Asia.

4 TREE PIPIT *Anthus trivialis* 15cm Forages on ground, walks with more deliberate gait than Meadow Pipit; when disturbed usually flies to settle in a nearby tree. **V** Rapid *chikchikchik chia-chia-chia-wich-wich-wich... tsee-tsee-tsee-tsee...* latter usually given as bird descends during display flight. Calls include drawn-out *tseep* or *teez* and high-pitched *seet-seet-seet*. **H** Grassy slopes with scattered trees or bushes, open forests, groves, cultivations, open country with scattered trees and stubble fields. **D** Breeds widely in Eurasia. Winters India and Africa.

5 OLIVE-BACKED PIPIT *Anthus hodgsoni* 14–15cm Forages on ground; when disturbed flies to nearby tree. **V** Repeated trilled phrases and dry rattles. Calls include loud *teaze*, thin *teez* or *tseep*. **H** Grassy and bracken-covered slopes, rocky ground, clearings in open forests, abandoned cultivation, scrub with isolated trees, plantations, groves. **D** Breeds in C and E Asia. Winters from S Asia to Philippines.

6 PECHORA PIPIT *Anthus gustavi* 15cm Mouse-like skulker; if flushed, flies, often without calling, a short distance then drops into cover. **V** Usually given in long lark-like hovering display flight; distinctive with short buzzy or clear notes, alternated with drawn-out very fast buzzy rattles. Call *tsip* or repeated *tsip-tsip-tsip-tsip*. **H** Wet shrubby tundra, marshes with tall sedge or reeds; on migration visits wet meadows, edges of marshes and paddies. **D** Breeds NC and NE Asia. Winters from Lesser Sunda Islands to Philippines.

7 ROSY PIPIT *Anthus roseatus* 15–16cm Forages on ground, usually in pairs; forms small loose flocks after breeding. Song is mainly given during display flight. **V** Monotonous *tree-tree-tree* mixed with some drawn-out notes, given as bird rises, and a *tsuli-tsuli-tsuli...* uttered during descent. Calls include *tzeep* or *tzeep-tzeep-tzeep*. **H** Breeds on alpine meadows and boulder-strewn grassy slopes; winters on short grassland, marshy areas, paddyfields. **D** Uzbekistan and Afghanistan to E and S China.

8 RED-THROATED PIPIT *Anthus cervinus* 14–15cm Forages on ground, often encountered in small flocks. Extent of red on head variable, sometimes absent. **V** Calls include short *tew* and longer, high-pitched *pseeeeu*. **H** Marshes, grasslands and stubble fields. **D** Breeds from N Europe to NW Alaska. Winters in Africa and from SE Asia to Philippines.

9 BUFF-BELLIED PIPIT *Anthus rubescens* 16cm Actions much like Water Pipit, although tends to have quicker, lighter gait and more often found in flocks. **V** Given from perch or display flight, series of fast, high notes repeated in phrases; *chwee-chwee, tsip-tsip-tsiru, treeu-treeu-treeu* or *pleetrr-pleetrr-pleetrr*. Calls include high *sipit, sip, tsweep* or *si-si-si-si-si* when flushed. **H** Rocky alpine and subalpine tundra; in winter moist grassland, marshland, other wetland edges. **D** Breeds Canada, Alaska and NE Asia. Winters S Asia, S USA and Central America.

10 WATER PIPIT *Anthus spinoletta* 16cm Usually more wary than Eurasian Rock Pipit, flying some distance when flushed and more likely to perch in trees. **V** Very similar to Eurasian Rock Pipit's but with additional buzzing *zeeer-zeeeer-zeeeer* phrase. Call a sharp *peezp*. **H** Mountain slopes with short grass and scattered boulders and bushes, usually above treeline. On migration and in winter visits sewage-farms, lakesides, marshes, wet meadows; rarely found on seashores. **D** Breeds C, W and S Eurasia. Winters N Africa.

11 EURASIAN ROCK PIPIT *Anthus petrosus* 16cm Usually relatively confiding, flying short distances when flushed. **V** Series of sharp repeated notes with various theme changes, usually given in display flight: *zru-zru-zru-zru-zru-zre-zre-zre-zre-zre-zre-zre-sui-sui-sui-sui-zre-zre....* Call *weest*, usually repeated several times but not as rapidly as Meadow Pipit. **H** Rocky coasts; in winter, occasionally inland lake and river shores. **D** Coasts in W Europe.

12 NILGIRI PIPIT *Anthus nilghiriensis* 17cm Forages on ground; when disturbed usually flies to settle on nearby bush or tree. Distinguished from Richard's, Paddyfield and Blyth's Pipit (all Plate 170) by dark lores and streaked flanks. **V** Feeble, hesitant, accelerating *tsip tsip tsip-tsip-sip-sip-sipsipsipsipsipsip*. Calls include *dzeep*, weak *see-see* and repeated *tzip-tzip-tzip-tzip*. **H** Hilltops and downs with short grass. **D** SW India.

13 UPLAND PIPIT *Anthus sylvanus* 17cm Forages on ground; when agitated stands upright, often on a rock, and twitches or flicks tail. **V** Repeated *seetyu-seetyu* or *tyu-see-tyu-ee*; also more monotonous *weeeee-tch-weeeeee-tch* or *wichee-wiche-wichee*. Call, infrequently given, a sparrow-like *chirp*. **H** Steep rocky and grassy slopes with scattered bushes and boulders, abandoned terrace cultivations, open pine forest with abundant grass or clearings. **D** N Pakistan to SE China.

14 BERTHELOT'S PIPIT *Anthus berthelotii* 14cm Often confiding. When wary will run rather than fly away. **V** Given from a perch or during undulating display flight, a series of cheerful energetic notes *tschilp-tschilp-tschilp-tschilp* or *tsiree-tsiree-tsiree-tsiree*. Call *tsri* or soft *chup*. **H** Various dry open areas. **D** Madeira, Canary Islands.

15 STRIPED PIPIT *Anthus lineiventris* 17cm Well-streaked, especially below. Note connection of cheek with mantle and greenish edges of wing feathers. **V** High, loud, rich song in short strophes. **H** Rocky, wooded slopes, often near water. **D** Rwanda and S Kenya to E South Africa, W Angola.

16 AFRICAN ROCK PIPIT *Anthus crenatus* 17cm Note very erect stance, dark underparts, yellowish edges to wing (visible at close range). **V** High, fluted *feeeEEpiurrrr* (second part descending). **H** Steep, rocky slopes. **D** South Africa, Lesotho, Swaziland.

17 SHORT-TAILED PIPIT *Anthus brachyurus* 12cm Skulks in grass, reluctant to be flushed. Note thin, short tail and indistinct eyebrow; outer-tail feathers greyish brown. **V** Call sounds like high, partly grating *fu-Weeh fu-Weeh*. **H** Open woodland, grassy hillsides. **D** SC Africa.

18 BUSHVELD PIPIT *Anthus caffer* 13cm From Short-tailed Pipit by less dark upperparts, white outer-tail feathers, absence of malar stripe, streaking below restricted to breast. **V** High, rather nasal *meetjeh meetjeh meetjeh meetjeh...* **H** Open woodland and bushland. **D** E and S Africa.

19 SOKOKE PIPIT *Anthus sokokensis* 14cm Note beautiful contrasting colouring with clear white wing stripes. **V** Song, given in flight, consists of repeated high-pitched phrases, *ee-see*; calls include *sweer*. **H** Open forests and woodland. **D** E Kenya, NE Tanzania.

20 MALINDI PIPIT *Anthus melindae* 15cm Most distinctive feature is dark breast streaks contrasting with white throat. No white or buff in tail. **V** Song, given from raised perch, is a jangling series of *kwee* notes; calls include *sweep* and *tweet-tweet*. **H** Grasslands near Tana River and seashore. **D** Coastal Somalia and Kenya.

21 YELLOW-BREASTED PIPIT *Anthus chloris* 17cm Breeding adult resembles longclaws due to yellow underparts, but differs by absence of distinct breast band. Separated from immature Yellow-throated Longclaw (Plate 170) by horizontal stance and isolated malar stripe. In non-breeding plumage, separated from other pipits by buffy breast and more scalloped than streaked upperparts. **V** High, liquid *tiew tiew tui tiew*. **H** Montane grasslands. **D** E South Africa.

22 ALPINE PIPIT *Anthus gutturalis* 18cm Large, lightly marked pipit with almost plain light buff underside. **V** Song series of high, clear, thin notes, becoming a trill; calls include *tsip* or *tseep*. **H** High alpine grassland with short turf. **D** New Guinea.

23 MADANGA *Anthus ruficollis* 13cm Little information. Last sightings were of birds in a mixed-species foraging flock; also recorded searching for food by climbing tree trunks like a nuthatch. DNA tests show this bird is related to pipits and not white-eyes as once thought. **V** Unrecorded. **H** Montane forest. **D** Buru, S Maluku Islands.

24 SPRAGUE'S PIPIT *Anthus spragueii* 17cm Solitary and secretive, usually keeps well hidden in grass or other short vegetation. Does not bob tail. In flight, shows white outer-tail feathers. **V** In flight, gives sharp, high-pitched *squeet* or *squeet squeet*, occasionally repeated several times in quick succession. **H** Open grassy areas, weedy fields. **D** Breeds SC Canada and NC USA. Winters S USA and Mexico.

25 YELLOWISH PIPIT *Anthus lutescens* 13cm Very small pipit. **V** Song a series of *tzit* notes in ascending flight, changing to *seee* when gliding down. **H** Open grassland near water. **D** W Panama and widely in South America.

26 PERUVIAN PIPIT *Anthus peruvianus* 13cm Above dark brown streaked buff. Below yellowish with speckled brown streaking on upper breast in fresh plumage, becoming plain whitish with wear. Brownish tail with white outer feathers. **V** Song, given in flight, begins with repeated *sip* notes then a characteristic buzzing *dzzeeeeeeeeeeoooooow* during long butterfly-like gliding descent, falling in pitch. **H** Short grass, often around lakes and dams. **D** W Peru to N Chile.

27 SHORT-BILLED PIPIT *Anthus furcatus* 14cm Note mottling on mantle; distinct breast streaking, more diffuse on flanks, and narrow malar streak. **V** Song, given in high flight, a short, fast chatter, a short buzz then another chatter. **H** Pastures and fields with short grass. **D** SE Brazil to Uruguay and Argentina.

28 PUNA PIPIT *Anthus brevirostris* 15cm Pink legs, short hind claw. Light brown above, streaked darker. Breast and flanks buff, breast with broad dark streaking, flanks and whitish belly unmarked. Outer-tail feathers buff-white, rest of tail brownish. **V** Flight song buzzing *teee-cliclicliclicli... teee-cliclicliclicli...*, performed very high, with short glides. **H** Short grass in lowlands, taller puna grasses in Andes. **D** C Peru to W Bolivia and C Argentina.

29 PAMPAS PIPIT *Anthus chacoensis* 13cm Not safely separable from Yellowish Pipit except by its trilling song, which is given flying around above its territory. **V** Song, given in flight, is a very prolonged series of more than 50 short, high-pitched *klee* or *pew* notes running together. **H** Pastures with taller grass, wheat fields. **D** Paraguay to EC Argentina.

30 CORRENDERA PIPIT *Anthus correndera* 15cm Not reliably separable from non-breeding Red-throated Pipit but their occurrence together is unlikely. **V** Short, fast, sharp twitter, sung singly or in short series given when rising and hovering high in the air. **H** Pastures, agricultural land, bogs, wetland edges. **D** WC South America and Southern Cone.

31 SOUTH GEORGIA PIPIT *Anthus antarcticus* 16cm The darkest pipit, almost olive-buff. Heavily streaked back with pale lines on scapulars as in Correndera Pipit. **H** Shorelines and grassy areas. **D** South Georgia, S Atlantic.

32 OCHRE-BREASTED PIPIT *Anthus nattereri* 14.5cm Separable from Yellowish Pipit by richer colouring and streaked flanks. **V** Short, hasty twitter when rising, concluded with hoarse *eeur* (repeated), when dropping back to ground. **H** Campo grasslands. **D** SE Brazil, S Paraguay, NE Argentina and Uruguay.

33 HELLMAYR'S PIPIT *Anthus hellmayri* 14.5cm Separable from Short-billed Pipit by distinct back streaking, and from Paramo Pipit by flank streaking. **V** Short, sharp twitter, often with embedded nasal *eeur* given in flight or from perch. **H** Grassland. **D** SE Peru to S Chile.

34 PARAMO PIPIT *Anthus bogotensis* 15cm Sparsely marked below except for fine streaking to upper breast and lower flanks. **V** Song resembles that of Hellmayr's Pipit but with less frequent *eeur* note. **H** Páramo, puna grassland, pastures. **D** NW Venezuela to NW Argentina.

1
2
3
4
5
6
7
8
9
10
11
12
13
14
15
16
17
18
19
20
21
22
23
24
25
26
27
28
29
30
31
32
33
34

PRZEVALSKI'S FINCH UROCYNCHRAMIDAE

1 PRZEVALSKI'S FINCH *Urocynchramus pylzowi* 16cm Feeds on ground and in bushes, perches on bush tops. Occurs mainly singly or in pairs in breeding season; in winter in small groups of 5–10. **V** Song hurried, chattering *chitri-chitr-tri* or *chitri-chitri-chitri-chitri*; call clear, ringing *kvuit-kvuit*. **H** Bushes, scrub and alpine thickets usually near water. **D** Tibet to C China.

FINCHES AND EUPHONIAS FRINGILLIDAE

2 COMMON CHAFFINCH *Fringilla coelebs* 15cm Feeds on ground and in tree foliage, also fly-catches. Gregarious in winter, flocks with other seed-eaters. **V** Song a short accelerating and descending twitter, ending with fast flourish; call *pink* or *pink-pink*, loud *whit* and wheezy *eeese*; *tsup* in flight. **H** Woodland, heath, farmland with hedges, parks, gardens, stubble fields in winter. **D** Europe, W and C Asia, N Africa, Canary Islands.

3 TENERIFE BLUE CHAFFINCH *Fringilla teydea* 17cm Habits similar to Common Chaffinch. **V** Song like Common Chaffinch, but slower, lacking end flourish; calls *chirp* or *chirp-chirp*, *che-wir*, sometimes sharp *sipp* in flight. **H** Pine forest, usually with rich undergrowth. **D** Tenerife (Canary Islands).

4 GRAN CANARIA BLUE CHAFFINCH *Fringilla polatzeki* 17cm Male primarily dusky blue-grey, female buff. Formerly considered conspecific with Tenerife Blue Chaffinch. Habits similar to Common Chaffinch. **V** Song descending chatter, like that of Common Chaffinch, end note downslurred; call rather weak *uit*, sometimes sharp *sip*. **H** Pine forest, usually with rich undergrowth. **D** Gran Canaria (Canary Islands).

5 BRAMBLING *Fringilla montifringilla* 15cm Behaviour much like Common Chaffinch. May form very large flocks in winter. In flight, shows white rump. **V** Song harsh, monotonous *zweeeeur* interspersed with weaker fluty notes and sometimes a rattling trill; call nasal *tsweek* or *zweee*; a *chuk-chuk* in flight. **H** Mixed forests, birch woods, edges and clearings, scrub, sometimes more open ground in winter. **D** Widespread in Eurasia. In winter, reaches N Africa.

6 BLACK-AND-YELLOW GROSBEAK *Mycerobas icterioides* 22cm Feeds mainly in treetop foliage, in pairs or loose flocks. **V** Song rich, clear *prr-trweeet-a-troweeet*; also *tookiyu-tuukiyu*; call high-pitched *pi-riu... pir-riu... pir-riu* or *tit-te-tew... tit-te-tew*; contact call short *chuck*. **H** Mainly conifer and deodar cedar forest, sometimes forest-edge oak and scrub. **D** Himalayas.

7 COLLARED GROSBEAK *Mycerobas affinis* 22cm Forages in tree canopy, in bushes or low vegetation and on ground, usually in pairs or small loose parties; in larger flocks in winter. Male similar to Black-and-yellow Grosbeak, but yellow more golden, thighs yellow. **V** Song loud, clear, piping *ti-di-li-ti-di-li-umm*, interspersed with creaky phrases and musical notes; calls include rapid, mellow *pip-pip-pip-pip-pip-pip-ugh*, alarm call sharp *kurr*. **H** Oak, rhododendron or mixed conifer and deciduous forest, occasionally in dwarf scrub above treeline. **D** Himalayas.

8 SPOT-WINGED GROSBEAK *Mycerobas melanozanthos* 22cm Shy. Forages in treetops, lower down or on ground; generally in pairs or flocks of 50 or so. **V** Song loud, melodious *tew-tew... teeeu* with mellow *tyop-tiu* or *tyu-tio* whistles; call rattling *krrr* or *charrarauk*; feeding flocks chatter constantly. **H** Mixed conifer and broadleaved forest. **D** NE Pakistan to N Vietnam.

9 WHITE-WINGED GROSBEAK *Mycerobas carnipes* 22cm Forages mainly in treetops, also in scrub and undergrowth, in pairs or small groups, with bigger flocks in winter. **V** Song piping *add-a-dit... un-di-di-di-dit* or *dja-dji-dji-dju*; calls include nasal *shwenk*, strident *wit*, grating *goink* and rapid, harsh *chet-et-et-et*. **H** Dwarf juniper forest above treeline, mixed forest near treeline and at lower elevations. **D** C and SC Asia.

10 EVENING GROSBEAK *Hesperiphona vespertina* 20cm Usually gregarious. Feeds in trees. **V** Song rambling, erratic warble ending with whistle; call loud *clee-ip* or *cleer*. **H** Coniferous and mixed forests, woods and copses; in winter, often visits parks and suburban gardens. **D** Canada to SW Mexico.

11 HOODED GROSBEAK *Hesperiphona abeillei* 18cm Black head diagnostic. **V** Song a long, jumbled series of rolling and ringing notes; calls include metallic, ringing *beenk beenk* and nasal *jerr*. **H** Forest interior and edges. **D** Mexico to Guatemala.

12 HAWFINCH *Coccothraustes coccothraustes* 17cm Forages in trees or on ground, in small scattered flocks of occasionally up to 30. **V** Calls include abrupt *tick* or *tzik*, also thin *seep* or *sreee*. **H** Woodland, wild olive forest, orchards. **D** Widespread in Eurasia, also N Africa.

13 CHINESE GROSBEAK *Eophona migratoria* 15–8cm Forages in trees, bushes and on ground; often remains concealed in foliage. Generally in pairs with larger parties post-breeding. **V** Song loud series of whistles, *chee chee choree kirichoo*; call loud *tek-tek*. **H** Edges and clearings in mixed or deciduous forests, wooded hills, river valleys, marsh edge, cultivations, orchards, parks, gardens. **D** Breeds E Asia. Winters SE Asia.

14 JAPANESE GROSBEAK *Eophona personata* 23cm Best located by call. Forages in canopy, occasionally lower down, usually seen in pairs or small flocks. **V** Calls include short, hard *tak tak* and high-pitched *kik* or *kick*. **H** Mixed and deciduous forests and woodlands, well-wooded hills and river valleys, cultivation edge, parks, gardens. **D** E Asia.

15 PINE GROSBEAK *Pinicola enucleator* 22cm Unobtrusive as it clambers about in foliage. Often more conspicuous in winter when small flocks feed on berries. **V** Song loud, musical, fluty warble; calls include fluty *teu-teu-teu* or *pee-lee-jeh*; also *pui-pui-pui* or *quid-quid-quid* given in flight. **H** Coniferous woods, moves to mainly deciduous trees in winter. **D** N North America and N Eurasia.

16 BROWN BULLFINCH *Pyrrhula nipalensis* 16–17cm Confiding, usually in pairs or small parties foraging in tops of trees or bushes. White rump-band shows during fast, direct flight. **V** Song hastily repeated, mellow *her-dee-a-duuee*; calls include mellow *per-lee* and soft whistling twitter while feeding. **H** Dense undergrowth or thick forests; favours oak, rhododendron or fir. **D** Himalayas to SE China, Taiwan and W Malaysia.

17 ORANGE BULLFINCH *Pyrrhula aurantiaca* 14cm Forages mainly on ground below trees, flying into trees when disturbed; white rump shows in flight. Usually quiet and unobtrusive, often sitting motionless for long periods. **V** Song loud *tew* followed by rapid, repeated metallic *tyatlinka-tlinka*; call soft, clear *tew*. **H** Open fir, birch or mixed forest. **D** Himalayas.

18 RED-HEADED BULLFINCH *Pyrrhula erythrocephala* 17cm Generally seen in pairs or small parties foraging on ground or low in bushes; often sits motionless in bushes for long periods. In flight shows prominent white rump. **V** Song low, mellow *terp-terp-tee*; call soft, plaintive *pew-pew*. **H** Dense cedar, pine or juniper forests; also mixed forest. **D** Himalayas.

19 GREY-HEADED BULLFINCH *Pyrrhula erythaca* 17cm White rump striking in flight. Forages low in bushes or on ground in pairs or small parties; can be approachable. **V** Song descending, then rising, mellow whistled warble mixed with long slurred and short creaky notes; call slow *soo-ee* or *poo-ee*, frequently repeated or given as triple whistle. **H** Conifer and rhododendron forests, willow and buckthorn thickets. **D** Himalayas to C China and Taiwan.

20 WHITE-CHEEKED BULLFINCH *Pyrrhula leucogenis* 15–17cm Noisy and conspicuous; in pairs or small groups, larger groups post-breeding. **V** Calls include ringing *pee-yuu*, loud, musical *chuck-a peeee yuuuu* and harsh, insect-like *zrrreeep*. **H** Montane forest, especially mossy forest and forest edges. **D** Philippines.

21 EURASIAN BULLFINCH *Pyrrhula pyrrhula* 16cm Generally wary; often first sign is call or flash of white rump as bird flies into cover. Feeds in bushes, shrubs or trees and on plant seed-heads, also on ground. Usually in pairs or small, loose flocks. **V** Song weak, scratchy warble interspersed with soft whistles; call low, melancholic, piping *peeu*, *pew* or *due-due*. **H** Woodland with dense undergrowth, hedgerows, thickets, orchards, parks, gardens. **D** Widespread in Eurasia.

22 AZORES BULLFINCH *Pyrrhula murina* 17cm Like pale, desaturated Eurasian Bullfinch; sexes similar. **V** Song and calls like those of Eurasian Bullfinch. **H** Native laurel forest, forest edges. **D** E São Miguel (Azores).

23 ASIAN CRIMSON-WINGED FINCH *Rhodopechys sanguineus* 16cm Forages mainly on ground. In flight shows much pink on wings and rump. **V** Calls include soft, musical *wee-tll-ee* or *wee-tell-er* and harsh *chilip*, often uttered in flight; other flight notes include fluty *dy-lit-dy-lit* and soft *chee-up*. **H** Bare mountain slopes, boulder fields, cliffs and gorges with or without sparse scrub. **D** Turkey to W China, Tajikistan and Afghanistan.

24 AFRICAN CRIMSON-WINGED FINCH *Rhodopechys alienus* 17cm Stocky, sandy-coloured finch with bright (male) or pale (female) pink wing markings. **V** Song soft, sparrow-like chirruping; calls *chilip* or *chee-rup*, *che-wir*; also fluting notes. **H** Bare or sparsely vegetated, rocky montane areas, semi-desert. **D** Morocco, Algeria.

25 TRUMPETER FINCH *Bucanetes githagineus* 15cm Gregarious, with larger flocks in winter; forages on ground, visits small pools to drink. **V** Song distinctive drawn-out, nasal, buzzing *cheeeeeu* interspersed with twittering, clicks and whistles. Calls include abrupt *chee* or *chit* and soft *weechp* or *dzit* given in flight. **H** Desert or semi-desert, including stony plains, rocky hills. **D** SC Asia, N Africa.

26 MONGOLIAN FINCH *Bucanetes mongolicus* 15cm Forages on ground, alone or in pairs or small parties; forms larger flocks in winter; makes evening or morning flights to drink at desert springs. **V** Song slow *do-mi-sol-mi* or *towit-too whit-tu-tu-churrrh*; calls include soft *djou-voud* or *djouddjou*; feeding flocks twitter constantly. **H** Arid or semi-arid mountain areas, with crags, ravines and rocky slopes with low vegetation. **D** Iran to S Russia, Mongolia and N China.

27 BLANFORD'S ROSEFINCH *Agraphospiza rubescens* 15cm Little known. Sings from tree tops; forages on ground, generally in pairs, with larger parties in winter. **V** Song loud musical warble, rising and falling in pitch, with downslurred last note; calls include high, thin *sip* and short rising and falling *pitch-ew*, *pitch-it*, *chit-it* or *chit-ew* notes. **H** Open areas in coniferous or mixed conifer and birch forest. **D** Himalayas to WC China.

28 SPECTACLED FINCH *Callacanthis burtoni* 17cm Unobtrusive ground feeder, eating mainly deodar cedar seeds. Usually in pairs during breeding season, often in flocks of 12 or more out of breeding season. **V** Song loud, trilling *il-til-til...*; also monotonous, repeated single note; call loud *pweee* or *chew-we*, often followed by *pweeu*, *pweuweu* or *chipeweu*, also light *chip* and rising *uh-eh* or *twee-yeh*. **H** Subalpine forests; oak and hemlock forests in winter. **D** Himalayas.

29 GOLDEN-NAPED FINCH *Pyrrhoplectes epauletta* 15cm Secretive. Forages on ground or in bushes and undergrowth; in winter forms small flocks, often with rosefinches. **V** Song rapid, high-pitched *pi-pi-pi-pi*, also low piping; calls include thin, repeated *teeu*, *tseu* or *peeuu*, *pur-lee* and *plee-e-e*. **H** High-altitude oak and rhododendron forests. **D** Himalayas.

30 DARK-BREASTED ROSEFINCH *Procarduelis nipalensis* 15–16cm Shy; feeds on ground and in or under bushes, in pairs or small parties. Often occurs in large, single-sex flocks with other rosefinches. **V** Song monotonous chirping; calls include plaintive, wailed double whistle, twittering and *cha-a-rr* alarm call. **H** Mixed scrubby forests, also weedy areas above treeline; edges and clearings in winter. **D** Himalayas to WC China and NE Myanmar.

31 PLAIN MOUNTAIN FINCH *Leucosticte nemoricola* 15cm Forages on ground, flying into treetops when disturbed; gregarious, often flying in large wheeling flocks. **V** Song sharp twittering *rick-pi-vitt* or *dui-dip-dip-dip* interspersed with trills and warbling notes; calls include soft twittering *chi-chi-chi-chi* and double-noted shrill whistle. **H** Mountains, hillsides, alpine meadows; descends in winter to terraced fields, open forested slopes, cultivations. **D** C Asia.

32 BRANDT'S MOUNTAIN FINCH *Leucosticte brandti* 16–19cm Forages on ground, often at edge of snow-melt or lakesides, perches on rocks and bushes. Usually encountered in pairs or small parties, with much larger, compact and active flocks in winter. **V** Song short trill; calls include loud *twit-twitt*, *twee-ti-ti* or *peek-peek* and harsh *churr*. **H** High-altitude cliffs and crags, barren stony mountaintops. **D** Himalayas.

33 ASIAN ROSY FINCH *Leucosticte arctoa* 16cm Tame. Ground feeder, in pairs or small parties; forms much larger winter flocks. **V** Song slow descending series of *chew* notes, given from ground or in display flight; calls *chew* or *cheew* repeated continuously, dry *pert* and high-pitched *chirp*. **H** High mountains; lower areas in winter. **D** C and NE Asia.

34 GREY-CROWNED ROSY FINCH *Leucosticte tephrocotis* 14–18cm Variable. Forages on ground, in pairs or small parties. **V** Song and calls like those of Asian Rosy Finch. **H** Tundra, grassy maritime plains, mountains, scree slopes. Winters in foothills, valleys, coastal plains, farmland, suburban areas. **D** E Russia and Alaska to W Canada and USA.

35 BLACK ROSY FINCH *Leucosticte atrata* 14–16cm Female duller. Similar to other rosy finches in behaviour. **V** Similar to Asian Rosy Finch. **H** Montane and sub-montane tundra, open scree slopes, mountain lakesides. Winters at lower levels in mountains usually near the snowline, also cultivation, forest edge. **D** WC USA.

36 BROWN-CAPPED ROSY FINCH *Leucosticte australis* 14–16.5cm Behaviour and habits like those of other rosy finches. **V** Similar to Asian Rosy Finch. **H** Montane tundra above treeline, rocky areas and scree slopes. Winters in montane meadows, grassy valleys, forest edges. **D** WC USA.

1
2
3
4
5
6
7
8
9
10
11
12
13
14
15
16
17
18
19
20
21
22
23
24
25
26
27
28
29
30
31
32
33
34
35
36

FINCHES AND EUPHONIAS *CONTINUED*

1 COMMON ROSEFINCH *Carpodacus erythrinus* 15cm Skulking, forages on ground or in low vegetation, bushes and trees, generally alone, in pairs or small parties; in winter often in larger flocks mixed with other seed-eaters. **V** Song cheery *twee-twee-tweeou* or *ti-dew-di-dew* or similar. Calls include rising *ooeet*, *ueet* or *too-ee*, and when alarmed sharp *chay-eeee*. **H** Streamside willows, rock and scrub, bushy slopes, open conifer forest; winters in open wooded country, scrub, bushes, cultivation. **D** Widespread in Eurasia.

2 SCARLET FINCH *Carpodacus sipahi* 18cm Feeds in trees and bushes and on ground. Often perches prominently. Forms single-sex flocks in winter. **V** Song clear, liquid *par-ree-reeeeee*; call loud *too-eee* or *pleeau* and *kwee-i-iu* or *chew-we-ah*. **H** Edges and clearings in montane coniferous forest, also in oak and bamboo forests in winter. **D** Himalayas.

3 STREAKED ROSEFINCH *Carpodacus rubicilloides* 19cm Generally shy. Perches on rocks, bushes and trees, forages on ground; flicks wings and tail when agitated. May form large roosts in willow groves. **V** Song slow descending *tsee-tsee-soo-soo-soo*, usually repeated; calls include *twink*, *pink* or *sink*, soft *sip* and melancholic *dooid-dooid*. **H** High-altitude rocky slopes, scree, plateaux and hillsides with scrub; winters in thickets. **D** Himalayas and China.

4 GREAT ROSEFINCH *Carpodacus rubicilla* 19–21cm Forages on ground or in low bushes; wary; generally alone or in pairs; in winter forms small flocks, often with other rosefinches. **V** Song low *wreep* and series of low chuckles; calls include rasping *jink*, soft *jeweet* and short twittering. **H** High-altitude areas with boulders and sparse vegetation; in winter also in hillside scrub, fields. **D** E Caucasus through C Asia to NW China.

5 BLYTH'S ROSEFINCH *Carpodacus grandis* 17cm Stout rosefinch with proportionately large head. Forages on ground, usually shy. Sings from tree-top. **V** Song a series of weak chirping and twittering notes with occasional squeaky whistles; calls include *kwee* or *skwee*, with a soft twitter given in flight. **H** High-altitude rocky slopes, scree, plateaux and hillsides with scrub; winters in thickets. **D** NW Afghanistan to W Himalayas.

6 RED-MANTLED ROSEFINCH *Carpodacus rhodochlamys* 18cm Usually secretive; forages on ground or low in bushes, generally in pairs; in winter alone or in small parties. **V** Song feeble series of short, wheezy *chirp* and *twit* notes, with occasional squeaky whistles; calls include plaintive, buzzing *kwee* or *squee* and sharp *wir*. **H** High-altitude juniper, briar and scrub; winters at lower levels in bushes, thorny scrub, cultivations, gardens. **D** Uzbekistan and Afghanistan to S Russia and Mongolia.

7 HIMALAYAN BEAUTIFUL ROSEFINCH *Carpodacus pulcherrimus* 15cm In flight shows pink rump. Forages on ground or low in bushes, usually in pairs or small flocks. **V** Calls include subdued *trip*, *trilp* or *trillip*, a tit-like twitter; in flight utters harsh *chaaannn*. **H** Steep hillsides with rhododendron and other bushes, near or above treeline; winters on open scrub-covered hillsides and terraced cultivation with nearby bushes. **D** C Himalayas to S Mongolia and WC and NC China.

8 CHINESE BEAUTIFUL ROSEFINCH *Carpodacus davidianus* 14cm Smallish, long-tailed rosefinch. Male's face and upper breast intensely crimson, underparts otherwise paler red-pink. **V** Song not known; calls include sparrow-like chirps and a soft trill. **H** Montane and submontane forest edges, scrub above treeline. **D** EC China.

9 PINK-RUMPED ROSEFINCH *Carpodacus waltoni* 15cm Forages on ground, often around edges of trees or bushes, usually in pairs or small parties. Bigger flocks occur post breeding, often mixed with other species. **V** Calls include sharp *pink* or *tink*, *tsip* or *tsick*, a thin rattle and harsh *pip-rit*. **H** Forest edge, secondary growth, scrub, cultivation borders. **D** E Himalayas to WC China.

10 PINK-BROWED ROSEFINCH *Carpodacus rodochroa* 14–15cm Unobtrusive ground forager, flying up to perch in bushes if disturbed; usually in pairs or small loose flocks. **V** Song loud, lilting *toowhi toowhi*; calls include loud *per-lee* or *chew-wee* and canary-like *sweet*. **H** Undergrowth in mixed fir and birch forest, willow bushes, rhododendrons and dwarf juniper; in winter also occurs in scrub jungle, open hillsides, grassy slopes, gardens. **D** Himalayas.

11 DARK-RUMPED ROSEFINCH *Carpodacus edwardsii* 16cm Paler-chinned than similar Vinaceous Rosefinch. Skulking, forages on ground under bushes, usually alone or in small groups; may form larger flocks post breeding. **V** Calls include metallic *twink* and rasping *che-wee*. **H** Breeds in rhododendron and silver fir forest; winters in open rhododendron or birch forest and mountainside bamboo and scrub. **D** Himalayas.

12 SPOT-WINGED ROSEFINCH *Carpodacus rodopeplus* 15cm Generally shy, although often perches prominently on bush tops; forages on ground. **V** Usually silent, but occasionally calls *chirp* or upslurred *churr-weee*. **H** Rhododendron scrub and bushes on alpine slopes and meadows; bamboo thickets, bushes and mixed forest in winter. **D** Himalayas from N India to S Tibet.

13 SHARPE'S ROSEFINCH *Carpodacus verreauxii* 17–20cm Generally shy, although will perch prominently on bush tops; feeds on ground. **V** Like that of Spot-winged Rosefinch. **H** Like that of Spot-winged Rosefinch. **D** SW and W China, N and NE Myanmar.

14 VINACEOUS ROSEFINCH *Carpodacus vinaceus* 15cm Shows distinct whitish tips to tertials. Forages on ground, in dense vegetation or in low bushes; usually in pairs or small parties. **V** Song simple *pee-de... be... do-do*. Calls include whiplash-like *pwit* or *zieh*, high *tip* and low *pink* or *zick*. **H** Moist mixed or bamboo forests, scrubby open hillsides. **D** Nepal and N India to S and C China and N Myanmar.

15 TAIWAN ROSEFINCH *Carpodacus formosanus* 15cm Male deep crimson with pinkish-brown wings; female rather uniform dark brown. Mainly forages on ground; usually seen in pairs or small groups. **V** Song a repeated, simple two-note phrase; call a brief *zip*. **H** High-altitude forest edges, bamboo, thickets. **D** Taiwan.

16 SINAI ROSEFINCH *Carpodacus synoicus* 15cm Shy, usually in small ground-feeding parties. Often larger parties at waterholes. **V** Not well documented; song said to be musical jumble, often containing buzzing notes; calls *cheeup* or *chip*, high *touit* or similar, and weak *stip*. **H** Arid mountains, foothills, rocky desert, wadis, cliffs and gorges. **D** NE Egypt, Israel, Jordan and W Saudi Arabia.

17 PALE ROSEFINCH *Carpodacus stoliczkae* 15cm A pale sandy-coloured, almost unmarked rosefinch; male with bright crimson on face only. Ground-feeder; may form large flocks post breeding. **V** Song little known but appears to combine buzzy and more melodious notes; calls include *tiu* and shorter *zik*, flight call *trizip*. **H** Arid mountains, hillsides, edges of desert and of cultivation. **D** C Afghanistan to C China.

18 TIBETAN ROSEFINCH *Carpodacus roborowskii* 17cm Ground feeder, shuffling gait. Little else known. **V** Generally silent, apart from short, plaintive whistle, often repeated as trill. **H** Barren rocky and alpine steppes; 4,500–5,400m, slightly lower in winter. **D** Himalayas to WC China.

19 SILLEM'S MOUNTAIN FINCH *Carpodacus sillemi* 15–17cm Rump and outer-tail feathers pale buff-white. Behaviour probably much as other *Carpodacus* finches. **V** Unknown. **H** High, barren plateaux. **D** N India and SW China.

20 LONG-TAILED ROSEFINCH *Carpodacus sibiricus* 16–17cm Forages on ground and in undergrowth, agile while picking at seedheads; usually alone, in pairs or small family parties. **V** Various pleasant rippling trills, liquid *pee-you-een* or *su-we-su-wee-sweeeoo-cheweeoo*; also rising *sit-it-it*. **H** Dense willow and birch thickets, grasslands, reedbeds, riverine woods, tall vegetation in ditches, wet meadows. **D** E and C Asia.

21 PALLAS'S ROSEFINCH *Carpodacus roseus* 16cm Feeds in trees, bushes or on ground, usually in pairs or small parties with larger winter flocks. **V** Song series of quiet, rising and falling notes; call a low, short whistle. **H** Conifer, cedar and birch forests, alpine meadows, shrub thickets and sparse scrub on mountains; cedar groves, scrub-covered hillsides in winter. **D** E Asia.

22 THREE-BANDED ROSEFINCH *Carpodacus trifasciatus* 18cm Lethargic, often sitting immobile for long periods hidden in bushes or trees. Mainly ground feeder, but also in bushes and trees; especially fond of crab apples. **V** Generally silent. **H** Undergrowth and thickets in light coniferous forest; 1,800–3,000m. Winters at lower elevations in farmland hedges, orchards, bushes. **D** W China.

23 HIMALAYAN WHITE-BROWED ROSEFINCH *Carpodacus thura* 17–18cm Forages on ground, usually in pairs or small parties, occasionally with other rosefinches or grosbeaks. **V** Song series of loud, short whistles followed by 3–4 short warbled notes ending with several longer whistles *drit-drit-drit-drit quip-quip-quip-quip dreep-dreep-dreep-dreep* or shorter *pew-pew-pew chit-chit naaar naar nah nah nah*; calls include buzzing *deep-deep deep-de-de-de-de*, bleating *veh ve ve ve ve ve ve*, rapid piping and loud *pwit-pwit*. **H** High-altitude open forest or forest edge, rhododendron, juniper and bamboo scrub above the treeline; winters on open hillsides with bushes and scrub. **D** Himalayas.

24 CHINESE WHITE-BROWED ROSEFINCH *Carpodacus dubius* 17cm Both sexes show whitish eyebrow, most prominent in brown, heavily streaked female. Usually confiding. **V** Song is a short and simple series of 3–5 rather nasal, bleating notes; calls include a fast chatter and nasal buzzing notes. **H** Montane forest clearings and edges, hillside scrub. **D** E Tibet, N, W and SW China.

25 RED-FRONTED ROSEFINCH *Carpodacus puniceus* 20cm Forages on ground among boulders, bushes or by melting snow, alone, in pairs or small parties. **V** Song short *twiddle-le-de* with various melodious downslurred whistles; calls include loud, cheery *are-you-quite ready*, cat-like *maaau* and *chirp* or *jeelp* in flight. **H** High-altitude boulder fields, rocky screes and slopes. **D** Himalayas.

26 CRIMSON-BROWED FINCH *Carpodacus subhimachalus* 19–20cm Unobtrusive, forages on ground or in bushes and low trees; usually in pairs or small parties. **V** Song bright variable warble, which may include *ter-ter-ter* or *terp terp tee* phrase; call a sparrow-like *chirp*. **H** Dense high-altitude scrub; winters in thick undergrowth in forests. **D** Himalayas.

27 OAHU ALAUAHIO *Paroreomyza maculata* 12cm Probably now extinct; said to have been very rare by 1930s and not seen with certainty since 1960s. Olive-green (male brighter, with yellow breast), with pale buff wing-bars in female; bill straighter than that of similar Oahu Amakihi (Plate 274). **V** Not known. **H** Forest. **D** Oahu (Hawaii).

28 MAUI ALAUAHIO *Paroreomyza montana* 11cm Male very bright yellow, without mask. Female uniform green with yellow throat and lores. **V** Song fast, bouncing warbling; call soft *dzip*. **H** Rainforest and plantations. **D** Hawaiian Islands.

29 AKIKIKI *Oreomystis bairdi* 13cm Greyish with pink bill. **V** Song short five-noted, descending series; call very high, soft, yet staccato *twit*. **H** Forest. **D** Hawaiian Islands.

30 LAYSAN FINCH *Telespiza cantans* 19cm No similar bird in its tiny range. Note very heavy bill. **V** High *chihchiteereh* and other carefree chirping phrases. **H** Grassland and shrubs. **D** NW Hawaiian Islands.

31 NIHOA FINCH *Telespiza ultima* 17cm No similar bird in its tiny range. Note heavy bill and grey plumage tones. **V** Short strophes; *chipchip-chéeroh* or *tuweréeh* or chirping series. **H** Grassland and shrubs. **D** NW Hawaiian Islands.

32 PALILA *Loxioides bailleui* 19cm Plumage pattern of yellow and grey tones diagnostic. **V** Song series of modest, high *teetjuw* and other notes with many short repetitions; call *ptweet*, *ohweet* and variations. **H** Dry forest. **D** Hawaiian Islands.

33 IIWI *Drepanis coccinea* 15cm Unmistakable by bill shape and, in adult, plumage colour. **V** Song calm series of short trills, nasal notes and liquid fragments; call ascending *weeh-ih* or *seeeh*. **H** Variety of forest types. **D** Hawaiian Islands.

34 AKOHEKOHE *Palmeria dolei* 18cm Unmistakable; no similar bird in area. **V** Song slow series of low frog-like notes, some with liquid quality; calls include very high, slightly upslurred *weeeh*. **H** Rainforest. **D** Hawaiian Islands.

35 APAPANE *Himatione sanguinea* 13cm Unmistakable. Note white vent and wing pattern. **V** Song and calls include nasal, liquid notes with canary-like quality. **H** Forest. **D** Hawaiian Islands.

1
2
3
4
5
6
7
8
9
10
11
12
13
14
15
16
17
18
19
20
21
22
23
24
25
26
27
28
29
30
31
32
33
34
35

FINCHES AND EUPHONIAS *CONTINUED*

1 KAUAI NUKUPUU *Hemignathus hanapepe* 14cm Unmistakable in range. Note curved lower mandible. **V** Song a sweet, short warbled phrase; call a whistled *keewit*. **H** Dense forest. **D** Kauai (Hawaii).

2 MAUI NUKUPUU *Hemignathus affinis* 14cm Critically endangered. Generally very slightly smaller than other nukupuu. Upper mandible long and downcurved, lower mandible short and straight. **V** Song a warbled phrase; contact call *keewit* given while foraging. **H** Varied forest and shrubland habitats. **D** Maui (Hawaii).

3 AKIAPOLAAU *Hemignathus wilsoni* 14cm Separated from nukupuus by range and straight lower mandible. **V** Song loud *wutwutwididjur*, last part louder and higher in pitch; call *tuwjur* and *tuweéh*. **H** Forest, parklands. **D** Hawaii (Hawaii).

4 MAUI PARROTBILL *Pseudonestor xanthophrys* 14cm Unmistakable by heavy bill. **V** Song short, descending, double-fluted series; call soft *twit* or *tweeh* or *turuweeh*. **H** Forest. **D** Maui (Hawaii).

5 ANIANIAU *Magumma parva* 10cm Anianiau, Oahu Amakihi, Hawaii Amakihi and Kauai Amakihi all very similar, differing mainly in shape and colour of bill. Range diagnostic in most cases. Pink legs and bill diagnostic. **V** Song variable short, high, almost level warblings with tit-like variations; call upslurred *tweet* or very high, short *pfit*. **H** Native forest. **D** Kauai (Hawaii).

6 HAWAII CREEPER *Loxops mana* 11cm Note drab green plumage and dark mask. **V** Song fast, warbling, slightly descending series; call very high *tweeh*. **H** Forest. **D** Hawaii (Hawaii).

7 AKEKEE *Loxops caeruleirostris* 11cm Note dark lores. **V** Song high, hurried, descending series *tjuttjut-wutwutwut* (3 secs); call very high *weeé* or upslurred *wuut*. **H** Forest. **D** Kauai (Hawaii).

8 HAWAII AKEPA *Loxops coccineus* 10cm Male unmistakable; female drab green, but note absence of darker lores. **V** Song very high, short, hurried warblings or longer, descending, slightly canary-like series; call very high, fast *tiritihi* or *tpeeh*. **H** Forest. **D** Hawaii (Hawaii).

9 MAUI AKEPA *Loxops ochraceus* 14cm Critically endangered. Small honeycreeper, with laterally curved lower mandible, a trait unique to its genus; very like Hawaii Akepa. **V** Presumed to be similar to that of Hawaii Akepa. **H** Presumed to be similar to that of Hawaii Akepa. **D** Maui (Hawaii).

10 HAWAII AMAKIHI *Chlorodrepanis virens* 11cm Note curved, dusky bill. Females vary by race; some much greener, others much greyer). **V** Song high, rhythmic, descending *tweet tweet... tweet* and similar with repetition of same note or phrase. **H** Rather dry forest. **D** Hawaiian Islands.

11 OAHU AMAKIHI *Chlorodrepanis flava* 11cm Short, almost straight bicoloured bill. Note wingbars of female. **V** Song a brief trill of fast *chee* notes; calls include *sweet* or disyllabic *ts-weet*, generally a little lower-pitched and buzzier than in other amakihis. **H** Native forest, also plantations and trees in urban areas. **D** Oahu (Hawaii).

12 KAUAI AMAKIHI *Chlorodrepanis stejnegeri* 11cm Note rather heavy bill (not pink as in Anianiau). **V** Song high, three-note, slightly descending *tjeutjewtjew* or reed warbler-like series; call *weeéh* or nasal *wew*. **H** Forest. **D** Kauai (Hawaii).

13 PURPLE FINCH *Haemorhous purpureus* 13.5–14.5cm Forages on ground or in trees and bushes. In winter, often with other finches. **V** Song rising and falling warble of rich bubbling notes; calls include sharp *pik*, *tick* or *pink* and musical *char-lee*, *chee-wee* or *whit whewe*. **H** Open conifer forests, mixed woodland, parks, gardens. **D** N and W North America, NW Mexico.

14 CASSIN'S FINCH *Haemorhous cassinii* 14.5–16.5cm Forages on ground and in treetops. In winter often associates with other finches. **V** Song high-pitched varied, jumbled warble; calls include dry *giddy-up*, *tee-dee-yip*, *kee-yup* or *cheedly-up*, all often given in flight. **H** Montane open conifer forests; in winter in conifer forests at lower levels. **D** W North America, N Mexico.

15 HOUSE FINCH *Haemorhous mexicanus* 13–15cm Forages mainly on ground. Yellow morph commoner in SW of range. **V** Song jumble of musical notes, ending with *wheeer*; calls include *cheet* or *queet* and *chirp*. **H** Towns and villages, farmland, scrub, orchards. **D** Canada, USA and Mexico.

16 EUROPEAN GREENFINCH *Chloris chloris* 15cm Yellow wing and tail patches conspicuous in flight. Often visits garden feeding stations. Generally sociable, in winter in larger groups, often associating with other seed-eaters. Sings from high perch or during slow, stiff-winged display flight. **V** Song variable, often starts with dry nasal trill then rising *teu-teu-teu-teu* interspersed with *tswee* note, sometimes just repetition of *tswee* note; call rapid *chichichichichit* or *chill-ill-ill-ill*, single *chit* or *teu*, *swee-it* or *tsooeet* when alarmed. **H** Various, including woodland edge, copses, thickets, hedgerows, parks, gardens. In winter, often in stubble fields and weedy areas. **D** Europe, SW and SC Asia.

17 GREY-CAPPED GREENFINCH *Chloris sinica* 14cm Usually in pairs or small groups, forages in trees, shrubs and on ground. Often sings in stiff-winged, slow display flight. **V** Song mixture of chattering and coarse notes interspersed with call notes; calls include distinctive *dzi-dzi-dzi-i* and nasal *dzwee*. **H** Woodlands, hedgerows, cultivations, parks, gardens. **D** E China to Vietnam.

18 YELLOW-BREASTED GREENFINCH *Chloris spinoides* 14cm Forages in treetops, bushes or on ground, usually in pairs or small family parties, with larger flocks after breeding. **V** Song long series of rapid twitters interspersed with short *chip* and *tew* notes; calls include distinct *swee-tu-tu* and twittering followed by harsh *dzwee*. **H** Open forest, forest edge, scrub, cultivations. **D** Pakistan to Myanmar.

19 VIETNAMESE GREENFINCH *Chloris monguilloti* 13–14cm Forages in pairs or small flocks, in trees and bushes. **V** Song slowly rising *seeuuu-seeuuu-seeuuu* or *teoo-teoo-teoo*, followed by nasal *weeeee* or *chweee*; calls include twittering *chi-chi-chi...* and a dry nasal *zweee*. **H** Open pine forest, forest edge, cultivations, gardens. **D** Vietnam.

20 BLACK-HEADED GREENFINCH *Chloris ambigua* 14cm Forages in bushes, low vegetation and on ground; in winter often in large flocks. **V** Song drawn-out, wheezy *wheeeeeeuu wheeeeeeuu* mixed with metallic notes, trills and harsh *screee* or *treeee-tetrah*; calls include wheezy *twzyee* and a twittering, often interspersed with harsh notes. **H** Open coniferous or deciduous forest, forest edge, forest clearings, scrub, cultivations. **D** Himalayas to N Thailand.

21 DESERT FINCH *Rhodospiza obsoleta* 15cm Forages on ground, but regularly perches on bushes, trees and wires. Generally in pairs; also small flocks post breeding. **V** Song repetitive jumble of twittering call notes, harsh trills and rolls. Calls include soft purring, harsh *turr* and in flight sharp *shreep*. **H** Dry plains with scattered trees and bushes. **D** Middle East to Mongolia and N China.

22 SOCOTRA GOLDEN-WINGED GROSBEAK *Rhynchostruthus socotranus* 15cm Feeds on seeds, buds and fruit. Usually occurs singly or in pairs, and in small parties of about 20 in winter. Often inactive, sitting unobtrusively in trees or bushes. Female slightly duller than male, lacks black face. **V** Song liquid, musical and jingling phrases such as *whit-whee-oo* or *tvit-te-vyt-te-viit*, interspersed with clear, fluty notes; call very variable, *wip* or *tzee*, rippling *tut-tut-tut* or *did-did-ee*, and soft, often repeated *tlyit*; also rapid *dy-dy-dy* followed by dry trill. **H** Hills and wadis with euphorbia, acacia or juniper. **D** Socotra, Gulf of Aden.

23 ARABIAN GOLDEN-WINGED GROSBEAK *Rhynchostruthus percivali* 15cm Feeds on seeds, buds and fruit. Usually alone or in pairs, with small parties of about 20 in winter. Often inactive, sitting unobtrusively in cover. Female lacks black face and is slightly duller than male. **V** Like that of Socotra Golden-winged Grosbeak. **H** Like that of Socotra Golden-winged Grosbeak. **D** SW Arabian Peninsula.

24 SOMALI GOLDEN-WINGED GROSBEAK *Rhynchostruthus louisae* 14cm Stocky, thick-billed finch, grey with black bib and lores, bright yellow wing and tail markings. Behaviour little known; may forage singly, in pairs or small groups. **V** Not known; presumed similar to others in its genus. **H** Scrub-covered rocky outcrops and wadis with giant euphorbias. **D** Somalia.

25 ORIOLE FINCH *Linurgus olivaceus* 13cm Separated from all other small yellow birds by orange bill and black nape. **V** Song comprises fast and slow trills interspersed with drawn-out *seeeeee* notes; calls include high-pitched, wheezy *tsip*, *tswee* and similar notes. **H** Undergrowth at montane forest edges. **D** WC, C and E Africa.

26 PRINCIPE SEEDEATER *Crithagra rufobrunnea* 12cm Strikingly short-tailed. No similar bird in its range. **V** Song is a slow series of warbled, twittered and chattering phrases finishing with loud upslurred *twee*; calls include clear *tsweet* and nasal *zwee*. **H** Forest edges, woodland, plantations, gardens. **D** São Tomé and Príncipe, Gulf of Guinea.

27 SAO TOME GROSBEAK *Crithagra concolor* 19cm One of the rarest birds in the world, few sightings. No similar bird occurs in its habitat and range. **V** Song is a repeated two-note phrase, second note at higher pitch than first; call four or five short whistled notes. **H** Tall forest trees. **D** E São Tomé, Gulf of Guinea.

28 AFRICAN CITRIL *Crithagra citrinelloides* 12cm Longer, sharper bill than Black-faced Canary (with which range slightly overlaps), and with more striped flanks. **V** Song comprises short phrases of steady high-pitched, often downslurred notes with short pauses in between; calls quiet, include *chit*, *cheep* and a soft twitter. **H** Forest edges, bamboo, woodland, lake sides, cultivation, gardens. **D** S Sudan, W and C Ethiopia, W Kenya.

29 WESTERN CITRIL *Crithagra frontalis* 11.5cm Separated from Black-faced Canary by greenish tone on cheek merging into sides of neck and mantle. Formerly considered a race of African Citril. **V** Song is a steady, sweet jingling phrase; call *chit-chit*. **H** Forest edges, bamboo, woodland, lake sides, cultivation, gardens. **D** E DR Congo to Uganda, W Tanzania and NE Zambia.

30 SOUTHERN CITRIL *Crithagra hyposticta* 12cm Note lack of well-defined eyebrow. Bill sharply pointed. Formerly treated as race of African Citril. **V** Song is a fast tinkling series of sweet and buzzy notes, sometimes a slower four-note phrase; calls include a quiet twitter given in flight. **H** Forest edges, bamboo, woodland, lake sides, cultivation, gardens. **D** E and SE Africa.

31 BLACK-FACED CANARY *Crithagra capistrata* 12cm Female separated from female African Citril by stubbier bill, less streaked underparts. **V** Song a fast and prolonged jumble of trills, buzzing notes and whistles; call *swee* or variants thereof. **H** Forest edges and clearings, swamps. **D** C Africa.

32 PAPYRUS CANARY *Crithagra koliensis* 12cm Note stubby bill and rather ill-defined wing bars. Restricted range and habitat. **V** Song a series of shrill, whistled notes and phrases, sometimes including toneless rattled trills; calls include *wee-titi-woy* but generally quiet. **H** Papyrus swamps. **D** C Uganda to E DR Congo, Burundi, NW Tanzania and W Kenya.

33 FOREST CANARY *Crithagra scotops* 13cm Note bright yellow eyebrow and upper throat, and heavy streaking on underparts. **V** Extremely high, hurried siffling with some hoarse, slightly lower notes. **H** Canopy of forests and nearby areas. **D** South Africa.

34 WHITE-RUMPED SEEDEATER *Crithagra leucopygia* 10cm Note white rump and pale colouring. **V** Song is a lengthy but quiet series of tuneful twitters and warbles; calls include *kwee* and *chu-wee*, also nasal *djerr*. **H** Wooded and bushy, natural and cultivated areas, including gardens. **D** W, C and E Africa.

35 BLACK-THROATED CANARY *Crithagra atrogularis* 11cm Blackish throat patch very variable, sometimes absent. Note conspicuous yellow rump in flight. **V** Very high, mumbled, continuous flow of warbles, flutes, trills and rolls. **H** Woodland, thornveld, farmland. **D** E and S Africa.

1
2
3
4
5
6
7
8
9
10
11
12
yellow morph
13
14
15
16
17
18
19
20
21
22
23
24
25
26
27
28
29
30
31
32
33
34
35

FINCHES AND EUPHONIAS *CONTINUED*

1 **YELLOW-RUMPED SEEDEATER** *Crithagra xanthopygia* 10cm Note yellow rump and dark-mottled throat. **V** Mid-high, rapid, warbled medley. **H** Wooded and bushy areas. **D** Ethiopia, Eritrea.

2 **REICHENOW'S SEEDEATER** *Crithagra reichenowi* 11cm Note yellow rump, striped chest and distinct malar stripe. **V** Song long, complex and pleasant series of fast trills, whistles and warbles; calls include *chewee*, *chit* and *chit-wee*. **H** Wooded and bushy areas. **D** Djibouti, Ethiopia and Somalia to S Tanzania.

3 **ARABIAN SERIN** *Crithagra rothschildi* 12cm Gently flicks tail. **V** Song slow, rising trill, occasionally followed by a varied musical jingle; call quiet *tsit-tsit* and a short ripple. **H** Rocky hills and wadis with bushes and trees, also orchards and gardens. **D** W and SW Arabian Peninsula.

4 **YELLOW-THROATED SEEDEATER** *Crithagra flavigula* 11cm Note yellow rump and throat. **V** Song consists of a monotonous three-note phrase, long-short-long; calls include *sip* and *swee*. **H** Wooded and bushy areas with some rough grass. **D** Ethiopia.

5 **SALVADORI'S SEEDEATER** *Crithagra xantholaema* 14cm Note yellow rump. Yellow throat divided by black line. **V** Song is a short series of whistled phrases, rising in pitch at end; call a brief *tsip*. **H** Dry rocky areas with some trees and shrubs. **D** Ethiopia.

6 **LEMON-BREASTED CANARY** *Crithagra citrinipectus* 11cm Light parts in complicated face pattern may be white or yellow. Often in flocks with Yellow-fronted Canary. **V** Song high, rather short, hurried strophes with basic *sri* sound. **H** Dry woodland and scrubland, grassy areas with palms, cultivation. **D** S Malawi to E South Africa.

7 **YELLOW-FRONTED CANARY** *Crithagra mozambica* 12cm Note sharply demarcated eyebrow and cheek, small, pointed bill, yellow rump. **V** Song very high, hurried, fluted variations on a short strophe, *pri-pree-pirre-pieeh-weeh*. **H** Open forest, woodland, bushveld, suburbs. **D** W, C, E and S Africa.

8 **WHITE-BELLIED CANARY** *Crithagra dorsostriata* 13cm Note dark (not black) malar stripe. Belly centre and under rail-coverts white. **V** Song mostly pleasant, tuneful trills and whistles with occasional harsher nasal or grating sounds; calls include *su-weee* and *zwee*. **H** Dry, more or less wooded and bushy mainly natural habitats. **D** E Africa.

9 **ANKOBER SERIN** *Crithagra ankoberensis* 11cm Overall ground colour almost white. **V** Song musical chirps and chirruping notes; calls include various single notes, similar in tone to song. **H** High, steep cliffs with some shrub. **D** Ethiopia.

10 **YEMEN SERIN** *Crithagra menachensis* 12cm Behaviour typical of genus. Tends to flock more than Arabian Serin. **V** Song *chew-chee-chee-chwee*; call *teee-oo*, also various flight notes, *dweep*, *twi-twi-twi-twi* and *chirrip-chirrip*. **H** Rocky hillsides with patches of cultivation or bushes, villages, towns. **D** SW Arabian Peninsula.

11 **CAPE SISKIN** *Crithagra totta* 13cm Note white tail tip and trailing edge to wing. Separated by distribution from Drakensberg Siskin. **V** Song high, rather hesitant flow of flutes, trills and chatters with basic *tiuh* sound. **H** Montane forest edges, pine plantations. Also at lower altitudes. **D** South Africa.

12 **DRAKENSBERG SISKIN** *Crithagra symonsi* 13cm Separated from Black-faced Canary (Plate 274) by more greenish head, streaked mantle and all-black tail and wings. **V** Song continuous flow of hurried sifflings, flutes and rolls with inclusions of very high *wut* notes and short rattles *djik-djik-djik-djik*. **H** Montane, grassy areas. **D** South Africa.

13 **NORTHERN GROSBEAK-CANARY** *Crithagra donaldsoni* 15cm Separated from Southern Grosbeak-Canary by white rather than yellow vent. Note heavy bill. **V** Song a fast series of up to 20 simple *siu* notes; calls rarely heard but include *seep* and *suweer*. **H** Dry wooded and bushy areas. **D** EC Ethiopia and Somalia to C Kenya.

14 **SOUTHERN GROSBEAK-CANARY** *Crithagra buchanani* 15cm Stout-billed finch. Very like Northern Grosbeak-Canary; separated by yellow rather than white vent. **V** Song combines short chipping notes with trills; calls include a three-syllable whistle and single upslurred note. **H** Dry thorn-scrub, thickets, bushy savannah areas. **D** S Kenya, N Tanzania.

15 **YELLOW CANARY** *Crithagra flaviventris* 13cm Colour of rump varies by race. Separated from Brimstone Canary by yellow forehead and whitish edges to tertials. **V** Song high, hurried flow of warbled flutes and trills around basic *wru* sound. **H** From montane bush to coastal scrub. **D** S Africa.

16 **BRIMSTONE CANARY** *Crithagra sulphurata* 15cm Note large bill and green rump. **V** Very high, short, warbled medleys of up-and-down flutes, trills and rolls. **H** Open woodland, bushveld, forest edges. **D** E, SE and S Africa.

17 **REICHARD'S SEEDEATER** *Crithagra reichardi* 15cm Separated from Streaky-headed Seedeater by different range. Often treated as race of Streaky-headed Seedeater. **V** Song combines sweet trills with harsher rattling sounds, and may include some mimicry; calls high-pitched *szee*, *zu-weeoo* and similar.. **H** As Streaky-headed Seedeater. **D** E Africa.

18 **STREAKY-HEADED SEEDEATER** *Crithagra gularis* 15cm Adult separated from Reichard's Seedeater by unmarked underparts. **V** Song sequence of high, short, up-and-down rattles, flutes and trills. **H** Wooded and bushy areas, cultivation, gardens. **D** SC and S Africa.

19 **WEST AFRICAN SEEDEATER** *Crithagra canicapilla* 14cm Forehead and crown striped dark brown and white. Adult separated from Reichard's Seedeater by unmarked underparts. **V** Song loud, varied, canary-like combination of rattles, trills and buzzes with pace shifting throughout series; may include mimicry. Calls various high-pitched and thin tweets and chirrups. **H** Dry woodland, miombo and gardens. **D** WC and C Africa.

20 **BLACK-EARED SEEDEATER** *Crithagra mennelli* 13cm Female separated from male by dark brown (not black) sides to face. **V** Song sustained flow of three short phrases, together *wuhwuh-feeh-tjeh*. **H** Miombo and other types of woodland. **D** Tanzania to E Angola, NE Botswana, Zimbabwe and S Mozambique.

21 **BROWN-RUMPED SEEDEATER** *Crithagra tristriata* 13cm Note narrow white eyebrow. **V** Song a simple, weak four-note phrase, *sip-sip twis-twis*; typical calls are brief notes such as *sip* or *chip*. **H** Forest edges. **D** Eritrea, Ethiopia and N Somalia.

22 **WHITE-THROATED CANARY** *Crithagra albogularis* 14cm Rump colour varies by race. Separated from Black-throated Canary (Plate 274) by larger bill and white chin, from Thick-billed Seedeater, Protea Canary and Streaky-Headed Seedeater by green or yellow rump. **V** Song very high, fast series of fluted rolls and warbles with returning *puwee* sound. **H** Karoo scrub, dry grassland, coastal dunes. **D** SW and S Africa.

23 **THICK-BILLED SEEDEATER** *Crithagra burtoni* 15cm Large with very heavy bill. **V** Song rarely heard, a quiet and varied series of churrs, trills and high-pitched warbles; call a soft *pleet* or *seewsew*. **H** Undergrowth of montane forest, bamboo. **D** C and EC Africa.

24 **STREAKY SEEDEATER** *Crithagra striolata* 15cm Heavily streaked; note characteristic face pattern. Often shows a greenish panel in closed wing. **V** Song varied in length and composition, includes downslurred whistles and pleasant trills; calls include high-pitched *sooeee* and *sooo-ip*, as well as lower-pitched nasal notes. **H** Montane forest edges. **D** E Africa.

25 **YELLOW-BROWED SEEDEATER** *Crithagra whytii* 14cm Streaked plumage and yellow 'underlayer' of head diagnostic. **V** Song and calls similar to those of Streaky Seedeater. **H** Like that of Kipengere Seedeater. **D** S Tanzania and N Malawi.

26 **KIPENGERE SEEDEATER** *Crithagra melanochroa* 15cm Dusky grey, stout-billed finch with heavily streaked underparts. Forages unobtrusively, mainly in mid-levels of trees and bushes. **V** Song is a long series of rattling, rambling notes. **H** Montane forest and bushland. **D** Tanzania.

27 **PROTEA CANARY** *Crithagra leucoptera* 16cm Note absence of yellow colouring, indistinct eyebrow, variable white throat patch. **V** Song differs from those of all other canaries by nasal quality and rattling repetitions, *tri-tri-tweet-weet-weet*. **H** Montane protea scrub. **D** South Africa.

28 **TWITE** *Linaria flavirostris* 13cm Often quite wary. Ground feeder. Forms large winter flocks. **V** Song like that of Common Linnet but harder and interspersed with call notes; call much as Common Linnet but with characteristic *tweee*, *chweee* or *chwaiie*. **H** Grassy, rocky mountain and highland areas, moorland and open hillsides. In winter, lowland pastures, cultivation, coastal salt marshes and waste ground. **D** Widespread in N Eurasia.

29 **COMMON LINNET** *Linaria cannabina* 13cm White in wings shows well in flight. Often in large winter flocks. Sings from prominent perch or during display flight. **V** Song lively, musical tinkling or twittering interspersed with short trills, drawn-out whistles and twanging notes; call rapid *tett-tett-terrett* or similar, and a soft *hoooi* or *tsooeet*. **H** Heaths and commons with scattered trees, farmland with hedges and thickets, parks, gardens. In winter, also stubble fields, various weedy areas, coastal marsh and shoreline. **D** Widespread in Eurasia, also NW Africa and Canary Islands.

30 **YEMEN LINNET** *Linaria yemenensis* 12cm White wing and tail patches conspicuous in flight. Winter flocks often include Yemen Serin. **V** Song rapid, melodious twittering; call musical *terrrit* or *wid-lee-lee* and a soft *vliet*. **H** Highlands, hillsides and wadis, usually with scattered trees or scrub, also cultivated areas. **D** SW Arabian Peninsula.

31 **WARSANGLI LINNET** *Linaria johannis* 13cm Bright orange-rufous upper rump with white patch below. **V** Song a prolonged and muddled combination of dry buzzing notes, trills and twitters; calls include *tweek*, *sis-sis-sis* and *swee-tee*. **H** Dry, stony and rocky areas with some trees and scrub. **D** NE Somalia.

32 **COMMON REDPOLL** *Acanthis flammea* 13cm Forages on ground or acrobatically in trees, in pairs or small parties. **V** Calls include metallic, twittering *chuch-uch-uch-uch*, plaintive *teu-teu-teu-teu*, and *tooee*. **H** Coastal and lowland open woods. **D** Widespread in Eurasia and North America.

33 **LESSER REDPOLL** *Acanthis cabaret* 12cm Behaviour similar to Common Redpoll. **V** As Common Redpoll, although flight note said to be higher pitched. **H** Coniferous, willow, birch and alder woodlands, hawthorn thickets and large gardens. **D** Britain to C Europe.

34 **ARCTIC REDPOLL** *Acanthis hornemanni* 13cm Behaviour similar to Common Redpoll, with which it often associates. **V** Similar to Common Redpoll, calls said to be more metallic or coarser. **H** Stunted trees or bushes on tundra, birch and willow thickets. **D** N Eurasia and North America.

1
2
3
4
5
6
♂
♀
7
8
♂
♀
9
10
11
♂
♀
12
♂
♀
13
♂
14
♂
♀
15
♂
♀
16
17
18
19
20
21
22
23
24
25
26
27
28
29
♀
30
♀
31
32
♂
♀
33
♂
34
♂
♀
♀
♂

FINCHES AND EUPHONIAS *CONTINUED*

1 PARROT CROSSBILL *Loxia pytyopsittacus* 17cm Tends to break off cones, extracting seeds while holding cone with a foot; otherwise behaviour similar to Red Crossbill. **V** Song similar to Red Crossbill but deeper and slower; calls much as Red Crossbill but deeper, *choop-choop, chok-chok; tsu-tsu-tsu-tsu* and hard *cherk-cherk* when alarmed. **H** Coniferous forests, especially Scots Pine. **D** N Europe to EC Russia.

2 SCOTTISH CROSSBILL *Loxia scotica* 16cm Actions and habits as Red Crossbill. **V** Very similar to Red Crossbill. **H** Scots pines in Caledonian pine forest. **D** Scotland.

3 RED CROSSBILL *Loxia curvirostra* 16–17cm Sociable, usually in small parties; regularly drinks at small pools. Often feeds acrobatically to extract seeds from pine cones. Commonly sings from topmost branch of a conifer. **V** Song consists of call notes that run into a *cheeree-cheeree-choop-chip-chip-chip-cheeree* combined with various trills, twitters and more call notes; call *chip-chip* or quieter *chuk-chuk*, often given in flight. **H** Conifer forests. **D** Widespread in Eurasia, North America and Central America.

4 CASSIA CROSSBILL *Loxia sinesciuris* 17cm Extremely similar to North American populations of Red Crossbill; separated by spectrographic analysis of call. Sedentary. **V** Most vocalisations indistinguishable from those of Red Crossbill, but song contains more buzzy notes, and flight call differs. **H** Forest containing mature Lodgepole Pines. **D** Idaho, USA.

5 TWO-BARRED CROSSBILL *Loxia leucoptera* 17cm Habits and many actions similar to Red Crossbill. As well as pine seeds, feeds on berries and insects. Often mixes with Red Crossbills and Pine and Evening grosbeaks (Plate 272). **V** Song rich and variable, with buzzing trills and harsh rattles; calls include metallic *glip-glip, kip-kip* or *chiff-chiff*, flocks often utter *chut-chut* or *chuch-chuch*. **H** Coniferous forests, especially larch, hemlock and spruce. **D** N Eurasia and North America.

6 HISPANIOLAN CROSSBILL *Loxia megaplaga* 15cm Usually secretive, although sometimes in noisy flocks; feeds acrobatically to extract pine seeds. Recently split from Two-barred Crossbill. **V** Call high-pitched, repeated *chu-chu-chu-chu*, usually given by feeding flocks. **H** Pine forests in high mountains. **D** Hispaniola.

7 MOUNTAIN SERIN *Chrysocorythus estherae* 12cm Shy and retiring, spending much time sitting quietly in low bushes or on the ground, usually alone or in small groups. **V** Calls include dull, metallic chittering note; in flight utters short tinkling. **H** Alpine and subalpine grassland and heather-dominated meadows with scattered bushes or scrub; on Mindanao, occurs in montane rainforest or dwarf ericaceous forest. **D** Sumatra, Java, Philippines, Sulawesi.

8 EUROPEAN GOLDFINCH *Carduelis carduelis* 14cm Feeds, often acrobatically, on plant seed-heads. Often in small groups with larger flocks in winter. **V** Song rapid tinkling *tsswit-witt-witt*, combined with various twittering or buzzing *zee-zee* notes; call liquid, tinkling *tickelit, tsee-yu* or *pee-uu*; also harsh *zeez*. **H** Woodland edges, scrub, hedgerows, waste ground, orchards, parks, gardens. **D** Europe and C and S Asia.

9 CITRIL FINCH *Carduelis citrinella* 12cm Ground feeder on seeds of weeds and conifers. Perches on trees, bushes and overhead wires. **V** Song has European Siskin or European Goldfinch-like short musical phrases interspersed with harsh twittering; call metallic *tiyie* or *tsi-ew* and *check* or *chwick*. **H** Montane conifer forest edge, rocky slopes with scattered conifers. Descends in winter. **D** Spain to Austria and Slovenia.

10 CORSICAN FINCH *Carduelis corsicana* 12cm Actions similar to Citril Finch, of which it is often thought to be a race. **V** Song like that of Citril Finch but more segmented. Also differs by having alternative song of purring, fluty, trumpeting notes. Call similar to that of Citril Finch. **H** Dry montane scrub, woodland and open vegetation. **D** Sardinia, Corsica.

11 RED-FRONTED SERIN *Serinus pusillus* 12cm Forages on ground, usually in small flocks, often with other finches. Female duller, with less red on forehead. **V** Song series of long twittering phrases mixed with trills and wheezy notes; calls include rapid, ringing trill, soft *dueet* and twittering *bri-ihihihihi*. **H** Rocky mountain slopes with grass and scrub, scattered junipers, stands of conifers and low rhododendrons. **D** Turkey to W China and Bhutan.

12 EUROPEAN SERIN *Serinus serinus* 11cm Frequently in small flocks. **V** Song jingling mixture of notes given from prominent perch or during stiff-winged display flight; call rapid, high-pitched trill, *tirrillilit* or *titteree*; also *tirrup* and *tsooee* or *tsswee*. **H** Woodland edges, pine woods, copses, parks, orchards, cultivations, large gardens. **D** C and S Europe to N Africa and the Middle East.

13 SYRIAN SERIN *Serinus syriacus* 12.5cm Behaviour similar to European Serin. Forms large, noisy winter flocks. **V** Song fast, linnet-like twittering, sometimes with buzzing sounds; call dry, rolling *pe-re-ret* or *tree-der-dee*, thin *shkeep* and dry *tearrrh* or *tsirr*. **H** Open bushy slopes and light woodland on mountains. In winter on cultivated land, vegetated wadis, desert or semi-desert scrub. **D** Syria and Iraq to Egypt.

14 ATLANTIC CANARY *Serinus canaria* 13cm Gregarious, often forming large flocks, frequently with other finches. Mainly ground feeder. **V** Song given from perch or display flight, similar to domesticated canary: melodic fluty whistles and trills with some twitters and churrs; call high-pitched *sooee, tsooeeet* or *sweee*, often accompanied by light trill. **H** Almost anywhere with trees or bushes. **D** Canary Islands, Azores, Madeira.

15 CAPE CANARY *Serinus canicollis* 13cm Long-tailed finch, slim looking, and lacking wingbars. **V** Song very high, very rapid sustained fluted warbling with basic *fiu* sound. **H** From montane grassland to coastal scrub, including gardens. **D** S Africa.

16 YELLOW-CROWNED CANARY *Serinus flavivertex* 12cm All-yellow head of male with faint greyer crown and cheeks distinctive; both sexes have bold wing pattern. **V** High, cheerful, unstructured twittering. **H** Wide range of montane habitats, from forest clearings to marshes, roadsides and bushy grassland. **D** NE and E Africa, C Angola.

17 ETHIOPIAN SISKIN *Serinus nigriceps* 12cm Small, with all-dark head in male, lighter grey in female. **V** Song a series of fast, sweet-tuned chuckling and trilling notes; calls include *trrr* and *chit-chit-chit*. **H** Grassland and moorland with giant heath patches. **D** Ethiopia.

18 BLACK-HEADED CANARY *Serinus alario* 13cm Some races interbreed, resulting in intermediate plumages. Male shows long or short black stripes on breast. Note rufous in wings of female. **V** Song jumbled, rather tuneless series of buzzing and nasal sounds with some fluted notes; calls include harsh *chay* and *tsweet*. **H** Dry, montane areas, Karoo scrub, bushveld, cultivation. **D** South Africa.

19 TIBETAN SERIN *Spinus thibetanus* 12cm Usually in small flocks, forages in treetops or on ground under bushes. **V** Song nasal, buzzy twittering interspersed with trills and *ti-ti tweeoo* phrases; calls include dry twittering, wheezy twang; in flight utters short twitters, trills and *chut-chut-chut*. **H** Open forest, forest edges. **D** Nepal to SW China and N Myanmar.

20 LAWRENCE'S GOLDFINCH *Spinus lawrencei* 12cm Habits much as Lesser Goldfinch, although spends much time on the ground. **V** Song tinkling twitter, incorporating mimicry of other bird songs; calls include *tink-oo* or *tink-il* and sharp *kee-yerr*. **H** Chaparral, dry grassy slopes, oak savannah. **D** SW USA, NW Mexico.

21 AMERICAN GOLDFINCH *Spinus tristis* 12cm Behaviour similar to Lesser Goldfinch. **V** Song lively series of twitters, trills and *swee* notes; calls include thin *toweeeowee* or *tweeee*, soft *tihoo*; in flight utters soft, descending *ti di di di*. **H** Open woodlands, thickets, weedy fields, orchards, gardens. **D** Canada to Mexico.

22 LESSER GOLDFINCH *Spinus psaltria* 11cm Forages in trees, bushes and on ground, usually in pairs or small flocks. In flight, male shows white patch on inner primaries and at base of outer-tail feathers. **V** Song rising twittering with paired phrases and imitations of other bird songs; call rising *pee-yee, cheeo* or *choo-ii*; also hoarse *chig chig chig* given in flight. **H** Woodland edges, dry bush, plantations, orchards, gardens. **D** W USA to Peru.

23 EURASIAN SISKIN *Spinus spinus* 12cm Acrobatic forager, mainly in trees; usually encountered in small parties. **V** Song given from treetop or butterfly-like display-flight, consists of a rapid, undulating series of twittering phrases interspersed with trills and wheezy notes, ending in a rasping *kreee*. Calls include plaintive *dlu-ee*, dry *tet* or *tet-tet*, and *twilit, tirrillili* or *titteree* in flight. **H** Open forest, forest edges and secondary growth. **D** Breeds W and N Europe to E Russia and NE China. Winters N Africa, the Middle East and S China.

24 ANTILLEAN SISKIN *Spinus dominicensis* 11cm Usually found in flocks, actively searching for seeds in trees, shrubs and on ground. **V** Song low twittering trill. In flight, utters low *chut-chut* or high-pitched *swee-ee*; also gives *seee-ip* and *chit chit chee-ee-o*. **H** Mountain forest edge and pine forests close to grassy clearings; after breeding visits agricultural areas with nearby scrub forest. **D** Hispaniola.

25 PINE SISKIN *Spinus pinus* 13cm Actions similar to Common Redpoll. Some are yellower on wing coverts and underparts. **V** Song rambling series of nasal and chattering notes, ending with rising *zzzhreeee*; calls include hoarse *tee-ee, clee-ip* or *chlee-it*, rising *sweeeet* and short *tit* or *twit-it-tit*. **H** Conifer and mixed woods, alder thickets, ornamental suburban trees. **D** Canada to Guatemala.

26 BLACK-CAPPED SISKIN *Spinus atriceps* 12cm Unmistakable. **V** Very high fast slurred song, typical of genus. **H** Humid montane woodland, open areas with scattered trees. **D** S Mexico and Guatemala.

27 BLACK-HEADED SISKIN *Spinus notatus* 11cm Black mask contrasts sharply with yellow collar and underparts. **V** Song is a prolonged and varied medley of twitters and ringing, twanging and nasal notes; calls include long *tsiuuu* and nasal *ti-chee*. **H** Woodland, open areas with scattered pines. **D** Mexico to Nicaragua.

28 BLACK-CHINNED SISKIN *Spinus barbatus* 13cm Black bib of male is diagnostic; note also black (male) or dark olivaceous (female) cap. **V** Song like that of Hooded Siskin. **H** Tall forest, brush, open country with scattered trees, gardens, suburbs. **D** S Chile, S Argentina and Falkland Islands.

29 YELLOW-BELLIED SISKIN *Spinus xanthogastrus* 11.5cm Distinctively patterned (compare to Black Siskin). Note dark olivaceous plumage of female. **V** Song like that of Hooded Siskin. **H** Edges and clearings in humid forest and woodland, plantations, pastures. **D** Costa Rica to Bolivia.

30 OLIVACEOUS SISKIN *Spinus olivaceus* 10.5cm Underparts show distinct olivaceous wash. **V** Song like that of Hooded Siskin. **H** Canopy, edges and clearings in montane forest. **D** E Ecuador to N Bolivia.

31 HOODED SISKIN *Spinus magellanicus* 11cm The only siskin in E and SE of its range. Probably indistinguishable from Olivaceous Siskin; separable from Saffron Siskin by lack of saffron tinge to mantle. Female variable; grey of head might be more sharply defined. **V** Song hurried, partly nasal warbling. **H** Woodland, savannah with scattered trees, plantations, cultivation, parks, suburbs. **D** Widespread in South America.

32 SAFFRON SISKIN *Spinus siemiradzkii* 10cm Lowland species; note saffron-tinged mantle and neck. **V** Song like that of Hooded Siskin. **H** Weedy growth at edges of dry forest and secondary growth, roadsides, scrub, suburbs. **D** SW Ecuador, NW Peru.

33 YELLOW-FACED SISKIN *Spinus yarrellii* 11cm No other siskin occurs in its range except Hooded Siskin, female of which is very similar and probably inseparable. **V** Hurried, high, nasal twittering. **H** Borders of humid forest, Caatinga, plantations, suburbs. **D** N Venezuela and NE Brazil.

34 RED SISKIN *Spinus cucullatus* 10cm Usually encountered in small flocks, feeding on seeds of trees, shrubs and grasses. **V** Song rambling, complicated semi-musical twitters, trills and chattering; calls include high-pitched twitter and sharp *chi-tit* or *chut-chut*. **H** Thick scrub on dry hills. **D** NE Colombia, N and W Venezuela, S Guyana.

35 BLACK SISKIN *Spinus atratus* 13cm Unmistakable; note habitat. **V** Hurried nasal warbling. **H** Rocky slopes with sparse vegetation, woodland edges, *Polylepis* woodland. **D** S Peru to N Chile and W Argentina.

36 YELLOW-RUMPED SISKIN *Spinus uropygialis* 13cm Separable from male Yellow-bellied Siskin by yellow rump and black blotching on flanks. **V** Song like that of Hooded Siskin. **H** Shrubby slopes, *Polylepis* woodland, desert scrub. **D** S Peru to Chile and W Argentina.

1
2
3
4
5
6
7
8
9
10
11
12
13
14
15
16
17
18
19
20
21
22
23
24
25
26
27
28
29
30
31
32
33
34
35
36

FINCHES AND EUPHONIAS *CONTINUED*

1 THICK-BILLED SISKIN *Spinus crassirostris* 13cm Note grey flanks of male and grey head of female. **V** Very high, sharp twittering. **H** Montane grassland, *Polylepis* woodland, shrubland. **D** S South America.

2 ANDEAN SISKIN *Spinus spinescens* 11cm Range differs from that of Yellow-faced Siskin, Saffron Siskin, Olivaceous Siskin and Black-chinned Siskin (all Plate 276). Female darker than female Hooded Siskin (Plate 276). **V** Hurried, very high twittering, introduced by inhaled, nasal note. **H** Low bush and scrub at treeline, borders and clearings in woodland and forest, páramo. **D** Colombia, Venezuela.

3 JAMAICAN EUPHONIA *Euphonia jamaica* 11.5cm Arboreal, feeds on mistletoe berries, small fruits, buds and flowers; often seen in small parties. **V** Song pleasant, squeaky whistle. Call staccato *chur-chur-chur...*; occasionally ends with a rising *chip.* **H** Woodlands, forest edge, shrubbery, gardens. **D** Jamaica.

4 PLUMBEOUS EUPHONIA *Euphonia plumbea* 9cm Grey tones are diagnostic. **V** Brief, high twittering. **H** Open woodland, savannah with scattered trees, scrubby vegetation, humid forest borders, mainly in lowlands. **D** N South America.

5 SCRUB EUPHONIA *Euphonia affinis* 10cm Note restricted yellow on forehead and small bill. See also White-vented Euphonia. **V** Very high sharp *deedeedee* and very fast canary-like twittering. **H** Forest, secondary growth, tree stands. **D** Mexico to Costa Rica.

6 YELLOW-CROWNED EUPHONIA *Euphonia luteicapilla* 10cm Separated from other euphonias by extent of yellow on crown. Female rich yellow, difficult to separate from other female euphonias. **V** Extremely high drawn-out *weeeet-weet.* **H** Forest edge, open woodland, open areas with scrub, hedges, tree stands. **D** Nicaragua to Panama.

7 PURPLE-THROATED EUPHONIA *Euphonia chlorotica* 10cm Note black throat, central white undertail area edged in black, and yellow front and underparts of male; separable from Trinidad Euphonia by less extensive yellow front, though this feature varies by race. Female (difficult to separate from other female euphonias) varies according to race. **V** Very high *tsee* (repeated up to three times). **H** Borders and clearings in forest, open woodland, riverine belts, trees in agricultural areas, suburbs. **D** SC South America, N Amazonia.

8 TRINIDAD EUPHONIA *Euphonia trinitatis* 9.5cm Male separable from male Purple-throated Euphonia by larger crown patch. Front of female always green (not yellow as in some races of Purple-throated). **V** Like that of Purple-throated Euphonia. **H** Forest, open woodland, riverine belts, cultivation, suburbs. **D** E Colombia and C Venezuela through the Guianas to N Brazil.

9 VELVET-FRONTED EUPHONIA *Euphonia concinna* 10cm Note lack of white in undertail of male, and short yellow eyebrow of female. **V** Series of high *wee-wee* phrases. **H** Dry open woodland, pastures with scattered trees. **D** Colombia.

10 ORANGE-CROWNED EUPHONIA *Euphonia saturata* 10cm Note extension of yellow on head, and orange wash to lower belly and vent. Female has a smaller bill than female Thick-billed Euphonia. **H** Lightly wooded habitats, including open woodland, forest edges, clearings. **D** Colombia to Peru.

11 FINSCH'S EUPHONIA *Euphonia finschi* 10cm Note lack of white in undertail of male. Female not reliably separable from female Violaceous Euphonia. **V** Series of high *wee-wee-wee* phrases. **H** Riverine belts, savannah with scattered trees, borders of humid forest. **D** Venezuela, the Guianas.

12 VIOLACEOUS EUPHONIA *Euphonia violacea* 11.5cm Separable from Thick-billed Euphonia by smaller bill. **V** Slow series of thin, scratchy *tets* notes. **H** Humid forest, secondary growth, shrubby clearings, orchards, suburbs. **D** N and SC South America.

13 THICK-BILLED EUPHONIA *Euphonia laniirostris* 11.5cm Note bill size, all-yellow underparts and crown patch size. Some races have black underside of tail. **V** Series of scratchy and nasal notes mixed with ultra-high *tsee.* **H** Semi-open habitats; clearings with scattered trees, riverine belts, plantations, habitation. **D** Costa Rica to Amazonia.

14 YELLOW-THROATED EUPHONIA *Euphonia hirundinacea* 11cm Separated from Thick-billed Euphonia by thinner bill and mainly different range. Female whitish below in centre of underparts. **V** Very high hurried warble (2 secs), with sharp *fjeet* and nasal *njèèh* notes. **H** Forest, secondary growth, gardens, semi-open areas with trees, forest edges. **D** E Mexico to Panama.

15 GREEN-CHINNED EUPHONIA *Euphonia chalybea* 11.5cm Note narrow black chin in male, and yellow face with grey sides of throat of female. **V** Series of low, nasal twittering notes. **H** Borders and clearings in tall forest and secondary growth. **D** E Paraguay, NE Argentina, SE Brazil.

16 ELEGANT EUPHONIA *Euphonia elegantissima* 11cm Unmistakable. **V** Song rapid and rambling series of trills, chirps, bubbling sounds and whistles; calls varied, with some sharp, others soft and melancholic. **H** Forest, tree stands. **D** Mexico to W Panama.

17 GOLDEN-RUMPED EUPHONIA *Euphonia cyanocephala* 11cm Blue cap is diagnostic. **V** Short, muttered twittering, punctuated with some dry, nasal notes. **H** Forest borders, secondary growth, plantations. **D** N Colombia to Trinidad and the Guianas, south to N Argentina, SE Brazil and E Paraguay.

18 ANTILLEAN EUPHONIA *Euphonia musica* 12cm Feeds in canopy, primarily on mistletoe berries; best located by calls. **V** Song jumbled tinkling, interspersed with explosive notes; calls include rapid, tinkling *ti-tit*, hard *chi-chink* and plaintive *whee.* **H** Dense forests. **D** West Indies.

19 FULVOUS-VENTED EUPHONIA *Euphonia fulvicrissa* 10cm Fulvous vent of male diagnostic. Female separable from female White-lored Euphonia and female Orange-bellied Euphonia by pale rufous (not yellow) vent. **V** Fast twittering, punctuated with short trills and drawn-out notes. **H** Interior and borders of humid forest, overgrown clearings, secondary growth. **D** Panama to Ecuador.

20 SPOT-CROWNED EUPHONIA *Euphonia imitans* 10cm Crown spots diagnostic, but difficult to see. Differs from other euphonias in its range by extent of yellow. **V** Song a single note repeated several times then changed; may be sweet or harsh and dry. Calls mostly dry and rattling. **H** Lowland forest and forest edges. **D** Costa Rica and Panama.

21 OLIVE-BACKED EUPHONIA *Euphonia gouldi* 10cm No similar bird in its range. Note dark-looking chestnut belly. **V** Calls include very high disyllabic slow *piuuw-swirrr.* **H** Forest. **D** SE Mexico to Panama.

22 WHITE-LORED EUPHONIA *Euphonia chrysopasta* 10cm Note striking dark-bordered white lores. **V** Hurried mixture of very short, high and low twitters. **H** Canopy and borders of várzea and terra firme forest, tall secondary growth. **D** W and N Amazonia.

23 BRONZE-GREEN EUPHONIA *Euphonia mesochrysa* 10cm Green upperparts with yellow front of male distinctive. Female shows greenish throat and white belly. **V** Mixture of clear rolls and very short, high twitters. **H** Clearings in cloud forest and roadsides with small trees and shrubs. **D** Colombia to Bolivia.

24 WHITE-VENTED EUPHONIA *Euphonia minuta* 9.5cm White vent is diagnostic; also note yellow breast band of female. **V** Very short, hurried, nasal twitters. **H** Borders of wet lowland forest. **D** Mexico to Ecuador, the Guianas to C Bolivia.

25 TAWNY-CAPPED EUPHONIA *Euphonia anneae* 11cm Separable from Orange-bellied Euphonia by whitish (not yellow) vent. **V** Mixture of very high *tsee-tsee-tsee* notes with low rolls and nasal notes. **H** Borders and clearings in dense wet forest and secondary growth. **D** Costa Rica to Colombia.

26 ORANGE-BELLIED EUPHONIA *Euphonia xanthogaster* 11cm Yellow crown patch of nominate race is larger than that of Violaceous Euphonia, Thick-billed Euphonia, and Green-chinned Euphonia. **V** Ultra-high *tsee*, mixed with high, loud, calm, warbled phrases. **H** Borders, gaps and clearings in humid and wet forest. **D** Panama to Bolivia and E Brazil.

27 RUFOUS-BELLIED EUPHONIA *Euphonia rufiventris* 11.5cm Note all-black head of male. Female shows pale buff-orange vent. **V** Little known; possible song combines *zit* and *weet* notes; usual call a harsh, dry rattle. **H** Borders and canopy of terra firme and várzea forest. **D** W and N Amazonia.

28 CHESTNUT-BELLIED EUPHONIA *Euphonia pectoralis* 11.5cm Separable from Golden-sided Euphonia by range. Note chestnut vent of female. **V** Short, sharp, rattled chatters. **H** Canopy, borders and clearings with scattered trees of humid forest and woodland. **D** E Paraguay, C and SE Brazil, NE Argentina.

29 GOLDEN-SIDED EUPHONIA *Euphonia cayennensis* 11.5cm Male unmistakable. Note pale greyish central underparts from chin to vent of female. **V** Short, dry, tuneless rattles. **H** Canopy and borders, secondary growth, forest patches in savannah. **D** N Amazonia.

30 YELLOW-COLLARED CHLOROPHONIA *Chlorophonia flavirostris* 10cm Unmistakable due to pale eyes and pink bill and legs. **V** High *peee* note, sometimes sung in very slow series. **H** Canopy and borders of wet forest. **D** Colombia, Ecuador.

31 BLUE-NAPED CHLOROPHONIA *Chlorophonia cyanea* 11.5cm Unmistakable due to blue-rimmed black eyes. Races vary in extent of blue on upperparts and lack or presence of yellow frontal patch. **V** High, drawn-out *tiuuue* notes. **H** Mainly in borders of humid forest and secondary growth. **D** N, W and SE South America.

32 CHESTNUT-BREASTED CHLOROPHONIA *Chlorophonia pyrrhophrys* 11.5cm Unmistakable due to blue cap and black eyebrows. **V** Nasal *tee-tee-teerowee* phrase. **H** Cloud forest and older secondary growth. **D** Venezuela to Peru.

33 BLUE-CROWNED CHLOROPHONIA *Chlorophonia occipitalis* 13cm Not present in range of Golden-browed or Yellow-collared clorophonias. **V** High descending pure whistled *plew* (2 secs), repeated frequently. **H** Forest. **D** Mexico to Nicaragua.

34 GOLDEN-BROWED CHLOROPHONIA *Chlorophonia callophrys* 13cm Yellow neck and (in female) eyebrow diagnostic. **V** High unstructured *pfuuuh pruit...* **H** Montane forest and adjacent more open areas. **D** Costa Rica, Panama.

1
2
3
4
5
6
7
8
9
10
11
12
13
14
15
16
17
18
19
20
21
22
23
24
25
26
27
28
29
30
31
32
33
34

LONGSPURS AND SNOW BUNTINGS CALCARIIDAE

1 MCCOWN'S LONGSPUR *Rhynchophanes mccownii* 15cm Tail mostly white, central feathers and tips black. Forms winter flocks with other seed-eaters. **V** Song energetic jumble of warbles and twitters, often given in display flight. Calls include *kittip* and popping *poik*. **H** Short grass prairie. Winters on barren fields, dry lake beds and areas of short grass. **D** Breeds SC Canada, NC USA. Winters N Mexico.

2 LAPLAND LONGSPUR *Calcarius lapponicus* 15cm Gregarious outside breeding season, often in mixed flocks with larks, pipits and snow buntings. **V** Song short, jangling warble, *djüü-tiiah-preeyu-chirio-twi-trii-tri-leeoh* or similar, often given during display flight; calls include melodious *tee-uu* and in flight high *jeeb*. **H** Tundra, mountains with low shrubbery. Winters in pastures, coastal marshes. **D** Breeds N North America and Eurasia. Winters S USA, C Eurasia.

3 SMITH'S LONGSPUR *Calcarius pictus* 15cm Outermost two pairs of tail feathers mainly white. Secretive. **V** Song sweet high warble *sew seeyu wee tee tee dzee tzeeyu* or similar. Flight call clear, descending rattle. **H** Wet sedge meadows in tundra area; grassy areas in mountains. In winter, grassy or stubble fields. **D** Breeds N and SE Alaska, Canada. Winters SC USA.

4 CHESTNUT-COLLARED LONGSPUR *Calcarius ornatus* 14cm Tail white with black central triangle. Forms large winter flocks. **V** Song pleasant warble, starting high, ending low and buzzy, often given in flight. Flight call *cheedle cheedle*. **H** Short grass plains, dense grasslands and wastelands. Winters in grassy areas. **D** Breeds SC Canada, NC USA. Winters Mexico.

5 SNOW BUNTING *Plectrophenax nivalis* 16.5cm In winter forms small ground-feeding flocks, often with larks, pipits and other buntings. **V** Song variable series of repeated phrases such as *terere-dzüü-weewa-tererere-dzüü-tsee-tsee...*, often given during display flight; calls include harsh *djee* and rippling *tirirririt* often followed by ringing *pyu*. **H** Barren tundra, rocky mountain tops, scree, sea cliffs, human settlements. Winters in open country and on beaches, coastal marshes. **D** Breeds N North America, N Eurasia. Winters S USA, Mediterranean.

6 MCKAY'S BUNTING *Plectrophenax hyperboreus* 18cm Generally whiter than Snow Bunting. Gregarious after breeding. **V** Similar to Snow Bunting. **H** Beaches and inland tundra. **D** W Alaska.

THRUSH-TANAGER RHODINOCICHLIDAE

7 ROSY THRUSH-TANAGER *Rhodinocichla rosea* 20cm Distinctive colour pattern but difficult to see in its shady, tangled habitat. **V** Calm series of well-separated, mellow, single, double and triple notes of different pitches. **H** Dense undergrowth of forest, secondary growth and woodland. **D** W Mexico to Venezuela.

BUNTINGS EMBERIZIDAE

8 CRESTED BUNTING *Emberiza lathami* 16–17cm Forages on the ground, regularly by paths and near cereal crops. Frequently perches on rocks or bushes. Sings from prominent elevated perch. In winter forms loose small flocks. **V** Song begins with a few subdued notes followed by low, mellow notes and finishing with 2–3 descending notes, *tzit dzit dzit see-see-suee* or similar; call soft *tip* or *tup*, uttered more emphatically in flight. **H** Dry rocky or grassy hillsides and terraced cultivation with rocky outcrops and scattered bushes. **D** Himalayas to SC China and N Vietnam.

9 SLATY BUNTING *Emberiza siemsseni* 13cm Forages on ground, often in cover; forms small parties post breeding. **V** Song variable, including high-pitched *zii-ziiiu-tzitzitzitzi hee, ze-ze-ze-ze swee twiitwit* and *tze-tze-tze teez teez teez*. Call sharp *zick*, often repeated. **H** Subtropical valley forests in hills and mountains, bamboo thickets in secondary forest, scrubby cover of degraded forest. In winter often seen in parks and around villages. **D** C China.

10 CORN BUNTING *Emberiza calandra* 17–18cm Forages on ground, regularly perches on bushes, posts or overhead wires; generally occurs in flocks post breeding. **V** Song starts with chipping notes followed by harsh jangling; call hard, dry *tuk* or *bitt*, often rapidly repeated. **H** Arable land and waste ground. **D** Europe, W and SC Asia, N Africa.

11 YELLOWHAMMER *Emberiza citrinella* 16.5cm Usually gregarious, forages on ground and regularly retreats to perch on bushes if disturbed. Hybridises with Pine Bunting, hybrids usually showing some yellow on throat or breast. **V** Song short rattle ending with long *dzeee* note; calls include metallic *tsit*, thin *see* and clicking *tit-tit-tit-tit*. **H** Upland agricultural land. **D** Breeds Europe, W and C Asia. Winters N Africa.

12 PINE BUNTING *Emberiza leucocephalos* 17cm Forages on ground, usually in small to large flocks, often with other buntings and finches. Non-breeding males have head pattern and back obscured by pale fringes. **V** Calls similar to those of Yellowhammer. **H** Grassy slopes with bushes and fallow or stubble fields. **D** Breeds C and E Asia. Winters S and SE Asia.

13 ROCK BUNTING *Emberiza cia* 16cm Forages on ground, among rocks or low vegetation, usually in pairs or small parties post breeding. **V** Song variable; starts hesitantly then accelerates into clear twittering phase, transcribed as *tsiritt churr chu-chut chirrirri tsirr chu-tsirriritt*; calls include sharp *tsii*, high-pitched *tseee* and when disturbed rolling *trrr*; also short *tup* in flight. **H** Rocky mountain and hill slopes with bushes and scattered trees, juniper and open pine forest, cultivation edges; in winter also in lowland fallow cultivation. **D** SE Europe, SW and SC Asia.

14 GODLEWSKI'S BUNTING *Emberiza godlewskii* 17cm Forages mainly on ground, but often perches in trees; in winter occurs in small flocks. **V** Song similar to Rock Bunting but starting with higher-pitched notes; calls include thin *tzii* and hard *pett-pett*. **H** Bushy and rocky mountainsides, often near forests; regularly visits fields in winter. **D** C Asia.

15 MEADOW BUNTING *Emberiza cioides* 16cm Forages mainly on ground, hopping with hunched posture. Winter flocks may include other seed-eaters, often seen at water-holes in early morning or late afternoon. Sings from bush top or other elevated perch. **V** Song short, simple *chi-hu chee-tsweet-tsweet-tuee*; call thin, repeated *zit zit zit*. **H** Lightly wooded rolling hills and mountain foothills; also shrub thickets. **D** E Asia.

16 WHITE-CAPPED BUNTING *Emberiza stewarti* 15cm Forages on ground and in bushes; in winter forms small flocks. Non-breeding males have chestnut and black plumage areas obscured by pale feather edges. **V** Song buzzy, metallic rattle ending with single high-pitched note; calls include high-pitched *tit* and squeaky, stacatto *tchirit*. **H** Mainly grassy and rocky slopes with or without bushes and trees; in winter dry foothills, scrub jungle, agricultural land, plain edges with scattered bushes. **D** Himalayas to C India.

17 JANKOWSKI'S BUNTING *Emberiza jankowskii* 16cm Little recorded. Forages on ground. Sings from tops of small bushes. **V** Song simple *chu-chu-cha-cha-chee* or *hsuii-dzja-dzja-dzjeee*; calls include *tsiit* or *tsiit-tsiit*; explosive *sstlitt* or thin *hsiu* when alarmed. **H** Dry, overgrown sand-dunes with little ground cover and low bushes. **D** NE China, N North Korea, extreme SE Russia.

18 GREY-NECKED BUNTING *Emberiza buchanani* 15cm Forages on ground in sparse vegetation, usually in pairs or small parties; more gregarious on migration and in winter. **V** Song *dzee-zeee-zeee-zee-zee-zee-deo*; calls include *tcheup* or *chep* and soft *tsip* given in flight. **H** Bare mountain slopes and foothills with sparse vegetation; in winter dry, stony cultivation, euphorbia-covered dry slopes. **D** Breeds SC Asia. Winters India.

19 CINEREOUS BUNTING *Emberiza cineracea* 16cm Wary. Migrants often join with Ortolan and Cretzschmar's buntings. **V** Song rapid, ringing *drip-drip-drip-drip-drie-drieh, dzuu-zuu-zuu-zuu-zee-uie* or similar; call soft *tsik* and descending *tieu*. **H** Rocky slopes with scattered bushes and scrub. In winter, dry lowland areas, including stubble fields. **D** Breeds SC Asia. Winters NE Africa.

20 ORTOLAN BUNTING *Emberiza hortulana* 16cm Feeds on ground and in trees. Gregarious. **V** Song variable, *dzii-dzii-dzii-dzii-hüü-hüü, witt-witt-witt-witt-hüü-hüü* or *zree-zree-zree-züü*; calls include dry *plet*, clear, metallic *ziie*, often followed by short *tew*. **H** Open cultivated country with scattered trees and bushes, bare mountainsides and gullies with scattered trees and bushes. **D** Breeds E Europe to SC Russia, W Mongolia, Iran and Turkey. Winters N Africa, S Asia.

21 CRETZSCHMAR'S BUNTING *Emberiza caesia* 16cm Terrestrial, rarely perches in bushes unless alarmed or singing. **V** Song variable, thin, low-pitched or buzzy with drawn-out final note, *dzree-dzree-dzreee, ziii-ziii-ziii-ziiiii*; calls include *tchipp, plet* and *tchu*. **H** Dry rocky hillsides with scattered bushes. **D** Breeds Greece and W and S Turkey to Israel and Jordan. Winters NE Africa.

22 CIRL BUNTING *Emberiza cirlus* 16cm Less conspicuous, unless singing, than Yellowhammer. **V** Song rattled trill, *sre'sre'sre'sre'sre'sre...* or similar; calls include sharp *zitt*, descending *zeee* and rapid clicking. **H** Lightly wooded country, farmland with tall hedgerows, orchards, vineyards, large gardens. In non-breeding season, in stubble or weedy fields. **D** S Europe, NW Africa.

23 STRIOLATED BUNTING *Emberiza striolata* 13cm Forages on ground, generally in pairs or small parties; in dry areas regularly drinks at water holes in morning. **V** Song simple, repetitive *trip trip te-tree-cha tre-tree-cha*; calls include nasal *dschu* and sparrow-like *chielp* flight note. **H** Dry rocky hills with sparse thorn or euphorbia scrub; in winter on sandy plains, tamarisk scrub, grassy areas by canals. **D** NE Africa to SW Asia and NW India.

24 HOUSE BUNTING *Emberiza sahari* 14cm After breeding forms parties with other buntings. **V** Song simple, repetitive *chippy-chiwy-chiwy-chiwy*; calls include nasal *dschu*, sparrow-like *tchiele* flight note. **H** Dry, desolate, rocky hills, sometimes human habitations. **D** NW Africa.

25 LARK-LIKE BUNTING *Emberiza impetuani* 13cm Separated from most true larks by longer tail, smaller bill and more horizontal stance. Note typical bunting form of bill with larger lower mandible. **V** Mid-high *tjuder-weeet-weet*. **H** Dry rocky areas with sparse scrub cover. **D** SW Africa.

26 CINNAMON-BREASTED BUNTING *Emberiza tahapisi* 14cm Usually occurs singly, in pairs or in small parties, sometimes in larger flocks in winter. **V** Song short *dzit-dzit-dzirera* or *try-tri... tre-rerir*, last two notes higher in pitch, occasionally ending with some scratchy notes; calls include subdued *dwee*, thin *tsiii-i* and nasal *per-we-e*. **H** Dry, rocky and stony hillsides with scattered vegetation. **D** E and S Africa, SW Arabian Peninsula.

27 GOSLING'S BUNTING *Emberiza goslingi* 14cm Very like Cinnamon-breasted Bunting in habits and appearance (and recently split from it); has more extensive rufous coloration in wings, and both sexes have grey throat. **V** Like that of Cinnamon-breasted Bunting. **H** Rocky and stony hillsides and gullies, sandy areas, old pastures and other similar open areas. **D** Mauritania and Senegal to SW Sudan and NE DR Congo.

28 SOCOTRA BUNTING *Emberiza socotrana* 15cm Very much like Cape Bunting, which is not present on Socotra. **V** Probable song a clear four- or five-note phrase, *tsee-too-tee-too* or similar. **H** Uplands. **D** Socotra, Gulf of Aden.

29 CAPE BUNTING *Emberiza capensis* 15cm Note extensively rusty-red wings. **V** Mid-high, hesitant *wit wit wit weeh wit* or *weeder-wit weeder-wit*. **H** Dry, rocky, mountainous and hilly areas, coastal scrub, miombo, suburbs. **D** S Africa.

30 VINCENT'S BUNTING *Emberiza vincenti* 15cm Differs from Cape Bunting by grey, darker upper- and underparts. **V** Very similar to Cape Bunting. **H** Rocky areas in highlands. **D** SE Zambia, S Malawi, N Mozambique, SW Tanzania.

31 TRISTRAM'S BUNTING *Emberiza tristrami* 15cm Forages on forest floor, usually in small flocks. **V** Call explosive *tzick*. **H** Undergrowth in forest, forest edge, secondary growth. **D** Breeds SE Russia, NE China. Winters S China, SE Asia.

32 CHESTNUT-EARED BUNTING *Emberiza fucata* 16cm Forages on ground, usually in pairs or small parties; after breeding often roosts in marshy reedbeds. **V** Song rapid, twittering *zwee-zwizewezwizizi-triip-triip* or similar; calls include explosive *pzick*, high-pitched *zii* or *zii-zii* and lower-pitched *chutt*. **H** Hillsides with bushes and scrub, especially near rivers; in winter wet stubbles, marshes, grassland and bushes. **D** Breeds C and E Asia. Winters SE Asia.

33 LITTLE BUNTING *Emberiza pusilla* 12–14cm Forages on ground and low in bushes and trees, often with other seed-eaters. **V** Song metallic *zree zree zree tsutsutsutsutzriiitu, tzru tzru tzru zee-zee-zee-zee zriiiiiru* or similar; calls include hard *tzik* or *pwick*. **H** Open and semi-open areas, including forest edge, scrubby hillsides, crop, stubble, paddyfields, orchards, gardens. **D** Breeds N Finland to E Russia. Winters SE Asia.

34 YELLOW-BROWED BUNTING *Emberiza chrysophrys* 15cm Secretive. Behaviour typical of genus. **V** Song *chuee-swii-swii-chew-chew* or *chueee-tzrrii-tzrrii-wee-wee-wee-tueei*; contact call short *ziit*. **H** Mixed forests with many low conifers, especially along rivers; prefers clearings and margins. **D** Breeds S and SC Russia. Winters E and S China.

35 RUSTIC BUNTING *Emberiza rustica* 14cm Behaviour as others of genus. Forms small to large flocks in winter. **V** Song hurried, mellow warble, *deduleu-dewee-deweea-weeu*; call sharp *tzik*, often repeated, also high-pitched *tsiee*. **H** Margins between damp coniferous or birch forest, willow scrub by rivers and fens, also boggy areas. In winter, woodland, cultivation, open country. **D** Breeds N Europe to E Russia. Winters E China.

36 YELLOW-THROATED BUNTING *Emberiza elegans* 15cm Forages on ground; forms small winter flocks, usually of up to 25; larger parties occur on migration. **V** Song long, monotonous twitter, *tswit tsu ri tu tswee witt tsuri weee dee tswit tsuri tu...*; call sharp *tzik*. **H** Deciduous and mixed deciduous–conifer forest, forest edge, clearings, in winter also in orchards, riverside vegetation. **D** Breeds E Asia. Winters SE Asia.

37 YELLOW-BREASTED BUNTING *Emberiza aureola* 15cm Forages on ground, when disturbed retreats to nearby bushes or trees; usually in flocks of up to 200. Non-breeding males have back and head pattern obscured by pale feather edges. **V** Song sweet, twittering *tswit tsu ri tu tswee witt tsuri weee dee tswit tsuri tu*; calls include sharp *tsik* and abrupt *chup* when flushed. **H** Cultivation, grassland, hedgerows, gardens. **D** Breeds N and E Asia. Winters SE Asia.

1
2
3
4
5
6
7
8
9
10
11
12
13
14
15
16
17
18
19
20
21
22
23
24
25
26
27
28
29
30
31
32
33
34
35
36
37

BUNTINGS *CONTINUED*

1 **SOMALI BUNTING** *Emberiza poliopleura* 14cm Note white underparts. Mantle streaked cream, rufous and black. **V** Very similar to Golden-breasted Bunting, but song phrase shorter. **H** Dry open bushy and wooded areas. **D** Ethiopia and Somalia to N Tanzania.

2 **GOLDEN-BREASTED BUNTING** *Emberiza flaviventris* 15cm White streak below eye diagnostic. **V** Very high, siffled *tueeh-tueeh-tueeh*. **H** Woodland, thornveld, plantations, gardens. **D** Widespread in Africa.

3 **BROWN-RUMPED BUNTING** *Emberiza affinis* 14cm Separated from Golden-breasted Bunting by lack of white in wings. **V** Song a brief trilling buzz; calls include short *chip* given in flight, and a more liquid note. **H** Dry open woodland. **D** W, C and E Africa.

4 **CABANIS'S BUNTING** *Emberiza cabanisi* 16cm Separated from Golden-breasted Bunting by absence of white stripe directly under eye, and more grey than rufous upperparts. **V** Extremely high, tit-like, fast *fijuut-fijuur-fijuur-fijuu-tuderleet-tuderleet*. **H** Miombo, other woodland and bushland. **D** W, C, E and SE Africa.

5 **CHESTNUT BUNTING** *Emberiza rutila* 14cm Forages on the ground, retreats into nearby bushes or trees when disturbed; usually in small flocks. Non-breeding adults have chestnut areas obscured by variable amounts of pale scaling. **V** Song high-pitched *wiie-wiie-wiie tzree-tzrree-tzrree zizizitt...*; calls include *zick* and thin, high *teseep*. **H** Open forest, woodland edge and clearings, cultivation, rice stubbles. **D** Breeds SC and SE Russia, Mongolia and NE China. Winters S China and SE Asia.

6 **TIBETAN BUNTING** *Emberiza koslowi* 16cm Forms small flocks in non-breeding season. Little else recorded, presumably behaviour similar to that of other buntings. **V** Song twittering *cheep-chüüp-tererep-cheechûû* or *chep-chip-chip-chiriree-chip-chee*; call thin *seee*, flight call *tsip-tsip*. **H** High-altitude cotoneaster and other low bushes on rocky slopes and ridges. **D** Himalayas to WC China.

7 **BLACK-HEADED BUNTING** *Emberiza melanocephala* 16–18cm Forages mainly on ground, retreats into nearby bushes or trees when disturbed; often in large flocks with Red-headed buntings. In winter head and back colours of male obscured by pale fringes. **V** Song low-pitched *zrit zrit srutt srutt-sruttsutteri-sutt sutterrih* or similar; calls include *cheep* or *chlip*, *dzuu* and in flight *chuhp*. **H** Cereal cultivations. **D** Breeds SC Europe to Iran and SW Pakistan. Winters India.

8 **RED-HEADED BUNTING** *Emberiza bruniceps* 16cm Behaviour similar to that of Black-headed Bunting. In winter head and back colours of male are obscured by pale fringes. **V** Song very similar to Black-headed Bunting's; calls include *chip*, *chuupp*, *zrit*, sharp *tsit* and a series of clicks, all very similar to those of Black-headed Bunting. **H** Cereal cultivations. **D** Breeds Kazakhstan to Iran and Afghanistan. Winters India.

9 **YELLOW BUNTING** *Emberiza sulphurata* 14cm Unobtrusive. Forms small flocks in non-breeding season. **V** Song alternating twittering phrases, *twee-twee-tsit-prewprew-zrii*, *ziriritt-zee-zee* and *psew-zereret-zeetew*; call *tsip-tsip*. **H** Deciduous and mixed forest edge in foothills and lower slopes of high mountains. Winters in weedy and bushy areas. **D** Breeds Japan, Korea. Winters SE China, Philippines.

10 **BLACK-FACED BUNTING** *Emberiza spodocephala* 15cm Forages mainly on ground. Face pattern of non-breeding males partly obscured by pale fringes. **V** Song variable series of ringing trills and chirps; calls include sharp *tzit* or *tzii*. **H** Breeds in shrubby and tall grassy areas with trees near water, mixed and moist conifer forest, open areas in sparse woodland and forest edge; in winter also in scrub, grass and cultivations, often near water. **D** Breeds E Asia. Winters SE Asia.

11 **GREY BUNTING** *Emberiza variabilis* 14–17cm Secretive, forages on ground, alone or in pairs. **V** Song simple, slow *swee swee chi chi chi*, *hsuuu twis-twis-twis* or *hsuuu tsisisisisis*; call sharp *tzii*. **H** Dense vegetation, usually in evergreen forest undergrowth, near streams; also in open cultivations, parks, gardens. **D** E Asia.

12 **PALLAS'S REED BUNTING** *Emberiza pallasi* 14cm Forages on ground, on grass-heads and in bushes; after breeding usually seen in flocks. Non-breeding male has browner head markings, obscured by pale fringes; also shows a faint supercilium and generally buffier breast and neck patch. **V** Calls include sparrow-like *chleep* or *tsilip* and slurred *dziu*. **H** Reedbeds, grassy fields, paddyfields, arable land, shrubs near watercourses. **D** E Asia.

13 **JAPANESE REED BUNTING** *Emberiza yessoensis* 14.5cm Behaviour similar to that of Common Reed Bunting. **V** Song brief twittering, *chuwi-chiwu-sii-psere-dsee* or *chui-tsui-chirin*; calls include short *tick*, flight call *bschet*. **H** Reedbeds, shrubby marshland, wet meadows in highlands. Winters in coastal marshes. **D** E Asia.

14 **COMMON REED BUNTING** *Emberiza schoeniclus* 14.5cm Unobtrusive, except when singing from reed stem or bush top. Forms small winter flocks. **V** Song short, simple series, *zritt-zreet-zreet-zreet-zritt-zriüüü* or similar; call *seeoo* used as contact note or when alarmed; migrating birds give hoarse *brzee*. **H** Marshy areas with scrub, reeds or other tall herbage. In non-breeding season often fields, woodland clearings, large gardens, not necessarily near water. **D** Breeds widely in Eurasia. Winters N Africa, S Asia.

NEW WORLD SPARROWS PASSERELLIDAE

15 **LARK BUNTING** *Calamospiza melanocorys* 18–19cm Terrestrial. Forms large winter flocks. **V** Song varied whistles and trills; call soft *hoo-ee*. **H** Dry grassland, farmland, semi-arid areas. **D** Breeds SC Canada to SC USA. Winters NC Mexico.

16 **RED FOX SPARROW** *Passerella iliaca* 18cm Reddish-toned with streaked back, shorter tail than other fox sparrows. **V** Song melodious notes, slowly rising then falling, often interspersed with buzzy trills; calls include *stsssp* and various *chips*. **H** Woodland undergrowth, thickets, chaparral. **D** Breeds NW and C Alaska to E Canada. Winters SC and SE USA.

17 **SOOTY FOX SPARROW** *Passerella unalaschcensis* 18cm Plumage predominantly dark slaty grey or ashy grey, underparts heavily marked. **V** Similar to that of Red Fox Sparrow. **H** Mostly deciduous thickets, riversides. **D** Breeds Aleutians and SW Alaska to NW British Columbia. Winters British Columbia to S California.

18 **SLATE-COLORED FOX SPARROW** *Passerella schistacea* 18cm Rather plain grey on head and mantle, shading to warm brown on wings and tail. **V** Similar to that of Red Fox Sparrow. **H** Riverside thickets and other shrubby areas. **D** Breeds SW Canada to WC USA. Winters California to N Baja California and S Arizona.

19 **THICK-BILLED FOX SPARROW** *Passerella megarhyncha* 18cm Grey head and mantle, otherwise mostly mid-brown; has a heavier bill and more lightly marked underparts than other fox sparrows. **V** Similar to that of Red Fox Sparrow. **H** Montane chaparral, fire-damaged forest. **D** Breeds SW Oregon and California. Winters S California and NW Baja California.

20 **SONG SPARROW** *Melospiza melodia* 16cm Variable; some races pale, others very dark. **V** Song clear notes followed by buzzing rattle and trill; calls include nasal *tchep* and a thin *tsee*. **H** Open shrubby areas, waterside thickets, urban gardens. **D** Widespread through North America to C Mexico.

21 **LINCOLN'S SPARROW** *Melospiza lincolnii* 15cm Secretive, forages on ground usually near cover. Pumps tail in flight. **V** Song continuous jumble of husky, chirping trills with many pitch changes; calls include nasal *tschup*, repeated when agitated; flight call buzzy *zeeet*. **H** Clearings in moist highland thickets, forest borders, coastal thickets. **D** Breeds W and N North America. Winters Central America.

22 **SWAMP SPARROW** *Melospiza georgiana* 15cm Shy and nervous; forages on ground or in low vegetation and also in shallow water. Pumps tail in flight. **V** Song slow, musical, one-pitched trill; calls include hard *chip*; in flight, gives buzzy *zeeet*. **H** Marshy and bushy areas. **D** Breeds NC and NE North America. Winters NE Mexico.

23 **RUFOUS-COLLARED SPARROW** *Zonotrichia capensis* 15–16.5cm Shy, forages on ground, usually in pair. **V** Song an accelerating trill, *whis-whis-whis-whis-whiswhisu-whiswhis*; call sharp *chip*. **H** Mountain forest edges, streamside thickets, undergrowth of pine forests. **D** S Mexico to S Argentina, also Hispaniola.

24 **HARRIS'S SPARROW** *Zonotrichia querula* 19cm Unmistakable. **V** Song high, clear *seeeeeeee seee seee*; call harsh *cheek*. **H** Stunted boreal forest; winters in thickets and open woodland. **D** Breeds NC and C Canada. Winters SC USA.

25 **WHITE-CROWNED SPARROW** *Zonotrichia leucophrys* 16cm Ground feeder, both in undergrowth cover and in open. Perches on bushes. Raises crown feathers when agitated. **V** Song sad *more-wet-wetter-chee-zee*; calls include metallic *pink* or *chink*, also thin *tseep*. **H** Various bushy areas, woodland edge, thickets, stunted conifers, mountain shrubbery, parks, gardens. **D** Breeds N North America. Winters C Mexico.

26 **WHITE-THROATED SPARROW** *Zonotrichia albicollis* 16cm Some have ochre eyebrow stripe. **V** Song whistled *dee-dee diddla-diddla-diddla*; call high *tseet* and sharp *pink*. **H** Open conifer or mixed woodland, bushy clearings, scrub, gardens. **D** Breeds NW and W Canada, NE and NC USA. Winters N Mexico.

27 **GOLDEN-CROWNED SPARROW** *Zonotrichia atricapilla* 16cm Shy. Ground feeder, usually near cover. **V** Song three melancholy, flute-like notes, *I'm-so-weary* or *oh-dear-me... three-bind-mice*; call sharp, loud *chink*, also thin *seet*. **H** Mountain and tundra, alpine meadows with bushes or small trees, scrubby hillsides, waterside scrub. **D** Breeds W Alaska to SW Canada. Winters NW Mexico.

28 **VOLCANO JUNCO** *Junco vulcani* 16cm No similar pale-eyed bird in its range and habitat. **V** Very high short warbling phrase *splipseewipseeklip* repeated. **H** Open ground, low and dense bamboo. **D** Costa Rica, Panama.

29 **DARK-EYED JUNCO** *Junco hyemalis* 16cm Distinctive, but very variable. **V** Song fast musical trill, interspersed with warbles, twitters and *chips*; calls similar to Yellow-eyed Junco. **H** Clearings and edges in coniferous and mixed woodland, weedy fields, gardens. **D** Breeds N and W North America. Winters S USA to Mexico.

30 **GUADALUPE JUNCO** *Junco insularis* 15cm No similar bird on Guadalupe. **V** Song is a varied series of chipping, trilling and buzzing sounds; call a metallic *sik*. **H** Steep woodland, scrub. **D** Guadalupe, off NW Mexico.

31 **YELLOW-EYED JUNCO** *Junco phaeonotus* 16cm Forages on ground and in bushes. **V** Song three-part series of whistles and trills; call high twittering and sharp *dit*. **H** Open coniferous or pine–oak forest. **D** SW USA to Guatemala.

32 **BAIRD'S JUNCO** *Junco bairdi* 15cm No similar bird in its range. **V** Song a sweet trilling warble lasting about 2 secs; call a harsh *tsik* or *tsk*. **H** Open woodland, savannah with scattered brush. **D** S Baja California (Mexico).

33 **SAVANNAH SPARROW** *Passerculus sandwichensis* 14–16cm Variable. **V** Song series of *chips*, then buzzy trill followed by low trill; call thin *seep*. **H** Grasslands, coastal salt marsh, sand dunes. **D** Canada to Guatemala.

34 **SEASIDE SPARROW** *Ammodramus maritimus* 15cm Variable. **V** Song muffled *tup teetle-zhrrrr*; call low *tup*. **H** Coastal marshes. **D** E and S coasts of USA.

35 **NELSON'S SPARROW** *Ammodramus nelsoni* 13cm Shy, forages in dense grass and bushes. **V** Song soft, fading *pl-tesh hhhhh-ush*; call hard *tek*. **H** Grassy marshes or meadows. **D** Breeds Canada and N USA. Winters SE USA.

36 **SALTMARSH SPARROW** *Ammodramus caudacutus* 13cm Shy, forages in dense grass and bushes. **V** Song rapid sequence of gurgling notes; call hard *tek*. **H** Grassy marshes or meadows. **D** Breeds E and NE USA. Winters SE USA.

37 **LECONTE'S SPARROW** *Ammodramus leconteii* 12cm Note silvery lores and cheeks, patterned upperparts with white in mantle. **V** Song two or three short chipping notes followed by very short trill; call a high-pitched, sharp *tsip*. **H** Grassland, fields. **D** Breeds C Canada to NC USA. Winters SE USA.

1
2
3
4
5
6
7
8
9
10
11
12
13
14
15
16
17
18
19
20
21
22
23
24
25
26
27
28
29
30
31
32
33
34
35
36
37

NEW WORLD SPARROWS *CONTINUED*

1 BAIRD'S SPARROW *Ammodramus bairdii* 14cm Very shy, best observed whilst singing from bush or plant stem. **V** Song formed of *zip* notes, then warble followed by short trill; calls include high *chip* and low *tr-r-i-p*. **H** Short grass prairie with scattered bushes. **D** Breeds SC Canada and NC USA. Winters N Mexico.

2 HENSLOW'S SPARROW *Ammodramus henslowii* 13cm Forages in dense cover. **V** Song insect-like *tsillik*; call sharp *tsik*. **H** Weedy fields. **D** Breeds C and EC North America. Winters SE USA.

3 GRASSHOPPER SPARROW *Ammodramus savannarum* 13cm Secretive, forages on ground; more conspicuous when singing from grass stem or low perch. **V** Song thin, insect-like buzz that starts with two sharp *tik* notes; also utters high-pitched tinkling; call high, thin *tip* or *titip*, also rising *tswees* flight call. **H** Weedy fields with tall grass. **D** Widespread across North America through Central America to Colombia. Resident on Jamaica, Hispaniola and Puerto Rico.

4 GRASSLAND SPARROW *Ammodramus humeralis* 13cm Note yellow lores. **V** Song short series of 4–5 inhaled and vibrating, melodious notes. **H** Grassy habitats, including Cerrado, pastures and fields. **D** Widespread in South America.

5 YELLOW-BROWED SPARROW *Ammodramus aurifrons* 13cm Yellow foreface is diagnostic. **V** Song short series of 2–4 cricket-like buzzy notes. **H** Grassy habitats. **D** Colombia through Amazonia.

6 SIERRA MADRE SPARROW *Xenospiza baileyi* 12cm Note notched tail, brown wings, tawny sides; yellow wing band in flight. **V** Vocalisations include *pjiew* and simple partly inhaled, partly nasal *tjiet-weeh-weh*... **H** Grassy, open pine woodland. **D** C Mexico.

7 AMERICAN TREE SPARROW *Spizelloides arborea* 16–17cm Mainly terrestrial. Gregarious during winter. **V** Song formed of clear *seet* notes followed by rapid, variable warble; calls include *tseet* and musical *teedle-eet-teedle-eet*. **H** Open scrub on the edge of tundra, winters in weedy areas. **D** Breeds N Canada, Alaska. Winters USA.

8 CHIPPING SPARROW *Spizella passerina* 13cm Feeds on ground or in bushes. Associates with other seed-eaters during winter. **V** Song monotonous, dry trill; call *chip* or *tsip*; flight note rising *tsisi*. **H** Open conifer or pine–oak woods, woodland edge, urban bushy areas. Winters in open grassy areas, thickets and gardens. **D** Widespread in North America and Central America.

9 FIELD SPARROW *Spizella pusilla* 14–15cm Pinkish bill. Forages mainly on ground. **V** Song clear whistles accelerating into a short trill; calls include *tsip* and *tseew* flight note. **H** Overgrown grassy areas with scattered bushes. **D** Breeds SE Canada to C and E USA. Winters NE Mexico.

10 WORTHEN'S SPARROW *Spizella wortheni* 15cm Forages on ground. Often considered a race of Field Sparrow, differs in song and habitat. **V** Slow monotonous trill. **H** Mesquite–juniper grassland. **D** N Mexico.

11 BLACK-CHINNED SPARROW *Spizella atrogularis* 15cm Secretive. Forages on ground or in bushes. **V** Song sharp, slurred notes that run into rapid trill; calls include weak *tsip* and *ssip* flight note. **H** Arid scrub, bushy hillsides and chaparral. **D** SW USA, Mexico.

12 CLAY-COLORED SPARROW *Spizella pallida* 14cm Forages on ground and in bushes. In winter often mixes with Chipping Sparrows. **V** Song short series of insect-like buzzes; calls include weak *chip* and rising *swit* flight note. **H** Scrubby grassland, thickets. Winters in open areas with scattered scrub. **D** Breeds C Canada to NC USA. Winters W Mexico.

13 BREWER'S SPARROW *Spizella breweri* 14cm Forages on ground and in bushes. Gregarious in winter, regularly associating with other sparrow species. **V** Song variable series of buzzes and trills; calls include sharp *tsip* and rising *swit* flight note. **H** Sagebrush and open scrubby areas. Winters in open, semi-arid areas and grassland. **D** Breeds WC North America. Winters C Mexico.

14 VESPER SPARROW *Pooecetes gramineus* 16cm Forages on ground. In flight shows white outer-tail feathers. In worn plumage (autumn), streaking heavier. **V** Song melodious, two long slurred notes, two higher notes then a short series of descending trills; calls include loud *hisp* and buzzy *seet* flight note. **H** Semi-arid scrub, weedy fields and open grasslands. **D** Breeds W and N North America. Winters SE USA and Mexico.

15 LARK SPARROW *Chondestes grammacus* 15cm Forages on ground. In flight, shows a large amount of white on outer-tail feathers and tips of inner tail feathers. **V** Song starts with two loud, clear notes followed by series of melodious notes, trills and unmusical buzzes; call metallic *tink*, also high *tsewp*, often repeated when alarmed. **H** Semi-arid, open areas with scattered bushes. **D** Breeds W and N North America. Winters SE USA to Guatemala.

16 FIVE-STRIPED SPARROW *Amphispiza quinquestriata* 15cm Forages on ground and in bushes. **V** Song short, repeated, tinkling phrases; calls include husky *terp*, soft *tink* and high *tip*. **H** Dense bush and scrub areas on grassy, rocky hillsides. **D** NW and W Mexico.

17 BLACK-THROATED SPARROW *Amphispiza bilineata* 14cm Forages on ground and in bushes, calling constantly and regularly flicking tail. In flight, shows white outer-tail feathers. **V** Song rapid, starting with two bell-like notes followed by trill; call high-pitched, tinkling *tip*. **H** Arid areas with scattered scrub. **D** W USA to C Mexico.

18 SAGEBRUSH SPARROW *Artemisiospiza nevadensis* 13.5cm Smallish, long-tailed sparrow with rather plain grey and sandy-brown plumage; dark lores and tail distinctive. **V** Song is a series of tuneful buzzing phrases; calls include bell-like *tink* and harsher *quid-quid*. **H** Open shrubby habitats, with sagebrush. **D** Breeds WC USA. Winters SW USA, N Mexico.

19 BELL'S SPARROW *Artemisiospiza belli* 16cm Often runs with tail cocked. **V** Song jumbled series of rising and falling phrases; calls include high twittering. **H** Chaparral, sagebrush and saltbush. **D** SW USA to Baja California.

20 TUMBES SPARROW *Rhynchospiza stolzmanni* 14.5cm Distinctive in its range; note rufous shoulders. **V** Series of 4–5 *tjud* notes. **H** Arid scrub, semi-desert, riverine scrub. **D** Ecuador, Peru.

21 STRIPE-CAPPED SPARROW *Rhynchospiza strigiceps* 16.5cm Head pattern is distinctive. **V** Sharp, slightly accelerating series of high *chee* notes, often preceded or followed by a few other notes. **H** Chaco woodlands, scrubby edges in grassy habitats. **D** SC South America.

22 STRIPE-HEADED SPARROW *Peucaea ruficauda* 17cm Head pattern with white stripe through black crown distinctive. **V** High hurried scratchy *petsútsjetsjúse wróng*. **H** Dry rocky areas with patches of dry scrub. **D** W Mexico to Costa Rica.

23 BLACK-CHESTED SPARROW *Peucaea humeralis* 16cm Black breast distinctive. **V** Ultra-high *pwchit-tsiwiwiwi*..., developing into rattle. **H** Dry scrub, woodland. **D** W Mexico.

24 BRIDLED SPARROW *Peucaea mystacalis* 16cm Resembles Black-chested Sparrow, but note black bib and different patterned back. **V** Ultra-high *ts-tsiet-tswi-twet* rattle. **H** Semi-open arid scrub, thickets, overgrown grassy clearings. **D** C Mexico.

25 CINNAMON-TAILED SPARROW *Peucaea sumichrasti* 16cm Cinnamon tail distinctive. Restricted range. **V** Ultra-high sharp hurried *tsjitsji*... or *tsitsitrrrrrr*. **H** Dry scrub. **D** SW Mexico.

26 RUFOUS-WINGED SPARROW *Peucaea carpalis* 14.5cm Forages on ground and in bushes. **V** Song variable; *chip* notes accelerating into a trill; call high *seep*. **H** Desert grassland with mesquite. **D** SW USA, NW Mexico.

27 CASSIN'S SPARROW *Peucaea cassinii* 15cm Shy except when singing. Forages on ground, often under cover. **V** Song, generally given during display flight, *tsisi seeeeeeee ssootssiit*; calls include high *teep* and loud *chip*. **H** Arid grassland with scattered brush or scrub. **D** SC and SW USA, N Mexico.

28 BACHMAN'S SPARROW *Peucaea aestivalis* 15cm Forages mainly on ground. **V** Song simple clear whistle followed by musical trill; call high *tsip* notes. **H** Open pine woods with patches of brush or grass. **D** SE USA.

29 BOTTERI'S SPARROW *Peucaea botterii* 14cm Separated from Cassin's Sparrow by unmarked sides of breast and flanks, and lack of malar stripe. **V** Ultra-high *tsjuh* or *fjuweet weet weetweetwerit* (almost rattling). **H** Dry open woodland, savannah, fields. **D** SW USA to Costa Rica.

30 RUFOUS-CROWNED SPARROW *Aimophila ruficeps* 15cm Forages on ground. **V** Song jumbled series of *chip* notes; calls include nasal *chur chur chur*, chatter and high *zeeet*. **H** Brush and grassy areas on rocky slopes. **D** SC and SW USA, Mexico.

31 RUSTY SPARROW *Aimophila rufescens* 18cm Bolder face pattern than Oaxaca Sparrow. **V** Varied, includes very high rapid *tsjech-tehtehtjuw*. **H** Forest edge, dry woodland, scrubby areas, overgrown fields. **D** NW Mexico to NW Costa Rica.

32 OAXACA SPARROW *Aimophila notosticta* 16cm Separated from Rufous-crowned Sparrow by warmer upperparts, from larger Rusty Sparrow by all-black bill. **V** Ultra-high sharp *tsjéeh-wehwehwehweehweeh*. **H** Dry scrub, open woodland. **D** S Mexico.

33 ZAPATA SPARROW *Torreornis inexpectata* 16.5cm Forages on or near ground. **V** Song buzzing trill, interspersed with chattering notes; calls include buzzing *zeee* and thin *tsip* or *tsip-tsip*. **H** Sawgrass with scattered bushes in the Zapata Swamp (race *inexpectata*), dry vegetation in arid areas east of Guantanamo Bay (race *sigmani*), woods and swampy areas on Cayo Coco (race *varonai*). **D** Cuba.

34 STRIPED SPARROW *Oriturus superciliosus* 17cm Note large size, thickset appearance, black bill, dark cheek, narrow white streak through brown crown. **V** Ultra-high rattling *tit-tjtjtjtjtj*... **H** Grassland with shrubby areas, open pine woodland. **D** W and C Mexico.

35 GREEN-TAILED TOWHEE *Pipilo chlorurus* 18cm Forages on or near ground. **V** Song begins with drawn-out *weet-chur*, then whistles followed by rasping trill; call *meewe*; flight call buzzy *zeereesh*. **H** Chaparral, thickets, dense brush. **D** Breeds W USA. Winters C Mexico.

36 COLLARED TOWHEE *Pipilo ocai* 20cm Dark above with rufous crown. Not in range of Orange-billed Sparrow (Plate 281). **V** Ultra-high descending sharp *fjieeeeeh* or strong *wrat-wrat*. **H** Montane forest. **D** Mexico.

37 SPOTTED TOWHEE *Pipilo maculatus* 20cm Often lumped with Eastern Towhee, and called 'Rufous-sided Towhee'. Behaviour similar to that of Eastern Towhee. **V** Song *che che che che che zhreee* or similar; call harsh *zhreee* or *grreeer*; in flight, *zeeeeweee*. **H** Dense brush, woodland edge, gardens. **D** SW Canada to W Guatemala.

38 EASTERN TOWHEE *Pipilo erythrophthalmus* 20cm Ground feeder, usually under cover. **V** Song transcribed as *drink-your-tea-ee-ee-ee-ee*; call rising *tow-whee* or *chee-wink*; also buzzy *zeeeeweee* flight note. **H** Undergrowth, woodland edge, gardens. **D** Breeds S Canada, C and E USA. Winters N Mexico.

1
2
3
4
5
6
7
8
9
10
11
12
13
14
15
16
17
18
19
20
21
22
23
24
25
26
27
28
29
30
31
32
33
34
35
36
37
38
♀
♀
♂
♂

NEW WORLD SPARROWS *CONTINUED*

1 WHITE-THROATED TOWHEE *Melozone albicollis* 20cm Note face pattern, narrow necklace and white breast. **V** Very high warbling song, strident in places, accelerating and rising in pitch in phrases, includes *pirrup-pirrup-pirrup*. **H** Dry scrub, woodland undergrowth. **D** Mexico.

2 CANYON TOWHEE *Melozone fusca* 19–24cm Forages by scratching in leaf litter. Originally considered conspecific with California Towhee and known as 'Brown Towhee'. **V** Song simple trill, introduced by call note, a nasal *klidl*. Calls also include dry *ch-ch-ch-ch* and buzzy *zeeee* flight note. **H** Primarily brush and arid scrub. **D** W USA, Mexico.

3 CALIFORNIA TOWHEE *Melozone crissalis* 21–22cm Habits similar to those of Canyon Towhee. **V** Song accelerating series of *teek* notes; calls include metallic *teek* and a buzzy *zeeee* given in flight. **H** Brush, arid scrub. **D** W USA, Baja California.

4 ABERT'S TOWHEE *Melozone aberti* 24cm Habits like those of Canyon Towhee. **V** Song variably pitched chipping trill; calls include sharp *teek*, high *seeeep* and buzzy *zeeoeeet* flight note. **H** Dense scrub, especially near water. **D** SW USA, NW Mexico.

5 RUSTY-CROWNED GROUND SPARROW *Melozone kieneri* 16cm Does not occur in range of Prevost's Ground Sparrow. **V** Very high, simple, calm and unstructured *tjitit-tjitit-tututut*. **H** Dry woodland, thornbush, semi-open areas. **D** Mexico.

6 PREVOST'S GROUND SPARROW *Melozone biarcuata* 16cm Note face pattern. **V** Song a series of chipping notes followed by a sweet double whistle; calls include various nasal, metallic chirps and chatters. **H** Weedy and scrubby undergrowth, hedgerows, plantations. **D** SE Mexico to El Salvador.

7 CABANIS'S GROUND SPARROW *Melozone cabanisi* 15cm Complex black, white and rufous head pattern, otherwise mostly greyish appearance. **V** Song combines fast and slow trills, buzzing notes and whistles; calls include high *tsit* and sharper *pseee psee psee*. **H** Undergrowth of forest edge, tall secondary growth, plantations. **D** C Costa Rica.

8 WHITE-EARED GROUND SPARROW *Melozone leucotis* 18cm Complicated head pattern with broken yellow collar diagnostic. **V** Extremely high, thin *seeweehtititit*. **H** Dense weedy vegetation at forest edges, thickets, tree stands. **D** S Mexico to Costa Rica.

9 OLIVE SPARROW *Arremonops rufivirgatus* 15cm Appearance varies by race. Separated from Green-backed Sparrow and Black-striped Sparrow by buff undertail coverts and brown head stripes. Black head stripes absent except in race *verticalis* (Yucatán). **V** Very high, bouncing phrase, fluted at the end: *tseet tseet tjut tjewd tjeeterrrreweet*. **H** Dense scrub, dry woodland, thickets, riverine belts. **D** S Texas to Costa Rica.

10 TOCUYO SPARROW *Arremonops tocuyensis* 13cm Separable from larger Black-striped Sparrow by whiter eyebrows, and generally found in drier habitats. **V** Similar to that of Black-striped Sparrow, but quieter and more tuneful. **H** Shrubby edges of dry forest. **D** Colombia, Venezuela.

11 GREEN-BACKED SPARROW *Arremonops chloronotus* 16cm Separated from Olive Sparrow race *verticalis* by overall brighter colouring with yellow undertail coverts. **V** Song a series of chipping notes (up to eight); calls include sharp *tsik* and thin, drawn-out *swiiirr*. **H** Forest edges, thornbush, dry scrub. **D** SE Mexico to Honduras.

12 BLACK-STRIPED SPARROW *Arremonops conirostris* 16cm Note distinctly marked head with narrow black stripes. Compare to Tocuyo Sparrow. **V** Calm series of musical, mellow *tue* and *wuh* notes. **H** Thickets, weedy areas, gardens, woodland edges. **D** Honduras to N Brazil.

13 PECTORAL SPARROW *Arremon taciturnus* 15cm Range differs from other yellow-billed sparrows. Bill is black except in race *axillaris*. **V** Slow, barely audible *see-see-see*. **H** Undergrowth of humid forest and secondary growth, usually lowlands but higher on tepuis. **D** NW South America, Amazonia.

14 HALF-COLLARED SPARROW *Arremon semitorquatus* 15cm Compare with Pectoral Sparrow (with which it has no or little overlap in range). **V** Extremely high, slightly descending series of four *see* notes. **H** Lower storeys of montane forest, secondary growth, gardens, neglected fields. **D** SE Brazil.

15 SAO FRANCISCO SPARROW *Arremon franciscanus* 15cm Has no overlap in range with Pectoral Sparrow or with Half-collared Sparrow. **V** Extremely high *see-see* rattle. **H** Caatinga scrub. **D** EC Brazil.

16 SAFFRON-BILLED SPARROW *Arremon flavirostris* 15.5cm Compare with Pectoral Sparrow; also similar to Half-collared Sparrow, but that species shows distinctly broken breast-band. **V** Slightly decelerating series of 4–8 very high *see* notes. **H** Undergrowth and edges of woodland and forest. **D** SE and SC South America.

17 ORANGE-BILLED SPARROW *Arremon aurantiirostris* 15cm Note complete breast-band and orangey-pink bill. **V** Series of extremely high *see-sisi-see* notes, and variations on this. **H** Undergrowth and edges of dense forest, secondary growth, thickets. **D** SE Mexico to NE Peru.

18 GOLDEN-WINGED SPARROW *Arremon schlegeli* 16cm Lack of white eyebrows is diagnostic. **V** Extremely high *see* notes, just beyond many people's hearing range. **H** Open woodland, secondary growth, clearings. **D** Colombia, Venezuela.

19 BLACK-CAPPED SPARROW *Arremon abeillei* 15cm Note largish head with crest often raised, lack of green (or only a very faint wash), narrow wingbars, and also distribution. Has some overlap with Orange-billed Sparrow, but separated by bill colour. **V** Very high, short series of *tee-tjuw* notes and short rattles. **H** Undergrowth of dry forest and woodland. **D** Ecuador, Peru.

20 CHESTNUT-CAPPED BRUSHFINCH *Arremon brunneinucha* 18cm Note distinctive chestnut cap. Race *frontalis* shows a black breast-band; this is lacking in race *allinornatus*. **V** Extremely high, barely audible *tee-tee-tititi*. **H** Undergrowth of dense forest and woodland. **D** Mexico to Ecuador.

21 GREEN-STRIPED BRUSHFINCH *Arremon virenticeps* 19cm Separated from Grey-browed Brushfinch by buff-green, not grey, eye-stripe and by different range. **V** Song is a long sequence of thin *ssi* notes given in uneven rhythm; call a single *ssi*. **H** Weedy and brushy forest edges, cultivations. **D** W and C Mexico.

22 BLACK-HEADED BRUSHFINCH *Arremon atricapillus* 19cm Note entirely black head. **V** Extremely high sizzling sound, just within or just beyond average human hearing. **H** Undergrowth and interior of dense forest and secondary growth. **D** E Panama, Colombian Andes.

23 COSTA RICAN BRUSHFINCH *Arremon costaricensis* 18cm Feeds on ground. Dark olive upperparts and grey underparts, white chin and black head with narrow pale grey eyebrow. **V** Song is a simple series of thin, high-pitched notes; calls include various sharp metallic notes. **H** Undergrowth of dense moist forest, sometimes forest edges, bamboo thickets. **D** Costa Rica, W Panama.

24 WHITE-BROWED BRUSHFINCH *Arremon torquatus* 19cm Broad white eyebrow and range are diagnostic. Some races have a distinct black breast-band. **V** Well-separated series of very high to extremely high notes, e.g. *psee*. **H** Undergrowth and borders of moist forest and secondary growth. **D** S Peru, Bolivia, Argentina.

25 SIERRA NEVADA BRUSHFINCH *Arremon basilicus* 19cm Very similar to Perija, Caracas and Paria Brushfinches; check range and length and colour of eyebrow. **V** Unstructured series of notes with an extremely high *tsee-tjef* phrase. **H** Undergrowth and edges of humid forest. **D** Sierra Nevada de Santa Marta, N Colombia.

26 PERIJA BRUSHFINCH *Arremon perijanus* 19cm Compare with Sierra Nevada Brushfinch. **V** Extremely high series of notes, including *tee* and *tjef* notes. **H** Undergrowth and edges of humid forest. **D** Serranía del Perijá, N Colombia and NW Venezuela.

27 GREY-BROWED BRUSHFINCH *Arremon assimilis* 19cm Eyebrow grey, and there is no breast-band. **V** Unstructured series of well-separated *tjee*, *see* and *fjut* notes. **H** Humid forest and older secondary growth. **D** Andes of Venezuela, Colombia, Ecuador, much of Peru.

28 CARACAS BRUSHFINCH *Arremon phaeopleurus* 19cm Compare with Sierra Nevada Brushfinch. **V** Song extremely high-pitched phrase of 3–5 notes, descending in pitch at end. **H** Undergrowth and edges of humid forest. **D** Cordillera de la Costa, N Venezuela.

29 PARIA BRUSHFINCH *Arremon phygas* 19cm Compare with Sierra Nevada Brushfinch. **V** Song is a long rambling series of high-pitched sibilant notes. **H** Undergrowth and edges of humid forest. **D** Cordillera de la Costa Oriental, NE Venezuela.

30 SOOTY-FACED FINCH *Arremon crassirostris* 16cm Separable from Olive Finch by more distinctive malar stripes. **V** Song a repeated double note, high and thin but emphatic; typical call a sharp whistling *tu-deee*. **H** Dense undergrowth of montane forest. **D** Costa Rica, Panama.

31 OLIVE FINCH *Arremon castaneiceps* 16cm Rather dark overall with grey head and rufous crown stripe. Compare with Sooty-faced Finch. **V** Extremely high *seet-sitsit*, given in slow series. **H** Dense undergrowth of humid forest, especially near streams. **D** Colombia to Peru.

32 LARGE-FOOTED FINCH *Pezopetes capitalis* 20cm Large size distinctive; very dark with large head. **V** Extremely high fluted *fjueet fijeh* (1/4 sec), each note very short. **H** Montane areas with forest, secondary growth, overgrown clearings, bamboo. **D** Costa Rica, Panama.

33 RUFOUS-CAPPED BRUSHFINCH *Atlapetes pileatus* 16cm Unmistakable due to rich yellow underparts and rufous crown. **V** Song begins with two thin high notes, followed by rapid short trill of chipping sounds; call *chik* or *chi-chi-chi*. **H** Weedy and brushy forest edges, plantations, cultivations. **D** Mexico.

34 MOUSTACHED BRUSHFINCH *Atlapetes albofrenatus* 18cm Separable from Yellow-breasted Brushfinch (Plate 282) by striking white whiskers. **V** Loose series of inhaled *see* notes. **H** Undergrowth and borders of humid forest, clearings, roadsides. **D** N Colombia.

35 MERIDA BRUSHFINCH *Atlapetes meridae* 17.5cm Coloration similar to that of Moustached Brushfinch (from which it was recently split), but more subdued, lacks black moustachial stripe. **V** Similar to that of Moustached Brushfinch; song begins with musical notes and finishes with more emphatic series of clattering notes. **H** Shrubby undergrowth and mid-storey of humid forest edges and clearings. **D** NW Venezuela.

36 OCHRE-BREASTED BRUSHFINCH *Atlapetes semirufus* 17cm The only rufous-headed finch in its range. **V** Forceful series of short phrases, with rattles, level chatters and *see-see-see* notes. **H** Shrubby undergrowth and borders of forest and secondary growth. **D** Venezuela, Colombia.

37 TEPUI BRUSHFINCH *Atlapetes personatus* 17cm The only brushfinch to live on tepuis. **V** Forceful series of strong rattles and loud, inhaled *fjeu* notes. **H** Shrubby borders, undergrowth and clearings in forest on tepuis. **D** Tepuis in Venezuela and Brazil.

1
2
3
4
5
6
7
8
9
10
11
12
13
14
15
16
17
18
19
20
21
22
23
24
25
26
27
28
29
30
31
32
33
34
35
36
37

NEW WORLD SPARROWS *CONTINUED*

1 WHITE-NAPED BRUSHFINCH *Atlapetes albinucha* 19cm Those in Mexico have yellow belly and breast; those further south have grey underparts with yellow restricted to throat. Note white crown stripe through black hood. **V** Very high sustained *tsjeet-tjutsjeet-tjeet* (*tju* phrase extremely high). **H** Forest, secondary growth. **D** Mexico to Colombia.

2 SANTA MARTA BRUSHFINCH *Atlapetes melanocephalus* 17cm Note silvery cheek patch. No similar finch occurs in its range. **V** Calm series of notes with single chirps. **H** Shrubby borders and undergrowth of forest, secondary growth, overgrown pastures. **D** N Colombia.

3 PALE-NAPED BRUSHFINCH *Atlapetes pallidinucha* 18cm Separable from White-naped Brushfinch by yellow-orange crown stripe and narrow malar stripes. **V** Repetitive, melodious, rolling warbling. **H** Borders and undergrowth of humid forest near the treeline. **D** Venezuela to Peru.

4 YELLOW-HEADED BRUSHFINCH *Atlapetes flaviceps* 17cm Yellowish plumage is distinctive. **V** Rapid, descending series of notes that gradually change in intonation. **H** Shrubby borders and clearings in moist forest. **D** Colombia.

5 DUSKY-HEADED BRUSHFINCH *Atlapetes fuscoolivaceus* 17cm No similar finch occurs in its range. Dark overall with narrow malar stripes. **V** Short, very high phrase with *see* and *teejuw* notes (one note given every 2 secs). **H** Borders and clearings in montane forest and secondary growth. **D** Colombia.

6 CHOCO BRUSHFINCH *Atlapetes crassus* 17cm Similar to Tricolored Brushfinch (and formerly considered conspecific) but generally darker. **V** Like that of Tricolored Brushfinch, but song phrases have longer spaces in between. **H** As for Tricolored Brushfinch. **D** Colombia, Ecuador.

7 TRICOLORED BRUSHFINCH *Atlapetes tricolor* 17cm Note yellow or brown crown stripe. **V** Song in two or three parts, beginning with two downslurred notes then proceeding to faster high-pitched phrases; calls include *zit-zit* and higher-pitched short *ti*. **H** Borders, undergrowth and shrubby clearings in humid forest. **D** Peru.

8 WHITE-RIMMED BRUSHFINCH *Atlapetes leucopis* 18cm Facial pattern and dark plumage are distinctive. **V** Short phrase with 5–10 sharp notes of different pitch. **H** Dense undergrowth of wet forest. **D** Colombia, Ecuador.

9 YELLOW-BREASTED BRUSHFINCH *Atlapetes latinuchus* 17cm Compare Moustached Brushfinch (Plate 281). Note rufous crown with yellow breast. **V** Song begins with 1–3 downslurred notes followed by a short trill; calls include *tink* and *tweet*. **H** Forest borders, secondary growth, roadsides. **D** Colombia to Peru.

10 BLACK-FRONTED BRUSHFINCH *Atlapetes nigrifrons* 17cm Formerly considered conspecific with Yellow-breasted Brushfinch; similar to that species but has more extensive black on face. **V** Song a long trilled phrase (up to 22 secs); also gives a short, simple five-note phrase, first note separated and downslurred. Calls include buzzes and simple *chip* notes. **H** Bushy undergrowth at forest edges, overgrown pastures. **D** Perijá mountains in extreme N Colombia, NW Venezuela.

11 ANTIOQUIA BRUSHFINCH *Atlapetes blancae* 17cm Note whitish underparts and wing patch, and rufous crown. **V** Not known. **H** Forest edges. **D** NC Colombia.

12 RUFOUS-EARED BRUSHFINCH *Atlapetes rufigenis* 18cm Facial pattern, lacking black around eyes, is distinctive. **V** Song comprises short, high phrases, such as *tee-tee-tjudud*. **H** Dense undergrowth of *Polylepis* and other woodland patches in highlands. **D** Peru.

13 APURIMAC BRUSHFINCH *Atlapetes forbesi* 18cm Black on face is only in front of eyes. **V** Song comprises short phrases, such as *tee*, *tjuwtuw*, *teetee* and *teeteetjuwtuw*. **H** *Polylepis* forest and scrub. **D** Peru.

14 BLACK-SPECTACLED BRUSHFINCH *Atlapetes melanopsis* 18cm Compare Rufous-eared and Apurimac Brushfinches. **V** Has varied vocalisations, including a short, rapid warble. **H** Montane scrub, forest edges, dry bushy areas. **D** Peru.

15 SLATY BRUSHFINCH *Atlapetes schistaceus* 18cm White wing patch present in nominate race. Note dark grey underparts. **V** Varied songs, such as *tjuwtuwtuw*. **H** Undergrowth and borders of highland humid forest and woodland, páramo, elfin forest and *Polylepis* woodland. **D** Venezuela to Peru.

16 WHITE-WINGED BRUSHFINCH *Atlapetes leucopterus* 16cm Large white wing patch is distinctive in all races. **V** Song comprises a couple of slow introductory notes followed by one or two trilled phrases; calls include *tink*, *tsew* and *seet*. **H** Dry to humid shrub and woodland, gardens, villages. **D** Ecuador, Peru.

17 WHITE-HEADED BRUSHFINCH *Atlapetes albiceps* 16cm Compare Pale-headed Brushfinch. Note lack of rufous on head. **V** Rapid, short nasal rattles. **H** Undergrowth of dry woodland, often along streams. **D** Ecuador, Peru.

18 PALE-HEADED BRUSHFINCH *Atlapetes pallidiceps* 16cm Separable from White-headed Brushfinch by different head pattern, which lacks black. **V** Short, rapid, slightly descending warbling. **H** Low, arid, scrubby woodland. **D** Ecuador.

19 BAY-CROWNED BRUSHFINCH *Atlapetes seebohmi* 16.5cm Note lack of white wing patch, and distinct malar stripes starting at chin. **V** Slow series of very short twitters and rolls. **H** Montane woodland and scrub. **D** Ecuador, Peru.

20 RUSTY-BELLIED BRUSHFINCH *Atlapetes nationi* 17cm No similar brushfinch with buff underparts occurs in its range. **V** Calls include *tchip* and *ti*. **H** Woodland patches. **D** Peru.

21 CUZCO BRUSHFINCH *Atlapetes canigenis* 18cm Very dark plumage with dark brown crown is distinctive. **V** Song a *wee-tju* phrase. **H** Humid montane forest and scrub. **D** Peru.

22 VILCABAMBA BRUSHFINCH *Atlapetes terborghi* 16.5cm Yellow throat with very short malar stripes is distinctive. **V** Not known. **H** Elfin forest, cloud forest. **D** Peru.

23 GREY-EARED BRUSHFINCH *Atlapetes melanolaemus* 16.5cm Compare with Bolivian Brushfinch. Rather dark overall with blackish throat. **V** Song comprises short phrases such as *tuweet-tju*. **H** Forest borders, secondary growth. **D** Peru.

24 BOLIVIAN BRUSHFINCH *Atlapetes rufinucha* 17cm Resembles Grey-eared Brushfinch but has yellow throat. **V** Song comprises thin, short phrases, including high, sharp *wee-tj-wee*. **H** Edges of montane forest, overgrown pastures, weedy places. **D** Bolivia.

25 FULVOUS-HEADED BRUSHFINCH *Atlapetes fulviceps* 17cm No other brushfinch with a fulvous head occurs in its range. **V** Song comprises hurried, very short phrases, some ending in a mini-rattle. **H** Undergrowth and edges of highland forest and woodland, often near streams, moves to lower elevations in winter. **D** Bolivia, NW Argentina.

26 YELLOW-STRIPED BRUSHFINCH *Atlapetes citrinellus* 17cm Head pattern diagnostic. **V** Song comprises short, hurried phrases including *tj-chachacha*. **H** Undergrowth and borders of woodland, forest and secondary growth. **D** NW Argentina.

27 YELLOW-THIGHED FINCH *Pselliophorus tibialis* 18cm Unmistakable by yellow thighs. **V** Very high staccato phrases of jangling sibilant notes and high trills. **H** Undergrowth at forest edge; >1,500m. **D** Costa Rica, Panama.

28 YELLOW-GREEN FINCH *Pselliophorus luteoviridis* 18cm Separated from Yellow-thighed Finch by body colouring. **V** Song is a rapid series of jumbled notes including long upslurred buzzes; call is a dry *peenk*. **H** Forest undergrowth, wet brush. **D** Panama.

29 TANAGER FINCH *Oreothraupis arremonops* 20.5cm Unmistakable due to facial pattern and rufous-brown body plumage. **V** Rapid, very high rattle, starting with barely audible *see* notes. **H** Undergrowth and borders, mainly in cloud forest. **D** Colombia, Ecuador.

30 COMMON BUSH TANAGER *Chlorospingus flavopectus* 14cm A complex species with more than 20 races recognised. Present species comprises numerous former species; at least five from Mexico alone. White spot behind eye diagnostic. **V** Extremely high hurried descending irregular *titi-tititru*. **H** Forest interior and edges. **D** SW Mexico to NW Argentina.

31 TACARCUNA BUSH TANAGER *Chlorospingus tacarcunae* 14cm The only bush tanager in its range. **V** Only known call is a high-pitched, emphatic *tseeu*. **H** Wet elfin forest and other forest habitats. **D** Panama, Colombia.

32 PIRRE BUSH TANAGER *Chlorospingus inornatus* 15cm Striking white eye in black mask, yellow underparts. **V** Calls include *tsip* and *chuweet*. **H** Montane forest. **D** Panama.

33 DUSKY BUSH TANAGER *Chlorospingus semifuscus* 14.5cm Note lack of yellow in plumage. **V** Calm series of raspy chirps. **H** Wet montane forest. **D** Colombia, Ecuador.

34 SOOTY-CAPPED BUSH TANAGER *Chlorospingus pileatus* 14cm Black-capped and brown-capped morphs occur. Larger than Bananaquit (Plate 301), with stubby bill, dark upperparts. **V** High-pitched *tjuk*; extremely high muttering. **H** Montane forest. **D** Costa Rica, W Panama.

35 YELLOW-WHISKERED BUSH TANAGER *Chlorospingus parvirostris* 14.5cm Note clear yellow whiskers. **V** Calls include *tsip*, *tsreet* and *seep*. **H** Wet montane forest. **D** Colombia to Bolivia.

36 YELLOW-THROATED BUSH TANAGER *Chlorospingus flavigularis* 15cm Yellow extends to chin, throat and cheeks. **V** Long, calm series of well-separated *weet-tjuw* phrases. **H** Borders and clearings in forest and woodland. **D** Panama to Peru.

37 ASHY-THROATED BUSH TANAGER *Chlorospingus canigularis* 14cm Note dark eyes, yellow-green frontal collar and unmarked throat. **V** Unstructured, very high to extremely high, sizzling notes and rattles. **H** Canopy and borders of tall forest and secondary growth. **D** Costa Rica to Peru.

1
2
3
4
5
6
7
8
9
10
11
12
13
14
15
16
17
18
19
20
21
22
23
24
25
26
27
28
29
30
31
32
33
34
35
36
37
black-capped morph
brown-capped morph

CHAT-TANAGERS CALYPTOPHILIDAE

1 WESTERN CHAT-TANAGER *Calyptophilus tertius* 20cm Habits similar to those of Eastern Chat-Tanager. **V** Similar to that of Eastern Chat-Tanager. **H** Dense undergrowth along watercourses in moist mountain broadleaf forests; the race on Gonâve Island occurs in semi-arid scrub. **D** Hispaniola.

2 EASTERN CHAT-TANAGER *Calyptophilus frugivorus* 17cm Mainly terrestrial and generally very secretive; forages among leaf litter in dense undergrowth. Until recently, combined with Western Chat-Tanager and called Chat Tanager; there still seem to be mixed opinions concerning this split. **V** Sharp *chick*. Song loud and clear whistling. **H** Thick undergrowth along streams in moist mountain broadleaf forest. **D** Hispaniola.

HISPANIOLAN TANAGERS PHAENICOPHILIDAE

3 BLACK-CROWNED TANAGER *Phaenicophilus palmarum* 18cm Slow, deliberate forager in thick cover, often part of mixed-species feeding flocks. Feeds on insects, seeds and fruit. **V** Low *chep* and nasal *pe-u*. **H** Any habitats with trees, forests, woodlands and thickets, also urban areas. **D** Hispaniola.

4 GREY-CROWNED TANAGER *Phaenicophilus poliocephalus* 18cm Habits similar to those of Black-crowned Tanager. **V** Short *peu*. Canary-like whispering song given during breeding season. **H** Forest and woodland, from sea level to mountains. **D** Hispaniola's southern peninsula.

5 GREEN-TAILED WARBLER *Microligea palustris* 12–14.5cm Forages, often in small groups, in dense undergrowth and thickets. **V** Song transcribed as *sip-sip-sip*. Call regularly repeated short rasping note. **H** Thickets and dense undergrowth in montane forest; in the Dominican Republic also found in semi-arid areas. **D** Hispaniola.

6 WHITE-WINGED WARBLER *Xenoligea montana* 13.5–14.5cm Classed as vulnerable. Actively forages for seeds and insects, from undergrowth to understorey. After breeding season becomes regular member of mixed-species feeding flocks. **V** Song short series of squeaky notes, delivered slowly and sometimes accelerating at the end; call thin *tseep*, also low chattering *suit suit suit chir suit suit suit chir chi...* **H** Humid broadleaf forest with dense understorey, scrub, thickets and wet shrubs in mountain areas. **D** Hispaniola.

PUERTO RICAN TANAGER NESOSPINGIDAE

7 PUERTO RICAN TANAGER *Nesospingus speculiferus* 18–20cm Regularly occurs in noisy flocks, often with other species, foraging in forest canopy. **V** Harsh, loud *chewp*, *chuck* or *chi-chi-chit*. During breeding season, utters soft warble. **H** Mountain forests. **D** Puerto Rico.

SPINDALISES SPINDALIDAE

8 WESTERN SPINDALIS *Spindalis zena* 15cm This and the following three species were formerly combined as Stripe-headed Tanager. Feeds on plants and fruit. **V** Song series of thin, high notes leading to buzzier phrases; calls include descending, high-pitched *see see see see...*, strong *seee* and sharp *tit*. **H** Wooded areas from coast to mountains. **D** Bahamas, Cuba, Cayman Islands, Cozumel Island.

9 HISPANIOLAN SPINDALIS *Spindalis dominicensis* 17cm Feeds on various fruits and plant parts. **V** Song thin, high-pitched whistle; call high-pitched *thseep*. **H** Pine, hardwood, mixed and mangrove forests. **D** Hispaniola.

10 PUERTO RICAN SPINDALIS *Spindalis portoricensis* 17cm Feeds on fruit and buds. **V** Song thin, high-pitched whistle *zée-tit-zée-tittit-zée*; call soft *teweep*, also a thin trill and short twitter. **H** Forests, woodlands. **D** Puerto Rico.

11 JAMAICAN SPINDALIS *Spindalis nigricephala* 18cm Feeds on fruit and plant parts. **V** Soft *seep* and a rapid *chi-chi-chi...* **H** Forests, woods, bushy areas. **D** Jamaica.

WRENTHRUSH ZELEDONIIDAE

12 WRENTHRUSH *Zeledonia coronata* 12cm Plump, long-legged, short-tailed, very dark. Note orange crown patch. **V** Very high drawn-up *pseeeér*; extremely high very fast *wseepuweeh*. **H** Dense undergrowth of montane forest. **D** Costa Rica, Panama.

CUBAN WARBLERS TERETISTRIDAE

13 YELLOW-HEADED WARBLER *Teretistris fernandinae* 13cm Usually forages in small noisy flocks, probing and gleaning for prey such as insects, spiders, caterpillars; also takes small fruits. Regularly forms part of mixed-species feeding flocks, often with migrant warblers and others. **V** Shrill, chattering *tsi-tsi-tsi...* All calls are similar to those of Oriente Warbler. **H** All forests with good understorey, scrubby thickets and occasionally open forests. **D** W and C Cuba.

14 ORIENTE WARBLER *Teretistris fornsi* 13cm Regularly forages in noisy flocks, probing bark crevices and gleaning vegetation for insects; often joins mixed-species flocks. **V** Sharp *tchip* and shrill *tsi-tsi-tsi...* All calls similar to those of Yellow-headed Warbler. **H** Forest scrub and swamp borders. **D** E Cuba.

YELLOW-BREASTED CHAT ICTERIIDAE

15 YELLOW-BREASTED CHAT *Icteria virens* 19cm Shy and retiring, forages in low dense cover, searching for invertebrates or fruit. **V** Song consists of loud jumble of rattles, cackles, squeals and whistles. Calls include harsh grating *chack*, nasal *cheewb* and soft *tuk*. **H** Dense thickets, scrub and woodland edge. **D** Breeds widely over North America and Central America. Winters Panama.

OROPENDOLAS, ORIOLES AND BLACKBIRDS ICTERIDAE

16 YELLOW-HEADED BLACKBIRD *Xanthocephalus xanthocephalus* 24cm In flight, male shows distinct white patches on primary coverts of upperwing, female only shows a pale crescent. Forages mainly on ground. **V** Song consists of low, hoarse rasping notes, ending with a buzz; call croaking *kruck*, *kack* or *ktuk*. **H** Marshes and reedbeds, farmland, pasture. **D** Breeds SW and SC Canada to SW USA and N Mexico. Winters S Mexico.

17 BOBOLINK *Dolichonyx oryzivorus* 18cm Forages on or near ground, among grasses and weeds, generally in flocks. Non-breeding male similar to breeding female although generally more yellowish buff. **V** Song bubbling, jangling warble; call soft, low *chuk*; in flight utters musical *pink*, *bink* or *bwink*. **H** Arable fields, open grassland, weedy areas, rich fields. **D** Breeds S Canada and N USA. Winters South America.

18 WESTERN MEADOWLARK *Sturnella neglecta* 23cm Very similar to Eastern Meadowlark, best identified by voice. **V** Variable series of bubbling, gurgling and flute-like notes that accelerates at the end; calls include bell-like *pluk* and dull rattle. **H** Grasslands, usually prefers drier areas than Eastern Meadowlark. **D** SW and SC Canada through west side of USA to C Mexico.

19 EASTERN MEADOWLARK *Sturnella magna* 23cm Forages on ground; typically rests and sings from posts and overhead wires. In flight, shows much white in outer-tail feathers. Non-breeding birds have less distinct breast-band. **V** Song consists of simple slurred whistles *seeeeooaaa seeeeadoo* or similar; call loud, harsh *dziit*. **H** Open grasslands, savannahs, marshes, pastures with scattered bushes or trees. **D** SE Canada to NE Brazil.

20 RED-BREASTED BLACKBIRD *Leistes militaris* 18.5cm Male very distinctively patterned. Female strongly striped, with characteristic barring of tail. **V** Very high *tzit* or low *nnnêh*, given singly or in combination. **H** Open natural and cultivated areas, often near water. **D** Costa Rica through C South America.

21 WHITE-BROWED BLACKBIRD *Leistes superciliaris* 18cm Male separable from male Red-breasted Blackbird by white eyebrow and stronger bill. Female not reliably separable from female Red-breasted Blackbird. **V** Short, fast combinations of high *tseeé* and low *zzêh* notes. **H** Wet and flooded grassland, fields, pastures. **D** N Bolivia and SW Brazil to C Argentina and Uruguay, E Brazil.

22 PERUVIAN MEADOWLARK *Leistes bellicosus* 20.5cm Separable from Pampas Meadowlark and Long-tailed Meadowlark by range. **V** Rapid series of short warbles, and high, fluted and wheezy notes. **H** Shrubs, meadows, fields. **D** Ecuador to N Chile.

23 LONG-TAILED MEADOWLARK *Leistes loyca* 24cm Compare with Pampas Meadowlark. **V** Series of high, sharp and warbled notes, followed by longer, wheezier note. **H** Grassland, pastures. **D** Southern Cone in South America.

24 PAMPAS MEADOWLARK *Leistes defilippii* 21cm Very similar to Long-tailed Meadowlark but darker on upperparts and with shorter tail and bill. **V** Song, given in flight, comprises loud musical whistles, ending with softer buzzy notes; call *peet*. **H** Natural and agricultural grassland. **D** EC Argentina.

25 YELLOW-BILLED CACIQUE *Amblycercus holosericeus* 20–21cm Compare with Ecuadorian Cacique (Plate 284). **V** Repertoire of repeated fluted notes. **H** Dense undergrowth, often in bamboo stands. **D** Mexico to N Bolivia.

26 MEXICAN CACIQUE *Cassiculus melanicterus* 30cm Unmistakable. Separated from Yellow-rumped Cacique (Plate 284) by thinner bill and black eye. **V** Song has a squeaky tone and includes hollow rattles, *krrow-krrow-kshara-chee*; calls very varied, including upslurred *whik*. **H** Forest edges, open woodland, gardens. **D** W Mexico to SE Guatemala.

27 CHESTNUT-HEADED OROPENDOLA *Psarocolius wagleri* male 35cm, female 27.5cm Note strong frontal shield of bill. Compare tail pattern of Crested Oropendola and Band-tailed Oropendola (Plate 184). **V** Nasal and mewing mechanical sounds, with occasional hurried, gurgling phrase. **H** Forest, plantations, parks, suburbs. **D** Mexico to Ecuador.

28 RUSSET-BACKED OROPENDOLA *Psarocolius angustifrons* male 46cm, female 36cm Races vary in bill colour and forehead colour. Note uniformly coloured bill. **V** Series of 2–3 rising, staccato yet gurgling notes. **H** Várzea and other types of wet forest. **D** W and N South America.

29 DUSKY-GREEN OROPENDOLA *Psarocolius atrovirens* male 42cm, female 33cm Note restricted yellow in tail. Yellow-fronted variant separable from similar form of Russet-backed Oropendola by dark green body plumage and dark (not pale) chestnut vent. **V** Rising series of 4–6 gurgling notes, nearly always accompanied by dry rattles and other mechanical sounds. **H** Montane forest, at settlements or near orchards. **D** Peru, Bolivia.

30 CRESTED OROPENDOLA *Psarocolius decumanus* male 46cm, female 37cm Black with dark chestnut rump and vent. Note tail pattern; compare with Chestnut-headed Oropendola, and Band-tailed Oropendola (Plate 284). **V** Amazing and rich repertoire of scratches, xylophone-like rattles and 'running', gurgling phrases. **H** Forest, secondary growth, riverine belts, tall trees in savannah, plantations. **D** Costa Rica, Panama, N and SC South America.

31 GREEN OROPENDOLA *Psarocolius viridis* male 47cm, female 40cm Unmistakable due to bill colour pattern and blue eyes. **V** Gurgling, meandering rattle accompanied by mechanical sounds. **H** Tall várzea forest, riverine belts. **D** Amazonia.

32 OLIVE OROPENDOLA *Psarocolius bifasciatus* male 50cm, female 44cm Black-headed and green-headed races exist. Unmistakable due to body and bill colour pattern, and by pink cheek patch; compare with Baudo Oropendola (Plate 284) which has a different range. **V** Fast-rising gurgling with accompanying mechanical sounds. **H** Terra firme forest. **D** E and S Amazonia.

1
2
3
4
5
6
7
8
9
10
11
12
13
14
15
16
17
18
19
20
21
22
23
24
25
26
27
28
29
30
31
32

OROPENDOLAS, ORIOLES AND BLACKBIRDS *CONTINUED*

1 MONTEZUMA OROPENDOLA *Psarocolius montezuma* 50cm Separated from Black Oropendola by mainly rufous body. Ranges do not overlap. **V** Low hurried bubbling *orreurreorre* together with nasal rhythmic hissing ending in *whéw* flutes. **H** Forest, tall trees. **D** S Mexico to Panama.

2 BLACK OROPENDOLA *Psarocolius guatimozinus* male 46cm, female 40cm Note bare blue cheeks and restricted chestnut in plumage. **V** Rapidly rising gurgling with accompanying mechanical sounds. **H** Forest, plantations. **D** Panama, Colombia.

3 BAUDO OROPENDOLA *Psarocolius cassini* male 46cm, female 40cm Separable from Olive Oropendola (Plate 283) by range. **V** Rapidly rising gurgling, ending in a 'full stop'. **H** Lowland forest. **D** NW Colombia.

4 SOLITARY CACIQUE *Cacicus solitarius* male 27cm, female 23cm Separable from all other dark-eyed black birds in range by sharp, concave-pointed white bill. **V** Rich repertoire of nasal shrieks. **H** Dense undergrowth in várzea and riverine belts. **D** C Amazonia to N Argentina.

5 GOLDEN-WINGED CACIQUE *Cacicus chrysopterus* male 20cm, female 18cm Compare with Northern Mountain Cacique. **V** Gurgled phrases in combination with loud shrieks. **H** Borders of humid montane forest, riverine belts. **D** C Bolivia to NC Argentina, S Brazil, E Paraguay, NE Argentina and Uruguay.

6 SELVA CACIQUE *Cacicus koepckeae* 23cm Compare with Southern Mountain Cacique. **V** Loud *weétweét-yoyo* call. **H** Riverine belts at small rivers and creeks. **D** Peru.

7 ECUADORIAN CACIQUE *Cacicus sclateri* male 23cm, female 20cm Separable from Yellow-billed Cacique (Plate 283) by pale bluish-grey (not yellow) bill and habitat (not in mountains). **V** Forceful *peét-yo*, sung as single or double phrases, or in rapid series. **H** Forests, often at water, mostly in lowlands. **D** Ecuador, Peru.

8 YELLOW-RUMPED CACIQUE *Cacicus cela* male 28cm, female 24cm Note yellow in rear body parts and tail. **V** Mixture of nasal chattering, hoarse shrieks and dry rattles. **H** Forest edges, riverine belts, large trees in savannah, pastures and clearings. **D** Panama to SE South America.

9 SCARLET-RUMPED CACIQUE *Cacicus microrhynchus* male 23cm, female 20cm Note restricted red on rump. Normally occurs at higher elevations than Red-rumped Cacique. **V** Aside from hoarse shrieks, also utters a rapid, yodelling series of notes. **H** Canopy and borders of humid forest. **D** Honduras to Ecuador.

10 SUBTROPICAL CACIQUE *Cacicus uropygialis* male 29cm, female 25cm Formerly treated as conspecific with Scarlet-rumped Cacique; larger, longer-tailed and shorter-winged than that species. **V** Song a series of whistles; calls very varied, including sharp *keap* and downslurred *gerrrrr*. **H** Humid montane forest, edges and clearings. **D** Panama to Peru.

11 NORTHERN MOUNTAIN CACIQUE *Cacicus leucoramphus* male 28cm, female 25cm Similar to Southern Mountain Cacique but has golden-yellow shoulders as well as rump. **V** Highly varied, including crow-like and hawk-like calls as well as squeals and rattles. **H** Humid montane forest. **D** Venezuela to Ecuador, N Peru.

12 SOUTHERN MOUNTAIN CACIQUE *Cacicus chrysonotus* male 30cm, female 25cm Separable from smaller Selva Cacique by habitat and range. **V** Nasal and falsetto shrieks, as single notes or in short phrases. **H** Humid montane forest. **D** Peru, N Bolivia.

13 BAND-TAILED OROPENDOLA *Cacicus latirostris* male 33cm, female 25cm Note tail pattern; compare with Chestnut-headed Oropendola and Crested Oropendola (both Plate 283). **V** Mixture of short gurgling and crow-like *chá* notes. **H** Mainly in várzea forest. **D** W Amazonia.

14 CASQUED OROPENDOLA *Cacicus oseryi* male 37cm, female 29cm Casqued bill and body colours are distinctive. **V** Descending, rolling shrieks, accompanied by gurgling. **H** Várzea and terra firme forest. **D** W Amazonia.

15 RED-RUMPED CACIQUE *Cacicus haemorrhous* male 28cm, female 23cm Note extensive red on rump and long wings. **V** Mixture of hoarse and nasal shrieks. **H** Interior, edges and clearings in forest and riverine belts. **D** Colombia to E Brazil and N Argentina.

16 SCOTT'S ORIOLE *Icterus parisorum* 23cm The black face on female is variable, can be more or less extensive than shown. **V** Song rich, mellow, fluty warble; calls include nasal *cheh-cheh*... or *chuhk* and quiet *huit*. **H** Dry hillsides, with yucca. **D** SC USA, Mexico.

17 YELLOW-BACKED ORIOLE *Icterus chrysater* 21cm Note yellow back and all-black wings and tail. **V** Voice like that of Yellow-tailed Oriole. **H** Borders and clearings in woodland and forest. **D** S Mexico to Venezuela and Ecuador.

18 AUDUBON'S ORIOLE *Icterus graduacauda* 24cm Shy and retiring, generally seen in pairs foraging in trees, often in shady densely vegetated parts. **V** Song melancholy, slurred *hooooo heeeowee heeew heweee*; calls include whistled *tooo* or *oooeh* and husky, rising *jeeek jeeek*... **H** Edges of dense forests and riparian thickets. **D** S Texas, Mexico.

19 JAMAICAN ORIOLE *Icterus leucopteryx* 21cm Arboreal, gleans, probes bromeliads and peels bark in search for insects; also feeds on fruit and nectar. **V** Song series of rapid whistles, repeated over and over; calls variously transcribed as *you cheat you cheat*, *cheat-you*, and *Auntie Katie*. **H** Montane subtropical forest, lowland forest, forest edges, wooded cultivated areas, gardens. **D** Jamaica, Cayman Islands.

20 ORANGE ORIOLE *Icterus auratus* 19cm Orange mantle diagnostic. **V** High *tiutju tjuh* or rattling staccato *titjutjutju*. **H** Dry open woodland. **D** SE Mexico.

21 ALTAMIRA ORIOLE *Icterus gularis* 25cm Arboreal, usually forages in pairs, feeding on insects, seeds, fruit and nectar. **V** Song series of loud, clear whistles; calls include whistled *teeu*, or similar; in flight gives hoarse, rising *griink*. **H** Open arid woodlands. **D** S Texas to Honduras.

22 YELLOW ORIOLE *Icterus nigrogularis* 20cm Separable from Yellow-backed Oriole by narrow black bib and white edging to wing feathers. **V** Rich repertoire of fluted, often melodious and sometimes hesitating phrases. **H** Open woodland, arid scrub, riverine belts, gardens. **D** N South America.

23 BULLOCK'S ORIOLE *Icterus bullockii* 23cm Sometimes combined as a single species with Baltimore Oriole, when known as Northern Oriole. Hybridises with Baltimore Oriole where ranges meet. **V** Short and lively, transcribed as *cut cut cudut whee up chooup*, less melodic than Baltimore Oriole; calls include short rattle and soft *chuk*. **H** Deciduous open woodland, woodland edges, urban parks. **D** Breeds SW Canada through W USA to Mexico. Winters Mexico to Nicaragua.

24 STREAK-BACKED ORIOLE *Icterus pustulatus* 20cm Wings patterned with grey, not with white. Note streaked and spotted mantle. **V** Very high sharp-fluted *itchee-itchee-itch* or tit-like calls. **H** Bushy land, dry woodland, scrub. **D** Widespread in Central America.

25 BLACK-BACKED ORIOLE *Icterus abeillei* 19cm Head pattern and extensive white in wing distinctive. **V** Yelping and chirping *whetwhet-ictariteeteet*. **H** Forest, dry woodland, parks and gardens in suburban regions. **D** SC Mexico.

26 BALTIMORE ORIOLE *Icterus galbula* 22cm Arboreal, attracted to trees with colourful flowers or dense foliage. **V** Song variable, consisting of clear whistled notes, transcribed as *pidoo tewdl tewdl yewdi tew tidew* or a simple series of *hew-li* notes; call rattling *cher-r-r-r-r-r*; also tinny *veeet* given in flight. **H** Open woodland, orchards, parks, gardens. **D** Breeds C and SE Canada to S and SE USA. Winters Colombia and Venezuela.

27 YELLOW-TAILED ORIOLE *Icterus mesomelas* 21cm Outer three pairs of tail feathers are yellow. **V** Very melodious, repeated phrases with notes of different pitch. **H** Borders and clearings in humid and swamp forest, plantations. **D** Mexico to Peru.

28 SPOT-BREASTED ORIOLE *Icterus pectoralis* 23–25cm Arboreal, forages in pairs or small groups; attracted to fruiting and flowering trees. **V** Song rich melodious warbled whistling; calls include nasal *nyeh*, sharp *whip* and chattering *ptcheck*. **H** Urban areas with flowering trees and shrubs. **D** Widespread in Central America.

29 WHITE-EDGED ORIOLE *Icterus graceannae* 20cm Note white-edged grey outer-tail feathers. Separable from Yellow-tailed Oriole by white wing patch. **V** Slow and rapid chattering and twittering. **H** Dry and arid scrub and woodland. **D** Ecuador, Peru.

30 CAMPO TROUPIAL *Icterus jamacaii* 23cm Separable from Venezuelan Troupial by orange (not white) shoulder and by smaller area of naked blue skin around eye. **V** Slow melodies of fluted notes at different pitches and variable tempo. **H** Borders and edges in forest and Caatinga; also seen in small towns. **D** E Brazil.

31 VENEZUELAN TROUPIAL *Icterus icterus* 25cm Usually seen singly or in pairs foraging in trees or bushes, will pick fallen fruit from the ground; often sings from top of bush or cactus. **V** Song repetitive series of whistles *troup troup troup*, *troup-ial troup-ial troup-ial* or *cheer taw cheer*; calls include mellow whistles and nasal sounds. **H** Primarily arid scrubland. **D** Colombia, Venezuela.

1
2
3
4
5
6
7
8
9
10
11
12
13
14
15
16
♀
♂
17
18
19
20
21
22
23
♂
♀
24
♀
♂
25
♀
♂
26
♀
♂
27
28
29
30
31

OROPENDOLAS, ORIOLES AND BLACKBIRDS *CONTINUED*

1 ORANGE-BACKED TROUPIAL *Icterus croconotus* 23cm Orange crown and pale eyes are diagnostic. **V** Slow series of high, fluted notes at random pitch. **H** At rivers in forest, clearings and secondary growth. **D** N, W and SW Amazonia to N Argentina.

2 BAR-WINGED ORIOLE *Icterus maculialatus* 20cm Separated from Scott's Oriole (Plate 284) by different range and deep yellow shoulder bar. **V** Very high meandering fluting *fjeefjeewuweehwee*. **H** Woodland, semi-open country with scattered trees. **D** S Mexico to El Salvador.

3 BLACK-VENTED ORIOLE *Icterus wagleri* 20–23cm Arboreal, favours flowering trees or bushes. **V** Song gurgling warble interspersed with nasal and squeaky notes; calls include *coo-nyah-ra* and nasal *nyeh*, which is often repeated when alarmed. **H** Dry scrub, open areas with scattered trees and riparian shrubs. **D** Mexico to Nicaragua.

4 HOODED ORIOLE *Icterus cucullatus* 20cm Arboreal, with a liking for palms. First-summer male like female but with black throat and bib, very similar to first-summer male Orchard Oriole. **V** Song rapid series of throaty whistles, trills and rattles; calls include hard *chairr*, rising *wheet* and sharp *veek* flight note. **H** Dry open woodland, parks, gardens. **D** SW USA, Mexico, Belize.

5 BLACK-COWLED ORIOLE *Icterus prosthemelas* 20cm Black on breast extends further down than in Black-vented Oriole, with pure yellow underparts. Note black wings and mask of female. **V** Very high sharp-fluted cautious *sweetohsweeterrohweeet*. **H** Forest edges and clearings, semi-open areas with scattered trees. **D** S Mexico to Panama.

6 ORCHARD ORIOLE *Icterus spurius* 18cm Arboreal, forages in trees and shrubs for insects and small fruits. **V** Song lively warbling with distinctive, ringing *plit titi zheeeer* ending; calls include clear, whistled *tweeo*, soft *chut*, rasping *jarrsh* and low, soft *yeeep* given in flight. **H** Woodlands, thickets, gardens. **D** SE Canada to C Mexico.

7 OCHRE ORIOLE *Icterus fuertesi* 16cm Paler than Orchard Oriole. With Orchard Oriole it is the only black-headed oriole with chestnut/ochre rather than yellow body. **V** Very high fluted sometimes sharp *ictustweereweetwet*. **H** Forest edges, orchards, suburban regions. **D** NE Mexico.

8 CUBAN ORIOLE *Icterus melanopsis* 20cm Long-tailed, black oriole with bright yellow shoulders, rump and thighs. Feeds mainly by gleaning insects from foliage; also takes nectar. **V** Song consists of clear whistled notes which may take long (11–12 notes) or short (5–6 notes) forms; calls include sharp *chip* and nasal *wheenk*. **H** All well-vegetated habitats, including gardens. **D** Cuba.

9 BAHAMA ORIOLE *Icterus northropi* 21cm Critically endangered. Bright yellow and black oriole with long tail. Mainly insectivorous but takes some nectar; will visit hummingbird feeders. **V** Sexes duet in clear, whistled song; calls include whining and whistling notes and abrupt *chit*. **H** Woodlands and wetlands of all kinds, urban areas. **D** Bahamas.

10 MARTINIQUE ORIOLE *Icterus bonana* 18–21cm Forages in family groups, pairs or singly, mainly in forest canopy where insects and fruit are main food items. **V** Song described as variable soft warbling; series of clear whistles, and shrill like that of a Carib Grackle (Plate 286); call harsh *cheeo* or *cheeo-cheo*. **H** Mangroves, dry forest, humid forest, forest edges, dense scrub, plantations, urban areas. **D** Martinique.

11 PUERTO RICAN ORIOLE *Icterus portoricensis* 22cm Black with bright yellow shoulders, rump, undertail and thighs. Mainly insectivorous; takes some small vertebrates. **V** Song comprises rising and falling clear whistles, also buzzes and warbles; call harsh *chk*. **H** Forest edges, palm groves, urban areas. **D** Puerto Rico.

12 MONTSERRAT ORIOLE *Icterus oberi* 20–22cm Generally occurs singly or in pairs; forages in canopy of moist montane forest searching for insects. **V** Song, only heard during the breeding season, a loud series of melodious whistles; calls include sharp *chic* or sharper *chuck* and scolding *chuur*. **H** Moist montane forest. **D** Montserrat.

13 ST. LUCIA ORIOLE *Icterus laudabilis* 20–22cm Usually occurs in pairs or small parties; forages in trees looking for insects and fruit. **V** Song consists of short series of sweet, varied whistles that are repeated a number of times; calls include harsh *chwee* and soft *chup*. **H** Humid forest, dry coastal scrub forest and adjoining mangroves. **D** St Lucia.

14 HISPANIOLAN ORIOLE *Icterus dominicensis* 20–22cm Usually encountered in pairs, foraging, sometimes acrobatically in palms and trees in search of insects, fruit and nectar. Split from Black-cowled Oriole. **V** Song a clear and pleasant whistle, although it is sometimes described as weak; call *churr-churr-r*. **H** Forests, forest edges, woodlands and gardens, usually near palms. **D** Hispaniola.

15 ORANGE-CROWNED ORIOLE *Icterus auricapillus* 20cm Note yellow epaulet and lack of white edging in wings. **H** Forest edges, riverine belts, scattered trees in cultivation. **D** Panama to Venezuela.

16 VARIABLE ORIOLE *Icterus pyrrhopterus* 20cm Races vary in colour of epaulet, from chestnut to yellow or intermediate between the two. **V** Rapid, musical series with nasal notes, short rattles and twittering. **H** Forest edges, palm savannah, swamp, Caatinga, Chaco woodland, suburbs. **D** SC South America.

17 EPAULET ORIOLE *Icterus cayanensis* 20.5cm Rarely (in Peruvian birds) shows some yellow in thighs. **V** Calm series of soft, drawn-out shrieks, mini-rattles and fluted notes. **H** Canopy, borders and clearings in tall forest. **D** N and C South America.

18 JAMAICAN BLACKBIRD *Nesopsar nigerrimus* 18cm Arboreal, seeks food among epiphytes and tree ferns. **V** Song buzzy *zwheezoo-whezoo whee* or similar; call *check* or *dzik*; when alarmed, gives thin, high *seee seee* or *chet-chet-chet*... **H** Wet mountain forests, occasionally in humid woodlands at lower elevations. **D** Jamaica.

19 YELLOW-SHOULDERED BLACKBIRD *Agelaius xanthomus* 19–22cm Classed as endangered. Usually forages in upper or mid-levels of mangrove or other woodland, probing into epiphytes, bark crevices or holes. Feeds mostly on arthropods and fruit when available, will descend to forage on ground where it eats grain or seeds. **V** Song a nasal rasp, transcribed as *nhyaaaaaaaaa* or *ttnyyaaa*; calls *check* and nasal *chwip*; when alarmed, gives *cut-zee*. **H** Mainly mangroves and arid scrublands. **D** Puerto Rico, most regular on Mona Island.

20 TAWNY-SHOULDERED BLACKBIRD *Agelaius humeralis* 19–22cm Usually occurs in flocks, especially in non-breeding season. Forages on ground and in trees, feeding on seeds, rice and nectar. In parts of Cuba very tame, scavenging on scraps in and around restaurants. **V** Song buzzy, drawn-out note, sometimes preceded by a higher-pitched buzz, transcribed as *preeee-whaaaaaaa*; calls include loud, short *chic-chic* or *chup-chup*, nasal *whaap* or *nhyaap* and high *pleeet* or *tweeep*. **H** Woodlands, farmland with scattered trees, mangroves, rich fields. **D** Cuba, Hispaniola (Haiti).

21 TRICOLORED BLACKBIRD *Agelaius tricolor* 22–24cm Regularly occurs in large flocks, often mixes with other blackbirds. Forages primarily on ground. **V** Song harsh *on-kee-kaaangh*; calls like those of Red-winged Blackbird but lower-pitched. **H** Marshes and dense thickets; winters in open agricultural areas. **D** California, N Baja California

22 RED-WINGED BLACKBIRD *Agelaius phoeniceus* 19–23cm Often encountered in large flocks foraging in marshes where they feed on seeds, fruit, insects and small vertebrates. Females of the endemic race on the Bahamas have a whiter throat than those from North America. **V** Song repeated, bubbling and shrill *ok-a-lee*; calls include sharp, throaty *check* and whistled *cheer* or *peet* given in alarm. **H** Swamps and marshes, fields, gardens. **D** Widespread in North America and Central America.

23 RED-SHOULDERED BLACKBIRD *Agelaius assimilis* 19–23cm Occasionally occurs in large flocks, especially in non-breeding season; forages in marshes and agricultural fields and pastures. **V** Song shrill, non-melodious *o-wi-hiiii*; call short *cheep*, *chek* or *chek-chek-chek*. **H** Swamps and marshes. **D** W Cuba.

24 SCREAMING COWBIRD *Molothrus rufoaxillaris* 19cm Adult separable from adult Shiny Cowbird by shorter bill and less shiny plumage. **V** Unstructured series of single scratchy notes. **H** Agricultural areas, parks, towns, grassland near woodland. **D** S Bolivia and S Brazil to EC Argentina.

25 GIANT COWBIRD *Molothrus oryzivorus* male 37cm, female 32cm Note large size; eyes might be brown or yellow. Male (with diagnostic ruff around neck) is much larger than female. **V** Series of single nasal notes at long intervals. **H** Forest borders and adjacent open grassland, around pastures and lawns. **D** E Mexico through N South America.

26 SHINY COWBIRD *Molothrus bonariensis* 18–20cm Forages on ground, often seen around livestock. Roosts communally, especially in non-breeding season, often in marshland. **V** Song consists of several liquid purrs, followed by a high whistle; call a rolling rattle. **H** Woodland edges, open country and agricultural areas. **D** SE USA through C South America.

27 BRONZED COWBIRD *Molothrus aeneus* 20cm Note ruff on neck, red eyes, shiny wings; female overall very dark brown (without gloss) and with red eyes. **V** Extremely high inhaled flutes with dry rattles. **H** Open woodlands, grasslands, fields, scrubland, suburban regions. **D** W USA to Panama.

28 BRONZE-BROWN COWBIRD *Molothrus armenti* male 20cm, female 18cm Note red eyes; short, heavy bill; more or less blue in wings; and 'thick' neck. **V** Like that of Bronzed Cowbird. **H** Open habitats such as dry scrub, pastures, roadsides, parks. **D** Colombia.

29 BROWN-HEADED COWBIRD *Molothrus ater* 19cm Forages on ground. Female has streaked underparts. **V** Song bubbly *glug-glug-gleeee*, followed by thin, slurred whistles; call harsh *chuk* and squeaky *weee-titi* often given in flight. **H** Farmland, open woodland, parks, gardens. **D** S Canada to S Mexico.

30 SCRUB BLACKBIRD *Dives warczewiczi* 21.5cm No similar blackbird occurs in its range. **V** Flowing series of gurgled and high notes, shrieks and rattles. **H** Dry open woodland and scrub, orchards, cultivation, suburbs, cities. **D** Ecuador, Peru.

1
2
3
4
5
6
7
8
9
10
11
12
13
14
15
16
17
18
19
20
21
22
23
24
25
26
27
28
29
30

OROPENDOLAS, ORIOLES AND BLACKBIRDS *CONTINUED*

1 MELODIOUS BLACKBIRD *Dives dives* 25cm Frequently flicks tail up. Separated from similar all-black skulking Black Catbird (Plate 237) by different habitat. **V** High melodious *whut whée-ut* or drawn-up *trrrr-uweeeét*. **H** Open areas with scattered tree stands, bush, hedges. **D** Mexico to Nicaragua.

2 CUBAN BLACKBIRD *Ptiloxena atroviolacea* 25–28cm Regularly forages on ground, on buildings and even on cattle. Gathers in flocks, sometimes in the company of Greater Antillean Grackles and Tawny-shouldered Blackbirds (Plate 285). **V** Song variable, consisting of repeated single sharp note or mellow phrases and a nasal call similar to that of a bleating sheep; call loud, repeated *to-teee* or *ti-o*; also various whistles and mews. **H** Woodlands, agricultural and open areas, parks, gardens. **D** Cuba.

3 RUSTY BLACKBIRD *Euphagus carolinus* 23cm Forages on ground, often with grackles or other blackbirds. If disturbed, retreats to nearby trees or bushes. **V** Song squeaky *kush-a-lee* or *chuck-la-weeeee*; call soft *chuck*. **H** Wooded swamps, lake shores, open agricultural fields. **D** Breeds N North America. Winters SE USA.

4 BREWER'S BLACKBIRD *Euphagus cyanocephalus* 23cm Forages on ground. Female usually dark-eyed; generally unmarked, dull greyish. **V** Song short, buzzy, cackling *t-kzzzz* or *t-zherr*; call short *ket*, *chak* or *chuk*. **H** Agricultural fields, grasslands, alpine meadows, beaches, urban areas. **D** Breeds SW and SC Canada through W USA. Winters S Mexico.

5 COMMON GRACKLE *Quiscalus quiscula* 32cm Usually winters in large flocks, often alongside blackbirds. Variable; female duller and smaller-tailed. **V** Song screechy *readle-eek*, *re-lick* or *scudle-eek*; call loud *chack* or *chuck*. **H** Almost any open area, including woodland, bogs, farmland and urban parks and gardens. **D** N, EC and SE North America.

6 NICARAGUAN GRACKLE *Quiscalus nicaraguensis* 30cm Much shorter-tailed than Great-tailed Grackle; note grey face in female. **V** Extremely high shrieks, shivers and drawn-up notes. **H** Wet grassland, fields, marshland. **D** Nicaragua, N Costa Rica.

7 CARIB GRACKLE *Quiscalus lugubris* 24–28cm Typically occurs in noisy flocks; can become very bold and tame in urban areas. **V** Song series of three- to seven-syllable squeaky notes, varying from island to island, ending with ringing bell-like note; call *chuck* and various whistles. **H** Lowland open situations and residential areas. **D** Lesser Antilles, N South America.

8 GREATER ANTILLEAN GRACKLE *Quiscalus niger* 25–30cm Typically encountered in flocks; forages on ground, often around livestock; also scavenges scraps in urban areas. **V** Song variable four-syllable phrase; calls include loud *chak-chak* and *chin-chin-chi-lin*. **H** Any type of open terrain, including farmland, mangrove edges and urban areas. **D** Greater Antilles.

9 BOAT-TAILED GRACKLE *Quiscalus major* male 42–44cm, female 37–39cm Gregarious at all times. A yellow-eyed race occurs on the Atlantic coast from New York to N Florida. **V** Very varied. Song consists of high, ringing notes; also harsh notes mixed with rustling sounds and various rattling, trilling and whistled notes; common call of male is deep *chuk*; also *kle-teet*, while female gives low *chenk* or *chuup*. **H** Saltwater and freshwater marshes. May spread to farmland in winter. **D** E and SE USA.

10 GREAT-TAILED GRACKLE *Quiscalus mexicanus* male 46–48cm, female 38–40cm Eastern females dark greyish-brown below. Often in large groups, especially after breeding. **V** Very varied, song starts with harsh notes followed by undulating *cheweechewe*, more harsh notes then finally several loud *cha-wee* calls; commonest call is low, hard *chuk*. **H** Open areas with scattered trees or bushes, including pastures, grassland, parkland, urban gardens. **D** S USA through to W and N South America.

11 RED-BELLIED GRACKLE *Hypopyrrhus pyrohypogaster* male 31cm, female 27cm Unmistakable due to rather large size, red-rimmed pale eyes, and red belly and vent. **V** Short, very high, nasal series of notes with tempo changes. **H** Canopy and borders of montane forest. **D** Colombia.

12 VELVET-FRONTED GRACKLE *Lampropsar tanagrinus* male 22cm, female 20cm Note relatively small, slender bill, flat crown and unobtrusive black velvet front. **V** High, partly hurried, slightly metallic series of 4–5 *gli* notes. **H** At water in forest, secondary growth, riverine belts, plantations, mangroves. **D** W and SW Amazonia, tepuis in N South America.

13 ORIOLE BLACKBIRD *Gymnomystax mexicanus* male 30cm, female 27cm Unmistakable due to contrasting plumage and facial pattern. **V** Slow series of rasping rattles with high *peec* (sung in a duet). **H** Scattered trees in savannah, tall grassland, marshes. **D** Amazonia.

14 COLOMBIAN MOUNTAIN GRACKLE *Macroagelaius subalaris* 27cm Compare with Red-winged Blackbird (Plate 285); chestnut underwing coverts are diagnostic. **V** Stuttered nasal chattering; excited chattering when in groups. **H** Canopy and borders of montane forest. **D** Colombia.

15 GOLDEN-TUFTED MOUNTAIN GRACKLE *Macroagelaius imthurni* male 28cm, female 25cm Note yellow tufts in armpits. **V** Mixture of high shrieks and musical notes. **H** Canopy and borders of montane forest on tepuis. **D** Tepuis of Venezuela, Guyana and Brazil.

16 AUSTRAL BLACKBIRD *Curaeus curaeus* male 26cm, female 24cm Note long, sharp-pointed bill. Compare smaller Shiny Cowbird (Plate 285), which has a smaller bill and shorter tail. In flocks, the difference between male and female is less obvious than in White-edged Oriole (Plate 284). **V** Series of notes with nasal shrieks, rattles and sharp flutes, changing in tempo. **H** Scrub and woodland, plantations, fields. **D** Southern Cone, South America.

17 SCARLET-HEADED BLACKBIRD *Amblyramphus holosericeus* 24cm Colour pattern is distinctive. **V** Series of high *dzee* or *pieu* notes and short chatters and trills, all at 1–2-sec intervals. **H** Marshes with tall vegetation. **D** C Bolivia to S Brazil, Uruguay and E Argentina.

18 FORBES'S BLACKBIRD *Anumara forbesi* male 24cm, female 21cm Note pointed, slender bill, slightly longer than that of Chopi Blackbird. **V** Sharp, level chatters. **H** Forest at lakes and marshes. **D** E Brazil.

19 CHOPI BLACKBIRD *Gnorimopsar chopi* 23cm Difficult to separate from Forbes's Blackbird, but note groove in curved maxilla and lanceolated neck feathers. **V** Unstructured series of loud, sharp *tuew* notes, combined with occasional small rattles. **H** Clearings, Chaco and open woodland, palm groves, plantations, cultivation, suburbs. **D** E, SE and SC South America.

20 BOLIVIAN BLACKBIRD *Oreopsar bolivianus* 23cm Note brown flight feathers and strong bill with curved maxilla. Separable from smaller Shiny Cowbird (Plate 285) by brown in wings and less iridescence. Lacks groove in maxilla of Chopi Blackbird, which occurs at lower elevations. **V** Rapid series of undulating phrases such as *tjow-ti-tjow*. **H** Dry woodland and scrub. **D** Bolivia.

21 GREYISH BAYWING *Agelaioides badius* 18.5cm Grey with black lores and rufous wings; very like Pale Baywing but ranges do not overlap. **V** Like that of Pale Baywing. **H** Open woodland, scrub, savannah, agricultural areas, gardens, suburbs. **D** SC South America.

22 PALE BAYWING *Agelaioides fringillarius* 18.5cm Distinctive with grey plumage with black lores and rufous wings, but compare with Screaming Cowbird (Plate 285) and with recently split Greyish Baywing (ranges do not overlap). **V** Stream of unstructured, rapid, sharp warbling and dry rattles. **H** Scrub, savannah, agricultural areas, gardens, suburbs. **D** NE Brazil.

23 YELLOW-WINGED BLACKBIRD *Agelasticus thilius* 18cm Yellow shoulder is often partly concealed. Male separable from male Variable Oriole and male Epaulet Oriole (both Plate 285) by different shape and manner. Female pale brownish grey with striped upperparts and underparts; note striking eyebrows. **V** Strong rattles, twitters and strange nasal, drawn-out notes. **H** Marshes and adjacent grassland. **D** SW and SC South America.

24 PALE-EYED BLACKBIRD *Agelasticus xanthophthalmus* 20.5cm Separable from all other blackbirds by pale eyes. **V** Series of loud staccato or downslurred notes and high rattles, most elements given 3–5 times. **H** Wetlands. **D** Ecuador, Peru.

25 UNICOLORED BLACKBIRD *Agelasticus cyanopus* 19cm Male all black with a rather long, pointed bill. Female varies. Note habitat. **V** Sharp, loud trills and rattles of different pitch and intonation. **H** Marshes with emergent and floating vegetation, and adjacent grassland and fields; <500m. **D** SC and E Brazil.

26 CHESTNUT-CAPPED BLACKBIRD *Chrysomus ruficapillus* 18.5cm male unmistakable due to head pattern. Note plain, pale buff throat of female and lack of eyebrows. **V** High, repetitive twittering. **H** Marshes, rice fields, wet grassland. **D** NE and SC South America.

27 YELLOW-HOODED BLACKBIRD *Chrysomus icterocephalus* 18cm Male unmistakable due to yellow head. Note striped grey-brown lower belly and vent of female. **V** Inhaled phrases and trills. **H** Marshes, rice fields, wet grassland. **D** N and C South America.

28 SAFFRON-COWLED BLACKBIRD *Xanthopsar flavus* 20cm Male unmistakable due to colour pattern. Female separable from female Yellow-hooded Blackbird by all-yellow underparts. **V** Low, unstructured chattering and high twittering. **H** Wet grassland, marshes, campo, rice fields. **D** E Paraguay, S Brazil, NE Argentina, Uruguay.

29 BROWN-AND-YELLOW MARSHBIRD *Pseudoleistes virescens* 24cm Note extensive dark flanks. **V** Unstructured twittering and chattering. **H** Marshes and wet grassland. **D** NE Argentina, S Brazil, Uruguay.

30 YELLOW-RUMPED MARSHBIRD *Pseudoleistes guirahuro* 23cm Note brown-and-yellow (not black-and-yellow) plumage. Separable from Brown-and-yellow Marshbird by yellow rump and flanks. **V** Fast series of shrieks and high twittering. **H** Marshes, adjacent wet grassland, flooded savannah. **D** E Paraguay, S Brazil, NE Argentina, Uruguay.

1
2
3
4
5
6
7
8
9
10
11
12
13
14
15
16
17
18
19
20
21
22
23
24
25
26
27
28
29
30

NEW WORLD WARBLERS PARULIDAE

1 OVENBIRD *Seiurus aurocapilla* 15cm Forages among forest floor leaf litter. Walks in a jerky manner. **V** Song emphatic *teecher-teecher-teecher*, rising in pitch and volume; call *chuk* or *tsuk*, often repeated. **H** Mature deciduous and mixed forest, with dense understorey. **D** Breeds N, C and E North America. Winters Mexico, Central America, West Indies, N South America.

2 WORM-EATING WARBLER *Helmitheros vermivorum* 14cm Forages in undergrowth and in trees, among dead leaf clumps. **V** Song monotonous even-pitched trill; calls include sharp *tchip* and buzzy *zeet-zeet*. **H** Wooded hillsides and ravines, with dense undergrowth, often near streams. **D** Breeds E USA. Winters Mexico, Central America, West Indies.

3 LOUISIANA WATERTHRUSH *Parkesia motacilla* 14.5–16cm Constantly bobs rear end as it walks on ground searching for prey among leaves and on logs. **V** Song loud, consists of short series of descending notes followed by warbling twitter; call high-pitched *chink*. **H** Primarily at edges of flowing water, mainly at higher elevations. **D** Breeds E USA. Winters Mexico, Central America, West Indies, N South America.

4 NORTHERN WATERTHRUSH *Parkesia noveboracensis* 12.5–15cm Behaviour similar that of Louisiana Waterthrush. Long-distance migrant. **V** Song transcribed as *swee-swee-chit-weedleo*, last note downslurred; call loud, metallic *chink*. **H** Watery borders of mangroves and coastal scrub. **D** Breeds Alaska through Canada to NW and NE USA. Winters Mexico, Central America, West Indies, N South America..

5 BACHMAN'S WARBLER *Vermivora bachmanii* 12cm Probably extinct. Often feeds high in treetops, gleaning insects from clumps of leaves or twigs. **V** Song buzzy, pulsating trill; only call recorded is low, hissing *zee-e-eep*. **H** Undergrowth in moist woods, forest edge near swamps and canebrakes. **D** Breeds SE USA. Winters Cuba, Bahamas.

6 GOLDEN-WINGED WARBLER *Vermivora chrysoptera* 12cm Very agile, often feeding tit-like in bushes and trees. Probes dead leaf clumps for insects. **V** Song soft, buzzy *zee-bee-bee-bee*, occasionally more trilling; call short *tchip*. **H** High forests, woodlands, gardens. **D** Breeds NC and NE USA, SE Canada. Winters Guatemala to N South America.

7 BLUE-WINGED WARBLER *Vermivora cyanoptera* 12cm Female duller than male. Agile acrobatic forager, usually in mid-level. **V** Song wheezy *beee-bzzz*; also longer *tsi tsi tsi tsi tsi zweeeeeeezt zt zt zt*; call sharp, dry *nik* or *chik*. **H** Brushy fields, woodland edges, streamside thickets. **D** Breeds C and NE USA, S Canada. Winters Mexico, Central America.

8 BLACK-AND-WHITE WARBLER *Mniotilta varia* 13cm Feeds nuthatch-like, probing bark for insects. **V** Song thin, high-pitched *see wee-see wee-see wee-see wee-see*; call sharp *tick* and thin *tzeet* or *tsip*. **H** Various types of deciduous and mixed woodlands, preferring those in moist areas. **D** Breeds C Canada to SE Canada and NE USA. Winters Mexico to Peru, and West Indies.

9 PROTHONOTARY WARBLER *Protonotaria citrea* 14cm Forages from low to mid-levels, behaviour much like that of Black-and-white Warbler. **V** Song series of ringing *zweet* notes; calls include ringing *tsip*, soft *psit* and thin *seet* flight note. **H** Swampy, mature woodlands. **D** Breeds C, E and SE USA. Winters Mexico, Central America, West Indies, N South America.

10 SWAINSON'S WARBLER *Limnothlypis swainsonii* 14cm Forages on ground and in low bushes and on logs. **V** Song transcribed as *wee wee wee wee-tu-weeu*; call strong, long *sship*. **H** Leaf litter in forests, woodland, thickets. **D** Breeds SE USA. Winters Mexico, Central America, West Indies.

11 CRESCENT-CHESTED WARBLER *Oreothlypis superciliosa* 11cm Forages at mid- to high levels. **V** Song short buzzy trill; call high-pitched *tchip*. **H** Montane forest. **D** Mexico to Nicaragua.

12 FLAME-THROATED WARBLER *Oreothlypis gutturalis* 12cm Unmistakable by general colour pattern. Forages in canopy. **V** Extremely high nasal *pssss*; very high *puWir-puWir-puWir*. **H** Montane forest and nearby trees. **D** Costa Rica, Panama.

13 TENNESSEE WARBLER *Leiothlypis peregrina* 12cm Usually forages high in tree canopy, although will descend to feed in bushes; agile and active. **V** Song series of staccato double and single notes, ending in a trill; calls include sharp *tsit* and thin *see*. **H** Conifer and mixed woodland; on migration occurs in open woodland and thickets. **D** Breeds SE Alaska and NW Canada to SE Canada and NE USA. Winters Mexico to N South America.

14 ORANGE-CROWNED WARBLER *Leiothlypis celata* 13cm Feeds from low to high levels on insects and small berries. **V** Song high-pitched trill followed by a lower, slower trill; calls include sharp *chet* and *see* flight note. **H** Open woodland, forest edges, thickets. **D** Breeds W and N North America. Winters Mexico, Central America.

15 COLIMA WARBLER *Leiothlypis crissalis* 15cm Forages, with slow deliberate actions, mainly in undergrowth. **V** Song short, monotonous chattering trill; call sharp, metallic *psit*. **H** Oak woodland. **D** Breeds SW Texas to C Mexico. Winters Mexico.

16 LUCY'S WARBLER *Leiothlypis luciae* 11cm Rufous rump. Forages at low to mid-levels. **V** Song short twittering trill followed by lower whistled notes *weeta weeta weeta che che che*; call high-pitched, husky *tzip*. **H** Mesquite woodland and scrub, often near water. **D** Breeds SW USA, NW Mexico. Winters W Mexico.

17 NASHVILLE WARBLER *Leiothlypis ruficapilla* 12cm Found mainly at low levels feeding on insects, nectar and berries. **V** Song consists of series of high-pitched *tsee* notes followed by low trill; call metallic *tink* or *spink*; in flight, gives high, clear *see* or *swit*. **H** Pine woodland and scrub woodland in highlands; also wooded areas in coastal lowlands. **D** Breeds W and N North America. Winters Mexico, Central America.

18 VIRGINIA'S WARBLER *Leiothlypis virginiae* 12cm Behaviour similar to that of Nashville Warbler. **V** Song series of rapid, accelerating, thin notes ending with several lower notes; calls similar, although a little rougher, to those of Nashville Warbler. **H** Chaparral and pinyon–juniper, yellow pine and scrub oak woodland. **D** Breeds SW USA. Winters Mexico.

19 SEMPER'S WARBLER *Leucopeza semperi* 14.5cm Critically endangered, possibly extinct. Forages on or close to ground. **V** Only call recorded is chattering *tuck-tick-tick-tuck*, given when alarmed. **H** Thick undergrowth in moist, mid-elevation forests, mountain thickets and dwarf forests. **D** St Lucia.

20 CONNECTICUT WARBLER *Oporornis agilis* 13–15cm Shy and skulking, forages on ground or in low bushes. Walks much in the manner of Ovenbird. **V** Song loud *wee-cher-cher wee-cher-cher wee-cher-cher wee*; calls include metallic *plink* and a high-pitched buzzy *zee* flight note. **H** Moist woodland understorey, usually near water. **D** Breeds SC Canada, NC USA. Winters C South America.

21 GREY-CROWNED YELLOWTHROAT *Geothlypis poliocephala* 14.5cm Frequently pumps tail up and down or twitches it from side to side. **V** Song varied, halting warble; call loud, slapping *chack*. **H** Damp fields, hedgerows, bushy savannahs. **D** Widespread in Mexico and Central America.

22 MASKED YELLOWTHROAT *Geothlypis aequinoctialis* 14cm Black mask of male bordered above by grey, and yellowish coloration runs from chin to vent. **V** Short, descending, warbling song. **H** Damp grassland, marshes, dense undergrowth at edges and in clearings in forest. **D** Venezuela and Colombia to N Amazonian Brazil, Trinidad.

23 CHIRIQUI YELLOWTHROAT *Geothlypis chiriquensis* 13cm Formerly treated as conspecific with Masked Yellowthroat. Black mask of male broader than in Masked Yellowthroat. **V** Short, descending, warbling song, faster than Masked Yellowthroat, and repeated more frequently, also given in flight. **H** Damp grassland, marshes, dense undergrowth at forest edges and clearings. **D** SW Costa Rica, Panama.

24 BLACK-LORED YELLOWTHROAT *Geothlypis auricularis* 13cm Formerly treated as conspecific with Masked Yellowthroat. Black mask of male much smaller than in Masked Yellowthroat. **V** Song similar to that of Masked Yellowthroat. **H** Grassy undergrowth at edges and in clearings in open, dry forest. **D** W Ecuador, W Peru.

25 SOUTHERN YELLOWTHROAT *Geothlypis velata* 13cm Formerly treated as conspecific with Masked Yellowthroat. Black mask of male narrower than in that species, with broader grey border. **V** Song similar to that of Masked Yellowthroat but longer and faster. **H** Damp grassland, marshes, scrub. **D** SE Peru, N Bolivia and S Amazonian Brazil to Argentina and Uruguay.

26 MACGILLIVRAY'S WARBLER *Geothlypis tolmiei* 13cm Broken white eye-ring. Feeds in low, dense cover. **V** Song consists of short series of buzzy notes followed by two or three lower *teeoo* notes; call dry *shik* or *twik*. **H** Open forest, forest edges, mountainside scrub. **D** Breeds W North America to N Mexico. Winters Mexico, Central America.

27 MOURNING WARBLER *Geothlypis philadelphia* 13cm Skulking, forages mainly on ground, where it hops, or in low, dense undergrowth. **V** Song rich, churring *churree churree churree turi turi*; calls include flat *chip* and *svit* or *zee* flight note. **H** Secondary growth, dense understorey of forest edges and clearings, often near damp situations. **D** Breeds C Canada to SE Canada and NE USA. Winters Nicaragua to N South America.

28 KENTUCKY WARBLER *Geothlypis formosa* 13cm Shy and skulking, forages on the ground, where it hops, and in low bushes. **V** A low, sharp *tship* and a buzzy *zeep*. Song consists of a series of loud, whistled *churree* notes. **H** Mature deciduous forests with dense undergrowth, often in damp situations. **D** Breeds C, E and SE USA. Winters Mexico to N South America.

29 OLIVE-CROWNED YELLOWTHROAT *Geothlypis semiflava* 13.5cm Black mask of male is not bordered by white or grey and extends over crown. **V** Song simple series of high, sharp notes, starting with almost level *tee-tjeet-weet*, and with more warbling towards the end (3–7 secs). **H** Tall grass in edges and clearings in forest, shrubby pastures, scrub. **D** Honduras to Ecuador.

30 BLACK-POLLED YELLOWTHROAT *Geothlypis speciosa* 13cm Note saffron wash over underparts and black crown. **V** Song very high descending *weetweetjetjuotjuh* or rattling *weetweet...*; call *trits*. **H** Marsh, reeds at lakes. **D** Mexico.

31 BELDING'S YELLOWTHROAT *Geothlypis beldingi* 14cm Separated from smaller Common Yellowthroat by stronger bill and brighter colours. Restricted range. **V** Song a short series of rich chuckling notes, similar to that of Common Yellowthroat; calls include dry *tdjip*, also rattling notes. **H** Marshland. **D** Baja California.

32 BAHAMA YELLOWTHROAT *Geothlypis rostrata* 15cm Behaviour similar to that of Common Yellowthroat, although a little less sprightly. **V** Song loud *wichity-wichity-wichit*, similar to song of Common Yellowthroat; call sharp *tchit*. **H** Dense low scrub and shrubs, also pine woods with thatch palm understorey. **D** N Bahamas.

33 ALTAMIRA YELLOWTHROAT *Geothlypis flavovelata* 13cm Yellow peaked crown diagnostic. **V** Song very high hurried up-and-down *titsjiputi...* **H** Marshland. **D** E Mexico.

34 COMMON YELLOWTHROAT *Geothlypis trichas* 13cm Skulking forager in thick vegetation, usually in damp areas. Often cocks tail. **V** Song variable, usually transcribed as *wichity-wichity-wichity-wich*; call dry *chep* or *tchuk*. **H** Bushes and thick vegetation at edge of various water bodies. **D** Breeds widely in North and Central America. Winters W South America.

35 HOODED YELLOWTHROAT *Geothlypis nelsoni* 12cm Separated from Common Yellowthroat by habitat, grey band at edge of mask. **V** Song *tsjúchohtsjúchot*. **H** Dry scrub. **D** Mexico.

36 WHISTLING WARBLER *Catharopeza bishopi* 14.5cm Generally shy and secretive. Forages purposely and fairly actively from low levels to tree canopy. **V** Song starts with series of rich, rising notes, ending with two or three emphatic notes; call soft *tuk* or *tchurk*. **H** Rainforest, palm brakes, humid secondary forest, elfin forest, forest edges. **D** St Vincent.

37 PLUMBEOUS WARBLER *Setophaga plumbea* 13cm Generally tame; constantly flicks tail as it forages, mainly in understorey. **V** Song simple, melodic *pa-pi-a* or *de-de-diu*; calls include a rattle and short *chek*. **H** Moist mountain forests, occasionally scrub forests and mangroves. **D** Lesser Antilles.

38 ELFIN WOODS WARBLER *Setophaga angelae* 11–13.5cm Very active, constantly flicks tail as it forages high in canopy, often with other species, searching for insects. **V** Song made up of a number of rapid, unmusical notes, on one pitch, that increases in volume and ends with a short series of double notes; call metallic *chip*. **H** Humid montane and lower montane elfin forests. **D** Puerto Rico.

39 ARROWHEAD WARBLER *Setophaga pharetra* 13cm Forages at high levels; constantly flicks tail downwards. Generally sings at dawn. **V** Song high-pitched *sww-sw-swee sww-sw-swee sww-sw-swee-swee-swee*; call metallic *tic*. **H** Moist humid forests. **D** Jamaica.

40 HOODED WARBLER *Setophaga citrina* 13cm Active forager, often makes fly-catching sallies; constantly flicks wings and spreads tail to reveal white in outer feathers. **V** Song loud *too-ee too-ee too-ee tee-ch*; call sharp *tchip* or *tchink* and buzzy *zrr*. **H** Moist forest undergrowth and mangrove swamps. **D** Breeds C, E and SE USA. Winters Mexico, Central America.

1
2
3
4
5
6
7
8
9
10
11
12
13
14
15
16
17
18
19
20
21
22
23
24
25
26
27
28
29
30
31
32
33
34
35
36
37
38
39
40

NEW WORLD WARBLERS *CONTINUED*

1 AMERICAN REDSTART *Setophaga ruticilla* 13cm Very active. Regularly fans wings and tail. **V** Song variable, usually high-pitched series ending with emphatic low note; calls include sweet *chip* and high, rising *sweet* flight note. **H** Open woodland, woodland edges, clearings, tall brush. **D** Breeds W Canada to SE Canada and E USA. Winters Mexico, Central America, West Indies, N South America.

2 KIRTLAND'S WARBLER *Setophaga kirtlandii* 15cm Solitary, generally forages low down in dense scrub. **V** Song emphatic *flip lip lip-lip-tip-tip-CHIDIP*; call loud *tchip*. **H** Mainly understorey, broadleaved scrub, thickets. **D** Breeds NC USA (Michigan). Winters Bahamas.

3 CAPE MAY WARBLER *Setophaga tigrina* 13cm Active forager in treetops, gleaning and occasionally using sallies to catch flying insects. **V** Song high *zi-zi-zi-zi-zi*; calls include very high *tsip* and slightly descending *tsee-tsee*, often given in flight. **H** Coniferous and mixed forest; all types of woodland frequented during migration. **D** Breeds C and SE Canada to NE and NC USA. Winters West Indies.

4 CERULEAN WARBLER *Setophaga cerulea* 12cm Active and agile; feeds mainly in canopy. **V** Song consists of short series of buzzy notes ending in high buzzing trill; call sharp *chip* and a loud *zzee* flight note. **H** Forests, also low bushes and small trees. **D** Breeds SE Canada to NC and NE USA. Winters Colombia to Peru.

5 NORTHERN PARULA *Setophaga americana* 11cm Very agile, often hangs upside-down whilst foraging in tree canopy. **V** Song ascending buzzing trill, ending with abrupt *tship*; calls include sharp *chip* and weak *tsif* given in flight. **H** Mainly lowland dry forests and scrub, but also damp mountain forests. **D** Breeds SC and SE Canada, C and E USA. Winters Mexico, Central America, West Indies.

6 TROPICAL PARULA *Setophaga pitiayumi* 11cm Gleans and hovers to collect insects in canopy. **V** Song accelerating buzzy trill, preceded by several high-pitched notes. **H** Deciduous forest, forest edges, clearings. **D** Widespread from S Texas to South America.

7 MAGNOLIA WARBLER *Setophaga magnolia* 13cm Active and agile forager in tree foliage at low to mid-levels. **V** Song short, musical *weety-weety-wee* or *weety-weety-weety-wee*, last note occasionally higher; calls include full *tship* or *dzip*, harsh *tshekk* and buzzy *zee* flight note. **H** Young conifer stands. On migration in other woods and tall scrub. **D** Breeds C and SE Canada to NE and NC USA. Winters Mexico, Central America, West Indies.

8 BAY-BREASTED WARBLER *Setophaga castanea* 14cm A deliberate forager from middle to high levels. **V** Song series of high *see* notes, all on one pitch; calls include high-pitched *sip* or *see*, and occasionally loud *chip*. **H** Forest edge, woodlands, open areas with scattered trees, gardens. **D** Breeds S and SE Canada to NE USA. Winters NW South America.

9 BLACKBURNIAN WARBLER *Setophaga fusca* 13cm Tends to forage high in canopy. **V** Song variable, tends to consist of series of *swee* notes followed by a high-pitched trill; call high, sharp *tsip* and thin *seet* given in flight. **H** Conifers, high trees, woodlands, tall bushes. **D** Breeds SC and SE Canada to NC, NE and E USA. Winters Costa Rica to Bolivia.

10 AMERICAN YELLOW WARBLER *Setophaga aestiva* 13cm Agile, active feeder in trees, bushes and on ground. **V** Song high-pitched *sweet-sweet-sweet-I'm-so-sweet*; calls include musical *tship* and buzzy *zzee*. **H** Riparian thickets, bushy areas including gardens. **D** Breeds widely in North America to central plateau of Mexico. Winters S North America to Amazonia.

11 MANGROVE WARBLER *Setophaga petechia* 13cm Split from American Yellow Warbler. Agile, active feeder in trees, bushes and on ground. Other races vary in intensity of rufous on crown. **V** Song loud, clear and rapid *sweet-sweet-sweet-ti-ti-ti-weet*; calls include loud *tship* and high *zzee*, usually given in flight. **H** Mangroves, coastal scrub and also in mountain forest on Martinique. Breeds from March to July. **D** Extreme S of USA through West Indies and coastal Mexico, Central America and South America.

12 CHESTNUT-SIDED WARBLER *Setophaga pensylvanica* 13cm Agile forager at low to middle levels in shrubs and lower branches of trees. **V** Song transcribed as *pleased-pleased-pleased-to-meecha*; call low, flat *tchip* and, in flight, utters rough *zeet*. **H** Open woodlands and gardens with trees. **D** Breeds SC and SE Canada to NC, NE and E USA. Winters Guatemala to N South America.

13 BLACKPOLL WARBLER *Setophaga striata* 14cm Constantly on the move. **V** Song high-pitched *sit-sit-sit-sit-sit-sit*...; notes all on one pitch with middle ones more emphatic; call loud *smack* and high *seet*. **H** In native N America: coniferous forest; in non-breeding season uses wider range of woodland, also thickets and scrubby or bushy areas. **D** Breeds Alaska to E and SE Canada, NE USA. Winters N and C South America.

14 BLACK-THROATED BLUE WARBLER *Setophaga caerulescens* 13cm Active. Forages from low levels to canopy, although primarily in understorey. **V** Song husky *zweea-zweea-zweea-zwee*; calls include dull *stip* or *chup* and metallic *twik* flight note. **H** Mature deciduous and mixed woodland with rich undergrowth; also woodland clearings and logged areas. **D** Breeds NE North America. Winters Bahamas to N South America.

15 PALM WARBLER *Setophaga palmarum* 14cm Forages low in vegetation or on ground; hops and wags tail. **V** Song rising and accelerating buzzy trill; call husky *chik* or *sup*; in flight utters high *seep*. **H** Damp areas of conifer forests. After breeding visits weedy fields, marshes, urban areas. **D** Breeds N North America. Winters S USA through Central America.

16 OLIVE-CAPPED WARBLER *Setophaga pityophila* 13cm Feeds from ground level to treetops. **V** Song series of shrill, whistled notes; call *tsip*. **H** Primarily pine forests. **D** Cuba, Bahamas (Grand Bahama and Abaco).

17 PINE WARBLER *Setophaga pinus* 14cm Often occurs in small parties, foraging from ground level to treetops. **V** Song simple, even-pitched trill; call sharp *chip* and *zeet* given in flight. **H** Mature pine forests. **D** Breeds E North America. Winters S USA, Bahamas, Hispaniola.

18 MYRTLE WARBLER *Setophaga coronata* 14cm Feeds in low vegetation and bushes as well as treetops. **V** Song slow trill, *uwee-tuwee-tuwee-tuwee-tuwee*..., which often changes pitch at end; calls include sharp *chek* and thin *tsee* flight note. **H** Open conifer and mixed woodland; after breeding, frequents hedgerows, thickets, gardens. **D** Breeds Alaska, NW, SC and SE Canada and NC and NE USA. Winters S North America, Central America.

19 AUDUBON'S WARBLER *Setophaga auduboni* 14cm Very close relative of Myrtle Warbler, with which it hybridises where their ranges meet. **V** Like that of Myrtle Warbler; calls typically softer. **H** As for Myrtle Warbler. **D** Breeds N, W and C Mexico. Winters S North America, Central America.

20 GOLDMAN'S WARBLER *Setophaga goldmani* 14cm Similar to Myrtle Warbler but grey parts of plumage more blackish. **V** Like that of Myrtle Warbler. **H** Like that of Myrtle Warbler. **D** S Chiapas (SE Mexico), W Guatemala.

21 YELLOW-THROATED WARBLER *Setophaga dominica* 13cm Generally feeds high in trees, although does forage in bushes. **V** Song consists of descending series of whistles, ending in flourish; call sharp *chip* and a high *see* given in flight. **H** Pine forests, lowland forests, palms, gardens. **D** Breeds E and SE USA. Winters Mexico, Central America, Greater Antilles.

22 BAHAMA WARBLER *Setophaga flavescens* 13cm Close relative of Yellow-throated Warbler. Has strikingly long bill and entirely yellow underparts. **V** Song consists of loud, clear whistles, ascending in pitch; call soft *tsip*. **H** Caribbean pine forest. **D** Bahamas.

23 VITELLINE WARBLER *Setophaga vitellina* 13cm Forages from ground to canopy. Quite tame on Little Cayman. **V** Song wheezy *szwee-szwee-szwee-zee*. **H** Dry woodland, scrub, urban areas. **D** Cayman Islands, Swan Islands.

24 PRAIRIE WARBLER *Setophaga discolor* 12cm Forages from ground to middle levels; often joins mixed-species feeding flocks. **V** Song series of rising buzzy notes; call low, sharp *tchip* or *tsup*. **H** Dry coastal forest, thickets, agricultural areas with scattered trees, mangroves, gardens. **D** Breeds E and SE USA. Winters West Indies.

25 ADELAIDE'S WARBLER *Setophaga adelaidae* 12.5cm Forages mainly at high levels. Often forms part of mixed-species flocks. **V** Song variable trill; call *chick*. **H** Mainly dry coastal scrub and thickets. **D** W Puerto Rico.

26 BARBUDA WARBLER *Setophaga subita* 12.5cm The only bird species endemic to Barbuda. Has yellow underside and eyebrow, light grey upperside. **V** Little known; song a sweet, rapid, descending series of whistles. **H** Found in almost all habitats on the island, especially xeric scrub and dry forest. **D** Barbuda (Leeward Islands).

27 ST. LUCIA WARBLER *Setophaga delicata* 12.5cm Striking grey-and-yellow warbler, with black-outlined yellow crescent below eye, and double whitish wing-bar. Forages mainly at high levels, fly-catching and gleaning for insect prey. **V** Song protracted but variable trill; call *chick*. **H** Most forest types on the island. **D** St Lucia.

28 GRACE'S WARBLER *Setophaga graciae* 13cm Forages in treetops, creeps along branches in search of insects. **V** Song series of downslurred whistles, quickening and rising towards end; calls include sweet *chirp* and thin *tss* flight note. **H** Pine–oak forests, especially those with ponderosa or yellow pine. **D** Breeds SW USA and Mexico to Nicaragua. Winters N and C Central America.

29 BLACK-THROATED GREY WARBLER *Setophaga nigrescens* 13cm Forages mainly in understorey. **V** Song buzzy *weezy-weezy-weezy-weezy WEE-too*; calls include *thick* or *tup* and high-pitched *see* or *sip* flight note. **H** Dry open woodlands with brushy understorey and chaparral. Occurs in any woodland or scrub during migration. **D** Breeds W North America. Winters Mexico.

30 TOWNSEND'S WARBLER *Setophaga townsendi* 13cm Gleans and fly-catches insects from low level to canopy. Often part of mixed-species flocks. **V** Song consists of series of high *zee* notes followed by two or three high-pitched buzzy notes; calls include metallic *tick* or *tip*. **H** Montane coniferous or mixed forest. **D** Breeds SE Alaska, W Canada and NW USA. Winters Mexico, Central America.

31 HERMIT WARBLER *Setophaga occidentalis* 13cm Active acrobatic forager in treetops. **V** Song variable, includes buzzy *ze ze ze ze ze ze zee sitew*, other variants are longer; calls like those of Townsend's Warbler. **H** Mature conifer forests. **D** Breeds W USA. Winters Mexico, Central America.

32 GOLDEN-CHEEKED WARBLER *Setophaga chrysoparia* 14cm Forages mainly at middle to high levels, gleaning or fly-catching for insects. **V** Song buzzy *bzzzz layzee dazzee*; call high *tchip*. **H** Open scrubby woodland with dense stands of juniper. During migration, frequents mountain woods and forests. **D** Breeds SC Texas. Winters Mexico to Nicaragua.

33 BLACK-THROATED GREEN WARBLER *Setophaga virens* 13cm Forages primarily from middle to high levels, feeding on insects and caterpillars by gleaning, hovering to pick from vegetation or by fly-catching. **V** Song lisping *zee zee zee zo zee*; calls very similar to those of Townsend's Warbler. **H** Low- to middle-elevation forests, shady coffee plantations, woodlands, gardens. **D** Breeds C and SE Canada to NE, E and NC USA. Winters Mexico to N South America, West Indies.

34 CITRINE WARBLER *Myiothlypis luteoviridis* 14cm Note striking broad eyebrows with black lores. **V** Short, high, hurried, sizzled warbling. **H** Humid forest with dense undergrowth. **D** Venezuela to N Bolivia.

35 SANTA MARTA WARBLER *Myiothlypis basilica* 14cm Compare with Three-striped Warbler (Plate 289). **V** High, nasal warbling (lasting more than 10 secs). **H** Stunted forest with dense undergrowth. **D** NE Colombia.

36 WHITE-STRIPED WARBLER *Myiothlypis leucophrys* 14.5cm Note grey crown and striking long white eyebrow. **V** Song short *teéteé-tjeu*. **H** Dense undergrowth of riverine belts. **D** SC Brazil.

37 FLAVESCENT WARBLER *Myiothlypis flaveola* 14.5cm Bright olive above and bright yellow below; no black in face. Forages at ground level. **V** Series of three extremely high *fee* notes, changing to four staccato *cha* notes. **H** Lowland forest, secondary growth. **D** Colombia and Venezuela to NE Argentina.

38 WHITE-RIMMED WARBLER *Myiothlypis leucoblephara* 14.5cm Separable from White-striped Warbler by different facial pattern and by range. **H** Dense understorey of forest, woodland and riverine belts. **D** SE South America.

39 PALE-LEGGED WARBLER *Myiothlypis signata* 13.5cm Separable from Citrine Warbler by shorter eyebrows and yellower crescents below eyes. **V** Long, very high, meandering rattle (3–5 secs). **H** Along streams in undergrowth of humid forest. **D** Peru to NW Argentina.

1
2
3
4
5
6
7
8
9
10
11
12
13
14
15
16
17
18
19
20
21
22
23
24
25
26
27
28
29
30
31
32
33
34
35
36
37
38
39

NEW WORLD WARBLERS *CONTINUED*

1 BLACK-CRESTED WARBLER *Myiothlypis nigrocristata* 13.5cm Black crown stripe diagnostic. **V** Song descending, accelerating series of notes that change in intonation. **H** Dense undergrowth at edges and in clearings in montane forest. **D** Venezuela to Peru.

2 BUFF-RUMPED WARBLER *Myiothlypis fulvicauda* 13.5cm Buff rump and basal half of tail diagnostic. Forages at ground level. **V** Song forceful, decelerating, high rattle (2 secs). **H** Along streams in forest and woodland. **D** Honduras to N Bolivia.

3 RIVERBANK WARBLER *Myiothlypis rivularis* 13.5cm Black lateral crown stripes are lacking in birds in SW of range, while those in NE of range are more buff, especially in face. **V** Song starts with 2–4 downslurred *fiu* notes and continues with forceful rattle-like phrase (3 secs). **H** Along streams in lowland forest and in swampy areas. **D** N, SE and SC South America.

4 TWO-BANDED WARBLER *Myiothlypis bivittata* 13.5cm Note white around eyes. **V** Song pleasant, vigorous, high, hurried, up-and-down babbling, slightly rising at end. **H** Dense undergrowth with bamboo thickets in humid forest. **D** Peru to NW Argentina.

5 RORAIMAN WARBLER *Myiothlypis roraimae* 13.5cm Very closely related to Two-banded Warbler, considered conspecific by some authorities. Dusky-olive with black eye-stripe and border to crown. **V** Song a sequence of short notes followed by upslurred buzz, also has slower, more hesitant song type. **H** Understorey of tall, humid montane forest. **D** SE Venezuela, W Guyana, N Brazil.

6 CUZCO WARBLER *Myiothlypis chrysogaster* 13cm Separable from Two-banded Warbler by yellow (not olive) eyebrows. **V** Song a series of 10 or so sweet *tew* notes, increasing in volume; call *tek* or *tzek*. **H** Dense understorey of humid forest and secondary growth. **D** Peru.

7 CHOCO WARBLER *Myiothlypis chlorophrys* 13cm Almost identical to Cuzco Warbler; head markings a little darker. **V** Song high and very fast staccato buzzing trill. **H** Lowland humid forest and secondary growth with dense understorey. **D** Colombia, Ecuador.

8 WHITE-LORED WARBLER *Myiothlypis conspicillata* 13.5cm Note white eye-rings and supralorals, and dark grey (not black) crown stripes. **V** Song very short, very high twitter. **H** Humid forest, secondary growth, plantations. **D** N Colombia.

9 GREY-THROATED WARBLER *Myiothlypis cinereicollis* 14cm Note lack of black in face and along yellow crown stripe. **V** Song simple, short, extremely high, sharp phrase. **H** Undisturbed undergrowth of humid forest. **D** Venezuela, Colombia.

10 GREY-AND-GOLD WARBLER *Myiothlypis fraseri* 14cm Distinctive in its range. Note all-grey upperparts with white supralorals. **V** High, nasal series of 6–7 *dzee* notes. **H** Dry (when breeding) and humid (after breeding) forest, woodland and scrub. **D** Ecuador, Peru.

11 RUSSET-CROWNED WARBLER *Myiothlypis coronata* 14cm Note striking central orange crown stripe and long stripe behind eyes. **V** Song hurried, randomly meandering phrase of rather pure, fluted notes, repeated in a continuous series (10 secs). **H** Dense understorey of humid forest and secondary growth. **D** Venezuela to C Bolivia.

12 GREY-HEADED WARBLER *Myiothlypis griseiceps* 14cm No other grey-headed warbler occurs in its range. **V** Song a loud series of tuneful slurred notes; calls include *tsek, chak* and *tseng*. **H** Undisturbed montane forest. **D** N Venezuela.

13 FAN-TAILED WARBLER *Basileuterus lachrymosus* 10cm Difficult to see well when feeding, much easier when singing, either from exposed perch or undulating song-flight. **V** Song simple, persistently repeated *zit-zit-zit-zit-zit...*; call usually *chip* or *plip*, also *tsipp-tsipp-tsipp*. **H** Tall grass, reeds, sedges, grassy wasteland, cultivations, salt-marsh vegetation. **D** Mexico to Nicaragua.

14 RUFOUS-CAPPED WARBLER *Basileuterus rufifrons* 13cm Several races occur, differing in extent of yellow on underparts. Whitish belly and finely black-marked white streak under russet cheek diagnostic. **V** Song extremely high warble (4 secs), with ending recalling Winter Wren (Plate 234), *wit weetweetweet*. **H** Semi-open areas with trees, scrub, hedges, open woodland. **D** Mexico and Guatemala south to Venezuela.

15 BLACK-CHEEKED WARBLER *Basileuterus melanogenys* 14cm Darker than Three-striped Warbler and not in same range. **V** Song extremely high rapid tinkling (3–7 secs). **H** Montane forest (undergrowth) and adjacent areas. **D** Costa Rica, Panama.

16 PIRRE WARBLER *Basileuterus ignotus* 13cm Note rufous coronal stripe and blackish cheeks. Dusky-looking, with striking eyebrows. **V** Call a compressed, penetrating *tseeit*. **H** Elfin forest. **D** Panama, Colombia.

17 GOLDEN-BROWED WARBLER *Basileuterus belli* 13cm No white in plumage. **V** Song high *tjutjutitisfiweehweeh* (*-sfi-* phrase extremely high). **H** Forest interior and edges. **D** Mexico to Honduras.

18 GOLDEN-CROWNED WARBLER *Basileuterus culicivorus* 12.5cm Often flicks wings and cocks tail. Race *hypoleucus* has white rather than yellow belly. **V** Song variable, normally short series of slurred whistles; call hard, dry *tek* and a loose rattle. **H** Sub-montane forest and forest edges. **D** Widespread in Central and South America.

19 BLACK-EARED WARBLER *Basileuterus melanotis* 13cm Distinct face pattern with central crown stripe grey, not rufous. **V** Song extremely high staccato *treetreetreetruwreet*. **H** Montane forest undergrowth. **D** Costa Rica, W Panama.

20 TACARCUNA WARBLER *Basileuterus tacarcunae* 13cm Little-known, rather drab warbler, closely related to Three-banded Warbler. **V** Song a free-form but agitated-sounding jumble of trills, twitters and buzzing notes; call is a sharp *tchp*, sometimes repeated in a rapid chatter. **H** Lower montane humid forest, edges and mature secondary growth. **D** E Panama, NW Colombia.

21 THREE-BANDED WARBLER *Basileuterus trifasciatus* 12.5cm No similar bird occurs in its range. **V** Song short, extremely high warbling (1 sec). **H** Rainforest, banks of streams in dry forest, secondary growth. **D** Ecuador, Peru.

22 YUNGAS WARBLER *Basileuterus punctipectus* 13cm Drab olive-toned warbler with dark ear-coverts and crown, and pale eyebrow and central crown stripe. Close relative of Three-striped Warbler. **V** Song like that of Tacarcuna Warbler. **H** Lower montane humid forest and secondary growth. **D** SE Peru to SC Bolivia.

23 THREE-STRIPED WARBLER *Basileuterus tristriatus* 13cm Separable from Santa Marta Warbler (Plate 288) by range and by slightly different face pattern. **V** Song like that of Golden-crowned Warbler. **H** Humid forest and secondary growth. **D** Venezuela to C Peru.

24 CANADA WARBLER *Cardellina canadensis* 13cm Feeds on insects by fly-catching or gleaning from vegetation in the understorey; active, often cocks tail. **V** Song variable warble; call sharp *chick* or *tyup*; flight call recorded both as low *plik* or high *zzee*. **H** Open vegetation among scattered trees, often near water or swamps. **D** Breeds C and SE Canada to NE, E and NC USA. Winters Panama to N South America.

25 WILSON'S WARBLER *Cardellina pusilla* 12cm Active, constantly flicks wings and tail. Forages in thick undergrowth. **V** Song staccato *chi-chi-chi-chi-chi-chet-chet*; call loud, low *chet*, hard *tik* and downslurred *tsip* flight note. **H** Various woodland types with thick undergrowth. **D** Breeds W and N North America. Winters Mexico, Central America.

26 RED-FACED WARBLER *Cardellina rubrifrons* 14cm Unmistakable. Usually forages on higher branches. **V** Song thin *towee towee towee tsew tsew wetoo weeeeew*; call sharp *tuk* or *tship*. **H** Firs and maples in mountain canyons. **D** Breeds SW USA, NW Mexico. Winters Mexico, Central America.

27 RED WARBLER *Cardellina rubra* 13cm Unmistakable. **V** Song very high rattling simple *wheetwheetsrrrrrrrohweet*. **H** Forest, woodland. **D** Mexico.

28 PINK-HEADED WARBLER *Cardellina versicolor* 13cm Unmistakable. **V** Song extremely high slightly rattling *tjuhtjuhtitjitjit*. **H** Forest interior, edges. **D** S Mexico, Guatemala.

29 PAINTED WHITESTART *Myioborus pictus* 13–13.5cm Active and acrobatic, constantly flicks and spreads wings and tail. **V** Song soft, musical warble; calls include *chidi-ew, chweee, tseeoo* and *bdeeyu*. **H** Pine–oak and oak forests. **D** SW USA to Nicaragua.

30 SLATE-THROATED WHITESTART *Myioborus miniatus* 13cm Several races occur, differing in colouring of underparts (from yellow to red) and barring of undertail coverts. Rufous-russet crown and slate-black wings diagnostic. Forages in middle levels. **V** Song very high *tritriwéehwéehwéeh*. **H** Montane forest and woodland. **D** Mexico to Bolivia and Guyana.

31 BROWN-CAPPED WHITESTART *Myioborus brunniceps* 13cm Separable from Tepui Whitestart by range. **V** Extremely high, level series of *see* notes (1 sec). **H** Montane and sub-montane forest and woodland; 1,400–3,200m, locally lower or higher. **D** Bolivia, NW and WC Argentina.

32 YELLOW-CROWNED WHITESTART *Myioborus flavivertex* 13.5cm Note that crown patch is yellow and mantle is olive-green. **V** Song like that of White-fronted Whitestart. **H** Humid montane forest. **D** NE Colombia.

33 WHITE-FRONTED WHITESTART *Myioborus albifrons* 13.5cm Note conspicuous white spectacles and lores. **V** Song very high, sharp warbling (5 secs). **H** Montane forest. **D** W Venezuela.

34 GOLDEN-FRONTED WHITESTART *Myioborus ornatus* 13.5cm Note yellow forecrown. No similar whitestart occurs in its range. **V** Song very high, unstructured warbling (5 secs). **H** Montane forest. **D** Colombia, Venezuela.

35 SPECTACLED WHITESTART *Myioborus melanocephalus* 13.5cm Has conspicuous yellow spectacles. **V** Unobtrusive, very high, repetitive warbling (5–10 secs). **H** Humid montane forest and adjacent scrub. **D** S Colombia to N Bolivia.

36 COLLARED WHITESTART *Myioborus torquatus* 13cm Unmistakable. **V** Song extremely high unstructured *tsjutsjurotjuwhotsjuh...*; call extremely high *tsjip*; **H** Montane forest, adjacent more open areas. **D** Costa Rica, Panama.

37 PARIA WHITESTART *Myioborus pariae* 13cm Separable from Brown-capped Whitestart and Tepui Whitestart by yellow (not white) lores and eye-rings. **V** Song short, high, sharp sizzling (shorter than 1 sec). **H** Edges and clearings in forest. **D** NE Venezuela.

38 WHITE-FACED WHITESTART *Myioborus albifacies* 13cm White face-sides and black cap are diagnostic. **V** Not known. **H** Wet forest. **D** S Venezuela.

39 GUAIQUINIMA WHITESTART *Myioborus cardonai* 13cm No similar whitestart occurs in its range. Note dark lores. **V** Not known. **H** Humid forest, riverine belts, scrub. **D** SE Venezuela.

40 TEPUI WHITESTART *Myioborus castaneocapilla* 13cm Compare with Brown-capped Whitestart. **V** Song very high, descending, rattle-like (1 sec). **H** Edges and clearings in montane forest. **D** Tepuis in SE Venezuela, W Guyana and N Brazil.

1
2
3
4
5
6
7
8
9
10
11
12
13
14
15
16
17
18
19
20
21
22
23
24
♀
♂
25
♂
♀
26
27
28
29
30
31
32
33
34
35
36
37
38
39
40

MITROSPINGID TANAGERS MITROSPINGIDAE

1 DUSKY-FACED TANAGER *Mitrospingus cassinii* 18cm Note distinctive whitish eye in black mask. **V** Song unmusical, unstructured, sharp, mid-high twittering. **H** Dense shrub at forest borders and in gaps, secondary growth and dense woodland, often along streams. **D** Costa Rica to Ecuador.

2 OLIVE-BACKED TANAGER *Mitrospingus oleagineus* 18.5cm No similar bird with grey mask and green plumage occurs in its range. **V** Song consists of single strained *zwee-eet* notes, rising and falling in pitch; calls include *seep, tic-tic-tic* and buzzing *pzzzz*. **H** Interior of forest on slopes of tepuis. **D** Tepuis in SE Venezuela, W Guyana and N Brazil.

3 OLIVE-GREEN TANAGER *Orthogonys chloricterus* 19cm Note rather bright yellow throat and slender, curved bill. **V** Song continuous fast twitters and rattles. **H** Canopy and borders of tall humid forest. **D** SE Brazil.

4 RED-BILLED PIED TANAGER *Lamprospiza melanoleuca* 17cm Note red bill and breast pattern. Mainly in canopy. **V** Song hurried, very high, nasal twittering, interrupted by *weeeh* and other longer notes. **H** Tall forest. **D** Amazonia.

CARDINALS, GROSBEAKS AND (TANAGER) ALLIES CARDINALIDAE

5 FLAME-COLORED TANAGER *Piranga bidentata* 18–19cm White tail corners. Forages in middle to upper levels. **V** Song fluty, powerful *chik churree chuwee* or *churee chiree ch-ree chiwee...*; calls include hard *ch-t-ruk, p-terruk* and short *ch-duk*. **H** Oak and pine–oak forest. **D** W Mexico to Panama.

6 TOOTH-BILLED TANAGER *Piranga lutea* 18–19cm A close relative of (and very similar to) Hepatic Tanager, and formerly considered conspecific. **V** Like that of Hepatic Tanager. **H** Montane forest edges. **D** Costa Rica to Bolivia.

7 RED TANAGER *Piranga flava* 18–20cm Similar in appearance and behaviour to Hepatic Tanager, from which it was recently split. Lacks dusky ear-coverts of Hepatic Tanager. **V** Like that of Hepatic Tanager. **H** Open woodlands. **D** S Guyana through E Brazil to Argentina.

8 HEPATIC TANAGER *Piranga hepatica* 18–20cm Usually unobtrusive, forages in upper branches of tall trees. **V** Song consists of sweet rising and falling phrases; call low *chup*. **H** Montane pine–oak forests. **D** W USA to Nicaragua.

9 SUMMER TANAGER *Piranga rubra* 19cm Mainly arboreal, usually seen singly or in pairs, regularly encountered with migrating vireos, warblers and thrushes. **V** Song thrush-like series of sweet, clear notes; calls include *chick* and chattering *pit-a-chuck piki-i-tuck* or *piki-i-tuck-i-tuck*. **H** Woodlands, forest edges, gardens. **D** Breeds S USA, N Mexico. Winters Amazonia, Bolivia.

10 ROSE-THROATED TANAGER *Piranga roseogularis* 16cm Red of crown, tail and wings sharply demarcated from greyish body. Note broken eye ring. **V** Mewing *titithere; teet tjut tittjuut*. **H** Forest interior and edges. **D** Mexico to Guatemala.

11 SCARLET TANAGER *Piranga olivacea* 18cm Forages primarily in tops of trees, but will descend to feed on ground. **V** Song raspy *querit-queer-query-querit-queer*; call hoarse *chip-burr*; in flight, utters clear *puwi*. **H** Mature deciduous woodland. **D** Breeds SC and SE Canada to SE USA. Winters Panama to Amazonia.

12 WESTERN TANAGER *Piranga ludoviciana* 18cm Arboreal. Non-breeding male has orange confined to face. **V** Song similar to that of Scarlet Tanager; calls include soft, rising rattle and soft whistled *howee* or *weet* flight note. **H** Coniferous mountain forests. **D** Breeds W and C Canada through W USA to extreme NW Mexico. Winters NW Mexico and Central America.

13 WHITE-WINGED TANAGER *Piranga leucoptera* 14cm Separated from Scarlet Tanager by white wing bars and black lores. **V** Very high sandpiper-like *wéeweezweeh*. **H** Montane forests, plantations. **D** E Mexico to Bolivia.

14 RED-HEADED TANAGER *Piranga erythrocephala* 15cm Unmistakable. Note white belly of female. **V** Song a series of throaty single or double *churr* notes; calls include *tic-c-c* and *spik*. **H** Forest, plantations. **D** Mexico.

15 RED-HOODED TANAGER *Piranga rubriceps* 18cm Unmistakable due to colour pattern. **V** Very high, sharp, partly nasal twittering, punctuated with short trills. **H** Humid and wet montane forest. **D** Colombia to Peru.

16 RED-CROWNED ANT TANAGER *Habia rubica* 18cm Dark red-brown overall. Narrow red crown bordered black. **V** Very high *peeteh-peeteh* or very high fluting *truhperritruhperritruh*, slightly crescendoing. **H** Forest, secondary growth. **D** Widespread in Central America and South America.

17 RED-THROATED ANT TANAGER *Habia fuscicauda* 19cm Deeper red than Red-crowned Ant Tanager with more concealed red crown. Female differs from Red-crowned Ant Tanager by yellow chin and lack of yellow crown stripe. **V** Low *scratch*, continued in chattering mellow up-and-down *tjuppoh tjuppoh tjupperoh*. **H** Undergrowth at forest edge. **D** Mexico to Colombia.

18 BLACK-CHEEKED ANT TANAGER *Habia atrimaxillaris* 18cm Black cheeks diagnostic. **V** Song comprises low piping notes, *pew pe-du pew dee* or similar; calls include nasal *chet* and a harsh scraping sound. **H** Forest understorey. **D** Costa Rica.

19 SOOTY ANT TANAGER *Habia gutturalis* 19cm Unmistakable due to sooty plumage and contrasting pink crown and throat. **V** Slow series of well-separated, high, fluted, staccato single notes. **H** Borders of humid forest and secondary growth. **D** Colombia.

20 CRESTED ANT TANAGER *Habia cristata* 19cm Rosy crest (often erect) is diagnostic. **V** Unstructured series of grating *kr-veék* notes. **H** Dense lower storeys and borders of montane forest. **D** Colombia.

21 CARMIOL'S TANAGER *Chlorothraupis carmioli* 17cm Uniform olive-green with yellower throat. **V** Rapid, high, nasal series of *tjew* notes, increasing in strength. **H** Undergrowth of humid and wet forest and tall secondary growth. **D** Nicaragua to Colombia.

22 OLIVE TANAGER *Chlorothraupis frenata* 17cm Almost uniform dusky olive-green with brighter throat and lower belly. Note striking black bill. **V** Song high, pleasant, musical and thrush-like, rising and falling *weoweeoweeoweeo-ps-wéeoo ps-wéeooo*. **H** Lower forest storeys. **D** W Amazonia.

23 LEMON-SPECTACLED TANAGER *Chlorothraupis olivacea* 17cm Note distinctive yellow eye-rings and lores. **V** Song accelerating and decelerating, loud series of *tjuw* notes (5–15 secs). **H** Lower storeys of wet forest and secondary growth. **D** Panama to Ecuador.

24 OCHRE-BREASTED TANAGER *Chlorothraupis stolzmanni* 18cm Note rather pale eyes and heavy bill. **V** Song hurried, warbled chattering, punctuated with extremely high *tsee* notes and short trills and rolls. **H** Lower storeys and borders of wet forest. **D** Colombia, Ecuador.

25 YELLOW GROSBEAK *Pheucticus chrysopeplus* 21–24cm In flight shows extensive white corners on tail; male has large, white patch at base of primaries, female shows much smaller patch. **V** Song variable, short, rich warble, transcribed as *chee wee chee-r weer weeuh* or *toodi todi toweeoo*; calls include sharp *piik* and soft *hu-oi* or *whoi*. **H** Oak or waterside woodlands. **D** Mexico, Guatemala.

26 BLACK-THIGHED GROSBEAK *Pheucticus tibialis* 20cm Large, heavy bill and small face mask distinctive. **V** High cheerful *wie er méh gaat wéht ick nog niet*. **H** Forest and adjacent more open country. **D** Costa Rica, Panama.

27 GOLDEN GROSBEAK *Pheucticus chrysogaster* 21cm Unmistakable due to yellow plumage and black wings with distinct white markings. **V** Slightly hurried, stuttering series of high double notes. **H** Open woodland, arid scrub, forest borders. **D** Venezuela to Peru.

28 BLACK-BACKED GROSBEAK *Pheucticus aureoventris* 22cm Male separable from male Golden Grosbeak by black head; female by streaked underparts. **V** Song unstructured series that jumps up and down, with high, sharp and musical notes, and incidental small melodious rattle. **H** Dry open woodland, gardens. **D** W and SC South America.

29 ROSE-BREASTED GROSBEAK *Pheucticus ludovicianus* 19cm Forages in trees and bushes, often in small parties mixed with other species. In flight, male shows rosy underwing coverts; female's buffy-yellow. **V** Song thrush-like slow, whistled warble; calls include squeaky *iik* or *eek* and soft, wheezy *wheek* flight call. **H** Forest edge, woodlands, shade coffee plantations, scrub, gardens. **D** Breeds C and SE Canada, E and C USA. Winters Mexico to N South America.

30 BLACK-HEADED GROSBEAK *Pheucticus melanocephalus* 21cm Behaviour like that of Rose-breasted Grosbeak. In flight, male shows orangeish upper rump, white primary bases and tail corners; underwing of male and female yellow. **V** Song whistled warble, higher and faster than that of Rose-breasted Grosbeak; call sharp *pik*. **H** Open woodlands, forest edges. **D** Breeds W North America. Winters Mexico.

31 RED-BREASTED CHAT *Granatellus venustus* 15cm Male and female darker and with more white in tail than Grey-throated Chat. White throat of male diagnostic. **V** Extremely high *tuwéetuwéesjeesjeeweeweeh; tsjt*. **H** Woodland. **D** Mexico.

32 GREY-THROATED CHAT *Granatellus sallaei* 13cm Unmistakable. **V** Very high short descending *fifjiwijuwiweeh; tutrrrrr*. **H** Thickets, woodland, forest. **D** Mexico, Guatemala, Belize.

33 ROSE-BREASTED CHAT *Granatellus pelzelni* 12.5cm Male unmistakable. Note pink vent of female. **V** Level series of 5–6 high, loud notes, which may accelerate, decelerate or change to a rattle. **H** Higher forest levels, woodland and shrubbery near water. **D** Amazonia.

34 NORTHERN CARDINAL *Cardinalis cardinalis* 22cm Unmistakable. Forages on ground and in trees and shrubs. **V** Song series of high, clear and slurred whistles, very variable, including *whoit whoit whoit what cheer what cheer wheat wheat* and *purty purty purty*; call high, hard *tik* and soft, rising *twik*. **H** Woodland margins, thickets, gardens. **D** E, S and SW USA, Mexico.

35 VERMILION CARDINAL *Cardinalis phoeniceus* 18.5cm Unmistakable due to long, erect crest. **V** Calm series of 3–8 *seet* or *sweet* notes. **H** Arid scrub. **D** Colombia, Venezuela.

36 PYRRHULOXIA *Cardinalis sinuatus* 22cm Unmistakable. Habits like those of Northern Cardinal. **V** Song loud, rich, repeated whistles; calls include low *spik* and chattering *pikpikpikpikpik*. **H** Arid or semi-arid scrub. **D** SW USA, Mexico.

1
2
3
4
5
6
7
8
9
10
11
12
13
14
15
16
17
18
19
20
21
22
23
24
25
26
27
28
29
30
31
32
33
34
35
36

CARDINALS, GROSBEAKS AND (TANAGER) ALLIES *CONTINUED*

1 BLACK-FACED GROSBEAK *Caryothraustes poliogaster* 18cm Distinguished from Yellow-green Grosbeak by grey rump and whitish belly, and by range. **V** Very high decisive *tji-tjuw-tjuw-tjuw-tji*. **H** Forest interior and edge, semi-open growth. **D** Mexico to Panama.

2 YELLOW-GREEN GROSBEAK *Caryothraustes canadensis* 18cm Unmistakable, with black mask contrasting with yellow plumage. **V** Short *dzeet*, followed by *chetchet*. **H** Borders and interior of humid forest. **D** E Panama to N and E Brazil.

3 CRIMSON-COLLARED GROSBEAK *Rhodothraupis celaeno* 21–22cm Skulking. In flight, underwing coverts are red in the male and yellowish in the female. **V** Song is a husky warble ending in an upslurred *weeee*. Calls include a clear *ssseeuu* and a piercing *seeip seeeiyu*. **H** Brushy woodland. **D** NE Mexico.

4 RED-AND-BLACK GROSBEAK *Periporphyrus erythromelas* 20.5cm Male especially is unmistakable. **V** Series of leisurely delivered phrases of full, rich, rising and descending *tjeeh-towhee* notes. **H** Lower storeys of mature humid forest. **D** NE South America, from SE Venezuela through the Guianas to NE Brazil.

5 CABANIS'S SEEDEATER *Amaurospiza concolor* 13cm Males of all four *Amaurospiza* seedeaters are predominantly dark blue, and females are warm brown. Distinguished from Indigo Bunting by uniform dull blue, not pure blue, plumage; from Blue-black Grassquit (Plate 299) by blunt bill and different habitat. **V** Warble, followed by splutter, *seet see-wee-see si-si-wee-su*. **H** Forest edge, woodland, shrubbery, bamboo. **D** S Mexico, Belize, Honduras to Panama.

6 ECUADORIAN SEEDEATER *Amaurospiza aequatorialis* 12.5cm Very like Cabanis's, Carrizal and Blackish-blue seedeaters, best separated by range. **V** Short *seet-seet-suweet-seet* phrase. **H** Borders and undergrowth with bamboo in forest and secondary growth. **D** SW Colombia to NW Peru.

7 BLACKISH-BLUE SEEDEATER *Amaurospiza moesta* 12.5cm Male is almost black on face and breast. As in the other *Amaurospiza* seedeaters, both sexes have white underwing coverts. **V** Song resembles that of Ecuadorian Seedeater, but sharper. **H** Forest edge with bamboo. **D** SE Brazil, E Paraguay, NE Argentina.

8 CARRIZAL SEEDEATER *Amaurospiza carrizalensis* 12.5cm Best separated from Cabanis's, Ecuadorian and Blackish-blue seedeaters by range. **V** Sings as Ecuadorian Seedeater. **H** Bamboo undergrowth within forest. **D** SE Venezuela.

9 DICKCISSEL *Spiza americana* 15–18cm Gregarious, often mixes with other species, foraging mostly on the ground. Non-breeding male's bib greyish. **V** Song is a staccato, insect-like *dik-dik-serr-si-si* or *dick- dick-ciss-ciss-ciss*. Calls include a dry *chek* or *pwik*, and a buzzing *dzzrrrt* given in flight. **H** Prairie, open grassland and weedy areas with scattered trees. **D** Breeds Midwest USA. Winters Mexico to Venezuela and Colombia.

10 GLAUCOUS-BLUE GROSBEAK *Cyanoloxia glaucocaerulea* 14cm Male is brighter blue than male Blue-black Grosbeak and male Ultramarine Grosbeak. **V** High, repetitive warbling (3 secs). **H** Undergrowth and borders of low forest, marshes, secondary growth. **D** SE Brazil, E Paraguay, NE Argentina.

11 BLUE-BLACK GROSBEAK *Cyanocompsa cyanoides* 16cm Generally not in same range as Rothschild's or Ultramarine Grosbeaks. Male is deep blue; female is uniform dark brown. **V** Song starts with a few full, fluted notes then disappears in lowered mumbling. **H** Dense lower growth in humid forest and woodland. **D** SE Mexico to NW Venezuela and NW Peru.

12 ROTHSCHILD'S GROSBEAK *Cyanocompsa rothschildii* 16cm Very similar to closely related Blue-black Grosbeak. Male is deep blue; female is uniform dark brown. **V** Song series of sweet but weak chiming notes with lilting rhythm; call dry *tchit*. **H** Understorey of humid forest, forest edges and secondary growth. **D** W Amazonia.

13 ULTRAMARINE GROSBEAK *Cyanocompsa brissonii* 15cm Similar to Blue-black Grosbeak, but male is brighter blue, with paler blue above eye, on malar area and on shoulder patch. **V** Song is a slightly lowered, loud, musical warbling. **H** Dense undergrowth in open woodland. **D** Colombia to E Brazil and N Argentina.

14 BLUE BUNTING *Cyanocompsa parellina* 14cm Heavy bill. Secretive, forages on or near the ground in cover. **V** Song is a variable sweet warble, often beginning with two separate notes and fading at the end. Call is a metallic *chik*. **H** Shrubby thickets, brushy forest, woodland and woodland edge. **D** Mexico to Nicaragua.

15 BLUE GROSBEAK *Passerina caerulea* 17cm Note double brown wing-bar in both sexes. Forages in trees, shrubs and on the ground. Regularly twitches and flares tail. **V** Song is a rapid warble with short rising and falling phrases. Calls include a metallic *chink* and a harsh buzzing flight call. **H** Forest edge, Casuarina groves, rice fields, seeding grass areas near woodland, gardens. **D** Breeds USA (from N California, South Dakota, New Jersey southwards) to Costa Rica. Winters Mexico to Panama.

16 INDIGO BUNTING *Passerina cyanea* 14cm Male all blue; female has buff wing-bars. Feeds in trees, bushes and on the ground, often in flocks. **V** Song is a high-pitched *sweet-sweet where-where here-here see-it-see-it*. Typical call is a sharp *tsick*. **H** Rich fields, pasture borders, grassy areas near thickets, woodland, dry scrub. **D** Breeds SC and SE Canada to SW, SC and SE USA. Winters Mexico to Venezuela.

17 LAZULI BUNTING *Passerina amoena* 14cm Male unmistakable, with spectacular plumage. Feeds on or near the ground in deep cover. **V** Rapid *see-see-sweert-sweert-sweert-zee-sweet-zeer-see-see*. Calls include a sharp *tzip* or *pit* and a dry buzz. **H** Open woodland, thickets (especially near water), overgrown fields, cultivation. **D** Breeds SW and SC Canada through W USA to extreme NW Mexico. Winters W Mexico.

18 VARIED BUNTING *Passerina versicolor* 14cm In dull light male can appear black. Shy, forages on or near the ground in cover. **V** Thin warbled notes. Call, a dry *spik*. **H** Thickets in canyons and washes. **D** SW USA to Guatemala.

19 PAINTED BUNTING *Passerina ciris* 13cm Male's plumage is a dramatic combination of blue, green and red. Forages primarily on or near the ground in thick vegetation. **V** Song is a sweet, continuous warble. Call is a loud *chip* or *pwich*. **H** Thickets, brush and grassy places in semi-arid areas, usually near water. **D** Breeds S, SE USA. Winters Mexico to Panama, Caribbean.

20 ROSE-BELLIED BUNTING *Passerina rositae* 14cm Note bright red belly of male and narrow eye-ring of both sexes, in female set in grey face side. **V** Ultra-high *tsiet*; very high descending *srietsriet-wuhsriet*. **H** Dry woodland, riverine belts. **D** S Mexico.

21 ORANGE-BREASTED BUNTING *Passerina leclancherii* 12cm Unmistakable. Note orange tone to breast of female. **V** High *wreetohwreetwitwetwit*. **H** Thornbush, dry woodland, arid scrub. **D** S Mexico.

TANAGERS AND ALLIES THRAUPIDAE

22 BROWN TANAGER *Orchesticus abeillei* 18cm Separable from foliage-gleaners (especially Buff-fronted Foliage-gleaner, Plate 130) by stubby bill. **V** Ultra-high twittering just within/beyond human hearing. **H** Humid forest, open woodland, secondary growth, settlements. **D** SE Brazil.

23 YELLOW CARDINAL *Gubernatrix cristata* 20cm Unmistakable, with long crest and distinctive facial pattern. **V** Song is a repeated phrase, *tff-tjee-tof* (repeated 3–4 times). **H** Large trees in thorny woodland. **D** N Argentina, Uruguay, extreme SE Brazil.

24 RED-CRESTED CARDINAL *Paroaria coronata* 19cm Combination of grey and white, with dramatic red crest, is distinctive. **V** Slow series of level or downslurred *tiur* notes. **H** Forest edge, Chaco and other woodland, parks, agricultural areas. **D** Bolivia, S Brazil, Uruguay to C Argentina.

25 RED-COWLED CARDINAL *Paroaria dominicana* 18.5cm No other cardinal occurs in its range. Note grey–black–white pattern on back and wings. **V** Slow series of alternating *tiuw* and *tuweet* notes. **H** Caatinga, scrub and other semi-open habitats. **D** NE Brazil.

26 RED-CAPPED CARDINAL *Paroaria gularis* 16.5cm Note black throat. Separable from Yellow-billed Cardinal by bicoloured bill and grey legs. **V** Slow series of alternating, sharp *tiuw* and *tuweet* notes. **H** Shrubbery and dense vegetation near waterbodies. **D** Amazonia.

27 MASKED CARDINAL *Paroaria nigrogenis* 21.5cm No other *Paroaria* cardinal occurs so far north in South America. Extensive red crown and nape is separated from lower cheeks and throat by black mask. **V** Song comprises well-spaced upslurred and downslurred chirping notes. Calls include sharp *tchep*. **H** Montane forest. **D** From E Colombia through N Venezuela to Trinidad.

28 CRIMSON-FRONTED CARDINAL *Paroaria baeri* 16.5cm Separable from Masked Cardinal by different range, black neck sides, and black-bordered red bib (reduced to red malar stripes in one race). **V** Song is a series of powerful, harsh descending whistles. Calls include a harsh sparrow-like chirrup. **H** Shrubby vegetation at riversides. **D** C Brazil.

29 YELLOW-BILLED CARDINAL *Paroaria capitata* 16.5cm Yellowish-pink bill and legs are diagnostic. **V** Very high series of alternating, piercing *tiuw* and *tuweet* notes. **H** Marshes and woody vegetation along rivers. **D** C and E Bolivia through SW Brazil and Paraguay to W Uruguay, NE Argentina.

30 CINNAMON TANAGER *Schistochlamys ruficapillus* 18cm Colour pattern and narrow black mask are distinctive. **V** Song resembles that of Black-Faced Tanager but slightly more rolling. **H** Semi-open habitats with scattered trees, bushes and secondary growth, clearings, gardens. **D** E and S Brazil to E Paraguay, NE Argentina.

31 BLACK-FACED TANAGER *Schistochlamys melanopis* 18cm All-grey plumage with black face and throat is unmistakable. **V** Fairly long, melodious phrase containing drawn-out *weee* and *tjuh* notes (3–5 secs). **H** Grassy areas with scattered shrub and tree stands, Cerrado, restinga. **D** Colombia to S Peru, SE Brazil.

32 MAGPIE TANAGER *Cissopis leverianus* 27cm Pied pattern and long tail are diagnostic. **V** Rapid mixture of high, nasal twittering, inhaled notes and small rattles in a hurriedly repeated phrase (7–9 secs). **H** Variety of semi-open habitats like forest clearings, river islands, plantations, gardens, open secondary growth. **D** From Colombia through Amazonia to SE Brazil, NE Argentina.

33 BLACK-AND-WHITE TANAGER *Conothraupis speculigera* 16cm Note grey rump and white shoulder patch of male. Breast of greenish female is faintly streaked. **V** Sustained series of alternating single, strident *tjiw* and soft, low *tjeuu* notes. **H** Woodland, shrubby clearings. **D** Ecuador, Peru. Breeds on west slopes of Andes. Winters east of Andes.

34 CONE-BILLED TANAGER *Conothraupis mesoleuca* 16cm Note large, pale bill, darker in brown female. Critically endangered. **V** Strong, bouncing-down rattle, starting with an *ee* and changing to a *u*, then suddenly rising to a hoarse, inhaled *seee*. **H** Riverine belts. **D** WC Brazil.

35 SCARLET-THROATED TANAGER *Compsothraupis loricata* 21.5cm Male is glossy blue-black apart from red throat and breast. Female all black. **V** Level, unstructured, nasal and mewing chatters. **H** Caatinga, riverine belts, semi-open habitats at waterbodies. **D** E Brazil.

36 WHITE-CAPPED TANAGER *Sericossypha albocristata* 23.5cm Large, thrush-sized tanager with a distinctive white crest. **V** Series of high to ultra-high, piercing shrieks. **H** Mainly interior of montane forest, but also forest edge. **D** Venezuela to Peru.

37 HOODED TANAGER *Nemosia pileata* 13cm Head pattern of male is distinctive; female has a bicoloured bill. **V** Ultra-high, thin trill. **H** Forest edge, riverine belts, clearings, Cerrado, mangrove. **D** South America, widespread as far south as N Argentina.

38 CHERRY-THROATED TANAGER *Nemosia rourei* 13.5cm No similar bird occurs in its restricted range. Critically endangered. **V** Short, hurried series of 2–4 very high *tseeeh* notes. **H** Canopy and borders of montane forest. **D** SE Brazil.

1
2
3
4
5
6
7
8
9
10
11
12
13
14
15
16
17
18
19
20
21
22
23
24
25
26
27
28
29
30
31
32
33
34
35
36
37
38

TANAGERS AND ALLIES *CONTINUED*

1 RUFOUS-CRESTED TANAGER *Creurgops verticalis* 16cm Grey upperparts, rufous underparts and dark eye area are diagnostic. Male has a semi-concealed rufous crown patch. **V** Ultra-high, barely audible *see* notes. **H** Mainly cloud forest. **D** Venezuela to Peru.

2 SLATY TANAGER *Creurgops dentatus* 15.5cm Male slate-grey with conspicuous chestnut cap. Female has grey upperparts, white supercilium, rufous breast and flanks. **V** Very high to ultra-high, thin twittering. **H** Mainly cloud forest. **D** SE Peru, Bolivia.

3 BLACK-CAPPED HEMISPINGUS *Hemispingus atropileus* 17cm Like all *Hemispingus* tanagers, is insectivorous and has a slender bill. Separable from smaller Superciliaried Hemispingus by more extensive black on crown and cheeks. **V** Mixture of meandering twitters. **H** Dense undergrowth and borders of montane forest. **D** Venezuela to Ecuador.

4 WHITE-BROWED HEMISPINGUS *Hemispingus auricularis* 16cm Closely related (and sometimes lumped with) Black-capped Hemispingus; has whiter eyebrow and more richly coloured underparts. **V** Song is a slow and steady series of *chew* notes alternating with higher-pitched *zeeet* notes. **H** Humid and wet montane forest and forest edges with dense undergrowth. **D** Peru.

5 ORANGE-BROWED HEMISPINGUS *Hemispingus calophrys* 16cm Note distinctive orange supercilium and throat. Separable from Superciliaried Hemispingus by solid black ear-streak and orange tinge to throat. **V** Ultra-high, thin, rhythmic ticking, like a clock, *tzic-tzac*. **H** Undergrowth with bamboo in humid forest. **D** Bolivia, S Peru.

6 PARODI'S HEMISPINGUS *Hemispingus parodii* 16cm Separable from similar Citrine Warbler (Plate 288; sometimes in same flock) by stouter, paler bill and greyish (not pink) bill. **V** Ultra-high twittering. **H** Bamboo in undergrowth of forest at the treeline. **D** SC Peru.

7 SUPERCILIARIED HEMISPINGUS *Hemispingus superciliaris* 14cm Black forehead and lores fade into green of crown and cheeks. White supercilium is distinctive. **V** Ultra-high rolls and twitters. **H** Canopy and borders of humid forest and secondary growth, *Polylepis* woodland. **D** Venezuela to N Bolivia.

8 GREY-CAPPED HEMISPINGUS *Hemispingus reyi* 14cm Combination of grey cap and lack of supercilium is diagnostic. **V** Fast, ultra-high twittering. **H** Bamboo in undergrowth of wet montane forest and woodland. **D** W Venezuela.

9 OLEAGINOUS HEMISPINGUS *Hemispingus frontalis* 14cm Nondescript tanager, with less black bill and greyer legs than *Basileuterus* warblers (Plate 289). **V** Very high, hurried, nasal twittering. **H** Thick undergrowth of humid montane forest. **D** Venezuela to Peru.

10 BLACK-EARED HEMISPINGUS *Hemispingus melanotis* 14cm Note grey crown, olive-grey upperparts, black mask, inconspicuous white arc below eye, cinnamon throat and breast. **V** Very high, nasal, thin twittering. **H** Undergrowth and borders of humid montane forest. **D** W Venezuela to WC Bolivia.

11 WESTERN HEMISPINGUS *Hemispingus ochraceus* 14cm Related to Black-eared Hemispingus, but plumage much drabber than that species. **V** Like that of Black-eared Hemispingus. **H** Moist to humid forest and mature secondary growth. **D** SW Colombia, W Ecuador.

12 PIURA HEMISPINGUS *Hemispingus piurae* 14cm Related to Black-eared Hemispingus, but has black head with striking white eyebrow, less grey on upperparts, and rich cinnamon-rufous underparts. **V** Song similar to that of Black-eared Hemispingus but with slower delivery. **H** Humid montane forest with bamboo understorey. **D** W Venezuela to WC Bolivia.

13 SLATY-BACKED HEMISPINGUS *Hemispingus goeringi* 14cm Distinctively patterned; note dark grey back. **V** Pairs produce a duet, one bird giving a series of fast staccato notes while the other produces a tinkling *chi-ti-tee chi-ti-tee*. **H** Dense borders of montane and elfin forest. **D** SW Venezuela.

14 RUFOUS-BROWED HEMISPINGUS *Hemispingus rufosuperciliaris* 15cm Strikingly patterned, with mostly grey upperparts, black ear-coverts and crown, tawny-red supercilium and breast. No similar bird occurs in its range. **V** Very high chirping to ultra-high twittering. **H** Bamboo and shrubbery in undergrowth of wet montane forest. **D** N and C Peru, east slopes of Andes.

15 BLACK-HEADED HEMISPINGUS *Hemispingus verticalis* 14cm Black head with pale eye and pale stripe over crown is distinctive. **V** Mixture of high, nasal and ultra-high, thin twittering. **H** Canopy and borders of dense wet montane forest, including elfin forest. **D** Colombia to N Peru.

16 DRAB HEMISPINGUS *Hemispingus xanthophthalmus* 13.5cm Pale eye and grey-white plumage are diagnostic. Note thin bill. **V** Song is a stuttering series of mostly high-pitched notes, sometimes given as a duet. Calls include *it-sit* and *tseee*. **H** Humid montane forest, scrubby woodland. **D** N Peru to NW Bolivia, east slopes of Andes.

17 THREE-STRIPED HEMISPINGUS *Hemispingus trifasciatus* 14cm Separable from similar hemispinguses by pale orange-yellow underparts and pale wing-bars. **V** Endlessly repeated, ultra-high *tzit*. **H** Treeline of forest and woodland. **D** Peru to C Bolivia, east slopes of Andes.

18 GREY-HOODED BUSH TANAGER *Cnemoscopus rubrirostris* 15cm Grey hood, contrasting with green back and yellow underparts, is diagnostic. Constantly wags tail. **V** Series of *tzit* and *rep* notes. **H** Canopy and borders of montane forest. **D** Venezuela to Bolivia.

19 FULVOUS-HEADED TANAGER *Thlypopsis fulviceps* 12.5cm Solid orange or brownish head is sharply demarcated from grey body. The *Thlypopsis* tanagers are small and insectivorous, with little or no overlap in their ranges. **V** Song is a series of abrupt stuttering notes, becoming faster then slower. Calls include brief *chip, tsit* and similar notes. **H** Dense canopy and borders of humid forest, secondary growth, shrubby gardens. **D** Colombia, Venezuela.

20 RUFOUS-CHESTED TANAGER *Thlypopsis ornata* 12.5cm Separable from Buff-bellied Tanager by uniformly coloured face sides. **V** High, nasal twittering. **H** Drier shrubby forest borders, secondary growth, woodland. **D** SW Colombia to Peru.

21 BROWN-FLANKED TANAGER *Thlypopsis pectoralis* 12.5cm Rufous head and breast contrast with olive-grey back. Flanks slightly browner than in other *Thlypopsis* tanagers. **V** Ultra-high twittering. **H** Open, wet elfin woodland. **D** C Peru.

22 ORANGE-HEADED TANAGER *Thlypopsis sordida* 13cm Head is two-toned: yellow around eye and throat, and orange on crown, nape and neck. **V** Calm, high to ultra-high twittering. **H** Open woodland, riverine belts, Cerrado. **D** C Venezuela, E Peru, Bolivia, W, C and E Brazil south to Paraguay and N Argentina.

23 BUFF-BELLIED TANAGER *Thlypopsis inornata* 12.5cm Male has crown and nape rich cinnamon-rufous, face yellowish, underparts pale buffy yellow. Female has a brownish-grey nape and neck. **V** Song is a fast series of short, high-pitched, slightly squeaky or nasal notes. Calls include soft *sip*. **H** Dry woodland and secondary growth, brushy clearings. **D** Extreme SE Ecuador to N Peru.

24 RUST-AND-YELLOW TANAGER *Thlypopsis ruficeps* 12.5cm The southernmost of the *Thlypopsis* tanagers. Male has bright rufous head contrasting with bright olive-green upperparts and yellow underparts. Note orange tinge to eye-stripe and cheeks of female. **V** Very high to ultra-high twittering, often in repeated phrases. **H** Montane and *Polylepis* woodland. **D** S Peru to NW Argentina.

25 CHESTNUT-HEADED TANAGER *Pyrrhocoma ruficeps* 14cm Male is distinctive, with grey body, deep chestnut head, small black mask. Female is grey with a green-yellow tinge, contrasting with yellowish-buff head. **V** Ultra-high, piercing series of 3–4 drawn-out *seee* notes. **H** Undergrowth and borders of forest and secondary growth. **D** E Paraguay, NE Argentina, S Brazil.

26 WHITE-RUMPED TANAGER *Cypsnagra hirundinacea* 16cm Distinctively patterned, with black upperparts, paler underparts, white wing-bars, and a conspicuous white rump. **V** Duet of strong *tjuw-tjuw* notes (female) and continuous low churring (male). **H** Grasslands with scattered trees, Cerrado, campina. **D** Suriname, French Guiana; NE Bolivia to Paraguay and SE Brazil.

27 PARDUSCO *Nephelornis oneilli* 13cm Rather drab, insectivorous tanager; note large head and pinkish bill and legs. **V** Call is a very high, thin *seep*, given in irregular twittering. Song not well documented. **H** Elfin woodland, low trees and bushes, treeline. **D** C Peru.

28 BLACK-GOGGLED TANAGER *Trichothraupis melanops* 16.5cm Plumage pattern of male is distinctive; female less so, but in flight both sexes show tawny undertail coverts and a white wing-bar. **V** Series of piercing, soft *swip* notes, separated by short, nasal three-syllable twitters. **H** Humid forest borders, secondary growth, riverine belts. **D** Disjunct populations: east slopes of Andes from N Peru to NW Argentina; E Brazil, E Paraguay, NE Argentina.

29 GREY-HEADED TANAGER *Eucometis penicillata* 17cm Grey head with shaggy crest is distinctive. **V** Loose series of well-separated *wits* notes. **H** Undergrowth of forest and woodland, riverine belts, secondary growth. **D** SE Mexico to SC Brazil and N Paraguay.

30 FLAME-CRESTED TANAGER *Tachyphonus cristatus* 16cm Male separable from male Fulvous-crested Tanager by all-black flanks. **V** Rapid phrase of four ultra-high *tsee* notes. **H** Higher levels and borders of humid forest. **D** Disjunct populations: Amazonia; E Brazil.

31 YELLOW-CRESTED TANAGER *Tachyphonus rufiventer* 15cm Male is distinctive. Note orange tinge to vent of female. **V** Poorly known. Possible song a series of high-pitched, upslurred buzzing notes. Calls may include a similar, single note. **H** Higher levels of humid forest. **D** SW Amazonia, from Peru to W Brazil, NW Bolivia.

32 FULVOUS-CRESTED TANAGER *Tachyphonus surinamus* 16cm White flanks of male are diagnostic. Yellow-buff crown patch may be hard to see. Note yellow-green eye-ring of female. **V** Call is a low, dry *tjurk*. **H** Interior and borders of humid forest and woodland. **D** N, W and C Amazonia.

33 WHITE-SHOULDERED TANAGER *Tachyphonus luctuosus* 13.5cm Broad white wing-bar of male is diagnostic. Underparts of female are uniform yellow. **V** Ultra-high, thin *seet-seet-seet*. **H** Higher levels and borders of forest, woodland, plantations. **D** From Honduras through Amazonia to Bolivia, C Brazil.

34 TAWNY-CRESTED TANAGER *Tachyphonus delatrii* 14.5cm Note bright crest of male; female is dark rufous brown, darker on wings and tail. **V** Ultra-high, thin twittering. **H** Interior and dense borders of humid forest. **D** Nicaragua to Ecuador.

35 RUBY-CROWNED TANAGER *Tachyphonus coronatus* 18cm Red crown of male (normally concealed) is diagnostic; in flight, shows white underwings. Head of female is duller than back. **V** Slow series of single staccato *tjip* and *tjirp* notes. **H** Open woodland, forest borders, orchards, gardens. **D** E Paraguay, NE Argentina, S Brazil.

36 WHITE-LINED TANAGER *Tachyphonus rufus* 18cm Male's white shoulders are normally visible only in flight. Female is uniformly rufous. **V** Series of *tjip* and *tjurp* notes. **H** Clearings, gardens. **D** Costa Rica to Peru; Colombia through the Guianas to E Brazil, Paraguay, NE Argentina.

37 RED-SHOULDERED TANAGER *Tachyphonus phoenicius* 16cm Red shoulder patch of male (normally concealed) is diagnostic. Female is distinctive. **V** Song is a tuneful series of rising then falling chirping notes, *twee twidoo twee twidoo*. Calls include brief *chup* and *tsit*. **H** Woodland and thickets on poor sandy soils. **D** Amazonia.

1
2
3
4
5
6
7
8
9
10
11
12
13
14
15
16
17
18
19
20
21
22
23
24
25
26
27
28
29
30
31
32
33
34
35
36
37

TANAGERS AND ALLIES *CONTINUED*

1 BLACK-THROATED SHRIKE-TANAGER *Lanio aurantius* 20cm Resembles New World orioles (Plates 284–285), but note wholly black head and yellow mantle of male. Female has grey head and olive-brown upperparts. **V** High *pjeet-tut* or *faldereeh-falderah*. **H** Middle levels of forest. **D** S Mexico to Honduras.

2 WHITE-THROATED SHRIKE-TANAGER *Lanio leucothorax* 20cm As Black-throated Shrike Tanager, but with white throat. Female has saffron-ochre throat. **V** High *weetjuh weetjuh* or very high *tjíeeuw*. **H** Middle forest levels. **D** Honduras to Panama.

3 FULVOUS SHRIKE-TANAGER *Lanio fulvus* 17cm Male unmistakable in its range. Female separable from female *Tachyphonus* tanagers (Plate 292) by larger bill. **V** Song short, with well-spaced phrases of slightly buzzy downslurred then upslurred notes. Most frequent call is a loud, sharp descending *tieuw*. **H** Higher levels and borders of humid forest. **D** Amazonia, mainly north of the Amazon, from Ecuador to the Guianas and N Brazil.

4 WHITE-WINGED SHRIKE-TANAGER *Lanio versicolor* 15cm Male unmistakable, with yellow, black and white plumage. Female has yellowish underparts. **V** High, rather sharp, downslurred *tseeuw* notes in a loose series. **H** Higher levels of várzea and terra firme forest. **D** Amazonia, south of the Amazon.

5 CRIMSON-COLLARED TANAGER *Ramphocelus sanguinolentus* 19cm Unmistakable, unlikely to be confused with other species in its range. **V** Song is a well-spaced series of alternating downslurred and upslurred notes. Calls include thin *tswee*, alarm call grating *chuck*. **H** Forest, secondary growth. **D** S Mexico to Panama.

6 MASKED CRIMSON TANAGER *Ramphocelus nigrogularis* 18cm Very distinctive. Female separable from male by brown (not black) belly. **V** Loose series of notes with well-spaced, short shrieks. **H** Shrubby borders. **D** W Amazonia.

7 CRIMSON-BACKED TANAGER *Ramphocelus dimidiatus* 17cm Note blackish throat and dark red belly of male. Female resembles male but less deep red and with a grey bill. **V** Simple series of *tset* notes. **H** Shrubby clearings, forest borders, gardens, cultivation. **D** Panama to Colombia, Venezuela.

8 HUALLAGA TANAGER *Ramphocelus melanogaster* 17cm Black belly of male is diagnostic. Note rosy fore-face of female. **V** Song very like that of Silver-beaked Tanager; call simple *chip*. **H** Secondary growth, shrubby growth, cultivation, gardens. **D** NC Peru.

9 SILVER-BEAKED TANAGER *Ramphocelus carbo* 18cm Note male's blackish-crimson head and underparts, darker upperparts, silvery-white lower mandible. Female is a dull dark reddish brown with a brighter red rump. **V** Unstructured series of *tset* and *tsuwee* notes. **H** Wooded habitats (but not forest interior), often near water. **D** From Venezuela and Colombia through Amazonia to Paraguay.

10 BRAZILIAN TANAGER *Ramphocelus bresilia* 18cm Male is unmistakable. Female has a grey throat, contrasting with reddish-brown belly. **V** Loose series of strong *tset* notes. **H** Atlantic forest borders, parks, swampy woodland, often near water. **D** E Brazil, extreme NE Argentina.

11 PASSERINI'S TANAGER *Ramphocelus passerinii* 16cm Best distinguished from Flame-rumped Tanager by different range. **V** Song is a series of upslurred *towee* notes interspersed with dry *chi-chi* notes. Calls include brief *ik* and *wah* notes. **H** Forest edge, thickets, scrub, suburban regions. **D** S Mexico to Panama.

12 CHERRIE'S TANAGER *Ramphocelus costaricensis* 16cm Formerly treated as conspecific with Passerini's Tanager. Males are similar, females differ in colour of rump and upper breast. **V** Very similar to that of Passerini's Tanager. **H** Forest edge, thickets, scrub, suburban regions. **D** W Costa Rica.

13 FLAME-RUMPED TANAGER *Ramphocelus flammigerus* 18cm Unmistakable in its range. Male differs from Passerini's and Cherrie's tanagers in having red of rump less deep red but more extensive. Female has blackish-brown upperparts, yellow underparts with orange-red breast patch, orange rump. **V** Call is a nasal *tjew*. **H** Borders, secondary growth, scrubby habitats. **D** WC Colombia.

14 LEMON-RUMPED TANAGER *Ramphocelus icteronotus* 18cm Formerly treated as a race of Flame-rumped Tanager. No similar black bird with yellow rump in the area. **V** Very high rattling staccato song consisting mainly of trills with whipping *swip* notes. Also a sparrow-like *kep*. **H** Forest edge, thickets, scrub, suburban regions. **D** Panama to Ecuador.

15 BLUE-GREY TANAGER *Thraupis episcopus* 17cm One of the most widespread birds of the Neotropics. Range differs from that of Sayaca and Azure-shouldered tanagers; separable from Glaucous Tanager by bluish-grey (not grey) head. **V** Loose series of notes, including nasal *tju* and hurried twittering. **H** Forest edge, savannah with scattered trees, plantations, city parks, suburban areas. **D** S Mexico to Bolivia, Amazonia.

16 SAYACA TANAGER *Thraupis sayaca* 16.5cm Pale blue-grey with sky-blue wings. Lacks the Azure-shouldered Tanager's azure shoulder patches. **V** Loose series of notes with sharp *swee* and hoarse, sizzled twittering. **H** Variety of forested and cultivated habitats, from open woodland and savannah with scattered trees to plantations, city parks, suburban areas. **D** S Amazonia to Uruguay, N Argentina.

17 GLAUCOUS TANAGER *Thraupis glaucocolpa* 16cm Separable from Blue-grey Tanager, Sayaca Tanager and Azure-shouldered Tanager by grey (not pale blue) head. **V** Unstructured series of downslurred shrieks. **H** Dry woodland, scrub, plantations, gardens. **D** B Colombia, Venezuela.

18 AZURE-SHOULDERED TANAGER *Thraupis cyanoptera* 18cm Azure shoulders and blackish lores are diagnostic. Note heavy bill. **V** Unmusical, grating twittering. **H** Canopy and borders of humid Atlantic forest and secondary growth. **D** SE Brazil.

19 GOLDEN-CHEVRONED TANAGER *Thraupis ornata* 18cm Median wing coverts bright golden-yellow. No similar tanager occurs in its range. **V** Song similar to Azure-shouldered Tanager, a mixture of squeaky and grating notes. **H** Canopy and borders of forest, clearings, gardens. **D** SE Brazil.

20 YELLOW-WINGED TANAGER *Thraupis abbas* 17cm Note lilac-blue head, blackish loral area, stout bill, bright yellow in wings. **V** Song is a reedy descending trill, preceded by one or two introductory notes. **H** Forest edge, plantations, areas with trees, hedges. **D** E Mexico to Nicaragua.

21 PALM TANAGER *Thraupis palmarum* 17.5cm Generally rather drab, but wing pattern, with pale coverts and dark flight feathers, is distinctive. **V** Hurried, very high twittering. **H** Wooded habitats, including suburban areas. **D** Guatemala to Paraguay, S Brazil, NE Argentina.

22 BLUE-CAPPED TANAGER *Thraupis cyanocephala* 17cm Black-masked blue head contrasts with green upperparts. **V** Mixture of very high warbling and twittering. **H** Montane forest borders, secondary growth. **D** N Venezuela to N Bolivia.

23 BLUE-AND-YELLOW TANAGER *Thraupis bonariensis* 17cm Distinctive. Male has blue head contrasting with greenish or blackish back and yellow underparts. Female is similarly patterned, but duller. **V** Hurried, repetitive, very high to ultra-high twittering. **H** Open woodland, scrub, gardens, agricultural areas. **D** Argentina, Uruguay, SE Brazil, Paraguay, and along Andean chain from N Chile to Ecuador.

24 VERMILION TANAGER *Calochaetes coccineus* 17.5cm Unmistakable. Scarlet, with black mask, throat, wings and tail. Sexes similar. **V** Song little known; may be quiet, flat trill of chipping notes. Calls include a weak *tsit*. **H** Humid montane forest. **D** Andes, from Colombia to Peru.

25 BLUE-BACKED TANAGER *Cyanicterus cyanicterus* 17cm Distinctive blue and yellow plumage and heavy slightly curved bill. **V** Series of very high, very sharp *seeet* and *seeet-seet* phrases. **H** Canopy of humid forest, tall trees in adjacent clearings. **D** The Guianas, N Brazil.

26 YELLOW-GREEN TANAGER *Bangsia flavovirens* 14.5cm Rather uniform green, duskier around eyes. **V** Short, ultra-high, warbled phrases. **H** Wet forest and clearings with scattered tall trees. **D** Colombia, Ecuador.

27 BLUE-AND-GOLD TANAGER *Bangsia arcaei* 15cm Upperparts dark blue to blackish, underparts yellow, with striking red eye. No similar tanager in the area. **V** Very high sharp *tch-please teh-please*. **H** Upper forest levels. **D** Costa Rica, Panama.

28 BLACK-AND-GOLD TANAGER *Bangsia melanochlamys* 16cm Note blue wing coverts and uppertail coverts. **V** Quick, barely audible *tsee-tsee-tsee*. **H** Humid Andean forest. **D** Colombia.

29 GOLDEN-CHESTED TANAGER *Bangsia rothschildi* 16cm Separable from Black-and-gold Tanager by lack of blue in the wings and less extensive yellow on underparts. **V** Very high, unstructured series of well-separated, twittered, single or double notes. **H** Undergrowth of humid Andean forest. **D** Colombia, extreme NW Ecuador.

30 MOSS-BACKED TANAGER *Bangsia edwardsi* 15cm Pale bill (dusky above), blue face sides and yellow breast patch are distinctive. **V** Long, very high, meandering rattle. **H** Cloud forest. **D** Colombia, Ecuador.

31 GOLD-RINGED TANAGER *Bangsia aureocincta* 16cm Black and yellow head pattern is diagnostic. **V** Song resembles that of Moss-backed Tanager but rattle is probably shorter. **H** Cloud forest. **D** Colombia.

32 ORANGE-THROATED TANAGER *Wetmorethraupis sterrhopteron* 17cm Unmistakable, with orange throat and plain buff-yellow underparts. **V** Unstructured stream of sharp, nasal, partly inhaled notes. **H** Mature forest. **D** N Peru, S Ecuador.

33 HOODED MOUNTAIN TANAGER *Buthraupis montana* 22cm Note black hood and throat, red eyes and blue band between yellow belly and vent. **V** Calm stream of thin, ultra-high notes. **H** Humid montane forest, tall secondary growth, scrub woodland. **D** S Venezuela to N Bolivia.

34 BLACK-CHESTED MOUNTAIN TANAGER *Buthraupis eximia* 21cm Note green (not blue) upperparts. **V** Song resembles that of Hooded Mountain Tanager, but probably with slightly more repeated elements. **H** Forest and woodland near the treeline. **D** Colombia to N Peru.

35 GOLDEN-BACKED MOUNTAIN TANAGER *Buthraupis aureodorsalis* 22cm Orange-brown arrowhead markings on flanks and yellow rump are diagnostic. **V** Song is a long series of chattering and more drawn-out upslurred notes. Calls include *chit*, *sweet* and *heet*. **H** Montane forest near the treeline. **D** Peru.

36 MASKED MOUNTAIN TANAGER *Buthraupis wetmorei* 20.5cm Yellow-bordered black mask is diagnostic. **V** Hurried, level series of ultra-high *see* notes. **H** High-altitude stunted forest and elfin woodland. **D** Colombia to N Peru.

1
2
3
4
5
6
7
8
9
10
11
12
13
14
15
16
17
18
19
20
21
22
23
24
25
26
27
28
29
30
31
32
33
34
35
36

TANAGERS AND ALLIES *CONTINUED*

1 SANTA MARTA MOUNTAIN TANAGER *Anisognathus melanogenys* 18cm Note all-blue upperparts and yellow mark below eye. Very restricted range. **V** Ultra-high sizzling. **H** Montane forest borders, secondary growth. **D** Sierra Nevada de Santa Marta, N Colombia.

2 LACRIMOSE MOUNTAIN TANAGER *Anisognathus lacrymosus* 17cm Separable from Santa Marta Mountain Tanager by wing pattern, and by range. **V** Ultra-high sizzling. **H** Wet montane forest, stunted forest, *Polylepis* woodland. **D** Colombia and Venezuela to Peru.

3 SCARLET-BELLIED MOUNTAIN TANAGER *Anisognathus igniventris* 16cm Unmistakable. Predominantly black, with bright blue rump and wing coverts, and red ear-coverts and belly. **V** Very high, repetitive warbling. **H** Interior and borders of high Andean forest, secondary growth, hedgerows, agricultural areas. **D** Extreme W Venezuela to Bolivia.

4 BLUE-WINGED MOUNTAIN TANAGER *Anisognathus somptuosus* 18cm Note yellow crown, and wing pattern with black greater coverts and blue shoulders. **V** Series of similar phrases that start very high and soft, but change to loud and forceful (5–6 secs). **H** Humid forest, tall secondary growth. **D** Venezuela to Bolivia.

5 BLACK-CHINNED MOUNTAIN TANAGER *Anisognathus notabilis* 18cm Note strong pointed bill and black head with narrow yellow crown-stripe. **V** Ultra-high sizzling. **H** Wet montane forest on west slopes of Andes. **D** Colombia, Ecuador.

6 GRASS-GREEN TANAGER *Chlorornis riefferii* 20.5cm Unmistakable, with brilliant green plumage and rufous mask and undertail coverts. Bill and legs are orange-red. **V** Short, low rattle, introduced with a dry *tuc*. **H** Wet, often stunted forest. **D** Colombia to Bolivia.

7 RUFOUS-BELLIED MOUNTAIN TANAGER *Pseudosaltator rufiventris* 22cm Once thought to be related to saltators (Plate 299), but distinguished by rufous underparts, stubby bill and very different voice. **V** Song is a short, subdued, evenly pitched *cheree-cheree-cheree-chew*. **H** *Polylepis* woodland, montane scrub. **D** Bolivia, NW Argentina.

8 BUFF-BREASTED MOUNTAIN TANAGER *Dubusia taeniata* 19cm Shape and pattern of long eyebrow are diagnostic. Differs from Chestnut-bellied Mountain Tanager in yellow (not rufous) belly. **V** Slow series of 3–4 very high, piercing, pressed-out *feee* notes. **H** Undergrowth and borders of montane forest and woodland. **D** Venezuela to Peru.

9 CHESTNUT-BELLIED MOUNTAIN TANAGER *Delothraupis castaneoventris* 17cm All-blue upperparts, chestnut underparts and black mask are distinctive. **V** Song consists of 1–3 very high, sharp, sizzled *seees* notes. **H** Canopy and borders of high-altitude montane forest and woodland. **D** Peru, Bolivia.

10 DIADEMED TANAGER *Stephanophorus diadematus* 19cm Predominantly blue-violet with a distinctive white and red crown patch, but often appears dull blackish. **V** High, sharp, melodious warbling (3–4 secs). **H** Open forest, shrubbery, parks and gardens. **D** SE Brazil, E Paraguay, Uruguay, NE Argentina.

11 PURPLISH-MANTLED TANAGER *Iridosornis porphyrocephalus* 16cm Mainly dark blue with a yellow throat and rufous vent. **V** Unstructured, very high to ultra-high mixture of notes, including a thin, downslurred *feeeh*. **H** Undergrowth of wet forest. **D** Colombia, NW Ecuador.

12 YELLOW-THROATED TANAGER *Iridosornis analis* 16cm Formerly considered conspecific with Purplish-mantled Tanager. Separable by greenish upperparts and dull green-tinged tawny-buff underparts, and by more southerly range. **V** Downslurred, very high *tseeeuw*. **H** Dense undergrowth and borders of wet forest. **D** SW Colombia to Peru.

13 GOLDEN-COLLARED TANAGER *Iridosornis jelskii* 14.5cm Unmistakable, with yellow band over crown and neck sides, black face, bright blue upperparts, deep tawny underparts. **V** Simple, high phrases, *tseeuw*, sung singly or lengthened to *tsee-oh-wee*. **H** Elfin woodland, scrub near the treeline. **D** Peru, Bolivia.

14 GOLDEN-CROWNED TANAGER *Iridosornis rufivertex* 15.5cm Yellow crown-stripe contrasting with otherwise dark plumage (violet-blue body, black head, rufous vent) is diagnostic. **V** Generally quiet. Song very high, thin, stuttering twitter; calls include *sit* and *seep*. **H** Undergrowth and borders of montane forest and woodland. **D** W Venezuela to N Peru.

15 YELLOW-SCARFED TANAGER *Iridosornis reinhardti* 16.5cm Note yellow band across nape and ear-coverts, and blue (not rufous) undertail coverts. **V** Quiet. Possible song series of clear *tsee* notes. Calls include *tzi-zi* and *zip*. **H** Montane forest, elfin woodland. **D** Peru.

16 FAWN-BREASTED TANAGER *Pipraeidea melanonota* 14cm General colour pattern (blue above, fawn below) with black 'Zorro' mask is diagnostic. **V** Ultra-high, rapid *tsi-tsi-tsi-tsi*. **H** Borders and clearings in forest, cultivated areas with large trees. **D** Two disjunct populations: Venezuela to Bolivia and NW Argentina; S Brazil, Paraguay, NE Argentina and Uruguay.

17 SHRIKE-LIKE TANAGER *Neothraupis fasciata* 16cm Face mask and wing pattern are distinctive. **V** High series of notes, including *tjiw* and *tjeuu* (6–8 secs). **H** Cerrado, open woodland. **D** NE Bolivia and NE Paraguay to E Brazil.

18 GLISTENING-GREEN TANAGER *Chlorochrysa phoenicotis* 12.5cm Glittering green plumage is unmistakable; note small red ear patches. **V** Unstructured, ultra-high twittering. **H** Cloud forest, older secondary growth, adjacent clearings. **D** Colombia, Ecuador.

19 ORANGE-EARED TANAGER *Chlorochrysa calliparaea* 12cm Predominantly green plumage, with an orange or rufous rump and orange-red neck sides. Extent of blue underparts in males varies, and two races have a black throat patch. **V** Generally quiet; voice also varies by region. Song high-pitched, warbling or wheezy; calls include *zzit*, *seep* and *see-up*. **H** Montane forest and woodland. **D** Colombia to Bolivia.

20 MULTICOLORED TANAGER *Chlorochrysa nitidissima* 12.5cm Note male's yellow face and throat, green nape, black and chestnut patch on neck sides, yellow mantle, green wings and tail, bright blue underparts with black on breast and belly. Females are similar but duller, without the yellow mantle or any black on the underparts. No similar-looking bird occurs in South America. **V** Song little known, possibly a short series of *see* notes. Calls include simple brief *swit*. **H** Montane forest and woodland. **D** Colombia.

21 PLAIN-COLORED TANAGER *Tangara inornata* 12cm Note the three beautiful shades of grey. In flight, shows blue shoulders. **V** Song extremely high, monotone rapid chattering trill; calls very short and simple, including *tsip* and *tswee* notes. **H** Forest, clearings, plantations, gardens, suburbs. **D** Costa Rica to Colombia.

22 CABANIS'S TANAGER *Tangara cabanisi* 14cm Greenish-blue all over, with distinct dark markings. **V** Very high sibilant *pfsweeeeo* and high trill, *spspsprrrrsp*. **H** Humid broadleaf forest. **D** Pacific slopes of S Mexico, Guatemala.

23 GREY-AND-GOLD TANAGER *Tangara palmeri* 14.5cm Unmistakable due to grey plumage with irregular black markings. **V** Noisy, with a wide range of calls that attract other flock-forming species. Sounds include *chup-chup-sweet*, *sweet* and *chit* notes. **H** Canopy and borders of wet forest. **D** Panama, W Colombia, W Ecuador.

24 TURQUOISE TANAGER *Tangara mexicana* 13.5cm Note metallic turquoise-blue face and breast, and black-edged flank feathers. Belly varies from white to yellow. Sexes similar. **V** Song extremely high, fast series of thin *zing* notes; call a single *zing*. **H** Humid forest, secondary growth, semi-open habitats. **D** Amazonia, SE Brazil.

25 PARADISE TANAGER *Tangara chilensis* 14cm Unmistakable multicoloured tanager, with green mask, black upperparts, scarlet or yellow rump, bright blue underparts. **V** Song is a simple combination of very high-pitched *chik* and *zeet* notes; calls include a sharp *chak*. **H** Canopy and borders of humid forest. **D** Amazonia.

26 SEVEN-COLORED TANAGER *Tangara fastuosa* 13.5cm Note turquoise-green head, bright blue underparts, black back, bright orange lower back and rump. **V** Calls include *ik-ik-Ik* and repeated *hwit* notes. **H** Forest borders, secondary growth. **D** E Brazil.

27 GREEN-HEADED TANAGER *Tangara seledon* 13cm Multicoloured tanager (turquoise, yellow-orange, lime green, violet and black), unmistakable in its range. **V** Song is a steady series of *chee* and *hwit* notes; calls include a buzzy, upslurred *zweet*. **H** Canopy and borders of humid forest, trees in small clearings, parks, gardens. **D** S and E Brazil, E Paraguay, NE Argentina.

28 RED-NECKED TANAGER *Tangara cyanocephala* 13cm Combination of blue crown, red 'scarf' and green underparts is distinctive. **V** Song very high-pitched, thin twittering; calls include sharp *seet* and softer *sip*. **H** Borders and canopy of forest, secondary growth, open woodland, Caatinga. **D** E Brazil, E Paraguay, NE Argentina.

29 BRASSY-BREASTED TANAGER *Tangara desmaresti* 13.5cm Brassy breast and neck are distinctive. Separable from Gilt-edged Tanager by green cheeks and greenish underparts. **V** Gives varied, very high-pitched sharp notes and twitters. **H** Humid montane forest. **D** SE Brazil.

30 GILT-EDGED TANAGER *Tangara cyanoventris* 13.5cm Note yellow head and black-streaked yellow mantle, contrasting with turquoise-blue breast and green belly. **V** Similar to that of Brassy-breasted Tanager. **H** Humid lowland forest, secondary growth, overgrown clearings. **D** E Brazil.

31 BLUE-WHISKERED TANAGER *Tangara johannae* 13.5cm Black mask with blue malar patch is diagnostic. **V** Typical call is a sharp, abrupt, high-pitched *tzit*. **H** Canopy and borders of forest and secondary growth. **D** W Colombia, NW Ecuador.

32 GREEN-AND-GOLD TANAGER *Tangara schrankii* 12cm Note facial pattern of blue lores, black cheeks and yellow crown. **V** Long, very high series of *ting-ting* notes. **H** Humid lowland forest. **D** Amazonia.

33 EMERALD TANAGER *Tangara florida* 13cm Separable from Golden Tanager by distinctive green and black (not yellow and black) striped plumage. **V** Probable song described as a series of high *zeeee* notes; calls include a harsh *dzeee*. **H** Humid forest and secondary growth. **D** Costa Rica to NW Ecuador.

34 GOLDEN TANAGER *Tangara arthus* 13.5cm Unmistakable, with predominantly yellow plumage. Back is black-streaked, wings and tail are mostly black, and ear-coverts are black. Amount of rufous varies. **V** Calls include various sharp and staccato, high-pitched and penetrating notes. **H** Humid montane forest. **D** Venezuela to Bolivia.

35 SILVER-THROATED TANAGER *Tangara icterocephala* 13cm White throat and narrow collar are diagnostic. **V** Song and calls both combine harsh, buzzing or hissing *tcheee* notes. **H** Wet mossy forest and secondary growth. **D** Costa Rica to Ecuador.

36 SAFFRON-CROWNED TANAGER *Tangara xanthocephala* 13cm A bluish-green tanager with black chin and mask, black streaks on the back, and a distinctive yellow head. **V** Possible song a short, squeaky warble; calls include a brief, sharp *chit*. **H** Humid montane forest, secondary growth, scattered trees in clearings. **D** Venezuela to Bolivia.

1
2
3
4
5
6
7
8
9
10
11
12
13
14
15
♂
16
♀
17
18
19
♀
♂
20
♀
♂
21
22
23
24
25
26
27
28
29
30
31
32
33
34
35
36

TANAGERS AND ALLIES ***CONTINUED***

1 GOLDEN-EARED TANAGER *Tangara chrysotis* 14cm Golden cheeks and long black malar stripes are distinctive. **V** Calls relatively low-pitched, including *tsuck* and *chup*. **H** Humid Andean forest, secondary growth. **D** Colombia to NW Bolivia.

2 FLAME-FACED TANAGER *Tangara parzudakii* 14.5cm Jet-black back and yellow, orange and black face pattern are distinctive. Note also pale shoulders. **V** Little known. Rapid series of *ti-ti-ti* notes has been reported. **H** Cloud forest and older secondary growth. **D** W Venezuela to Peru.

3 YELLOW-BELLIED TANAGER *Tangara xanthogastra* 12cm Feathers of mantle are blue-edged, as are wing feathers. Underparts are yellow-green (not white). **V** A high, buzzy song of chatters and wheezes has been noted; calls include weak *chit* and *seet*. **H** Canopy and borders of humid forest. **D** N and W Amazonia, from Venezuela and Guyana to Bolivia.

4 SPOTTED TANAGER *Tangara punctata* 12cm Similar to Speckled Tanager and Yellow-bellied Tanager, but bluish-grey head is diagnostic. **V** Song is a very high-pitched spluttering trill of *chit* notes; call a single *chip*. **H** Wet forest borders, secondary growth. **D** Amazonia from S Venezuela, the Guianas and N Brazil; Andes from Ecuador to Bolivia.

5 SPECKLED TANAGER *Tangara guttata* 13.5cm Separable from Spotted Tanager by yellow-green (not bluish) head. Ground colour of breast and belly is white. **V** Song is a fairly short rapid trill, downslurring and increasing in volume; calls include weak *chit* and *tsip*. **H** Canopy and borders of montane forest, woodland and adjacent clearings. **D** Costa Rica to Venezuela and NW Brazil.

6 DOTTED TANAGER *Tangara varia* 11.5cm Note pale blue-edged wing and mantle feathers. **V** Song is a short series of *chit* notes followed by a few drawn-out *see* notes; calls include a high *tsip*. **H** Humid lowland forest, secondary growth, plantations. **D** N and W Amazonia: S Venezuela to N Brazil; C Peru.

7 RUFOUS-THROATED TANAGER *Tangara rufigula* 12cm Dull rufous throat, contrasting with black face and black-spotted breast, is distinctive. **V** Song and calls involve series of high-pitched *tik* notes. **H** Mossy forest, broken forest. **D** Pacific slopes of Andes in Colombia, N Ecuador.

8 BAY-HEADED TANAGER *Tangara gyrola* 13cm Chestnut-red head is diagnostic. Underparts vary from blue to green, depending on race. **V** High, hurried, nasal twittering. **H** Canopy and borders of humid lowland forest, woodland, plantations, mature secondary growth. **D** Nicaragua to W Ecuador; N Colombia and N Venezuela through N Amazonia to Bolivia.

9 RUFOUS-WINGED TANAGER *Tangara lavinia* 13cm Rufous head and primaries are diagnostic. **V** Song is a rapidly accelerating series of *tit* notes that run into a trill; calls include *ti* and *tit*. **H** Borders and interior of lowland forest, secondary growth and open woodland, plantations, parks. **D** Guatemala to Ecuador.

10 BURNISHED-BUFF TANAGER *Tangara cayana* 13.5cm Dull buff plumage contrasting with black mask and blue-green wings and tail is diagnostic. **V** Song is a series of squeaky *tsweek* notes, accelerating into a long trill; calls include a single *tsweek* and *tsit*. **H** Open and semi-open habitats, savannah woodland, riverine belts, wasteland, gardens. **D** E Colombia through the Guianas to N Brazil; E and C Brazil to E Bolivia, S Paraguay, NE Argentina.

11 LESSER ANTILLEAN TANAGER *Tangara cucullata* 15cm Note greenish wings, dark mask and dark chestnut to rufous cap. Attracted to fruiting trees, usually in pairs or small flocks. **V** Weak *weet-weet-weet-witwitwitwit*. **H** Moist and dry forests, secondary growth, gardens. **D** Grenada and St Vincent (Lesser Antilles).

12 BLACK-BACKED TANAGER *Tangara peruviana* 14cm Male separable from male Chestnut-backed Tanager by black mantle. **V** Calls include a thin, rather hissing *sweeeek*. **H** Coastal forest, restinga, woodland borders, gardens. **D** SE Brazil.

13 CHESTNUT-BACKED TANAGER *Tangara preciosa* 14.5cm Very like Black-backed Tanager (and formerly considered conspecific), but male's back is golden-rufous, matching colour of crown. Females not reliably separable. **V** Very like Black-backed Tanager. **H** Forest borders, gardens, shrubby areas. **D** S Brazil, E Paraguay, Uruguay, NE Argentina.

14 SCRUB TANAGER *Tangara vitriolina* 14cm Note silvery-green plumage, black mask and sharply defined orange crown. **V** Song begins with high-pitched single *slip* notes that build to rapid trilling *si-si-si-si-si*; call a buzzy *ziit*. **H** Dry scrub, agricultural areas, habitation. **D** Colombia, Ecuador.

15 GREEN-CAPPED TANAGER *Tangara meyerdeschauenseei* 14cm Female separable from female Black-backed and Chestnut-backed tanagers by yellowish cap. Mask is less pronounced than in Burnished-buff Tanager. **V** Call is a flat *chut*. **H** Forest edges, scrub, gardens. **D** SE Peru, W Bolivia.

16 RUFOUS-CHEEKED TANAGER *Tangara rufigenis* 13cm Opalescent green with dull rufous cheeks and flight feathers. **V** Call is a weak *tsit*, sometimes repeated in a fast series. **H** Wet and humid forest and secondary growth. **D** N Venezuela.

17 GOLDEN-NAPED TANAGER *Tangara ruficervix* 13cm Note irregular marking along cheeks and lower flanks; golden-yellow crown patch is diagnostic. **V** Song begins with single *tik* notes, which fall in pitch and accelerate into a short trill; call a simple *tit* or *seee*. **H** Montane forest and woodland. **D** Andes, from Colombia to NW Bolivia.

18 METALLIC-GREEN TANAGER *Tangara labradorides* 13cm Note small mask, narrow black crown-stripe and green-blue flanks. **V** Calls varied, include *jiit*, *tip* and *nyah*. **H** Borders of humid montane forest, secondary growth, clearings. **D** Andes, from Colombia to Peru.

19 BLUE-BROWED TANAGER *Tangara cyanotis* 12cm Striking blue-green supercilium is distinctive. Note also buff-rufous vent. **V** Song a wheezy chatter; calls include high-pitched *tip*. **H** Montane forest and clearings. **D** Andes, from Colombia to Bolivia.

20 BLUE-NECKED TANAGER *Tangara cyanicollis* 13cm Brilliant blue head contrasts with black back and breast. Note also yellowish shoulders. **V** Little known. Possible song a fast chitter, ending with a few more drawn out buzzing notes. Calls include *chep*, *tsit* and *seet*. **H** Borders and clearings, overgrown pastures, riverine belts. **D** Disjunct populations: Andean foothills from W Venezuela to C Bolivia; Amazonia of C Brazil.

21 GOLDEN-HOODED TANAGER *Tangara larvata* 12cm Golden hood, sharply defined from black mantle and breast, is diagnostic. **V** Soft, dry rattles starting with a stutter. **H** Forest edge, secondary growth, clearings with scattered trees, gardens. **D** S Mexico to W Colombia, W Ecuador.

22 MASKED TANAGER *Tangara nigrocincta* 13cm Has no yellow in plumage; also separable from Blue-necked Tanager by white lower belly and vent. **V** Song comprises three long, slightly buzzy notes, with last note downslurring sharply. Calls combine *chit* and *tsit* notes, given singly or in a chatter. **H** Canopy and borders of humid forest, clearings with scattered trees. **D** N and W Amazonia.

23 SPANGLE-CHEEKED TANAGER *Tangara dowii* 13cm Distinctively coloured and patterned. Not in range of Green-naped Tanager. **V** Probable song a cheerful short chitter; calls include *tsip* and *seek*. **H** Forest and nearby areas. **D** Costa Rica, Panama.

24 GREEN-NAPED TANAGER *Tangara fucosa* 13cm Distinguished from Beryl-spangled Tanager by more extensive black mask and paler underparts. **V** Calls include *tsit* and a longer *tseeet*. **H** Borders and interior of montane forest. **D** Extreme E Panama.

25 BERYL-SPANGLED TANAGER *Tangara nigroviridis* 13cm Spangled blue-green and black appearance is unmistakable. **V** Typical call is a sharp, whip-cracking *hwit*. **H** High-altitude wet forest, secondary growth, trees in pastures. **D** Venezuela to Bolivia.

26 BLUE-AND-BLACK TANAGER *Tangara vassorii* 13cm Primarily dark cobalt-blue with black mask. Note also small bill and thick-set appearance. **V** Song is an accelerating series of high-pitched, upslurred notes; calls include an emphatic *tsit*. **H** High-altitude humid forest, secondary growth and adjacent clearings. **D** Venezuela to Bolivia.

27 BLACK-CAPPED TANAGER *Tangara heinei* 13cm Male is distinctive, predominantly grey-blue with greenish cheeks and a black cap. Female may be hard to distinguish from female Black-headed Tanager, but shows blue edging of primaries, more striping on lower throat and yellower flanks. **V** Ultra-high, hurried *tja-tja-tja*. **H** Humid forest, secondary growth, shrubby clearings. **D** NW Venezuela to N Ecuador.

28 SIRA TANAGER *Tangara phillipsi* 13cm Has a small range. Male separable from male Silver-backed Tanager by green-blue throat. Female not reliably separable from female Straw-backed Tanager. **V** Calls include an abrupt *tsip*, given singly or in a sequence (up to six). **H** Canopy and borders of tall humid forest. **D** Cerros del Sira, C Peru.

29 SILVER-BACKED TANAGER *Tangara viridicollis* 13cm Both male and female have distinctive plumage; note coppery-orange cheeks of male and chestnut cap of female. **V** Song is a very long sequence of rather buzzy notes, descending slightly, recalling a pipit. Calls include a downslurred wheezing *tseeer* and a thin *pit*. **H** Forest, woodland, shrubby clearings. **D** Ecuador, Peru, Bolivia.

30 STRAW-BACKED TANAGER *Tangara argyrofenges* 13cm Male unmistakable. Note green cap of female. **V** Song is a mixture of *prrt* and high *seeet* notes; calls include *tsew* and a dry *tip*. **H** Humid montane forest. **D** Ecuador, Peru, Bolivia.

31 BLACK-HEADED TANAGER *Tangara cyanoptera* 13.5cm Colour pattern of male is unmistakable. Female is predominantly greenish with a greyer head – similar to female Black-capped Tanager, but with limited range overlap. **V** Song a series of high, thin and buzzy notes that fall in pitch; calls include *djeet*. **H** Montane forest and woodland. **D** Colombia to Guyana.

32 OPAL-RUMPED TANAGER *Tangara velia* 14cm Resembles Opal-crowned Tanager, but with opal rump, rufous lower belly and vent, and different facial pattern. **V** Song a simple twittering of *tiz* notes; calls include thin *sit* or *siz*. **H** Borders and canopy of humid forest, secondary growth, trees in clearings. **D** Disjunct populations in Amazonia and SE Brazil.

33 OPAL-CROWNED TANAGER *Tangara callophrys* 14.5cm Note striking opal forecrown and brow. Sexes alike. **V** Probable song a high squeaky chatter; calls include thin *zit*, very high *pseet* and lower, rich *chup*. **H** Tall forest. **D** W Amazonia, mainly in Colombia, Ecuador, Peru.

34 SWALLOW TANAGER *Tersina viridis* 15cm Male is turquoise-blue, with black on forehead and throat. Female is dull green with buffy-yellow underparts. Arboreal, often perches on open branches within the canopy; regularly makes fly-catching sallies. Flight swallow-like. **V** Song is a squeaky twitter. Typical call is a high thin *tsee*. **H** Moist forest, light woodland, forest and woodland edge. **D** Panama to NE Argentina.

1
2
3
4
5
6
7
8
9
10
11
12
13
14
15
16
17
18
19
20
21
22
23
24
25
26
27
28
29
30
31
32
33
34

TANAGERS AND ALLIES *CONTINUED*

1 WHITE-BELLIED DACNIS *Dacnis albiventris* 11.5cm Male cobalt blue (not turquoise like male Black-faced Dacnis and Turquoise Dacnis) with black mask and white belly; note irregularly spotted flanks. Female rather greenish. **V** Hurried, ultra-high sizzling. **H** Canopy and borders of tall humid forest. **D** W Amazonia, from SW Venezuela to N Peru.

2 BLACK-FACED DACNIS *Dacnis lineata* 11.5cm Note black mask of male extending to nape and back. Wings are mainly black. White belly separates this species from Yellow-tufted Dacnis. Female rather brown. **V** Song rarely heard, said to be a lengthy series of buzzy notes; most frequent call is a short trilling *tsrrrrip*. **H** Canopy and borders of tall humid forest, scattered trees in clearings, riverine belts. **D** N and W Amazonia.

3 YELLOW-TUFTED DACNIS *Dacnis egregia* 11cm Very like formerly conspecific Black-faced Dacnis. Male has yellow belly and shoulder. Female more grey-toned than brown. **V** Little known. Call high-pitched *srrip* note. **H** Like that of Black-faced Dacnis. **D** Colombia, Ecuador.

4 YELLOW-BELLIED DACNIS *Dacnis flaviventer* 12.5cm Male unmistakable, mostly black and yellow with olive-green crown. Female is not as brown as female Black-faced Dacnis. **V** Song described as a two-note phrase, *whu-zeeh*, sometimes with additional *pzeeIPP* notes. Calls include *zeet* and *zreet*. **H** Várzea forest and secondary growth. **D** W and C Amazonia.

5 TURQUOISE DACNIS *Dacnis hartlaubi* 11cm Male separable from male Black-faced Dacnis and Yellow-tufted Dacnis by uniform blue underparts and black chin. **V** Song possibly a series of three distinct notes followed by buzzing short trill. Calls include high-pitched *tik* and *sic*. **H** Humid forest, secondary growth, habitats with scattered trees. **D** Colombia.

6 BLACK-LEGGED DACNIS *Dacnis nigripes* 11cm Male differs from male Blue Dacnis in having black (not red) legs and all-black primaries. **V** Calls include a high *sweet* and a lower, nasal *diuh*. **H** Borders and clearings in humid forest and secondary growth. **D** SE Brazil.

7 SCARLET-THIGHED DACNIS *Dacnis venusta* 12cm Note red eyes. Male unmistakable due to black belly and turquoise cap. Bluish head, scapulars and rump of female are distinctive. Scarlet thighs are rarely visible. **V** Song consists of two or three short scratchy trills. Calls include a metallic *tzirp* and various scratchy notes. **H** Canopy and borders of wet forest. **D** Costa Rica to Ecuador.

8 BLUE DACNIS *Dacnis cayana* 12cm Male is bright turquoise blue with a black throat, back and tail; female is mostly green with a blue crown. Note pink legs. Separable from Short-billed Honeycreeper by black mantle, and from Red-legged Honeycreeper by black bib and shorter bill. **V** Song is a slow series of buzzy *tswee* notes; calls include thin *tsip* and high *chit*. **H** Variety of forested and wooded habitats, including várzea and terra firme forest, riverine belts and gardens. **D** From Honduras through Amazonia to NE Argentina.

9 VIRIDIAN DACNIS *Dacnis viguieri* 11.5cm Has a restricted range. Note pale eyes. Green upperparts of female merge gradually into pale underparts, not sharply defined as in female White-bellied Dacnis and female Turquoise Dacnis. **V** Not known but presumably similar to that of other dacnis species. **H** Humid forest, scrubby secondary growth. **D** E Panama, NW Colombia.

10 SCARLET-BREASTED DACNIS *Dacnis berlepschi* 12cm No similar bird occurs in its small range. Both sexes distinctive. **V** Song is a fast, steady series of high-pitched single notes, recalling a pipit; call *tze*, often repeated. **H** Canopy and clearings with scattered trees in humid forest. **D** SW Colombia, extreme NW Ecuador.

11 SHORT-BILLED HONEYCREEPER *Cyanerpes nitidus* 9.5cm Short bill (much shorter than in other honeycreepers) and pink or reddish legs are diagnostic. **V** Usual call a brief, high *tsee*. **H** Interior and borders of humid forest. **D** W Amazonia.

12 SHINING HONEYCREEPER *Cyanerpes lucidus* 10cm Note bright yellow legs of male (rather duller in female). Male not reliably separable from male Purple Honeycreeper, but note that bib is larger, with a more rounded lower edge. **V** Song may be a simple series of chittering notes; calls short and simple, *tic* or *tsip*. **H** Canopy and borders of humid forest, clearings with scattered trees. **D** SE Mexico to NW Colombia.

13 PURPLE HONEYCREEPER *Cyanerpes caeruleus* 11cm Very similar to Shining Honeycreeper, and ranges overlap. Female has buff face sides. **V** Calls include a high, downslurring *tseeu* note and a fast, dry chipping. **H** Canopy and borders of tall humid forest, secondary growth, plantations, riverine belts, gardens. **D** E Panama to Trinidad and the Guianas, Ecuador to Bolivia, W and C Amazonia.

14 RED-LEGGED HONEYCREEPER *Cyanerpes cyaneus* 11.5cm Male unmistakable due to size, pale cap, red legs and yellow visible in wings in flight. Female has faint eyebrow and no blue in face. **V** Song seldom heard, *tsip-tsip-cha*, repeated. Calls include various weak buzzy or thin and high-pitched notes. **H** Humid and drier forest, open areas with scattered trees, older secondary growth, open woodland, habitation. **D** Widespread, from Mexico to C Bolivia and E Brazil.

15 GREEN HONEYCREEPER *Chlorophanes spiza* 13.5cm Males are blue-green with a broad black mask and a slightly downcurved, mostly yellow bill. Females are duller, more uniform green. **V** Song seldom heard, a scratchy combination of *tst* and louder *chip* notes. Calls include *tsup* and high-pitched *tseet*. **H** Forest, woodland, gardens. **D** Widespread, from Mexico to C Bolivia and E Brazil.

16 GOLDEN-COLLARED HONEYCREEPER *Iridophanes pulcherrimus* 11cm Male is unmistakable, with black hood and upper back separated by yellow-buff collar. Collar is fainter in female. **V** Call a long, thin *tzeeeeeeeu* that rises and falls. **H** Canopy of humid Andean forest, forest edge, secondary growth. **D** S Colombia to SE Peru.

17 SULPHUR-RUMPED TANAGER *Heterospingus rubrifrons* 16cm Yellow rump and undertail coverts diagnostic. **V** Calls include *tsip*, *tseeet* and buzzy *dzeet*. **H** Forest (middle levels) and adjacent semi-open areas of Caribbean slopes. **D** Costa Rica, Panama.

18 SCARLET-BROWED TANAGER *Heterospingus xanthopygius* 17.5cm Male is unmistakable, with bright red post-ocular tuft. Female is leaden-grey with a yellow rump. **V** Rapid, slightly stuttering, partly nasal twittering. **H** Tall humid forest and secondary growth. **D** E Panama to W Ecuador.

19 GUIRA TANAGER *Hemithraupis guira* 13cm Male unmistakable, with black face and throat outlined by a narrow yellowish border. Female similar to female Rufous-headed Tanager. **V** Very high, descending, twittered phrase. **H** Canopy of humid to dry forest, secondary growth, plantations, scattered trees in clearings, Cerrado, mangrove. **D** Widespread from N Colombia to S Brazil, NE Argentina.

20 RUFOUS-HEADED TANAGER *Hemithraupis ruficapilla* 13cm Male unmistakable, with rufous head and yellow neck; female similar to female Guira Tanager. **V** Strange mixture of single nasal notes with short, high twitters. **H** Forest borders, open woodland, plantations, gardens. **D** E and SE Brazil.

21 YELLOW-BACKED TANAGER *Hemithraupis flavicollis* 13cm Male unmistakable, with yellow rump contrasting with black upperparts; female separable from female Guira Tanager and female Rufous-headed Tanager by yellow-edged wing feathers. **V** Song is a high, toneless and rapid *tsi-si-tzi-tzi-tzi*… Calls include high-pitched *tsik* and *tsew*. **H** Tall humid forest, scattered trees in clearings. **D** Panama to SE Brazil and SE Peru.

22 BLACK-AND-YELLOW TANAGER *Chrysothlypis chrysomelas* 12cm Male is very striking, bright yellow apart from narrow black eye-ring and jet-black upper back, wings and tail. Female shows uniform yellow underparts. **V** Call is a thin hummingbird-like *tsee* or *tseeeut*. **H** Canopy and borders of humid forest and adjacent secondary growth. **D** Costa Rica to extreme NW Colombia.

23 SCARLET-AND-WHITE TANAGER *Chrysothlypis salmoni* 13cm Male is unmistakable; female has little or no yellow below. **V** Possible song a series of jumbled, rather weak high-pitched *ti-ti* and similar notes, finishing with a stronger *heeet*. Calls include a weak and slurring *scipp*. **H** Stunted secondary growth at borders and in openings of tall forest. **D** Colombia, Ecuador.

24 TIT-LIKE DACNIS *Xenodacnis parina* 11cm Note very distinctive stubby bill, thick-set appearance with long legs, and dark blue plumage of male. **V** Rapid series of 3–5 liquid *weet* notes, followed by a very high *see-ee*. **H** Stunted and *Polylepis* woodland at the treeline. **D** SW Ecuador, Peru.

25 CHESTNUT-VENTED CONEBILL *Conirostrum speciosum* 11cm Rufous vent of male, contrasting with grey plumage, is distinctive. Female shows a grey head and olive upperparts; separable from female Capped Conebill by buff (not green) underparts. **V** Unstructured, very high chirping. **H** Canopy and borders of forest, secondary growth, dry open woodland. **D** Widespread from Venezuela to N Argentina.

26 WHITE-EARED CONEBILL *Conirostrum leucogenys* 9.5cm Facial pattern of male is diagnostic. Female is bicoloured buff and grey. Note small white rump patch on both male and female. **V** Very high twittering. **H** Open woodland, plantations, riverine belts, secondary growth. **D** E Panama to Venezuela.

27 BICOLORED CONEBILL *Conirostrum bicolor* 11.5cm Not obviously bicoloured, despite the name. Greyish-blue upperparts, greyish-buff underparts. Sexes similar. **V** Song varied, a series of squeaky, rollicking notes; calls include various soft, sibilant notes. **H** Shrubby riverside vegetation, river islands, mangroves. **D** Along the Amazon and its tributaries; coastal South America, from Venezuela to SE Brazil.

28 PEARLY-BREASTED CONEBILL *Conirostrum margaritae* 11.5cm Not easily separable from Bicolored Conebill. Less sharp contrast between upperparts and underparts; less buff on underparts. **V** Hurried, very high, partly nasal, repetitive twittering. **H** Riverside vegetation, especially *Cecropia* trees. **D** Amazon and its tributaries, in W Brazil, Peru, N Bolivia.

29 CINEREOUS CONEBILL *Conirostrum cinereum* 12cm Note forked wing mark, sharp bill and distinct supercilium. **V** Very high, level sizzling. **H** Lightly wooded areas, bushy borders, stunted shrubs, gardens; from sea level up to 4,500m. **D** S Colombia to N Chile.

30 TAMARUGO CONEBILL *Conirostrum tamarugense* 12.5cm Rufous supercilium, throat and vent are distinctive. Sexes similar, but female slightly duller than male. **V** Very high, hurried sizzling. **H** Low scrubby woodland, plantations, riverine scrub, agricultural areas. **D** Breeds N Chile. Spreads into S Peru in non-breeding season.

31 WHITE-BROWED CONEBILL *Conirostrum ferrugineiventre* 12cm Bluish-grey upperparts, rufous underparts, prominent white supercilium. Note also indistinct white malar stripe. **V** Very high, hurried twittering. **H** Stunted forest, *Polylepis* woodland, shrubbery. **D** Peru and Bolivia, east slopes of the Andes.

32 RUFOUS-BROWED CONEBILL *Conirostrum rufum* 12.5cm Beautifully and distinctively coloured, with deep rufous face and underparts. **V** Very high, hurried, repetitive sizzling. **H** Bushy and shrubby habitats, from elfin woodland to gardens and parks. **D** Colombia.

33 BLUE-BACKED CONEBILL *Conirostrum sitticolor* 13cm Distinctively patterned, with black head and throat, blue back, rich rufous belly. Sexes alike. **V** High twittering, punctuated with small rattles. **H** Humid montane forest, elfin woodland. **D** W Venezuela to C Bolivia.

34 CAPPED CONEBILL *Conirostrum albifrons* 13cm Note bluish forehead and cap of female. Male's cap varies from white to violet-blue, depending on race. **V** Very high, repetitive sizzling. **H** High, wet montane forest, tall secondary growth. **D** Venezuela to Bolivia.

35 GIANT CONEBILL *Oreomanes fraseri* 15.5cm Unmistakable facial pattern and long, sharp bill. Extent of white on face varies. Sexes alike. Forages by probing the bark of *Polylepis* trees. **V** High, stammered twittering. **H** Restricted to high-altitude *Polylepis* woodland. **D** Colombia to Bolivia, N Chile, NW Argentina.

1
2
3
4
5
6
7
8
9
10
11
12
13
14
15
16
17
18
19
20
21
22
23
24
25
26
27
28
29
30
31
32
33
34
35

TANAGERS AND ALLIES *CONTINUED*

1 CINNAMON-BELLIED FLOWERPIERCER *Diglossa baritula* 11cm Male is blue-grey above and chestnut below, with a strikingly pale base to the lower mandible. Female is largely olive-brown. Note the distinctive hooked tip of the flowerpiercer bill, adapted for piercing the base of a flower to feed on nectar. **V** Very high short nasal fast twittering *witwitwitweetfjeetwheet*. **H** Flowering trees and shrub in montane forest, gardens. **D** Mexico to Nicaragua.

2 SLATY FLOWERPIERCER *Diglossa plumbea* 10cm Plain grey (male) or brown (female) flowerpiercer, with hook-tipped slightly upturned bill. **V** Ultra-high *sreesreesreeruhsreeh*. **H** Flowering trees and shrubs in or near montane forest. **D** Costa Rica, W Panama.

3 RUSTY FLOWERPIERCER *Diglossa sittoides* 11.5cm Male is distinctive. Note lack of white plumes in female (compare female White-sided Flowerpiercer). **V** Short bursts of very high twittering. **H** Shrubby areas, gardens, open woodland. **D** Venezuela to NW Argentina.

4 CHESTNUT-BELLIED FLOWERPIERCER *Diglossa gloriosissima* 14.5cm Separable from Black-throated Flowerpiercer by all-black head and upper breast, and from Merida Flowerpiercer by rufous flanks and different range. **V** High, sharp, rapid twittering. **H** Wet scrub and elfin woodland at the treeline. **D** Very restricted range in W Colombian Andes.

5 GLOSSY FLOWERPIERCER *Diglossa lafresnayii* 14.5cm Glossy black with pale blue-grey on wing coverts. Sexes alike. Not reliably separable from Black Flowerpiercer where their ranges overlap in Venezuela. **V** High, stuttered twittering. **H** Shrub and stunted forest at the treeline, overgrown pastures, gardens. **D** Venezuela to Peru.

6 MOUSTACHED FLOWERPIERCER *Diglossa mystacalis* 14.5cm Note black breast and belly, and distinctive 'moustache', which varies from rufous to buffish white. **V** High twittering. **H** Stunted forest, elfin forest, *Polylepis* woodland, scrubby woodland. **D** Peru, Bolivia.

7 MERIDA FLOWERPIERCER *Diglossa gloriosa* 13.5cm Range differs from that of Black-throated Flowerpiercer. Variant is also shown. **V** Probable song a short, jumbled series of sweet, fast warbles, fading at end. Calls include a short, thin note given in flight. **H** Dry scrub, gardens, stunted thickets, cultivation. **D** W Venezuela.

8 BLACK FLOWERPIERCER *Diglossa humeralis* 13.5cm Smaller than Glossy Flowerpiercer, not as glossy, and with smaller shoulder patch (but birds in W Venezuela have more blue-grey on the shoulder and are easily confused with that species). **V** Very high, hurried twittering in rather short bursts. **H** Shrubbery, gardens, stunted forest, parks, gardens, suburbs. **D** W Venezuela to N Peru.

9 BLACK-THROATED FLOWERPIERCER *Diglossa brunneiventris* 14cm Black chin and rufous whiskers are diagnostic. **V** High, fast, rattled twittering. **H** Montane scrub, elfin forest, páramo, cultivated areas, gardens. **D** Colombia; Peru to N Chile.

10 GREY-BELLIED FLOWERPIERCER *Diglossa carbonaria* 14cm Note pale blue shoulders and black crown (not grey as in Greater Flowerpiercer). Variable; a variant is shown. **V** High, raspy twittering in rather short bursts. **H** Scrub, woodland patches, gardens. **D** Bolivia, NW Argentina.

11 VENEZUELAN FLOWERPIERCER *Diglossa venezuelensis* 12.5cm Male is black with a white patch on the flanks. Female is greenish. Range differs from that of White-sided Flowerpiercer. **V** Song is a rapid and complex, but rather repetitive, series of soft, low-pitched notes. **H** Borders of humid forest and secondary growth. **D** NE Venezuela.

12 WHITE-SIDED FLOWERPIERCER *Diglossa albilatera* 12cm Male resembles Venezuelan Flowerpiercer. Female has olive-brown upperparts and cinnamon to yellowish-buff (not greenish) underparts. **V** One or two high notes followed by a trill, *swee-ti ti'ti'ti'ti'ti'ti*. **H** Shrubby borders and clearings in montane forest. **D** NW Venezuela to Peru.

13 SCALED FLOWERPIERCER *Diglossa duidae* 14cm Note pale horn-coloured base of maxilla, and more or less distinct spotting to underparts. **V** No information. **H** Scrubby forest on tepuis. **D** S Venezuela, extreme N Brazil.

14 GREATER FLOWERPIERCER *Diglossa major* 16.5cm Mostly slate-blue. Note narrow or very narrow malar stripe and rufous undertail coverts. **V** Unmusical scratchy rattling introduced by a series of tinkling notes. Recalls camera shutter or radio static. **H** Higher levels and borders of humid and wet montane forest on tepuis, secondary growth. **D** SE Venezuela, S Guyana, extreme N Brazil.

15 INDIGO FLOWERPIERCER *Diglossa indigotica* 11.5cm Separable from larger Masked Flowerpiercer by smaller mask and less fierce red eyes. **V** Ultra-high, slowly descending *sisisi* rattle. **H** Cloud forest, mossy forest borders, secondary growth. **D** Colombia, Ecuador.

16 GOLDEN-EYED FLOWERPIERCER *Diglossa glauca* 12cm Striking yellow eyes are diagnostic. Plumage almost entirely deep blue. **V** Barely audible, descending *si-si-si* series of notes. **H** Cloud forest, mossy forest borders. **D** Colombia to Bolivia.

17 BLUISH FLOWERPIERCER *Diglossa caerulescens* 13.5cm Dull blue and grey with a small, poorly defined mask. **V** Song a thin, high, descending warble with slight flourish at end. Calls include metallic *tink*. **H** Humid forest, secondary growth. **D** N Venezuela to N Bolivia.

18 MASKED FLOWERPIERCER *Diglossa cyanea* 14.5cm Note striking black mask and red eyes. **V** Rapid, ultra-thin, sizzled twittering. **H** Cloud forest, elfin forest, shrubby growth. **D** N Venezuela to C Bolivia.

19 BLACK-BACKED BUSH TANAGER *Urothraupis stolzmanni* 15cm Black above, with bright white throat and mottled greyish underparts. Sexes similar. **V** Song a very prolonged, fast-accelerating twitter. Calls include simple *see* and *tik* notes. **H** High-altitude dense wet forest and woodland, up to the treeline. **D** Colombia, Ecuador.

20 COAL-CRESTED FINCH *Charitospiza eucosma* 11.5cm Male is unmistakable due to pattern and crest (often laid flat). Female separable from seedeaters (Plates 299–300) by crest and white in tail. **V** Song consists of hurried, high phrases. **H** Cerrado, Caatinga, grassy savannah. **D** C and E Brazil, NE Bolivia, N Argentina.

21 BLACK-MASKED FINCH *Coryphaspiza melanotis* 13.5cm Male unmistakable, with bicoloured bill and striking facial pattern. Note yellow carpal patch of female. **V** Song is an ultra-high, thin *tji-ti*. **H** Cerrado with tall grass. **D** SE Peru, E Bolivia, E Brazil, Paraguay, NE Argentina. Also Ilha de Marajó, NE Brazil.

22 GREY PILEATED FINCH *Coryphospingus pileatus* 13cm Red crown of male, combined with grey plumage, is diagnostic. Female has brownish-grey crown. **V** Calm series of 5–10 *tui-tjee-tjaw* chirps. **H** Arid open woodland, dry scrub, forest edges, roadsides. **D** Colombia, Venezuela, EC Brazil.

23 RED PILEATED FINCH *Coryphospingus cucullatus* 13.5cm Reddish plumage with white eye-ring is diagnostic. **V** Calm series of 8–12 high chirps. **H** Dry scrub, grassland, villages, Chaco woodland. **D** The Guianas and NE Brazil; S Ecuador and Peru; C and S Brazil through Bolivia, Paraguay, Uruguay to N Argentina.

24 CRIMSON-BREASTED FINCH *Rhodospingus cruentus* 11cm Male unmistakable. Female less so, dull brown with paler underparts. Note rather long, slender bill. **V** Very high to ultra-high *tsee-tzi*. **H** Tall grass habitats in or near dry forest, low woodland and agricultural areas. **D** W Ecuador, NW Peru.

25 BLACK-HOODED SIERRA FINCH *Phrygilus atriceps* 15.5cm Note solid black hood of male and ochraceous back of female. **V** Slow, level chirping (4–8 notes). **H** Montane scrub, *Polylepis* woodland, settlements. **D** S Peru through SW Bolivia to N Chile, NW Argentina.

26 PERUVIAN SIERRA FINCH *Phrygilus punensis* 15.5cm Separable from Black-hooded Sierra Finch by paler hood and more olivaceous mantle. **V** Slow, unstructured, sharp chirping. **H** Scrubby mountain slopes, settlements, *Polylepis* woodland. **D** Peru, W Bolivia.

27 GREY-HOODED SIERRA FINCH *Phrygilus gayi* 16cm Separable from Patagonian Sierra Finch by habitat. Note olive back of male. **V** Slow *tjiw* or hurried *ti-tjerp* notes in series. **H** Open shrubby and scrubby habitats. **D** Patagonia, south to Tierra del Fuego.

28 PATAGONIAN SIERRA FINCH *Phrygilus patagonicus* 14.5cm Note ochraceous mantle of male. Female difficult to separate from male Grey-hooded Sierra Finch (which has an ochraceous wash to underparts). **V** Slow, high, unstructured chirping. **H** Openings and clearings in forest. **D** W Patagonia, south to Tierra del Fuego.

29 MOURNING SIERRA FINCH *Phrygilus fruticeti* 18cm Note striking white wing-bars of both sexes. Female shows a cinnamon cheek patch. **V** Series of 4–5 notes, the second lower, nasal and stretched. **H** Open and semi-open arid habitats, including agricultural land. **D** Peru to S Chile, S Argentina.

30 PLUMBEOUS SIERRA FINCH *Phrygilus unicolor* 15cm Male is uniform grey, including lores. Female separable from female Ash-breasted Sierra Finch by narrower streaking below. **V** Song series of five tuneful upslurred notes followed by one sharply downslurred note. Calls include quiet *chp* and high-pitched *zee*. **H** Open grassy habitats. **D** Andean chain, from Venezuela to Tierra del Fuego.

31 RED-BACKED SIERRA FINCH *Phrygilus dorsalis* 18cm Unmistakable, with grey head and rufous back. Sexes alike. **V** Calls include a nasal *phew* and a brief *wheenk*. **H** Dry rocky slopes, cushion bogs, puna grassland. **D** N Chile, S Bolivia, NW Argentina.

32 WHITE-THROATED SIERRA FINCH *Phrygilus erythronotus* 18cm Closely related to Red-backed Sierra Finch. Compare White-winged Diuca Finch (Plate 298), which shows white in wings. **V** Calls include a brief *weenk* and a short, sharp *tsip*. **H** Cushion bogs, rocky slopes. **D** S Peru, SW Bolivia, N Chile.

33 ASH-BREASTED SIERRA FINCH *Phrygilus plebejus* 12cm Note mantle streaking on male. Female similar to female Plumbeous Sierra Finch. **V** Series of 4–5 *tjew* notes, preceded by a buzzy first note. **H** Puna grassland, rocky slopes, cushion bogs. **D** Ecuador to W Chile, N Argentina.

34 CARBONATED SIERRA FINCH *Phrygilus carbonarius* 14.5cm Separable from Mourning Sierra Finch by lack of wing-bars, by black belly, and in female by grey-brown (not cinnamon) cheeks. **V** Fast series of very high *tjerre* notes, reaching a crescendo. **H** Open shrubby habitats. **D** C Argentina.

35 BAND-TAILED SIERRA FINCH *Phrygilus alaudinus* 14cm Note yellow bill of male. Diagnostic white tail-band is visible only in flight. **V** Repeated variations of a high two- or three-syllable phrase, the penultimate note sharper and higher-pitched, and the last one lower. **H** Semi-open arid shrubby areas. **D** Ecuador to N Argentina, Chile.

36 BLUE FINCH *Porphyrospiza caerulescens* 12.5cm Combination of blue plumage and yellow bill of male is diagnostic. Female's bill is duller yellowish. **V** Two very high, sharp notes, followed by a downslurred note. **H** Cerrado, rocky savannah. **D** NE Bolivia to E Brazil.

37 WHITE-BRIDLED FINCH *Melanodera melanodera* 15cm Note distinctive facial markings of male, and yellow in tail of both sexes. **V** Calm series of high, piercing notes with various intonations. **H** Tussock grassland, heathland. **D** S Chile, S Argentina, Falkland Islands.

38 YELLOW-BRIDLED FINCH *Melanodera xanthogramma* 16cm Note facial pattern. Has less yellow in wings and tail than White-bridled Finch. **V** Melodious notes, delivered slowly. **H** Grassy, rocky habitats above the treeline. **D** S Chile, SW Argentina.

39 SLATY FINCH *Haplospiza rustica* 12.5cm Separable from Uniform Finch by range. Resembles Blue-black Grassquit (Plate 299). **V** Ultra-high, hurried sizzling. **H** Bamboo in humid forest and adjacent grassy areas. **D** S Mexico to Bolivia; S Venezuela.

40 UNIFORM FINCH *Haplospiza unicolor* 12.5cm Compare Slaty Finch. No other grey finch occurs in its range. **V** Like that of Slaty Finch. **H** Borders and undergrowth of humid forest with bamboo. **D** S Brazil, SE Paraguay, NE Argentina.

1
2
3
4
5
variant
6
7
8
9
variant
10
11
12
13
14
15
16
17
18
19
20
21
22
23
24
25
26
27
28
29
30
31
32
33
34
35
36
37
38
39
40

TANAGERS AND ALLIES *CONTINUED*

1 PEG-BILLED FINCH *Acanthidops bairdi* 14cm Note distinctive bill shape and pale lower mandible. Male is very dark; female is brownish olive with pale supercilium and two wing-bars. Feeds on the ground. **V** Very high buzzing and insect-like *spss spss spss*, unstructured, but mostly in twos and threes. Also ultra-high single whistle. **H** Montane forest edge and clearings, bamboo. **D** Costa Rica, Panama.

2 BLACK-CRESTED FINCH *Lophospingus pusillus* 14cm Unmistakable, with striking black and white facial pattern and pointed crest. **V** Fast, nasal chattering. **H** Chaco scrub, low woodland, pastures. **D** SE Bolivia, W Paraguay to C Argentina.

3 GREY-CRESTED FINCH *Lophospingus griseocristatus* 14cm Like a Black-crested Finch with an all-grey head. White tail corners show well in flight. **V** Series of monotonous, rolling double notes. **H** Desert, semi-desert and other dry habitats with sparse vegetation. **D** C Bolivia to NW Argentina.

4 LONG-TAILED REED FINCH *Donacospiza albifrons* 15cm Facial pattern and long tail are distinctive. **V** High, nasal twittering. **H** Marshes, reedbeds, shrubbery near water. **D** C Bolivia; E Paraguay, S Brazil, Uruguay, NE Argentina.

5 GOUGH FINCH *Rowettia goughensis* 25cm Large slim-billed finch. Male leaf-green with black bib and lores, female greyer with faint streaking. Omnivorous ground-feeder, even takes carrion on occasion. **V** Male has high-pitched whistling song; female responds with chattering call. **H** Most habitats, but especially upland tussocky grassland. **D** Gough Island, S Atlantic.

6 INACCESSIBLE ISLAND FINCH *Nesospiza acunhae* 19cm Large dull-green finch. Male brighter than female but shows considerable individual variation. **V** Song simple, repeated three- or four-note phrase; calls include sharp *chip* contact note. **H** Varies by subspecies; includes coastal lowlands, cliffs, woodland and tussocky grassland. **D** Inaccessible Island (Tristan da Cunha, S Atlantic).

7 NIGHTINGALE ISLAND FINCH *Nesospiza questi* 17cm Medium-sized, yellow-grey finch with faint streaking on upperside and underside; little sexual dimorphism. Feeds mainly on ground, taking seeds, berries and a few invertebrates. **V** Similar to that of Inaccessible Island Finch. **H** Found in all habitat types on the island. **D** Nightingale Island (Tristan da Cunha, S Atlantic).

8 WILKINS'S FINCH *Nesospiza wilkinsi* 21cm Large, stocky and very thick-billed finch with streaked yellow-green plumage. Feeds on ground and in trees, using bill to break into tough woody tree fruits. **V** Similar to that of Inaccessible Island Finch but lower-pitched. **H** Woodland with myrtles, adjacent grasslands. **D** Nightingale Island (Tristan da Cunha, S Atlantic).

9 WHITE-WINGED DIUCA FINCH *Diuca speculifera* 19cm White in wing, below eye and on throat is distinctive. **V** Ultra-high, thin twitters. **H** High Andean puna grassland. **D** W Peru to N Chile, W Bolivia.

10 COMMON DIUCA FINCH *Diuca diuca* 16.5cm Unmistakable in its range. Note white throat and belly, and rufous lower flanks. **V** Unstructured stream of well-separated, staccato, chirping notes. **H** Open and semi-open areas, including natural and agricultural habitats. **D** Chile, Argentina, extreme S Bolivia.

11 SHORT-TAILED FINCH *Idiopsar brachyurus* 18cm Note rather heavy, long bill and white grizzling below eyes. **V** Song is a series of high-pitched, rasping notes. Calls include rising *shrii* and *seet*. **H** High-altitude fields with rocks and boulders, puna grassland. **D** S Peru, Bolivia, N Argentina.

12 CINEREOUS FINCH *Piezorina cinerea* 16.5cm Note massive yellow bill and pale grey plumage. **V** Calm, unstructured series of nasal chirps. **H** Coastal plains, open terrain with scarce grass and scattered trees. **D** W Peru.

13 SLENDER-BILLED FINCH *Xenospingus concolor* 15cm Mostly grey, with black lores. Slender yellow bill is diagnostic. **V** Unstructured series of chirps of varying pitch and intonation. **H** Dense shrubbery, dry woodland patches. **D** S Peru, NW Chile.

14 GREAT INCA FINCH *Incaspiza pulchra* 16.5cm Mainly brown above, grey below, with black mask and yellowish bill and legs. **V** Short, ultra-high, shivering trill, the last note lower-pitched. **H** Arid slopes and ravines of the western Andes. **D** W Peru.

15 RUFOUS-BACKED INCA FINCH *Incaspiza personata* 16.5cm Similar to Great Inca Finch, but with rufous of wings extending over back, more black on face, less black on throat. **V** Series of sharp, very high *seet-suweet* notes. **H** Dry slopes with cacti and *Agave*. **D** W Peru, especially Marañón River valley.

16 GREY-WINGED INCA FINCH *Incaspiza ortizi* 16.5cm Lacks rufous in plumage. **V** Ultra-high *sueet-sweetie*. **H** Desert scrub with bromeliads and cacti. **D** NW Peru.

17 BUFF-BRIDLED INCA FINCH *Incaspiza laeta* 14.5cm Note white or buffish malar patch. **V** Calls include a high-pitched *tseet* and a dry scolding note. **H** Dry woodland, thorn scrub. **D** Marañón River valley, Peru.

18 LITTLE INCA FINCH *Incaspiza watkinsi* 13cm Note small size and faintly streaked mantle. **V** Calls include a simple high-pitched *tseet* and a rasping *tchew* note. **H** Desert scrub with bromeliads; 400–900m. **D** Very restricted range within the Marañón River valley, Peru.

19 BAY-CHESTED WARBLING FINCH *Poospiza thoracica* 13.5cm Note grey head without white supercilium, striking rufous breast-band and grey rump. **V** Mixture of very high *tsi* and *tsu* notes (three per sec). **H** Edges of montane forest, clearings with scattered trees, open woodland. **D** SE Brazil.

20 BOLIVIAN WARBLING FINCH *Poospiza boliviana* 16cm Note white supercilium, white throat with no dark malar stripe, rufous breast. **V** Call is a sharp high-pitched *twipp*. **H** Dry and semi-humid habitats, open woodland, *Polylepis*, riverine thickets, dry scrub. **D** Bolivia, NW Argentina.

21 PLAIN-TAILED WARBLING FINCH *Poospiza alticola* 16cm No similar bird occurs in its range. Wings are plain, except for white edges to outer flight feathers. **V** Calm series of nasal *wit-wee* notes. **H** *Polylepis* and other woodland, montane scrub and shrubbery. **D** C Peru.

22 RUFOUS-SIDED WARBLING FINCH *Poospiza hypocondria* 16cm Note narrow dark malar stripe and pale greyish breast contrasting with rufous flanks. **V** Song is a rushed and jumbled series of rather unmelodic notes. Typical call is a high-pitched *zwee*. **H** Shrub and hedges on hillsides and in agricultural areas. **D** W Bolivia, NW Argentina.

23 RUSTY-BROWED WARBLING FINCH *Poospiza erythrophrys* 14cm Note rufous supercilium and white in wings. **V** Calm series of *tjiw* and sharp *tee* notes. **H** Edges of humid forest and secondary growth. **D** C Bolivia, NW Argentina, east slopes of the Andes.

24 CINNAMON WARBLING FINCH *Poospiza ornata* 13cm Note distinctive long supercilium and much white in wings. **V** Series of short, rapid, twittering phrases. **H** Arid woodland. **D** W Argentina.

25 BLACK-AND-RUFOUS WARBLING FINCH *Poospiza nigrorufa* 15cm Striking, mostly white supercilium and plain wings are diagnostic. **V** Series of *tee-tee-tjuw* notes, the last one lower-pitched. **H** Woodlands, shrubby habitats, suburban areas. **D** E Paraguay and S Brazil through Uruguay to E Argentina.

26 BLACK-AND-CHESTNUT WARBLING FINCH *Poospiza whitii* 15cm Formerly considered conspecific with Black-and-rufous Warbling Finch. Back is greyer, breast is darker chestnut, and tail has more white – but best separated by range. **V** Song variable, complex, with tuneful notes and trilling, quavering phrases. Calls include a high-pitched *tziip*. **H** Semi-arid scrub and woodland, forest edge, highland agricultural areas. **D** C Bolivia to NW Argentina.

27 BUFF-THROATED WARBLING FINCH *Poospiza lateralis* 15cm Note rufous rump, buff throat and grey cheeks. **V** Calm, unstructured series of sharp, very high to ultra-high *tjiw* notes. **H** Moist montane forest. **D** E Brazil.

28 GREY-THROATED WARBLING FINCH *Poospiza cabanisi* 15cm Separable from Buff-throated Warbling Finch by range and greyer throat. **V** Calm series of piercing *tjiw* and lower *tjuw* notes. **H** Riverine belts, forest edges, open woodland. **D** SE Brazil, E Paraguay, Uruguay, NE Argentina.

29 RUFOUS-BREASTED WARBLING FINCH *Poospiza rubecula* 15cm Grey upperparts with rufous forecrown and short rufous supercilium, rufous underparts. No similar bird occurs in its range, but compare some tanagers with grey and rufous in their plumage. **V** Unstructured series of *chiff-chaff-sree* phrases. **H** Montane scrub and woodland. **D** W Peru.

30 COLLARED WARBLING FINCH *Poospiza hispaniolensis* 13.5cm Note distinctive facial pattern; shows striking white inner webs of outer-tail feathers in flight. **V** Song is a single or double *wee-tjuw*. **H** Dense desert scrub, agricultural areas with scattered bush and low trees. **D** SW Ecuador, W Peru.

31 RINGED WARBLING FINCH *Poospiza torquata* 13.5cm Distinctive in its range; note white streak in closed wings, and rusty undertail coverts. **V** Series of repetitive warbling, incorporating *wee-tjuw* notes. **H** Chaco, *Polylepis* and other woodland, dry forest. **D** Bolivia, N and C Argentina.

32 BLACK-CAPPED WARBLING FINCH *Poospiza melanoleuca* 13cm Note distinctive contrasting monochrome plumage. Eyes may appear orange. **V** Unstructured series of high-pitched notes. **H** Chaco woodland, savannah, dense shrub. **D** C Bolivia through Paraguay to NC Argentina, Uruguay.

33 CINEREOUS WARBLING FINCH *Poospiza cinerea* 13cm Separable from Black-capped Warbling Finch by range and by different head pattern (not all black). **V** Fast, very high warbling. **H** Cerrado and other dry woodland, agricultural areas. **D** SC Brazil.

34 CHESTNUT-BREASTED MOUNTAIN FINCH *Poospiza caesar* 18.5cm Note long white supercilium and white throat. No similar bird occurs in its range. **V** Short, warbling phrase. **H** Montane scrub, woodland, secondary growth. **D** SC Peru.

35 COCHABAMBA MOUNTAIN FINCH *Compsospiza garleppi* 18cm Note large size and distinctive grey and orange-rufous plumage. **V** Short series of warbling phrases, incorporating a few nasal notes. **H** Shrubby canyons and slopes, *Polylepis* woodland. **D** C Bolivia.

36 TUCUMAN MOUNTAIN FINCH *Compsospiza baeri* 18cm Differs from Cochabamba Mountain Finch in having greyish breast and belly, and in different range. **V** Calm, unstructured series of high, sharp, staccato notes. **H** Puna grassland with woodland patches, gardens. **D** High Andes of NW Argentina.

37 STRIPE-TAILED YELLOW FINCH *Sicalis citrina* 11cm Male separable from other male yellow finches by streaking on mantle (except greenish Grassland Yellow Finch, Plate 299) and white terminal halves of outer-tail feathers. Note streaking on head of female. **V** Series of short, high, sharp, warbled phrases, with the last note rapidly repeated 3–5 times. **H** Grassy and open habitats, Cerrado. **D** Disjunct populations: Colombia, Venezuela, the Guianas, N Brazil; EC Brazil; SE Peru to NW Argentina.

38 PUNA YELLOW FINCH *Sicalis lutea* 11.5cm No other yellow finch in its high Andes habitat has such uniform yellow plumage. Female only slightly duller-plumaged than male. **V** Song pleasant and tuneful conversational chattering; calls include rich *tchew* and *meluu*. **H** Rocky slopes, dry grassland. **D** SC Peru to NW Argentina.

39 BRIGHT-RUMPED YELLOW FINCH *Sicalis uropigyalis* 14cm In addition to grey wings, note grey cheeks and flanks (less extensive or missing in N and C Peru birds). **V** Song a loud twittering series, *twit-twit-twit-twit-teu-teu-teu-teu*... Calls include a *we-week* given in flight, and a nasal *kuik*. **H** Puna grassland, habitations. **D** Peru to N Chile, NW Argentina.

40 CITRON-HEADED YELLOW FINCH *Sicalis luteocephala* 14cm Separable from Bright-rumped Yellow Finch by yellow face contrasting with grey neck and crown; grey on cheeks is absent. **V** Long, hurried series of nasal notes and rollers, each one doubled. **H** Grassy highland areas with scattered patches of shrub. **D** Bolivia, NW Argentina.

1
2
3
4
5
6
7
8
9
10
11
12
13
14
15
16
17
18
19
20
21
22
23
24
25
26
27
28
29
30
31
32
33
34
35
36
37
38
39
40

TANAGERS AND ALLIES *CONTINUED*

1 GREATER YELLOW FINCH *Sicalis auriventris* 14.5cm No similar yellow-finch with such grey wings occurs in its range. **V** Hurried, sharp, partly nasal twittering. **H** High, harsh alpine habitats; in lower and damper areas in Patagonia. **D** Chile, Argentina.

2 GREENISH YELLOW FINCH *Sicalis olivascens* 14cm Compare Puna Yellow Finch (Plate 298) (brighter yellow) and Greater Yellow Finch (greyer wings). **V** Rapid, level rattle, changing in intonation (2 secs). **H** Shrubby puna, cultivated areas. **D** Peru to Bolivia, N Chile, N Argentina.

3 MONTE YELLOW FINCH *Sicalis mendozae* 14cm Male bright yellow, female mostly grey with some yellow on throat, centre of belly and vent. **V** Song is a complex combination of rhythmic rolling notes combined with mimicry. Calls include *tweep* and brief *tik*. **H** Arid, shrubby areas. **D** W Argentina.

4 PATAGONIAN YELLOW FINCH *Sicalis lebruni* 14cm Has more grey plumage than yellow. Grey particularly extensive in the female. **V** Very high, rapid, tinkling warbling. **H** Dry steppe. **D** S Argentina, S Chile.

5 ORANGE-FRONTED YELLOW FINCH *Sicalis columbiana* 11.5cm Male is separable from Saffron Finch by dusky lores and olive nape and mantle. **V** Song is a series of sparrow-like chirrups; typical calls are similar single notes. **H** Shrubby and grassy areas close to water. **D** Colombia and Venezuela, C Brazil, E Brazil.

6 SAFFRON FINCH *Sicalis flaveola* 14cm Male's mantle is only slightly less yellow than underparts. Female much streakier. Feeds on the ground, on grass seeds. **V** Song is a melodious, slightly harsh *chit chit chit chit-chit*, varying in length. Calls include a soft or loud *pink* and a whistled *wheat*. **H** Open and semi-open savannah, thorn scrub, agricultural land, gardens, urban lawns. **D** Colombia, Venezuela, Trinidad, the Guianas; Ecuador and NW Peru; Bolivia to E Brazil and N Argentina.

7 GRASSLAND YELLOW FINCH *Sicalis luteola* 12cm Male has streaky olive crown and much grey on the face. Female duller. Feeds in grass or on the ground. **V** Song includes a buzzy trill. **H** Open grassy areas. **D** Widespread from S Mexico to C Argentina. Southernmost populations migrate north for the winter.

8 RAIMONDI'S YELLOW FINCH *Sicalis raimondii* 11.5cm Greyer than Grassland Yellow Finch, with less yellow on lore and eye-ring. Sexes similar, but female slightly duller. **V** Song is a sweet, reedy warble; calls include dry *djerr-djerr* and upslurred *kewee*. **H** Rocky slopes, dry grassy habitats. **D** W Peru.

9 SULPHUR-THROATED FINCH *Sicalis taczanowskii* 12cm Note heavy bill, short tail and pale yellow throat. Sexes alike. **V** Calls include various squeaky and dry scratchy notes. **H** Desert scrub and nearly barren areas. **D** SW Ecuador, NW Peru.

10 WEDGE-TAILED GRASS FINCH *Emberizoides herbicola* 19cm Note long tail, white throat and green outer wings. **V** Series of well-separated short melodious phrases, *tjer-tjer*. **H** Grassland with patches of tall grass, savannah. **D** Costa Rica to N Peru, N Brazil; Bolivia to E Brazil, N Argentina.

11 LESSER GRASS FINCH *Emberizoides ypiranganus* 17.5cm Similar to Wedge-tailed Grass Finch, best separated by song. Smaller than Duida Grass Finch and with more contrasting mantle streaking. **V** Series of rapid, level, raspy phrases, almost like rattles (once every 4 secs). **H** Wet and drier grassy habitats. **D** E Paraguay, S Brazil, Uruguay, NE Argentina.

12 DUIDA GRASS FINCH *Emberizoides duidae* 21cm Separable from Wedge-tailed Grass Finch by very restricted range, darker plumage and proportionally longer tail. **V** Not known. **H** Dry savannah. **D** Cerro Duida, S Venezuela.

13 PAMPA FINCH *Embernagra platensis* 22cm Note pale yellow-orange bill, green mantle and long tail. **V** Short, very high, rapid, muttered phrases, often with partly inhaled notes. **H** Grassland with some scrub, often at the edge of lakes. **D** Bolivia through Paraguay, SE Brazil, Uruguay to C Argentina.

14 SERRA FINCH *Embernagra longicauda* 21.5cm Separable from Pampa Finch by different head pattern. Note that mantle is unstreaked and rather grey. **V** Short, high, twittered phrases. **H** Dry savannah, grassy scrub with scattered palms. **D** E Brazil.

15 MANY-COLORED CHACO FINCH *Saltatricula multicolor* 18cm Unmistakable in its habitat. Note brown crown, black and grey face with bold white posterior supercilium. Sexes similar. **V** Song is a persistently given short warbled phrase. Calls include a low *chup* and a higher-pitched *swiip*. **H** Open Chaco woodland, semi-open country, field edges, roadsides. **D** E Bolivia and W Paraguay to W Uruguay, C Argentina.

16 SLATE-COLORED GROSBEAK *Saltator grossus* 20cm Note distinctive red bill and white throat. **V** High, pure, fluted notes in short phrases. **H** Várzea and terra firme forest, and mature secondary growth. **D** Honduras to Ecuador, Amazonia.

17 BLACK-THROATED GROSBEAK *Saltator fuliginosus* 22cm Separable from Slate-colored Grosbeak by range and black throat. **V** Song like Slate-colored Grosbeak. **H** Humid forest and woodland. **D** E Paraguay, NE Argentina, S and SE Brazil.

18 BLACK-HEADED SALTATOR *Saltator atriceps* 25cm Blackish head with contrasting white chin and eye stripe distinctive. Note also buff-rufous undertail coverts. **V** Cracking powerful nasal *tjih tjihtihwee* with a dry rattle. **H** Forest edge, secondary growth, scrub. **D** E Mexico to Panama.

19 BUFF-THROATED SALTATOR *Saltator maximus* 21cm Separable from Greyish Saltator by green back, and from Green-winged and Thick-billed saltators by green tail. **V** Slow series of well-separated groups of warbled musical notes and short phrases. **H** Borders of humid forest, secondary growth, plantations, suburbs. **D** E Mexico to Ecuador, Venezuela and through Amazonia to E Brazil.

20 BLACK-WINGED SALTATOR *Saltator atripennis* 20.5cm Unmistakable, with black crown and sides of head, and conspicuous white supercilium. Bright olive back contrasts with black wings and tail. **V** Slow series of well-separated groups of short, melodious phrases and rising or downslurred notes. **H** Rather wet forest, secondary growth. **D** C Colombia to W Ecuador.

21 GREEN-WINGED SALTATOR *Saltator similis* 20.5cm Note green wings and long eye-stripe. **V** Slow, calm phrases of well-enunciated musical notes. **H** Woodland, forest borders, riverine belts. **D** E Bolivia and C Brazil to Paraguay, NE Argentina, Uruguay.

22 GREYISH SALTATOR *Saltator coerulescens* 21cm Grey upperparts are diagnostic. Note also black malar stripe, white throat, cinnamon undertail coverts. **V** Varied repertoire of calm, melodious phrases of repeated musical notes. **H** Edges and clearings in forest, riverine belts, dry scrub. **D** Mexico to Argentina.

23 ORINOCO SALTATOR *Saltator orenocensis* 18.5cm Unmistakable in its range, with black head sides, broad white supercilium, mostly white underparts with buffy flanks and undertail. **V** Rather hurried, high, loud twittering. **H** Dry woodland and brush, suburbs, riverine belts. **D** NE Colombia, N Venezuela.

24 THICK-BILLED SALTATOR *Saltator maxillosus* 21cm Note short, heavy bill, normally with at least some orange at base. **V** Calm phrases of 4–7 musical notes. **H** Canopy and borders of humid forest. **D** NE Argentina, SE Brazil.

25 BLACK-COWLED SALTATOR *Saltator nigriceps* 22cm Very distinctive in its restricted range; mostly grey with black hood and stout orange-red bill. **V** Slow series of single or calm, short phrases of high, sharp notes. **H** Montane humid forest and scrub on west slopes of the Andes. **D** SW Ecuador, NW Peru.

26 GOLDEN-BILLED SALTATOR *Saltator aurantiirostris* 20cm No other saltator has a white eyebrow starting above eye (not at bill). **V** Pleasant yet sharp, short phrases at 2–3-sec intervals. **H** Montane scrub and woodland, agricultural areas, gardens. **D** Peru through Bolivia and Paraguay to N and WC Argentina.

27 MASKED SALTATOR *Saltator cinctus* 21.5cm Extensive black mask and black breast-band are conspicuous. **V** Song is a simple *tjew-tjew-tjew-tjuwee*. **H** Montane forest. **D** Colombia, Ecuador, Peru.

28 BLACK-THROATED SALTATOR *Saltator atricollis* 20.5cm Mostly dull upperparts, with distinctive diffusely bordered black face and throat, and red bill. **V** Short, hurried, sharp phrases, ending as a *tuwee*. **H** Cerrado and Caatinga shrub. **D** E Brazil to E Bolivia, E Paraguay.

29 LESSER ANTILLEAN SALTATOR *Saltator albicollis* 22cm Arboreal, feeds primarily on fruit and seeds. Formerly treated as a race of Streaked Saltator. **V** A series of rising and falling harsh, loud notes. **H** Forest-edge undergrowth, secondary growth, thickets, dry scrub. **D** Lesser Antilles.

30 STREAKED SALTATOR *Saltator striatipectus* 20cm No other saltator in its range has streaking below. **V** Calm series of short, high, fluted phrases. **H** Regenerating wooded habitats, gardens. **D** Costa Rica to Peru and Trinidad.

31 BLUE-BLACK GRASSQUIT *Volatinia jacarina* 11cm Forages in low vegetation in search of seeds and insects. In display, male perches on an exposed branch and jumps up and down with tail spread and wings open, showing white axillaries. **V** A loud *eee-slick*; during display jump utters a wheezing *jweee*. **H** Shrubby fields, farmland, scrubby secondary growth and roadsides. **D** Widespread from Mexico to Argentina.

32 BUFFY-FRONTED SEEDEATER *Sporophila frontalis* 13cm Note heavy bill, wing stripes and wooded habitat. **V** Rapid phrases of 4–5 level, staccato *tjub* notes. **H** Undergrowth, edges and adjacent overgrown clearings in humid forest. **D** SE Brazil, E Paraguay, NE Argentina.

33 TEMMINCK'S SEEDEATER *Sporophila falcirostris* 11cm Male almost entirely grey with yellow bill, female olive-brown. May have some white on face and/or throat. **V** Short, very high, thin trill (once every 4–6 secs). **H** Bamboo-rich undergrowth, forest edge, secondary growth. **D** SE Brazil, E Paraguay, NE Argentina.

34 SLATE-COLORED SEEDEATER *Sporophila schistacea* 11cm Male has white malar stripe that widens onto neck side. Habitat normally differs from that of Grey Seedeater. **V** Very high to ultra-high, hurried trills, ending *tue-tue*. **H** Bamboo-rich borders of humid forest, secondary growth, plantations. **D** Scattered populations, S Mexico to Peru, Bolivia, Brazil.

35 PLUMBEOUS SEEDEATER *Sporophila plumbea* 11cm Note dark bill in both male and female. **V** Rich, repetitive warbling. **H** Savannah, grassland, Cerrado with tall grass. **D** Colombia, Venezuela, the Guianas, N Brazil; C and S Brazil, SE Peru, Bolivia, NE Argentina.

36 TROPEIRO SEEDEATER *Sporophila beltoni* 12cm Large bill, yellowish in both sexes. Male smoky-grey, female greenish-brown with pale underparts. **V** Song elaborate with much mimicry; calls similar to those of Plumbeous Seedeater. **H** Dry grassland with tall shrubs. **D** S Brazil (Rio Grande do Sul, Santa Catarina, Paraná).

37 VARIABLE SEEDEATER *Sporophila corvina* 11cm The only seedeater in its range with black upperparts. Extent of black in male varies greatly, with some birds almost entirely black. **V** Rapid series of repeated notes and small rattles. **H** Scrub, agricultural areas, habitation. **D** E Mexico to Peru.

38 GREY SEEDEATER *Sporophila intermedia* 11cm Male mostly grey, with white wing panel. Female very like most other female seedeaters. **V** Hurried, nasal twittering, each element repeated 2–4 times. **H** Grassland, shrubby clearings, wasteland in suburbs. **D** Colombia to Guyana, N Brazil, Trinidad.

1
2
3
4
5
6
7
8
9
10
11
12
13
14
15
16
17
18
19
20
21
22
23
24
25
26
27
28
29
30
31
32
33
34
35
36
37
variant
37
38

TANAGERS AND ALLIES *CONTINUED*

1 WING-BARRED SEEDEATER *Sporophila americana* 11.5cm Male separable from Caqueta Seedeater by more distinct wing-bars. Note white collar, interrupted at neck. **V** Hurried, sharp warbling. **H** Forest edges and secondary growth on coastal plains. **D** N and E Amazonia, from NE Venezuela to N Brazil.

2 CAQUETA SEEDEATER *Sporophila murallae* 11cm Very similar to Wing-barred Seedeater, and formerly treated as conspecific. **V** Simple, rather monotonous warbling that mounts slightly in strength. **H** Riverine scrub, forest edges, shrubby areas. **D** W Amazonia, from SE Colombia to E Ecuador, NE Peru, W Brazil.

3 CINNAMON-RUMPED SEEDEATER *Sporophila torqueola* 11.5cm Note male's blackish head, white neck side, black breast-band, cinnamon underparts. Sings from fences, trees and wires. **V** Song is a series of sweet whistles, *sweet sweet tew tew tew tew sit*. Calls include a husky *quit*, a nasal *cheh* and a loud *seeu*. **H** Dense grass near tall cane, weedy fields and secondary growth. **D** W Mexico.

4 MORELET'S SEEDEATER *Sporophila morelleti* 11cm Male has distinctive, bold black-and-white pattern. Female plain brown, darker on wings with two narrow whitish wing-bars. Feeds mainly on grass stems. **V** Song is a melodious whistled phrase, *wee wee wee* followed by a dry trill; calls include *tik* or *che*. **H** Open grassland, often near water. **D** S Texas, E Mexico to Panama.

5 RUSTY-COLLARED SEEDEATER *Sporophila collaris* 12cm No similar seedeater occurs in its range. Note white marks above and below eyes in male. Female more distinctly patterned than other female seedeaters. **V** Hurried, repetitive nasal warbling. **H** Wet grassland, shrubbery near water. **D** Bolivia to SC Brazil, Paraguay, Uruguay, N Argentina.

6 LESSON'S SEEDEATER *Sporophila bouvronides* 11cm Note male's black head with triangular white patch between throat and moustachial area. Separable from male Lined Seedeater by all-black crown. **V** A rattle, changing to a trill. **H** Shrubby and grassy areas near water. **D** Breeds N Colombia to the Guianas (Apr–Nov). Spreads through N Amazonia in non-breeding season.

7 LINED SEEDEATER *Sporophila lineola* 11cm Southern equivalent of Lesson's Seedeater. Male separable by white crown-stripe. Female indistinguishable in the field. **V** Short, unvaried warbling, the main section comprising a strong rattle. **H** Grassy and shrubby habitats, Chaco woodland, Caatinga. **D** Breeds E Brazil, Paraguay, N Argentina, S Brazil (Nov–Apr). Spreads through Amazonia in non-breeding season.

8 BLACK-AND-WHITE SEEDEATER *Sporophila luctuosa* 11cm Note male's white crescent below eye, and prominent wing patch. Female looks much like most other female seedeaters. **V** Nasal twittering with embedded short rolls and rattles. **H** Grassy and shrubby forest edges and roadsides. **D** Venezuela to Bolivia.

9 YELLOW-BELLIED SEEDEATER *Sporophila nigricollis* 10.5cm Male has black hood and olive mantle, but belly is at best only pale yellow. Like all seedeaters, forages for seeds and insects in low vegetation. **V** Song is a short warble followed by buzzy notes. Calls include a short *cheep* and a chittering when excited. **H** Forest edge, thickets, roadsides, shrubby fields and field edges. **D** Costa Rica to Peru, Bolivia, Venezuela, the Guianas, Trinidad, E Brazil, N Argentina.

10 DUBOIS'S SEEDEATER *Sporophila ardesiaca* 11cm Male whiter below than male Yellow-bellied Seedeater (and sometimes considered a race of that species). **V** Short, very high phrase of 5–6 notes, starting with *tjew-tjew*. **H** Shrubby clearings and roadsides. **D** SE Brazil.

11 DOUBLE-COLLARED SEEDEATER *Sporophila caerulescens* 11cm Head pattern of male is distinctive. Female is a typical nondescript female seedeater. **V** Song loud and varied, tailing off at the end. **H** Shrubby areas and neglected farmland. **D** E Bolivia, S and E Brazil through Paraguay and Uruguay to N Argentina. Some northward movement into Amazonia in non-breeding season.

12 WHITE-THROATED SEEDEATER *Sporophila albogularis* 11cm Note dark breast-band in male, separating white throat from white underparts. **V** Short phrase of pretty warbling with an embedded *tjup-tjup-tjup*. **H** Caatinga scrub and woodland. **D** NE Brazil.

13 WHITE-BELLIED SEEDEATER *Sporophila leucoptera* 12cm Note male's bicoloured plumage and pinkish bill. Resembles White-throated Seedeater but lacks dark breast-band. **V** Series of 5–7 drawn-out, rising and falling *tuweet* notes. **H** Thickets, forest edges near grassy places, water and marshes. **D** Suriname, NE, E, C and S Brazil, SE Peru, Bolivia, Paraguay, N Argentina.

14 PARROT-BILLED SEEDEATER *Sporophila peruviana* 11.5cm Heavy bill is distinctive, and makes females more easily identifiable than most female seedeaters. Note also black(ish) chin. **V** Series of hoarse and twittered notes. **H** Arid scrub, agricultural areas, villages. **D** SW Ecuador, W Peru.

15 DRAB SEEDEATER *Sporophila simplex* 11cm Dull-coloured plumage and narrow wing-bars are distinctive. **V** Short, rapid, gently descending, slightly hoarse series of 3–4 *tjuw* notes, introduced by two soft *mew* notes. **H** Dry scrubby areas, fields. **D** SW Ecuador, W Peru.

16 BLACK-AND-TAWNY SEEDEATER *Sporophila nigrorufa* 10cm Black cap, dark grey or black mantle and tawny-cinnamon underparts of male are diagnostic. **V** Calm mixture of 5–15 drawn-out, rising or falling *tjuw* notes. **H** Cerrado, tall grass, floodplains. **D** E Bolivia, SW Brazil.

17 COPPER SEEDEATER *Sporophila bouvreuil* 10cm Male is mostly cinnamon, with black wings and tail and an isolated black crown. **V** Short, sharp, warbled phrases, often preceded by *tjuw-tjee*. **H** Cerrado, savannah, tall grass. **D** S Suriname, E and SE Brazil.

18 PEARLY-BELLIED SEEDEATER *Sporophila pileata* 10cm Pale brown seedeater, much lighter than formerly conspecific Copper Seedeater; male has neatly delineated black crown and whitish cheeks. **V** Like that of Copper Seedeater. **H** Cerrado, savannah, tall grasslands, pastures. **D** NE Bolivia, C and S Brazil, Paraguay, NE Argentina, Uruguay.

19 RUDDY-BREASTED SEEDEATER *Sporophila minuta* 10cm Rufous underparts of male contrast with grey head and grey-brown upperparts. Separable from Tawny-bellied Seedeater by range, grey cheeks and small white mark at gape. **V** Short, unstructured series of notes, including *tjuw* and *tjee*. **H** Shrubby roadsides, parks, pastures, farms. **D** SW Mexico to Colombia, Venezuela, the Guianas, N Brazil.

20 TAWNY-BELLIED SEEDEATER *Sporophila hypoxantha* 10cm Similar to Ruddy-breasted Seedeater, but occurs further south. Generally paler than Rufous-rumped Seedeater, especially on throat. **V** Short series of 2–10 high, gliding notes. **H** Tall grass, roadsides. **D** E Bolivia to SE Brazil, NE Argentina, W Uruguay.

21 RUFOUS-RUMPED SEEDEATER *Sporophila hypochroma* 10cm Very similar to Tawny-bellied Seedeater, but with more saturated colouring. **V** Slow series of sharply fluted notes with incidental small rolls and rattles. **H** Tall grassland, often adjacent to marsh. **D** Breeds during austral spring and summer, Paraguay, NE Argentina. Winters north into Bolivia, SC Brazil.

22 DARK-THROATED SEEDEATER *Sporophila ruficollis* 10cm Black(ish) throat of male is diagnostic. **V** Short series of 2–10 high, fluted notes. **H** Tall grass, roadsides, scrub, Cerrado. **D** Breeds during austral spring and summer, SE Bolivia, SW Brazil to NE Argentina and Uruguay. Winters north into Bolivia, C Brazil.

23 MARSH SEEDEATER *Sporophila palustris* 10cm Separable from other similarly plumaged seedeaters by white throat of male and strong preference for marshy habitats. **V** Short series of 2–10 high, fluted notes. **H** Damp grassland, marsh. **D** Breeds during austral spring and summer, NE Argentina, Uruguay, S Brazil. Winters SC Brazil.

24 CHESTNUT-BELLIED SEEDEATER *Sporophila castaneiventris* 10cm Male has extensive blue-grey flanks (matching colour of mantle) and rufous median underparts. **V** Warbled phrases, introduced by a high, sharp *pseet* (repeated 1–2 times). **H** Grassy clearings, fields and pastures, habitation, edges of lakes and rivers. **D** Amazonia, resident.

25 CHESTNUT SEEDEATER *Sporophila cinnamomea* 10cm Note contrast between male's chestnut plumage (including neck and mantle) and pale grey crown. Female not separable from other seedeater species. **V** Slow series of sharply fluted notes. **H** Tall grass next to wet areas, floodplains. **D** Breeds during austral spring and summer, S Paraguay, NE Argentina, S Brazil, Uruguay. Winters SC Brazil.

26 BLACK-BELLIED SEEDEATER *Sporophila melanogaster* 10cm Male distinctly patterned with black and grey. **V** Series of *fiu* and *tsi-tsi* notes, sung at a very slow tempo. **H** Damp grassland. **D** Breeds S Brazil during austral spring and summer. Winters north to C Brazil.

27 CHESTNUT-THROATED SEEDEATER *Sporophila telasco* 10cm Note chestnut chin of male. Colour morph '*insulata*', formerly considered to be a separate species, is separable from Chestnut Seedeater by grey neck. **V** Short, unstructured, twittering phrases. **H** At shrubby edges of grassland, fields, hedgerows, villages. **D** Ecuador, W Chile.

28 THICK-BILLED SEED FINCH *Oryzoborus funereus* 12.5cm The only seed finch to occur west of the Andes. **V** Pretty warbling with short trills and a few inhaled notes. **H** Scrub at forest edges, fields, shrubby clearings. **D** SE Mexico to W Ecuador.

29 CHESTNUT-BELLIED SEED FINCH *Oryzoborus angolensis* 12.5cm Chestnut belly is diagnostic. **V** Song like that of Thick-billed Seed Finch, musical whistled notes interspersed with chattering, often ending in a trill. **H** Shrubby clearings and edges of woodland and forest. **D** Amazonian lowlands, from Colombia to S Brazil.

30 NICARAGUAN SEED FINCH *Oryzoborus nuttingi* 15cm Massive bill (yellowish pink in the male) is diagnostic in its range. **V** High irregular *petertjuwpetertjuwwuw*. **H** Marshy, grassy and weedy areas. **D** Nicaragua to Panama.

31 LARGE-BILLED SEED FINCH *Oryzoborus crassirostris* 14.5cm Note pale bluish bill. **V** Pretty warbling with many nasal notes. **H** Shrubby areas, tall weeds and grass, usually near water or marsh. **D** Colombia to Peru and N Brazil.

32 GREAT-BILLED SEED FINCH *Oryzoborus maximiliani* 16cm Note pinkish or chalk-white bill. **V** Pretty warbling with many *tue* notes. **H** Shrub near marsh, tall weedy vegetation near water. **D** Venezuela to N Brazil; E Bolivia to EC Brazil.

33 BLACK-BILLED SEED FINCH *Oryzoborus atrirostris* 16cm Both male and female have a black bill. **V** Song is a rich and leisurely warbling, similar to Thick-billed Seed Finch. **H** Tall weedy vegetation near lakes, marshes, riverine belts. **D** Colombia to Bolivia.

34 CUBAN BULLFINCH *Melopyrrha nigra* 14–15cm Male black with a conspicuous white wing patch; female duller. Forages from near the ground to the canopy, usually in small flocks; often mixes with other species. **V** A buzzing *chip*. The song is a long, descending then ascending trill. **H** Forests, woods, bushes, undergrowth in pine woods, mangroves. **D** Cuba, Cayman Islands (Grand Cayman).

35 WHITE-NAPED SEEDEATER *Dolospingus fringilloides* 13.5cm Head of male is distinctly patterned; note his sharp-pointed bill and black chin. Female is mainly warm brown. **V** Short, energetic series in which every note is repeated three times. **H** Edges and clearings in forest, grassy savannah. **D** N and NW Amazonia, in Colombia, Venezuela, N Brazil.

36 BAND-TAILED SEEDEATER *Catamenia analis* 12.5cm Separable from Plain-colored Seed Eater and Paramo Seedeater by white band across tail (in flight) and yellow bill. **V** Fast, rolling *diddle diddle diddle didee* song. **H** Shrubby grassland, edges of fields, pastures and villages. **D** Andes, from N Colombia to C Argentina.

37 PLAIN-COLORED SEEDEATER *Catamenia inornata* 14cm Bill is pinkish. Male uniform grey (paler that Paramo Seedeater); note rufous vent (similar in Band-tailed Seedeater and Paramo Seedeater). Streaking on female less dark than in female Paramo Seedeater. **V** Song is a faint buzz (1 sec). **H** Shrubby slopes, open grassland, *Polylepis* woodland. **D** Andes, from N Colombia and Venezuela to C Argentina.

38 PARAMO SEEDEATER *Catamenia homochroa* 13.5cm The darkest of the *Catamenia* seedeaters. Compare Band-tailed Seedeater and Plain-colored Seedeater. **V** High, piped *eeee*, followed by a higher, short buzz. **H** Shrubbery and bamboo at the treeline. **D** N Andes, from Colombia to Bolivia; N Colombia; tepuis of S Venezuela.

1
2
3
4
5
6
7
8
9
10
11
12
13
14
15
16
17
18
19
20
21
22
23
24
25
26
27
27
28
29
30
31
32
33
34
35
36
37
38
'insulata'

TANAGERS AND ALLIES *CONTINUED*

1 BANANAQUIT *Coereba flaveola* 10–12.5cm Generally dark grey or olive above, pale rump, prominent pale supercilium, underparts bright yellow, but very variable, with over 40 races recognised. Feeds on nectar from flowering trees and other plants. **V** Song variable, generally consisting of short series of high-pitched unmusical buzzes. Typical call a metallic *tsip*. **H** Almost anywhere there are flowering plants. **D** Mexico and Caribbean, south to Paraguay and N Argentina.

2 CUBAN GRASSQUIT *Tiaris canorus* 11.5cm Black mask and breast patch of male contrast with yellow collar. Forages on the ground, searching out small seeds, generally in flocks. **V** A simple *chip* and a shrill, rasping *chiri-wichi-wichi chibiri-wichi-wichi*. **H** Semi-arid country near the coast, pine undergrowth, woodland edge, plantations, farmland. **D** Cuba.

3 YELLOW-FACED GRASSQUIT *Tiaris olivaceus* 10cm Male unmistakable due to facial pattern. The same pattern is detectable on the female, but much fainter. **V** Short, ultra-high, thin rattles. **H** Scrubby and grassy habitats. **D** Mexico and Caribbean to Ecuador, Venezuela.

4 DULL-COLORED GRASSQUIT *Tiaris obscurus* 11cm Best identified by its bicoloured bill and its song. **V** Bursts of ultra-high, fast twittering, preceded by a nasal *eu*. **H** Shrubby habitats, thickets, dry scrub. **D** W Venezuela to Bolivia, N Argentina.

5 BLACK-FACED GRASSQUIT *Tiaris bicolor* 11.5cm Forages in pairs or small flocks, searching for small seeds and occasionally insects. In display, male flies a short distance with rapidly beating wings, giving a buzzing call. **V** Song is a loud, buzzing *dik-zeezeezee*. Typical call a soft, musical *tsip*. **H** Almost any open area with grasses and shrubs. **D** Caribbean, N Venezuela.

6 SOOTY GRASSQUIT *Tiaris fuliginosus* 11.5cm Plumage entirely dark, sooty black in male, dark brown in female. Note pinkish gape of male, yellowish in female. **V** Very high, repetitive twittering. **H** Shrubby clearings, woodland borders. **D** Scattered populations, Venezuela and Trinidad to E Bolivia, NE and C Brazil, NE Argentina.

7 YELLOW-SHOULDERED GRASSQUIT *Loxipasser anoxanthus* 10cm Charcoal-black head of male contrasts with yellow wing coverts. Forages in small parties in shrubs and trees, searching for fruits and seeds. **V** Descending, insect-like *zwee-ze-ze-ze-ze*. **H** Forest-edge shrubs, woodlands, gardens near wooded areas. **D** Jamaica.

8 PUERTO RICAN BULLFINCH *Loxigilla portoricensis* 16.5cm Very black, with reddish markings on crown, throat and undertail coverts. Sexes similar. Secretive, more often heard than seen. **V** Song is a rising whistle of 2–10 notes. Calls include a soft *tseet* or *check*. **H** Dense mountain forest, coffee plantations, coastal thickets, mangroves. **D** Puerto Rico.

9 GREATER ANTILLEAN BULLFINCH *Loxigilla violacea* 15–18cm Note reddish supercilium. Female similar to male, but rather duller. Keeps to dense cover, where it feeds on fruit, seeds and plant parts. **V** Calls include an insect-like *zeet* and a thin *split* given in alarm. Song is a repetition of the *zeet* call note. **H** Dense thickets and undergrowth, from coastal scrub to wet mountain forests. **D** Bahamas, Hispaniola, Jamaica.

10 LESSER ANTILLEAN BULLFINCH *Loxigilla noctis* 14–15.5cm Red throat, small red patch above the eye, and (in some races) red vent. Conspicuous and fairly tame; feeds on fruit and seeds. **V** A harsh *chuck*, a thin *tseep tseep*, a crisp trill and a lengthy twitter. **H** Shrubbery, thickets, forest understorey, gardens. **D** Lesser Antilles.

11 BARBADOS BULLFINCH *Loxigilla barbadensis* 15cm Formerly treated as a race of Lesser Antillean Bullfinch. Very plain plumage, and very tame. The only bird endemic to Barbados. **V** A simple twittering and a sharp trill. **H** Undergrowth, shrubbery, gardens. **D** Barbados.

12 ORANGEQUIT *Euneornis campestris* 14cm Note short downcurved bill. Forages at low to medium levels, searching for nectar, fruits and sap oozing from holes made by Yellow-bellied sapsuckers (Plate 105). **V** A thin, high-pitched *tseet*, *swee* or *fi-swee*. **H** Humid forest and woodland, mainly at mid-elevations. **D** Jamaica.

13 ST. LUCIA BLACK FINCH *Melanospiza richardsoni* 13–14cm Male uniformly black. Note pink legs of both sexes. Forages on or near the ground, primarily among leaf litter. **V** Burry *tick-zwee-swisiwis-you*, with accents on the second and last note. **H** Moist and semi-arid forests. **D** St Lucia (Lesser Antilles).

14 LARGE GROUND FINCH *Geospiza magnirostris* 16cm Largest of the 18 species of Darwin's finches, with a massive bill. **V** Song consists of loud couplets in groups of 2–3. **H** Arid scrub. **D** Galápagos (widespread).

15 ESPANOLA CACTUS FINCH *Geospiza conirostris* 14.5cm Male's black bill becomes orange-yellow outside the breeding season. **V** Varied phrases of repeated nasal *tjiew* or *tjiuw* notes. **H** Dry scrub and areas with *Opuntia*. **D** Galápagos (Española).

16 MEDIUM GROUND FINCH *Geospiza fortis* 12cm Of the three ground finches that are widespread across Galápagos, this is the middle-sized one. Variation in bill size with ecological niche has been the subject of much study. **V** Short phrases of 2–5 notes, the first few of which are level, nasal and voiceless. **H** During the warm/wet season, occurs in coastal and transition zones; at other times, moves into highlands. **D** Galápagos (widespread).

17 SMALL GROUND FINCH *Geospiza fuliginosa* 10cm The smallest of the widespread ground finches, with a dainty bill. **V** Song is a two-note rather raspy phrase, *tchew-tchew*. **H** All habitats, most common in arid and transition zones; moves into highlands in non-breeding season. **D** Galápagos (widespread).

18 SHARP-BEAKED GROUND FINCH *Geospiza difficilis* 12.5cm A small ground finch with a small pointed bill. Male's bill is black in breeding season. **V** Similar to that of others in its genus. Calls include a very high-pitched *tzeeeeuw*. **H** Dense montane and arid scrub, inside seabird colonies. **D** Galápagos (Pinta, Fernandina, Santiago).

19 GENOVESA GROUND FINCH *Geospiza acutirostris* 12cm Dark, very short-tailed ground finch with rather long and slightly downcurved bill. **V** Song is a series of well-spaced buzzy, low-pitched chirruping notes. **H** Arid scrub. **D** Galápagos (Genovesa).

20 VAMPIRE GROUND FINCH *Geospiza septentrionalis* 12cm Dark ground finch with strikingly long, sharply pointed and downcurved bill. Seedeater but in dry season feeds on seabird eggs and will drink blood from their wounds. **V** Song varies by island; on Darwin gives a phrase of buzzy, low-pitched notes, while on Wolf the song is a more chirruped, musical double note. **H** Low-lying grassy scrub. **D** Galápagos (Darwin, Wolf).

21 COMMON CACTUS FINCH *Geospiza scandens* 14cm Among the Darwin's finches, the two cactus finches have relatively long bills, an adaptation for feeding on *Opuntia* (prickly pear) seeds. **V** Loud, rapid, level phrases of 5–10 *djep* notes. **H** Mainly in the arid zone. **D** Galápagos (widespread).

22 GENOVESA CACTUS FINCH *Geospiza propinqua* 14cm Cactus finch with long and thick-based bill. Feeds on all parts of *Opuntia* (prickly pear); also takes insects. **V** Song rapid, rolling phrases; calls include shrill *tzeeeer*. **H** Arid scrub with grasses and *Opuntia*. **D** Galápagos (Genovesa).

23 VEGETARIAN FINCH *Platyspiza crassirostris* 16cm Note black hood and otherwise streaked appearance of male. Feeds mainly on leaves, flowers and fruits, and is often seen with a leaf or bud in its bill. **V** Strange nasal phrases of about three notes, with the middle one low and stretched. **H** Transition zone between arid and humid zones of montane slopes. **D** Galápagos (widespread).

24 LARGE TREE FINCH *Camarhynchus psittacula* 13cm Largest of the three imaginatively named tree finches, with the largest bill. Blackish hood of male extends to throat and breast. **V** Series of four high *treet* notes. **H** Highland forest, moving to lower elevations outside the warm/wet season. **D** Galápagos (widespread).

25 MEDIUM TREE FINCH *Camarhynchus pauper* 13cm Intermediate in size, but similar in appearance to Large and Small tree finches. Critically endangered. **V** Song is a high-pitched *tzee* followed by a lower *tew-tew-tew-tew*. Calls include nasal *phwee* and higher-pitched *zwee*. **H** Forest at high and middle levels, moving to lower elevations outside the warm/wet season. **D** Galápagos (Floreana).

26 SMALL TREE FINCH *Camarhynchus parvulus* 11cm Males on San Cristóbal lack the dark hood, and resemble females. **V** Rapid series of 5–6 staccato *tjaw* notes. **H** Highlands, but also in the arid zone. **D** Galápagos (widespread).

27 WOODPECKER FINCH *Camarhynchus pallidus* 15cm One of the very few bird species that use tools, in this case a thin twig to ease insects out of crevices. **V** Short series of rapid, separated three-syllable phrases. **H** Mostly in the transition zone and in the higher humid forest zone. **D** Galápagos (widespread).

28 MANGROVE FINCH *Camarhynchus heliobates* 14cm The most habitat-specific of all of the Darwin's finches. **V** Probably a rapid, high series of three *kreh* notes. **H** Dense mangrove stands. **D** Galápagos (Isabela, Fernandina).

29 GREEN WARBLER-FINCH *Certhidea olivacea* 10cm The two warbler-finches are the smallest of the Darwin's finches, with the finest bills. They were formerly considered a single species, but differ in habitat and range. **V** Very high series of well-separated *tsi* notes, sung singly, as couplets or as triplets. **H** All habitats, but most common in humid highlands of larger islands. **D** Galápagos (several islands).

30 GREY WARBLER-FINCH *Certhidea fusca* 10cm Separable from Green Warbler-Finch by range and habitat. **V** Song is like that of Green Warbler-Finch. **H** Arid lowlands of smaller islands, even at the vegetation edge on beaches. **D** Galápagos (several islands).

31 COCOS FINCH *Pinaroloxias inornata* 12cm An outlier of the Darwin's finch group, the only finch on Cocos Island. A generalist, occupying a wide ecological niche. Feeds on fruit, nectar, arthropods, grass seeds. **V** Song like other Darwin's finches, a buzzy double note, *chz-zhweeeuu* or *phfft-zheeuu*. **H** Forest, woodland, open areas. **D** Cocos Island, EC Pacific.

32 PLUSHCAP *Catamblyrhynchus diadema* 14cm Unmistakable, with dark bluish-grey upperparts, rufous underparts, and golden-yellow forecrown. Note stubby bill. **V** Very long, rapid, silvery twittering, slowing at the end as if getting tired. **H** Bamboo in undergrowth and borders of montane forest and woodland. **D** Venezuela to NW Argentina.

33 YELLOW-SHOULDERED GROSBEAK *Parkerthraustes humeralis* 16cm Distinctive, with grey crown and underparts, black mask, olive-green upperparts with yellow shoulder patch, yellow undertail coverts, red eye. **V** Repeated phrase of ultra-high twittering, punctuated with a high, short rattle. **H** Canopy and borders of terra firme forest. **D** W and S Amazonia.

1
2
3
4
5
6
7
8
9
10
11
12
13
14
15
16
17
18
19
20
21
22
23
24
25
26
27
28
29
30
31
32
33

English Name Index

A

C

D

E

F

G

H

I

J

K

L

M

N

O

P

Q

R

S

T

U

V

W

X

Y

Scientific Name Index

A

B

C

D

E

F

G

H

I

J

K

L

M

N

O

P

Q

R